More Than Just a Te

Internet Resources

StudentWorks™ *Plus* Online This interactive **eBook** includes the complete Student Edition with audio, Math in Motion, Personal Tutor, Self-Check Quizzes, and much more – all at point of use!

Step 1 **Connect to** Math Online glencoe.com

Step 2 **Connect to resources by using simple and convenient** QuickPass **codes.**

A14801c1

"A1" for "Algebra 1"

This edition, ISBN 978-0-07-888480-1

Enter the appropriate chapter number. c1 = Chapter 1

For Students

Connect to the Student Edition **eBook** that contains all of the following online resources. You don't need to take your textbook home every night.

- Personal Tutor
- Self-Check Quizzes
- Chapter Readiness Quizzes
- Math in Motion: Animation
- Math in Motion: BrainPOP®
- Math in Motion: Interactive Lab
- Extra Examples
- Chapter Test Practice
- Standardized Test Practice
- Study to Go
- Vocabulary Review Games
- Graphing Calculator Keystrokes
- Multilingual eGlossary
- Scavenger Hunts
- Workbooks
- Hotmath Math Homework Help Homework Help

For Teachers

- Teaching Today
- **Advance Tracker**
  - Diagnostic, formative, and summative assessment
  - Progress reports
  - Differentiated instruction
- State Resources
- Professional Development at www.mhpd.com
  - Video Clips
  - Online Credit Courses
- Research
  - White Papers
  - Efficacy Studies

For Parents

Connect to www.glencoe.com to access **StudentWorks *Plus* Online** and all of the resources for students and teachers listed above.

Glencoe McGraw-Hill

Algebra 1

Authors
Carter • Cuevas • Day • Malloy • Holliday • Luchin

Mc Graw Hill Glencoe

About the Authors

Glencoe/McGraw-Hill K–12 Mathematics Lead Authors

Our lead authors ensure that Glencoe/McGraw-Hill mathematics programs are truly vertically aligned by beginning with the end in mind – success in Algebra 1 and beyond. By "backmapping" the content from the high school programs, all of our mathematics programs are well articulated in their scope and sequence, ensuring that the content in each program provides a solid foundation for moving forward. These authors also worked closely with the entire K–12 author team to ensure vertical alignment of the instructional approach and visual design.

Dr. John A. Carter, Ph.D.
Assistant Principal for Teaching and Learning
Adlai E. Stevenson High School
Lincolnshire, IL

Areas of Expertise: Using technology and manipulatives to visualize concepts, mathematics achievement of English-language learners

Dr. Gilbert J. Cuevas, Ph.D.
Professor of Mathematics Education
Texas State University
San Marcos, TX

Areas of Expertise: Applying concepts and skills in mathematically rich contexts, mathematical representations

Dr. Roger Day, Ph.D., NBCT
Mathematics Department Chairperson
Pontiac Township High School
Pontiac, IL

Areas of Expertise: Understanding and applying probability and statistics, mathematics teacher education

Dr. Carol Malloy, Ph.D.
Associate Professor
University of North Carolina at Chapel Hill
Chapel Hill, NC

Areas of Expertise: Representations and critical thinking, student success in Algebra 1

Glencoe McGraw-Hill

Algebra 1

Authors

Carter • Cuevas • Day • Malloy • Holliday • Luchin

Mc Graw Hill Glencoe

About the Cover

On your mark, get set, GO! Math is everywhere, even on a track. The lanes on a track never cross or intersect. Lines that continue infinitely and never intersect are called *parallel lines*. You will learn more about parallel lines, intersecting lines, and the equations that represent them in Chapter 4.

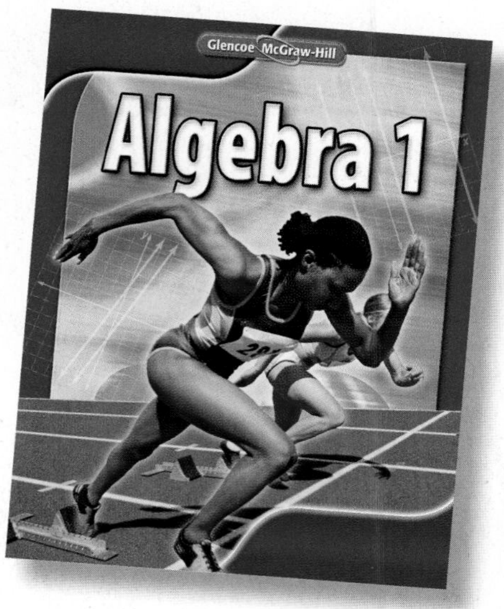

The McGraw-Hill Companies

Glencoe

Send all inquiries to:
Glencoe/McGraw-Hill
8787 Orion Place
Columbus, OH 43240-4027

ISBN: 978-0-07-888480-1
MHID: 0-07-888480-2

Printed in the United States of America.

2 3 4 5 6 7 8 9 10 079/043 17 16 15 14 13 12 11 10 09

CONTENTS IN BRIEF

About the Authors

Glencoe/McGraw-Hill K–12 Mathematics Lead Authors

Our lead authors ensure that Glencoe/McGraw-Hill mathematics programs are truly vertically aligned by beginning with the end in mind – success in Algebra 1 and beyond. By "backmapping" the content from the high school programs, all of our mathematics programs are well articulated in their scope and sequence, ensuring that the content in each program provides a solid foundation for moving forward. These authors also worked closely with the entire K–12 author team to ensure vertical alignment of the instructional approach and visual design.

Dr. John A. Carter, Ph.D.
Assistant Principal for Teaching and Learning
Adlai E. Stevenson High School
Lincolnshire, IL

Areas of Expertise: Using technology and manipulatives to visualize concepts, mathematics achievement of English-language learners

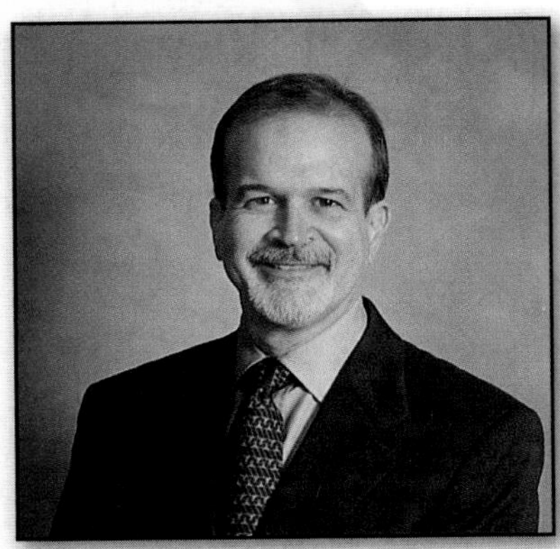

Dr. Gilbert J. Cuevas, Ph.D.
Professor of Mathematics Education
Texas State University
San Marcos, TX

Areas of Expertise: Applying concepts and skills in mathematically rich contexts, mathematical representations

Dr. Roger Day, Ph.D., NBCT
Mathematics Department Chairperson
Pontiac Township High School
Pontiac, IL

Areas of Expertise: Understanding and applying probability and statistics, mathematics teacher education

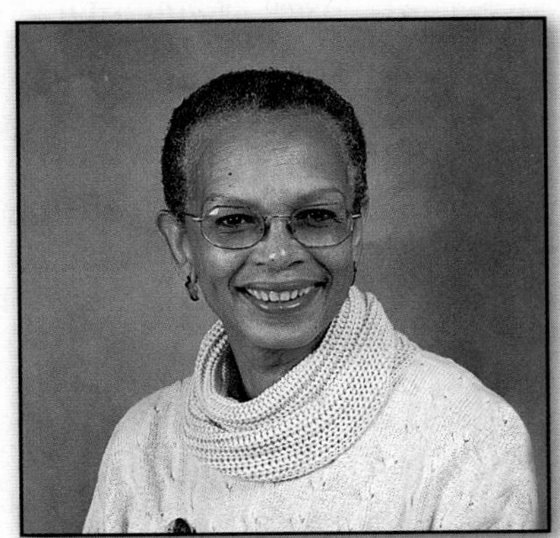

Dr. Carol Malloy, Ph.D.
Associate Professor
University of North Carolina at Chapel Hill
Chapel Hill, NC

Areas of Expertise: Representations and critical thinking, student success in Algebra 1

Additional Algebra 1 Authors

The entire Algebra 1 author team strives to create a program that can be used by all types of Algebra 1 teachers with all types of Algebra 1 students. Each author brings their special expertise to making a program that will contribute to the success of every student who uses this instructional resource.

Dr. Berchie Holliday, Ed.D.
National Mathematics Consultant
Silver Spring, MD
Areas of Expertise: Using mathematics to model and understand real-world data, the effect of graphics on the mathematical understanding

Beatrice Luchin
Mathematics Consultant
League City, TX
Areas of Expertise: Using reading strategies to aid in mathematical understanding

Contributing Author

This program is the beneficiary of the imagination of Dinah Zike through the contribution of the Foldables Study Organizers.

Dinah Zike
Educational Consultant
Dinah-Might Activities, Inc.
San Antonio, TX

Math Online Meet the authors at **glencoe.com**.

Consultants

Glencoe/McGraw-Hill wishes to thank the following professionals for their feedback. They were instrumental in providing valuable input toward the development of this program in these specific areas.

Mathematical Content

Viken Hovsepian
Professor of Mathematics
Rio Hondo College
Whittier, California

Grant A. Fraser, Ph.D.
Professor of Mathematics
California State University, Los Angeles
Los Angeles, California

Arthur K. Wayman, Ph.D.
Professor of Mathematics Emeritus
California State University, Long Beach
Long Beach, California

Gifted and Talented

Shelbi K. Cole
Research Assistant
University of Connecticut
Storrs, Connecticut

College Readiness

Robert Lee Kimball, Jr.
Department Head, Math and Physics
Wake Technical Community College
Raleigh, North Carolina

Differentiation for English-Language Learners

Susana Davidenko
State University of New York
Cortland, New York

Alfredo Gómez
Mathematics/ESL Teacher
George W. Fowler High School
Syracuse, New York

Graphing Calculator

Ruth M. Casey
T^3 National Instructor
Frankfort, Kentucky

Jerry Cummins
Former President
National Council of Supervisors of Mathematics
Western Springs, Illinois

Mathematical Fluency

Robert M. Capraro
Associate Professor
Texas A&M University
College Station, Texas

Pre-AP

Dixie Ross
Lead Teacher for Advanced Placement Mathematics
Pflugerville High School
Pflugerville, Texas

Reading and Writing

ReLeah Cossett Lent
Author and Educational Consultant
Morganton, Georgia

Lynn T. Havens
Director of Project CRISS
Kalispell, Montana

Teacher Reviewers

Each Reviewer reviewed at least two chapters of the Student Edition, giving feedback and suggestions for improving the effectiveness of the mathematics instruction.

Sherri Abel
Mathematics Teacher
Eastside High School
Taylors, South Carolina

Kelli Ball, NBCT
Mathematics Teacher
Owasso 7th Grade Center
Owasso, Oklahoma

Cynthia A. Burke
Mathematics Teacher
Sherrard Junior High School
Wheeling, West Virginia

Patrick M. Cain, Sr.
Assistant Principal
Stanhope Elmore High School
Millbrook, Alabama

Robert D. Cherry
Mathematics Instructor
Wheaton Warrenville South High School
Wheaton, Illinois

Tammy Cisco
8th Grade Mathematics/Algebra Teacher
Celina Middle School
Celina, Ohio

Amber L. Contrano
High School Teacher
Naperville Central High School
Naperville, Illinois

Catherine Creteau
Mathematics Department
Delaware Valley Regional High School
Frenchtown, New Jersey

Glenna L. Crockett
Mathematics Department Chair
Fairland High School
Fairland, Oklahoma

Jami L. Cullen
Mathematics Teacher/Leader
Hilltonia Middle School
Columbus, Ohio

Franco DiPasqua
Director of K-12 Mathematics
West Seneca Central Schools
West Seneca, New York

Kendrick Fearson
Mathematics Department Chair
Amos P. Godby High School
Tallahassee, Florida

Lisa K. Gleason
Mathematics Teacher
Gaylord High School
Gaylord, Michigan

Debra Harley
Director of Math & Science
East Meadow School District
Westbury, NewYork

Tracie A. Harwood
Mathematics Teacher
Braden River High School
Bradenton, Florida

Bonnie C. Hill
Mathematics Department Chair
Triad High School
Troy, Illinois

Clayton Hutsler
Teacher
Goodwyn Junior High School
Montgomery, Alabama

Gureet Kaur
7th Grade Mathematics Teacher
Quail Hollow Middle School
Charlotte, North Carolina

Rima Seals Kelley, NBCT
Mathematics Teacher/Department Chair
Deerlake Middle School
Tallahassee, Florida

Holly W. Loftis
8th Grade Mathematics Teacher
Greer Middle School
Lyman, South Carolina

Katherine Lohrman
Teacher, Math Specialist, New Teacher Mentor
John Marshall High School
Rochester, New York

Carol Y. Lumpkin
Mathematics Educator
Crayton Middle School
Columbia, South Carolina

Ron Mezzadri
Supervisor of Mathematics K–12
Fair Lawn Public Schools
Fair Lawn, New Jersey

Bonnye C. Newton
SOL Resource Specialist
Amherst County Public Schools
Amherst, Virginia

Kevin Olsen
Mathematics Teacher
River Ridge High School
New Port Richey, Florida

Kara Painter
Mathematics Teacher
Downers Grove South High School
Downers Grove, Illinois

Sheila L. Ruddle, NBCT
Mathematics Teacher, Grades 7 and 8
Pendleton County Middle/High School
Franklin, West Virginia

Angela H. Slate
Mathematics Teacher/Grade 7, Pre-Algebra, Algebra
LeRoy Martin Middle School
Raleigh, North Carolina

Cathy Stellern
Mathematics Teacher
West High School
Knoxville, Tennessee

Dr. Maria J. Vlahos
Mathematics Division Head for Grades 6–12
Barrington High School
Barrington, Illinois

Susan S. Wesson
Mathematics Consultant/Teacher (Retired)
Pilot Butte Middle School
Bend, Oregon

Mary Beth Zinn
High School Mathematics Teacher
Chippewa Valley High Schools
Clinton Township, Michigan

CHAPTER 0

Preparing for Algebra

Chapter 0 Support

Helping You Learn

Math Online

Unit 1
Foundations for Functions

CHAPTER 1

Expressions, Equations, and Functions

Chapter 1 Support

Helping You Learn

Math Online

Preparing for Testing

Unit 1
Foundations for Functions

CHAPTER 2

Linear Equations

Bob Daemmrich/PhotoEdit

Chapter 2 Support

Helping You Learn

Math Online

Preparing for Testing

Unit 2
Linear Functions and Relations

CHAPTER 3

Linear Functions

Wolfgang Kaehler

Chapter 3 Support

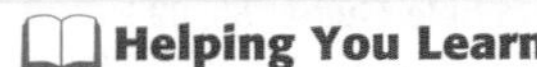

Helping You Learn

- **New Vocabulary** 153, 161, 170, 180, 187, 195
- **Key Concepts** 153, 161, 170, 173, 181, 187, 195
- **Check Your Progress** 153, 154, 155, 156, 162, 163, 170, 171, 172, 173, 174, 180, 181, 182, 188, 189, 190, 196, 197
- **Check Your Understanding** 157, 164, 175, 183, 191, 198
- **Multiple Representations** 159, 185, 192
- **H.O.T. Problems** 159, 165, 177, 185, 192, 199
- **Skills Review** 160, 166, 178, 186, 193, 200

Math Online

- **Math in Motion: Animation** 156, 169
- **Graphing Technology Personal Tutor** 167
- **Personal Tutor** 153, 154, 155, 156, 161, 162, 163, 170, 171, 172, 173, 174, 180, 181, 182, 187, 188, 189, 190, 195, 196, 197
- **Self-Check Quizzes** 153, 161, 170, 180, 187, 195
- **Extra Examples** 153, 161, 170, 180, 187, 195
- **Homework Help** 153, 161, 170, 180, 187, 195

Preparing for Testing

- **Extended Response** 166
- **Multiple Choice** 154, 160, 166, 178, 179, 186, 193, 200
- **Short/Gridded Response** 160, 178, 186, 193, 200
- **Worked-Out Example** 154

Alex Segre/Alamy

Linear Functions and Relations

Chapter 4 Support

Helping You Learn

- **New Vocabulary** 214, 224, 231, 237, 245, 253
- **Key Concepts** 214, 231, 232, 240, 245, 246, 261, 262, 264
- **Check Your Progress** 214, 215, 216, 217, 224, 225, 226, 231, 232, 233, 237, 238, 239, 245, 247, 254, 255, 261, 262, 263
- **Check Your Understanding** 217, 227, 233, 240, 248, 256, 264
- **Multiple Representations** 228, 242, 267
- **H.O.T. Problems** 220, 229, 235, 242, 250, 259, 267
- **Skills Review** 221, 230, 236, 243, 251, 260, 268

Math Online

- **Math in Motion: BrainPOP®** 214
- **Math in Motion: Interactive Labs** 246
- **Graphing Technology Personal Tutor** 222, 269
- **Personal Tutor** 214, 215, 216, 217, 224, 225, 226, 231, 232, 233, 237, 238, 239, 245, 247, 254, 255
- **Self-Check Quizzes** 214, 224, 231, 237, 245, 253, 261
- **Extra Examples** 214, 224, 231, 237, 245, 253, 261
- **Homework Help** 214, 224, 231, 237, 245, 253, 261

Preparing for Testing

- **Extended Response** 221
- **Multiple Choice** 216, 221, 230, 236, 243, 251, 260, 268
- **Short/Gridded Response** 230, 236, 243, 251, 268
- **Worked-Out Example** 216

Unit 2
Linear Functions and Relations

CHAPTER 5

Linear Inequalities

Chapter 5 Support

Helping You Learn

- **New Vocabulary** 283, 304, 315
- **Key Concepts** 283, 284, 285, 290, 292, 315
- **Check Your Progress** 283, 284, 285, 291, 292, 296, 297, 298, 305, 306, 310, 311, 316, 317
- **Check Your Understanding** 286, 293, 298, 306, 312, 318
- **Multiple Representations** 287, 294, 308, 313, 319
- **H.O.T. Problems** 287, 294, 300, 308, 313, 319
- **Skills Review** 288, 295, 301, 309, 314, 320

Math Online

- **Math in Motion: Animation** 289, 305, 311
- **Math in Motion: BrainPOP®** 297
- **Math in Motion: Interactive Labs** 298
- **Graphing Technology Personal Tutor** 321
- **Personal Tutor** 283, 284, 285, 290, 296, 297, 298, 304, 305, 306, 310, 311, 315, 316, 317
- **Self-Check Quizzes** 283, 290, 296, 304, 310, 315
- **Extra Examples** 283, 290, 296, 304, 310, 315
- **Homework Help** 283, 290, 296, 304, 310, 315

Preparing for Testing

- **Extended Response** 320
- **Multiple Choice** 284, 288, 295, 301, 302, 309, 314, 320, 325
- **Short/Gridded Response** 288, 295, 301, 309, 314
- **Worked-Out Example** 284

Unit 2
Linear Functions and Relations

CHAPTER 6

Systems of Linear Equations and Inequalities

Chapter 6 Support

Masterfile

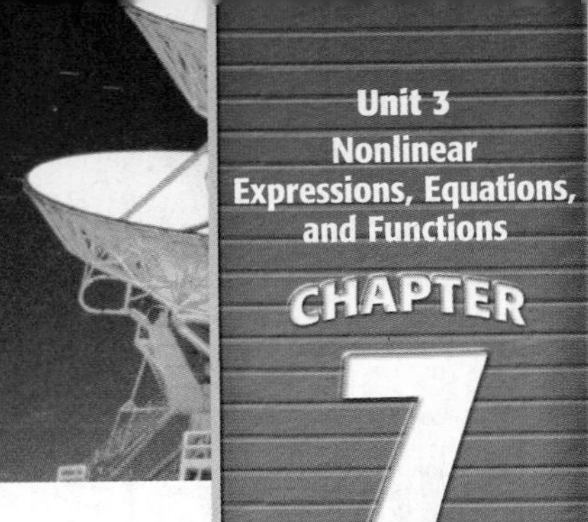

Polynomials

Roger Ressmeyer/CORBIS

Chapter 7 Support

Helping You Learn

Math Online

Preparing for Testing

Unit 3
Nonlinear Expressions, Equations, and Functions

CHAPTER 8

Factoring and Quadratic Equations

Jon Arnold Images Ltd/Alamy

Chapter 8 Support

Unit 3
Nonlinear Expressions, Equations, and Functions

CHAPTER 9

Quadratic and Exponential Functions

Stephen Chernin/Getty Images

Chapter 9 Support

Helping You Learn

Math Online

Preparing for Testing

Unit 4
Advanced Functions and Equations

CHAPTER 10

Radical Functions and Geometry

Chapter 10 Support

Helping You Learn

Math Online

Preparing for Testing

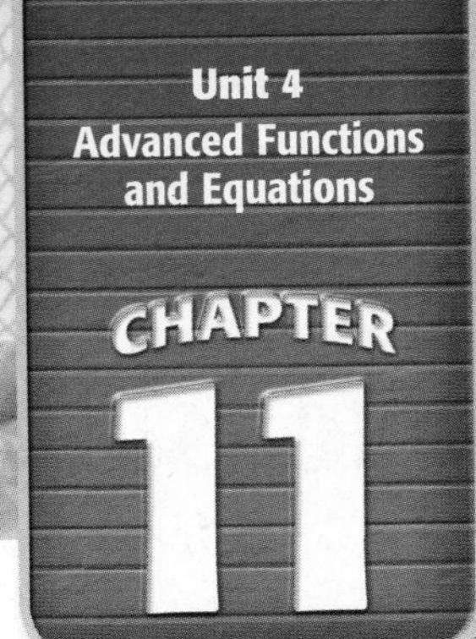

Doug Pensinger/Getty Images

Rational Functions and Equations

Chapter 11 Support

Helping You Learn

Math Online

Preparing for Testing

Dex Image/Alamy

Unit 5
Data Analysis

CHAPTER 12

Statistics and Probability

Chapter 12 Support

Helping You Learn

Math Online

Preparing for Testing

Student Handbook

Contents

CHAPTER 0

Preparing for Algebra

Chapter 0 contains lessons on topics from previous courses. You can use this chapter in various ways.

- Begin the school year by taking the Pretest. If you need additional review, complete the lessons in this chapter. To verify that you have successfully reviewed the topics, take the Posttest.
- As you work through the text, you may find that there are topics you need to review. When this happens, complete the individual lessons that you need.
- Use this chapter for reference. When you have questions about any of these topics, flip back to this chapter to review definitions or key concepts.

Get Started on Chapter 0

You will review several concepts, skills, and vocabulary terms as you study Chapter 0. To get ready, identify important terms and organize your resources.

FOLDABLES® Study Organizer

Throughout this text, you will be invited to use Foldables to organize your notes.

Why should you use them?

- They help you organize, display, and arrange information.
- They make great study guides, specifically designed for you.
- You can use them as your math journal for recording main ideas, problem-solving strategies, examples, or questions you may have.
- They give you a chance to improve your math vocabulary.

How should you use them?

- Write general information – titles, vocabulary terms, concepts, questions, and main ideas – on the front tabs of your Foldable.
- Write specific information – ideas, your thoughts, answers to questions, steps, notes, and definitions – under the tabs.
- Use the tabs for:
  - math concepts in parts, like types of triangles,
  - steps to follow, or
  - parts of a problem, like *compare* and *contrast* (2 parts) or *what*, *where*, *when*, *why*, and *how* (5 parts).
- You may want to store your Foldables in a plastic zipper bag that you have three-hole punched to fit in your notebook.

When should you use them?

- Set up your Foldable as you begin a chapter, or when you start learning a new concept.
- Write in your Foldable every day.
- Use your Foldable to review for homework, quizzes, and tests.

New Vocabulary

English		Español
integer	• p. P7 •	entero
absolute value	• p. P11 •	valor absolute
opposites	• p. P11 •	opuestos
reciprocal	• p. P18 •	recíproco
perimeter	• p. P23 •	perímetro
circle	• p. P24 •	círculo
diameter	• p. P24 •	diámetro
center	• p. P24 •	centro
circumference	• p. P24 •	circunferencia
radius	• p. P24 •	radio
area	• p. P26 •	area
volume	• p. P29 •	volumen
surface area	• p. P31 •	area de superficie
probability	• p. P33 •	probabilidad
sample space	• p. P33 •	espacio muestral
complements	• p. P33 •	complementos
tree diagram	• p. P34 •	diagrama de árbol
odds	• p. P35 •	probabilidades
mean	• p. P37 •	media
median	• p. P37 •	mediana
mode	• p. P37 •	moda
range	• p. P38 •	rango
quartile	• p. P38 •	cuartil
lower quartile	• p. P38 •	cuartil inferior
upper quartile	• p. P38 •	cuartil superior
bar graph	• p. P40 •	gráfica de barras
histogram	• p. P40 •	histograma
line graph	• p. P41 •	gráfica lineal
circle graph	• p. P41 •	gráfica circular
outliers	• p. P42 •	valores atípicos

Multilingual eGlossary glencoe.com

Math Online glencoe.com

- Study the chapter online
- Explore **Math in Motion**
- Get extra help from your own **Personal Tutor**
- Use **Extra Examples** for additional help
- Take a **Self-Check Quiz**
- **Review Vocabulary** in fun ways

CHAPTER 0 Pretest

Determine whether you need an estimate or an exact answer. Then solve.

1. **SHOPPING** Addison paid \$1.29 for gum and \$0.89 for a package of notebook paper. She gave the cashier a \$5 bill. If the tax was \$0.14, how much change should Addison receive?

2. **DISTANCE** Luis rode his bike 1.2 miles to his friend's house, then 0.7 mile to the video store, then 1.9 miles to the library. If he rode the same route back home, about how far did he travel in all?

Find each sum or difference.

3. $20 + (-7)$
4. $-15 + 6$
5. $-9 - 22$
6. $18.4 - (-3.2)$
7. $23.1 + (-9.81)$
8. $-5.6 + (-30.7)$

Find each product or quotient.

9. $11(-8)$
10. $-15(-2)$
11. $63 \div (-9)$
12. $-22 \div 11$

Replace each ● with <, >, or = to make a true sentence.

13. $\frac{7}{20}$ ● $\frac{2}{5}$
14. 0.15 ● $\frac{1}{8}$

15. Order 0.5, $-\frac{1}{7}$, -0.2, and $\frac{1}{3}$ from least to greatest.

Find each sum or difference. Write in simplest form.

16. $\frac{5}{6} + \frac{2}{3}$
17. $\frac{11}{12} - \frac{3}{4}$
18. $\frac{1}{2} + \frac{4}{9}$
19. $-\frac{3}{5} + \left(-\frac{1}{5}\right)$

Find each product or quotient.

20. $2.4(-0.7)$
21. $-40.5 \div (-8.1)$

Name the reciprocal of each number.

22. $\frac{4}{11}$
23. $-\frac{3}{7}$

Find each product or quotient. Write in simplest form.

24. $\frac{2}{21} \div \frac{1}{3}$
25. $\frac{1}{5} \cdot \frac{3}{20}$
26. $\frac{6}{25} \div \left(-\frac{3}{5}\right)$
27. $\frac{1}{9} \cdot \frac{3}{4}$
28. $-\frac{2}{21} \div \left(-\frac{2}{15}\right)$
29. $2\frac{1}{2} \cdot \frac{2}{15}$

Express each percent as a fraction in simplest form.

30. 20%
31. 7.5%

Use the percent proportion to find each number.

32. 18 is what percent of 72?
33. 35 is what percent of 200?
34. 24 is 60% of what number?

35. **TEST SCORES** James answered 14 items correctly on a 16-item quiz. What percent did he answer correctly?

36. **BASKETBALL** Emily made 75% of the baskets that she attempted. If she made 9 baskets, how many attempts did she make?

Find the perimeter and area of each figure.

37.

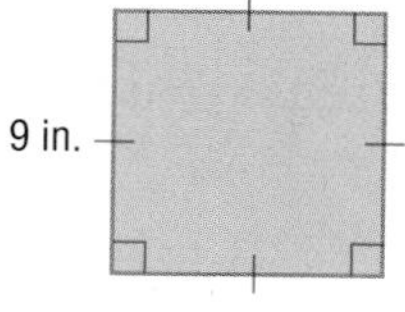

38.

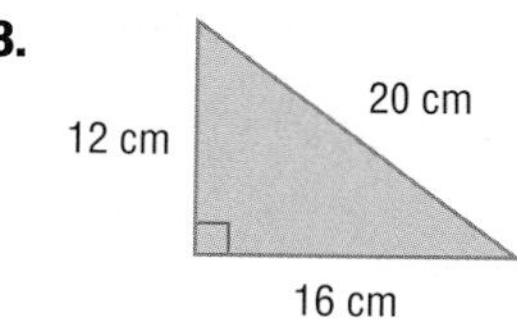

39. A parallelogram has side lengths of 7 inches and 11 inches. Find the perimeter.

40. **GARDENS** Find the perimeter of the garden.

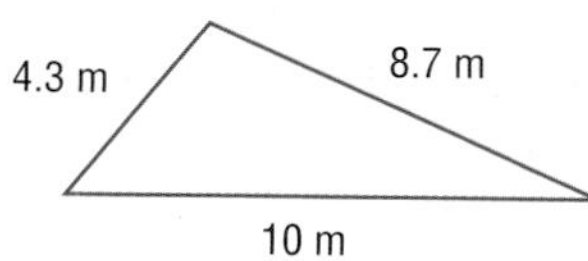

Find the circumference and area of each circle. Round to the nearest tenth.

41.

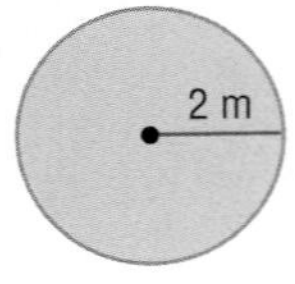

42.

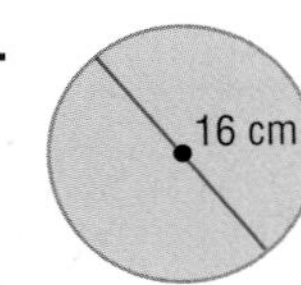

43. **BIRDS** The floor of a birdcage is a circle with a circumference of about 47.1 inches. What is the diameter of the birdcage floor? Round to the nearest inch.

Find the volume and surface area of each rectangular prism given the measurements below.

44. $\ell = 3$ cm, $w = 1$ cm, $h = 3$ cm

45. $\ell = 6$ ft, $w = 2$ ft, $h = 5$ ft

46. Find the volume and surface area of the rectangular prism.

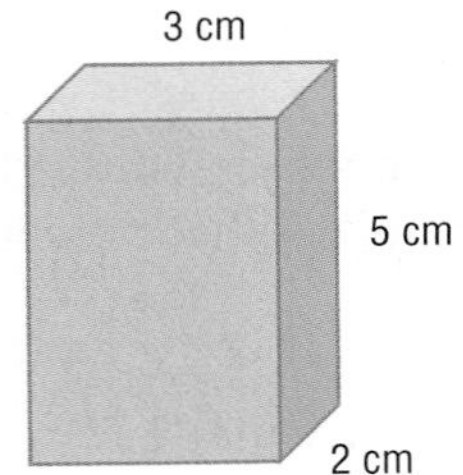

One pencil is randomly selected from a case containing 3 red, 4 green, 2 black, and 6 blue pencils. Find each probability.

47. *P*(green)

48. *P*(red or blue)

49. Use a tree diagram to find the sample space for the event *a die is rolled, and a coin is tossed*. State the number of possible outcomes.

One coin is randomly selected from a jar containing 20 pennies, 15 nickels, 3 dimes, and 12 quarters. Find the odds of each outcome. Write in simplest form.

50. a penny

51. a penny or nickel

52. A coin is tossed 50 times. The results are shown in the table. Find the experimental probability of heads. Write as a fraction in simplest form.

Lands Face-Up	Number of Times
head	22
tails	28

Find the mean, median, and mode for each set of data.

53. {10, 11, 18, 24, 30}

54. {4, 8, 9, 9, 10, 14, 16}

55. Find the range, median, lower quartile, and upper quartile for {16, 19, 21, 24, 25, 31, 35}.

56. **SCHOOL** Devonte's scores on his first four Spanish tests are 92, 85, 90, and 92. What test score must Devonte earn on the fifth test so that the mean will be exactly 90?

57. **MUSIC** The table shows the results of a survey in which students were asked to choose which of four instruments they would like to learn. Make a bar graph of the data.

Favorite Instrument	
Instrument	**Number of Students**
drums	8
guitar	12
piano	5
trumpet	7

58. Make a stem-and-leaf plot of the data: 42, 50, 38, 59, 50, 44, 46, 62, 47, 35, 55, and 56.

59. **EXPENSES** The table shows how Dylan spent his money at the fair. Make a circle graph of the data.

Money Spent at the Fair	
How Spent	**Amount ($)**
rides	6
food	10
games	4

0-1 Plan for Problem Solving

Using the **four-step problem-solving plan** can help you solve any word problem.

Objective

Use the four-step problem-solving plan.

New Vocabulary

four-step problem-solving plan
defining a variable

Key Concept — Four-Step Problem-Solving Plan

Step 1 Understand the problem.
Step 2 Plan the solution.
Step 3 Solve the problem.
Step 4 Check the solution.

Each step of the plan is important.

Step 1 Understand the Problem

To solve a verbal problem, first read the problem carefully and explore what the problem is about.

- Identify what information is given.
- Identify what you need to find.

Step 2 Plan the Solution

One strategy you can use is to write an equation. Choose a variable to represent one of the unspecified numbers in the problem. This is called **defining a variable**. Then use the variable to write expressions for the other unspecified numbers in the problem.

Step 3 Solve the Problem

Use the strategy you chose in Step 2 to solve the problem.

Step 4 Check the Solution

Check your answer in the context of the original problem.

- Does your answer make sense?
- Does it fit the information in the problem?

EXAMPLE 1

FLOORS Ling's hallway is 10 feet long and 4 feet wide. He paid $200 to tile his hallway floor. How much did Ling pay per square foot for the tile?

Understand We are given the measurements of the hallway and the total cost of the tile. We are asked to find the cost of each square foot of tile.

Plan Write an equation. Let f represent the cost of each square foot of tile. The area of the hallway is 10×4 or 40 ft^2.

40	times	the cost per square foot	equals	200
40	$\cdot$	f	$=$	200

Solve $40 \cdot f = 200$. Find f mentally by asking, "What number times 40 is 200?"

$f = 5$

The tile cost $5 per square foot.

Check If the tile costs $5 per square foot, then 40 square feet of tile costs $5 \cdot 40$ or $200. The answer makes sense.

When an exact value is needed, you can use estimation to check your answer.

EXAMPLE 2

TRAVEL Emily's family drove 254.6 miles. Their car used 19 gallons of gasoline. Describe the car's gas mileage.

Understand We are given the total miles driven and how much gasoline was used. We are asked to find the gas mileage of the car.

Plan Write an equation. Let G represent the car's gas mileage.

gas mileage = number of miles ÷ number of gallons used

$G = 254.6 \div 19$

Solve $G = 254.6 \div 19$

$= 13.4$ mi/gal

The car's gas mileage is 13.4 miles per gallon.

Check Use estimation to check your solution.

260 mi ÷ 20 gal = 13 mi/gal

Since the solution 13.4 is close to the estimate, the answer is reasonable.

Exercises

Real-World Link

In a recent year, an average of $2.9 billion was spent on grooming and boarding dogs in the United States.

Source: American Pet Products Manufacturers Association

Determine whether you need an estimate or an exact answer. Then use the four step problem-solving plan to solve.

1. **DRIVING** While on vacation, the Jacobson family drove 312.8 miles the first day, 177.2 miles the second day, and 209 miles the third day. About how many miles did they travel in all?
2. **PETS** Ms. Hernandez boarded her dog at a kennel for 4 days. It cost $18.90 per day, and she had a coupon for $5 off. What was the final cost for boarding her dog?
3. **MEASUREMENT** William is using a 1.75-liter container to fill a 14-liter container of water. About how many times will he need to fill the smaller container?
4. **SEWING** Fabric costs $5.15 per yard. The drama department needs 18 yards of the fabric for their new play. About how much should they expect to pay?
5. **FINANCIAL LITERACY** The table shows donations to help purchase a new tree for the school. How much money did the students donate in all?

Number of Students	Amount of Each Donation
20	$2.50
15	$3.25

6. **SHOPPING** Is $12 enough to buy a half gallon of milk for $2.30, a bag of apples for $3.99, and four cups of yogurt that cost $0.79 each? Explain.

0-2 Real Numbers

Objective

Classify and use real numbers.

New Vocabulary

positive number
negative number
natural number
whole number
integer
rational number
square root
perfect square
irrational number
real number
graph
coordinate

A number line can be used to show the sets of natural numbers, whole numbers, integers, and rational numbers. Values greater than 0, or **positive numbers**, are listed to the right of 0, and values less than 0, or **negative numbers**, are listed to the left of 0.

natural numbers: 1, 2, 3, …

whole numbers: 0, 1, 2, 3, …

integers: …, −3, −2, −1, 0, 1, 2, 3, …

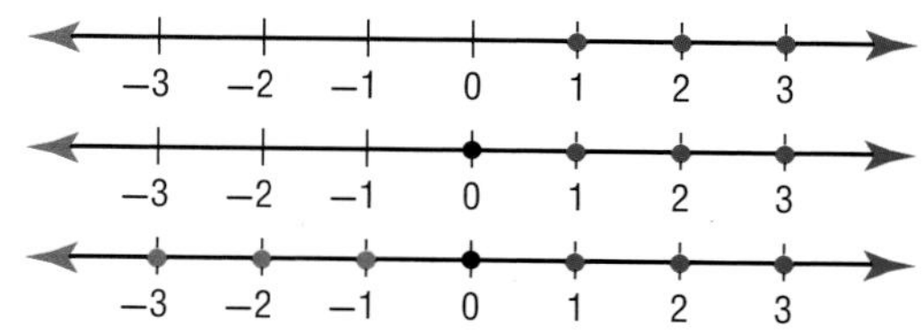

rational numbers: numbers that can be expressed in the form $\frac{a}{b}$, where a and b are integers and $b \neq 0$

A **square root** is one of two equal factors of a number. For example, one square root of 64, written as $\sqrt{64}$, is 8 since $8 \cdot 8$ or 8^2 is 64. Another square root of 64 is −8 since $(-8) \cdot (-8)$ or $(-8)^2$ is also 64. A number like 64, with a square root that is a rational number, is called a **perfect square**. The square roots of a perfect square are rational numbers.

A number such as $\sqrt{3}$ is the square root of a number that is not a perfect square. It cannot be expressed as a terminating or repeating decimal; $\sqrt{3} \approx 1.73205\ldots$. Numbers that cannot be expressed as terminating or repeating decimals, or in the form $\frac{a}{b}$, where a and b are integers and $b \neq 0$, are called **irrational numbers**. Irrational numbers and rational numbers together form the set of **real numbers**.

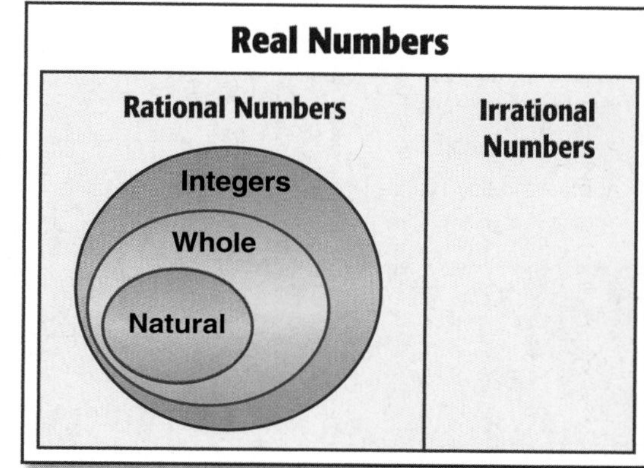

EXAMPLE 1 Classify Real Numbers

Math *in Motion*, Animation glencoe.com

Name the set or sets of numbers to which each real number belongs.

a. $\frac{5}{22}$

Because 5 and 22 are integers and $5 \div 22 = 0.2272727\ldots$ or $0.2\overline{27}$, which is a repeating decimal, this number is a rational number.

b. $\sqrt{81}$

Because $\sqrt{81} = 9$, this number is a natural number, a whole number, an integer, and a rational number.

c. $\sqrt{56}$

Because $\sqrt{56} = 7.48331477\ldots$, which is not a repeating or terminating decimal, this number is irrational.

To graph a set of numbers means to draw, or plot, the points named by those numbers on a number line. The number that corresponds to a point on a number line is called the coordinate of that point. The rational numbers and the irrational numbers complete the number line.

EXAMPLE 2 Graph Real Numbers

Graph each set of numbers on a number line.

a. $\left\{-\frac{4}{3}, -\frac{1}{3}, \frac{2}{3}, \frac{5}{3}\right\}$

$-\frac{5}{3}$ $-\frac{4}{3}$ -1 $-\frac{2}{3}$ $-\frac{1}{3}$ 0 $\frac{1}{3}$ $\frac{2}{3}$ 1 $\frac{4}{3}$ $\frac{5}{3}$ 2 $\frac{7}{3}$

b. $x > -2$

−3 −2 −1 0 1 2 3 4 5 6 7 8

The heavy arrow indicates that all numbers to the right of −2 are included in the graph. The circle at −2 indicates that −2 is not included in the graph.

c. $b \leq 4.5$

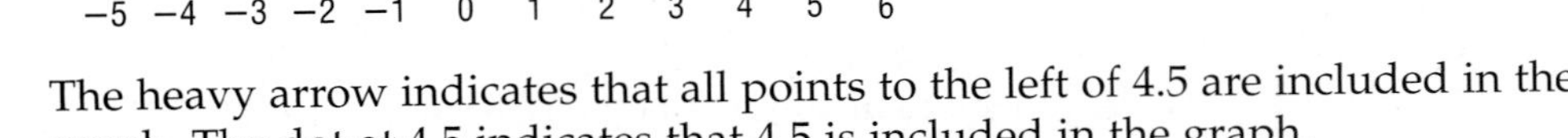

−5 −4 −3 −2 −1 0 1 2 3 4 5 6

The heavy arrow indicates that all points to the left of 4.5 are included in the graph. The dot at 4.5 indicates that 4.5 is included in the graph.

d. $h \geq -3\frac{2}{5}$

−5 −4 −3 −2 −1 0 1 2 3 4 5

The heavy arrow indicates that all points to the right of $-3\frac{2}{5}$ are included in the graph. The dot at $-3\frac{2}{5}$ indicates that $-3\frac{2}{5}$ is included in the graph.

StudyTip

Finite and Infinite The set of numbers in Example 2a is *finite*, that is, the elements can be counted. However, the set of numbers in Example 2b is *infinite*. Its graph includes all values in the range including integers like 3 and −1, as well as rational numbers like $\frac{3}{8}$ and $-\frac{12}{13}$ and irrational numbers like $\sqrt{10}$ and π.

Any repeating decimal can be written as a fraction.

EXAMPLE 3 Write Repeating Decimals as Fractions

Write $0.\overline{7}$ as a fraction in simplest form.

Step 1

$N = 0.777\ldots$	**Let *N* represent the repeating decimal.**
$10N = 10(0.777\ldots)$	**Since only one digit repeats, multiply each side by 10.**
$10N = 7.777\ldots$	**Simplify.**

Step 2 Subtract N from $10N$ to eliminate the part of the number that repeats.

$$10N = 7.777\ldots$$
$$\underline{-(N = 0.777\ldots)}$$

$9N = 7$	**Subtract.**
$\frac{9N}{9} = \frac{7}{9}$	**Divide each side by 9.**
$N = \frac{7}{9}$	**Simplify.**

Perfect squares can also be used to simplify square roots of rational numbers.

Key Concept — Perfect Square

For Your FOLDABLE

Words Rational numbers with square roots that are rational numbers.

Examples 25 is a perfect square since $\sqrt{25} = 5$.

144 is a perfect square since $\sqrt{144} = 12$.

EXAMPLE 4 Square Roots

Simplify each square root.

a. $-\sqrt{\frac{49}{256}}$

$-\sqrt{\frac{49}{256}} = -\sqrt{\left(\frac{7}{16}\right)^2}$ $\quad$ **$7^2 = 49$ and $16^2 = 256$**

$= -\frac{7}{16}$ $\quad$ **Simplify.**

b. $\sqrt{\frac{4}{121}}$

$\sqrt{\frac{4}{121}} = \sqrt{\left(\frac{2}{11}\right)^2}$ $\quad$ **$2^2 = 4$ and $11^2 = 121$**

$= \frac{2}{11}$ $\quad$ **Simplify.**

You can estimate square roots of numbers that are not perfect squares.

EXAMPLE 5 Estimate Square Roots

Estimate each square root to the nearest whole number.

a. $\sqrt{15}$

Find the two perfect squares closest to 15. List some perfect squares.

1, 4, 9, 16, 25, 36, …

$9 < 15 < 16$ $\quad$ **Write an inequality.**

$\sqrt{9} < \sqrt{15} < \sqrt{16}$ $\quad$ **Take the square root of each number.**

$3 < \sqrt{15} < 4$ $\quad$ **Simplify.**

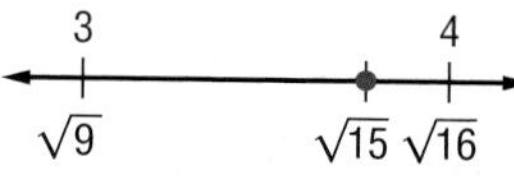

Since 15 is closer to 16 than 9, the best whole-number estimate for $\sqrt{15}$ is 4.

b. $\sqrt{130}$

Find the two perfect squares closest to 130. List some perfect squares.

81, 100, 121, 144, ...

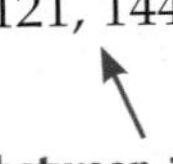

130 is between 121 and 144.

$121 < 130 < 144$	**Write an inequality.**
$\sqrt{121} < \sqrt{130} < \sqrt{144}$	**Take the square root of each number.**
$11 < \sqrt{130} < 12$	**Simplify.**

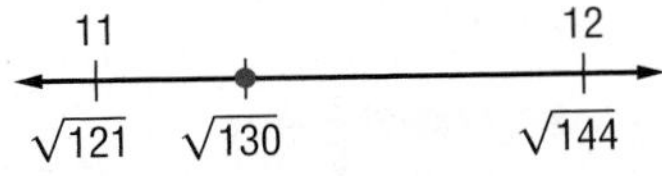

> **StudyTip**
>
> **Draw a Diagram** Graphing points on a number line can help you analyze your estimate for accuracy.

Since 130 is closer to 121 than 144, the best whole-number estimate for $\sqrt{130}$ is 11.

CHECK $\sqrt{130} \approx 11.4018...$ **Use a calculator.**

Rounded to the nearest whole number, $\sqrt{130}$ is 11. So our estimate is valid.

Exercises

Name the set or sets of numbers to which each real number belongs.

1. $-\sqrt{64}$ **2.** $\frac{8}{3}$ **3.** $\sqrt{28}$ **4.** $\frac{56}{7}$

5. $-\sqrt{22}$ **6.** $\frac{36}{6}$ **7.** $-\frac{5}{12}$ **8.** $\frac{18}{3}$

9. $\sqrt{10.24}$ **10.** $\frac{-54}{19}$ **11.** $\sqrt{\frac{82}{20}}$ **12.** $-\frac{72}{8}$

Graph each set of numbers.

13. $\{-4, -2, 1, 5, 7\}$ **14.** $x < -3.5$ **15.** $x \geq -7$

16. $\{-4, -2, -1, 1, 3\}$ **17.** $\{..., -2, 0, 2, 4, 6\}$ **18.** $x > -12$

Write each repeating decimal as a fraction in simplest form.

19. $0.\overline{5}$ **20.** $0.\overline{4}$

21. $0.\overline{13}$ **22.** $0.\overline{21}$

Simplify each square root.

23. $-\sqrt{25}$ **24.** $\sqrt{1.44}$ **25.** $\pm\sqrt{\frac{16}{49}}$

26. $\sqrt{361}$ **27.** $\sqrt{49}$ **28.** $\pm\sqrt{0.64}$

29. $-\sqrt{6.25}$ **30.** $\sqrt{\frac{169}{196}}$ **31.** $\sqrt{\frac{25}{324}}$

Estimate each square root to the nearest whole number.

32. $\sqrt{31}$ **33.** $\sqrt{24}$ **34.** $\sqrt{112}$ **35.** $\sqrt{152}$

0-3 Operations with Integers

Objective

To add, subtract, multiply, and divide integers.

New Vocabulary

absolute value
opposites
additive inverses

An integer is any number from the set {…, −3, −2, −1, 0, 1, 2, 3, …}. You can use a number line to add integers.

EXAMPLE 1

Use a number line to find −3 + (−4).

Step 1 Draw an arrow from 0 to −3.

Step 2 Draw a second arrow 4 units to the left to represent adding −4.

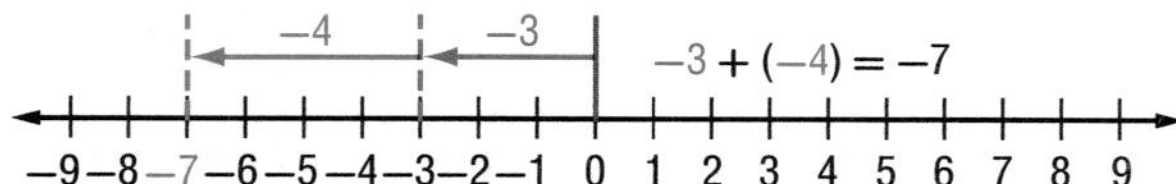

The second arrow ends at −7. So, −3 + (−4) = −7.

You can also use absolute value to add integers. The **absolute value** of a number is its distance from 0 on the number line.

Same Signs (+ + or − −)		Different Signs (+ − or − +)	
3 + 5 = 8	3 and 5 are positive. Their sum is positive.	3 + (−5) = −2	−5 has the greater absolute value. Their sum is negative.
−3 + (−5) = −8	−3 and −5 are negative. Their sum is negative.	−3 + 5 = 2	5 has the greater absolute value. Their sum is positive.

EXAMPLE 2

Find −11 + (−7).

$-11 + (-7) = -(|-11| + |-7|)$ **Add the absolute values. Both numbers are negative, so the sum is negative.**

$= -(11 + 7)$ **Absolute values of nonzero numbers are always positive.**

$= -18$ **Simplify.**

Every positive integer can be paired with a negative integer. These pairs are called **opposites**. A number and its opposite are **additive inverses**. Additive inverses can be used when you subtract integers.

EXAMPLE 3

Find 18 − 23.

$18 - 23 = 18 + (-23)$ **To subtract 23, add its inverse.**

$= -(|-23| - |18|)$ **Subtract the absolute values. Because |−23| is greater than |18|, the result is negative.**

$= -(23 - 18)$ **Absolute values of nonzero numbers are always positive.**

$= -5$ **Simplify.**

StudyTip

Products and Quotients The product or quotient of two numbers having the *same sign* is positive. The product or quotient of two numbers having *different signs* is negative.

Same Signs ($+ +$ or $- -$)		Different Signs ($+ -$ or $- +$)	
$3(5) = 15$	3 and 5 are positive. Their product is positive.	$3(-5) = -15$	3 and −5 have different signs. Their product is negative.
$-3(-5) = 15$	−3 and −5 are negative. Their product is positive.	$-3(5) = -15$	−3 and 5 have different signs. Their product is negative.

EXAMPLE 4

Find each product or quotient.

a. $\mathbf{4(-5)}$
$4(-5) = -20$ different signs ⟶ negative product

b. $\mathbf{-51 \div (-3)}$
$-51 \div (-3) = 17$ same sign ⟶ positive quotient

c. $\mathbf{-12(-14)}$
$-12(-14) = 168$ same sign ⟶ positive product

d. $\mathbf{-63 \div 7}$
$-63 \div 7 = -9$ different signs ⟶ negative quotient

Exercises

Find each sum or difference.

1. $-8 + 13$
2. $11 + (-19)$
3. $-19 - 8$
4. $-77 + (-46)$
5. $12 - 34$
6. $41 + (-56)$
7. $50 - 82$
8. $-47 - 13$
9. $-80 + 102$

Find each product or quotient.

10. $5(18)$
11. $60 \div 12$
12. $-12(15)$
13. $-64 \div (-8)$
14. $8(-22)$
15. $54 \div (-6)$
16. $30(14)$
17. $-23(5)$
18. $-200 \div 2$

19. WEATHER The outside temperature was −4°F in the morning and 13°F in the afternoon. By how much did the temperature increase?

20. DOLPHINS A dolphin swimming 24 feet below the ocean's surface dives 18 feet straight down. How many feet below the ocean's surface is the dolphin now?

21. MOVIES A movie theater gave out 50 coupons for $3 off each movie. What is the total amount of discounts provided by the theater?

22. WAGES Emilio earns $11 per hour. He works 14 hours a week. His employer withholds $32 from each paycheck for taxes. If he is paid weekly, what is the amount of his paycheck?

23. FINANCIAL LITERACY Talia is working on a monthly budget. Her monthly income is $500. She has allocated $200 for savings, $100 for vehicle expenses, and $75 for clothing. How much is available to spend on entertainment?

0-4

Adding and Subtracting Rational Numbers

Objective

Add and subtract rational numbers.

You can use different methods to compare rational numbers. One way is to compare two fractions with common denominators. Another way is to compare decimals.

EXAMPLE 1

Replace ● with <, >, or = to make $\frac{2}{3}$ ● $\frac{5}{6}$ a true sentence.

Method 1 Write the fractions with the same denominator.

The least common denominator of $\frac{2}{3}$ and $\frac{5}{6}$ is 6.

$\frac{2}{3} = \frac{4}{6}$

$\frac{5}{6} = \frac{5}{6}$

Since $\frac{4}{6} < \frac{5}{6}$, $\frac{2}{3} < \frac{5}{6}$.

Method 2 Write as decimals.

Write $\frac{2}{3}$ and $\frac{5}{6}$ as decimals. You may want to use a calculator.

2 [÷] 3 [ENTER] .6666666667

so, $\frac{2}{3} = 0.\overline{6}$

5 [÷] 6 [ENTER] .8333333333

so, $\frac{5}{6} = 0.8\overline{3}$

Since $0.\overline{6} < 0.8\overline{3}$, $\frac{2}{3} < \frac{5}{6}$.

You can order rational numbers by writing all of the fractions as decimals.

EXAMPLE 2

Order $5\frac{2}{9}$, $5\frac{3}{8}$, 4.9, and $-5\frac{3}{5}$ from least to greatest.

$5\frac{2}{9} = 5.\overline{2}$ $\quad$ $5\frac{3}{8} = 5.375$

$4.9 = 4.9$ $\quad$ $-5\frac{3}{5} = -5.6$

$-5.6 < 4.9 < 5.\overline{2} < 5.375$. So, from least to greatest, the numbers are $-5\frac{3}{5}$, 4.9, $5\frac{2}{9}$, and $5\frac{3}{8}$.

To add or subtract fractions with the same denominator, add or subtract the numerators and write the sum or difference over the denominator.

EXAMPLE 3

Find each sum or difference. Write in simplest form.

a. $\frac{3}{5} + \frac{1}{5}$

$\frac{3}{5} + \frac{1}{5} = \frac{3+1}{5}$ **The denominators are the same. Add the numerators.**

$= \frac{4}{5}$ **Simplify.**

b. $\frac{7}{16} - \frac{1}{16}$

$\frac{7}{16} - \frac{1}{16} = \frac{7-1}{16}$ **The denominators are the same. Subtract the numerators.**

$= \frac{6}{16}$ **Simplify.**

$= \frac{3}{8}$ **Rename the fraction.**

c. $\frac{4}{9} - \frac{7}{9}$

$\frac{4}{9} - \frac{7}{9} = \frac{4-7}{9}$ **The denominators are the same. Subtract the numerators.**

$= -\frac{3}{9}$ **Simplify.**

$= -\frac{1}{3}$ **Rename the fraction.**

StudyTip

Mental Math If the denominators of the fractions are the same, you can use mental math to determine the sum or difference.

To add or subtract fractions with unlike denominators, first find the least common denominator (LCD). Rename each fraction with the LCD, and then add or subtract. Simplify if possible.

EXAMPLE 4

Find each sum or difference. Write in simplest form.

a. $\frac{1}{2} + \frac{2}{3}$

$\frac{1}{2} + \frac{2}{3} = \frac{3}{6} + \frac{4}{6}$ **The LCD for 2 and 3 is 6. Rename $\frac{1}{2}$ as $\frac{3}{6}$ and $\frac{2}{3}$ as $\frac{4}{6}$.**

$= \frac{3+4}{6}$ **Add the numerators.**

$= \frac{7}{6}$ or $1\frac{1}{6}$ **Simplify.**

b. $\frac{3}{8} - \frac{1}{3}$

$\frac{3}{8} - \frac{1}{3} = \frac{9}{24} - \frac{8}{24}$ **The LCD for 8 and 3 is 24. Rename $\frac{3}{8}$ as $\frac{9}{24}$ and $\frac{1}{3}$ as $\frac{8}{24}$.**

$= \frac{9-8}{24}$ **Subtract the numerators.**

$= \frac{1}{24}$ **Simplify.**

c. $\frac{2}{5} - \frac{3}{4}$

$\frac{2}{5} - \frac{3}{4} = \frac{8}{20} - \frac{15}{20}$ **The LCD for 5 and 4 is 20. Rename $\frac{2}{5}$ as $\frac{8}{20}$ and $\frac{3}{4}$ as $\frac{15}{20}$.**

$= \frac{8-15}{20}$ **Subtract the numerators.**

$= -\frac{7}{20}$ **Simplify.**

StudyTip

Number Line To use a number line, put your pencil at the first number. Then move left to find the difference. To find the sum, move your pencil to the right.

You can use a number line to add rational numbers.

EXAMPLE 5

Use a number line to find 2.5 + (−3.5).

Step 1 Draw an arrow from 0 to 2.5.

Step 2 Draw a second arrow 3.5 units to the left.

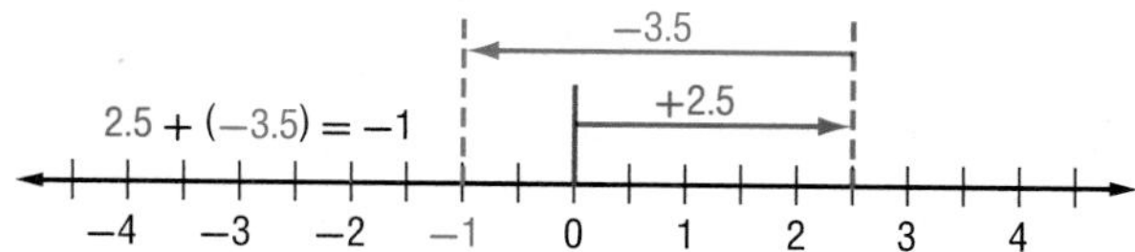

The second arrow ends at −1.

So, 2.5 + (−3.5) = −1.

You can also use absolute value to add rational numbers.

Same Signs (+ + or − −)		Different Signs (+ − or − +)	
3.1 + 2.5 = 5.6	3.1 and 2.5 are positive, so the sum is positive.	3.1 + (−2.5) = 0.6	3.1 has the greater absolute value, so the sum is positive.
−3.1 + (−2.5) = −5.6	−3.1 and −2.5 are negative, so the sum is negative.	−3.1 + 2.5 = −0.6	−3.1 has the greater absolute value, so the sum is negative.

EXAMPLE 6

Find each sum.

a. −13.12 + (−8.6)

$-13.12 + (-8.6) = -(|-13.12| + |-8.6|)$ **Both numbers are negative, so the sum is negative.**

$= -(13.12 + 8.6)$ **Absolute values of nonzero numbers are always positive.**

$= -21.72$ **Simplify.**

b. $\frac{7}{16} + \left(-\frac{3}{8}\right)$

$\frac{7}{16} + \left(-\frac{3}{8}\right) = \frac{7}{16} + \left(-\frac{6}{16}\right)$ **The LCD is 16. Replace $-\frac{3}{8}$ with $-\frac{6}{16}$.**

$= \left(\left|\frac{7}{16}\right| - \left|-\frac{6}{16}\right|\right)$ **Subtract the absolute values. Because $\left|\frac{7}{16}\right|$ is greater than $\left|-\frac{6}{16}\right|$, the result is positive.**

$= \frac{7}{16} - \frac{6}{16}$ **Absolute values of nonzero numbers are always positive.**

$= \frac{1}{16}$ **Simplify.**

To subtract a negative rational number, add its inverse.

EXAMPLE 7

Find $-32.25 - (-42.5)$.

$-32.25 - (-42.5) = -32.25 + 42.5$	**To subtract −42.5, add its inverse.**
$= \|42.5\| - \|-32.25\|$	**Subtract the absolute values. Because |42.5| is greater than |−32.25|, the result is positive.**
$= 42.5 - 32.25$	**Absolute values of nonzero numbers are always positive.**
$= 10.25$	**Simplify.**

Exercises

Replace each ● with <, >, or = to make a true sentence.

1. $-\frac{5}{8} \bullet \frac{3}{8}$ **2.** $\frac{4}{5} \bullet 0.71$ **3.** $\frac{5}{6} \bullet 0.875$

4. $1.2 \bullet 1\frac{2}{9}$ **5.** $\frac{8}{15} \bullet 0.5\overline{3}$ **6.** $-\frac{7}{11} \bullet -\frac{2}{3}$

Order each set of rational numbers from least to greatest.

7. $3.8, 3.06, 3\frac{1}{6}, 3\frac{3}{4}$ **8.** $2\frac{1}{4}, 1\frac{7}{8}, 1.75, 2.4$

9. $0.11, -\frac{1}{9}, -0.5, \frac{1}{10}$ **10.** $-4\frac{3}{5}, -3\frac{2}{5}, -4.65, -4.09$

Find each sum or difference. Write in simplest form.

11. $\frac{2}{5} + \frac{1}{5}$ **12.** $\frac{3}{9} + \frac{4}{9}$ **13.** $\frac{5}{16} - \frac{4}{16}$

14. $\frac{6}{7} - \frac{3}{7}$ **15.** $\frac{2}{3} + \frac{1}{3}$ **16.** $\frac{5}{8} + \frac{7}{8}$

17. $\frac{4}{3} + \frac{4}{3}$ **18.** $\frac{7}{15} - \frac{2}{15}$ **19.** $\frac{1}{3} - \frac{2}{9}$

20. $\frac{1}{2} + \frac{1}{4}$ **21.** $\frac{1}{2} - \frac{1}{3}$ **22.** $\frac{3}{7} + \frac{5}{14}$

23. $\frac{7}{10} - \frac{2}{15}$ **24.** $\frac{3}{8} + \frac{1}{6}$ **25.** $\frac{13}{20} - \frac{2}{5}$

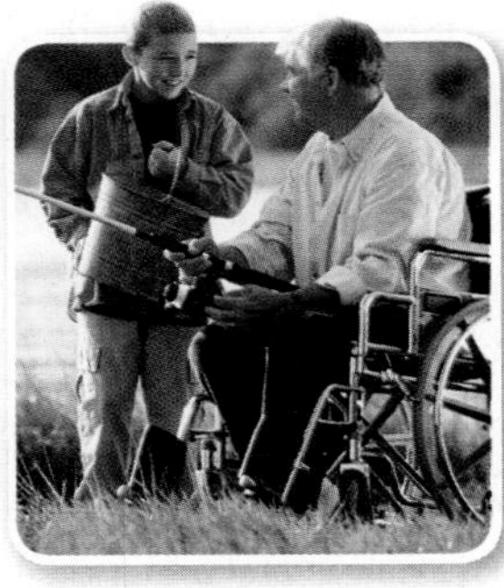

Real-World Link

About 97% of the water is saltwater from oceans. Of the freshwater, only 1% is on the surface from lakes, rivers, and swamps.

Source: U.S. Geological Survey

Find each sum or difference. Write in simplest form if necessary.

26. $-1.6 + (-3.8)$ **27.** $-32.4 + (-4.5)$ **28.** $-38.9 + 24.2$

29. $-9.16 - 10.17$ **30.** $26.37 + (-61.1)$ **31.** $72.5 - (-81.3)$

32. $43.2 + (-27.9)$ **33.** $79.3 - (-14)$ **34.** $1.34 - (-0.458)$

35. $-\frac{1}{6} - \frac{2}{3}$ **36.** $\frac{1}{2} - \frac{4}{5}$ **37.** $-\frac{2}{5} + \frac{17}{20}$

38. $-\frac{4}{5} + \left(-\frac{1}{3}\right)$ **39.** $-\frac{1}{12} - \left(-\frac{3}{4}\right)$ **40.** $-\frac{7}{8} - \left(-\frac{3}{16}\right)$

41. GEOGRAPHY About $\frac{7}{10}$ of the surface of Earth is covered by water. The rest of the surface is covered by land. How much of Earth's surface is covered by land?

0-5 Multiplying and Dividing Rational Numbers

Objective

Multiply and divide rational numbers.

New Vocabulary

multiplicative inverses
reciprocals

The product or quotient of two rational numbers having the *same sign* is positive. The product or quotient of two rational numbers having *different signs* is negative.

EXAMPLE 1

Find each product or quotient.

a. $7.2(-0.2)$

different signs ⟶ negative product

$7.2(-0.2) = -1.44$

b. $-23.94 \div (-10.5)$

same sign ⟶ positive quotient

$-23.94 \div (-10.5) = 2.28$

To multiply fractions, multiply the numerators and multiply the denominators. If the numerators and denominators have common factors, you can simplify before you multiply by canceling.

EXAMPLE 2

Find each product.

a. $\frac{2}{5} \cdot \frac{1}{3}$

$\frac{2}{5} \cdot \frac{1}{3} = \frac{2 \cdot 1}{5 \cdot 3}$ — Multiply the numerators. Multiply the denominators.

$= \frac{2}{15}$ — Simplify.

b. $\frac{3}{5} \cdot 1\frac{1}{2}$

$\frac{3}{5} \cdot 1\frac{1}{2} = \frac{3}{5} \cdot \frac{3}{2}$ — Write $1\frac{1}{2}$ as an improper fraction.

$= \frac{3 \cdot 3}{5 \cdot 2}$ — Multiply the numerators. Multiply the denominators.

$= \frac{9}{10}$ — Simplify.

c. $\frac{1}{4} \cdot \frac{2}{9}$

$\frac{1}{4} \cdot \frac{2}{9} = \frac{1}{\cancel{4}_2} \cdot \frac{\cancel{2}^1}{9}$ — Divide by the GCF, 2.

$= \frac{1 \cdot 1}{2 \cdot 9}$ or $\frac{1}{18}$ — Multiply the numerators. Multiply the denominators and simplify.

EXAMPLE 3

Find $-\left(\frac{3}{4}\right)\left(\frac{3}{8}\right)$.

$\left(-\frac{3}{4}\right)\left(\frac{3}{8}\right) = -\left(\frac{3}{4} \cdot \frac{3}{8}\right)$ — different signs ⟶ negative product

$= -\left(\frac{3 \cdot 3}{4 \cdot 8}\right)$ or $\frac{9}{32}$ — Multiply the numerators. Multiply the denominators and simplify.

Two numbers whose product is 1 are called **multiplicative inverses** or **reciprocals**.

EXAMPLE 4

Name the reciprocal of each number.

a. $\frac{3}{8}$

$\frac{3}{8} \cdot \frac{8}{3} = 1$ — **The product is 1.**

The reciprocal of $\frac{3}{8}$ is $\frac{8}{3}$.

b. $2\frac{4}{5}$

$2\frac{4}{5} = \frac{14}{5}$ — **Write $2\frac{4}{5}$ as $\frac{14}{5}$.**

$\frac{14}{5} \cdot \frac{5}{14} = 1$ — **The product is 1.**

The reciprocal of $2\frac{4}{5}$ is $\frac{5}{14}$.

To divide one fraction by another fraction, multiply the dividend by the reciprocal of the divisor.

EXAMPLE 5

Find each quotient.

a. $\frac{1}{3} \div \frac{1}{2}$

$\frac{1}{3} \div \frac{1}{2} = \frac{1}{3} \cdot \frac{2}{1}$ — **Multiply $\frac{1}{3}$ by $\frac{2}{1}$, the reciprocal of $\frac{1}{2}$.**

$= \frac{2}{3}$ — **Simplify.**

b. $\frac{3}{8} \div \frac{2}{3}$

$\frac{3}{8} \div \frac{2}{3} = \frac{3}{8} \cdot \frac{3}{2}$ — **Multiply $\frac{3}{8}$ by $\frac{3}{2}$, the reciprocal of $\frac{2}{3}$.**

$= \frac{9}{16}$ — **Simplify.**

c. $\frac{3}{4} \div 2\frac{1}{2}$

$\frac{3}{4} \div 2\frac{1}{2} = \frac{3}{4} \div \frac{5}{2}$ — **Write $2\frac{1}{2}$ as a mixed number.**

$= \frac{3}{4} \cdot \frac{2}{5}$ — **Multiply $\frac{3}{4}$ by $\frac{2}{5}$, the reciprocal of $2\frac{1}{2}$.**

$= \frac{6}{20}$ or $\frac{3}{10}$ — **Simplify.**

d. $-\frac{1}{5} \div \left(-\frac{3}{10}\right)$

$-\frac{1}{5} \div \left(-\frac{3}{10}\right) = -\frac{1}{5} \cdot \left(-\frac{10}{3}\right)$ — **Multiply $-\frac{1}{5}$ by $-\frac{10}{3}$, the reciprocal of $-\frac{3}{10}$.**

$= \frac{10}{15}$ or $\frac{2}{3}$ — **Same sign ⟶ positive quotient; simplify.**

StudyTip

Use Estimation You can justify your answer by using estimation. $\frac{3}{8}$ is close to $\frac{1}{2}$ and $\frac{2}{3}$ is close to 1. So, the quotient is close to $\frac{1}{2}$ divided by 1 or $\frac{1}{2}$.

StudyTip

Negative Fractions A negative fraction can be written as $-\frac{1}{2}$ or $\frac{-1}{2}$.

Exercises

Find each product or quotient. Round to the nearest hundredth if necessary.

1. $6.5(0.13)$ **2.** $-5.8(2.3)$ **3.** $42.3 \div (-6)$

4. $-14.1(-2.9)$ **5.** $-78 \div (-1.3)$ **6.** $108 \div (-0.9)$

7. $0.75(-6.4)$ **8.** $-23.94 \div 10.5$ **9.** $-32.4 \div 21.3$

Find each product. Simplify before multiplying if possible.

10. $\frac{3}{4} \cdot \frac{1}{5}$ **11.** $\frac{2}{5} \cdot \frac{3}{7}$ **12.** $-\frac{1}{3} \cdot \frac{2}{5}$

13. $-\frac{2}{3} \cdot \left(-\frac{1}{11}\right)$ **14.** $2\frac{1}{2} \cdot \left(-\frac{1}{4}\right)$ **15.** $3\frac{1}{2} \cdot 1\frac{1}{2}$

16. $\frac{2}{9} \cdot \frac{1}{2}$ **17.** $\frac{3}{2} \cdot \left(-\frac{1}{3}\right)$ **18.** $\frac{1}{3} \cdot \frac{6}{5}$

19. $-\frac{9}{4} \cdot \frac{1}{18}$ **20.** $\frac{11}{3} \cdot \frac{9}{44}$ **21.** $\left(-\frac{30}{11}\right) \cdot \left(-\frac{1}{3}\right)$

22. $-\frac{3}{5} \cdot \frac{5}{6}$ **23.** $\left(-\frac{1}{3}\right)\left(-7\frac{1}{2}\right)$ **24.** $\frac{2}{7} \cdot 4\frac{2}{3}$

Name the reciprocal of each number.

25. $\frac{6}{7}$ **26.** $\frac{1}{22}$

27. $-\frac{14}{23}$ **28.** $2\frac{3}{4}$

29. $-5\frac{1}{3}$ **30.** $3\frac{3}{4}$

Find each quotient.

31. $\frac{2}{3} \div \frac{1}{3}$ **32.** $\frac{16}{9} \div \frac{4}{9}$

33. $\frac{3}{2} \div \frac{1}{2}$ **34.** $\frac{3}{7} \div \left(-\frac{1}{5}\right)$

35. $-\frac{9}{10} \div 3$ **36.** $\frac{1}{2} \div \frac{3}{5}$

37. $2\frac{1}{4} \div \frac{1}{2}$ **38.** $-1\frac{1}{3} \div \frac{2}{3}$

39. $\frac{11}{12} \div 1\frac{2}{3}$ **40.** $4 \div \left(-\frac{2}{7}\right)$

41. $-\frac{1}{3} \div \left(-1\frac{1}{5}\right)$ **42.** $\frac{3}{25} \div \frac{2}{15}$

43. PIZZA A large pizza at Pizza Shack has 12 slices. If Bobby ate $\frac{1}{4}$ of the pizza, how many slices of pizza did he eat?

44. MUSIC Samantha practices the flute for $4\frac{1}{2}$ hours each week. How many hours does she practice in a month?

45. BAND How many band uniforms can be made with $131\frac{3}{4}$ yards of fabric if each uniform requires $3\frac{7}{8}$ yards?

46. CARPENTRY How many boards, each 2 feet 8 inches long, can be cut from a board 16 feet long if there is no waste?

47. SEWING How many 9-inch ribbons can be cut from $1\frac{1}{2}$ yards of ribbon?

0-6 The Percent Proportion

Objective

Use and apply the percent proportion.

New Vocabulary

percent

percent proportion

A **percent** is a ratio that compares a number to 100. To write a percent as a fraction, express the ratio as a fraction with a denominator of 100. Fractions should be expressed in simplest form.

EXAMPLE 1

Express each percent as a fraction or mixed number.

a. 79%

$79\% = \frac{79}{100}$ **Definition of percent**

b. 107%

$107\% = \frac{107}{100}$ **Definition of percent**

$= 1\frac{7}{100}$ **Simplify.**

c. 0.5%

$0.5\% = \frac{0.5}{100}$ **Definition of percent**

$= \frac{5}{1000}$ **Multiply the numerator and denominator by 10 to eliminate the decimal.**

$= \frac{1}{200}$ **Simplify.**

In the **percent proportion**, the ratio of a part of something to the whole (base) is equal to the percent written as a fraction.

part → $\frac{a}{b} = \frac{p}{100}$ ← percent
whole →

Example: 25% (percent) of 40 (whole) is 10 (part).

You can use the percent proportion to find the part.

EXAMPLE 2

40% of 30 is what number?

$\frac{a}{b} = \frac{p}{100}$ **The percent is 40, and the base is 30. Let *a* represent the part.**

$\frac{a}{30} = \frac{40}{100}$ **Replace *b* with 30 and *p* with 40.**

$100a = 30(40)$ **Find the cross products.**

$100a = 1200$ **Simplify.**

$\frac{100a}{100} = \frac{1200}{100}$ **Divide each side by 100.**

$a = 12$ **Simplify.**

The part is 12. So, 40% of 30 is 12.

You can also use the percent proportion to find the percent of the base.

EXAMPLE 3

SURVEYS Kelsey took a survey of students in her lunch period. 42 out of the 70 students Kelsey surveyed said their family had a pet. What percent of the students had pets?

$\frac{a}{b} = \frac{p}{100}$ **The part is 42, and the base is 70. Let *p* represent the percent.**

$\frac{42}{70} = \frac{p}{100}$ **Replace *a* with 42 and *b* with 70.**

$4200 = 70p$ **Find the cross products.**

$\frac{4200}{70} = \frac{70p}{70}$ **Divide each side by 70.**

$60 = p$ **Simplify.**

The percent is 60, so $\frac{60}{100}$ or 60% of the students had pets.

StudyTip

Percent Proportion In percent problems, the whole, or base usually follows the word *of*.

EXAMPLE 4

67.5 is 75% of what number?

$\frac{a}{b} = \frac{p}{100}$ **The percent is 75, and the part is 67.5. Let *b* represent the base.**

$\frac{67.5}{b} = \frac{75}{100}$ **Replace *a* with 67.5 and *p* with 75.**

$6750 = 75b$ **Find the cross products.**

$\frac{6750}{75} = \frac{75b}{75}$ **Divide each side by 75.**

$90 = b$ **Simplify.**

The base is 90, so 67.5 is 75% of 90.

Exercises

Express each percent as a fraction or mixed number in simplest form.

1. 5%
2. 60%
3. 11%
4. 120%
5. 78%
6. 2.5%
7. 0.6%
8. 0.4%
9. 1400%

Use the percent proportion to find each number.

10. 25 is what percent of 125?
11. 16 is what percent of 40?
12. 14 is 20% of what number?
13. 50% of what number is 80?
14. What number is 25% of 18?
15. Find 10% of 95.
16. What percent of 48 is 30?
17. What number is 150% of 32?
18. 5% of what number is 3.5?
19. 1 is what percent of 400?
20. Find 0.5% of 250.
21. 49 is 200% of what number?
22. 15 is what percent of 12?
23. 36 is what percent of 24?

24. **BASKETBALL** Madeline usually makes 85% of her shots in basketball. If she attempts 20, how many will she likely make?

25. **TEST SCORES** Brian answered 36 items correctly on a 40-item test. What percent did he answer correctly?

26. **CARD GAMES** Juanita told her dad that she won 80% of the card games she played yesterday. If she won 4 games, how many games did she play?

StudyTip

Word Problems When a problem starts with the result and asks for something that happened earlier, work backward.

27. **SOLUTIONS** A glucose solution is prepared by dissolving 6 milliliters of glucose in 120 milliliters of pure solution. What is the percent of glucose in the resulting solution?

28. **DRIVER'S ED** Kara needs to get a 75% on her driving education test in order to get her license. If there are 35 questions on the test, how many does she need to answer correctly?

29. **HEALTH** The U.S. Food and Drug Administration require food manufacturers to label their products with a nutritional label. The label shows the information from a package of macaroni and cheese.

Nutrition Facts		
Serving Size 1 cup (228g)		
Servings per container 2		
Amount per serving		
Calories 250	Calories from Fat 110	
		%Daily value*
Total Fat 12g		18%
Saturated Fat 3g		15%
Cholesterol 30mg		10%
Sodium 470mg		20%
Total Carbohydrate 31g		10%
Dietary Fiber 0g		0%
Sugars 5g		
Protein 5g		
Vitamin A 4%	•	Vitamin C 2%
Calcium 20%	•	Iron 4%

a. The label states that a serving contains 3 grams of saturated fat, which is 15% of the daily value recommended for a 2000-Calorie diet. How many grams of saturated fat are recommended for a 2000-Calorie diet?

b. The 470 milligrams of sodium (salt) in the macaroni and cheese is 20% of the recommended daily value. What is the recommended daily value of sodium?

c. For a healthy diet, the National Research Council recommends that no more than 30 percent of the total Calories come from fat. What percent of the Calories in a serving of this macaroni and cheese come from fat?

30. **TEST SCORES** The table shows the number of points each student in Will's study group earned on a recent math test. There were 88 points possible on the test. Express all answers to the nearest tenth of a percent.

Name	Will	Penny	Cheng	Minowa	Rob
Score	72	68	81	87	75

a. Find Will's percent correct on the test.

b. Find Cheng's percent correct on the test.

c. Find Rob's percent correct on the test.

d. What was the highest percentage? The lowest?

31. **PET STORE** In a pet store, 15% of the animals are hamsters. If the store has 40 animals, how many of them are hamsters?

0-7 Perimeter

Objective

Find the perimeter of two-dimensional figures.

New Vocabulary

perimeter
circle
diameter
circumference
center
radius

Perimeter is the distance around a figure. Perimeter is measured in linear units.

Rectangle

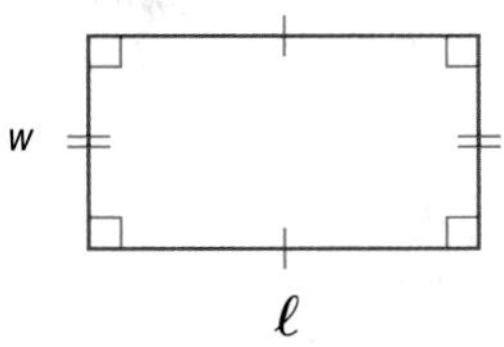

$P = 2(\ell + w)$ or
$P = 2\ell + 2w$

Parallelogram

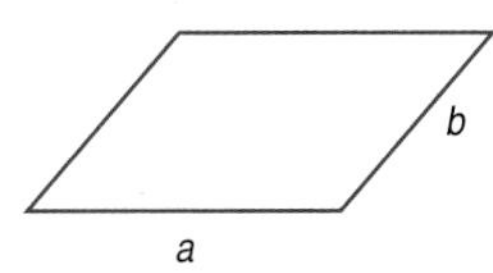

$P = 2(a + b)$ or
$P = 2a + 2b$

Square

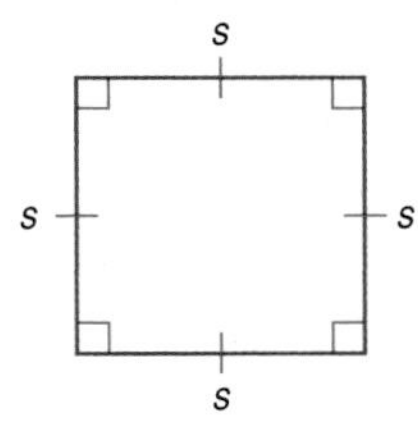

$P = 4s$

Triangle

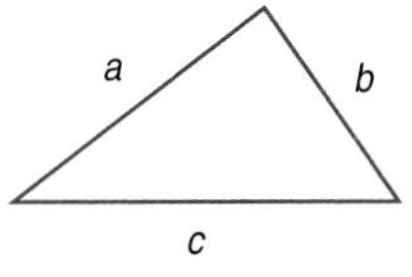

$P = a + b + c$

EXAMPLE 1

Find the perimeter of each figure.

a. a rectangle with a length of 5 inches and a width of 1 inch

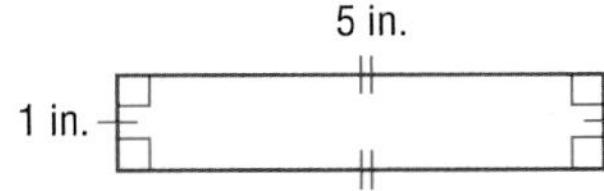

$P = 2(\ell + w)$	**Perimeter formula**
$= 2(5 + 1)$	**$\ell = 5$, $w = 1$**
$= 2(6)$	**Add.**
$= 12$	The perimeter is 12 inches.

b. a square with a side length of 7 centimeters

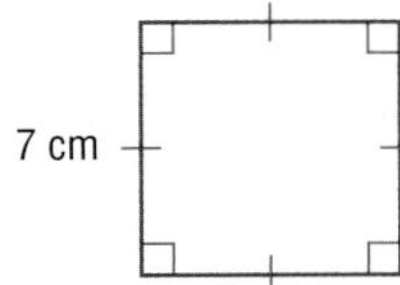

$P = 4s$	**Perimeter formula**
$= 4(7)$	**Replace s with 7.**
$= 28$	The perimeter is 28 centimeters.

EXAMPLE 2

Find the perimeter of each figure.

a.

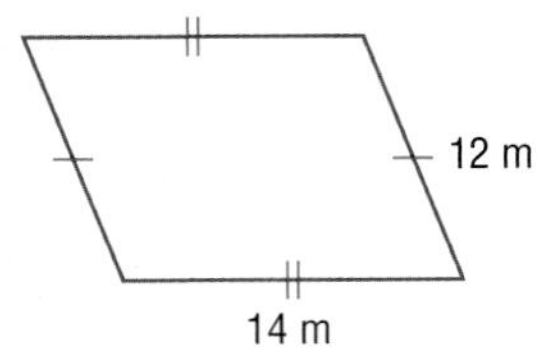

$P = 2(a + b)$ **Perimeter formula**

$= 2(14 + 12)$ **$a = 14$, $b = 12$**

$= 2(26)$ **Add.**

$= 52$ **Multiply.**

The perimeter of the parallelogram is 52 meters.

b.

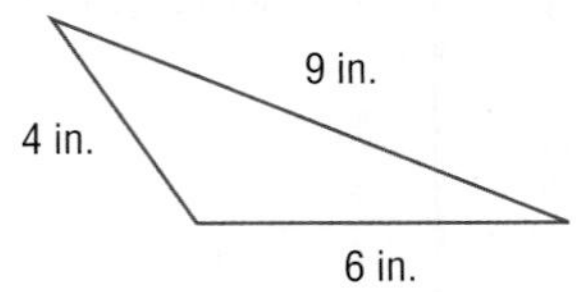

$P = a + b + c$ **Perimeter formula**

$= 4 + 6 + 9$ **$a = 4$, $b = 6$, $c = 9$**

$= 19$ **Add.**

The perimeter of the triangle is 19 inches.

StudyTip

Pi To perform a calculation that involves π, use a calculator.

A **circle** is the set of all points in a plane that are the same distance from a given point.

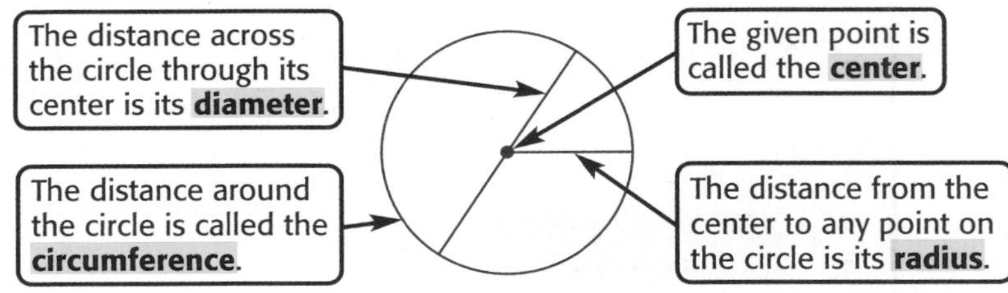

The formula for the circumference of a circle is $C = \pi d$ or $C = 2\pi r$.

EXAMPLE 3

Find each circumference to the nearest tenth.

a. The radius is 4 feet.

$C = 2\pi r$ **Circumference formula**

$= 2\pi(4)$ **Replace r with 4.**

$= 8\pi$ **Simplify.**

The exact circumference is 8π feet.

8 [π] [ENTER] 25.13274123

The circumference is about 25.1 feet.

b. The diameter is 15 centimeters.

$C = \pi d$ **Circumference formula**

$= \pi(15)$ **Replace d with 15.**

$= 15\pi$ **Simplify.**

≈ 47.1 **Use a calculator to evaluate 15π.**

The circumference is about 47.1 centimeters.

c. (circle with radius 3 m)

$C = 2\pi r$ **Circumference formula**

$= 2\pi(3)$ **Replace r with 3.**

$= 6\pi$ **Simplify.**

≈ 18.8 **Use a calculator to evaluate 6π.**

The circumference is about 18.8 meters.

Exercises

Find the perimeter of each figure.

StudyTip

Congruent Marks The hash marks on the figures indicate sides that have the same length.

1.

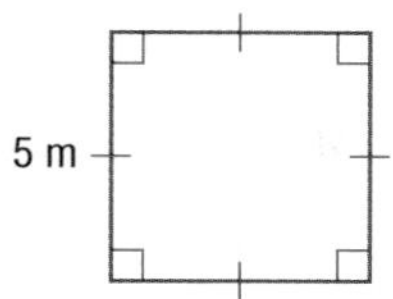

2.

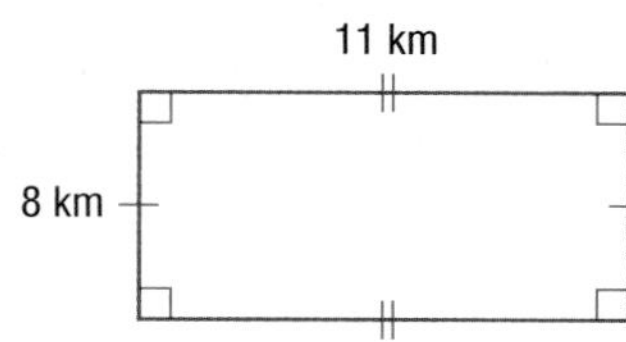

3.

4.

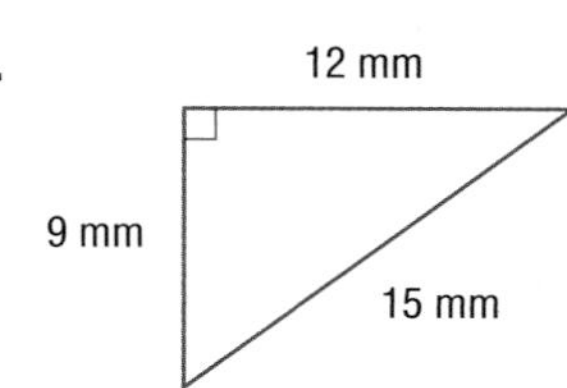

5. a square with side length 8 inches

6. a rectangle with length 9 centimeters and width 3 centimeters

7. a triangle with sides 4 feet, 13 feet, and 12 feet

8. a parallelogram with side lengths $6\frac{1}{4}$ inches and 5 inches

9. a quarter-circle with a radius of 7 inches

Find the circumference of each circle. Round to the nearest tenth.

10.

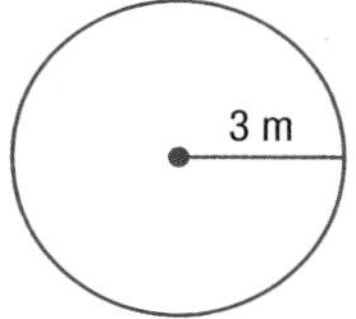

11.

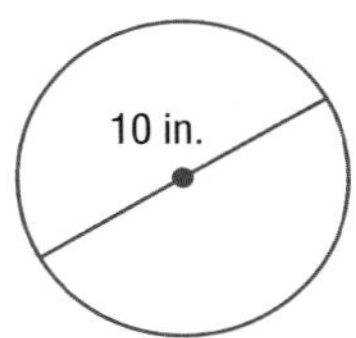

12.

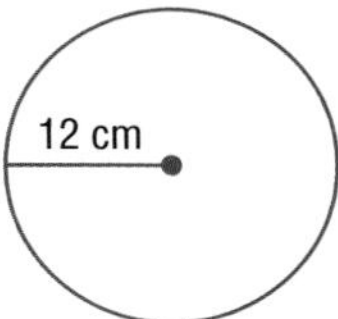

13. GARDENS A square garden has a side length of 5.8 meters. What is the perimeter of the garden?

14. ROOMS A rectangular room is $12\frac{1}{2}$ feet wide and 14 feet long. What is the perimeter of the room?

15. CYCLING The tire for a 10-speed bicycle has a diameter of 27 inches. Find the distance traveled in 10 rotations of the tire. Round to the nearest tenth.

16. GEOGRAPHY Earth's circumference is approximately 25,000 miles. If you could dig a tunnel to the center of the Earth, how long would the tunnel be? Round to the nearest tenth mile.

Find the perimeter of each figure. Round to the nearest tenth.

17.

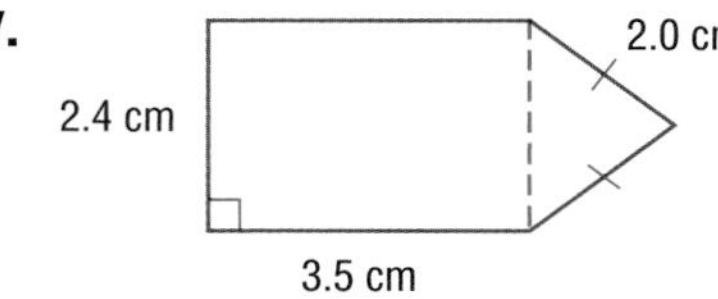

18.

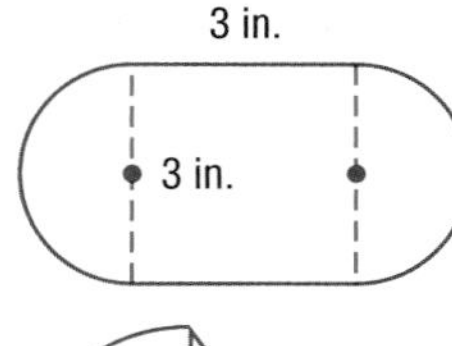

19.

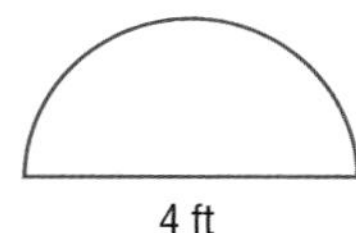

20.

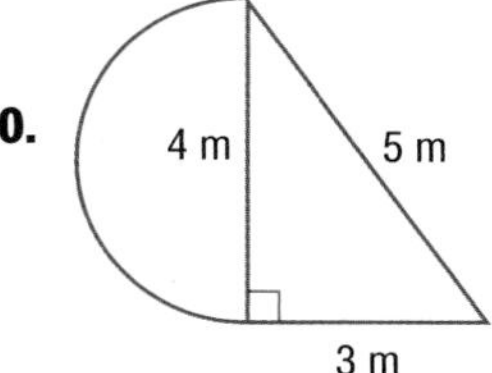

0-8

Area

Objective

Find the area of two-dimensional figures.

New Vocabulary

area

Area is the number of square units needed to cover a surface. Area is measured in square units.

Rectangle

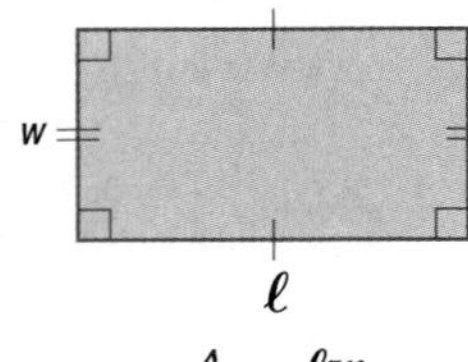

$A = \ell w$

Parallelogram

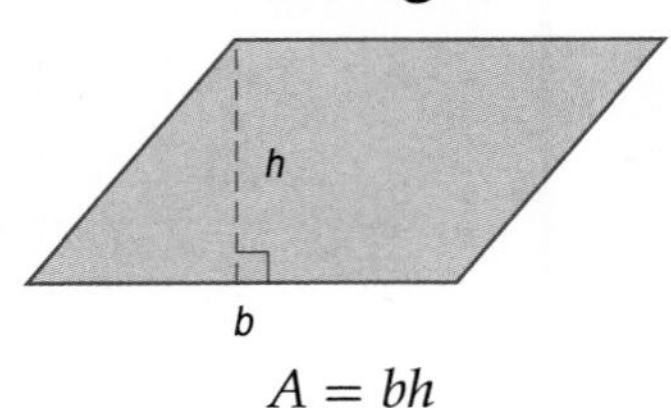

$A = bh$

Square

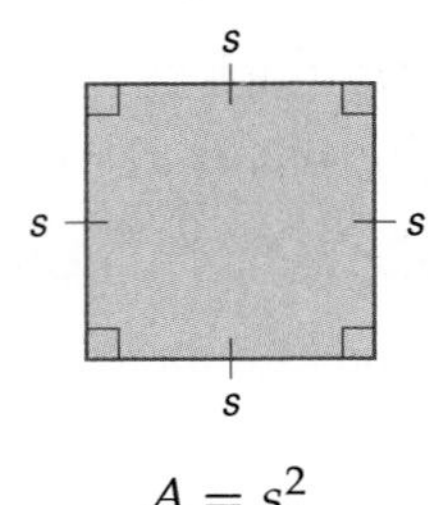

$A = s^2$

Triangle

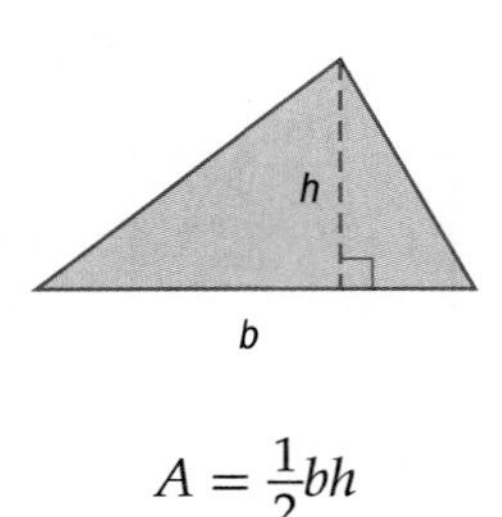

$A = \frac{1}{2}bh$

EXAMPLE 1

Find the area of each figure.

a. a rectangle that has a length of 7 yards and a width of 1 yard

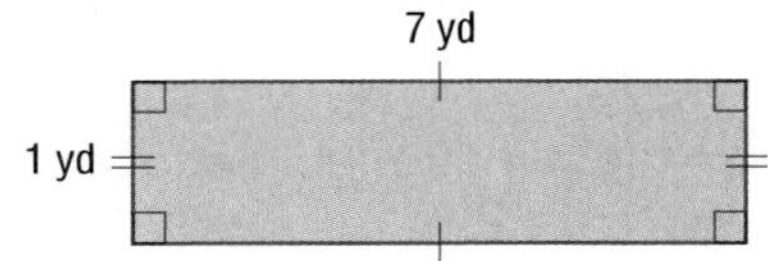

$A = \ell w$	**Area formula**
$= 7(1)$	**$\ell = 7, w = 1$**
$= 7$	The area of the rectangle is 7 square yards.

b. a square that has a side length of 2 meters

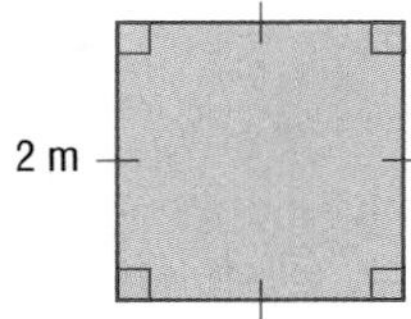

$A = s^2$	**Area formula**
$= 2^2$	**$s = 2$**
$= 4$	The area is 4 square meters.

EXAMPLE 2

Find the area of each figure.

a. a parallelogram that has a base of 11 feet and a height of 9 feet

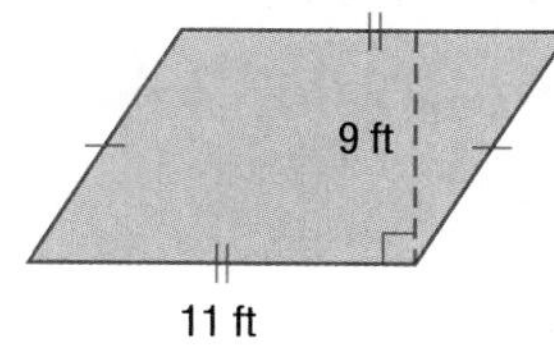

$A = bh$ **Area formula**

$= 11(9)$ **$b = 11, h = 9$**

$= 99$ **Multiply.**

The area is 99 square feet.

b. a triangle that has a base of 12 millimeters and a height of 5 millimeters

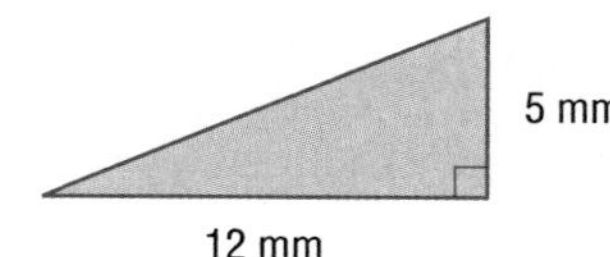

$A = \frac{1}{2}bh$ **Area formula**

$= \frac{1}{2}(12)(5)$ **$b = 12, h = 5$**

$= 30$ **Multiply.**

The area is 30 square millimeters.

The formula for the area of a circle is $A = \pi r^2$.

EXAMPLE 3

Find the area of each circle to the nearest tenth.

StudyTip

Mental Math You can use mental math to check your solutions. Square the radius and then multiply by 3.

a. The radius is 3 centimeters.

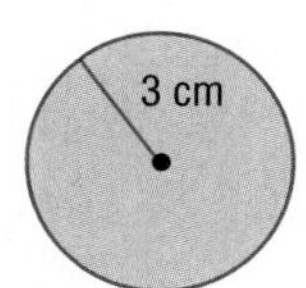

$A = \pi r^2$ **Area formula**

$= \pi(3)^2$ **Replace r with 3.**

$= 9\pi$ **Simplify.**

≈ 28.3 **Use a calculator to evaluate 9π.**

The area is about 28.3 square centimeters.

b. The diameter is 21 meters.

21 m

$A = \pi r^2$ **Area formula**

$= \pi(10.5)^2$ **Replace r with 10.5.**

$= 110.25\pi$ **Simplify.**

≈ 346.4 **Use a calculator to evaluate 110.25π.**

The area is about 346.4 square meters.

Exercises

Find the area of each figure.

1.

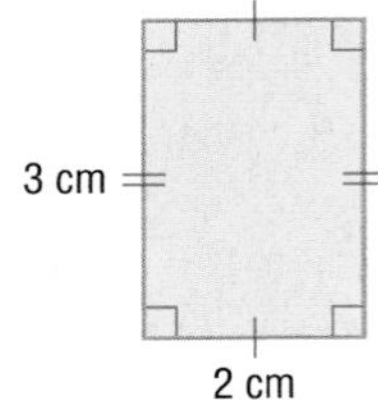

2.

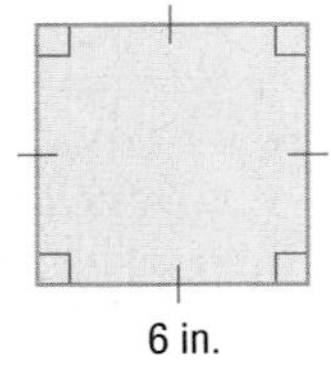

3.

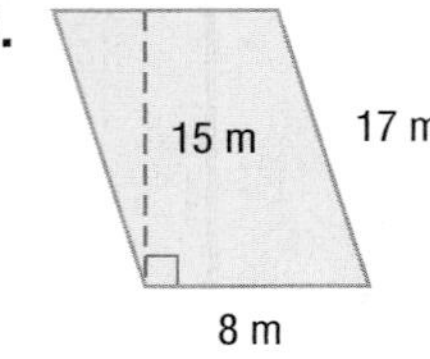

Find the area of each figure. Round to the nearest tenth if necessary.

4. a triangle with a base 12 millimeters and height 11 millimeters

5. a square with side length 9 feet

6. a rectangle with length 8 centimeters and width 2 centimeters

7. a triangle with a base 6 feet and height 3 feet

8. a quarter-circle with a diameter of 4 meters

9. a semi-circle with a radius of 3 inches

Find the area of each circle. Round to the nearest tenth.

10.

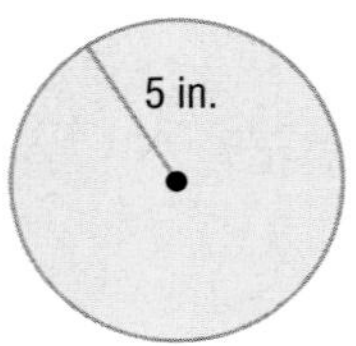

11.

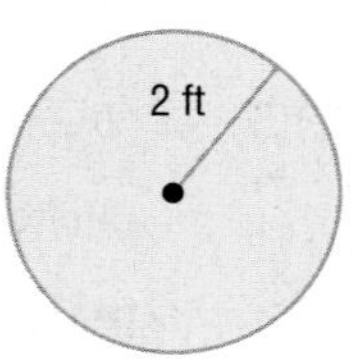

12.

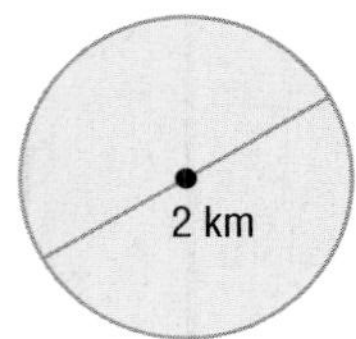

13. The radius is 4 centimeters.

14. The radius is 7.2 millimeters.

15. The diameter is 16 inches.

16. The diameter is 25 feet.

17. **RECREATION** The Granville Parks and Recreation Department uses an empty city lot for a community vegetable garden. Each participant is allotted a space of 18 feet by 90 feet for a garden. What is the area of each plot?

18. **CAMPING** The square floor of a tent has an area of 49 square feet. What is the side length of the tent?

19. **PUBLIC SAFETY** The sound emitted from the siren of a tornado warning system can be heard for a 2.5-mile radius. Find the area of the region that hears the siren. Round to the nearest tenth square mile.

20. **HISTORY** Stonehenge is an ancient monument in Wiltshire, England. The giant stones of Stonehenge are arranged in a circle 30 meters in diameter. Find the area of the circle. Round to the nearest tenth square meter.

Real-World Link

It is estimated that Stonehenge was built around 2300 B.C. The construction of the monument took place in three phases.

Source: Window on Britain

Find the area of each figure. Round to the nearest tenth.

21.

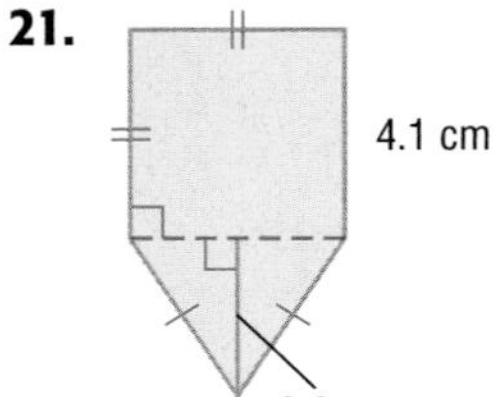

22.

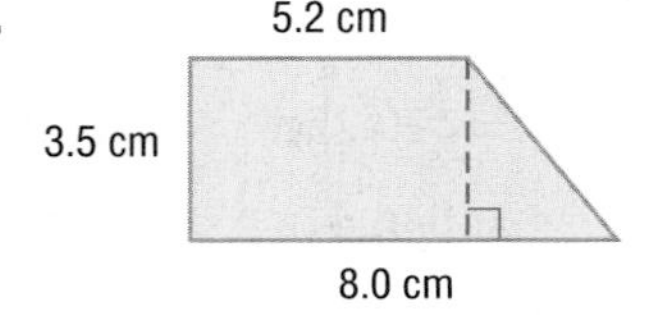

23.

0-9 Volume

Objective

Find the volume of rectangular prisms.

New Vocabulary

volume

Volume is the measure of space occupied by a solid. Volume is measured in cubic units.

To find the volume of a rectangular prism, multiply the length times the width times the height. The formula for the volume of a rectangular prism is shown below.

$w = 2$

$h = 3$

$\ell = 2$

$$V = \ell \cdot w \cdot h$$

The prism at the right has a volume of $2 \cdot 2 \cdot 3$ or 12 cubic units.

EXAMPLE 1

Find the volume of each rectangular prism.

a. The length is 8 centimeters, the width is 1 centimeter, and the height is 5 centimeters.

$V = \ell \cdot w \cdot h$	**Volume formula**
$= 8 \cdot 1 \cdot 5$	**Replace ℓ with 8, w with 1, and h with 5.**
$= 40$	**Simplify.**

The volume is 40 cubic centimeters.

b.

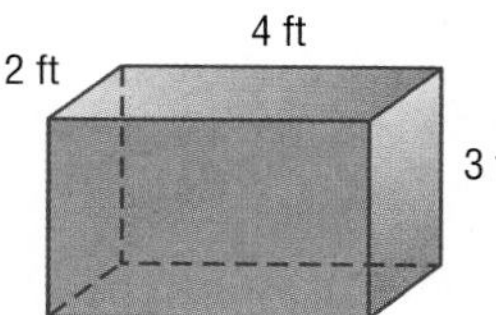

The prism has a length of 4 feet, width of 2 feet, and height of 3 feet.

$V = \ell \cdot w \cdot h$	**Volume formula**
$= 4 \cdot 2 \cdot 3$	**Replace ℓ with 4, w with 2, and h with 3.**
$= 24$	**Simplify.**

The volume is 24 cubic feet.

The volume of a solid is the product of the area of the base and the height of the solid. For a cylinder, the area of the base is πr^2. So the volume is $V = \pi r^2 h$.

EXAMPLE 2

Find the volume of the cylinder.

$V = \pi r^2 h$	**Volume of a cylinder**
$= \pi(3^2)6$	**$r = 3$, $h = 6$**
$= 54\pi$	**Simplify.**
≈ 169.6	**Use a calculator.**

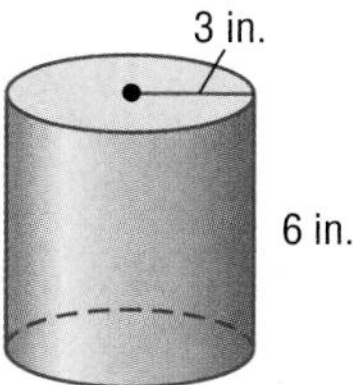

The volume is 169.6 cubic inches.

Exercises

Find the volume of each rectangular prism given the length, width, and height.

1. $\ell = 5$ cm, $w = 3$ cm, $h = 2$ cm
2. $\ell = 10$ m, $w = 10$ m, $h = 1$ m
3. $\ell = 6$ yd, $w = 2$ yd, $h = 4$ yd
4. $\ell = 2$ in., $w = 5$ in., $h = 12$ in.
5. $\ell = 13$ ft, $w = 9$ ft, $h = 12$ ft
6. $\ell = 7.8$ mm, $w = 0.6$ mm, $h = 8$ mm

Find the volume of each rectangular prism.

7.
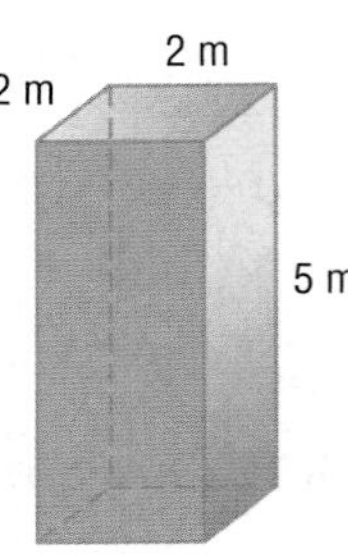

8.
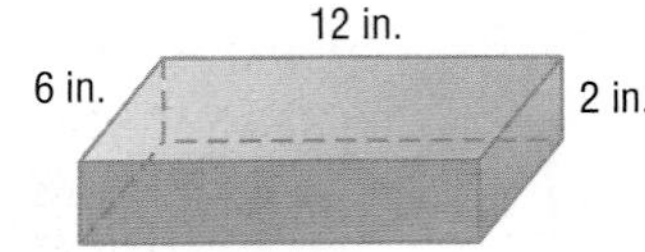

9. **GEOMETRY** A cube measures 3 meters on a side. What is its volume?

10. **AQUARIUMS** An aquarium is 8 feet long, 5 feet wide, and 5.5 feet deep. What is the volume of the tank?

> **StudyTip**
> **Draw a Diagram** Draw a diagram to organize the information given in the problem.

11. **COOKING** What is the volume of a microwave oven that is 18 inches wide by 10 inches long with a depth of $11\frac{1}{2}$ inches?

12. **BOXES** A cardboard box is 32 inches long, 22 inches wide, and 16 inches tall. What is the volume of the box?

13. **SWIMMING POOLS** A children's rectangular pool holds 480 cubic feet of water. What is the depth of the pool if its length is 30 feet and its width is 16 feet?

14. **BAKING** A rectangular cake pan has a volume of 234 cubic inches. If the length of the pan is 9 inches and the width is 13 inches, what is the height of the pan?

15. **GEOMETRY** The volume of the rectangular prism at the right is 440 cubic centimeters. What is the width?

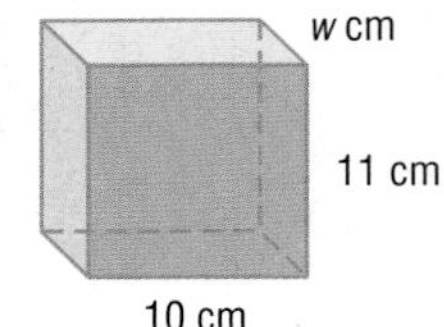

Find the volume of each cylinder. Round to the nearest tenth.

16.
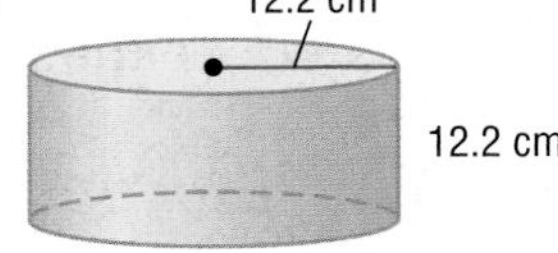

17.
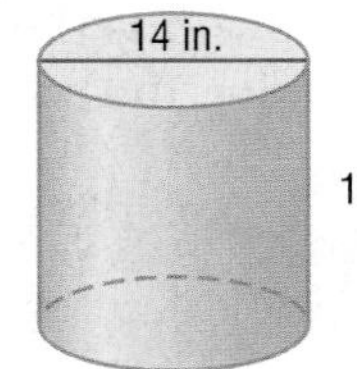

18.
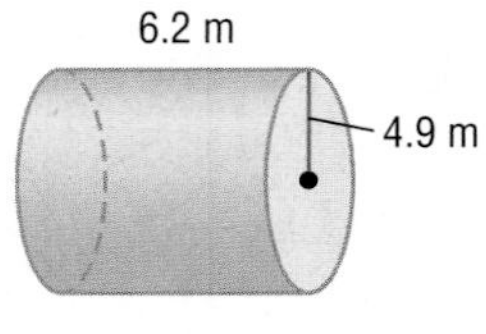

19. **FIREWOOD** Firewood is usually sold by a measure known as a *cord*. A full cord may be a stack $8 \times 4 \times 4$ feet or a stack $8 \times 8 \times 2$ feet.
    a. What is the volume of a full cord of firewood?
    b. A "short cord" of wood is $8 \times 4 \times$ the length of the logs. What is the volume of a short cord of $2\frac{1}{2}$-foot logs?
    c. If you have an area that is 12 feet long and 2 feet wide in which to store your firewood, how high will the stack be if it is a full cord of wood?

0-10 Surface Area

Objective

Find the surface area of rectangular prisms.

New Vocabulary

surface area

Surface area is the sum of the areas of all the surfaces, or faces, of a solid. Surface area is measured in square units.

Key Concept Surface Area — For Your FOLDABLE

Prism

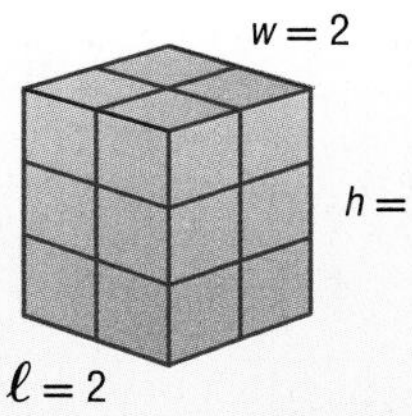

$S = 2\ell w + 2\ell h + 2wh$

Cylinder

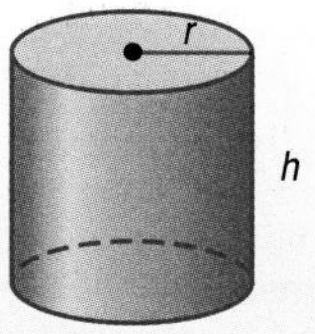

$S = 2\pi rh + 2\pi r^2$

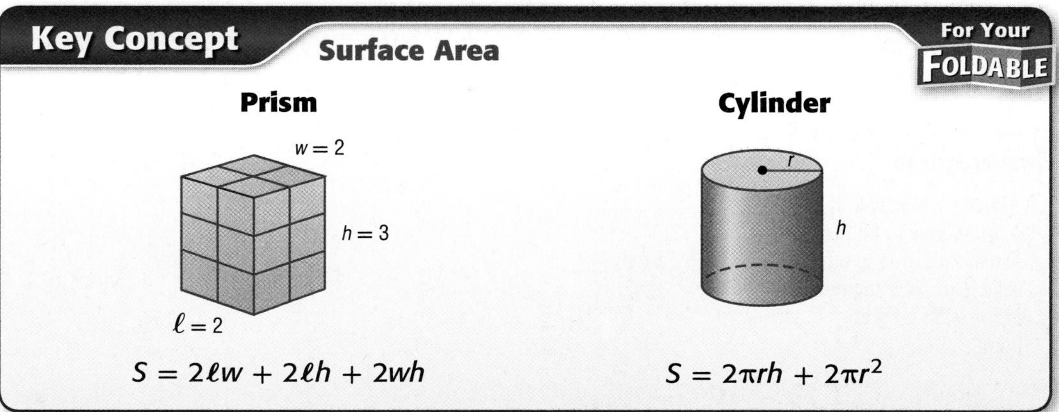

EXAMPLE

Find the surface area of each solid. Round to the nearest tenth if necessary.

a.

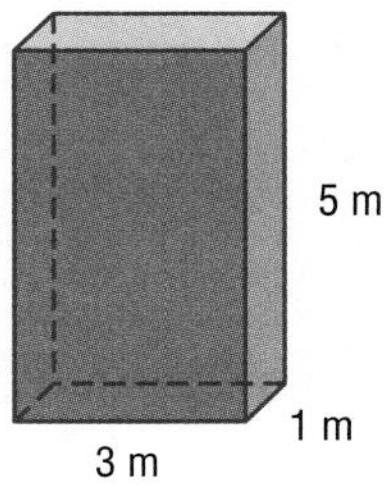

The prism has a length of 3 meters, width of 1 meter, and height of 5 meters.

$S = 2\ell w + 2\ell h + 2wh$	**Surface area formula**
$= 2(3)(1) + 2(3)(5) + 2(1)(5)$	**$\ell = 3, w = 1, h = 5$**
$= 6 + 30 + 10$	**Multiply.**
$= 46$	**Add.**

The surface area is 46 square meters.

b.

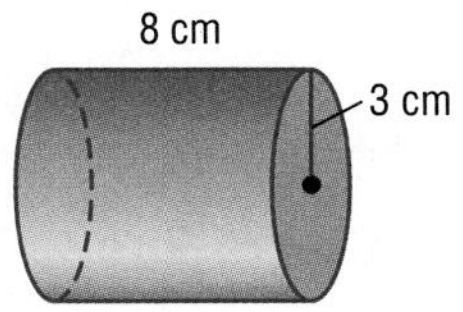

The height is 8 centimeters and the radius of the base is 3 centimeters. The surface area is the sum of the area of each base, $2\pi r^2$, and the area of the side, given by the circumference of the base times the height or $2\pi rh$.

$S = 2\pi rh + 2\pi r^2$	**Formula for surface area of a cylinder.**
$= 2\pi(3)(8) + 2\pi(3^2)$	**$r = 3, h = 8$**
$= 48\pi + 18\pi$	**Simplify.**
$\approx 207.3\ \text{cm}^2$	**Use a calculator.**

Exercises

Find the surface area of each rectangular prism given the measurements below.

1. $\ell = 6$ in., $w = 1$ in., $h = 4$ in

2. $\ell = 8$ m, $w = 2$ m, $h = 2$ m

3. $\ell = 10$ mm, $w = 4$ mm, $h = 5$ mm

4. $\ell = 6.2$ cm, $w = 1$ cm, $h = 3$ cm

5. $\ell = 7$ ft, $w = 2$ ft, $h = \frac{1}{2}$ ft

6. $\ell = 7.8$ m, $w = 3.4$ m, $h = 9$ m

> **StudyTip**
>
> **Alternate Method** Another way to find the surface area of a solid is to draw the net of the solid on grid paper.

Find the surface area of each solid.

7.

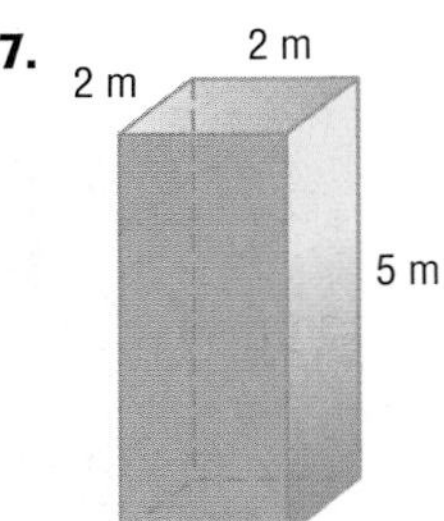

8.

2 ft
4 ft
3 ft

9.

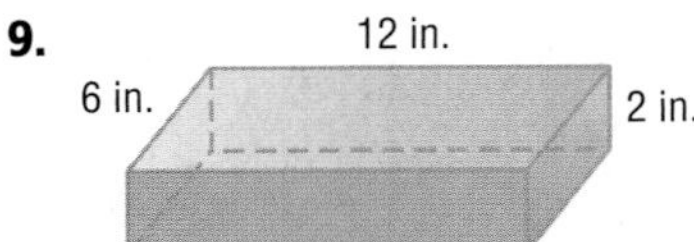

10.

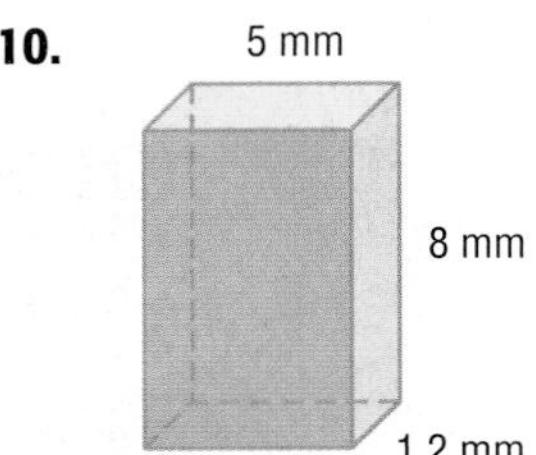

11.

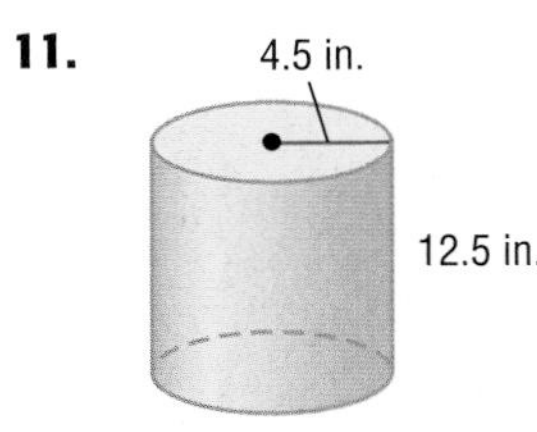

12.

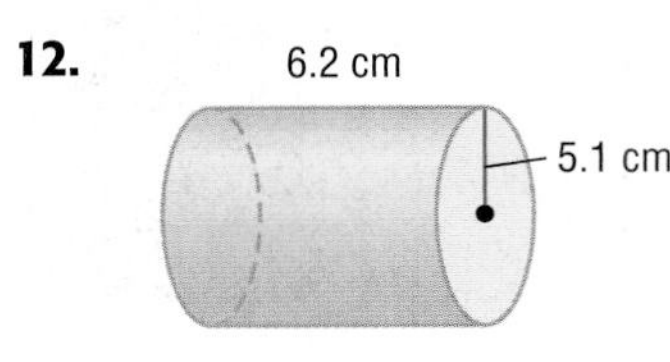

13. **GEOMETRY** What is the surface area of a cube with a side length of 2 meters?

14. **GIFTS** A gift box is a rectangular prism 14 inches long, 5 inches wide, and 4 inches high. If the box is to be covered in fabric, how much fabric is needed if there is no overlap?

15. **BOXES** A new refrigerator is shipped in a box 34 inches deep, 66 inches high, and $33\frac{1}{4}$ inches wide. What is the surface area of the box in square feet? Round to the nearest square foot. (*Hint:* 1 ft^2 = 144 in^2)

16. **PAINTING** A cabinet is 6 feet high, 3 feet wide, and 2 feet long. The entire outside surface of the cabinet is being painted except for the bottom. What is the surface area of the cabinet that is being painted?

17. **SOUP** A soup can is 4 inches tall and has a diameter of $3\frac{1}{4}$ inches. How much paper is needed for the label on the can? Round your answer to the nearest tenth.

18. **CRAFTS** For a craft project, Sarah is covering all the sides of a box with stickers. The length of the box is 8 inches, the width is 6 inches, and the height is 4 inches. If each sticker has a length of 2 inches and a width of 4 inches, how many stickers does she need to cover the box?

0-11 Simple Probability and Odds

Objective

Find the probability and odds of simple events.

New Vocabulary

probability
sample space
equally likely
tree diagram
odds
complements
theoretical probability
experimental probability

The **probability** of an event is the ratio of the number of favorable outcomes for the event to the total number of possible outcomes. When you roll a die, there are six possible outcomes: 1, 2, 3, 4, 5, or 6. This list of all possible outcomes is called the **sample space**.

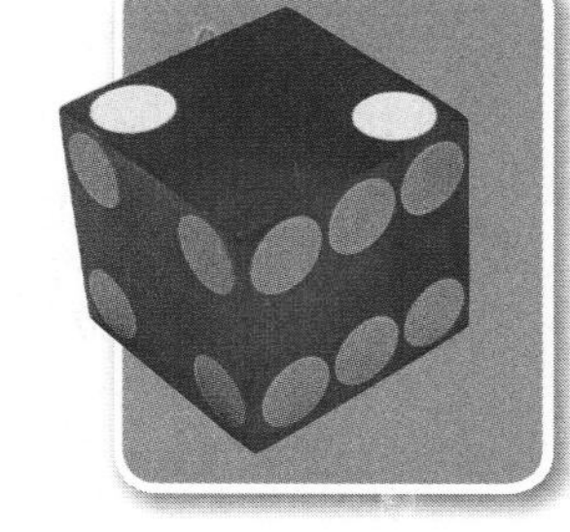

When there are n outcomes and the probability of each one is $\frac{1}{n}$, we say that the outcomes are **equally likely**. For example, when you roll a die, the 6 possible outcomes are equally likely because each outcome has a probability of $\frac{1}{6}$. The probabilty of an event is always between 0 and 1, inclusive. The closer a probability is to 1, the more likely it is to occur.

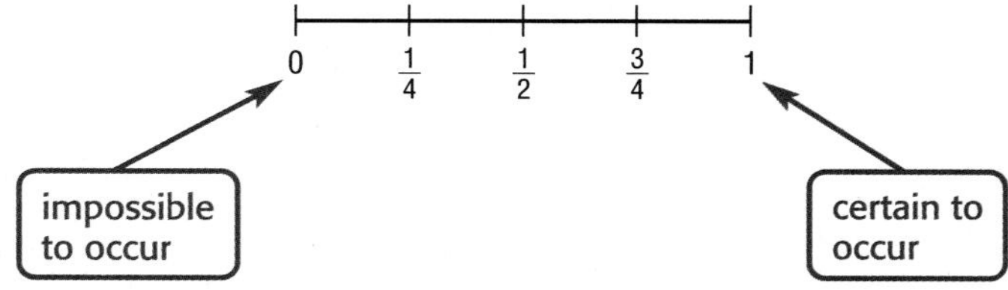

EXAMPLE 1

A die is rolled. Find each probability.

a. rolling a 1 or 5.

There are six possible outcomes. There are two favorable outcomes, 1 and 5.

$$\text{probability} = \frac{\text{number of favorable outcomes}}{\text{total number of possible outcomes}} = \frac{2}{6}$$

So, $P(1 \text{ or } 5) = \frac{2}{6}$ or $\frac{1}{3}$.

b. rolling an even number

Three of the six outcomes are even numbers. So, there are three favorable outcomes.

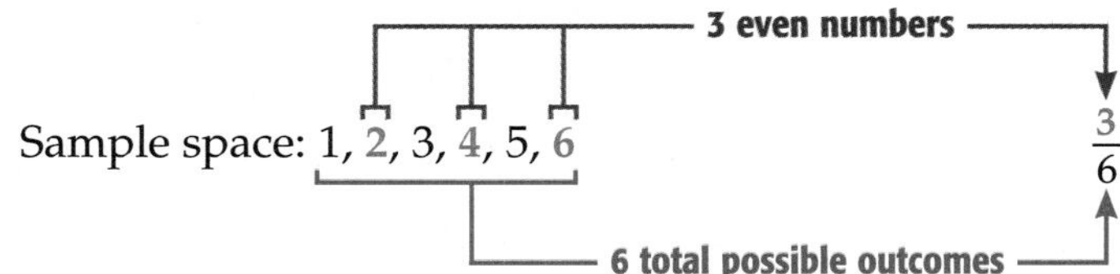

So, $P(\text{even number}) = \frac{3}{6}$ or $\frac{1}{2}$.

The events for rolling a 1 and for *not* rolling a 1 are called **complements**.

P(1) P(not 1) P(sum of probabilities)

$$\frac{1}{6} + \frac{5}{6} = \frac{6}{6} \text{ or } 1$$

The sum of the probabilities for any two complementary events is always 1.

EXAMPLE 2

A bowl contains 5 red chips, 7 blue chips, 6 yellow chips, and 10 green chips. One chip is randomly drawn. Find each probability.

a. blue

There are 7 blue chips and 28 total chips.

$$P(\text{blue chip}) = \frac{7}{28} \quad \begin{array}{l} \leftarrow \text{number of favorable outcomes} \\ \leftarrow \text{number of possible outcomes} \end{array}$$

$$= \frac{1}{4}$$

The probability can be stated as $\frac{1}{4}$, 0.25, or 25%.

b. red or yellow

There are 5 + 6 or 11 chips that are red or yellow.

$$P(\text{red or yellow}) = \frac{11}{28} \quad \begin{array}{l} \leftarrow \text{number of favorable outcomes} \\ \leftarrow \text{number of possible outcomes} \end{array}$$

$$\approx 0.39$$

The probability can be stated as $\frac{11}{28}$, about 0.39, or about 39%.

c. not green

There are 5 + 7 + 6 or 18 chips that are not green.

$$P(\text{not green}) = \frac{18}{28} \quad \begin{array}{l} \leftarrow \text{number of favorable outcomes} \\ \leftarrow \text{number of possible outcomes} \end{array}$$

$$= \frac{9}{14} \text{ or about } 0.64$$

The probability can be stated as $\frac{9}{14}$, about 0.64, or about 64%.

StudyTip

Alternate Method A chip drawn will either be green or not green. So, another method for finding *P*(not green) is to find *P*(green) and subtract that probability from 1.

One method used for counting the number of possible outcomes is to draw a **tree diagram**. The last column of a tree diagram shows all of the possible outcomes.

EXAMPLE 3

School baseball caps come in blue, yellow, or white. The caps have either the school mascot or the school's initials. Use a tree diagram to determine the number of different caps possible.

Color	Design	Outcomes
blue	mascot	blue, mascot
	initials	blue, initials
yellow	mascot	yellow, mascot
	initials	yellow, initials
white	mascot	white, mascot
	initials	white, initials

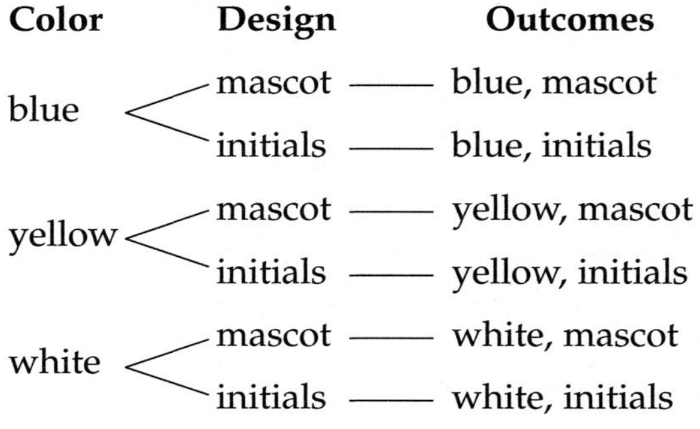

The tree diagram shows that there are 6 different caps possible.

This example is an illustration of the **Fundamental Counting Principle**, which relates the number of outcomes to the number of choices.

Key Concept — Fundamental Counting Principle

For Your FOLDABLE

Words If event M can occur in m ways and is followed by event N that can occur in n ways, then the event M followed by N can occur in $m \cdot n$ ways.

Example If there are 4 possible sizes for fish tanks and 3 possible shapes, then there are $4 \cdot 3$ or 12 possible fish tanks.

EXAMPLE 4

a. An ice cream shop offers one, two, or three scoops of ice cream from among 12 different flavors. The ice cream can be served in a wafer cone, a sugar cone, or in a cup. Use the Fundamental Counting Principle to determine the number of choices possible.

There are 3 ways the ice cream is served, 3 different servings, and there are 12 different flavors of ice cream.

Use the Fundamental Counting Principle to find the number of possible choices.

number of scoops		number of flavors		number of serving options		number of choices of ordering ice cream
3	·	12	·	3	=	108

So, there are 108 different ways to order ice cream.

b. Jimmy needs to make a 3-digit password for his log-on name on a Web site. The password can include any digit from 0-9, but the digits may not repeat. How many possible 3-digit passwords are there?

If the first digit is a 4, then the next digit cannot be a 4.

We can use the Fundamental Counting Principle to find the number of possible passwords.

1st digit		2nd digit		3rd digit		number of passwords
10	·	9	·	8	=	720

So, there are 720 possible 3-digit passwords.

The **odds** of an event occurring is the ratio that compares the number of ways an event can occur (successes) to the number of ways it cannot occur (failures).

StudyTip

Odds The sum of the number of successes and the number of failures equals the size of the sample space, or the number of possible outcomes.

EXAMPLE 5

Find the odds of rolling a number less than 3.

There are six possible outcomes; 2 are successes and 4 are failures.

So, the odds of rolling a number less than 3 are $\frac{1}{2}$ or 1:2.

Exercises

One coin is randomly selected from a jar containing 70 nickels, 100 dimes, 80 quarters, and 50 one-dollar coins. Find each probability.

1. P(quarter)
2. P(dime)
3. P(quarter or nickel)
4. P(value greater than \$0.10)
5. P(value less than \$1)
6. P(value at most \$1)

One of the polygons below is chosen at random. Find each probability.

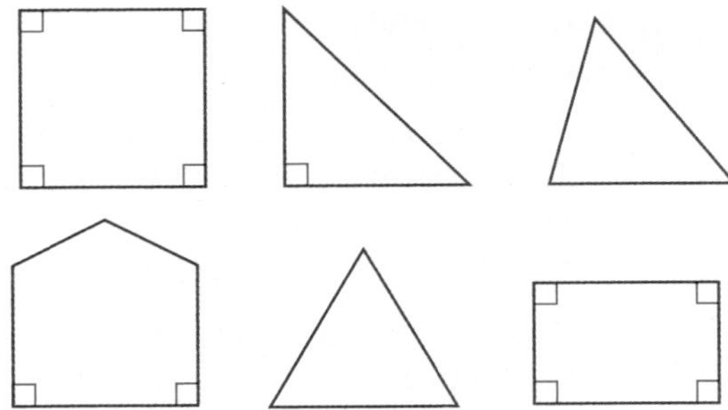

7. P(triangle)
8. P(pentagon)
9. P(not a quadrilateral)
10. P(more than 2 right angles)

StudyTip

Counting Outcomes When counting possible outcomes, make a column in your tree diagram for each part of the event.

Use a tree diagram to find the sample space for each event. State the number of possible outcomes.

11. The spinner at the right is spun and two coins are tossed.
12. At a restaurant, you have several choices of sides to have with breakfast. You can choose white or whole wheat toast. You can choose sausage links, sausage patties, or bacon.
13. How many different 3-character codes are there using A, B, or C for the first character, 8 or 9 for the second character, and 0 or 1 for the third character?

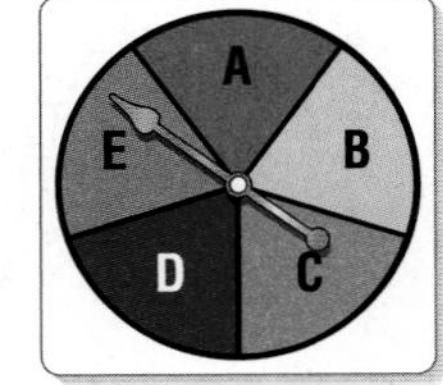

A bag is full of different colored marbles. The probability of randomly selecting a red marble from the bag is $\frac{1}{8}$. The probability of selecting a blue marble is $\frac{13}{24}$. Find each probability.

14. P(not red)
15. P(not blue)

Find the odds of each outcome if a computer randomly picks a letter in the name THE UNITED STATES OF AMERICA.

16. the letter A
17. the letter T
18. a vowel
19. a consonant

Margaret wants to order a sub at the local deli.

20. Find the number of possible orders of a sub with one topping and one dressing option.
21. Find the number of possible ham subs with mayonnaise, any combination of toppings or no toppings at all.
22. Find the number of possible orders of a sub with any combination of dressing and/or toppings.

Subs	
ham, salami, roast beef, turkey, bologna, pepperoni	
Dressing	**Toppings**
mayonnaise, mustard, vinegar, oil	lettuce, onions, peppers, olives

0-12 Mean, Median, Mode, Range, and Quartiles

Objective

Calculate the measures of central tendency of a set of data.

New Vocabulary

measures of central tendency
mean
median
mode
measures of variation
range
quartiles
lower quartile
upper quartile

Measures of central tendency are numbers used to represent a set of data. Three types of measures of central tendency are mean, median, and mode. The **mean** is the sum of the numbers in a set of data divided by the number of items.

EXAMPLE 1

Katherine has a lemonade stand. She made a profit of \$3.50 on Tuesday, \$4.00 on Wednesday, \$5.00 on Thursday, and \$4.50 on Friday. What was her mean daily profit?

$$\text{mean} = \frac{\text{sum of daily profits}}{\text{number of days}}$$

$$= \frac{\$3.50 + \$4.00 + \$5.00 + \$4.50}{4}$$

$$= \frac{\$17.00}{4} \text{ or } \$4.25$$

Katherine's mean daily profit was \$4.25.

The **median** is the middle number in a set of data when the data are arranged in numerical order. If there is an even number of data, the median is the mean of the two middle numbers.

EXAMPLE 2

The table shows the number of hits Marcus made for his team. Find the median of the data.

Team Played	Number of Hits by Marcus
Badgers	3
Hornets	6
Bulldogs	5
Vikings	2
Rangers	3
Panthers	7

To find the median, order the numbers from least to greatest. The median is in the middle.

2, 3, 3, 5, 6, 7

$$\frac{3 + 5}{2} = 4$$

There is an even number of items. Find the mean of the middle two.

The median number of hits is 4.

The **mode** is the number or numbers that appear most often in a set of data. If no item appears most often, the set has no mode.

EXAMPLE 3

The table shows the heights in inches of the members of a college men's basketball team. What is the mode of the heights?

Men's Basketball Team				
74	78	79	80	78
72	81	83	76	78
76	75	77	79	72

78 occurs three times. 72, 76, and 79 each occur twice. All the other heights occur once.

Since 78 occurs most frequently, the mode height is 78.

You can use measures of central tendency to solve problems.

EXAMPLE 4

SCHOOL On her first five history tests, Yoko received the following scores: 82, 96, 92, 83, and 91. What must she earn on the sixth test to have an average (mean) of 90?

$$\text{mean} = \frac{\text{sum of the first five scores} + \text{sixth score}}{\text{total number of tests}}$$ **Write an equation.**

$$90 = \frac{82 + 96 + 92 + 83 + 91 + x}{6}$$ **Use x to represent the sixth score.**

$$90 = \frac{444 + x}{6}$$ **Simplify.**

$$540 = 444 + x$$ **Multiply each side by 6.**

$$96 = x$$ **Subtract 444 from each side.**

To have an average score of 90, Yoko must earn a 96 on the sixth test.

Measures of variation are used to describe the distribution of the data. One measure, the difference between the greatest and the least data values, is called the **range**.

StudyTip

Describing Data The measures of variation including range describe how the data in a set vary. This is another way to describe data.

EXAMPLE 5

The times in minutes it took Olivia to walk to school each day this week are 18, 15, 15, 12, and 14. Find the range of the times.

range = greatest value − least value **Write an equation.**

= 18 − 12 or 6 **The greatest value is 18, and the least value is 12.**

The range of the times is 6 minutes.

In a set of data, the **quartiles** are values that separate the data into four equal subsets, each containing one fourth of the data. Q_1, Q_2, and Q_3 are used to represent the three quartiles. Q_1 is the **lower quartile**. It divides the lower half of the data into two equal parts. Q_2 is the median since it separates the data into two equal parts. Q_3 is the **upper quartile**. It divides the upper half of the data into two equal parts.

EXAMPLE 6

Find the median, lower quartile, and upper quartile of the data shown below.

22, 16, 35, 26, 14, 17, 28, 29, 21, 17, 20

Order the data from least to greatest. Then use the list to determine the quartiles.

14, 16, 17, 17, 20, 21, 22, 26, 28, 29, 35

Q_1 (17) Q_2 (21) Q_3 (28)

The median (Q_2) is 21, the lower quartile (Q_1) is 17, and the upper quartile (Q_3) is 28.

Exercises

StudyTip

Calculating Measures of Central Tendency Remember to order the data from least to greatest before finding the median of the data.

Find the mean, median, and mode for each set of data.

1. {1, 2, 3, 5, 5, 6, 13}
2. {3, 5, 8, 1, 4, 11, 3}
3. {52, 53, 53, 53, 55, 55, 57}
4. {8, 7, 5, 19}
5. {3, 11, 26, 4, 1}
6. {201, 201, 200, 199, 199}
7. {4, 5, 6, 7, 8}
8. {3, 7, 21, 23, 63, 27, 29, 95, 23}

Find the range, median, lower quartile, and upper quartile for each set of data.

9. {4, 7, 11, 19, 26, 26, 32}
10. {62, 65, 67, 68, 73, 80, 81, 83, 99}
11. {17, 9, 10, 17, 18, 5, 2}
12. {33, 38, 29, 25, 41, 40}
13. {10, 9, 8, 7, 6, 5, 4}
14. {111, 109, 112, 114, 119, 112}

15. **SCHOOL** The table shows the cost of some school supplies. Find the mean, median, and mode costs.

Cost of School Supplies	
Supply	**Cost**
pencils	\$0.50
pens	\$2.00
paper	\$2.00
pocket Folder	\$1.25
calculator	\$5.25
notebook	\$3.00
eraser	\$2.50
markers	\$3.50

16. **NUTRITION** The table shows the number of servings of fruits and vegetables that Cole eats one week. Find the range, median, lower quartile, and upper quartile.

Fruit and Vegetable Servings	
Day	**Number of Servings**
Monday	5
Tuesday	7
Wednesday	5
Thursday	4
Friday	3
Saturday	3
Sunday	8

17. **SCHOOL** Bill's scores on his first four science tests are 86, 90, 84, and 91. What must Bill earn on the fifth test so that his average (mean) will be exactly 88?

18. **BOWLING** Sue's average for 9 games of bowling is 108. What is the lowest score she can receive for the tenth game to have an average of 110?

19. **SCHOOL** Olivia has an average score of 92 on five French tests. If she earns a score of 96 on the sixth test, what will her new average score be?

20. **JOBS** The number of hours Maria and her friends each work at their part-time jobs is 20, 10, 8, 5, 25, 12 and 10 hours. Find the average amount of time Maria and her friends work at their jobs to the nearest hour.

21. **MOVIES** At a movie theater, ten movies are playing and their lengths are 105, 95, 115, 120, 150, 130, 100, 125, 110, and 135 minutes. Find the average length of a movie playing at this theater to the nearest tenth.

22. **BASKETBALL** The heights of players of a girls' basketball team are shown. Find the average height of the team to the nearest tenth.

Height of Players (in.)			
72	71	69	66
62	70	64	69
67	65	65	70

0-13

Representing Data

Data can be displayed and organized by different methods. In a **frequency table**, you use tally marks to record and display the frequency of events. A **bar graph** compares categories of data with bars representing the frequency.

Objective

To represent sets of data using different visual displays.

New Vocabulary

frequency table
bar graph
histogram
line graph
stem-and-leaf plot
circle graph
box-and-whisker plot
interquartile range
outliers

EXAMPLE 1

The frequency table shows the results of a survey of students' favorite sports. Make a bar graph to display the data.

Sport	Tally	Frequency
basketball	𝍸 𝍸 𝍸	15
football	𝍸 𝍸 𝍸 𝍸 𝍸	25
soccer	𝍸 𝍸 𝍸 |||	18
baseball	𝍸 𝍸 𝍸 𝍸		21
tennis	𝍸 𝍸 𝍸		16

Step 1 Draw a horizontal axis and a vertical axis. Label the axes as shown. Add a title.

Step 2 Draw a bar to represent each sport. The vertical scale is the number of students who chose each sport. The horizontal scale identifies the sport.

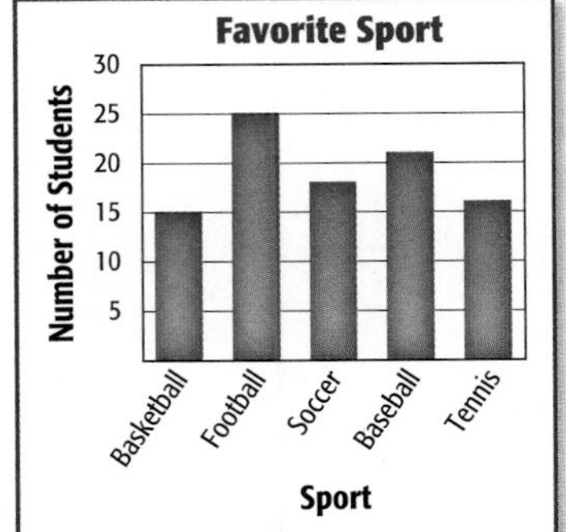

A **histogram** is a type of bar graph used to display numerical data that have been organized into equal intervals.

EXAMPLE 2

The frequency table shows the heights of students in a class. Make a histogram of the data.

Heights of Students		
Height (cm)	Tally	Frequency
131–140	||||	4
141–150	𝍸		6
151–160	𝍸 |||	8
161–170	𝍸	5
171–180	|||	3

Step 1 Draw and label a horizontal and vertical axis. Include a title.

Step 2 Show the intervals from the frequency table on the horizontal axis.

Step 3 For each height interval, draw a bar whose height is given by the frequencies. There is no space between the bars.

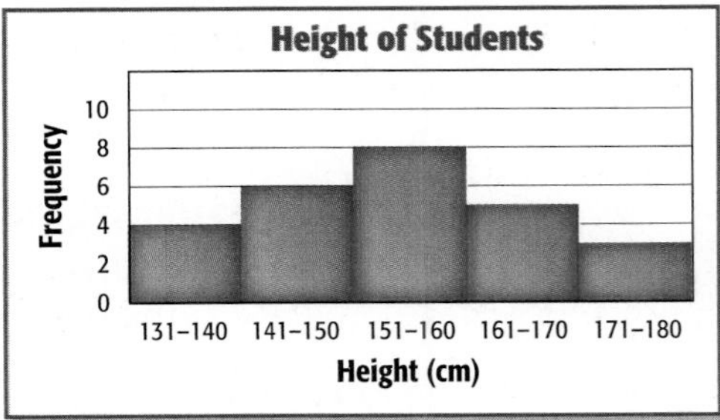

Another way to represent data is by using a line graph. A **line graph** usually shows how data change over a period of time.

EXAMPLE 3

Sales at the Marshall High School Store are shown in the table. Make a line graph of the data.

School Store Sales Amounts					
September	$670	December	$168	March	$412
October	$229	January	$290	April	$309
November	$300	February	$388	May	$198

Step 1 Draw a horizontal axis and a vertical axis and label them as shown. Include a title.

Step 2 Plot the points.

Step 3 Draw a line connecting each pair of consecutive points.

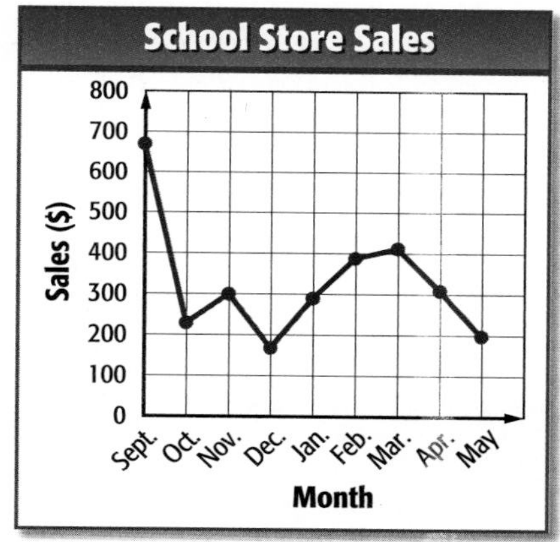

Data can also be organized and displayed by using a stem-and-leaf plot. In a **stem-and-leaf plot**, the greatest common place value is used for the *stems*. The numbers in the next greatest place value are used to form the *leaves*.

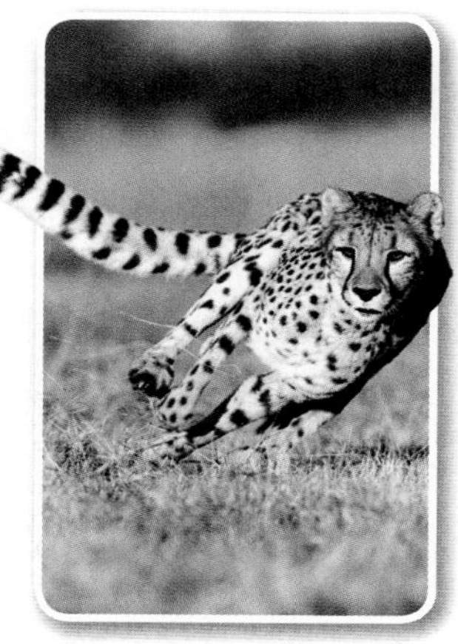

Real-World Link

The fastest animal on land is the cheetah. Cheetahs can run at speeds up to 60 miles per hour.

Source: Infoplease

EXAMPLE 4

The speeds (mph) of 20 of the fastest land animals are listed at the right. Use the data to make a stem-and-leaf plot.

42	40	40	35	50
32	50	36	50	40
45	70	43	45	32
40	35	61	48	35

Source: *The World Almanac*

The greatest place value is tens. So, 32 miles per hour would have a stem of 3 and a leaf of 2.

Stem	Leaf
3	2 2 5 5 5 6
4	0 0 0 0 2 3 5 5 8
5	0 0 0
6	1
7	0

Key: 3|2 = 32

A **circle graph** is a graph that shows the relationship between parts of the data and the whole. The circle represents all of the data.

EXAMPLE 5

The table shows how Lily spent 8 hours of one day at summer camp.

Summer Camp	
Activity	**Hours**
canoeing	3
crafts	1
eating	2
hiking	2

First, find the ratio that compares the number of hours for each activity to 8. Then multiply each ratio by 360° to find the number of degrees for each section of the graph.

Canoeing: $\frac{3}{8} \cdot 360° = 135°$

Crafts: $\frac{1}{8} \cdot 360° = 45°$

Eating: $\frac{2}{8} \cdot 360° = 90°$

Hiking: $\frac{2}{8} \cdot 360° = 90°$

Summer Camp

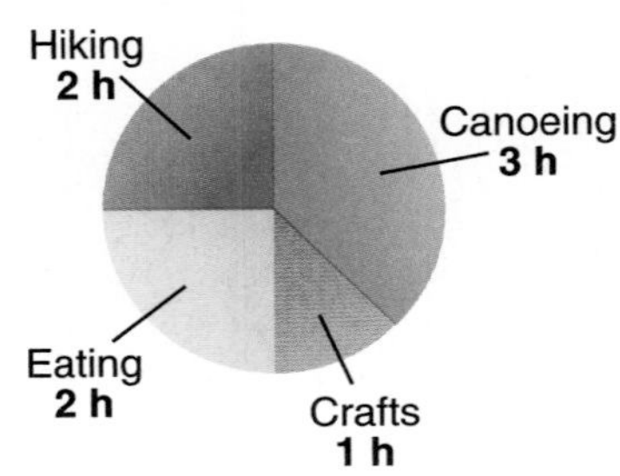

StudyTip

Interquartile Range When the interquartile range is a small value, the data in the set are close together. A large interquartile range means that the data are spread out.

Data can be organized and displayed by dividing a set of data into four parts using the median and quartiles. This is a **box-and-whisker plot**. The box in a box-and-whisker plot represents the interquartile range. The **interquartile range** is the difference between the upper and lower quartiles. Data that are more than 1.5 times the value of the interquartile range beyond the quartiles are called **outliers**.

EXAMPLE 6

Draw a box-and-whisker plot for these data. Describe how the outlier affects the quartile points.

14, 30, 16, 20, 18, 16, 20, 18, 22, 13, 8

Step 1 Order the data from least to greatest. Then determine the quartiles.

8, 13, **14**, 16, 16, **18**, 18, 20, **20**, 22, 30

↑ Q_1 ↑ Q_2 ↑ Q_3

Determine the interquartile range.

$IQR = Q_3 - Q_1$

$= 20 - 14$ or 6

Check to see if there are any outliers.

$14 - 1.5(6) = 5 \qquad 20 + 1.5(6) = 29$

Numbers less than 5 or greater than 29 are outliers.

The only outlier is 30.

Step 2 Draw a number line that includes the least and greatest numbers in the data. Place dots above the number line to represent the three quartile points, any outliers, the least number that is not an outlier, and the greatest number that is not an outlier.

Step 3 Draw the box and the whiskers. The vertical rules go through the quartiles. The outliers are not connected to the whiskers.

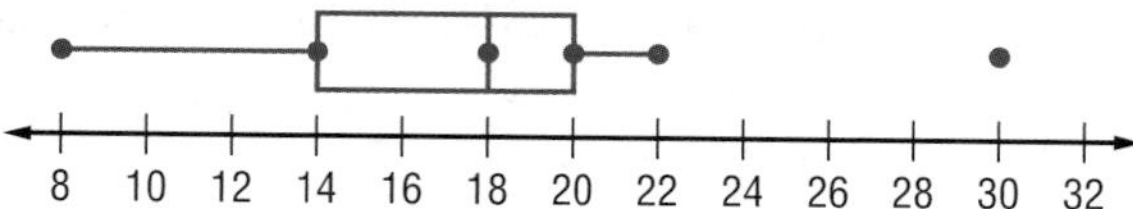

Step 4 Omit 30 from the data. Repeat Step 1 to determine Q_1, Q_2, and Q_3.
8, 13, **14,** 16, **16, 18,** 18, **20,** 20, 22

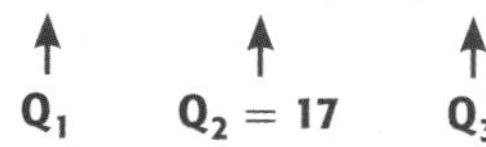

Removing the outlier does not affect Q_1 or Q_2 and thus does not affect the interquartile range. The value of Q_2 changes from 18 to 17.

Exercises

StudyTip

Best Representation A set of data can be represented by several different displays. There is usually one type of graph that is best for displaying the data.

1. **SURVEYS** Alana surveyed several students to find the number of hours of sleep they typically get each night. The results are shown in the table. Make a bar graph of the data.

Hours of Sleep					
Alana	8	Kwam	7.5	Tomas	7.75
Nick	8.25	Kate	7.25	Sharla	8.5

2. **PLAYS** The frequency table at the right shows the ages of people attending a high school play. Make a histogram to display the data.

Age	Tally	Frequency
0–19	𝍸 𝍸 𝍸 𝍸 𝍸 𝍸 𝍸 𝍸 𝍸 II	47
20–39	𝍸 𝍸 𝍸 𝍸 𝍸 𝍸 𝍸 𝍸 III	43
40–59	𝍸 𝍸 𝍸 𝍸 𝍸 𝍸 I	31
60–79	𝍸 III	8

3. **LAWN CARE** Marcus started a lawn care service. The chart shows how much money he made over summer break. Make a line graph of the data.

Lawn Care Profits ($)								
Week	1	2	3	4	5	6	7	8
Profit	25	40	45	50	75	85	95	95

Use each set of data to make a stem-and-leaf plot and a box-and-whisker plot. Describe how the outliers affects the quartile points.

4. {65, 63, 69, 71, 73, 59, 60, 70, 72, 66, 71, 58}

5. {31, 30, 28, 26, 22, 34, 26, 31, 47, 32, 18, 33, 26, 23, 18}

6. **FINANCIAL LITERACY** The table shows how Ping spent his allowance of $40. Make a circle graph of the data.

Allowance	
How Spent	**Amount ($)**
savings	15
downloaded music	8
snacks	5
T-shirt	12

7. **JOGGING** The table shows the number of miles Hannah jogged each day for 10 days. Make a line graph of the data.

Day	1	2	3	4	5	6	7	8	9	10
Miles Jogged	2	2	3	3.5	4	4.5	2.5	3	4	5

CHAPTER 0

Posttest

Determine whether you need an estimate or an exact answer. Then use the four-step problem-solving plan to solve.

1. **DISTANCE** Fabio rode his scooter 2.3 miles to his friend's house, then 0.7 mile to the video store, then 2.1 miles to the library. If he rode the same route back home, about how far did he travel in all?

2. **SHOPPING** The regular price of a T-shirt is \$9.99. It is on sale for 15% off. Sales tax is 6%. If you give the cashier a \$10 bill, how much change will you receive?

Find each sum or difference.

3. $-31 + (-4)$
4. $48 - 55$
5. $-71 - (-10)$
6. $31 - 42.9$
7. $-11.5 + 8.1$
8. $-0.38 - (-1.06)$

Find each product or quotient.

9. $-21(-5)$
10. $-81 \div (-3)$
11. $-120 \div 8$
12. $-39 \div -3$

Replace each ● with <, >, or = to make a true sentence.

13. -0.62 ● $-\frac{6}{7}$
14. $\frac{12}{44}$ ● $\frac{8}{11}$
15. Order $4\frac{4}{5}$, 4.85, $2\frac{5}{8}$, and 2.6 from least to greatest.

Find each sum or difference. Write in simplest form.

16. $\frac{1}{7} + \frac{5}{7}$
17. $\frac{7}{8} - \frac{1}{8}$
18. $\frac{1}{6} + \left(-\frac{1}{2}\right)$
19. $-\frac{1}{12} - \left(-\frac{3}{4}\right)$

Find each product or quotient.

20. $-1.2(9.3)$
21. $-20.93 \div (-2.3)$
22. $10.5 \div (-1.2)$
23. $(-3.4)(-2.8)$

Name the reciprocal of each number.

24. 6
25. $1\frac{2}{5}$
26. $-2\frac{3}{7}$
27. $-\frac{1}{2}$
28. $\frac{4}{3}$
29. $5\frac{1}{3}$

Find each product or quotient. Write in simplest form.

30. $\frac{2}{5} \cdot \frac{5}{9}$
31. $\frac{4}{5} \div \frac{1}{5}$
32. $-\frac{7}{8} \cdot 2$
33. $\frac{1}{3} \div 2\frac{1}{4}$
34. $-6 \cdot \left(-\frac{3}{4}\right)$
35. $\frac{7}{18} \div \left(-\frac{14}{15}\right)$
36. **PICNIC** Joseph is mixing $5\frac{1}{2}$ gallons of orange drink for his class picnic. Every $\frac{1}{2}$ gallon requires 1 packet of orange drink mix. How many packets of orange drink mix does Joseph need?

Express each percent as a fraction in simplest form.

37. 6%
38. 140%

Use the percent proportion to find each number.

39. 50% of what number is 31?
40. What number is 110% of 51?
41. Find 8% of 95.
42. **SOLUTIONS** A solution is prepared by dissolving 24 milliliters of saline in 150 milliliters of pure solution. What is the percent of saline in the pure solution?
43. **SHOPPING** Marta got 60% off a pair of shoes. If the shoes cost \$9.75 (before sales tax), what was the original price of the shoes?

Find the perimeter and area of each figure.

44.

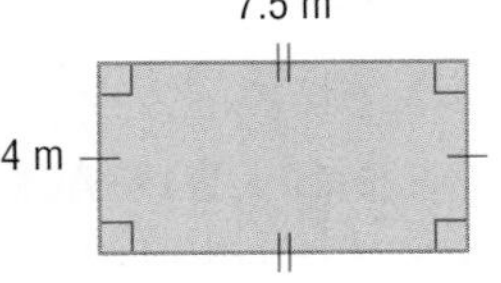

45.

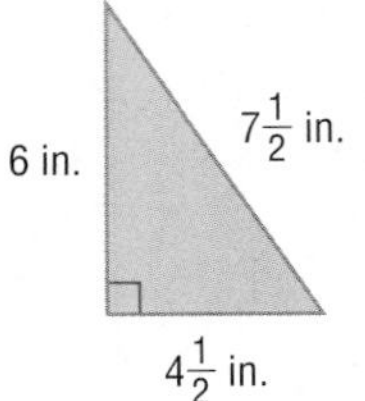

46. A parallelogram has a base of 20 millimeters and a height of 6 millimeters. Find the area.
47. **GARDENS** Find the perimeter of the garden.

6.0 m
3.5 m
4.0 m

Find the circumference and area of each circle. Round to the nearest tenth.

48.

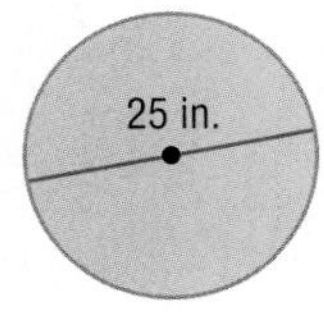

49.

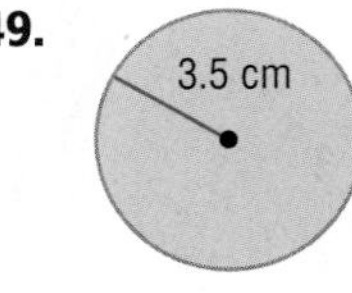

50. **PARKS** A park has a circular area for a fountain that has a circumference of about 16 feet. What is the radius of the circular area? Round to the nearest tenth.

Find the volume and surface area of each rectangular prism given the measurements below.

51. $\ell = 1.5$ m, $w = 3$ m, $h = 2$ m

52. $\ell = 4$ in., $w = 1$ in., $h = \frac{1}{2}$ in.

53. Find the volume and surface area of the rectangular prism.

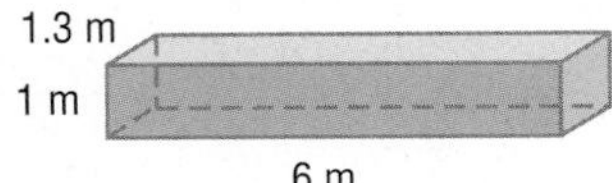

One marble is randomly selected from a jar containing 3 red, 4 green, 2 black, and 6 blue marbles. Find each probability.

54. P(red or blue)

55. P(green or red)

56. P(not black)

57. P(not blue)

58. A movie theater is offering snack specials. You can choose a small, medium, large, or jumbo popcorn with or without butter, and soda or bottled water. Use a tree diagram to find the sample space for the event. State the number of possible outcomes.

One coin is randomly selected from a jar containing 20 pennies, 15 nickels, 3 dimes, and 12 quarters. Find the odds of each outcome. Write in simplest form.

59. a dime

60. a value less than $0.25

61. a value greater than $0.10

62. a value less than $0.05

63. **SCHOOL** In a science class, each student must choose a lab project from a list of 15, write a paper on one of 6 topics, and give a presentation about one of 8 subjects. How many ways can students choose to do their assignments?

64. **GAMES** Marcos has been dealt seven different cards. How many different ways can he play his cards if he is required to play one card at a time?

Find the mean, median, and mode for each set of data.

65. {99, 88, 88, 92, 100}

66. {30, 22, 38, 41, 33, 41, 30, 24}

67. Find the range, median, lower quartile, and upper quartile for {77, 75, 72, 70, 79, 77, 70, 76}.

68. **TESTS** Kevin's scores on the first four science tests are 88, 92, 82, and 94. What score must he earn on the fifth test so that the mean will be 90?

69. **FOOD** The table shows the results of a survey in which students were asked to choose their favorite food. Make a bar graph of the data.

Favorite Foods	
Food	**Number of Students**
pizza	15
chicken nuggets	10
cheesy potatoes	8
ice cream	5

70. Make a box-and-whisker plot of the following data: 26, 18, 26, 29, 18, 20, 35, 32, 31, 24, 26, and 22.

71. **BUDGET** The table shows how Kat spends her allowance. Make a circle graph of the data.

Category	Amount ($)
Savings	25
Clothes	10
Entertainment	15

CHAPTER 1

Expressions, Equations, and Functions

Then

You have learned how to perform operations on whole numbers.

Now

In Chapter 1, you will:

- Write algebraic expressions.
- Use the order of operations.
- Solve equations.
- Represent relations and functions.
- Use conditional statements and counterexamples.

Why?

SCUBA DIVING A scuba diving store rents air tanks and wet suits. An algebraic expression can be written to represent the total cost to rent this equipment. This expression can be evaluated to determine the total cost for a group of people to rent the equipment.

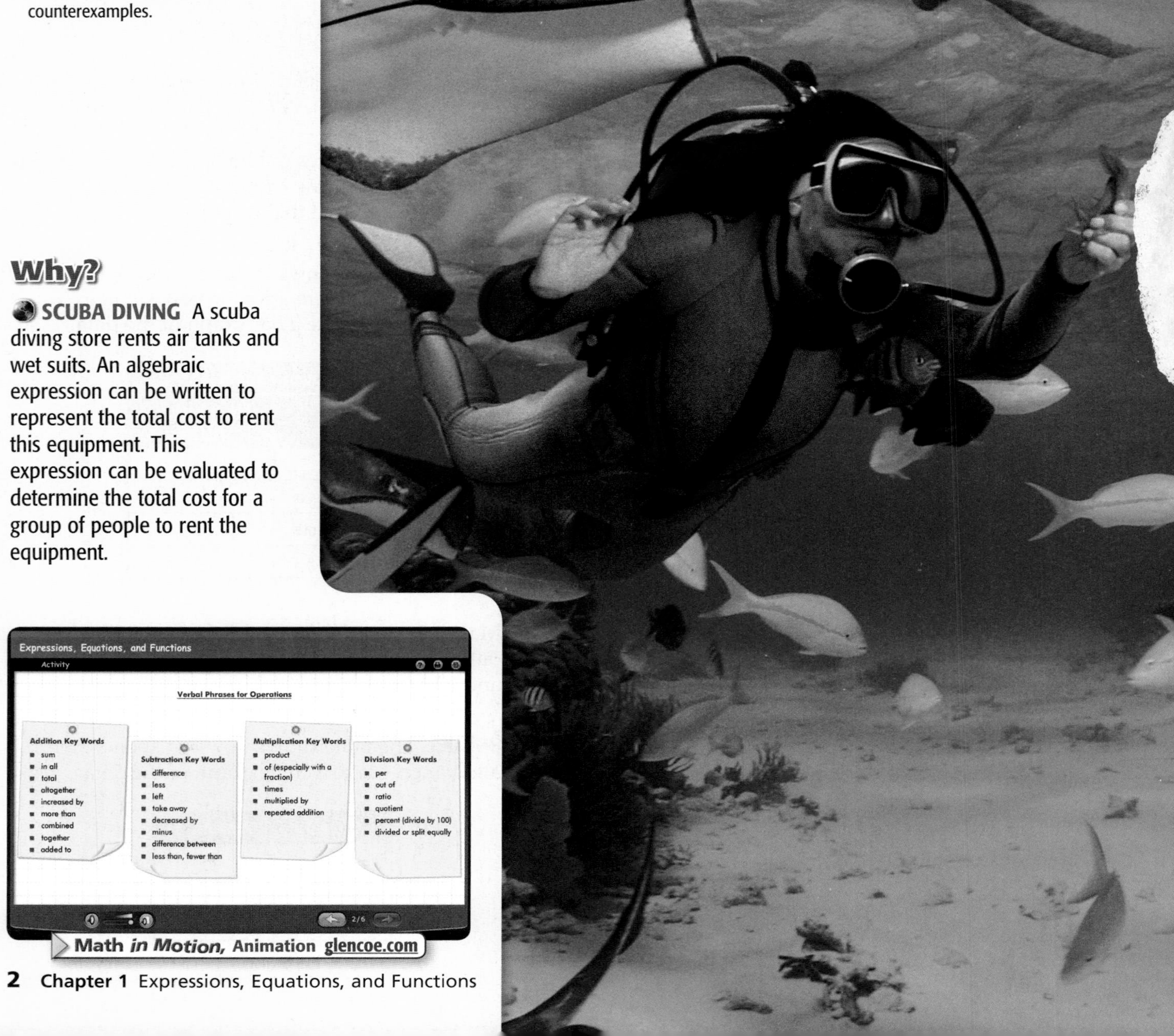

Math *in Motion,* Animation glencoe.com

Get Ready for Chapter 1

Diagnose Readiness You have two options for checking Prerequisite Skills.

Text Option

Take the Quick Check below. Refer to the Quick Review for help.

*Quick*Check

Write each fraction in simplest form. If the fraction is already in simplest form, write *simplest form*. (Lesson 0-4)

1. $\frac{24}{36}$
2. $\frac{34}{85}$
3. $\frac{36}{12}$
4. $\frac{27}{45}$
5. $\frac{11}{18}$
6. $\frac{5}{65}$
7. $\frac{19}{1}$
8. $\frac{16}{44}$
9. $\frac{64}{88}$
10. **ICE CREAM** Fifty-four out of 180 customers said that cookie dough ice cream was their favorite flavor. What fraction of customers was this? (Lesson 0-5)

Find the perimeter of each figure. (Lesson 0-7)

11.

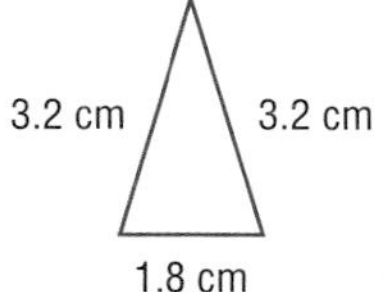

12.

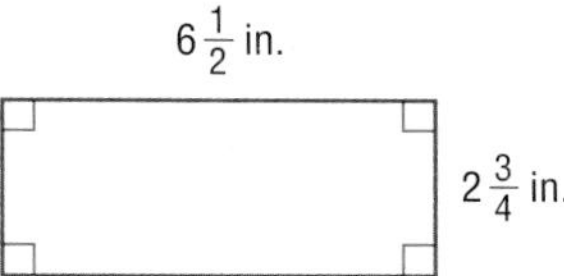

13. **FENCING** Jolon needs to fence a garden. The dimensions of the garden are 6 meters by 4 meters. How much fencing does Jolon need to purchase?

Evaluate. (Lesson 0-5)

14. $6 \cdot \frac{2}{3}$
15. $4.2 \cdot 8.1$
16. $\frac{3}{8} \div \frac{1}{4}$
17. $5.13 \div 2.7$
18. $3\frac{1}{5} \cdot \frac{3}{4}$
19. $2.8 \cdot 0.2$
20. **CONSTRUCTION** A board measuring 7.2 feet must be cut into three equal pieces. Find the length of each piece.

*Quick*Review

EXAMPLE 1

Write $\frac{24}{40}$ in simplest form.

Find the greatest common factor (GCF) of 24 and 40.

factors of 24: 1, 2, 3, 4, 6, 8, 12, 24
factors of 40: 1, 2, 4, 5, 8, 10, 20, 40

The GCF of 24 and 40 is 8.

$\frac{24 \div 8}{40 \div 8} = \frac{3}{5}$ **Divide the numerator and denominator by their GCF, 8.**

EXAMPLE 2

Find the perimeter.

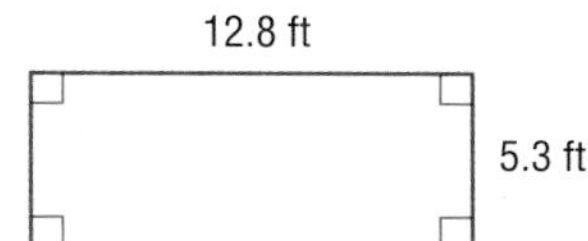

$P = 2\ell + 2w$

$= 2(12.8) + 2(5.3)$ **$\ell = 12.8$ and $w = 5.3$**

$= 25.6 + 10.6$ or 36.2 **Simplify.**

The perimeter is 36.2 feet.

EXAMPLE 3

Find $2\frac{1}{4} \div 1\frac{1}{2}$.

$2\frac{1}{4} \div 1\frac{1}{2} = \frac{9}{4} \div \frac{3}{2}$ **Write mixed numbers as improper fractions.**

$= \frac{9}{4}\left(\frac{2}{3}\right)$ **Multiply by the reciprocal.**

$= \frac{18}{12}$ or $1\frac{1}{2}$ **Simplify.**

Online Option

Math Online Take a self-check Chapter Readiness Quiz at glencoe.com.

Get Started on Chapter 1

You will learn several new concepts, skills, and vocabulary terms as you study Chapter 1. To get ready, identify important terms and organize your resources. You may wish to refer to **Chapter 0** to review prerequisite skills.

FOLDABLES® Study Organizer

Expressions, Equations, and Functions Make this Foldable to help you organize your Chapter 1 notes about expressions, equations, and functions. Begin with five sheets of grid paper.

1. **Fold** each sheet of grid paper in half along the width. Then cut along the crease.

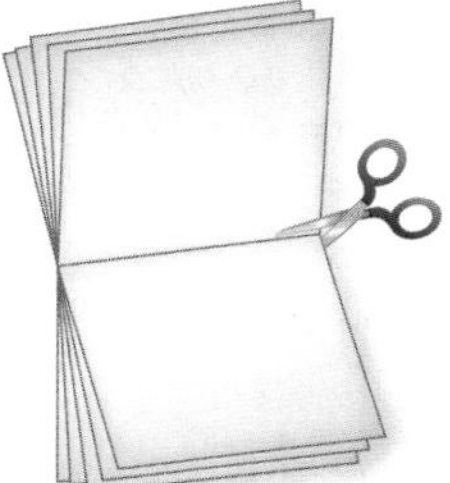

2. **Staple** the ten half-sheets together to form a booklet.

3. **Cut** nine lines from the bottom of the top sheet, eight lines from the second sheet, and so on.

4. **Label** each of the tabs with a lesson number. The ninth tab is for the properties and the last tab is for the vocabulary.

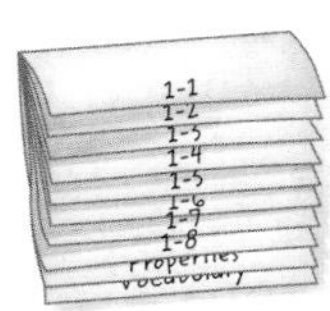

Math Online glencoe.com

- Study the chapter online
- Explore **Math in Motion**
- Get extra help from your own **Personal Tutor**
- Use **Extra Examples** for additional help
- Take a **Self-Check Quiz**
- **Review Vocabulary** in fun ways

New Vocabulary

English		Español
algebraic expression	p. 5	expression algebraica
variable	p. 5	variable
term	p. 5	término
power	p. 5	potencia
coefficient	p. 26	coeficiente
equation	p. 31	ecuación
solution	p. 31	solución
identity	p. 33	identidad
relation	p. 38	relacíon
domain	p. 38	domino
range	p. 38	rango
independent variable	p. 40	variable independiente
dependent variable	p. 40	variable dependiente
function	p. 45	función
nonlinear function	p. 48	función no lineal
deductive reasoning	p. 55	razonamiento deductivo
counterexample	p. 56	contraejemplo

Review Vocabulary

additive inverse • p. P11 • inverso de la adición a number and its opposite

multiplicative inverse • p. P18 • inverso multiplicativo two numbers with a product of 1

perimeter • p. P23 • perímetro the distance around a geometric figure

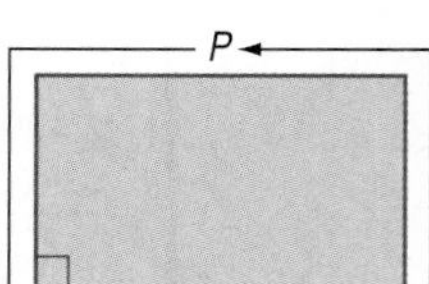

Multilingual eGlossary glencoe.com

1-1

Variables and Expressions

Why?

Cassie and her friends are at a baseball game. The stadium is running a Dime-A-Dog promotion where hot dogs are \$0.10 each. Suppose d represents the number of hot dogs Cassie and her friends eat. Then $0.10d$ represents the cost of the hot dogs they eat.

Then

You performed operations on integers. (Lesson 0-3)

Now

- Write verbal expressions for algebraic expressions.
- Write algebraic expressions for verbal expressions.

New Vocabulary

algebraic expression
variable
term
factor
product
power
exponent
base

Math Online

glencoe.com

- Extra Examples
- Personal Tutor
- Self-Check Quiz
- Homework Help

Write Verbal Expressions An **algebraic expression** consists of sums and/or products of numbers and variables. In the algebraic expression $0.10d$, the letter d is called a variable. In algebra, **variables** are symbols used to represent unspecified numbers or values. Any letter may be used as a variable.

$$0.10d \qquad 2x + 4 \qquad 3 + \frac{z}{6} \qquad p \cdot q \qquad 4cd \div 3mn$$

A **term** of an expression may be a number, a variable, or a product or quotient of numbers and variables. For example, $0.10d$, $2x$ and 4 are each terms.

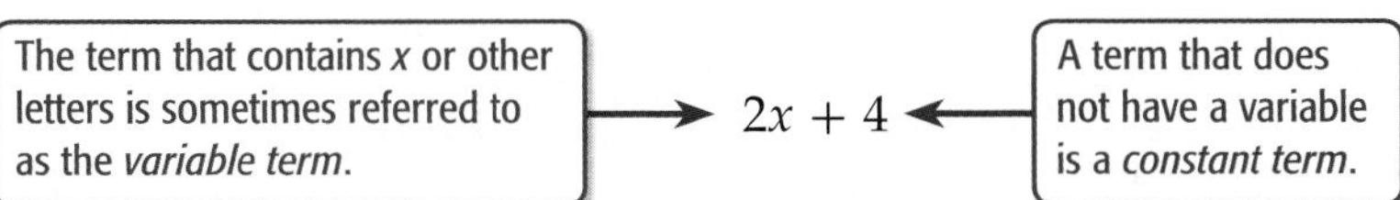

In a multiplication expression, the quantities being multiplied are **factors**, and the result is the **product**. A raised dot or set of parentheses are often used to indicate a product. Here are several ways to represent the product of x and y.

$$xy \qquad x \cdot y \qquad x(y) \qquad (x)y \qquad (x)(y)$$

An expression like x^n is called a **power**. The word *power* can also refer to the exponent. The **exponent** indicates the number of times the base is used as a factor. In an expression of the form x^n, the **base** is x. The expression x^n is read "x to the nth power." When no exponent is shown, it is understood to be 1. For example, $a = a^1$.

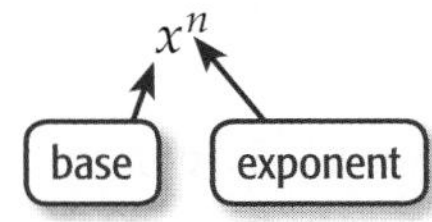

EXAMPLE 1 Write Verbal Expressions

Write a verbal expression for each algebraic expression.

a. $3x^4$
three times x to the fourth power

b. $5z^2 + 16$
5 times z to the second power plus sixteen

Check Your Progress

1A. $16u^2 - 3$

1B. $\frac{1}{2}a + \frac{6b}{7}$

Personal Tutor glencoe.com

StudyTip

Order of Operations Remember to follow the order of operations when writing a sentence to represent an algebraic expression.

Write Algebraic Expressions Another important skill is translating verbal expressions into algebraic expressions.

Key Concept For Your FOLDABLE

Translating Verbal to Algebraic Expressions

Operation	Verbal Phrases
Addition	more than, sum, plus, increased by, added to
Subtraction	less than, subtracted from, difference, decreased by, minus
Multiplication	product of, multiplied by, times, of
Division	quotient of, divided by

EXAMPLE 2 **Write Algebraic Expressions**

Write an algebraic expression for each verbal expression.

a. a number t more than 6

The words *more than* suggest addition.
Thus, the algebraic expression is $6 + t$ or $t + 6$.

b. 10 less than the product of 7 and f

Less than implies subtraction, and *product* suggests multiplication.
So the expression is written as $7f - 10$.

c. two thirds of the volume v

The word *of* with a fraction implies that you should multiply.
The expression could be written as $\frac{2}{3}v$ or $\frac{2v}{3}$.

Check Your Progress

2A. the product of p and 6

2B. one third of the area a

▷ Personal Tutor glencoe.com

Variables can represent quantities that are known and quantities that are unknown. They are also used in formulas, expressions, and equations.

Real-World Career

Sports Marketing Sports marketers promote and manage athletes, teams, facilities and sports-related businesses and organizations. A minimum of a bachelor's degree in sports management or business administration is preferred.

Real-World EXAMPLE 3 **Write an Expression**

SPORTS MARKETING Mr. Martinez orders 250 key chains printed with his athletic team's logo and 500 pencils printed with their Web address. Write an algebraic expression that represents the cost of the order.

Let k be the cost of each key chain and p be the cost of each pencil. Then the cost of the key chains is $250k$ and the cost of the pencils is $500p$. The cost of the order is represented by $250k + 500p$.

Check Your Progress

3. COFFEE SHOP Katie estimates that $\frac{1}{8}$ of the people who order beverages also order pastries. Write an algebraic expression to represent this situation.

▷ Personal Tutor glencoe.com

1-1

Variables and Expressions

Why?

Cassie and her friends are at a baseball game. The stadium is running a Dime-A-Dog promotion where hot dogs are \$0.10 each. Suppose d represents the number of hot dogs Cassie and her friends eat. Then $0.10d$ represents the cost of the hot dogs they eat.

Then

You performed operations on integers. (Lesson 0-3)

Now

- Write verbal expressions for algebraic expressions.
- Write algebraic expressions for verbal expressions.

New Vocabulary

algebraic expression
variable
term
factor
product
power
exponent
base

Math Online

glencoe.com

- Extra Examples
- Personal Tutor
- Self-Check Quiz
- Homework Help

Write Verbal Expressions An **algebraic expression** consists of sums and/or products of numbers and variables. In the algebraic expression $0.10d$, the letter d is called a variable. In algebra, **variables** are symbols used to represent unspecified numbers or values. Any letter may be used as a variable.

$$0.10d \qquad 2x + 4 \qquad 3 + \frac{z}{6} \qquad p \cdot q \qquad 4cd \div 3mn$$

A **term** of an expression may be a number, a variable, or a product or quotient of numbers and variables. For example, $0.10d$, $2x$ and 4 are each terms.

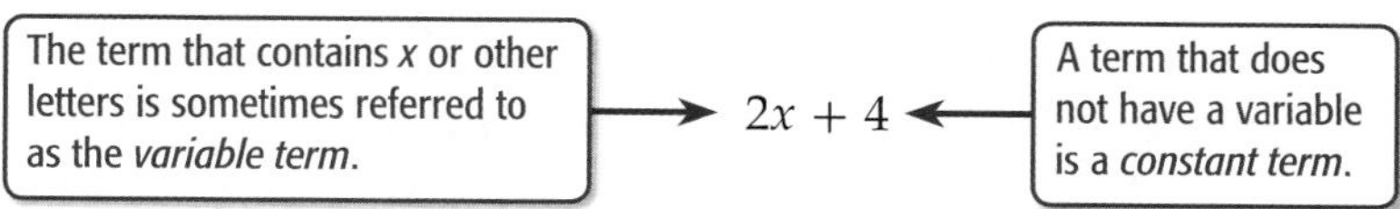

In a multiplication expression, the quantities being multiplied are **factors**, and the result is the **product**. A raised dot or set of parentheses are often used to indicate a product. Here are several ways to represent the product of x and y.

$$xy \qquad x \cdot y \qquad x(y) \qquad (x)y \qquad (x)(y)$$

An expression like x^n is called a **power**. The word *power* can also refer to the exponent. The **exponent** indicates the number of times the base is used as a factor. In an expression of the form x^n, the **base** is x. The expression x^n is read "x to the nth power." When no exponent is shown, it is understood to be 1. For example, $a = a^1$.

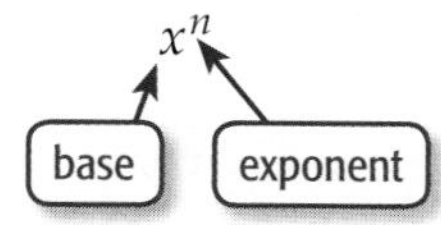

EXAMPLE 1 Write Verbal Expressions

Write a verbal expression for each algebraic expression.

a. $3x^4$
three times x to the fourth power

b. $5z^2 + 16$
5 times z to the second power plus sixteen

Check Your Progress

1A. $16u^2 - 3$

1B. $\frac{1}{2}a + \frac{6b}{7}$

Personal Tutor glencoe.com

StudyTip

Order of Operations Remember to follow the order of operations when writing a sentence to represent an algebraic expression.

Write Algebraic Expressions Another important skill is translating verbal expressions into algebraic expressions.

Key Concept For Your FOLDABLE

Translating Verbal to Algebraic Expressions

Operation	Verbal Phrases
Addition	more than, sum, plus, increased by, added to
Subtraction	less than, subtracted from, difference, decreased by, minus
Multiplication	product of, multiplied by, times, of
Division	quotient of, divided by

EXAMPLE 2 Write Algebraic Expressions

Write an algebraic expression for each verbal expression.

a. a number t more than 6

The words *more than* suggest addition.
Thus, the algebraic expression is $6 + t$ or $t + 6$.

b. 10 less than the product of 7 and f

Less than implies subtraction, and *product* suggests multiplication.
So the expression is written as $7f - 10$.

c. two thirds of the volume v

The word *of* with a fraction implies that you should multiply.
The expression could be written as $\frac{2}{3}v$ or $\frac{2v}{3}$.

Check Your Progress

2A. the product of p and 6

2B. one third of the area a

Personal Tutor glencoe.com

Variables can represent quantities that are known and quantities that are unknown. They are also used in formulas, expressions, and equations.

Real-World Career

Sports Marketing Sports marketers promote and manage athletes, teams, facilities and sports-related businesses and organizations. A minimum of a bachelor's degree in sports management or business administration is preferred.

Real-World EXAMPLE 3 Write an Expression

SPORTS MARKETING Mr. Martinez orders 250 key chains printed with his athletic team's logo and 500 pencils printed with their Web address. Write an algebraic expression that represents the cost of the order.

Let k be the cost of each key chain and p be the cost of each pencil. Then the cost of the key chains is $250k$ and the cost of the pencils is $500p$. The cost of the order is represented by $250k + 500p$.

Check Your Progress

3. COFFEE SHOP Katie estimates that $\frac{1}{8}$ of the people who order beverages also order pastries. Write an algebraic expression to represent this situation.

Personal Tutor glencoe.com

Check Your Understanding

Example 1 p. 5

Write a verbal expression for each algebraic expression.

1. $2m$
2. $\frac{2}{3}r^4$
3. $a^2 - 18b$

Example 2 p. 6

Write an algebraic expression for each verbal expression.

4. the sum of a number and 14
5. 6 less a number t
6. 7 more than 11 times a number
7. 1 minus the quotient of r and 7
8. two fifths of a number j squared
9. n cubed increased by 5

Example 3 p. 6

10. **GROCERIES** Mr. Bailey purchased some groceries that cost d dollars. He paid with a \$50 bill. Write an expression for the amount of change he will receive.

Practice and Problem Solving

 = **Step-by-Step Solutions** begin on page R12.
Extra Practice begins on page 815.

Example 1 p. 5

Write a verbal expression for each algebraic expression.

11. $4q$
12. $\frac{1}{8}y$
13. $15 + r$
14. $w - 24$
15. $3x^2$
16. $\frac{r^4}{9}$
17. $2a + 6$
18. $r^4 \cdot t^3$

Example 2 p. 6

Write an algebraic expression for each verbal expression.

19. x more than 7
20. a number less 35
21. 5 times a number
22. one third of a number
23. f divided by 10
24. the quotient of 45 and r
25. three times a number plus 16
26. 18 decreased by 3 times d
27. k squared minus 11
28. 20 divided by t to the fifth power

Example 3 p. 6

29. **GEOMETRY** The volume of a cylinder is π times the radius r squared multiplied by the height h. Write an expression for the volume.

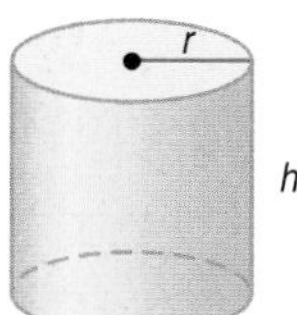

30. **FINANCIAL LITERACY** Jocelyn makes x dollars per hour working at the grocery store and n dollars per hour babysitting. Write an expression that describes her earnings if she babysat for 25 hours and worked at the grocery store for 15 hours.

Write a verbal expression for each algebraic expression.

31. $25 + 6x^2$
32. $6f^2 + 5f$
33. $\frac{3a^5}{2}$

34. **HEALTH** If the body mass index (BMI) is 25 or higher, then you are at a higher risk for heart disease. The BMI is the product of 703 and the quotient of the weight in pounds and the square of the height in inches.
    a. Write an expression that describes how to calculate the BMI.
    b. Calculate the BMI for a 140-pound person who is 65 inches tall.
    c. Calculate the BMI for a 155-pound person who is 5 feet 8 inches tall.

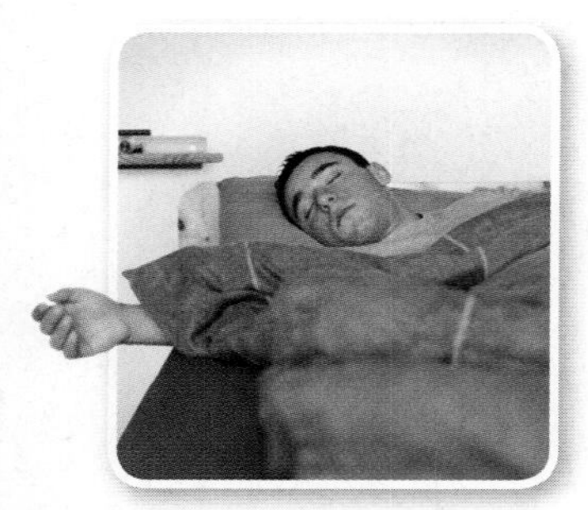

Real-World Link

About 20% of our dreams are about animals. About $\frac{3}{4}$ of our dreams involve people that we know.

Source: National Dream Hotline

35. DREAMS Refer to the information at the left.

a. Write an expression to describe the number of dreams that feature people you know if you have *d* dreams.

b. Use the expression you wrote to predict the number of dreams that include people you know out of 28 dreams.

36. SPORTS In football, a touchdown is awarded 6 points and the team can then try for a point after a touchdown.

a. Write an expression that describes the number of points scored on touchdowns and points after touchdowns by one team in a game.

b. If a team wins a football game 27-0, write an equation to represent the possible number of touchdowns and points after touchdowns by the winning team.

c. If a team wins a football game 7-21, how many possible number of touchdowns and points after touchdowns were scored during the game by both teams?

37. MULTIPLE REPRESENTATIONS In this problem, you will explore the multiplication of powers with like bases.

a. **TABULAR** Copy and complete the table.

10^2	×	10^1	=	10 × 10 × 10	=	10^3
10^2	×	10^2	=	10 × 10 × 10 × 10	=	10^4
10^2	×	10^3	=	10 × 10 × 10 × 10 × 10	=	?
10^2	×	10^4	=	?	=	?

b. **ALGEBRAIC** Write an equation for the pattern in the table.

c. **VERBAL** Make a conjecture about the exponent of the product of two powers.

H.O.T. Problems

Use Higher-Order Thinking Skills

38. REASONING Explain the differences between an algebraic expression and a verbal expression.

39. OPEN ENDED Define a variable to represent a real-life quantity, such as time in minutes or distance in feet. Then use the variable to write an algebraic expression to represent one of your daily activities. Describe in words what your expression represents, and explain your reasoning.

40. FIND THE ERROR Consuelo and James are writing an algebraic expression for *three times the sum of n squared and 3*. Is either of them correct? Explain your reasoning.

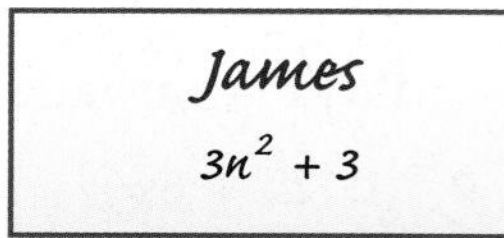

41. CHALLENGE For the cube, *x* represents a positive whole number. Find the value of *x* such that the volume of the cube and 6 times the area of one of its faces have the same value.

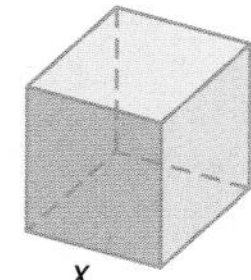

42. WRITING IN MATH Describe how to write an algebraic expression from a real-world situation. Include a definition of algebraic expression in your own words.

Standardized Test Practice

43. Which expression best represents the volume of the cube?

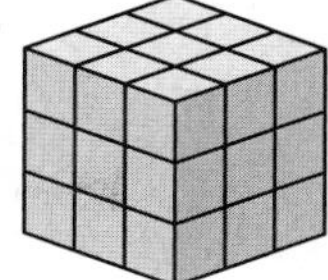

A the product of three and five
B three to the fifth power
C three squared
D three cubed

44. Which expression best represents the perimeter of the rectangle?

ℓ

w

F $2\ell w$
G $\ell + w$
H $2\ell + 2w$
J $4(\ell + w)$

45. SHORT RESPONSE The yards of fabric needed to make curtains is 3 times the length of a window in inches, divided by 36. Write an expression that represents the yards of fabric needed in terms of the length of the window ℓ.

46. GEOMETRY Find the area of the rectangle.

A 14 square meters
B 16 square meters
C 50 square meters
D 60 square meters

Spiral Review

47. AMUSEMENT PARKS A roller coaster enthusiast club took a poll to see what each member's favorite ride was. Make a bar graph of the results. (Lesson 0-13)

Our Favorite Rides							
Ride	Big Plunge	Twisting Time	The Shiner	Raging Bull	The Bat	Teaser	The Adventure
Number of Votes	5	22	16	9	25	6	12

48. SPORTS The results for an annual 5K race are shown at the right. Make a box-and-whisker plot for the data. Write a sentence describing what the length of the box-and-whisker plot tells about the times for the race. (Lesson 0–13)

Annual 5-K Race Results			
Joe	14:48	Carissa	19:58
Jessica	19:27	Jordan	14:58
Lupe	15:06	Taylor	20:47
Dante	20:39	Mi-Ling	15:48
Tia	15:54	Winona	21:35
Amber	20:49	Angel	16:10
Amanda	16:30	Catalina	20:21

Find the mean, median, and mode for each set of data. (Lesson 0–12)

49. {7, 6, 5, 7, 4, 8, 2, 2, 7, 8}

50. {−1, 0, 5, 2, −2, 0 ,−1, 2, −1, 0}

51. {17, 24, 16, 3, 12, 11, 24, 15}

52. SPORTS Lisa has a rectangular trampoline that is 6 feet long and 12 feet wide. What is the area of her trampoline in square feet? (Lesson 0–8)

Find each product or quotient. (Lesson 0–5)

53. $\frac{3}{5} \cdot \frac{7}{11}$

54. $\frac{4}{3} \div \frac{7}{6}$

55. $\frac{5}{6} \cdot \frac{8}{3}$

Skills Review

Evaluate each expression. (Lesson 0–4)

56. $\frac{3}{5} + \frac{4}{9}$

57. $5.67 - 4.21$

58. $\frac{5}{6} - \frac{8}{3}$

59. $10.34 + 14.27$

60. $\frac{11}{12} + \frac{5}{36}$

61. $37.02 - 15.86$

1-2 Order of Operations

Then

You expressed algebraic expressions verbally. (Lesson 1-1)

Now

- Evaluate numerical expressions by using the order of operations.
- Evaluate algebraic expressions by using the order of operations.

New Vocabulary

evaluate
order of operations

Math Online

glencoe.com

- Extra Examples
- Personal Tutor
- Self-Check Quiz
- Homework Help

Why?

The admission prices for SeaWorld Adventure Park in Orlando, Florida, are shown in the table. If four adults and three children go the park, the expression below represents the cost of admission for the group.

$$4(64.95) + 3(53.95)$$

Ticket	Price ($)
Adult	64.95
Child	53.95

Evaluate Numerical Expressions To find the cost of admission, the expression $4(64.95) + 3(53.95)$ must be evaluated. To **evaluate** an expression means to find its value.

EXAMPLE 1 Evaluate Expressions

Evaluate 3^5.

$3^5 = 3 \cdot 3 \cdot 3 \cdot 3 \cdot 3$	**Use 3 as a factor 5 times.**
$= 243$	**Multiply.**

Check Your Progress

1A. 2^4 **1B.** 4^5 **1C.** 7^3

Personal Tutor glencoe.com

The numerical expression that represents the cost of admission contains more than one operation. The rule that lets you know which operation to perform first is called the **order of operations**.

Key Concept Order of Operations

For Your FOLDABLE

Step 1 Evaluate expressions inside grouping symbols.

Step 2 Evaluate all powers.

Step 3 Multiply and/or divide from left to right.

Step 4 Add and/or subtract from left to right.

EXAMPLE 2 Use Order of Operations

Evaluate $16 - 8 \div 2^2 + 14$.

$16 - 8 \div 2^2 + 14 = 16 - 8 \div 4 + 14$	**Evaluate powers.**
$= 16 - 2 + 14$	**Divide 8 by 4.**
$= 14 + 14$	**Subtract 2 from 16.**
$= 28$	**Add 14 and 14.**

Check Your Progress Evaluate each expression.

2A. $3 + 42 \cdot 2 - 5$ **2B.** $20 - 7 + 8^2 - 7 \cdot 11$

Personal Tutor glencoe.com

StudyTip

Grouping Symbols Grouping symbols such as parentheses ( ), brackets [ ], and braces { } are used to clarify or change the order of operations.

When one or more grouping symbols are used, evaluate within the innermost grouping symbols first.

EXAMPLE 3 Expressions with Grouping Symbols

Evaluate each expression.

a. $4 \div 2 + 5(10 - 6)$

$4 \div 2 + 5(10 - 6) = 4 \div 2 + 5(4)$	**Evaluate inside parentheses.**
$= 2 + 5(4)$	**Divide 4 by 2.**
$= 2 + 20$	**Multiply 5 by 4.**
$= 22$	**Add 2 to 20.**

b. $6\left[32 - (2 + 3)^2\right]$

$6\left[32 - (2 + 3)^2\right] = 6\left[32 - (5)^2\right]$	**Evaluate innermost expression first.**
$= 6[32 - 25]$	**Evaluate power.**
$= 6[7]$	**Subtract 25 from 32.**
$= 42$	**Multiply.**

c. $\dfrac{2^3 - 5}{15 + 9}$

$\dfrac{2^3 - 5}{15 + 9} = \dfrac{8 - 5}{15 + 9}$	**Evaluate the power in the numerator.**
$= \dfrac{3}{15 + 9}$	**Subtract 5 from 8 in the numerator.**
$= \dfrac{3}{24}$ or $\dfrac{1}{8}$	**Add 15 and 9 in denominator, and simplify.**

StudyTip

Grouping Symbols A fraction bar is considered a grouping symbol. So, evaluate expressions in the numerator and denominator before completing the division.

Check Your Progress

3A. $5 \cdot 4(10 - 8) + 20$ **3B.** $15 - \left[10 + (3 - 2)^2\right] + 6$ **3C.** $\dfrac{(4 + 5)^2}{3(7 - 4)}$

Personal Tutor glencoe.com

Evaluate Algebraic Expressions To evaluate an algebraic expression, replace the variables with their values. Then find the value of the numerical expression using the order of operations.

EXAMPLE 4 Evaluate an Algebraic Expression

Evaluate $3x^2 + \left(2y + z^3\right)$ if $x = 4$, $y = 5$, $z = 3$.

$3x^2 + \left(2y + z^3\right)$

$= 3(4)^2 + \left(2 \cdot 5 + 3^3\right)$	**Replace *x* with 4, *y* with 5, and *z* with 3.**
$= 3(4)^2 + (2 \cdot 5 + 27)$	**Evaluate 3^3.**
$= 3(4)^2 + (10 + 27)$	**Multiply 2 by 10.**
$= 3(4)^2 + (37)$	**Add 10 to 27.**
$= 3(16) + 37$	**Evaluate 4^2.**
$= 48 + 37$	**Multiply 3 by 16.**
$= 85$	**Add 48 to 37.**

Check Your Progress

Evaluate each expression.

4A. $a^2(3b + 5) \div c$ if $a = 2, b = 6, c = 4$ **4B.** $5d + (6f - g)$ if $d = 4, f = 3, g = 12$

Personal Tutor glencoe.com

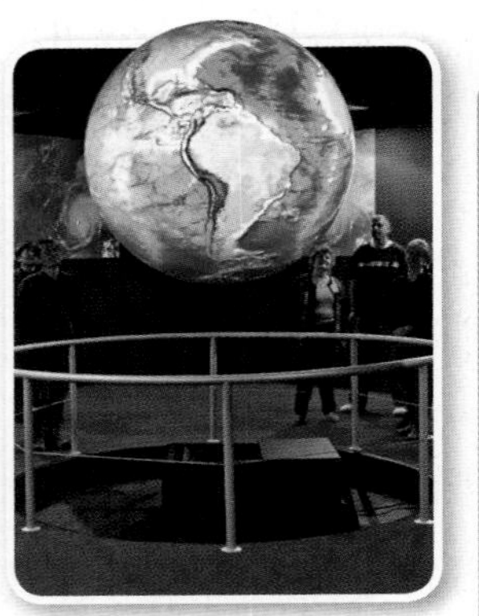

Real-World Link

The National Oceanic & Atmospheric Administration (NOAA) developed the Science on a Sphere system to educate people about Earth's processes. There are five computers and four video projectors that power the sphere.

Source: NOAA

Real-World EXAMPLE 5 Write and Evaluate an Expression

ENVIRONMENTAL STUDIES Science on a Sphere (SOS)® demonstrates the effects of atmospheric storms, climate changes, and ocean temperature on the environment. The volume of a sphere is four thirds of π multiplied by the radius r to the third power.

a. Write an expression that represents the volume of a sphere.

Words	four thirds	of	π multiplied by radius to the third power
Variable	Let r = radius.		
Expression	$\frac{4}{3}$	$\times$	πr^3 or $\frac{4}{3}\pi r^3$

b. Find the volume of the 3-foot radius sphere used for SOS.

$V = \frac{4}{3}\pi r^3$	**Volume of a sphere**
$= \frac{4}{3}\pi(3)^3$	**Replace r with 3.**
$= \left(\frac{4}{3}\right)\pi(27)$	**Evaluate $3^3 = 27$.**
$= 36\pi$	**Multiply $\frac{4}{3}$ by 27.**

The volume of the sphere is 36π cubic feet.

Check Your Progress

5. FOREST FIRES According to the California Department of Forestry, an average of 539.2 fires each year are started by burning debris, while campfires are responsible for an average of 129.1 each year.

A. Write an algebraic expression that represents the number of fires, on average, in d years of debris burning and c years of campfires.

B. How many fires would there be in 5 years?

Personal Tutor glencoe.com

Check Your Understanding

Examples 1–3 pp. 10–11

Evaluate each expression.

1. 9^2
2. 4^4
3. 3^5
4. $30 - 14 \div 2$
5. $5 \cdot 5 - 1 \cdot 3$
6. $(2 + 5)4$
7. $[8(2) - 4^2] + 7(4)$
8. $\frac{11 - 8}{1 + 7 \cdot 2}$
9. $\frac{(4 \cdot 3)^2}{9 + 3}$

Example 4 p. 11

Evaluate each expression if $a = 4$, $b = 6$, and $c = 8$.

10. $8b - a$
11. $2a + (b^2 \div 3)$
12. $\frac{b(9 - c)}{a^2}$

Example 5 p. 12

13. **BOOKS** Akira bought one new book for $20 and three used books for $4.95 each. Write and evaluate an expression to find how much money the books cost.

14. **FOOD** Koto purchased food for herself and her friends. She bought 4 cheeseburgers for $2.25 each, 3 French fries for $1.25 each, and 4 drinks for $4.00. Write and evaluate an expression to find how much the food cost.

Practice and Problem Solving

● = **Step-by-Step Solutions** begin on page R12.
Extra Practice begins on page 815.

Examples 1–3 pp. 10–11

Evaluate each expression.

15. 7^2
16. 14^3
17. 2^6
18. $35 - 3 \cdot 8$
19. $18 \div 9 + 2 \cdot 6$
20. $10 + 8^3 \div 16$
21. $24 \div 6 + 2^3 \cdot 4$
22. $(11 \cdot 7) - 9 \cdot 8$
23. $29 - 3(9 - 4)$
24. $(12 - 6) \cdot 5^2$
25. $3^5 - (1 + 10^2)$
26. $108 \div [3(9 + 3^2)]$
27. $[(6^3 - 9) \div 23]4$
28. $\dfrac{8 + 3^3}{12 - 7}$
29. $\dfrac{(1 + 6)9}{5^2 - 4}$

Example 4 p. 11

Evaluate each expression if $g = 2$, $r = 3$, and $t = 11$.

30. $g + 6t$
31. $7 - gr$
32. $r^2 + (g^3 - 8)^5$
33. $(2t + 3g) \div 4$
34. $t^2 + 8rt + r^2$
35. $3g(g + r)^2 - 1$

36. **GEOMETRY** Write an algebraic expression to represent the area of the triangle. Then evaluate it to find the area when $h = 12$ inches.

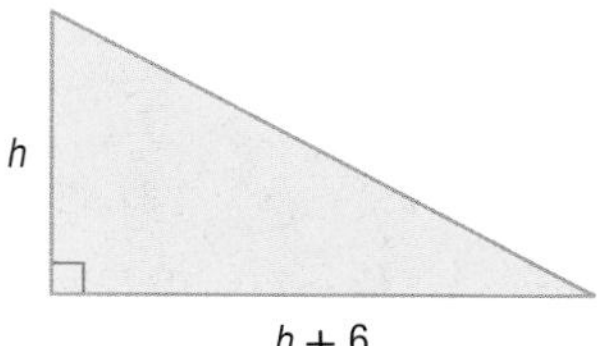

37. **AMUSEMENT PARKS** In 1997, there were 3344 amusement parks and arcades. This decreased by 148 by 2002. Write and evaluate an expression to find the number of amusement parks and arcades in 2002.

38. **SPORTS** Kamilah works at the Duke University Athletic Ticket Office. One week she sold 15 preferred season tickets, 45 blue zone tickets, and 55 general admission tickets. Write and evaluate an expression to find the amount of money Kamilah processed.

Duke University Football Ticket Prices	
Preferred Season Ticket	$100
Blue Zone	$80
General Admission	$70

Source: Duke University

Evaluate each expression.

39. 4^2
40. 12^3
41. 3^6
42. 11^5
43. $(3 - 4^2)^2 + 8$
44. $23 - 2(17 + 3^3)$
45. $3[4 - 8 + 4^2(2 + 5)]$
46. $\dfrac{2 \cdot 8^2 - 2^2 \cdot 8}{2 \cdot 8}$
47. $25 + \left[(16 - 3 \cdot 5) + \dfrac{12 + 3}{5}\right]$
48. $7^3 - \frac{2}{3}(13 \cdot 6 + 9)4$

Evaluate each expression if $a = 8$, $b = 4$, and $c = 16$.

49. $a^2bc - b^2$
50. $\dfrac{c^2}{b^2} + \dfrac{b^2}{a^2}$
51. $\dfrac{2b + 3c^2}{4a^2 - 2b}$
52. $\dfrac{3ab + c^2}{a}$
53. $\left(\dfrac{a}{b}\right)^2 - \dfrac{c}{a - b}$
54. $\dfrac{2a - b^2}{ab} + \dfrac{c - a}{b^2}$

55. **SALES** One day, 28 small and 12 large merchant spaces were rented. Another day, 30 small and 15 large spaces were rented. Write and evaluate an expression to show the total rent collected.

THE FLEA MARKET
MERCHANT SPACE RENTALS
Small space $7.00/day
Large space $9.75/day
Open Daily from 9:00–6:00

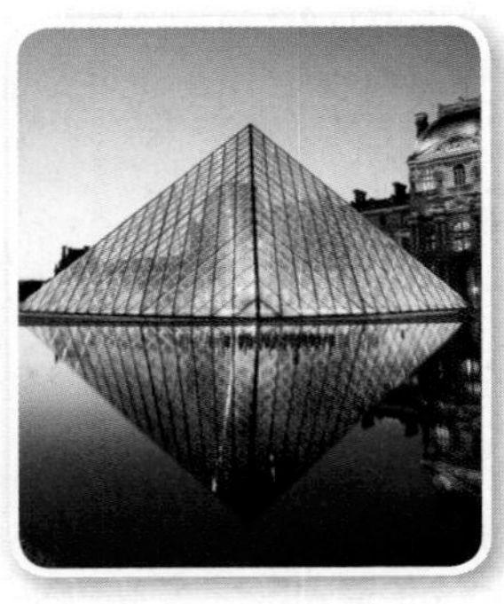

Real-World Link

The pyramid at the Louvre in Paris was designed by famed architect I.M. Pei. He also designed the Rock & Roll Hall of Fame in Cleveland, Ohio.

Source: Infoplease

56. **SHOPPING** Evelina is shopping for back-to-school clothes. She bought 3 skirts, 2 pairs of jeans, and 4 sweaters. Write and evaluate an expression to find out how much money Evelina spent on clothes, without including sales tax.

Clothing	
skirt	\$25.99
jeans	\$39.99
sweater	\$22.99

57. **PYRAMIDS** The pyramid at the Louvre has a square base with a side of 35.42 meters and a height of 21.64 meters. The Great Pyramid in Egypt has a square base with a side of 230 meters and a height of 146.5 meters. The expression for the volume of a pyramid is $\frac{1}{3}Bh$, where B is the area of the base and h is the height.
    a. Draw both pyramids and label the dimensions.
    b. Write a verbal expression for the difference in volume of the two pyramids.
    c. Write an algebraic expression for the difference in volume of the two pyramids. Find the difference in volume.

58. **FINANCIAL LITERACY** A sales representative receives an annual salary s, an average commission each month c, and a bonus b for each sales goal that she reaches.
    a. Write an algebraic expression to represent her total earnings in one year if she receives four equal bonuses.
    b. Suppose her annual salary is \$52,000 and her average commission is \$1225 per month. If each of the four bonuses equals \$1150, what does she earn annually?

H.O.T. Problems Use Higher-Order Thinking Skills

59. **FIND THE ERROR** Tara and Curtis are simplifying $[4(10) - 3^2] + 6(4)$. Is either of them correct? Explain your reasoning.

Tara	Curtis
$= [4(10) - 9] + 6(4)$	$= [4(10) - 9] + 6(4)$
$= 4(1) + 6(4)$	$= (40 - 9) + 6(4)$
$= 4 + 6(4)$	$= 31 + 6(4)$
$= 4 + 24$	$= 31 + 24$
$= 28$	$= 55$

60. **REASONING** Explain how to evaluate $a[(b - c) \div d] - f$ if you were given values for a, b, c, d, and f. How would you evaluate the expression differently if the expression was $a \cdot b - c \div d - f$?

61. **CHALLENGE** Write an expression using the whole numbers 1 to 5 using all five digits and addition and/or subtraction to create a numeric expression with a value of 3.

62. **OPEN ENDED** Write an expression that uses exponents, at least three different operations and two sets of parentheses. Explain the steps you would take to evaluate the expression.

63. **WRITING IN MATH** Choose a geometric formula and explain how the order of operations applies when using the formula.

64. **WRITING IN MATH** Equivalent expression have the same value. Are the expressions $(30 + 17) \times 10$ and $10 \times 30 + 10 \times 17$ equivalent? Explain why or why not.

Standardized Test Practice

65. Let m represent the number of miles. Which algebraic expression represents the number of feet in m miles?

A $5280m$

B $\frac{5280}{m}$

C $m + 5280$

D $5280 - m$

66. SHORT RESPONSE
Simplify: $[10 + 15(2^3)] \div [7(2^2) - 2]$

Step 1 $[10 + 15(8)] \div [7(4) - 2]$

Step 2 $[10 + 120] \div [28 - 2]$

Step 3 $130 \div 26$

Step 4 $\frac{1}{5}$

Which is the first *incorrect* step? Explain the error.

67. EXTENDED RESPONSE A local movie theater has advertised that one out of every four customers will receive a free popcorn with the purchase of a movie ticket. So far, 25 of the first 80 customers have won.

Part A Based on the results so far, what is the experimental probability that a customer will win?

Part B What is the theoretical probability that a customer will win?

Part C Explain the difference between theoretical and experimental probabilities.

68. GEOMETRY What is the perimeter of the triangle if $a = 9$ and $b = 10$?

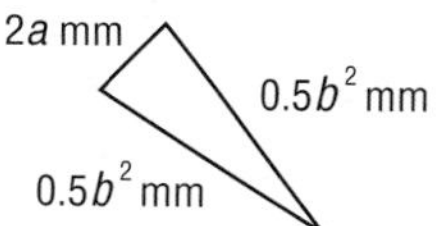

F 164 mm

G 118 mm

H 28 mm

J 4 mm

Spiral Review

Write a verbal expression for each algebraic expression. (Lesson 1-1)

69. $14 - 9c$

70. $k^3 + 13$

71. $\frac{4 - v}{w}$

72. MONEY Destiny earns $8 per hour babysitting and $15 for each lawn she mows. Write an expression to show the amount of money she earns babysitting h hours and mowing m lawns. (Lesson 1-1)

Find the area of each figure. (Lesson 0-7)

73.

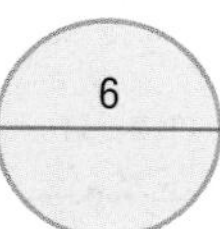

74.

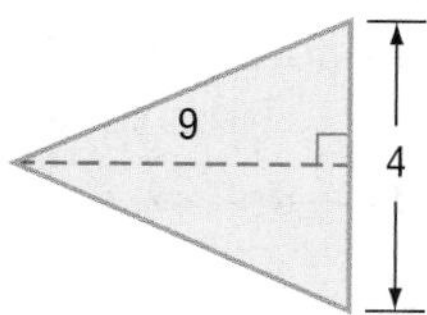

75.

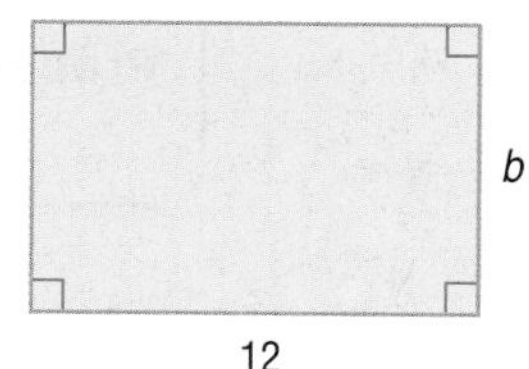

76. SCHOOL Aaron correctly answered 27 out of 30 questions on his last biology test. What percent of the questions did he answer correctly? (Lesson 0-5)

Skills Review

Find the value of each expression. (Lessons 0-4 and 0-5)

77. $5.65 - 3.08$

78. $6 \div \frac{4}{5}$

79. $4.85(2.72)$

80. $1\frac{1}{12} + 3\frac{2}{3}$

81. $\frac{4}{9} \cdot \frac{3}{2}$

82. $7\frac{3}{4} - 4\frac{7}{10}$

1-3 Properties of Numbers

Then

You used the order of operations to simplify expressions. (Lesson 1-2)

Now

- Recognize the properties of equality and identity.
- Recognize the Commutative and Associative Properties.

New Vocabulary

equivalent expressions
additive identity
multiplicative identity
multiplicative inverse
reciprocal

Math Online

glencoe.com

- Extra Examples
- Personal Tutor
- Self-Check Quiz
- Homework Help
- Math in Motion

Why?

Nate lives 32 miles away from the mall. The distance from his house to the mall is the same as the distance from the mall to his house. This is an example of the *Reflexive Property*.

Properties of Equality and Identity The expressions $4k + 8k$ and $12k$ are called **equivalent expressions** because they represent the same number. The properties below allow you to write an equivalent expression for a given expression.

Key Concept — Properties of Equality (For Your FOLDABLE)

Property	Words	Symbols	Examples
Reflexive Property	Any quantity is equal to itself.	For any number a, $a = a$.	$5 = 5$ $4 + 7 = 4 + 7$
Symmetric Property	If one quantity equals a second quantity, then the second quantity equals the first.	For any numbers a and b, if $a = b$, then $b = a$.	If $8 = 2 + 6$, then $2 + 6 = 8$.
Transitive Property	If one quantity equals a second quantity and the second quantity equals a third quantity, then the first quantity equals the third quantity.	For any numbers a, b, and c, if $a = b$ and $b = c$, then $a = c$.	If $6 + 9 = 3 + 12$ and $3 + 12 = 15$, then $6 + 9 = 15$.
Substitution Property	A quantity may be substituted for its equal in any expression.	If $a = b$, then a may be replaced by b in any expression.	If $n = 11$, then $4n = 4 \cdot 11$

The sum of any number and 0 is equal to the number. Thus, 0 is called the **additive identity**.

Key Concept — Addition Properties (For Your FOLDABLE)

Property	Words	Symbols	Examples
Additive Identity	For any number a, the sum of a and 0 is a.	$a + 0 = 0 + a = a$	$2 + 0 = 2$ $0 + 2 = 2$
Additive Inverse	A number and its opposite are additive inverses of each other.	$a + (-a) = 0$	$3 + (-3) = 0$ $4 - 4 = 0$

There are also special properties associated with multiplication. Consider the following equations.

$4 \cdot n = 4$

The solution of the equation is 1. Since the product of any number and 1 is equal to the number, 1 is called the **multiplicative identity**.

$6 \cdot m = 0$

The solution of the equation is 0. The product of any number and 0 is equal to 0. This is called the **Multiplicative Property of Zero**.

Two numbers whose product is 1 are called **multiplicative inverses** or **reciprocals**. Zero has no reciprocal because any number times 0 is 0.

StudyTip

Properties and Identities These properties are true for all real numbers. They are also referred to as *field properties*.

Key Concept Multiplication Properties

For Your FOLDABLE

Property	Words	Symbols	Example
Multiplicative Identity	For any number a, the product of a and 1 is a.	$a \cdot 1 = 1$ $1 \cdot a = a$	$14 \cdot 1 = 14$ $1 \cdot 14 = 14$
Multiplicative Property of Zero	For any number a, the product of a and 0 is 0.	$a \cdot 0 = 0$ $0 \cdot a = 0$	$9 \cdot 0 = 0$ $0 \cdot 9 = 0$
Multiplicative Inverse	For every number $\frac{a}{b}$, where a, $b \neq 0$, there is exactly one number $\frac{b}{a}$ such that the product of $\frac{a}{b}$ and $\frac{b}{a}$ is 1.	$\frac{a}{b} \cdot \frac{b}{a} = 1$ $\frac{b}{a} \cdot \frac{a}{b} = 1$	$\frac{4}{5} \cdot \frac{5}{4} = \frac{20}{20}$ or 1 $\frac{5}{4} \cdot \frac{4}{5} = \frac{20}{20}$ or 1

EXAMPLE 1 Evaluate Using Properties

Evaluate $7(4 - 3) - 1 + 5 \cdot \frac{1}{5}$. Name the property used in each step.

$7(4 - 3) - 1 + 5 \cdot \frac{1}{5} = 7(1) - 1 + 5 \cdot \frac{1}{5}$ Substitution: $4 - 3 = 1$

$= 7 - 1 + 5 \cdot \frac{1}{5}$ Multiplicative Identity: $7 \cdot 1 = 7$

$= 7 - 1 + 1$ Multiplicative Inverse: $5 \cdot \frac{1}{5} = 1$

$= 6 + 1$ Substitution: $7 - 1 = 6$

$= 7$ Substitution: $6 + 1 = 7$

Check Your Progress

Name the property used in each step.

1A. $2 \cdot 3 + (4 \cdot 2 - 8)$
$= 2 \cdot 3 + (8 - 8)$ ___?___
$= 2 \cdot 3 + (0)$ ___?___
$= 6 + 0$ ___?___
$= 6$ ___?___

1B. $7 \cdot \frac{1}{7} + 6(15 \div 3 - 5)$
$= 7 \cdot \frac{1}{7} + 6(5 - 5)$ ___?___
$= 7 \cdot \frac{1}{7} + 6(0)$ ___?___
$= 1 + 6(0)$ ___?___
$= 1 + 0$ ___?___
$= 1$ ___?___

Personal Tutor glencoe.com

Use Commutative and Associate Properties Nikki walks 2 blocks to her friend Sierra's house. They walk another 4 blocks to school. At the end of the day, Nikki and Sierra walk back to Sierra's house, and then Nikki walks home.

The distance from Nikki's house to school	equals	the distance from the school to Nikki's house.
$2 + 4$	$=$	$4 + 2$

This is an example of the **Commutative Property** for addition.

Key Concept — Commutative Property

For Your FOLDABLE

Words The order in which you add or multiply numbers does not change their sum or product.

Symbols For any numbers a and b, $a + b = b + a$ and $a \cdot b = b \cdot a$.

Examples $4 + 8 = 8 + 4$ $\quad 7 \cdot 11 = 11 \cdot 7$

Math *in Motion,* BrainPOP® glencoe.com

An easy way to find the sum or product of numbers is to group, or associate, the numbers using the **Associative Property**.

Key Concept — Associative Property

For Your FOLDABLE

Words The way you group three or more numbers when adding or multiplying does not change their sum or product.

Symbols For any numbers a, b, and c,
$(a + b) + c = a + (b + c)$ and $(ab)c = a(bc)$.

Examples $(3 + 5) + 7 = 3 + (5 + 7)$ $\quad (2 \cdot 6) \cdot 9 = 2 \cdot (6 \cdot 9)$

Math *in Motion,* BrainPOP® glencoe.com

Real-World Link

A child's birthday party may cost about $200 depending on the number of children invited.

Source: Family Corner

Real-World EXAMPLE 2 Apply Properties of Numbers

PARTY PLANNING Eric makes a list of items that he needs to buy for a party and their costs. Find the total cost of these items.

Party Supplies	
Item	**Cost ($)**
balloons	6.75
decorations	14.00
food	23.25
beverages	20.50

Balloons		Decorations		Food		Beverages
6.75	+	14.00	+	23.25	+	20.50

$= 6.75 + 23.25 + 14.00 + 20.50$ Commutative (+)
$= (6.75 + 23.25) + (14.00 + 20.50)$ Associative (+)
$= 30.00 + 34.50$ Substitution
$= 64.50$ Substitution

The total cost is $64.50.

Check Your Progress

2. FURNITURE Rafael is buying furnishings for his first apartment. He buys a couch for $300, lamps for $30.50, a rug for $25.50, and a table for $50. Find the total cost of these items.

Personal Tutor glencoe.com

EXAMPLE 3 Use Multiplication Properties

Evaluate $5 \cdot 7 \cdot 4 \cdot 2$ using the properties of numbers. Name the property used in each step.

$$
\begin{aligned}
5 \cdot 7 \cdot 4 \cdot 2 &= 5 \cdot 2 \cdot 7 \cdot 4 && \text{Commutative } (\times)\\
&= (5 \cdot 2) \cdot (7 \cdot 4) && \text{Associative } (\times)\\
&= 10 \cdot 28 && \text{Substitution}\\
&= 280 && \text{Substitution}
\end{aligned}
$$

Check Your Progress

Evaluate each expression using the properties of numbers. Name the property used in each step.

3A. $2.9 \cdot 4 \cdot 10$

3B. $\frac{5}{3} \cdot 25 \cdot 3 \cdot 2$

Personal Tutor glencoe.com

Check Your Understanding

Example 1 p. 17

Evaluate each expression. Name the property used in each step.

1. $(1 \div 5)5 \cdot 14$

2. $6 + 4(19 - 15)$

3. $5(14 - 5) + 6(3 + 7)$

4. FINANCIAL LITERACY Carolyn has 9 quarters, 4 dimes, 7 nickels, and 2 pennies, which can be represented as $9(25) + 4(10) + 7(5) + 2$. Evaluate the expression to find how much money she has. Name the property used in each step.

Examples 2 and 3 pp. 18–19

Evaluate each expression using the properties of numbers. Name the property used in each step.

5. $23 + 42 + 37$

6. $2.75 + 3.5 + 4.25 + 1.5$

7. $3 \cdot 7 \cdot 10 \cdot 2$

8. $\frac{1}{4} \cdot 24 \cdot \frac{2}{3}$

Practice and Problem Solving

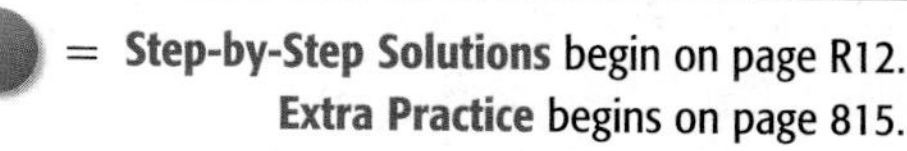
= **Step-by-Step Solutions** begins on page R12.
Extra Practice begins on page 815.

Example 1 p. 17

Evaluate each expression. Name the property used in each step.

9. $3(22 - 3 \cdot 7)$

10. $7 + (9 - 3^2)$

11. $\frac{3}{4}[4 \div (7 - 4)]$

12. $[3 \div (2 \cdot 1)]\frac{2}{3}$

13. $2(3 \cdot 2 - 5) + 3 \cdot \frac{1}{3}$

14. $6 \cdot \frac{1}{6} + 5(12 \div 4 - 3)$

Example 2 p. 18

15. GEOMETRY The expression $2 \cdot \frac{22}{7} \cdot 14^2 + 2 \cdot \frac{22}{7} \cdot 14 \cdot 7$ represents the approximate surface area of the cylinder at the right. Evaluate this expression to find the approximate surface area. Name the property used in each step.

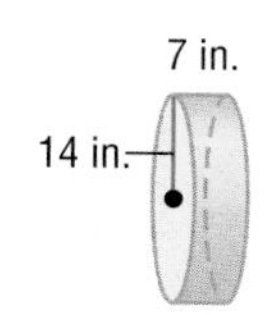

16. HOTEL RATES A traveler checks into a hotel on Friday and checks out the following Tuesday morning. Use the table to find the total cost of the room including tax.

Hotel Rates Per Day

Day	Room Charge	Sales Tax
Monday–Friday	$72	$5.40
Saturday–Sunday	$63	$5.10

Examples 2 and 3
pp. 18–19

Evaluate each expression using properties of numbers. Name the property used in each step.

17. $25 + 14 + 15 + 36$

18. $11 + 7 + 5 + 13$

19. $3\frac{2}{3} + 4 + 5\frac{1}{3}$

20. $4\frac{4}{9} + 7\frac{2}{9}$

21. $4.3 + 2.4 + 3.6 + 9.7$

22. $3.25 + 2.2 + 5.4 + 10.75$

23. $12 \cdot 2 \cdot 6 \cdot 5$

24. $2 \cdot 8 \cdot 10 \cdot 2$

25. $0.2 \cdot 4.6 \cdot 5$

26. $3.5 \cdot 3 \cdot 6$

27. $1\frac{5}{6} \cdot 24 \cdot 3\frac{1}{11}$

28. $2\frac{3}{4} \cdot 1\frac{1}{8} \cdot 32$

29. SCUBA DIVING The sign shows the equipment rented or sold by a scuba diving store.

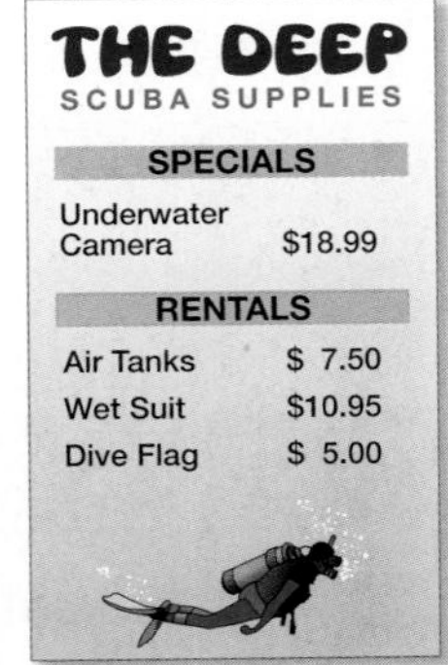

a. Write two expressions to represent the total sales to rent 2 wet suits, 3 air tanks, 2 dive flags, and selling 5 underwater cameras.

b. What are the total sales?

30. COOKIES Bobby baked 2 dozen chocolate chip cookies, 3 dozen sugar cookies, and a dozen oatmeal raisin cookies. How many total cookies did he bake?

Evaluate each expression if $a = -1$, $b = 4$, and $c = 6$.

31 $4a + 9b - 2c$

32. $-10c + 3a + a$

33. $a - b + 5a - 2b$

34. $8a + 5b - 11a - 7b$

35. $3c^2 + 2c + 2c^2$

36. $3a - 4a^2 + 2a$

37. FOOTBALL A football team is on the 35-yard line. The quarterback is sacked at the line of scrimmage. The team gains 0 yards, so they are still at the 35-yard line. Which identity or property does this represent? Explain.

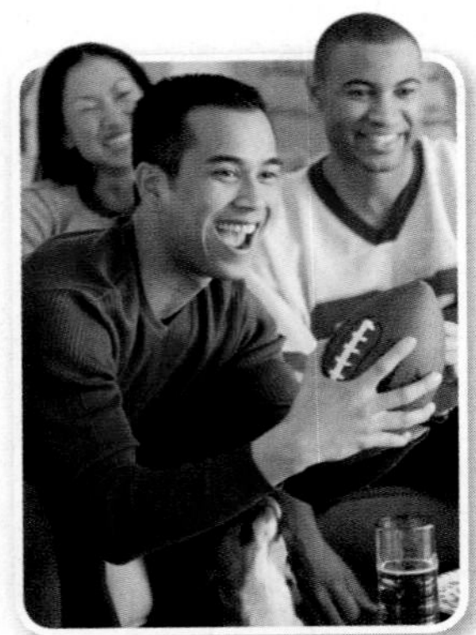

Real-World Link

Fantasy football has about 10 million participants per year. Fans participate in drafts to pick their teams and then compete throughout the season.

Source: CNN

Find the value of x. Then name the property used.

38. $8 = 8 + x$

39. $3.2 + x = 3.2$

40. $10x = 10$

41. $\frac{1}{2} \cdot x = \frac{1}{2} \cdot 7$

42. $x + 0 = 5$

43. $1 \cdot x = 3$

44. $5 \cdot \frac{1}{5} = x$

45. $2 + 8 = 8 + x$

46. $x + \frac{3}{4} = 3 + \frac{3}{4}$

47. $\frac{1}{3} \cdot x = 1$

48. GEOMETRY Write an expression to represent the perimeter of the triangle. Then find the perimeter if $x = 2$ and $y = 7$.

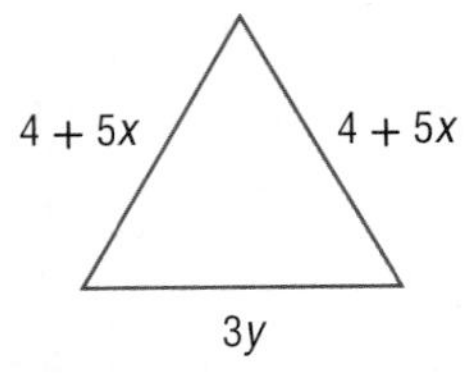

49. SPORTS Tickets to a baseball game cost $25 each plus a $4.50 handling charge per ticket. If Sharon has a coupon for $10 off and orders 4 tickets, how much will she be charged?

50. RETAIL The table shows prices on children's clothing.

Shorts	Shirts	Tank Tops
$7.99	$8.99	$6.99
$5.99	$4.99	$2.99

a. Write three different expressions that represent 8 pairs of shorts and 8 tops.

b. Evaluate the three expressions in part **a** to find the costs of the 16 items. What do you notice about all the total costs?

c. If you buy 8 shorts and 8 tops, you receive a discount of 15%. Find the greatest and least amount of money you can spend on the 16 items at the sale.

51. **GEOMETRY** A regular octagon measures $(3x + 5)$ units on each side. What is the perimeter if $x = 2$?

52. **MULTIPLE REPRESENTATIONS** You can use *algebra tiles* to model and explore algebraic expressions. The rectangular tile has an area of x, with dimensions 1 by x. The small square tile has an area of 1, with dimensions 1 by 1.

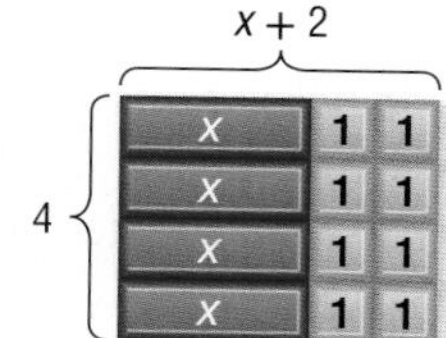

a. **CONCRETE** Make a rectangle with algebra tiles to model the expression $4(x + 2)$ as shown above. What are the dimensions of this rectangle? What is its area?

b. **ANALYTICAL** What are the areas of the green region and of the yellow region?

c. **VERBAL** Complete this statement: $4(x + 2) =$ __?__. Write a convincing argument to justify your statement.

53. **GEOMETRY** It is given that $\overline{AB} \cong \overline{CD}$, $\overline{AB} \cong \overline{BD}$, and $\overline{AB} \cong \overline{AC}$. Pedro wants to prove $\triangle ADB \cong \triangle ADC$. To do this, he must show that $\overline{AD} \cong \overline{AD}$, $\overline{AB} \cong \overline{DC}$ and $\overline{BD} \cong \overline{AC}$.

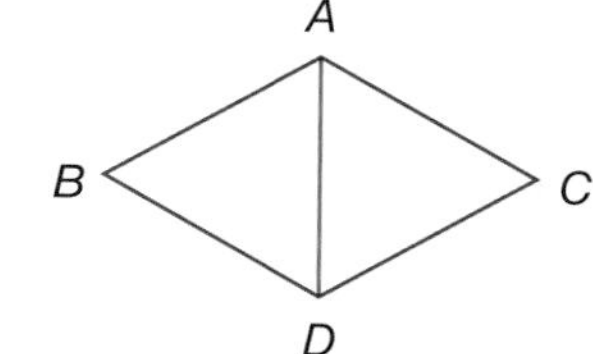

a. Copy the figure and label on your drawing that $\overline{AB} \cong \overline{CD}$, $\overline{AB} \cong \overline{BD}$, and $\overline{AB} \cong \overline{AC}$.

b. Explain how he can use the Reflexive and Transitive Properties to prove $\triangle ADB \cong \triangle ADC$.

c. If the length of $\overline{AC}$ is x cm, write an equation for the perimeter of the quadrilateral $ACDB$.

StudyTip

Proof A **proof** is a logical argument in which each statement you make is supported by a statement that is accepted as true.

H.O.T. Problems

Use Higher-Order Thinking Skills

54. **OPEN ENDED** Write two equations showing the Transitive Property of Equality. Justify your reasoning.

55. **REASONING** Explain why 0 has no multiplicative inverse.

56. **REASONING** The sum of any two whole numbers is always a whole number. So, the set of whole numbers {0, 1, 2, 3, 4, … } is said to be closed under addition. This is an example of the **Closure Property**. State whether each statement is *true* or *false*. If false, justify your reasoning.

a. The set of whole numbers is closed under subtraction.

b. The set of whole numbers is closed under multiplication.

c. The set of whole numbers is closed under division.

57. **CHALLENGE** Does the Commutative Property *sometimes*, *always* or *never* hold for subtraction? Explain your reasoning.

58. **REASONING** Explain whether 1 can be an additive identity. Give an example to justify your answer.

59. **WHICH ONE DOESN'T BELONG?** Identify the sentence that does not belong with the other three. Explain your reasoning.

$x + 12 = 12 + x$	$7h = h \cdot 7$	$1 + a = a + 1$	$(2j)k = 2(jk)$

60. **WRITING IN MATH** Determine whether the Commutative Property applies to division. Justify your answer.

Standardized Test Practice

61. A deck is shaped like a rectangle with a width of 12 feet and a length of 15 feet. What is the area of the deck?

A 3 ft^2

B 27 ft^2

C 108 ft^2

D 180 ft^2

62. GEOMETRY A box in the shape of a rectangular prism has a volume of 56 cubic inches. If the length of each side is multiplied by 2, what will be the approximate volume of the box?

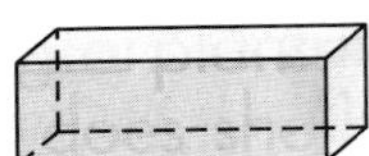

F 112 in^3 **H** 336 in^3

G 224 in^3 **J** 448 in^3

63. $27 \div 3 + (12 - 4) =$

A $\frac{-11}{5}$ **C** 17

B $\frac{27}{11}$ **D** 25

64. GRIDDED RESPONSE Ms. Beal had 1 bran muffin, 16 ounces of orange juice, 3 ounces of sunflower seeds, 2 slices of turkey, and half a cup of spinach. Find the total number of grams of protein she consumed.

Protein Content

Food	Protein (g)
bran muffin (1)	3
orange juice (8 0z)	2
sunflower seeds (1 oz)	2
turkey (1 slice)	12
spinach (1 c)	5

Spiral Review

Evaluate each expression. (Lesson 1-2)

65. $3 \cdot 5 + 1 - 2$

66. $14 \div 2 \cdot 6 - 5^2$

67. $\frac{3 \cdot 9^2 - 3^2 \cdot 9}{3 \cdot 9}$

68. GEOMETRY Write an expression for the perimeter of the figure. (Lesson 1-1)

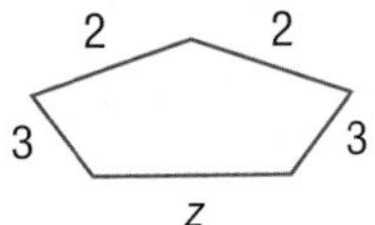

Find the perimeter and area of each figure. (Lessons 0-7 and 0-8)

69. a rectangle with length 5 feet and width 8 feet

70. a square with length 4.5 inches

71. SURVEY Andrew took a survey of his friends to find out their favorite type of music. Of the 34 friends surveyed, 22 said they liked rock music the best. What percent like rock music the best? (Lesson 0-6)

Name the reciprocal of each number. (Lesson 0-5)

72. $\frac{6}{17}$

73. $\frac{2}{23}$

74. $3\frac{4}{5}$

Skills Review

Find each product. Express in simplest form. (Lesson 0-5)

75. $\frac{12}{15} \cdot \frac{3}{14}$

76. $\frac{5}{7} \cdot \left(-\frac{4}{5}\right)$

77. $\frac{10}{11} \cdot \frac{21}{35}$

78. $\frac{63}{65} \cdot \frac{120}{126}$

79. $-\frac{4}{3} \cdot \left(-\frac{9}{2}\right)$

80. $\frac{1}{3} \cdot \frac{2}{5}$

1-4 The Distributive Property

Then

You explored Associative and Commutative Properties. (Lesson 1-3)

Now

- Use the Distributive Property to evaluate expressions.
- Use the Distributive Property to simplify expressions.

New Vocabulary

like terms
simplest form
coefficient

Math Online

glencoe.com

- Extra Examples
- Personal Tutor
- Self-Check Quiz
- Homework Help
- Math in Motion

Why?

John burns approximately 420 Calories per hour by inline skating. The chart below shows the time he spent inline skating in one week.

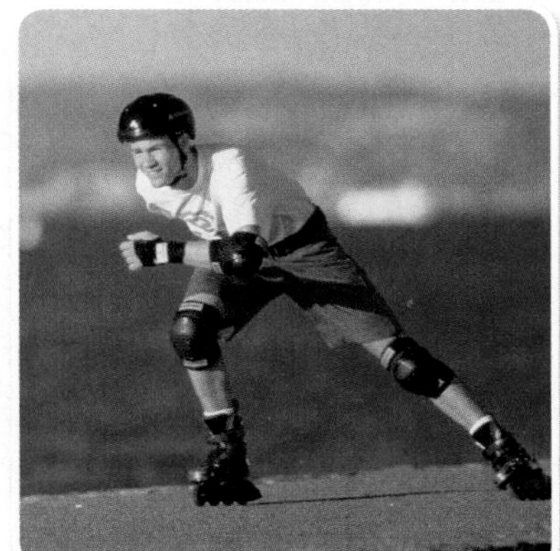

Day	Mon	Tue	Wed	Thu	Fri	Sat	Sun
Time (h)	1	$\frac{1}{2}$	0	1	0	2	$2\frac{1}{2}$

To determine the total number of Calories that he burned inline skating that week, you can use the Distributive Property.

Evaluate Expressions There are two methods you could use to calculate the number of Calories John burned inline skating. You could find the total time spent inline skating and then multiply by the Calories burned per hour. Or you could find the number of Calories burned each day and then add to find the total.

Method 1 Rate Times Total Time

$$420\left(1 + \frac{1}{2} + 1 + 2 + 2\frac{1}{2}\right)$$
$$= 420(7)$$
$$= 2940$$

Method 2 Sum of Daily Calories Burned

$$420(1) + 420\left(\frac{1}{2}\right) + 420(1) + 420(2) + 420\left(2\frac{1}{2}\right)$$
$$= 420 + 210 + 420 + 840 + 1050$$
$$= 2940$$

Either method gives the same total of 2940 Calories burned. This is an example of the **Distributive Property**.

Key Concept For Your FOLDABLE

Distributive Property

Symbol For any numbers a, b, and c,
$a(b + c) = ab + ac$ and $(b + c)a = ba + ca$ and
$a(b - c) = ab - ac$ and $(b - c)a = ba - ca$.

Examples

$3(2 + 5) = 3 \cdot 2 + 3 \cdot 5$
$3(7) = 6 + 15$
$21 = 21$

$4(9 - 7) = 4 \cdot 9 - 4 \cdot 7$
$4(2) = 36 - 28$
$8 = 8$

Math *in Motion*, BrainPOP® glencoe.com

The Symmetric Property of Equality allows the Distributive Property to be written as follows.

If $a(b + c) = ab + ac$, then $ab + ac = a(b + c)$.

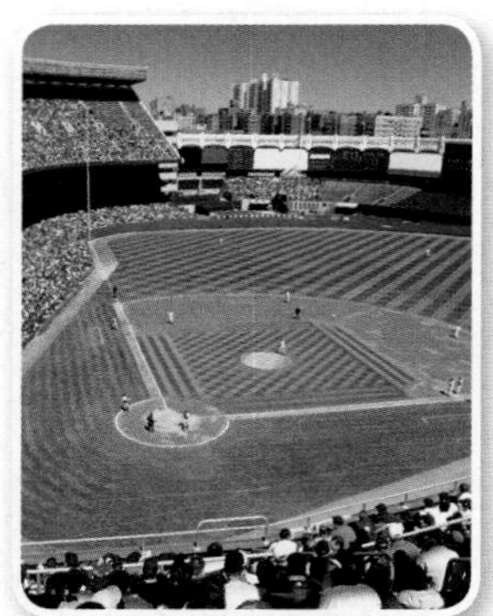

Real-World Link

The record attendance for a single baseball game was set in 1959. There were 92,706 spectators at a game between the Los Angeles Dodgers and the Chicago White Sox.

Source: *Baseball Almanac*

Real-World EXAMPLE 1 Distribute Over Addition

SPORTS A group of 7 adults and 6 children are going to a University of South Florida Bulls baseball game. Use the Distributive Property to write and evaluate an expression for the total ticket cost.

USF Bulls Baseball Tickets	
Ticket	Cost ($)
Adult Single Game	5
Children Single Game (12 and under)	3
Groups of 10 or more Single Game	2
Senior Single Game (65 and over)	3

Source: USF

Understand You need to find the cost of each ticket and then find the total cost.

Plan 7 + 6 or 13 people are going to the game, so the tickets are $2 each.

Solve Write an expression that shows the product of the cost of each ticket and the sum of adult tickets and children's tickets.

$2(7 + 6) = 2(7) + 2(6)$ **Distributive Property**

$= 14 + 12$ **Multiply.**

$= 26$ **Add.**

The total cost is $26.

Check The total number of tickets needed is 13 and they cost $2 each. Multiply 13 by 2 to get 26. Therefore, the total cost of tickets is $26.

Check Your Progress

1. **SPORTS** A group of 3 adults, an 11-year old, and 2 children under 10 years old are going to a baseball game. Write and evaluate an expression to determine the cost of tickets for the group.

Personal Tutor glencoe.com

You can use the Distributive Property to make mental math easier.

EXAMPLE 2 Mental Math

Use the Distributive Property to rewrite 7 • 49. Then evaluate.

$7 \cdot 49 = 7(50 - 1)$ **Think: 49 = 50 − 1**

$= 7(50) - 7(1)$ **Distributive Property**

$= 350 - 7$ **Multiply.**

$= 343$ **Subtract.**

Check Your Progress

Use the Distributive Property to rewrite each expression. Then evaluate.

2A. 304(15)

2B. $44 \cdot 2\frac{1}{2}$

2C. 210(5)

2D. 52(17)

Personal Tutor glencoe.com

Simplify Expressions You can use algebra tiles to investigate how the Distributive Property relates to algebraic expressions.

Problem-Solving Tip

Make a Model It can be helpful to visualize a problem using algebra tiles or folded paper.

The rectangle at the right has 3 x-tiles and 6 1-tiles. The area of the rectangle is $x + 1 + 1 + x + 1 + 1 + x + 1 + 1$ or $3x + 6$. Therefore, $3(x + 2) = 3x + 6$.

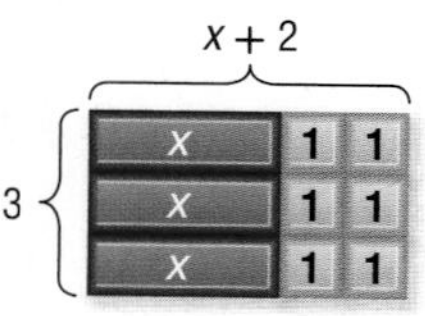

EXAMPLE 3 Algebraic Expressions

Rewrite each expression using the Distributive Property. Then simplify.

a. $7(3w - 5)$

$7(3w - 5) = 7 \cdot 3w - 7 \cdot 5$ **Distributive Property**

$= 21w - 35$ **Multiply.**

b. $(6v^2 + v - 3)4$

$(6v^2 + v - 3)4 = 6v^2(4) + v(4) - 3(4)$ **Distributive Property**

$= 24v^2 + 4v - 12$ **Multiply.**

Check Your Progress

3A. $(8 + 4n)2$

3B. $-6(r + 3g - t)$

3C. $(2 - 5q)(-3)$

3D. $-4(-8 - 3m)$

Personal Tutor glencoe.com

Review Vocabulary

term a number, a variable, or a product or quotient of numbers and variables (Lesson 1-1)

Like terms are terms that contain the same variables, with corresponding variables having the same power.

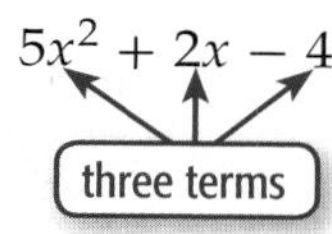

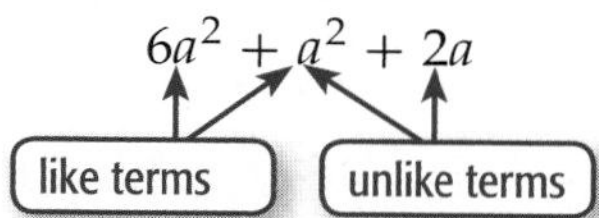

The Distributive Property and the properties of equality can be used to show that $4k + 8k = 12k$. In this expression, $4k$ and $8k$ are like terms.

$4k + 8k = (4 + 8)k$ **Distributive Property**

$= 12k$ **Substitution**

An expression is in **simplest form** when it contains no like terms or parentheses.

EXAMPLE 4 Combine Like Terms

a. Simplify $17u + 25u$.

$17u + 25u = (17 + 25)u$ **Distributive Property**

$= 42u$ **Substitution**

b. Simplify $6t^2 + 3t - t$.

$6t^2 + 3t - t = 6t^2 + (3 - 1)t$ **Distributive Property**

$= 6t^2 + 2t$ **Substitution**

Check Your Progress

Simplify each expression. If not possible, write *simplified*.

4A. $6n - 4n$

4B. $b^2 + 13b + 13$

4C. $4y^3 + 2y - 8y + 5$

4D. $7a + 4 - 6a^2 - 2a$

Personal Tutor glencoe.com

EXAMPLE 5 Write and Simplify Expressions

Use the expression *twice the difference of 3x and y increased by five times the sum of x and 2y*.

a. Write an algebraic expression for the verbal expression.

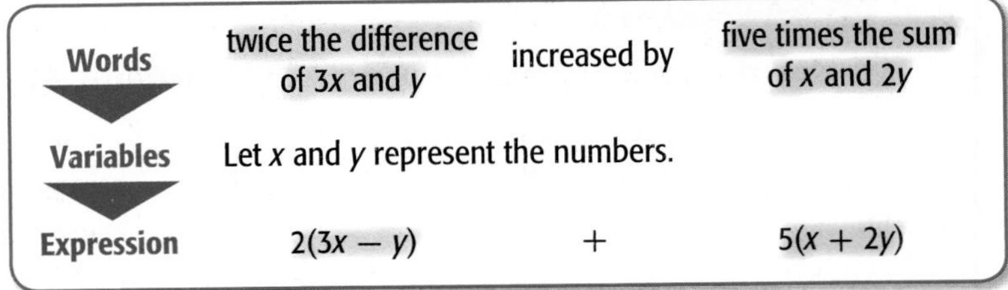

b. Simplify the expression, and indicate the properties used.

$2(3x - y) + 5(x + 2y) = 2(3x) - 2(y) + 5(x) + 5(2y)$	**Distributive Property**
$= 6x - 2y + 5x + 10y$	**Multiply.**
$= 6x + 5x - 2y + 10y$	**Commutative (+)**
$= (6 + 5)x + (-2 + 10)y$	**Distributive Property**
$= 11x + 8y$	**Substitution**

Check Your Progress

5. Write an algebraic expression *5 times the difference of q squared and r plus 8 times the sum of 3q and 2r*.

A. Write an algebraic expression for the verbal expression.

B. Simplify the expression, and indicate the properties used.

Personal Tutor glencoe.com

The **coefficient** of a term is the numerical factor. For example, in $6ab$, the coefficient is 6, and in $\frac{x^2}{3}$, the coefficient is $\frac{1}{3}$. In the term y, the coefficient is 1 since $1 \cdot y = y$ by the Multiplicative Identity Property.

StudyTip

Like Terms *Like terms* could be defined as terms with the same variables to the same powers.

Concept Summary

For Your FOLDABLE

Properties of Numbers

The following properties are true for any numbers a, b, and c.

Properties	Addition	Multiplication
Commutative	$a + b = b + a$	$ab = ba$
Associative	$(a + b) + c = a + (b + c)$	$(ab)c = a(bc)$
Identity	0 is the identity. $a + 0 = 0 + a = a$	1 is the identity. $a \cdot 1 = 1 \cdot a = a$
Zero	—	$a \cdot 0 = 0 \cdot a = 0$
Distributive	$a(b + c) = ab + ac$ and $(b + c)a = ba + ca$	
Substitution	If $a = b$, then a may be substituted for b.	

Check Your Understanding

Example 1 p. 24

1. **PILOT** A pilot at an air show charges \$25 per passenger for rides. If 12 adults and 15 children ride in one day, write and evaluate an expression to describe the situation.

Example 2 p. 24

Use the Distributive Property to rewrite each expression. Then evaluate.

2. $14(51)$
3. $6\frac{1}{9}(9)$

Example 3 p. 25

Use the Distributive Property to rewrite each expression. Then simplify.

4. $2(4 + t)$
5. $(g - 9)5$

Example 4 p. 25

Simplify each expression. If not possible, write *simplified*.

6. $15m + m$
7. $3x^3 + 5y^3 + 14$
8. $(5m + 2m)10$

Example 5 p. 26

Write an algebraic expression for each verbal expression. Then simplify, indicating the properties used.

9. 4 times the sum of 2 times x and six
10. one half of 4 times y plus the quantity of y and 3

Practice and Problem Solving

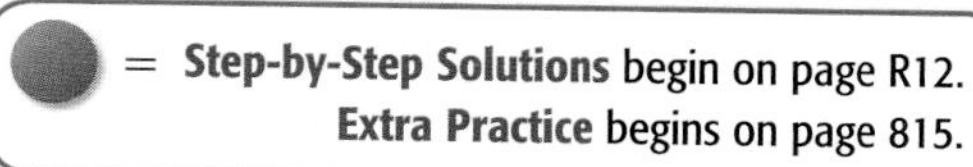
= **Step-by-Step Solutions** begin on page R12.
Extra Practice begins on page 815.

Example 1 p. 24

11. **TIME MANAGEMENT** Margo uses dots to track her activities on a calendar. Red dots represent homework, yellow dots represent work, and green dots represent track practice. In a typical week, she uses 5 red dots, 3 yellow dots, and 4 green dots. How many activities does Margo do in 4 weeks?

12. **BLOOD SUPPLY** The Red Cross is holding blood drives in two locations. In one day, Center 1 collected 715 pints and Center 2 collected 1035 pints. Write and evaluate an expression to estimate the total number of pints of blood donated over a 3-day period.

Example 2 p. 24

Use the Distributive Property to rewrite each expression. Then evaluate.

13. $(4 + 5)6$
14. $7(13 + 12)$
15. $6(6 - 1)$
16. $(3 + 8)15$
17. $14(8 - 5)$
18. $(9 - 4)19$
19. $4(7 - 2)$
20. $7(2 + 1)$
21. $7 \cdot 497$
22. $6(525)$
23. $36 \cdot 3\frac{1}{4}$
24. $\left(4\frac{2}{7}\right)21$

Example 3 p. 25

Use the Distributive Property to rewrite each expression. Then simplify.

25. $2(x + 4)$
26. $(5 + n)3$
27. $(4 - 3m)8$
28. $-3(2x - 6)$

Example 4 p. 25

Simplify each expression. If not possible, write *simplified*.

29. $13r + 5r$
30. $3x^3 - 2x^2$
31. $7m + 7 - 5m$
32. $5z^2 + 3z + 8z^2$
33. $(2 - 4n)17$
34. $11(4d + 6)$
35. $7m + 2m + 5p + 4m$
36. $3x + 7(3x + 4)$
37. $4(fg + 3g) + 5g$

Example 5 p. 26

Write an algebraic expression for each verbal expression. Then simplify, indicating the properties used.

38. the product of 5 and m squared, increased by the sum of the square of m and 5
39. 7 times the sum of a squared and b minus 4 times the sum of a squared and b

Math History Link

Kambei Mori
(c. 1600–1628)
Kambei Mori was a Japanese scholar who popularized the abacus. He changed the focus of mathematics from philosophy to computation.

40. **GEOMETRY** Find the perimeter of an isosceles triangle with side lengths of $5 + x$, $5 + x$, and xy. Write in simplest form.

41. **GEOMETRY** A regular hexagon measures $3x + 5$ units on each side. What is the perimeter in simplest form?

Simplify each expression.

42. $6x + 4y + 5x$ **43.** $3m + 5g + 6g + 11m$ **44.** $4a + 5a^2 + 2a^2 + a^2$

45. $5k + 3k^3 + 7k + 9k^3$ **46.** $6d + 4(3d + 5)$ **47.** $2(6x + 4) + 7x$

48. **FOOD** Kenji is picking up take-out food for his study group.

a. Write and evaluate an expression to find the total cost of four sandwiches, three soups, three salads, and five drinks.

b. How much would it cost if Kenji bought four of each item on the menu?

Menu

Item	Cost ($)
sandwich	2.49
cup of soup	1.29
side salad	0.99
drink	1.49

Use the Distributive Property to rewrite each expression. Then simplify.

49. $\left(\frac{1}{3} - 2b\right)27$ **50.** $4(8p + 4q - 7r)$ **51.** $6(2c - cd^2 + d)$

Simplify each expression. If not possible, write *simplified.*

52. $6x^2 + 14x - 9x$ **53.** $4y^3 + 3y^3 + y^4$ **54.** $a + \frac{a}{5} + \frac{2}{5}a$

55. **MULTIPLE REPRESENTATIONS** The area of the model is $2(x - 4)$ or $2x - 8$. The expression $2(x - 4)$ is in *factored form.*

a. **GEOMETRIC** Use algebra tiles to form a rectangle with area $2x + 6$. Use the result to write $2x + 6$ in factored form.

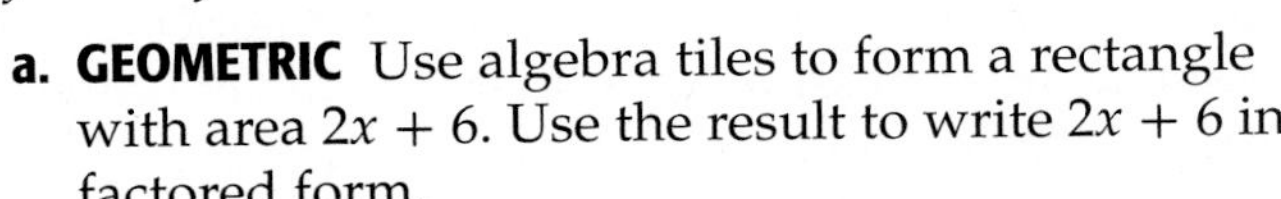

b. **TABULAR** Use algebra tiles to form rectangles to represent each area in the table. Record the factored form of each expression.

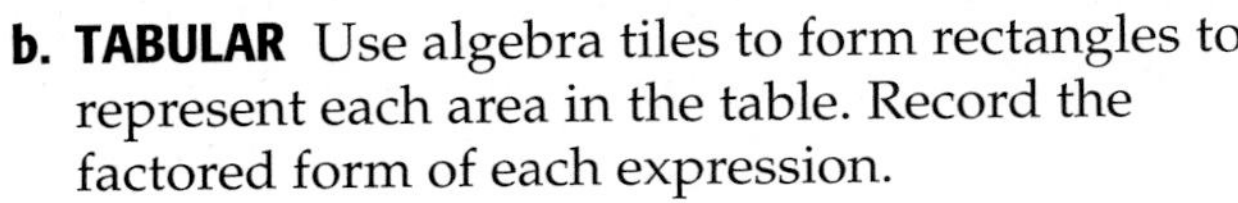

c. **VERBAL** Explain how you could find the factored form of an expression.

Area	Factored Form
$2x + 6$	
$3x + 3$	
$3x - 12$	
$5x + 10$	

H.O.T. Problems Use Higher-Order Thinking Skills

56. **CHALLENGE** Use the Distributive Property to simplify $6x^2[(3x - 4) + (4x + 2)]$.

57. **REASONING** Should the Distributive Property be a property of multiplication, addition, or both? Explain your answer.

58. **OPEN ENDED** Write a real-life example in which the Distributive Property would be useful. Write an expression that demonstrates the example.

59. **WRITING IN MATH** Use the data about skating on page 23 to explain how the Distributive Property can be used to calculate quickly. Also, compare the two methods of finding the total Calories burned.

Standardized Test Practice

60. Which illustrates the Symmetric Property of Equality?

A If $a = b$, then $b = a$.
B If $a = b$, and $b = c$, then $a = c$.
C If $a = b$, then $b = c$.
D If $a = a$, then $a + 0 = a$.

61. Anna is three years younger than her sister Emily. Which expression represents Anna's age if we express Emily's age as y years?

F $y + 3$
G $y - 3$
H $3y$
J $\frac{3}{y}$

62. Which property is used below?
If $4xy^2 = 8y^2$ and $8y^2 = 72$, then $4xy^2 = 72$.

A Reflexive Property
B Substitution Property
C Symmetric Property
D Transitive Property

63. SHORT RESPONSE A drawer contains the socks in the chart. What is the probability that a randomly chosen sock is blue?

Color	Number
white	16
blue	12
black	8

Spiral Review

Evaluate each expression. Name the property used in each step. (Lesson 1-3)

64. $14 + 23 + 8 + 15$

65. $0.24 \cdot 8 \cdot 7.05$

66. $1\frac{1}{4} \cdot 9 \cdot \frac{5}{6}$

67. SPORTS Braden runs 6 times a week for 30 minutes and lifts weights 3 times a week for 20 minutes. Write and evaluate an expression for the number of hours Braden works out in 4 weeks. (Lesson 1-2)

SPORTS Refer to the table showing Blanca's cross-country times for the first 8 meets of the season. Round answers to the nearest second. (Lesson 0-12)

68. Find the mean of the data.

69. Find the median of the data.

70. Find the mode of the data.

Cross Country	
Meet	Time
1	22:31
2	22:21
3	21:48
4	22:01
5	21:48
6	20:56
7	20:34
8	20:15

71. SURFACE AREA What is the surface area of the cube? (Lesson 0-10)

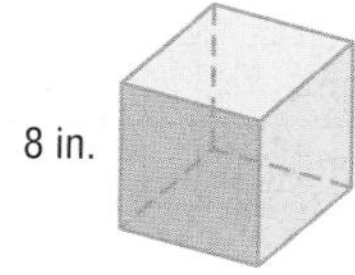

Skills Review

Evaluate each expression. (Lesson 1-2)

72. $12(7 + 2)$

73. $11(5) - 8(5)$

74. $(13 - 9) \cdot 4$

75. $3(6) + 7(6)$

76. $(1 + 19) \cdot 8$

77. $16(5 + 7)$

CHAPTER 1

Mid-Chapter Quiz

Lessons 1-1 through 1-4

Math Online glencoe.com
Math *in Motion,* Animation

Write a verbal expression for each algebraic expression. (Lesson 1-1)

1. $21 - x^3$
2. $3m^5 + 9$

Write an algebraic expression for each verbal expression. (Lesson 1-1)

3. five more than s squared
4. four times y to the fourth power
5. **CAR RENTAL** The XYZ Car Rental Agency charges a flat rate of \$29 per day plus \$0.32 per mile driven. Write an algebraic expression for the rental cost of a car for x days that is driven y miles. (Lesson 1-1)

Evaluate each expression. (Lesson 1-2)

6. $24 \div 3 - 2 \cdot 3$
7. $5 + 2^2$
8. $4(3 + 9)$
9. $36 - 2(1 + 3)^2$
10. $\frac{40 - 2^3}{4 + 3(2^2)}$
11. **AMUSEMENT PARK** The costs of tickets to a local amusement park are shown. Write and evaluate an expression to find the total cost for 5 adults and 8 children. (Lesson 1-2)

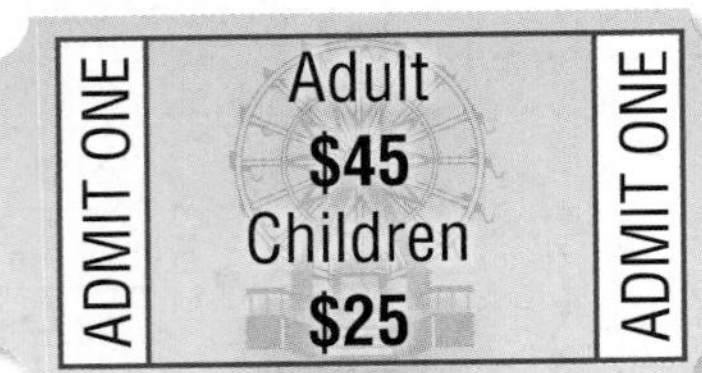

12. **MULTIPLE CHOICE** Write an algebraic expression to represent the perimeter of the rectangle shown below. Then evaluate it to find the perimeter when $w = 8$ cm. (Lesson 1-2)

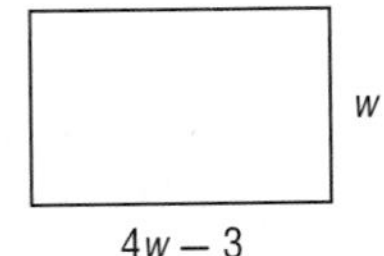

A 37 cm
B 232 cm
C 74 cm
D 45 cm

Evaluate each expression. Name the property used in each step. (Lesson 1-3)

13. $(8 - 2^3) + 21$
14. $3(1 \div 3) \cdot 9$
15. $[5 \div (3 \cdot 1)]\frac{3}{5}$
16. $18 + 35 + 32 + 15$
17. $0.25 \cdot 7 \cdot 4$

Use the Distributive Property to rewrite each expression. Then evaluate. (Lesson 1-4)

18. $3(5 + 2)$
19. $(9 - 6)12$
20. $8(7 - 4)$

Use the Distributive Property to rewrite each expression. Then simplify. (Lesson 1-4)

21. $4(x + 3)$
22. $(6 - 2y)7$
23. $-5(3m - 2)$
24. **DVD SALES** A video store chain has three locations. Use the information in the table below to write and evaluate an expression to estimate the total number of DVDs sold over a 4-day period. (Lesson 1-4)

Location	Daily Sales Numbers
Location 1	145
Location 2	211
Location 3	184

25. **MULTIPLE CHOICE** Rewrite the expression $(8 - 3p)(-2)$ using the Distributive Property. (Lesson 1-4)

F $16 - 6p$
G $-10p$
H $-16 + 6p$
J $10p$

1-5 Equations

Why?

Mark's baseball team scored 3 runs in the first inning. At the top of the third inning, their score was 4. The open sentence below represents the change in their score.

$$3 + r = 4$$

The solution is 1. The team got 1 run in the second inning.

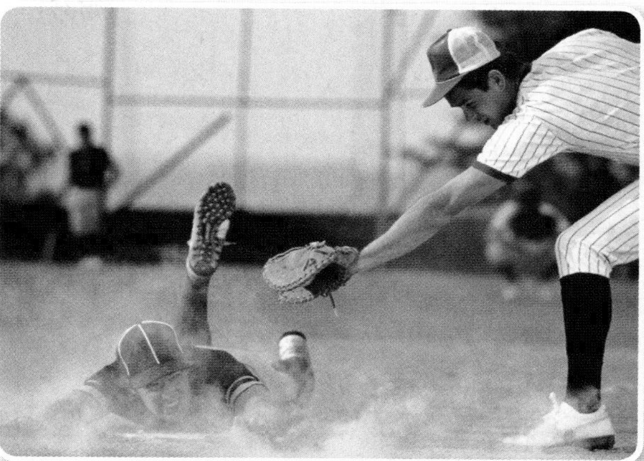

Then

You simplified expressions. (Lesson 1-1 through 1-4)

Now

- Solve equations with one variable.
- Solve equations with two variables.

New Vocabulary

open sentence
equation
solving
solution
replacement set
set
element
solution set
identity

Math Online

glencoe.com

- Extra Examples
- Personal Tutor
- Self-Check Quiz
- Homework Help
- Math in Motion

Solve Equations A mathematical statement that contains algebraic expressions and symbols is an **open sentence**. A sentence that contains an equals sign, =, is an **equation**.

expression → $3x + 7$ $3x + 7 = 13$ ← equation

Finding a value for a variable that makes a sentence true is called **solving** the open sentence. This replacement value is a **solution**.

A set of numbers from which replacements for a variable may be chosen is called a **replacement set**. A **set** is a collection of objects or numbers that is often shown using braces. Each object or number in the set is called an **element**, or member. A **solution set** is the set of elements from the replacement set that make an open sentence true.

EXAMPLE 1 Use a Replacement Set

Find the solution set of the equation $2q + 5 = 13$ if the replacement set is {2, 3, 4, 5, 6}.

Use a table to solve. Replace q in $2q + 5 = 13$ with each value in the replacement set.

q	$2q + 5 = 13$	True or False?
2	$2(2) + 5 = 13$	False
3	$2(3) + 5 = 13$	False
4	$2(4) + 5 = 13$	True
5	$2(5) + 5 = 13$	False
6	$2(6) + 5 = 13$	False

Since the equation is true when $q = 4$, the solution of $2q + 5 = 13$ is $q = 4$.

The solution set is {4}.

Check Your Progress

Find the solution set for each equation if the replacement set is {0, 1, 2, 3}.

1A. $8m - 7 = 17$

1B. $28 = 4(1 + 3d)$

Personal Tutor glencoe.com

You can often solve an equation by applying the order of operations.

Test-Taking Tip

Rewrite the Equation If you are allowed to write in your testing booklet, it can be helpful to rewrite the equation with simplified terms.

STANDARDIZED TEST EXAMPLE 2

Solve $6 + (5^2 - 5) \div 2 = p$.

A 3 **B** 6 **C** 13 **D** 16

Read the Test Item

You need to apply the order of operations to the expression in order to solve for p.

Solve the Test Item

$6 + (5^2 - 5) \div 2 = p$	**Original equation**
$6 + (25 - 5) \div 2 = p$	**Evaluate powers.**
$6 + 20 \div 2 = p$	**Subtract 5 from 25.**
$6 + 10 = p$	**Divide 20 by 2.**
$16 = p$	Add. The correct answer is D.

Check Your Progress

2. Solve $t = 9^2 \div (5 - 2)$.

F 3 **G** 6 **H** 14.2 **J** 27

Personal Tutor glencoe.com

Some equations have a unique solution. Other equations do not have a solution.

EXAMPLE 3 Solutions of Equations

Solve each equation.

a. $7 - (4^2 - 10) + n = 10$

Simplify the equation first and then look for a solution.

$7 - (4^2 - 10) + n = 10$	**Original equation**
$7 - (16 - 10) + n = 10$	**Evaluate powers.**
$7 - 6 + n = 10$	**Subtract 10 from 16.**
$1 + n = 10$	**Subtract 6 from 7.**

The only value for n that makes the equation true is 9. Therefore, this equation has a unique solution of 9.

Math *in Motion*, Interactive Lab glencoe.com

b. $n(3 + 2) + 6 = 5n + (10 - 3)$

$n(3 + 2) + 6 = 5n + (10 - 3)$	**Original equation**
$n(5) + 6 = 5n + (10 - 3)$	**Add 3 + 2.**
$n(5) + 6 = 5n + 7$	**Subtract 3 from 10.**
$5n + 6 = 5n + 7$	**Commutative (×)**

No matter what real value is substituted for n, the left side of the equation will always be one less than the right side. So, the equation will never be true. Therefore, there is no solution of this equation.

Check Your Progress **Solve each equation.**

3A. $(18 + 4) + m = (5 - 3)m$

3B. $8 \cdot 4 \cdot k + 9 \cdot 5 = (36 - 4)k - (2 \cdot 5)$

Personal Tutor glencoe.com

An equation that is true for every value of the variable is called an **identity**.

ReadingMath

Reading Math
Identities An identity is an equation that shows that a number or expression is equivalent to itself.

EXAMPLE 4 Identities

Solve $(2 \cdot 5 - 8)(3h + 6) = [(2h + h) + 6]2$.

$(2 \cdot 5 - 8)(3h + 6) = [(2h + h) + 6]2$	**Original Equation**
$(10 - 8)(3h + 6) = [(2h + h) + 6]2$	**Multiply $2 \cdot 5$.**
$2(3h + 6) = [(2h + h) + 6]2$	**Subtract 8 from 10.**
$6h + 12 = [(2h + h) + 6]2$	**Distributive Property**
$6h + 12 = [3h + 6]2$	**Add $2h + h$.**
$6h + 12 = 6h + 12$	**Distributive Property**

No matter what value is substituted for h, the left side of the equation will always be equal to the right side. So, the equation will always be true. Therefore, the solution of this equation could be any real number.

Check Your Progress

Solve each equation.

4A. $12(10 - 7) + 9g = g(2^2 + 5) + 36$

4B. $2d + (2^3 - 5) = 10(5 - 2) + d(12 \div 6)$

4C. $3(b + 1) - 5 = 3b - 2$

4D. $5 - \frac{1}{2}(c - 6) = 4$

Personal Tutor glencoe.com

Solve Equations with Two Variables Some equations contain two variables. It is often useful to make a table of values and use substitution to find the corresponding values of the second variable.

EXAMPLE 5 Equations Involving Two Variables

MOVIE RENTALS Mr. Hernandez pays $10 each month for movies delivered by mail. He can also rent movies in the store for $1.50 per title. Write and solve an equation to find the total amount Mr. Hernandez spends this month if he rents 3 movies from the store.

The cost of the movie plan is a flat rate. The variable is the number of movies he rents from the store. The total cost is the price of the plan plus $1.50 times the number of movies from the store. Let C be the total cost and m be the number of movies.

$C = 1.50m + 10$	**Original equation**
$= 1.50(3) + 10$	**Substitute 3 for m.**
$= 4.50 + 10$	**Multiply.**
$= 14.50$	

Mr. Hernandez spends $14.50 on movie rentals in one month.

Check Your Progress

5. TRAVEL Amelia drives an average of 65 miles per hour. Write and solve an equation to find the time it will take her to drive 36 miles.

Personal Tutor glencoe.com

Check Your Understanding

Example 1 p. 31

Find the solution set for each equation if the replacement set is {11, 12, 13, 14, 15}.

1. $n + 10 = 23$
2. $7 = \frac{c}{2}$
3. $29 = 3x - 7$
4. $(k - 8)12 = 84$

Example 2 p. 32

5. **MULTIPLE CHOICE** Solve $\frac{d + 5}{10} = 2$.

A 10　　B 15　　C 20　　D 25

Examples 3 and 4 pp. 32–33

Solve each equation.

6. $x = 4(6) + 3$
7. $14 - 82 = w$
8. $5 + 22a = 2 + 10 \div 2$
9. $(2 \cdot 5) + \frac{c^3}{3} = c^3 \div (1^5 + 2) + 10$

Example 5 p. 33

10. **RECYCLING** San Francisco has a recycling facility that accepts unused paint. Volunteers blend and mix the paint and give it away in 5-gallon buckets. Write and solve an equation to find the number of buckets of paint given away from the 30,000 gallons that are donated.

Practice and Problem Solving

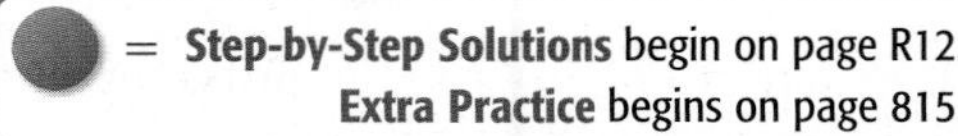
= **Step-by-Step Solutions** begin on page R12.
Extra Practice begins on page 815.

Example 1 p. 31

Find the solution set of each equation if the replacement sets are y: {1, 3, 5, 7, 9} and z: {10, 12, 14, 16, 18}.

11. $z + 10 = 22$
12. $52 = 4z$
13. $\frac{15}{y} = 3$
14. $17 = 24 - y$
15. $2z - 5 = 27$
16. $4(y + 1) = 40$
17. $22 = \frac{60}{y} + 2$
18. $111 = z^2 + 11$

Examples 2–4 pp. 32–33

Solve each equation.

19. $a = 32 - 9(2)$
20. $w = 56 \div (2^2 + 3)$
21. $\frac{27 + 5}{16} = g$
22. $\frac{12 \cdot 5}{15 - 3} = y$
23. $r = \frac{9(6)}{(8 + 1)3}$
24. $a = \frac{4(14 - 1)}{3(6) - 5} + 7$
25. $(4 - 2^2 + 5)w = 25$
26. $7 + x - (3 + 32 \div 8) = 3$
27. $3^2 - 2 \cdot 3 + u = (3^3 - 3 \cdot 8)(2) + u$
28. $(3 \cdot 6 \div 2)v + 10 = 3^2v + 9$
29. $6k + (3 \cdot 10 - 8) = (2 \cdot 3)k + 22$
30. $(3 \cdot 5)t + (21 - 12) = 15t + 3^2$
31. $(2^4 - 3 \cdot 5)q + 13 = (2 \cdot 9 - 4^2)q + \left(\frac{3 \cdot 4}{12} - 1\right)$
32. $\frac{3 \cdot 22}{18 + 4}r - \left(\frac{4^2}{9 + 7} - 1\right) = r + \left(\frac{8 \cdot 9}{3} \div 3\right)$
33. **SCHOOL** A conference room can seat a maximum of 85 people. The principal and two counselors need to meet with the school's juniors to discuss college admissions. If each student must bring a parent with them, how many students can attend each meeting? Assume that each student has a unique set of parents.
34. **GEOMETRY** The perimeter of a regular octagon is 128 inches. Find the length of each side.

Example 5 p. 33

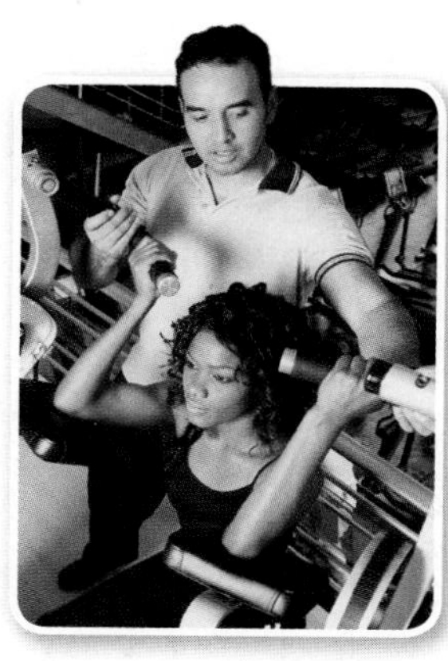

Real-World Link

Athletes in training should have a specific blend of sources for their Calories.
69% carbohydrates
20% fats
11% protein

Source: *Food and Sport*

35. **SPORTS** A 200-pound athlete who trains for four hours per day requires 2836 Calories for basic energy requirements. During training, the same athlete requires 3091 Calories for extra energy requirements. Write an equation to find C, the total daily Calorie requirement for this athlete. Then solve the equation.

36. **ENERGY** An electric generator can power 3550 watts of electricity. Write and solve an equation to find how many 75-watt light bulbs a generator could power.

Make a table of values for each equation if the replacement set is {−2, −1, 0, 1, 2}.

37. $y = 3x - 2$

38. $3.25x + 0.75 = y$

Solve each equation using the given replacement set.

39. $t - 13 = 7$; {10, 13, 17, 20}

40. $14(x + 5) = 126$; {3, 4, 5, 6, 7}

41. $22 = \frac{n}{3}$; {62, 64, 66, 68, 70}

42. $35 = \frac{g - 8}{2}$; {78, 79, 80, 81}

Solve each equation.

43. $\frac{3(9) - 2}{1 + 4} = d$

44. $j = 15 \div 3 \cdot 5 - 4^2$

45. $c + (3^2 - 3) = 21$

46. $(3^3 - 3 \cdot 9) + (7 - 2^2)b = 24b$

47. **HEALTH** Blood flow rate can be expressed as $F = \frac{p_1 - p_2}{r}$, where F is the flow rate, p_1 and p_2 are the initial and final pressure exerted against the blood vessel's walls, respectively, and r is the resistance created by the size of the vessel.

   a. Write and solve an equation to determine the resistance of the blood vessel for an initial pressure of 100 millimeters of mercury, a final pressure of 0 millimeters of mercury, and a flow rate of 5 liters per minute.

   b. Use the equation to complete the table below.

Initial Pressure p_1 (mm Hg)	Final Pressure p_2 (mm Hg)	Resistance r (mm Hg/L/min)	Blood Flow Rate F (L/min)
100	0		5
100	0	30	
	5	40	4
90		10	6

Determine whether the given number is a solution of the equation.

48. $x + 6 = 15$; 9

49. $12 + y = 26$; 14

50. $2t - 10 = 4$; 3

51. $3r + 7 = -5$; 2

52. $6 + 4m = 18$; 3

53. $-5 + 2p = -11$; −3

54. $\frac{q}{2} = 20$; 10

55. $\frac{w - 4}{5} = -3$; −11

56. $\frac{g}{3} - 4 = 12$; 48

Make a table of values for each equation if the replacement set is {−2, −1, 0, 1, 2}.

57. $y = 3x + 5$

58. $-2x - 3 = y$

59. $y = \frac{1}{2}x + 2$

60. $4.2x - 1.6 = y$

61. **GEOMETRY** The length of a rectangle is 2 inches greater than the width. The length of the base of an isosceles triangle is 12 inches, and the lengths of the other two sides are 1 inch greater than the width of the rectangle.

   a. Draw a picture of each figure and label the dimensions.

   b. Write two expressions to find the perimeters of the rectangle and triangle.

   c. Find the width of the rectangle if the perimeters of the figures are equal.

Real-World Link

In 2007, Chicago had three "supertall" skyscrapers (over 1000 feet) under construction that used a new method of steel construction. A web of supports stretch from the center to the outside wall supports. This method allows buildings to be built taller and with more features than ever before.

62. **CONSTRUCTION** The construction of a building requires 10 tons of steel per story.
    a. Define a variable and write an equation for the number of tons of steel required if the building has 15 stories.
    b. How many tons of steel are needed?

63. **MULTIPLE REPRESENTATIONS** In this problem, you will further explore writing equations.
    a. **CONCRETE** Use centimeter cubes to build a tower similar to the one shown at the right.

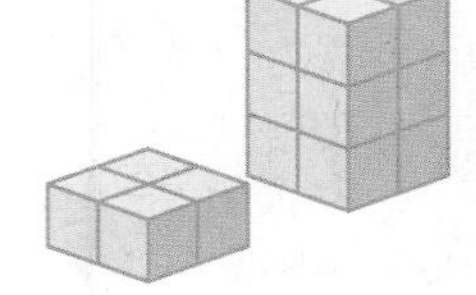

    b. **TABULAR** Copy and complete the table shown below. Record the number of layers in the tower and the number of cubes used in the table.

Layers	1	2	3	4	5	6	7
Cubes	?	?	?	?	?	?	?

    c. **ANALYTICAL** As the number of layers in the tower increases, how does the number of cubes in the tower change?
    d. **ALGEBRAIC** Write a rule that gives the number of cubes in terms of the number of layers in the tower.

H.O.T. Problems

Use **H**igher-**O**rder **T**hinking Skills

64. **REASONING** Compare and contrast an expression and an equation.

65. **OPEN ENDED** Write an equation that is an identity.

66. **REASONING** Explain why an open sentence always has at least one variable.

67. **FIND THE ERROR** Tom and Li-Cheng are solving the equation $x = 4(3 - 2) + 6 \div 8$. Is either of them correct? Explain your reasoning.

Tom

$x = 4(3 - 2) + 6 \div 8$
$= 4(1) + 6 \div 8$
$= 4 + 6 \div 8$
$= 4 + \frac{6}{8}$
$= 4\frac{3}{4}$

Li-Cheng

$x = 4(3 - 2) + 6 \div 8$
$= 4(1) + 6 \div 8$
$= 4 + 6 \div 8$
$= 10 \div 8$
$= \frac{5}{4}$

68. **CHALLENGE** Find all of the solutions of $x^2 + 5 = 30$.

69. **OPEN ENDED** Write an equation that involves two or more operations with a solution of -7.

70. **WRITING IN MATH** Explain how you can determine that an equation has no real numbers as a solution. How can you determine that an equation has all real numbers as solutions?

Standardized Test Practice

71. **STATISTICS** A researcher wants to find out how often teens in her town exercise. Which sample group should she survey to get results that best represent all the teens in the town?

A a summer baseball league
B her nieces and nephews
C high school students chosen at random
D the teens at the mall one Saturday afternoon

72. **SHORT RESPONSE** The expected attendance for the Drama Club production is 65% of the student body. If the student body consists of 300 students, how many students are expected to attend?

73. **GEOMETRY** A speedboat and a sailboat take off from the same port. The diagram shows their travel. What is the distance between the boats?

F 12 mi
G 15 mi
H 18 mi
J 24 mi

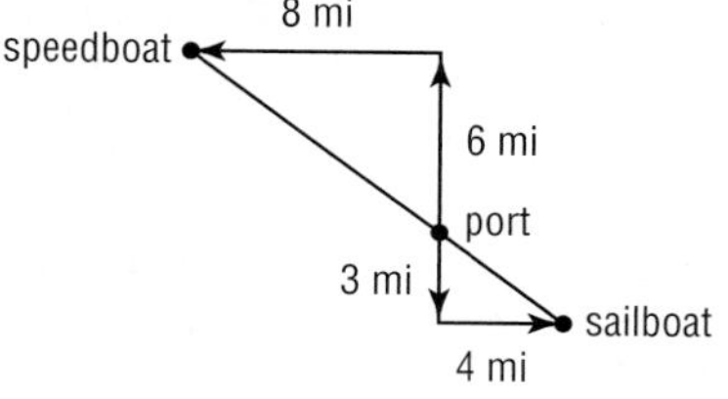

74. Michelle can read 1.5 pages per minute. How many pages can she read in two hours?

A 90 pages
B 150 pages
C 120 pages
D 180 pages

Spiral Review

75. **ZOO** A zoo has about 500 children and 750 adults visit each day. Write an expression to represent about how many visitors the zoo will have over a month. (Lesson 1-4)

Find the value of p in each equation. Then name the property that is used. (Lesson 1-3)

76. $7.3 + p = 7.3$
77. $12p = 1$
78. $1p = 4$

79. **MOVING BOXES** The figure shows the dimensions of the boxes Steve uses to pack. How many cubic inches can each box hold? (Lesson 0-9)

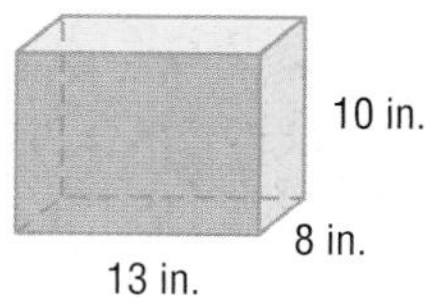

Express each percent as a fraction. (Lesson 0-6)

80. 35%
81. 15%
82. 28%

For each problem, determine whether you need an estimate or an exact answer. Then solve. (Lessons 0-6 and 0-1)

83. **TRAVEL** The distance from Raleigh, North Carolina, to Philadelphia, Pennsylvania, is approximately 428 miles. The average gas mileage of José's car is 45 miles per gallon. About how many gallons of gas will be needed to make the trip?

84. **PART-TIME JOB** An employer pays $8.50 per hour. If 20% of pay is withheld for taxes, what are the take-home earnings from 28 hours of work?

Skills Review

Find each sum or difference. (Lesson 0-4)

85. $1.14 + 5.6$
86. $4.28 - 2.4$
87. $8 - 6.35$
88. $\frac{4}{5} + \frac{1}{6}$
89. $\frac{2}{7} + \frac{3}{4}$
90. $\frac{6}{8} - \frac{1}{2}$

1-6 Relations

Then

You solved equations with one or two variables. (Lesson 1-5)

Now

- Represent relations.
- Interpret graphs of relations.

New Vocabulary

coordinate system
x- and y-axes
origin
ordered pair
x- and y-coordinates
relation
domain
range
independent variable
dependent variable

Math Online

glencoe.com

- Extra Examples
- Personal Tutor
- Self-Check Quiz
- Homework Help

Why?

The deeper in the ocean you are, the greater pressure is on your body. This is because there is more water over you. The force of gravity pulls the water weight down, creating a greater pressure.

The equation that relates the total pressure of the water to the depth is $P = rgh$, where
P = the pressure,
r = the density of water,
g = the acceleration due to gravity, and
h = the height of water above you.

Represent a Relation This relationship between the depth and the pressure exerted can be represented by a line on a coordinate grid.

A **coordinate system** is formed by the intersection of two number lines, the *horizontal axis* and the *vertical axis.*

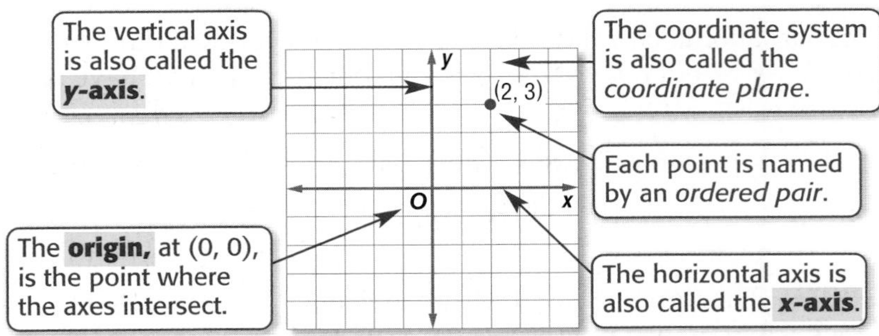

A point is represented on a graph using ordered pairs.

- An **ordered pair** is a set of numbers, or *coordinates*, written in the form (x, y).
- The x-value, called the ***x*-coordinate**, represents the horizontal placement of the point.
- The y-value, or ***y*-coordinate**, represents the vertical placement of the point.

A set of ordered pairs is called a **relation**. A relation can be depicted in several different ways. An equation can represent a relation as well as graphs, tables, and mappings.

A **mapping** illustrates how each element of the *domain* is paired with an element in the *range*. The set of the first numbers of the ordered pairs is the **domain**. The set of second numbers of the ordered pairs is the **range** of the relation. This mapping represents the ordered pairs (−2, 4), (−1, 4), (0, 6) (1, 8), and (2, 8).

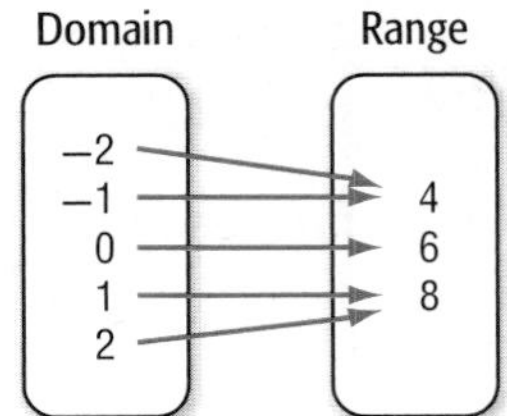

StudyTip

Multiple Representations Each representation of the same relation serves a different purpose. Graphing the points can show the pattern between the points. A mapping shows you at a glance if elements are paired with the same element.

Study the different representations of the same relation below.

Ordered Pairs

(1, 2)
(−2, 4)
(0, −3)

Table

x	y
1	2
−2	4
0	−3

Graph

Mapping

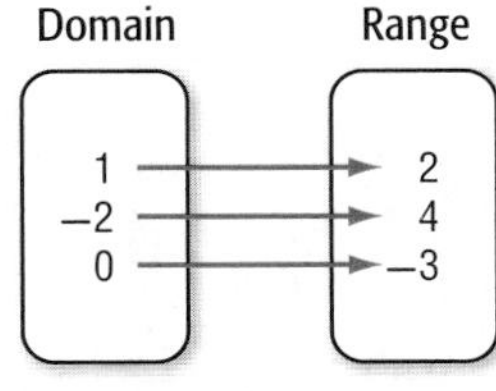

The x-values of a relation are members of the domain and the y-values of a relation are members of the range. In the relation above, the domain is {−2, 1, 0} and the range is {−3, 2, 4}.

EXAMPLE 1 Representations of a Relation

a. Express {(2, 5), (−2, 3), (5, −2), (−1, −2)} as a table, a graph, and a mapping.

Table
Place the x-coordinates into the first column of the table. Place the corresponding y-coordinates in the second column of the table.

x	y
2	5
−2	3
5	−2
−1	−2

Graph
Graph each ordered pair on a coordinate plane.

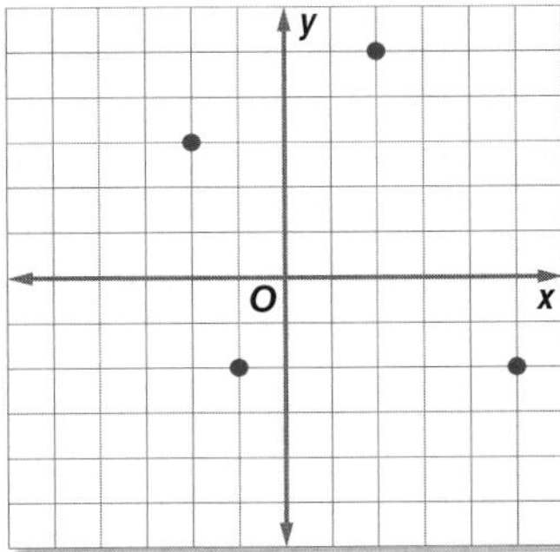

Mapping
List the x-values in the domain and the y-values in the range. Draw arrows from the x-values in the domain to the corresponding y-values in the range.

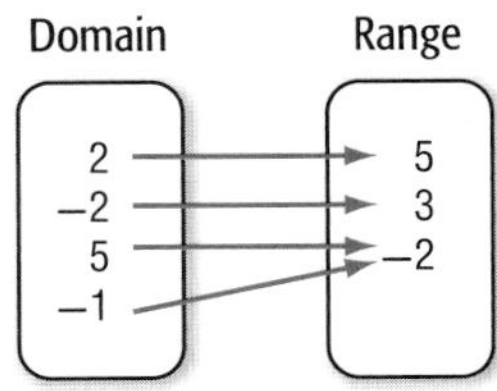

b. Determine the domain and the range of the relation.

The domain of the relation is {2, −2, 5, −1}. The range of the relation is {5, 3, −2}.

Check Your Progress

1A. Express {(4, −3), (3, 2), (−4, 1), (0, −3)} as a table, graph, and mapping.

1B. Determine the domain and range.

Personal Tutor glencoe.com

In a relation, the value of the variable that determines the output is called the **independent variable**. The variable with a value that is dependent on the value of the independent variable is called the **dependent variable**. The domain contains values of the independent variable. The range contains the values of the dependent variable.

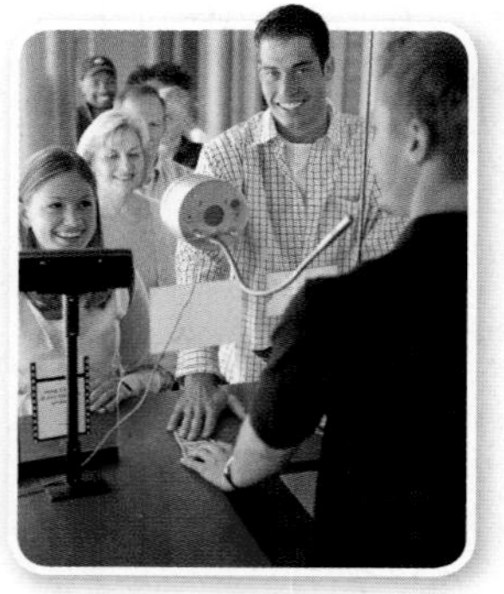

Real-World Link

In 1948, a movie ticket cost $0.36. In 2006, the average ticket price in the United States was $6.50.

Source: National Association of Theatre Owners

Real-World EXAMPLE 2 Independent and Dependent Variables

Identify the independent and dependent variables for each relation.

a. DANCE The dance committee is selling tickets to the Fall Ball. The more tickets that they sell, the greater the amount of money they can spend for decorations.

The number of tickets sold is the independent variable because it is unaffected by the money spent on decorations. The money spent on decorations is the dependent variable because it depends on the number of tickets sold.

b. MOVIES Generally, the average price of going to the movies has steadily increased over time.

Time is the independent variable because it is unaffected by the cost of attending the movies. The price of going to the movies is the dependent variable because it is affected by time.

Check Your Progress

Identify the independent and dependent variables for each relation.

2A. The air pressure inside a tire increases with the temperature.

2B. As the amount of rain decreases, so does the water level of the river.

Personal Tutor glencoe.com

Graphs of a Relation A relation can be graphed without a scale on either axis. These graphs can be interpreted by analyzing their shape.

EXAMPLE 3 Analyze Graphs

The graph represents the distance Francesca has ridden on her bike. Describe what happens in the graph.

Bike Ride

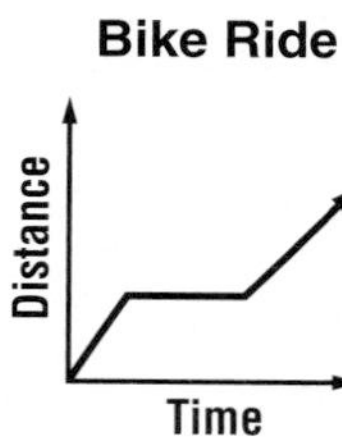

As time increases, the distance increases until the graph becomes a horizontal line.

So, time is increasing but the distance remains constant. At this section Francesca stopped. Then she continued to ride her bike.

Check Your Progress

Describe what is happening in each graph.

3A. Driving to School

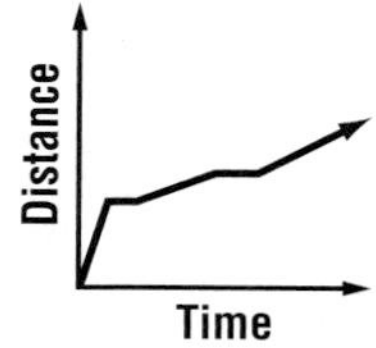

3B. Change in Income

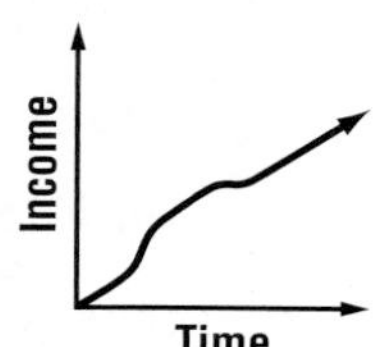

Personal Tutor glencoe.com

Check Your Understanding

Example 1 p. 39 **Express each relation as a table, a graph, and a mapping. Then determine the domain and range.**

1. $\{(4, 3), (-2, 2), (5, -6)\}$
2. $\{(5, -7), (-1, 4), (0, -5), (-2, 3)\}$

Example 2 p. 40 **Identify the independent and dependent variables for each relation.**

3. Increasing the temperature of a compound inside a sealed container increases the pressure inside a sealed container.
4. Mike's cell phone is part of a family plan. If he uses more minutes than his share, then there are fewer minutes available for the rest of his family.
5. Julian is buying concert tickets for him and his friends. The more concert tickets he buys the greater the cost.
6. A store is having a sale over Labor Day weekend. The more purchases, the greater the profits.

Example 3 p. 40 **Describe what is happening in each graph.**

7. The graph represents the distance the track team runs during a practice.

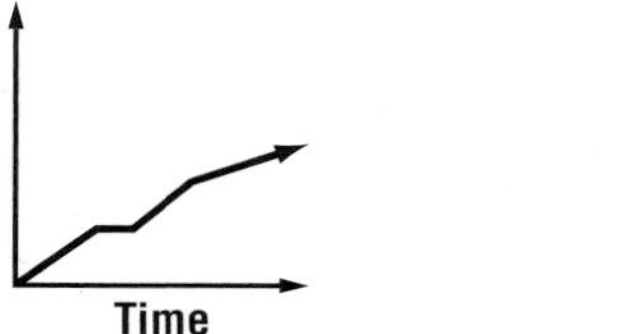

8. The graph represents revenues generated through an online store.

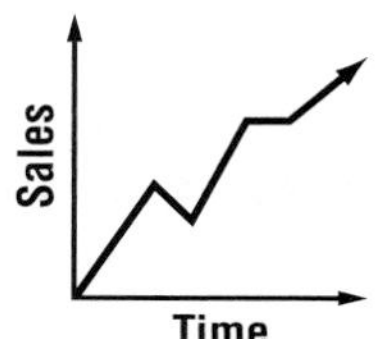

Practice and Problem Solving

● = **Step-by-Step Solutions** begin on page R12. **Extra Practice** begins on page 815.

Example 1 p. 39 **Express each relation as a table, a graph, and a mapping. Then determine the domain and range.**

9. $\{(0, 0), (-3, 2), (6, 4), (-1, 1)\}$
10. $\{(5, 2), (5, 6), (3, -2), (0, -2)\}$
11. $\{(6, 1), (4, -3), (3, 2), (-1, -3)\}$
12. $\{(-1, 3), (3, -6), (-1, -8), (-3, -7)\}$
13. $\{(6, 7), (3, -2), (8, 8), (-6, 2), (2, -6)\}$
14. $\{(4, -3), (1, 3), (7, -2), (2, -2), (1, 5)\}$

Example 2 p. 40 **Identify the independent and dependent variables for each relation.**

15. The Spanish classes are having a fiesta lunch. Each student that attends is to bring a Spanish side dish or dessert. The more students that attend, the more food there will be.
16. The faster you drive your car, the longer it will take to come to a complete stop.

Example 3 p. 40 **Describe what is happening in each graph.**

17. The graph represents the height of a bungee jumper.

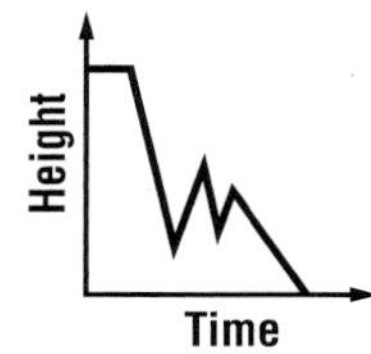

18. The graph represents the sales of lawn mowers.

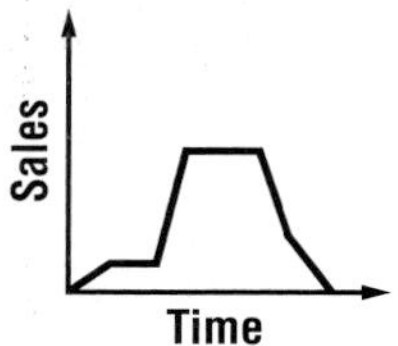

Describe what is happening in each graph.

19. The graph represents the value of a rare baseball card.

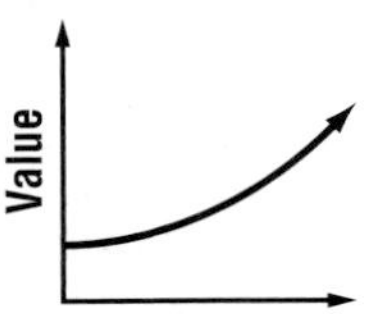

20. The graph represents the distance covered on an extended car ride.

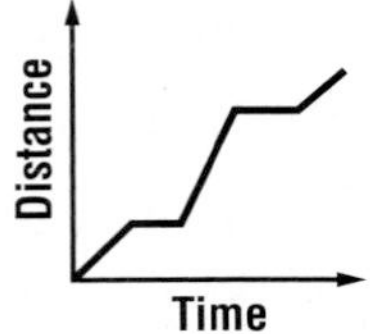

For Exercises 21–23, use the graph at the right.

21. Name the ordered pair at point *A* and explain what it represents.

22. Name the ordered pair at point *B* and explain what it represents.

23. Identify the independent and dependent variables for the relation.

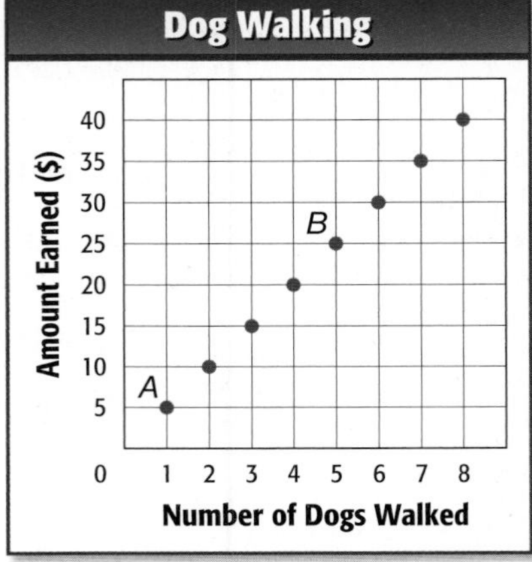

For Exercises 24–26, use the graph at the right.

24. Name the ordered pair at point *C* and explain what it represents.

25. Name the ordered pair at point *D* and explain what it represents.

26. Identify the independent and dependent variables.

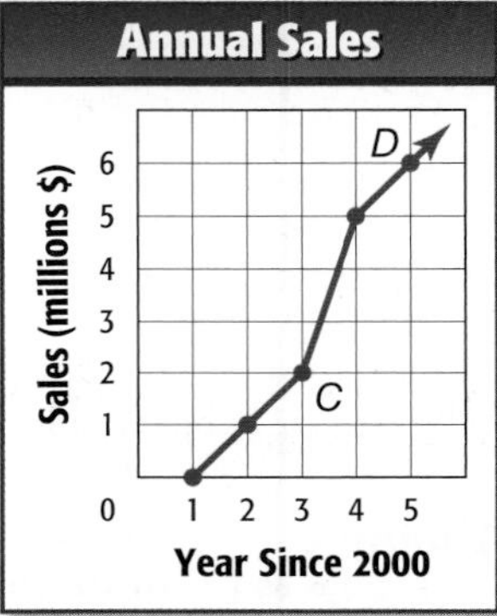

Real-World Link

Allow one gallon of water for each inch of fish you have in the tank.

Source: Tim's Tropicals

Express each relation as a set of ordered pairs. Describe the domain and range.

27.

Buying Aquarium Fish	
Number of Fish	Total Cost
1	$2.50
2	$5.50
5	$10.00
8	$18.75

28.

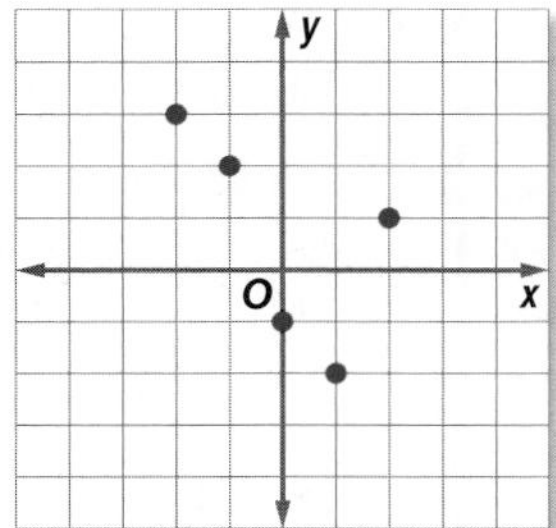

Express the relation in each table, mapping, or graph as a set of ordered pairs.

29.

x	y
4	−1
8	9
−2	−6
7	−3

30.

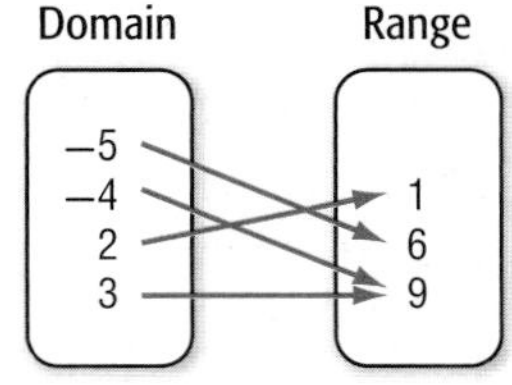

31.

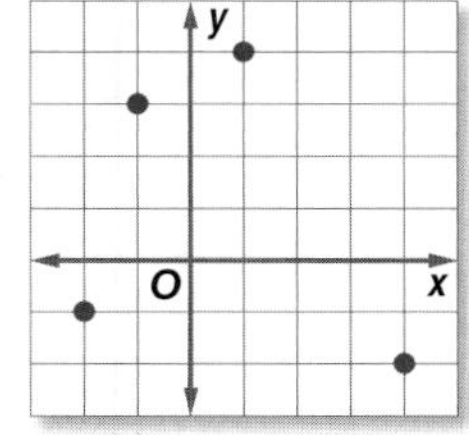

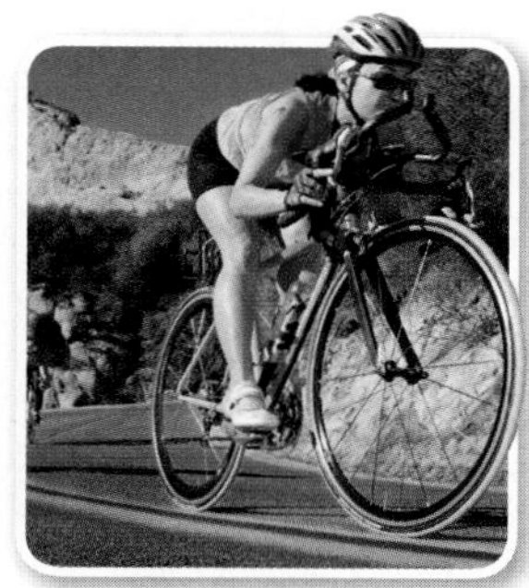

Real-World Link

A triathlon is a competitive sport in which participants swim 2.4 miles, bicycle 112 miles, and then run 26.2 miles. The athlete's total time includes transitioning from one activity to the next.

Source: Ironman World Championship

32. COMPETITIVE SPORTS Refer to the information at the left. Which of the following graphs best represents a participant in a triathlon? Explain.

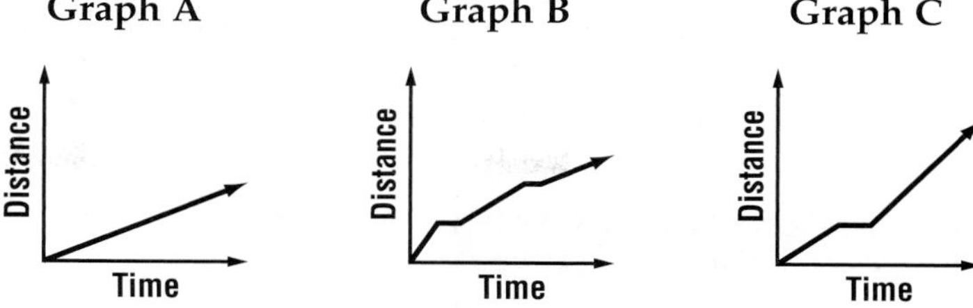

Draw a graph to represent each situation.

33. ANTIQUES A grandfather clock that is over 100 years old has increased in value rapidly from when it was first purchased.

34. CAR A car depreciates in value. The value decreases quickly in the first few years.

35. REAL ESTATE A house typically increases in value over time.

36. EXERCISE An athlete alternates between running and walking during a workout.

37. PHYSIOLOGY A typical adult has about 2 pounds of water for each 3 pounds of body weight. This can be represented by the equation $w = 2\left(\frac{b}{3}\right)$, where w is the weight of water in pounds and b is the body weight in pounds.

a. Make a table to show the relation between body and water weight for people weighing 100, 105, 110, 115, 120, 125, and 130 pounds. Round to the nearest tenth if necessary.

b. What are the independent and dependent variables?

c. State the domain and range, and then graph the relation.

d. Reverse the independent and dependent variables. Graph this relation. Explain what the graph indicates in this circumstance.

H.O.T. Problems

Use Higher-Order Thinking Skills

38. OPEN ENDED Describe a real-life situation that can be represented using a relation and discuss how one of the quantities in the relation depends on the other. Then represent the relation in three different ways.

39. CHALLENGE Describe a real-world situation where it is reasonable to have a negative number included in the domain or range.

40. REASONING Compare and contrast dependent and independent variables.

41. CHALLENGE The table presents a relation. Graph the ordered pairs. Then reverse the y-coordinate and the x-coordinate in each ordered pair. Graph these ordered pairs on the same coordinate plane. Graph the line $y = x$. Describe the relationship between the two sets of ordered pairs.

x	y
0	1
1	3
2	5
3	7

42. WRITING MATH Use the data about the pressure of water on page 38 to explain the difference between dependent and independent variables.

Standardized Test Practice

43. A school's cafeteria employees surveyed 250 students asking what beverage they drank with lunch. They used the data to create the table below.

Beverage	Number of Students
milk	38
chocolate milk	112
juice	75
water	25

What percent of the students surveyed preferred drinking juice with lunch?

A 25% **C** 35%

B 30% **D** 40%

44. Which of the following is equivalent to $6(3 - g) + 2(11 - g)$?

F $2(20 - g)$ **H** $8(5 - g)$

G $8(14 - g)$ **J** $40 - g$

45. SHORT RESPONSE Grant and Hector want to build a clubhouse at the midpoint between their houses. If Grant's house is at point G and Hector's house is at point H, what will be the coordinates of the clubhouse?

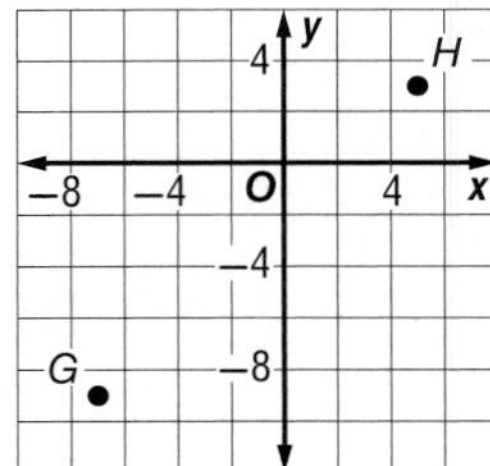

46. If $3b = 2b$, which of the following is true?

A $b = 0$

B $b = \frac{2}{3}$

C $b = 1$

D $b = \frac{3}{2}$

Spiral Review

Solve each equation. (Lesson 1-5)

47. $6(a + 5) = 42$

48. $92 = k + 11$

49. $17 = \frac{45}{w} + 2$

50. HOT-AIR BALLOON A hot-air balloon owner charges \$150 for a one-hour ride. If he gave 6 rides on Saturday and 5 rides on Sunday, write and evaluate an expression to describe his total income for the weekend. (Lesson 1-4)

51. LOLLIPOPS A bag of lollipops contains 19 cherry, 13 grape, 8 sour apple, 15 strawberry, and 9 orange flavored lollipops. What is the probability of drawing a sour apple flavored lollipop? (Lesson 0-11)

Find the perimeter of each figure. (Lesson 0-7)

52.

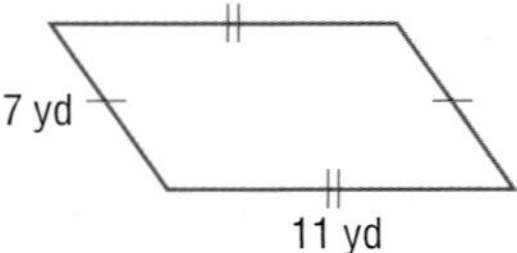

53.

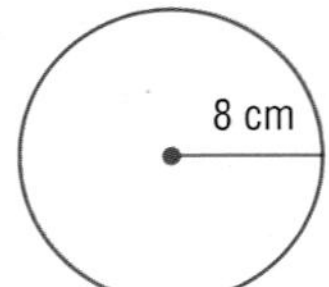

54.

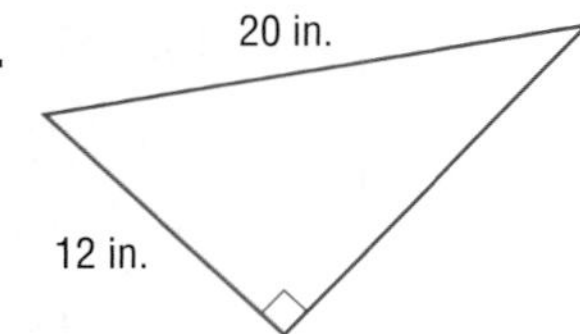

Skills Review

Evaluate each expression. (Lesson 1-2)

55. 8^2

56. $(-6)^2$

57. $(2.5)^2$

58. $(-1.8)^2$

59. $(3 + 4)^2$

60. $(1 - 4)^2$

1-7

Functions

Then

You solved equations with elements from a replacement set. (Lesson 1-5)

Now

- Determine whether a relation is a function.
- Find function values.

New Vocabulary

function
discrete function
continuous function
vertical line test
non linear function

Math Online

glencoe.com

- Extra Examples
- Personal Tutor
- Self-Check Quiz
- Homework Help

Why?

The distance a car travels from when the brakes are applied to the car's complete stop is the stopping distance. This includes time for the driver to react. The faster a car is traveling, the longer the stopping distance. The stopping distance is a function of the speed of the car.

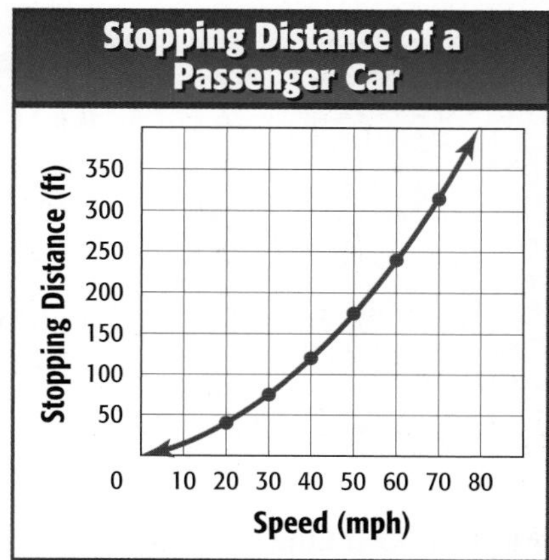

Identify Functions A **function** is a relationship between input and output. In a function, there is exactly one output for each input.

Key Concept — Function

For Your FOLDABLE

Words A function is a relation in which each element of the domain is paired with *exactly* one element of the range.

Examples

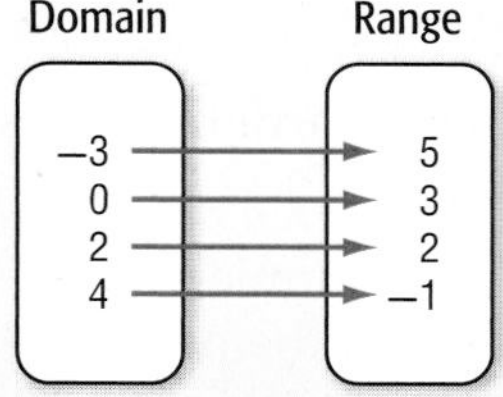

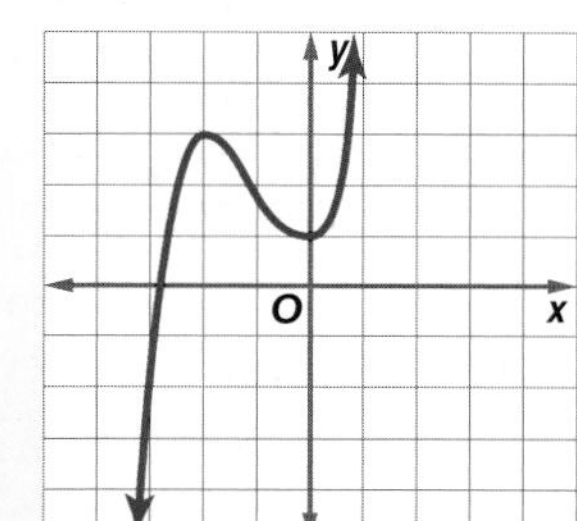

EXAMPLE 1 Identify Functions

Determine whether each relation is a function. Explain.

a. Domain: −2, 0, 3, 4; Range: −3, 6, 9

For each member of the domain, there is only one member of the range. So this mapping represents a function. It does not matter if more than one element of the domain is paired with one element of the range.

b.

Domain	1	3	5	1
Range	4	2	4	−4

The element 1 in the domain is paired with both 4 and −4 in the range. So, when x equals 1 there is more than one possible value for y. This relation is not a function.

Check Your Progress

1. {(2, 1), (3, −2), (3, 1), (2, −2)}

Personal Tutor glencoe.com

A graph that consists of points that are not connected is a **discrete function**. A function graphed with a line or smooth curve is a **continuous function**.

Real-World Link

The Icehotel, located in the Arctic Circle in Sweden, is a hotel made out of ice. The ice insulates the igloo-like hotel so the temperature is at least −8°C.

Source: Icehotel

EXAMPLE 2 Draw Graphs

ICE SCULPTING At an ice sculpting competition, each sculpture's height was measured to make sure that it was within the regulated height range of 0 to 6 feet. The measurements were as follows: Team 1, 4 feet; Team 2, 4.5 feet; Team 3, 3.2 feet; Team 4, 5.1 feet; Team 5, 4.8 feet.

a. Make a table of values showing the relation between the ice sculpting team and the height of their sculpture.

Team Number	1	2	3	4	5
Height (ft)	4	4.5	3.2	5.1	4.8

b. Determine the domain and range of the function.

The domain of the function is {1, 2, 3, 4, 5} because this set represents values of the independent variable. It is unaffected by the heights.

The range of the function is {4, 4.5, 3.2, 5.1, 4.8} because this set represents values of the dependent variable. This value depends on the team number.

c. Write the data as a set of ordered pairs. Then graph the data.

Use the table. The team number is the independent variable and the height of the sculpture is the dependent variable. Therefore, the ordered pairs are (1, 4), (2, 4.5), (3, 3.2), (4, 5.1), and (5, 4.8).

Because the team numbers and their corresponding heights cannot be between the points given, the points should not be connected.

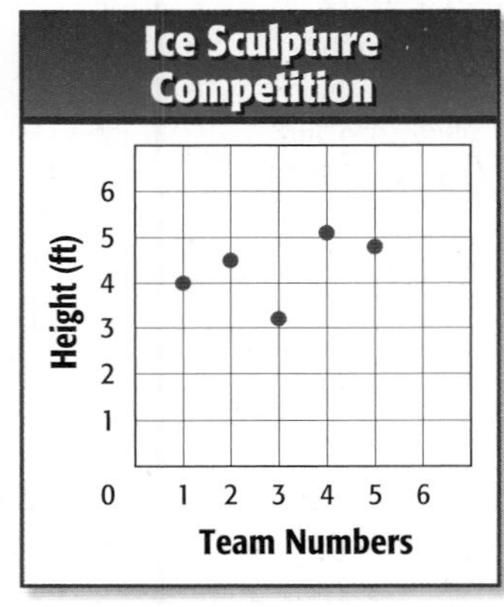

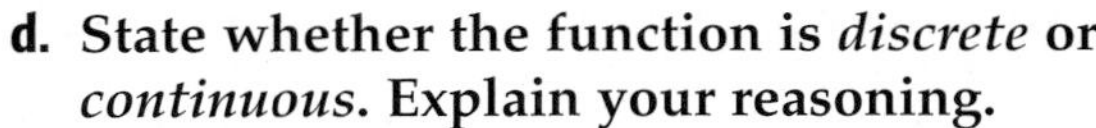

d. State whether the function is *discrete* or *continuous*. Explain your reasoning.

Because the points are not connected, the function is discrete.

Check Your Progress

2. A bird feeder will hold up to 3 quarts of seed. The feeder weighs 2.3 pounds when empty and 13.4 pounds when full.

A. Make a table that shows the bird feeder with 0, 1, 2, and 3 quarts of seed in it weighing 2.3, 6, 9.7, 13.4 pounds respectively.

B. Determine the domain and range of the function.

C. Write the data as a set of ordered pairs. Then graph the data.

D. State whether the function is *discrete* or *continuous*. Explain your reasoning.

Personal Tutor glencoe.com

You can use the **vertical line test** to see if a graph represents a function. If a vertical line intersects the graph more than once, then the graph is not a function. Otherwise, the relation is a function.

Function

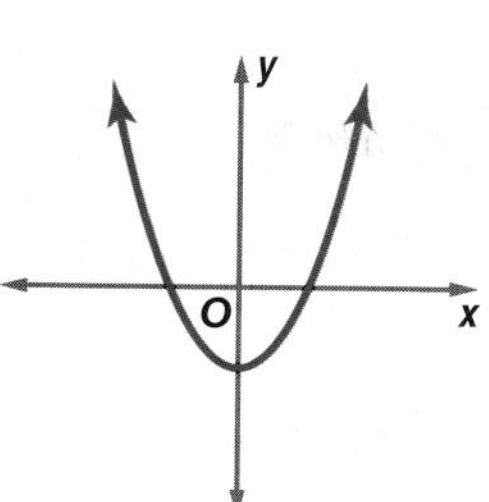

Not a Function

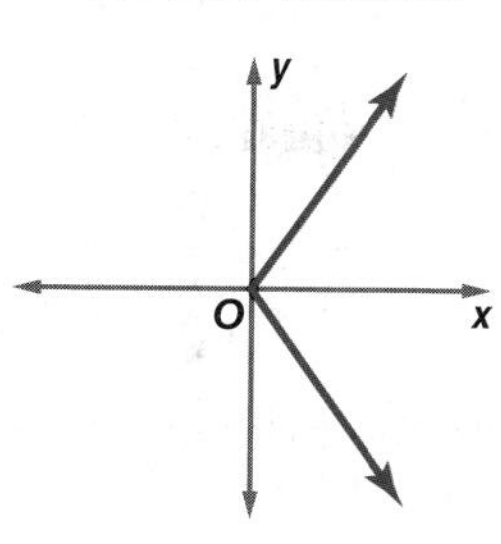

Function

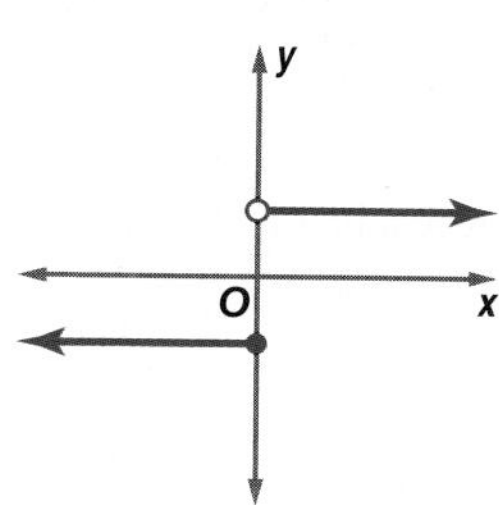

Recall from Lesson 1-6 that an equation is a representation of a relation. If the relation is a function, then the equation represents a function.

EXAMPLE 3 Equations as Functions

Determine whether $-3x + y = 8$ represents a function.

First make a table of values. Then graph the equation.

x	-1	0	1	2
y	5	8	11	14

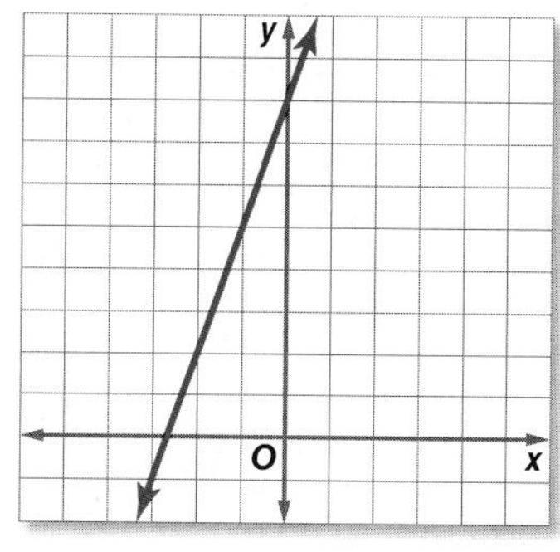

The graph is a line. Place a pencil at the left of the graph to represent a vertical line. Slowly move the pencil across the graph.

For any value of x, the vertical line passes through no more than one point on the graph. So, the graph and the equation represent a function.

StudyTip

Vertical Line Test One way to perform the vertical line test is to use a pencil. Place your pencil vertically on the graph and move from left to right. If the pencil passes over the graph in only one place, then the graph represents a function.

Check Your Progress

Determine if each of the equations represents a function.

3A. $4x = 8$

3B. $4x = y + 8$

Personal Tutor glencoe.com

A function can be represented in different ways.

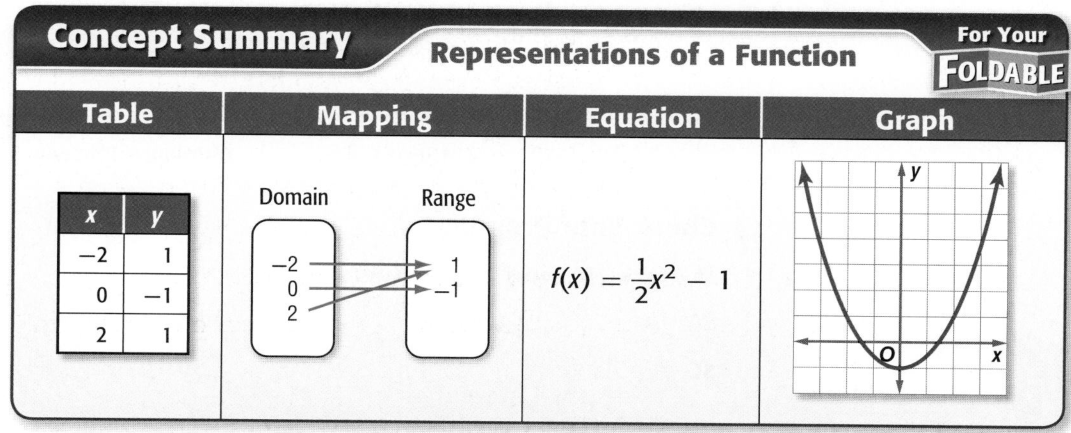

x	y
-2	1
0	-1
2	1

StudyTip

Function Notation Functions are indicated by the symbol $f(x)$. This is read f of x. Other letters, such as g or h, can be used to represent functions.

Find Function Values Equations that are functions can be written in a form called **function notation**. For example, consider $y = 3x - 8$.

Equation	Function Notation
$y = 3x - 8$	$f(x) = 3x - 8$

In a function, x represents the elements of the domain, and $f(x)$ represents the elements of the range. Suppose you want to find the value in the range that corresponds to the element 5 in the domain. This is written $f(5)$ and is read "f of 5." The value $f(5)$ is found by substituting 5 for x in the equation.

EXAMPLE 4 Function Values

For $f(x) = -4x + 7$, find each value.

a. $f(2)$

$f(2) = -4(2) + 7$ **$x = 2$**

$= -8 + 7$ **Multiply.**

$= -1$ **Add.**

b. $f(-3) + 1$

$f(-3) + 1 = [-4(-3) + 7] + 1$ **$x = -3$**

$= 19 + 1$ **Simplify.**

$= 20$ **Add.**

Check Your Progress

For $f(x) = 2x - 3$, find each value.

4A. $f(1)$

4B. $6 - f(5)$

4C. $f(-2)$

4D. $f(-1) + f(2)$

Personal Tutor glencoe.com

A function with a variable term that has an exponent other than 1 forms a **nonlinear function** and the graph is not a line.

EXAMPLE 5 Nonlinear Function Values

If $h(t) = -16t^2 + 68t + 2$, find each value.

a. $h(4)$

$h(4) = -16(4)^2 + 68(4) + 2$ **Replace t with 4.**

$= -256 + 272 + 2$ **Multiply.**

$= 18$ **Add.**

b. $2[h(g)]$

$2[h(g)] = 2[-16(g)^2 + 68(g) + 2]$ **Replace t with g.**

$= 2(-16g^2 + 68g + 2)$ **Simplify.**

$= -32g^2 + 136g + 4$ **Distributive Property**

Check Your Progress

If $f(t) = 2t^3$, find each value.

5A. $f(4)$

5B. $3[f(t)] + 2$

5C. $f(-5)$

5D. $f(-3) - f(1)$

Personal Tutor glencoe.com

Check Your Understanding

Examples 1 and 3 pp. 45, 47

Determine whether each relation is a function. Explain.

1. 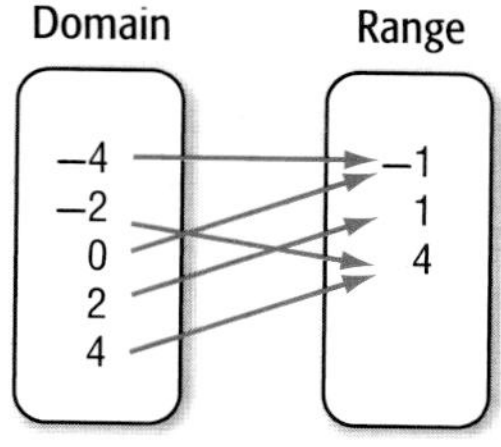

2.

Domain	Range
2	6
5	7
6	9
6	10

3. $\{(2, 2), (-1, 5), (5, 2), (2, -4)\}$

4. $y = \frac{1}{2}x - 6$

5. 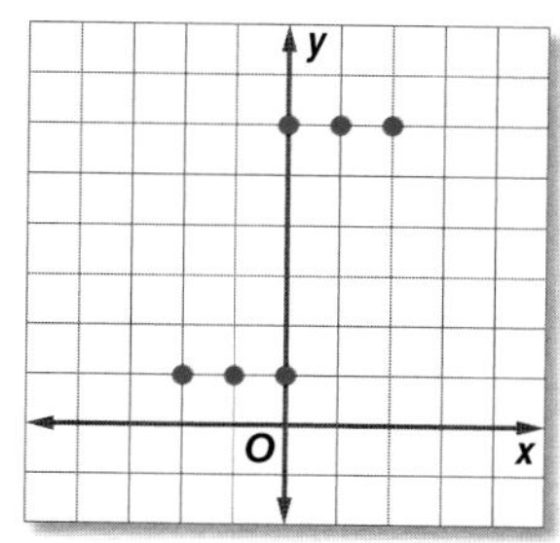

6. 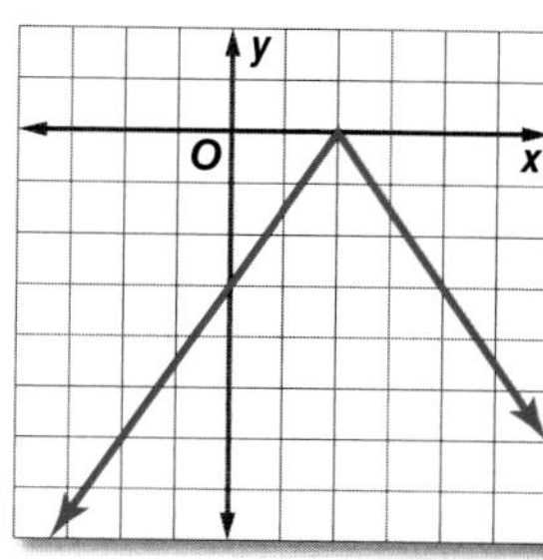

7. 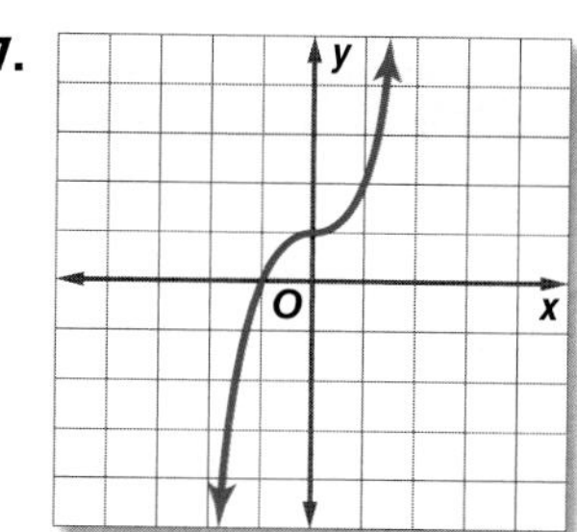

8. 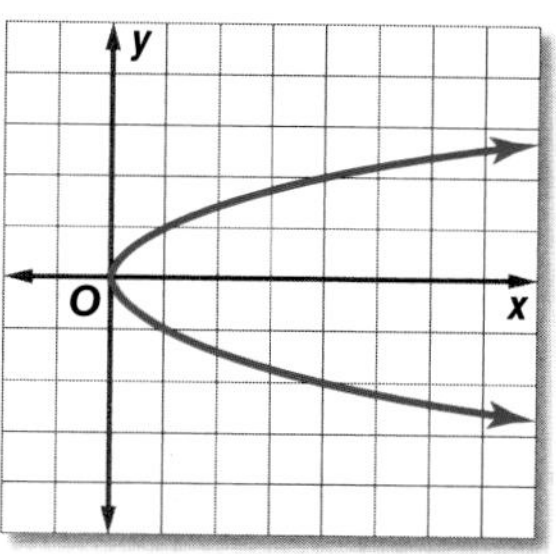

Example 2 p. 46

9. SCHOOL ENROLLMENT The table shows the total enrollment in U.S. public schools.

School Year	2004–05	2005–06	2006–07	2007–08
Enrollment (in thousands)	48,560	48,710	48,948	49,091

Source: *The World Almanac*

a. Write a set of ordered pairs representing the data in the table if x is the number of school years since 2004–2005.

b. Draw a graph showing the relationship between the year and enrollment.

c. Describe the domain and range of the data.

10. CELL PHONES The cost of sending cell phone pictures is given by $y = 0.25x$, where x is the number of pictures that you send. Write the equation in function notation and then find $f(5)$ and $f(12)$. What do these values represent? Determine the domain and range of this function.

Examples 4 and 5 p. 48

If $f(x) = 6x + 7$ and $g(x) = x^2 - 4$, find each value.

11. $f(-3)$

12. $f(m)$

13. $f(r - 2)$

14. $g(5)$

15. $g(a) + 9$

16. $g(-4t)$

17. $f(q + 1)$

18. $f(2) + g(-2)$

19. $g(-b)$

Standardized Test Practice

56. Which point on the number line represents a number whose square is less than itself?

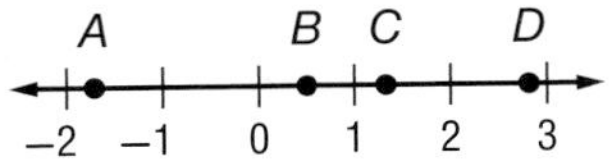

A A **C** C

B B **D** D

57. Determine which of the following relations is a function.

F $\{(-3, 2), (4, 1), (-3, 5)\}$
G $\{(2, -1), (4, -1), (2, 6)\}$
H $\{(-3, -4), (-3, 6), (8, -2)\}$
J $\{(5, -1), (3, -2), (-2, -2)\}$

58. GEOMETRY What is the value of x?

A 3 in.
B 4 in.
C 5 in.
D 6 in.

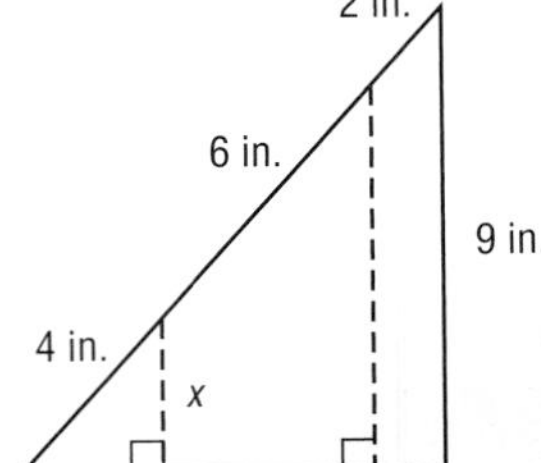

59. SHORT RESPONSE Camille made 16 out of 19 of her serves during her first volleyball game. She made 13 out of 16 of her serves during her second game. During which game did she make a greater percent of her serves?

Spiral Review

Solve each equation. (Lesson 1-5)

60. $x = \frac{27 + 3}{10}$ **61.** $m = \frac{3^2 + 4}{7 - 5}$ **62.** $z = 32 + 4(-3)$

63. SCHOOL SUPPLIES The table shows the prices of some items Tom needs. If he needs 4 glue sticks, 10 pencils, and 4 notebooks, write and solve an equation to determine whether Tom can get them for under \$10. Describe what the variables represent. (Lesson 1-6)

School Supplies Prices	
glue stick	\$1.99
pencil	\$0.25
notebook	\$1.85

Write a verbal expression for each algebraic expression. (Lesson 1-1)

64. $4y + 2$ **65.** $\frac{2}{3}x$ **66.** $a^2b + 5$

Find the volume of each rectangular prism. (Lesson 0-9)

67.

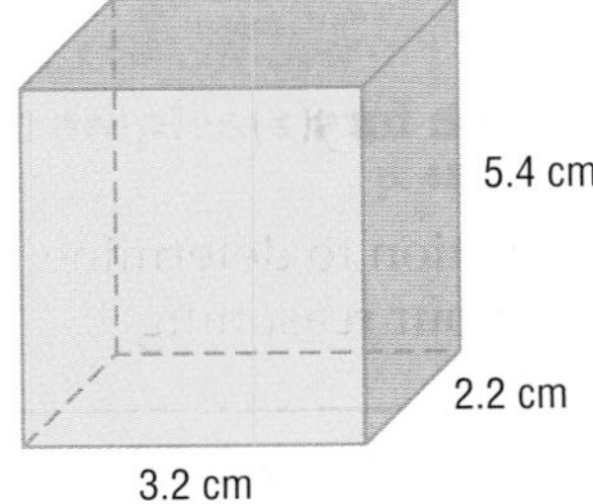

68.

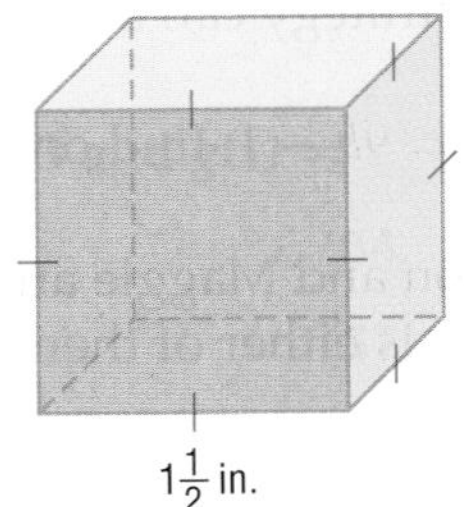

69.

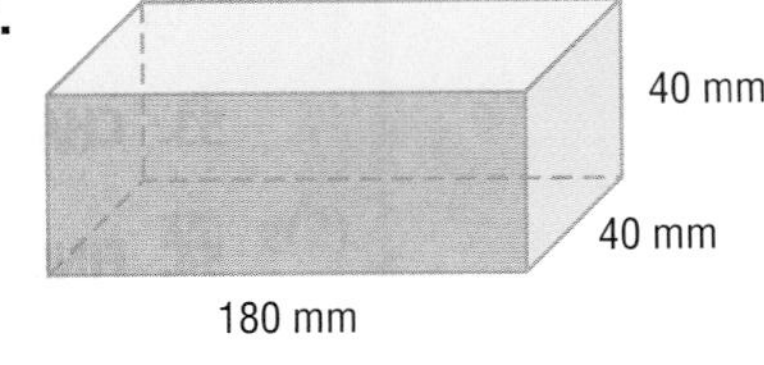

Skills Review

Evaluate each expression. (Lesson 1-2)

70. If $x = 3$, then $6x - 5 = \underline{?}$. **71.** If $n = -1$, then $2n + 1 = \underline{?}$. **72.** If $p = 4$, then $3p + 4 = \underline{?}$.

73. If $q = 7$, then $7q - 9 = \underline{?}$ **74.** If $k = -11$, then $4k + 6 = \underline{?}$ **75.** If $y = 10$, then $8y - 15 = \underline{?}$

Graphing Technology Lab
Representing Functions

Math Online glencoe.com
- Other Calculator Keystrokes
- Graphing Technology Personal Tutor

You can use TI-Nspire™ or TI-Nspire™ CAS technology to explore the different ways to represent a function.

ACTIVITY

Graph $f(x) = 2x + 3$ on the TI-Nspire graphing calculator.

Step 1 From the Home screen, select Graphs & Geometry.

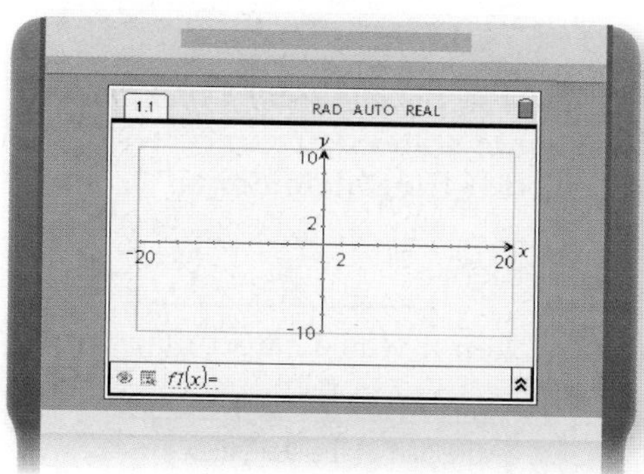

Step 2 Type $2x + 3$ (enter) in the entry line.

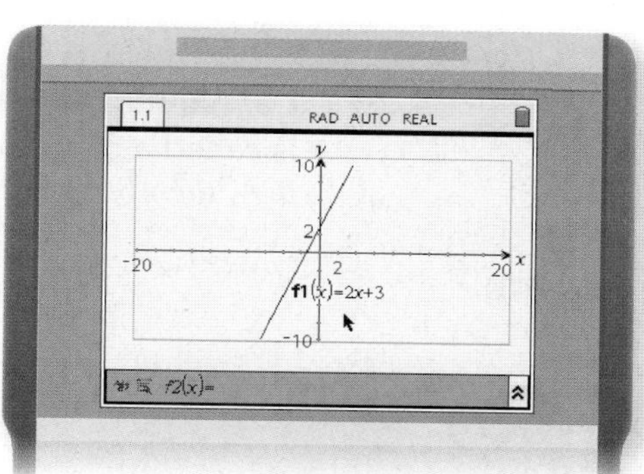

Represent the function as a table.

Step 3 Press (menu). Choose View, then Add Function Table. Then press (enter) or the click button.

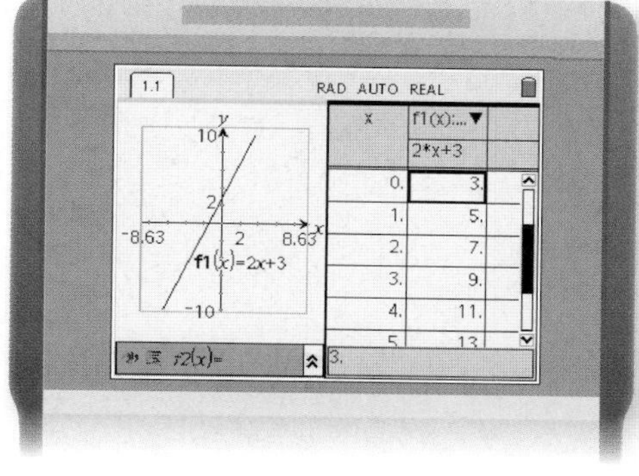

Step 4 Press (ctrl) + (tab) to toggle from the table to the graph. Press (tab) until an arrow appears on the graph. Use the click button to grab the line and move it. Notice how the values in the table change.

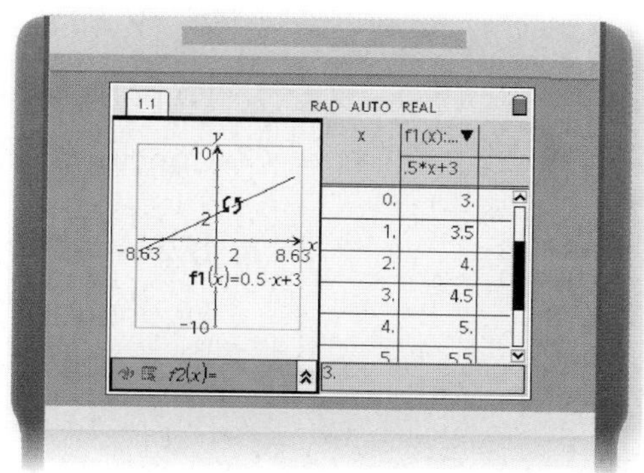

Analyze the Results

Graph each function. Make a table of five ordered pairs that also represents the function.

1. $g(x) = -x - 3$
2. $h(x) = \frac{1}{3}x + 3$
3. $f(x) = -\frac{1}{2}x - 5$
4. $f(x) = 3x - \frac{1}{2}$
5. $g(x) = -2x + 5$
6. $h(x) = \frac{1}{5}x + 4$

1-8 Logical Reasoning and Counterexamples

Then

You applied the properties of real numbers. (Lesson 1-3)

Now

- Identify the hypothesis and conclusion in a conditional statement.
- Use a counterexample to show that an assertion is false.

New Vocabulary

conditional statement
if-then statements
hypothesis
conclusion
deductive reasoning
counterexample

Math Online

glencoe.com

- Extra Examples
- Personal Tutor
- Self-Check Quiz
- Homework Help

Why?

The Butterfly Gardens is a conservatory in British Columbia, Canada, with over 50 species of butterflies. There is also an Emerging Room where you can see caterpillars change into butterflies.

Conditional Statements The statement *If an insect is a butterfly, then it was a caterpillar* is called a conditional statement. A **conditional statement** can be written in the form *If A, then B*. Statements in this form are called **if-then statements**.

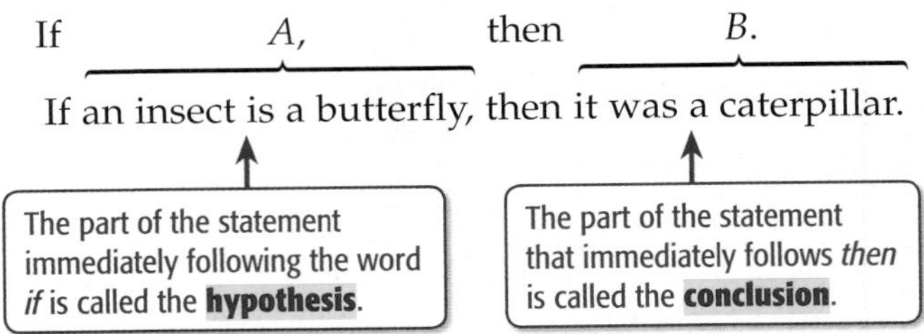

EXAMPLE 1 Identify Hypothesis and Conclusion

Identify the hypothesis and conclusion of each statement.

a. CELEBRATION If it is the Fourth of July, then we will see fireworks.

The hypothesis follows the word *if* and the conclusion follows the *then*.

Hypothesis: it is the 4th of July

Conclusion: we will see fireworks

b. If $2x - 10 = 0$, then $x = 5$.

Hypothesis: $2x - 10 = 0$

Conclusion: $x = 5$

Check Your Progress

1A. If we have enough sugar, then we will make cookies.

1B. If $16z - 5 = 43$, then $z = 3$.

Personal Tutor glencoe.com

Sometimes a conditional statement does not contain the words *if* and *then*. But a conditional statement can always be rewritten in if-then form.

StudyTip

Conditional Statements If a conditional statement is true, the hypothesis need not always be true. For example, if Daniel plays air hockey, then he is at an arcade. But just because Daniel is at an arcade does not mean that he plays air hockey.

EXAMPLE 2 Write a Conditional in If-Then Form

Identify the hypothesis and conclusion of each statement. Then write each statement in if-then form.

a. Chen gets chocolate chip ice cream when she is at the ice cream parlor.

Hypothesis: Chen is at the ice cream parlor

Conclusion: she will get chocolate chip ice cream

If-Then Form: If Chen is at the ice cream parlor, then she will get chocolate chip ice cream.

b. For the equation $3y + 4 = 25$, $y = 7$.

Hypothesis: $3y + 4 = 25$

Conclusion: $y = 7$

If-Then Form: If $3y + 4 = 25$, then $y = 7$.

Check Your Progress

2A. The neon light is on when the store is open.

2B. A circle with a radius of $w - 4$ has a circumference of $2\pi(w - 4)$.

Personal Tutor glencoe.com

Deductive Reasoning and Counterexamples The process of using facts, rules, definitions, or properties to reach a valid conclusion is called **deductive reasoning**. If you know that the hypothesis of a true conditional is true for a given case, deductive reasoning allows you to say that the conclusion is true for that case.

ReadingMath

If-Then Statements Note that *if* is not part of the hypothesis, and *then* is not part of the conclusion.

EXAMPLE 3 Deductive Reasoning

Determine a valid conclusion that follows from the statement below for each condition. If a valid conclusion does not follow, write *no valid conclusion* and explain why.

If one number is odd and another is even, then their product must be even.

a. The numbers are 5 and 8.

5 is odd and 8 is even, so the hypothesis is true.

Their product is 40, which is even, so the conclusion is also true.

b. The product is 24.

The product is part of the conclusion. The product is even, so the conclusion is true.

The hypothesis is also true for numbers such as 3 and 8. However, for numbers such as 4 and 6 the hypothesis is not true. So, there is no valid conclusion.

Check Your Progress

Determine a valid conclusion that follows from the statement *If one number is negative and another is positive, then their product must be negative.* If a valid conclusion does not follow, write *no valid conclusion* and explain why.

3A. The numbers are -3 and 4.

3B. The product is 10.

Personal Tutor glencoe.com

To show that a conditional is false, we can use a counterexample. A **counterexample** is a specific case in which the hypothesis is true and the conclusion is false.

StudyTip

Counterexamples It takes only one counterexample to show that a statement is false.

EXAMPLE 4 Counterexamples

Find a counterexample for each conditional statement.

a. If $a + b > c$, then $b > c$.

One counterexample is when $a = 7$, $b = 3$, and $c = 9$. The hypothesis is true, $7 + 3 > 9$. However, the conclusion $3 > 9$ is false.

b. If the leaves on the tree are brown, then it is fall.

If the leaves are brown then the tree could have died. So, the conclusion is not necessarily true.

Check Your Progress

4A. If $ab > 0$, then a and b are greater than 0.

4B. If a clothing store is selling wool coats, then it must be December.

Personal Tutor glencoe.com

Check Your Understanding

Example 1 p. 54

Identify the hypothesis and conclusion of each statement.

1. If the game is on Saturday, then Eduardo will play.
2. If the chicken burns, then it was left in the oven too long.
3. If $52 - 4x = 28$, then $x = 6$.

Example 2 p. 55

Identify the hypothesis and conclusion of each statement. Then write each statement in if-then form.

4. Alisa plays with her dog in the yard when the weather is nice.
5. Two lines that are perpendicular form right angles.
6. A prime number is only divisible by one and itself.

Example 3 p. 55

Determine a valid conclusion that follows from the statement below for each given condition. If a valid conclusion does not follow, write *no valid conclusion* and explain why.

If a number is a multiple of 10, then the number is divisible by 5.

7. The number is divisible by 5.
8. The number is 5010.
9. The number is 955.

Example 4 p. 56

Find a counterexample for each conditional statement.

10. If Jack is at the park, then he is flying a kite.
11. If a teacher assigns a writing project, then it must be more than two pages long.
12. If $|x| = 7$, then $x = 7$.
13. If a number y is multiplied by $\frac{1}{3}$, then $\frac{1}{3}y < y$.

Practice and Problem Solving

= **Step-by-Step Solutions** begin on page R12.
Extra Practice begins on page 815.

Example 1 p. 54

Identify the hypothesis and conclusion of each statement.

14. If a team is playing at home, then they wear their white uniforms.

15 If you are in a grocery store, then you will buy food.

16. If $2n - 7 > 25$, then $n > 16$.

17. If x equals y and y equals z, then x equals z.

18. If it is not raining outside, we will walk the dogs.

19. If you play basketball, then you are tall.

Example 2 p. 55

Identify the hypothesis and conclusion of each statement. Then write each statement in if-then form.

20. Lamar's third-period class is art.

21. Joe will go to the mall after class.

22. For $x = 4$, $6x - 10 = 14$.

23. $5m - 8 < 52$ when $m < 12$.

24. A rectangle with sides of equal length is a square.

25. The sum of two even numbers is an even number.

26. August has 31 days.

27. Science teachers like to conduct experiments.

Example 3 p. 55

Determine whether a valid conclusion follows from the statement below for each given condition. If a valid conclusion does not follow, write *no valid conclusion* and explain why.

If Belinda scores higher than 90% on the exam, then she will receive an A for the course.

28. Belinda scores a 91% on the exam.

29. Belinda scores an 89% on the exam.

30. Belinda receives an A for the course.

31. Belinda receives a B for the course.

Example 4 p. 56

Find a counterexample for each conditional statement.

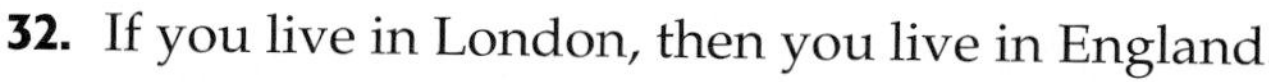

32. If you live in London, then you live in England.

33. If you attend the banquet, then you will eat the food.

34. If the four sides of a quadrilateral are congruent, then the shape is a square.

35. If a number is divisible by 3, then the number is odd.

36. If $3x + 17 \leq 53$, then $x < 12$.

37. If $x^2 = 1$, then x must equal 1.

38. If an animal has spots, then it is a Dalmatian.

39. If a number is prime, then it is an odd number.

40. If an animal cannot fly, then the animal is not a bird.

Real-World Link

The Old Farmer's Almanac uses a formula devised in 1792 to predict weather patterns. It claims 80% accuracy in its forecasts.

41. RESEARCH Use the Internet or some other resource to research the weather predictions and actual weather for your region for the past five years. Summarize your data as examples and counterexamples.

42. Determine whether a valid conclusion follows from the statement below for each given condition. If a valid conclusion does not follow, write *no valid conclusion* and explain why.

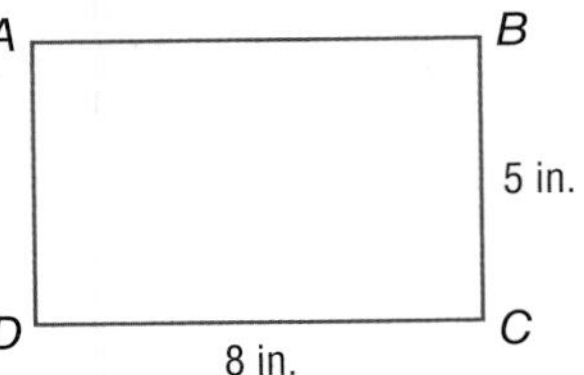

If the dimensions of rectangle ABCD are doubled, then the perimeter is doubled.

a. The new rectangle measures 16 inches by 10 inches.

b. The perimeter of the new rectangle is 52 inches.

43. **GEOMETRY** Use the following statement.

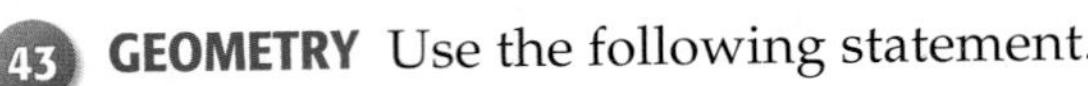

If there are three line segments $\overline{AB}$, $\overline{BC}$, and $\overline{CD}$, then they form a triangle.

a. Draw a diagram to provide an example for the conditional statement.

b. Draw a diagram to provide a counterexample for the conditional statement.

Real-World Link

Ground Hog's Day began with Pennsylvania's early settlers. They based it on the legend of Candlemas Day which states,"For as the sun shines on Candlemas Day, so far will the snow swirl in May...". The first official trek to Gobbler's Knob in Punxsutawney, Pennsylvania was on February 2, 1887.

Source: Punxsutawney Groundhog Club

44. **GROUNDHOG DAY** On Groundhog Day, some people say that if a groundhog sees its shadow, then there will be 6 more weeks of winter. If it does not see its shadow, then there will be an early spring.

a. The most famous groundhog, Punxsutawney Phil in Pennsylvania, sees his shadow 85% of the time. Write an algebraic expression to represent how many times he sees his shadow in y years.

b. The table lists each possible scenario. From the given conditional statement, determine whether this is *true* or *false*.

Sees His Shadow or Not	6 More Weeks of Winter or an Early Spring	True or False
shadow	Winter	true
shadow	Spring	?
no shadow	Winter	?
no shadow	Spring	?

c. Of the situations listed in the table, explain which situation could be considered a counterexample to the original statement.

H.O.T. Problems

Use Higher-Order Thinking Skills

45. **CHALLENGE** Determine whether the following statement is always true. If not, provide a counterexample.

$$\text{If } 2(b + c) = 2b + 2c, \text{ then } 2 + (b \cdot c) = (2 + b)(2 + c).$$

46. **CHALLENGE** For what values of n is the opposite of n greater than n? For what values of n is the opposite of n less than n? For what values is n equal to its opposite?

47. **OPEN ENDED** Write a conditional statement. Label the hypothesis and conclusion.

48. **REASONING** Determine whether this statement is true or false. *If the length of a rectangle is doubled, then the area of the rectangle is doubled.* Justify your answer.

49. **OPEN ENDED** Write a conditional statement. Write a counterexample to the statement. Explain your reasoning.

50. **WRITING IN MATH** Explain how deductive reasoning is used to show that a conditional is true or false.

Standardized Test Practice

51. Which value of b serves as a counterexample to the statement $2b < 3b$?

A -4
B $\frac{1}{4}$
C $\frac{1}{2}$
D 4

52. SHORT RESPONSE A deli serves boxed lunches with a sandwich, fruit, and a dessert. The sandwich choices are turkey, roast beef, or ham. The fruit choices are an orange or an apple. The dessert choices are a cookie or a brownie. How many different boxed lunches does the deli serve?

53. Which illustrates the Transitive Property of Equality?

F If $c = 1$, then $c \cdot \frac{1}{c} = 1$.
G If $c = d$ and $d = f$, then $c = f$.
H If $c = d$, then $d = c$.
J If $c = d$ and $d = c$, then $c = 1$.

54. Simplify the expression $5d(7 - 3) - 16d + 3 \cdot 2d$.

A $10d$
B $14d$
C $21d$
D $25d$

Spiral Review

Determine whether each relation is a function. (Lesson 1-7)

55.

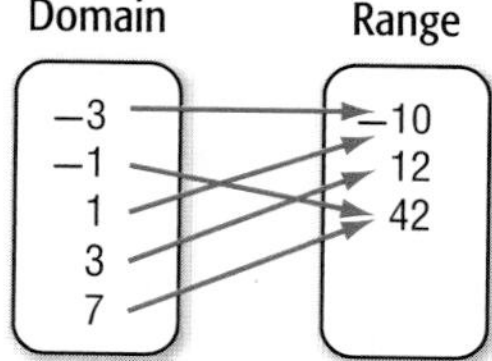

56. $\{(0, 2), (3, 5), (0, -1), (-2, 4)\}$

57.

x	y
17	6
18	6
19	5
20	4

58. GEOMETRY Express the relation in the graph as a set of ordered pairs and describe the domain and range. (Lesson 1-6)

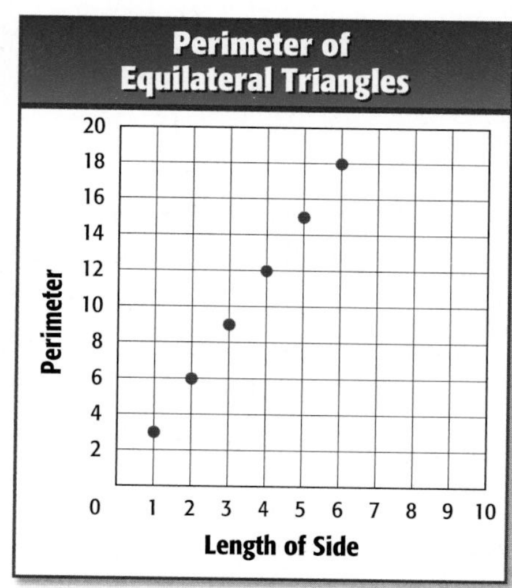

59. CLOTHING Robert has 30 socks in his sock drawer. 16 of the socks are white, 6 are black, 2 are red, and 6 are yellow. What is the probability that he randomly pulls out a black sock? (Lesson 0-9)

Find the perimeter of each figure. (Lesson 0-7)

60.

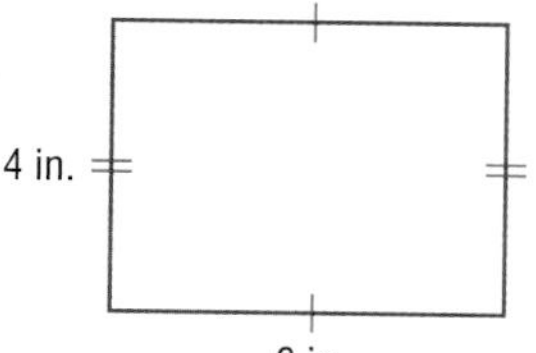

61.

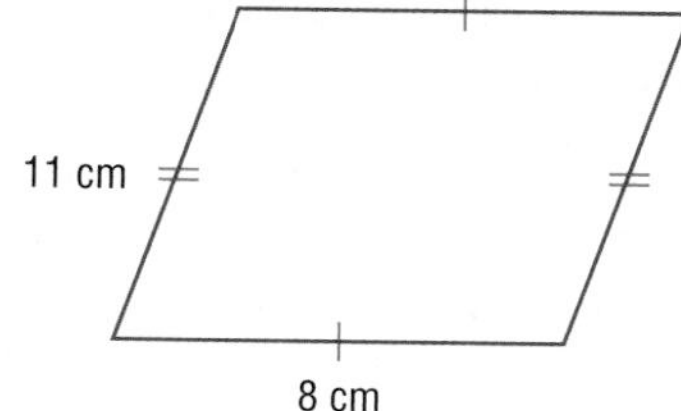

Skills Review

Evaluate each expression. (Lesson 1-2)

62. 7^2

63. $(-9)^2$

64. 2.7^2

65. $(-12.25)^2$

66. 5^2

67. 25^2

EXTEND 1-8

Algebra Lab
Sets

Math Online glencoe.com
Math *in Motion*, Animation

A **set** is any collection of objects. The set that contains all objects is called the **universal set**, or the **universe**, usually labeled U. Each object is called a **member** or **element** of the set.

ACTIVITY 1

Step 1 Cut 6 pieces of paper for each color shown. Draw the shapes shown at the right.

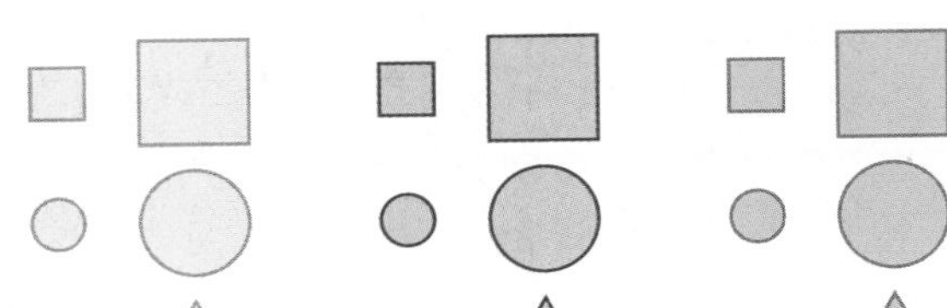

Step 2 Place the shapes inside a loop of string. Label the space inside of the loop U.

Step 3 Arrange the shapes and string as shown. Call the set of squares A.

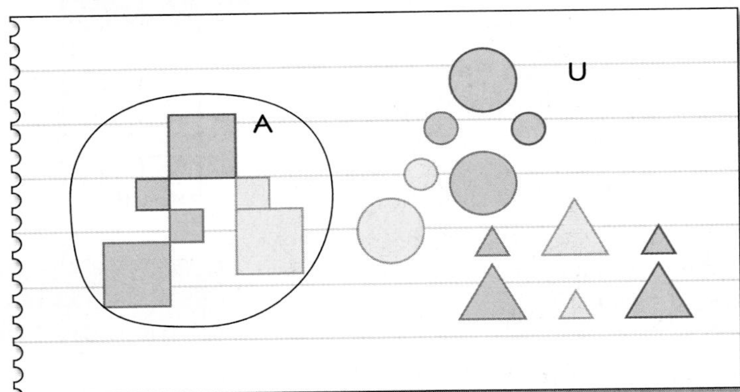

The set of squares is a **subset** of U. The **empty set**, denoted by { } or $\varnothing$, is a set with no objects. It is a subset of any set. A set is also a subset of itself. In math notation, we can write $A \subseteq U$, $A \subseteq A$, and $\varnothing \subseteq U$.

Step 4 We can identify a set by writing a description in brackets, such as {squares}. Put a loop around B = {circles}. Label it B. Notice that $B \subseteq U$.

Step 5 If A = {squares}, then the **complement** of A, written $\overline{A}$, is every object in U that is not in A. $\overline{A}$ = {circles and triangles}, or {nonsquares}. Draw the elements in $\overline{B}$. Write a description of $\overline{B}$ in brackets.

Model and Analyze

1. Let C = {triangles}. Write a description of the complement of set C in brackets.
2. Let R = {yellow shapes}. Write a description of the complement of set R in brackets.
3. Let U = {squares}. Subsets of U can have 0, 1, 2, 3, 4, 5, or 6 elements. How many subsets of U have exactly two elements? How many subsets are there total?

ACTIVITY 2

You can perform operations on two or more numbers, such as addition, subtraction, multiplication, and division. Finding the complement of a set is an operation on one set. You can also perform operations on two or more sets at a time.

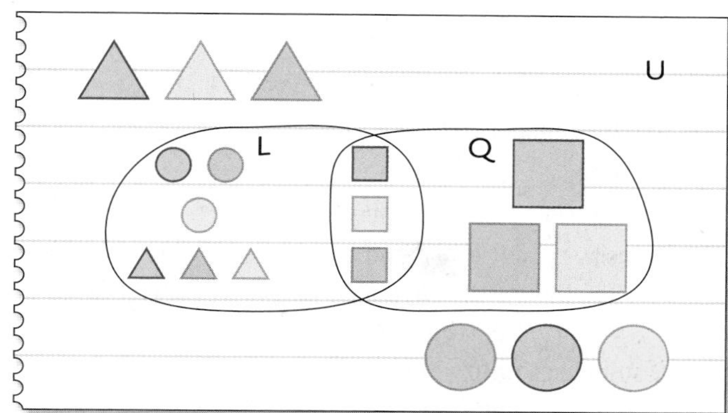

Step 1 Use U from Step 2 in Activity 1. Arrange the shapes as shown. Label the sets.

Write a description of L in brackets. Write a description of Q in brackets.

Step 2 In the diagram in Step 1, the region where L and Q overlap is shaded. Describe the shapes in the shaded region.

Step 3 The **intersection** of two sets is the set of elements common to both. The symbol for this operation is $\cap$. Intersection means that an element is in L and Q. Draw the elements in $L \cap Q$.

Step 4 The **union** of two sets is the set of elements in one set or the other set. The symbol for this operation is $\cup$. You might think of this operation as *adding up* or *combining* all elements in two or more sets. Draw the elements in the set $L \cup Q$.

Step 5 Recall that finding the complement is an operation on only one set. Draw the elements in $\overline{L \cap Q}$.

Step 6 Draw the elements in $\overline{L \cup Q}$.

Exercises

Refer to the Venn diagram shown at the right. Write a description of the shapes in each set.

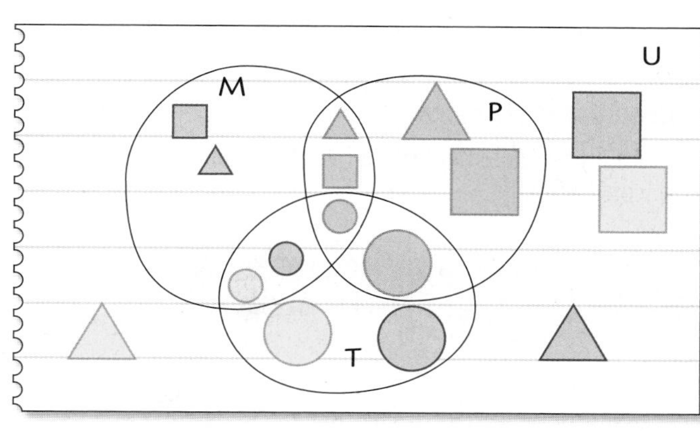

4. M

5. P

6. T

7. $M \cap P$

8. $M \cap T$

9. $P \cap T$

10. $M \cup P$

11. $M \cup T$

12. $P \cup T$

13. $M \cap P \cap T$

14. $M \cup P \cup T$

15. $\overline{M \cup P \cup T}$

16. CHALLENGE Use U from Step 2, Activity 1. Find two sets W and Z such that $W \cap Z = \emptyset$. Draw a diagram with W, Z, and U labeled and all shapes shown. Write a description of W, Z, and $\overline{W \cup Z}$ in brackets.

Study Guide and Review

Math Online glencoe.com
- STUDY TO GO
- Vocabulary Review

Chapter Summary

Key Concepts

Order of Operations (Lesson 1-2)

- Evalute expressions inside grouping symbols.
- Evaluate all powers.
- Multiply and/or divide in order from left to right.
- Add or subtract in order from left to right.

Properties of Equality (Lessons 1-3 and 1-4)

- For any numbers a, b, and c:

Reflexive: $a = a$
Symmetric: If $a = b$, then $b = a$.
Transitive: If $a = b$ and $b = c$, then $a = c$.
Substitution: If $a = b$, then a may be replaced by b in any expression.
Distributive: $a(b + c) = ab + ac$ and $a(b - c) = ab - ac$
Commutative: $a + b = b + a$ and $ab = ba$
Associative: $(a + b) + c = a + (b + c)$ and $(ab)c = a(bc)$

Solving Equations (Lesson 1-5)

- Apply order of operations and the properties of real numbers to solve equations.

Relations (Lesson 1-6)

- Relations can be represent by ordered pairs, a table, a mapping, or a graph.

Functions (Lesson 1-7)

- Use the vertical line test to determine if a relation is a function.

Conditional Statements (Lesson 1-8)

- An if-then statement has a hypothesis and a conclusion.

FOLDABLES® Study Organizer

Be sure the Key Concepts are noted in your Foldable.

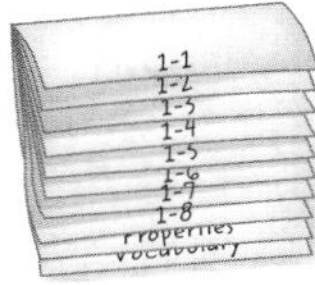

Key Vocabulary

algebraic expression (p. 5)
base (p. 5)
coefficient (p. 26)
conclusion (p. 54)
conditional statement (p. 54)
coordinate system (p. 38)
counterexample (p. 56)
deductive reasoning (p. 55)
dependent variable (p. 40)
domain (p. 38)
equation (p. 31)
exponent (p. 5)
function (p. 45)
hypothesis (p. 54)
independent variable (p. 40)
like terms (p. 25)
mapping (p. 38)
ordered pair (p. 38)
order of operations (p. 10)
origin (p. 38)
power (p. 5)
range (p. 38)
reciprocal (p. 17)
relation (p. 38)
replacement set (p. 31)
simplest form (p. 25)
solution (p. 31)
term (p. 5)
variables (p. 5)
vertical line test (p. 47)

Vocabulary Check

State whether each sentence is *true* or *false*. If *false*, replace the underlined term to make a true sentence.

1. A coordinate system is formed by two intersecting number lines.
2. An exponent indicates the number of times the base is to be used as a factor.
3. An expression is in simplest form when it contains like terms and parentheses.
4. In an expression involving multiplication, the quantities being multiplied are called factors.
5. In a function, there is exactly one output for each input.
6. Order of operations tells us to perform multiplication before subtraction.
7. Since the product of any number and 1 is equal to the number, 1 is called the multiplicative inverse.

Lesson-by-Lesson Review

1-1 Variables and Expressions (pp. 5–9)

Write a verbal expression for each algebraic expression.

8. $h - 7$ **9.** $3x^2$ **10.** $5 + 6m^3$

Write an algebraic expression for each verbal expression.

11. a number increased by 9

12. two thirds of a number d to the third power

13. 5 less than four times a number

Evaluate each expression.

14. 2^5 **15.** 6^3 **16.** 4^4

17. BOWLING Fantastic Pins Bowling Alley charges \$2.50 for shoe rental plus \$3.25 for each game. Write an expression representing the cost to rent shoes and bowl g games.

EXAMPLE 1

Write a verbal expression for $4x + 9$.

nine more than four times a number x

EXAMPLE 2

Write an algebraic expression for *the difference of twelve and two times a number cubed.*

Variable Let x represent the number.

Expression $12 - 2x^3$

EXAMPLE 3

Evaluate 3^4.

The base is 3 and the exponent is 4.

$3^4 = 3 \cdot 3 \cdot 3 \cdot 3$ **Use 3 as a factor 4 times.**

$= 81$ **Multiply.**

1-2 Order of Operations (pp. 10–15)

Evaluate each expression.

18. $24 - 4 \cdot 5$ **19.** $15 + 3^2 - 6$

20. $7 + 2(9 - 3)$ **21.** $8 \cdot 4 - 6 \cdot 5$

22. $\left[(2^5 - 5) \div 9\right]11$ **23.** $\dfrac{11 + 4^2}{5^2 - 4^2}$

Evaluate each expression if $a = 4$, $b = 3$, and $c = 9$.

24. $c + 3a$

25. $5b^2 \div c$

26. $(a^2 + 2bc) \div 7$

27. ICE CREAM The cost of a one-scoop sundae is \$2.75, and the cost of a two-scoop sundae is \$4.25. Write and evaluate an expression to find the total cost of 3 one-scoop sundaes and 2 two-scoop sundaes.

EXAMPLE 4

Evaluate the expression $3(9 - 5)^2 \div 8$.

$3(9 - 5)^2 \div 8 = 3(4)^2 \div 8$ **Work inside parentheses.**

$= 3(16) \div 8$ **Evaluate 4^2.**

$= 48 \div 8$ **Multiply.**

$= 6$ **Divide.**

EXAMPLE 5

Evaluate the expression $(5m - 2n) \div p^2$ if $m = 8$, $n = 4$, $p = 2$.

$(5m - 2n) \div p^2$

$= (5 \cdot 8 - 2 \cdot 4) \div 2^2$ **Replace m with 8, n with 4, and p with 2.**

$= (40 - 8) \div 2^2$ **Multiply.**

$= 32 \div 2^2$ **Subtract.**

$= 32 \div 4$ **Evaluate 2^2.**

$= 8$ **Divide.**

Study Guide and Review

1-3 Properties of Numbers (pp. 16–22)

Evaluate each expression using properties of numbers. Name the property used in each step.

28. $18 \cdot 3(1 \div 3)$

29. $[5 \div (8 - 6)]\frac{2}{5}$

30. $(16 - 4^2) + 9$

31. $2 \cdot \frac{1}{2} + 4(4 \cdot 2 - 7)$

32. $18 + 41 + 32 + 9$

33. $7\frac{2}{5} + 5 + 2\frac{3}{5}$

34. $8 \cdot 0.5 \cdot 5$

35. $5.3 + 2.8 + 3.7 + 6.2$

36. SCHOOL SUPPLIES Monica needs to purchase a binder, a textbook, a calculator, and a workbook for her algebra class. The binder costs $9.25, the textbook $32.50, the calculator $18.75, and the workbook $15.00. Find the total cost for Monica's algebra supplies.

EXAMPLE 6

Evaluate $6(4 \cdot 2 - 7) + 5 \cdot \frac{1}{5}$. Name the property used in each step.

$6(4 \cdot 2 - 7) + 5 \cdot \frac{1}{5}$

$= 6(8 - 7) + 5 \cdot \frac{1}{5}$ **Substitution**

$= 6(1) + 5 \cdot \frac{1}{5}$ **Substitution**

$= 6 + 5 \cdot \frac{1}{5}$ **Multiplicative Identity**

$= 6 + 1$ **Multiplicative Inverse**

$= 7$ **Substitution**

1-4 The Distributive Property (pp. 23–29)

Use the Distributive Property to rewrite each expression. Then evaluate.

37. $(2 + 3)6$

38. $5(18 + 12)$

39. $8(6 - 2)$

40. $(11 - 4)3$

41. $-2(5 - 3)$

42. $(8 - 3)4$

Rewrite each expression using the Distributive Property. Then simplify.

43. $3(x + 2)$

44. $(m + 8)4$

45. $6(d - 3)$

46. $-4(5 - 2t)$

47. $(9y - 6)(-3)$

48. $-6(4z + 3)$

49. TUTORING Write and evaluate an expression for the number of tutoring lessons Mrs. Green gives in 4 weeks.

Tutoring Schedule	
Day	**Students**
Monday	3
Tuesday	5
Wednesday	4

EXAMPLE 7

Use the Distributive Property to rewrite the expression 5(3 + 8). Then evaluate.

$5(3 + 8) = 5(3) + 5(8)$ **Distributive Property**

$= 15 + 40$ **Multiply.**

$= 55$ **Simplify.**

EXAMPLE 8

Rewrite the expression $6(x + 4)$ using the Distributive Property. Then simplify.

$6(x + 4) = 6 \cdot x + 6 \cdot 4$ **Distributive Property**

$= 6x + 24$ **Simplify.**

EXAMPLE 9

Rewrite the expression $(3x - 2)(-5)$ using the Distributive Property. Then simplify.

$(3x - 2)(-5)$

$= (3x)(-5) - (2)(-5)$ **Distributive Property**

$= -15x + 10$ **Simplify.**

MIXED PROBLEM SOLVING
For mixed problem-solving practice, see page 845.

1-5 Equations (pp. 31–37)

Find the solution of each equation if the replacement sets are x: {1, 3, 5, 7, 9} and y: {6, 8, 10, 12, 14}

50. $y - 9 = 3$

51. $14 + x = 21$

52. $4y = 32$

53. $3x - 11 = 16$

54. $\frac{42}{y} = 7$

55. $2(x - 1) = 8$

Solve each equation.

56. $a = 24 - 7(3)$

57. $z = 63 \div (3^2 - 2)$

58. AGE Shandra's age is four more than three times Sherita's age. Write an equation for Shandra's age. Solve if Sherita is 3 years old.

EXAMPLE 10

Solve the equation $5w - 19 = 11$ if the replacement set is w: {2, 4, 6, 8, 10}.

Replace w in $5w - 19 = 11$ with each value in the replacement set.

w	$5w - 19 = 11$	True or False?
2	$5(2) - 19 = 11$	False
4	$5(4) - 19 = 11$	False
6	$5(6) - 19 = 11$	True
8	$5(8) - 19 = 11$	False
10	$5(10) - 19 = 11$	False

Since the equation is true when $w = 6$, the solution of $5w - 19 = 11$ is $w = 6$.

1-6 Representing Relations (pp. 38–44)

Express each relation as a table, a graph, and a mapping. Then determine the domain and range.

59. {(1, 3), (2, 4), (3, 5), (4, 6)}

60. {(−1, 1), (0, −2), (3, 1), (4, −1)}

61. {(−2, 4), (−1, 3), (0, 2), (−1, 2)}

Express the relation shown in each table, mapping, or graph as a set of ordered pairs.

62.

x	y
5	3
3	−1
1	2
−1	0

63.

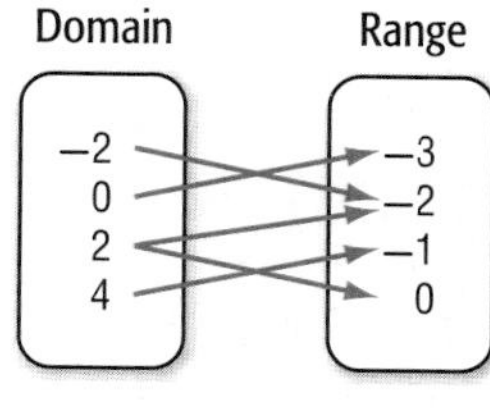

64. GARDENING On average, 7 plants grow for every 10 seeds of a certain type planted. Make a table to show the relation between seeds planted and plants growing for 50, 100, 150, and 200 seeds. Then state the domain and range and graph the relation.

EXAMPLE 11

Express the relation {(−3, 4), (1, −2), (0, 1), (3, −1)} as a table, a graph, and a mapping.

Table

Place the x-coordinates into the first column. Place the corresponding y-coordinates in the second column.

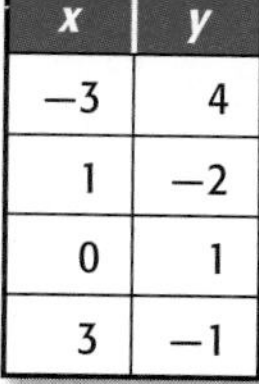

x	y
−3	4
1	−2
0	1
3	−1

Graph

Graph each ordered pair on a coordinate plane.

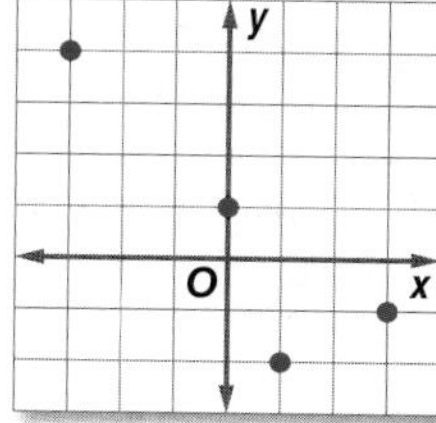

Mapping

List the x-values in the domain and the y-values in the range. Draw arrows from the x-values in set X to the corresponding y-values in set Y.

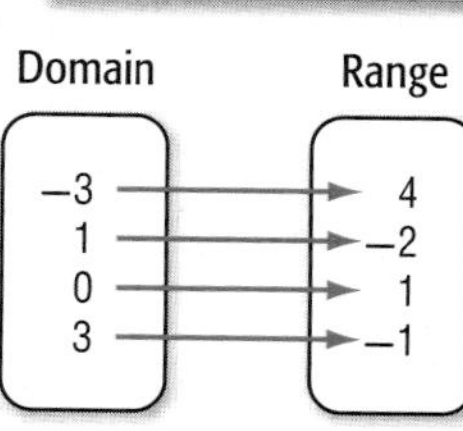

CHAPTER 1 Study Guide and Review

1-7 Representing Functions (pp. 45–52)

Determine whether each relation is a function.

65.

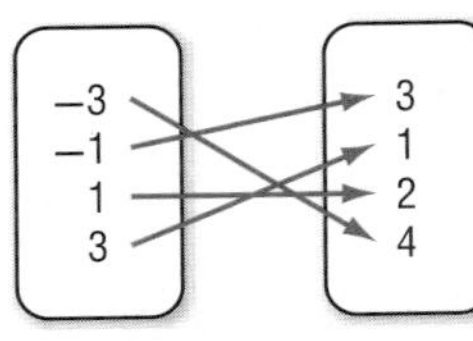

66.

x	y
−4	3
2	0
1	−2
2	1

67.

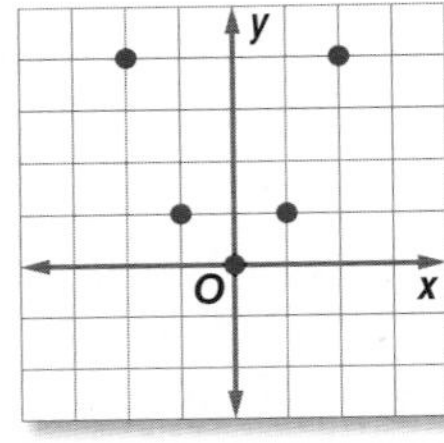

68. $\{(8, 4), (6, 3), (4, 2), (2, 1), (0, 0)\}$

If $f(x) = 2x + 4$ and $g(x) = x^2 - 3$, find each value.

69. $f(-3)$ **70.** $g(2)$ **71.** $f(0)$

72. $g(-4)$ **73.** $f(m + 2)$ **74.** $g(3p)$

75. GRADES A teacher claims that the relationship between number of hours studied for a test and test score can be described by $g(x) = 45 + 9x$, where x represents the number of hours studied. Graph this function.

EXAMPLE 12

Determine whether the relation shown below is a function.

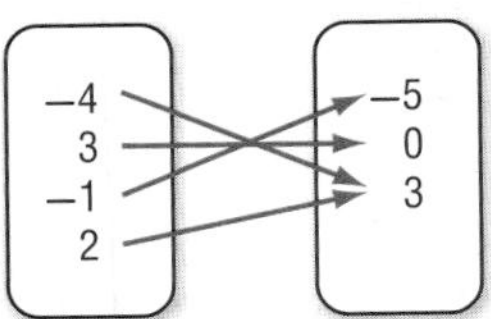

For each member of the domain, there is only one member of the range that corresponds to it. So this mapping represents a function. It does not matter that more than one element of the domain is paired with one element of the range.

EXAMPLE 13

Determine whether $2x - y = 1$ represents a function.

First make a table of values. Then graph the equation.

x	y
−1	−3
0	−1
1	1
2	3
3	5

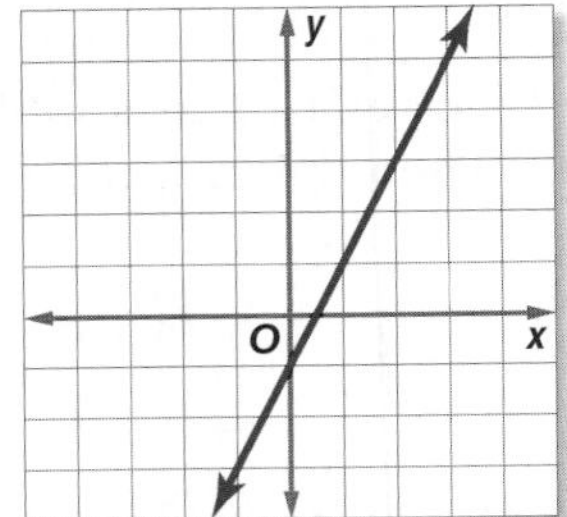

Using the vertical line test, it can be shown that $2x - y = 1$ does represent a function.

1-8 Logical Reasoning and Counterexamples (pp. 54–59)

Identify the hypothesis and conclusion of each statement.

76. If Orlando practices the piano, then he will perform well at his recital.

77. If $2x + 7 > 31$, then $x > 12$.

Find a counterexample for each conditional statement.

78. If it is raining outside, then you will get wet.

79. If $4x - 11 = 53$, then $x < 16$.

EXAMPLE 14

Identify the hypothesis and the conclusion for the statement "If the football team wins their last game, then they will win the championship."

The hypothesis follows the word *if*, and the conclusion follows the word *then*.

Hypothesis: the football team wins their last game

Conclusion: they will win the championship

CHAPTER 1

Practice Test

Math Online glencoe.com
Chapter Test

Write an algebraic expression for each verbal expression.

1. six more than a number
2. twelve less than the product of three and a number
3. four divided by the difference between a number and seven

Evaluate each expression.

4. $32 \div 4 + 2^3 - 3$
5. $\frac{(2 \cdot 4)^2}{7 + 3^2}$

6. **MULTIPLE CHOICE** Find the value of the expression $a^2 + 2ab + b^2$ if $a = 6$ and $b = 4$.

   A 68

   B 92

   C 100

   D 121

Evaluate each expression. Name the property used in each step.

7. $13 + (16 - 4^2)$
8. $\frac{2}{9}[9 \div (7 - 5)]$
9. $37 + 29 + 13 + 21$

Rewrite each expression using the Distributive Property. Then simplify.

10. $4(x + 3)$
11. $(5p - 2)(-3)$

12. **MOVIE TICKETS** A company operates three movie theaters. The chart shows the typical number of tickets sold each week at the three locations. Write and evaluate an expression for the total typical number of tickets sold by all three locations in four weeks.

Location	Tickets Sold
A	438
B	374
C	512

Find the solution of each equation if the replacement sets are x: {1, 3, 5, 7, 9} and y: {2, 4, 6, 8, 10}.

13. $3x - 9 = 12$
14. $y^2 - 5y - 11 = 13$

15. **CELL PHONES** The ABC Cell Phone Company offers a plan that includes a flat fee of \$29 per month plus a \$0.12 charge per minute. Write an equation to find C, the total monthly cost for m minutes. Then solve the equation for $m = 50$.

Express the relation shown in each table, mapping, or graph as a set of ordered pairs.

16.

x	y
−2	4
1	2
3	0
4	−2

17. Domain: −3, −1, 1, 3 — Range: −2, 0, 2, 4

18. **MULTIPLE CHOICE** Determine the domain and range for the relation {(2, 5), (−1, 3), (0, −1), (3, 3), (−4, −2)}.

   F D: {2, −1, 0, 3, −4}, R: {5, 3, −1, 3, −2}

   G D: {5, 3, −1, 3, −2}, R: {2, −1, 0, 3, 4}

   H D: {0, 1, 2, 3, 4}, R: {−4, −3, −2, −1, 0}

   J D: {2, −1, 0, 3, −4}, R: {2, −1, 0, 3, 4}

19. Determine whether the relation {(2, 3), (−1, 3), (0, 4), (3, 2), (−2, 3)} is a function.

If $f(x) = 5 - 2x$ and $g(x) = x^2 + 7x$, find each value.

20. $g(3)$
21. $f(-6y)$

Identify the hypothesis and conclusion of each statement.

22. If the temperature goes below 32°F, it will snow outside.
23. If Ivan breaks his arm, he will need to go to the hospital.

Find a counterexample for each conditional statement.

24. If you go to the pool, you will get wet.
25. If a quadrilateral has one pair of sides that are parallel, then it is a square.

Eliminate Unreasonable Answers

You can eliminate unreasonable answers to help you find the correct one when solving multiple choice test items. Doing so will save you time by narrowing down the list of possible correct answers.

Strategies for Eliminating Unreasonable Answers

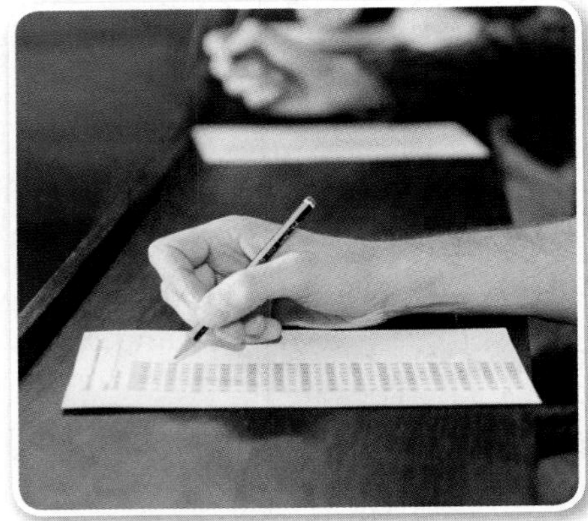

Step 1

Read the problem statement carefully to determine exactly what you are being asked to find.

Ask yourself:

- What am I being asked to solve?
- What format (i.e., fraction, number, decimal, percent, type of graph) will the correct answer be?
- What units (if any) will the correct answer have?

Step 2

Carefully look over each possible answer choice and evaluate for reasonableness.

- Identify any answer choices that are clearly incorrect and eliminate them.
- Eliminate any answer choices that are not in the proper format.
- Eliminate any answer choices that do not have the correct units.

Step 3

Solve the problem and choose the correct answer from those remaining. Check your answer.

EXAMPLE

Read each problem. Eliminate any unreasonable answers. Then use the information in the problem to solve.

Jason earns 8.5% commission on his weekly sales at an electronics retail store. Last week he had $4200 in sales. What was his commission for the week?

A $332

B $357

C $425

D $441

Using mental math, you know that 10% of \$4200 is \$420. Since 8.5% is less than 10%, you know that Jason earned less than \$420 in commission for his weekly sales. So, answer choices C and D can be eliminated because they are greater than \$420. The correct answer is either A or B.

$\$4200 \times 0.085 = \$357$

So, the correct answer is B.

Exercises

Read each problem. Eliminate any unreasonable answers. Then use the information in the problem to solve.

1. Coach Roberts expects 35% of the student body to turn out for a pep rally. If there are 560 students, how many does Coach Roberts expect to attend the pep rally?

A 184

B 196

C 214

D 390

2. Jorge and Sally leave school at the same time. Jorge walks 300 yards north and then 400 yards east. Sally rides her bike 600 yards south and then 800 yards west. What is the distance between the two students?

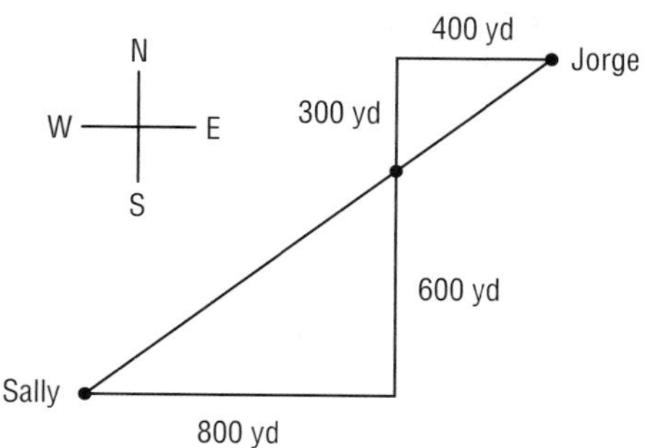

F 500 yd

G 750 yd

H 1,200 yd

J 1,500 yd

3. What is the range of the relation below?

$\{(1, 2), (3, 4), (5, 6), (7, 8)\}$

A all real numbers

B all even numbers

C $\{2, 4, 6, 8\}$

D $\{1, 3, 5, 7\}$

4. The expression $3n + 1$ gives the total number of squares needed to make each figure of the pattern where n is the figure number. How many squares will be needed to make Figure 9?

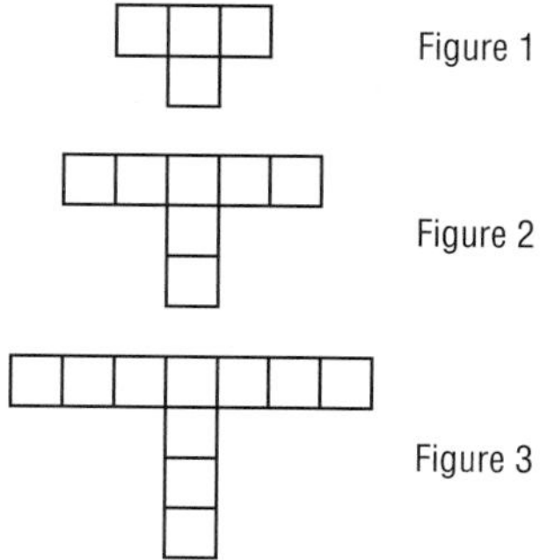

F 28 squares

G 32.5 squares

H 56 squares

J 88.5 squares

5. The expression $3x - (2x + 4x - 6)$ is equivalent to

A $-3x - 6$

B $-3x + 6$

C $3x + 6$

D $3x - 6$

CHAPTER 1 Standardized Test Practice

Chapter 1

Multiple Choice

Read each question. Then fill in the correct answer on the answer document provided by your teacher or on a sheet of paper.

1. Evaluate the expression 2^6.

A 12

B 32

C 64

D 128

2. Monica claims: *If you are in the drama club, then you are also on the academic team.* Which student is a counterexample to this statement?

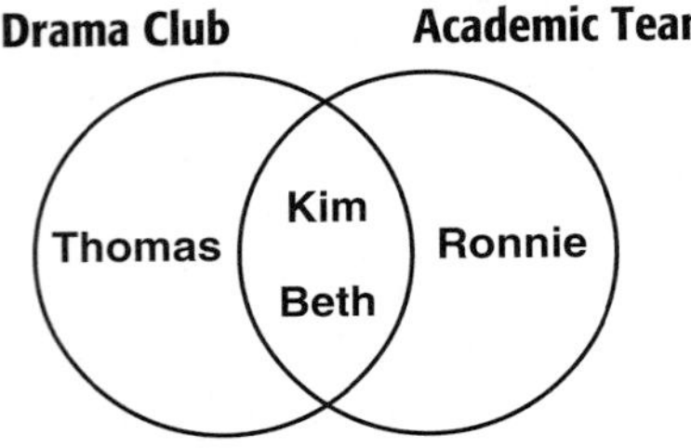

F Beth

G Kim

H Ronnie

J Thomas

3. Let y represent the number of yards. Which algebraic expression represents the number of feet in y?

A $y - 3$

B $y + 3$

C $3y$

D $\frac{3}{y}$

4. What is the domain of the following relation?

$$\{(1, 3), (-6, 4), (8, 5)\}$$

F $\{3, 4, 5\}$

G $\{-6, 1, 8\}$

H $\{-6, 1, 3, 4, 5, 8\}$

J $\{1, 3, 4, 5, 8\}$

5. The table shows the number of some of the items sold at the concession stand at the first day of a soccer tournament. Estimate how many items were sold from the concession stand through out the four days of the tournament.

Concession Sales Day 1 Results	
Item	Number Sold
Popcorn	78
Hot Dogs	80
Chip	48
Sodas	51
Bottled Water	92

A 1350 items

B 1400 items

C 1450 items

D 1500 items

6. There are 24 more cars for sale at a dealership than twice the number of trucks. If there are 100 cars for sale, how many trucks are there for sale at the dealership?

F 28

G 32

H 34

J 38

7. Refer to the relation in the table below. Which of the following values would result in the relation *not* being a function?

x	−6	−2	0	?	3	5
y	−1	8	3	−3	4	0

A −1

B 3

C 7

D 8

Test-TakingTip

Question 2 A *counterexample* is a specific case in which the hypothesis of a conditional statement is true, but the conclusion is false.

Short Response/Gridded Response

Record your answers on the answer sheet provided by your teacher or on a sheet of paper.

8. The edge of each box below is 1 unit long.

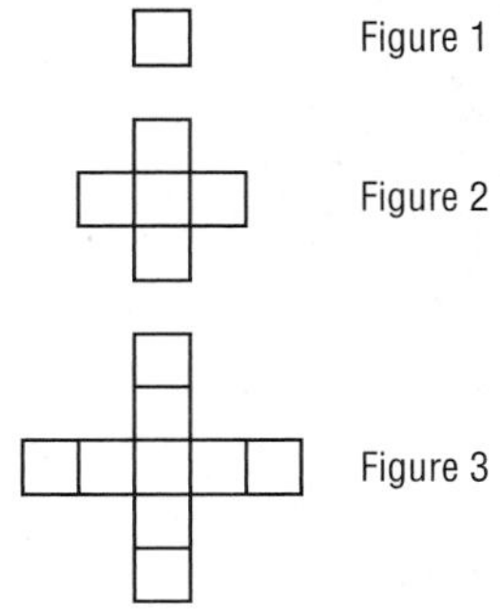

a. Make a table showing the perimeters of the first 3 figures in the pattern.

b. Look for a pattern in the perimeters of the shapes. Write an algebraic expression for the perimeter of Figure n.

c. What would be the perimeter of Figure 10 in the pattern?

9. The table shows the costs of certain items at a corner hardware store.

Item	Cost
box of nails	\$3.80
box of screws	\$5.25
claw hammer	\$12.95
electric drill	\$42.50

a. Write two expressions to represent the total cost of 3 boxes of nails, 2 boxes of screws, 2 hammers, and 1 electric drill.

b. What is the total cost of the items purchased?

10. GRIDDED RESPONSE Evaluate the expression below.

$$\frac{5^3 \cdot 4^2 - 5^2 \cdot 4^3}{5 \cdot 4}$$

11. Use the equation $y = 2(4 + x)$ to answer each question.

a. Complete the table for each value of x.

x	y
1	
2	
3	
4	
5	
6	

b. Plot the points from the table on a coordinate grid. What do you notice about the points?

c. Make a conjecture about the relationship between the change in x and the change in y.

Extended Response

Record your answers on a sheet of paper. Show your work.

12. The volume of a sphere is four-thirds the product of π and the radius cubed.

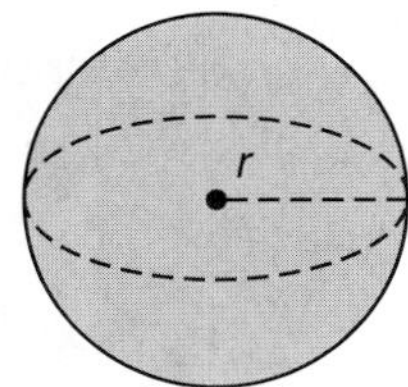

a. Write an expression for the volume of a sphere with radius r.

b. Find the volume of a sphere with a radius of 6 centimeters. Describe how you found your answer.

Need Extra Help?												
If you missed Question...	1	2	3	4	5	6	7	8	9	10	11	12
Go to Lesson or Page...	1-1	1-8	1-2	1-6	1-4	1-5	1-7	1-5	1-3	1-2	1-4	1-1

CHAPTER 2

Linear Equations

Then

In Chapter 1 you learned to simplify algebraic expressions.

Now

In Chapter 2, you will:

- Solve equations by using the four basic operations.
- Solve equations by using multiple steps.
- Solve proportions.
- Use formulas to solve real-world problems.

Why?

SHOPPING In recent years, the percent of change in sales per year at shopping malls in the U.S. averaged 5%. A store manager can use this data to set a sales goal for the upcoming year.

Linear Equations

Activity

If sales for Cyber Monday continue to increase at the same percentage, how much money would you expect to be spent on Cyber Monday in 2008?

First find the percent of increase from 2006 to 2007 for Cyber Monday.

$$\text{Percent of change} = \frac{\text{Amount of change}}{\text{Original amount}}$$

$$\text{Percent of change} = \frac{\square - \square}{\square}$$

Holiday Retail Season

Retail Date	2006 (millions $)	2007 (millions $)
Holiday Season Nov 1-Dec 31	24,570	29,170
Thanksgiving Day	210	270
Black Friday	430	530
Cyber Monday	610	730
Green Monday	660	880

source: comScore, Inc.

6/7

Math *in Motion*, Animation glencoe.com

Get Ready for Chapter 2

Diagnose Readiness You have two options for checking Prerequisite Skills.

Text Option

Take the Quick Check below. Refer to the Quick Review for help.

*Quick*Check

Write an algebraic expression for each verbal expression. (Lesson 1-1)

1. four less than three times a number n
2. a number d cubed less seven
3. the difference between two times b and eleven

Evaluate each expression. (Lesson 1-2)

4. $(9 - 4)^2 + 3$
5. $\frac{3 \cdot 8 - 12 \div 2}{3^2}$
6. $5(8 - 2) \div 3$
7. $\frac{1}{3}(21) + \frac{1}{8}(32)$
8. $72 \div 9 + 3 \cdot 2^3$
9. $\frac{11 - 3}{2} + 7$
10. $2[(5 - 3)^2 + 8] + (3 - 1) \div 2$
11. **BAKERY** Sue buys 1 carrot cake for \$14, 6 large chocolate chip cookies for \$1.50 each, and a dozen doughnuts for \$0.45 each. How much money did Sue spend at the bakery?

Find each percent. (Lesson 0-6)

12. What percent of 400 is 260?
13. Twelve is what percent of 60?
14. What percent of 25 is 75?
15. **ICE CREAM** What percent of the people surveyed prefer strawberry ice cream?

Favorite Flavor	Number of Responses
vanilla	82
chocolate	76
strawberry	42

*Quick*Review

EXAMPLE 1

Write an algebraic expression for the phrase *the product of eight and w increased by nine.*

the product of eight and w increased by nine

$8 \quad \cdot \quad w \qquad + \qquad 9$

The expression is $8w + 9$.

EXAMPLE 2

Evaluate $9 - \left[\frac{8 + 2^2}{2} - 2(5 \times 2 - 8)\right]$.

$9 - \left[\frac{8 + 2^2}{2} - 2(5 \times 2 - 8)\right]$	**Original expression**
$= 9 - \left[\frac{8 + 2^2}{2} - 2(2)\right]$	**Evaluate inside the parentheses.**
$= 9 - \left(\frac{8 + 2^2}{2} - 4\right)$	**Multiply.**
$= 9 - \left(\frac{8 + 4}{2} - 4\right)$	**Evaluate the power.**
$= 9 - (6 - 4)$	**Add and then divide.**
$= 7$	**Simplify.**

EXAMPLE 3

32 is what percent of 40?

$\frac{a}{b} = \frac{p}{100}$	**Use the percent proportion.**
$\frac{32}{40} = \frac{p}{100}$	**Replace *a* with 32 and *b* with 40.**
$32(100) = 40p$	**Find the cross products.**
$3200 = 40p$	**Multiply.**
$80 = p$	**Divide each side by 40.**

32 is 80% of 40.

Online Option

Math Online Take a self-check Chapter Readiness Quiz at glencoe.com.

Get Started on Chapter 2

You will learn several new concepts, skills, and vocabulary terms as you study Chapter 2. To get ready, identify important terms and organize your resources. You may wish to refer to **Chapter 0** to review prerequisite skills.

FOLDABLES® Study Organizer

Linear Functions Make this Foldable to help you organize your Chapter 2 notes about linear equations. Begin with 5 sheets of grid paper.

1. **Fold** each sheet in half along the width.

2. **Unfold** each sheet and tape to form one long piece.

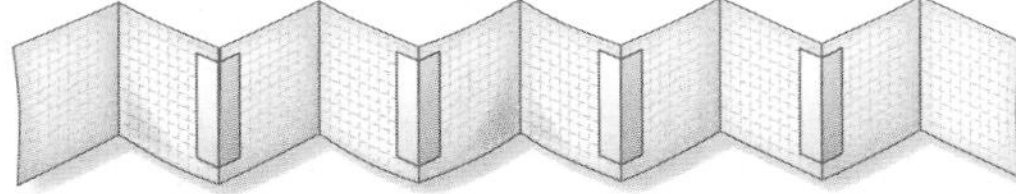

3. **Label** each page with the lesson number as shown. Refold to form a booklet.

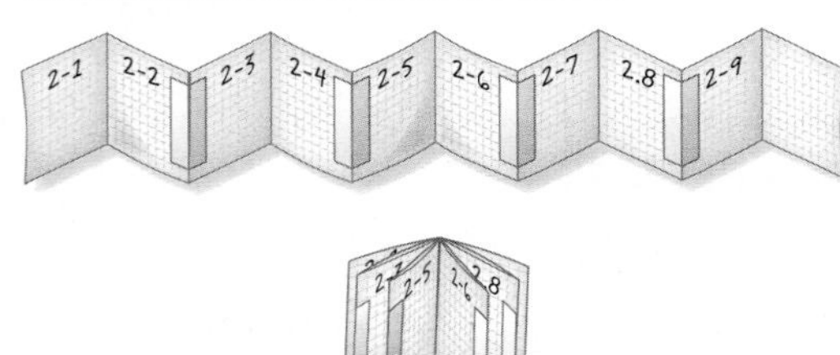

Math Online glencoe.com

- Study the chapter online
- Explore **Math in Motion**
- Get extra help from your own **Personal Tutor**
- Use **Extra Examples** for additional help
- Take a **Self-Check Quiz**
- **Review Vocabulary** in fun ways

New Vocabulary

English		Español
formula	• p. 76 •	fórmula
solve an equation	• p. 83 •	resolver una ecuación
equivalent equations	• p. 83 •	ecuaciones equivalentes
multi-step equation	• p. 91 •	ecuación de varios pasos
identity	• p. 98 •	identidad
ratio	• p. 111 •	razón
proportion	• p. 111 •	proporción
rate	• p. 113 •	tasa
unit rate	• p. 113 •	tasa unitaria
scale model	• p. 114 •	modelo de escala
percent of change	• p. 119 •	porcentaje de cambio
literal equation	• p. 127 •	ecuación literal
dimensional analysis	• p. 128 •	análisis dimensional
weighted average	• p. 132 •	promedio ponderado

Review Vocabulary

algebraic expression • p. 5 • expresion algebraica
an expression consisting of one or more numbers and variables along with one or more arithmetic operations

coordinate system • p. 38 • sistema de coordenedas
the grid formed by the intersection of two number lines, the horizontal axis and the vertical axis

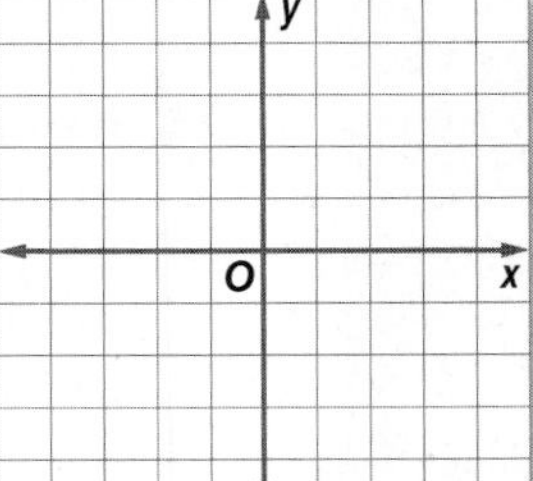

function • p. 45 • función
a relation in which each element of the domain is paired with exactly one element of the range

Multilingual eGlossary glencoe.com

2-1 Writing Equations

Then

You evaluated and simplified algebraic expressions. (Lesson 1-2)

Now

- Translate sentences into equations.
- Translate equations into sentences.

New Vocabulary

formula

Math Online

glencoe.com

- Extra Examples
- Personal Tutor
- Self-Check Quiz
- Homework Help

Why?

The Daytona 500 is widely considered to be the most important event of the NASCAR circuit. The distance around the track is 2.5 miles, and the race is a total of 500 miles. We can write an equation to determine how many laps it takes to finish the race.

Write Verbal Expressions To write an equation, identify the unknown for which you are looking and assign a variable to it. Then, write the sentence as an equation. Look for key words such as *is*, *is as much as*, *is the same as*, or *is identical to* that indicate where you should place the equals sign.

Consider the Daytona 500 example above.

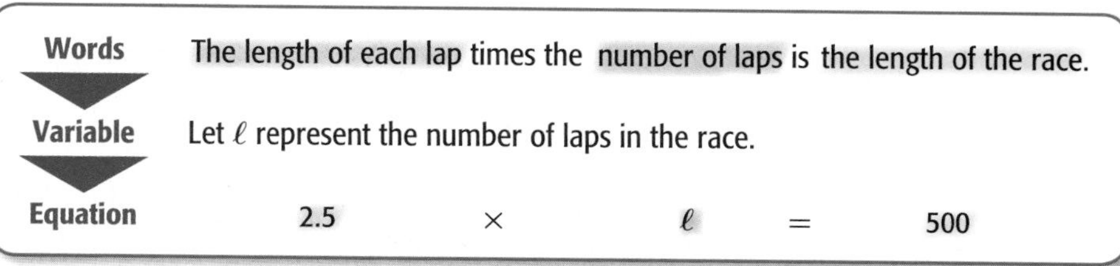

EXAMPLE 1 Translate Sentences into Equations

Translate each sentence into an equation.

a. Seven times a number squared is five times the difference of k and m.

Seven	times	*n* squared	is	five	times	the difference of *k* and *m*.
7	$\cdot$	n^2	$=$	5	$\cdot$	$(k - m)$

The equation is $7n^2 = 5(k - m)$.

b. Fifteen times a number subtracted from 80 is 25.

You can rewrite the verbal sentence so it is easier to translate. *Fifteen times a number subtracted from 80* is the same as *80 minus 15 times a number is 25*. Let n represent the number.

80	minus	15	times	a number	is	25.
80	$-$	15	$\cdot$	n	$=$	25

The equation is $80 - 15n = 25$.

Check Your Progress

1A. Two plus the quotient of a number and 8 is the same as 16.

1B. Twenty-seven times k is h squared decreased by 9.

Personal Tutor glencoe.com

Translating sentences to algebraic expressions and equations is a valuable skill in solving real-world problems.

Real-World Link

In 1919, Britain and France offered a flight that carried two passengers at a time. Now there are approximately 45,000 flights each day in the U.S., carrying hundreds of passengers on each flight.

Source: *Flightaware*

Real-World EXAMPLE 2 Use the Four-Step Problem-Solving Plan

AIR TRAVEL Refer to the information at the left. In how many days will 180,000 flights have occurred in the United States?

Understand The information given in the problem is that there are approximately 45,000 flights per day in the United States. We are asked to find how many days it will take for 180,000 flights to have occurred.

Plan Write an equation. Let d represent the number of days needed.

45,000	times	the number of days	equals	180,000.
45,000	$\cdot$	d	$=$	180,000

Solve $45{,}000d = 180{,}000$ **Find *d* by asking, "What number times 45,000 is 180,000?"**

$d = 4$

Check Check your answer by substituting 4 for d in the equation.

$45{,}000(4) \stackrel{?}{=} 180{,}000$ **Substitute 4 for *d*.**

$180{,}000 = 180{,}000$ ✓ **Multiply.**

The answer makes sense and works for the original problem.

Check Your Progress

2. GOVERNMENT There are 50 members in the North Carolina Senate. This is 70 fewer than the number in the North Carolina House of Representatives. How many members are in the North Carolina House of Representatives?

Personal Tutor glencoe.com

A rule for the relationship between certain quantities is called a **formula**. These equations use variables to represent numbers and form general rules.

EXAMPLE 3 Write a Formula

GEOMETRY Translate the sentence into a formula.

The area of a triangle equals the product of $\frac{1}{2}$ the length of the base and the height.

Words	The	area of a triangle	equals	the product of $\frac{1}{2}$ the length of the base and the height.
Variables	Let A = area, b = base, and h = height.			
Equation		A	$=$	$\frac{1}{2}bh$

The formula for the area of a triangle is $A = \frac{1}{2}bh$.

Check Your Progress

3. GEOMETRY Translate the sentence into a formula.
In a right triangle, the square of the measure of the hypotenuse c is equal to the sum of the squares of the measures of the legs, a and b.

Personal Tutor glencoe.com

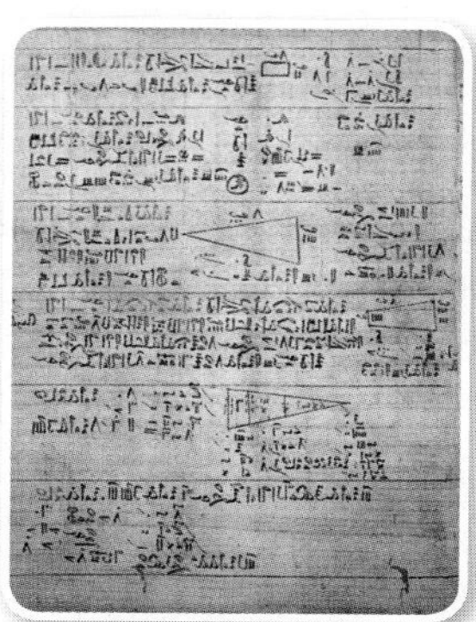

Math History Link

Ahmes
(about 1680–1620 B.C.)
Ahmes was the Egyptian mathematician and scribe who copied the Rhind Mathematical Papyrus. The papyrus contains 87 algebra problems of the same type. The first set of problems asks how to divide n loaves of bread between 10 people.

Write Sentences from Equations If you are given an equation, you can write a sentence or create your own word problem.

EXAMPLE 4 Translate Equations into Sentences

Translate each equation into a sentence.

a. $6z - 15 = 45$

$6z$	$-$	15	$=$	45
Six times z	minus	fifteen	equals	forty-five.

b. $y^2 + 3x = w$

y^2	$+$	$3x$	$=$	w
The sum of y squared	and	three times x	is	w.

Check Your Progress

4A. $15 = 25u^2 + 2$

4B. $\frac{3}{2}r - t^3 = 132$

Personal Tutor glencoe.com

When given a set of information, you can create a problem that relates a story.

EXAMPLE 5 Write a Problem

Write a problem based on the given information.

t = the time that Maxine drove; $t + 4$ = the time that Tia drove; $2t + (t + 4) = 28$

Sample problem:

Maxine and Tia went on a trip, and they took turns driving. During her turn, Tia drove 4 hours more than Maxine. Maxine took 2 turns, and Tia took 1 turn. Together they drove for 28 hours. How many hours did Maxine drive?

Check Your Progress

5. p = Beth's salary; $0.1p$ = bonus; $p + 0.1p = 525$

Personal Tutor glencoe.com

Check Your Understanding

Example 1 p. 75

Translate each sentence into an equation.

1. Three times r less than 15 equals 6.
2. The sum of q and four times t is equal to 29.
3. A number n squared plus 12 is the same as the quotient of p and 4.
4. Half of j minus 5 is the sum of k and 13.
5. The sum of 8 and three times k equals the difference of 5 times k and 3.
6. Three fourths of w plus 5 is one half of w increased by nine.
7. The quotient of 25 and t plus 6 is the same as twice t plus 1.
8. Thirty-two divided by y is equal to the product of three and y minus four.

Example 2 p. 76

9. **FINANCIAL LITERACY** Samuel has \$1900 in the bank. He wishes to increase his account to a total of \$2500 by depositing \$30 per week from his paycheck. Write and solve an equation to find how many weeks he needs to reach his goal.

10. **PAINTING** Miguel is earning extra money by painting houses. He charges a \$200 fee plus \$12 per can of paint needed to complete the job. Write and use an equation to find how many cans of paint he needs for a \$260 job.

Example 3 p. 76

Translate each sentence into a formula.

11. The perimeter of a regular pentagon is 5 times the length of each side.
12. The area of a circle is the product of π and the radius r squared.
13. Four times π times the radius squared is the surface area of a sphere.
14. One third the product of the length of the side squared and the height is the volume of a pyramid with a square base.

Example 4 p. 77

Translate each equation into a sentence.

15. $7m - q = 23$
16. $6 + 9k + 5j = 54$
17. $3(g + 8) = 4h - 10$
18. $6d^2 - 7f = 8d + f^2$

Example 5 p. 77

Write a problem based on the given information.

19. g = gymnasts on a team; $3g = 45$
20. c = cost of a notebook; $0.25c$ = markup; $c + 0.25c = 3.75$

Practice and Problem Solving

 = **Step-by-Step Solutions** begin on page R12. **Extra Practice** begins on page 815.

Example 1 p. 75

Translate each sentence into an equation.

21. The difference of f and five times g is the same as 25 minus f.
22. Three times b less than 100 is equal to the product of 6 and b.
23. Four times the sum of 14 and c is a squared.

Example 2 p. 76

24. **MUSIC** A piano has 52 white keys. Write and use an equation to find the number of octaves on a piano keyboard.

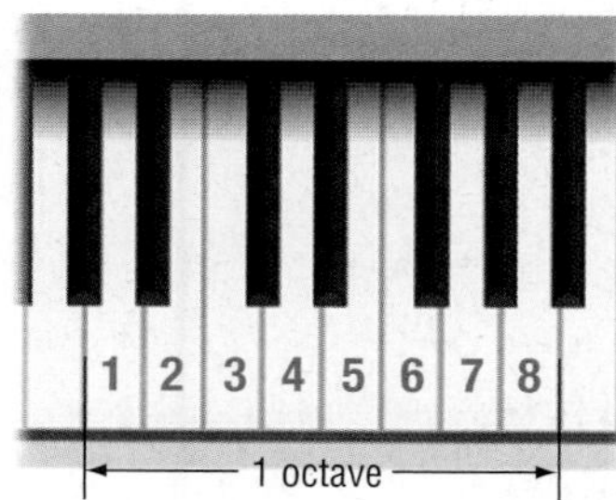

25. **GARDENING** A flat of plants contains 12 plants. Yoshi wants a garden that has three rows with 10 plants per row. Write and solve an equation for the number of flats Yoshi should buy.

Example 3 p. 76

Translate each sentence into a formula.

26. The perimeter of a rectangle is equal to 2 times the length plus twice the width.
27. Celsius temperature C is five ninths times the difference of the Fahrenheit temperature F and 32.
28. The density of an object is the quotient of its mass and its volume.
29. Simple interest is computed by finding the product of the principal amount p, the interest rate r, and the time t.

Example 4 p. 77

Translate each equation into a sentence.

30. $j + 16 = 35$
31. $4m = 52$
32. $7(p + 23) = 102$
33. $r^2 - 15 = t + 19$
34. $\frac{2}{5}v + \frac{3}{4} = \frac{2}{3}x^2$
35. $\frac{1}{3} - \frac{4}{5}z = \frac{4}{3}y^3$

Example 5 p. 77

Write a problem based on the given information.

36. q = quarts of strawberries; $2.50q = 10$

37. p = the principal amount; $0.12p$ = the interest charged; $p + 0.12p = 224$

38. m = number of movies rented; $10 + 1.50m = 14.50$

39. p = the number of players in the game; $5p + 7$ = number of cards in a deck

For Exercises 40–43, match each sentence with an equation.

A. $g^2 = 2(g - 10)$ **C.** $g^3 = 24g + 4$

B. $\frac{1}{2}g + 32 = 15 + 6g$ **D.** $3g^2 = 30 + 9g$

40. One half of g plus thirty-two is as much as the sum of fifteen and six times g.

41. A number g to the third power is the same as the product of 24 and g plus 4.

42. The square of g is the same as two times the difference of g and 10.

43. The product of 3 and the square of g equals the sum of thirty and the product of nine and g.

44. FINANCIAL LITERACY Tim's bank contains quarters, dimes, and nickels. He has three more dimes than quarters and 6 fewer nickels than quarters. If he has 63 coins, write and solve an equation to find how many quarters Tim has.

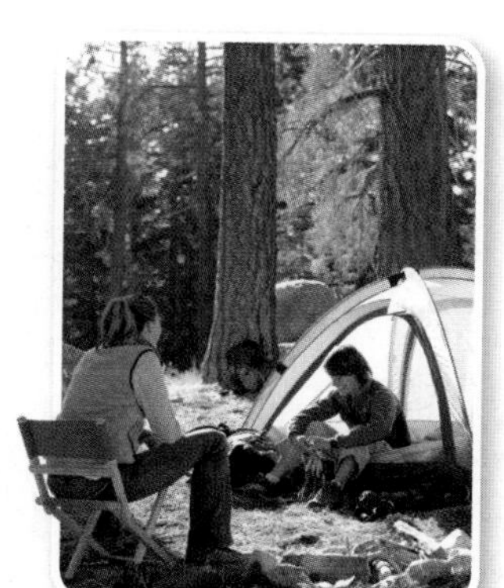

Real-World Link

There are more than 16,000 commercial and public campgrounds nationwide. Camping is the number one outdoor vacation activity in America.

Source: Travel Industry Association of America

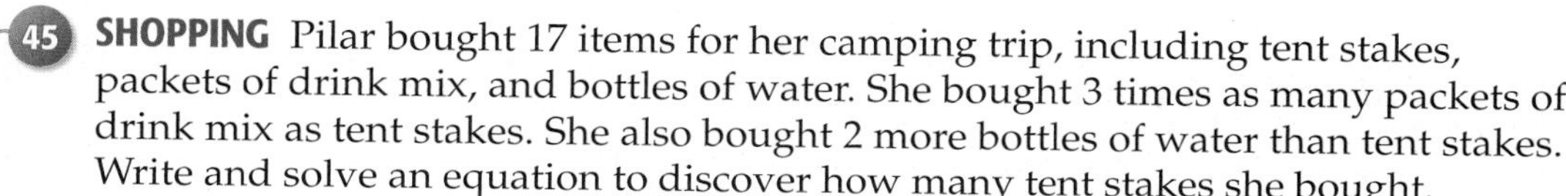

45. SHOPPING Pilar bought 17 items for her camping trip, including tent stakes, packets of drink mix, and bottles of water. She bought 3 times as many packets of drink mix as tent stakes. She also bought 2 more bottles of water than tent stakes. Write and solve an equation to discover how many tent stakes she bought.

46. MULTIPLE REPRESENTATIONS In this problem, you will explore how to translate relations with powers.

x	2	3	4	5	6
y	5	10	17	26	37

a. VERBAL Write a sentence to describe the relationship between x and y in the table.

b. ALGEBRAIC Write an equation that represents the data in the table.

c. GRAPHICAL Graph each ordered pair and draw the function. Describe the graph as discrete or continuous.

H.O.T. Problems

Use Higher-Order Thinking Skills

47. OPEN ENDED Write a problem about your favorite television show that uses the equation $x + 8 = 30$.

48. REASONING The surface area of a three-dimensional object is the sum of the areas of the faces. If ℓ represents the length of the side of a cube, write a formula for the surface area of the cube.

49. CHALLENGE Given the perimeter P and width w of a rectangle, write a formula to find the length ℓ.

50. WRITING IN MATH Explain how to translate a verbal sentence into an algebraic equation. Include any tips that you may have for your fellow students.

Standardized Test Practice

51. Which equation *best* represents the relationship between the number of hours an electrician works h and the total charges c?

Cost of Electrician	
Emergency House Call	$30 one time fee
Rate	$55/hour

A $c = 30 + 55$
B $c = 30h + 55$
C $c = 30 + 55h$
D $c = 30h + 55h$

52. A car traveled at 55 miles per hour for 2.5 hours and then at 65 miles per hour for 3 hours. How far did the car travel in all?

F 300.5 mi
G 305 mi
H 330 mi
J 332.5 mi

53. SHORT RESPONSE Suppose each dimension of rectangle $ABCD$ is doubled. What is the perimeter of the new $ABCD$?

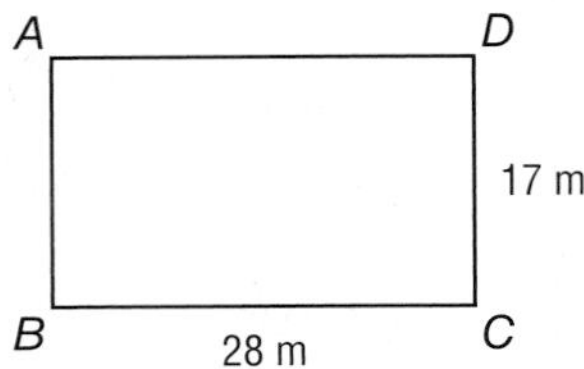

54. STATISTICS Stacy's first five science test scores were 95, 86, 83, 95, and 99. Which of the following is a true statement?

A The mode is the same as the median.
B The median is the same as the mean.
C The range is the same as the mode.
D The mode is the same as the mean.

Spiral Review

Write a counterexample for each conditional statement. (Lesson 1-8)

55. If you were born in Florida, then you live in Florida.

56. If the product of two numbers is an even number, then both factors must be even numbers.

57. If a number is divisible by 2, then it is divisible by 4.

58. SHOPPING Cuties is having a sale on earrings (Lesson 1-7)

a. Make a table that shows the cost of buying 1 to 5 pairs of earrings.

b. Write the data as a set of ordered pairs.

c. Graph the data.

SALE
Earrings $29.00 each pair
Buy 2 pairs Get 1 pair FREE

59. GEOMETRY Refer to the table below. (Lesson 1-6)

Polygon	triangle	quadrilateral	pentagon	hexagon	heptagon
Number of Sides	3	4	5	6	7
Interior Angle Sum	180	360	540	720	900

a. Identify the independent and dependent variables.

b. Identify the domain and range for this situation.

c. State whether the function is *discrete* or *continuous*. Explain.

Skills Review

Evaluate each expression. (Lesson 1-1)

60. 9^2

61. 10^6

62. 3^5

63. 5^3

Algebra Lab
Solving Equations

Math Online glencoe.com
Math *in Motion,* Animation

You can use **algebra tiles** to model solving equations. To **solve an equation** means to find the value of the variable that makes the equation true. An x tile represents the variable x. The 1 tile represents a positive 1. The -1 tile represents a negative 1. And, the $-x$ tile represents the variable negative x. The goal is to get the x-tile by itself on one side of the mat by using the rules stated below.

Rules for Equation Models When Adding or Subtracting	
You can remove or add the same number of identical algebra tiles to each side of the mat without changing the equation.	1 1 1 = 1 1 1
One positive tile and one negative tile of the same unit are called a zero pair. Since $1 + (-1) = 0$, you can remove or add zero pairs to either side of the equation mat without changing the equation.	−1 1 1 = 1

ACTIVITY 1 Addition Equation

Use an equation model to solve $x + 3 = -4$.

Step 1 Model the equation. Place 1 x-tile and 3 positive 1-tiles on one side of the mat. Place 4 negative 1-tiles on the other side of the mat.

Step 2 Isolate the x-term. Add 3 negative 1-tiles to each side. The resulting equation is $x = -7$.

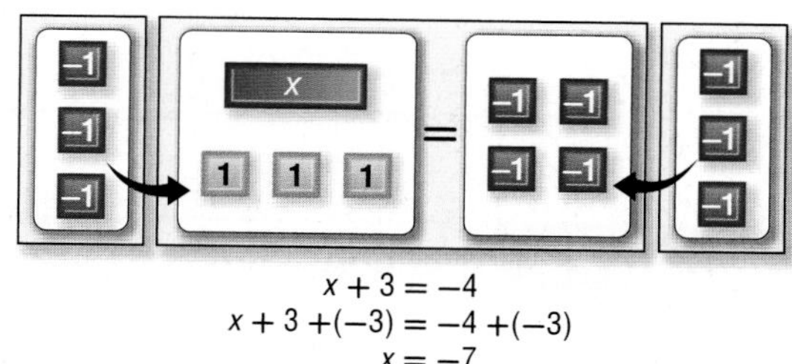

$x + 3 = -4$
$x + 3 + (-3) = -4 + (-3)$
$x = -7$

ACTIVITY 2 Subtraction Equation

Use an equation model to solve $x - 2 = 1$.

Step 1

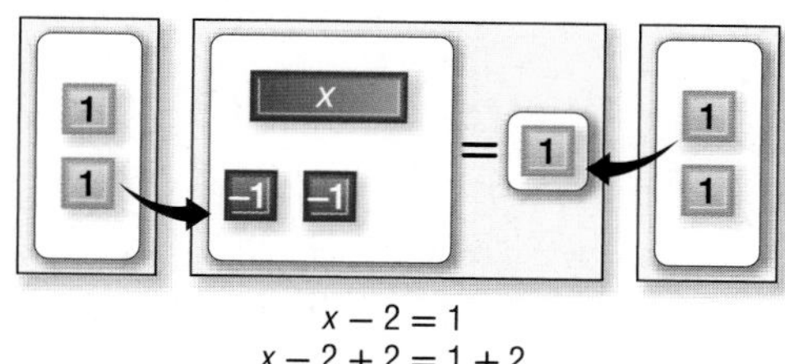

$x - 2 = 1$
$x - 2 + 2 = 1 + 2$

Place 1 x-tile and 2 negative 1-tiles on one side of the mat. Place 1 positive 1-tile on the other side of the mat. Then add 2 positive 1-tiles to each side.

Step 2

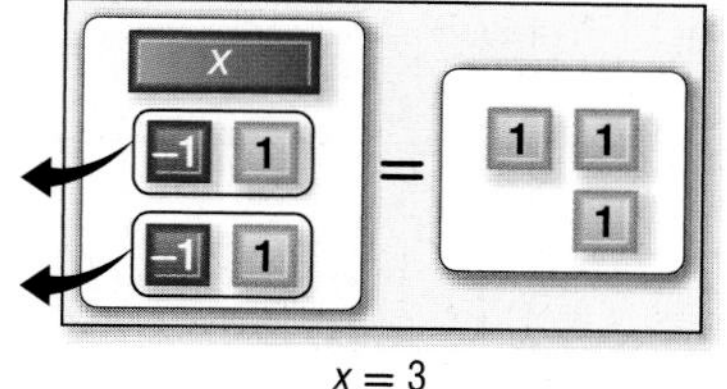

$x = 3$

Group the tiles to form zero pairs. Then remove all the zero pairs. The resulting equation is $x = 3$.

Model and Analyze

Use algebra tiles to solve each equation.

1. $x + 4 = 9$
2. $x + (-3) = -4$
3. $x + 7 = -2$
4. $x + (-2) = 11$
5. **WRITING IN MATH** If $a = b$, what can you say about $a + c$ and $b + c$? about $a - c$ and $b - c$?

When solving multiplication equations, the goal is still to get the x-tile by itself on one side of the mat by using the rules for dividing.

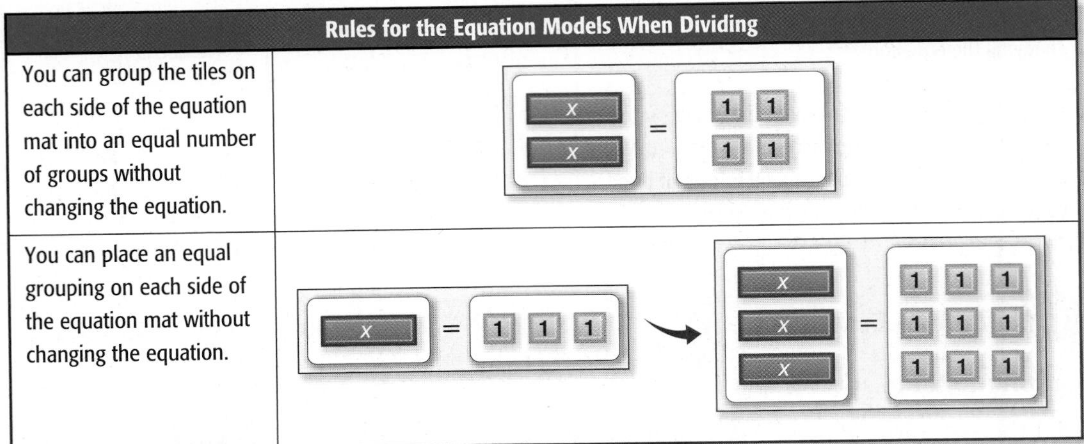

Rules for the Equation Models When Dividing	
You can group the tiles on each side of the equation mat into an equal number of groups without changing the equation.	
You can place an equal grouping on each side of the equation mat without changing the equation.	

ACTIVITY 3 Multiplication Equation

Use an equation model to solve $3x = 12$.

Step 1 Model the equation. Place 3 x-tiles on one side of the mat. Place 12 positive 1-tiles on the other side of the mat.

Step 2 Isolate the x-term. Separate the tiles into 3 equal groups to match the 3 x-tiles. Each x-tile is paired with 4 positive 1-tiles. The resulting equation is $x = 4$.

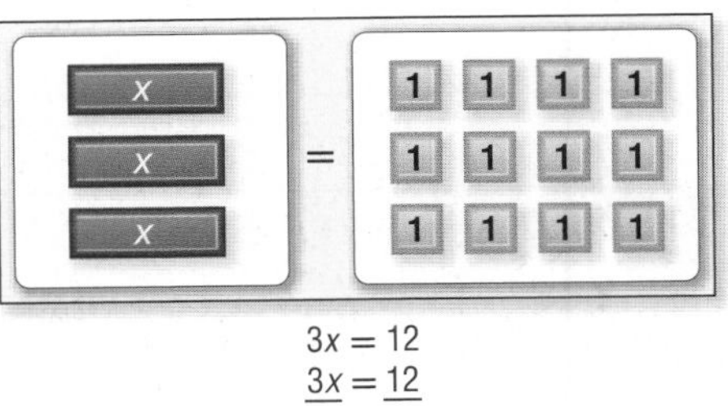

$$3x = 12$$
$$\frac{3x}{3} = \frac{12}{3}$$
$$x = 4$$

Model and Analyze

Use algebra tiles to solve each equation.

6. $5x = -15$
7. $-3x = -9$
8. $4x = 8$
9. $-6x = 18$
10. **MAKE A CONJECTURE** How would you use algebra tiles to solve $\frac{x}{4} = 5$? Discuss the steps you would take to solve this equation algebraically.

2-2 Solving One-Step Equations

Then

You translated sentences into equations. (Lesson 2-1)

Now

- Solve equations by using addition and subtraction.
- Solve equations by using multiplication and division.

New Vocabulary

solve an equation
equivalent equations

Math Online

glencoe.com

- Extra Examples
- Personal Tutor
- Self-Check Quiz
- Homework Help

Why?

A record for the most snow angels made at one time was set in Michigan when 3784 people participated. North Dakota had 8910 people register to break the record. To determine how many more people North Dakota had than Michigan, solve the equation $3784 + x = 8910$.

Solve Equations Using Addition or Subtraction In an equation, the variable represents the number that satisfies the equation. To **solve an equation** means to find the value of the variable that makes the equation true.

The process of solving an equation involves isolating the variable (with a coefficient of 1) on one side of the equation. Each step in this process results in equivalent equations. **Equivalent equations** have the same solution.

Key Concept — Addition Property of Equality

For Your FOLDABLE

Words If an equation is true and the same number is added to each side of the equation, the resulting equivalent equation is also true.

Symbols For any real numbers a, b, and c, if $a = b$, then $a + c = b + c$.

Examples

$$14 = 14$$
$$14 + 3 = 14 + 3$$
$$17 = 17$$

$$-3 = -3$$
$$+9 = +9$$
$$6 = 6$$

EXAMPLE 1 Solve by Adding

Solve $c - 22 = 54$.

Horizontal Method

$c - 22 = 54$	Original equation
$c - 22 + 22 = 54 + 22$	Add 22 to each side.
$c = 76$	Simplify.

Vertical Method

$$c - 22 = 54$$
$$+22 = +22$$
$$c = 76$$

To check that 76 is the solution, substitute 76 for c in the original equation.

CHECK		
	$c - 22 = 54$	Original equation
	$76 - 22 \stackrel{?}{=} 54$	Substitute 76 for c.
	$54 = 54$ ✓	Subtract.

Check Your Progress Solve each equation.

1A. $113 = g - 25$

1B. $j - 87 = -3$

Personal Tutor glencoe.com

Similar to the Addition Property of Equality, the **Subtraction Property of Equality** can also be used to solve equations.

StudyTip

Subtraction Subtracting a value is equivalent to adding the opposite of the value.

Key Concept Subtraction Property of Equality — For Your FOLDABLE

Words If an equation is true and the same number is subtracted from each side of the equation, the resulting equivalent equation is also true.

Symbols For any real numbers a, b, and c, if $a = b$, then $a - c = b - c$.

Examples

$$87 = 87$$
$$87 - 17 = 87 - 17$$
$$70 = 70$$

$$\begin{aligned} 13 &= 13 \\ -28 &= -28 \\ \hline -15 &= -15 \end{aligned}$$

StudyTip

Solving an Equation When solving equations you can use either the horizontal method or the vertical method. Both methods will produce the same result.

EXAMPLE 2 Solve by Subtracting

Solve $63 + m = 79$.

Horizontal Method

$63 + m = 79$	Original equation	
$63 - 63 + m = 79 - 63$	Subtract 63 from each side.	
$m = 16$	Simplify.	

Vertical Method

$$\begin{aligned} 63 + m &= 79 \\ -63 \quad &= -63 \\ \hline m &= 16 \end{aligned}$$

To check that 16 is the solution, replace m with 16 in the original equation.

CHECK

$63 + m = 79$	Original equation
$63 + 16 \stackrel{?}{=} 79$	Substitution, $m = 16$
$79 = 79$ ✓	Simplify.

Check Your Progress Solve each equation.

2A. $27 + k = 30$

2B. $-12 = p + 16$

Personal Tutor glencoe.com

Solve Equations Using Multiplication or Division In the equation $\frac{x}{3} = 9$, the variable x is divided by 3. To solve for x, undo the division by multiplying each side by 3. This is an example of the **Multiplication Property of Equality**.

Key Concept Multiplication Property of Equality — For Your FOLDABLE

Words If an equation is true and each side is multiplied by the same nonzero number, the resulting equation is equivalent.

Symbols For any real numbers a, b, and c, $c \neq 0$, if $a = b$, then $ac = bc$.

Example If $x = 5$, then $3x = 15$.

Division Property of Equality

Words If an equation is true and each side is divided by the same nonzero number, the resulting equation is equivalent.

Symbols For any real numbers a, b, and c, $c \neq 0$, if $a = b$, then $\frac{a}{c} = \frac{b}{c}$.

Example If $x = -20$, then $\frac{x}{5} = \frac{-20}{5}$ or -4.

The reciprocal of a number can be used to solve equations.

Review Vocabulary

reciprocal the multiplicative inverse of a number (Lesson 1-3)

EXAMPLE 3 Solve by Multiplying and Dividing

Solve each equation.

a. $\frac{2}{3}q = \frac{1}{2}$

$\frac{2}{3}q = \frac{1}{2}$	**Original equation**
$\frac{3}{2}\left(\frac{2}{3}\right)q = \frac{3}{2}\left(\frac{1}{2}\right)$	**Multiply each side by $\frac{3}{2}$, the reciprocal of $\frac{2}{3}$.**
$q = \frac{3}{4}$	**Check the result.**

b. $39 = -3r$

$39 = -3r$	**Original equation**
$\frac{39}{-3} = \frac{-3r}{-3}$	**Divide each side by −3.**
$-13 = r$	**Check the result.**

Check Your Progress

3A. $\frac{3}{5}k = 6$

3B. $-\frac{1}{4} = \frac{2}{3}b$

Personal Tutor glencoe.com

We can also use reciprocals and properties of equality to solve real-world problems.

Real-World EXAMPLE 4 Solve by Multiplying

SURVEYS Of a group of 13- to 15-year-old girls surveyed, 225, or about $\frac{9}{20}$ said they talk on the telephone while they watch television. About how many girls were surveyed?

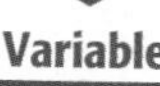

Words	Nine twentieths times those surveyed	is	225.
Variable	Let g = the number of girls surveyed.		
Equation	$\frac{9}{20}g$	=	225

$\frac{9}{20}g = 225$	**Original equation**
$\left(\frac{20}{9}\right)\frac{9}{20}g = \left(\frac{20}{9}\right)225$	**Multiply each side by $\frac{20}{9}$.**
$g = \frac{4500}{9}$	$\left(\frac{20}{9}\right)\left(\frac{9}{20}\right) = 1$
$g = 500$	**Simplify.**

About 500 girls were surveyed.

Real-World Link

Almost half of 10- to 18-year-olds in the U.S. use a cell phone. Of those, 53% play games on their phones, more than 33% download games, 52% use the calendar/organizer, and nearly all teens with camera phones snap pictures.

Source: Lexdon Business Library

Check Your Progress

4. STAINED GLASS Allison is making a stained glass window. Her pattern requires that one fifth of the glass should be blue. She has 288 square inches of blue glass. If she intends to use all of her blue glass, how much glass will she need for the entire project?

Personal Tutor glencoe.com

Check Your Understanding

Examples 1 and 3 pp. 83–85

Solve each equation. Check your solution.

1. $g + 5 = 33$
2. $104 = y - 67$
3. $\frac{2}{3} + w = 1\frac{1}{2}$
4. $-4 + t = -7$
5. $a + 26 = 35$
6. $-6 + c = 32$
7. $1.5 = y - (-5.6)$
8. $3 + g = \frac{1}{4}$
9. $x + 4 = \frac{3}{4}$
10. $\frac{t}{7} = -5$
11. $\frac{a}{36} = \frac{4}{9}$
12. $\frac{2}{3}n = 10$
13. $\frac{8}{9} = \frac{4}{5}k$
14. $12 = \frac{x}{-3}$
15. $-\frac{r}{4} = \frac{1}{7}$

Example 4 p. 85

16. FUNDRAISING The television show "Idol Gives Back" raised money for relief organizations. During this show, viewers could call in and vote for their favorite performer. The parent company contributed $5 million for the 50 million votes cast. What did they pay for each vote?

17. SHOPPING Hana decides to buy her cat a bed from an online fund that gives $\frac{7}{8}$ of her purchase to shelters that care for animals. How much of Hana's money went to the animal shelter?

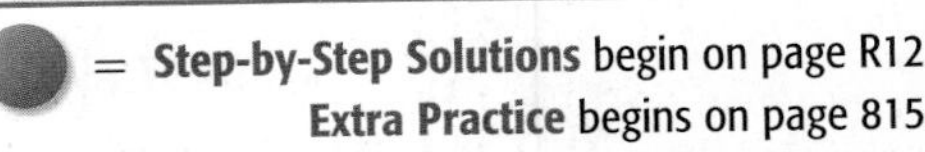

Practice and Problem Solving

= Step-by-Step Solutions begin on page R12.
Extra Practice begins on page 815.

Examples 1 and 3 pp. 83–85

Solve each equation. Check your solution.

18. $v - 9 = 14$
19. $44 = t - 72$
20. $-61 = d + (-18)$
21. $18 + z = 40$
22. $-4a = 48$
23. $12t = -132$
24. $18 - (-f) = 91$
25 $-16 - (-t) = -45$
26. $\frac{1}{3}v = -5$
27. $\frac{u}{8} = -4$
28. $\frac{a}{6} = -9$
29. $-\frac{k}{5} = \frac{7}{5}$
30. $\frac{3}{4} = w + \frac{2}{5}$
31. $-\frac{1}{2} + a = \frac{5}{8}$
32. $-\frac{t}{7} = \frac{1}{15}$
33. $-\frac{5}{7} = y - 2$
34. $v + 914 = -23$
35. $447 + x = -261$
36. $-\frac{1}{7}c = 21$
37. $-\frac{2}{3}h = -22$
38. $\frac{3}{5}q = -15$
39. $\frac{n}{8} = -\frac{1}{4}$
40. $\frac{c}{4} = -\frac{9}{8}$
41. $\frac{2}{3} + r = -\frac{4}{9}$

Example 4 p. 85

42. CATS A domestic cat can run at speeds of 27.5 miles per hour when chasing prey. A cheetah can run 42.5 miles per hour faster when chasing prey. How fast can the cheetah go?

43. CARS The average time t it takes to manufacture a car in the United States is 24.9 hours. This is 8.1 hours longer than the average time it takes to manufacture a car in Japan. Write and solve an equation to find the average time in Japan.

Solve each equation. Check your solution.

44. $\frac{x}{9} = 10$

45. $\frac{b}{7} = -11$

46. $\frac{3}{4} = \frac{c}{24}$

47. $\frac{2}{3} = \frac{1}{8}y$

48. $\frac{2}{3}n = 14$

49. $\frac{3}{5}g = -6$

50. $4\frac{1}{5} = 3p$

51. $-5 = 3\frac{1}{2}x$

52. $6 = -\frac{1}{2}n$

53. $-\frac{2}{5} = -\frac{z}{45}$

54. $-\frac{g}{24} = \frac{5}{12}$

55. $-\frac{v}{5} = -45$

Write an equation for each sentence. Then solve the equation.

56. Six times a number is 132.

57. Two thirds equals negative eight times a number.

58. Five elevenths times a number is 55.

59. Four fifths is equal to ten sixteenths of a number.

60. Three and two thirds times a number equals two ninths.

61 Four and four fifths times a number is one and one fifth.

62. SHOPPING Adelina is comparing prices for two brands of health and energy bars at the local grocery store. She wants to get the best price for each bar.

a. Write an equation to find the price for each bar of the Feel Great brand.

b. Write an equation to find the price of each bar for the Super Power brand.

c. Which bar should Adelina buy? Explain.

Feel Great energy bars 12 bars — $18.00

Super Power energy bars 15 bars — $21.75

63. MEDIA The world's largest passenger plane, the Airbus A380, was first used by Singapore Airlines in 2005. The following description appeared on a news Web site after the plane was introduced.

"That airline will see the A380 transporting some 555 passengers, 139 more than a similarly set-up 747."

How many passengers will a similarly set-up 747 transport?

Real-World Link

Ethanol is produced from corn and is considered energy efficient because it yields 25% more energy than the process to create it.

Source: U.S. Department of Energy

64. FUEL In 2004, approximately 5 million cars and trucks were classified as flex-fuel, which means they could run on gasoline or ethanol. In 2006, that number increased to 7.5 million. How many more cars and trucks were flex-fuel in 2006?

65. CHEERLEADING At a certain cheerleading competition, the maximum time per team, including the set up, is 3 minutes. The Ridgeview High School squad's performance time is 2 minutes and 34 seconds. How much time does the squad have left for their set up?

66. COMIC BOOKS An X-Men #1 comic book in mint condition recently sold for $45,000. An Action Comics #63 (Mile High), also in mint condition, sold for $15,000. How much more did the X-Men comic book sell for than the Action Comics book?

67. MOVIES A certain movie made $1.6 million in ticket sales. Its sequel made $0.8 million in ticket sales. How much more did the first movie make than the sequel?

68. CAMERAS An electronics store sells a certain digital camera for $126. This is $\frac{2}{3}$ of the price that a photography store charges. What is the cost of the camera at the photography store?

69. **BLOGS** In 2006, 57 million American adults read online blogs. However, 45 million fewer American adults say that they maintain their own blog. How many American adults maintain a blog?

70. **SCIENCE CAREERS** According to the Bureau of Labor and Statistics, approximately 65,000,000 women were employed in the United States in 2004.

a. The number of women in the computer science fields times 26 is the number of working women. Write an equation to represent the number of women employed in the computer sciences in 2004. Then solve the equation.

b. The number of women in natural science fields is 2,266,000 less than the number of women in computer science fields. How many women are in natural science fields?

Real-World Link

Schools have begun using an online voting system that allows students to log in and vote for homecoming king and queen.

Source: NewBay Media

71. **DANCES** Student Council has a budget of \$1000 for the homecoming dance. So far, they have spent \$350 dollars for music.

a. Write an equation to represent the amount of money left to spend. Then solve the equation.

b. They then spent \$225 on decorations. Write an equation to represent the amount of money left.

c. If the Student Council spent their entire budget, write an equation to represent how many \$6 tickets they must sell to make a profit.

H.O.T. Problems

Use **H**igher-**O**rder **T**hinking Skills

72. **WHICH ONE DOESN'T BELONG?** Identify the equation that does not belong with the other three. Explain your reasoning.

$n + 14 = 27$ | $12 + n = 25$ | $n - 16 = 29$ | $n - 4 = 9$

73. **OPEN ENDED** Write an equation involving addition and demonstrate two ways to solve it.

74. **REASONING** For which triangle is the height not $4\frac{1}{2}b$, where b is the length of the base?

Triangle	Base (cm)	Height (cm)
$\triangle ABC$	3.8	17.1
$\triangle MQP$	5.4	24.3
$\triangle RST$	6.3	28.5
$\triangle TRW$	1.6	7.2

75. **CHALLENGE** Determine whether each sentence is *sometimes*, *always*, or *never true*. Explain your reasoning.

a. $x + x = x$ **b.** $x + 0 = x$

76. **REASONING** Determine the value for each statement below.

a. If $x - 7 = 14$, what is the value of $x - 2$?

b. If $t + 8 = -12$, what is the value of $t + 1$?

77. **CHALLENGE** Discuss why $\frac{2}{3}b = 16$ and $48 = 2c$ have the same solution.

78. **WRITING IN MATH** Consider the Multiplication Property of Equality and the Division Property of Equality. Explain why they can be considered the same property. Which one do you think is easier to use?

Standardized Test Practice

79. Which of the following best represents the equation $w - 15 = 33$?

A Jake added w ounces of water to his bottle, which originally contained 33 ounces of water. How much water did he add?

B Jake added 15 ounces of water to his bottle, for a total of 33 ounces. How much water w was originally in the bottle?

C Jake drank 15 ounces of water from his bottle and 33 ounces were left. How much water w was originally in the bottle?

D Jake drank 15 ounces of water from his water bottle, which originally contained 33 ounces. How much water w was left?

80. SHORT RESPONSE Charlie's company pays him for every mile that he drives on his trip. When he drives 50 miles, he is paid \$30. To the nearest tenth, how many miles did he drive if he was paid \$275?

81. The table shows the results of a survey given to 500 international travelers. Based on the data, which statement is true?

Vacation Plans	
Destination	**Percent**
The Tropics	37
Europe	19
Asia	17
Other	17
No Vacation	10

F Fifty have no vacation plans.

G Fifteen are going to Asia.

H One third are going to the tropics.

J One hundred are going to Europe.

82. GEOMETRY The amount of water needed to fill a pool represents the pool's ____.

A volume

B surface area

C circumference

D perimeter

Spiral Review

Translate each sentence into an equation. (Lesson 2-1)

83. The sum of twice r and three times k is identical to thirteen.

84. The quotient of t and forty is the same as twelve minus half of u.

85. The square of m minus the cube of p is sixteen.

86. Two times z is equal to two times the sum of v and x.

Write each statement in if-then form. (Lesson 1-8)

87. The trash is picked up on Monday.

88. Vito will call after school.

89. For $x = 8$, $x^2 - 3x = 40$.

90. $4q + 6 > 42$ when $q > 9$.

Skills Review

91. COMMUNICATION Sato communicates with his friends for a math project. In a week, he averages 5 hours using e-mail, 18 hours on the phone, and 12 hours meeting with them in person. Write and evaluate an expression to predict how many hours he will spend communicating with his friends over the next 12 weeks. (Lesson 1-4)

92. PETS The Poochie Pet supply store has the following items on sale. Write and evaluate an expression to find the total cost of purchasing 1 collar, 2 T-shirts, 3 kerchiefs, 1 leash, and 4 flying disks. (Lesson 1-4)

Item	Cost (\$)
studded collar	4.50
kerchief	3.00
doggy T-shirt	6.25
leash	5.50
flying disk	3.25

EXPLORE 2-3

Algebra Lab

Solving Multi-Step Equations

Math Online glencoe.com

Math *in Motion,* Animation

You can use algebra tiles to model solving multi-step equations.

ACTIVITY **Use an equation model to solve $4x + 3 = -5$.**

Step 1 Model the equation.

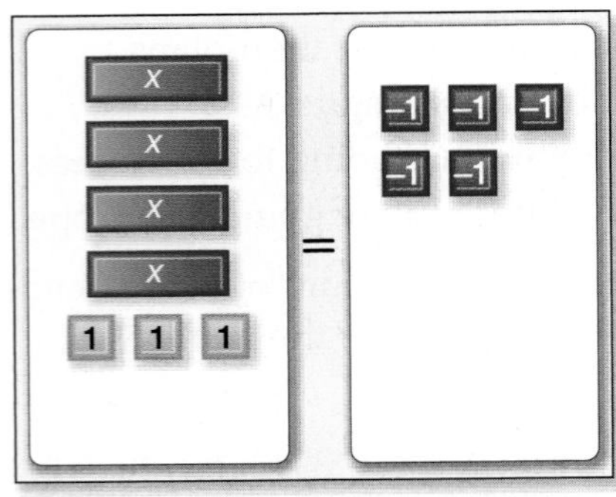

$4x + 3 = -5$

Place 4 x-tiles and 3 positive 1-tiles on one side of the mat. Place 5 negative 1-tiles on the other side.

Step 2 Isolate the x-term.

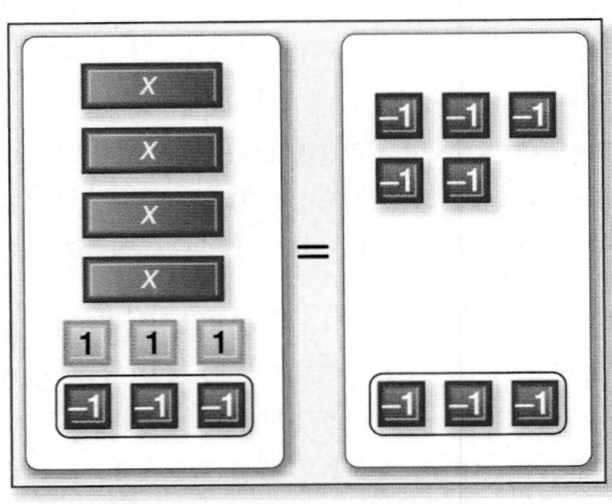

$4x + 3 - 3 = -5 - 3$

Since there are 3 positive 1-tiles with the x-tiles, add 3 negative 1-tiles to each side to form zero pairs.

Step 3 Remove zero pairs.

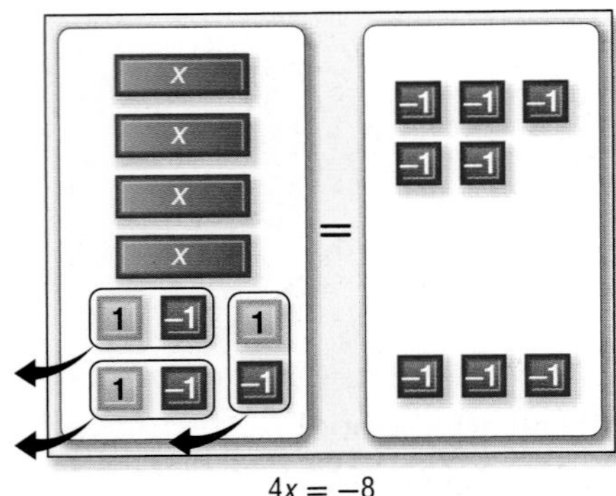

$4x = -8$

Group the tiles to form zero pairs and remove the zero pairs.

Step 4 Group the tiles.

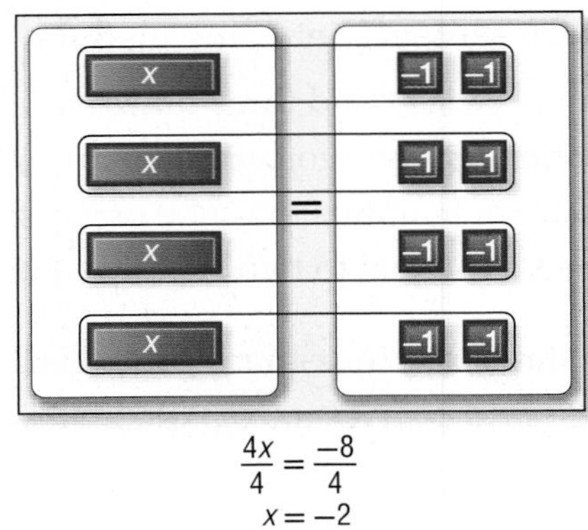

$$\frac{4x}{4} = \frac{-8}{4}$$
$$x = -2$$

Separate the remaining tiles into 4 equal groups to match the 4 x-tiles. Each x-tile is paired with 2 negative 1-tiles. The resulting equation is $x = -2$.

Model **Use algebra tiles to solve each equation.**

1. $3x - 7 = -10$ **2.** $2x + 5 = 9$ **3.** $5x - 7 = 8$ **4.** $-7 = 3x + 8$

5. $5 + 4x = -11$ **6.** $3x + 1 = 7$ **7.** $11 = 2x - 5$ **8.** $7 + 6x = -11$

9. What would be your first step in solving $8x - 29 = 67$?

10. What steps would you use to solve $9x + 14 = -49$?

2-3 Solving Multi-Step Equations

Then

You solved single-step equations. (Lesson 2-2)

Now

- Solve equations involving more than one operation.
- Solve equations involving consecutive integers.

New Vocabulary

multi-step equation
consecutive integers
number theory

Math Online

glencoe.com

- Extra Examples
- Personal Tutor
- Self-Check Quiz
- Homework Help

Why?

The Tour de France is the premier cycling event in the world. The map shows the 2007 Tour de France course. If the length of the shortest portion of the race can be represented by k, the expression $4k + 20$ is the length of the longest stage or 236 kilometers.

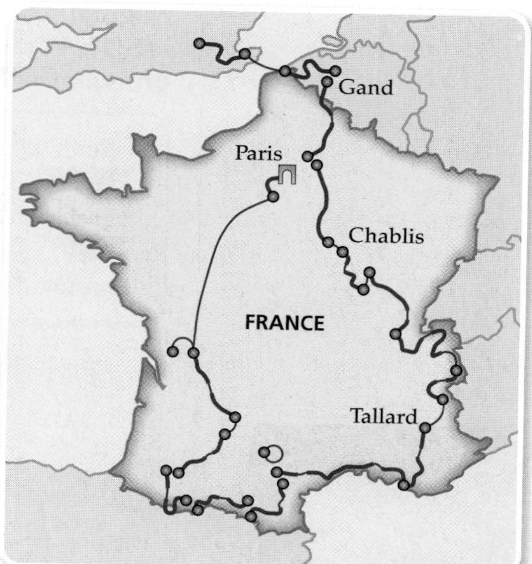

Solve Multi-Step Equations The situation above can be described by the equation $4k + 20 = 236$. Because this equation requires more than one step to solve, it is called a **multi-step equation**. To solve this equation, we must undo each operation by working backward.

EXAMPLE 1 Solve Multi-Step Equations

Solve each equation.

a. $11x - 4 = 29$

$11x - 4 = 29$	**Original equation**
$11x - 4 + 4 = 29 + 4$	**Add 4 to each side.**
$11x = 33$	**Simplify.**
$\frac{11x}{11} = \frac{33}{11}$	**Divide each side by 11.**
$x = 3$	**Simplify.**

b. $\frac{a+7}{8} = 5$

$\frac{a+7}{8} = 5$	**Original equation**
$8\left(\frac{a+7}{8}\right) = 8(5)$	**Multiply each side by 8.**
$a + 7 = 40$	**Simplify.**
$-7 = -7$	**Subtract 7 from each side.**
$a = 33$	**Simplify.**

You can check your solutions by substituting the results back into the original equations.

✓ Check Your Progress

Solve each equation. Check your solution.

1A. $2a - 6 = 4$

1B. $\frac{n+1}{-2} = 15$

Personal Tutor glencoe.com

Real-World EXAMPLE 2 Write and Solve a Multi-Step Equation

SHOPPING **Hiroshi is buying a pair of water skis that are on sale for $\frac{2}{3}$ of the original price. After he uses a \$25 gift certificate, the total cost before taxes is \$115. What was the original price of the skis? Write an equation for the problem. Then solve the equation.**

Words	Two thirds	of	the price	minus	25	is	115.
Variable	Let p = original price of the skis.						
Equation	$\frac{2}{3}$	$\cdot$	p	$-$	25	$=$	115

$\frac{2}{3}p - 25 = 115$ **Original equation**

$\frac{2}{3}p - 25 + 25 = 115 + 25$ **Add 25 to each side.**

$\frac{2}{3}p = 140$ **Simplify.**

$\frac{3}{2}\left(\frac{2}{3}p\right) = \frac{3}{2}(140)$ **Multiply each side by $\frac{3}{2}$.**

$p = 210$ **Simplify.**

The original price of the skis was \$210.

Real-World Link

Shoppers in Shanghai, China, can pay for purchased items at a terminal that can match the buyers' fingerprints with their bank accounts.

Source: *Shanghai Daily*

Check Your Progress

2A. **RETAIL** A music store has sold $\frac{3}{5}$ of their hip-hop CDs, but 10 were returned. Now the store has 62 hip-hop CDs. How many were there originally?

2B. **READING** Len read $\frac{3}{4}$ of a graphic novel over the weekend. Monday, he read 22 more pages. If he has read 220 pages, how many pages does the book have?

Personal Tutor glencoe.com

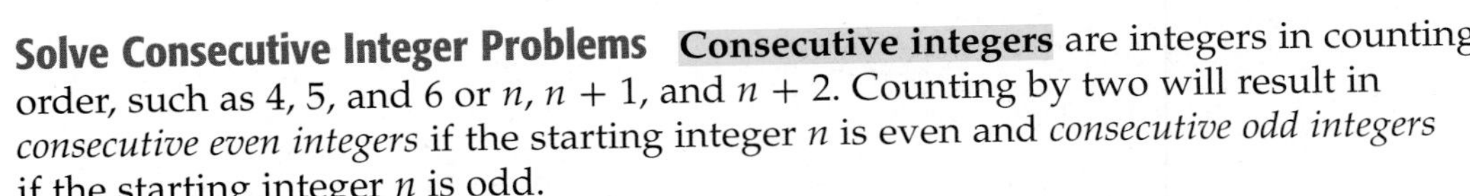

Solve Consecutive Integer Problems **Consecutive integers** are integers in counting order, such as 4, 5, and 6 or n, $n + 1$, and $n + 2$. Counting by two will result in *consecutive even integers* if the starting integer n is even and *consecutive odd integers* if the starting integer n is odd.

Concept Summary — Consecutive Integers

For Your FOLDABLE

Type	Words	Symbols	Example
Consecutive Integers	Integers that come in counting order.	$n, n + 1, n + 2, \ldots$	…, −2, −1, 0, 1, 2, …
Consecutive Even Integers	Even integer followed by the next even integer.	$n, n + 2, n + 4, \ldots$	…, −2, 0, 2, 4, …
Consective Odd Integers	Odd integer followed by the next even integer.	$n, n + 2, n + 4, \ldots$	…, −1, 1, −3, 5, …

Number theory is the study of numbers and the relationships between them.

EXAMPLE 3 Solve a Consecutive Integer Problem

NUMBER THEORY Write an equation for the following problem. Then solve the equation and answer the problem.

Find three consecutive odd integers with a sum of −51.

Let n = the least odd integer.

Then $n + 2$ = the next greater odd integer, and $n + 4$ = the greatest of the three integers.

Words	The sum of three consecutive odd integers	is	−51.
Equation	$n + (n + 2) + (n + 4)$	$=$	-51

$$n + (n + 2) + (n + 4) = -51 \quad \textbf{Original equation}$$
$$3n + 6 = -51 \quad \textbf{Simplify.}$$
$$-6 = -6 \quad \textbf{Subtract 6 from each side.}$$
$$3n = -57 \quad \textbf{Simplify.}$$
$$\frac{3n}{3} = \frac{-57}{3} \quad \textbf{Divide each side by 3.}$$
$$n = -19 \quad \textbf{Simplify.}$$

$n + 2 = -19 + 2$ or -17 $\quad n + 4 = -19 + 4$ or -15

The consecutive odd integers are −19, −17, and −15.

CHECK −19, −17, and −15 are consecutive odd integers.
$-19 + (-17) + (-15) = -51$ ✓

StudyTip

Representing Consecutive Integers You can use the same expressions to represent either consecutive even integers or consecutive odd integers. It is the value of n (odd or even) that differs between the two expressions.

Check Your Progress

3. Write an equation for the following problem. Then solve the equation and answer the problem.

Find three consecutive integers with a sum of 21.

Personal Tutor glencoe.com

Check Your Understanding

Example 1 p. 91

Solve each equation. Check your solution.

1. $3m + 4 = -11$
2. $12 = -7f - 9$
3. $-3 = 2 + \frac{a}{11}$
4. $\frac{3}{2}a - 8 = 11$
5. $8 = \frac{x - 5}{7}$
6. $\frac{c + 1}{-3} = -21$

Example 2 p. 92

7. **NUMBER THEORY** Twelve decreased by twice a number equals −34. Write an equation for this situation and then find the number.
8. **BASEBALL** Among the career home run leaders for Major League Baseball, Hank Aaron has 175 fewer than twice the number that Dave Winfield has. Hank Aaron hit 755 home runs. Write an equation for this situation. How many home runs did Dave Winfield hit in his career?

Example 3 p. 93

Write an equation and solve each problem.

9. Find three consecutive odd integers with a sum of 75.
10. Find three consecutive integers with a sum of −36.

Practice and Problem Solving

= **Step-by-Step Solutions** begin on page R12.
Extra Practice begins on page 815.

Example 1 p. 91

Solve each equation. Check your solution.

11. $3t + 7 = -8$

12. $8 = 16 + 8n$

13. $-34 = 6m - 4$

14. $9x + 27 = -72$

15. $\frac{y}{5} - 6 = 8$

16. $\frac{f}{-7} - 8 = 2$

17. $1 + \frac{r}{9} = 4$

18. $\frac{k}{3} + 4 = -16$

19. $\frac{n-2}{7} = 2$

20. $14 = \frac{6+z}{-2}$

21. $-11 = \frac{a-5}{6}$

22. $\frac{22-w}{3} = -7$

Example 2 p. 92

23 **FINANCIAL LITERACY** The Cell+ Cellular Phone store offers the plans shown in the table. Raul chose the business plan and has budgeted \$100 per month. Write an equation for this situation, and determine how many minutes per month he can use the phone and stay within budget.

Plan	Flat Monthly Fee	Anytime Minutes	Cost per Minute After Anytime Minutes
personal	\$29.99	250	\$0.20
business	\$49.99	650	\$0.15
executive	\$59.99	1200	\$0.10

Example 3 p. 93

Write an equation and solve each problem.

24. Fourteen less than three fourths of a number is negative eight. Find the number.

25. Seventeen is thirteen subtracted from six times a number. What is the number?

26. Find three consecutive even integers with the sum of −84.

27. Find three consecutive odd integers with the sum of 141.

28. Find four consecutive integers with the sum of 54.

29. Find four consecutive integers with the sum of −142.

Solve each equation. Check your solution.

30. $-6m - 8 = 24$

31. $45 = 7 - 5n$

32. $\frac{2b}{3} + 6 = 24$

33. $\frac{5x}{9} - 11 = -51$

34. $65 = \frac{3}{4}c - 7$

35. $9 + \frac{2}{3}x = 81$

36. $-\frac{5}{2} = \frac{3}{4}z + \frac{1}{2}$

37. $\frac{5}{6}k + \frac{2}{3} = \frac{4}{3}$

38. $-\frac{1}{5} - \frac{4}{9}a = \frac{2}{15}$

39. $-\frac{3}{7} = \frac{3}{4} - \frac{b}{2}$

Write an equation and solve each problem.

40. **FAMILY** The ages of three brothers are consecutive integers with the sum of 96. How old are the brothers?

41. **VOLCANOES** Moving lava can build up and form beaches at the coast of an island. The growth of an island in a seaward direction may be modeled as $8y + 2$ centimeters, where y represents the number of years that the lava flows. An island has expanded 60 centimeters seaward. How long has the lava flowed?

Solve each equation. Check your solution.

42. $-5x - 4.8 = 6.7$

43. $3.7q + 26.2 = 111.67$

44. $0.6a + 9 = 14.4$

45. $\frac{c}{2} - 4.3 = 11.5$

46. $9 = \frac{-6p - (-3)}{-8}$

47. $3.6 - 2.4m = 12$

48. If $7m - 3 = 53$, what is the value of $11m + 2$?

49. If $13y + 25 = 64$, what is the value of $4y - 7$?

50. If $-5c + 6 = -69$, what is the value of $6c - 15$?

Real-World Link

The top amusement parks in the world for attendance are:

1. Magic Kingdom – Orlando, FL
2. Disneyland – Anaheim, CA
3. Tokyo Disneyland – Urayasu, Chiba, Japan
4. Tokyo Disney Sea – Urayasu, Chiba, Japan
5. Disneyland Paris – Cedex, Marne La Valle, France

51. **AMUSEMENT PARKS** An amusement park offers a yearly membership of \$275 that allows for free parking and admission to the park. Members can also use the water park for an additional \$5 per day. Nonmembers pay \$6 for parking, \$15 for admission, and \$9 for the water park.

 a. Write and solve an equation to find the number of visits it would take for the total cost to be the same for a member and a nonmember if they both use the water park at each visit.

 b. Make a table for the costs of members and nonmembers after 3, 6, 9, 12, and 15 visits to the park.

 c. Plot these points on a coordinate graph and describe things you notice from the graph.

52. **SHOPPING** At The Family Farm, you can pick your own fruits and vegetables.

 a. The cost of a bag of potatoes is \$1.50 less than $\frac{1}{2}$ of the price of apples. Write and solve an equation to find the cost of potatoes.

 b. The price of each zucchini is 3 times the price of winter squash minus \$7. Write and solve an equation to find the cost of zucchini.

 c. Write an equation to represent the cost of a pumpkin using the cost of the blueberries.

The Family Farm

Fruit	Price ($)
Apples	6.99/bag
Pumpkins	5.00 each
Blueberries	2.99/qt
Winter squash	2.99 each

H.O.T. Problems Use Higher-Order Thinking Skills

53. **OPEN ENDED** Write a problem that can be modeled by the equation $2x + 40 = 60$. Then solve the equation and explain the solution in the context of the problem.

54. **REASONING** Describe the steps you can use to solve $\frac{w + 3}{5} - 4 = 6$.

55. **CHALLENGE** To find the measure of an interior angle of a regular polygon, you can use the formula $m = \frac{180(n - 2)}{n}$, where m represents the measure of each angle and n represents the number of sides in the polygon. If $m = 156$, how many sides does the polygon have?

56. **CHALLENGE** Determine whether the following statement is *sometimes*, *always*, or *never* true. Explain your reasoning.

 The sum of three consecutive odd integers equals an even integer.

57. **WRITING IN MATH** Write a paragraph explaining the order of the steps that you would take to solve a multi-step equation.

Standardized Test Practice

58. Which is the best estimate for the number of minutes on the calling card advertised below?

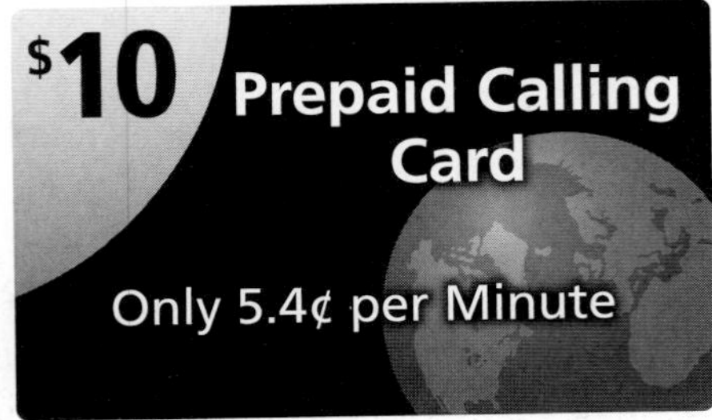

A 10 min

B 20 min

C 50 min

D 200 min

59. GRIDDED RESPONSE The scale factor for two similar triangles is 2 : 3. The perimeter of the smaller triangle is 56 cm. What is the perimeter of the larger triangle in centimeters?

60. Mr. Morrison is draining his cylindrical pool. The pool has a radius of 10 feet and a standard height of 4.5 feet. If the pool water is pumped out at a constant rate of 5 gallons per minute, about how long will it take to drain the pool? (1 ft^3 = 7.5 gal)

F 37.8 min

G 7 h

H 25.4 h

J 35.3 h

61. STATISTICS Look at the golf scores for the five players in the table.

Player	1	2	3	4	5
Score	80	91	103	79	78

Which of these is the range of the golf scores?

A 10

B 25

C 35

D 40

Spiral Review

62. GAS MILEAGE A midsize car with a 4-cylinder engine travels 34 miles on a gallon of gas. This is 10 miles more than a luxury car with an 8-cylinder engine travels on a gallon of gas. How many miles does a luxury car travel on a gallon of gas? (Lesson 2-2)

63. DEER In a recent year, 1286 female deer were born in Clark County. That is 93 fewer than the number of male deer born. How many male deer were born that year? (Lesson 2-2)

Translate each equation into a verbal sentence. (Lesson 2-1)

64. $f - 15 = 6$

65. $3h + 7 = 20$

66. $k^2 + 18 = 54 - m$

67. $3p = 8p - r$

68. $\frac{3}{5}t + \frac{1}{3} = t$

69. $\frac{1}{2}v = \frac{2}{3}v + 4$

70. GEOGRAPHY The Pacific Ocean covers about 46% of Earth. If P represents the surface area of the Pacific Ocean and E represents the surface area of Earth, write an equation for this situation. (Lesson 2-1)

Find the value of n in each equation. Then name the property that is used. (Lesson 1-3)

71. $1.5 + n = 1.5$

72. $8n = 1$

73. $4 - n = 0$

74. $1 = 2n$

Skills Review

Evaluate each expression. (Lesson 1-2)

75. $5 + 3(4^2)$

76. $\frac{38 - 12}{2 \cdot 13}$

77. $[5(1 + 1)]^3$

78. $[8(2) - 4^2] + 7(4)$

2-4 Solving Equations with the Variable on Each Side

Then
You solved multi-step equations. (Lesson 2-3)

Now
- Solve equations with the variable on each side.
- Solve equations involving grouping symbols.

New Vocabulary
identity

Math Online
glencoe.com
- Extra Examples
- Personal Tutor
- Self-Check Quiz
- Homework Help
- Math in Motion

Why?

The equation $y = 1.3x + 19$ represents the number of times Americans eat in their cars each year, where x is the number of years since 1985, and y is the number of times that they eat in their car. The equation $y = -1.3x + 93$ represents the number of times Americans eat in restaurants each year, where x is the number of years since 1985, and y is the number of times that they eat in a restaurant.

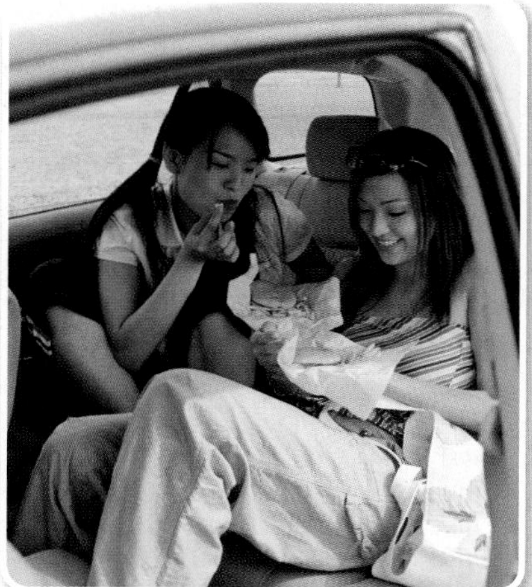

The equation $1.3x + 19 = -1.3x + 93$ represents the year when the number of times Americans eat in their cars will equal the number of times Americans eat in restaurants.

Variables on Each Side To solve an equation that has variables on each side, use the Addition or Subtraction Property of Equality to write an equivalent equation with the variable terms on one side.

EXAMPLE 1 Solve an Equation with Variables on Each Side

Solve $2 + 5k = 3k - 6$. Check your solution.

$2 + 5k = 3k - 6$	Original equation
$-3k = -3k$	Subtract $3k$ from each side.
$2 + 2k = -6$	Simplify.
$-2 \quad = -2$	Subtract 2 from each side.
$2k = -8$	Simplify.
$\frac{2k}{2} = \frac{-8}{2}$	Divide each side by 2.
$k = -4$	Simplify.
CHECK $2 + 5k = 3k - 6$	Original equation
$2 + 5(-4) \stackrel{?}{=} 3(-4) - 6$	Substitution, $k = -4$
$2 + -20 \stackrel{?}{=} -12 - 6$	Multiply.
$-18 = -18$ ✓	Simplify.

Check Your Progress

Solve each equation. Check your solution.

1A. $3w + 2 = 7w$

1B. $5a + 2 = 6 - 7a$

1C. $\frac{x}{2} + 1 = \frac{1}{4}x - 6$

1D. $1.3c = 3.3c + 2.8$

Personal Tutor glencoe.com

Grouping Symbols If equations contain grouping symbols such as parentheses or brackets, use the Distributive Property first to remove the grouping symbols.

StudyTip

Solving an Equation You may want to eliminate the terms with a variable from one side before eliminating a constant.

EXAMPLE 2 Solve an Equation with Grouping Symbols

Solve $6(5m - 3) = \frac{1}{3}(24m + 12)$.

$6(5m - 3) = \frac{1}{3}(24m + 12)$	**Original equation**
$30m - 18 = 8m + 4$	**Distributive Property**
$30m - 18 - 8m = 8m + 4 - 8m$	**Subtract $8m$ from each side.**
$22m - 18 = 4$	**Simplify.**
$22m - 18 + 18 = 4 + 18$	**Add 18 to each side.**
$22m = 22$	**Simplify.**
$\frac{22m}{22} = \frac{22}{22}$	**Divide each side by 22.**
$m = 1$	**Simplify.**

Check Your Progress

Solve each equation. Check your solution.

2A. $8s - 10 = 3(6 - 2s)$

2B. $7(n - 1) = -2(3 + n)$

Personal Tutor glencoe.com

Some equations may have no solution. That is, there is no value of the variable that will result in a true equation. Some equations are true for all values of the variables. These are called **identities**.

ReadingMath

No Solution The symbol that represents no solution is Ø.

EXAMPLE 3 Find Special Solutions

Solve each equation.

a. $5x + 5 = 3(5x - 4) - 10x$

$5x + 5 = 3(5x - 4) - 10x$	**Original equation**
$5x + 5 = 15x - 12 - 10x$	**Distributive Property**
$5x + 5 = 5x - 12$	**Simplify.**
$-5x \quad = -5x$	**Subtract $5x$ from each side.**
$5 \neq -12$	

Since $5 \neq -12$, this equation has no solution.

b. $3(2b - 1) - 7 = 6b - 10$

$3(2b - 1) - 7 = 6b - 10$	**Original equation**
$6b - 3 - 7 = 6b - 10$	**Distributive Property**
$6b - 10 = 6b - 10$	**Simplify.**
$0 = 0$	**Subtract $6b - 10$ from each side.**

Since the expressions on each side of the equation are the same, this equation is an identity. It is true for all values of b.

Check Your Progress

3A. $7x + 5(x - 1) = -5 + 12x$

3B. $6(y - 5) = 2(10 + 3y)$

Personal Tutor glencoe.com

The steps for solving an equation can be summarized as follows.

Concept Summary — **Steps for Solving Equations** — For Your FOLDABLE

Step 1 Simplify the expressions on each side. Use the Distributive Property as needed.

Step 2 Use the Addition and/or Subtraction Properties of Equality to get the variables on one side and the numbers without variables on the other side. Simplify.

Step 3 Use the Multiplication or Division Property of Equality to solve.

Math *in Motion*, BrainPOP® glencoe.com

There are many situations in which variables are on both sides of the equation.

STANDARDIZED TEST EXAMPLE 4

Find the value of x so that the figures have the same area.

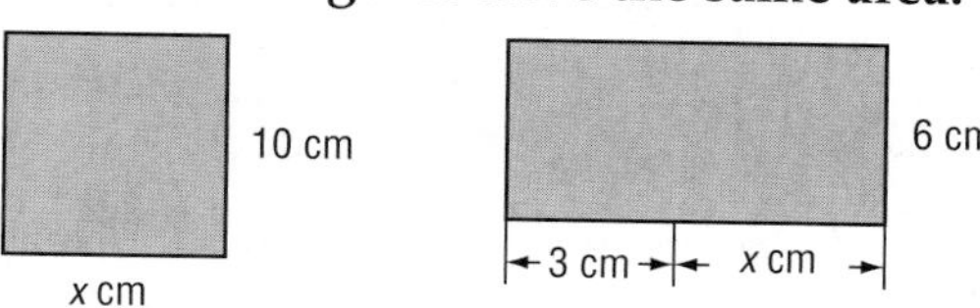

A 3

B 4.5

C 6.5

D 7

Read the Test Item

The area of the first rectangle is $10x$, and the area of the second is $6(3 + x)$. The equation $10x = 6(3 + x)$ represents this situation.

Solve the Test Item

A
$$10x = 6(3 + x)$$
$$10(3) \stackrel{?}{=} 6(3 + 3)$$
$$30 \stackrel{?}{=} 6(6)$$
$$30 \neq 36 \text{ ✗}$$

B
$$10x = 6(3 + x)$$
$$10(4.5) \stackrel{?}{=} 6(3 + 4.5)$$
$$45 \stackrel{?}{=} 6(7.5)$$
$$45 = 45 \text{ ✓}$$

Since the value 4.5 results in a true statement, you do not need to check 6.5 and 7. The answer is B.

Test-TakingTip

Choose a Method There is often more than one way to solve a problem. In this example, you can write an algebraic equation and solve for x. Or you can substitute each answer choice into the formulas to find the correct answer.

Check Your Progress

4. **Find the value of x so that the figures have the same perimeter.**

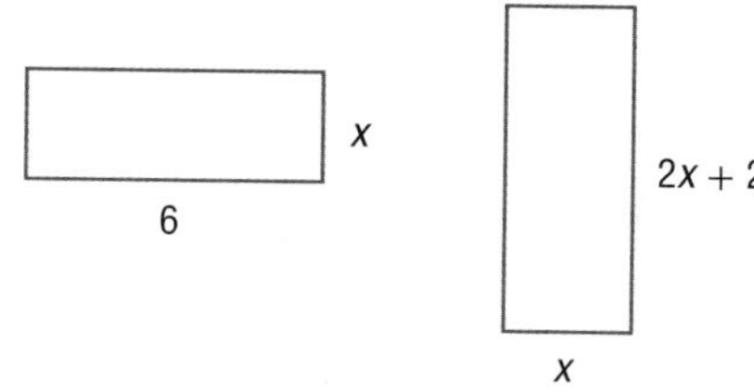

F 1.5 **G** 2 **H** 3.2 **J** 4

Personal Tutor glencoe.com

Check Your Understanding

Examples 1–3 pp. 97–98

Solve each equation. Check your solution.

1. $13x + 2 = 4x + 38$
2. $\frac{2}{3} + \frac{1}{6}q = \frac{5}{6}q + \frac{1}{3}$
3. $6(n + 4) = -18$
4. $7 = -11 + 3(b + 5)$
5. $5 + 2(n + 1) = 2n$
6. $7 - 3r = r - 4(2 + r)$
7. $14v + 6 = 2(5 + 7v) - 4$
8. $5h - 7 = 5(h - 2) + 3$

Example 4 p. 99

9. **MULTIPLE CHOICE** Find the value of x so that the figures have the same perimeter.

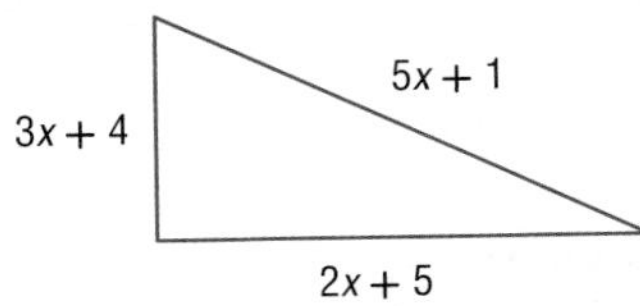

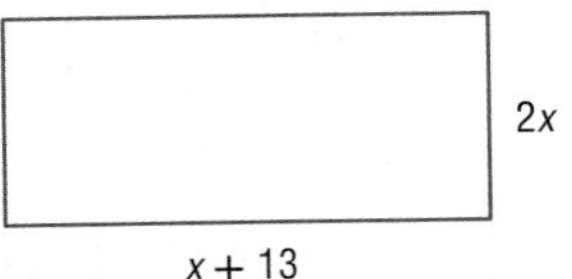

A 4 **B** 5 **C** 6 **D** 7

Practice and Problem Solving

● = **Step-by-Step Solutions** begin on page R12.
Extra Practice begins on page 815.

Examples 1–3 pp. 97–98

Solve each equation. Check your solution.

10. $7c + 12 = -4c + 78$
11. $2m - 13 = -8m + 27$
12. $9x - 4 = 2x + 3$
13. $6 + 3t = 8t - 14$
14. $\frac{b - 4}{6} = \frac{b}{2}$
15. $\frac{5v - 4}{10} = \frac{4}{5}$
16. $8 = 4(r + 4)$
17. $6(n + 5) = 66$
18. $5(g + 8) - 7 = 103$
19. $12 - \frac{4}{5}(x + 15) = 4$
20. $3(3m - 2) = 2(3m + 3)$
21. $6(3a + 1) - 30 = 3(2a - 4)$

Example 4 p. 99

22. **GEOMETRY** Find the value of x so the rectangles have the same area.

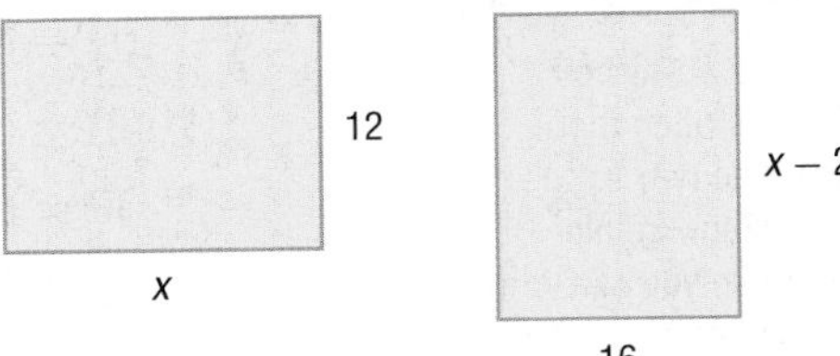

23. **NUMBER THEORY** Four times the lesser of two consecutive even integers is 12 less than twice the greater number. Find the integers.

24. **NUMBER THEORY** Two times the least of three consecutive odd integers exceeds three times the greatest by 15. What are the integers?

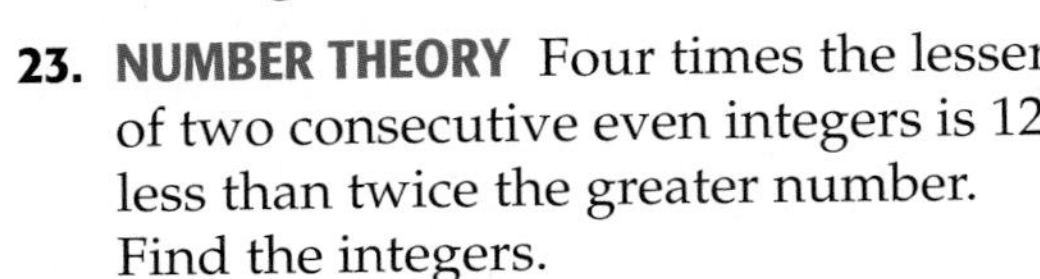

Solve each equation. Check your solution.

25. $2x = 2(x - 3)$
26. $\frac{2}{5}h - 7 = \frac{12}{5}h - 2h + 3$
27. $-5(3 - q) + 4 = 5q - 11$
28. $2(4r + 6) = \frac{2}{3}(12r + 18)$
29. $\frac{3}{5}f + 24 = 4 - \frac{1}{5}f$
30. $\frac{1}{12} + \frac{3}{8}y = \frac{5}{12} + \frac{5}{8}y$
31. $\frac{2m}{5} = \frac{1}{3}(2m - 12)$
32. $\frac{1}{8}(3d - 2) = \frac{1}{4}(d + 5)$
33. $6.78j - 5.2 = 4.33j + 2.15$
34. $14.2t - 25.2 = 3.8t + 26.8$
35. $3.2k - 4.3 = 12.6k + 14.5$
36. $5[2p - 4(p + 5)] = 25$

Real-World Link

About 25% of American adults have Internet access and give time or money to charity. Of those, 56% have never visited the Web site of a charity, and only 7% say they have given online.

Source: Craver, Mathews, Smith & Co.

37. **NUMBER THEORY** Three times the lesser of two consecutive even integers is 6 less than six times the greater number. Find the integers.

38. **MONEY** Chris has saved twice the number of quarters that Nora saved plus 6. The number of quarters Chris saved is also five times the difference of the number of quarters and 3 that Nora has saved. Write and solve an equation to find the number of quarters they each have saved.

39. **DVD** A company that replicates DVDs spends \$1500 per day in building overhead plus \$0.80 per DVD in supplies and labor. If the DVDs sell for \$1.59 per disk, how many DVDs must the company sell each day before it makes a profit?

40. **INTERNET ACCESS** The table shows the percent of households that have broadband Internet access and the average growth rates for two age groups. How long will it take for the percents to be the same?

Age Group	Percent with Broadband in Their Homes in 2006	Growth Rate Percentage per Year
18–49	52.5	42
50+	25.5	52

Source: Pew Internet & American Life Project

41. **MULTIPLE REPRESENTATIONS** In this problem, you will explore $2x + 4 = -x - 2$.

   a. **GRAPHICAL** Make a table of values with five points for $y = 2x + 4$ and $y = -x - 2$. Graph the points from the tables.

   b. **ALGEBRAIC** Solve $2x + 4 = -x - 2$.

   c. **VERBAL** Explain how the solution you found in part **b** is related to the intersection point of the graphs in part **a**.

H.O.T. Problems

Use **H**igher-**O**rder **T**hinking Skills

42. **REASONING** Solve the equation below. Describe each step.

$$t = 2 - 2[2t - 3(1 - t)]$$

43. **CHALLENGE** Write an equation with the variable on each side of the equals sign, at least one fractional coefficient, and a solution of −6. Discuss the steps you used.

44. **OPEN ENDED** Create an equation with at least two grouping symbols for which there is no solution.

45. **REASONING** Determine whether each solution is correct. If the solution is not correct, describe the error and give the correct solution.

   a.
$$\begin{aligned} 2(g + 5) &= 22 \\ 2g + 5 &= 22 \\ 2g + 5 - 5 &= 22 \\ 2g &= 17 \\ 2g &= 8.5 \end{aligned}$$

   b.
$$\begin{aligned} 5d &= 2d - 18 \\ 5d - 2d &= 2d - 18 - 2d \\ 3d &= -18 \\ d &= -6 \end{aligned}$$

   c.
$$\begin{aligned} -6z + 13 &= 7z \\ -6z + 13 - 6z &= 7z - 6z \\ 13 &= z \end{aligned}$$

46. **CHALLENGE** Find the value of k for which each equation is an identity.

   a. $k(3x - 2) = 4 - 6x$

   b. $15y - 10 + k = 2(ky - 1) - y$

47. **WRITING IN MATH** Compare and contrast solving equations with variables on both sides of the equation to solving one-step or multi-step equations with a variable on one side of the equation.

Standardized Test Practice

48. A hang glider 25 meters above the ground starts to descend at a constant rate of 2 meters per second. Which equation shows the height h after t seconds of descent?

A $h = 25t + 2t$
B $h = -25t + 2$
C $h = 2t + 25$
D $h = -2t + 25$

49. GEOMETRY Two rectangular walls each with a length of 12 feet and a width of 23 feet need to be painted. It costs \$0.08 per square foot for paint. How much will it cost to paint the walls?

F \$22.08
G \$23.04
H \$34.50
J \$44.16

50. SHORT RESPONSE Maddie works at Game Exchange. They are having a sale as shown.

Item	Price	Special
video games	\$20	Buy 2 get 1 Free
DVDs	\$15	Buy 1 get 1 Free

She purchases four video games and uses her employee discount of 15%. If sales tax is 7.25%, how much does she spend on the games?

51. Solve $\frac{4}{5}x + 7 = \frac{3}{15}x - 3$.

A $-16\frac{2}{3}$
B $-14\frac{4}{9}$
C $-6\frac{2}{3}$
D -10

Spiral Review

Solve each equation. Check your solution. (Lesson 2-3)

52. $5n + 6 = -4$
53. $-1 = 7 + 3c$
54. $\frac{1}{2}z + 7 = 16 - \frac{3}{5}z$
55. $\frac{2}{5}x + 6 = \frac{2}{3}x + 10$
56. $\frac{a}{7} - 3 = -2$
57. $9 + \frac{y}{5} = 6$

58. WORLD RECORDS In 1998, Winchell's House of Donuts in Pasadena, California, made the world's largest donut. It weighed 5000 pounds and had a circumference of 298.3 feet. What was the donut's diameter to the nearest tenth? (*Hint:* $C = \pi d$) (Lesson 2-2)

59. ZOO At a zoo, the cost of admission is posted on the sign. Find the cost of admission for two adults and two children. (Lesson 1-3)

ZOO ADMISSION
Adults.........\$9.75
Children......\$7.25

Find the value of n. Then name the property used in each step. (Lesson 1-3)

60. $25n = 25$
61. $n \cdot 1 = 2$
62. $12 \cdot n = 12 \cdot 6$
63. $n + 0 = \frac{2}{3}$
64. $4 \cdot \frac{1}{4} = n$
65. $(10 - 8)(7) = 2(n)$

Skills Review

Translate each sentence into an equation. (Lesson 2-1)

66. Twice a number t decreased by eight equals seventy.

67. Five times the sum of m and k is the same as seven times k.

68. Half of p is the same as p minus 3.

Evaluate each expression. (Lesson 0-3)

69. $-9 - (-14)$
70. $-10 + (20)$
71. $-15 - 9$
72. $5(14)$
73. $-55 \div (-5)$
74. $-25(-5)$

2-5 Solving Equations Involving Absolute Value

Then

You solved equations with the variable on each side. (Lesson 2-5)

Now

- Evaluate absolute value expressions.
- Solve absolute value equations.

Math Online

glencoe.com

- Extra Examples
- Personal Tutor
- Self-Check Quiz
- Homework Help

Why?

In 2007, a telephone poll was conducted to determine the reading habits of Americans. People in this survey were allowed to select more than one type of book.

The survey had a margin of error of 3 percentage points. This means that the results could be three points higher or lower. So, the percent of people who read religious material could be as high as 69% or as low as 63%.

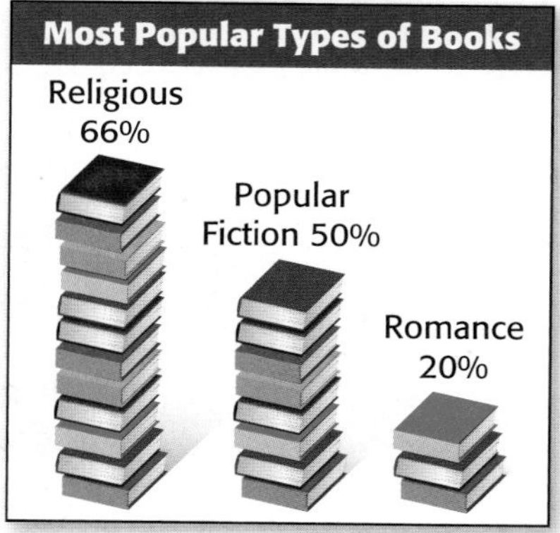

Source: CNN

Absolute Value Expressions Expressions with absolute values define an upper and lower range in which a value must lie. Expressions involving absolute value can be evaluated using the given value for the variable.

EXAMPLE 1 Expressions with Absolute Value

Evaluate $|m + 6| - 14$ if $m = 4$.

$|m + 6| - 14 = |4 + 6| - 14$	Replace m with 4.
$= |10| - 14$	$4 + 6 = 10$
$= 10 - 14$	$|10| = 10$
$= -4$	Simplify.

Check Your Progress

1. Evaluate $23 - |3 - 4x|$ if $x = 2$.

Personal Tutor glencoe.com

Absolute Value Equations Looking at the example at the top of the page, we notice that the margin of error in the bar graph is an example of absolute value. The distance between 66 and 69 on a number line is the same as the distance between 63 and 66.

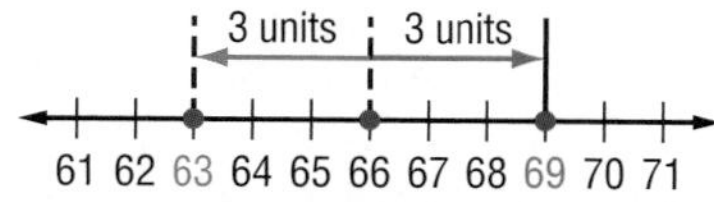

There are three types of open sentences involving absolute value, $|x| = n$, $|x| < n$, and $|x| > n$. In this lesson, we will consider only the first type. Look at the equation $|x| = 4$. This means that the distance between 0 and x is 4.

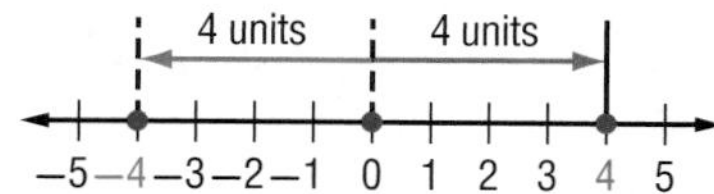

If $|x| = 4$, then $x = -4$ or $x = 4$. Thus, the solution set is $\{-4, 4\}$.

For each absolute value equation, we must consider both cases. To solve an absolute value equation, first isolate the absolute value on one side of the equals sign if it is not already by itself.

ReadingMath

Absolute Values The expression $|f + 5|$ is read the *absolute value of the quantity f plus 5.*

Key Concept: Absolute Value Equations

Words When solving equations that involve absolute values, there are two cases to consider.

Case 1: The expression inside the absolute value symbol is positive or zero.

Case 2: The expression inside the absolute value symbol is negative.

Symbols For any real numbers a and b, if $|a| = b$, then $a = b$ or $a = -b$.

Example $|d| = 10$, so $d = 10$ or $d = -10$.

EXAMPLE 2 Solve Absolute Value Equations

Solve each equation. Then graph the solution set.

a. $|f + 5| = 17$

$|f + 5| = 17$ **Original equation**

Case 1

$f + 5 = 17$

$f + 5 - 5 = 17 - 5$ **Subtract 5 from each side.**

$f = 12$ **Simplify.**

Case 2

$f + 5 = -17$

$f + 5 - 5 = -17 - 5$

$f = -22$

−25 −20 −15 −10 −5 0 5 10 15 20 25

b. $|b - 1| = -3$

$|b - 1| = -3$ means the distance between b and 1 is −3. Since distance cannot be negative, the solution is the empty set ∅.

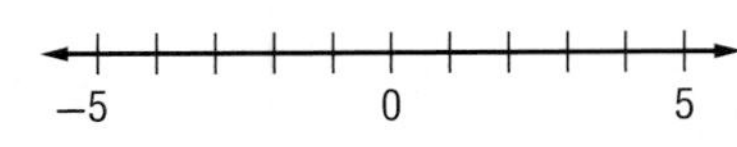

Check Your Progress

2A. $|y + 2| = 4$

2B. $|3n - 4| = -1$

Personal Tutor glencoe.com

Absolute value equations occur in real-world situations that describe a range within which a value must lie.

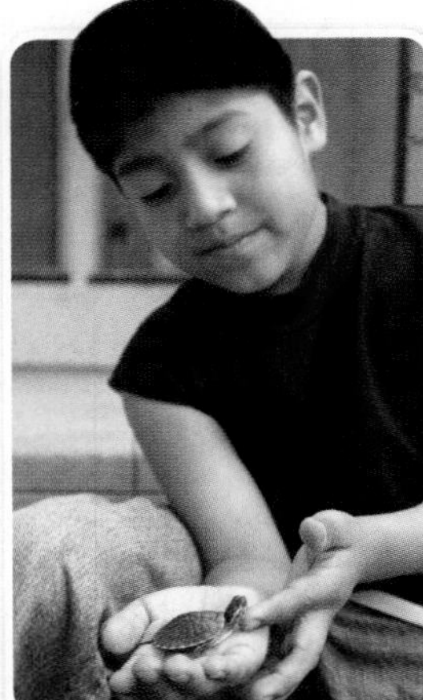

Real-World Link

In 2001, the number of households in the U.S. that had either a turtle, snake, lizard, or other reptile as a pet was 1,678,000.

Source: American Veterinary Medical Association

Real-World EXAMPLE 3 Solve an Absolute Value Equation

SNAKES The temperature of an enclosure for a pet snake should be about 80°F, give or take 5°. Find the maximum and minimum temperatures.

You can use a number line to solve.

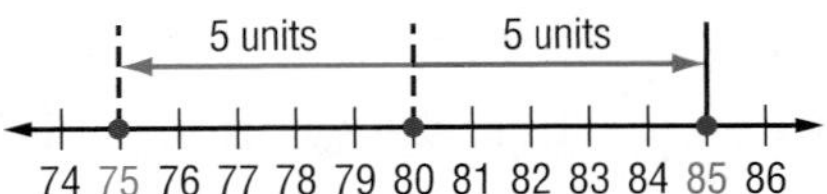

The distance from 80 to 75 is 5 units.
The distance from 80 to 85 is 5 units.

The solution set is {75, 85}. The maximum and minimum temperatures are 85° and 75°.

Check Your Progress

3. ICE CREAM Ice cream should be stored at 5°F with an allowance for 5°. Write and solve an equation to find the maximum and minimum temperatures at which the ice cream should be stored.

Personal Tutor glencoe.com

When given two points on a graph, you can write an absolute value equation for the graph.

> **StudyTip**
>
> **Find the Midpoint** To find the point midway between two points, add the values together and divide by 2. For Example 4, $11 + 19 = 30$, $30 \div 2 = 15$. So 15 is the point halfway between 11 and 19.

EXAMPLE 4 Write an Absolute Value Equation

Write an equation involving absolute value for the graph.

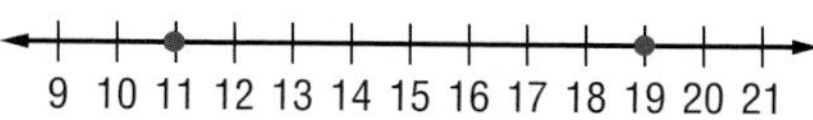

Find the point that is the same distance from 11 and from 19. This is the midpoint between 11 and 19, which is 15.

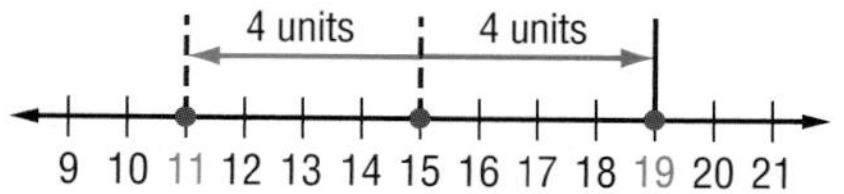

The distance from 15 to 11 is 4 units.
The distance from 15 to 19 is 4 units.

So an equation is $|x - 15| = 4$.

Check Your Progress

4. Write an equation involving absolute value for the graph.

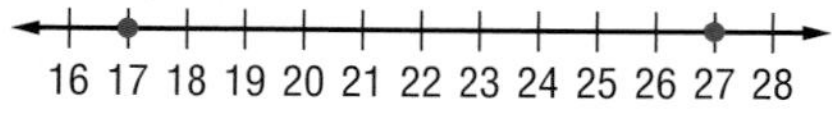

Personal Tutor glencoe.com

Check Your Understanding

Example 1 p. 103

Evaluate each expression if $f = 3$, $g = -4$, and $h = 5$.

1. $|3 - h| + 13$ **2.** $16 - |g + 9|$ **3.** $|f + g| - h$

Example 2 p. 104

Solve each equation. Then graph the solution set.

4. $|n + 7| = 5$ **5.** $|3z - 3| = 9$ **6.** $|4n - 1| = -6$

7. $|b + 4| = 2$ **8.** $|2t - 4| = 8$ **9.** $|5h + 2| = -8$

Example 3 p. 104

10. FINANCIAL LITERACY For a company to invest in a product, they must believe they will receive a 12% return on investment (ROI) plus or minus 3%. Write an equation to find the least and the greatest ROI they believe they will receive.

Example 4 p. 105

Write an equation involving absolute value for each graph.

11.

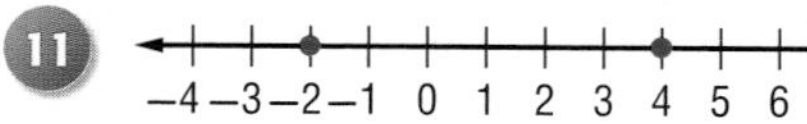

12. −10 −8 −6 −4 −2 0 2 4 6

Practice and Problem Solving

= **Step-by-Step Solutions** begin on page R12.
Extra Practice begins on page 815.

Example 1 p. 103

Evaluate each expression if $a = -2$, $b = -3$, $c = 2$, $x = 2.1$, $y = 3$, and $z = -4.2$.

13. $|2x + z| + 2y$
14. $4a - |3b + 2c|$
15. $-|5a + c| + |3y + 2z|$
16. $-a + |2x - a|$
17. $|y - 2z| - 3$
18. $3|3b - 8c| - 3$
19. $|2x - z| + 6b$
20. $-3|z| + 2(a + y)$
21. $-4|c - 3| + 2|z - a|$

Example 2 p. 104

Solve each equation. Then graph the solution set.

22. $|n - 3| = 5$
23. $|f + 10| = 1$
24. $|v - 2| = -5$
25. $|4t - 8| = 20$
26. $|8w + 5| = 21$
27. $|6y - 7| = -1$
28. $\left|\frac{1}{2}x + 5\right| = -3$
29. $|-2y + 6| = 6$
30. $\left|\frac{3}{4}a - 3\right| = 9$

Example 3 p. 104

31. **SURVEY** The circle graph at the right shows the results of a survey that asked, "How likely is it that you will be rich some day?" If the margin of error is ±4%, what is the range of the percent of teens who say it is very likely that they will be rich?

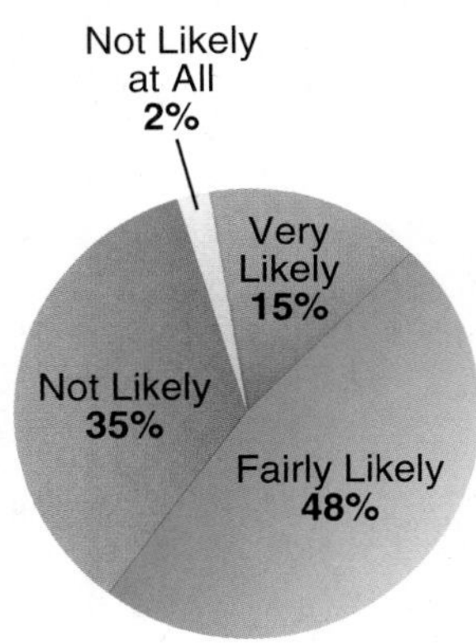

32. **CHEERLEADING** For competition, the cheerleading team is preparing a dance routine that must last 4 minutes, with a variation of ±5 seconds.
    a. Find the least and greatest possible times for the routine in minutes and seconds.
    b. Find the least and greatest possible times in seconds.

Example 4 p. 105

Write an equation involving absolute value for each graph.

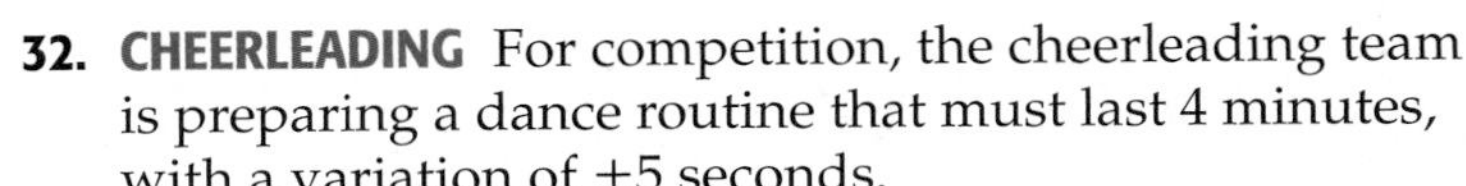

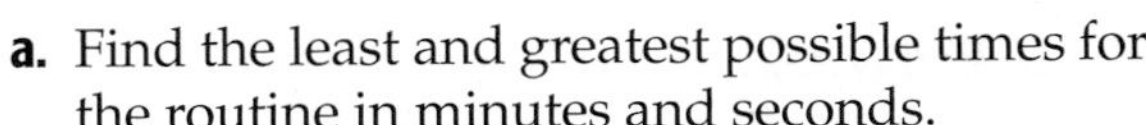

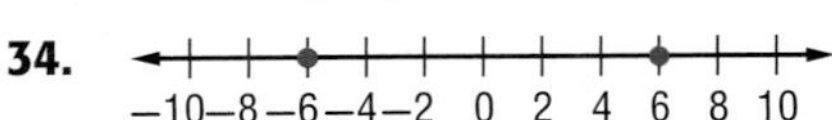

Solve each equation. Then graph the solution set.

37. $\left|-\frac{1}{2}b - 2\right| = 10$
38. $|-4d + 6| = 12$
39. $|5f - 3| = 12$
40. $2|h| - 3 = 8$
41. $4 - 3|q| = 10$
42. $\frac{4}{|p|} + 12 = 14$

43. **TRACK** The 4×400 relay is a race where 4 runners take turns running 400 meters, or one lap around the track.
    a. If a runner runs the first leg in 52 seconds plus or minus 2 seconds, write an equation to find the fastest and slowest times.
    b. If the runners of the second and third legs run their laps in 53 seconds plus or minus 1 second, write an equation to find the fastest and slowest times.
    c. Suppose the runner of the fourth leg is the fastest on the team. If he runs an average of 50.5 seconds plus or minus 1.5 seconds, what are the team's fastest and slowest times?

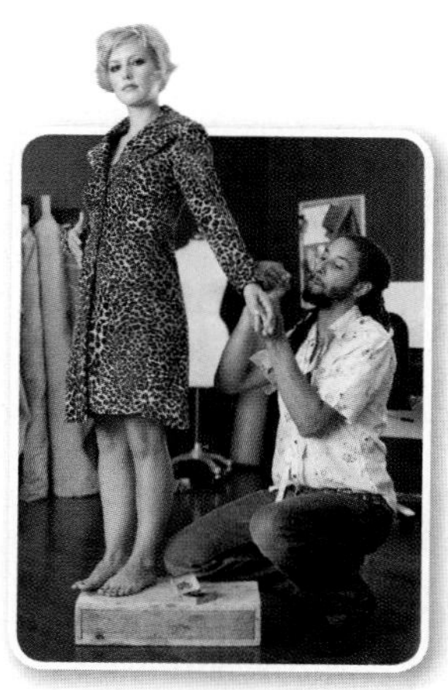

Real-World Link

An increasing number of fashion designers are using computer-aided design (CAD). CAD allows designers to view designs on virtual models.

Source: Bureau of Labor Statistics

44. FASHION To allow for a model's height, a designer is willing to use models that require him to change hems either up or down 2 inches. The length of the skirts is 20 inches.

a. Write an absolute value equation that represents the length of the skirts.

b. What is the range of the lengths of the skirts?

c. If a 20-inch skirt was fitted for a model that is 5 feet 9 inches tall, will the designer use a 6-foot-tall model?

45. CARS Speedometer accuracy can be affected by many details such as tire diameter and axle ratio. For example, there is variation of ± 3 miles per hour when calibrated at 50 miles per hour.

a. What is the range of actual speeds of the car if calibrated at 50 miles per hour?

b. A speedometer calibrated at 45 miles per hour has an accepted variation of ± 1 mile per hour. What can we conclude from this?

Write an equation involving absolute value for each graph.

46. 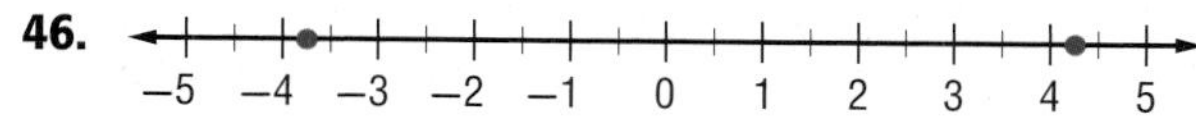

47. 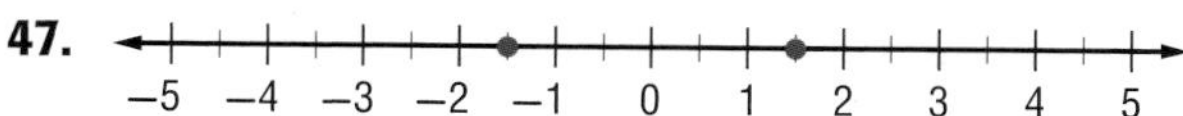

48. −5 −4 −3 −2 −1 0 1 2 3 4 5

49. −2 −1 0 1 2

50. −3 −2 −1 0 1 2 3

51. −3 −2 −1 0 1 2 3

52. MUSIC A CD will record an hour and a half of music plus or minus 3 minutes for time to change tracks.

a. Write an absolute value equation that represents the recording time.

b. What is the range of time in minutes that the CD could run?

c. Graph the possible times on a number line.

53. ACOUSTICS The Red Rocks Amphitheater located in the Red Rock Park near Denver, Colorado, is the only naturally occurring amphitheater. The acoustic qualities here are such that a maximum of 20,000 people, plus or minus 1000, can hear natural voices clearly.

a. Write an equation involving an absolute value that represents the number of people that can hear natural voices at Red Rocks Amphitheater.

b. Find the maximum and minimum number of people that can hear natural voices clearly in the amphitheater.

c. What is the range of people in part **b**?

54. **BOOK CLUB** The members of a book club agree to read within ten pages of the last page of the chapter. The chapter ends on page 203.

   a. Write an absolute value equation that represents the pages where club members could stop reading.

   b. Write the range of the pages where the club members could stop reading.

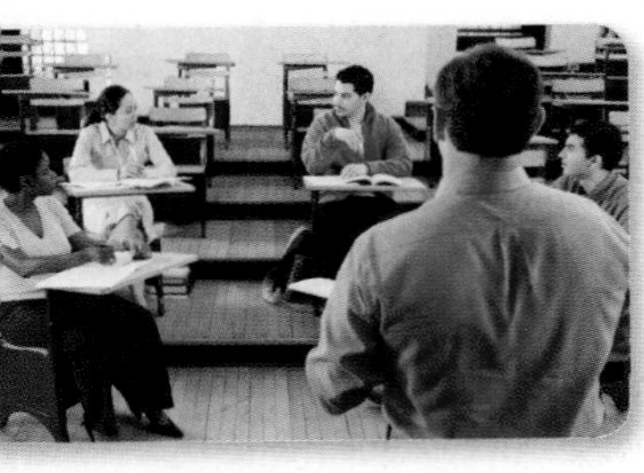

Real-World Link

Each spring since 1997, Gallaudet University holds its National Academic Bowl for Deaf and Hard of Hearing High School Students. Here, teams of high school students compete in local, regional, and national events throughout the U.S.

Source: Galluadet University

55. **SCHOOL** Washington High School and McKinley High School are competing in an academic challenge. The team with a correct response is awarded 10 points. An incorrect response has a point value of -10. There are 5 mathematics questions.

   a. Write an equation that represents the scoring for the challenge.

   b. Make a table of values for the possible points that a school could receive during the mathematics portion of the challenge.

   c. Write about how absolute values can be used in classes other than math.

H.O.T. Problems — Use Higher-Order Thinking Skills

56. **OPEN ENDED** Describe a real-world situation that could be represented by the absolute value equation $|x - 4| = 10$.

REASONING Determine whether the following statements are *sometimes*, *always*, or *never* true, if c is an integer. Explain your reasoning.

57. The value of $|x + 1|$ is greater than zero.

58. The solution of $|x + c| = 0$ is greater than 0.

59. The inequality $|x| + c < 0$ has no solution.

60. The value of $|x + c| + c$ is greater than zero.

61. **REASONING** Explain why an absolute value can never be negative.

62. **CHALLENGE** Use the sentence $x = 7 \pm 4.6$.

   a. Describe the values of x that make the sentence true.

   b. Translate the sentence into an equation involving absolute value.

63. **ERROR ANALYSIS** Alex and Wesley are solving $|x + 5| = -3$. Is either of them correct? Explain your reasoning.

Alex

$|x+5| = 3$ or $|x+5| = -3$

$x + 5 = 3$ $x + 5 = -3$

$-5 \quad -5$ $-5 \quad -5$

$x = -2$ $x = -8$

Wesley

$|x + 5| = -3$

The solution is $\varnothing$.

64. **WRITING IN MATH** Explain why there are either two, one, or no solutions for absolute value equations. Demonstrate an example of each possibility.

Standardized Test Practice

65. Which equation represents the second step of the solution process?

Step 1: $4(2x + 7) - 6 = 3x$
Step 2: ____________
Step 3: $5x + 28 - 6 = 0$
Step 4: $5x = -22$
Step 5: $x = -4.4$

A $4(2x - 6) + 7 = 3x$
B $4(2x + 1) = 3x$
C $8x + 7 - 6 = 3x$
D $8x + 28 - 6 = 3x$

66. GEOMETRY The area of a circle is 25π square centimeters. What is the circumference?

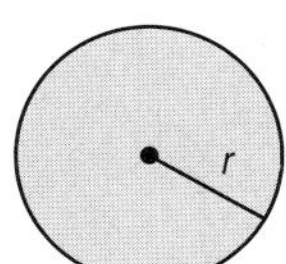

F 625π cm
G 50π cm
H 5π cm
J 10π cm

67. Tanya makes \$5 an hour and 15% commission of the total dollar value on cosmetics she sells. Suppose Tanya's commission is increased to 17%. How much money will she make if she sells \$300 worth of product and works 30 hours?

A \$201 C \$255
B \$226 D \$283

68. EXTENDED RESPONSE John's mother has agreed to take him driving every day for two weeks. On the first day, John drives for 20 minutes. Each day after that, John drives 5 minutes more than the day before.

a. Write an expression for the nth term. Explain how you found the expression.

b. For how many minutes will John drive on the last day? Show your work.

c. John's driver's education teacher requires that each student drive for 30 hours with an adult outside of class. Will John's sessions with his mother fulfill this requirement?

Spiral Review

Write and solve an equation for each sentence. (Lesson 2-4)

69. One half of a number increased by 16 is four less than two thirds of the number.

70. The sum of one half of a number and 6 equals one third of the number.

71. SHOE If ℓ represents the length of a man's foot in inches, the expression $2\ell - 12$ can be used to estimate his shoe size. What is the approximate length of a man's foot if he wears a size 8? (Lesson 2-3)

Skills Review

Write an equation for each problem. Then solve the equation. (Lesson 2-2)

72. Seven times a number equals −84. What is the number?

73. Two fifths of a number equals −24. Find the number.

74. Negative 117 is nine times a number. Find the number.

75. Twelve is one fifth of a number. What is the number?

CHAPTER 2

Mid-Chapter Quiz

Lessons 2-1 through 2-5

Math Online glencoe.com
Math *in Motion*, Animation

Translate each sentence into an equation. (Lesson 2-1)

1. The sum of three times a and four is the same as five times a.
2. One fourth of m minus six is equal to two times the sum of m and 9.
3. The product of five and w is the same as w to the third power.
4. **MARBLES** Drew has 50 red, green, and blue marbles. He has six more red marbles than blue marbles and four fewer green marbles than blue marbles. Write and solve an equation to determine how many blue marbles Drew has. (Lesson 2-1)

Solve each equation. Check your solution. (Lesson 2-2)

5. $p + 8 = 13$
6. $-26 = b - 3$
7. $\frac{t}{6} = 3$
8. **MULTIPLE CHOICE** Solve the equation $\frac{3}{5}a = \frac{1}{4}$. (Lesson 2-2)
   A $\frac{3}{20}$
   B 2
   C $\frac{5}{12}$
   D -3

Solve each equation. Check your solution. (Lesson 2-3)

9. $2x + 5 = 13$
10. $-21 = 7 - 4y$
11. $\frac{m}{6} - 3 = 8$
12. $-4 = \frac{d + 3}{5}$
13. **FISH** The average length of a yellow-banded angelfish is 12 inches. This is 4.8 times as long as an average common goldfish. (Lesson 2-3)
   a. Write an equation you could use to find the length of the average common goldfish.
   b. What is the length of an average common goldfish?

Write an equation and solve each problem. (Lesson 2-3)

14. Three less than three fourths of a number is negative 9. Find the number.
15. Thirty is twelve added to six times a number. What is the number?
16. Find four consecutive integers with a sum of 106.

Solve each equation. Check your solution. (Lesson 2-4)

17. $8p + 3 = 5p + 9$
18. $\frac{3}{4}w + 6 = 9 - \frac{1}{4}w$
19. $\frac{z + 6}{3} = \frac{2z}{4}$
20. **PERIMETER** Find the value of x so that the triangles have the same perimeter. (Lesson 2-4)

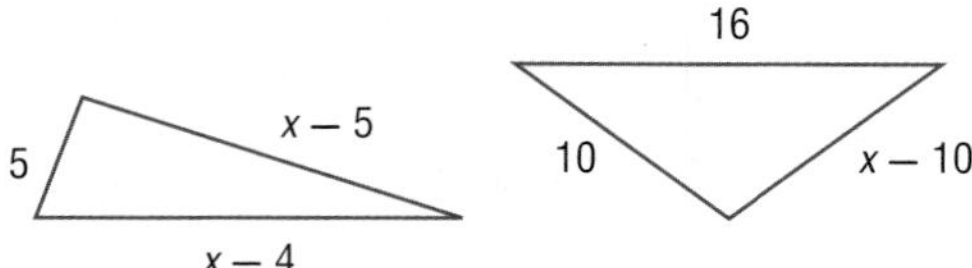

21. **PRODUCTION** ABC Sporting Goods Company produces baseball gloves. Their fixed monthly production cost is \$8000 with a per glove cost of \$5. XYZ Sporting Goods Company also produces baseball gloves. Their fixed monthly production cost is \$10,000 with a per glove cost of \$3. Find the value of x, the number of gloves produced monthly, so that the total monthly production cost is the same for both companies. (Lesson 2-4)

Evaluate each expression if $x = -4$, $y = 7$, and $z = -9$. (Lesson 2-5)

22. $|3x - 2| + 2y$
23. $|-4y + 2z| - 7z$
24. **MULTIPLE CHOICE Solve $|6m - 3| = 9$.** (Lesson 2-5)
   F $\{2\}$
   G $\{-1, 2\}$
   H $\{-3, 6\}$
   J $\{-3, 3\}$
25. **COFFEE** Some say to brew an excellent cup of coffee, you must have a brewing temperature of 200° F, plus or minus 5 degrees. Write and solve an equation describing the maximum and minimum brewing temperatures for an excellent cup of coffee.

2-6 Ratios and Proportions

Then

You evaluated percents by using a proportion. (Lesson 0-5)

Now

- Compare ratios.
- Solve proportions.

New Vocabulary

ratio
proportion
means
extremes
rate
unit rate
scale
scale model

Math Online

glencoe.com

- Extra Examples
- Personal Tutor
- Self-Check Quiz
- Homework Help
- Math in Motion

Why?

Ratios allow us to compare many items by using a common reference. The table below shows the number of a certain popular fast food restaurants, per 10,000 people, in the United States as well as other countries. This allows us to compare the number of these restaurants using an equal reference.

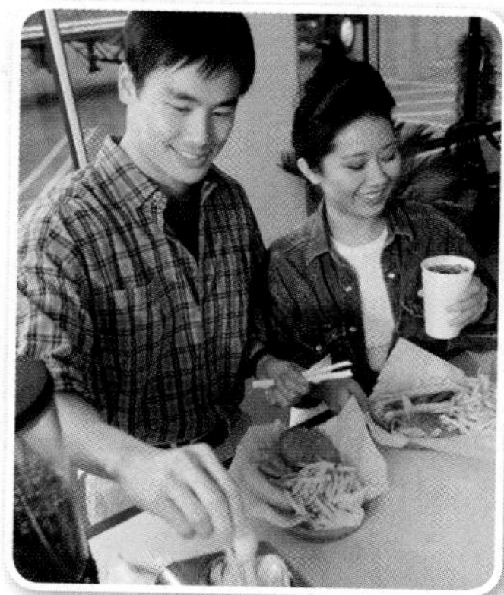

Countries	United States	New Zealand	Canada	Australia	Japan	Singapore
Number of Restaurants per 10,000 People	0.433	0.369	0.352	0.349	0.282	0.273

Ratios and Proportions The comparison between the number of restaurants and the number of people is a ratio. A **ratio** is a comparison of two numbers by division. The ratio of x to y can be expressed in the following ways.

$$x \text{ to } y \qquad x{:}y \qquad \frac{x}{y}$$

Suppose you wanted to determine the number of restaurants per 100,000 people in Australia. Notice that this ratio is equal to the original ratio.

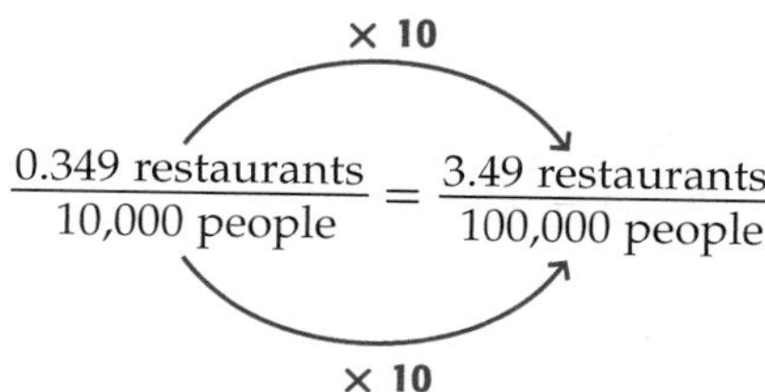

$$\frac{0.349 \text{ restaurants}}{10{,}000 \text{ people}} = \frac{3.49 \text{ restaurants}}{100{,}000 \text{ people}}$$

(× 10 above and below)

An equation stating that two ratios are equal is called a **proportion**. So, we can state that $\frac{0.349}{10{,}000} = \frac{3.49}{100{,}000}$ is a proportion.

EXAMPLE 1 Determine Whether Ratios Are Equivalent

Determine whether $\frac{2}{3}$ and $\frac{16}{24}$ are equivalent ratios. Write *yes* or *no*. Justify your answer.

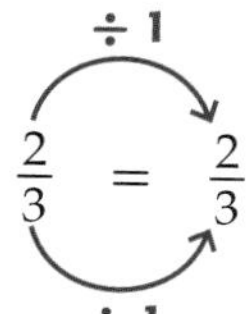

$\frac{2}{3} = \frac{2}{3}$ (÷ 1)

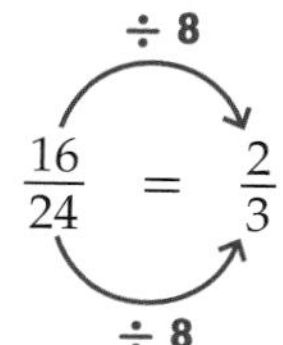

$\frac{16}{24} = \frac{2}{3}$ (÷ 8)

When expressed in simplest form, the ratios are equivalent.

Check Your Progress

Determine whether each pair of ratios are equivalent ratios. Write *yes* or *no*. Justify your answer.

1A. $\frac{6}{10}, \frac{2}{5}$

1B. $\frac{1}{6}, \frac{5}{30}$

Personal Tutor glencoe.com

StudyTip

Means and Extremes To solve a proportion using cross products, write an equation that sets the product of the extremes equal to the product of the means.

There are special names for the terms in a proportion.

1.5 and 1.2 are called the **means**. They are the middle terms of the proportion.

$$0.2 : 1.5 = 1.2 : 9.0$$

0.2 and 9.0 are called the **extremes**. They are the first and last terms of the proportion.

Key Concept

For Your FOLDABLE

Means-Extremes Property of Proportion

Words In a proportion, the product of the extremes is equal to the product of the means.

Symbols If $\frac{a}{b} = \frac{c}{d}$ and $b, d \neq 0$, then $ad = bc$.

Example Since $\frac{2}{4} = \frac{1}{2}$, $2(2) = 4(1)$ or $4 = 4$.

Another way to determine whether two ratios form a proportion is to use cross products. If the cross products are equal, then the ratios form a proportion.

This is the same as multiplying the means, and multiplying the extremes.

EXAMPLE 2 Cross Products

Use cross products to determine whether each pair of ratios forms a proportion.

a. $\frac{2}{3.5}, \frac{8}{14}$

$\frac{2}{3.5} \stackrel{?}{=} \frac{8}{14}$	**Original proportion**
$2(14) \stackrel{?}{=} 3.5(8)$	**Cross products**
$28 = 28$ ✓	**Simplify.**

The cross products are equal, so the ratios form a proportion.

b. $\frac{0.3}{1.5}, \frac{0.5}{2.0}$

$\frac{0.3}{1.5} \stackrel{?}{=} \frac{0.5}{2.0}$	**Original proportion**
$0.3(2.0) \stackrel{?}{=} 1.5(0.5)$	**Cross products**
$0.6 \neq 0.75$ ✗	**Simplify.**

The cross products are not equal, so the ratios do not form a proportion.

Check Your Progress

2A. $\frac{0.2}{1.8}, \frac{1}{0.9}$

2B. $\frac{15}{36}, \frac{35}{42}$

Personal Tutor glencoe.com

StudyTip

Cross Products When you find cross products, you are said to be *cross multiplying*.

Math *in Motion*, BrainPOP® glencoe.com

Solve Proportions To solve proportions, use cross products.

EXAMPLE 3 Solve a Proportion

Solve each proportion. If necessary, round to the nearest hundredth.

a. $\frac{x}{10} = \frac{3}{5}$

$\frac{x}{10} = \frac{3}{5}$	Original proportion
$x(5) = 10(3)$	Find the cross products.
$5x = 30$	Simplify.
$\frac{5x}{5} = \frac{30}{5}$	Divide each side by 5.
$x = 6$	Simplify.

b. $\frac{x-2}{14} = \frac{2}{7}$

$\frac{x-2}{14} = \frac{2}{7}$	Original proportion
$(x-2)7 = 14(2)$	Find the cross products.
$7x - 14 = 28$	Simplify.
$7x = 42$	Add 14 to each side.
$x = 6$	Divide each side by 7.

Check Your Progress

3A. $\frac{r}{8} = \frac{25}{40}$

3B. $\frac{x+4}{5} = \frac{3}{8}$

Personal Tutor glencoe.com

The ratio of two measurements having different units of measure is called a **rate**. For example, a price of $9.99 per 10 songs is a rate. A rate that tells how many of one item is being compared to 1 of another item is called a **unit rate**.

Real-World Career

Retail Buyer
A retail buyer purchases goods for stores, primarily from wholesalers, for resale to the general public. Buyers use math to determine the amount of each product to order. A bachelor's degree with an emphasis on business studies is usually required.

Real-World EXAMPLE 4 Rate of Growth

RETAIL In the past two years, a retailer has opened 232 stores. If the rate of growth remains constant, how many stores will the retailer open in the next 3 years?

Understand Let r represent the number of retail stores.

Plan Write a proportion for the problem.

$$\frac{232 \text{ retail stores}}{2 \text{ years}} = \frac{r \text{ retail stores}}{3 \text{ years}}$$

Solve

$\frac{232}{2} = \frac{r}{3}$	Original proportion
$232(3) = 2r$	Find the cross products.
$696 = 2r$	Simplify.
$\frac{696}{2} = \frac{2r}{2}$	Divide each side by 2.
$348 = r$	Simplify.

It will open 348 stores in 3 years.

Check If the clothing retailer continues to open 232 stores every 2 years, then in the next 3 years, it will open 348 stores.

Check Your Progress

4. **EXERCISE** It takes 7 minutes for Isabella to walk around the gym track twice. At this rate, how many times can Isabella walk around the track in a half hour?

Personal Tutor glencoe.com

A **scale** is used when making a model of something that is too large or too small to be convenient at actual size. The scale compares the model to the actual size of the object using a proportion. A **scale model** is a three-dimensional reproduction of an item that has been reduced or increased in size proportionally.

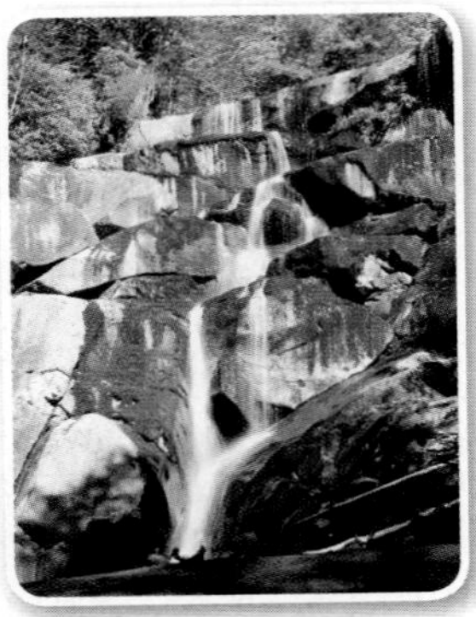

Real-World Link

The Great Smoky Mountains National Park in Tennessee is home to several waterfalls. The Ramsey Cascades is 100 feet tall. It is the tallest in the park.

Source: National Park Service

Math in Motion, Interactive Lab glencoe.com

Real-World EXAMPLE 5 Scale and Scale Models

MOUNTAIN TRAIL The scale on a map of the Great Smoky Mountains National Park is 3 inches = 10 miles. The length of the Ramsey Cascades Trail is about $1\frac{1}{8}$ inches on the map. What is the actual length of the trail?

Let ℓ represent the actual length.

scale → $\frac{3}{10} = \frac{1\frac{1}{8}}{\ell}$ ← scale
actual → ← actual

$3(\ell) = 1\frac{1}{8}(10)$ **Find the cross products.**

$3\ell = \frac{45}{4}$ **Simplify.**

$3\ell \div 3 = \frac{45}{4} \div 3$ **Divide each side by 3.**

$\ell = \frac{15}{4}$ or $3\frac{3}{4}$ **Simplify.**

The actual length is about $3\frac{3}{4}$ miles.

Check Your Progress

5. **AIRPLANES** On a model airplane, the scale is 5 centimeters = 2 meters. If the wingspan of the scale model is 28.5 centimeters, what is the actual wingspan?

Personal Tutor glencoe.com

Check Your Understanding

Examples 1 and 2 pp. 111–112

Determine whether each pair of ratios are equivalent ratios. Write *yes* or *no*.

1. $\frac{3}{7}, \frac{9}{14}$
2. $\frac{7}{8}, \frac{42}{48}$
3. $\frac{2.8}{4.4}, \frac{1.4}{2.1}$

Examples 3 p. 113

Solve each proportion. If necessary, round to nearest hundredth.

4. $\frac{n}{9} = \frac{6}{27}$
5. $\frac{4}{u} = \frac{28}{35}$
6. $\frac{3}{8} = \frac{b}{10}$

Example 4 p. 113

7. **RACE** Jennie ran the first 6 miles of a marathon in 58 minutes. If she is able to maintain the same pace, how long will it take her to finish the 26.2 miles?

Example 5 p. 114

8. **MAPS** On a map of North Carolina, Raleigh and Asheville are about 8 inches apart. If the scale is 1 inch = 12 miles, how far apart are the cities?

Practice and Problem Solving

= **Step-by-Step Solutions** begin on page R12.
Extra Practice begins on page 815.

Examples 1 and 2 pp. 111–112

Determine whether each pair of ratios are equivalent ratios. Write *yes* or *no*.

9. $\frac{9}{11}, \frac{81}{99}$
10. $\frac{3}{7}, \frac{18}{42}$
11. $\frac{8.4}{9.2}, \frac{8.8}{9.6}$
12. $\frac{4}{3}, \frac{6}{8}$
13. $\frac{29.2}{10.4}, \frac{7.3}{2.6}$
14. $\frac{39.68}{60.14}, \frac{6.4}{9.7}$

Example 3 p. 113

Solve each proportion. If necessary, round to the nearest hundredth.

15. $\frac{3}{8} = \frac{15}{a}$
16. $\frac{t}{2} = \frac{6}{12}$
17. $\frac{4}{9} = \frac{13}{q}$
18. $\frac{15}{35} = \frac{g}{7}$
19. $\frac{7}{10} = \frac{m}{14}$
20. $\frac{8}{13} = \frac{v}{21}$
21. $\frac{w}{2} = \frac{4.5}{6.8}$
22. $\frac{1}{0.19} = \frac{12}{n}$
23. $\frac{2}{0.21} = \frac{8}{n}$
24. $\frac{2.4}{3.6} = \frac{k}{1.8}$
25. $\frac{t}{0.3} = \frac{1.7}{0.9}$
26. $\frac{7}{1.066} = \frac{z}{9.65}$
27. $\frac{x-3}{5} = \frac{6}{10}$
28. $\frac{7}{x+9} = \frac{21}{36}$
29. $\frac{10}{15} = \frac{4}{x-5}$

Example 4 p. 113

30. **CAR WASH** The B-Clean Car Wash washed 128 cars in 3 hours. At that rate, how many cars can they wash in 8 hours?

Example 5 p. 114

31. **MENU** On Monday, a restaurant made $545 from selling 110 hamburgers. If they sold 53 hamburgers on Tuesday, how much did they make?

32. **MODELS** An artist used interlocking building blocks to build a scale model of Kennedy Space Center, Florida. In the model, 1 inch equals 1.67 feet of an actual space shuttle. The model is 110.3 inches tall. How tall is the actual space shuttle? Round to the nearest tenth.

33. **GEOGRAPHY** On a map of Florida, the distance between Jacksonville and Tallahassee is 7.5 centimeters. If 2 centimeters = 40 miles, what is the distance between the two cities?

Solve each proportion. If necessary, round to the nearest hundredth.

34. $\frac{6}{14} = \frac{7}{x-3}$
35. $\frac{7}{4} = \frac{f-4}{8}$
36. $\frac{3-y}{4} = \frac{1}{9}$
37. $\frac{4v+7}{15} = \frac{6v+2}{10}$
38. $\frac{9b-3}{9} = \frac{5b+5}{3}$
39. $\frac{2n-4}{5} = \frac{3n+3}{10}$

40. **ATHLETES** At Piedmont High School, 3 out of every 8 students are athletes. If there are 1280 students at the school, how many are not athletes?

41. **BRACES** Two out of five students in the ninth grade have braces. If there are 325 students in the ninth grade, how many have braces?

42. **PAINT** Joel used a half gallon of paint to cover 84 square feet of wall. He has 932 square feet of wall to paint. How many gallons of paint should he purchase?

Real-World Link

At a drive-in theater, the film is projected onto a large outdoor screen while the moviegoers sit in their cars. The sound for the film is broadcast on a radio station. In 1987, there were over 2000 drive-in theaters. By 1989, half of them had closed.

Source: North American Theater Owners

43. **MOVIE THEATERS** Use the table at the right.

a. Write a ratio of the number of indoor theaters to the total number of theaters for each year.

b. Do any two of the ratios you wrote for part a form a proportion? If so, explain the real-world meaning of the proportion.

Year	Indoor	Drive-In	Total
2000	35,567	683	36,250
2001	34,490	683	35,173
2002	35,170	666	35,836
2003	35,361	634	35,995
2004	36,012	640	36,652
2005	37,092	648	37,740
2006	37,776	649	38,425

Source: North American Theater Owners

44. **DIARIES** In a survey, 36% of the students said that they kept an electronic diary. There were 900 students who kept an electronic diary. How many students were in the survey?

45. **MULTIPLE REPRESENTATIONS** In this problem, you will explore how changing the lengths of the sides of a shape by a factor changes the perimeter of that shape.

a. **GEOMETRIC** Draw a square *ABCD*. Measure and label the sides. Draw a second square *MNPQ* with sides twice as long as *ABCD*. Draw a third square *FGHJ* with sides half as long as *ABCD*.

b. **TABULAR** Complete the table below using the appropriate measures.

ABCD		*MNPQ*		*FGHJ*	
Side length		Side length		Side length	
Perimeter		Perimeter		Perimeter	

c. **VERBAL** Make a conjecture about the change in the perimeter of a square if the side length is increased or decreased by a factor.

H.O.T. Problems

Use Higher-Order Thinking Skills

46. **OPEN ENDED** Write a real-life example of a ratio.

47. **REASONING** Compare and contrast ratios and rates.

48. **CHALLENGE** If $\frac{a+1}{b-1} = \frac{5}{1}$ and $\frac{a-1}{b+1} = \frac{1}{1}$, find the value of $\frac{b}{a}$. (*Hint:* Choose different values of a and b for which the proportions are true and evaluate the expression $\frac{b}{a}$.)

49. **FIND THE ERROR** Tim and Aisha are solving the following problem. Is either of them correct? Explain.

Two years ago, 78 women were enrolled in a dance class, while 162 men were enrolled. This year 193 men enrolled, while the ratio of women to men did not change. How many women enrolled this year?

Tim

$$\frac{78}{162} = \frac{193}{x}$$
$$78x = (162)(193)$$
$$78x = 31{,}266$$
$$x \approx 400.8$$

Aisha

$$\frac{162}{78} = \frac{x}{193}$$
$$162(193) = 78x$$
$$31{,}266 = 78x$$
$$x = 400.8$$

50. **WRITING IN MATH** Describe how businesses can use ratios. Write about a real-world situation in which a business would use a ratio.

Standardized Test Practice

51. In the figure, $x : y = 2 : 3$ and $y : z = 3 : 5$. If $x = 10$, find the value of z.

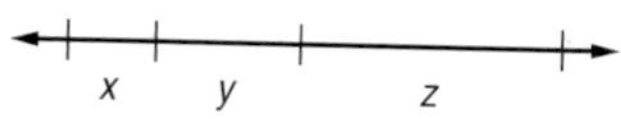

A 15
B 20
C 25
D 30

52. GRIDDED RESPONSE A race car driver records the finishing times for recent practice trials.

Trial	Time (seconds)
1	5.09
2	5.10
3	4.95
4	4.91
5	5.05

What is the mean time, in seconds, for the trials?

53. GEOMETRY If $\angle LMN$ is similar to $\angle LPO$, what is z?

F 240
G 140
H 120
J 70

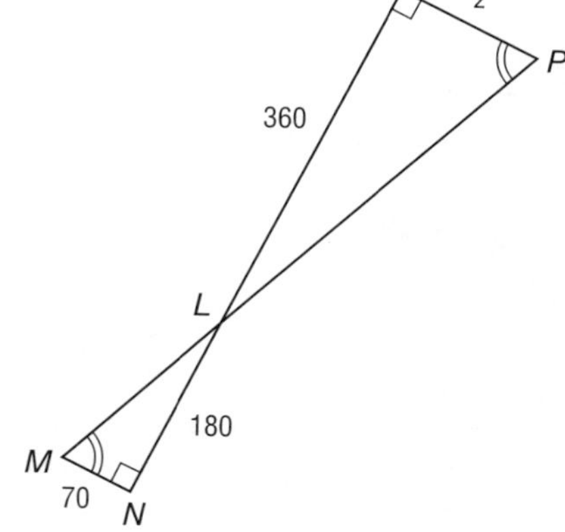

54. Which equation below illustrates the Commutative Property?

A $(3x + 4y) + 2z = 3x + (4y + 2z)$
B $7(x + y) = 7x + 7y$
C $xyz = yxz$
D $x + 0 = x$

Spiral Review

Solve each equation. (Lesson 2-5)

55. $|x + 5| = -8$

56. $|b + 9| = 2$

57. $|2p - 3| = 17$

58. $|5c - 8| = 12$

59. HEALTH When exercising, a person's pulse rate should not exceed a certain limit. This maximum rate is represented by the expression $0.8(220 - a)$, where a is age in years. Find the age of a person whose maximum pulse rate is 152. (Lesson 2-4)

Solve each equation. Check your solution. (Lesson 2-3)

60. $15 = 4a - 5$

61. $7g - 14 = -63$

62. $9 + \frac{y}{5} = 6$

63. $\frac{t}{8} - 6 = -12$

64. GEOMETRY Find the area of $\triangle ABC$ if each small triangle has a base of 5.2 inches and a height of 4.5 inches. (Lesson 1-3)

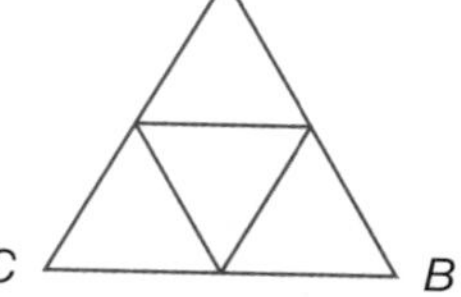

Evaluate each expression. (Lesson 1-2)

65. $3 + 16 \div 8 \cdot 5$

66. $4^2 \cdot 3 - 5(6 + 3)$

Skills Review

Solve each equation. (Lesson 2-2)

67. $4p = 22$

68. $5h = 33$

69. $1.25y = 4.375$

70. $9.8m = 30.87$

Spreadsheet Lab
Financial Ratios

Math Online glencoe.com
Graphing Technology Personal Tutor

You can use a spreadsheet to investigate the debt-to-income ratio in mortgage lending.

ACTIVITY

Dorrie is thinking about buying a house. She has the following expenses: rent of $650, credit card monthly bills of $320, a car payment of $410, and a student loan payment of $115. Dorrie has a yearly salary of $46,500. You can use a spreadsheet to find Dorrie's debt-to-income ratio.

Step 1 Enter Dorrie's debts in column B.

Step 2 Add her debts using a function in cell B6. Go to Insert and then Function. Then choose Sum. The resulting sum of 1495 should appear in B6.

Step 3 Now insert Dorrie's salary in column C. Remember to find her monthly salary by dividing the yearly salary by 12.

A mortgage company will use her debt-to-income ratio in part to determine if Dorrie qualifies for a mortgage loan. The **debt-to-income ratio** is calculated as *how much she owes per month* divided by *how much she earns each month.*

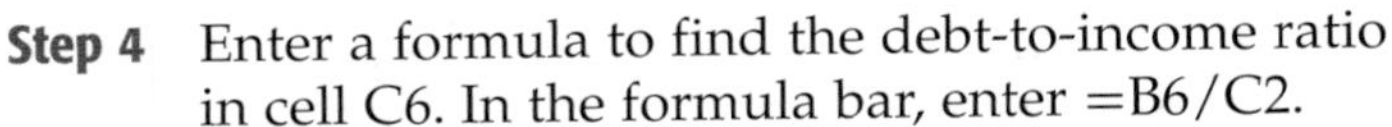

Step 4 Enter a formula to find the debt-to-income ratio in cell C6. In the formula bar, enter =B6/C2.

The ratio of about 0.39 appears. An ideal ratio would be 0.36 or less. A ratio higher than 0.36 would cause an increased interest rate or may require a higher down payment.

The spreadsheet shows a debt-to-income ratio of about 0.39. Dorrie should try to eliminate or reduce some debts or try to earn more money in order to lower her debt-to-income ratio.

Lab 2-6 B Spreadsheet.xls

◇	A	B	C
1	Type of Debt	Expenses	Salary
2	Rent	650	3875
3	Credit Cards	320	
4	Car Payment	410	
5	Student Loan	115	
6		1495	0.385806
7			

Sheet 1 Sheet 2 Sheet 3

Exercises

1. If Dorrie waits until she pays off her credit card bills to buy a house, what would be her new debt-to-income ratio?
2. Dorrie decides to reduce her monthly credit card payments to $160 per month, and she sells her car. How would her debt-to-income ratio change?
3. How could Dorrie improve her debt-to-income ratio?
4. How would your spreadsheet be different if Dorrie had income other than her monthly salary?

2-7

Percent of Change

Then

You solved proportions. (Lesson 2-6)

Now

- Find the percent of change.
- Solve problems involving percent of change.

New Vocabulary

percent of change
percent of increase
percent of decrease

Math Online

glencoe.com

- Extra Examples
- Personal Tutor
- Self-Check Quiz
- Homework Help

Why?

Every year, millions of people volunteer their time to improve their community. The difference in the number of volunteers from one year to the next can be used to determine a percent to represent the increase or decrease in volunteers.

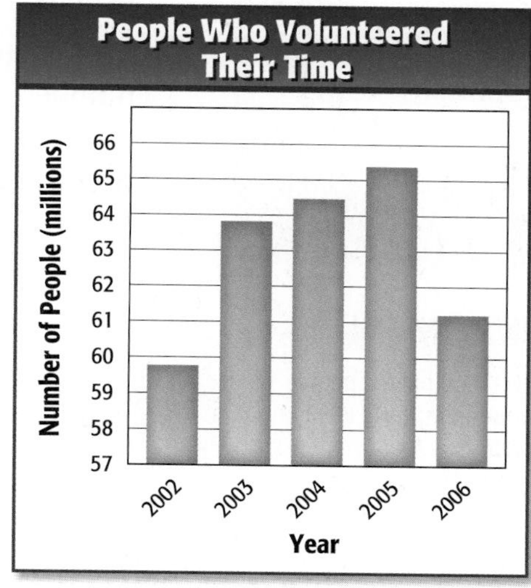

Source: Bureau of Labor and Statistics

Percent of Change

Percent of change is the ratio of the change in an amount to the original amount expressed as a percent. If the new number is greater than the original number, the percent of change is a **percent of increase.** If the new number is less than the original number, the percent of change is a **percent of decrease.**

EXAMPLE 1 Percent of Change

Determine whether each percent of change is a percent of *increase* or a percent of *decrease*. Then find the percent of change.

a. original: 20
final: 23

Subtract the original amount from the final amount to find the amount of change: $23 - 20 = 3$.

Since the new amount is greater than the original, this is a percent of increase.

Use the original number, 20, as the base.

change →, original amount →

$$\frac{3}{20} = \frac{r}{100}$$

$$3(100) = r(20)$$

$$300 = 20r$$

$$\frac{300}{20} = \frac{20r}{20}$$

$$15 = r$$

The percent of increase is 15%.

b. original: 25
final: 17

Subtract the original amount from the final amount to find the amount of change: $17 - 25 = -8$.

Since the new amount is less than the original, this is a percent of decrease.

Use the original number, 25, as the base.

change →, original amount →

$$\frac{-8}{25} = \frac{r}{100}$$

$$-8(100) = r(25)$$

$$-800 = 25r$$

$$\frac{-800}{25} = \frac{25r}{25}$$

$$-32 = r$$

The percent of decrease is 32%.

Check Your Progress

1A. original: 66
new: 30

1B. original: 9.8
new: 12.1

1C. original: 24
new: 40

1D. original: 500
new: 131

Personal Tutor glencoe.com

Real-World Link

Over 11.8 million people around the world went on a cruise in 2006.

Source: M&L Research, Inc.

Real-World EXAMPLE 2 Percent of Change

CRUISE The number of cruise ships in North America increased 18% from 2000 to 2005. If there were 192 ships in 2005, how many were there in 2000?

Let c = the number of cruise ships in 2000. Since 18% is a percent of increase, the number of cruise ships in 2000 is less than the number of ships in 2005.

change → original amount →	$\frac{192 - c}{c} = \frac{18}{100}$	**Percent proportion**
	$(192 - c)100 = 18c$	**Find the cross products.**
	$19{,}200 - 100c = 18c$	**Distibutive Property**
	$19{,}200 - 100c + 100c = 18c + 100c$	**Add 100c to each side.**
	$19{,}200 = 118c$	**Simplify.**
	$\frac{19{,}200}{118} = \frac{118c}{118}$	**Divide each side by 118.**
	$163 \approx c$	**Simplify.**

There were approximately 163 cruise ships in 2000.

Check Your Progress

2. TUITION A recent percent of increase in tuition at Northwestern University, in Evanston, Illinois, was 5.4%. If the new cost is $33,408 per year, find the original cost per year.

Personal Tutor glencoe.com

Solve Problems Two applications of percent of change are sales tax and discounts. Sales tax is an example of a percent of increase. Discount is an example of a percent of decrease.

EXAMPLE 3 Sales Tax

SHOPPING Marta is purchasing wire and beads to make jewelry. Her merchandise is $28.62 before tax. If the tax is 7.25% of the total sales, what is the cost?

Step 1 Find the tax.

The tax is 7.25% of the price of the merchandise.

7.25% of \$28.62 = 0.0725 × 28.62	**7.25% = 0.0725**
= 2.07495	**Use a calculator.**

Step 2 Find the cost with tax.

Round $2.07495 to $2.07 since tax is always rounded to the nearest cent. Add this amount to the original price: $28.62 + $2.07 = $30.69.

The total cost of Marta's jewelry supplies is $30.69.

Check Your Progress

3. SHOPPING A new DVD costs $24.99. If the sales tax is 6.85%, what is the total cost?

Personal Tutor glencoe.com

To find a discounted amount, you will follow similar steps to those for sales tax.

StudyTip

Key Words When translating a problem from word sentences to math sentences, the word "is" translates to =, and the word "of" translates to ×.

EXAMPLE 4 Discounts

DISCOUNT Since Tyrell has earned good grades in school, he qualifies for the Good Student Discount on his car insurance. His monthly payment without the discount is $85. If the discount is 20%, what will he pay each month?

Step 1 Find the discount.

The discount is 20% of the original payment.

20% of \$85 = 0.20 × 85 — **20% = 0.20**

= 17 — **Use a calculator.**

Step 2 Find the cost after discount.

Subtract \$17 from the original payment: \$85 − \$17 = \$68.

With the Good Student Discount, Tyrell will pay \$68 per month.

Check Your Progress

4. SALES A picture frame originally priced at \$14.89 is on sale for 40% off. What is the discounted price?

Personal Tutor glencoe.com

Check Your Understanding

Example 1 p. 119

State whether each percent of change is a percent of *increase* or a percent of *decrease*. Then find the percent of change. Round to the nearest whole percent.

1. original: 78
   new: 125
2. original: 41
   new: 24
3. original: 6 candles
   new: 8 candles
4. original: 35 computers
   new: 32 computers

Example 2 p. 120

5. **GEOGRAPHY** The distance from Phoenix to Tucson is 120 miles. The distance from Phoenix to Flagstaff is about 21.7% longer. To the nearest mile, what is the distance from Phoenix to Flagstaff?

Example 3 p. 120

Find the total price of each item.

6. dress: \$22.50
   sales tax: 7.5%
7. video game: \$35.99
   sales tax: 6.75%
8. **PROM** A limo costs \$85 to rent for 3 hours plus a 7% sales tax. What is the total cost to rent a limo for 6 hours?
9. **GAMES** A computer game costs \$49.95 plus a 6.25% sales tax. What is the total cost of the game?

Example 4 p. 121

Find the discounted price of each item.

10. guitar: \$95.00
    discount: 15%
11. DVD: \$22.95
    discount: 25%
12. **SKATEBOARD** A skateboard costs \$99.99. If you have a coupon for 20% off, how much will you save?
13. **TICKETS** Tickets to the county fair are \$8 for an adult and \$5 for a child. If you have a 15% discount card, how much will 2 adult tickets and 2 child tickets cost?

Practice and Problem Solving

● = **Step-by-Step Solutions** begin on page R12.
Extra Practice begins on page 815.

Example 1 p. 119

State whether each percent of change is a percent of *increase* or a percent of *decrease*. Then find the percent of change. Round to the nearest whole percent.

14. original: 35
new: 40

15 original: 16
new: 10

16. original: 27
new: 73

17. original: 92
new: 21

18. original: 21.2 grams
new: 10.8 grams

19. original: 11 feet
new: 25 feet

20. original: \$68
new: \$76

21. original: 21 hours
new: 40 hours

Example 2 p. 120

22. GASOLINE The average cost of regular gasoline in North Carolina increased by 73% from 2006 to 2007. If the average cost of a gallon of gas in 2006 was \$2.069, what was the average cost in 2007? Round to the nearest cent.

23. CARS Beng is shopping for a car. The cost of a new car is \$15,500. This is 25% greater than the cost of a used car. What is the cost of the used car?

Example 3 p. 120

Find the total price of each item.

24. messenger bag: \$28.00
tax: 7.25%

25. software: \$45.00
tax: 5.5%

26. vase: \$5.50
tax: 6.25%

27. book: \$25.95
tax: 5.25%

28. magazine: \$3.50
tax: 5.75%

29. pillow: \$9.99
tax: 6.75%

Example 4 p. 121

Find the discounted price of each item.

30. computer: \$1099.00
discount: 25%

31. CD player: \$89.99
discount: 15%

32. athletic shoes: \$59.99
discount: 40%

33. jeans: \$24.50
discount: 33%

34. jacket: \$125.00
discount: 25%

35. belt: \$14.99
discount: 20%

Find the final price of each item.

36. sweater: \$14.99
discount: 12%
tax: 6.25%

37. printer: \$60.00
discount: 25%
tax: 6.75%

38. board game: \$25.00
discount: 15%
tax: 7.5%

39. CONSUMER PRICE INDEX An *index* measures the percent change of a value from a base year. An index of 115 means that there was a 15% increase from the base year. In 2000, the consumer price index of dairy products was 160.7. In 2005, it was 182.4. Determine the percent of change.

40. FINANCIAL LITERACY The current price of each share of a technology company is \$135. If this represents a 16.2% increase over the past year, what was the price per share a year ago?

41. SHOPPING A group of girls are shopping for dresses to wear to the spring dance. One finds a dress priced \$75 with a 20% discount. A second girl finds a dress priced \$85 with a 30% discount.

a. Find the amount of discount for each dress.

b. Which girl is getting the better price for the dress?

42. RECREATIONAL SPORTS In 1995, there were 73,567 youth softball teams. By 2007, there were 86,049. Determine the percent of increase.

Real-World Link

Softball became an Olympic event in 1996. The American women, winning the gold medal in 2004, capped a 79-game winning streak.

Source: *USA TODAY*

43 **GROCERIES** Which grocery item had the greatest percent increase in cost from 2000 to 2005?

Average Retail Prices of Selected Grocery Items		
Grocery Item	Cost in 2000 ($ per pound)	Cost in 2005 ($ per pound)
milk (gallon)	2.79	3.24
eggs (dozen)	0.96	1.35
chicken (whole)	1.08	1.06
ground beef	1.63	2.30
apples	0.82	0.97
iceberg lettuce	0.85	0.85
peanut butter	1.96	1.70

Source: Statistical Abstract of the United States

44. **MULTIPLE REPRESENTATIONS** In this problem, you will explore patterns in percentages.

a. **TABULAR** Copy and complete the following table.

1% of	500	is 5.	100% of		is 20.		% of 80 is 20.
2% of		is 5.	50% of		is 20.		% of 40 is 20.
4% of		is 5.	25% of		is 20.		% of 20 is 20.
8% of		is 5.	12.5% of		is 20.		% of 10 is 20.

b. **VERBAL** Describe the patterns in the second and fifth columns.

c. **ANALYTICAL** Use the patterns to write the fifth row of the table.

H.O.T. Problems

Use Higher-Order Thinking Skills

45. **OPEN ENDED** Write a real-world problem to find the total price of an item including sales tax.

Real-World Link

In a recent year, the amount of sales tax varied across the United States from 0% to 7.25%.

Source: Federation of Tax Administrators

46. **REASONING** If you have 75% of a number n, what percent of decrease is it from the number n? If you have 40% of a number a, what percent of decrease do you have from the number a? What pattern do you notice? Is this always true?

47. **FIND THE ERROR** Maddie and Xavier are solving for the percent change if the original amount was \$25 and the new amount is \$28. Is either of them correct? Explain your reasoning.

Maddie

$$\frac{3}{28} = \frac{r}{100}$$
$$3(100) = 28r$$
$$300 = 28r$$
$$10.7 = r$$

Xavier

$$\frac{3}{25} = \frac{r}{100}$$
$$3(100) = 25r$$
$$300 = 25r$$
$$12 = r$$

48. **CHALLENGE** Determine whether the following statement is *sometimes*, *always*, or *never* true. *The percent of change is less than 100%.*

49. **WRITING IN MATH** Explain how to find a percent of change between two values and how to determine whether the change is a percent of increase or decrease.

Standardized Test Practice

50. **GEOMETRY** The rectangle has a perimeter of P centimeters. Which equation could be used to find the length ℓ of the rectangle?

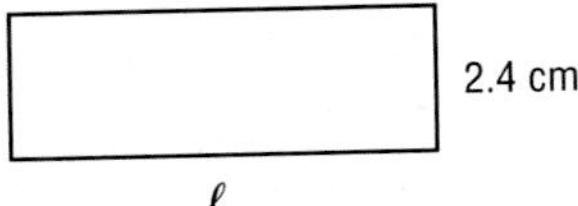

A $P = 2.4\ell$ C $P = 2.4 + 2\ell$
B $P = 4.8 + \ell$ D $P = 4.8 + 2\ell$

51. **SHORT RESPONSE** Henry is painting a room with four walls that are 12 feet by 14 feet. A gallon of paint costs $18 and covers 350 square feet. If he uses two coats of paint, how much will it cost him to paint the room?

52. The number of students at Franklin High School increased from 840 to 910 over a 5-year period. What was the percent of increase?

F 8.3%
G 14.0%
H 18.5%
J 92.3%

53. **PROBABILITY** Two dice are rolled. What is the probability that the sum is 10?

A $\frac{1}{6}$ B $\frac{1}{3}$ C $\frac{1}{12}$ D $\frac{1}{36}$

Spiral Review

54. **TRAVEL** The Chan's minivan requires 5 gallons of gasoline to travel 120 miles. How many gallons of gasoline will they need to travel 360 miles? (Lesson 2-6)

Evaluate each expression if $x = -2$, $y = 6$, and $z = 4$. (Lesson 2-5)

55. $|3 - x| + 7$
56. $12 - |z + 9|$
57. $|y + x| - z + 4$

Solve each equation. Round to the nearest hundredth. Check your solution. (Lesson 2-4)

58. $1.03p - 4 = -2.15p + 8.72$
59. $18 - 3.8t = 7.36 - 1.9t$
60. $5.4w + 8.2 = 9.8w - 2.8$
61. $2[d + 3(d - 1)] = 18$

Solve each equation. Check your solution. (Lesson 2-3)

62. $5n + 6 = -4$
63. $-11 = 7 + 3c$
64. $15 = 4a - 5$
65. $-14 + 7g = -63$

66. **RIVERS** The Congo River in Africa is 2900 miles long. That is 310 miles longer than the Niger River, which is also in Africa. (Lesson 2-2)
   a. Write an equation you could use to find the length of the Niger River.
   b. What is the length of the Niger River?

67. **GEOMETRY** Two perpendicular lines meet to form four right angles. Write two different if-then statements for this definition. (Lesson 1-8)

Skills Review

Translate each equation into a sentence. (Lesson 2-1)

68. $d - 14 = 5$
69. $2f + 6 = 19$
70. $y - 12 = y + 8$
71. $3a + 5 = 27 - 2a$
72. $-6c^2 - 4c = 25$
73. $d^4 + 64 = 3d^3 + 77$

EXTEND 2-7

Algebra Lab
Percentiles

Objective
You will use percentiles to represent data.

A **percentile** is a measure that is often used to report test data, such as standardized test scores. It tells us what percent of the total scores were below a given score.

- Percentiles measure rank from the bottom.
- There is no 0 percentile rank. The lowest score is at the 1st percentile.
- There is no 100th percentile rank. The highest score is at the 99th percentile.

ACTIVITY

A talent show was held for the fifteen finalists in the Teen Idol contest. Each performer received a score from 0 through 30 with 30 being the highest.

Name	Score	Name	Score
Arnold	17	Malik	10
Benito	9	Natalie	26
Carmen	21	Pearl	4
Delia	29	Twyla	6
Fernando	15	Victor	28
Horatio	5	Warren	22
Ingrid	11	Yolanda	18
Ishi	27		

Step 1 Write one score on each of 15 slips of paper.

Step 2 Arrange the slips vertically from greatest to least score.

Step 3 Find Victor's percentile rank.

Victor had a score of 28. There are 13 scores below his score. To find his percentile rank, use the following formula:

$$\frac{\text{number of scores below 28}}{\text{total number of scores}} \cdot 100 = \frac{13}{15} \cdot 100 \text{ or about } 87.$$

Victor scored at the 87th percentile in the contest.

Analyze the Results

1. Find the median, lower quartile, and upper quartile of the scores.
2. Which performer was at the 50th percentile? Which performer was at the 25th percentile? the 75th percentile?
3. Compare and contrast the values for the median, lower quartile, and upper quartile and the scores for the 25th, 50th, and 75th percentiles.
4. While Victor scored at the 87th percentile, what percent of the 30 possible points did he score?
5. Compare and contrast the percentile rank and the percent score.
6. Are there any outliers in the data that could alter the results of our computations?

2-8 Literal Equations and Dimensional Analysis

Then

You solved equations with variables on each side. (Lesson 2-4)

Now

- Solve equations for given variables.
- Use formulas to solve real-world problems.

New Vocabulary

literal equation
dimensional analysis
unit analysis

Math Online

glencoe.com

- Extra Examples
- Personal Tutor
- Self-Check Quiz
- Homework Help

Why?

Each year, more people use credit cards to make everyday purchases. If the entire balance is not paid by the due date, compound interest is applied. The formula for computing the balance of an account with compound interest added annually is $A = P(1 + r)$.

- A represents the amount of money in the account including the interest,
- P is the amount in the account before interest is added,
- r is the interest rate written as a decimal.

Solve for a Specific Variable Some equations such as the one above contain more than one variable. At times, you will need to solve these equations for one of the variables.

EXAMPLE 1 Solve for a Specific Variable

Solve $4m - 3n = 8$ for m.

$4m - 3n = 8$	Original equation
$4m - 3n + 3n = 8 + 3n$	Add $3n$ to each side.
$4m = 8 + 3n$	Simplify.
$\frac{4m}{4} = \frac{8 + 3n}{4}$	Divide each side by 4.
$m = \frac{8}{4} + \frac{3}{4}n$	Simplify.
$m = 2 + \frac{3}{4}n$	Simplify.

Check Your Progress

Solve each equation for the variable indicated.

1A. $15 = 3n + 6p$, for n

1B. $\frac{k - 2}{5} = 11j$, for k

1C. $28 = t(r + 4)$, for t

1D. $a(q - 8) = 23$, for q

Personal Tutor glencoe.com

Sometimes we need to solve equations for a variable that is on both sides of the equation. When this happens, you must get all terms with that variable onto one side of the equation. It is then helpful to use the Distributive Property to isolate the variable for which you are solving.

StudyTip

Solving for a Specific Variable When an equation has more than one variable, it can be helpful to highlight the variable for which you are solving on your paper.

EXAMPLE 2 Solve for a Specific Variable

Solve $3x - 2y = xz + 5$ for x.

$3x - 2y = xz + 5$	**Original equation**
$3x - 2y + 2y = xz + 5 + 2y$	**Add $2y$ to each side.**
$3x - xz = xz - xz + 5 + 2y$	**Subtract xz from each side.**
$3x - xz = 5 + 2y$	**Simplify.**
$x(3 - z) = 5 + 2y$	**Distributive Property**
$\frac{x(3 - z)}{3 - z} = \frac{5 + 2y}{3 - z}$	**Divide each side by $3 - z$.**
$x = \frac{5 + 2y}{3 - z}$	**Simplify.**

Since division by 0 is undefined, $3 - z \neq 0$ so $z \neq 3$.

Check Your Progress

Solve each equation for the variable indicated.

2A. $d + 5c = 3d - 1$, for d

2B. $6q - 18 = qr + t$, for q

Personal Tutor glencoe.com

Use Formulas A formula or equation that involves several variables is called a **literal equation**. To solve a literal equation, apply the process of solving for a specific variable.

Real-World Link

The largest yo-yo in the world is 32.7 feet in circumference. It was launched by crane from a height of 189 feet.

Source: *Guinness Book of World Records*

Real-World EXAMPLE 3 Use Literal Equations

YO-YOS Use the information about the largest yo-yo at the left. The formula for the circumference of a circle is $C = 2\pi r$, where C represents circumference and r represents radius.

a. Solve the formula for r.

$C = 2\pi r$	**Formula for circumference**
$\frac{C}{2\pi} = \frac{2\pi r}{2\pi}$	**Divide each side by 2π.**
$\frac{C}{2\pi} = r$	**Simplify.**

b. Find the radius of the yo-yo.

$\frac{C}{2\pi} = r$	**Formula for radius**
$\frac{32.7}{2\pi} = r$	**$C = 32.7$**
$5.2 \approx r$	**Use a calculator.**

The yo-yo has a radius of about 5.2 feet.

Check Your Progress

3. GEOMETRY The formula for the volume of a rectangular prism is $V = \ell wh$, where ℓ is the length, w is the width, and h is the height.

A. Solve the formula for w.

B. Find the width of a rectangular prism that has a volume of 79.04 cubic centimeters, a length of 5.2 centimeters, and a height of 4 centimeters.

Personal Tutor glencoe.com

When using formulas, you may want to use dimensional analysis. **Dimensional analysis** or **unit analysis** is the process of carrying units throughout a computation.

StudyTip

Rewriting Equations with Exponents When rewriting an equation to solve for a specific variable, the exponents of a variable are moved with that variable. They are not separated.

EXAMPLE 4 Use Dimensional Analysis

RUNNING A 10K run is 10 kilometers long. If 1 meter = 1.094 yards, use dimensional analysis to find the length of the race in miles. (*Hint*: 1 mi = 1760 yd)

Since the given conversion relates meters to yards, first convert 10 kilometers to meters. Then multiply by the conversion factor such that the unit meters are divided out. To convert from yards to miles, multiply by $\frac{1 \text{ mi}}{1760 \text{ yd}}$.

length of run	×	kilometers to meters	×	meters to yards	×	yards to miles
10 km	×	$\frac{1000 \text{ m}}{1 \text{ km}}$	×	$\frac{1.094 \text{ yd}}{1 \text{ m}}$	×	$\frac{1 \text{ mi}}{1760 \text{ yd}}$

Notice how the units cancel, leaving the unit to which you are converting.

$$10 \cancel{\text{km}} \times \frac{1000 \cancel{\text{m}}}{1 \cancel{\text{km}}} \times \frac{1.094 \cancel{\text{yd}}}{1 \cancel{\text{m}}} \times \frac{1 \text{ mi}}{1760 \cancel{\text{yd}}} = \frac{10940}{1760} \text{ mi}$$

$$= \frac{10,940}{1760}$$

$$\approx 6.2 \text{ mi}$$

A 10K race is approximately 6.2 miles.

Check Your Progress

4. A car travels a distance of 100 feet in about 2.8 seconds. What is the velocity of the car in miles per hour? Round to the nearest whole number.

Personal Tutor glencoe.com

Check Your Understanding

Examples 1 and 2 pp. 126–127

Solve each equation or formula for the variable indicated.

1. $5a + c = -8a$, for a

2. $7h + f = 2h + g$, for g

3. $\frac{k + m}{-7} = n$, for k

4. $q = p(r + s)$, for p

Example 3 p. 127

5. PACKAGING A soap company wants to use a cylindrical container to hold their new liquid soap.

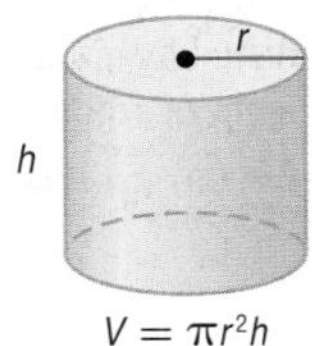

$V = \pi r^2 h$

a. Solve the formula for h.

b. What is the height of a container if the volume is 56.52 cubic inches and the radius is 1.5 inches? Round to the nearest tenth.

Example 4 p. 128

6. SHOPPING Scott found a rare video game on an online auction site priced at 35 Australian dollars. If the exchange rate is $1 U.S. = $1.24 Australian, find the cost of the game in United States dollars. Round to the nearest cent.

7. PHOTOGRAPHY A fisheye lens has a minimum focus range of 13.5 centimeters. If 1 centimeter is equal in length to about 0.39 inches, what is the minimum focus range of the lens in feet?

Practice and Problem Solving

= Step-by-Step Solutions begin on page R12. Extra Practice begins on page 815.

Examples 1 and 2 pp. 126–127

Solve each equation or formula for the variable indicated.

8. $u = vw + z$, for v

9. $x = b - cd$, for c

10. $fg - 9h = 10j$, for g

11. $10m - p = -n$, for m

12. $r = \frac{2}{3}t + v$, for t

13. $\frac{5}{9}v + w = z$, for v

14. $\frac{10ac - x}{11} = -3$, for a

15. $\frac{df + 10}{6} = g$, for f

Example 3 p. 127

16. FITNESS The formula to compute a person's body mass index is $B = 703 \cdot \frac{w}{h^2}$. B represents the body mass index, w is the person's weight in pounds, and h represents the person's height in inches.

a. Solve the formula for w.

b. What is the weight to the nearest pound of a person who is 64 inches tall and has a body mass index of 21.45?

17. PHYSICS Acceleration is the measure of how fast a velocity is changing. The formula for acceleration is $a = \frac{v_f - v_i}{t}$. a represents the acceleration rate, v_f is the final velocity, v_i is the initial velocity, and t represents the time in seconds.

a. Solve the formula for v_f.

b. What is the final velocity of a runner who is accelerating at 2 feet per second squared for 3 seconds with an initial velocity of 4 feet per second?

Example 4 p. 128

18. SWIMMING If each lap in a pool is 100 meters long, how many laps equal one mile? Round to the nearest tenth. (*Hint*: 1 foot ≈ 0.3048 meter)

19. GASOLINE How many liters of gasoline are needed to fill a 13.2-gallon tank? There are about 1.06 quarts per 1 liter. Round to the nearest tenth.

Solve each equation or formula for the variable indicated.

20. $-14n + q = rt - 4n$, for n

21. $18t + 11v = w - 13t$, for t

22. $ax + z = aw - y$, for a

23. $10c - f = -13 + cd$, for c

Select an appropriate unit from the choices below and convert the rate to that unit.

ft/s	mph	mm/s	km/s

24. a car traveling at 36 ft/s

25. a snail moving at 3.6 m/h

26. a person walking at 3.4 mph

27. a satellite moving at 234,000 m/min

28. DANCING The formula $P = \frac{1.2W}{H^2}$ represents the amount of pressure exerted on the floor by a ballroom dancer's heel. In this formula, P is the pressure in pounds per square inch, W is the weight of a person wearing the shoe in pounds, and H is the width of the heel of the shoe in inches.

a. Solve the formula for W.

b. Find the weight of the dancer if the heel is 3 inches wide and the pressure exerted is 30 pounds per square inch.

Real-World Link

In most dance competitions, dancers compete as couples, or formation teams. At "Jack & Jill" competitions, individuals are paired by random draw.

Source: USA Dance

Write an equation and solve for the variable indicated.

29. Seven less than a number t equals another number r plus 6. Solve for t.

30. Ten plus eight times a number a equals eleven times another number d minus six. Solve for a.

31. Nine tenths of a number g is the same as seven plus two thirds of another number k. Solve for k.

32. Three fourths of a number p less two is five sixths of another number r plus five. Solve for r.

33 **GIFTS** Ashley has 214 square inches of paper to wrap a gift box. The surface area S of the box can be found by using the formula $S = 2w(\ell + h) + 2\ell h$, where w is the width of the box, ℓ is the length of the box, and h is the height. If the length of the box is 7 inches and the width is 6 inches, how tall can Ashley's box be?

34. **MULTIPLE REPRESENTATIONS** In this problem, you will investigate cylinders. The surface area of cylinder can be found by the formula $S = 2\pi rh + 2\pi r^2$.

a. ALGEBRAIC Solve for h. Rewrite the solution using 3.14 for π and 2500 for S.

b. TABULAR Make a table of values using your new formula to find h if $r = 20$, 15, 10, 5, and 0. Round to the nearest hundredth.

c. VERBAL What do we know about the domain (possible values of r)?

H.O.T. Problems

Use Higher-Order Thinking Skills

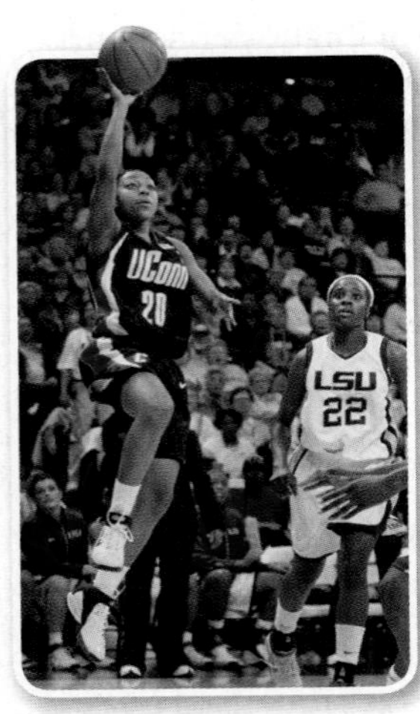

Real-World Link

In 2007, Louisiana State University lost to the University of Connecticut by a score of 71 to 72, ending a 43-game winning streak, an NCAA Women's Basketball record.

Source: National Collegiate Athletic Association

35. **CHALLENGE** The circumference of an NCAA women's basketball is 29 inches, and the rubber coating is $\frac{3}{16}$ inch thick. Use the formula $v = \frac{4}{3}\pi r^3$, where v represents the volume and r is the radius of the inside of the ball, to determine the volume of the air inside the ball. Round to the nearest whole number.

36. **REASONING** Select an appropriate unit to describe the highway speed of a car and the speed of a caterpillar crawling on a tree. Can the same unit be used for both situations? Explain.

37. **FIND THE ERROR** Sandrea and Fernando are solving $4a - 5b = 7$ for b. Is either of them correct? Explain.

Sandrea	Fernando
$4a - 5b = 7$	$4a - 5b = 7$
$-5b = 7 - 4a$	$5b = 7 - 4a$
$\frac{-5b}{-5} = \frac{7 - 4a}{-5}$	$\frac{5b}{5} = \frac{7 - 4a}{5}$
$b = \frac{7 - 4a}{-5}$	$b = \frac{7 - 4a}{5}$

38. **OPEN ENDED** Write a formula for A, the area of a geometric figure such as a triangle or rectangle. Then solve the formula for a variable other than A.

39. **CHALLENGE** Solve each equation or formula for the variable indicated.

a. $n = \frac{x + y - 1}{xy}$ for x

b. $\frac{x + y}{x - y} = \frac{1}{2}$ for y

40. **WRITING IN MATH** Explain what a literal equation is and how to solve one.

Standardized Test Practice

41. Eula is investing \$6000, part at 4.5% interest and the rest at 6% interest. If d represents the amount invested at 4.5%, which expression represents the amount of interest earned in one year by the amount paying 6%?

A $0.06d$
B $0.06(d - 6000)$
C $0.06(d + 6000)$
D $0.06(6000 - d)$

42. Todd drove from Boston to Cleveland, a distance of 616 miles. His breaks, gasoline, and food stops took 2 hours. If his trip took 16 hours altogether, what was Todd's average speed?

F 38.5 mph
G 40 mph
H 44 mph
J 47.5 mph

43. SHORT RESPONSE Brian has 3 more books than Erika. Jasmine has triple the number of books that Brian has. Altogether Brian, Erika, and Jasmine have 22 books. How many books does Jasmine have?

44. GEOMETRY Which of the following best describes a plane?

A a location having neither size nor shape
B a flat surface made up of points having no depth
C made up of points and has no thickness or width
D a boundless, three-dimensional set of all points

Spiral Review

Find the final price of each item. (Lesson 2-7)

45. lamp: \$120.00
discount: 20%
tax: 6%

46. dress: \$70.00
discount: 30%
tax: 7%

47. camera: \$58.00
discount: 25%
tax: 6.5%

48. jacket: \$82.00
discount: 15%
tax: 6%

49. comforter: \$67.00
discount: 20%
tax: 6.25%

50. lawnmower: \$720.00
discount: 35%
tax: 7%

Solve each proportion. If necessary, round to the nearest hundredth. (Lesson 2-6)

51. $\frac{3}{4.5} = \frac{x}{2.5}$

52. $\frac{2}{0.36} = \frac{7}{p}$

53. $\frac{m}{9} = \frac{2.8}{4.9}$

54. JOBS Laurie mows lawns to earn extra money. She can mow at most 30 lawns in one week. She profits \$15 on each lawn she mows. Identify a reasonable domain and range for this situation and draw a graph. (Lesson 1-6)

55. ENTERTAINMENT Each member of the pit orchestra is selling tickets for the school musical. The trombone section sold 50 floor tickets and 90 balcony tickets. Write and evaluate an expression to find how much money the trombone section collected. (Lesson 1-2)

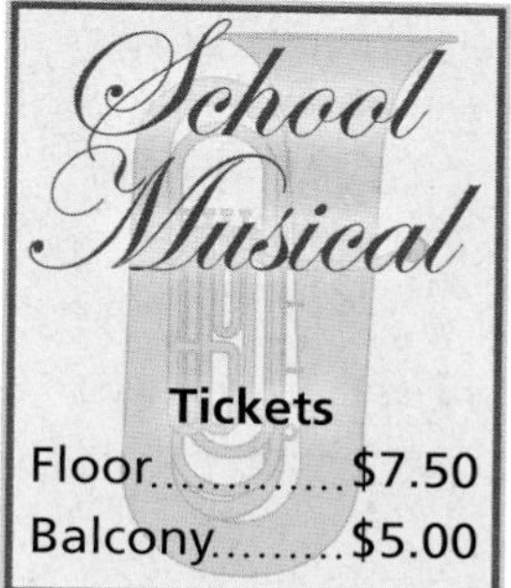

Skills Review

Solve each equation. (Lesson 2-5)

56. $8k + 9 = 7k + 6$

57. $3 - 4q = 10q + 10$

58. $\frac{3}{4}n + 16 = 2 - \frac{1}{8}n$

59. $\frac{1}{4} - \frac{2}{3}y = \frac{3}{4} - \frac{1}{3}y$

60. $4(2a - 1) = -10(a - 5)$

61. $2(w - 3) + 5 = 3(w - 1)$

2-9 Weighted Averages

Why?

Baseball players' performance is measured in large part by statistics. Slugging average (SLG) is a weighted average that measures the power of a hitter. The slugging average is calculated by using the following formula.

$$\text{SLG} = \frac{1B + (2 \times 2B) + (3 \times 3B) + (4 \times HR)}{\text{at bats}}$$

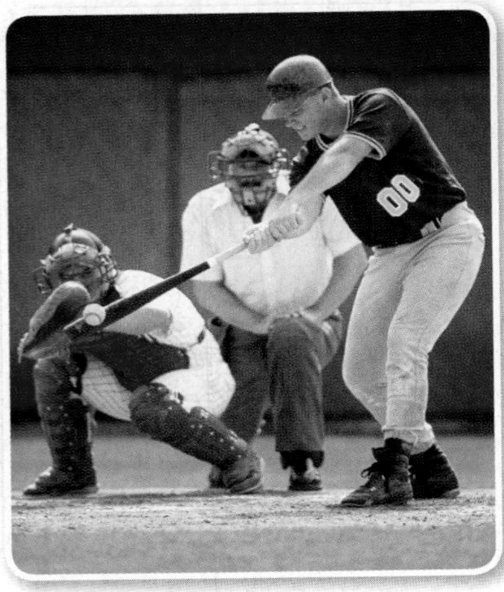

Then

You translated sentences into equations. (Lesson 2-1)

Now

- Solve mixture problems.
- Solve uniform motion problems.

New Vocabulary

weighted average
mixture problem
uniform motion problem
rate problem

Math Online

glencoe.com

- Extra Examples
- Personal Tutor
- Self-Check Quiz
- Homework Help

Weighted Averages The batter's slugging percentage is an example of a weighted average. The **weighted average** M of a set of data is the sum of the product of the number of units and the value per unit divided by the sum of the number of units.

Mixture problems are problems in which two or more parts are combined into a whole. They are solved using weighted averages. In a mixture problem, the units are usually the number of gallons or pounds and the value is the cost, value, or concentration per unit.

Real-World EXAMPLE 1 Mixture Problem

RETAIL A tea company sells blended tea for $25 per pound. To make blackberry tea, dried blackberries that cost $10.50 per pound are blended with black tea that costs $35 per pound. How many pounds of black tea should be added to 5 pounds of dried blackberries to make blackberry tea?

Step 1 Let w be the weight of the black tea. Make a table to organize the information.

	Number of Units (lb)	Price per Unit ($)	Total Price (price)(units)
Dried Blackberries	5	10.50	10.50(5)
Black Tea	w	35	$35w$
Blackberry Tea	$5 + w$	25	$25(5 + w)$

Write an equation using the information in the table.

Price of blackberries	plus	price of tea	equals	price of blackberry tea.
$10.50(5)$	$+$	$35w$	$=$	$25(5 + w)$

Step 2 Solve the equation.

$10.50(5) + 35w = 25(5 + w)$ **Original equation**

$52.5 + 35w = 125 + 25w$ **Distributive Property**

$52.5 + 35w - 25w = 125 + 25w - 25w$ **Subtract 25w from each side.**

$52.5 + 10w = 125$ **Simplify.**

$52.5 - 52.5 + 10w = 125 - 52.5$ **Subtract 52.5 from each side.**

$10w = 72.5$ **Simplify.**

$w = 7.25$ **Divide each side by 10.**

StudyTip

Mixture Problems When you organize the information in mixture problems, remember that the final mixture must contain the sum of the parts in the correct quantities and at the correct percents.

To make the blackberry tea, 7.25 pounds of black tea will need to be added to the dried blackberries.

Check Your Progress

1. **COFFEE** How many pounds of Premium coffee beans should be mixed with 2 pounds of Supreme coffee to make the Blend coffee?

Personal Tutor glencoe.com

Sometimes mixture problems are expressed in terms of percents.

Real-World EXAMPLE 2 Percent Mixture Problem

FRUIT PUNCH Mrs. Matthews has 16 cups of punch that is 3% pineapple juice. She also has a punch that is 33% pineapple juice. How many cups of the 33% punch will she need to add to the 3% punch to obtain a punch that is 20% pineapple juice?

Problem-SolvingTip

Make a Table Using a table is a great way to organize the given information. It also helps you understand how to write an equation to solve for the missing value.

Step 1 Let x = the amount of 33% solution to be added. Make a table.

	Amount of Punch (cups)	Amount of Pineapple Juice
3% Punch	16	0.03(16)
33% Punch	x	$0.33x$
20% Punch	$16 + x$	$0.20(16 + x)$

Write an equation using the information in the table.

Amount of pineapple juice in 3% punch	plus	amount of pineapple juice in 33% punch	equals	amount of pineapple juice in 20% punch.
$0.03(16)$	$+$	$0.33x$	$=$	$0.20(16 + x)$

Step 2 Solve the equation.

$$0.03(16) + 0.33x = 0.20(16 + x)$$ **Original equation**

$$0.48 + 0.33x = 3.2 + 0.20x$$ **Simplify.**

$$0.48 + 0.33x - 0.20x = 3.2 + 0.20x - 0.20x$$ **Subtract 0.20x from each side.**

$$0.48 + 0.13x = 3.2$$ **Simplify.**

$$0.48 - 0.48 + 0.13x = 3.2 - 0.48$$ **Subtract 0.48 from each side.**

$$0.13x = 2.72$$ **Simplify.**

$$\frac{0.13x}{0.13} = \frac{2.72}{0.13}$$ **Divide each side by 0.13.**

$$x \approx 20.9$$ **Round to the nearest tenth.**

Mrs. Matthews should add about 20.9 cups of the 33% punch to the 16 cups of the 3% punch.

Check Your Progress

2. **ANTIFREEZE** One type of antifreeze is 40% glycol, and another type of antifreeze is 60% glycol. How much of each kind should be used to make 100 gallons of antifreeze that is 48% glycol?

Personal Tutor glencoe.com

Uniform Motion Problems **Uniform motion problems** or **rate problems** are problems in which an object moves at a certain speed or rate. The formula $d = rt$ is used to solve these problems. In the formula, d represents distance, r represents rate, and t represents time.

Real-World EXAMPLE 3 Speed of One Vehicle

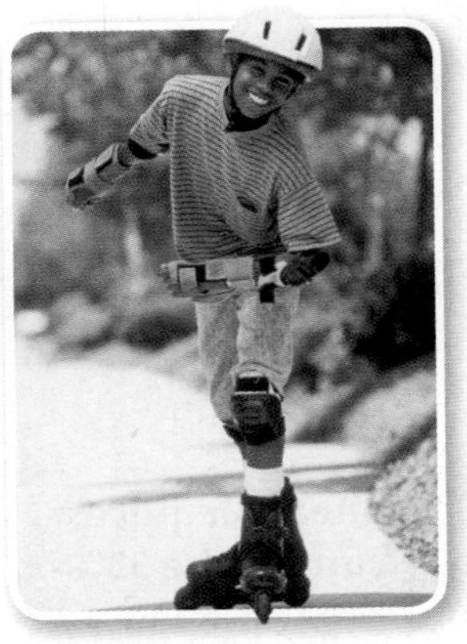

Real-World Link

In-line skating is the fourth most popular recreational activity in the U.S.

Source: *Statistical Abstract of the United States*

INLINE SKATING It took Travis and Tony 40 minutes to skate 5 miles. The return trip took them 30 minutes. What was their average speed for the trip?

Understand We know that the boys did not travel the same amount of time on each portion of their trip. So, we will need to find the weighted average of their speeds. We are asked to find their average speed for both portions of the trip.

Plan First find the rate of the going portion, and then the return portion of the trip. Because the rate is in miles per hour we convert 40 minutes to about 0.667 hours and 30 minutes to 0.5 hours.

Going

$r = \frac{d}{t}$ **Formula for rate**

$\approx \frac{5 \text{ miles}}{0.667 \text{ hour}}$ or about 7.5 miles per hour **Substitution $d = 5$ mi, $t = 0.667$ h**

Return

$r = \frac{d}{t}$ **Formula for rate**

$= \frac{5 \text{ miles}}{0.5 \text{ hour}}$ or 10 miles per hour **Substitution $d = 5$ mi, $t = 0.5$ h**

Because we are looking for a weighted average we cannot just average their speeds. We need to find the weighted average for the round trip.

Solve

$$M = \frac{(\text{rate of going})(\text{time of going}) + (\text{rate of return})(\text{time of return})}{\text{time of going} + \text{time of return}}$$

$\approx \frac{(7.5)(0.667) + (10)(0.5)}{0.667 + 0.5}$ **Substitution**

$\approx \frac{10.0025}{1.167}$ or about 8.6 **Simplify.**

Their average speed was about 8.6 miles per hour.

Check Our solution of 8.6 miles per hour is between the going portion rate, 7.5 miles per hour, and the return rate, 10 miles per hour. So, we know that our answer is reasonable.

Check Your Progress

3. **EXERCISE** Austin jogged 2.5 miles in 16 minutes and then walked 1 mile in 10 minutes. What was his average speed?

Personal Tutor glencoe.com

Math *in Motion*, Animation glencoe.com

The formula $d = rt$ can also be used to solve real-world problems involving two vehicles in motion.

StudyTip

Draw a Diagram Drawing a diagram is not just for geometry problems. You can use diagrams to visualize many problem situations that can be represented by equations.

Real-World EXAMPLE 4 Speeds of Two Vehicles

FREIGHT TRAINS Two trains are 550 miles apart heading toward each other on parallel tracks. Train A is traveling east at 35 miles per hour, while Train B travels west at 45 miles per hour. When will the trains pass each other?

Step 1 Draw a diagram.

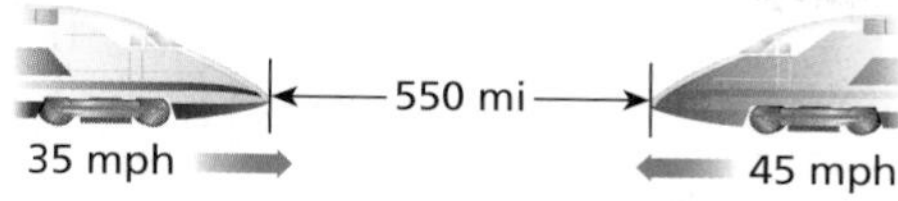

Step 2 Let t = the number of hours until the trains pass each other. Make a table.

	r	t	$d = rt$
Train A	35	t	$35t$
Train B	45	t	$45t$

Step 3 Write and solve an equation.

Distance traveled by Train A	plus	distance traveled by Train B	equals	550 miles.
$35t$	$+$	$45t$	$=$	550

$35t + 45t = 550$ **Original equation**

$80t = 550$ **Simplify.**

$\frac{80t}{80} = \frac{550}{80}$ **Divide each side by 80.**

$t = 6.875$ **Simplify.**

The trains will pass each other in about 6.875 hours.

Check Your Progress

4. CYCLING Two cyclists begin traveling in opposite directions on a circular bike trail that is 5 miles long. One cyclist travels 12 miles per hour, and the other travels 18 miles per hour. How long will it be before they meet?

Personal Tutor glencoe.com

Check Your Understanding

Example 1 p. 132

1. FOOD Tasha ordered soup and salad for lunch. If Tasha ordered 10 ounces of soup for lunch and the total cost was $3.30, how many ounces of salad did Tasha order?

15¢/ounce

20¢/ounce

Example 2 p. 133

2. CHEMISTRY Margo has 40 milliliters of 25% solution. How many milliliters of 60% solution should she add to obtain the required 30% solution?

Example 3 p. 134

3. TRAVEL A boat travels 16 miles due north in 2 hours and 24 miles due west in 2 hours. What is the average speed of the boat?

4. EXERCISE Felisa jogged 3 miles in 25 minutes and then jogged 3 more miles in 30 minutes. What was her average speed in miles per minute?

Example 4 p. 135

5. CYCLING A cyclist begins traveling 18 miles per hour. At the same time and at the same starting point, an inline skater follows the cyclist's path and begins traveling 6 miles per hour. After how much time will they be 24 miles apart?

Practice and Problem Solving

= **Step-by-Step Solutions** begin on page R12.
Extra Practice begins on page 815.

Example 1 p. 132

6. **CANDY** A candy store wants to create a mix using two hard candies. One is priced at \$5.45 per pound, and the other is priced at \$7.33 per pound. How many pounds of the \$7.33 candy should be mixed with 11 pounds of the \$5.45 candy to sell the mixture for \$6.14 per pound?

7. **BUSINESS** Party Supplies Inc. sells metallic balloons for \$2 each and helium balloons for \$3.50 per bunch. Yesterday, they sold 36 more metallic balloons than the number of bunches of helium balloons. The total sales for both types of balloons were \$281. Let b represent the number of metallic balloons sold.

a. Copy and complete the table representing the problem.

	Number	Price	Total Price
Metallic Balloons	b		
Bunches of Helium Balloons	$b - 36$		

b. Write an equation to represent the problem.

c. How many metallic balloons were sold?

d. How many bunches of helium balloons were sold?

8. **FINANCIAL LITERACY** Lakeisha spent \$4.57 on color and black-and-white copies for her project. She made 7 more black-and-white copies than color copies. How many color copies did she make?

Type of Copy	Cost per Page
color	\$0.44
black-and-white	\$0.07

Example 2 p. 133

9. **FISH** Rosamaria is setting up a 20-gallon saltwater fish tank that needs to have a salt content of 3.5%. If Rosamaria has water that has 2.5% salt and water that has 3.7% salt, how many gallons of the water with 3.7% salt content should Rosamaria use?

10. **CHEMISTRY** Hector is performing a chemistry experiment that requires 160 milliliters of 40% sulfuric acid solution. He has a 25% sulfuric acid solution and a 50% sulfuric acid solution. How many millimeters of each solution should he mix to obtain the needed solution?

Example 3 p. 134

11. **TRAVEL** A boat travels 36 miles in 1.5 hours and then 14 miles in 0.75 hour. What is the average speed of the boat?

12. **RUNNING** A runner ran 1.5 miles in 28 minutes and then 1.2 more miles in 10 minutes. What was the average speed in miles per minute?

13. **AIRLINERS** Two airliners are 1600 miles apart and heading toward each other at different altitudes. The first plane is traveling north at 620 miles per hour, while the second is traveling south at 780 miles per hour. When will the planes pass each other?

Example 4 p. 135

14. **SAILING** A ship is sailing due east at 20 miles per hour when it passes the lighthouse. At the same time a ship is sailing due west at 15 miles per hour when it passes a point. The lighthouse and the point are 175 miles apart. When will these ships pass each other?

15. **CHEMISTRY** A lab technician has 40 gallons of a 15% iodine solution. How many gallons of a 40% iodine solution must he add to make a 20% iodine solution?

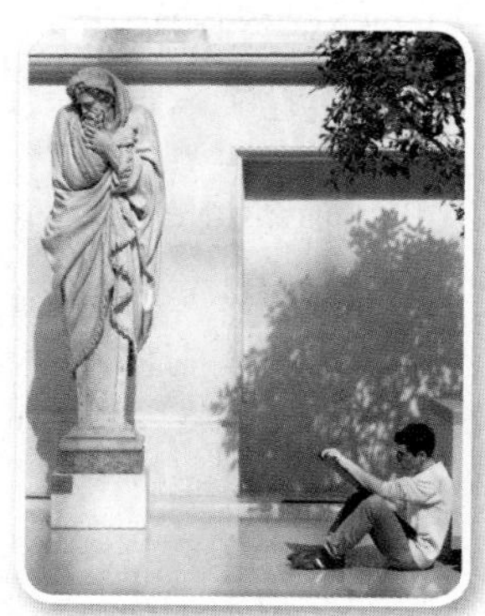

Real-World Link

Different countries have individual grading scales. For example, French schools give number grades ranging from 0 to 20, rather than letter grades like those in the U.S.

Source: Morris

16. **GRADES** At Westbridge High School, a student's grade point average (GPA) is based on the student's grade and the class credit rating. Brittany's grades for this quarter are shown. Find Brittany's GPA if a grade of A equals 4 and a B equals 3.

Class	Credit Rating	Grade
Algebra 1	1	A
Science	1	A
English	1	B
Spanish	1	A
Music	$\frac{1}{2}$	B

17. **SPORTS** In a triathlon, Steve swam 0.5 mile in 15 minutes, biked 20 miles in 90 minutes, and ran 4 miles in 30 minutes. What was Steve's average speed for the triathlon in miles per hour?

18. **MUSIC** Amalia has 10 songs on her MP3 player. If 3 songs are 5 minutes long, 3 are 4 minutes long, 2 are 2 minutes long, and 2 are 3.5 minutes long, what is the average length of the songs?

19. **DISTANCE** Garcia is driving to Florida for vacation. The trip is a total of 625 miles.
   a. How far can he drive in 6 hours at 65 miles per hour?
   b. If Garcia maintains a speed of 65 miles per hour, how long will it take him to drive to Florida?

20. **TRAVEL** Two buses leave Smithville at the same time, one traveling north and the other traveling south. The northbound bus travels at 50 miles per hour, and the southbound bus travels at 65 miles per hour. Let t represent the amount of time since their departure.
   a. Copy and complete the table representing the situation.

	r	t	$d = rt$
Northbound bus	?	?	?
Southbound bus	?	?	?

   b. Write an equation to find when the buses will be 345 miles apart.
   c. Solve the equation. Explain how you found your answer.

21. **TRAVEL** A subway travels 60 miles per hour from Glendale to Midtown. Another subway, traveling at 45 miles per hour, takes 11 minutes longer for the same trip. How far apart are Glendale and Midtown?

H.O.T. Problems

Use Higher-Order Thinking Skills

22. **OPEN ENDED** Write a problem that depicts motion in opposite directions.

23. **REASONING** Describe the conditions so that adding a 50% solution to a 100% solution would produce a 75% solution.

24. **CHALLENGE** Find five consecutive odd integers from least to greatest in which the sum of the first and the fifth is one less than three times the fourth.

25. **CHALLENGE** Describe a situation involving mixtures that could be represented by $1.00x + 0.15(36) = 0.50(x + 36)$.

26. **WRITING IN MATH** Describe how a gallon of 25% solution is added to an unknown amount of 10% solution to get a 15% solution.

Standardized Test Practice

27. If $2x + y = 5$, what is the value of $4x$?

A $10 - y$

B $10 - 2y$

C $\frac{5 - y}{2}$

D $\frac{10 - y}{2}$

28. Which expression is equivalent to $7x^2 3x^{-4}$?

F $21x^{-8}$

G $21x^2$

H $21x^{-6}$

J $21x^{-2}$

29. GEOMETRY What is the base of the triangle if the area is 56 square meters?

A 4 m

B 8 m

C 16 m

D 28 m

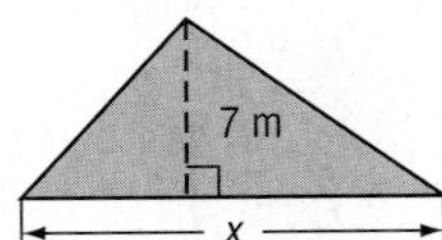

30. SHORT RESPONSE Brianne makes blankets for a baby store. She works on the blankets 30 hours per week. The store pays her \$9.50 per hour plus 30% of the profit. If her hourly rate is increased by \$0.75 and her commission is raised to 40%, how much will she earn for a \$300 profit?

Spiral Review

Solve each equation or formula for x. (Lesson 2-8)

31. $2bx - b = -5$

32. $3x - r = r(-3 + x)$

33. $A = 2\pi r^2 + 2\pi rx$

34. SKIING Yuji is registering for ski camp. The cost of the camp is \$1254, but there is a sales tax of 7%. What is the total cost of the camp including tax? (Lesson 2-7)

Translate each equation into a sentence. (Lesson 2-1)

35. $\frac{n}{-6} = 2n + 1$

36. $18 - 5h = 13h$

37. $2x^2 + 3 = 21$

Refer to the graph.

38. Name the ordered pair at point A and explain what it represents. (Lesson 1-6)

39. Name the ordered pair at point B and explain what it represents. (Lesson 1-6)

40. Identify the independent and dependent variables for the function. (Lesson 1-6)

41. BASEBALL Tickets to a baseball game cost \$18.95, \$12.95, or \$9.95. A hot dog and soda combo costs \$5.50. The Madison family is having a reunion. They buy 10 tickets in each price category and plan to buy 30 combos. What is the total cost for the tickets and meals? (Lesson 1-3)

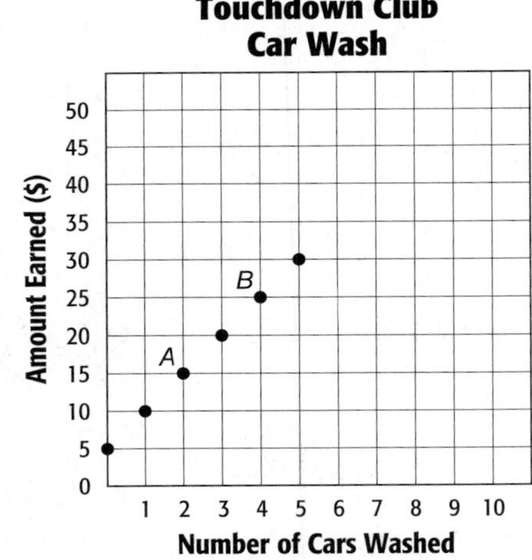

Skills Review

Solve each equation. (Lesson 2-4)

42. $a - 8 = 15$

43. $9m - 11 = -29$

44. $18 - 2k = 24$

45. $5 - 8y = 61$

46. $7 = \frac{h}{2} + 3$

47. $\frac{n}{6} + 1 = 5$

Study Guide and Review

Math Online glencoe.com
• STUDY TO GO
• Vocabulary Review

Chapter Summary

Key Concepts

Writing Equations (Lesson 2-1)

- Identify the unknown you are looking for and assign a variable to it. Then, write the sentence as an equation.

Solving Equations (Lessons 2-2 to 2-4)

- Addition and Subtraction Properties of Equality: If an equation is true and the same number is added to or subtracted from each side, the resulting equation is true.
- Multiplication and Division Properties of Equality: If an equation is true and each side is multiplied or divided by the same nonzero number, the resulting equation is true.
- Steps for Solving Equations:

  Step 1 Simplify the expression on each side. Use the Distributive Property as needed.

  Step 2 Use the Addition and/or Subtraction Properties of Equality to get the variables on one side and the numbers without variables on the other side.

  Step 3 Use the Multiplication or Division Property of Equality to solve.

Absolute Value Equations (Lesson 2-5)

- For any real numbers a and b, if $|a| = b$, then $a = b$ or $a = -b$.

Ratios and Proportions (Lesson 2-6)

- The Means-Extremes Property of Proportion states that in a proportion, the product of the extremes is equal to the product of the means.

FOLDABLES® Study Organizer

Be sure the Key Concepts are noted in your Foldable.

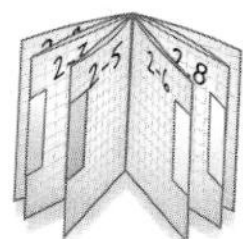

Key Vocabulary

consecutive integers (p. 92)
dimensional analysis (p. 128)
equivalent equations (p. 83)
extremes (p. 112)
formula (p. 76)
identity (p. 98)
literal equation (p. 127)
means (p. 112)
multi-step equations (p. 91)
number theory (p. 92)
percent of change (p. 119)
percent of decrease (p. 119)
percent of increase (p. 119)
proportion (p. 111)
rate (p. 113)
ratio (p. 111)
scale (p. 114)
scale model (p. 114)
solve an equation (p. 83)
unit analysis (p. 128)
unit rate (p. 113)
weighted average (p. 132)

Vocabulary Check

State whether each sentence is *true* or *false*. If *false*, replace the underlined term to make a true sentence.

1. In order to write an equation to solve a problem, identify the unknown for which you are looking and assign a(n) <u>number</u> to it.
2. To <u>solve an equation</u> means to find the value of the variable that makes the equation true.
3. The numbers 10, 12, and 14 are an example of <u>consecutive even integers</u>.
4. The <u>absolute value</u> of any number is simply the distance the number is away from zero on a number line.
5. A(n) <u>equation</u> is a comparison of two numbers by division.
6. An equation stating that two ratios are equal is called a(n) <u>proportion</u>.
7. If the new number is less than the original number, the percent of change is a percent of <u>increase</u>.
8. The <u>weighted average</u> of a set of data is the sum of the product of the number of units and the value per unit divided by the sum of the number of units.

CHAPTER 2

Study Guide and Review

Lesson-by-Lesson Review

2-1 Writing Equations (pp. 75–80)

Translate each sentence into an equation.

9. The sum of five times a number x and three is the same as fifteen.
10. Four times the difference of b and six is equal to b squared.
11. One half of m cubed is the same as four times m minus nine.

Translate each equation into a sentence.

12. $3p + 8 = 20$
13. $h^2 - 5h + 6 = 0$
14. $\frac{3}{4}w^2 + \frac{2}{3}w - \frac{1}{5} = 2$
15. **FENCING** Adrianne wants to create an outdoor rectangular kennel. The length will be three feet more than twice the width. Write and use an equation to find the length and the width of the kennel if Adrianne has 54 feet of fencing.

EXAMPLE 1

Translate the following sentence into an equation.

Six times the sum of a number n and four is the same as the difference between two times n to the second power and ten.

$6(n + 4) = 2n^2 - 10$

EXAMPLE 2

Translate $3d^2 - 9d + 8 = 4(d + 2)$ into a sentence.

Three times a number d squared minus nine times d increased by eight is equal to four times the sum of d and two.

2-2 Solving One-Step Equations (pp. 83–89)

Solve each equation. Check your solution.

16. $x - 9 = 4$
17. $-6 + g = -11$
18. $\frac{5}{9} + w = \frac{7}{9}$
19. $3.8 = m + 1.7$
20. $\frac{a}{12} = 5$
21. $8y = 48$
22. $\frac{2}{5}b = -4$
23. $-\frac{t}{16} = -\frac{7}{8}$
24. **AGE** Max is four years younger than his sister Brenda. The total of their ages is 16. Write and solve an equation to find their ages.

EXAMPLE 3

Solve $x - 13 = 9$. Check your solution.

$x - 13 = 9$	**Original equation**
$x - 13 + 13 = 9 + 13$	**Add 13 to each side.**
$x = 22$	**$-13 + 13 = 0$ and $9 + 13 = 22$**

To check that 22 is the solution, substitute 22 for x in the original equation.

CHECK	$x - 13 = 9$	**Original equation**
	$22 - 13 \stackrel{?}{=} 9$	**Substitute 22 for x.**
	$9 = 9$ ✓	**Subtract.**

2-3 Solving Multi-Step Equations (pp. 91–96)

Solve each equation. Check your solution.

25. $2d - 4 = 8$
26. $-9 = 3t + 6$
27. $14 = -8 - 2k$
28. $\frac{n}{4} - 7 = -2$
29. $\frac{r + 4}{3} = 7$
30. $-18 = \frac{9 - a}{2}$
31. $6g - 3.5 = 8.5$
32. $0.2c + 4 = 6$
33. $\frac{f}{3} - 9.2 = 3.5$
34. $4 = \frac{-3u - (-7)}{-8}$
35. **CONSECUTIVE INTEGERS** Find three consecutive odd integers with a sum of 63.
36. **CONSECUTIVE INTEGERS** Find three consecutive integers with a sum of –39.

EXAMPLE 4

Solve $7y - 9 = 33$. Check your solution.

	$7y - 9 = 33$	**Original equation**
	$7y - 9 + 9 = 33 + 9$	**Add 9 to each side.**
	$7y = 42$	**Simplify.**
	$\frac{7y}{7} = \frac{42}{7}$	**Divide each side by 7.**
	$y = 6$	**Simplify.**
CHECK	$7y - 9 = 33$	**Original equation**
	$7(6) - 9 \stackrel{?}{=} 33$	**Substitute 6 for *y*.**
	$42 - 9 \stackrel{?}{=} 33$	**Multiply.**
	$33 = 33$ ✓	**Subtract.**

2-4 Solving Equations with the Variable on Each Side (pp. 97–102)

Solve each equation. Check your solution.

37. $8m + 7 = 5m + 16$
38. $2h - 14 = -5h$
39. $21 + 3j = 9 - 3j$
40. $\frac{x - 3}{4} = \frac{x}{2}$
41. $\frac{6r - 7}{10} = \frac{r}{4}$
42. $3(p + 4) = 33$
43. $-2(b - 3) - 4 = 18$
44. $4(3w - 2) = 8(2w + 3)$

Write an equation and solve each problem.

45. Find the sum of three consecutive odd integers if the sum of the first two integers is equal to twenty-four less than four times the third integer.
46. **TRAVEL** Mr. Jones drove 480 miles to a business meeting. His travel time to the meeting was 8 hours and from the meeting was 7.5 hours. Find his rate of travel for each leg of the trip.

EXAMPLE 5

Solve $9w - 24 = 6w + 18$.

$9w - 24 = 6w + 18$	**Original equation**
$9w - 24 - 6w = 6w + 18 - 6w$	**Subtract 6*w* from each side.**
$3w - 24 = 18$	**Simplify.**
$3w - 24 + 24 = 18 + 24$	**Add 24 to each side.**
$3w = 42$	**Simplify.**
$\frac{3w}{3} = \frac{42}{3}$	**Divide each side by 3.**
$w = 14$	**Simplify.**

EXAMPLE 6

Write an equation to find three consecutive integers such that three times the sum of the first two integers is the same as thirteen more than four times the third integer.

Let x, $x + 1$, and $x + 2$ represent the three consecutive integers.

$3(x + x + 1) = 4(x + 2) + 13$

CHAPTER 2

Study Guide and Review

2-5 Solving Equations Involving Absolute Value (pp. 103–109)

Evaluate each expression if $m = -8$, $n = 4$, and $p = -12$.

47. $|3m - n|$

48. $|-2p + m| - 3n$

49. $-3|6n - 2p|$

50. $4|7m + 3p| + 4n$

Solve each equation. Then graph the solution set.

51. $|x - 6| = 11$

52. $|-4w + 2| = 14$

53. $\left|\frac{1}{3}d - 6\right| = 15$

54. $\left|\frac{2b}{3} + 8\right| = 20$

EXAMPLE 7

Solve $|y - 9| = 16$. Then graph the solution set.

Case 1

$y - 9 = 16$	**Original equation**
$y - 9 + 9 = 16 + 9$	**Add 9 to each side.**
$y = 25$	**Simplify.**

Case 2

$y - 9 = -16$	**Original equation**
$y - 9 + 9 = -16 + 9$	**Add 9 to each side.**
$y = -7$	**Simplify.**

The solution set is $\{-7, 25\}$.

Graph the points on a number line.

2-6 Ratios and Proportions (pp. 111–117)

Determine whether each pair of ratios are equivalent ratios. Write *yes* or *no*.

55. $\frac{27}{45}, \frac{3}{5}$

56. $\frac{18}{32}, \frac{3}{4}$

Solve each proportion. If necessary, round to the nearest hundredth.

57. $\frac{4}{9} = \frac{a}{45}$

58. $\frac{3}{8} = \frac{21}{t}$

59. $\frac{9}{12} = \frac{g}{16}$

60. CONSTRUCTION A new gym is being built at Greenfield Middle School. The length of the gym as shown on the builder's blueprints is 12 inches. Find the actual length of the new gym.

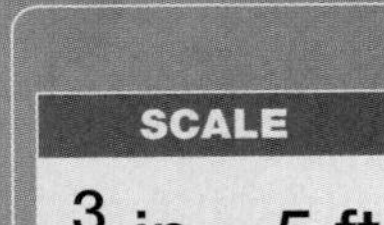

EXAMPLE 8

Determine whether $\frac{7}{9}$ and $\frac{42}{54}$ are equivalent ratios. Write *yes* or *no*. Justify your answer.

First, simplify each ratio. $\frac{7}{9}$ is already in simplest form.

$$\frac{42}{54} = \frac{42 \div 6}{54 \div 6} = \frac{7}{9}$$

When expressed in simplest form, the ratios are equivalent. The answer is yes.

EXAMPLE 9

Solve $\frac{r}{8} = \frac{3}{4}$. If necessary, round to the nearest hundredth.

$\frac{r}{8} = \frac{3}{4}$	**Original equation**
$r(4) = 3(8)$	**Find the cross products.**
$4r = 24$	**Simplify.**
$\frac{4r}{4} = \frac{24}{4}$	**Divide each side by 4.**
$r = 6$	**Simplify.**

MIXED PROBLEM SOLVING
For mixed problem-solving practice, see page 846.

2-7 Percent of Change (pp. 119–124)

State whether each percent of change is a percent of *increase* or a percent of *decrease*. Then find the percent of change. Round to the nearest whole percent.

61. original: 40, new: 50

62. original: 36, new: 24

63. original: \$72, new: \$60

Find the total price of each item.

64. boots: \$64, tax: 7%

65. video game: \$49, tax: 6.5%

66. hockey skates: \$199, tax: 5.25%

Find the discounted price of each item.

67. MP3 player: \$69.00, discount: 20%

68. jacket: \$129, discount: 15%

69. backpack: \$45, discount: 25%

70. **ATTENDANCE** An amusement park recorded attendance of 825,000 one year. The next year, the attendance increased to 975,000. Determine the percent of increase in attendance.

EXAMPLE 10

State whether the percent of change is a percent of *increase* or a percent of *decrease*. Then find the percent of change. Round to the nearest whole percent.

original: 80
new: 60

Since the new amount is less than the original, this is a percent of decrease. Subtract to find the amount of change: $80 - 60 = 20$.

Use the original number, 80, as the base.

$$\frac{\text{change}}{\text{original amount}} \rightarrow \frac{20}{80} = \frac{r}{100}$$

$$20(100) = r(80)$$

$$2000 = 80r$$

$$\frac{2000}{80} = \frac{80r}{80}$$

$$25 = r$$

The percent of decrease is 25%.

2-8 Literal Equations and Dimensional Analysis (pp. 126–131)

Solve each equation or formula for the variable indicated.

71. $3x + 2y = 9$, for y

72. $P = 2\ell + 2w$, for ℓ

73. $-5m + 9n = 15$, for m

74. $14w + 15x = y - 21w$, for w

75. $m = \frac{2}{5}y + n$, for y

76. $7d - 3c = f + 2d$, for d

77. **GEOMETRY** The formula for the area of a trapezoid is $A = \frac{1}{2}h(a + b)$, where h represents the height and a and b represent the lengths of the bases. Solve for h.

EXAMPLE 11

Solve $6p - 8n = 12$ for p.

$6p - 8n = 12$	**Original equation**
$6p - 8n + 8n = 12 + 8n$	**Add 8*n* to each side.**
$6p = 12 + 8n$	**Simplify.**
$\frac{6p}{6} = \frac{12 + 8n}{6}$	**Divide each side by 6.**
$\frac{6p}{6} = \frac{12}{6} + \frac{8}{6}n$	**Simplify.**
$p = 2 + \frac{4}{3}n$	**Simplify.**

2-9 Weighted Averages (pp. 132–138)

78. CANDY Michael is mixing two types of candy for a party. The chocolate pieces cost \$0.40 per ounce, and the hard candy costs \$0.20 per ounce. Michael purchases 20 ounces of the chocolate pieces, and the total cost of his candy was \$11. How many ounces of hard candy did he purchase?

79. TRAVEL A car travels 100 miles east in 2 hours and 30 miles north in half an hour. What is the average speed of the car?

80. FINANCIAL LITERACY A candle supply store sells votive wax and low-shrink wax. How many pounds of low-shrink wax should be mixed with 8 pounds of votive wax to obtain a blend that sells for \$0.98 a pound?

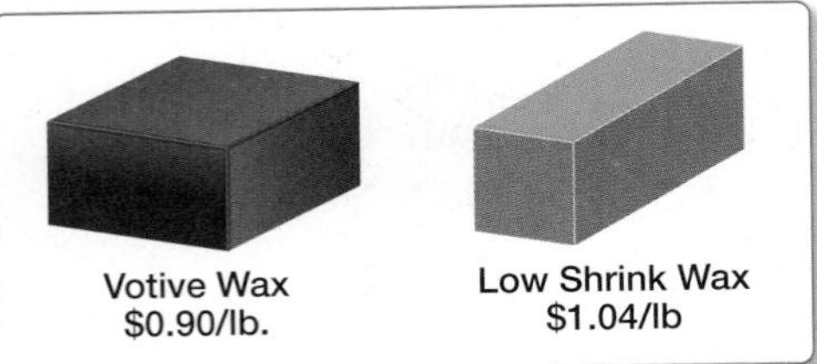

EXAMPLE 12

METALS An alloy of metals is 25% copper. Another alloy is 50% copper. How much of each should be used to make 1000 grams of an alloy that is 45% copper?

Let x = the amount of the 25% copper alloy. Write and solve an equation.

$0.25x + 0.50(1000 - x) = 0.45(1000)$	**Original Equation**
$0.25x + 500 - 0.50x = 450$	**Distributive Property**
$-0.25x + 500 = 450$	**Simplify.**
$-0.25x + 500 - 500 = 450 - 500$	**Subtract 500 from each side.**
$-0.25x = -50$	**Simplify.**
$\frac{-0.25x}{-0.25} = \frac{-50}{-0.25}$	**Divide each side by −0.25.**
$x = 200$	**Simplify.**

200 grams of the 25% alloy and 800 grams of the 50% alloy should be used.

Practice Test

Math Online glencoe.com
Chapter Test

Translate each sentence into an equation.

1. The sum of six and four times d is the same as d minus nine.
2. Three times the difference of two times m and five is equal to eight times m to the second power increased by four.

Solve each equation. Check your solutions.

3. $x - 5 = -11$
4. $\frac{2}{3} = w + \frac{1}{4}$
5. $\frac{t}{6} = -3$

Solve each equation. Check your solution.

6. $2a - 5 = 13$
7. $\frac{p}{4} - 3 = 9$
8. **MULTIPLE CHOICE** At Mama Mia Pizza, the price of a large pizza is determined by $P = 9 + 1.5x$, where x represents the number of toppings added to a cheese pizza. Daniel spent $13.50 on a large pizza. How many toppings did he get?

   A 0

   B 1

   C 3

   D 5

Solve each equation. Check your solution.

9. $5y - 4 = 9y + 8$
10. $3(2k - 2) = -2(4k - 11)$
11. **GEOMETRY** Find the value of x so that the figures have the same perimeter.

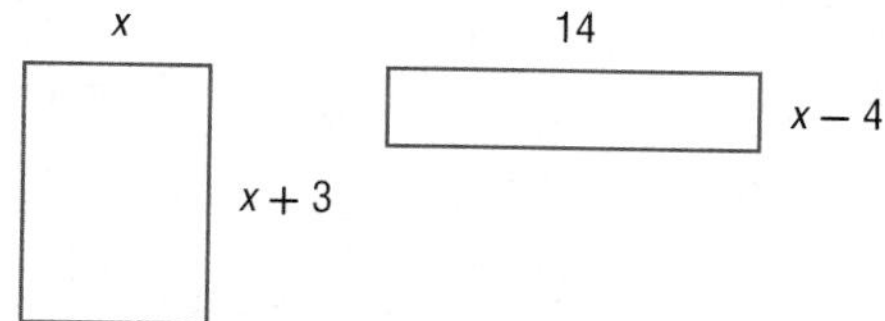

12. Evaluate the expression $|3t - 2u| + 5v$ if $t = 2$, $u = -5$, and $v = -3$.

Solve each equation. Then graph the solution set.

13. $|p - 4| = 6$
14. $|2b + 5| = 9$

Solve each proportion. If necessary, round to the nearest hundredth.

15. $\frac{a}{3} = \frac{16}{24}$
16. $\frac{9}{k + 3} = \frac{3}{5}$
17. **MULTIPLE CHOICE** Akiko uses 2 feet of thread for every three squares that she sews for her quilt. How many squares can she sew if she has 38 feet of thread?

    F 19

    G 57

    H 76

    J 228

18. State whether the percent of change is a percent of *increase* or a percent of *decrease*. Then find the percent of change. Round to the nearest whole percent.

    original: 54 new: 45

19. Find the total price of a sweatshirt that is priced at $48 and taxed at 6.5%.
20. **SHOPPING** Kirk wants to purchase a wide-screen TV. He sees an advertisement for a TV that was originally priced at $3200 and is 20% off. Find the discounted price of the TV.
21. Solve $5x - 3y = 9$ for y.
22. Solve $A = \frac{1}{2}bh$ for h.
23. **CHEMISTRY** Deon has 12 milliliters of a 5% solution. He also has a solution that has a concentration of 30%. How many milliliters of the 30% solution does Deon need to add to the 5% solution to obtain a 20% solution?
24. **BICYCLING** Shanee bikes 5 miles to the park in 30 minutes and 3 miles to the library in 45 minutes. What was her average speed?
25. **MAPS** On a map of North Carolina, the distance between Charlotte and Wilmington is 14.75 inches. If 2 inches equals 24 miles, what is the approximate distance between the two cities?

Gridded Response Questions

In addition to multiple-choice, short-answer, and extended-response questions, you will likely encounter gridded-response questions on standardized tests. For gridded-response questions, you must print your answer on an answer sheet and mark in the correct circles on the grid to match your answer.

Strategies for Solving Gridded Response Questions

Step 1

Read the problem carefully.

- **Ask yourself:** "What information is given?" "What do I need to find?" "How do I solve this type of problem?"
- **Solve the Problem:** Use the information given in the problem to solve.
- **Check your answer:** If time permits, check your answer to make sure you have solved the problem correctly.

Step 2

Write your answer in the answer boxes.

- Print only one digit or symbol in each answer box.
- Do not write any digits or symbols outside the answer boxes.
- You may write your answer with the first digit in the left answer box, or with the last digit in the right answer box. You may leave blank any boxes you do not need on the right or the left side of your answer.

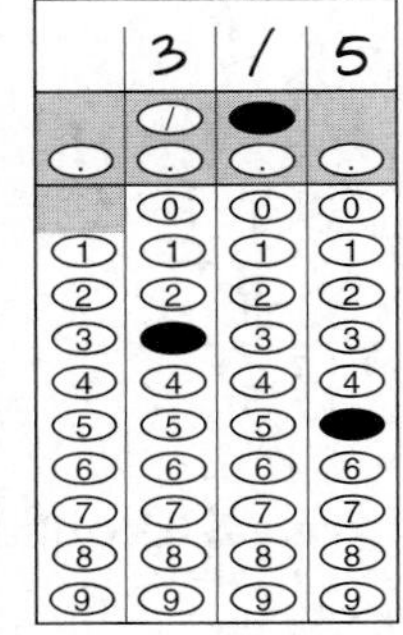

Step 3

Fill in the grid.

- Fill in only one bubble for every answer box that you have written in. Be sure not to fill in a bubble under a blank answer box.
- Fill in each bubble completely and clearly.

EXAMPLE

Read the problem. Identify what you need to know. Then use the information in the problem to solve.

GRIDDED RESPONSE Ashley is 3 years older than her sister, Tina. Combined, the sum of their ages is 27 years. How old is Ashley?

Read the problem carefully. You are told that Ashley is 3 years older than her sister and that their ages combined equal 27 years. You need to find Ashley's age.

Solve the Problem

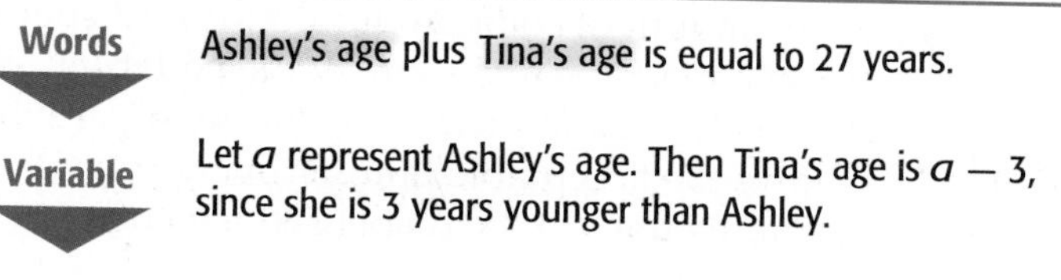

Solve the equation for a.

$a + (a - 3) = 27$	**Original equation.**
$2a - 3 = 27$	**Add like terms.**
$2a = 30$	**Add 3 to each side.**
$a = 15$	**Divide each side by 2.**

Since we let a represent Ashley's age, we know that she is 15 years old.

Fill in the Grid

1 5

Exercises

Read each problem. Identify what you need to know. Then use the information in the problem to solve. Copy and complete an answer grid on your paper.

1. Orlando has \$1350 in the bank. He wants to increase his balance to a total of \$2550 by depositing \$40 each week from his paycheck. How many weeks will he need to save in order to reach his goal?

2. Fourteen less than three times a number is equal to 40. Find the number.

3. The table shows the regular prices and sale prices of certain items at a department store this week. What is the percent of discount during the sale?

Item	Regular Price (\$)	Sale Price (\$)
pillows	25	20
sweaters	30	24
entertainment center	125	100

4. Maureen is driving from Raleigh, North Carolina, to Charlotte, North Carolina, to visit her brother at college. If she averages 65 miles per hour on the trip, then the equation $\frac{d}{2.65} = 65$ can be solved for the distance d. What is the distance to the nearest mile from Raleigh to Charlotte?

5. Find the value of x so that the figures below have the same area.

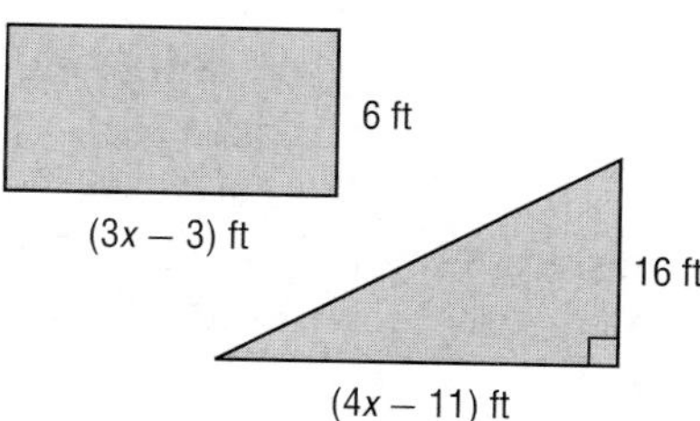

6. The sum of three consecutive whole numbers is 18. What is the greatest of the numbers?

CHAPTER 2 Standardized Test Practice

Cumulative, Chapters 1 and 2

Multiple Choice

Read each question. Then fill in the correct answer on the answer document provided by your teacher or on a sheet of paper.

1. Which point on the number line best represents the position of $\sqrt{8}$? (Lesson 0-2)

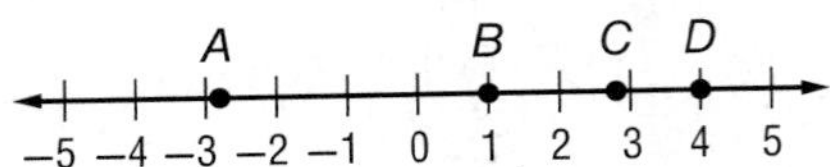

A −2.8
B 1
C 2.8
D 4

2. Find the value of x so that the figures have the same area. (Lesson 2-5)

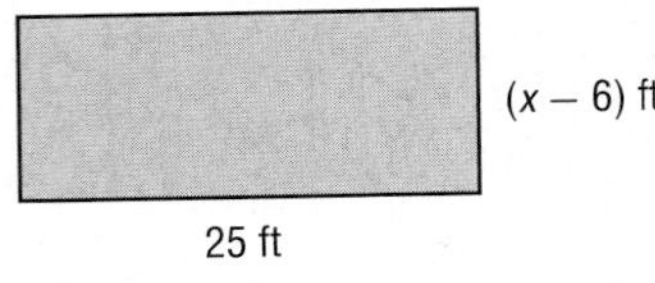

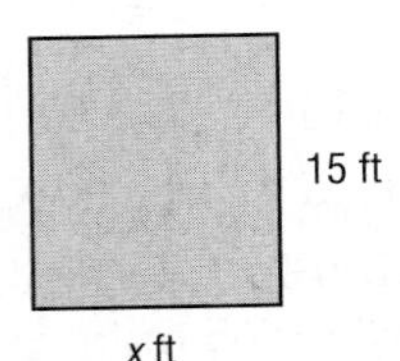

A 10
B 12
C 13
D 15

3. The elevation of Black Mountain is 27 feet more than 16 times the lowest point in the state. If the elevation of the lowest point in the state is 257 feet, what is the elevation of Black Mountain? (Lesson 2-2)

A 4,085 feet
B 4,103 feet
C 4,139 feet
D 4,215 feet

4. The expression $(3x^2 + 5x - 12) - 2(x^2 + 4x + 9)$ is equivalent to which of the following? (Lesson 1-4)

A $x^2 - 3x - 30$
B $x^2 + 13x + 6$
C $5x^2 + x - 18$
D $x^2 + 3x - 21$

5. The amount of soda, in fluid ounces, dispensed from a machine must satisfy the equation $|a - 0.4| = 20$. Which of the following graphs shows the acceptable minimum and maximum amounts that can be dispensed from the machine? (Lesson 2-5)

A
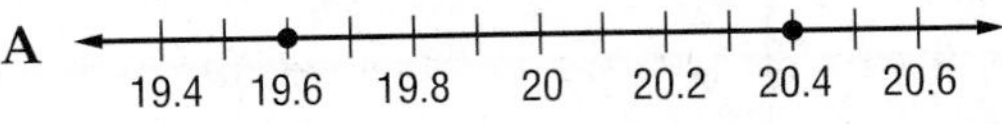

B
19.4 19.6 19.8 20 20.2 20.4 20.6

C
19.4 19.6 19.8 20 20.2 20.4 20.6

D
19.4 19.6 19.8 20 20.2 20.4 20.6

6. If a and b represent integers, $ab = ba$ is an example of which property? (Lesson 1-3)

A Associative Property
B Commutative Property
C Distributive Property
D Closure Property

7. The sum of one fifth of a number and three is equal to half of the number. What is the number? (Lesson 2-4)

A 5
B 10
C 15
D 20

8. Aaron charges \$15 to mow the lawn and \$10 per hour for other gardening work. Which expression represents his earnings? (Lesson 1-1)

A $10h$
B $15h$
C $15h + 10$
D $15 + 10h$

Test-TakingTip

Question 2 Use the figures and the formula for area to set up an equation. The product of the length and width of each figure should be equal.

Short Response/Gridded Response

Record your answers on the answer sheet provided by your teacher or on a sheet of paper.

9. The formula for the lateral area of a cylinder is $A = 2\pi rh$, where r is the radius and h is the height. Solve the equation for h. (Lesson 2-9)

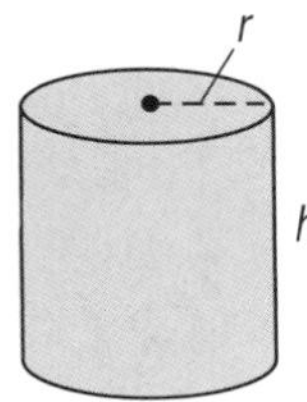

10. GRIDDED RESPONSE Solve the proportion $\frac{x}{18} = \frac{7}{21}$. (Lesson 2-6)

11. GRIDDED RESPONSE The table shows the cost of renting a moving van. If Miguel budgeted $75, how many miles could he drive the van and maintain his budget? (Lesson 2-3)

Moving Van Rentals	
Flat Fee	$50 for up to 300 miles
Variable Fee	$0.20 per mile over 300

12. Find the height of a soup can if the area of the label is 302 square centimeters and the radius of the can is 4 centimeters. Round to the nearest whole number. (Lesson 2-8)

13. GRIDDED RESPONSE Lara's car needed a particular part that costs $75. The mechanic charges $50 per hour to install the part. If the total cost was $350, how many hours did it take to install the part?

14. Lucinda is buying a set of patio furniture that is on sale for $\frac{4}{5}$ of the original price. After she uses a $50 gift certificate, the total cost before sales tax is $222. What was the original price of the patio furniture? (Lesson 2-3)

Extended Response

Record your answers on a sheet of paper. Show your work.

15. The city zoo offers a yearly membership that costs $120. A yearly membership includes free parking. Members can also purchase a ride pass for an additional $2 per day that allows them unlimited access to the rides in the park. Nonmembers pay $12 for admission to the park, $5 for parking, and $5 for a ride pass. (Lesson 2-4)

a. Write an equation that could be solved for the number of visits it would take for the total cost to be the same for a member and a nonmember if they both purchase a ride pass each day. Solve the equation.

b. What would the total cost be for members and nonmembers after this number of visits?

c. Georgena is deciding whether or not to purchase a yearly membership. Explain how she could use the results above to help make her decision.

Need Extra Help?															
If you missed Question...	1	2	3	4	5	6	7	8	9	10	11	12	13	14	15
Go to Lesson or Page...	0-2	2-5	2-2	1-4	2-5	1-3	2-4	1-1	2-8	2-6	2-3	2-8	2-3	2-3	2-4

CHAPTER 3 Linear Functions

Then

In Chapter 2, you solved linear equations algebraically.

Now

In Chapter 3, you will:

- Identify linear equations, intercepts, and zeros.
- Graph and write linear equations.
- Use rate of change to solve problems.

Why?

AMUSEMENT PARKS The Magic Kingdom in Orlando, Florida, is one of the most popular amusement parks in the world. Yearly attendance figures increase steadily each year. Quantities like populations that change with respect to time can be described using rate of change. Often you can represent these situations with linear functions.

Linear Functions

Activity

On their vacation, Tabitha and Jim are staying at the Fort Wilderness Cabins, little houses at Walt Disney World complete with everything, even kitchens.

Jim is washing dishes in the sink which is 7 inches deep. We will look at the depth of the water compared to time.

5/7

Math *in Motion,* Animation glencoe.com

3-1 Graphing Linear Equations

Why?

Recycling one ton of waste paper saves an average of 17 trees, 7000 gallons of water, 3 barrels of oil, and about 3.3 cubic yards of landfill space.

The relationship between the amount of paper recycled and the number of trees saved can be expressed with the equation $y = 17x$, where y represents the number of trees and x represents the tons of paper recycled.

Then

You represented relationships among quantities using equations. (Lesson 2-1)

Now

- Identify linear equations, intercepts, and zeros.
- Graph linear equations.

New Vocabulary

linear equation
standard form
constant
x-intercept
y-intercept

Math Online

glencoe.com

- Extra Examples
- Personal Tutor
- Self-Check Quiz
- Homework Help
- Math in Motion

Linear Equations and Intercepts A **linear equation** is an equation that forms a line when it is graphed. Linear equations are often written in the form $Ax + By = C$. This is called the **standard form** of a linear equation. In this equation, C is called a **constant**, or a number. Ax and By are variable terms.

Key Concept — Standard Form of a Linear Equation (For Your FOLDABLE)

Words The standard form of a linear equation is $Ax + By = C$, where $A \geq 0$, A and B are not both zero, and A, B, and C are integers with a greatest common factor of 1.

Examples In $3x + 2y = 5$, $A = 3$, $B = 2$, and $C = 5$.

In $x = -7$, $A = 1$, $B = 0$, and $C = -7$.

EXAMPLE 1 Identify Linear Equations

Determine whether each equation is a linear equation. Write the equation in standard form.

a. $y = 4 - 3x$

Rewrite the equation so that it appears in standard form.

$y = 4 - 3x$	**Original equation**
$y + 3x = 4 - 3x + 3x$	**Add $3x$ to each side.**
$3x + y = 4$	**Simplify.**

The equation is now in standard form where $A = 3$, $B = 1$, and $C = 4$. This is a linear equation.

b. $6x - xy = 4$

Since the term xy has two variables, the equation cannot be written in the form $Ax + By = C$. Therefore, this is not a linear equation.

Check Your Progress

1A. $\frac{1}{3}y = -1$

1B. $y = x^2 - 4$

Personal Tutor glencoe.com

A linear equation can be represented on a coordinate graph. The x-coordinate of the point at which the graph of an equation crosses the x-axis is an **x-intercept**. The y-coordinate of the point at which the graph crosses the y-axis is called a **y-intercept**.

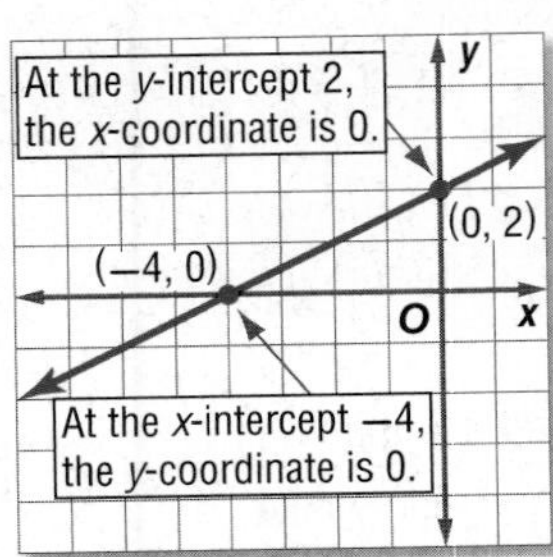

The graph of a linear equation has at most one x-intercept and one y-intercept, unless it is the equation $x = 0$ or $y = 0$, in which case every number is a y-intercept or an x-intercept, respectively.

STANDARDIZED TEST EXAMPLE 2

Find the x- and y-intercepts of the line graphed at the right.

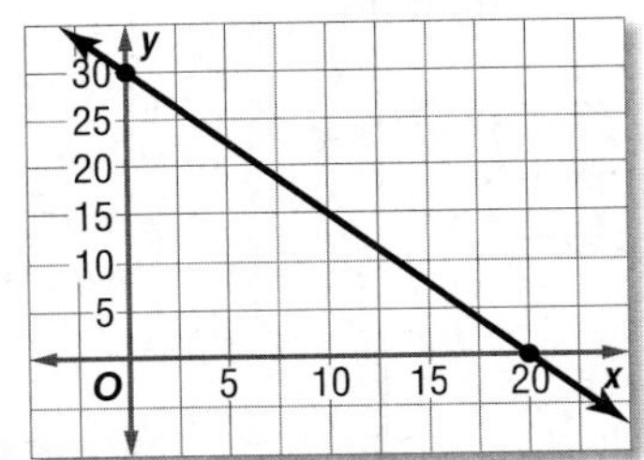

A x-intercept is 0; y-intercept is 30.

B x-intercept is 20; y-intercept is 30.

C x-intercept is 20; y-intercept is 0.

D x-intercept is 30; y-intercept is 20.

ReadingMath

Intercepts Usually, the individual coordinates are called the x-intercept and the y-intercept. The x-intercept 20 is located at (20, 0). The y-intercept 60 is located at (0, 60).

Read the Test Item

We need to determine the x- and y-intercepts of the line in the graph.

Solve the Test Item

Step 1 Find the x-intercept. Look for the point where the line crosses the x-axis.

The line crosses at (20, 0). The x-intercept is 20 because it is the x-coordinate of the point where the line crosses the x-axis.

Step 2 Find the y-intercept. Look for the point where the line crosses the y-axis.

The line crosses the y-axis at (0, 30). The y-intercept is 30 because it is the y-coordinate of the point where the line crosses the y-axis.

Thus, the answer is B.

Check Your Progress

2. **HEALTH** Find the x- and y-intercepts of the graph.

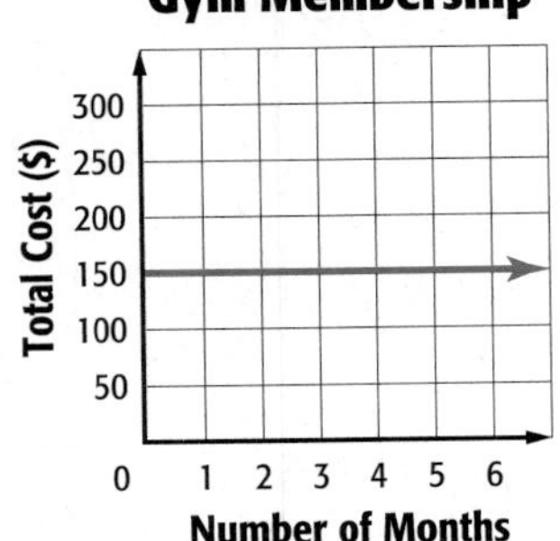

F x-intercept is 0; y-intercept is 150.

G x-intercept is 150; y-intercept is 0.

H x-intercept is 150; no y-intercept.

J No x-intercept; y-intercept is 150.

Personal Tutor glencoe.com

When equations represent a real-world situation, the x- and y-intercepts have a real-world meaning.

StudyTip

Defining Variables In Example 3, time is the *independent* variable, and volume of water is the *dependent* variable.

Real-World EXAMPLE 3 Find Intercepts

SWIMMING POOL A swimming pool is being drained at a rate of 720 gallons per hour. The table shows the function relating the volume of water in a pool and the time in hours that the pool has been draining.

Draining a Pool	
Time (h)	Volume (gal)
x	y
0	10,080
2	8640
6	5760
10	2880
12	1440
14	0

a. Find the x- and y-intercepts of the graph of the function.

x-intercept $= 14$ **14 is the value of x when $y = 0$.**

y-intercept $= 10{,}080$ **10,080 is the value of y when $x = 0$.**

b. Describe what the intercepts mean in this situation.

The x-intercept 14 means that after 14 hours, the water has a volume of 0 gallons, or the pool is completely drained.

The y-intercept 10,080 means that the pool contained 10,080 gallons of water at time 0, or before it started to drain. This is shown in the graph.

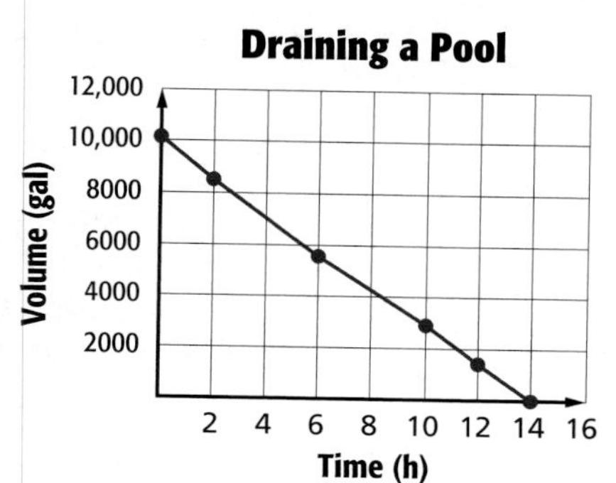

Check Your Progress

3. DRIVING The table shows the function relating the distance to an amusement park in miles and the time in hours the Torres family has driven. Find the x- and y-intercepts. Describe what the intercepts mean in this situation.

Time (h)	Distance (mi)
0	248
1	186
2	124
3	62
4	0

Personal Tutor glencoe.com

StudyTip

Equivalent Equations Rewriting equations by solving for y may make it easier to find values for y. $4x + y = -3$ → $y = -4x - 3$

Graph Linear Equations By first finding the x- and y-intercepts, you have two ordered pairs of two points through which the graph of the linear equation passes. This information can be used to graph the line because only two points are needed to graph a line.

EXAMPLE 4 Graph by Using Intercepts

Graph $2x + 4y = 16$ by using the x- and y-intercepts.

To find the x-intercept, let $y = 0$.

$2x + 4y = 16$ **Original equation**

$2x + 4(0) = 16$ **Replace y with 0.**

$2x = 16$ **Simplify.**

$x = 8$ **Divide each side by 2.**

The x-intercept is 8. This means that the graph intersects the x-axis at (8, 0).

(continued on the next page)

StudyTip

Intercepts The x-intercept is where the graph crosses the x-axis. So the y-value is always 0. The y-intercept is where the graph crosses the y-axis. So, the x-value is always 0.

To find the y-intercept, let $x = 0$.

$2x + 4y = 16$	**Original equation**
$2(0) + 4y = 16$	**Replace x with 0.**
$4y = 16$	**Simplify.**
$y = 4$	**Divide each side by 4.**

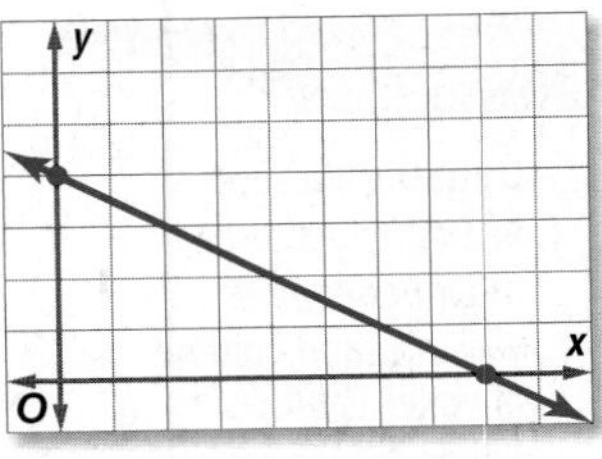

The y-intercept is 4. This means the graph intersects the y-axis at (0, 4).

Plot these two points and then draw a line through them.

Check Your Progress

Graph each equation by using the x- and y-intercepts.

4A. $-x + 2y = 3$ **4B.** $y = -x - 5$

Personal Tutor glencoe.com

Note that the equation in Example 4 has both an x- and a y-intercept. Some lines have an x-intercept and no y-intercept or vice versa. The graph of $y = b$ is a horizontal line that only has a y-intercept (unless $b = 0$). The intercept occurs at $(0, b)$. The graph of $x = a$ is a vertical line that only has an x-intercept (unless $a = 0$). The intercept occurs at $(a, 0)$.

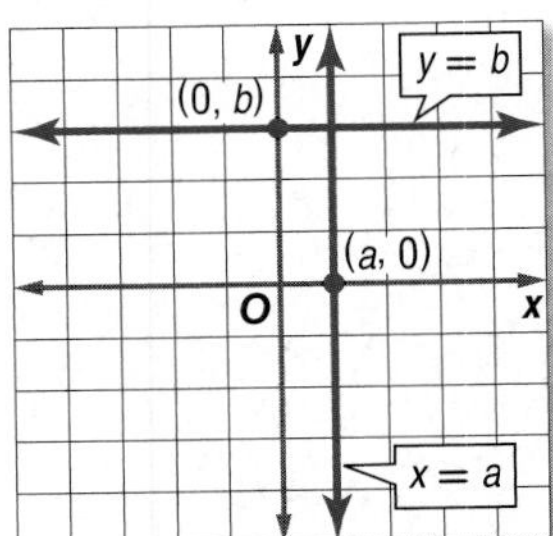

Every ordered pair that makes an equation true represents a point on the graph. So, the graph of an equation represents all of its solutions. Any ordered pair that does not make the equation true represents a point that is not on the line.

EXAMPLE 5 Graph by Making a Table

Graph $y = \frac{1}{3}x + 2$.

The domain is all real numbers. Select values from the domain and make a table. When the x-coefficient is a fraction, select a number from the domain that is a multiple of the denominator. Create ordered pairs and graph them.

x	$\frac{1}{3}x + 2$	y	(x, y)
−3	$\frac{1}{3}(-3) + 2$	1	(−3, 1)
0	$\frac{1}{3}(0) + 2$	2	(0, 2)
3	$\frac{1}{3}(3) + 2$	3	(3, 3)
6	$\frac{1}{3}(6) + 2$	4	(6, 4)

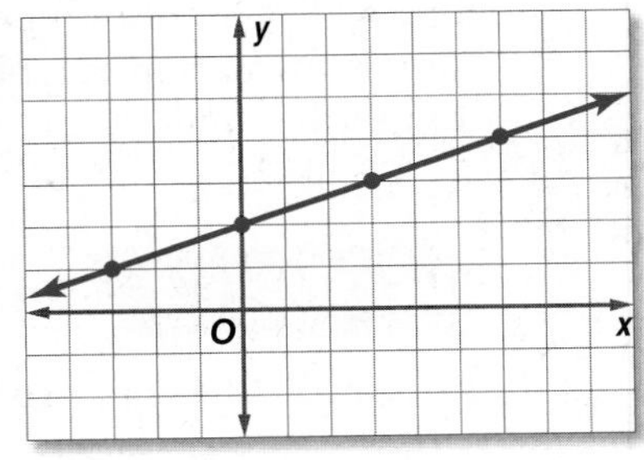

Math in Motion, Animation glencoe.com

Check Your Progress

Graph each equation by making a table.

5A. $2x - y = 2$ **5B.** $x = 3$ **5C.** $y = -2$

Personal Tutor glencoe.com

Check Your Understanding

Example 1 p. 153

Determine whether each equation is a linear equation. Write *yes* or *no*. If yes, write the equation in standard form.

1. $x = y - 5$ **2.** $-2x - 3 = y$ **3.** $-4y + 6 = 2$ **4.** $\frac{2}{3}x - \frac{1}{3}y = 2$

Examples 2 and 3 pp. 154–155

Find the x- and y-intercepts of each linear function. Describe what the intercepts mean.

5.

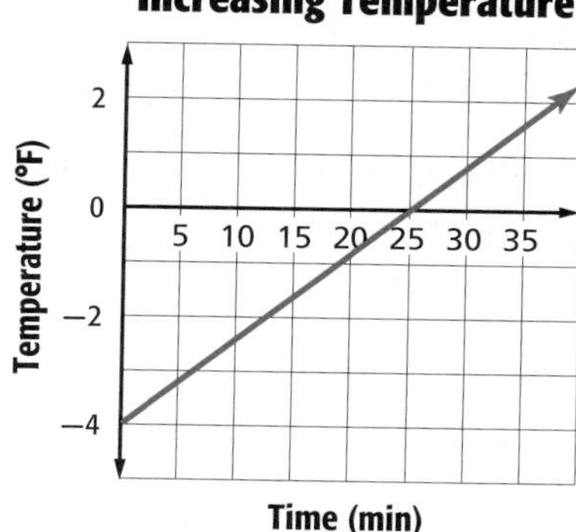

6.

Position of Scuba Diver	
Time (s)	Depth (m)
x	y
0	−24
3	−18
6	−12
9	−6
12	0

Example 4 p. 155

Graph each equation by using the x- and y-intercepts.

7. $y = 4 + x$ **8.** $2x - 5y = 1$

Example 5 p. 156

Graph each equation by making a table.

9. $x + 2y = 4$ **10.** $-3 + 2y = -5$ **11.** $y = 3$

12. RODEOS The equation $5x + 10y = 60$ represents the number of children x and adults y who can attend the rodeo for \$60.

a. Use the x- and y-intercepts to graph the equation.

b. Describe what these values mean.

Practice and Problem Solving

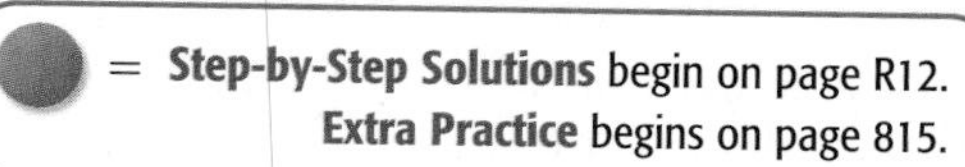

Example 1 p. 153

Determine whether each equation is a linear equation. Write *yes* or *no*. If yes, write the equation in standard form.

13. $5x + y^2 = 25$ **14.** $8 + y = 4x$ **15.** $9xy - 6x = 7$

16. $4y^2 + 9 = -4$ **17.** $12x = 7y - 10y$ **18.** $y = 4x + x$

Examples 2 and 3 pp. 154–155

Find the x- and y-intercepts of the graph of each linear function.

19.

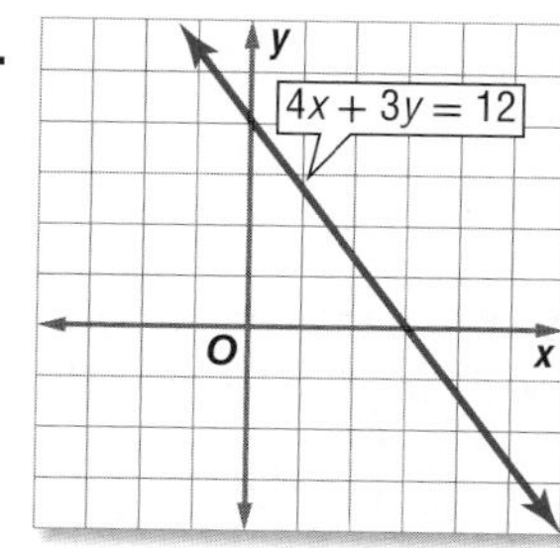

20.

x	y
−3	−1
−2	0
−1	1
0	2
1	3

Examples 2 and 3
pp. 154–155

Find the x- and y-intercepts of each linear function. Describe what the intercepts mean.

21.

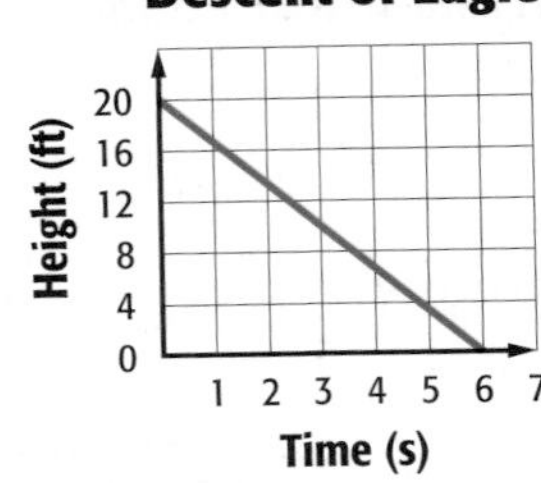

22.

Eva's Distance from Home	
Time (min)	Distance (mi)
x	y
0	4
2	3
4	2
6	1
8	0

Example 4
p. 155

Graph each equation by using the x- and y-intercepts.

23. $y = 4 + 2x$ **24.** $5 - y = -3x$ **25.** $x = 5y + 5$

26. $x + y = 4$ **27.** $x - y = -3$ **28.** $y = 8 - 6x$

Example 5
p. 156

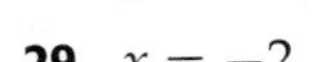

Graph each equation by making a table.

29. $x = -2$ **30.** $y = -4$ **31.** $y = -8x$

32. $3x = y$ **33.** $y - 8 = -x$ **34.** $x = 10 - y$

Real-World Link

In a recent survey, the top five uses for a DVR were: to skip commercials easily (53%); to be able to watch one show while recording another (47%); to pause live TV (32%); to use the on-screen program guide (31%); to record all the episodes of a given show (31%).

Source: MIT

35 **TV RATINGS** The number of people who watch a singing competition can be given by $p = 0.15v$, where p represents the number of people in millions who saw the show and v is the number of potential viewers in millions.

a. Make a table of values for the points (v, p).

b. Graph the equation.

c. Use the graph to estimate the number of people who saw the show if there are 14 million potential viewers.

d. Explain why it would not make sense for v to be a negative number.

Determine whether each equation is a linear equation. Write *yes* or *no*. If yes, write the equation in standard form.

36. $x + \frac{1}{y} = 7$ **37.** $\frac{x}{2} = 10 + \frac{2y}{3}$

38. $7n - 8m = 4 - 2m$ **39.** $3a + b - 2 = b$

40. $2r - 3rt + 5t = 1$ **41.** $\frac{3m}{4} = \frac{2n}{3} - 5$

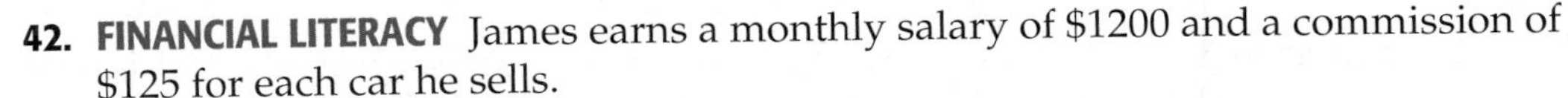

42. **FINANCIAL LITERACY** James earns a monthly salary of \$1200 and a commission of \$125 for each car he sells.

a. Graph an equation that represents how much James earns in a month in which he sells x cars.

b. Use the graph to estimate the number of cars James needs to sell in order to earn \$5000.

Graph each equation.

43. $2.5x - 4 = y$ **44.** $1.25x + 7.5 = y$ **45.** $y + \frac{1}{5}x = 3$

46. $\frac{2}{3}x + y = -7$ **47.** $2x - 3 = 4y + 6$ **48.** $3y - 7 = 4x + 1$

49. **VACATION** Mrs. Johnson is renting a car for vacation and plans to drive a total of 800 miles. A rental car company charges \$153 for the week including 700 miles and \$0.23 for each additional mile. If Mrs. Johnson has only \$160 to spend on the rental car, can she afford to rent a car? Explain your reasoning.

Real-World Link

Attendance at America's theme parks outnumber attendance to all NHL, NFL, NBA, and MLB games combined each year.

Source: *Pricewater Coopers*

50. **AMUSEMENT PARKS** An amusement park charges $50 for admission before 6 P.M. and $20 for admission after 6 P.M. On Saturday, the park took in a total of $20,000.
    a. Write an equation that represents the number of admissions that may have been sold. Let x represent the admissions sold before 6 P.M., and let y represent the admissions sold after 6 P.M.
    b. Graph the equation.
    c. Find the x- and y-intercepts of the graph. What does each intercept represent?

Find the x-intercept and y-intercept of the graph of each equation.

51. $5x + 3y = 15$
52. $2x - 7y = 14$
53. $2x - 3y = 5$
54. $6x + 2y = 8$
55. $y = \frac{1}{4}x - 3$
56. $y = \frac{2}{3}x + 1$

57. **ONLINE GAMES** The percent of teens who play online games can be modeled by $p = \frac{15}{4}t + 66$. p is the percent of students and t represents time in years since 2000.
    a. Graph the equation.
    b. Use the graph to estimate the percent of students playing the games in 2008.

58. **MULTIPLE REPRESENTATIONS** In this problem, you will explore x- and y-intercepts of graphs of linear equations.
    a. **GRAPHICAL** If possible, use a straightedge to draw a line with each of the following characteristics.

x- and y-intercept	x-intercept, no y-intercept	exactly 2 x-intercepts	no x-intercept, y-intercept	exactly 2 y-intercepts

    b. **ANALYTICAL** For which characteristics were you able to create a line and for which characteristics were you unable to create a line? Explain.
    c. **VERBAL** What must be true of the x- and y-intercepts of a line?

H.O.T. Problems

Use Higher-Order Thinking Skills

59. **CHALLENGE** Copy and complete each table. State whether any of the tables show a linear relationship. Explain.

Perimeter of a Square	
Side Length	Perimeter
1	
2	
3	
4	

Area of a Square	
Side Length	Area
1	
2	
3	
4	

Volume of a Cube	
Side Length	Volume
1	
2	
3	
4	

60. **REASONING** Compare and contrast the graphs of $y = 2x + 1$ with the domain $\{1, 2, 3, 4\}$ and $y = 2x + 1$ with the domain of all real numbers.

OPEN ENDED Give an example of a linear equation of the form $Ax + By = C$ for each condition. Then describe the graph of the equation.

61. $A = 0$
62. $B = 0$
63. $C = 0$

64. **WRITING IN MATH** Explain how to find the x-intercept and y-intercept of a graph and summarize how to graph a linear equation.

Standardized Test Practice

65. Sancho can ride 8 miles on his bicycle in 30 minutes. At this rate, about how long would it take him to ride 30 miles?

A 8 hours

B 6 hours 32 minutes

C 2 hours

D 1 hour 53 minutes

66. GEOMETRY Which is a true statement about the relation graphed?

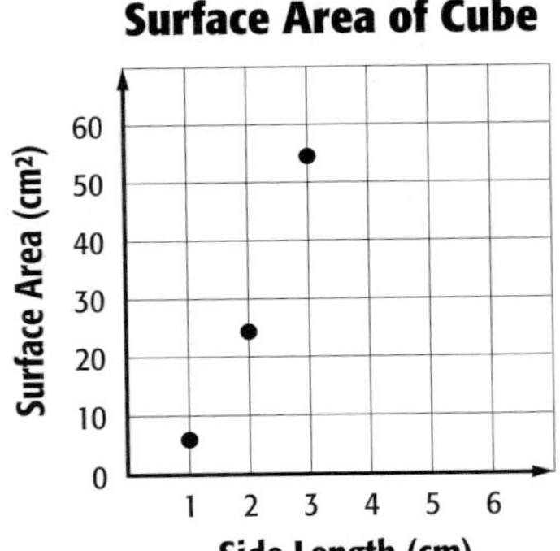

F The relation is not a function.

G Surface area is the independent quantity.

H The surface area of a cube is a function of the side length.

J As the side length of a cube increases, the surface area decreases.

67. SHORT RESPONSE Selena deposited \$2000 into a savings account that pays 1.5% interest compounded annually. If she does not deposit any more money into her account, how much will she earn in interest at the end of one year?

68. A candle burns as shown in the graph.

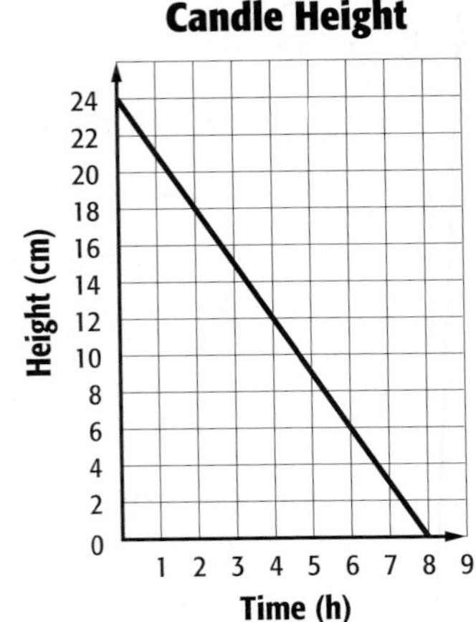

If the height of the candle is 8 centimeters, approximately how long has the candle been burning?

A 0 hours

B 24 minutes

C 64 minutes

D $5\frac{1}{2}$ hours

Spiral Review

69. FUNDRAISING The Madison High School Marching Band sold solid-color gift wrap for \$4 and print gift wrap for \$6 per roll. The total number of rolls sold was 480, and the total amount of money collected was \$2,340. How many rolls of each kind of gift wrap were sold? (Lesson 2-9)

Solve each equation or formula for the variable specified. (Lesson 2-8)

70. $S = \frac{n}{2}(A + t)$, for A

71. $2g - m = 5 - gh$, for g

72. $\frac{y + a}{3} = c$, for y

73. $4z + b = 2z + c$, for z

Skills Review

Evaluate each expression if $x = 2$, $y = 5$, and $z = 7$. (Lesson 1-2)

74. $3x^2 - 4y$

75. $\frac{x - y^2}{2z}$

76. $\left(\frac{y}{z}\right)^2 + \frac{xy}{2}$

77. $z^2 - y^3 + 5x^2$

3-2 Solving Linear Equations by Graphing

Then

You graphed linear equations by using tables and finding roots, zeros, and intercepts. (Lesson 3-1)

Now

- Solve equations by graphing.
- Estimate solutions to an equation by graphing.

New Vocabulary

linear function
parent function
family of graphs
root
zeros

Math Online

glencoe.com

- Extra Examples
- Personal Tutor
- Self-Check Quiz
- Homework Help

Why?

The cost of braces can vary widely. The graph shows the balance of the cost of treatments as payments are made. This is modeled by the function $b = -85p + 5100$, where p represents the number of \$85 payments made, and b is the remaining balance.

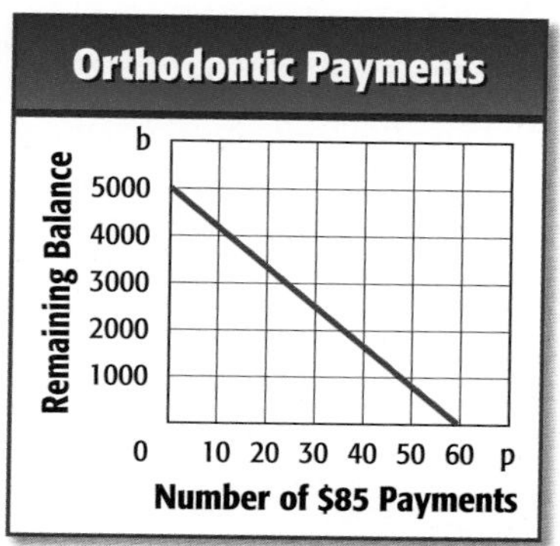

Solve by Graphing A **linear function** is a function for which the graph is a line. The simplest linear function is $f(x) = x$ and is called the **parent function** of the family of linear functions. A **family of graphs** is a group of graphs with one or more similar characteristics.

Key Concept — Linear Function

For Your FOLDABLE

Parent function: $f(x) = x$

Type of graph: line

Domain: all real numbers

Range: all real numbers

The solution or **root** of an equation is any value that makes the equation true. A linear equation has at most one root. You can find the root of an equation by graphing its related function. To write the related function for an equation, replace 0 with $f(x)$.

Linear Equation	Related Function
$2x - 8 = 0$	$f(x) = 2x - 8$ or $y = 2x - 8$

Values of x for which $f(x) = 0$ are called **zeros** of the function f. The zero of a function is located at the x-intercept of the function. The root of an equation is the value of the x-intercept. So:

- 4 is the x-intercept of $2x - 8 = 0$.
- 4 is the solution of $2x - 8 = 0$.
- 4 is the root of $2x - 8 = 0$.
- 4 is the zero of $f(x) = 2x - 8$.

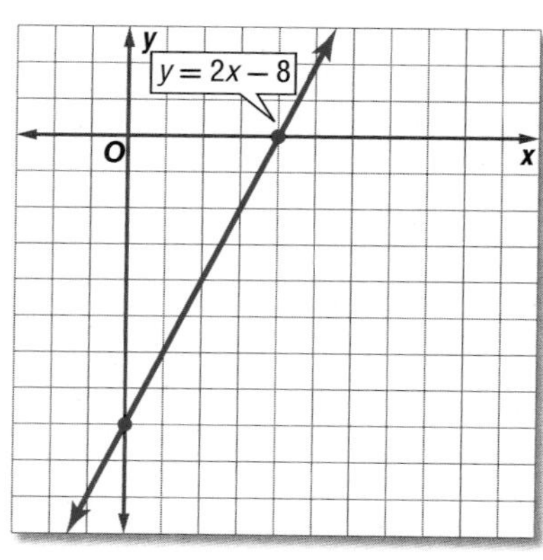

EXAMPLE 1 Solve an Equation with One Root

Solve each equation.

a. $0 = \frac{1}{3}x - 2$

Method 1 Solve algebraically.

$0 = \frac{1}{3}x - 2$	**Original equation**
$0 + 2 = \frac{1}{3}x - 2 + 2$	**Add 2 to each side.**
$3(2) = 3\left(\frac{1}{3}x\right)$	**Multiply each side by 3.**
$6 = x$	**Solve.**

The solution is 6.

b. $3x + 1 = -2$

Method 2 Solve by graphing.

Find the related function. Rewrite the equation with 0 on the right side.

$3x + 1 = -2$	**Original equation**
$3x + 1 + 2 = -2 + 2$	**Add 2 to each side.**
$3x + 3 = 0$	**Simplify.**

The related function is $f(x) = 3x + 3$. To graph the function, make a table.

x	$f(x) = 3x + 3$	$f(x)$	$(x, f(x))$
−2	$f(-2) = 3(-2) + 3$	−3	(−2, −3)
1	$f(1) = 3(1) + 3$	6	(1, 6)

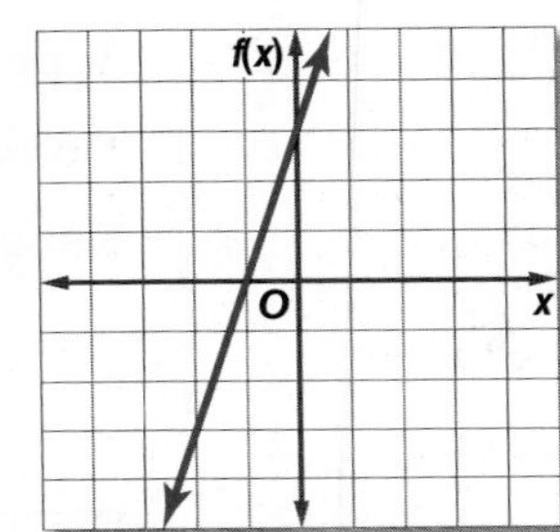

The graph intersects the x-axis at −1.
So, the solution is −1.

StudyTip

Zeros from tables The zero is located at the x-intercept, so the value of y will equal 0. When looking at a table, the zero is the x-value when $y = 0$.

Check Your Progress

1A. $0 = \frac{2}{5}x + 6$

1B. $-1.25x + 3 = 0$

Personal Tutor glencoe.com

For equations with the same variable on each side of the equation, use addition or subtraction to get the terms with variables on one side. Then solve.

EXAMPLE 2 Solve an Equation with No Solution

Solve each equation.

a. $3x + 7 = 3x + 1$

Method 1 Solve algebraically.

$3x + 7 = 3x + 1$	**Original equation**
$3x + 7 - 1 = 3x + 1 - 1$	**Subtract 1 from each side.**
$3x + 6 = 3x$	**Simplify.**
$3x - 3x + 6 = 3x - 3x$	**Subtract $3x$ from each side.**
$6 = 0$	**Simplify.**

The related function is $f(x) = 6$. The root of a linear equation is the value of x when $f(x) = 0$. Since $f(x)$ is always equal to 6, this equation has no solution.

b. $2x - 4 = 2x - 6$

Method 2 Solve by graphing.

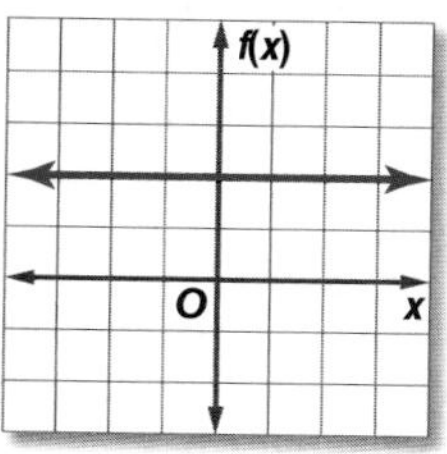

$2x - 4 = 2x - 6$	**Original equation**
$2x - 4 + 6 = 2x - 6 + 6$	**Add 6 to each side.**
$2x + 2 = 2x$	**Simplify.**
$2x - 2x + 2 = 2x - 2x$	**Subtract $2x$ from each side.**
$2 = 0$	**Simplify.**

Graph the related function, which is $f(x) = 2$. The graph does not intersect the x-axis. Thus, there is no solution.

Check Your Progress

2A. $4x + 3 = 4x - 5$

2B. $2 - 3x = 6 - 3x$

Personal Tutor glencoe.com

Estimate Solutions by Graphing Graphing may provide only an estimate. In these cases, solve algebraically to find the exact solution.

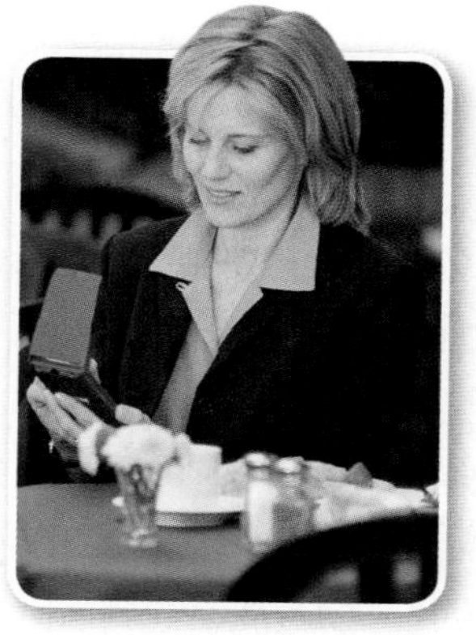

Real-World Career

Entertainment Manager An entertainment manager supervises tech tests, calls show cues, schedules performances and performers, coaches employees and guest talent, and manages expenses. Entertainment managers need a college degree in a field such as communication or theater.

Real-World EXAMPLE 3 Estimate by Graphing

AMUSEMENT PARKS Emily is going to a local carnival. The function $m = 20 - 0.75r$ represents the amount of money m she has left after r rides. Find the zero of this function. Describe what this value means in this context.

Carnival

Amount of Money ($): 0, 5, 10, 15, 20, 25, 30, 35, 40, 45 (m)

Number of Rides: 5, 10, 15, 20, 25, 30, 35, 40, 45 (r)

Make a table of values.

r	$m = 20 - 0.75r$	m	(r, m)
0	$m = 20 - 0.75(0)$	20	(0, 20)
5	$m = 20 - 0.75(5)$	16.25	(5, 16.25)

The graph appears to intersect the r-axis at 27.

Next, solve algebraically to check.

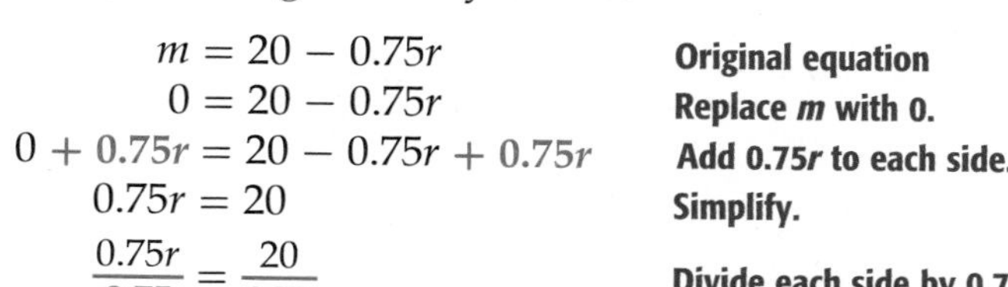

$m = 20 - 0.75r$	**Original equation**
$0 = 20 - 0.75r$	**Replace m with 0.**
$0 + 0.75r = 20 - 0.75r + 0.75r$	**Add $0.75r$ to each side.**
$0.75r = 20$	**Simplify.**
$\frac{0.75r}{0.75} = \frac{20}{0.75}$	**Divide each side by 0.75.**
$r \approx 26.67$	**Simplify and round to the nearest hundredth.**

The zero of this function is about 26.67. Since Emily cannot ride part of a ride, she can ride 26 rides before she will run out of money.

Check Your Progress

3. FINANCIAL LITERACY Antoine's class is selling candy to raise money for a class trip. They paid \$45 for the candy, and they are selling each candy bar for \$1.50. The function $y = 1.50x - 45$ represents their profit y when they sell x candy bars. Find the zero and describe what it means in the context of this situation.

Personal Tutor glencoe.com

Check Your Understanding

Examples 1 and 2 pp. 162–163

Solve each equation.

1. $-2x + 6 = 0$

2. $-x - 3 = 0$

3. $4x - 2 = 0$

4. $9x + 3 = 0$

5. $2x - 5 = 2x + 8$

6. $4x + 11 = 4x - 24$

7. $3x - 5 = 3x - 10$

8. $-6x + 3 = -6x + 5$

Example 3 p. 163

9. **NEWSPAPERS** The function $w = 30 - \frac{3}{4}n$ represents the weight w in pounds of the papers in Tyrone's newspaper delivery bag after he delivers n newspapers. Find the zero and explain what it means in the context of this situation.

Practice and Problem Solving

● = **Step-by-Step Solutions** begin on page R12. **Extra Practice** begins on page 815

Examples 1 and 2 pp. 162–163

Solve each equation.

10. $0 = x - 5$

11. $0 = x + 3$

12. $5 - 8x = 16 - 8x$

13. $3x - 10 = 21 + 3x$

14. $4x - 36 = 0$

15. $0 = 7x + 10$

16. $2x + 22 = 0$

17. $5x - 5 = 5x + 2$

18. $-7x + 35 = 20 - 7x$

19. $-4x - 28 = 3 - 4x$

20. $0 = 6x - 8$

21. $12x + 132 = 12x - 100$

Example 3 p. 163

22. **TEXT MESSAGING** Sean is sending text messages to his friends. The function $y = 160 - x$ represents the number of characters y the message can hold after he has typed x characters. Find the zero and explain what it means in the context of this situation.

Real-World Link

In 2006, 158 billion text messages were sent nationwide, nearly double the amount in 2005.

Source: The Wireless Association

23. **GIFT CARDS** For her birthday Kwan receives a \$50 gift card to download songs. The function $m = -0.50d + 50$ represents the amount of money m that remains on the card after a number of songs d are downloaded. Find the zero and explain what it means in the context of this situation.

Solve each equation.

24. $-7 = 4x + 1$

25. $4 - 2x = 20$

26. $2 - 5x = -23$

27. $10 - 3x = 0$

28. $15 + 6x = 0$

29. $0 = 13x + 34$

30. $0 = 22x - 10$

31. $25x - 17 = 0$

32. $0 = \frac{1}{2} + \frac{2}{3}x$

33. $0 = \frac{3}{4} - \frac{2}{5}x$

34. $13x + 117 = 0$

35. $24x - 72 = 0$

36. **SEA LEVEL** Parts of New Orleans lie 0.5 meter below sea level. After d days of rain the equation $w = 0.3d - 0.5$ represents the water level w in meters. Find the zero, and explain what it means in the context of this situation.

37. **ICE SCULPTURE** An artist completed an ice sculpture when the temperature was $-10°$C. The equation $t = 1.25h - 10$ shows the temperature h hours after the sculpture's completion. If the artist completed the sculpture at 8:00 A.M., at what time will it begin to melt?

Solve each equation by graphing. Verify your answer algebraically.

38. $7 - 3x = 8 - 4x$

39. $19 + 3x = 13 + x$

40. $16x + 6 = 14x + 10$

41. $15x - 30 = 5x - 50$

42. $\frac{1}{2}x - 5 = 3x - 10$

43. $3x - 11 = \frac{1}{3}x - 8$

44. HAIR PRODUCTS Chemical hair straightening makes curly hair straight and smooth. The percent of the process left to complete is modeled by $p = -12.5t + 100$, where t is the time in minutes that the solution is left on the hair, and p represents the percent of the process left to complete.

a. Find the zero of this function.

b. Make a graph of this situation.

c. Explain what the zero represents in this context.

d. State the possible domain and range of this function.

StudyTip

Zero of a Function The zero of a function is also called the root or x-intercept.

45. MUSIC DOWNLOADS In this problem, you will investigate the change between two quantities.

a. Copy and complete the table.

Number of Songs Downloaded	Total Cost ($)	$\frac{\text{Total Cost}}{\text{Number of Songs Downloaded}}$
2	4	
4	8	
6	12	

b. As the number of songs downloaded increases, how does the total cost change?

c. Interpret the value of the total cost divided by the number of songs downloaded.

H.O.T. Problems

Use Higher-Order Thinking Skills

46. FIND THE ERROR Clarissa and Koko solve $3x + 5 = 2x + 4$ by graphing the related function. Is either of them correct? Explain your reasoning.

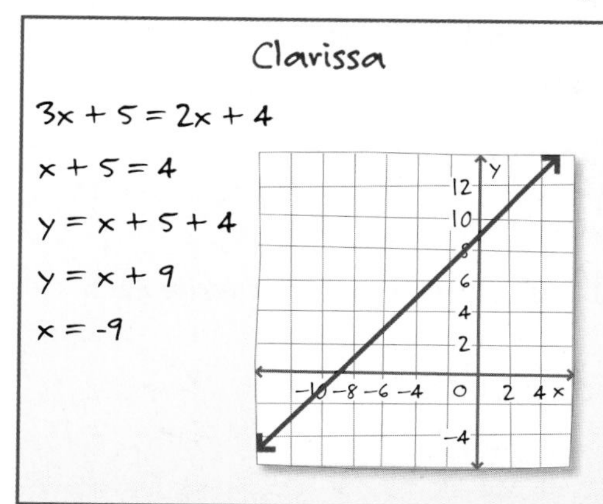

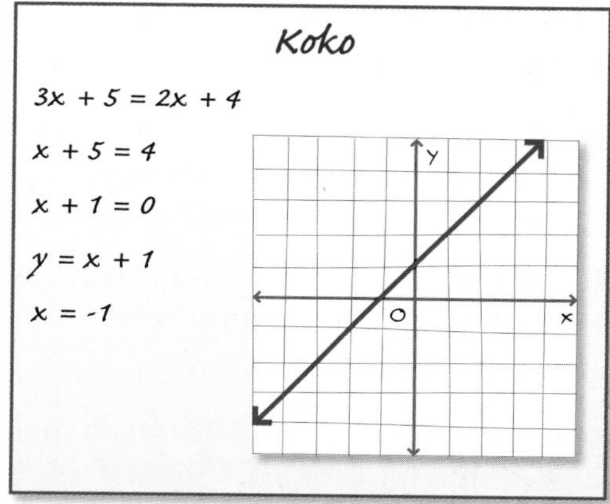

47. CHALLENGE Use a graphing calculator to find the solution of $\frac{2}{3}(x + 3) = \frac{1}{2}(x + 5)$. Verify your solution algebraically.

48. REASONING Explain when it is better to solve an equation using algebraic methods and when it is better to solve by graphing.

49. OPEN ENDED Write a linear equation that has a root of $-\frac{3}{4}$. Write its related function.

50. WRITING IN MATH Summarize how to solve a linear equation algebraically and graphically.

Standardized Test Practice

51. What are the x- and y-intercepts of the graph of the function?

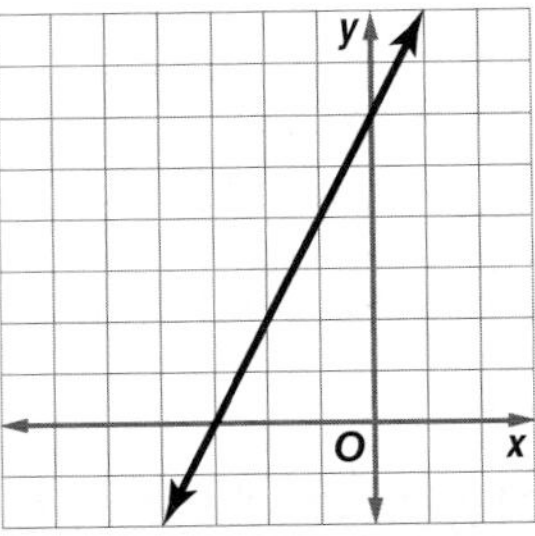

A $-3, 6$
B $6, -3$
C $3, -6$
D $-6, 3$

52. The table shows the cost C of renting a pontoon boat for h hours.

Hours	1	2	3
Cost ($)	7.25	14.5	21.75

Which equation best represents the data?

F $C = 7.25h$
G $C = h + 7.25$
H $C = 21.75 - 7.25h$
J $C = 7.25h + 21.75$

53. Which is the best estimate for the x-intercept of the graph of the linear function represented in the table?

x	y
0	5
1	3
2	1
3	−1
4	−3

A between 0 and 1
B between 2 and 3
C between 1 and 2
D between 3 and 4

54. EXTENDED RESPONSE Mr. Kauffmann has the following options for a backyard pool.

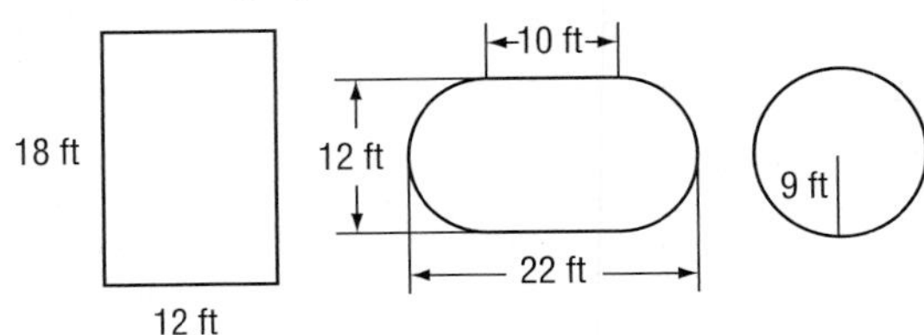

Which pool would give the greatest area to swim? Explain your reasoning.

Spiral Review

Find the x- and y-intercepts of the graph of each linear equation. (Lesson 3-1)

55. $y = 2x + 10$

56. $3y = 6x - 9$

57. $4x - 14y = 28$

58. FOOD If 2% milk contains 2% butterfat and whipping cream contains 9% butterfat, how much whipping cream and 2% milk should be mixed to obtain 35 gallons of milk with 4% butterfat? (Lesson 2-9)

Identify the hypothesis and conclusion of each statement. Then write each statement in if-then form. (Lesson 1-8)

59. A number that is divisible by 10 is also divisible by 5.

60. A rectangle is a quadrilateral with four right angles.

Skills Review

Simplify. (Lesson 0-3)

61. $\frac{25}{10}$

62. $\frac{-4}{-12}$

63. $\frac{6}{-12}$

64. $\frac{-36}{8}$

Evaluate $\frac{a-b}{c-d}$ for the given values. (Lesson 1-2)

65. $a = 6, b = 2, c = 9, d = 3$

66. $a = -8, b = 4, c = 5, d = -3$

67. $a = 4, b = -7, c = -1, d = -2$

EXTEND 3-2

Graphing Technology Lab
Graphing Linear Functions

Math Online glencoe.com
- Other Calculator Keystrokes
- Graphing Technology Personal Tutor

The power of a graphing calculator is the ability to graph different types of equations accurately and quickly. By entering one or more equations in the calculator you can view features of a graph, such as the x-intercept, y-intercept, the origin, intersections, and the coordinates of specific points.

Often linear equations are graphed in the **standard viewing window**, which is $[-10, 10]$ by $[-10, 10]$ with a scale of 1 on each axis. To quickly choose the standard viewing window on a TI-83/84 Plus, press [Zoom] 6.

ACTIVITY 1 Graph a Linear Equation

Graph $3x - y = 4$.

Step 1 Enter the equation in the Y= list.

- The Y= list shows the equation or equations that you will graph.
- Equations must be entered with the y isolated on one side of the equation. Solve the equation for y, then enter it into the calculator.

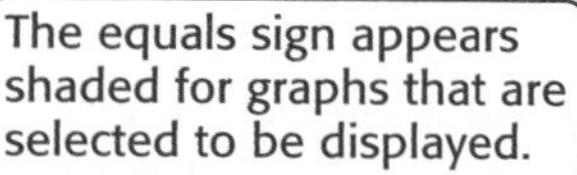

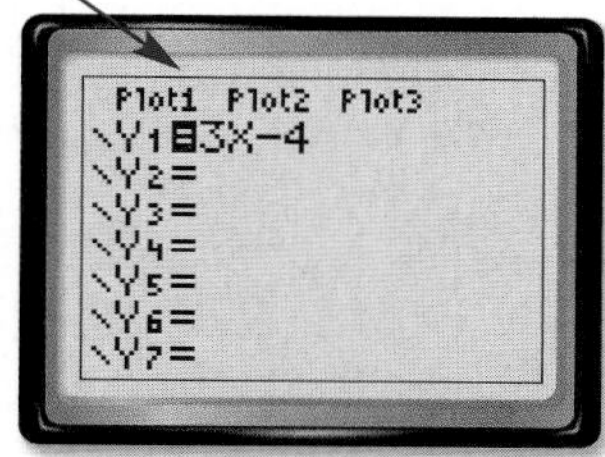

$3x - y = 4$	**Original equation**
$3x - y - 3x = 4 - 3x$	**Subtract $3x$ from each side.**
$-y = -3x + 4$	**Simplify.**
$y = 3x - 4$	**Multiply each side by -1.**

KEYSTROKES: [Y=] 3 [X,T,θ,n] [−] 4

Step 2 Graph the equation in the standard viewing window.

- Graph the selected equation.

KEYSTROKES: [Zoom] 6

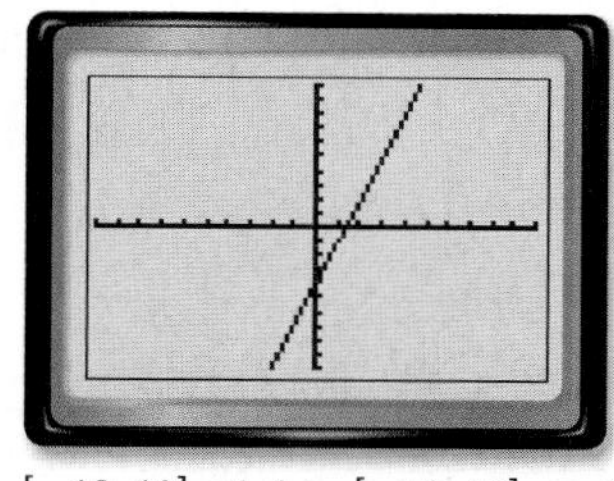

$[-10, 10]$ scl: 1 by $[-10, 10]$ scl: 1

Sometimes a complete graph is not displayed using the standard viewing window. A **complete graph** includes all of the important characteristics of the graph on the screen including the origin and the x- and y-intercepts. Note that the graph above is a complete graph because all of these points are visible.

When a complete graph is not displayed using the standard viewing window, you will need to change the viewing window to accommodate these important features. Use what you have learned about intercepts to help you choose an appropriate viewing window.

ACTIVITY 2 Graph a Complete Graph

Graph $y = 5x - 14$.

Step 1 Enter the equation in the Y= list and graph in the standard viewing window.

- Clear the previous equation from the Y= list. Then enter the new equation and graph.

KEYSTROKES: [Y=] [CLEAR] 5 [X,T,θ,n] [−] 14 [Zoom] 6

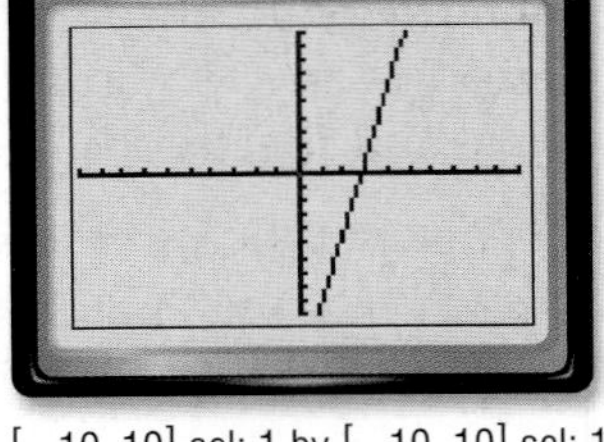

[−10, 10] scl: 1 by [−10, 10] scl: 1

Step 2 Modify the viewing window and graph again.

- The origin and the x-intercept are displayed in the standard viewing window. But notice that the y-intercept is outside of the viewing window.

Find the y-intercept.

$y = 5x - 14$	**Original equation**
$= 5(0) - 14$	**Replace x with 0.**
$= -14$	**Simplify.**

Since the y-intercept is -14, choose a viewing window that includes a number less than -14. The window $[-10, 10]$ by $[-20, 5]$ with a scale of 1 on each axis is a good choice.

This window allows the complete graph, including the y-intercept, to be displayed.

[−10, 10] scl: 1 by [−20, 5] scl: 1

KEYSTROKES: [WINDOW] -10 [ENTER] 10 [ENTER] 1 [ENTER] -20 [ENTER] 5 [ENTER] 1 [GRAPH]

Exercises

Use a graphing calculator to graph each equation in the standard viewing window. Sketch the result.

1. $y = x + 5$

2. $y = 5x + 6$

3. $y = 9 - 4x$

4. $3x + y = 5$

5. $x + y = -4$

6. $x - 3y = 6$

Graph each equation in the standard viewing window. Determine whether the graph is complete. If the graph is not complete, adjust the viewing window and graph the equation again.

7. $y = 4x + 7$

8. $y = 9x - 5$

9. $y = 2x - 11$

10. $4x - y = 16$

11. $6x + 2y = 23$

12. $x + 4y = -36$

Consider the linear equation $y = 3x + b$.

13. Choose several different positive and negative values for b. Graph each equation in the standard viewing window.

14. For which values of b is the complete graph in the standard viewing window?

15. How is the value of b related to the y-intercept of the graph of $y = 3x + b$?

EXPLORE
3-3

Algebra Lab
Rate of Change of a Linear Function

Math Online glencoe.com

Math *in Motion*, Animation

Objective
Investigate the steepness of a line using concrete models.

In mathematics, you can measure the steepness of a line using a ratio.

Set Up the Lab

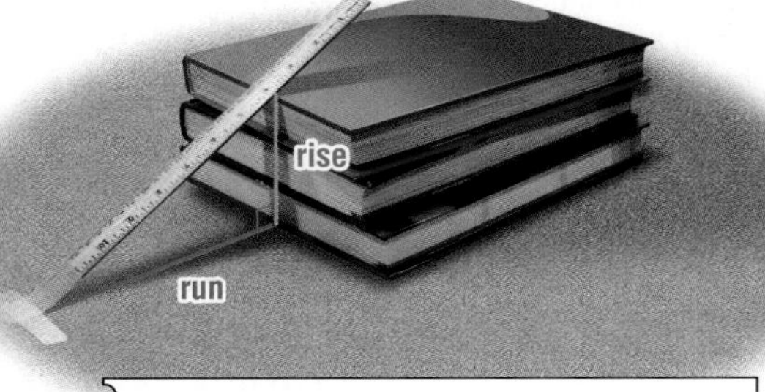

- Stack three books on your desk.
- Lean a ruler on the books to create a ramp.
- Tape the ruler to the desk.
- Measure the **rise** and the **run**. Record your data in a table like the one at the right.
- Calculate and record the ratio $\frac{\text{rise}}{\text{run}}$.

rise	run	$\frac{\text{rise}}{\text{run}}$

Step 1

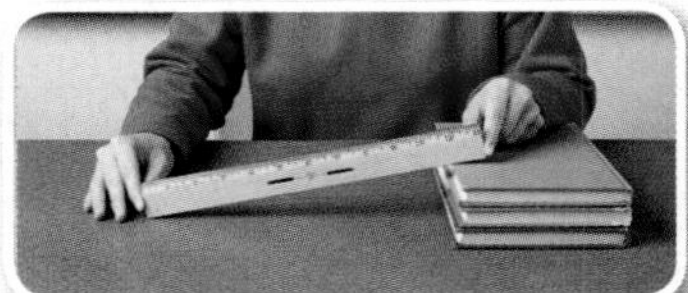

Move the books to make the ramp steeper. Measure and record the **rise** and the **run**. Calculate and record $\frac{\text{rise}}{\text{run}}$.

Step 2

Add books to the stack to make the ramp even steeper. Measure, calculate, and record your data in the table.

Analyze the Results

1. Examine the ratios you recorded. How did they change as the ramp became steeper?

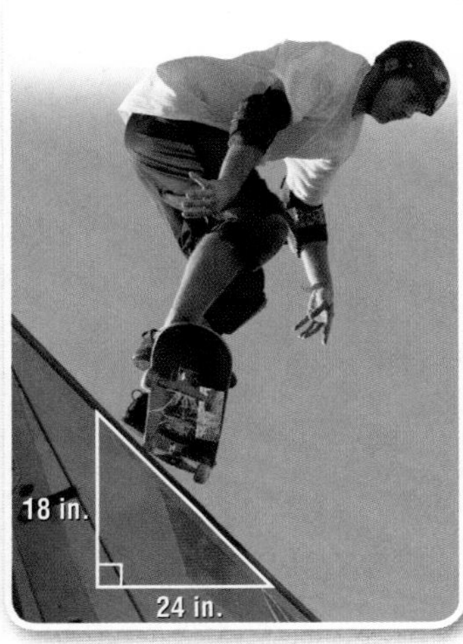

$m = \frac{18}{24} = \frac{3}{4}$

2. **MAKE A PREDICTION** Suppose you want to construct a skateboard ramp that is not as steep as the one shown at the left. List three different sets of $\frac{\text{rise}}{\text{run}}$ measurements that will result in a less steep ramp. Verify your predictions by calculating the ratio $\frac{\text{rise}}{\text{run}}$ for each ramp.

3. Copy the coordinate graph and draw a line through the origin with a $\frac{\text{rise}}{\text{run}}$ ratio greater than the original line. Then draw a line through the origin with a ratio less than that of the original line. Explain using the words *rise* and *run* why the lines you drew have a ratio greater or less than the original line.

4. We have seen what happens on the graph as the $\frac{\text{rise}}{\text{run}}$ ratio gets closer to zero. What would you predict will happen when the ratio is zero? Explain your reasoning. Give an example to support your prediction.

3-3 Rate of Change and Slope

Then

You graphed ordered pairs in the coordinate plane. (Lesson 1-6)

Now

- Use rate of change to solve problems.
- Find the slope of a line.

New Vocabulary

rate of change
slope

Math Online

glencoe.com

- Extra Examples
- Personal Tutor
- Self-Check Quiz
- Homework Help

Why?

The Daredevil Drop at Wet 'n Wild Emerald Pointe in Greensboro, North Carolina, is a thrilling ride that drops you 76 feet down a steep water chute. The *rate of change* of the ride describes how far a rider goes over the course of time on the ride.

Rate of Change **Rate of change** is a ratio that describes, on average, how much one quantity changes with respect to a change in another quantity.

Key Concept — Rate of Change

For Your FOLDABLE

If x is the independent variable and y is the dependent variable, then

$$\text{rate of change} = \frac{\text{change in } y}{\text{change in } x}.$$

Real-World EXAMPLE 1 — Find Rate of Change

ENTERTAINMENT Use the table to find the rate of change. Then explain its meaning.

Number of Computer Games	Total Cost ($)
x	y
2	78
4	156
6	234

$$\text{rate of change} = \frac{\text{change in } y}{\text{change in } x} \begin{matrix}\leftarrow \text{dollars}\\ \leftarrow \text{games}\end{matrix}$$

$$= \frac{\text{change in cost}}{\text{change in number of games}}$$

$$= \frac{156 - 78}{4 - 2}$$

$$= \frac{78}{2} \text{ or } \frac{39}{1}$$

The rate of change is $\frac{39}{1}$. This means that each game costs $39.

Check Your Progress

1. **REMODELING** The table shows how the tiled surface area changes with the number of floor tiles.

   A. Find the rate of change.

   B. Explain the meaning of the rate of change.

Number of Floor Tiles	Area of Tiled Surface (in^2)
x	y
3	48
6	96
9	144

Personal Tutor glencoe.com

StudyTip

Rate A positive rate of change indicates an increase over time. A negative rate of change indicates that a quantity is decreasing.

So far, you have seen rates of change that are *constant*. Many real-world situations involve rates of change that are not constant.

Real-World EXAMPLE 2 Variable Rate of Change

AMUSEMENT PARKS The graph shows the number of people who visited U.S. theme parks in recent years.

a. Find the rates of change for 2000–2002 and 2002–2004.

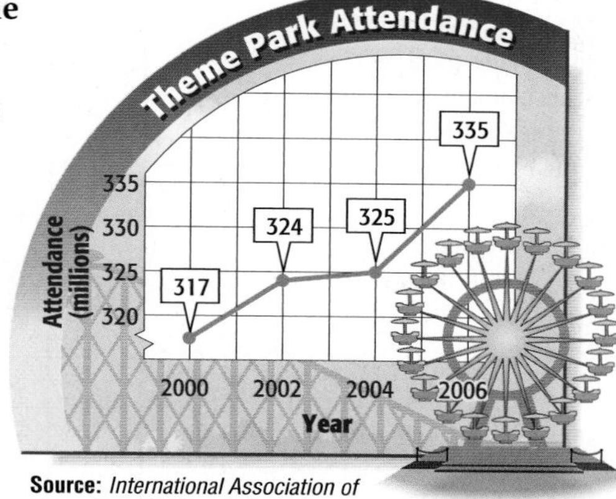

Source: *International Association of Amusement Parks and Attractions*

2000–2002:

$$\frac{\text{change in attendance}}{\text{change in time}} = \frac{324 - 317}{2002 - 2000} \begin{matrix} \leftarrow \text{people} \\ \leftarrow \text{years} \end{matrix} \qquad \text{Substitute.}$$

$$= \frac{7}{2} \text{ or } 3.5 \qquad \text{Simplify.}$$

Over this 2-year period, attendance increased by 7 million, for a rate of change of 3.5 million per year.

2002–2004:

$$\frac{\text{change in attendance}}{\text{change in time}} = \frac{328 - 324}{2004 - 2002} \qquad \text{Substitute.}$$

$$= \frac{4}{2} \text{ or } 2 \qquad \text{Simplify.}$$

Over this 2-year period, attendance increased by 4 million, for a rate of change of 2 million per year.

b. Explain the meaning of the rate of change in each case.

For 2000–2002, on average, 3.5 million more people went to a theme park each year than the last.

For 2002–2004, on average, 2 million more people attended theme parks each year than the last.

c. How are the different rates of change shown on the graph?

There is a greater vertical change for 2000–2002 than for 2002–2004. Therefore, the section of the graph for 2000–2002 is steeper.

Check Your Progress

2. Refer to the graph above. Without calculating, find the 2-year period that has the least rate of change. Then calculate to verify your answer.

Personal Tutor glencoe.com

A rate of change is constant for a function when the rate of change is the same between any pair of points on the graph of the function. Linear functions have a constant rate of change.

StudyTip

Linear or Nonlinear Function? Notice that the changes in x and y are not the same. For the rate of change to be linear, the change in x-values must be constant and the change in y-values must be constant.

EXAMPLE 3 Constant Rates of Change

Determine whether each function is linear. Explain.

a.

x	y
1	−6
4	−8
7	−10
10	−12
13	−14

x	y	rate of change
1	−6	
4	−8	$\frac{-8-(-6)}{4-1}$ or $-\frac{2}{3}$
7	−10	$\frac{-10-(-8)}{7-4}$ or $-\frac{2}{3}$
10	−12	$\frac{-12-(-10)}{10-7}$ or $-\frac{2}{3}$
13	−14	$\frac{-14-(-12)}{13-10}$ or $-\frac{2}{3}$

The rate of change is constant. Thus, the function is linear.

b.

x	y
−3	10
−1	12
1	16
3	18
5	22

x	y	rate of change
−3	10	
−1	12	$\frac{12-10}{-1-(-3)}$ or 1
1	16	$\frac{16-12}{1-(-1)}$ or 2
3	18	$\frac{18-16}{3-1}$ or 1
5	22	$\frac{22-18}{5-3}$ or 2

This rate of change is not constant. Thus, the function is not linear.

Check Your Progress

3A.

x	y
−3	11
−2	15
−1	19
1	23
2	27

3B.

x	y
12	−4
9	1
6	6
3	11
0	16

Personal Tutor glencoe.com

Find Slope The **slope** of a nonvertical line is the ratio of the change in the y-coordinates (rise) to the change in the x-coordinates (run) as you move from one point to another.

It can be used to describe a rate of change. Slope describes how steep a line is. The greater the absolute value of the slope, the steeper the line.

The graph shows a line that passes through (−1, 3) and (2, −2).

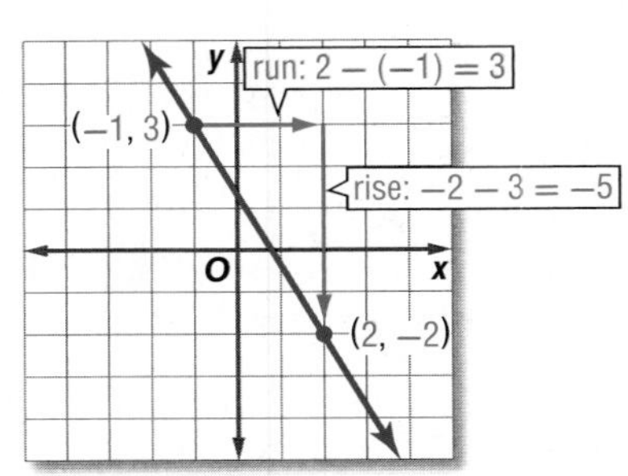

$$\textbf{slope} = \frac{\text{rise}}{\text{run}}$$

$$= \frac{\text{change in } y\text{-coordinates}}{\text{change in } x\text{-coordinates}}$$

$$= \frac{-2-3}{2-(-1)} \text{ or } -\frac{5}{3}$$

So, the slope of the line is $-\frac{5}{3}$.

Because a linear function has a constant rate of change, any two points on a nonvertical line can be used to determine its slope.

ReadingMath

Subscripts y_1 is read as *y sub one* and x_2 is read as *x sub two*. The 1 and 2 are subscripts and refer to the first and second point to which the *x*- and *y*-values correspond.

Key Concept — Slope

For Your FOLDABLE

Words The slope of a nonvertical line is the ratio of the rise to the run.

Symbols The slope m of a nonvertical line through any two points, (x_1, y_1) and (x_2, y_2), can be found as follows.

$$m = \frac{y_2 - y_1}{x_2 - x_1} \begin{array}{l} \leftarrow \text{change in } y \\ \leftarrow \text{change in } x \end{array}$$

Graph

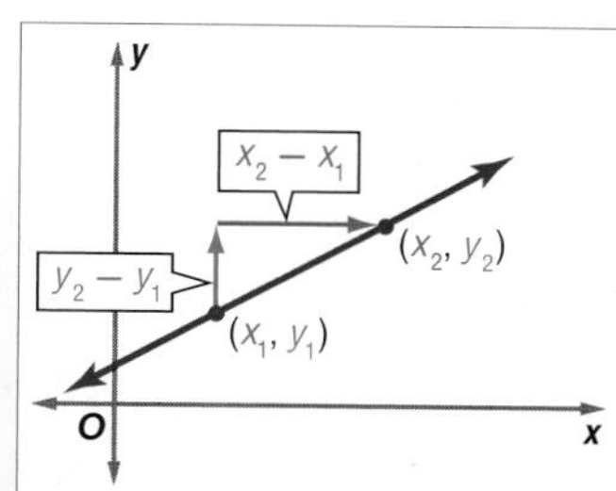

The slope of a line can be positive, negative, zero, or undefined. If the line is not horizontal or vertical, then the slope is either positive or negative.

EXAMPLE 4 Positive, Negative, and Zero Slope

Find the slope of a line that passes through each pair of points.

a. (−2, 0) and (1, 5)

$m = \frac{y_2 - y_1}{x_2 - x_1}$ **rise/run**

$= \frac{5 - 0}{1 - (-2)}$ **$(-2, 0) = (x_1, y_1)$ and $(1, 5) = (x_2, y_2)$**

$= \frac{5}{3}$ **Simplify.**

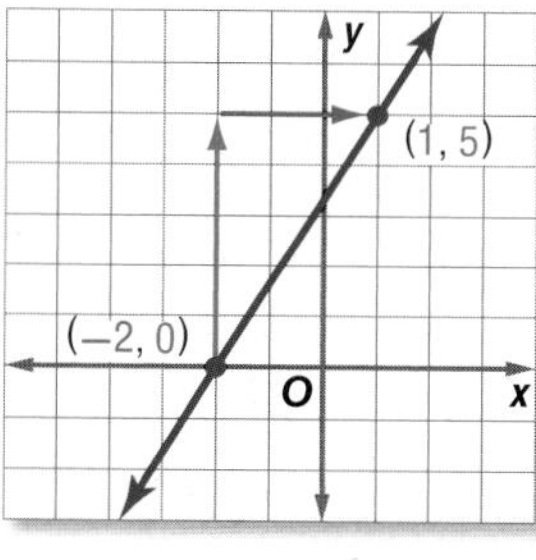

b. (−3, 4) and (2, −3)

$m = \frac{y_2 - y_1}{x_2 - x_1}$ **rise/run**

$= \frac{-3 - 4}{2 - (-3)}$ **$(-3, 4) = (x_1, y_1)$ and $(2, -3) = (x_2, y_2)$**

$= \frac{-7}{5}$ or $-\frac{7}{5}$ **Simplify.**

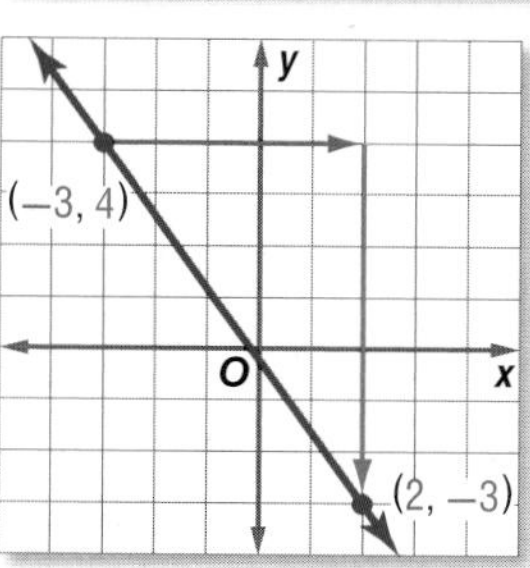

c. (−3, −1) and (2, −1)

$m = \frac{y_2 - y_1}{x_2 - x_1}$ **rise/run**

$= \frac{-1 - (-1)}{2 - (-3)}$ **Substitute.**

$= \frac{0}{2}$ or 0 **Simplify.**

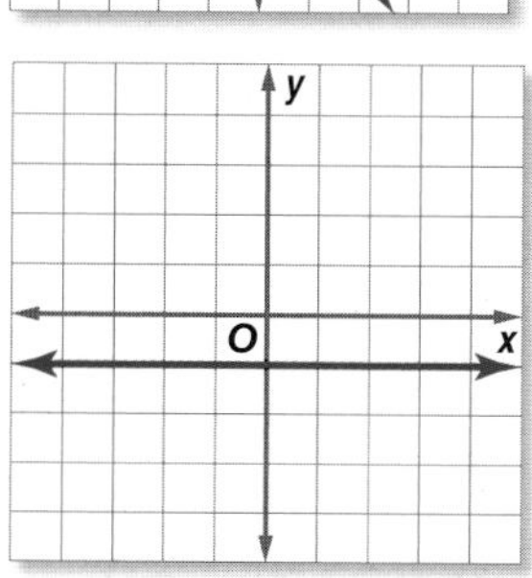

Watch Out!

Order Be careful not to transpose the order of the *x*-values or the *y*-values.

Check Your Progress

Find the slope of the line that passes through each pair of points.

4A. (3, 6), (4, 8) **4B.** (−4, −2), (0, −2) **4C.** (−4, 2), (−2, 10)

4D. (6, 7), (−2, 7) **4E.** (−2, 2), (−6, 4) **4F.** (4, 3), (−1, 11)

Personal Tutor glencoe.com

StudyTip

Zero and Undefined Slopes If the change in y-values is 0, then the graph of the line is horizontal. If the change in x-values is 0, then the slope is undefined. This graph is a vertical line.

EXAMPLE 5 Undefined Slope

Find the slope of the line that passes through (−2, 4) and (−2, −3).

$m = \frac{y_2 - y_1}{x_2 - x_1}$ — **$\frac{\text{rise}}{\text{run}}$**

$= \frac{-3 - 4}{-2 - (-2)}$ — **Substitute.**

$= \frac{-7}{0}$ or undefined — **Simplify.**

Check Your Progress

Find the slope of the line that passes through each pair of points.

5A. (6, 3), (6, 7) **5B.** (−3, 2), (−3, −1)

Personal Tutor glencoe.com

The graphs of lines with different slopes are summarized below.

Concept Summary Slope

For Your FOLDABLE

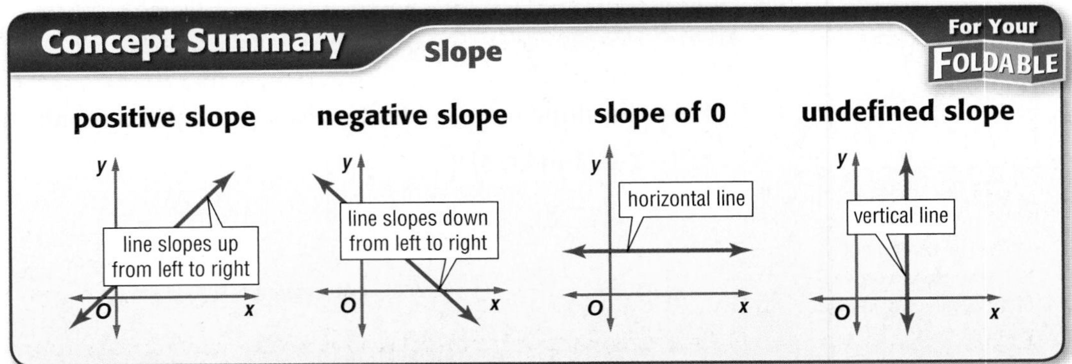

Sometimes you are given the slope and must find a missing coordinate.

EXAMPLE 6 Find Coordinates Given the Slope

Find the value of r so that the line through (1, 4) and (−5, r) has a slope of $\frac{1}{3}$.

$m = \frac{y_2 - y_1}{x_2 - x_1}$ — **Slope Formula**

$\frac{1}{3} = \frac{r - 4}{-5 - 1}$ — **Let (1, 4) = (x_1, y_1) and (−5, r) = (x_2, y_2).**

$\frac{1}{3} = \frac{r - 4}{-6}$ — **Subtract.**

$3(r - 4) = 1(-6)$ — **Find the cross products.**

$3r - 12 = -6$ — **Distributive Property.**

$3r = 6$ — **Add 12 to each side and simplify.**

$r = 2$ — **Divide each side by 3 and simplify.**

So, the line goes through (−5, 2).

Check Your Progress

Find the value of r so the line that passes through each pair of points has the given slope.

6A. (−2, 6), (r, −4); $m = -5$ **6B.** (r, −6), (5, −8); $m = -8$

Personal Tutor glencoe.com

Check Your Understanding

Example 1 p. 170

Find the rate of change represented in each table or graph.

1. 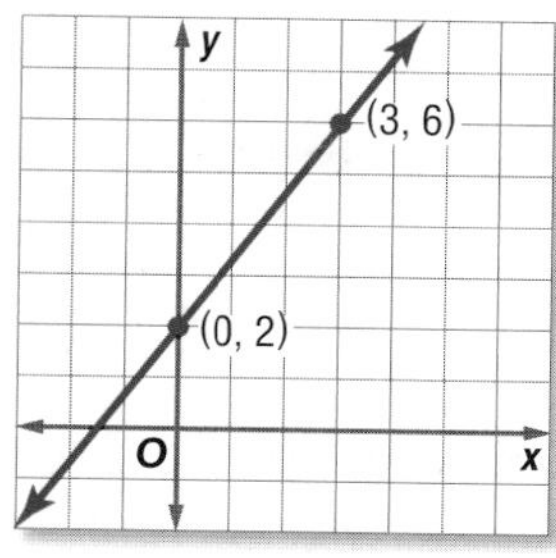

2.

x	y
3	−6
5	2
7	10
9	18
11	26

Example 2 p. 171

3. **SPORTS** Refer to the graph at the right.
   a. Find the rate of change of prices from 2002 to 2004. Explain the meaning of the rate of change.
   b. Without calculating, find a two-year period that had a greater rate of change than 2002–2004. Explain.
   c. Between which years would you guess the new stadium was built? Explain your reasoning.

Source: *Team Marketing Report*

Example 3 p. 172

Determine whether each function is linear. Write *yes* or *no*. Explain.

4.

x	−7	−4	−1	2	5
y	5	4	3	2	1

5.

x	8	12	16	20	24
y	7	5	3	0	−2

Examples 4 and 5 pp. 173–174

Find the slope of the line that passes through each pair of points.

6. (5, 3), (6, 9)
7. (−4, 3), (−2, 1)
8. (6, −2), (8, 3)
9. (1, 10), (−8, 3)
10. (−3, 7), (−3, 4)
11. (5, 2), (−6, 2)

Example 6 p. 174

Find the value of r so the line that passes through each pair of points has the given slope.

12. $(-4, r), (-8, 3), m = -5$
13. $(5, 2), (-7, r), m = \frac{5}{6}$

Practice and Problem Solving

● = **Step-by-Step Solutions** begin on page R12.
Extra Practice begins on page 815.

Example 1 p. 170

Find the rate of change represented in each table or graph.

14.

x	y
5	2
10	3
15	4
20	5

15.

x	y
1	15
2	9
3	3
4	−3

Example 1 p. 170

Find the rate of change represented in each table or graph.

16.

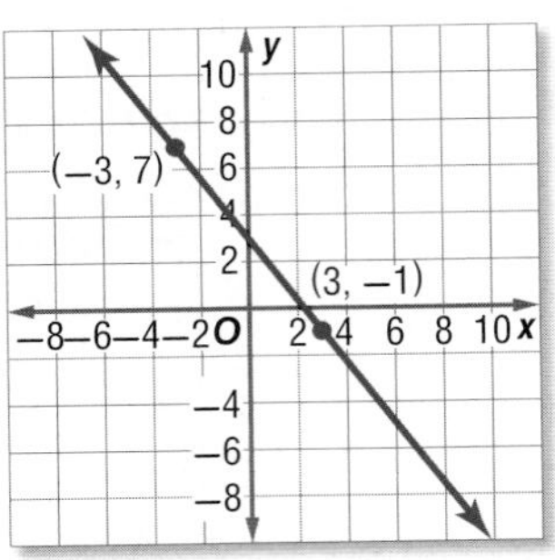

17.

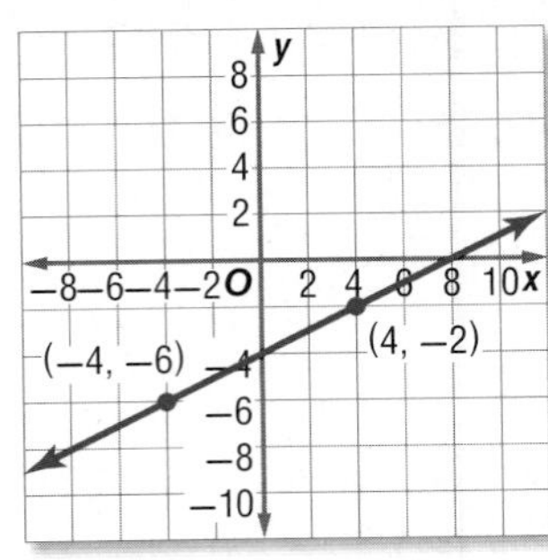

Example 2 p. 171

18. SPORTS What was the annual rate of change from 1995 to 2003 for women competing in triathlons? Explain the meaning of the rate of change.

Year	Number of Women
1995	4600
2003	19,100

19. RETAIL The average retail price in the spring of 2008 for a used car is shown in the table at the right.

Age (years)	Value ($)
2	15,924.96
3	14,113.29

a. Write a linear function to model the price of the car with respect to age.

b. Interpret the meaning of the slope of the line.

c. Assuming a constant rate of change predict the average retail price for a 7-year-old car.

Example 3 p. 172

Determine whether each function is linear. Write *yes* or *no*. Explain.

20.

x	4	2	0	−2	−4
y	−1	1	3	5	7

21.

x	−7	−5	−3	−1	0
y	11	14	17	20	23

22.

x	−0.2	0	0.2	0.4	0.6
y	0.7	0.4	0.1	0.3	0.6

23.

x	$\frac{1}{2}$	$\frac{3}{2}$	$\frac{5}{2}$	$\frac{7}{2}$	$\frac{9}{2}$
y	$\frac{1}{2}$	1	$\frac{3}{2}$	2	$\frac{5}{2}$

Examples 4 and 5 pp. 173–174

Find the slope of the line that passes through each pair of points.

24. (4, 3), (−1, 6) **25.** (8, −2), (1, 1) **26.** (2, 2), (−2, −2)

27. (6, −10), (6, 14) **28.** (5, −4), (9, −4) **29.** (11, 7), (−6, 2)

30. (−3, 5), (3, 6) **31.** (−3, 2), (7, 2) **32.** (8, 10), (−4, −6)

33. (−8, 6), (−8, 4) **34.** (−12, 15), (18, −13) **35.** (−8, −15), (−2, 5)

Example 6 p. 174

Find the value of *r* so the line that passes through each pair of points has the given slope.

36. $(12, 10), (-2, r), m = -4$ **37.** $(r, -5), (3, 13), m = 8$

38. $(3, 5), (-3, r), m = \frac{3}{4}$ **39.** $(-2, 8), (r, 4), m = -\frac{1}{2}$

ESTIMATION **Use a ruler to estimate the slope of each object.**

40.

41.

42. **DRIVING** When driving up a certain hill, you rise 15 feet for every 1000 feet you drive forward. What is the slope of the road?

Find the slope of the line that passes through each pair of points.

43.

x	y
4.5	−1
5.3	2

44.

x	y
0.75	1
0.75	−1

45.

x	y
$2\frac{1}{2}$	$-1\frac{1}{2}$
$-\frac{1}{2}$	$\frac{1}{2}$

46. **GROWTH RATE** May's hair was 8 inches long. In three months, it grew another inch at a steady rate. Assume that her hair growth continues at the same rate.

  a. Make a table that shows May's hair length for each of the three months and for the next three months.

  b. Draw a graph showing the relationship between May's hair length and time in months.

  c. What is the slope of the graph? What does it represent?

Real-World Link

You have about 100,000 hairs on your head. Often hair grows faster in warm weather.

Source: HairBoutique

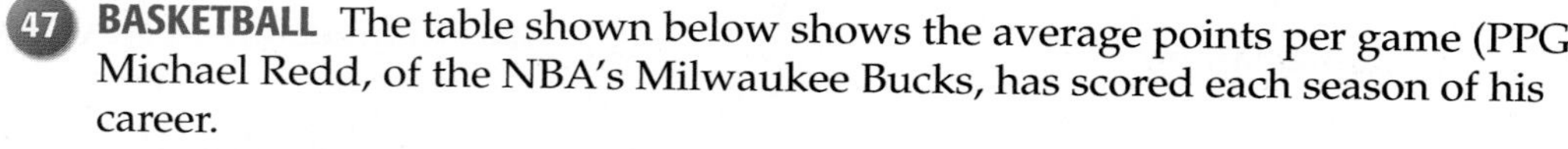

47. **BASKETBALL** The table shown below shows the average points per game (PPG) Michael Redd, of the NBA's Milwaukee Bucks, has scored each season of his career.

Season	2000–01	2001–02	2002–03	2003–04	2004–05	2005–06	2006–07
PPG	2.2	11.4	15.1	21.7	23.0	25.4	26.7

  a. Make a graph of the data. Connect each pair of adjacent points with a line.

  b. Use the graph to determine in which period Michael Redd's PPG increased the fastest. Explain your reasoning.

  c. Discuss the difference in the rate of change from the 2000–01 through the 2003–04 seasons and from the 2003–04 through the 2006–07 seasons.

H.O.T. Problems

Use Higher-Order Thinking Skills

48. **REASONING** Why does the Slope Formula not work for vertical lines? Explain.

49. **OPEN ENDED** Use what you know about rate of change to describe the function represented by the table.

Time (wk)	Height of Plant (in.)
4	9.0
6	13.5
8	18.0

50. **CHALLENGE** Find the value of d so the line that passes through (a, b) and (c, d) has a slope of $\frac{1}{2}$.

51. **WRITING IN MATH** Explain how the rate of change and slope are related and how to find the slope of a line.

52. **FIND THE ERROR** Kyle and Luna are finding the value of x so the line that passes through $(10, x)$ and $(-2, 8)$ has a slope of $\frac{1}{4}$. Is either of them correct? Explain.

Kyle

$$\frac{-2-10}{8-x} = \frac{1}{4}$$

$$1(8-x) = 4(-12)$$

$$8 - x = -48$$

$$x = 56$$

Luna

$$\frac{8-x}{-2-10} = \frac{1}{4}$$

$$4(8-x) = 1(-12)$$

$$32 - 4x = -12$$

$$x = 11$$

Standardized Test Practice

53. The cost of prints from an online photo processor is given by $C(p) = 29.99 + 0.13p$. \$29.99 is the cost of the membership, and p is the number of 4-inch by 6-inch prints. What does the slope represent?

A cost per print
B cost of the membership
C cost of the membership and 1 print
D number of prints

54. Danita bought a computer for \$1200 and its value depreciated linearly. After 2 years, the value was \$250. What was the amount of yearly depreciation?

F \$950
G \$475
H \$250
J \$225

55. SHORT RESPONSE The graph represents how much the Wright Brothers National Monument charges visitors. How much does the park charge each visitor?

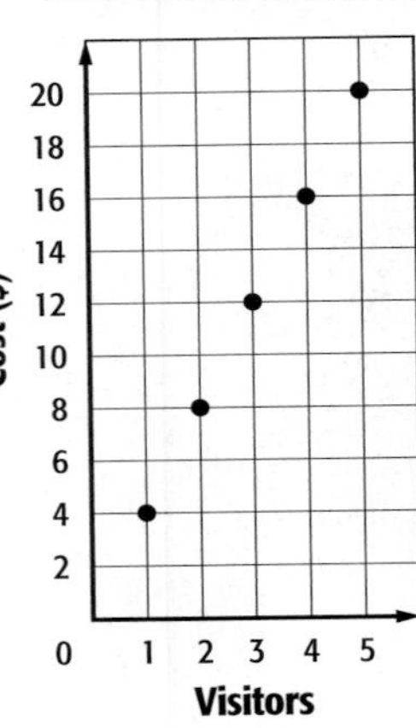

56. PROBABILITY At a gymnastics camp, 1 gymnast is chosen at random from each team. The Flipstars Gymnastics Team consists of 5 eleven-year-olds, 7 twelve-year-olds, 10 thirteen-year-olds, and 8 fourteen-year-olds. What is the probability that the age of the gymnast chosen is an odd number?

A $\frac{1}{30}$ **B** $\frac{1}{15}$ **C** $\frac{1}{2}$ **D** $\frac{3}{5}$

Spiral Review

Solve each equation by graphing. (Lesson 3-2)

57. $3x + 18 = 0$ **58.** $8x - 32 = 0$ **59.** $0 = 12x - 48$

Find the x- and y-intercepts of the graph of each linear function. (Lesson 3-1)

60.

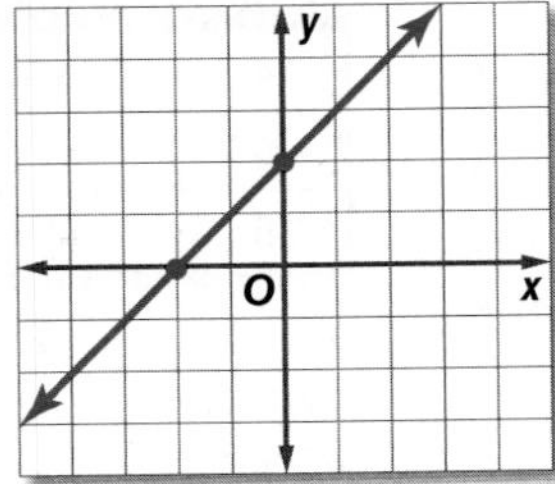

61.

x	y
−3	−4
−2	−2
−1	0
0	2
1	4

62. HOMECOMING Dance tickets are \$9 for one person and \$15 for two people. If a group of seven students wishes to go to the dance, write and solve an equation that would represent the least expensive price p of their tickets. (Lesson 1-3)

Skills Review

Find each quotient. (Lesson 0-5)

63. $8 \div \frac{2}{3}$ **64.** $\frac{3}{8} \div \frac{1}{4}$ **65.** $\frac{5}{8} \div 2$ **66.** $\frac{12 \cdot 6}{9}$ **67.** $\frac{2 \cdot 15}{6}$ **68.** $\frac{18 \cdot 5}{15}$

Mid-Chapter Quiz

Lessons 3-1 through 3-3

Determine whether each equation is a linear equation. Write *yes* or *no*. If yes, write the equation in standard form. (Lesson 3-1)

1. $y = -4x + 3$
2. $x^2 + 3y = 8$
3. $\frac{1}{4}x - \frac{3}{4}y = -1$

Graph each equation using the *x*- and *y*-intercepts. (Lesson 3-1)

4. $y = 3x - 6$
5. $2x + 5y = 10$

Graph each equation by making a table. (Lesson 3-1)

6. $y = -2x$
7. $x = 8 - y$

8. **BOOK SALES** The equation $5x + 12y = 240$ describes the total amount of money collected when selling x paperback books at \$5 per book and y hardback books at \$12 per book. Graph the equation using the x- and y-intercepts. (Lesson 3-1)

Find the root of each equation. (Lesson 3-2)

9. $x + 8 = 0$
10. $4x - 24 = 0$
11. $18 + 8x = 0$
12. $\frac{3}{5}x - \frac{1}{2} = 0$

Solve each equation by graphing. (Lesson 3-2)

13. $-5x + 35 = 0$
14. $14x - 84 = 0$
15. $118 + 11x = -3$

16. **MULTIPLE CHOICE** The function $y = -15 + 3x$ represents the outside temperature, in degrees Fahrenheit, in a small Alaskan town where x represents the number of hours after midnight. The function is accurate for x values representing midnight through 4:00 P.M. Find the zero of this function. (Lesson 3-2)

    A 0
    B 3
    C 5
    D −15

17. Find the rate of change represented in the table. (Lesson 3-3)

x	y
1	2
4	6
7	10
10	14

Find the slope of the line that passes through each pair of points. (Lesson 3-3)

18. (2, 6), (4, 12)
19. (1, 5), (3, 8)
20. (−3, 4), (2, −6)
21. $\left(\frac{1}{3}, \frac{3}{4}\right), \left(\frac{2}{3}, \frac{1}{4}\right)$

22. **MULTIPLE CHOICE** Find the value of r so the line that passes through the pair of points has the given slope. (Lesson 3-3)

$$(-4, 8), (r, 12), m = \frac{4}{3}$$

    F −4
    G −1
    H 0
    J 3

23. Find the slope of the line that passes through the pair of points. (Lesson 3-3)

x	y
2.6	−2
3.1	4

24. **POPULATION GROWTH** The graph shows the population growth in Leesburg, Florida, since 2000. (Lesson 3-3)

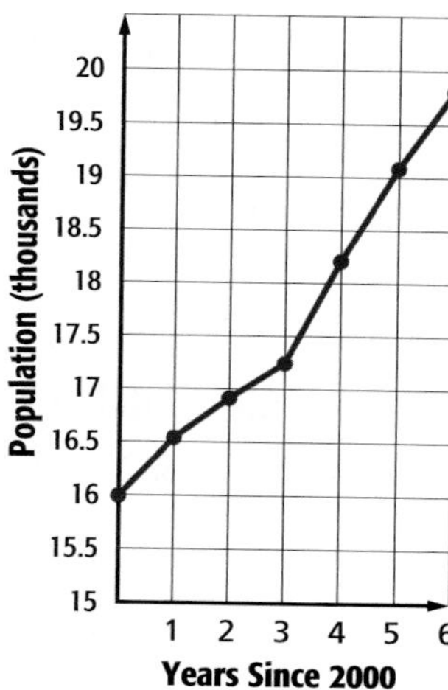

a. For which time period is the rate of change the greatest?

b. Explain the meaning of the slope from 2000 to 2006.

3-4 Direct Variation

Then

You found rates of change of linear functions. (Lesson 3-3)

Now

- Write and graph direct variation equations.
- Solve problems involving direct variation.

New Vocabulary

direct variation
constant of variation
constant of proportionality

Math Online

glencoe.com

- Extra Examples
- Personal Tutor
- Self-Check Quiz
- Homework Help

Why?

Bianca is saving her money to buy a designer purse that costs \$295. To help raise the money, she charges \$12.50 per hour to babysit her neighbors' two children. The slope of the line that represents the amount of money Bianca earns is 12.5, and the rate of change is constant.

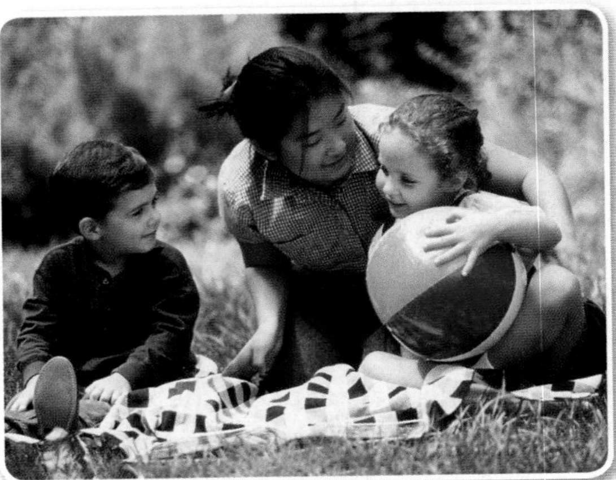

Direct Variation Equations A **direct variation** is described by an equation of the form $y = kx$, where $k \neq 0$. The equation $y = kx$ illustrates a constant rate of change, and k is the **constant of variation**, also called the **constant of proportionality**.

EXAMPLE 1 Slope and Constant of Variation

Name the constant of variation for each equation. Then find the slope of the line that passes through each pair of points.

a.

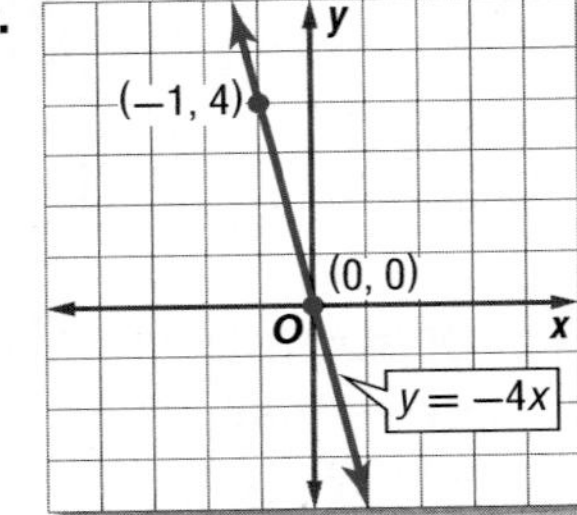

The constant of variation is −4.

$m = \frac{y_2 - y_1}{x_2 - x_1}$ **Slope Formula**

$= \frac{4 - 0}{-1 - 0}$ **$(x_1, y_1) = (0, 0)$** **$(x_2, y_2) = (-1, 4)$**

$= -4$ **The slope is −4.**

b.

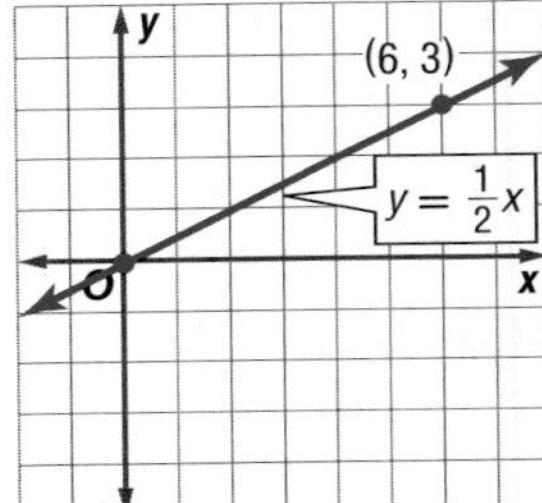

The constant of variation is $\frac{1}{2}$.

$m = \frac{y_2 - y_1}{x_2 - x_1}$ **Slope Formula**

$= \frac{3 - 0}{6 - 0}$ **$(x_1, y_1) = (0, 0)$** **$(x_2, y_2) = (6, 3)$**

$= \frac{1}{2}$ **The slope is $\frac{1}{2}$.**

Check Your Progress

1A. Name the constant of variation for $y = \frac{1}{4}x$. Then find the slope of the line that passes through (0, 0) and (4, 1), two points on the line.

1B. Name the constant of variation for $y = -2x$. Then find the slope of the line that passes through (0, 0) and (1, −2), two points on the line.

Personal Tutor glencoe.com

The slope of the graph of $y = kx$ is k. Since $0 = k(0)$, the graph of $y = kx$ always passes through the origin. Therefore the x- and y-intercepts are zero.

StudyTip

Constant of Variation A line with a positive constant of variation will go up from left to right and a line with a negative constant of variation will go down from left to right.

EXAMPLE 2 Graph a Direct Variation

Graph $y = -6x$.

Step 1 Write the slope as a ratio.

$-6 = \frac{-6}{1}$ $\frac{\text{rise}}{\text{run}}$

Step 2 Graph (0, 0).

Step 3 From the point (0, 0), move down 6 units and right 1 unit. Draw a dot.

Step 4 Draw a line containing the points.

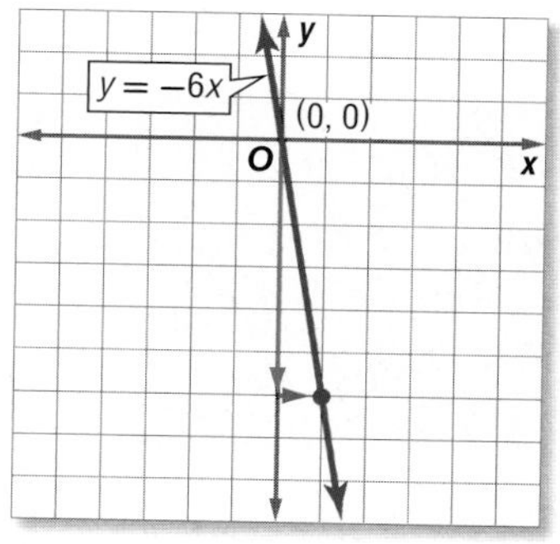

Check Your Progress

2A. $y = 6x$ **2B.** $y = \frac{2}{3}x$ **2C.** $y = -5x$ **2D.** $y = -\frac{3}{4}x$

Personal Tutor glencoe.com

The graphs of all direct variation equations share some common characteristics.

Concept Summary Direct Variation Graphs

For Your FOLDABLE

- Direct variation equations are of the form $y = kx$, where $k \neq 0$.
- The graph of $y = kx$ always passes through the origin.
- The slope is positive if $k > 0$.

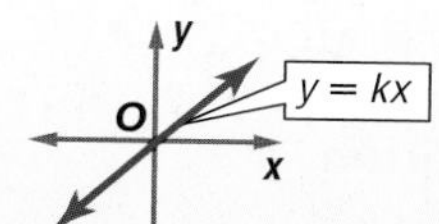

- The slope is negative if $k < 0$.

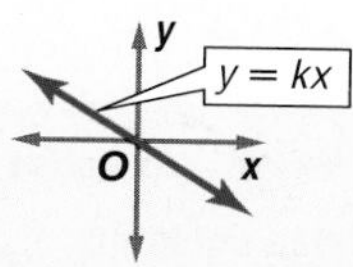

Math *in Motion*, Animation glencoe.com

If the relationship between the values of y and x can be described by a direct variation equation, then we say that y varies directly as x.

EXAMPLE 3 Write and Solve a Direct Variation Equation

Suppose y varies directly as x, and $y = 72$ when $x = 8$.

a. Write a direct variation equation that relates x and y.

$y = kx$	**Direct variation formula**
$72 = k(8)$	**Replace y with 72 and x with 8.**
$9 = k$	**Divide each side by 8.**

Therefore, the direct variation equation is $y = 9x$.

b. Use the direct variation equation to find x when $y = 63$.

$y = 9x$	**Direct variation formula**
$63 = 9x$	**Replace y with 63.**
$7 = x$	**Divide each side by 9.**

Therefore, $x = 7$ when $y = 63$.

Check Your Progress

3. Suppose y varies directly as x, and $y = 98$ when $x = 14$. Write a direct variation equation that relates x and y. Then find y when $x = -4$.

Personal Tutor glencoe.com

Direct Variation Problems One of the most common applications of direct variation is the formula $d = rt$. Distance d varies directly as time t, and the rate r is the constant of variation.

Real-World Link

In 2006, domestic airlines transported over 660 million passengers an average distance of 724 miles per flight.

Source: Bureau of Transportation Statistics

Real-World EXAMPLE 4 Estimate Using Direct Variation

TRAVEL The distance a jet travels varies directly as the number of hours it flies. A jet traveled 3420 miles in 6 hours.

a. Write a direct variation equation for the distance d flown in time t.

Words	Distance	equals	rate	times	time.
Variable	Let r = rate.				
Equation	3420	=	r	×	6

Solve for the rate.

$3420 = r(6)$ **Original equation**

$\frac{3420}{6} = \frac{r(6)}{6}$ **Divide each side by 6.**

$570 = r$ **Simplify.**

Therefore, the direct variation equation is $d = 570t$. The airliner flew at a rate of 570 miles per hour.

b. Graph the equation.

The graph of $d = 570t$ passes through the origin with slope 570.

$m = \frac{570}{1}$ **$\frac{\text{rise}}{\text{run}}$**

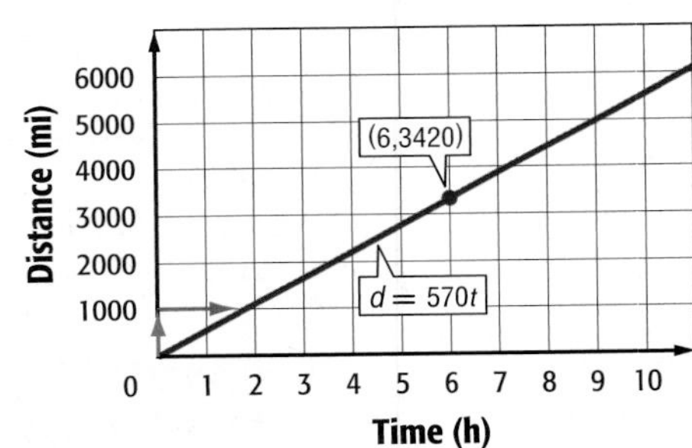

c. Estimate how many hours it will take for an airliner to fly 6500 miles.

$d = 570t$ **Original equation**

$6500 = 570t$ **Replace d with 6500.**

$\frac{6500}{570} = \frac{570t}{570}$ **Divide each side by 570.**

$t \approx 11.4$ **Simplify.**

It would take the airliner approximately 11.4 hours to fly 6500 miles.

Problem-Solving Tip

Write an Equation Sometimes the best way to solve a problem is to write an equation from the given information and solve.

Check Your Progress

4. HOT-AIR BALLOONS A hot-air balloon's height varies directly as the balloon's ascent time in minutes.

350 ft. in 5 min.

A. Write a direct variation for the distance d ascended in time t.

B. Graph the equation.

C. Estimate how many minutes it would take to ascend 2100 feet.

D. About how many minutes would it take to ascend 3500 feet?

Personal Tutor glencoe.com

Check Your Understanding

Example 1 p. 180

Name the constant of variation for each equation. Then find the slope of the line that passes through each pair of points.

1.

2. 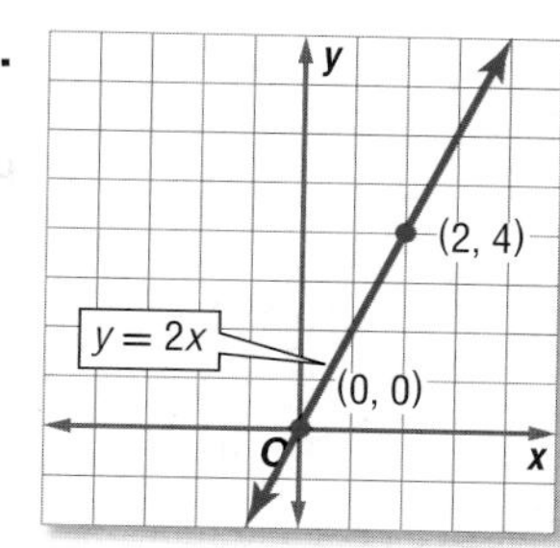

Example 2 p. 181

Graph each equation.

3. $y = -x$

4. $y = \frac{3}{4}x$

5. $y = -8x$

6. $y = -\frac{8}{5}x$

Example 3 p. 181

Suppose y varies directly as x. Write a direct variation equation that relates x and y. Then solve.

7. If $y = 15$ when $x = 12$, find y when $x = 32$.

8. If $y = -11$ when $x = 6$, find x when $y = 44$.

Example 4 p. 182

9. **MESSAGE BOARDS** You find that the number of messages you receive on your message board varies directly as the number of messages you post. When you post 5 messages, you receive 12 messages in return.

   a. Write a direct variation equation relating your posts to the messages received. Then graph the equation.

   b. Find the number of messages you need to post to receive 96 messages.

Practice and Problem Solving

● = **Step-by-Step Solutions** begin on page R12.
Extra Practice begins on page 815.

Example 1 p. 180

Name the constant of variation for each equation. Then find the slope of the line that passes through each pair of points.

10.

11. 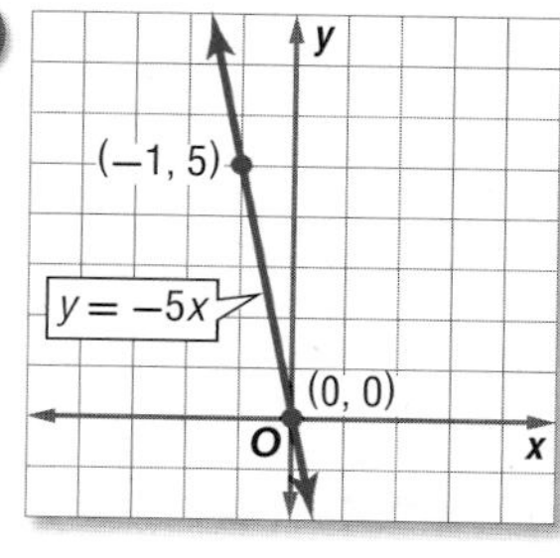

12. 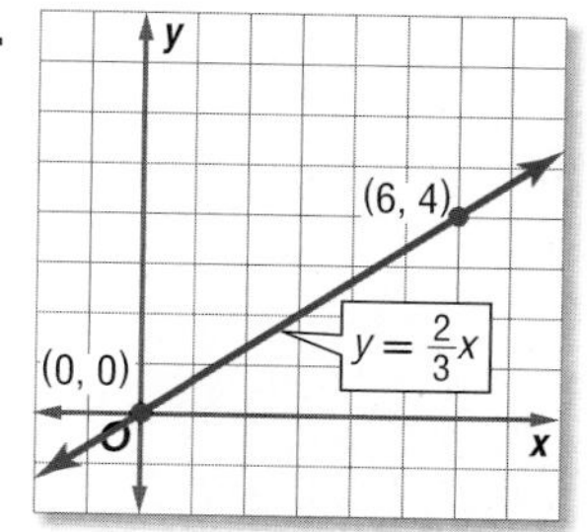

13.

14. 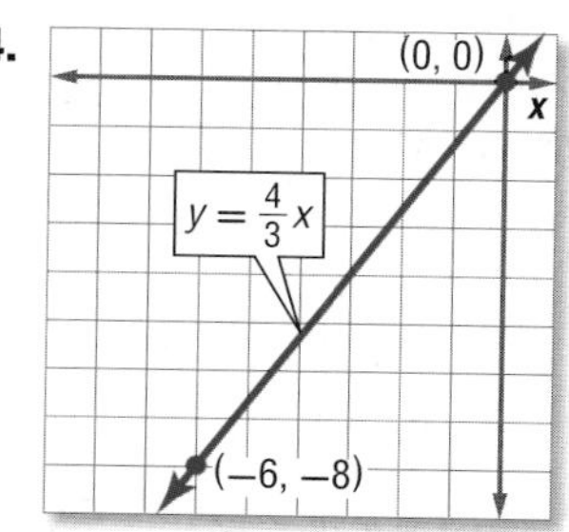

15. 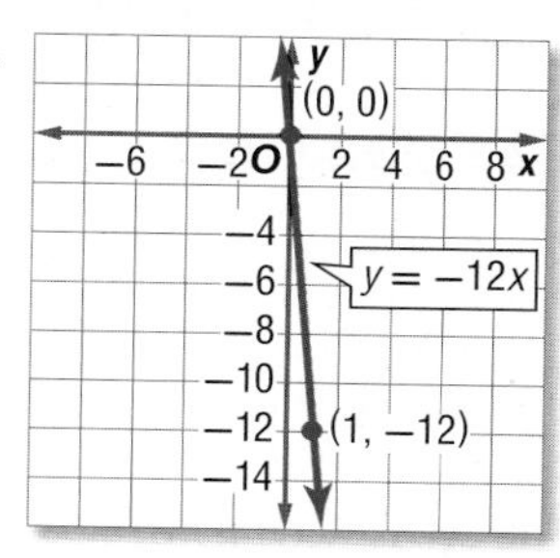

Example 2 p. 181

Graph each equation.

16. $y = 10x$
17. $y = -7x$
18. $y = x$
19. $y = \frac{7}{6}x$
20. $y = \frac{1}{6}x$
21. $y = \frac{2}{9}x$
22. $y = \frac{6}{5}x$
23. $y = -\frac{5}{4}x$

Example 3 p. 181

Suppose y varies directly as x. Write a direct variation equation that relates x and y. Then solve.

24. If $y = 6$ when $x = 10$, find x when $y = 18$.
25. If $y = 22$ when $x = 8$, find y when $x = -16$.
26. If $y = 4\frac{1}{4}$ when $x = \frac{3}{4}$, find y when $x = 4\frac{1}{2}$.
27. If $y = 12$ when $x = \frac{6}{7}$, find x when $y = 16$.

Example 4 p. 182

28. **SPORTS** The distance a golf ball travels at an altitude of 7000 feet varies directly with the distance the ball travels at sea level, as shown.

Hitting a Golf Ball		
Altitude (ft)	0 (sea level)	7000
Distance (yd)	200	210

  a. Write and graph an equation that relates the distance a golf ball travels at an altitude of 7000 feet y with the distance at sea level x.
  b. What would be a person's average driving distance at 7000 feet if his average driving distance at sea level is 180 yards?

29. **FINANCIAL LITERACY** Depreciation is the decline in a car's value over the course of time. The table below shows the values of a car with an average depreciation.

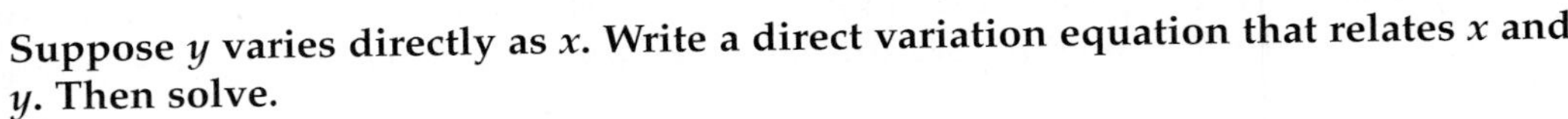

Age of Car (years)	1	2	3	4	5
Value (dollars)	12,000	10,200	8400	6600	4800

  a. Write an equation that relates the age x of the car to the value y that it lost after each year.
  b. Find the age of the car if the value is $300.

Suppose y varies directly as x. Write a direct variation equation that relates x and y. Then solve.

30. If $y = 3.2$ when $x = 1.6$, find y when $x = 19$.
31. If $y = 15$ when $x = \frac{3}{4}$, find x when $y = 25$.
32. If $y = 4.5$ when $x = 2.5$, find y when $x = 12$.
33. If $y = -6$ when $x = 1.6$, find y when $x = 8$.

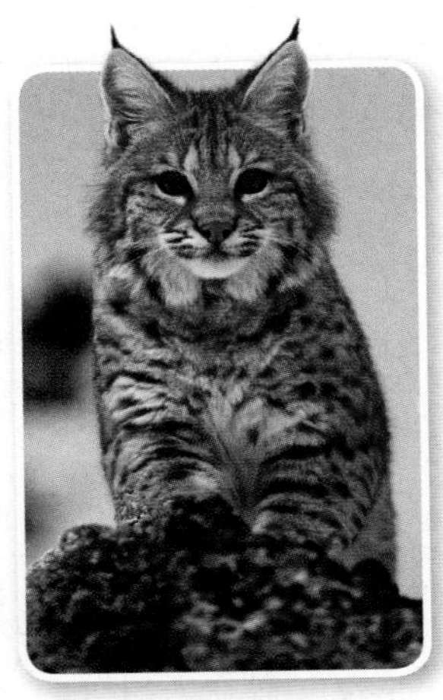

Real-World Link

The name Lynx comes from the Greek word "to shine." It may be in reference to the reflective ability of the cat's eyes.

Source: Lions, Tigers, and Bears

ENDANGERED SPECIES Certain endangered species experience cycles in their populations as shown in the graph at the right. Match each animal below to one of the colored lines in the graph.

34. red grouse, 8 years per cycle
35. voles, 3 years per cycle
36. lemmings, 4 years per cycle

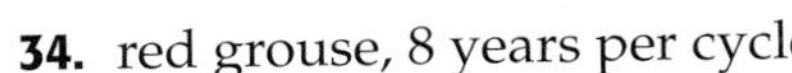

37. lynx, 10 years per cycle.

Population Cycles of Endangered Species

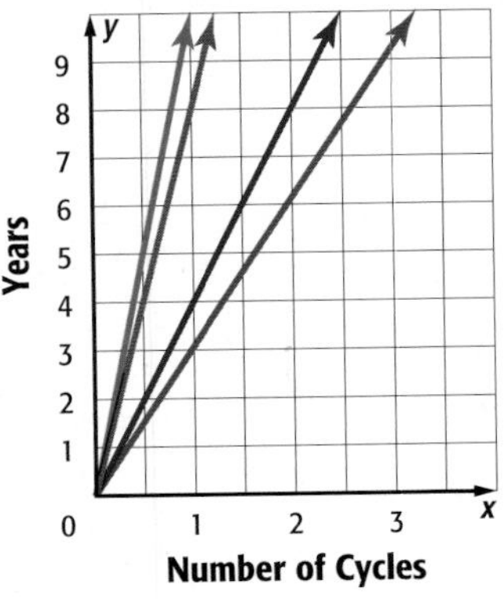

In Exercises 38–40, write and graph a direct variation equation that relates the variables.

38. PHYSICAL SCIENCE The weight W of an object is $9.8\ \text{m/s}^2$ times the mass of the object m.

39. MUSIC Music downloads are $0.99 per song. The total cost of d songs is T.

40. GEOMETRY The circumference of a circle C is approximately 3.14 times the diameter d.

41. MULTIPLE REPRESENTATIONS In this problem, you will investigate the family of direct variation functions.

a. GRAPHICAL Graph $y = x$, $y = 3x$, and $y = 5x$ on the same coordinate plane.

b. ALGEBRAIC Describe the relationship among the constant of variation, the slope of the line, and the rate of change of the graph.

c. VERBAL Make a conjecture about how you can determine without graphing which of two direct variation equations has the steeper graph.

42. TRAVEL A map of North Carolina is scaled so that 3 inches represents 93 miles. How far apart are Raleigh and Charlotte if they are 1.8 inches apart on the map?

43. INTERNET A company will design and maintain a Web site for your company for $9.95 per month. Write a direct variation equation to find the total cost C for having a Web page for n months.

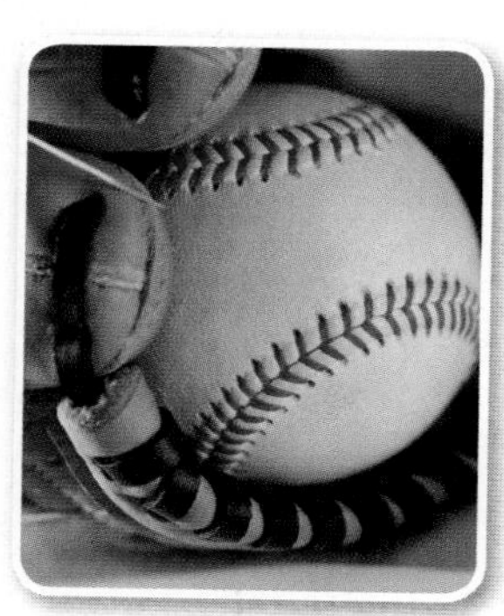

Real-World Link

More than 41 million fans attended Minor League Baseball games in 2006. Minor League Baseball draws more fans than the NBA or NFL.

Source: Minor League Basebal

44. BASEBALL Before their first game, high school student Todd McCormick warmed all 5200 seats in a new minor league stadium. He started at 11:50 A.M. and finished around 3 P.M.

a. Write a direct variation equation relating the number of seats to time. What is the meaning of the constant of variation in this situation?

b. About how many seats had Todd sat in by 1:00 P.M.?

c. How long would you expect it to take Todd to sit in all of the seats at a major league stadium with more than 40,000 seats?

H.O.T. Problems

Use Higher-Order Thinking Skills

45. WHICH ONE DOESN'T BELONG? Identify the equation that does not belong. Explain.

$9 = rt$	$9a = 0$	$z = \frac{1}{9}x$	$w = \frac{9}{t}$

46. REASONING How are the constant of variation and the slope related in a direct variation equation? Explain your reasoning.

47. OPEN ENDED Model a real-world situation using a direct variation equation. Graph the equation and describe the rate of change.

48. CHALLENGE Suppose y varies directly as x. If the value of x is doubled, then the value of y is also *always, sometimes* or *never* doubled. Explain your reasoning.

49. FIND THE ERROR Eddy says the slope between any two points on the graph of a direct variation equation $y = kx$ is $\frac{1}{k}$. Adelle says the slope depends on the points chosen. Is either of them correct? Explain.

50. WRITING IN MATH Describe the graph of a direct variation equation.

Standardized Test Practice

51. Patricia pays \$1.19 each to download songs to her MP3 player. If n is the number of downloaded songs, which equation represents the cost C in dollars?

A $C = 1.19n$
B $n = 1.19C$
C $C = 1.19 \div n$
D $C = n + 1.19$

52. Suppose that y varies directly as x, and $y = 8$ when $x = 6$. What is the value of y when $x = 8$?

F 6
G 12
H $10\frac{2}{3}$
J 16

53. What is the relationship between the input (x) and output (y)?

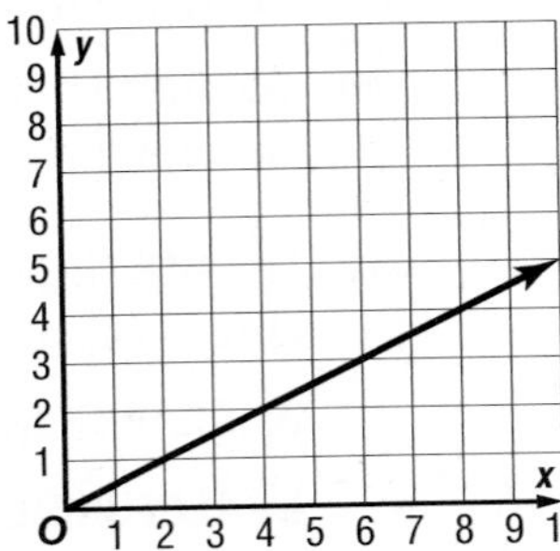

A The output is two more than the input.
B The output is two less than the input.
C The output is twice the input.
D The output is half the input.

54. SHORT RESPONSE A telephone company charges \$40 per month plus \$0.07 per minute. How much would a month of service cost a customer if the customer talked for 200 minutes?

Spiral Review

55. TELEVISION The graph shows the average number of television channels American households receive. What was the annual rate of change from 2004 to 2006? Explain the meaning of the rate of change. (Lesson 3-3)

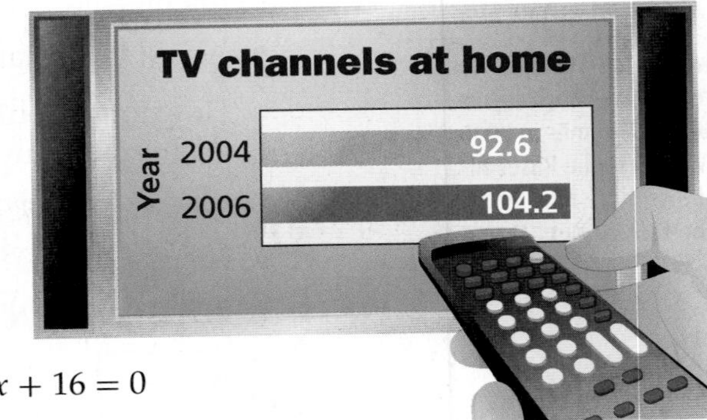

Solve each equation by graphing. (Lesson 3-2)

56. $0 = 18 - 9x$ **57.** $2x + 14 = 0$ **58.** $-4x + 16 = 0$

59. $-5x - 20 = 0$ **60.** $8x - 24 = 0$ **61.** $12x - 144 = 0$

Evaluate each expression if $a = 4$, $b = -2$, and $c = -4$. (Lesson 2-5)

62. $|2a + c| + 1$ **63.** $4a - |3b + 2|$ **64.** $-|a + 1| + |3c|$

65. $-a + |2 - a|$ **66.** $|c - 2b| - 3$ **67.** $-2|3b - 8|$

Skills Review

Find each difference. (Lesson 0-3)

68. $13 - (-1)$ **69.** $4 - 16$ **70.** $-3 - 3$

71. $-8 - (-2)$ **72.** $16 - (-10)$ **73.** $-8 - 4$

3-5 Arithmetic Sequences as Linear Functions

Then

You indentified linear functions. (Lesson 3-1)

Now

- Recognize arithmetic sequences.
- Relate arithmetic sequences to linear functions.

New Vocabulary

sequence
terms of the sequence
arithmetic sequence
common difference

Math Online

glencoe.com

- Extra Examples
- Personal Tutor
- Self-Check Quiz
- Homework Help

Why?

During a 2000-meter race, the coach of a women's crew team recorded the team's times at several intervals.

- At 400 meters, the time was 1 minute 32 seconds.
- At 800 meters, it was 3 minutes 4 seconds.
- At 1200 meters, it was 4 minutes 36 seconds.
- At 1600 meters, it was 6 minutes 8 seconds.

They completed the race with a time of 7 minutes 40 seconds.

Recognize Arithmetic Sequences You can relate the pattern of team times to linear functions. A **sequence** is a set of numbers, called the **terms of the sequence**, in a specific order. Look for a pattern in the information given for the women's crew team. Make a table to analyze the data.

Distance (m)	400	800	1200	1600	2000
Time (min : sec)	1:32	3:04	4:36	6:08	7:40

+ 1:32 + 1:32 + 1:32 + 1:32

As the distance increases in regular intervals, the time increases by 1 minute 32 seconds. Since the difference between successive terms is constant, this is an **arithmetic sequence**. The difference between the terms is called the **common difference** d.

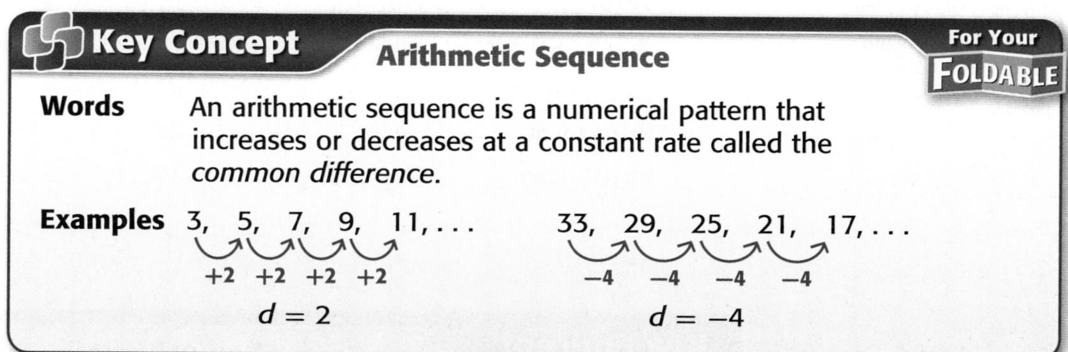

Key Concept Arithmetic Sequence

For Your FOLDABLE

Words An arithmetic sequence is a numerical pattern that increases or decreases at a constant rate called the *common difference*.

Examples 3, 5, 7, 9, 11, . . . (+2 +2 +2 +2) $d = 2$

33, 29, 25, 21, 17, . . . (−4 −4 −4 −4) $d = -4$

The three dots used with sequences are called an *ellipsis*. The ellipsis indicates that there are more terms in the sequence that are not listed.

StudyTip

Common Difference If the terms of an arithmetic sequence are increasing, the common difference is positive. If the terms are decreasing, the common difference is negative.

EXAMPLE 1 Identify Arithmetic Sequences

Determine whether each sequence is an arithmetic sequence. Explain.

a. $-4, -2, 0, 2, \ldots$

$-4 \quad -2 \quad 0 \quad 2$

$+2 \quad +2 \quad +2$

The difference between terms in the sequence is constant. Therefore, this sequence is arithmetic.

b. $\frac{1}{2}, \frac{5}{8}, \frac{3}{4}, \frac{13}{16}, \ldots$

$\frac{1}{2} \quad \frac{5}{8} \quad \frac{3}{4} \quad \frac{13}{16}$

$+\frac{1}{8} \quad +\frac{1}{8} \quad +\frac{1}{16}$

This is not an arithmetic sequence. The difference between terms is not constant.

Check Your Progress

1A. $-26, -22, -18, -14, \ldots$

1B. $1, 4, 9, 25, \ldots$

Personal Tutor glencoe.com

You can use the common difference of an arithmetic sequence to find the next term in the sequence.

Math History Link

Mina Rees (1902–1997) Rees received the first award for Distinguished Service to Mathematics from the Mathematical Association of America. Her work in analyzing patterns is still inspiring young women to study mathematics today.

EXAMPLE 2 Find the Next Term

Find the next three terms of the arithmetic sequence 15, 9, 3, −3, … .

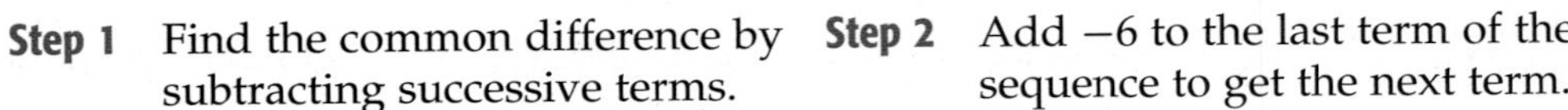

Step 1 Find the common difference by subtracting successive terms.

$15 \quad 9 \quad 3 \quad -3$

$-6 \quad -6 \quad -6$

The common difference is −6.

Step 2 Add −6 to the last term of the sequence to get the next term.

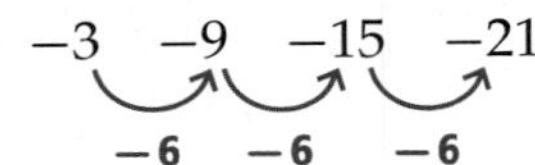

$-3 \quad -9 \quad -15 \quad -21$

$-6 \quad -6 \quad -6$

The next three terms in the sequence are −9, −15, and −21.

Check Your Progress

2. Find the next four terms of the arithmetic sequence 9.5, 11.0, 12.5, 14.0, … .

Personal Tutor glencoe.com

Each term in an arithmetic sequence can be expressed in terms of the first term a_1 and the common difference d.

Term	Symbol	In Terms of a_1 and d	Numbers
first term	a_1	a_1	8
second term	a_2	$a_1 + d$	$8 + 1(3) = 11$
third term	a_3	$a_1 + 2d$	$8 + 2(3) = 14$
fourth term	a_4	$a_1 + 3d$	$8 + 3(3) = 17$
⋮	⋮	⋮	⋮
nth term	a_n	$a_1 + (n-1)d$	$8 + (n-1)(3)$

Key Concept ***n*th Term of an Arithmetic Sequence** For Your FOLDABLE

The nth term of an arithmetic sequence with first term a_1 and common difference d is given by $a_n = a_1 + (n-1)d$, where n is a positive integer.

StudyTip

***n*th Terms** Since n represents the number of the term, the inputs for n are the counting numbers.

EXAMPLE 3 Find the *n*th Term

a. Write an equation for the *n*th term of the arithmetic sequence −12, −8, −4, 0, … .

Step 1 Find the common difference.

$$-12 \quad -8 \quad -4 \quad 0$$

+4 +4 +4 **The common difference is 4.**

Step 2 Write an equation.

$a_n = a_1 + (n - 1)d$	**Formula for the *n*th term**
$= -12 + (n - 1)4$	**$a_1 = -12$ and $d = 4$**
$= -12 + 4n - 4$	**Distributive Property**
$= 4n - 16$	**Simplify.**

b. Find the 9th term of the sequence.

Substitute 9 for n in the formula for the nth term.

$a_n = 4n - 16$	**Formula for the *n*th term**
$a_9 = 4(9) - 16$	**$n = 9$**
$a_9 = 36 - 16$	**Multiply.**
$a_9 = 20$	**Simplify.**

c. Graph the first five terms of the sequence.

n	$4n - 16$	a_n	(n, a_n)
1	4(1) − 16	−12	(1, −12)
2	4(2) − 16	−8	(2, −8)
3	4(3) − 16	−4	(3, −4)
4	4(4) − 16	0	(4, 0)
5	4(5) − 16	4	(5, 4)

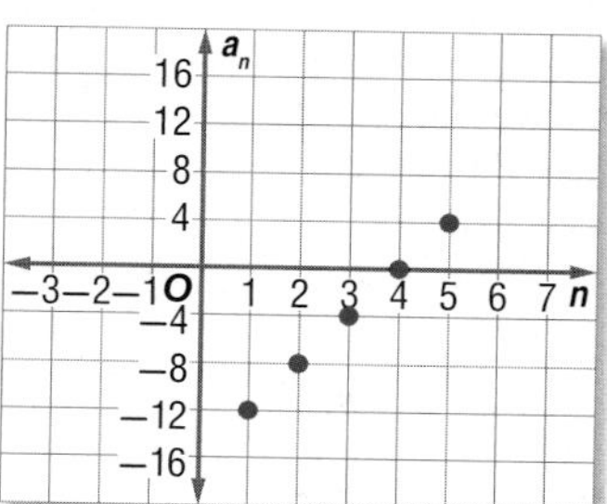

d. Which term of the sequence is 32?

In the formula for the nth term, substitute 32 for a_n.

$a_n = 4n - 16$	**Formula for the *n*th term**
$32 = 4n - 16$	**$a_n = 32$**
$32 + 16 = 4n - 16 + 16$	**Add 16 to each side.**
$48 = 4n$	**Simplify.**
$12 = n$	**Divide each side by 4.**

Check Your Progress

Consider the arithmetic sequence 3, −10, −23, −36, … .

3A. Write an equation for the nth term of the sequence.

3B. Find the 15th term in the sequence.

3C. Graph the first five terms of the sequence.

3D. Which term of the sequence is −114?

Personal Tutor glencoe.com

Arithmetic Sequences and Functions As you can see from Example 3, the graph of the first five terms of the arithmetic sequence lie on a line. An arithmetic sequence is a linear function in which n is the independent variable, a_n is the dependent variable, and d is the slope. The formula can be rewritten as the function $f(n) = (n - 1)d + a_1$, where n is a counting number.

While the domain of most linear functions are all real numbers, in Example 3 the domain of the function is the set of counting numbers and the range of the function is the set of integers on the line.

Real-World Link

When a Latina turns 15, her family may host a quinceañera for her birthday. The quinceañera is a traditional Hispanic ceremony and reception that signifies the transition from childhood to adulthood.

Source: Quince Girl

Real-World EXAMPLE 4 Arithmetic Sequences as Functions

INVITATIONS Marisol is mailing invitations to her quinceañera. The arithmetic sequence \$0.42, \$0.84, \$1.26, \$1.68, ... represents the cost of postage.

a. Write a function to represent this sequence.

The first term, a_1, is 0.42. Find the common difference.

0.42   0.84   1.26   1.68

+0.42 +0.42 +0.42

The common difference is 0.41.

$a_n = a_1 + (n - 1)d$	**Formula for the nth term**
$= 0.42 + (n - 1)0.42$	**$a_1 = 0.42$ and $d = 0.42$**
$= 0.42 + 0.42n - 0.42$	**Distributive Property**
$= 0.42n$	**Simplify.**

The function is $f(n) = 0.42n$.

b. Graph the function and determine the domain.

The rate of change of the function is 0.42. Make a table and plot points.

n	$f(n)$
1	0.42
2	0.84
3	1.26
4	1.68
5	2.10

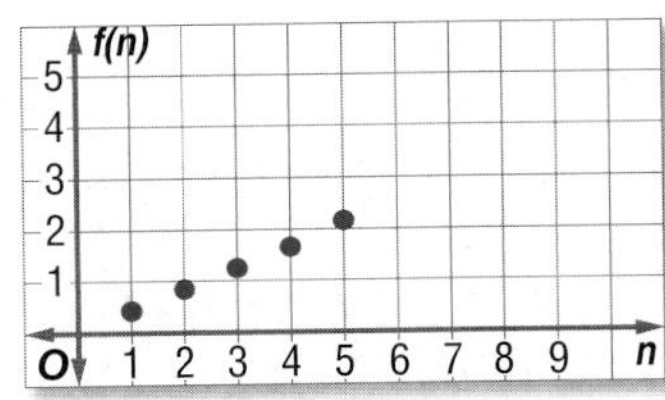

The domain of a function is the number of invitations Marisol mails. So, the domain is {0, 1, 2, 3, ...}.

Check Your Progress

4. TRACK The chart below shows the length of Martin's long jumps.

Jump	1	2	3	4
Length (ft)	8	9.5	11	12.5

A. Write a function to represent this arithmetic sequence.

B. Then graph the function.

Personal Tutor glencoe.com

Check Your Understanding

Example 1 p. 188

Determine whether each sequence is an arithmetic sequence. Write *yes* or *no*. Explain.

1. 18, 16, 15, 13, …

2. 4, 9, 14, 19, …

Example 2 p. 188

Find the next three terms of each arithmetic sequence.

3. 12, 9, 6, 3, …

4. $-2, 2, 6, 10, \ldots$

Example 3 p. 189

Write an equation for the *n*th term of each arithmetic sequence. Then graph the first five terms of the sequence.

5. 15, 13, 11, 9, …

6. $-1, -0.5, 0, 0.5, \ldots$

Example 4 p. 190

7. **SAVINGS** Kaia has \$525 in a savings account. After one month she has \$580 in the account. The next month the balance is \$635. The balance after the third month is \$690. Write a function to represent the arithmetic sequence. Then graph the function.

Practice and Problem Solving

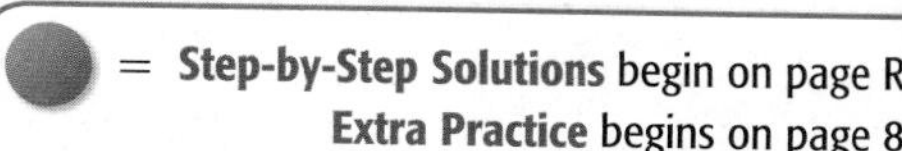
= Step-by-Step Solutions begin on page R12.
Extra Practice begins on page 815.

Example 1 p. 188

Determine whether each sequence is an arithmetic sequence. Write *yes* or *no*. Explain.

8. $-3, 1, 5, 9, \ldots$

9. $\frac{1}{2}, \frac{3}{4}, \frac{5}{8}, \frac{7}{16}, \ldots$

10. $-10, -7, -4, 1, \ldots$

11. $-12.3, -9.7, -7.1, -4.5, \ldots$

Example 2 p. 188

Find the next three terms of each arithmetic sequence.

12. 0.02, 1.08, 2.14, 3.2, …

13. 6, 12, 18, 24, …

14. 21, 19, 17, 15, …

15. $-\frac{1}{2}, 0, \frac{1}{2}, 1, \ldots$

16. $2\frac{1}{3}, 2\frac{2}{3}, 3, 3\frac{1}{3}, \ldots$

17. $\frac{7}{12}, 1\frac{1}{3}, 2\frac{1}{12}, 2\frac{5}{6}, \ldots$

Example 3 p. 189

Write an equation for the *n*th term of the arithmetic sequence. Then graph the first five terms in the sequence.

18. $-3, -8, -13, -18, \ldots$

19. $-2, 3, 8, 13, \ldots$

20. $-11, -15, -19, -23, \ldots$

21. $-0.75, -0.5, -0.25, 0, \ldots$

Example 4 p. 190

22. **AMUSEMENT PARKS** Shiloh and her friends spent the day at an amusement park. In the first hour, they rode two rides. After 2 hours, they had ridden 4 rides. They had ridden 6 rides after 3 hours.

a. Write a function to represent the arithmetic sequence.

b. Graph the function and determine the domain.

23. **JOBS** The table shows how Ryan is paid at his lumber yard job.

Linear Feet of 2×4 Planks Cut	10	20	30	40	50	60	70
Amount Paid in Commission (\$)	8	16	24	32	40	48	56

a. Write a function to represent Ryan's commission.

b. Graph the function and determine the domain.

24. The graph is a representation of an arithmetic sequence.

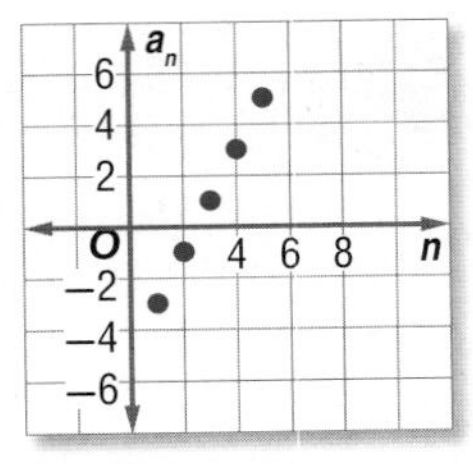

a. List the first five terms.

b. Write the formula for the nth term.

c. Write the function.

25. **NEWSPAPERS** A local newspaper charges by the number of words for advertising. Write a function to represent the advertising costs.

DAILY NEWS ADVERTISING	
10 words $7.50	20 words $10.00
15 words $8.75	25 words $11.25

26. The fourth term of an arithmetic sequence is 8. If the common difference is 2, what is the first term?

27. The common difference of an arithmetic sequence is −5. If a_{12} is 22, what is a_1?

28. The first four terms of an arithmetic sequence are 28, 20, 12, and 4. Which term of the sequence is −36?

29. **CARS** Jamal's odometer of his car reads 24,521. If Jamal drives 45 miles every day, what will the odometer reading be in 25 days?

Real-World Link

The first high school yearbook, called *The Evergreen,* was published in 1845 in Waterville, New York.

Source: Brownielocks

30. **YEARBOOKS** The yearbook staff is unpacking a box of school yearbooks. The arithmetic sequence 281, 270, 259, 248 ... represents the total number of ounces that the box weighs as each yearbook is taken out of the box.

a. Write a function to represent this sequence.

b. Determine the weight of each yearbook.

c. If the box weighs at least 11 ounces empty and 292 ounces when it is full, how many yearbooks were in the box?

31. **MULTIPLE REPRESENTATIONS** The Fibonacci sequence can be defined by a recursive formula. This means each term after the first two terms comes from one or more previous terms. The first six terms are 1, 1, 2, 3, 5, 8 ... .

a. **LOGICAL** Determine the relationship between the terms of the sequence. What are the next five terms in the sequence?

b. **ALGEBRAIC** Write a formula for the nth term if $n \geq 3$.

c. **ALGEBRAIC** Find the 15th term.

d. **ANALYTICAL** Explain why the Fibonacci sequence is not an arithmetic sequence.

H.O.T. Problems Use Higher-Order Thinking Skills

32. **OPEN ENDED** Create an arithmetic sequence with a common difference of −10.

33. **CHALLENGE** Find the value of x that makes $x + 8$, $4x + 6$, and $3x$ the first three terms of an arithmetic sequence.

34. **REASONING** Compare and contrast the domain and range of the linear functions described by $Ax + By = C$ and $a_n = a_1 + (n - 1)d$.

35. **CHALLENGE** Determine whether each sequence is an arithmetic sequence. Write *yes* or *no*. Explain. If yes, find the common difference and the next three terms.

a. $2x + 1, 3x + 1, 4x + 1 \ldots$

b. $2x, 4x, 8x, \ldots$

36. **WRITING IN MATH** Explain how to find a certain term of an arithmetic sequence and how an arithmetic sequence is related to a linear function.

Standardized Test Practice

37. GRIDDED RESPONSE The population of Westerville is about 35,000. Each year the population increases by about 400. This can be represented by the following equation, where n represents the number of years from now and p represents the population.

$$p = 35{,}000 + 400n$$

In how many years will the Westerville population be about 38,200?

38. Which relation is a function?

A $\{(-5, 6), (4, -3), (2, -1), (4, 2)\}$
B $\{(3, -1), (3, -5), (3, 4), (3, 6)\}$
C $\{(-2, 3), (0, 3), (-2, -1), (-1, 2)\}$
D $\{(-5, 6), (4, -3), (2, -1), (0, 2)\}$

39. Find the formula for the nth term of the arithmetic sequence.

$$-7, -4, -1, 2, \ldots$$

F $a_n = 3n - 4$
G $a_n = -7n + 10$
H $a_n = 3n - 10$
J $a_n = -7n + 4$

40. STATISTICS A class received the following scores on the ACT. What is the difference between the median and the mode in the scores?

18, 26, 20, 30, 25, 21, 32, 19, 22, 29, 29, 27, 24

A 1
B 2
C 3
D 4

Spiral Review

Name the constant of variation for each direct variation. Then find the slope of the line that passes through each pair of points. (Lesson 3-4)

41.

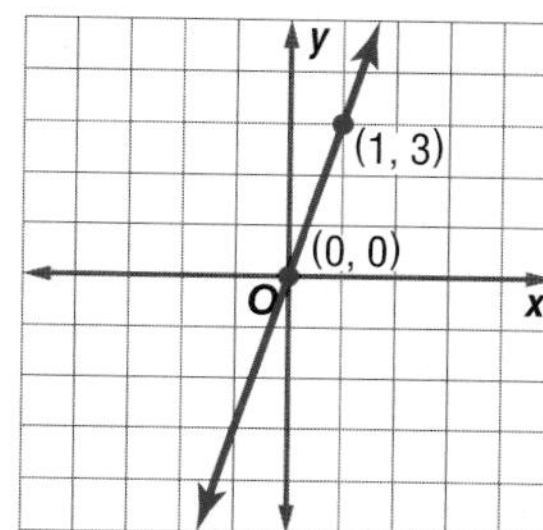

42.

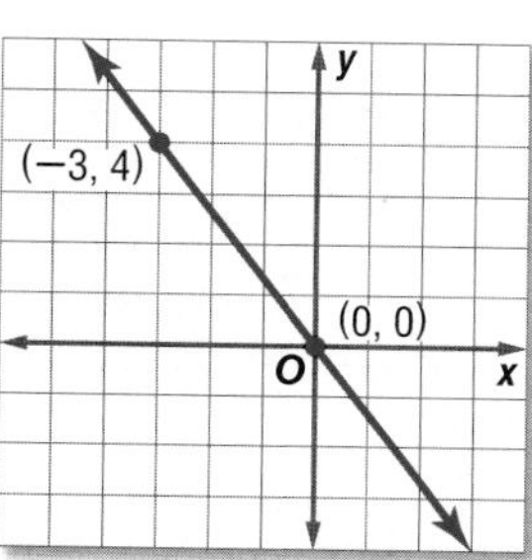

Find the slope of the line that passes through each pair of points. (Lesson 3-3)

43. $(5, 3), (-2, 6)$

44. $(9, 2), (-3, -1)$

45. $(2, 8), (-2, -4)$

Solve each equation. Check your solution. (Lesson 2-4)

46. $5x + 7 = -8$

47. $8 = 2 + 3n$

48. $12 = \frac{c - 6}{2}$

49. SPORTS The most popular sports for high school girls are basketball and softball. Write and use an equation to find how many more girls play on basketball teams than on softball teams. (Lesson 2-1)

Basketball
453,000 girls

Softball
369,000 girls

Skills Review

Graph each point on the same coordinate plane.

50. $A(2, 5)$

51. $B(-2, 1)$

52. $C(-3, -1)$

53. $D(0, 4)$

54. $F(5, -3)$

55. $G(-5, 0)$

EXTEND 3-5

Algebra Lab
Inductive and Deductive Reasoning

Objective
Investigate inductive and deductive reasoning.

If Jolene is not feeling well, she may go to a doctor. The doctor will ask her questions about how she is feeling and possibly run other tests. Based on her symptoms, the doctor can diagnose Jolene's illness. This is an example of inductive reasoning. **Inductive reasoning** is used to derive a general rule after observing many events.

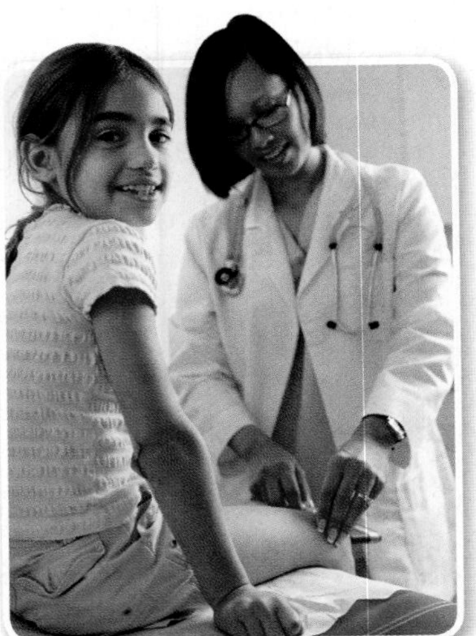

To use inductive reasoning:

Step 1 Observe many examples.

Step 2 Look for a pattern.

Step 3 Make a conjecture.

Step 4 Check the conjecture.

Step 5 Discover a likely conclusion.

With **deductive reasoning**, you come to a conclusion by accepting facts. The results of the tests ordered by the doctor may support the original diagnosis or lead to a different conclusion. This is an example of deductive reasoning. There is no conjecturing involved. Consider the two statements below.

1) If the strep test is positive, then the patient has strep throat.

2) Jolene tested positive for strep.

If these two statements are accepted as facts, then the obvious conclusion is that Jolene has strep throat. This is an example of deductive reasoning.

Exercises

1. Explain the difference between *inductive* and *deductive* reasoning. Then give an example of each.
2. When a detective reaches a conclusion about the height of a suspect from the distance between footprints, what kind of reasoning is being used? Explain.
3. When you examine a finite number of terms in a sequence of numbers and decide that it is an arithmetic sequence, what kind of reasoning are you using? Explain.
4. Suppose you have found the common difference for an arithmetic sequence based on analyzing a finite number of terms, what kind of reasoning do you use to find the 100th term in the sequence?
5. **a.** Copy and complete the table.

3^1	3^2	3^3	3^4	3^5	3^6	3^7	3^8	3^9
3	9	27						

   b. Write the sequence of numbers representing the numbers in the ones place.

   c. Find the number in the ones place for the value of 3^{100}. Explain your reasoning. State the type of reasoning that you used.

3-6 Proportional and Nonproportional Relationships

Then

You recognized arithmetic sequences and related them to linear functions. (Lesson 3-5)

Now

- Write an equation for a proportional relationship.
- Write an equation for a nonproportional relationship.

Math Online

glencoe.com

- Extra Examples
- Personal Tutor
- Self-Check Quiz
- Homework Help

Why?

Heather is planting flats of flowers. The table shows the number of flowers that she has planted and the amount of time that she has been working in the garden.

Number of flowers planted (p)	1	6	12	18
Number of minutes working (t)	5	30	60	90

The relationship between the flowers planted and the time that Heather worked in minutes can be graphed. Let p represent the number of flowers planted. Let t represent the number of minutes that Heather has worked.

When the ordered pairs are graphed, they form a linear pattern. This pattern can be described by an equation.

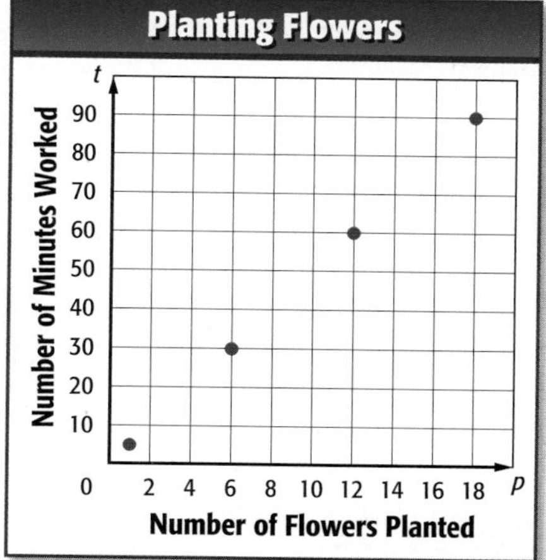

Proportional Relationships If the relationship between the domain and range of a relation is linear, the relationship can be described by a linear equation. If the equation is of the form $y = kx$, then the relationship is proportional. In a proportional relationship, the graph will pass through (0, 0). So, direct variations are proportional relationships.

Key Concept — **Proportional Relationship** — For Your FOLDABLE

Words A relationship is proportional if its equation is of the form $y = kx$, $k \neq 0$. The graph passes through (0, 0).

Example $y = 3x$

x	0	1	2	3	4
y	0	3	6	9	12

The ratio of the value of x to the value of y is constant when $x \neq 0$.

Real-World Link

Attendance at fitness clubs has steadily grown over the past fifteen years. Members' ages are expanding to a range of 15–34 on average.

Source: International Health, Raquet, and Sportsclub Association

Real-World EXAMPLE 1 Proportional Relationships

BONUS PAY **Marcos is a personal trainer at a gym. In addition to his salary, he receives a bonus for each client he sees.**

Number of Clients	1	2	3	4	5
Bonus Pay ($)	45	90	135	180	225

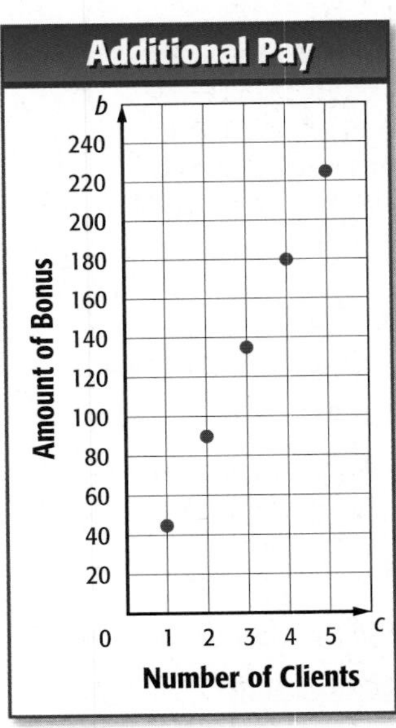

a. Graph the data. What can you deduce from the pattern about the relationship between the number of clients and the bonus pay?

The graph demonstrates a linear relationship between the number of clients and the bonus pay.

The graph also passes through the point (0, 0) because when Marcos sees 0 clients, he does not receive any bonus money. Therefore, the relationship is proportional.

b. Write an equation to describe this relationship.

Look for a pattern that can be described in an equation.

+1 +1 +1 +1

Number of Clients	1	2	3	4	5
Bonus Pay ($)	45	90	135	180	225

+45 +45 +45 +45

The difference between the values for the number of clients c is 1. The difference in the values for the bonus pay b is 45. This suggests that the k-value is $\frac{45}{1}$ or 45. So the equation is $b = 45c$. You can check this equation by substituting values for c into the equation.

CHECK If $c = 1$, then $b = 45(1)$ or 45. ✓
If $c = 5$, then $b = 45(5)$ or 225. ✓

c. Use this equation to predict the amount of Marcos' bonus if he sees 8 clients.

$b = 45c$ **Original equation**
$= 45(8)$ or 360 $c = 8$

Marcos will receive a bonus of $360 if he sees 8 clients.

StudyTip

Patterns Look for a pattern that shows a constant rate of change between the terms.

Check Your Progress

1. **CHARITY** A professional soccer team is donating money to a local charity for each goal they score.

Number of Goals	1	2	3	4	5
Donation ($)	75	150	225	300	375

A. Graph the data. What can you deduce from the pattern about the relationship between the number of goals and the money donated?

B. Write an equation to describe this relationship.

C. Use this equation to predict how much money will be donated for 12 goals.

Personal Tutor glencoe.com

StudyTip

A value added to or subtracted from one side of the equation $y = ax$ will cause a shift along the y-axis for the graph of the line.

Nonproportional Relationships Some linear equations can represent a nonproportional relationship. If the ratio of the value of x to the value of y is different for select ordered pairs that are on the line, the equation is nonproportional and the graph will not pass through (0, 0).

EXAMPLE 2 Nonproportional Relationships

Write an equation in function notation for the graph.

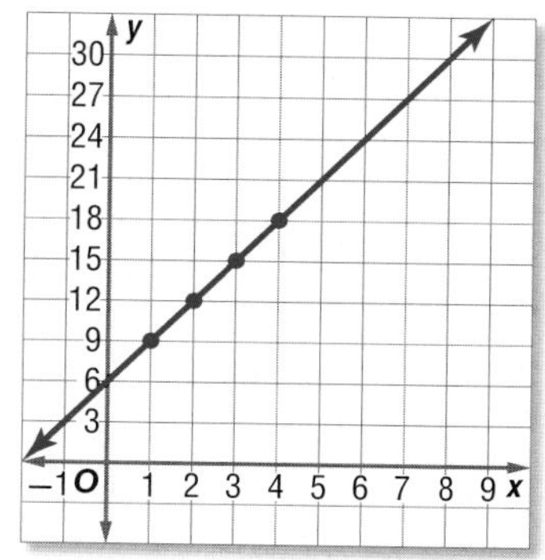

Understand You are asked to write an equation of the relation that is graphed in function notation.

Plan Find the difference between the x-values and the difference between the y-values.

Solve Select points from the graph and place them in a table.

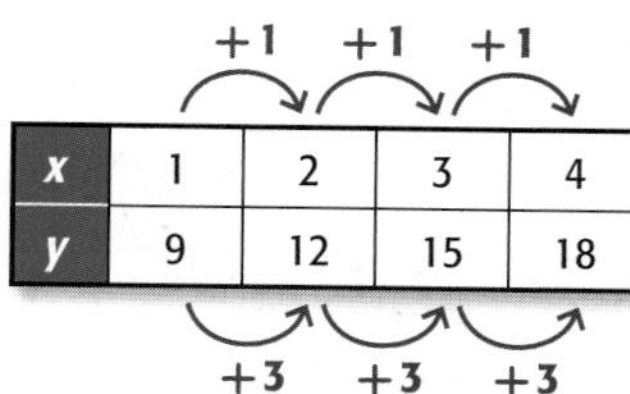

x	1	2	3	4
y	9	12	15	18

Notice that $\frac{1}{9} \neq \frac{2}{12} \neq \frac{3}{15} \neq \frac{4}{18}$.

The difference between the x-values is 1, while the difference between the y-values is 3. This suggests that $y = 3x$ or $f(x) = 3x + 6$.

If $x = 1$, then $y = 3(1)$ or 3. But the y-value for $x = 1$ is 9. Let's try some other values and see if we can detect a pattern.

x	1	2	3	4
$3x$	3	6	9	12
y	9	12	15	18

***y* is always 6 more than 3*x*.**

This pattern shows that 6 should be added to one side of the equation. Thus, the equation is $y = 3x + 6$ or $f(x) = 3x + 6$.

Check Compare the ordered pairs from the table to the graph. The points correspond. ✓

Check Your Progress

2. Write an equation in function notation for the relation shown in the table.

A.

x	1	2	3	4
y	3	2	1	0

B. Write an equation in function notation for the graph.

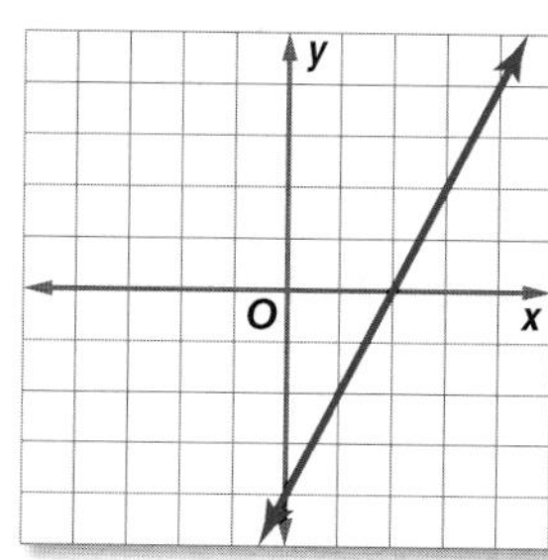

Personal Tutor glencoe.com

Check Your Understanding

Example 1 p. 196

1. **GEOMETRY** The table shows the perimeter of a square with sides of a given length.

Side Length (in.)	1	2	3	4	5
Perimeter (in.)	4	8	12	16	20

a. Graph the data.

b. Write an equation to describe the relationship.

c. What conclusion can you make regarding the relationship between the side and the perimeter?

Example 2 p. 197

Write an equation in function notation for each relation.

2. 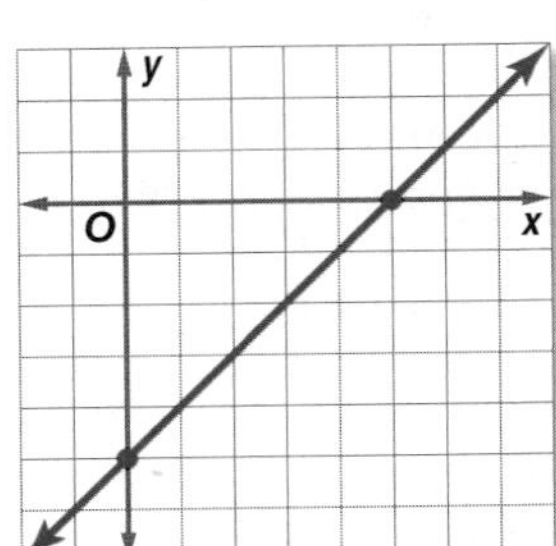

3. 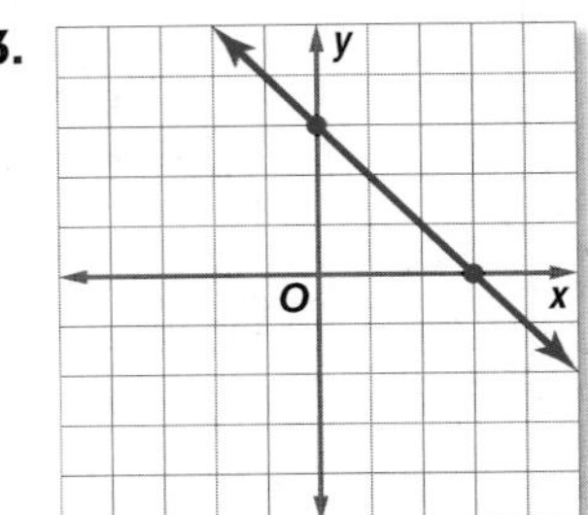

Practice and Problem Solving

● = **Step-by-Step Solutions** begin on page R12.
Extra Practice begins on page 815.

Example 1 p. 196

4. The table shows the pages of comic books read.

Books Read	1	2	3	4	5
Pages Read	35	70	105	140	175

a. Graph the data.

b. Write an equation to describe the relationship.

c. Find the number of pages read if 8 comic books were read.

Example 2 p. 197

Write an equation in function notation for each relation.

5. 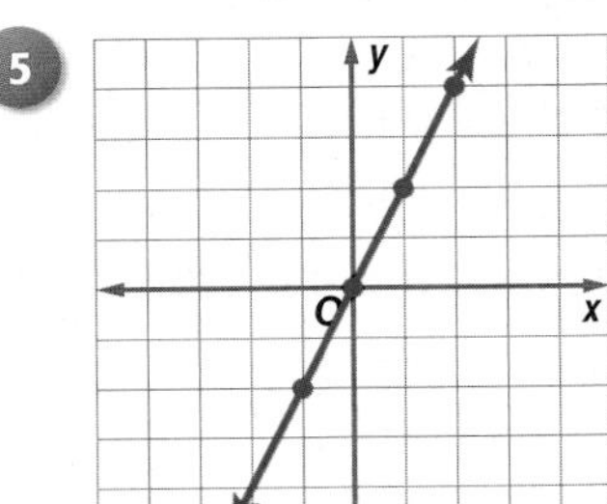

6. 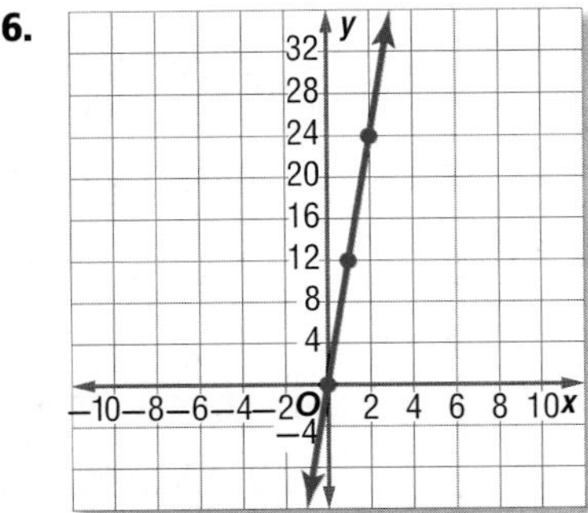

7. 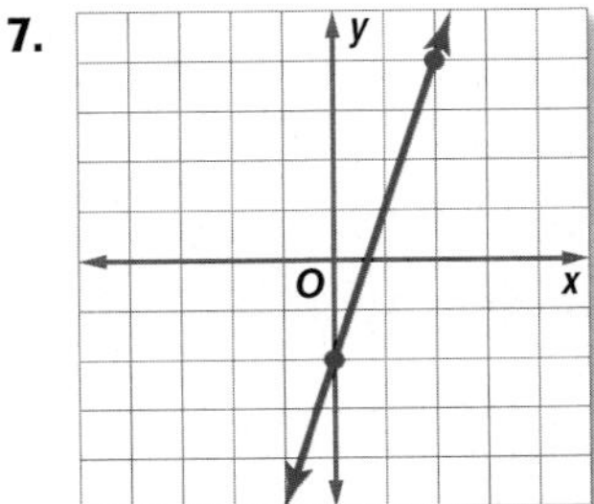

8. 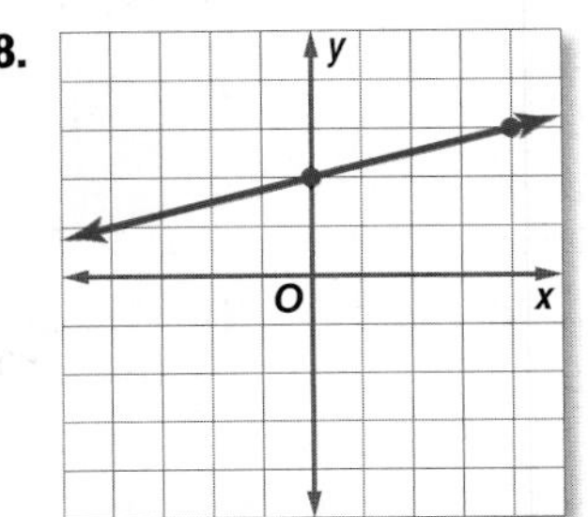

Real-World Link

The ski resort town of Mt. Baker, Washington, set the world record for the most snowfall in one year at 93 feet 8 inches.

Source: *USA TODAY*

For each arithmetic sequence, determine the related function. Then determine if the function is *proportional* or *nonproportional*. Explain.

9. 0, 3, 6, …

10. −4, 0, 4, …

11. **PHOTOGRAPH** Marielle wants to enlarge a picture of her family. The store charges $2.50 to develop the picture, and the table shows the list of prices for enlarging photographs. Write an equation to represent the total price y of the photograph with an enlargement of x size.

Size	Price ($)
3×5	2.25
4×6	4.50
5×7	6.75
8×10	9

12. **SNOWFALL** The total snowfall each hour of a winter snowstorm is shown in the table below.

Hour	1	2	3	4
Inches of Snowfall	1.65	3.30	4.95	6.60

a. Write an equation to fit the data in the table.

b. Describe the relationship between the hour and inches of snowfall.

13 **FUNDRAISER** The Cougar Pep Squad wants to sell T-shirts in the bookstore for the spring dance. The cost in dollars to order T-shirts in their school colors is represented by the equation $C = 2t + 3$.

a. Make a table of values that represents this relationship.

b. Rewrite the equation in function notation.

c. Graph the function.

d. Describe the relationship between the number of T-shirts and the cost.

H.O.T. Problems

Use **H**igher-**O**rder **T**hinking Skills

14. **ERROR ANALYSIS** Quentin and Claudia are writing an equation to describe the following relationship. Is either of them correct? Explain.

x	y
2	1
4	2
6	3

Quentin

Since the difference in the y-values is half as much as the difference in the x-values, the equation is $y = \frac{1}{2}x$.

Claudia

Since the difference in the x-values is half as much as the difference in the y-values, the equation is $y = 2x$.

15. **OPEN ENDED** Create an arithmetic sequence in which the first term is 4. Explain the pattern that you used. Write an equation that represents your sequence.

16. **CHALLENGE** Describe how inductive reasoning can be used to write an equation from a pattern.

17. **REASONING** Provide a counterexample to the following statement. *The related function of an arithmetic sequence is always proportional.* Explain why the counterexample is true.

18. **WRITING IN MATH** Compare and contrast proportional relationships with nonproportional relationships.

Standardized Test Practice

19. What is the slope of a line that contains the point $(1, -5)$ and has the same y-intercept as $2x - y = 9$?

A -9 **C** 2

B -7 **D** 4

20. SHORT RESPONSE $\triangle FGR$ is an isosceles triangle. What is the measure of $\angle G$?

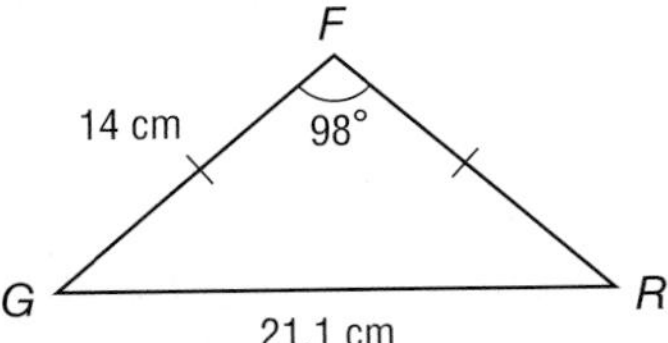

21. Luis deposits \$25 each week into a savings account from his part-time job. If he has \$350 in savings now, how much will he have in 12 weeks?

F \$600 **H** \$650

G \$625 **J** \$675

22. GEOMETRY Omar and Mackenzie want to build a zip-line by attaching one end of a rope to their 8-foot-tall tree house and anchoring the other end to the ground 28 feet away from the base of the tree house. How long, to the nearest foot, does the piece of rope need to be?

A 26 ft **C** 28 ft

B 27 ft **D** 29 ft

Spiral Review

Find the next three terms in each sequence. (Lesson 3-5)

23. 3, 13, 23, 33, …

24. $-2, -1.4, -0.8, -0.2, \ldots$

25. $\frac{3}{4}, \frac{7}{8}, 1, \frac{9}{8}, \ldots$

Suppose y varies directly as x. Write a direct variation equation that relates x and y. Then solve. (Lesson 3-4)

26. If $y = 45$ when $x = 9$, find y when $x = 7$.

27. If $y = -7$ when $x = -1$, find x when $y = -84$.

28. GENETICS About $\frac{2}{25}$ of the male population in the world cannot distinguish red from green. If there are 14 boys in the ninth grade who cannot distinguish red from green, about how many ninth-grade boys are there in all? Write and solve an equation to find the answer. (Lesson 2-3)

29. GEOMETRY The volume V of a cone equals one third times the product of π, the square of the radius r of the base, and the height h. (Lesson 2-1)

a. Write the formula for the volume of a cone.

b. Find the volume of a cone if r is 10 centimeters and h is 30 centimeters.

Skills Review

Solve each equation for y. (Lesson 2-8)

30. $3x = y + 7$

31. $2y = 6x - 10$

32. $9y + 2x = 12$

Graph each equation. (Lesson 3-1)

33. $y = x - 8$

34. $x - y = -4$

35. $2x + 4y = 8$

Study Guide and Review

Math Online glencoe.com
• STUDY TO GO
• Vocabulary Review

Chapter Summary

Key Concepts

Graphing Linear Equations (Lesson 3-1)

- The standard form of a linear equation is $Ax + By = C$, where $A \geq 0$, A and B are not both zero, and A, B, and C are integers whose greatest common factor is 1.

Solving Linear Equations by Graphing (Lesson 3-2)

- Values of x for which $f(x) = 0$ are called zeros of the function f. A zero of a function is located at an x-intercept of the graph of the function.

Rate of Change and Slope (Lesson 3-3)

- If x is the independent variable and y is the dependent variable, then rate of change equals

$$\frac{\text{change in } y}{\text{change in } x}.$$

- The slope of a line is the ratio of the rise to the run.

$$m = \frac{y_2 - y_1}{x_2 - x_1}$$

Direct Variation (Lesson 3-4)

- A direct variation is described by an equation of the form $y = kx$, where $k \neq 0$.

Arithmetic Sequences (Lesson 3-5)

- The nth term a_n of an arithmetic sequence with first term a_1 and common difference d is given by $a_n = a_1 + (n - 1)d$, where n is a positive integer.

Proportional and Nonproportional Relationships (Lesson 3-6)

- In a proportional relationship, the graph will pass through (0, 0).
- In a nonproportional relationship, the graph will *not* pass through (0, 0).

FOLDABLES® Study Organizer

Be sure the Key Concepts are noted in your Foldable.

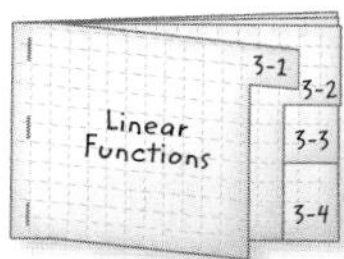

Key Vocabulary

arithmetic sequence (p. 187)
common difference (p. 187)
constant (p. 153)
constant of variation (p. 180)
direct variation (p. 180)
inductive reasoning (p. 194)
linear equation (p. 153)
linear function (p. 161)
rate of change (p. 170)
root (p. 161)
sequence (p. 187)
slope (p. 172)
standard form (p. 153)
terms of the sequence (p. 187)
x-intercept (p. 154)
y-intercept (p. 154)
zero of a function (p. 161)

Vocabulary Check

State whether each sentence is *true* or *false*. If *false*, replace the underlined word or number to make a true sentence.

1. The x-coordinate of the point at which the graph of an equation crosses the x-axis is an <u>x-intercept</u>.
2. A <u>linear equation</u> is an equation of a line.
3. The difference between successive terms of an arithmetic sequence is the <u>constant of variation</u>.
4. The <u>regular form</u> of a linear equation is $Ax + By = C$.
5. Values of x for which $f(x) = 0$ are called <u>zeros</u> of the function f.
6. Any two points on a nonvertical line can be used to determine the <u>slope</u>.
7. The slope of the line $y = 5$ is <u>5</u>.
8. The graph of any direct variation equation passes through <u>(0, 1)</u>.
9. A ratio that describes, on average, how much one quantity changes with respect to a change in another quantity is a <u>rate of change</u>.
10. In the linear equation $4x + 3y = 12$, the constant term is <u>12</u>.

CHAPTER 3 Study Guide and Review

Lesson-by-Lesson Review

3-1 Graphing Linear Equations (pp. 153–160)

Find the x- intercept and y- intercept of the graph of each linear function.

11.

x	y
−8	0
−4	3
0	6
4	9
8	12

12.

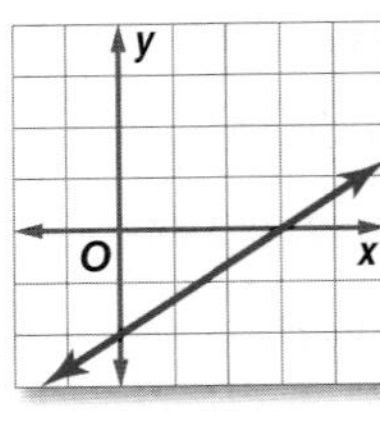

Graph each equation.

13. $y = -x + 2$

14. $x + 5y = 4$

15. $2x - 3y = 6$

16. $5x + 2y = 10$

17. SOUND The distance d in kilometers that sound waves travel through water is given by $d = 1.6t$, where t is the time in seconds.

a. Make a table of values and graph the equation.

b. Use the graph to estimate how far sound can travel through water in 7 seconds.

EXAMPLE 1

Graph $3x - y = 4$ by using the x- and y-intercepts.

Find the x-intercept.

$3x - y = 4$

$3x - 0 = 4$ **Let $y = 0$.**

$3x = 4$

$x = \frac{4}{3}$

Find the y-intercept.

$3x - y = 4$

$3(0) - y = 4$ **Let $x = 0$.**

$-y = 4$

$y = -4$

x-intercept: $\frac{4}{3}$

y-intercept: -4

The graph intersects the x-axis at $\left(\frac{4}{3}, 0\right)$ and the y-axis at $(0, -4)$. Plot these points. Then draw the line through them.

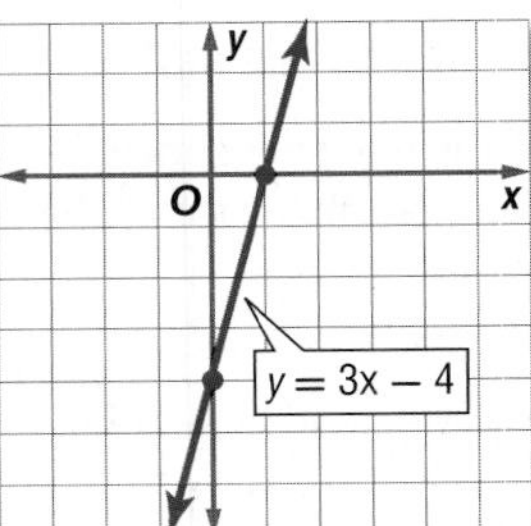

3-2 Solving Linear Equations by Graphing (pp. 161–168)

Find the root of each equation.

18. $0 = 2x + 8$

19. $0 = 4x - 24$

20. $3x - 5 = 0$

21. $6x + 3 = 0$

Solve each equation by graphing.

22. $0 = 16 - 8x$

23. $0 = 21 + 3x$

24. $-4x - 28 = 0$

25. $25x - 225 = 0$

26. FUNDRAISING Sean's class is selling boxes of popcorn to raise money for a class trip. Sean's class paid \$85 for the popcorn, and they are selling each box for \$1. The function $y = x - 85$ represents their profit y for each box of popcorn sold x. Find the zero and describe what it means in this situation.

EXAMPLE 2

Solve $3x + 1 = -2$ by graphing.

The first step is to find the related function.

$3x + 1 = -2$ **Original equation**

$3x + 1 + 2 = -2 + 2$ **Add 2 to each side.**

$3x + 3 = 0$ **Simplify.**

The related function is $y = 3x + 3$.

The graph intersects the x-axis at -1. So, the solution is -1.

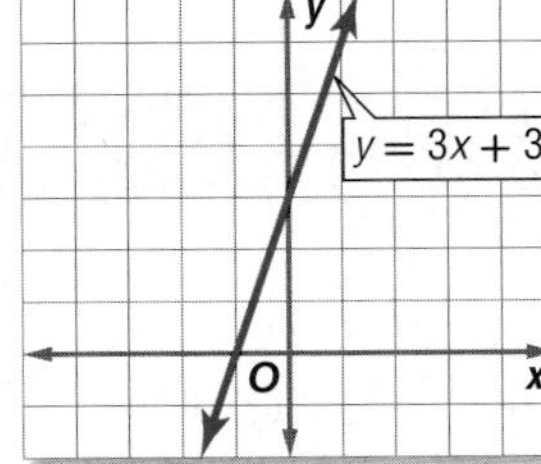

MIXED PROBLEM SOLVING
For mixed problem-solving practice, see page 847.

3-3 Rate of Change and Slope (pp. 169–178)

Find the rate of change represented in each table or graph.

27.

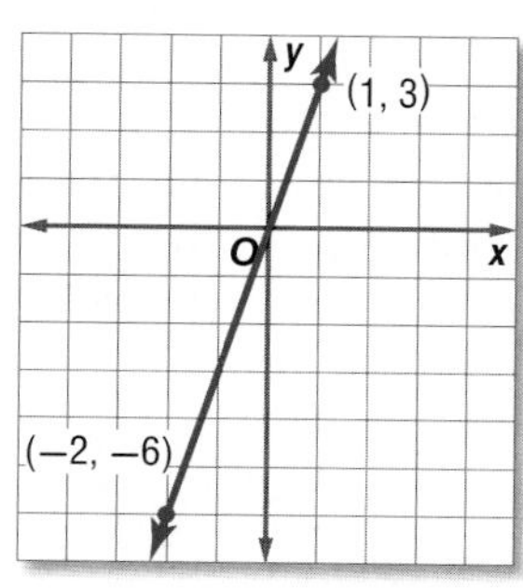

28.

x	y
−2	−3
0	−3
4	−3
12	−3

Find the slope of the line that passes through each pair of points.

29. (0, 5), (6, 2)

30. (−6, 4), (−6, −2)

31. PHOTOS The average cost of online photos decreased from $0.50 per print to $0.27 per print between 2002 and 2007. Find the average rate of change in the cost. Explain what it means.

EXAMPLE 3

Find the slope of the line that passes through (0, −4) and (3, 2).

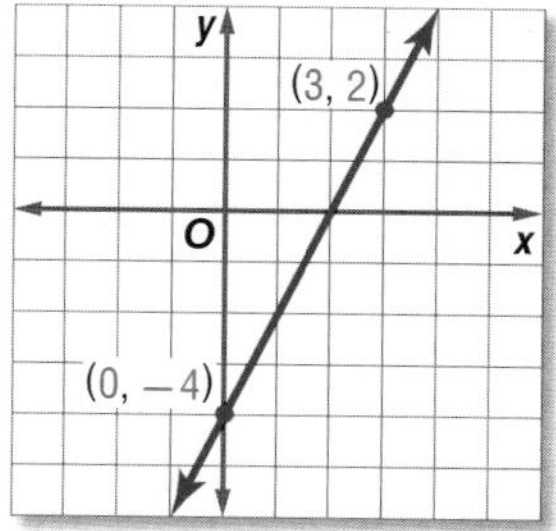

Let (0, −4) = (x_1, y_1) and (3, 2) = (x_2, y_2).

$m = \frac{y_2 - y_1}{x_2 - x_1}$ **Slope formula**

$= \frac{2 - (-4)}{3 - 0}$ **$x_1 = 0, x_2 = 3, y_1 = -4, y_2 = 2$**

$= \frac{6}{3}$ or 2 **Simplify.**

3-4 Direct Variation (pp. 180–186)

Graph each equation.

32. $y = x$

33. $y = \frac{4}{3}x$

34. $y = -2x$

Suppose y varies directly as x. Write a direct variation equation that relates x and y. Then solve.

35. If $y = 15$ when $x = 2$, find y when $x = 8$.

36. If $y = -6$ when $x = 9$, find x when $y = -3$.

37. If $y = 4$ when $x = -4$, find y when $x = 7$.

38. JOBS Suppose you earn $127 for working 20 hours.

a. Write a direct variation equation relating your earnings to the number of hours worked.

b. How much would you earn for working 35 hours?

EXAMPLE 4

Suppose y varies directly as x, and $y = -24$ when $x = 8$.

a. Write a direct variation equation that relates x and y.

$y = kx$ **Direct variation equation**

$-24 = k(8)$ **Substitute −24 for y and 8 for x.**

$\frac{-24}{8} = \frac{k(8)}{8}$ **Divide each side by 8.**

$-3 = k$ **Simplify.**

So, the direct variation equation is $y = -3x$.

b. Use the direct variation equation to find x when $y = -18$.

$y = -3x$ **Direct variation equation**

$-18 = -3x$ **Replace y with −18.**

$\frac{-18}{-3} = \frac{-3x}{-3}$ **Divide each side by −3.**

$6 = x$ **Simplify.**

Therefore, $x = 6$ when $y = -18$.

Study Guide and Review

3-5 Arithmetic Sequences as Linear Functions (pp. 187–194)

Find the next three terms of each arithmetic sequence.

39. 6, 11, 16, 21, …

40. 1.4, 1.2, 1.0, …

Write an equation for the *n*th term of each arithmetic sequence.

41. $a_1 = 6, d = 5$

42. 28, 25, 22, 19, …

43. **SCIENCE** The table shows the distance traveled by sound in water. Write an equation for this sequence. Then find the time for sound to travel 72,300 feet.

Time (s)	1	2	3	4
Distance (ft)	4820	9640	14,460	19,280

EXAMPLE 5

Find the next three terms of the arithmetic sequence 10, 23, 36, 49, … .

Find the common difference.

10 23 36 49
+13 +13 +13

So, $d = 13$.

Add 13 to the last term of the sequence. Continue adding 13 until the next three terms are found.

49 62 75 88
+13 +13 +13

The next three terms are 62, 75, and 88.

3-6 Proportional and Nonproportional Relationships (pp. 195–200)

44. Write an equation in function notation for this relation.

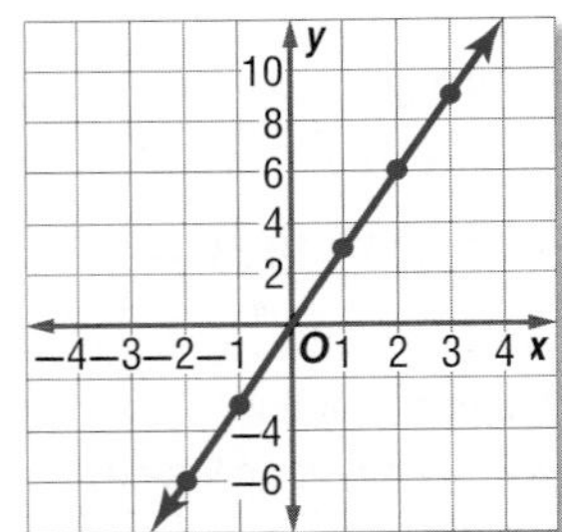

45. **ANALYZE TABLES** The table shows the cost of picking your own strawberries at a farm.

Number of Pounds	1	2	3	4
Total Cost ($)	1.25	2.50	3.75	5.00

a. Graph the data.

b. Write an equation in function notation to describe this relationship.

c. How much would it cost to pick 6 pounds of strawberries?

EXAMPLE 6

Write an equation in function notation for this relation.

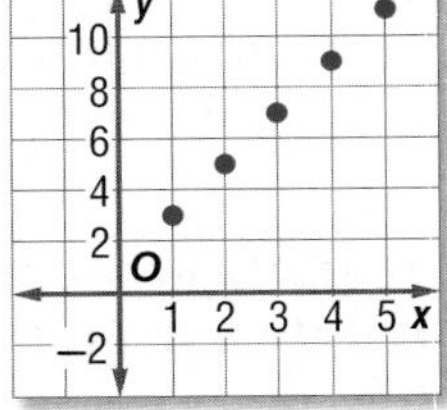

Make a table of ordered pairs for several points on the graph.

x	1	2	3	4	5
y	3	5	7	9	11

The difference in y-values is twice the difference of x values. This suggests that $y = 2x$. However, $3 \neq 2(1)$. Compare the values of y to the values of $2x$.

x	1	2	3	4	5
2x	2	4	6	8	10
y	3	5	7	9	11

The difference between y and $2x$ is always 1. So the equation is $y = 2x + 1$. Since this relation is also a function, it can be written as $f(x) = 2x + 1$.

Practice Test

Math Online glencoe.com
Chapter Test

1. **TEMPERATURE** The equation to convert Celsius temperature C to Kelvin temperature K is shown.

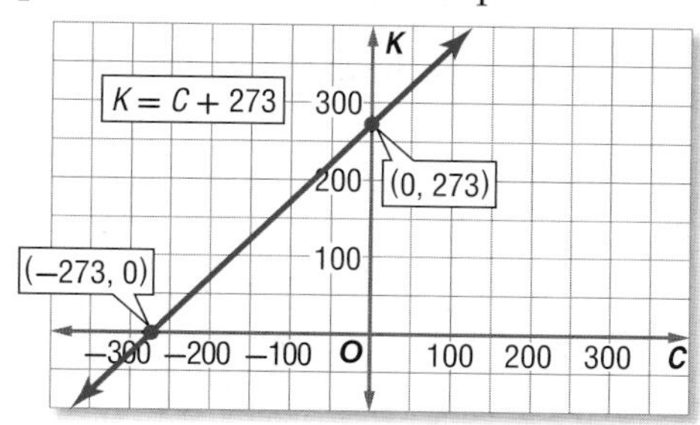

   a. State the independent and dependent variables. Explain.

   b. Determine the C- and K-intercepts and describe what the intercepts mean in this situation.

Graph each equation.

2. $y = x + 2$
3. $y = 4x$
4. $x + 2y = -1$
5. $-3x = 5 - y$

Solve each equation by graphing.

6. $4x + 2 = 0$
7. $0 = 6 - 3x$
8. $5x + 2 = -3$
9. $12x = 4x + 16$

Find the slope of the line that passes through each pair of points.

10. $(5, 8), (-3, 7)$
11. $(5, -2), (3, -2)$
12. $(-4, 7), (8, -1)$
13. $(6, -3), (6, 4)$

14. **MULTIPLE CHOICE** Which is the slope of the linear function shown in the graph?

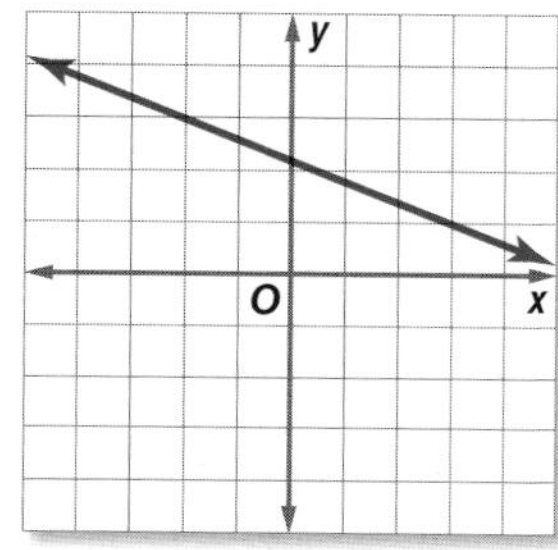

   A $-\frac{5}{2}$
   C $\frac{5}{2}$
   B $-\frac{2}{5}$
   D $\frac{2}{5}$

Suppose y varies directly as x. Write a direct variation equation that relates x and y. Then solve.

15. If $y = 6$ when $x = 9$, find x when $y = 12$.
16. When $y = -8$, $x = 8$. What is x when $y = -6$?
17. If $y = -5$ when $x = -2$, what is y when $x = 14$?
18. If $y = 2$ when $x = -12$, find y when $x = -4$.

19. **BIOLOGY** The number of pints of blood in a human body varies directly with the person's weight. A person who weighs 120 pounds has about 8.4 pints of blood in his or her body.

   a. Write and graph an equation relating weight and amount of blood in a person's body.

   b. Predict the weight of a person whose body holds 12 pints of blood.

Find the next three terms in each sequence.

20. 5, −10, 15, −20, 25, …
21. 5, 5, 6, 8, 11, 15, …

Determine whether each sequence is an arithmetic sequence. If it is, state the common difference.

22. −40, −32, −24, −16, …
23. 0.75, 1.5, 3, 6, 12, …
24. 5, 17, 29, 41, …

25. **MULTIPLE CHOICE** In each figure, only one side of each regular pentagon is shared with another pentagon. Each side of each pentagon is 1 centimeter. If the pattern continues, what is the perimeter of a figure that has 6 pentagons?

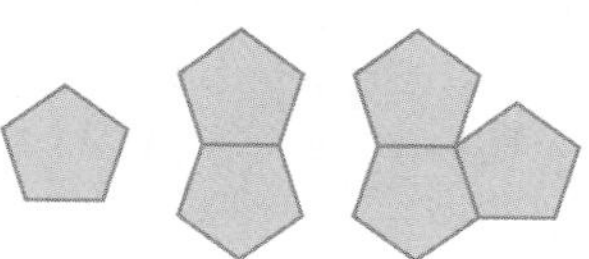

   F 15 cm
   H 20 cm
   G 25 cm
   J 30 cm

CHAPTER 3 Preparing for Standardized Tests

Reading Math Problems

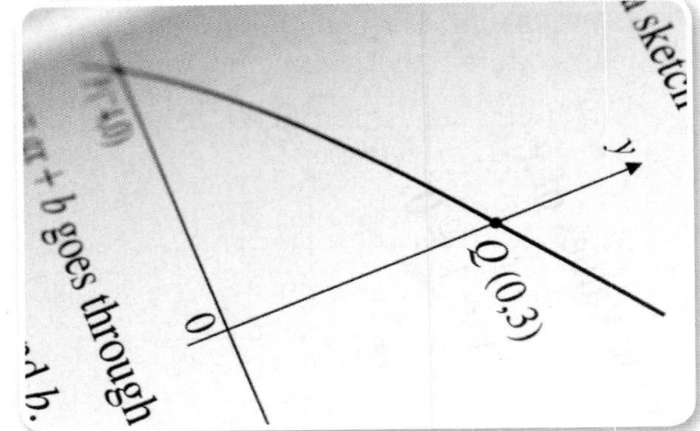

The first step to solving any math problem is to read the problem. When reading a math problem to get the information you need to solve, it is helpful to use special reading strategies.

Strategies for Reading Math Problems

Step 1

Read the problem quickly to gain a general understanding of it.

- **Ask yourself:** "What do I know?" "What do I need to find out?"
- **Think:** "Is there enough information to solve the problem? Is there extra information?"
- **Highlight:** If you are allowed to write in your test booklet, underline or highlight important information. Cross out any information you don't need.

Step 2

Reread the problem to identify relevant facts.

- **Analyze:** Determine how the facts are related.
- **Key Words:** Look for keywords to solve the problem.
- **Vocabulary:** Identify mathematical terms. Think about the concepts and how they are related.
- **Plan:** Make a plan to solve the problem.
- **Estimate:** Quickly estimate the answer.

Step 3

Identify any obvious wrong answers.

- **Eliminate:** Eliminate any choices that are very different from your estimate.
- **Units of Measure:** Identify choices that are possible answers based on the units of measure in the question. For example, if the question asks for area, only answers in square units will work.

Step 4

Look back after solving the problem.

Check: Make sure you have answered the question.

EXAMPLE

Read the problem. Identify what you need to know. Then use the information in the problem to solve.

> Jamal, Gina, Lisa, and Renaldo are renting a car for a road trip. The cost of renting the car is given by the function $C = 12.5 + 21d$, where C is the total cost for renting the car for d days. What does the slope of the function represent?
>
> **A** number of people
>
> **B** cost per day
>
> **C** number of days
>
> **D** miles per gallon

Read the problem carefully. The number of people going on the trip is not needed information. You need to know what the slope of the function represents.

Slope is a ratio. The word "per" in answers B and D imply that they are both ratios. Since choices A and C are not ratios, eliminate them.

The problem says that C represents the cost of renting the car. So the slope cannot represent the miles per gallon of the car. The slope must represent the cost per day.

The correct answer is B.

Exercises

Read each problem. Identify what you need to know. Then use the information in the problem to solve.

1. What does the x-intercept mean in the context of the situation given below?

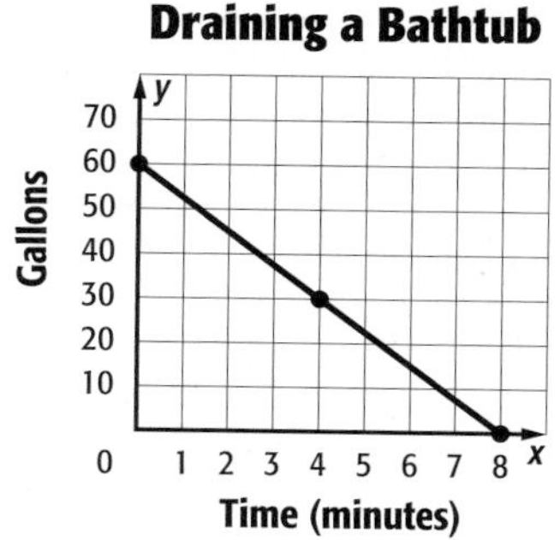

A amount of time needed to drain the bathtub

B number of gallons in the tub when the drain plug is pulled

C number of gallons in the tub after x minutes

D amount of water drained each minute

2. The amount of money raised by a charity carwash varies directly as the number of cars washed. When 11 cars are washed, $79.75 is raised. How many cars must be washed to raise $174.00?

F 10 cars

G 16 cars

H 22 cars

J 24 cars

3. The function $C = 25 + 0.45(x - 450)$ represents the cost of a monthly cell phone bill, when x minutes are used. Which statement best represents the formula for the cost of the bill?

A The cost consists of a flat fee of $0.45 and $25 for each minute used over 450.

B The cost consists of a flat fee of $450 and $0.45 for each minute used over 25.

C The cost consists of a flat fee of $25 and $0.45 for each minute used over 450.

D The cost consists of a flat fee of $25 and $0.45 for each minute used.

Standardized Test Practice

Cumulative, Chapters 1 through 3

Multiple Choice

Read each question. Then fill in the correct answer on the answer document provided by your teacher or on a sheet of paper.

1. Horatio is purchasing a computer cable for \$15.49. If the sales tax rate in his state is 5.25%, what is the total cost of the purchase?

   A \$16.42
   B \$16.30
   C \$15.73
   D \$15.62

2. What is the value of the expression below?

   $$3^2 + 5^3 - 2^5$$

   F 14
   G 34
   H 102
   J 166

3. What is the slope of the linear function graphed below?

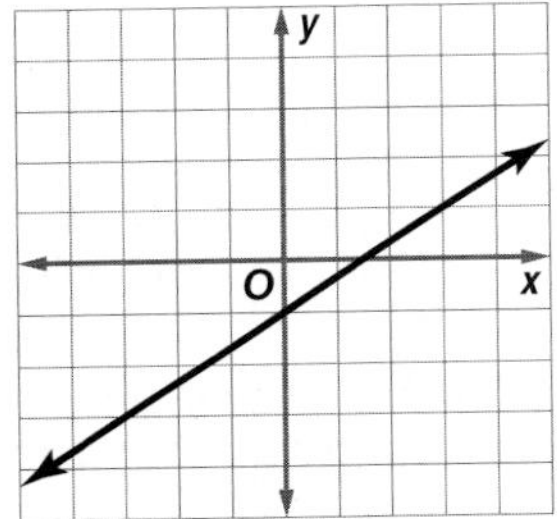

   A $-\frac{1}{3}$
   B $\frac{1}{2}$
   C $\frac{2}{3}$
   D $\frac{3}{2}$

4. Find the rate of change for the linear function represented in the table.

Hours Worked	1	2	3	4
Money Earned (\$)	5.50	11.00	16.50	22.00

   F increase \$6.50/h
   G increase \$5.50/h
   H decrease \$5.50/h
   J decrease \$6.50/h

5. Suppose that y varies directly as x, and $y = 14$ when $x = 4$. What is the value of y when $x = 9$?

   A 25.5
   B 27.5
   C 29.5
   D 31.5

6. Write an equation for the nth term of the arithmetic sequence shown below.

   $$-2, 1, 4, 7, 10, 13, \ldots$$

   F $a_n = 2n - 1$
   G $a_n = 2n + 4$
   H $a_n = 3n + 2$
   J $a_n = 3n - 5$

7. The table shows the labor charges of an electrician for jobs of different lengths.

Number of Hours (n)	Labor Charges (C)
1	\$60
2	\$85
3	\$110
4	\$135

   Which function represents the situation?

   A $C(n) = 25n + 35$
   B $C(n) = 25n + 30$
   C $C(n) = 35n + 25$
   D $C(n) = 35n + 40$

8. Find the value of x so that the figures have the same area.

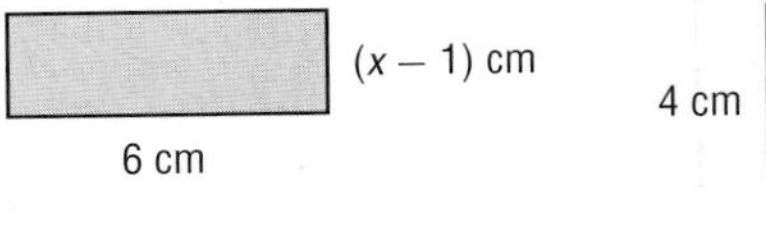

   F 3
   G 4
   H 5
   J 6

9. The table shows the total amount of rain during a storm. Write a formula to find out how much rain will fall after a given hour.

Hour (h)	1	2	3	4
Inches (n)	0.45	0.9	1.35	1.8

   A $h = 0.45n$
   B $n = 0.45h$
   C $h = 0.9n$
   D $h = 1.8n$

Test-TakingTip

Question 3 You can *eliminate unreasonable answers* to multiple choice items. The line slopes up from left to right, so the slope is positive. Answer choice A can be eliminated.

Short Response/Gridded Response

Record your answers on the answer sheet provided by your teacher or on a sheet of paper.

10. The scale on a map is 1.5 inches = 6 miles. If two cities are 4 inches apart on the map, what is the actual distance between the cities?

11. Write a direct variation equation to represent the graph below.

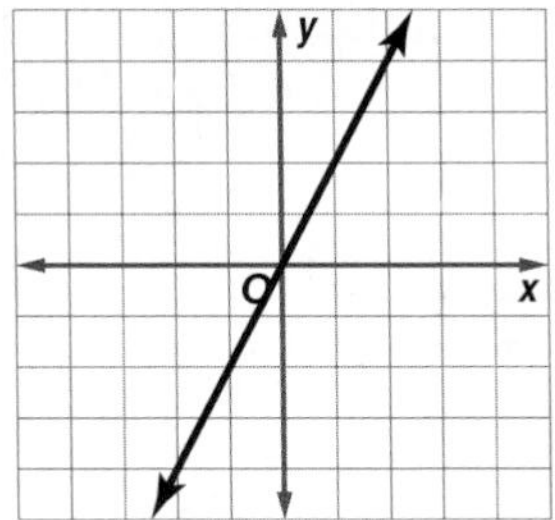

12. Justine bought a car for $18,500 and its value depreciated linearly. After 3 years, the value was $14,150. What is the amount of yearly depreciation?

13. GRIDDED RESPONSE Use the graph to determine the solution to the equation $-\frac{1}{3}x + 1 = 0$?

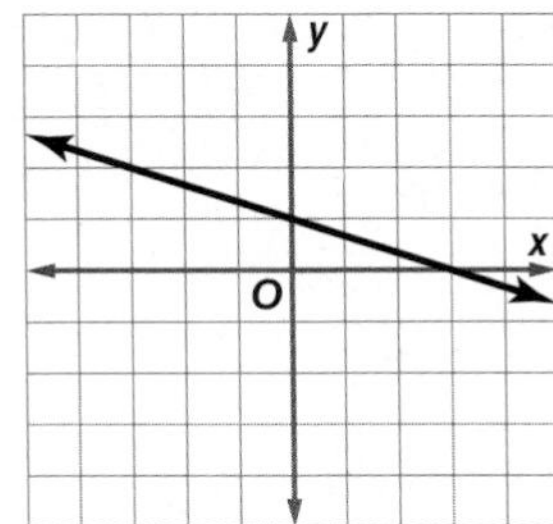

14. Write an expression that represents the total surface area (including the top and bottom) of a tower of n cubes each having a side length of s. (Do not include faces that cover each other.)

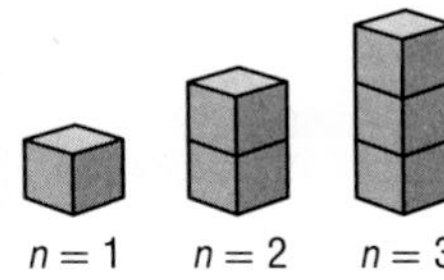

15. GRIDDED RESPONSE There are 120 members in the North Carolina House of Representatives. This is 70 more than the number of members in the North Carolina Senate. How many members are in the North Carolina Senate?

Extended Response

Record your answers on a sheet of paper. Show your work.

16. A hot air balloon was at a height of 60 feet above the ground when it began to ascend. The balloon climbed at a rate of 15 feet per minute.

a. Make a table that shows the height of the hot air balloon after climbing for 1, 2, 3, and 4 minutes.

b. Let t represent the time in minutes since the balloon began climbing. Write an algebraic equation for a sequence that can be used to find the height, h, of the balloon after t minutes.

c. Use your equation from part b to find the height, in feet, of the hot air balloon after climbing for 8 minutes.

Need Extra Help?																
If you missed Question...	1	2	3	4	5	6	7	8	9	10	11	12	13	14	15	16
Go to Lesson or Page...	2-7	1-2	3-3	3-3	3-4	3-5	2-1	0-8	3-5	2-8	3-4	3-3	3-2	0-10	2-1	3-5

CHAPTER 4 Linear Functions and Relations

Then

In Chapter 3, you graphed linear functions.

Now

In Chapter 4, you will:

- Write and graph linear equations in various forms.
- Use scatter plots and lines of fit, and write equations of best-fit lines using linear regression.
- Identify and graph special functions.

Why?

TRAVEL The number of trips people take changes from year to year. From the yearly data, patterns emerge. Rate of change can be applied to these data to determine a linear model. This can be used to predict the number of trips taken in future years.

Linear Functions and Relations

Activity

Use the slope formula to find the slope of the line containing the two points.

Drag the coordinates into the formula.

1991 | 19.1 | 2006 | 11.9

$m = \frac{y_2 - y_1}{x_2 - x_1}$

Check Answer

Math *in Motion,* Animation glencoe.com

Get Ready for Chapter 4

Diagnose Readiness You have two options for checking Prerequisite Skills.

Text Option

Take the Quick Check below. Refer to the Quick Review for help.

QuickCheck

Evaluate $3a^2 - 2ab + c$ for the values given. (Lesson 1-5)

1. $a = 2, b = 1, c = 5$
2. $a = -3, b = -2, c = 3$
3. $a = -1, b = 0, c = 11$
4. $a = 5, b = -3, c = -9$
5. **CAR RENTAL** The cost of renting a car is given by $49x + 0.3y$. Let x represent the number of days rented, and let y represent the number of miles driven. Find the cost for a five-day rental over 125 miles.

Solve each equation for the given variable. (Lesson 2-8)

6. $x + y = 5$ for y
7. $2x - 4y = 6$ for x
8. $y - 2 = x + 3$ for y
9. $4x - 3y = 12$ for x
10. **GEOMETRY** The formula for the perimeter of a rectangle is $P = 2w + 2\ell$, where w represents width and ℓ represents length. Solve for w.

Write the ordered pair for each point. (Lesson 1-6)

11. A
12. B
13. C
14. D
15. E
16. F

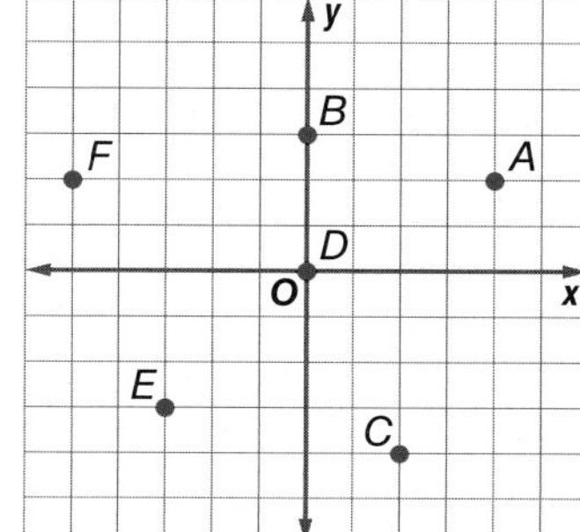

QuickReview

EXAMPLE 1

Evaluate $2(m - n)^2 + 3p$ for $m = 5$, $n = 2$, and $p = -3$.

$2(m - n)^2 + 3p$	**Original expression**
$= 2(5 - 2)^2 + 3(-3)$	**Substitute.**
$= 2(3)^2 + 3(-3)$	**Subtract.**
$= 2(9) + 3(-3)$	**Evaluate power.**
$= 18 + (-9)$	**Multiply.**
$= 9$	**Add.**

EXAMPLE 2

Solve $5x + 15y = 9$ for x.

$5x + 15y = 9$	**Original equation**
$5x + 15y - 15y = 9 - 15y$	**Subtract 15y from each side.**
$5x = 9 - 15y$	**Simplify.**
$\frac{5x}{5} = \frac{9 - 15y}{5}$	**Divide each side by 5.**
$x = \frac{9}{5} - 3y$	**Simplify.**

EXAMPLE 3

Write the ordered pair for A.

Step 1 Begin at point A.

Step 2 Follow along a vertical line to the x-axis. The x-coordinate is -4.

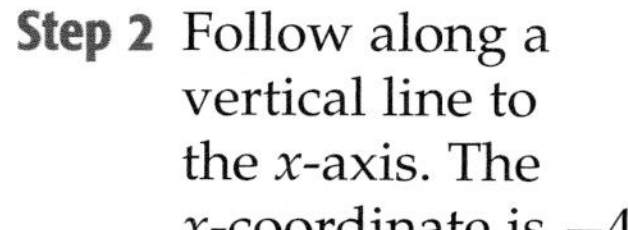

Step 3 Follow along a horizontal line to the y-axis. The y-coordinate is 2.

The ordered pair for point A is $(-4, 2)$.

Online Option

Math Online Take a self-check Chapter Readiness Quiz at glencoe.com.

Get Started on Chapter 4

You will learn several new concepts, skills, and vocabulary terms as you study Chapter 4. To get ready, identify important terms and organize your resources. You may wish to refer to **Chapter 0** to review prerequisite skills.

FOLDABLES® Study Organizer

Linear Functions Make this Foldable to help you organize your Chapter 4 notes about linear functions. Begin with one sheet of 11" by 17" paper.

1. **Fold** each end of the paper in about 2 inches.

2. **Fold** along the width and the length. Unfold. Cut along the fold line from the top to the center.

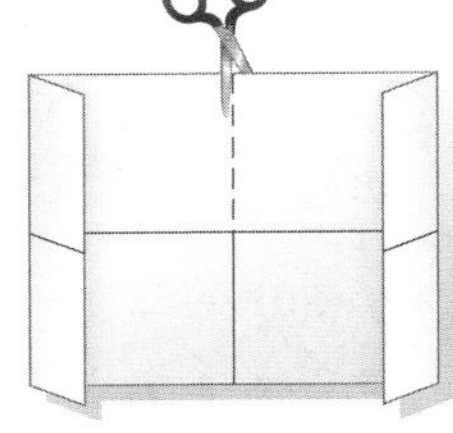

3. **Fold** the top flaps down. Then fold in half and turn to form a folder. Staple the flaps down to form pockets.

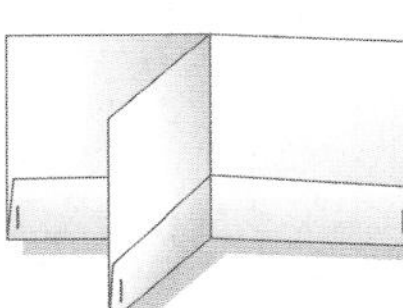

4. **Label** the front with the chapter title.

New Vocabulary

English		Español
slope-intercept form	p. 214	forma pendiente-intersección
linear extrapolation	p. 226	extrapolación lineal
point-slope form	p. 231	forma punto-pendiente
parallel lines	p. 237	rectas paralelas
perpendicular lines	p. 238	rectas perpendiculares
scatter plot	p. 245	gráfica de dispersión
line of fit	p. 246	recta de ajuste
linear interpolation	p. 247	interpolación lineal
best-fit line	p. 253	recta de ajuste óptimo
linear regression	p. 253	retroceso lineal
correlation coefficient	p. 253	coeficiente de correlación
median-fit line	p. 255	línea de mediana-ataque
step function	p. 261	función etapa
piecewise function	p. 261	función a intervalos
greatest integer function	p. 261	función del máximo entero

Review Vocabulary

coefficient • p. 26 • coeficiente the numerical factor of a term

function • p. 45 • función a relation in which each element of the domain is paired with exactly one element of the range

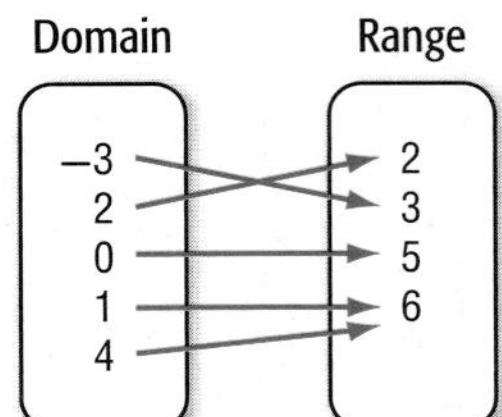

ratio • p. 111 • razon a comparison of two numbers by division

Multilingual eGlossary glencoe.com

Math Online glencoe.com

- Study the chapter online
- Explore **Math in Motion**
- Get extra help from your own **Personal Tutor**
- Use **Extra Examples** for additional help
- Take a **Self-Check Quiz**
- **Review Vocabulary** in fun ways

EXPLORE 4-1

Graphing Technology Lab
Investigating Slope-Intercept Form

Objective
Use a graphing calculator to collect data and investigate slope-intercept form.

Set Up the Lab

- Cut a small hole in a top corner of a plastic sandwich bag. Hang the bag from the end of the force sensor.
- Connect the force sensor to your data collection device.

ACTIVITY

Step 1 Use the sensor to collect the weight with 0 washers in the bag. Record the data pair in the calculator.

Step 2 Place one washer in the plastic bag. Wait for the bag to stop swinging, then measure and record the weight.

Step 3 Repeat the experiment, adding different numbers of washers to the bag. Each time, record the number of washers and the weight.

Analyze the Results

1. The domain contains values of the independent variable, number of washers. The range contains values of the dependent variable, weight. Use the graphing calculator to create a scatter plot using the ordered pairs (washers, weight).
2. Write a sentence that describes the points on the graph.
3. Describe the position of the point on the graph that represents the trial with no washers in the bag.
4. The rate of change can be found by using the formula for slope.

$$\frac{\text{rise}}{\text{run}} = \frac{\text{change in weight}}{\text{change in number of washers}}$$

   Find the rate of change in the weight as more washers are added.
5. Explain how the rate of change is shown on the graph.

Make a Conjecture

The graph shows sample data from a washer experiment. Describe the graph for each situation.

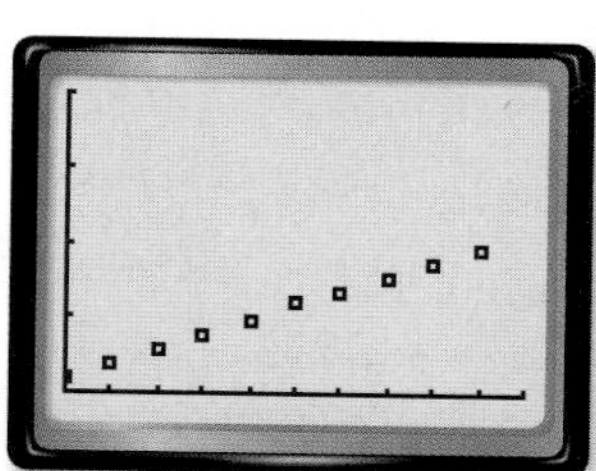

[0, 20] scl: 2 by [0, 1] scl: 0.25

6. a bag that hangs weighs 0.8 N when empty and increases in weight at the rate of the sample
7. a bag that has the same weight when empty as the sample and increases in weight at a faster rate
8. a bag that has the same weight when empty as the sample and increases in weight at a slower rate

4-1 Graphing Equations in Slope-Intercept Form

Then

You found rates of change and slopes. (Lesson 3-3)

Now

- Write and graph linear equations in slope-intercept from.
- Model real-world data with equations in slope-intercept form.

New Vocabulary

slope-intercept form

Math Online

glencoe.com

- Extra Examples
- Personal Tutor
- Self-Check Quiz
- Homework Help
- Math in Motion

Why?

Jamil has 500 songs on his MP3 player. He joins a music club that lets him download 30 songs per month for a monthly fee. The number of songs that Jamil could eventually have in his player if he does not delete any songs is represented by $y = 30x + 500$.

Slope-Intercept Form An equation of the form $y = mx + b$, where m is the slope and b is the y-intercept, is in **slope-intercept form**. The variables m and b are called *parameters* of the equation. Changing either value changes the equation's graph.

Key Concept Slope-Intercept Form

For Your FOLDABLE

Words The slope-intercept form of a linear equation is $y = mx + b$, where m is the slope and b is the y-intercept.

Example

$y = mx + b$

$y = 2x + 6$

slope (2) ↑ ↑ y-intercept (6)

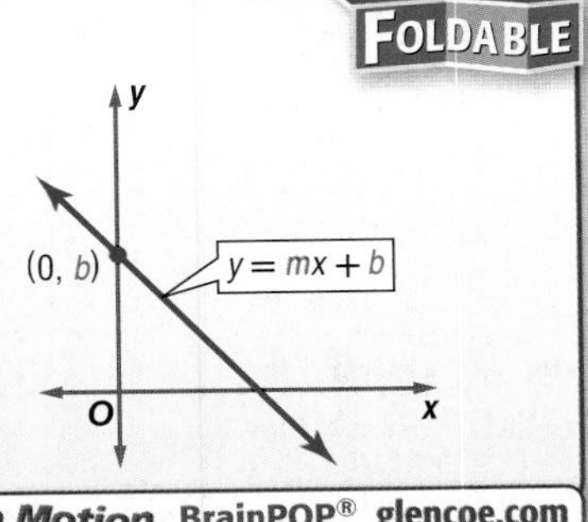

Math *in Motion*, BrainPOP® glencoe.com

EXAMPLE 1 Write and Graph an Equation

Write an equation in slope-intercept form for the line with a slope of $\frac{3}{4}$ and a y-intercept of -2. Then graph the equation.

$y = mx + b$ **Slope-intercept form**

$y = \frac{3}{4}x + (-2)$ **Replace m with $\frac{3}{4}$ and b with -2.**

$y = \frac{3}{4}x - 2$ **Simplify.**

Now graph the equation.

$y = \frac{3}{4}x - 2$

Step 1 Plot the y-intercept $(0, -2)$.

Step 2 The slope is $\frac{\text{rise}}{\text{run}} = \frac{3}{4}$. From $(0, -2)$, move up 3 units and right 4 units. Plot the point.

Step 3 Draw a line through the two points.

Check Your Progress

Write an equation of a line in slope intercept form with the given slope and y-intercept. Then graph the equation.

1A. slope: $-\frac{1}{2}$, y-intercept: 3

1B. slope: -3, y-intercept: -8

Personal Tutor glencoe.com

When an equation is not written in slope-intercept form, it may be easier to rewrite it before graphing.

StudyTip

Dependent Variables y is the *dependent* variable because it *depends* on changes in x.

EXAMPLE 2 Graph Linear Equations

Graph $3x + 2y = 6$.

Rewrite the equation in slope-intercept form.

$3x + 2y = 6$	**Original equation**
$3x + 2y - 3x = 6 - 3x$	**Subtract $3x$ from each side.**
$2y = 6 - 3x$	**Simplify.**
$2y = -3x + 6$	**$6 - 3x = 6 + (-3x)$ or $-3x + 6$**
$\frac{2y}{2} = \frac{-3x + 6}{2}$	**Divide each side by 2.**
$y = -\frac{3}{2}x + 3$	**Slope-intercept form**

Now graph the equation. The slope is $-\frac{3}{2}$, and the y-intercept is 3.

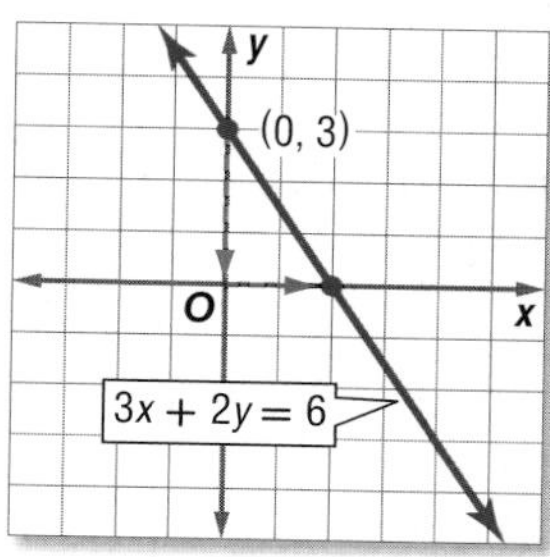

Step 1 Plot the y-intercept (0, 3).

Step 2 The slope is $\frac{\text{rise}}{\text{run}} = -\frac{3}{2}$. From (0, 3), move down 3 units and right 2 units. Plot the point.

Step 3 Draw a line through the two points.

StudyTip

Counting and Direction When counting rise and run, a negative sign may be associated with the value in the numerator or denominator. If with the numerator, begin by counting down for the rise. If with the denominator, count left when counting the run. The resulting line will be the same.

Check Your Progress

Graph each equation.

2A. $3x - 4y = 12$

2B. $-2x + 5y = 10$

Personal Tutor glencoe.com

Horizontal lines have a slope of 0. So, equations of horizontal lines can be written in slope intercept form as $y = 0x + b$ or $y = b$. Vertical lines have no slope. So, equations of vertical lines cannot be written in slope-intercept form.

EXAMPLE 3 Graph Linear Equations

Graph $y = -3$.

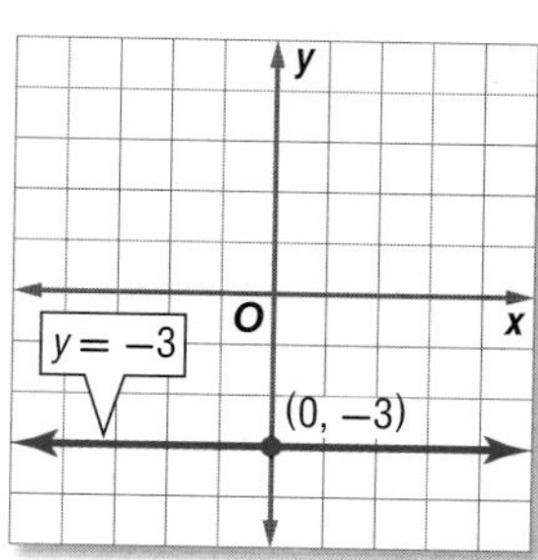

Step 1 Plot the y-intercept (0, −3).

Step 2 The slope is 0. Draw a line through the points with y-coordinate −3.

Check Your Progress

Graph each equation.

3A. $y = 5$

3B. $2y = 1$

Personal Tutor glencoe.com

Notice that the equations of horizontal lines do not have an x variable. The graph of a horizontal line does not cross the x-axis. The equation $y = 0$ lies on the x-axis.

There are times when you will need to write an equation when given a graph. To do this, locate the y-intercept and use the rise and run to find another point on the graph. Then write the equation in slope-intercept form.

Test-TakingTip

Eliminating Choices Analyze the graph to determine the slope and the y-intercept. Then you can save time by eliminating answer choices that do not match the graph.

STANDARDIZED TEST EXAMPLE 4

Which of the following is an equation in slope-intercept form for the line shown?

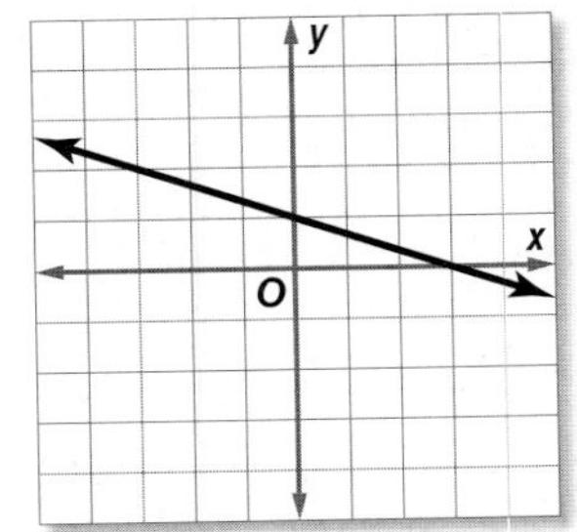

A $y = -3x + 1$

B $y = -3x + 3$

C $y = -\frac{1}{3}x + 1$

D $y = -\frac{1}{3}x + 3$

Read the Test Item

You need to find the slope and y-intercept of the line to write the equation.

Solve the Test Item

Step 1 The line crosses the y-axis at (0, 1), so the y-intercept is 1. The answer is either A or C.

Step 2 To get from (0, 1) to (3, 0), go down 1 unit and 3 units to the right. The slope is $-\frac{1}{3}$.

Step 3 Write the equation.

$y = mx + b$

$y = -\frac{1}{3}x + 1$

CHECK The graph also passes through (−3, 2). If the equation is correct, this should be a solution.

$y = -\frac{1}{3}x + 1$

$2 \stackrel{?}{=} -\frac{1}{3}(-3) + 1$

$2 \stackrel{?}{=} 1 + 1$

$2 = 2$ ✓ The answer is C.

Check Your Progress

4. Which of the following is an equation in slope-intercept form for the line shown?

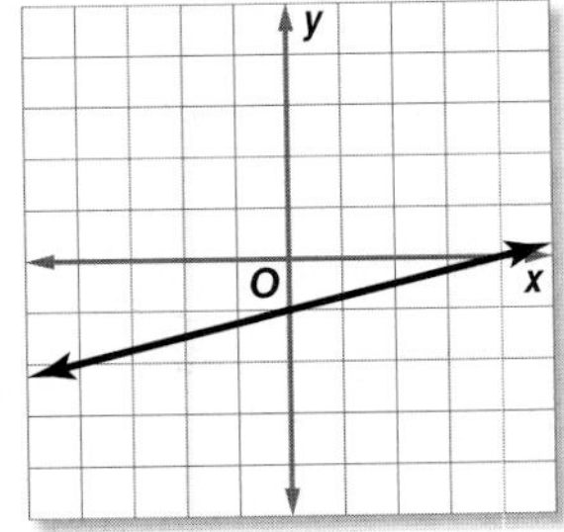

F $y = \frac{1}{4}x - 1$

G $y = \frac{1}{4}x + 4$

H $y = 4x - 1$

J $y = 4x + 4$

Personal Tutor glencoe.com

Modeling Real-World Data Real-world data can be modeled by a linear equation if there is a constant rate of change. The rate of change represents the slope. The y-intercept is the point where the value of the independent variable is 0.

Standardized Test Practice

67. A music store has x CDs in stock. If 350 are sold and $3y$ are added to stock, which expression represents the number of CDs in stock?

A $350 + 3y - x$ C $x + 350 + 3y$
B $x - 350 + 3y$ D $3y - 350 - x$

68. PROBABILITY The table shows the result of a survey of favorite activities. What is the probability that a student's favorite activity is sports or drama club?

Extracurricular Activity	Students
art club	24
band	134
choir	37
drama club	46
mock trial	19
school paper	26
sports	314

F $\frac{3}{8}$ G $\frac{4}{9}$ H $\frac{3}{5}$ J $\frac{2}{3}$

69. A recipe for fruit punch calls for 2 ounces of orange juice for every 8 ounces of lemonade. If Jennifer uses 64 ounces of lemonade, which proportion can she use to find x, the number of ounces of orange juice needed?

A $\frac{2}{x} = \frac{64}{6}$ C $\frac{2}{8} = \frac{x}{64}$
B $\frac{8}{x} = \frac{64}{2}$ D $\frac{6}{2} = \frac{x}{64}$

70. EXTENDED RESPONSE The table shows the results of a canned food drive. 1225 cans were collected, and the 12th-grade class collected 55 more cans than the 10th-grade class. How many cans each did the 10th- and 12th-grade classes collect? Show your work.

Grade	Cans
9	340
10	x
11	280
12	y

Spiral Review

For each arithmetic sequence, determine the related function. Then determine if the function is *proportional* or *nonproportional*. (Lesson 3-6)

71. 3, 7, 11, ... **72.** 8, 6, 4, ... **73.** 0, 3, 6, ... **74.** 1, 2, 3, ...

75. GAME SHOWS Contestants on a game show win money by answering 10 questions. (Lesson 3-5)

a. Find the value of the 10th question.

b. If all questions are answered correctly, how much are the winnings?

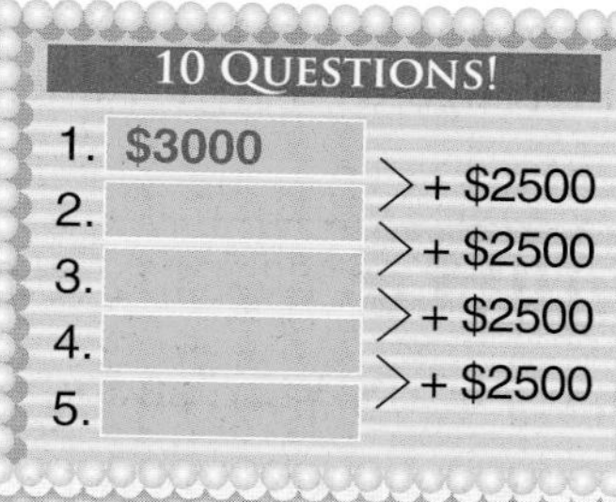

Suppose y varies directly as x. Write a direct variation equation that relates x and y. Then solve. (Lesson 3-4)

76. If $y = 10$ when $x = 5$, find y when $x = 6$.

77. If $y = -16$ when $x = 4$, find x when $y = 20$.

78. If $y = 6$ when $x = 18$, find y when $x = -12$.

79. If $y = 12$ when $x = 15$, find x when $y = -6$.

Skills Review

Find the slope of the line that passes through each pair of points. (Lesson 3-3)

80. (2, 3), (9, 7) **81.** (−3, 6), (2, 4) **82.** (2, 6), (−1, 3) **83.** (−3, 3), (1, 3)

EXTEND 4-1

Graphing Technology Lab The Family of Linear Graphs

Math Online glencoe.com
- Other Calculator Keystrokes
- Graphing Technology Personal Tutor

A family of people is related by birth, marriage, or adoption. Often people in families share characteristics. The graphs in a family share at least one characteristic. Graphs in the linear family are all lines, with the simplest graph in the family being that of the parent function $y = x$.

You can use a graphing calculator to investigate how changing the parameters m and b in $y = mx + b$ affects the graphs in the family of linear functions.

Parent Graph

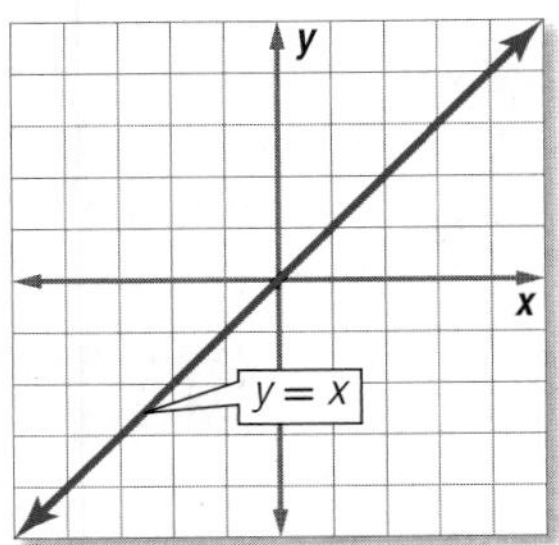

ACTIVITY 1 Changing b in $y = mx + b$

Graph $y = x$, $y = x + 4$, and $y = x - 2$ in the standard viewing window.

Enter the equations in the Y= list as Y1, Y2, and Y3. Then graph the equations.

KEYSTROKES: *Review graphing on pages 167 and 168.*

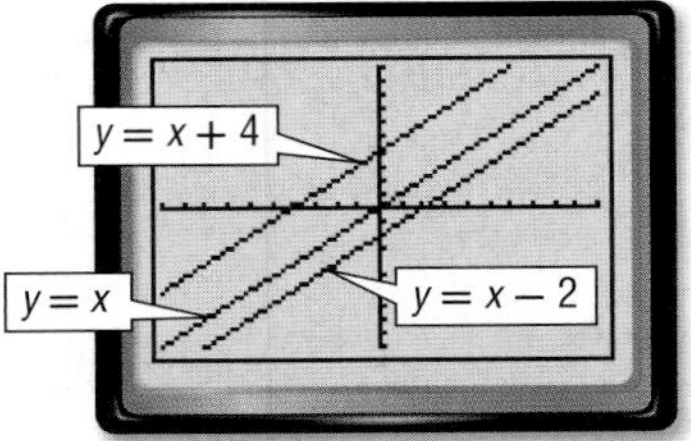

[−10, 10] scl: 1 by [−10, 10] scl: 1

1A. How do the slopes of the graphs compare?

1B. Compare the graph of $y = x + 4$ and the graph of $y = x$. How would you obtain the graph of $y = x + 4$ from the graph of $y = x$?

1C. How would you obtain the graph of $y = x - 2$ from the graph of $y = x$?

Changing m in $y = mx + b$ affects the graphs in a different way than changing b. First, investigate positive values of m.

ACTIVITY 2 Changing m in $y = mx + b$, Positive Values

Graph $y = x$, $y = 2x$, and $y = \frac{1}{3}x$ in the standard viewing window.

Enter the equations in the Y= list and graph.

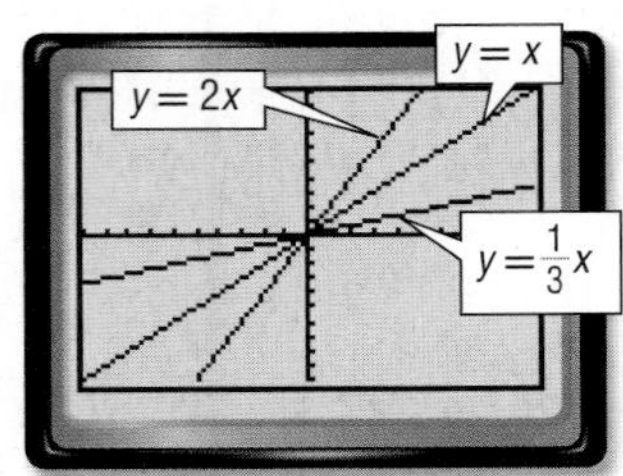

[−10, 10] scl: 1 by [−10, 10] scl: 1

2A. How do the y-intercepts of the graphs compare?

2B. Compare the graph of $y = 2x$ and the graph of $y = x$.

2C. Which is steeper, the graph of $y = \frac{1}{3}x$ or the graph of $y = x$?

Does changing m to a negative value affect the graph differently than changing it to a positive value?

ACTIVITY 3 Changing m in $y = mx + b$, Negative Values

Graph $y = x$, $y = -x$, $y = -3x$, and $y = -\frac{1}{2}x$ in the standard viewing window.

Enter the equations in the Y= list and graph.

3A. How are the graphs with negative values of m different than graphs with a positive m?

3B. Compare the graphs of $y = -x$, $y = -3x$, and $y = -\frac{1}{2}x$. Which is steepest?

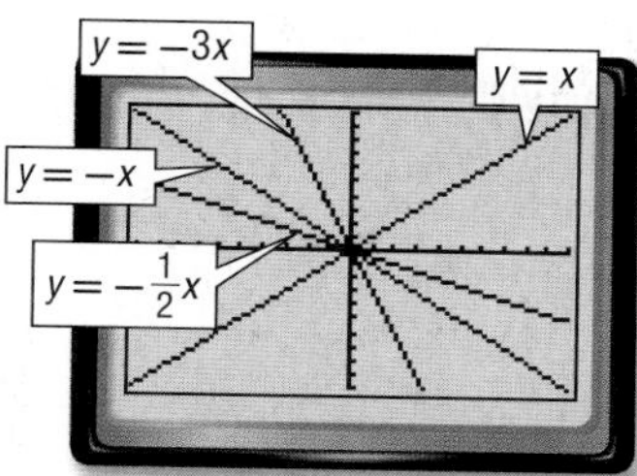

[−10, 10] scl: 1 by [−10, 10] scl: 1

Analyze the Results

Graph each set of equations on the same screen. Describe the similarities or differences.

1. $y = 2x$
$y = 2x + 3$
$y = 2x - 7$

2. $y = x + 1$
$y = 2x + 1$
$y = \frac{1}{4}x + 1$

3. $y = x + 4$
$y = 2x + 4$
$y = \frac{3}{4}x + 4$

4. $y = 0.5x + 2$
$y = 0.5x - 5$
$y = 0.5x + 4$

5. $y = -2x - 2$
$y = -4.2x - 2$
$y = -\frac{1}{3}x - 2$

6. $y = 3x$
$y = 3x + 6$
$y = 3x - 7$

7. Families of graphs have common characteristics. What do the graphs of all equations of the form $y = mx + b$ have in common?

8. How does the value of b affect the graph of $y = mx + b$?

9. What is the result of changing the value of m on the graph of $y = mx + b$ if m is positive?

10. How can you determine which graph is steepest by examining the following equations?
$y = 3x$, $y = -4x - 7$, $y = \frac{1}{2}x + 4$

11. Explain how knowing about the effects of m and b can help you sketch the graph of an equation.

12. The equation $y = k$ can also be a parent graph. Graph $y = 5$, $y = 2$, and $y = -4$ on the same screen. Describe the similarities or differences among the graphs.

Extension

Nonlinear functions can also be defined in terms of a family of graphs. Graph each set of equations on the same screen. Describe the similarities or differences.

13. $y = x^2$
$y = -3x^2$
$y = (-3x)^2$

14. $y = x^2$
$y = x^2 + 3$
$y = (x - 2)^2$

15. $y = x^2$
$y = 2x^2 + 4$
$y = (3x)^2 - 5$

16. Describe the similarities and differences in the classes of functions $f(x) = x^2 + c$ and $f(x) = (x + c)^2$, where c is any real number.

4-2

Writing Equations in Slope-Intercept Form

Then

You graphed lines given the slope and the y-intercept. (Lesson 4-1)

Now

- Write an equation of a line in slope-intercept form given the slope and one point.
- Write an equation of a line in slope-intercept form given two points.

New Vocabulary

linear extrapolation

Math Online

glencoe.com

- Extra Examples
- Personal Tutor
- Self-Check Quiz
- Homework Help

Why?

In 2000, Americans took 337.1 million vacations. In 2004, Americans took 375.4 million vacations. You can find the average rate of change for these data. Then you can write an equation that would model the average number of vacations taken per year.

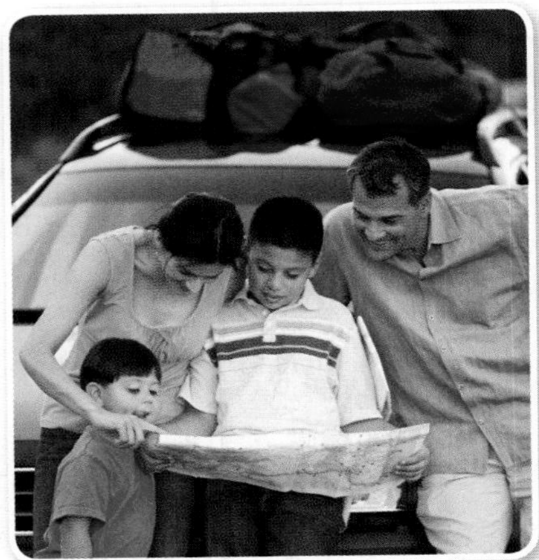

Write an Equation Given the Slope and a Point The next example shows how to write an equation of a line if you are given a slope and a point other than the y-intercept.

EXAMPLE 1 Write an Equation Given the Slope and a Point

Write an equation of the line that passes through (2, 1) with a slope of 3.

You are given the slope but not the y-intercept.

Step 1 Find the y-intercept.

$y = mx + b$	**Slope-intercept form**
$1 = 3(2) + b$	**Replace m with 3, y with 1, and x with 2.**
$1 = 6 + b$	**Simplify.**
$1 - 6 = 6 + b - 6$	**Subtract 6 from each side.**
$-5 = b$	**Simplify.**

Step 2 Write the equation in slope-intercept form.

$y = mx + b$	**Slope-intercept form**
$y = 3x - 5$	**Replace m with 3 and b with −5.**

Therefore, the equation of the line is $y = 3x - 5$.

Check Your Progress

Write an equation of a line that passes through the given point and has the given slope.

1A. $(-2, 5)$, slope 3

1B. $(4, -7)$, slope -1

Personal Tutor glencoe.com

Write an Equation Given Two Points If you are given two points through which a line passes, you can use them to find the slope first. Then follow the steps in Example 1 to write the equation.

EXAMPLE 2 Write an Equation Given Two Points

Write an equation of the line that passes through each pair of points.

a. (3, 1) and (2, 4)

Step 1 Find the slope of the line containing the given points.

$m = \frac{y_2 - y_1}{x_2 - x_1}$	**Slope Formula**
$= \frac{4 - 1}{2 - 3}$	**$(x_1, y_1) = (3, 1)$ and $(x_2, y_2) = (2, 4)$**
$= \frac{3}{-1}$ or -3	**Simplify.**

Step 2 Use either point to find the y-intercept.

$y = mx + b$	**Slope-intercept form**
$4 = (-3)(2) + b$	**Replace m with -3, x with 2, and y with 4.**
$4 = -6 + b$	**Simplify.**
$4 - (-6) = -6 + b - (-6)$	**Subtract -6 from each side.**
$10 = b$	**Simplify.**

Step 3 Write the equation in slope-intercept form.

$y = mx + b$	**Slope-intercept form**
$y = -3x + 10$	**Replace m with -3 and b with 10.**

Therefore, the equation is $y = -3x + 10$.

b. (−4, −2) and (−5, −6)

Step 1 Find the slope of the line containing the given points.

$m = \frac{y_2 - y_1}{x_2 - x_1}$	**Slope Formula**
$= \frac{-6 - (-2)}{-5 - (-4)}$	**$(x_1, y_1) = (-4, -2)$ and $(x_2, y_2) = (-5, -6)$**
$= \frac{-4}{-1}$ or 4	**Simplify.**

Step 2 Use either point to find the y-intercept.

$y = mx + b$	**Slope-intercept form**
$-2 = 4(-4) + b$	**Replace m with 4, x with -4, and y with -2.**
$-2 = -16 + b$	**Simplify.**
$-2 - (-16) = -16 + b - (-16)$	**Subtract -16 from each side.**
$14 = b$	**Simplify.**

Step 3 Write the equation in slope-intercept form.

$y = mx + b$	**Slope-intercept form**
$y = 4x + 14$	**Replace m with 4 and b with 14.**

Therefore, the equation is $y = 4x + 14$.

Check Your Progress

Write an equation of the line that passes through each pair of points.

2A. $(-1, 12), (4, -8)$

2B. $(5, -8), (-7, 0)$

Personal Tutor glencoe.com

StudyTip

Choosing a point Given two points on a line, you may select either point to be (x_1, y_1). Be sure to remain consistent throughout the problem.

StudyTip

Slope If the (x_1, y_1) coordinates are negative, be sure to account for both the negative signs and the subtraction symbols in the Slope Formula.

Real-World Career

Baggage Handler
Airline ground crew responsibilities include checking tickets, helping passengers with luggage, and making sure that baggage is secure. This job usually requires a high school diploma or GED.

Source: Airline Jobs

Real-World EXAMPLE 3 Use Slope-Intercept Form

AIR FARES The table shows the average fares for domestic flights in certain years. Write an equation that could be used to predict air fares if fares continue to increase at this rate.

Year	Cost ($)
2004	354
2005	366
2006	378
2007	390

Understand You know the air fares for the years listed.

Plan Let x represent the number of years since 2000, and let y represent the air fare. Write an equation of the line that passes through (5, 366) and (6, 378).

Solve Find the slope.

$m = \frac{y_2 - y_1}{x_2 - x_1}$ — **Slope Formula**

$= \frac{378 - 366}{6 - 5}$ — **Let $(x_1, y_1) = (5, 366)$ and $(x_2, y_2) = (6, 378)$.**

$= \frac{12}{1}$ or 12 — **Simplify.**

Choose (6, 378) and find the y-intercept of the line.

$y = mx + b$ — **Slope-intercept form**

$378 = 12(6) + b$ — **Replace m with 12, x with 6, and y with 378.**

$378 = 72 + b$ — **Simplify.**

$306 = b$ — **Subtract 72 from each side.**

Write the equation using $m = 12$ and $b = 306$.

$y = mx + b$ — **Slope-intercept form**

$y = 12x + 306$ — **Replace m with 12 and b with 306.**

Check Check your result by using the coordinates of the other point.

$y = 12x + 306$ — **Original equation**

$366 \stackrel{?}{=} 12(5) + 306$ — **Replace x with 5 and y with 366.**

$366 = 366$ ✓ — **Simplify.**

Check Your Progress

3. FINANCIAL LITERACY In addition to his weekly salary, Ethan is paid $16 per delivery. Last week, he made 5 deliveries, and his total pay was $215. Write a linear equation to find Ethan's total weekly pay T if he makes d deliveries.

Personal Tutor glencoe.com

You can use a linear equation to make predictions about values that are beyond the range of the data. This process is called **linear extrapolation**.

Real-World EXAMPLE 4 Predict from Slope-Intercept Form

AIR FARES Use the equation from Example 3 to estimate the cost of airfares in 2010.

$y = 12x + 306$ — **Original equation**

$= 12(10) + 306$ — **Replace x with 10.**

$= 426$ — An estimate of the average air fares is $426.

Check Your Progress

4. MONEY Use the equation in Check Your Progress 3 to predict how much money Ethan will earn in a week if he makes 8 deliveries.

Personal Tutor glencoe.com

Check Your Understanding

Example 1 p. 224

Write an equation of the line that passes through the given point and has the given slope.

1. (3, −3), slope 3
2. (2, 4), slope 2
3. (1, 5), slope −1
4. (−4, 6), slope −2

Example 2 p. 225

Write an equation of the line that passes through each pair of points.

5. (4, −3), (2, 3)
6. (−7, −3), (−3, 5)
7. (−1, 3), (0, 8)
8. (−2, 6), (0, 0)

Examples 3 and 4 p. 226

9. WHITEWATER RAFTING Ten people from a local youth group went to Black Hills Whitewater Rafting Tour Company for a one-day rafting trip. The group paid $425.

a. Write an equation in slope-intercept form to find the total cost C for p people.

b. How much would it cost for 15 people?

Practice and Problem Solving

 = **Step-by-Step Solutions** begin on page R12.
Extra Practice begins on page 815.

Example 1 p. 224

Write an equation of the line that passes through the given point and has the given slope.

10. (3, 1), slope 2
11. (−1, 4), slope −1
12. (1, 0), slope 1
13. (7, 1), slope 8
14. (2, 5), slope −2
15. (2, 6), slope 2

Example 2 p. 225

Write an equation of the line that passes through each pair of points.

16. (9, −2), (4, 3)
17. (−2, 5), (5, −2)
18. (−5, 3), (0, −7)
19. (3, 5), (2, −2)
20. (−1, −3), (−2, 3)
21. (−2, −4), (2, 4)

Examples 3 and 4 p. 226

22. RC CAR Greg is driving a remote control car at a constant speed. He starts the timer when the car is 5 feet away. After 2 seconds the car is 35 feet away.

a. Write a linear equation to find the distance d of the car from Greg.

b. Estimate the distance the car has traveled after 10 seconds.

Problem-SolvingTip

Determine Reasonable Answers Deciding whether an answer is reasonable is useful when an exact answer is not neccessary.

23. TRAVEL Refer to the beginning of the lesson.

a. Write a linear equation to find the number of vacations (in millions) y after x years. Let x be the number of years since 2000.

b. Estimate the number of vacations that will be taken in 2012.

24. BOOKS In 1904, a dictionary cost 30¢. Since then the cost of a dictionary has risen an average of 6¢ per year.

a. Write a linear equation to find the cost C of a dictionary y years after 2004.

b. If this trend continues, what will the cost of a dictionary be in 2020?

Write an equation of the line that passes through the given point and has the given slope.

25. (4, 2), slope $\frac{1}{2}$
26. (3, −2), slope $\frac{1}{3}$
27. (6, 4), slope $-\frac{3}{4}$
28. (2, −3), slope $\frac{2}{3}$
29. (2, −2), slope $\frac{2}{7}$
30. (−4, −2), slope $-\frac{3}{5}$

Real-World Link

There are approximately 73 million dogs kept as pets in the United States. Thirty-nine percent of households in the United States own at least one dog.

Source: The Humane Society of the United States

31. DOGS In 2001, there were about 56.1 thousand golden retrievers registered in the United States. In 2002, the number was 62.5 thousand.

a. Write a linear equation to find the number of golden retrievers G that will be registered in year t, where $t = 0$ is the year 2000.

b. Graph the equation.

c. Estimate the number of golden retrievers that will be registered in 2012.

32. GYM MEMBERSHIPS A local recreation center offers a yearly membership for $265. The center offers aerobics classes for an additional $5 per class.

a. Write an equation that represents the total cost of the membership.

b. Carly spent $500 one year. How many aerobics classes did she take?

33. SUBSCRIPTION A magazine offers an online subscription that allows you to view up to 25 archived articles free. To view 30 archived articles, you pay $49.15. To view 33 archived articles, you pay $57.40.

a. What is the cost of each archived article for which you pay a fee?

b. What is the cost of the magazine subscription?

Write an equation of the line that passes through the given points.

34. $(5, -2), (7, 1)$ **35** $(5, -3), (2, 5)$ **36.** $\left(\frac{5}{4}, 1\right), \left(-\frac{1}{4}, \frac{3}{4}\right)$ **37.** $\left(\frac{5}{12}, -1\right), \left(-\frac{3}{4}, \frac{1}{6}\right)$

Determine whether the given point is on the line. Explain why or why not.

38. $(3, -1); y = \frac{1}{3}x + 5$ **39.** $(6, -2); y = \frac{1}{2}x - 5$

For Exercises 40–42, determine which equation best represents each situation. Explain the meaning of each variable.

A $y = -\frac{1}{3}x + 72$	**B** $y = 2x + 225$	**C** $y = 8x + 4$

40. CONCERTS Tickets to a concert cost $8 each plus a processing fee of $4 per order.

41. FUNDRAISING The freshman class has $225. They sell raffle tickets at $2 each to raise money for a field trip.

42. POOLS The current water level of a swimming pool in Tucson, Arizona, is 6 feet. The rate of evaporation is $\frac{1}{3}$ inch per day.

43. ENVIRONMENT A manufacturer implemented a program to reduce waste. In 1998 they sent 946 tons of waste to landfills. Each year after that, they reduced their waste by an average 28.4 tons.

a. How many tons were sent to the landfill in 2010?

b. In what year will it become impossible for this trend to continue? Explain.

44. MULTIPLE REPRESENTATIONS In this problem, you will explore the slopes of perpendicular lines.

a. GRAPHICAL On a coordinate plane, graph $y = \frac{3}{4}x + 1$.

b. PICTORIAL Use a straightedge and a protractor to draw a line that is perpendicular to the line you graphed.

c. ALGEBRAIC Find the equation of the line that is perpendicular to the original line. Describe which method you used to write the equation.

d. ANALYTICAL Compare the slopes of the lines. Describe the relationship, if any, between the two values.

Real-World Link

55% of people who buy tickets for events use online ticket agents.

Source: Pew Internet & American Life Project

45. **CONCERT TICKETS** Jackson is ordering tickets for a concert online. There is a processing fee for each order, and the tickets are \$52 each. Jackson ordered 5 tickets and the cost was \$275.

 a. Determine the processing fee. Write a linear equation to represent the total cost C for t tickets.

 b. Make a table of values for at least three other numbers of tickets.

 c. Graph this equation. Predict the cost of 8 tickets.

46. **MUSIC** A music store is offering a Frequent Buyers Club membership. The membership costs \$22 per year, and then a member can buy CDs at a reduced price. If a member buys 17 CDs in one year, the cost is \$111.25.

 a. Determine the cost of each CD for a member.

 b. Write a linear equation to represent the total cost y of a one year membership, if x CDs are purchased.

 c. Graph this equation.

H.O.T. Problems

Use **H**igher-**O**rder **T**hinking Skills

47. **FIND THE ERROR** Tess and Jacinta are writing an equation of the line through $(3, -2)$ and $(6, 4)$. Is either of them correct? Explain your reasoning.

Tess

$m = \frac{4-(-2)}{6-3} = \frac{6}{3}$ or 2

$y = mx + b$

$6 = 2(4) + b$

$6 = 8 + b$

$-2 = b$

$y = 2x - 2$

Jacinta

$m = \frac{4-(-2)}{6-3} = \frac{6}{3}$ or 2

$y = mx + b$

$-2 = 2(3) + b$

$-2 = 6 + b$

$-8 = b$

$y = 2x - 8$

48. **CHALLENGE** Consider three points, $(3, 7)$, $(-6, 1)$ and $(9, p)$, on the same line. Find the value of p and explain your steps.

49. **REASONING** Consider the standard form of a linear equation, $Ax + By = C$.

 a. Rewrite the equation in slope-intercept form.

 b. What is the slope?

 c. What is the y-intercept?

 d. Is this true for all real values of A, B, and C?

50. **OPEN ENDED** Create a real-world situation that fits the graph at the right. Define the two quantities and describe the functional relationship between them. Write an equation to represent this relationship and describe what the slope and y-intercept mean.

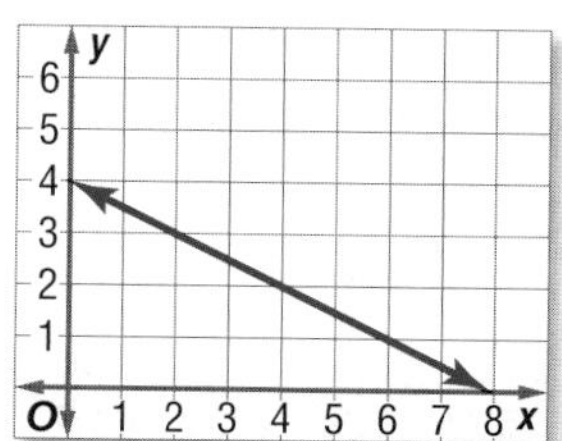

51. **WRITING IN MATH** Linear equations are useful in predicting future events. Describe some factors in real-world situations that might affect the reliability of the graph in making any predictions.

52. **WRITING IN MATH** What information is needed to write the equation of a line? Explain.

Standardized Test Practice

53. Which equation *best* represents the graph?

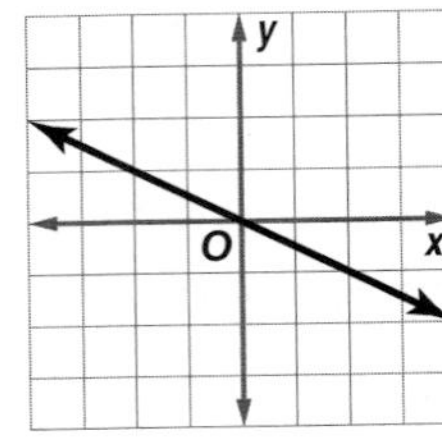

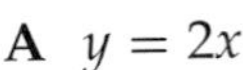

A $y = 2x$

B $y = -2x$

C $y = \frac{1}{2}x$

D $y = -\frac{1}{2}x$

54. Roberto receives an employee discount of 12%. If he buys a \$355 item at the store, what is his discount to the nearest dollar?

F \$3

G \$4

H \$30

J \$43

55. **GEOMETRY** The midpoints of the sides of the large square are joined to form a smaller square. What is the area of the smaller square?

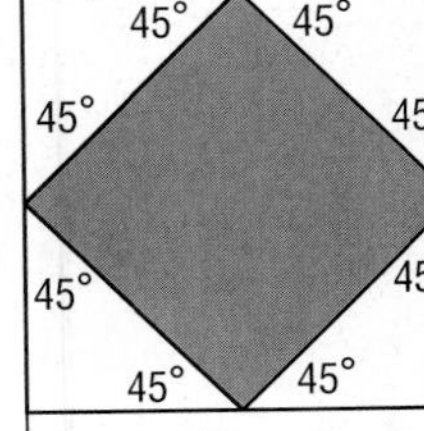

A 64 cm^2

B 128 cm^2

C 248 cm^2

D 256 cm^2

56. **SHORT RESPONSE** If $\frac{5(x + 4)}{2} + 7 = 37$, what is the value of $3x - 9$?

Spiral Review

Graph each equation. (Lesson 4-1)

57. $y = 3x + 2$

58. $y = -4x + 2$

59. $3y = 2x + 6$

60. $y = \frac{1}{2}x + 6$

61. $3x + y = -1$

62. $2x + 3y = 6$

Write an equation in function notation for each relation. (Lesson 3-6)

63.

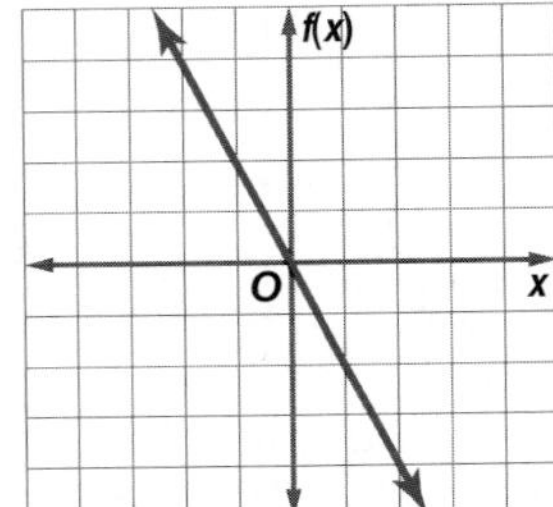

64.

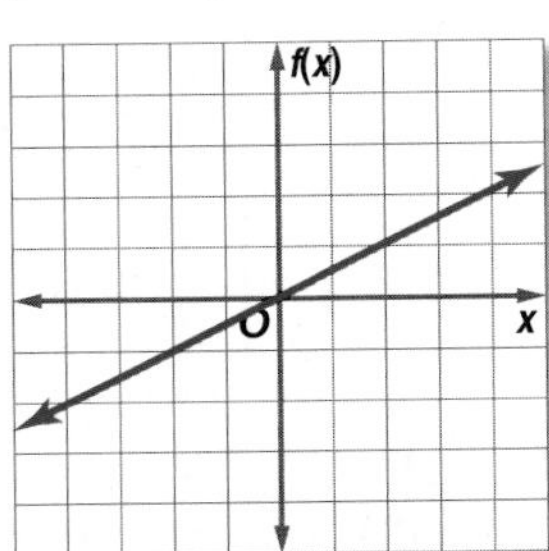

65. **METEOROLOGY** The distance d in miles that the sound of thunder travels in t seconds is given by the equation $d = 0.21t$. (Lesson 3-1)

a. Graph the equation.

b. Use the graph to estimate how long it will take you to hear thunder from a storm 3 miles away.

Solve each equation. Check your solution. (Lesson 2-3)

66. $-5t - 2.2 = -2.9$

67. $-5.5a - 43.9 = 77.1$

68. $4.2r + 7.14 = 12.6$

69. $-14 - \frac{n}{9} = 9$

70. $\frac{-8b - (-9)}{-10} = 17$

71. $9.5x + 11 - 7.5x = 14$

Skills Review

Find the value of r so the line through each pair of points has the given slope. (Lesson 3-3)

72. $(6, -2), (r, -6), m = 4$

73. $(8, 10), (r, 4), m = 6$

74. $(7, -10), (r, 4), m = -3$

75. $(6, 2), (9, r), m = -1$

76. $(9, r), (6, 3), m = -\frac{1}{3}$

77. $(5, r), (2, -3), m = \frac{4}{3}$

4-3 Writing Equations in Point-Slope Form

Then

You wrote linear equations given either one point and the slope or two points. (Lesson 4-2)

Now

- Write equations of lines in point-slope form.
- Write linear equations in different forms.

New Vocabulary

point-slope form

Math Online

glencoe.com

- Extra Examples
- Personal Tutor
- Self-Check Quiz
- Homework Help

Why?

Most humane societies have foster homes for newborn puppies, kittens, and injured or ill animals. During the spring and summer, a large shelter can place 3000 animals in homes each month.

If a shelter had 200 animals in foster homes at the beginning of spring, the number of animals in foster homes at the end of the summer could be represented by $y = 3000x + 200$, where x is the number of months and y is the number of animals.

Point-Slope Form An equation of a line can be written in **point-slope form** when given the coordinates of one known point on a line and the slope of that line.

Key Concept — Point-Slope Form

For Your FOLDABLE

Words The linear equation $y - y_1 = m(x - x_1)$ is written in point-slope form, where (x_1, y_1) is a given point on a nonvertical line and m is the slope of the line.

Symbols $y - y_1 = m(x - x_1)$

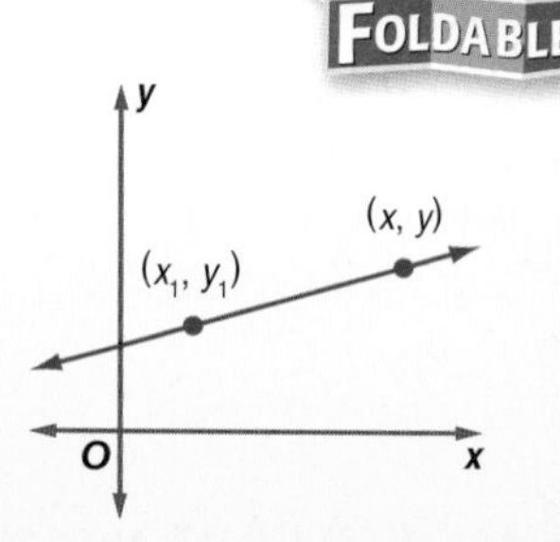

EXAMPLE 1 Write and Graph an Equation in Point-Slope Form

Write an equation in point-slope form for the line that passes through (3, −2) with a slope of $\frac{1}{4}$. Then graph the equation.

$y - y_1 = m(x - x_1)$ **Point-slope form**

$y - (-2) = \frac{1}{4}(x - 3)$ $(x_1, y_1) = (3, -2), m = \frac{1}{4}$

$y + 2 = \frac{1}{4}(x - 3)$ **Simplify.**

Plot the point at (3, −2) and use the slope to find another point on the line. Draw a line through the two points.

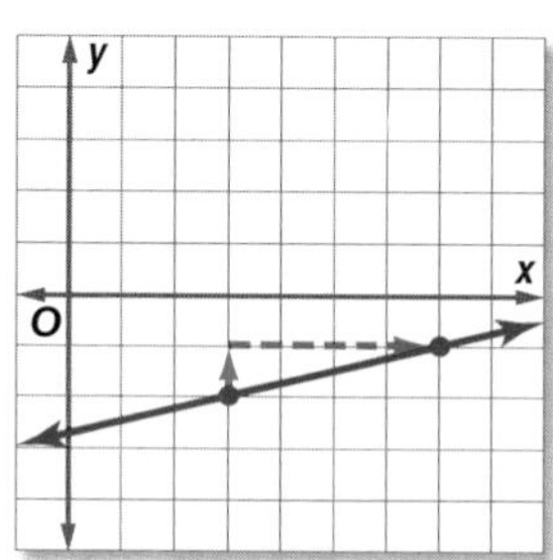

Check Your Progress

1. Write an equation in point-slope form for the line that passes through (−2, 1) with a slope of −6. Then graph the equation.

Personal Tutor glencoe.com

Forms of Linear Equations If you are given the slope and the coordinates of one or two points, you can write the linear equation in the following ways.

StudyTip

Slope The slope of the line remains unchanged throughout the line. You can go in either direction along the line using the same rise over run and you will always end at a point on the line.

Concept Summary **Writing Equations** For Your FOLDABLE

Given the Slope and One Point

Step 1 Substitute the value of m and let the x and y coordinates be (x_1, y_1). Or, substitute the values of m, x, and y into the slope-intercept form and solve for b.

Step 2 Rewrite the equation in the needed form.

Given Two Points

Step 1 Find the slope.

Step 2 Choose one of the two points to use.

Step 3 Follow the steps for writing an equation given the slope and one point.

Review Vocabulary

standard form of a linear equation $Ax + By = C$, where $A \geq 0$, A and B are not both zero, and A, B, and C are integers with a greatest common factor of 1 (Lesson 3-1)

EXAMPLE 2 **Standard Form**

Write $y - 1 = -\frac{2}{3}(x - 5)$ in standard form.

$y - 1 = -\frac{2}{3}(x - 5)$ **Original equation**

$3(y - 1) = 3\left(-\frac{2}{3}\right)(x - 5)$ **Multiply each side by 3 to eliminate the fraction.**

$3(y - 1) = -2(x - 5)$ **Simplify.**

$3y - 3 = -2x + 10$ **Distributive Property**

$3y = -2x + 13$ **Add 3 to each side.**

$2x + 3y = 13$ **Add $2x$ to each side.**

Check Your Progress

2. Write $y - 1 = 7(x + 5)$ in standard form.

Personal Tutor glencoe.com

To find the y-intercept of an equation, rewrite the equation in slope-intercept form.

EXAMPLE 3 **Slope-Intercept Form**

Write $y + 3 = \frac{3}{2}(x + 1)$ in slope-intercept form.

$y + 3 = \frac{3}{2}(x + 1)$ **Original equation**

$y + 3 = \frac{3}{2}x + \frac{3}{2}$ **Distributive Property**

$y = \frac{3}{2}x - \frac{3}{2}$ **Subtract 3 from each side.**

Check Your Progress

3. Write $y + 6 = -3(x - 4)$ in slope-intercept form.

Personal Tutor glencoe.com

Being able to use a variety of forms of linear equations can be useful in other subjects as well.

StudyTip

Slopes in Squares Nonvertical opposite sides of a square have equal slopes. If the coordinates for one of the vertices are unavailable, use the slope of the opposite side.

EXAMPLE 4 Point-Slope Form and Standard Form

GEOMETRY The figure shows square *RSTU*.

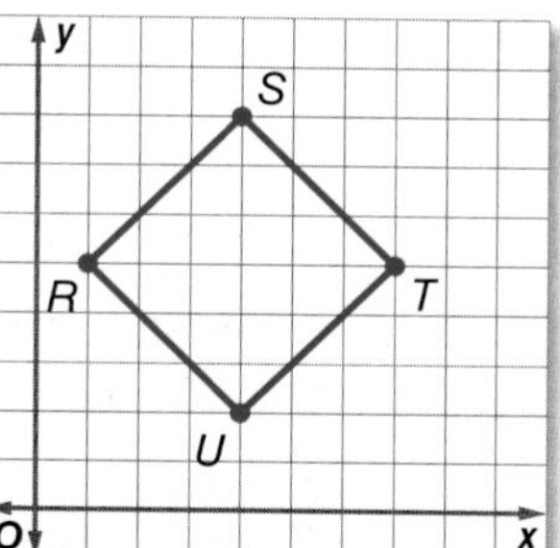

a. Write an equation in point-slope form for the line containing side $\overline{TU}$.

Step 1 Find the slope of $\overline{TU}$.

$m = \frac{y_2 - y_1}{x_2 - x_1}$ **Slope Formula**

$= \frac{5-2}{7-4}$ or 1 **$(x_1, y_1) = (4, 2)$ and $(x_2, y_2) = (7, 5)$**

Step 2 You can select either point for (x_1, y_1) in the point-slope form.

$y - y_1 = m(x - x_1)$ **Point-slope form**

$y - 2 = 1(x - 4)$ **$(x_1, y_1) = (4, 2)$**

$y - 5 = 1(x - 7)$ **$(x_1, y_1) = (7, 5)$**

b. Write an equation in standard form for the same line.

$y - 2 = 1(x - 4)$	**Original equation**	$y - 5 = 1(x - 7)$
$y - 2 = 1x - 4$	**Distributive Property**	$y - 5 = 1x - 7$
$y = 1x - 2$	**Add to each side.**	$y = 1x - 2$
$-1x + y = -2$	**Subtract 1*x* from each side.**	$-1x + y = -2$
$x - y = 2$	**Multiply each side by −1.**	$x - y = 2$

Check Your Progress

4A. Write an equation in point-slope form of the line containing side $\overline{ST}$.

4B. Write an equation in standard form of the line containing $\overline{ST}$.

Personal Tutor glencoe.com

Check Your Understanding

Example 1 p. 231

Write an equation in point-slope form for the line that passes through the given point with the slope provided. Then graph the equation.

1. $(-2, 5)$, slope -6
2. $(-2, -8)$, slope $\frac{5}{6}$
3. $(4, 3)$, slope $-\frac{1}{2}$

Example 2 p. 232

Write each equation in standard form.

4. $y + 2 = \frac{7}{8}(x - 3)$
5. $y + 7 = -5(x + 3)$
6. $y + 2 = \frac{5}{3}(x + 6)$

Example 3 p. 232

Write each equation in slope-intercept form.

7. $y - 10 = 4(x + 6)$
8. $y - 7 = -\frac{3}{4}(x + 5)$
9. $y - 9 = x + 4$

Example 4
p. 233

10. **GEOMETRY** Use right triangle FGH.

    a. Write an equation in point-slope form for the line containing $\overline{GH}$.

    b. Write the standard form of the line containing $\overline{GH}$.

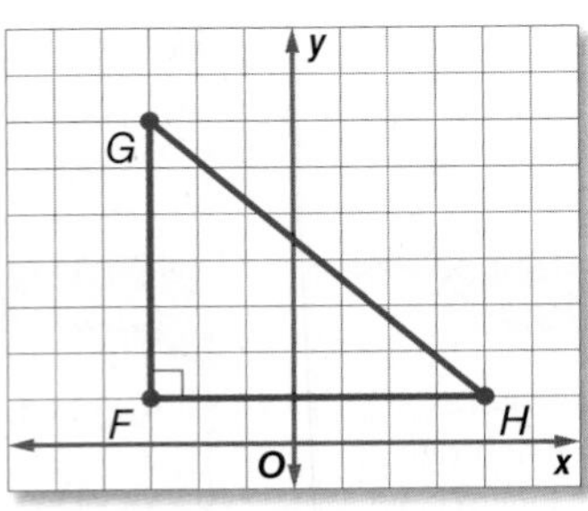

Practice and Problem Solving

● = **Step-by-Step Solutions** begin on page R12.
Extra Practice begins on page 815.

Example 1
p. 231

Write an equation in point-slope form for the line that passes through each point with the given slope. Then graph the equation.

11. $(5, 3), m = 7$
12. $(2, -1), m = -3$
13. $(-6, -3), m = -1$
14. $(-7, 6), m = 0$
15. $(-2, 11), m = \frac{4}{3}$
16. $(-7, -8), m = -\frac{3}{7}$
17. Write an equation in point-slope form for a line that passes through $(-2, -9)$ with a slope of $-\frac{7}{5}$.
18. Write an equation in point-slope form for a horizontal line that passes through $(-6, 0)$.

Example 2
p. 232

Write each equation in standard form.

19. $y - 10 = 2(x - 8)$
20. $y - 6 = -3(x + 2)$
21. $y - 9 = -6(x + 9)$
22. $y + 4 = \frac{2}{3}(x + 7)$
23. $y + 7 = \frac{9}{10}(x + 3)$
24. $y + 7 = -\frac{3}{2}(x + 1)$
25. $2y + 3 = -\frac{1}{3}(x - 2)$
26. $4y - 5x = 3(4x - 2y + 1)$

Example 3
p. 232

Write each equation in slope-intercept form.

27. $y - 6 = -2(x - 7)$
28. $y - 11 = 3(x + 4)$
29. $y + 5 = -6(x + 7)$
30. $y - 1 = \frac{4}{5}(x + 5)$
31. $y + 2 = \frac{1}{6}(x - 4)$
32. $y + 6 = -\frac{3}{4}(x + 8)$
33. $y + 3 = -\frac{1}{3}(2x + 6)$
34. $y + 4 = 3(3x + 3)$

Example 4
p. 233

35. **MOVIE RENTALS** The number of copies of a movie rented at a video store decreased at a constant rate of 5 copies per week. The 6th week after the movie was released, 4 copies were rented. How many copies were rented during the second week?

36. **CABLE** A company offers premium cable for \$39.95 per month plus a one-time setup fee. The total cost for setup and 6 months of service is \$264.70.

    a. Write an equation in point-slope form to find the total price y for any number of months x. (*Hint*: The point (6, 264.70) is a solution to the equation.)

    b. Write the equation in slope-intercept form.

    c. What is the setup fee?

Write each equation in standard form.

37. $y + 8 = -\frac{11}{12}(x - 14)$
38. $y - 3 = 2.5(x + 1)$
39. $y + 2.1 = 1.4(x - 5)$

Write an equation in point-slope form for each line.

40.

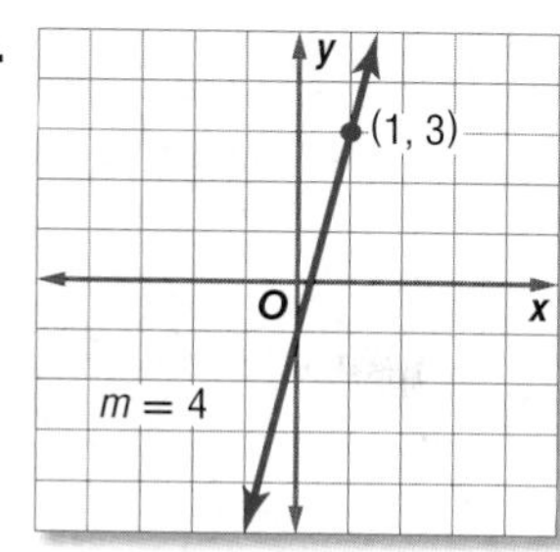

41.

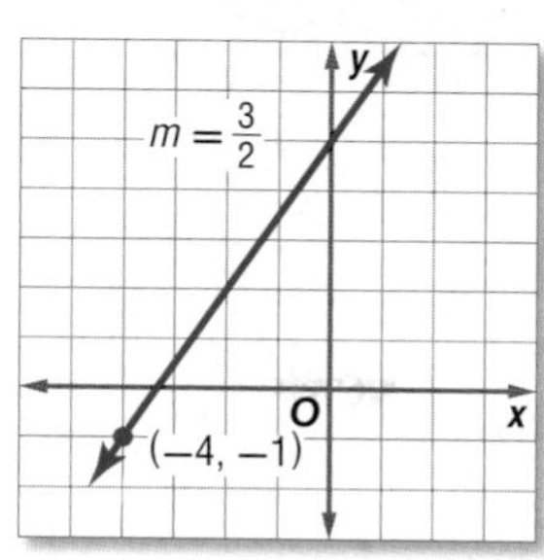

42.

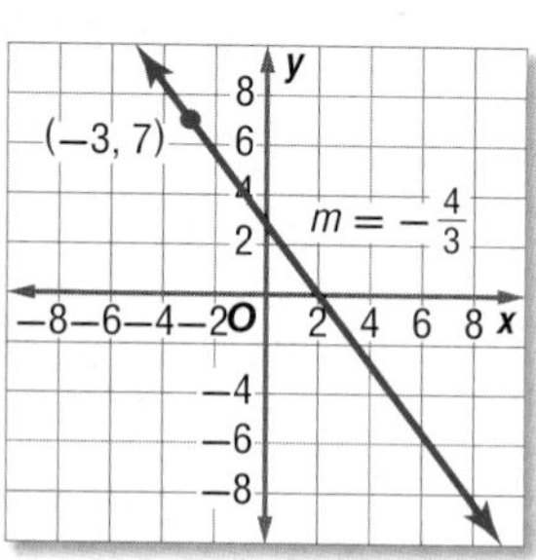

Write each equation in slope-intercept form.

43. $y + \frac{3}{5} = x - \frac{2}{5}$ **44.** $y - \frac{7}{2} = \frac{1}{2}(x - 4)$ **45.** $y + \frac{1}{3} = \frac{5}{6}(x + \frac{2}{5})$

46. Write an equation in point-slope form, slope-intercept form, and standard form for a line that passes through $(-2, 8)$ with slope $\frac{8}{5}$.

47. Line ℓ passes through $(-9, 4)$ with slope $\frac{4}{7}$. Write an equation in point-slope form, slope-intercept form, and standard form for line ℓ.

48. WEATHER Barometric pressure is a linear function of altitude. The barometric pressure is 598 millimeters of mercury (mmHg) at an altitude of 1.8 kilometers. The pressure is 577 millimeters of mercury at 2.1 kilometers.

a. Write a formula for the barometric pressure as a function of the altitude.

b. What is the altitude if the pressure is 657 millimeters of mercury?

Real-World Link

At higher altitudes, air is at a lower pressure and contains less oxygen. Prolonged exposure to low air pressure causes altitude sickness. Experienced mountain climbers take precautions to prevent altitude sickness.

Source: Altitude Physiology Expeditions Charity

H.O.T. Problems

Use **H**igher-**O**rder **T**hinking Skills

49. WHICH ONE DOESN'T BELONG? Identify the equation that does not belong. Explain your reasoning.

$y - 5 = 3(x - 1)$	$y + 1 = 3(x + 1)$	$y + 4 = 3(x + 1)$	$y - 8 = 3(x - 2)$

50. FIND THE ERROR Juana and Sabrina wrote an equation in point-slope form for the line that passes through $(3, -7)$ and $(-6, 4)$. Is either of them correct? Explain.

Juana	Sabrina
$y - 7 = -\frac{11}{9}(x + 3)$	$y - 4 = -\frac{9}{11}(x + 6)$

51. OPEN ENDED Describe a real-life scenario that has a constant rate of change and a value of y for a particular value of x. Represent this situation using an equation in point-slope form and an equation in slope-intercept form.

52. REASONING Write an equation for the line that passes through $(-4, 8)$ and $(3, -7)$. What is the slope? Where does the line intersect the x-axis? the y-axis?

53. CHALLENGE Write an equation in point-slope form for the line that passes through the points (f, g) and (h, j).

54. WRITING IN MATH Demonstrate how you can use the Slope Formula to write the point-slope form of an equation of a line.

Standardized Test Practice

55. Which statement is *most* strongly supported by the graph?

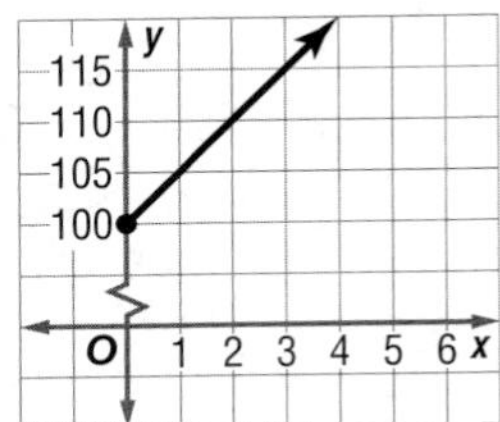

A You have \$100 and spend \$5 weekly.
B You have \$100 and save \$5 weekly.
C You need \$100 for a new CD player and save \$5 weekly.
D You need \$100 for a new CD player and spend \$5 weekly.

56. SHORT RESPONSE A store offers customers a \$5 gift certificate for every \$75 they spend. How much would a customer have to spend to earn \$35 worth of gift certificates?

57. GEOMETRY Which triangle is similar to $\triangle ABC$?

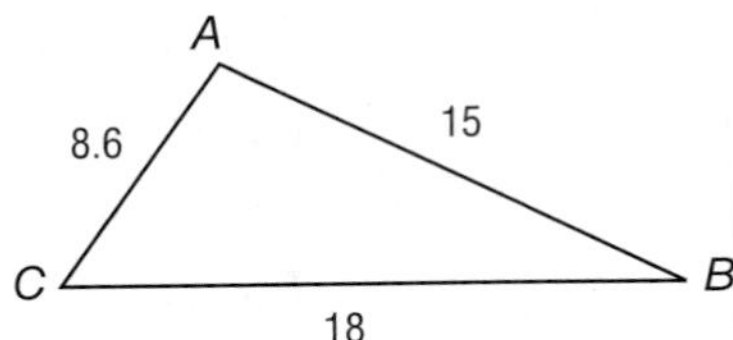

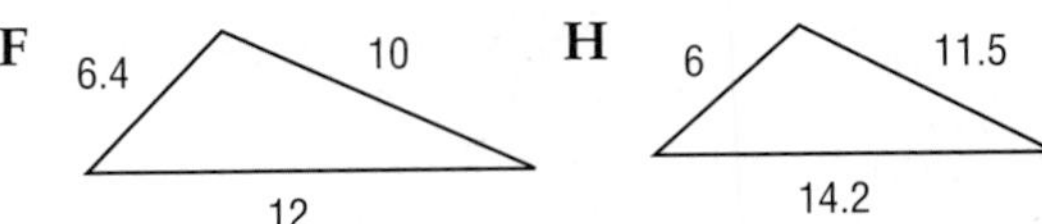

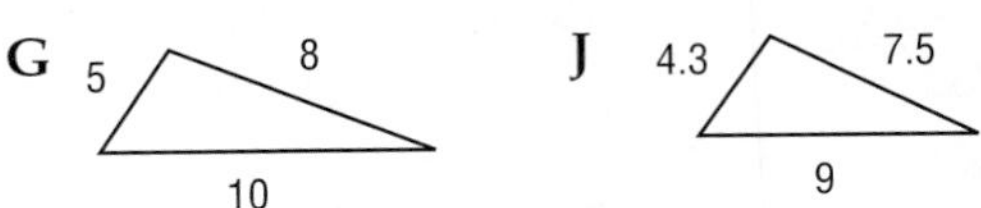

58. In a class of 25 students, 6 have blue eyes, 15 have brown hair, and 3 have blue eyes and brown hair. How many students have neither blue eyes nor brown hair?

A 4
B 7
C 10
D 22

Spiral Review

Write an equation of the line that passes through each pair of points. (Lesson 4-2)

59. (4, 2), (−2, −4)
60. (3, −2), (6, 4)
61. (−1, 3), (2, −3)
62. (2, −2), (3, 2)
63. (7, −2), (−4, −2)
64. (0, 5), (−3, 5)

Write an equation in slope-intercept form of the line with the given slope and *y*-intercept. (Lesson 4-1)

65. slope: −2, *y*-intercept: 6
66. slope: 3, *y*-intercept: −5
67. slope: $\frac{1}{2}$, *y*-intercept: 3
68. slope: $-\frac{3}{5}$, *y*-intercept: 12
69. slope: 0, *y*-intercept: 3
70. slope: −1, *y*-intercept: 0

71. THEATER The Coral Gables Actors' Playhouse has 7 rows of seats in the orchestra section. The number of seats in the rows forms an arithmetic sequence, as shown in the table. On opening night, 368 tickets were sold for the orchestra section. Was the section oversold? (Lesson 3-5)

Rows	Number of Seats
7	76
6	68
5	60

Skills Review

Solve each equation or formula for the variable specified. (Lesson 2-7)

72. $y = mx + b$, for m
73. $v = r + at$, for a
74. $km + 5x = 6y$, for m
75. $4b - 5 = -t$, for b

4-4 Parallel and Perpendicular Lines

Then

You wrote equations in point-slope form. (Lesson 4-3)

Now

- Write an equation of the line that passes through a given point, parallel to a given line.
- Write an equation of the line that passes through a given point, perpendicular to a given line.

New Vocabulary

parallel lines
perpendicular lines

Math Online
glencoe.com
- Extra Examples
- Personal Tutor
- Self-Check Quiz
- Homework Help

Why?

Notice the squares, rectangles and lines in the piece of art shown at the right. Some of the lines intersect forming right angles. Other lines do not intersect at all.

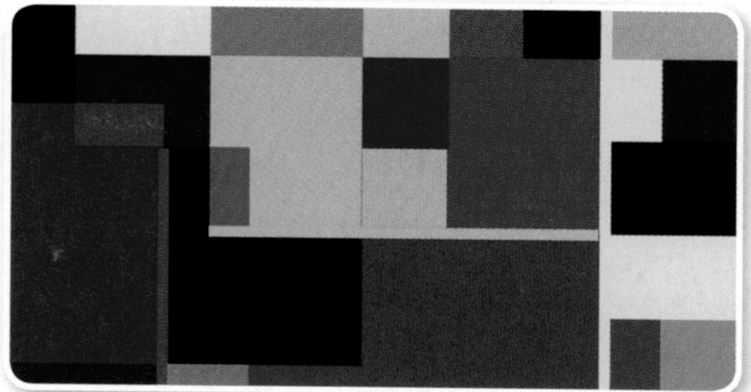

Parallel Lines Lines in the same plane that do not intersect are called **parallel lines**. Parallel lines have the same slope.

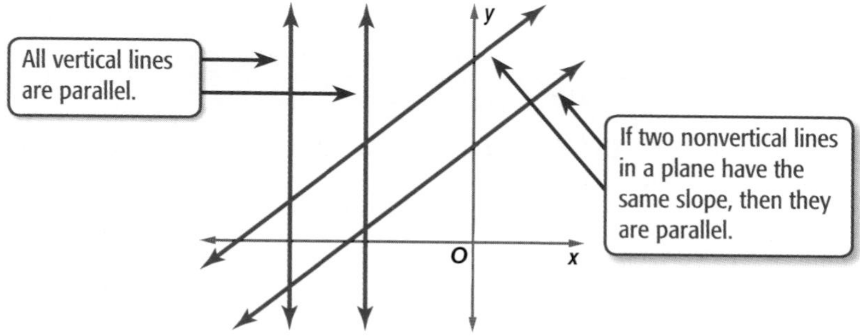

You can write an equation of a line parallel to a given line if you know a point on the line and an equation of the given line. First find the slope of the given line. Then, substitute the point provided and the slope from the given line into the point-slope form.

EXAMPLE 1 Parallel Line Through a Given Point

Write an equation in slope-intercept form for the line that passes through (−3, 5) and is parallel to the graph of $y = 2x - 4$.

Step 1 The slope of the line with equation $y = 2x - 4$ is 2. The line parallel to $y = 2x - 4$ has the same slope, 2.

Step 2 Find the equation in slope-intercept form.

$y - y_1 = m(x - x_1)$	**Point-slope form**
$y - 5 = 2[x - (-3)]$	**Replace m with 2 and (x_1, y_1) with (−3, 5).**
$y - 5 = 2(x + 3)$	**Simplify.**
$y - 5 = 2x + 6$	**Distributive Property**
$y - 5 + 5 = 2x + 6 + 5$	**Add 5 to each side.**
$y = 2x + 11$	**Write the equation in slope-intercept form.**

Check Your Progress

1. Write an equation in point-slope form for the line that passes through (4, −1) and is parallel to the graph of $y = \frac{1}{4}x + 7$.

Personal Tutor glencoe.com

Review Vocabulary

opposite reciprocals The opposite reciprocal of $\frac{a}{b}$ is $-\frac{b}{a}$. Their product is -1.

Perpendicular Lines Lines that intersect at right angles are called **perpendicular lines**. The slopes of perpendicular lines are opposite reciprocals. That is, if the slope of a line is 4, the slope of the line perpendicular to it is $-\frac{1}{4}$.

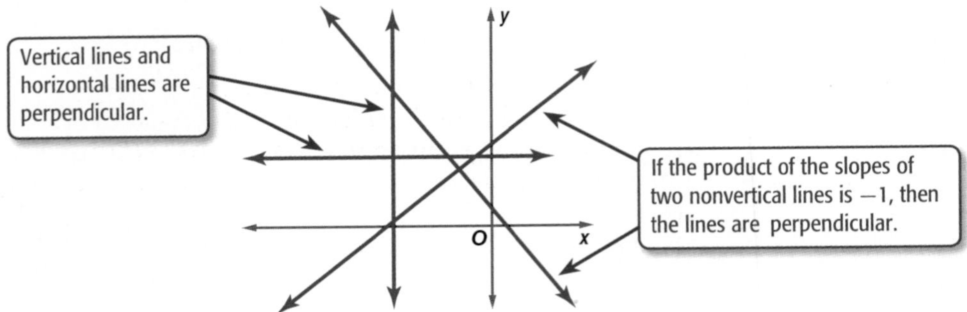

You can use slope to determine whether two lines are perpendicular.

Real-World EXAMPLE 2 Slopes of Perpendicular Lines

DESIGN The outline of a company's new logo is shown on a coordinate plane.

a. Is ∠*DFE* a right angle in the logo?

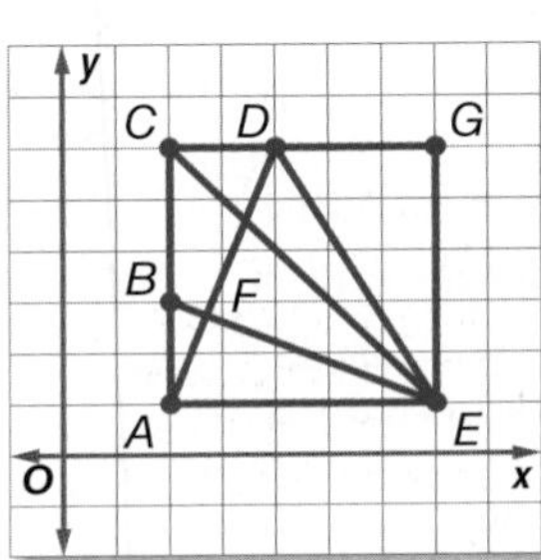

If $\overline{BE}$ and $\overline{AD}$ are perpendicular, then $\angle DFE$ is a right angle. Find the slopes of $\overline{BE}$ and $\overline{AD}$.

slope of $\overline{BE}$: $m = \frac{1-3}{7-2}$ or $-\frac{2}{5}$

slope of $\overline{AD}$: $m = \frac{6-1}{4-2}$ or $\frac{5}{2}$

The line segments are perpendicular because $-\frac{2}{5} \times \frac{5}{2} = -1$. Therefore, $\angle DFE$ is a right angle.

b. Is each pair of opposite sides parallel?

If a pair of opposite sides are parallel, then they have the same slope.

slope of $\overline{AC}$: $m = \frac{6-1}{2-2}$ or undefined

Since $\overline{AC}$ and $\overline{GE}$ are both parallel to the y-axis, they are vertical and are therefore parallel.

slope of $\overline{CG}$: $m = \frac{6-6}{7-2}$ or 0

Since $\overline{CG}$ and $\overline{AE}$ are both parallel to the x-axis, they are horizontal and are therefore parallel.

Check Your Progress

2. **CONSTRUCTION** On the plans for a treehouse, a beam represented by $\overline{QR}$ has endpoints $Q(-6, 2)$ and $R(-1, 8)$. A connecting beam represented by $\overline{ST}$ has endpoints $S(-3, 6)$ and $T(-8, 5)$. Are the beams perpendicular? Explain.

Personal Tutor glencoe.com

Real-World Link

The oldest treehouse still in existence was built in England in the 1700s.

Source: The Treehouse Company

You can determine whether the graphs of two linear equations are parallel or perpendicular by comparing the slopes of the lines.

ReadingMath

Parallel and Perpendicular Lines The symbol for parallel is ||. The symbol for perpendicular is ⊥.

EXAMPLE 3 Parallel or Perpendicular Lines

Determine whether the graphs of $y = 5$, $x = 3$, and $y = -2x + 1$ are *parallel* or *perpendicular*. Explain.

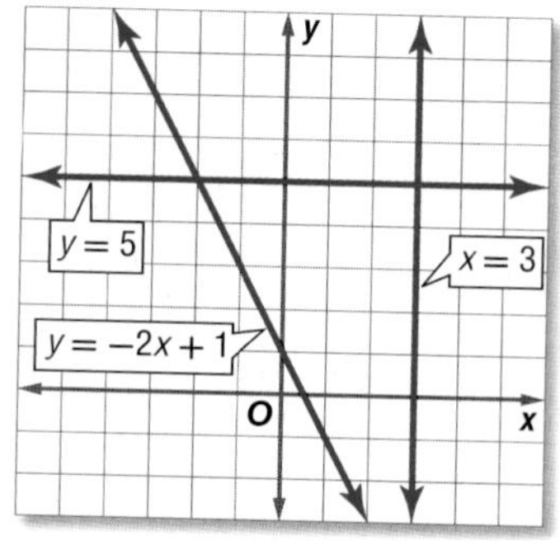

Graph each line on a coordinate plane.

From the graph, you can see that $y = 5$ is parallel to the x-axis and $x = 3$ is parallel to the y-axis. Therefore, they are perpendicular. None of the lines are parallel.

Check Your Progress

3. Determine whether the graphs of $6x - 2y = -2$, $y = 3x - 4$, and $y = 4$ are *parallel* or *perpendicular*. Explain.

Personal Tutor glencoe.com

You can write the equation of a line perpendicular to a given line if you know a point on the line and the equation of the given line.

StudyTip

Visualize the Problem Graph the given equation on a coordinate grid and plot the given point. Using a ruler, draw a line perpendicular to the given line that passes through the point.

EXAMPLE 4 Perpendicular Line Through a Given Point

Write an equation in slope-intercept form for the line that passes through $(-4, 6)$ and is perpendicular to the graph of $2x + 3y = 12$.

Step 1 Find the slope of the given line by solving the equation for y.

$2x + 3y = 12$	**Original equation**
$2x - 2x + 3y = -2x + 12$	**Subtract 2*x* from each side.**
$3y = -2x + 12$	**Simplify.**
$\frac{3y}{3} = \frac{-2x + 12}{3}$	**Divide each side by 3.**
$y = -\frac{2}{3}x + 4$	**Simplify.**

The slope is $-\frac{2}{3}$.

Step 2 The slope of the perpendicular line is the opposite reciprocal of $-\frac{2}{3}$ or $\frac{3}{2}$. Find the equation of the perpendicular line.

$y - y_1 = m(x - x_1)$	**Point-slope form**
$y - 6 = \frac{3}{2}(x - (-4))$	**$(x_1, y_1) = (-4, 6)$ and $m = \frac{3}{2}$**
$y - 6 = \frac{3}{2}(x + 4)$	**Simplify.**
$y - 6 = \frac{3}{2}x + 6$	**Distributive Property**
$y - 6 + 6 = \frac{3}{2}x + 6 + 6$	**Add 6 to each side.**
$y = \frac{3}{2}x + 12$	**Simplify.**

Check Your Progress

4. Write an equation in slope-intercept form for the line that passes through $(4, 7)$ and is perpendicular to the graph of $y = \frac{2}{3}x - 1$.

Personal Tutor glencoe.com

Concept Summary — Parallel and Perpendicular Lines

For Your FOLDABLE

	Parallel Lines	Perpendicular Lines
Words	Two nonvertical lines are parallel if they have the same slope.	Two nonvertical lines are perpendicular if the product of their slopes is -1.
Symbols	$\overline{AB} \parallel \overline{CD}$	$\overline{EF} \perp \overline{GH}$
Models		

Check Your Understanding

Example 1 p. 237

Write an equation in slope-intercept form for the line that passes through the given point and is parallel to the graph of the given equation.

1. $(-1, 2), y = \frac{1}{2}x - 3$

2. $(0, 4), y = -4x + 5$

Example 2 p. 238

3. **GARDENS** A garden is in the shape of a quadrilateral with vertices $A(-2, 1)$, $B(3, -3)$, $C(5, 7)$, and $D(-3, 4)$. Two paths represented by $\overline{AC}$ and $\overline{BD}$ cut across the garden. Are the paths perpendicular? Explain.

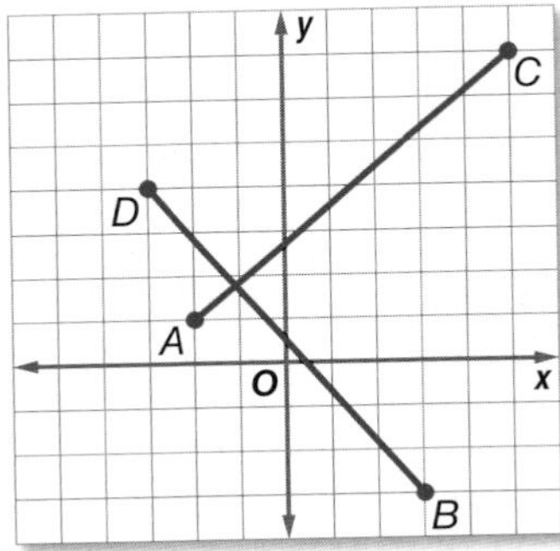

4. **GEOMETRY** A square is a quadrilateral that has opposite sides parallel, consecutive sides that are perpendicular, and diagonals that are perpendicular. Determine whether the quadrilateral is a square. Explain.

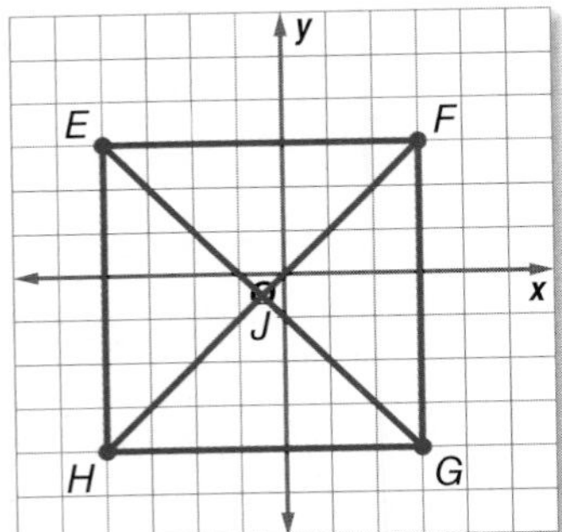

Example 3 p. 239

Determine whether the graphs of the following equations are *parallel* or *perpendicular*. Explain.

5. $y = -2x, 2y = x, 4y = 2x + 4$

6. $y = \frac{1}{2}x, 3y = x, y = -\frac{1}{2}x$

Example 4 p. 239

Write an equation in slope-intercept form for the line that passes through the given point and is perpendicular to the graph of the equation.

7. $(-2, 3), y = -\frac{1}{2}x - 4$

8. $(-1, 4), y = 3x + 5$

9. $(2, 3), 2x + 3y = 4$

10. $(3, 6), 3x - 4y = -2$

Practice and Problem Solving

● = **Step-by-Step Solutions** begin on page R12.
Extra Practice begins on page 815.

Example 1 p. 237

Write an equation in slope-intercept form for the line that passes through the given point and is parallel to the graph of the given equation.

11. $(3, -2), y = x + 4$ **12.** $(4, -3), y = 3x - 5$ **13.** $(0, 2), y = -5x + 8$

14. $(-4, 2), y = -\frac{1}{2}x + 6$ **15.** $(-2, 3), y = -\frac{3}{4}x + 4$ **16.** $(9, 12), y = 13x - 4$

Example 2 p. 238

17. GEOMETRY A trapezoid is a quadrilateral that has exactly one pair of parallel opposite sides. Is *ABCD* a trapezoid? Explain your reasoning.

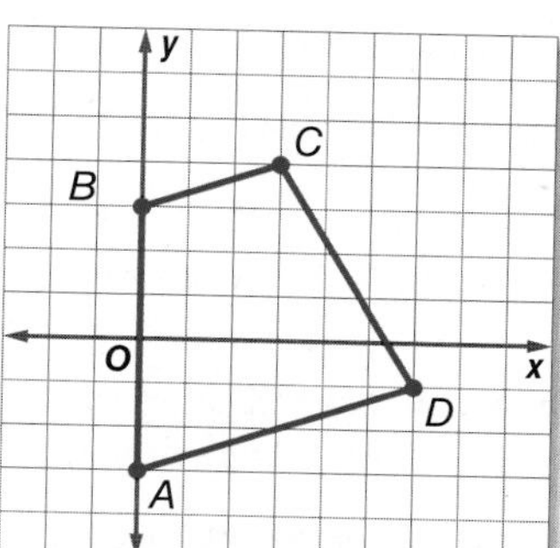

18. GEOMETRY *CDEF* is a kite. Are the diagonals of the kite perpendicular? Explain your reasoning.

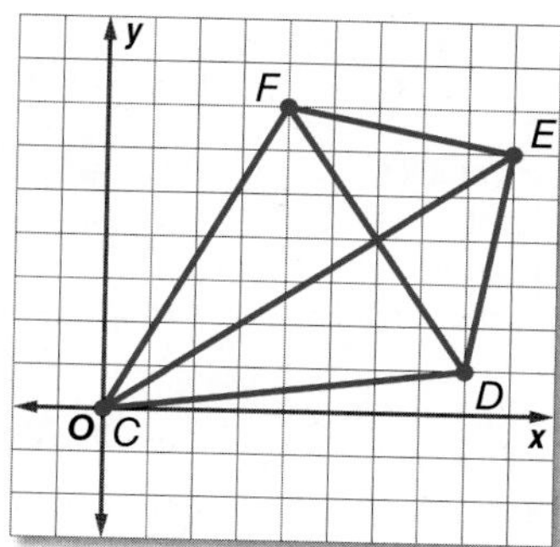

19. Determine whether the graphs of $y = -6x + 4$ and $y = \frac{1}{6}x$ are perpendicular. Explain.

20. MAPS On a map, Elmwood Drive passes through $R(4, -11)$ and $S(0, -9)$, and Taylor Road passes through $J(6, -2)$ and $K(4, -5)$. If they are straight lines, are the two streets perpendicular? Explain.

Example 3 p. 239

Determine whether the graphs of the following equations are *parallel* or *perpendicular*. Explain.

21. $2x - 8y = -24, 4x + y = -2, x - 4y = 4$

22. $3x - 9y = 9, 3y = x + 12, 2x - 6y = 12$

Example 4 p. 239

Write an equation in slope-intercept form for the line that passes through the given point and is perpendicular to the graph of the equation.

23 $(-3, -2), y = -2x + 4$ **24.** $(-5, 2), y = \frac{1}{2}x - 3$ **25.** $(-4, 5), y = \frac{1}{3}x + 6$

26. $(2, 6), y = -\frac{1}{4}x + 3$ **27.** $(3, 8), y = 5x - 3$ **28.** $(4, -2), y = 3x + 5$

Write an equation in slope-intercept form for a line perpendicular to the graph of the equation that passes through the *x*-intercept of that line.

29. $y = -\frac{1}{2}x - 4$ **30.** $y = \frac{2}{3}x - 6$ **31.** $y = 5x + 3$

32. Write an equation in slope-intercept form for the line that is perpendicular to the graph of $3x + 2y = 8$ and passes through the *y*-intercept of that line.

Determine whether the graphs of each pair of equations are *parallel*, *perpendicular*, or *neither*.

33. $y = 4x + 3$
$4x + y = 3$

34. $y = -2x$
$2x + y = 3$

35. $3x + 5y = 10$
$5x - 3y = -6$

36. $-3x + 4y = 8$
$-4x + 3y = -6$

37. $2x + 5y = 15$
$3x + 5y = 15$

38. $2x + 7y = -35$
$4x + 14y = -42$

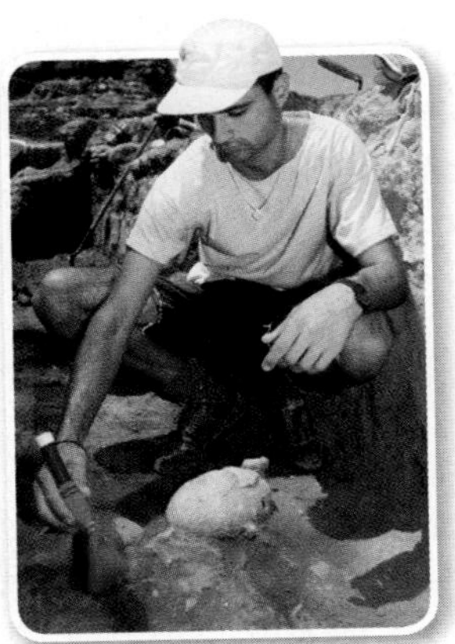

Real-World Career

Archaeologist
An archaeologist studies artifacts of ancient civilizations to piece together information about ancient societies. Archaeology is considered a branch of anthropology. Most archaeologists have a master's degree at least.

39. Write an equation of the line that is parallel to the graph of $y = 7x - 3$ and passes through the origin.

40. **EXCAVATION** Scientists excavating a dinosaur mapped the site on a coordinate plane. If one bone lies from $(-5, 8)$ to $(10, -1)$ and a second bone lies from $(-10, -3)$ to $(-5, -6)$, are the bones parallel? Explain.

41. **ARCHAEOLOGY** In the ruins of an ancient civilization, an archaeologist found pottery at $(2, 6)$ and hair accessories at $(4, -1)$. A pole is found with one end at $(7, 10)$ and the other end at $(14, 12)$. Is the pole perpendicular to the line through the pottery and the hair accessories? Explain.

42. **GRAPHICS** To create a design on a computer, Andeana must enter the coordinates for points on the design. One line segment she drew has endpoints of $(-2, 1)$ and $(4, 3)$. The other coordinates that Andeana entered are $(2, -7)$ and $(8, -3)$. Could these points be the vertices of a rectangle? Explain.

43. **MULTIPLE REPRESENTATIONS** In this problem, you will explore parallel and perpendicular lines.
    a. **GRAPHICAL** Graph the points $A(-3, 3)$, $B(3, 5)$, and $C(-4, 0)$ on a coordinate plane.
    b. **ANALYTICAL** Determine the coordinates of a fourth point D that would form a parallelogram. Explain your reasoning.
    c. **ANALYTICAL** What is the minimum number of points that could be moved to make the parallelogram a rectangle? Describe which points should be moved, and explain why.

H.O.T. Problems Use Higher-Order Thinking Skills

44. **CHALLENGE** If the line through $(-2, 4)$ and $(5, d)$ is parallel to the graph of $y = 3x + 4$, what is the value of d?

45. **REASONING** Is a horizontal line perpendicular to a vertical line *sometimes*, *always*, or *never*? Explain your reasoning.

46. **OPEN ENDED** Graph a line that is parallel and a line that is perpendicular to $y = 2x - 1$.

47. **FIND THE ERROR** Carmen and Chase are finding an equation of the line that is perpendicular to the graph of $y = \frac{1}{3}x + 2$ and passes through the point $(-3, 5)$. Is either of them correct? Explain your reasoning.

Carmen	Chase
$y - 5 = -3[x - (-3)]$	$y - 5 = 3[x - (-3)]$
$y - 5 = -3(x + 3)$	$y - 5 = 3(x + 3)$
$y = -3x - 9 + 5$	$y = 3x + 9 + 5$
$y = -3x - 4$	$y = -3x + 14$

48. **WRITING IN MATH** Illustrate how you can determine whether two lines are parallel or perpendicular. Write an equation for the graph that is parallel and an equation for the graph that is perpendicular to the line shown. Explain your reasoning.

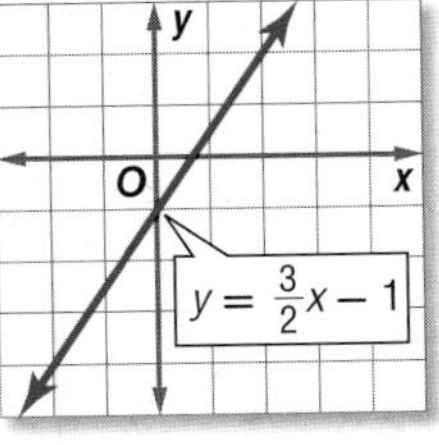

Standardized Test Practice

49. Which of the following is an algebraic translation of the following phrase?

5 less than the quotient of a number and 8

A $5 - \frac{n}{8}$

B $\frac{n}{8} - 5$

C $5 - \frac{8}{n}$

D $\frac{8}{n} - 5$

50. A line through which two points would be parallel to a line with a slope of $\frac{3}{4}$?

F (0, 5) and (−4, 2)

G (0, 2) and (−4, 1)

H (0, 0) and (0, −2)

J (0, −2) and (−4, −2)

51. Which equation best fits the data in the table?

A $y = x + 4$

B $y = 2x + 3$

C $y = 7$

D $y = 4x - 5$

x	y
1	5
2	7
3	9
4	11

52. SHORT RESPONSE Tyler is filling his 6000-gallon pool at a constant rate. After 4 hours, the pool contained 800 gal. How many total hours will it take to completely fill the pool?

Spiral Review

Write each equation in standard form. (Lesson 4-3)

53. $y - 13 = 4(x - 2)$

54. $y - 5 = -2(x + 2)$

55. $y + 3 = -5(x + 1)$

56. $y + 7 = \frac{1}{2}(x + 2)$

57. $y - 1 = \frac{5}{6}(x - 4)$

58. $y - 2 = -\frac{2}{5}(x - 8)$

59. CANOE RENTAL Latanya and her friends rented a canoe for 3 hours and paid a total of $45. (Lesson 4-2)

a. Write a linear equation to find the total cost C of renting the canoe for h hours.

b. How much would it cost to rent the canoe for 8 hours?

Canoe Rentals
Daily rates
plus $10 per hour

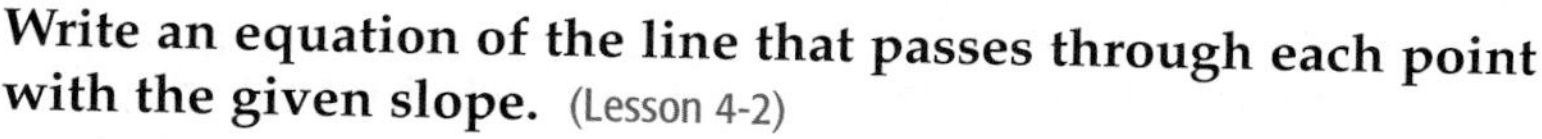

Write an equation of the line that passes through each point with the given slope. (Lesson 4-2)

60. $(5, -2), m = 3$

61. $(-5, 4), m = -5$

62. $(3, 0), m = -2$

63. $(3, 5), m = 2$

64. $(-3, -1), m = -3$

65. $(-2, 4), m = -5$

Simplify each expression. If not possible, write *simplified*. (Lesson 1-4)

66. $13m + m$

67. $14a^2 + 13b^2 + 27$

68. $3(x + 2x)$

69. FINANCIAL LITERACY At a Farmers' Market, merchants can rent a small table for $5.00 and a large table for $8.50. One time, 25 small and 10 large tables were rented. Another time, 35 small and 12 large were rented. (Lesson 1-2)

a. Write an expression to show the total amount of money collected.

b. Evaluate the expression.

Skills Review

Express each relation as a graph. Then determine the domain and range. (Lesson 1-6)

70. $\{(3, 8), (3, 7), (2, -9), (1, -9), (-5, -3)\}$

71. $\{(3, 4), (4, 3), (2, 2), (5, -4), (-4, 5)\}$

72. $\{(0, 2), (-5, 1), (0, 6), (-1, 9), (-4, -5)\}$

73. $\{(7, 6), (3, 4), (4, 5), (-2, 6), (-3, 2)\}$

Mid-Chapter Quiz

Lessons 4-1 through 4-4

Math Online glencoe.com
Math *in Motion,* Animation

Write an equation in slope-intercept form for each graph shown. (Lesson 4-1)

1.
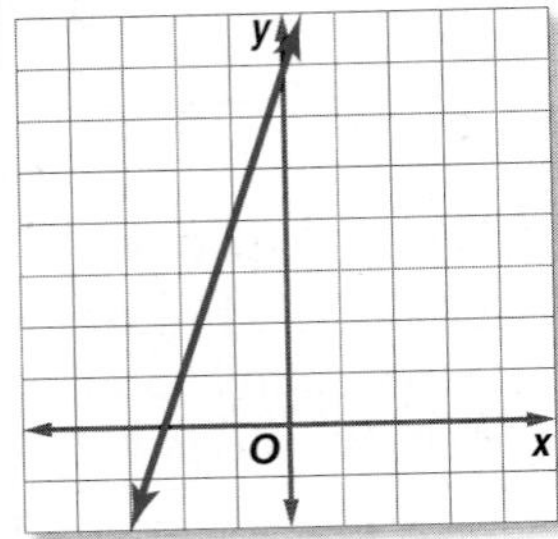

2.
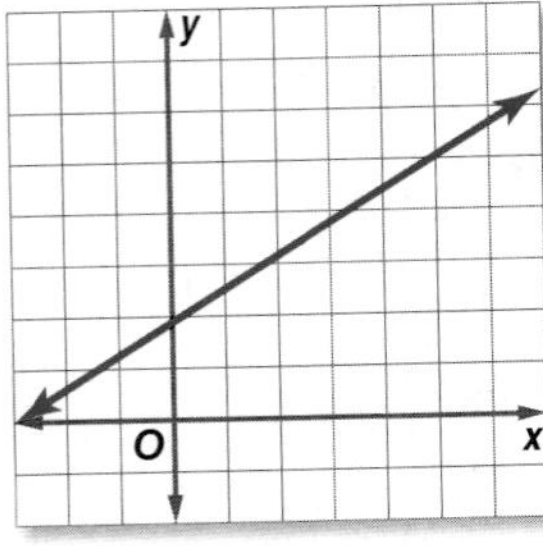

Graph each equation. (Lesson 4-1)

3. $y = 2x + 3$

4. $y = \frac{1}{3}x - 2$

5. **BOATS** Write an equation in slope-intercept form for the total rental cost C for a pontoon boat used for t hours. (Lesson 4-1)

Write an equation of the line with the given conditions. (Lesson 4-2)

6. (2, 5); slope 3

7. (−3, −1), slope $\frac{1}{2}$

8. (−3, 4), (1, 12)

9. (−1, 6), (2, 4)

10. (2, 1), slope 0

11. **MULTIPLE CHOICE** Write an equation of the line that passes through the point (0, 0) and has slope −4. (Lesson 4-2)

 A $y = x - 4$

 B $y = x + 4$

 C $y = -4x$

 D $y = 4 - x$

Write an equation in point-slope form for the line that passes through each point with the given slope. (Lesson 4-3)

12. (1, 4), $m = 6$

13. (−2, −1), $m = -3$

14. Write an equation in point-slope form for the line that passes through the point (8, 3), $m = -2$. (Lesson 4-3)

15. Write $y + 3 = \frac{1}{2}(x - 5)$ in standard form. (Lesson 4-3)

16. Write $y + 4 = -7(x - 3)$ in slope-intercept form. (Lesson 4-3)

Write each equation in standard form. (Lesson 4-3)

17. $y - 5 = -2(x - 3)$

18. $y + 4 = \frac{2}{3}(x - 3)$

Write each equation in slope-intercept form. (Lesson 4-3)

19. $y - 3 = 4(x + 3)$

20. $y + 1 = \frac{1}{2}(x - 8)$

21. **MULTIPLE CHOICE** Determine whether the graphs of the pair of equations are *parallel*, *perpendicular*, or *neither*. (Lesson 4-4)

$$y = -6x + 8$$
$$3x + \frac{1}{2}y = -3$$

 F parallel

 G perpendicular

 H neither

 J not enough information

Write an equation in slope-intercept form for the line that passes through the given point and is perpendicular to the graph of the equation. (Lesson 4-4)

22. (3, −4); $y = -\frac{1}{3}x - 5$

23. (0, −3); $y = -2x + 4$

24. (−4, −5); $-4x + 5y = -6$

25. (−1, −4); $-x - 2y = 0$

4-5 Scatter Plots and Lines of Fit

Then

You wrote linear equations given a point and the slope. (Lesson 4-3)

Now

- Investigate relationships between quantities by using points on scatter plots.
- Use lines of fit to make and evaluate predictions.

New Vocabulary

bivariate data
scatter plot
line of fit
linear interpolation

Math Online

glencoe.com

- Extra Examples
- Personal Tutor
- Self-Check Quiz
- Homework Help
- Math in Motion

Why?

The graph shows the number of people from the United States who travel to other countries. The points do not all lie on the same line; however, you may be able to draw a line that is close to all of the points. That line would show a linear relationship between the year x and the number of travelers each year y. Generally, international travel has increased.

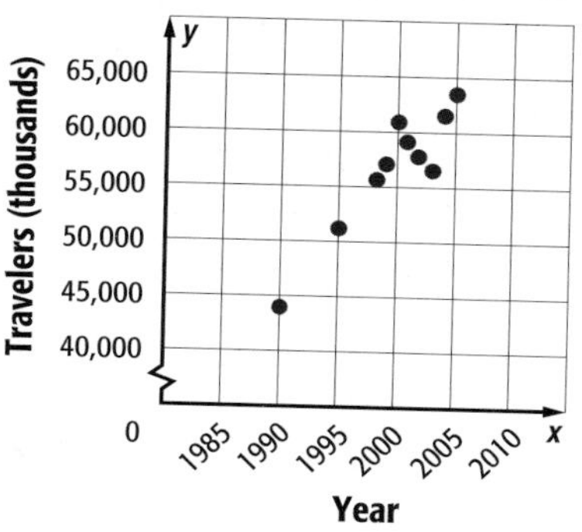

Source: *Statistical Abstract of the United States*

Investigate Relationships Using Scatter Plots Data with two variables are called **bivariate data**. A **scatter plot** is a graph in which two sets of data are plotted as ordered pairs in a coordinate plane. Scatter plots are used to investigate a relationship between two quantities.

Concept Summary — Scatter Plots

For Your FOLDABLE

Positive Correlation	Negative Correlation	No Correlation

Real-World EXAMPLE 1 — Evaluate a Correlation

WAGES Determine whether the graph shows a *positive*, *negative*, or *no* correlation. If there is a positive or negative correlation, describe its meaning in the situation.

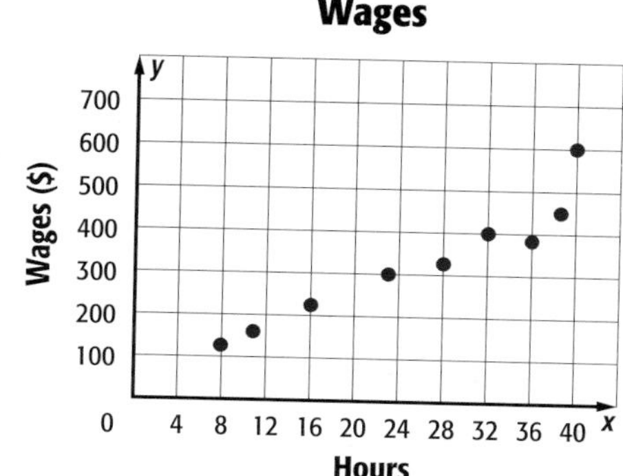

The graph shows a positive correlation. As the number of hours worked increases, the wages usually increase.

Check Your Progress

1. Refer to the graph on international travel. Determine whether the graph shows a *positive*, *negative*, or *no* correlation. If there is a positive or negative correlation, describe its meaning.

Personal Tutor glencoe.com

StudyTip

Correlations
positive: as x increases, y increases
negative: as x increases, y decreases
no correlation: no relationship between x and y

Use Lines of Fit Scatter plots can show whether there is a trend in a set of data. When the data points all lie close to a line, a **line of fit** or *trend line* can model the trend.

Key Concept Using a Linear Function to Model Data

For Your FOLDABLE

Step 1 Make a scatter plot. Determine whether any relationship exists in the data.

Step 2 Draw a line that seems to pass close to most of the data points.

Step 3 Use two points on the line of fit to write an equation for the line.

Step 4 Use the line of fit to make predictions.

Math *in Motion*, Interactive Lab glencoe.com

Real-World Link

The Kingda Ka roller coaster at Six Flags Great Adventure in Jackson, New Jersey, has broken three records: tallest roller coaster at 456 feet, fastest at 128 miles per hour, and largest vertical drop of 418 feet.

Source: Ultimate Roller Coaster

Real-World EXAMPLE 2 Write a Line of Fit

ROLLER COASTERS The table shows the largest vertical drops of nine roller coasters in the United States and the number of years after 1988 that they were opened. Identify the independent and the dependant variables. Is there a relationship in the data? If so, predict the vertical drop in a roller coaster built 25 years after 1988.

Years Since 1988	1	3	5	8	12	12	12	13	15
Vertical Drop (ft)	151	155	225	230	306	300	255	255	400

Source: Ultimate Roller Coaster

Step 1 Make a scatter plot.

The independent variable is the year, and the dependent variable is the vertical drop. As the number of years increases, the vertical drop of roller coasters increases. There is a positive correlation between the two variables.

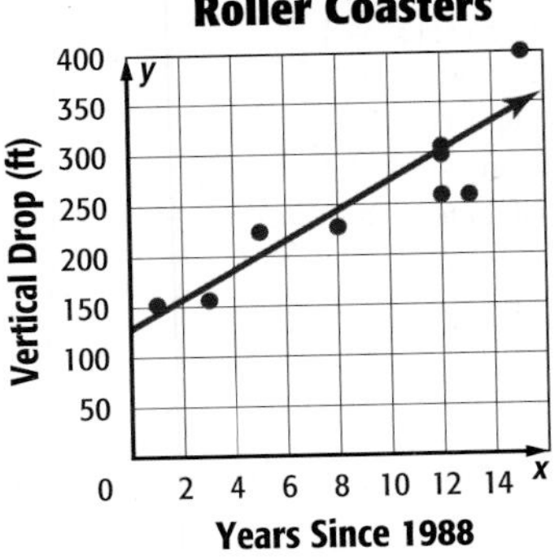

Step 2 Draw a line of fit.

No one line will pass through all of the data points. Draw a line that passes close to the points. A line of fit is shown.

Step 3 Write the slope-intercept form of an equation for the line of fit.

The line of fit passes close to (2, 150) and the data point (12, 300).

Find the slope.

$$m = \frac{y_2 - y_1}{x_2 - x_1} \qquad (x_1, y_1) = (2, 150), (x_2, y_2) = (12, 300)$$

$$= \frac{300 - 150}{12 - 2}$$

$$= \frac{150}{10} \text{ or } 15$$

Use $m = 15$ and either the point-slope form or the slope-intercept form to write the equation of the line of fit.

$$y - y_1 = m(x - x_1)$$

$$y - 150 = 15(x - 2)$$

$$y - 150 = 15x - 30$$

$$y = 15x + 120$$

A slope of 15 means that the vertical drops increased an average of 15 feet per year. To predict the vertical drop of a roller coaster built 25 years after 1988, substitute 25 for x in the equation. The vertical drop is $15(25) + 120$ or 495 feet.

Check Your Progress

2. MUSIC The table shows the dollar value in millions for the sales of CDs for the year. Make a scatter plot and determine what relationship exists, if any.

Year	2000	2001	2002	2003	2004	2005
Sales	13,215	12,909	12,044	11,233	11,447	10,520

Personal Tutor glencoe.com

ReadingMath

Interpolation and Extrapolation The Latin prefix *inter-* means between, and the Latin prefix *extra-* means beyond.

In Lesson 4-2, you learned that linear extrapolation is used to predict values *outside* the range of the data. You can also use a linear equation to predict values *inside* the range of the data. This is called **linear interpolation**.

Real-World EXAMPLE 3 Use Interpolation or Extrapolation

TRAVEL Use the scatter plot to find the approximate number of United States travelers to international countries in 1996.

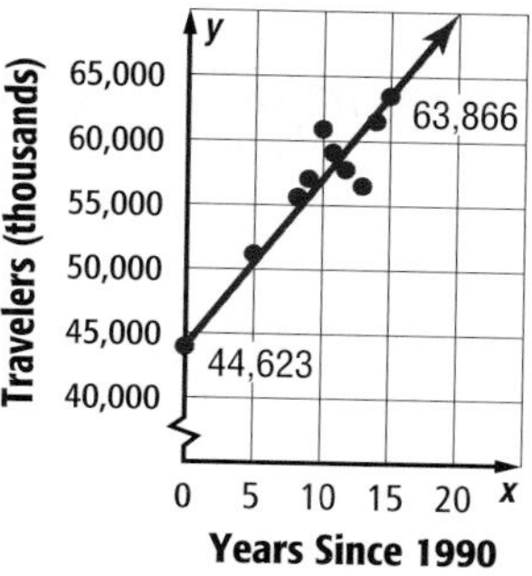

Source: *Statistical Abstract of the United States*

Step 1 Draw a line of fit. The line should be as close to as many points as possible.

Step 2 Write the slope-intercept form of the equation. The line of fit passes through (0, 44,623) and (15, 63,866).

Find the slope.

$$m = \frac{y_2 - y_1}{x_2 - x_1}$$ **Slope Formula**

$$= \frac{63{,}866 - 44{,}623}{15 - 0}$$ **$(x_1, y_1) = (0, 44{,}623)$, $(x_2, y_2) = (15, 63{,}866)$**

$$= \frac{19{,}243}{15}$$ **Simplify.**

Use $m = \frac{19{,}243}{15}$ and either the point-slope form or the slope-intercept form to write the equation of the line of fit.

$$y - y_1 = m(x - x_1)$$

$$y - 44{,}623 = \frac{19{,}243}{15}(x - 0)$$

$$y - 44{,}623 = \frac{19{,}243}{15}x$$

$$y = \frac{19{,}243}{15}x + 44{,}623$$

Step 3 Evaluate the function for $x = 1996 - 1990$ or 6.

$$y = \frac{19{,}243}{15}x + 44{,}623$$ **Equation of best-fit line**

$$= \frac{19{,}243}{15}(6) + 44{,}623$$ **$x = 6$**

$$= 7697\frac{1}{5} + 44{,}623 \text{ or } 52{,}320\frac{1}{5}$$ **Add.**

In 1996, there were approximately 52,320 thousand or 52,320,000 people who traveled from the United States to international countries.

Check Your Progress

3. MUSIC Use the equation for the line of fit in Check Your Progress 2 to estimate CD sales in 2015.

Personal Tutor glencoe.com

Check Your Understanding

Example 1
p. 245

Determine whether each graph shows a *positive, negative,* or *no* correlation. If there is a positive or negative correlation, describe its meaning in the situation.

1.

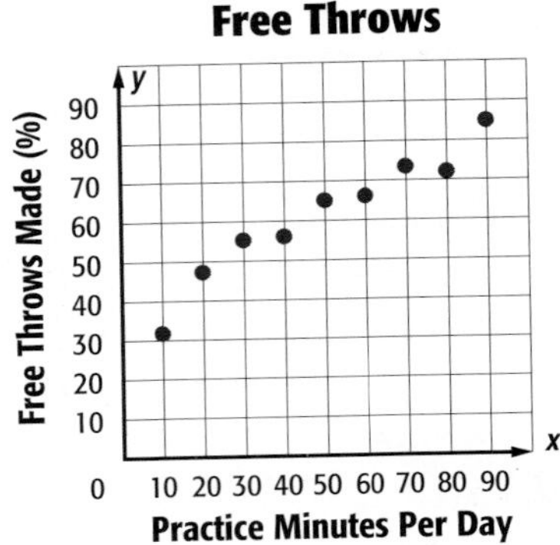

2.

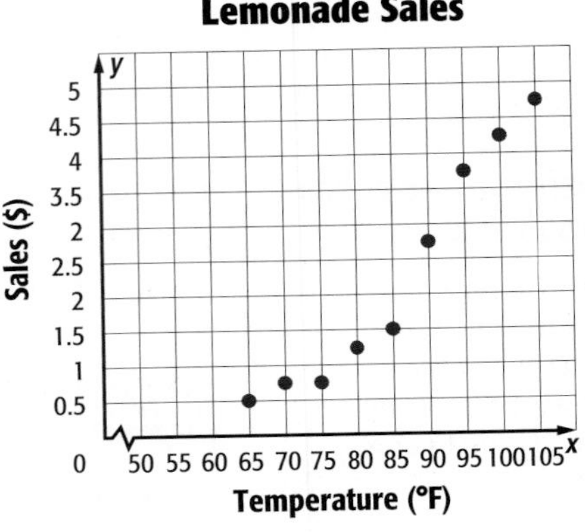

Example 2
p. 246

3. MARRIAGE The table shows the median age of females when they were first married.

Year	Age
1996	24.8
1997	25.0
1998	25.0
1999	25.1
2000	25.1
2001	25.1
2002	25.3
2003	25.3
2005	25.5
2006	25.9

Source: U.S. Bureau of Census

a. Make a scatter plot and determine what relationship exists, if any, in the data. Identify the independent and the dependant variables.

b. Draw a line of fit for the scatter plot.

c. Write an equation in slope-intercept form for the line of fit.

Example 3
p. 247

d. Predict what the median age of females when they are first married will be in 2016.

e. Do you think the equation can give a reasonable estimate for the year 2056? Explain.

Practice and Problem Solving

● = **Step-by-Step Solutions** begin on page R12.
Extra Practice begins on page 815.

Example 1
p. 245

Determine whether each graph shows a *positive, negative,* or *no* correlation. If there is a positive or negative correlation, describe its meaning in the situation.

4. **Game Tickets at the Fair**

Prizes (y): 0, 2, 4, 6, 8, 10
Tickets (x): 2, 4, 6, 8, 10, 12, 14, 16, 18

5. **NBA 3-Point Percentage**

3-Point Percentage Made (y): 0, 0.2, 0.3, 0.4, 0.5
Height (in.) (x): 73, 75, 77, 79, 81, 83, 85

6. 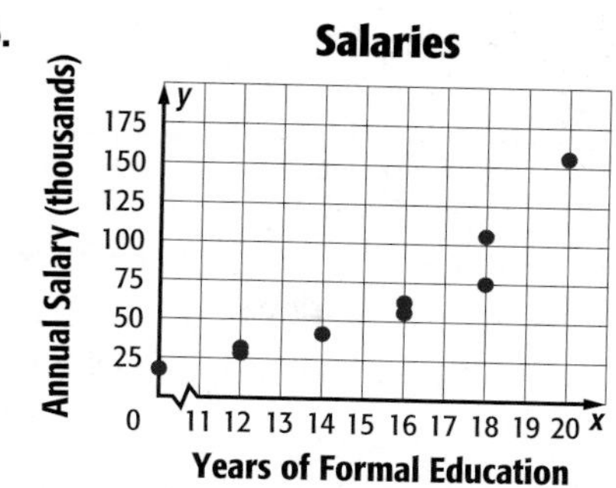

7 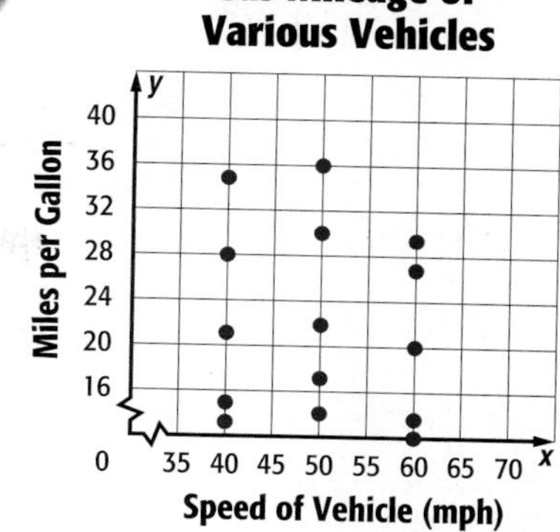

Examples 2 and 3
pp. 246-247

8. **MILK** Refer to the scatter plot of gallons of milk consumption per person for selected years.

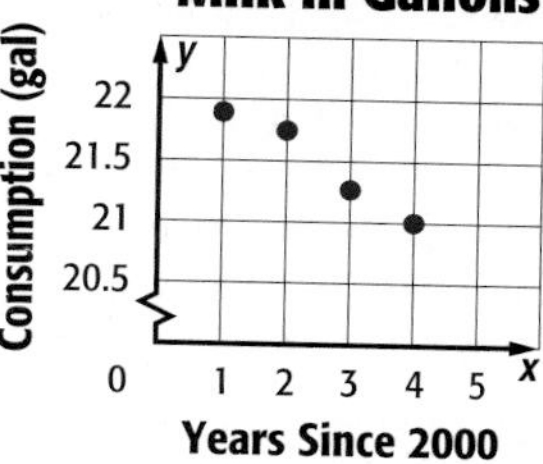

   a. Use the points (2, 21.75) and (4, 21) to write the slope-intercept form of an equation for the line of fit.
   b. Predict the milk consumption in 2015.
   c. Predict in what year milk consumption will be 10 gallons.
   d. Is it reasonable to use the equation to estimate the consumption of milk for any year? Explain.

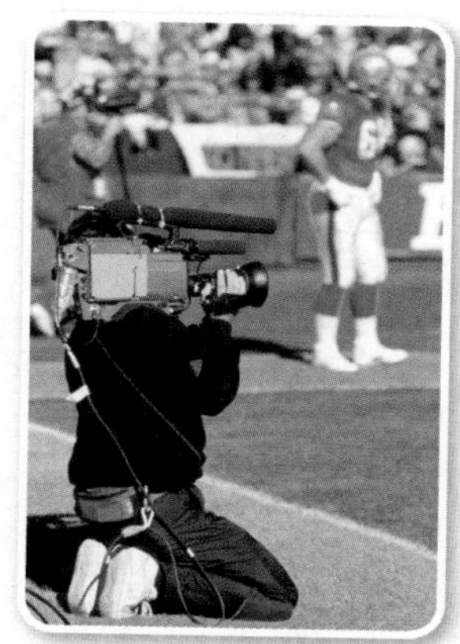

Real-World Career

Camera Operator
A camera operator is responsible for physically operating a camera and framing a scene. This job requires on-the-job training, while the educational requirements range from a high school diploma to a college degree.

9. **FOOTBALL** Use the scatter plot.

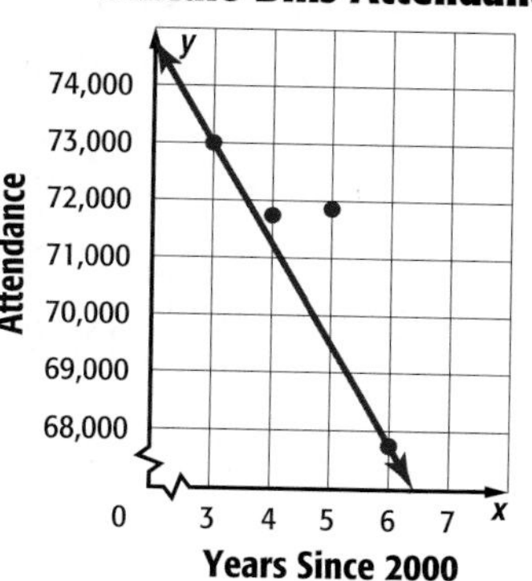

   a. Use the points (3, 73,000) and (6, 67,650) to write the slope-intercept form of an equation for the line of fit shown in the scatter plot.
   b. Predict the average attendance at a game in 2012.
   c. Can you use the equation to make a decision about the average attendance in any given year in the future? Explain.

10. **WEIGHT** The Body Mass Index (BMI) is a measure of body fat using height and weight. The heights and weights of twelve men with normal BMI are given in the table at the right.
   a. Make a scatter plot comparing the height in inches to the weight in pounds.
   b. Draw a line of fit for the data.
   c. Write the slope-intercept form of an equation for the line of fit.
   d. Predict the normal weight for a man who is 84 inches tall.
   e. A man's weight is 188 pounds. Use the equation of the line of fit to predict the height of the man.

Height (in.)	Weight (lb)
62	115
63	124
65	120
67	134
67	140
68	138
68	144
68	152
69	147
72	155
73	168
73	166

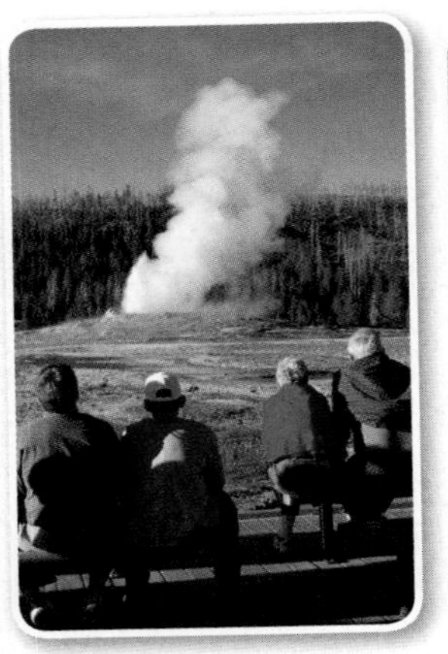

Real-World Link

Old Faithful is located in Yellowstone National Park. The Washburn Expedition of 1870 named the geyser because of its size and frequency.

Source: The Geyser Observation and Study Association

11. **GEYSERS** The time to the next eruption of Old Faithful can be predicted by using the duration of the current eruption.

Duration (min)	1.5	2	2.5	3	3.5	4	4.5	5
Interval (min)	48	55	70	72	74	82	93	100

a. Identify the independent and the dependent variables. Make a scatter plot and determine what relationship, if any, exists in the data. Draw a line of fit for the scatter plot.

b. Let x represent the duration of the previous interval. Let y represent the time between eruptions. Write the slope-intercept form of the equation for the line of fit. Predict the interval after a 7.5-minute eruption.

c. Make a critical judgment about using the equation to predict the duration of the next eruption. Would the equation be a useful model?

12. **COLLECT DATA** Use a tape measure to measure both the foot size and the height in inches of ten individuals.

a. Record your data in a table.

b. Make a scatter plot and draw a line of fit for the data.

c. Write an equation for the line of fit.

d. Make a conjecture about the relationship between foot size and height.

H.O.T. Problems

Use Higher-Order Thinking Skills

13. **OPEN ENDED** Describe a real-life situation that can be modeled using a scatter plot. Decide whether there is a *positive*, *negative*, or *no* correlation. Explain what this correlation means.

14. **WHICH ONE DOESN'T BELONG?** Analyze the following situations and determine which one does not belong.

hours worked and amount of money earned	height of an athlete and favorite color
seedlings that grow an average of 2 centimeters each week	number of photos stored on a camera and capacity of camera

15. **CHALLENGE** Determine which line of fit is better for the scatter plot. Explain your reasoning.

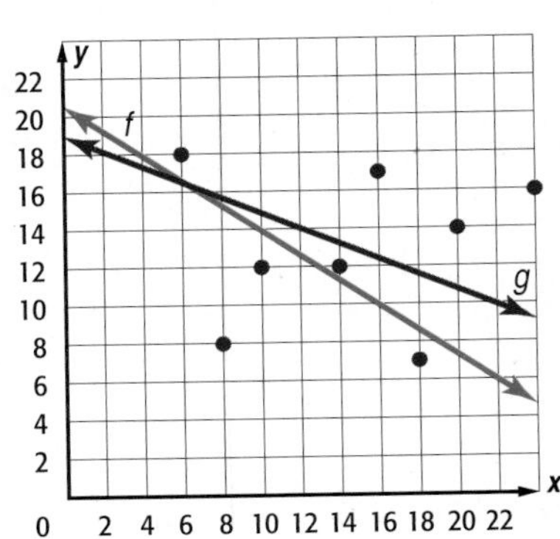

16. **REASONING** What can make a scatter plot and line of fit more useful for accurate predictions? Does an accurate line of fit always predict what will happen in the future? Explain.

17. **WRITING IN MATH** Make a scatter plot that shows the height of a person and age. Explain how you could use the scatter plot to predict the age of a person given his or her height. How can the information from a scatter plot be used to identify trends and make decisions?

Standardized Test Practice

18. Which equation best describes the relationship between the values of x and y in the table?

A $y = x - 5$
B $y = 2x - 5$
C $y = 3x - 7$
D $y = 4x - 7$

x	y
−1	−7
0	−5
2	−1
4	3

19. STATISTICS Mr. Hernandez collected data on the heights and average stride lengths of a random sample of high school students. He then made a scatter plot. What kind of correlation did he most likely see?

F positive
G constant
H negative
J no

20. GEOMETRY Mrs. Aguilar's rectangular bedroom measures 13 feet by 11 feet. She wants to purchase carpet for the bedroom that costs \$2.95 per square foot, including tax. How much will the carpet cost?

A \$70.80
B \$141.60
C \$145.95
D \$421.85

21. SHORT RESPONSE Nikia bought a one-month membership to a fitness center for \$35. Each time she goes, she rents a locker for \$0.25. If she spent \$40.50 at the fitness center last month, how many days did she go?

Spiral Review

Determine whether the graphs of each pair of equations are *parallel*, *perpendicular*, or *neither*. (Lesson 4-4)

22. $y = -2x + 11$
$y + 2x = 23$

23. $3y = 2x + 14$
$2x + 3y = 2$

24. $y = -5x$
$y = 5x - 18$

25. $y = 3x + 2$
$y = -\frac{1}{3}x - 2$

Write each equation in standard form. (Lesson 4-3)

26. $y - 13 = 4(x - 2)$

27. $y - 5 = -2(x + 2)$

28. $y + 3 = -5(x + 1)$

29. $y + 7 = \frac{1}{2}(x + 2)$

30. $y - 1 = \frac{5}{6}(x - 4)$

31. $y - 2 = -\frac{2}{5}(x - 8)$

Graph each equation. (Lesson 4-1)

32. $y = 2x + 3$

33. $4x + y = -1$

34. $3x + 4y = 7$

Find the slope of the line that passes through each pair of points. (Lesson 3-3)

35. (3, 4), (10, 8)

36. (−4, 7), (3, 5)

37. (3, 7), (−2, 4)

38. (−3, 2), (−3, 4)

39. (−2, −6), (−1, 10)

40. (1, −5), (−3, −5)

41. DRIVING Latisha drove 248 miles in 4 hours. At that rate, how long will it take her to drive an additional 93 miles? (Lesson 2-6)

Skills Review

Express each relation as a graph. Then determine the domain and range. (Lesson 1-6)

42. {(4, 5), (5, 4), (−2, −2), (4, −5), (−5, 4)}

43. {(7, 6), (3, 4), (4, 5), (−2, 6), (−3, 2)}

EXTEND
4-5

Algebra Lab
Correlation and Causation

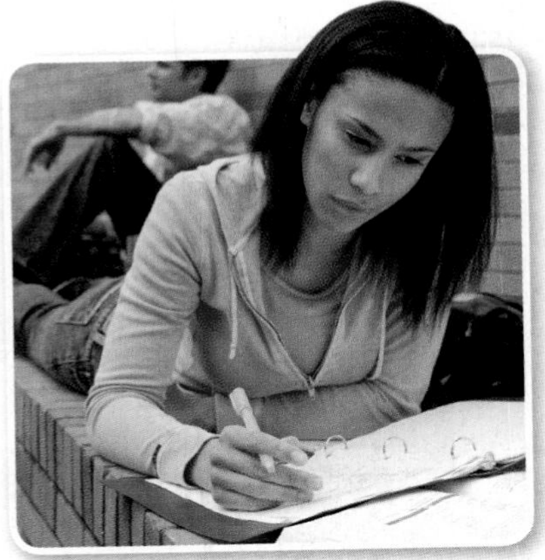

You may be considering attending a college or technical school in the future. What factors cause tuition to rise—increased building costs, higher employee salaries, or the amount of bottled water consumed?

Let's see how bottled water and college tuition are related. The table shows the average college tuition and fees for public colleges and the per person consumption of bottled water per year for 2001 through 2005.

Year	2001	2002	2003	2004	2005
Water Consumed (gallons)	18.8	20.9	22.4	24.0	26.1
Tuition ($)	3725	4081	4694	5132	5491

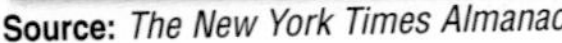

Source: *The New York Times Almanac*

ACTIVITY

Follow the steps to learn about correlation and causation.

Step 1 Graph the ordered pairs (gallons, tuition) to create a scatter plot. For example, one ordered pair is (18.8, 2562). Describe the graph.

Step 2 Is the correlation *positive* or *negative*? Explain.

Step 3 Do you think drinking more bottled water *causes* college tuition costs to rise? Explain.

Step 4 **Causation** occurs when a change in one variable produces a change in another variable. Correlation can be observed between many variables, but causation can only be determined from data collected from a controlled experiment. Describe an experiment that could illustrate causation.

Exercises

For each exercise, determine whether each situation illustrates *correlation* or *causation*. Explain your reasoning, including other factors that might be involved.

1. A survey showed that sleeping with the light on was positively correlated to nearsightedness.
2. A controlled experiment showed a positive correlation between the number of cigarettes smoked and the probability of developing lung cancer.
3. A random sample of students found that owning a cell phone had a negative correlation with riding the bus to school.
4. A controlled experiment showed a positive correlation between the number of hours using headphones when listening to music and the level of hearing loss.
5. DeQuan read in the newspaper that shark attacks are positively correlated with monthly ice cream sales.

4-6 Regression and Median-Fit Lines

Then

You used lines of fit and scatter plots to evaluate trends and make predictions. (Lesson 4-5)

Now

- Write equations of best-fit lines using linear regression.
- Write equations of median-fit lines.

New Vocabulary

best-fit line
linear regression
correlation coefficient
median-fit line

Math Online
glencoe.com

- Extra Examples
- Personal Tutor
- Self-Check Quiz
- Homework Help

Why?

Spider-Man comics have been around since 1962. Since then Spider-Man has made appearances in magazines, books, and movies.

The table shows the number of comic books and appearances that he has made. We can estimate how many appearances he will make in 2012.

Spider-Man Comics and Appearances

Year	Number
1962	1
1972	39
1982	110
1992	164
2002	278
2012	?

Best-Fit Lines You have learned how to find and write equations for lines of fit by hand. Many calculators use complex algorithms that find a more precise line of fit called the **best-fit line**. One algorithm is called **linear regression**.

Your calculator may also compute a number called the **correlation coefficient**. This number will tell you if your correlation is positive or negative and how closely the equation is modeling the data. The closer the correlation coefficient is to 1 or −1, the more closely the equation models the data.

Real-World EXAMPLE 1 Best-Fit Line

MOVIES The table shows the amount of money made by movies in the United States. Use a graphing calculator to write an equation for the best-fit line for that data.

Year	2000	2001	2002	2003	2004	2005	2006
Income ($ billion)	7.66	8.41	9.52	9.49	9.54	8.99	9.49

Before you begin, make sure that your Diagnostic setting is on. You can find this under the CATALOG menu. Press D and then scroll down and click DiagnosticOn. Then press [ENTER].

Step 1 Enter the data by pressing [STAT] and selecting the Edit option. Let the year 2000 be represented by 0. Enter the years since 2000 into List 1 (L_1). These will represent the x-values. Enter the income ($ billion) into List 2 (L_2). These will represent the y-values.

Step 2 Perform the regression by pressing [STAT] and selecting the CALC option. Scroll down to LinReg (ax+b) and press [ENTER].

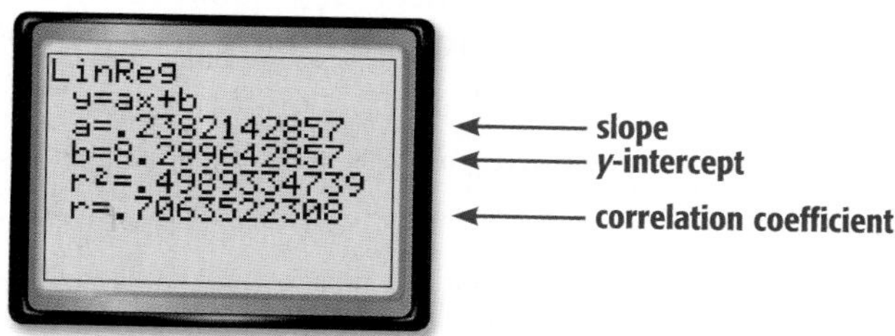

Real-World Link

In 1994, Minnesota became the first state to sanction girls' ice hockey as a high school varsity sport.

Source: ESPNET SportsZone

Step 3 Write the equation of the regression line by rounding the a and b values. The form that we chose was $ax + b$, so the equation is $y = 0.24x + 8.30$. The correlation coefficient is about 0.7064, which means that the equation models the data fairly well.

Check Your Progress

Write an equation of the best-fit line for the data in each table. Name the correlation coefficient. Round to the nearest ten-thousandth. Let x be the number of years since 2003.

1A. HOCKEY The table shows the number of goals of leading scorers.

Mustang Girls Hockey Leading Scorers								
Year	2003	2004	2005	2006	2007	2008	2009	2010
Goals	30	23	41	35	31	43	33	45

1B. HOCKEY The table gives the number of goals scored by the team each season.

Mustang Girls Hockey Team Goals								
Year	2003	2004	2005	2006	2007	2008	2009	2010
Goals	63	44	55	63	81	85	93	84

Personal Tutor glencoe.com

We can use points on the best-fit line to estimate values that are not in the data. Recall that when we estimate values that are between known values, this is called *linear interpolation*. When we estimate a number outside of the range of the data, it is called *linear extrapolation*.

Real-World EXAMPLE 2 Use Interpolation and Extrapolation

PAINTBALL The table shows the points received by the top ten paintball teams at a tournament. How many points did the 20th-ranked team receive?

Top Ten Teams										
Rank	1	2	3	4	5	6	7	8	9	10
Score	100	89	96	99	97	98	78	70	64	80

Write an equation of the best-fit line for the data. Then extrapolate to find the missing value.

Step 1 Enter the data from the table into the lists as you did before. Let the ranks be the x-values and the scores be the y-values. Then graph the scatter plot.

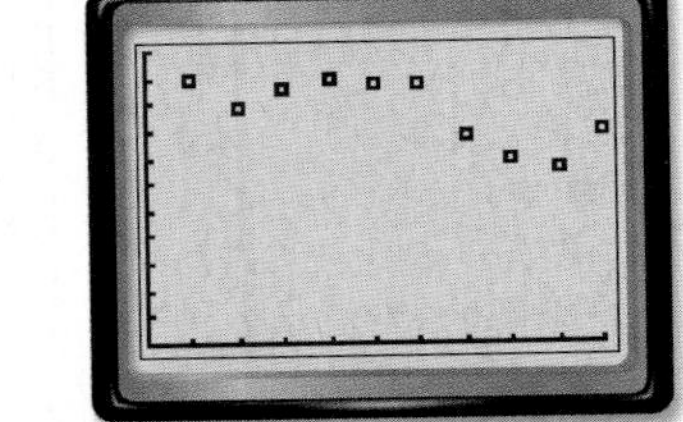

[0, 10] scl: 1 by [0, 110] scl: 10

Step 2 Perform the linear regression using the data in the lists. Find the equation of the best-fit line.

The equation is $y = -3.32x + 105.3$.

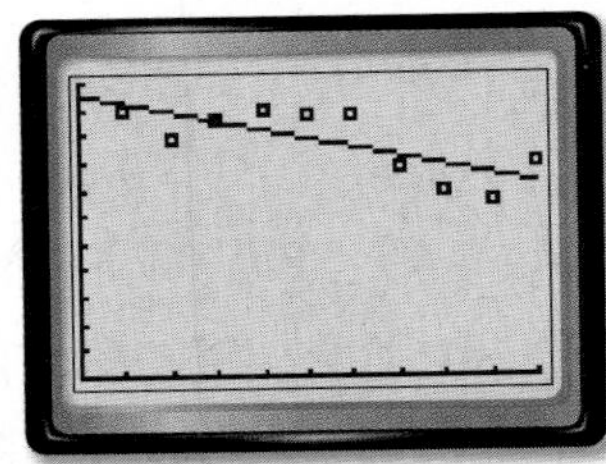

[0, 10] scl: 1 by [0, 110] scl: 10

StudyTip

Median-Fit Line The median-fit line is computed using a different algorithm than linear regression.

Step 3 Graph the best-fit line. Press [2nd] [CALC] [ENTER] to find that when $x = 20$, $y \approx 39$.

It is estimated that the 20th ranked team received 39 points.

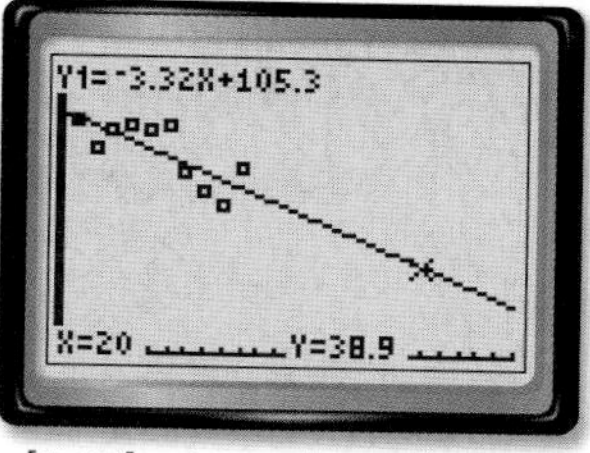

[0, 25] scl: 1 by [0, 125] scl: 10

Check Your Progress

ONLINE GAMES Use linear interpolation to find the percent of Americans that play online games for the following ages.

Percent of Americans Who Play Online Games					
Age	15	20	30	40	50
Percent	81	54	37	29	25

Source: Pew Internet & American Life Survey

2A. 35 years

2B. 18 years

Personal Tutor glencoe.com

Median-Fit Lines A second type of fit line that can be found using a graphing calculator is a **median-fit line**. The equation of a median-fit line is calculated using the medians of the coordinates of the data points.

EXAMPLE 3 Median-Fit Line

PAINTBALL Find and graph the equation of a median-fit line for the data in Example 2. Then predict the score of the 15th ranked team.

Step 1 The data should be in the lists. Graph the scatter plot.

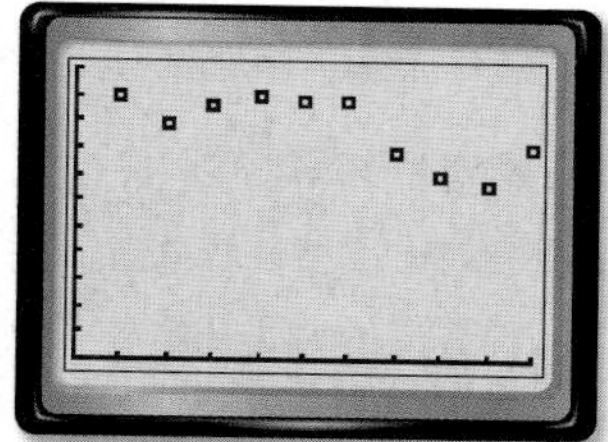

[0, 10] scl: 1 by [0, 110] scl: 10

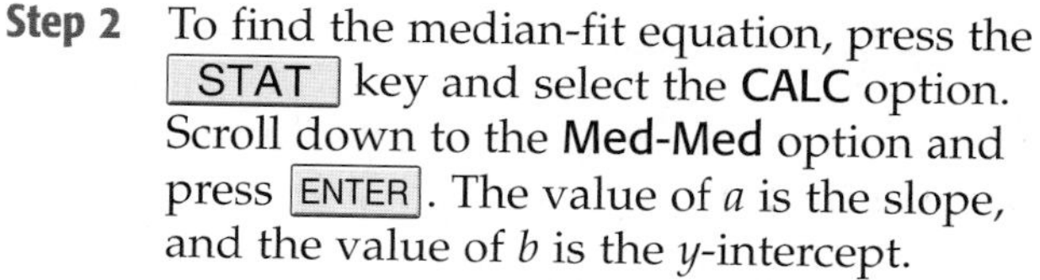

Step 2 To find the median-fit equation, press the [STAT] key and select the CALC option. Scroll down to the Med-Med option and press [ENTER]. The value of a is the slope, and the value of b is the y-intercept.

The equation for the median-fit line is $y = -3.71x + 108.26$.

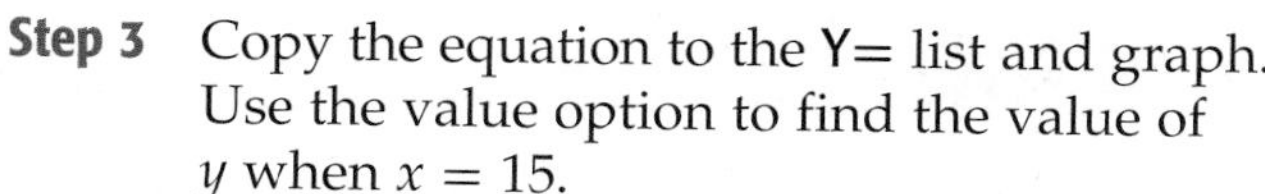

Step 3 Copy the equation to the Y= list and graph. Use the value option to find the value of y when $x = 15$.

The 15th place team scored about 53 points.

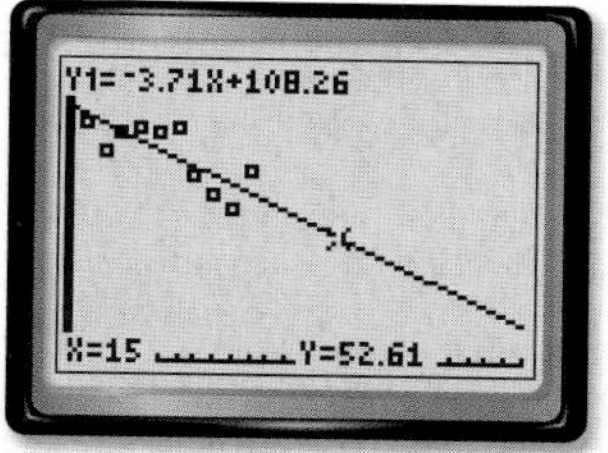

[0, 25] scl: 1 by [0, 125] scl: 10

Notice that the equations for the regression line and the median-fit line are very similar.

Real-World Link

Paintball is more popular with 12- to 17-year-olds than any other age group. In 2002, 3,649,000 teenagers participated in paintball while 2,195,000 18- to 24-year-olds participated.

Source: *Statistical Abstract of the United States*

Check Your Progress

3. Use the data from Check Your Progress 2 and a median-fit line to estimate the numbers of 18- and 35-year-olds who play online games. Compare these values with the answers from the regression line.

Personal Tutor glencoe.com

Check Your Understanding

Example 1 p. 253

1. **POTTERY** A local university is keeping track of the number of art students who use the pottery studio each day. Write an equation of the regression line and find the correlation coefficient.

Students Throwing Pottery							
Day	1	2	3	4	5	6	7
Students	10	15	18	15	13	19	20

Example 2 p. 254

2. **COMPUTERS** The table below shows the percent of Americans with a broadband connection at home in a recent year. Use linear extrapolation and a regression equation to estimate the percentage of 60-year-olds with broadband at home.

Percentage of Americans with Broadband At Home						
Age	25	30	35	40	45	50
Percent	40	42	36	35	36	32

Example 3 p. 255

3. **VACATION** The Smiths want to rent a house on the lake that sleeps eight people. The cost of the house per night is based on how close it is to the water.

Rental Properties							
Distance from Lake (mi)	0.0 (houseboat)	0.3	0.5	1.0	1.25	1.5	2.0
Price/Night ($)	785	325	250	200	150	140	100

a. Find and graph an equation for the median-fit line.

b. What would you estimate is the cost of a rental 1.75 miles from the lake?

Practice and Problem Solving

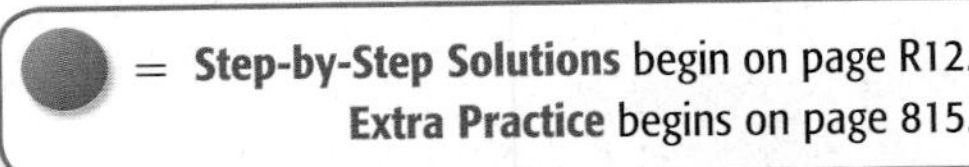
= **Step-by-Step Solutions** begin on page R12. **Extra Practice** begins on page 815.

Examples 1 and 2 pp. 253–254

Write an equation of the regression line for the data in each table. Then find the correlation coefficient.

4. **SKYSCRAPERS** The table ranks the ten tallest buildings in the world.

Tallest Buildings and Stories										
Rank	1	2	3	4	5	6	7	8	9	10
Stories	101	88	110	88	88	80	69	102	78	70

5. **MUSIC** The table gives the number of annual violin auditions held by a youth symphony each year since 2000. Let x be the number of years since 2000.

Youth Symphony Violin Auditions							
Year	2000	2001	2002	2003	2004	2005	2006
Auditions	22	19	25	37	32	35	42

6. **RETAIL** The table gives the sales of jeans at a clothing chain since 2004. Let x be the number of years since 2004.

Jeans Sales By Year						
Year	2004	2005	2006	2007	2008	2009
Sales (Millions of Dollars)	6.84	7.6	10.9	15.4	17.6	21.2

Example 3 p. 255

7. **MARATHON** The number of entrants in the Boston Marathon every five years since 1975 is shown. Let x be the number of years since 1975.

Year	1975	1980	1985	1990	1995	2000	2005
Entrants	2395	5417	5594	9412	9416	17,813	20,453

Source: Boston Athletic Association

a. Find an equation for the median-fit line.

b. According to the equation, how many entrants were there in 2003?

8. **CAMPING** A campground keeps a record of the number of campsites rented the week of July 4 for several years. Let x be the number of years since 2000.

Campsites Rented July 4th Week						
Year	2002	2003	2004	2005	2006	2007
Sites Rented	34	45	42	53	58	47

a. Find an equation for the regression line.

b. Predict the number of campsites that will be rented in 2010.

c. Predict the number of campsites that will be rented in 2020.

Real-World Link

The average American eats 20 quarts of ice cream per year.

Source: *Real Simple Magazine*

9. **ICE CREAM** An ice cream company keeps a count of the tubs of cookie dough ice cream delivered to each of their stores in a particular area.

Store Delivery of Cookie Dough Ice Cream					
Store Size (ft^2)	2100	2225	3135	3569	4587
Tubs (hundreds)	110	102	215	312	265

a. Find an equation for the median-fit line.

b. Graph the points and the median-fit line.

c. How many tubs would be delivered to a 1500-square-foot store? a 5000-square-foot store?

10. **COLLEGE TESTING** The ACT is an exam that evaluates students' readiness to perform college-level work. The table below shows the number of participants who took the test in the given years. Let x be the number of years since 1990.

Years	1990	1995	2000	2001	2002	2003	2004	2005	2006
Participants (thousands)	817	945	1065	1070	1116	1175	1171	1186	1206

Source: ACT

a. Find an equation for the regression line.

b. According to the equation, how many participants were there in 1998?

c. How many students would you predict will participate in 2011?

11. **FINANCIAL LITERACY** The prices of the eight top-selling brands of jeans at Jeanie's Jeans are given in the table below.

Sales Rank	1	2	3	4	5	6	7	8
Price ($)	43	44	50	61	64	135	108	78

a. Find the equation for the regression line.

b. According to the equation, what would be the price of a pair of the 12th best-selling brand?

c. Is this a reasonable prediction? Explain.

Math History Link

Florence Nightingale David (1909–1993)
A renowned statistician born in Ivington, England, Florence Nightingale David received the first Elizabeth L. Scott Award for her "efforts in opening the door to women in statistics; … for research contributions to … statistical methods …; and her spirit as a lecturer and as a role model."

12. **STATE FAIRS** Opening day attendance at the North Carolina State Fair for 2000 was 53,331, for 2001 it was 47,940, for 2002 it was 54,036, for 2003 it was 61,364, for 2004 it was 61,289, and for 2005 it was 52,201.
   a. Construct a table for the given data.
   b. Graph the points and the regression line.
   c. Predict the attendance on opening day in 2012.

13. **MUSIC** For the following data, find the equation of the regression line and give the correlation coefficient to the nearest ten-thousandth place.

Battle of the Bands Concessions									
Year	1998	1999	2000	2001	2002	2003	2004	2005	2006
Amount Raised ($)	1236	1560	1423	1740	2230	2563	3215	4517	4219

14. **FIREFIGHTERS** The table shows statistics from the U.S. Fire Administration.
   a. Find an equation for the median-fit line.
   b. Graph the points and the median-fit line.
   c. Does the median-fit line give you an accurate picture of the number of firefighters? Explain.

Age	Number of Firefighters
18	40,919
25	245,516
35	330,516
45	296,665
55	167,087
65	54,559

15. **ATHLETICS** The table shows the number of participants in high school athletics.

Year Since 1970	1	10	20	30	35
Athletes	3,960,932	5,356,913	5,298,671	6,705,223	7,159,904

   a. Find an equation for the regression line.
   b. According to the equation, how many participated in 1988?

16. **ART** A count was kept on the number of paintings sold at an auction by the year in which they were painted. Let x be the number of years since 1950.

Paintings Sorted by Year of Execution						
Year Painted	1950	1955	1960	1965	1970	1975
Paintings Solds	8	5	25	21	9	22

   a. Find the equation for the linear regression line.
   b. How many paintings were sold that were painted in 1961?
   c. Is the linear regression equation an accurate model of the data? Explain why or why not.

17. **SCHOOL** The table shows the average cost of a technical school between 2000 and 2006. Let x be 0 for the school year 2000–2001.

Average Cost of Public 2-Year Institution						
Year	2000–2001	2001–2002	2002–2003	2003–2004	2004–2005	2005–2006
Cost ($)	4839	5137	5601	6020	6375	6492

Source: U.S. Department of Education

   a. Find the equation of the median-fit line.
   b. Graph the points and the median-fit line.
   c. What would you estimate the cost to be in 2020–2021?

18. **SPACE EXPLORATION** As of 2006, the names of the space shuttles launched each year since 1993 are given below. Let x be the year.

Space Shuttle Launches	
Year	**Launches**
1993	STS-61, STS-58, STS-51, STS-57, STS-55, STS-56, STS-54
1994	STS-66, STS-68, STS-64, STS-65, STS-59, STS-62, STS-60
1995	STS-74, STS-73, STS-69, STS-70, STS-71, STS-67, STS-63
1996	STS-80, STS-79, STS-78, STS-77, STS-76, STS-75, TSTS-72
1997	STS-87, STS-86, STS-85, STS-94, STS-84, STS-83, STS-82, STS-81
1998	STS-88, STS-95, STS-91, STS-90, STS-89
1999	STS-103, STS-93, STS-96
2000	STS-97, STS-92, STS-106, STS-101, STS-99
2001	STS-108, STS-105, STS-104, STS-100, STS-102, STS-98
2002	STS-113, STS-112, STS-111, STS-110, STS-109
2003	STS-107
2004	
2005	STS-114
2006	STS-116, STS-115, STS-121

a. Construct a table of values for the data that could be used for graphing.

b. Find an equation for the regression line.

c. Predict the number of launches there will be in 2015. Is your prediction reasonable? What influences might cause the actual number of flights to be different from your prediction? Explain.

H.O.T. Problems Use Higher-Order Thinking Skills

19. **CHALLENGE** Below are the results of the World Superpipe Championships in 2008.

Men	Score	Rank	Women	Score
Shaun White	93.00	1	Torah Bright	96.67
Mason Aguirre	90.33	2	Kelly Clark	93.00
Janne Korpi	85.33	3	Soko Yamaoka	85.00
Luke Mitrani	85.00	4	Ellery Hollingsworth	79.33
Keir Dillion	81.33	5	Sophie Rodriguez	71.00

Find an equation of the regression line for each, and graph them on the same coordinate plane. Compare and contrast the men's and women's graphs.

Real-World Link

The World Superpipe Championship is held in Park City, Utah. This is home to one of the largest half-pipes in the world, with walls 22 feet high.

Source: Park City Mountain Resort

20. **REASONING** For a class project, the scores that 10 randomly selected students earned on the first 8 tests of the school year are given. Explain how to find a line of best fit. Could it be used to predict the scores of other students? Explain your reasoning.

21. **OPEN ENDED** For 10 different people, measure their heights and the lengths of their heads from chin to top. Use these data to generate a linear regression equation and a median-fit equation. Make a prediction using both of the equations.

22. **WRITING IN MATH** Using the data at the beginning of the lesson, describe the steps you would take to determine the number of appearances Spiderman will make in 2012.

Standardized Test Practice

23. **GEOMETRY** Sam is putting a border around a poster. x represents the poster's width, and y represents the poster's length. Which equation represents how much border Sam will use if he doubles the length and the width?

A $4xy$
B $(x + y)^4$
C $4(x + y)$
D $16(x + y)$

24. **SHORT RESPONSE** Tatiana wants to run 5 miles at an average pace of 9 minutes per mile. After 4 miles, her average pace is 9 minutes 10 seconds. In how many minutes must she complete the final mile to reach her goal?

25. What is the slope of the line that passes through (1, 3) and (−3, 1)?

F -2
G $-\frac{1}{2}$
H $\frac{1}{2}$
J 2

26. What is an equation of the line that passes through (0, 1) and has a slope of 3?

A $y = 3x - 1$
B $y = 3x - 2$
C $y = 3x + 4$
D $y = 3x + 1$

Spiral Review

27. **USED CARS** Gianna wants to buy a specific make and model of a used car. She researched prices from dealers and private sellers and made the graph shown. (Lesson 4-5)

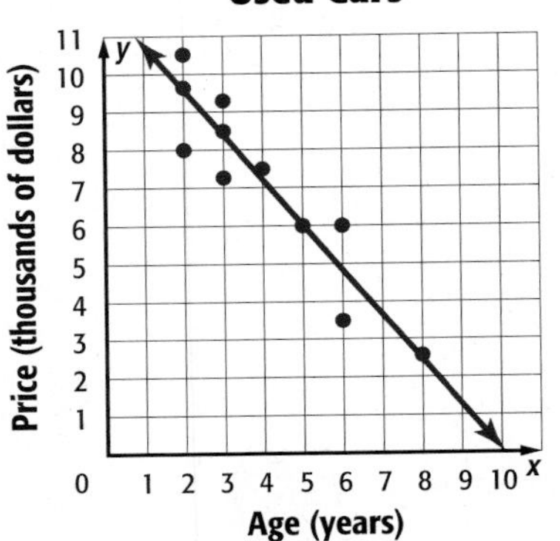

a. Describe the relationship in the data.

b. Use the line of fit to predict the price of a car that is 7 years old.

c. Is it reasonable to use this line of fit to predict the price of a 10-year-old car? Explain.

28. **GEOMETRY** A quadrilateral has sides with equations $y = -2x$, $2x + y = 6$, $y = \frac{1}{2}x + 6$, and $x - 2y = 9$. Is the figure a rectangle? Explain your reasoning. (Lesson 4-4)

Write each equation in standard form. (Lesson 4-3)

29. $y - 2 = 3(x - 1)$

30. $y - 5 = 6(x + 1)$

31. $y + 2 = -2(x - 5)$

32. $y + 3 = \frac{1}{2}(x + 4)$

33. $y - 1 = \frac{2}{3}(x + 9)$

34. $y + 3 = -\frac{1}{4}(x + 2)$

Find the slope of the line that passes through each pair of points. (Lesson 3-3)

35. (3, 4), (10, 8)

36. (−4, 7), (3, 5)

37. (3, 7), (−2, 4)

38. (−3, 2), (−3, 4)

Skills Review

If $f(x) = x^2 - x + 1$, find each value. (Lesson 1-7)

39. $f(-1)$

40. $f(5) - 3$

41. $f(a)$

42. $f(b + 2)$

Graph each equation. (Lesson 3-1)

43. $y = x + 2$

44. $x + 5y = 4$

45. $2x - 3y = 6$

46. $5x + 2y = 6$

4-7 Special Functions

Then

You wrote equations of lines that pass through a particular point, either parallel or perpendicular to a given line. (Lesson 4-4)

Now

- Identify and graph step functions.
- Identify and graph absolute value and piecewise-defined functions.

New Vocabulary

step function
piecewise-linear function
greatest integer function
absolute value function
piecewise-defined function

Math Online
glencoe.com
- Extra Examples
- Personal Tutor
- Self-Check Quiz
- Homework Help

Why?

Kim is ordering books online. The site charges for shipping based on the amount of the order. If the order is less than \$10, shipping costs \$3. If the order is more than \$10 but less than \$20, it will cost \$5 to ship it.

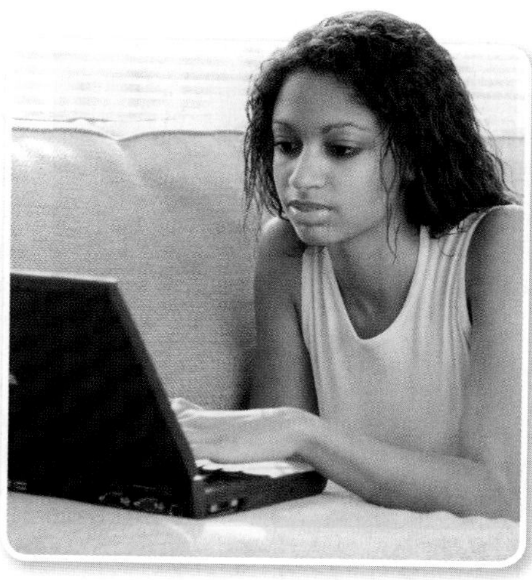

Step Functions The graph of a **step function** is a series of line segments. Because each part of a step function is linear, this type of function is called a **piecewise-linear function**.

One example of a step function is the **greatest integer function**, written as $f(x) = [\![x]\!]$, where $f(x)$ is the greatest integer not greater than x. For example, $[6.8] = 6$ because 6 is the greatest integer that is not greater than 6.8.

Key Concept — Greatest Integer Function

For Your FOLDABLE

Parent function: $f(x) = [\![x]\!]$
Type of graph: disjointed line segments
Domain: all real numbers
Range: all integers

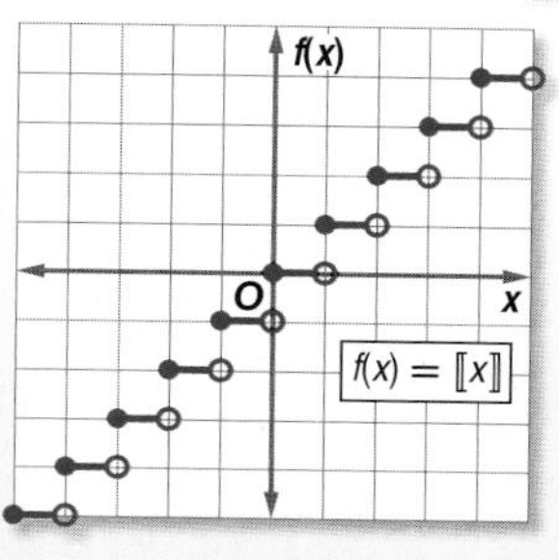

EXAMPLE 1 Greatest Integer Function

Graph $f(x) = [\![x + 2]\!]$. State the domain and range.

First, make a table. Select a few values between integers. On the graph, dots represent included points. Circles represent points not included.

x	$x + 2$	$[\![x + 2]\!]$
0	2	2
0.25	2.25	2
0.5	2.5	2
1	3	3
1.25	3.25	3
1.5	3.5	3
2	4	4
2.25	4.25	4

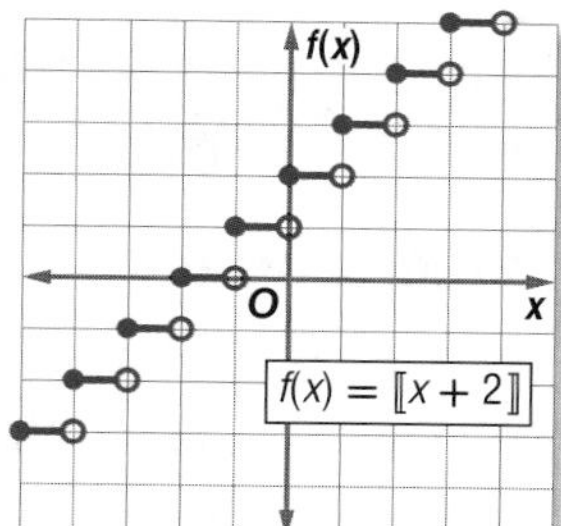

Note that this is the graph of $f(x) = [\![x]\!]$ shifted 2 units to the left.

Because the dots and circles overlap, the domain is all real numbers. The range is all integers.

Check Your Progress

1. Graph $g(x) = 2[\![x]\!]$. State the domain and range.

Personal Tutor glencoe.com

Step functions can be used to represent many real-world situations involving money.

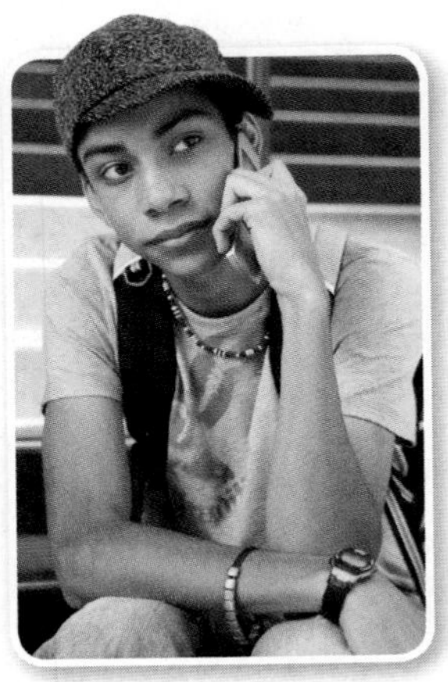

Real-World Link

North Americans are the most likely to have cell phones with 93.2% of the population owning phones.

Source: IT Facts

Real-World EXAMPLE 2 Step Function

CELL PHONE PLANS Cell phone companies charge by the minute, not by the second. A cell phone company charges $0.45 per minute or any fraction thereof for exceeding the number of minutes allotted on each plan. Draw a graph that represents this situation.

The total cost for the extra minutes will be a multiple of $0.45, and the graph will be a step function. If the time is greater than 0 but less than or equal to 1 minute, the charge will be $0.45. If the time is greater than 2 but is less than or equal to 3 minutes, you will be charged for 3 minutes or $1.35.

x	$f(x)$
$0 < x \leq 1$	0.45
$1 < x \leq 2$	0.90
$2 < x \leq 3$	1.35
$3 < x \leq 4$	1.80
$4 < x \leq 5$	2.25
$5 < x \leq 6$	2.70
$6 < x \leq 7$	3.15

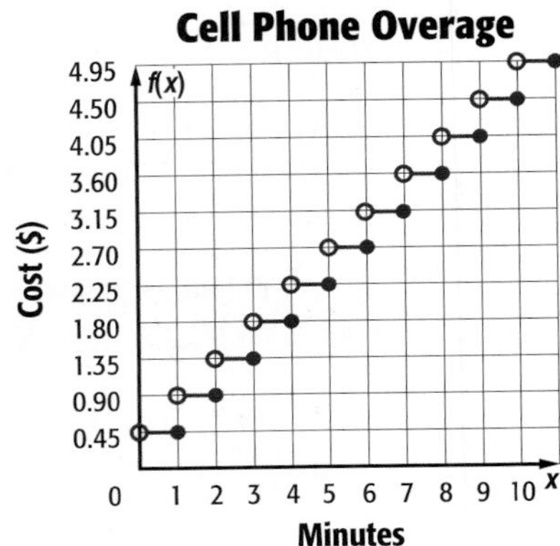

Check Your Progress

2. PARKING A garage charges $4 for the first hour and $1 for each additional hour. Draw a graph that represents this situation.

Personal Tutor glencoe.com

Review Vocabulary

absolute value the distance a number is from zero on a number line; written $|n|$ (Lesson 2-5)

Absolute Value Functions Another type of piecewise-linear function is the **absolute value function**. Recall that the absolute value of a number is always nonnegative. So in the absolute value parent function, written as $f(x) = |x|$, all of the values of the range are nonnegative.

Key Concept Absolute Value Function

For Your FOLDABLE

Parent function: $f(x) = |x|$, defined as

$$f(x) = \begin{cases} x \text{ if } x > 0 \\ 0 \text{ if } x = 0 \\ -x \text{ if } x < 0 \end{cases}$$

Type of graph: V-shaped

Domain: all real numbers

Range: all nonnegative real numbers

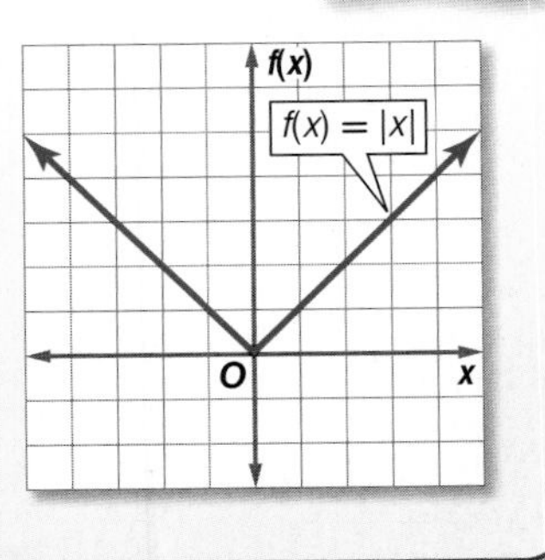

The absolute value function is called a **piecewise-defined function** because it is defined using two or more expressions.

EXAMPLE 3 Absolute Value Function

Graph $f(x) = |x - 4|$. State the domain and range.

Since $f(x)$ cannot be negative, the minimum point of the graph is where $f(x) = 0$.

$f(x) = \|x - 4\|$	**Original function**
$0 = x - 4$	**Replace $f(x)$ with 0 and $\|x - 4\|$ with $x - 4$.**
$4 = x$	**Add 4 to each side.**

Next make a table of values. Include values for $x > 4$ and $x < 4$.

$f(x) = \|x - 4\|$	
x	$f(x)$
−2	6
0	4
2	2
4	0
5	1
6	2
7	3
8	4

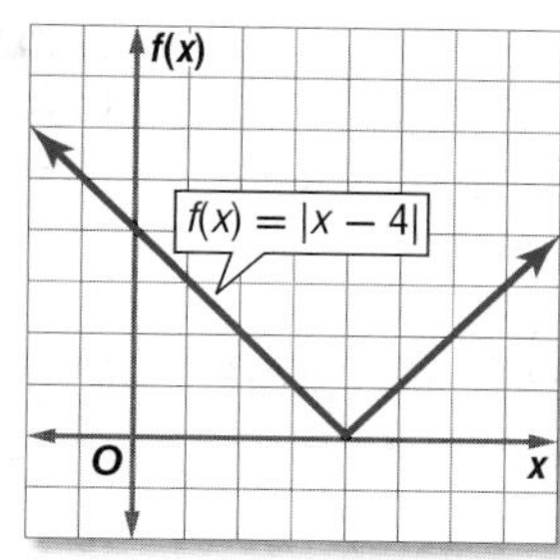

The domain is all real numbers. The range is all real numbers greater than or equal to 0. Note that this is the graph of $f(x) = |x|$ shifted 4 units to the right.

Check Your Progress

3. Graph $f(x) = |2x + 1|$. State the domain and range.

Personal Tutor glencoe.com

Not all piecewise-defined functions are absolute value functions. Step functions are also piecewise-defined functions. In fact, all piecewise-linear functions are piecewise-defined.

StudyTip

Piecewise Functions To graph a piecewise-defined function, graph each "piece" separately. There should be a dot or line that contains each member of the domain.

EXAMPLE 4 Piecewise-Defined Function

Graph $f(x) = \begin{cases} -2x \text{ if } x > 1 \\ x + 3 \text{ if } x \leq 1 \end{cases}$. State the domain and range.

Graph the first expression. Create a table of values for when $x > 1$, $f(x) = -2x$ and draw the graph. Since x is not equal to 1, place a circle at $(1, -2)$.

Next, graph the second expression. Create a table of values for when $x \leq 1$, $f(x) = x + 3$ and draw the graph. If $x = 1$, then $f(x) = 4$; place a dot at $(1, 4)$.

The domain is all real numbers. The range is $y \leq 4$.

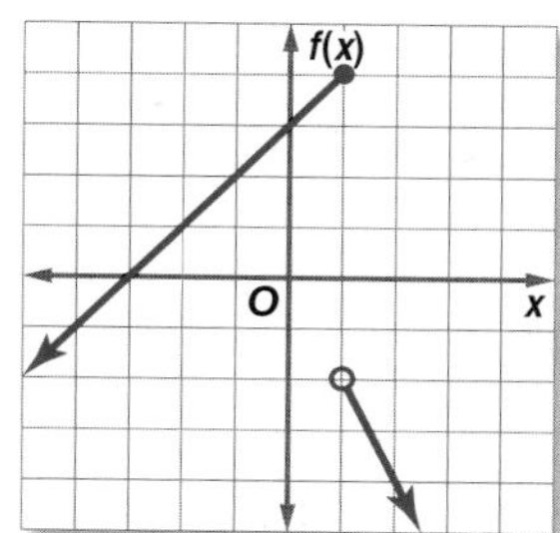

Check Your Progress

4. Graph $f(x) = \begin{cases} 2x + 1 \text{ if } x > 0 \\ 3 \text{ if } x \leq 0 \end{cases}$. State the domain and range.

Personal Tutor glencoe.com

Concept Summary — Special Functions

For Your FOLDABLE

Step Function | **Absolute Value Function** | **Piecewise-Defined Function**

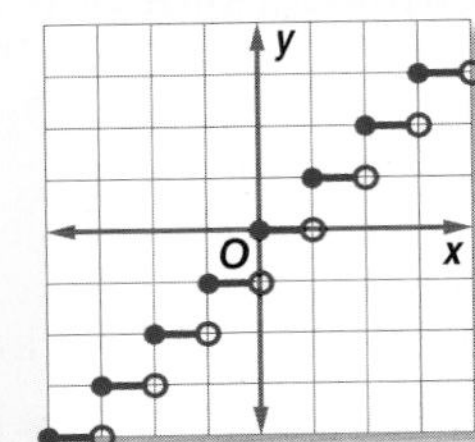

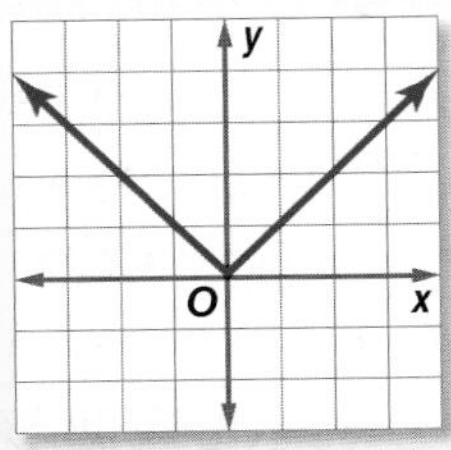

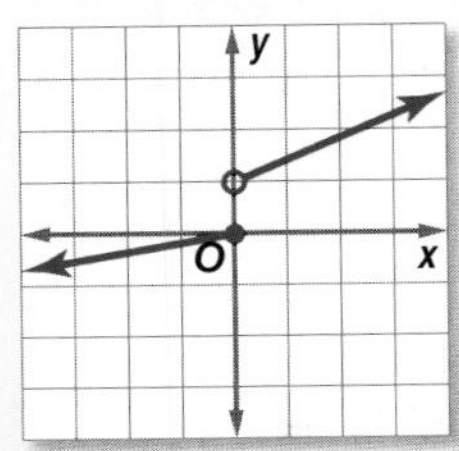

Check Your Understanding

Example 1 p. 261

Graph each function. State the domain and range.

1. $f(x) = \frac{1}{2}[\![x]\!]$
2. $g(x) = -[\![x]\!]$
3. $h(x) = [\![2x]\!]$

Example 2 p. 262

4. **SHIPPING** Elan is ordering a gift for his dad online. The table shows the shipping rates. Graph the step function.

Order Total ($)	Shipping Cost ($)
0–15	3.99
15.01–30	5.99
30.01–50	6.99
50.01–75	7.99
75.01–100	8.99
Over $100	9.99

Examples 3 and 4 p. 263

Graph each function. State the domain and range.

5. $f(x) = |x - 3|$
6. $g(x) = |2x + 4|$
7. $f(x) = \begin{cases} 2x - 1 \text{ if } x > -1 \\ -x \text{ if } x \leq -1 \end{cases}$
8. $g(x) = \begin{cases} -3x - 2 \text{ if } x > -2 \\ -x + 1 \text{ if } x \leq -2 \end{cases}$

Practice and Problem Solving

● = **Step-by-Step Solutions** begin on page R12.
Extra Practice begins on page 815.

Example 1 p. 261

Graph each function. State the domain and range.

9. $f(x) = 3[\![x]\!]$
10. $f(x) = [\![-x]\!]$
11. $g(x) = -2[\![x]\!]$
12. $g(x) = [\![x]\!] + 3$
13. $h(x) = [\![x]\!] - 1$
14. $h(x) = \frac{1}{2}[\![x]\!] + 1$

Example 2 p. 262

15. **CAB FARES** Lauren wants to take a taxi from a hotel to a friend's house. The rate is \$3 plus \$1.50 per mile after the first mile. Every fraction of a mile is rounded up to the next mile.

    a. Draw a graph to represent the cost of using a taxi cab.

    b. What is the cost if the trip is 8.5 miles long?

16. **POSTAGE** The United States Postal Service increases the rate of postage periodically. The table shows the cost to mail a letter weighing 1 ounce or less from 1988 through 2007. Draw a step graph to represent the data.

Year	1988	1991	1995	1999	2001	2002	2006	2007
Cost ($)	0.25	0.29	0.32	0.33	0.34	0.37	0.39	0.41

Examples 3 and 4
p. 263

Graph each function. State the domain and range.

17. $f(x) = |2x - 1|$

18. $f(x) = |x + 5|$

19. $g(x) = |-3x - 5|$

20. $g(x) = |-x - 3|$

21. $f(x) = \left|\frac{1}{2}x - 2\right|$

22. $f(x) = \left|\frac{1}{3}x + 2\right|$

23. $g(x) = |x + 2| + 3$

24. $g(x) = |2x - 3| + 1$

25. $f(x) = \begin{cases} \frac{1}{2}x - 1 \text{ if } x > 3 \\ -2x + 3 \text{ if } x \leq 3 \end{cases}$

26. $f(x) = \begin{cases} 2x - 5 \text{ if } x > 1 \\ 4x - 3 \text{ if } x \leq 1 \end{cases}$

27. $f(x) = \begin{cases} 2x + 3 \text{ if } x \geq -3 \\ -\frac{1}{3}x + 1 \text{ if } x < -3 \end{cases}$

28. $f(x) = \begin{cases} 3x + 4 \text{ if } x \geq 1 \\ x + 3 \text{ if } x < 1 \end{cases}$

29. $f(x) = \begin{cases} 3x + 2 \text{ if } x > -1 \\ -\frac{1}{2}x - 3 \text{ if } x \leq -1 \end{cases}$

30. $f(x) = \begin{cases} 2x + 1 \text{ if } x < -2 \\ -3x - 1 \text{ if } x \geq -2 \end{cases}$

Determine the domain and range of each function.

StudyTip

Nonlinear Functions The greatest integer function, absolute value function, and piecewise defined functions are examples of nonlinear functions.

31.

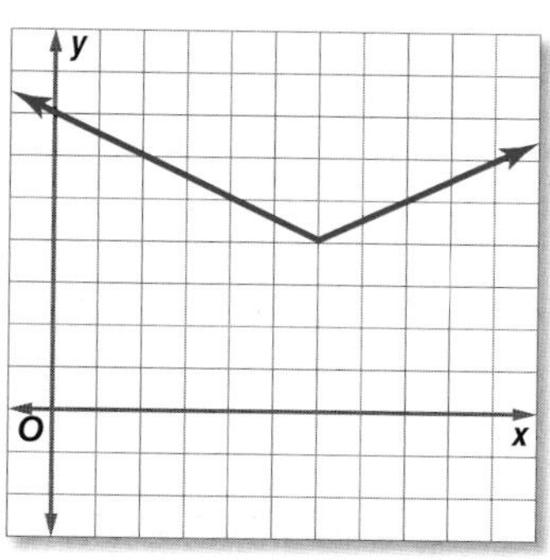

32.

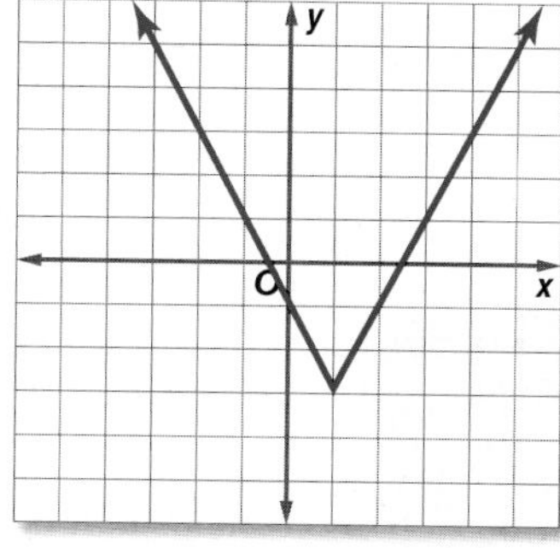

33.

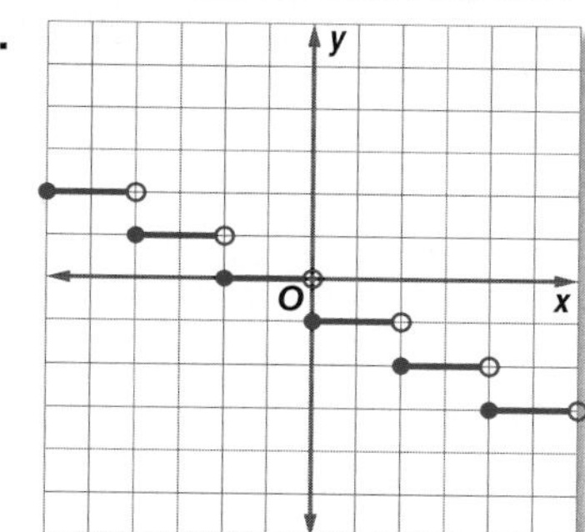

34.

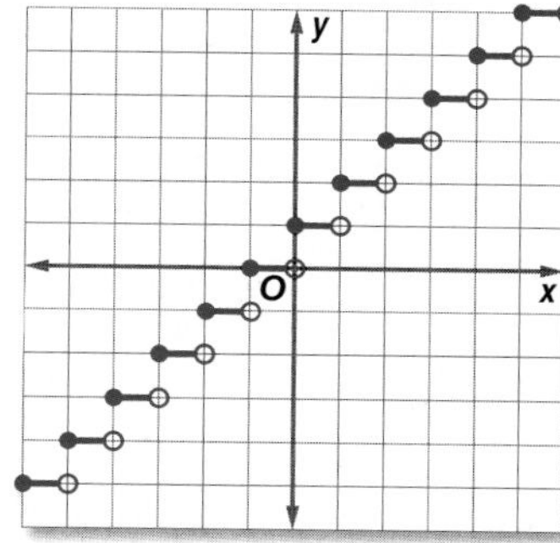

35.

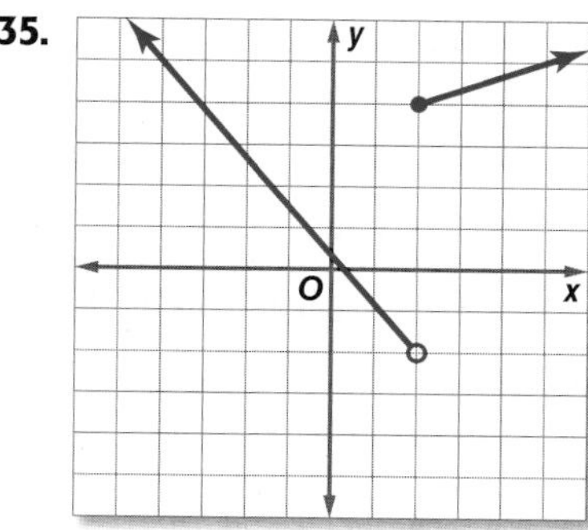

36.

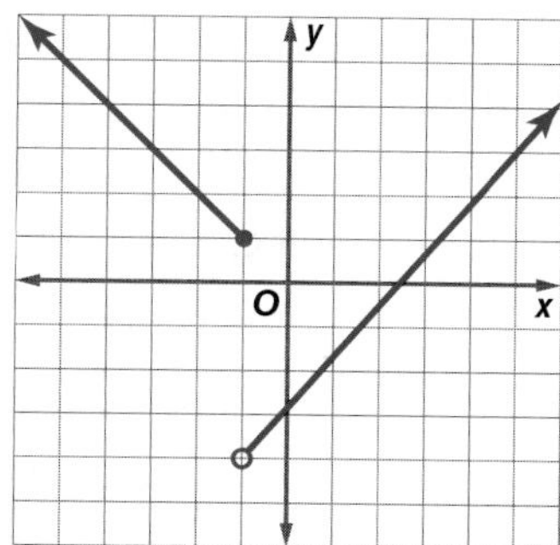

37. **BOATING** According to Boat Minnesota, the maximum number of people that can safely ride in a boat is determined by the boat's length and width. The table shows some guidelines for the length of a boat that is 6 feet wide. Graph this relation.

Length of Boat (ft)	18–19	20–22	23–24
Number of People	7	8	9

For Exercises 38–41, match each graph to one of the following equations.

A	B	C	D
$y = 2x - 1$	$y = [\![2x]\!] - 1$	$y = \lvert 2x \rvert - 1$	$y = \begin{cases} 2x + 1 & \text{if } x > 0 \\ -2x + 1 & \text{if } x \leq 0 \end{cases}$

38. 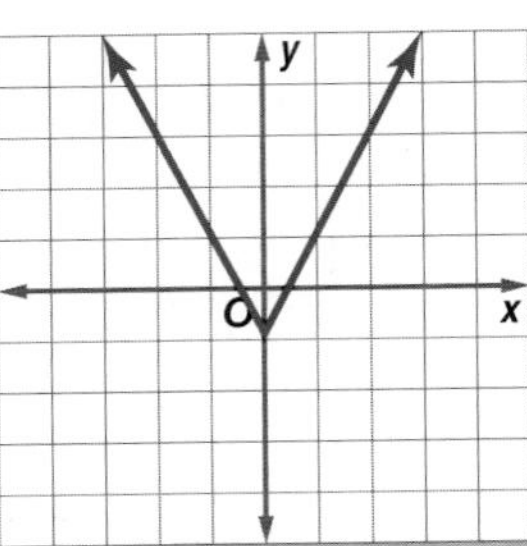

39. 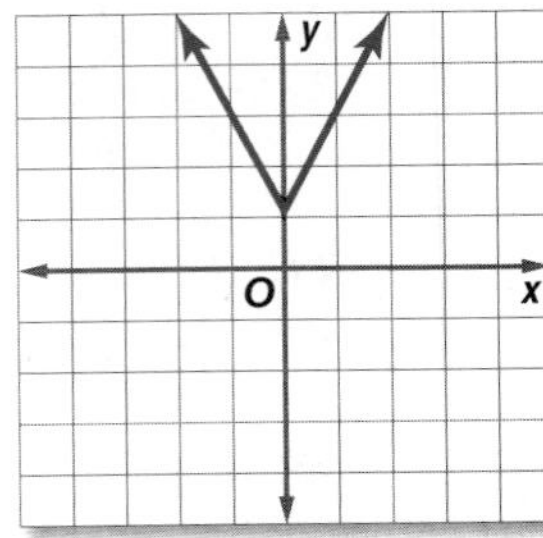

40. 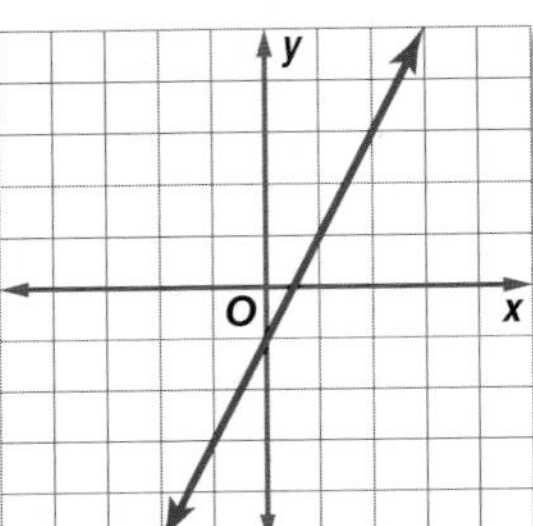

41. 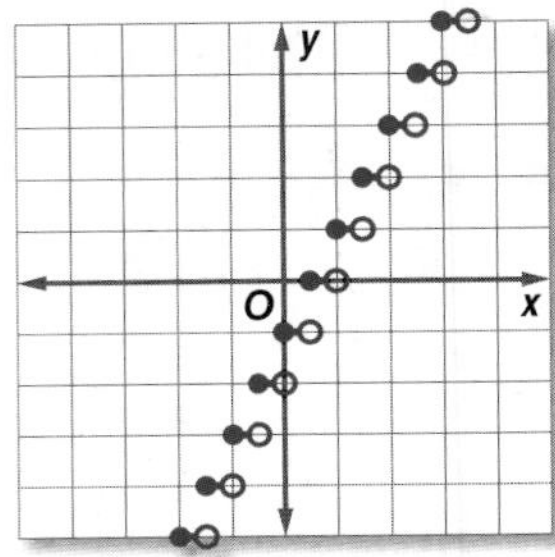

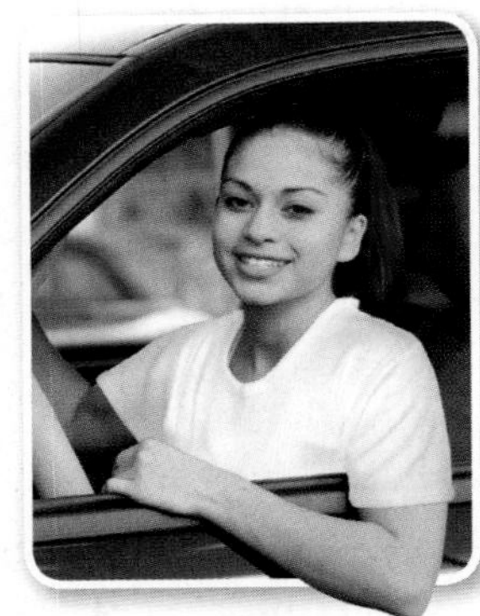

Real-World Link

Leasing a car differs from buying a car. When leasing a car, you may pay a lower monthly cost; however, you do not own the car at the end of the lease term. Most leasing agreements limit the number of miles you can drive the car before incurring additional costs.

Source: Federal Reserve

42. CAR LEASE As part of Marcus' leasing agreement, he will be charged \$0.20 per mile for each mile over 12,000. Any fraction of a mile is rounded up to the next mile. Make a step graph to represent the cost of going over the mileage.

43. BASEBALL A baseball team is ordering T-shirts with the team logo on the front and the players' names on the back. A graphic design store charges \$10 to set up the artwork plus \$10 per shirt, \$4 each for the team logo, and \$2 to print the last name for an order of 10 shirts or less. For orders of 11–20 shirts, a 5% discount is given. For orders of more than 20 shirts, a 10% discount is given.

a. Organize the information into a table. Include a column showing the total order price for each size order.

b. Write an equation representing the total price for an order of x shirts.

c. Graph the piecewise relation.

44. Consider the function $f(x) = \lvert 2x + 3 \rvert$.

a. Make a table of values where x is all integers from -5 to 5, inclusive.

b. Plot the points on a coordinate grid.

c. Graph the function.

45. Consider the function $f(x) = \lvert 2x \rvert + 3$.

a. Make a table of values where x is all integers from -5 to 5, inclusive.

b. Plot the points on a coordinate grid.

c. Graph the function.

d. Describe how this graph is different from the graph in Exercise 44.

Real-World Link

In addition to traditional dance classes like ballet and tap, more studios are offering classes in hip-hop, jazz/hip-hip, and cardio dance.

Source: Ariel Dance Studio

46. DANCE A local studio must have at least 5 students enrolled in a class, or else the class will be canceled. Once 10 students are enrolled, a second class is started. Draw a graph for this situation.

47. THEATERS A certain theater will not have a show unless it has sold 50 tickets for that show. Once the capacity of 250 seats are sold, the theater begins selling tickets for the next show. Draw a graph that describes this situation.

Graph each function.

48. $f(x) = \frac{1}{2}|x| + 2$ **49** $g(x) = \frac{1}{3}|x| + 4$ **50.** $h(x) = -2|x - 3| + 2$

51. $f(x) = -4|x + 2| - 3$ **52.** $g(x) = -\frac{2}{3}|x + 6| - 1$ **53.** $h(x) = -\frac{3}{4}|x - 8| + 1$

54. MULTIPLE REPRESENTATIONS In this problem, you will explore piecewise-defined functions.

a. TABULAR Copy and complete the table of values for $f(x) = |[\![x]\!]|$ and $g(x) = [\![|x|]\!]$.

x	$[\![x]\!]$	$f(x) = \lvert[\![x]\!]\rvert$	$\lvert x\rvert$	$g(x) = [\![\lvert x\rvert]\!]$
−3	−3	3	3	3
−2.5				
−2				
0				
0.5				
1				
1.5				

b. GRAPHICAL Graph each function on a coordinate plane.

c. ANALYTICAL Compare and contrast the graphs of $f(x)$ and $g(x)$.

H.O.T. Problems Use Higher-Order Thinking Skills

55. REASONING Does the piecewise relation below represent a function? Why or why not?

$$y = \begin{cases} -2x + 4 \text{ if } x \geq 2 \\ -\frac{1}{2}x - 1 \text{ if } x \leq 4 \end{cases}$$

CHALLENGE Refer to the graph.

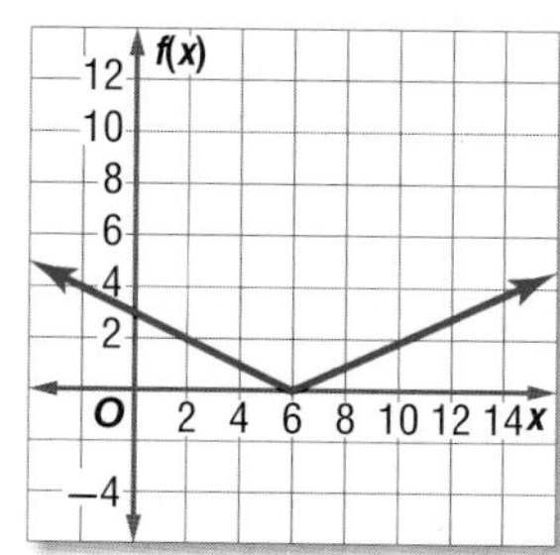

56. Write an absolute value function that represents the graph.

57. Write a piecewise function to represent the graph.

58. What are the domain and range?

59. WRITING IN MATH Refer to the information on cell phone plans in Example 2. Explain why the graph of this description is called a *step graph*.

60. CHALLENGE A bicyclist travels up and down a hill. The hill has a vertical cross section that can be modeled by the equation $y = -\frac{1}{4}|x - 300| + 400$ where x and y are measured in feet.

a. If $0 \leq x \leq 600$, find the slope for the uphill portion of the trip and then the downhill portion of the trip.

b. Graph this function.

c. What are the domain and range of the graph?

Standardized Test Practice

61. Which equation represents a line that is perpendicular to the graph and passes through the point at (2, 0)?

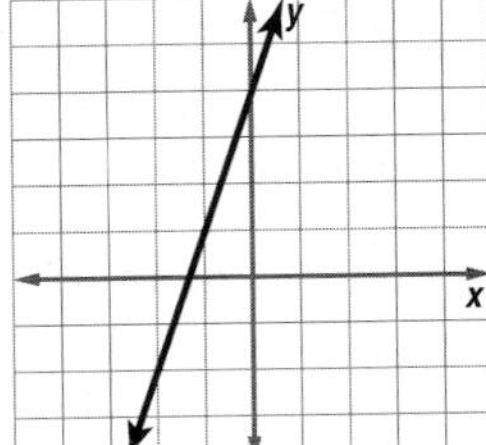

A $y = 3x - 6$

B $y = -3x + 6$

C $y = -\frac{1}{3}x + \frac{2}{3}$

D $y = \frac{1}{3}x - \frac{2}{3}$

62. A giant tortoise travels at a rate of 0.17 mile per hour. Which equation models the time t it would take the giant tortoise to travel 0.8 mile?

F $t = \frac{0.8}{0.17}$

G $t = (0.17)(0.8)$

H $t = \frac{0.17}{0.8}$

J $0.8 = \frac{0.17}{t}$

63. GEOMETRY If $\triangle JKL$ is similar to $\triangle JNM$ what is the value a?

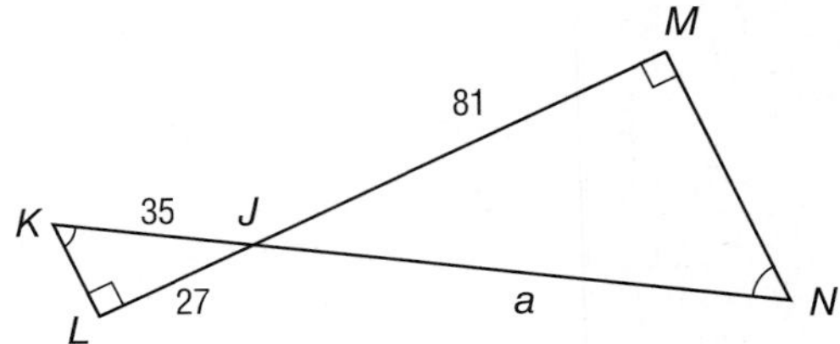

A 62.5

B 105

C 125

D 155.5

64. GRIDDED RESPONSE What is the difference in the value of $2.1(x + 3.2)$, when $x = 5$ and when $x = 3$?

Spiral Review

Write an equation of the regression line for the data in each table. (Lesson 4-6)

65.

x	1	3	5	7	9
y	3	8	15	18	21

66.

x	3	5	7	9	11
y	7.2	23.5	41.2	56.4	73.1

67.

x	1	2	3	4	5
y	21	33	39	54	64

68.

x	2	4	6	8	10
y	1.4	2.4	2.9	3.3	4.2

69. TESTS Determine whether the graph at the right shows a *positive*, *negative*, or *no* correlation. If there is a correlation, describe its meaning. (Lesson 4-5)

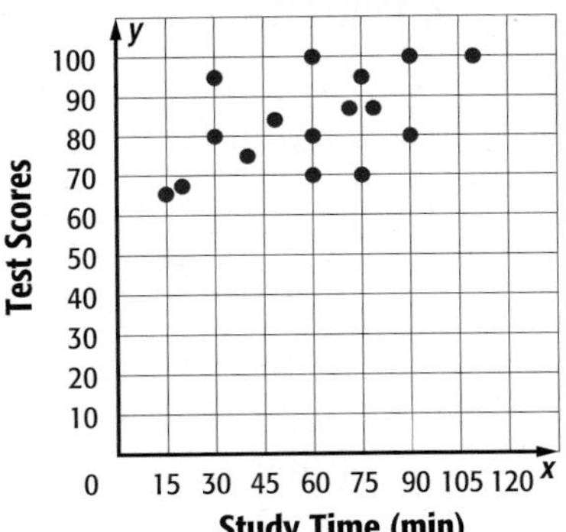

Suppose y varies directly as x. (Lesson 3-4)

70. If $y = 2.5$ when $x = 0.5$, find y when $x = 20$.

71. If $y = -6.6$ when $x = 9.9$, find y when $x = 6.6$.

72. If $y = 2.6$ when $x = 0.25$, find y when $x = 1.125$.

73. If $y = 6$ when $x = 0.6$, find x when $y = 12$.

Skills Review

Solve each equation. (Lesson 2-2)

74. $104 = k - 67$

75. $-4 + x = -7$

76. $\frac{m}{7} = -11$

77. $\frac{2}{3}p = 14$

78. $-82 = n + 18$

79. $\frac{9}{t} = -27$

EXTEND 4-7

Graphing Technology Lab: Piecewise-Linear Functions

Math Online glencoe.com
- Other Calculator Keystrokes
- Graphing Technology Personal Tutor

You can use a graphing calculator to graph and analyze various piecewise functions, including greatest integer functions and absolute value functions.

ACTIVITY 1 Greatest Integer Functions

Graph $f(x) = [\![x]\!]$ in the standard viewing window.

The calculator may need to be changed to dot mode for the function to graph correctly. Press MODE then use the arrow and ENTER keys to select DOT.

Enter the equation in the Y= list. Then graph the equation.

KEYSTROKES: Y= MATH ▶ 5 X,T,θ,n) Zoom 6

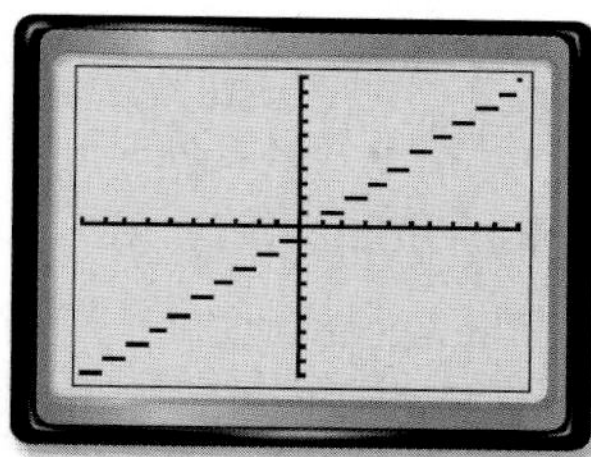

[−10, 10] scl: 1 by [−10, 10] scl: 1

1A. How does the graph of $f(x) = [\![x]\!]$ compare to the graph of $f(x) = x$?

1B. What are the domain and range of the function $f(x) = [\![x]\!]$? Explain.

The graphs of piecewise functions are affected by changes in parameters.

ACTIVITY 2 Absolute Value Functions

Graph $y = |x| - 3$ and $y = |x| + 1$ in the standard viewing window.

Enter the equations in the Y= list. Then graph.

KEYSTROKES: Y= MATH ▶ 1 X,T,θ,n) − 3 ENTER MATH ▶ 1 X,T,θ,n) + 1 Zoom 6

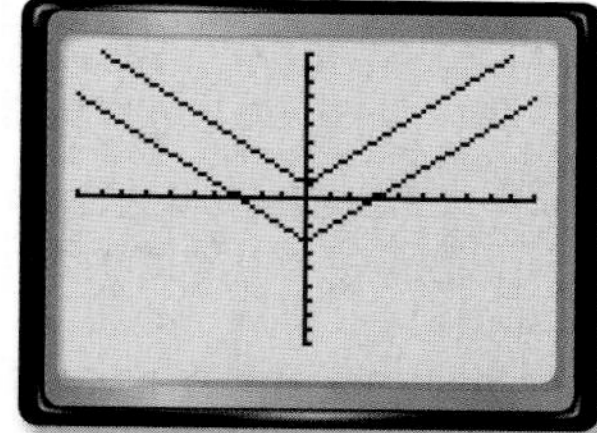

[−10, 10] scl: 1 by [−10, 10] scl: 1

2A. Compare and contrast the graphs to the graph of $y = |x|$.

2B. How does the value of c affect the graph of $y = |x| + c$?

Analyze The Results

1. A parking garage charges $4 for every hour or fraction of an hour. Is this situation modeled by a *linear* function or a *step* function? Explain your reasoning.
2. A maintenance technician is testing an elevator system. The technician starts the elevator at the fifth floor. It is sent to the ground floor, then back to the fifth floor. Assume the elevator travels at a constant rate. Should the height of the elevator be modeled by a step function or an absolute value function? Explain.
3. **MAKE A CONJECTURE** Explain why the greatest integer function is sometimes called the *floor function*.
4. Graph $y = -|x|$ in the standard viewing window. How is this graph to related to the graph of $y = |x|$?
5. **MAKE A CONJECTURE** Describe the transformation of the parent graph to $y = |x + c|$. Use a graphing calculator with different values of c to test your conjecture.

Study Guide and Review

Math Online glencoe.com
- STUDY TO GO
- Vocabulary Review

Chapter Summary

Key Concepts

Slope-Intercept Form (Lessons 4-1 and 4-2)

- The slope-intercept form of a linear equation is $y = mx + b$, where m is the slope and b is the y-intercept.
- If you are given two points through which a line passes, use them to find the slope first.

Point-Slope Form (Lesson 4-3)

- The linear equation $y - y_1 = m(x - x_1)$ is written in point-slope form, where (x_1, y_1) is a given point on a nonvertical line and m is the slope of the line.

Parallel and Perpendicular Lines (Lesson 4-4)

- Nonvertical parallel lines have the same slope.
- Lines that intersect at right angles are called perpendicular lines. The slopes of perpendicular lines are opposite reciprocals.

Scatter Plots and Lines of Fit (Lesson 4-5)

- Data with two variables are called bivariate data.
- A scatter plot is a graph in which two sets of data are plotted as ordered pairs in a coordinate plane.

Regression and Median-Fit Lines (Lesson 4-6)

- A graphing calculator can be used to find regression lines and median-fit lines.

Special Functions (Lesson 4-7)

- The greatest integer function is written as $f(x) = [\![x]\!]$, where $f(x)$ is the greatest integer not greater than x.
- The absolute value function is written as $f(x) = |x|$, where $f(x)$ is the distance between x and 0 on a number line.

FOLDABLES® Study Organizer

Be sure the Key Concepts are noted in your Foldable.

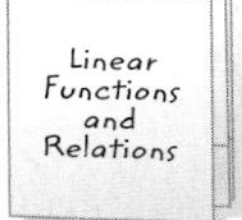

Key Vocabulary

absolute value function (p. 262)
best-fit line (p. 253)
bivariate data (p. 245)
correlation coefficient (p. 253)
greatest integer function (p. 261)
linear extrapolation (p. 226)
linear interpolation (p. 247)
linear regression (p. 253)
line of fit (p. 246)
median-fit line (p. 255)
parallel lines (p. 237)
perpendicular lines (p. 238)
piecewise-defined function (p. 262)
piecewise-linear function (p. 261)
point-slope form (p. 231)
scatter plot (p. 245)
slope-intercept form (p. 214)
step function (p. 261)

Vocabulary Check

State whether each sentence is *true* or *false*. If *false*, replace the underlined term to make a true sentence.

1. The <u>y-intercept</u> is the y-coordinate of the point where the graph crosses the y-axis.
2. The process of using a linear equation to make predictions about values that are beyond the range of the data is called <u>linear regression</u>.
3. A graph in which two sets of data are plotted as ordered pairs in a coordinate plane is called a <u>step function</u>.
4. The <u>correlation coefficient</u> describes whether the correlation between the variables is positive or negative and how closely the regression equation is modeling the data.
5. Lines in the same plane that do not intersect are called <u>parallel</u> lines.
6. Lines that intersect at <u>acute</u> angles are called perpendicular lines.
7. A function that is defined differently for different parts of its domain is called a <u>piecewise-defined function</u>.
8. The <u>range</u> of the greatest integer function is the set of all real numbers.
9. A <u>piecewise-linear function</u> is also called a step function.

Lesson-by-Lesson Review

4-1 Graphing Equations in Slope-Intercept Form (pp. 214–221)

Write an equation of a line in slope-intercept form with the given slope and y-intercept. Then graph the equation.

10. slope: 3, y-intercept: 5

11. slope: -2, y-intercept: -9

12. slope: $\frac{2}{3}$, y-intercept: 3

13. slope: $-\frac{5}{8}$, y-intercept: -2

Graph each equation.

14. $y = 4x - 2$

15. $y = -3x + 5$

16. $y = \frac{1}{2}x + 1$

17. $3x + 4y = 8$

18. SKI RENTAL Write an equation in slope-intercept form for the total cost of skiing for h hours with one lift ticket.

Slippery Slope
Ski Lodge
Lift Ticket $15/day
Ski Rental $5/hour

EXAMPLE 1

Write an equation of a line in slope-intercept form with slope -5 and y-intercept -3. Then graph the equation.

$y = mx + b$	**Slope-intercept form**
$y = -5x + (-3)$	**$m = -5$ and $b = -3$**
$y = -5x - 3$	**Simplify.**

To graph the equation, plot the y-intercept $(0, -3)$. Then move up 5 units and left 1 unit. Plot the point. Draw a line through the two points.

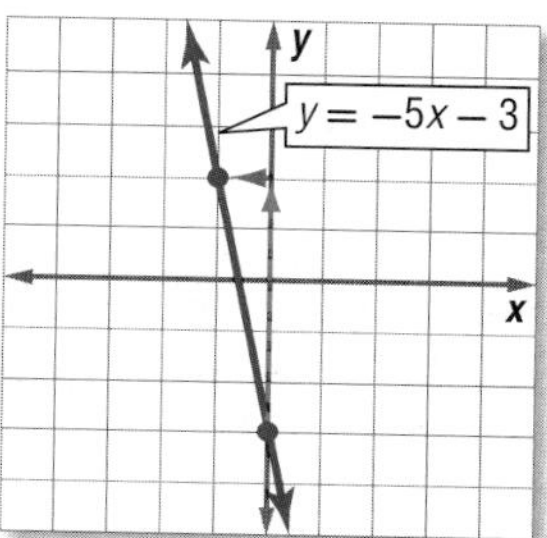

4-2 Writing Equations in Slope-Intercept Form (pp. 224–230)

Write an equation of the line that passes through the given point and has the given slope.

19. (1, 2), slope 3

20. (2, −6), slope −4

21. (−3, −1), slope $\frac{2}{5}$

22. (5, −2), slope $-\frac{1}{3}$

Write an equation of the line that passes through the given points.

23. (2, −1), (5, 2)

24. (−4, 3), (1, 13)

25. (3, 5), (5, 6)

26. (2, 4), (7, 2)

27. CAMP In 2000, a camp had 450 campers. Five years later, the number of campers rose to 750. Write a linear equation that represents the number of campers that attend camp.

EXAMPLE 2

Write an equation of the line that passes through (3, 2) with a slope of 5.

Step 1 Find the y-intercept.

$y = mx + b$	**Slope-intercept form**
$2 = 5(3) + b$	**$m = 5$, $y = 2$, and $x = 3$**
$2 = 15 + b$	**Simplify.**
$-13 = b$	**Subtract 15 from each side.**

Step 2 Write the equation in slope-intercept form.

$y = mx + b$	**Slope-intercept form**
$y = 5x - 13$	**$m = 5$ and $b = -13$**

Study Guide and Review

4-3 Point-Slope Form (pp. 231–236)

Write an equation in point-slope form for the line that passes through the given point with the slope provided.

28. (6, 3), slope 5

29. (−2, 1), slope −3

30. (−4, 2), slope 0

Write each equation in standard form.

31. $y - 3 = 5(x - 2)$

32. $y - 7 = -3(x + 1)$

33. $y + 4 = \frac{1}{2}(x - 3)$

34. $y - 9 = -\frac{4}{5}(x + 2)$

Write each equation in slope-intercept form.

35. $y - 2 = 3(x - 5)$

36. $y - 12 = -2(x - 3)$

37. $y + 3 = 5(x + 1)$

38. $y - 4 = \frac{1}{2}(x + 2)$

EXAMPLE 3

Write an equation in point-slope form for the line that passes through (3, 4) with a slope of −2.

$y - y_1 = m(x - x_1)$	**Point-slope form**
$y - 4 = -2(x - 3)$	**Replace m with −2 and (x_1, y_1) with (3, 4).**

EXAMPLE 4

Write $y + 6 = -4(x - 3)$ in standard form.

$y + 6 = -4(x - 3)$	**Original equation**
$y + 6 = -4x + 12$	**Simplify.**
$4x + y + 6 = 12$	**Add $4x$ to each side.**
$4x + y = 6$	**Subtract 6 from each side.**

4-4 Parallel and Perpendicular Lines (pp. 237–243)

Write an equation in slope-intercept form for the line that passes through the given point and is parallel to the graph of each equation.

39. (2, 5), $y = x - 3$

40. (0, 3), $y = 3x + 5$

41. (−4, 1), $y = -2x - 6$

42. (−5, −2), $y = -\frac{1}{2}x + 4$

Write an equation in slope-intercept form for the line that passes through the given point and is perpendicular to the graph of the given equation.

43. (2, 4), $y = 3x + 1$

44. (1, 3), $y = -2x - 4$

45. (−5, 2), $y = \frac{1}{3}x + 4$

46. (3, 0), $y = -\frac{1}{2}x$

EXAMPLE 5

Write an equation in slope-intercept form for the line that passes through (−2, 4) and is parallel to the graph of $y = 6x - 3$.

The slope of the line with equation $y = 6x - 3$ is 6. The line parallel to $y = 6x - 3$ has the same slope, 6.

$y - y_1 = m(x - x_1)$	**Point-slope form**
$y - 4 = 6(x - (-2))$	**Substitute.**
$y - 4 = 6(x + 2)$	**Simplify.**
$y - 4 = 6x + 12$	**Distributive Property**
$y = 6x + 16$	**Add 4 to each side.**

4-5 Scatter Plots and Lines of Fit (pp. 245–252)

47. Determine whether the graph shows a *positive*, *negative*, or *no* correlation. If there is a positive or negative correlation, describe its meaning.

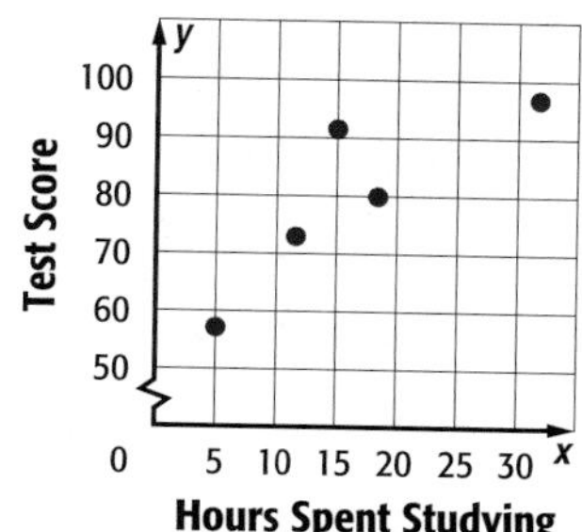

48. ATTENDANCE A scatter plot of data compares the number of years since a business has opened and its annual number of sales. It contains the ordered pairs (2, 650) and (5, 1280). Write an equation in slope-intercept form for the line of fit for this situation.

EXAMPLE 6

The scatter plot displays the number of text messages and the number of phone calls made daily. Write an equation for the line of fit.

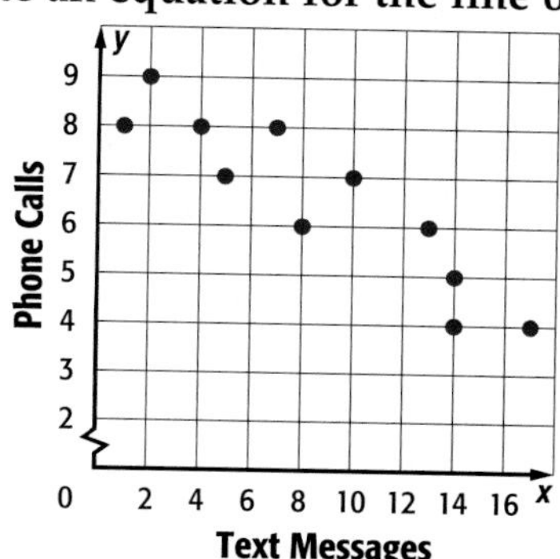

First, find the slope using (2, 9) and (17, 4).

$m = \frac{4 - 9}{17 - 2} = \frac{-5}{15}$ or $-\frac{1}{3}$ **Substitute and simplify.**

Then find the y-intercept.

$9 = -\frac{1}{3}(2) + b$ **Substitute.**

$9\frac{2}{3} = b$ **Add $\frac{2}{3}$ to each side.**

Write the equation. $y = -\frac{1}{3}x + 9\frac{2}{3}$

4-6 Regression and Median-Fit Lines (pp. 253–260)

49. SALE The table shows the number of sales made at an outerwear store during a sale. Write an equation of the regression line. Then estimate the daily sales on day 10 of the sale.

Days Since Sale Began	1	2	3	4	5	6	7
Daily Sales ($)	15	21	32	30	40	38	51

50. MOVIES The table shows ticket sales during the first week. Write an equation of the regression line. Then estimate the daily ticket sales on the 15th day after the movie opens.

Days Since Movie Opened	1	2	3	4	5	6	7
Daily Ticket Sales ($)	85	92	89	78	65	68	55

EXAMPLE 7

ATTENDANCE The table shows the annual attendance at an amusement park. Write an equation of the regression line for the data.

Year (since 2000)	0	1	2	3	4	5	6
Attendance (thousands)	75	80	72	68	65	60	53

Step 1 Enter the data by pressing STAT and selecting the **Edit** option.

Step 2 Perform the regression by pressing STAT and selecting the **CALC** option. Scroll down to **LinReg (ax + b)** and press ENTER.

Step 3 Write the equation of the regression line by rounding the a- and b-values on the screen. $y = -4.04x + 79.68$

4-7 Special Functions (pp. 261–268)

Graph each function. State the domain and range.

51. $f(x) = [\![x]\!]$

52. $f(x) = [\![2x]\!]$

53. $f(x) = |x|$

54. $f(x) = |2x - 2|$

55. $f(x) = \begin{cases} x - 2 \text{ if } x < 1 \\ 3x \text{ if } x \geq 1 \end{cases}$

56. $f(x) = \begin{cases} 2x - 3 \text{ if } x \leq 2 \\ x + 1 \text{ if } x > 2 \end{cases}$

57. Determine the domain and range of the function graphed below.

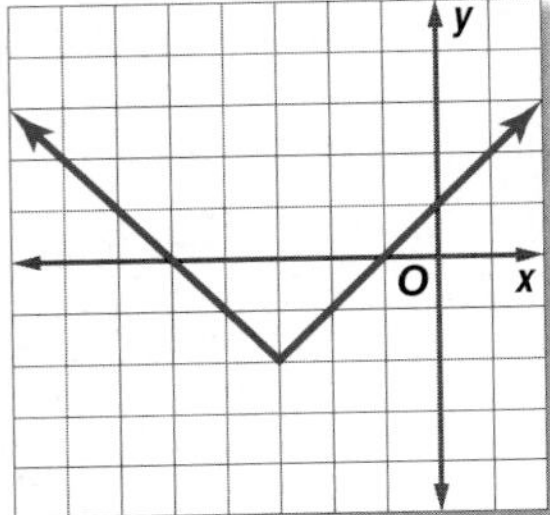

EXAMPLE 8

Graph $f(x) = |x + 3|$. State the domain and range.

Since $f(x)$ cannot be negative, the minimum point of the graph is where $f(x) = 0$.

$f(x) = \lvert x + 3 \rvert$	**Original function**
$0 = x + 3$	**Replace $f(x)$ with 0.**
$-3 = x$	**Subtract 3 from each side.**

Next, make a table of values. Include values for $x > -3$ and $x < -3$.

x	−5	−4	−3	−2	−1
$f(x)$	2	1	0	1	2

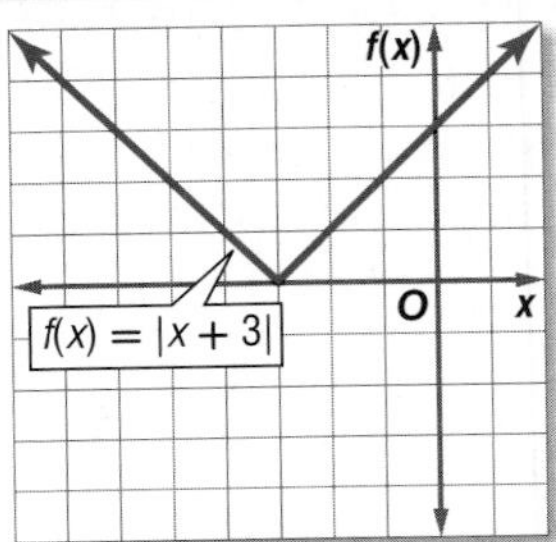

The domain is all real numbers, and the range is $f(x) \geq 0$.

CHAPTER 4

Practice Test

Math Online glencoe.com
Chapter Test

1. Graph $y = 2x - 3$.

2. **MULTIPLE CHOICE** A popular pizza parlor charges \$12 for a large cheese pizza plus \$1.50 for each additional topping. Write an equation in slope-intercept form for the total cost C of a pizza with t toppings.

   A $C = 12t + 1.50$

   B $C = 13.50t$

   C $C = 12 + 1.50t$

   D $C = 1.50t - 12$

Write an equation of a line in slope-intercept form that passes through the given point and has the given slope.

3. $(-4, 2)$; slope -3

4. $(3, -5)$; slope $\frac{2}{3}$

Write an equation of the line in slope-intercept form that passes through the given points.

5. $(1, 4), (3, 10)$

6. $(2, 5), (-2, 8)$

7. $(0, 4), (-3, 0)$

8. $(7, -1), (9, -4)$

9. **PAINTING** The data in the table show the size of a room in square feet and the time it takes to paint the room in minutes.

Room Size	100	150	200	400	500
Painting Time	160	220	270	500	680

   a. Use the points (100, 160) and (500, 680) to write an equation in slope-intercept form.

   b. Predict the amount of time required to paint a room measuring 750 square feet.

10. **SALARY** The table shows the relationship between years of experience and teacher salary.

Years Experience	1	5	10	15	20
Salary (thousands of dollars)	28	31	42	49	64

   a. Write an equation for the best-fit line.

   b. Find the correlation coefficient and explain what it tells us about the relationship between experience and salary.

Write an equation in slope-intercept form for the line that passes through the given point and is parallel to the graph of each equation.

11. $(2, -3), y = 4x - 9$

12. $(-5, 1), y = -3x + 2$

Write an equation in slope-intercept form for the line that passes through the given point and is perpendicular to the graph of the equation.

13. $(1, 4), y = -2x + 5$

14. $(-3, 6), y = \frac{1}{4}x + 2$

15. **MULTIPLE CHOICE** The graph shows the relationship between outside temperature and daily ice cream cone sales. What type of correlation is shown?

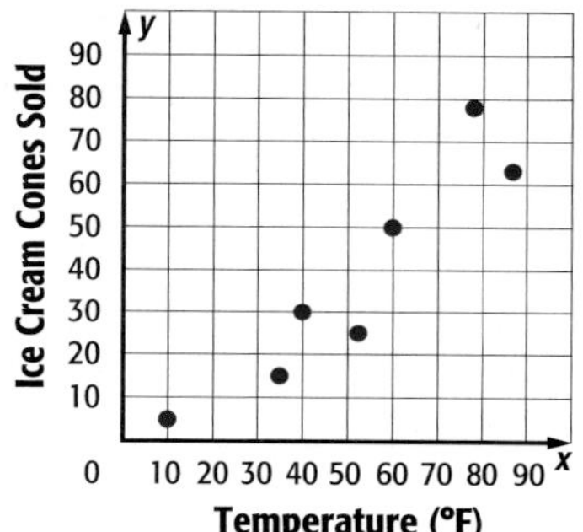

   F positive correlation

   G negative correlation

   H no correlation

   J not enough information

Graph each function.

16. $f(x) = |x - 1|$

17. $f(x) = -|2x|$

18. $f(x) = [\![x]\!]$

19. $f(x) = \begin{cases} 2x - 1 \text{ if } x < 2 \\ x - 3 \text{ if } x \geq 2 \end{cases}$

20. The table shows the number of children from Russia adopted by U.S. citizens.

Years Since 2000	0	1	2	3	4
Number of Children	4269	4279	4939	5209	6936

   a. Write the slope-intercept form of the equation for the line of fit.

   b. Predict the number of children from Russia who will be adopted in 2025.

CHAPTER 4

Preparing for Standardized Tests

Short Answer Questions

Short answer questions require you to provide a solution to the problem, along with a method, explanation, and/or justification used to arrive at the solution.

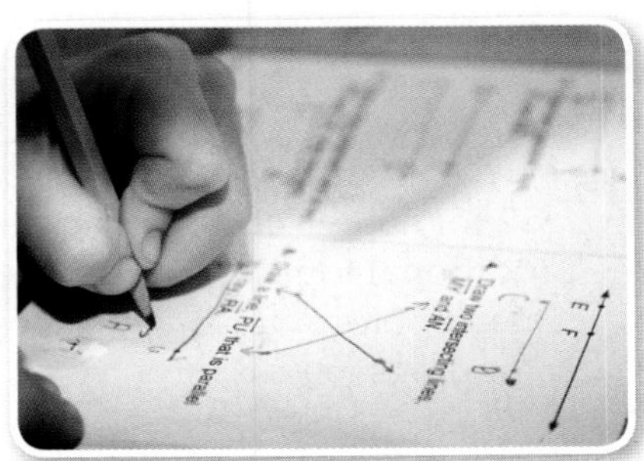

Strategies for Solving Short Answer Questions

Step 1

Short answer questions are typically graded using a **rubric**, or a scoring guide. The following is an example of a short answer question scoring rubric.

Scoring Rubric	
Criteria	**Score**
Full Credit: The answer is correct and a full explanation is provided that shows each step.	2
Partial Credit: • The answer is correct, but the explanation is incomplete. • The answer is incorrect, but the explanation is correct.	1
No Credit: Either an answer is not provided or the answer does not make sense.	0

Step 2

In solving short answer questions, remember to...

- explain your reasoning or state your approach to solving the problem.
- show all of your work or steps.
- check your answer if time permits.

EXAMPLE

Read the problem. Identify what you need to know. Then use the information in the problem to solve. Show your work.

The table shows production costs for building different numbers of skateboards. Determine the missing value, x, that will result in a linear model.

Skateboards Built	Production Costs
14	\$325
28	\$500
x	\$375
22	\$425

Read the problem carefully. You are given several data points and asked to find the missing value that results in a linear model.

Example of a 2-point response:

Set up a coordinate grid and plot the three given points: (14, 325), (28, 500), (22, 425).

Then draw a straight line through them and find the x-value that produces a y-value of 375.

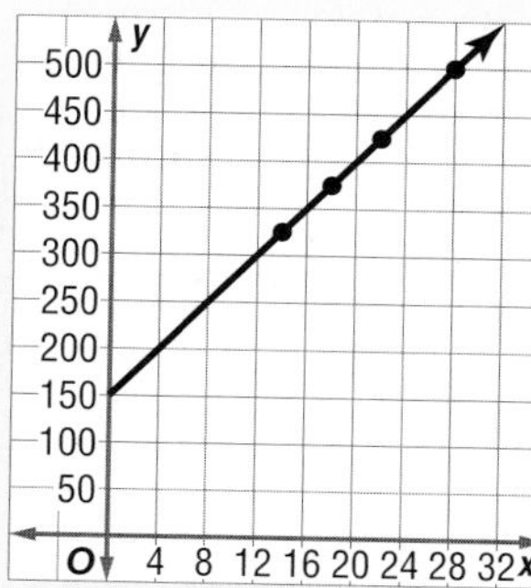

So, building 18 skateboards would result in production costs of $375. These data form a linear model.

The steps, calculations, and reasoning are clearly stated. The student also arrives at the correct answer. So, this response is worth the full 2 points.

Exercises

Read each problem. Identify what you need to know. Then use the information in the problem to solve. Show your work.

1. Given points $M(-1, 7)$, $N(3, -5)$, $O(6, 1)$, and $P(-3, -2)$, determine two segments that are perpendicular to each other.

2. Write the equation of a line that is parallel to $4x + 2y = 8$ and has a y-intercept of 5.

3. Three vertices of a quadrilateral are shown on the coordinate grid. Determine a fourth vertex that would result in a trapezoid.

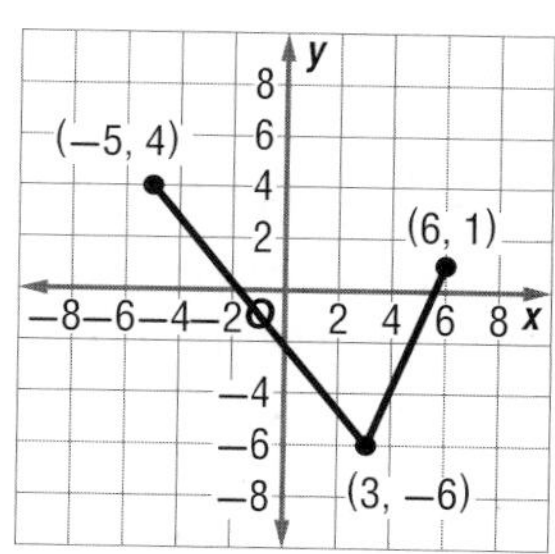

CHAPTER 4

Standardized Test Practice

Cumulative, Chapters 1 through 4

Multiple Choice

Read each question. Then fill in the correct answer on the answer document provided by your teacher or on a sheet of paper.

1. A shipping company charges \$8.50 to ship packages that weigh up to 1 pound and \$7.25 for each additional pound. Which of the following piecewise-defined functions represents the cost C of shipping a package that weighs p pounds?

A $C(p) = \begin{cases} 8.50 & p \leq 1 \\ 8.50 + 7.25p & p > 1 \end{cases}$

B $C(p) = \begin{cases} 7.25 & p \leq 1 \\ 7.25 + 8.50p & p > 1 \end{cases}$

C $C(p) = \begin{cases} 8.50 & p \leq 1 \\ 8.50p + 7.25(p - 1) & p > 1 \end{cases}$

D $C(p) = \begin{cases} 8.50 & p \leq 1 \\ 8.50 + 7.25(p - 1) & p > 1 \end{cases}$

2. Refer to the information given in Exercise 2. How much would it cost a customer to ship a package that weighs 5 pounds 11 ounces? (Assume that partial pounds are rounded up to the nearest whole pound.)

F \$42.35 H \$48.20

G \$44.75 J \$52.00

3. Jaime bought a car in 2005 for \$28,500. By 2008, the car was worth \$23,700. Based on a linear model, what will the value of the car be in 2012?

A \$17,300 C. \$18,100

B \$17,550 D. \$18,475

4. If the graph of a line has a positive slope and a negative y-intercept, what happens to the x-intercept if the slope and the y-intercept are doubled?

F The x-intercept becomes four times larger.

G The x-intercept becomes twice as large.

H The x-intercept becomes one-fourth as large.

J The x-intercept remains the same.

5. Which absolute value equation has the graph below as its solution?

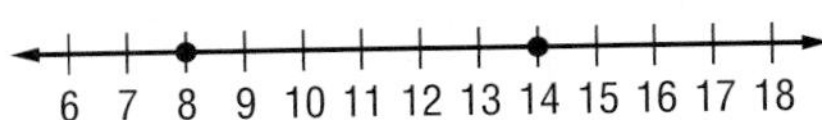

A $|x - 3| = 11$

B $|x - 4| = 12$

C $|x - 11| = 3$

D $|x - 12| = 4$

6. The table below shows the relationship between certain temperatures in degrees Fahrenheit and degrees Celsius. Which of the following linear equations correctly models this relationship?

F $F = \frac{8}{5}C + 35$

G $F = \frac{4}{5}C + 42$

H $F = \frac{9}{5}C + 32$

J $F = \frac{12}{5}C + 26$

Celsius (C)	Fahrenheit (F)
10°	50°
15°	59°
20°	68°
25°	77°
30°	86°

Test-TakingTip

Question 3 Find the average annual depreciation between 2005 and 2008. Then extend the pattern to find the car's value in 2012.

Short Response/Gridded Response

Record your answers on the answer sheet provided by your teacher or on a sheet of paper.

7. What is the equation of the line graphed below?

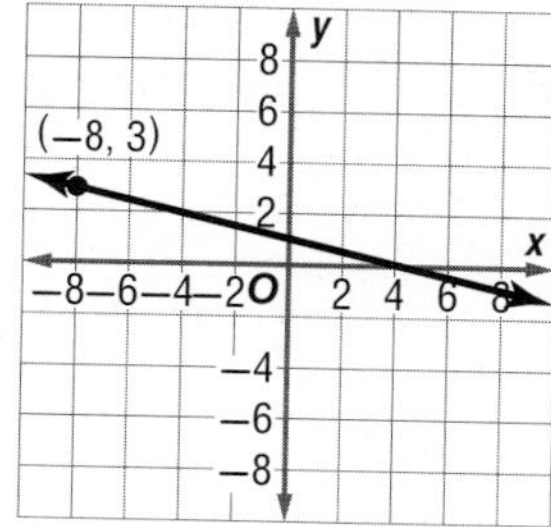

Express your answer in point slope form using the point (−8, 3).

8. GRIDDED RESPONSE The linear equation below is a best fit model for the peak depth of the Mad River when x inches of rain fall. What would you expect the peak depth of the river to be after a storm that produces $1\frac{3}{4}$ inches of rain? Round your answer to the nearest tenth of a foot if necessary.

$$y = 2.5x + 14.8$$

9. Jacob formed an advertising company in 1992. Initially, the company only had 14 employees. In 2008, the company had grown to a total of 63 employees. Find the percent of change in the number of employees working at Jacob's company. Round to the nearest tenth of a percent if necessary.

10. The table shows the total amount of rain during a storm.

Hour	1	2	3	4
Inches	0.45	0.9	1.35	1.8

a. Write an equation to fit the data in the table.

b. Describe the relationship between the hour and the amount of rain received.

11. An electrician charges a $25 consultation fee plus $35 per hour for labor.

a. Copy and complete the following table showing the charges for jobs that take 1, 2, 3, 4, or 5 hours.

Hours, h	Total Cost, C
1	
2	
3	
4	
5	

b. Write an equation in slope-intercept form for the total cost of a job that takes h hours.

c. If the electrician bills in quarter hours, how much would it cost for a job that takes 3 hours 15 minutes to complete?

Extended Response

Record your answer on a sheet of paper. Show your work.

12. Explain how you can determine whether two lines are parallel or perpendicular.

Need Extra Help?												
If you missed Question...	1	2	3	4	5	6	7	8	9	10	11	12
Go to Lesson or Page...	4-7	4-7	4-5	3-1	2-5	4-2	4-3	4-5	2-7	3-6	4-2	4-4

CHAPTER 5 Linear Inequalities

Then

In Chapter 2, you solved equations.

Now

In Chapter 5, you will:

- Solve one-step and multi-step inequalities.
- Solve compound inequalities and inequalities involving absolute value.
- Graph inequalities in two variables.

Why?

PETS In the United States, about 75 million dogs are kept as pets. Approximately 16% of these were adopted from animal shelters. About 14% of dog owners own more than 3 dogs.

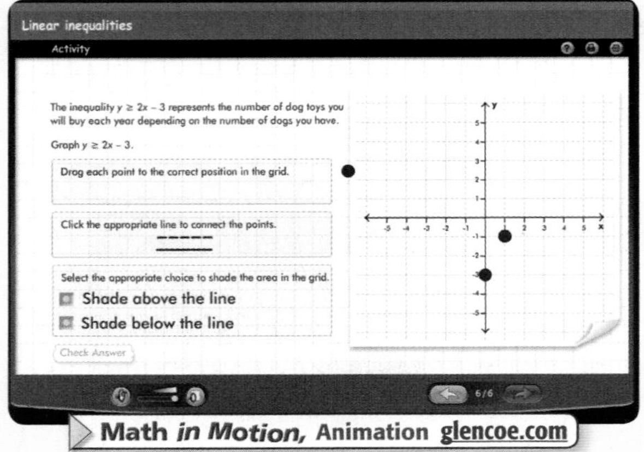

Math *in Motion,* Animation glencoe.com

Get Ready for Chapter 5

Diagnose Readiness You have two options for checking Prerequisite Skills.

Text Option

Take the Quick Check below. Refer to the Quick Review for help.

QuickCheck

Evaluate each expression for the given values. (Lesson 1-2)

1. $3x + y$ if $x = -4$ and $y = 2$
2. $-2m + 3k$ if $m = -8$ and $k = 3$
3. **CARS** The expression $\frac{m \text{ mi}}{g \text{ gal}}$ represents the gas mileage of a car. Find the gas mileage of a car that goes 295 miles on 12 gallons of gasoline. Round to the nearest tenth.

Solve each equation. (Lesson 2-2)

4. $x - 4 = 9$
5. $x + 8 = -3$
6. $4x = -16$
7. $\frac{x}{3} = 7$
8. $2x + 1 = 9$
9. $4x - 5 = 15$
10. $9x + 2 = 3x - 10$
11. $3(x - 2) = -2(x + 13)$
12. **FINANCIAL LITERACY** Claudia opened a savings account with \$325. She saves \$100 per month. Write an equation to determine how much money d, she has after m months. (Lesson 2-1)

Solve each equation. (Lesson 2-5)

13. $|x + 11| = 18$
14. $|3x - 2| = 16$
15. **SURVEYS** In a survey, 32% of the people chose pizza as their favorite food. The results were reported to within 2% accuracy. What is the maximum and minimum percent of people who chose pizza? (Lesson 2-5)

QuickReview

EXAMPLE 1

Evaluate $-3x^2 + 4x - 6$ if $x = -2$.

$-3x^2 + 4x - 6$	**Original expression**
$= -3(-2)^2 + 4(-2) - 6$	**Replace x with -2.**
$= -3(4) + 4(-2) - 6$	**Evaluate the power.**
$= -12 + (-8) - 6$	**Multiply.**
$= -26$	**Add and subtract.**

EXAMPLE 2

Solve $-2(x - 4) = 7x - 19$.

$-2(x - 4) = 7x - 19$	**Original equation**
$-2x + 8 = 7x - 19$	**Distributive Property**
$-2x + 8 + 2x = 7x - 19 + 2x$	**Add $2x$.**
$8 = 9x - 19$	**Simplify.**
$8 + 19 = 9x - 19 + 19$	**Add 19.**
$27 = 9x$	**Simplify.**
$3 = x$	**Divide by 3.**

EXAMPLE 3

Solve $|x - 4| = 9$.

If $|x - 4| = 9$, then $x - 4 = 9$ or $x - 4 = -9$.

$x - 4 = 9$	or	$x - 4 = -9$
$x - 4 + 4 = 9 + 4$		$x - 4 + 4 = -9 + 4$
$x = 13$		$x = -5$

So, the solution set is $\{-5, 13\}$.

Online Option

Math Online Take a self-check Chapter Readiness Quiz at glencoe.com.

Get Started on Chapter 5

You will learn several new concepts, skills, and vocabulary terms as you study Chapter 5. To get ready, identify important terms and organize your resources. You may wish to refer to **Chapter 0** to review prerequisite skills.

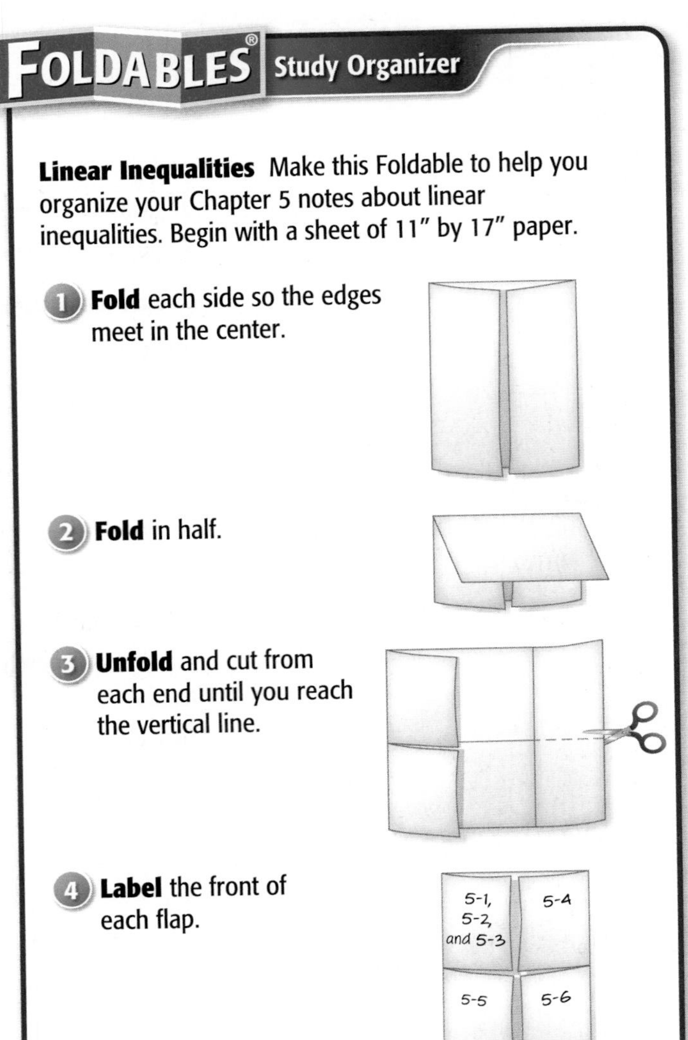

New Vocabulary

English		Español
set-builder notation	• p. 284 •	notación de construcción de conjuntos
compound inequality	• p. 304 •	desigualdad compuesta
intersection	• p. 304 •	intersección
union	• p. 305 •	unión
boundary	• p. 315 •	frontera
half-plane	• p. 315 •	semiplano
closed half-plane	• p. 315 •	semiplano cerrada
open half-plane	• p. 315 •	semiplano abierto

Review Vocabulary

equivalent equations • p. 83 • ecuaciones equivalentes equations that have the same solution

linear equation • p. 153 • ecuación lineal an equation in the form $Ax + By = C$, with a graph consisting of points on a straight line

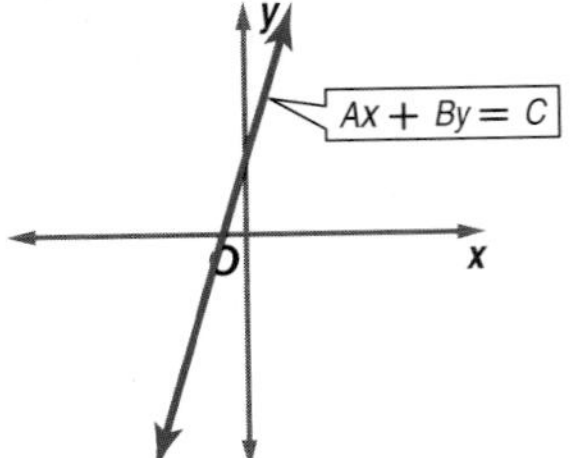

solution set • p. 31 • conjunto solución the set of elements from the replacement set that makes an open sentence true

Multilingual eGlossary glencoe.com

Math Online glencoe.com

- Study the chapter online
- Explore **Math in Motion**
- Get extra help from your own **Personal Tutor**
- Use **Extra Examples** for additional help
- Take a **Self-Check Quiz**
- **Review Vocabulary** in fun ways

5-1

Solving Inequalities by Addition and Subtraction

Then

You solved equations by using addition and subtraction. (Lesson 2-2)

Now

- Solve linear inequalities by using addition.
- Solve linear inequalities by using subtraction.

New Vocabulary

set-builder notation

Math Online

glencoe.com

- Extra Examples
- Personal Tutor
- Self-Check Quiz
- Homework Help

Why?

The data in the table show that the recommended daily allowance of Calories for girls 11–14 years old is less than that of girls between 15–18 years old.

Calories	
Girls 11–14 Years	Girls 15–18
1845	2110

Source: *Vital Health Zone*

$$1845 < 2110$$

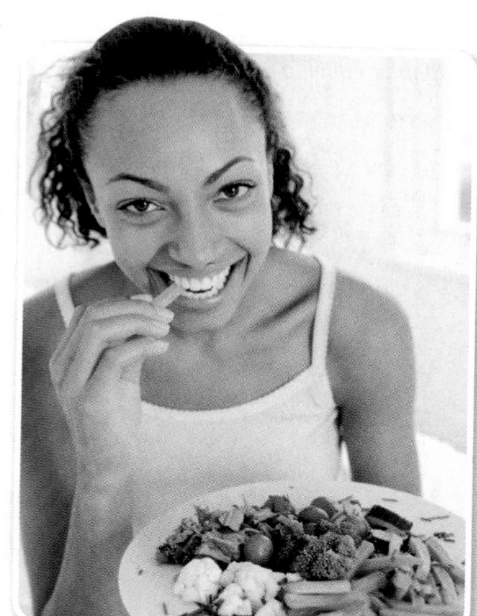

If a 13-year-old girl and a 16-year-old girl each eat 150 more Calories in a day than is suggested, the 16-year-old will still eat more Calories.

$$1845 + 150 \; \underline{?} \; 2110 + 150$$
$$1995 < 2260$$

Solve Inequalities by Addition This example illustrates the Addition Property of Inequalities.

Key Concept — **Addition Property of Inequalities** *For Your FOLDABLE*

Words If the same number is added to each side of a true inequality, the resulting inequality is also true.

Symbols For all numbers a, b, and c, the following are true.

1. If $a > b$, then $a + c > b + c$.
2. If $a < b$, then $a + c < b + c$.

This property is also true for $\geq$ and $\leq$.

EXAMPLE 1 Solve by Adding

Solve $x - 12 \geq 8$. Check your solution.

$x - 12 \geq 8$	**Original inequality**
$x - 12 + 12 \geq 8 + 12$	**Add 12 to each side.**
$x \geq 20$	**Simplify.**

The solution is the set {all numbers greater than or equal to 20}.

CHECK To check, substitute three different values into the original inequality: 20, a number less than 20, and a number greater than 20.

Check Your Progress

Solve each inequality. Check your solution.

1A. $22 > m - 8$

1B. $d - 14 \geq -19$

Personal Tutor glencoe.com

ReadingMath

set-builder notation $\{x \mid x \geq 20\}$ is read *the set of all numbers x such that x is greater than or equal to 20.*

A more concise way of writing a solution set is to use **set-builder notation**. In set-builder notation, the solution set in Example 1 is $\{x \mid x \geq 20\}$.

This solution set can be graphed on a number line. Be sure to check if the endpoint of the graph of an inequality should be a circle or a dot. If the endpoint is not included in the graph, use a circle, otherwise use a dot.

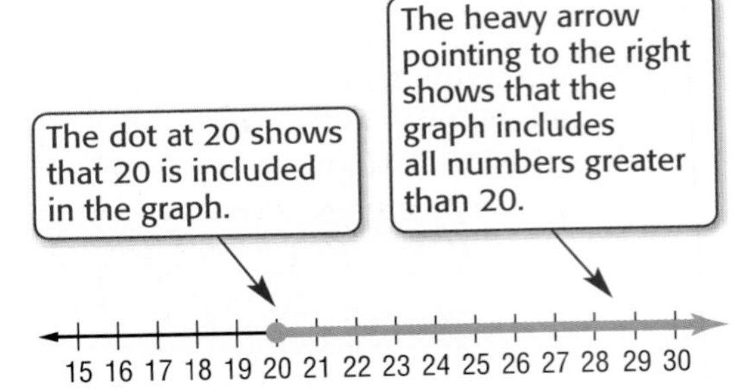

Solve Inequalities by Subtraction Subtraction can also be used to solve inequalities.

Key Concept **Subtraction Property of Inequalities** For Your FOLDABLE

Words If the same number is subtracted from each side of a true inequality, the resulting inequality is also true.

Symbols For all numbers a, b, and c, the following are true.

1. If $a > b$, then $a - c > b - c$.

2. If $a < b$, then $a - c < b - c$.

This property is also true for $\geq$ and $\leq$.

Test-TakingTip

Isolating the Variable When solving inequalities, the goal is to isolate the variable on one side of the inequality. This is the same as with solving equations.

STANDARDIZED TEST EXAMPLE 2

Solve $m + 19 > 56$.

A $\{m \mid m < 41\}$ **B** $\{m \mid m < 37\}$ **C** $\{m \mid m > 37\}$ **D** $\{m \mid m > 41\}$

Read the Test Item

You need to find the solution set for the inequality.

Solve the Test Item

Step 1 Solve the inequality.

$m + 19 > 56$ **Original inequality**

$m + 19 - 19 > 56 - 19$ **Subtract 19 from each side.**

$m > 37$ **Simplify.**

Step 2 Write in set-builder notation: $\{m \mid m > 37\}$. The answer is C.

Check Your Progress

2. Solve $p + 8 \leq 18$.

F $\{p \mid p \geq 10\}$ **G** $\{p \mid p \leq 10\}$ **H** $\{p \mid p \leq 26\}$ **J** $\{p \mid p \geq 126\}$

Personal Tutor glencoe.com

StudyTip

Writing Inequalities Simplifying the inequality so that the variable is on the left side, as in $a \geq 6$, prepares you to write the solution set in set-builder notation.

Terms that are constants are not the only terms that can be subtracted. Terms with variables can also be subtracted from each side to solve inequalities.

EXAMPLE 3 Variables on Each Side

Solve $3a + 6 \leq 4a$. Then graph the solution set on a number line.

$3a + 6 \leq 4a$	**Original inequality**
$3a - 3a + 6 \leq 4a - 3a$	**Subtract $3a$ from each side.**
$6 \leq a$	**Simplify.**

Since $6 \leq a$ is the same as $a \geq 6$, the solution set is $\{a \mid a \geq 6\}$.

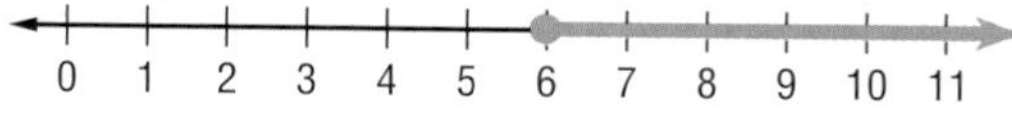

Check Your Progress

Solve each inequality. Then graph the solution set on a number line.

3A. $9n - 1 < 10n$

3B. $5h \leq 12 + 4h$

Personal Tutor glencoe.com

Verbal problems containing phrases like *greater than* or *less than* can be solved by using inequalities. The chart shows some other phrases that indicate inequalities.

Concept Summary Phrases for Inequalities

For Your FOLDABLE

<	>	≤	≥
less than fewer than	greater than more than	at most, no more than, less than or equal to	at least, no less than, greater than or equal to

Real-World EXAMPLE 4 Use an Inequality to Solve a Problem

PETS **Felipe needs for the temperature of his leopard gecko's basking spot to be at least 82°F. Currently the basking spot is 62.5°F. How much warmer does the basking spot need to be?**

Words	The current temperature needs to be at least 82°F.		
Variable	Let t = the number of degrees that the temperature needs to rise.		
Inequality	$62.5 + t$	$\geq$	82

$62.5 + t \geq 82$	**Original inequality**
$62.5 + t - 62.5 \geq 82 - 62.5$	**Subtract 62.5 from each side.**
$t \geq 19.5$	**Simplify.**

Felipe needs to raise the temperature of the basking spot 19.5°F or more.

Real-World Link

Leopard geckos are commonly yellow and white with black spots. They are nocturnal and easy to tame. They do not have toe pads like other geckos, so they do not climb.

Source: Exotic Pets

Check Your Progress

4. **SHOPPING** Sanjay has \$65 to spend at the mall. He bought a T-shirt for \$18 and a belt for \$14. If Sanjay wants a pair of jeans, how much can he spend?

Personal Tutor glencoe.com

Check Your Understanding

Examples 1 and 2 pp. 283–284

Solve each inequality. Then graph the solution set on a number line.

1. $x - 3 > 7$
2. $5 \geq 7 + y$
3. $g + 6 < 2$
4. $11 \leq p + 4$
5. $10 > n - 1$
6. $k + 24 > -5$
7. $8r + 6 < 9r$
8. $8n \geq 7n - 3$

Example 3 p. 285

Define a variable, write an inequality, and solve each problem. Check your solution.

9. A number increased by 4 is at least 10.
10. Three more than a number is less than twice the number.

Example 4 p. 285

11. **AMUSEMENT** A thrill ride swings passengers back and forth, a little higher each time up to 137 feet. Suppose the height of the swing after 30 seconds is 45 feet. How much higher will the ride swing?

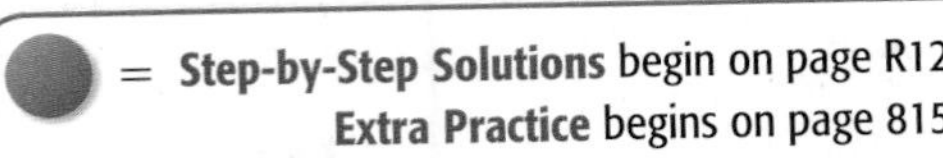

= **Step-by-Step Solutions** begin on page R12. **Extra Practice** begins on page 815.

Practice and Problem Solving

Examples 1 and 2 pp. 283–284

Solve each inequality. Then graph the solution set on a number line.

12. $m - 4 < 3$
13. $p - 6 \geq 3$
14. $r - 8 \leq 7$
15. $t - 3 > -8$
16. $b + 2 \geq 4$
17. $13 > 18 + r$
18. $5 + c \leq 1$
19. $-23 \geq q - 30$
20. $11 + m \geq 15$
21. $h - 26 < 4$
22. $8 \leq r - 14$
23. $-7 > 20 + c$
24. $2a \leq -4 + a$
25. $z + 4 \geq 2z$
26. $w - 5 \leq 2w$
27. $3y + 6 \leq 2y$
28. $6x + 5 \geq 7x$
29. $-9 + 2a < 3a$

Example 3 p. 285

Define a variable, write an inequality, and solve each problem. Check your solution.

30. The sum of a number and -4 is at least 8.
31. A number decreased by 8 is less than 21.
32. Twice a number is more than the sum of that number and 9.
33. The sum of twice a number and 5 is at most 3 less than the number.

Example 4 p. 285

Define a variable, write an inequality, and solve each problem. Then interpret your solution.

34. **FINANCIAL LITERACY** Keisha is babysitting at $8 per hour to earn money for a car. So far she has saved $1300. The car that Keisha wants to buy costs at least $5440. How much money does Keisha still need to earn to buy the car?
35. **TECHNOLOGY** A recent survey found that more than 21 million people between the ages of 12 and 17 use the Internet. Of those, about 16 million said they use the Internet at school. How many teens that are online do not use the Internet at school?
36. **MUSIC** A DJ added 20 more songs to his MP3 player, making the total more than 61. How many songs were originally on the player?

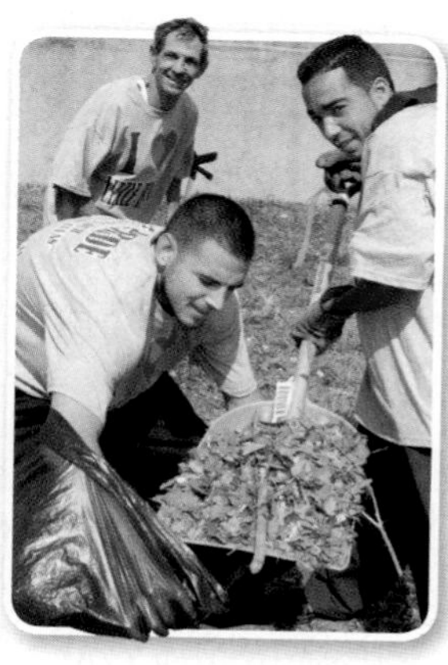

Real-World Link

In a recent year, 55% of American teenagers said they volunteered within the last year, nearly double the percentage of adults.

Source: Corporation for National and Community Service

37. **TEMPERATURE** The water temperature in a swimming pool increased 4°F this morning. The temperature is now less than 81°F. What was the water temperature this morning?

38. **BASKETBALL** A player's goal was to score at least 150 points this season. So far, she has scored 123 points. If there is one game left, how many points must she score to reach her goal?

39. **SPAS** Samantha received a $75 gift card for a local day spa for her birthday. She plans to get a haircut and a manicure. How much money will be left on her gift card after her visit?

Service	Cost ($)
haircut	at least 32
manicure	at least 26

40. **VOLUNTEER** Kono knows that he can only volunteer up to 25 hours per week. If he has volunteered for the times recorded at the right, how much more time can Kono volunteer this week?

Center	Time (h)
Shelter	3 h 15 min
Kitchen	2 h 20 min

Solve each inequality. Check your solution, and then graph it on a number line.

41. $c + (-1.4) \geq 2.3$

42. $9.1g + 4.5 < 10.1g$

43. $k + \frac{3}{4} > \frac{1}{3}$

44. $\frac{3}{2}p - \frac{2}{3} \leq \frac{4}{9} + \frac{1}{2}p$

45. **MULTIPLE REPRESENTATIONS** In this problem, you will explore multiplication and division in inequalities.

a. **GEOMETRIC** Suppose a balance has 12 pounds on the left side and 18 pounds on the right side. Draw a picture to represent this situation.

b. **NUMERICAL** Write an inequality to represent the situation.

c. **TABULAR** Create a table showing the result of doubling, tripling, or quadrupling the weight on each side of the balance. Create a second table showing the result of reducing the weight on each side of the balance by a factor of $\frac{1}{2}$, $\frac{1}{3}$, or $\frac{1}{4}$. Include a column in each table for the inequality representing each situation.

d. **VERBAL** Describe the effect multiplying or dividing each side of an inequality by the same positive value has on the inequality.

If $m + 7 \geq 24$, then complete each inequality.

46. $m \geq \underline{?}$

47. $m + \underline{?} \geq 27$

48. $m - 5 \geq \underline{?}$

49. $m - \underline{?} \geq 14$

50. $m - 19 \geq \underline{?}$

51. $m + \underline{?} \geq 43$

H.O.T. Problems

Use Higher-Order Thinking Skills

52. **REASONING** Compare and contrast the graphs of $a < 4$ and $a \leq 4$.

53. **CHALLENGE** Suppose $b > d + \frac{1}{3}$, $c + 1 < a - 4$, and $d + \frac{5}{8} > a + 2$. Order a, b, c, and d from least to greatest.

54. **OPEN ENDED** Write three linear inequalities that are equivalent to $y < -3$.

55. **WRITING IN MATH** Summarize the process of solving and graphing linear inequalities.

56. **WRITING IN MATH** Explain why $x - 2 > 5$ has the same solution set as $x > 7$.

Standardized Test Practice

57. Which equation represents the relationship shown?

x	y
1	1
2	9
3	17
4	25
5	33
6	41

A $y = 7x - 8$
B $y = 7x + 8$
C $y = 8x - 7$
D $y = 8x + 7$

58. What is the solution set of the inequality $7 + x < 5$?

F $\{x \mid x < 2\}$
G $\{x \mid x > 2\}$
H $\{x \mid x < -2\}$
J $\{x \mid x > -2\}$

59. Francisco has \$3 more than $\frac{1}{4}$ the number of dollars that Kayla has. Which expression represents how much money Francisco has?

A $3\left(\frac{1}{4}k\right)$
B $\frac{1}{4}k + 3$
C $3 - \frac{1}{4}k$
D $\frac{1}{4} + 3k$

60. GRIDDED RESPONSE The mean score for 10 students on the chemistry final exam was 178. However, the teacher had made a mistake and recorded one student's score as ten points less than the actual score. What should the mean score be?

Spiral Review

Graph each function. (Lesson 4-7)

61. $f(x) = |3x + 2|$

62. $f(x) = \begin{cases} x - 2 \text{ if } x > -1 \\ x + 3 \text{ if } x \leq -1 \end{cases}$

63. $f(x) = [\![x + 1]\!]$

64. $f(x) = \left|\frac{1}{4}x - 1\right|$

Write the slope-intercept form of an equation for the line that passes through the given point and is perpendicular to the graph of each equation. (Lesson 4-4)

65. $(-2, 0), y = x - 6$

66. $(-3, 1), y = -3x + 7$

67. $(1, -3), y = \frac{1}{2}x + 4$

68. $(-2, 7), 2x - 5y = 3$

69. TRAVEL On an island cruise in Hawaii, each passenger is given a lei. A crew member hands out 3 red, 3 blue, and 3 green leis in that order. If this pattern is repeated, what color lei will the 50th person receive? (Lesson 3-6)

Find the *n*th term of each arithmetic sequence described. (Lesson 3-5)

70. $a_1 = 52, d = 12, n = 102$

71. $-9, -7, -5, -3, \ldots$ for $n = 18$

72. $0.5, 1, 1.5, 2, \ldots$ for $n = 50$

73. JOBS Refer to the time card shown. Write a direct variation equation relating your pay to the hours worked and find your pay if you work 30 hours. (Lesson 3-4)

Weekly Time Card

Day	Hours
FRIDAY	2.0
SATURDAY	3.5
SUNDAY	2.0
TOTAL HOURS	7.5
PAY	\$52.50

Skills Review

Solve each equation. (Lesson 2-2)

74. $8y = 56$

75. $4p = -120$

76. $-3a = -21$

77. $2c = \frac{1}{5}$

78. $\frac{r}{2} = 21$

79. $-\frac{3}{4}g = -12$

80. $\frac{2}{5}w = -4$

81. $-6x = \frac{2}{3}$

Algebra Lab
Solving Inequalities

Math Online glencoe.com
Math *in Motion*, Animation

You can use algebra tiles to solve inequalities.

ACTIVITY Solve Inequalities

Solve $-2x \le 4$.

Step 1 Use a self-adhesive note to cover the equals sign on the equation mat. Then write a $\le$ symbol on the note. Model the inequality.

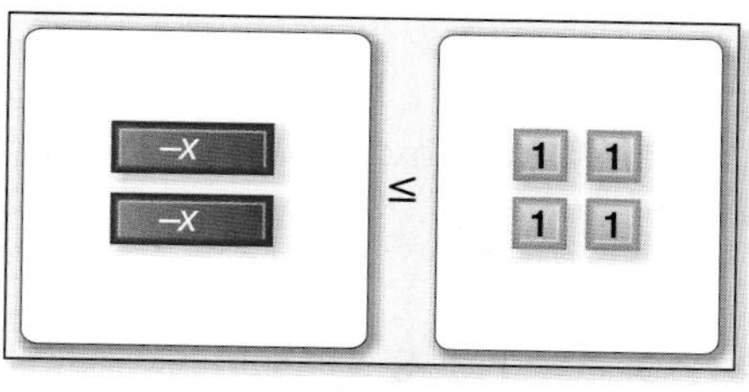

$-2x \le 4$

Step 2 Since you do not want to solve for a negative x-tile, eliminate the negative x-tiles by adding 2 positive x-tiles to each side. Remove the zero pairs.

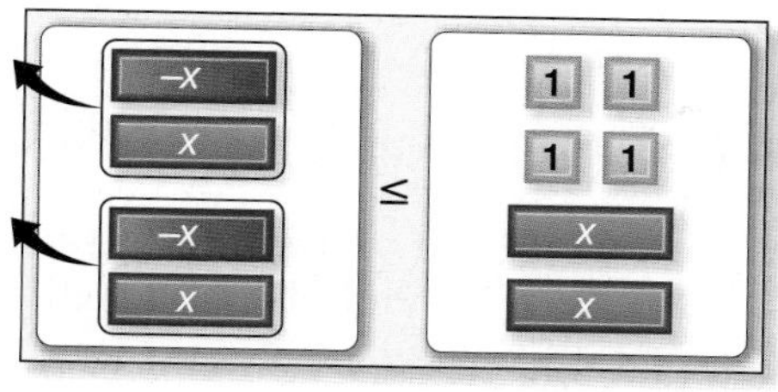

$-2x + 2x \le 4 + 2x$

Step 3 Add 4 negative 1-tiles to each side to isolate the x-tiles. Remove the zero pairs.

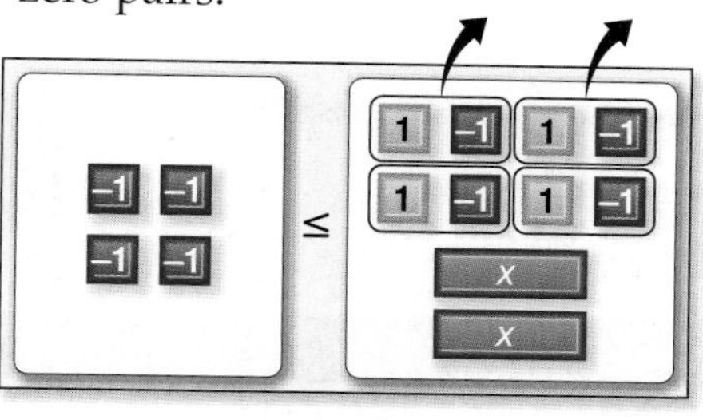

$-4 \le 2x$

Step 4 Separate the tiles into 2 groups.

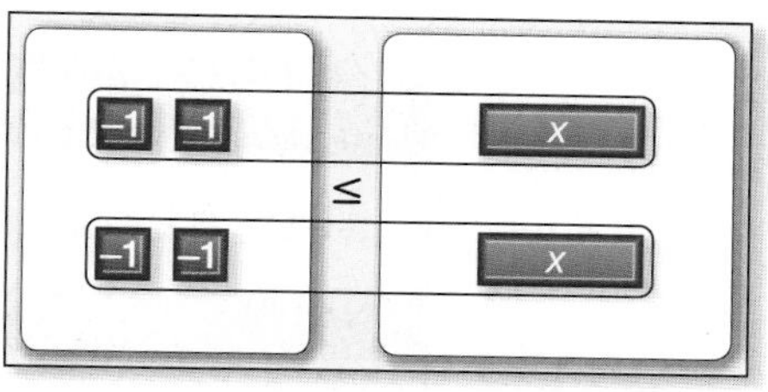

$-2 \le x$ or $x \ge -2$

Model and Analyze

Use algebra tiles to solve each inequality.

1. $-3x < 9$
2. $-4x > -4$
3. $-5x \ge 15$
4. $-6x \le -12$
5. In Exercises 1–4, is the coefficient of x in each inequality positive or negative?
6. Compare the inequality symbols and locations of the variable in Exercises 1–4 with those in their solutions. What do you find?
7. Model the solution for $3x \le 12$. How is this different from solving $-3x \le 12$?
8. Write a rule for solving inequalities involving multiplication and division. (*Hint:* Remember that dividing by a number is the same as multiplying by its reciprocal.)

5-2 Solving Inequalities by Multiplication and Division

Then

You solved equations by using multiplication and division. (Lesson 2-3)

Now

- Solve linear inequalities by using multiplication.
- Solve linear inequalities by using division.

Math Online

glencoe.com

- Extra Examples
- Personal Tutor
- Self-Check Quiz
- Homework Help

Why?

Terrell received a gift card for \$20 of music downloads. If each download costs \$0.89, the number of downloads he can purchase can be represented by the inequality $0.89d \leq 20$.

Solve Inequalities by Multiplication If you multiply each side of an inequality by a positive number, then the inequality remains true.

$4 > 2$	**Original inequality**
$4(3) \overset{?}{\_} 2(3)$	**Multiply each side by 3.**
$12 > 6$	**Simplify.**

Notice that the direction of the inequality remains the same.

If you multiply each side of an inequality by a negative number, the inequality symbol changes direction.

$7 < 9$	**Original inequality**
$7(-2) \overset{?}{\_} 9(-2)$	**Multiply each side by −2.**
$-14 > -18$	**Simplify.**

These examples demonstrate the **Multiplication Property of Inequalities**.

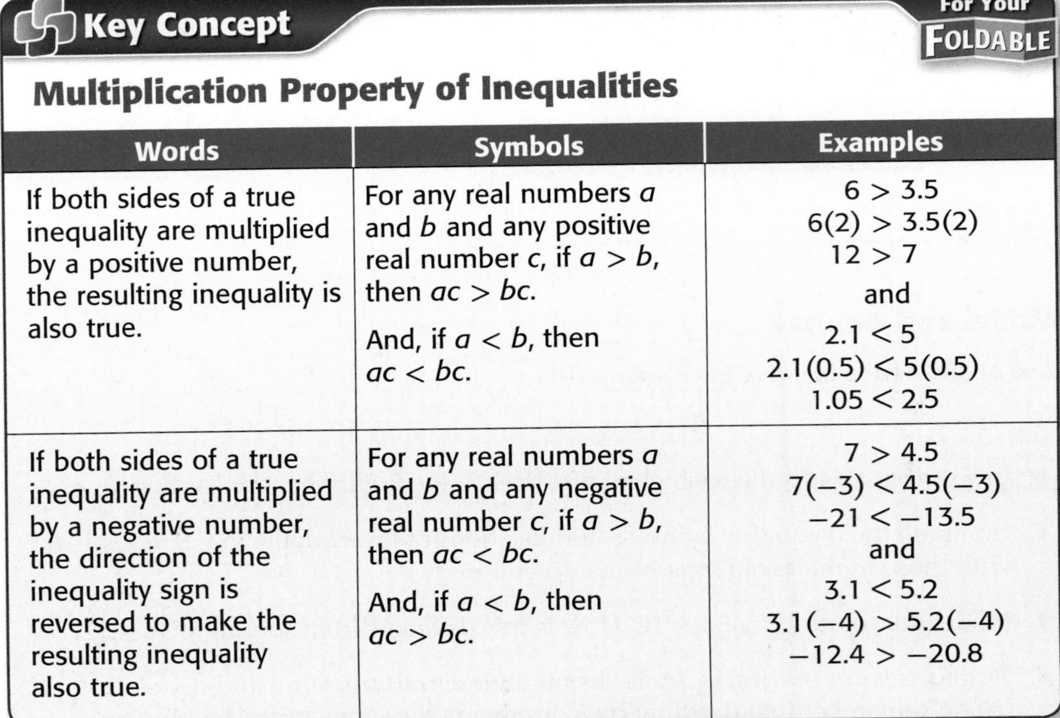

Key Concept For Your FOLDABLE

Multiplication Property of Inequalities

Words	Symbols	Examples
If both sides of a true inequality are multiplied by a positive number, the resulting inequality is also true.	For any real numbers a and b and any positive real number c, if $a > b$, then $ac > bc$. And, if $a < b$, then $ac < bc$.	$6 > 3.5$ $6(2) > 3.5(2)$ $12 > 7$ and $2.1 < 5$ $2.1(0.5) < 5(0.5)$ $1.05 < 2.5$
If both sides of a true inequality are multiplied by a negative number, the direction of the inequality sign is reversed to make the resulting inequality also true.	For any real numbers a and b and any negative real number c, if $a > b$, then $ac < bc$. And, if $a < b$, then $ac > bc$.	$7 > 4.5$ $7(-3) < 4.5(-3)$ $-21 < -13.5$ and $3.1 < 5.2$ $3.1(-4) > 5.2(-4)$ $-12.4 > -20.8$

This property also holds for inequalities involving $\leq$ and $\geq$.

StudyTip

Checking Solutions In Example 1, you could also check the solution by substituting a number greater than 672 and verifying that the resulting inequality is false.

Real-World EXAMPLE 1 Write and Solve an Inequality

SURVEYS Of the students surveyed at Madison High School, fewer than eighty-four said they have never purchased an item online. This is about one eighth of those surveyed. How many students were surveyed?

Understand You know the number of students who have never purchased an item online and the portion this is of the number of students surveyed.

Plan Let n = the number of students surveyed. Write an open sentence that represents this situation.

Words	One eighth	times	the number of students surveyed	is less than	84.
Inequality	$\frac{1}{8}$	$\cdot$	n	$<$	84.

Solve Solve for n.

$\frac{1}{8}n < 84$ **Original inequality**

$(8)\frac{1}{8}n < (8)84$ **Multiply each side by 8.**

$n < 672$ **Simplify.**

Check To check this answer, substitute a number less than 672 into the original inequality. If $n = 80$, then $\frac{1}{8}(80)$ or $10 < 84$, so the solution checks.

The solution set is $\{n \mid n < 672\}$, so fewer than 672 students were surveyed at Madison High School.

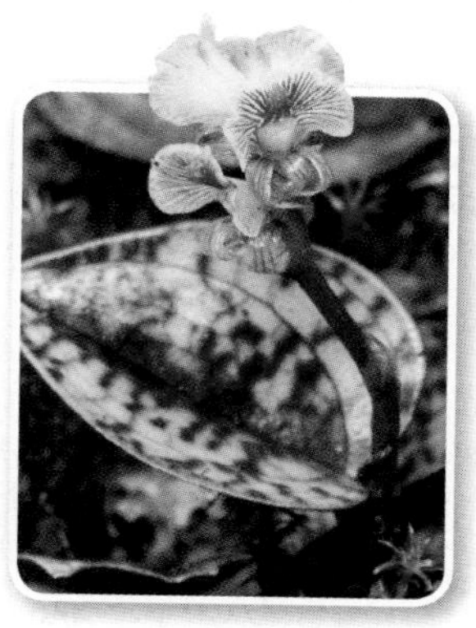

Real-World Link

More than 30,000 different orchid species flower in the wild on every continent except Antarctica.

Source: Aloha Orchid Nursery

Check Your Progress

1. **BIOLOGY** Mount Kinabalue in Malaysia has the greatest concentration of wild orchids on Earth. It contains more than 750 species, or about one fourth of all orchid species in Malaysia. How many orchid species are there in Malaysia?

Personal Tutor glencoe.com

You can also use multiplicative inverses with the Multiplication Property of Inequalities to solve an inequality.

EXAMPLE 2 Solve by Multiplying

Solve $-\frac{3}{7}r < 21$. Check your solution.

$-\frac{3}{7}r < 21$ **Original inequality**

$\left(-\frac{7}{3}\right)\left(-\frac{3}{7}r\right) > 21\left(-\frac{7}{3}\right)$ **Multiply each side by $-\frac{7}{3}$. Reverse the inequality symbol.**

$r > -49$ **Simplify. Check by substituting values.**

The solution set is $\{r \mid r > -49\}$.

Check Your Progress **Solve each inequality. Check your solution.**

2A. $-\frac{n}{6} \leq 8$

2B. $-\frac{4}{3}p > -10$

2C. $\frac{1}{5}m \geq -3$

2D. $\frac{3}{8}t < 5$

Personal Tutor glencoe.com

> **Watch Out!**
>
> **Negatives** A negative sign in an inequality does not necessarily mean that the direction of the inequality should change. For example, when solving $\frac{x}{6} > -3$, do not change the direction of the inequality.

Solve Inequalities by Division If you divide each side of an inequality by a positive number, then the inequality remains true.

$-10 < -5$	**Original inequality**
$\frac{-10}{5} \; ? \; \frac{-5}{5}$	**Divide each side by −5.**
$-2 < -1$	**Simplify.**

Notice that the direction of the inequality remains the same. If you divide each side of an inequality by a negative number, the inequality symbol changes direction.

$15 < 18$	**Original inequality**
$\frac{15}{-3} \; ? \; \frac{18}{-3}$	**Divide each side by −3.**
$-5 > -6$	**Simplify.**

These examples demonstrate the **Division Property of Inequalities.**

Key Concept

For Your FOLDABLE

Division Property of Inequalities

Words	Symbols	Examples
If both sides of a true inequality are divided by a positive number, the resulting inequality is also true.	For any real numbers a and b and any positive real number c, if $a > b$, then $\frac{a}{c} > \frac{b}{c}$. And, if $a < b$, then $\frac{a}{c} < \frac{b}{c}$.	$4.5 > 2.1$, $\frac{4.5}{3} > \frac{2.1}{3}$, $1.5 > 0.7$ and $1.5 < 5$, $\frac{1.5}{0.5} < \frac{5}{0.5}$, $3 < 10$
If both sides of a true inequality are divided by a negative number, the direction of the inequality sign is reversed to make the resulting inequality also true.	For any real numbers a and b and any negative real number c, if $a > b$, then $\frac{a}{c} < \frac{b}{c}$. And, if $a < b$, then $\frac{a}{c} > \frac{b}{c}$.	$6 > 2.4$, $\frac{6}{-6} < \frac{2.4}{-6}$, $-1 < -0.4$ and $-1.8 < 3.6$, $\frac{-1.8}{-9} < \frac{3.6}{-9}$, $0.2 > -0.4$

This property also holds for inequalities involving ≤ and ≥.

EXAMPLE 3 Divide to Solve an Inequality

Solve each inequality. Check your solution.

a. $60t > 8$

$60t > 8$	**Original inequality**
$\frac{60t}{60} > \frac{8}{60}$	**Divide each side by 60.**
$t > \frac{2}{15}$	**Simplify.**

b. $-7d \leq 147$

$-7d \leq 147$	**Original inequality**
$\frac{-7d}{-7} \geq \frac{147}{-7}$	**Divide each side by −7.**
$d \geq -21$	**Simplify.**

3A. $8p < 58$ **3B.** $-42 > 6r$ **3C.** $-12h > 15$ **3D.** $-\frac{1}{2}n < 6$

Personal Tutor glencoe.com

Check Your Understanding

Example 1 p. 291

1. **FUNDRAISING** The Jefferson Band Boosters raised more than \$5500 from sales of their \$15 band DVD. Define a variable, and write an inequality to represent the number of DVDs they sold. Solve the inequality and interpret your solution.

Examples 2 and 3 pp. 291–292

Solve each inequality. Check your solution.

2. $30 > \frac{1}{2}n$
3. $-\frac{3}{4}r \leq -6$
4. $-\frac{c}{6} \geq 7$
5. $\frac{h}{2} < -5$
6. $9t > 108$
7. $-84 < 7v$
8. $-28 \leq -6x$
9. $40 \geq -5z$

Practice and Problem Solving

● = **Step-by-Step Solutions** begin on page R12.
Extra Practice begins on page 815.

Example 1 p. 291

Define a variable, write an inequality, and solve each problem. Then interpret your solution.

10. **CELL PHONE PLAN** Mario purchases a prepaid phone plan for \$50 at \$0.13 per minute. How many minutes can Mario talk on this plan?

11. **FINANCIAL LITERACY** Rodrigo needs at least \$560 to pay for his spring break expenses, and he is saving \$25 from each of his weekly paychecks. How long will it be before he can pay for his trip?

Examples 2 and 3 pp. 291–292

Solve each inequality. Check your solution.

12. $\frac{1}{4}m \leq -17$
13. $\frac{1}{2}a < 20$
14. $-11 > -\frac{c}{11}$
15. $-2 \geq -\frac{d}{34}$
16. $-10 \leq \frac{x}{-2}$
17. $-72 < \frac{f}{-6}$
18. $\frac{2}{3}h > 14$
19. $-\frac{3}{4}j \geq 12$
20. $-\frac{1}{6}n \leq -18$
21. $6p \leq 96$
22. $4r < 64$
23. $32 > -2y$
24. $-26 < 26t$
25. $-6v > -72$
26. $-33 \geq -3z$
27. $4b \leq -3$
28. $-2d < 5$
29. $-7f > 5$

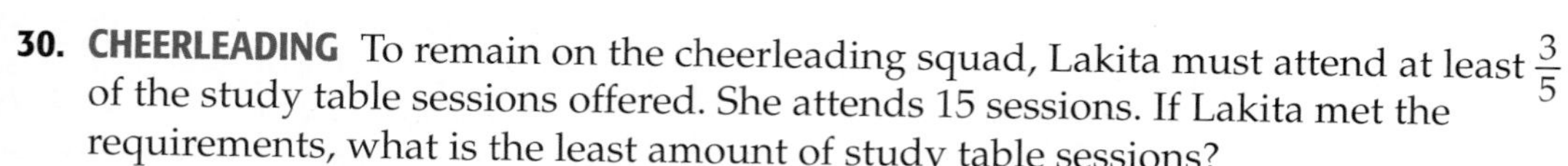

30. **CHEERLEADING** To remain on the cheerleading squad, Lakita must attend at least $\frac{3}{5}$ of the study table sessions offered. She attends 15 sessions. If Lakita met the requirements, what is the least amount of study table sessions?

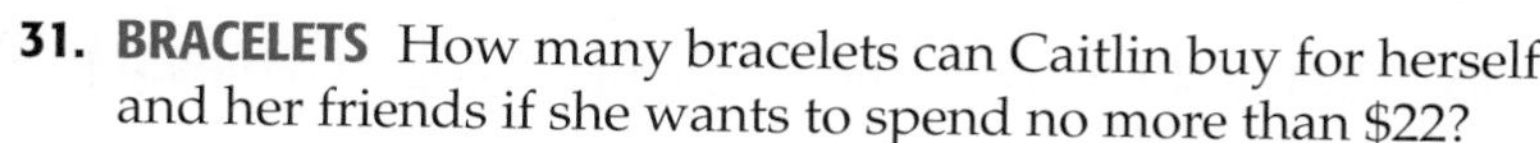

31. **BRACELETS** How many bracelets can Caitlin buy for herself and her friends if she wants to spend no more than \$22?

Math History Link

Thomas Harriot (1560–1621)

Harriot was a prolific astronomer. He was the first to map the moon's surface and to see sunspots. Harriot is best known for his work in algebra.

32. **CHARITY** The National Honor Society at Pleasantville High School wants to raise at least \$500 for a local charity. Each student earns \$0.50 for every quarter of a mile walked in a walk-a-thon. How many miles will the students need to walk?

33. **MUSEUM** The American history classes are planning a trip to a local museum. Admission is \$8 per person. Determine how many people can go for \$260.

34. **GASOLINE** If gasoline costs \$3.15 per gallon, how many gallons of gasoline, to the nearest tenth, can Jan buy for \$24?

Real-World Link

At the FedNor Pavillion Royal Winter Fair in Toronto, Canada, the Northwest Fudge Factory made the world's largest slab of fudge. The fudge weighed 3010 pounds.

Source: *Guinness World Records*

Match each inequality to the graph of its solution.

35. $-\frac{2}{3}h \leq 9$ **36.** $25j \geq 8$ **37.** $3.6p < -4.5$ **38.** $2.3 < -5t$

a. 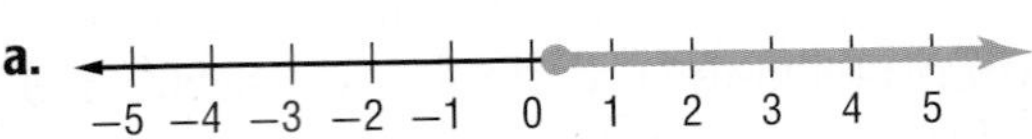

b. 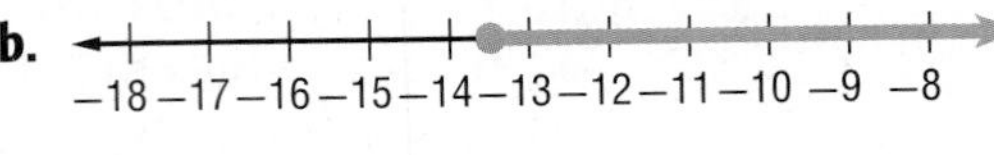

c.

d.

39 **CANDY** Fewer than 42 employees at a factory stated that they preferred fudge over fruit candy. This is about two thirds of the employees. How many employees are there?

40. **TRAVEL** A certain travel agency employs more than 275 people at all of its branches. Approximately three fifths of all the people are employed at the west branch. How many people work at the west branch?

41. **MULTIPLE REPRESENTATIONS** The equation for the volume of a pyramid is $\frac{1}{3}$ the area of the base times the height.

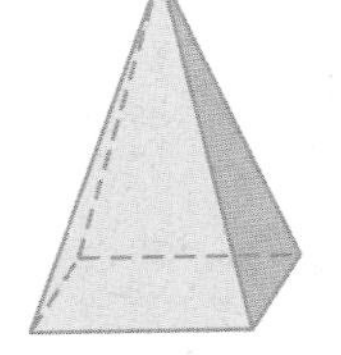

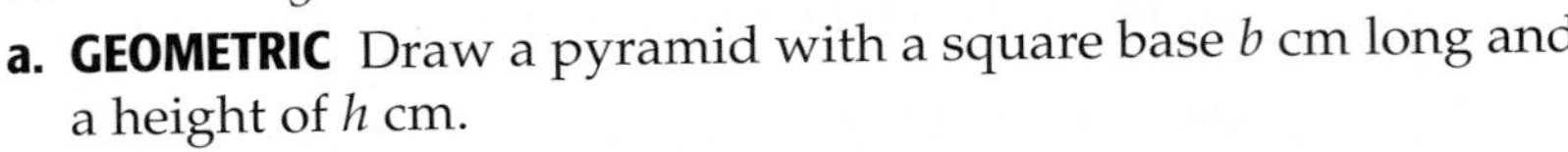

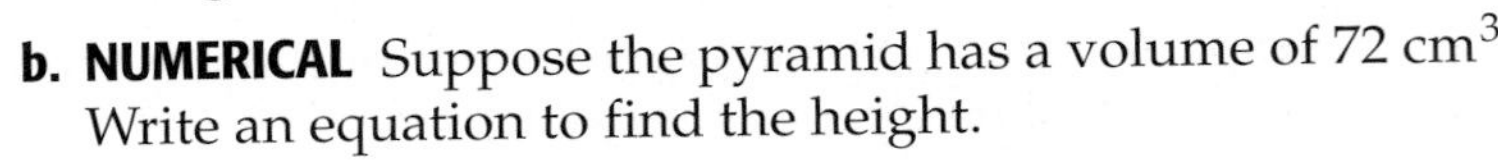

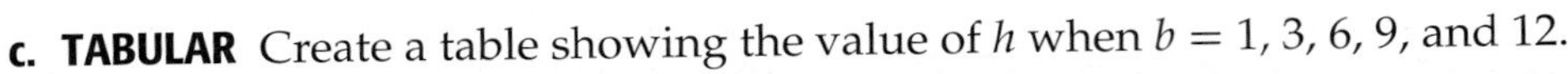

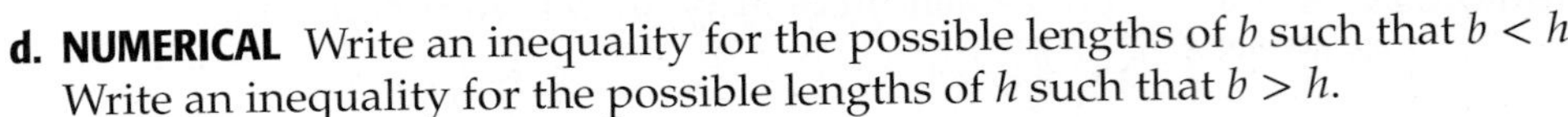

a. **GEOMETRIC** Draw a pyramid with a square base b cm long and a height of h cm.

b. **NUMERICAL** Suppose the pyramid has a volume of 72 cm^3. Write an equation to find the height.

c. **TABULAR** Create a table showing the value of h when $b = 1, 3, 6, 9,$ and 12.

d. **NUMERICAL** Write an inequality for the possible lengths of b such that $b < h$. Write an inequality for the possible lengths of h such that $b > h$.

H.O.T. Problems

Use **H**igher-**O**rder **T**hinking Skills

42. **FIND THE ERROR** Taro and Jamie are solving $6d \geq -84$. Is either of them correct? Explain your reasoning.

Taro	Jamie
$6d \geq -84$	$6d \geq -84$
$\frac{6d}{6} \geq \frac{-84}{6}$	$\frac{6d}{6} \leq \frac{-84}{6}$
$d \geq -14$	$d \leq -14$

43. **CHALLENGE** Solve $-96c < 12d$ for c using two methods. Show your work.

44. **CHALLENGE** Determine whether $x^2 > 1$ and $x > 1$ are equivalent. Explain.

45. **REASONING** Explain whether the statement *If* $a > b$, *then* $\frac{1}{a} > \frac{1}{b}$ is *sometimes, always,* or *never* true.

46. **OPEN ENDED** Create a real-world situation to represent the inequality $-\frac{5}{8} \geq x$.

47. **WRITING IN MATH** Explain the circumstances under which the inequality symbol changes directions. Use examples to support your explanation.

Standardized Test Practice

48. Juan's long-distance phone company charges 9¢ for each minute. Which inequality can be used to find how long he can talk to a friend if he does not want to spend more than \$2.50 on the call?

A $0.09 \geq 2.50m$

B $0.09 \leq 2.50m$

C $0.09m \geq 2.50$

D $0.09m \leq 2.50$

49. **SHORT RESPONSE** Find the value of x.

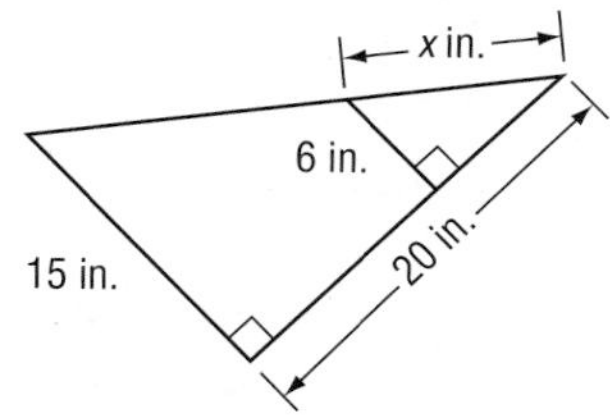

50. What is the greatest rate of decrease of this function?

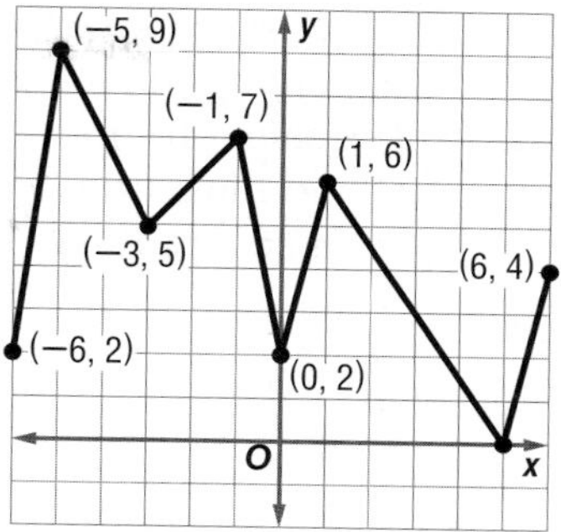

F -5

G -3

H -2

J 1

51. What is the value of x if $4x - 3 = -2x$?

A -2

B $-\frac{1}{2}$

C $\frac{1}{2}$

D 2

Spiral Review

Solve each inequality. Check your solution, and then graph it on a number line. (Lesson 5-1)

52. $-8 + 4a < 6a$

53. $2y + 11 \geq -24y$

54. $7 - 2b > 12b$

Determine the domain and range for each function. (Lesson 4-7)

55. $f(x) = |2x - 5|$

56. $h(x) = |x - 1|$

57. $g(x) = \begin{cases} -3x + 4 \text{ if } x > 2 \\ x - 1 \text{ if } x \leq 2 \end{cases}$

58. **HOME DECOR** Pam is having blinds installed at her home. The cost c of installation for any number of blinds b can be described by $c = 25 + 6.5b$. Graph the equation and determine how much it would cost if Pam has 8 blinds installed. (Lesson 3-1)

59. **RESCUE** A boater radioed for a helicopter to pick up a sick crew member. At that time, the boat and the helicopter were at the positions shown. How long will it take for the helicopter to reach the boat? (Lesson 2-8)

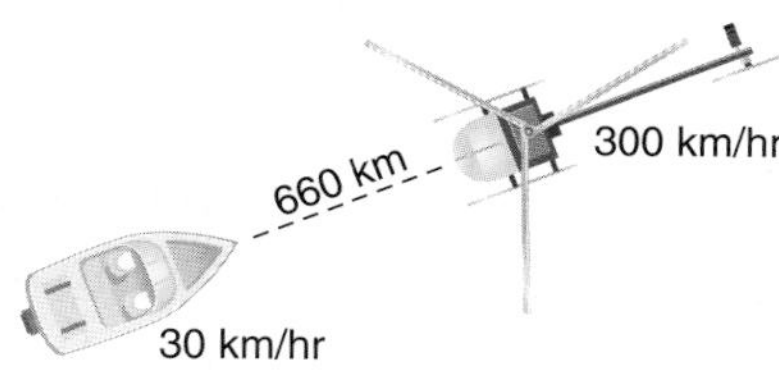

Solve each open sentence. (Lesson 2-5)

60. $|x + 3| = 10$

61. $|2x - 8| = 6$

62. $|3x + 1| = -2$

Skills Review

Solve each equation. (Lessons 2-3 and 2-4)

63. $4y + 11 = 19$

64. $2x - 7 = 9 + 4x$

65. $\frac{1}{4} + 2x = 4x - 8$

66. $\frac{1}{3}(6w - 3) = 3w + 12$

67. $\frac{7r + 5}{2} = 13$

68. $\frac{1}{2}a = \frac{a - 3}{4}$

5-3 Solving Multi-Step Inequalities

Then

You solved multi-step equations. (Lesson 2-3)

Now

- Solve linear inequalities involving more than one operation.
- Solve linear inequalities involving the Distributive Property.

Math Online

glencoe.com

- Extra Examples
- Personal Tutor
- Self-Check Quiz
- Homework Help
- Math in Motion

Why?

A salesperson may make a base monthly salary and earn a commission on each of her sales. To find the number of sales she needs to make to pay her monthly bills, you can use a multi-step inequality.

Solve Multi-Step Inequalities Multi–step inequalities can be solved by undoing the operations in the same way you would solve a multi-step equation.

Real-World EXAMPLE 1 Solve a Multi-Step Inequality

SALES Write and solve an inequality to find the sales Mrs. Jones needs if she earns a salary of \$2000 plus a 10% commission on her sales. Her goal is to make at least \$4000 per month. What sales does she need to meet her goal?

base salary + (commission × sales) ≥ income needed

$2000 + 0.10x \geq 4000$	**Substitution**
$0.10x \geq 2000$	**Subtract 2000 from each side.**
$x \geq 20{,}000$	**Divide each side by 0.10.**

She must make at least \$20,000 in sales to meet her monthly goal.

Check Your Progress

1. **FINANCIAL LITERACY** The Print Shop advertises a special to print 400 flyers for less than the competition. The price includes a \$3.50 set-up fee. If the competition charges \$35.50, what does the Print Shop charge for each flyer?

Personal Tutor glencoe.com

When multiplying or dividing by a negative number, the direction of the inequality symbol changes. This holds true for multi-step inequalities.

EXAMPLE 2 Inequality Involving a Negative Coefficient

Solve $-11y - 13 > 42$.

$-11y - 13 > 42$	**Original inequality**
$-11y > 55$	**Add 13 to each side and simplify.**
$\frac{-11y}{-11} < \frac{55}{-11}$	**Divide each side by −11, and reverse the inequality.**
$y < -5$	**Simplify.**

Check Your Progress Solve each inequality.

2A. $23 \geq 10 - 2w$

2B. $43 > -4y + 11$

Personal Tutor glencoe.com

Math *in Motion*, BrainPOP® glencoe.com

You can translate sentences into multi-step inequalities and then solve them using the Properties of Inequalities.

Math *in Motion*,
Interactive Lab
glencoe.com

EXAMPLE 3 Write and Solve an Inequality

Define a variable, write an inequality, and solve the problem.

Five minus 6 times a number is more than four times the number plus 45.

Five	minus	six times a number	is more	four times a number	plus	forty-five.
5	$-$	$6n$	$>$	$4n$	$+$	45

$5 - 10n > 45$ **Subtract $4n$ from each side and simplify.**

$-10n > 40$ **Subtract 5 from each side and simplify.**

$\frac{-10n}{-10} < \frac{40}{-10}$ **Divide each side by -10, and reverse the inequality.**

$n < -4$ **Simplify.**

The solution set is $\{n \mid n < -4\}$.

Check Your Progress

3. *Two more than half of a number is greater than twenty-seven.*

Personal Tutor glencoe.com

Review Vocabulary

order of operations
1. Evaluate expressions inside grouping symbols.
2. Evaluate all powers.
3. Multiply and/or divide from left to right.
4. Add and/or subtract from left to right.

(Lesson 1-2)

Solve Inequalities Involving the Distributive Property When solving inequalities that contain grouping symbols, use the Distributive Property to remove the grouping symbols first. Then use the order of operations to simplify the resulting inequality.

EXAMPLE 4 Distributive Property

Solve $4(3t - 5) + 7 \geq 8t + 3$.

$4(3t - 5) + 7 \geq 8t + 3$ **Original inequality**

$12t - 20 + 7 \geq 8t + 3$ **Distributive Property**

$12t - 13 \geq 8t + 3$ **Combine like terms.**

$4t - 13 \geq 3$ **Subtract $8t$ from each side and simplify.**

$4t \geq 16$ **Add 13 to each side.**

$\frac{4t}{4} \geq \frac{16}{4}$ **Divide each side by 4.**

$t \geq 4$ **Simplify.**

The solution set is $\{t \mid t \geq 4\}$.

Check Your Progress

Solve each inequality. Check your solution.

4A. $6(5z - 3) \leq 36z$

4B. $2(h + 6) > -3(8 - h)$

Personal Tutor glencoe.com

Watch Out!

Distributive Property If a negative number is multiplied by a sum or difference, remember to distribute the negative sign along with the number to each term inside the parentheses.

If solving an inequality results in a statement that is always true, the solution set is the set of all real numbers. This solution set is written as $\{x \mid x \text{ is a real number.}\}$. If solving an inequality results in a statement that is never true, the solution set is the empty set, which is written as the symbol $\varnothing$. The empty set has no members.

StudyTip

Empty Set Set-builder notation is not required when the solution set is the empty set. Instead, the solution is written as the symbol ∅.

EXAMPLE 5 Empty Set and All Reals

Solve each inequality. Check your solution.

a. $9t - 5(t - 5) \leq 4(t - 3)$

$9t - 5(t - 5) \leq 4(t - 3)$	**Original inequality**
$9t - 5t + 25 \leq 4t - 12$	**Distributive Property**
$4t + 25 \leq 4t - 12$	**Combine like terms.**
$4t + 25 - 4t \leq 4t - 12 - 4t$	**Subtract 4*t* from each side.**
$25 \leq -12$	**Simplify.**

Since the inequality results in a false statement, the solution set is the empty set, ∅.

b. $3(4m + 6) \leq 42 + 6(2m - 4)$

$3(4m + 6) \leq 42 + 6(2m - 4)$	**Original inequality**
$12m + 18 \leq 42 + 12m - 24$	**Distributive Property**
$12m + 18 \leq 12m + 18$	**Combine like terms.**
$12m + 18 - 12m \leq 12m + 18 - 12m$	**Subtract 12*m* from each side.**
$18 \leq 18$	**Simplify.**

All values of m make the inequality true. All real numbers are solutions.

Check Your Progress

Solve each inequality. Check your solution.

5A. $18 - 3(8c + 4) \geq -6(4c - 1)$

5B. $46 \leq 8m - 4(2m + 5)$

Personal Tutor glencoe.com

Check Your Understanding

Example 1 p. 296

1. **CANOEING** If four people plan to use the canoe with 60 pounds of supplies, write and solve an inequality to find the allowable average weight per person.

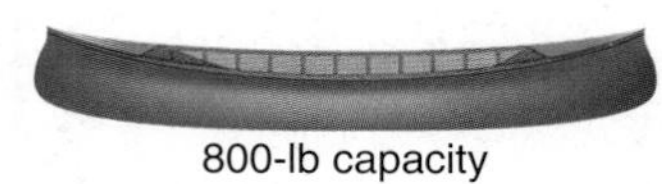

800-lb capacity

2. **SHOPPING** Rita is ordering a movie for \$11.95 and a few CDs. She has \$50 to spend. Shipping and sales tax will be \$10. If each CD costs \$9.99, write and solve an inequality to find the greatest number of CDs that she can buy.

Example 2 p. 296

Solve each inequality. Check your solution.

3. $6h - 10 \geq 32$
4. $-3 \leq \frac{2}{3}r + 9$
5. $-3x + 7 > 43$
6. $4m - 17 < 6m + 25$

Example 3 p. 297

Define a variable, write an inequality, and solve each problem. Then check your solution.

7. Four times a number minus 6 is greater than eight plus two times the number.
8. Negative three times a number plus 4 is less than five times the number plus 8.

Examples 4 and 5 pp. 297–298

Solve each inequality. Check your solution.

9. $-6 \leq 3(5v - 2)$
10. $-5(g + 4) > 3(g - 4)$
11. $3 - 8x \geq 9 + 2(1 - 4x)$

Practice and Problem Solving

= Step-by-Step Solutions begin on page R12.
Extra Practice begins on page 815.

Examples 1 and 2 p. 296

Solve each inequality. Check your solution.

12. $5b - 1 \geq -11$

13. $21 > 15 + 2a$

14. $-9 \geq \frac{2}{5}m + 7$

15. $\frac{w}{8} - 13 > -6$

16. $-a + 6 \leq 5$

17. $37 < 7 - 10w$

18. $8 - \frac{z}{3} \geq 11$

19. $-\frac{5}{4}p + 6 < 12$

20. $3b - 6 \geq 15 + 24b$

21. $15h + 30 < 10h - 45$

Example 3 p. 297

Define a variable, write an inequality, and solve each problem. Check your solution.

22. Three fourths of a number decreased by nine is at least forty-two.

23. Two thirds of a number added to six is at least twenty-two.

24. Seven tenths of a number plus 14 is less than forty-nine.

25. Eight times a number minus twenty-seven is no more than the negative of that number plus eighteen.

26. Ten is no more than 4 times the sum of twice a number and three.

27. Three times the sum of a number and seven is greater than five times the number less thirteen.

28. The sum of nine times a number and fifteen is less than or equal to the sum of twenty-four and ten times the number.

Examples 4 and 5 pp. 297–298

Solve each inequality. Check your solution.

29. $-3(7n + 3) < 6n$

30. $21 \geq 3(a - 7) + 9$

31. $2y + 4 > 2(3 + y)$

32. $3(2 - b) < 10 - 3(b - 6)$

33. $7 + t \leq 2(t + 3) + 2$

34. $8a + 2(1 - 5a) \leq 20$

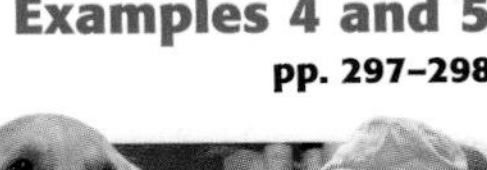

Real-World Career

Veterinarian
Veterinarians take care of sick and injured animals. Vets can work anywhere from a zoo to a research facility to owning their own practice. Vets need to earn a bachelor's degree, attend vet college for 4 years, and take a test to get licensed.

Define a variable, write an inequality, and solve each problem. Then interpret your solution.

35. **CARS** A car salesperson is paid a base salary of \$35,000 a year plus 8% of sales. What are the sales needed to have an annual income greater than \$65,000?

36. **ANIMALS** Keith's dog weighs 90 pounds. A healthy weight for his dog would be less than 75 pounds. If Keith's dog can lose an average of 1.25 pounds per week on a certain diet, how long until the dog reaches a healthy weight?

37. Solve $6(m - 3) > 5(2m + 4)$. Show each step and justify your work.

38. Solve $8(a - 2) \leq 10(a + 2)$. Show each step and justify your work.

39. **MUSICAL** A high school drama club is performing a musical to benefit a local charity. Tickets are \$5 each. They also received donations of \$565. They want to raise at least \$1500.
    a. Write an inequality that describes this situation. Then solve the inequality.
    b. Graph the solution.

40. **ICE CREAM** Benito has \$6 to spend. A sundae costs \$3.25 plus \$0.65 per topping. Write and solve an inequality to find how many toppings he can order.

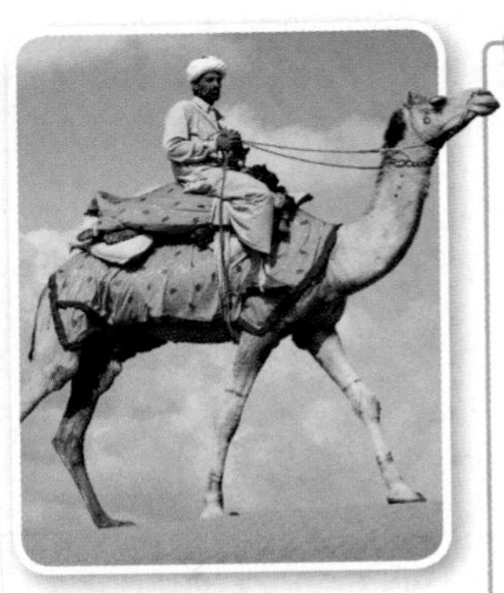

Real-World Link

Unlike most animals, camels move both legs on one side of their body at the same time as they walk.

Source: National Zoo

41. **SCIENCE** The normal body temperature of a camel is 97.7°F in the morning. If it has had no water by noon, its body temperature can be greater than 104°F.

   a. Write an inequality that represents a camel's body temperature at noon if the camel had no water.

   b. If C represents degrees Celsius, then $F = \frac{9}{5}C + 32$. Write and solve an inequality to find the camel's body temperature at noon in degrees Celsius.

42. **NUMBER THEORY** Find all sets of three consecutive positive even integers with a sum no greater than 36.

43. **NUMBER THEORY** Find all sets of four consecutive positive odd integers whose sum is less than 42.

Solve each inequality. Check your solution.

44. $2(x - 4) \leq 2 + 3(x - 6)$

45. $\frac{2x - 4}{6} \geq -5x + 2$

46. $5.6z + 1.5 < 2.5z - 4.7$

47. $0.7(2m - 5) \geq 21.7$

GRAPHING CALCULATOR **Use a graphing calculator to solve each inequality.**

48. $3x + 7 > 4x + 9$

49. $13x - 11 \leq 7x + 37$

50. $2(x - 3) < 3(2x + 2)$

51. $\frac{1}{2}x - 9 < 2x$

52. $2x - \frac{2}{3} \geq x - 22$

53. $\frac{1}{3}(4x + 3) \geq \frac{2}{3}x + 2$

54. **MULTIPLE REPRESENTATIONS** In this problem, you will solve compound inequalities. A number x is greater than 4, and the same number is less than 9.

   a. **NUMERICAL** Write two separate inequalities for the statement.

   b. **GRAPHICAL** Graph the solution set for the first inequality in red. Graph the solution set for the second inequality in blue. Highlight the portion of the graph in which the red and blue overlap.

   c. **TABULAR** Make a table using ten points from your number line, including points from each section. Use one column for each inequality and a third column titled "Both are True." Complete the table by writing true or false.

   d. **VERBAL** Describe the relationship between the colored regions of the graph and the chart.

   e. **LOGICAL** Make a prediction of what the graph of $4 < x < 9$ looks like.

H.O.T. Problems

Use Higher-Order Thinking Skills

55. **REASONING** Explain how you could solve $-3p + 7 \geq -2$ without multiplying or dividing each side by a negative number.

56. **CHALLENGE** If $ax + b < ax + c$ has infinitely many solutions, what will be the solution of $ax + b > ax + c$? Explain how you know.

57. **OPEN ENDED** Write two different multi-step inequalities that have the same graph.

58. **WHICH ONE DOESN'T BELONG?** Name the inequality that does not belong. Explain.

$4y + 9 > -3$	$3y - 4 > 5$	$-2y + 1 < -5$	$-5y + 2 < -13$

59. **WRITING IN MATH** Explain when the solution set of an inequality will be the empty set or the set of all real numbers. Show an example of each.

Standardized Test Practice

60. What is the solution set of the inequality $4t + 2 < 8t - (6t - 10)$?

A $\{t \mid t < -6.5\}$
B $\{t \mid t > -6.5\}$
C $\{t \mid t < 4\}$
D $\{t \mid t > 4\}$

61. GEOMETRY The section of Liberty Ave. between 5th St. and King Ave. is temporarily closed. Traffic is being detoured right on 5th St., left on King Ave. and then back on Liberty Ave. How long is the closed section of Liberty Ave.?

F 100 ft
G 120 ft
H 144 ft
J 180 ft

Liberty Ave.
72 ft
5th St.
King Ave.
96 ft

62. SHORT RESPONSE Rhiannon is paid \$52 for working 4 hours. At this rate, how many hours will it take her to earn \$845?

63. GEOMETRY Classify the triangle.

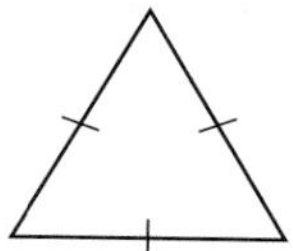

A right
B parallel
C obtuse
D equilateral

Spiral Review

Solve each inequality. Check your solution. (Lesson 5-2)

64. $\frac{y}{2} \leq -5$

65. $12b > -48$

66. $-\frac{2}{3}t \leq -30$

Solve each inequality. Check your solution, and graph it on a number line. (Lesson 5-1)

67. $6 - h > -8$

68. $p - 9 < 2$

69. $3 \geq 4 - m$

Solve each equation by graphing. Verify your answer algebraically. (Lesson 3-2)

70. $2x - 7 = 4x + 9$

71. $5 + 3x = 7x - 11$

72. $2(x - 3) = 5x + 12$

73. THEME PARKS In 2006, attendance at the top 20 theme parks in North America was 119.8 million. That represents an increase of about 1.5% from 2005. What was the approximate attendance in 2005? (Lesson 2-7)

If $f(x) = 4x - 3$ and $g(x) = 2x^2 + 5$, find each value. (Lesson 1-7)

74. $f(-2)$

75. $g(2) - 5$

76. $f(c + 3)$

77. COSMETOLOGY On average, a barber received a tip of \$4 for each of 12 haircuts. Write and evaluate an expression to determine the total amount that she earned. (Lesson 1-4)

Skills Review

Graph each set of numbers on a number line.

78. $\{-4, -2, 2, 4\}$

79. $\{-3, 0, 1, 5\}$

80. {integers less than 3}

81. {integers greater than or equal to -2}

82. {integers between -3 and 4}

83. {integers less than -1}

CHAPTER 5

Mid-Chapter Quiz

Lessons 5-1 through 5-3

Math Online glencoe.com
Math *in Motion,* Animation

Solve each inequality. Then graph it on a number line. (Lesson 5-1)

1. $x - 8 > 4$
2. $m + 2 \geq 6$
3. $p - 4 < -7$
4. $12 \leq t - 9$

5. **CONCERTS** Lupe's allowance for the month is $60. She wants to go to a concert for which a ticket costs $45. (Lesson 5-1)

   a. Write and solve an inequality that shows how much money she can spend that month after buying a concert ticket.

   b. She spends $9.99 on music downloads and $2 on lunch in the cafeteria. Write and solve an inequality that shows how much she can spend after these purchases and the concert ticket.

Define a variable, write an inequality, and solve each problem. Check your solution. (Lesson 5-2)

6. The sum of a number and -2 is no more than 6.
7. A number decreased by 4 is more than -1.
8. Twice a number increased by 3 is less than the number decreased by 4.

9. **MULTIPLE CHOICE** Jane is saving money to buy a new cell phone that costs no more than $90. So far, she has saved $52. How much more money does Jane need to save? (Lesson 5-2)

   A $38

   B more than $38

   C no more than $38

   D at least $38

Solve each inequality. Check your solution. (Lesson 5-2)

10. $\frac{1}{3}y \geq 5$
11. $4 < \frac{c}{5}$
12. $-8x > 24$
13. $2m \leq -10$
14. $\frac{x}{2} < \frac{5}{8}$
15. $-9a \geq -45$
16. $\frac{w}{6} > -3$
17. $\frac{k}{7} < -2$

18. **ANIMALS** The world's heaviest flying bird is the great bustard. A male bustard can be up to 4 feet long and weigh up to 40 pounds. (Lesson 5-2)

    a. Write inequalities to describe the ranges of lengths and weights of male bustards.

    b. Male bustards are usually about four times as heavy as females. Write and solve an inequality that describes the range of weights of female bustards.

19. **GARDENING** Bill is building a fence around a square garden to keep deer out. He has 60 feet of fencing. Find the maximum length of a side of the garden. (Lesson 5-3)

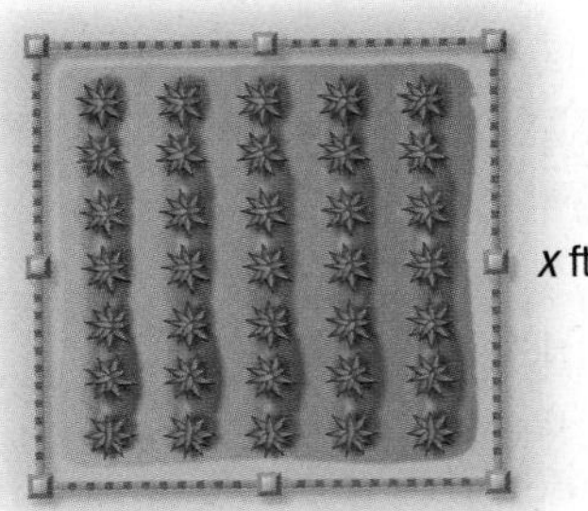

Solve each inequality. Check your solution. (Lesson 5-3)

20. $4a - 2 > 14$
21. $2x + 11 \leq 5x - 10$
22. $-p + 4 < -9$
23. $\frac{d}{4} + 1 \geq -3$
24. $-2(4b + 1) < -3b + 8$

Define a variable, write an inequality, and solve each problem. Check your solution. (Lesson 5-3)

25. Three times a number increased by 8 is no more than the number decreased by 4.
26. Two thirds of a number plus 5 is greater than 17.
27. **MULTIPLE CHOICE** Shoe rental costs $2, and each game bowled costs $3. How many games can Kyle bowl without spending more than $15? (Lesson 5-3)

    F 2

    G 3

    H 4

    J 5

EXPLORE 5-4

Algebra Lab

Reading Compound Statements

A compound statement is made up of two simple statements connected by the word *and* or *or*. Before you can determine whether a compound statement is true or false, you must understand what the words *and* and *or* mean.

A spider has eight legs, *and* a dog has five legs.

For a compound statement connected by the word *and* to be true, both simple statements must be true.

A spider has eight legs. ⟶ true

A dog has five legs. ⟶ false

Since one of the statements is false, the compound statement is false.

A compound statement connected by the word *or* may be *exclusive* or *inclusive*. For example, the statement "With your lunch, you may have milk *or* juice," is exclusive. In everyday language, *or* means one or the other, but not both. However, in mathematics, *or* is inclusive. It means one or the other or both.

A spider has eight legs, *or* a dog has five legs.

For a compound statement connected by the word *or* to be true, at least one of the simple statements must be true. Since it is true that a spider has eight legs, the compound statement is true.

Exercises

Is each compound statement *true* or *false*? Explain.

1. Most top 20 movies in 2004 were rated PG-13, *or* most top 20 movies in 2002 were rated G.
2. In 2005 more top 20 movies were rated PG than were rated G, *and* more were rated PG than rated PG-13.
3. For the years shown most top 20 movies are rated PG-13, *and* no top 20 movies in 2002 were rated R.
4. No top 20 movies in 2005 were rated G, *or* most top 20 movies in 2005 were *not* rated PG.
5. $11 < 5$ or $9 < 7$
6. $-2 > 0$ and $3 < 7$
7. $5 > 0$ and $-3 < 0$
8. $-2 > -3$ or $0 = 0$
9. $8 \neq 8$ or $-2 > -5$
10. $5 > 10$ and $4 > -2$

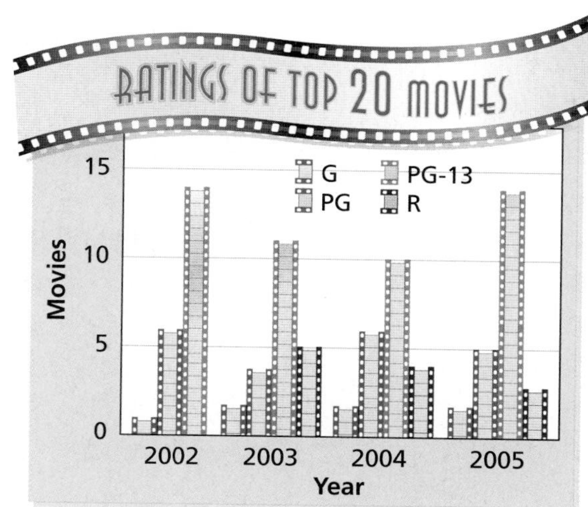

Source: National Association of Theater Owners

5-4 Solving Compound Inequalities

Then

You solved absolute value equations with two cases. (Lesson 2-5)

Now

- Solve compound inequalities containing the word *and* and graph their solution set.
- Solve compound inequalities containing the word *or* and graph their solution set.

New Vocabulary

compound inequality
intersection
union

Math Online

glencoe.com

- Extra Examples
- Personal Tutor
- Self-Check Quiz
- Homework Help
- Math in Motion

Why?

To ride the Mind Eraser roller coaster at Six Flags in Baltimore, Maryland, you must be at least 52 inches tall, and your height cannot exceed 72 inches. If h represents the height of a rider, we can write two inequalities to represent this.

at least 52 inches	cannot exceed 72 inches
$h \geq 52$	$h \leq 72$

The inequalities $h \geq 52$ and $h \leq 72$ can be combined and written without using *and* as $52 \leq h \leq 72$.

Inequalities Containing *and* When considered together, two inequalities such as $h \geq 52$ and $h \leq 72$ form a **compound inequality**. A compound inequality containing *and* is only true if both inequalities are true. Its graph is where the graphs of the two inequalities overlap. This is called the **intersection** of the two graphs.

The intersection can be found by graphing each inequality and then determining where the graphs intersect.

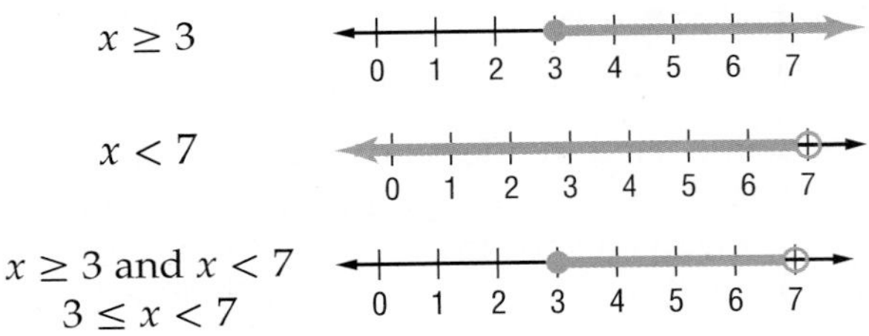

The statement $3 \leq x < 7$ can be read as x *is greater than or equal to 3 and less than 7* or x *is between 3 and 7 including 3.*

EXAMPLE 1 Solve and Graph an Intersection

Solve $-2 \leq x - 3 < 4$. Then graph the solution set.

First, express $-2 \leq x - 3 < 4$ using *and*. Then solve each inequality.

$-2 \leq x - 3$	**and**	$x - 3 < 4$	**Write the inequalities.**
$-2 + 3 \leq x - 3 + 3$		$x - 3 + 3 < 4 + 3$	**Add 3 to each side.**
$1 \leq x$		$x < 7$	**Simplify.**

The solution set is $\{x \mid 1 \leq x < 7\}$. Now graph the solution set.

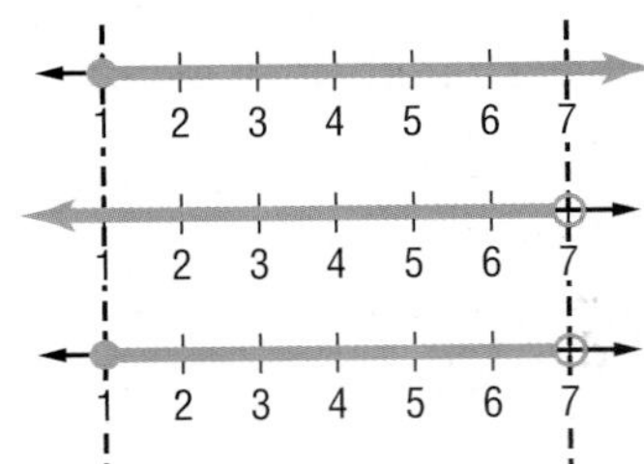

Graph $1 \leq x$ or $x \geq 1$.

Graph $x < 7$.

Find the intersection of the graphs.

Check Your Progress

Solve each compound inequality. Then graph the solution set.

1A. $y - 3 \geq -11$ and $y - 3 \leq -8$ **1B.** $6 \leq r + 7 < 10$

Personal Tutor glencoe.com

Inequalities Containing *or* Another type of compound inequality contains the word *or*. A compound inequality containing *or* is true if at least one of the inequalities is true. Its graph is the **union** of the graphs of two inequalities.

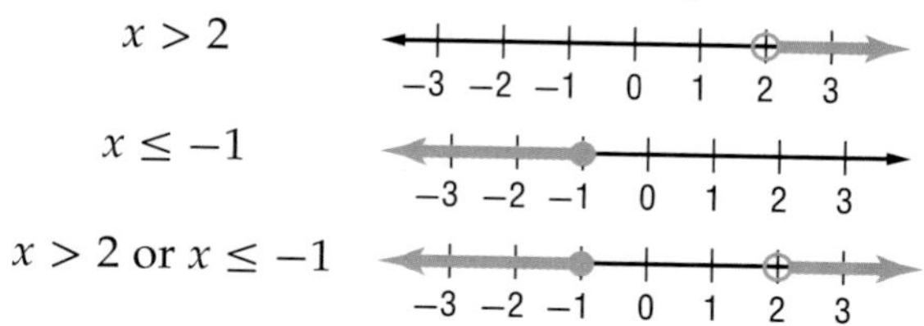

When solving problems involving inequalities, *within* is meant to be inclusive, so use $\geq$ or $\leq$. *Between* is meant to be exclusive, so use $<$ or $>$.

Real-World EXAMPLE 2 Write and Graph a Compound Inequality

SOUND The human ear can only detect sounds between the frequencies 20 Hertz and 20,000 Hertz. Write and graph a compound inequality that describes the frequency of sounds humans cannot hear.

The problem states that humans can hear the frequencies between 20 Hz and 20,000 Hz. We are asked to find the frequencies humans cannot hear.

Words	The frequency	is at most	20 Hertz	or	The frequency	is at least	20,000 Hertz.
Variable	Let f be the frequency.						
Inequality	f	$<$	20	or	f	$>$	20,000

Now, graph the solution set.

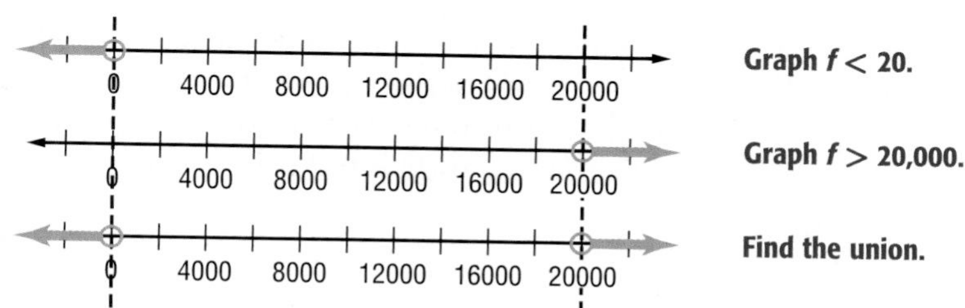

Notice that the graphs do not intersect. Humans cannot hear sounds at a frequency less than 20 Hertz or greater than 20,000 Hertz. The compound inequality is $\{f \mid f < 20 \text{ or } f > 20{,}000\}$.

Check Your Progress

2. MANUFACTURING A company is manufacturing an action figure that must be at least 11.2 centimeters and at most 11.4 centimeters tall. Write and graph a compound inequality that describes how tall the action figure can be.

Personal Tutor glencoe.com

ReadingMath

At Most The phrase *at most* in Example 2 indicates $\leq$. It could also have been phrased as *no more than* or *less than or equal to.*

Math *in Motion,* Animation glencoe.com

EXAMPLE 3 Solve and Graph a Union

Solve $-2m + 7 \leq 13$ or $5m + 12 > 37$. Then graph the solution set.

$-2m + 7 \leq 13$	**or**	$5m + 12 > 37$
$-2m + 7 - 7 \leq 13 - 7$	**Subtract.**	$5m + 12 - 12 > 37 - 12$
$-2m \leq 6$	**Simplify.**	$5m > 25$
$\frac{-2m}{-2} \geq \frac{6}{-2}$	**Divide.**	$\frac{5m}{5} > \frac{25}{5}$
$m \geq -3$	**Simplify.**	$m > 5$

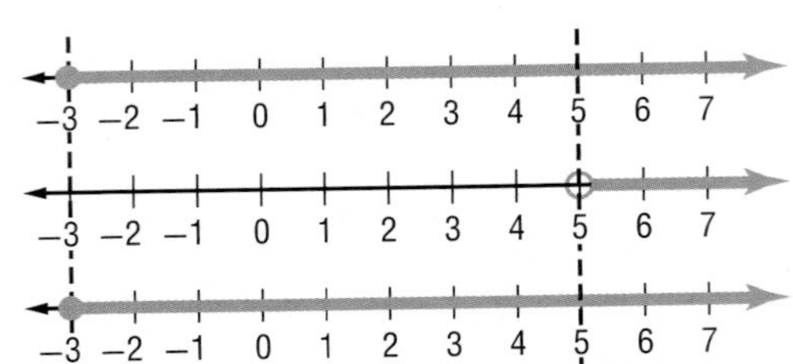

Notice that the graph of $m \geq -3$ contains every point in the graph of $m > 5$. So, the union is the graph of $m \geq -3$. The solution set is $\{m \mid m \geq -3\}$.

StudyTip

Intersections and Unions The graphs of compound inequalities containing *and* will be an intersection. The graphs of compound inequalities containing *or* will be a union.

Check Your Progress

Solve each compound inequality. Then graph the solution set.

3A. $a + 1 < 4$ or $a - 1 \geq 3$

3B. $x \leq 9$ or $2 + 4x < 10$

Personal Tutor glencoe.com

Check Your Understanding

Examples 1 and 3 pp. 304, 306

Solve each compound inequality. Then graph the solution set.

1. $4 \leq p - 8$ and $p - 14 \leq 2$
2. $r + 6 < -8$ or $r - 3 > -10$
3. $4a + 7 \geq 31$ or $a > 5$
4. $2 \leq g + 4 < 7$

Example 2 p. 305

5. **BIKES** The recommended air pressure for the tires of a mountain bike is at least 35 pounds per square inch (psi), but no more than 80 pounds per square inch. If a bike's tires have 24 pounds per square inch, what is the recommended range of air that should be put into the tires?

Practice and Problem Solving

= **Step-by-Step Solutions** begin on page R12.
Extra Practice begins on page 815.

Examples 1 and 3 pp. 304, 306

Solve each compound inequality. Then graph the solution set.

6. $f - 6 < 5$ and $f - 4 \geq 2$
7. $n + 2 \leq -5$ and $n + 6 \geq -6$
8. $y - 1 \geq 7$ or $y + 3 < -1$
9. $t + 14 \geq 15$ or $t - 9 < -10$
10. $-5 < 3p + 7 \leq 22$
11. $-3 \leq 7c + 4 < 18$
12. $5h + -4 \geq 6$ and $7h + 11 < 32$
13. $22 \geq 4m - 2$ or $5 - 3m \leq -13$
14. $-4a + 13 \geq 29$ and $10 < 6a - 14$
15. $-y + 5 \geq 9$ or $3y + 4 < -5$

Example 2
p. 305

16. **SPEED** The posted speed limit on an interstate highway is shown. Write an inequality that represents the sign. Graph the inequality.

SPEED
LIMIT
70
MINIMUM
40

17. **NUMBER THEORY** Find all sets of two consecutive positive odd integers with a sum that is at least 8 and less than 24.

Write a compound inequality for each graph.

18. −2 −1 0 1 2 3 4

19. −4 −3 −2 −1 0 1 2

20. −1 0 1 2 3 4

21. −6 −5 −4 −3 −2 −1 0

22. 1 2 3 4 5 6 7

23. −4 −3 −2 −1 0 1 2

Solve each compound inequality. Then graph the solution set.

24. $3b + 2 < 5b - 6 \leq 2b + 9$

25. $-2a + 3 \geq 6a - 1 > 3a - 10$

26. $10m - 7 < 17m$ or $-6m > 36$

27. $5n - 1 < -16$ or $-3n - 1 < 8$

28. **COUPON** Juanita has a coupon for 10% off any digital camera at a local electronics store. She is looking at digital cameras that range in price from \$100 to \$250.
   a. How much are the cameras after the coupon is used?
   b. If the tax amount is 6.5%, how much should Juanita expect to spend?

Define a variable, write an inequality, and solve each problem. Then check your solution.

29. Eight less than a number is no more than 14 and no less than 5.
30. The sum of 3 times a number and 4 is between −8 and 10.
31. The product of −5 and a number is greater than 35 or less than 10.
32. One half a number is greater than 0 and less than or equal to 1.

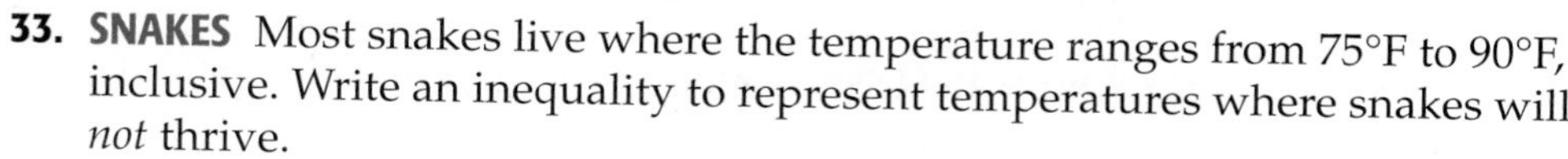

33. **SNAKES** Most snakes live where the temperature ranges from 75°F to 90°F, inclusive. Write an inequality to represent temperatures where snakes will *not* thrive.

34. **FUNDRAISING** Yumas is selling gift cards to raise money for a class trip. He can earn prizes depending on how many cards he sells. So far, he has sold 34 cards. How many more does he need to sell to earn a prize in category 4?

Cards	Prize
1–15	1
16–30	2
31–45	3
46–60	4
+61	5

Real-World Link

The organization Field Trip Earth monitors the locations of Atlantic sea turtles. This data allows the scientists to track migration patterns as part of a wildlife conservation project.

Source: Field Trip Earth

35. **TURTLES** Atlantic sea turtle eggs that incubate below 23°C or above 33°C rarely hatch. Write an inequality for the temperatures at which the eggs should be incubated.

36. **GEOMETRY** The *Triangle Inequality Theorem* states that the sum of the measures of any two sides of a triangle is greater than the measure of the third side.
   a. Write and solve three inequalities to express the relationships among the measures of the sides of the triangle shown at the right.
   b. What are four possible lengths for the third side of the triangle?
   c. Write a compound inequality for the possible values of x.

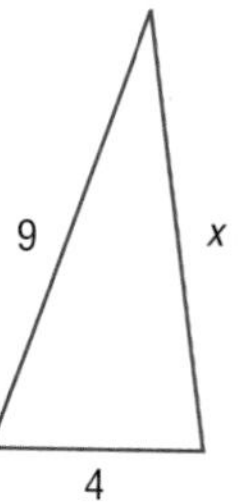

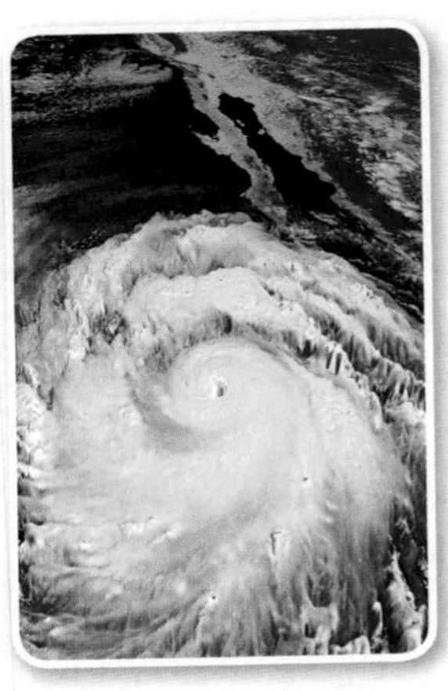

Real-World Link

Between 2001 and 2004, a total of nine hurricanes struck the mainland United States. Of these, three were classified as Category 3, 4, or 5.

Source: National Weather Service

37. **HURRICANES** The Saffir-Simpson Hurricane Scale rates hurricanes on a scale from 1 to 5 based on their wind speed.

Category	Wind Speed (mph)	Example (year)
1	74–95	Gaston (2004)
2	96–110	Frances (2004)
3	111–130	Ivan (2004)
4	131–155	Charley (2004)
5	> 155	Andrew (1992)

a. Write a compound inequality for the wind speeds of a category 3 and a category 4 hurricane.

b. What is the intersection of the two graphs of the inequalities you found in part **a**?

38. **MULTIPLE REPRESENTATIONS** In this problem, you will investigate measurements. The **absolute error** of a measurement is equal to one half the unit of measure. The **relative error** of a measure is the ratio of the absolute error to the expected measure.

a. **TABULAR** Copy and complete the table.

Measure	Absolute Error	Relative Error
14.3 cm	$\frac{1}{2}(0.1) = 0.05$ cm	$\frac{\text{absolute error}}{\text{expected measure}} = \frac{0.05 \text{ cm}}{14.3 \text{ cm}}$ ≈ 0.0035 or 0.4%
1.85 cm		
61.2 cm		
237 cm		

b. **ANALYTICAL** You measured a length of 12.8 centimeters. Compute the absolute error and then write the range of possible measures.

c. **LOGICAL** To what precision would you have to measure a length in centimeters to have an absolute error of less than 0.05 centimeter?

d. **ANALYTICAL** To find the relative error of an area or volume calculation, add the relative errors of each linear measure. If the measures of the sides of a rectangular box are 6.5 centimeters, 7.2 centimeters, and 10.25 centimeters, what is the relative error of the volume of the box?

H.O.T. Problems

Use Higher-Order Thinking Skills

39. **ERROR ANALYSIS** Chloe and Jonas are solving $3 < 2x - 5 < 7$. Is either of them correct? Explain your reasoning.

Chloe	Jonas
$3 < 2x - 5 < 7$	$3 < 2x - 5 < 7$
$3 < 2x < 12$	$8 < 2x < 7$
$\frac{3}{2} < x < 6$	$4 < x < \frac{7}{2}$

40. **REASONING** Write a compound inequality for which the graph is the empty set and one for which the graph is the set of all real numbers.

41. **OPEN ENDED** Create an example of a compound inequality containing *or* that has infinitely many solutions.

42. **CHALLENGE** Determine whether the following statement is *always, sometimes,* or *never* true. Explain. *The graph of a compound inequality that involves an* or *statement is bounded on the left and right by two values of x.*

43. **WRITING IN MATH** Give an example of a compound inequality you might encounter at an amusement park. Does the example represent an intersection or a union?

Standardized Test Practice

44. What is the solution set of the inequality $-7 < x + 2 < 4$?

A $\{x \mid -5 < x < 6\}$
B $\{x \mid -5 < x < 2\}$
C $\{x \mid -9 < x < 2\}$
D $\{x \mid -9 < x < 6\}$

45. **GEOMETRY** What is the surface area of the rectangular solid?

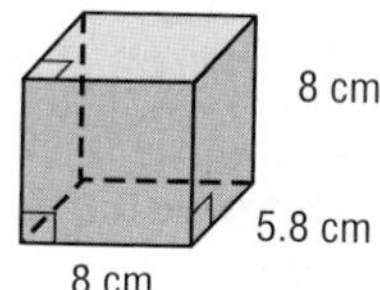

F 249.6 cm^2
G 278.4 cm^2
H 313.6 cm^2
J 371.2 cm^2

46. **GRIDDED RESPONSE** What is the next term in the sequence?

$$\frac{13}{2}, \frac{18}{5}, \frac{23}{8}, \frac{28}{11}, \frac{33}{14}, \ldots$$

47. After paying a \$15 membership fee, members of a video club can rent movies for \$2. Nonmembers can rent movies for \$4. What is the least number of movies which must be rented for it to be less expensive for members?

A 9
B 8
C 7
D 6

Spiral Review

48. **BABYSITTING** Marilyn earns \$150 per month delivering newspapers plus \$7 an hour babysitting. If she wants to earn at least \$300 this month, how many hours will she have to babysit? (Lesson 5-3)

49. **MAGAZINES** Carlos has earned more than \$260 selling magazine subscriptions. Each subscription was sold for \$12. How many did Carlos sell? (Lesson 5-2)

50. **PUNCH** Raquel is mixing lemon-lime soda and a fruit juice blend that is 45% juice. If she uses 3 quarts of soda, how many quarts of fruit juice must be added to produce punch that is 30% juice? (Lesson 2-9)

Solve each proportion. If necessary, round to the nearest hundredth. (Lesson 2-6)

51. $\frac{14}{x} = \frac{20}{8}$
52. $\frac{0.47}{6} = \frac{1.41}{m}$
53. $\frac{16}{7} = \frac{9}{b}$
54. $\frac{2 + y}{5} = \frac{10}{3}$
55. $\frac{8}{9} = \frac{2r - 3}{4}$
56. $\frac{6 - 2y}{8} = \frac{2}{18}$

Determine whether a valid conclusion follows from the statement below for each given condition. If a valid conclusion does not follow, write *no valid conclusion* and explain why. (Lesson 1-8)

If a DVD box set costs less than \$70, then Ian will buy one.

57. A DVD box set costs \$59.
58. A DVD box set costs \$89.
59. Ian will not buy a DVD box set.
60. Ian bought 2 DVD box sets.

Evaluate each expression. Name the property used in each step. (Lesson 1-2)

61. $5 + (4 - 2^2)$
62. $\frac{3}{8}[8 \div (7 - 4)]$
63. $2(4 \cdot 9 - 3) + 5 \cdot \frac{1}{5}$

Skills Review

Solve each equation. (Lesson 2-3)

64. $4p - 2 = -6$
65. $18 = 5p + 3$
66. $9 = 1 + \frac{m}{7}$
67. $1.5a - 8 = 11$
68. $20 = -4c - 8$
69. $\frac{b + 4}{-2} = -17$
70. $\frac{n - 3}{8} = 20$
71. $6y - 16 = 44$
72. $130 = 11k + 9$

5-5

Inequalities Involving Absolute Value

Then

You solved equations involving absolute value. (Lesson 2-5)

Now

- Solve and graph absolute value inequalities (<).
- Solve and graph absolute value inequalities (>).

Math Online

glencoe.com

- Extra Examples
- Personal Tutor
- Self-Check Quiz
- Homework Help
- Math in Motion

Why?

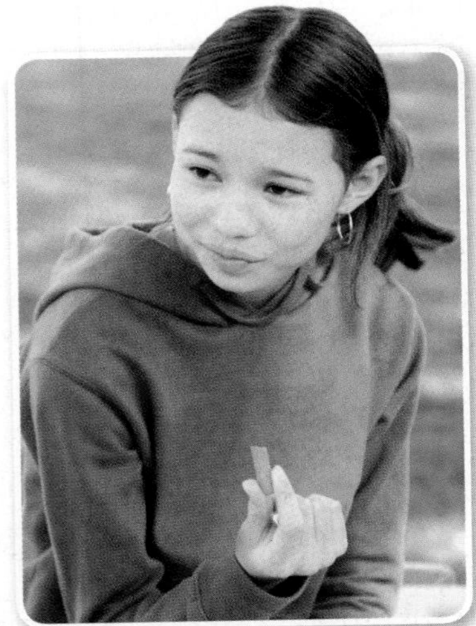

Some companies use absolute value inequalities to control the quality of their product. To make baby carrots, long carrots are sliced into 3-inch sections and peeled. If the machine is accurate to within $\frac{1}{8}$ of an inch, the length ranges from $2\frac{7}{8}$ inches to $3\frac{1}{8}$ inches.

Absolute Value Inequalities (<) The inequality $|x| < 3$ means that the distance between x and 0 is less than 3.

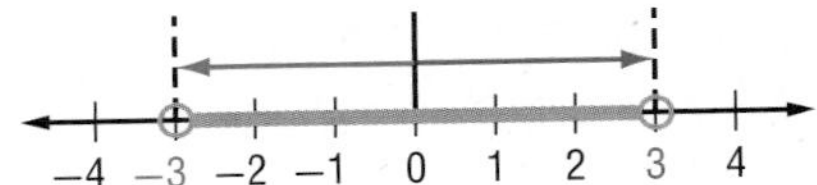

So, $x > -3$ and $x < 3$. The solution set is $\{x \mid -3 < x < 3\}$.

When solving absolute value inequalities, there are two cases to consider.

Case 1 The expression inside the absolute value symbols is nonnegative.

Case 2 The expression inside the absolute value symbols is negative.

The solution is the union of the solutions of these two cases.

EXAMPLE 1 Solve Absolute Value Inequalities (<)

Solve each inequality. Then graph the solution set.

a. $|m + 2| < 11$

Rewrite $|m + 2| < 11$ for Case 1 *and* Case 2.

Case 1 $m + 2$ is nonnegative. **and** **Case 2** $m + 2$ is negative.

Case 1:

$$m + 2 < 11$$
$$m + 2 - 2 < 11 - 2$$
$$m < 9$$

Case 2:

$$-(m + 2) < 11$$
$$m + 2 > -11$$
$$m + 2 - 2 > -11 - 2$$
$$m > -13$$

So, $m < 9$ and $m > -13$. The solution set is $\{m \mid -13 < m < 9\}$.

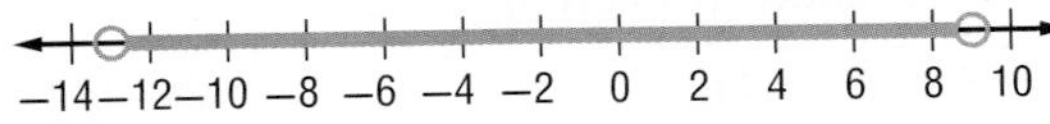

b. $|y - 1| < -2$

$|y - 1|$ cannot be negative. So it is not possible for $|y - 1|$ to be less than -2. Therefore, there is no solution, and the solution set is the empty set, Ø.

Math *in Motion*, Animation glencoe.com

Check Your Progress

1A. $|n - 8| \leq 2$

1B. $|2c - 5| < -3$

Personal Tutor glencoe.com

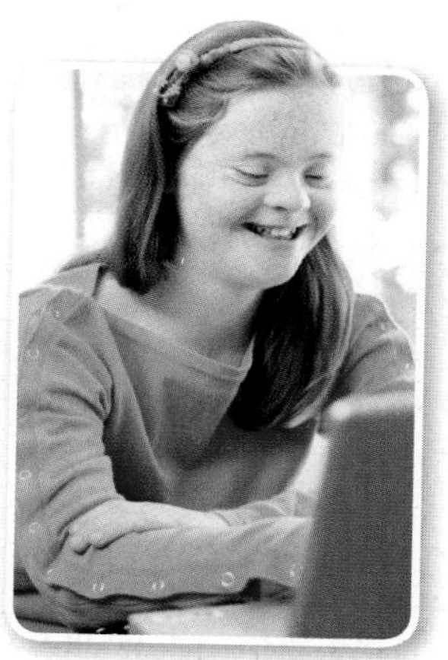

Real-World Link

One in five Americans use the Internet to view videos. Young adults tend to watch funny videos, while other age groups tend to watch the news.

Source: Pew Internet and American Life Project

Real-World EXAMPLE 2 Apply Absolute Value Inequalities

INTERNET A recent survey showed that 65% of young adults watched online video clips. The margin of error was within 3 percentage points. Find the range of young adults who use video sharing sites.

The difference between the actual number of viewers and the number from the survey is less than or equal to 3. Let x be the actual number of viewers. Then $|x - 65| \leq 3$.

Solve each case of the inequality.

Case 1 $x - 65$ is nonnegative.

$$x - 65 \leq 3$$
$$x - 65 + 65 \leq 3 + 65$$
$$x \leq 68$$

and

Case 2 $x - 65$ is negative.

$$-(x - 65) \leq 3$$
$$x - 65 \geq -3$$
$$x \geq 62$$

The range of young adults who use video sharing sites is $\{x \mid 62 \leq x \leq 68\}$.

Check Your Progress

2. CHEMISTRY The melting point of ice is 0°C. During a chemistry experiment, Jill observed ice melting within 2°C of this measurement. Write the range of temperatures that Jill observed.

Personal Tutor glencoe.com

Absolute Value Inequalities (>) The inequality $|x| > 3$ means that the distance between x and 0 is greater than 3.

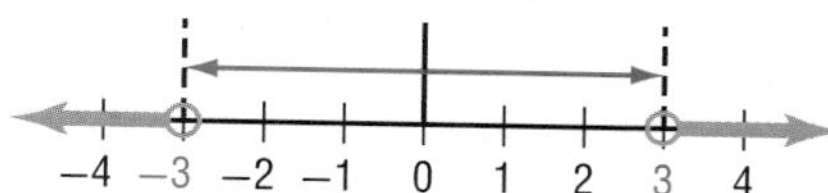

So, $x < -3$ or $x > 3$. The solution set is $\{x \mid x < -3 \text{ or } x > 3\}$.

As in the previous example, we must consider both cases.

Case 1 The expression inside the absolute value symbols is nonnegative.

Case 2 The expression inside the absolute value symbols is negative.

StudyTip

Absolute Value For $|a| \geq b$, where a is any linear expression in one variable and b is a negative number, the solution set will always be the set of all real numbers. Since $|a|$ is always greater than or equal to zero, $|a|$ is always greater than or equal to b.

EXAMPLE 3 Solve Absolute Value Inequalities (>)

Solve $|3n + 6| \geq 12$. Then graph the solution set.

Rewrite $|3n + 6| \geq 12$ for Case 1 *or* Case 2.

Case 1 $3n + 6$ is nonnegative.

$$3n + 6 \geq 12$$
$$3n + 6 - 6 \geq 12 - 6$$
$$3n \geq 6$$
$$n \geq 2$$

or

Case 2 $3n + 6$ is negative.

$$-(3n + 6) \geq 12$$
$$3n + 6 \leq -12$$
$$3n \leq -18$$
$$n \leq -6$$

So, $n \geq 2$ or $n \leq -6$. The solution set is $\{n \mid n \geq 2 \text{ or } n \leq -6\}$.

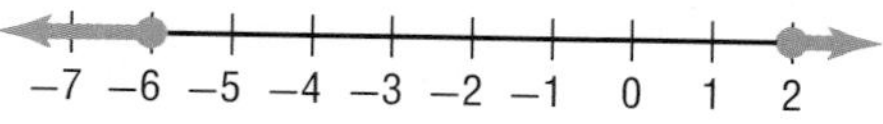

Check Your Progress

Solve each inequality. Then graph the solution set.

3A. $|2k + 1| > 7$

3B. $|r - 6| \geq -5$

Personal Tutor glencoe.com

Check Your Understanding

Examples 1 and 3 pp. 310–311

Solve each inequality. Then graph the solution set.

1. $|a - 5| < 3$

2. $|u + 3| < 7$

3. $|t + 4| \leq -2$

4. $|c + 2| > -2$

5. $|n + 5| \geq 3$

6. $|p - 2| \geq 8$

Example 2 p. 311

7. FINANCIAL LITERACY Jerome bought stock in his favorite fast-food restaurant chain at \$70.85. However, it has fluctuated up to \$0.75 in a day. Find the range of prices for which the stock could trade in a day.

Practice and Problem Solving

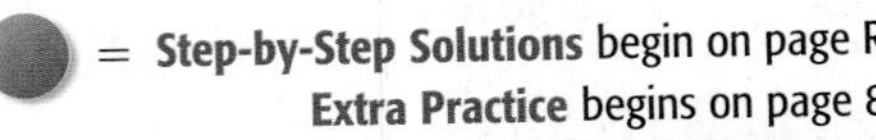
= **Step-by-Step Solutions** begin on page R12.
Extra Practice begins on page 815.

Examples 1 and 3 pp. 310–311

Solve each inequality. Then graph the solution set.

8. $|x + 8| < 16$

9. $|r + 1| \leq 2$

10. $|2c - 1| \leq 7$

11. $|3h - 3| < 12$

12. $|m + 4| < -2$

13. $|w + 5| < -8$

14. $|r + 2| > 6$

15. $|k - 4| > 3$

16. $|2h - 3| \geq 9$

17. $|4p + 2| \geq 10$

18. $|5v + 3| > -9$

19. $|-2c - 3| > -4$

Example 2 p. 311

20. SCUBA DIVING The pressure of a scuba tank should be within 500 pounds per square inch (psi) of 2500 psi. Write the range of optimum pressures.

Solve each inequality. Then graph the solution set.

21. $|4n + 3| \geq 18$

22. $|5t - 2| \leq 6$

23. $\left|\frac{3h + 1}{2}\right| < 8$

24. $\left|\frac{2p - 8}{4}\right| \geq 9$

25. $\left|\frac{7c + 3}{2}\right| \leq -5$

26. $\left|\frac{2g + 3}{2}\right| > -7$

27. $|-6r - 4| < 8$

28. $|-3p - 7| > 5$

29. $|-h + 1.5| < 3$

30. MUSIC DOWNLOADS Kareem is allowed to download \$10 worth of music each month. This month he has spent within \$3 of his allowance.

a. What is the range of money he has spent on music downloads this month?

b. Graph the range of the money that he spent.

31. CHEMISTRY Water can be present in our atmosphere as a solid, liquid, or gas. Water freezes at 32°F and vaporizes at 212°F.

a. Write the range of temperatures in which water is not a liquid.

b. Graph this range.

c. Write the absolute value inequality that describes this situation.

Write an open sentence involving absolute value for each graph.

32. −5 −4 −3 −2 −1 0 1 2 3 4 5

33. −6 −5 −4 −3 −2 −1 0 1 2 3 4

34. −6 −5 −4 −3 −2 −1 0 1 2 3 4

35. 0 1 2 3 4 5 6 7 8 9 10 11

Real-World Link

In the 1930s, there were about 20,000 miniature golf courses in the United States. Today there are about 4000 courses.

Source: Professional Miniature Golf Association

36. **ANIMALS** A sheep's normal body temperature is 39°C. However, a healthy sheep may have body temperatures 1°C above or below this temperature. What is the range of body temperatures for a sheep?

37. **MINIATURE GOLF** Ginger's score was within 5 strokes of her average score of 52. Determine the range of scores for Ginger's game.

Express each statement using an inequality involving absolute value. Do *not* solve.

38. The pH of a swimming pool must be within 0.3 of a pH of 7.5.
39. The temperature inside a refrigerator should be within 1.5 degrees of 38°F.
40. Ramona's bowling score was within 6 points of her average score of 98.
41. The cruise control of a car should keep the speed within 3 miles per hour of 55.

42. **MULTIPLE REPRESENTATIONS** In this problem, you will investigate the graphs of absolute value inequalities on a coordinate plane.
    a. **TABULAR** Copy and complete the table. Substitute the x and $f(x)$ values for each point into each inequality. Mark whether the resulting statement is *true* or *false*.

Point	$f(x) \geq \lvert x - 1 \rvert$	true/false	$f(x) \leq \lvert x - 1 \rvert$	true/false
(−4, 2)				
(−2, 2)				
(0, 2)				
(2, 2)				
(4, 2)				

    b. **GRAPHICAL** Graph $f(x) = \lvert x - 1 \rvert$.
    c. **GRAPHICAL** Plot each point that made $f(x) \geq \lvert x - 1 \rvert$ a true statement on the graph in red. Plot each point that made $f(x) \leq \lvert x - 1 \rvert$ on the graph in blue.
    d. **LOGICAL** Make a conjecture about what the graphs of $f(x) \geq \lvert x - 1 \rvert$ and $f(x) \leq \lvert x - 1 \rvert$ look like. Complete the table with other points to verify your conjecture.
    e. **GRAPHICAL** Use what you discovered to graph $f(x) \geq \lvert x - 3 \rvert$.

H.O.T. Problems

Use Higher-Order Thinking Skills

43. **FIND THE ERROR** Lucita sketched a graph of her solution to $\lvert 2a - 3 \rvert > 1$. Is she correct? Explain your reasoning.

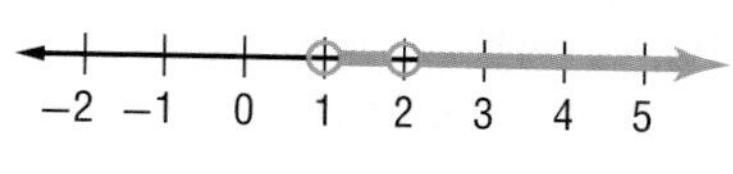

44. **REASONING** The graph of an absolute value inequality is *sometimes, always,* or *never* the union of two graphs. Explain.

45. **CHALLENGE** Demonstrate why the solution of $\lvert t \rvert > 0$ is not all real numbers. Explain your reasoning.

46. **OPEN ENDED** Write an absolute value inequality to represent a real-world situation. Interpret the solution.

47. **WRITING IN MATH** Explain how to determine whether an absolute value inequality uses a compound inequality with *and* or a compound inequality with *or*. Then summarize how to solve absolute value inequalities.

Standardized Test Practice

48. The formula for acceleration in a circle is $a = \frac{v^2}{r}$. Which of the following shows the equation solved for v?

A $v = ar$

B $v = \sqrt{ar}$

C $v^2 = ar$

D $v = \frac{\sqrt{a}}{r}$

49. An engraver charges a \$3 set-up fee and \$0.25 per word. Which table shows the total price p for w words?

F

w	p
15	\$3
20	\$4.25
25	\$5.50
30	\$7.75

G

w	p
15	\$6.75
20	\$7
25	\$7.25
30	\$7.50

H

w	p
15	\$3.75
20	\$5
25	\$6.25
30	\$8.50

J

w	p
15	\$6.75
20	\$8
25	\$9.25
30	\$10.50

50. **SHORT RESPONSE** The table shows the items in stock at the school store the first day of class. What is the probability that an item chosen at random was a notebook?

Item	Number Purchased
pencil	57
pen	38
eraser	6
folder	25
notebook	18

51. Solve for n.

$$|2n - 3| = 5$$

A $\{-4, -1\}$

B $\{-1, 4\}$

C $\{1, 1\}$

D $\{4, 4\}$

Spiral Review

Solve each compound inequality. Then graph the solution set. (Lesson 5-4)

52. $b + 3 < 11$ and $b + 2 > -3$

53. $6 \leq 2t - 4 \leq 8$

54. $2c - 3 \geq 5$ or $3c + 7 \leq -5$

55. **FINANCIAL LITERACY** In a recent year, the sum of the number of \$2 bills and \$50 bills in circulation was 1,857,573,945. The number of \$50 bills was 494,264,809 more than the number of \$2 bills. How many of each type of bill was in circulation? (Lesson 5-3)

56. **GEOMETRY** One angle of a triangle measures 10° more than the second. The measure of the third angle is twice the sum of the measure of the first two angles. Find the measure of each angle. (Lesson 2-4)

Solve each equation. Then check your solution. (Lesson 2-2)

57. $c - 7 = 11$

58. $2w = 24$

59. $9 + p = -11$

60. $\frac{t}{5} = 20$

Skills Review

Graph each equation. (Lesson 3-1)

61. $y = 4x - 1$

62. $y - x = 3$

63. $2x - y = -4$

64. $3y + 2x = 6$

65. $4y = 4x - 16$

66. $2y - 2x = 8$

67. $-9 = -3x - y$

68. $-10 = 5y - 2x$

5-6 Graphing Inequalities in Two Variables

Then

You graphed linear equations. (Lesson 3-1)

Now

- Graph linear inequalities on the coordinate plane.
- Solve inequalities by graphing.

New Vocabulary

boundary
half-plane
closed half-plane
open half-plane

Math Online

glencoe.com

- Extra Examples
- Personal Tutor
- Self-Check Quiz
- Homework Help

Why?

Hannah has budgeted \$35 every three months for car maintenance. From this she must buy oil costing \$3 and filters that cost \$7 each. How much oil and how many filters can Hannah buy and stay within her budget?

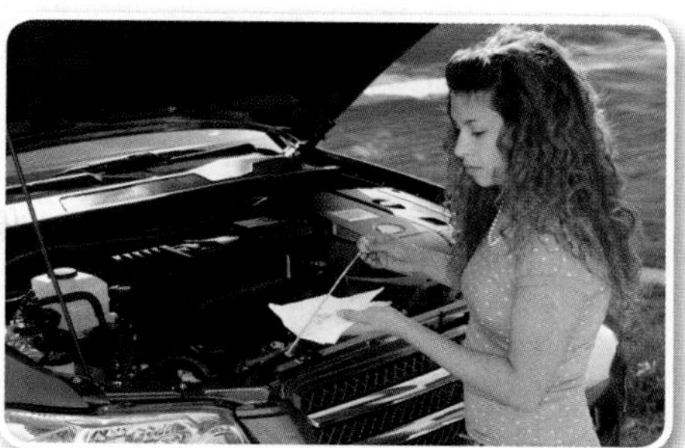

Graph Linear Inequalities The graph of a linear inequality is the set of points that represent all of the possible solutions of that inequality. An equation defines a **boundary**, which divides the coordinate plane into two **half-planes**.

The boundary may or may not be included in the graph of an inequality. When it is included, the solution is a **closed half-plane**. When not included, the solution is an **open half-plane**.

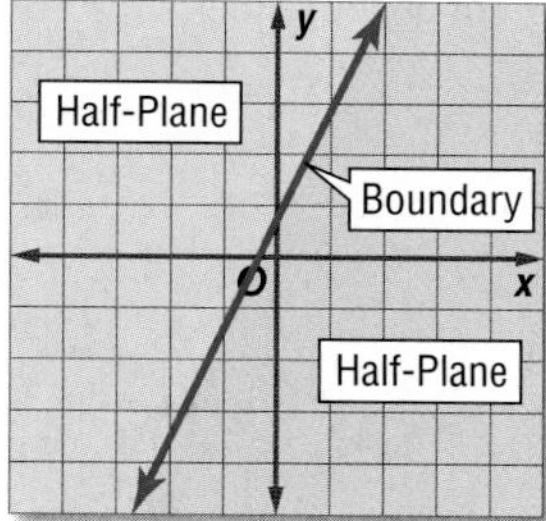

Key Concept — Graphing Linear Inequalities (For Your FOLDABLE)

Step 1 Graph the boundary. Use a solid line when the inequality contains $\leq$ or $\geq$. Use a dashed line when the inequality contains $<$ or $>$.

Step 2 Use a test point to determine which half-plane should be shaded.

Step 3 Shade the half-plane that contains the solution.

EXAMPLE 1 Graph an Inequality ($<$ or $>$)

Graph $3x - y < 2$.

Step 1 First, solve for y in terms of x.

$$3x - y < 2$$
$$-y < -3x + 2$$
$$y > 3x - 2$$

Then, graph $y = 3x - 2$. Because the inequality involves $>$, graph the boundary with a dashed line.

Step 2 Select a test point in either half-plane. A simple choice is (0, 0).

$3x - y < 2$	**Original inequality**
$3(0) - 0 < 2$	**$x = 0$ and $y = 0$**
$0 < 2$	**true**

Step 3 So, the half-plane containing the origin is the solution. Shade this half-plane.

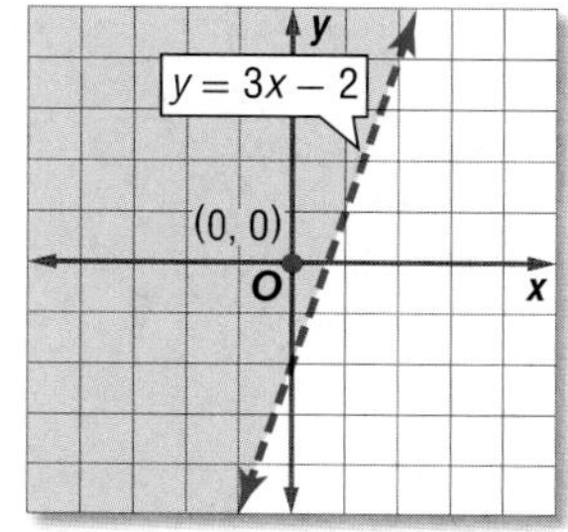

Check Your Progress Graph each inequality.

1A. $y > \frac{1}{2}x + 3$

1B. $x - 1 > y$

Personal Tutor glencoe.com

EXAMPLE 2 Graph an Inequality ($\leq$ or $\geq$)

Graph $x + 5y \leq 10$.

Step 1 Solve for y in terms of x.

$x + 5y \leq 10$	**Original inequality**
$5y \leq -x + 10$	**Subtract x from each side and simplify.**
$y \leq -\frac{1}{5}x + 2$	**Divide each side by 5.**

Graph $y = -\frac{1}{5}x + 2$. Because the inequality symbol is $\leq$, graph the boundary with a solid line.

Step 2 Select a test point. Let's use (3, 3). Substitute the values into the original inequality.

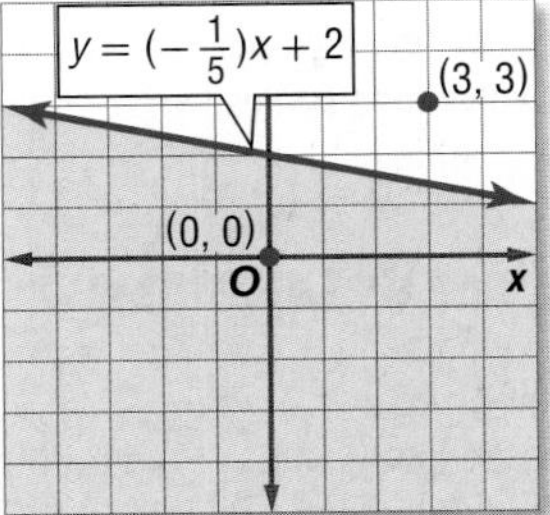

$x + 5y \leq 10$	**Original inequality**
$3 + 5(3) \leq 10$	**$x = 3$ and $y = 3$**
$18 \nleq 10$	**Simplify.**

Step 3 Since this statement is false, shade the other half-plane.

Selecting a Test Point When selecting a test point, a standard choice is the origin because it offers easy calculations. However, if it lies on the border, you must choose another point that is not on the border.

(StudyTip)

Check Your Progress Graph each inequality.

2A. $x - y \leq 3$

2B. $2x + 3y \geq 18$

Personal Tutor glencoe.com

Solve Linear Inequalities We can use a coordinate plane to solve inequalities with one variable.

EXAMPLE 3 Solve Inequalities From Graphs

Use a graph to solve $3x + 5 < 14$.

Step 1 First graph the boundary, which is the related equation. Replace the inequality sign with an equals sign, and solve for x.

$3x + 5 < 14$	**Original inequality**
$3x + 5 = 14$	**Change $<$ to $=$.**
$3x = 9$	**Subtract 5 from each side and simplify.**
$x = 3$	**Divide each side by 3.**

Graph $x = 3$ with a dashed line.

Step 2 Choose (0, 0) as a test point. These values in the original inequality give us $5 < 14$.

Step 3 Since this statement is true, shade the half-plane that contains the point (0, 0).

Notice that the x-intercept of the graph is at 3. Since the half-plane to the left of the x-intercept is shaded, the solution is $x < 3$.

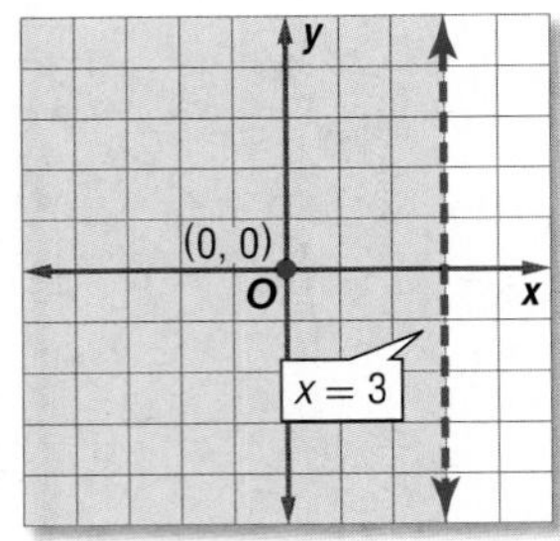

Check Your Progress Use a graph to solve each inequality.

3A. $4x - 3 \geq 17$

3B. $-2x + 6 > 12$

Personal Tutor glencoe.com

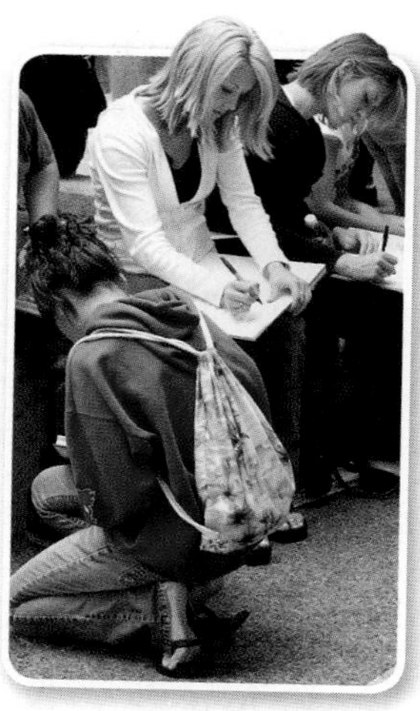

Real-World Link

As a supplement to traditional yearbooks, many schools are producing digital versions. They include features that allow you to click on a picture and see a short video clip.

Source: *eSchool News*

When using inequalities to solve real-world problems, the domain and the range are often restricted to nonnegative or whole numbers.

Real-World EXAMPLE 4 Write and Solve an Inequality

CLASS PICNIC A yearbook company promises to give the junior class a picnic if they spend at least \$28,000 on yearbooks and class rings. Each yearbook costs \$35, and each class ring costs \$140. How many yearbooks and class rings must the junior class buy to get their picnic?

Understand You know the cost of each item and the minimum amount the class needs to spend.

Plan Let $x =$ the number of yearbooks and $y =$ the number of class rings the class must buy. Write an inequality.

\$35	times	the number of yearbooks	plus	\$140	times	the number of rings	is at least	\$28,000.
35	$\cdot$	x	$+$	140	$\cdot$	y	$\geq$	28,000

Solve Solve for y in terms of x.

$35x + 140y - 35x \geq 28{,}000 - 35x$ **Subtract 35x from each side.**

$140y \geq -35x + 28{,}000$ **Divide each side by 140.**

$\frac{140y}{140} \geq \frac{-35x}{140} + \frac{28000}{140}$ **Simplify.**

$y \geq -0.25x + 200$ **Simplify.**

Because the yearbook company cannot sell a negative number of items, the domain and range must be nonnegative numbers. Graph the boundary with a solid line. If we test (0, 0), the result is $0 \geq 28{,}000$, which is false. Shade the closed half-plane that does not include the origin. One solution is (500, 100), or 500 yearbooks and 100 class rings.

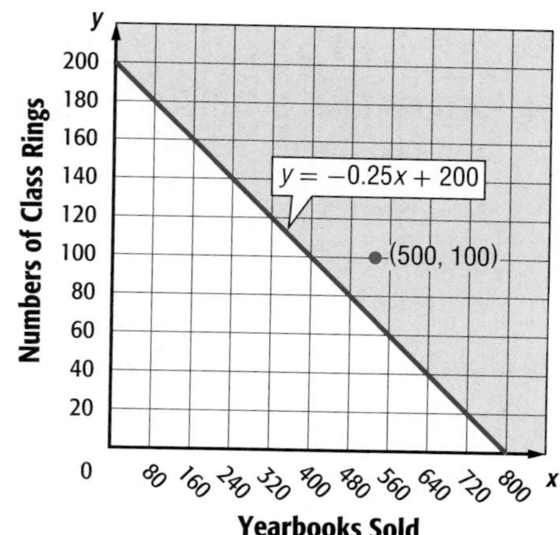

Check If we test (500, 100), the result is $100 \geq 75$, which is true. Because the company cannot sell a fraction of an item, only points with whole-number coordinates can be solutions.

Problem-SolvingTip

Use a Graph You can use a graph to visualize data, analyze trends, and make predictions.

Check Your Progress

4. **MARATHONS** Neil wants to run a marathon at a pace of at least 6 miles per hour. Write and graph an inequality for the miles y he will run in x hours.

Personal Tutor glencoe.com

Check Your Understanding

Examples 1 and 2 pp. 315–316

Graph each inequality.

1. $y > x + 3$
2. $y \geq -8$
3. $x + y > 1$
4. $y \leq x - 6$
5. $y < 2x - 4$
6. $x - y \leq 4$

Example 3 p. 316

Use a graph to solve each inequality.

7. $7x + 1 < 15$
8. $-3x - 2 \geq 11$
9. $3y - 5 \leq 34$
10. $4y - 21 > 1$

Example 4 p. 317

11. **FINANCIAL LITERACY** The surf shop has a weekly overhead of $2300.
    a. Write an inequality to describe this situation.
    b. How many skimboards and longboards must the shop sell each week to make a profit?

Practice and Problem Solving

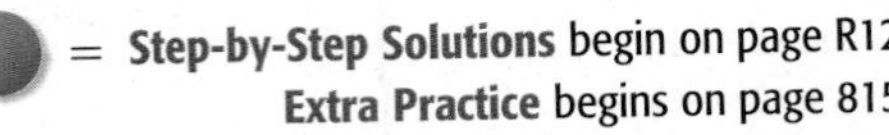

Examples 1 and 2 pp. 315–316

Graph each inequality.

12. $y < x - 3$
13. $y > x + 12$
14. $y \geq 3x - 1$
15. $y \leq -4x + 12$
16. $6x + 3y > 12$
17. $2x + 2y < 18$
18. $5x + y > 10$
19. $2x + y < -3$
20. $-2x + y \geq -4$
21. $8x + y \leq 6$
22. $10x + 2y \leq 14$
23. $-24x + 8y \geq -48$

Example 3 p. 316

Use a graph to solve each inequality.

24. $10x - 8 < 22$
25. $20x - 5 > 35$
26. $4y - 77 \geq 23$
27. $5y + 8 \leq 33$
28. $35x + 25 < 6$
29. $14x - 12 > -31$

Example 4 p. 317

30. **DECORATING** Sybrina is decorating her bedroom. She has $300 to spend on paint and bed linens. A gallon of paint costs $14, while a set of bed linens costs $60.
    a. Write an inequality for this situation.
    b. How many gallons of paint and bed linen sets can Sybrina buy and stay within her budget?

Use a graph to solve each inequality.

31. $3x + 2 < 0$
32. $4x - 1 > 3$
33. $-6x - 8 \geq -4$
34. $-5x + 1 < 3$
35. $-7x + 13 < 10$
36. $-4x - 4 \leq -6$

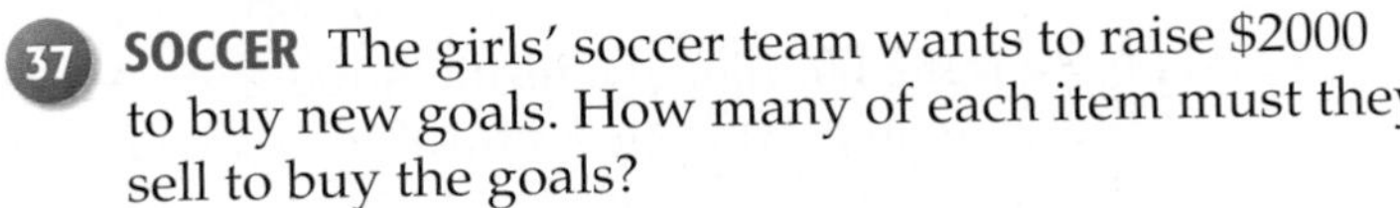

37. **SOCCER** The girls' soccer team wants to raise $2000 to buy new goals. How many of each item must they sell to buy the goals?
    a. Write an inequality that represents this situation.
    b. Graph this inequality.
    c. Make a table of values that shows at least five possible solutions.
    d. Plot the solutions from part c.

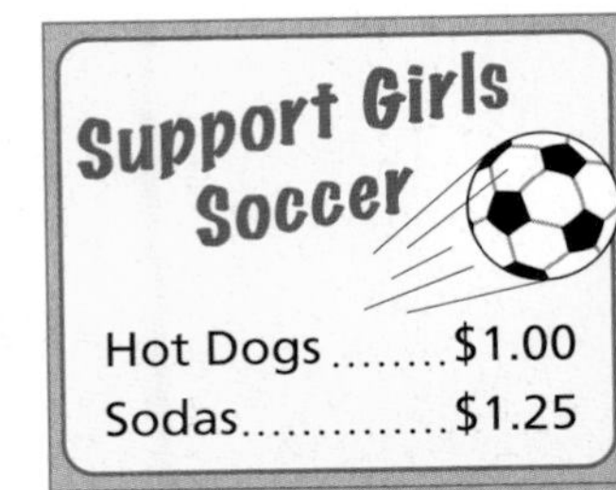

Graph each inequality. Determine which of the ordered pairs are part of the solution set for each inequality.

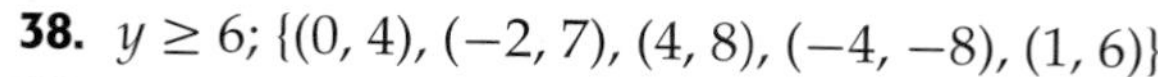

38. $y \geq 6$; {(0, 4), (−2, 7), (4, 8), (−4, −8), (1, 6)}

39. $x < -4$; {(2, 1), (−3, 0), (0, −3), (−5, −5), (−4, 2)}

40. $2x - 3y \leq 1$; {(2, 3), (3, 1), (0, 0), (0, −1), (5, 3)}

41. $5x + 7y \geq 10$; {(−2, −2), (1, −1), (1, 1), (2, 5), (6, 0)}

42. $-3x + 5y < 10$; {(3, −1), (1, 1), (0, 8), (−2, 0), (0, 2)}

43. $2x - 2y \geq 4$; {(0, 0), (0, 7), (7, 5), (5, 3), (2, −5)}

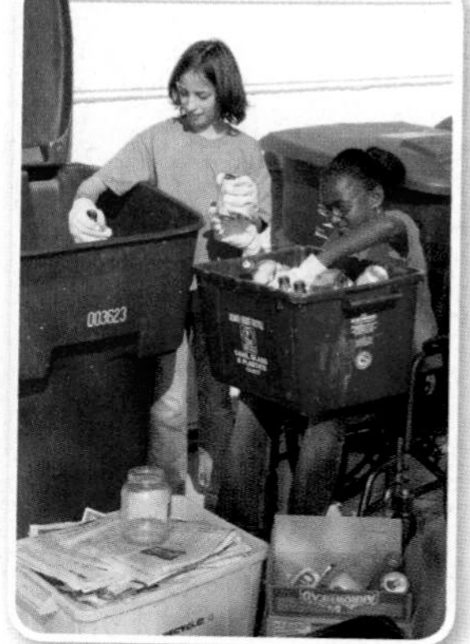

Real-World Link

The energy saved when recycling one glass bottle is enough to light a traditional light bulb for four hours.

Source: PlanetPals

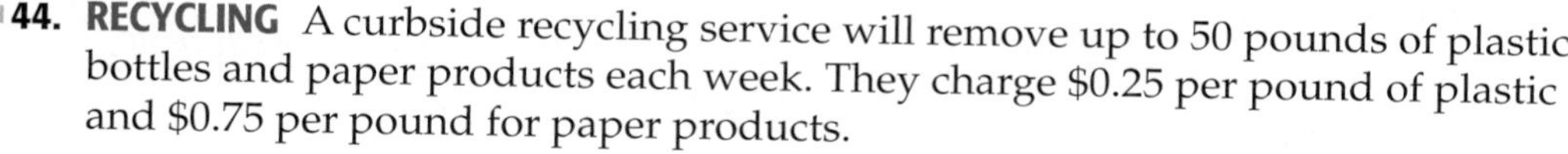

44. RECYCLING A curbside recycling service will remove up to 50 pounds of plastic bottles and paper products each week. They charge \$0.25 per pound of plastic and \$0.75 per pound for paper products.

a. Write an inequality that describes the pounds of each kind of product that can be included in the curbside service.

b. Write an inequality that describes the charge.

c. Graph each inequality.

d. Compare the two graphs.

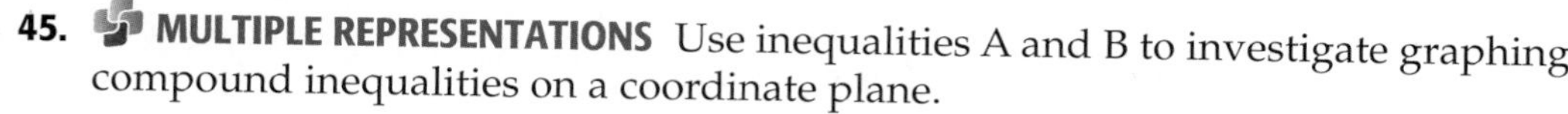

45. MULTIPLE REPRESENTATIONS Use inequalities A and B to investigate graphing compound inequalities on a coordinate plane.

A. $7(y + 6) \leq 21x + 14$ **B.** $-3y \leq 3x - 12$

a. NUMERICAL Solve each inequality for y.

b. GRAPHICAL Graph both inequalities on one graph. Shade the half-plane that makes A true in red. Shade the half-plane that makes B true in blue.

c. VERBAL What does the overlapping region represent?

H.O.T. Problems

Use **H**igher-**O**rder **T**hinking Skills

46. FIND THE ERROR Reiko and Kristin are solving $4y \leq \frac{8}{3}x$ by graphing. Is either of them correct? Explain your reasoning.

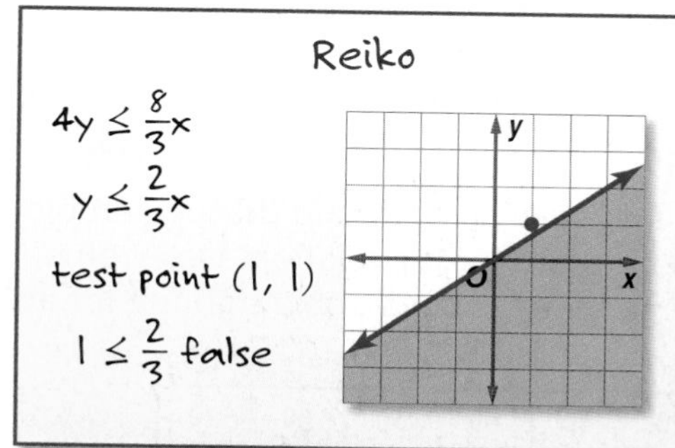

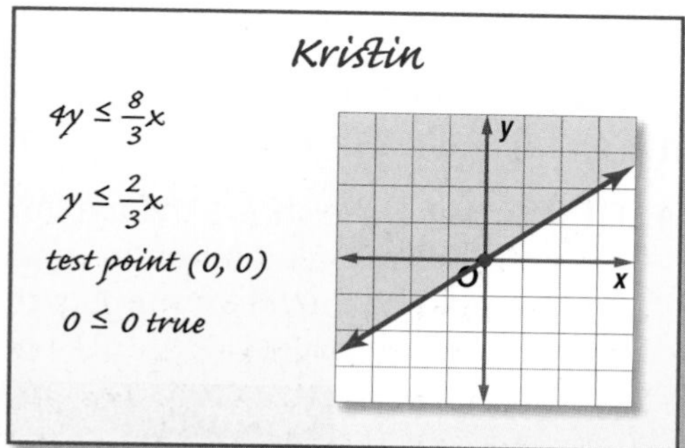

47. CHALLENGE Graph $y > |x + 5|$.

48. REASONING Explain why a point on the boundary should not be used as a test point.

49. OPEN ENDED Write a two-variable inequality with a restricted domain and range to represent a real-world situation. Give the domain and range, and explain why they are restricted.

50. WRITING IN MATH Summarize the steps to graph an inequality in two variables.

Standardized Test Practice

51. What is the domain of this function?

A $\{x \mid 0 \leq x \leq 3\}$
B $\{x \mid 0 \leq x \leq 9\}$
C $\{y \mid 0 \leq y \leq 9\}$
D $\{y \mid 0 \leq y \leq 3\}$

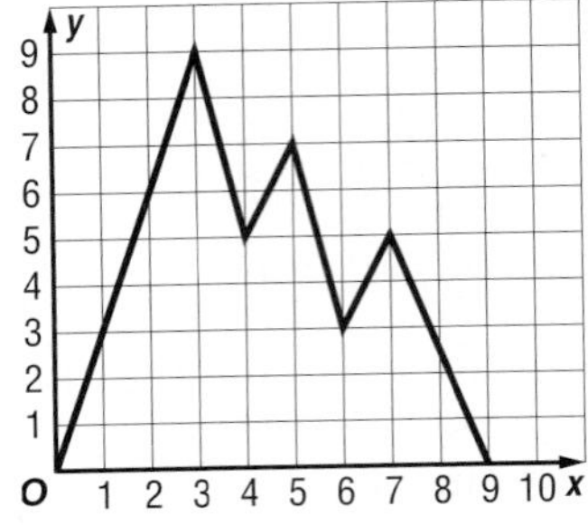

52. EXTENDED RESPONSE An arboretum will close for the winter when all of the trees have lost their leaves. The table shows the number of trees each day that still have leaves.

Day	5	10	15	20
Trees with Leaves	325	260	195	130

a. Write an equation that represents the number of trees with leaves y after d days.

b. Find the y-intercept. What does it mean in the context of this problem?

c. After how many days will the arboretum close? Explain how you got your answer.

53. Which inequality best represents the statement below?

A jar contains 832 gumballs. Ebony's guess was within 46 pieces.

F $|g - 832| \leq 46$
G $|g + 832| \leq 46$
H $|g - 832| \geq 46$
J $|g + 832| \geq 46$

54. GEOMETRY If the rectangular prism has a volume of 10,080 cm^3, what is the value of x?

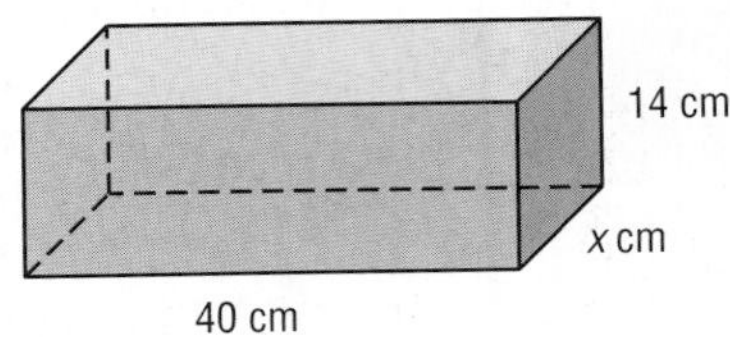

A 12
B 14
C 16
D 18

Spiral Review

Solve each open sentence. (Lesson 5-5)

55. $|y - 2| > 4$ **56.** $|t - 6| \leq 5$ **57.** $|3 + d| < -4$

Solve each compound inequality. (Lesson 5-4)

58. $4c - 4 < 8c - 16 < 6c - 6$ **59.** $5 < \frac{1}{2}p + 3 < 8$ **60.** $0.5n \geq -7$ or $2.5n + 2 \leq 9$

Write an equation of the line that passes through each pair of points. (Lesson 4-2)

61. $(1, -3)$ and $(2, 5)$ **62.** $(-2, -4)$ and $(-7, 3)$ **63.** $(-6, -8)$ and $(-8, -5)$

64. FITNESS The table shows the maximum heart rate to maintain during aerobic activities. Write an equation in function notation for the relation. Determine what would be the maximum heart rate to maintain in aerobic training for an 80-year-old. (Lesson 3-5)

Age (yr)	20	30	40	50	60	70
Pulse rate (beats/min)	175	166	157	148	139	130

Skills Review

65. WORK The formula $s = \frac{w - 10r}{m}$ is used to find keyboarding speeds. In the formula, s represents the speed in words per minute, w the number of words typed, r the number of errors, and m the number of minutes typed. Solve for r. (Lesson 2-8)

Graphing Technology Lab
Graphing Inequalities

Math Online glencoe.com
- Other Calculator Keystrokes
- Graphing Technology Personal Tutor

You can use a graphing calculator to investigate the graphs of inequalities.

ACTIVITY 1 Less Than

Graph $y \leq 2x + 5$.

Clear all functions from the Y= list.

KEYSTROKES: [Y=] [CLEAR]

Graph $y \leq 2x + 5$ in the standard window.

KEYSTROKES: 2 [X,T,θ,n] [+] 5 [◄] [◄] [◄] [◄] [◄] [◄] [ENTER] [ENTER] [ENTER] [Zoom] 6

All ordered pairs for which y is *less than or equal to* $2x + 5$ lie *below or on* the line and are solutions.

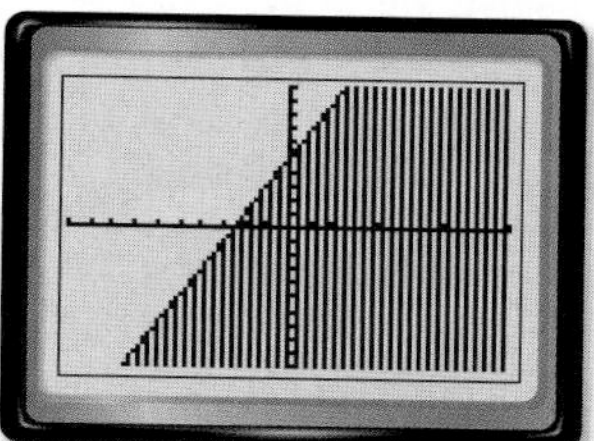

[−10, 10] scl: 1 by [−10, 10] scl: 1

ACTIVITY 2 Greater Than

Graph $y - 2x \geq 5$.

Clear the graph that is currently displayed.

KEYSTROKES: [Y=] [CLEAR]

Rewrite $y - 2x \geq 5$ as $y \geq 2x + 5$ and graph it.

KEYSTROKES: 2 [X,T,θ,n] [+] 5 [◄] [◄] [◄] [◄] [◄] [◄] [ENTER] [ENTER] [Zoom] 6

All ordered pairs for which y is *greater than or equal to* $2x + 5$ lie *above or on* the line and are solutions.

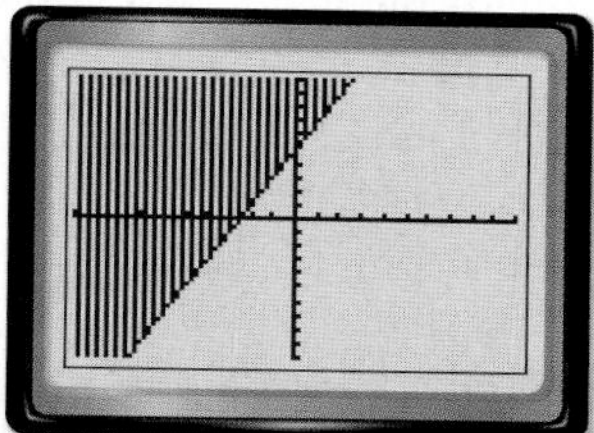

[−10, 10] scl: 1 by [−10, 10] scl: 1

Exercises

1. Compare and contrast the two graphs shown above.
2. Graph $y \geq -3x + 1$ in the standard viewing window. Using your graph, name four solutions of the inequality.
3. Suppose student water park tickets cost $16, and adult water park tickets cost $20. You would like to buy at least 10 tickets but spend no more than $200.
   - **a.** Let x = number of student tickets and y = number of adult tickets. Write two inequalities, one representing the total number of tickets and the other representing the total cost of the tickets.
   - **b.** Graph the inequalities. Use the viewing window [0, 20] scl: 1 by [0, 20] scl: 1.
   - **c.** Name four possible combinations of student and adult tickets.

Study Guide and Review

Math Online glencoe.com
- STUDY TO GO
- Vocabulary Review

Chapter Summary

Key Concepts

Solving One-Step Inequalities (Lessons 5-1 and 5-2)

For all numbers a, b, and c, the following are true.

- If $a > b$ and c is positive, $ac > bc$.
- If $a > b$ and c is negative, $ac < bc$.

Multi-Step and Compound Inequalities (Lessons 5-3 and 5-4)

- Multi-step inequalities can be solved by undoing the operations in the same way you would solve a multi-step equation.
- A compound inequality containing *and* is only true if both inequalities are true.
- A compound inequality containing *or* is true if at least one of the inequalities is true.

Absolute Value Inequalities (Lesson 5-5)

- The absolute value of any number x is its distance from zero on a number line and is written as $|x|$. If $x \geq 0$, then $|x| = x$. If $x < 0$, then $|x| = -x$.
- If $|x| < n$ and $n > 0$, then $-n < x < n$.
- If $|x| > n$ and $n > 0$, then $x > n$ or $x < -n$.

Inequalities in Two Variables (Lesson 5-6)

To graph an inequality:

Step 1 Graph the boundary. Use a solid line when the inequality contains $\leq$ or $\geq$. Use a dashed line when the inequality contains $<$ or $>$.

Step 2 Use a test point to determine which half-plane should be shaded.

Step 3 Shade the half-plane.

FOLDABLES® Study Organizer

Be sure the Key Concepts are noted in your Foldable.

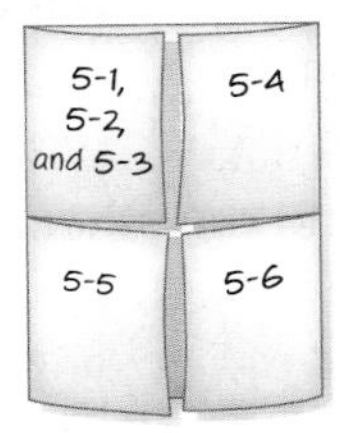

Key Vocabulary

boundary (p. 315)

closed half-plane (p. 315)

compound inequality (p. 304)

half plane (p. 315)

intersection (p. 304)

open half-plane (p. 315)

set-builder notation (p. 284)

union (p. 305)

Vocabulary Check

State whether each sentence is *true* or *false*. If *false*, replace the underlined term to make a true sentence.

1. Set-builder notation is a <u>less</u> concise way of writing a solution set.
2. There are <u>two</u> types of compound inequalities.
3. The graph of a compound inequality containing *and* shows the <u>union</u> of the individual graphs.
4. A compound inequality containing *or* is true if one or both of the inequalities is true. Its graph is the <u>union</u> of the graphs of the two inequalities.
5. The graph of an inequality of the form $y < ax + b$ is a region on the coordinate plane called a <u>half-plane</u>.
6. A <u>point</u> defines the boundary of an open half-plane.
7. The <u>boundary</u> is the graph of the equation of the line that defines the edge of each half-plane.
8. The solution set to the inequality $y \geq x$ includes the <u>boundary</u>.
9. When solving an inequality, <u>multiplying</u> each side by a negative number reverses the inequality symbol.
10. The graph of a compound inequality that contains <u>*and*</u> is the intersection of the graphs of the two inequalities.

Lesson-by-Lesson Review

5-1 Solving Inequalities by Addition and Subtraction (pp. 283–288)

Solve each inequality. Then graph it on a number line.

11. $w - 4 > 9$

12. $x + 8 \leq 3$

13. $6 + h < 1$

14. $-5 < a + 2$

15. $13 - p \geq 15$

16. $y + 1 \leq 8$

17. FIELD TRIP A bus can hold 44 people. If there are 35 students in Samantha's class, how many more people can ride on the bus?

EXAMPLE 1

Solve $x - 9 < -4$. Then graph it on a number line.

$x - 9 < -4$	**Original inequality**
$x - 9 + 9 < -4 + 9$	**Add 9 to each side.**
$x < 5$	**Simplify.**

The solution set is $\{x \mid x < 5\}$.

−5 −4 −3 −2 −1 0 1 2 3 4 5

5-2 Solving Inequalities by Multiplication and Division (pp. 290–295)

Solve each inequality. Check your solution.

18. $\frac{1}{3}x > 6$

19. $\frac{1}{5}g \geq -4$

20. $4p < 32$

21. $-55 \leq -5w$

22. $-2m > 100$

23. $\frac{2}{3}t < -48$

24. MOVIE RENTAL Jack has no more than $24 to spend on DVDs for a party. Each DVD rents for $4. Find the maximum number of DVDs Jack can rent for his party.

EXAMPLE 2

Solve $-14h < 56$. Check your solution.

$-14h < 56$	**Original inequality**
$\frac{-14h}{-14} > \frac{56}{-14}$	**Divide each side by −14.**
$h > -4$	**Simplify.**

CHECK To check, substitute three different values into the original inequality: −4, a number less than −4, and a number greater than −4.

5-3 Solving Multi-Step Inequalities (pp. 296–301)

Solve each inequality. Check your solution.

25. $3h - 7 < 14$

26. $4 + 5b > 34$

27. $18 \leq -2x + 8$

28. $\frac{t}{3} - 6 > -4$

29. Four times a number decreased by 6 is less than −2. Define a variable, write an inequality, and solve for the number.

30. TICKET SALES The drama club collected $160 from ticket sales for the spring play. They need to collect at least $400 to pay for new lighting for the stage. If tickets sell for $3 each, how many more tickets need to be sold?

EXAMPLE 3

Solve $-6y - 13 > 29$. Check your solution.

$-6y - 13 > 29$	**Original inequality**
$-6y - 13 + 13 > 29 + 13$	**Add 13 to each side.**
$-6y > 42$	**Simplify.**
$\frac{-6y}{-6} < \frac{42}{-6}$	**Divide each side by −6 and change > to <.**
$y < -7$	**Simplify.**

The solution set is $\{y \mid y < -7\}$.

CHECK	$-6y - 13 > 29$	**Original inequality**
	$-6(-10) - 13 \stackrel{?}{>} 29$	**Substitute −10 for *y*.**
	$47 > 29$ ✓	**Simplify.**

5-4 Solving Compound Inequalities (pp. 304–309)

Solve each compound inequality. Then graph the solution set.

31. $m - 3 < 6$ and $m + 2 > 4$

32. $-4 < 2t - 6 < 8$

33. $3x + 2 \leq 11$ or $5x - 8 > 22$

34. **KITES** A kite can be flown in wind speeds no less than 7 miles per hour and no more than 16 miles per hour. Write an inequality for the wind speeds at which the kite can fly.

EXAMPLE 4

Solve $-3w + 4 > -8$ and $2w - 11 > -19$. Then graph the solution set.

$-3w + 4 > -8$ and $2w - 11 > -19$

$w < 4$ $\qquad$ $w > -4$

To graph the solution set, graph $w < 4$ and graph $w > -4$. Then find the intersection.

−5 −4 −3 −2 −1 0 1 2 3 4 5

5-5 Inequalities Involving Absolute Value (pp. 310–314)

Solve each inequality. Then graph the solution set.

35. $|x - 4| < 9$

36. $|p + 2| > 7$

37. $|2c + 3| \leq 11$

38. $|f - 9| \geq 2$

39. $|3d - 1| \leq 8$

40. $\left|\frac{4b - 2}{3}\right| < 12$

41. $\left|\frac{2t + 6}{2}\right| > 10$

42. $|-4y - 3| < 13$

43. $|m + 19| \leq 1$

44. $|-k - 7| \geq 4$

EXAMPLE 5

Solve $|x - 6| < 9$. Then graph the solution set.

Case 1 $x - 6$ is nonnegative.

$x - 6 < 9$

$x < 15$

Case 2 $x - 6$ is negative.

$-(x - 6) < 9$

$x > -3$

The solution set is $\{x \mid -3 < x < 15\}$.

−4 −2 0 2 4 6 8 10 12 14 16 18

5-6 Graphing Inequalities in Two Variables (pp. 315–320)

Graph each inequality.

45. $y > x - 3$

46. $y < 2x + 1$

47. $3x - y \leq 4$

48. $y \geq -2x + 6$

49. $5x - 2y < 10$

50. $3x + 4y > 12$

Graph each inequality. Determine which of the ordered pairs are part of the solution set for each inequality.

51. $y \leq 4$; $\{(3, 6), (1, 2), (-4, 8), (3, -2), (1, 7)\}$

52. $-2x + 3y \geq 12$; $\{(-2, 2), (-1, 1), (0, 4), (2, 2)\}$

53. **BAKERY** Ben has $24 to spend on cookies and cupcakes. Write and graph an inequality that represents what Ben can buy.

$2

$3

EXAMPLE 6

Graph $2x - y > 3$.

Solve for y in terms of x.

$2x - y > 3$	**Original inequality**
$-y > -2x + 3$	**Subtract 2x from each side.**
$y < 2x - 3$	**Multiply each side by −1.**

Graph the boundary using a dashed line. Choose (0, 0) as a test point.

$2(0) - 0 \stackrel{?}{>} 3$

$0 \ngtr 3$

Since 0 is not greater than 3, shade the plane that does not contain (0, 0).

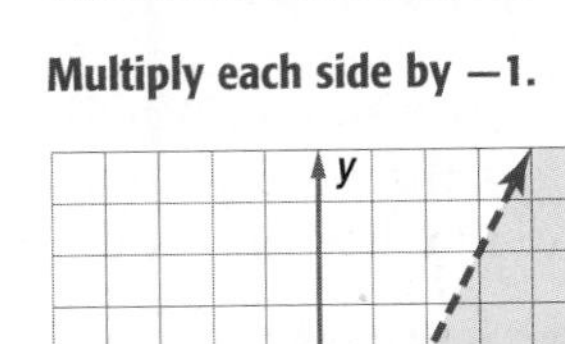
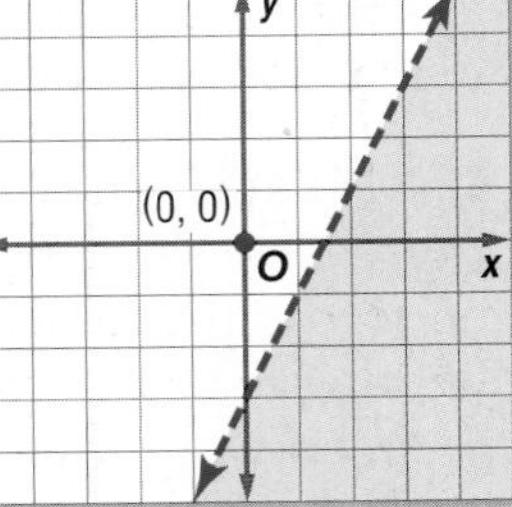

CHAPTER 5 Practice Test

Math Online glencoe.com
Chapter Test

Solve each inequality. Then graph it on a number line.

1. $x - 9 < -4$
2. $6p \geq 5p - 3$

3. **MULTIPLE CHOICE** Drew currently has 31 comic books in his collection. His friend Connor has 58 comic books. How many more comic books does Drew need to add to his collection in order to have a larger collection than Connor?

   A no more than 21

   B 27

   C at least 28

   D more than 30

Solve each inequality. Check your solution.

4. $\frac{1}{5}h > 3$
5. $7w \leq -42$
6. $-\frac{2}{3}t \geq 24$
7. $-9m < -36$
8. $3c - 7 < 11$
9. $\frac{g}{4} + 3 \leq -9$
10. $-2(x - 4) > 5x - 13$

11. **ZOO** The 8th grade science class is going to the zoo. The class can spend up to $300 on admission.

Zoo Admission	
Visitor	**Cost**
Student	$8
Adult	$10

   a. Write an inequality for this situation.

   b. If there are 32 students in the class and 1 adult will attend for every 8 students, how much will admission be?

Solve each compound inequality. Then graph the solution set.

12. $y - 8 < -3$ or $y + 5 > 19$
13. $-11 \leq 2h - 5 \leq 13$
14. $3z - 2 > -5$ and $7z + 4 < -17$

Define a variable, write an inequality, and solve the problem. Check your solution.

15. The difference of a number and 4 is no more than 8.

16. Nine times a number decreased by four is at least twenty-three.

17. **MULTIPLE CHOICE** Write a compound inequality for the graph shown below.

−5 −4 −3 −2 −1 0 1 2 3 4 5

   F $-2 \leq x < 3$

   G $x \leq -2$ or $x \geq 3$

   H $x < -2$ or $x \geq 3$

   J $-2 < x \leq 3$

Solve each inequality. Then graph the solution set.

18. $|p - 5| < 3$
19. $|2f + 7| \geq 21$
20. $|-4m + 3| \leq 15$
21. $\left|\frac{x - 3}{4}\right| > 5$

22. **RETAIL** A sporting goods store is offering a $15 coupon on any pair of shoes.

   a. The most and least expensive pairs of shoes are $149.95 and $24.95. What is the range of costs for customers with coupons?

   b. When buying a pair of $109.95 shoes, you can use a coupon or a 15% discount. Which option is best?

Graph each inequality.

23. $y < 4x - 1$
24. $2x + 3y \geq 12$

25. Graph $y > -2x + 5$. Then determine which of the ordered pairs in $\{(-2, 0), (-1, 5), (2, 3), (7, 3)\}$ are in the solution set.

26. **PRESCHOOL** Mrs. Jones is buying new books and puzzles for her preschool classroom. Each book costs $6, and each puzzle costs $4. Write and graph an inequality to determine how many books and puzzles she can buy for $96.

CHAPTER 5 Preparing for Standardized Tests

Write and Solve an Inequality

Many multiple-choice items will require writing and solving inequalities. Follow the steps below to help you successfully solve these types of problems.

Strategies for Writing and Solving Inequalities

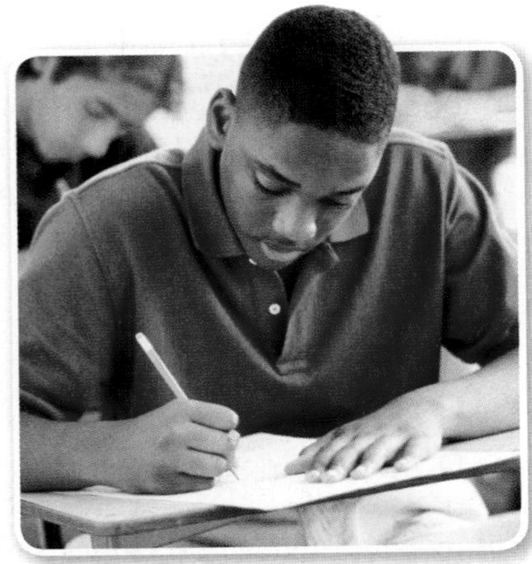

Step 1

Read the problem statement carefully.

Ask yourself:

- What am I being asked to solve?
- What information is given in the problem?
- What are the unknowns for which I need to solve?

Step 2

Translate the problem statement into an inequality.

- Assign variables to the unknown(s).
- Write the word sentence as a mathematical number sentence looking for words such as *greater than*, *less than*, *no more than*, *up to*, or *at least to* indicate the type of inequality as well as where to place the inequality sign.

Step 3

Solve the inequality.

- Solve for the unknowns in the inequality.
- Remember that multiplying or dividing each side by a negative number reverses the direction of the inequality.
- Check your answer to make sure it makes sense.

EXAMPLE

Read the problem. Identify what you need to know. Then use the information in the problem to solve. Show your work.

Pedro has earned scores of 89, 74, 79, 85, and 88 on his tests this semester. He needs a test average of at least 85 in order to earn an A for the semester. There will be one more test given this semester.

A Write an inequality to model the situation.

B What score must he have on his final test to earn an A for the semester?

Read the problem carefully. You are given Pedro's first 5 test scores and told that he needs an average of *at least* 85 after his next test to earn an A for the semester.

a. Write the inequality.

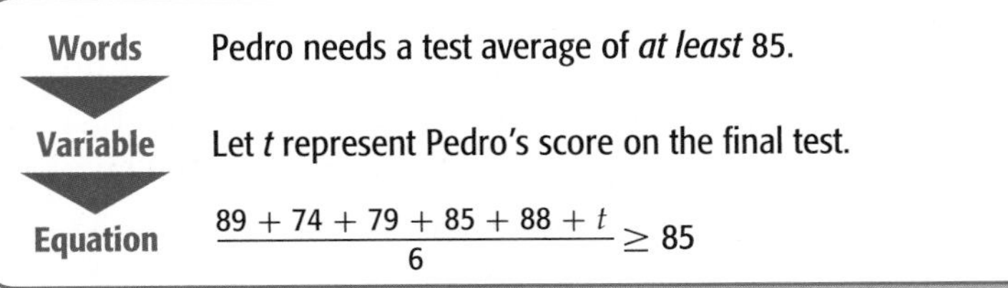

b. Solve the inequality for t.

$$\frac{89 + 74 + 79 + 85 + 88 + t}{6} \geq 85$$

$$89 + 74 + 79 + 85 + 88 + t \geq 85(6)$$

$$415 + t \geq 510$$

$$t \geq 95$$

So, Pedro's final test score must be greater than or equal to 95 in order for him to earn an A for the semester.

Exercises

Read each problem. Identify what you need to know. Then use the information in the problem to solve.

1. Craig has \$20 to order a pizza. The pizza costs \$12.50 plus \$0.95 per topping. If there is also a \$3 deliver fee, how many toppings can Craig order?

2. To join an archery club, Nina had to pay an initiation fee of \$75, plus \$40 per year in membership dues.
   a. Write an equation to model the total cost, y, of belonging to the club for x years.
   b. How many years will it take her to spend more than \$400 to belong to the club?

3. The area of the triangle below is no more than 84 square millimeters. What is the height of the triangle?

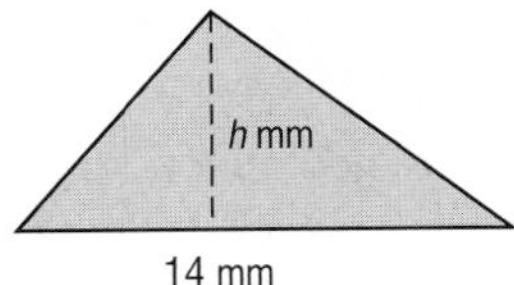

4. Rosa earns \$200 a month delivering newspapers, plus an average of \$11 per hour babysitting. If her goal is to earn at least \$295 this month, how many hours will she have to babysit?

5. To earn money for a new bike, Ethan is selling some of his baseball cards. He has saved \$245. If the bike costs \$1400, and he can sell 154 cards, for how much money will he need to sell each card to reach his goal?

6. In a certain lacrosse league, there can be no more than 22 players on each team, and no more than 10 teams per age group. There are 6 age groups.
   a. Write an inequality to represent this situation.
   b. What is the greatest number of players that can play lacrosse in this league?

7. Sarah has \$120 to shop for herself and to buy some gifts for 6 of her friends. She has purchased a shirt for herself for \$32. What is the maximum that she can spend on each friend?

Standardized Test Practice

Cumulative, Chapters 1 through 5

Multiple Choice

Read each question. Then fill in the correct answer on the answer document provided by your teacher or on a sheet of paper.

1. Miguel received a $100 gift certificate for a graduation gift. He wants to buy a CD player that costs $38 and CDs that cost $12 each. Which of the following inequalities represents how many CDs Miguel can buy?

A $n \leq 6$

B $n \geq 5$

C $n < 5$

D $n \leq 5$

2. Craig is paid time-and-a-half for any additional hours over 40 that he works.

Time	Pay Rate
Up to 40 hours	$12.80/hr
Additional hours worked over 40	$19.20/hr

If Craig's goal is to earn at least $600 next week, what is the minimum number of hours he needs to work?

F 43 hours **H** 44 hours

G 45 hours **J** 46 hours

3. Which equation has a slope of $-\frac{2}{3}$ and a y-intercept of 6?

A $y = 6x + \frac{2}{3}$ **C** $y = -\frac{2}{3}x + 6$

B $y = -\frac{2}{3}x - 6$ **D** $y = 6x - \frac{2}{3}$

4. The highest score that is on record on a video game is 10,219 points. The lowest score on record is 257 points. Which of the following inequalities best shows the range of scores recorded on the game?

F $x \leq 10{,}219$

G $x \geq 257$

H $257 < x < 10{,}219$

J $257 \leq x \leq 10{,}219$

5. Kyle scored 14 points in his last basketball game, bringing his total points for the season to over 100. Which number line represents the number of points Kyle had scored prior to the last game?

A
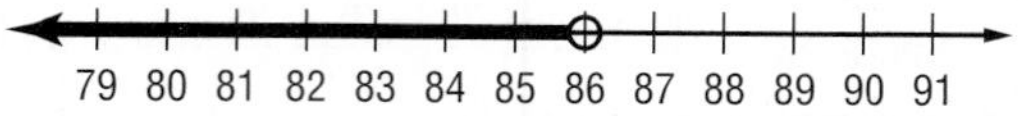

B
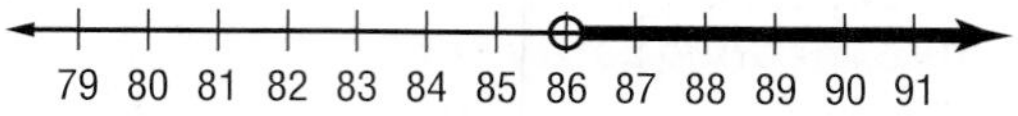

C
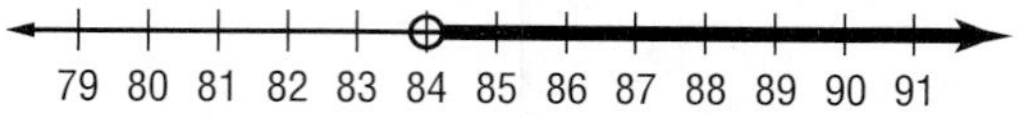

D
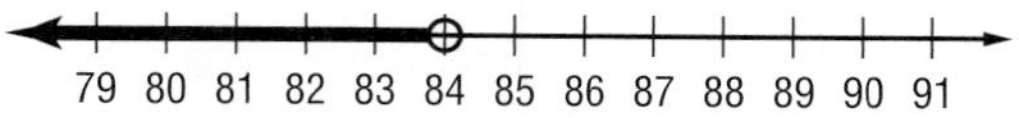

6. The girls' volleyball team is selling T-shirts and pennants to raise money for new uniforms. The team hopes to raise more than $250.

Item	Price
T-shirt	$10
Pennant	$4

Which of the following combinations of items sold would meet this goal?

F 16 T-shirts and 20 pennants

G 20 T-shirts and 12 pennants

H 18 T-shirts and 18 pennants

J 15 T-shirts and 20 pennants

7. What type of line does not have a defined slope?

A horizontal **C** perpendicular

B parallel **D** vertical

8. Which expression below illustrates the Associative Property?

F $abc = bac$

G $2(x - 3) = 2x - 6$

H $(p + 3) - t = p(3 - t)$

J $5 + (-5) = 0$

Test-TakingTip

Question 2 You can check your answer by finding Craig's earnings for the hours worked.

Short Response/Gridded Response

Record your answers on the answer sheet provided by your teacher or on a sheet of paper.

9. Solve $-4 < 3x + 8 \leq 23$.

10. GRIDDED RESPONSE Tien is saving money for a new television. She needs to save at least $720 to pay for her expenses. Each week Tien saves $50 toward her new television. How many weeks will it take so she can pay for the television?

11. Write an inequality that best represents the graph.

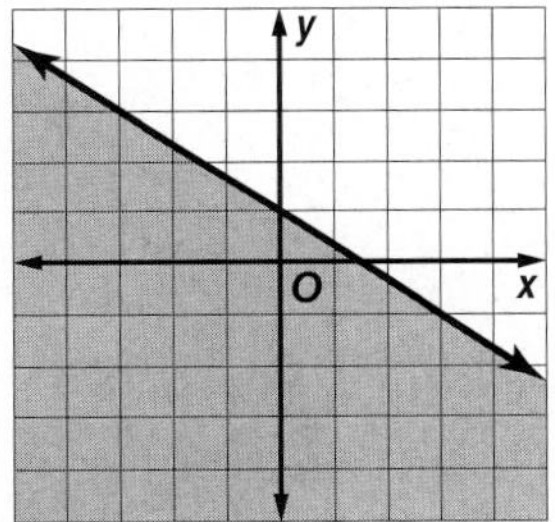

12. Solve $|x - 4| < 2$.

13. GRIDDED RESPONSE Daniel wants to ship a set of golf clubs and several boxes of golf balls in a box that can hold up to 20 pounds. If the set of clubs weighs 9 pounds and each box of golf balls weighs 12 ounces, how many boxes of golf balls can Daniel ship?

14. Graph the solution set for the inequality $3x - 6 \leq 4x - 4 \leq 3x + 1$.

15. Write an equation that represents the data in the table.

x	y
3	12.5
4	16
5	19.5
6	23
7	26.5

16. A sporting goods company near the beach rents bicycles for $10 plus $5 per hour. Write an equation in slope-intercept form that shows the total cost, y, of renting a bicycle for x hours. How much would it cost Emily to rent a bicycle for 6 hours?

Extended Response

Record your answers on a sheet of paper. Show your work.

17. Theresa is saving money for a vacation. She needs to save at least $640 to pay for her expenses. Each week, she puts $35 towards her vacation savings.

a. Let w represent the number of weeks Theresa saves money. Write an inequality to model the situation.

b. Solve the inequality from part a. What is the minimum number of weeks Theresa must save money in order to reach her goal?

c. If Theresa were to save $45 each week instead, by how many weeks would the minimum savings time be decreased?

Need Extra Help?

If you missed Question...	1	2	3	4	5	6	7	8	9	10	11	12	13	14	15	16	17
Go to Lesson or Page...	5-3	5-2	4-1	5-4	5-1	5-6	3-3	1-3	5-4	5-2	5-6	5-5	5-3	5-4	2-1	4-2	5-2

CHAPTER 6 Systems of Linear Equations and Inequalities

Then

In Chapter 2, you solved linear equations in one variable.

Now

In Chapter 6, you will:

- Solve systems of linear equations by graphing, substitution, and elimination.
- Solve systems of linear inequalities by graphing.

Why?

MUSIC \$1500 worth of tickets were sold for a marching band competition. Adult tickets were \$12 each, and student tickets were \$8 each. If you knew how many total tickets were sold, you could use a system of equations to determine how many adult tickets and how many student tickets were sold.

Systems of Linear Equations and Inequalities

Introduction

\$1500 worth of tickets were sold for a marching band competition. Adult tickets were \$12 each, and student tickets were \$8 each. If you want to know how many tickets were sold, you could use a system of equations to determine how many adult tickets and how many student tickets were sold.

Adult Ticket \$12 (Admit One)

Student Ticket \$8 (Admit One)

1/5

Math *in Motion,* Animation glencoe.com

Get Ready for Chapter 6

Diagnose Readiness You have two options for checking Prerequisite Skills.

Text Option Take the Quick Check below. Refer to the Quick Review for help.

*Quick*Check

Name the ordered pair for each point on the coordinate plane. (Lesson 1-6)

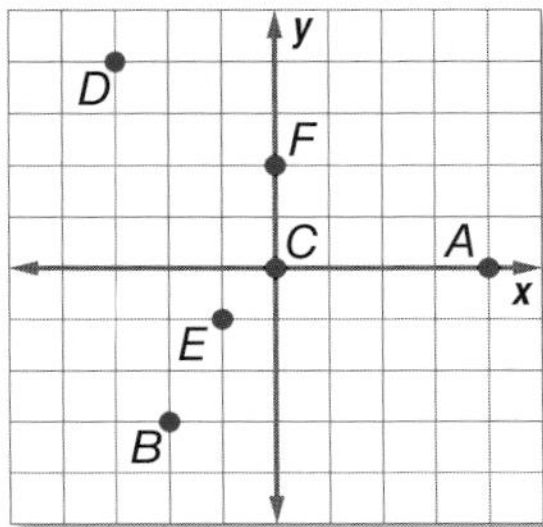

1. A
2. D
3. B
4. C
5. E
6. F

Solve each equation or formula for the variable specified. (Lesson 2-8)

7. $2x + 4y = 12$, for x
8. $x = 3y - 9$, for y
9. $m - 2n = 6$, for m
10. $y = mx + b$, for x
11. $P = 2\ell + 2w$, for ℓ
12. $5x - 10y = 40$, for y
13. **GEOMETRY** The formula for the area of a triangle is $A = \frac{1}{2}bh$, where A represents the area, b is the base, and h is the height of the triangle. Solve the equation for b.

*Quick*Review

EXAMPLE 1

Name the ordered pair for Q on the coordinate plane.

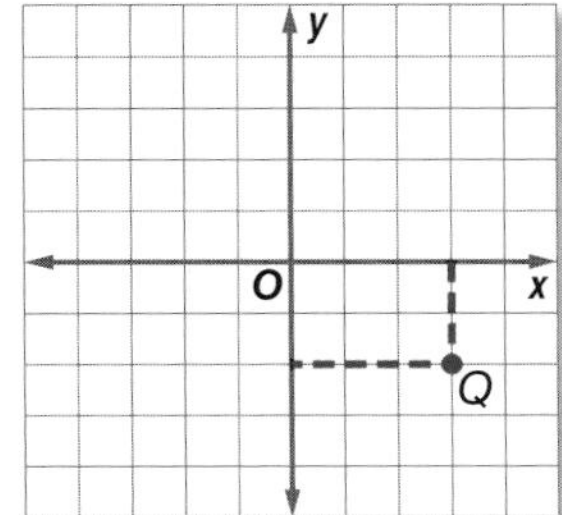

Follow a vertical line from the point to the x-axis. This gives the x-coordinate, 3.

Follow a horizontal line from the point to the y-axis. This gives the y-coordinate, -2.

The ordered pair is $(3, -2)$.

EXAMPLE 2

Solve $12x + 3y = 36$ for y.

$12x + 3y = 36$	**Original equation**
$12x + 3y - 12x = 36 - 12x$	**Subtract 12*x* from each side.**
$3y = 36 - 12x$	**Simplify.**
$\frac{3y}{3} = \frac{36 - 12x}{3}$	**Divide each side by 3.**
$y = 12 - 4x$	**Simplify.**

Online Option Math Online Take a self-check Chapter Readiness Quiz at glencoe.com.

Get Started on Chapter 6

You will learn several new concepts, skills, and vocabulary terms as you study Chapter 6. To get ready, identify important terms and organize your resources. You may wish to refer to **Chapter 0** to review prerequisite skills.

FOLDABLES® Study Organizer

Linear Functions Make this Foldable to help you organize your Chapter 6 notes about solving systems of equations and inequalities. Begin with a sheet of notebook paper.

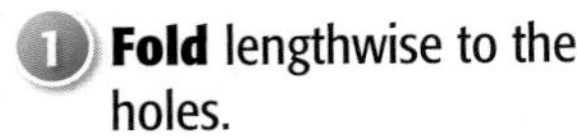

1. **Fold** lengthwise to the holes.

2. **Cut** 8 tabs.

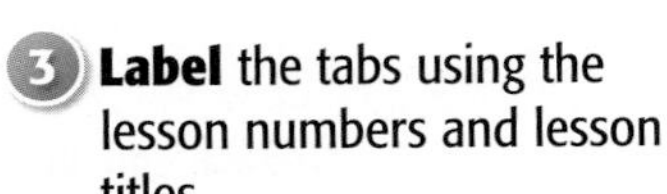

3. **Label** the tabs using the lesson numbers and lesson titles.

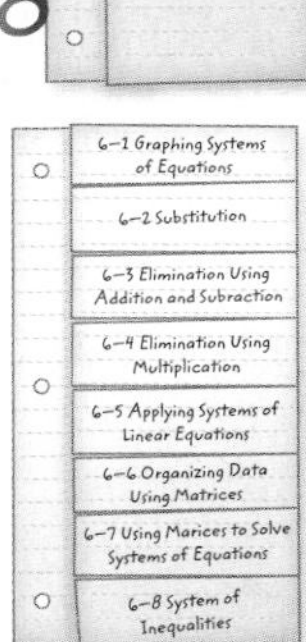

Math Online glencoe.com

- Study the chapter online
- Explore **Math in Motion**
- Get extra help from your own **Personal Tutor**
- Use **Extra Examples** for additional help
- Take a **Self-Check Quiz**
- **Review Vocabulary** in fun ways

New Vocabulary

English		Español
system of equations	p. 333	sistema de ecuaciones
consistent	p. 333	consistente
independent	p. 333	independiente
dependent	p. 333	dependiente
inconsistent	p. 333	inconsistente
substitution	p. 342	sustitución
elimination	p. 348	eliminación
matrix	p. 369	matriz
element	p. 369	elemento
dimension	p. 369	dimensión
scalar	p. 371	escalar
scalar multiplication	p. 371	multiplicación por escalares
augmented matrix	p. 376	matriz ampliada
row reduction	p. 377	reducción de fila
identity matrix	p. 377	matriz
system of inequalities	p. 382	sistema de desigualdades

Review Vocabulary

domain • p. 39 • dominio the set of the first numbers of the ordered pairs in a relation

intersection • p. 304 • intersección the graph of a compound inequality containing *and*; the solution is the set of elements common to both graphs

proportion • p. 111 • proporción an equation stating that two ratios are equal

Proportion

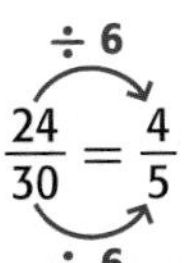

Multilingual eGlossary glencoe.com

6-1 Graphing Systems of Equations

Then

You graphed linear equations. (Lesson 3-1)

Now

- Determine the number of solutions a system of linear equations has.
- Solve systems of linear equations by graphing.

New Vocabulary

system of equations
consistent
independent
dependent
inconsistent

Math Online

glencoe.com

- Extra Examples
- Personal Tutor
- Self-Check Quiz
- Homework Help
- Math in Motion

Why?

The cost to begin production on a band's CD is \$1500. Each CD costs \$4 to produce and will sell for \$10. The band wants to know how many CDs they will have to sell to earn a profit.

Graphing a system can show when a company makes a profit. The cost of producing the CD can be modeled by the equation $y = 4x + 1500$, where y represents the cost of production and x is the number of CDs produced.

The income from the CDs sold can be modeled by the equation $y = 10x$, where y represents the total income of selling the CDs, and x is the number of CDs sold.

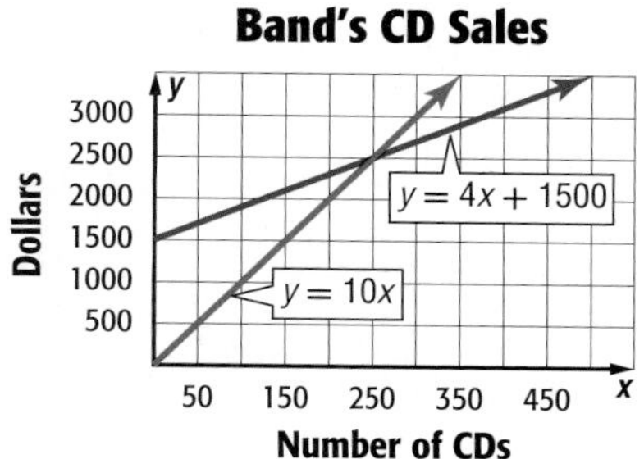

If we graph these equations, we can see at which point the band begins making a profit. The point where the two graphs intersect is where the band breaks even. This happens when the band sells 250 CDs. If the band sells more than 250 CDs, they will make a profit.

Possible Number of Solutions The two equations, $y = 4x + 1500$ and $y = 10x$, form a **system of equations**. The ordered pair that is a solution of both equations is the solution of the system. A system of two linear equations can have one solution, an infinite number of solutions, or no solution.

- If a system has at least one solution, it is said to be **consistent**. The graphs intersect at one point or are the same line.
- If a consistent system has exactly one solution, it is said to be **independent**. If it has an infinite number of solutions, it is **dependent**. This means that there are unlimited solutions that satisfy both equations.
- If a system has no solution, it is said to be **inconsistent**. The graphs are parallel.

Concept Summary: Possible Solutions

For Your FOLDABLE

Number of Solutions	exactly one	infinite	no solution
Terminology	consistent and independent	consistent and dependent	inconsistent
Graph			

StudyTip

Number of Solutions When both equations are of the form $y = mx + b$, the values of m and b can determine the number of solutions.

Compare m and b	Number of Solutions
different m values	one
same m value, but different b values	none
same m value, and same b value	infinite

Math *in Motion*, Animation glencoe.com

EXAMPLE 1 Number of Solutions

Use the graph at the right to determine whether each system is *consistent* or *inconsistent* and if it is *independent* or *dependent*.

a. $y = -2x + 3$
$y = x - 5$

Since the graphs of these two lines intersect at one point, there is exactly one solution. Therefore, the system is consistent and independent.

b. $y = -2x - 5$
$y = -2x + 3$

Since the graphs of these two lines are parallel, there is no solution of the system. Therefore, the system is inconsistent.

Check Your Progress

1A. $y = 2x + 3$
$y = -2x - 5$

1B. $y = x - 5$
$y = -2x - 5$

Personal Tutor glencoe.com

Solve by Graphing One method of solving a system of equations is to graph the equations carefully on the same coordinate grid and find their point of intersection. This point is the solution of the system.

EXAMPLE 2 Solve by Graphing

Graph each system and determine the number of solutions that it has. If it has one solution, name it.

a. $y = -3x + 10$
$y = x - 2$

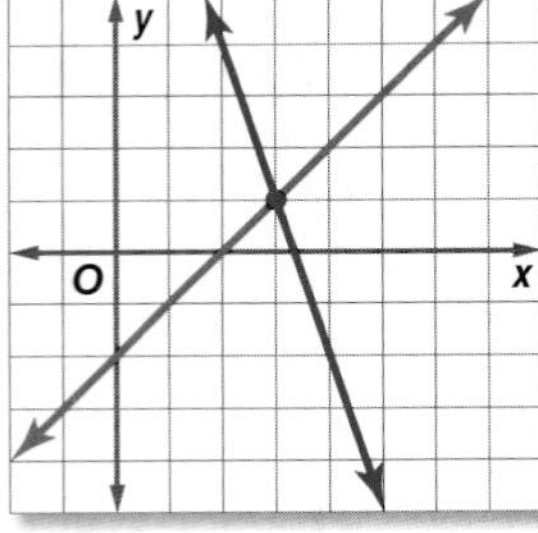

The graphs appear to intersect at the point (3, 1). You can check this by substituting 3 for x and 1 for y.

CHECK

$y = -3x + 10$	**Original equation**
$1 \stackrel{?}{=} -3(3) + 10$	**Substitution**
$1 \stackrel{?}{=} -9 + 10$	**Multiply.**
$1 = 1$ ✓	
$y = x - 2$	**Original equation**
$1 \stackrel{?}{=} 3 - 2$	**Substitution**
$1 = 1$ ✓	**Multiply.**

The solution is (3, 1).

b. $2x - y = -1$
$4x - 2y = 6$

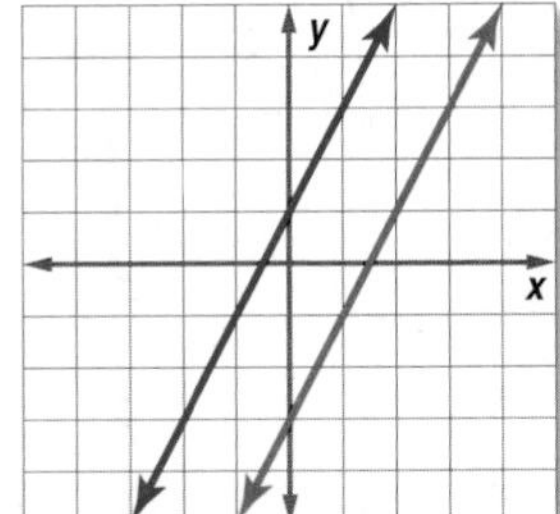

The lines have the same slope but different y-intercepts, so the lines are parallel. Since they do not intersect, there is no solution of this system. The system is inconsistent.

Review Vocabulary

parallel lines never intersect and have the same slope (Lesson 4-4)

Check Your Progress

Graph each system and determine the number of solutions that it has. If it has one solution, name it.

2A. $x - y = 2$
$3y + 2x = 9$

2B. $y = -2x - 3$
$6x + 3y = -9$

Personal Tutor glencoe.com

We can use what we know about systems of equations to solve many real-world problems that involve two or more different functions.

Real-World Link

In 2004, 2.9 million girls participated in high school sports. This was an all-time high for female participation.

Source: National Federation of State High School Associations

Math *in Motion*, Interactive Lab glencoe.com

Real-World EXAMPLE 3 Write and Solve a System of Equations

SPORTS The number of girls participating in high school soccer and track and field has steadily increased over the past few years. Use the information in the table to predict the approximate year when the number of girls participating in these two sports will be the same.

High School Sport	Number of Girls Participating in 2004 (thousands)	Average rate of increase (thousands per year)
soccer	309	8
track and field	418	3

Source: National Federation of State High School Associations

Words	Number of girls participating	equals	rate of increase	times	number of years after 2004	plus	number participating in 2004.
Variables	Let y = number of girls competing. Let x = number of years after 2004.						
Equations	Soccer: y	$=$	8	$\cdot$	x	$+$	309
	Track and field: y	$=$	3	$\cdot$	x	$+$	418

Graph $y = 8x + 309$ and $y = 3x + 418$. The graphs appear to intersect at approximately (22, 485).

CHECK Use substitution to check this answer.

$y = 8x + 309$ $y = 3x + 418$

$485 \stackrel{?}{=} 8(22) + 309$ $485 \stackrel{?}{=} 3(22) + 418$

$485 = 485$ ✓ $485 \approx 484$ ✓

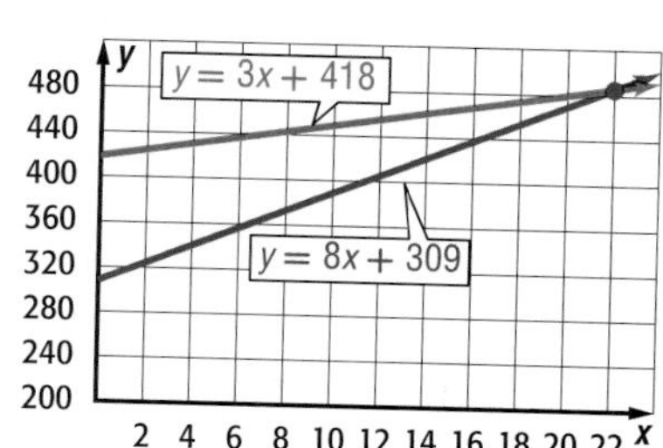

The solution means that approximately 22 years after 2004, or in 2026, the number of girls participating in high school soccer and track and field will be the same, about 485,000.

Check Your Progress

3. VIDEO GAMES Joe and Josh each want to buy a video game. Joe has $14 and saves $10 a week. Josh has $26 and saves $7 a week. In how many weeks will they have the same amount?

Personal Tutor glencoe.com

Check Your Understanding

Example 1 p. 334

Use the graph at the right to determine whether each system is *consistent* or *inconsistent* and if it is *independent* or *dependent*.

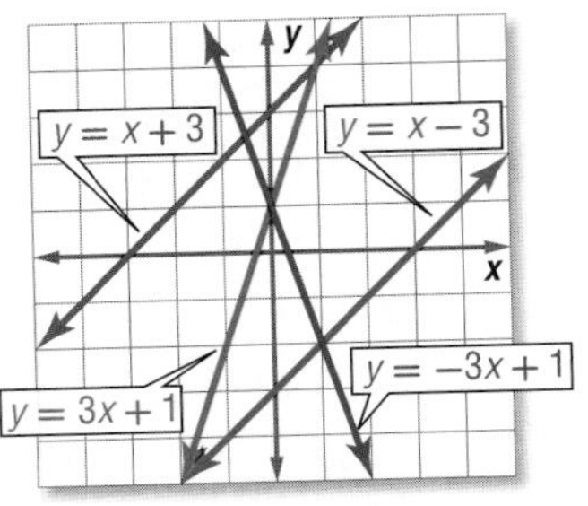

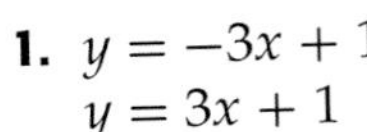

1. $y = -3x + 1$; $y = 3x + 1$
2. $y = 3x + 1$; $y = x - 3$
3. $y = x - 3$; $y = x + 3$
4. $y = x + 3$; $x - y = -3$
5. $x - y = -3$; $y = -3x + 1$
6. $y = -3x + 1$; $y = x - 3$

Example 2 p. 334

Graph each system and determine the number of solutions that it has. If it has one solution, name it.

7. $y = x + 4$; $y = -x - 4$
8. $y = x + 3$; $y = 2x + 4$

Example 3 p. 335

9. **READING** Alberto and Ashanti are reading a graphic novel.

a. Write an equation to represent the pages each boy has read.

b. Graph each equation.

c. How long will it be before Alberto has read more pages than Ashanti? Check and interpret your solution.

Practice and Problem Solving

 = **Step-by-Step Solutions** begin on page R12.
Extra Practice begins on page 815.

Example 1 p. 334

Use the graph at the right to determine whether each system is *consistent* or *inconsistent* and if it is *independent* or *dependent*.

10. $y = 6$; $y = 3x + 4$
11. $y = 3x + 4$; $y = -3x + 4$
12. $y = -3x + 4$; $y = -3x - 4$
13. $y = -3x - 4$; $y = 3x - 4$
14. $3x - y = -4$; $y = -3x + 4$
15. $3x - y = 4$; $3x + y = 4$

Example 2 p. 334

Graph each system and determine the number of solutions that it has. If it has one solution, name it.

16. $y = -3$; $y = x - 3$
17. $y = 4x + 2$; $y = -2x - 3$
18. $y = x - 6$; $y = x + 2$
19. $x + y = 4$; $3x + 3y = 12$
20. $x - y = -2$; $-x + y = 2$
21. $x + 2y = 3$; $x = 5$
22. $2x + 3y = 12$; $2x - y = 4$
23. $2x + y = -4$; $y + 2x = 3$
24. $2x + 2y = 6$; $5y + 5x = 15$

Example 3 p. 335

25. **SCHOOL DANCE** Akira and Jen are competing to see who can sell the most tickets for the Winter Dance. On Monday, Akira sold 22 and then sold 30 per day after that. Jen sold 53 one Monday and then sold 20 per day after that.

 a. Write equations for the number of tickets each person has sold.

 b. Graph each equation.

 c. Solve the system of equations. Check and interpret your solution.

26. **TRAVEL** If x is the number of years since 2000 and y is the percent of people using travel services, the following equations represent the percent of people using travel agents and the percent of the people using the Internet to plan travel.

 Travel agents: $y = -2x + 30$ Internet: $y = 6x + 41$

 a. Graph the system of equations.

 b. Estimate the year travel agents and the Internet were used equally.

Graph each system and determine the number of solutions that it has. If it has one solution, name it.

27. $y = \frac{1}{2}x$, $y = x + 2$

28. $y = 6x + 6$, $y = 3x + 6$

29. $y = 2x - 17$, $y = x - 10$

30. $8x - 4y = 16$, $-5x - 5y = 5$

31. $3x + 5y = 30$, $3x + y = 18$

32. $-3x + 4y = 24$, $4x - y = 7$

33. $2x - 8y = 6$, $x - 4y = 3$

34. $4x - 6y = 12$, $-2x + 3y = -6$

35. $2x + 3y = 10$, $4x + 6y = 12$

36. $3x + 2y = 10$, $2x + 3y = 10$

37. $3y - x = -2$, $y - \frac{1}{3}x = 2$

38. $\frac{8}{5}y = \frac{2}{5}x + 1$, $\frac{2}{5}y = \frac{1}{10}x + \frac{1}{4}$

39. $\frac{1}{3}x + \frac{1}{3}y = 1$, $x + y = 1$

40. $\frac{3}{4}x + \frac{1}{2}y = \frac{1}{4}$, $\frac{2}{3}x + \frac{1}{6}y = \frac{1}{2}$

41. $\frac{5}{6}x + \frac{2}{3}y = \frac{1}{2}$, $\frac{2}{5}x + \frac{1}{5}y = \frac{3}{5}$

Real-World Link

Many photographers feel that the most useful aspect of digital cameras is the ability to see photos instantly. The second most important aspect is the ability to manipulate and correct digital photos.

Source: Photography

42. **PHOTOGRAPHY** Suppose x represents the number of cameras sold and y represents the number of years since 2000. Then the number of digital cameras sold each year since 2000, in millions, can be modeled by the equation $y = 12.5x + 10.9$. The number of film cameras sold each year since 2000, in millions, can be modeled by the equation $y = -9.1x + 78.8$.

 a. Graph each equation.

 b. In which year did digital camera sales surpass film camera sales?

 c. In what year will film cameras stop selling altogether?

 d. What are the domain and range of each of the functions in this situation?

Graph each system and determine the number of solutions that it has. If it has one solution, name it.

43. $2y = 1.2x - 10$, $4y = 2.4x$

44. $x = 6 - \frac{3}{8}y$, $4 = \frac{2}{3}x + \frac{1}{4}y$

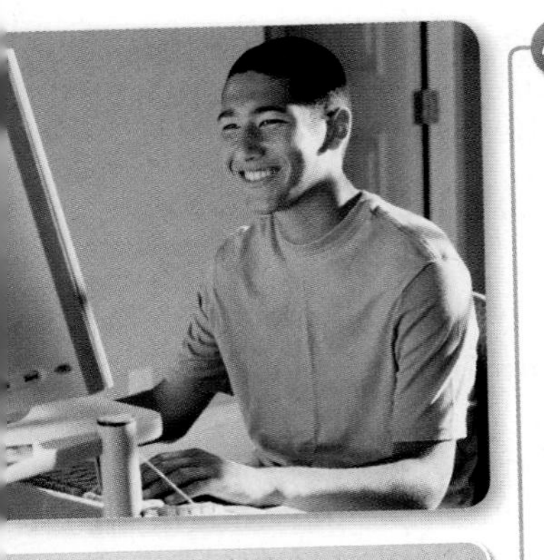

Real-World Link

Social networking sites have grown in popularity in recent years. Approximately 2 out of every 3 people online visit a social networking site.

Source: Read/Write Web

45. **WEB SITES** Personal publishing site *Lookatme* had 2.5 million visitors in 2005. Each year after that, the number of visitors rose by 13.1 million. Online auction site *Buyourstuff* had 59 million visitors in 2005, but each year after that the number of visitors fell by 2 million.

   a. Write an equation for each of the companies.

   b. Make a table of values for 5 years for each of the companies.

   c. Graph each equation.

   d. When will *Lookatme* and *Buyourstuff's* sites have the same number of visitors?

   e. Name the domain and range of these functions in this situation.

46. **MULTIPLE REPRESENTATIONS** In this problem, you will explore different methods for finding the intersection of the graphs of two linear equations.

   a. ALGEBRAIC Use algebra to solve the equation $\frac{1}{2}x + 3 = -x + 12$.

   b. GRAPHICAL Use a graph to solve $y = \frac{1}{2}x + 3$ and $y = -x + 12$.

   c. ANALYTICAL How is the equation in part **a** related to the system in part **b**?

   d. VERBAL Explain how to use the graph in part b to solve the equation in part a.

H.O.T. Problems

Use Higher-Order Thinking Skills

47. **FIND THE ERROR** Store A is offering a 10% discount on the purchase of all electronics in their store. Store B is offering $10 off all the electronics in their store. Francisca and Alan are deciding which offer will save them more money. Is either of them correct? Explain your reasoning.

Francisca	Alan
You can't determine which store has the better offer unless you know the price of the items you want to buy.	*Store A has the better offer because 10% of the sale price is a greater discount than $10.*

48. **CHALLENGE** Use graphing to find the solution of the system of equations $2x + 3y = 5$, $3x + 4y = 6$, and $4x + 5y = 7$.

49. **REASONING** Determine whether a system of two linear equations with (0, 0) and (2, 2) as solutions *sometimes, always,* or *never* has other solutions. Explain.

50. **WHICH ONE DOESN'T BELONG?** Which one of the following systems of equations doesn't belong with the other three? Explain your reasoning.

$4x - y = 5$ $-2x + y = -1$	$-x + 4y = 8$ $3x - 6y = 6$	$4x + 2y = 14$ $12x + 6y = 18$	$3x - 2y = 1$ $2x + 3y = 18$

51. **OPEN ENDED** Write three equations such that they form three systems of equations with $y = 5x - 3$. The three systems should be inconsistent, consistent and independent, and consistent and dependent, respectively.

52. **WRITING IN MATH** Describe the advantages and disadvantages to solving systems of equations by graphing.

Standardized Test Practice

53. SHORT RESPONSE Certain bacteria can reproduce every 20 minutes, doubling the population. If there are 450,000 bacteria in a population at 9:00 A.M., how many bacteria will be in the population at 2:00 P.M.?

54. GEOMETRY An 84-centimeter piece of wire is cut into equal segments and then attached at the ends to form the edges of a cube. What is the volume of the cube?

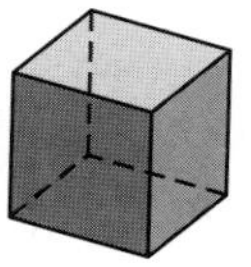

A 294 cm^3 C 1158 cm^3
B 343 cm^3 D 2744 cm^3

55. What is the solution of the inequality $-9 < 2x + 3 < 15$?

F $-x \geq 0$ H $-6 < x < 6$
G $x \leq 0$ J $-5 < x < 5$

56. What is the solution of the system of equations?

$$x + 2y = -1$$
$$2x + 4y = -2$$

A $(-1, -1)$ C no solution
B $(2, 1)$ D infinitely many solutions

Spiral Review

Graph each inequality. (Lesson 5-6)

57. $3x + 6y > 0$
58. $4x - 2y < 0$
59. $3y - x \leq 9$
60. $4y - 3x \geq 12$
61. $y < -4x - 8$
62. $3x - 1 > y$

63. LIBRARY To get a grant from the city's historical society, the number of history books must be within 25 of 1500. What is the range of the number of historical books that must be in the library? (Lesson 5-5)

64. SCHOOL Camilla's scores on three math tests are shown in the table. The fourth and final test of the grading period is tomorrow. She needs an average of at least 92 to receive an A for the grading period. (Lesson 5-3)

Test	Score
1	91
2	95
3	88

a. If m represents her score on the fourth math test, write an inequality to represent this situation.

b. If Camilla wants an A in math, what must she score on the test?

c. Is your solution reasonable? Explain.

Write the slope-intercept form of an equation for the line that passes through the given point and is perpendicular to the graph of the equation. (Lesson 4-4)

65. $(-3, 1), y = \frac{1}{3}x + 2$
66. $(6, -2), y = \frac{3}{5}x - 4$
67. $(2, -2), 2x + y = 5$
68. $(-3, -3), -3x + y = 6$

Skills Review

Find the solution of each equation using the given replacement set. (Lesson 1-5)

69. $f - 14 = 8$; {12, 15, 19, 22}
70. $15(n + 6) = 165$; {3, 4, 5, 6, 7}
71. $23 = \frac{d}{4}$; {91, 92, 93, 94, 95}
72. $36 = \frac{t-9}{2}$; {78, 79, 80, 81}

Evaluate each expression if $a = 2$, $b = -3$, and $c = 11$. (Lesson 1-2)

73. $a + 6b$
74. $7 - ab$
75. $(2c + 3a) \div 4$
76. $b^2 + (a^3 - 8)5$

EXTEND 6-1

Graphing Technology Lab

Systems of Equations

Math Online glencoe.com

- **Other Calculator Keystrokes**
- **Graphing Technology Personal Tutor**

You can use a graphing calculator to graph and solve a system of equations.

ACTIVITY 1 Solve a System of Equations

Solve the system of equations. State the decimal solution to the nearest hundredth.

$$5.23x + y = 7.48$$
$$6.42x - y = 2.11$$

Step 1 Solve each equation for y to enter them into the calculator.

$5.23x + y = 7.48$	**First equation**
$5.23x + y - 5.23x = 7.48 - 5.23x$	**Subtract 5.23*x* from each side.**
$y = 7.48 - 5.23x$	**Simplify.**
$6.42x - y = 2.11$	**Second equation**
$6.42x - y - 6.42x = 2.11 - 6.42x$	**Subtract 6.42*x* from each side.**
$-y = 2.11 - 6.42x$	**Simplify.**
$(-1)(-y) = (-1)(2.11 - 6.42x)$	**Multiply each side by −1.**
$y = -2.11 + 6.42x$	**Simplify.**

Step 2 Enter these equations in the Y= list and graph.

KEYSTROKES: *Review on pages 167–168.*

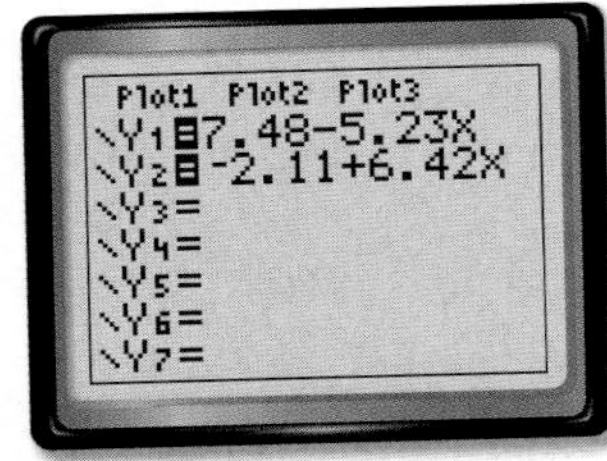

Step 3 Use the CALC menu to find the point of intersection.

KEYSTROKES: 2nd [CALC] 5 ENTER ENTER ENTER

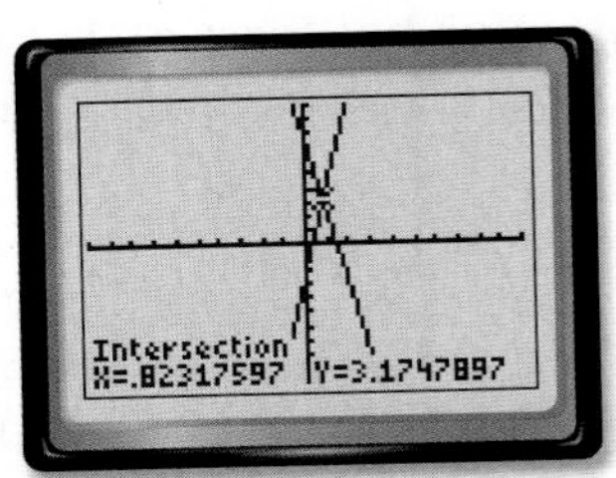

[−10, 10] scl: 1 by [−10, 10] scl: 1

The solution is approximately (0.82, 3.17).

One method you can use to solve an equation with one variable is by graphing and solving a system of equations based on the equation. To do this, write a system using both sides of the equation. Then use a graphing calculator to solve the system.

ACTIVITY 2 Use a System to Solve a Linear Equation

Use a system of equations to solve $5x + 6 = -4$.

Step 1 Write a system of equations.
Set each side of the equation equal to y.

$y = 5x + 6$ **First equation**

$y = -4$ **Second equation**

Step 2 Enter these equations in the **Y=** list and graph.

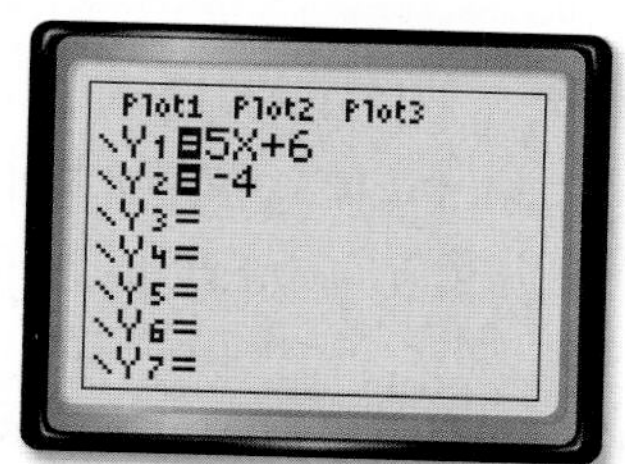

Step 3 Use the **CALC** menu to find the point of intersection.

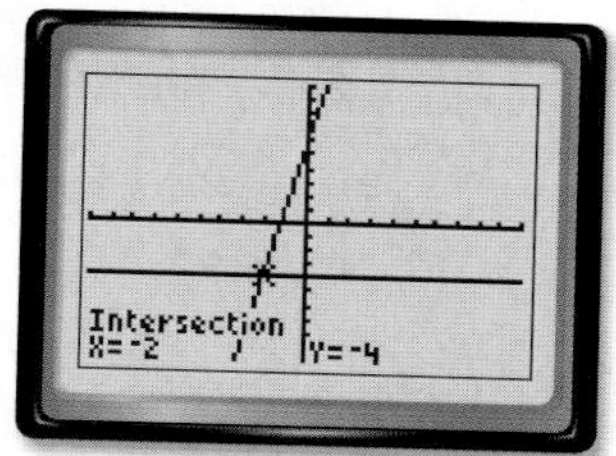

[−10, 10] scl: 1 by [−10, 10] scl: 1

The solution is −2.

Exercises

Use a graphing calculator to solve each system of equations. Write decimal solutions to the nearest hundredth.

1. $y = 2x - 3$
$y = -0.4x + 5$

2. $y = 6x + 1$
$y = -3.2x - 4$

3. $x + y = 9.35$
$5x - y = 8.75$

4. $2.32x - y = 6.12$
$4.5x + y = -6.05$

5. $5.2x - y = 4.1$
$1.5x + y = 6.7$

6. $1.8 = 5.4x - y$
$y = -3.8 - 6.2x$

7. $7x - 2y = 16$
$11x + 6y = 32.3$

8. $3x + 2y = 16$
$5x + y = 9$

9. $0.62x + 0.35y = 1.60$
$-1.38x + y = 8.24$

10. $75x - 100y = 400$
$33x - 10y = 70$

Use a graphing calculator to solve each equation. Write decimal solutions to the nearest hundredth.

11. $4x - 2 = -6$

12. $3 = 1 + \frac{x}{2}$

13. $\frac{x + 4}{-2} = -1$

14. $\frac{x}{7} - 3 = -2$

15. $-9 = 7 + 3x$

16. $-2 + 10x = 8x - 1$

17. WRITING IN MATH Explain why you can solve an equation like $r = ax + b$ by solving the system of equations $y = r$ and $y = ax + b$.

6-2 Substitution

Why?

Two movies were released at the same time. Movie A earned $31 million in its opening week, but fell to $15 million the following week. Movie B opened earning $21 million and fell to $11 million the following week. If the earnings for each movie continue to decrease at the same rate, when will they earn the same amount?

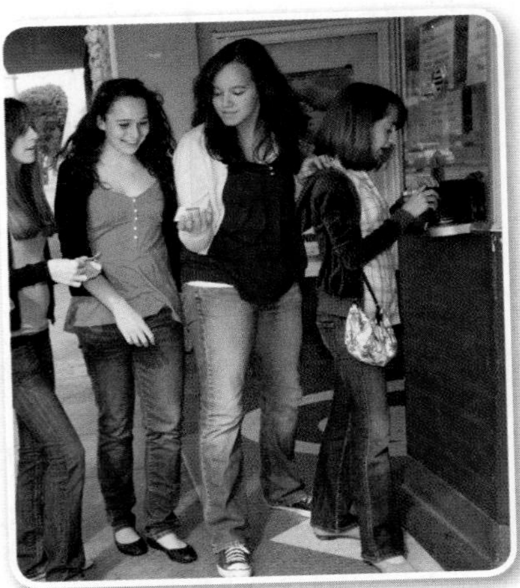

Then

You solved systems of equations by graphing. (Lesson 6-1)

Now

- Solve systems of equations by using substitution.
- Solve real-world problems involving systems of equations by using substitution.

New Vocabulary

substitution

Math Online

glencoe.com

- Extra Examples
- Personal Tutor
- Self-Check Quiz
- Homework Help

Solve by Substitution You can use a system of equations to find when the movie earnings are the same. One method of finding an exact solution of a system of equations is called **substitution**.

Key Concept — Solving by Substitution

For Your FOLDABLE

Step 1 When necessary, solve at least one equation for one variable.

Step 2 Substitute the resulting expression from Step 1 into the other equation to replace the variable. Then solve the equation.

Step 3 Substitute the value from Step 2 into either equation, and solve for the other variable. Write the solution as an ordered pair.

EXAMPLE 1 Solve a System by Substitution

Use substitution to solve the system of equations.

$y = 2x + 1$ ← **Step 1** One equation is already solved for y.
$3x + y = -9$

Step 2 Substitute $2x + 1$ for y in the second equation.

$3x + y = -9$	**Second equation**
$3x + 2x + 1 = -9$	**Substitute 2x + 1 for y.**
$5x + 1 = -9$	**Combine like terms.**
$5x = -10$	**Subtract 1 from each side.**
$x = -2$	**Divide each side by 5.**

Step 3 Substitute -2 for x in either equation to find y.

$y = 2x + 1$	**First equation**
$= 2(-2) + 1$	**Substitute −2 for x.**
$= -3$	**Simplify.**

The solution is $(-2, -3)$.

CHECK You can check your solution by graphing.

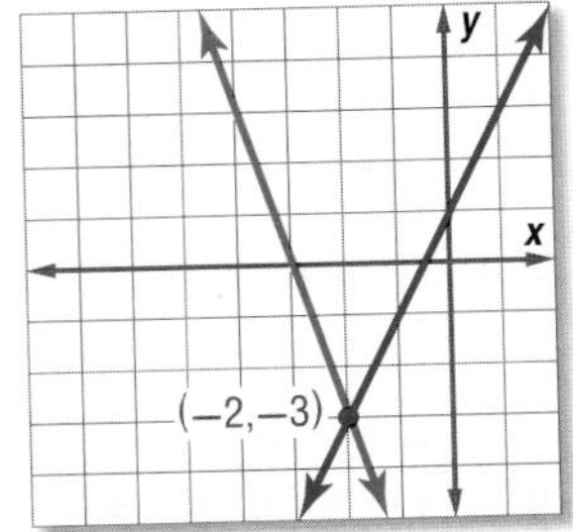

Check Your Progress

1A. $y = 4x - 6$
$5x + 3y = -1$

1B. $2x + 5y = -1$
$y = 3x + 10$

Personal Tutor glencoe.com

If a variable is not isolated in one of the equations in a system, solve an equation for a variable first. Then you can use substitution to solve the system.

StudyTip

Slope-Intercept Form If both equations are in the form $y = mx + b$, they can simply be set equal to each other and then solved for x. The solution for x can then be used to find the value of y.

EXAMPLE 2 Solve and then Substitute

Use substitution to solve the system of equations.

$\mathbf{x + 2y = 6}$

$\mathbf{3x - 4y = 28}$

Step 1 Solve the first equation for x since the coefficient is 1.

$x + 2y = 6$	**First equation**
$x + 2y - 2y = 6 - 2y$	**Subtract $2y$ from each side.**
$x = 6 - 2y$	**Simplify.**

Step 2 Substitute $6 - 2y$ for x in the second equation to find the value of y.

$3x - 4y = 28$	**Second equation**
$3(6 - 2y) - 4y = 28$	**Substitute $6 - 2y$ for x.**
$18 - 6y - 4y = 28$	**Distributive Property**
$18 - 10y = 28$	**Combine like terms.**
$18 - 10y - 18 = 28 - 18$	**Subtract 18 from each side.**
$-10y = 10$	**Simplify.**
$y = -1$	**Divide each side by -10.**

Step 3 Find the value of x.

$x + 2y = 6$	**First equation**
$x + 2(-1) = 6$	**Substitute -1 for y.**
$x - 2 = 6$	**Simplify.**
$x = 8$	**Add 2 to each side.**

The solution is $(8, -1)$.

Check Your Progress

2A. $4x + 5y = 11$
$y - 3x = -13$

2B. $x - 3y = -9$
$5x - 2y = 7$

Personal Tutor glencoe.com

Generally, if you solve a system of equations and the result is a false statement such as $3 = -2$, there is no solution. If the result is an identity, such as $3 = 3$, then there are an infinite number of solutions.

StudyTip

Dependent Systems There are infinitely many solutions of the system in Example 3 because the equations in slope-intercept form are equivalent, and they have the same graph.

EXAMPLE 3 No Solution or Infinitely Many Solutions

Use substitution to solve the system of equations.

$\mathbf{y = 2x - 4}$

$\mathbf{-6x + 3y = -12}$

Substitute $2x - 4$ for y in the second equation.

$-6x + 3y = -12$	**Second equation**
$-6x + 3(2x - 4) = -12$	**Substitute $2x - 4$ for y.**
$-6x + 6x - 12 = -12$	**Distributive Property**
$-12 = -12$	**Combine like terms.**

This statement is an identity. Thus, there are an infinite number of solutions.

 Check Your Progress

Use substitution to solve each system of equations.

3A. $2x - y = 8$
$y = 2x - 3$

3B. $4x - 3y = 1$
$6y - 8x = -2$

Personal Tutor glencoe.com

Solve Real-World Problems You can use substitution to find the solution of a real-world problem involving a system of equations.

Real-World Career

Sound Engineering Technician
Sound engineering technicians record, synchronize, mix, and reproduce music, voices, and sound effects in recording studios, sporting arenas, and theater, movie, or video productions. They need to have at least a 2-year associate's degree in electronics.

Real-World EXAMPLE 4 Write and Solve a System of Equations

MUSIC A store sold a total of 125 car stereo systems and speakers in one week. The stereo systems sold for \$104.95, and the speakers sold for \$18.95. The sales from these two items totaled \$6926.75. How many of each item were sold?

Let c = the number of car stereo systems sold, and let t = the number of speakers sold.

Number of Units Sold	c	t	125
Sales (\$)	$104.95c$	$18.95t$	6926.75

So, the two equations are $c + t = 125$ and $104.95c + 18.95t = 6926.75$.

Step 1 Solve the first equation for c.

$c + t = 125$ **First equation**

$c + t - t = 125 - t$ **Subtract t from each side.**

$c = 125 - t$ **Simplify.**

Step 2 Substitute $125 - t$ for c in the second equation.

$104.95c + 18.95t = 6926.75$ **Second equation**

$104.95(125 - t) + 18.95t = 6926.75$ **Substitute $125 - t$ for c.**

$13{,}118.75 - 104.95t + 18.95t = 6926.75$ **Distributive Property**

$13{,}118.75 - 86t = 6926.75$ **Combine like terms.**

$-86t = -6192$ **Subtract 13118.75 from each side.**

$t = 72$ **Divide each side by −86.**

Step 3 Substitute 72 for t in either equation to find the value of c.

$c + t = 125$ **First equation**

$c + 72 = 125$ **Substitute 72 for t.**

$c = 53$ **Subtract 72 from each side.**

The store sold 53 car stereo systems and 72 speakers.

 Check Your Progress

4. BASEBALL As of 2007, the New York Yankees and the Cincinnati Reds together had won a total of 31 World Series. The Yankees had won 5.2 times as many as the Reds. How many World Series had each team won?

Personal Tutor glencoe.com

Check Your Understanding

Examples 1–3 pp. 342–343

Use substitution to solve each system of equations.

1. $y = x + 5$
   $3x + y = 25$
2. $x = y - 2$
   $4x + y = 2$
3. $3x + y = 6$
   $4x + 2y = 8$
4. $2x + 3y = 4$
   $4x + 6y = 9$
5. $x - y = 1$
   $3x = 3y + 3$
6. $2x - y = 6$
   $-3y = -6x + 18$

Example 4 p. 344

7. **GEOMETRY** The sum of the measures of angles X and Y is 180°. The measure of angle X is 24° greater than the measure of angle Y.
   a. Define the variables, and write equations for this situation.
   b. Find the measure of each angle.

Practice and Problem Solving

● = **Step-by-Step Solutions** begin on page R12.
Extra Practice begins on page 815.

Examples 1–3 pp. 342–343

Use substitution to solve each system of equations.

8. $y = 5x + 1$
   $4x + y = 10$
9. $y = 4x + 5$
   $2x + y = 17$
10. $y = 3x - 34$
    $y = 2x - 5$
11. $y = 3x - 2$
    $y = 2x - 5$
12. $2x + y = 3$
    $4x + 4y = 8$
13. $3x + 4y = -3$
    $x + 2y = -1$
14. $y = -3x + 4$
    $-6x - 2y = -8$
15. $-1 = 2x - y$
    $8x - 4y = -4$
16. $x = y - 1$
    $-x + y = -1$
17. $y = -4x + 11$
    $3x + y = 9$
18. $y = -3x + 1$
    $2x + y = 1$
19. $3x + y = -5$
    $6x + 2y = 10$
20. $5x - y = 5$
    $-x + 3y = 13$
21. $2x + y = 4$
    $-2x + y = -4$
22. $-5x + 4y = 20$
    $10x - 8y = -40$

Example 4 p. 344

23. **ECONOMICS** In 2000, the demand for nurses was 2,000,000, while the supply was only 1,890,000. The projected demand for nurses in 2010 is 2,820,000, while the supply is only projected to be 1,810,000.
    a. Define the variables, and write equations to represent these situations.
    b. Use substitution to determine during which year the supply of nurses was equal to the demand.

Real-World Link

As the population's age increases, the need for nursing and home care is also increasing. Employment among RNs is expected to grow faster than the average for all occupations.

24. **TOURISM** The table shows the approximate number of tourists in two areas of the world during a recent year and the average rates of change in tourism.

Destination	Number of Tourists	Average Rates of Change in Tourists (millions per year)
South America and the Caribbean	40.3 million	increase of 0.8
Middle East	17.0 million	increase of 1.8

   a. Define the variables, and write an equation for each region's tourism rate.
   b. If the trends continue, in how many years would you expect the number of tourists in the regions to be equal?

25 **SPORTS** The table shows the winning times for the Triathlon World Championship.

Year	Men's	Women's
2000	1:51:39	1:54:43
2005	1:49:31	1:58:03

a. The times are in hours, minutes, and seconds. Rewrite the times rounded to the nearest minute.

b. Let the year 2000 be 0. Assume that the rate of change remains the same for years after 2000. Write an equation to represent each of the men's and women's winning times y in any year x.

c. If the trend continues, when would you expect the men's and women's winning times to be the same? Explain your reasoning.

Real-World Link

Recent marketing surveys reveal that two thirds of concert attendees are female, and one third are male.

Source: Concert Promotions Company

26. **CONCERT TICKETS** Booker is buying tickets online for a concert. He finds tickets for himself and his friends for $65 each plus a one-time fee of $10. Paula is looking for tickets to the same concert. She finds them at another Web site for $69 and a one-time fee of $13.60.

a. Define the variables, and write equations to represent this situation.

b. Create a table of values for 1 to 5 tickets for each person's purchase.

c. Graph each of these equations.

d. Analyze the graph. How many solutions are there? Explain why.

H.O.T. Problems

Use **H**igher-**O**rder **T**hinking Skills

27. **FIND THE ERROR** In the system $a + b = 7$ and $1.29a + 0.49b = 6.63$, a represents pounds of apples and b represents pounds of bananas. Guillermo and Cara are finding and interpreting the solution. Is either of them correct? Explain.

Guillermo

$$1.29a + 0.49b = 6.63$$
$$1.29a + 0.49(a + 7) = 6.63$$
$$1.29 + 0.49a + 3.43 = 6.63$$
$$0.49a = 3.2$$
$$a = 1.9$$

$a + b = 7$, so $b = 5$. The solution (2, 5) means that 2 pounds of apples and 5 pounds of bananas were bought.

Cara

$$1.29a + 0.49b = 6.63$$
$$1.29(7 - b) + 0.49b = 6.63$$
$$9.03 - 1.29b + 0.49b = 6.63$$
$$-0.8b = -2.4$$
$$b = 3$$

The solution $b = 3$ means that 3 pounds of apples and 3 pounds of bananas were bought.

28. **CHALLENGE** A local charity has 60 volunteers. The ratio of boys to girls is 7:5. Find the number of boy and the number of girl volunteers.

29. **REASONING** Compare and contrast the solution of a system found by graphing and the solution of the same system found by substitution.

30. **OPEN ENDED** Create a system of equations that has one solution. Illustrate how the system could represent a real-world situation and describe the significance of the solution in the context of the situation.

31. **WRITING IN MATH** Explain how to determine what to substitute when using the substitution method of solving systems of equations.

Standardized Test Practice

32. The debate team plans to make and sell trail mix. They can spend \$34.

Item	Cost Per Pound
sunflower seeds	\$4.00
raisins	\$1.50

The pounds of raisins in the mix are to be 3 times the pounds of sunflower seeds. Which system can be used to find r, the pounds of raisins, and p, pounds of sunflower seeds, they should buy?

A $3p = r$
$4p + 1.5r = 34$

B $3p = r$
$4r + 1.5p = 34$

C $3r = p$
$4p + 1.5r = 34$

D $3r = p$
$4r + 1.5p = 34$

33. GRIDDED RESPONSE The perimeters of two similar polygons are 250 centimeters and 300 centimeters, respectively. What is the scale factor between the two polygons?

34. Based on the graph, which statement is true?

Sports Drinks Supply

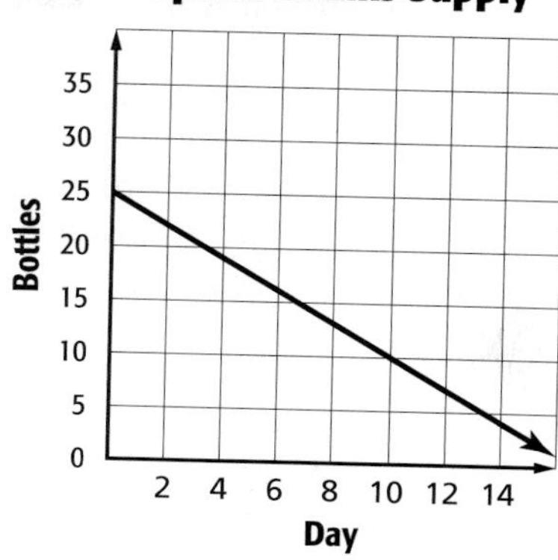

F Mary started with 30 bottles.
G On day 10, Mary will have 10 bottles left.
H Mary will be out of sports drinks on day 14.
J Mary drank 5 bottles the first two days.

35. If p is an integer, which of the following is the solution set for $2|p| = 16$?

A $\{0, 8\}$
B $\{-8, 0\}$
C $\{-8, 8\}$
D $\{-8, 0, 8\}$

Spiral Review

Graph each system of equations. Then determine whether the system has *no* solution, *one* solution, or *infinitely many* solutions. If the system has one solution, name it. (Lesson 6-1)

36. $y = -5$
$3x + y = 1$

37. $x = 1$
$2x - y = 7$

38. $y = x + 5$
$y = x - 2$

39. $x + y = 1$
$3y + 3x = 3$

40. ENTERTAINMENT Coach Ross wants to take the soccer team out for pizza after their game. Her budget is at most \$70. (Lesson 5-6)

a. Using the sign, write an inequality that represents this situation.

b. Are there any restrictions on the domain or range? Explain.

Welcome to Rini's Pizza
Large Pizza \$12
Pitcher of Soft Drinks \$2

Solve each inequality. Check your solution. (Lesson 5-3)

41. $6v + 1 \geq -11$

42. $24 > 18 + 2n$

43. $-11 \geq \frac{2}{5}q + 5$

44. $\frac{a}{8} - 10 > -3$

45. $-3t + 9 \leq 0$

46. $54 > -10 - 8n$

Skills Review

Rewrite each product using the Distributive Property. Then simplify. (Lesson 1-4)

47. $10b + 5(3 + 9b)$

48. $5(3t^2 + 4) - 8t$

49. $7h^2 + 4(3h + h^2)$

50. $-2(7a + 5b) + 5(2a - 7b)$

6-3

Elimination Using Addition and Subtraction

Then

You solved systems of equations by using substitution. (Lesson 6-2)

Now

- Solve systems of equations by using elimination with addition.
- Solve systems of equations by using elimination with subtraction.

New Vocabulary

elimination

Math Online

glencoe.com

- Extra Examples
- Personal Tutor
- Self-Check Quiz
- Homework Help

Why?

In Chicago, Illinois, there are two more months a when the mean high temperature is below 70°F than there are months b when it is above 70°F. The system of equations, $a + b = 12$ and $a - b = 2$, represents this situation.

Elimination Using Addition If you add these equations, the variable b will be eliminated. Using addition or subtraction to solve a system is called **elimination**.

Key Concept — Solving by Elimination

For Your FOLDABLE

Step 1 Write the system so like terms with the same or opposite coefficients are aligned.

Step 2 Add or subtract the equations, eliminating one variable. Then solve the equation.

Step 3 Substitute the value from Step 2 into one of the equations and solve for the other variable. Write the solution as an ordered pair.

EXAMPLE 1 Elimination Using Addition

Use elimination to solve the system of equations.

$\mathbf{4x + 6y = 32}$
$\mathbf{3x - 6y = 3}$ ← **Step 1** $6y$ and $-6y$ have opposite coefficients.

Step 2 Add the equations.

$$
\begin{array}{rll}
4x + 6y = 32 & \\
(+)\ 3x - 6y = 3 & \\
\hline
7x \quad = 35 & \text{The variable } y \text{ is eliminated.}\\
\frac{7x}{7} = \frac{35}{7} & \text{Divide each side by 7.}\\
x = 5 & \text{Simplify.}
\end{array}
$$

Step 3 Substitute 5 for x in either equation to find the value of y.

$4x + 6y = 32$	**First equation**
$4(5) + 6y = 32$	**Replace x with 5.**
$20 + 6y = 32$	**Multiply.**
$20 + 6y - 20 = 32 - 20$	**Subtract 20 from each side.**
$6y = 12$	**Simplify.**
$\frac{6y}{6} = \frac{12}{6}$	**Divide each side by 6.**
$y = 2$	**Simplify.**

The solution is (5, 2).

Check Your Progress

1A. $-4x + 3y = -3$
$4x - 5y = 5$

1B. $4y + 3x = 22$
$3x - 4y = 14$

Personal Tutor glencoe.com

ReadingMath

Elimination When adding or subtracting two equations causes the coefficients of a variable to result in 0, the variable is said to be *eliminated.*

We can use elimination to find specific numbers that are described as being related to each other.

EXAMPLE 2 Write and Solve a System of Equations

Negative three times one number plus five times another number is −11. Three times the first number plus seven times the other number is −1. Find the numbers.

Negative three times one number	plus	five times another number	is	−11.
$-3x$	$+$	$5y$	$=$	-11
Three times the first number	plus	seven times the other number	is	−1.
$3x$	$+$	$7y$	$=$	-1

Steps 1 and 2 Write the equations vertically and add.

$$\begin{aligned} -3x + 5y &= -11 \\ (+)\ 3x + 7y &= -1 \\ \hline 12y &= -12 \end{aligned}$$ **The variable *x* is eliminated.**

$$\frac{12y}{12} = \frac{-12}{12}$$ **Divide each side by 12.**

$$y = -1$$ **Simplify.**

Step 3 Substitute −1 for *y* in either equation to find the value of *x*.

$$3x + 7y = -1$$ **Second equation**

$$3x + 7(-1) = -1$$ **Replace *y* with −1.**

$$3x + (-7) = -1$$ **Simplify.**

$$3x + (-7) + 7 = -1 + 7$$ **Add 7 to each side.**

$$3x = 6$$ **Simplify.**

$$\frac{3x}{3} = \frac{6}{3}$$ **Divide each side by 3.**

$$x = 2$$ **Simplify.**

The numbers are 2 and −1.

CHECK

$$-3x + 5y = -11$$ **First equation**

$$-3(2) + 5(-1) \stackrel{?}{=} -11$$ **Substitute 2 for *x* and −1 for *y*.**

$$-11 = -11 \checkmark$$ **Simplify.**

$$3x + 7y = -1$$ **Second equation**

$$3(2) + 7(-1) \stackrel{?}{=} -1$$ **Substitute 2 for *x* and −1 for *y*.**

$$-1 = -1 \checkmark$$ **Simplify.**

StudyTip

Coefficients When the coefficients of a variable are the same, subtracting the equations will eliminate the variable. When the coefficients are opposites, adding the equations will eliminate the variable.

Check Your Progress

2. The sum of two numbers is −10. Negative three times the first number minus the second number equals 2. Find the numbers.

Personal Tutor glencoe.com

Elimination Using Subtraction Sometimes we can eliminate a variable by subtracting one equation from another.

STANDARDIZED TEST EXAMPLE 3 Elimination Using Subtraction

Solve the system of equations. $2t + 5r = 6$
$9r + 2t = 22$

A $(-7, 15)$ **B** $\left(7, \frac{8}{9}\right)$ **C** $(4, -7)$ **D** $\left(4, -\frac{2}{5}\right)$

Read the Test Item

Since both equations contain $2t$, use elimination by subtraction.

Solve the Test Item

Step 1 Subtract the equations.

$$\begin{array}{rrl} & 5r + 2t = & 6 \\ (-) & 9r + 2t = & 22 \\ \hline & -4r \quad = & -16 \\ & r = & 4 \end{array}$$

Write the system so like terms are aligned. The variable t is eliminated. Simplify.

Step 2 Substitute 4 for r in either equation to find the value of t.

$5r + 2t = 6$	**First equation**
$5(4) + 2t = 6$	$r = 4$
$20 + 2t = 6$	**Simplify.**
$20 + 2t - 20 = 6 - 20$	**Subtract 20 from each side.**
$2t = -14$	**Simplify.**
$t = -7$	**Simplify.**

The solution is $(4, -7)$. The correct answer is C .

Check Your Progress

3. Solve the system of equations. $8b + 3c = 11$
$8b + 7c = 7$

F $(1.5, -1)$ **G** $(1.75, -1)$ **H** $(1.75, 1)$ **J** $(1.5, 1)$

Personal Tutor glencoe.com

Real-World Link

The five most dangerous jobs for teenagers are: delivery and other driving jobs, working alone in cash-based businesses, traveling youth crews, cooking, and construction.

Source: National Consumers League

Real-World EXAMPLE 4 Write and Solve a System of Equations

JOBS Cheryl and Jackie work at an ice cream shop. Cheryl earns \$8.50 per hour and Jackie earns \$7.50 per hour. During a typical week, Cheryl and Jackie earn \$299.50 together. One week, Jackie doubles her work hours, and the girls earn \$412. How many hours does each girl work during a typical week?

Understand You know how much Cheryl and Jackie each earn per hour and how much they earned together.

Plan Let c = Cheryl's hours and j = Jackie's hours.

Cheryl's pay	plus	Jackie's pay	equals	\$299.50.
$8.50c$	$+$	$7.50j$	$=$	299.50
Cheryl's pay	**plus**	**Jackie's pay**	**equals**	**\$412.**
$8.50c$	$+$	$7.50(2)j$	$=$	412

Solve Subtract the equations to eliminate one of the variables. Then solve for the other variable.

$$\begin{array}{r} 8.50c + 7.50j = 299.50 \\ (-)\ 8.50c + 7.50(2)j = 412 \\ \hline \end{array}$$ **Write the equations vertically.**

$$\begin{array}{r} 8.50c + 7.50j = 299.50 \\ (-)\ 8.50c + 15j = 412 \\ \hline \end{array}$$ **Simplify.**

$$-7.50j = -112.50$$ **Subtract. The variable c is eliminated.**

$$\frac{-7.50j}{-7.50} = \frac{-112.50}{-7.50}$$ **Divide each side by −7.50.**

$$j = 15$$ **Simplify.**

StudyTip

Another Method Instead of subtracting the equations, you could also multiply one equation by −1 and then add the equations.

Now substitute 15 for j in either equation to find the value of c.

$$8.50c + 7.50j = 299.50$$ **First equation**

$$8.50c + 7.50(15) = 299.50$$ **Substitute 15 for j.**

$$8.50c + 112.50 = 299.50$$ **Simplify.**

$$8.50c = 187$$ **Subtract 112.50 from each side.**

$$c = 22$$ **Divide each side by 8.50.**

Check Substitute both values into the other equation to see if the equation holds true. If $c = 22$ and $j = 15$, then $8.50(22) + 15(15)$ or 412.

Cheryl works 22 hours, while Jackie works 15 hours during a typical week.

Check Your Progress

4. PARTIES Tamera and Adelina are throwing a birthday party for their friend. Tamera invited 5 fewer friends than Adelina. Together they invited 47 guests. How many guests did each girl invite?

Personal Tutor glencoe.com

Check Your Understanding

Examples 1 and 3 pp. 348 and 350

Use elimination to solve each system of equations.

1. $5m - p = 7$
$7m - p = 11$

2. $8x + 5y = 38$
$-8x + 2y = 4$

3. $7f + 3g = -6$
$7f - 2g = -31$

4. $6a - 3b = 27$
$2a - 3b = 11$

Example 2 p. 349

5. The sum of two numbers is 24. Five times the first number minus the second number is 12. What are the two numbers?

Example 4 pp. 350–351

6. RECYCLING The recycling and reuse industry employs approximately 1,025,000 more workers than the waste management industry. Together they provide 1,275,000 jobs. How many jobs does each industry provide?

Practice and Problem Solving

= **Step-by-Step Solutions** begin on page R12.
Extra Practice begins on page 815.

Examples 1 and 3
pp. 348 and 350

Use elimination to solve each system of equations.

7. $-v + w = 7$
$v + w = 1$

8. $y + z = 4$
$y - z = 8$

9. $-4x + 5y = 17$
$4x + 6y = -6$

10. $5m - 2p = 24$
$3m + 2p = 24$

11. $a + 4b = -4$
$a + 10b = -16$

12. $6r - 6t = 6$
$3r - 6t = 15$

13. $6c - 9d = 111$
$5c - 9d = 103$

14. $11f + 14g = 13$
$11f + 10g = 25$

15. $9x + 6y = 78$
$3x - 6y = -30$

16. $3j + 4k = 23.5$
$8j - 4k = 4$

17. $-3x - 8y = -24$
$3x - 5y = 4.5$

18. $6x - 2y = 1$
$10x - 2y = 5$

Example 2
p. 349

19. The sum of two numbers is 22, and their difference is 12. What are the numbers?

20. Find the two numbers with a sum of 41 and a difference of 9.

21 Three times a number minus another number is −3. The sum of the numbers is 11. Find the numbers.

22. A number minus twice another number is 4. Three times the first number plus two times the second number is 12. What are the numbers?

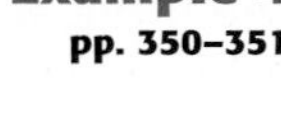

Example 4
pp. 350–351

23. TOURS The Blackwells and Joneses are going to Hershey's Really Big 3D Show in Pennsylvania. Find the adult price and the children's price of the show.

Family	Number of Adults	Number of Children	Total Cost
Blackwell	2	5	$31.65
Jones	2	3	$23.75

Use elimination to solve each system of equations.

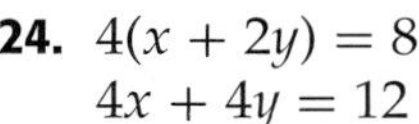

24. $4(x + 2y) = 8$
$4x + 4y = 12$

25. $3x - 5y = 11$
$5(x + y) = 5$

26. $4x + 3y = 6$
$3x + 3y = 7$

27. $6x - 7y = -26$
$6x + 5y = 10$

28. $\frac{1}{2}x + \frac{2}{3}y = 2\frac{3}{4}$
$\frac{1}{4}x - \frac{2}{3}y = 6\frac{1}{4}$

29. $\frac{3}{5}x + \frac{1}{2}y = 8\frac{1}{3}$
$-\frac{3}{5}x + \frac{3}{4}y = 8\frac{1}{3}$

30. ARCHITECTURE The total height of an office building b and the granite statue that stands on top of it g is 326.6 feet. The difference in heights between the building and the statue is 295.4 feet.

a. How tall is the statue?

b. How tall is the building?

Real-World Link

Cross-country mountain bike racing became an Olympic event in 1996. Dutch cyclist Bart Brentjens earned the first gold medal in the event with a time of 2:17:36.

Source: Cycling News

31. BIKE RACING Professional Mountain Bike Racing currently has 66 teams. The number of non-U.S. teams is 30 more than the number of U.S. teams.

a. Let x represent the number of non-U.S. teams and y represent the number of U.S. teams. Write a system of equations that represents the number of U.S. teams and non-U.S. teams.

b. Use elimination to find the solution of the system of equations.

c. Interpret the solution in the context of the situation.

d. Graph the system of equations to check your solution.

Math History Link

Leonardo Pisano (1170–1250)

Leonardo Pisano is better known by his nickname *Fibonacci*. His book introduced the Hindu-Arabic place-valued decimal system. Systems of linear equations are studied in this work.

32. ONLINE CATALOGS Let x represent the number of years since 2004 and y represent the number of catalogs.

Catalogs	Number in 2004	Growth Rate (number per year)
online	7440	1293
print	3805	−1364

Source: MediaPost Publications

a. Write a system of equations to represent this situation.

b. Use elimination to find the solution to the system of equations.

c. Analyze the solution in terms of the situation. Determine the reasonableness of the solution.

33 **MULTIPLE REPRESENTATIONS** Collect 9 pennies and 9 paper clips. For this game, you may use a maximum of 9 objects to create a certain required number of points. Each paper clip is worth 1 point and each penny is worth 3 points. Let p represent a penny and c represent a paper clip.

9 points = (2 pennies) + (3 paper clips) $= 2p + 3c$

a. CONCRETE You must have exactly 15 points using at least one of each piece. Compare your pattern to other students.

b. ANALYTICAL Write and solve a system of equations to find the number of paper clips and pennies used.

c. TABULAR Make a table showing the number of paper clips used and the total number of points when the number of pennies is 0, 1, 2, 3, 4, or 5.

d. VERBAL Does the result in the table match the results in part **b**? Explain.

H.O.T. Problems

Use **H**igher-**O**rder **T**hinking Skills

34. REASONING Describe the solution of a system of equations if after you added two equations the result was $0 = 0$.

35. REASONING What is the solution of a system of equations if the sum of the equations is $0 = 2$?

36. OPEN ENDED Create a system of equations that can be solved by using addition to eliminate one variable. Formulate a general rule for creating such systems.

37. REASONING The solution of a system of equations is $(-3, 2)$. One equation in the system is $x + 4y = 5$. Find a second equation for the system. Explain how you derived this equation.

38. CHALLENGE If a number is multiplied by 7, the result is 182. The sum of that number's two digits is 8. Define the variables and write the system of equations that you would use to find the number. Then solve the system and find the number.

39. WRITING IN MATH Describe when it would be most beneficial to use elimination to solve a system of equations.

Standardized Test Practice

40. SHORT RESPONSE Martina is on a train traveling at a speed of 188 mph between two cities 1128 miles apart. If the train has been traveling for an hour, how many more hours is her train ride?

41. GEOMETRY Ms. Miller wants to tile her rectangular kitchen floor. She knows the dimensions of the floor. Which formula should she use to find the area?

A $A = \ell w$

B $V = Bh$

C $P = 2\ell + 2w$

D $c^2 = a^2 + b^2$

42. If the pattern continues, what is the 8th number in the sequence?

$$2, 3, \frac{9}{2}, \frac{27}{4}, \frac{81}{8}, \ldots$$

F $\frac{2187}{64}$

G $\frac{2245}{64}$

H $\frac{2281}{64}$

J $\frac{2445}{64}$

43. What is the solution of this system of equations?

$$x + 4y = 1$$
$$2x - 3y = -9$$

A $(2, -8)$

B $(-3, 1)$

C no solution

D infinitely many solutions

Spiral Review

Use substitution to solve each system of equations. If the system does not have exactly one solution, state whether it has no solution or infinitely many solutions. (Lesson 6-2)

44. $y = 6x$
$2x + 3y = 40$

45. $x = 3y$
$2x + 3y = 45$

46. $x = 5y + 6$
$x = 3y - 2$

47. $y = 3x + 2$
$y = 4x - 1$

48. $3c = 4d + 2$
$c = d - 1$

49. $z = v + 4$
$2z - v = 6$

50. FINANCIAL LITERACY Gregorio and Javier each want to buy a bicycle. Gregorio has already saved \$35 and plans to save \$10 per week. Javier has \$26 and plans to save \$13 per week. (Lesson 6-1)

a. In how many weeks will Gregorio and Javier have saved the same amount of money?

b. How much will each person have saved at that time?

51. GEOMETRY A *parallelogram* is a quadrilateral in which opposite sides are parallel. Determine whether $ABCD$ is parallelogram. Explain your reasoning. (Lesson 4-4)

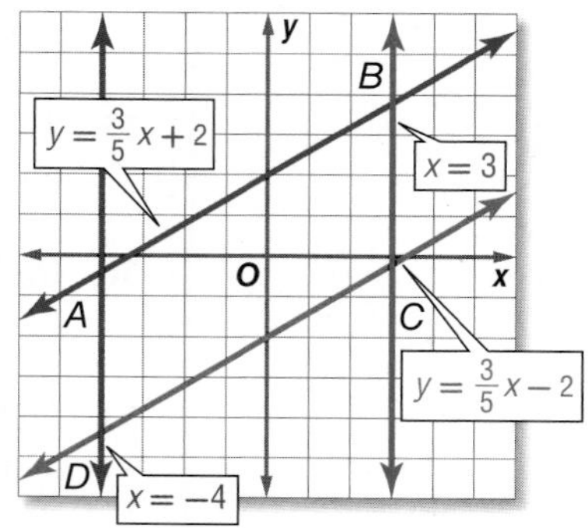

Solve each equation. Check your solution. (Lesson 2-2)

52. $6u = -48$

53. $75 = -15p$

54. $\frac{2}{3}a = 8$

55. $-\frac{3}{4}d = 15$

Skills Review

Simplify each expression. If not possible, write *simplified*. (Lesson 1-4)

56. $6q - 3 + 7q + 1$

57. $7w^2 - 9w + 4w^2$

58. $10(2 + r) + 3r$

59. $5y - 7(y + 5)$

6-4 Elimination Using Multiplication

Then

You used elimination with addition and subtraction to solve systems of equations. (Lesson 6-3)

Now

- Solve systems of equations by using elimination with multiplication.
- Solve real-world problems involving systems of equations.

Math Online

glencoe.com

- Extra Examples
- Personal Tutor
- Self-Check Quiz
- Homework Help

Why?

The table shows the number of cars at Scott's Auto Repair Shop for each type of service.

Item	Repairs	Maintenance
body	3	4
engine	2	2

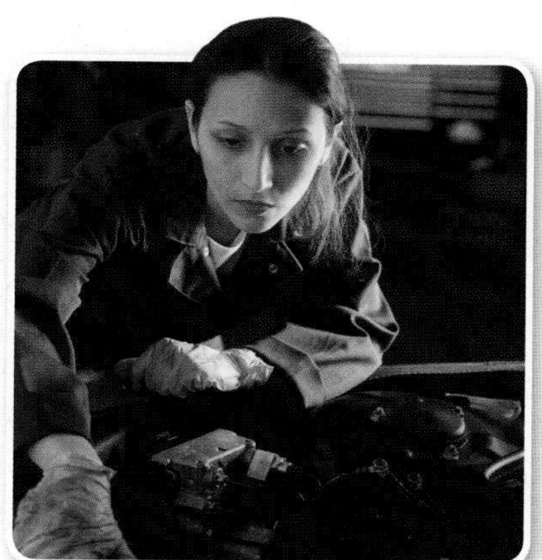

The manager has allotted 1110 minutes for body work and 570 minutes for engine work. The system $3r + 4m = 1110$ and $2r + 2m = 570$ can be used to find the average time for each service.

Elimination Using Multiplication In the system above, neither variable can be eliminated by adding or subtracting. You can use multiplication to solve.

Key Concept — Solving by Elimination

For Your FOLDABLE

Step 1 Multiply at least one equation by a constant to get two equations that contain opposite terms.

Step 2 Add or subtract the equations, eliminating one variable. Then solve the equation.

Step 3 Substitute the value from Step 2 into one of the equations and solve for the other variable. Write the solution as an ordered pair.

EXAMPLE 1 Multiply One Equation to Eliminate a Variable

Use elimination to solve the system of equations.

$5x + 6y = -8$

$2x + 3y = -5$

Steps 1 and 2

$5x + 6y = -8$

$2x + 3y = -5$ **Multiply each term by −2.** →

$$\begin{aligned} 5x + 6y &= -8 \\ (+)\ -4x - 6y &= 10 \quad \textbf{Add.} \\ \hline x \qquad &= 2 \quad \textbf{y is eliminated.} \end{aligned}$$

Step 3

$2x + 3y = -5$	**Second equation**
$2(2) + 3y = -5$	**Substitution, $x = 2$**
$4 + 3y = -5$	**Simplify.**
$3y = -9$	**Subtract 4 from each side and simplify.**
$y = -3$	**Divide each side by 3 and simplify.**

The solution is (2, −3).

Check Your Progress

1A. $6x - 2y = 10$
$3x - 7y = -19$

1B. $9r + q = 13$
$3r + 2q = -4$

Personal Tutor glencoe.com

StudyTip

Choosing a Variable to Eliminate Unless the problem is asking for the value of a specific variable, you may use multiplication to eliminate either variable.

Sometimes you have to multiply each equation by a different number in order to solve the system.

EXAMPLE 2 Multiply Both Equations to Eliminate a Variable

Use elimination to solve the system of equations.

$4x + 2y = 8$

$3x + 3y = 9$

Method 1 Eliminate x.

$4x + 2y = 8$ **Multiply by 3.** → $12x + 6y = 24$

$3x + 3y = 9$ **Multiply by −4.** → $(+)\ -12x - 12y = -36$

$-6y = -12$	**Add equations. x is eliminated.**
$\frac{-6y}{-6} = \frac{-12}{-6}$	**Divide each side by −6.**
$y = 2$	**Simplify.**

Now substitute 2 for y in either equation to find the value of x.

$3x + 3y = 9$	**Second equation**
$3x + 3(2) = 9$	**Substitute 2 for y.**
$3x + 6 = 9$	**Simplify.**
$3x = 3$	**Subtract 6 from each side and simplify.**
$\frac{3x}{3} = \frac{3}{3}$	**Divide each side by 3.**
$x = 1$	**The solution is (1, 2).**

Method 2 Eliminate y.

$4x + 2y = 8$ **Multiply by 3.** → $12x + 6y = 24$

$3x + 3y = 9$ **Multiply by −2.** → $(+)\ -6x - 6y = -18$

$6x = 6$	**Add equations. y is eliminated.**
$\frac{6x}{6} = \frac{6}{6}$	**Divide each side by 6.**
$x = 1$	**Simplify.**

Now substitute 1 for x in either equation to find the value of y.

$3x + 3y = 9$	**Second equation**
$3(1) + 3y = 9$	**Substitute 1 for x.**
$3 + 3y = 9$	**Simplify.**
$3y = 6$	**Subtract 3 from each side and simplify.**
$\frac{3y}{3} = \frac{6}{3}$	**Divide each side by 3.**
$y = 2$	**Simplify.**

The solution is (1, 2), which matches the result obtained with Method 1.

CHECK Substitute 1 for x and 2 for y in the first equation.

$4x + 2y = 8$	**Original equation**
$4(1) + 2(2) \stackrel{?}{=} 8$	**Substitute (1, 2) for (x, y).**
$4 + 4 \stackrel{?}{=} 8$	**Multiply.**
$8 = 8$ ✓	**Add.**

Check Your Progress

2A. $5x - 3y = 6$
$2x + 5y = -10$

2B. $6a + 2b = 2$
$4a + 3b = 8$

Personal Tutor glencoe.com

Solve Real-World Problems Sometimes it is necessary to use multiplication before elimination in real-world problem solving too.

Real-World EXAMPLE 3 Solve a System of Equations

FLIGHT A personal aircraft traveling with the wind flies 520 miles in 4 hours. On the return trip, the airplane takes 5 hours to travel the same distance. Find the speed of the airplane if the air is still.

You are asked to find the speed of the airplane in still air.

Let a = the rate of the airplane if the air is still.
Let w = the rate of the wind.

	r	t	d	$r \cdot t = d$
With the Wind	$a + w$	4	520	$(a + w)4 = 520$
Against the Wind	$a - w$	5	520	$(a - w)5 = 520$

So, our two equations are $4a + 4w = 520$ and $5a - 5w = 520$.

$4a + 4w = 520$ **Multiply by 5.** → $20a + 20w = 2600$

$5a - 5w = 520$ **Multiply by 4.** → $(+)\ 20a - 20w = 2080$

$40a = 4680$ — ***w* is eliminated.**

$\frac{40a}{40} = \frac{4680}{40}$ — **Divide each side by 40.**

$a = 117$ — **Simplify.**

The rate of the airplane in still air is 117 miles per hour.

Check Your Progress

3. CANOEING A canoeist travels 4 miles downstream in 1 hour. The return trip takes the canoeist 1.5 hours. Find the rate of the boat in still water.

Personal Tutor glencoe.com

Check Your Understanding

Examples 1 and 2 pp. 355–356

Use elimination to solve each system of equations.

1. $2x - y = 4$
$7x + 3y = 27$

2. $2x + 7y = 1$
$x + 5y = 2$

3. $4x + 2y = -14$
$5x + 3y = -17$

4. $9a - 2b = -8$
$-7a + 3b = 12$

Example 3 p. 357

5. KAYAKING A kayaking group with a guide travels 16 miles downstream, stops for a meal, and then travels 16 miles upstream. The speed of the current remains constant throughout the trip. Find the speed of the kayak in still water.

Leave	10:00 A.M.
Stop for meal	12:00 noon
Return	1:00 P.M.
Finish	5:00 P.M.

6. PODCASTS Steve subscribed to 10 podcasts for a total of 340 minutes. He used his two favorite tags, Hobbies and Recreation and Soliloquies. Each of the Hobbies and Recreation episodes lasted about 32 minutes. Each Soliloquies episode lasted 42 minutes. To how many of each tag did Steve subscribe?

Practice and Problem Solving

= **Step-by-Step Solutions** begin on page R12.
Extra Practice begins on page 815.

Examples 1 and 2 pp. 355–356

Use elimination to solve each system of equations.

7. $x + y = 2$
$-3x + 4y = 15$

8. $x - y = -8$
$7x + 5y = 16$

9. $x + 5y = 17$
$-4x + 3y = 24$

10. $6x + y = -39$
$3x + 2y = -15$

11. $2x + 5y = 11$
$4x + 3y = 1$

12. $3x - 3y = -6$
$-5x + 6y = 12$

13. $3x + 4y = 29$
$6x + 5y = 43$

14. $8x + 3y = 4$
$-7x + 5y = -34$

15. $8x + 3y = -7$
$7x + 2y = -3$

16. $4x + 7y = -80$
$3x + 5y = -58$

17. $12x - 3y = -3$
$6x + y = 1$

18. $-4x + 2y = 0$
$10x + 3y = 8$

Example 3 p. 357

19. **NUMBER THEORY** Seven times a number plus three times another number equals negative one. The sum of the two numbers is negative three. What are the numbers?

20. **FOOTBALL** A field goal is 3 points and the extra point after a touchdown is 1 point. In a recent post-season, Adam Vinatieri of the Indianapolis Colts made a total of 21 field goals and extra point kicks for 49 points. Find the number of field goals and extra points that he made.

Use elimination to solve each system of equations.

21. $2.2x + 3y = 15.25$
$4.6x + 2.1y = 18.325$

22. $-0.4x + 0.25y = -2.175$
$2x + y = 7.5$

23. $\frac{1}{4}x + 4y = 2\frac{3}{4}$
$3x + \frac{1}{2}y = 9\frac{1}{4}$

24. $\frac{2}{5}x + 6y = 24\frac{1}{5}$
$3x + \frac{1}{2}y = 3\frac{1}{2}$

Real-World Link

TOBOR, *robot* spelled backward, is a robot that delivers medications directly from the pharmacy to a patient's room. TOBOR can talk, detect when someone is blocking its path, and interface with an elevator.

Source: *U.S. Medicine Magazine*

25. **ROBOTS** TOBOR saves 120 minutes of a nurse's time n and 180 minutes of support staff time s each day. Another robot that aids stroke patients' limbs is estimated to save 90 minutes of nursing time and 120 minutes of support staff time each day.

a. To be cost effective, TOBOR must save a total of 1500 minutes per day. Write an equation that represents this relationship.

b. To make the stroke assistant cost effective, it must save a total of 1050 minutes per day. Write an equation that represents this relationship.

c. Solve the system of equations, and interpret the solution in the context of the situation.

26. **GEOMETRY** The graphs of $x + 2y = 6$ and $2x + y = 9$ contain two of the sides of a triangle. A vertex of the triangle is at the intersection of the graphs.

a. What are the coordinates of the vertex?

b. Draw the graph of the two lines. Identify the vertex of the triangle.

c. The line that forms the third side of the triangle is the line $x - y = -3$. Draw this line on the previous graph.

d. Name the other two vertices of the triangle.

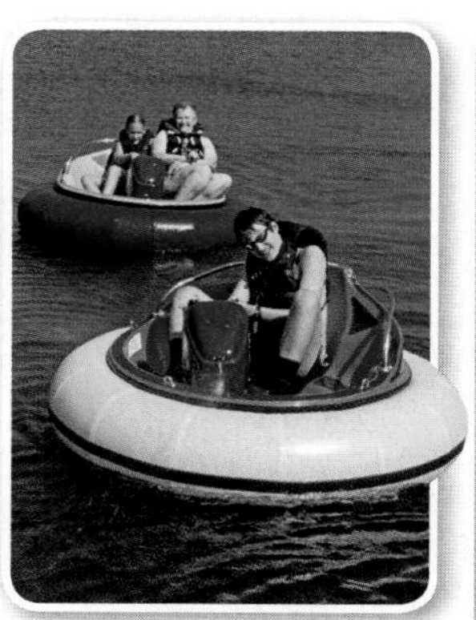

Real-World Link

In addition to batting cages and miniature golf, many entertainment centers offer go-karts, bumper boats, and a video arcade.

Source: Camelot Park

27. **ENTERTAINMENT** At an entertainment center, two groups of people bought batting tokens and miniature golf games, as shown in the table.

Group	Number of Batting Tokens	Number of Miniature Golf Games	Total Cost
A	16	3	$30
B	22	5	$43

a. Define the variables, and write a system of linear equations from this situation.

b. Solve the system of equations, and explain what the solution represents.

28. **TESTS** Mrs. Henderson discovered that she had accidentally reversed the digits of a test score and did not give a student 36 points. Mrs. Henderson told the student that the sum of the digits was 14 and agreed to give the student his correct score plus extra credit if he could determine his actual score. What was his correct score?

H.O.T. Problems

Use Higher-Order Thinking Skills

29. **REASONING** Explain how you could recognize a system of linear equations with infinitely many solutions.

30. **FIND THE ERROR** Jason and Daniela are solving a system of equations. Is either of them correct? Explain your reasoning.

Jason

$2r + 7t = 11$

$r - 9t = -7$

$2r + 7t = 11$

$(-)\ 2r - 18t = -14$

$25t = 25$

$t = 1$

$2r + 7t = 11$

$2r + 7(1) = 11$

$2r + 7 = 11$

$2r = 4$

$\frac{2r}{2} = \frac{4}{2}$

$r = 2$

The solution is (2, 1).

Daniela

$2r + 7t = 11$

$(-)\ r - 9t = -7$

$r = 18$

$2r + 7t = 11$

$2(18) + 7t = 11$

$36 + 7t = 11$

$7t = -25$

$\frac{7t}{7} = -\frac{25}{7}$

$t = -3.6$

The solution is (18, -3.6).

31. **OPEN ENDED** Write a system of equations that can be solved by multiplying one equation by -3 and then adding the two equations together.

32. **CHALLENGE** The solution of the system $4x + 5y = 2$ and $6x - 2y = b$ is $(3, a)$. Find the values of a and b. Discuss the steps that you used.

33. **WRITING IN MATH** Explain how to decide which variable to eliminate when using multiplication.

Standardized Test Practice

34. What is the solution of this system of equations?

$2x - 3y = -9$
$-x + 3y = 6$

A (3, 3) **C** (−3, 1)
B (−3, 3) **D** (1, −3)

35. A buffet has one price for adults and another for children. The Taylor family has two adults and three children, and their bill was \$40.50. The Wong family has three adults and one child. Their bill was \$38. Which system of equations could be used to determine the price for an adult and for a child?

F $x + y = 40.50$, $x + y = 38$
H $2x + 3y = 40.50$, $x + 3y = 38$
G $2x + 3y = 40.50$, $3x + y = 38$
J $2x + 2y = 40.50$, $3x + y = 38$

36. SHORT RESPONSE A customer at the paint store has ordered 3 gallons of ivy green paint. Melissa mixes the paint in a ratio of 3 parts blue to one part yellow. How many quarts of blue paint does she use?

37. PROBABILITY The table shows the results of a number cube being rolled. What is the experimental probability of rolling a 3?

Outcome	Frequency
1	4
2	8
3	2
4	0
5	5
6	1

A $\frac{2}{3}$ **B** $\frac{1}{3}$ **C** 0.2 **D** 0.1

Spiral Review

Use elimination to solve each system of equations. (Lesson 6-3)

38. $f + g = -3$, $f - g = 1$

39. $6g + h = -7$, $6g + 3h = -9$

40. $5j + 3k = -9$, $3j + 3k = -3$

41. $2x - 4z = 6$, $x - 4z = -3$

42. $-5c - 3v = 9$, $5c + 2v = -6$

43. $4b - 6n = -36$, $3b - 6n = -36$

44. JOBS Brandy and Adriana work at an after-school child care center. Together they cared for 32 children this week. Brandy cared for 0.6 times as many children as Adriana. How many children did each girl care for? (Lesson 6-2)

Solve each inequality. Then graph the solution set. (Lesson 5-5)

45. $|m - 5| \leq 8$

46. $|q + 11| < 5$

47. $|2w + 9| > 11$

48. $|2r + 1| \geq 9$

Skills Review

Translate each sentence into a formula. (Lesson 2-1)

49. The area A of a triangle equals one half times the base b times the height h.

50. The circumference C of a circle equals the product of 2, π, and the radius r.

51. The volume V of a rectangular box is the length ℓ times the width w multiplied by the height h.

52. The volume of a cylinder V is the same as the product of π and the radius r to the second power multiplied by the height h.

53. The area of a circle A equals the product of π and the radius r squared.

54. Acceleration A equals the increase in speed s divided by time t in seconds.

Mid-Chapter Quiz

Lessons 6-1 through 6-4

Use the graph to determine whether each system is *consistent* or *inconsistent* and if it is *independent* or *dependent*.

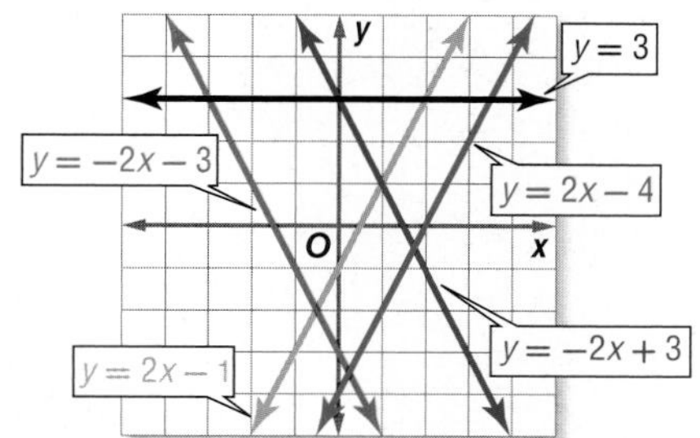

1. $y = 2x - 1$
   $y = -2x + 3$

2. $y = -2x + 3$
   $y = -2x - 3$

Graph each system and determine the number of solutions that it has. If it has one solution, name it.

3. $y = 2x - 3$
   $y = x + 4$

4. $x + y = 6$
   $x - y = 4$

5. $x + y = 8$
   $3x + 3y = 24$

6. $x - 4y = -6$
   $y = -1$

7. $3x + 2y = 12$
   $3x + 2y = 6$

8. $2x + y = -4$
   $5x + 3y = -6$

Use substitution to solve each system of equations.

9. $y = x + 4$
   $2x + y = 16$

10. $y = -2x - 3$
    $x + y = 9$

11. $x + y = 6$
    $x - y = 8$

12. $y = -4x$
    $6x - y = 30$

13. **FOOD** The cost of two meals at a restaurant is shown in the table below.

Meal	Total Cost
3 tacos, 2 burritos	$7.40
4 tacos, 1 burrito	$6.45

a. Define variables to represent the cost of a taco and the cost of a burrito.

b. Write a system of equations to find the cost of a single taco and a single burrito.

c. Solve the systems of equations, and explain what the solution means.

d. How much would a customer pay for 2 tacos and 2 burritos?

14. **AMUSEMENT PARKS** The cost of two groups going to an amusement park is shown in the table.

Group	Total Cost
4 adults, 2 children	$184
4 adults, 3 children	$200

a. Define variables to represent the cost of an adult ticket and the cost of a child ticket.

b. Write a system of equations to find the cost of an adult ticket and a child ticket.

c. Solve the system of equations, and explain what the solution means.

d. How much will a group of 3 adults and 5 children be charged for admission?

15. **MULTIPLE CHOICE** Angelina needs to buy 12 pieces of candy to take to a meeting. She has $16. Each chocolate bar costs $2, and each lollipop costs $1. Determine how many of each she can buy.

A 6 chocolate bars, 6 lollipops

B 4 chocolate bars, 8 lollipops

C 7 chocolate bars, 5 lollipops

D 3 chocolate bars, 9 lollipops

Use elimination to solve each system of equations.

16. $x + y = 9$
    $x - y = -3$

17. $x + 3y = 11$
    $x + 7y = 19$

18. $9x - 24y = -6$
    $3x + 4y = 10$

19. $-5x + 2y = -11$
    $5x - 7y = 1$

20. **MULTIPLE CHOICE** The Blue Mountain High School Drama Club is selling tickets to their spring musical. Adult tickets are $4 and student tickets are $1. A total of 285 tickets are sold for $765. How many of each type of ticket are sold?

F 145 adult, 140 student

G 120 adult, 165 student

H 180 adult, 105 student

J 160 adult, 125 student

6-5 Applying Systems of Linear Equations

Then

You solved systems of equations by using substitution and elimination.
(Lessons 6-2, 6-3, and 6-4)

Now

- Determine the best method for solving systems of equations.
- Apply systems of equations.

Math Online

glencoe.com

- Extra Examples
- Personal Tutor
- Self-Check Quiz
- Homework Help

Why?

In speed skating, competitors race two at a time on a double track. Indoor speed skating rinks have two track sizes for race events: an official track and a short track.

Speed Skating Tracks	
official track	x
short track	y

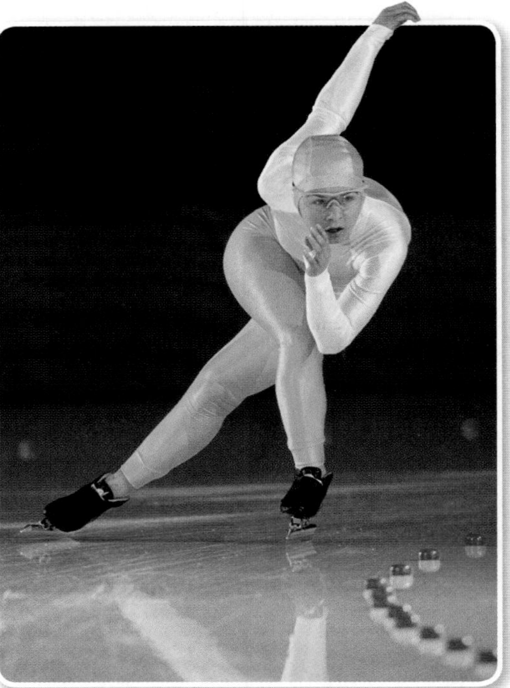

The total length of the two tracks is 511 meters. The official track is 44 meters less than four times the short track. The total length is represented by $x + y = 511$. The length of the official track is represented by $x = 4y - 44$.

You can solve the system of equations to find the length of each track.

Determine the Best Method You have learned five methods for solving systems of linear equations. The table summarizes the methods and the types of systems for which each method works best.

Concept Summary — Solving Systems of Equations — For Your FOLDABLE

Method	The Best Time to Use
Graphing	To estimate solutions, since graphing usually does not give an exact solution.
Substitution	If one of the variables in either equation has a coefficient of 1 or -1.
Elimination Using Addition	If one of the variables has opposite coefficients in the two equations.
Elimination Using Subtraction	If one of the variables has the same coefficient in the two equations.
Elimination Using Multiplication	If none of the coefficients are 1 or -1 and neither of the variables can be eliminated by simply adding or subtracting the equations.

Substitution and elimination are algebraic methods for solving systems of equations. An algebraic method is best for an exact solution. Graphing, with or without technology, is a good way to estimate a solution.

EXAMPLE 1 Choose the Best Method

Determine the best method to solve the system of equations. Then solve the system.

$4x - 4y = 8$
$-8x + y = 19$

StudyTip

Alternate Method The system of equations in Example 1 can also be solved by using elimination with multiplication. You can multiply the first equation by 2 and then add to eliminate the x-term.

Understand To determine the best method to solve the system of equations, look closely at the coefficients of each term.

Plan Neither the coefficients of x nor y are the same or additive inverses, so you cannot add or subtract to eliminate a variable. Since the coefficient of y in the second equation is 1, you can use substitution.

Solve First, solve the second equation for y.

$-8x + y = 19$	**Second equation**
$-8x + y + 8x = 19 + 8x$	**Add 8x to each side.**
$y = 19 + 8x$	**Simplify.**

Next, substitute $19 + 8x$ for y in the first equation.

$4x - 4y = 8$	**First equation**
$4x - 4(19 + 8x) = 8$	**Substitution**
$4x - 76 - 32x = 8$	**Distributive Property**
$-28x - 76 = 8$	**Simplify.**
$-28x - 76 + 76 = 8 + 76$	**Add 76 to each side.**
$-28x = 84$	**Simplify.**
$\frac{-28x}{-28} = \frac{84}{-28}$	**Divide each side by −28.**
$x = -3$	**Simplify.**

Last, substitute -3 for x in the second equation.

$-8x + y = 19$	**Second equation**
$-8(-3) + y = 19$	**$x = -3$**
$y = -5$	**Simplify.**

The solution of the system of equations is $(-3, -5)$.

Check Use a graphing calculator to check your solution. If your algebraic solution is correct, then the graphs will intersect at $(-3, -5)$.

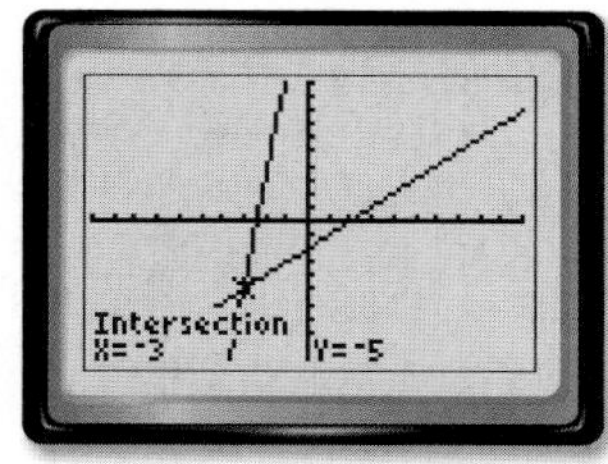

[−10, 10] scl: 1 [−10, 10] scl: 1

Check Your Progress

1A. $5x + 7y = 2$
$-2x + 7y = 9$

1B. $3x - 4y = -10$
$5x + 8y = -2$

1C. $x - y = 9$
$7x + y = 7$

1D. $5x - y = 17$
$3x + 2y = 5$

Personal Tutor glencoe.com

Apply Systems of Linear Equations When applying systems of linear equations to problems, it is important to analyze each solution in the context of the situation.

Real-World Link

Four species of penguins are on the endangered species list. The colonies of penguins that live closest to human inhabitants are the most at risk for extinction.

Source: PBS

Real-World EXAMPLE 2 Apply Systems of Linear Equations

PENGUINS Of the 17 species of penguins in the world, the largest species is the emperor penguin. One of the smallest is the Galapagos penguin. The total height of the two penguins is 169 centimeters. The emperor penguin is 22 centimeters more than twice the height of the Galapagos penguin. Find the height of each penguin.

The total height of the two species can be represented by $p + g = 169$, where p represents the height of the emperor penguin and g the height of the Galapagos penguin. Next write an equation to represent the height of the emperor penguin.

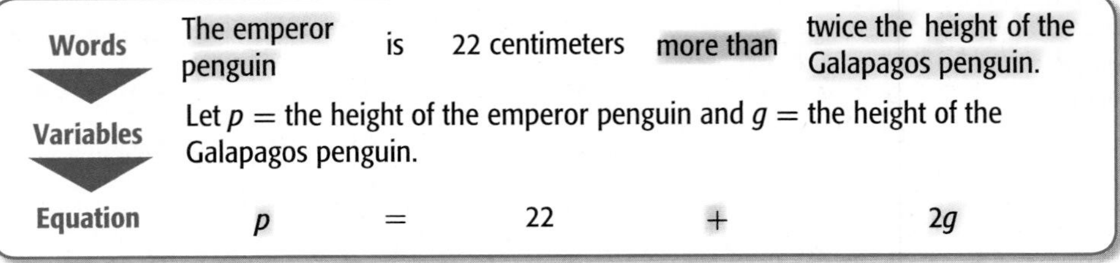

First rewrite the second equation.

$$
\begin{aligned}
p &= 22 + 2g && \textbf{Second equation} \\
p - 2g &= 22 && \textbf{Subtract } 2g \textbf{ from each side.}
\end{aligned}
$$

You can use elimination by subtraction to solve this system of equations.

$$
\begin{aligned}
p + g &= 169 && \textbf{First equation} \\
(-)\ p - 2g &= \ \ 22 && \textbf{Subtract the second equation.} \\
3g &= 147 && \textbf{Eliminate } p. \\
\frac{3g}{3} &= \frac{147}{3} && \textbf{Divide each side by 3.} \\
g &= 49 && \textbf{Simplify.}
\end{aligned}
$$

Next substitute 49 for g in one of the equations.

$$
\begin{aligned}
p &= 22 + 2g && \textbf{Second equation} \\
&= 22 + 2(49) && g = 49 \\
&= 120 && \textbf{Simplify.}
\end{aligned}
$$

The height of the emperor penguin is 120 centimeters, and the height of the Galapagos penguin is 49 centimeters.

Does the solution make sense in the context of the problem?
Check by verifying the given information. The penguins' heights added together would be 120 + 49 or 169 centimeters and 22 + 2(49) is 120 centimeters.

Check Your Progress

2. **VOLUNTEERING** Jared has volunteered 50 hours and plans to volunteer 3 hours in each coming week. Clementine is a new volunteer who plans to volunteer 5 hours each week. Write and solve a system of equations to find how long it will be before they will have volunteered the same number of hours.

Personal Tutor glencoe.com

Check Your Understanding

Example 1 p. 363

Determine the best method to solve each system of equations. Then solve the system.

1. $2x + 3y = -11$
   $-8x - 5y = 9$
2. $3x + 4y = 11$
   $2x + y = -1$
3. $3x - 4y = -5$
   $-3x + 2y = 3$
4. $3x + 7y = 4$
   $5x - 7y = -12$

Example 2 p. 364

5. **SHOPPING** At a sale, Salazar bought 4 T-shirts and 3 pairs of jeans for \$181. At the same store, Jenna bought 1 T-shirt and 2 pairs of jeans for \$94. The T-shirts were all the same price, and the jeans were all the same price.
   a. Write a system of equations that can be used to represent this situation.
   b. Determine the best method to solve the system of equations.
   c. Solve the system.

Practice and Problem Solving

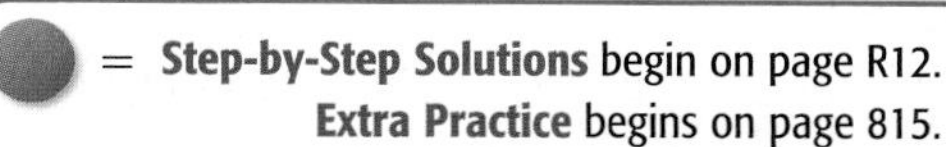
= **Step-by-Step Solutions** begin on page R12.
Extra Practice begins on page 815.

Example 1 p. 363

Determine the best method to solve each system of equations. Then solve the system.

6. $-3x + y = -3$
   $4x + 2y = 14$
7. $2x + 6y = -8$
   $x - 3y = 8$
8. $3x - 4y = -5$
   $-3x - 6y = -5$
9. $5x + 8y = 1$
   $-2x + 8y = -6$
10. $y + 4x = 3$
    $y = -4x - 1$
11. $-5x + 4y = 7$
    $-5x - 3y = -14$

Example 2 p. 364

12. **FINANCIAL LITERACY** For a Future Teachers of America fundraiser, Denzell sold food as shown in the table. He sold 11 more subs than pizzas and earned a total of \$233. Write and solve a system of equations to represent this situation. Then describe what the solution means.

Item	Selling Price
pizza	\$5.00
sub	\$3.00

13. **DVDs** Manuela has a total of 40 DVDs of movies and television shows. The number of movies is 4 less than 3 times the number of television shows. Write and solve a system of equations to find the numbers of movies and television shows that she has on DVD.

14. **CAVES** The Caverns of Sonora have two different tours: the Crystal Palace tour and the Horseshoe Lake tour. The total length of both tours is 3.25 miles. The Crystal Palace tour is a half-mile less than twice the distance of the Horseshoe Lake tour. Determine the length of each tour.

15. **YEARBOOKS** The *break-even point* is the point at which income equals expenses. Ridgemont High School is paying \$13,200 for the writing and research of their yearbook plus a printing fee of \$25 per book. If they sell the books for \$40 each, how many will they have to sell to break even? Explain.

Real-World Link

Anyone can create their own yearbook using the Internet and layout software. The yearbook is all in color, and you can order exactly as many as you need and order more anytime you want.

Source: dotPhoto Inc

16. **PAINTBALL** Clara and her friends are planning a trip to a paintball park. Find the cost of lunch and the cost of each paintball. What would be the cost for 400 paintballs and lunch?

17 **RECYCLING** Mara and Ling each recycled aluminum cans and newspaper, as shown in the table. Mara earned \$3.77, and Ling earned \$4.65.

Materials	Pounds Recycled	
	Mara	Ling
aluminum cans	9	9
newspaper	26	114

a. Define variables and write a system of linear equations from this situation.

b. What was the price per pound of aluminum? Determine the reasonableness of your solution.

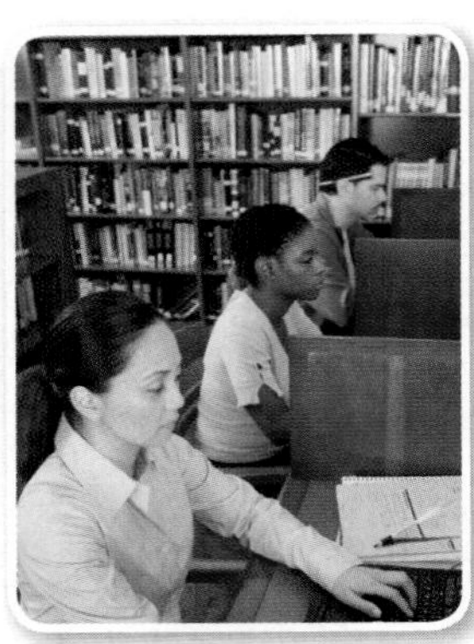

Real-World Link

There are many things for teenagers at a library besides books. Public libraries may have DVDs of new movies, current music, free Internet access, magazines, free online homework help, and SAT workshops.

Source: MSN Encarta

18. **BOOKS** The library is having a book sale. Hardcover books sell for \$4 each, and paperback books are \$2 each. If Connie spends \$26 for 8 books, how many hardcover books did she buy?

19. **MUSIC** An online music club offers individual songs for one price or entire albums for another. Kendrick pays \$14.90 to download 5 individual songs and 1 album. Geoffrey pays \$21.75 to download 3 individual songs and 2 albums.

a. How much does the music club charge to download a song?

b. How much does the music club charge to download an entire album?

20. **DRIVING** Malik drove his car for 45 miles at an average speed of r miles per hour. On the return trip, traffic has increased, and Malik's average speed is $\frac{3}{4}r$. The round trip took a total of 1 hour and 45 minutes. Find the average speed for each portion of the trip.

H.O.T. Problems

Use Higher-Order Thinking Skills

21. **OPEN ENDED** Formulate a system of equations that represents a situation in your school. Describe the method that you would use to solve the system. Then solve the system and explain what the solution means.

22. **REASONING** In a system of equations, x represents the time spent riding a bike, and y represents the distance traveled. You determine the solution to be $(-1, 7)$. Use this problem to discuss the importance of analyzing solutions in the context of real-world problems.

23. **CHALLENGE** Solve the following system of equations by using three different methods. Show your work.

$$4x + y = 13$$
$$6x - y = 7$$

24. **WRITE A QUESTION** A classmate says that elimination is the best way to solve a system of equations. Write a question to challenge his conjecture.

25. **WHICH ONE DOESN'T BELONG?** Which system is different? Explain.

$x - y = 3$ $x + \frac{1}{2}y = 1$	$-x + y = 0$ $5x = 2y$	$y = x - 4$ $y = \frac{2}{x}$	$y = x + 1$ $y = 3x$

26. **WRITING IN MATH** Explain when graphing would be the best method of solving a system of equations. When would solving a system of equations algebraically be the best method?

Standardized Test Practice

27. If $5x + 3y = 12$ and $4x - 5y = 17$, what is y?

A -1 B 3 C $(-1, 3)$ D $(3, -1)$

28. STATISTICS The scatter plot shows the number of hay bales used on the Bostwick farm during the last year.

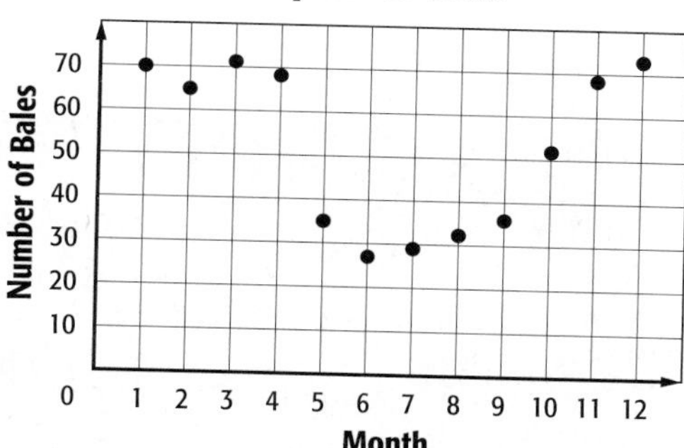

Which is an invalid conclusion?

F The Bostwicks used less hay in the summer than they did in the winter.

G The Bostwicks used about 629 bales of hay during the year.

H On average, the Bostwicks used about 52 bales each month.

J The Bostwicks used the most hay in February.

29. SHORT RESPONSE At noon, Cesar cast a shadow 0.15 foot long. Next to him a streetlight cast a shadow 0.25 foot long. If Cesar is 6 feet tall, how tall is the streetlight?

30. The graph shows the solution to which of the following systems of equations?

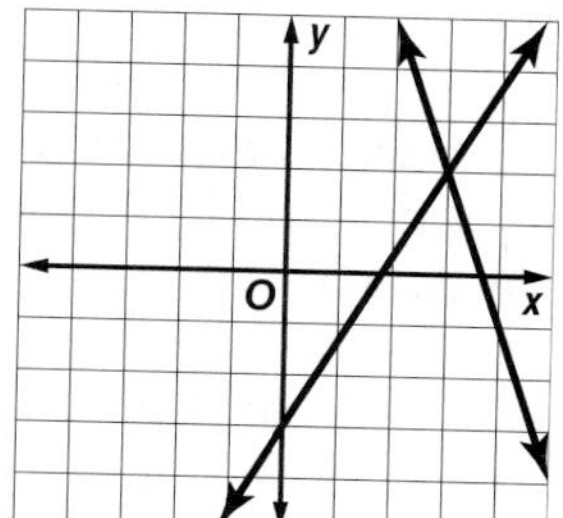

A $y = -3x + 11$
$3y = 5x - 9$

B $y = 5x - 15$
$2y = x + 7$

C $y = -3x + 11$
$2y = 4x - 5$

D $y = 5x - 15$
$3y = 2x + 18$

Spiral Review

Use elimination to solve each system of equations. (Lesson 6-4)

31. $x + y = 3$
$3x - 4y = -12$

32. $-4x + 2y = 0$
$2x - 3y = 16$

33. $4x + 2y = 10$
$5x - 3y = 7$

34. TRAVELING A youth group is traveling in two vans to visit an aquarium. The number of people in each van and the cost of admission for that van are shown. What are the adult and student prices? (Lesson 6-3)

Van	Number of Adults	Number of Students	Total Cost
A	2	5	\$77
B	2	7	\$95

Graph each inequality. (Lesson 5-6)

35. $y < 4$

36. $x \geq 3$

37. $7x + 12y > 0$

38. $y - 3x \leq 4$

Skills Review

Find each sum or difference. (Lesson 0-4)

39. $(-3.81) + (-8.5)$

40. $12.625 + (-5.23)$

41. $21.65 + (-15.05)$

42. $(-4.27) + 1.77$

43. $(-78.94) - 14.25$

44. $(-97.623) - (-25.14)$

EXTEND 6-5

Spreadsheet Lab
Credit Cards and Cash

Math Online glencoe.com
• Graphing Technology Personal Tutor

You can use a spreadsheet to compare the advantages and disadvantages of using cash versus using a credit card for a purchase.

ACTIVITY

Jun wants to purchase a car for $4000. He can save $350 per month toward the purchase of the car. Or he can use a credit card that charges 15% interest and pay $200 a month on the card. How much money will Jun save on his purchase if he waits and pays cash?

Part 1 To find out how long it will take Jun to pay cash for the car, divide $4000 by $350. This is about 11.4 months. So Jun would need to save for 12 months to pay cash for the car.

Part 2 If Jun uses his credit card to pay for the car, he would have the car right away, but he would have to pay interest. You can use a spreadsheet to find the costs by month.

Column A: List months from 0–30.

Column B: List each of the $200 payments.

Column C: Place the remaining balance. Begin with 4000 in C2.

Column D: Find the amount of interest paid each month by using the formula =C2•0.15/12.

Column E: Find the principal paid by subtracting the interest from 200 using the formula =200−D2.

Spreadsheet.xls

◇	A	B	C	D	E
1	Month	Payment	Remaining Balance	Interest Paid	Principal Paid
2	0	200	4000	50	150
3	1	200	3850.00	48.13	151.875
4	2	200	3698.13	46.23	153.77
5	3	200	3544.35	44.30	155.70
6	4	200	3388.66	42.36	157.64
21	19	200	805.48	10.07	189.93
22	20	200	615.55	7.69	192.31
23	21	200	423.25		
24	22	200			

Sheet 1 | Sheet 2 | Sheet 3

Analyze the Results

1. How long will it take Jun to pay for his car using his credit card?
2. What is the amount of Jun's last payment?
3. How can you find how much Jun pays in interest as he pays back his credit card?
4. How much total interest did Jun pay?
5. What are the benefits of using cash to pay for the car instead of using a credit card?

6-6

Organizing Data Using Matrices

Then

You represented data using statistical graphs. (Lesson 0-13)

Now

- Organize data in matrices.
- Perform matrix operations.

New Vocabulary

matrix
element
dimension
scalar
scalar multiplication

Math Online

glencoe.com

- Extra Examples
- Personal Tutor
- Self-Check Quiz
- Homework Help

Why?

The table shows high school participation in various sports.

Sport	Girls		Boys	
	Schools	Participation	Schools	Participation
basketball	17,175	456,543	17,482	545,497
cross country	12,345	170,450	12,727	201,719
lacrosse	1270	48,086	1334	59,993
tennis	9646	169,292	9426	148,530

Source: The National Federation of State High School Associations

These data can be organized into two matrices with the figures for girls and boys.

Organize Data Using Matrices A **matrix** is a rectangular arrangement of numbers in rows and columns enclosed in brackets. Each number in a matrix is called an **element**. A matrix is usually named using an uppercase letter. A matrix can be described by its **dimensions** or the number of rows and columns in the matrix. A matrix with m rows and n columns is an $m \times n$ matrix (read "m by n").

$$A = \begin{bmatrix} 7 & -9 & 5 & 3 \\ -1 & 3 & -3 & 6 \\ 0 & -4 & 8 & 2 \end{bmatrix}$$

3 rows

4 columns

The element −1 is in Row 2, Column 1.

The element 2 is in Row 3, Column 4.

Matrix A above is a 3×4 matrix because it has 3 rows and 4 columns.

EXAMPLE 1 Dimensions of a Matrix

State the dimensions of each matrix. Then identify the position of the circled element in each matrix.

a. $A = \begin{bmatrix} 1 & \textcircled{7} \\ -4 & 0 \\ 2 & -5 \end{bmatrix}$

$$\begin{bmatrix} 1 & \textcircled{7} \\ -4 & 0 \\ 2 & -5 \end{bmatrix}$$

3 rows

2 columns

Matrix A has 3 rows and 2 columns. Therefore, it is a 3×2 matrix. The circled element is in the first row and the second column.

b. $B = \begin{bmatrix} -5 & 10 & \textcircled{2} & -3 \end{bmatrix}$

$$\begin{bmatrix} -5 & 10 & \textcircled{2} & -3 \end{bmatrix}$$

1 row

4 columns

Matrix B has 1 row and 4 columns. Therefore, it is a 1×4 matrix. The circled element is in the first row and the third column.

Check Your Progress

1A. $C = \begin{bmatrix} -1 & 7 & 12 \\ 4 & \textcircled{-5} & -2 \end{bmatrix}$

1B. $D = \begin{bmatrix} -3 & 6 \\ 4 & \textcircled{-8} \end{bmatrix}$

Personal Tutor glencoe.com

Real-World EXAMPLE 2 Organize Data into a Matrix

SWIMMING **At a meet, 10 points were awarded for each first-place finish, 8 points for each second-place finish, and 5 points for each third-place finish. Use a matrix to organize each team's points. Which school had the most first-place finishes?**

School	Freestyle	Backstroke	Breaststroke	Butterfly
North	10	8	8	10
South	5	5	10	8
Jefferson	8	10	5	5

Organize the points awarded into labeled columns and rows.

$$\begin{array}{r} \\ \text{North} \\ \text{South} \\ \text{Jefferson} \end{array} \begin{array}{c} \text{Freestyle} \quad \text{Backstroke} \quad \text{Breaststroke} \quad \text{Butterfly} \\ \begin{bmatrix} 10 & 8 & 8 & 10 \\ 5 & 5 & 10 & 8 \\ 8 & 10 & 5 & 5 \end{bmatrix} \end{array}$$

North High School earned 10 points in both the freestyle event and the butterfly event, so they had the most first-place finishes.

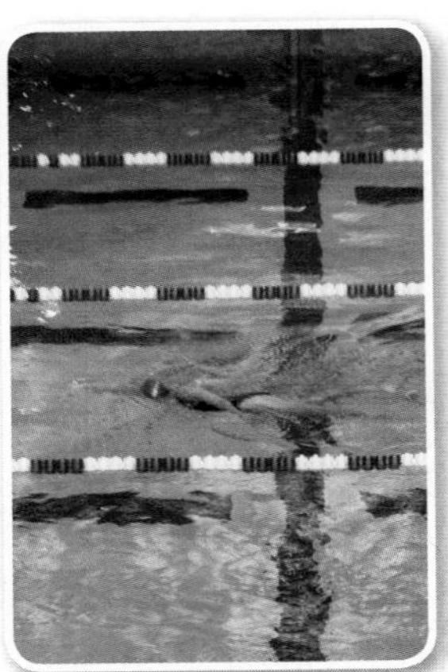

Real-World Link

In June 2007, the Projecte Home Balear set a world record by having the most participants swim one pool length each in a 24-hour relay. A total of 3,168 people participated in the relay at the Municipal Sports Complex in Palma de Mallorca, Spain.

Source: Guinness World Records

Check Your Progress

2. MOVIES For a matinee, a movie theater charges \$5.25 for an adult and \$4.50 for a child. Evening admission is \$8.75 for an adult and \$5.75 for a child. Organize the prices into a matrix. What are the dimensions of the matrix?

Personal Tutor glencoe.com

Matrix Operations If two matrices have the same dimensions, they can be added together. You add matrices by adding the corresponding elements of the matrices.

StudyTip

Corresponding Elements *Corresponding* means that the elements are in the exact same position in each matrix.

EXAMPLE 3 Add Matrices

Find each sum for $A = \begin{bmatrix} 12 & 2 \\ -9 & 15 \end{bmatrix}$, $B = \begin{bmatrix} -4 & 4 \\ -3 & -10 \end{bmatrix}$, **and** $C = \begin{bmatrix} -2 \\ 8 \end{bmatrix}$.

a. $A + B$

$A + B = \begin{bmatrix} 12 & 2 \\ -9 & 15 \end{bmatrix} + \begin{bmatrix} -4 & 4 \\ -3 & -10 \end{bmatrix}$ **Substitution**

$= \begin{bmatrix} 12 + (-4) & 2 + 4 \\ -9 + (-3) & 15 + (-10) \end{bmatrix}$ or $\begin{bmatrix} 8 & 6 \\ -12 & 5 \end{bmatrix}$ **Simplify.**

b. $B + C$

$B + C = \begin{bmatrix} -4 & 4 \\ -3 & -10 \end{bmatrix} + \begin{bmatrix} -2 \\ 8 \end{bmatrix}$ **Substitution**

Matrix B is a 2×2 matrix, and matrix C is a 2×1 matrix. Since the matrices do not have the same dimensions, it is not possible to add these matrices.

Check Your Progress

3A. $\begin{bmatrix} 7 & -2 & 11 \\ -14 & 8 & 1 \end{bmatrix} + \begin{bmatrix} -3 & 4 & 5 \\ -4 & 6 & -1 \end{bmatrix}$

3B. $\begin{bmatrix} 12 & 8 \\ -3 & -7 \\ -6 & 9 \end{bmatrix} + \begin{bmatrix} -2 & 4 \\ 1 & -1 \\ -9 & 5 \end{bmatrix}$

Personal Tutor glencoe.com

If two matrices have the same dimensions, then they can also be subtracted. You subtract matrices by subtracting the corresponding elements of the matrices.

EXAMPLE 4 Subtract Matrices

Find each difference for $A = \begin{bmatrix} 4 & -2 & 8 \\ -17 & 10 & 6 \end{bmatrix}$, $B = \begin{bmatrix} 6 & -14 & 2 \end{bmatrix}$, and $C = \begin{bmatrix} 1 & -3 & 5 \\ -10 & 8 & 7 \end{bmatrix}$. If the difference does not exist, write *impossible*.

a. $B - A$

$$B - A = \begin{bmatrix} 6 & -14 & 2 \end{bmatrix} - \begin{bmatrix} 4 & -2 & 8 \\ -7 & 10 & 6 \end{bmatrix}$$ **Substitution**

Matrix B is a 1×3 matrix, while matrix A is a 2×3 matrix. Since the dimensions are not the same, it is not possible to subtract these matrices.

b. $A - C$

$$A - C = \begin{bmatrix} 4 & -2 & 8 \\ -17 & 10 & 6 \end{bmatrix} - \begin{bmatrix} 1 & -3 & 5 \\ -10 & 8 & 7 \end{bmatrix}$$ **Substitution**

$$= \begin{bmatrix} 4 - 1 & -2 - (-3) & 8 - 5 \\ -17 - (-10) & 10 - 8 & 6 - 7 \end{bmatrix} \text{ or } \begin{bmatrix} 3 & 1 & 3 \\ -7 & 2 & -1 \end{bmatrix}$$ **Simplify.**

> **Watch Out!**
>
> **Dimensions** When working with matrices, remember that the dimensions are always given as rows by columns.

Check Your Progress

4A. $\begin{bmatrix} 2 & -5 \\ 8 & 12 \end{bmatrix} - \begin{bmatrix} 6 & -3 \\ -9 & 1 \end{bmatrix}$

4B. $\begin{bmatrix} 16 & -6 & 1 \\ -2 & 5 & -1 \end{bmatrix} - \begin{bmatrix} 21 & 3 & -6 \\ -12 & -2 & 1 \end{bmatrix}$

Personal Tutor glencoe.com

You can multiply any matrix by a constant called a **scalar**. This operation is called **scalar multiplication** and is done by multiplying each element of the matrix by the scalar.

> **StudyTip**
>
> **Matrix Operations** The order of operations for matrices is similar to that of real numbers. You would perform scalar multiplication before matrix addition and subtraction.

EXAMPLE 5 Multiply a Matrix by a Scalar

If $A = \begin{bmatrix} 2 & -4 & -7 & 9 \\ 1 & -10 & 8 & 6 \end{bmatrix}$, find $4A$.

$$4A = 4\begin{bmatrix} 2 & -4 & -7 & 9 \\ 1 & -10 & 8 & 6 \end{bmatrix}$$ **Substitution**

$$= \begin{bmatrix} 4(2) & 4(-4) & 4(-7) & 4(9) \\ 4(1) & 4(-10) & 4(8) & 4(6) \end{bmatrix}$$ **Definition of scalar multiplication**

$$= \begin{bmatrix} 8 & -16 & -28 & 36 \\ 4 & -40 & 32 & 24 \end{bmatrix}$$ **Simplify.**

Check Your Progress

5A. If $A = \begin{bmatrix} -9 & 3 \\ 5 & -11 \\ -2 & 7 \end{bmatrix}$, find $-3A$.

5B. If $B = \begin{bmatrix} -12 & 8 \\ -3 & 0 \end{bmatrix}$, find $2B$.

Personal Tutor glencoe.com

Check Your Understanding

Example 1 p. 369 — **State the dimensions of each matrix. Then identify the position of the circled element in each matrix.**

1. $\begin{bmatrix} 8 & -2 & 1 & -3 \\ \textcircled{0} & 5 & 7 & -11 \end{bmatrix}$

2. $\begin{bmatrix} 6 \\ -7 \\ 2 \\ \textcircled{1} \end{bmatrix}$

3. $\begin{bmatrix} 9 & -12 & \textcircled{6} & 2 \end{bmatrix}$

4. $\begin{bmatrix} 6 & -8 & 12 & 9 & -1 \\ 2 & 9 & 7 & \textcircled{11} & -5 \\ 5 & 0 & 1 & 3 & 4 \end{bmatrix}$

Example 2 p. 370

5. HOTELS The costs for an overnight stay at a hotel are listed in the table at the right.

Room	Weekday	Weekend
single	\$69	\$89
double	\$79	\$109
suite	\$99	\$139

a. Write a matrix to organize the costs of an overnight stay at the hotel.

b. What are the dimensions of the matrix?

c. Which room and night is the most expensive? least expensive?

Examples 3–5 pp. 370–371 — **Perform the indicated matrix operations. If the matrix does not exist, write *impossible*.**

6. $\begin{bmatrix} 5 & -2 \\ 7 & -6 \end{bmatrix} + \begin{bmatrix} -5 & -8 \\ 3 & 1 \end{bmatrix}$

7. $\begin{bmatrix} 8 & 11 & 5 \\ -3 & 7 & 8 \\ -1 & -2 & 0 \end{bmatrix} - \begin{bmatrix} -4 & 10 & -9 \\ 6 & -12 & -1 \\ 7 & 5 & 3 \end{bmatrix}$

8. $-2\begin{bmatrix} 7 & -2 & 0 & 1 \\ -8 & 11 & -9 & 3 \\ -4 & -7 & 5 & 6 \end{bmatrix}$

9. $\begin{bmatrix} 15 \\ -8 \\ 4 \end{bmatrix} - \begin{bmatrix} -3 & -2 & 7 \end{bmatrix}$

Practice and Problem Solving

 = **Step-by-Step Solutions** begin on page R12.
Extra Practice begins on page 815.

Example 1 p. 369 — **State the dimensions of each matrix. Then identify the position of the circled element in each matrix.**

10. $\begin{bmatrix} 6 & 8 & -2 & 3 \\ -7 & -12 & 58 & 1 \\ 86 & \textcircled{12} & 7 & -9 \\ 0 & -6 & 21 & 79 \end{bmatrix}$

11. $\begin{bmatrix} 2 & 9 \\ \textcircled{-3} & -5 \\ 7 & -8 \\ 1 & -1 \\ -2 & 3 \end{bmatrix}$

12. $\begin{bmatrix} 8 & -10 & 4 & 6 & -2 \\ \textcircled{3} & 7 & 9 & 5 & -1 \end{bmatrix}$

13. $\begin{bmatrix} 8 & 2 & -1 & 4 & 3 & -7 \\ 9 & 10 & -17 & 0 & 1 & -8 \\ -1 & 5 & -2 & \textcircled{7} & -3 & 0 \\ -9 & 7 & 5 & 3 & -6 & 6 \end{bmatrix}$

14. $\begin{bmatrix} 1 & -20 & -16 \\ -5 & 0 & 7 \\ 9 & -13 & 12 \\ 8 & \textcircled{-9} & 2 \\ -3 & 5 & 10 \\ 6 & -14 & 25 \end{bmatrix}$

15. $\begin{bmatrix} 3 & 2 & 7 & 0 \\ 4 & 9 & 10 & \textcircled{4} \\ -1 & 7 & 6 & 5 \\ 0 & -3 & 12 & -5 \\ 8 & -5 & -10 & -8 \\ -2 & 4 & 11 & -2 \end{bmatrix}$

16. $\begin{bmatrix} 8 & 6 & -4 & 2 & 1 & 3 \\ -8 & -4 & 0 & \textcircled{9} & -5 & 6 \\ 2 & 3 & -1 & 7 & -9 & 0 \end{bmatrix}$

Example 2 p. 370

17 **GEOGRAPHY** The land area in square miles and the number of people per square mile in 2000 are shown.

State	Land Area	People per Square Mile
Ohio	40,948	277.3
Florida	53,926	296.4
New York	47,213	401.9
North Carolina	48,710	165.2

a. Write a matrix to organize the given data.

b. What are the dimensions of the matrix?

c. Which state has the most people per square mile? the fewest people per square mile?

Examples 3–5 pp. 370–371

Perform the indicated matrix operations. If the matrix does not exist, write ***impossible.***

18. $\begin{bmatrix} 8 & -5 & 1 \\ 3 & -7 & -4 \end{bmatrix} - \begin{bmatrix} 6 & -2 & -7 & 9 \\ 10 & -3 & 1 & -4 \end{bmatrix}$

19. $\begin{bmatrix} -9 & 5 & 1 \\ 14 & -6 & 7 \end{bmatrix} + \begin{bmatrix} 3 & -4 & -1 \\ -7 & -2 & 8 \end{bmatrix}$

20. $-3\begin{bmatrix} 6 & -8 & 9 & 1 & -3 \\ 0 & 7 & -2 & -4 & 5 \end{bmatrix}$

21. $5\begin{bmatrix} 2 & -1 & 0 \\ 1 & -3 & 5 \\ 7 & 10 & -11 \\ 8 & -9 & -4 \end{bmatrix}$

22. $\begin{bmatrix} 17 & 10 \\ -5 & 1 \\ 7 & 6 \\ -8 & -2 \\ 3 & 8 \end{bmatrix} - \begin{bmatrix} 20 & 6 \\ -4 & -5 \\ -9 & 0 \\ -1 & 9 \\ 6 & -12 \end{bmatrix}$

23. $\begin{bmatrix} 6 & 8 & -4 & -2 \end{bmatrix} - \begin{bmatrix} 9 & -4 & 7 & 8 \end{bmatrix}$

24. **VOTING** The results of a recent poll are organized in the matrix shown at the right.

$$\begin{array}{l} \\ \text{Propositon 1} \\ \text{Propositon 2} \\ \text{Propositon 3} \end{array} \begin{array}{c} \begin{array}{cc} \text{For} & \text{Against} \end{array} \\ \begin{bmatrix} 562 & 1025 \\ 789 & 921 \\ 1255 & 301 \end{bmatrix} \end{array}$$

a. How many people voted for Proposition 1?

b. How many more people voted against Proposition 2 than for Proposition 2?

c. How many votes were cast against the propositions?

25. **SALES** The manager of The Donut Delight Shop keeps records of the types of donuts sold each day. Two days of sales are shown.

Day	Store	Sales of Each Type of Donut ($)			
		Chocolate	Glazed	Powdered	Lemon Filled
Saturday	Main St.	95	205	70	51
	Elm St.	105	245	79	49
Sunday	Main St.	167	295	99	79
	Elm St.	159	289	107	88

a. Describe what 245 represents.

b. Write a matrix that represents the sales for each day.

c. How much did each store make in sales over the two days for each type of donut?

d. Which donut made the company the most money?

Real-World Link

Some Dutch settlers brought donuts to Colonial America. Since ovens were not always available, people began frying the dough. Then sugar and spices were added.

Source: Oracle Education Foundation

Real-World Link

The SAT test consists of three parts: math, critical reading, and writing. Each part is worth 800 points, and 2400 is the best possible score. The average SAT score is about 1518.

Source: *The Washington Post*

26. **SCORES** The average SAT scores for males and females are shown.

Year	Verbal Score		Mathematical Score	
	Male	Female	Male	Female
1998	509	502	531	496
2000	507	504	533	498
2002	507	502	534	500
2004	512	504	537	501
2005	513	505	538	504

a. Organize the verbal scores and mathematical scores into two matrices.

b. Find the total score that males and females earned on the SATs each year.

c. Express the difference between the verbal and mathematical scores in a matrix.

Perform the indicated matrix operations. If an operation cannot be performed, write *impossible*.

27. $2\begin{bmatrix} -5 & 8 & 2 \end{bmatrix} + \begin{bmatrix} -6 & 9 & 5 \end{bmatrix}$

28. $\begin{bmatrix} 9 & -5 \\ -3 & 4 \end{bmatrix} + (-9)\begin{bmatrix} 2 & -1 \\ 0 & -7 \end{bmatrix}$

29. $-3\begin{bmatrix} 7 \\ -4 \\ -2 \\ 1 \end{bmatrix} - 2\begin{bmatrix} 7 \\ -5 \\ -3 \end{bmatrix}$

30. $-1\begin{bmatrix} 5 & -8 & 14 \\ 12 & -7 & -3 \\ 1 & 0 & 8 \end{bmatrix} + \begin{bmatrix} 7 & 8 & 2 \\ -7 & -2 & 6 \\ -6 & 3 & -1 \end{bmatrix}$

31. $\begin{bmatrix} -5 & 2 \\ 12 & -11 \\ 9 & 0 \\ -1 & 7 \\ 6 & 5 \\ -4 & 2 \end{bmatrix} + 4\begin{bmatrix} 10 & 4 \\ -1 & -3 \\ 5 & -8 \\ -9 & 0 \\ 1 & 4 \\ -3 & 2 \end{bmatrix}$

32. $2\begin{bmatrix} 4 & 0 \\ -1 & 7 \\ -3 & 2 \end{bmatrix} - 3\begin{bmatrix} 1 & -4 \\ -8 & 9 \\ 10 & 7 \end{bmatrix} + \begin{bmatrix} 6 & 3 \\ 8 & -4 \\ -1 & -2 \end{bmatrix}$

H.O.T. Problems

Use **H**igher-**O**rder **T**hinking Skills

33. **OPEN ENDED** Write two matrices with a difference of $\begin{bmatrix} 5 & -2 & 7 \\ -3 & 1 & 0 \end{bmatrix}$.

34. **CHALLENGE** Write three matrices with a sum of $\begin{bmatrix} 5 & -1 \\ 9 & 7 \\ -5 & 8 \end{bmatrix}$ if at least one of the addends is shown as the product of a scalar and a matrix.

35. **REASONING** For matrix $A = \begin{bmatrix} 1 & 2 \\ 3 & 4 \end{bmatrix}$, the *transpose* of A is $A^T = \begin{bmatrix} 1 & 3 \\ 2 & 4 \end{bmatrix}$. Write a matrix B that is equal to its transpose B^T.

36. **REASONING** Is it possible to add a 3×2 matrix and a 2×3 matrix? Explain. Include an example or counterexample to support your answer.

37. **OPEN ENDED** Describe a real-world situation that can be modeled by using a matrix. Then write a matrix to model the situation.

38. **WRITING IN MATH** Summarize how to perform matrix operations on matrices.

Standardized Test Practice

39. If $A = \begin{bmatrix} 5 & -8 & 1 \\ 7 & -3 & 4 \end{bmatrix}$ and $B = \begin{bmatrix} 9 & -5 & 2 \\ -1 & -7 & 6 \end{bmatrix}$, find $A + B$.

A $\begin{bmatrix} 14 & -13 & 3 \\ 7 & -3 & 4 \end{bmatrix}$

B $\begin{bmatrix} 5 & -8 & 1 \\ 6 & -10 & 10 \end{bmatrix}$

C $\begin{bmatrix} 14 & -13 & 3 \\ 6 & -10 & 10 \end{bmatrix}$

D $\begin{bmatrix} -4 & -3 & -1 \\ 8 & 4 & -2 \end{bmatrix}$

40. SHORT RESPONSE The difference between the length and width of a rectangle is 9 inches. Find the dimensions of the rectangle if its perimeter is 52 inches.

41. At a movie theater, the costs for various amounts of popcorn and hot dogs are shown.

Hot Dogs	Boxes of Popcorn	Total Cost
1	1	$8.50
2	4	$21.60

Which pair of equations can be used to find p, the cost of a box of popcorn, and h, the cost of a hot dog?

F $p + h = 8.5$
$p + 2h = 10.8$

G $p + h = 8.5$
$2h + 4p = 21.6$

H $p + h = 8.5$
$2p + 4h = 21.6$

J $p + h = 8.5$
$2p + 2h = 21.6$

42. What is the solution set for $9 + x \geq 3$?

A $\{x \mid x \geq -6\}$

B $\{x \mid x \geq 6\}$

C $\{x \mid x \leq -6\}$

D $\{x \mid x \leq 6\}$

Spiral Review

43. CHEMISTRY Orion Labs needs to make 500 gallons of 34% acid solution. The only solutions available are a 25% acid solution and a 50% acid solution. Write and solve a system of equations to find the number of gallons of each solution that should be mixed to make the 34% solution. (Lesson 6-5)

Use elimination to solve each system of equations. (Lesson 6-4)

44. $x + y = 7$
$2x + y = 11$

45. $a - b = 9$
$7a + b = 7$

46. $q + 4r = -8$
$3q + 2r = 6$

47. SALES Marissa wants to make at least $75 selling caramel apples at the school carnival. She plans to sell each apple for $1.50. Write and solve an inequality to find the number of apples a she needs to make and sell to reach her goal if it costs her $0.30 per apple. (Lesson 5-3)

Find the next three terms of each arithmetic sequence. (Lesson 3-5)

48. 4, 7, 10, 13, . . .

49. 18, 24, 30, 36, . . .

50. −66, −70, −74, −78, . . .

51. CRAFTS Mandy makes baby blankets and stuffed rabbits to sell at craft fairs. She sells blankets for $28 and rabbits for $18. Write and evaluate an expression to find her total amount of sales if she sells 25 blankets and 25 rabbits. (Lesson 1-4)

Skills Review

Solve each equation. (Lesson 2-3)

52. $5 = 4t - 7$

53. $-3x + 10 = 19$

54. $\frac{c}{-4} - 2 = -36$

55. $6 + \frac{y}{3} = -45$

56. $9 = \frac{d + 5}{8}$

57. $\frac{r + 1}{3} = 8$

6-7 Using Matrices to Solve Systems of Equations

Then

You solved systems of equations by graphing, using substitution, and using elimination. (Lesson 6-1 through 6-4)

Now

- Write systems of equations as augmented matrices.
- Solve systems of equations by using elementary row operations.

New Vocabulary

augmented matrix
row reduction
identity matrix

Math Online
glencoe.com

- Extra Examples
- Personal Tutor
- Self-Check Quiz
- Homework Help

Why?

The 30 members of the Washington High School's Ski Club went on a one-day ski trip. Members can rent skis for \$22 per day or snowboards for \$24 per day. The club paid a total of \$700 for rental equipment.

The resort can use this information to find how many members rented each type of equipment.

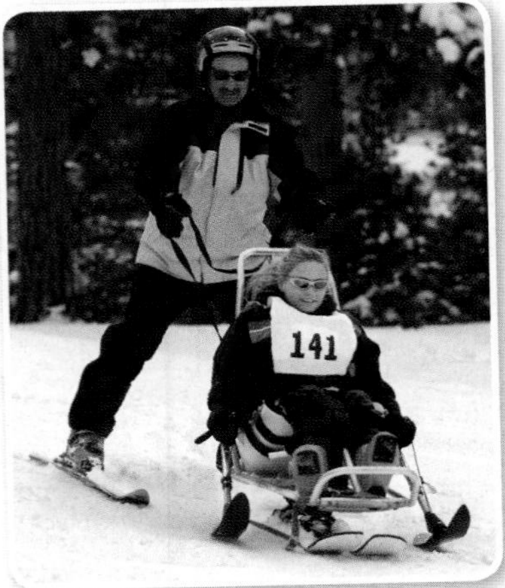

Augmented Matrices You can use a matrix called an **augmented matrix** to solve a system of equations. An augmented matrix consists of the coefficients and the constant terms of a system of equations. The coefficients and constant terms are usually separated by a dashed line.

Linear System	Augmented Matrix
$x - 3y = 8$ $-9x + 2y = -4$	$\left[\begin{array}{cc:c} 1 & -3 & 8 \\ -9 & 2 & -4 \end{array}\right]$

Make sure that the coefficients of the x-terms are listed in one column, the coefficients of the y-terms are in another column, and the constant terms are in a third column.

EXAMPLE 1 Write an Augmented Matrix

Write an augmented matrix for each system of equations.

a. $-2x + 7y = 11$
$6x - 4y = 2$

Place the coefficients of the equations and the constant terms into a matrix.

$$\begin{aligned} -2x + 7y &= 11 \\ 6x - 4y &= 2 \end{aligned} \longrightarrow \left[\begin{array}{cc:c} -2 & 7 & 11 \\ 6 & -4 & 2 \end{array}\right]$$

b. $x - 2y = 5$
$y = -4$

$$\begin{aligned} x - 2y &= 5 \\ y &= -4 \end{aligned} \longrightarrow \left[\begin{array}{cc:c} 1 & -2 & 5 \\ 0 & 1 & -4 \end{array}\right]$$

Check Your Progress

1A. $6x - 8y = -10$
$-5x = -20$

1B. $3x - 2y = 6$
$2x + 3y = 12$

Personal Tutor glencoe.com

Solve Systems of Equations You can solve a system of equations by using an augmented matrix. By performing row operations, you can change the form of the matrix.

Key Concept — Elementary Row Operations

For Your FOLDABLE

The following operations can be performed on an augmented matrix.

- Interchange any two rows.
- Multiply all elements in a row by a nonzero constant.
- Replace one row with the sum of that row and a multiple of another row.

Row reduction is the process of performing elementary row operations on an augmented matrix to solve a system. The goal is to get the coefficients portion of the matrix to have the form $\begin{bmatrix} 1 & 0 \\ 0 & 1 \end{bmatrix}$, also known as the **identity matrix**.

The first row will give you the solution for x, because the coefficient of y is 0 and the coefficient of x is 1. The second row will give you the solution for y, because the coefficient of x is 0 and the coefficient of y is 1.

EXAMPLE 2 Use Row Operations to Solve a System

Use an augmented matrix to solve the system of equations.

$-5x + 3y = 6$

$x - y = 4$

Step 1 Write the augmented matrix: $\left[\begin{array}{cc|c} -5 & 3 & 6 \\ 1 & -1 & 4 \end{array}\right]$.

Step 2 Notice that the first element in the second row is 1. Interchange the rows so 1 can be in the upper left-hand corner.

$$\left[\begin{array}{cc|c} -5 & 3 & 6 \\ 1 & -1 & 4 \end{array}\right] \xrightarrow{\text{Interchange } R_1 \text{ and } R_2.} \left[\begin{array}{cc|c} 1 & -1 & 4 \\ -5 & 3 & 6 \end{array}\right]$$

Step 3 To make the first element in the second row a 0, multiply the first row by 5 and add the result to row 2.

$$\left[\begin{array}{cc|c} 1 & -1 & 4 \\ -5 & 3 & 6 \end{array}\right] \xrightarrow{5R_1 + R_2} \left[\begin{array}{cc|c} 1 & -1 & 4 \\ 0 & -2 & 26 \end{array}\right]$$

The result is placed in row 2.

Step 4 To make the second element in the second row a 1, multiply the second row by $-\frac{1}{2}$.

$$\left[\begin{array}{cc|c} 1 & -1 & 4 \\ 0 & -2 & 26 \end{array}\right] \xrightarrow{-\frac{1}{2}R_2} \left[\begin{array}{cc|c} 1 & -1 & 4 \\ 0 & 1 & -13 \end{array}\right]$$

The result is placed in row 2.

Step 5 To make the second element in the first row a 0, add the rows together.

$$\left[\begin{array}{cc|c} 1 & -1 & 4 \\ 0 & 1 & -13 \end{array}\right] \xrightarrow{R_1 + R_2} \left[\begin{array}{cc|c} 1 & 0 & -9 \\ 0 & 1 & -13 \end{array}\right]$$

The result is placed in row 1.

The solution is $(-9, -13)$.

Check Your Progress

2A. $x + 2y = 6$
$2x + y = 9$

2B. $2x - 3y = 3$
$x + y = 14$

Personal Tutor glencoe.com

StudyTip

Alternate Method Row operations can be performed in different orders to arrive at the same result. In Example 2, you could have started by multiplying the first row, R_1, by $-\frac{1}{5}$ instead of interchanging the rows.

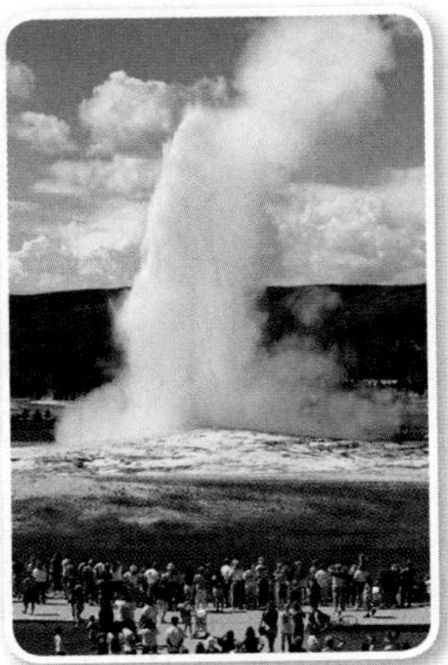

Real-World Link

Yellowstone National Park is America's first national park. It is located in Wyoming, Montana, and Idaho. The park was established in 1872. The most popular geyser in the world, Old Faithful, is located there.

Source: National Park Service

Matrices are useful for solving real-world problems. First, model the situation with a system of equations and then write the augmented matrix.

Real-World EXAMPLE 3

PARKS A youth group traveling in two vans visited Yellowstone National Park. The number of people in each van and the park fees are shown in the table.

Van	Number of Adults	Number of Students	Total Cost
A	2	6	\$102
B	2	7	\$114

a. Write a system of linear equations to model the situation. Then write the augmented matrix.

Let a represent the adult fee, and let s represent the student fee.

$$\begin{aligned} 2a + 6s &= 102 \\ 2a + 7s &= 114 \end{aligned} \rightarrow \left[\begin{array}{cc|c} 2 & 6 & 102 \\ 2 & 7 & 114 \end{array}\right]$$

b. Find the entrance fee for an adult and a student.

Step 1 To make the first element in the first row a 1, multiply the first row by $\frac{1}{2}$.

$$\left[\begin{array}{cc|c} 2 & 6 & 102 \\ 2 & 7 & 114 \end{array}\right] \xrightarrow{\frac{1}{2}R_1} \left[\begin{array}{cc|c} 1 & 3 & 51 \\ 2 & 7 & 114 \end{array}\right]$$

Step 2 To make the first element in the second row a 0, multiply the first row by −2 and add the result to row 2.

$$\left[\begin{array}{cc|c} 1 & 3 & 51 \\ 2 & 7 & 114 \end{array}\right] \xrightarrow{-2R_1 + R_2} \left[\begin{array}{cc|c} 1 & 3 & 51 \\ 0 & 1 & 12 \end{array}\right]$$

Step 3 To make the second element in the first row a 0, multiply the second row by −3 and add the result to row 1.

$$\left[\begin{array}{cc|c} 1 & 3 & 51 \\ 0 & 1 & 12 \end{array}\right] \xrightarrow{-3R_2 + R_1} \left[\begin{array}{cc|c} 1 & 0 & 15 \\ 0 & 1 & 12 \end{array}\right]$$

The solution is (15, 12). The adult fee is \$15, and the student fee is \$12.

Check Your Progress

3. CARNIVAL At a carnival, 44 tickets are required for 4 meals and 8 rides, and 58 tickets are required for 6 meals and 10 rides. How many tickets are required for each item?

Personal Tutor glencoe.com

Check Your Understanding

Example 1 p. 376

Write an augmented matrix for each system of equations.

1. $-x + 3y = -10$
$5x - 2y = 7$

2. $x - 4y = 5$
$-2x + 8y = 1$

3. $x + 2y = -1$
$2x - 2y = -9$

4. $-4x + 6y = 2$
$-x - 8y = 0$

5. $3x + 4y = -5$
$2x - y = 6$

6. $-x + 3y = 8$
$6x - 3y = -3$

Example 2 p. 377

Use an augmented matrix to solve each system of equations.

7. $x + y = -3$
$x - y = 1$

8. $x - y = -2$
$2x + 2y = 12$

9. $3x - 4y = -27$
$x + 2y = 11$

10. $x + 4y = -6$
$2x - 5y = 1$

Example 3 p. 378

11. **SHOPPING** Darnell and Sandra went shopping for graphic novels. The store charges one price for all new books and another for all old books.

   a. Write a system of linear equations to model the situation. Let n represent new books, and let b represent old books.

   b. Write the augmented matrix.

   c. What is the price for each type of book?

Practice and Problem Solving

 = **Step-by-Step Solutions** begin on page R12.
Extra Practice begins on page 815.

Example 1 p. 376

Write an augmented matrix for each system of equations.

12. $x + 2y = -3$
$3x - y = 2$

13. $-4x - 3y = -8$
$x + y = -12$

14. $2x + y = 1$
$x + 4y = -5$

15. $-6x + y = -15$
$x - 2y = 13$

16. $3x + y = -6$
$x - 5y = -7$

17. $x - y = 7$
$9x - 5y = 23$

Example 2 p. 377

Use an augmented matrix to solve each system of equations.

18. $x - 3y = -2$
$4x + y = 31$

19. $x + 2y = 3$
$-3x + 3y = 27$

20. $2x - 3y = -20$
$x + 2y = 11$

21. $x - y = 2$
$3x - 2y = 2$

22. $x - 2y = -15$
$2x + 5y = 78$

23. $4x + 3y = -9$
$x + 4y = -25$

24. $-2x - 2y = 8$
$5x + 2y = -2$

25. $3x - 6y = 36$
$2x + 4y = -40$

26. $2x + 3y = 11$
$3x - y = -11$

27. $4x - 3y = 24$
$2x + 5y = -14$

28. $3x + 2y = 6$
$x + 2y = 8$

29. $4x - 3y = 5$
$2x + 9y = 6$

Example 3 p. 378

30. **CHEERLEADING** If the Franklin High School cheerleaders replace 8 uniforms and 6 poms, the cost is $378. If they replace 6 uniforms and 9 poms, the cost is $333.

   a. Write a system of linear equations to model the situation. Let u represent uniforms, and let p represent poms.

   b. Write the augmented matrix.

   c. What is the cost of each uniform and each pom?

Write a system of equations for each augmented matrix.

31. $\left[\begin{array}{cc|c} 1 & 0 & 16 \\ 0 & 1 & -2 \end{array}\right]$

32. $\left[\begin{array}{cc|c} 1 & 7 & 3 \\ -5 & 10 & 4 \end{array}\right]$

33. $\left[\begin{array}{cc|c} 3 & 2 & 7 \\ -1 & -4 & 5 \end{array}\right]$

34. $\left[\begin{array}{cc|c} 1 & 0 & -10 \\ 0 & 4 & 8 \end{array}\right]$

35. $\left[\begin{array}{cc|c} -1 & 9 & 12 \\ 2 & 3 & -7 \end{array}\right]$

36. $\left[\begin{array}{cc|c} 6 & -6 & 5 \\ -8 & 11 & 0 \end{array}\right]$

Real-World Link

According to a survey of girls between the ages of 8 and 15, watching videos is a popular slumber party activity.

Source: Blockbuster

37. **SCHOOL STORE** Nari is checking items being shipped to the school store. The shipment contains notebooks that cost \$22 per box and mugs that cost \$40 per box. She counts 16 boxes, and the invoice states that the order totals \$460. How many boxes of each item were received?

38. **PARTIES** Mel is having a few friends over, and she is buying subs and cans of sodas for them. Mel bought 28 items. If Mel spent \$56.70, how many subs did she buy? How many sodas did she buy?

39. **RENTALS** Makya and his three sisters rented 2 items each at the video store. Members can rent movies for \$4 and video games for \$4.50. If Makya and his sisters spent \$33.50, how many movies did they rent? How many games did they rent?

Use an augmented matrix to solve each system of equations.

40. $2x + y = -1$
$-2x + y = -4$

41. $x + 2y = 3$
$2x + 4y = 6$

42. $3x - y = 1$
$-12x + 4y = 3$

43. $3x - 9y = 12$
$-2x + 6y = 9$

44. $4x - 3y = 1$
$-8x + 6y = -2$

45. $6x - 2y = -4$
$-3x + y = 2$

46. **MULTIPLE REPRESENTATIONS** In this problem, you will investigate the different representations of the following problem and their impact on the solution.

Paloma exercises every morning for 40 minutes. She does a combination of aerobics, which burns about 11 Calories per minute, and stretching, which burns about 4 Calories per minute. Her goal is to burn 335 Calories during her routine. How long should she do each activity to burn 335 Calories?

a. **VERBAL** List the representations that would be appropriate to solve the problem.

b. **ALGEBRAIC** Select a representation and solve the problem.

c. **ALGEBRAIC** Select a different representation and solve the problem.

d. **VERBAL** Write about the relationship between the representations of the problem. How did each affect your solution?

H.O.T. Problems

Use Higher-Order Thinking Skills

47. **REASONING** Explain why the system represented by $\left[\begin{array}{cc|c} 6 & 2 & 3 \\ 6 & 2 & -5 \end{array}\right]$ has no solution.

Problem-Solving Tip

Look for a Pattern Looking for a pattern can help to identify a function and write an equation.

48. **WRITING IN MATH** Describe the advantages and disadvantages of using an augmented matrix to solve a system of equations.

49. **CHALLENGE** For $a \neq 0$, what is the solution of the system represented by $\left[\begin{array}{cc|c} a & 2 & 4 \\ a & -3 & -6 \end{array}\right]$?

50. **OPEN ENDED** Write a word problem for the system represented by $\left[\begin{array}{cc|c} 3 & 1 & 13 \\ 2 & 1 & 9.5 \end{array}\right]$. Solve the system, and explain its meaning in this situation.

51. **WRITING IN MATH** Summarize how to write and use an augmented matrix to solve a system of linear equations.

Standardized Test Practice

52. SHORT RESPONSE Tonisha paid \$25.75 for 3 games of miniature golf and 2 rides on go-karts. Trevor paid \$35.75 for 4 games of miniature golf and 3 rides on go-karts. How much did each activity cost?

53. What is the solution of this system of equations?

$$0.5x - 2y = 17$$
$$2x + y = 104$$

A (50, 4)
B (4, 50)
C no solution
D infinitely many solutions

54. PROBABILITY Lexis scored 88, 95, 77, and 93 on her first four tests. What grade must she get on her fifth test to earn an average of 90 for all five tests?

F 85 **H** 96
G 90 **J** 97

55. Pablo's Pizza Place estimates that 42% of their annual sales go toward paying employees. If the pizza place makes \$4156.50 on Friday, approximately how much went for paying employees?

A \$98.96 **C** \$174.57
B \$1745.73 **D** \$17457.30

Spiral Review

Perform the indicated matrix operations. If an operation cannot be performed, write *impossible*. (Lesson 6-6)

56. $\begin{bmatrix} 5 & -2 & 4 \\ -3 & 7 & 9 \\ 12 & -1 & 8 \end{bmatrix} + \begin{bmatrix} -6 & 7 & -12 \\ 15 & -8 & 1 \\ 9 & 3 & -5 \end{bmatrix}$

57. $\begin{bmatrix} 8 & -13 \\ 4 & 2 \\ -4 & -6 \end{bmatrix} - \begin{bmatrix} -3 & 5 \\ 10 & -16 \\ 8 & -2 \end{bmatrix}$

58. $-3\begin{bmatrix} -7 & -1 & 3 & 0 \\ -1 & 5 & 7 & 9 \end{bmatrix}$

59. $\begin{bmatrix} 5 \\ -8 \\ 3 \end{bmatrix} + \begin{bmatrix} 9 & -4 & 2 \end{bmatrix}$

60. SPORTS In the 2006 Winter Olympic Games, the total number of gold and silver medals won by the U.S. was 18. The total points scored for gold and silver medals was 45. Write and solve a system of equations to find how many gold and silver medals were won by the U.S. (Lesson 6-5)

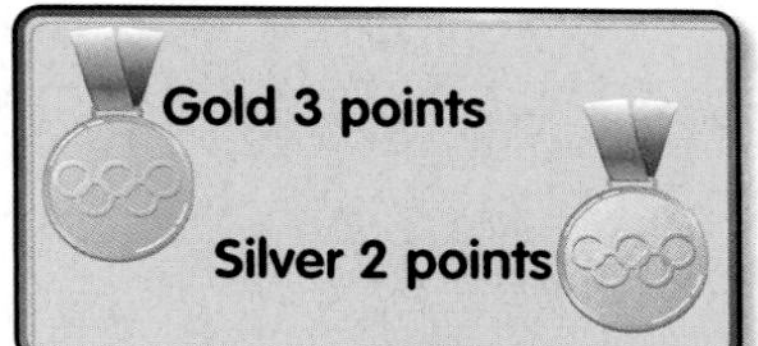

61. DRIVING Tires should be kept within 2 pounds per square inch (psi) of the manufacturer's recommended tire pressure. If the recommendation for a tire is 30 psi, what is the range of acceptable pressures? (Lesson 5-5)

Write an equation in slope-intercept form for the line that passes through the given point and is parallel to the graph of each equation. (Lesson 4-4)

62. $(-3, 2); y = x - 6$

63. $(2, -1); y = 2x + 2$

64. $(-5, -4); y = \frac{1}{2}x + 1$

65. $(3, 3); y = \frac{2}{3}x - 1$

66. $(-4, -3); y = -\frac{1}{3}x + 3$

67. $(-1, 2); y = -\frac{1}{2}x - 4$

Skills Review

Simplify each expression. (Lesson 1-4)

68. $2(7p + 4)$

69. $-3(3 - 8x)$

70. $-3y + 5(4y)$

71. $-2(4m - 6 + 8m)$

72. $5g - 8g + 2(-4g)$

73. $12(4c + 3b)$

6-8 Systems of Inequalities

Then

You graphed and solved linear inequalities. (Lesson 5-6)

Now

- Graph systems of linear inequalities.
- Solve systems of linear inequalities by graphing.

New Vocabulary

system of inequalities

Math Online

glencoe.com

- Extra Examples
- Personal Tutor
- Self-Check Quiz
- Homework Help
- Math in Motion

Why?

Jacui is beginning an exercise program that involves an intense cardiovascular workout. Her trainer recommends that for a person her age, her heart rate should stay within the following range as she exercises.

- It should be higher than 102 beats per minute.
- It should not exceed 174 beats per minute.

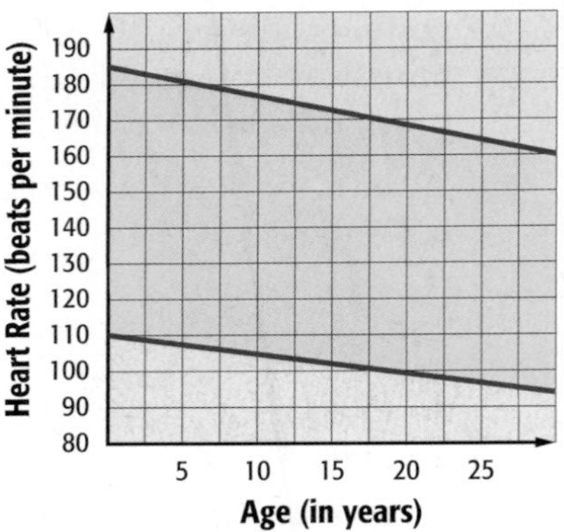

The graph shows the maximum and minimum target heart rate for people ages 0 to 30 as they exercise. If the preferred range is in light green, how old do you think Jacui is?

Systems of Inequalities The graph above is a graph of two inequalities. A set of two or more inequalities with the same variables is called a **system of inequalities**.

The solution of a system of inequalities with two variables is the set of ordered pairs that satisfy all of the inequalities in the system. The solution set is represented by the overlap, or intersection, of the graphs of the inequalities.

EXAMPLE 1 Solve by Graphing

Solve the system of inequalities by graphing.

$y > -2x + 1$
$y \leq x + 3$

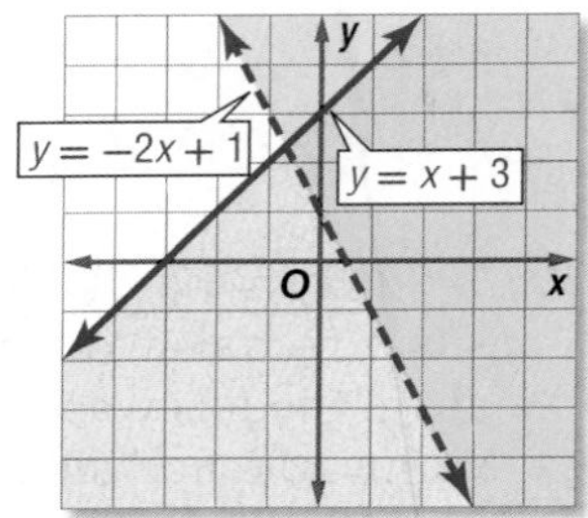

The graph of $y = -2x + 1$ is dashed and is not included in the graph of the solution. The graph of $y = x + 3$ is solid and is included in the graph of the solution.

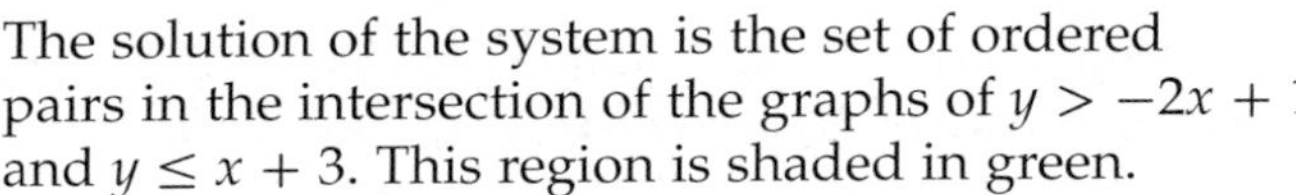

The solution of the system is the set of ordered pairs in the intersection of the graphs of $y > -2x + 1$ and $y \leq x + 3$. This region is shaded in green.

When graphing more than one region, it is helpful to use two different colored pencils or two different patterns for each region. This will make it easier to see where the regions intersect and find possible solutions.

Check Your Progress

1A. $y \leq 3$
$x + y \geq 1$

1B. $2x + y \geq 2$
$2x + y < 4$

1C. $y \geq -4$
$3x + y \leq 2$

1D. $x + y > 2$
$-4x + 2y < 8$

Personal Tutor glencoe.com

Sometimes the regions never intersect. When this happens, there is no solution because there are no points in common.

StudyTip

Parallel Boundaries A system of equations that are parallel lines does not have a solution. However, a system of inequalities with parallel boundaries can have a solution. For example:

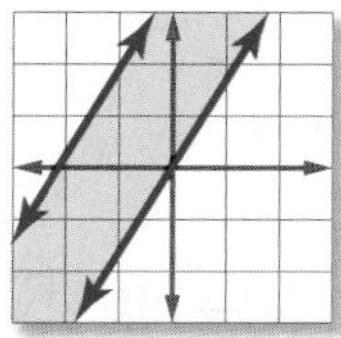

Math *in Motion*, Animation glencoe.com

Real-World Link

Student government might be a good activity for you if you like to bring about change, plan events, and work with others.

EXAMPLE 2 No Solution

Solve the system of inequalities by graphing.

$3x - y \geq 2$
$3x - y < -5$

The graphs of $3x - y = 2$ and $3x - y = -5$ are parallel lines. The two regions do not intersect at any point, so the system has no solution.

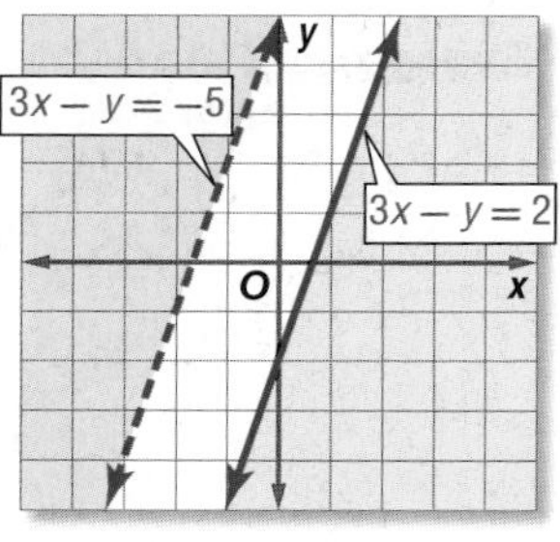

Check Your Progress

2A. $y > 3$
$y < 1$

2B. $x + 6y \leq 2$
$y \geq -\frac{1}{6}x + 7$

Personal Tutor glencoe.com

When using a system of inequalities in a real-world problem, sometimes only whole-number solutions will make sense.

Real-World EXAMPLE 3 Whole-Number Solutions

ELECTIONS Monifa is running for student council. The election rules say that for the election to be valid, at least 80% of the 900 students must vote. Monifa knows that she needs more than 330 votes to win.

a. Define the variables, and write a system of inequalities to represent this situation. Then graph the system.

Let r = the number of votes required by the election rules; 80% of 900 students is 720 students. So $r \geq 720$.

Let v = the number of votes that Monifa needs to win. So $v > 330$.

The system of inequalities is $r \geq 720$ and $v > 330$.

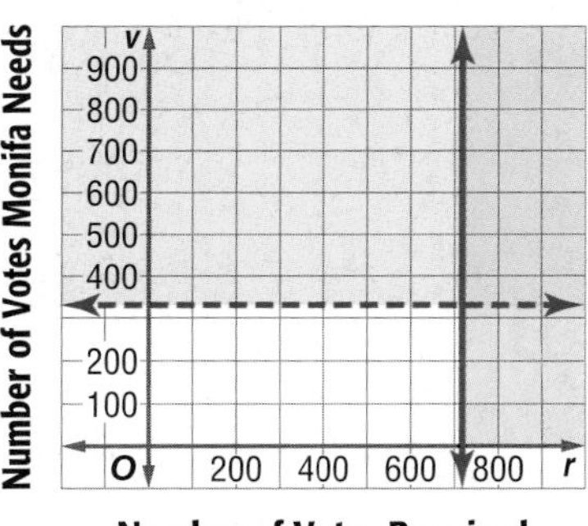

b. Name one possible solution.

Only whole-number solutions make sense in this problem. One possible solution is (800, 400); 800 students voted and Monifa received 400 votes.

Check Your Progress

3. FUNDRAISING The Theater Club is selling shirts. They have only enough supplies to print 120 shirts. They will sell sweatshirts for \$22 and T-shirts for \$15, with a goal of at least \$2000 in sales.

A. Define the variables, and write a system of inequalities to represent this situation.

B. Then graph the system.

C. Name one possible solution.

D. Is (45, 30) a solution? Explain.

Personal Tutor glencoe.com

Check Your Understanding

Examples 1 and 2 pp. 382–383

Solve each system of inequalities by graphing.

1. $x \geq 4$
$y \leq x - 3$

2. $y > -2$
$y \leq x + 9$

3. $y < 3x + 8$
$y \geq 4x$

4. $3x - y \geq -1$
$2x + y \geq 5$

5. $y \leq 2x - 7$
$y \geq 2x + 7$

6. $y > -2x + 5$
$y \geq -2x + 10$

7. $2x + y \leq 5$
$2x + y \leq 7$

8. $5x - y < -2$
$5x - y > 6$

Example 3 p. 383

9. AUTO RACING At a racecar driving school there are safety requirements.

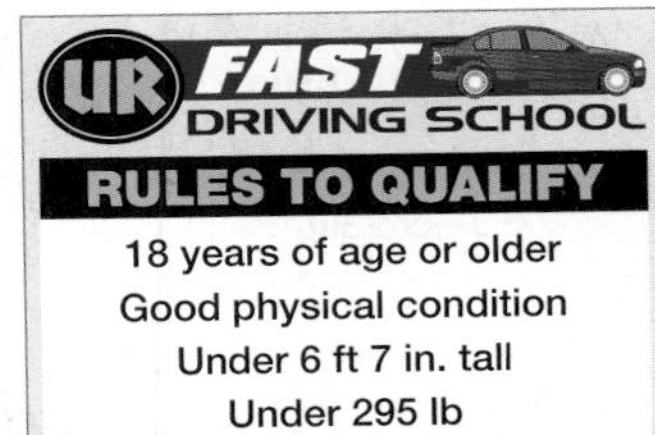

a. Define the variables, and write a system of inequalities to represent the height and weight requirements in this situation. Then graph the system.

b. Name one possible solution.

c. Is (50, 180) a solution? Explain.

Practice and Problem Solving

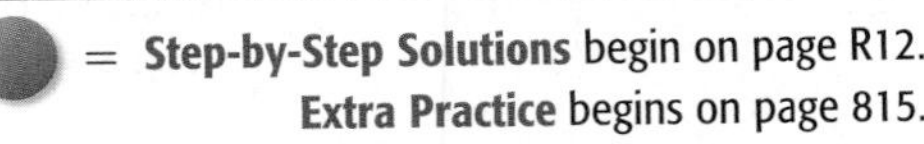

Examples 1 and 2 pp. 382–383

Solve each system of inequalities by graphing.

10. $y < 6$
$y > x + 3$

11. $y \geq 0$
$y \leq x - 5$

12. $y \leq x + 10$
$y > 6x + 2$

13. $y < 5x - 2$
$y > -6x + 2$

14. $2x - y \leq 6$
$x - y \geq -1$

15. $3x - y > -5$
$5x - y < 9$

16. $y \geq x + 10$
$y \leq x - 3$

17. $y < 5x - 5$
$y > 5x + 9$

18. $y \geq 3x - 5$
$3x - y > -4$

19. $4x + y > -1$
$y < -4x + 1$

20. $3x - y \geq -2$
$y < 3x + 4$

21. $y > 2x - 3$
$2x - y \geq 1$

22. $5x - y < -6$
$3x - y \geq 4$

23. $x - y \leq 8$
$y < 3x$

24. $4x + y < -2$
$y > -4x$

Example 3 p. 383

25. ICE RINKS Ice resurfacers are used for rinks of at least 1000 square feet and up to 17,000 square feet. The price ranges from as little as \$10,000 to as much as \$150,000.

a. Define the variables, and write a system of inequalities to represent this situation. Then graph the system.

b. Name one possible solution.

c. Is (15,000, 30,000) a solution? Explain.

26. PIZZERIA Josefina works between 10 and 30 hours per week at a pizzeria. She earns \$6.50 an hour, but can earn tips when she delivers pizzas.

a. Write a system of inequalities to represent the dollars d she could earn for working h hours in a week.

b. Graph this system.

c. If Josefina received \$17.50 in tips and earned a total of \$180 for the week, how many hours did she work?

Solve each system of inequalities by graphing.

27. $x + y \geq 1$
$x + y \leq 2$

28. $3x - y < -2$
$3x - y < 1$

29. $2x - y \leq -11$
$3x - y \geq 12$

30. $y < 4x + 13$
$4x - y \geq 1$

31. $4x - y < -3$
$y \geq 4x - 6$

32. $y \leq 2x + 7$
$y < 2x - 3$

33. $y > -12x + 1$
$y \leq 9x + 2$

34. $2y \geq x$
$x - 3y > -6$

35. $x - 5y > -15$
$5y \geq x - 5$

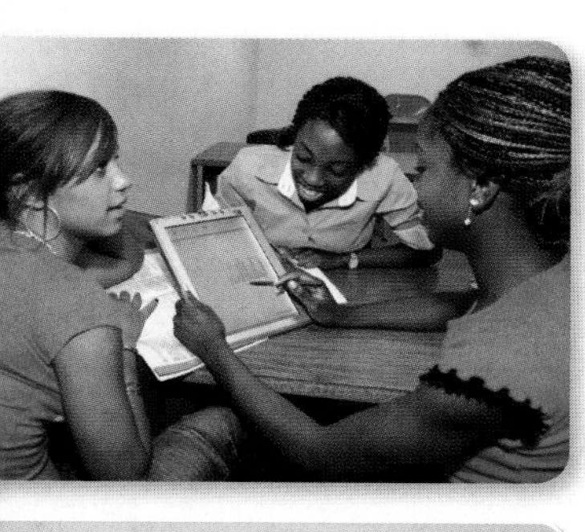

Real-World Link

Junior Achievement is a worldwide organization of volunteers that teaches students about entrepreneurship, work readiness, and finance through hands-on programs.

Source: *Junior Achievement*

36. **CLASS PROJECT** An economics class formed a company to sell school supplies. They would like to sell at least 20 notebooks and 50 pens per week, with a goal of earning at least $60 per week.

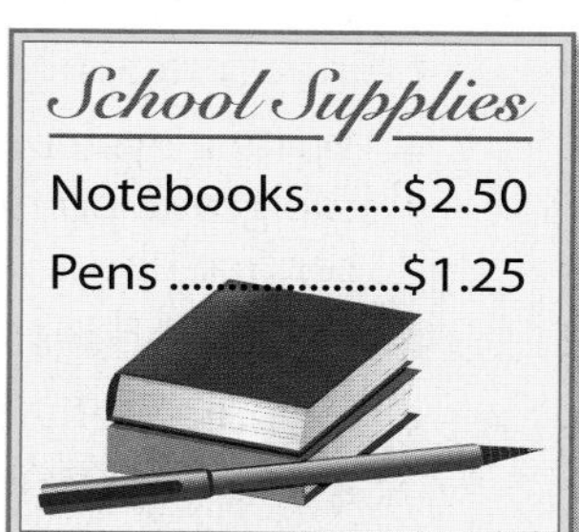

a. Define the variables, and write a system of inequalities to represent this situation.

b. Graph the system.

c. Name one possible solution.

37. **FINANCIAL LITERACY** Opal makes $15 per hour working for a photographer. She also coaches a competitive soccer team for $10 per hour. Opal needs to earn at least $90 per week, but she does not want to work more than 20 hours per week.

a. Define the variables, and write a system of inequalities to represent this situation.

b. Graph this system.

c. Give two possible solutions to describe how Opal can meet her goals.

d. Is (2, 2) a solution? Explain.

H.O.T. Problems — Use Higher-Order Thinking Skills

38. **CHALLENGE** Create a system of inequalities equivalent to $|x| \leq 4$.

39. **REASONING** State whether the following statement is *sometimes*, *always*, or *never* true. Explain your answer with an example or counterexample.

Systems of inequalities with parallel boundaries have no solutions.

40. **REASONING** Describe the graph of the solution of this system without graphing.
$6x - 3y \leq -5$
$6x - 3y \geq -5$

41. **OPEN ENDED** One inequality in a system is $3x - y > 4$. Write a second inequality so that the system will have no solution.

42. **CHALLENGE** Graph the system of inequalities. Estimate the area of the solution.
$y \geq 1$
$y \leq x + 4$
$y \leq -x + 4$

43. **WRITING IN MATH** Refer to the beginning of the lesson. Explain what each colored region of the graph represents. Explain how shading in various colors can help to clearly show the solution set of a system of inequalities.

Standardized Test Practice

44. EXTENDED RESPONSE To apply for a scholarship, you must have a minimum of 20 hours of community service and a grade-point average of at least 3.75. Another scholarship requires at least 40 hours of community service and a minimum grade-point average of 3.0.

a. Write a system of inequalities to represent the grade point average g and community service hours c you must have to apply.

b. Graph the system of inequalities.

c. If you are eligible for both scholarships, give one possible solution.

45. GEOMETRY What is the measure of $\angle 1$?

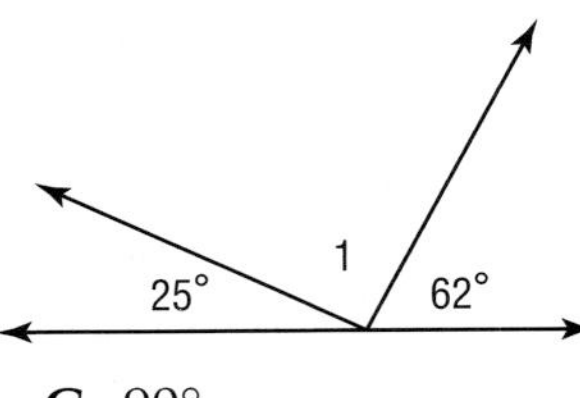

A 83°
B 87°
C 90°
D 93°

46. GEOMETRY What is the volume of the triangular prism?

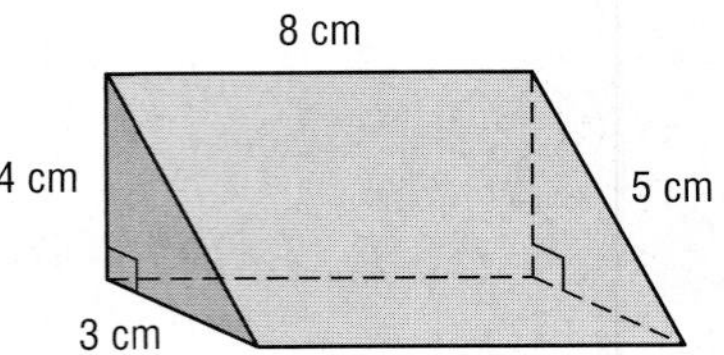

F 120 cm^3
G 96 cm^3
H 48 cm^3
J 30 cm^3

47. Ten pounds of fresh tomatoes make about 15 cups of cooked tomatoes. How many cups of cooked tomatoes does one pound of fresh tomatoes make?

A $1\frac{1}{2}$ cups
B 3 cups
C 4 cups
D 5 cups

Spiral Review

Solve each system by using matrices. (Lesson 6-7)

48. $3x + 2y = 12$
$-2x - 2y = -10$

49. $-2x + 5y = -9$
$2x - 4y = 6$

50. $2x + y = 2$
$5x + 2y = 7$

51. $-3x - 6y = -42$
$x + 4y = 30$

52. $-5x + 6y = 41$
$3x - 4y = -27$

53. $-3x - 4y = -46$
$7x - 6y = 0$

If $A = \begin{bmatrix} 4 & 5 & 6 \\ 0 & -1 & -3 \\ 2 & -3 & 7 \end{bmatrix}$ and $B = \begin{bmatrix} 3 & 9 & 1 \\ -3 & 5 & -3 \\ 1 & 2 & 6 \end{bmatrix}$, **find each sum, difference, or product.** (Lesson 6-6)

54. $A + B$

55. $A - B$

56. $3A$

57. $B + A$

58. $B - A$

59. $-2B$

60. ENTERTAINMENT A group of 11 adults and children bought tickets for the baseball game. If the total cost was $156, how many of each type of ticket did they buy? (Lesson 6-4)

TICKET Adult............$15 Children......$12 TICKET

Graph each inequality. (Lesson 5-6)

61. $4x - 2 \geq 2y$

62. $9x - 3y < 0$

63. $2y \leq -4x - 6$

Skills Review

Evaluate each expression. (Lesson 1-1)

64. 3^3

65. 2^4

66. $(-4)^3$

EXTEND 6-8

Graphing Technology Lab
Systems of Inequalities

Math Online glencoe.com
• Graphing Technology Personal Tutor

You can use a TI-Nspire™ or TI-Nspire™ CAS technology to explore systems of inequalities. To prepare your calculator, select New Document from the Home screen. Then select **Add Graphs & Geometry**.

ACTIVITY Graph Systems of Inequalities

Mr. Jackson owns a car washing and detailing business. It takes 20 minutes to wash a car and 60 minutes to detail a car. He works at most 8 hours per day and does at most 4 details per day. Write a system of linear inequalities to represent this situation.

First, write a linear inequality that represents the time it takes for car washing and car detailing. Let x represent the number of car washes, and let y represent the number of car details. Then $20x + 60y \leq 480$.

To graph this using a graphing calculator, solve for y.

$20x + 60y \leq 480$ **Original inequality**

$60y \leq -20x + 480$ **Subtract $20x$ from each side and simplify.**

$y \leq -\frac{1}{3}x + 8$ **Divide each side by 60 and simplify.**

Mr. Jackson does at most 4 details per day. This means that $y \leq 4$.

Step 1 Graph $y \leq 4$. Press (menu) Window; Window Settings (enter) −4, 30, −2, 10 (enter). Press **clear** to delete = and then type (<) (=) 4 (enter).

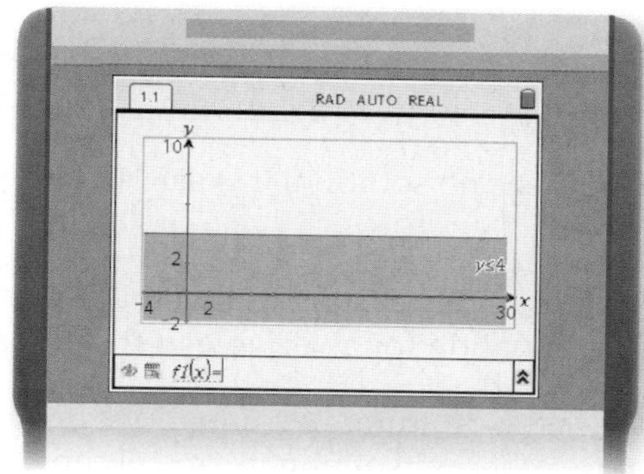

Step 2 Graph $y \leq -\frac{1}{3}x + 8$. Press **clear** once, delete =, and then type (<)(=)((−)) (1 ÷ 3) $x + 8$ (enter).

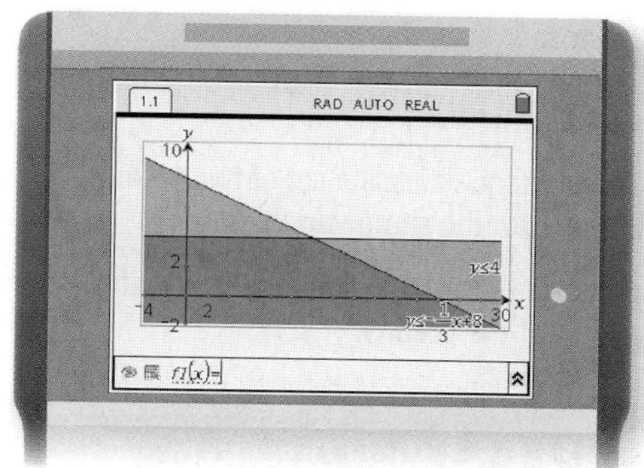

The darkest shaded half-plane of the graph represents the solutions.

Analyze the Results

1. If Mr. Jackson charges $75 for each car he details and $25 for each car wash, what is the maximum amount of money he could earn in one day?
2. What is the greatest number of car washes that Mr. Jackson could do in a day? Explain your reasoning.

Study Guide and Review

Math Online glencoe.com
- STUDY TO GO
- Vocabulary Review

Chapter Summary

Key Concepts

Systems of Equations (Lessons 6-1 through 6-4)

- A system with a graph of two intersecting lines has one solution and is *consistent and independent*.
- Graphing a system of equations can only provide approximate solutions. For exact solutions, you must use algebraic methods.
- In the substitution method, one equation is solved for a variable and the expression substituted to find the value of another variable.
- In the elimination method, one variable is eliminated by adding or subtracting the equations. Sometimes multiplying one equation by a constant makes it easier to use the elimination method.

Matrices (Lessons 6-6 and 6-7)

- Matrices can be added or subtracted only if they have the same dimensions. Add or subtract corresponding elements.
- To multiply a matrix by a scalar *k*, multiply each element in the matrix by *k*.
- An augmented matrix can be used to solve a system of equations.

Systems of Inequalities (Lesson 6-8)

- A system of inequalities is a set of two or more inequalities with the same variables.
- The solution of a system of inequalities is the intersection of the graphs.

FOLDABLES® Study Organizer

Be sure the Key Concepts are noted in your Foldable.

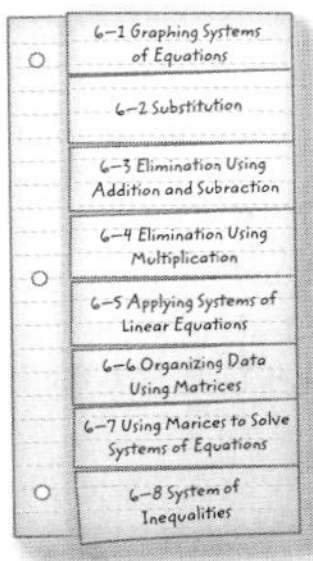

Key Vocabulary

augmented matrix (p. 376)
consistent (p. 333)
dependent (p. 333)
dimension (p. 369)
element (p. 369)
elimination (p. 348)
inconsistent (p. 333)
independent (p. 333)
matrix (p. 369)
scalar (p. 371)
scalar multiplication (p. 371)
substitution (p. 342)
system of equations (p. 333)
system of inequalities (p. 382)

Vocabulary Check

State whether each sentence is *true* or *false*. If *false*, replace the underlined term to make a true sentence.

1. If a system has at least one solution, it is said to be consistent.
2. If a consistent system has exactly two solution(s), it is said to be independent.
3. If a consistent system has an infinite number of solutions, it is said to be inconsistent.
4. If a system has no solution, it is said to be inconsistent.
5. Substitution involves substituting an expression from one equation for a variable in the other.
6. In some cases, dividing two equations in a system together will eliminate one of the variables. This process is called elimination.
7. Each number in a matrix is called a(n) dimension.
8. A constant by which you multiply a matrix is called a scalar.
9. An augmented matrix consists of the coefficients and the constant terms of a system of equations.
10. A set of two or more inequalities with the same variables is called a system of equations.

Lesson-by-Lesson Review

6-1 Graphing Systems of Equations (pp. 333–339)

Graph each system and determine the number of solutions that it has. If it has one solution, name it.

11. $x - y = 1$
$x + y = 5$

12. $y = 2x - 4$
$4x + y = 2$

13. $2x - 3y = -6$
$y = -3x + 2$

14. $-3x + y = -3$
$y = x - 3$

15. $x + 2y = 6$
$3x + 6y = 8$

16. $3x + y = 5$
$6x = 10 - 2y$

17. MAGIC NUMBERS Sean is trying to find two numbers with a sum of 14 and a difference of 4. Define two variables, write a system of equations, and solve by graphing.

EXAMPLE 1

Graph the system and determine the number of solutions it has. If it has one solution, name it.

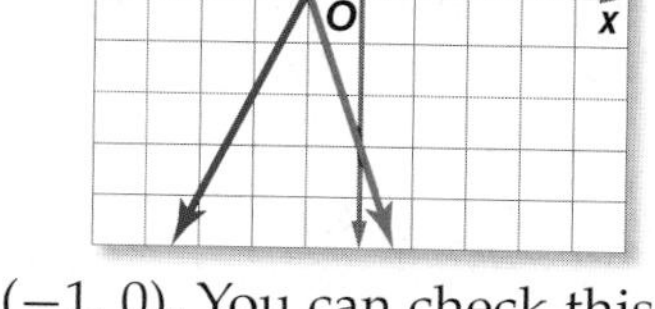

$\mathbf{y = 2x + 2}$
$\mathbf{y = -3x - 3}$

The lines appear to intersect at the point $(-1, 0)$. You can check this by substituting -1 for x and 0 for y.

CHECK

$y = 2x + 2$	**Original equation**
$0 \stackrel{?}{=} 2(-1) + 2$	**Substitution**
$0 \stackrel{?}{=} -2 + 2$	**Multiply.**
$0 = 0$ ✓	
$y = -3x - 3$	**Original equation**
$0 \stackrel{?}{=} -3(-1) - 3$	**Substitution**
$0 \stackrel{?}{=} 3 - 3$	**Multiply.**
$0 = 0$ ✓	

The solution is $(-1, 0)$.

6-2 Substitution (pp. 342–347)

Use substitution to solve each system of equations.

18. $x + y = 3$
$x = 2y$

19. $x + 3y = -28$
$y = -5x$

20. $3x + 2y = 16$
$x = 3y - 2$

21. $x - y = 8$
$y = -3x$

22. $y = 5x - 3$
$x + 2y = 27$

23. $x + 3y = 9$
$x + y = 1$

24. GEOMETRY The perimeter of a rectangle is 48 inches. The length is 6 inches greater than the width. Define the variables, and write equations to represent this situation. Solve the system by using substitution.

EXAMPLE 2

Use substitution to solve the system.

$\mathbf{3x - y = 18}$
$\mathbf{y = x - 4}$

$3x - y = 18$	**First equation**
$3x - (x - 4) = 18$	**Substitute $x - 4$ for y.**
$2x + 4 = 18$	**Simplify.**
$2x = 14$	**Subtract 4 from each side.**
$x = 7$	**Divide each side by 2.**

Use the value of x and either equation to find the value for y.

$y = x - 4$	**Second equation**
$= 7 - 4$ or 3	**Substitute and simplify.**

The solution is $(7, 3)$.

CHAPTER 6

Study Guide and Review

6-3 Elimination Using Addition and Subtraction (pp. 348–354)

Use elimination to solve each system of equations.

25. $x + y = 13$
$x - y = 5$

26. $-3x + 4y = 21$
$3x + 3y = 14$

27. $x + 4y = -4$
$x + 10y = -16$

28. $2x + y = -5$
$x - y = 2$

29. $6x + y = 9$
$-6x + 3y = 15$

30. $x - 4y = 2$
$3x + 4y = 38$

31. $2x + 2y = 4$
$2x - 8y = -46$

32. $3x + 2y = 8$
$x + 2y = 2$

33. BASEBALL CARDS Cristiano bought 24 baseball cards for \$50. One type cost \$1 per card, and the other cost \$3 per card. Define the variables, and write equations to find the number of each type of card he bought. Solve by using elimination.

EXAMPLE 3

Use elimination to solve the system of equations.

$\mathbf{3x - 5y = 11}$
$\mathbf{x + 5y = -3}$

$$\begin{array}{rrl} & 3x - 5y = 11 & \\ (+) & x + 5y = -3 & \\ \hline & 4x \quad = 8 & \text{The variable } y \text{ is eliminated.} \\ & x = 2 & \text{Divide each side by 4.} \end{array}$$

Now, substitute 2 for x in either equation to find the value of y.

$3x - 5y = 11$	First equation
$3(2) - 5y = 11$	Substitute.
$6 - 5y = 11$	Multiply.
$-5y = 5$	Subtract 6 from each side.
$y = -1$	Divide each side by -5.

The solution is $(2, -1)$.

6-4 Elimination Using Multiplication (pp. 355–360)

Use elimination to solve each system of equations.

34. $x + y = 4$
$-2x + 3y = 7$

35. $x - y = -2$
$2x + 4y = 38$

36. $3x + 4y = 1$
$5x + 2y = 11$

37. $-9x + 3y = -3$
$3x - 2y = -4$

38. $8x - 3y = -35$
$3x + 4y = 33$

39. $2x + 9y = 3$
$5x + 4y = 26$

40. $-7x + 3y = 12$
$2x - 8y = -32$

41. $8x - 5y = 18$
$6x + 6y = -6$

42. BAKE SALE On the first day, a total of 40 items were sold for \$356. Define the variables, and write a system of equations to find the number of cakes and pies sold. Solve by using elimination.

EXAMPLE 4

Use elimination to solve the system of equations.
$\mathbf{3x + 6y = 6}$
$\mathbf{2x + 3y = 5}$

Notice that if you multiply the second equation by -2, the coefficients of the y-terms are additive inverses.

$3x + 6y = 6$
$2x + 3y = 5$ **Multiply by −2.** →

$$\begin{array}{rr} & 3x + 6y = 6 \\ (+) & -4x - 6y = -10 \\ \hline & -x \quad = -4 \\ & x = 4 \end{array}$$

Now, substitute 4 for x in either equation to find the value of y.

$2x + 3y = 5$	Second equation
$2(4) + 3y = 5$	Substitution
$8 + 3y = 5$	Multiply.
$3y = -3$	Subtract 8 from both sides.
$y = -1$	Divide each side by 3.

The solution is $(4, -1)$.

MIXED PROBLEM SOLVING
For mixed problem-solving practice, see page 850.

6-5 Applying Systems of Linear Equations (pp. 362–367)

Determine the best method to solve each system of equations. Then solve the system.

43. $y = x - 8$
$y = -3x$

44. $y = -x$
$y = 2x$

45. $x + 3y = 12$
$x = -6y$

46. $x + y = 10$
$x - y = 18$

47. $3x + 2y = -4$
$5x + 2y = -8$

48. $6x + 5y = 9$
$-2x + 4y = 14$

49. $3x + 4y = 26$
$2x + 3y = 19$

50. $11x - 6y = 3$
$5x - 8y = -25$

51. COINS Tionna has saved dimes and quarters in her piggy bank. Define the variables, and write a system of equations to determine the number of dimes and quarters. Then solve the system using the best method for the situation.

EXAMPLE 5

Determine the best method to solve the system of equations. Then solve the system.

$3x + 5y = 4$
$4x + y = -6$

Solve the second equation for y.

$4x + y = -6$	**Second equation**
$y = -6 - 4x$	**Subtract 4x from each side.**

Substitute $-6 - 4x$ for y in the first equation.

$3x + 5(-6 - 4x) = 4$	**Substitute.**
$3x - 30 - 20x = 4$	**Distributive Property**
$-17x - 30 = 4$	**Simplify.**
$-17x = 34$	**Add 30 to each side.**
$x = -2$	**Divide by −17.**

Last, substitute -2 for x in either equation to find y.

$4x + y = -6$	**Second equation**
$4(-2) + y = -6$	**Substitute.**
$-8 + y = -6$	**Multiply.**
$y = 2$	**Add 8 to each side.**

The solution is $(-2, 2)$.

6-6 Organizing Data Using Matrices (pp. 369–375)

Perform the indicated matrix operation.

52. $\begin{bmatrix} 5 & -6 \\ -9 & 4 \end{bmatrix} + \begin{bmatrix} 2 & -3 \\ -4 & 2 \end{bmatrix}$

53. $\begin{bmatrix} 5 & -8 & 4 \\ -6 & 2 & -9 \end{bmatrix} - \begin{bmatrix} 6 & -4 & -3 \\ 8 & 2 & 1 \end{bmatrix}$

54. $3\begin{bmatrix} 5 & -3 & -1 & 4 \\ 7 & -6 & 8 & 1 \end{bmatrix}$

55. POLLS The results of a poll are shown. How many votes were cast against both tax levies?

	For	Against
Tax Levy 1	68	105
Tax Levy 2	69	71

EXAMPLE 6

Find $\begin{bmatrix} 3 & 9 & -6 \\ 7 & 2 & 1 \end{bmatrix} + \begin{bmatrix} -2 & -1 & 6 \\ -6 & 4 & -5 \end{bmatrix}$.

To add the 2×3 matrices, add the corresponding elements.

$$\begin{bmatrix} 3 & 9 & -6 \\ 7 & 2 & 1 \end{bmatrix} + \begin{bmatrix} -2 & -1 & 6 \\ -6 & 4 & -5 \end{bmatrix}$$

$$= \begin{bmatrix} 3 + (-2) & 9 + (-1) & -6 + 6 \\ 7 + (-6) & 2 + 4 & 1 + (-5) \end{bmatrix}$$

$$= \begin{bmatrix} 1 & 8 & 0 \\ 1 & 6 & -4 \end{bmatrix}$$

Study Guide and Review

6-7 Using Matrices to Solve Systems of Equations (pp. 376–381)

Use an augmented matrix to solve each system of equations.

56. $x - 2y = -5$
$2x + 3y = 4$

57. $x - y = 5$
$3x + 3y = 3$

58. $2x + 3y = 8$
$x - 5y = -22$

59. $2x + 4y = -16$
$x - 2y = 0$

60. $2x - 4y = 10$
$x + 2y = 5$

61. $x - 4y = 11$
$3x + 5y = -1$

62. $x - 2y = -26$
$2x - 3y = -42$

63. $x - 3y = 24$
$5x - 2y = 29$

64. $x - 2y = -16$
$3x + 4y = 72$

65. $x + 3y = 12$
$2x - 4y = 14$

66. ART Mateo spent a total of $36 on 10 art supply items. How many bottles of paint and paint brushes did Mateo buy?

EXAMPLE 7

Use an augmented matrix to solve the system of equations.

$\mathbf{x + 3y = 11}$
$\mathbf{3x + 4y = 18}$

Step 1 Write the augmented matrix.

$$\left[\begin{array}{cc|c} 1 & 3 & 11 \\ 3 & 4 & 18 \end{array}\right]$$

Step 2 Multiply the first row by -3 and add the result to row 2.

$$\left[\begin{array}{cc|c} 1 & 3 & 11 \\ 3 & 4 & 18 \end{array}\right] \xrightarrow{-3R_1 + R_2} \left[\begin{array}{cc|c} 1 & 3 & 11 \\ 0 & -5 & -15 \end{array}\right]$$

Step 3 Multiply the second row by $-\frac{1}{5}$.

$$\left[\begin{array}{cc|c} 1 & 3 & 11 \\ 0 & -5 & -15 \end{array}\right] \xrightarrow{-\frac{1}{5}R_2} \left[\begin{array}{cc|c} 1 & 3 & 11 \\ 0 & 1 & 3 \end{array}\right]$$

Step 4 Multiply the second row by -3 and add the result to row 1.

$$\left[\begin{array}{cc|c} 1 & 3 & 11 \\ 0 & 1 & 3 \end{array}\right] \xrightarrow{-3R_2 + R_1} \left[\begin{array}{cc|c} 1 & 0 & 2 \\ 0 & 1 & 3 \end{array}\right]$$

The solution is (2, 3).

6-8 Systems of Inequalities (pp. 382–387)

Solve each system of inequalities by graphing.

67. $x > 3$
$y < x + 2$

68. $y \leq 5$
$y > x - 4$

69. $y < 3x - 1$
$y \geq -2x + 4$

70. $y \leq -x - 3$
$y \geq 3x - 2$

71. JOBS Kishi makes $7 an hour working at the grocery store and $10 an hour delivering newspapers. She cannot work more than 20 hours per week. Graph two inequalities that Kishi can use to determine how many hours she needs to work at each job if she wants to earn at least $90 per week.

EXAMPLE 8

Solve the system of inequalities by graphing.

$\mathbf{y < 3x + 1}$
$\mathbf{y \geq -2x + 3}$

The solution set of the system is the set of ordered pairs in the intersection of the two graphs. This portion is shaded in the graph below.

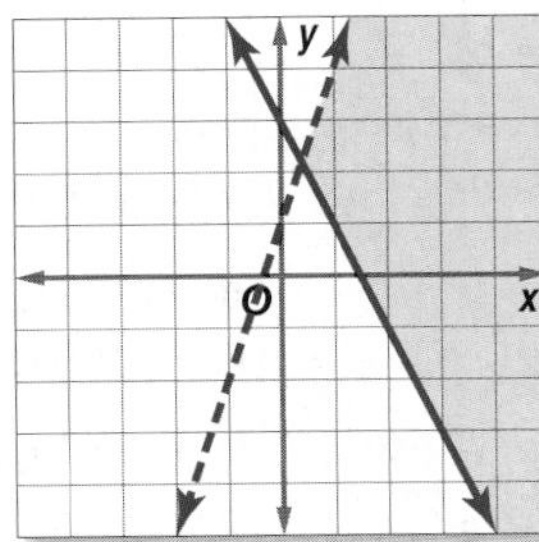

Practice Test

Math Online glencoe.com
Chapter Test

Graph each system and determine the number of solutions that it has. If it has one solution, name it.

1. $y = 2x$
   $y = 6 - x$
2. $y = x - 3$
   $y = -2x + 9$
3. $x - y = 4$
   $x + y = 10$
4. $2x + 3y = 4$
   $2x + 3y = -1$

Use substitution to solve each system of equations.

5. $y = x + 8$
   $2x + y = -10$
6. $x = -4y - 3$
   $3x - 2y = 5$
7. **GARDENING** Corey has 42 feet of fencing around his garden. The garden is rectangular in shape, and its length is equal to twice the width minus 3 feet. Define the variables, and write a system of equations to find the length and width of the garden. Solve the system by using substitution.

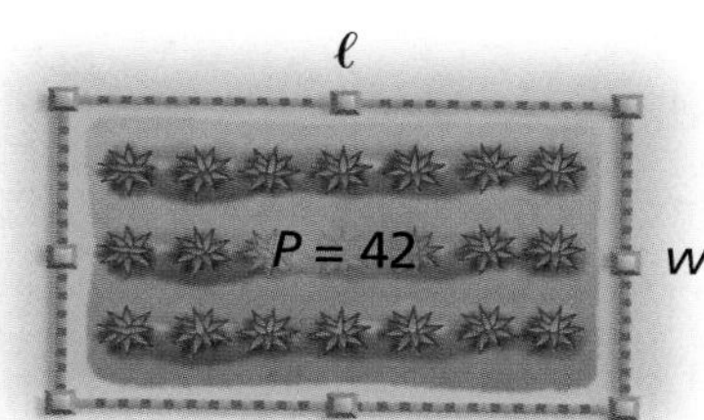

8. **MULTIPLE CHOICE** Use elimination to solve the system.

$$6x - 4y = 6$$
$$-6x + 3y = 0$$

A (5, 6)

B (−3, −6)

C (1, 0)

D (4, −8)

9. **SHOPPING** Shelly has $175 to shop for jeans and sweaters. Each pair of jeans costs $25, each sweater costs $20, and she buys 8 items. Determine the number of pairs of jeans and sweaters Shelly bought.

Use elimination to solve each system of equations.

10. $x + y = 13$
    $x - y = 5$
11. $3x + 7y = 2$
    $3x - 4y = 13$
12. $x + y = 8$
    $x - 3y = -4$
13. $2x + 6y = 18$
    $3x + 2y = 13$
14. **MAGAZINES** Julie subscribes to a sports magazine and a fashion magazine. She received 24 issues this year. The number of fashion issues is 6 less than twice the number of sports issues. Define the variables, and write a system of equations to find the number of issues of each magazine.

Perform the indicated matrix operations. If an operation cannot be performed, write *impossible*.

15. $\begin{bmatrix} 3 & -5 & 2 \end{bmatrix} + \begin{bmatrix} 12 & -7 & -6 \end{bmatrix}$
16. $\begin{bmatrix} 1 & 8 & -4 \\ -2 & 6 & 7 \\ 0 & 9 & -6 \end{bmatrix} - \begin{bmatrix} -4 & 2 & 0 \\ -5 & 3 & 8 \\ -9 & -7 & 1 \end{bmatrix}$
17. $-2\begin{bmatrix} -2 & 7 \\ 6 & -4 \\ 8 & 5 \end{bmatrix}$

Use an augmented matrix to solve each system of equations.

18. $y = 3x$
    $x + 2y = 21$
19. $x + y = 12$
    $y = x - 4$
20. $x + y = 15$
    $x - y = 9$
21. $3x + 5y = 7$
    $2x - 3y = 11$

Solve each system of inequalities by graphing.

22. $x > 2$
    $y < 4$
23. $x + y \leq 5$
    $y \geq x + 2$
24. $3x - y > 9$
    $y > -2x$
25. $y \geq 2x + 3$
    $-4x - 3y > 12$

Guess and Check

It is very important to pace yourself and keep track of how much time you have when taking a standardized test. If time is running short, or if you are unsure how to solve a problem, the guess and check strategy may help you determine the correct answer quickly.

Strategies for Guessing and Checking

Step 1

Carefully look over each possible answer choice, and evaluate for reasonableness. Eliminate unreasonable answers.

Ask yourself:

- Are there any answer choices that are clearly incorrect?
- Are there any answer choices that are not in the proper format?
- Are there any answer choices that do not have the proper units for the correct answer?

Step 2

For the remaining answer choices, use the guess and check method.

- **Equations:** If you are solving an equation, substitute the answer choice for the variable and see if this results in a true number sentence.
- **Inequalities:** Likewise, you can substitute the answer choice for the variable and see if it satisfies the inequality.
- **System of Equations:** Find the answer choice that satisfies both equations of the system.

Step 3

Choose an answer choice and see if it satisfies the constraints of the problem statement. Identify the correct answer.

- If the answer choice you are testing does not satisfy the problem, move on to the next reasonable guess and check it.
- When you find the correct answer choice, stop. You do not have to check the other answer choices.

EXAMPLE

Read the problem. Identify what you need to know. Then use the information in the problem to solve.

Solve $\begin{cases} 4x - 8y = 20 \\ -3x + 5y = -14 \end{cases}$.

A (5, 0) **C** (3, –1)

B (4, –2) **D** (–6, –5)

The solution of a system of equations is an ordered pair, (x, y). Since all four answer choices are of this form, they are all possible correct answers and must be checked. Begin with the first answer choice and substitute it in each equation. Continue until you find the ordered pair that satisfies both equations of the system.

	First Equation	Second Equation
Guess: (5, 0)	$4x - 8y = 20$ $4(5) - 8(0) = 20$ ✓	$-3x + 5y = -14$ $-3(5) + 5(0) \neq -14$ ✗

	First Equation	Second Equation
Guess: (4, –2)	$4x - 8y = 20$ $4(4) - 8(-2) \neq 20$ ✗	$-3x + 5y = -14$ $-3(4) + 5(-2) \neq -14$ ✗

	First Equation	Second Equation
Guess: (3, –1)	$4x - 8y = 20$ $4(3) - 8(-1) = 20$ ✓	$-3x + 5y = -14$ $-3(3) + 5(-1) = -14$ ✓

The ordered pair (3, –1) satisfies both equations of the system. So, the correct answer is C.

Exercises

Read each problem. Eliminate any unreasonable answers. Then use the information in the problem to solve.

1. Gina bought 5 hot dogs and 3 soft drinks at the ball game for $11.50. Renaldo bought 4 hot dogs and 2 soft drinks for $8.50. How much does a single hot dog and a single drink cost?

A hot dogs: $1.25
soft drinks: $1.50

B hot dogs: $1.25
soft drinks: $1.75

C hot dogs: $1.50
soft drinks: $1.25

D hot dogs: $1.50
soft drinks: $1.75

2. The bookstore hopes to sell at least 30 binders and calculators each week. The store also hopes to have sales revenue of at least $200 in binders and calculators. How many binders and calculators could be sold to meet both of these sales goals?

Store Prices	
Item	**Price**
binders	$3.65
calculators	$14.80

F 25 binders, 5 calculators

G 12 binders, 15 calculators

H 22 binders, 9 calculators

J 28 binders, 6 calculators

CHAPTER 6

Standardized Test Practice

Cumulative, Chapters 1 through 6

Multiple Choice

Read each question. Then fill in the correct answer on the answer document provided by your teacher or on a sheet of paper.

1. Which of the following terms ***best*** describes the system of equations shown in the graph?

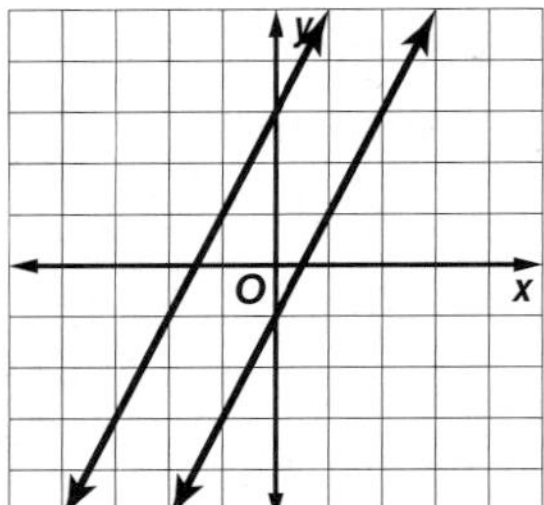

A consistent

B consistent and dependent

C consistent and independent

D inconsistent

2. Use substitution to solve the system of equations below.

$$\begin{cases} y = 4x - 7 \\ 3x - 2y = -1 \end{cases}$$

F (3, 5)

G (4, −1)

H (5, −2)

J (−6, 2)

3. Which ordered pair is the solution of the system of linear equations shown below?

$$\begin{cases} 3x - 8y = -50 \\ 3x - 5y = -38 \end{cases}$$

A $\left(\frac{5}{8}, \frac{3}{2}\right)$

B (4, −9)

C $\left(-\frac{2}{7}, \frac{4}{9}\right)$

D (−6, 4)

4. A home goods store received \$881 from the sale of 4 table saws and 9 electric drills. If the receipts from the saws exceeded the receipts from the drills by \$71, what is the price of an electric drill?

F \$45

G \$59

H \$108

J \$119

5. A region is defined by this system.

$$y > -\frac{1}{2}x - 1$$
$$y > -x + 3$$

In which quadrant(s) of the coordinate plane is the region located?

A I and IV only

B III only

C I, II, and IV only

D II and III only

6. Which of the following terms ***best*** describes the system of equations shown in the graph?

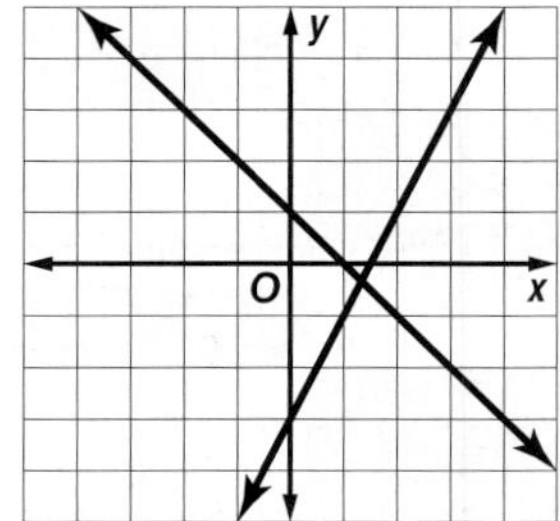

F consistent

G consistent and independent

H consistent and dependent

J inconsistent

7. Use elimination to solve the system of equations below.

$$3x + 2y = -2$$
$$2x - 2y = -18$$

A (1, 3)

B (7, −4)

C (−2, −3)

D (−4, 5)

8. What is the solution of the following system of equations?

$$\begin{cases} y = 6x - 1 \\ y = 6x + 4 \end{cases}$$

F (2, 11)

G (−3, −14)

H (7, 5)

J no solution

Test-TakingTip

Question 8 You can subtract the second equation from the first equation to eliminate the *x*-variable. Then solve for *y*.

Short Response/Gridded Response

Record your answers on the answer sheet provided by your teacher or on a sheet of paper.

9. **GRIDDED RESPONSE** Angie and her sister have \$15 to spend on pizza. A medium pizza costs \$11.50 plus \$0.75 per topping. What is the maximum number of toppings Angie and her sister can get on their pizza?

10. Write an inequality for the graph below.

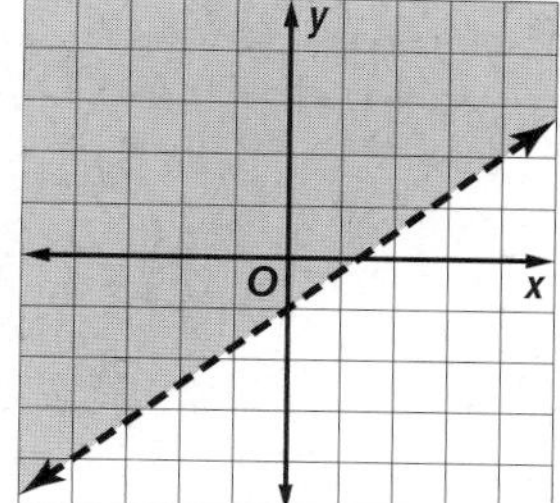

11. **GRIDDED RESPONSE** Christy is taking a road trip. After she drives 12 more miles, she will have driven at least half of the 108-mile trip. What is the least number of miles she has driven so far?

12 Write an equation in slope-intercept form with a slope of $-\frac{2}{3}$ and a y-intercept of 6.

13. A rental company charges \$9.50 per hour for a scooter plus a \$15 fee. Write an equation in slope-intercept form for the total rental cost C of renting a scooter for h hours.

14. **GRIDDED RESPONSE** A computer supplies store is having a storewide sale this weekend. An inkjet printer that normally sells for \$179.00 is on sale for \$143.20. What is the percent discount of the sale price?

15. In 1980, the population of Kentucky was about 3.66 million people. By 2000, this number had grown to about 4.04 million people. What was the annual rate of change in population from 1980 to 2000?

16. Joseph's cell phone service charges him \$0.15 per each text message sent. Write an equation that represents the cost C of his cell phone service for text messages t sent each month.

17. A store is offering a \$15 mail-in-rebate on all printers. If Mark is looking at printers that range from \$45 to \$89, how much can he expect to pay?

Extended Response

Record your answers on a sheet of paper. Show your work.

18. The table shows how many canned goods were collected during the first day of a charity food drive.

Food Drive Day 1 Results	
Class	**Number Collected**
10th graders	78
11th graders	80
12th graders	92

a. Estimate how many canned goods will be collected during the 5-day food drive. Explain your answer.

b. Is this estimate a reasonable expectation? Explain.

Need Extra Help?																		
If you missed Question...	1	2	3	4	5	6	7	8	9	10	11	12	13	14	15	16	17	18
Go to Lesson or Page...	6-1	6-2	6-3	6-3	6-8	6-1	6-3	6-3	5-3	5-6	5-3	4-2	4-2	3-3	2-7	2-1	5-4	1-4

CHAPTER 7

Polynomials

Then

In Chapter 1, you performed operations on expressions with exponents.

Now

In Chapter 7, you will:

- Simplify expressions involving monomials.
- Use scientific notation.
- Find degrees of polynomials, write polynomials in standard form, and add, subtract, and multiply polynomials.

Why?

SPACE The Very Large Array is an arrangement of 27 radio antennas in a Y pattern. The data the antennas collect is used by astronomers around the world to study the planets and stars. Astrophysicists use and apply properties of exponents to model the distance and orbit of celestial bodies.

Polynomials

Introduction

Space

The Very Large Array is an arrangement of 27 radio antennas in a Y pattern. The data the antennas collect is used by astronomers around the world to study the planets and stars. Astrophysicists use and apply properties of exponents to model the distance and orbit of celestial bodies.

1/6

Math *in Motion*, Animation glencoe.com

Get Ready for Chapter 7

Diagnose Readiness You have two options for checking Prerequisite Skills.

Text Option

Take the Quick Check below. Refer to the Quick Review for help.

QuickCheck

Write each expression using exponents. (Lesson 1-1)

1. $4 \cdot 4 \cdot 4 \cdot 4 \cdot 4$
2. $y \cdot y \cdot y$
3. $6 \cdot 6$
4. $2 \cdot 2 \cdot 2 \cdot 2 \cdot 2 \cdot 2 \cdot 2 \cdot 2 \cdot 2$
5. $b \cdot b \cdot b \cdot b \cdot b \cdot b$
6. $m \cdot m \cdot m \cdot p \cdot p \cdot p \cdot p \cdot p \cdot p$
7. $\frac{1}{3} \cdot \frac{1}{3} \cdot \frac{1}{3} \cdot \frac{1}{3} \cdot \frac{1}{3} \cdot \frac{1}{3} \cdot \frac{1}{3} \cdot \frac{1}{3}$
8. $\frac{x}{y} \cdot \frac{x}{y} \cdot \frac{x}{y} \cdot \frac{x}{y} \cdot \frac{w}{z} \cdot \frac{w}{z}$

Evaluate each expression. (Lesson 1-2)

9. 2^3
10. $(-5)^2$
11. 3^3
12. $(-4)^3$
13. $\left(\frac{2}{3}\right)^2$
14. $\left(\frac{1}{2}\right)^4$
15. **SCHOOL** The probability of guessing correctly on 5 true-false questions is $\left(\frac{1}{2}\right)^5$. Express this probability as a fraction without exponents.

Find the area or volume of each figure. (Lessons 0-8 and 0-9)

16. 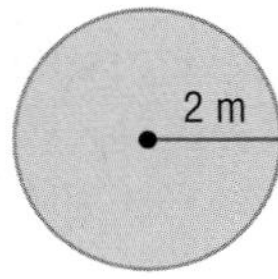

17. 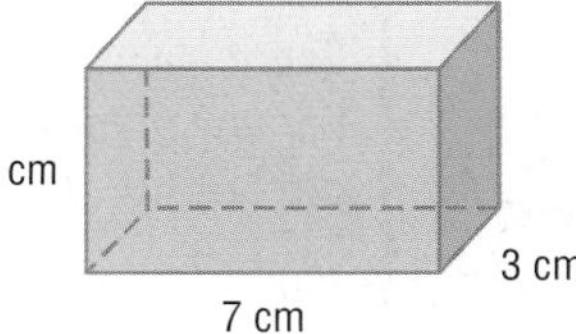

18. **PHOTOGRAPHY** A photo is 4 inches by 6 inches. What is the area of the photo?

QuickReview

EXAMPLE 1

Write $5 \cdot 5 \cdot 5 \cdot 5 + x \cdot x \cdot x$ using exponents.

4 factors of 5 is 5^4.

3 factors of x is x^3.

So, $5 \cdot 5 \cdot 5 \cdot 5 + x \cdot x \cdot x = 5^4 + x^3$.

EXAMPLE 2

Evaluate $\left(\frac{5}{7}\right)^2$.

$\left(\frac{5}{7}\right)^2 = \frac{5^2}{7^2}$ **Power of a Quotient**

$= \frac{25}{49}$ **Simplify.**

EXAMPLE 3

Find the volume of the figure.

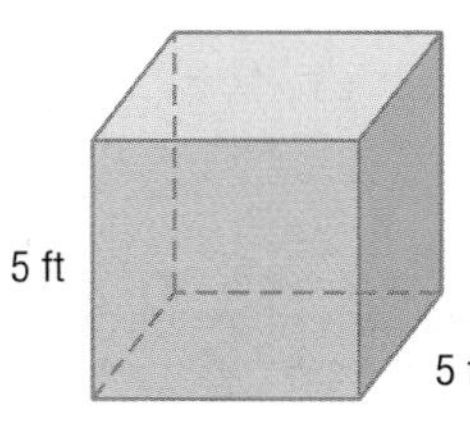

$V = \ell wh$ **Volume of a rectangular prism**

$= 5 \cdot 5 \cdot 5$ or 125 **$\ell = 5$, $w = 5$, and $h = 5$**

The volume is 125 cubic feet.

Online Option

Math Online Take a self-check Chapter Readiness Quiz at glencoe.com.

Get Started on Chapter 7

You will learn several new concepts, skills, and vocabulary terms as you study Chapter 7. To get ready, identify important terms and organize your resources. You may wish to refer to **Chapter 0** to review prerequisite skills.

FOLDABLES® Study Organizer

Polynomials Make this Foldable to help you organize your Chapter 7 notes about polynomials. Begin with nine sheets of notebook paper.

1. **Arrange** the paper into a stack.

2. **Staple** along the left side. Starting with the second sheet of paper, cut along the right side to form tabs.

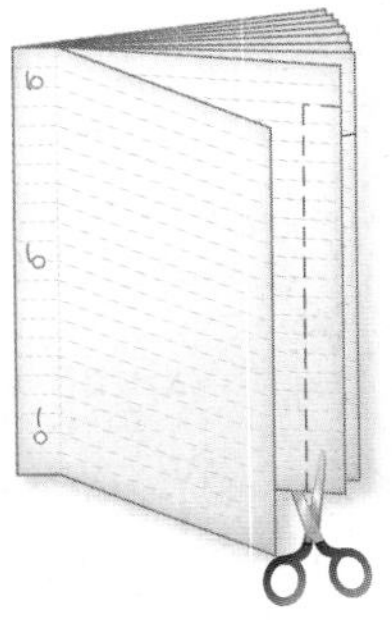

3. **Label** the cover sheet "Polynomials" and label each tab with a lesson number.

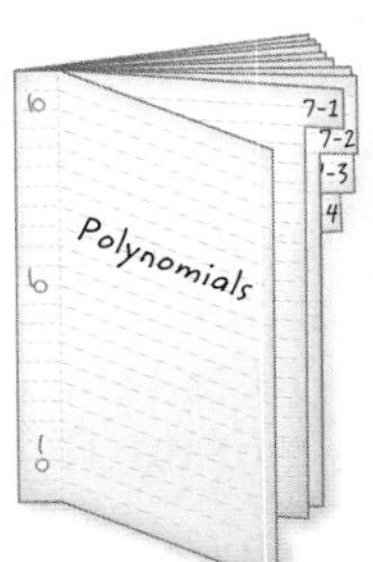

Math Online glencoe.com

- Study the chapter online
- Explore **Math in Motion**
- Get extra help from your own **Personal Tutor**
- Use **Extra Examples** for additional help
- Take a **Self-Check Quiz**
- **Review Vocabulary** in fun ways

New Vocabulary

English		Español
constant	p. 401	constante
monomial	p. 401	monomio
negative exponent	p. 410	exponente negativo
zero exponent	p. 410	cero exponente
order of magnitude	p. 411	ordenar de magnitud
scientific notation	p. 416	notación científica
binomial	p. 424	binomio
degree of a monomial	p. 424	grado de un monomio
degree of a polynomial	p. 424	grado de un polinomio
polynomial	p. 424	polinomio
trinomial	p. 424	trinomio
leading coefficient	p. 425	coeficiente líder
standard form of a polynomial	p. 425	forma estándar de polinomio
FOIL method	p. 448	método foil
quadratic expression	p. 448	expresion cuadrática

Review Vocabulary

base • p. 5 • base In an expression of the form x^n, the base is x.

Distributive Property • p. 23 • Propiedad distributiva For any numbers a, b, and c, $a(b + c) = ab + ac$ and $a(b - c) = ab - ac$.

exponent • p. 5 • exponente In an expression of the form x^n, the exponent is n. It indicates the number of times x is used as a factor.

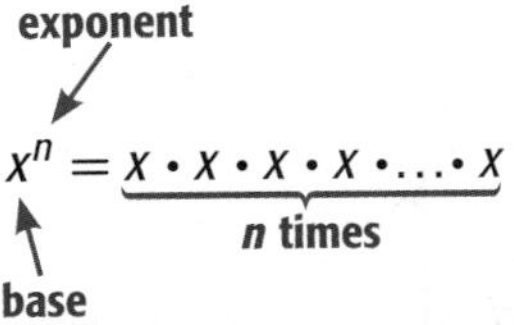

Multilingual eGlossary glencoe.com

7-1 Multiplying Monomials

Why?

Then

You performed operations on expressions with exponents. (Lesson 1-1)

Now

- Multiply monomials.
- Simplify expressions involving monomials.

New Vocabulary

monomial
constant

Math Online

glencoe.com

- Extra Examples
- Personal Tutor
- Self-Check Quiz
- Homework Help

Many formulas contain *monomials*. For example, the formula for the horsepower of a car is $H = w\left(\frac{v}{234}\right)^3$. H represents the horsepower produced by the engine, w equals the weight of the car with passengers, and v is the velocity of the car at the end of a quarter of a mile. As the velocity increases, the horsepower increases.

Monomials A **monomial** is a number, a variable, or the product of a number and one or more variables with nonnegative integer exponents. It has only one term. In the formula to calculate the horsepower of a car, the term $w\left(\frac{v}{234}\right)^3$ is a monomial. An expression that involves division by a variable, like $\frac{ab}{c}$, is not a monomial.

A **constant** is a monomial that is a real number. The monomial $3x$ is an example of a *linear expression* since the exponent of x is 1. The monomial $2x^2$ is a *nonlinear expression* since the exponent is a positive number other than 1.

EXAMPLE 1 Identify Monomials

Determine whether each expression is a monomial. Write *yes* or *no*. Explain your reasoning.

a. 10 — Yes; this is a constant, so it is a monomial.

b. $f + 24$ — No; this expression has addition, so it has more than one term.

c. h^2 — Yes; this expression is a product of variables.

d. j — Yes; single variables are monomials.

Check Your Progress

1A. $-x + 5$

1B. $23abcd^2$

1C. $\frac{xyz^2}{2}$

1D. $\frac{mp}{n}$

Personal Tutor glencoe.com

Recall that an expression of the form x^n is called a *power* and represents the result of multiplying x by itself n times. x is the *base*, and n is the *exponent*. The word *power* is also used sometimes to refer to the exponent.

$$3^4 = \underbrace{3 \cdot 3 \cdot 3 \cdot 3}_{\text{4 factors}} = 81$$

exponent: 4; base: 3

By applying the definition of a power, you can find the product of powers. Look for a pattern in the exponents.

$$2^2 \cdot 2^4 = \overbrace{2 \cdot 2}^{\text{2 factors}} \cdot \overbrace{2 \cdot 2 \cdot 2 \cdot 2}^{\text{4 factors}} \quad (2 + 4 = 6 \text{ factors})$$

$$4^3 \cdot 4^2 = \overbrace{4 \cdot 4 \cdot 4}^{\text{3 factors}} \cdot \overbrace{4 \cdot 4}^{\text{2 factors}} \quad (3 + 2 = 5 \text{ factors})$$

These examples demonstrate the property for the product of powers.

Key Concept — Product of Powers

For Your FOLDABLE

Words To multiply two powers that have the same base, add their exponents.

Symbols For any real number a and any integers m and p, $a^m \cdot a^p = a^{m+p}$.

Examples $b^3 \cdot b^5 = b^{3+5}$ or b^8 $\quad g^4 \cdot g^6 = g^{4+6}$ or g^{10}

EXAMPLE 2 Product of Powers

Simplify each expression.

a. $(6n^3)(2n^7)$

$(6n^3)(2n^7) = (6 \cdot 2)(n^3 \cdot n^7)$ Group the coefficients and the variables.

$= (6 \cdot 2)(n^{3+7})$ Product of Powers

$= 12n^{10}$ Simplify.

b. $(3pt^3)(p^3t^4)$

$(3pt^3)(p^3t^4) = (3 \cdot 1)(p \cdot p^3)(t^3 \cdot t^4)$ Group the coefficients and the variables.

$= (3 \cdot 1)(p^{1+3})(t^{3+4})$ Product of Powers

$= 3p^4t^7$ Simplify.

StudyTip

Coefficients and Powers of 1 A variable with no exponent or coefficient shown can be assumed to have an exponent and coefficient of 1. For example, $x = 1x^1$.

Check Your Progress

2A. $(3y^4)(7y^5)$

2B. $(-4rx^2t^3)(-6r^5x^2t)$

Personal Tutor glencoe.com

We can use the Product of Powers Property to find the power of a power. In the following examples, look for a pattern in the exponents.

$$(3^2)^4 = \overbrace{(3^2)(3^2)(3^2)(3^2)}^{\text{4 factors}} = 3^{2+2+2+2} = 3^8$$

$$(r^4)^3 = \overbrace{(r^4)(r^4)(r^4)}^{\text{3 factors}} = r^{4+4+4} = r^{12}$$

These examples demonstrate the property for the power of a power.

Key Concept — Power of a Power

For Your FOLDABLE

Words To find the power of a power, multiply the exponents.

Symbols For any real number a and any integers m and p, $(a^m)^p = a^{m \cdot p}$.

Examples $(b^3)^5 = b^{3 \cdot 5}$ or b^{15} $\quad (g^6)^7 = g^{6 \cdot 7}$ or g^{42}

StudyTip

Power Rules If you are unsure about when to multiply the exponents and when to add the exponents, write the expression in expanded form.

EXAMPLE 3 Power of a Power

Simplify $[(2^3)^2]^4$.

$[(2^3)^2]^4 = (2^{3 \cdot 2})^4$	Power of a Power
$= (2^6)^4$	Simplify.
$= 2^{6 \cdot 4}$	Power of a Power
$= 2^{24}$ or 16,777,216	Simplify.

Check Your Progress

Simplify each expression.

3A. $[(2^2)^2]^4$

3B. $[(3^2)^3]^2$

Personal Tutor glencoe.com

We can use the Product of Powers Property and the Power of a Power Property to find the power of a product. In the following examples, look for a pattern in the exponents.

$$(tw)^3 = \underbrace{(tw)(tw)(tw)}_{\text{3 factors}} = (t \cdot t \cdot t)(w \cdot w \cdot w) = t^3w^3$$

$$(2yz^2)^3 = \underbrace{(2yz^2)(2yz^2)(2yz^2)}_{\text{3 factors}} = (2 \cdot 2 \cdot 2)(y \cdot y \cdot y)(z^2 \cdot z^2 \cdot z^2) = 2^3y^3z^6 \text{ or } 8y^3z^6$$

These examples demonstrate the property for the power of a product.

Key Concept Power of a Product

For Your

Words To find the power of a product, find the power of each factor and multiply.

Symbols For any real numbers a and b and any integer m, $(ab)^m = a^mb^m$.

Example $(-2xy^3)^5 = (-2)^5x^5y^{15}$ or $-32x^5y^{15}$

EXAMPLE 4 Power of a Product

GEOMETRY Express the area of the circle as a monomial.

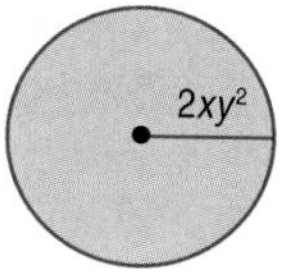

Area $= \pi r^2$	Formula for the area of a circle
$= \pi(2xy^2)^2$	Replace r with $2xy^2$.
$= \pi(2^2x^2y^4)$	Power of a Product
$= 4x^2y^4\pi$	Simplify.

The area of the circle is $4x^2y^4\pi$ square units.

Check Your Progress

4A. Express the area of a square with sides of length $3xy^2$ as a monomial.

4B. Express the area of a triangle with a height of $4a$ and a base of $5ab^2$ as a monomial.

Personal Tutor glencoe.com

StudyTip

Simplify When simplifying expressions with multiple grouping symbols, begin at the innermost expression and work outward.

Simplify Expressions We can combine and use these properties to simplify expressions involving monomials.

Key Concept **Simplify Expressions** For Your FOLDABLE

To simplify a monomial expression, write an equivalent expression in which:

- each variable base appears exactly once,
- there are no powers of powers, and
- all fractions are in simplest form.

EXAMPLE 5 **Simplify Expressions**

Simplify $(3xy^4)^2[(-2y)^2]^3$.

$(3xy^4)^2[(-2y)^2]^3 = (3xy^4)^2(-2y)^6$	**Power of a Power**
$=(3)^2x^2(y^4)^2(-2)^6y^6$	**Power of a Product**
$= 9x^2y^8(64)y^6$	**Power of a Power**
$= 9(64)x^2 \cdot y^8 \cdot y^6$	**Commutative**
$= 576x^2y^{14}$	**Product of Powers**

Check Your Progress

5. Simplify $\left(\frac{1}{2}a^2b^2\right)^3[(-4b)^2]^2$.

Personal Tutor glencoe.com

Check Your Understanding

Example 1 p. 401

Determine whether each expression is a monomial. Write *yes* or *no*. Explain your reasoning.

1. 15

2. $2 - 3a$

3. $\frac{5c}{d}$

4. $-15g^2$

5. $\frac{r}{2}$

6. $7b + 9$

Examples 2 and 3 pp. 402–403

Simplify each expression.

7. $k(k^3)$

8. $m^4(m^2)$

9. $2q^2(9q^4)$

10. $(5u^4v)(7u^4v^3)$

11. $[(3^2)^2]^2$

12. $(xy^4)^6$

13. $(4a^4b^9c)^2$

14. $(-2f^2g^3h^2)^3$

15. $(-3p^5t^6)^4$

Example 4 p. 403

16. **GEOMETRY** The formula for the surface area of a cube is $SA = 6s^2$, where SA is the surface area and s is the length of any side.

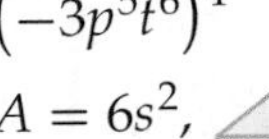

a. Express the surface area of the cube as a monomial.

b. What is the surface area of the cube if $a = 3$ and $b = 4$?

Example 5 p. 404

Simplify each expression.

17. $(5x^2y)^2(2xy^3z)^3(4xyz)$

18. $(-3d^2f^3g)^2[(-3d^2f)^3]^2$

19. $(-2g^3h)(-3gj^4)^2(-ghj)^2$

20. $(-7ab^4c)^3[(2a^2c)^2]^3$

Practice and Problem Solving

= Step-by-Step Solutions begin on page R12.
Extra Practice begins on page 815.

Example 1 p. 401

Determine whether each expression is a monomial. Write *yes* or *no*. Explain your reasoning.

21. 122
22. $3a^4$
23. $2c + 2$
24. $\frac{-2g}{4h}$
25. $\frac{5k}{10}$
26. $6m + 3n$

Examples 2 and 3 pp. 402–403

Simplify each expression.

27. $(q^2)(2q^4)$
28. $(-2u^2)(6u^6)$
29. $(9w^2x^8)(w^6x^4)$
30. $(y^6z^9)(6y^4z^2)$
31. $(b^8c^6d^5)(7b^6c^2d)$
32. $(14fg^2h^2)(-3f^4g^2h^2)$
33. $(j^5k^7)^4$
34. $(n^3p)^4$
35. $[(2^2)^2]^2$
36. $[(3^2)^2]^4$
37. $[(4r^2t)^3]^2$
38. $[(-2xy^2)^3]^2$

Example 4 p. 403

GEOMETRY Express the area of each triangle as a monomial.

39.

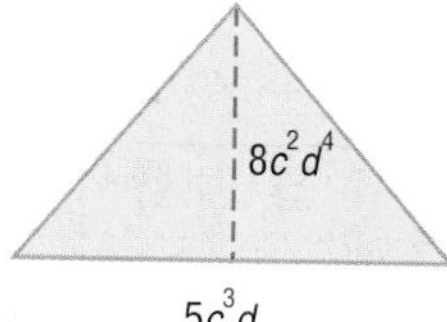

40.

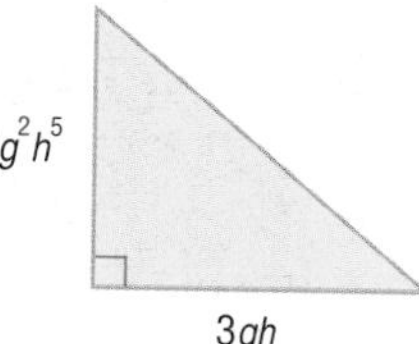

Example 5 p. 404

Simplify each expression.

41. $(2a^3)^4(a^3)^3$
42. $(c^3)^2(-3c^5)^2$
43. $(2gh^4)^3[(-2g^4h)^3]^2$
44. $(5k^2m)^3[(4km^4)^2]^2$
45. $(p^5r^2)^4(-7p^3r^4)^2(6pr^3)$
46. $(5x^2y)^2(2xy^3z)^3(4xyz)$
47. $(5a^2b^3c^4)(6a^3b^4c^2)$
48. $(10xy^5z^3)(3x^4y^6z^3)$
49. $(0.5x^3)^2$
50. $(0.4h^5)^3$
51. $\left(-\frac{3}{4}c\right)^3$
52. $\left(\frac{4}{5}a^2\right)^2$
53. $(8y^3)(-3x^2y^2)\left(\frac{3}{8}xy^4\right)$
54. $\left(\frac{4}{7}m\right)^2(49m)(17p)\left(\frac{1}{34}p^5\right)$
55. $(-3r^3w^4)^3(2rw)^2(-3r^2)^3(4rw^2)^3(2r^2w^3)^4$
56. $(3ab^2c)^2(-2a^2b^4)^2(a^4c^2)^3(a^2b^4c^5)^2(2a^3b^2c^4)^3$

Real-World Link

84% of teens have some money saved. The average teen has saved $1044.

Source: Charles Schwab Teens & Money Survey

57. **FINANCIAL LITERACY** Cleavon has money in an account that earns 3% simple interest. The formula for computing simple interest is $I = Prt$, where I is the interest earned, P represents the principal that he put into the account, r is the interest rate (in decimal form), and t represents time in years.
    a. Cleavon makes a deposit of $\$2c$ and leaves it for 2 years. Write a monomial that represents the interest earned.
    b. If c represents a birthday gift of \$250, how much will Cleavon have in this account after 2 years?

GEOMETRY Express the volume of each solid as a monomial.

58.

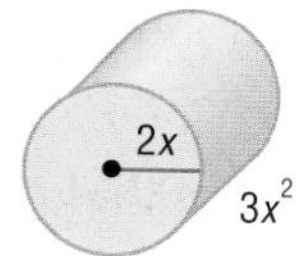

59.

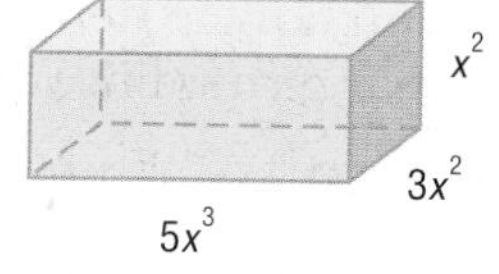

60.

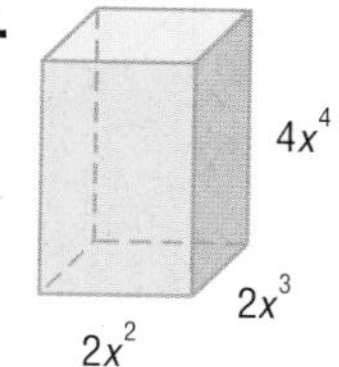

Math History Link

Albert Einstein (1879–1955)

Albert Einstein is perhaps the most well-known scientist of the 20th century. His formula $E = mc^2$, where E represents the energy, m is the mass of the material, and c is the speed of light, shows that if mass is accelerated enough, it could be converted into usable energy.

61. **PACKAGING** For a commercial art class, Aiko must design a new container for individually wrapped pieces of candy. The shape that she chose is a cylinder. The formula for the volume of a cylinder is $V = \pi r^2 h$.
    a. The radius that Aiko would like to use is $2p^3$, and the height is $4p^3$. Write a monomial that represents the volume of her container.
    b. Make a table of values for five possible radius widths and heights if the volume is to remain the same.
    c. What is the volume of Aiko's container if the height is doubled?

62. **ENERGY** Matter can be converted completely into energy by using the formula at the left. Energy is measured in joules, mass in kilograms, and the speed of light is about 300 million meters per second.
    a. Complete the calculations to convert 3 kilograms of gasoline completely into energy.
    b. What happens to the energy if the amount of gasoline is doubled?

63. **MULTIPLE REPRESENTATIONS** In this problem, you will explore exponents.
    a. **TABULAR** Copy and use a calculator to complete the table.

Power	3^4	3^3	3^2	3^1	3^0	3^{-1}	3^{-2}	3^{-3}	3^{-4}
Value						$\frac{1}{3}$	$\frac{1}{9}$	$\frac{1}{27}$	$\frac{1}{81}$

    b. **ANALYTICAL** What do you think the values of 5^0 and 5^{-1} are? Verify your conjecture using a calculator.
    c. **ANALYTICAL** Complete: For any nonzero number a and any integer n, $a^{-n} =$ ______.
    d. **VERBAL** Describe the value of a nonzero number raised to the zero power.

H.O.T. Problems

Use Higher-Order Thinking Skills

64. **CHALLENGE** For any nonzero real numbers a and b and any integers m and t, simplify the expression $\left(-\frac{a^m}{b^t}\right)^{2t}$ and describe each step.

65. **REASONING** Copy the table below.

Equation	Related Expression	Power of x	Linear or Nonlinear
$y = x$			
$y = x^2$			
$y = x^3$			

    a. For each equation, write the related expression and record the power of x.
    b. Graph each equation using a graphing calculator.
    c. Classify each graph as *linear* or *nonlinear*.
    d. Explain how to determine whether an equation, or its related expression, is linear or nonlinear without graphing.

66. **OPEN ENDED** Write three different expressions that can be simplified to x^6.

67. **WRITING IN MATH** Write two formulas that have monomial expressions in them. Explain how each is used in a real-world situation.

Standardized Test Practice

68. Which of the following is not a monomial?

A $-6xy$

B $\frac{1}{2}a^2$

C $-\frac{1}{2b^3}$

D $5gh^4$

69. GEOMETRY The accompanying diagram shows the transformation of $\triangle XYZ$ to $\triangle X'Y'Z'$.

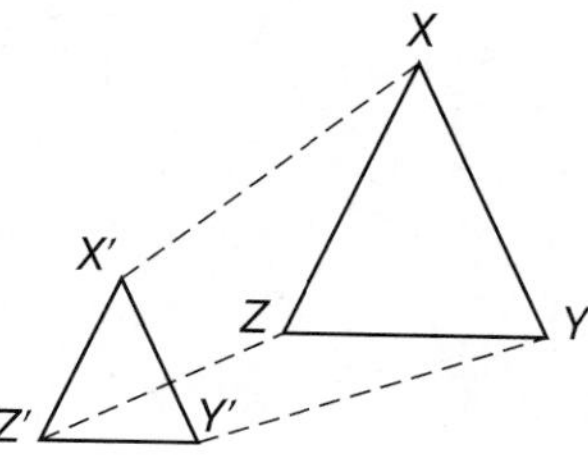

This transformation is an example of a

F dilation
G line reflection
H rotation
J translation

70. CARS In 1994, the average price of a new domestic car was \$16,930. In 2002, the average price was \$19,126. Based on a linear model, what is the predicted average price for 2010?

A \$22,969

B \$21,322

C \$20,773

D \$18,577

71. SHORT RESPONSE If a line has a positive slope and a negative y-intercept, what happens to the x-intercept if the slope and the y-intercept are both doubled?

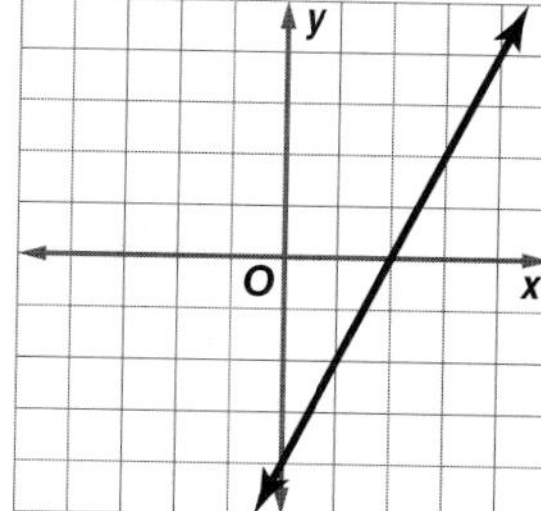

Spiral Review

Solve each system of inequalities by graphing. (Lesson 6-8)

72. $y < 4x$
$2x + 3y \geq -21$

73. $y \geq 2$
$2y + 2x \leq 4$

74. $y > -2x - 1$
$2y \leq 3x + 2$

75. $3x + 2y < 10$
$2x + 12y < -6$

Perform the indicated matrix operations. If an operation cannot be performed, write *impossible*. (Lesson 6-7)

76. $\begin{bmatrix} 2 & 5 & 3 \\ -5 & -1 & 10 \\ 4 & -4 & 0 \end{bmatrix} + \begin{bmatrix} -8 & 2 & -6 \\ 3 & 6 & -1 \\ -6 & -10 & 6 \end{bmatrix}$

77. $\begin{bmatrix} 11 & 0 & 7 \\ 8 & 11 & -10 \end{bmatrix} - \begin{bmatrix} -3 & 0 & 4 \end{bmatrix}$

78. $\begin{bmatrix} -5 & 2 & -11 \\ 2 & -2 & 1 \end{bmatrix} + \begin{bmatrix} 2 & 5 \\ 3 & -9 \end{bmatrix}$

79. $\begin{bmatrix} 2 & -5 & -7 \\ -1 & 11 & 1 \\ 6 & -3 & 4 \end{bmatrix} + \begin{bmatrix} -4 & 0 & -9 \\ 12 & -12 & 8 \\ 12 & 0 & 8 \end{bmatrix}$

80. BABYSITTING Alexis charges \$10 plus \$4 per hour to babysit. Alexis needs at least \$40 more to buy a television for which she is saving. Write an inequality for this situation. Will she be able to get her television if she babysits for 5 hours? (Lesson 5-6)

Skills Review

Find each quotient. (Lesson 0-3)

81. $-64 \div (-8)$

82. $-78 \div 1.3$

83. $42.3 \div (-6)$

84. $-23.94 \div 10.5$

85. $-32.5 \div (-2.5)$

86. $-98.44 \div 4.6$

7-2 Dividing Monomials

Then

You multiplied monomials. (Lesson 7-1)

Now

- Find the quotient of two monomials.
- Simplify expressions containing negative and zero exponents.

New Vocabulary

zero exponents
negative exponent
order of magnitude

Math Online

glencoe.com

- Extra Examples
- Personal Tutor
- Self-Check Quiz
- Homework Help
- Math in Motion

Why?

The tallest redwood tree is 112 feet or about 10^2 meters tall. The average height of a woman in the United States is 1.62 meters. The closest power of ten to 1.62 is 10^0, so a woman is about 10^0 meters tall. The ratio of the tree's height to the woman's height is $\frac{10^2}{10^0}$ or 10^2. This means the tallest redwood tree is approximately 100 times as tall as the average woman.

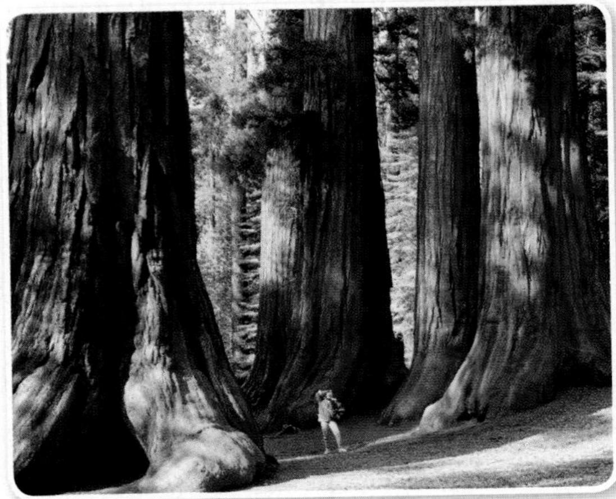

Quotients of Monomials We can use the principles for reducing fractions to find quotients of monomials like $\frac{10^2}{10^0}$. In the following examples, look for a pattern in the exponents.

$$\frac{2^7}{2^4} = \frac{\overbrace{\cancel{2}\cdot\cancel{2}\cdot\cancel{2}\cdot\cancel{2}\cdot 2\cdot 2\cdot 2}^{7\text{ factors}}}{\underbrace{\cancel{2}\cdot\cancel{2}\cdot\cancel{2}\cdot\cancel{2}}_{4\text{ factors}}} = 2\cdot 2\cdot 2 \text{ or } 2^3$$

$$\frac{t^4}{t^3} = \frac{\overbrace{\cancel{t}\cdot\cancel{t}\cdot\cancel{t}\cdot t}^{4\text{ factors}}}{\underbrace{\cancel{t}\cdot\cancel{t}\cdot\cancel{t}}_{3\text{ factors}}} = t$$

These examples demonstrate the Quotient of Powers Rule.

Key Concept — **Quotient of Powers** (For Your FOLDABLE)

Words To divide two powers with the same base, subtract the exponents.

Symbols For any nonzero number a, and any integers m and p, $\frac{a^m}{a^p} = a^{m-p}$.

Examples $\frac{c^{11}}{c^8} = c^{11-8}$ or c^3 $\qquad \frac{r^5}{r^2} = r^{5-2} = r^3$

Math *in Motion*, BrainPOP® glencoe.com

EXAMPLE 1 Quotient of Powers

Simplify $\frac{g^3h^5}{gh^2}$. Assume that no denominator equals zero.

$\frac{g^3h^5}{gh^2} = \left(\frac{g^3}{g}\right)\left(\frac{h^5}{h^2}\right)$ **Group powers with the same base.**

$= (g^{3-1})(h^{5-2})$ **Quotient of Powers**

$= g^2h^3$ **Simplify.**

Check Your Progress

Simplify each expression. Assume that no denominator equals zero.

1A. $\frac{x^3y^4}{x^2y}$ **1B.** $\frac{k^7m^{10}p}{k^5m^3p}$

Personal Tutor glencoe.com

We can use the Product of Powers Rule to find the powers of quotients for monomials. In the following example, look for a pattern in the exponents.

$$\left(\frac{3}{4}\right)^3 = \overbrace{\left(\frac{3}{4}\right)\left(\frac{3}{4}\right)\left(\frac{3}{4}\right)}^{3\text{ factors}} = \frac{\overbrace{3 \cdot 3 \cdot 3}^{3\text{ factors}}}{\underbrace{4 \cdot 4 \cdot 4}_{3\text{ factors}}} = \frac{3^3}{4^3}$$

$$\left(\frac{c}{d}\right)^2 = \overbrace{\left(\frac{c}{d}\right)\left(\frac{c}{d}\right)}^{2\text{ factors}} = \frac{\overbrace{c \cdot c}^{2\text{ factors}}}{\underbrace{d \cdot d}_{2\text{ factors}}} = \frac{c^2}{d^2}$$

StudyTip

Power Rules with Variables
The power rules apply to variables as well as numbers. For example, $\left(\frac{3a}{4b}\right)^3 = \frac{(3a)^3}{(4b)^3}$ or $\frac{27a^3}{64b^3}$.

Key Concept Power of a Quotient — For Your FOLDABLE

Words To find the power of a quotient, find the power of the numerator and the power of the denominator.

Symbols For any real numbers a and $b \neq 0$, and any integer m, $\left(\frac{a}{b}\right)^m = \frac{a^m}{b^m}$.

Examples $\left(\frac{3}{5}\right)^4 = \frac{3^4}{5^4}$ $\left(\frac{r}{t}\right)^5 = \frac{r^5}{t^5}$

EXAMPLE 2 Power of a Quotient

Simplify $\left(\frac{3p^3}{7}\right)^2$.

$\left(\frac{3p^3}{7}\right)^2 = \frac{(3p^3)^2}{7^2}$ **Power of a Quotient**

$= \frac{3^2(p^3)^2}{7^2}$ **Power of a Product**

$= \frac{9p^6}{49}$ **Power of a Power**

Check Your Progress

Simplify each expression.

2A. $\left(\frac{3x^4}{4}\right)^3$ **2B.** $\left(\frac{5x^5y}{6}\right)^2$ **2C.** $\left(\frac{2y^2}{3z^3}\right)^2$ **2D.** $\left(\frac{4x^3}{5y^4}\right)^3$

Personal Tutor glencoe.com

A calculator can be used to explore expressions with 0 as the exponent. There are two methods to explain why a calculator gives a value of 1 for 3^0.

Method 1

$\frac{3^5}{3^5} = 3^{5-5}$ Quotient of Powers

$= 3^0$ Simplify.

Method 2

$\frac{3^5}{3^5} = \frac{\cancel{3} \cdot \cancel{3} \cdot \cancel{3} \cdot \cancel{3} \cdot \cancel{3}}{\cancel{3} \cdot \cancel{3} \cdot \cancel{3} \cdot \cancel{3} \cdot \cancel{3}}$ Definition of powers

$= 1$ Simplify.

Since $\frac{3^5}{3^5}$ can only have one value, we can conclude that $3^0 = 1$.

Key Concept — Zero Exponent Property

For Your FOLDABLE

Words Any nonzero number raised to the zero power is equal to 1.

Symbols For any nonzero number a, $a^0 = 1$.

Examples $15^0 = 1$ $\left(\frac{b}{c}\right)^0 = 1$ $\left(\frac{2}{7}\right)^0 = 1$

EXAMPLE 3 Zero Exponent

Simplify each expression. Assume that no denominator equals zero.

a. $\left(-\frac{4n^2q^5r^2}{9n^3q^2r}\right)^0$

$\left(-\frac{4n^2q^5r^2}{9n^3q^2r}\right)^0 = 1$ $a^0 = 1$

b. $\frac{x^5y^0}{x^3}$

$\frac{x^5y^0}{x^3} = \frac{x^5(1)}{x^3}$ $a^0 = 1$

$= x^2$ Quotient of Powers

Check Your Progress

3A. $\frac{b^4c^2d^0}{b^2c}$

3B. $\left(\frac{2f^4g^7h^3}{15f^3g^9h^6}\right)^0$

Personal Tutor glencoe.com

StudyTip

Zero Exponent Be careful of parentheses. The expression $(5x)^0$ is 1 but $5x^0 = 5$.

Negative Exponents To investigate the meaning of a negative exponent, we can simplify expressions like $\frac{c^2}{c^5}$ using two methods.

Method 1

$\frac{c^2}{c^5} = c^{2-5}$ Quotient of Powers

$= c^{-3}$ Simplify.

Method 2

$\frac{c^2}{c^5} = \frac{\cancel{c} \cdot \cancel{c}}{\cancel{c} \cdot \cancel{c} \cdot c \cdot c \cdot c}$ Definition of powers

$= \frac{1}{c^3}$ Simplify.

Since $\frac{c^2}{c^5}$ can only have one value, we can conclude that $c^{-3} = \frac{1}{c^3}$.

Key Concept — Negative Exponent Property

For Your FOLDABLE

Words For any nonzero number a and any integer n, a^{-n} is the reciprocal of a^n. Also, the reciprocal of a^{-n} is a^n.

Symbols For any nonzero number a and any integer n, $a^{-n} = \frac{1}{a^n}$ and $\frac{1}{a^{-n}} = a^n$.

Examples $2^{-4} = \frac{1}{2^4} = \frac{1}{16}$ $\frac{1}{j^{-4}} = j^4$

StudyTip

Negative Signs Be aware of where a negative sign is placed. $5^{-1} = \frac{1}{5}$, while $-5^1 \neq \frac{1}{5}$.

An expression is considered simplified when it contains only positive exponents, each base appears exactly once, there are no powers of powers, and all fractions are in simplest form.

EXAMPLE 4 Negative Exponents

Simplify each expression. Assume that no denominator equals zero.

a. $\frac{n^{-5}p^4}{r^{-2}}$

$$\frac{n^{-5}p^4}{r^{-2}} = \left(\frac{n^{-5}}{1}\right)\left(\frac{p^4}{1}\right)\left(\frac{1}{r^{-2}}\right) \quad \text{Write as a product of fractions.}$$

$$= \left(\frac{1}{n^5}\right)\left(\frac{p^4}{1}\right)\left(\frac{r^2}{1}\right) \quad a^{-n} = \frac{1}{a^n} \text{ and } \frac{1}{a^{-n}} = a^n$$

$$= \frac{p^4r^2}{n^5} \quad \text{Multiply.}$$

b. $\frac{5r^{-3}t^4}{-20r^2t^7u^{-5}}$

$$\frac{5r^{-3}t^4}{-20r^2t^7u^{-5}} = \left(\frac{5}{-20}\right)\left(\frac{r^{-3}}{r^2}\right)\left(\frac{t^4}{t^7}\right)\left(\frac{1}{u^{-5}}\right) \quad \text{Group powers with the same base.}$$

$$= \left(-\frac{1}{4}\right)(r^{-3-2})(t^{4-7})(u^5) \quad \text{Quotient of Powers and Negative Exponents Property}$$

$$= -\frac{1}{4}r^{-5}t^{-3}u^5 \quad \text{Simplify.}$$

$$= -\frac{1}{4}\left(\frac{1}{r^5}\right)\left(\frac{1}{t^3}\right)(u^5) \quad \text{Negative Exponent Property}$$

$$= -\frac{u^5}{4r^5t^3} \quad \text{Multiply.}$$

c. $\frac{2a^2b^3c^{-5}}{10a^{-3}b^{-1}c^{-4}}$

$$\frac{2a^2b^3c^{-5}}{10a^{-3}b^{-1}c^{-4}} = \left(\frac{2}{10}\right)\left(\frac{a^2}{a^{-3}}\right)\left(\frac{b^3}{b^{-1}}\right)\left(\frac{c^{-5}}{c^{-4}}\right) \quad \text{Group powers with the same base.}$$

$$= \left(\frac{1}{5}\right)(a^{2-(-3)})(b^{3-(-1)})(c^{-5-(-4)}) \quad \text{Quotient of Powers and Negative Exponents Property}$$

$$= \frac{1}{5}a^5b^4c^{-1} \quad \text{Simplify.}$$

$$= \frac{1}{5}(a^5)(b^4)\left(\frac{1}{c}\right) \quad \text{Negative Exponent Property}$$

$$= \frac{a^5b^4}{5c} \quad \text{Multiply.}$$

Simplify each expression. Assume that no denominator equals zero.

4A. $\frac{v^{-3}wx^2}{wy^{-6}}$ **4B.** $\frac{32a^{-8}b^3c^{-4}}{4a^3b^5c^{-2}}$ **4C.** $\frac{5j^{-3}k^2m^{-6}}{25k^{-4}m^{-2}}$

Personal Tutor glencoe.com

Real-World Link

An adult human weighs about 70 kilograms and an adult dairy cow weighs about 700 kilograms. Their weights differ by 1 order of magnitude.

Order of magnitude is used to compare measures and to estimate and perform rough calculations. The **order of magnitude** of a quantity is the number rounded to the nearest power of 10. For example, the power of 10 closest to 95,000,000,000 is 10^{11}, or 100,000,000,000. So the order of magnitude of 95,000,000,000 is 10^{11}.

Real-World Link

There are over 14,000 species of ants living all over the world. Some ants can carry objects that are 50 times their own weight.

Source: Maine Animal Coalition

Real-World EXAMPLE 5 Apply Properties of Exponents

HEIGHT Suppose the average height of a man is about 1.7 meters, and the average height of an ant is 0.0008 meter. How many orders of magnitude as tall as an ant is a man?

Understand We must find the order of magnitude of the heights of the man and ant. Then find the ratio of the orders of magnitude of the man's height to that of the ant's height.

Plan Round each height to the nearest power of ten. Then find the ratio of the height of the man to the height of the ant.

Solve The average height of a man is close to 1 meter. So, the order of magnitude is 10^0 meter. The average height of an ant is about 0.001 meter. So, the order of magnitude is 10^{-3} meters.

The ratio of the height of a man to the height of an ant is about $\frac{10^0}{10^{-3}}$.

$\frac{10^0}{10^{-3}} = 10^{0-(-3)}$ **Quotient of Powers**

$= 10^3$ **$0 - (-3) = 0 + 3$ or 3**

$= 1000$ **Simplify.**

So, a man is approximately 1000 times as tall as an ant, or a man is 3 orders of magnitude as tall as an ant.

Check The ratio of the man's height to the ant's height is $\frac{1.7}{0.0008} = 2125$. The order of magnitude of 2125 is 10^3. ✓

Check Your Progress

5. **ASTRONOMY** The order of magnitude of the mass of Earth is about 10^{27}. The order of magnitude of the Milky Way galaxy is about 10^{44}. How many orders of magnitude as big is the Milky Way galaxy as Earth?

Personal Tutor glencoe.com

Check Your Understanding

Examples 1–4 pp. 408–411

Simplify each expression. Assume that no denominator equals zero.

1. $\frac{t^5u^4}{t^2u}$
2. $\frac{a^6b^4c^{10}}{a^3b^2c}$
3. $\frac{m^6r^5p^3}{m^5r^2p^3}$
4. $\frac{b^4c^6f^8}{b^4c^3f^5}$
5. $\frac{g^8h^2m}{hg^7}$
6. $\frac{r^4t^7v^2}{t^7v^2}$
7. $\frac{x^3y^2z^6}{z^5x^2y}$
8. $\frac{n^4q^4w^6}{q^2n^3w}$
9. $\left(\frac{2a^3b^5}{3}\right)^2$
10. $\frac{r^3v^{-2}}{t^{-7}}$
11. $\left(\frac{2c^3d^5}{5g^2}\right)^5$
12. $\left(-\frac{3xy^4z^2}{x^3yz^4}\right)^0$
13. $\left(\frac{3f^4gh^4}{32f^3g^4h}\right)^0$
14. $\frac{4r^2v^0t^5}{2rt^3}$
15. $\frac{f^{-3}g^2}{h^{-4}}$
16. $\frac{-8x^2y^8z^{-5}}{12x^4y^{-7}z^7}$
17. $\frac{2a^2b^{-7}c^{10}}{6a^{-3}b^2c^{-3}}$

Example 5 p. 412

18. **FINANCIAL LITERACY** The gross domestic product (GDP) for the United States in 2006 was $13.06 trillion, and the GDP per person was $43,800. Use order of magnitude to approximate the population of the United States in 2006.

Practice and Problem Solving

● = **Step-by-Step Solutions** begin on page R12.
Extra Practice begins on page 815.

Examples 1–4 pp. 408–411

Simplify each expression. Assume that no denominator equals zero.

19. $\frac{m^4p^2}{m^2p}$
20. $\frac{p^{12}t^3r}{p^2tr}$
21. $\frac{3m^{-3}r^4p^2}{12t^4}$
22. $\frac{c^4d^4f^3}{c^2d^4f^3}$
23. $\left(\frac{3xy^4}{5z^2}\right)^2$
24. $\left(\frac{3t^6u^2v^5}{9tuv^{21}}\right)^0$
25. $\left(\frac{p^2t^7}{10}\right)^3$
26. $\frac{x^{-4}y^9}{z^{-2}}$
27. $\frac{a^7b^8c^8}{a^5bc^7}$
28. $\left(\frac{3np^3}{7q^2}\right)^2$
29. $\left(\frac{2r^3t^6}{5u^9}\right)^4$
30. $\left(\frac{3m^5r^3}{4p^8}\right)^4$
31. $\left(-\frac{5f^9g^4h^2}{fg^2h^3}\right)^0$
32. $\frac{p^{12}t^7r^2}{p^2t^7r}$
33. $\frac{p^4t^{-3}}{r^{-2}}$
34. $-\frac{5c^2d^5}{8cd^5f^0}$
35. $\frac{-2f^3g^2h^0}{8f^2g^2}$
36. $\frac{12m^{-4}p^2}{-15m^3p^{-9}}$
37. $\frac{k^4m^3p^2}{k^2m^2}$
38. $\frac{14f^{-3}g^2h^{-7}}{21k^3}$
39. $\frac{39t^4uv^{-2}}{13t^{-3}u^7}$
40. $\left(\frac{a^{-2}b^4c^5}{a^{-4}b^{-4}c^3}\right)^2$
41. $\frac{r^3t^{-1}x^{-5}}{tx^5}$
42. $\frac{g^0h^7j^{-2}}{g^{-5}h^0j^{-2}}$

Example 5 p. 412

43. **INTERNET** In a recent year, there were approximately 3.95 million Internet hosts. Suppose there were 208 million Internet users. Determine the order of magnitude for the Internet hosts and Internet users. Using the orders of magnitude, how many Internet users were there compared to Internet hosts?

44. **PROBABILITY** The probability of rolling a die and getting an even number is $\frac{1}{2}$. If you roll the die twice, the probability of getting an even number both times is $\left(\frac{1}{2}\right)\left(\frac{1}{2}\right)$ or $\left(\frac{1}{2}\right)^2$. Write an expression to represent the probability of rolling a die d times and getting an even number every time. Write the expression as a power of 2.

Simplify each expression. Assume that no denominator equals zero.

45. $\frac{-4w^{12}}{12w^3}$
46. $\frac{13r^7}{39r^4}$
47. $\frac{(4k^3m^2)^3}{(5k^2m^{-3})^{-2}}$
48. $\frac{3wy^{-2}}{(w^{-1}y)^3}$
49. $\frac{20qr^{-2}t^{-5}}{4q^0r^4t^{-2}}$
50. $\frac{-12c^3d^0f^{-2}}{6c^5d^{-3}f^4}$
51. $\frac{(2g^3h^{-2})^2}{(g^2h^0)^{-3}}$
52. $\frac{(5pr^{-2})^{-2}}{(3p^{-1}r)^3}$
53. $\left(\frac{-3x^{-6}y^{-1}z^{-2}}{6x^{-2}yz^{-5}}\right)^{-2}$
54. $\left(\frac{2a^{-2}b^4c^2}{-4a^{-2}b^{-5}c^{-7}}\right)^{-1}$
55. $\frac{(16x^2y^{-1})^0}{(4x^0y^{-4}z)^{-2}}$
56. $\left(\frac{4^0c^2d^3f}{2c^{-4}d^{-5}}\right)^{-3}$

57. **COMPUTERS** In 1993, the processing speed of a desktop computer was about 10^8 instructions per second. By 2004, it had increased to 10^{10} instructions per second. The newer computer is how many times as fast as the older one?

Real-World Career

Astronomer
An astronomer studies the universe and analyzes space travel and satellite communications. To be a technician or research assistant, a bachelor's degree is required.

58. ASTRONOMY The brightness of a star is measured in magnitudes. The lower the magnitude, the brighter the star. A magnitude 9 star is 2.51 times as bright as a magnitude 10 star. A magnitude 8 star is $2.51 \cdot 2.51$ or 2.51^2 times as bright as a magnitude 10 star.

a. How many times as bright is a magnitude 3 star as a magnitude 10 star?

b. Write an expression to compare a magnitude m star to a magnitude 10 star.

c. Magnitudes can be measured in negative numbers. Does your expression hold true? Give an example or counterexample.

59. PROBABILITY The probability of rolling a die and getting a 3 is $\frac{1}{6}$. If you roll the die twice, the probability of getting a 3 both times is $\frac{1}{6} \cdot \frac{1}{6}$ or $\left(\frac{1}{6}\right)^2$.

a. Write an expression to represent the probability of rolling a die d times and getting a 3 each time.

b. Write the expression as a power of 6.

60. MULTIPLE REPRESENTATIONS To find the area of a circle, use $A = \pi r^2$. The formula for the area of a square is $A = s^2$.

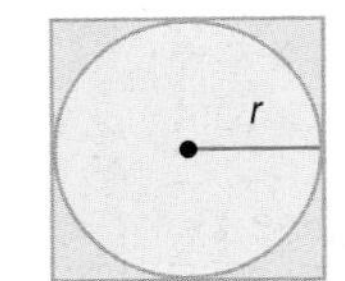

a. ALGEBRAIC Find the ratio of the area of the circle to the area of the square.

b. ALGEBRAIC If the radius of the circle and the length of each side of the square are doubled, find the ratio of the area of the circle to the square.

c. TABULAR Copy and complete the table.

Radius	Area of Circle	Area of Square	Ratio
r			
$2r$			
$3r$			
$4r$			
$5r$			
$6r$			

d. ANALYTICAL What conclusion can be drawn from this?

H.O.T. Problems Use Higher-Order Thinking Skills

61. REASONING Is $x^y \cdot x^z = x^{yz}$ *sometimes*, *always*, or *never* true? Explain.

62. OPEN ENDED Name two monomials with a quotient of $24a^2b^3$.

63. CHALLENGE Use the Quotient of Powers Property to explain why $x^{-n} = \frac{1}{x^n}$.

64. REASONING Write a convincing argument to show why $3^0 = 1$ using the following pattern: $3^5 = 243$, $3^4 = 81$, $3^3 = 27$, $3^2 = 9$.

65. WRITING IN MATH Explain how to use the Quotient of Powers property and the Power of a Quotient property.

Standardized Test Practice

66. GEOMETRY What is the perimeter of the figure in meters?

A $40x$

B $80x$

C $160x$

D $400x$

8x

12x

20x

67. In researching her science project, Leigh learned that light travels at a constant rate and that it takes 500 seconds for light to travel the 93 million miles from the Sun to Earth. Mars is 142 million miles from the Sun. About how many seconds will it take for light to travel from the Sun to Mars?

F 235 seconds

G 327 seconds

H 642 seconds

J 763 seconds

68. EXTENDED RESPONSE Jessie and Jonas are playing a game using the spinners below. Each spinner is equally likely to stop on any of the four numbers. In the game, a player spins both spinners and calculates the product of the two numbers on which the spinners have stopped.

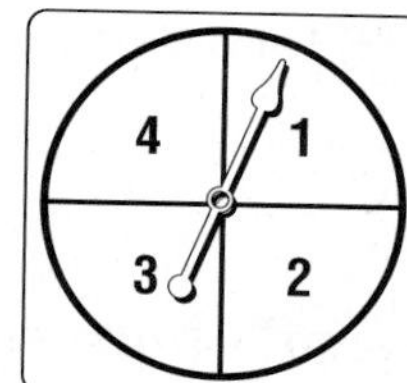

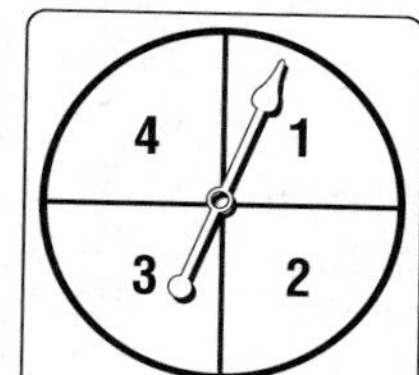

a. What product has the greatest probability of occurring?

b. What is the probability of that product occurring?

69. Simplify $(4^{-2} \cdot 5^0 \cdot 64)^3$.

A $\frac{1}{64}$

B 64

C 320

D 1024

Spiral Review

70. GEOLOGY The seismic waves of a magnitude 6 earthquake are 10^2 times as great as a magnitude 4 earthquake. The seismic waves of a magnitude 4 earthquake are 10 times as great as a magnitude 3 earthquake. How many times as great are the seismic waves of a magnitude 6 earthquake as those of a magnitude 3 earthquake? (Lesson 7-1)

Solve each system of inequalities by graphing. (Lesson 6-8)

71. $y \geq 1$, $x < -1$

72. $y \geq -3$, $y - x < 1$

73. $y < 3x + 2$, $y \geq -2x + 4$

74. $y - 2x < 2$, $y - 2x > 4$

Solve each inequality. Check your solution. (Lesson 5-3)

75. $5(2h - 6) > 4h$

76. $22 \geq 4(b - 8) + 10$

77. $5(u - 8) \leq 3(u + 10)$

78. $8 + t \leq 3(t + 4) + 2$

79. $9n + 3(1 - 6n) \leq 21$

80. $-6(b + 5) > 3(b - 5)$

81. GRADES In a high school science class, a test is worth three times as much as a quiz. What is the student's average grade? (Lesson 2-9)

Science Grades

Tests	Quizzes
85	82
92	75
	95

Skills Review

Evaluate each expression. (Lesson 1-1)

82. 9^2

83. 11^2

84. 10^6

85. 10^4

86. 3^5

87. 5^3

88. 12^3

89. 4^6

7-3 Scientific Notation

Then

You found products and quotients of monomials. (Lessons 7-1 and 7-2)

Now

- Express numbers in scientific notation.
- Find products and quotients of numbers expressed in scientific notation.

New Vocabulary

scientific notation

Math Online

glencoe.com

- Extra Examples
- Personal Tutor
- Self-Check Quiz
- Homework Help

Why?

Space tourism is a multibillion dollar industry. For a price of \$20 million, a civilian can travel on a rocket or shuttle and visit the International Space Station (ISS) for a week.

Scientific Notation Very large and very small numbers such as \$20 million can be cumbersome to use in calculations. For this reason, numbers are often expressed in scientific notation. A number written in **scientific notation** is of the form $a \times 10^n$, where $1 \leq a < 10$ and n is an integer.

Key Concept — Standard Form to Scientific Notation

For Your FOLDABLE

Step 1 Move the decimal point until it is to the right of the first nonzero digit. The result is a real number a.

Step 2 Note the number of places n and the direction that you moved the decimal point.

Step 3 If the decimal point is moved left, write the number as $a \times 10^n$. If the decimal point is moved right, write the number as $a \times 10^{-n}$.

Step 4 Remove the unnecessary zeros.

EXAMPLE 1 Standard Form to Scientific Notation

Express each number in scientific notation.

a. 201,000,000

Step 1 201,000,000 ⟶ 2.01000000 $\quad$ $a = 2.01000000$

Step 2 The decimal point moved 8 places to the left, so $n = 8$.

Step 3 $201{,}000{,}000 = 2.01000000 \times 10^8$

Step 4 2.01×10^8

b. 0.000051

Step 1 0.000051 ⟶ 00005.1 $\quad$ $a = 00005.1$

Step 2 The decimal point moved 5 places to the right, so $n = 5$.

Step 3 $0.000051 = 00005.1 \times 10^{-5}$

Step 4 5.1×10^{-5}

Check Your Progress

1A. 68,700,000,000 $\qquad$ **1B.** 0.0000725

Personal Tutor glencoe.com

Watch Out!

Negative Signs Be careful about the placement of negative signs. A negative sign in the exponent means that the number is between 0 and 1. A negative sign before the number means that it is less than 0.

You can also rewrite numbers in scientific notation in standard form.

Key Concept **Scientific Notation to Standard Form** For Your FOLDABLE

Step 1 In $a \times 10^n$, note whether $n > 0$ or $n < 0$.

Step 2 If $n > 0$, move the decimal point n places right.
If $n < 0$, move the decimal point $-n$ places left.

Step 3 Insert zeros, decimal point, and commas as needed for place value.

EXAMPLE 2 Scientific Notation to Standard Form

Express each number in standard form.

a. 6.32×10^9

Step 1 The exponent is 9, so $n = 9$.

Step 2 Since $n > 0$, move the decimal point 9 places to the right.
$6.32 \times 10^9 \longrightarrow 6320000000$

Step 3 $6.32 \times 10^9 = 6{,}320{,}000{,}000$ **Rewrite; insert commas.**

b. 4×10^{-7}

Step 1 The exponent is -7, so $n = -7$.

Step 2 Since $n < 0$, move the decimal point 7 places to the left.
$4 \times 10^{-7} \longrightarrow 0000004$

Step 3 $4 \times 10^{-7} = 0.0000004$ **Rewrite; insert a 0 before the decimal point.**

Check Your Progress

2A. 3.201×10^6

2B. 9.03×10^{-5}

Personal Tutor glencoe.com

Product and Quotients in Scientific Notation You can use scientific notation to simplify multiplying and dividing very large and very small numbers.

Problem-SolvingTip

Estimate Reasonable Answers Estimating an answer before computing the solution can help you determine if your answer is reasonable.

EXAMPLE 3 Multiply with Scientific Notation

Evaluate $(3.5 \times 10^{-3})(7 \times 10^5)$. Express the result in both scientific notation and standard form.

$(3.5 \times 10^{-3})(7 \times 10^5)$ **Original expression**

$= (3.5 \times 7)(10^{-3} \times 10^5)$ **Commutative and Associative Properties**

$= 24.5 \times 10^2$ **Product of Powers**

$= (2.45 \times 10^1) \times 10^2$ **$24.5 = 2.45 \times 10$**

$= 2.45 \times 10^3$ **Product of Powers**

$= 2450$ **Standard form**

Check Your Progress

Evaluate each product. Express the results in both scientific notation and standard form.

3A. $(6.5 \times 10^{12})(8.7 \times 10^{-15})$

3B. $(1.95 \times 10^{-8})(7.8 \times 10^{-2})$

Personal Tutor glencoe.com

StudyTip

Quotient of Powers Recall that the Quotient of Powers Property is only valid for powers that have the same base. Since 10^8 and 10^3 have the same base, the property applies.

EXAMPLE 4 Divide with Scientific Notation

Evaluate $\frac{3.066 \times 10^8}{7.3 \times 10^3}$. Express the result in both scientific notation and standard form.

$$\frac{3.066 \times 10^8}{7.3 \times 10^3} = \left(\frac{3.066}{7.3}\right)\left(\frac{10^8}{10^3}\right) \quad \textbf{Product rule for fractions}$$
$$= 0.42 \times 10^5 \quad \textbf{Quotient of Powers}$$
$$= 4.2 \times 10^{-1} \times 10^5 \quad \mathbf{0.42 = 4.2 \times 10^{-1}}$$
$$= 4.2 \times 10^4 \quad \textbf{Product of Powers}$$
$$= 42{,}000 \quad \textbf{Standard form}$$

Check Your Progress

Evaluate each quotient. Express the results in both scientific notation and standard form.

4A. $\frac{2.3958 \times 10^3}{1.98 \times 10^8}$

4B. $\frac{1.305 \times 10^3}{1.45 \times 10^{-4}}$

Personal Tutor glencoe.com

Real-World Link

The platinum award was created in 1976. In 2004, the criteria for the award was extended to digital sales. The top-selling artist of all time is the Beatles with 170 million units sold.

Source: Recording Industry Association of America

Real-World EXAMPLE 5 Use Scientific Notation

MUSIC In the United States, a CD reaches gold status once 500 thousand copies are sold. A CD reaches platinum status once 1 million or more copies are sold.

a. Express the number of copies of CDs that need to be sold to reach each status in standard notation.

gold status: 500 thousand = 500,000; platinum status: 1 million = 1,000,000

b. Write each number in scientific notation.

gold status: $500{,}000 = 5 \times 10^5$; platinum status: $1{,}000{,}000 = 1 \times 10^6$

c. How many copies of a CD have sold if it has gone platinum 13 times? Write your answer in scientific notation and standard form.

A CD reaches platinum status once it sells 1 million records. Since the CD has gone platinum 13 times, we need to multiply by 13.

$$(13)(1 \times 10^6) \quad \textbf{Original expression}$$
$$= (13 \times 1)(10^6) \quad \textbf{Associative Property}$$
$$= 13 \times 10^6 \quad \mathbf{13 \times 1 = 13}$$
$$= (1.3 \times 10^1) \times 10^6 \quad \mathbf{13 = 1.3 \times 10}$$
$$= 1.3 \times 10^7 \quad \textbf{Product of Powers}$$
$$= 13{,}000{,}000 \quad \textbf{Standard form}$$

Check Your Progress

5. SATELLITE RADIO Suppose a satellite radio company earned \$125.4 million in one year.

A. Write this number in standard form.

B. Write this number in scientific notation.

C. If the following year the company earned 2.5 times the amount earned the previous year, determine the amount earned. Write your answer in scientific notation and standard form.

Personal Tutor glencoe.com

Check Your Understanding

Example 1 p. 416

Express each number in scientific notation.

1. 185,000,000
2. 1,902,500,000
3. 0.000564
4. 0.00000804

MONEY Express each number in scientific notation.

5. Teenagers spend \$13 billion annually on clothing.
6. Teenagers have an influence on their families' spending habit. They control about \$1.5 billion of discretionary income.

Example 2 p. 417

Express each number in standard form.

7. 1.98×10^7
8. 4.052×10^6
9. 3.405×10^{-8}
10. 6.8×10^{-5}

Example 3 p. 417

Evaluate each product. Express the results in both scientific notation and standard form.

11. $(1.2 \times 10^3)(1.45 \times 10^{12})$
12. $(7.08 \times 10^{14})(5 \times 10^{-9})$
13. $(5.18 \times 10^2)(9.1 \times 10^{-5})$
14. $(2.9 \times 10^{-2})(5.2 \times 10^{-9})$

Example 4 p. 418

Evaluate each quotient. Express the results in both scientific notation and standard form.

15. $\frac{1.035 \times 10^8}{2.3 \times 10^4}$
16. $\frac{2.542 \times 10^5}{4.1 \times 10^{-10}}$
17. $\frac{1.445 \times 10^{-7}}{1.7 \times 10^5}$
18. $\frac{2.05 \times 10^{-8}}{4 \times 10^{-2}}$

Example 5 p. 418

19. **AIR FILTERS** Salvador bought an air purifier to help him deal with his allergies. The filter in the purifier will stop particles as small as one hundredth of a micron. A micron is one millionth of a millimeter.
    a. Write one hundredth and one micron in standard form.
    b. Write one hundredth and one micron in scientific notation.
    c. What is the smallest size particle in meters that the filter will stop? Write the result in both standard form and scientific notation.

Practice and Problem Solving

● = **Step-by-Step Solutions** begin on page R12.
Extra Practice begins on page 815.

Example 1 p. 416

Express each number in scientific notation.

20. 1,220,000
 21. 58,600,000
22. 1,405,000,000,000
23. 0.0000013
24. 0.000056
25. 0.000000000709

E-MAIL Express each number in scientific notation.

26. Approximately 100 million e-mails sent to the President are put into the National Archives.
27. By 2010, the e-mail security market will generate \$5.5 billion.

Example 2 p. 417

Express each number in standard form.

28. 1×10^{12}
29. 9.4×10^7
30. 8.1×10^{-3}
31. 5×10^{-4}
32. 8.73×10^{11}
33. 6.22×10^{-6}

Example 2 p. 417

INTERNET Express each number in standard form.

34. About 2.1×10^7 people, aged 12 to 17, use the Internet.

35. Approximately 1.1×10^7 teens go online daily.

Examples 3 and 4 pp. 417–418

Evaluate each product or quotient. Express the results in both scientific notation and standard form.

36. $(3.807 \times 10^3)(5 \times 10^2)$

37. $\dfrac{9.6 \times 10^3}{1.2 \times 10^{-4}}$

38. $\dfrac{2.88 \times 10^3}{1.2 \times 10^{-5}}$

39. $(6.5 \times 10^7)(7.2 \times 10^{-2})$

40. $(9.5 \times 10^{-18})(9 \times 10^9)$

41. $\dfrac{8.8 \times 10^3}{4 \times 10^{-4}}$

42. $\dfrac{9.15 \times 10^{-3}}{6.1 \times 10}$

43. $(2.01 \times 10^{-4})(8.9 \times 10^{-3})$

44. $(2.58 \times 10^2)(3.6 \times 10^6)$

45. $\dfrac{5.6498 \times 10^{10}}{8.2 \times 10^4}$

46. $\dfrac{1.363 \times 10^{16}}{2.9 \times 10^6}$

47. $(9.04 \times 10^6)(5.2 \times 10^{-4})$

48. $(1.6 \times 10^{-5})(2.3 \times 10^{-3})$

49. $\dfrac{6.25 \times 10^{-4}}{1.25 \times 10^2}$

50. $\dfrac{3.75 \times 10^{-9}}{1.5 \times 10^{-4}}$

51. $(3.4 \times 10^4)(7.2 \times 10^{-15})$

52. $\dfrac{8.6 \times 10^4}{2 \times 10^{-6}}$

53. $(6.3 \times 10^{-2})(3.5 \times 10^{-4})$

Example 5 p. 418

54. **ASTRONOMY** The distance between Earth and the Sun varies throughout the year. Earth is closest to the Sun in January when the distance is 91.4 million miles. In July, the distance is greatest at 94.4 million miles.
   a. Write 91.4 million in both standard form and in scientific notation.
   b. Write 94.4 million in both standard form and in scientific notation.
   c. What is the percent increase in distance from January to July? Round to the nearest tenth of a percent.

Real-World Link

The distance from Earth to the Sun does not determine the seasons. The seasons are determined by the tilt of the Earth's axis and the elliptical orbit around the Sun.

Source: University of British Columbia Okanagan

Evaluate each product or quotient. Express the results in both scientific notation and standard form.

55. $(4.65 \times 10^{-2})(5 \times 10^6)$

56. $\dfrac{2.548 \times 10^5}{2.8 \times 10^{-2}}$

57. $\dfrac{2.135 \times 10^5}{3.5 \times 10^{12}}$

58. $(4.8 \times 10^5)(3.16 \times 10^{-5})$

59. $(4.3 \times 10^{-3})(4.5 \times 10^4)$

60. $\dfrac{5.184 \times 10^{-5}}{7.2 \times 10^3}$

61. $(5 \times 10^3)(1.8 \times 10^{-7})$

62. $\dfrac{1.032 \times 10^{-4}}{8.6 \times 10^{-5}}$

LIGHT The speed of light is approximately 3×10^8 meters per second.

63. Write an expression to represent the speed of light in kilometers per second.

64. Write an expression to represent the speed of light in kilometers per hour.

65. Make a table to show how many kilometers light travels in a day, a week, a 30-day month, and a 365-day year. Express your results in scientific notation.

66. The distance from Earth to the Moon is approximately 3.844×10^5 kilometers. How long would it take light to travel from Earth to the Moon?

Real-World Link

The longest river on Earth is the Nile River in Africa. It is 4,160 miles long, beginning in Burundi to its mouth at the Mediterranean Sea. The Nile River basin covers an area of 1.293×10^6 square miles.

Source: *Encarta Encyclopedia*

67. **EARTH** The population of Earth is about 6.623×10^9. The land surface of Earth is 1.483×10^8 square kilometers. What is the population density for the land surface area of Earth?

68. **RIVERS** A drainage basin separated from adjacent basins by a ridge, hill, or mountain is known as a watershed. The watershed of the Amazon River is 2,300,000 square miles. The watershed of the Mississippi River is 1,200,000 square miles.

    a. Write each of these numbers in scientific notation.

    b. How many times as large is the Amazon River watershed as the Mississippi River watershed?

69. **AGRICULTURE** In a recent year, farmers planted approximately 92.9 million acres of corn. They also planted 64.1 million acres of soybeans and 11.1 million acres of cotton.

    a. Write each of these numbers in scientific notation and in standard form.

    b. How many times as much corn was planted as soybeans? Write your results in standard form and in scientific notation. Round your answer to four decimal places.

    c. How many times as much corn was planted as cotton? Write your results in standard form and in scientific notation. Round your answer to four decimal places.

H.O.T. Problems

Use Higher-Order Thinking Skills

70. **REASONING** Which is greater, 100^{10} or 10^{100}? Explain your reasoning.

71. **FIND THE ERROR** Syreeta and Pete are solving a division problem with scientific notation. Is either of them correct? Explain your reasoning.

Syreeta	Pete
$\frac{3.65 \times 10^{-12}}{5 \times 10^5} = 0.73 \times 10^{-17}$	$\frac{3.65 \times 10^{-12}}{5 \times 10^5} = 0.73 \times 10^{-17}$
$= 7.3 \times 10^{-16}$	$= 7.3 \times 10^{-18}$

72. **CHALLENGE** Order these numbers from least to greatest without converting them to standard form.

    5.46×10^{-3}, 6.54×10^3, 4.56×10^{-4}, -5.64×10^4, -4.65×10^5

73. **REASONING** Determine whether the statement is *always, sometimes,* or *never* true. Give examples or a counterexample to verify your reasoning.

    When multiplying two numbers written in scientific notation, the resulting number can have no more than two digits to the left of the decimal point.

74. **OPEN ENDED** Write two numbers in scientific notation with a product of 1.3×10^{-3}. Then name two numbers in scientific notation with a quotient of 1.3×10^{-3}.

75. **WRITING IN MATH** Write the steps that you would use to divide two numbers written in scientific notation. Then describe how you would write the results in standard form.

Standardized Test Practice

76. Which number represents 0.05604×10^8 written in standard form?

A 0.0000000005604 **C** 5,604,000

B 560,400 **D** 50,604,000

77. Toni left school and rode her bike home. The graph below shows the relationship between her distance from the school and time.

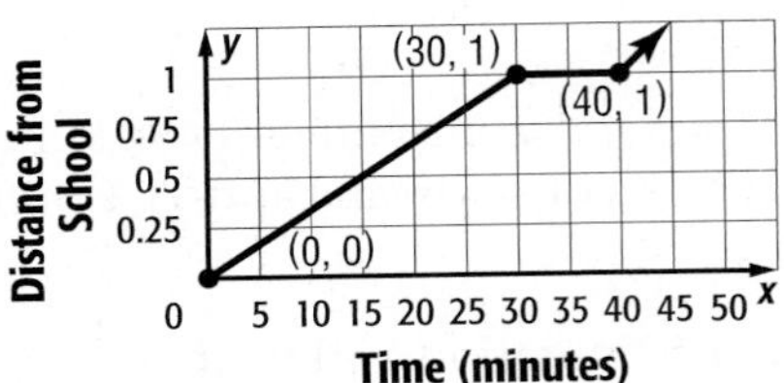

Which explanation could account for the section of the graph from $t = 30$ to $t = 40$?

F Toni rode her bike down a hill.

G Toni ran all the way home.

H Toni stopped at a friend's house on her way home.

J Toni returned to school to get her mathematics book.

78. SHORT RESPONSE In his first four years of coaching football, Coach Delgato's team won 5 games the first year, 10 games the second year, 8 games the third year, and 7 games the fourth year. How many games does the team need to win during the fifth year to have an average of 8 wins per year?

79. The table shows the relationship between Calories and grams of fat contained in an order of fried chicken from various restaurants.

Calories	305	410	320	500	510	440
Fat (g)	28	34	28	41	42	38

Assuming that the data can best be described by a linear model, about how many grams of fat would you expect to be in a 275-Calorie order of fried chicken?

A 22

B 25

C 27

D 28

Spiral Review

Simplify. Assume that no denominator is equal to zero. (Lesson 7-2)

80. $\frac{8^9}{8^6}$

81. $\frac{6^5}{6^3}$

82. $\frac{r^8t^{12}}{r^2t^7}$

83. $\left(\frac{3a^4b^4}{8c^2}\right)^4$

84. $\left(\frac{5d^3g^2}{3h^4}\right)^2$

85. $\left(\frac{4n^2p^4}{8p^3}\right)^3$

86. CHEMISTRY Lemon juice is 10^2 times as acidic as tomato juice. Tomato juice is 10^3 times as acidic as egg whites. How many times as acidic is lemon juice as egg whites? (Lesson 7-1)

Write each equation in slope-intercept form. (Lesson 4-2)

87. $y - 2 = 3(x - 1)$

88. $y - 5 = 6(x + 1)$

89. $y + 2 = -2(x + 5)$

90. $y + 3 = \frac{1}{2}(x + 4)$

91. $y - 1 = \frac{2}{3}(x + 9)$

92. $y + 3 = -\frac{1}{4}(x + 2)$

Skills Review

Simplify each expression. If not possible, write *simplified*. (Lesson 1-4)

93. $3u + 10u$

94. $5a - 2 + 6a$

95. $6m^2 - 8m$

96. $4w^2 + w + 15w^2$

97. $13(5 + 4a)$

98. $(4t - 6)16$

Algebra Lab
Polynomials

Math Online glencoe.com
Math in Motion, Animation

Algebra tiles can be used to model polynomials. A polynomial is a monomial or the sum of monomials. The diagram at the right shows the models.

Polynomial Models	
Polynomials are modeled using three types of tiles.	1 x x^2
Each tile has an opposite.	−1 −x −x^2

ACTIVITY **Use algebra tiles to model each polynomial.**

- $5x$

  To model this polynomial, you will need 5 green x-tiles.

- $3x^2 - 1$

  To model this polynomial, you will need 3 blue x^2-tiles and 1 red -1-tile.

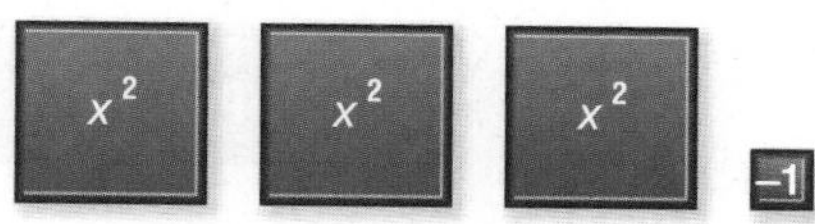

- $-2x^2 + x + 3$

  To model this polynomial, you will need 2 red $-x^2$-tiles, 1 green x-tile, and 3 yellow 1-tiles.

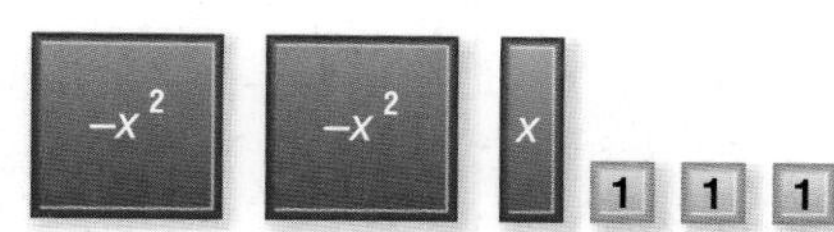

Model and Analyze

Use algebra tiles to model each polynomial. Then draw a diagram of your model.

1. $-4x^2$

2. $3x - 5$

3. $2x^2 - 3x$

4. $x^2 + 2x + 1$

Write an algebraic expression for each model.

5.

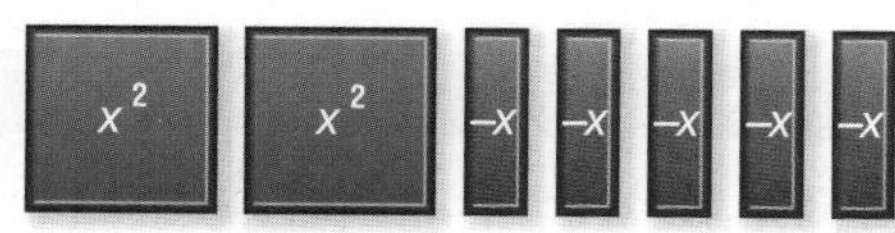

6.

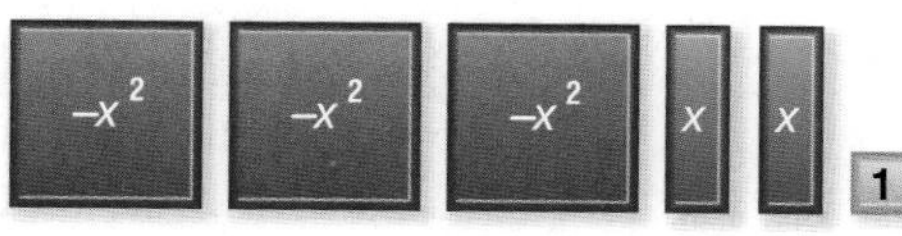

7.

8.

9. MAKE A CONJECTURE Write a sentence or two explaining why algebra tiles are sometimes called *area tiles*.

7-4 Polynomials

Why?

In 2011, sales of digital audio players are expected to reach record numbers. The sales data can be modeled by the equation $U = -2.7t^2 + 49.4t + 128.7$, where U is the number of units shipped in millions and t is the number of years since 2005.

The expression $-2.7t^2 + 49.4t + 128.7$ is an example of a polynomial. Polynomials can be used to model situations.

Then

You identified monomials and their characteristics. (Lesson 7-1)

Now

- Find the degree of a polynomial.
- Write polynomials in standard form.

New Vocabulary

polynomial
binomial
trinomial
degree of a monomial
degree of a polynomial
standard form of a polynomial
leading coefficient

Math Online

glencoe.com

- Extra Examples
- Personal Tutor
- Self-Check Quiz
- Homework Help
- Math in Motion

Degree of a Polynomial A **polynomial** is a monomial or the sum of monomials, each called a *term* of the polynomial. Some polynomials have special names. A **binomial** is the sum of *two* monomials, and a **trinomial** is the sum of *three* monomials.

EXAMPLE 1 Identify Polynomials

Determine whether each expression is a polynomial. If so, identify the polynomial as a *monomial, binomial,* or *trinomial.*

Expression	Is it a polynomial?	Monomial, binomial, or trinomial?
a. $4y - 5xz$	Yes; $4y - 5xz$ is the sum of the two monomials $4y$ and $-5xz$.	binomial
b. -6.5	Yes; -6.5 is a real number.	monomial
c. $7a^{-3} + 9b$	No; $7a^{-3} = \frac{7}{a^3}$, which is not a monomial.	none of these
d. $6x^3 + 4x + x + 3$	Yes; $6x^3 + 4x + x + 3 = 6x^3 + 5x + 3$, the sum of three monomials.	trinomial

Check Your Progress

1A. x

1B. $-3y^2 - 2y + 4y - 1$

1C. $5rx + 7tuv$

1D. $10x^{-4} - 8x^a$

Personal Tutor glencoe.com

The **degree of a monomial** is the sum of the exponents of all its variables. A nonzero constant has degree 0. Zero has no degree.

The **degree of a polynomial** is the greatest degree of any term in the polynomial. To find the degree of a polynomial, you must find the degree of each term. Some polynomials have special names based on their degree.

Degree	Name
0	constant
1	linear
2	quadratic
3	cubic
4	quartic
5	quintic
6 or more	6th degree, 7th degree, and so on

ReadingMath

Prefixes The prefixes *mono*, *bi*, and *tri* mean one, two, and three, respectively. Hence, a monomial has one term, a binomial has two terms, and a trinomial has three terms.

Math *in Motion*, BrainPOP® glencoe.com

EXAMPLE 2 Degree of a Polynomial

Find the degree of each polynomial.

a. $3a^2b^3 + 6$

Step 1 Find the degree of each term.

$3a^2b^3$: degree $= 2 + 3$ or 5 6: degree 0

Step 2 The degree of the polynomial is the greatest degree, 5.

b. $2d^3 - 5c^5d - 7$

$2d^3$: degree $= 3$ $-5c^5d$: degree $= 5 + 1$ or 6

-7: degree 0 The degree of the polynomial is 6.

Check Your Progress

2A. $7xy^5z$

2B. $2rt - 3rt^2 - 7r^2t^2 - 13$

Personal Tutor glencoe.com

Polynomials in Standard Form The terms of a polynomial may be written in any order. Polynomials written in only one variable are usually written in standard form.

The **standard form of a polynomial** is written with the terms in order from greatest degree to least degree. When a polynomial is written in standard form, the coefficient of the first term is called the **leading coefficient**.

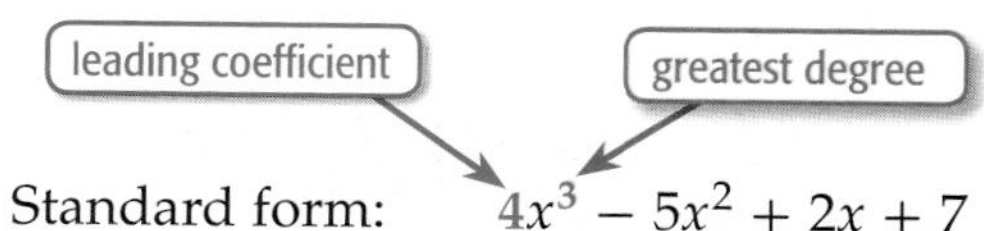

Standard form: $4x^3 - 5x^2 + 2x + 7$

EXAMPLE 3 Standard Form of a Polynomial

Write each polynomial in standard form. Identify the leading coefficient.

a. $3x^2 + 4x^5 - 7x$

Step 1 Find the degree of each term.

Degree: 2, 5, 1

Polynomial: $3x^2 + 4x^5 - 7x$

Step 2 Write the terms in descending order: $4x^5 + 3x^2 - 7x$.

The leading coefficient is 4.

Math *in Motion*, Interactive Lab glencoe.com

b. $5y - 9 - 2y^4 - 6y^3$

Step 1 Degree: 1, 0, 4, 3

Polynomial: $5y - 9 - 2y^4 - 6y^3$

Step 2 $-2y^4 - 6y^3 + 5y - 9$ The leading coefficient is -2.

Check Your Progress

3A. $8 - 2x^2 + 4x^4 - 3x$

3B. $y + 5y^3 - 2y^2 - 7y^6 + 10$

Personal Tutor glencoe.com

We can use polynomials to estimate values between two points. We can also use them to predict values of events before they occur.

Real-World Link

The world's biggest skateboard was built in 1996. The skateboard is 10 feet long, 4 feet wide, 3 feet tall and is fully functional. It is on display in San Diego, California.

Source: Foundation Skateboard Company

Real-World EXAMPLE 4 Use a Polynomial

BUSINESS From 2000 through 2006, the number U of skateboards (in thousands) produced at a manufacturing plant can be modeled by the equation $U = 3t^2 - 2t + 10$, where t is the number of years since 2000. How many skateboards were produced in 2002?

Find the value of t, and substitute the value of t to find the number of skateboards produced.

Since t is the number of years since 2000, t equals 2002 – 2000 or 2.

$U = 3t^2 - 2t + 10$	**Original equation**
$= 3(2)^2 - 2(2) + 10$	**$t = 2$**
$= 3(4) - 4 + 10$	**Simplify.**
$= 12 - 4 + 10$	**Multiply.**
$= 18$	**Simplify.**

Since U is in thousands, the number of skateboards produced was 18 thousand or 18,000.

Check Your Progress

4A. How many skateboards were produced in 2005?

4B. If this trend continues, how many skateboards will be produced in 2015?

Personal Tutor glencoe.com

Check Your Understanding

Example 1 p. 424

Determine whether each expression is a polynomial. If so, identify the polynomial as a *monomial*, *binomial*, or *trinomial*.

1. $7ab + 6b^2 - 2a^3$

2. $2y - 5 + 3y^2$

3. $3x^2$

4. $\frac{4m}{3p}$

5. $5m^2p^3 + 6$

6. $5q^{-4} + 6q$

Example 2 p. 425

Find the degree of each polynomial.

7. -3

8. $6p^3 - p^4$

9. $-7z$

10. $\frac{3}{4}$

11 $12 - 7q^2t + 8r$

12. $2a^2b^5 + 5 - ab$

13. $6df^3 + 3d^2f^2 + 2d + 1$

14. $9hjk - 4h^2j^3 + 5j^2k^2 - h^3k^3$

Example 3 p. 425

Write each polynomial in standard form. Identify the leading coefficient.

15. $2x^5 - 12 + 3x$

16. $-y^3 + 3y - 3y^2 + 2$

17. $4z - 2z^2 - 5z^4$

18. $2a + 4a^3 - 5a^2 - 1$

Example 4 p. 426

19. ENROLLMENT Suppose the number N (in hundreds) of students projected to attend a high school from 1998 to 2007 can be modeled by the equation $N = t^2 + 1.5t + 0.5$, where t is the number of years since 1998.

a. How many students were enrolled in the high school in 2003?

b. How many students were enrolled in the high school in 2005?

Practice and Problem Solving

● = **Step-by-Step Solutions** begin on page R12.
Extra Practice begins on page 815.

Example 1 p. 424

Determine whether each expression is a polynomial. If so, identify the polynomial as a *monomial, binomial,* or *trinomial.*

20. $\frac{5y^3}{x^2} + 4x$

21. 21

22. $c^4 - 2c^2 + 1$

23. $d + 3d^{-c}$

24. $a - a^2$

25. $5n^3 + nq^3$

Example 2 p. 425

Find the degree of each polynomial.

26. $13 - 4ab + 5a^3b$

27. $3x - 8$

28. -4

29. $17g^2h$

30. $10 + 2cd^4 - 6d^2g$

31. $2z^2y^2 - 7 + 5y^3w^4$

Example 3 p. 425

Write each polynomial in standard form. Identify the leading coefficient.

32. $5x^2 - 2 + 3x$

33. $8y + 7y^3$

34. $4 - 3c - 5c^2$

35 $-4d^4 + 1 - d^2$

36. $11t + 2t^2 - 3 + t^5$

37. $2 + r - r^3$

38. $\frac{1}{2}x - 3x^4 + 7$

39. $-9b^2 + 10b - b^6$

Example 4 p. 426

40. FIREWORKS A firework shell is launched two feet from the ground at a speed of 150 feet per second. The height H of the firework shell is modeled by the equation $H = -16t^2 + 150t + 2$, where t is time in seconds.

a. How high will the firework be after 3 seconds?

b. How high will the firework be after 5 seconds?

Classify each polynomial according to its degree and number of terms.

41. $4x - 3x^2 + 5$

42. $11z^3$

43. $9 + y^4$

44. $3x^3 - 7$

45. $-2z^5 - x^2 + 5x - 8$

46. $10t - 4t^2 + 6t^3$

47. ICE CREAM An ice cream shop is changing the size of their cone.

a. If the volume of a cone is the product of $\frac{1}{3}$, π, the square of the radius r, and the height h, write a polynomial that represents the volume.

b. How much will the cone hold if the radius is 1.5 inches and the height is 4 inches?

c. If the volume of the cone must be 63 cubic inches and the radius of the cone is 3 inches, how tall is the cone?

48. GEOMETRY Write two expressions for the perimeter and area of the rectangle.

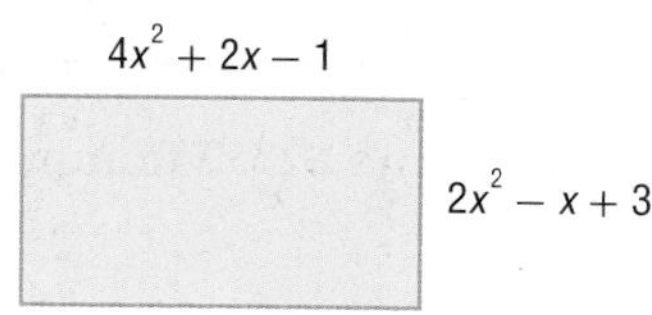

49. GEOMETRY Write a polynomial for the area of the shaded region shown.

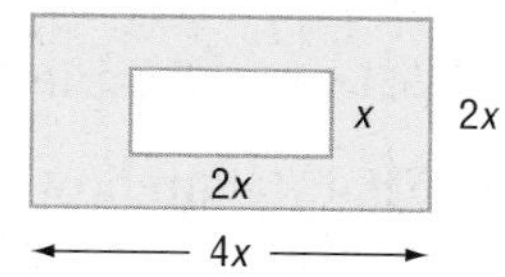

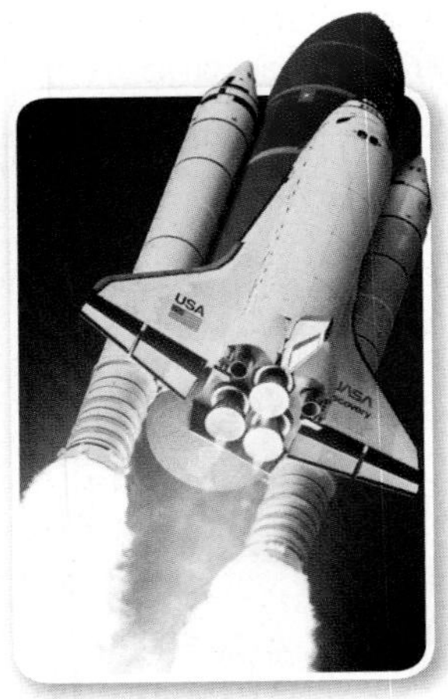

Real-World Link

A space shuttle has three parts: the orbiter, an external tank, and two solid rocket boosters. Including fuel, the shuttle weighs a total of 4.4 million pounds at launch.

Source: How Stuff Works

50. PROJECT Rocky and Arturo are designing a rocket for a competition. The top must be cone-shaped and the body of the rocket must be cylindrical. The volume of a cone is the product of $\frac{1}{3}$, π, the height h, and the square of the radius r. The volume of a cylinder is the product of π, the height t, and the square of the radius r.

a. Write a polynomial that represents the volume of the rocket.

b. If the height of the body of the rocket is 8 inches, the height of the top is 6 inches, and the radius is 3 inches, find the volume of the rocket.

c. If the height of the body of the rocket is 9 inches, the height of the top is 5 inches, and the radius is 4 inches, find the volume of the rocket.

51. MULTIPLE REPRESENTATIONS In this problem, you will explore perimeter and area.

a. GEOMETRIC Draw three rectangles that each have a perimeter of 400 feet.

b. TABULAR Record the width and length of each rectangle in a table like the one shown below. Find the area of each rectangle.

Rectangle	Length	Width	Area
1	100 ft		
2	50 ft		
3	75 ft		
4	x ft		

c. GRAPHICAL On a coordinate system, graph the area of rectangle 4 in terms of the length, x. Use the graph to determine the largest area possible.

d. ANALYTICAL Determine the length and width that produce the largest area.

H.O.T. Problems

Use **H**igher-**O**rder **T**hinking Skills

52. FIND THE ERROR Chuck and Claudio are writing $2x^2 - 3 + 5x$ in standard form. Is either of them correct? Explain your reasoning.

Chuck	Claudio
$2x^2$: degree 2	$2x^2$: degree 2
-3 : degree 0	-3 : degree 0
5x : degree 1	5x : degree 1
$2x^2 - 5x + 3$	$2x^2 + 5x - 3$

53. CHALLENGE Write a polynomial that represents any odd integer if x is an integer. Explain.

54. REASONING Is the following statement *sometimes, always,* or *never* true? Explain.

A binomial can have a degree of zero.

55. OPEN ENDED Write an example of a cubic trinomial.

56. WRITING IN MATH Explain how to write a polynomial in standard form and how to identify the leading coefficient.

Standardized Test Practice

57. Matrices P and Q are given below.

$$P = \begin{bmatrix} 3 & 2 \\ 6 & 9 \\ 1 & 0 \end{bmatrix} \qquad Q = \begin{bmatrix} -3 & -2 \\ -6 & -9 \\ 4 & 0 \end{bmatrix}$$

What is $P - Q$?

A $\begin{bmatrix} -6 & -9 \\ -8 & -15 \\ 3 & 0 \end{bmatrix}$ **C** $\begin{bmatrix} 0 & -5 \\ 4 & 3 \\ 5 & 0 \end{bmatrix}$

B $\begin{bmatrix} 0 & 5 \\ -4 & -3 \\ -5 & 0 \end{bmatrix}$ **D** $\begin{bmatrix} 6 & 4 \\ 12 & 18 \\ -3 & 0 \end{bmatrix}$

58. You have a coupon from The Really Quick Lube Shop for an \$8 off oil change this month. An oil change costs \$19.95, and a new oil filter costs \$4.95. You use the coupon for an oil change and filter. Before adding tax, how much should you pay?

F \$11.95
G \$16.90
H \$24.90
J \$27.95

59. SHORT RESPONSE In a recent poll, 3000 people were asked to pick their favorite baseball team. The accompanying circle graph shows the results of that poll. How many people polled picked the Black Sox as their favorite team?

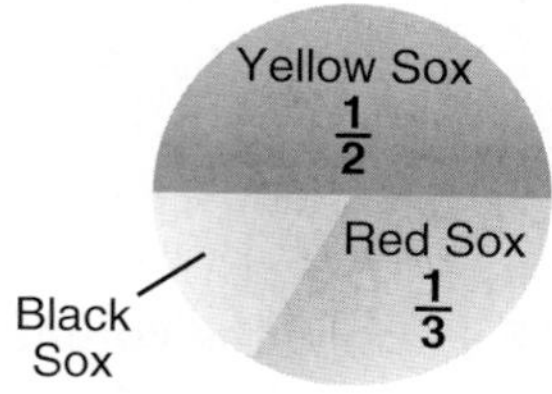

60. What value for y satisfies the system of equations below?

$$2x + y = 19$$
$$4x - 6y = -2$$

A 5
B 7
C 8
D 10

Spiral Review

Express each number in standard notation. (Lesson 7-3)

61. 6×10^{-7} **62.** 7.2×10^{-10} **63.** 8.1×10^{5}

64. 7×10^{6} **65.** 0.132×10^{-6} **66.** 1.88×10^{0}

Simplify. Assume that no denominator is equal to zero. (Lesson 7-2)

67. $a^0(a^4)(a^{-8})$ **68.** $\dfrac{(4m^{-3}c^6)^0}{mc}$ **69.** $\dfrac{(3f^2g^6)^0}{(18f^6g^2)^0}$

70. 12^{-1} **71.** $\dfrac{k^{-4}}{m^2p^{-8}}$ **72.** $\dfrac{(nq^{-1})^3}{(n^4q^8)^{-1}}$

73. FINANCIAL LITERACY The owners of a new restaurant have hired enough servers to handle 17 tables of customers. The fire marshal has approved the restaurant for a limit of 56 customers. How many two-seat tables and how many four-seat tables should the owners buy? (Lesson 6-4)

Skills Review

Simplify each expression. If not possible, write *simplified*. (Lesson 1-5)

74. $7b^2 + 14b - 10b$ **75.** $5t + 12t^2 - 8t$ **76.** $3y^4 + 2y^4 + 2y^5$

77. $7h^5 - 7j^5 + 8k^5$ **78.** $n + \frac{n}{3} + \frac{2}{3}n$ **79.** $2u + \frac{u}{2} + u^2$

Mid-Chapter Quiz

Lessons 7-1 through 7-4

Math Online glencoe.com
Math *in Motion,* Animation

Simplify each expression. (Lesson 7-1)

1. $(x^3)(4x^5)$
2. $(m^2p^5)^3$
3. $[(2xy^3)^2]^3$
4. $(6ab^3c^4)(-3a^2b^3c)$

5. **MULTIPLE CHOICE** Express the volume of the solid as a monomial. (Lesson 7-1)

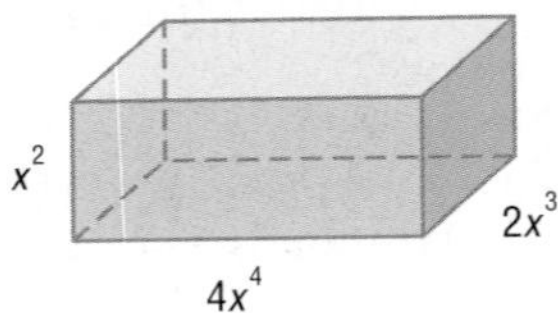

A $6x^9$

B $8x^9$

C $8x^{24}$

D $7x^{24}$

Simplify each expression. Assume that no denominator equals 0. (Lesson 7-2)

6. $\left(\frac{2a^4b^3}{c^6}\right)^3$
7. $\frac{2xy^0}{6x}$
8. $\frac{m^7n^4p}{m^3n^3p}$
9. $\frac{p^4t^{-2}}{r^{-5}}$

10. **ASTRONOMY** Physicists estimate that the number of stars in the universe has an order of magnitude of 10^{21}. The number of stars in the Milky Way galaxy is around 100 billion. Using orders of magnitude, how many times as many stars are there in the universe as the Milky Way? (Lesson 7-2)

Express each number in scientific notation. (Lesson 7-3)

11. 0.00000054
12. 0.0042
13. 234,000
14. 418,000,000

Express each number in standard form. (Lesson 7-3)

15. 4.1×10^{-3}
16. 2.74×10^5
17. 3×10^9
18. 9.1×10^{-5}

Evaluate each product or quotient. Express the results in scientific notation. (Lesson 7-3)

19. $(2.13 \times 10^2)(3 \times 10^5)$
20. $(7.5 \times 10^6)(2.5 \times 10^{-2})$
21. $\frac{7.5 \times 10^8}{2.5 \times 10^4}$
22. $\frac{6.6 \times 10^5}{2 \times 10^{-3}}$

Determine whether each expression is a polynomial. If so, identify the polynomial as a *monomial*, *binomial*, or *trinomial*. (Lesson 7-4)

23. $3y^2 - 2$
24. $4t^5 + 3t^2 + t$
25. $\frac{3x}{5y}$
26. ax^{-3}
27. $3b^2$
28. $2x^{-3} - 4x + 1$

29. **POPULATION** The table shows the population density for Nevada for various years. (Lesson 7-4)

Year	Years Since 1930	People/ Square Mile
1930	0	0.8
1960	30	2.6
1980	50	7.3
1990	60	10.9
2000	70	18.2

a. The population density d of Nevada from 1930 to 2000 can be modeled by $d = 0.005y^2 - 0.127y + 1$, where y represents the number of years since 1930. Identify the type of polynomial for $0.005y^2 - 0.127y + 1$.

b. What is the degree of the polynomial?

c. Predict the population density of Nevada for 2020. Explain your method.

d. Predict the population density of Nevada for 2030. Explain your method.

Algebra Lab
Adding and Subtracting Polynomials

Math Online glencoe.com
Math *in Motion*, Animation

Monomials such as $3x$ and $-2x$ are called *like terms* because they have the same variable to the same power. When you use algebra tiles, you can recognize like terms because the individual tiles have the same size and shape.

Polynomial Models	
Like terms are represented by tiles that have the same shape and size.	x x $-x$ like terms
A *zero pair* may be formed by pairing one tile with its opposite. You can remove or add zero pairs without changing the polynomial.	x $-x$ → 0

ACTIVITY 1 Add Polynomials

Use algebra tiles to find $(2x^2 - 3x + 5) + (x^2 + 6x - 4)$.

Step 1 Model each polynomial.

$2x^2 - 3x + 5 \longrightarrow$ $2x^2$ + $-3x$ + 5

$x^2 + 6x - 4 \longrightarrow$ x^2 + $6x$ + -4

Step 2 Combine like terms and remove zero pairs.

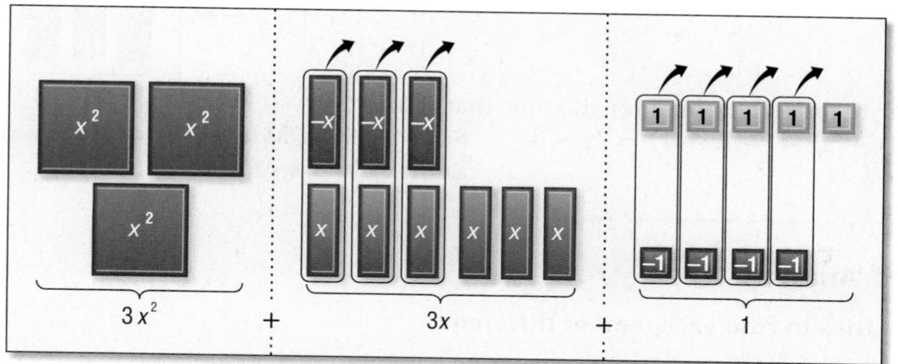

Step 3 Write the polynomial for the tiles that remain.

$(2x^2 - 3x + 5) + (x^2 + 6x - 4) = 3x^2 + 3x + 1$

ACTIVITY 2 Subtract Polynomials

Use algebra tiles to find $(4x + 5) - (-3x + 1)$.

Step 1 Model the polynomial $4x + 5$.

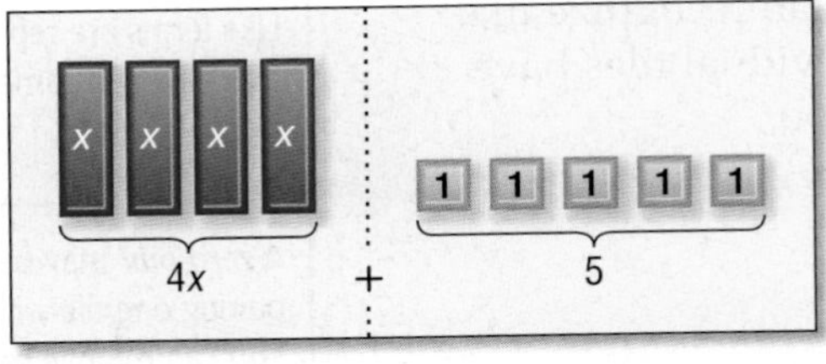

Step 2 To subtract $-3x + 1$, you must remove 3 red $-x$-tiles and 1 yellow 1-tile. You can remove the yellow 1-tile, but there are no red $-x$-tiles. Add 3 zero pairs of x-tiles. Then remove the 3 red $-x$-tiles.

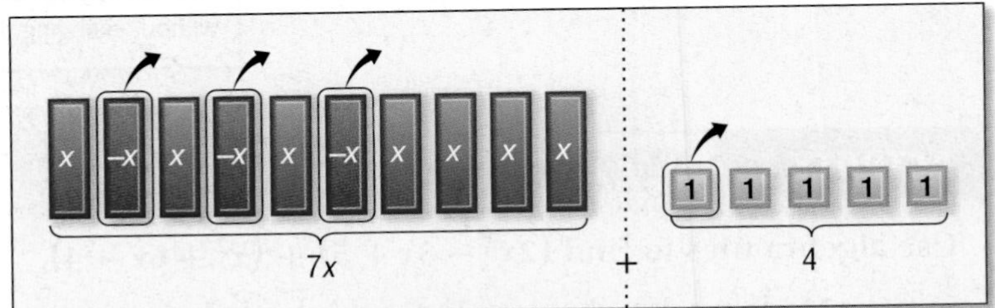

Step 3 Write the polynomial for the tiles that remain. $(4x + 5) - (-3x + 1) = 7x + 4$

Recall that you can subtract a number by adding its additive inverse or opposite. Similarly, you can subtract a polynomial by adding its opposite.

ACTIVITY 3 Subtract Polynomials Using Additive Inverse

Use algebra tiles to find $(4x + 5) - (-3x + 1)$.

Step 1 To find the difference of $4x + 5$ and $-3x + 1$, add $4x + 5$ and the opposite of $-3x + 1$.

$4x + 5$ →

The opposite of $-3x + 1$ is $3x - 1$. →

4x + 4

3x + −1

Step 2 Write the polynomial for the tiles that remain. $(4x + 5) - (-3x + 1) = 7x + 4$.

Notice that this is the same answer as in Activity 2.

Model and Analyze

Use algebra tiles to find each sum or difference.

1. $(x^2 + 5x - 2) + (3x^2 - 2x + 6)$
2. $(2x^2 + 8x + 1) - (x^2 - 4x - 2)$
3. $(-4x^2 + x) - (x^2 + 5x)$
4. **WRITING IN MATH** Find $(4x^2 - x + 3) - (2x + 1)$ using each method from Activity 2 and Activity 3. Illustrate with drawings, and explain in writing how zero pairs are used in each case.

7-5

Adding and Subtracting Polynomials

Then

You wrote polynomials in standard form. (Lesson 7-4)

Now

- Add polynomials.
- Subtract polynomials.

Math Online

glencoe.com

- Extra Examples
- Personal Tutor
- Self-Check Quiz
- Homework Help

Why?

From 2000 to 2003, sales (in millions of dollars) of rap/hip-hop music R and country music C in the United States can be modeled by the following equations, where t is the number of years since 2000.

$R = -132.3t^3 + 624.7t^2 - 773.6t + 1847.7$

$C = -3.4t^3 + 8.6t^2 - 95t + 1532.6$

The total music sales T of rap/hip-hop music and country music is $R + C$.

Add Polynomials Adding polynomials involves adding like terms. You can group like terms by using a horizontal or vertical format.

EXAMPLE 1 Add Polynomials

Find each sum.

a. $(2x^2 + 5x - 7) + (3 - 4x^2 + 6x)$

Horizontal Method

$(2x^2 + 5x - 7) + (3 - 4x^2 + 6x)$

$= [2x^2 + (-4x^2)] + [5x + 6x] + [-7 + 3]$ — Group like terms.

$= -2x^2 + 11x - 4$ — Combine like terms.

Vertical Method

Align like terms in columns and combine.

$$\begin{array}{r} 2x^2 + 5x - 7 \\ (+)\ -4x^2 + 6x + 3 \\ \hline -2x^2 + 11x - 4 \end{array}$$

b. $(3y + y^3 - 5) + (4y^2 - 4y + 2y^3 + 8)$

Horizontal Method

$(3y + y^3 - 5) + (4y^2 - 4y + 2y^3 + 8)$

$= [y^3 + 2y^3] + 4y^2 + [3y + (-4y)] + [(-5) + 8]$ — Group like terms.

$= 3y^3 + 4y^2 - y + 3$ — Combine like terms.

Vertical Method

$$\begin{array}{r} y^3 + 0y^2 + 3y - 5 \\ (+)\ 2y^3 + 4y^2 - 4y + 8 \\ \hline 3y^3 + 4y^2 - y + 3 \end{array}$$

Insert a placeholder to help align the terms.
Align and combine like terms.

Check Your Progress

1A. Find $(5x^2 - 3x + 4) + (6x - 3x^2 - 3)$.

1B. Find $(y^4 - 3y + 7) + (2y^3 + 2y - 2y^4 - 11)$.

Personal Tutor glencoe.com

StudyTip

Additive Inverse When finding the additive inverse of a polynomial, you are multiplying every term by -1.

Subtract Polynomials Recall that you can subtract a real number by adding its opposite or additive inverse. Similarly, you can subtract a polynomial by adding its additive inverse.

To find the additive inverse of a polynomial, write the opposite of each term in the polynomial.

$$-(3x^2 + 2x - 6) = \underbrace{-3x^2 - 2x + 6}_{\text{Additive Inverse}}$$

EXAMPLE 2 Subtract Polynomials

Find each difference.

a. $(3 - 2x + 2x^2) - (4x - 5 + 3x^2)$

Horizontal Method

Subtract $4x - 5 + 3x^2$ by adding its additive inverse.

$(3 - 2x + 2x^2) - (4x - 5 + 3x^2)$

$= (3 - 2x + 2x^2) + (-4x + 5 - 3x^2)$ — **The additive inverse of $4x - 5 + 3x^2$ is $-4x + 5 - 3x^2$.**

$= [2x^2 + (-3x^2)] + [(-2x) + (-4x)] + [3 + 5]$ — **Group like terms.**

$= -x^2 - 6x + 8$ — **Combine like terms.**

Vertical Method

Align like terms in columns and subtract by adding the additive inverse.

$$\begin{array}{rr} & 2x^2 - 2x + 3 \\ (-) & 3x^2 + 4x - 5 \\ \hline \end{array} \quad \textbf{Add the opposite.} \quad \begin{array}{rr} & 2x^2 - 2x + 3 \\ (+) & -3x^2 - 4x + 5 \\ \hline & -x^2 - 6x + 8 \end{array}$$

Thus, $(3 - 2x + 2x^2) - (4x - 5 + 3x^2) = -x^2 - 6x + 8$.

StudyTip

Vertical Method Notice that the polynomials are written in standard form with like terms aligned.

b. $(7p + 4p^3 - 8) - (3p^2 + 2 - 9p)$

Horizontal Method

Subtract $3p^2 + 2 - 9p$ by adding its additive inverse.

$(7p + 4p^3 - 8) - (3p^2 + 2 - 9p)$

$= (7p + 4p^3 - 8) + (-3p^2 - 2 + 9p)$ — **The additive inverse of $3p^2 + 2 - 9p$ is $-3p^2 - 2 + 9p$.**

$= [7p + 9p] + 4p^3 + (-3p^2) + [(-8) + (-2)]$ — **Group like terms.**

$= 4p^3 - 3p^2 + 16p - 10$ — **Combine like terms.**

Vertical Method

Align like terms in columns and subtract by adding the additive inverse.

$$\begin{array}{rr} & 4p^3 + 0p^2 + 7p - 8 \\ (-) & 3p^2 - 9p + 2 \\ \hline \end{array} \quad \textbf{Add the opposite.} \quad \begin{array}{rr} & 4p^3 + 0p^2 + 7p - 8 \\ (+) & -3p^2 + 9p - 2 \\ \hline & 4p^3 - 3p^2 + 16p - 10 \end{array}$$

Thus, $(7p + 4p^3 - 8) - (3p^2 + 2 - 9p) = 4p^3 - 3p^2 + 16p - 10$.

Check Your Progress

2A. Find $(4x^3 - 3x^2 + 6x - 4) - (-2x^3 + x^2 - 2)$.

2B. Find $(8y - 10 + 5y^2) - (7 - y^3 + 12y)$.

Personal Tutor glencoe.com

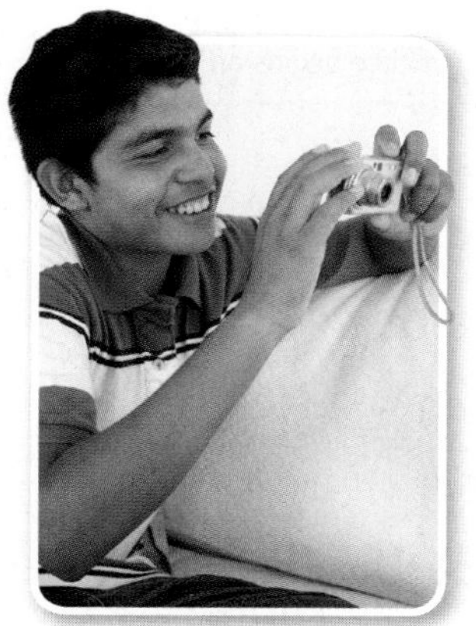

Real-World Link

Sales of digital cameras recently increased by 42% in one year. Sales are expected to increase by at least 15% each year as consumers upgrade their cameras.

Source: Big Planet Marketing Company

Real-World EXAMPLE 3 Add and Subtract Polynomials

CONSUMER ELECTRONICS An electronics store is starting to track sales of cell phones and digital cameras. The equations below represent the number of cell phones P and the number of digital cameras C sold in m months.

$$P = 7m + 137 \qquad C = 4m + 78$$

a. Write an equation for the monthly sales T of phones and cameras.

Add the polynomial for P with the polynomial for C.

total sales = cell phone sales + digital camera sales

$T = 7m + 137 + 4m + 78$ **Substitution**

$= 11m + 215$ **Combine like terms.**

An equation is $T = 11m + 215$.

b. Use the equation to predict the number of cell phones and digital cameras sold in 10 months.

$T = 11(10) + 215$ **Substitute 10 for m.**

$= 110 + 215$ **Simplify.**

$= 325$

Thus, a total of 325 cell phones and digital cameras will be sold in 10 months.

Check Your Progress

3. Use the information above to write an equation that represents the difference in the monthly sales of cell phones and the monthly sales of digital cameras. Use the equation to predict the difference in monthly sales in 24 months.

Personal Tutor glencoe.com

Check Your Understanding

Examples 1 and 2 pp. 433–434

Find each sum or difference.

1. $(6x^3 - 4) + (-2x^3 + 9)$

2. $(g^3 - 2g^2 + 5g + 6) - (g^2 + 2g)$

3. $(4 + 2a^2 - 2a) - (3a^2 - 8a + 7)$

4. $(8y - 4y^2) + (3y - 9y^2)$

5. $(-4z^3 - 2z + 8) - (4z^3 + 3z^2 - 5)$

6. $(-3d^2 - 8 + 2d) + (4d - 12 + d^2)$

7 $(2c^2 + 6c + 4) + (5c^2 - 7)$

8. $(3n^3 - 5n + n^2) - (-8n^2 + 3n^3)$

Example 3 p. 435

9. VACATION The total number of students T who traveled for spring break consists of two groups: students who flew to their destinations F and students who drove to their destination D. The number (in thousands) of students who flew and the total number of students who flew or drove can be modeled by the following equations, where n is the number of years since 1995.

$$T = 14n + 21 \qquad F = 8n + 7$$

a. Write an equation that models the number of students who drove to their destination for this time period.

b. Predict the number of students who will drive to their destination in 2012.

c. How many students will drive or fly to their destination in 2015?

Practice and Problem Solving

● = Step-by-Step Solutions begin on page R12. Extra Practice begins on page 815.

Examples 1 and 2 pp. 433–434

Find each sum or difference.

10. $(y + 5) + (2y + 4y^2 - 2)$
11. $(2x + 3x^2) - (7 - 8x^2)$
12. $(3c^3 - c + 11) - (c^2 + 2c + 8)$
13. $(z^2 + z) + (z^2 - 11)$
14. $(2x - 2y + 1) - (3y + 4x)$
15. $(4a - 5b^2 + 3) + (6 - 2a + 3b^2)$
16. $(x^2y - 3x^2 + y) + (3y - 2x^2y)$
17. $(-8xy + 3x^2 - 5y) + (4x^2 - 2y + 6xy)$
18. $(5n - 2p^2 + 2np) - (4p^2 + 4n)$
19. $(4rxt - 8r^2x + x^2) - (6rx^2 + 5rxt - 2x^2)$
20. $(6ab^2 + 2ab) + (3a^2b - 4ab + ab^2)$
21. $(cd^2 + 2cd - 4) + (-6 + 4cd - 2cd^2)$

Example 3 p. 435

22. **PETS** From 1997 through 2007, the number of dogs D and the number of cats C (in hundreds) adopted from animal shelters in the United States are modeled by the following equations, where n is the number of years since 1997.

$$D = 2n + 3 \qquad C = n + 4$$

   a. Write an equation that models the total number T of dogs and cats adopted in hundreds for this time period.

   b. If this trend continues, how many dogs and cats will be adopted in 2011?

Find each sum or difference.

23. $(4x + 2y - 6z) + (5y - 2z + 7x) + (-9z - 2x - 3y)$
24. $(5a^2 - 4) + (a^2 - 2a + 12) + (4a^2 - 6a + 8)$
25. $(3c^2 - 7) + (4c + 7) - (c^2 + 5c - 8)$
26. $(3n^3 + 3n - 10) - (4n^2 - 5n) + (4n^3 - 3n^2 - 9n + 4)$

27. **GEOMETRY** Write a polynomial that represents the perimeter of the figure at the right.

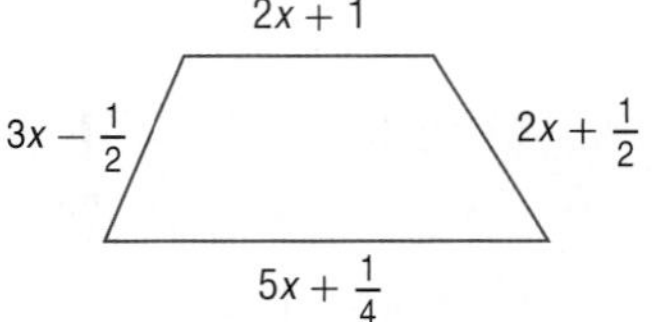

28. **PAINTING** Kin is painting two walls of her bedroom. The area of one wall can be modeled by $3x^2 + 14$, and the area of the other wall can be modeled by $2x - 3$. What is the total area of the two walls?

29. **GEOMETRY** The perimeter of the figure at the right is represented by the expression $3x^2 - 7x + 2$. Write a polynomial that represents the measure of the third side.

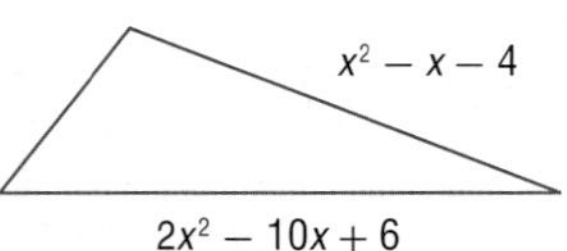

Real-World Link

The Pro Football Hall of Fame located in Canton, Ohio, is approximately 83,000 square feet and has had more than eight million visitors.

Source: Pro Football Hall of Fame

30. **FOOTBALL** The National Football League is divided into two conferences, the American A and the National N. From 1996 through 2004, the total attendance T (in thousands) for both conferences and for the American Conference games are modeled by the following equations, where y is the number of years since 1996.

$$T = 35y^3 + 27y^2 + 1899 \qquad A = 16y^3 + 13y^2 + 2y + 905$$

Determine how many people attended a National Conference football game in 2002.

31. **GEOMETRY** The width of a rectangle is represented by $5x + 2y$, and the length is represented by $6y - 2x$. Write a polynomial that represents the perimeter.

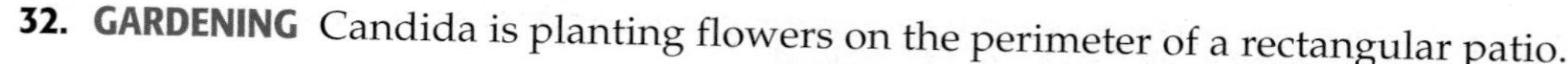

32. GARDENING Candida is planting flowers on the perimeter of a rectangular patio.

a. If the perimeter of the patio is $210x$ and one side measures $32x$, find the length of the other side.

b. Write a polynomial that represents the area of the rectangular patio.

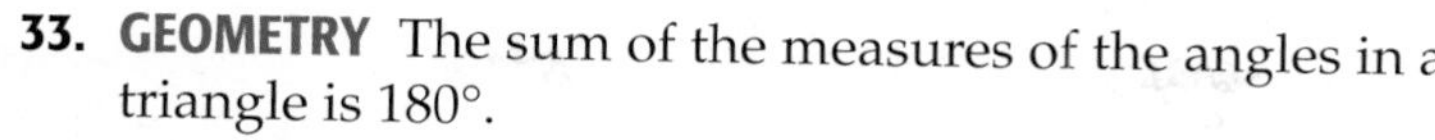

33. GEOMETRY The sum of the measures of the angles in a triangle is 180°.

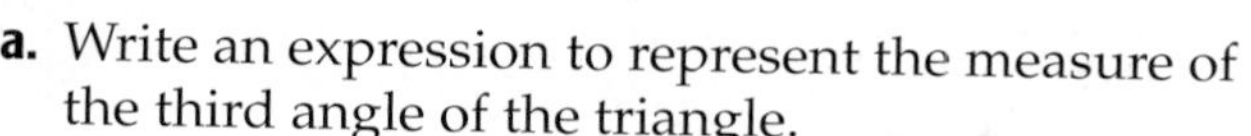

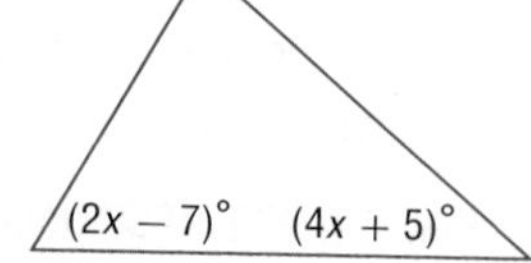

a. Write an expression to represent the measure of the third angle of the triangle.

b. If $x = 23$, find the measures of the three angles.

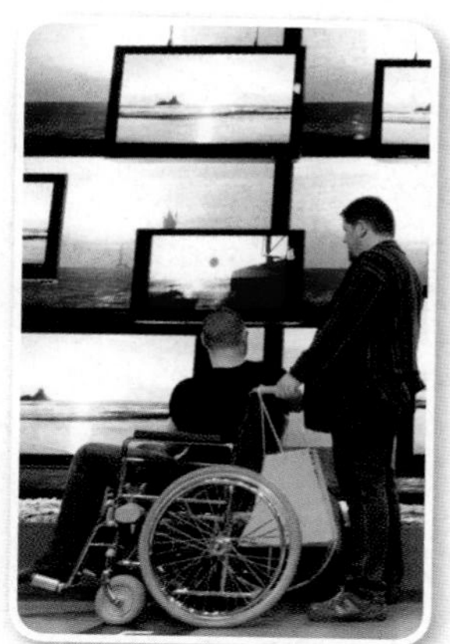

Real-World Link

On average, an LCD television lasts about 60,000 hours. This means the lifespan of an LCD television can be 20 years or more if the television is used less than 8 hours per day.

Source: LCD TV Buying Guide

34. SALES An electronics store estimates that the cost, in dollars, of selling t units of LCD televisions is given by the expression $0.002t^2 + 4t + 400$. The revenue from the sales of t LCD televisions is $8t$.

a. Write a polynomial that represents the profit of selling t units.

b. If 750 LCD televisions are sold, how much did the store earn?

c. If 575 LCD televisions are sold, how much did the store earn?

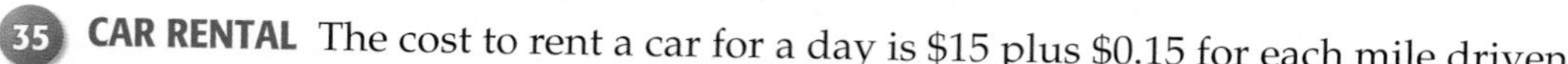

35. CAR RENTAL The cost to rent a car for a day is \$15 plus \$0.15 for each mile driven.

a. Write a polynomial that represents the cost of renting a car for m miles.

b. If a car is driven 145 miles, how much would it cost to rent?

c. If a car is driven 105 miles each day for four days, how much would it cost to rent a car?

d. If a car is driven 220 miles each day for seven days, how much would it cost to rent a car?

H.O.T. Problems

Use Higher-Order Thinking Skills

36. FIND THE ERROR Cheyenne and Sebastian are finding $(2x^2 - x) - (3x + 3x^2 - 2)$. Is either of them correct? Explain your reasoning.

Cheyenne	Sebastian
$(2x^2 - x) - (3x + 3x^2 - 2)$	$(2x^2 - x) - (3x + 3x^2 - 2)$
$= (2x^2 - x) + (-3x + 3x^2 - 2)$	$= (2x^2 - x) + (-3x - 3x^2 - 2)$
$= 5x^2 - 4x - 2$	$= -x^2 - 4x - 2$

37. OPEN ENDED Write two trinomials with a difference of $2x^3 - 7x + 8$.

38. CHALLENGE Write a polynomial that represents the sum of an odd integer $2n + 1$ and the next two consecutive odd integers.

39. REASONING Find a counterexample to the following statement.

The order in which polynomials are subtracted does not matter.

40. OPEN ENDED Write three trinomials with a sum of $4x^4 + 3x^2$.

41. WRITING IN MATH Describe how to add and subtract polynomials using both the vertical and horizontal formats. Which one do you think is easier? Why?

Standardized Test Practice

42. Three consecutive integers can be represented by x, $x + 1$, and $x + 2$. What is the sum of these three integers?

A $x(x + 1)(x + 2)$ **C** $3x + 3$

B $x^3 + 3$ **D** $x + 3$

43. SHORT RESPONSE What is the perimeter of a square with sides that measure $2x + 3$ units?

44. Jim cuts a board in the shape of a regular hexagon and pounds in a nail at each vertex, as shown. How many rubber bands will he need to stretch a rubber band across every possible pair of nails?

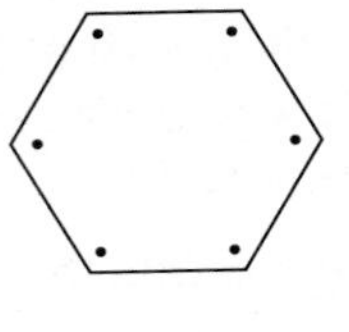

F 15 **G** 12 **H** 14 **J** 9

45. Which ordered pair is in the solution set of the system of inequalities shown in the graph?

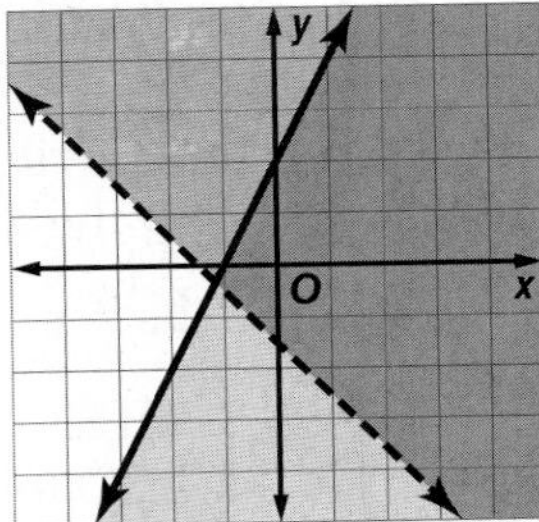

A $(-3, 0)$ **C** $(5, 0)$

B $(0, -3)$ **D** $(0, 5)$

Spiral Review

Find the degree of each polynomial. (Lesson 7-4)

46. $6b^4$

47. $10t$

48. $5g^2h$

49. $7np^4$

50. 25

51. $t^3 + 6u$

52. $2 + 3ab^3 - a^2b + 4a^6$

53. $6 - v^4 + v^2z^3 + 6v^3$

54. POPULATION The 2005 population of North Carolina's Beaufort County was approximately 46,000. Express this number in scientific notation. (Lesson 7-3)

55. JOBS Kimi received an offer for a new job. She wants to compare the offer with her current job. What is total amount of sales that Kimi must get each month to make the same income at either job? (Lesson 6-2)

New Offer
\$600/mo 2% commission

Current Job
\$1000/mo 1.5% commission

Determine whether each sequence is an arithmetic sequence. If it is, state the common difference. (Lesson 3-5)

56. 24, 16, 8, 0, …

57. $3\frac{1}{4}$, $6\frac{1}{2}$, 13, 26, …

58. 7, 6, 5, 4, …

59. 10, 12, 15, 18, …

60. −15, −11, −7, −3, …

61. −0.3, 0.2, 0.7, 1.2, …

Skills Review

Simplify. (Lesson 7-1)

62. $t(t^5)(t^7)$

63. $n^3(n^2)(-2n^3)$

64. $(5t^5v^2)(10t^3v^4)$

65. $(-8u^4z^5)(5uz^4)$

66. $[(3)^2]^3$

67. $[(2)^3]^2$

68. $(2m^4k^3)^2(-3mk^2)^3$

69. $(6xy^2)^2(2x^2y^2z^2)^3$

7-6 Multiplying a Polynomial by a Monomial

Then

You multiplied monomials. (Lesson 7-1)

Now

- Multiply a polynomial by a monomial.
- Solve equations involving the products of monomials and polynomials.

Math Online

glencoe.com

- Extra Examples
- Personal Tutor
- Self-Check Quiz
- Homework Help

Why?

Charmaine Brooks is opening a fitness club. She tells the contractor that the length of the fitness room should be three times the width plus 8 feet.

To cover the floor with mats for exercise classes, Ms. Brooks needs to know the area of the floor. So she multiplies the width times the length, $w(3w + 8)$.

Polynomial Multiplied by Monomial To find the product of a polynomial and a monomial, you can use the Distributive Property.

EXAMPLE 1 Multiply a Polynomial by a Monomial

Find $-3x^2(7x^2 - x + 4)$.

Horizontal Method

$$\begin{aligned} &-3x^2(7x^2 - x + 4) && \textbf{Original expression} \\ &= -3x^2(7x^2) - (-3x^2)(x) + (-3x^2)(4) && \textbf{Distributive Property} \\ &= -21x^4 - (-3x^3) + (-12x^2) && \textbf{Multiply.} \\ &= -21x^4 + 3x^3 - 12x^2 && \textbf{Simplify.} \end{aligned}$$

Vertical Method

$$\begin{array}{rl} 7x^2 - x + 4 & \\ (\times) \qquad\quad -3x^2 & \textbf{Distributive Property} \\ \hline -21x^4 + 3x^3 - 12x^2 & \textbf{Multiply.} \end{array}$$

Check Your Progress

Find each product.

1A. $5a^2(-4a^2 + 2a - 7)$

1B. $-6d^3(3d^4 - 2d^3 - d + 9)$

Personal Tutor glencoe.com

We can use this same method more than once to simplify large expressions.

EXAMPLE 2 Simplify Expressions

Simplify $2p(-4p^2 + 5p) - 5(2p^2 + 20)$.

$$\begin{aligned} &2p(-4p^2 + 5p) - 5(2p^2 + 20) && \textbf{Original expression} \\ &= (2p)(-4p^2) + (2p)(5p) + (-5)(2p^2) + (-5)(20) && \textbf{Distributive Property} \\ &= -8p^3 + 10p^2 - 10p^2 - 100 && \textbf{Multiply.} \\ &= -8p^3 + (10p^2 - 10p^2) - 100 && \textbf{Commutative and Associative Properties} \\ &= -8p^3 - 100 && \textbf{Combine like terms.} \end{aligned}$$

Check Your Progress

Simplify each expression.

2A. $3(5x^2 + 2x - 4) - x(7x^2 + 2x - 3)$ **2B.** $15t(10y^3t^5 + 5y^2t) - 2y(yt^2 + 4y^2)$

Personal Tutor glencoe.com

We can use the Distributive Property to multiply monomials by polynomials and solve real world problems.

Test-TakingTip

Formulas Many standardized tests provide formula sheets with commonly used formulas. If you are unsure of the correct formula, check the sheet before beginning to solve the problem.

STANDARDIZED TEST EXAMPLE 3

GRIDDED RESPONSE The theme for a school dance is "Solid Gold." For one decoration, Kana is covering a trapezoid-shaped piece of poster board with metallic gold paper to look like a bar of gold. If the height of the poster board is 18 inches, how much metallic paper will Kana need in square inches?

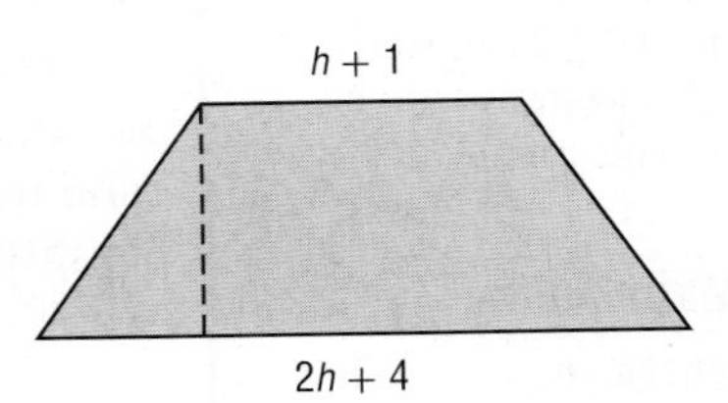

Read the Test Item

The question is asking you to find the area of the trapezoid with a height of h and bases of $h + 1$ and $2h + 4$.

Solve the Test Item

Write an equation to represent the area of the trapezoid.
Let $b_1 = h + 1$, let $b_2 = 2h + 4$ and let h = height of the trapezoid.

$A = \frac{1}{2}h(b_1 + b_2)$	**Area of a trapezoid**
$= \frac{1}{2}h[(h + 1) + (2h + 4)]$	$b_1 = h + 1$ **and** $b_2 = 2h + 4$
$= \frac{1}{2}h(3h + 5)$	**Add and simplify.**
$= \frac{3}{2}h^2 + \frac{5}{2}h$	**Distributive Property**
$= \frac{3}{2}(18)^2 + \frac{5}{2}(18)$	$h = 18$
$= 531$	**Simplify.**

5	3	1		

Kana will need 531 square inches of metallic paper. Grid in your response of 531.

Real-World Link

In a recent year, the pet supply business hit an estimated $7.05 billion in sales. This business ranges from gourmet food to rhinestone tiaras, pearl collars, and cashmere coats.

Source: *Entrepreneur Magazine*

Check Your Progress

3. Kachima is making triangular bandanas for the dogs and cats in her pet club. The base of the bandana is the length of the collar with 4 inches added to each end to tie it on. The height is $\frac{1}{2}$ of the collar length.

A. If Kachima's dog has a collar length of 12 inches, how much fabric does she need in square inches?

B. If Kachima makes a bandana for her friend's cat with a 6-inch collar, how much fabric does Kachima need in square inches?

Personal Tutor glencoe.com

StudyTip

Combining Like Terms When simplifying a long expression, it may be helpful to put a circle around one set of like terms, a rectangle around another set, a triangle around another set, and so on.

Solve Equations with Polynomial Expressions We can use the Distributive Property to solve equations that involve the products of monomials and polynomials.

EXAMPLE 4 Equations with Polynomials on Both Sides

Solve $2a(5a - 2) + 3a(2a + 6) + 8 = a(4a + 1) + 2a(6a - 4) + 50$.

$2a(5a - 2) + 3a(2a + 6) + 8 = a(4a + 1) + 2a(6a - 4) + 50$	**Original equation**
$10a^2 - 4a + 6a^2 + 18a + 8 = 4a^2 + a + 12a^2 - 8a + 50$	**Distributive Property**
$16a^2 + 14a + 8 = 16a^2 - 7a + 50$	**Combine like terms.**
$14a + 8 = -7a + 50$	**Subtract $16a^2$ from each side.**
$21a + 8 = 50$	**Add $7a$ to each side.**
$21a = 42$	**Subtract 8 from each side.**
$a = 2$	**Divide each side by 21.**

CHECK

$2a(5a - 2) + 3a(2a + 6) + 8 = a(4a + 1) + 2a(6a - 4) + 50$	
$2(2)[5(2) - 2] + 3(2)[2(2) + 6] + 8 \stackrel{?}{=} 2[4(2) + 1] + 2(2)[6(2) - 4] + 50$	
$4(8) + 6(10) + 8 \stackrel{?}{=} 2(9) + 4(8) + 50$	**Simplify.**
$32 + 60 + 8 \stackrel{?}{=} 18 + 32 + 50$	**Multiply.**
$100 = 100$ ✓	**Add and subtract.**

Solve each equation.

4A. $2x(x + 4) + 7 = (x + 8) + 2x(x + 1) + 12$

4B. $d(d + 3) - d(d - 4) = 9d - 16$

Personal Tutor glencoe.com

Check Your Understanding

Example 1 p. 439

Find each product.

1. $5w(-3w^2 + 2w - 4)$

2. $6g^2(3g^3 + 4g^2 + 10g - 1)$

3. $4km^2(8km^2 + 2k^2m + 5k)$

4. $-3p^4r^3(2p^2r^4 - 6p^6r^3 - 5)$

5 $2ab(7a^4b^2 + a^5b - 2a)$

6. $c^2d^3(5cd^7 - 3c^3d^2 - 4d^3)$

Example 2 p. 439

Simplify each expression.

7. $t(4t^2 + 15t + 4) - 4(3t - 1)$

8. $x(3x^2 + 4) + 2(7x - 3)$

9. $-2d(d^3c^2 - 4dc^2 + 2d^2c) + c^2(dc^2 - 3d^4)$

10. $-5w^2(8w^2x - 11wx^2) + 6x(9wx^4 - 4w - 3x^2)$

Example 3 p. 440

11. GRIDDED RESPONSE Marlene is buying a new plasma television. The height of the screen of the television is one half the width plus 5 inches. The width is 30 inches. Find the height of the screen in inches.

Example 4 p. 441

Solve each equation.

12. $-6(11 - 2c) = 7(-2 - 2c)$

13. $t(2t + 3) + 20 = 2t(t - 3)$

14. $-2(w + 1) + w = 7 - 4w$

15. $3(y - 2) + 2y = 4y + 14$

16. $a(a + 3) + a(a - 6) + 35 = a(a - 5) + a(a + 7)$

17. $n(n - 4) + n(n + 8) = n(n - 13) + n(n + 1) + 16$

Practice and Problem Solving

● = Step-by-Step Solutions begin on page R12.
Extra Practice begins on page 815.

Example 1 p. 439

Find each product.

18. $b(b^2 - 12b + 1)$

19. $f(f^2 + 2f + 25)$

20. $-3m^3(2m^3 - 12m^2 + 2m + 25)$

21. $2j^2(5j^3 - 15j^2 + 2j + 2)$

22. $2pr^2(2pr + 5p^2r - 15p)$

23. $4t^3u(2t^2u^2 - 10tu^4 + 2)$

Example 2 p. 439

Simplify each expression.

24. $-3(5x^2 + 2x + 9) + x(2x - 3)$

25. $a(-8a^2 + 2a + 4) + 3(6a^2 - 4)$

26. $-4d(5d^2 - 12) + 7(d + 5)$

27. $-9g(-2g + g^2) + 3(g^2 + 4)$

28. $2j(7j^2k^2 + jk^2 + 5k) - 9k(-2j^2k^2 + 2k^2 + 3j)$

29. $4n(2n^3p^2 - 3np^2 + 5n) + 4p(6n^2p - 2np^2 + 3p)$

Example 3 p. 440

30. **DAMS** A new dam being built has the shape of a trapezoid. The base at the bottom of the dam is 2 times the height. The base at the top of the dam is $\frac{1}{5}$ times the height minus 30 feet.

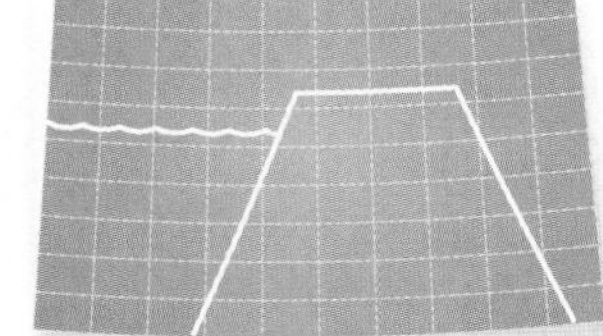

   a. Write an expression to find the area of the trapezoidal cross section of the dam.

   b. If the height of the dam is 180 feet, find the area of this cross section.

Example 4 p. 441

Solve each equation.

31. $7(t^2 + 5t - 9) + t = t(7t - 2) + 13$

32. $w(4w + 6) + 2w = 2(2w^2 + 7w - 3)$

33. $5(4z + 6) - 2(z - 4) = 7z(z + 4) - z(7z - 2) - 48$

34. $9c(c - 11) + 10(5c - 3) = 3c(c + 5) + c(6c - 3) - 30$

35. $2f(5f - 2) - 10(f^2 - 3f + 6) = -8f(f + 4) + 4(2f^2 - 7f)$

36. $2k(-3k + 4) + 6(k^2 + 10) = k(4k + 8) - 2k(2k + 5)$

Simplify each expression.

37. $\frac{2}{3}np^2(30p^2 + 9n^2p - 12)$

38. $\frac{3}{5}r^2t(10r^3 + 5rt^3 + 15t^2)$

39. $-5q^2w^3(4q + 7w) + 4qw^2(7q^2w + 2q) - 3qw(3q^2w^2 + 9)$

40. $-x^2z(2z^2 + 4xz^3) + xz^2(xz + 5x^3z) + x^2z^3(3x^2z + 4xz)$

41. **PARKING** A parking garage charges \$30 per month plus \$0.50 per daytime hour and \$0.25 per hour during nights and weekends. Suppose Trent parks in the garage for 47 hours in January and h of those are night and weekend hours.

   a. Find an expression for Trent's January bill.

   b. Find the cost if Trent had 12 hours of night and weekend hours.

42. **PETS** Che is building a dog house for his new puppy. The upper face of the dog house is a trapezoid. If the height of the trapezoid is 12 inches, find the area of the face of this piece of the dog house.

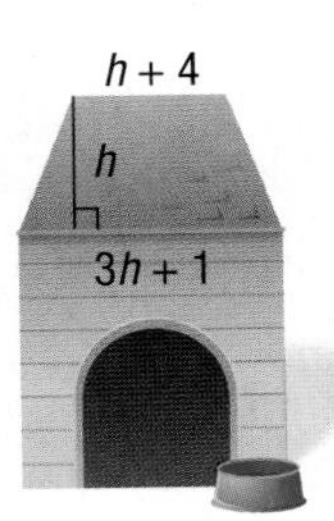

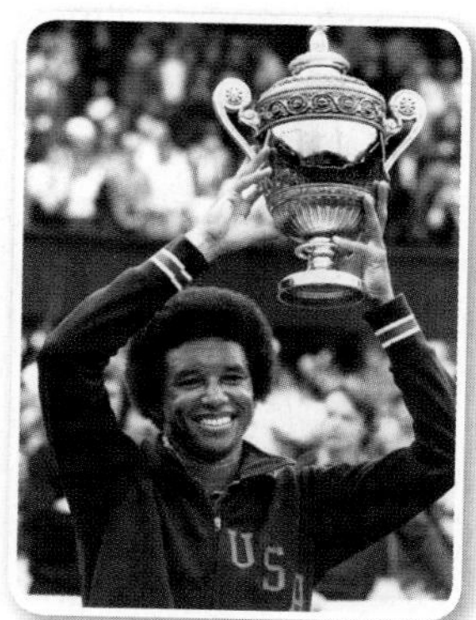

Real-World Link

Arthur Ashe is the only African-American male tennis player to win both Wimbledon and the U.S. Open and the first and only African-American player to be ranked the #1 tennis player in the world.

Source: CMG Worldwide

43. **TENNIS** The tennis club is building a new tennis court with a path around it.

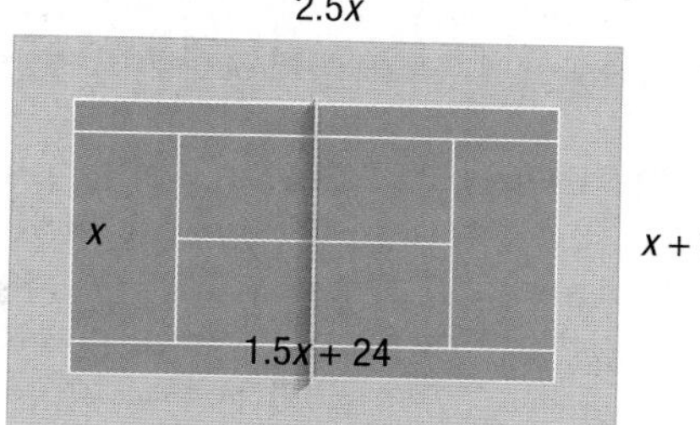

a. Write an expression for the area of the tennis court.

b. Write an expression for the area of the path.

c. Every three feet around the court, there will be a stepping stone placed in the path. If $x = 36$, how many stones will there be?

44. **MULTIPLE REPRESENTATIONS** In this problem, you will investigate the degree of the product of a monomial and a polynomial.

a. TABULAR Write three monomials of different degrees and three polynomials of different degrees. Determine the degree of each monomial and polynomial. Multiply the monomials by the polynomials. Determine the degree of each product. Record your results in a table like the one shown below.

Monomial	Degree	Polynomial	Degree	Product of Monomial and Polynomial	Degree

b. VERBAL Make a conjecture about the degree of the product of a monomial and a polynomial. What is the degree of the product of a monomial of degree a and a polynomial of degree b?

H.O.T. Problems

Use **H**igher-**O**rder **T**hinking Skills

45. **FIND THE ERROR** Pearl and Ted both worked on this problem. Is either of them correct? Explain your reasoning.

Pearl	Ted
$2x^2(3x^2 + 4x + 2)$	$2x^2(3x^2 + 4x + 2)$
$6x^4 + 8x^2 + 4x^2$	$6x^4 + 8x^3 + 4x^2$
$6x^4 + 12x^2$	

46. **CHALLENGE** Find p such that $3x^p(4x^{2p+3} + 2x^{3p-2}) = 12x^{12} + 6x^{10}$.

47. **CHALLENGE** Simplify $4x^{-3}y^2(2x^5y^{-4} + 6x^{-7}y^6 - 4x^0y^{-2})$.

48. **REASONING** Is there a value for x that makes the statement $(x + 2)^2 = x^2 + 2^2$ true? If so, find a value for x. Explain your reasoning.

49. **OPEN ENDED** Write a monomial and a polynomial using n as the variable. Find their product.

50. **WRITING IN MATH** Describe the steps to multiply a polynomial by a monomial.

Standardized Test Practice

51. Every week a store sells j jeans and t T-shirts. The store makes \$8 for each T-shirt and \$12 for each pair of jeans. Which of the following expressions represents the total amount of money, in dollars, the store makes every week?

A $8j + 12t$
B $12j + 8t$
C $20(j + t)$
D $96jt$

52. If $a = 5x + 7y$ and $b = 2y - 3x$, what is $a + b$?

F $2x - 9y$
G $3y + 4x$
H $2x + 9y$
J $2x - 5y$

53. GEOMETRY A triangle has sides of length 5 inches and 8.5 inches. Which of the following cannot be the length of the third side?

A 3.5 inches
B 4 inches
C 5.5 inches
D 12 inches

54. SHORT RESPONSE Write an equation in which x varies directly as the cube of y and inversely as the square of z.

Spiral Review

Find each sum or difference. (Lesson 7-5)

55. $(2x^2 - 7) + (8 - 5x^2)$
56. $(3z^2 + 2z - 1) + (z^2 - 6)$
57. $(2a - 4a^2 + 1) - (5a^2 - 2a - 6)$
58. $(a^3 - 3a^2 + 4) - (4a^2 + 7)$
59. $(2ab - 3a + 4b) + (5a + 4ab)$
60. $(8c^3 - 3c^2 + c - 2) - (3c^3 + 9)$

Find the degree of each polynomial. (Lesson 7-4)

61. $12y$
62. -10
63. $2x^2 - 5$
64. $9a - 8a^3 + 6$
65. $7b^2c^3$
66. $-3p^4r^5t^2$

67. TRAVEL In 1990, about 3.6 million people took cruises. Between 1990 and 2000, the number increased by about 300,000 each year. Write the point-slope form of an equation to find the total number of people y taking a cruise for any year x. Estimate the number of people who will take a cruise in 2010. (Lesson 4-3)

Write an equation in function notation for each relation. (Lesson 3-6)

68.

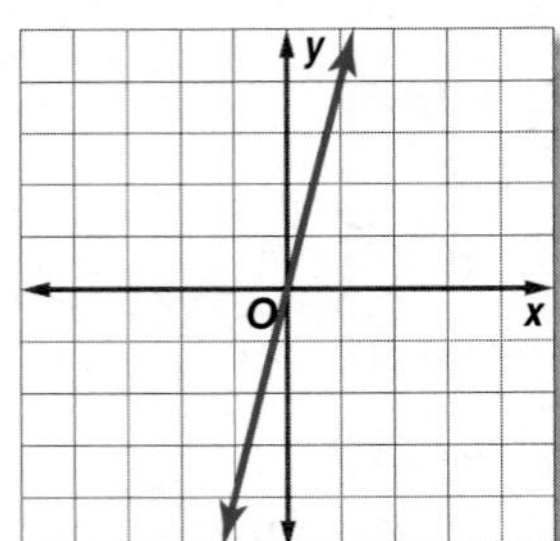

69.

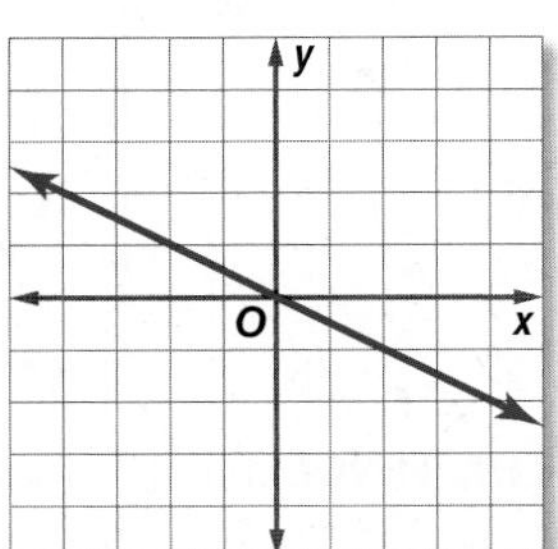

Skills Review

Simplify. (Lesson 7-1)

70. $b(b^2)(b^3)$
71. $2y(3y^2)$
72. $-y^4(-2y^3)$
73. $-3z^3(-5z^4 + 2z)$
74. $2m(-4m^4) - 3(-5m^3)$
75. $4p^2(-2p^3) + 2p^4(5p^6)$

EXPLORE 7-7

Algebra Lab: Multiplying Polynomials

Math Online glencoe.com
Math in Motion, Animation

You can use algebra tiles to find the product of two binomials.

ACTIVITY 1 Multiply Binomials

Use algebra tiles to find $(x + 3)(x + 4)$.

The rectangle will have a width of $x + 3$ and a length of $x + 4$. Use algebra tiles to mark off the dimensions on a product mat. Then complete the rectangle with algebra tiles.

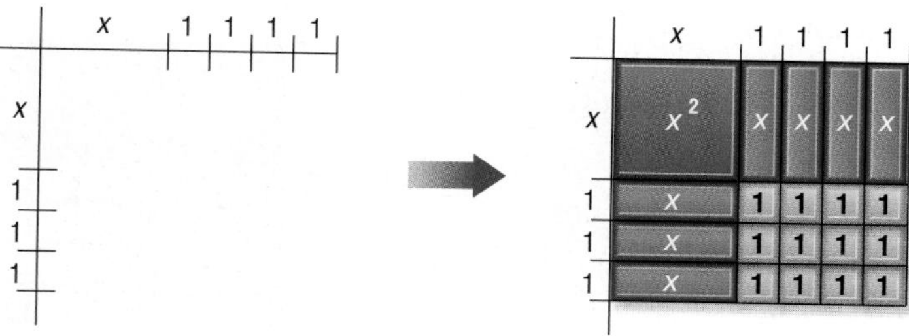

The rectangle consists of 1 blue x^2-tile, 7 green x-tiles, and 12 yellow 1-tiles. The area of the rectangle is $x^2 + 7x + 12$. So, $(x + 3)(x + 4) = x^2 + 7x + 12$.

ACTIVITY 2 Multiply Binomials

Use algebra tiles to find $(x - 2)(x - 5)$.

Step 1 The rectangle will have a width of $x - 2$ and a length of $x - 5$. Use algebra tiles to mark off the dimensions on a product mat. Then begin to make the rectangle with algebra tiles.

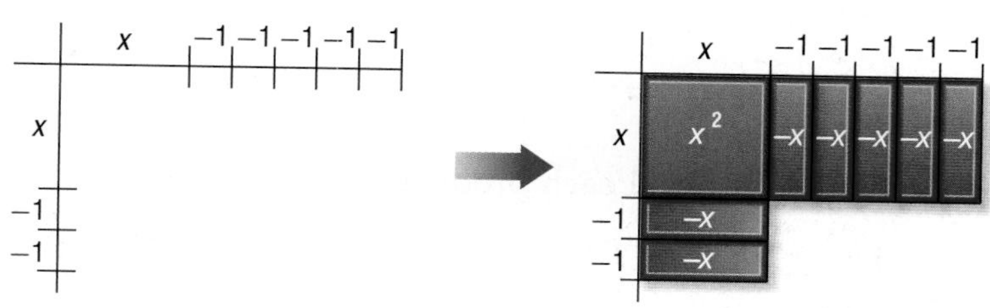

Step 2 Determine whether to use 10 yellow 1-tiles or 10 red -1-tiles to complete the rectangle. The area of each yellow tile is the product of -1 and -1. Fill in the space with 10 yellow 1-tiles to complete the rectangle.

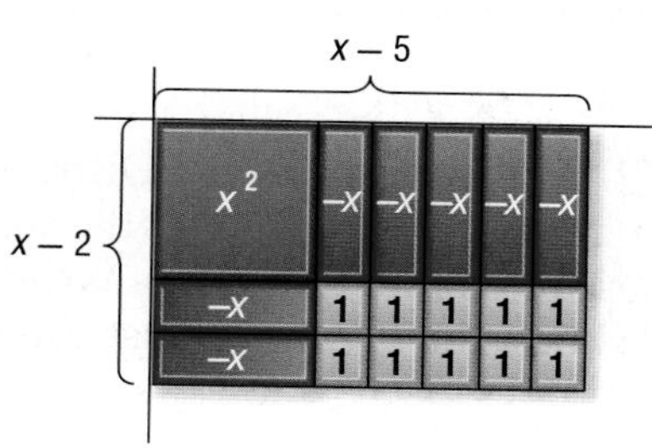

The rectangle consists of 1 blue x^2-tile, 7 red $-x$-tiles, and 10 yellow 1-tiles. The area of the rectangle is $x^2 - 7x + 10$. So, $(x - 2)(x - 5) = x^2 - 7x + 10$.

ACTIVITY 3 Multiply Binomials

Use algebra tiles to find $(x - 4)(2x + 3)$.

Step 1 The rectangle will have a width of $x - 4$ and a length of $2x + 3$. Use algebra tiles to mark off the dimensions on a product mat. Then begin to make the rectangle with algebra tiles.

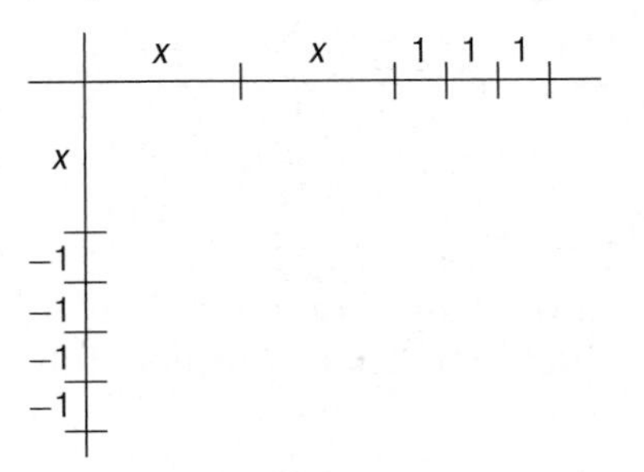

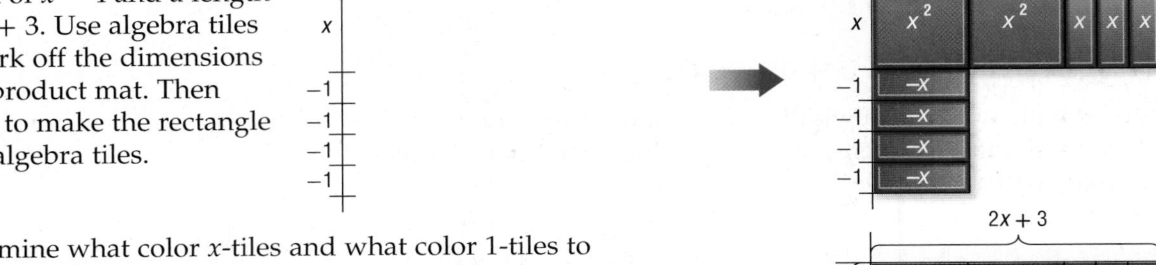

Step 2 Determine what color x-tiles and what color 1-tiles to use to complete the rectangle. The area of each red x-tile is the product of x and -1. The area of each red -1-tile is represented by the product of 1 and -1 or 1.

Complete the rectangle with 4 red x-tiles and 12 red -1-tiles.

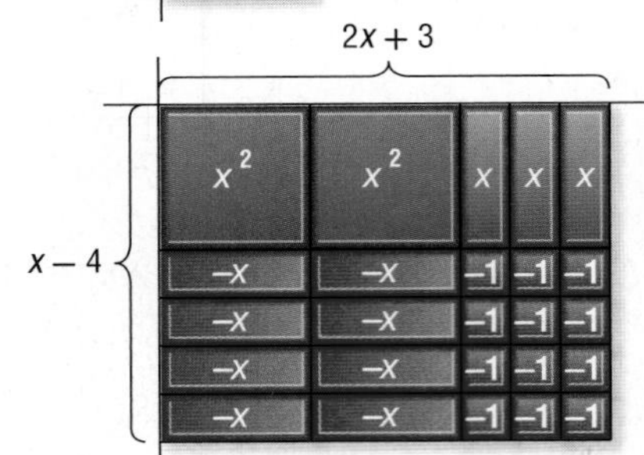

Step 3 Rearrange the tiles to simplify the polynomial you have formed. Notice that a 3 zero pair are formed by three positive and three negative x-tiles.

There are 2 blue x^2-tiles, 5 red $-x$-tiles, and 12 red -1-tiles left. In simplest form, $(x - 4)(2x + 3) = 2x^2 - 5x - 12$.

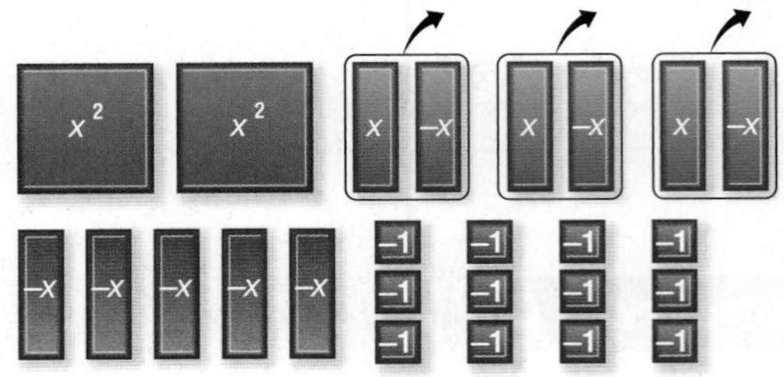

Model and Analyze

Use algebra tiles to find each product.

1. $(x + 1)(x + 4)$

2. $(x - 3)(x - 2)$

3. $(x + 5)(x - 1)$

4. $(x + 2)(2x + 3)$

5. $(x - 1)(2x - 1)$

6. $(x + 4)(2x - 5)$

Is each statement *true* or *false*? Justify your answer with a drawing of algebra tiles.

7. $(x - 4)(x - 2) = x^2 - 6x + 8$

8. $(x + 3)(x + 5) = x^2 + 15$

9. WRITING IN MATH You can also use the Distributive Property to find the product of two binomials. The figure at the right shows the model for $(x + 4)(x + 5)$ separated into four parts. Write a sentence or two explaining how this model shows the use of the Distributive Property.

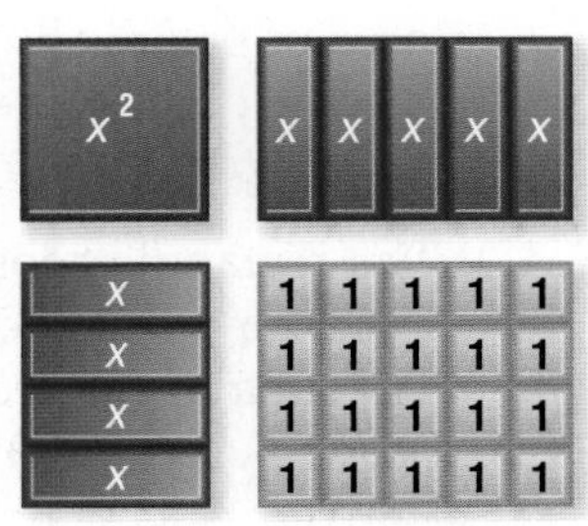

7-7

Multiplying Polynomials

Why?

Bodyboards, which are used to ride waves, are made of foam and are more rectangular than surfboards. A bodyboard's dimensions are determined by the height and skill level of the user.

The length of Ann's bodyboard should be Ann's height h minus 32 inches or $h - 32$. The board's width should be half of Ann's height plus 11 inches or $\frac{1}{2}h + 11$. To approximate the area of the bodyboard, you need to find $(h - 32)\left(\frac{1}{2}h + 11\right)$.

Then

You multiplied polynomials by monomials. (Lesson 7-6)

Now

- Multiply polynomials by using the Distributive Property.
- Multiply binomials by using the FOIL method.

New Vocabulary

FOIL method
quadratic expression

Math Online

- Extra Examples
- Personal Tutor
- Self-Check Quiz
- Homework Help
- Math in Motion

Multiply Binomials To multiply two binomials such as $h - 32$ and $\frac{1}{2}h + 11$, the Distributive Property is used. Binomials can be multiplied horizontally or vertically.

EXAMPLE 1 The Distributive Property

Find each product.

a. $(2x + 3)(x + 5)$

Vertical Method

Multiply by 5.	Multiply by x.	Combine like terms.
$\begin{array}{r} 2x + 3 \\ (\times)\ x + 5 \\ \hline 10x + 15 \end{array}$	$\begin{array}{r} 2x + 3 \\ (\times)\ x + 5 \\ \hline 10x + 15 \\ 2x^2 + 3x \\ \hline \end{array}$	$\begin{array}{r} 2x + 3 \\ (\times)\ x + 5 \\ \hline 10x + 15 \\ 2x^2 + 3x \\ \hline 2x^2 + 13x + 15 \end{array}$
$5(2x + 3) = 10x + 15$	$x(2x + 3) = 2x^2 + 3x$	

Horizontal Method

$$\begin{aligned} (2x + 3)(x + 5) &= 2x(x + 5) + 3(x + 5) && \textbf{Rewrite as the sum of two products.} \\ &= 2x^2 + 10x + 3x + 15 && \textbf{Distributive Property} \\ &= 2x^2 + 13x + 15 && \textbf{Combine like terms.} \end{aligned}$$

b. $(x - 2)(3x + 4)$

Vertical Method

Multiply by 4.	Multiply by $3x$.	Combine like terms.
$\begin{array}{r} x - 2 \\ (\times)\ 3x + 4 \\ \hline 4x - 8 \end{array}$	$\begin{array}{r} x - 2 \\ (\times)\ 3x + 4 \\ \hline 4x - 8 \\ 3x^2 - 6x \\ \hline \end{array}$	$\begin{array}{r} x - 2 \\ (\times)\ 3x + 4 \\ \hline 4x - 8 \\ 3x^2 - 6x \\ \hline 3x^2 - 2x - 8 \end{array}$
$4(x - 2) = 4x - 8$	$3x(x - 2) = 3x^2 - 6x$	

Horizontal Method

$$\begin{aligned} (x - 2)(3x + 4) &= x(3x + 4) - 2(3x + 4) && \textbf{Rewrite as the difference of two products.} \\ &= 3x^2 + 4x - 6x - 8 && \textbf{Distributive Property} \\ &= 3x^2 - 2x - 8 && \textbf{Combine like terms.} \end{aligned}$$

Check Your Progress

1A. $(3m + 4)(m + 5)$

1B. $(5y - 2)(y + 8)$

Personal Tutor glencoe.com

A shortcut version of the Distributive Property for multiplying binomials is called the **FOIL method**.

Key Concept

For Your FOLDABLE

FOIL Method

Words To multiply two binomials, find the sum of the products of **F** the *First* terms, **O** the *Outer* terms, **I** the *Inner* terms, **L** and the *Last* terms.

Example

		Product of First Terms		Product of Outer Terms		Product of Inner Terms		Product of Last Terms
$(x + 4)(x - 2)$	$=$	$(x)(x)$	$+$	$(x)(-2)$	$+$	$(4)(x)$	$+$	$(4)(-2)$

$$= x^2 - 2x + 4x - 8$$

$$= x^2 + 2x - 8$$

Math *in Motion*, Animation glencoe.com

ReadingMath

Polynomials as Factors

The expression $(x + 4)(x - 2)$ is read *the quantity x plus 4 times the quantity x minus 2.*

EXAMPLE 2 **FOIL Method**

Find each product.

a. $(2y - 7)(3y + 5)$

$(2y - 7)(3y + 5) = (2y)(3y) + (2y)(5) + (-7)(3y) + (-7)(5)$ **FOIL method**

$= 6y^2 + 10y - 21y - 35$ **Multiply.**

$= 6y^2 - 11y - 35$ **Combine like terms.**

b. $(4a - 5)(2a - 9)$

$(4a - 5)(2a - 9)$

$= (4a)(2a) + (4a)(-9) + (-5)(2a) + (-5)(-9)$ **FOIL method**

$= 8a^2 - 36a - 10a + 45$ **Multiply.**

$= 8a^2 - 46a + 45$ **Combine like terms.**

Check Your Progress

2A. $(x + 3)(x - 4)$

2B. $(4b - 5)(3b + 2)$

2C. $(2y - 5)(y - 6)$

2D. $(5a + 2)(3a - 4)$

Personal Tutor glencoe.com

Notice that when two linear expressions are multiplied, the result is a quadratic expression. A **quadratic expression** is an expression in one variable with a degree of 2. When three linear expressions are multiplied, the result has a degree of 3.

The FOIL method can be used to find an expression that represents the area of a rectangular object when the lengths of the sides are given as binomials.

Real-World Link

The cost of a swimming pool depends on many factors, including the size of the pool, whether the pool is an above-ground or an in-ground pool, and the material used.

Source: American Dream Homes

Real-World EXAMPLE 3 FOIL Method

SWIMMING POOL A contractor is building a deck around a rectangular swimming pool. The deck is x feet from every side of the pool. Write an expression for the total area of the pool and deck.

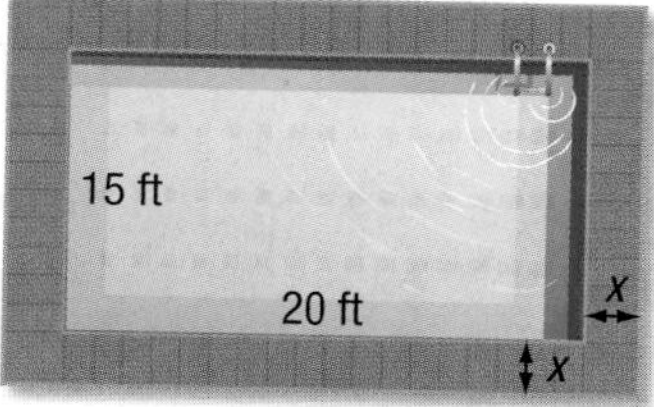

Understand We need to find an expression for the total area of the pool and deck.

Plan Use the formula for the area of a rectangle and determine the length and width of the pool with the deck.

Solve Since the deck is the same distance from every side of the pool, the length and width of the pool are $2x$ longer. So, the length can be represented by $2x + 20$ and the width can be represented by $2x + 15$.

Area = length • width	**Area of a rectangle**
$= (2x + 20)(2x + 15)$	**Substitution**
$= (2x)(2x) + (2x)(15) + (20)(2x) + (20)(15)$	**FOIL Method**
$= 4x^2 + 30x + 40x + 300$	**Multiply.**
$= 4x^2 + 70x + 300$	**Combine like terms.**

So, the total area of the deck and pool is $4x^2 + 70x + 300$.

Check Choose a value for x. Substitute this value into $(2x + 20)(2x + 15)$ and $4x^2 + 70x + 300$. The result should be the same for both expressions.

Check Your Progress

3. If the pool is 25 feet long and 20 feet wide, find the area of the pool and deck.

Personal Tutor glencoe.com

Multiply Polynomials The Distributive Property can also be used to multiply any two polynomials.

EXAMPLE 4 The Distributive Property

Find each product.

a. $(6x + 5)(2x^2 - 3x - 5)$

$(6x + 5)(2x^2 - 3x - 5)$	
$= 6x(2x^2 - 3x - 5) + 5(2x^2 - 3x - 5)$	**Distributive Property**
$= 12x^3 - 18x^2 - 30x + 10x^2 - 15x - 25$	**Multiply.**
$= 12x^3 - 8x^2 - 45x - 25$	**Combine like terms.**

b. $(2y^2 + 3y - 1)(3y^2 - 5y + 2)$

$(2y^2 + 3y - 1)(3y^2 - 5y + 2)$	
$= 2y^2(3y^2 - 5y + 2) + 3y(3y^2 - 5y + 2) - 1(3y^2 - 5y + 2)$	**Distributive Property**
$= 6y^4 - 10y^3 + 4y^2 + 9y^3 - 15y^2 + 6y - 3y^2 + 5y - 2$	**Multiply.**
$= 6y^4 - y^3 - 14y^2 + 11y - 2$	**Combine like terms.**

StudyTip

Multiplying Polynomials If a polynomial with c terms and a polynomial with d terms are multiplied together, there will be $c \cdot d$ terms before simplifying. In Example 4a, there are 2 • 3 or 6 terms before simplifying.

Check Your Progress

4A. $(3x - 5)(2x^2 + 7x - 8)$

4B. $(m^2 + 2m - 3)(4m^2 - 7m + 5)$

Personal Tutor glencoe.com

Check Your Understanding

Examples 1 and 2 pp. 447–448

Find each product.

1. $(x + 5)(x + 2)$
2. $(y - 2)(y + 4)$
3. $(b - 7)(b + 3)$
4. $(4n + 3)(n + 9)$
5. $(8h - 1)(2h - 3)$
6. $(2a + 9)(5a - 6)$

Example 3 p. 449

7. **FRAME** Hugo is designing a frame to surround the picture shown at the right. The frame is the same distance all the way around. Write an expression that represents the total area of the picture and frame.

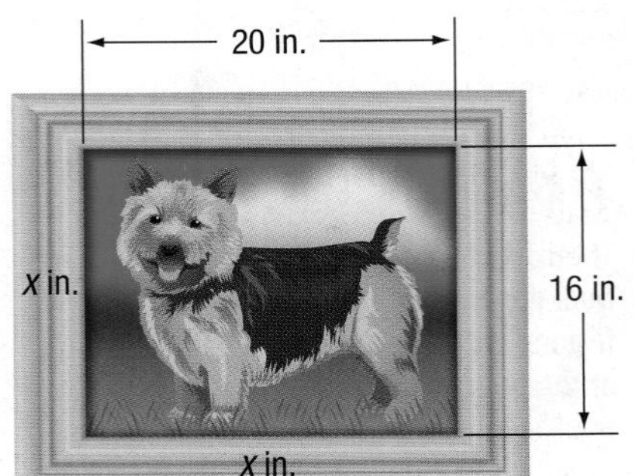

Example 4 p. 449

Find each product.

8. $(2a - 9)(3a^2 + 4a - 4)$
9. $(4y^2 - 3)(4y^2 + 7y + 2)$
10. $(x^2 - 4x + 5)(5x^2 + 3x - 4)$
11. $(2n^2 + 3n - 6)(5n^2 - 2n - 8)$

Practice and Problem Solving

= **Step-by-Step Solutions** begin on page R12. **Extra Practice** begins on page 815.

Examples 1 and 2 pp. 447–448

Find each product.

12. $(3c - 5)(c + 3)$
13. $(g + 10)(2g - 5)$
14. $(6a + 5)(5a + 3)$
15. $(4x + 1)(6x + 3)$
16. $(5y - 4)(3y - 1)$
17. $(6d - 5)(4d - 7)$
18. $(3m + 5)(2m + 3)$
19. $(7n - 6)(7n - 6)$
20. $(12t - 5)(12t + 5)$
21. $(5r + 7)(5r - 7)$
22. $(8w + 4x)(5w - 6x)$
23. $(11z - 5y)(3z + 2y)$

Example 3 p. 449

24. **GARDEN** A walkway surrounds a rectangular garden. The width of the garden is 8 feet, and the length is 6 feet. The width x of the walkway around the garden is the same on every side. Write an expression that represents the total area of the garden and walkway.

Example 4 p. 449

Find each product.

25. $(2y - 11)(y^2 - 3y + 2)$
26. $(4a + 7)(9a^2 + 2a - 7)$
27. $(m^2 - 5m + 4)(m^2 + 7m - 3)$
28. $(x^2 + 5x - 1)(5x^2 - 6x + 1)$
29. $(3b^3 - 4b - 7)(2b^2 - b - 9)$
30. $(6z^2 - 5z - 2)(3z^3 - 2z - 4)$

Simplify.

31. $(m + 2)[(m^2 + 3m - 6) + (m^2 - 2m + 4)]$
32. $[(t^2 + 3t - 8) - (t^2 - 2t + 6)](t - 4)$

GEOMETRY Find an expression to represent the area of each shaded region.

33.

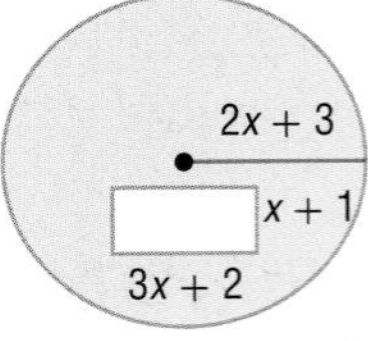

34.

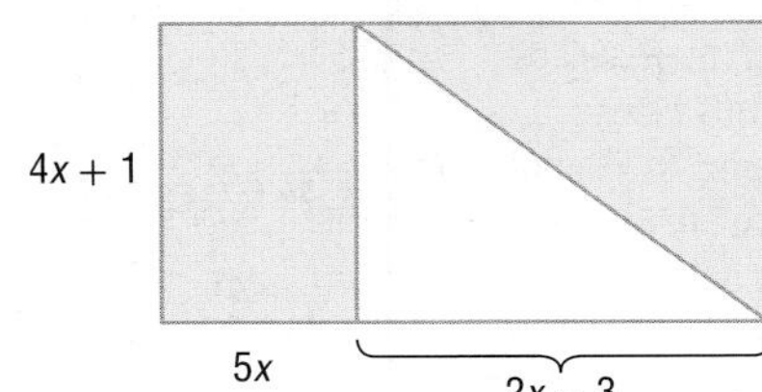

Real-World Link

On May 20, 2007, Misty May-Treanor won her 73rd professional beach volleyball title. May-Treanor has more wins than any other woman.

Source: Association of Volleyball Professionals

35. **VOLLEYBALL** The dimensions of a sand volleyball court are represented by a width of $6y - 5$ feet and a length of $3y + 4$ feet.

a. Write an expression that represents the area of the court.

b. The length of a sand volleyball court is 31 feet. Find the area of the court.

36. **GEOMETRY** Write an expression for the area of a triangle with a base of $2x + 3$ and a height of $3x - 1$.

Find each product.

37. $(a - 2b)^2$

38. $(3c + 4d)^2$

39. $(x - 5y)^2$

40. $(2r - 3t)^3$

41. $(5g + 2h)^3$

42. $(4y + 3z)(4y - 3z)^2$

43. **CONSTRUCTION** A sandbox kit allows you to build a square sandbox or a rectangular sandbox as shown.

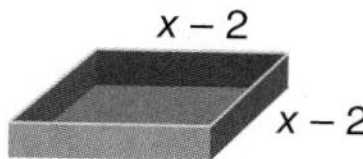

a. What are the possible values of x? Explain.

b. Which shape has the greater area?

c. What is the difference in areas between the two?

44. **MULTIPLE REPRESENTATIONS** In this problem, you will investigate the square of a sum.

a. **TABULAR** Copy and complete the table for each sum.

Expression	(Expression)^2
$x + 5$	
$3y + 1$	
$z + q$	

b. **VERBAL** Make a conjecture about the terms of the square of a sum.

c. **SYMBOLIC** For a sum of the form $a + b$, write an expression for the square of the sum.

H.O.T. Problems

Use Higher-Order Thinking Skills

45. **REASONING** Determine if the following statement is *sometimes*, *always*, or *never* true. Explain your reasoning.

The FOIL method can be used to multiply a binomial and a trinomial.

46. **CHALLENGE** Find $(x^m + x^p)(x^{m-1} - x^{1-p} + x^p)$.

47. **OPEN ENDED** Write a binomial and a trinomial involving a single variable. Then find their product.

48. **REASONING** Compare and contrast the procedure used to multiply a trinomial by a binomial using the vertical method with the procedure used to multiply a three-digit number by a two-digit number.

49. **WRITING IN MATH** Summarize the methods that can be used to multiply polynomials.

Standardized Test Practice

50. What is the product of $2x - 5$ and $3x + 4$?

A $5x - 1$
B $6x^2 - 7x - 20$
C $6x^2 - 20$
D $6x^2 + 7x - 20$

51. Which statement is correct about the symmetry of this design?

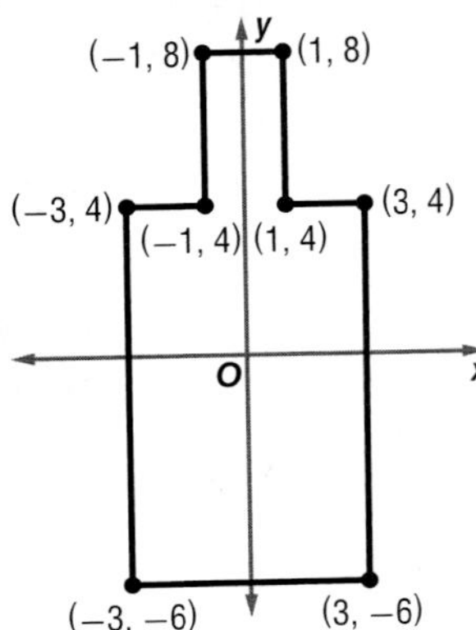

F The design is symmetrical only about the y-axis.
G The design is symmetrical only about the x-axis.
H The design is symmetrical about both the y- and the x-axes.
J The design has no symmetry.

52. Which point on the number line represents a number that, when cubed, will result in a number greater than itself?

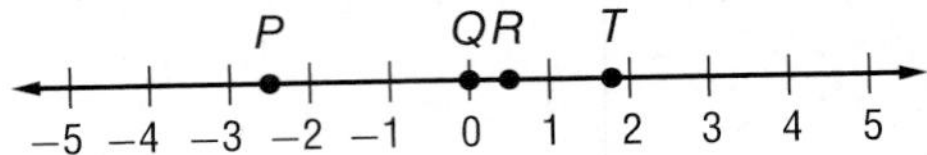

A P
B Q
C R
D T

53. SHORT RESPONSE For a science project, Jodi selected three bean plants of equal height. Then, for five days, she measured their heights in centimeters and plotted the values on the graph below.

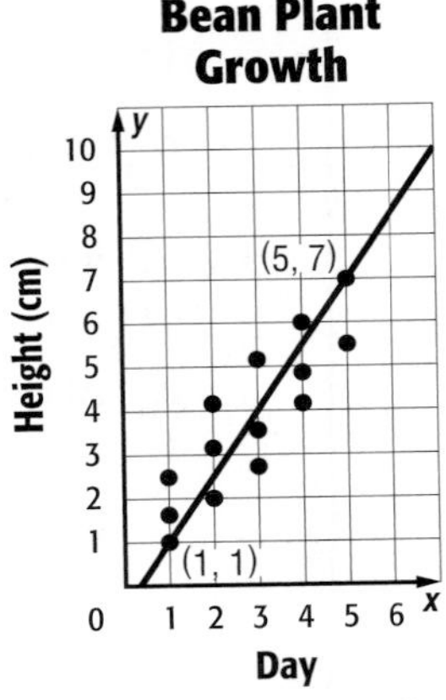

She drew a line of best fit on the graph. What is the slope of the line that she drew?

Spiral Review

54. SAVINGS Carrie has \$6000 to invest. She puts x dollars of this money into a savings account that earns 2% interest per year. She uses the rest of the money to purchase a certificate of deposit that earns 4% interest. Write an equation for the amount of money that Carrie will have in one year. (Lesson 7-6)

Find each sum or difference. (Lesson 7-5)

55. $(7a^2 - 5) + (-3a^2 + 10)$

56. $(8n - 2n^2) + (4n - 6n^2)$

57. $(4 + n^3 + 3n^2) + (2n^3 - 9n^2 + 6)$

58. $(-4u^2 - 9 + 2u) + (6u + 14 + 2u^2)$

59. $(b + 4) + (c + 3b - 2)$

60. $(3a^3 - 6a) - (3a^3 + 5a)$

61. $(-4m^3 - m + 10) - (3m^3 + 3m^2 - 7)$

62. $(3a + 4ab + 3b) - (2b + 5a + 8ab)$

Skills Review

Simplify. (Lesson 7-1)

63. $(-2t^4)^3 - 3(-2t^3)^4$

64. $(-3h^2)^3 - 2(-h^3)^2$

65. $2(-5y^3)^2 + (-3y^3)^3$

66. $3(-6n^4)^2 + (-2n^2)^2$

7-8 Special Products

Then

You multiplied binomials by using the FOIL method. (Lesson 7-7)

Now

- Find squares of sums and differences.
- Find the product of a sum and a difference.

Math Online

glencoe.com

- Extra Examples
- Personal Tutor
- Self-Check Quiz
- Homework Help
- Math in Motion

Why?

Colby wants to attach a dartboard to a square piece of corkboard. If the radius of the dartboard is $r + 12$, how large does the square corkboard need to be?

Colby knows that the diameter of the dartboard is $2(r + 12)$ or $2r + 24$. Each side of the square also measures $2r + 24$. To find how much corkboard is needed, Colby must find the area of the square: $A = (2r + 24)^2$.

Squares of Sums and Differences Some pairs of binomials, such as squares like $(2r + 24)^2$, have products that follow a specific pattern. Using the pattern can make multiplying easier. The square of a sum, $(a + b)^2$ or $(a + b)(a + b)$, is one of those products.

$$(a + b)^2 = a^2 + ab + ab + b^2$$

Key Concept — Square of a Sum

For Your FOLDABLE

Words The square of $a + b$ is the square of a plus twice the product of a and b plus the square of b.

Symbols $(a + b)^2 = (a + b)(a + b)$
$= a^2 + 2ab + b^2$

Example $(x + 4)^2 = (x + 4)(x + 4)$
$= x^2 + 8x + 16$

Math *in Motion*, Animation glencoe.com

EXAMPLE 1 Square of a Sum

Find $(3x + 5)^2$.

$(a + b)^2 = a^2 + 2ab + b^2$ **Square of a sum**

$(3x + 5)^2 = (3x)^2 + 2(3x)(5) + 5^2$ **$a = 3x$, $b = 5$**

$= 9x^2 + 30x + 25$ **Simplify. Use FOIL to check your solution.**

Check Your Progress

Find each product.

1A. $(8c + 3d)^2$

1B. $(3x + 4y)^2$

Personal Tutor glencoe.com

There is also a pattern for the *square of a difference*. Write $a - b$ as $a + (-b)$ and square it using the square of a sum pattern.

$$(a - b)^2 = [a + (-b)]^2$$
$$= a^2 + 2(a)(-b) + (-b)^2 \quad \text{Square of a sum}$$
$$= a^2 - 2ab + b^2 \quad \text{Simplify.}$$

Key Concept — Square of a Difference

For Your FOLDABLE

Words The square of $a - b$ is the square of a minus twice the product of a and b plus the square of b.

Symbols $(a + b)^2 = (a + b)(a + b) = a^2 + 2ab + b^2$

Example $(x - 3)^2 = (x - 3)(x - 3) = x^2 - 6x + 9$

Watch Out!

Square of a Difference Remember that $(x - 7)^2$ does not equal $x^2 - 7^2$, or $x^2 - 49$.

$(x - 7)^2$
$= (x - 7)(x - 7)$
$= x^2 - 14x + 49$

EXAMPLE 2 Square of a Difference

Find $(2x - 5y)^2$.

$(a - b)^2 = a^2 - 2ab + b^2$ — **Square of a difference**

$(2x - 5y)^2 = (2x)^2 - 2(2x)(5y) + (5y)^2$ — **$a = 2x$ and $b = 5y$**

$= 4x^2 - 20xy + 25y^2$ — **Simplify.**

✓ Check Your Progress

Find each product.

2A. $(6p - 1)^2$

2B. $(a - 2b)^2$

▷ **Personal Tutor glencoe.com**

The product of the square of a sum or the square of a difference is called a *perfect square trinomial*. We can use these to find patterns to solve real-world problems.

Real-World EXAMPLE 3 Square of a Difference

PHYSICAL SCIENCE Each edge of a cube of aluminum is 4 centimeters less than each edge of a cube of copper. Write an equation to model the surface area of the aluminum cube.

Let c = the length of each edge of the cube of copper.
So, each edge of the cube of aluminum is $c - 4$.

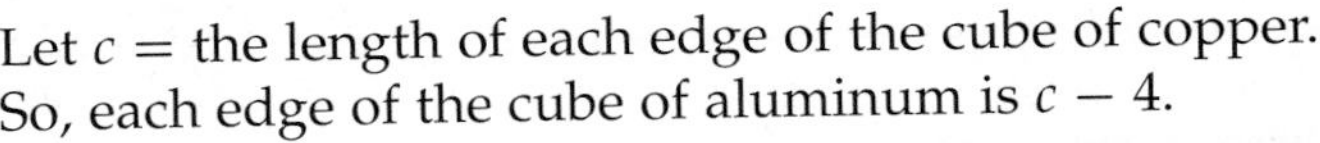

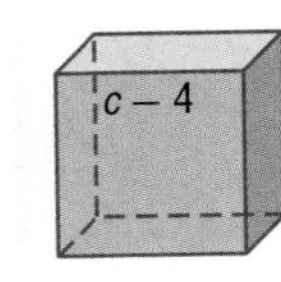

$SA = 6s^2$ — **Formula for surface area of a cube**

$SA = 6(c - 4)^2$ — **Replace s with $c - 4$.**

$SA = 6[c^2 - 2(4)(c) + 4^2]$ — **Square of a difference**

$SA = 6(c^2 - 8c + 16)$ — **Simplify.**

✓ Check Your Progress

3. GARDENING Alano has a garden that is g feet long and g feet wide. He wants to add 3 feet to the length and the width.

A. Show how the new area of the garden can be modeled by the square of a binomial.

B. Find the square of this binomial.

▷ **Personal Tutor glencoe.com**

StudyTip

Patterns When using any of these patterns, a and b can be numbers, variables, or expressions with numbers and variables.

Product of a Sum and a Difference Now we will see what the result is when we multiply a sum and a difference, or $(a + b)(a - b)$. Recall that $a - b$ can be written as $a + (-b)$.

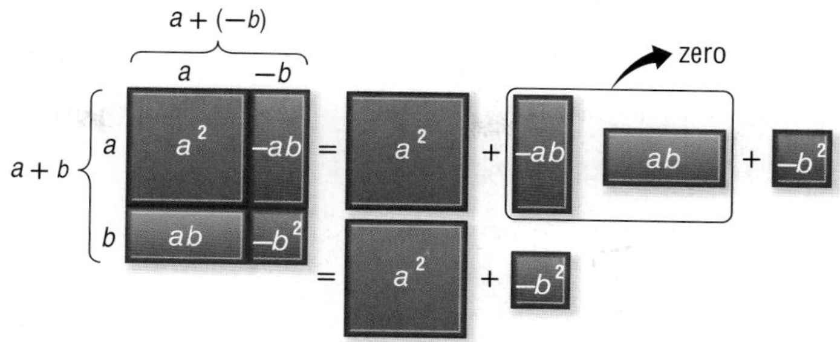

Notice that the middle terms are opposites and add to a zero pair. So $(a + b)(a - b) = a^2 - ab + ab - b^2 = a^2 - b^2$.

Key Concept Product of a Sum and a Difference — For Your FOLDABLE

Words The product of $a + b$ and $a - b$ is the square of a minus the square of b.

Symbols $(a + b)(a - b) = (a - b)(a + b)$
$= a^2 - b^2$

Math *in Motion*, Animation glencoe.com

EXAMPLE 4 Product of a Sum and a Difference

Find $(2x^2 + 3)(2x^2 - 3)$.

$(a + b)(a - b) = a^2 - b^2$	Product of a sum and difference
$(2x^2 + 3)(2x^2 - 3) = (2x^2)^2 - (3)^2$	$a = 2x^2$ and $b = 3$
$= 4x^4 - 9$	Simplify.

Check Your Progress

Find each product.

4A. $(3n + 2)(3n - 2)$

4B. $(4c - 7d)(4c + 7d)$

Personal Tutor glencoe.com

Check Your Understanding

Examples 1 and 2 pp. 453–454

Find each product.

1. $(x + 5)^2$

2. $(11 - a)^2$

3. $(2x + 7y)^2$

4. $(3m - 4)(3m - 4)$

5. $(g - 4h)(g - 4h)$

6. $(3c + 6d)^2$

Example 3 p. 454

7. GENETICS The color of a Labrador retriever's fur is genetic. Dark genes D are dominant over yellow genes y. A dog with genes DD or Dy will have dark fur. A dog with genes yy will have yellow fur. Pepper's genes for fur color are Dy, and Ramiro's are yy.

	D	y
D	DD	Dy
y	Dy	yy

a. Write an expression for the possible fur colors of Pepper's and Ramiro's puppies.

b. What is the probability that a puppy will have yellow fur?

Example 4
p. 455

Find each product.

8. $(a - 3)(a + 3)$

9. $(x + 5)(x - 5)$

10. $(6y - 7)(6y + 7)$

11. $(9t + 6)(9t - 6)$

Practice and Problem Solving

= **Step-by-Step Solutions** begin on page R12.
Extra Practice begins on page 815.

Examples 1 and 2
pp. 453–454

Find each product.

12. $(a + 10)(a + 10)$

13. $(b - 6)(b - 6)$

14. $(h + 7)^2$

15. $(x + 6)^2$

16. $(8 - m)^2$

17. $(9 - 2y)^2$

18. $(2b + 3)^2$

19. $(5t - 2)^2$

20. $(8h - 4n)^2$

Example 3
p. 454

21. GENETICS The ability to roll your tongue is inherited genetically from parents if either parent has the dominant trait T. Children of two parents without the trait will not be able to roll their tongues.

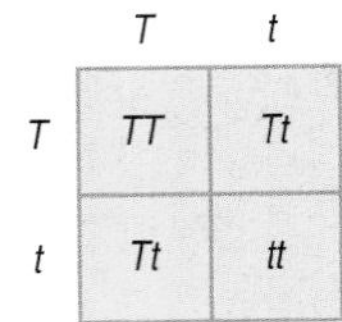

a. Show how the combinations can be modeled by the square of a sum.

b. Predict the percent of children that will have both dominant genes, one dominant gene, and both recessive genes.

Example 4
p. 455

Find each product.

22. $(u + 3)(u - 3)$

23 $(b + 7)(b - 7)$

24. $(2 + x)(2 - x)$

25. $(4 - x)(4 + x)$

26. $(2q + 5r)(2q - 5r)$

27. $(3a^2 + 7b)(3a^2 - 7b)$

28. $(5y + 7)^2$

29. $(8 - 10a)^2$

30. $(10x - 2)(10x + 2)$

31. $(3t + 12)(3t - 12)$

32. $(a + 4b)^2$

33. $(3q - 5r)^2$

34. $(2c - 9d)^2$

35. $(g + 5h)^2$

36. $(6y - 13)(6y + 13)$

37. $(3a^4 - b)(3a^4 + b)$

38. $(5x^2 - y^2)^2$

39. $(8a^2 - 9b^3)(8a^2 + 9b^3)$

40. $\left(\frac{3}{4}k + 8\right)^2$

41. $\left(\frac{2}{5}y - 4\right)^2$

42. $(7z^2 + 5y^2)(7z^2 - 5y^2)$

43. $(2m + 3)(2m - 3)(m + 4)$

44. $(r + 2)(r - 5)(r - 2)(r + 5)$

45. GEOMETRY Write a polynomial that represents the area of the figure at the right.

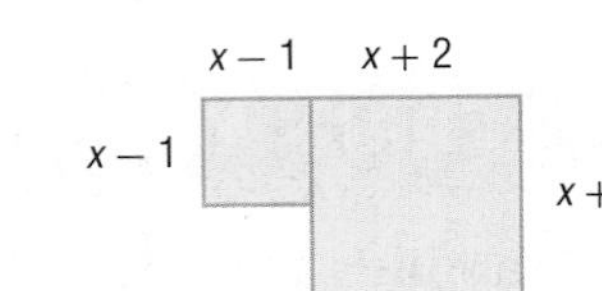

46. FLYING DISKS A flying disk shaped like a circle has a radius of $x + 3$ inches.

a. Write an expression representing the area of the flying disk.

b. If the diameter of the flying disk is 8 inches, what is its area?

Real-World Link

In the 1870s, a baker named William Frisbie put his name on the bottom of tin pie pans. In the 1940s, students from Yale University began throwing the pie pans through the air. Eventually, the pie tins became known as Frisbees.

Source: Idea Finder

GEOMETRY Find the area of each shaded region.

47.

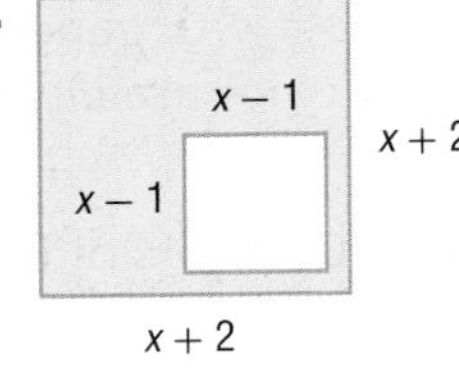

48.

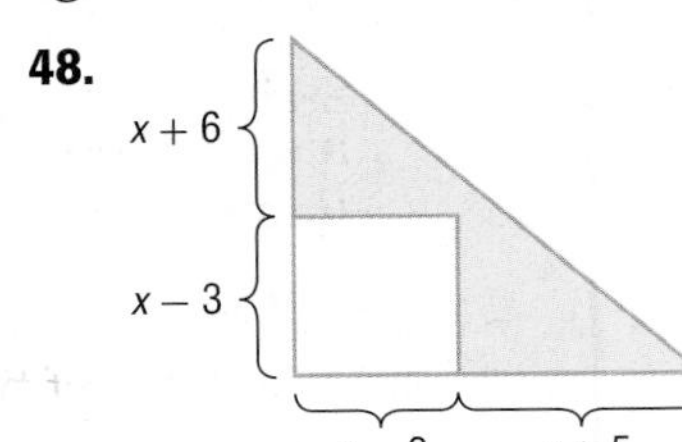

Real-World Link

Cael Sanderson of Iowa State University is the only wrestler in NCAA Division I history to be undefeated for four years. He compiled a 159-0 record from 1999–2002.

Source: Team Sanderson

Find each product.

49. $(c + d)(c + d)(c + d)$

50. $(2a - b)^3$

51. $(f + g)(f - g)(f + g)$

52. $(k - m)(k + m)(k - m)$

53. $(n - p)^2(n + p)$

54. $(q + r)^2(q - r)$

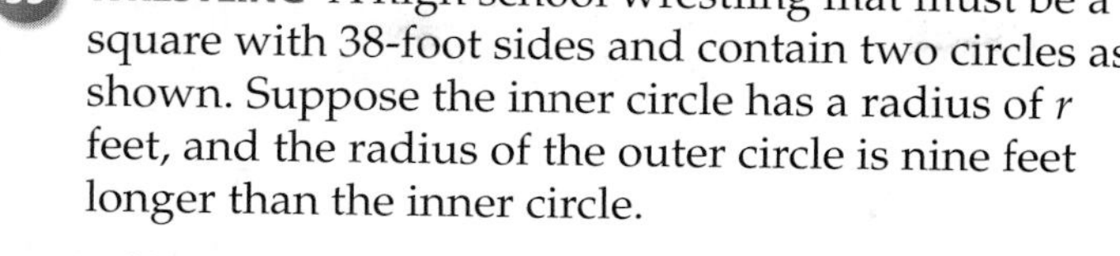

55. WRESTLING A high school wrestling mat must be a square with 38-foot sides and contain two circles as shown. Suppose the inner circle has a radius of r feet, and the radius of the outer circle is nine feet longer than the inner circle.

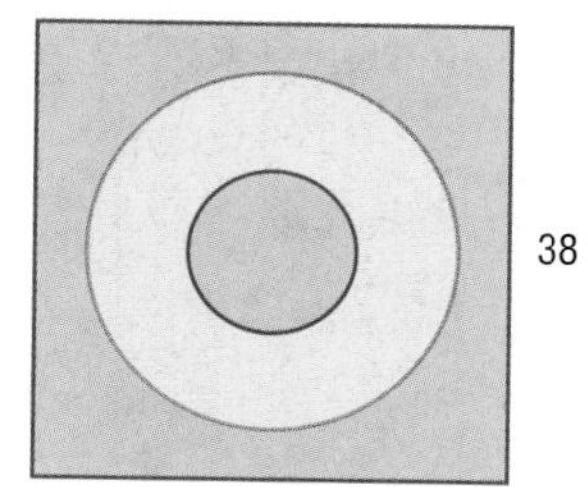

a. Write an expression for the area of the larger circle.

b. Write an expression for the area of the portion of the square outside the larger circle.

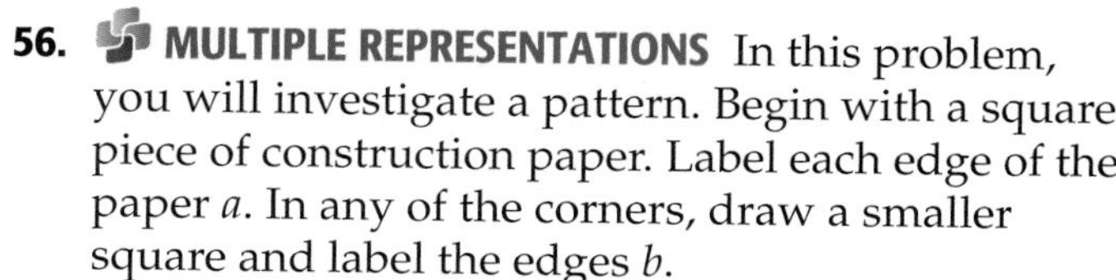

56. MULTIPLE REPRESENTATIONS In this problem, you will investigate a pattern. Begin with a square piece of construction paper. Label each edge of the paper a. In any of the corners, draw a smaller square and label the edges b.

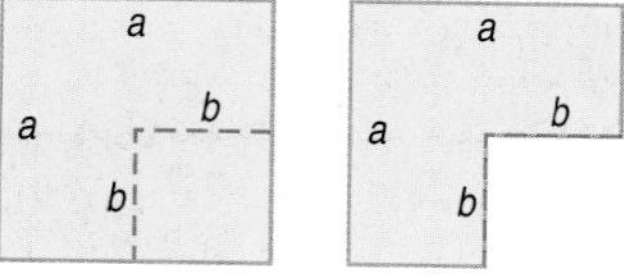

a. **NUMERICAL** Find the area of each of the squares.

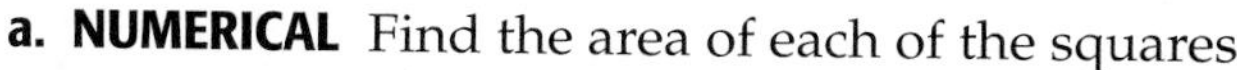

b. **CONCRETE** Cut the smaller square out of the corner. What is the area of the shape?

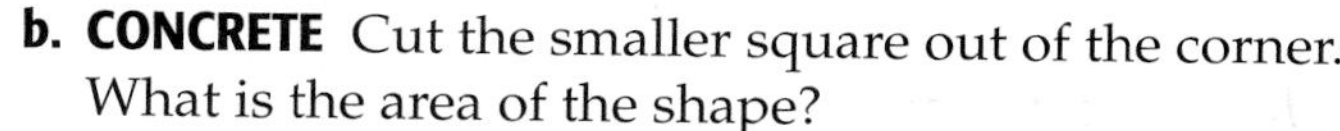

c. **ANALYTICAL** Remove the smaller rectangle on the bottom. Turn it and slide it next to the top rectangle. What is the length of the new arrangement? What is the width? What is the area?

d. **ANALYTICAL** What pattern does this verify?

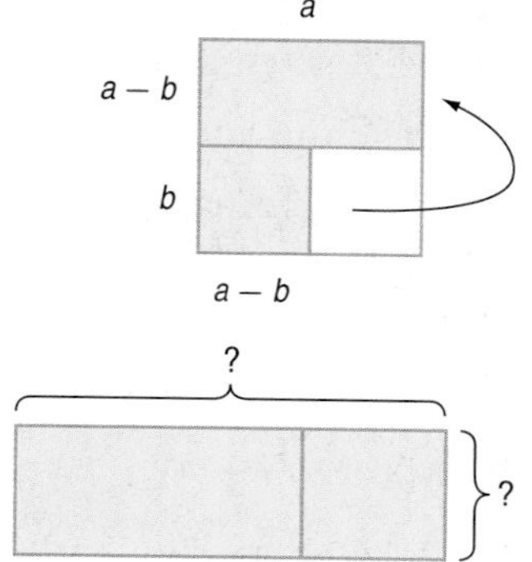

H.O.T. Problems

Use Higher-Order Thinking Skills

57. WHICH ONE DOESN'T BELONG? Which expression does not belong? Explain.

$(2c - d)(2c - d)$	$(2c + d)(2c - d)$	$(2c + d)(2c + d)$	$(c + d)(c + d)$

58. CHALLENGE Does a pattern exist for the cube of the sum $(a + b)^3$?

a. Investigate this question by finding the product $(a + b)(a + b)(a + b)$.

b. Use the pattern you discovered in part a to find $(x + 2)^3$.

c. Draw a diagram of a geometric model for the cube of a sum.

d. What is the pattern for the cube of a difference, $(a - b)^3$?

59. REASONING Find c that makes $25x^2 - 90x + c$ a perfect square trinomial.

60. OPEN ENDED Write two binomials with a product that is a binomial. Then write two binomials with a product that is not a binomial.

61. WRITING IN MATH Describe how to square the sum of two quantities, square the difference of two quantities, and how to find the product of a sum of two quantities and a difference of two quantities.

Standardized Test Practice

62. GRIDDED RESPONSE In the right triangle, $\overline{DB}$ bisects $\angle B$. What is the measure of $\angle ADB$ in degrees?

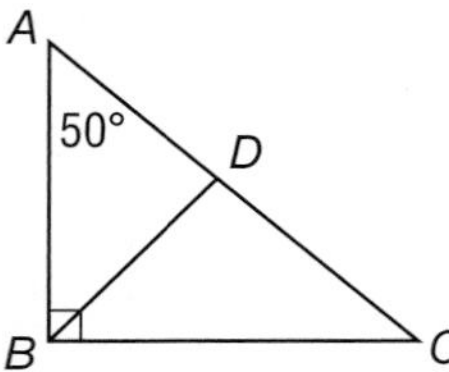

63. What is the product of $(2a - 3)$ and $(2a - 3)$?

A $4a^2 + 12a + 9$
B $4a^2 + 9$
C $4a^2 - 12a - 9$
D $4a^2 - 12a + 9$

64. Myron can drive 4 miles in m minutes. At this rate, how many minutes will it take him to drive 19 miles?

F $76m$
G $\frac{19m}{4}$
H $\frac{4m}{19}$
J $\frac{4}{19m}$

65. What property is illustrated by the equation $2x + 0 = 2x$?

A Commutative Property of Addition
B Additive Inverse Property
C Additive Identity Property
D Associative Property of Addition

Spiral Review

Find each product. (Lesson 7-7)

66. $(y - 4)(y - 2)$
67. $(2c - 1)(c + 3)$
68. $(d - 9)(d + 5)$
69. $(4h - 3)(2h - 7)$
70. $(3x + 5)(2x + 3)$
71. $(5m + 4)(8m + 3)$

Simplify. (Lesson 7-6)

72. $x(2x - 7) + 5x$
73. $c(c - 8) + 2c(c + 3)$
74. $8y(-3y + 7) - 11y^2$
75. $-2d(5d) - 3d(d + 6)$
76. $5m(2m^3 + m^2 + 8) + 4m$
77. $3p(6p - 4) + 2\left(\frac{1}{2}p^2 - 3p\right)$

Use substitution to solve each system of equations. (Lesson 6-2)

78. $4c = 3d + 3$
$c = d - 1$

79. $c - 5d = 2$
$2c + d = 4$

80. $5r - t = 5$
$-4r + 5t = 17$

81. BIOLOGY Each type of fish thrives in a specific range of temperatures. The best temperatures for sharks range from 18°C to 22°C, inclusive. Write a compound inequality to represent temperatures where sharks will not thrive. (Lesson 6-2)

Write an equation of the line that passes through each pair of points. (Lesson 4-2)

82. (1, 1), (7, 4)
83. (5, 7), (0, 6)
84. (5, 1), (8, −2)

85. COFFEE A coffee store wants to create a mix using two coffees. How many pounds of coffee A should be mixed with 9 pounds of coffee B to get a mixture that can sell for $6.95 per pound? (Lesson 2-9)

Skills Review

Find the prime factorization of each number. (Concepts and Skills Bank Lesson 3)

86. 40
87. 120
88. 900
89. 165

Study Guide and Review

Math Online glencoe.com
- STUDY TO GO
- Vocabulary Review

Chapter Summary

Key Concepts

For any nonzero real numbers a and b and any integers m, n, and p, the following are true.

Multiplying Monomials (Lesson 7-1)

- Product of Powers: $a^m \cdot a^n = a^{m+n}$
- Power of a Power: $(a^m)^n = a^{m \cdot n}$
- Power of a Product: $(ab)^m = a^m b^m$

Dividing Monomials (Lesson 7-2)

- Quotient of Powers: $\frac{a^m}{a^p} = a^{m-p}$
- Power of a Quotient: $\left(\frac{a}{b}\right)^m = \frac{a^m}{b^m}$
- Zero Exponent: $a^0 = 1$
- Negative Exponent: $a^{-n} = \frac{1}{a^n}$ and $\frac{1}{a^{-n}} = a^n$

Scientific Notation (Lesson 7-3)

- A number is in scientific notation if it is in the form $a \times 10^n$, where $1 \le a < 10$.
- To write in standard form:
  - If $n > 0$, move the decimal n places right.
  - If $n < 0$, move the decimal n places left.

Operations with Polynomials (Lessons 7-5 through 7-8)

- To add or subtract polynomials, add or subtract like terms. To multiply polynomials, use the Distributive Property.
- Special products: $(a + b)^2 = a^2 + 2ab + b^2$
  $(a - b)^2 = a^2 - 2ab + b^2$
  $(a + b)(a - b) = a^2 - b^2$

FOLDABLES® Study Organizer

Be sure the Key Concepts are noted in your Foldable.

Key Vocabulary

binomial (p. 424)
constant (p. 401)
degree of a monomial (p. 424)
degree of a polynomial (p. 424)
FOIL method (p. 448)
leading coefficient (p. 425)
monomial (p. 401)
order of magnitude (p. 411)
polynomial (p. 424)
quadratic expression (p. 448)
scientific notation (p. 416)
standard form of a polynomial (p. 425)
trinomial (p. 424)

Vocabulary Check

Choose a term from the Key Vocabulary list above that best describes each expression or equation.

1. $x^2 + 1$
2. $5^0 = 1$
3. $x^2 - 3x + 2$
4. $(xy^3)(x^2y^4) = x^3y^7$
5. $(a^7)^3 = a^{21}$
6. $5^{-2} = \frac{1}{5^2}$
7. 6.2×10^5
8. $(x + 2)(x - 5) = x^2 - 3x - 10$
9. $x^3 + 2x^2 - 3x - 1$
10. $7xy^4$

CHAPTER 7

Study Guide and Review

Lesson-by-Lesson Review

7-1 Multiplying Monomials (pp. 401–407)

Simplify each expression.

11. $x \cdot x^3 \cdot x^5$
12. $(2xy)(-3x^2y^5)$
13. $(-4ab^4)(-5a^5b^2)$
14. $(6x^3y^2)^2$
15. $[(2r^3t)^3]^2$
16. $(-2u^3)(5u)$
17. $(2x^2)^3(x^3)^3$
18. $\frac{1}{2}(2x^3)^3$
19. **GEOMETRY** Use the formula $V = \pi r^2 h$ to find the volume of the cylinder.

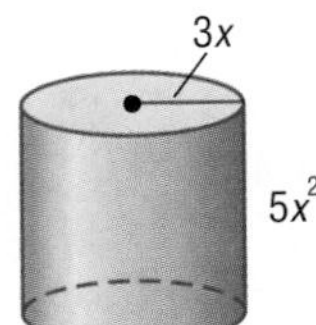

EXAMPLE 1

Simplify $(5x^2y^3)(2x^4y)$.

$(5x^2y^3)(2x^4y)$

$= (5 \cdot 2)(x^2 \cdot x^4)(y^3 \cdot y)$ — **Commutative Property**

$= 10x^6y^4$ — **Product of Powers**

EXAMPLE 2

Simplify $(3a^2b^4)^3$.

$(3a^2b^4)^3 = 3^3(a^2)^3(b^4)^3$ — **Power of a Product**

$= 27a^6b^{12}$ — **Simplify.**

7-2 Dividing Monomials (pp. 408–415)

Simplify each expression. Assume that no denominator equals zero.

20. $\frac{(3x)^0}{2a}$
21. $\left(\frac{3xy^3}{2z}\right)^3$
22. $\frac{12y^{-4}}{3y^{-5}}$
23. $a^{-3}b^0c^6$
24. $\frac{-15x^7y^8z^4}{-45x^3y^5z^3}$
25. $\frac{(3x^{-1})^{-2}}{(3x^2)^{-2}}$
26. $\left(\frac{6xy^{11}z^9}{48x^6yz^{-7}}\right)^0$
27. $\left(\frac{12}{2}\right)\left(\frac{x}{y^5}\right)\left(\frac{y^4}{x^4}\right)$
28. **GEOMETRY** The area of a rectangle is $25x^2y^4$ square feet. The width of the rectangle is $5xy$ feet. What is the length of the rectangle?

EXAMPLE 3

Simplify $\frac{2k^4m^3}{4k^2m}$. Assume that no denominator equals zero.

$\frac{2k^4m^3}{4k^2m} = \left(\frac{2}{4}\right)\left(\frac{k^4}{k^2}\right)\left(\frac{m^3}{m}\right)$ — **Group powers with the same base.**

$= \left(\frac{1}{2}\right)k^{4-2}m^{3-1}$ — **Quotient of Powers**

$= \frac{k^2m^2}{2}$ — **Simplify.**

EXAMPLE 4

Simplify $\frac{t^4uv^{-2}}{t^{-3}u^7}$. Assume that no denominator equals zero.

$\frac{t^4uv^{-2}}{t^{-3}u^7} = \left(\frac{t^4}{t^{-3}}\right)\left(\frac{u}{u^7}\right)(v^{-2})$ — **Group the powers with the same base.**

$= (t^{4+3})(u^{1-7})(v^{-2})$ — **Quotient of Powers**

$= t^7u^{-6}v^{-2}$ — **Simplify.**

$= \frac{t^7}{u^6v^2}$ — **Simplify.**

MIXED PROBLEM SOLVING
For mixed problem-solving practice, see page 851.

7-3 Scientific Notation (pp. 416–422)

Express each number in scientific notation.

29. 2,300,000

30. 0.0000543

31. ASTRONOMY Earth has a diameter of about 8000 miles. Jupiter has a diameter of about 88,000 miles. Write in scientific notation the ratio of Earth's diameter to Jupiter's diameter.

EXAMPLE 5

Express 300,000,000 in scientific notation.

Step 1 300,000,000 ⟶ 3.00000000

Step 2 The decimal point moved 8 places to the left, so $n = 8$.

Step 3 $300{,}000{,}000 = 3 \times 10^8$

7-4 Polynomials (pp. 424–429)

Write each polynomial in standard form.

32. $x + 2 + 3x^2$

33. $1 - x^4$

34. $2 + 3x + x^2$

35. $3x^5 - 2 + 6x - 2x^2 + x^3$

36. GEOMETRY Write a polynomial that represents the perimeter of the figure.

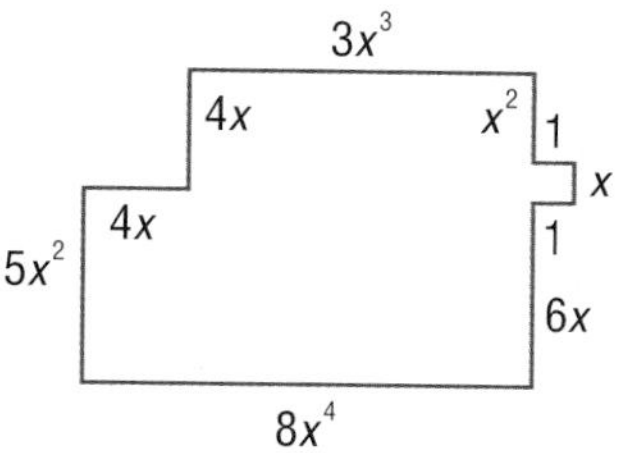

EXAMPLE 6

Write $3 - x^2 + 4x$ in standard form.

Step 1 Find the degree of each term.

3: degree 0

$-x^2$: degree 2

$4x$: degree 1

Step 2 Write the terms in descending order of degree.

$3 - x^2 + 4x = -x^2 + 4x + 3$

7-5 Adding and Subtracting Polynomials (pp. 433–438)

Find each sum or difference.

37. $(x^3 + 2) + (-3x^3 - 5)$

38. $a^2 + 5a - 3 - (2a^2 - 4a + 3)$

39. $(4x - 3x^2 + 5) + (2x^2 - 5x + 1)$

40. $(6ab + 3b^2) - (3ab - 2b^2)$

41. PICTURE FRAMES Jean is framing a painting that is a rectangle. What is the perimeter of the frame?

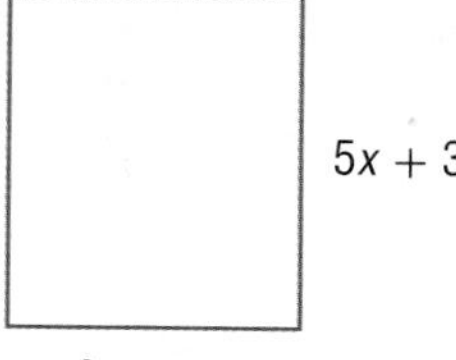

EXAMPLE 7

Find $(8r^2 + 3r) - (10r^2 - 5)$.

$(8r^2 + 3r) - (10r^2 - 5)$

$= (8r^2 + 3r) + (-10r^2 + 5)$ **Use the additive inverse.**

$= (8r^2 - 10r^2) + 3r + 5$ **Group like terms.**

$= -2r^2 + 3r + 5$ **Add like terms.**

CHAPTER 7

Study Guide and Review

7-6 Multiplying a Polynomial by a Monomial (pp. 439–444)

Solve each equation.

42. $x^2(x + 2) = x(x^2 + 2x + 1)$

43. $2x(x + 3) = 2(x^2 + 3)$

44. $2(4w + w^2) - 6 = 2w(w - 4) + 10$

45. $6k(k + 2) = 6(k^2 + 4)$

46. GEOMETRY Find the area of the rectangle.

$3x$

$x^2 + x - 7$

EXAMPLE 8

Solve $m(2m - 5) + m = 2m(m - 6) + 16$.

$$m(2m - 5) + m = 2m(m - 6) + 16$$
$$2m^2 - 5m + m = 2m^2 - 12m + 16$$
$$2m^2 - 4m = 2m^2 - 12m + 16$$
$$-4m = -12m + 16$$
$$8m = 16$$
$$m = 2$$

7-7 Multiplying Polynomials (pp. 447–452)

Find each product.

47. $(x - 3)(x + 7)$

48. $(3a - 2)(6a + 5)$

49. $(3r - 7t)(2r + 5t)$

50. $(2x + 5)(5x + 2)$

51. PARKING LOT The parking lot shown is to be paved. What is the area to be paved?

$2x + 3$

$5x - 4$

EXAMPLE 9

Find $(6x - 5)(x + 4)$.

$(6x - 5)(x + 4)$

F O I L

$= (6x)(x) + (6x)(4) + (-5)(x) + (-5)(4)$

$= 6x^2 + 24x - 5x - 20$ Multiply.

$= 6x^2 + 19x - 20$ Combine like terms.

7-8 Special Products (pp. 453–458)

Find each product. See margin

52. $(x + 5)(x - 5)$

53. $(3x - 2)^2$

54. $(5x + 4)^2$

55. $(2x - 3)(2x + 3)$

56. $(2r + 5t)^2$

57. $(3m - 2)(3m + 2)$

58. GEOMETRY Write an expression to represent the area of the shaded region.

$2x + 5$

$x + 2$

$x - 2$

$2x - 5$

EXAMPLE 10

Find $(x - 7)^2$.

$(a - b)^2 = a^2 - 2ab + b^2$ Square of a Difference

$(x - 7)^2 = x^2 - 2(x)(7) + (-7)^2$ $a = x$ and $b = 7$

$= x^2 - 14x + 49$ Simplify.

EXAMPLE 11

Find $(5a - 4)(5a + 4)$.

$(a + b)(a - b) = a^2 - b^2$ Product of a Sum and Difference

$(5a - 4)(5a + 4) = (5a)^2 - (4)^2$ $a = 5a$ and $b = 4$

$= 25a^2 - 16$ Simplify.

Practice Test

Math Online glencoe.com
Chapter Test

Simplify each expression.

1. $(x^2)(7x^8)$
2. $(5a^7bc^2)(-6a^2bc^5)$
3. **MULTIPLE CHOICE** Express the volume of the solid as a monomial.

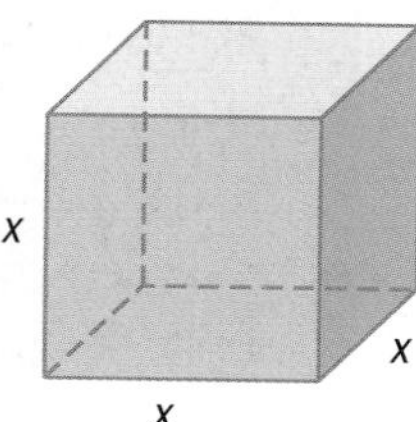

A x^3 C $6x^3$

B $6x$ D x^6

Simplify each expression. Assume that no denominator equals 0.

4. $\frac{x^6y^8}{x^2}$
5. $\left(\frac{2a^4b^3}{c^6}\right)^0$
6. $\frac{2xy^{-7}}{8x}$

Express each number in scientific notation. (Lesson 7-3)

7. 0.00021
8. 58,000

Express each number in standard form.

9. 2.9×10^{-5}
10. 9.1×10^6

Evaluate each product or quotient. Express the results in scientific notation.

11. $(2.5 \times 10^3)(3 \times 10^4)$
12. $\frac{8.8 \times 10^2}{4 \times 10^{-4}}$
13. **ASTRONOMY** The average distance from Mercury to the Sun is 35,980,000 miles. Express this distance in scientific notation.

Find each sum or difference.

14. $(x + 5) + (x^2 - 3x + 7)$
15. $(7m - 8n^2 + 3n) - (-2n^2 + 4m - 3n)$
16. **MULTIPLE CHOICE** Antonia is carpeting two of the rooms in her house. The dimensions are shown. What is the total area to be carpeted?

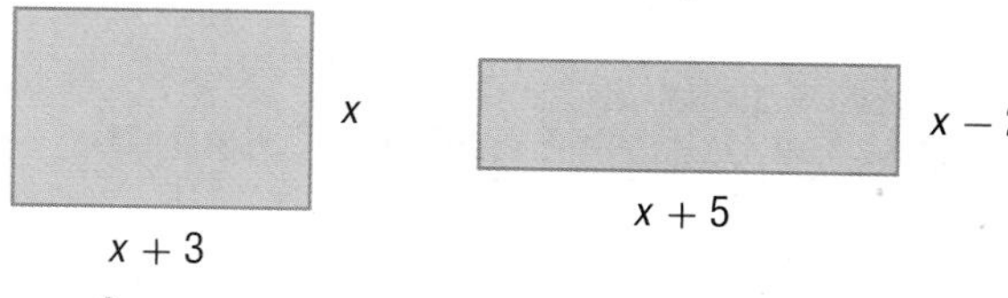

F $x^2 + 3x$ H $x^2 + 3x - 5$

G $2x^2 + 6x - 10$ J $8x + 12$

Find each product.

17. $a(a^2 + 2a - 10)$
18. $(2a - 5)(3a + 5)$
19. $(x - 3)(x^2 + 5x - 6)$
20. $(x + 3)^2$
21. $(2b - 5)(2b + 5)$
22. **GEOMETRY** A rectangular prism has dimensions x, $x + 3$, and $2x + 5$.
    a. Find the volume of the prism in terms of x.
    b. Choose two values for x. How do the volumes compare?

Solve each equation.

23. $5(t^2 - 3t + 2) = t(5t - 2)$
24. $3x(x + 2) = 3(x^2 - 2)$
25. **FINANCIAL LITERACY** Money invested in a certificate of deposit (CD) earns interest once per year. Suppose you invest \$4000 in a 2-year CD.
    a. If the interest rate is 5% per year, the expression $4000(1 + 0.05)^2$ can be evaluated to find the total amount of money after two years. Explain the numbers in this expression.
    b. Find the amount at the end of two years.
    c. Suppose you invest \$10,000 in a CD for 4 years at an annual rate of 6.25%. What is the total amount of money you will have after 4 years?

CHAPTER 7 Preparing for Standardized Tests

Using a Scientific Calculator

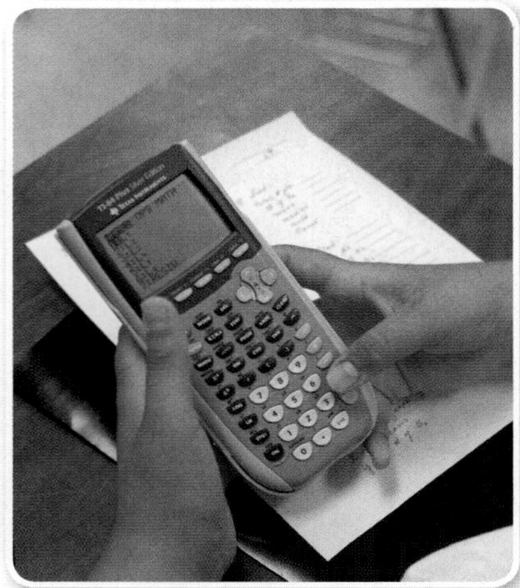

Scientific calculators are powerful problem-solving tools. There are times when using a scientific calculator can be used to make computations faster and easier, such as computations with very large numbers. However, there are times when using a scientific calculator is necessary, like the estimation of irrational numbers.

Strategies for Using a Scientific Calculator

Step 1

Familiarize yourself with the various functions of a scientific calculator as well as when they should be used:

- **Exponents** scientific notation, calculating with large or small numbers
- **Pi** solving circle problems, like circumference and area
- **Square roots** distance on a coordinate plane, Pythagorean theorem
- **Graphs** analyzing paired data in a scatter plot, graphing functions, finding roots of equations

Step 2

Use your scientific or graphing calculator to solve the problem.

- Remember to work as efficiently as possible. Some steps may be done mentally or by hand, while others should be completed using your calculator.
- If time permits, check your answer.

EXAMPLE

Read the problem. Identify what you need to know. Then use the information in the problem to solve.

The distance from the Sun to Jupiter is approximately 7.786×10^{11} meters. If the speed of light is about 3×10^{8} meters per second, how long does it take for light from the Sun to reach Jupiter? Round to the nearest minute.

A about 43 minutes

B about 51 minutes

C about 1876 minutes

D about 2595 minutes

Read the problem carefully. You are given the approximate distance from the Sun to Jupiter as well as the speed of light. Both quantities are given in scientific notation. You are asked to find how many minutes it takes for light from the Sun to reach Jupiter. Use the relationship distance = rate × time to find the amount of time.

$d = r \times t$

$\frac{d}{r} = t$

To find the amount of time, divide the distance by the rate. Notice, however, that the units for time will be seconds.

$\frac{7.786 \times 10^{11} \text{ m}}{3 \times 10^{8} \text{ m/s}} = t$ seconds

Use a scientific calculator to quickly find the quotient. On most scientific calculators, the EE key is used to enter numbers in scientific notation.

KEYSTROKES: (7.786 [2nd] [EE] 11) / (3 [2nd] [EE] 8)

The result is 2595.33333333 seconds. To convert this number to minutes, use your calculator to divide the result by 60. This gives an answer of about 43.2555 minutes. The answer is A.

Exercises

Read each problem. Identify what you need to know. Then use the information in the problem to solve.

1. Since its creation 5 years ago, approximately 2.504×10^7 items have been sold or traded on a popular online website. What is the average daily number of items sold or traded over the 5-year period?

   A about 9640 items per day

   B about 13,720 items per day

   C about 1,025,000 items per day

   D about 5,008,000 items per day

2. Evaluate $\sqrt{ab}$ if $a = 121$ and $b = 23$.

   F about 5.26

   G about 9.90

   H about 12

   J about 52.75

3. The population of the United States is about 3.034×10^8 people. The land area of the country is about 3.54×10^6 square miles. What is the average *population density* (number of people per square mile) of the United States?

   A about 136.3 people per square mile

   B about 112.5 people per square mile

   C about 94.3 people per square mile

   D about 85.7 people per square mile

4. Eleece is making a cover for the marching band's bass drum. The drum has a diameter of 20 inches. Estimate the area of the face of the bass drum.

   F 31.41 square inches

   G 62.83 square inches

   H 78.54 square inches

   J 314.16 square inches

CHAPTER 7

Standardized Test Practice

Cumulative, Chapters 1 through 7

Multiple Choice

Read each question. Then fill in the correct answer on the answer document provided by your teacher or on a sheet of paper.

1. Express the area of the triangle below as a monomial.

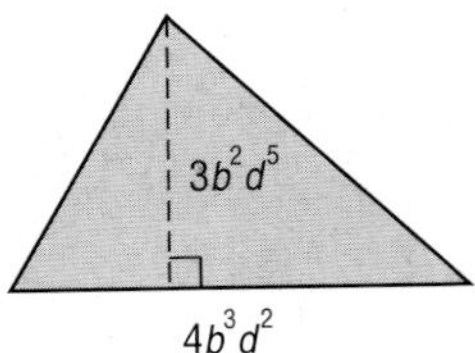

A $12b^5d^7$

B $12b^6d^{10}$

C $6b^6d^{10}$

D $6b^5d^7$

2. Simplify the following expression.

$$\left(\frac{2w^2z^5}{3y^4}\right)^3$$

F $\frac{2w^5z^8}{3y^7}$

G $\frac{8w^6z^{15}}{27y^{12}}$

H $\frac{8w^5z^8}{27y^7}$

J $\frac{2w^6z^{15}}{3y^{12}}$

3. Which equation of a line is perpendicular to $y = \frac{3}{5}x - 3$?

A $y = -\frac{5}{3}x + 2$

B $y = -\frac{3}{5}x + 2$

C $y = \frac{5}{3}x - 2$

D $y = \frac{3}{5}x - 2$

Test-TakingTip

Question 2 Use the laws of exponents to simplify the expression. Remember, to find the power of a power, multiply the exponents.

4. Express the perimeter of the rectangle below as a polynomial.

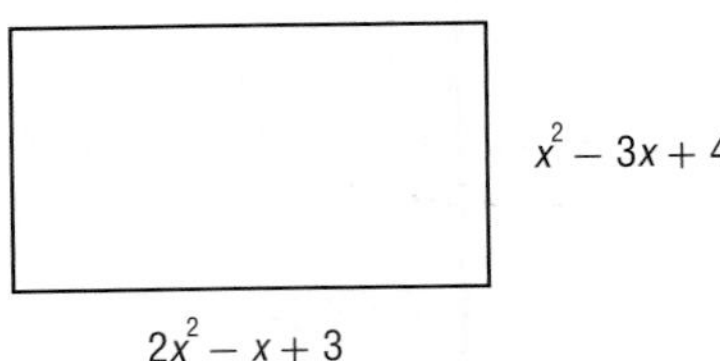

F $3x^2 - 4x + 7$

G $3x^2 + x + 7$

H $6x^2 - 8x + 14$

J $6x^2 - 4x + 7$

5. Subtract the polynomials below.

$(7a^2 + 6a - 2) - (-4a^3 + 3a^2 + 5)$

A $4a^3 + 4a^2 + 6a - 7$

B $11a^2 + 3a - 7$

C $4a^3 + 10a^2 + 6a + 3$

D $4a^3 + 7a^3 - 3a$

6. Which inequality is shown in the graph?

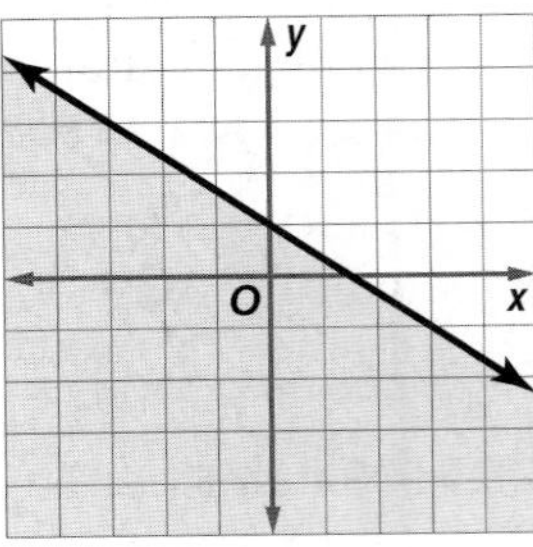

F $y \leq -\frac{2}{3}x - 1$

G $y \leq -\frac{3}{4}x - 1$

H $y \leq -\frac{2}{3}x + 1$

J $y \leq -\frac{3}{4}x + 1$

Short Response/Gridded Response

7. Mickey has 180 feet of fencing that she wants to use to enclose a play area for her puppy. She will use her house as one of the sides of the region.

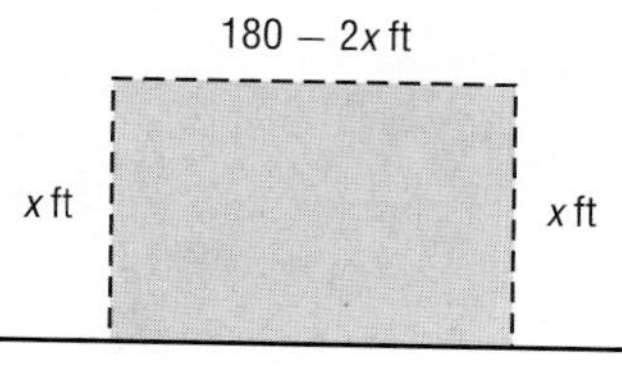

a. If she makes the play area x feet deep as shown in the figure, write a polynomial in standard form to represent the area of the region.

b. How many square feet of area will the puppy have to play in if Mickey makes it 40 feet deep?

8. Identify the expression below that does not belong with the other two. Explain.

$(3m - 2n)(3m + 2n)$

$(3m + 2n)(3m + 2n)$

$(3m + 2n)(3m - 2n)$

9. What is the solution to the following system of equations? Show your work.

$$\begin{cases} y = 6x - 1 \\ y = 6x + 4 \end{cases}$$

10. **GRIDDED RESPONSE** At a family fun center, the Wilson and Sanchez families each bought video game tokens and batting cage tokens as shown in the table.

Family	Wilson	Sanchez
Number of Video Game Tokens	25	30
Number of Batting Cage Tokens	8	6
Total Cost	$26.50	$25.50

What is the cost in dollars of a batting cage token at the family fun center?

Extended Response

Record your answers on a sheet of paper. Show your work.

11. The table below shows the distances from the Sun to Mercury, Earth, Mars, and Saturn. Use the data to answer each question.

Planet	Distance from Sun (km)
Mercury	5.79×10^7
Earth	1.50×10^8
Mars	2.28×10^8
Saturn	1.43×10^9

a. Of the planets listed, which one is the closest to the Sun?

b. About how many times as far from the Sun is Mars as Earth?

Need Extra Help?											
If you missed Question...	1	2	3	4	5	6	7	8	9	10	11
Go to Lesson or Page...	7-1	7-2	4-4	7-5	7-5	5-6	7-6	7-8	6-1	6-4	3-5

CHAPTER 8 Factoring and Quadratic Equations

Then

In Chapter 7, you multiplied monomials and polynomials.

Now

In Chapter 8, you will:

- Factor monomials.
- Factor trinomials.
- Factor differences of squares.
- Solve quadratic equations.

Why?

ARCHITECTURE Quadratic equations can be used to model the shape of architectural structures such as the tallest memorial in the United States, the Gateway Arch in St. Louis, Missouri.

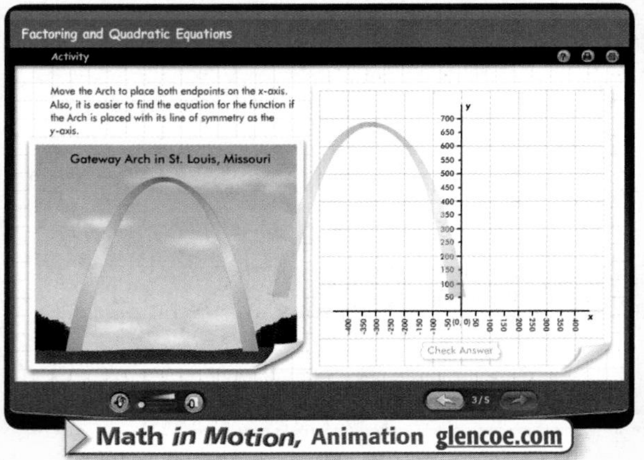

Math *in Motion,* Animation glencoe.com

Get Ready for Chapter 8

Diagnose Readiness You have two options for checking Prerequisite Skills.

Text Option

Take the Quick Check below. Refer to the Quick Review for help.

QuickCheck

Rewrite each expression using the Distributive Property. Then simplify. (Lesson 1-4)

1. $a(a + 5)$
2. $2(3 + x)$
3. $n(n - 3n^2 + 2)$
4. $-6(x^2 - 5x + 6)$
5. **FINANCIAL LITERACY** Five friends will pay \$9 per ticket, \$3 per drink, and \$6 per popcorn at the movies. Write an expression that could be used to determine the cost for them to go to the movies.

Find each product. (Lesson 7-7)

6. $(x + 2)(x - 5)$
7. $(x + 4)(x - 1)$
8. $(2a - 3)(5a + 4)$
9. $(3x - 4)(x + 5)$
10. $(x + 4)(x + 7)$
11. $(6a - 2b)(9a + b)$
12. **TABLECLOTH** The dimensions of a tablecloth are represented by a width of $2x + 3$ and a length of $x + 1$. Find an expression for the area of the tablecloth.

Find each product. (Lesson 7-8)

13. $(3 - a)^2$
14. $(x + 5)^2$
15. $(3x - 2y)^2$
16. $(2x + 5y)(2x - 5y)$
17. **PHOTOGRAPHY** A photo is $x + 6$ inches by $x - 6$ inches. What is the area of the photo?

QuickReview

EXAMPLE 1

Rewrite $6x(-3x - 5x - 5x^2 + x^3)$ using the Distributive Property. Then simplify.

$6x(-3x - 5x - 5x^2 + x^3)$

$= 6x(-3x) + 6x(-5x) + 6x(-5x^2) + 6x(x^3)$

$= -18x^2 - 30x^2 - 30x^3 + 6x^4$

$= -48x^2 - 30x^3 + 6x^4$

EXAMPLE 2

Find $(x + 3)(2x - 1)$.

$(x + 3)(2x - 1)$	**Original expression**
$= x(2x) + x(-1) + 3(2x) + 3(-1)$	**FOIL method**
$= 2x^2 - x + 6x - 3$	**Multiply.**
$= 2x^2 + 5x - 3$	**Combine like terms.**

EXAMPLE 3

Find $(y + 8)^2$.

$(a + b)^2 = a^2 + 2ab + b^2$	**Square of a sum**
$(y + 8)^2 = (y)^2 + 2(y)(8) + 8^2$	**$a = y, b = 8$**
$= y^2 + 16y + 64$	**Simplify.**

Online Option

Math Online Take a self-check Chapter Readiness Quiz at glencoe.com.

Get Started on Chapter 8

You will learn several new concepts, skills, and vocabulary terms as you study Chapter 8. To get ready, identify important terms and organize your resources. You may wish to refer to **Chapter 0** to review prerequisite skills.

FOLDABLES® Study Organizer

Factoring and Quadratic Equations Make this Foldable to help you organize your Chapter 8 notes about factoring and quadratic equations. Begin with four sheets of grid paper.

1. **Fold** in half along the width. On the first two sheets, cut 5 centimeters along the fold at the ends. On the second two sheets cut in the center, stopping 5 centimeters from the ends.

First Sheets

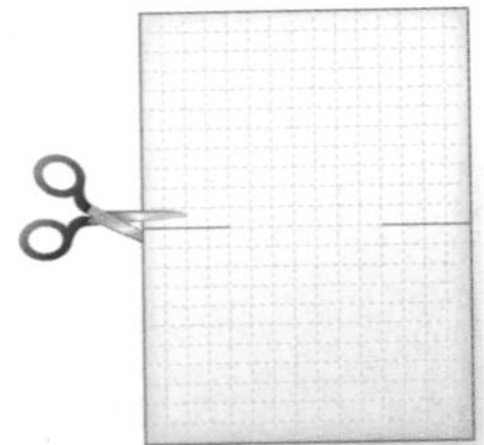

Second Sheets

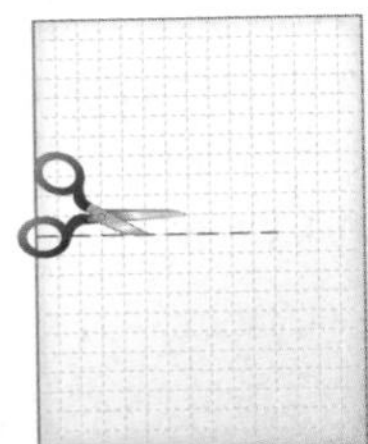

2. **Insert** the first sheets through the second sheets and align the folds. Label the front Chapter 8, Factoring and Quadratic Equations. Label the pages with lesson numbers and the last page with vocabulary.

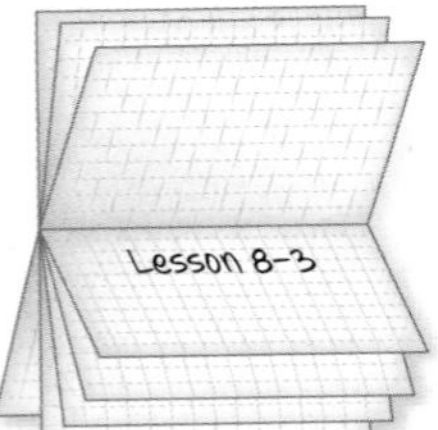

Math Online glencoe.com

- Study the chapter online
- Explore **Math in Motion**
- Get extra help from your own **Personal Tutor**
- Use **Extra Examples** for additional help
- Take a **Self-Check Quiz**
- **Review Vocabulary** in fun ways

New Vocabulary

English		Español
factored form	• p. 471 •	forma reducida
greatest common factor (GCF)	• p. 471 •	máximo común divisor (MCD)
factoring	• p. 476 •	factorización
factoring by grouping	• p. 477 •	factorización por agrupamiento
Zero Product Property	• p. 478 •	propiedad del producto de cero
quadratic equation	• p. 488 •	ecuación cuadrática
prime polynomial	• p. 495 •	polinomio primo
difference of two squares	• p. 499 •	diferencia de cuadrados
perfect square trinomial	• p. 505 •	trinomio cuadrado perfecto
Square Root Property	• p. 508 •	Propeidad de la raíz cuadrada

Review Vocabulary

absolute value • p. 103 • valor absoluto the absolute value of any number n is the distance the number is from zero on a number line and is written $|n|$

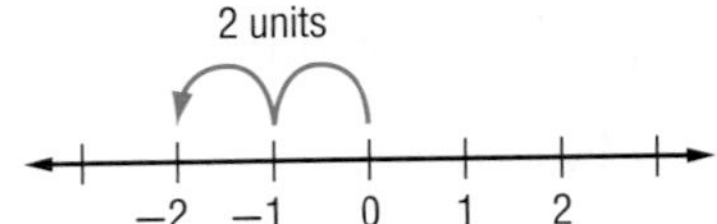

The absolute value of −2 is 2 because it is 2 units from 0.

perfect square • p. P7 • cuadrado perfecto a number with a square root that is a rational number

prime number • p. 861 • numero primo a whole number, greater than 1, with the only factor being 1 and itself

Multilingual eGlossary glencoe.com

8-1 Monomials and Factoring

Then

You multiplied monomials and divided a polynomial by a monomial. (Lesson 7-1 and 7-2)

Now

- Factor monomials.
- Find the greatest common factors of monomials.

New Vocabulary

factored form
greatest common factor (GCF)

Math Online

glencoe.com

- Extra Examples
- Personal Tutor
- Self-Check Quiz
- Homework Help
- Math in Motion

Why?

Susie is making beaded bracelets for extra money. She has 60 gemstone beads and 15 glass beads. She wants each bracelet to have only one type of bead and all of the bracelets to have the same number of beads. Susie needs to determine the *greatest common factor* of 60 and 15.

Factor Monomials Factoring a monomial is similar to factoring a whole number. A monomial is in **factored form** when it is expressed as the product of prime numbers and variables, and no variable has an exponent greater than 1.

EXAMPLE 1 Monomial in Factored Form

Factor $-20x^3y^2$ completely.

$-20x^3y^2 = -1 \cdot 20x^3y^2$ — Express -20 as $-1 \cdot 20$.

$= -1 \cdot 2 \cdot 10 \cdot x \cdot x \cdot x \cdot y \cdot y$ — $20 = 2 \cdot 10$, $x^3 = x \cdot x \cdot x$, and $y^2 = y \cdot y$

$= -1 \cdot 2 \cdot 2 \cdot 5 \cdot x \cdot x \cdot x \cdot y \cdot y$ — $10 = 2 \cdot 5$

Thus, $-20x^3y^2$ in factored form is $-1 \cdot 2 \cdot 2 \cdot 5 \cdot x \cdot x \cdot x \cdot y \cdot y$.

Check Your Progress

Factor each monomial completely.

1A. $34x^4y^3$

1B. $-52a^2b$

Personal Tutor glencoe.com

Greatest Common Factor Two or more whole numbers may have some common prime factors. The product of the common prime factors is called their greatest common factor. The **greatest common factor (GCF)** is the greatest number that is a factor of both original numbers. The GCF of two or more monomials can be found in a similar way.

EXAMPLE 2 GCF of a Set of Monomials

Find the GCF of $12a^2b^2c$ and $18ab^3$.

$12a^2b^2c = (2) \cdot 2 \cdot (3) \cdot (a) \cdot a \cdot (b) \cdot (b) \cdot c$ — Factor each number, and write all powers of variables as products.

$18ab^3 = (2) \cdot (3) \cdot 3 \cdot (a) \cdot (b) \cdot (b) \cdot b$ — Circle the common prime factors.

The GCF of $12a^2b^2c$ and $18ab^2$ is $2 \cdot 3 \cdot a \cdot b \cdot b$ or $6ab^2$.

Check Your Progress

Find the GCF of each pair of monomials.

2A. $6xy^3$, $18yz$

2B. $11a^2b$, $21ab^2$

2C. $30q^3r^2t$, $50q^2rt$

Personal Tutor glencoe.com

Real-World EXAMPLE 3 Find a GCF

FLOWERS A florist has 20 roses and 30 tulips to make bouquets. What is the greatest number of identical bouquets she can make without having any flowers left over? How many of each kind of flower will be in each bouquet?

Find the GCF of 20 and 30.

$20 = 2^2 \cdot 5$ **Write the prime factorization of each number.**

$30 = 2 \cdot 3 \cdot 5$ **The common prime factors are 2 and 5 or 10.**

The GCF of 20 and 30 is 10. So, the florist can make 10 bouquets. Since $2 \times 10 = 20$ and $3 \times 10 = 30$, each bouquet will have 2 roses and 3 tulips.

Math *in Motion*, Animation glencoe.com

Check Your Progress

3. What is the greatest possible value for the widths of two rectangles if their areas are 84 square inches and 70 square inches, respectively, and the length and width are whole numbers?

Personal Tutor glencoe.com

Check Your Understanding

Example 1 p. 471

Factor each monomial completely.

1. $12g^2h^4$

2. $-38rp^2t^2$

3. $-17x^3y^2z$

4. $23ab^3$

Examples 2 and 3 pp. 471–472

Find the GCF of each pair of monomials.

5. $24cd^3, 48c^2d$

6. $7gh, 11mp$

7. $8x^2y^5, 31xy^3$

8. $10ab, 25a$

9. GEOMETRY The areas of two rectangles are 15 square inches and 16 square inches. The length and width of both figures are whole numbers. If the rectangles have the same width, what is the greatest possible value for their widths?

Practice and Problem Solving

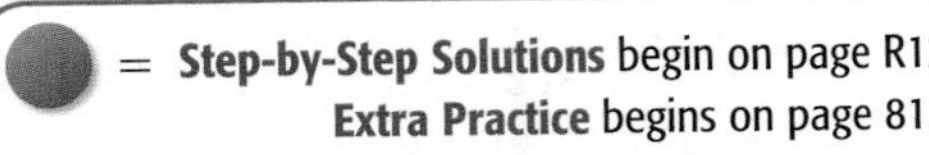

Example 1 p. 471

Factor each monomial completely.

10. $95xy^2$

11. $-35a^3c^2$

12. $42g^3h^3$

13. $81n^5p$

14. $-100q^4r$

15. $121abc^3$

Examples 2 and 3 pp. 471–472

Find the GCF of each set of monomials.

16. $25x^3, 45x^4, 65x^2$

17. $26z^2, 32z, 44z^4$

18. $30gh^2, 42g^2h, 66g$

19. $12qr, 8r^2, 16rt$

20. $42a^2b, 6a^2, 18a^3$

21. $15r^2t, 35t^2, 70rt$

22. BAKING Delsin wants to package the same number of cookies in each bag, and each bag should have every type of cookie. If he puts the greatest possible number of cookies in each bag, how many bags can he make?

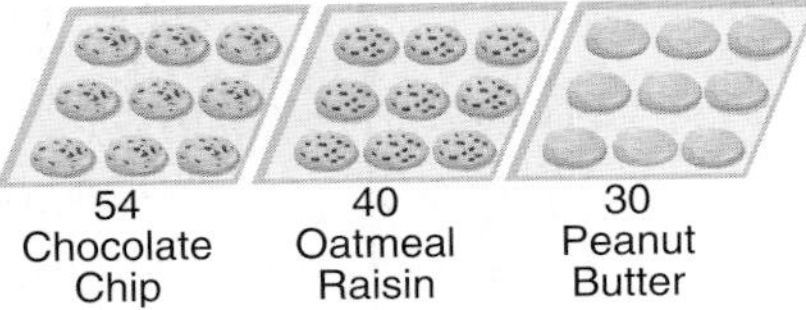

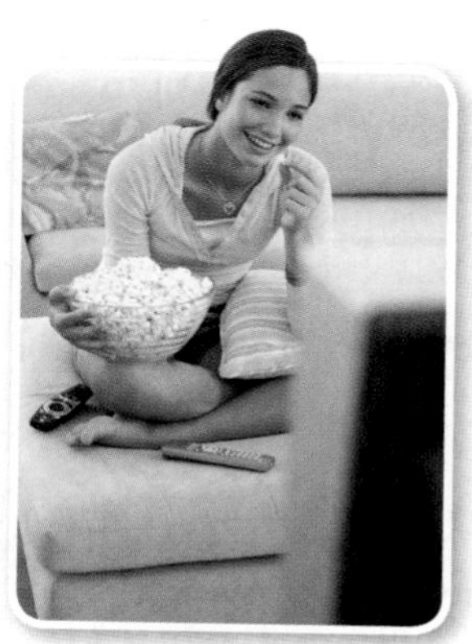

Real-World Link

About 77% of 18- to 25-year-olds say their favorite way to watch a movie at home is watching a DVD or video, while only 17% say they watch movies that are on television.

Source: 2006 Gen Next Survey

23. **GEOMETRY** The area of a triangle is 28 square inches. What are possible whole-number dimensions for the base and height of the triangle?

24. **MUSIC** In what ways can Clara organize her 36 CDs so that she has the same number of CDs on each shelf, at least 4 per shelf, and at least 2 shelves of CDs?

25. **MOVIES** In what ways can Shannon arrange her 80 DVDs so that she has at least 4 shelves of DVDS, the same number on each shelf, and at least 5 on each shelf?

26. **VOLUNTEER** Denzell is donating packages of school supplies to an elementary school where he volunteers. He bought 200 pencils, 150 glue sticks, and 120 folders. How many packages can Denzell make using an equal number of each item? How many items of each type will each package contain?

27. **NUMBER THEORY** *Twin primes* are two consecutive odd numbers that are prime. The first two pairs of twin primes are 3 and 5 and 5 and 7. List the next five pairs.

28. **MULTIPLE REPRESENTATIONS** In this problem, you will investigate a method of factoring a number.

    a. **ANALYTICAL** Copy the ladder diagram shown at the right six times and record six whole numbers, two of which are prime, in the top right portion of the diagrams.

    b. **ANALYTICAL** Choose a prime factor of one of your numbers. Record the factor on the left of the number in the diagram. Divide the two numbers. Keep dividing by prime factors until the quotient is 1. Add to or subtract boxes from the diagram as necessary. Repeat this process with all of your numbers.

    c. **VERBAL** What is the prime factorization of your six numbers?

3 | 12
  2 | 4
    2 | 2
      | 1

So, the prime factorization of 12 is $2^2 \cdot 3$.

H.O.T. Problems

Use Higher-Order Thinking Skills

29. **CHALLENGE** Find the least pair of numbers that satisfies the following conditions. The GCF of the numbers is 11. One number is even and the other number is odd. One number is not a multiple of the other.

30. **REASONING** The *least common multiple* (LCM) of two or more numbers is the least number that is a multiple of each number. Compare and contrast the GCF and LCM of two or more numbers.

31. **REASONING** Determine whether the following statement is *true* or *false*. Provide an example or counterexample.

    Two monomials always have a greatest common factor that is not equal to 1.

32. **CHALLENGE** Two or more integers or monomials with a GCF of 1 are said to be *relatively prime*. Copy and complete the chart to determine which pairs of monomials are relatively prime.

Monomial	Prime Factorization
$15a^2bc^3$	
$6b^3c^3d$	
$12cd^2f$	
$22d^3fg^2$	
$30f^2gh^2$	

33. **OPEN ENDED** Name three monomials with a GCF of $6y^3$. Explain your answer.

34. **WRITING IN MATH** Define *prime factorization* in your own words. Explain how to find the prime factorization of a monomial, and how a prime factorization helps you determine the GCF of two or more monomials.

Standardized Test Practice

35. Abigail surveyed 320 of her classmates about what type of movie they prefer. The results of the survey are shown below. What percent of her classmates enjoyed action movies?

Type of Movie	Number of Responses
comedy	160
drama	25
science fiction	55
action	80

A 25%
C 75%
B 50%
D 95%

36. What is the value of c in the equation $4c - 27 = 19 + 2c$?

F -4
G 4
H 23
J 46

37. Which equation best represents a line parallel to the line shown below?

A $y = 2x + 4$
B $y = -2x - 5$
C $y = \frac{1}{2}x - 6$
D $y = -\frac{1}{2}x + 3$

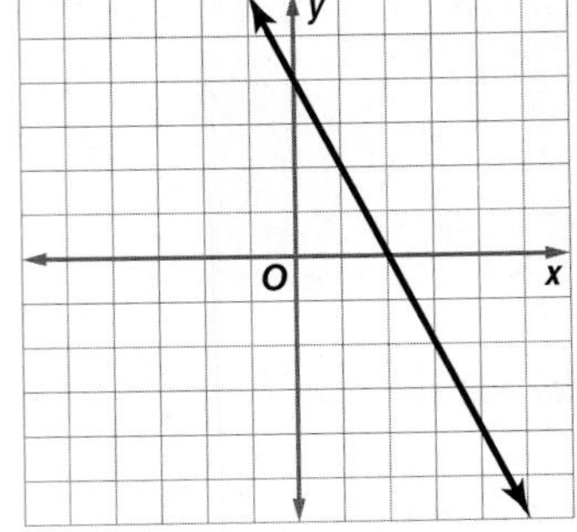

38. SHORT RESPONSE The table shows a five-day forecast indicating high (H) and low (L) temperatures. Organize the temperatures in a matrix.

	Mon	Tue	Wed	Thurs	Fri
H	92	87	85	88	90
L	68	64	62	65	66

Spiral Review

Find each product. (Lesson 7-8)

39. $(a - 4)^2$
40. $(c + 6)^2$
41. $(z - 5)^2$
42. $(n - 3)(n + 3)$
43. $(y + 2)^2$
44. $(d - 7)(d + 7)$

Find each product. (Lesson 7-7)

45. $(2m - 3)(m + 4)$
46. $(h - 2)(3h - 5)$
47. $(t + 2)(t + 9)$
48. $(8r - 1)(r - 6)$
49. $(p + 3q)(p + 3q)$
50. $(n - 4)(n + 2)(n + 1)$

Write an augmented matrix to solve each system of equations. (Lesson 6-7)

51. $y = 2x + 3$
$y = 4x - 1$

52. $8x + 2y = 13$
$4x + y = 11$

53. $-x + \frac{1}{3}y = 5$
$2x + 3y = 1$

54. FINANCIAL LITERACY Suppose you have already saved $50 toward the cost of a new television. You plan to save $5 more each week. Write and graph an equation for the total amount T that you will have w weeks from now. (Lesson 4-1)

Skills Review

Use the Distributive Property to rewrite each expression. (Lesson 1-4)

55. $2(4x - 7)$
56. $\frac{1}{2}d(2d + 6)$
57. $-h(6h - 1)$
58. $9m - 9p$
59. $5y - 10$
60. $3z - 6x$

Algebra Lab: Factoring Using the Distributive Property

Math Online glencoe.com
Math *in Motion*, Animation

When two or more numbers are multiplied, these numbers are *factors* of the product. Sometimes you know the product of binomials and are asked to find the factors. This is called factoring. You can use algebra tiles and a product mat to factor binomials.

ACTIVITY 1 Use algebra tiles to factor $2x - 8$.

Step 1 Model $2x - 8$.

Step 2 Arrange the tiles into a rectangle. The total area of the rectangle represents the product, and its length and width represent the factors.

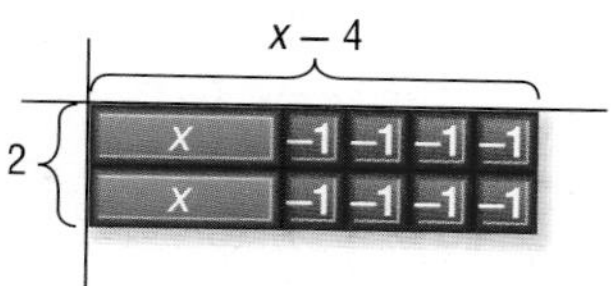

The rectangle has a width of 2 and a length of $x - 4$. Therefore, $2x - 8 = 2(x - 4)$.

ACTIVITY 2 Use algebra tiles to factor $x^2 + 3x$.

Step 1 Model $x^2 + 3x$.

Step 2 Arrange the tiles into a rectangle.

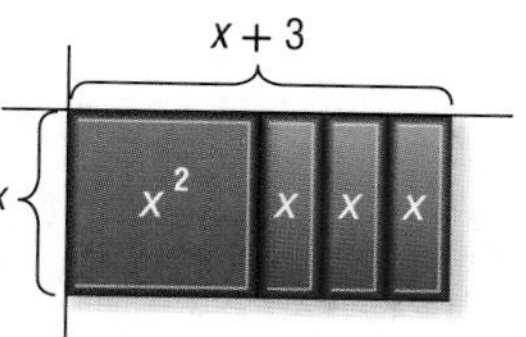

The rectangle has a width of x and a length of $x + 3$. Therefore, $x^2 + 3x = x(x + 3)$.

Model and Analyze

Use algebra tiles to factor each binomial.

1. $4x + 12$ **2.** $4x - 6$ **3.** $3x^2 + 4x$ **4.** $10 - 2x$

Determine whether each binomial can be factored. Justify your answer with a drawing.

5. $6x - 9$ **6.** $5x - 4$ **7.** $4x^2 + 7$ **8.** $x^2 + 3x$

9. WRITING IN MATH Write a paragraph that explains how you can use algebra tiles to determine whether a binomial can be factored. Include an example of one binomial that can be factored and one that cannot.

8-2 Using the Distributive Property

Then
You found the GCF of a set of monomials. (Lesson 8-1)

Now
- Use the Distributive Property to factor polynomials.
- Solve quadratic equations of the form $ax^2 + bx = 0$.

New Vocabulary
factoring
factoring by grouping
Zero Product Property

Math Online
glencoe.com
- Extra Examples
- Personal Tutor
- Self-Check Quiz
- Homework Help

Why?

The cost of rent for Ms. Cole's store is determined by the square footage of the space. The area of the store can be modeled by the equation $A = 1.6w^2 + 6w$, where w is the width of the store in feet. We can use factoring and the Zero Product Property to find possible dimensions of the store.

Use the Distributive Property to Factor In Chapter 7, the Distributive Property was used to multiply a monomial by a polynomial.

$$5z(4z + 7) = 5z(4z) + 5z(7)$$
$$= 20z^2 + 35z$$

You can work backward to express a polynomial as a product of a monomial factor and a polynomial factor.

$$1.6w^2 + 6w = 1.6w(w) + 6(w)$$
$$= w(1.6w + 6)$$

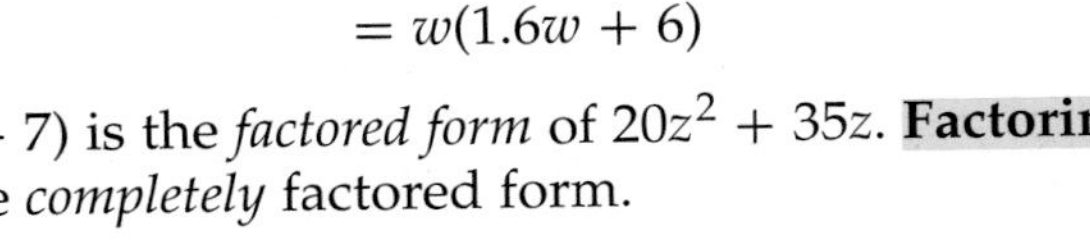

So, $5z(4z + 7)$ is the *factored form* of $20z^2 + 35z$. **Factoring** a polynomial involves finding the *completely* factored form.

EXAMPLE 1 Use the Distributive Property

Use the Distributive Property to factor each polynomial.

a. $\mathbf{27y^2 + 18y}$

Find the GCF of each term.

$27y^2 = 3 \cdot 3 \cdot 3 \cdot y \cdot y$ — Factor each term.

$18y = 2 \cdot 3 \cdot 3 \cdot y$ — Circle common factors.

$\text{GCF} = 3 \cdot 3 \cdot y$ or $9y$

Write each term as the product of the GCF and the remaining factors. Use the Distributive Property to *factor out* the GCF.

$27y^2 + 18y = 9y(3y) + 9y(2)$ — Rewrite each term using the GCF.

$= 9y(3y + 2)$ — Distributive Property

b. $\mathbf{-4a^2b - 8ab^2 + 2ab}$

$-4a^2b = -1 \cdot 2 \cdot 2 \cdot a \cdot a \cdot b$ — Factor each term.

$-8ab^2 = -1 \cdot 2 \cdot 2 \cdot 2 \cdot a \cdot b \cdot b$ — Circle common factors.

$2ab = 2 \cdot a \cdot b$

$\text{GCF} = 2 \cdot a \cdot b$ or $2ab$

$-4a^2b - 8ab^2 + 2ab = 2ab(-2a) - 2ab(4b) + 2ab(1)$ — Rewrite each term using the GCF.

$= 2ab(-2a - 4b + 1)$ — Distributive Property

Check Your Progress

1A. $15w - 3v$

1B. $7u^2t^2 + 21ut^2 - ut$

Personal Tutor glencoe.com

Using the Distributive Property to factor polynomials with four or more terms is called **factoring by grouping** because terms are put into groups and then factored. The Distributive Property is then applied to a common binomial factor.

Key Concept Factoring by Grouping

For Your FOLDABLE

Words A polynomial can be factored by grouping only if all of the following conditions exist.

- There are four or more terms.
- Terms have common factors that can be grouped together.
- There are two common factors that are identical or additive inverses of each other.

Symbols
$$ax + bx + ay + by = (ax + bx) + (ay + by)$$
$$= x(a + b) + y(a + b)$$
$$= (x + y)(a + b)$$

EXAMPLE 2 Factor by Grouping

Factor $4qr + 8r + 3q + 6$.

$4qr + 8r + 3q + 6$ — **Original expression**

$= (4qr + 8r) + (3q + 6)$ — **Group terms with common factors.**

$= 4r(q + 2) + 3(q + 2)$ — **Factor the GCF from each group.**

Notice that $(q + 2)$ is common in both groups, so it becomes the GCF.

$= (4r + 3)(q + 2)$ — **Distributive Property**

Check Your Progress

Factor each polynomial.

2A. $rn + 5n - r - 5$

2B. $3np + 15p - 4n - 20$

Personal Tutor glencoe.com

It can be helpful to recognize when binomials are additive inverses of each other. For example $6 - a = -1(a - 6)$.

StudyTip

Check To check your factored answers, multiply your factors out. You should get your original expression as a result.

EXAMPLE 3 Factor by Grouping with Additive Inverses

Factor $2mk - 12m + 42 - 7k$.

$2mk - 12m + 42 - 7k$

$= (2mk - 12m) + (42 - 7k)$ — **Group terms with common factors.**

$= 2m(k - 6) + 7(6 - k)$ — **Factor the GCF from each group.**

$= 2m(k - 6) + 7[(-1)(k - 6)]$ — **$6 - k = -1(k - 6)$**

$= 2m(k - 6) - 7(k - 6)$ — **Associative Property**

$= (2m - 7)(k - 6)$ — **Distributive Property**

Check Your Progress

Factor each polynomial.

3A. $c - 2cd + 8d - 4$

3B. $3p - 2p^2 - 18p + 27$

Personal Tutor glencoe.com

Solve Equations by Factoring Some equations can be solved by factoring. Consider the following.

$3(0) = 0$ $\quad 0(2 - 2) = 0$ $\quad -312(0) = 0$ $\quad 0(0.25) = 0$

Notice that in each case, at least one of the factors is 0. These examples are demonstrations of the **Zero Product Property**.

Key Concept **Zero Product Property** For Your FOLDABLE

Words If the product of two factors is 0, then at least one of the factors must be 0.

Symbols For any real numbers a and b, if $ab = 0$, then $a = 0$, $b = 0$, or both a and b equal zero.

Recall from Lesson 3-2 that a solution or root of an equation is any value that makes the equation true.

Watch Out!

Unknown Value It may be tempting to solve an equation by dividing each side by the variable. However, the variable has an unknown value, so you may be dividing by 0, which is undefined.

EXAMPLE 4 Solve Equations

Solve each equation. Check your solutions.

a. $(2d + 6)(3d - 15) = 0$

$(2d + 6)(3d - 15) = 0$ — **Original equation**

$2d + 6 = 0$ or $3d - 15 = 0$ — **Zero Product Property**

$2d = -6$ $\quad 3d = 15$ — **Solve each equation.**

$d = -3$ $\quad d = 5$ — **Divide.**

The roots are −3 and 5.

CHECK Substitute −3 and 5 for d in the original equation.

$$(2d + 6)(3d - 15) = 0 \qquad (2d + 6)(3d - 15) = 0$$
$$[2(-3) + 6][3(-3) - 15] \stackrel{?}{=} 0 \qquad [2(5) + 6][3(5) - 15] \stackrel{?}{=} 0$$
$$(-6 + 6)(-9 - 15) \stackrel{?}{=} 0 \qquad (10 + 6)(15 - 15) \stackrel{?}{=} 0$$
$$(0)(-24) \stackrel{?}{=} 0 \qquad 16(0) \stackrel{?}{=} 0$$
$$0 = 0 \checkmark \qquad 0 = 0 \checkmark$$

b. $c^2 = 3c$

$c^2 = 3c$ — **Original equation**

$c^2 - 3c = 0$ — **Subtract $3c$ from each side to get 0 on one side of the equation.**

$c(c - 3) = 0$ — **Factor by using the GCF to get the form $ab = 0$.**

$c = 0$ or $c - 3 = 0$ — **Zero Product Property**

$c = 3$ — **Solve each equation.**

The roots are 0 and 3. — **Check by substituting 0 and 3 for c.**

Check Your Progress

4A. $3n(n + 2) = 0$ $\quad$ **4B.** $8b^2 - 40b = 0$ $\quad$ **4C.** $x^2 = -10x$

Personal Tutor glencoe.com

Real-World Link

Dog agility tests a person's skills as a trainer and handler. Competitors race through an obstacle course that includes hurdles, tunnels, a see-saw, and line poles.

Source: United States Dog Agility Association

Real-World EXAMPLE 5 Use Factoring

AGILITY Penny is a Labrador Retriever who competes with her trainer in the agility course. Within the course, Penny must leap over a hurdle. Penny's jump can be modeled by the equation $h = -16t^2 + 20t$, where h is the height of the leap in inches at t seconds. Find the values of t when $h = 0$.

$h = -16t^2 + 20t$	Original equation
$0 = -16t^2 + 20t$	Substitution, $h = 0$
$0 = 4t(-4t + 5)$	Factor by using the GCF.
$4t = 0$ or $-4t + 5 = 0$	Zero Product Property
$t = 0$ $\quad -4t = -5$	Solve each equation.
$t = \frac{5}{4}$ or 1.25	Divide each side by -4.

Penny's height is 0 inches at 0 seconds and 1.25 seconds into the jump.

Check Your Progress

5. **KANGAROOS** The hop of a kangaroo can be modeled by $h = 24t - 16t^2$ where h represents the height of the hop in meters and t is the time in seconds. Find the values of t when $h = 0$.

Personal Tutor glencoe.com

Check Your Understanding

Example 1 p. 476

Use the Distributive Property to factor each polynomial.

1. $21b - 15a$
2. $14c^2 + 2c$
3. $10g^2h^2 + 9gh^2 - g^2h$
4. $12jk^2 + 6j^2k + 2j^2k^2$

Examples 2 and 3 p. 477

Factor each polynomial.

5. $np + 2n + 8p + 16$
6. $xy - 7x + 7y - 49$
7. $3bc - 2b - 10 + 15c$
8. $9fg - 45f - 7g + 35$

Example 4 p. 478

Solve each equation. Check your solutions.

9. $3k(k + 10) = 0$
10. $(4m + 2)(3m - 9) = 0$
11. $20p^2 - 15p = 0$
12. $r^2 = 14r$

Example 5 p. 479

13. **SPIDERS** Jumping spiders can commonly be found in homes and barns throughout the United States. A jumping spider's jump can be modeled by the equation $h = 33.3t - 16t^2$, where t represents the time in seconds and h is the height in feet.
    a. When is the spider's height at 0 feet?
    b. What is the spider's height after 1 second? after 2 seconds?

14. **ROCKETS** At a Fourth of July celebration, a rocket is launched straight up with an initial velocity of 125 feet per second. The height h of the rocket in feet above sea level is modeled by the formula $h = 125t - 16t^2$, where t is the time in seconds after the rocket is launched.
    a. What is the height of the rocket when it returns to the ground?
    b. Let $h = 0$ in the equation and solve for t.
    c. How many seconds will it take for the rocket to return to the ground?

Practice and Problem Solving

= **Step-by-Step Solutions** begin on page R12.
Extra Practice begins on page 815.

Example 1 p. 476

Use the Distributive Property to factor each polynomial.

15. $16t - 40y$
16. $30v + 50x$
17. $2k^2 + 4k$
18. $5z^2 + 10z$
19. $4a^2b^2 + 2a^2b - 10ab^2$
20. $5c^2v - 15c^2v^2 + 5c^2v^3$

Examples 2 and 3 p. 477

Factor each polynomial.

21. $fg - 5g + 4f - 20$
22. $a^2 - 4a - 24 + 6a$
23. $hj - 2h + 5j - 10$
24. $xy - 2x - 2 + y$
25. $45pq - 27q - 50p + 30$
26. $24ty - 18t + 4y - 3$
27. $3dt - 21d + 35 - 5t$
28. $8r^2 + 12r$
29. $21th - 3t - 35h + 5$
30. $vp + 12v + 8p + 96$
31. $5br - 25b + 2r - 10$
32. $2nu - 8u + 3n - 12$
33. $5gf^2 + g^2f + 15gf$
34. $rp - 9r + 9p - 81$
35. $27cd^2 - 18c^2d^2 + 3cd$
36. $18r^3t^2 + 12r^2t^2 - 6r^2t$
37. $48tu - 90t + 32u - 60$
38. $16gh + 24g - 2h - 3$

Example 4 p. 478

Solve each equation. Check your solutions.

39. $3b(9b - 27) = 0$
40. $2n(3n + 3) = 0$
41. $(8z + 4)(5z + 10) = 0$
42. $(7x + 3)(2x - 6) = 0$
43. $b^2 = -3b$
44. $a^2 = 4a$

Example 5 p. 479

45. **GEOMETRY** Use the drawing at the right.

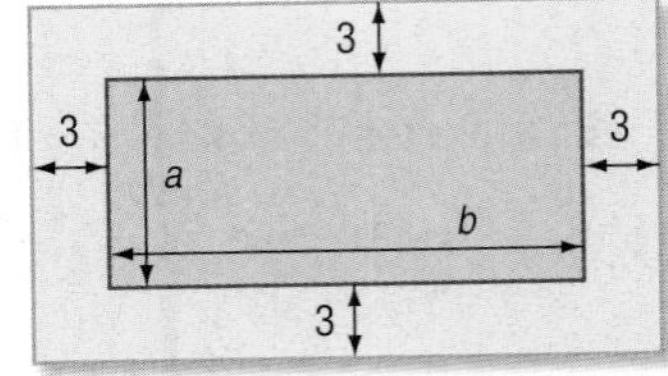

   a. Write an expression in factored form to represent the area of the blue section.
   b. Write an expression in factored form to represent the area of the region formed by the outer edge.
   c. Write an expression in factored form to represent the orange region.

46. **FIREWORKS** A ten-inch fireworks shell is fired from ground level. The height of the shell in feet is given by the formula $h = 263t - 16t^2$, where t is the time in seconds after launch.
   a. Write the expression that represents the height in factored form.
   b. At what time will the height be 0? Is this answer practical? Explain.
   c. What is the height of the shell 8 seconds and 10 seconds after being fired?
   d. At 10 seconds, is the shell rising or falling?

47. **ARCHITECTURE** The frame of a doorway is an arch that can be modeled by the graph of the equation $y = -3x^2 + 12x$, where x and y are measured in feet. On a coordinate plane, the floor is represented by the x-axis.
   a. Make a table of values for the height of the arch if $x = 0, 1, 2, 3,$ and 4 feet.
   b. Plot the points from the table on a coordinate plane and connect the points to form a smooth curve to represent the arch.
   c. How high is the doorway?

Real-World Link

The Xtreme Skyflyer at Virginia's Kings Dominion uses a free-fall/cable swing to hoist people about 160 feet in the air. Then the riders experience a free-fall followed by a period of swinging back and forth.

Source: Theme Park Insider

48. **RIDES** Suppose the height of a rider after being dropped can be modeled by $h = -16t^2 - 96t + 160$, where h is the height in feet and t is time in seconds.
    a. Write an expression to represent the height in factored form.
    b. From what height is the rider initially dropped?
    c. At what height will the rider be after 3 seconds of falling? Is this possible? Explain.

49. **ARCHERY** The height h in feet of an arrow can be modeled by the equation $h = 64t - 16t^2$, where t is time in seconds. Ignoring the height of the archer, how long after the arrow is released does it hit the ground?

50. **TENNIS** A tennis player hits a tennis ball upward with an initial velocity of 80 feet per second. The height h in feet of the tennis ball can be modeled by the equation $h = 80t - 16t^2$, where t is time in seconds. Ignoring the height of the tennis player, how long does it take the ball to hit the ground?

51. **MULTIPLE REPRESENTATIONS** In this problem, you will explore the *box method* of factoring. To factor $x^2 + x - 6$, write the first term in the top left-hand corner of the box, and then write the last term in the lower right-hand corner.

	?	?
?	x^2	?
?	?	-6

    a. **ANALYTICAL** Determine which two factors have a product of -6 and a sum of 1.
    b. **SYMBOLIC** Write each factor in an empty square in the box. Include the positive or negative sign and variable.
    c. **ANALYTICAL** Find the factor for each row and column of the box. What are the factors of $x^2 + x - 6$?
    d. **VERBAL** Describe how you would use the box method to factor $x^2 - 3x - 40$.

H.O.T. Problems

Use **H**igher-**O**rder **T**hinking Skills

52. **FIND THE ERROR** Hernando and Rachel are solving $2m^2 = 4m$. Is either of them correct? Explain your reasoning.

Hernando

$2m^2 = 4m$

$\frac{2m^2}{m} = \frac{4m^2}{2m}$

$2m = 2$

$m = 1$

Rachel

$2m^2 = 4m$

$2m^2 - 4m = 0$

$2m(m - 2) = 0$

$2m = 0$ or $m - 2 = 0$

$m = 0$ or 2

53. **CHALLENGE** Given the equation $(ax + b)(ax - b) = 0$, solve for x. What do we know about the values of a and b?

54. **OPEN ENDED** Write a four-term polynomial that can be factored by grouping. Then factor the polynomial.

55. **REASONING** Given the equation $c = a^2 - ab$, for what values of a and b does $c = 0$?

56. **WRITING IN MATH** Explain how to solve a quadratic equation by using the Zero Product Property.

Standardized Test Practice

57. Which is a factor of $6z^2 - 3z - 2 + 4z$?

A $2z + 1$
B $3z - 2$
C $z + 2$
D $2z - 1$

58. PROBABILITY Hailey has 10 blocks: 2 red, 4 blue, 3 yellow, and 1 green. What is the probability that Hailey chooses either a red or a yellow block?

F $\frac{3}{10}$
G $\frac{1}{5}$
H $\frac{1}{2}$
J $\frac{7}{10}$

59. GRIDDED RESPONSE Cho is making a 140-inch by 160-inch quilt with quilt squares that measure 8 inches on each side. How many will be needed to make the quilt?

60. GEOMETRY The area of the right triangle shown below is $5h$ square centimeters. What is the height of the triangle?

A 2 cm
B 5 cm
C 8 cm
D 10 cm

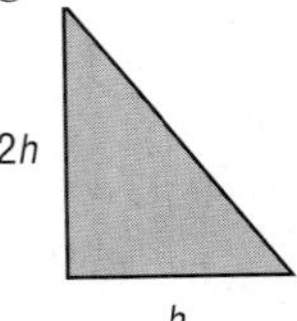

Spiral Review

Find the GCF of each set of monomials. (Lesson 8-1)

61. 15, 25
62. 40, 100
63. $16x, 24x^2$
64. $30a^2, 50ab^2$
65. $8c^2d^3, 16c^3d$
66. $4y, 18y^2, 6y^3$

67. GENETICS Brown genes B are dominant over blue genes b. A person with genes BB or Bb has brown eyes. Someone with genes bb has blue eyes. Elisa has brown eyes with Bb genes, and Bob has blue eyes. Write an expression for the possible eye coloring of Elisa and Bob's children. Determine the probability that their child would have blue eyes. (Lesson 7-8)

Simplify. (Lesson 7-1)

68. $(ab^4)(ab^2)$
69. $(p^5r^4)(p^2r)$
70. $(-7c^3d^4)(4cd^3)$
71. $(9xy^7)^2$
72. $[(3^2)^4]^2$
73. $[(4^2)^3]^2$

74. BASKETBALL In basketball, a free throw is 1 point and a field goal is either 2 or 3 points. In a season, Tim Duncan of the San Antonio Spurs scored a total of 1342 points. The total number of 2-point field goals and 3-point field goals was 517, and he made 305 of the 455 free throws that he attempted. Find the number of 2-point field goals and 3-point field goals Duncan made that season. (Lesson 6-4)

Solve each inequality. Check your solution. (Lesson 5-3)

75. $3y - 4 > -37$
76. $-5q + 9 > 24$
77. $-2k + 12 < 30$
78. $5q + 7 \leq 3(q + 1)$
79. $\frac{z}{4} + 7 \geq -5$
80. $8c - (c - 5) > c + 17$

Skills Review

Find each product. (Lesson 7-7)

81. $(a + 2)(a + 5)$
82. $(d + 4)(d + 10)$
83. $(z - 1)(z - 8)$
84. $(c + 9)(c - 3)$
85. $(x - 7)(x - 6)$
86. $(g - 2)(g + 11)$

Algebra Lab: Factoring Trinomials

Math Online glencoe.com
Math *in Motion*, Animation

You can use algebra tiles to factor trinomials. If a polynomial represents the area of a rectangle formed by algebra tiles, then the rectangle's length and width are *factors* of the area. If a rectangle cannot be formed to represent the trinomial, then the trinomial is not factorable.

ACTIVITY 1 Factor $x^2 + bx + c$

Use algebra tiles to factor $x^2 + 4x + 3$.

Step 1 Model $x^2 + 4x + 3$.

Step 2 Place the x^2-tile at the corner of the product mat. Arrange the 1-tiles into a rectangular array. Because 3 is prime, the 3 tiles can be arranged in a rectangle in one way, a 1-by-3 rectangle.

Step 3 Complete the rectangle with the x-tiles.

The rectangle has a width of $x + 1$ and a length of $x + 3$.

Therefore, $x^2 + 4x + 3 = (x + 1)(x + 3)$.

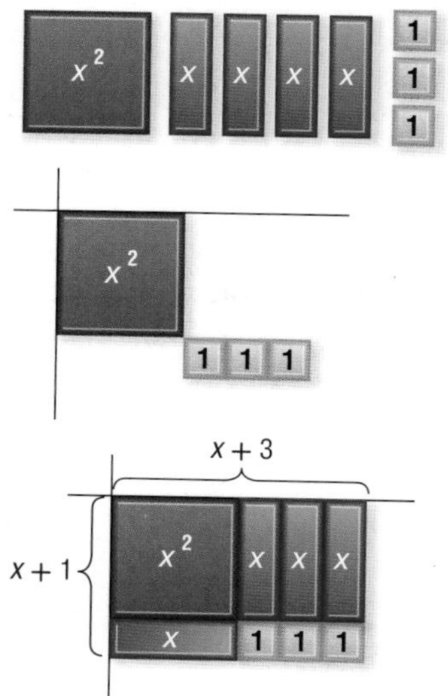

ACTIVITY 2 Factor $x^2 + bx + c$

Use algebra tiles to factor $x^2 + 8x + 12$.

Step 1 Model $x^2 + 8x + 12$.

Step 2 Place the x^2-tile at the corner of the product mat. Arrange the 1-tiles into a rectangular array. Since $12 = 3 \times 4$, try a 3-by-4 rectangle. Try to complete the rectangle. Notice that there is an extra x-tile.

Step 3 Arrange the 1-tiles into a 2-by-6 rectangular array. This time you can complete the rectangle with the x-tiles.

The rectangle has a width of $x + 2$ and a length of $x + 6$.

Therefore, $x^2 + 8x + 12 = (x + 2)(x + 6)$.

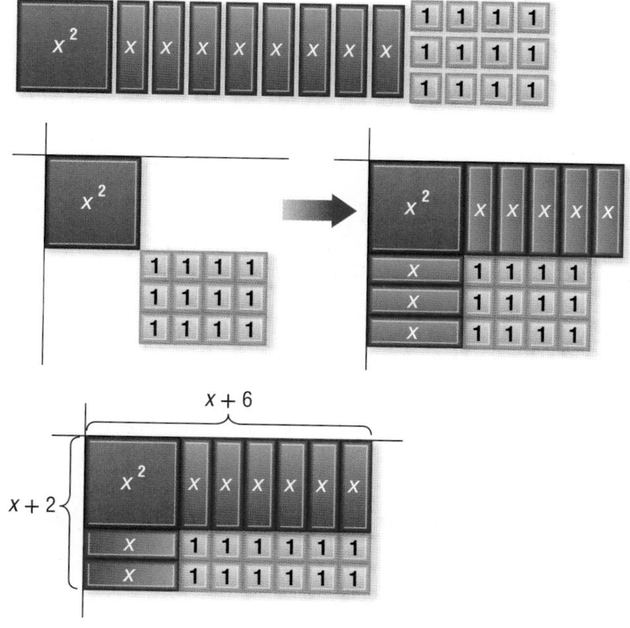

ACTIVITY 3 Factor $x^2 - bx + c$

Use algebra tiles to factor $x^2 - 5x + 6$.

Step 1 Model $x^2 - 5x + 6$.

Step 2 Place the x^2-tile at the corner of the product mat. Arrange the 1-tiles into a 2-by-3 rectangular array as shown.

Step 3 Complete the rectangle with the x-tiles. The rectangle has a width of $x - 2$ and a length of $x - 3$.

Therefore, $x^2 - 5x + 6 = (x - 2)(x - 3)$.

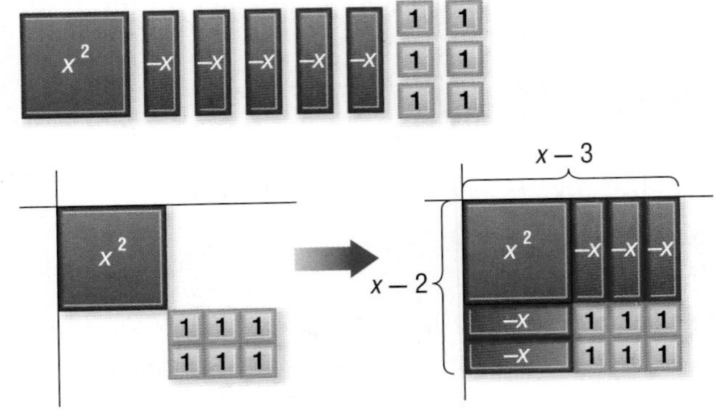

ACTIVITY 4 Factor $x^2 - bx - c$

Use algebra tiles to factor $x^2 - 4x - 5$.

Step 1 Model $x^2 - 4x - 5$.

Step 2 Place the x^2-tile at the corner of the product mat. Arrange the 1-tiles into a 1-by-5 rectangular array as shown.

Step 3 Place the x-tile as shown. Recall that you can add zero pairs without changing the value of the polynomial. In this case, add a zero pair of x-tiles.

The rectangle has a width of $x + 1$ and a length of $x - 5$.

Therefore, $x^2 - 4x - 5 = (x + 1)(x - 5)$.

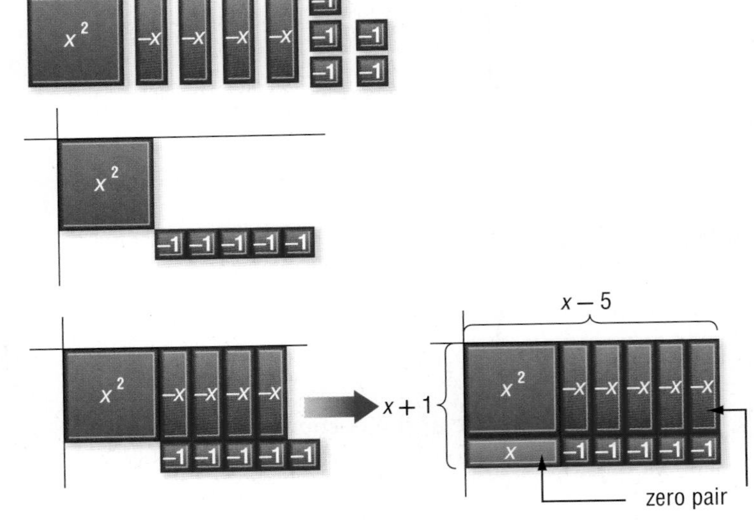

Model and Analyze

Use algebra tiles to factor each trinomial.

1. $x^2 + 3x + 2$
2. $x^2 + 6x + 8$
3. $x^2 + 3x - 4$
4. $x^2 - 7x + 12$
5. $x^2 + 7x + 10$
6. $x^2 - 2x + 1$
7. $x^2 + x - 12$
8. $x^2 - 8x + 15$

Tell whether each trinomial can be factored. Justify your answer with a drawing.

9. $x^2 + 3x + 6$
10. $x^2 - 5x - 6$
11. $x^2 - x - 4$
12. $x^2 - 4$

13. **WRITING IN MATH** How can you use algebra tiles to determine whether a trinomial can be factored?

8-3 Quadratic Equations: $x^2 + bx + c = 0$

Then

You multiplied binomials by using the FOIL method. (Lesson 7-7)

Now

- Factor trinomials of the form $x^2 + bx + c$.
- Solve equations of the form $x^2 + bx + c = 0$.

New Vocabulary

quadratic equation

Math Online

glencoe.com

- Extra Examples
- Personal Tutor
- Self-Check Quiz
- Homework Help

Why?

Diana is having a rectangular in-ground swimming pool installed and she wants to include a 24-foot fence around the pool. The pool requires a space of 36 square feet. What dimensions should the pool have?

To solve this problem, the landscape architect needs to find two numbers that have a product of 36 and a sum of 12, half the perimeter of the pool.

Factor $x^2 + bx + c$ In Lesson 7-7, you learned how to multiply two binomials by using the FOIL method. Each of the binomials was a factor of the product. The pattern for multiplying two binomials can be used to factor certain types of trinomials.

$$(x + 3)(x + 4) = x^2 + 4x + 3x + 3 \cdot 4 \quad \text{Use the FOIL method.}$$
$$= x^2 + (4 + 3)x + 3 \cdot 4 \quad \text{Distributive Property}$$
$$= x^2 + 7x + 12 \quad \text{Simplify.}$$

Notice that the coefficient of the middle term, $7x$, is the sum of 3 and 4, and the last term, 12, is the product of 3 and 4.

Observe the following pattern in this multiplication.

$$(x + 3)(x + 4) = x^2 + (4 + 3)x + (3 \cdot 4)$$
$$(x + m)(x + p) = x^2 + (p + m)x + mp \quad \text{Let } 3 = m \text{ and } 4 = p.$$
$$= x^2 + \underbrace{(m + p)x}_{bx} + \underbrace{mp}_{c} \quad \text{Commutative (+)}$$
$$x^2 + bx + c \quad b = m + p \text{ and } c = mp$$

Notice that the coefficient of the middle term is the sum of m and p, and the last term is the product of m and p. This pattern can be used to factor trinomials of the form $x^2 + bx + c$.

Key Concept Factoring $x^2 + bx + c$ — For Your FOLDABLE

Words To factor trinomials in the form $x^2 + bx + c$, find two integers, m and p, with a sum of b and a product of c. Then write $x^2 + bx + c$ as $(x + m)(x + p)$.

Symbols $x^2 + bx + c = (x + m)(x + p)$ when $m + p = b$ and $mp = c$

Example $x^2 + 6x + 8 = (x + 2)(x + 4)$, because $2 + 4 = 6$ and $2 \cdot 4 = 8$.

When c is positive, its factors have the same signs. Both of the factors are positive or negative based upon the sign of b. If b is positive, the factors are positive. If b is negative, the factors are negative.

Problem-SolvingTip

Guess and Check When factoring a trinomial, make an educated guess, check for reasonableness, and then adjust the guess until the correct answer is found.

EXAMPLE 1 *b* and *c* Are Positive

Factor $x^2 + 9x + 20$.

In this trinomial, $b = 9$ and $c = 20$. Since c is positive and b is positive, you need to find two positive factors with a sum of 9 and a product of 20. Make an organized list of the factors of 20, and look for the pair of factors with a sum of 9.

Factors of 20	Sum of Factors
1, 20	21
2, 10	12
4, 5	9

The correct factors are 4 and 5.

$x^2 + 9x + 20 = (x + m)(x + p)$ **Write the pattern.**

$= (x + 4)(x + 5)$ **$m = 4$ and $p = 5$**

CHECK You can check this result by multiplying the two factors. The product should be equal to the original expression.

$(x + 4)(x + 5) = x^2 + 5x + 4x + 20$ **FOIL Method**

$= x^2 + 9x + 20$ ✓ **Simplify.**

Check Your Progress **Factor each polynomial.**

1A. $d^2 + 11x + 24$ **1B.** $9 + 10t + t^2$

Personal Tutor glencoe.com

When factoring a trinomial in which b is negative and c is positive, use what you know about the product of binomials to narrow the list of possible factors.

StudyTip

Finding Factors Once the correct factors are found, it is not necessary to test any other factors. In Example 2, -2 and -6 are the correct factors, so -3 and -4 do not need to be tested.

EXAMPLE 2 *b* Is Negative and *c* Is Positive

Factor $x^2 - 8x + 12$.

In this trinomial, $b = -8$ and $c = 12$. Since c is positive and b is negative, you need to find two negative factors with a sum of -8 and a product of 12.

Factors of 12	Sum of Factors
−1, −12	−13
−2, −6	−8
−3, −4	−7

The correct factors are −2 and −6.

$x^2 - 8x + 12 = (x + m)(x + p)$ **Write the pattern.**

$= (x - 2)(x - 6)$ **$m = -2$ and $p = -6$**

CHECK Graph $y = x^2 - 8x + 12$ and $y = (x - 2)(x - 6)$ on the same screen. Since only one graph appears, the two graphs must coincide. Therefore, the trinomial has been factored correctly. ✓

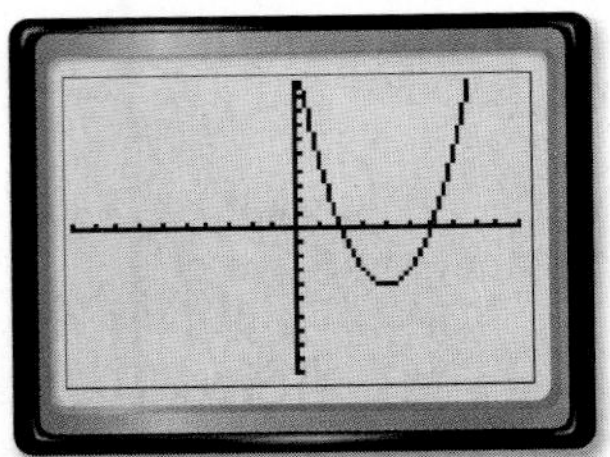

[−10, 10] scl: 1 by [−10, 10] scl: 1

Check Your Progress **Factor each polynomial.**

2A. $21 - 22m + m^2$ **2B.** $w^2 - 11w + 28$

Personal Tutor glencoe.com

Review Vocabulary

absolute value the distance a number is from zero on a number line, written $|n|$ (Lesson 0-2)

When c is negative, its factors have opposite signs. To determine which factor is positive and which is negative, look at the sign of b. The factor with the greater absolute value has the same sign as b.

EXAMPLE 3 c is Negative

Factor each polynomial.

a. $x^2 + 2x - 15$

In this trinomial, $b = 2$ and $c = -15$. Since c is negative, the factors m and p have opposite signs. So either m or p is negative, but not both. Since b is positive, the factor with the greater absolute value is also positive.

List the factors of -15, where one factor of each pair is negative. Look for the pair of factors with a sum of 2.

Factors of -15	Sum of Factors
$-1, 15$	14
$-3, 5$	2

The correct factors are -3 and 5.

$x^2 + 2x - 15 = (x + m)(x + p)$ **Write the pattern.**

$= (x - 3)(x + 5)$ **$m = -3$ and $p = 5$**

CHECK $(x - 3)(x + 5) = x^2 + 5x - 3x - 15$ **FOIL Method**

$= x^2 + 2x - 15$ ✓ **Simplify.**

b. $x^2 - 7x - 18$

In this trinomial, $b = -7$ and $c = -18$. Either m or p is negative, but not both. Since b is negative, the factor with the greater absolute value is also negative.

List the factors of -18, where one factor of each pair is negative. Look for the pair of factors with a sum of -7.

Factors of -18	Sum of Factors
$1, -18$	-17
$2, -9$	-7
$3, -6$	-3

The correct factors are 2 and -9.

$x^2 - 7x - 18 = (x + m)(x + p)$ **Write the pattern.**

$= (x + 2)(x - 9)$ **$m = 2$ and $p = -9$**

CHECK Graph $y = x^2 - 7x - 18$ and $y = (x + 2)(x - 9)$ on the same screen.

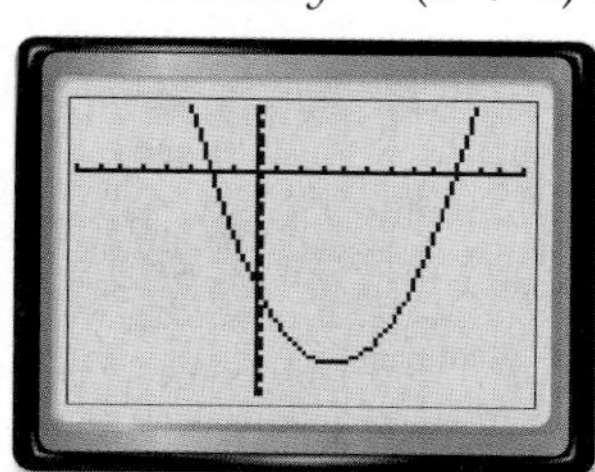

[−10, 15] scl: 1 by [−40, 20] scl: 1

The graphs coincide. Therefore, the trinomial has been factored correctly. ✓

Check Your Progress

3A. $y^2 + 13y - 48$

3B. $r^2 - 2r - 24$

Personal Tutor glencoe.com

Solve Equations by Factoring A **quadratic equation** can be written in the standard form $ax^2 + bx + c = 0$, where $a \neq 0$. Some equations of the form $x^2 + bx + c = 0$ can be solved by factoring and then using the Zero Product Property.

EXAMPLE 4 Solve an Equation by Factoring

Solve $x^2 + 6x = 27$. Check your solutions.

$x^2 + 6x = 27$	**Original equation**
$x^2 + 6x - 27 = 0$	**Subtract 27 from each side.**
$(x - 3)(x + 9) = 0$	**Factor.**
$x - 3 = 0$ or $x + 9 = 0$	**Zero Product Property**
$x = 3$ $\qquad x = -9$	**Solve each equation.**

The roots are 3 and −9.

CHECK Substitute 3 and −9 for x in the original equation.

$$\begin{aligned} x^2 + 6x &= 27 \\ (3)^2 + 6(3) &\stackrel{?}{=} 27 \\ 9 + 18 &\stackrel{?}{=} 27 \\ 27 &= 27 \checkmark \end{aligned} \qquad \begin{aligned} x^2 + 6x &= 27 \\ (-9)^2 + 6(-9) &\stackrel{?}{=} 27 \\ 81 - 54 &\stackrel{?}{=} 27 \\ 27 &= 27 \checkmark \end{aligned}$$

Check Your Progress **Solve each equation. Check your solutions.**

4A. $z^2 - 3z = 70$

4B. $x^2 + 3x - 18 = 0$

Personal Tutor glencoe.com

Factoring can be useful when solving real-world problems.

Real-World Link

A company that produces event signs recommends foamcore boards for event signs that will be used only once. For signs used more than once, use a stronger type of foamcore board.

Source: MegaPrint Inc.

Real-World EXAMPLE 5 Solve a Problem by Factoring

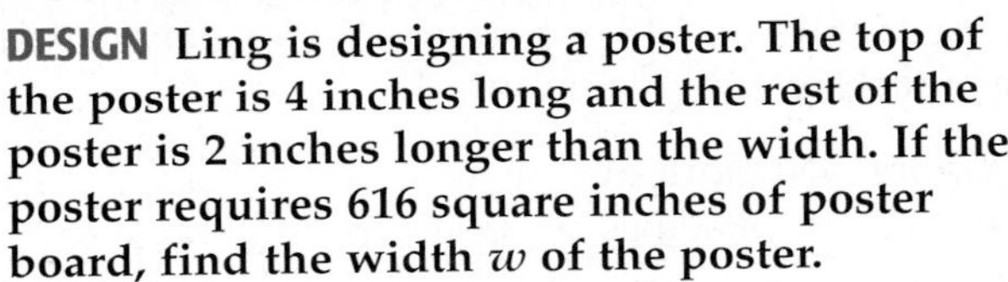

DESIGN Ling is designing a poster. The top of the poster is 4 inches long and the rest of the poster is 2 inches longer than the width. If the poster requires 616 square inches of poster board, find the width w of the poster.

Understand You want to find the width of the poster.

Plan Since the poster is a rectangle, width • length = area.

Solve Let w = the width of the poster. The length is $w + 2 + 4$ or $w + 6$.

$w(w + 6) = 616$	**Write the equation.**
$w^2 + 6w = 616$	**Multiply.**
$w^2 + 6w - 616 = 0$	**Subtract 616 from each side.**
$(w + 28)(w - 22) = 0$	**Factor.**
$w + 28 = 0$ or $w - 22 = 0$	**Zero Product Property**
$w = -28$ $\qquad w = 22$	**Solve each equation.**

Since dimensions cannot be negative, the width is 22 inches.

Check If the width is 22 inches, then the area of the poster is $22 \cdot (22 + 6)$ or 616 square inches, which is the amount the poster requires. ✓

Check Your Progress

5. GEOMETRY The height of a parallelogram is 18 centimeters less than its base. If the area is 175 square centimeters, what is its height?

Personal Tutor glencoe.com

Check Your Understanding

Examples 1–3 pp. 486–487

Factor each polynomial.

1. $x^2 + 14x + 24$

2. $y^2 - 7y - 30$

3. $n^2 + 4n - 21$

4. $m^2 - 15m + 50$

Example 4 p. 488

Solve each equation. Check your solutions.

5. $x^2 - 4x - 21 = 0$

6. $n^2 - 3n + 2 = 0$

7. $x^2 - 15x + 54 = 0$

8. $x^2 + 12x = -32$

9. $x^2 - x - 72 = 0$

10. $x^2 - 10x = -24$

Example 5 p. 488

11. **FRAMING** Tina bought a frame for a photo, but the photo is too big for the frame. Tina needs to reduce the width and length of the photo by the same amount. The area of the photo should be reduced to half the original area. If the original photo is 12 inches by 16 inches, what will be the dimensions of the smaller photo?

Practice and Problem Solving

= **Step-by-Step Solutions** begin on page R12.
Extra Practice begins on page 815.

Examples 1–3 pp. 486–487

Factor each polynomial.

12. $x^2 + 17x + 42$

13. $y^2 - 17y + 72$

14. $a^2 + 8a - 48$

15. $n^2 - 2n - 35$

16. $44 + 15h + h^2$

17. $40 - 22x + x^2$

18. $-24 - 10x + x^2$

19. $-42 - m + m^2$

Example 4 p. 488

Solve each equation. Check your solutions.

20. $x^2 - 7x + 12 = 0$

21. $y^2 + y = 20$

22. $x^2 - 6x = 27$

23. $a^2 + 11a = -18$

24. $c^2 + 10c + 9 = 0$

25. $x^2 - 18x = -32$

26. $n^2 - 120 = 7n$

27. $d^2 + 56 = -18d$

28. $y^2 - 90 = 13y$

29. $h^2 + 48 = 16h$

Example 5 p. 488

30. **GEOMETRY** A triangle has an area of 36 square feet. If the height of the triangle is 6 feet more than its base, what are its height and base?

31. **GEOMETRY** A rectangle has an area represented by $x^2 - 4x - 12$ square feet. If the length is $x + 2$ feet, what is the width of the rectangle?

32. **SOCCER** The width of a high school soccer field is 45 yards shorter than its length.

a. Define a variable, and write an expression for the area of the field.

b. The area of the field is 9000 square yards. Find the dimensions.

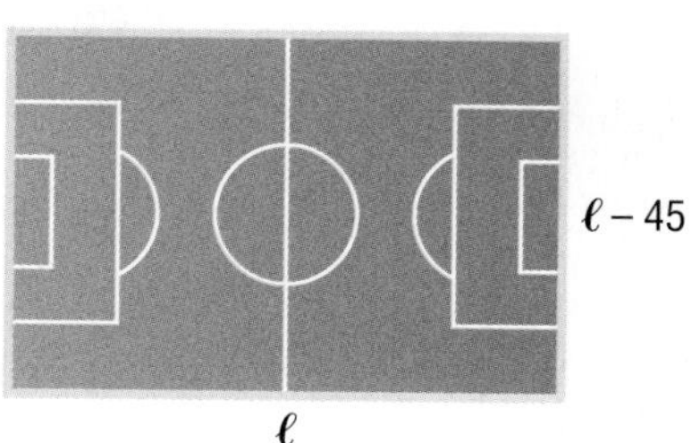

Factor each polynomial.

33. $q^2 + 11qr + 18r^2$

34. $x^2 - 14xy - 51y^2$

35. $x^2 - 6xy + 5y^2$

36. $a^2 + 10ab - 39b^2$

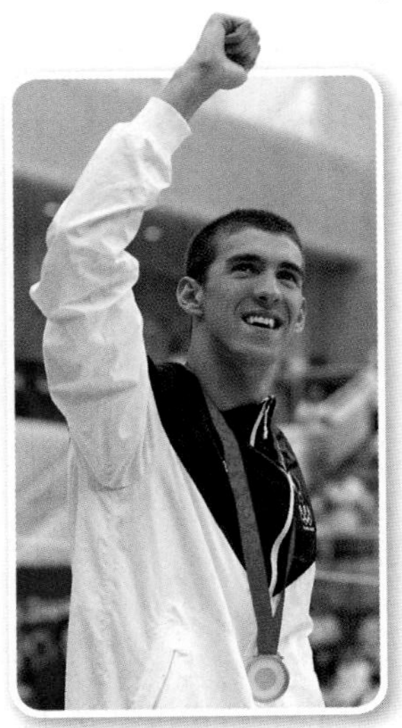

Real-World Link

At the 2008 Olympics in Beijing, Michael Phelps won eight gold medals out of eight swimming events, the highest total for any athlete in a non-boycotted Games. He broke seven world records.

Source: *NBC Olympics*

37. **SWIMMING** The length of a rectangular swimming pool is 20 feet greater than its width. The area of the pool is 525 square feet.

 a. Define a variable and write an equation for the area of the pool.

 b. Solve the equation.

 c. Interpret the solutions. Do both solutions make sense? Explain.

GEOMETRY Find an expression for the perimeter of a rectangle with the given area.

38. $A = x^2 + 24x - 81$

39. $A = x^2 + 13x - 90$

40. **MULTIPLE REPRESENTATIONS** In this problem, you will explore factoring when the leading coefficient is not 1.

 a. TABULAR Copy and complete the table below.

Product of Two Binomials	$ax^2 + mx + px + c$	$ax^2 + bx + c$	$m \times p$	$a \times c$
$(2x + 3)(x + 4)$	$2x^2 + 8x + 3x + 12$	$2x^2 + 11x + 12$	24	24
$(x + 1)(3x + 5)$				
$(2x - 1)(4x + 1)$				
$(3x + 5)(4x - 2)$				

 b. ANALYTICAL How are m and p related to a and c?

 c. ANALYTICAL How are m and p related to b?

 d. VERBAL Describe a process you can use for factoring a polynomial of the form $ax^2 + bx + c$.

H.O.T. Problems

Use Higher-Order Thinking Skills

41. **FIND THE ERROR** Jerome and Charles have factored $x^2 + 6x - 16$. Is either of them correct? Explain your reasoning.

Jerome	Charles
$x^2 + 6x - 16 = (x + 2)(x - 8)$	$x^2 + 6x - 16 = (x - 2)(x + 8)$

CHALLENGE Find all values of k so that each polynomial can be factored using integers.

42. $x^2 + kx - 19$

43. $x^2 + kx + 14$

44. $x^2 - 8x + k, k > 0$

45. $x^2 - 5x + k, k > 0$

46. **REASONING** For any factorable trinomial, $x^2 + bx + c$, will the absolute value of b *sometimes*, *always*, or *never* be less than the absolute value of c? Explain.

47. **OPEN ENDED** Give an example of a trinomial that can be factored using the factoring techniques presented in this lesson. Then factor the trinomial.

48. **CHALLENGE** Factor $(4y - 5)^2 + 3(4y - 5) - 70$.

49. **WRITING IN MATH** Explain how to factor trinomials of the form $x^2 + bx + c$ and how to determine the signs of the factors of c.

Standardized Test Practice

50. Which inequality is shown in the graph below?

A $y \leq -\frac{3}{4}x + 3$

B $y < -\frac{3}{4}x + 3$

C $y > -\frac{3}{4}x + 3$

D $y \geq -\frac{3}{4}x + 3$

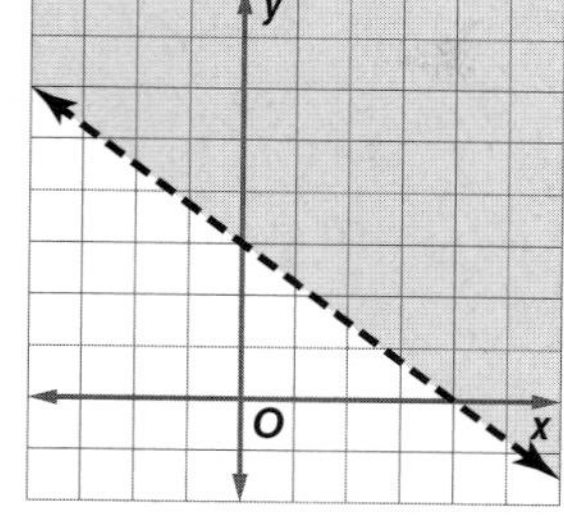

51. SHORT RESPONSE Olivia must earn more than \$254 from selling candy bars in order to go on a trip with the National Honor Society. If each candy bar is sold for \$1.25, what is the fewest candy bars she must sell?

52. GEOMETRY Which expression represents the length of the rectangle?

$A = x^2 - 3x - 18$; $x + 3$

F $x + 5$

G $x + 6$

H $x - 6$

J $x - 5$

53. The difference of 21 and a number n is 6. Which equation shows the relationship?

A $21 - n = 6$

B $21 + n = 6$

C $21n = 6$

D $6n = -21$

Spiral Review

Factor each polynomial. (Lesson 8-2)

54. $10a^2 + 40a$

55. $11x + 44x^2y$

56. $2m^3p^2 - 16mp^2 + 8mp$

57. $2ax + 6xc + ba + 3bc$

58. $8ac - 2ad + 4bc - bd$

59. $x^2 - xy - xy + y^2$

60. FLOORING Emma is replacing her dining room floor, which is 10 feet by 12 feet. The flooring comes in pieces 1 foot by 1 foot, 2 foot by 2 foot, 3 foot by 3 foot, and 2 foot by 3 foot. Without cutting the pieces, which of the four sizes of flooring can Emma use? Explain. (Lesson 8-1)

Perform the indicated matrix operations. If an operation cannot be performed, write *impossible*. (Lesson 6-6)

61. $\begin{bmatrix} 10 & 0 & 8 \\ 2 & -5 & -7 \end{bmatrix} + \begin{bmatrix} 2 & -9 & 4 \\ -1 & 2 & -3 \end{bmatrix}$

62. $\begin{bmatrix} 7 & 3 & 1 \\ 9 & -2 & -4 \\ 4 & -8 & 7 \end{bmatrix} - \begin{bmatrix} -7 & 4 & 2 \\ 6 & 3 & -3 \\ 11 & -16 & 5 \end{bmatrix}$

63. $\begin{bmatrix} 25 \\ -8 \\ -23 \end{bmatrix} + \begin{bmatrix} -9 & 21 & -2 & 7 \end{bmatrix}$

64. $3\begin{bmatrix} -5 & 3 \\ 4 & 0 \\ -1 & 8 \end{bmatrix}$

65. LANDSCAPING Kendrick is planning a circular flower garden with a low fence around the border. He has 38 feet of fence. What is the radius of the largest garden he can make? (*Hint:* $C = 2\pi r$) (Lesson 5-2)

Skills Review

Factor each polynomial. (Lesson 8-2)

66. $6mx - 4m + 3rx - 2r$

67. $3ax - 6bx + 8b - 4a$

68. $2d^2g + 2fg + 4d^2h + 4fh$

Mid-Chapter Quiz

Lessons 8-1 through 8-3

Math Online glencoe.com
Math *in Motion*, Animation

Factor each monomial completely. (Lesson 8-1)

1. $16x^3y^2$
2. $35ab^4$
3. $-20m^5n^2$
4. $-13xy^3$

5. **ROOM DESIGN** The area of a rectangular room is 120 square feet. What are the possible whole-number dimensions for the length and width of the room? (Lesson 8-1)

Find the GCF of each set of monomials. (Lesson 8-1)

6. $10a$, $20a^2$, $25a$
7. $13c$, $25d$
8. $21ab$, $35a$, $56ab^3$

9. **FASHION** A sales clerk is organizing 24 pairs of shoes for a sales display. In what ways can she organize the shoes so that she has the same number of shoes on each shelf, at least 4 pairs of shoes per shelf, and at least 2 shelves of shoes?

Use the Distributive Property to factor each polynomial. (Lesson 8-2)

10. $3xy - 9x$
11. $6ab + 12ab^2 + 18b$

12. **MULTIPLE CHOICE** The area of the rectangle is $3x^2 + 6x - 12$ square units. What is the width of the rectangle? (Lesson 8-2)

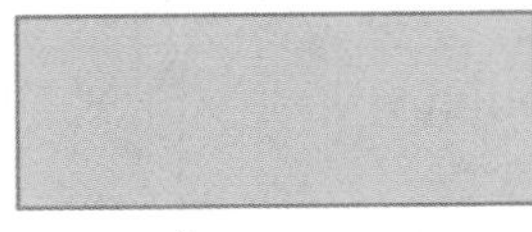

$x^2 + 2x - 4$

A 2 units

B 3 units

C 4 units

D 6 units

Factor each polynomial. (Lesson 8-2)

13. $5h + 40g$
14. $3x^2 + 6x + x + 2$
15. $5a^2 - 25a - a + 5$

Solve each equation. Check your solutions. (Lesson 8-2)

16. $2x(x - 5) = 0$
17. $6p^2 - 3p = 0$
18. $a^2 = 15a$

19. **ARCHITECTURE** The curve of the archway under a bridge can be modeled by the equation $y = -\frac{1}{5}x^2 + 6x$, where x and y are measured in feet. Copy and complete the table for each value of x. (Lesson 8-2)

x	y
0	
10	
15	
20	
30	

Factor each polynomial. (Lesson 8-3)

20. $x^2 - 4x - 21$
21. $x^2 - 10x + 24$
22. $x^2 + 4x - 21$

Solve each equation. Check your solutions. (Lesson 8-3)

23. $x^2 - 5x = 14$
24. $x^2 - 3x - 18 = 0$
25. $24 + x^2 = 10x$

26. **MULTIPLE CHOICE** A rectangle has a length that is 2 inches longer than its width. The area of the rectangle is 48 square inches. What is the length of the rectangle? (Lesson 8-3)

F 48 in.

G 8 in.

H 6 in.

J 2 in.

8-4 Quadratic Equations: $ax^2 + bx + c = 0$

Then

You factored trinomials of the form $x^2 + bx + c$. (Lesson 8-3)

Now

- Factor trinomials of the form $ax^2 + bx + c$.
- Solve equations of the form $ax^2 + bx + c = 0$.

New Vocabulary

prime polynomial

Math Online

glencoe.com

- Extra Examples
- Personal Tutor
- Self-Check Quiz
- Homework Help

Why?

At amusement parks around the country, the paths of riders can be modeled by the expression $16t^2 - 5t + 120$.

Factoring this expression can help the ride operators determine how long a rider rides on the initial swing.

Factor $ax^2 + bx + c$ In the last lesson, you factored quadratic expressions of the form $ax^2 + bx + c$, where $a = 1$. In this lesson, you will apply the factoring methods to quadratic expressions in which a is not 1.

The dimensions of the rectangle formed by the algebra tiles are the factors of $2x^2 + 5x + 3$. The factors of $2x^2 + 5x + 3$ are $x + 1$ and $2x + 3$.

You can also use the method of factoring by grouping to solve this expression.

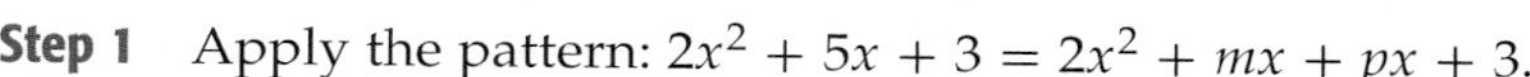

Step 1 Apply the pattern: $2x^2 + 5x + 3 = 2x^2 + mx + px + 3$.

Step 2 Find two numbers that have a product of $2 \cdot 3$ or 6 and a sum of 5.

Factors of 6	Sum of Factors
1, 6	7
2, 3	5

Step 3 Use grouping to find the factors.

$2x^2 + 5x + 3 = 2x^2 + mx + px + 3$	**Write the pattern.**
$= 2x^2 + 2x + 3x + 3$	**$m = 2$ and $p = 3$**
$= (2x^2 + 2x) + (3x + 3)$	**Group terms with common factors.**
$= 2x(x + 1) + 3(x + 1)$	**Factor the GCF.**
$= (2x + 3)(x + 1)$	**$x + 1$ is the common factor.**

Therefore, $2x^2 + 5x + 3 = (2x + 3)(x + 1)$.

Key Concept — Factoring $ax^2 + bx + c$ (For Your FOLDABLE)

Words To factor trinomials of the form $ax^2 + bx + c$, find two integers, m and p, with a sum of b and a product of ac. Then write $ax^2 + bx + c$ as $ax^2 + mx + px + c$, and factor by grouping.

Example $5x^2 - 13x + 6 = 5x^2 - 10x - 3x + 6$ **$m = -10$ and $p = -3$**

$= 5x(x - 2) + (-3)(x - 2)$

$= (5x - 3)(x - 2)$

StudyTip

Greatest Common Factor Always look for a GCF of the terms of a polynomial before you factor.

EXAMPLE 1 Factor $ax^2 + bx + c$

Factor each trinomial.

a. $7x^2 + 29x + 4$

In this trinomial, $a = 7$, $b = 29$, and $c = 4$. You need to find two numbers with a sum of 29 and a product of $7 \cdot 4$ or 28. Make a list of the factors of 28 and look for the pair of factors with the sum of 29.

Factors of 28	Sum of Factors
1, 28	29

The correct factors are 1 and 28.

$$\begin{aligned} 7x^2 + 29x + 4 &= 7x^2 + mx + px + 4 && \text{Write the pattern.} \\ &= 7x^2 + 1x + 28x + 4 && m = 1 \text{ and } p = 28 \\ &= (7x^2 + 1x) + (28x + 4) && \text{Group terms with common factors.} \\ &= x(7x + 1) + 4(7x + 1) && \text{Factor the GCF.} \\ &= (x + 4)(7x + 1) && 7x + 1 \text{ is the common factor.} \end{aligned}$$

b. $3x^2 + 15x + 18$

The GCF of the terms $3x^2$, $15x$, and 18 is 3. Factor this first.

$$\begin{aligned} 3x^2 + 15x + 18 &= 3(x^2 + 5x + 6) && \text{Distributive Property} \\ &= 3(x + 3)(x + 2) && \text{Find two factors of 6 with a sum of 5.} \end{aligned}$$

Check Your Progress

1A. $5x^2 + 13x + 6$

1B. $6x^2 + 22x - 8$

Personal Tutor glencoe.com

Sometimes the coefficient of the x-term is negative.

EXAMPLE 2 Factor $ax^2 - bx + c$

Factor $3x^2 - 17x + 20$.

In this trinomial, $a = 3$, $b = -17$, and $c = 20$. Since b is negative, $m + p$ will be negative. Since c is positive, mp will be positive.

To determine m and p, list the negative factors of ac or 60. The sum of m and p should be -17.

Factors of 60	Sum of Factors
−2, −30	−32
−3, −20	−23
−4, −15	−19
−5, −12	−17

The correct factors are −5 and −12.

$$\begin{aligned} 3x^2 - 17x + 20 &= 3x^2 - 12x - 5x + 20 && m = -12 \text{ and } p = -5 \\ &= (3x^2 - 12x) + (-5x + 20) && \text{Group terms with common factors.} \\ &= 3x(x - 4) + (-5)(x - 4) && \text{Factor the GCF.} \\ &= (3x - 5)(x - 4) && \text{Distributive Property} \end{aligned}$$

Check Your Progress

2A. $2n^2 - n - 1$

2B. $10y^2 - 35y + 30$

Personal Tutor glencoe.com

A polynomial that cannot be written as a product of two polynomials with integral coefficients is called a **prime polynomial**.

EXAMPLE 3 Determine Whether a Polynomial is Prime

Factor $4x^2 - 3x + 5$, if possible. If the polynomial cannot be factored using integers, write *prime*.

In this trinomial, $a = 4$, $b = -3$, and $c = 5$. Since b is negative, $m + p$ is negative. Since c is positive, mp is positive. So, m and p are both negative. Next, list the factors of 20. Look for the pair with a sum of -3.

Factors of 20	Sum of Factors
$-20, -1$	-21
$-4, -5$	-9
$-2, -10$	-12

There are no factors with a sum of -3. So the quadratic expression cannot be factored using integers. Therefore, $4x^2 - 3x + 5$ is prime.

Check Your Progress

Factor each polynomial, if possible. If the polynomial cannot be factored using integers, write *prime*.

3A. $4r^2 - r + 7$

3B. $2x^2 + 3x - 5$

Personal Tutor glencoe.com

Solve Equations by Factoring A model for the height of a projectile is given by $h = -16t^2 + vt + h_0$, where h is the height in feet, t is the time in seconds, v is the initial velocity in feet per second, and h_0 is the initial height in feet. Equations of the form $ax^2 + bx + c = 0$ can be solved by factoring and by using the Zero Product Property.

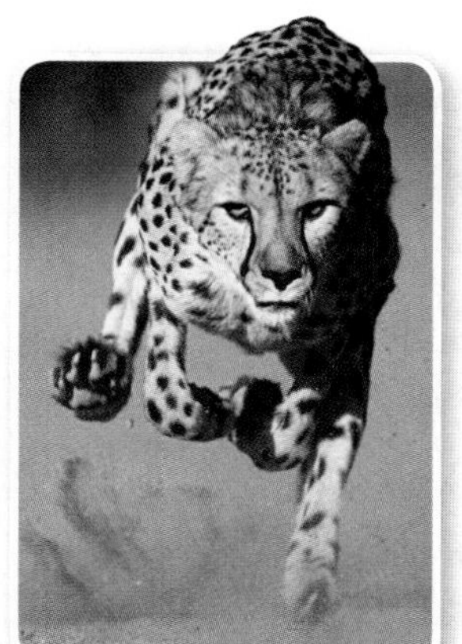

Real-World Link

Cheetahs are the fastest land animals in the world, reaching speeds of up to 70 mph. It can accelerate from 0 to 40 mph in 3 strides. It takes just seconds for the cheetah to reach the full speed of 70 mph.

Source: Cheetah Conservation Fund

Real-World EXAMPLE 4 Solve Equations by Factoring

WILDLIFE Suppose a cheetah pouncing on an antelope leaps with an initial velocity of 49 feet per second. How long is the cheetah in the air if it lands on the antelope's hind quarter, 3 feet from the ground?

$h = -16t^2 + vt + h_0$	**Equation for height**
$3 = -16t^2 + 49t + 0$	**$h = 3$, $v = 49$, and $s = 0$**
$0 = -16t^2 + 49t - 3$	**Subtract 3 from each side.**
$0 = 16t^2 - 49t + 3$	**Multiply each side by -1.**
$0 = (16t - 1)(t - 3)$	**Factor $16t^2 - 49t + 3$.**
$16t - 1 = 0$ or $t - 3 = 0$	**Zero Product Property**
$16t = 1$ $\quad t = 3$	**Solve each equation.**
$t = \frac{1}{16}$	

The solutions are $\frac{1}{16}$ and 3 seconds. It takes the cheetah $\frac{1}{16}$ second to reach a height of 3 feet on his way up. It takes the cheetah 3 seconds to reach a height of 3 feet on his way down. So, the cheetah is in the air 3 seconds before he catches the antelope.

Watch Out!

Keep the -1 Do not forget to carry the -1 that was factored out through the rest of the steps or multiply both sides by -1.

Check Your Progress

4. PHYSICAL SCIENCE A person throws a ball upward from a 506-foot tall building. The ball's height h in feet after t seconds is given by the equation $h = -16t^2 + 48t + 506$. The ball lands on a balcony that is 218 feet above the ground. How many seconds was it in the air?

Personal Tutor glencoe.com

Check Your Understanding

Examples 1–3
pp. 494–495

Factor each polynomial, if possible. If the polynomial cannot be factored using integers, write *prime*.

1. $3x^2 + 17x + 10$
2. $2x^2 + 22x + 56$
3. $5x^2 - 3x + 4$
4. $3x^2 - 11x - 20$

Example 4
p. 495

Solve each equation. Check your solutions.

5. $2x^2 + 9x + 9 = 0$
6. $3x^2 + 17x + 20 = 0$
7. $3x^2 - 10x + 8 = 0$
8. $2x^2 - 17x + 30 = 0$

9. **DISCUS** Ken throws the discus at a school meet.
   a. What is the initial height of the discus?
   b. After how many seconds does the discus hit the ground?

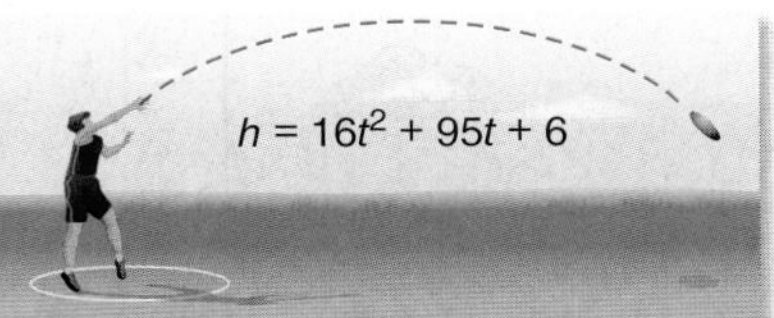

Practice and Problem Solving

 = **Step-by-Step Solutions** begin on page R12.
Extra Practice begins on page 815.

Examples 1–3
pp. 494–495

Factor each polynomial, if possible. If the polynomial cannot be factored using integers, write *prime*.

10. $5x^2 + 34x + 24$
11. $2x^2 + 19x + 24$
12. $4x^2 + 22x + 10$
13. $4x^2 + 38x + 70$
14. $2x^2 - 3x - 9$
15. $4x^2 - 13x + 10$
16. $2x^2 + 3x + 6$
17. $5x^2 + 3x + 4$
18. $12x^2 + 69x + 45$
19. $4x^2 - 5x + 7$
20. $5x^2 + 23x + 24$
21. $3x^2 - 8x + 15$

Example 4
p. 495

22. **SHOT PUT** An athlete throws a shot put with an initial velocity of 29 feet per second and from an initial height of 6 feet.
    a. Write an equation that models the height of the shot put in feet with respect to time in seconds.
    b. After how many seconds will the shot put hit the ground?

Solve each equation. Check your solutions.

23. $2x^2 + 9x - 18 = 0$
24. $4x^2 + 17x + 15 = 0$
25. $-3x^2 + 26x = 16$
26. $-2x^2 + 13x = 15$
27. $-3x^2 + 5x = -2$
28. $-4x^2 + 19x = -30$

29. **BASKETBALL** When Jerald shoots a free throw, the ball is 6 feet from the floor and has an initial velocity of 20 feet per second. The hoop is 10 feet from the floor.
    a. Use the vertical motion model to determine an equation that models Jerald's free throw.
    b. How long is the basketball in the air before it reaches the hoop?
    c. Raymond shoots a free throw that is 5 foot 9 inches from the floor with the same initial velocity. Will the ball be in the air more or less time? Explain.

30. **DIVING** Ben dives from a 10-foot platform. The equation $h = -16t^2 + 27t + 10$ models the dive. How long will it take Ben to reach the water?

Real-World Career

Urban Planner
Urban planners design the layout of an area. They take into consideration the available land and geographical and environmental factors to design an area that benefits the community the most. City planners have a bachelor's degree in planning and almost half have a master's degree.

31. **NUMBER THEORY** Six times the square of a number x plus 11 times the number equals 2. What are possible values of x?

Factor each polynomial, if possible. If the polynomial cannot be factored using integers, write *prime*.

32. $-6x^2 - 23x - 20$ **33.** $-4x^2 - 15x - 14$ **34.** $-5x^2 + 18x + 8$

35. $-6x^2 + 31x - 35$ **36.** $-4x^2 + 5x - 12$ **37.** $-12x^2 + x + 20$

38. **URBAN PLANNING** The city has commissioned the building of a new park. The area of the park can be expressed as $660x^2 + 524x + 85$. Factor this expression to find binomials with integer coefficients that represent possible dimensions of the park. If $x = 8$, what is the perimeter of the park?

39. **MULTIPLE REPRESENTATIONS** In this problem, you will explore factoring a special type of polynomial.

a. GEOMETRIC Draw a square and label the sides a. Within this square, draw a smaller square that shares a vertex with the first square. Label the sides b. What are the areas of the two squares?

b. GEOMETRIC Cut and remove the small square. What is the area of the remaining region?

c. ANALYTICAL Draw a diagonal line between the inside corner and outside corner of the figure, and cut along this line to make two congruent pieces. Then rearrange the two pieces to form a rectangle. What are the dimensions?

d. ANALYTICAL Write the area of the rectangle as the product of two binomials.

e. VERBAL Complete this statement: $a^2 - b^2 = \ldots$ Why is this statement true?

H.O.T. Problems

Use Higher-Order Thinking Skills

40. **FIND THE ERROR** Zachary and Samantha are solving $6x^2 - x = 12$. Is either of them correct? Explain your reasoning.

Zachary	Samantha
$6x^2 - x = 12$	$6x^2 - x = 12$
$x(6x - 1) = 12$	$6x^2 - x - 12 = 0$
$x = 12$ or $6x - 1 = 12$	$(2x - 3)(3x + 4) = 0$
$6x = 13$	$2x - 3 = 0$ or $3x + 4 = 0$
$x = \frac{13}{6}$	$x = \frac{3}{2}$ $x = -\frac{4}{3}$

41. **REASONING** A square has an area of $9x^2 + 30xy + 25y^2$ square inches. The dimensions are binomials with positive integer coefficients. What is the perimeter of the square? Explain.

42. **CHALLENGE** Find all values of k so that $2x^2 + kx + 12$ can be factored as two binomials using integers.

43. **OPEN ENDED** Write a quadratic equation with integer coefficients that has $\frac{1}{2}$ and $-\frac{3}{5}$ as solutions. Explain your reasoning.

44. **WRITING IN MATH** Explain how to determine which values should be chosen for m and n when factoring a polynomial of the form $ax^2 + bx + c$.

Standardized Test Practice

45. GRIDDED RESPONSE Savannah has two sisters. One sister is 8 years older than her and the other sister is 2 years younger than her. The product of Savannah's sisters' ages is 56. How old is Savannah?

46. What is the product of $\frac{2}{3}a^3b^5$ and $\frac{3}{5}a^5b^2$?

A $\frac{2}{5}a^8b^7$

B $\frac{2}{5}a^2b^3$

C $\frac{2}{5}a^8b^3$

D $\frac{2}{5}a^2b^7$

47. What is the solution set of $x^2 + 2x - 24 = 0$?

F $\{-4, 6\}$ H $\{-3, 8\}$

G $\{3, -8\}$ J $\{4, -6\}$

48. Which is the solution set of $x \geq -2$?

A
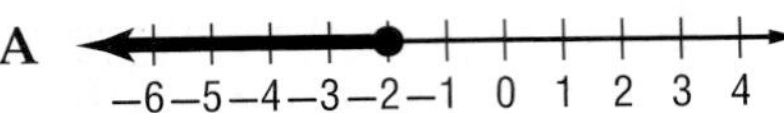

B
−6 −5 −4 −3 −2 −1 0 1 2 3 4

C
−6 −5 −4 −3 −2 −1 0 1 2 3 4

D
−6 −5 −4 −3 −2 −1 0 1 2 3 4

Spiral Review

Factor each polynomial. (Lesson 8-3)

49. $x^2 - 9x + 14$ **50.** $n^2 - 8n + 15$ **51.** $x^2 - 5x - 24$

52. $z^2 + 15z + 36$ **53.** $r^2 + 3r - 40$ **54.** $v^2 + 16v + 63$

Solve each equation. Check your solutions. (Lesson 8-2)

55. $a(a - 9) = 0$ **56.** $(2y + 6)(y - 1) = 0$ **57.** $10x^2 - 20x = 0$

58. $8b^2 - 12b = 0$ **59.** $15a^2 = 60a$ **60.** $33x^2 = -22x$

61. ART A painter has 32 units of yellow dye and 54 units of blue dye to make two shades of green. The units needed to make a gallon of light green and a gallon of dark green are shown. Make a graph showing the numbers of gallons of the two greens she can make, and list three possible solutions. (Lesson 6-8)

Color	Units of Yellow Dye	Units of Blue Dye
light green	4	1
dark green	1	6

Solve each compound inequality. Then graph the solution set. (Lesson 5-4)

62. $k + 2 > 12$ and $k + 2 \leq 18$ **63.** $d - 4 > 3$ or $d - 4 \leq 1$ **64.** $3 < 2x - 3 < 15$

65. $3t - 7 \geq 5$ and $2t + 6 \leq 12$ **66.** $h - 10 < -21$ or $h + 3 < 2$ **67.** $4 < 2y - 2 < 10$

68. FINANCIAL LITERACY A home security company provides security systems for \$5 per week, plus an installation fee. The total cost for installation and 12 weeks of service is \$210. Write the point-slope form of an equation to find the total fee y for any number of weeks x. What is the installation fee? (Lesson 4-3)

Skills Review

Find the principal square root of each number. (Lesson 0-2)

69. 16 **70.** 36 **71.** 64

72. 81 **73.** 121 **74.** 100

8-5 Quadratic Equations: Differences of Squares

Then

You factored trinomials into two binomials. (Lesson 8-3, 8-4)

Now

- Factor binomials that are the difference of squares.
- Use the difference of squares to solve equations.

New Vocabulary

difference of two squares

Math Online

glencoe.com

- Extra Examples
- Personal Tutor
- Self-Check Quiz
- Homework Help
- Math in Motion

Why?

Computer graphics designers use a combination of art and mathematics skills to design images and videos. They use equations to form shapes and lines on computers. Factoring can help to determine the dimensions and shapes of the figures.

Factor Differences of Squares Recall that in Lesson 7-8, you learned about the product of the sum and difference of two quantities. This resulting product is referred to as the **difference of two squares**. So, the factored form of the difference of squares is called the product of the sum and difference of the two quantities.

Key Concept — Difference of Squares

For Your FOLDABLE

Symbols $a^2 - b^2 = (a + b)(a - b)$ or $(a - b)(a + b)$

Examples $x^2 - 25 = (x + 5)(x - 5)$ or $(x - 5)(x + 5)$

$t^2 - 64 = (t + 8)(t - 8)$ or $(t - 8)(t + 8)$

EXAMPLE 1 Factor Differences of Squares

Factor each polynomial.

a. $16h^2 - 9a^2$

$16h^2 - 9a^2 = (4h)^2 - (3a)^2$ — **Write in the form of $a^2 - b^2$.**

$= (4h + 3a)(4h - 3a)$ — **Factor the difference of squares.**

b. $121 - 4b^2$

$121 - 4b^2 = (11)^2 - (2b)^2$ — **Write in the form of $a^2 - b^2$.**

$= (11 - 2b)(11 + 2b)$ — **Factor the difference of squares.**

c. $27g^3 - 3g$

Because the terms have a common factor, factor out the GCF first. Then proceed with other factoring techniques.

$27g^3 - 3g = 3g(9g^2 - 1)$ — **Factor out the GCF of $3g$.**

$= 3g[(3g)^2 - (1)^2]$ — **Write in the form $a^2 - b^2$.**

$= 3g(3g - 1)(3g + 1)$ — **Factor the difference of squares.**

Check Your Progress

1A. $81 - c^2$

1B. $64g^2 - h^2$

1C. $9x^3 - 4x$

1D. $-4y^3 + 9y$

Personal Tutor glencoe.com

Watch Out!

Sum of Squares The sum of squares, $a^2 + b^2$, does not factor into $(a + b)(a + b)$. The sum of squares is a prime polynomial and cannot be factored.

Math *in Motion*, Animation glencoe.com

To factor a polynomial completely, a technique may need to be applied more than once. This also applies to the difference of squares pattern.

EXAMPLE 2 Apply a Technique More than Once

Factor each polynomial.

a. $b^4 - 16$

$b^4 - 16 = (b^2)^2 - (4)^2$ — Write $b^4 - 16$ in $a^2 - b^2$ form.

$= (b^2 + 4)(b^2 - 4)$ — Factor the difference of squares.

Notice that the factor $b^2 - 4$ is also the difference of squares.

$= (b^2 + 4)(b^2 - 2^2)$ — Write $b^2 - 4$ in $a^2 - b^2$ form.

$= (b^2 + 4)(b + 2)(b - 2)$ — Factor the difference of squares.

b. $625 - x^4$

$625 - x^4 = (25)^2 - (x^2)^2$ — Write $625 - x^4$ in $a^2 - b^2$ form.

$= (25 + x^2)(25 - x^2)$ — Factor the difference of squares.

$= (25 + x^2)(5^2 - x^2)$ — Write $25 - x^2$ in $a^2 - b^2$ form.

$= (25 + x^2)(5 - x)(5 + x)$ — Factor the difference of squares.

Check Your Progress

Factor each polynomial.

2A. $y^4 - 1$

2B. $4a^4 - b^4$

2C. $81 - x^4$

2D. $16y^4 - 1$

Personal Tutor glencoe.com

Sometimes more than one factoring technique needs to be applied to ensure that a polynomial is factored completely.

EXAMPLE 3 Apply Different Techniques

Factor each polynomial.

a. $5x^5 - 45x$

$5x^5 - 45x = 5x(x^4 - 9)$ — Factor out GCF.

$= 5x[(x^2)^2 - (3)^2]$ — Write $x^4 - 9$ in the form $a^2 - b^2$.

$= 5x(x^2 - 3)(x^2 + 3)$ — Factor the difference of squares.

Notice that the factor $x^2 - 3$ is not the difference of squares because 3 is not a perfect square.

b. $7x^3 + 21x^2 - 7x - 21$

$7x^3 + 21x^2 - 7x - 21$ — Original expression

$= 7(x^3 + 3x^2 - x - 3)$ — Factor out GCF.

$= 7[(x^3 + 3x^2) - (x + 3)]$ — Group terms with common factors.

$= 7[x^2(x + 3) - 1(x + 3)]$ — Factor each grouping.

$= 7(x + 3)(x^2 - 1)$ — $x + 3$ is the common factor.

$= 7(x + 3)(x + 1)(x - 1)$ — Factor the difference of squares.

Check Your Progress

Factor each polynomial.

3A. $2y^4 - 50$

3B. $6x^4 - 96$

3C. $2m^3 + m^2 - 50m - 25$

3D. $r^3 + 6r^2 + 11r + 66$

Personal Tutor glencoe.com

Solve Equations by Factoring After factoring, you can apply the Zero Product Property to an equation that is written as the product of factors set equal to 0.

STANDARDIZED TEST EXAMPLE 4

In the equation $y = x^2 - \frac{9}{16}$, which is a value of x when $y = 0$?

A $-\frac{9}{4}$ **B** 0 **C** $\frac{3}{4}$ **D** $\frac{9}{4}$

Read the Test Item

Replace y with 0 and then solve.

Solve the Test Item

$y = x^2 - \frac{9}{16}$ **Original equation**

$0 = x^2 - \frac{9}{16}$ **Replace y with 0.**

$0 = x^2 - \left(\frac{3}{4}\right)^2$ **Write in the form $a^2 - b^2$.**

$0 = \left(x + \frac{3}{4}\right)\left(x - \frac{3}{4}\right)$ **Factor the difference of squares.**

$0 = x + \frac{3}{4}$ or $0 = x - \frac{3}{4}$ **Zero Product Property**

$x = -\frac{3}{4}$ $x = \frac{3}{4}$ The correct answer is C.

Test-TakingTip

Use Another Method Another method that can be used to solve this equation is to substitute each answer choice into the equation.

Check Your Progress

4. Which are the solutions of $18x^3 = 50x$?

F $0, \frac{5}{3}$ **G** $-\frac{5}{3}, \frac{5}{3}$ **H** $-\frac{5}{3}, \frac{5}{3}, 0$ **J** $-\frac{5}{3}, \frac{5}{3}, 1$

Personal Tutor glencoe.com

Check Your Understanding

Examples 1–3 pp. 499–500

Factor each polynomial.

1. $x^2 - 9$ **2.** $4a^2 - 25$

3. $9m^2 - 144$ **4.** $2p^3 - 162p$

5. $u^4 - 81$ **6.** $2d^4 - 32f^4$

7 $20r^4 - 45n^4$ **8.** $256n^4 - c^4$

9. $2c^3 + 3c^2 - 2c - 3$ **10.** $f^3 - 4f^2 - 9f + 36$

11. $3t^3 + 2t^2 - 48t - 32$ **12.** $w^3 - 3w^2 - 9w + 27$

Example 4 p. 501

EXTENDED RESPONSE After an accident, skid marks may result from sudden breaking. The formula $\frac{1}{24}s^2 = d$ approximates a vehicle's speed s in miles per hour given the length d in feet of the skid marks on dry concrete.

13. If skid marks on dry concrete are 54 feet long, how fast was the car traveling when the brakes were applied?

14. If the skid marks on dry concrete are 150 feet long, how fast was the car traveling when the brakes were applied?

Practice and Problem Solving

● = **Step-by-Step Solutions** begin on page R12.
Extra Practice begins on page 815.

Examples 1–3
pp. 499–500

Factor each polynomial.

15. $q^2 - 121$

16. $r^4 - k^4$

17. $6n^4 - 6$

18. $w^4 - 625$

19. $r^2 - 9t^2$

20. $2c^2 - 32d^2$

21. $h^3 - 100h$

22. $h^4 - 256$

23. $2x^3 - x^2 - 162x + 81$

24. $x^2 - 4y^2$

25. $7h^4 - 7p^4$

26. $3c^3 + 2c^2 - 147c - 98$

27. $6k^2h^4 - 54k^4$

28. $5a^3 - 20a$

29. $f^3 + 2f^2 - 64f - 128$

30. $3r^3 - 192r$

31. $10q^3 - 1210q$

32. $3xn^4 - 27x^3$

33. $p^3r^5 - p^3r$

34. $8c^3 - 8c$

35. $r^3 - 5r^2 - 100r + 500$

36. $3t^3 - 7t^2 - 3t + 7$

37. $a^2 - 49$

38. $4m^3 + 9m^2 - 36m - 81$

39. $3m^4 + 243$

40. $3x^3 + x^2 - 75x - 25$

41. $12a^3 + 2a^2 - 192a - 32$

42. $x^4 + 6x^3 - 36x^2 - 216x$

43. $15m^3 + 12m^2 - 375m - 300$

Example 4
p. 501

44. GEOMETRY The drawing at the right is a square with a square cut out of it.

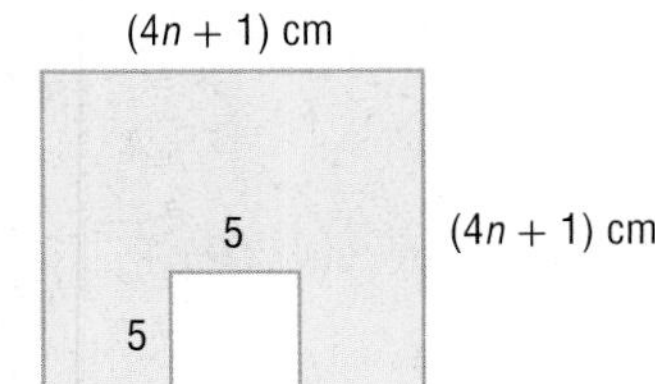

a. Write an expression that represents the area of the shaded region.

b. Find the dimensions of a rectangle with the same area as the shaded region in the drawing. Assume that the dimensions of the rectangle must be represented by binomials with integral coefficients.

45. DECORATIONS An arch decorated with balloons was used to decorate the gym for the spring dance. The shape of the arch can be modeled by the equation $y = -0.5x^2 + 4.5x$, where x and y are measured in feet and the x-axis represents the floor.

a. Write the expression that represents the height of the arch in factored form.

b. How far apart are the two points where the arch touches the floor?

c. Graph this equation on your calculator. What is the highest point of the arch?

46. DECKS Zelda is building a deck in her backyard. The plans for the deck show that it is to be 24 feet by 24 feet. Zelda wants to reduce one dimension by a number of feet and increase the other dimension by the same number of feet. If the area of the reduced deck is 512 square feet, what are the dimensions of the deck?

Real-World Link

Teens were asked "If you were given $20 today, what would you most likely spend it on?" Of the teens surveyed, 32% would spend it on a music CD.

Source: *USA TODAY*

47. SALES The sales of a particular CD can be modeled by the equation $S = -25m^2 + 125m$, where S is the number of CDs sold in thousands, and m is the number of months that it is on the market.

a. In what month should the music store expect the CD to stop selling?

b. In what month will CD sales peak?

c. How many copies will the CD sell at its peak?

StudyTip

Solving an Equation By Factoring Remember to get 0 on one side of the equation before factoring.

Solve each equation by factoring. Check your solutions.

48. $36w^2 = 121$

49. $100 = 25x^2$

50. $64x^2 - 1 = 0$

51. $4y^2 - \frac{9}{16} = 0$

52. $\frac{1}{4}b^2 = 16$

53. $81 - \frac{1}{25}x^2 = 0$

54. $9d^2 - 81 = 0$

55. $4a^2 = \frac{9}{64}$

56. **MULTIPLE REPRESENTATIONS** In this problem, you will investigate perfect square trinomials.

a. **TABULAR** Copy and complete the table below by factoring each polynomial. Then write the first and last terms of the given polynomials as perfect squares.

Polynomial	Factored Polynomial	First Term	Last Term	Middle Term
$4x^2 + 12x + 9$	$(2x + 3)(2x + 3)$	$4x^2 = (2x)^2$	$9 = 3^2$	
$9x^2 - 24x + 16$				
$4x^2 - 20x + 25$				
$16x^2 + 24x + 9$				
$25x^2 + 20x + 4$				

b. **ANALYTICAL** Write the middle term of each polynomial using the square roots of the perfect squares of the first and last terms.

c. **ALGEBRAIC** Write the pattern for a perfect square trinomial.

d. **VERBAL** What conditions must be met for a trinomial to be classified as a perfect square trinomial?

H.O.T. Problems

Use Higher-Order Thinking Skills

57. **FIND THE ERROR** Elizabeth and Lorenzo are factoring an expression. Is either of them correct? Explain your reasoning.

Elizabeth	Lorenzo
$16x^4 - 25y^2 =$	$16x^4 - 25y^2 =$
$(4x - 5y)(4x + 5y)$	$(4x^2 - 5y)(4x^2 + 5y)$

58. **CHALLENGE** Factor and simplify $9 - (k + 3)^2$, a difference of squares.

59. **CHALLENGE** Factor $x^{16} - 81$.

60. **REASONING** Write and factor a binomial that is the difference of two perfect squares and that has a greatest common factor of $5mk$.

61. **REASONING** Determine whether the following statement is *true* or *false*. Give an example or counterexample to justify your answer.

All binomials that have a perfect square in each of the two terms can be factored.

62. **OPEN ENDED** Write a binomial in which the difference of squares pattern must be repeated to factor it completely. Then factor the binomial.

63. **WRITING IN MATH** Describe why the difference of squares pattern has no middle term with a variable.

Standardized Test Practice

64. One of the roots of $2x^2 + 13x = 24$ is -8. What is the other root?

A $-\frac{3}{2}$ C $\frac{2}{3}$

B $\frac{3}{2}$ D $-\frac{2}{3}$

65. Which of the following is the sum of both solutions of the equation $x^2 + 3x = 54$?

F -21 H 3

G -3 J 21

66. What are the x-intercepts of the graph of $y = -3x^2 + 7x + 20$?

A $\frac{5}{3}, -4$ C $-\frac{5}{3}, 4$

B $-\frac{5}{3}, -4$ D $\frac{5}{3}, 4$

67. EXTENDED RESPONSE Two cars leave Cleveland at the same time from different parts of the city and both drive to Cincinnati. The distance in miles of the cars from the center of Cleveland can be represented by the two equations below, where t represents the time in hours.

Car A: $65t + 15$ Car B: $60t + 25$

a. Which car is faster? Explain.

b. Find an expression that models the distance between the two cars.

c. How far apart are the cars after $2\frac{1}{2}$ hours?

Spiral Review

Factor each trinomial, if possible. If the trinomial cannot be factored using integers, write *prime*. (Lesson 8-4)

68. $5x^2 - 17x + 14$ **69.** $5a^2 - 3a + 15$ **70.** $10x^2 - 20xy + 10y^2$

Solve each equation. Check your solutions. (Lesson 8-3)

71. $n^2 - 9n = -18$ **72.** $10 + a^2 = -7a$ **73.** $22x - x^2 = 96$

74. SAVINGS Victoria and Trey each want to buy a scooter. In how many weeks will Victoria and Trey have saved the same amount of money, and how much will each of them have saved? (Lesson 6-1)

Solve each inequality. Graph the solution set on a number line. (Lesson 5-1)

75. $t + 14 \geq 18$ **76.** $d + 5 \leq 7$ **77.** $-5 + k > -1$

78. $5 < 3 + g$ **79.** $2 \leq -1 + m$ **80.** $2y > -8 + y$

81. FITNESS Silvia is beginning an exercise program that calls for 20 minutes of walking each day for the first week. Each week thereafter, she has to increase her daily walking for a week by 7 minutes. In which week will she first walk over an hour a day? (Lesson 3-5)

Skills Review

Find each product. (Lesson 7-8)

82. $(x - 6)^2$ **83.** $(x - 2)(x - 2)$ **84.** $(x + 3)(x + 3)$

85. $(2x - 5)^2$ **86.** $(6x - 1)^2$ **87.** $(4x + 5)(4x + 5)$

8-6 Quadratic Equations: Perfect Squares

Then

You found the product of a sum and difference. (Lesson 7-8)

Now

- Factor perfect square trinomials.
- Solve equations involving perfect squares.

New Vocabulary

perfect square trinomial

Math Online

glencoe.com

- Extra Examples
- Personal Tutor
- Self-Check Quiz
- Homework Help
- Math in Motion

Why?

In a vacuum, a feather and a piano would fall at the same speed, or velocity. To find about how long it takes an object to hit the ground if it is dropped from an initial height of h_0 feet above ground, you would need to solve the equation $0 = -16t^2 + h_0$, where t is time in seconds after the object is dropped.

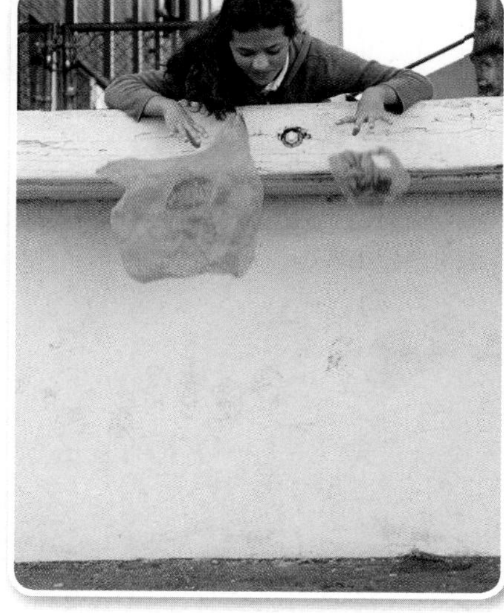

Factor Perfect Square Trinomials In Lesson 7-8, you learned the patterns for the products of the binomials $(a + b)^2$ and $(a - b)^2$. Recall that these are special products that follow specific patterns.

$$\begin{aligned}(a + b)^2 &= (a + b)(a + b)\\ &= a^2 + ab + ab + b^2\\ &= a^2 + 2ab + b^2\end{aligned}$$

$$\begin{aligned}(a - b)^2 &= (a - b)(a - b)\\ &= a^2 - ab - ab + b^2\\ &= a^2 - 2ab + b^2\end{aligned}$$

These products are called **perfect square trinomials**, because they are the squares of binomials. The above patterns can help you factor perfect square trinomials.

For a trinomial to be factorable as a perfect square, the first and last terms must be perfect squares and the middle term must be two times the square roots of the first and last terms.

The trinomial $16x^2 + 24x + 9$ is a perfect square trinomial, as illustrated below.

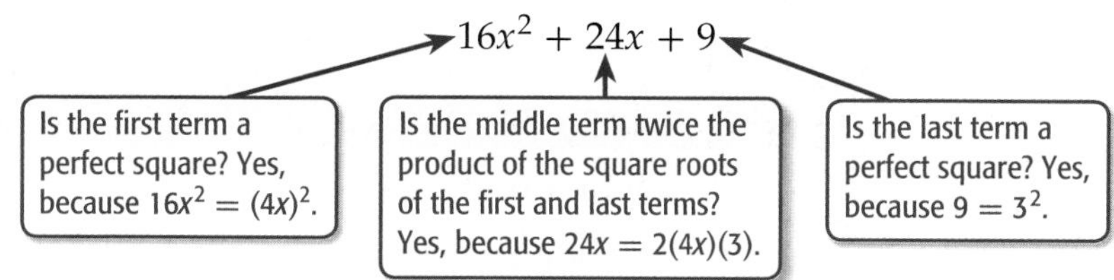

Key Concept Factoring Perfect Square Trinomials — For Your FOLDABLE

Symbols $a^2 + 2ab + b^2 = (a + b)(a + b) = (a + b)^2$

$a^2 - 2ab + b^2 = (a - b)(a - b) = (a - b)^2$

Examples $x^2 + 8x + 16 = (x + 4)(x + 4)$ or $(x + 4)^2$

$x^2 - 6x + 9 = (x - 3)(x - 3)$ or $(x - 3)^2$

StudyTip

Recognizing Perfect Square Trinomials If the constant term of the trinomial is negative, the trinomial is not a perfect square trinomial, so it is not necessary to check the other conditions.

EXAMPLE 1 Recognize and Factor Perfect Square Trinomials

Determine whether each trinomial is a perfect square trinomial. Write *yes* or *no*. If so, factor it.

a. $4y^2 + 12y + 9$

1. Is the first term a perfect square? Yes, $4y^2 = (2y)^2$.
2. Is the last term a perfect square? Yes, $9 = 3^2$.
3. Is the middle term equal to $2(2y)(3)$? Yes, $12y = 2(2y)(3)$

Since all three conditions are satisfied, $4y^2 + 12y + 9$ is a perfect square trinomial.

$4y^2 + 12y + 9 = (2y)^2 + 2(2y)(3) + 3^2$ **Write as $a^2 + 2ab + b^2$.**

$= (2y + 3)^2$ **Factor using the pattern.**

b. $9x^2 - 6x + 4$

1. Is the first term a perfect square? Yes, $9x^2 = (3x)^2$.
2. Is the last term a perfect square? Yes, $4 = 2^2$.
3. Is the middle term equal to $-2(3x)(2)$? No, $-6x \neq -2(3x)(2)$.

Since the middle term does not satisfy the required condition, $9x^2 - 6x + 4$ is not a perfect square trinomial.

Check Your Progress

1A. $9y^2 + 24y + 16$

1B. $2a^2 + 10a + 25$

Personal Tutor glencoe.com

A polynomial is completely factored when it is written as a product of prime polynomials. More than one method might be needed to factor a polynomial completely. When completely factoring a polynomial, the Concept Summary can help you decide where to start.

Remember, if the polynomial does not fit any pattern or cannot be factored, the polynomial is prime.

Concept Summary Factoring Methods

For Your FOLDABLE

Steps		Number of Terms	Examples
Step 1	Factor out the GCF.	any	$4x^3 + 2x^2 - 6x = 2x(2x^2 + x - 3)$
Step 2	Check for a difference of squares or a perfect square trinomial.	2 or 3	$9x^2 - 16 = (3x + 4)(3x - 4)$ $16x^2 + 24x + 9 = (4x + 3)^2$
Step 3	Apply the factoring patterns for $x^2 + bx + c$ or $ax^2 + bx + c$ (general trinomials), or factor by grouping.	3 or 4	$x^2 - 8x + 12 = (x - 2)(x - 6)$ $2x^2 + 13x + 6 = (2x + 1)(x + 6)$ $12y^2 + 9y + 8y + 6$ $= (12y^2 + 9y) + (8y + 6)$ $= 3y(4y + 3) + 2(4y + 3)$ $= (4y + 3)(3y + 2)$

EXAMPLE 2 Factor Completely

Factor each polynomial, if possible. If the polynomial cannot be factored, write *prime*.

a. $5x^2 - 80$

Step 1 The GCF of $5x^2$ and -80 is 5, so factor it out.

Step 2 Since there are two terms, check for a difference of squares.

$$5x^2 - 80 = 5(x^2 - 16) \quad \text{5 is the GCF of the terms.}$$
$$= 5(x^2 - 4^2) \quad x^2 = x \cdot x \text{ and } 16 = 4 \cdot 4$$
$$= 5(x - 4)(x + 4) \quad \text{Factor the difference of squares.}$$

b. $9x^2 - 6x - 35$

Step 1 The GCF of $9x^2$, $-6x$, and -35 is 1.

Step 2 Since 35 is not a perfect square, this is not a perfect square trinomial.

Step 3 Factor using the pattern $ax^2 + bx + c$. Are there two numbers with a product of $9(-35)$ or -315 and a sum of -6? Yes, the product of 15 and -21 is -315, and the sum is -6.

$$9x^2 - 6x - 35 = 9x^2 + mx + nx - 35 \quad \text{Write the pattern.}$$
$$= 9x^2 + 15x - 21x - 35 \quad m = 15 \text{ and } n = -21$$
$$= (9x^2 + 15x) + (-21x - 35) \quad \text{Group terms with common factors.}$$
$$= 3x(3x + 5) - 7(3x + 5) \quad \text{Factor out the GCF from each grouping.}$$
$$= (3x + 5)(3x - 7) \quad 3x + 5 \text{ is the common factor.}$$

StudyTip

Check Your Answer You can check your answer by:
- Using the FOIL method.
- Using the Distributive Property.
- Graphing the original expression and factored expression and comparing the graphs.

If the product of the factors does not match the original expression exactly, the answer is incorrect.

Check Your Progress

2A. $2x^2 - 32$

2B. $12x^2 + 5x - 25$

Personal Tutor glencoe.com

Solve Equations with Perfect Squares When solving equations involving repeated factors, it is only necessary to set one of the repeated factors equal to zero.

EXAMPLE 3 Solve Equations with Repeated Factors

Solve $9x^2 - 48x = -64$.

$9x^2 - 48x = -64$	Original equation
$9x^2 - 48x + 64 = 0$	Add 64 to each side.
$(3x)^2 - 2(3x)(8) + (8)^2 = 0$	Recognize $9x^2 - 48x + 64$ as a perfect square trinomial.
$(3x - 8)^2 = 0$	Factor the perfect square trinomial.
$(3x - 8)(3x - 8) = 0$	Write $(3x - 8)^2$ as two factors.
$3x - 8 = 0$	Set the repeated factor equal to zero.
$3x = 8$	Add 8 to each side.
$x = \frac{8}{3}$	Divide each side by 3.

Check Your Progress

Solve each equation. Check your solutions.

3A. $a^2 + 12a + 36 = 0$

3B. $y^2 - \frac{4}{3}y + \frac{4}{9} = 0$

Personal Tutor glencoe.com

You have solved equations like $x^2 - 16 = 0$ by factoring. You can also use the definition of a square root to solve the equation.

Reading Math

Square Root Solutions $\pm\sqrt{16}$ is read as *plus or minus the square root of 16.*

$x^2 - 16 = 0$	**Original equation**
$x^2 = 16$	**Add 16 to each side.**
$x = \pm\sqrt{16}$	**Take the square root of each side.**

Remember that there are two square roots of 16, namely 4 and −4. Therefore, the solution set is $\{-4, 4\}$. You can express this as $\{\pm 4\}$.

Key Concept — Square Root Property

For Your FOLDABLE

Words To solve a quadratic equation in the form $x^2 = n$, take the square root of each side.

Symbols For any number $n \geq 0$, if $x^2 = n$, then $x = \pm\sqrt{n}$.

Example $x^2 = 25$

$x = \pm\sqrt{25}$ or ± 5

In the equation $x^2 = n$, if n is not a perfect square, you need to approximate the square root. Use a calculator to find an approximation. If n is a perfect square, you will have an exact answer.

EXAMPLE 4 Use the Square Root Property

Solve each equation. Check your solutions.

a. $(y - 6)^2 = 81$

$(y - 6)^2 = 81$	**Original equation**
$y - 6 = \pm\sqrt{81}$	**Square Root Property**
$y - 6 = \pm 9$	**81 = 9 • 9**
$y = 6 \pm 9$	**Add 6 to each side.**
$y = 6 + 9$ or $y = 6 - 9$	**Separate into two equations.**
$= 15$ $\quad = -3$	**Simplify.**
The roots are 15 and −3.	**Check in the original equation.**

b. $(x + 6)^2 = 12$

$(x + 6)^2 = 12$	**Original equation**
$x + 6 = \pm\sqrt{12}$	**Square Root Property**
$x = -6 \pm \sqrt{12}$	**Subtract 6 from each side.**

The roots are $-6 \pm \sqrt{12}$ or $-6 + \sqrt{12}$ and $-6 - \sqrt{12}$.

Using a calculator, $-6 + \sqrt{12} \approx -2.54$ and $-6 - \sqrt{12} \approx -9.46$.

Math in Motion, Interactive Lab glencoe.com

Check Your Progress

4A. $(a - 10)^2 = 121$

4B. $(z + 3)^2 = 26$

Personal Tutor glencoe.com

Math History Link

Galileo Galilei (1564–1642)
Galileo was the first person to prove that objects of different weights fall at the same velocity by dropping two objects of different weights from the top of the Leaning Tower of Pisa in 1589.

Real-World EXAMPLE 5 Solve an Equation

PHYSICAL SCIENCE During an experiment, a ball is dropped from a height of 205 feet. The formula $h = -16t^2 + h_0$ can be used to approximate the number of seconds t it takes for the ball to reach height h from an initial height of h_0 in feet. Find the time it takes the ball to reach the ground.

At ground level, $h = 0$ and the initial height is 205, so $h_0 = 205$.

$h = -16t^2 + h_0$	**Original Formula**
$0 = -16t^2 + 205$	**Replace h with 0 and h_0 with 205.**
$-205 = -16t^2$	**Subtract 205 from each side.**
$12.8125 = t^2$	**Divide each side by −16.**
$\pm 3.6 \approx t$	**Use the Square Root Property.**

Since a negative number does not make sense in this situation, the solution is 3.6. It takes about 3.6 seconds for the ball to reach the ground.

Check Your Progress

5. Find the time it takes a ball to reach the ground if it is dropped from a bridge that is half as high as the one described above.

Personal Tutor glencoe.com

Check Your Understanding

Example 1 p. 506

Determine whether each trinomial is a perfect square trinomial. Write *yes* or *no*. If so, factor it.

1. $25x^2 + 60x + 36$

2. $6x^2 + 30x + 36$

Example 2 p. 507

Factor each polynomial, if possible. If the polynomial cannot be factored, write *prime*.

3. $2x^2 - x - 28$

4. $6x^2 - 34x + 48$

5. $4x^2 + 64$

6. $4x^2 + 9x - 16$

Examples 3 and 4 pp. 507–508

Solve each equation. Check your solutions.

7. $4x^2 = 36$

8. $25a^2 - 40a = -16$

9. $64y^2 - 48y + 18 = 9$

10. $(z + 5)^2 = 47$

Example 5 p. 509

11. PAINT While painting his bedroom, Nick drops his paintbrush off his ladder from a height of 6 feet. Use the formula $h = -16t^2 + h_0$ to approximate the number of seconds it takes for the paintbrush to hit the floor.

Practice and Problem Solving

● = **Step-by-Step Solutions** begin on page R12.
Extra Practice begins on page 815.

Example 1 p. 506

Determine whether each trinomial is a perfect square trinomial. Write *yes* or *no*. If so, factor it.

12. $4x^2 - 42x + 110$

13. $16x^2 - 56x + 49$

14. $81x^2 - 90x + 25$

15 $x^2 + 26x + 168$

Example 2 p. 507

Factor each polynomial, if possible. If the polynomial cannot be factored, write *prime*.

16. $24d^2 + 39d - 18$

17. $8x^2 + 10x - 21$

18. $2b^2 + 12b - 24$

19. $8y^2 - 200z^2$

20. $16a^2 - 121b^2$

21. $12m^3 - 22m^2 - 70m$

22. $8c^2 - 88c + 242$

23. $12x^2 - 84x + 147$

24. $w^4 - w^2$

25. $12p^3 - 3p$

26. $16q^3 - 48q^2 + 36q$

27. $4t^3 + 10t^2 - 84t$

28. $x^3 + 2x^2y - 4x - 8y$

29. $2a^2b^2 - 2a^2 - 2ab^3 + 2ab$

30. $2r^3 - r^2 - 72r + 36$

31. $3k^3 - 24k^2 + 48k$

32. $4c^4d - 10c^3d + 4c^2d^3 - 10cd^3$

33. $g^2 + 2g - 3h^2 + 4h$

Examples 3 and 4 pp. 507–508

Solve each equation. Check the solutions.

34. $4m^2 - 24m + 36 = 0$

35 $(y - 4)^2 = 7$

36. $a^2 + \frac{10}{7}a + \frac{25}{49} = 0$

37. $x^2 - \frac{3}{2}x + \frac{9}{16} = 0$

38. $x^2 + 8x + 16 = 25$

39. $5x^2 - 60x = -180$

40. $4x^2 = 80x - 400$

41. $9 - 54x = -81x^2$

42. $4c^2 + 4c + 1 = 15$

43. $x^2 - 16x + 64 = 6$

44. **PHYSICAL SCIENCE** For an experiment in physics class, a water balloon is dropped from the window of the school building. The window is 40 feet high. How long does it take until the balloon hits the ground? Round to the nearest hundredth.

45. **SCREENS** The area A in square feet of a projected picture on a movie screen can be modeled by the equation $A = 0.25d^2$, where d represents the distance from a projector to a movie screen. At what distance will the projected picture have an area of 100 square feet?

Example 5 p. 509

46. **GEOMETRY** The area of a square is represented by $9x^2 - 42x + 49$. Find the length of each side.

47. **GEOMETRY** The area of a square is represented by $16x^2 + 40x + 25$. Find the length of each side.

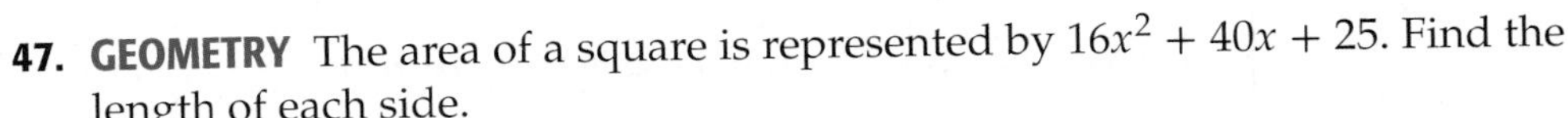

48. **ELECTION** For the student council elections, Franco is building the voting box shown with a volume of 672 cubic inches.

Vote for Student Council

h

$h - 2$

$h + 6$

a. Write a polynomial that represents the volume of the box.

b. What are the dimensions of the voting box?

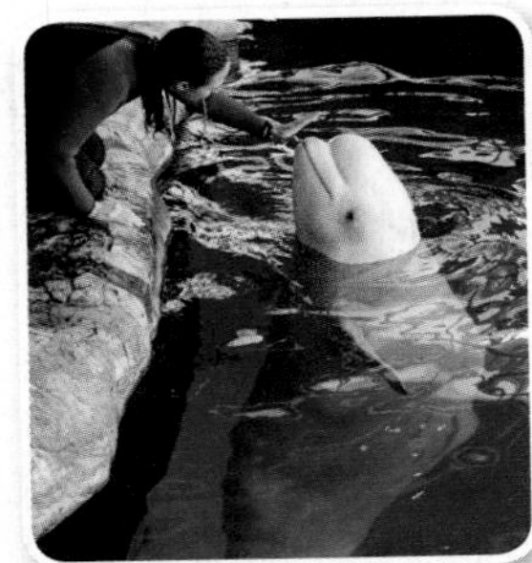

Real-World Link

The Shedd Aquarium in Chicago, Illinois, is home to over 1500 species of fish and mammals.

Source: *Chicago Traveler*

49. **AQUARIUM** Dexter has a fish tank shaped like a rectangular prism. It has a volume of 480 cubic inches. The height of the tank is 8 inches taller than the width, and the length of the tank is 6 inches longer than the width.

a. Write a polynomial that represents the volume of the fish tank.

b. What are the dimensions of the fish tank?

Real-World Link

In 2005, 219,000 above-ground swimming pools were sold. This brought the total number of above-ground swimming pools in the United States to 3.6 million.

Source: Association of Pool & Spa Professionals

50. **GEOMETRY** The volume of a rectangular prism is represented by the expression $8y^3 + 40y^2 + 50y$. Find the possible dimensions of the prism if the dimensions are represented by polynomials with integer coefficients.

51. **POOLS** Ichiro wants to buy an above-ground swimming pool for his yard. Model A is 42 inches deep and holds 1750 cubic feet of water. The length of the rectangular pool is 5 feet more than the width.
    - **a.** What is the surface area of the water?
    - **b.** What are the dimensions of the pool?
    - **c.** Model B pool holds twice as much water as Model A. What are some possible dimensions for this pool?
    - **d.** Model C has length and width that are both twice as long as Model A, but the height is the same. What is the ratio of the volume of Model A to Model C?

52. **GEOMETRY** Use the rectangular prism at the right.
    - **a.** Write an expression for the height and width of the prism in terms of the length, ℓ.
    - **b.** Write a polynomial for the volume of the prism in terms of the length.

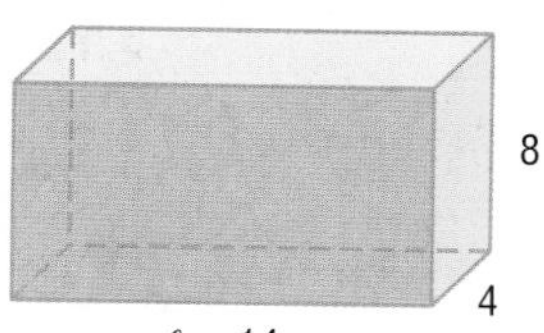

H.O.T. Problems

Use **H**igher-**O**rder **T**hinking Skills

53. **FIND THE ERROR** Debbie and Adriano are factoring the expression $x^8 - x^4$ completely. Is either of them correct? Explain your reasoning.

Debbie	Adriano
$x^8 - x^4 = x^4(x^2 + 1)(x^2 - 1)$	$x^8 - x^4 = x^4(x^2 + 1)(x - 1)(x + 1)$

54. **CHALLENGE** Factor $x^{n+6} + x^{n+2} + x^n$ completely.

55. **OPEN ENDED** Write a perfect square trinomial equation in which the coefficient of the middle term is negative and the last term is a fraction. Solve the equation.

56. **REASONING** Find a counterexample to the following statement.

    A polynomial equation of degree three always has three real solutions.

57. **WRITING IN MATH** Explain how to factor a polynomial completely.

58. **WHICH ONE DOESN'T BELONG?** Identify the trinomial that does not belong. Explain.

$4x^2 - 36x + 81$	$25x^2 + 10x + 1$	$4x^2 + 10x + 4$	$9x^2 - 24x + 16$

59. **OPEN ENDED** Write a binomial that can be factored using the difference of two squares twice. Set your binomial equal to zero and solve the equation.

60. **WRITING IN MATH** Explain how to determine whether a trinomial is a perfect square trinomial.

Standardized Test Practice

61. What is the solution set for the equation $(x - 3)^2 = 25$?

A $\{-8, 2\}$
B $\{-2, 8\}$
C $\{4, 14\}$
D $\{-4, 14\}$

62. SHORT RESPONSE Write an equation in slope-intercept form for the graph shown below.

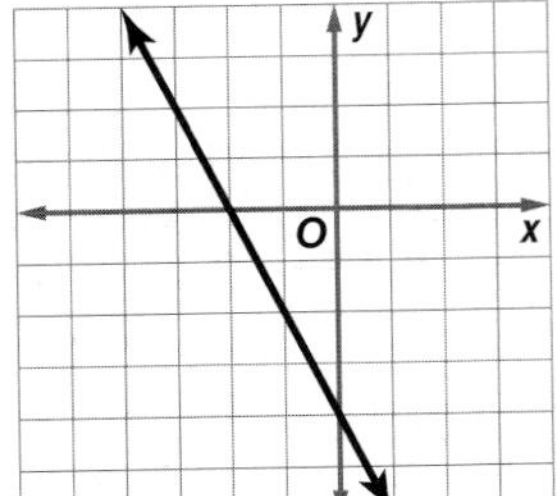

63. At an amphitheater, the price of 2 lawn seats and 2 pavilion seats is \$120. The price of 3 lawn seats and 4 pavilion seats is \$225. How much do lawn and pavilion seats cost?

F \$20 and \$41.25
G \$10 and \$50
H \$15 and \$45
J \$30 and \$30

64. GEOMETRY The circumference of a circle is $\frac{6\pi}{5}$ units. What is the area of the circle?

A $\frac{3\pi}{5}$ units^2
B $\frac{12\pi}{5}$ units^2
C $\frac{9\pi}{25}$ units^2
D $\frac{30\pi}{25}$ units^2

Spiral Review

Factor each polynomial, if possible. If the polynomial cannot be factored, write *prime*. (Lesson 8-5)

65. $x^2 - 16$
66. $4x^2 - 81y^2$
67. $1 - 100p^2$
68. $3a^2 - 20$
69. $25n^2 - 1$
70. $36 - 9c^2$

Solve each equation. Check your solutions. (Lesson 8-4)

71. $4x^2 - 8x - 32 = 0$
72. $6x^2 - 48x + 90 = 0$
73. $14x^2 + 14x = 28$
74. $2x^2 - 10x = 48$
75. $5x^2 - 25x = -30$
76. $8x^2 - 16x = 192$

SOUND The intensity of sound can be measured in watts per square meter. The table gives the watts per square meter for some common sounds. (Lesson 7-2)

Watts Per Square Meter	Common Sounds
10^{-11}	rustling leaves
10^{-10}	whisper
10^{-6}	normal conversation
10^{-5}	busy street traffic
10^{-4}	vacuum cleaner
10^{-1}	front rows of rock concert
10^{1}	threshold of pain
10^{2}	military jet takeoff

77. How many times more intense is the sound from busy street traffic than sound from normal conversation?

78. Which sound is 10,000 times as loud as a busy street traffic?

79. How does the intensity of a whisper compare to that of normal conversation?

Skills Review

Find the slope of the line that passes through each pair of points. (Lesson 3-3)

80. (5, 7), (−2, −3)
81. (2, −1), (5, −3)
82. (−4, −1), (−3, −3)
83. (−3, −4), (5, −1)
84. (−2, 3), (8, 3)
85. (−5, 4), (−5, −1)

Math Online glencoe.com
• STUDY TO GO
• Vocabulary Review

Chapter Summary

Key Concepts

Monomials and Factoring (Lesson 8-1)

- The greatest common factor (GCF) of two or more monomials is the product of their common prime factors.

Factoring Using the Distributive Property (Lesson 8-2)

- Using the Distributive Property to factor polynomials with four or more terms is called factoring by grouping.
  $ax + bx + ay + by = x(a + b) + y(a + b)$
  $= (a + b)(x + y)$
- Factoring can be used to solve some equations. According to the Zero Product Property, for any real numbers a and b, if $ab = 0$, then either $a = 0$, $b = 0$, or both a and b equal zero.

Factoring Trinomials and Differences of Squares (Lessons 8-3 through 8-5)

- To factor $x^2 + bx + c$, find m and p with a sum of b and a product of c. Then write $x^2 + bx + c$ as $(x + m)(x + p)$.
- To factor $ax^2 + bx + c$, find m and p with a sum of b and a product of ac. Then write as $ax^2 + mx + px + c$ and factor by grouping.
- $a^2 - b^2 = (a - b)(a + b)$

Perfect Squares and Factoring (Lesson 8-6)

- For a trinomial to be a perfect square, the first and last terms must be perfect squares, and the middle term must be twice the product of the square roots of the first and last terms.
- For any number $n \geq 0$, if $x^2 = n$, then $x = \pm\sqrt{n}$.

FOLDABLES® Study Organizer

Be sure the Key Concepts are noted in your Foldable.

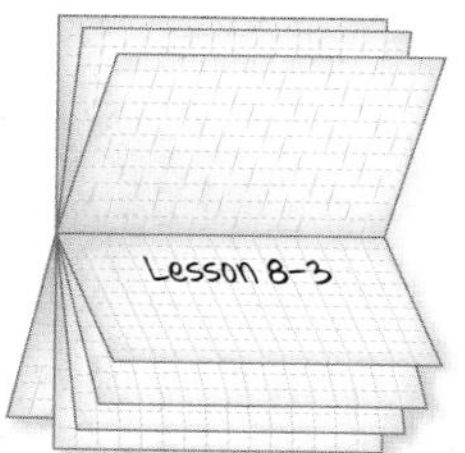

Key Vocabulary

difference of two squares (p. 499)
factored form (p. 471)
factoring (p. 476)
factoring by grouping (p. 477)
greatest common factor (GCF) (p. 471)
perfect square trinomial (p. 505)
prime polynomial (p. 495)
quadratic equation (p. 488)
Square Root Property (p. 508)
Zero Product Property (p. 478)

Vocabulary Check

State whether each sentence is *true* or *false*. If *false*, replace the underlined phrase or expression to make a true sentence.

1. $\underline{x^2 + 5x + 6}$ is an example of a prime polynomial.
2. $\underline{(x + 5)(x - 5)}$ is the factorization of a difference of squares.
3. $\underline{5x}$ is the greatest common factor of $10x$ and $15xy^2$.
4. $\underline{(x + 5)(x - 2)}$ is the factored form of $x^2 - 3x - 10$.
5. Expressions with four or more unlike terms can sometimes be <u>factored by grouping</u>.
6. The Zero Product Property states that if <u>$ab = 1$, then a or b is 1</u>.
7. $x^2 - 12x + 36$ is an example of a <u>perfect square trinomial</u>.
8. $\underline{x - 2 = 0}$ is a quadratic equation.
9. $x^2 - 16$ is an example of a <u>perfect square trinomial</u>.
10. The greatest common factor of $8x$ and $4x^2$ is $\underline{4x}$.

Study Guide and Review

Lesson-by-Lesson Review

8-1 Monomials and Factoring (pp. 471–474)

Factor each monomial completely.

11. $28x^3$
12. $-33x^2y^3$
13. $68cd^3$
14. $120mq$

Find the greatest common factor of each set of monomials.

15. $22b, 33c$
16. $21xy, 28x^2y, 42xy^2$
17. $6ab, 24ab^4$
18. $10ab, 30a, 40a^2b$

19. **HOME IMPROVEMENT** A landscape architect is designing a stone path 36 inches wide and 120 inches long. What is the maximum size square stone that can be used so that none of the stones have to be cut?

EXAMPLE 1

Factor $24a^2b^3$ completely.

$24a^2b^3 = 4 \cdot 6 \cdot a \cdot a \cdot b \cdot b \cdot b$

$= 2 \cdot 2 \cdot 2 \cdot 3 \cdot a \cdot a \cdot b \cdot b \cdot b$

EXAMPLE 2

Find the greatest common factor of $12xy$ and $8xy^2$.

$12xy = (2) \cdot (2) \cdot 3 \cdot (x) \cdot (y)$

$8xy^2 = (2) \cdot (2) \cdot 2 \cdot (x) \cdot (y) \cdot y$

Factor each monomial. Circle the common prime factors.

The greatest common factor is $2 \cdot 2 \cdot x \cdot y$ or $4xy$.

8-2 Using the Distributive Property (pp. 476–482)

Use the Distributive Property to factor each polynomial.

20. $12x + 24y$
21. $14x^2y - 21xy + 35xy^2$
22. $8xy - 16x^3y + 10y$
23. $a^2 - 4ac + ab - 4bc$
24. $2x^2 - 3xz - 2xy + 3yz$
25. $24am - 9an + 40bm - 15bn$

Solve each equation. Check your solutions.

26. $x(3x - 6) = 0$
27. $6x^2 = 12x$
28. $x^2 = 3x$
29. $3x^2 = 5x$

30. **GEOMETRY** The area of the rectangle shown is $x^3 - 2x^2 + 5x$ square units. What is the length?

EXAMPLE 3

Factor $12y^2 + 9y + 8y + 6$.

$12y^2 + 9y + 8y + 6$

$= (12y^2 + 9y) + (8y + 6)$ — **Group terms with common factors.**

$= 3y(4y + 3) + 2(4y + 3)$ — **Factor the GCF from each group.**

$= (4y + 3)(3y + 2)$ — **Distributive Property**

EXAMPLE 4

Solve $x^2 - 6x = 0$. Check your solutions.

Write the equation so that it is of the form $ab = 0$.

$x^2 - 6x = 0$ — **Original equation**

$x(x - 6) = 0$ — **Factor by using the GCF.**

$x = 0$ or $x - 6 = 0$ — **Zero Product Property**

$x = 6$ — **Solve.**

The roots are 0 and 6. Check by substituting 0 and 6 for x in the original equation.

MIXED PROBLEM SOLVING
For mixed problem-solving practice, see page 852.

8-3 Quadratic Equations: $x^2 + bx + c = 0$ (pp. 485–491)

Factor each trinomial.

31. $x^2 - 8x + 15$

32. $x^2 + 9x + 20$

33. $x^2 - 5x - 6$

34. $x^2 + 3x - 18$

Solve each equation. Check your solutions.

35. $x^2 + 5x - 50 = 0$

36. $x^2 - 6x + 8 = 0$

37. $x^2 + 12x + 32 = 0$

38. $x^2 - 2x - 48 = 0$

39. $x^2 + 11x + 10 = 0$

40. ART An artist is working on a painting that is 3 inches longer than it is wide. The area of the painting is 154 square inches. What is the length of the painting?

EXAMPLE 5

Factor $x^2 + 10x + 21$

$b = 10$ and $c = 21$, so $m + p$ is positive and mp is positive. Therefore, m and p must both be positive. List the positive factors of 21, and look for the pair of factors with a sum of 10.

Factors of 21	Sum of 10
1, 21	22
3, 7	10

The correct factors are 3 and 7.

$x^2 + 10x + 21 = (x + m)(x + p)$ **Write the pattern.**

$= (x + 3)(x + 7)$ **$m = 3$ and $p = 7$**

8-4 Quadratic Equations: $ax^2 + bx + c = 0$ (pp. 493–498)

Factor each trinomial, if possible. If the trinomial cannot be factored, write *prime*.

41. $12x^2 + 22x - 14$

42. $2y^2 - 9y + 3$

43. $3x^2 - 6x - 45$

44. $2a^2 + 13a - 24$

Solve each equation. Check your solutions.

45. $40x^2 + 2x = 24$

46. $2x^2 - 3x - 20 = 0$

47. $-16t^2 + 36t - 8 = 0$

48. $6x^2 - 7x - 5 = 0$

49. GEOMETRY The area of the rectangle shown is $6x^2 + 11x - 7$ square units. What is the width of the rectangle?

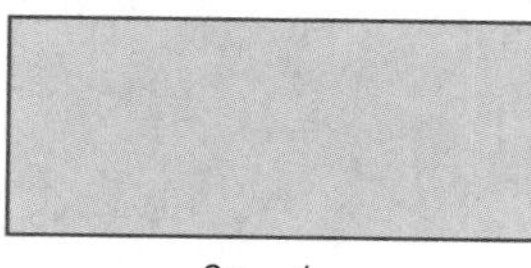

EXAMPLE 6

Factor $12a^2 + 17a + 6$

$a = 12$, $b = 17$, and $c = 6$. Since b is positive, $m + p$ is positive. Since c is positive, mp is positive. So, m and p are both positive. List the factors of 12(6) or 72, where both factors are positive.

Factors of 72	Sum of 17
1, 72	73
2, 36	38
3, 24	27
4, 18	22
6, 12	18
8, 9	17

The correct factors are 8 and 9.

$$\begin{aligned} 12a^2 + 17a + 6 &= 12a^2 + ma + pa + 6 \\ &= 12a^2 + 8a + 9a + 6 \\ &= (12a^2 + 8a) + (9a + 6) \\ &= 4a(3a + 2) + 3(3a + 2) \\ &= (3a + 2)(4a + 3) \end{aligned}$$

So, $12a^2 + 17a + 6 = (3a + 2)(4a + 3)$.

CHAPTER 8

Study Guide and Review

8-5 Quadratic Equations: Differences of Squares (pp. 499–504)

Factor each polynomial.

50. $y^2 - 81$

51. $64 - 25x^2$

52. $16a^2 - 21b^2$

53. $3x^2 - 3$

Solve each equation by factoring. Check your solutions.

54. $a^2 - 25 = 0$

55. $9x^2 - 25 = 0$

56. $81 - y^2 = 0$

57. $x^2 - 5 = 20$

58. **EROSION** A boulder falls down a mountain into water 64 feet below. The distance d that the boulder falls in t seconds is given by the equation $d = 16t^2$. How long does it take the boulder to hit the water?

EXAMPLE 7

Solve $x^2 - 4 = 12$ by factoring.

$x^2 - 4 = 12$	**Original equation**
$x^2 - 16 = 0$	**Subtract 12 from each side.**
$x^2 - (4)^2 = 0$	**$16 = 4^2$**
$(x + 4)(x - 4) = 0$	**Factor the difference of squares.**
$x + 4 = 0$ or $x - 4 = 0$	**Zero Product Property**
$x = -4$ $\quad$ $x = 4$	**Solve each equation.**

The solutions are −4 and 4.

8-6 Quadratic Equations: Perfect Squares (pp. 505–512)

Factor each polynomial, if possible. If the polynomial cannot be factored write *prime*.

59. $x^2 + 12x + 36$

60. $x^2 + 5x + 25$

61. $9y^2 - 12y + 4$

62. $4 - 28a + 49a^2$

63. $x^4 - 1$

64. $x^4 - 16x^2$

Solve each equation. Check your solutions.

65. $(x - 5)^2 = 121$

66. $4c^2 + 4c + 1 = 9$

67. $4y^2 = 64$

68. $16d^2 + 40d + 25 = 9$

69. **LANDSCAPING** A sidewalk of equal width is being built around a square yard. What is the width of the sidewalk?

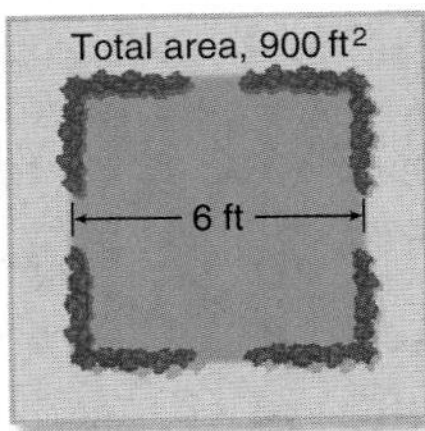

EXAMPLE 8

Solve $(x - 9)^2 = 144$.

$(x - 9)^2 = 144$	**Original equation**
$x - 9 = \pm\sqrt{144}$	**Square Root Property**
$x - 9 = \pm 12$	**144 = 12 • 12**
$x = 9 \pm 12$	**Add 9 to each side.**
$x = 9 + 12$ or $x = 9 - 12$	**Zero Product Property**
$x = 21$ $\quad$ $x = -3$	**Solve.**

CHECK

$(x - 9)^2 = 144$	$(x - 9)^2 = 144$
$(21 - 9)^2 \stackrel{?}{=} 144$	$(-3 - 9)^2 \stackrel{?}{=} 144$
$(12)^2 \stackrel{?}{=} 144$	$(-12)^2 \stackrel{?}{=} 144$
$144 = 144$ ✓	$144 = 144$ ✓

Factor each monomial completely.

1. $25x^2y^4$
2. $17ab^2$
3. $-18c^5d^3$
4. **GARDENING** Conrado is planting 140 pumpkins in a rectangular arrangement in his garden. In what ways can he arrange them so that he has at least 4 rows of pumpkins, the same number of pumpkins in each row, and at least 6 pumpkins in each row?

Find the greatest common factor of each set of monomials.

5. $2a, 8a^2, 16a^3$
6. $7c, 24d$
7. $50g^2h, 120gh^2$
8. $8q^2r^2, 36qr$
9. **MULTIPLE CHOICE** The area of the rectangle shown below is $2x^2 - x - 15$ square units. What is the width of the rectangle?

A $x - 5$

B $x + 3$

C $x - 3$

D $2x - 3$

Use the Distributive Property to factor each polynomial.

10. $5xy - 10x$
11. $7ab + 14ab^2 + 21a^2b$

Factor each polynomial.

12. $4x^2 + 8x + x + 2$
13. $10a^2 - 50a - a + 5$

Solve each equation. Check your solutions.

14. $y(y - 14) = 0$
15. $3x(x + 6) = 0$
16. $a^2 = 12a$
17. **MULTIPLE CHOICE** Chantel is carpeting a room that has an area of $x^2 - 100$ square feet. If the width of the room is $x - 10$ feet, what is the length of the room?

F $x - 10$ ft

G $x + 10$ ft

H $x - 100$ ft

J 10 ft

Factor each trinomial.

18. $x^2 + 7x + 6$
19. $x^2 - 3x - 28$
20. $10x^2 - x - 3$
21. $15x^2 + 7x - 2$
22. $x^2 - 25$
23. $4x^2 - 81$
24. $9x^2 - 12x + 4$
25. $16x^2 + 40x + 25$

Solve each equation. Check your solutions.

26. $x^2 - 4x = 21$
27. $x^2 - 2x - 24 = 0$
28. $6x^2 - 5x - 6 = 0$
29. $2x^2 - 13x + 20 = 0$
30. **MULTIPLE CHOICE** Which choice is a factor of $x^4 - 1$ when it is factored completely?

A $x^2 - 1$

B $x - 1$

C x

D 1

Solve Multi-Step Problems

Some problems that you will encounter on standardized tests require you to solve multiple parts in order to come up with the final solution. Use this lesson to practice these types of problems.

Strategies for Solving Multi-Step Problems

Step 1

Read the problem statement carefully.

Ask yourself:

- What am I being asked to solve? What information is given?
- Are there any intermediate steps that need to be completed before I can solve the problem?

Step 2

Organize your approach.

- List the steps you will need to complete in order to solve the problem.
- Remember that there may be more than one possible way to solve the problem.

Step 3

Solve and check.

- Work as efficiently as possible to complete each step and solve.
- If time permits, check your answer.

EXAMPLE

Read the problem. Identify what you need to know. Then use the information in the problem to solve.

A florist has 80 roses, 50 tulips, and 20 lilies that he wants to use to create bouquets. He wants to create the maximum number of bouquets possible and use all of the flowers. Each bouquet should have the same number of each type of flower. How many roses will be in each bouquet?

A 4 roses

B 8 roses

C 10 roses

D 15 roses

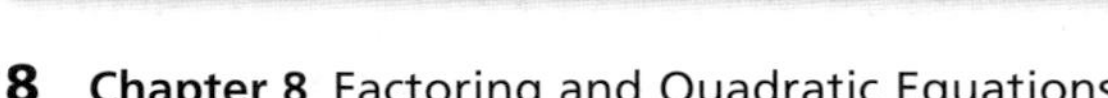

Read the problem carefully. You are given the number of roses, tulips, and lilies and told that bouquets will be made using the same number of flowers in each. You need to find the number of roses that will be in each bouquet.

Step 1 Find the GCF of the number of roses, tulips, and lilies.

Step 2 Use the GCF to determine how many bouquets will be made.

Step 3 Divide the total number of roses by the number of bouquets.

Step 1 Write the prime factorization of each number of flowers to find the GCF.

$80 = \mathbf{2} \cdot 2 \cdot 2 \cdot 2 \cdot \mathbf{5}$

$50 = \mathbf{2} \cdot \mathbf{5} \cdot 5$

$20 = \mathbf{2} \cdot 2 \cdot \mathbf{5}$

$\text{GCF} = 2 \cdot 5 = 10$

Step 2 The GCF of the number of roses, tulips, and lilies tells you how many bouquets can be made because each bouquet will contain the same number of flowers. So, the florist can make a total of 10 bouquets.

Step 3 Divide the number of roses by the number of bouquets to find the number of roses in each bouquet.

$\frac{80}{10} = 8$

So, there will be 8 roses in each bouquet. The answer is B.

Exercises

Read each problem. Identify what you need to know. Then use the information in the problem to solve.

1. Which of the following values is not a solution to $x^3 - 3x^2 - 25x + 75 = 0$?

   A $x = 5$
   B $x = 3$
   C $x = -3$
   D $x = -5$

2. There are 12 teachers, 90 students, and 36 parent volunteers going on a field trip. Mrs. Bartholomew wants to divide everyone into equal groups with the same number of teachers, students, and parents in each group. If she makes as many groups as possible, how many students will be in each group?

   F 6
   G 9
   H 12
   J 15

3. What is the area of the square?

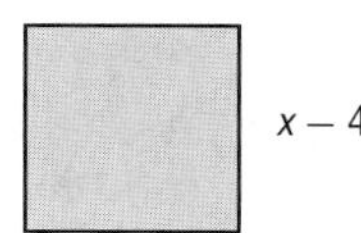

   A $x^2 + 16$
   B $4x - 16$
   C $x^2 - 8x - 16$
   D $x^2 - 8x + 16$

4. Students are selling magazines to raise money for a field trip. They make $2.75 for each magazine they sell. If they want to raise $600, what is the least amount of magazines they need to sell?

   F 121
   G 177
   H 202
   J 219

CHAPTER 8

Standardized Test Practice

Cumulative, Chapters 1 through 8

Multiple Choice

Read each question. Then fill in the correct answer on the answer document provided by your teacher or on a sheet of paper.

1. A baker has the number of cookies shown in the table below available to put into gift baskets. She wants to put the same number of each cookie into each basket, and each basket should have each type of cookie. If she puts the greatest possible number of cookies in each basket, how many baskets can she make?

Type of Cookie	Number
Chocolate Chip	54
Peanut Butter	45
Oatmeal Raisin	36
Sugar	60

A 18

B 16

C 12

D 10

2. Refer to the information given in Exercise 1. How many of each type of cookie will be in each of the baker's gift baskets?

F 2

G 3

H 4

J 6

3. Factor the $mn + 5m - 3n - 15$.

A $(mn - 3)(5)$

B $(n - 3)(m + 5)$

C $(m - 5)(n + 3)$

D $(m - 3)(n + 5)$

4. Which of the following is a solution to $x^2 + 6x - 112 = 0$?

F -14

G -8

H 6

J 12

5. Which of the following polynomials is prime? (Lesson 8-4)

A $5x^2 + 34x + 24$

B $4x^2 + 22x + 10$

C $4x^2 + 38x + 70$

D $5x^2 + 3x + 4$

6. Which of the following is not a factor of the polynomial $45a^2 - 80b^2$?

F 5

G $3a - 4b$

H $2a - 5b$

J $3a + 4b$

7. A rectangular gift box has dimensions that can be represented as shown in the figure. The volume of the box is $56w$ cubic inches. Which of the following is *not* a dimension of the box?

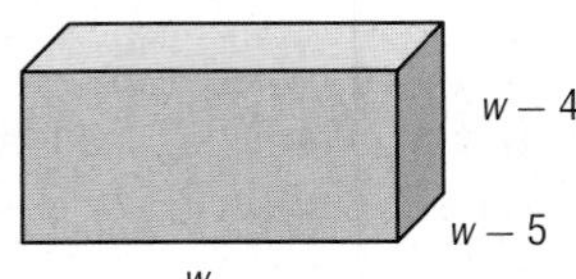

A 6 in.

B 7 in.

C 8 in.

D 12 in.

8. Factor the polynomial $y^2 - 9y + 20$.

F $(y - 2)(y - 10)$

G $(y - 4)(y - 5)$

H $(y - 2)(y - 7)$

J $(y - 5)(y + 2)$

9. Which of the following numbers is less than zero?

A 1.03×10^{-21}

B 7.5×10^2

C 8.21543×10^{10}

D none of the above

Test-Taking Tip

Question 4 If time permits, be sure to check your answer. Substitute it into the equation to see if you get a true number sentence.

Short Response/Gridded Response

Record your answers on the answer sheet provided by your teacher or on a sheet of paper.

10. **GRIDDED RESPONSE** Mr. Branson bought a total of 9 tickets to the zoo. He bought children tickets at the rate of $6.50 and adult tickets for $9.25 each. If he spent $69.50 altogether, how many adult tickets did Mr. Branson purchase?

11. What is the domain of the following relation? $\{(2, -1), (4, 3), (7, 6)\}$

12. Lawrence just added 15 more songs to his MP3 player, making the total number of songs more than 84. Draw a number line that represents the original number of songs he had on his MP3 player?

13. **GRIDDED RESPONSE** Carlos bought a rare painting in 1995 for $14,200. By 2003, the painting was worth $17,120. Write an equation in slope-intercept form that represents the value V of the painting after t years.

14. The equation $h = -16t^2 + 40t + 3$ models the height h in feet of a soccer ball after t seconds. What is the height of the ball after 2 seconds?

15. Marcel spent $24.50 on peanuts and walnuts for a party. He bought 1.5 pounds more peanuts than walnuts. How many pounds of peanuts and walnuts did he buy?

Product	Price per pound
Peanuts p	$3.80
Cashews c	$6.90
Walnuts w	$5.60

16. **GRIDDED RESPONSE** The amount of money that Humberto earns varies directly as the number of hours that he works as shown in the graph. How much money will he earn for working 40 hours next week? Express your answer in dollars.

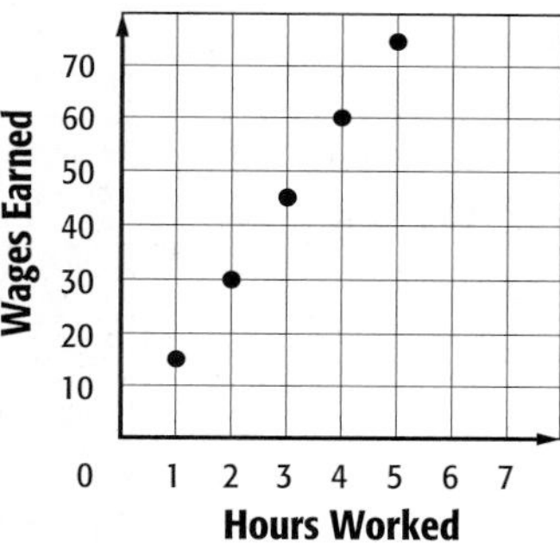

Extended Response

Record your answers on a sheet of paper. Show your work.

17. The height in feet of a model rocket t seconds after being launched into the air is given by the function $h(t) = -16t^2 + 200t$.
    a. Write the expression that shows the height of the rocket in factored form.
    b. At what time(s) is the height of the rocket equal to zero feet above the ground? Describe the real world meaning of your answer.
    c. What is the greatest height reached by the model rocket? When does this occur?

Need Extra Help?

If you missed Question...	1	2	3	4	5	6	7	8	9	10	11	12	13	14	15	16	17
Go to Lesson or Page...	8-1	8-1	8-2	8-3	8-4	8-5	8-6	8-3	7-3	6-5	1-6	5-1	4-2	8-4	2-9	3-4	8-2

CHAPTER 9
Quadratic and Exponential Functions

Then

In Chapter 8, you solved quadratic equations by factoring and by using the Square Root Property.

Now

In Chapter 9, you will:

- Solve quadratic equations by graphing, completing the square, and using the Quadratic Formula.
- Graph exponential functions.
- Identify geometric sequences.

Why?

FINANCE The value of a certain company's stock can be modeled by the function $f(x) = x^2 - 12x + 75$. By graphing this quadratic function, we can make an educated guess as to how the stock will perform in the near future.

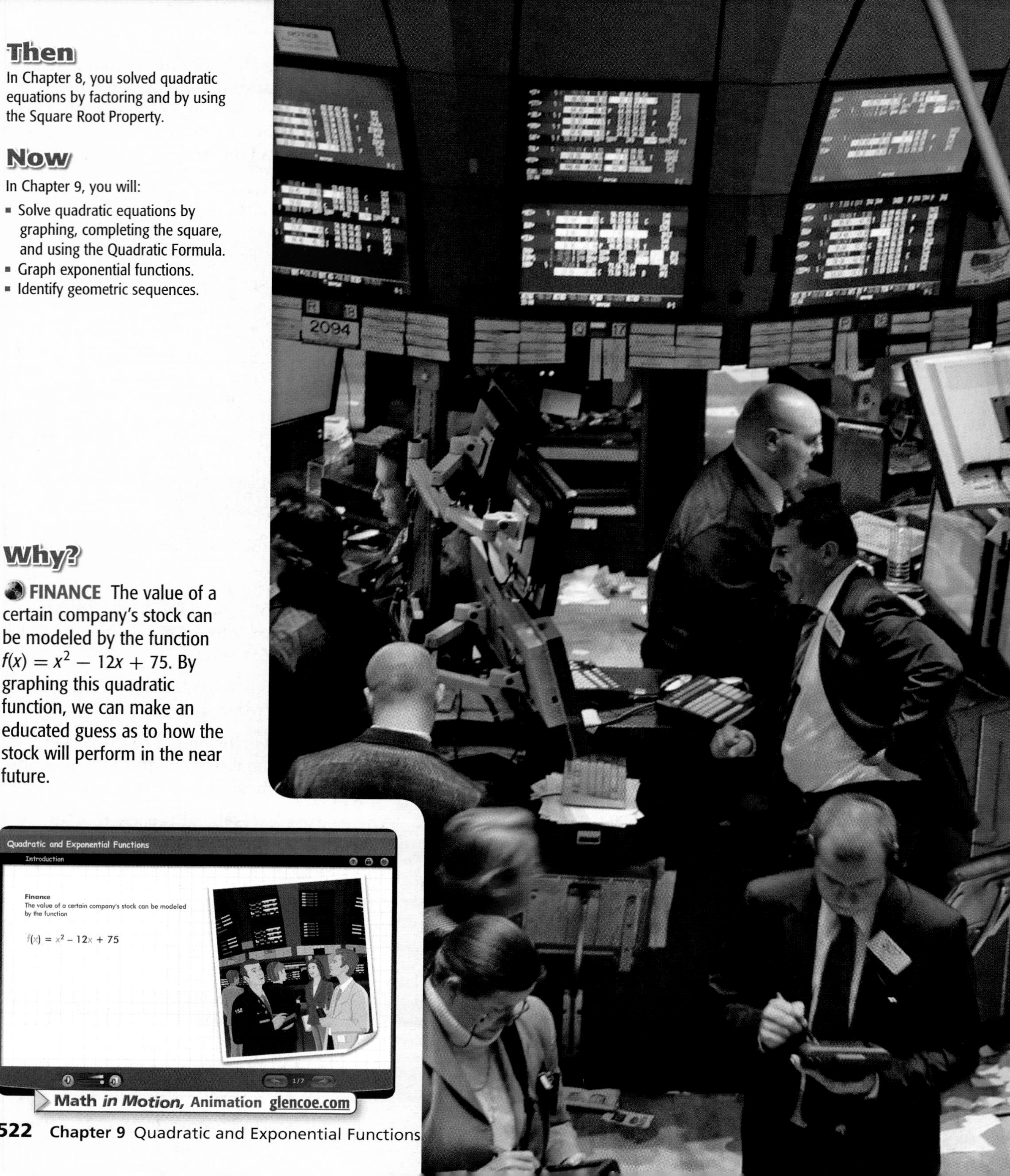

Math *in Motion*, Animation glencoe.com

Get Ready for Chapter 9

Diagnose Readiness You have two options for checking Prerequisite Skills.

Text Option

Take the Quick Check below. Refer to the Quick Review for help.

*Quick*Check

Use a table of values to graph each equation. (Lesson 3-1)

1. $y = x + 3$
2. $y = 2x + 2$
3. $y = -2x - 3$
4. $y = 0.5x - 1$
5. $4x - 3y = 12$
6. $3y = 6 + 9x$
7. **SAVINGS** Jack has $100 to buy a game system. He plans to save $10 each week. Graph an equation to show the total amount T Jack will have in w weeks.

Determine whether each trinomial is a perfect square trinomial. Write *yes* or *no*. If so, factor it. (Lesson 8-6)

8. $a^2 + 12a + 36$
9. $x^2 + 5x + 25$
10. $x^2 - 12x + 32$
11. $x^2 + 20x + 100$
12. $4x^2 + 28x + 49$
13. $k^2 - 16k + 64$
14. $a^2 - 22a + 121$
15. $5t^2 - 12t + 25$

Find the next three terms of each arithmetic sequence. (Lessons 3-5)

16. 16, 4, −8, −20, …
17. 2, 10, 18, 26, …
18. −5, −2, 1, 4, …
19. 3, 5, 7, 9, …
20. **GEOMETRY** Write a formula that can be used to find the perimeter of a figure containing n squares.

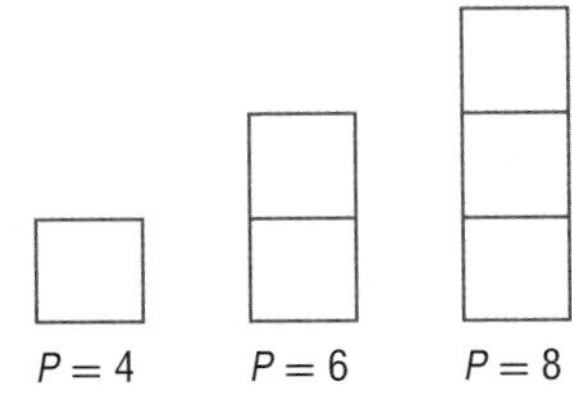

*Quick*Review

EXAMPLE 1

Use a table of values to graph $y = 3x + 1$.

x	$y = 3x + 1$	y
−1	3(−1) + 1	−2
0	3(0) + 1	1
1	3(1) + 1	4
2	3(2) + 1	7

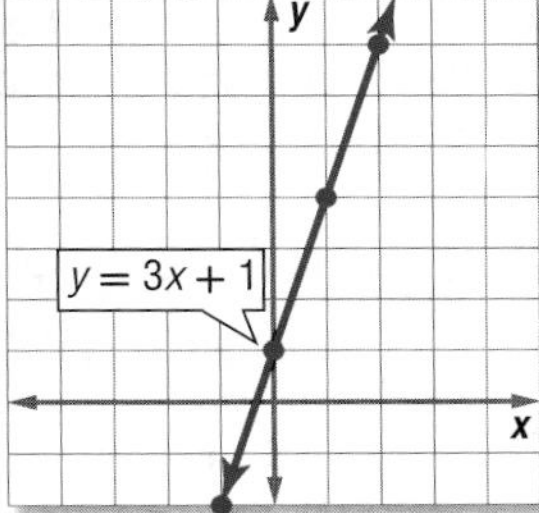

EXAMPLE 2

Determine whether $x^2 - 10x + 25$ is a perfect square trinomial. Write *yes* or *no*. If so, factor it.

1. Is the first term a perfect square? **yes**
2. Is the last term a perfect square? **yes**
3. Is the middle term equal to $-2(1x)(5)$? **yes**

$x^2 - 10x + 25 = (x - 5)^2$

EXAMPLE 3

Find the next three terms of the arithmetic sequence 5, 9, 13, 17, … .

Find the common difference by subtracting a term from the next term.

$$9 - 5 = 4$$

Add to find the next three terms.

$17 + 4 = 21,\ 21 + 4 = 25,\ 25 + 4 = 29$

The next three terms are 21, 25, 29.

Online Option

Math Online Take a self-check Chapter Readiness Quiz at glencoe.com.

Get Started on Chapter 9

You will learn several new concepts, skills, and vocabulary terms as you study Chapter 9. To get ready, identify important terms and organize your resources. You may wish to refer to **Chapter 0** to review prerequisite skills.

FOLDABLES® Study Organizer

Quadratic and Exponential Functions Make this Foldable to help you organize your Chapter 9 notes about quadratic functions. Begin with a sheet of notebook paper.

1. **Fold** the sheet of paper along the length so that the edge of the paper aligns with the margin rule on the paper.

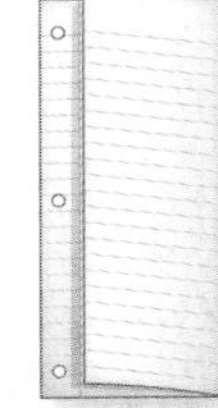

2. **Fold** the sheet twice widthwise to form four sections.

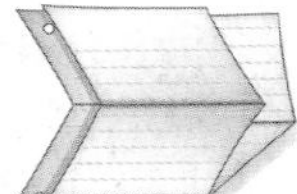

3. **Unfold** the sheet, and cut along the folds on the front flap only.

4. **Label** each section as shown.

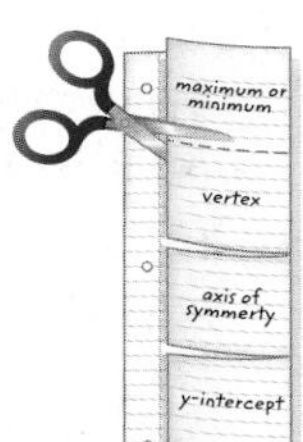

Math Online glencoe.com

- Study the chapter online
- Explore **Math in Motion**
- Get extra help from your own **Personal Tutor**
- Use **Extra Examples** for additional help
- Take a **Self-Check Quiz**
- **Review Vocabulary** in fun ways

New Vocabulary

English		Español
axis of symmetry	• p. 525 •	eje de simetría
maximum	• p. 525 •	máximo
minimum	• p. 525 •	mínimo
nonlinear function	• p. 525 •	función no lineal
parabola	• p. 525 •	parábola
quadratic function	• p. 525 •	función cuadrática
vertex	• p. 525 •	vértice
double root	• p. 538 •	doble raíz
transformation	• p. 544 •	transformación
completing the square	• p. 552 •	completar el cuadrado
Quadratic Formula	• p. 558 •	Formula cuadrática
discriminant	• p. 561 •	discriminante
exponential function	• p. 567 •	función exponencial
compound interest	• p. 574 •	interés es compuesta
common ratio	• p. 578 •	proporción común
geometric sequence	• p. 578 •	secuencia geométrica

Review Vocabulary

domain • p. 38 • dominio all the possible values of the independent variable, x

leading coefficient • p. 425 • coeficiente delantero the coefficient of the first term of a polynomial written in standard form

range • p. 38 • rango all the possible values of the dependent variable, y

In the function represented by the table, the domain is {0, 2, 4, 6}, and the range is {3, 5, 7, 9}.

x	y
0	3
2	5
4	7
6	9

Multilingual eGlossary glencoe.com

9-1

Graphing Quadratic Functions

Then

You graphed linear functions. (Lesson 3-2)

Now

- Analyze the characteristics of graphs of quadratic functions.
- Graph quadratic functions.

New Vocabulary

nonlinear function
quadratic function
standard form
parabola
axis of symmetry
vertex
minimum
maximum
symmetry

Math Online

glencoe.com

- Extra Examples
- Personal Tutor
- Self-Check Quiz
- Homework Help

Why?

The Innovention Fountain in Epcot's Futureworld in Orlando, Florida, is an elaborate display of water, light, and music. The sprayers shoot water in shapes that can be modeled by quadratic equations. You can use the graph of this equation to show the path of the water.

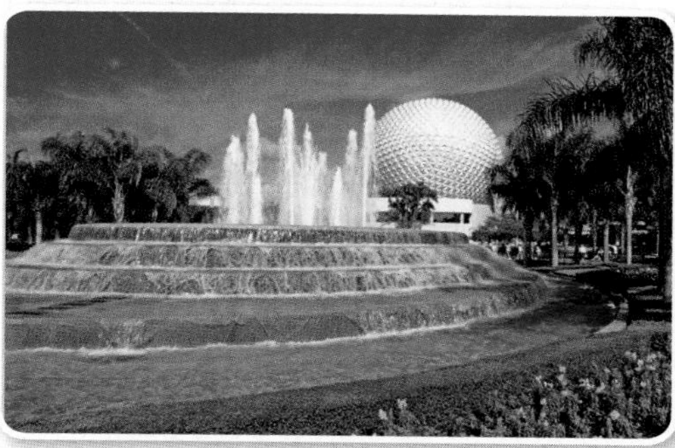

Characteristics of Quadratic Functions You have studied linear functions. There are also **nonlinear functions** with graphs of other shapes. **Quadratic functions** are nonlinear and can be written in the form $f(x) = ax^2 + bx + c$, where $a \neq 0$. This form is called the **standard form** of a quadratic function.

The shape of the graph of a quadratic function is called a **parabola**. Parabolas are symmetric about a central line called the **axis of symmetry**. The axis of symmetry intersects a parabola at only one point, called the **vertex**.

Key Concept | **Quadratic Functions** | For Your FOLDABLE

Parent Function: $f(x) = x^2$

Standard Form: $f(x) = ax^2 + bx + c$

Type of Graph: parabola

Axis of Symmetry: $x = -\frac{b}{2a}$

y-intercept: c

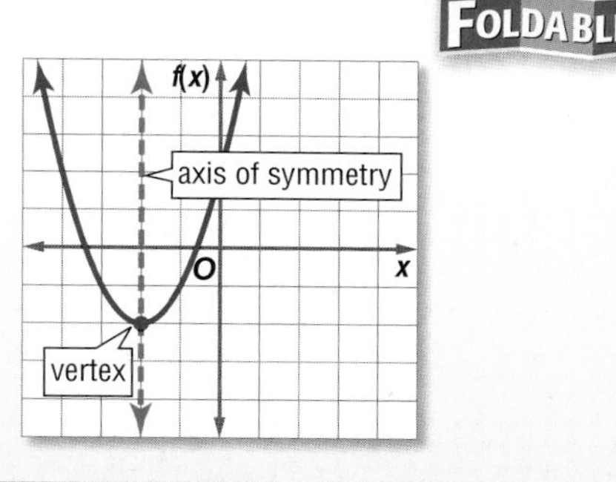

When $a > 0$, the graph of $y = ax^2 + bx + c$ opens upward. The lowest point on the graph is the **minimum**. When $a < 0$, the graph of $y = ax^2 + bx + c$ opens downward. The highest point on the graph is the **maximum**. The maximum or minimum is the vertex.

EXAMPLE 1 Graph a Parabola

Use a table of values to graph $y = 3x^2 + 6x - 4$. State the domain and range.

x	y
1	5
0	−4
−1	−7
−2	−4
−3	5

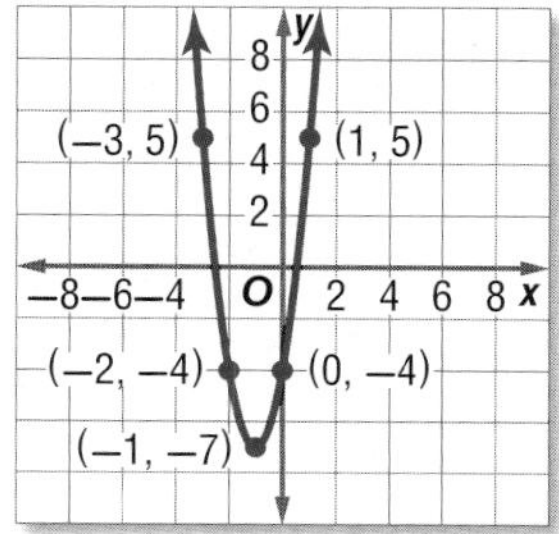

Graph the ordered pairs, and connect them to create a smooth curve. The parabola extends to infinity. The domain is all real numbers. The range is $\{y \mid y \geq -7\}$, because −7 is the minimum.

Check Your Progress

1. Use a table of values to graph $y = x^2 + 3$. State the domain and range.

Personal Tutor glencoe.com

Review Vocabulary

Domain and Range The domain is the set of all of the possible values of the independent variable x. The range is the set of all of the possible values of the dependent variable y.

Figures that possess **symmetry** are those in which each half of the figure matches exactly.

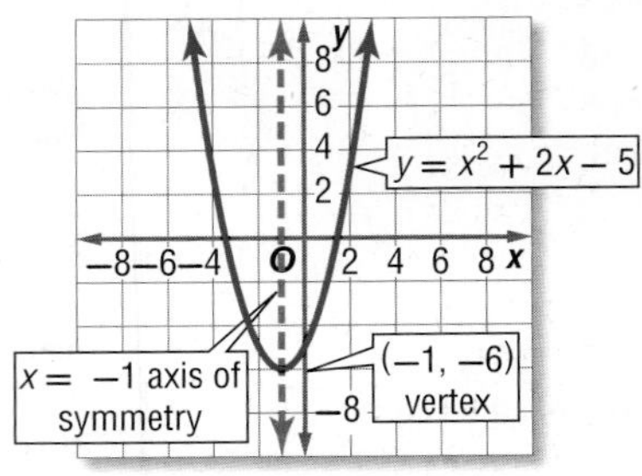

A parabola is symmetric about the axis of symmetry. Every point on the parabola to the left of the axis of symmetry has a corresponding point on the other half.

When identifying characteristics from a graph, it is often easiest to locate the vertex first. It is either the maximum or minimum point of the graph.

EXAMPLE 2 Identify Characteristics from Graphs

Find the vertex, the equation of the axis of symmetry, and the y-intercept of each graph.

a.

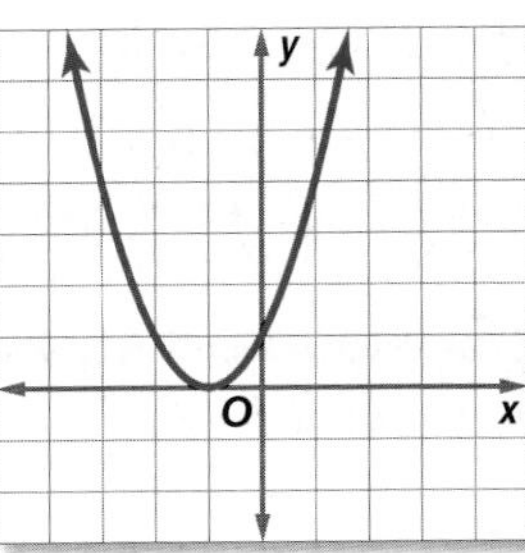

Step 1 Find the vertex.
Because the parabola opens upward, the vertex is located at the minimum point of the parabola. It is located at $(-1, 0)$.

Step 2 Find the axis of symmetry.
The axis of symmetry is the line that goes through the vertex and divides the parabola into congruent halves. It is located at $x = -1$.

Step 3 Find the y-intercept.
The y-intercept is the point where the graph intersects the y-axis. It is located at $(0, -1)$, so the y-intercept is -1.

b.

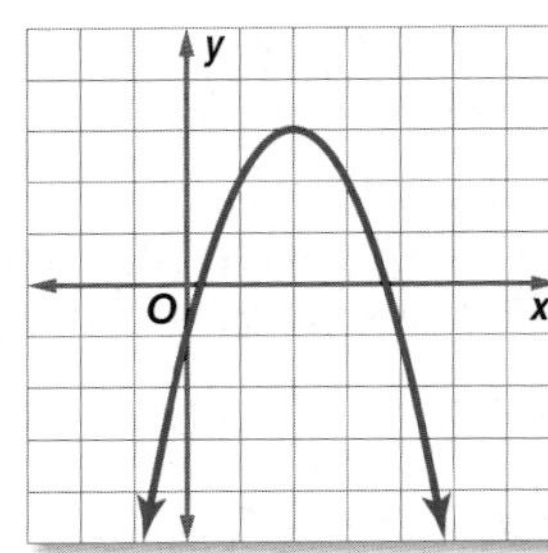

Step 1 Find the vertex.
The parabola opens downward, so the vertex is located at its maximum point, $(2, 3)$.

Step 2 Find the axis of symmetry.
The axis of symmetry is located at $x = 2$.

Step 3 Find the y-intercept.
The y-intercept is where the parabola crosses the y-axis. It is located at $(0, -1)$, so the y-intercept is -1.

Check Your Progress

2A.

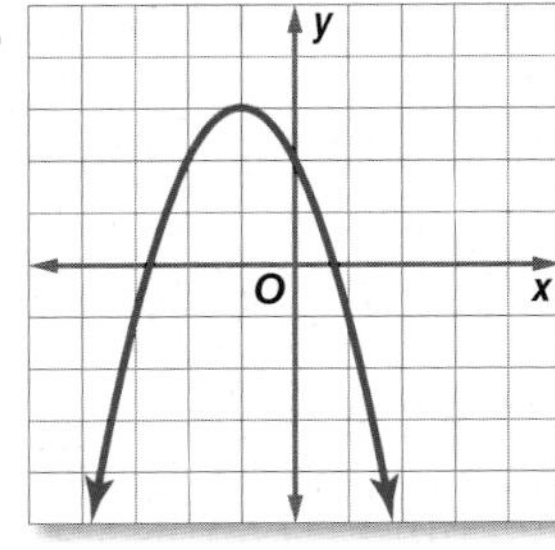

2B.

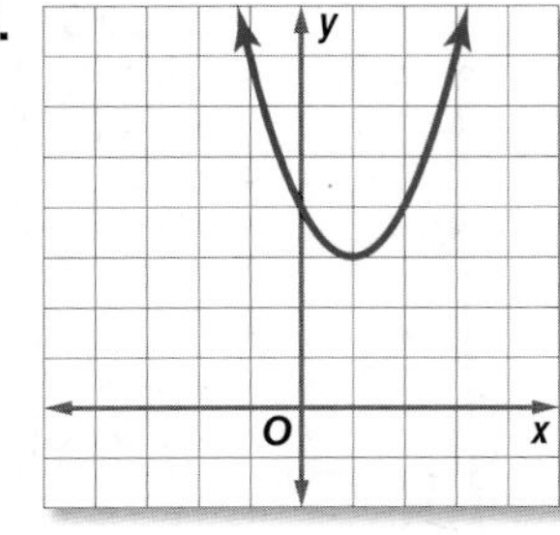

Personal Tutor glencoe.com

StudyTip

Function Characteristics When identifying characteristics of a function, it is often easiest to locate the axis of symmetry first.

EXAMPLE 3 Identify Characteristics from Functions

Find the vertex, the equation of the axis of symmetry, and the y-intercept of each function.

a. $y = 2x^2 + 4x - 3$

$x = -\frac{b}{2a}$ **Formula for the equation of the axis of symmetry**

$x = -\frac{4}{2 \cdot 2}$ **$a = 2$ and $b = 4$**

$x = -1$ **Simplify.**

The equation for the axis of symmetry is $x = -1$.

To find the vertex, use the value you found for the axis of symmetry as the x-coordinate of the vertex. To find the y-coordinate, substitute that value for x in the original equation.

$y = 2x^2 + 4x - 3$ **Original equation**

$= 2(-1)^2 + 4(-1) - 3$ **$x = -1$**

$= -5$ **Simplify.**

The vertex is at $(-1, -5)$.

The y-intercept always occurs at $(0, c)$. So, the y-intercept is -3.

b. $y = -x^2 + 6x + 4$

$x = -\frac{b}{2a}$ **Formula for the equation of the axis of symmetry**

$x = -\frac{6}{2(-1)}$ **$a = -1$ and $b = 6$**

$x = 3$ **Simplify.**

The equation of the axis of symmetry is $x = 3$.

$y = -x^2 + 6x + 4$ **Original equation**

$= -(3)^2 + 6(3) + 4$ **$x = 3$**

$= 13$ **Simplify.**

The vertex is at $(3, 13)$.

The y-intercept is 4.

StudyTip

***y*-intercept** The y-coordinate of the y-intercept is also the constant term (c) of the quadratic function in standard form.

Check Your Progress

3A. $y = -3x^2 + 6x - 5$

3B. $y = 2x^2 + 2x + 2$

Personal Tutor glencoe.com

There are general differences between linear functions and quadratic functions.

	Linear Functions	Quadratic Functions
Standard Form	$y = ax + b$	$y = ax^2 + bx + c; a \neq 0$
Degree	1; Notice that all of the variables are to the first power.	2; Notice that the independent variable, x, is squared in the first term. The coefficient a can not equal 0, or the equation would be linear.
Example	$y = 2x + 6$	$y = 3x^2 + 5x - 4$
Graph	line	parabola

Next you will learn how to identify whether the parabola opens up or down and whether the vertex is a maximum or a minimum point.

Key Concept — Maximum and Minimum Values

For Your FOLDABLE

Words The graph of $f(x) = ax^2 + bx + c$, where $a \neq 0$:

- opens upward and has a minimum value when $a > 0$, and
- opens downward and has a maximum value when $a < 0$.
- The range of a quadratic function is all real numbers greater than or equal to the minimum, or all real numbers less than or equal to the maximum.

Examples a is positive. a is negative.

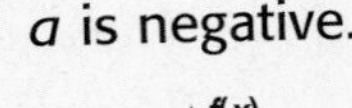

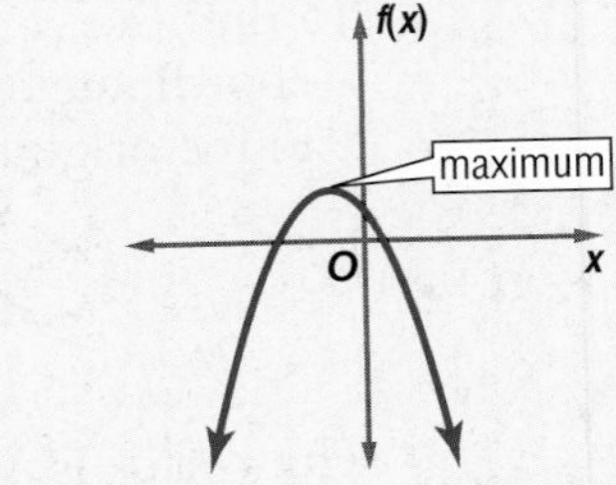

> **Watch Out!**
>
> **Minimum and Maximum Values** Don't forget to find both coordinates of the vertex (x, y). The minimum or maximum value is the y-coordinate.

EXAMPLE 4 Maximum and Minimum Values

Consider $f(x) = -2x^2 - 4x + 6$.

a. Determine whether the function has a *maximum* or *minimum* value.

For $f(x) = -2x^2 - 4x + 6$, $a = -2$, $b = -4$, and $c = 6$.
Because a is negative the graph opens down, so the function has a maximum value.

b. State the maximum or minimum value of the function.

The maximum value is the y-coordinate of the vertex.

The x-coordinate of the vertex is $\frac{-b}{2a}$ or $\frac{4}{2(-2)}$ or -1.

$f(x) = -2x^2 - 4x + 6$	**Original function**
$f(-1) = -2(-1)^2 - 4(-1) + 6$	$x = -1$
$f(-1) = 8$	**Simplify.**

The maximum value is 8.

c. State the domain and range of the function.

The domain is all real numbers. The range is all real numbers less than or equal to the maximum value, or $\{y \mid y \leq 8\}$.

Check Your Progress

Consider $g(x) = 2x^2 - 4x - 1$.

4A. Determine whether the function has a *maximum* or *minimum* value.

4B. State the maximum or minimum value.

4C. State the domain and range of the function.

Personal Tutor glencoe.com

Math *in Motion,* Animation glencoe.com

Graph Quadratic Functions You have learned how to find several important characteristics of quadratic functions.

Key Concept — Graph Quadratic Functions

For Your FOLDABLE

Step 1 Find the equation of the axis of symmetry.

Step 2 Find the vertex, and determine whether it is a maximum or minimum.

Step 3 Find the y-intercept.

Step 4 Use symmetry to find additional points on the graph, if necessary.

Step 5 Connect the points with a smooth curve.

EXAMPLE 5 Graph Quadratic Functions

Graph $f(x) = x^2 + 4x + 3$.

Step 1 Find the equation of the axis of symmetry.

$x = \frac{-b}{2a}$ **Formula for the equation of the axis of symmetry**

$x = \frac{-4}{2 \cdot 1}$ **$a = 1$ and $b = 4$**

$x = -2$ **Simplify.**

Step 2 Find the vertex, and determine whether it is a maximum or minimum.

$y = x^2 + 4x + 3$ **Original equation**

$= (-2)^2 + 4(-2) + 3$ **$x = -2$**

$= -1$ **Simplify.**

The vertex lies at $(-2, -1)$. Because a is positive the graph opens up, and the vertex is a minimum.

Step 3 Find the y-intercept.

$y = x^2 + 4x + 3$ **Original equation**

$= (0)^2 + 4(0) + 3$ **$x = 0$**

$= 3$ **Simplify.**

The y-intercept is 3.

Step 4 The axis of symmetry divides the parabola into two equal parts. So if there is a point on one side, there is a corresponding point on the other side that is the same distance from the axis of symmetry and has the same y-value.

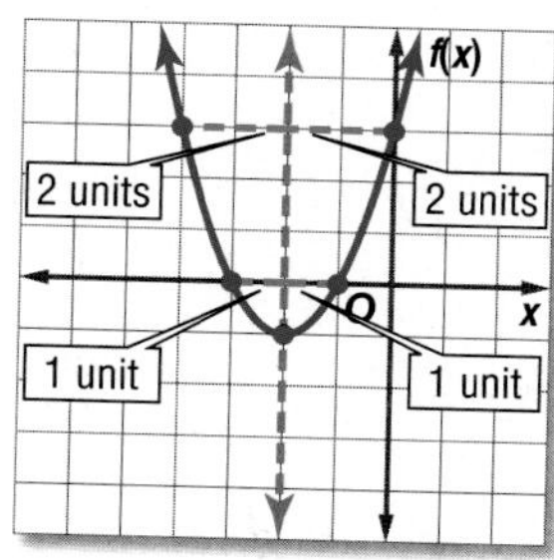

Step 5 Connect the points with a smooth curve.

StudyTip

Symmetry and Points When locating points that are on opposite sides of the axis of symmetry, they are not only the same distance to the left and right of the axis of symmetry. They are also the same number of spaces up or down from the vertex.

 **Check Your Progress**

Graph each function.

5A. $f(x) = -2x^2 + 2x - 1$

5B. $f(x) = 3x^2 - 6x + 2$

Personal Tutor glencoe.com

You have used what you know about quadratic functions, parabolas, and symmetry to create graphs. You can analyze these graphs to solve real-world problems.

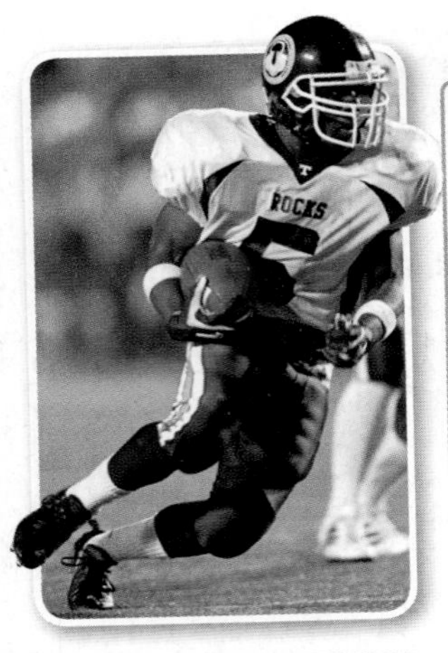

Real-World Link

About 1 in 17 high school seniors playing football will go on to play football at an NCAA school.

Source: National Collegiate Athletic Association

Real-World EXAMPLE 6 Use a Graph of a Quadratic Function

SCHOOL SPIRIT The cheerleaders at Lake High School launch T-shirts into the crowd every time the Lakers score a touchdown. The height of the T-shirt can be modeled by the function $h(x) = -16x^2 + 48x + 6$, where $h(x)$ represents the height in feet of the T-shirt after x seconds.

a. Graph the function.

$x = -\frac{b}{2a}$ — **Equation of the axis of symmetry**

$x = -\frac{48}{2(-16)}$ or $\frac{3}{2}$ — **$a = -16$ and $b = 48$**

The equation of the axis of symmetry is $x = \frac{3}{2}$. Thus, the x-coordinate for the vertex is $\frac{3}{2}$.

$y = -16x^2 + 48x + 6$ — **Original equation**

$= -16\left(\frac{3}{2}\right)^2 + 48\left(\frac{3}{2}\right) + 6$ — **$x = \frac{3}{2}$**

$= -16\left(\frac{9}{4}\right) + 48\left(\frac{3}{2}\right) + 6$ — **$\left(\frac{3}{2}\right)^2 = \frac{9}{4}$**

$= -36 + 72 + 6$ or 42 — **Simplify.**

The vertex is at $\left(\frac{3}{2}, 42\right)$.

Let's find another point. Choose an x-value of 0 and substitute. Our new point is at (0, 6). The point paired with it on the other side of the axis of symmetry is (3, 6).

Repeat this and choose an x-value of 1 to get (1, 38) and its corresponding point (2, 38). Connect these points and create a smooth curve.

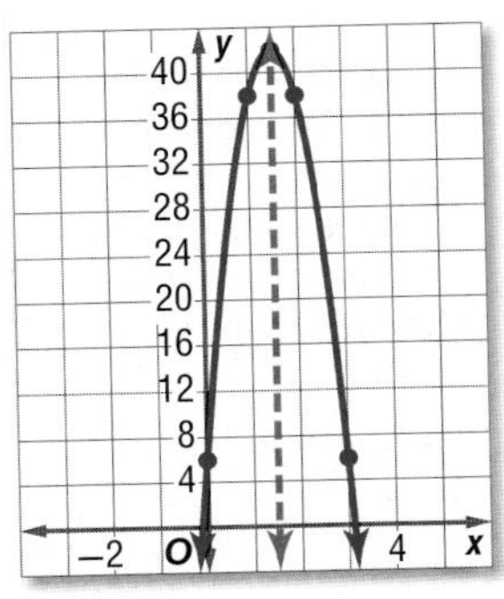

b. At what height was the T-shirt launched?
The T-shirt is launched when time equals 0, or at the y-intercept. So, the T-shirt was launched 6 feet from the ground.

c. What is the maximum height of the T-shirt? When was the maximum height reached?
The maximum height of the T-shirt occurs at the vertex. So the T-shirt reaches a maximum height of 42 feet. The time was $\frac{3}{2}$ or 1.5 seconds after launch.

Check Your Progress

6. TRACK Emilio is competing in the javelin throw. The height of the javelin can be modeled by the equation $y = -16x^2 + 64x + 6$, where y represents the height in feet of the javelin after x seconds.

A. Graph the path of the javelin.

B. At what height is the javelin thrown?

C. What is the maximum height of the javelin?

Personal Tutor glencoe.com

Check Your Understanding

Example 1 p. 525 — Use a table of values to graph each equation. State the domain and range.

1. $y = 2x^2 + 4x - 6$
2. $y = x^2 + 2x - 1$
3. $y = x^2 - 6x - 3$
4. $y = 3x^2 - 6x - 5$

Example 2 p. 526 — Find the vertex, the equation of the axis of symmetry, and the y-intercept of each graph.

5. 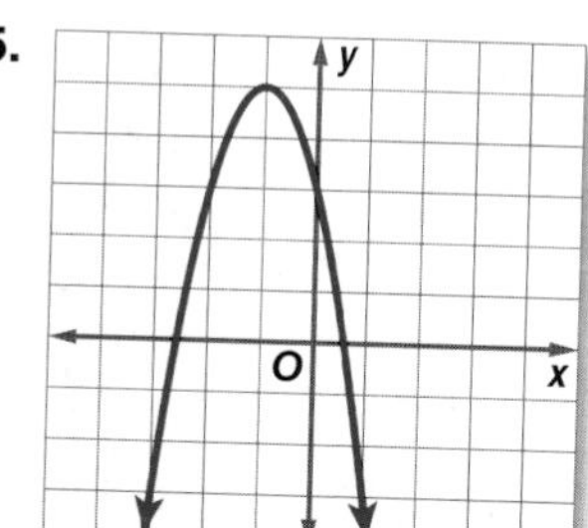

6. 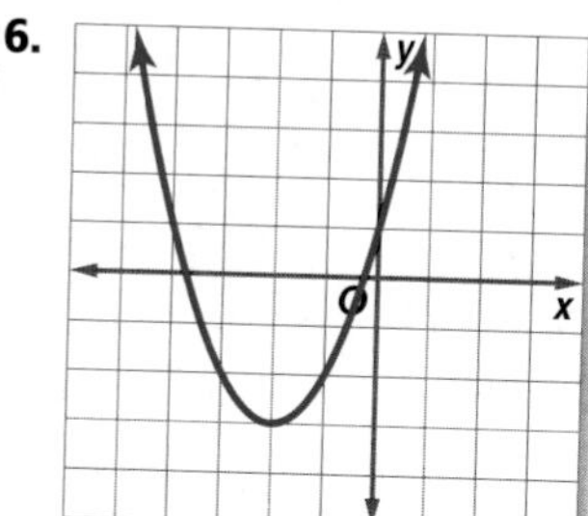

7. 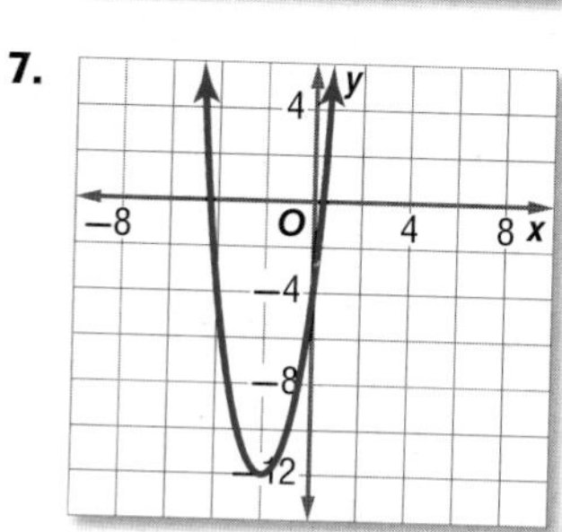

8. 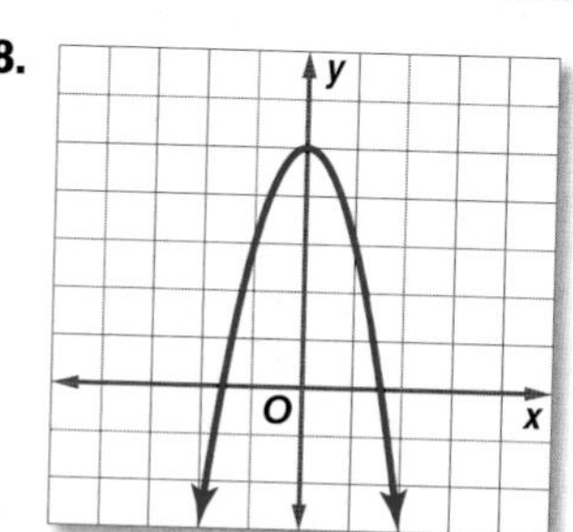

Example 3 p. 527 — Find the vertex, the equation of the axis of symmetry, and the y-intercept of the graph of each function.

9. $y = -3x^2 + 6x - 1$
10. $y = -x^2 + 2x + 1$
11. $y = x^2 - 4x + 5$
12. $y = 4x^2 - 8x + 9$

Example 4 p. 528 — Consider each function.

a. Determine whether the function has *maximum* or *minimum* value.

b. State the maximum or minimum value.

c. What are the domain and range of the function?

13. $y = -x^2 + 4x - 3$
14. $y = -x^2 - 2x + 2$
15. $y = -3x^2 + 6x + 3$
16. $y = -2x^2 + 8x - 6$

Example 5 p. 529 — Graph each function.

17. $f(x) = -3x^2 + 6x + 3$
18. $f(x) = -2x^2 + 4x + 1$
19. $f(x) = 2x^2 - 8x - 4$
20. $f(x) = 3x^2 - 6x - 1$

Example 6 p. 530

21. **JUGGLING** A juggler is tossing a ball into the air. The height of the ball in feet can be modeled by the equation $y = -16x^2 + 16x + 5$, where y represents the height of the ball at x seconds.

    a. Graph this equation.

    b. At what height is the ball thrown?

    c. What is the maximum height of the ball?

Practice and Problem Solving

= **Step-by-Step Solutions** begin on page R12.
Extra Practice begins on page 815.

Example 1 p. 525

Use a table of values to graph each equation. State the domain and range.

22. $y = x^2 + 4x + 6$
23. $y = 2x^2 + 4x + 7$
24. $y = 2x^2 - 8x - 5$
25. $y = 3x^2 + 12x + 5$
26. $y = 3x^2 - 6x - 2$
27. $y = x^2 - 2x - 1$

Example 2 p. 526

Find the vertex, the equation of the axis of symmetry, and the y-intercept of each graph.

28. 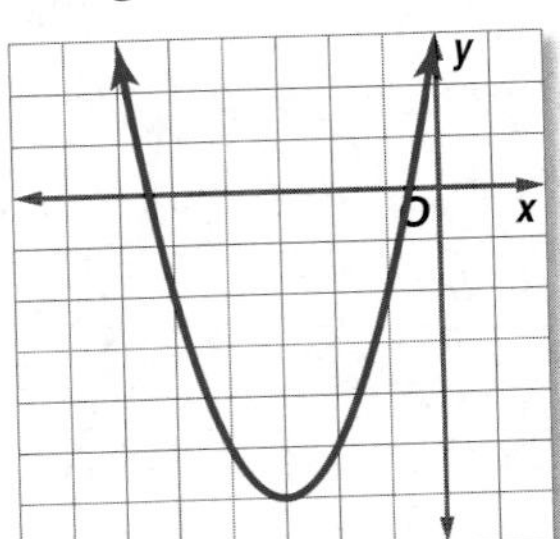

29. 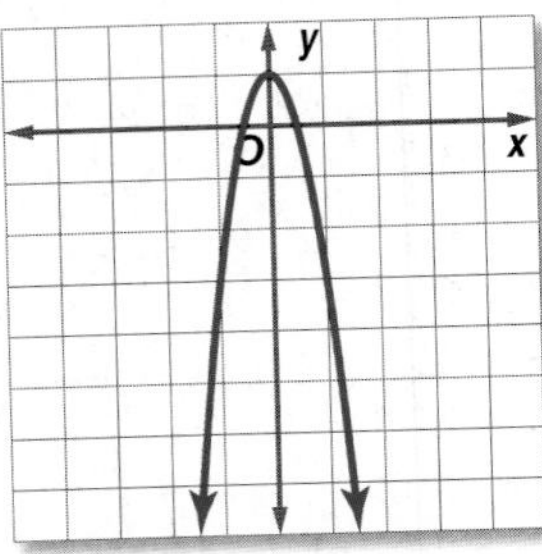

30. 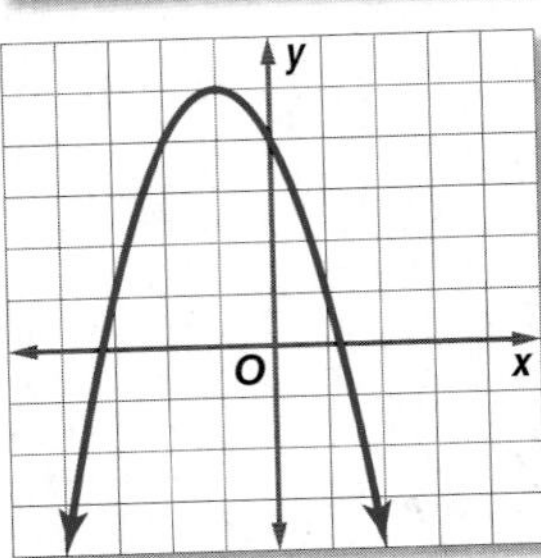

31. 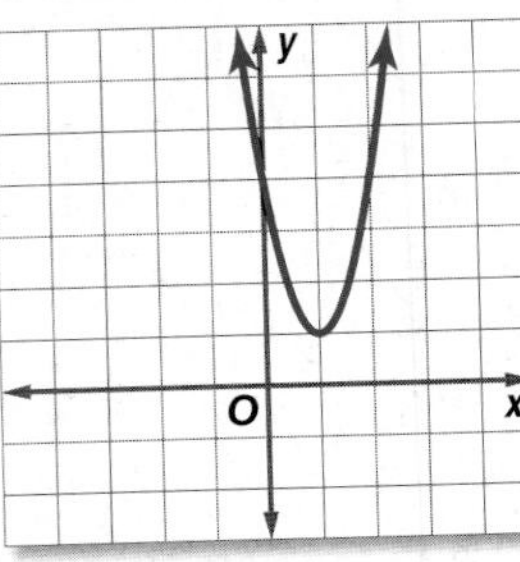

32. 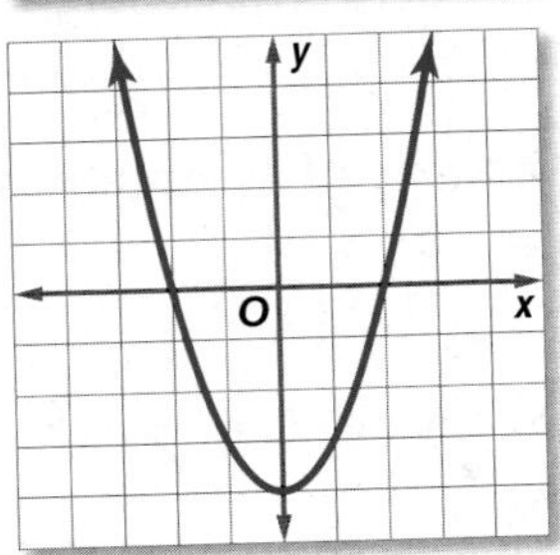

33. 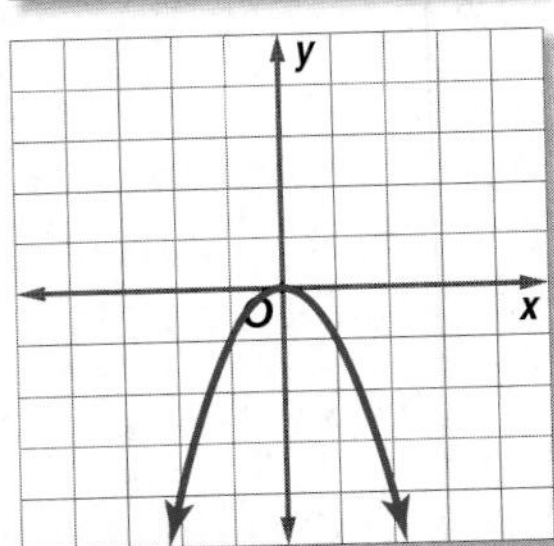

Example 3 p. 527

Find the vertex, the equation of the axis of symmetry, and the y-intercept of each function.

34. $y = x^2 + 8x + 10$
35. $y = 2x^2 + 12x + 10$
36. $y = -3x^2 - 6x + 7$
37. $y = -x^2 - 6x - 5$
38. $y = 5x^2 + 20x + 10$
39. $y = 7x^2 - 28x + 14$
40. $y = 2x^2 - 12x + 6$
41. $y = -3x^2 + 6x - 18$
42. $y = -x^2 + 10x - 13$

Example 4 p. 528

Consider each function.

a. Determine whether the function has a *maximum* or *minimum* value.
b. State the maximum or minimum value.
c. What are the domain and range of the function?

43. $y = -2x^2 - 8x + 1$
44. $y = x^2 + 4x - 5$
45. $y = 3x^2 + 18x - 21$
46. $y = -2x^2 - 16x + 18$
47. $y = -x^2 - 14x - 16$
48. $y = 4x^2 + 40x + 44$
49. $y = -x^2 - 6x - 5$
50. $y = 2x^2 + 4x + 6$
51. $y = -3x^2 - 12x - 9$

Example 5 p. 529

Graph each function.

52. $y = -3x^2 + 6x - 4$
53. $y = -2x^2 - 4x - 3$
54. $y = -2x^2 - 8x + 2$
55. $y = x^2 + 6x - 6$
56. $y = x^2 - 2x + 2$
57. $y = 3x^2 - 12x + 5$

Example 6 p. 530

58. BOATING Miranda has her boat docked on the west side of Casper Point. She is boating over to the Casper Marina. The distance traveled by Miranda over time can be modeled by the equation $d = -16t^2 + 66t$, where d is the number of feet she travels in t minutes.

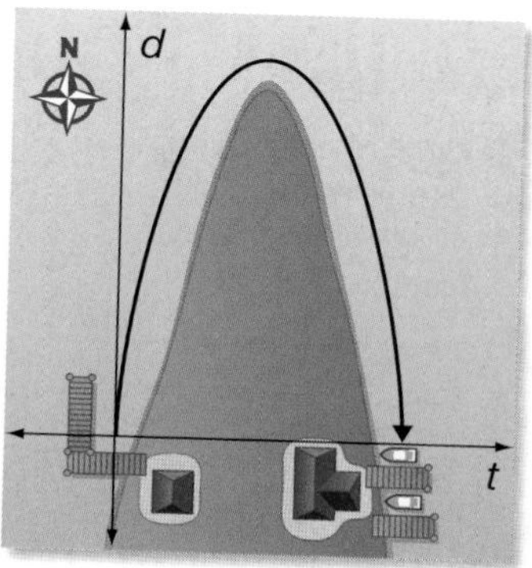

a. Graph this equation.

b. What is the maximum number of feet north that she traveled?

c. How long did it take her to reach Casper Marina?

GRAPHING CALCULATOR Graph each equation. Use the TRACE feature to find the vertex on the graph. Round to the nearest thousandth if necessary.

59. $y = 4x^2 + 10x + 6$

60. $y = 8x^2 - 8x + 8$

61. $y = -5x^2 - 3x - 8$

62. $y = -7x^2 + 12x - 10$

63. GOLF The average amateur golfer can hit a ball with an initial velocity of 31.3 meters per second. If the ball is hit straight up, the height can be modeled by the equation $h = -4.9t^2 + 31.3t$, where h is the height of the ball, in meters, after t seconds.

a. Graph this equation.

b. At what height is the ball hit?

c. What is the maximum height of the ball?

d. How long did it take for the ball to hit the ground?

e. State a reasonable range and domain for this situation.

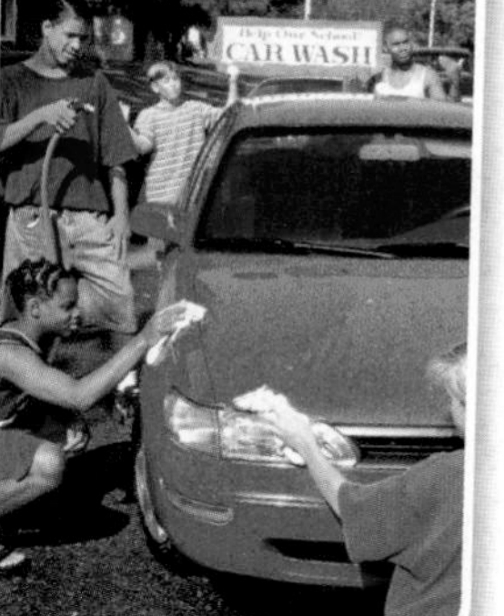

Real-World Link

School fundraisers provide revenue for extracurricular activities that are not part of a school budget.

64. FUNDRAISING The marching band is selling poinsettias to buy new uniforms. Last year the band charged \$5 each, and they sold 150. They want to increase the price this year, and they expect to lose 10 sales for each \$1 increase. The sales revenue R, in dollars, generated by selling the poinsettias is predicted by the function $R = (5 + p)(150 - 10p)$, where p is the number of \$1 price increases.

a. Write the function in standard form.

b. Find the maximum value of the function.

c. At what price should the poinsettias be sold to generate the most sales revenue? Explain your reasoning.

65 FOOTBALL A football is kicked up from ground level at an initial upward velocity of 90 feet per second. The equation $h = -16t^2 + 90t$ gives the height h of the football after t seconds.

a. What is the height of the ball after one second?

b. When is the ball 126 feet high?

c. When is the height of the ball 0 feet? What do these points represent in the context of the situation?

66. REASONING Let $f(x) = x^2 - 9$.

a. What is the domain of $f(x)$?

b. What is the range of $f(x)$?

c. For what values of x is $f(x)$ negative?

d. When x is a real number, what are the domain and range of $f(x) = \sqrt{x^2 - 9}$?

StudyTip

Zeros The number of zeros is equal to the degree of the related function.

67. **MULTIPLE REPRESENTATIONS** In this problem, you will investigate solving quadratic equations using tables.

a. ALGEBRAIC Determine the related function for each equation. Copy and complete the table below.

Equation	Related Function	Zeros	y-Values
$x^2 - x = 12$	?	?	?
$x^2 + 8x = 9$	?	?	?
$x^2 = 14x - 24$	?	?	?
$x^2 + 16x = -28$	?	?	?

b. GRAPHICAL Graph each related function with a graphing calculator.

c. ANALYTICAL Use the table feature on your calculator to determine the zeros of each related function. Record the zeros in the table above. Also record the values of the function one unit less than and one unit more than each zero.

d. VERBAL Compare the signs of the function values for x-values just before and just after a zero. What happens to the sign of the function value before and after a zero?

H.O.T. Problems

Use **H**igher-**O**rder **T**hinking Skills

68. **OPEN ENDED** Write and graph a quadratic function for which the graph has the axis of symmetry $x = -\frac{3}{8}$. Summarize your steps.

69. **FIND THE ERROR** Chase and Jade are finding the axis of symmetry of a parabola. Is either of them correct? Explain your reasoning.

Chase

$y = -x^2 - 4x + 6$

$x = -\frac{b}{2a}$

$x = -\frac{4}{2(-1)}$

$x = 2$

Jade

$y = -x^2 - 4x + 6$

$x = -\frac{b}{2a}$

$x = -\frac{-4}{2(-1)}$

$x = -2$

70. **CHALLENGE** Using the axis of symmetry and one x-intercept, write an equation for the graph shown.

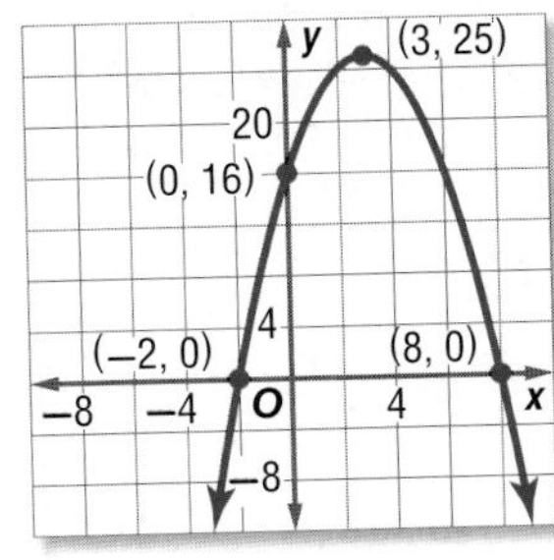

71. **REASONING** The graph of a quadratic function has a vertex at (2, 0). One point on the graph is (5, 9). Find another point on the graph. Explain how you found it.

72. **OPEN ENDED** Describe a real-world situation that involves a quadratic equation. Explain what the vertex represents.

73. **REASONING** Provide a counterexample to the following statement. *The vertex of a parabola is always the minimum of the graph.*

74. **WRITING IN MATH** Explain how to find the axis of symmetry from an equation for a quadratic function. Then explain what other characteristics of the graph you can derive from this, and how you would do that.

Standardized Test Practice

75. Which of the following is an equation for the line that passes through $(2, -5)$ and is perpendicular to $2x + 4y = 8$?

A $y = 2x + 10$
B $y = -\frac{1}{2}x - 4$
C $y = 2x - 9$
D $y = -2x - 1$

76. GEOMETRY The area of the circle is 36π square units. If the radius is doubled, what is the area of the new circle?

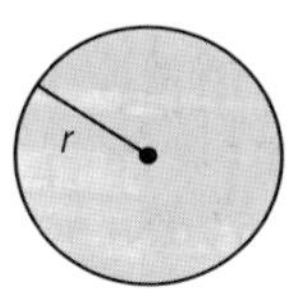

$A = 36\pi$

F 72π units^2
G 144π units^2
H 1296π units^2
J 9π units^2

77. What is the range of the function $f(x) = -4x^2 - \frac{1}{2}$?

A $\left\{\text{all integers less than or equal to } \frac{1}{2}\right\}$
B {all nonnegative integers}
C {all real numbers}
D $\left\{\text{all real numbers less than or equal to } -\frac{1}{2}\right\}$

78. SHORT RESPONSE Dylan delivers newspapers for extra money. He starts delivering the newspapers at 3:15 P.M. and finishes at 5:05 P.M. How long does it take Dylan to complete his route?

Spiral Review

Determine whether each trinomial is a perfect square trinomial. Write *yes* or *no*. If so, factor it. (Lesson 8-6)

79. $4x^2 + 4x + 1$
80. $4x^2 - 20x + 25$
81. $9x^2 + 8x + 16$

Factor each polynomial if possible. If the polynomial cannot be factored, write *prime*. (Lesson 8-5)

82. $n^2 - 16$
83. $x^2 + 25$
84. $9 - 4a^2$

Find each product. (Lesson 7-7)

85. $(b - 7)(b + 3)$
86. $(c - 6)(c - 5)$
87. $(2x - 1)(x + 9)$

88. MULTIPLE BIRTHS The number of quadruplet births Q in the United States in recent years can be modeled by $Q = -0.5t^3 + 11.7t^2 - 21.5t + 218.6$, where t represents the number of years since 1992. For what values of t does this model no longer allow for realistic predictions? Explain your reasoning. (Lesson 7-4)

Use elimination to solve each system of equations. (Lesson 6-4)

89. $2x + y = 5$
$3x - 2y = 4$

90. $4x - 3y = 12$
$x + 2y = 14$

91. $2x - 3y = 2$
$5x + 4y = 28$

92. HEALTH About 20% of the time you sleep is spent in rapid eye movement (REM), which is associated with dreaming. If an adult sleeps 7 to 8 hours, how much time is spent in REM sleep? (Lesson 5-4)

Skills Review

Find the x-intercept of the graph of each equation. (Lesson 3-1)

93. $x + 2y = 10$
94. $2x - 3y = 12$
95. $3x - y = -18$

EXTEND
9-1

Algebra Lab

Rate of Change of a Quadratic Function

Math Online glencoe.com
Math *in Motion,* Animation

Objective
Investigate the rate of change for a quadratic function.

A model rocket is launched from the ground with an upward velocity of 144 feet per second. The function $y = -16x^2 + 144x$ models the height y of the rocket in feet after x seconds. Using this function, we can investigate the rate of change of a quadratic function.

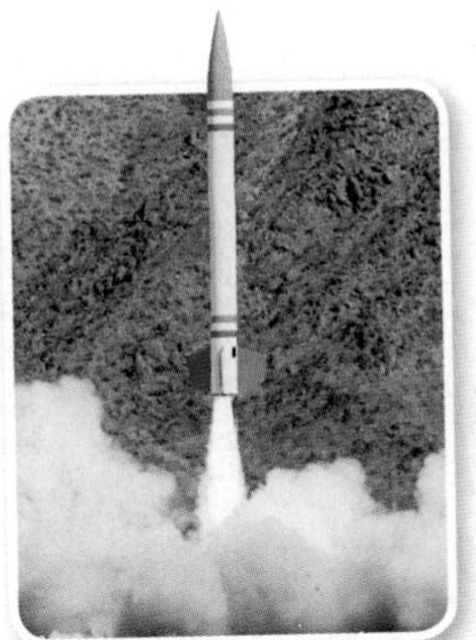

ACTIVITY

Step 1 Copy the table below.

x	0	0.5	1.0	1.5	...	9.0
y	0					
Rate of Change	–					

Step 2 Find the value of y for each value of x from 0 through 9.

Step 3 Graph the ordered pairs (x, y) on grid paper. Connect the points with a smooth curve. Notice that the function *increases* when $0 < x < 4.5$ and *decreases* when $4.5 < x < 9$.

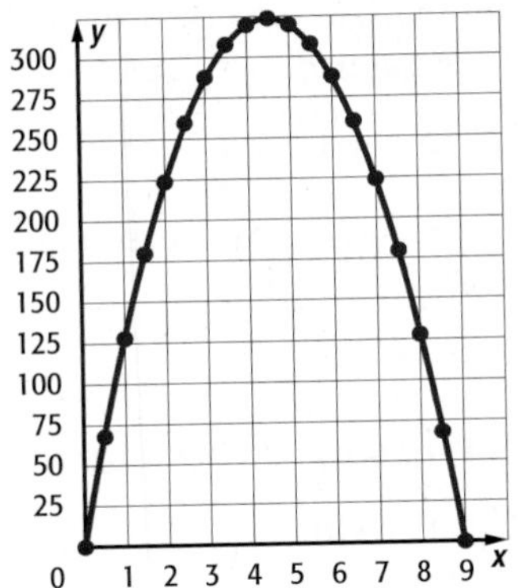

Step 4 Recall that the *rate of change* is the change in y divided by the change in x. Find the rate of change for each half second interval of x and y.

Exercises

Use the quadratic function $y = x^2$.

1. Make a table, similar to the one in the Activity, for the function using $x = -4, -3, -2, -1, 0, 1, 2, 3$, and 4. Find the values of y for each x-value.

2. Graph the ordered pairs on grid paper. Connect the points with a smooth curve. Describe where the function is increasing and where it is decreasing.

3. Find the rate of change for each column starting with $x = -3$. Compare the rates of change when the function is increasing and when it is decreasing.

4. **CHALLENGE** If an object is dropped from 100 feet in the air and air resistance is ignored, the object will fall at a rate that can be modeled by the equation $f(x) = -16x^2 + 100$, where $f(x)$ represents the object's height in feet after x seconds. Make a table like that in Exercise 1, selecting appropriate values for x. Fill in the x-values, the y-values, and rates of change. Compare the rates of change. Describe any patterns that you see.

9-2

Solving Quadratic Equations by Graphing

Then

You solved quadratic equations by factoring. (Lesson 8-3)

Now

- Solve quadratic equations by graphing.
- Estimate solutions of quadratic equations by graphing.

New Vocabulary

double root

Math Online

glencoe.com

- Extra Examples
- Personal Tutor
- Self-Check Quiz
- Homework Help
- Math in Motion

Why?

Dorton Arena at the state fairgrounds in Raleigh, North Carolina, has a shape created by two intersecting parabolas. The shape of one of the parabolas can be modeled by the equation $y = -x^2 + 127x$, where x represents the width of the parabola in feet, and y represents the length of the parabola in feet.

The x-intercepts of the graph of this function can be used to determine the distance between the points where the parabola meets the ground.

Solve by Graphing A quadratic equation can be written in the standard form $ax^2 + bc + c = 0$, where $a \neq 0$. To write a quadratic function as an equation, replace y or $f(x)$ with 0. Recall that the solutions or roots of an equation can be identified by finding the x-intercepts of the related graph. Quadratic equations may have two, one, or no solutions.

Key Concept Solutions of Quadratic Equations — For Your FOLDABLE

two unique real solutions	*one* unique real solution	*no* real solutions

Math *in Motion*, Animation glencoe.com

EXAMPLE 1 Two Roots

Solve $x^2 - 2x - 8 = 0$ by graphing.

Graph the related function $f(x) = x^2 - 2x - 8$.

The x-intercepts of the graph appear to be at -2 and 4, so the solutions are -2 and 4.

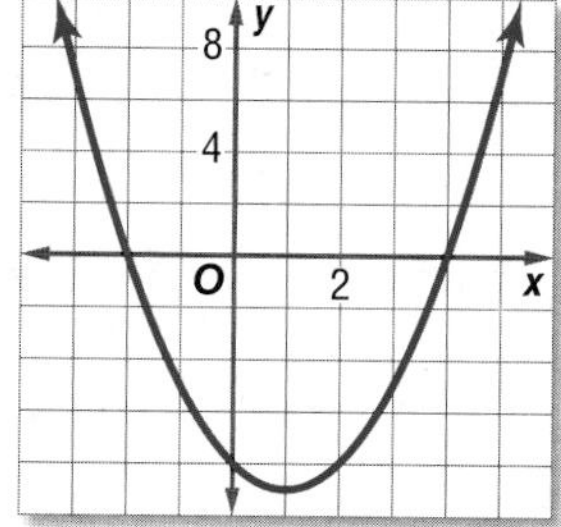

CHECK Check each solution in the original equation.

$x^2 - 2x - 8 = 0$	**Original equation**	$x^2 - 2x - 8 = 0$
$(-2)^2 - 2(-2) - 8 \stackrel{?}{=} 0$	**$x = -2$ or $x = 4$**	$(4)^2 - 2(4) - 8 \stackrel{?}{=} 0$
$0 = 0$ ✓	**Simplify.**	$0 = 0$ ✓

Check Your Progress Solve each equation by graphing.

1A. $-x^2 - 3x + 18 = 0$

1B. $x^2 - 4x + 3 = 0$

Personal Tutor glencoe.com

The solutions in Example 1 were two distinct numbers. Sometimes the two roots are the same number, called a **double root**.

EXAMPLE 2 Double Root

Solve $x^2 - 6x = -9$ by graphing.

Step 1 Rewrite the equation in standard form.

$x^2 - 6x = -9$ **Original equation**

$x^2 - 6x + 9 = 0$ **Add 9 to each side.**

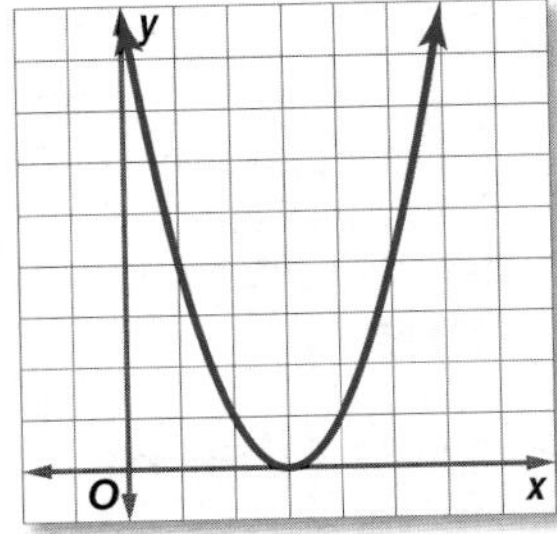

Step 2 Graph the related function $f(x) = x^2 - 6x + 9$.

Step 3 Locate the x-intercepts of the graph. Notice that the vertex of the parabola is the only x-intercept. Therefore, there is only one solution, 3.

CHECK Solve by factoring.

$x^2 - 6x + 9 = 0$ **Original equation**

$(x - 3)(x - 3) = 0$ **Factor.**

$x - 3 = 0$ or $x - 3 = 0$ **Zero Product Property**

$x = 3$ $\quad x = 3$ **Add 3 to each side.**

The only solution is 3.

Watch Out!

Exact Solutions Solutions found from the graph of an equation may appear to be exact, but you cannot be sure unless you can check them in the original equation.

Check Your Progress

Solve each equation by graphing.

2A. $x^2 + 25 = 10x$

2B. $x^2 = -8x - 16$

Personal Tutor glencoe.com

Sometimes the roots are not real numbers.

EXAMPLE 3 No Real Roots

Solve $2x^2 - 3x + 5 = 0$ by graphing.

Step 1 Rewrite the equation in standard form.

This equation is written in standard form.

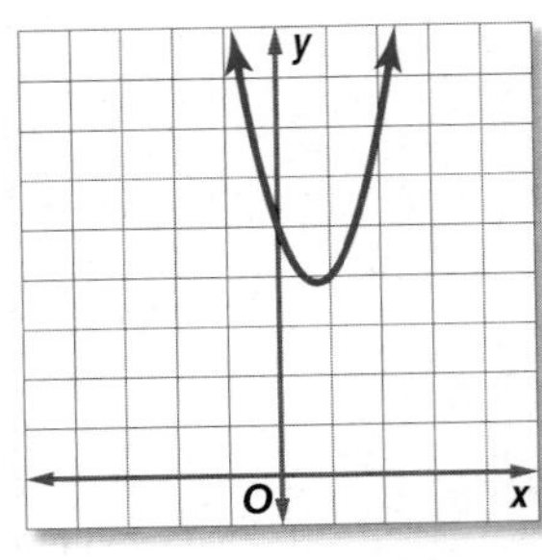

Step 2 Graph the related function $f(x) = 2x^2 - 3x + 5$.

Step 3 Locate the x-intercepts of the graph. This graph has no x-intercepts. Therefore, this equation has no real number solutions. The solution set is $\varnothing$.

CHECK Solve by factoring.

There are no factors of 10 that have a sum of -3, so the expression is not factorable. Thus, the equation has no real number solutions.

Check Your Progress

Solve each equation by graphing.

3A. $-x^2 - 3x = 5$

3B. $-2x^2 - 8 = 6x$

Personal Tutor glencoe.com

StudyTip

Location of Zeros Since quadratic functions are continuous, there must be a zero between two x-values for which the corresponding y-values have opposite signs.

Estimate Solutions The real roots found thus far have been integers. However, the roots of quadratic equations are usually not integers. In these cases, use estimation to approximate the roots of the equation.

EXAMPLE 4 Approximate Roots with a Table

Solve $x^2 + 6x + 6 = 0$ by graphing. If integral roots cannot be found, estimate the roots to the nearest tenth.

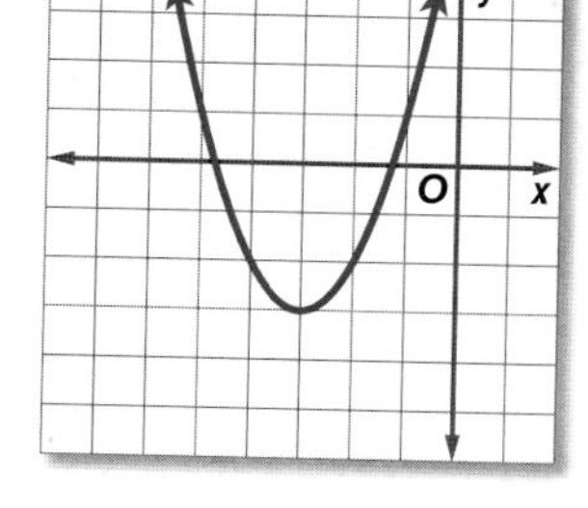

Graph the related function $f(x) = x^2 + 6x + 6$.

The x-intercepts are located between -5 and -4 and between -2 and -1.

Make a table using an increment of 0.1 for the x-values located between -5 and -4 and between -2 and -1.

Look for a change in the signs of the function values. The function value that is closest to zero is the best approximation for a zero of the function.

x	−4.9	−4.8	−4.7	−4.6	−4.5	−4.4	−4.3	−4.2	−4.1
y	0.61	0.24	−0.11	−0.44	−0.75	−1.04	−1.31	−1.56	−1.79

x	−1.9	−1.8	−1.7	−1.6	−1.5	−1.4	−1.3	−1.2	−1.1
y	−1.79	−1.56	−1.31	−1.04	−0.75	−0.44	−0.11	0.24	0.61

For each table, the function value that is closest to zero when the sign changes is -0.11. Thus, the roots are approximately -4.7 and -1.3.

Check Your Progress

4. Solve $2x^2 + 6x - 3 = 0$ by graphing. If integral roots cannot be found, estimate the roots to the nearest tenth.

Personal Tutor glencoe.com

Approximating the x-intercepts of graphs is helpful for real-world applications.

Real-World EXAMPLE 5 Approximate Roots with a Calculator

SOCCER A goalie kicks a soccer ball with an upward velocity of 65 feet per second, and her foot meets the ball 1 foot off the ground. The quadratic function $h = -16t^2 + 65t + 1$ represents the height of the ball h in feet after t seconds. Approximately how long is the ball in the air?

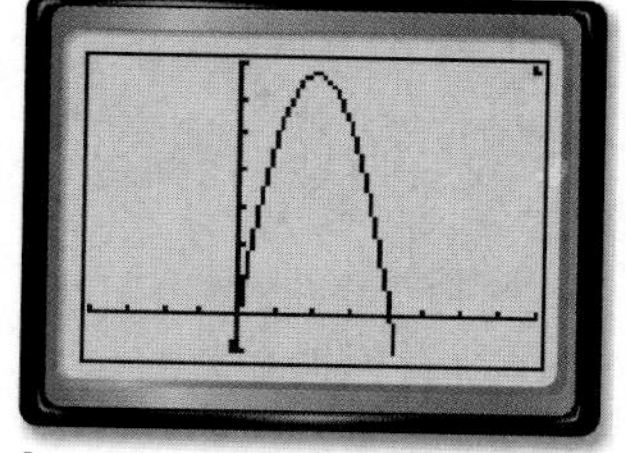

[−4, 7] scl: 1 by [−10, 70] scl: 10

You need to find the roots of the equation $-16t^2 + 65t + 1 = 0$. Use a graphing calculator to graph the related function $f(x) = -16t^2 + 65t + 1$.

The positive x-intercept of the graph is approximately 4. Therefore, the ball is in the air for approximately 4 seconds.

Real-World Link

The game of soccer, called "football" outside of North America, began in 1863 in Britain when the Football Association was founded. Soccer is played on every continent of the world.

Source: Sports Know How

Check Your Progress

5. If the goalie kicks the soccer ball with an upward velocity of 55 feet per second and his foot meets the ball 2 feet off the ground, approximately how long is the ball in the air?

Personal Tutor glencoe.com

Check Your Understanding

Examples 1–3 pp. 537–538

Solve each equation by graphing.

1. $x^2 + 3x - 10 = 0$
2. $2x^2 - 8x = 0$
3. $x^2 + 4x = -4$
4. $x^2 + 12 = -8x$

Example 4 p. 539

Solve each equation by graphing. If integral roots cannot be found, estimate the roots to the nearest tenth.

5. $-x^2 - 5x + 1 = 0$
6. $-9 = x^2$
7. $x^2 = 25$
8. $x^2 - 8x = -9$

Example 5 p. 539

9. **SCIENCE FAIR** Ricky built a model rocket. Its flight can be modeled by the equation shown, where h is the height of the rocket in feet after t seconds. About how long was Ricky's rocket in the air?

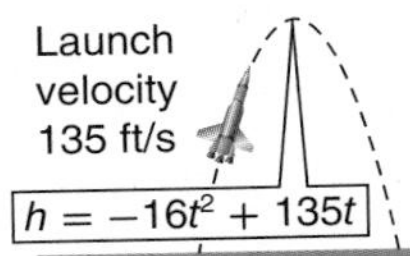

● = **Step-by-Step Solutions** begin on page R12.
Extra Practice begins on page 815.

Practice and Problem Solving

Examples 1–3 pp. 537–538

Solve each equation by graphing.

10. $x^2 + 7x + 14 = 0$
11. $x^2 + 2x - 24 = 0$
12. $x^2 - 16x + 64 = 0$
13. $x^2 - 5x + 12 = 0$
14. $x^2 + 14x = -49$
15. $x^2 = 2x - 1$
16. $x^2 - 10x = -16$
17. $-2x^2 - 8x = 13$
18. $2x^2 - 16x = -30$
19. $2x^2 = -24x - 72$
20. $-3x^2 + 2x = 15$
21. $x^2 = -2x + 80$

Example 4 p. 539

Solve each equation by graphing. If integral roots cannot be found, estimate the roots to the nearest tenth.

22. $x^2 + 2x - 9 = 0$
23. $x^2 - 4x = 20$
24. $x^2 + 3x = 18$
25. $2x^2 - 9x = -8$
26. $3x^2 = -2x + 7$
27. $5x = 25 - x^2$

Example 5 p. 539

28. **SOFTBALL** Sofia hits a softball straight up. The equation $h = -16t^2 + 90t$ models the height h, in feet, of the ball after t seconds. How long is the ball in the air?

29. **RIDES** A skyrocket roller coaster takes riders straight up and then returns straight down. The equation $h = -16t^2 + 185t$ models the height h, in feet, of the coaster after t seconds. How long is it until the coaster returns to the bottom?

Use factoring to determine how many times the graph of each function intersects the x-axis. Identify each zero.

30. $y = x^2 - 8x + 16$
31. $y = x^2 + 3x + 4$
32. $y = x^2 + 2x - 24$
33. $y = x^2 + 12x + 32$

34. **NUMBER THEORY** Use a quadratic equation to find two numbers that have a sum of 9 and a product of 20.

35. **NUMBER THEORY** Use a quadratic equation to find two numbers that have a sum of 1 and a product of -12.

36. **GOLF** The height of a golf ball in the air can be modeled by the equation $h = -16t^2 + 60t + 3$, where h is the height in feet of the ball after t seconds.

    a. How long was the ball in the air?
    b. What is the ball's maximum height?
    c. When will the ball reach its maximum height?

Real-World Link

In the 1998 Winter Games in Japan, snowboarding became an Olympic event for the first time. A total of four competitions were held that were divided into two categories: men's and women's halfpipe and men's and women's giant slalom.

Source: About, Inc.

37. **SNOWBOARDING** Stefanie is in a snowboarding competition. The equation $h = -16t^2 + 30t + 10$ models Stefanie's height h, in feet, in the air after t seconds.
   a. How long is Stefanie in the air?
   b. When will Stefanie reach a height of 15 feet?
   c. To earn bonus points in the competition, you must reach a height of 20 feet. Will Stefanie earn bonus points?

38. **MULTIPLE REPRESENTATIONS** In this problem, you will explore how to further interpret the relationship between quadratic functions and graphs.
   a. GRAPHICAL Graph $y = x^2$.
   b. ANALYTICAL Name the vertex and two other points on the graph.
   c. GRAPHICAL Graph $y = x^2 + 2$, $y = x^2 + 4$, and $y = x^2 + 6$ on the same coordinate plane as the previous graph.
   d. ANALYTICAL Name the vertex and two points from each of these graphs that have the same x-coordinates as the first graph.
   e. ANALYTICAL What conclusion can you draw from this?

GRAPHING CALCULATOR Approximate the zeros of each cubic function by graphing. If integral zeros cannot be found, estimate the zeros to the nearest tenth.

39. $f(x) = x^3 - 3x^2 - 6x + 8$

40. $g(x) = x^3 - 4x^2 + 5x - 12$

H.O.T. Problems

Use Higher-Order Thinking Skills

41. **FIND THE ERROR** Iku and Zachary are finding the number of real zeros of the function graphed at the right. Iku says that the function has no real zeros because there are no x-intercepts. Zachary says that the function has one real zero because the graph has a y-intercept. Is either of them correct? Explain your reasoning.

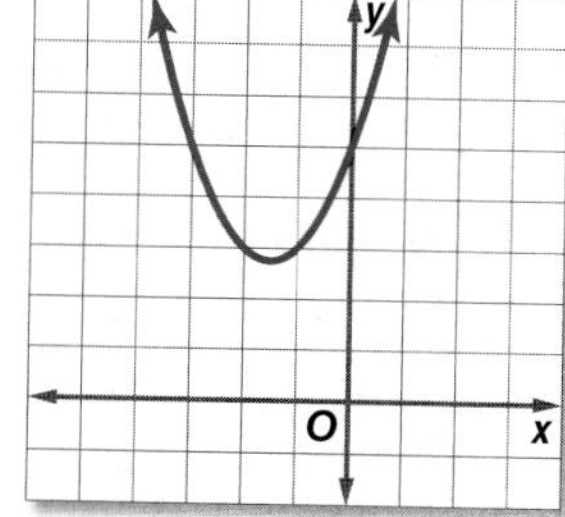

42. **OPEN ENDED** Describe a real-world situation in which a thrown object travels in the air. Write an equation that models the height of the object with respect to time, and determine how long the object travels in the air.

43. **REASONING** The graph shown is that of a *quadratic inequality*. Analyze the graph, and determine whether the y-value of a solution of the inequality is *sometimes*, *always*, or *never* greater than 2. Explain.

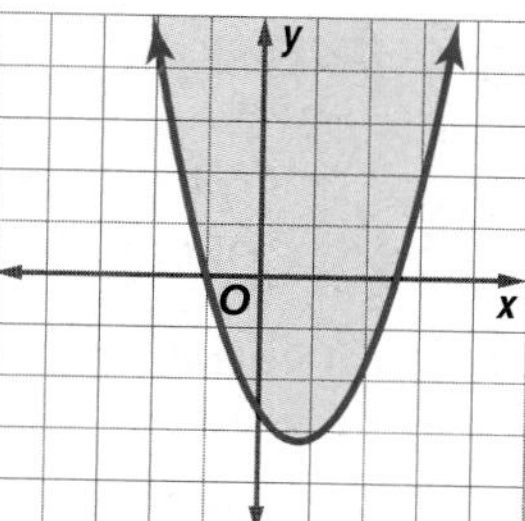

44. **CHALLENGE** Write a quadratic equation that has the roots described.
   a. one double root
   b. one rational (nonintegral) root and one integral root
   c. two distinct integral roots that are additive opposites.

45. **CHALLENGE** Find the roots of $x^2 = 2.25$ without using a calculator. Explain your strategy.

46. **WRITING IN MATH** Explain how to approximate the roots of a quadratic equation when the roots are not integers.

Standardized Test Practice

47. Adrahan earned 50 out of 80 points on a test. What percentage did Adrahan score on the test?

A 62.5% C 1.6%
B 6.25% D 16%

48. Ernesto needs to loosen a bolt. He needs a wrench that is smaller than a $\frac{7}{8}$-inch wrench, but larger than a $\frac{3}{4}$-inch wrench. Which of the following sizes should Ernesto use?

F $\frac{11}{16}$ inch H $\frac{13}{16}$ inch
G $\frac{5}{8}$ inch J $\frac{3}{8}$ inch

49. **EXTENDED RESPONSE** Two boats leave a dock. One boat travels 4 miles east and then 5 miles north. The second boat travels 12 miles south and 9 miles west. Draw a diagram that represents the paths traveled by the boats. How far apart are the boats in miles?

50. The formula $s = \frac{1}{2}at^2$ represents the distance s in meters that a free-falling object will fall near a planet or the Moon in a given time t in seconds. Solve the formula for a, the acceleration due to gravity.

A $a = \frac{1}{2}t^2 - s$ C $a = s - \frac{1}{2}t^2$
B $a = 2s - t^2$ D $a = \frac{2s}{t^2}$

Spiral Review

Write the equation of the axis of symmetry, and find the coordinates of the vertex of the graph of each function. Identify the vertex as a maximum or minimum. Then graph the function. (Lesson 9-1)

51. $y = 3x^2$
52. $y = -4x^2 - 5$
53. $y = -x^2 + 4x - 7$
54. $y = x^2 - 6x - 8$
55. $y = 3x^2 + 2x + 1$
56. $y = -4x^2 - 8x + 5$

Solve each equation. Check the solutions. (Lesson 8-6)

57. $2x^2 = 32$
58. $(x - 4)^2 = 25$
59. $4x^2 - 4x + 1 = 16$
60. $2x^2 + 16x = -32$
61. $(x + 3)^2 = 5$
62. $4x^2 - 12x = -9$

Find each sum or difference. (Lesson 7-5)

63. $(3n^2 - 3) + (4 + 4n^2)$
64. $(2d^2 - 7d - 3) - (4d^2 + 7)$
65. $(2b^3 - 4b^2 + 4) - (3b^4 + 5b^2 - 9)$
66. $(8 - 4h^2 + 6h^4) + (5h^2 - 3 + 2h^3)$

67. **GEOMETRY** Supplementary angles are two angles with measures that have a sum of 180°. For the supplementary angles in the figure, the measure of the larger angle is 24° greater than the measure of the smaller angle. Write and solve a system of equations to find these measures. (Lesson 6-5)

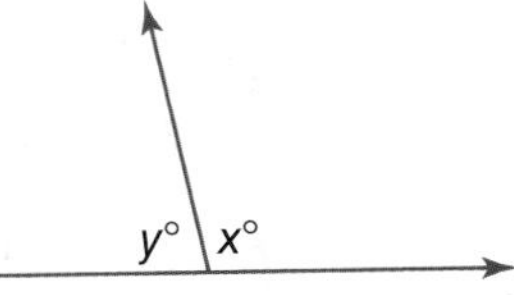

Write an equation in point-slope form for the line that passes through each point with the given slope. (Lesson 4-3)

68. $(2, 5), m = 3$
69. $(-3, 6), m = -7$
70. $(-1, -2), m = -\frac{1}{2}$

Skills Review

Graph each function. (Lesson 9-1)

71. $y = x^2 + 5$
72. $y = x^2 - 8$
73. $y = 2x^2 - 7$
74. $y = -x^2 + 2$
75. $y = -0.5x^2 - 3$
76. $y = (-x)^2 + 1$

EXTEND 9-2

Graphing Technology Lab
Quadratic Inequalities

Math Online glencoe.com
- Other Calculator Keystrokes
- Graphing Technology Personal Tutor

Recall that the graph of a linear inequality consists of the boundary and the shaded half plane. The solution set of the inequality lies in the shaded region of the graph. Graphing quadratic inequalities is similar to graphing linear inequalities.

ACTIVITY 1 Shade Inside a Parabola

Graph $y \geq x^2 - 5x + 4$ in the standard viewing window.

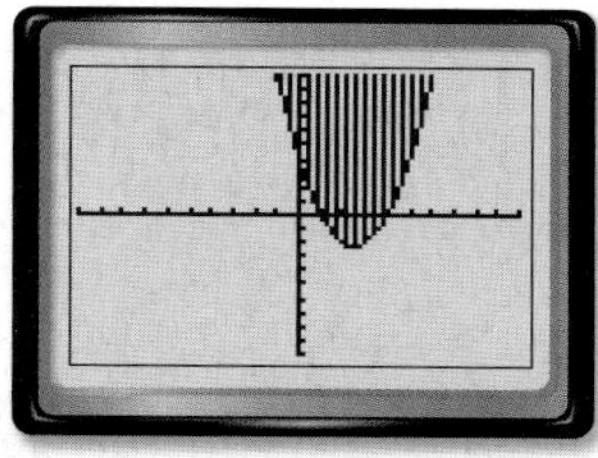

[−10, 10] scl: 1 by [−10, 10] scl: 1

First, clear all functions from the Y= list.

To graph $y \geq x^2 - 5x + 4$, enter the equation in the Y= list. Then use the left arrow to select =. Press ENTER until shading above the line is selected.

KEYSTROKES: ◀ ◀ ENTER ENTER ▶ ▶ X,T,θ,n x^2 − 5 X,T,θ,n + 4 GRAPH

All ordered pairs for which y is *greater than or equal* to $x^2 - 5x + 4$ lie *above or on* the line and are solutions.

A similar procedure will be used to graph an inequality in which the shading is outside of the parabola.

ACTIVITY 2 Shade Outside a Parabola

Graph $y - 4 \leq x^2 - 5x$ in the standard viewing window.

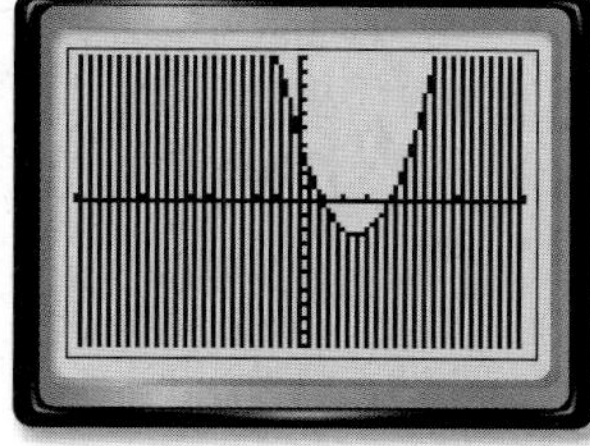

[−10, 10] scl: 1 by [−10, 10] scl: 1

First, clear the graph that is displayed.

KEYSTROKES: Y= CLEAR

Then rewrite $y - 4 \leq x^2 - 5x$ as $y \leq x^2 - 5x + 4$, and graph it.

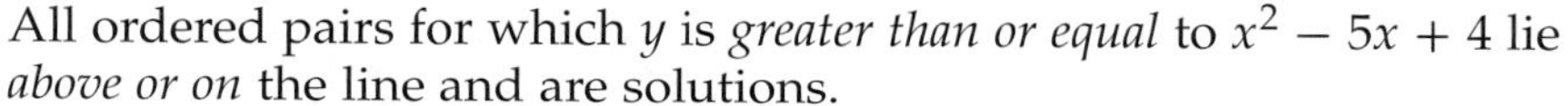

KEYSTROKES: ◀ ◀ ENTER ENTER ENTER ▶ ▶ X,T,θ,n x^2 − 5 X,T,θ,n + 4 GRAPH

All ordered pairs for which y is *less than or equal* to $x^2 - 5x + 4$ lie *below or on* the line and are solutions.

Exercises

1. Compare and contrast the two graphs shown above.
2. Graph $y - 2x + 6 \geq 5x^2$ in the standard viewing window. Name three solutions of the inequality.
3. Graph $y - 6x \leq -x^2 - 3$ in the standard viewing window. Name three solutions of the inequality.

9-3 Transformations of Quadratic Functions

Then

You graphed quadratic functions by using the vertex and axis of symmetry. (Lesson 9-1)

Now

- Apply translations of quadratic functions.
- Apply dilations and reflections to quadratic functions.

New Vocabulary

transformation
translation
dilation
reflection

Math Online

glencoe.com

- Extra Examples
- Personal Tutor
- Self-Check Quiz
- Homework Help
- Math in Motion

Why?

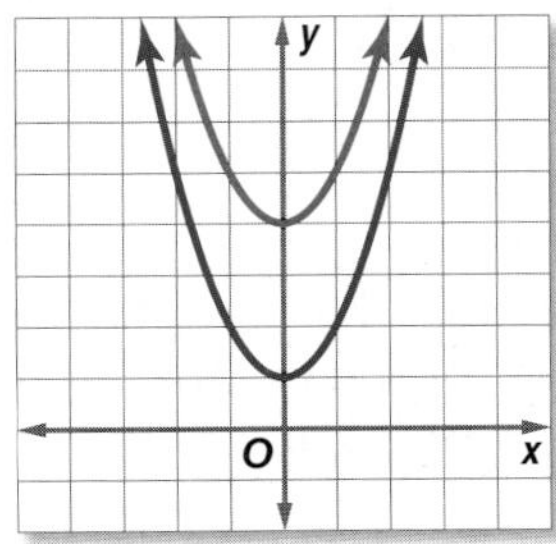

The graphs of the parabolas shown at the right are the same size and shape, but notice that the vertex of the red parabola is higher on the y-axis than the vertex of the blue parabola. Shifting a parabola up and down is an example of a transformation.

Translations A **transformation** changes the position or size of a figure. One type of transformation, a **translation**, moves a figure up, down, left, or right. When a constant c is added to or subtracted from the parent function, the graph of the resulting function $f(x) \pm c$ is the graph of the parent function translated up or down.

The parent function of the family of quadratics is $f(x) = x^2$. All other quadratic functions have graphs that are transformations of the graph of $f(x) = x^2$.

Key Concept — Vertical Translations

For Your FOLDABLE

The graph of $g(x) = x^2 + c$ is the graph of $f(x) = x^2$ translated vertically.

If $c > 0$, the graph of $f(x) = x^2$ is translated $|c|$ units **up**.

If $c < 0$, the graph of $f(x) = x^2$ is translated $|c|$ units **down**.

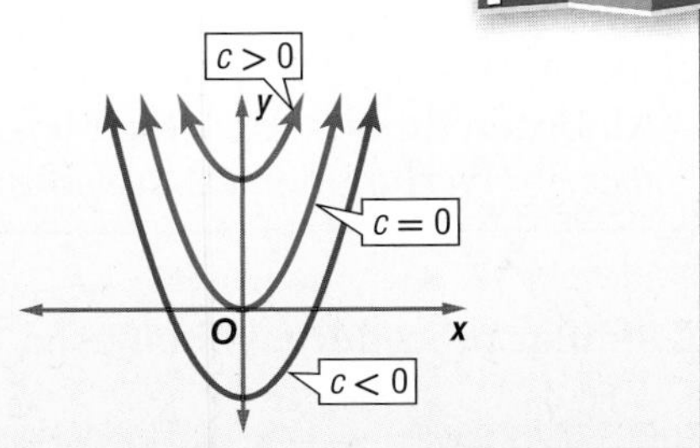

Math *in Motion*, Animation glencoe.com

EXAMPLE 1 Describe and Graph Translations

Describe how the graph of each function is related to the graph of $f(x) = x^2$.

a. $h(x) = x^2 + 3$

The value of c is 3, and $3 > 0$. Therefore, the graph of $y = x^2 + 3$ is a translation of the graph of $y = x^2$ up 3 units.

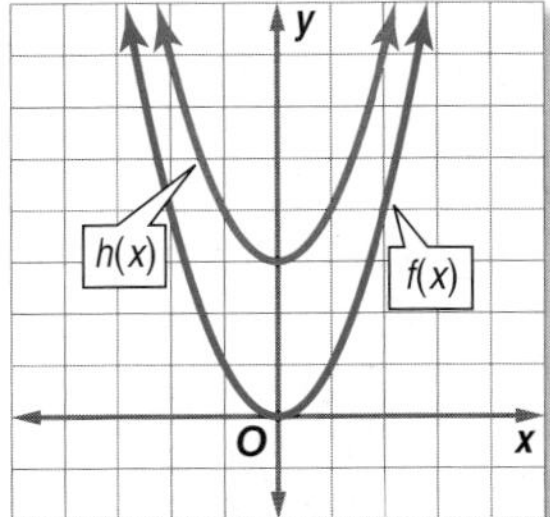

b. $g(x) = x^2 - 4$

The value of c is -4, and $-4 < 0$. Therefore, the graph of $y = x^2 - 4$ is a translation of the graph of $y = x^2$ down 4 units.

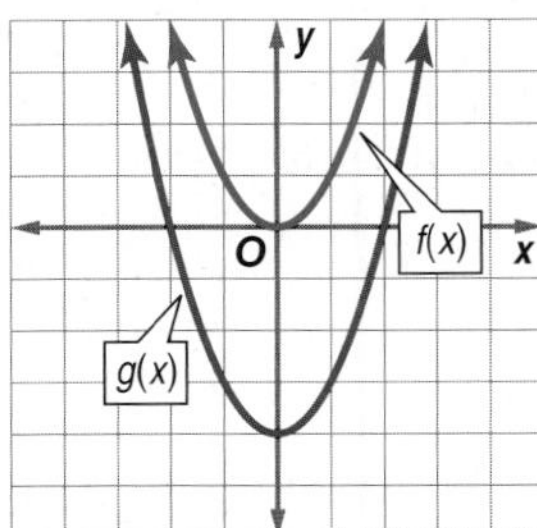

Check Your Progress

1A. $f(x) = x^2 - 7$ **1B.** $g(x) = 5 + x^2$ **1C.** $h(x) = -5 + x^2$ **1D.** $f(x) = x^2 + 1$

Personal Tutor glencoe.com

Dilations and Reflections Another type of transformation is a dilation. A **dilation** makes the graph narrower than the parent graph or wider than the parent graph. When the parent function $f(x) = x^2$ is multiplied by a constant a, the graph of the resulting function $f(x) = ax^2$ is either stretched or compressed vertically.

Key Concept — Dilations

For Your FOLDABLE

The graph of $g(x) = ax^2$ stretches or compresses the graph of $f(x) = x^2$ vertically.

If $|a| > 1$, the graph of $f(x) = x^2$ is stretched vertically.

If $0 < |a| < 1$, the graph of $f(x) = x^2$ is compressed vertically.

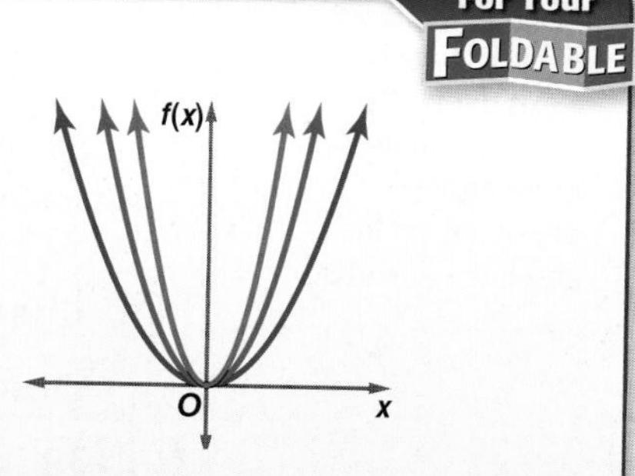

StudyTip

Compress or Stretch When the graph of a quadratic function is stretched vertically, the shape of the graph is narrower than that of the parent function. When it is compressed vertically, the graph is wider than the parent function.

EXAMPLE 2 Describe and Graph Dilations

Describe how the graph of each function is related to the graph of $f(x) = x^2$.

a. $h(x) = \frac{1}{2}x^2$

The function can be written $h(x) = ax^2$, where $a = \frac{1}{2}$. Since $0 < \frac{1}{2} < 1$, the graph of $y = \frac{1}{2}x^2$ is a dilation of the graph of $y = x^2$ that is compressed vertically.

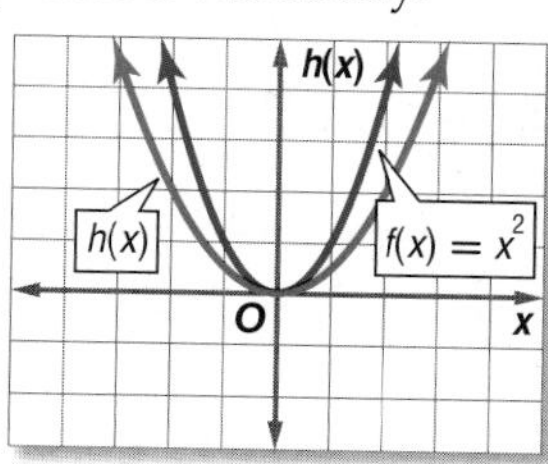

b. $g(x) = 3x^2 + 2$

The function $g(x) = ax^2 + c$, where $a = 3$ and $c = 2$. Since $2 > 0$ and $3 > 1$, the graph of $y = 3x^2 + 2$ translates the graph $y = x^2$ up 2 units and stretches it vertically.

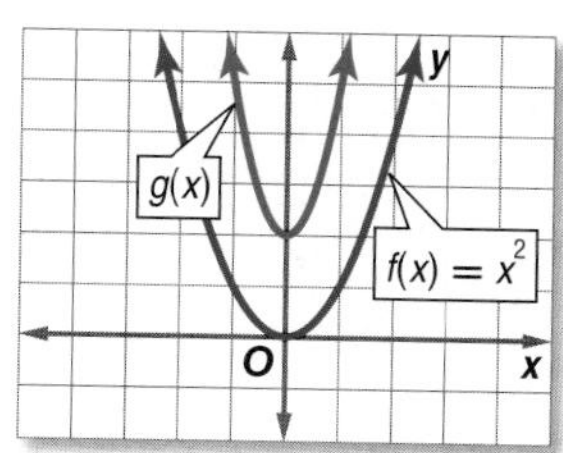

Check Your Progress

2A. $j(x) = 2x^2$

2B. $h(x) = 5x^2 - 2$

2C. $g(x) = \frac{1}{3}x^2 + 2$

Personal Tutor glencoe.com

StudyTip

Reflection A reflection of $f(x) = x^2$ over the y-axis results in the same function, because $f(-x) = (-x)^2 = x^2$.

A **reflection** flips a figure over a line. When $f(x) = x^2$ or the variable x is multiplied by -1, the graph is reflected over the x- or y-axis.

Key Concept — Reflections

For Your FOLDABLE

The graph of $-f(x)$ is the reflection of the graph of $f(x) = x^2$ across the x-axis.

The graph of $f(-x)$ is the reflection of the graph of $f(x) = x^2$ across the y-axis.

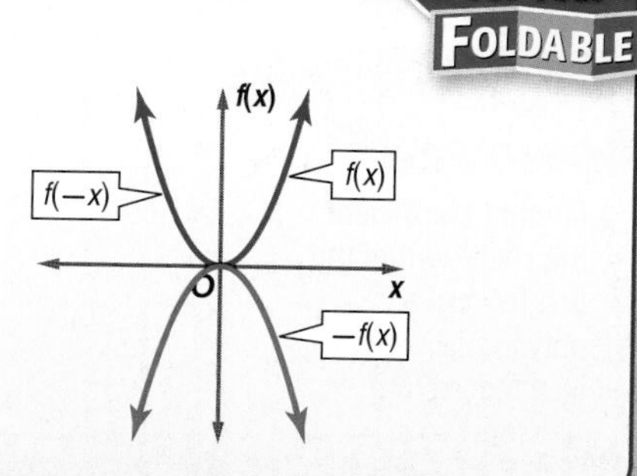

Watch Out!

Transformations The graph of $f(x) = -ax^2$ can result in two transformations of the graph of $f(x) = x^2$: a reflection over the x-axis if $a > 0$ and either a compression or expansion depending on the absolute value of a.

EXAMPLE 3 Describe and Graph Reflections

Describe how the graph of $g(x) = -2x^2 - 3$ is related to the graph of $f(x) = x^2$.

Three separate transformations are occurring. The negative sign of the coefficient of x^2 causes a reflection across the x-axis. Then a dilation occurs and finally a translation down 3 units.

So the graph of $y = -2x^2 - 3$ is reflected across the x-axis, compressed, and translated down 3 units.

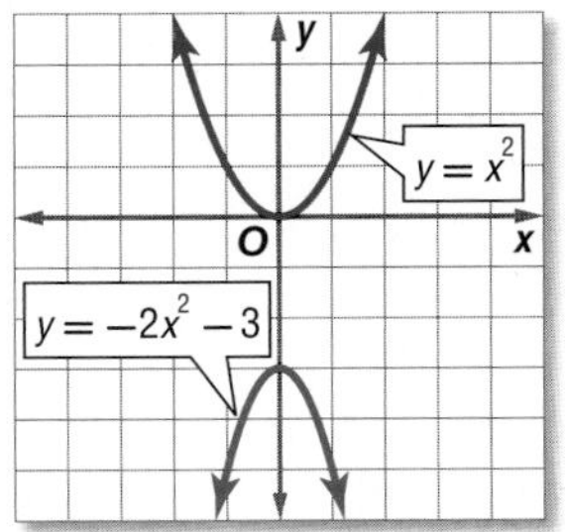

Check Your Progress

Describe how the graph of each function is related to the graph of $f(x) = x^2$.

3A. $h(x) = 2(-x)^2 - 9$

3B. $g(x) = -\frac{1}{5}x^2 + 3$

Personal Tutor glencoe.com

You can use what you know about the characteristics of graphs of quadratic equations to match an equation with a graph.

STANDARDIZED TEST EXAMPLE 4

Which is an equation for the function shown in the graph?

A $y = \frac{1}{2}x^2 - 5$

B $y = -2x^2 - 5$

C $y = -\frac{1}{2}x^2 + 5$

D $y = 2x^2 + 5$

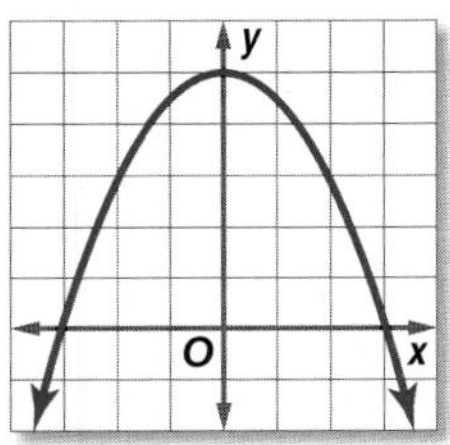

Read the Test Item

You are given the graph of a parabola. You need to find an equation of the graph.

Solve the Test Item

Notice that the graph opens downward. Therefore, the graph of $y = x^2$ has been reflected across the x-axis. The leading coefficient should be negative, so eliminate choices A and D.

The parabola is translated up 5 units, so $c = 5$. Look at the equations. Only choices C and D have $c = 5$. The answer is C.WS

Check Your Progress

4. Which is the graph of $y = -3x^2 + 1$?

F

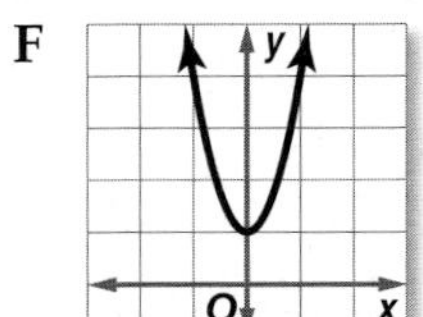

G

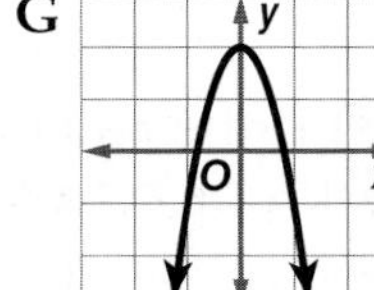

H

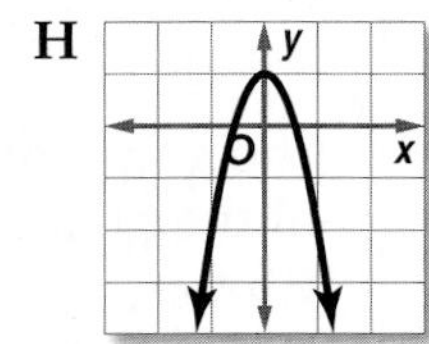

J

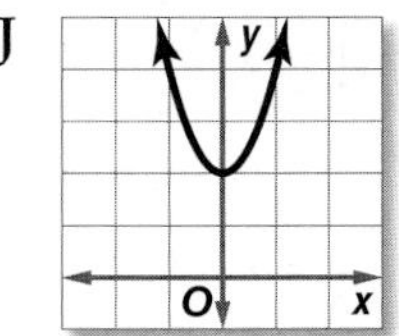

Personal Tutor glencoe.com

Review Vocabulary

leading coefficient the coefficient of the first term of a polynomial written in standard form (Lesson 7-4)

Check Your Understanding

Examples 1–3 pp. 544–546

Describe how the graph of each function is related to the graph of $f(x) = x^2$.

1. $g(x) = x^2 - 11$

2. $h(x) = \frac{1}{2}x^2$

3. $h(x) = -x^2 + 8$

4. $g(x) = x^2 + 6$

5. $g(x) = -4x^2$

6. $h(x) = -x^2 - 2$

Example 4 p. 546

7. MULTIPLE CHOICE Which is an equation for the function shown in the graph?

A $g(x) = \frac{1}{5}x^2 + 2$

B $g(x) = -5x^2 - 2$

C $g(x) = \frac{1}{5}x^2 - 2$

D $g(x) = -\frac{1}{5}x^2 - 2$

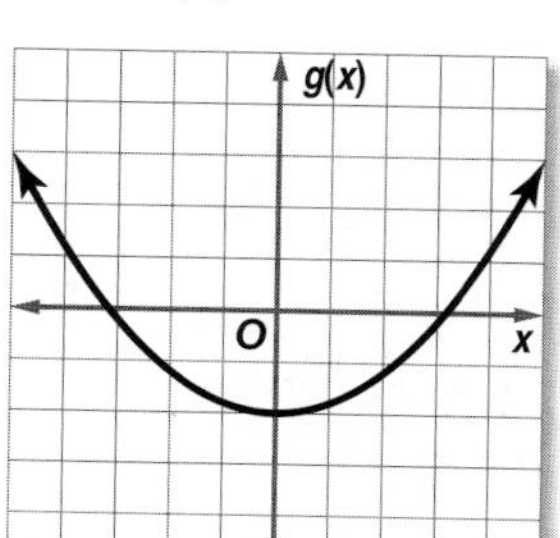

Practice and Problem Solving

● = **Step-by-Step Solutions** begin on page R12.
Extra Practice begins on page 815.

Examples 1–3 pp. 544–546

Describe how the graph of each function is related to the graph of $f(x) = x^2$.

8. $g(x) = -10 + x^2$

9 $h(x) = -7 - x^2$

10. $g(x) = 2x^2 + 8$

11. $h(x) = 6 + \frac{2}{3}x^2$

12. $g(x) = -5 - \frac{4}{3}x^2$

13. $h(x) = 3 + \frac{5}{2}x^2$

14. $g(x) = 0.25x^2 - 1.1$

15. $h(x) = 1.35x^2 + 2.6$

16. $g(x) = \frac{3}{4}x^2 + \frac{5}{6}$

17. $h(x) = 1.01x^2 - 6.5$

Example 4 p. 546

Match each equation to its graph.

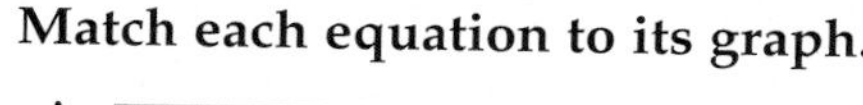

A

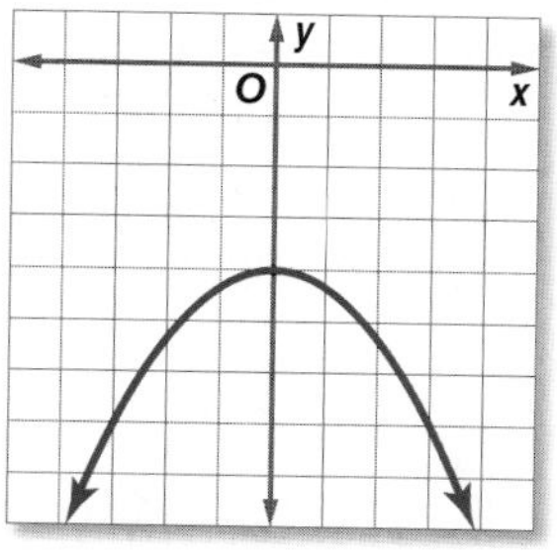

B

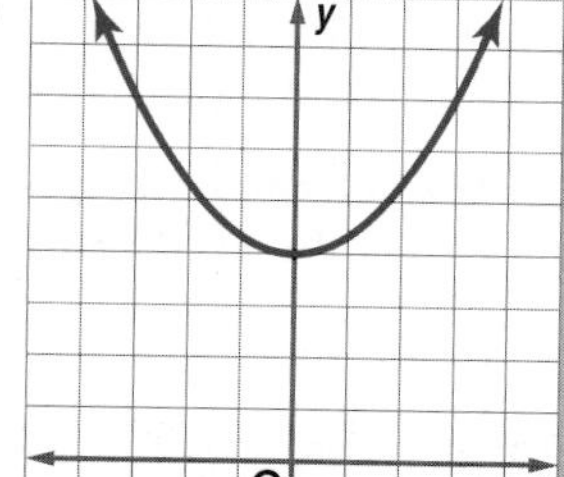

C

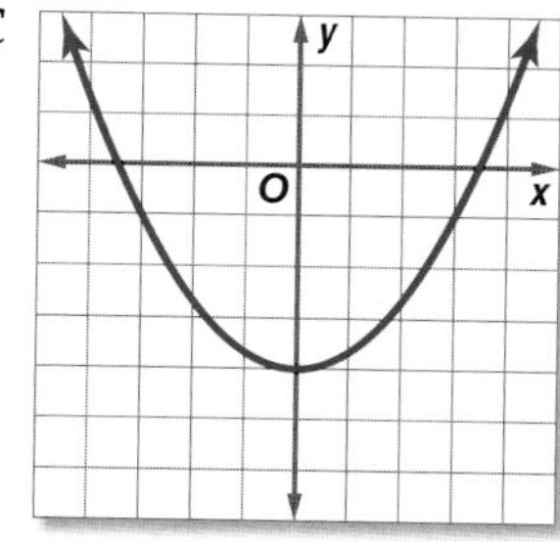

D

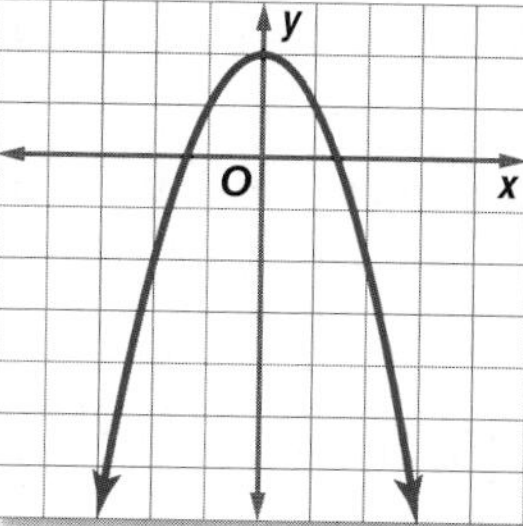

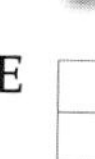

E

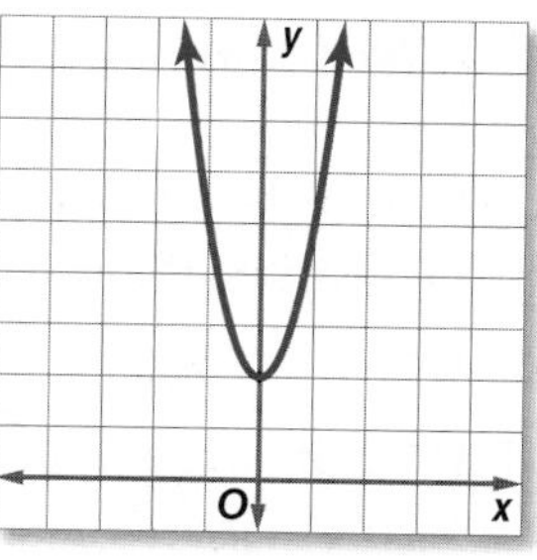

F

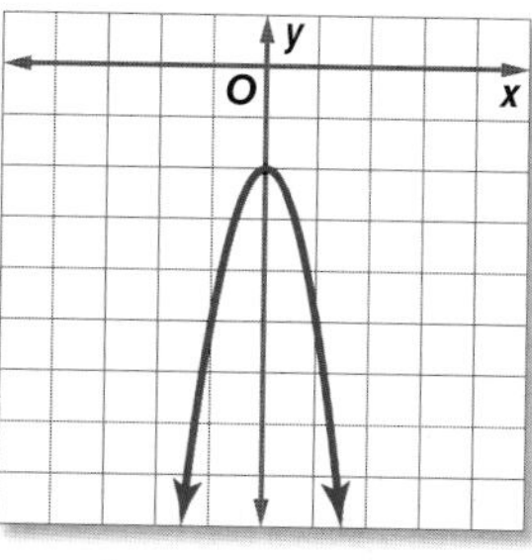

18. $y = \frac{1}{3}x^2 - 4$

19. $y = -\frac{1}{3}x^2 - 4$

20. $y = \frac{1}{3}x^2 + 4$

21. $y = -3x^2 - 2$

22. $y = -x^2 + 2$

23. $y = 3x^2 + 2$

Real-World Link

There are over 365 species of squirrels or squirrel-like mammals in the world. The three most common are the flying squirrel, the tree squirrel, and the ground squirrel.

Source: The Squirrel Place

24. SQUIRRELS A squirrel 12 feet above the ground drops an acorn from a tree. The function $h = -16t^2 + 12$ models the height of the acorn above the ground in feet after t seconds. Graph the function and compare this graph to the graph of its parent function.

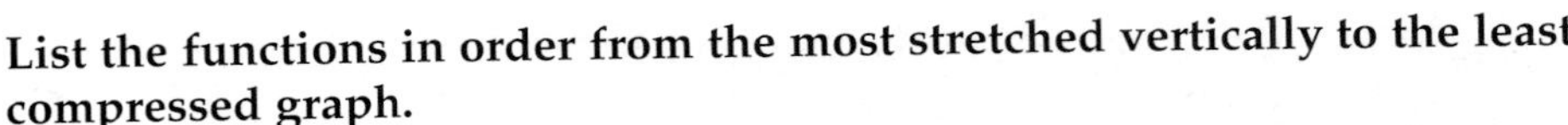

List the functions in order from the most stretched vertically to the least compressed graph.

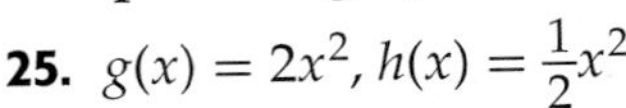

25. $g(x) = 2x^2, h(x) = \frac{1}{2}x^2$

26. $g(x) = -3x^2, h(x) = \frac{2}{3}x^2$

27. $g(x) = -4x^2, h(x) = 6x^2, f(x) = 0.3x^2$

28. $g(x) = -x^2, h(x) = \frac{5}{3}x^2, f(x) = -4.5x^2$

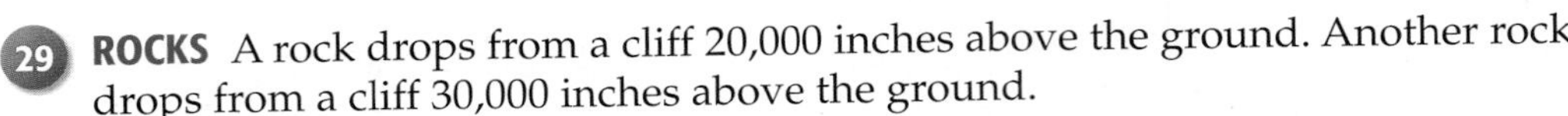

29 **ROCKS** A rock drops from a cliff 20,000 inches above the ground. Another rock drops from a cliff 30,000 inches above the ground.

a. Write two functions that model the heights h of the rocks after t seconds.

b. Which rock will reach the ground first?

Real-World Link

This fountain is in the East Concourse of Midfield Terminal in Detroit, Michigan. The streams are choreographed to create feelings ranging from tranquil to energetic.

Source: The WET Design

30. **SPRINKLERS** The path of water from a sprinkler can be modeled by quadratic functions. The following functions model paths for three different sprinklers.

Sprinkler A: $y = -0.35x^2 + 3.5$ Sprinkler B: $y = -0.21x^2 + 1.7$
Sprinkler C: $y = -0.08x^2 + 2.4$

a. Which sprinkler will send water the farthest? Explain.

b. Which sprinkler will send water the highest? Explain.

c. Which sprinkler will produce the narrowest path? Explain.

Describe the transformations to obtain the graph of $g(x)$ from the graph of $f(x)$.

31. $f(x) = x^2 + 3$
$g(x) = x^2 - 2$

32. $f(x) = x^2 - 4$
$g(x) = x^2 + 7$

33. $f(x) = -6x^2$
$g(x) = -3x^2$

34. **MULTIPLE REPRESENTATIONS** In this problem, you will investigate another type of transformation using your graphing calculator.

a. GRAPHICAL Graph the following family of equations: $y = x^2$, $y = (x - 2)^2$, $y = (x - 4)^2$, $y = (x + 3)^2$, and $y = (x + 5)^2$ on the same screen. Describe how the graphs of the functions are related to the graph of $f(x) = x^2$.

b. ALGEBRAIC Write a concept for quadratic functions, similar to the concept for vertical translations, to describe the effect of a value being added to or subtracted from x inside the parentheses.

c. ANALYTICAL Predict where the graphs of $y = (x - 7)^2$ and $y = (x + 4)^2$ will be located. Verify your answer by graphing each equation.

H.O.T. Problems

Use **H**igher-**O**rder **T**hinking Skills

35. **REASONING** Are the following statements *sometimes*, *always*, or *never* true? Explain.

a. The graph of $y = x^2 + c$ has its vertex at the origin.

b. The graphs of $y = ax^2$ and of $y = -ax^2$ are the same width.

c. The graph of $y = x^2 + c$ opens downward.

36. **CHALLENGE** Write a function of the form $y = ax^2 + c$ with a graph that passes through the points $(-2, 3)$ and $(4, 15)$.

37. **REASONING** Determine whether all quadratic functions that are reflected across the y-axis produce the same graph. Explain your answer.

38. **OPEN ENDED** Write a quadratic function that opens downward and is wider than the parent graph.

39. **WRITING IN MATH** Describe how the values of a and c affect the graphical and tabular representations for the functions $y = ax^2$, $y = x^2 + c$, and $y = ax^2 + c$.

Standardized Test Practice

40. SHORT RESPONSE A plumber charges a flat fee of \$55 and \$30 for each hour of work. Write a function that represents the total charge C, in terms of the number of hours h worked.

41. Which *best* describes the graph of $y = 2x^2$?

A a line with a y-intercept of (0, 2) and an x-intercept at the origin
B a parabola with a minimum point at (0, 0) and that is twice as wide as the graph of $y = x^2$ when $y = 2$
C a parabola with a maximum point at (0, 0) and that is half as wide as the graph of $y = x^2$ when $y = 2$
D a parabola with a minimum point at (0, 0) and that is half as wide as the graph of $y = x^2$ when $y = 2$

42. Candace is 5 feet tall. If 1 inch is about 2.54 centimeters, how tall is Candace to the nearest centimeter?

F 123 cm
G 26 cm
H 13 cm
J 152 cm

43. While in England, Imani spent 49.60 British pounds on a pair of jeans. If this is equivalent to \$100 in U.S. currency, how many British pounds would Imani have spent on a sweater that cost \$60?

A 8.26 pounds
B 29.76 pounds
C 2976 pounds
D 19.84 pounds

Spiral Review

Solve each equation by graphing. (Lesson 9-2)

44. $x^2 + 6 = 0$
45. $x^2 - 10x = -24$
46. $x^2 + 5x + 4 = 0$
47. $2x^2 - x = 3$
48. $2x^2 - x = 15$
49. $12x^2 = -11x + 15$

Find the vertex, the equation of the axis of symmetry, and the y-intercept of each graph. (Lesson 9-1)

50.

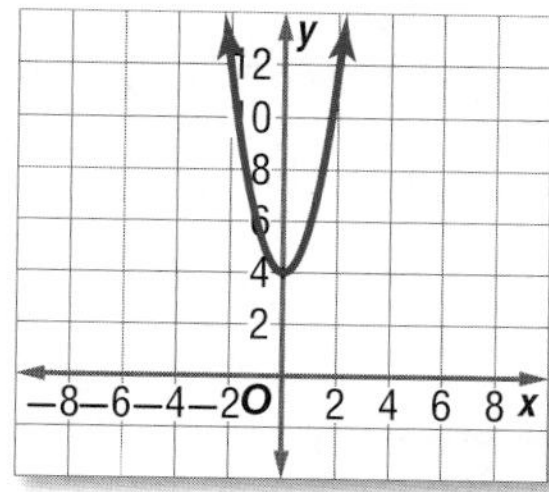

51.

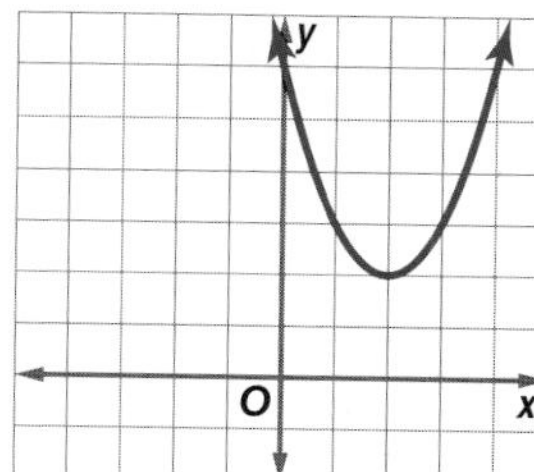

52.

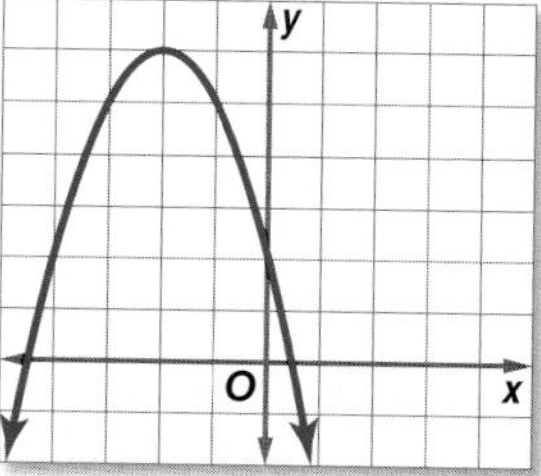

53. CLASS TRIP Mr. Wong's American History class will take taxis from their hotel in Washington, D.C., to the Lincoln Memorial. The fare is \$2.75 for the first mile and \$1.25 for each additional mile. If the distance is m miles and t taxis are needed, write an expression for the cost to transport the group. (Lesson 7-6)

Solve each inequality. Check your solution. (Lesson 5-3)

54. $-3t + 6 \leq -3$
55. $59 > -5 - 8f$
56. $-2 - \frac{d}{5} < 23$

Skills Check

Determine whether each trinomial is a perfect square trinomial. If so, factor it. (Lesson 8-6)

57. $16x^2 - 24x + 9$
58. $9x^2 + 6x + 1$
59. $25x^2 - 60x + 36$
60. $x^2 - 8x + 81$
61. $36x^2 - 84x + 49$
62. $4x^2 - 3x + 9$

EXTEND 9-3

Graphing Technology Lab

Systems of Linear and Quadratic Equations

Math Online glencoe.com
- Other Calculator Keystrokes
- Graphing Technology Personal Tutor

You can use a graphing calculator to solve systems involving linear and quadratic equations.

ACTIVITY 1

Use a graphing calculator to solve the system of equations.

$y = x^2 - x - 6$
$y = x - 3$

Step 1 Enter each equation in the Y= list. Enter the quadratic equation as Y1 and the linear equation as Y2.

KEYSTROKES: [X,T,θ,n] [x^2] [−] [X,T,θ,n] [−] 6 [ENTER] [X,T,θ,n] [−] 3

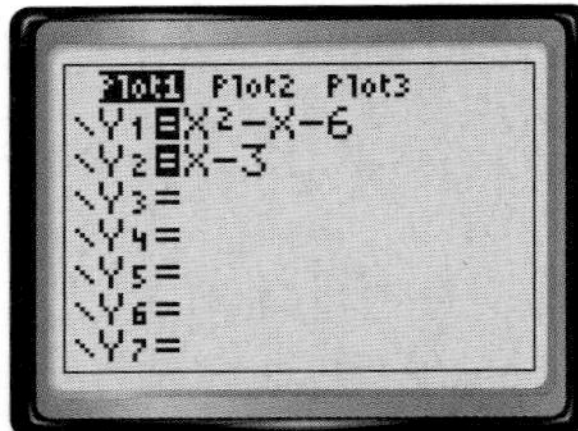

Step 2 Graph the system. **KEYSTROKE:** [Graph]

The solutions of the system are the intersection points. The graphs intersect at two points. So, there are two solutions.

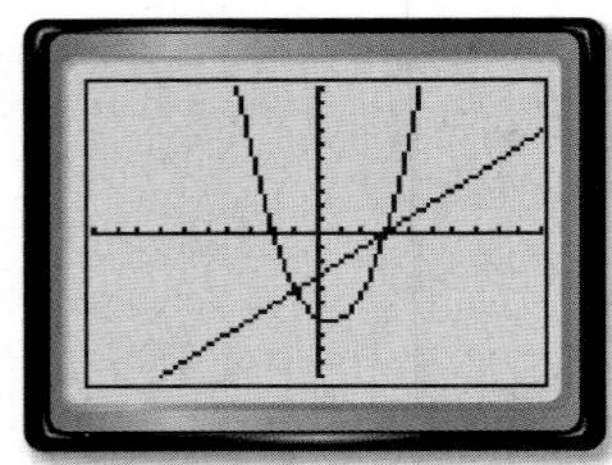

[−10, 10] scl: 1 by [−10, 10] scl: 1

Step 3 Find the first intersection of the graphs by using the CALC menu.

KEYSTROKES: [2nd] [CALC] 5

On the screen, notice the question "First Curve?" The cursor should be on the parabola. Press [ENTER].

Notice that the question changes to "Second curve?" and the cursor jumps to the line. Press [ENTER].

Use the arrow keys to move the cursor as close as possible to the intersection point in Quadrant III. Press [ENTER] again.

The intersection is the point at (−1, −4).

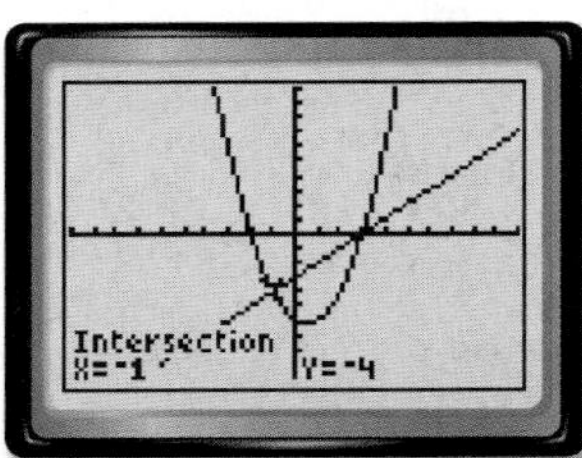

[−10, 10] scl: 1 by [−10, 10] scl: 1

Step 4 Move the cursor to the second intersection. Find the second intersection by repeating Step 3.

The intersection is at (3, 0).

Therefore, the solutions of the system of equations are (–1, −4) and (3, 0).

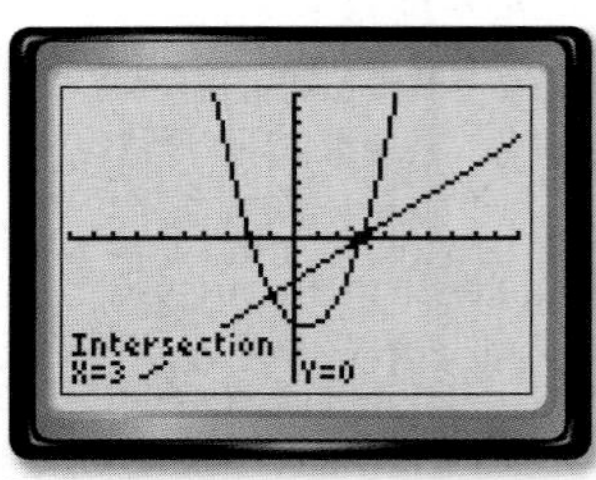

[−10, 10] scl: 1 by [−10, 10] scl: 1

ACTIVITY 2

Use a graphing calculator to solve the system of equations.

$y = x^2 - 8x + 19$
$y = 2x - 6$

Step 1 Enter each equation in the Y= list.

Enter the quadratic equation as Y1 and the linear equation as Y2.

Step 2 Graph the system.

In this case, the graphs of the equations intersect at only one point. Therefore, there is only one solution of this system of equations.

Step 3 Find the intersection of the graphs of the equations.

The intersection is the point at about (5, 4).

Thus, the solution of the system of equations is about (5, 4).

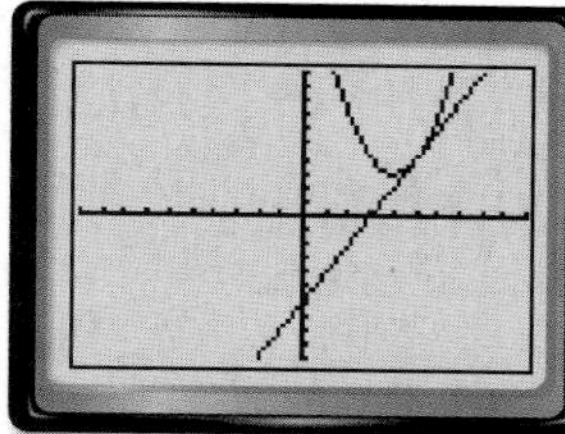

[−10, 10] scl: 1 by [−10, 10] scl: 1

ACTIVITY 3

Use a graphing calculator to solve the system of equations.

$y = -x^2 - 4x - 6$
$y = -\frac{1}{3}x + 4$

Step 1 Enter each equation in the Y= list.

Enter the quadratic equation as Y1 and the linear equation as Y2.

Step 2 Graph the system.

The graphs of the equations do not intersect. Thus, this system of equations has no solution.

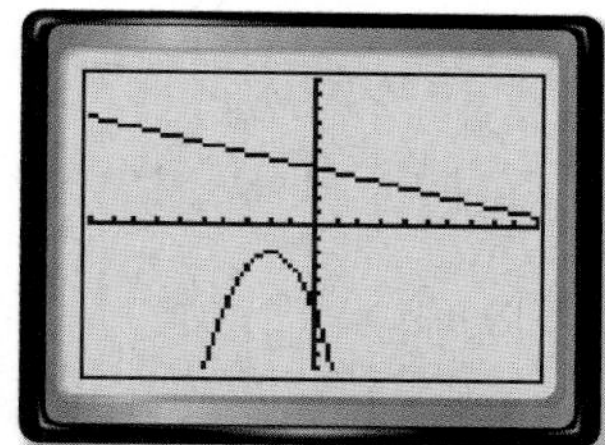

[−10, 10] scl: 1 by [−10, 10] scl: 1

Exercises

Use factoring to solve each system of equations. Then use a graphing calculator to check your solutions.

1. $y = x^2 + 7x + 12$
   $y = 2x + 8$
2. $y = x^2 - x - 20$
   $y = 3x + 12$
3. $y = 3x^2 - x - 2$
   $y = -2x + 2$

Use a graphing calculator to solve each system of equations.

4. $y = x^2$
   $y = 2x$
5. $y = -x^2 - 6x - 3$
   $y = 6$
6. $y = -x^2 + 4$
   $y = \frac{1}{2}x + 5$
7. $y = x^2 + 5x + 4$
   $y = -x - 8$
8. $y = \frac{1}{2}x^2 - 4$
   $y = 3x + 4$
9. $y = x^2$
   $y = -2x - 1$

9-4 Solving Quadratic Equations by Completing the Square

Then

You solved quadratic equations by using the square root property. (Lesson 8-6)

Now

- Complete the square to write perfect square trinomials.
- Solve quadratic equations by completing the square.

New Vocabulary

completing the square

Math Online

glencoe.com

- Extra Examples
- Personal Tutor
- Self-Check Quiz
- Homework Help
- Math in Motion

Why?

In competitions, skateboarders may launch themselves from a half pipe into the air to perform tricks. The equation $h = -16t^2 + 20t + 12$ can be used to model their height, in feet, after t seconds.

To find how long a skateboarder is in the air if he is 25 feet above the half pipe, you can solve $25 = -16t^2 + 20t + 12$ by using a method called completing the square.

Complete the Square In Lesson 8-6, you solved equations by taking the square root of each side. This method worked only because the expression on the left-hand side was a perfect square. In perfect square trinomials in which the leading coefficient is 1, there is a relationship between the coefficient of the x-term and the constant term.

$$(x + 5)^2 = x^2 + 2(5)(x) + 5^2$$
$$= x^2 + 10x + 25$$

Notice that $\left(\frac{10}{2}\right)^2 = 25$. To get the constant term, divide the coefficient of the x-term by 2 and square the result. Any quadratic expression in the form $x^2 + bx$ can be made into a perfect square by using a method called **completing the square**.

Key Concept — Completing the Square

For Your FOLDABLE

Words To complete the square for any quadratic expression of the form $x^2 + bx$, follow the steps below.

Step 1 Find one half of b, the coefficient of x.

Step 2 Square the result in Step 1.

Step 3 Add the result of Step 2 to $x^2 + bx$.

Symbols $x^2 + bx + \left(\frac{b}{2}\right)^2 = \left(x + \frac{b}{2}\right)^2$

Math *in Motion,* Animation glencoe.com

EXAMPLE 1 Complete the Square

Find the value of c that makes $x^2 + 4x + c$ a perfect square trinomial.

Method 1 Use algebra tiles.

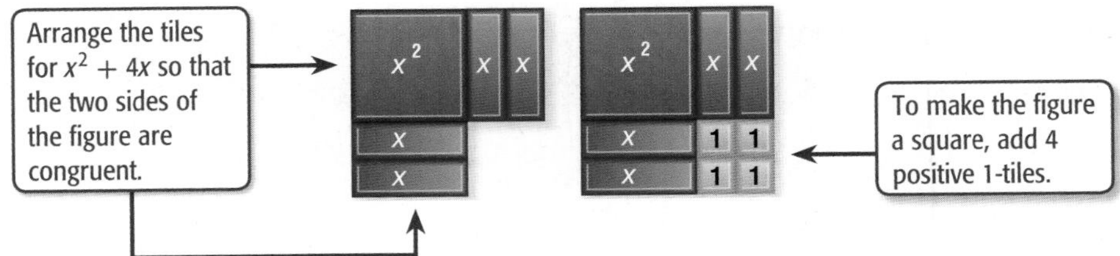

StudyTip

Algorithms An algorithm is a series of steps for carrying out a procedure or solving a problem.

Method 2 Use complete the square algorithm.

Step 1 Find $\frac{1}{2}$ of 4. $\frac{4}{2} = 2$

Step 2 Square the result in Step 1. $2^2 = 4$

Step 3 Add the result of Step 2 to $x^2 + 4x$. $x^2 + 4x + 4$

Thus, $c = 4$. Notice that $x^2 + 4x + 4 = (x + 2)^2$.

Check Your Progress

1. Find the value of c that makes $r^2 - 8r + c$ a perfect square trinomial.

Personal Tutor glencoe.com

Solve Equations by Completing the Square You can complete the square to solve quadratic equations. First, you must isolate the x^2- and bx-terms.

EXAMPLE 2 Solve an Equation by Completing the Square

Solve $x^2 - 6x + 12 = 19$ by completing the square.

$x^2 - 6x + 12 = 19$	**Original equation**
$x^2 - 6x = 7$	**Subtract 12 from each side.**
$x^2 - 6x + 9 = 7 + 9$	**Since $\left(\frac{-6}{2}\right)^2 = 9$, add 9 to each side.**
$(x - 3)^2 = 16$	**Factor $x^2 - 6x + 9$.**
$x - 3 = \pm 4$	**Take the square root of each side.**
$x = 3 \pm 4$	**Add 3 to each side.**
$x = 3 + 4$ or $x = 3 - 4$	**Separate the solutions.**
$= 7$ $\quad = -1$	The solutions are 7 and −1.

Check Your Progress

2. Solve $x^2 - 12x + 3 = 8$ by completing the square.

Personal Tutor glencoe.com

To solve a quadratic equation in which the leading coefficient is not 1, divide each term by the coefficient. Then isolate the x^2- and x-terms and complete the square.

Watch Out!

Leading Coefficient Remember that the leading coefficient has to be 1 before you can complete the square.

EXAMPLE 3 Equation with $a \neq 1$

Solve $-2x^2 + 8x - 18 = 0$ by completing the square.

$-2x^2 + 8x - 18 = 0$	**Original equation**
$\frac{-2x^2 + 8x - 18}{-2} = \frac{0}{-2}$	**Divide each side by −2.**
$x^2 - 4x + 9 = 0$	**Simplify.**
$x^2 - 4x = -9$	**Subtract 9 from each side.**
$x^2 - 4x + 4 = -9 + 4$	**Since $\left(\frac{-4}{2}\right)^2 = 4$, add 4 to each side.**
$(x - 2)^2 = -5$	**Factor $x^2 - 4x + 4$.**

No real number has a negative square. So, this equation has no real solutions.

Check Your Progress

3. Solve $3x^2 - 9x - 3 = 21$ by completing the square.

Personal Tutor glencoe.com

Real-World Link

The oldest public high school rivalry takes place between Wellesley High School and Needham Heights High School in Massachusetts. The first football game between them took place on Thanksgiving morning in 1882 in Needham.

Source: USA Football

Real-World EXAMPLE 4 Solve a Problem by Completing the Square

JERSEYS The senior class at Bay High School buys jerseys to wear to the football games. The cost of the jerseys can be modeled by the equation $C = 0.1x^2 + 2.4x + 25$, where C is the amount it costs to buy x jerseys. How many jerseys can they purchase for $430?

The seniors have $430, so set the equation equal to 430 and complete the square.

$0.1x^2 + 2.4x + 25 = 430$	**Original equation**
$\frac{0.1x^2 + 2.4x + 25}{0.1} = \frac{430}{0.1}$	**Divide each side by 0.1.**
$x^2 + 24x + 250 = 4300$	**Simplify.**
$x^2 + 24x + 250 - 250 = 4300 - 250$	**Subtract 250 from each side.**
$x^2 + 24x = 4050$	**Simplify.**
$x^2 + 24x + 144 = 4050 + 144$	**Since $\left(\frac{24}{2}\right)^2 = 144$, add 144 to each side.**
$x^2 + 24x + 144 = 4194$	**Simplify.**
$(x + 12)^2 = 4194$	**Factor $x^2 + 24x + 144$.**
$x + 12 = \pm\sqrt{4194}$	**Take the square root of each side.**
$x = -12 \pm\sqrt{4194}$	**Subtract 12 from each side.**

Use a calculator to approximate each value of x.

$x = -12 + \sqrt{4194}$ or $x = -12 - \sqrt{4194}$ **Separate the solutions.**

≈ 52.7 ≈ -76.7 **Evaluate.**

Since you cannot buy a negative number of jerseys, the negative solution is not reasonable. The seniors can afford to buy 52 jerseys.

Check Your Progress

4. If the senior class were able to raise $620, how many jerseys could they buy?

Personal Tutor glencoe.com

Check Your Understanding

Example 1 pp. 552–553

Find the value of c that makes each trinomial a perfect square.

1. $x^2 - 18x + c$
2. $x^2 + 22x + c$
3. $x^2 + 9x + c$
4. $x^2 - 7x + c$

Examples 2 and 3 p. 553

Solve each equation by completing the square. Round to the nearest tenth if necessary.

5. $x^2 + 4x = 6$
6. $x^2 - 8x = -9$
7. $4x^2 + 9x - 1 = 0$
8. $-2x^2 + 10x + 22 = 4$

Example 4 p. 554

9. **CONSTRUCTION** Collin is building a deck on the back of his family's house. He has enough lumber for the deck to be 144 square feet. The length should be 10 feet more than its width. What should the dimensions of the deck be?

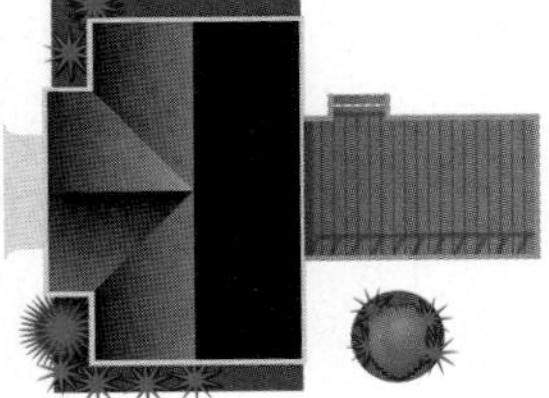

Practice and Problem Solving

● = Step-by-Step Solutions begin on page R12.
Extra Practice begins on page 815.

Example 1 pp. 552–553

Find the value of c that makes each trinomial a perfect square.

10. $x^2 + 26x + c$
11. $x^2 - 24x + c$
12. $x^2 - 19x + c$
13. $x^2 + 17x + c$
14. $x^2 + 5x + c$
15. $x^2 - 13x + c$
16. $x^2 - 22x + c$
17. $x^2 - 15x + c$
18. $x^2 + 24x + c$

Examples 2 and 3 p. 553

Solve each equation by completing the square. Round to the nearest tenth if necessary.

19. $x^2 + 6x - 16 = 0$
20. $x^2 - 2x - 14 = 0$
21. $x^2 - 8x - 1 = 8$
22. $x^2 + 3x + 21 = 22$
23. $x^2 - 11x + 3 = 5$
24. $5x^2 - 10x = 23$
25. $2x^2 - 2x + 7 = 5$
26. $3x^2 + 12x + 81 = 15$
27. $4x^2 + 6x = 12$
28. $4x^2 + 5 = 10x$
29. $-2x^2 + 10x = -14$
30. $-3x^2 - 12 = 14x$

Example 4 p. 554

31. **FINANCIAL LITERACY** The price p in dollars for a particular stock can be modeled by the quadratic equation $p = 3.5t - 0.05t^2$, where t represents the number of days after the stock is purchased. When is the stock worth $60?

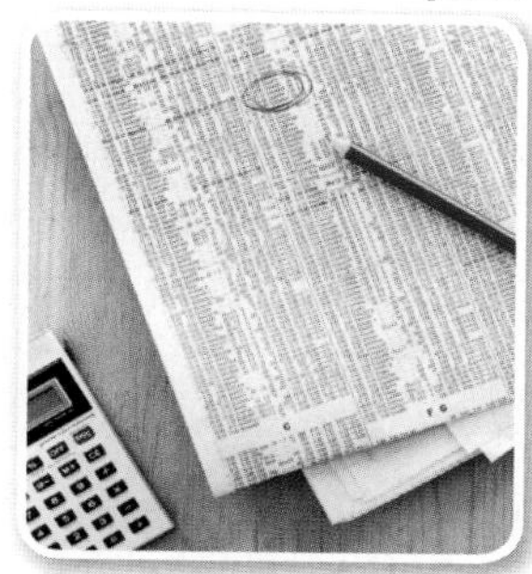

Real-World Link

According to a Junior Achievement survey, 25% of 13- and 14-year-olds own stocks in their own name. Most got their shares as gifts.

Source: *Money Magazine*

GEOMETRY **Find the value of x for each figure. Round to the nearest tenth if necessary.**

32. $A = 45\text{ in}^2$

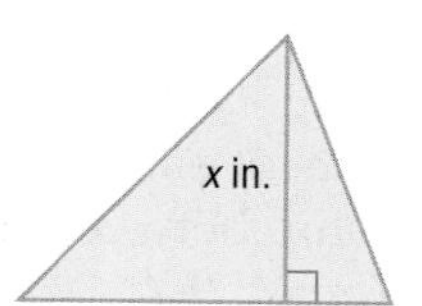

33. $A = 110\text{ ft}^2$

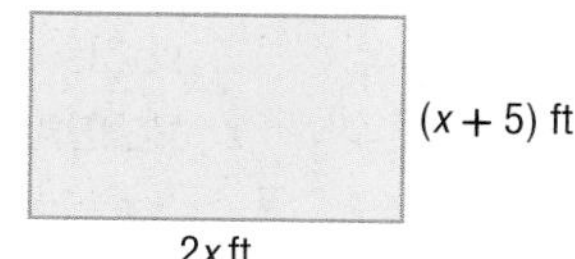

34. **NUMBER THEORY** The product of two consecutive even integers is 224. Find the integers.

35. **NUMBER THEORY** The product of two consecutive negative odd integers is 483. Find the integers.

36. **GEOMETRY** Find the area of the triangle below.

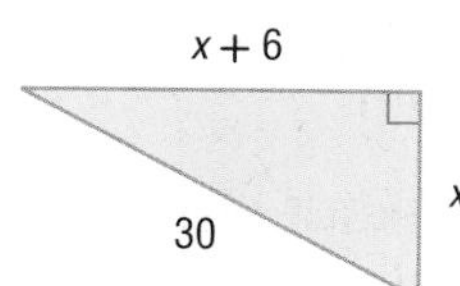

Solve each equation by completing the square. Round to the nearest tenth if necessary.

37. $0.2x^2 - 0.2x - 0.4 = 0$
38. $0.5x^2 = 2x - 0.3$
39. $2x^2 - \frac{11}{5}x = -\frac{3}{10}$
40. $\frac{2}{3}x^2 - \frac{4}{3}x = \frac{5}{6}$
41. $\frac{1}{4}x^2 + 2x = \frac{3}{8}$
42. $\frac{2}{5}x^2 + 2x = \frac{1}{5}$

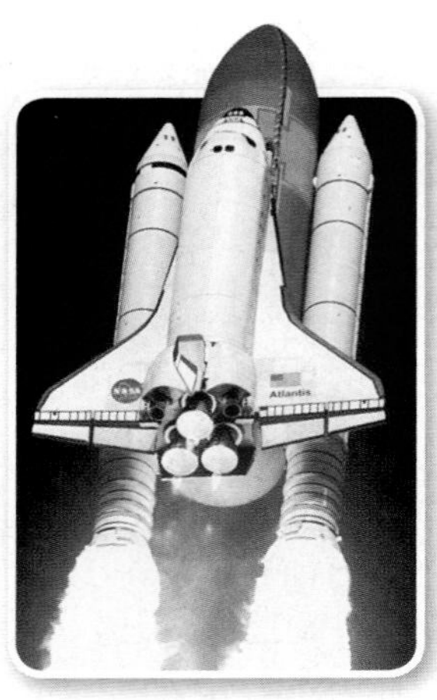

Real-World Link

The National Aeronautics and Space Administration (NASA) began operations on October 1, 1958. NASA employs about 8000 people and has an annual budget of $100 million.

Source: NASA

43. **ASTRONOMY** The height of an object t seconds after it is dropped is given by the equation $h = -\frac{1}{2}gt^2 + h_0$, where h_0 is the initial height and g is the acceleration due to gravity. The acceleration due to gravity near the surface of Mars is $3.73\ \text{m/s}^2$, while on Earth it is $9.8\ \text{m/s}^2$. Suppose an object is dropped from an initial height of 120 meters above the surface of each planet.
   a. On which planet would the object reach the ground first?
   b. How long would it take the object to reach the ground on each planet? Round each answer to the nearest tenth.
   c. Do the times that it takes the object to reach the ground seem reasonable? Explain your reasoning.

44. Find all values of c that make $x^2 + cx + 100$ a perfect square trinomial.

45. Find all values of c that make $x^2 + cx + 225$ a perfect square trinomial.

46. **PAINTING** Before she begins painting a picture, Donna stretches her canvas over a wood frame. The frame has a length of 60 inches and a width of 4 inches. She has enough canvas to cover 480 square inches. Donna decides to increase the dimensions of the frame. If the increase in the length is 10 times the increase in the width, what will the dimensions of the frame be?

47. **MULTIPLE REPRESENTATIONS** In this problem, you will investigate a property of quadratic equations.
   a. **TABULAR** Copy the table shown and complete the second column.
   b. **ALGEBRAIC** Set each trinomial equal to zero, and solve the equation by completing the square. Complete the last column of the table with the number of roots of each equation.

Trinomial	$b^2 - 4ac$	Number of Roots
$x^2 - 8x + 16$	0	1
$2x^2 - 11x + 3$		
$3x^2 + 6x + 9$		
$x^2 - 2x + 7$		
$x^2 + 10x + 25$		
$x^2 + 3x - 12$		

   c. **VERBAL** Compare the number of roots of each equation to the result in the $b^2 - 4ac$ column. Is there a relationship between these values? If so, describe it.
   d. **ANALYTICAL** Predict how many solutions $2x^2 - 9x + 15 = 0$ will have. Verify your prediction by solving the equation.

H.O.T. Problems

Use Higher-Order Thinking Skills

48. **CHALLENGE** Given $y = ax^2 + bx + c$ with $a \neq 0$, derive the equation for the axis of symmetry by completing the square and rewriting the equation in the form $y = a(x - h)^2 + k$.

49. **REASONING** Determine the number of solutions $x^2 + bx = c$ has if $c < -\left(\frac{b}{2}\right)^2$. Explain.

50. **WHICH ONE DOESN'T BELONG?** Identify the expression that does not belong with the other three. Explain your reasoning.

$n^2 - n + \frac{1}{4}$	$n^2 + n + \frac{1}{4}$	$n^2 - \frac{2}{3}n + \frac{1}{9}$	$n^2 + \frac{1}{3}n + \frac{1}{9}$

51. **OPEN ENDED** Write a quadratic equation for which the only solution is 4.

52. **WRITING IN MATH** Compare and contrast the following strategies for solving $x^2 - 5x - 7 = 0$: completing the square, graphing, and factoring.

Standardized Test Practice

53. The length of a rectangle is 3 times its width. The area of the rectangle is 75 square feet. Find the length of the rectangle in feet.

A 25 B 15 C 10 D 5

54. PROBABILITY At a festival, winners of a game draw a token for a prize. There is one token for each prize. The prizes include 9 movie passes, 8 stuffed animals, 5 hats, 10 jump ropes, and 4 glow necklaces. What is the probability that the first person to draw a token will win a movie pass?

F $\frac{9}{61}$ G $\frac{1}{9}$ H $\frac{1}{4}$ J $\frac{1}{36}$

55. GRIDDED RESPONSE The population of a town can be modeled by $P = 22{,}000 + 125t$, where P represents the population and t represents the number of years from 2000. How many years after 2000 will the population be 26,000?

56. Percy delivers pizzas for Pizza King. He is paid \$6 an hour plus \$2.50 for each pizza he delivers. Percy earned \$280 last week. If he worked a total of 30 hours, how many pizzas did he deliver?

A 250 pizzas
B 184 pizzas
C 40 pizzas
D 34 pizzas

Spiral Review

Describe how the graph of each function is related to the graph of $f(x) = x^2$. (Lesson 9-3)

57. $g(x) = -12 + x^2$ **58.** $h(x) = 2 - x^2$ **59.** $g(x) = 2x^2 + 5$

60. $h(x) = -6 + \frac{2}{3}x^2$ **61.** $g(x) = 6 + \frac{4}{3}x^2$ **62.** $h(x) = -1 - \frac{3}{2}x^2$

63. RIDES A popular amusement park ride whisks riders to the top of a 250-foot tower and drops them. A function for the height of a rider is $h = -16t^2 + 250$, where h is the height and t is the time in seconds. The ride stops the descent of the rider 40 feet above the ground. Write an equation that models the drop of the rider. How long does it take to fall from 250 feet to 40 feet? (Lesson 9-2)

Simplify. Assume that no denominator is equal to zero. (Lesson 7-2)

64. $\frac{a^6}{a^3}$ **65.** $\frac{4^7}{4^5}$ **66.** $\frac{c^3d^4}{cd^7}$

67. $\left(\frac{4h^{-2}g}{2g^5}\right)^0$ **68.** $\frac{5q^{-2}t^6}{10q^2t^{-4}}$ **69.** $b^3(m^{-3})(b^{-6})$

Solve each open sentence. (Lesson 5-5)

70. $|y - 2| > 7$ **71.** $|z + 5| < 3$ **72.** $|2b + 7| \leq -6$

73. $|3 - 2y| \geq 8$ **74.** $|9 - 4m| < -1$ **75.** $|5c - 2| \leq 13$

Skills Check

Evaluate $\sqrt{b^2 - 4ac}$ for each set of values. Round to the nearest tenth if necessary. (Lesson 1-2)

76. $a = 2, b = -5, c = 2$ **77.** $a = 1, b = 12, c = 11$ **78.** $a = -9, b = 10, c = -1$

79. $a = 1, b = 7, c = -3$ **80.** $a = 2, b = -4, c = -6$ **81.** $a = 3, b = 1, c = 2$

9-5 Solving Quadratic Equations by Using the Quadratic Formula

Then

You solved quadratic equations by completing the square. (Lesson 9-4)

Now

- Solve quadratic equations by using the Quadratic Formula.
- Use the discriminant to determine the number of solutions of a quadratic equation.

New Vocabulary

Quadratic Formula
discriminant

Math Online

glencoe.com

- Extra Examples
- Personal Tutor
- Self-Check Quiz
- Homework Help

Why?

For adult women, the normal systolic blood pressure P in millimeters of mercury (mm Hg) can be modeled by $P = 0.01a^2 + 0.05a + 107$, where a is age in years. This equation can be used to approximate the age of a woman with a certain systolic blood pressure. However, it would be difficult to solve by factoring, graphing, or completing the square.

Quadratic Formula Completing the square of the quadratic equation $ax^2 + bx + c = 0$ produces a formula that allows you to find the solutions of *any* quadratic equation that is written in standard form. This formula is called the **Quadratic Formula**.

Key Concept — The Quadratic Formula

For Your FOLDABLE

The solutions of a quadratic equation $ax^2 + bx + c = 0$, where $a \neq 0$, are given by the Quadratic Formula.

$$x = \frac{-b \pm \sqrt{b^2 - 4ac}}{2a}$$

You will derive this formula in Lesson 10-2.

EXAMPLE 1 Use the Quadratic Formula

Solve $x^2 - 12x = -20$ by using the Quadratic Formula.

Step 1 Rewrite the equation in standard form.

$x^2 - 12x = -20$ — **Original equation**

$x^2 - 12x + 20 = 0$ — **Add 20 to each side.**

Step 2 Apply the Quadratic Formula.

$x = \frac{-b \pm \sqrt{b^2 - 4ac}}{2a}$ — **Quadratic Formula**

$= \frac{-(-12) \pm \sqrt{(-12)^2 - 4(1)(20)}}{2(1)}$ — **$a = 1$, $b = -12$, and $c = 20$**

$= \frac{12 \pm \sqrt{144 - 80}}{2}$ — **Multiply.**

$= \frac{12 \pm \sqrt{64}}{2}$ or $\frac{12 \pm 8}{2}$ — **Subtract and take the square root.**

$x = \frac{12 - 8}{2}$ or $x = \frac{12 + 8}{2}$ — **Separate the solutions.**

$= 2 \qquad = 10$ — **Simplify.**

The solutions are 2 and 10.

Check Your Progress

1. Solve $2x^2 + 9x = 18$ by using the Quadratic Formula.

Personal Tutor glencoe.com

The solutions of quadratic equations are not always integers.

EXAMPLE 2 Use the Quadratic Formula

Solve each equation by using the Quadratic Formula. Round to the nearest tenth if necessary.

a. $3x^2 + 5x - 12 = 0$

For this equation, $a = 3$, $b = 5$, and $c = -12$.

$$x = \frac{-b \pm \sqrt{b^2 - 4ac}}{2a}$$ **Quadratic Formula**

$$= \frac{-(5) \pm \sqrt{(5)^2 - 4(3)(-12)}}{2(3)}$$ **$a = 3$, $b = 5$, and $c = -12$**

$$= \frac{-5 \pm \sqrt{25 + 144}}{6}$$ **Multiply.**

$$= \frac{-5 \pm \sqrt{169}}{6} \text{ or } \frac{-5 \pm 13}{6}$$ **Add and simplify.**

$$x = \frac{-5 - 13}{6} \text{ or } x = \frac{-5 + 13}{6}$$ **Separate the solutions.**

$$= -3 \qquad = \frac{4}{3}$$ **Simplify.**

The solutions are -3 and $\frac{4}{3}$.

b. $10x^2 - 5x = 25$

Step 1 Rewrite the equation in standard form.

$$10x^2 - 5x = 25$$ **Original equation**

$$10x^2 - 5x - 25 = 0$$ **Subtract 25 from each side.**

Step 2 Apply the Quadratic Formula.

$$x = \frac{-b \pm \sqrt{b^2 - 4ac}}{2a}$$ **Quadratic Formula**

$$= \frac{-(-5) \pm \sqrt{(-5)^2 - 4(10)(-25)}}{2(10)}$$ **$a = 10$, $b = -5$, and $c = -25$**

$$= \frac{5 \pm \sqrt{25 + 1000}}{20}$$ **Multiply.**

$$= \frac{5 \pm \sqrt{1025}}{20}$$ **Add.**

$$= \frac{5 - \sqrt{1025}}{20} \text{ or } \frac{5 + \sqrt{1025}}{20}$$ **Separate the solutions.**

$$\approx -1.4 \qquad \approx 1.9$$ **Simplify.**

The solutions are about -1.4 and 1.9.

StudyTip

Exact Answers In Example 2, the number $\sqrt{1025}$ is irrational, so the calculator can only give you an approximation of its value. So, the exact answer in Example 2 is  $\frac{5 \pm \sqrt{1025}}{20}$. The numbers -1.4 and 1.9 are approximations.

Check Your Progress

2A. $4x^2 - 24x + 35 = 0$

2B. $3x^2 - 2x - 9 = 0$

Personal Tutor glencoe.com

You can solve quadratic equations by using many different methods. No one way is always best.

EXAMPLE 3 Solve Quadratic Equations Using Different Methods

Solve $x^2 - 4x = 12$.

Method 1 Graphing

Rewrite the equation in standard form.

$x^2 - 4x = 12$ **Original equation**

$x^2 - 4x - 12 = 0$ **Subtract 12 from each side.**

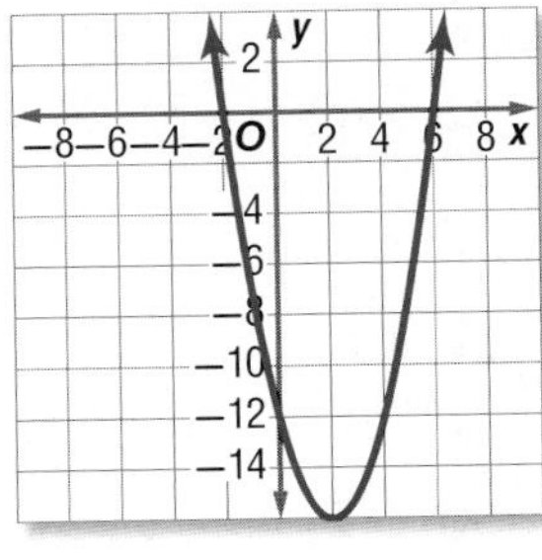

Graph the related function $f(x) = x^2 - 4x - 12$. Locate the x-intercepts of the graph. The solutions are −2 and 6.

Method 2 Factoring

$x^2 - 4x = 12$	**Original equation**
$x^2 - 4x - 12 = 0$	**Subtract 12 from each side.**
$(x - 6)(x + 2) = 0$	**Factor.**
$x - 6 = 0$ or $x + 2 = 0$	**Zero Product Property**
$x = 6 \qquad x = -2$	**Solve for x.**

Method 3 Completing the Square

The equation is in the correct form to complete the square, since the leading coefficient is 1 and the x^2 and x terms are isolated.

$x^2 - 4x = 12$	**Original equation**
$x^2 - 4x + 4 = 12 + 4$	**Since $\left(\frac{-4}{2}\right)^2 = 4$, add 4 to each side.**
$(x - 2)^2 = 16$	**Factor $x^2 - 4x + 4$.**
$x - 2 = \pm 4$	**Take the square root of each side.**
$x = 2 \pm 4$	**Add 2 to each side.**
$x = 2 + 4$ or $x = 2 - 4$	**Separate the solutions.**
$= 6 \qquad = -2$	**Simplify.**

> **Watch Out!**
>
> **Solutions** No matter what method is used to solve a quadratic equation, all of the methods should produce the same solution(s).

Method 4 Quadratic Formula

From Method 1, the standard form of the equation is $x^2 - 4x - 12 = 0$.

$x = \frac{-b \pm \sqrt{b^2 - 4ac}}{2a}$	**Quadratic Formula**
$= \frac{-(-4) \pm \sqrt{(-4)^2 - 4(1)(-12)}}{2(1)}$	**$a = 1$, $b = -4$, and $c = -12$**
$= \frac{4 \pm \sqrt{16 + 48}}{2}$	**Multiply.**
$= \frac{4 \pm \sqrt{64}}{2}$ or $\frac{4 \pm 8}{2}$	**Add and simplify.**
$x = \frac{4 - 8}{2}$ or $x = \frac{4 + 8}{2}$	**Separate the solutions.**
$= -2 \qquad = 6$	**Simplify.**

Check Your Progress

Solve each equation.

3A. $2x^2 - 17x + 8 = 0$

3B. $4x^2 - 4x - 11 = 0$

Personal Tutor glencoe.com

Concept Summary — Solving Quadratic Equations

For Your FOLDABLE

Method	When to Use
Factoring	Use when the constant term is 0 or if the factors are easily determined. Not all equations are factorable.
Graphing	Use when an approximate solution is sufficient.
Using Square Roots	Use when an equation can be written in the form $x^2 = n$. Can only be used if the equation has no x-term.
Completing the Square	Can be used for any equation $ax^2 + bx + c = 0$, but is simplest to apply when b is even and $a = 1$.
Quadratic Formula	Can be used for any equation $ax^2 + bx + c = 0$.

The Discriminant In the Quadratic Formula, the expression under the radical sign, $b^2 - 4ac$, is called the **discriminant**. The discriminant can be used to determine the number of real solutions of a quadratic equation.

Key Concept — Using the Discriminant

For Your FOLDABLE

Equation	$x^2 + 2x + 5 = 0$	$x^2 + 10x + 25 = 0$	$2x^2 - 7x + 2 = 0$
Discriminant	$b^2 - 4ac = -16$ negative	$b^2 - 4ac = 0$ zero	$b^2 - 4ac = 33$ positive
Graph of Related Function	0 x-intercepts	1 x-intercept	2 x-intercepts
Real Solutions	0	1	2

StudyTip

Discriminant Recall that when the left side of the standard form of an equation is a perfect square trinomial, there is only one solution. Therefore, the discriminant of a perfect square trinomial will always be zero.

EXAMPLE 4 Use the Discriminant

State the value of the discriminant of $4x^2 + 5x = -3$. Then determine the number of real solutions of the equation.

Step 1 Rewrite in standard form. $4x^2 - 5x = -3 \longrightarrow 4x^2 - 5x + 3 = 0$

Step 2 Find the discriminant.

$$b^2 - 4ac = (-5)^2 - 4(4)(3) \quad a = 4, b = -5, \text{ and } c = 3$$
$$= -23 \quad \text{Simplify.}$$

Since the discriminant is negative, the equation has no real solutions.

Check Your Progress

4A. $2x^2 + 11x + 15 = 0$

4B. $9x^2 - 30x + 25 = 0$

Personal Tutor glencoe.com

Check Your Understanding

Examples 1 and 2 pp. 558–559

Solve each equation by using the Quadratic Formula. Round to the nearest tenth if necessary.

1. $x^2 - 2x - 15 = 0$
2. $x^2 - 10x + 16 = 0$
3. $x^2 - 8x = -10$
4. $x^2 + 3x = 12$
5. $10x^2 - 31x + 15 = 0$
6. $5x^2 + 5 = -13x$

Example 3 p. 560

Solve each equation. State which method you used.

7. $2x^2 + 11x - 6 = 0$
8. $2x^2 - 3x - 6 = 0$
9. $9x^2 = 25$
10. $x^2 - 9x = -19$

Example 4 p. 561

State the value of the discriminant for each equation. Then determine the number of real solutions of the equation.

11. $x^2 - 9x + 21 = 0$
12. $2x^2 - 11x + 10 = 0$
13. $9x^2 + 24x = -16$
14. $3x^2 - x = 8$

15. **TRAMPOLINE** Eva is jumping on a trampoline. Her height h in feet can be modeled by the equation $h = -16t^2 + 2.4t + 6$, where t is time in seconds. Use the discriminant to determine if Eva will ever reach a height of 20 feet. Explain.

Practice and Problem Solving

 = **Step-by-Step Solutions** begin on page R12. **Extra Practice** begins on page 815.

Examples 1 and 2 pp. 558–559

Solve each equation by using the Quadratic Formula. Round to the nearest tenth if necessary.

16. $4x^2 + 5x - 6 = 0$
17. $x^2 + 16 = 0$
18. $6x^2 - 12x + 1 = 0$
19. $5x^2 - 8x = 6$
20. $2x^2 - 5x = -7$
21. $5x^2 + 21x = -18$
22. $81x^2 = 9$
23. $8x^2 + 12x = 8$
24. $4x^2 = -16x - 16$
25. $10x^2 = -7x + 6$
26. $-3x^2 = 8x - 12$
27. $2x^2 = 12x - 18$

28. **AMUSEMENT PARKS** The Demon Drop at Cedar Point in Ohio takes riders to the top of a tower and drops them 60 feet. A function that approximates this ride is $h = -16t^2 + 64t - 60$, where h is the height in feet and t is the time in seconds. About how many seconds does it take for riders to drop from 60 feet to 0 feet?

Example 3 p. 560

Solve each equation. State which method you used.

29. $2x^2 - 8x = 12$
30. $3x^2 - 24x = -36$
31. $x^2 - 3x = 10$
32. $4x^2 + 100 = 0$
33. $x^2 = -7x - 5$
34. $12 - 12x = -3x^2$

Example 4 p. 561

State the value of the discriminant for each equation. Then determine the number of real solutions of the equation.

35. $0.2x^2 - 1.5x + 2.9 = 0$
36. $2x^2 - 5x + 20 = 0$
37. $x^2 - \frac{4}{5}x = 3$
38. $0.5x^2 - 2x = -2$
39. $2.25x^2 - 3x = -1$
40. $2x^2 = \frac{5}{2}x + \frac{3}{2}$

41. **INTERNET** The percent of U.S. households with high-speed Internet h can be estimated by $h = -0.2n^2 + 7.2n + 1.5$, where n is the number of years since 1990.
    a. Use the Quadratic Formula to determine when 20% of the population will have high-speed Internet.
    b. Is a quadratic equation a good model for this information? Explain.

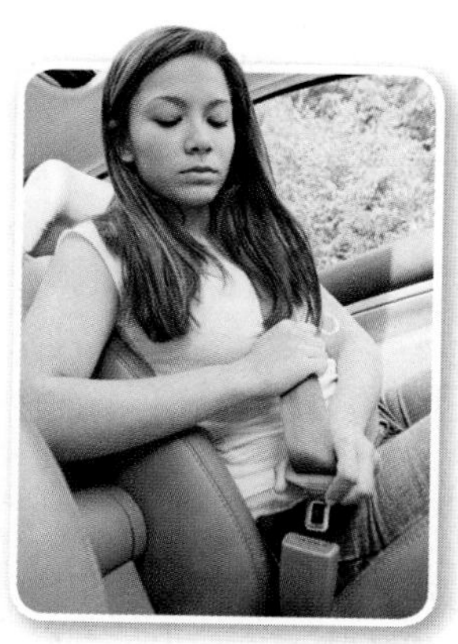

Real-World Link

The risk of motor vehicle crashes is higher among 16- to 19-year-olds than any other age group. Per mile driven, 16- to 19-year-olds are four times more likely than older drivers to crash.

Source: Centers for Disease Control and Prevention

42. TRAFFIC The equation $d = 0.05v^2 + 1.1v$ models the distance d in feet it takes a car traveling at a speed of v miles per hour to come to a complete stop. The speed limit on some highways is 65 miles per hour. If Hannah's car stopped after 250 feet, was she speeding? Explain your reasoning.

Without graphing, determine the number of x-intercepts of the graph of the related function for each function.

43. $4.25x + 3 = -3x^2$

44. $x^2 + \frac{2}{25} = \frac{3}{5}x$

45. $0.25x^2 + x = -1$

Solve each equation by using the Quadratic Formula. Round to the nearest tenth if necessary.

46. $-2x^2 - 7x = -1.5$

47. $2.3x^2 - 1.4x = 6.8$

48. $x^2 - 2x = 5$

49. POSTER Bartolo is making a poster for the dance. He wants to cover three fourths of the area with text.

a. Write an equation for the area of the section with text.

b. Solve the equation by using the Quadratic Formula.

c. What should be the margins of the poster?

50. MULTIPLE REPRESENTATIONS In this problem, you will investigate *exponential* functions.

a. TABULAR Copy and complete the table.

b. GRAPHICAL Construct a graph from the information given in the table using the points (time, number of bacteria). Is the graph linear or quadratic?

c. ANALYTICAL What happens to the number of bacteria after every hour? Write a function that models the pattern in the table.

Time (hours)	Number of Bacteria
0	$1 = 2^0$
1	$2 = 2^1$
2	$4 = 2^2$
3	
4	
5	
6	

H.O.T. Problems

Use **H**igher-**O**rder **T**hinking Skills

51. CHALLENGE Find all values of k such that $2x^2 - 3x + 5k = 0$ has two solutions.

52. REASONING Use factoring techniques to determine the number of real zeros of $f(x) = x^2 - 8x + 16$. Compare this method to using the discriminant.

REASONING Determine whether there are *two*, *one*, or *no* real solutions.

53. The graph of a quadratic function does not have an x-intercept.

54. The graph of a quadratic function is tangent at the x-axis.

55. The graph of a quadratic function intersects the x-axis twice.

56. Both a and b are greater than 0 and c is less than 0 in a quadratic equation.

57. OPEN ENDED Write a quadratic function that has a positive discriminant, one with a negative discriminant, and one with a zero discriminant.

58. WRITING IN MATH Describe the advantages and disadvantages of each method of solving quadratic equations. Which method do you prefer, and why?

Standardized Test Practice

59. If n is an even integer, which expression represents the product of three consecutive even integers?

A $n(n + 1)(n + 2)$

B $(n + 1)(n + 2)(n + 3)$

C $3n + 2$

D $n(n + 2)(n + 4)$

60. SHORT RESPONSE The triangle shown is an isosceles triangle. What is the value of x?

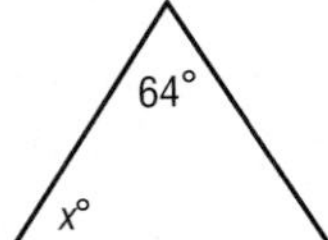

61. Which statement best describes the graph of $x = 5$?

F It is parallel to the x-axis.

G It is parallel to the y-axis.

H It passes through the point (2, 5).

J It has a y-intercept of 5.

62. What are the solutions of the quadratic equation $6h^2 + 6h = 72$?

A 3 or −4

B −3 or 4

C no solution

D 12 or −48

Spiral Review

Solve each equation by completing the square. Round to the nearest tenth if necessary. (Lesson 9-4)

63. $6x^2 - 17x + 12 = 0$

64. $x^2 - 9x = -12$

65. $4x^2 = 20x - 25$

Describe the transformations needed to obtain the graph of $g(x)$ from the graph of $f(x)$. (Lesson 9-3)

66. $f(x) = 4x^2$
$g(x) = 2x^2$

67. $f(x) = x^2 + 5$
$g(x) = x^2 - 1$

68. $f(x) = x^2 - 6$
$g(x) = x^2 + 3$

Determine whether each graph shows a *positive correlation*, a *negative correlation*, or *no correlation*. If there is a positive or negative correlation, describe its meaning in the situation. (Lesson 4-4)

69. **Electronic Tax Returns**

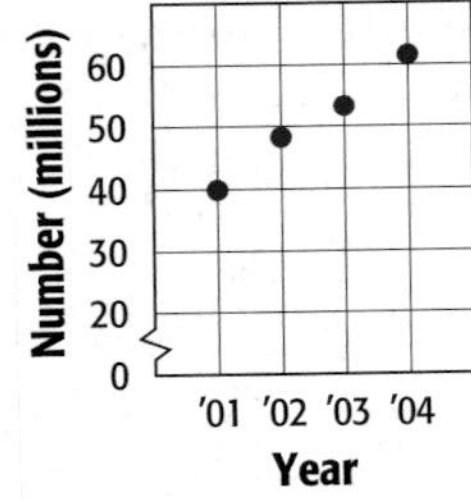

70. **Atlantic Hurricanes**

Number

10, 8, 6, 4, 2, 0

'92 '94 '96 '98 '00 '02 '04 '06

Year

71. ENTERTAINMENT Coach Washington wants to take her softball team out for pizza and soft drinks after the last game of the season. A large pizza costs \$12 and a pitcher of a soft drink costs \$3. She does not want to spend more than \$60. Write an inequality that represents this situation and graph the solution set. (Lesson 5-6)

Skills Check

Evaluate $a(b^x)$ for each of the given values. (Lesson 1-2)

72. $a = 1, b = 2, x = 4$

73. $a = 4, b = 1, x = 7$

74. $a = 5, b = 3, x = 0$

75. $a = 0, b = 6, x = 8$

76. $a = -2, b = 3, x = 1$

77. $a = -3, b = 5, x = 2$

Graphing Technology Lab: Cubic Functions

Math Online glencoe.com
- Other Calculator Keystrokes
- Graphing Technology Personal Tutor

You have studied linear functions and monomials. Some functions can be defined by the sums of monomials. One function that can be defined this way is a cubic function. A **cubic equation** has the form $ax^3 + bx^2 + cx + d = 0$, where $a \neq 0$. All cubic equations have at least one but no more than three real roots.

ACTIVITY

Solve $x^3 - 6x^2 + 3x + 10 = 0$ by graphing.

Step 1 Enter the related function in the Y= list.

KEYSTROKES: [Y=] [X,T,θ,n] [^] 3 [−] 6 [X,T,θ,n] [x^2] [+] 3 [X,T,θ,n] [+] 10

Step 2 Graph the function in the standard viewing window.

KEYSTROKES: [Zoom] 6

Step 3 Find the zeros of the function by determining where the graph crosses the x-axis. Notice that this graph crosses the x-axis three times. Therefore, there are 3 real solutions for the equation.

[−10, 10] scl: 1 by [−10, 10] scl: 1

KEYSTROKES: [2nd] [CALC] 2

Press the left arrow to move to the left of the intercept closest to the origin.

Press [ENTER].

Press the right arrow to move to the right of the intercept.

Do not go past another intercept. Press [ENTER].

Notice the arrows above the intercept. The intercept you are finding should be between these two arrows.

Press the left arrow to move as close as possible to the intercept. Press [ENTER].

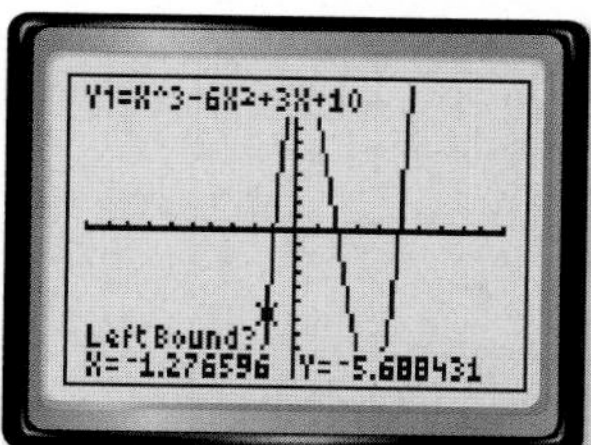

[−10, 10] scl:1 by [−10, 10] scl:1

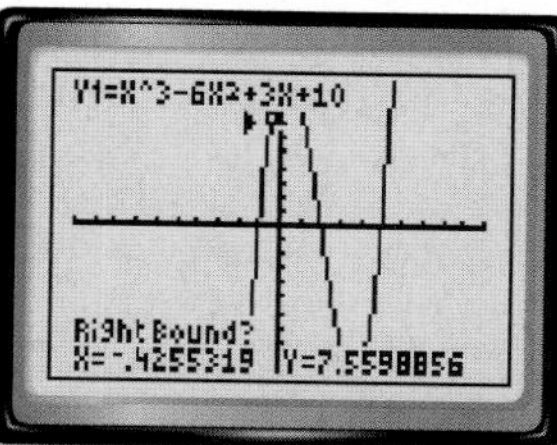

[−10, 10] scl:1 by [−10, 10] scl:1

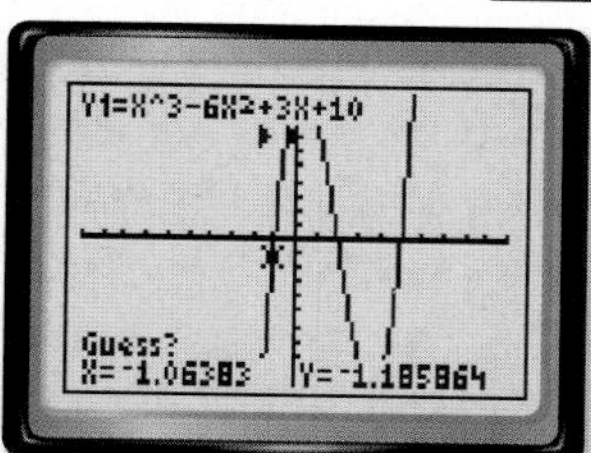

[−10, 10] scl:1 by [−10, 10] scl:1

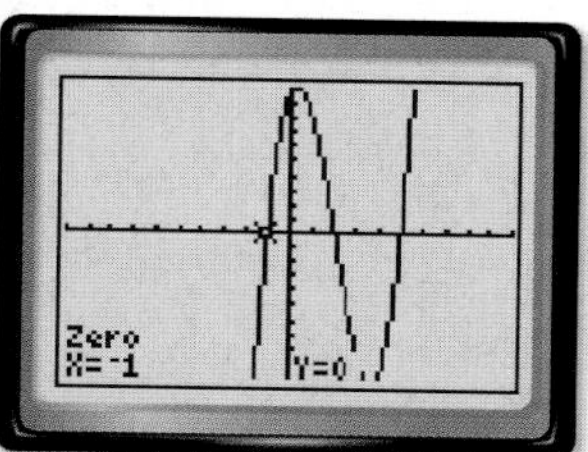

[−10, 10] scl:1 by [−10, 10] scl:1

One root is $x = -1$.

Step 4 Repeat Step 3 for each additional root.

The solutions for $x^3 - 6x^2 + 3x + 10 = 0$ are $x = -1$, 2, and 5.

Exercises

Solve each equation by graphing.

1. $x^3 - 4x^2 - 9x + 36 = 0$
2. $x^3 - 6x^2 - 6x - 7 = 0$
3. $x^3 + x^2 + x - 3 = 0$
4. $x^3 - 5x^2 - 2x + 24 = 0$

CHAPTER 9

Mid-Chapter Quiz

Lessons 9-1 through 9-5

Math Online glencoe.com
Math *in Motion,* Animation

Use a table of values to graph each equation. State the domain and range. (Lesson 9-1)

1. $y = x^2 + 3x + 1$
2. $y = 2x^2 - 4x + 3$
3. $y = -x^2 - 3x - 3$
4. $y = -3x^2 - x + 1$

Consider $y = x^2 - 5x + 4$. (Lesson 9-1)

5. Write the equation of the axis of symmetry.
6. Find the coordinates of the vertex. Is it a maximum or minimum point?
7. Graph the function.
8. **SOCCER** A soccer ball is kicked from ground level with an initial upward velocity of 90 feet per second. The equation $h = -16t^2 + 90t$ gives the height h of the ball after t seconds. (Lesson 9-1)
   a. What is the height of the ball after one second?
   b. How many seconds will it take for the ball to reach its maximum height?
   c. When is the height of the ball 0 feet? What do these points represent in this situation?

Solve each equation by graphing. If integral roots cannot be found, estimate the roots to the nearest tenth. (Lesson 9-2)

9. $x^2 + 5x + 6 = 0$
10. $x^2 + 8 = -6x$
11. $-x^2 + 3x - 1 = 0$
12. $x^2 = 12$
13. **BASEBALL** Juan hits a baseball. The equation $h = -16t^2 + 120t$ models the height h, in feet, of the ball after t seconds. How long is the ball in the air? (Lesson 9-2)
14. **CONSTRUCTION** Christopher is repairing the roof on a shed. He accidentally dropped a box of nails from a height of 14 feet. This is represented by the equation $h = -16t^2 + 14$, where h is the height in feet and t is the time in seconds. Describe how the graph is related to $h = t^2$. (Lesson 9-3)

Describe how the graph of each function is related to the graph of $f(x) = x^2$. (Lesson 9-3)

15. $g(x) = x^2 + 3$
16. $h(x) = 2x^2$
17. $g(x) = x^2 - 6$
18. **MULTIPLE CHOICE** Which is an equation for the function shown in the graph? (Lesson 9-3)

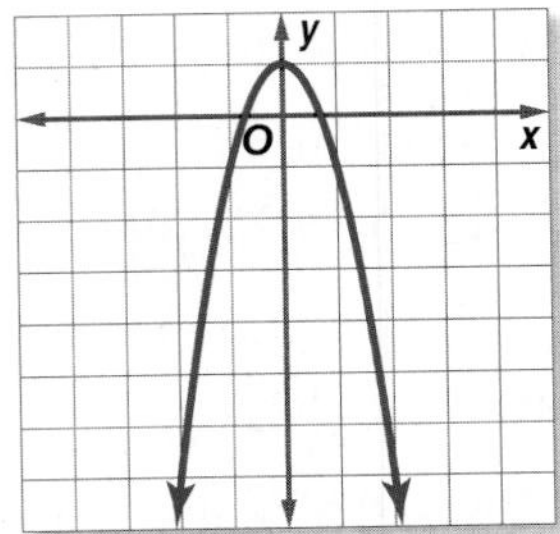

A $y = -2x^2$
B $y = 2x^2 + 1$
C $y = x^2 - 1$
D $y = -2x^2 + 1$

Solve each equation by completing the square. Round to the nearest tenth. (Lesson 9-4)

19. $x^2 + 4x + 2 = 0$
20. $x^2 - 2x - 10 = 0$
21. $2x^2 + 4x - 5 = 7$

Solve each equation by using the Quadratic Formula. Round to the nearest tenth if necessary. (Lesson 9-5)

22. $x^2 - 3x - 18 = 0$
23. $x^2 - 10x = -24$
24. $2x^2 + 5x - 3 = 0$
25. **PARTIES** Della's parents are throwing a Sweet 16 party for her. At 10:00, a ball will slide 25 feet down a pole and light up. A function that models the drop is $h = -t^2 + 5t + 25$, where h is height in feet of the ball after t seconds. How many seconds will it take for the ball to reach the bottom of the pole? (Lesson 9-5)

25 ft

9-6 Exponential Functions

Then

You simplified numerical expressions involving exponents. (Lesson 1-2)

Now

- Graph exponential functions.
- Identify data that display exponential behavior.

New Vocabulary

exponential function

Math Online

glencoe.com

- Extra Examples
- Personal Tutor
- Self-check Quiz
- Homework Help

Why?

Tarantulas can appear scary with their large hairy bodies and legs, but they are harmless to humans. The graph shows a tarantula spider population that increases over time. Notice that the graph is neither linear nor quadratic.

The graph represents the function $y = 3(2)^x$. This is an example of an *exponential* function.

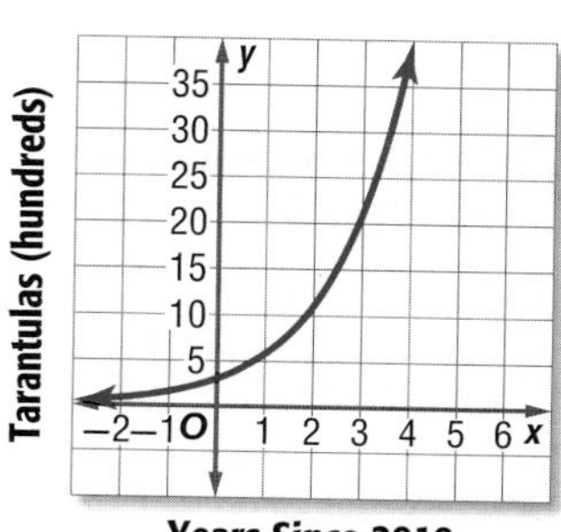

Graph Exponential Functions An **exponential function** is a function of the form $y = ab^x$, where $a \neq 0$, $b > 0$, and $b \neq 1$. Notice that the base is a constant and the exponent is a variable. Exponential functions are nonlinear and nonquadratic functions.

Key Concept Exponential Function

For Your FOLDABLE

Words An exponential function is a function that can be described by an equation of the form $y = ab^x$, where $a \neq 0$, $b > 0$, and $b \neq 1$.

Examples $y = 2(3)^x$ $y = 4^x$ $y = \left(\frac{1}{2}\right)^x$

EXAMPLE 1 Graph with $a > 0$ and $b > 1$

a. Graph $y = 3^x$. Find the y-intercept, and state the domain and range.

x	3^x	y
−2	3^{-2}	$\frac{1}{9}$
−1	3^{-1}	$\frac{1}{3}$
0	3^0	1
1	3^1	3
2	3^2	9

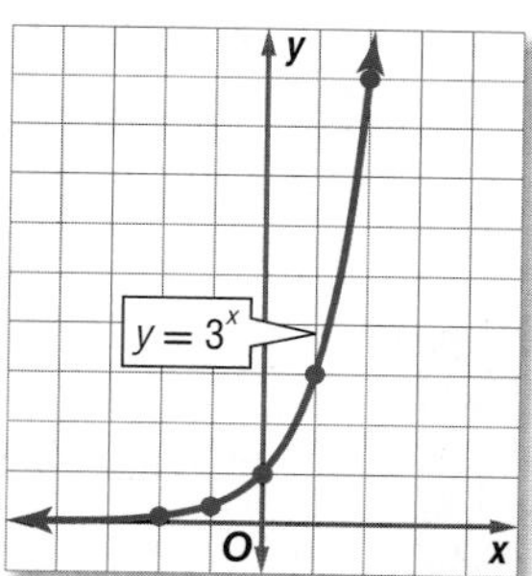

Graph the ordered pairs, and connect the points with a smooth curve.
The graph crosses the y-axis at 1, so the y-intercept is 1.
The domain is all real numbers, and the range is all positive real numbers.

b. Use the graph to approximate the value of $3^{0.7}$.

The graph represents all real values of x and their corresponding values of y for $y = 3^x$. So, when $x = 0.7$, y is about 2. Use a calculator to confirm this value: $3^{0.7} \approx 2.157669$.

Check Your Progress

1A. Graph $y = 7^x$. Find the y-intercept, and state the domain and range.

1B. Use the graph to approximate the value of $y = 7^{0.5}$ to the nearest tenth. Use a calculator to confirm the value.

Personal Tutor glencoe.com

The graphs of functions of the form $y = ab^x$, where $a > 0$ and $b > 1$, all have the same shape as the graph in Example 1. The greater the base or b-value, the faster the graph rises as you move from left to right on the graph. The graphs of functions of the form $y = ab^x$, where $a > 0$ and $0 < b < 1$, also have the same general shape.

StudyTip

$a < 0$ If the value of a is less than 0, the graph will be reflected across the x-axis.

EXAMPLE 2 Graph with $a > 0$ and $0 < b < 1$

a. Graph $y = \left(\frac{1}{3}\right)^x$. Find the y-intercept, and state the domain and range.

x	$\left(\frac{1}{3}\right)^x$	y
−2	$\left(\frac{1}{3}\right)^{-2}$	9
0	$\left(\frac{1}{3}\right)^{0}$	1
2	$\left(\frac{1}{3}\right)^{2}$	$\frac{1}{9}$

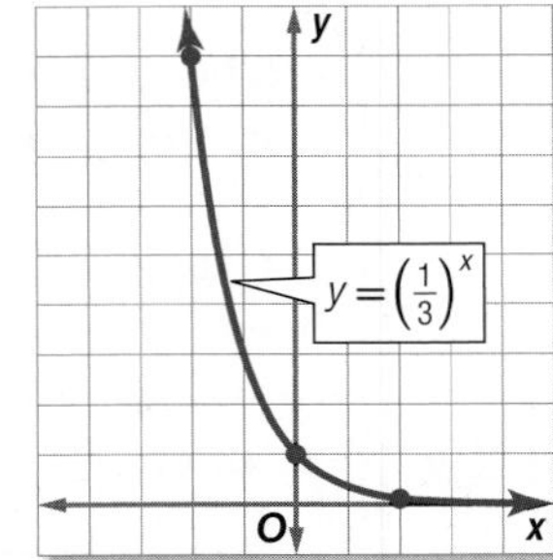

The y-intercept is 1. The domain is all real numbers, and the range is all positive real numbers. Notice that as x increases, the y-values decrease less rapidly.

b. Use the graph to approximate the value of $\left(\frac{1}{3}\right)^{-1.5}$.

When $x = -1.5$, the value of y is about 5. Use a calculator to confirm this value:

KEYSTROKES: [( ] 1 [÷] 3 [ )] [^] −1.5 [ENTER] 5.196152.

Check Your Progress

2A. Graph $y = \left(\frac{1}{2}\right)^x - 1$. Find the y-intercept, and state the domain and range.

2B. Use the graph to approximate the value of $\left(\frac{1}{2}\right)^{-2.5} - 1$ to the nearest tenth. Use a calculator to confirm the value.

Personal Tutor glencoe.com

Exponential functions occur in many real world situations.

Real-World Link

The United States is the largest soda consumer in the world. In a recent year, the United States accounted for one third of the world's total soda consumption.

Source: Worldwatch Institute

Real-World EXAMPLE 3 Use Exponential Functions to Solve Problems

SODA The consumption of soda has increased each year since 2000. The function $C = 179(1.029)^t$ models the amount of soda consumed in the world, where C is the amount consumed in billions of liters and t is the number of years since 2000.

a. Graph the function. What values of C and t are meaningful in the context of the problem?

Since t represents time, $t > 0$. At $t = 0$, the consumption is 179 billion liters. Therefore, in the context of this problem, $C > 179$ is meaningful.

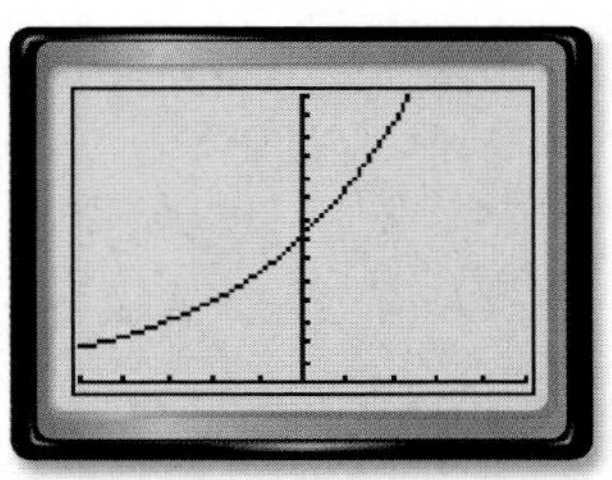

[−50, 50] scl: 10 by [0, 350] scl: 25

b. How much soda was consumed in 2005?

$C = 179(1.029)^t$ **Original equation**

$= 179(1.029)^5$ **$t = 5$**

≈ 206.5 **Use a calculator.**

The world soda consumption in 2005 was approximately 206.5 billion liters.

Check Your Progress

3. A certain bacteria population doubles every 20 minutes. Beginning with 10 cells in a culture, the population can be represented by the function $B = 10(2)^t$, where B is the number of bacteria cells and t is the time in 20 minute increments. How many will there be after 2 hours?

Personal Tutor glencoe.com

Identify Exponential Behavior Recall from Lesson 3-3 that linear functions have a constant rate of change. Exponential functions do not have constant rates of change, but they do have constant ratios.

EXAMPLE 4 Identify Exponential Behavior

Determine whether the set of data shown below displays exponential behavior. Write *yes* or *no*. Explain why or why not.

x	0	5	10	15	20	25
y	64	32	16	8	4	2

Method 1 Look for a pattern.

The domain values are at regular intervals of 5. Look for a common factor among the range values.

$$64 \quad 32 \quad 16 \quad 8 \quad 4 \quad 2$$
$$\times\tfrac{1}{2} \quad \times\tfrac{1}{2} \quad \times\tfrac{1}{2} \quad \times\tfrac{1}{2} \quad \times\tfrac{1}{2}$$

The range values differ by the common factor of $\frac{1}{2}$.

Since the domain values are at regular intervals and the range values differ by a positive common factor, the data are probably exponential. Its equation may involve $\left(\frac{1}{2}\right)^x$.

Method 2 Graph the data.

Plot the points and connect them with a smooth curve. The graph shows a rapidly decreasing value of y as x increases. This is a characteristic of exponential behavior in which the base is between 0 and 1.

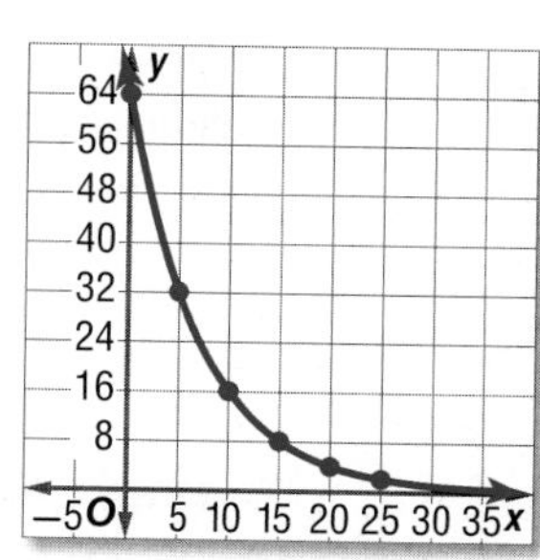

Check Your Progress

4. Determine whether the set of data shown below displays exponential behavior. Write *yes* or *no*. Explain why or why not.

x	0	3	6	9	12	15
y	12	16	20	24	28	32

Personal Tutor glencoe.com

Problem-Solving Tip

Make an Organized List Making an organized list of x-values and corresponding y-values is helpful in graphing the function. It can also help you identify patterns in the data.

Study Tip

Checking Answers The graph of an exponential function may resemble part of the graph of a quadratic function. Be sure to check for a pattern as well as to look at a graph.

Check Your Understanding

Examples 1 and 2
pp. 567–568

Graph each function. Find the y-intercept and state the domain and range. Then use the graph to determine the approximate value of the given expression to the nearest tenth. Use a calculator to confirm the value.

1. $y = 2^x; 2^{1.5}$

2. $y = -5^x; -5^{0.5}$

3. $y = -\left(\frac{1}{5}\right)^x; -\left(\frac{1}{5}\right)^{-0.5}$

4. $y = 3\left(\frac{1}{4}\right)^x; 3\left(\frac{1}{4}\right)^{0.5}$

Graph each function. Find the y-intercept, and state the domain and range.

5. $f(x) = 6^x + 3$

6. $f(x) = 2 - 2^x$

Example 3
pp. 568–569

7. **BIOLOGY** The function $f(t) = 100(1.05)^t$ models the growth of a fruit fly population, where $f(t)$ is the number of flies and t is time in days.

a. What values for the domain and range are reasonable in the context of this situation? Explain.

b. After two weeks, approximately how many flies are in this population?

Example 4
p. 569

Determine whether the set of data shown below displays exponential behavior. Write *yes* or *no*. Explain why or why not.

8.

x	1	2	3	4	5	6
y	−4	−2	0	2	4	6

9.

x	2	4	6	8	10	12
y	1	4	16	64	256	1024

Practice and Problem Solving

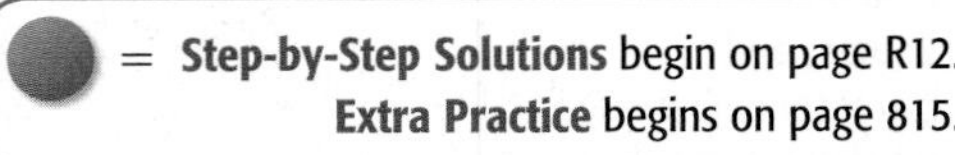
= **Step-by-Step Solutions** begins on page R12.
Extra Practice begins on page 815.

Examples 1 and 2
pp. 567–568

Graph each function. Find the y-intercept and state the domain and range. Then use the graph to determine the approximate value of the given expression to the nearest tenth. Use a calculator to confirm the value.

10. $y = 2 \cdot 8^x, 2(8)^{-0.5}$

11. $y = 2 \cdot \left(\frac{1}{6}\right)^x; 2\left(\frac{1}{6}\right)^{1.5}$

12. $y = \left(\frac{1}{12}\right)^x; \left(\frac{1}{12}\right)^{0.5}$

13. $y = -3 \cdot 9^x, -3(9)^{-0.5}$

14. $y = -4 \cdot 10^x, -4(10)^{-0.5}$

15. $y = 3 \cdot 11^x, 3(11)^{-0.2}$

Graph each function. Find the y-intercept and state the domain and range.

16. $y = 4^x + 3$

17. $y = \frac{1}{2}(2^x - 8)$

18. $y = 5(3^x) + 1$

19. $y = -2(3^x) + 5$

Example 3
pp. 568–569

20. **BIOLOGY** A population of bacteria in a culture increases according to the model $p = 300(2.7)^{0.02t}$, where t is the number of hours and $t = 0$ corresponds to 9:00 A.M.

a. Use this model to estimate the number of bacteria at 11 A.M.

b. Graph the function and name the p-intercept. Describe what the p-intercept represents, and describe a reasonable domain and range for this situation.

Example 4
p. 569

Determine whether the set of data shown below displays exponential behavior. Write *yes* or *no*. Explain why or why not.

21.

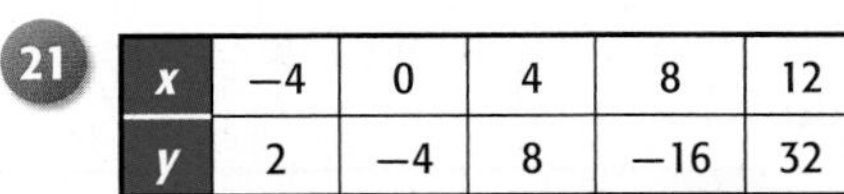

x	−4	0	4	8	12
y	2	−4	8	−16	32

22.

x	−6	−3	0	3
y	5	10	15	20

23.

x	−8	−6	−4	−2
y	0.25	0.5	1	2

24.

x	20	30	40	50	60
y	1	0.4	0.16	0.064	0.0256

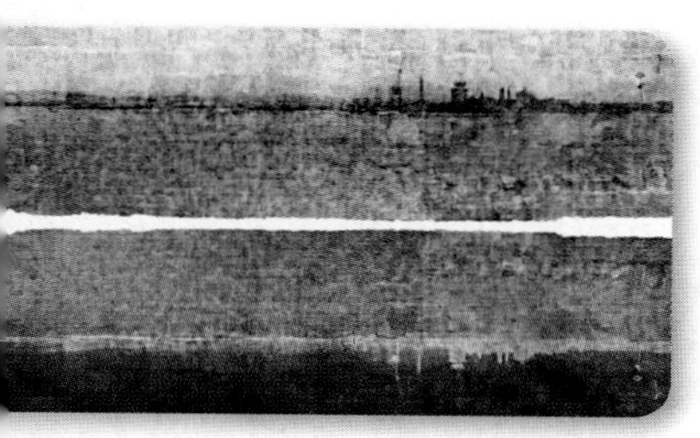

Real-World Link

The world's largest photograph, named The Great Picture, was created by a group of photographers known as The Legacy Project. The photograph has an area of 3375 square feet.

Source: Photoshop Support

25. **PHOTOGRAPHY** Jameka is enlarging a photograph to make a poster for school. She will enlarge the picture repeatedly at 150%. The function $P = 1.5^x$ models the new size of the picture being enlarged, where x is the number of enlargements. How many times as big is the picture after 4 enlargements?

26. **FINANCIAL LITERACY** Daniel invested \$500 into a savings account. The equation $A = 500(1.005)^{12t}$ models the value of Daniel's investment A after t years. How much will Daniel's investment be worth in 8 years?

Identify each function as *linear, quadratic,* or *exponential.*

27. 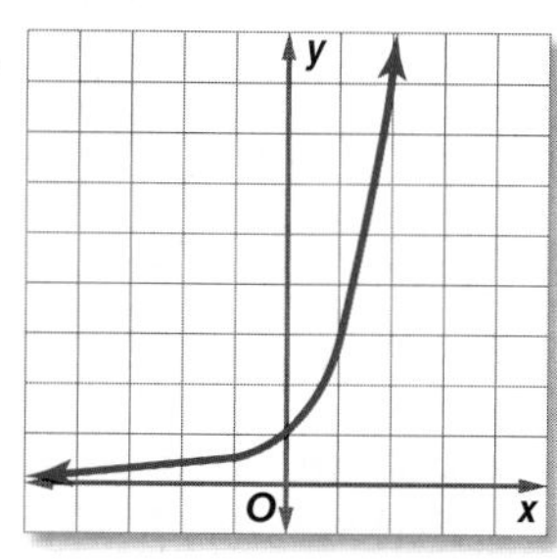

28. 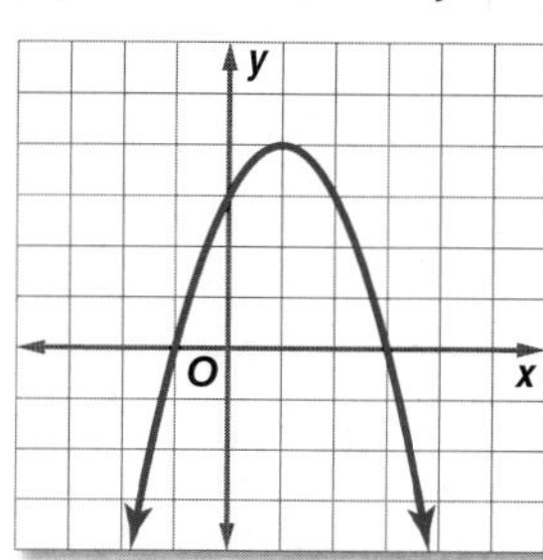

29. 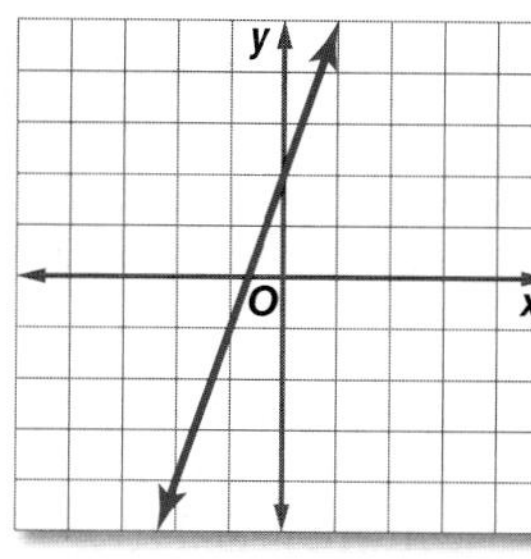

30. $y = 4^x$

31. $y = 2x(x - 1)$

32. $5x + y = 8$

33. **GRADUATION** The number of graduates at a high school has increased by a factor of 1.055 every year since 2001. In 2001, 110 students graduated. The function $N = 110(1.055)^t$ models N, the number of students expected to graduate t years after 2001. How many students will graduate in 2012?

Describe the graph of each equation as a transformation of the graph of $y = 2^x$.

34. $y = 2^x + 6$

35. $y = 3(2)^x$

36. $y = -\frac{1}{4}(2)^x$

37. $y = -3 + 2^x$

38. $y = \left(\frac{1}{2}\right)^x$

39. $y = -5(2)^x$

40. **DEER** The deer population at a national park doubles every year. In 2000, there were 25 deer in the park. The function $N = 25(2)^t$ models the number of deer N in the park t years after 2000. What will the deer population be in 2015?

H.O.T. Problems

Use **H**igher-**O**rder **T**hinking Skills

41. **CHALLENGE** Write an exponential function for which the graph passes through the points at (0, 3) and (1, 6).

42. **REASONING** Determine whether the graph of $y = ab^x$, where $a \neq 0$, $b > 0$, and $b \neq 1$, *sometimes, always,* or *never* has an x-intercept. Explain your reasoning.

43. **OPEN ENDED** Find an exponential function that represents a real-world situation, and graph the function. Analyze the graph.

44. **REASONING** Compare and contrast a function of the form $y = ab^x + c$, where $a \neq 0$, $b > 0$, and $b \neq 1$ and a quadratic function of the form $y = ax^2 + c$.

45. **WRITING IN MATH** Explain how to determine whether a set of data displays exponential behavior.

Standardized Test Practice

46. SHORT RESPONSE What are the zeros of the function graphed below?

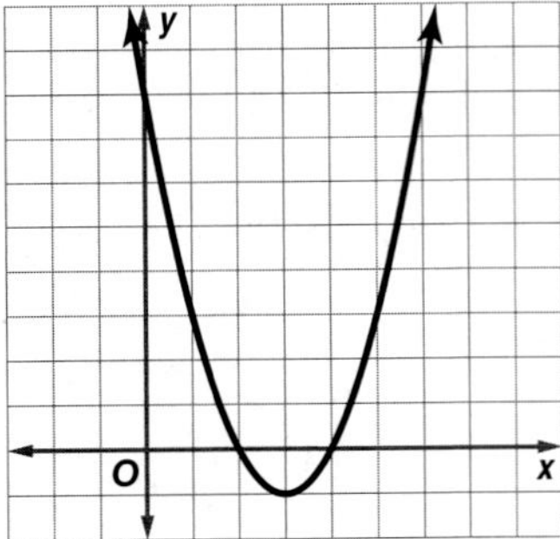

47. Hinto invested \$300 into a savings account. The equation $A = 300(1.005)^{12t}$ models the amount in Hinto's account A after t years. How much will be in Hinto's account after 7 years?

A \$25,326
B \$456.11
C \$385.01
D \$301.52

48. GEOMETRY Ayana placed a circular piece of paper on a square picture as shown below. If the picture extends 4 inches beyond the circle on each side, what is the perimeter of the square picture?

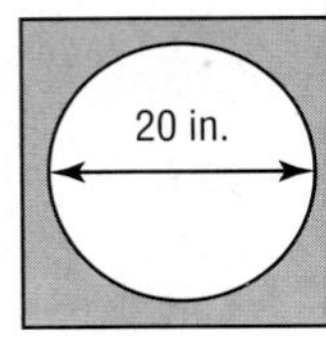

F 64 in.
G 80 in.
H 94 in.
J 112 in.

49. Which of the following shows $4x^2 - 8x - 12$ factored completely?

A $4(x - 3)(x + 1)$
B $4(x + 3)(x - 1)$
C $(4x + 12)(x - 1)$
D $(x - 3)(4x + 4)$

Spiral Review

Solve each equation by using the Quadratic Formula. Round to the nearest tenth if necessary. (Lesson 9-5)

50. $6x^2 - 3x - 30 = 0$
51. $4x^2 + 18x = 10$
52. $2x^2 + 6x = 7$

Solve each equation by taking the square root of each side. Round to the nearest tenth if necessary. (Lesson 9-4)

53. $x^2 = 25$
54. $x^2 + 6x + 9 = 16$
55. $x^2 - 14x + 49 = 15$

Evaluate each product. Express the results in both scientific notation and standard form. (Lesson 7-3)

56. $(1.9 \times 10^2)(4.7 \times 10^6)$
57. $(4.5 \times 10^{-3})(5.6 \times 10^4)$
58. $(3.8 \times 10^{-4})(6.4 \times 10^{-8})$

59. DEMOLITION DERBY When a car hits an object, the damage is measured by the collision impact. For a certain car the collision impact I is given by $I = 2v^2$, where v represents the speed in kilometers per minute. What is the collision impact if the speed of the car is 4 kilometers per minute? (Lesson 7-1)

Use elimination to solve each system of equations. (Lesson 6-3)

60. $x + y = -3$
$x - y = 1$

61. $3a + b = 5$
$2a + b = 10$

62. $3x - 5y = 16$
$-3x + 2y = -10$

Skills Review

Find the next three terms of each arithmetic sequence. (Lesson 3-5)

63. $1, 3, 5, 7, \ldots$
64. $-6, -4, -2, 0, \ldots$
65. $6.5, 9, 11.5, 14, \ldots$
66. $10, 3, -4, -11, \ldots$
67. $\frac{1}{2}, \frac{5}{4}, 2, \frac{11}{4}, \ldots$
68. $1, \frac{3}{4}, \frac{1}{2}, \frac{1}{4}, \ldots$

9-7 Growth and Decay

Then

You analyzed exponential functions. (Lesson 9-6)

Now

- Solve problems involving exponential growth.
- Solve problems involving exponential decay.

New Vocabulary

exponential growth
compound interest
exponential decay

Math Online

glencoe.com

- Extra Examples
- Personal Tutor
- Self-Check Quiz
- Homework Help

Why?

The number of Weblogs or blogs increased at a monthly rate of about 13.7% over 21 months. The average number of blogs per month can be modeled by $y = 1.1(1 + 0.137)^t$ or $y = 1.1(1.137)^t$, where y represents the total number of blogs in millions and t is the number of months since November 2003.

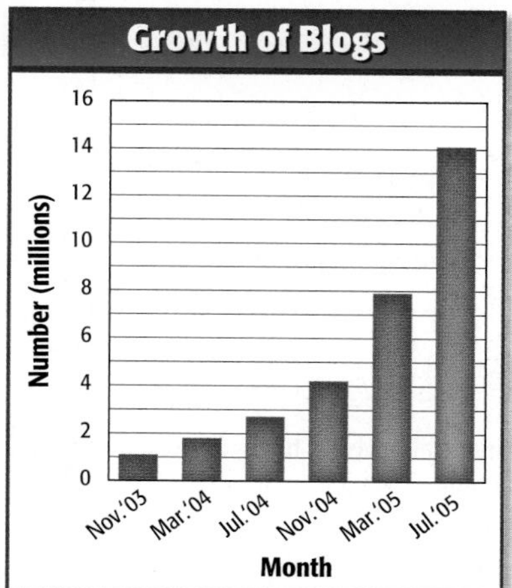

Exponential Growth The equation for the number of blogs is in the form $y = a(1 + r)^t$. This is the general equation for **exponential growth**.

Key Concept: Equation for Exponential Growth

For Your FOLDABLE

$$y = a(1 + r)^t$$

- a is the initial amount.
- t is time.
- y is the final amount.
- r is the rate of change expressed as a decimal, $r > 0$.

Real-World EXAMPLE 1 Exponential Growth

CONTEST A radio station is sponsoring a contest. The prize begins as a \$100 gift card. Once a day, the disc jockey announces a name, and the person has 15 minutes to call. If the person does not call within the allotted time, the prize increases by 2.5%.

a. Write an equation to represent the amount of the gift card in dollars after t days with no winners.

$y = a(1 + r)^t$	Equation for exponential growth
$y = 100(1 + 0.025)^t$	$a = 100$ and $r = 2.5\%$ or 0.025
$y = 100(1.025)^t$	Simplify.

In the equation $y = 100(1.025)^t$, y is the amount of the gift card and t is the number of days since the contest began.

b. How much will the gift card be worth if no one wins after 10 days?

$y = 100(1.025)^t$	Equation for amount of gift card
$= 100(1.025)^{10}$	$t = 10$
≈ 128.01	Use a calculator.

In 10 days, the gift card will be worth \$128.01.

Check Your Progress

1. **TUITION** A college's tuition has risen 5% each year since 2000. If the tuition in 2000 was \$10,850, write an equation for the amount of the tuition t years after 2000. Predict the cost of tuition for this college in 2015.

Personal Tutor glencoe.com

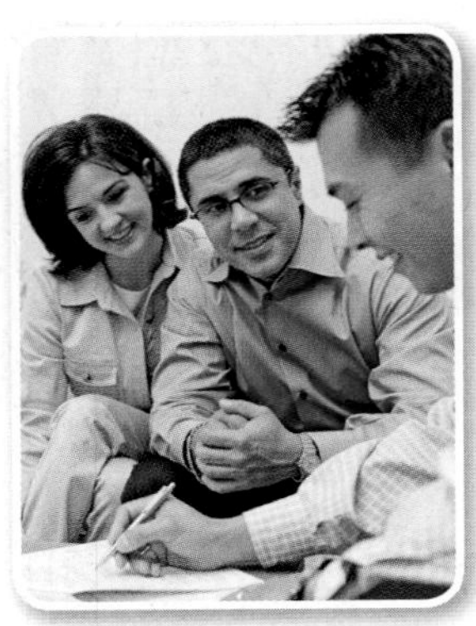

Real-World Career

Financial advisors help people plan their financial futures. A good financial advisor has mathematical, problem-solving, and communication skills. A bachelor's degree is strongly preferred but not required.

Compound interest is interest earned or paid on both the initial investment and previously earned interest. It is an application of exponential growth.

Key Concept: Equation for Compound Interest

For Your FOLDABLE

$$A = P\left(1 + \frac{r}{n}\right)^{nt}$$

- A is the current amount.
- P is the principal or initial amount.
- r is the annual interest rate expressed as a decimal, $r > 0$.
- n is the number of times the interest is compounded each year, and t is time in years.

Real-World EXAMPLE 2 Compound Interest

FINANCE Maria's parents invested \$14,000 at 6% per year compounded monthly. How much money will there be in the account after 10 years?

$A = P\left(1 + \frac{r}{n}\right)^{nt}$	Compound interest equation
$= 14{,}000\left(1 + \frac{0.06}{12}\right)^{12(10)}$	$P = 14{,}000$, $r = 6\%$ or 0.06, $n = 12$, and $t = 10$
$= 14{,}000(1.005)^{120}$	Simplify.
$\approx 25{,}471.55$	Use a calculator.

There will be about \$25,471.55 in 10 years.

Check Your Progress

2. **FINANCE** Determine the amount of an investment if \$300 is invested at an interest rate of 3.5% compounded monthly for 22 years.

Personal Tutor glencoe.com

Exponential Decay In **exponential decay**, the original amount decreases by the same percent over a period of time. A variation of the growth equation can be used as the general equation for exponential decay.

StudyTip

Growth and Decay Since r is added to 1, the value inside the parentheses will be greater than 1 for exponential growth functions. For exponential decay functions, this value will be less than 1 since r is subtracted from 1.

Key Concept: Equation for Exponential Decay

For Your FOLDABLE

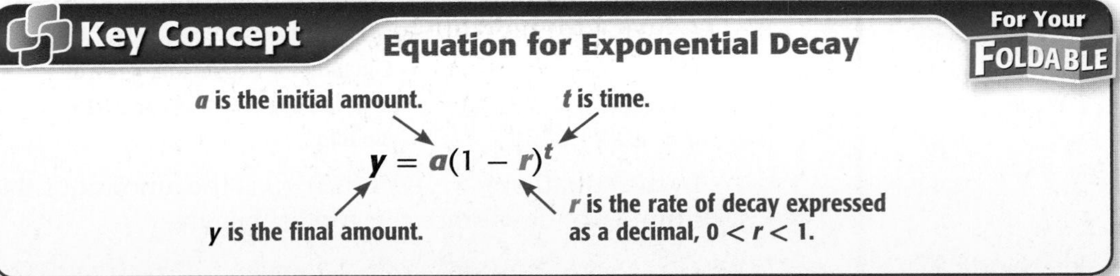

Real-World EXAMPLE 3 Exponential Decay

SWIMMING A fully inflated child's raft for a pool is losing 6.6% of its air every day. The raft originally contained 4500 cubic inches of air.

a. Write an equation to represent the loss of air.

$y = a(1 - r)^t$	Equation for exponential decay
$= 4500(1 - 0.066)^t$	$a = 4500$ and $r = 6.6\%$ or 0.066
$= 4500(0.934)^t$	Simplify.

$y = 4500(0.934)^t$, where y is the air in the raft in cubic inches after t days.

b. **Estimate the amount of air in the raft after 7 days.**

$y = 4500(0.934)^t$ — **Equation for air loss**

$= 4500(0.934)^7$ — $t = 7$

≈ 2790 — **Use a calculator.**

The amount of air in the raft after 7 days will be about 2790 cubic inches.

Check Your Progress

3. **POPULATION** The population of Campbell County, Kentucky, has been decreasing at an average rate of about 0.3% per year. In 2000, its population was 88,647. Write an equation to represent the population since 2000. If the trend continues, predict the population in 2010.

Personal Tutor glencoe.com

Check Your Understanding

Example 1 p. 573

1. **SALARY** Ms. Acosta received a job as a teacher with a starting salary of $34,000. According to her contract, she will receive a 1.5% increase in her salary every year. How much will Ms. Acosta earn in 7 years?

Example 2 p. 574

2. **MONEY** Paul invested $400 into an account with a 5.5% interest rate compounded monthly. How much will Paul's investment be worth in 8 years?

Example 3 p. 574

3. **ENROLLMENT** In 2000, 2200 students attended Polaris High School. The enrollment has been declining 2% annually.

a. Write an equation for the enrollment of Polaris High School t years after 2000.

b. If this trend continues, how many students will be enrolled in 2015?

Practice and Problem Solving

 = **Step-by-Step Solutions** begins on page R12. **Extra Practice** begins on page 815.

Example 1 p. 573

4. **MEMBERSHIPS** The Work-Out Gym sold 550 memberships in 2001. Since then the number of memberships sold has increased 3% annually.

a. Write an equation for the number of memberships sold at Work-Out Gym t years after 2001.

b. If this trend continues, predict how many memberships the gym will sell in 2020.

5. **COMPUTERS** The number of people who own computers has increased 23.2% annually since 1990. If half a million people owned a computer in 1990, predict how many people will own a computer in 2015.

6. **COINS** Camilo purchased a rare coin from a dealer for $300. The value of the coin increases 5% each year. Determine the value of the coin in 5 years.

Example 2 p. 574

7. **INVESTMENTS** Theo invested $6600 at an interest rate of 4.5% compounded monthly. Determine the value of his investment in 4 years.

8. **FINANCE** Paige invested $1200 at an interest rate of 5.75% compounded quarterly. Determine the value of her investment in 7 years.

9. **SAVINGS** Brooke is saving money for a trip to the Bahamas that costs $295.99. She puts $150 into a savings account that pays 7.25% interest compounded quarterly. Will she have enough money in the account after 4 years? Explain.

10. **INVESTMENTS** Jin's investment of $4500 has been losing its value at a rate of 2.5% each year. What will his investment be worth in 5 years?

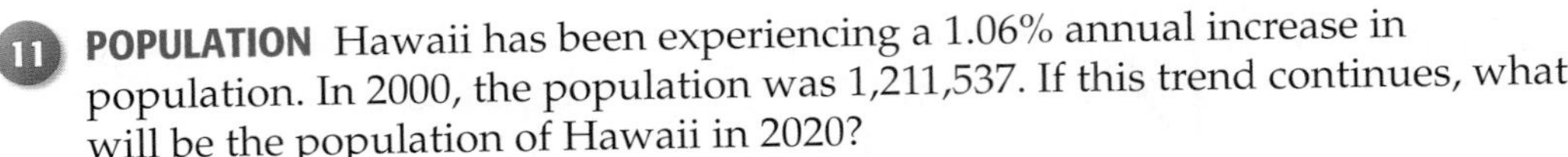

Example 3
p. 574

11. POPULATION Hawaii has been experiencing a 1.06% annual increase in population. In 2000, the population was 1,211,537. If this trend continues, what will be the population of Hawaii in 2020?

12. CARS Leonardo purchases a car for \$18,995. The car depreciates at a rate of 18% annually. After 6 years, Manuel offers to buy the car for \$4500. Should Leonardo sell the car? Explain.

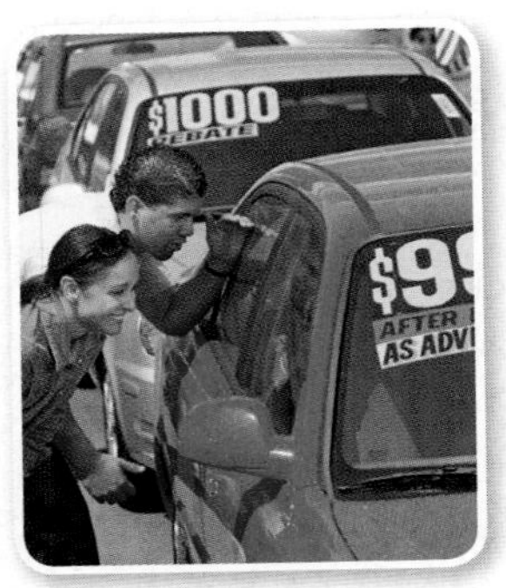

Real-World Link

A car loses 15% to 20% of its value each year. Brand, model, and the condition of the car all contribute to a used car's value. Some brands depreciate much slower than other brands.

Source: Bankrate

13. HOUSING The median house price in the United States increased an average of 8.6% each year between 2002 and 2004. Assume that this pattern continues.

a. Write an equation for the median house price for t years after 2004.

b. Predict the median house price in 2015.

Median House Price	
2002	\$187,600
2003	\$195,000
2004	\$221,000

Source: *Real Estate Journal*

14. ELEMENTS A radioactive element's half-life is the time it takes for one half of the element's quantity to decay. The half-life of Plutonium-241 is 14.4 years. The number of grams A of Plutonium-241 left after t years can be modeled by $A = p(0.5)^{\frac{t}{14.4}}$, where p is the original amount of the element.

a. How much of a 0.2-gram sample of Plutonium-241 remains after 72 years?

b. How much of a 5.4-gram sample of Plutonium-241 remains after 1095 days?

15. FINANCIAL LITERACY Marta is planning to buy a new car. She will finance \$16,000 at an annual interest rate of 7% over a period of 60 months. In the formula, P is the amount of each payment, r is the annual interest rate in decimal form, and t is the time in years of the loan.

$$\text{Amount financed} = P\left[\frac{1-\left(1+\frac{r}{12}\right)^{-12t}}{\frac{r}{12}}\right]$$

a. Use the formula to find her monthly payment.

b. Assuming that she does not pay ahead, what will she have paid on the car?

H.O.T. Problems Use Higher-Order Thinking Skills

16. REASONING Determine the growth rate (as a percent) of a population that quadruples every year. Explain.

17. CHALLENGE Santos invested \$1200 into an account with an interest rate of 8% compounded monthly. Use a calculator to approximate how long it will take for Santos' investment to reach \$2500.

18. REASONING The amount of water in a container doubles every minute. After 8 minutes, the container is full. After how many minutes was the container half full? Explain.

19. OPEN ENDED Create a real-world situation that can be modeled by $y = 200(1.05)^t$.

20. WRITING IN MATH Compare and contrast the exponential growth formula and the exponential decay formula.

Standardized Test Practice

21. GEOMETRY The parallelogram has an area of 35 square inches. Find the height h of the parallelogram.

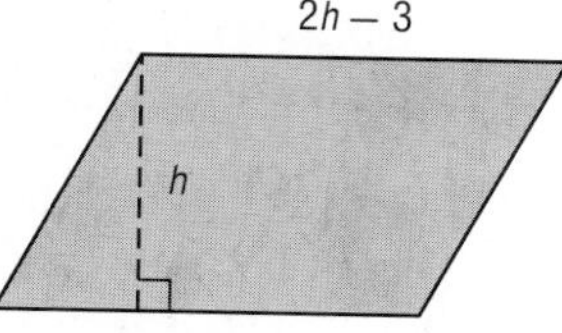

A 3.5 inches
B 4 inches
C 5 inches
D 7 inches

22. What are the roots of $x^2 + 2x = 48$?

F 6 and 8
G -6 and -8
H 6 and -8
J -6 and 8

23. Thi purchased a car for \$22,900. The car depreciated at an annual rate of 16%. Which of the following equations models the value of Thi's car after 5 years?

A $A = 22{,}900(1.16)^5$
B $A = 22{,}900(0.16)^5$
C $A = 16(22{,}900)^5$
D $A = 22{,}900(0.84)^5$

24. GRIDDED RESPONSE A deck measures 12 feet by 18 feet. If a painter charges \$2.65 per square foot, including tax, how much will it cost in dollars to have the deck painted?

Spiral Review

Graph each function. Find the y-intercept and state the domain and range. (Lesson 9-6)

25. $y = 3^x$
26. $y = \left(\frac{1}{2}\right)^x$
27. $y = 6^x$

Solve each equation by using the Quadratic Formula. Round to the nearest tenth if necessary. (Lesson 9-5)

28. $4x^2 + 15x = 25$
29. $3x^2 - 4x = 5$
30. $2x^2 = -2x + 11$
31. $4x^2 + 16x = -16$
32. $5x^2 + 5x = 60$
33. $2x^2 = 3x + 15$

34. EVENT PLANNING A hall does not charge a rental fee as long as at least \$4000 is spent on food. For the prom, the hall charges \$28.95 per person for a buffet. How many people must attend the prom to avoid a rental fee for the hall? (Lesson 5-2)

Determine whether the graphs of each pair of equations are *parallel*, *perpendicular*, or *neither*. (Lesson 4-4)

35. $y = -2x + 11$
$y + 2x = 23$

36. $3y = 2x + 14$
$-3x - 2y = 2$

37. $y = -5x$
$y = 5x - 18$

38. AGES The table shows equivalent ages for horses and humans. Write an equation that relates human age to horse age and find the equivalent horse age for a human who is 16 years old. (Lesson 3-4)

Horse age (x)	0	1	2	3	4	5
Human age (y)	0	3	6	9	12	15

Find the total price of each item. (Lesson 2-7)

39. umbrella: \$14.00
tax: 5.5%

40. sandals: \$29.99
tax: 5.75%

41. backpack: \$35.00
tax: 7%

Skills Check

Graph each set of ordered pairs. (Lesson 1-6)

42. (3, 0), (0, 1), (−4, −6)
43. (0, −2), (−1, −6), (3, 4)
44. (2, 2), (−2, −3), (−3, −6)

9-8 Geometric Sequences as Exponential Functions

Then

You related arithmetic sequences to linear functions. (Lesson 3-5)

Now

- Identify and generate geometric sequences.
- Relate geometric sequences to exponential functions.

New Vocabulary

geometric sequence
common ratio

Math Online

glencoe.com

- Extra Examples
- Personal Tutor
- Self-Check Quiz
- Homework Help

Why?

You send a chain e-mail to a friend who forwards the e-mail to five more people. Each of these five people forwards the e-mail to five more people. The number of new e-mails generated forms a geometric sequence.

Recognize Geometric Sequences The first person generates 5 e-mails. If each of these people sends the e-mail to 5 more people, 25 e-mails are generated. If each of the 25 people sends 5 e-mails, 125 e-mails are generated. The sequence of e-mails generated, 1, 5, 25, 125, … is an example of a **geometric sequence**.

In a geometric sequence, the first term is nonzero and each term after the first is found by multiplying the previous term by a nonzero constant r called the **common ratio**. The common ratio can be found by dividing any term by its previous term.

EXAMPLE 1 Identify Geometric Sequences

Determine whether each sequence is *arithmetic*, *geometric*, or *neither*. Explain.

a. 256, 128, 64, 32, …

Find the ratios of consecutive terms.

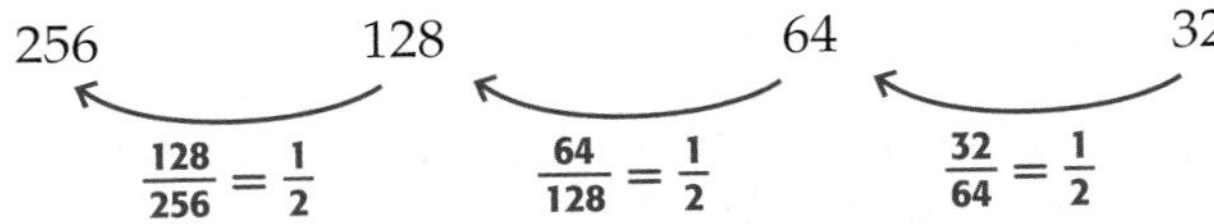

Since the ratios are constant, the sequence is geometric. The common ratio is $\frac{1}{2}$.

b. 4, 9, 12, 18, …

Find the ratios of consecutive terms.

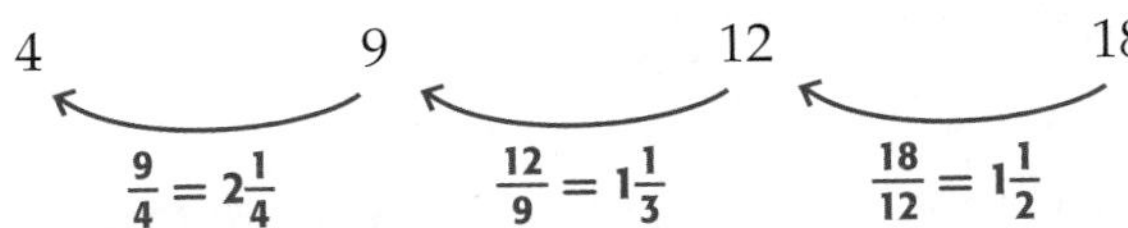

The ratios are not constant, so the sequence is not geometric.

Find the differences of consecutive terms.

4 9 12 18

$9 - 4 = 5$ $12 - 9 = 3$ $18 - 12 = 6$

There is no common difference, so the sequence is not arithmetic. Thus, the sequence is neither geometric nor arithmetic.

Check Your Progress

1A. 1, 3, 9, 27, … **1B.** $-20, -15, -10, -5, \ldots$ **1C.** 2, 8, 14, 22, …

Personal Tutor glencoe.com

Once the common ratio is known, more terms of a sequence can be generated. The recursive formula can be rewritten as $a_n = a_1r^{n-1}$, where n is a counting number and r is the common ratio.

StudyTip

Common Ratio If the terms of a geometric sequence alternate between positive and negative terms or vice versa, the common ratio is negative.

EXAMPLE 2 Find Terms of Geometric Sequences

Find the next three terms in each geometric sequence.

a. 1, −4, 16, −64, ...

Step 1 Find the common ratio.

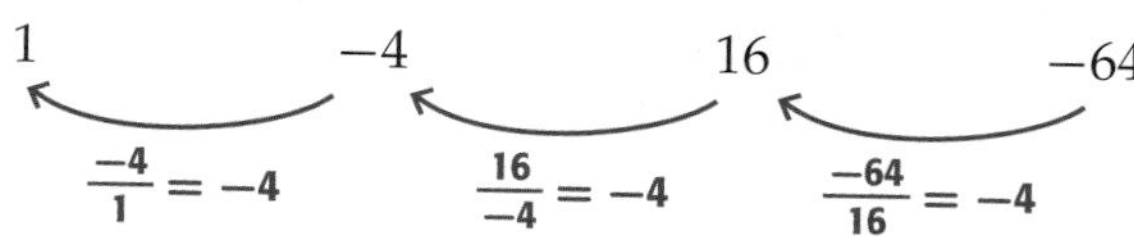

Step 2 Multiply each term by the common ratio to find the next three terms.

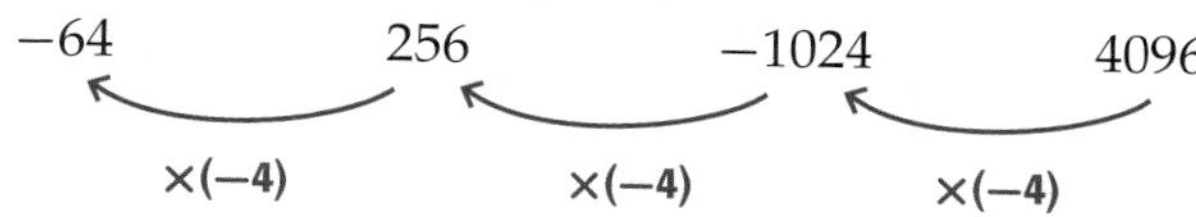

The next three terms are 256, −1024, and 4096.

b. $9, 3, 1, \frac{1}{3}$ ...

Step 1 Find the common ratio.

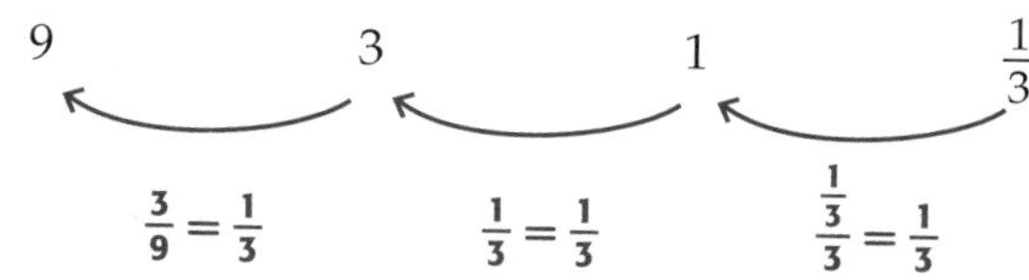

The value of r is $\frac{1}{3}$.

Step 2 Multiply each term by the common ratio to find the next three terms.

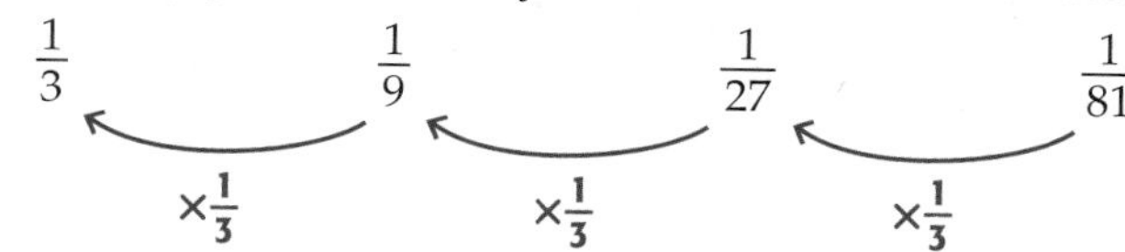

The next three terms are $\frac{1}{9}$, $\frac{1}{27}$, and $\frac{1}{81}$.

Check Your Progress

2A. −3, 15, −75, 375, ...

2B. 24, 36, 54, 81, ...

Personal Tutor glencoe.com

The Granger Collection, New York

Math History Link

Thomas Robert Malthus (1766–1834)
Malthus studied populations and had pessimistic views about the future population of the world. In his work, he stated: "Population increases in a geometric ratio, while the means of subsistence increases in an arithmetic ratio."

Geometric Sequences and Functions Finding the nth term of a geometric sequence would be tedious if we used the above method. The table below shows a rule for finding the nth term of a geometric sequence.

Position, n	1	2	3	4	...	n
Term, a_n	a_1	a_1r	a_1r^2	a_1r^3	...	a_1r^{n-1}

Notice that the common ratio between the terms is r. The table shows that to get the nth term, you multiply the first term by the common ratio r raised to the power $n - 1$. A geometric sequence can be defined by an exponential function in which n is the independent variable, a_n is the dependent variable, and r is the base. The domain is the counting numbers.

Key Concept — *n*th term of a Geometric Sequence

For Your FOLDABLE

The nth term a_n of a geometric sequence with first term a_1 and common ratio r is given by the following formula, where n is any positive integer and $a_1, r \neq 0$.

$$a_n = a_1 r^{n-1}$$

Watch Out!

Negative Common Ratio If the common ratio is negative, as in Example 3, make sure to enclose the common ratio in parentheses. $(-2)^8 \neq -2^8$

EXAMPLE 3 Find the *n*th Term of a Geometric Sequence

a. Write an equation for the nth term of the sequence −6, 12, −24, 48, … .

The first term of the sequence is −6. So, $a_1 = -6$. Now find the common ratio.

$-6 \quad 12 \quad -24 \quad 48$

$\frac{12}{-6} = -2 \qquad \frac{-24}{12} = -2 \qquad \frac{48}{-24} = -2$

The common ratio is −2.

$a_n = a_1 r^{n-1}$ — **Formula for *n*th term**

$a_n = -6(-2)^{n-1}$ — **$a_1 = -6$ and $r = 2$**

b. Find the ninth term of this sequence.

$a_n = a_1 r^{n-1}$ — **Formula for *n*th term**

$a_9 = -6(-2)^{9-1}$ — **For the *n*th term, $n = 9$.**

$= -6(-2)^8$ — **Simplify.**

$= -6(256)$ — **$(-2)^8 = 256$**

$= -1536$

Check Your Progress

3. Write an equation for the nth term of the geometric sequence 96, 48, 24, 12, … . Then find the tenth term of the sequence.

Personal Tutor glencoe.com

Real-World Link

The first NCAA Division I women's basketball tournament was held in 1982. The University of Tennessee has won the most national titles with 8 titles as of 2008.

Source: NCAA Sports

Real-World EXAMPLE 4 Graph a Geometric Sequence

BASKETBALL The NCAA women's basketball tournament begins with 64 teams. In each round, one half of the teams are left to compete, until only one team remains. Draw a graph to represent how many teams are left in each round.

Compared to the previous rounds, one half of the teams remain. So, $r = \frac{1}{2}$. Therefore, the geometric sequence that models this situation is 64, 32, 16, 8, 4, 2, 1. So in round two, 32 teams compete, in round three 16 teams compete and so forth. Use this information to draw a graph.

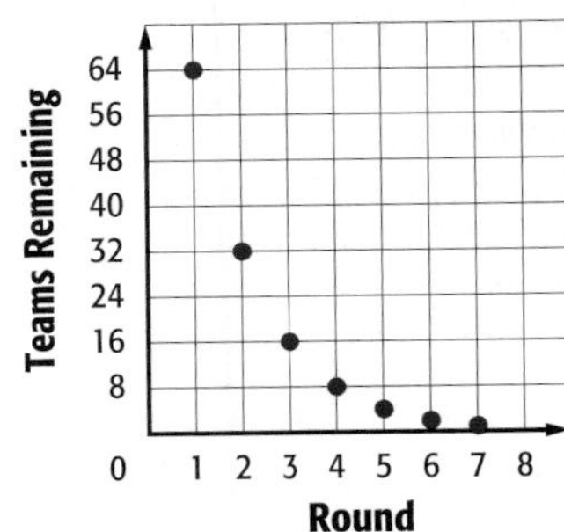

Check Your Progress

4. TENNIS A tennis ball is dropped from a height of 12 feet. Each time the ball bounces back to 80% of the height from which it fell. Draw a graph to represent the height of the ball after each bounce.

Personal Tutor glencoe.com

Check Your Understanding

Example 1 p. 580

Determine whether each sequence is *arithmetic, geometric,* or *neither*. Explain.

1. 200, 40, 8, …
2. 2, 4, 16, …
3. −6, −3, 0, 3, …
4. 1, −1, 1, −1, …

Example 2 p. 581

Find the next three terms in each geometric sequence.

5. 10, 20, 40, 80, …
6. 100, 50, 25, …
7. $4, -1, \frac{1}{4}, \ldots$
8. −7, 21, −63, …

Example 3 p. 582

Write an equation for the *n*th term of each geometric sequence, and find the indicated term.

9. the fifth term of −6, −24, −96, …
10. the seventh term of −1, 5, −25, …
11. the tenth term of 72, 48, 32, …
12. the ninth term of 112, 84, 63, …

Example 4 p. 582

13. **EXPERIMENT** In a physics class experiment, Diana drops a ball from a height of 16 feet. Each bounce has 70% the height of the previous bounce. Draw a graph to represent the height of the ball after each bounce.

Practice and Problem Solving

 = **Step-by-Step Solutions** begin on page R12. **Extra Practice** begins on page 815.

Example 1 p. 580

Determine whether each sequence is *arithmetic, geometric,* or *neither*. Explain.

14. 4, 1, 2, …
15. 10, 20, 30, 40, …
16. 4, 20, 100, …
17. 212, 106, 53, …
18. −10, −8, −6, −4, …
19. 5, −10, 20, 40, …

Example 2 p. 581

Find the next three terms in each geometric sequence.

20. 2, −10, 50, …
21. 36, 12, 4, …
22. 4, 12, 36, …
23. 400, 100, 25, …
24. −6, −42, −294, …
25. 1024, −128, 16, …

Example 3 p. 582

26. The first term of a geometric series is 1 and the common ratio is 9. What is the 8th term of the sequence?
27. The first term of a geometric series is 2 and the common ratio is 4. What is the 14th term of the sequence?
28. What is the 15th term of the geometric series −9, 27, −81, …?
29. What is the 10th term of the geometric series 6, −24, 96, …?

Example 4 p. 582

30. **PENDULUM** The first swing of a pendulum is shown. On each swing after that, the arc length is 60% of the length of the previous swing. Draw a graph that represents the arc length after each swing.

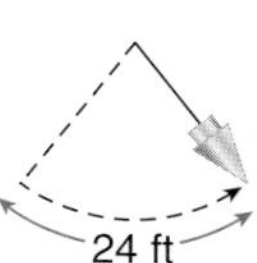

31. Find the eighth term of a geometric sequence for which $a_3 = 81$ and $r = 3$.
32. **MAPS** At an online mapping site, Mr. Mosley notices that when he clicks a spot on the map, the map zooms in on that spot. The magnification increases by 20% each time.
    a. Write a formula for the *n*th term of the geometric sequence that represents the magnification of each zoom level. (*Hint:* The common ratio is not just 0.2.)
    b. What is the fourth term of this sequence? What does it represent?

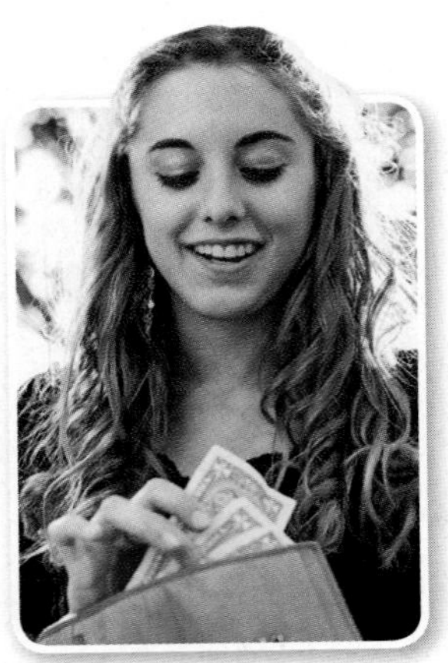

Real-World Link

The average American 9- to 14-year-old gets $9.15 each week for allowance.

Source: *Money Magazine*

33. **ALLOWANCE** Danielle's parents have offered her two different options to earn her allowance for a 9-week period over the summer. She can either get paid $30 each week or $1 the first week, $2 for the second week, $4 for the third week, and so on.

 a. Does the second option form a geometric sequence? Explain.

 b. Which option should Danielle choose? Explain.

34. **SIERPINSKI'S TRIANGLE** Consider the inscribed equilateral triangles at the right. The perimeter of each triangle is one half of the perimeter of the next larger triangle. What is the perimeter of the smallest triangle?

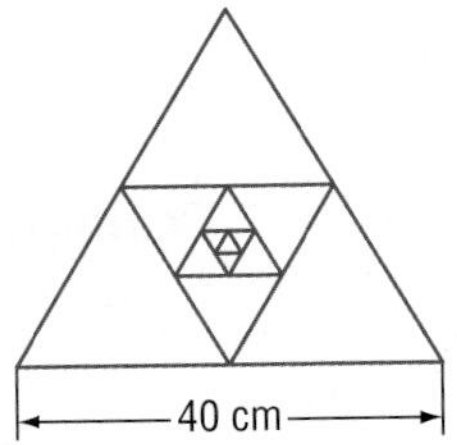

35. If the second term of a geometric sequence is 3 and the third term is 1, find the first and fourth terms of the sequence.

36. If the third term of a geometric sequence is -12 and the fourth term is 24, find the first and fifth terms of the sequence.

37. **EARTHQUAKES** The Richter scale is used to measure the force of an earthquake. The table shows the increase in magnitude for the values on the Richter scale.

Richter Number (x)	Increase in Magnitude (y)	Rate of Change (slope)
1	1	—
2	10	9
3	100	
4	1000	
5	10,000	

 a. Copy and complete the table. Remember that the rate of change is the change in y divided by the change in x.

 b. Plot the ordered pairs (Richter number, increase in magnitude).

 c. Describe the graph that you made of the Richter scale data. Is the rate of change between any two points the same?

 d. Write an exponential equation that represents the Richter scale.

H.O.T. Problems

Use **H**igher-**O**rder **T**hinking Skills

38. **CHALLENGE** Write a sequence that is both geometric and arithmetic. Explain your answer.

39. **FIND THE ERROR** Haro and Matthew are finding the ninth term of the geometric sequence $-5, 10, -20, \ldots$. Is either of them correct? Explain your reasoning.

Haro	Matthew
$r = \frac{10}{-5}$ or -2	$r = \frac{10}{-5}$ or -2
$a_9 = -5(-2)^{9-1}$	$a_9 = -5 \cdot (-2)^{9-1}$
$= -5(512)$	$= -5 \cdot -256$
$= -2560$	$= 1280$

40. **REASONING** Write a sequence of numbers that form a pattern but are neither arithmetic nor geometric. Explain the pattern.

41. **OPEN ENDED** Write a geometric sequence that has a common ratio of $\frac{3}{4}$.

42. **WRITING IN MATH** Summarize how to find a specific term of a geometric sequence.

Standardized Test Practice

43. Find the eleventh term of the sequence 3, −6, 12, −24, … .

A 1024
B 3072
C 33
D −6144

44. What is the total amount of the investment shown in the table below if interest is compounded monthly?

Principal	\$500
Length of Investment	4 years
Annual Interest Rate	5.25%

F \$613.56
G \$616.00
H \$616.56
J \$718.75

45. SHORT RESPONSE Gloria has \$6.50 in quarters and dimes. If she has 35 coins in total, how many of each coin does she have?

46. A sidewalk is being built along the inside edges of all four sides of a rectangular lawn. The lawn is 32 feet long and 24 feet wide. The remaining lawn will have an area of 425 square feet. How wide will the sidewalk be?

A 3.5 feet
B 17 feet
C 24.5 feet
D 25 feet

Spiral Review

Find the next three terms in each geometric sequence. (Lesson 9-7)

47. 2, 6, 18, 54, …
48. −5, −10, −20, −40, …
49. $1, -\frac{1}{2}, \frac{1}{4}, -\frac{1}{8}, \ldots$
50. −3, 1.5, −0.75, 0.375, …
51. 1, 0.6, 0.36, 0.216, …
52. 4, 6, 9, 13.5, …

Graph each function. Find the y-intercept and state the domain and range. (Lesson 9-6)

53. $y = \left(\frac{1}{4}\right)^x - 5$
54. $y = 2(4)^x$
55. $y = \frac{1}{2}(3^x)$

56. LANDSCAPING A blue spruce grows an average of 6 inches per year. A hemlock grows an average of 4 inches per year. If a blue spruce is 4 feet tall and a hemlock is 6 feet tall, write a system of equations to represent their growth. Find and interpret the solution in the context of the situation. (Lesson 6-2)

57. MONEY City Bank requires a minimum balance of \$1500 to maintain free checking services. If Mr. Hayashi is going to write checks for the amounts listed in the table, how much money should he start with in order to have free checking? (Lesson 5-1)

Check	Amount
750	\$1300
751	\$947

Write an equation in slope-intercept form of the line with the given slope and y-intercept. (Lesson 4-1)

58. slope: 4, y-intercept: 2
59. slope: −3, y-intercept: $-\frac{2}{3}$
60. slope: $-\frac{1}{4}$, y-intercept: −5
61. slope: $\frac{1}{2}$, y-intercept: −9
62. slope: $-\frac{2}{5}$, y-intercept: $\frac{3}{4}$
63. slope: −6, y-intercept: −7

Skills Check

Evaluate $a(1 + r)^t$ to the nearest hundredth for each of the given values. (Lesson 1-2)

64. $a = 20, r = 0.25, t = 5$
65. $a = 1000, r = 0.65, t = 4$
66. $a = 200, r = 0.35, t = 8$
67. $a = 60, r = 0.2, t = 10$
68. $a = 8, r = 0.5, t = 2$
69. $a = 500, r = 0.55, t = 12$

9-9

Analyzing Functions with Successive Differences

Then

You graphed linear, quadratic, and exponential functions. (Lessons 3-2, 9-1, 9-6)

Now

- Identify linear, quadratic, and exponential functions from given data.
- Write equations that model data.

Math Online

glencoe.com

- Extra Examples
- Personal Tutor
- Self-Check Quiz
- Homework Help

Why?

Every year the golf team sells candy to raise money for charity. By knowing what type of function models the sales of the candy, they can determine the best price of the candy.

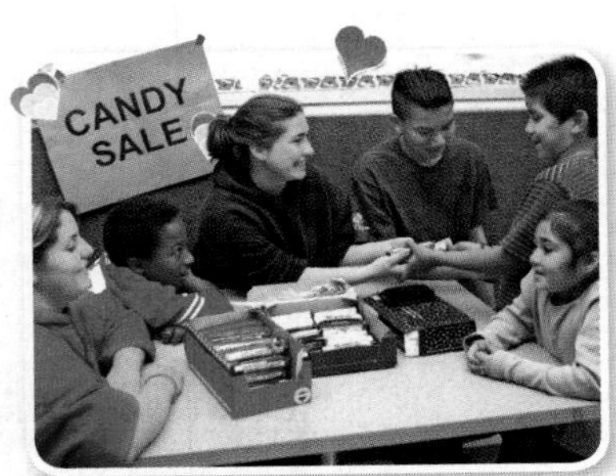

Identify Functions You can use linear functions, quadratic functions, and exponential functions to model data. The general forms of the equations and a graph of each function type are listed below.

Concept Summary — Linear and Nonlinear Functions — For Your FOLDABLE

Linear Function	Quadratic Function	Exponential Function
$y = mx + b$	$y = ax^2 + bx + c$	$y = ab^x$, when $b > 0$
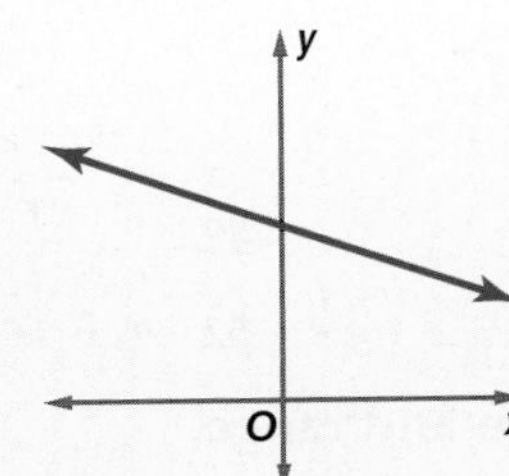	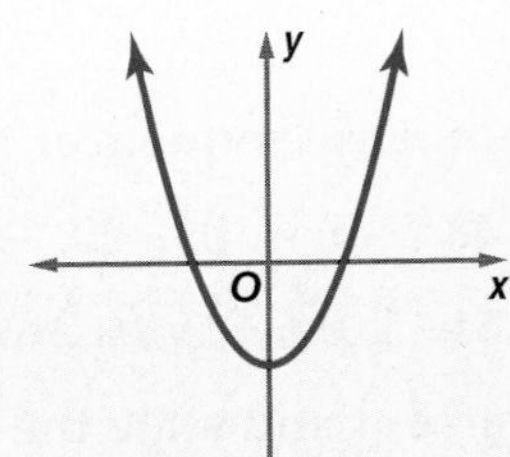	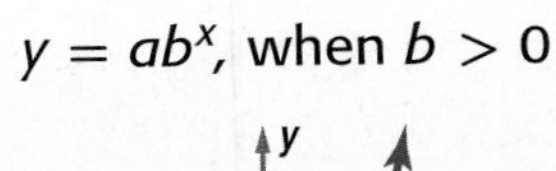 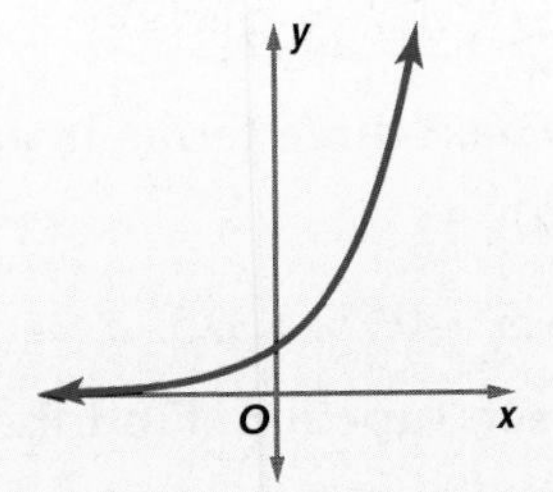

EXAMPLE 1 Choose a Model Using Graphs

Graph each set of ordered pairs. Determine whether the ordered pairs represent a *linear* function, a *quadratic* function, or an *exponential* function.

a. $\{(-2, 5), (-1, 2), (0, 1), (1, 2), (2, 5)\}$

The ordered pairs appear to represent a quadratic function.

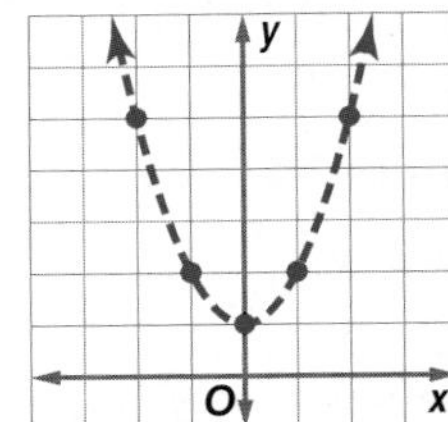

b. $\left\{\left(-2, \frac{1}{4}\right), \left(-1, \frac{1}{2}\right), (0, 1), (1, 2), (2, 4)\right\}$

The ordered pairs appear to represent an exponential function.

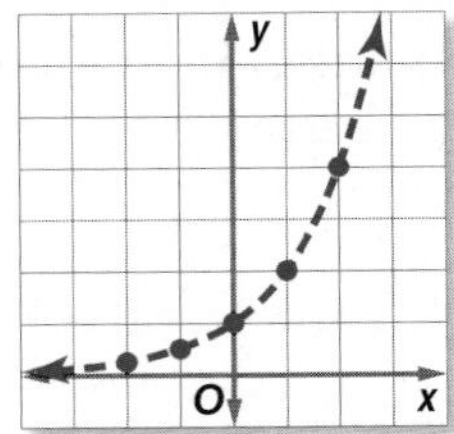

Check Your Progress

1A. $(-2, -3), (-1, -1), (0, 1), (1, 3)$

1B. $(-1, 0.25), (0, 1), (1, 4), (2, 16)$

Personal Tutor glencoe.com

Another way to determine which model best describes data is to use patterns. The differences of successive y-values are called *first differences*. The differences of successive first differences are called *second differences*.

- If the differences of successive y-values are all equal, the data represent a linear function.
- If the second differences are all equal, but the first differences are not equal, the data represent a quadratic function.
- If the ratios of successive y-values are all equal and $r \neq 1$, the data represent an exponential function.

> **Watch Out!**
>
> **x-Values** Before you check for successive differences or ratios, make sure the x-values are increasing by the same amount.

EXAMPLE 2 Choose a Model Using Differences or Ratios

Look for a pattern in each table of values to determine which kind of model best describes the data.

a.

x	−2	−1	0	1	2
y	−8	−3	2	7	12

First differences:

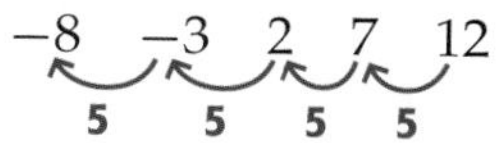

Since the first differences are all equal, the table of values represents a linear function.

b.

x	−1	0	1	2	3
y	8	4	2	1	0.5

First differences: 8 4 2 1 0.5
−4 −2 −1 −0.5

The first differences are not all equal. So, the table of values does not represent a linear function. Find the second differences and compare.

First differences: −4 −2 −1 −0.5

Second differences: **2 1 0.5**

The second differences are not all equal. So, the table of values does not represent a quadratic function. Find the ratios of the y-values and compare.

8 4 2 1 0.5

Ratios: $\frac{4}{8} = \frac{1}{2}$ $\frac{2}{4} = \frac{1}{2}$ $\frac{1}{2}$ $\frac{0.5}{1} = \frac{1}{2}$

The ratios of successive y-values are equal. Therefore, the table of values can be modeled by an exponential function.

Check Your Progress

2A.

x	−3	−2	−1	0	1
y	−3	−7	−9	−9	−7

2B.

x	−2	−1	0	1	2
y	−18	−13	−8	−3	2

Personal Tutor glencoe.com

Write Equations Once you find the model that best describes the data, you can write an equation for the function. For a quadratic function in this lesson, the equation will have the form $y = ax^2$.

EXAMPLE 3 Write an Equation

Determine which kind of model best describes the data. Then write an equation for the function that models the data.

x	−4	−3	−2	−1	0
y	32	18	8	2	0

Step 1 Determine which model fits the data.

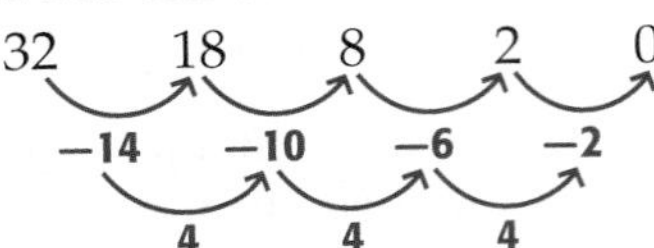

First differences: −14 −10 −6 −2

Second differences: 4 4 4

Since the second differences are equal, a quadratic function models the data.

Step 2 Write an equation for the function that models the data.

The equation has the form $y = ax^2$. Find the value of a by choosing one of the ordered pairs from the table of values. Let's use $(-1, 2)$.

$y = ax^2$ **Equation for quadratic function**

$2 = a(-1)^2$ **$x = -1$ and $y = 2$**

$2 = a$ An equation that models the data is $y = 2x^2$.

> **Watch Out!**
>
> **Finding *a***
> In Example 3, the point (0, 0) cannot be used to find the value of a. You will have to divide each side by 0, giving you an undefined value for a.

Check Your Progress

3A.

x	−2	−1	0	1	2
y	11	7	3	−1	−5

3B.

x	−3	−2	−1	0	1
y	0.375	0.75	1.5	3	6

Personal Tutor glencoe.com

Real-World EXAMPLE 4 Write an Equation for a Real-World Situation

BOOK CLUB The table shows the number of book club members for four consecutive years. Determine which model best represents the data. Then write a function that models the data.

Time (years)	0	1	2	3	4
Members	5	10	20	40	80

Understand We need to find a model for the data, and then write a function.

Plan Find a pattern using successive differences or ratios. Then use the general form of the equation to write a function.

Solve The constant ratio is 2. This is the value of the base. An exponential function of the form $y = ab^x$ models the data.

$y = ab^x$ **Equation for exponential function**

$5 = a(2)^0$ **$x = 0$, $y = 5$, and $b = 2$**

$5 = a$ The equation that models the data is $y = 5 \cdot 2^x$.

Check You used (0, 5) to write the function. Verify that every other ordered pair satisfies the equation.

> **Real-World Link**
>
> A poll by the National Education Association found that 87% of all teens polled found reading relaxing, 85% viewed reading as rewarding, and 79% found reading exciting.
>
> **Source:** *American Demographics*

Check Your Progress

4. ADVERTISING The table shows the cost of placing an ad in a newspaper. Determine a model that best represents the data and write a function that models the data.

No. of Lines	5	6	7	8
Total Cost ($)	14.50	16.60	18.70	20.80

Personal Tutor glencoe.com

Check Your Understanding

Example 1 p. 586

Graph each set of ordered pairs. Determine whether the ordered pairs represent a *linear* function, a *quadratic* function, or an *exponential* function.

1. (−2, 8), (−1, 5), (0, 2), (1, −1)
2. (−3, 7), (−2, 3), (−1, 1), (0, 1), (1, 3)
3. (−3, 8), (−2, 4), (−1, 2), (0, 1), (1, 0.5)
4. (0, 2), (1, 2.5), (2, 3), (3, 3.5)

Example 2 p. 587

Look for a pattern in each table of values to determine which kind of model best describes the data.

5.

x	0	1	2	3	4
y	5	8	17	32	53

6.

x	−3	−2	−1	0
y	−6.75	−7.5	−8.25	−9

7.

x	−1	0	1	2	3
y	3	6	12	24	48

8.

x	3	4	5	6	7
y	−1.5	0	2.5	6	10.5

Example 3 p. 588

Determine which kind of model best describes the data. Then write an equation for the function that models the data.

9.

x	−1	0	1	2	3
y	1	3	9	27	81

10.

x	−5	−4	−3	−2	−1
y	125	80	45	20	5

11.

x	−3	−2	−1	0	1
y	1	1.5	2	2.5	3

12.

x	−1	0	1	2
y	−1.25	−1	−0.75	−0.5

Examples 4 p. 588

13. **PLANTS** The table shows the height of a plant for four consecutive weeks. Determine which kind of function best models the height. Then write a function that models the data.

Week	0	1	2	3	4
Height (in.)	3	3.5	4	4.5	5

Practice and Problem Solving

● = **Step-by-Step Solutions** begin on page R12.
Extra Practice begins on page 815.

Example 1 p. 586

Graph each set of ordered pairs. Determine whether the ordered pairs represent a *linear* function, a *quadratic* function, or an *exponential* function.

14. (−1, 1), (0, −2), (1, −3), (2, −2), (3, 1)
15. (1, 2.75), (2, 2.5), (3, 2.25), (4, 2)
16. (−3, 0.25), (−2, 0.5), (−1, 1), (0, 2)
17. (−3, −11), (−2, −5), (−1, −3), (0, −5)
18. (−2, 6), (−1, 1), (0, −4), (1, −9)
19. (−1, 8), (0, 2), (1, 0.5), (2, 0.125)

Examples 2 and 3 pp. 587–588

Look for a pattern in each table of values to determine which kind of model best describes the data. Then write an equation for the function that models the data.

20.

x	−3	−2	−1	0
y	−8.8	−8.6	−8.4	−8.2

21.

x	−2	−1	0	1	2
y	10	2.5	0	2.5	10

22.

x	−1	0	1	2	3
y	0.75	3	12	48	192

23.

x	−2	−1	0	1	2
y	0.008	0.04	0.2	1	5

24.

x	0	1	2	3	4
y	0	4.2	16.8	37.8	67.2

25.

x	−3	−2	−1	0	1
y	14.75	9.75	4.75	−0.25	−5.25

Example 4
p. 588

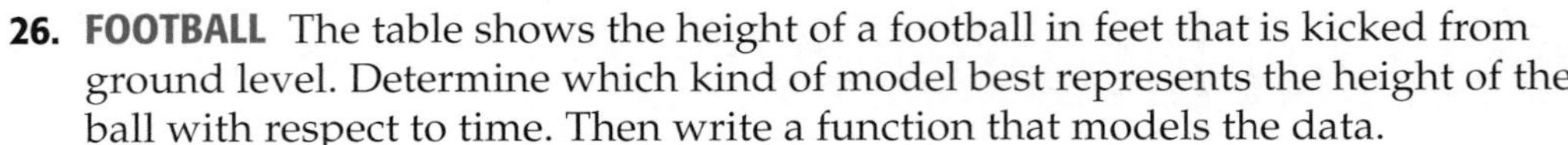

26. **FOOTBALL** The table shows the height of a football in feet that is kicked from ground level. Determine which kind of model best represents the height of the ball with respect to time. Then write a function that models the data.

Time (s)	0	1	2	3	4
Height	0	66	100	102	72

27. **LONG DISTANCE** The cost of a long-distance telephone call depends on the length of the call. The table shows the cost for up to 6 minutes.

Length of call (min)	1	2	3	4	5	6
Cost ($)	0.12	0.24	0.36	0.48	0.60	0.72

a. Graph the data and determine which kind of function best models the data.

b. Write an equation for the function that models the data.

c. Use your equation to determine how much a 10-minute call would cost.

Real-World Link

The top three forms of communication used by teenagers are e-mail, cell phones, and landline telephones.

Source: Harris Interactive

28. **DEPRECIATION** The value of a car depreciates over time. The table shows the value of a car over a period of time.

Year	0	1	2	3	4
Value ($)	18,500	15,910	13,682.60	11,767.04	10,119.65

a. Determine which kind of function best models the data.

b. Write an equation for the function that models the data.

c. Use your equation to determine how much the car is worth after 7 years.

29. **BACTERIA** A scientist estimates that a bacteria culture with an initial population of 12 will triple every hour.

a. Make a table to show the bacteria population for the first 4 hours.

b. Which kind of model best represents the data?

c. Write a function that models the data.

d. How many bacteria will there be after 8 hours?

30. **PRINTING** A printing company charges the fees shown to print flyers. Write a function that models the total cost of the flyers, and determine how much 30 flyers would cost.

H.O.T. Problems

Use **H**igher-**O**rder **T**hinking Skills

31. **CHALLENGE** Write a function that has constant second differences, first differences that are not constant, a y-intercept of -5, and passes through the point at (2, 3).

32. **REASONING** What type of function will have a constant third differences but not constant second differences? Explain.

33. **OPEN ENDED** Write a linear function that has a constant first difference of 4.

34. **REASONING** If data can be modeled by a quadratic function, what is the relationship between the coefficient of x^2 and the constant second difference?

35. **WRITING IN MATH** Summarize how to determine whether a given set of data is modeled by a *linear* function, a *quadratic* function, or an *exponential* function.

Standardized Test Practice

36. SHORT RESPONSE Write an equation that models the data in the table.

x	0	1	2	3	4
y	3	6	12	24	48

37. What is the equation of the line below?

A $y = \frac{2}{5}x + 2$

B $y = \frac{2}{5}x - 2$

C $y = \frac{5}{2}x + 2$

D $y = \frac{5}{2}x - 2$

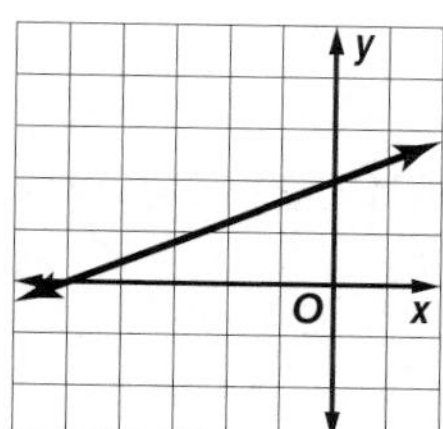

38. The point $(r, -4)$ lies on a line with an equation of $2x + 3y = -8$. Find the value of r.

F -10

G 0

H 2

J 8

39. GEOMETRY The rectangle has an area of 220 square feet. Find the length ℓ.

A 8 feet

B 10 feet

C 22 feet

D 34 feet

Spiral Review

40. INVESTMENTS Joey's investment of $2500 has been decreasing in value at a rate of 1.5% each year. What will his investment be worth in 5 years? (Lesson 9-8)

Write an equation for the nth term of each geometric sequence, and find the seventh term of each sequence. (Lesson 9-7)

41. 1, 2, 4, 8, …

42. $-20, -10, -5, \ldots$

43. $4, -12, 36, \ldots$

44. $99, -33, 11, \ldots$

45. 22, 44, 88, …

46. $\frac{2}{3}, \frac{1}{3}, \frac{1}{6}, \ldots$

Find each product. (Lesson 7-8)

47. $(x - 4)^2$

48. $(2y + 3)^2$

49. $(4x - 7)^2$

50. $(a - 5)(a + 5)$

51. $(5x - 6y)(5x + 6y)$

52. $(9c - 2d^2)(9c + 2d^2)$

53. CANOE RENTAL To rent a canoe, you must pay a daily rate plus $10 per hour. Ilia and her friends rented a canoe for 3 hours and paid $45. Write a linear equation for the cost C of renting the canoe for h hours, and determine how much it cost to rent the canoe for 8 hours. (Lesson 4-2)

Determine whether each equation is a linear equation. If so, write the equation in standard form. (Lesson 3-1)

54. $3x = 5y$

55. $6 - y = 2x$

56. $6xy + 3x = 4$

57. $y + 5 = 0$

58. $7y = 2x + 5x$

59. $y = 4x^2 - 1$

Skills Review

Graph each function. (Lesson 4-7)

60. $f(x) = |x - 2|$

61. $g(x) = |3x + 4|$

62. $f(x) = \left|\frac{1}{2}x + 5\right|$

Graphing Technology Lab: Curve Fitting

Math Online glencoe.com
• Other Calculator Keystrokes
• Graphing Technology Personal Tutor

If there is a constant increase or decrease in data values, there is a linear trend. If the values are increasing or decreasing more and more rapidly, there may be a quadratic or exponential trend.

Linear Trend

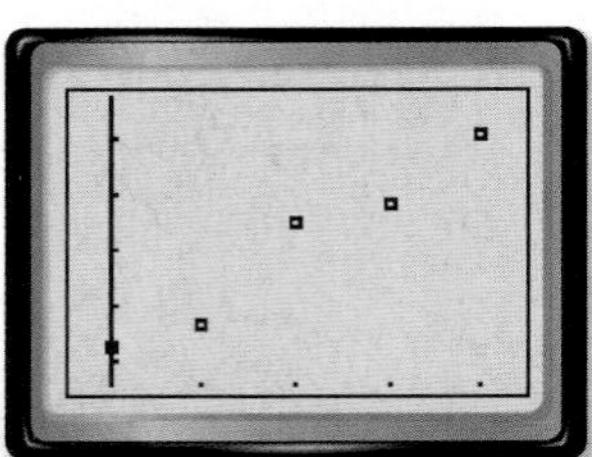

[0, 5] scl: 1 by [0, 6] scl: 1

Quadratic Trend

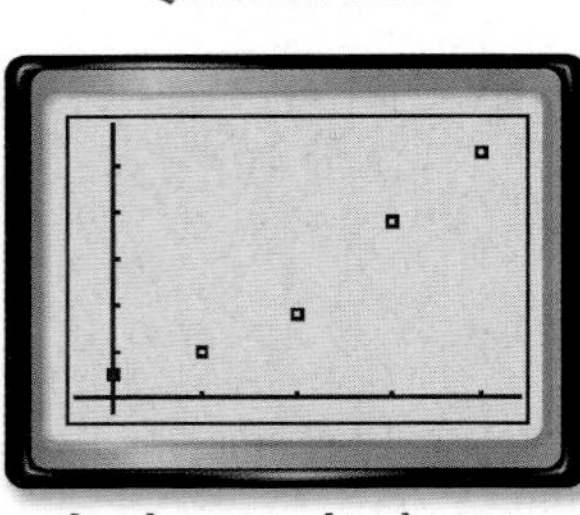

[0, 5] scl: 1 by [0, 6] scl: 1

Exponential Trend

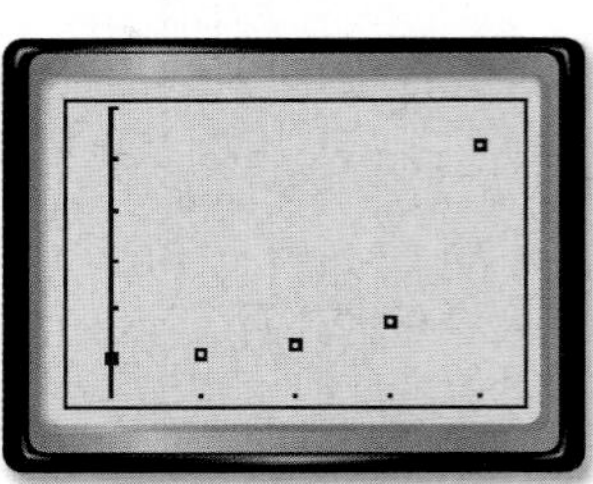

[0, 5] scl: 1 by [0, 6] scl: 1

With a graphing calculator, you can find the appropriate regression equation.

ACTIVITY

CHARTER AIRLINE The table shows the average monthly number of flights made each year by a charter airline that was founded in 2000.

Year	2000	2001	2002	2003	2004	2005	2006	2007
Flights	17	20	24	28	33	38	44	50

Step 1 Make a scatter plot.

- Enter the number of years since 2000 in L1 and the number of flights in L2.

KEYSTROKES: *Review entering a list on page 253.*

- Use STAT PLOT to graph the scatter plot.

KEYSTROKES: *Review statistical plots on page 254.*

Use [Zoom] 9 to graph.

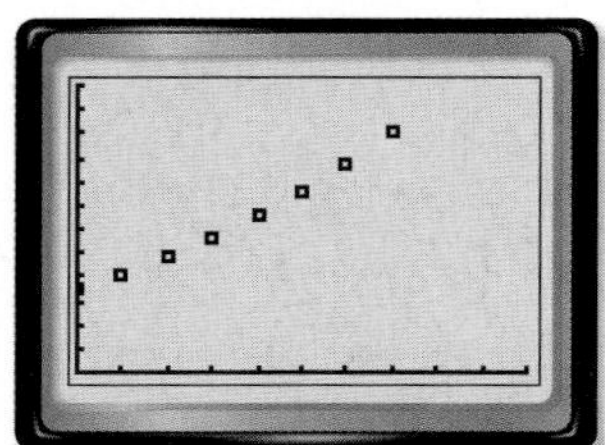

[0, 10] scl: 1 by [0, 60] scl: 5

From the scatter plot we can see that the data may have either a quadratic trend or an exponential trend.

Step 2 Find the regression equation.

We will check both trends by examining their regression equations.

- Select DiagnosticOn from the CATALOG.
- Select QuadReg on the [STAT] menu.

KEYSTROKES: [STAT] [▶] 5 [ENTER] [ENTER]

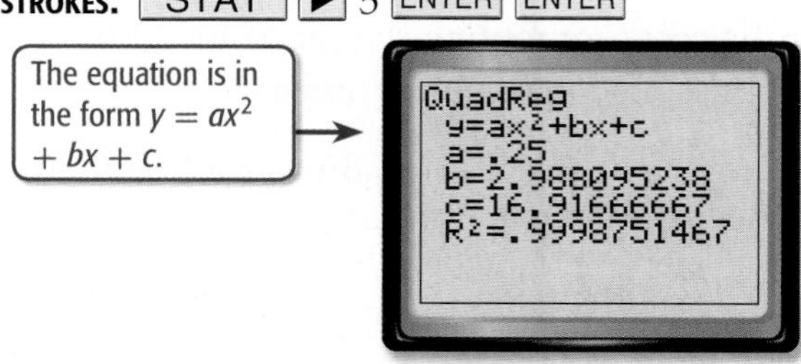

The equation is about $y = 0.25x^2 + 3x + 17$.

R^2 is the **coefficient of determination**. The closer R^2 is to 1, the better the model. To acquire the exponential equation select ExpReg on the [STAT] menu. To choose a quadratic or exponential model, fit both and use the one with the R^2 value closer to 1.

Step 3 Graph the quadratic regression equation.

- Copy the equation to the Y= list and graph.

KEYSTROKES: [Y=] [VARS] 5 [▶] [▶] 1 [Zoom] 9

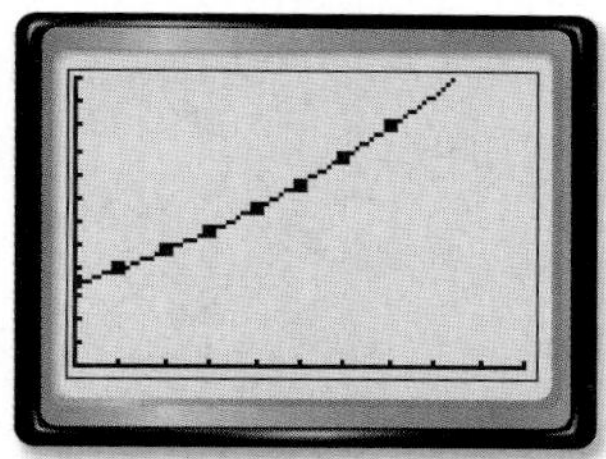

[0, 10] scl: 1 by [0, 60] scl: 5

Step 4 Predict using the equation.

If this trend continues, we can use the graph of our equation to predict the monthly number of flights the airline will make in a specific year. Let's check the year 2020. First adjust the window.

KEYSTROKES: [2nd] [CALC] 1 At x = enter 20 [ENTER].

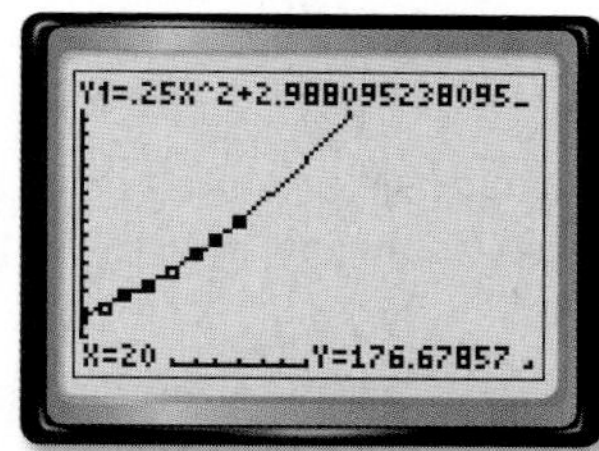

[0, 25] scl: 1 by [0, 200] scl: 5

There will be approximately 177 flights per month if this trend continues.

Exercises

Plot each set of data points. Determine whether to use a *linear, quadratic* or *exponential* regression equation. State the coefficient of determination.

1.

x	y
1	30
2	40
3	50
4	55
5	50
6	40

2.

x	y
0.0	12.1
0.1	9.6
0.2	6.3
0.3	5.5
0.4	4.8
0.5	1.9

3.

x	y
0	1.1
2	3.3
4	2.9
6	5.6
8	11.9
10	19.8

4.

x	y
1	1.67
5	2.59
9	4.37
13	6.12
17	5.48
21	3.12

5. **BAKING Alyssa baked a cake and is waiting for it to cool so she can ice it. The table shows the temperature of the cake every 5 minutes after Alyssa took it out of the oven.**

Time (min)	Temperature (°F)
0	350
5	244
10	178
15	137
20	112
25	96
30	89

   a. Make a scatter plot of the data.

   b. Which regression equation has an R^2 value closest to 1? Is this the equation that best fits the context of the problem? Explain your reasoning.

   c. Find an appropriate regression equation, and state the coefficient of determination. What is the domain and range?

   d. Alyssa will ice the cake when it reaches room temperature (70°F). Use the regression equation to predict when she can ice her cake.

Study Guide and Review

Math Online glencoe.com
- STUDY TO GO
- Vocabulary Review

Chapter Summary

Key Concepts

Graphing Quadratic Functions (Lesson 9-1)

- A quadratic function can be described by an equation of the form $y = ax^2 + bx + c$, where $a \neq 0$.
- The axis of symmetry for the graph of $y = ax^2 + bx + c$, where $a \neq 0$, is $x = -\frac{b}{2a}$.

Solving Quadratic Equations (Lessons 9-2, 9-4, and 9-5)

- Quadratic equations can be solved by graphing. The solutions are the x-intercepts or zeros of the related quadratic function.
- Quadratic equations can be solved by completing the square. To complete the square for $x^2 + bx$, find $\frac{1}{2}$ of b, square this result, and then add the result to $x^2 + bx$.
- Quadratic equations can be solved by using the Quadratic Formula, $x = \frac{-b \pm \sqrt{b^2 - 4ac}}{2a}$.

Transformations of Quadratic Functions (Lesson 9-3)

- $f(x) = x^2 + c$ translates the graph up or down.
- $f(x) = ax^2$ compresses or expands the graph vertically.

Exponential Functions (Lessons 9-6 and 9-7)

- An exponential function can be described by an equation of the form $y = ab^x$, where $a \neq 0$, $b > 0$ and $b \neq 1$.
- The general equation for exponential growth is $y = a(1 + r)^t$, where $r > 0$, and the general equation for exponential decay is $y = a(1 - r)^t$, where $0 < r < 1$. y represents the final amount, a is the initial amount, r represents the rate of change, and t is the time in years.

FOLDABLES® Study Organizer

Be sure the Key Concepts are noted in your Foldable.

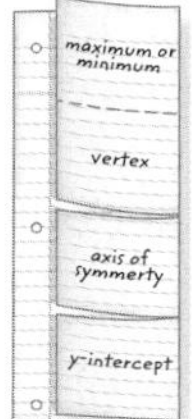

Key Vocabulary

axis of symmetry (p. 525)
common ratio (p. 578)
completing the square (p. 552)
compound interest (p. 574)
dilation (p. 545)
discriminant (p. 561)
double root (p. 538)
exponential decay (p. 574)
exponential function (p. 567)
exponential growth (p. 573)
geometric sequence (p. 578)
maximum (p. 525)
minimum (p. 525)
nonlinear function (p. 525)
parabola (p. 525)
Quadratic Formula (p. 558)
quadratic function (p. 525)
reflection (p. 545)
standard form (p. 525)
symmetry (p. 526)
transformation (p. 544)
translation (p. 544)
vertex (p. 525)

Vocabulary Check

State whether each sentence is *true* or *false*. If *false*, replace the underlined term to make a true sentence.

1. The axis of symmetry of a quadratic function can be found by using the equation $x = -\frac{b}{2a}$.
2. The vertex is the maximum or minimum point of a parabola.
3. The graph of a quadratic function is a straight line.
4. The graph of a quadratic function has a maximum if the coefficient of the x^2 is positive.
5. A quadratic equation with a graph that has two x-intercepts has one real root.
6. The expression $b^2 - 4ac$ is called the discriminant.
7. An example of an exponential function is $y = 3^x$.
8. The exponential growth equation is $y = C(1 - r)^t$.
9. The solutions of a quadratic equation are called roots.
10. The graph of the parent function is translated down to form the graph of $f(x) = x^2 + 5$.

Lesson-by-Lesson Review

9-1 Graphing Quadratic Functions (pp. 525–535)

Consider each equation.

a. **Determine whether the function has a *maximum* or *minimum* value.**

b. **State the maximum or minimum value.**

c. **What are the domain and range of the function?**

11. $y = x^2 - 4x + 4$

12. $y = -x^2 + 3x$

13. $y = x^2 - 2x - 3$

14. $y = -x^2 + 2.$

15. BASEBALL A baseball is thrown with an upward velocity of 32 feet per second. The equation $h = -16t^2 + 32t$ gives the height of the ball t seconds after it is thrown.

a. Determine whether the function has a *maximum* or *minimum* value.

b. State the maximum or minimum value.

c. State a reasonable domain and range for this situation.

EXAMPLE 1

Consider $f(x) = x^2 + 6x + 5$.

a. **Determine whether the function has a *maximum* or *minimum* value.**

For $f(x) = x^2 + 6x + 5$, $a = 1$, $b = 6$, and $c = 5$.

Because a is positive, the graph opens up, so the function has a minimum value.

b. **State the *minimum* or *maximum* value of the function.**

The minimum value is the y-coordinate of the vertex.

The x-coordinate of the vertex is $\frac{-b}{2a}$ or $\frac{-6}{2(1)}$ or -3.

$f(x) = x^2 + 6x + 5$	**Original function**
$f(-3) = (-3)^2 + 6(-3) + 5$	$x = -3$
$f(-3) = -4$	**Simplify.**

The minimum value is -4.

c. **State the domain and range of the function.**

The domain is all real numbers. The range is all real numbers greater than or equal to the minimum value, or $\{y \mid y \geq -4\}$.

9-2 Solving Quadratic Equations by Graphing (pp. 537–542)

Solve each equation by graphing. If integral roots cannot be found, estimate the roots to the nearest tenth.

16. $x^2 - 3x - 4 = 0$

17. $-x^2 + 6x - 9 = 0$

18. $x^2 - x - 12 = 0$

19. $x^2 + 4x - 3 = 0$

20. $x^2 - 10x = -21$

21. $6x^2 - 13x = 15$

22. NUMBER THEORY Find two numbers that have a sum of 2 and a product of -15.

EXAMPLE 2

Solve $x^2 - x - 6 = 0$ by graphing.

Graph the related function $f(x) = x^2 - x - 6$.

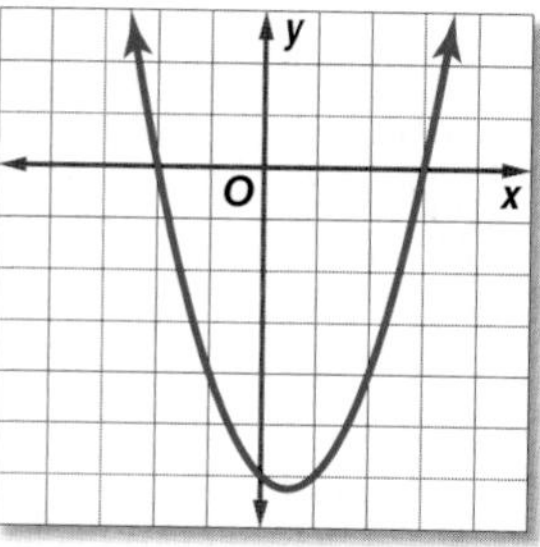

The x-intercepts of the graph appear to be at -2 and 3, so the solutions are -2 and 3.

CHAPTER 9 Study Guide and Review

9-3 Transformations of Quadratic Functions (pp. 544–549)

Describe how the graph of each function is related to the graph of $f(x) = x^2$.

23. $f(x) = x^2 + 8$
24. $f(x) = x^2 - 3$
25. $f(x) = 2x^2$
26. $f(x) = 4x^2 - 18$
27. $f(x) = \frac{1}{3}x^2$
28. $f(x) = \frac{1}{4}x^2$

29. Write an equation for the function shown in the graph.

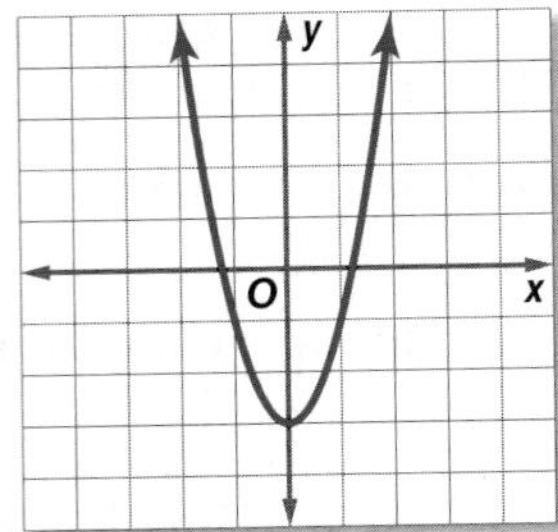

30. **PHYSICS** A ball is dropped off a cliff that is 100 feet high. The function $h = -16t^2 + 100$ models the height h of the ball after t seconds. Compare the graph of this function to the graph of $h = t^2$.

EXAMPLE 3

Describe how the graph of $f(x) = x^2 - 2$ is related to the graph of $f(x) = x^2$.

The graph of $f(x) = x^2 + c$ represents a translation up or down of the parent graph.

Since $c = -2$, the translation is down.

So, the graph is shifted down from the parent function.

EXAMPLE 4

Write an equation for the function shown in the graph.

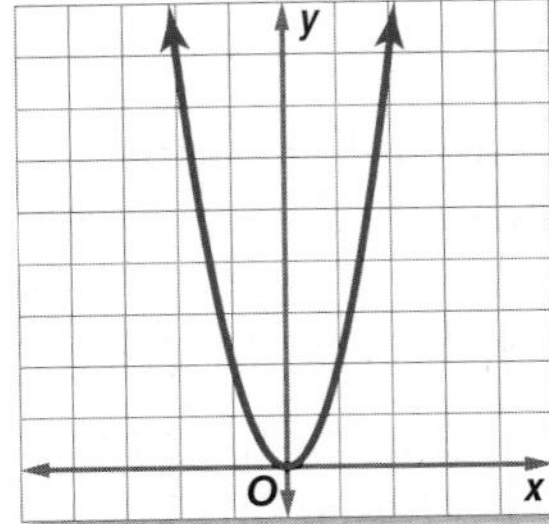

Since the graph opens upward, the leading coefficient must be positive. The parabola has not been translated up or down, so $c = 0$. Since the graph is stretched vertically, it must be of the form of $f(x) = ax^2$ where $a > 1$. The equation for the function is $y = 2x^2$.

9-4 Solving Quadratic Equations by Completing the Square (pp. 552–557)

Solve each equation by completing the square. Round to the nearest tenth if necessary.

31. $x^2 + 6x + 9 = 16$
32. $-a^2 - 10a + 25 = 25$
33. $y^2 - 8y + 16 = 36$
34. $y^2 - 6y + 2 = 0$
35. $n^2 - 7n = 5$
36. $-3x^2 + 4 = 0$

37. **NUMBER THEORY** Find two numbers that have a sum of -2 and a product of -48.

EXAMPLE 5

Solve $x^2 - 16x + 32 = 0$ by competing the square. Round to the nearest tenth if necessary.

Isolate the x^2- and x-terms. Then complete the square and solve.

$$x^2 - 16x + 32 = 0$$
$$x^2 - 16x = -32$$
$$x^2 - 16x + 64 = -32 + 64$$
$$(x - 8)^2 = 32$$
$$x - 8 = \pm\sqrt{32}$$
$$x = 8 \pm\sqrt{32}$$
$$x = 8 \pm 4\sqrt{2}$$

The solutions are about 2.3 and 13.7.

MIXED PROBLEM SOLVING
For mixed problem-solving practice, see page 845.

9-5 Solving Quadratic Equations by Using the Quadratic Formula (pp. 558–564)

Solve each equation by using the Quadratic Formula. Round to the nearest tenth if necessary.

38. $x^2 - 8x = 20$

39. $21x^2 + 5x - 7 = 0$

40. $d^2 - 5d + 6 = 0$

41. $2f^2 + 7f - 15 = 0$

42. $2h^2 + 8h + 3 = 3$

43. $4x^2 + 4x = 15$

44. **GEOMETRY** The area of a square can be quadrupled by increasing the side length and width by 4 inches. What is the side length?

EXAMPLE 6

Solve $x^2 + 10x + 9 = 0$ by using the Quadratic Formula.

$x = \frac{-b \pm \sqrt{b^2 - 4ac}}{2a}$ **Quadratic Formula**

$= \frac{-10 \pm \sqrt{10^2 - 4(1)(9)}}{2(1)}$ **$a = 1, b = 10, c = 9$**

$= \frac{-10 \pm \sqrt{64}}{2}$ **Simplify.**

$= \frac{-10 + 8}{2}$ or $\frac{-10 - 8}{2}$ **Separate the solutions.**

$= -1$ or -9 **Simplify.**

9-6 Exponential Functions (pp. 567–572)

Graph each function. Find the y-intercept, and state the domain and range.

45. $y = 2^x$

46. $y = 3^x + 1$

47. $y = 4^x + 2$

48. $y = 2^x - 3$

49. **BIOLOGY** The population of bacteria in a petri dish increases according to the model $p = 550(2.7)^{0.008t}$, where t is the number of hours and $t = 0$ corresponds to 1:00 P.M. Use this model to estimate the number of bacteria in the dish at 5:00 P.M.

EXAMPLE 7

Graph $y = 3^x + 6$. Find the y-intercept, and state the domain and range.

x	$3^x + 6$	y
−3	$3^{-3} + 6$	6.04
−2	$3^{-2} + 6$	6.11
−1	$3^{-1} + 6$	6.33
0	$3^{-0} + 6$	7
1	$3^{1} + 6$	9

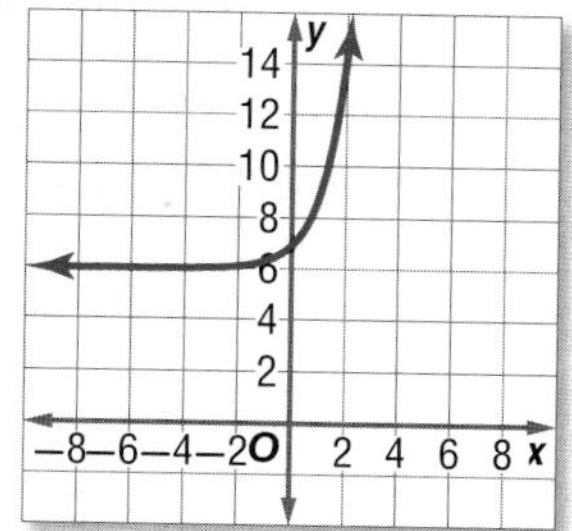

The y-intercept is (0, 7). The domain is all real numbers, and the range is all real numbers greater than 6.

9-7 Growth and Decay (pp. 573–579)

50. Find the final value of \$2500 invested at an interest rate of 2% compounded monthly for 10 years.

51. **COMPUTERS** Zita's computer is depreciating at a rate of 3% per year. She bought the computer for \$1200.

a. Write an equation to represent this situation.

b. What will the computer's value be after 5 years?

EXAMPLE 8

Find the final value of \$2000 invested at an interest rate of 3% compounded quarterly for 8 years.

$A = P\left(1 + \frac{r}{n}\right)^{nt}$ **Compound interest equation**

$= 2000\left(1 + \frac{0.03}{4}\right)^{4(8)}$ **$P = 2000, r = 0.03, n = 4$, and $t = 8$**

$\approx \$2540.22$ **Use a calculator.**

There will be about \$2540.22 in 8 years.

CHAPTER 9 Study Guide and Review

9-8 Geometric Sequences as Exponential Functions (pp. 578–583)

Find the next three terms in each geometric sequence.

52. $-1, 1, -1, 1, \ldots$

53. $3, 9, 27, \ldots$

54. $256, 128, 64, \ldots$

Write the equation for the *n*th term of each geometric sequence.

55. $-1, 1, -1, 1, \ldots$

56. $3, 9, 27, \ldots$

57. $256, 128, 64, \ldots$

58. SPORTS A basketball is dropped from a height of 20 feet. It bounces to $\frac{1}{2}$ its height after each bounce. Draw a graph to represent the situation.

EXAMPLE 9

Find the next three terms in the geometric sequence 2, 6, 18, … .

Step 1 Find the common ratio. Each number is 3 times the previous number, so $r = 3$.

Step 2 Multiply each term by the common ratio to find the next three terms.

$18 \times 3 = 54, 54 \times 3 = 162, 162 \times 3 = 486$

The next three terms are 54, 162, and 486.

EXAMPLE 10

Write the equation for the *n*th term of the geometric sequence −3, 12, −48, … .

The common ratio is −4. So $r = -4$.

$a_n = a_1 r^{n-1}$ **Formula for the *n*th term**

$a_n = -3(-4)^{n-1}$ **$a_1 = -3$ and $r = -4$**

9-9 Analyzing Functions with Successive Differences (pp. 584–589)

Look for a pattern in each table of values to determine which kind of model best describes the data. Then write an equation for the function that models the data.

59.

x	0	1	2	3	4
y	0	3	12	27	48

60.

x	0	1	2	3	4
y	1	2	4	8	16

61.

x	0	1	2	3	4
y	0	−1	−4	−9	−16

62.

x	0	1	2	3	4
y	3	6	9	12	15

63. SCHOOL SPIRIT The table shows the cost to purchase school-spirit posters. Determine which kind of model best describes the data. Then write the equation.

No. of posters	2	4	6	8
Cost	4	7	10	13

EXAMPLE 11

Determine which kind of model best describes the data. Then write an equation for the function that models the data.

x	0	1	2	3	4
y	3	4	5	6	7

Step 1 Determine which model fits the data.

First differences: 

Since the first differences are all equal, a linear function models the data.

Step 2 Write an equation for the function that models the data.

The equation has the form $y = mx + b$.

The slope is 1 and the y-intercept is 3, so the equation is $y = x + 3$.

Math Online glencoe.com
Chapter Test

Use a table of values to graph the following functions. State the domain and range.

1. $y = x^2 + 2x + 5$
2. $y = 2x^2 - 3x + 1$

Consider $y = x^2 - 7x + 6$.

3. Determine whether the function has a *maximum* or *minimum* value.
4. State the maximum or minimum value.
5. What are the domain and range?

Solve each equation by graphing. If integral roots cannot be found, estimate the roots to the nearest tenth.

6. $x^2 + 7x + 10 = 0$
7. $x^2 - 5 = -3x$

Describe how the graph of each function is related to the graph of $f(x) = x^2$.

8. $g(x) = x^2 - 5$
9. $g(x) = -3x^2$
10. $h(x) = \frac{1}{2}x^2 + 4$

11. **MULTIPLE CHOICE** Which is an equation for the function shown in the graph?

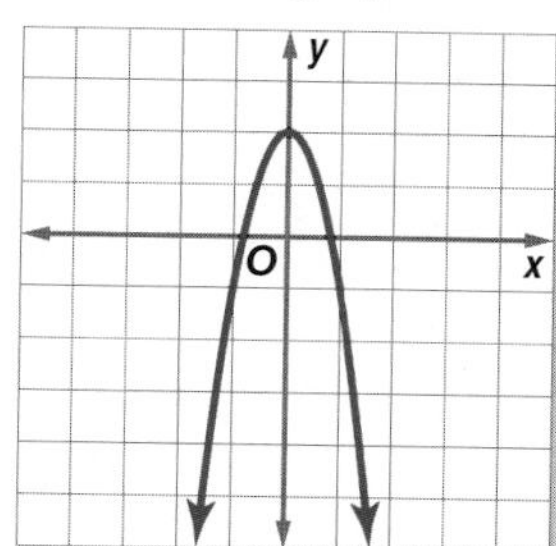

A $y = -3x^2$

B $y = 3x^2 + 1$

C $y = x^2 + 2$

D $y = -3x^2 + 2$

Solve each equation by completing the square.

12. $x^2 + 2x + 5 = 0$
13. $x^2 - x - 6 = 0$
14. $2x^2 - 36 = -6x$

Solve each equation by using the Quadratic Formula. Round to the nearest tenth if necessary.

15. $x^2 - x - 30 = 0$
16. $x^2 - 10x = -15$
17. $2x^2 + x - 15 = 0$

18. **BASEBALL** Elias hits a baseball into the air. The equation $h = -16t^2 + 60t + 3$ models the height h in feet of the ball after t seconds. How long is the ball in the air?

Graph each function. Find the y-intercept, and state the domain and range.

19. $y = 2(5)^x$
20. $y = -3(11)^x$
21. $y = 3x + 2$

Find the next three terms in each geometric sequence.

22. 2, −6, 18, …
23. 1000, 500, 250, …
24. 32, 8, 2, …

25. **MONEY** Lynne invested \$500 into an account with a 6.5% interest rate compounded monthly. How much will Lynne's investment be worth in 10 years?

F \$600.00 H \$956.09

G \$938.57 J \$957.02

26. **INVESTMENTS** Shelly's investment of \$3000 has been losing value at a rate of 3% each year. What will her investment be worth in 6 years?

27. Graph $\{(-2, 4), (-1, 1), (0, 0), (1, 1), (2, 4)\}$. Determine whether the ordered pairs represent a *linear function*, a *quadratic function*, or an *exponential function*.

28. Look for a pattern in the table to determine which kind of model best describes the data.

x	0	1	2	3	4
y	1	3	5	7	9

CHAPTER 9 Preparing for Standardized Tests

Use a Formula

A *formula* is an equation that shows a relationship among certain quantities. Many standardized test problems will require using a formula to solve them.

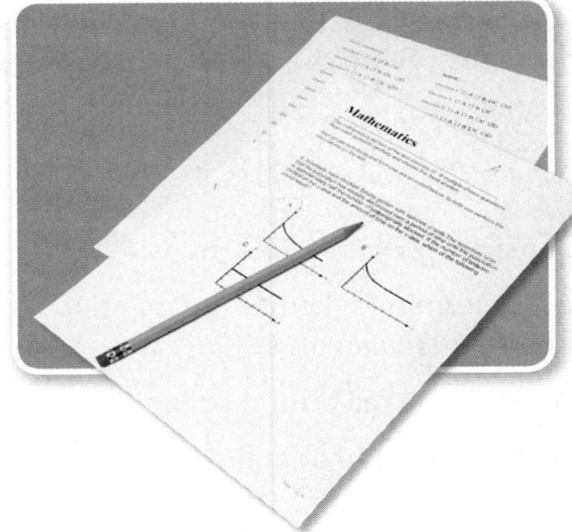

Strategies for Using a Formula

Step 1

Become familiar with common formulas and their uses. You may or may not be given access to a formula sheet to use during the test.

- **If given a formula sheet,** be sure to practice with the formulas on it before taking the test so you know how to apply them.
- **If not given a formula sheet,** study and practice with common formulas such as perimeter, area, and volume formulas, the Distance Formula, the Pythagorean Theorem, the Midpoint Formula, the Quadratic Formula, and others.

Step 2

Choose a formula and solve.

- **Ask Yourself:** What quantities are given in the problem statement?
- **Ask Yourself:** What quantities am I looking for?
- **Ask Yourself:** Is there a formula I know that relates these quantities?
- **Write:** Write the formula out that you have chosen each time.
- **Solve:** Substitute known quantities into the formula and solve for the unknown quantity.
- **Check:** Check your answer if time permits.

EXAMPLE

Read the problem. Identify what you need to know. Then use the information in the problem to solve.

Find the exact roots of the quadratic equation $-2x^2 + 6x + 5 = 0$.

A $\frac{3 \pm \sqrt{17}}{4}$

B $\frac{4 \pm \sqrt{17}}{3}$

C $\frac{3 \pm \sqrt{19}}{2}$

D $\frac{3 \pm \sqrt{19}}{4}$

Read the problem carefully. You are given a quadratic equation and asked to find the exact roots of the equation. Use the **Quadratic Formula** to find the roots.

$-2x^2 + 6x + 5 = 0$	**Original equation**
$a = -2, b = 6, c = 5$	**Identify the coefficients of the equation.**
$x = \frac{-b \pm \sqrt{b^2 - 4ac}}{2a}$	**Quadratic Formula**
$= \frac{-(6) \pm \sqrt{(6)^2 - 4(-2)(5)}}{2(-2)}$	**$a = -2$, $b = 6$, and $c = 5$**
$= \frac{-6 \pm \sqrt{36 - (-40)}}{-4}$	**Simplify.**
$= \frac{-6 \pm \sqrt{76}}{-4}$	**Subtract.**
$= \frac{-6 \pm 2\sqrt{19}}{-4}$	**$\sqrt{76} = \sqrt{4 \cdot 19}$ or $2\sqrt{19}$.**
$= \frac{-2(3 \pm \sqrt{19})}{-2(2)}$	**Factor out −2 from the numerator and denominator.**
$= \frac{3 \pm \sqrt{19}}{2}$	**Simplify.**

The roots of the equation are $\frac{3 + \sqrt{19}}{2}$ and $\frac{3 - \sqrt{19}}{2}$. The correct answer is C.

Exercises

Read each problem. Identify what you need to know. Then use the information in the problem to solve.

1. Find the exact roots of the quadratic equation $x^2 + 5x - 12 = 0$.

A $\frac{-5 \pm \sqrt{73}}{2}$

B $\frac{4 \pm \sqrt{61}}{3}$

C $\frac{-3 \pm \sqrt{73}}{4}$

D $\frac{-1 \pm \sqrt{61}}{2}$

2. The area of a triangle in which the length of the base is 4 centimeters greater than twice the height is 80 square centimeters. What is the length of the base of the triangle?

F −10

G 8

H 16

J 20

3. Find the volume of the figure below.

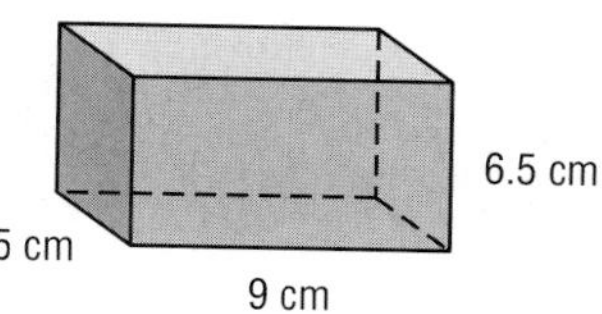

A 18.5 cm^2

B 91 cm^2

C 272 cm^2

D 292.5 cm^2

4. Myron is traveling 263.5 miles at an average rate of 62 miles per hour. How long will it take Myron to complete his trip?

F 5 h 25 min

G 4 h 15 min

H 5 h 10 min

J 4 h 25 min

Standardized Test Practice

Cumulative, Chapters 1 through 9

Multiple Choice

Read each question. Then fill in the correct answer on the answer document provided by your teacher or on a sheet of paper.

1. What is the vertex of the parabola graphed below?

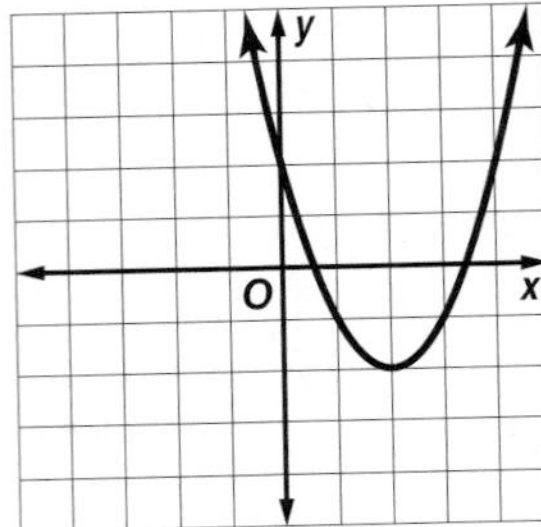

A (2, 0)

B (0, 2)

C (−2, 2)

D (2, −2)

2. Write an equation in slope-intercept form with a slope of $\frac{9}{10}$ and y-intercept of 3.

F $y = 3x + \frac{9}{10}$

G $y = \frac{9}{10}x + 3$

H $y = \frac{9}{10}x - 3$

J $y = 3x - \frac{9}{10}$

3. Use the Quadratic Formula to find the exact solutions of the equation $2x^2 - 6x + 3 = 0$.

A $\frac{3 \pm \sqrt{3}}{2}$

B $\frac{3 \pm \sqrt{2}}{4}$

C $\frac{2 \pm \sqrt{5}}{3}$

D $\frac{5 \pm \sqrt{2}}{2}$

4. Write an expression for the area of the rectangle below.

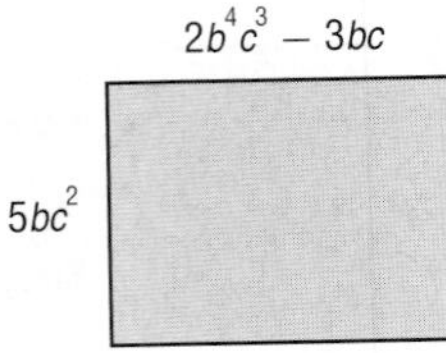

F $10b^5c^5 - 3bc$

G $10b^5c^5 - 15b^2c^3$

H $2b^5c^5 - 3b^2c^3$

J $10b^4c^6 - 15bc^2$

5. Solve the quadratic equation below by graphing.

$$x^2 - 2x - 15 = 0$$

A −1, 4

B −3, 5

C 3, −5

D ∅

6. Jason is playing games at a family fun center. So far he has won 38 prize tickets. How many more tickets would he need to win to place him in the gold prize category?

Number of Tickets	Prize Category
1–20	bronze
21–40	silver
41–60	gold
61–80	platinum

F $2 \leq t \leq 22$

G $3 \leq t \leq 22$

H $1 \leq t \leq 20$

J $3 \leq t \leq 20$

Test-Taking Tip

Question 5 If permitted, you can use a graphing calculator to quickly graph an equation and find its roots.

Short Response/Gridded Response

Record your answers on the answer sheet provided by your teacher or on a sheet of paper.

7. **GRIDDED RESPONSE** Misty purchased a car several years ago for $21,459. The value of the car depreciated at a rate of 15% annually. What was the value of the car after 5 years? Round your answer to the nearest whole dollar.

8. Use the graph of the quadratic equation shown below to answer each question.

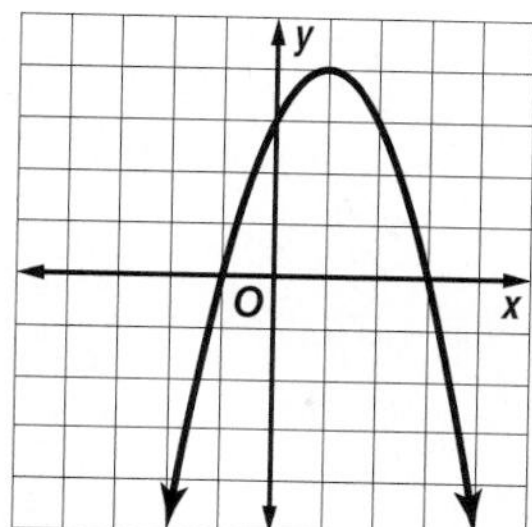

   a. What is the vertex?

   b. What is the y-intercept?

   c. What is the axis of symmetry?

   d. What are the roots of the corresponding quadratic equation?

9. The cost of 5 notebooks and 3 pens is $9.75. The cost of 4 notebooks and 6 pens is $10.50. Which of the following systems can be used to find the cost of a notebook n and a pen p?

   a. Write a system of equations to model the situation.

   b. Solve the system of equations. How much does each item cost?

10. The table shows the total cost of renting a canoe for n hours.

Number of Hours (n)	Rental Cost (C)
1	$15
2	$20
3	$25
4	$30

   a. Write a function to represent the situation.

   b. How much would it cost to rent the canoe for 7 hours?

Extended Response

Record your answers on a sheet of paper. Show your work.

11. Use the equation and its graph to answer each question.

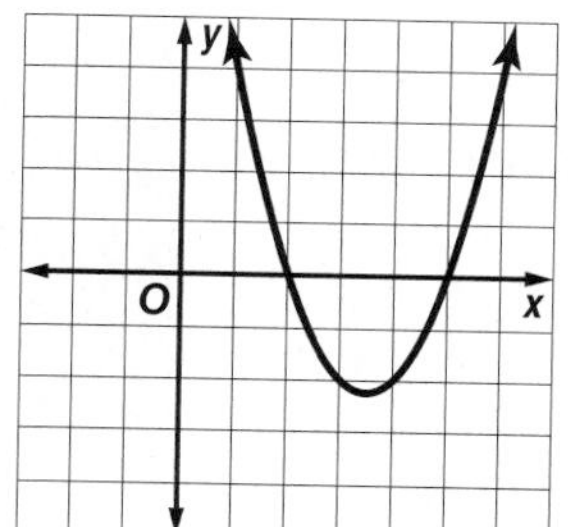

   a. Factor $x^2 - 7x + 10$.

   b. What are the solutions of $x^2 - 7x + 10 = 0$?

   c. What do you notice about the graph of the quadratic equation and where it crosses the x-axis? How do these values compare to the solutions of $x^2 - 7x + 10 = 0$? Explain.

Need Extra Help?											
If you missed Question...	1	2	3	4	5	6	7	8	9	10	11
Go to Lesson or Page...	9-1	4-2	9-5	7-6	9-2	5-1	9-7	9-1	6-4	3-5	8-3

CHAPTER 10 Radical Functions and Geometry

Then

In Chapters 8 and 9, you solved quadratic equations.

Now

In Chapter 10, you will:

- Graph and transform radical functions.
- Simplify, add, subtract, and multiply radical expressions.
- Solve radical equations.
- Use the Pythagorean Theorem.
- Find trigonometric ratios.

Why?

OCEANS Tsunamis, or large waves, are generated by undersea earthquakes. A radical equation can be used to find the speed of a tsunami in meters per second or the depth of the ocean in meters.

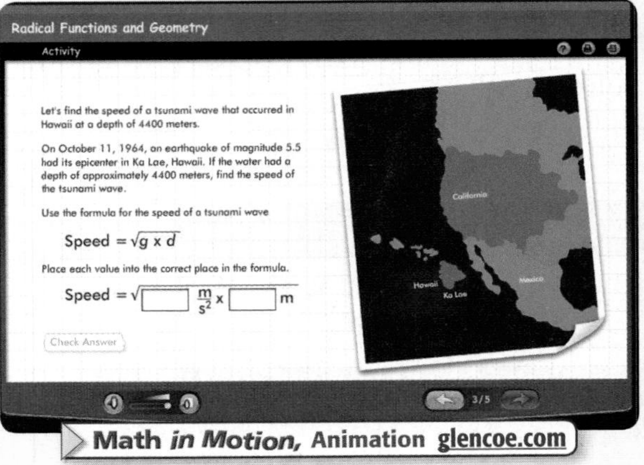

Math *in Motion,* Animation glencoe.com

Get Ready for Chapter 10

Diagnose Readiness You have two options for checking Prerequisite Skills.

Text Option Take the Quick Check below. Refer to the Quick Review for help.

*Quick*Check

Find each square root. If necessary, round to the nearest hundredth. (Lesson 0-2)

1. $\sqrt{82}$
2. $\sqrt{26}$
3. $\sqrt{15}$
4. $\sqrt{99}$
5. **SANDBOX** Isaac is making a square sandbox with an area of 100 square feet. How long is a side of the sandbox?

Simplify each expression. (Lesson 1-4)

6. $(21x + 15y) - (9x - 4y)$
7. $13x - 5y + 2y$
8. $(10a - 5b) + (6a + 5b)$
9. $6m + 5n + 4 - 3m - 2n + 6$
10. $x + y - 3x - 4y + 2x - 8y$

Solve each equation. (Lesson 8-4)

11. $2x^2 - 4x = 0$
12. $6x^2 - 5x - 4 = 0$
13. $x^2 - 7x + 10 = 0$
14. $2x^2 + 7x - 5 = -1$
15. **GEOMETRY** The area of the rectangle is 90 square feet. Find x. (Lesson 8-3)

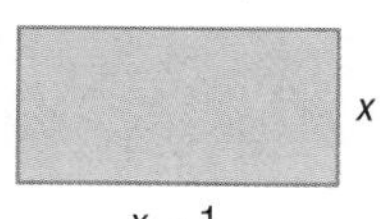

Use cross products to determine whether each pair of ratios forms a proportion. Write *yes* or *no*. (Lesson 2-6)

16. $\frac{2}{3}$ and $\frac{4}{9}$
17. $\frac{3}{4}$ and $\frac{15}{20}$
18. **MAPS** On a map, 1 inch = 10 miles. If the distance between cities is 50 miles, how many inches will it be on the map? (Lesson 2-6)

*Quick*Review

EXAMPLE 1

Find the square root of $\sqrt{50}$. If necessary, round to the nearest hundredth.

$\sqrt{50} = 7.071067812\ldots$ **Use a calculator.**

To the nearest hundredth, $\sqrt{50} = 7.07$.

EXAMPLE 2

Simplify $3x + 7y - 4x - 8y$.

$3x + 7y - 4x - 8y$

$= (3x - 4x) + (7y - 8y)$ **Combine like terms.**

$= -x - y$ **Simplify.**

EXAMPLE 3

Solve $x^2 - 5x + 6 = 0$.

$x^2 - 5x + 6 = 0$ **Original equation**

$(x - 3)(x - 2) = 0$ **Factor.**

$x - 3 = 0$ or $x - 2 = 0$ **Zero Product Property**

$x = 3$ $\quad x = 2$ **Solve each equation.**

EXAMPLE 4

Use cross products to determine whether $\frac{2}{3}$ and $\frac{8}{12}$ form a proportion.

$\frac{2}{3} \stackrel{?}{=} \frac{8}{12}$ **Write the equation.**

$2(12) \stackrel{?}{=} 3(8)$ **Find the cross products.**

$24 = 24$ ✓ **Simplify.**

They form a proportion.

Online Option **Math Online** Take a self-check Chapter Readiness Quiz at glencoe.com.

Get Started on Chapter 10

You will learn several new concepts, skills, and vocabulary terms as you study Chapter 10. To get ready, identify important terms and organize your resources. You may wish to refer to **Chapter 0** to review prerequisite skills.

FOLDABLES® Study Organizer

Radical Functions and Geometry Make this Foldable to help you organize your Chapter 10 notes about radical functions and geometry. Begin with four sheets of grid paper.

1. **Fold** in half along the width.

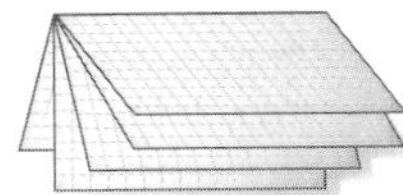

2. **Staple** along the fold.

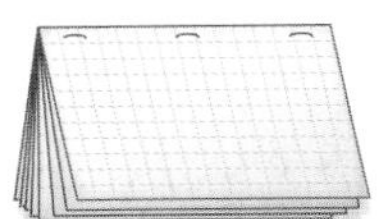

3. **Turn** the fold to the left and write the title of the chapter on the front. On each left-hand page of the booklet, write the title of a lesson from the chapter.

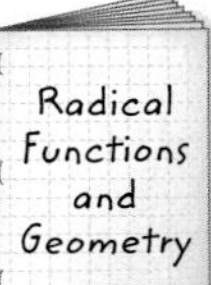

Math Online glencoe.com

- Study the chapter online
- Explore **Math in Motion**
- Get extra help from your own **Personal Tutor**
- Use **Extra Examples** for additional help
- Take a **Self-Check Quiz**
- **Review Vocabulary** in fun ways

New Vocabulary

English		Español
radicand	p. 605	radicando
radical function	p. 605	función radicales
conjugate	p. 614	conjugado
radical equations	p. 624	ecuaciones radicales
hypotenuse	p. 630	hipotenusa
legs	p. 630	catetos
converse	p. 631	recíproco
midpoint	p. 638	punto medio
similar triangles	p. 642	semejantes
cosine	p. 649	coseno
tangent	p. 649	tangente
trigonometry	p. 649	trigonometría
inverse cosine	p. 651	coseno inverso
inverse sine	p. 651	seno inverso
inverse tangent	p. 651	tangente inverse

Review Vocabulary

FOIL method • p. 448 • metodo FOIL to multiply two binomials, find the sum of the products of the First terms, Outer terms, Inner terms, and Last terms

perfect square • p. P7 • cuadrado perfecto a number with a square root that is a rational number

proportion • p. 111 • proporcion an equation of the form $\frac{a}{b} = \frac{c}{d}$ stating that two ratios are equivalent

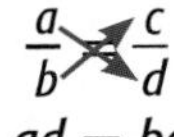

$ad = bc$

Multilingual eGlossary glencoe.com

10-1

Square Root Functions

Then

You solved quadratic equations by using the Quadratic Formula. (Lesson 9-5)

Now

- Graph and analyze dilations of radical functions.
- Graph and analyze reflections and translations of radical functions.

New Vocabulary

square root function
radical function
radicand

Math Online

glencoe.com

- Extra Examples
- Personal Tutor
- Self-Check Quiz
- Homework Help

Why?

Scientists use sounds of whales to track their movements. The distance to a whale can be found by relating time to the speed of sound in water.

The speed of sound in water can be described by the *square root function* $c = \sqrt{\frac{E}{d}}$, where E represents the bulk modulus elasticity of the water and d represents the density of the water.

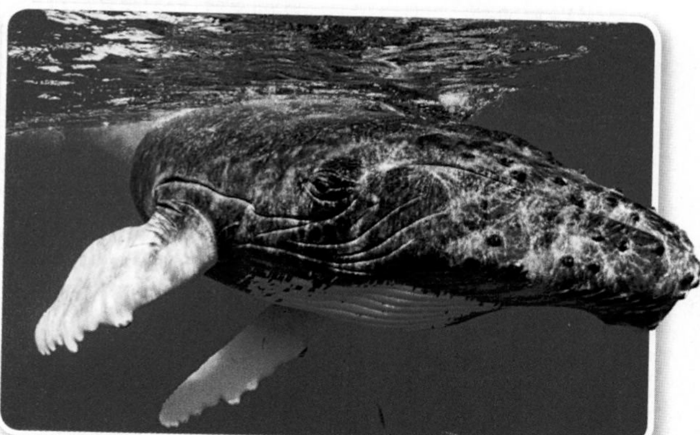

Dilations of Radical Functions A **square root function** contains the square root of a variable. Square root functions are a type of **radical function**. The expression under the radical sign is called the **radicand**. For a square root to be a real number, the radicand cannot be negative. Values that make the radicand negative are not included in the domain.

Key Concept — Square Root Function

For Your FOLDABLE

Parent function: $f(x) = \sqrt{x}$

Type of graph: curve

Domain: $\{x \mid x \geq 0\}$

Range: $\{y \mid y \geq 0\}$

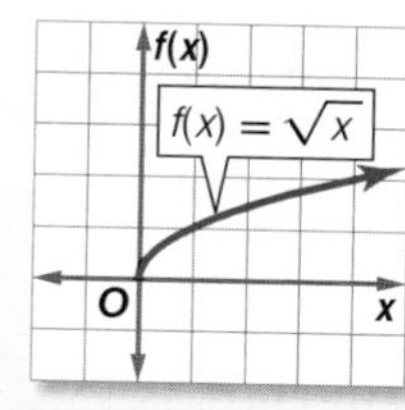

EXAMPLE 1 Dilation of the Square Root Function

Graph $f(x) = 2\sqrt{x}$. State the domain and range.

Step 1 Make a table.

x	0	0.5	1	2	3	4
$f(x)$	0	≈1.4	2	≈2.8	≈3.5	4

Step 2 Plot points. Draw a smooth curve.

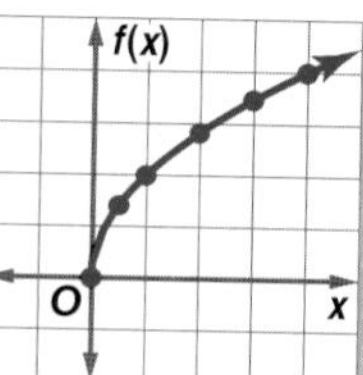

The domain is $\{x \mid x \geq 0\}$, and the range is $\{y \mid y \geq 0\}$.

Check Your Progress

1A. $g(x) = 4\sqrt{x}$

1B. $h(x) = 6\sqrt{x}$

Personal Tutor glencoe.com

Reflections and Translations of Radical Functions Recall that when the value of a is negative in the quadratic function $f(x) = ax^2$, the graph of the parent function is reflected across the x-axis and stretched vertically.

StudyTip

Graphing Radical Functions Choose perfect squares for x-values that will result in coordinates that are easy to plot.

Key Concept Graphing $y = a\sqrt{x + h} + c$ — For Your FOLDABLE

Step 1 Draw the graph of $y = a\sqrt{x}$. The graph starts at the origin and passes through $(1, a)$. If $a > 0$, the graph is in quadrant I. If $a < 0$, the graph is reflected across the x-axis and is in quadrant IV.

Step 2 Translate the graph $|c|$ units up if $c > 0$ and down if $c < 0$.

Step 3 Translate the graph $|h|$ units left if $h > 0$ and right if $h < 0$.

EXAMPLE 2 Reflection of the Square Root Function

Graph $y = -3\sqrt{x}$. Compare to the parent graph. State the domain and range.

Make a table of values. Then plot the points on a coordinate system and draw a smooth curve that connects them.

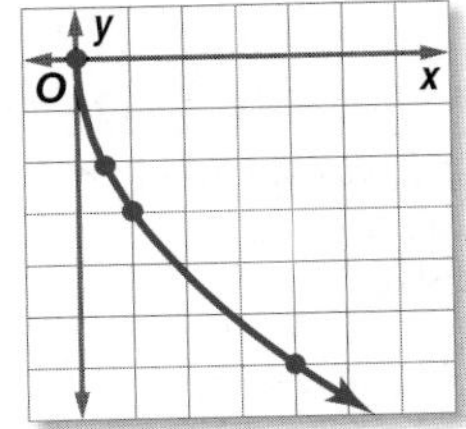

x	0	0.5	1	4
y	0	≈ -2.1	-3	-6

Notice that the graph is in the 4th quadrant. It is obtained by stretching the graph of $y = \sqrt{x}$ vertically and then reflecting across the x-axis. The domain is $\{x \mid x \geq 0\}$, and the range is $\{y \mid y \leq 0\}$.

Check Your Progress

2A. $y = -2\sqrt{x}$

2B. $y = -4\sqrt{x}$

Personal Tutor glencoe.com

StudyTip

Translating Radical Functions If $c > 0$, a radical function $f(x) = \sqrt{x - c}$ is a horizontal translation c units to the right. $f(x) = \sqrt{x + c}$ is a horizontal translation c units to the left.

EXAMPLE 3 Translation of the Square Root Function

Graph each function. Compare to the parent graph. State the domain and range.

a. $g(x) = \sqrt{x} + 1$

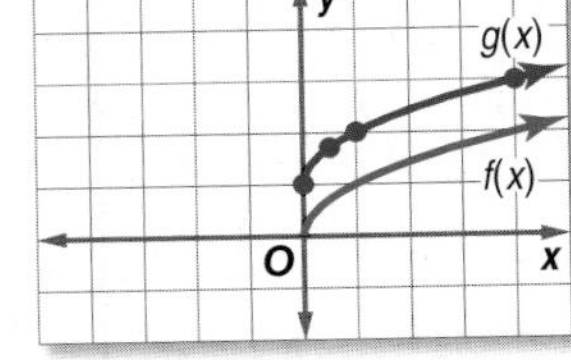

x	0	0.5	1	4	9
y	0	≈ 1.7	2	3	4

Notice that the values of $g(x)$ are 1 greater than those of $f(x) = \sqrt{x}$. This is a vertical translation 1 unit up from the parent function. The domain is $\{x \mid x \geq 0\}$, and the range is $\{y \mid y \geq 1\}$.

b. $h(x) = \sqrt{x - 2}$

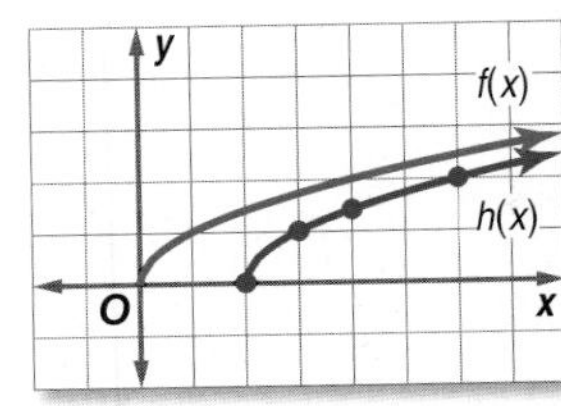

x	2	3	4	6
y	0	1	≈ 1.4	2

This is a horizontal translation 2 units to the right of the parent function. The domain is $\{x \mid x \geq 2\}$, and the range is $\{y \mid y \geq 0\}$.

Check Your Progress

3A. $g(x) = \sqrt{x} - 4$

3B. $h(x) = \sqrt{x + 3}$

Personal Tutor glencoe.com

Physical phenomena such as motion can be modeled by radical functions. Often these functions are transformations of the parent square root function.

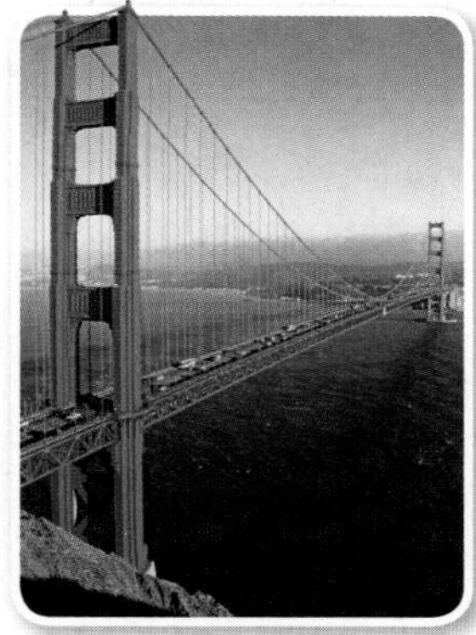

Real-World Link

Approximately 39 million cars cross the Golden Gate Bridge in San Francisco each year.

Source: San Francisco Convention and Visitors Bureau

Real-World EXAMPLE 4 Analyze a Radical Function

BRIDGES The Golden Gate Bridge is about 67 meters above the water. The velocity v of a freely falling object that has fallen h meters is given by $v = \sqrt{2gh}$, where g is the constant 9.8 meters per second squared. Graph the function. If an object is dropped from the bridge, what is its velocity when it hits the water?

Use a graphing calculator to graph the function. To find the velocity of the object, substitute 67 meters for h.

Y1=√(19.6X)

X=67 Y=36.238101

$v = \sqrt{2gh}$	**Original function**
$= \sqrt{2(9.8)(67)}$	$g = 9.8$ **and** $h = 67$
$= \sqrt{1313.2}$	**Simplify.**
≈ 36.2 m/s	**Use a calculator.**

The velocity of the object is about 36.2 meters per second after dropping 67 meters.

Check Your Progress

4. Use the graph above to estimate the initial height of an object if it is moving at 20 meters per second when it hits the water.

Personal Tutor glencoe.com

Transformations such as reflections, translations, and dilations can be combined in one equation.

EXAMPLE 5 Transformations of the Square Root Function

Graph $y = -2\sqrt{x} + 1$, and compare to the parent graph. State the domain and range.

x	0	1	4	9
y	1	−1	−3	−5

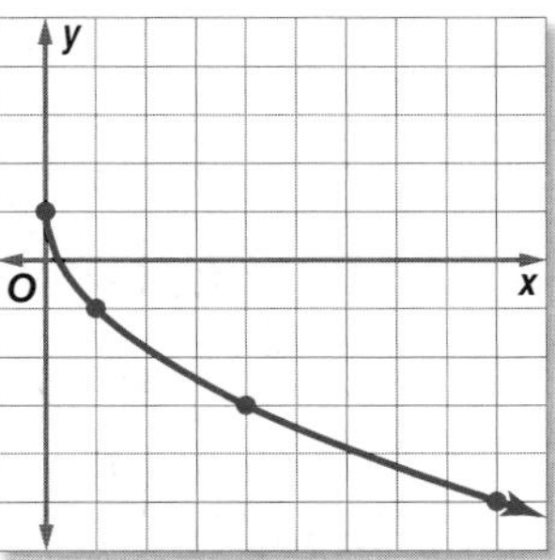

This graph is the result of a vertical stretch of the graph of $y = \sqrt{x}$ followed by a reflection across the x-axis, and then a translation 1 unit up. The domain is $\{x \mid x \geq 0\}$, and the range is $\{y \mid y \leq 1\}$.

Check Your Progress

5A. $y = \frac{1}{2}\sqrt{x} - 1$

5B. $y = -2\sqrt{x - 1}$

Personal Tutor glencoe.com

Check Your Understanding

Examples 1 and 3
pp. 605–606

Graph each function. Compare to the parent graph. State the domain and range.

1. $y = 3\sqrt{x}$

2. $y = -5\sqrt{x}$

3. $y = \frac{1}{3}\sqrt{x}$

4. $y = -\frac{1}{2}\sqrt{x}$

5. $y = \sqrt{x} + 3$

6. $y = \sqrt{x} - 2$

7. $y = \sqrt{x + 2}$

8. $y = \sqrt{x - 3}$

Example 4
p. 607

9. **FREE FALL** The time in seconds that it takes an object to fall a distance d is given by the function $t = \frac{1}{4}\sqrt{d}$ (assuming zero air resistance). Graph the function, and state the domain and range.

Example 5
p. 607

Graph each function, and compare to the parent graph. State the domain and range.

10. $y = \frac{1}{2}\sqrt{x} + 2$

11. $y = -\frac{1}{4}\sqrt{x} - 1$

12. $y = -2\sqrt{x + 1}$

13. $y = 3\sqrt{x - 2}$

Practice and Problem Solving

Examples 1 and 3
pp. 605–606

Graph each function. Compare to the parent graph. State the domain and range.

14. $y = 5\sqrt{x}$

15. $y = \frac{1}{2}\sqrt{x}$

16. $y = -\frac{1}{3}\sqrt{x}$

17. $y = 7\sqrt{x}$

18. $y = -\frac{1}{4}\sqrt{x}$

19. $y = -\sqrt{x}$

20. $y = -\frac{1}{5}\sqrt{x}$

21. $y = -7\sqrt{x}$

22. $y = \sqrt{x} + 2$

23. $y = \sqrt{x} + 4$

24. $y = \sqrt{x} - 1$

25. $y = \sqrt{x} - 3$

26. $y = \sqrt{x} + 1.5$

27. $y = \sqrt{x} - 2.5$

28. $y = \sqrt{x + 4}$

29. $y = \sqrt{x - 4}$

30. $y = \sqrt{x + 1}$

31. $y = \sqrt{x - 0.5}$

32. $y = \sqrt{x + 5}$

33. $y = \sqrt{x - 1.5}$

Example 4
p. 607

34. **GEOMETRY** The perimeter of a square is given by the function $P = 4\sqrt{A}$, where A is the area of the square.

a. Graph the function.

b. Determine the perimeter of a square with an area of 225 m^2.

c. When will the perimeter and the area be the same value?

Example 5
p. 607

Graph each function, and compare to the parent graph. State the domain and range.

35. $y = -2\sqrt{x} + 2$

36. $y = -3\sqrt{x} - 3$

37. $y = \frac{1}{2}\sqrt{x + 2}$

38. $y = -\sqrt{x - 1}$

39. $y = \frac{1}{4}\sqrt{x - 1} + 2$

40. $y = \frac{1}{2}\sqrt{x - 2} + 1$

41. **ENERGY** An object has kinetic energy when it is in motion. The velocity in meters per second of an object of mass m kilograms with an energy of E joules is given by the function $v = \sqrt{\frac{2E}{m}}$. Use a graphing calculator to graph the function that represents the velocity of a basketball with a mass of 0.6 kilogram.

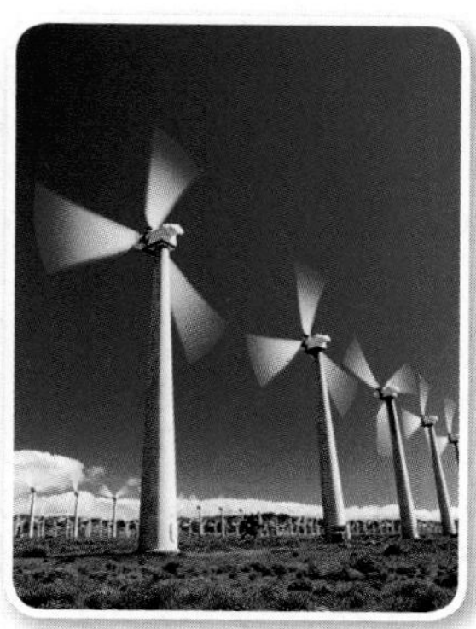

Real-World Link

Wind farms harness the kinetic energy of the wind and convert it to usable power. A turbine that is designed to power a house can have a rotor with a diameter of 50 feet.

Source: American Wind Energy Association

42. **GEOMETRY** The radius of a circle is given by $r = \sqrt{\frac{A}{\pi}}$, where A is the area of the circle.

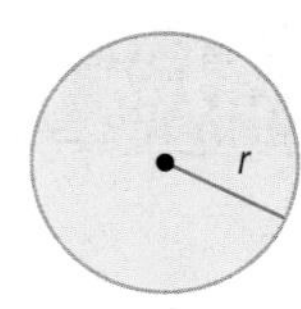

    a. Graph the function.

    b. Use a graphing calculator to determine the radius of a circle that has an area of 27 in^2.

43. **SPEED OF SOUND** The speed of sound in air is determined by the temperature of the air. The speed c in meters per second is given by $c = 331.5\sqrt{1 + \frac{t}{273.15}}$, where t is the temperature of the air in degrees Celsius.

    a. Use a graphing calculator to graph the function.

    b. How fast does sound travel when the temperature is 55°C?

    c. How is the speed of sound affected when the temperature increases by 10°?

44. **MULTIPLE REPRESENTATIONS** In this problem, you will explore the relationship between the graphs of square root functions and parabolas.

    a. **GRAPHICAL** Graph $y = x^2$ on a coordinate system.

    b. **ALGEBRAIC** Write a piecewise-defined function to describe the graph of $y^2 = x$ in each quadrant.

    c. **GRAPHICAL** On the same coordinate system, graph $y = \sqrt{x}$ and $y = -\sqrt{x}$.

    d. **GRAPHICAL** On the same coordinate system, graph $y = x$. Plot the points (2, 4), (4, 2), and (1, 1).

    e. **ANALYTICAL** Compare the graph of the parabola to the graphs of the square root functions.

H.O.T. Problems

Use Higher-Order Thinking Skills

CHALLENGE Determine whether each statement is *true* or *false*. Provide an example or counterexample to support your answer.

45. Numbers in the domain of a radical function will always be nonnegative.

46. Numbers in the range of a radical function will always be nonnegative.

47. **REASONING** Write a radical function that translates $y = \sqrt{x}$ four units to the right. Graph the function.

48. **CHALLENGE** Write a radical function with a domain of all real numbers greater than or equal to 2 and a range of all real numbers less than or equal to 5.

49. **WHICH DOES NOT BELONG?** Identify the equation that does not belong. Explain.

$y = 3\sqrt{x}$	$y = 0.7\sqrt{x}$	$y = \sqrt{x} + 3$	$y = \frac{\sqrt{x}}{6}$

50. **OPEN ENDED** Write a function that is a reflection, translation, and a dilation of the parent graph $y = \sqrt{x}$.

51. **REASONING** If the range of the function $y = a\sqrt{x}$ is $\{y \mid y \leq 0\}$, what can you conclude about the value of a? Explain your reasoning.

52. **WRITING IN MATH** Compare and contrast the graphs of $f(x) = \sqrt{x} + 2$ and $g(x) = \sqrt{x + 2}$.

Standardized Test Practice

53. 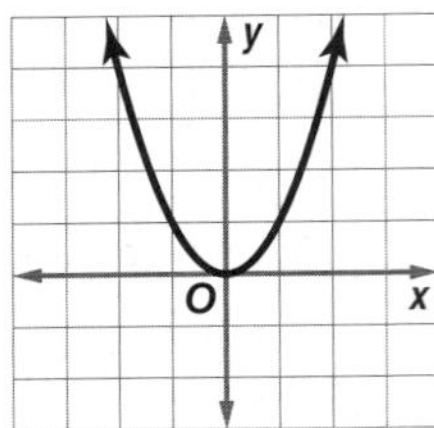

Which function *best* represents the graph?

A $y = x^2$ **C** $y = \sqrt{x}$

B $y = 2^x$ **D** $y = x$

54. The statement "$x < 10$ and $3x - 2 \geq 7$" is true when x is equal to what?

F 0 **H** 8

G 2 **J** 12

55. Which of the following is the equation of a line parallel to $y = -\frac{1}{2}x + 3$ and passing through $(-2, -1)$?

A $y = \frac{1}{2}x$ **C** $y = -\frac{1}{2}x + 2$

B $y = 2x + 3$ **D** $y = -\frac{1}{2}x - 2$

56. SHORT RESPONSE A landscaper needs to mulch 6 rectangular flower beds that are 8 feet by 4 feet and 4 circular flower beds each with a radius of 3 feet. One bag of mulch covers 25 square feet. How many bags of mulch are needed to cover the flower beds?

Spiral Review

Graph each set of ordered pairs. Determine whether the ordered pairs represent a *linear* function, a *quadratic* function, or an *exponential* function. (Lesson 9-9)

57. $\{(-2, 5), (-1, 3), (0, 1), (1, -1), (2, -3)\}$

58. $\{(0, 0), (1, 3), (2, 4), (3, 3), (4, 0)\}$

59. $\left\{\left(-2, \frac{1}{4}\right), (0, 1), (1, 2), (2, 4), (3, 8)\right\}$

60. $\{(-4, 4), (-2, 1), (0, 0), (2, 1), (4, 4)\}$

Find the next three terms in each geometric sequence. (Lesson 9-8)

61. 5, 20, 80, 320, …

62. $-4, 2, -1, \frac{1}{2}, \ldots$

63. $\frac{1}{8}, \frac{1}{4}, \frac{1}{2}, 1, \ldots$

64. HEALTH Aida exercises every day by walking and jogging at least 3 miles. Aida walks at a rate of 4 miles per hour and jogs at a rate of 8 miles per hour. Suppose she has exactly one half-hour to exercise today. (Lesson 6-8)

a. Draw a graph showing the possible amounts of time she can spend walking and jogging.

b. List three possible solutions.

65. NUTRITION Determine whether the graph shows a *positive* correlation, a *negative* correlation, or *no* correlation. If there is a positive or negative correlation, describe its meaning in the situation. (Lesson 4-5)

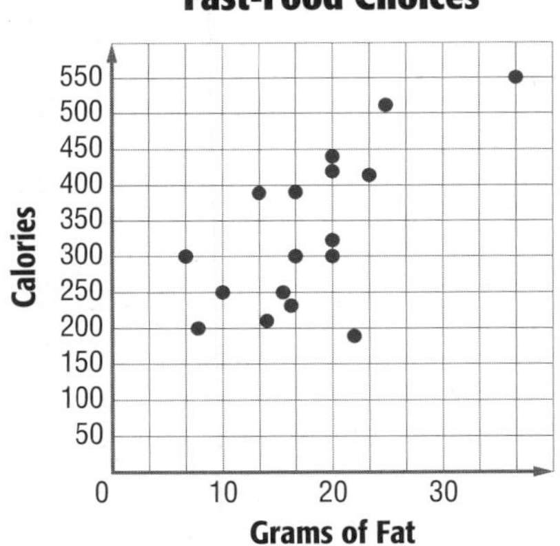

Skills Review

Factor each monomial completely. (Lesson 8-1)

66. $28n^3$

67. $-33a^2b$

68. $150rt$

69. $-378nq^2r^2$

70. $225a^3b^2c$

71. $-160x^2y^4$

EXTEND 10-1

Graphing Technology Lab
Graphing Square Root Functions

Math Online glencoe.com
- Other Calculator Keystrokes
- Graphing Technology Personal Tutor

For a square root to be a real number, the radicand cannot be negative. When graphing a radical function, determine when the radicand would be negative and exclude those values from the domain.

ACTIVITY 1 Parent Function

Graph $y = \sqrt{x}$.

Enter the equation in the Y= list.

KEYSTROKES: Y= 2nd [√] X,T,θ,n) GRAPH

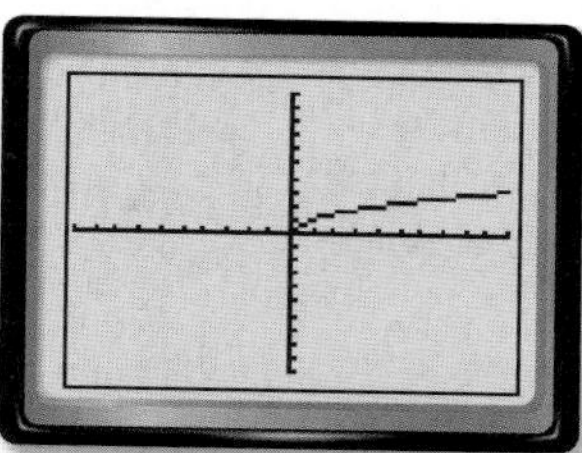

[−10, 10] scl: 1 by [−10, 10] scl: 1

1A. Examine the graph. What is the domain of the function?

1B. What is the range of the function?

ACTIVITY 2 Translation of Parent Function

Graph $y = \sqrt{x - 2}$.

Enter the equation in the Y= list.

KEYSTROKES: Y= 2nd [√] X,T,θ,n − 2) GRAPH

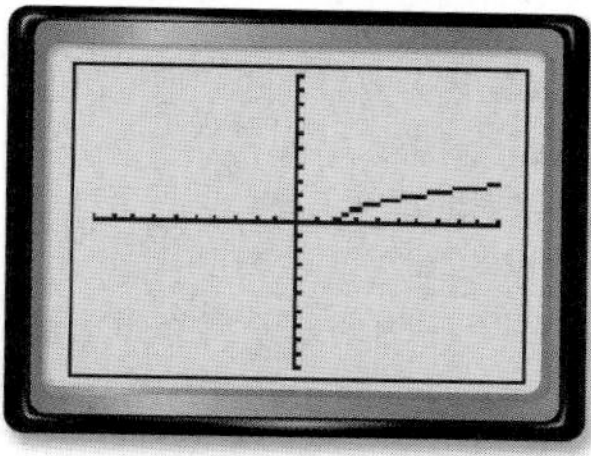

[−10, 10] scl: 1 by [−10, 10] scl: 1

2A. What are the domain and range of the function?

2B. How does the graph of $y = \sqrt{x - 2}$ compare to the graph of the parent function $y = \sqrt{x}$?

Exercises

Graph each equation, and sketch the graph on your paper. State the domain and range. Describe how the graph differs from that of the parent function $y = \sqrt{x}$.

1. $y = \sqrt{x - 1}$

2. $y = \sqrt{x + 3}$

3. $y = \sqrt{x} - 2$

4. $y = \sqrt{-x}$

5. $y = -\sqrt{x}$

6. $y = \sqrt{2x}$

7. $y = \sqrt{2 - x}$

8. $y = \sqrt{x - 3} + 2$

9. Does $x = y^2$ represent a function? Explain your reasoning.

10. Does $x^2 + y^2 = 4$ determine y as a function of x? Explain.

11. Does $x^2 + y^2 = 2$ determine y as a function of x? Explain.

Write a function with a graph that translates $y = \sqrt{x}$ in each way.

12. Shifted 4 units to the left

13. Shifted up 7 units

14. Shifted down 6 units

15. Shifted 5 units to the right and up 3 units

10-2 Simplifying Radical Expressions

Then

You simplified radicals. (Lesson 0-2)

Now

- Simplify radical expressions by using the Product Property of Square roots.
- Simplify radical expressions by using the Quotient Property of Square roots.

New Vocabulary

radical expression
rationalizing the denominator
conjugate

Math Online

glencoe.com

- Extra Examples
- Personal Tutor
- Self-Check Quiz
- Homework Help

Why?

The Sunshine Skyway Bridge across Tampa Bay in Florida, is supported by 21 steel cables, each 9 inches in diameter.

To find the diameter a steel cable should have to support a given weight, you can use the equation $d = \sqrt{\frac{w}{8}}$, where d is the diameter of the cable in inches and w is the weight in tons.

Product Property of Square Roots A **radical expression** contains a radical, such as a square root. Recall the expression under the radical sign is called the radicand. A radicand is in simplest form if the following three conditions are true.

- No radicands have perfect square factors other than 1.
- No radicands contain fractions.
- No radicals appear in the denominator of a fraction.

The following property can be used to simplify square roots.

Key Concept — Product Property of Square Roots — For Your FOLDABLE

Words For any nonnegative real numbers a and b, the square root of ab is equal to the square root of a times the square root of b.

Symbols $\sqrt{ab} = \sqrt{a} \cdot \sqrt{b}$, if $a \geq 0$ and $b \geq 0$

Examples $\sqrt{4 \cdot 9} = \sqrt{36}$ or 6 $\qquad \sqrt{4 \cdot 9} = \sqrt{4} \cdot \sqrt{9} = 2 \cdot 3$ or 6

EXAMPLE 1 Simplify Square Roots

Simplify $\sqrt{80}$.

$\sqrt{80} = \sqrt{2 \cdot 2 \cdot 2 \cdot 2 \cdot 5}$	Prime factorization of 80
$= \sqrt{2^2} \cdot \sqrt{2^2} \cdot \sqrt{5}$	Product Property of Square Roots
$= 2 \cdot 2 \cdot \sqrt{5}$ or $4\sqrt{5}$	Simplify.

Check Your Progress

1A. $\sqrt{54}$ **1B.** $\sqrt{180}$

Personal Tutor glencoe.com

EXAMPLE 2 Multiply Square Roots

Simplify $\sqrt{2} \cdot \sqrt{14}$.

$\sqrt{2} \cdot \sqrt{14} = \sqrt{2} \cdot \sqrt{2} \cdot \sqrt{7}$ **Product Property of Square Roots**

$= \sqrt{2^2} \cdot \sqrt{7}$ or $2\sqrt{7}$ **Product Property of Square Roots**

Check Your Progress

2A. $\sqrt{5} \cdot \sqrt{10}$

2B. $\sqrt{6} \cdot \sqrt{8}$

Personal Tutor glencoe.com

Consider the expression $\sqrt{x^2}$. It may seem that $x = \sqrt{x^2}$, but when finding the principal square root of an expression containing variables, you have to be sure that the result is not negative. Consider $x = -3$.

$\sqrt{x^2} \stackrel{?}{=} x$

$\sqrt{(-3)^2} \stackrel{?}{=} -3$ **Replace x with -3.**

$\sqrt{9} \stackrel{?}{=} -3$ **$(-3)^2 = 9$**

$3 \neq -3$ **$\sqrt{9} = 3$**

Notice in this case, if the right hand side of the equation were $|x|$, the equation would be true. For expressions where the exponent of the variable inside a radical is even and the simplified exponent is odd, you must use absolute value.

$\sqrt{x^2} = |x|$ $\quad \sqrt{x^3} = |x|\sqrt{x}$ $\quad \sqrt{x^4} = x^2$ $\quad \sqrt{x^6} = |x^3|$

EXAMPLE 3 Simplify a Square Root with Variables

Simplify $\sqrt{90x^3y^4z^5}$.

$\sqrt{90x^3y^4z^5} = \sqrt{2 \cdot 3^2 \cdot 5 \cdot x^3 \cdot y^4 \cdot z^5}$ **Prime factorization**

$= \sqrt{2} \cdot \sqrt{3^2} \cdot \sqrt{5} \cdot \sqrt{x^2} \cdot \sqrt{x} \cdot \sqrt{y^4} \cdot \sqrt{z^4} \cdot \sqrt{z}$ **Product Property**

$= \sqrt{2} \cdot 3 \cdot \sqrt{5} \cdot |x| \cdot \sqrt{x} \cdot y^2 \cdot z^2 \cdot \sqrt{z}$ **Simplify.**

$= 3y^2z^2|x|\sqrt{10xz}$ **Simplify.**

Check Your Progress

3A. $\sqrt{32r^2k^4t^5}$

3B. $\sqrt{56xy^{10}z^5}$

Personal Tutor glencoe.com

Quotient Property of Square Roots To divide square roots and simplify radical expressions, you can use the Quotient Property of Square Roots.

ReadingMath

Fractions in the Radicand The expression $\sqrt{\frac{a}{b}}$ is read *the square root of a over b, or the square root of the quantity of a over b.*

Key Concept Quotient Property of Square Roots

For Your FOLDABLE

Words For any real numbers a and b, where $a \geq 0$ and $b > 0$, the square root of $\frac{a}{b}$ is equal to the square root of a divided by the square root of b.

Symbols $\sqrt{\frac{a}{b}} = \frac{\sqrt{a}}{\sqrt{b}}$

You can use the properties of square roots to **rationalize the denominator** of a fraction with a radical. This involves multiplying the numerator and denominator by a factor that eliminates radicals in the denominator.

Test-TakingTip

Simplify Look at the radicand to see if it can be simplified first. This may make your computations simpler.

STANDARDIZED TEST EXAMPLE 4

Which expression is equivalent to $\sqrt{\frac{35}{15}}$?

A $\frac{5\sqrt{21}}{15}$ B $\frac{\sqrt{21}}{3}$ C $\frac{\sqrt{525}}{15}$ D $\frac{\sqrt{35}}{15}$

Read the Test Item

The radical expression needs to be simplified.

Solve the Test Item

$\sqrt{\frac{35}{15}} = \frac{\sqrt{35}}{\sqrt{15}}$ **Quotient Property of Square Roots**

$= \frac{\sqrt{35}}{\sqrt{15}} \cdot \frac{\sqrt{15}}{\sqrt{15}}$ **Multiply by $\frac{\sqrt{15}}{\sqrt{15}}$.**

$= \frac{\sqrt{525}}{15}$ **Product Property of Square Roots**

$= \frac{\sqrt{3 \cdot 5 \cdot 5 \cdot 7}}{15}$ **Prime factorization**

$= \frac{5\sqrt{21}}{15}$ or $\frac{\sqrt{21}}{3}$ The correct choice is B.

Check Your Progress

4. Simplify $\frac{\sqrt{6y}}{\sqrt{12}}$.

Personal Tutor glencoe.com

Binomials of the form $a\sqrt{b} + c\sqrt{d}$ and $a\sqrt{b} - c\sqrt{d}$, where a, b, c, and d are rational numbers, are called **conjugates**. For example, $2 + \sqrt{7}$ and $2 - \sqrt{7}$ are conjugates. The product of two conjugates is a rational number and can be found using the pattern for the difference of squares.

EXAMPLE 5 Use Conjugates to Rationalize a Denominator

Simplify $\frac{3}{5 + \sqrt{2}}$.

$\frac{3}{5 + \sqrt{2}} = \frac{3}{5 + \sqrt{2}} \cdot \frac{5 - \sqrt{2}}{5 - \sqrt{2}}$ **The conjugate of $5 + \sqrt{2}$ is $5 - \sqrt{2}$.**

$= \frac{3(5 - \sqrt{2})}{5^2 - (\sqrt{2})^2}$ **$(a - b)(a + b) = a^2 - b^2$**

$= \frac{15 - 3\sqrt{2}}{25 - 2}$ or $\frac{15 - 3\sqrt{2}}{23}$ **$(\sqrt{2})^2 = 2$**

Check Your Progress

Simplify each expression.

5A. $\frac{3}{2 + \sqrt{2}}$

5B. $\frac{7}{3 - \sqrt{7}}$

Personal Tutor glencoe.com

Check Your Understanding

Examples 1–3 pp. 612–613

Simplify each expression.

1. $\sqrt{24}$
2. $3\sqrt{16}$
3. $2\sqrt{25}$
4. $\sqrt{10} \cdot \sqrt{14}$
5. $\sqrt{3} \cdot \sqrt{18}$
6. $3\sqrt{10} \cdot 4\sqrt{10}$
7. $\sqrt{60x^4y^7}$
8. $\sqrt{88m^3p^2r^5}$
9. $\sqrt{99ab^5c^2}$

Example 4 p. 614

10. **MULTIPLE CHOICE** Which expression is equivalent to $\sqrt{\frac{45}{10}}$?

A $\frac{5\sqrt{2}}{10}$ B $\frac{\sqrt{450}}{10}$ C $\frac{\sqrt{50}}{10}$ D $\frac{3\sqrt{2}}{2}$

Example 5 p. 614

Simplify each expression.

11. $\frac{3}{3 + \sqrt{5}}$
12. $\frac{5}{2 - \sqrt{6}}$
13. $\frac{2}{1 - \sqrt{10}}$
14. $\frac{1}{4 + \sqrt{12}}$
15. $\frac{4}{6 - \sqrt{7}}$
16. $\frac{6}{5 + \sqrt{11}}$

Practice and Problem Solving

= **Step-by-Step Solutions** begin on page R12.
Extra Practice begins on page 815.

Examples 1 and 3 pp. 612–613

Simplify each expression.

17. $\sqrt{52}$
18. $\sqrt{56}$
19. $\sqrt{72}$
20. $3\sqrt{18}$
21. $\sqrt{243}$
22. $\sqrt{245}$
23. $\sqrt{5} \cdot \sqrt{10}$
24. $\sqrt{10} \cdot \sqrt{20}$
25. $3\sqrt{8} \cdot 2\sqrt{7}$
26. $4\sqrt{2} \cdot 5\sqrt{8}$
27. $3\sqrt{25t^2}$
28. $5\sqrt{81q^5}$
29. $\sqrt{28a^2b^3}$
30. $\sqrt{75qr^3}$
31. $7\sqrt{63m^3p}$
32. $4\sqrt{66g^2h^4}$
33. $\sqrt{2ab^2} \cdot \sqrt{10a^5b}$
34. $\sqrt{4c^3d^3} \cdot \sqrt{8c^3d}$

35. **ROLLER COASTER** The velocity v of a roller coaster in feet per second at the bottom of a hill can be approximated by $v = \sqrt{64h}$, where h is the height of the hill in feet.
   a. Simplify the equation.
   b. Determine the velocity of a roller coaster at the bottom of a 134-foot hill.

36. **FIREFIGHTING** When fighting a fire, the velocity v of water being pumped into the air is modeled by the function $v = \sqrt{2hg}$, where h represents the maximum height of the water and g represents the acceleration due to gravity (32 ft/s^2).
   a. Solve the function for h.
   b. The Hollowville Fire Department needs a pump that will propel water 80 feet into the air. Will a pump advertised to project water with a velocity of 70 feet per second meet their needs? Explain.
   c. The Jackson Fire Department must purchase a pump that will propel water 90 feet into the air. Will a pump that is advertised to project water with a velocity of 77 feet per second meet the fire department's need? Explain.

Real-World Link

In 1736, Benjamin Franklin founded the first volunteer fire organization, the Union Fire Company, in Philadelphia.

Source: *Firehouse Magazine*

Examples 4 and 5 p. 614

Simplify each expression.

37. $\sqrt{\frac{32}{t^4}}$

38. $\sqrt{\frac{27}{m^5}}$

39. $\frac{\sqrt{68ac^3}}{\sqrt{27a^2}}$

40. $\frac{\sqrt{h^3}}{\sqrt{8}}$

41. $\sqrt{\frac{3}{16}} \cdot \sqrt{\frac{9}{5}}$

42. $\sqrt{\frac{7}{2}} \cdot \sqrt{\frac{5}{3}}$

43. $\frac{7}{5 + \sqrt{3}}$

44. $\frac{9}{6 - \sqrt{8}}$

45. $\frac{3\sqrt{3}}{-2 + \sqrt{6}}$

46. $\frac{3}{\sqrt{7} - \sqrt{2}}$

47. $\frac{5}{\sqrt{6} + \sqrt{3}}$

48. $\frac{2\sqrt{5}}{2\sqrt{7} + 3\sqrt{3}}$

Real-World Link

The first hand-held hair dryer was sold in 1925 and dried hair with 100 watts of heat. Modern hair dryers may have 2000 watts.

Source: *Enotes Encyclopedia*

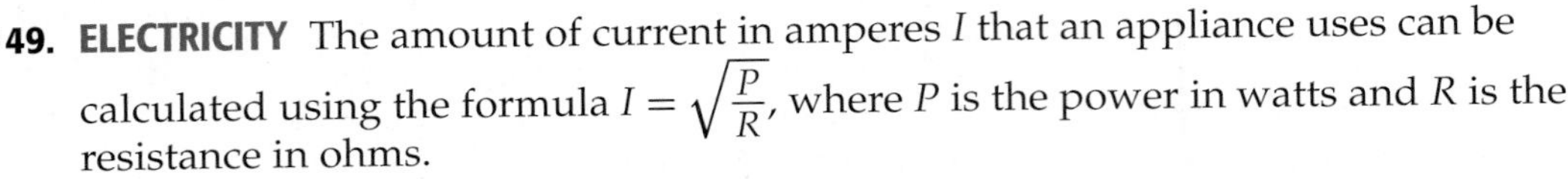

49. **ELECTRICITY** The amount of current in amperes I that an appliance uses can be calculated using the formula $I = \sqrt{\frac{P}{R}}$, where P is the power in watts and R is the resistance in ohms.

   a. Simplify the formula.

   b. How much current does an appliance use if the power used is 75 watts and the resistance is 5 ohms?

50. **KINETIC ENERGY** The speed v of a ball can be determined by the equation $v = \sqrt{\frac{2k}{m}}$, where k is the kinetic energy and m is the mass of the ball.

   a. Simplify the formula if the mass of the ball is 3 kilograms.

   b. If the ball is traveling 7 meters per second, what is the kinetic energy of the ball in Joules?

51. **SUBMARINES** The greatest distance d in miles that a lookout can see on a clear day is modeled by the formula shown. Determine how high the submarine would have to raise its periscope to see a ship, if the submarine is the given distances away from the ship.

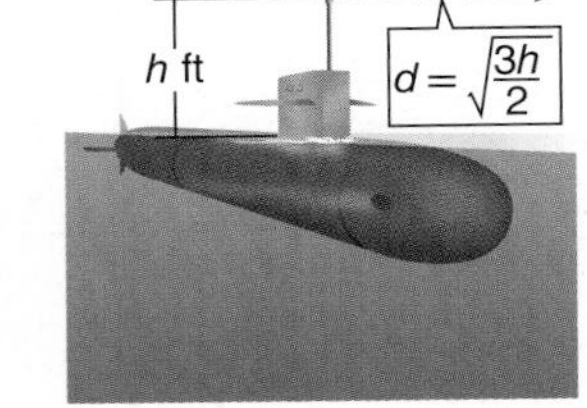

Distance	3	6	9	12	15
Height					

H.O.T. Problems

Use Higher-Order Thinking Skills

52. **REASONING** Explain how to solve $(3x - 2)^2 = (2x + 6)^2$.

53. **CHALLENGE** Solve $|y^3| = \frac{1}{3\sqrt{3}}$ for y.

54. **REASONING** Marge takes a number, subtracts 4, multiplies by 4, takes the square root, and takes the reciprocal to get $\frac{1}{2}$. What number did she start with? Write a formula to describe the process.

55. **OPEN ENDED** Write two binomials of the form $a\sqrt{b} + c\sqrt{f}$ and $a\sqrt{b} - c\sqrt{f}$. Then find their product.

56. **CHALLENGE** Use the Quotient Property of Square Roots to derive the Quadratic Formula by solving the quadratic equation $ax^2 + bx + c = 0$. (*Hint*: Begin by completing the square.)

57. **WRITING IN MATH** Summarize how to write a radical expression in simplest form.

Standardized Test Practice

58. Jerry's electric bill is \$23 less than his natural gas bill. The two bills are a total of \$109. Which of the following equations can be used to find the amount of his natural gas bill?

A $g + g = 109$

B $23 + 2g = 109$

C $g - 23 = 109$

D $2g - 23 = 109$

59. Solve $a^2 - 2a + 1 = 25$.

F $-4, -6$

G $4, -6$

H $-4, 6$

J $4, 6$

60. The expression $\sqrt{160x^2y^5}$ is equivalent to which of the following?

A $16|x|y^2\sqrt{10y}$

B $|x|y^2\sqrt{160y}$

C $4|x|y^2\sqrt{10y}$

D $10|x|y^2\sqrt{4y}$

61. **GRIDDED RESPONSE** Miki earns \$10 an hour and 10% commission on sales. If Miki worked 38 hours and had a total sales of \$1275 last week, how much did she make?

Spiral Review

Graph each function. Compare to the parent graph. State the domain and range. (Lesson 10-1)

62. $y = 2\sqrt{x} - 1$

63. $y = \frac{1}{2}\sqrt{x}$

64. $y = 2\sqrt{x + 2}$

65. $y = -\sqrt{x + 1}$

66. $y = -3\sqrt{x - 3}$

67. $y = -2\sqrt{x} + 1$

Look for a pattern in each table of values to determine which kind of model best describes the data. (Lesson 9-9)

68.

x	0	1	2	3	4
y	1	3	9	27	81

69.

x	−3	−2	−1	0	1
y	18	8	2	0	2

70.

x	1	2	3	4	5
y	1	3	5	7	9

71. **POPULATION** The country of Latvia has been experiencing a 1.1% annual decrease in population. In 2005, its population was 2,290,237. If the trend continues, predict Latvia's population in 2015. (Lesson 9-7)

Solve each equation by using the Quadratic Formula. Round to the nearest tenth if necessary. (Lesson 9-5)

72. $x^2 - 25 = 0$

73. $r^2 + 25 = 0$

74. $4w^2 + 100 = 40w$

75. $2r^2 + r - 14 = 0$

76. $5v^2 - 7v = 1$

77. $11z^2 - z = 3$

Factor each polynomial, if possible. If the polynomial cannot be factored, write *prime*. (Lesson 8-5)

78. $n^2 - 81$

79. $4 - 9a^2$

80. $2x^5 - 98x^3$

81. $32x^4 - 2y^4$

82. $4t^2 - 27$

83. $x^3 - 3x^2 - 9x + 27$

84. **GARDENING** Cleveland is planting 120 jalapeno pepper plants in a rectangular arrangement in his garden. In what ways can he arrange them so that he has at least 4 rows of plants, the same number of plants in each row, and at least 6 plants in each row? (Lesson 8-1)

Skills Review

Write the prime factorization of each number. (Concepts and Skills Bank Lesson 6)

85. 24

86. 88

87. 180

88. 31

89. 60

90. 90

EXTEND 10-2

Graphing Technology Lab
Rational Exponents

Math Online glencoe.com
- Other Calculator Keystrokes
- Graphing Technology Personal Tutor

You have studied the properties of exponents that are whole numbers. Some exponents are rational numbers or fractions. You can use a calculator to explore the meaning of rational exponents.

ACTIVITY Rational Exponents

Step 1 Evaluate $16^{\frac{1}{2}}$ and $\sqrt{16}$.

KEYSTROKES: 16 [^] [(] 1 [÷] 2 [)] [ENTER]

KEYSTROKES: [2nd] [$\sqrt{\ }$] 16 [ENTER]

Record the results in a table like the one at the right.

Step 2 Use a calculator to evaluate each expression. Record each result in your table. To find a root other than a square root, choose the $\sqrt[x]{\ }$ function from the [MATH] menu.

Expression	Value	Expression	Value
$16^{\frac{1}{2}}$	4	$\sqrt{16}$	4
$25^{\frac{1}{2}}$		$\sqrt{25}$	
$64^{\frac{1}{3}}$		$\sqrt[3]{64}$	
$125^{\frac{1}{3}}$		$\sqrt[3]{125}$	
$64^{\frac{2}{3}}$		$\sqrt[3]{64^2}$	
$81^{\frac{3}{4}}$		$\sqrt[4]{81^3}$	

1A. Study the table. What do you observe about the value of an expression of the form $a^{\frac{1}{n}}$?

1B. What do you observe about the value of an expression of the form $a^{\frac{m}{n}}$?

Exercises

1. Recall the Power of a Power Property. For any number a and all integers m and n, $(a^m)^n = (a^{m \cdot n})$. Assume that fractional exponents behave as whole number exponents and find the value of $\left(b^{\frac{1}{2}}\right)^2$.

$\left(b^{\frac{1}{2}}\right)^2 = b^{\frac{1}{2} \cdot 2}$ **Power of a Power Property**

$= b^1$ or b **Simplify.**

Thus, $b^{\frac{1}{2}}$ is a number whose square equals b. So it makes sense to define $b^{\frac{1}{2}} = \sqrt{b}$. Use a similar process to define $b^{\frac{1}{n}}$.

2. Define $b^{\frac{m}{n}}$. Justify your answer.

Write each root as an expression using a fractional exponent. Then evaluate the expression.

3. $\sqrt{36}$
4. $\sqrt{121}$
5. $\sqrt[4]{256}$
6. $\sqrt[5]{32}$
7. $\sqrt[3]{8^2}$
8. $\sqrt[4]{1296}$
9. $\sqrt[4]{16^3}$
10. $\sqrt[3]{8^3}$

10-3 Operations with Radical Expressions

Then

You simplified radical expressions. (Lesson 10-2)

Now

- Add and subtract radical expressions.
- Multiply radical expressions.

Math Online

glencoe.com

- Extra Examples
- Personal Tutor
- Self-Check Quiz
- Homework Help

Why?

Conchita is going to run in her neighborhood to get ready for the soccer season. She plans to run the course that she has laid out three times each day.

How far does Conchita have to run to complete the course that she laid out?

How far does she run every day?

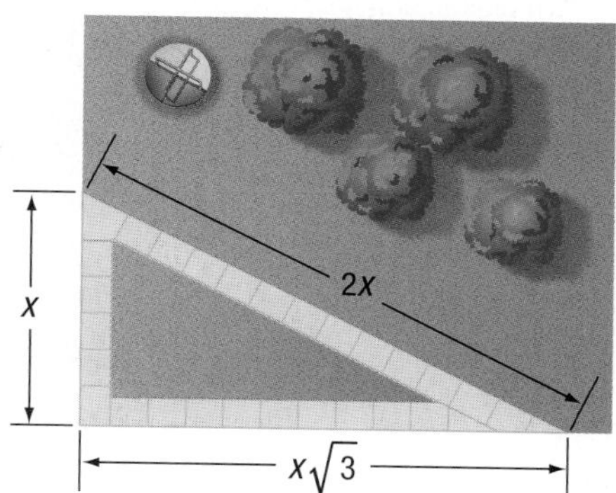

Add or Subtract Radical Expressions To add or subtract radical expressions, the radicands must be alike in the same way that monomial terms must be alike to add or subtract.

Monomials	Radical Expressions
$4a + 2a = (4 + 2)a$	$4\sqrt{5} + 2\sqrt{5} = (4 + 2)\sqrt{5}$
$= 6a$	$= 6\sqrt{5}$
$9b - 2b = (9 - 2)b$	$9\sqrt{3} - 2\sqrt{3} = (9 - 2)\sqrt{3}$
$= 7b$	$= 7\sqrt{3}$

Notice that when adding and subtracting radical expressions, the radicand does not change. This is the same as when adding or subtracting monomials.

EXAMPLE 1 Add and Subtract Expressions with Like Radicands

Simplify each expression.

a. $5\sqrt{2} + 7\sqrt{2} - 6\sqrt{2}$

$5\sqrt{2} + 7\sqrt{2} - 6\sqrt{2} = (5 + 7 - 6)\sqrt{2}$ **Distributive Property**

$= 6\sqrt{2}$ **Simplify.**

b. $10\sqrt{7} + 5\sqrt{11} + 4\sqrt{7} - 6\sqrt{11}$

$10\sqrt{7} + 5\sqrt{11} + 4\sqrt{7} - 6\sqrt{11} = (10 + 4)\sqrt{7} + (5 - 6)\sqrt{11}$ **Distributive Property**

$= 14\sqrt{7} - \sqrt{11}$ **Simplify.**

Check Your Progress

1A. $3\sqrt{2} - 5\sqrt{2} + 4\sqrt{2}$

1B. $6\sqrt{11} + 2\sqrt{11} - 9\sqrt{11}$

1C. $15\sqrt{3} - 14\sqrt{5} + 6\sqrt{5} - 11\sqrt{3}$

1D. $4\sqrt{3} + 3\sqrt{7} - 6\sqrt{3} + 3\sqrt{7}$

Personal Tutor glencoe.com

Not all radical expressions have like radicands. Simplifying the expressions may make it possible to have like radicands so that they can be added or subtracted.

StudyTip

Simplify First Simplify each radical term first. Then perform the operations needed.

EXAMPLE 2 Add and Subtract Expressions with Unlike Radicands

Simplify $2\sqrt{18} + 2\sqrt{32} + \sqrt{72}$.

$2\sqrt{18} + 2\sqrt{32} + \sqrt{72} = 2(\sqrt{3^2} \cdot \sqrt{2}) + 2(\sqrt{4^2} \cdot \sqrt{2}) + (\sqrt{6^2} \cdot \sqrt{2})$	**Product Property**
$= 2(3\sqrt{2}) + 2(4\sqrt{2}) + (6\sqrt{2})$	**Simplify.**
$= 6\sqrt{2} + 8\sqrt{2} + 6\sqrt{2}$	**Multiply.**
$= 20\sqrt{2}$	**Simplify.**

Check Your Progress

2A. $4\sqrt{54} + 2\sqrt{24}$

2B. $4\sqrt{12} - 6\sqrt{48}$

2C. $3\sqrt{45} + \sqrt{20} - \sqrt{245}$

2D. $\sqrt{24} - \sqrt{54} + \sqrt{96}$

Personal Tutor glencoe.com

Multiply Radical Expressions Multiplying radical expressions is similar to multiplying monomial algebraic expressions. Let $x \geq 0$.

Monomials	Radical Expressions
$(2x)(3x) = 2 \cdot 3 \cdot x \cdot x$	$(2\sqrt{x})(3\sqrt{x}) = 2 \cdot 3 \cdot \sqrt{x} \cdot \sqrt{x}$
$= 6x^2$	$= 6x$

You can also apply the Distributive Property to radical expressions.

Watch Out!

Multiplying Radicands Make sure that you multiply the radicands when multiplying radical expressions. A common mistake is to add the radicands rather than multiply.

EXAMPLE 3 Multiply Radical Expressions

Simplify each expression.

a. $3\sqrt{2} \cdot 2\sqrt{6}$

$3\sqrt{2} \cdot 2\sqrt{6} = (3 \cdot 2)(\sqrt{2} \cdot \sqrt{6})$	**Associative Property**
$= 6(\sqrt{12})$	**Multiply.**
$= 6(2\sqrt{3})$	**Simplify.**
$= 12\sqrt{3}$	**Multiply.**

b. $3\sqrt{5}(2\sqrt{5} + 5\sqrt{3})$

$3\sqrt{5}(2\sqrt{5} + 5\sqrt{3}) = (3\sqrt{5} \cdot 2\sqrt{5}) + (3\sqrt{5} \cdot 5\sqrt{3})$	**Distributive Property**
$= [(3 \cdot 2)(\sqrt{5} \cdot \sqrt{5})] + [(3 \cdot 5)(\sqrt{5} \cdot \sqrt{3})]$	**Associative Property**
$= [6(\sqrt{25})] + [15(\sqrt{15})]$	**Multiply.**
$= [6(5)] + [15(\sqrt{15})]$	**Simplify.**
$= 30 + 15\sqrt{15}$	**Multiply.**

Check Your Progress

3A. $2\sqrt{6} \cdot 7\sqrt{3}$

3B. $9\sqrt{5} \cdot 11\sqrt{15}$

3C. $3\sqrt{2}(4\sqrt{3} + 6\sqrt{2})$

3D. $5\sqrt{3}(3\sqrt{2} - \sqrt{3})$

Personal Tutor glencoe.com

You can also multiply radical expressions with more than one term in each factor. This is similar to multiplying two algebraic binomials with variables.

Real-World EXAMPLE 4 Multiply Radical Expressions

GEOMETRY **Find the area of the rectangle in simplest form.**

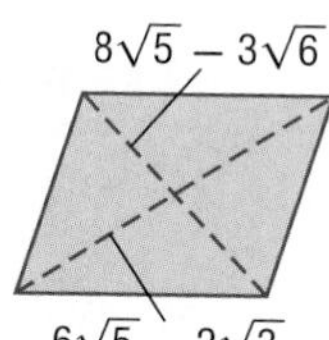

$A = (5\sqrt{2} - \sqrt{3})(\sqrt{5} + 4\sqrt{3})$ $\quad A = \ell \cdot w$

$$= \overbrace{(5\sqrt{2})(\sqrt{5})}^{\text{First Terms}} + \overbrace{(5\sqrt{2})(4\sqrt{3})}^{\text{Outer Terms}} + \overbrace{(-\sqrt{3})(\sqrt{5})}^{\text{Inner Terms}} + \overbrace{(-\sqrt{3})(4\sqrt{3})}^{\text{Last Terms}}$$

$= 5\sqrt{10} + 20\sqrt{6} - \sqrt{15} - 4\sqrt{9}$ **Multiply.**

$= 5\sqrt{10} + 20\sqrt{6} - \sqrt{15} - 12$ **Simplify.**

Review Vocabulary

FOIL Method Multiply two binomials by finding the sum of the products of the First terms, the Outer terms, the Inner terms, and the Last terms. (Lesson 7-7)

Check Your Progress

4. **GEOMETRY** The area A of a rhombus can be found using the equation $A = \frac{1}{2}d_1d_2$, where d_1 and d_2 are the lengths of the diagonals. What is the area of the rhombus at the right?

$8\sqrt{5} - 3\sqrt{6}$

$6\sqrt{5} - 2\sqrt{3}$

Personal Tutor glencoe.com

Concept Summary

For Your FOLDABLE

Operations with Radical Expressions

Operation	Symbols	Example
addition, $b \geq 0$	$a\sqrt{b} + c\sqrt{b} = (a + c)\sqrt{b}$ like radicands	$4\sqrt{3} + 6\sqrt{3} = (4 + 6)\sqrt{3} = 10\sqrt{3}$
subtraction, $b \geq 0$	$a\sqrt{b} - c\sqrt{b} = (a - c)\sqrt{b}$ like radicands	$12\sqrt{5} - 8\sqrt{5} = (12 - 8)\sqrt{5} = 4\sqrt{5}$
multiplication, $b \geq 0, g \geq 0$	$a\sqrt{b}(f\sqrt{g}) = af\sqrt{bg}$ Radicands do not have to be like radicands.	$3\sqrt{2}(5\sqrt{7}) = (3 \cdot 5)(\sqrt{2 \cdot 7}) = 15\sqrt{14}$

Check Your Understanding

Examples 1–3 pp. 619–620

Simplify each expression.

1. $3\sqrt{5} + 6\sqrt{5}$
2. $8\sqrt{3} + 5\sqrt{3}$
3. $\sqrt{7} - 6\sqrt{7}$
4. $10\sqrt{2} - 6\sqrt{2}$
5. $4\sqrt{5} + 2\sqrt{20}$
6. $\sqrt{12} - \sqrt{3}$
7. $\sqrt{8} + \sqrt{12} + \sqrt{18}$
8. $\sqrt{27} + 2\sqrt{3} - \sqrt{12}$
9. $9\sqrt{2}(4\sqrt{6})$
10. $4\sqrt{3}(8\sqrt{3})$
11. $\sqrt{3}(\sqrt{7} + 3\sqrt{2})$
12. $\sqrt{5}(\sqrt{2} + 4\sqrt{2})$

Example 4 p. 621

13. **GEOMETRY** The area A of a triangle can be found by using the formula $A = \frac{1}{2}bh$, where b represents the base and h is the height. What is the area of the triangle at the right?

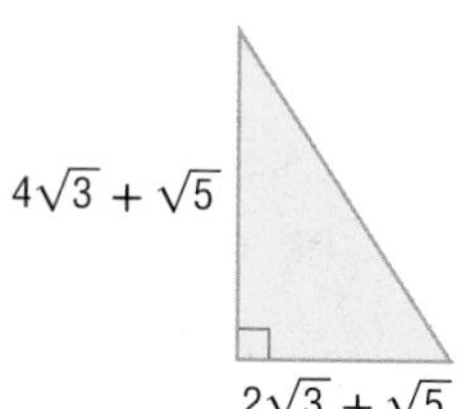

Practice and Problem Solving

= Step-by-Step Solutions begin on page R12.
Extra Practice begins on page 815.

Examples 1–3 pp. 619–620

Simplify each expression.

14. $7\sqrt{5} + 4\sqrt{5}$

15. $2\sqrt{6} + 9\sqrt{6}$

16. $3\sqrt{5} - 2\sqrt{20}$

17. $3\sqrt{50} - 3\sqrt{32}$

18. $7\sqrt{3} - 2\sqrt{2} + 3\sqrt{2} + 5\sqrt{3}$

19. $\sqrt{5}(\sqrt{2} + 4\sqrt{2})$

20. $\sqrt{6}(2\sqrt{10} + 3\sqrt{2})$

21. $4\sqrt{5}(3\sqrt{5} + 8\sqrt{2})$

22. $5\sqrt{3}(6\sqrt{10} - 6\sqrt{3})$

23. $(\sqrt{3} - \sqrt{2})(\sqrt{15} + \sqrt{12})$

24. $(3\sqrt{11} + 3\sqrt{15})(3\sqrt{3} - 2\sqrt{2})$

25. $(5\sqrt{2} + 3\sqrt{5})(2\sqrt{10} - 5)$

Example 4 p. 621

26. GEOMETRY Find the perimeter and area of a rectangle with a width of $2\sqrt{7} - 2\sqrt{5}$ and a length of $3\sqrt{7} + 3\sqrt{5}$.

Simplify each expression.

27. $\sqrt{\frac{1}{5}} - \sqrt{5}$

28. $\sqrt{\frac{2}{3}} + \sqrt{6}$

29. $2\sqrt{\frac{1}{2}} + 2\sqrt{2} - \sqrt{8}$

30. $8\sqrt{\frac{5}{4}} + 3\sqrt{20} - 10\sqrt{\frac{1}{5}}$

31. $(3 - \sqrt{5})^2$

32. $(\sqrt{2} + \sqrt{3})^2$

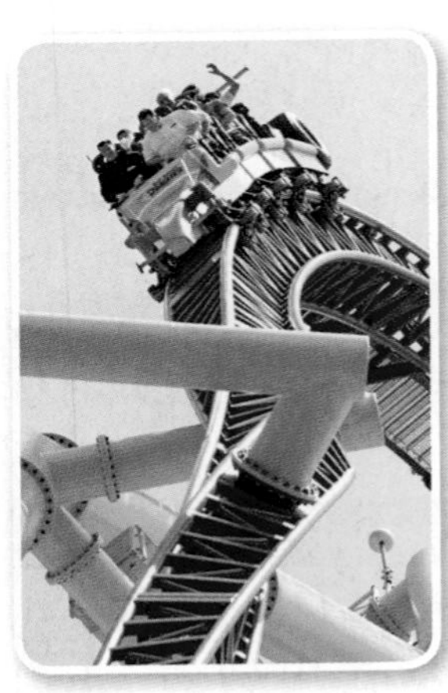

Real-World Link

The Top Thrill Dragster roller coaster at Ohio's Cedar Point is powered by a 10,000 horsepower motor–equivalent to about 10 Formula One racecars.

Source: *National Geographic*

33. ROLLER COASTERS The velocity v in feet per second of a roller coaster at the bottom of a hill is related to the vertical drop h in feet and the velocity v_0 of the coaster at the top of the hill by the formula $v_0 = \sqrt{v^2 - 64h}$.

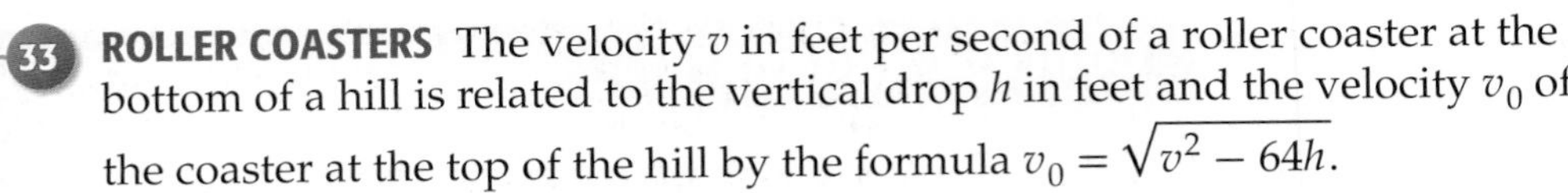

a. What velocity must a coaster have at the top of a 225-foot hill to achieve a velocity of 120 feet per second at the bottom?

b. Explain why $v_0 = v - 8\sqrt{h}$ is not equivalent to the formula given.

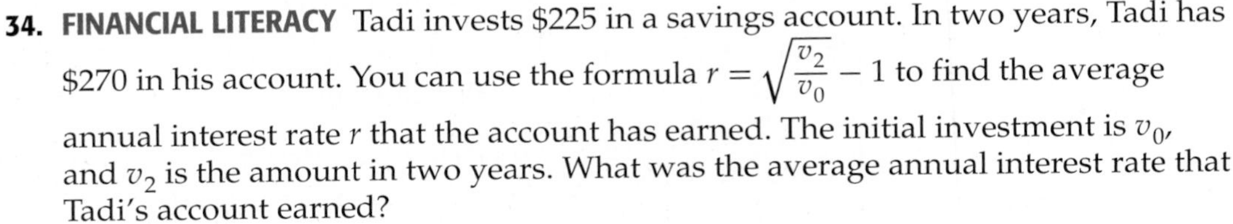

34. FINANCIAL LITERACY Tadi invests \$225 in a savings account. In two years, Tadi has \$270 in his account. You can use the formula $r = \sqrt{\frac{v_2}{v_0}} - 1$ to find the average annual interest rate r that the account has earned. The initial investment is v_0, and v_2 is the amount in two years. What was the average annual interest rate that Tadi's account earned?

35. ELECTRICITY Electricians can calculate the electrical current in amps A by using the formula $A = \frac{\sqrt{w}}{\sqrt{r}}$, where w is the power in watts and r the resistance in ohms. How much electrical current is running through a microwave oven that has 850 watts of power and 5 ohms of resistance? Write the number of amps in simplest radical form, and then estimate the amount of current to the nearest tenth.

H.O.T. Problems

Use **H**igher-**O**rder **T**hinking Skills

36. CHALLENGE Determine whether the following statement is *true* or *false*. Provide an example or counterexample to support your answer.

$$x + y > \sqrt{x^2 + y^2} \text{ when } x > 0 \text{ and } y > 0$$

37. REASONING Let a, b, c, d, and f be rational numbers. Show that if you multiply $a\sqrt{b} + c\sqrt{f}$ and $a\sqrt{b} - c\sqrt{f}$, the product has no radicals. Explain why this occurs.

38. OPEN ENDED Write an equation that shows a sum of two radicals with different radicands. Explain how you could combine these terms.

39. WRITING IN MATH Describe step by step how to multiply two radical expressions, each with two terms. Write an example to demonstrate your description.

Standardized Test Practice

40. SHORT RESPONSE The population of a town is 13,000 and is increasing by about 250 people per year. This can be represented by the equation $p = 13{,}000 + 250y$, where y is the number of years from now and p represents the population. In how many years will the population of the town be 14,500?

41. GEOMETRY Which expression represents the sum of the lengths of the 12 edges on this rectangular solid?

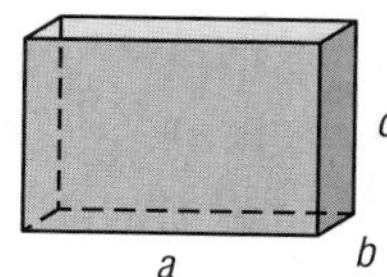

A $2(a + b + c)$
B $3(a + b + c)$
C $4(a + b + c)$
D $12(a + b + c)$

42. Which of the following is equivalent to $8(3 - y) + 5(3 - y)$?

F $39 - y$
G $13(3 - y)$
H $40(30 - y)$
J $13(6 - 2y)$

43. The current I in a simple electrical circuit is given by the formula $I = \frac{V}{R}$, where V is the voltage and R is the resistance of the circuit. If the voltage remains unchanged, what effect will doubling the resistance of the circuit have on the current?

A The current will remain the same.
B The current will double its previous value.
C The current will be half its previous value.
D The current will be two units more than its previous value.

Spiral Review

Simplify. (Lesson 10-2)

44. $\sqrt{18}$
45. $\sqrt{24}$
46. $\sqrt{60}$
47. $\sqrt{50a^3b^5}$
48. $\sqrt{169x^4y^7}$
49. $\sqrt{63c^3d^4f^5}$

Graph each function. Compare to the parent graph. State the domain and range. (Lesson 10-1)

50. $y = 2\sqrt{x}$
51. $y = -3\sqrt{x}$
52. $y = \sqrt{x + 1}$
53. $y = \sqrt{x - 4}$
54. $y = \sqrt{x} + 3$
55. $y = \sqrt{x} - 2$

56. FINANCIAL LITERACY Determine the value of an investment if \$400 is invested at an interest rate of 7.25% compounded quarterly for 7 years. (Lesson 9-7)

Factor each trinomial. (Lesson 8-2)

57. $x^2 + 12x + 27$
58. $y^2 + 13y + 30$
59. $p^2 - 17p + 72$
60. $x^2 + 6x - 7$
61. $y^2 - y - 42$
62. $-72 + 6w + w^2$

Skills Review

Solve each equation. Round each solution to the nearest tenth, if necessary. (Lesson 2-3)

63. $-4c - 1.2 = 0.8$
64. $-2.6q - 33.7 = 84.1$
65. $0.3m + 4 = 9.6$
66. $-10 - \frac{n}{5} = 6$
67. $\frac{-4h - (-5)}{-7} = 13$
68. $3.6t + 6 - 2.5t = 8$

10-4 Radical Equations

Why?

The waterline length of a sailboat is the length of the line made by the water's edge when the boat is full. A sailboat's hull speed is the fastest speed that it can travel.

You can estimate hull speed h by using the formula $h = 1.34\sqrt{\ell}$, where ℓ is the length of the sailboat's waterline.

Then

You added, subtracted, and multiplied radical expressions. (Lesson 10-3)

Now

- Solve radical equations.
- Solve radical equations with extraneous solutions.

New Vocabulary

radical equations
extraneous solutions

Math Online

glencoe.com

- Extra Examples
- Personal Tutor
- Self-Check Quiz
- Homework Help

Radical Equations Equations that contain variables in the radicand, like $h = 1.34\sqrt{\ell}$, are called **radical equations**. To solve, isolate the desired variable on one side of the equation first. Then square each side of the equation to eliminate the radical.

Key Concept — Power Property of Equality

For Your FOLDABLE

Words If you square both sides of a true equation, the resulting equation is still true.

Symbols If $a = b$, then $a^2 = b^2$.

Example If $\sqrt{x} = 4$, then $(\sqrt{x})^2 = 4^2$.

Real-World EXAMPLE 1 — Variable as a Radicand

SAILING Idris and Sebastian are sailing in a friend's sailboat. They measure the hull speed at 9 nautical miles per hour. Find the length of the sailboat's waterline. Round to the nearest foot.

Understand You know how fast the boat will travel and that it relates to the length.

Plan The boat travels at 9 nautical miles per hour. The formula for hull speed is $h = 1.34\sqrt{\ell}$.

Solve

$h = 1.34\sqrt{\ell}$	**Formula for hull speed**
$9 = 1.34\sqrt{\ell}$	**Substitute 9 for *h*.**
$\frac{9}{1.34} = \frac{1.34\sqrt{\ell}}{1.34}$	**Divide each side by 1.34.**
$6.72 \approx \sqrt{\ell}$	**Simplify.**
$(6.72)^2 \approx (\sqrt{\ell})^2$	**Square each side of the equation.**
$45.16 \approx \ell$	**Simplify.**

The sailboat's waterline length is about 45 feet.

Check Check by substituting the estimate into the original formula.

$h = 1.34\sqrt{\ell}$	**Formula for hull speed**
$9 \stackrel{?}{=} 1.34\sqrt{45}$	**$h = 9$ and $\ell = 45$**
$9 \approx 8.98899327$ ✓	**Multiply.**

Check Your Progress

1. **DRIVING** The equation $v = \sqrt{2.5r}$ represents the maximum velocity that a car can travel safely on an unbanked curve when v is the maximum velocity in miles per hour and r is the radius of the turn in feet. If a road is designed for a maximum speed of 65 miles per hour, what is the radius of the turn?

Personal Tutor glencoe.com

To solve a radical equation, isolate the radical first. Then square both sides of the equation.

Watch Out!

Squaring Each Side Remember that when you square each side of the equation, you must square the entire side of the equation, even if there is more than one term on the side.

EXAMPLE 2 Expression as a Radicand

Solve $\sqrt{a+5} + 7 = 12$.

$\sqrt{a+5} + 7 = 12$ — **Original equation**

$\sqrt{a+5} = 5$ — **Subtract 7 from each side.**

$\left(\sqrt{a+5}\right)^2 = 5^2$ — **Square each side.**

$a + 5 = 25$ — **Simplify.**

$a = 20$ — **Subtract 5 from each side.**

Check Your Progress Solve each equation.

2A. $\sqrt{c-3} - 2 = 4$

2B. $4 + \sqrt{h+1} = 14$

Personal Tutor glencoe.com

Extraneous Solutions Squaring each side of an equation sometimes produces a solution that is not a solution of the original equation. These are called **extraneous solutions**. Therefore, you must check all solutions in the original equation.

StudyTip

Extraneous Solutions When checking solutions for extraneous solutions, we are only interested in principal roots.

EXAMPLE 3 Variable on Each Side

Solve $\sqrt{k+1} = k - 1$. Check your solution.

$\sqrt{k+1} = k - 1$ — **Original equation**

$\left(\sqrt{k+1}\right)^2 = (k-1)^2$ — **Square each side.**

$k + 1 = k^2 - 2k + 1$ — **Simplify.**

$0 = k^2 - 3k$ — **Subtract k and 1 from each side.**

$0 = k(k-3)$ — **Factor.**

$k = 0$ or $k - 3 = 0$ — **Zero Product Property**

$k = 3$ — **Solve.**

CHECK

$\sqrt{k+1} = k - 1$	**Original equation**	$\sqrt{k+1} = k - 1$	**Original equation**
$\sqrt{0+1} \stackrel{?}{=} 0 - 1$	**$k = 0$**	$\sqrt{3+1} \stackrel{?}{=} 3 - 1$	**$k = 3$**
$\sqrt{1} \stackrel{?}{=} -1$	**Simplify.**	$\sqrt{4} \stackrel{?}{=} 2$	**Simplify.**
$1 \neq -1$ ✗	**False**	$2 = 2$ ✓	**True**

Since 0 does not satisfy the original equation, 3 is the only solution.

Check Your Progress

Solve each equation. Check your solution.

3A. $\sqrt{t+5} = t + 3$

3B. $x - 3 = \sqrt{x-1}$

Personal Tutor glencoe.com

Check Your Understanding

Example 1 p. 624

1. **GEOMETRY** The surface area of a basketball is x square inches. What is the radius of the basketball if the formula for the surface area of a sphere is $SA = 4\pi r^2$?

Examples 2 and 3 p. 625

Solve each equation. Check your solution.

2. $\sqrt{10h} + 1 = 21$
3. $\sqrt{7r + 2} + 3 = 7$
4. $5 + \sqrt{g - 3} = 6$
5. $\sqrt{3x - 5} = x - 5$
6. $\sqrt{2n + 3} = n$
7. $\sqrt{a - 2} + 4 = a$

Practice and Problem Solving

= **Step-by-Step Solutions** begin on page R12.
Extra Practice begins on page 815.

Example 1 p. 624

8. **EXERCISE** Suppose the function $S = \pi\sqrt{\frac{9.8\ell}{7}}$, where S represents speed in meters per second and ℓ is the leg length of a person in meters, can approximate the maximum speed that a person can run.
   a. What is the maximum running speed of a person with a leg length of 1.1 meters to the nearest tenth of a meter?
   b. What is the leg length of a person with a running speed of 2.7 meters per second to the nearest tenth of a meter?
   c. As leg length increases, does maximum speed increase or decrease? Explain.

Examples 2 and 3 p. 625

Solve each equation. Check your solution.

9. $\sqrt{a} + 11 = 21$
10. $\sqrt{t} - 4 = 7$
11. $\sqrt{n - 3} = 6$
12. $\sqrt{c + 10} = 4$
13. $\sqrt{h - 5} = 2\sqrt{3}$
14. $\sqrt{k + 7} = 3\sqrt{2}$
15. $y = \sqrt{12 - y}$
16. $\sqrt{u + 6} = u$
17. $\sqrt{r + 3} = r - 3$
18. $\sqrt{1 - 2t} = 1 + t$
19. $5\sqrt{a - 3} + 4 = 14$
20. $2\sqrt{x - 11} - 8 = 4$

21. **RIDES** The amount of time t, in seconds, that it takes a simple pendulum to complete a full swing is called the *period*. It is given by $t = 2\pi\sqrt{\frac{\ell}{32}}$, where ℓ is the length of the pendulum, in feet.
   a. The Giant Swing completes a period in about 8 seconds. About how long is the pendulum's arm? Round to the nearest foot.
   b. Does increasing the length of the pendulum increase or decrease the period? Explain.

Real-World Link

The Giant Swing at Silver Dollar City in Branson, Missouri, swings riders at 45 miles per hour and reaches a height of 7 stories.

Source: Silver Dollar City Amusement Park

Solve each equation. Check your solution.

22. $\sqrt{6a - 6} = a + 1$
23. $\sqrt{x^2 + 9x + 15} = x + 5$
24. $6\sqrt{\frac{5k}{4}} - 3 = 0$
25. $\sqrt{\frac{5y}{6}} - 10 = 4$
26. $\sqrt{2a^2 - 121} = a$
27. $\sqrt{5x^2 - 9} = 2x$

28. **GEOMETRY** The formula for the slant height c of a cone is $c = \sqrt{h^2 + r^2}$, where h is the height of the cone and r is the radius of its base. Find the height of the cone if the slant height is 4 and the radius is 2. Round to the nearest tenth.

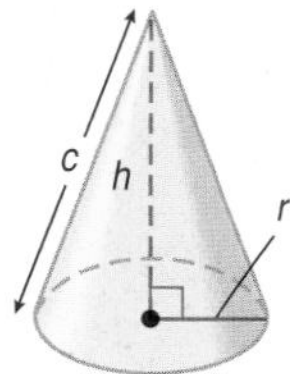

29. **MULTIPLE REPRESENTATIONS** Consider $\sqrt{2x - 7} = x - 7$.

a. **GRAPHICAL** Clear the Y= list. Enter the left side of the equation as Y1 $= \sqrt{2x - 7}$. Enter the right side of the equation as Y2 $= x - 7$. Press GRAPH.

b. **GRAPHICAL** Sketch what is shown on the screen.

c. **ANALYTICAL** Use the **intersect** feature on the CALC menu to find the point of intersection.

d. **ANALYTICAL** Solve the radical equation algebraically. How does your solution compare to the solution from the graph?

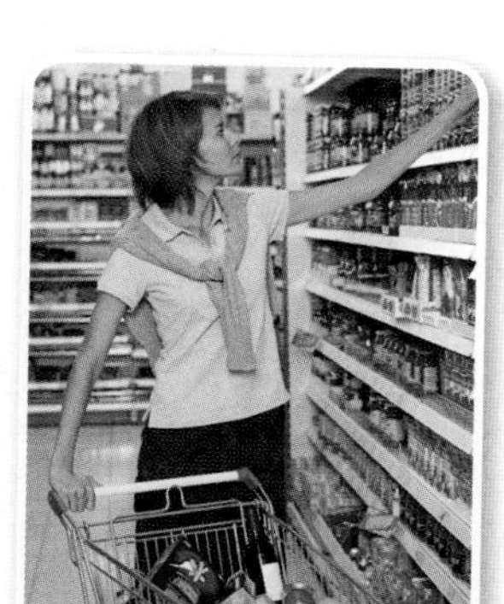

Real-World Link

Packaging has several objectives, including physical protection, information transmission, marketing, convenience, security, and portion control.

Source: *Packaging World*

30. **PACKAGING** A cylindrical container of chocolate drink mix has a volume of 162 cubic inches. The radius r of the container can be found by using the formula $r = \sqrt{\frac{V}{\pi h}}$, where V is the volume of the container and h is the height.

a. If the radius is 2.5 inches, find the height of the container. Round to the nearest hundredth.

b. If the height of the container is 10 inches, find the radius. Round to the nearest hundredth.

H.O.T. Problems

Use **H**igher-**O**rder **T**hinking Skills

31. **FIND THE ERROR** Jada and Fina solved $\sqrt{6 - b} = \sqrt{b + 10}$. Is either of them correct? Explain.

Jada

$\sqrt{6-b} = \sqrt{b+10}$

$(\sqrt{6-b})^2 = (\sqrt{b+10})^2$

$6 - b = b + 10$

$-2b = 4$

$b = -2$

Check $\sqrt{6-(-2)} \stackrel{?}{=} \sqrt{(-2)+10}$

$\sqrt{8} = \sqrt{8}$ ✓

Fina

$\sqrt{6-b} = \sqrt{b+10}$

$(\sqrt{6-b})^2 = (\sqrt{b+10})^2$

$6 - b = b + 10$

$2b = 4$

$b = 2$

Check $\sqrt{6-(2)} \stackrel{?}{=} \sqrt{(2)+10}$

$\sqrt{4} \neq \sqrt{12}$ X

no solution

32. **REASONING** Which equation has the same solution set as $\sqrt{4} = \sqrt{x + 2}$? Explain.

A. $\sqrt{4} = \sqrt{x} + \sqrt{2}$ **B.** $4 = x + 2$ **C.** $2 - \sqrt{2} = \sqrt{x}$

33. **REASONING** Explain how solving $5 = \sqrt{x} + 1$ is different from solving $5 = \sqrt{x + 1}$.

34. **OPEN ENDED** Write a radical equation with a variable on each side. Then solve the equation.

35. **REASONING** Is the following equation *sometimes, always* or *never* true? Explain.

$$\sqrt{(x - 2)^2} = x - 2$$

36. **CHALLENGE** Solve $\sqrt{x + 9} = \sqrt{3} + \sqrt{x}$.

37. **WRITING IN MATH** Write some general rules about how to solve radical equations. Demonstrate your rules by solving a radical equation.

Standardized Test Practice

38. SHORT RESPONSE Zack needs to drill a hole at A, B, C, D, and E on circle P.

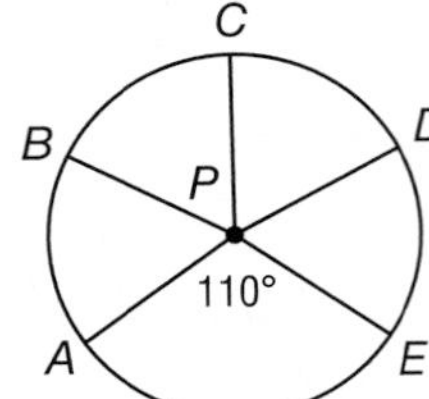

If Zack drills holes so that $m\angle APE = 110°$ and the other four angles are congruent, what is $m\angle CPD$?

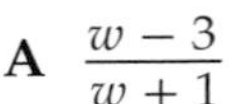

39. Which expression is undefined when $w = 3$?

A $\frac{w-3}{w+1}$

B $\frac{w^2-3w}{3w}$

C $\frac{w+1}{w^2-3w}$

D $\frac{3w}{3w^2}$

40. What is the slope of a line that is parallel to the line?

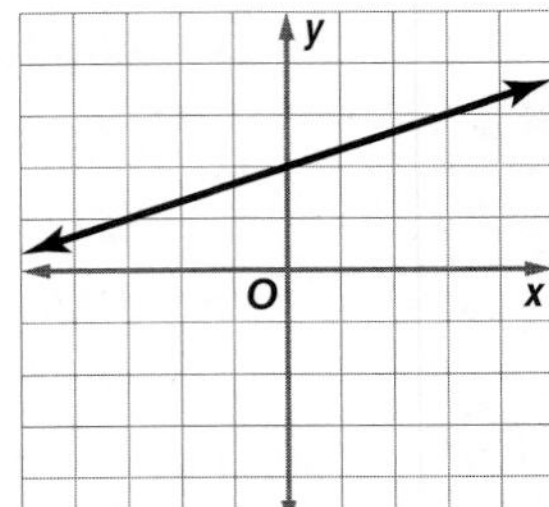

F -3

G $-\frac{1}{3}$

H 3

J $\frac{1}{3}$

41. What are the solutions of $\sqrt{x+3} - 1 = x - 4$?

A $1, 6$

B $-1, -6$

C 1

D 6

Spiral Review

42. ELECTRICITY The voltage V required for a circuit is given by $V = \sqrt{PR}$, where P is the power in watts and R is the resistance in ohms. How many more volts are needed to light a 100-watt light bulb than a 75-watt light bulb if the resistance of both is 110 ohms? (Lesson 10-3)

Simplify each expression. (Lesson 10-2)

43. $\sqrt{6} \cdot \sqrt{8}$

44. $\sqrt{3} \cdot \sqrt{6}$

45. $7\sqrt{3} \cdot 2\sqrt{6}$

46. $\sqrt{\frac{27}{a^2}}$

47. $\sqrt{\frac{5c^5}{4d^5}}$

48. $\frac{\sqrt{9x^3y}}{\sqrt{16x^2y^2}}$

49. PHYSICAL SCIENCE A projectile is shot straight up from ground level. Its height h, in feet, after t seconds is given by $h = 96t - 16t^2$. Find the value(s) of t when h is 96 feet. (Lesson 9-5)

Factor each trinomial, if possible. If the trinomial cannot be factored using integers, write *prime*. (Lesson 8-4)

50. $2x^2 + 7x + 5$

51. $6p^2 + 5p - 6$

52. $5d^2 + 6d - 8$

53. $8k^2 - 19k + 9$

54. $9g^2 - 12g + 4$

55. $2a^2 - 9a - 18$

Determine whether each expression is a monomial. Write *yes* or *no*. Explain. (Lesson 7-1)

56. 12

57. $4x^3$

58. $a - 2b$

59. $4n + 5p$

60. $\frac{x}{y^2}$

61. $\frac{1}{5}abc^{14}$

Skills Review

Simplify. (Lesson 1-1)

62. 9^2

63. 10^6

64. 4^5

65. $(8v)^2$

66. $\left(\frac{w^3}{9}\right)^2$

67. $(10y^2)^3$

Mid-Chapter Quiz

Lessons 10-1 through 10-4

Math Online glencoe.com
Math *in Motion*, Animation

Graph each function. Compare to the parent graph. State the domain and range. (Lesson 10-1)

1. $y = 2\sqrt{x}$
2. $y = -4\sqrt{x}$
3. $y = \frac{1}{2}\sqrt{x}$
4. $y = \sqrt{x} - 3$
5. $y = \sqrt{x - 1}$
6. $y = 2\sqrt{x - 2}$

7. **GEOMETRY** The length of the side of a square is given by the function $s = \sqrt{A}$, where A is the area of the square. What is the length of the side of a square that has an area of 121 square inches? (Lesson 10-1)

   A 121 inches　　C 44 inches
   B 11 inches　　D 10 inches

Simplify each expression. (Lesson 10-2)

8. $2\sqrt{25}$
9. $\sqrt{12} \cdot \sqrt{8}$
10. $\sqrt{72xy^5z^6}$
11. $\frac{3}{1 + \sqrt{5}}$
12. $\frac{1}{5 - \sqrt{7}}$

13. **SATELLITES** A satellite is launched into orbit 200 kilometers above Earth. The orbital velocity of a satellite is given by the formula $v = \sqrt{\frac{Gm_E}{r}}$. v is velocity in meters per second, G is a given constant, m_E is the mass of Earth, and r is the radius of the satellite's orbit in meters. (Lesson 10-2)

    a. The radius of Earth is 6,380,000 meters. What is the radius of the satellite's orbit in meters?

    b. The mass of Earth is 5.97×10^{24} kilograms, and the constant G is $6.67 \times 10^{-11}\ \text{N} \cdot \frac{\text{m}^2}{\text{kg}^2}$ where N is in Newtons. Use the formula to find the orbital velocity of the satellite in meters per second.

14. Which expression is equivalent to $\sqrt{\frac{16}{32}}$? (Lesson 10-2)

    F $\frac{1}{2}$

    G 2

    H $\frac{\sqrt{2}}{2}$

    J 4

Simplify each expression. (Lesson 10-3)

15. $3\sqrt{2} + 5\sqrt{2}$
16. $\sqrt{11} - 3\sqrt{11}$
17. $6\sqrt{2} + 4\sqrt{50}$
18. $\sqrt{27} - \sqrt{48}$
19. $4\sqrt{3}(2\sqrt{6})$
20. $3\sqrt{20}(2\sqrt{5})$
21. $(\sqrt{5} + \sqrt{7})(\sqrt{20} + \sqrt{3})$

22. **GEOMETRY** Find the area of the rectangle. (Lesson 10-3)

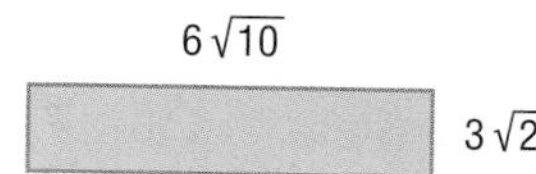

Solve each equation. Check your solution. (Lesson 10-4)

23. $\sqrt{5x} - 1 = 4$
24. $\sqrt{a - 2} = 6$
25. $\sqrt{15 - x} = 4$
26. $\sqrt{3x^2 - 32} = x$
27. $\sqrt{2x - 1} = 2x - 7$
28. $\sqrt{x + 1} + 2 = 4$

29. **GEOMETRY** The lateral surface area S of a cone can be found by using the formula $S = \pi r\sqrt{r^2 + h^2}$, where r is the radius of the base and h is the height of the cone. Find the height of the cone. (Lesson 10-4)

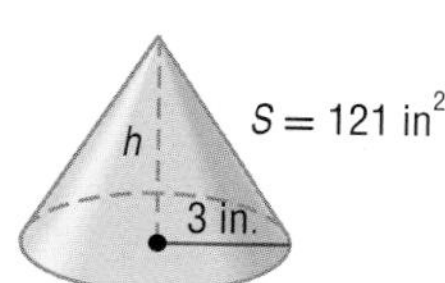

10-5 The Pythagorean Theorem

Why?

The designer television shown is made of black and white leather just like a real soccer ball. Televisions are measured along the diagonal of the screen. If the height and width of the screen is known, the Pythagorean Theorem can be used to find the measure of the diagonal.

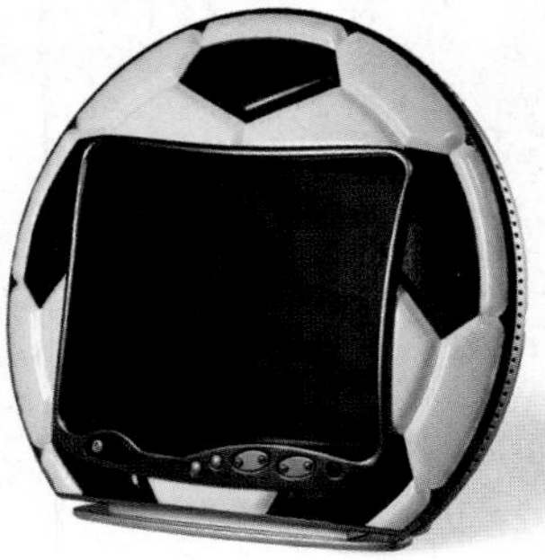

Then

You solved quadratic equations by using the Square Root Property. (Lesson 8-6)

Now

- Solve problems by using the Pythagorean Theorem.
- Determine whether a triangle is a right triangle.

New Vocabulary

hypotenuse
legs
converse
Pythagorean triple

Math Online

glencoe.com

- Extra Examples
- Personal Tutor
- Self-Check Quiz
- Homework Help

The Pythagorean Theorem In a right triangle, the side opposite the right angle is the **hypotenuse**. This side is always the longest. The other two sides are the **legs**.

Key Concept — The Pythagorean Theorem

For Your FOLDABLE

Words If a triangle is a right triangle, then the square of the length of the hypotenuse is equal to the sum of the squares of the lengths of the legs.

Symbols $c^2 = a^2 + b^2$

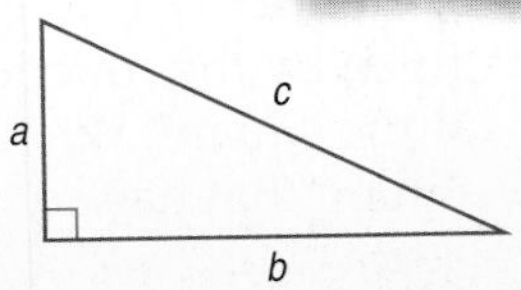

EXAMPLE 1 Find the Length of a Side

Find each missing length. If necessary, round to the nearest hundredth.

a.

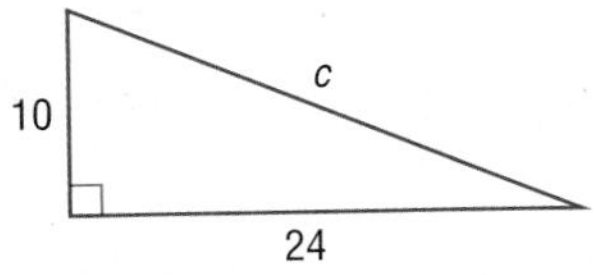

$c^2 = a^2 + b^2$	Pythagorean Theorem
$c^2 = 10^2 + 24^2$	$a = 10$ and $b = 24$
$c^2 = 100 + 576$	Evaluate squares.
$c^2 = 676$	Simplify.
$c = \pm\sqrt{676}$	Take the square root of each side.
$c = \pm 26$	$(\pm 26)^2 = 676$

A length cannot be negative. The missing length is 26 units.

b.

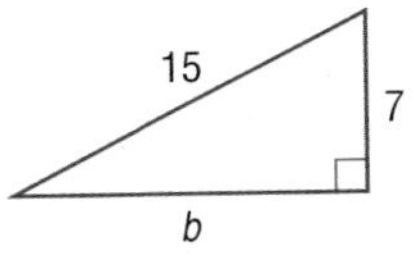

$c^2 = a^2 + b^2$	Pythagorean Theorem
$15^2 = 7^2 + b^2$	$a = 7$ and $c = 15$
$225 = 49 + b^2$	Evaluate squares.
$176 = b^2$	Subtract 49 from each side.
$\pm\sqrt{176} = b$	Take the square root of each side.
$\pm 13.27 \approx b$	Use a calculator to evaluate $\sqrt{176}$.

The missing length is 13.27 units.

Check Your Progress

1A.

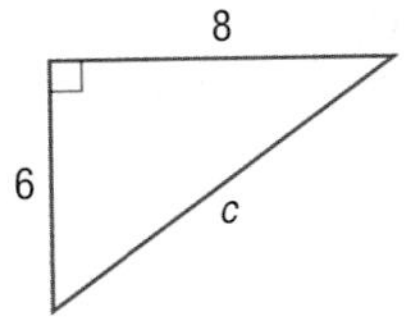

1B.

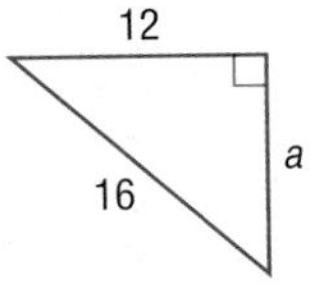

Personal Tutor glencoe.com

Real-World Link

A keelboat is a sailboat with a weighted keel, a vertical fin at the bottom of the boat. Keels are 20 to 30 inches in length.

Source: United States Sailing Association

Real-World EXAMPLE 2 Find the Length of a Side

SAILING The sail of a keelboat forms a right triangle as shown. Find the height of the sail.

$20^2 = h^2 + 10^2$	**Pythagorean Theorem**
$400 = h^2 + 100$	**Evaluate squares.**
$300 = h^2$	**Subtract 100 from each side.**
$\pm 17.32 \approx h$	**Take the square root of each side.**
$17.32 \approx h$	**Use the positive value.**

The sail is approximately 17.32 feet high.

Check Your Progress

2. Suppose the longest side of the sail is 30 feet long and the shortest side is 14 feet long. Find the height of the sail.

Personal Tutor glencoe.com

Right Triangles If you exchange the hypothesis and conclusion of an if-then statement, the result is the **converse** of the statement. The converse of the Pythagorean Theorem can be used to determine whether a triangle is a right triangle.

Key Concept Converse of the Pythagorean Theorem

For Your FOLDABLE

If a triangle has side lengths a, b, and c such that $c^2 = a^2 + b^2$, then the triangle is a right triangle. If $c^2 \neq a^2 + b^2$, then the triangle is not a right triangle.

A **Pythagorean triple** is a group of three counting numbers that satisfy the equation $c^2 = a^2 + b^2$, where c is the greatest number. Examples include (3, 4, 5) and (5, 12, 13). Multiples of Pythagorean triples also satisfy the converse of the Pythagorean Theorem, so (6, 8, 10) is also a Pythagorean triple.

EXAMPLE 3 Check for Right Triangles

Determine whether 9, 12, and 16 can be the lengths of the sides of a right triangle.

Since the measure of the longest side is 16, let $c = 16$, $a = 9$, and $b = 12$.

$c^2 = a^2 + b^2$	**Pythagorean Theorem**
$16^2 \stackrel{?}{=} 9^2 + 12^2$	**$a = 9$, $b = 12$, and $c = 16$**
$256 \stackrel{?}{=} 81 + 144$	**Evaluate squares.**
$256 \neq 225$	**Add.**

Since $c^2 \neq a^2 + b^2$, segments with these measures cannot form a right triangle.

Check Your Progress

Determine whether each set of measures can be the lengths of the sides of a right triangle.

3A. 30, 40, 50

3B. 6, 12, 18

Personal Tutor glencoe.com

Check Your Understanding

Example 1 p. 630

Find each missing length. If necessary, round to the nearest hundredth.

1. 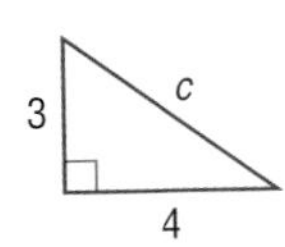

2. 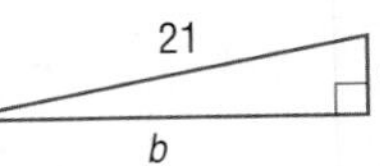

3. 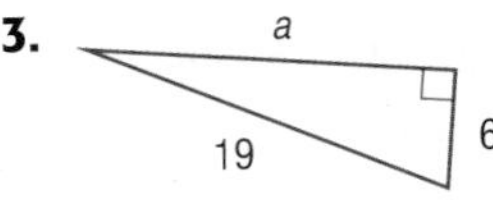

4. 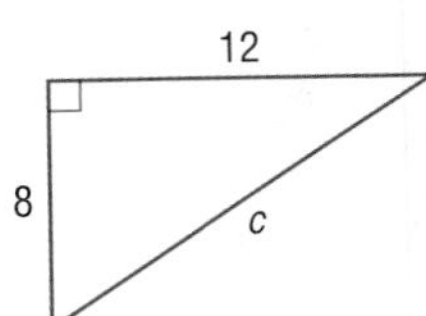

Example 2 p. 631

5. BASEBALL A baseball diamond is a square. The distance between consecutive bases is 90 feet.

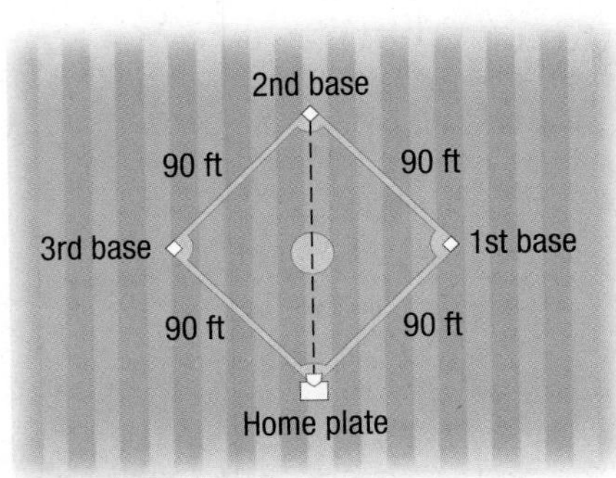

a. How far does a catcher have to throw the ball from home plate to second base?

b. How far does a third baseman have to throw the ball to the first baseman?

c. If the catcher is five feet behind home plate, how far does he have to throw the ball to second base?

Example 3 p. 631

Determine whether each set of measures can be the lengths of the sides of a right triangle.

6. 8, 12, 16

7. 28, 45, 53

8. 7, 24, 25

9. 15, 25, 45

Practice and Problem Solving

● = **Step-by-Step Solutions** begin on page R12.
Extra Practice begins on page 815.

Example 1 p. 630

Find each missing length. If necessary, round to the nearest hundredth.

10. 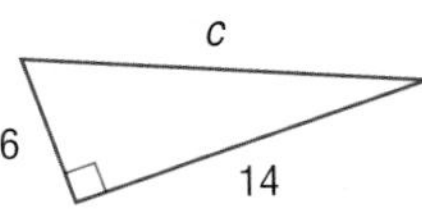

11 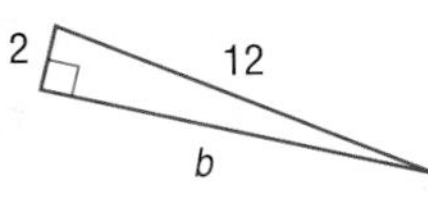

12. 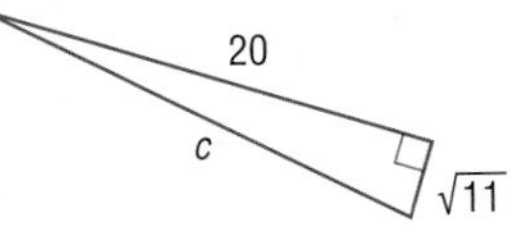

13. 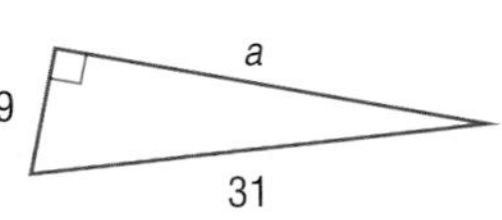

14. 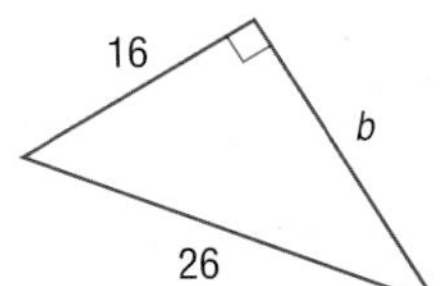

15. 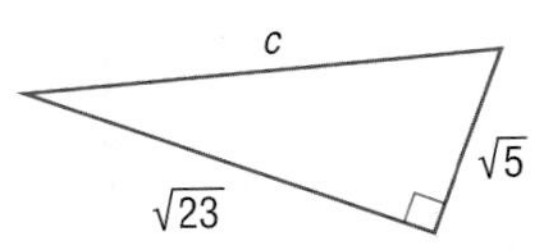

16. 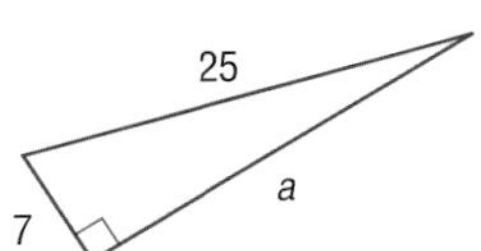

17. 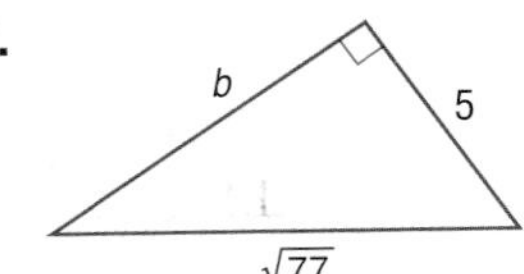

18. 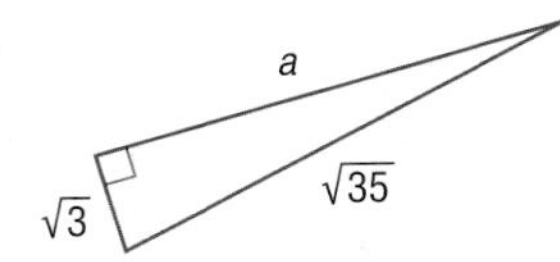

Example 2
p. 631

19. TELEVISION Larry is buying an entertainment stand for his television. The diagonal of his television is 27 inches. The space for the television measures 20 inches by 26 inches. Will Larry's television fit? Explain.

Example 3
p. 631

Determine whether each set of measures can be the lengths of the sides of a right triangle. Then determine whether they form a Pythagorean triple.

20. 9, 40, 41

21. $3, 2\sqrt{10}, \sqrt{41}$

22. $4, \sqrt{26}, 12$

23. $\sqrt{5}, 7, 14$

24. 8, 31.5, 32.5

25. $\sqrt{65}, 6\sqrt{2}, \sqrt{97}$

26. 18, 24, 30

27. 36, 77, 85

28. 17, 33, 98

29. GEOMETRY Refer to the triangle at the right.

a. What is a?

b. Find the area of the triangle.

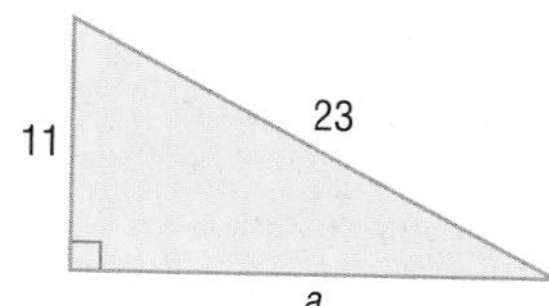

30. GARDENING Khaliah wants to plant flowers in a triangular plot. She has three lengths of plastic garden edging that measure 8 feet, 15 feet, and 17 feet. Determine whether these pieces form a right triangle. Explain.

31. LADDER Mr. Takeo is locked out of his house. The only open window is on the second floor. There is a bush along the edge of the house, so he places the neighbor's ladder 10 feet from the house. To the nearest foot, what length of ladder does he need to reach the window?

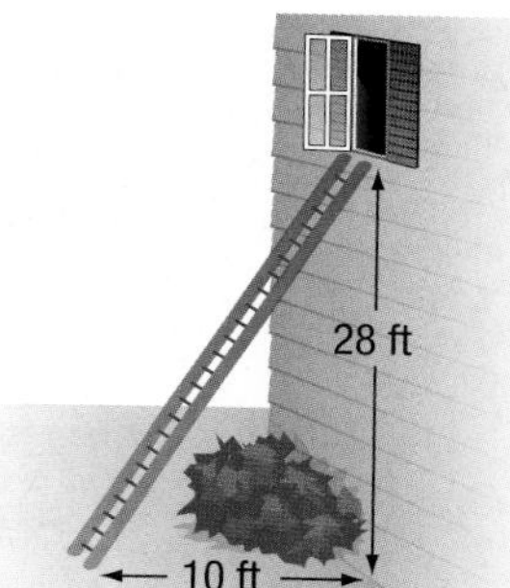

Math History Link

Pythagoras
(580–500 B.C.)
Greek mathematician Pythagoras founded the famous Pythagorean school for study of philosophy, mathematics, and natural science.

Find the length of the hypotenuse. Round to the nearest hundredth.

32.

33.

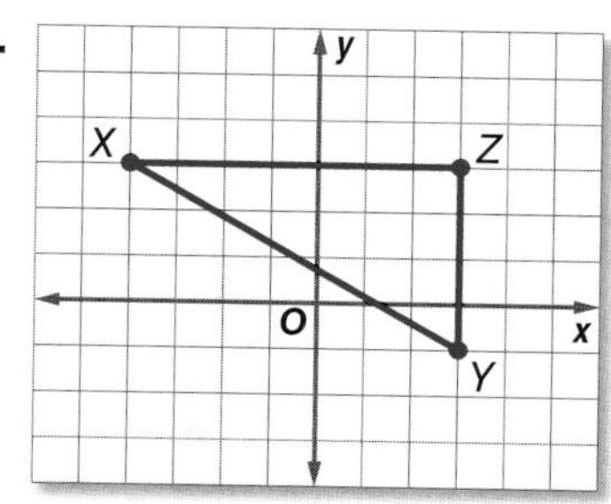

34. GEOMETRY A rectangle has a base of 5 feet and a height of 12 feet. What is the length of the diagonal?

35. GEOMETRY A square has a diagonal with length of 6 meters. Find the length of the sides of the square.

36. DOLLHOUSE Alonso is building a dollhouse for his sister's birthday. The house is 24 inches across and the slanted side is 16 inches long as shown. Find the height of the roof to the nearest tenth of an inch.

37. GEOMETRY Each side of a cube is 5 inches long. Find the length of a diagonal of the cube.

Real-World Link

The largest town square in the United States is Washington Square Park in New York. The park is 9.75 acres or 39,000 square meters.

Source: *The New York Times*

38. **TOWN SQUARES** The largest town square in the world is Tiananmen Square in Beijing, China, covering 98 acres.
    a. One square mile is 640 acres. Assuming that Tiananmen Square is a square, how many feet long is a side to the nearest foot?
    b. To the nearest foot, what is the diagonal distance across Tiananmen Square?

39. **TRUCKS** Violeta needs to construct a ramp to roll a cart of moving boxes from her garage into the back of her truck. How long does the ramp have to be?

6 ft
36 in.

40. **GEOMETRY** A square has an area of 242 square inches. Find the length of a diagonal.

41. **GEOMETRY** A rectangle has a width that is twice as long as its length and an area of 722 square inches. Find the length of a diagonal.

If c is the measure of the hypotenuse of a right triangle, find each missing measure. If necessary, round to the nearest hundredth.

42. $a = x, b = x + 41, c = 85$
43. $a = 8, b = x, c = x + 2$
44. $a = 12, b = x - 2, c = x$
45. $a = x, b = x + 7, c = 97$
46. $a = x - 47, b = x, c = x + 2$
47. $a = x - 32, b = x - 1, c = x$

48. **GEOMETRY** A right triangle has one leg that is 8 inches shorter than the other leg. The hypotenuse is 30 inches long. Find the length of each leg.

49. **GEOMETRY** A rectangle has a diagonal with length 8 centimeters. Its length is 4 centimeters greater than the width. Find the length and width of the rectangle.

H.O.T. Problems

Use **H**igher-**O**rder **T**hinking Skills

50. **FIND THE ERROR** Wyatt and Dario are determining whether 36, 77, and 85 form a Pythagorean triple. Is either of them correct? Explain your reasoning.

Wyatt

$36^2 + 77^2 \stackrel{?}{=} 85^2$

$1296 + 5929 \stackrel{?}{=} 7225$

$7225 = 7225$

yes

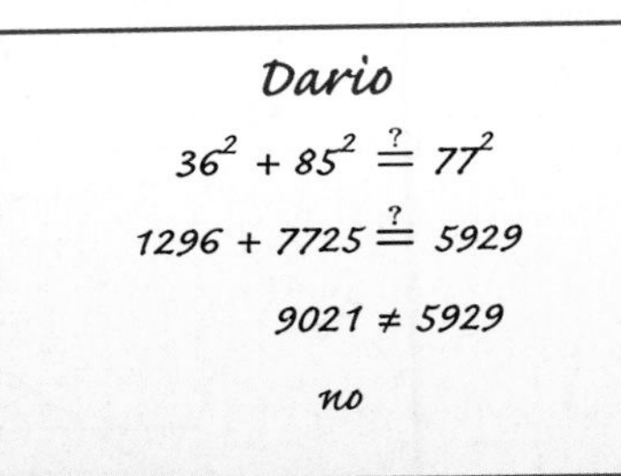

Dario

$36^2 + 85^2 \stackrel{?}{=} 77^2$

$1296 + 7725 \stackrel{?}{=} 5929$

$9021 \neq 5929$

no

51. **CHALLENGE** Find the value of x in the figure.

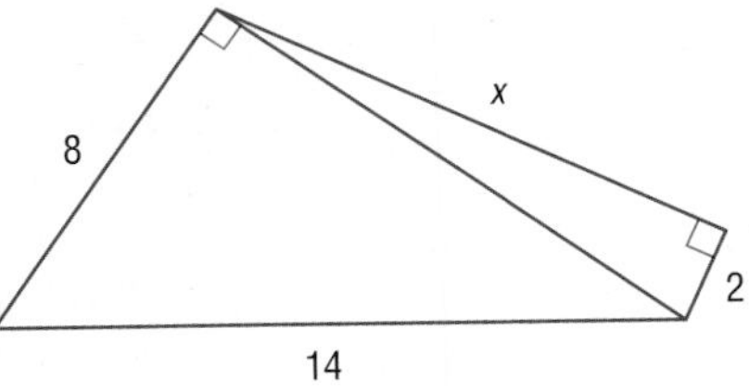

52. **REASONING** Provide a counterexample to the statement.
Any two right triangles with the same hypotenuse have the same area.

53. **OPEN ENDED** Draw a right triangle that has a hypotenuse of $\sqrt{72}$ units.

54. **WRITING IN MATH** Explain how to determine whether segments in three lengths could form a right triangle.

Standardized Test Practice

55. GEOMETRY Find the missing length.

A -17
B $-\sqrt{161}$
C $\sqrt{161}$
D 17

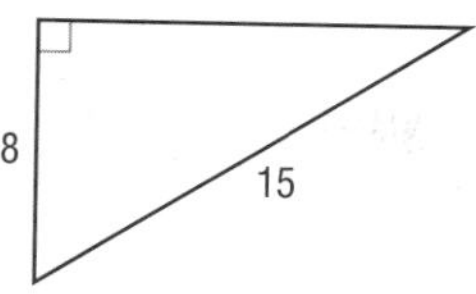

56. What is a solution of this equation?

$$x + 1 = \sqrt{x + 1}$$

F $0, 3$ H 0
G 3 J no solutions

57. SHORT RESPONSE A plumber charges \$40 for the first hour of each house call plus \$8 for each additional half hour. If the plumber works for 4 hours, how much does he charge?

58. Find the next term in the geometric sequence $4, 3, \frac{9}{4}, \frac{27}{16}, \ldots$.

A $\frac{4}{3}$ B $\frac{81}{64}$ C $\frac{64}{81}$ D $\frac{243}{64}$

Spiral Review

Solve each equation. Check your solution. (Lesson 10-4)

59. $\sqrt{x} = 16$
60. $\sqrt{4x} = 64$
61. $\sqrt{10x} = 10$
62. $\sqrt{8x} + 1 = 65$
63. $\sqrt{x + 1} + 2 = 4$
64. $\sqrt{x - 15} = 3 - \sqrt{x}$

Simplify each expression. (Lesson 10-3)

65. $2\sqrt{3} + 5\sqrt{3}$
66. $4\sqrt{5} - 2\sqrt{5}$
67. $6\sqrt{7} + 2\sqrt{28}$
68. $\sqrt{18} - 4\sqrt{2}$
69. $3\sqrt{5} - 5\sqrt{3} + 9\sqrt{5}$
70. $4\sqrt{3} + 6\sqrt{12}$

71. BUSINESS The amount of money spent at West Outlet Mall continues to increase. The total $T(x)$ in millions of dollars can be estimated by the function $T(x) = 12(1.12)^x$, where x is the number of years after it opened in 2005. Find the amount of sales in 2015, 2016, and 2017. (Lesson 9-6)

Describe how the graph of each function is related to the graph of $f(x) = x^2$. (Lesson 9-3)

72. $g(x) = x^2 - 8$
73. $h(x) = \frac{1}{4}x^2$
74. $h(x) = -x^2 + 5$
75. $g(x) = x^2 + 10$
76. $g(x) = -2x^2$
77. $h(x) = -x^2 - \frac{4}{3}$

78. ROCK CLIMBING While rock climbing, Damaris launches a grappling hook from a height of 6 feet with an initial upward velocity of 56 feet per second. The hook just misses the stone ledge that she wants to scale. As it falls, the hook anchors on a ledge 30 feet above the ground. How long was the hook in the air? (Lesson 8-4)

Find each product. (Lesson 7-7)

79. $(b + 8)(b + 2)$
80. $(x - 4)(x - 9)$
81. $(y + 4)(y - 8)$
82. $(p + 2)(p - 10)$
83. $(2w - 5)(w + 7)$
84. $(8d + 3)(5d + 2)$

Skills Review

Solve each proportion. (Lesson 2-6)

85. $\frac{x}{5} = \frac{12}{3}$
86. $\frac{12}{x} = \frac{3}{4}$
87. $\frac{5}{4} = \frac{10}{x}$
88. $\frac{3}{5} = \frac{12}{x + 8}$

10-6 The Distance and Midpoint Formulas

Then

You used the Pythagorean Theorem. (Lesson 10-5)

Now

- Find the distance between two points on a coordinate plane.
- Find the midpoint between two points on a coordinate plane.

New Vocabulary

Distance Formula
midpoint
Midpoint Formula

Math Online

glencoe.com

- Extra Examples
- Personal Tutor
- Self-Check Quiz
- Homework Help

Why?

Rescue helicopters use electronic Global Positioning Systems (GPS) to compute direct distances between two locations.

A rescue helicopter can fly 450 miles before it needs to refuel. A person needs to be flown from Washington, North Carolina, to Huntington, West Virginia. Sides of the grid squares are 50 miles long. Asheville, North Carolina, is at the origin, Huntington is at (0, 196), and Washington is at (310, 0). Can the helicopter make the trip without refueling?

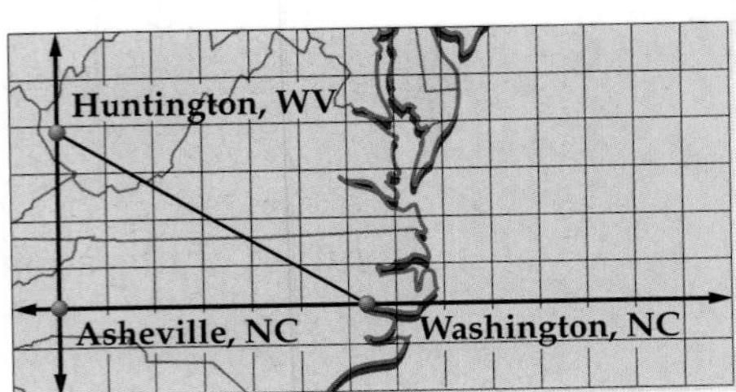

Distance Formula The GPS system calculates direct distances by using the **Distance Formula**, which is based on the Pythagorean Theorem.

Key Concept The Distance Formula

For Your FOLDABLE

Words The distance d between any two points with coordinates (x_1, y_1) and (x_2, y_2) is given by the following formula.

$$d = \sqrt{(x_2 - x_1)^2 + (y_2 - y_1)^2}$$

Model

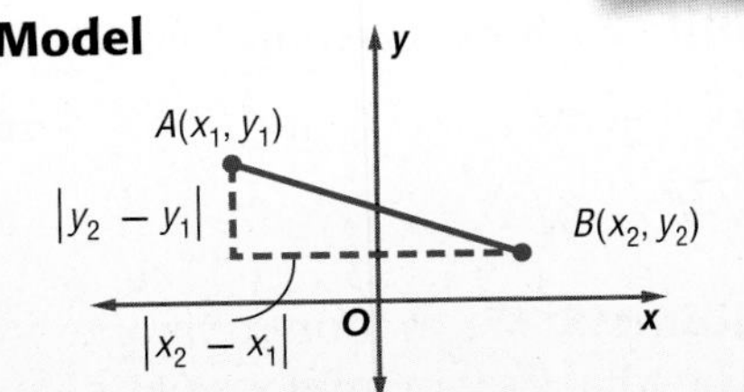

You can use the Distance Formula to find the distance between any two points on a coordinate plane.

EXAMPLE 1 Distance Between Two Points

Find the distance between points at (5, 3) and (1, −2).

$d = \sqrt{(x_2 - x_1)^2 + (y_2 - y_1)^2}$ Distance Formula

$= \sqrt{(1 - 5)^2 + (-2 - 3)^2}$ $(x_1, y_1) = (5, 3)$ and $(x_2, y_2) = (1, -2)$

$= \sqrt{(-4)^2 + (-5)^2}$ Simplify.

$= \sqrt{16 + 25}$ Evaluate squares.

$= \sqrt{41}$ or about 6.4 units Simplify.

Check Your Progress

Find the distance between points with the given coordinates.

1A. (4, 2) and (−3, −1).

1B. (−7, −2) and (−5, −8)

Personal Tutor glencoe.com

Real-World Link

About 22% of U.S. residents have a home theater. Technology installations is a $9.6 billion per year business.

Source: *Realty Times*

Real-World EXAMPLE 2 Use the Distance Formula

ENTERTAINMENT The Vaccaro Family is having a home theater system installed. The TV and the seating will be placed in opposite corners of the room. The manufacturer of the TV recommends that for the size of TV that they want, the seating should be placed at least 13 feet away. If each grid square is 1 foot long, is the Vaccaro's room large enough for the TV?

The front of the TV screen is located at (1, 11), and the front of the sofa is located at (7, 2).

$d = \sqrt{(x_2 - x_1)^2 + (y_2 - y_1)^2}$ **Distance Formula**

$= \sqrt{(7 - 1)^2 + (2 - 11)^2}$ **$(x_1, y_1) = (1, 11)$ and $(x_2, y_2) = (7, 2)$**

$= \sqrt{6^2 + (-9)^2}$ **Simplify.**

$= \sqrt{117}$ or about 10.8 feet

No, the room is not large enough for the TV.

Check Your Progress

2. The manufacturer of the speakers recommends that they be placed at least 8 feet from the seating. If one of the speakers is being placed at (0, 9), is the Vaccaros' family room large enough for the speakers? Explain.

Personal Tutor glencoe.com

When we know the distance and one of the points, we can use the Distance Formula to find the coordinates of the other point.

StudyTip

Two Possibilities When finding a missing coordinate, there will usually be two possibilities because the point could be the same distance in different directions.

EXAMPLE 3 Find a Missing Coordinate

Find the possible values for *a* if the distance between points at (4, 7) and (*a*, 3) are 5 units apart.

$d = \sqrt{(x_2 - x_1)^2 + (y_2 - y_1)^2}$ **Distance Formula**

$5 = \sqrt{(a - 4)^2 + (3 - 7)^2}$ **$(x_1, y_1) = (4, 7)$ and $(x_2, y_2) = (a, 3)$, and $d = 5$**

$5 = \sqrt{(a - 4)^2 + (-4)^2}$ **Simplify.**

$5 = \sqrt{a^2 - 8a + 32}$ **Evaluate squares and simplify.**

$25 = a^2 - 8a + 32$ **Square each side.**

$0 = a^2 - 8a + 7$ **Subtract 25 from each side.**

$0 = (a - 1)(a - 7)$ **Factor.**

$a - 1 = 0$ or $a - 7 = 0$ **Zero Product Property**

$a = 1$ $\quad$ $a = 7$ **Solve each equation.**

Check Your Progress

3. Find the possible values of a if the distance between points at $(2, a)$ and $(-6, 2)$ is 10 units.

Personal Tutor glencoe.com

Midpoint Formula The point on the segment that joins two points and is equidistant from the endpoints is called the **midpoint**. You can find the coordinates of the midpoint by using the **Midpoint Formula**.

Key Concept The Midpoint Formula

For Your FOLDABLE

Words The midpoint M of a line segment with endpoints at (x_1, y_1) and (x_2, y_2) is given by

$$M = \left(\frac{x_1 + x_2}{2}, \frac{y_1 + y_2}{2}\right).$$

Model

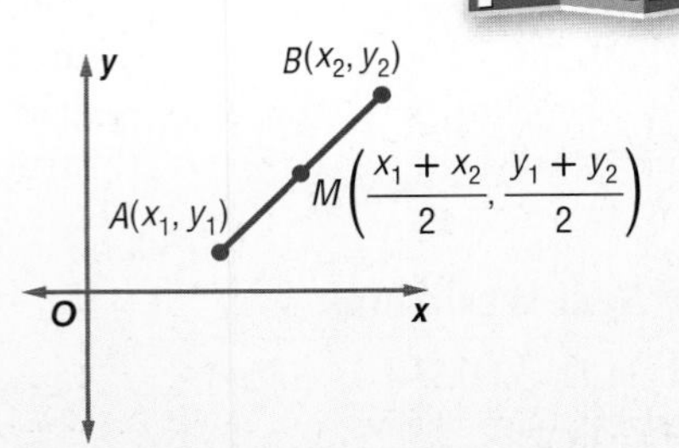

Watch Out!

Midpoint Formula Be careful to add, and not subtract, when using the Midpoint Formula.

EXAMPLE 4 Find the Midpoint

Find the coordinates of the midpoint of the segment with endpoints at (−1, −2) and (3, −4).

$M = \left(\frac{x_1 + x_2}{2}, \frac{y_1 + y_2}{2}\right)$ **Midpoint Formula**

$= \left(\frac{-1 + 3}{2}, \frac{-2 + (-4)}{2}\right)$ $(x_1, y_1) = (-1, -2)$ **and** $(x_2, y_2) = (3, -4)$

$= \left(\frac{2}{2}, \frac{-6}{2}\right)$ **Simplify the numerators.**

$= (1, -3)$ **Simplify.**

Check Your Progress

Find the coordinates of the midpoint of the segment with the given endpoints.

4A. (12, 3), (−8, 3) **4B.** (0, 0), (5, 12) **4C.** (6, 8), (3, 4)

Personal Tutor glencoe.com

Check Your Understanding

Example 1 p. 636

Find the distance between points with the given coordinates.

1. (6, −2), (12, 8)

2. (4, 8), (−3, −6)

3 (3, 0), (6, −2)

4. (−2, −4), (−5, −3)

Example 2 p. 637

5. GOLF Addison hit a golf ball from a tee to the point at (−3, 12), 12 feet past the hole and 3 feet to the left. The hole is located at the point at (0, 0). Her first putt traveled to the point at (1, 2), 2 feet above the hole and 1 foot to the right.

a. How far did the ball travel on her first putt?

b. How far was her first putt from the cup?

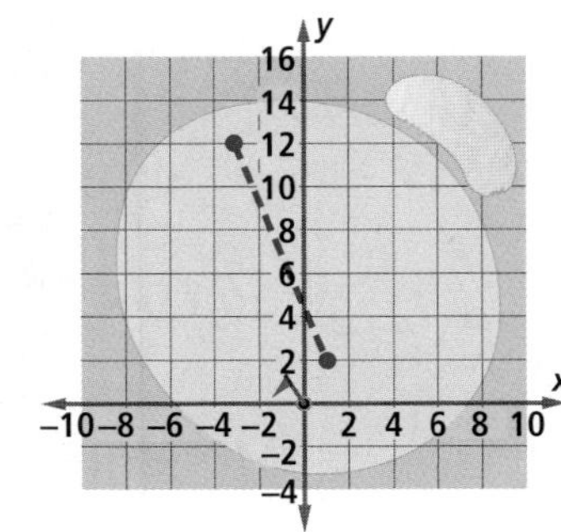

Example 3 p. 637

Find the possible values for a if the points with the given coordinates are the indicated distance apart.

6. $(-5, a), (3, 1); d = \sqrt{89}$

7. $(6, a), (5, 0); d = \sqrt{17}$

8. $(5, 8), (a, 2); d = 3\sqrt{5}$

9. $(a, 6), (-6, 2); d = 4\sqrt{10}$

Example 4 p. 638

Find the coordinates of the midpoint of the segment with the given endpoints.

10. $(5, -10), (5, 8)$

11. $(2, -2), (6, 2)$

12. $(5, 0), (0, 3)$

13. $(-4, 1), (3, -1)$

14. $(3, -17), (2, -8)$

15. $(-2, 2), (4, 10)$

16. $(3, 10), (3, 3)$

17. $(-17, 8), (-2, 20)$

Practice and Problem Solving

= **Step-by-Step Solutions** begin on page R12.
Extra Practice begins on page 815.

Example 1 p. 636

Find the distance between points with the given coordinates.

18. $(5, 8), (5, 7)$

19. $(6, -9), (9, -9)$

20. $(3, -3), (7, 2)$

21. $(5, 1), (0, 4)$

22. $(-5, 2), (4, -2)$

23. $(3, 5), (-6, 0)$

24. $(-7, 8), (3, 10)$

25. $(-11, 9), (3, -4)$

26. $(8, 6), (-13, -2)$

27. $(5, 2), (3, -3)$

28. $(4, 2), (5, 5)$

29. $(-3, 5), (5, -3)$

Example 2 p. 637

30. NAVIGATION Lawana and Ken are meeting at a restaurant in a marina. Ken takes his boat, while Lawana is driving her car. The sides of each grid square on the map represent 1 mile.

a. How far did Ken travel?

b. How far did Lawana travel?

c. How many times as great is the distance that Ken traveled as the distance that Lawana traveled?

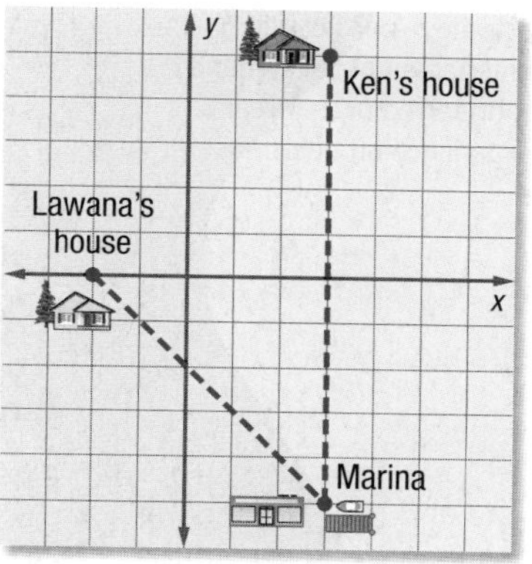

Example 3 p. 637

Find the possible values for a if the points with the given coordinates are the indicated distance apart.

31. $(-9, -2), (a, 5); d = 7$

32. $(a, -6), (-5, 2); d = 10$

33. $(a, 0), (3, 1); d = \sqrt{2}$

34. $(4, a), (8, 4); d = 2\sqrt{5}$

35. $(7, 5), (-9, a); d = 2\sqrt{65}$

36. $(-2, a), (6, 1); d = 4\sqrt{5}$

Example 4 p. 638

Find the coordinates of the midpoint of the segment with the given endpoints.

37. $(0, 2), (7, 3)$

38. $(5, -2), (3, -6)$

39. $(-4, 0), (0, 14)$

40. $(10, -3), (-8, -5)$

41 $(-5, 5), (3, -3)$

42. $(-16, -7), (-4, -3)$

Find the distance between points with the given coordinates.

43. $(4, 2), \left(6, -\frac{2}{3}\right)$

44. $\left(\frac{4}{5}, -1\right), \left(2, -\frac{1}{2}\right)$

45. $(4\sqrt{5}, 7), (6\sqrt{5}, 1)$

46. GEOMETRY Triangle ABC has vertices $A(1, 3)$, $B(-2, 5)$, and $C(8, 8)$. Find the perimeter of the triangle. Use a calculator to estimate the perimeter to the nearest tenth.

47. GEOMETRY Quadrilateral $JKLM$ has vertices $J(-3, -4)$, $K(-1, 4)$, $L(4, 5)$, and $M(6, -5)$. Find the perimeter of the quadrilateral to the nearest tenth.

48. **TEMPERATURE** The temperature dropped from 25° to −8° over a 12-hour period beginning at 12:00 noon as shown.

Time	12:00 noon	4:00	8:00	12:00 midnight
Temperature	25°	14°	3°	−8°

a. Plot these points on a coordinate plane with time on the x-axis and temperature on the y-axis. Let x represent the number of hours, and let 12:00 noon correspond to $x = 0$.

b. Draw a segment to connect the points. Find the midpoint of this segment. Interpret the meaning of the midpoint in this situation.

Real-World Career

Cruise Director
The cruise director is in charge of all onboard entertainment and activities. A professional entertainment background is preferred or 2–5 years experience on board.

49. **NAVIGATION** Two cruise ships are leaving St. Lucia Island at the same time. One travels 10 miles due east and then 8 miles north. The second ship travels 12 miles due north and then 6 miles west.

a. If St. Lucia is at the origin, how far is the first ship from St. Lucia?

b. How far is the second ship from St. Lucia?

c. How far apart are the ships?

50. **TOURING** Sasha is using the GPS system in her car to go from her hotel to the art museum, to a restaurant, and then to the theater. Sides of grid squares represent 500 feet. Round your answers to the nearest hundredth.

Art Museum (2.25, 3)
Hotel (0, 0)
Restaurant (4.5, 1)
Theater (3.5, −1)

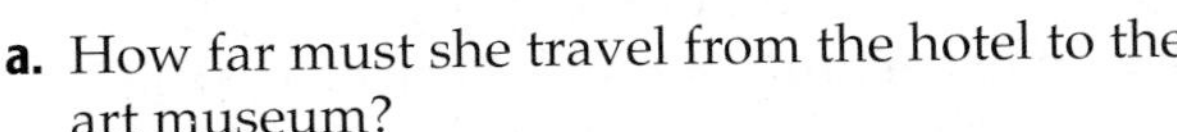

a. How far must she travel from the hotel to the art museum?

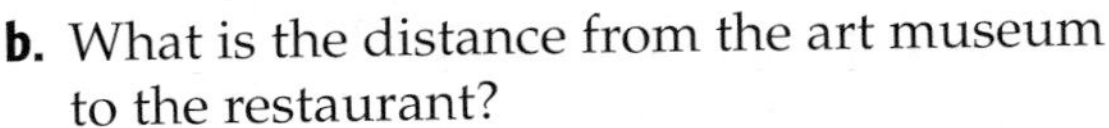

b. What is the distance from the art museum to the restaurant?

c. How far is it from the restaurant to the theater?

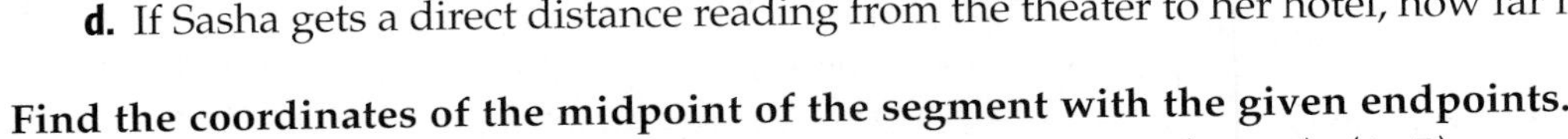

d. If Sasha gets a direct distance reading from the theater to her hotel, how far is it?

Find the coordinates of the midpoint of the segment with the given endpoints.

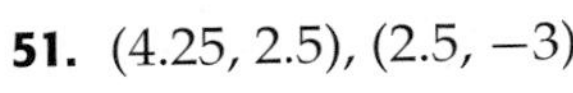

51. (4.25, 2.5), (2.5, −3)

52. $\left(5, -\frac{1}{2}\right), \left(-3, \frac{5}{2}\right)$

53. $\left(\frac{2}{5}, -\frac{1}{5}\right), \left(\frac{1}{3}, \frac{5}{2}\right)$

H.O.T. Problems

Use **H**igher-**O**rder **T**hinking Skills

54. **CHALLENGE** $A(-7, 3)$, $B(4, 0)$, and $C(-4, 4)$ are the vertices of a triangle. Discuss two different ways to determine whether $\triangle ABC$ is a right triangle.

55. **REASONING** Explain why there are usually two possible values when looking for a missing coordinate when you are given two sets of coordinates and the distance between the two points.

56. **REASONING** Is the following statement *true* or *false*? Explain your reasoning.

It matters which ordered pair is first when using the Distance Formula.

57. **OPEN ENDED** Plot two points on a coordinate plane and draw the segment between them. Find the coordinates of the midpoint.

58. **WRITING IN MATH** Explain how the Midpoint Formula is related to finding the mean.

Standardized Test Practice

59. SHORT RESPONSE Two sailboats leave Key Largo, Florida, at the same time. One travels east and then north. The other travels south and then west. How far apart are the boats?

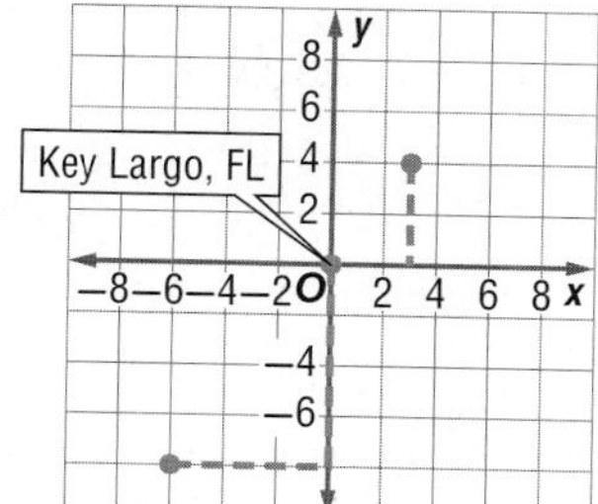

60. While in Tokyo, Callie spent 560 yen for a strand of pearls. The cost of the pearls was equivalent to $35 in U.S. currency. At the time of Callie's purchase, how many yen were equivalent to $20 in U.S. currency?

A 109 yen
B 320 yen
C 980 yen
D 2350 yen

61. SHORT RESPONSE L represents a lighthouse and B represents a buoy. A ship is at the midpoint between L and B. Which coordinates best represent the ship's position?

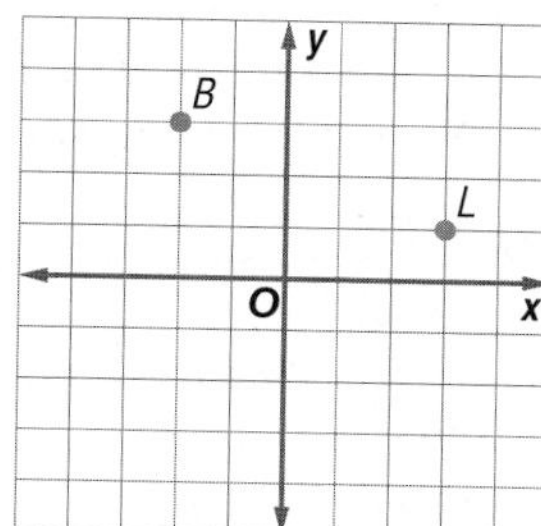

F $\left(2, \frac{1}{2}\right)$
G $\left(1, \frac{1}{2}\right)$
H $\left(\frac{1}{2}, 2\right)$
J $\left(\frac{1}{2}, 5\right)$

62. At a family reunion, Guido cut a slice of cheesecake that was about one sixteenth of the cake. If the entire cheesecake contained 4480 Calories, which is the closest to the number of Calories in Guido's slice?

A 280
B 373
C 498
D 560

Spiral Review

If c is the measure of the hypotenuse of a right triangle, find each missing measure. If necessary, round to the nearest hundredth. (Lesson 10-5)

63. $a = 16, b = 63, c = ?$

64. $b = 3, a = \sqrt{112}, c = ?$

65. $c = 14, a = 9, b = ?$

66. $a = 6, b = 3, c = ?$

67. $b = \sqrt{77}, c = 12, a = ?$

68. $a = 4, b = \sqrt{11}, c = ?$

69. AVIATION The relationship between a plane's length L in feet and the pounds P its wings can lift is described by $L = \sqrt{kP}$, where k is the constant of proportionality. Find k for this plane to the nearest hundredth. (Lesson 10-4)

Skills Review

Solve each proportion. If necessary, round to the nearest hundredth. (Lesson 2-6)

70. $\frac{4}{d} = \frac{2}{10}$

71. $\frac{6}{5} = \frac{f}{15}$

72. $\frac{20}{28} = \frac{h}{21}$

73. $\frac{6}{7} = \frac{7}{j}$

74. $\frac{16}{7} = \frac{9}{m}$

75. $\frac{p}{2} = \frac{45}{68}$

10-7 Similar Triangles

Then

You solved proportions. (Lesson 2-6)

Now

- Determine whether two triangles are similar.
- Find the unknown measures of sides of two similar triangles.

New Vocabulary

similar triangles

Math Online

glencoe.com

- Extra Examples
- Personal Tutor
- Self-Check Quiz
- Homework Help
- Math in Motion

Why?

Simona needs to measure the height of a Ferris wheel for a class project. Simona can measure her shadow and the shadow of the Ferris wheel. She can then use similar triangles and indirect measurement to find the height of the Ferris wheel.

Similar Triangles **Similar triangles** have the same shape, but not necessarily the same size. The symbol ~ is used to denote that two triangles are similar. The vertices of similar triangles are written in order to show the corresponding parts.

Key Concept — Similar Triangles

For Your FOLDABLE

Words If two triangles are similar, then the measures of their corresponding angles are equal, and the measures of their corresponding sides are proportional.

Example If $\triangle ABC \sim \triangle DEF$, then $m\angle A = m\angle D$, $m\angle B = m\angle E$, $m\angle C = m\angle F$, and $\frac{AB}{DE} = \frac{BC}{EF} = \frac{AC}{DF} = \frac{1}{2}$.

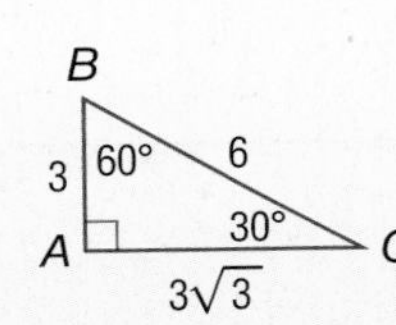

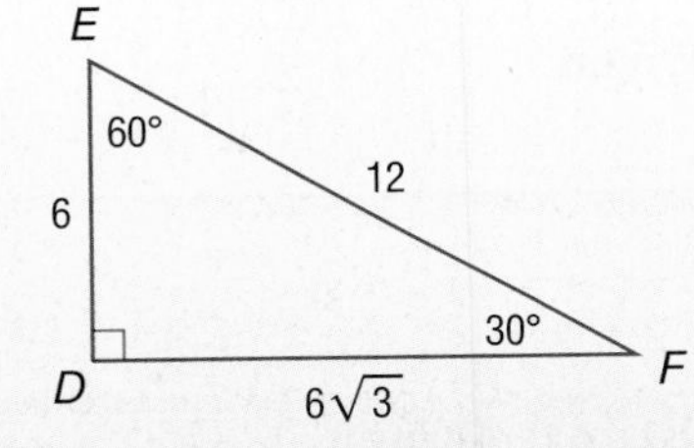

Math *in Motion*, Animation glencoe.com

EXAMPLE 1 Determine Whether Two Triangles are Similar

Determine whether the pair of triangles is similar. Justify your answer.

The measure of $\angle T$ is $180 - (57 + 57)$ or $66°$.

In $\triangle XYZ$, $\angle X$ and $\angle Z$ have the same measure.

Let x = the measure of $\angle X$ and $\angle Z$.

$x + x + 66 = 180$ **Sum of the angle measures is 180.**

$2x = 114$ **Subtract 66 from each side.**

$x = 57$ **Divide each side by 2.**

So, $m\angle X = 57°$ and $m\angle Z = 57°$.

Since the corresponding angles have equal measures, $\triangle XYZ \sim \triangle STQ$.

Check Your Progress

1. Determine whether $\triangle ABC$ with $m\angle A = 68°$ and $m\angle B = m\angle C$ is similar to $\triangle DEF$ with $m\angle E = m\angle F = 54°$. Justify your answer.

Personal Tutor glencoe.com

The ratios of the lengths of the corresponding sides can also be compared to show that two triangles are similar.

ReadingMath

Angle Measures The $m\angle A$ is read as *the measure of angle A.*

EXAMPLE 2 **Determine Whether Two Triangles are Similar**

Determine whether the pair of triangles is similar. Justify your answer.

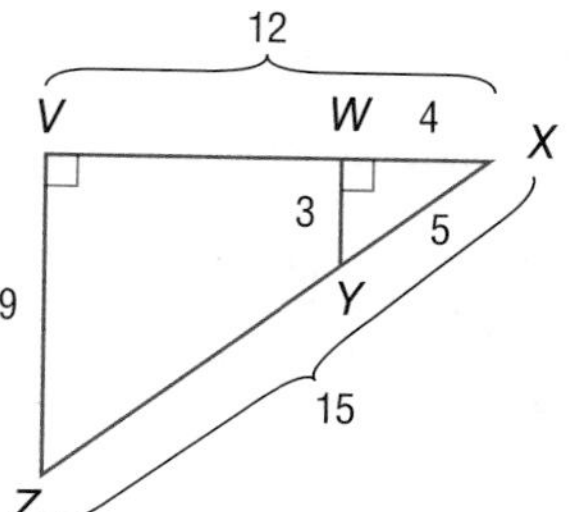

If $\triangle VXZ$ and $\triangle WXY$ are similar, then the measures of their corresponding sides are proportional.

$\frac{VX}{WX} = \frac{12}{4} = 3$ $\quad \frac{XZ}{XY} = \frac{15}{5} = 3$ $\quad \frac{VZ}{WY} = \frac{9}{3} = 3$

Since the corresponding sides are proportional, $\triangle VXZ \sim \triangle WXY$.

Check Your Progress

2. Determine whether $\triangle ABC$ with $AB = 6$, $BC = 16$, and $AC = 20$ is similar to $\triangle JKL$ with $JK = 3$, $KL = 8$, and $JL = 9$. Justify your answer.

Personal Tutor glencoe.com

Find Unknown Measures When some of the measurements of the sides of similar triangles are known, proportions can be used to find the missing measures.

EXAMPLE 3 **Find Missing Measures**

Find the missing measures for the pair of similar triangles.

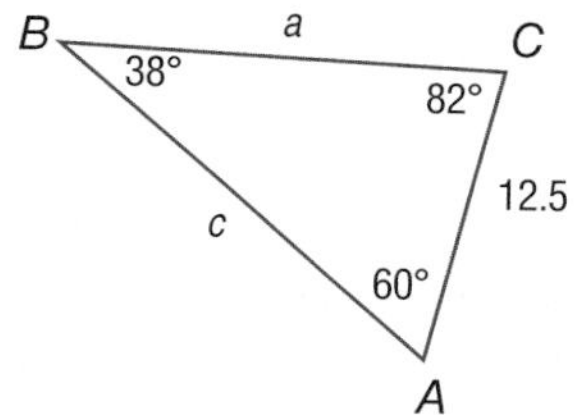

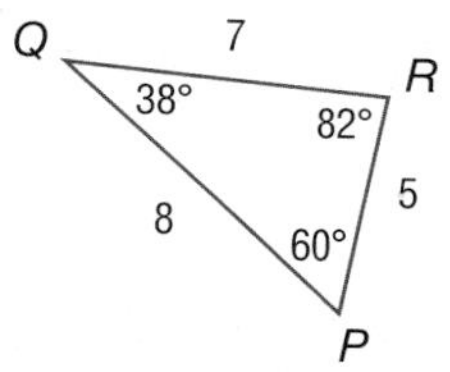

StudyTip

Overlapping Triangles If two triangles overlap, you may want to draw each triangle separately. Make sure the corresponding parts are in the same position and label the corresponding angles and/or sides.

$\frac{AB}{PQ} = \frac{AC}{PR}$	**Corresponding sides of similar triangles are proportional.**
$\frac{c}{8} = \frac{12.5}{5}$	**$AB = c$, $AC = 12.5$, $PQ = 8$, $PR = 5$**
$5c = 100$	**Find the cross products.**
$c = 20$	**Divide each side by 5.**

$\frac{BC}{QR} = \frac{AC}{PR}$	**Corresponding sides of similar triangles are proportional.**
$\frac{a}{7} = \frac{12.5}{5}$	**$BC = a$, $AC = 12.5$, $QR = 7$, $PR = 5$**
$5a = 87.5$	**Find the cross products.**
$a = 17.5$	**Divide each side by 5.**

The missing measures are 20 and 17.5.

Check Your Progress

3A.

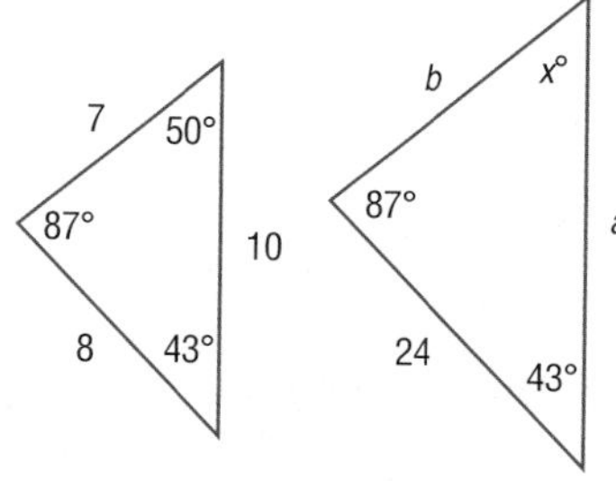

3B.

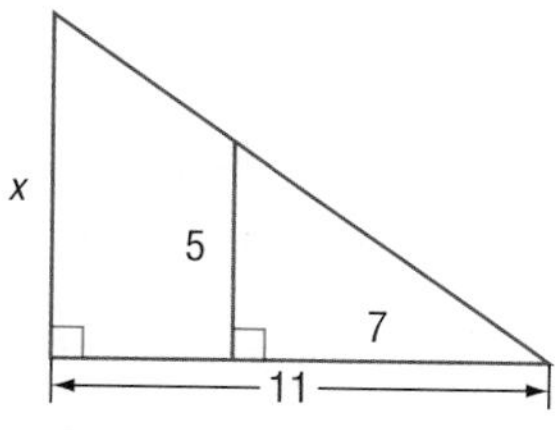

Personal Tutor glencoe.com

Real-World EXAMPLE 4 Indirect Measurement

SHADOWS Tori is 5 feet 6 inches tall, and her shadow is 2 feet 9 inches long. She is standing next to a flagpole. If the length of the shadow of the flagpole is 12 feet long, how tall is the flagpole?

x ft
5.5 ft
2.75 ft
12 ft

Math *in Motion*, BrainPOP® glencoe.com

Understand Find the height of the flagpole.

Plan Make a sketch of the situation.

Solve The Sun's rays form similar triangles. Write a proportion that compares the heights of the objects and the lengths of their shadows.

Let $x =$ the height of the flagpole.

height of flagpole → **Tori's height** → $\frac{x}{5.5} = \frac{12}{2.75}$ ← **length of the flagpole's shadow** ← **Tori's shadow**

$2.75x = 66$

$x = 24$ The height of the flagpole is 24 feet.

Check $\frac{24}{5.5} \stackrel{?}{=} \frac{12}{2.75}$ **Substitute 24 for *x*.**

$4.36 = 4.36$ ✓

Check Your Progress

4. TENTS The directions for pitching a tent include a scale drawing in which 1 inch represents 4.5 feet. In the drawing, the tent is $1\frac{3}{4}$ inches tall. How tall should the actual tent be?

Personal Tutor glencoe.com

Check Your Understanding

Examples 1 and 2 pp. 642–643

Determine whether each pair of triangles is similar. Justify your answer.

1.

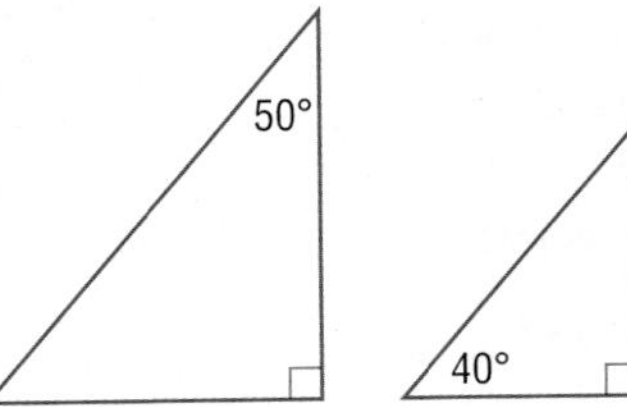

2.

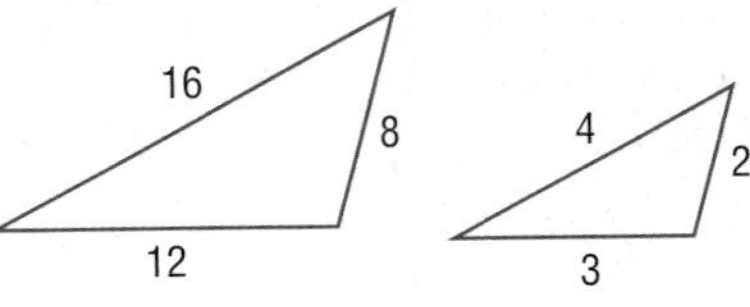

Example 3 p. 643

Find the missing measures for the pair of similar triangles if $\triangle ABC \sim \triangle XYZ$.

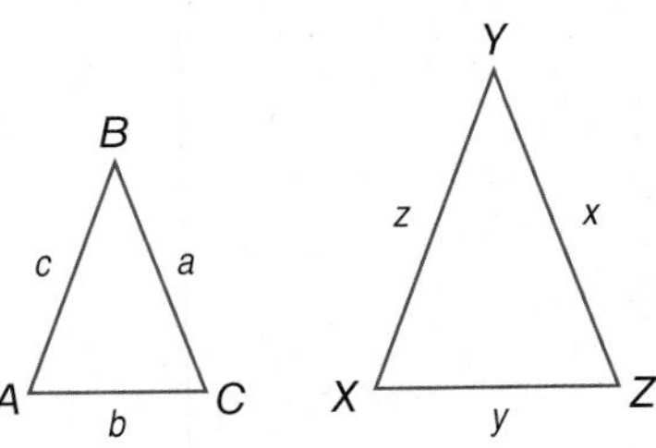

3. $a = 4, b = 6, c = 8, x = 6$

4. $x = 9, y = 15, z = 21, c = 7$

5. $a = 2, b = 5, x = 10, z = 30$

6. $b = 6, c = 10, x = 30, y = 15$

Example 4 p. 644

7. TREES Marla wants to know the height of the tree in her backyard. The tree casts a shadow 8 feet 6 inches long. Marla is 5 feet tall, and her shadow is 2 feet 6 inches long. How tall is the tree?

Practice and Problem Solving

= **Step-by-Step Solutions** begin on page R12.
Extra Practice begins on page 815.

Examples 1 and 2 pp. 642–643

Determine whether each pair of triangles is similar. Justify your answer.

8. 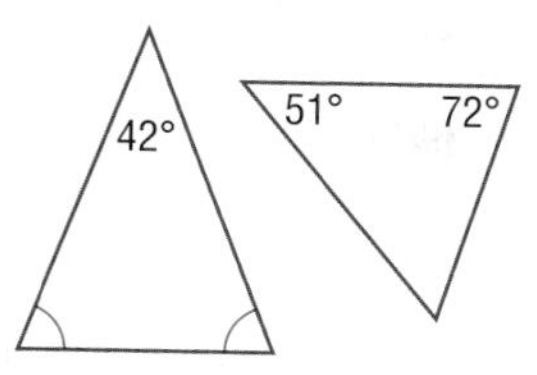

9. 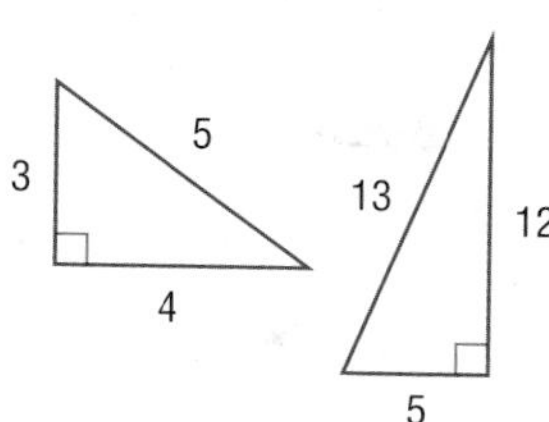

10. 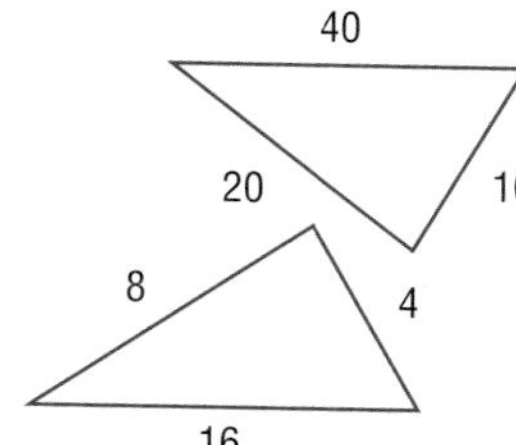

11. 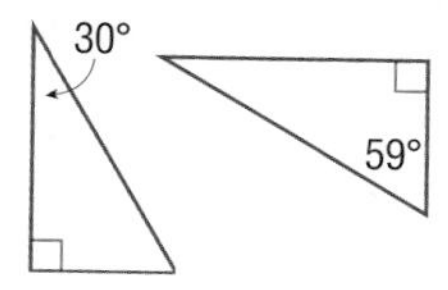

12. 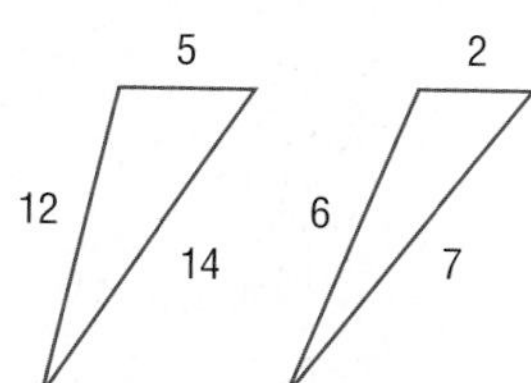

13. 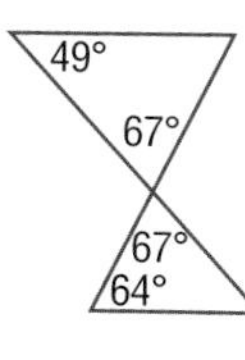

Example 3 p. 643

Find the missing measures for the pair of similar triangles if $\triangle HKM \sim \triangle PTR$.

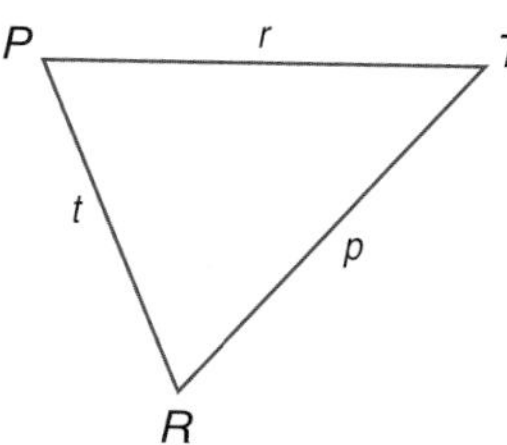

14. $m = 2, k = 7, h = 6, r = 4$

15. $r = 7.5, p = 15, t = 20, h = 6$

16. $m = 3.5, k = 9, t = 13.5, p = 9.75$

17. $m = 1.4, h = 2.8, p = 0.56, t = 0.84$

18. $m = \sqrt{7}, h = 2\sqrt{2}, t = 4\sqrt{3}, r = \sqrt{21}$

19. $m = \sqrt{2}, k = \sqrt{7}, t = \sqrt{14}, p = \sqrt{10}$

Example 4 p. 644

20. TOYS Diecast model cars use a scale of 1 inch : 2 feet of the real vehicle. The original vehicle has a window shaped like a right triangle. If the height of the window on the actual vehicle is 2.5 feet, what will the height of the window be on the model?

21. GOLF Beatriz is playing miniature golf on a hole like the one shown at the right. She wants to putt her ball U so that it will bank at T and travel into the hole at R. Use similar triangles to find where Beatriz's ball should strike the wall.

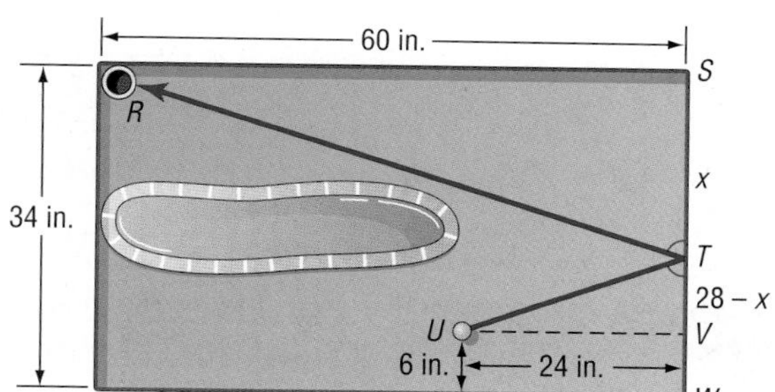

22. MAPS The scale on an Ohio map shows that 2.5 centimeters represents 100 miles. The distance on the map from Cleveland to Cincinnati is 5.5 centimeters. About how many miles apart are the two cities?

23. SCHOOL PROJECT For extra credit in his history class, Marquez plans to make a model of the Statue of Liberty in the scale 1 inch : 10 feet. If the height of the actual Statue of Liberty is 151 feet, what will be the height of the model?

24. TENNIS Andy wants to hit the ball just over the net so it will land 16 feet away from the base of the net. If Andy hits the ball 40 feet away from the net, how high does he have to hit the ball?

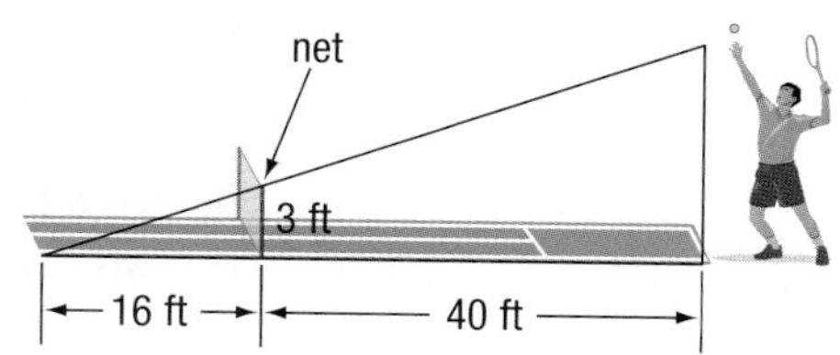

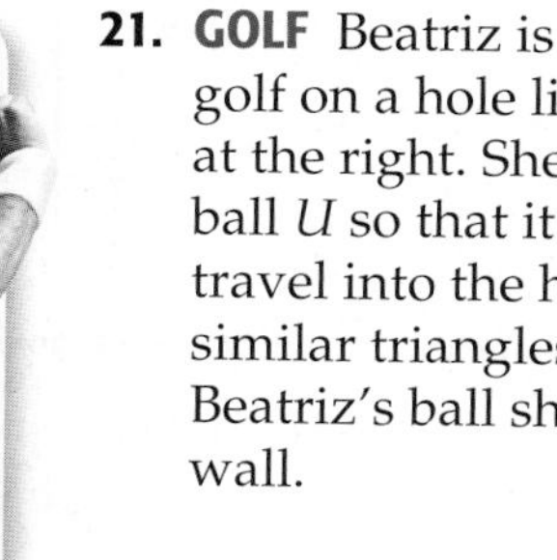

Real-World Link

Roger Federer was the number 1 ranked tennis player for four consecutive years beginning in 2004. He was the first player to reach all four Grand Slam finals in back-to-back years.

Source: ATP Tennis

25 **MULTIPLE REPRESENTATIONS** In this problem, you will compare the ratios of corresponding sides and perimeters of similar triangles.

a. **ALGEBRAIC** What is the ratio of the corresponding sides of each pair of similar triangles? Record your results in the table.

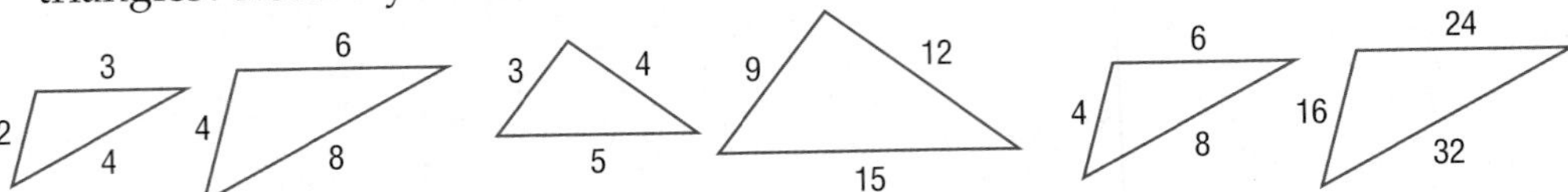

b. **TABULAR** Find the perimeter of each triangle. Then find the ratio of the perimeters for each pair of triangles. Record your results in the table.

Similar Triangles		Ratios of Sides	Perimeters	Ratios of Perimeters
Pair 1	smaller triangle			
	larger triangle			
Pair 2	smaller triangle			
	larger triangle			
Pair 3	smaller triangle			
	larger triangle			

c. **ANALYTICAL** How is the ratio of the perimeters related to the ratio of the lengths of corresponding sides for each pair of triangles?

d. **ANALYTICAL** If the ratio of the lengths of corresponding sides of two similar triangles is 1:6, what would be the ratio of their perimeters?

H.O.T. Problems Use Higher-Order Thinking Skills

26. **FIND THE ERROR** Kwam and Rosalinda are comparing the similar triangles. Is either of them correct? Explain.

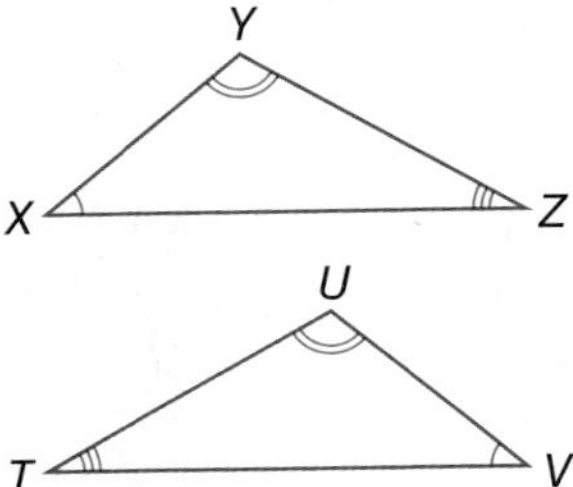

Kwam

$m\angle X = m\angle T$

$m\angle Y = m\angle U$

$m\angle Z = m\angle V$

$\triangle XYZ, \triangle TUV$

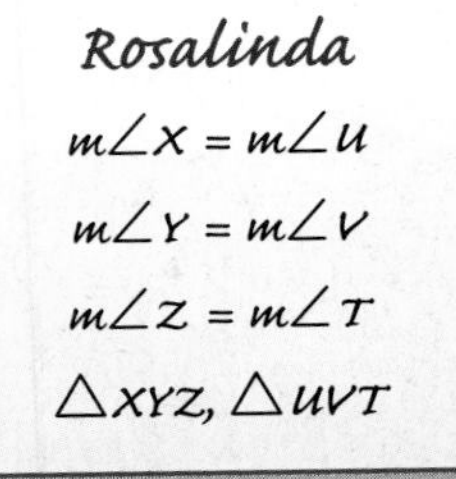

Rosalinda

$m\angle X = m\angle U$

$m\angle Y = m\angle V$

$m\angle Z = m\angle T$

$\triangle XYZ, \triangle UVT$

27. **CHALLENGE** Triangle XYZ is similar to the two triangles formed by the line segment from Z perpendicular to $\overline{XY}$, and these two triangles are similar to each other. Write three similarity statements about these triangles. Why are the triangles similar to each other?

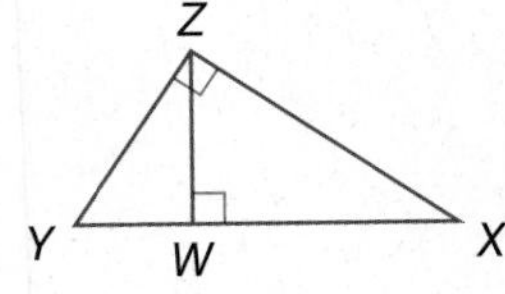

Problem-SolvingTip

Draw a Diagram When a problem involves spatial reasoning, or geometric figures, draw a diagram. For example, in Exercise 27, draw each triangle separately to help determine the answer.

28. **REASONING** Is the statement *sometimes*, *always*, or *never* true? Explain.

If the measures of the sides of a triangle are multiplied by 3, then the measures of the angles of the enlarged triangle will have the same measures as the angles of the original triangle.

29. **OPEN ENDED** Draw and label a triangle ABC. Then draw and label a similar triangle PQR so that the area of $\triangle PQR$ is four times the area of $\triangle ABC$. Explain your strategy.

30. **WRITING IN MATH** Summarize how to determine whether two triangles are similar to each other and how to find missing measures of similar triangles.

Standardized Test Practice

31. Find the distance between the points at (2, −4) and (−5, 8).

A 5
B 7
C $\sqrt{95}$
D $\sqrt{193}$

32. GEOMETRY Find the value of a if the two triangles are similar.

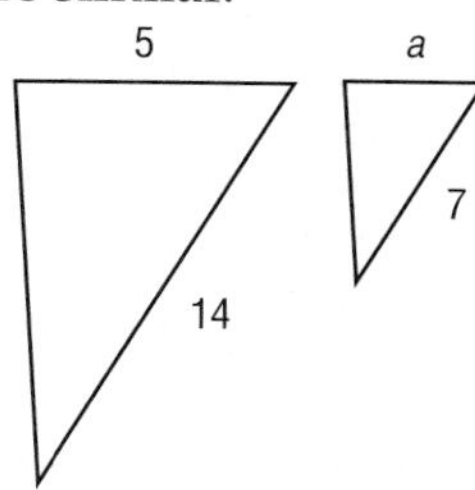

F 2.5 G 10 H 19 J 21

33. Which equation represents a line with a y-intercept of −4 and a slope of 6?

A $y = 6x - 4$
B $y = -4x + 6$
C $y = -6x - 4$
D $y = 6x + 4$

34. SHORT RESPONSE What are the x- and y-intercepts of the function graphed below?

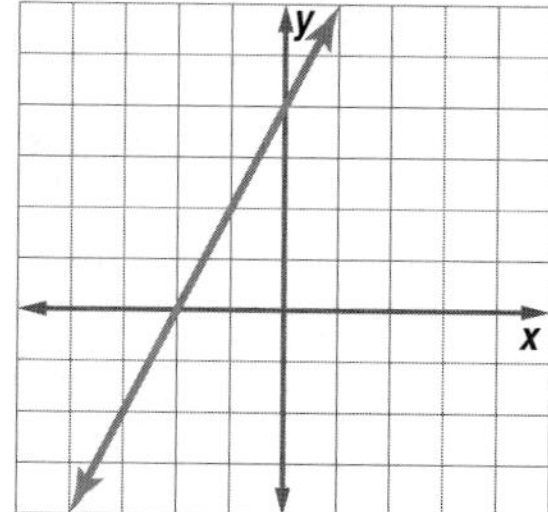

Spiral Review

Find the distance between the points with the given coordinates. (Lesson 10-6)

35. (0, 3), (1, 9)
36. (−2, 4), (5, 13)
37. (1, −5), (−1, −5)
38. (7, −2), (−2, 4)
39. (−6, −3), (−1, 2)
40. (−4, −3), (−7, −8)

Determine whether the measures can be the lengths of the sides of a right triangle. (Lesson 10-5)

41. 3, 4, 5
42. 8, 10, 12
43. 10, 24, 26
44. 5, 12, 13
45. 6, 9, 14
46. 4, 5, 6

47. NUTRITION In the function $y = 0.059x^2 - 7.423x + 362.1$, y represents the consumption of bread and cereal in pounds per person in the United States, and x represents the number of years since 1900. If this trend continues, in what future year will the average American consume 300 pounds of bread and cereal? (Lesson 9-4)

Factor each polynomial, if possible. If the polynomial cannot be factored, write *prime*. (Lesson 8-6)

48. $4k^2 - 100$
49. $4a^2 - 36b^2$
50. $x^2 + 6x - 9$
51. $50g^2 + 40g + 8$
52. $9t^3 + 66t^2 - 48t$
53. $20n^2 + 34n + 6$

54. DRIVING Average speed is calculated by dividing distance by time. If the speed limit on an interstate is 65 miles per hour, how far can a person travel legally in $1\frac{1}{2}$ hours? (Lesson 5-2)

Skills Review

Evaluate if $a = 3$, $b = -2$, and $c = 6$. (Lesson 1-2)

55. $\frac{b}{c}$
56. $\frac{2ab}{c}$
57. $\frac{ac}{-4b}$
58. $\frac{-3ac}{2b}$
59. $\frac{-2bc}{a}$

EXPLORE

10-8

Algebra Lab
Investigating Trigonometric Ratios

Math Online glencoe.com
Math *in Motion,* Animation

Objective
Investigate trigonometric ratios.

You can use paper triangles to investigate the ratios of the lengths of sides of right triangles.

Collect the Data

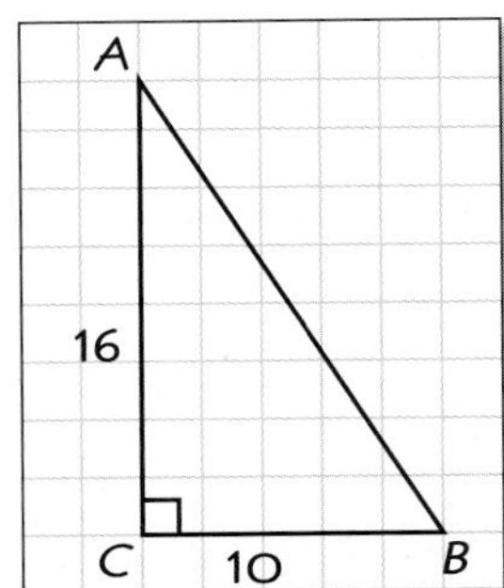

Step 1 Use a ruler and grid paper to draw several right triangles with legs in a ratio of 5:8. Include right triangles with the side lengths listed in the table below and several more right triangles similar to these three. Label the vertices of each triangle as *A*, *B*, and *C*, where *C* is at the right angle, *B* is opposite the longest leg, and *A* is opposite the shortest leg.

Step 2 Copy the table below. Complete the first three columns by measuring the hypotenuse (side $\overline{AB}$) in each right triangle you created and recording its length to the nearest tenth.

Step 3 Calculate and record the ratios in the middle two columns. Round to the nearest hundredth.

Step 4 Use a protractor to carefully measure angles *A* and *B* to the nearest degree in each right triangle. Record the angle measures in the table.

Side Lengths			Ratios		Angle Measures		
side *BC*	side *AC*	side *AB*	$\frac{BC}{AC}$	$\frac{BC}{AB}$	angle *A*	angle *B*	angle *C*
2.5	4						90°
5	8						90°
10	16						90°
							90°
							90°
							90°

Analyze the Results

1. Examine the measures and ratios in the table. What do you notice? Write a sentence or two to describe any patterns you see.

Make a Conjecture

2. For any right triangle similar to the ones you have drawn here, what will be the value of the ratio of the length of the shortest leg to the length of the longest leg?

3. If you draw a right triangle and calculate the ratio of the length of the shortest leg to the length of the hypotenuse to be approximately 0.53, what will be the measure of the larger acute angle in the right triangle?

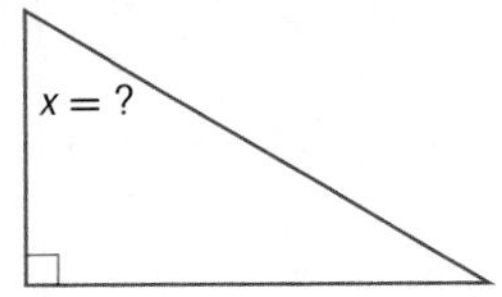

10-8 Trigonometric Ratios

Then

You used the Pythagorean Theorem (Lesson 10-5)

Now

- Find trigonometric ratios of angles.
- Use trigonometry to solve triangles.

New Vocabulary

trigonometry
trigonometric ratio
sine
cosine
tangent
solving the triangle

Math Online

glencoe.com

- Extra Examples
- Personal Tutor
- Self-Check Quiz
- Homework Help

Why?

If a road has a percent grade of 8%, this means the road rises or falls 8 feet over a horizontal distance of 100 feet. Trigonometric ratios can be used to determine the angle that the road rises or falls.

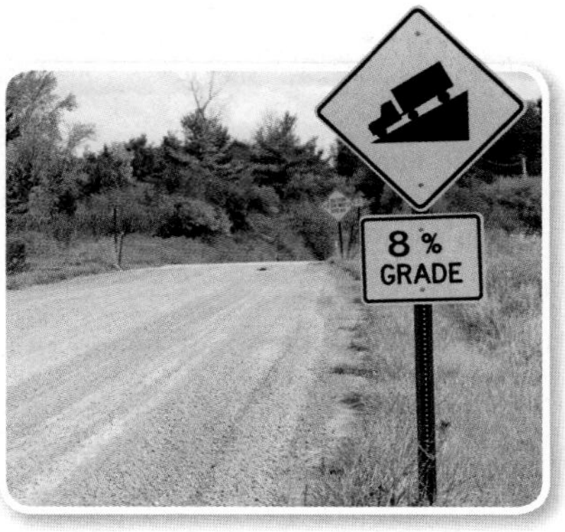

Trigonometric Ratios **Trigonometry** is the study of relationships among the angles and sides of triangles. A **trigonometric ratio** is a ratio that compares the side lengths of two sides of a right triangle. The three most common trigonometric ratios, **sine**, **cosine**, and **tangent**, are described below.

Key Concept — Trigonometric Ratios

For Your FOLDABLE

Words	Symbols	Model
sine of $\angle A = \dfrac{\text{leg opposite } \angle A}{\text{hypotenuse}}$	$\sin A = \dfrac{a}{c}$	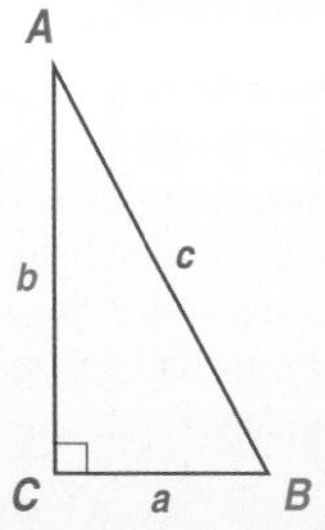
cosine of $\angle A = \dfrac{\text{leg adjacent to } \angle A}{\text{hypotenuse}}$	$\cos A = \dfrac{b}{c}$	
tangent of $\angle A = \dfrac{\text{leg opposite } \angle A}{\text{leg adjacent to } \angle A}$	$\tan A = \dfrac{a}{b}$	

Opposite, adjacent, and hypotenuse are abbreviated *opp*, *adj*, and *hyp*, respectively.

EXAMPLE 1 Find Sine, Cosine, and Tangent Ratios

Find the values of the three trigonometric ratios for angle A.

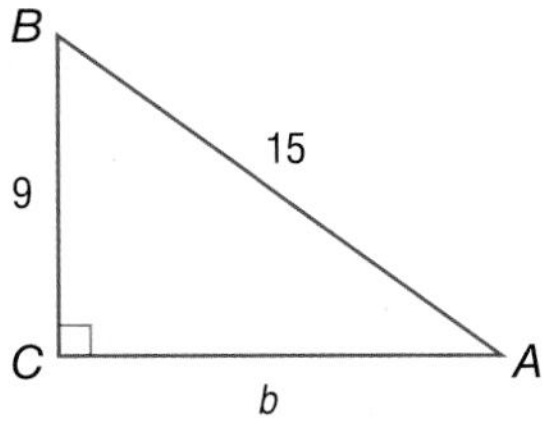

Step 1 Use the Pythagorean Theorem to find AC.

$a^2 + b^2 = c^2$ **Pythagorean Theorem**

$9^2 + b^2 = 15^2$ **$a = 9$ and $c = 15$**

$81 + b^2 = 225$ **Simplify.**

$b^2 = 144$ **Subtract 81 from each side.**

$b = 12$ **Take the square root of each side.**

Step 2 Use the side lengths to write the trigonometric ratios.

$$\sin A = \frac{\text{opp}}{\text{hyp}} = \frac{9}{15} = \frac{3}{5} \qquad \cos A = \frac{\text{adj}}{\text{hyp}} = \frac{12}{15} = \frac{4}{5} \qquad \tan A = \frac{\text{opp}}{\text{adj}} = \frac{9}{12} = \frac{3}{4}$$

Check Your Progress

1. Find the values of the three trigonometric ratios for angle B.

Personal Tutor glencoe.com

Watch Out!

Calculator Mode Make sure your graphing calculator is in degree mode.

EXAMPLE 2 Use a Calculator to Evaluate Expressions

Use a calculator to find cos 42° to the nearest ten-thousandth.

KEYSTROKES: [COS] 42 [)] [ENTER]

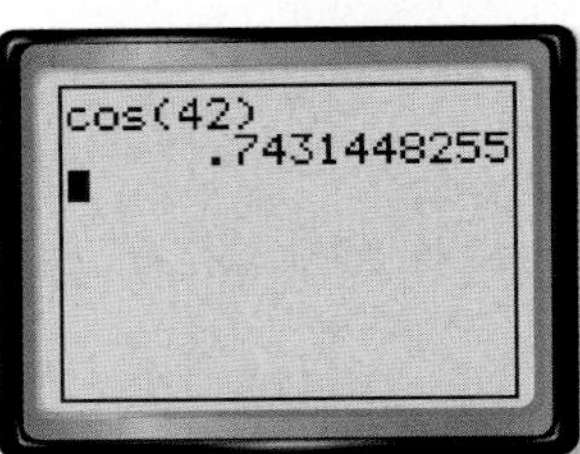

Rounded to the nearest ten-thousandth, $\cos 42° \approx 0.7431$.

Check Your Progress

2A. $\sin 31°$ **2B.** $\tan 76°$ **2C.** $\cos 55°$

Personal Tutor glencoe.com

Use Trigonometric Ratios When you find all unknown measures of the sides and angles of a right triangle, you are **solving the triangle**. You can find the missing measures if you know the measure of two sides of the triangle or the measure of one side and the measure of one acute angle.

StudyTip

Remembering Trigonometric Ratios SOH–CAH–TOA can be used to help you remember the ratios for sine, cosine, and tangent. Each letter represents a word.

$$\sin A = \frac{\text{opp}}{\text{hyp}}$$

$$\cos A = \frac{\text{adj}}{\text{hyp}}$$

$$\tan A = \frac{\text{opp}}{\text{adj}}$$

EXAMPLE 3 Solve a Triangle

Solve the right triangle. Round each side length to the nearest tenth.

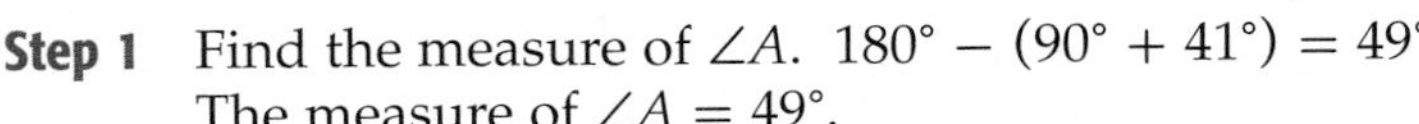

Step 1 Find the measure of $\angle A$. $180° - (90° + 41°) = 49°$
The measure of $\angle A = 49°$.

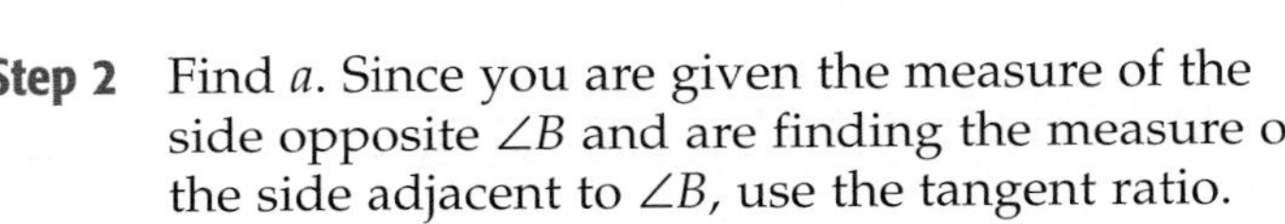

Step 2 Find a. Since you are given the measure of the side opposite $\angle B$ and are finding the measure of the side adjacent to $\angle B$, use the tangent ratio.

$\tan 41° = \frac{6}{a}$ **Definition of tangent**

$a \tan 41° = 6$ **Multiply each side by *a*.**

$a = \frac{6}{\tan 41°}$ or about 6.9 **Divide each side by tan 41°. Use a calculator.**

So the measure of a or $\overline{BC}$ is about 6.9.

Step 3 Find c. Since you are given the measure of the side opposite $\angle B$ and are finding the measure of the hypotenuse, use the sine ratio.

$\sin 41° = \frac{6}{c}$ **Definition of sine**

$c \sin 41° = 6$ **Multiply each side by *c*.**

$c = \frac{6}{\sin 41°}$ or about 9.1 **Divide each side by sin 41°. Use a calculator.**

So the measure of c or $\overline{AB}$ is about 9.1.

Check Your Progress

3A.

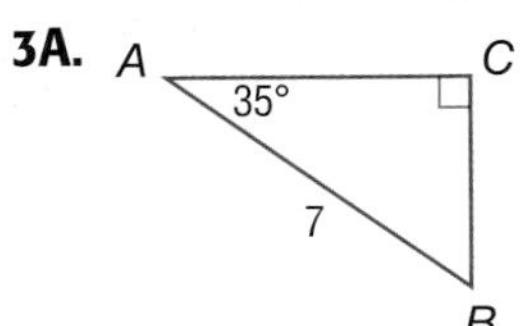

3B.

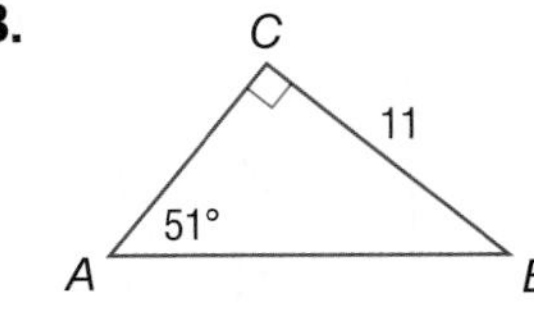

Personal Tutor glencoe.com

Real-World Link

For optimum health, all adults ages 18–65 should get at least 30 minutes of moderately intense activity five days per week.

Source: American Heart Association

Real-World EXAMPLE 4 Find a Missing Side Length

EXERCISE A trainer sets the incline on a treadmill to 10°. The walking surface of the treadmill is 5 feet long. About how many inches is the end of the treadmill from the floor?

$\sin 10° = \frac{h}{5}$ **Definition of sine**

$5 \cdot \sin 10° = h$ **Multiply each side by 5.**

$0.87 \approx h$ **Use a calculator.**

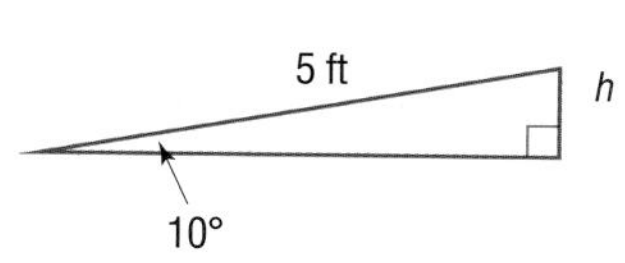

The value of h is in feet. Multiply 0.87 by 12 to convert feet to inches. The trainer raised the treadmill about 10.4 inches.

Check Your Progress

4. **SKATEBOARDING** The angle that a skateboarding ramp forms with the ground is 25° and the height of the ramp is 6 feet. Determine the length of the ramp.

Personal Tutor glencoe.com

A trigonometric function has a rule given by a trigonometric ratio. If you know the sine, cosine, or tangent of an acute angle, you can use the *inverse* of the trigonometric function to find the measure of the angle.

Key Concept Inverse Trigonometric Functions

For Your FOLDABLE

Words	If $\angle A$ is an acute angle and the sine of A is x, then the **inverse sine** of x is the measure of $\angle A$.
Symbols	If $\sin A = x$, then $\sin^{-1} x = m\angle A$.
Words	If $\angle A$ is an acute angle and the cosine of A is x, then the **inverse cosine** of x is the measure of $\angle A$.
Symbols	If $\cos A = x$, then $\cos^{-1} x = m\angle A$.
Words	If $\angle A$ is an acute angle and the tangent of A is x, then the **inverse tangent** of x is the measure of $\angle A$.
Symbols	If $\tan A = x$, then $\tan^{-1} x = m\angle A$.

EXAMPLE 5 Find a Missing Angle Measure

Find $m\angle Y$ to the nearest degree.

You know the measure of the side adjacent to $\angle Y$ and the measure of the hypotenuse. Use the cosine ratio.

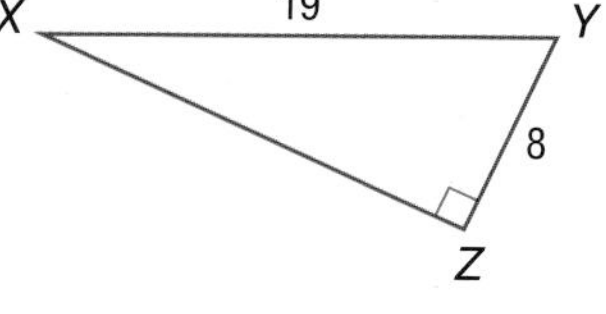

$\cos Y = \frac{8}{19}$ **Definition of cosine**

Use a calculator and the [cos^{-1}] function to find the measure of the angle.

KEYSTROKES: 2nd [cos^{-1}] 8 ÷ 19) ENTER 65.098937 So, $m\angle Y = 65°$.

Check Your Progress

5. Find $m\angle X$ to the nearest degree if $XY = 14$ and $YZ = 5$.

Personal Tutor glencoe.com

Check Your Understanding

Example 1 p. 649

Find the values of the three trigonometric ratios for angle A.

1.

2. 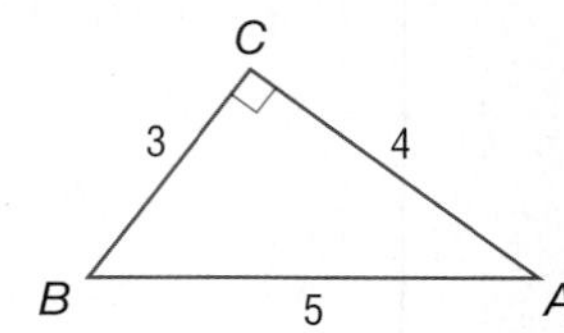

3. 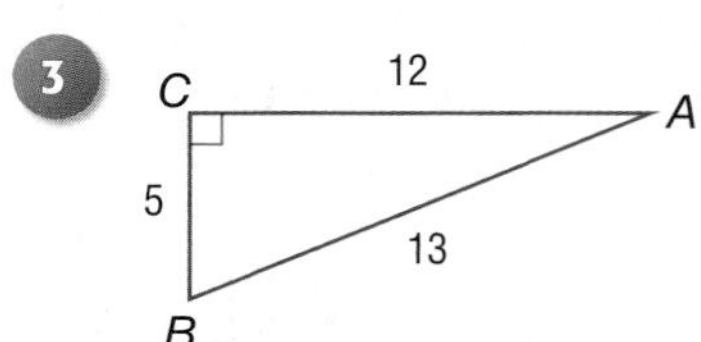

4. 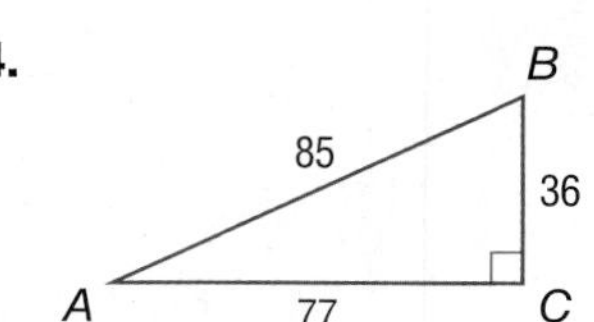

Example 2 p. 650

Use a calculator to find the value of each trigonometric ratio to the nearest ten-thousandth.

5. sin 37° **6.** cos 23° **7.** tan 14° **8.** cos 82°

Example 3 p. 650

Solve each right triangle. Round each side length to the nearest tenth.

9. 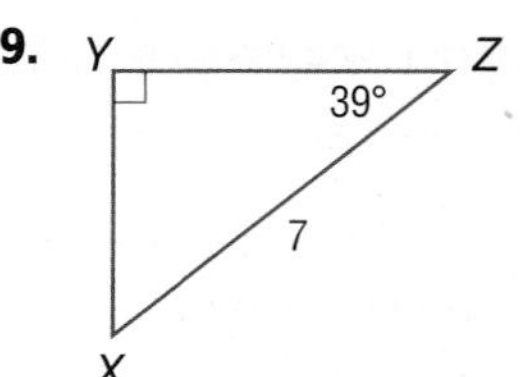

10. 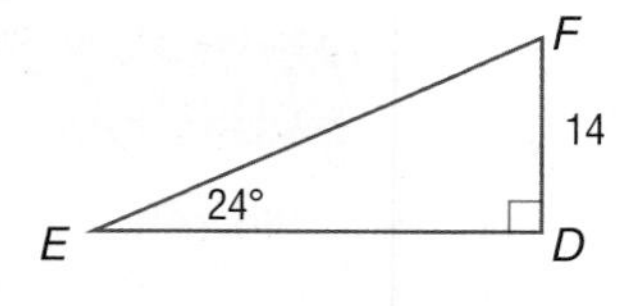

11. 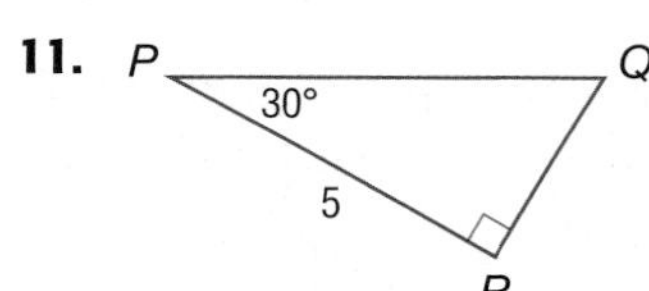

12. 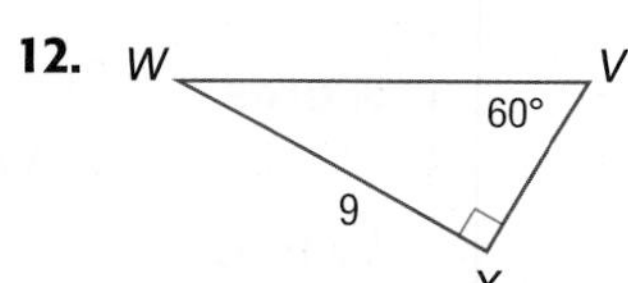

Example 4 p. 651

13. **SNOWBOARDING** A hill used for snowboarding has a vertical drop of 3500 feet. The angle the run makes with the ground is 18°. Estimate the length of r.

Example 5 p. 651

Find m∠X for each right triangle to the nearest degree.

14. 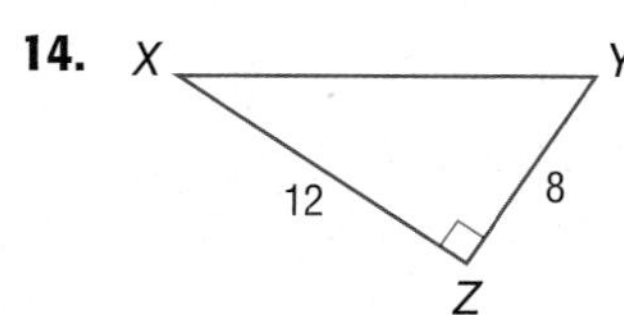

15. 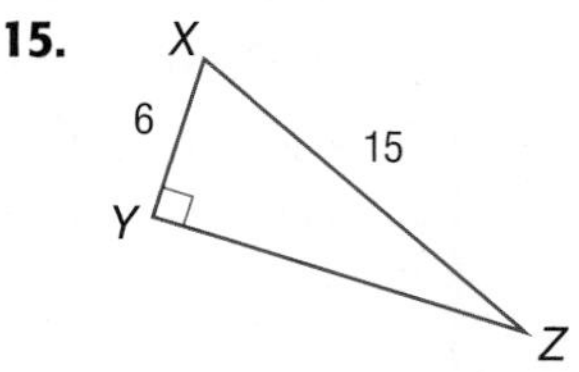

16. 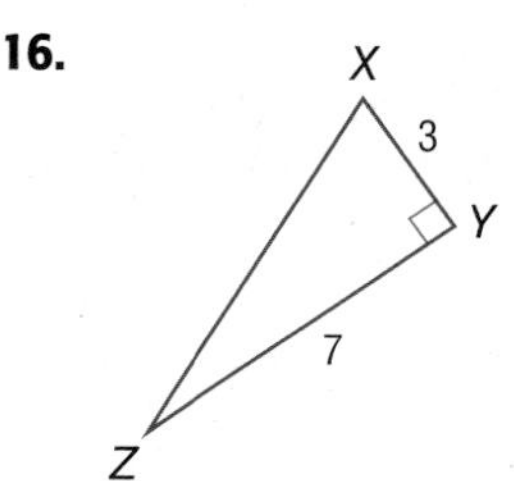

17. 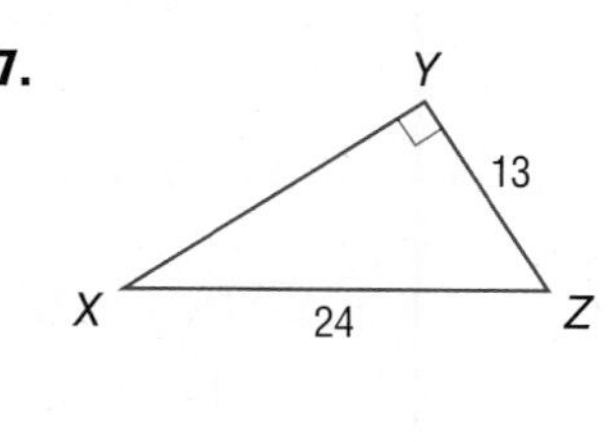

Practice and Problem Solving

● = **Step-by-Step Solutions** begin on page R12.
Extra Practice begins on page 815.

Example 1 p. 649

Find the values of the three trigonometric ratios for angle B.

18.

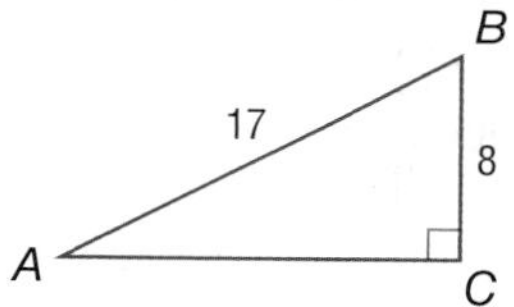

19.

20.

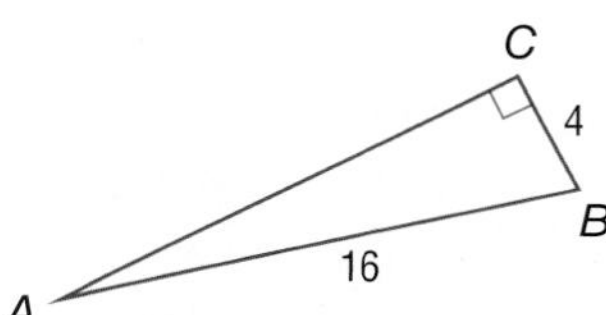

Example 2 p. 650

Use a calculator to find the value of each trigonometric ratio to the nearest ten-thousandth.

21. tan 2° **22.** sin 89° **23.** cos 44° **24.** tan 45°

25. sin 73° **26.** cos 90° **27.** sin 30° **28.** tan 60°

Example 3 p. 650

Solve each right triangle. Round each side length to the nearest tenth.

29.

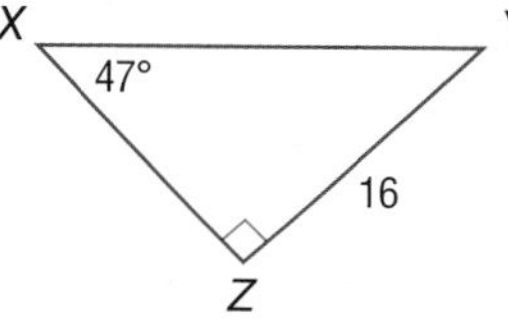

30.

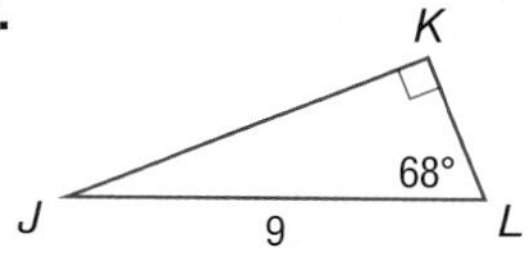

31.

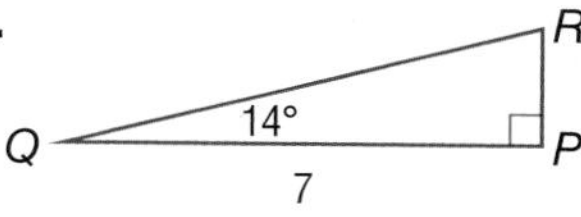

32.

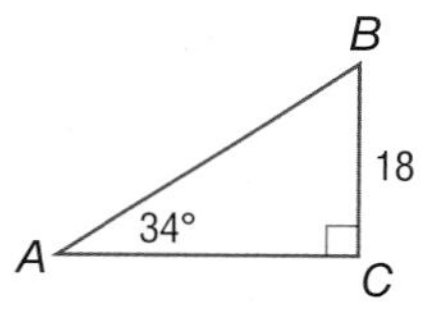

33.

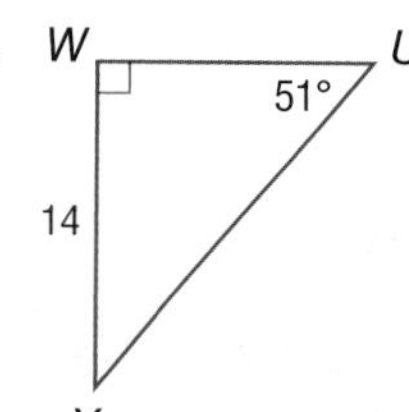

34.

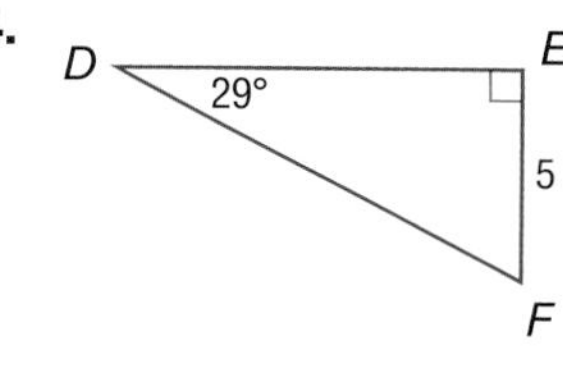

Example 4 p. 651

35. ESCALATORS At a local mall, an escalator is 110 feet long. The angle the escalator makes with the ground is 29°. Find the height reached by the escalator.

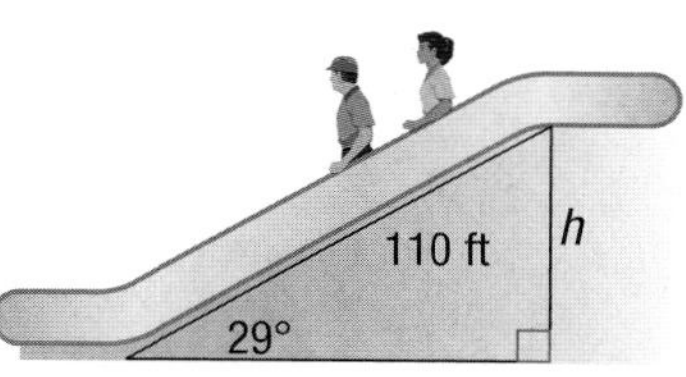

Example 5 p. 651

Find $m\angle J$ for each right triangle to the nearest degree.

36.

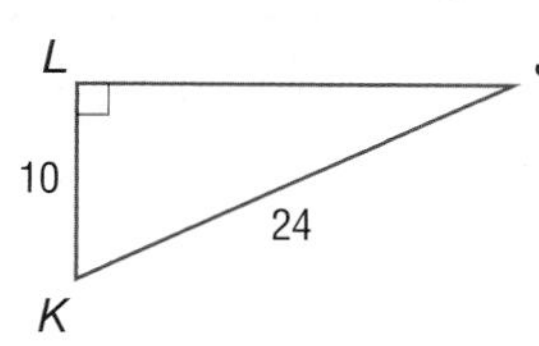

37.

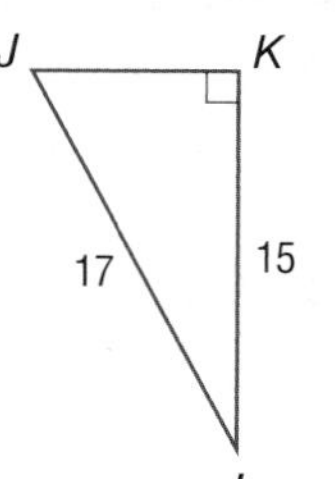

38.

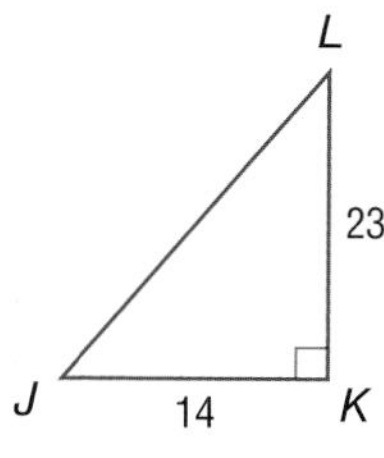

39.

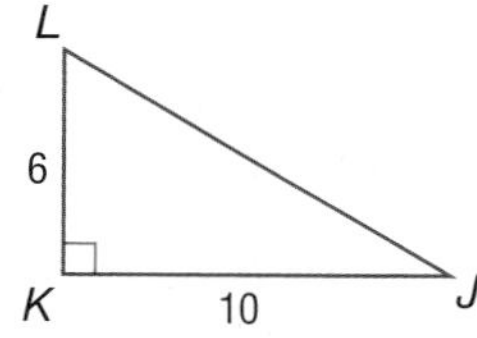

40.

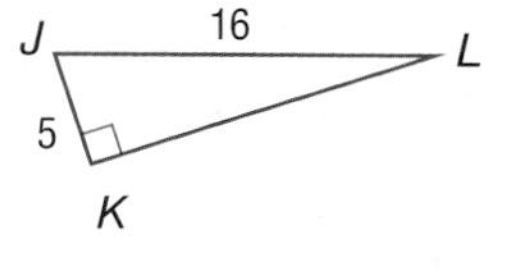

41.

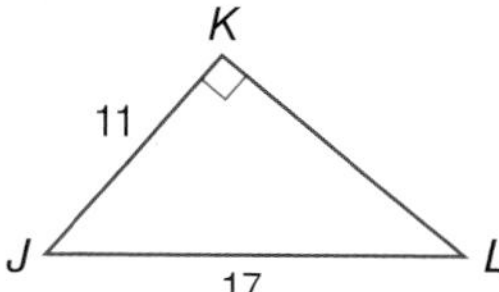

42. MONUMENTS The Lincoln Memorial building measures 204 feet long, 134 feet wide, and 99 feet tall. Chloe is looking at the top of the monument at an angle of 55°. How far away is she standing from the monument?

Real-World Link

The tallest living tree is a redwood called Hyperion. It measures 378.1 feet. There are about 135 redwood trees that are taller than 350 feet.

Source: SFGate

43. **AIRPLANES** Ella looks down at a city from an airplane window. The airplane is 5000 feet in the air, and she looks down at an angle of 8°. Determine the horizontal distance to the city.

44. **FORESTS** A forest ranger estimates the height of a tree is about 175 feet. If the forest ranger is standing 100 feet from the base of the tree, what is the measure of the angle formed by the ranger and the top of the tree?

Suppose $\angle A$ is an acute angle of right triangle ABC.

45. Find $\sin A$ and $\tan A$ if $\cos A = \frac{3}{4}$.

46. Find $\tan A$ and $\cos A$ if $\sin A = \frac{2}{7}$.

47. Find $\cos A$ and $\tan A$ if $\sin A = \frac{1}{4}$.

48. Find $\sin A$ and $\cos A$ if $\tan A = \frac{5}{3}$.

49. **SUBMARINES** A submarine descends into the ocean at an angle of 10° below the water line and travels 3 miles diagonally. How far beneath the surface of the water has the submarine reached?

50. **MULTIPLE REPRESENTATIONS** In this problem, you will explore a relationship between the sine and cosine functions.

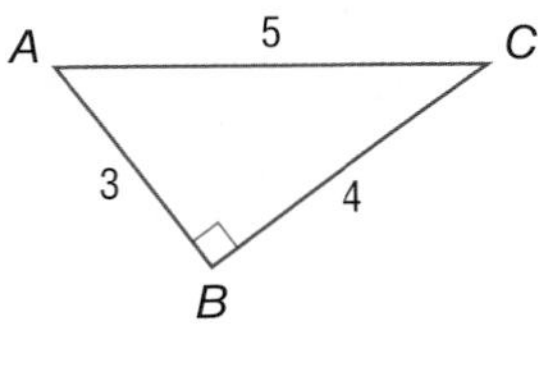

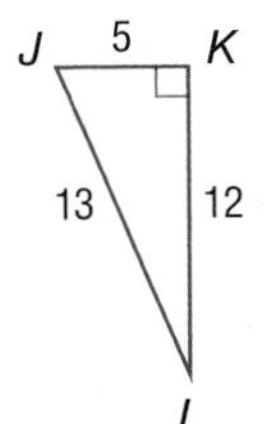

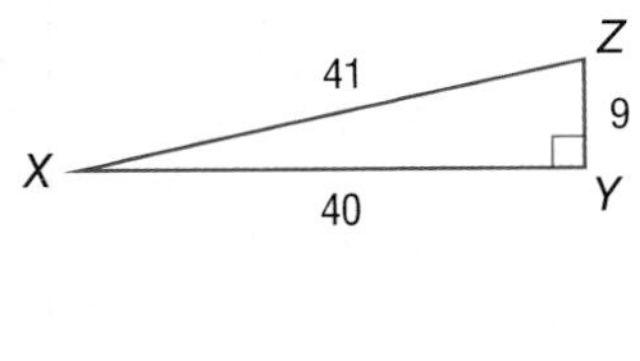

a. **TABULAR** Copy and complete the table using the triangles shown above.

Triangle	Trigonometric Ratios		$\sin^2$	$\cos^2$	$\sin^2 + \cos^2 =$
ABC	$\sin A =$	$\cos A =$	$\sin^2 A =$	$\cos^2 A =$	
	$\sin C =$	$\cos C =$	$\sin^2 C =$	$\cos^2 C =$	
JKL	$\sin J =$	$\cos J =$	$\sin^2 J =$	$\cos^2 J =$	
	$\sin L =$	$\cos L =$	$\sin^2 L =$	$\cos^2 L =$	
XYZ	$\sin X =$	$\cos X =$	$\sin^2 X =$	$\cos^2 X =$	
	$\sin Z =$	$\cos Z =$	$\sin^2 Z =$	$\cos^2 Z =$	

b. **VERBAL** Make a conjecture about the sum of the squares of the sine and cosine functions of an acute angle in a right triangle.

H.O.T. Problems

Use Higher-Order Thinking Skills

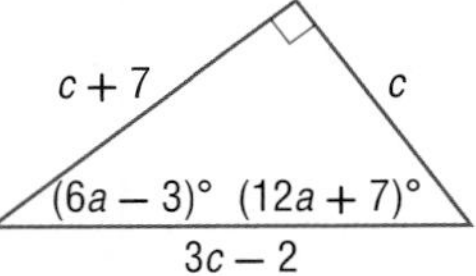

51. **CHALLENGE** Solve the triangle shown.

52. **REASONING** Use the definitions of the sine and cosine ratios to define the tangent ratio.

53. **OPEN ENDED** Write a problem that uses the cosine ratio to find the measure of an unknown angle in a triangle. Then solve the problem.

54. **REASONING** The sine and cosine of an acute angle in a right triangle are equal. What can you conclude about the triangle?

55. **WRITING IN MATH** Explain how to use trigonometric ratios to find the missing length of a side of a right triangle given the measure of one acute angle and the length of one side.

Standardized Test Practice

56. Which graph below represents the solution set for $-2 \leq x \leq 4$?

A 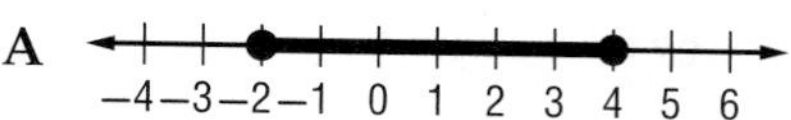

B −4 −3 −2 −1 0 1 2 3 4 5 6

C −4 −3 −2 −1 0 1 2 3 4 5 6

D −4 −3 −2 −1 0 1 2 3 4 5 6

57. PROBABILITY Suppose one chip is chosen from a bin with the chips shown. To the nearest tenth, what is the probability that a green chip is chosen?

Color	Number
yellow	7
blue	9
orange	3
green	5
red	6

F 0.2
G 0.5
H 0.6
J 0.8

58. In the graph, for what value(s) of x is $y = 0$?

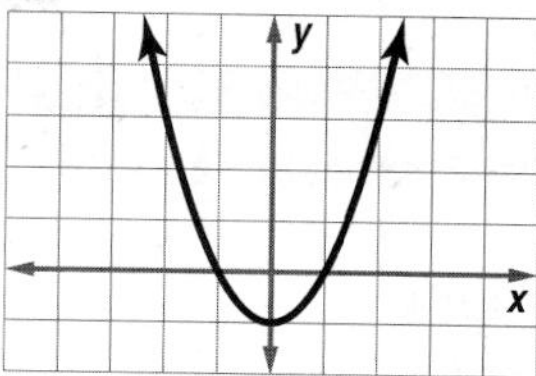

A 0
B −1
C 1
D 1 and −1

59. EXTENDED RESPONSE A 16-foot ladder is placed against the side of a house so that the bottom of the ladder is 8 feet from the base of the house.

a. If the bottom of the ladder is moved closer to the base of the house, does the height reached by the ladder increase or decrease?

b. What conclusion can you make about the distance between the bottom of the ladder and the base of the house and the height reached by the ladder?

c. How high does the ladder reach if the ladder is 3 feet from the base of the house?

Spiral Review

For each set of measures given, find the measures of the missing sides if $\triangle ABC \sim \triangle DFH$. (Lesson 10-7)

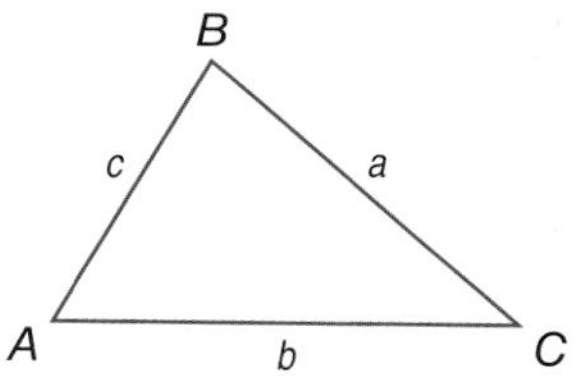

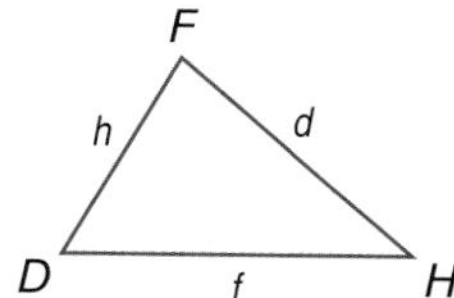

60. $a = 16, b = 12, c = 8, f = 6$

61. $d = 9, f = 6, h = 4, b = 18$

62. $a = 36, b = 21, h = 11, f = 14$

63. $c = 22.5, b = 20, h = 9, d = 2$

Find the coordinates of the midpoint of the segment with the given endpoints. (Lesson 10-6)

64. $(5, 3), (11, 9)$

65. $(8, 2), (6, 4)$

66. $(-1, 7), (13, -3)$

67. FINANCIAL LITERACY A salesperson is paid $32,000 a year plus 5% of the amount in sales made. What is the amount of sales needed to have an annual income greater than $45,000? (Lesson 5-3)

Skills Review

Solve each proportion. (Lesson 2-6)

68. $\frac{8}{9} = \frac{6}{z}$

69. $\frac{p}{6} = \frac{4}{3}$

70. $\frac{0.3}{r} = \frac{0.9}{1.7}$

71. $\frac{0.6}{1.1} = \frac{y}{8.47}$

Study Guide and Review

Math Online glencoe.com
• STUDY TO GO
• Vocabulary Review

Chapter Summary

Key Concepts

Simplifying Radical Expressions (Lesson 10-2)

- A radical expression is in simplest form when
  - no radicands have perfect square factors other than 1,
  - no radicals contain fractions,
  - and no radicals appear in the denominator of a fraction.

Operations with Radical Expressions and Equations (Lessons 10-3 and 10-4)

- Radical expressions with like radicals can be added or subtracted.
- Use the FOIL method to multiply radical expressions.

Pythagorean Theorem, Distance Formula, and Midpoint Formula (Lessons 10-5 and 10-6)

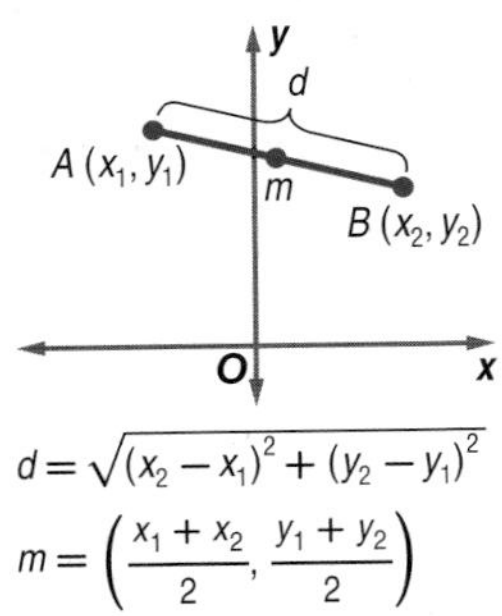

$$d = \sqrt{(x_2 - x_1)^2 + (y_2 - y_1)^2}$$

$$m = \left(\frac{x_1 + x_2}{2}, \frac{y_1 + y_2}{2}\right)$$

Similar Triangles (Lesson 10-7)

- Similar triangles have congruent corresponding angles and proportional corresponding sides.

  If $\triangle ABC \sim \triangle DEF$, then $\frac{AB}{DE} = \frac{BC}{EF} = \frac{AC}{DF}$.

FOLDABLES® Study Organizer

Be sure the Key Concepts are noted in your Foldable.

Radical Functions and Geometry

Key Vocabulary

conjugate (p. 614)
converse (p. 631)
cosine (p. 649)
Distance Formula (p. 636)
extraneous solutions (p. 625)
hypotenuse (p. 630)
inverse cosine (p. 651)
inverse sine (p. 651)
inverse tangent (p. 651)
legs (p. 630)
midpoint (p. 638)
Midpoint Formula (p. 638)
Pythagorean triple (p. 631)
radical equations (p. 624)
radical expression (p. 612)
radical function (p. 605)
radicand (p. 605)
rationalizing the denominator (p. 614)
similar triangles (p. 642)
sine (p. 649)
solving the triangle (p. 650)
square root function (p.605)
tangent (p. 649)
trigonometric ratio (p. 649)
trigonometry (p. 649)

Vocabulary Check

State whether each sentence is *true* or *false*. If *false*, replace the underlined word, phrase, expression, or number to make a true sentence.

1. A triangle with sides having measures of <u>3, 4, and 6</u> is a right triangle.
2. Two triangles are <u>congruent</u> if corresponding angles are congruent.
3. The expressions $2 + \sqrt{5}$ and <u>$2 - \sqrt{5}$</u> are conjugates.
4. In the expression $-5\sqrt{2}$, the radicand is <u>2</u>.
5. The <u>shortest</u> side of a right triangle is the hypotenuse.
6. The cosine of an angle is found by dividing the measure of the side <u>opposite</u> the angle by the hypotenuse.
7. The domain of the function $y = \sqrt{x}$ is <u>$\{x \mid x \leq 0\}$</u>.
8. After the first step in solving $\sqrt{2x + 4} = x + 5$, you would have <u>$2x + 4 = x^2 + 10x + 25$</u>
9. The converse of the Pythagorean Theorem is <u>true</u>.
10. The range of the function $y = \sqrt{x}$ is <u>$\{y \mid > 0\}$</u>.

Lesson-by-Lesson Review

10-1 Radical Functions (pp. 605–610)

Graph each function. Compare to the parent graph. State the domain and range.

11. $y = \sqrt{x} + 3$

12. $y = \sqrt{x} - 2$

13. $y = -5\sqrt{x}$

14. $y = \sqrt{x} - 6$

15. $y = \sqrt{x - 1}$

16. $y = \sqrt{x} + 5$

17. GEOMETRY The function $s = \sqrt{A}$ can be used to find the length of a side of a square given its area. Use this function to determine the length of a side of a square with an area of 90 square inches. Round to the nearest tenth if necessary.

EXAMPLE 1

Graph $y = -3\sqrt{x}$. Compare to the parent graph. State the domain and range.

Make a table. Choose nonnegative values for x.

x	0	1	2	3	4
y	0	−3	≈−4.2	≈−5.2	−6

Plot points and draw a smooth curve.

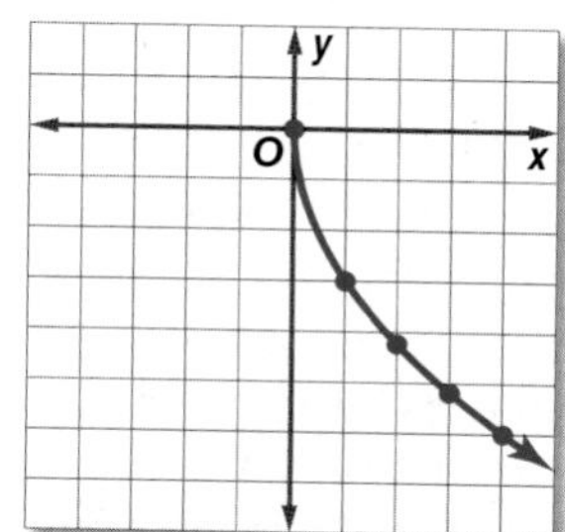

The graph of $y = \sqrt{x}$ is stretched vertically and is reflected across the x-axis.

The domain is $\{x \mid x \geq 0\}$.

The range is $\{y \mid y \leq 0\}$.

10-2 Simplifying Radical Expressions (pp. 612–617)

Simplify.

18. $\sqrt{36x^2y^7}$

19. $\sqrt{20ab^3}$

20. $\sqrt{3} \cdot \sqrt{6}$

21. $2\sqrt{3} \cdot 3\sqrt{12}$

22. $(4 - \sqrt{5})^2$

23. $(1 + \sqrt{2})^2$

24. $\sqrt{\frac{50}{a^2}}$

25. $\sqrt{\frac{2}{5}} \cdot \sqrt{\frac{3}{4}}$

26. $\frac{3}{2 - \sqrt{5}}$

27. $\frac{5}{\sqrt{7} + 6}$

28. WEATHER To estimate how long a thunderstorm will last, use $t = \sqrt{\frac{d^3}{216}}$, where t is the time in hours and d is the diameter of the storm in miles. A storm is 10 miles in diameter. How long will it last?

EXAMPLE 2

Simplify $\frac{2}{4 + \sqrt{3}}$.

$\frac{2}{4 + \sqrt{3}}$

$= \frac{2}{4 + \sqrt{3}} \cdot \frac{4 - \sqrt{3}}{4 - \sqrt{3}}$ **Rationalize the denominator.**

$= \frac{2(4) - 2\sqrt{3}}{4^2 - (\sqrt{3})^2}$ $(a - b)(a + b) = a^2 - b^2$

$= \frac{8 - 2\sqrt{3}}{16 - 3}$ $(\sqrt{3})^2 = 3$

$= \frac{8 - 2\sqrt{3}}{13}$ **Simplify.**

CHAPTER 10

Study Guide and Review

10-3 Operations with Radical Expressions (pp. 619–623)

Simplify each expression.

29. $\sqrt{6} - \sqrt{54} + 3\sqrt{12} + 5\sqrt{3}$

30. $2\sqrt{6} - \sqrt{48}$

31. $4\sqrt{3x} - 3\sqrt{3x} + 3\sqrt{3x}$

32. $\sqrt{50} + \sqrt{75}$

33. $\sqrt{2}(5 + 3\sqrt{3})$

34. $(2\sqrt{3} - \sqrt{5})(\sqrt{10} + 4\sqrt{6})$

35. $(6\sqrt{5} + 2)(4\sqrt{2} + \sqrt{3})$

36. **MOTION** The velocity of a dropped object when it hits the ground can be found using $v = \sqrt{2gd}$, where v is the velocity in feet per second, g is the acceleration due to gravity, and d is the distance in feet the object drops. Find the speed of a penny when it hits the ground, after being dropped from 984 feet. Use 32 feet per second squared for g.

EXAMPLE 3

Simplify $2\sqrt{6} - \sqrt{24}$.

$2\sqrt{6} - \sqrt{24} = 2\sqrt{6} - \sqrt{4 \cdot 6}$	**Product Property**
$= 2\sqrt{6} - 2\sqrt{6}$	**Simplify.**
$= 0$	**Simplify.**

EXAMPLE 4

Simplify $(\sqrt{3} - \sqrt{2})(\sqrt{3} + 2\sqrt{2})$.

$(\sqrt{3} - \sqrt{2})(\sqrt{3} + 2\sqrt{2})$

$= (\sqrt{3})(\sqrt{3}) + (\sqrt{3})(2\sqrt{2}) + (-\sqrt{2})(\sqrt{3}) + (\sqrt{2})(2\sqrt{2})$

$= 3 + 2\sqrt{6} - \sqrt{6} + 4$

$= 7 + \sqrt{6}$

10-4 Radical Equations (pp. 624–628)

Solve each equation. Check your solution.

37. $10 + 2\sqrt{x} = 0$

38. $\sqrt{5 - 4x} - 6 = 7$

39. $\sqrt{a + 4} = 6$

40. $\sqrt{3x} = 2$

41. $\sqrt{x + 4} = x - 8$

42. $\sqrt{3x - 14} + x = 6$

43. **FREE FALL** Assuming no air resistance, the time t in seconds that it takes an object to fall h feet can be determined by $t = \frac{\sqrt{h}}{4}$. If a skydiver jumps from an airplane and free falls for 10 seconds before opening the parachute, how many feet does she free fall?

EXAMPLE 5

Solve $\sqrt{7x + 4} - 18 = 5$.

$\sqrt{7x + 4} - 18 = 5$	**Original equation**
$\sqrt{7x + 4} = 23$	**Add 18 to each side.**
$(\sqrt{7x + 4})^2 = 23^2$	**Square each side.**
$7x + 4 = 529$	**Simplify.**
$7x = 525$	**Subtract 4 from each side.**
$x = 75$	**Divide each side by 7.**

CHECK

$\sqrt{7x + 4} - 18 = 5$	**Original equation**
$\sqrt{7(75) + 4} - 18 \stackrel{?}{=} 5$	**$x = 75$**
$\sqrt{525 + 4} - 18 \stackrel{?}{=} 5$	**Multiply.**
$\sqrt{529} - 18 \stackrel{?}{=} 5$	**Add.**
$23 - 18 \stackrel{?}{=} 5$	**Simplify.**
$5 = 5$ ✓	**True.**

MIXED PROBLEM SOLVING
For mixed problem-solving practice, see page 845.

10-5 The Pythagorean Theorem (pp. 630–635)

Determine whether each set of measures can be the lengths of the sides of a right triangle.

44. 6, 8, 10

45. 3, 4, 5

46. 12, 16, 21

47. 10, 12, 15

48. 2, 3, 4

49. 7, 24, 25

50. 5, 12, 13

51. 15, 19, 23

52. LADDER A ladder is leaning on a building. The base of the ladder is 10 feet from the building, and the ladder reaches up 15 feet on the building. How long is the ladder?

EXAMPLE 6

Determine whether the set of measures 12, 16, and 20 can be the lengths of the sides of a right triangle.

$a^2 + b^2 = c^2$	**Pythagorean Theorem**
$12^2 + 16^2 \stackrel{?}{=} 20^2$	**$a = 12$, $b = 16$, and $c = 20$**
$144 + 256 \stackrel{?}{=} 400$	**Multiply.**
$400 = 400$ ✓	**Add.**

The measures can be the lengths of the sides of a right triangle.

10-6 The Distance and Midpoint formulas (pp. 636–641)

Find the distance between points with the given coordinates and the midpoint of the segment with the given endpoints. Round to the nearest hundredth if necessary.

53. $(2, 4), (-3, 4)$

54. $(-1, -3), (3, 5)$

55. $(-6, 7), (0, 0)$

56. $(1, 5), (-4, -5)$

Find the possible values for a if the points with the given coordinates are the indicated distance apart.

57. $(5, -2), (a, -3); d = \sqrt{170}$

58. $(1, a), (-3, 2); d = 5$

59. PLAYGROUND How far apart in feet are the swings from the slide?

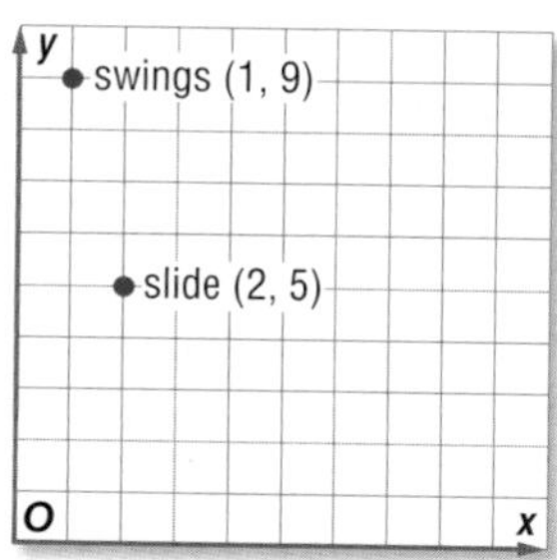

EXAMPLE 7

Find the distance between $(-3, 5)$ and $(-5, -3)$ and the midpoint of the segment with those endpoints. Round to the nearest tenth if necessary.

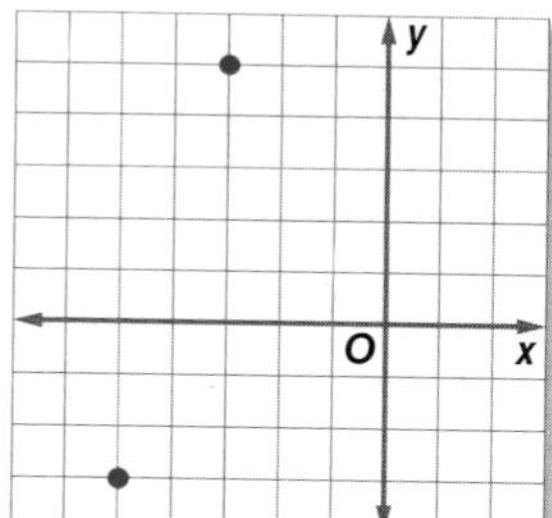

$d = \sqrt{(x_2 - x_1)^2 + (y_2 - y_1)^2}$	**Distance Formula**
$= \sqrt{[-3 - (-5)]^2 + [5 - (-3)]^2}$	**$(x_1, y_1) = (-5, -3)$, $(x_2, y_2) = (-3, 5)$**
$= \sqrt{2^2 + 8^2}$	**Simplify.**
$= \sqrt{4 + 64}$	**Evaluate squares.**
≈ 8.2	**Simplify.**
$M = \left(\frac{-5 + (-3)}{2}, \frac{-3 + 5}{2}\right)$	**Midpoint Formula**
$= (-4, 1)$	**Simplify.**

10-7 Similar Triangles (pp. 642–647)

Find the missing measures for the pair of similar triangles if $\triangle ABC \sim \triangle DEF$.

60. $a = 3, b = 4, c = 5, d = 12$

61. $a = 3, b = 4, c = 5, d = 4.5$

62. $a = 4, b = 8, c = 11, e = 4$

63. $a = 5, b = 7, c = 9, f = 18$

64. MODELS Kristin is making a model of the artwork shown in the scale of 1 inch = 2 feet. If the height of the artwork is 10 feet, what will the height of the model be?

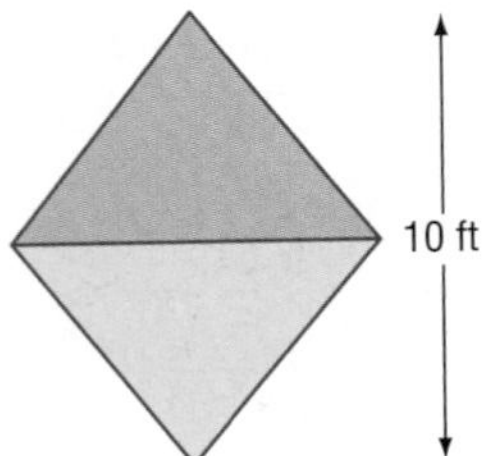

EXAMPLE 8

Find the missing length for the pair of triangles if $\triangle ABC \sim \triangle DFG$.

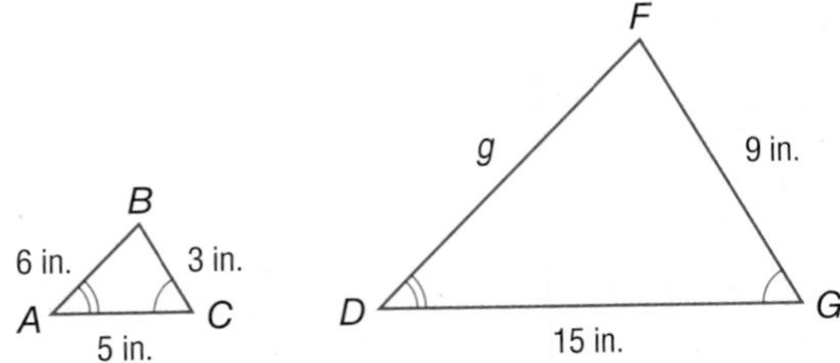

$\frac{AB}{DF} = \frac{BC}{FG}$ **Corresponding sides of similar triangles are proportional.**

$\frac{6}{g} = \frac{3}{9}$ **$AB = 6$, $BC = 3$, $DF = g$, and $FG = 9$**

$18 = g$ **Find the cross products and simplify.**

10-8 Trigonometric Ratios (pp. 649–655)

Find the values of the three trigonometric ratios for angle A.

65.

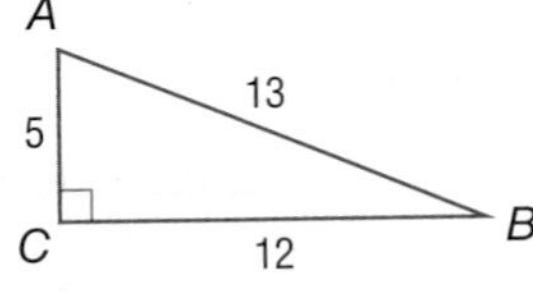

66.

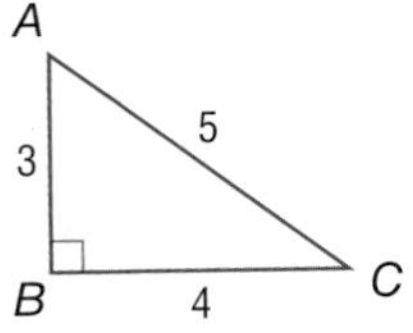

67. RAMPS How long is the ramp?

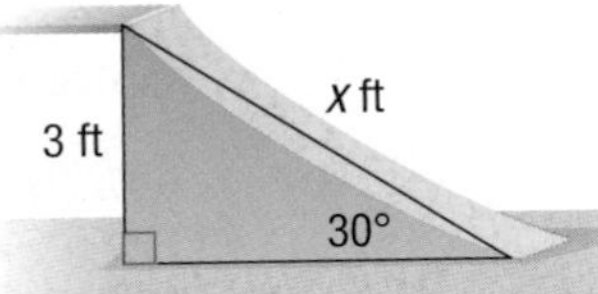

EXAMPLE 9

Find the values of the three trigonometric ratios for angle A.

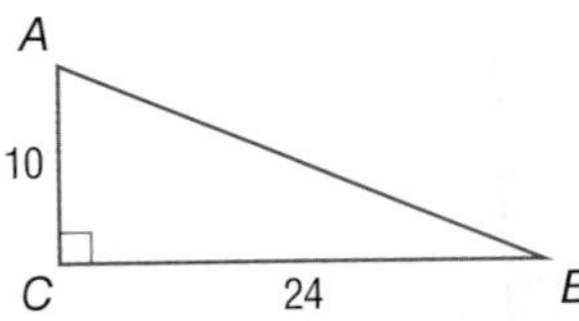

Find the hypotenuse: $c^2 = 10^2 + 24^2$, so $c = 26$.

$$\sin A = \frac{\text{leg opposite } \angle A}{\text{hypotenuse}} = \frac{24}{26} = \frac{12}{13}$$

$$\cos A = \frac{\text{leg adjacent } \angle A}{\text{hypotenuse}} = \frac{10}{26} = \frac{5}{13}$$

$$\tan A = \frac{\text{leg opposite } \angle A}{\text{leg adjacent } \angle A} = \frac{24}{10} = \frac{12}{5}$$

Practice Test

Math Online glencoe.com
Chapter Test

Graph each function, and compare to the parent graph. State the domain and range.

1. $y = -\sqrt{x}$
2. $y = \frac{1}{4}\sqrt{x}$
3. $y = \sqrt{x} + 5$
4. $y = \sqrt{x + 4}$

5. **GEOMETRY** The length of the side of a square is given by the function $s = \sqrt{A}$, where A is the area of the square. What is the perimeter of a square that has an area of 64 square inches?

   A 64 inches
   B 8 inches
   C 32 inches
   D 16 inches

Simplify each expression.

6. $5\sqrt{36}$
7. $\frac{3}{1 - \sqrt{2}}$
8. $2\sqrt{3} + 7\sqrt{3}$
9. $3\sqrt{6}(5\sqrt{2})$

10. **GEOMETRY** Find the area of the rectangle.

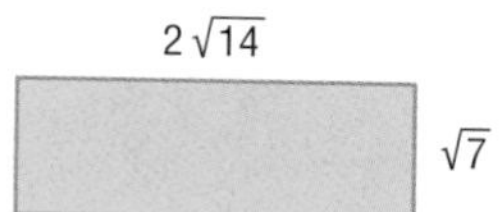

   F $14\sqrt{2}$
   G 14
   H $98\sqrt{2}$
   J $7\sqrt{2}$

Solve each equation. Check your solution.

11. $\sqrt{10x} = 20$
12. $\sqrt{4x - 3} = 6 - x$

13. **PACKAGING** A cylindrical container of chocolate drink mix has a volume of about 162 in^3. The radius of the container can be found by using the formula $r = \sqrt{\frac{V}{\pi h}}$, where r is the radius and h is the height. If the height is 8.25 inches, find the radius of the container.

Find each missing length. If necessary, round to the nearest tenth.

14.
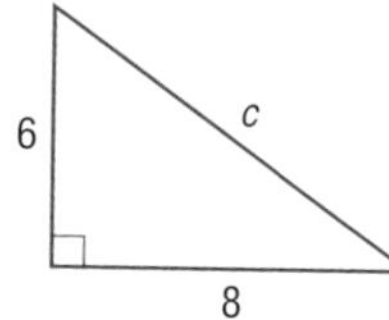

15.
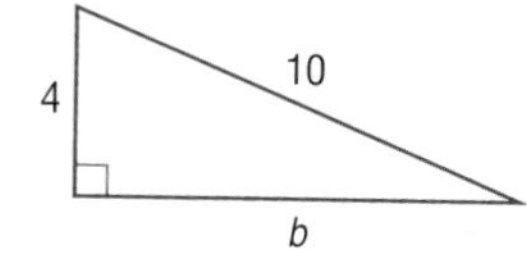

Find the distance between the points with the given coordinates.

16. (2, 3), (3, 5)
17. (−3, 4), (−2, −3)
18. (−1, −1), (3, 2)
19. (−4, −6), (−7, 1)

Find the coordinates of the midpoint of the segment with the given endpoints.

20. (2, 3), (3, 5)
21. (−3, 4), (−2, −3)
22. (−1, −1), (3, 2)
23. (−4, −8), (10, −6)

24. **PIZZA DELIVERY** The Pizza Place delivers to any location within a radius of 5 miles from the store for free. A delivery person drives 32 blocks north and then 45 blocks east to deliver a pizza. In this city, there are about 6 blocks per half mile.
   a. Should there be a charge for delivery? Explain.
   b. Describe two delivery situations that would result in about 5 miles.

25. Find the missing lengths if $\triangle ABC \sim \triangle XYZ$.

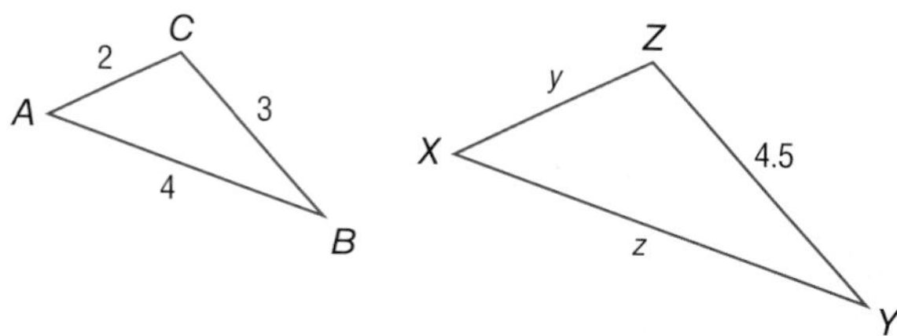

26. Find the values of the three trigonometric ratios for angle A.

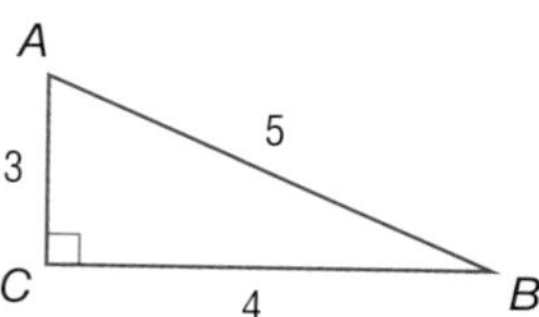

27. Find $m\angle X$ to the nearest degree.

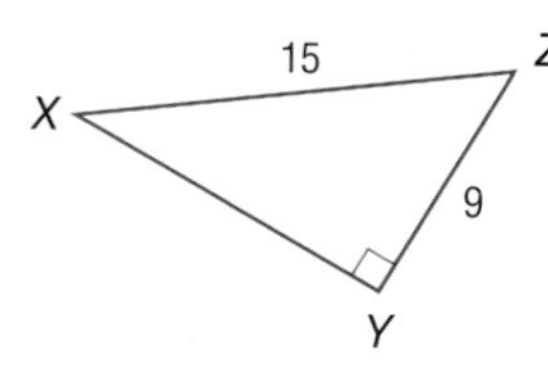

CHAPTER 10

Preparing for Standardized Tests

Draw a Picture

Sometimes it is easier to visualize how to solve a problem if you draw a picture first. You can sketch your picture on scrap paper or in your test booklet (if allowed). Be careful not make any marks on your answer sheet other than your answers.

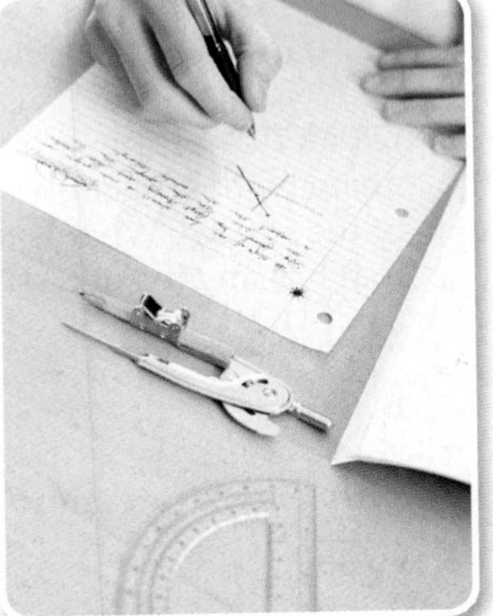

Strategies for Drawing a Picture

Step 1

Read the problem statement carefully.

Ask yourself:

- What am I being asked to solve?
- What information is given in the problem?
- What is the unknown quantity for which I need to solve?

Step 2

Sketch and label your picture.

- Draw your picture as clearly and accurately as possible.
- Label the picture carefully. Be sure to include all of the information given in the problem statement.

Step 3

Solve the problem.

- Use your picture to help you model the problem situation with an equation. Then solve the equation.
- Check your answer to make sure it is reasonable.

EXAMPLE

Read the problem. Identify what you need to know. Then use the information in the problem to solve. Show your work.

> One sunny day, a church steeple casts a shadow that is 24 feet 3 inches long. At the same time, Nicole casts a shadow that is 1 foot 9 inches long. If Nicole is 5 feet 3 inches tall, what is the height of the steeple?

Read the problem statement carefully. You know Nicole's height, her shadow length, and the shadow length of the steeple. You need to find the height of the steeple.

Scoring Rubric	
Criteria	**Score**
Full Credit: The answer is correct and a full explanation is provided that shows each step.	2
Partial Credit: • The answer is correct, but the explanation is incomplete. • The answer is incorrect, but the explanation is correct.	1
No Credit: Either an answer is not provided or the answer does not make sense.	0

Example of a 2-point response:

First convert all measurements to feet.

24 feet 3 inches $= 24\frac{3}{12}$ or 24.25 feet

1 foot 9 inches $= 1\frac{9}{12}$ or 1.75 feet

5 feet 3 inches $= 5\frac{3}{12}$ or 5.25 feet

Use similar triangles to find the height of the church steeple. Draw and label two triangles to represent the situation.

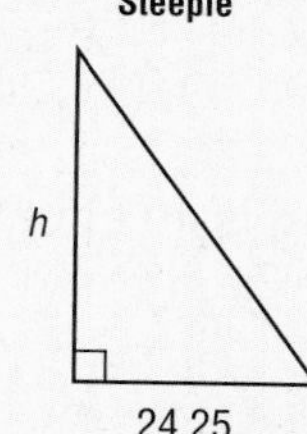

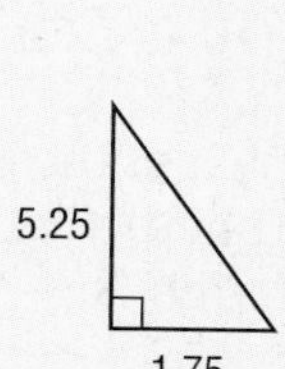

Use the similar triangles to set up and solve a proportion.

$$\frac{h}{24.25} = \frac{5.25}{1.75}$$

$$1.75h = (24.25)(5.25)$$

$$1.75h = 127.3125$$

$$h = 72.75$$

The height of the church steeple is 72.75 feet or 72 feet 9 inches.

Exercises

Read each problem. Identify what you need to know. Then use the information in the problem to solve. Show your work.

1. A building casts a 15-foot shadow, while a billboard casts a 4.5-foot shadow. If the billboard is 26 feet high, what is the height of the building? Round to the nearest tenth if necessary.

2. Jamey places a mirror on the ground at a distance of 56 feet from the base of a water tower. When he stands at a distance of 6 feet from the mirror, he can see the top of the water tower in the mirror's reflection and forms a pair of similar triangles. If Jamey is 5 feet 6 inches tall, what is the height of the water tower? Express your answer in feet and inches.

CHAPTER 10

Standardized Test Practice

Cumulative, Chapters 1 through 10

Multiple Choice

Read each question. Then fill in the correct answer on the answer document provided by your teacher on a sheet of paper.

1. What is the equation of the square root function graphed below?

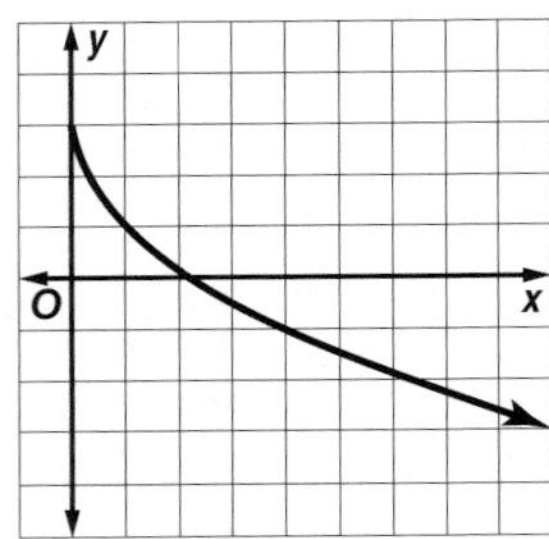

A $y = -2\sqrt{x} + 1$

B $y = -2\sqrt{x} + 3$

C $y = 2\sqrt{x} + 3$

D $y = 2\sqrt{x} + 1$

2. Simplify $\frac{1}{4 + \sqrt{2}}$.

A $\frac{4 + \sqrt{2}}{14}$

B $\frac{2 - \sqrt{2}}{7}$

C $\frac{4 - \sqrt{2}}{14}$

D $\frac{2 + \sqrt{2}}{7}$

3. What is the area of the triangle below?

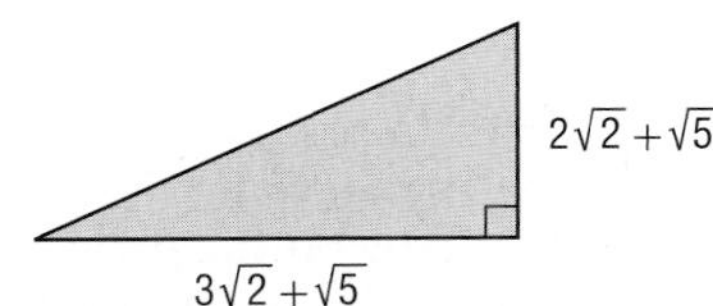

A $3\sqrt{2} + 10\sqrt{5}$

B $17 + 5\sqrt{10}$

C $12\sqrt{2} + 8\sqrt{5}$

D $8.5 + 2.5\sqrt{10}$

4. The formula for the slant height c of a cone is $c = \sqrt{h^2 + r^2}$, where h is the height of the cone and r is the radius of its base. What is the radius of the cone below? Round to the nearest tenth.

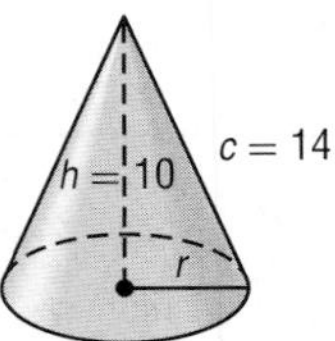

A 4.9

B 6.3

C 9.8

D 10.2

5. Which of the following sets of measures could not be the sides of a right triangle?

A (12, 16, 24)

B (10, 24, 26)

C (24, 45, 51)

D (18, 24, 30)

6. Which of the following is an equation of the line perpendicular to $4x - 2y = 6$ and passing through $(4, -4)$?

A $y = -\frac{3}{4}x + 3$

B $y = -\frac{3}{4}x - 1$

C $y = -\frac{1}{2}x - 4$

D $y = -\frac{1}{2}x - 2$

7. The scale on a map shows that 1.5 centimeters is equivalent to 40 miles. If the distance on the map between two cities is 8 centimeters, about how many miles apart are the cities?

A 178 miles

B 213 miles

C 224 miles

D 275 miles

Test-TakingTip

Question 4 Substitute for c and h in the formula. Then solve for r.

Short Response/Gridded Response

Record your answers on the answer sheet provided by your teacher or on a sheet of paper.

8. **GRIDDED RESPONSE** How many times does the graph of $y = x^2 - 4x + 10$ cross the x-axis?

9. Factor $2x^4 - 32$ completely.

10. **GRIDDED RESPONSE** In football, a field goal is worth 3 points, and the extra point after a touchdown is worth 1 point. During the 2006 season, John Kasay of the Carolina Panthers scored a total of 100 points for his team by making a total of 52 field goals and extra points. How many field goals did he make?

11. Shannon bought a satellite radio and a subscription to satellite radio. What is the total cost for his first year of service?

Item	Cost
radio	\$39.99
subscription	\$11.99 per month

12. **GRIDDED RESPONSE** The distance required for a car to stop is directly proportional to the square of its velocity. If a car can stop in 242 meters at 22 kilometers per hour, how many meters are needed to stop at 30 kilometers per hour?

13. The highest point in Kentucky is at an elevation of 4145 feet above sea level. The lowest point in the state is at an elevation of 257 feet above sea level. Which of the following inequalities best describes the elevations in the state of Kentucky?

14. Simplify the expression below. Show your work.

$$\left(\frac{-2r^{-2}q^5t^2}{5r^4q^2t^{-3}}\right)^{-2}$$

15. **GRIDDED RESPONSE** For the first home basketball game, 652 tickets were sold for a total revenue of \$5216. If each ticket costs the same, how much is the cost per ticket? State your answer in dollars.

Extended Response

Record your answers on a sheet of paper. Show your work.

16. Karen is making a map of her hometown using a coordinate grid. The scale of her map is 1 unit = 2.5 miles.

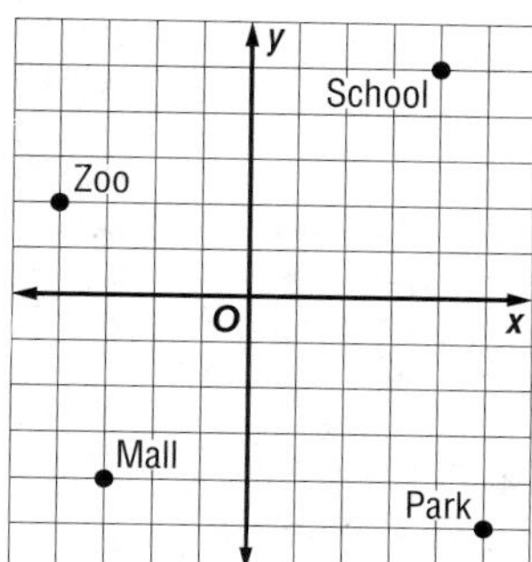

a. What is the actual distance between Karen's school and the park? Round to the nearest tenth of a mile if necessary.

b. Suppose Karen's house is located midway between the mall and the school. What coordinates represent her house? Show your work.

Need Extra Help?												
If you missed Question...	1	2	3	4	5	6	7	8	9	10	11	12
Go to Lesson or Page...	10-7	7-5	8-2	10-5	10-6	9-4	9-6	10-2	10-4	4-4	10-8	4-3

CHAPTER 11 Rational Functions and Equations

Then

In Chapter 7, you simplified expressions involving monomials and polynomials.

Now

In Chapter 11, you will:

- Identify and graph inverse variations.
- Identify excluded values of rational functions.
- Multiply, divide, and add rational expressions.
- Divide polynomials.
- Solve rational equations.

Why?

HOCKEY The time it will take for a puck hit from the blue line to reach the goal line is given by the rational expression $\frac{64}{x}$, where x is the speed of the puck in feet per seconds. If a player hits the puck at 100 miles per hour, the puck will reach the goal line in 0.34 second.

Math *in Motion,* Animation glencoe.com

Get Ready for Chapter 11

Diagnose Readiness You have two options for checking Prerequisite Skills.

Text Option

Take the Quick Check below. Refer to the Quick Review for help.

*Quick*Check

Solve each proportion. (Lesson 2-6)

1. $\frac{y}{3} = \frac{8}{9}$

2. $\frac{5}{12} = \frac{x}{36}$

3. $\frac{7}{2} = \frac{y}{3}$

4. $\frac{5}{x} = \frac{10}{4}$

5. DRAWING Rosie is making a scale drawing. She is using the scale 1 inch = 3 feet. How many inches will represent 10 feet?

*Quick*Review

EXAMPLE 1

Solve $\frac{3}{5} = \frac{x}{12}$.

$\frac{3}{5} = \frac{x}{12}$	**Original equation**
$3 \cdot 12 = 5 \cdot x$	**Cross products**
$36 = 5x$	**Simplify.**
$\frac{36}{5} = \frac{5x}{5}$	**Divide each side by 5.**
$\frac{36}{5} = x$	**Simplify.**

Find the GCF of each pair of monomials. (Lesson 8-1)

6. $12ab$, $18b$

7. $15cd^2$, $25c^2d$

8. $60r^2$, $45r^3$

9. $12xy$, $16x^2y$

10. GAMES Fifty girls and 75 boys attend a sports club. For a game, boys and girls are going to split into groups. The number in each group has to be the same. How large can the groups be?

EXAMPLE 2

Find the greatest common factor of 30 and 42.

$2 \cdot 3 \cdot 5$	**Prime factorization of 30**
$2 \cdot 3 \cdot 7$	**Prime factorization of 42**
$2 \cdot 3 = 6$	**Product of the common factors**

The greatest common factor of 30 and 42 is 6.

Factor each polynomial. (Lessons 8-2 and 8-4)

11. $2x^2 - 4x$

12. $6x^2 - 5x - 4$

13. $6xy + 15x$

14. $2c^2d - 4c^2d^2$

15. AREA The area of a rectangle is $x^2 + 5x + 6$. What binomial expressions represent the side lengths of the rectangle?

$A = x^2 + 5x + 6$

EXAMPLE 3

Factor $x^2 + 4x - 45$.

In this trinomial, $b = 4$ and $c = -45$. Find factors of -45 with a sum of 4. The correct factors are -5 and 9.

$x^2 + 4x - 45$	**Original expression**
$= (x + m)(x + p)$	**Write the pattern.**
$= (x - 5)(x + 9)$	**$m = -5$ and $p = 9$**

Online Option

Math Online Take a self-check Chapter Readiness Quiz at glencoe.com.

Get Started on Chapter 11

You will learn several new concepts, skills, and vocabulary terms as you study Chapter 11. To get ready, identify important terms and organize your resources. You may wish to refer to **Chapter 0** to review prerequisite skills.

FOLDABLES® Study Organizer

Rational Functions and Equations Make this Foldable to help you organize your Chapter 11 notes about rational functions and equations. Begin with 3 sheets of notebook paper.

1. **Take** one sheet of paper and fold in half along the width. Cut 1 inch slits on each side of the paper.

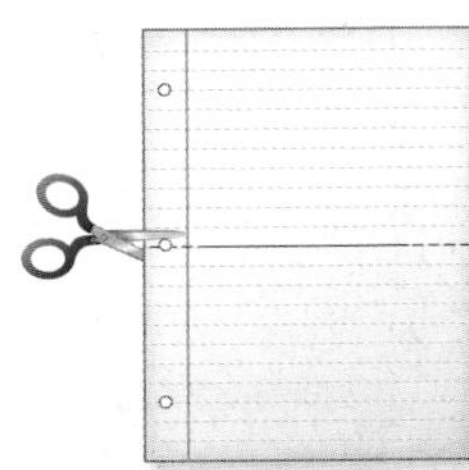

2. **Stack** the two sheets of paper and fold in half along the width. Cut a slit through the center stopping 1 inch from each side.

3. **Insert** the first sheet through the second sheets and align the folds to form a booklet. Label the cover with the chapter title.

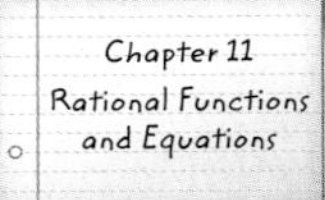

New Vocabulary

English		Español
inverse variation	• p. 670 •	variación inversa
product rule	• p. 671 •	regla del producto
excluded value	• p. 678 •	valores excluidos
rational function	• p. 678 •	función racional
asymptote	• p. 679 •	asíntota
rational expression	• p. 684 •	expresión racional
least common multiple (LCM)	• p. 707 •	mínimo común múltiplo (mcm)
least common denominator (LCD)	• p. 708 •	mínimo común denominador (mcd)
complex fraction	• p. 714 •	fracción compleja
mixed expression	• p. 714 •	expresión mixta
rational equation	• p. 720 •	ecuacion racional
extraneous solutions	• p. 721 •	soluciones extrañas
work problems	• p. 722 •	problemas de trabajo
rate problems	• p. 723 •	problemas de tasas

Review Vocabulary

direct variation • p. 180 • variación directa an equation of the form $y = kx$, where $k \neq 0$

Quotient of Powers • p. 408 • cociente de potencia
$\frac{a^m}{a^n} = a^{m-n}$

$$\frac{x^5}{x^3} = \frac{x \cdot x \cdot x \cdot x \cdot x}{x \cdot x \cdot x} = x \cdot x \text{ or } x^2$$
$$\frac{x^5}{x^3} = x^{5-3} \text{ or } x^2$$

Zero Product Property • p. 478 • propiedad del producto de cero if the product of two factors is 0, then at least one of the factors must be 0

Multilingual eGlossary glencoe.com

Math Online glencoe.com

- Study the chapter online
- Explore **Math in Motion**
- Get extra help from your own **Personal Tutor**
- Use **Extra Examples** for additional help
- Take a **Self-Check Quiz**
- **Review Vocabulary** in fun ways

EXPLORE 11-1

Graphing Technology Lab
Inverse Variation

You can use a data collection device to investigate the relationship between volume and pressure.

Set Up The Lab

- Connect a syringe to the gas pressure sensor. Then connect the data collection device to both the sensor and the calculator as shown.
- Start the collection program and select the sensor.

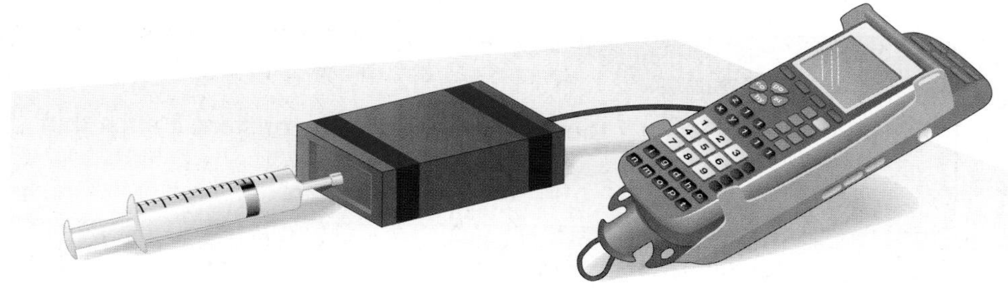

ACTIVITY Collect Data

Step 1 Open the valve between the atmosphere and the syringe. Set the inside ring of the syringe to 20 mL and close the valve. This ensures that the amount of air inside the syringe will be constant throughout the experiment.

Step 2 Press the plunger of the syringe to the 5 mL mark. Wait for the pressure gauge to stop changing, then take the data reading. Enter 5 as the volume in the calculator. The pressure is measured in atmospheres (atm).

Step 3 Repeat step 2, pressing the plunger to 7.5 mL, 10.0 mL, 12.5 mL, 15.0 mL, 17.5 mL, and 20.0 mL. Record the volume from each data reading.

Step 4 After taking the last data reading, use **STAT PLOT** to create a line graph.

Exercises

1. Does the pressure vary directly as the volume? Explain.
2. As the volume changes from 10 to 20 mL, what happens to the pressure?
3. Predict what the pressure of the gas in the syringe would be if the volume increased to 40 mL.
4. Add a column to the data table to find the product of the volume and the pressure for each data reading. What pattern do you observe?
5. **MAKE A CONJECTURE** The relationship between the pressure and volume of a gas is called Boyle's Law. Write an equation relating the volume v in milliliters and pressure p in atmospheres in your experiment. Compare your conjecture to those of two classmates. Formulate mathematical questions about their conjectures.

11-1 Inverse Variation

Then

You solved problems involving direct variation. (Lesson 3-4)

Now

- Identify and use inverse variations.
- Graph inverse variations.

New Vocabulary

inverse variation
product rule

Math Online

glencoe.com

- Extra Examples
- Personal Tutor
- Self-Check Quiz
- Homework Help

Why?

The time it takes a runner to finish a race is inversely proportional to the average pace of the runner.

Identify and Use Inverse Variations In the situation above, the runner's time decreases as the pace of the runner increases. So, these quantities are *inversely proportional*. An **inverse variation** can be represented by the equation $y = \frac{k}{x}$ or $xy = k$.

Key Concept — Inverse Variation (For Your FOLDABLE)

y varies inversely as x if there is some nonzero constant k such that $y = \frac{k}{x}$ or $xy = k$, where $x \neq 0$ and $y \neq 0$.

In an inverse variation, the product of two values remains constant. Recall that a relationship of the form $y = kx$ is a *direct variation*. For either a direct or indirect variations, the constant k is called the *constant of variation* or the *constant of proportionality*.

EXAMPLE 1 Identify Inverse and Direct Variations

Determine whether each table or equation represents an *inverse* or a *direct variation*. Explain.

a.

x	y
1	16
2	8
4	4

In an inverse variation, xy equals a constant k. Find xy for each ordered pair in the table.

$1 \cdot 16 = 16$
$2 \cdot 8 = 16$
$4 \cdot 4 = 16$

The product is constant, so the table represents an inverse variation.

b.

x	y
1	3
2	6
3	9

Notice that xy is not constant. So, the table does not represent an indirect variation.

$3 = k(1)$ $6 = k(2)$ $9 = k(3)$
$3 = k$ $3 = k$ $3 = k$

The table of values represents the direct variation $y = 3x$.

c. $x = 2y$

The equation can be written as $y = \frac{1}{2}x$. Therefore, it represents a direct variation.

d. $2xy = 10$

$2xy = 10$ **Write the equation.**
$xy = 5$ **Divide each side by 2.**

The equation represents an inverse variation.

Check Your Progress

1A.

x	1	2	5
y	10	5	2

1B. $-2x = y$

Personal Tutor glencoe.com

ReadingMath

Variation Equations For direct variation equations, you say that *y varies directly* as *x*. For inverse variation equations, you say that *y varies inversely* as *x*.

You can use $xy = k$ to write an inverse variation equation that relates x and y.

EXAMPLE 2 Write an Inverse Variation

Assume that y varies inversely as x. If $y = 18$ when $x = 2$, write an inverse variation equation that relates x and y.

$xy = k$	**Inverse variation equation**
$2(18) = k$	**$x = 2$ and $y = 18$**
$36 = k$	**Simplify.**

The constant of variation is 36. So, an equation that relates x and y is $xy = 36$ or $y = \frac{36}{x}$.

Check Your Progress

2. Assume that y varies inversely as x. If $y = 5$ when $x = -4$, write an inverse variation equation that relates x and y.

Personal Tutor glencoe.com

If (x_1, y_1) and (x_2, y_2) are solutions of an inverse variation, then $x_1y_1 = k$ and $x_2y_2 = k$.

$x_1y_1 = k$ and $x_2y_2 = k$	
$x_1y_1 = x_2y_2$	**Substitute x_2y_2 for k.**

The equation $x_1y_1 = x_2y_2$ is called the **product rule** for inverse variations.

Key Concept Product Rule for Inverse Variations

For Your FOLDABLE

Words If (x_1, y_1) and (x_2, y_2) are solutions of an inverse variation, then the products x_1y_1 and x_2y_2 are equal.

Symbols $x_1y_1 = x_2y_2$ or $\frac{x_1}{x_2} = \frac{y_2}{y_1}$

EXAMPLE 3 Solve for x or y

Assume that y varies inversely as x. If $y = 3$ when $x = 12$, find x when $y = 4$.

$x_1y_1 = x_2y_2$	**Product rule for inverse variations**
$12 \cdot 3 = x_2 \cdot 4$	**$x_1 = 12$, $y_1 = 3$, and $y_2 = 4$**
$36 = x_2 \cdot 4$	**Simplify.**
$\frac{36}{4} = x_2$	**Divide each side by 4.**
$9 = x_2$	**Simplify.**

So, when $y = 4$, $x = 9$.

Check Your Progress

3. If y varies inversely as x and $y = 4$ when $x = -8$, find y when $x = -4$.

Personal Tutor glencoe.com

The product rule for inverse variations can be used to write an equation to solve real-world problems.

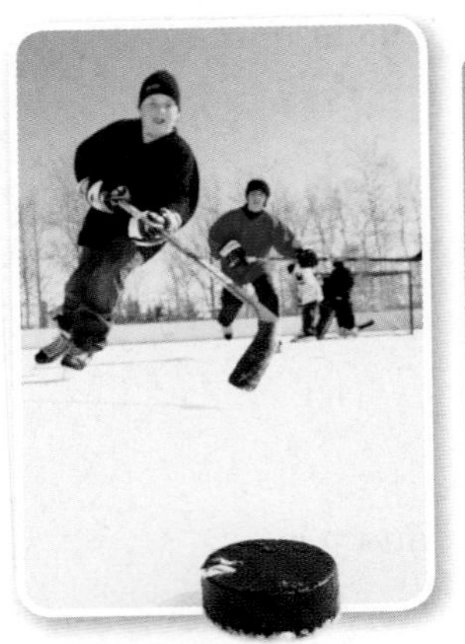

Real-World Link

A standard hockey puck is 1 inch thick and 3 inches in diameter. Its mass is between approximately 156 and 170 grams.

Source: *NHL Rulebook*

Real-World EXAMPLE 4 Use Inverse Variations

PHYSICS The acceleration a of a hockey puck is inversely proportional to its mass m. Suppose a hockey puck with a mass of 164 grams is hit so that it accelerates 122 m/s^2. Find the acceleration of a 158-gram hockey puck if the same amount of force is applied.

Make a table to organize the information.

Let $m_1 = 164$, $a_1 = 122$, and $m_2 = 164$. Solve for a_2.

Puck	Mass	Acceleration
1	164 g	122 m/s^2
2	158 g	a_2

$m_1a_1 = m_2a_2$ — **Use the product rule to write an equation.**

$164 \cdot 122 = 158a_2$ — **$m_1 = 164$, $a_1 = 122$, and $m_2 = 158$**

$20{,}008 = 158a_2$ — **Simplify.**

$126.6 \approx a_2$ — **Divide each side by 158 and simplify.**

The 158-gram puck has an acceleration of approximately 126.6 m/s^2.

Check Your Progress

4. RACING Manuel runs an average of 8 miles per hour and finishes a race in 0.39 hour. Dyani finished the race in 0.35 hour. What was her average pace?

Personal Tutor glencoe.com

Graph Inverse Variations The graph of an inverse variation is not a straight line like the graph of a direct variation.

EXAMPLE 5 Graph an Inverse Variation

Graph an inverse variation equation in which $y = 8$ when $x = 3$.

Problem-Solving Tip

Solve a Simpler Problem Sometimes it is necessary to break a problem into parts, solve each part, and then combine them to find the solution to the problem.

Step 1 Write an inverse variation equation.

$xy = k$ — **Inverse variation equation**

$3(8) = k$ — **$x = 3$, $y = 8$**

$24 = k$ — **Simplify.**

The inverse variation equation is $xy = 24$ or $y = \frac{24}{x}$.

Step 2 Choose values for x and y that have a product of 24.

Step 3 Plot each point and draw a smooth curve that connects the points.

x	y
−12	−2
−8	−3
−4	−6
−2	−12
0	undefined
2	12
3	8
6	4
12	2

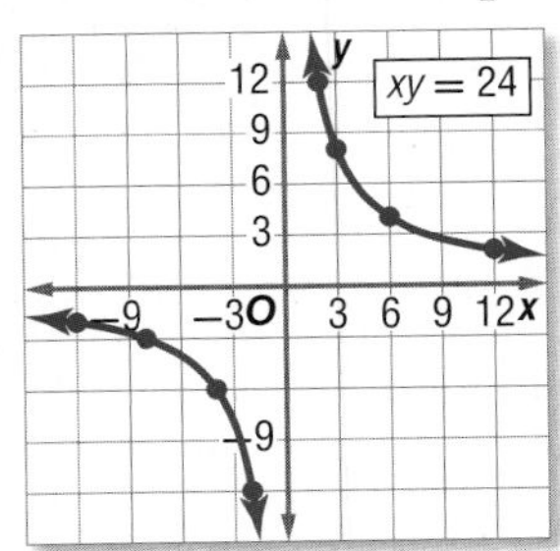

Notice that since y is undefined when $x = 0$, there is no point on the graph when $x = 0$. This graph is called a hyperbola.

Check Your Progress

5. Graph an inverse variation equation in which $y = 16$ when $x = 4$.

Personal Tutor glencoe.com

Concept Summary — Direct and Inverse Variations

Direct Variation

$k > 0$

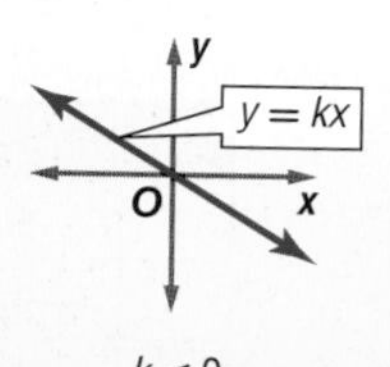

$k < 0$

Inverse Variation

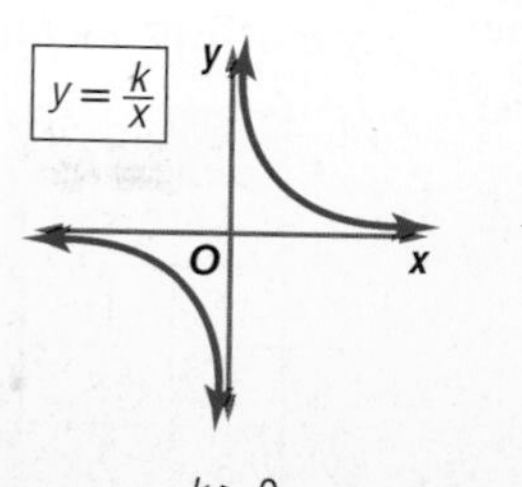

$k > 0$

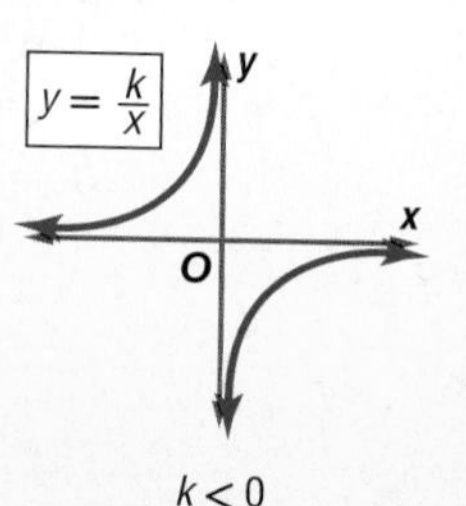

$k < 0$

- $y = kx$
- y varies directly as x.
- The ratio $\frac{y}{x}$ is a constant.

- $y = \frac{k}{x}$
- y varies inversely as x.
- The product xy is a constant.

Check Your Understanding

Example 1 p. 670

Determine whether each table or equation represents an *inverse* or a *direct* variation. Explain.

1.

x	1	4	8	12
y	2	8	16	24

2.

x	1	2	3	4
y	24	12	8	6

3. $xy = 4$

4. $y = \frac{x}{10}$

Examples 2 and 5 pp. 671–672

Assume that y varies inversely as x. Write an inverse variation equation that relates x and y. Then graph the equation.

5. $y = 8$ when $x = 6$

6. $y = 2$ when $x = 5$

7. $y = 3$ when $x = -10$

8. $y = -1$ when $x = -12$

Example 3 p. 671

Solve. Assume that y varies inversely as x.

9. If $y = 8$ when $x = 4$, find x when $y = 2$.

10. If $y = 7$ when $x = 6$, find y when $x = -21$.

11. If $y = -5$ when $x = 9$, find y when $x = 6$.

Example 4 p. 672

12. RACING The time it takes to complete a go-cart race course is inversely proportional to the average speed of the go-cart. One rider has an average speed of 73.3 feet per second and completes the course in 30 seconds. Another rider completes the course in 25 seconds. What was the average speed of the second rider?

13. OPTOMETRY When a person does not have clear vision, an optometrist can prescribe lenses to correct the condition. The power P of a lens, in a unit called diopters, is equal to 1 divided by the focal length f, in meters, of the lens.

a. Graph the inverse variation $P = \frac{1}{f}$.

b. Find the powers of lenses with focal lengths +0.2 to −0.4 meters.

Practice and Problem Solving

= **Step-by-Step Solutions** begin on page R12.
Extra Practice begins on page 815.

Example 1 p. 670

Determine whether each table or equation represents an *inverse* or a *direct* variation. Explain.

14.

x	y
1	30
2	15
5	6
6	5

15.

x	y
2	−6
3	−9
4	−12
5	−15

16.

x	y
−4	−2
−2	−1
2	1
4	2

17.

x	y
−5	8
−2	20
4	−10
8	−5

18. $5x - y = 0$

19. $xy = \frac{1}{4}$

20. $x = 14y$

21. $\frac{y}{x} = 9$

Examples 2 and 5 pp. 671–672

Assume that y varies inversely as x. Write an inverse variation equation that relates x and y. Then graph the equation.

22. $y = 2$ when $x = 20$

23. $y = 18$ when $x = 4$

24. $y = -6$ when $x = -3$

25. $y = -4$ when $x = -3$

26. $y = -4$ when $x = 16$

27. $y = 12$ when $x = -9$

Example 3 p. 671

Solve. Assume that y varies inversely as x.

28. If $y = 12$ when $x = 3$, find x when $y = 6$.

29. If $y = 5$ when $x = 6$, find x when $y = 2$.

30. If $y = 4$ when $x = 14$, find x when $y = -5$.

31. If $y = 9$ when $x = 9$, find y when $x = -27$.

32. If $y = 15$ when $x = -2$, find y when $x = 3$.

33. If $y = -8$ when $x = -12$, find y when $x = 10$.

Example 4 p. 672

34. **EARTH SCIENCE** The water level in a river varies inversely with air temperature. When the air temperature was 90° Fahrenheit, the water level was 11 feet. If the air temperature was 110° Fahrenheit, what was the level of water in the river?

Real-World Link

A medium-sized piano has about 230 strings with a combined tension of 15 to 20 tons. A concert grand piano may have a combined string tension of up to 30 tons.

Source: Piano World

35. **MUSIC** When under equal tension, the frequency of a vibrating string in a piano varies inversely with the string length. If a string that is 420 millimeters in length vibrates at a frequency of 523 cycles a second, at what frequency will a 707-millimeter string vibrate?

Determine whether each situation is an example of an *inverse* or a *direct* variation. Justify your reasoning.

36. The drama club can afford to purchase 10 wigs at \$2 each or 5 wigs at \$4 each.

37. The Spring family buys several lemonades for \$1.50 each.

38. Nicole earns \$14 for babysitting 2 hours, and \$21 for babysitting 3 hours.

39. Thirty video game tokens are divided evenly among a group of friends.

Determine whether each table or graph represents an *inverse* or a *direct* variation. Explain.

40.

x	y
5	1
8	1.6
11	2.2

41.

x	y
−3	−7
−2	−10.5
4	5.25

42.

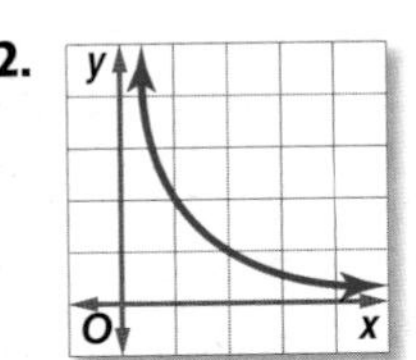

43.

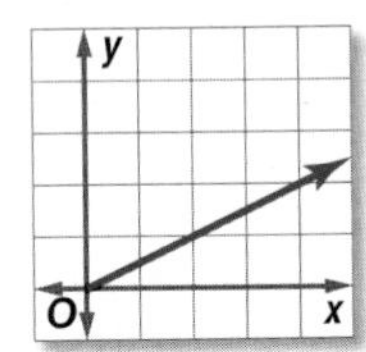

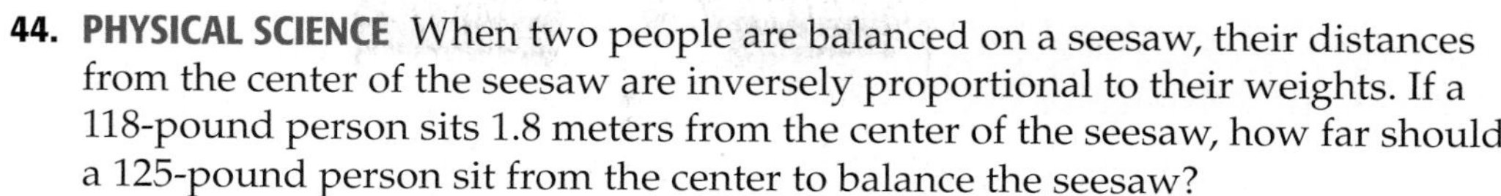

44. PHYSICAL SCIENCE When two people are balanced on a seesaw, their distances from the center of the seesaw are inversely proportional to their weights. If a 118-pound person sits 1.8 meters from the center of the seesaw, how far should a 125-pound person sit from the center to balance the seesaw?

Solve. Assume that y varies inversely as x.

45. If $y = 9.2$ when $x = 6$, find x when $y = 3$.

46. If $y = 3.8$ when $x = 1.5$, find x when $y = 0.3$.

47. If $y = \frac{1}{5}$ when $x = -20$, find y when $x = -\frac{8}{5}$.

48. If $y = -6.3$ when $x = \frac{2}{3}$, find y when $x = 8$.

49. SWIMMING Logan and Brianna each bought a pool membership. Their average cost per day is inversely proportional to the number of days that they go to the pool. Logan went to the pool 25 days for an average cost per day of $5.60. Brianna went to the pool 35 days. What was her average cost per day?

50. PHYSICAL SCIENCE The amount of force required to do a certain amount of work in moving an object is inversely proportional to the distance that the object is moved. Suppose 90 N of force is required to move an object 10 feet. Find the force needed to move another object 15 feet if the same amount of work is done.

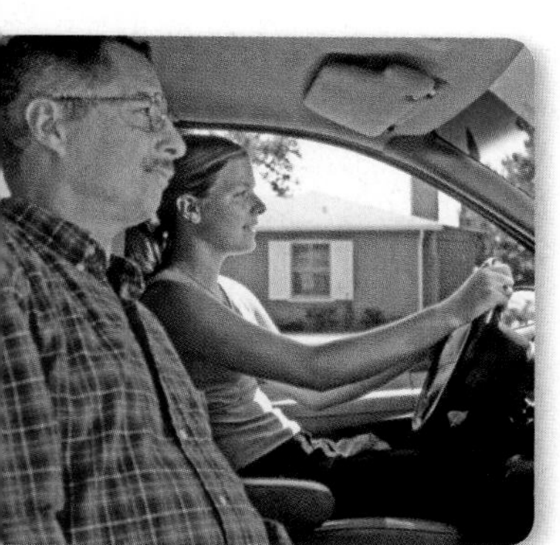

Real-World Link

In 2007, Illinois began the Operation Teen Safe Driving program. The program challenges high schools to compete against each other to develop the most comprehensive and creative safe driving community.

Source: Ford Motor Company

51. DRIVING Lina must practice driving 40 hours with a parent or guardian before she is allowed to take the test to get her driver's license. She plans to practice the same number of hours each week.

a. Let h represent the number of hours per week that she practices driving. Make a table showing the number of weeks w that she will need to practice for the following values of h: 1, 2, 4, 5, 8, and 10.

b. Describe how the number of weeks changes as the number of hours per week increases.

c. Write and graph an equation that shows the relationship between h and w.

H.O.T. Problems

Use **H**igher-**O**rder **T**hinking Skills

52. FIND THE ERROR Christian and Trevor found an equation such that x and y vary inversely, and $y = 10$ when $x = 5$. Is either of them correct? Explain.

Christian

$k = \frac{y}{x}$

$= \frac{10}{2}$ or 5

$y = 5x$

Trevor

$k = xy$

$= (5)(10)$ or 50

$y = \frac{50}{x}$

53. CHALLENGE Suppose f varies inversely with g, and g varies inversely with h. What is the relationship between f and h?

54. REASONING Does $xy = -k$ represent an inverse variation when $k \neq 0$? Explain.

55. OPEN ENDED Give a real-world situation or phenomena that can be modeled by an inverse variation equation. Use the correct terminology to describe your example and explain why this situation is an inverse variation.

56. WRITING IN MATH Compare and contrast direct and inverse variation. Include a description of the relationship between slope and the graphs of a direct and inverse variation.

Standardized Test Practice

57. Given a constant force, the acceleration of an object varies inversely with its mass. Assume that a constant force is acting on an object with a mass of 6 pounds resulting in an acceleration of 10 ft/s^2. The same force acts on another object with a mass of 12 pounds. What would be the resulting acceleration?

A 4 ft/s^2
B 5 ft/s^2
C 6 ft/s^2
D 7 ft/s^2

58. Fiona had an average of 56% on her first seven tests. What would she have to make on her eighth test to average 60% on 8 tests?

F 82%
G 88%
H 100%
J 98%

59. Anthony takes a picture of a 1-meter snake beside a brick wall. When he develops the pictures, the 1-meter snake is 2 centimeters long and the wall is 4.5 centimeters high. What was the actual height of the brick wall?

A 2.25 cm
B 22.5 cm
C 225 cm
D 0.225 cm

60. SHORT RESPONSE Find the area of the rectangle.

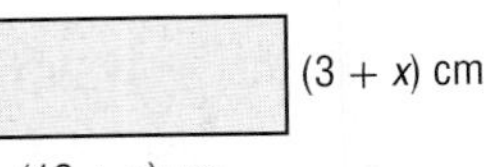

Spiral Review

For each triangle, find sin A, cos A, and tan A to the nearest ten-thousandth. (Lesson 10-8)

61.

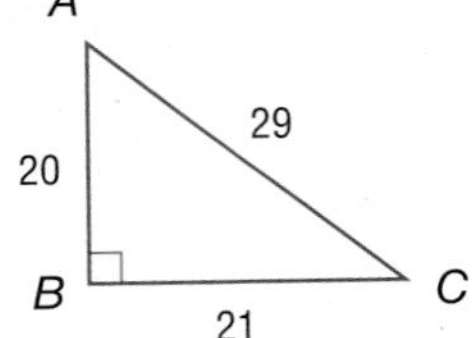

62.

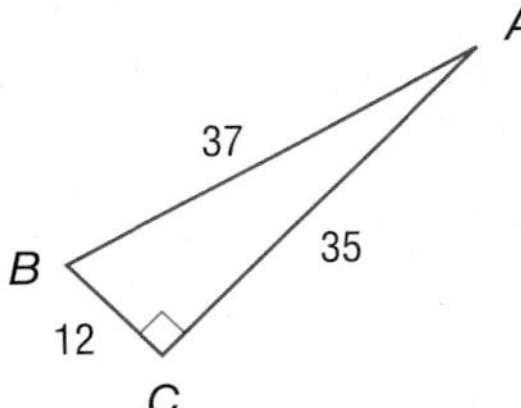

63.

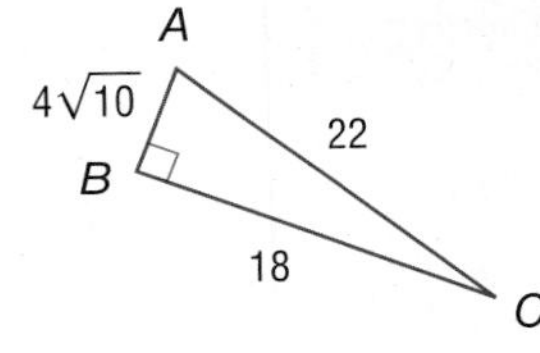

64. CRAFTS Jane is making a stained glass window using several triangular pieces of glass that are similar to the one shown. If two of the sides measure 4 inches, what is the length of the third side? (Lesson 10-7)

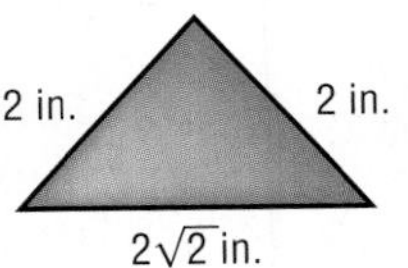

Solve each equation. (Lesson 10-4)

65. $\sqrt{10c} + 2 = 5$

66. $\sqrt{9h + 19} = 9$

67. $\sqrt{7k + 2} + 2 = 5$

68. $\sqrt{5r - 1} = r - 5$

69. $6 + \sqrt{2x + 11} = -x$

70. $4 + \sqrt{4t - 4} = t$

Skills Review

Simplify. Assume that no denominator is equal to zero. (Lesson 7-2)

71. $\frac{7^8}{7^6}$

72. $\frac{x^8y^{12}}{x^2y^7}$

73. $\frac{5pq^7}{10p^6q^3}$

74. $\left(\frac{2c^3d}{7z^2}\right)^3$

75. $\left(\frac{4a^2b}{2c^3}\right)^2$

76. $y^0(y^5)(y^{-9})$

77. $\frac{(4m^{-3}n^5)^0}{mn}$

78. $\frac{(3x^2y^5)^0}{(21x^5y^2)^0}$

EXPLORE 11-2

Algebra Lab
Reading Rational Expressions

Several concepts need to be applied when reading rational expressions.

A fraction bar acts as a grouping symbol, where the entire numerator is divided by the entire denominator.

ACTIVITY 1

Read the expression $\frac{4y + 6}{14}$.

It is correct to read the expression as *the quantity four y plus six divided by fourteen.*

It is incorrect to read the expression as *four y plus six divided by fourteen or four y divided by fourteen plus six.*

If a fraction consists of two or more terms divided by a monomial or one-term denominator, the denominator divides each term.

ACTIVITY 2

Simplify $\frac{4y + 6}{14}$.

It is correct to write $\frac{4y + 6}{14} = \frac{4y}{14} + \frac{6}{14}$.

$$= \frac{2y}{7} + \frac{3}{7} \text{ or } \frac{2y + 3}{7}$$

It is also correct to write $\frac{4y + 6}{14} = \frac{2(2y + 3)}{2 \cdot 7}$

$$= \frac{\cancel{2}(2y + 3)}{\cancel{2} \cdot 7} \text{ or } \frac{2y + 3}{7}$$

It is incorrect to write $\frac{4y + 6}{14} = \frac{\overset{2y}{\cancel{4y}} + 6}{\underset{7}{\cancel{14}}} = \frac{2y + 6}{7}$.

Exercises

Write the verbal translation of each rational expression.

1. $\frac{x - 3}{5}$
2. $\frac{2x}{x + 3}$
3. $\frac{c + 3}{c^2 - 4}$
4. $\frac{b^2 - 9}{b - 3}$
5. $\frac{n^2 + 2n - 8}{n - 4}$
6. $\frac{h^2 - 6h + 1}{h^2 + h + 5}$

Simplify each expression.

7. $\frac{2x + 4}{10}$
8. $\frac{4m + 12}{16}$
9. $\frac{2y^2 - 4y}{16y}$
10. $\frac{g - 9}{g^2 - 81}$
11. $\frac{2p - 5}{4p^2 - 20p + 25}$
12. $\frac{2d - 7}{2d^2 + d - 28}$

11-2 Rational Functions

Why?

Trina is reading a 300-page book. The average number of pages she reads each day y is given by $y = \frac{300}{x}$, where x is the number of days that she reads.

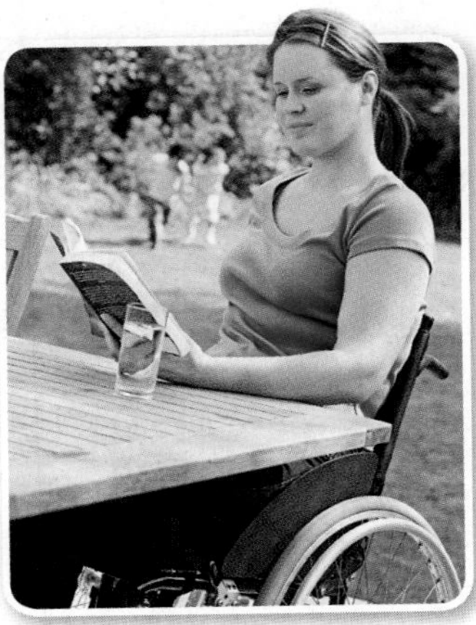

Identify Excluded Values The function $y = \frac{300}{x}$ is an example of a **rational function**. This function is *nonlinear*.

Then
You wrote inverse variation equations. (Lesson 11-1)

Now
- Identify excluded values.
- Identify and use asymptotes to graph rational functions.

New Vocabulary
rational function
excluded value
asymptote

Math Online
glencoe.com
- Extra Examples
- Personal Tutor
- Self-Check Quiz
- Homework Help

Key Concept — Rational Functions

Words A rational function can be described by an equation of the form $y = \frac{p}{q}$, where p and q are polynomials and $q \neq 0$.

Parent function:	$f(x) = \frac{1}{x}$
Type of graph:	hyperbola
Domain:	$\{x \mid x \neq 0\}$
Range:	$\{y \mid y \neq 0\}$

Graph

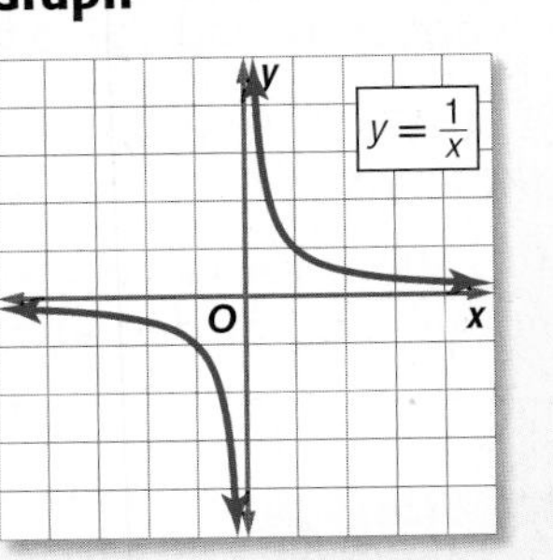

Since division by zero is undefined, any value of a variable that results in a denominator of zero in a rational function is excluded from the domain of the function. These are called **excluded values** for the rational function.

EXAMPLE 1 Find Excluded Values

State the excluded value for each function.

a. $y = -\frac{2}{x}$

The denominator cannot equal 0. So, the excluded value is $x = 0$.

b. $y = \frac{2}{x+1}$

$x + 1 = 0$ **Set the denominator equal to 0.**

$x = -1$ **Subtract 1 from each side.**

The excluded value is $x = -1$.

c. $y = \frac{5}{4x-8}$

$4x - 8 = 0$ **Set the denominator equal to 0.**

$4x = 8$ **Add 8 to each side.**

$x = 2$ **Divide each side by 4.**

The excluded value is $x = 2$.

Check Your Progress

1A. $y = \frac{5}{2x}$ **1B.** $y = \frac{x}{x-7}$ **1C.** $y = \frac{4}{3x+9}$

Personal Tutor glencoe.com

Depending on the real-world situation, in addition to excluding x-values that make a denominator zero from the domain of a rational function, additional values might have to be excluded from the domain as well.

Real-World Link

As the temperature of the gas inside a hot air balloon increases, the density of the gas decreases. A hot air balloon rises because the density of the air inside it is less than the density of the air outside.

Source: Goddard Space Flight Center

Real-World EXAMPLE 2 Graph Real-Life Rational Functions

BALLOONS If there are x people in the basket of a hot air balloon, the function $y = \frac{20}{x}$ represents the average number of square feet y per person. Graph this function.

Since the number of people cannot be zero, it is reasonable to exclude negative values and only use positive values for x.

Number of People x	2	4	5	10
Square Feet per Person y	10	5	4	2

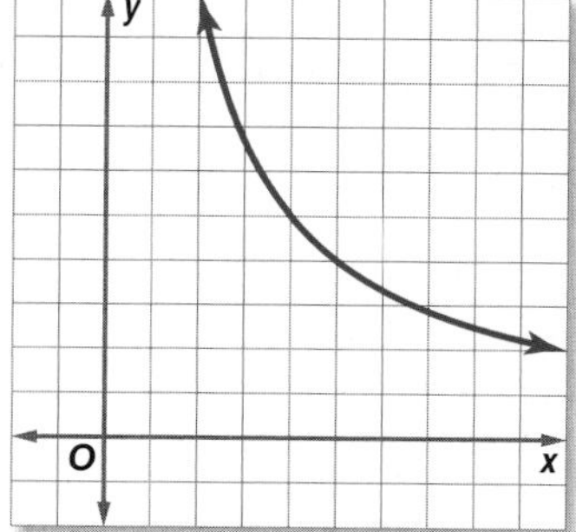

Check Your Progress

2. GEOMETRY A rectangle has an area of 18 square inches. The function $\ell = \frac{18}{w}$ shows the relationship between the length and width. Graph the function.

Personal Tutor glencoe.com

Identify and Use Asymptotes In Example 2, an excluded value is $x = 0$. Notice that the graph approaches the vertical line $x = 0$, but never touches it.

The graph also approaches but never touches the horizontal line $y = 0$. The lines $x = 0$ and $y = 0$ are called *asymptotes*. An **asymptote** is a line that the graph of a function approaches.

StudyTip

Use Asymptotes Asymptotes are helpful for graphing rational functions. However, they are not part of the graph.

Key Concept Asymptotes

For Your FOLDABLE

Words A rational function in the form $y = \frac{a}{x - b} + c$, $a \neq 0$, has a vertical asymptote at the x-value that makes the denominator equal zero, $x = b$. It has a horizontal asymptote at $y = c$.

Model **Example**

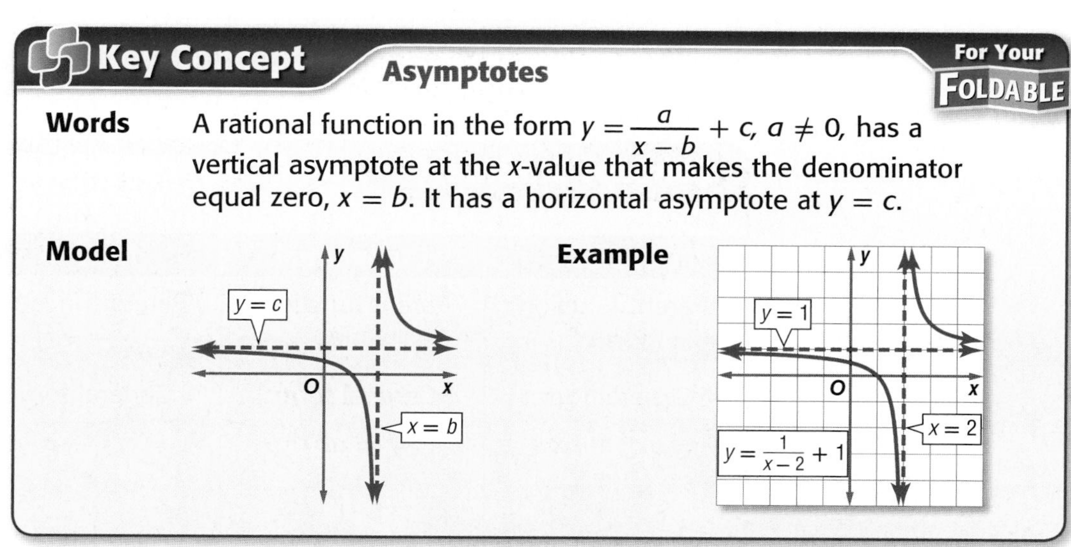

The domain of $y = \frac{a}{x - b} + c$ is all real numbers except $x = b$. The range is all real numbers except $y = c$. Rational functions cannot be traced with a pencil that never leaves the paper, so choose x-values on both sides of the vertical asymptote to graph both portions of the function.

Math History Link

Evelyn Boyd Granville (1924–)
Granville majored in mathematics and physics at Smith College in 1945, where she graduated summa cum laude. She earned an M.A. in mathematics and physics and a Ph.D. in mathematics from Yale University. Granville's doctoral work focused on functional analysis.

EXAMPLE 3 Identify and Use Asymptotes to Graph Functions

Identify the asymptotes of each function. Then graph the function.

a. $y = \frac{2}{x} - 4$

Step 1 Identify and graph the asymptotes using dashed lines.

vertical asymptote: $x = 0$
horizontal asymptote: $y = -4$

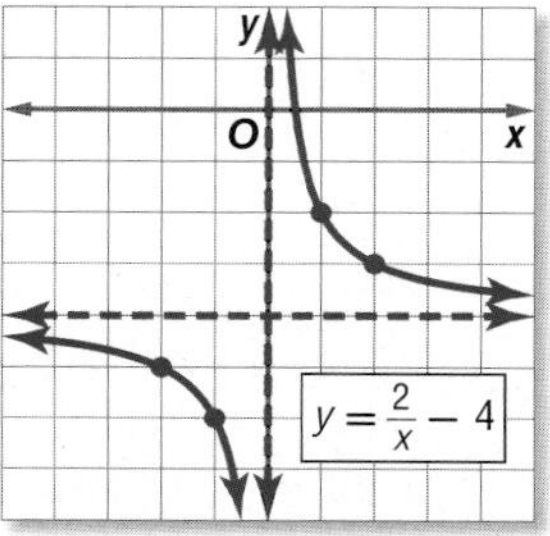

Step 2 Make a table of values and plot the points. Then connect them.

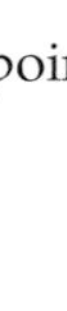

x	−2	−1	1	2
y	−5	−6	−2	−3

b. $y = \frac{1}{x + 1}$

Step 1 To find the vertical asymptote, find the excluded value.

$x + 1 = 0$ **Set the denominator equal to 0.**

$x = -1$ **Subtract 1 from each side.**

vertical asymptote: $x = -1$
horizontal asymptote: $y = 0$

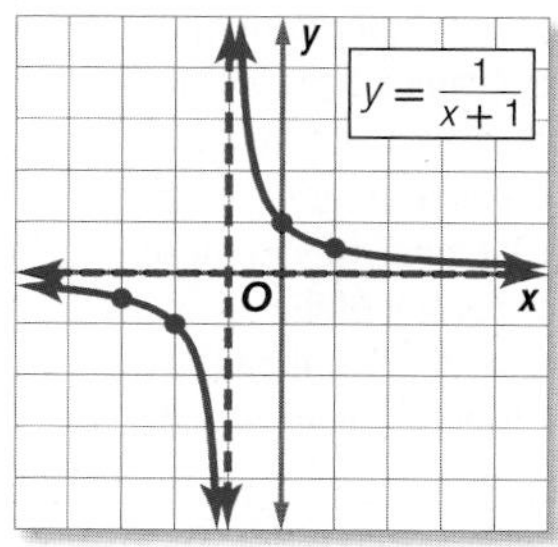

Step 2

x	−3	−2	0	1
y	−0.5	−1	1	0.5

Check Your Progress

3A. $y = -\frac{6}{x}$ **3B.** $y = \frac{1}{x - 3}$ **3C.** $y = \frac{2}{x + 2} + 1$

Personal Tutor glencoe.com

Four types of nonlinear functions are shown below.

Concept Summary Families of Functions

For Your FOLDABLE

Quadratic	Exponential	Radical	Rational
Parent function: $y = x^2$	Parent function: varies	Parent function: $y = \sqrt{x}$	Parent function: $y = \frac{1}{x}$
General form: $y = ax^2 + bx + c$	General form: $y = ab^x$	General form: $y = \sqrt{x - b} + c$	General form: $y = \frac{a}{x - b} + c$

Check Your Understanding

Example 1 p. 678

State the excluded value for each function.

1. $y = \frac{5}{x}$
2. $y = \frac{1}{x + 3}$
3. $y = \frac{x + 2}{x - 1}$
4. $y = \frac{x}{2x - 8}$

Example 2 p. 679

5. **PARTY PLANNING** The cost of decorations for a party is \$32. This is split among a group of friends. The amount each person pays y is given by $y = \frac{32}{x}$, where x is the number of people. Graph the function.

Example 3 p. 680

Identify the asymptotes of each function. Then graph the function.

6. $y = \frac{2}{x}$
7. $y = \frac{3}{x} - 1$
8. $y = \frac{1}{x - 2}$
9. $y = \frac{-4}{x + 2}$
10. $y = \frac{3}{x - 1} + 2$
11. $y = \frac{2}{x + 1} - 5$

Practice and Problem Solving

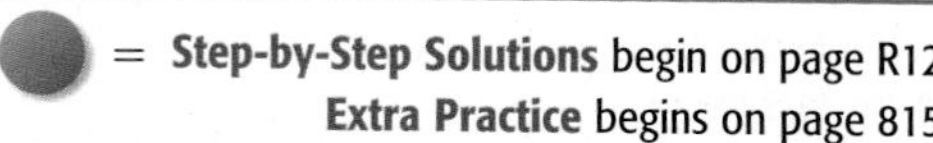
= **Step-by-Step Solutions** begin on page R12.
Extra Practice begins on page 815.

Example 1 p. 678

State the excluded value for each function.

12. $y = \frac{-1}{x}$
13. $y = \frac{8}{x - 8}$
14. $y = \frac{x}{x + 2}$
15. $y = \frac{4}{x + 6}$
16. $y = \frac{x + 1}{x - 3}$
17. $y = \frac{2x + 5}{x + 5}$
18. $y = \frac{7}{5x - 10}$
19. $y = \frac{x}{2x + 14}$

Example 2 p. 679

20. **ANTELOPES** A pronghorn antelope can run 40 miles without stopping. The average speed is given by $y = \frac{40}{x}$, where x is the time it takes to run the distance.
    a. Graph $y = \frac{40}{x}$.
    b. Describe the asymptotes.
21. **CYCLING** A cyclist rides 10 miles each morning. Her average speed y is given by $y = \frac{10}{x}$, where x is the time it takes her to ride 10 miles. Graph the function.

Example 3 p. 680

Identify the asymptotes of each function. Then graph the function.

22. $y = \frac{5}{x}$
23. $y = \frac{-3}{x}$
24. $y = \frac{2}{x} + 3$
25. $y = \frac{1}{x} - 2$
26. $y = \frac{1}{x + 3}$
27. $y = \frac{1}{x - 2}$
28. $y = \frac{-2}{x + 1}$
29. $y = \frac{4}{x - 1}$
30. $y = \frac{1}{x - 2} + 1$
31. $y = \frac{3}{x - 1} - 2$
32. $y = \frac{2}{x + 1} - 4$
33. $y = \frac{-1}{x + 4} + 3$

34. **READING** Refer to the application at the beginning of the lesson.
    a. Graph the function.
    b. Choose a point on the graph, and describe what it means in the context of the situation.
35. The graph shows a translation of the graph of $y = \frac{1}{x}$.
    a. Describe the asymptotes.
    b. Write a function for the graph.

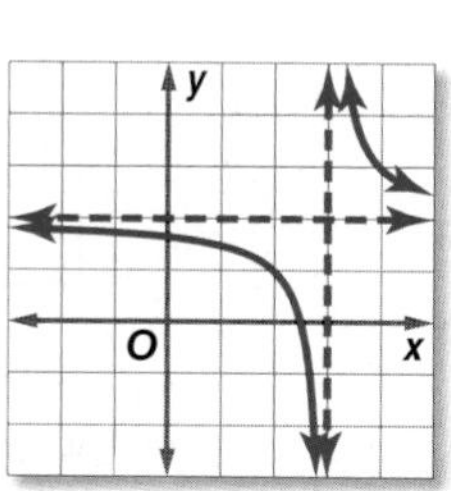

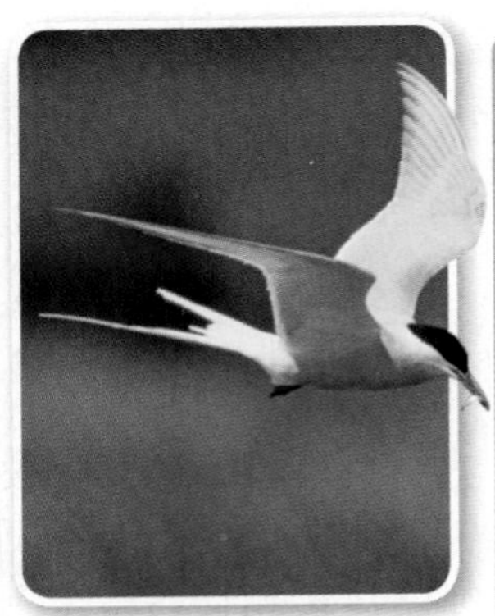

Real-World Link

The bird with the longest migration is the arctic tern. The arctic tern flies up to 20,000 miles per year, from the Arctic to the Antarctic and back.

Source: University of Wisconsin

36. BIRDS A long-tailed jaeger is a sea bird that can migrate 5000 miles or more each year. The average rate in miles per hour r can be given by the function $r = \frac{5000}{t}$, where t is the time in hours. Use the function to determine the average rate of the bird if it spends 250 hours flying.

37. CLASS TRIP The freshmen class is going to a science museum. As part of the trip, each person in the class is also contributing an equal amount of money to name a star.

a. Write a verbal description for the cost per person.

b. Write an equation to represent the total cost y per person if p people go to the museum.

c. Use a graphing calculator to graph the equation.

d. Estimate the number of people needed for the total cost of the trip to be about \$15.

Graph each function. Identify the asymptotes.

38. $y = \frac{4x + 3}{2x - 4}$

39. $y = \frac{x^2}{x^2 - 1}$

40. $y = \frac{x}{x^2 - 9}$

41 GEOMETRY The equation $h = \frac{2(64)}{b_1 + 8}$ represents the height h of a quadrilateral with an area of 64 square units. The quadrilateral has two opposite sides that are parallel and h units apart; one is b_1 units long and another is 8 units long.

a. Describe a reasonable domain and range for the function.

b. Graph the function in the first quadrant.

c. Use the graph to estimate the value of h when $b_1 = 10$.

H.O.T. Problems

Use **H**igher-**O**rder **T**hinking Skills

42. CHALLENGE Graph $y = \frac{1}{x^2 - 4}$. State the domain and the range of the function.

43. REASONING Without graphing, describe the transformation that takes place between the graph of $y = \frac{1}{x}$ and the graph of $y = \frac{1}{x + 5} - 2$.

44. OPEN ENDED Write a rational function if the asymptotes of the graph are at $x = 3$ and $y = 1$. Explain how you found the function.

45. REASONING Is the following statement *true* or *false*? If false, give a counterexample.

The graph of a rational function will have at least one intercept.

46. WHICH ONE DOESN'T BELONG Identify the function that does not belong with the other three. Explain your reasoning.

$y = \frac{4}{x}$	$y = \frac{6}{x + 1}$	$y = \frac{8}{x} + 1$	$y = \frac{10}{2x}$

47. WRITING IN MATH Write a rule to find the vertical asymptotes of a rational function.

Standardized Test Practice

48. Simplify $\frac{2a^2d}{3bc} \cdot \frac{9b^2c}{16ad^2}$.

A $\frac{abd}{c}$

B $\frac{ab}{d}$

C $\frac{6a}{4bd}$

D $\frac{3ab}{8d}$

49. SHORT RESPONSE One day Lola ran 100 meters in 15 seconds, 200 meters in 45 seconds, and 200 meters over low hurdles in one and a half minutes. How many more seconds did it take her to run 200 meters over low hurdles than the 200-meter dash?

50. Scott and Ian started a T-shirt printing business. The total start-up costs were $450. It costs $5.50 to print one T-shirt. Write a rational function $A(x)$ for the average cost of producing x T-shirts.

F $A(x) = \frac{450 + 5.5x}{x}$

G $A(x) = \frac{450}{x} + 5.5$

H $A(x) = 450x + 5.5$

J $A(x) = 450 + 5.5x$

51. GEOMETRY Which of the following is a quadrilateral with exactly one pair of parallel sides?

A parallelogram

B rectangle

C square

D trapezoid

Spiral Review

52. TRAVEL The Brooks family can drive to the beach, which is 220 miles away, in 4 hours if they drive 55 miles per hour. Kendra says that they would save at least a half an hour if they were to drive 65 miles per hour. Is Kendra correct? Explain. (Lesson 11-1)

Use a calculator to find the measure of each angle to the nearest degree. (Lesson 10-8)

53. $\sin C = 0.9781$

54. $\tan H = 0.6473$

55. $\cos K = 0.7658$

56. $\tan Y = 3.6541$

57. $\cos U = 0.5000$

58. $\sin N = 0.3832$

If c is the measure of the hypotenuse of a right triangle, find each missing measure. If necessary round to the nearest hundredth. (Lesson 10-5)

59. $a = 15, b = 60, c = ?$

60. $a = 17, c = 35, b = ?$

61. $a = \sqrt{110}, b = 1, c = ?$

62. $a = \sqrt{17}, b = \sqrt{12}, c = ?$

63. $a = 6, c = 11, b = ?$

64. $a = 9, b = 6, c = ?$

65. SIGHT The formula $d = \sqrt{\frac{3h}{2}}$ represents the distance d in miles that a person h feet high can see. Irene is standing on a cliff that is 310 feet above sea level. How far can Irene see from the cliff? Write a simplified radical expression and a decimal approximation. (Lesson 10-3)

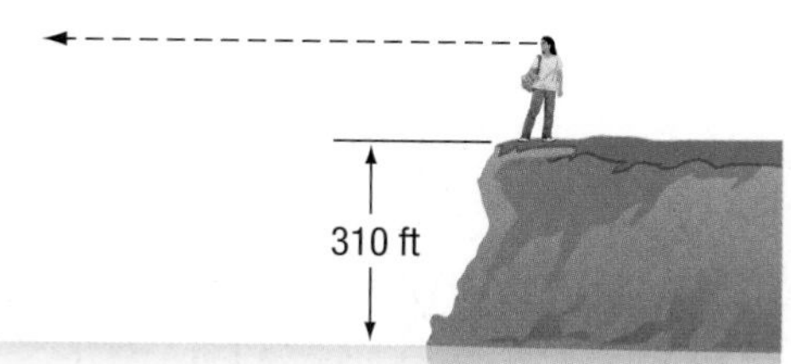

Skills Review

Factor each trinomial. (Lessons 8-3 and 8-4)

66. $x^2 + 11x + 24$

67. $w^2 + 13w - 48$

68. $p^2 - 2p - 35$

69. $72 + 27a + a^2$

70. $c^2 + 12c + 35$

71. $d^2 - 7d + 10$

72. $g^2 - 19g + 60$

73. $n^2 + 3n - 54$

74. $5x^2 + 27x + 10$

75. $24b^2 - 14b - 3$

76. $12a^2 - 13a - 35$

77. $6x^2 - 14x - 12$

11-3 Simplifying Rational Expressions

Then

You simplified expressions involving the quotient of monomials. (Lesson 7-2)

Now

- Identify values excluded from the domain of a rational expression.
- Simplify rational expressions.

New Vocabulary

rational expression

Math Online

glencoe.com

- Extra Examples
- Personal Tutor
- Self-Check Quiz
- Homework Help
- Math in Motion

Why?

Big-O is a "hubless" Ferris wheel in Tokyo, Japan. The *centripetal force*, or the force acting toward the center, is given by $\frac{mv^2}{r}$, where m is the mass of the Ferris wheel, v is the velocity, and r is the radius.

Identify Excluded Values The expression $\frac{mv^2}{r}$ is an example of a rational expression. A **rational expression** is an algebraic fraction whose numerator and denominator are polynomials. Since division by zero is undefined, the polynomial in the denominator cannot be 0.

EXAMPLE 1 Find Excluded Values

State the excluded values for each rational expression.

a. $\frac{-8}{r^2 - 36}$

Exclude the values for which $r^2 - 36 = 0$.

$r^2 - 36 = 0$	**The denominator cannot be zero.**
$(r - 6)(r + 6) = 0$	**Factor.**
$r - 6 = 0$ or $r + 6 = 0$	**Zero Product Property**
$r = 6$ $\quad r = -6$	

Therefore, r cannot equal 6 or -6.

b. $\frac{n^2}{n^2 + 4n - 5}$

Exclude the values for which $n^2 + 4n - 5 = 0$.

$n^2 + 4n - 5 = 0$	**The denominator cannot be zero.**
$(n - 1)(n + 5) = 0$	**Factor.**
$n - 1 = 0$ or $n + 5 = 0$	**Zero Product Property**
$n = 1$ $\quad n = -5$	

Therefore, n cannot equal 1 or -5.

Check Your Progress

1A. $\frac{5x}{x^2 - 81}$

1B. $\frac{3a - 2}{a^2 + 6a + 8}$

Personal Tutor glencoe.com

Real-World EXAMPLE 2 Use Rational Expressions

GEOMETRY Find the height of a cylinder that has a volume of 821 cubic inches and a radius of 7 inches. Round to the nearest tenth.

Understand You have a rational expression with two variables, V and r.

Plan Substitute 821 for V and 7 for r and simplify.

Solve $\frac{V}{\pi r^2} = \frac{821}{\pi(7)^2}$ **Replace V with 821 and r with 7.**

≈ 5.3 **The height of the cylinder is about 5.3 inches.**

Check Use estimation to determine whether the answer is reasonable.

$\frac{800}{3(50)} \approx 5$ ✓ **The solution is reasonable.**

Check Your Progress

2. Find the height of the cylinder that has a volume of 710 cubic inches and a diameter of 18 inches.

Personal Tutor glencoe.com

StudyTip

Cylinder The volume of a cylinder is $V = \pi r^2 h$. The height of a cylinder with volume V and radius r is given by $\frac{V}{\pi r^2}$.

Simplify Expressions A rational expression is in simplest form when the numerator and denominator have no common factors except 1. To simplify a rational expression, divide out any common factors of the numerator and denominator.

Key Concept Simplifying Rational Expressions

For Your FOLDABLE

Words Let a, b, and c, be polynomials with $a \neq 0$ and $c \neq 0$.

Symbols $\frac{ba}{ca} = \frac{b \cdot a}{c \cdot a} = \frac{b}{c}$

Example $\frac{3x - 9}{4x - 12} = \frac{3(x - 3)}{4(x - 3)} = \frac{3}{4}$

Math *in Motion*, Interactive Lab glencoe.com

STANDARDIZED TEST EXAMPLE 3

Which expression is equivalent to $\frac{(-3x^2)(4x^5)}{9x^6}$?

A $\frac{4}{3}x$ **B** $\frac{4}{3x}$ **C** $-\frac{4}{3x}$ **D** $-\frac{4}{3}x$

Read the Test Item The expression represents the product of two monomials and the division of that product by another monomial.

Solve the Test Item

Step 1 Factor the numerator and denominator, using their GCF. $\frac{(3x^6)(-4x)}{(3x^6)(3)}$

Step 2 Simplify. The correct answer is D. $\frac{\cancel{(3x^6)}(-4x)}{\cancel{(3x^6)}(3)}$ or $-\frac{4}{3}x$

Test-TakingTip

Eliminate Possibilities Since there is only one negative factor in the expression in Example 3, the simplified expression should have a negative sign. So, you can eliminate choices A and B.

Check Your Progress

3. Which expression is equivalent to $\frac{16c^2b^4}{8c^3b}$?

F $\frac{2b^3}{c}$ **G** $\frac{b^3}{2c}$ **H** $\frac{1}{2b^3c}$ **J** $2b^3c$

Personal Tutor glencoe.com

You can use the same procedure to simplify a rational expression in which the numerator and denominator are polynomials.

EXAMPLE 4 Simplify Rational Expressions

Simplify $\frac{2r + 18}{r^2 + 8r - 9}$. State the excluded values of r.

$\frac{2r + 18}{r^2 + 8r - 9} = \frac{2(r + 9)}{(r + 9)(r - 1)}$ **Factor.**

$= \frac{2\cancel{(r + 9)}^{1}}{\cancel{(r + 9)}_{1}(r - 1)}$ or $\frac{2}{r - 1}$ **Divide the numerator and denominator by the GCF, $r + 9$.**

Exclude the values for which $r^2 + 8r - 9$ equals 0.

$r^2 + 8r - 9 = 0$ **The denominator cannot equal zero.**

$(r + 9)(r - 1) = 0$ **Factor.**

$r = -9$ or $r = 1$ **Zero Product Property**

So, $r \neq -9$ and $r \neq 1$.

StudyTip

Excluded Values Determine the excluded values using the *original* expression rather than the simplified expression.

Check Your Progress

Simplify each rational expression. State the excluded values of the variables.

4A. $\frac{n + 3}{n^2 + 10n + 21}$

4B. $\frac{y^2 + 9y - 10}{2y + 20}$

Personal Tutor glencoe.com

When simplifying rational expressions, look for binomials that are opposites. For example, $5 - x$ and $x - 5$ are opposites because $5 - x = -1(x - 5)$. So, you can write $\frac{x - 5}{5 - x}$ as $\frac{x - 5}{-1(x - 5)}$.

EXAMPLE 5 Recognize Opposites

Simplify $\frac{36 - t^2}{5t - 30}$. State the excluded values of t.

$\frac{36 - t^2}{5t - 30} = \frac{(6 - t)(6 + t)}{5(t - 6)}$ **Factor.**

$= \frac{-1(t - 6)(6 + t)}{5(t - 6)}$ **Rewrite $6 - t$ as $-1(t - 6)$.**

$= \frac{-1\cancel{(t - 6)}^{1}(6 + t)}{5\cancel{(t - 6)}_{1}}$ or $-\frac{6 + t}{5}$ **Divide out the common factor, $t - 6$.**

Exclude the values for which $5t - 30$ equals 0.

$5t - 30 = 0$ **The denominator cannot equal zero.**

$5t = 30$ **Add 30 to each side.**

$t = 6$ **Zero Product Property**

So, $t \neq 6$.

Check Your Progress

Simplify each expression. State the excluded values of x.

5A. $\frac{12x + 36}{x^2 - x - 12}$

5B. $\frac{x^2 - 2x - 35}{x^2 - 9x + 14}$

Personal Tutor glencoe.com

StudyTip

Zeros Once simplified, a rational function has zeros at the values that make the numerator 0.

Recall that to find the zeros of a quadratic function, you need to find the values of x when $f(x) = 0$. The zeros of a rational function are found in the same way.

EXAMPLE 6 Rational Functions

Find the zeros of $f(x) = \frac{x^2 + 3x - 18}{x - 3}$.

$f(x) = \frac{x^2 + 3x - 18}{x - 3}$ — Original function

$0 = \frac{x^2 + 3x - 18}{x - 3}$ — $f(x) = 0$

$0 = \frac{(x + 6)(x - 3)}{x - 3}$ — Factor.

$0 = \frac{(x + 6)\cancel{(x - 3)}^{1}}{\cancel{x - 3}_{1}}$ — Divide out common factors.

$0 = x + 6$ — Simplify.

When $x = -6$, the numerator becomes 0, so $f(x) = 0$. Therefore, the zero of the function is -6.

Check Your Progress

Find the zeros of each function.

6A. $f(x) = \frac{x^2 + 2x - 15}{x + 1}$

6B. $f(x) = \frac{x^2 + 6x + 8}{x^2 + x - 2}$

Personal Tutor glencoe.com

Check Your Understanding

Example 1 p. 684

State the excluded values for each rational expression.

1. $\frac{8}{x^2 - 16}$

2. $\frac{3m}{m^2 - 6m + 5}$

Example 2 p. 685

3. PHYSICAL SCIENCE A 0.16-kilogram ball attached to a string is being spun in a circle 7.26 meters per second. The expression $\frac{mv^2}{r}$, where m is the mass of the ball, v is the velocity, and r is the radius, can be used to find the force that keeps the ball spinning in a circle. If the circle has a radius of 0.5 meter, find the force that must be exerted to keep the ball spinning. Round to the nearest tenth.

Examples 3–5 pp. 685–686

Simplify each expression. State the excluded values of the variables.

4. $\frac{28ab^3}{16a^2b}$

5 $\frac{(-3r)(10r^4)}{6r^5}$

6. $\frac{5d + 15}{d^2 - d - 12}$

7. $\frac{x^2 + 11x + 28}{x + 4}$

8. $\frac{2r - 12}{r^2 - 36}$

9. $\frac{3y - 27}{81 - y^2}$

Example 6 p. 687

Find the zeros of each function.

10. $f(x) = \frac{x^2 - x - 12}{x - 2}$

11. $f(x) = \frac{x^2 - x - 6}{x^2 + 8x + 12}$

Practice and Problem Solving

= **Step-by-Step Solutions** begin on page R12.
Extra Practice begins on page 815.

Example 1 p. 684

State the excluded values for each rational expression.

12. $\frac{-n}{n^2 - 49}$

13. $\frac{5x + 1}{x^2 - 1}$

14. $\frac{12a}{a^2 - 3a - 10}$

15. $\frac{k^2 - 4}{k^2 + 5k - 24}$

Example 2 p. 685

16. GEOMETRY The volume of a rectangular prism is $3x^3 + 34x^2 + 72x - 64$. If the height is $x + 4$, what is the area of the base of the prism?

17. GEOMETRY Use the circle at the right to write the ratio $\frac{\text{circumference}}{\text{area}}$. Then simplify. State the excluded value of the variable.

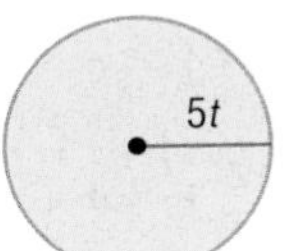

Examples 3–5 pp. 685–686

Simplify each expression. State the excluded values of the variables.

18. $\frac{15x^4y^2}{40x^3y^3}$

19. $\frac{32n^2p}{2n^4p}$

20. $\frac{(4t^3)(2t)}{20t^2}$

21. $\frac{(7c^2)(-6c^3)}{21c^4}$

22. $\frac{4x - 24}{x^2 - 12x + 36}$

23. $\frac{a^2 + 3a}{a^2 - 3a - 18}$

24. $\frac{n^2 + 7n - 18}{n - 2}$

25. $\frac{x^2 + 4x - 32}{x + 8}$

26. $\frac{x^2 - 25}{x^2 + 5x}$

27. $\frac{2p^2 - 14p}{p^2 - 49}$

28. $\frac{2x - 10}{25 - x^2}$

29. $\frac{64 - c^2}{c^2 - 7c - 8}$

Example 6 p. 687

Find the zeros of each function.

30. $f(x) = \frac{x^2 - x - 12}{x^2 + 2x - 35}$

31 $f(x) = \frac{x^2 + 3x - 4}{x^2 + 9x + 20}$

32. $f(x) = \frac{2x^2 + 11x - 40}{2x + 5}$

33. $f(x) = \frac{3x^2 - 18x + 24}{x - 6}$

34. $f(x) = \frac{x^3 + x^2 - 6x}{x - 1}$

35. $f(x) = \frac{x^3 - 4x^2 - 12x}{x + 2}$

36. PYRAMIDS The perimeter of the base of the Pyramid of the Sun is 4π times the height. The perimeter of the base of the Great Pyramid of Giza is 2π times the height. Write and simplify each ratio comparing the base perimeters.

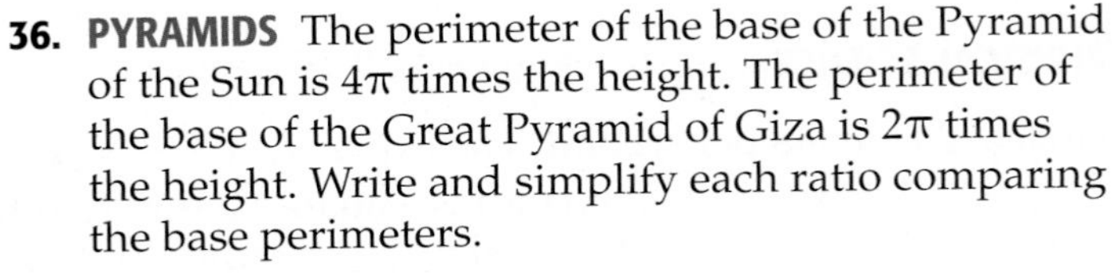

a. Pyramid of the Sun to the Great Pyramid

b. Great Pyramid to the Pyramid of the Sun

Pyramid	Height (ft)
Pyramid of the Sun (Mexico)	233.5
Great Pyramid (Egypt)	481.4

Source: Nexus Network Journal

Real-World Link

George Ferris built the first Ferris wheel for the World's Columbian Exposition in 1893. It had a diameter of 250 feet.

Source: *The New York Times*

37. FERRIS WHEELS Refer to the Real-World Link.

a. To find the speed traveled by a car located on the wheel, you can find the circumference of a circle and divide by the time it takes for one rotation. Write a rational expression for the speed of a car rotating in time t.

b. Suppose the first Ferris wheel rotated once every 5 minutes. What was the speed of a car on the circumference in feet per minute?

Simplify each expression. State the excluded values of the variables.

38. $\frac{3a^2b^4 + 9a^3b - 6a^5b}{3a^2b}$

39. $\frac{8x^5 - 10xy^2}{2xy^3}$

40. $\frac{x + 5}{3x^2 + 14x - 5}$

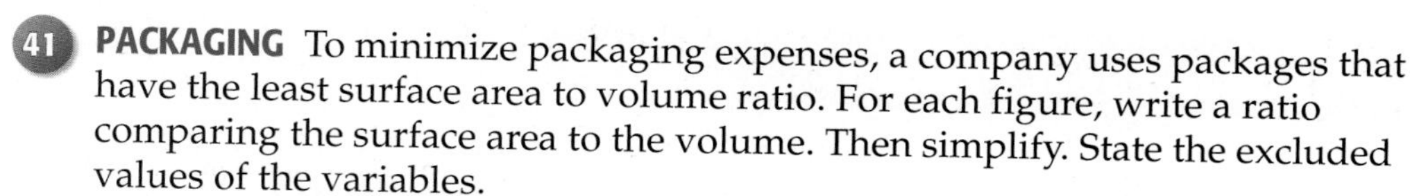
41 **PACKAGING** To minimize packaging expenses, a company uses packages that have the least surface area to volume ratio. For each figure, write a ratio comparing the surface area to the volume. Then simplify. State the excluded values of the variables.

a.
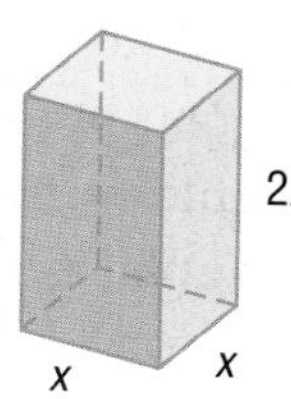

b.
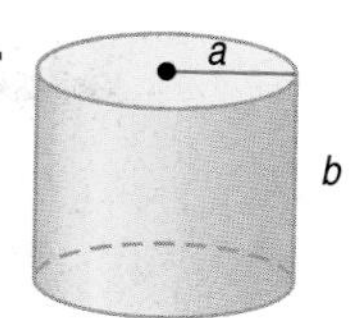

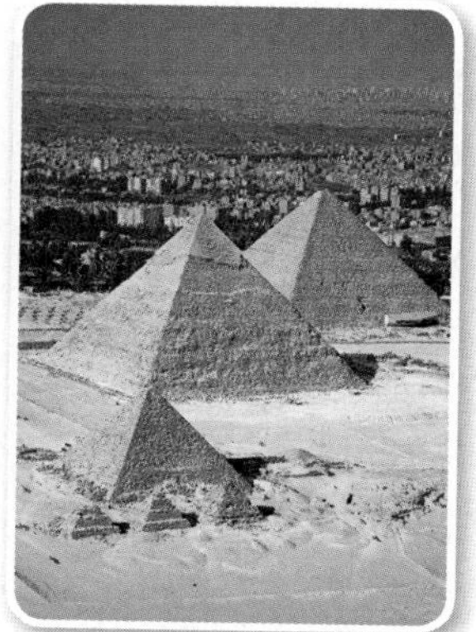

Real-World Link

The ancient Egyptians probably used levers to help them maneuver the giant blocks they used to build the pyramids. An estimated 20,000 to 30,000 workers built the Pyramids at Giza over 80 years.

Source: *National Geographic*

42. **HISTORY** The diagram shows how a lever may have been used to move blocks.

a. The mechanical advantage of a lever is $\frac{L_A}{L_R}$, where L_A is the length of the effort arm and L_R is the length of the resistance arm. Find the mechanical advantage of the lever shown.

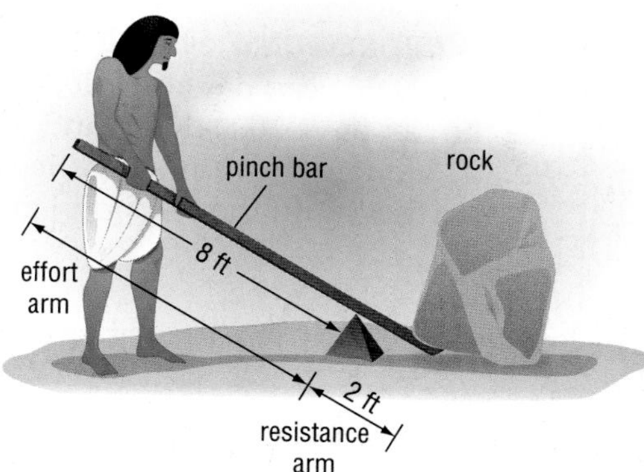

b. The force placed on the rock is the product of the mechanical advantage and the force applied to the end of the lever. If the Egyptian worker can apply a force of 180 pounds, what is the greatest weight he can lift with the lever?

c. To lift a 535-pound rock using a 7-foot lever with the fulcrum 2 feet from the rock, how much force will have to be used?

H.O.T. Problems

Use Higher-Order Thinking Skills

43. **FIND THE ERROR** Colleen and Sanson examined $\frac{12x + 36}{x^2 - x - 12}$ and found the excluded value(s). Is either of them correct? Explain.

Colleen

$\frac{12x + 36}{x^2 - x - 12} = \frac{12(x + 3)}{(x - 4)(x + 3)}$

The excluded values are 4 and -3.

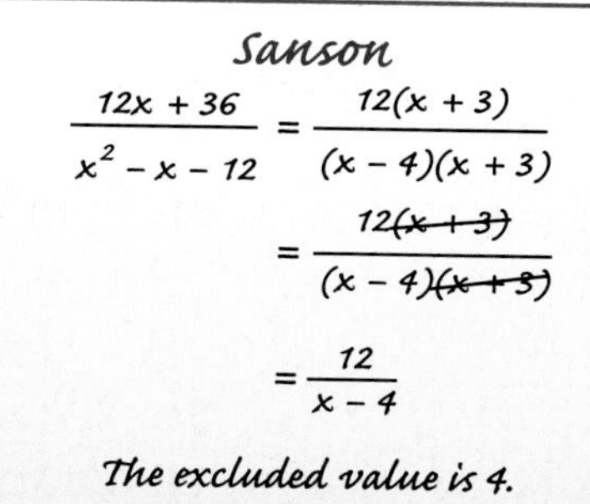
Sanson

$\frac{12x + 36}{x^2 - x - 12} = \frac{12(x + 3)}{(x - 4)(x + 3)}$

$= \frac{12\cancel{(x + 3)}}{(x - 4)\cancel{(x + 3)}}$

$= \frac{12}{x - 4}$

The excluded value is 4.

44. **CHALLENGE** Compare and contrast the graphs of $y = x - 2$ and $y = \frac{x^2 + 5x - 14}{x + 7}$.

45. **REASONING** Explain why every polynomial is also a rational expression.

46. **OPEN ENDED** Write a rational expression with excluded values −2 and 2. Explain how you found the expression.

47. **REASONING** Is $\frac{2x^2 - 4x}{x - 2}$ in simplest form? Justify your answer.

48. **WRITING IN MATH** List the steps you would use to simplify $\frac{x^2 + x - 20}{x + 5}$. State the excluded value.

Standardized Test Practice

49. Simplify $\frac{2x+4}{2}$.

A $x + 1$
B x
C $x + 2$
D $\frac{x}{2}$

50. **SHORT RESPONSE** Shiro is buying a car for \$5800. He can pay the full amount in cash, or he can pay \$1000 down and \$230 a month for 24 months. How much more would he pay for the car on the second plan?

51. **GEOMETRY** What is the name of the figure?

F triangular pyramid
G triangular prism
H rectangular prism
J triangulon

52. A rectangle has a length of 10 inches and a width of 5 inches. Another rectangle has the same area as the first rectangle but its width is 2 inches. Find the length of the second rectangle.

A 30 in.
B 60 in.
C 20 in.
D 25 in.

Spiral Review

State the excluded value for each function. (Lesson 11-2)

53. $y = \frac{6}{x}$
54. $y = \frac{2}{x-5}$
55. $y = \frac{x-4}{x-3}$
56. $y = \frac{3x}{2x+6}$

Solve. Assume that y varies inversely as x. (Lesson 11-1)

57. If $y = 10$ when $x = 4$, find x when $y = 2$.
58. If $y = 12$ when $x = 3$, find x when $y = 6$.
59. If $y = -5$ when $x = 3$, find x when $y = -3$.
60. If $y = 21$ when $x = -6$, find x when $y = 7$.

61. **CRAFTS** Melinda is working on a quilt using the pattern shown. She has several triangular pieces of material with two sides that measure 6 inches. If these pieces are similar to the pattern shown, what is the length of the third side? (Lesson 10-7)

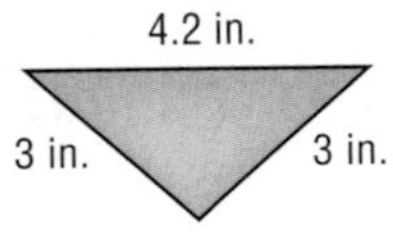

Find the distance between each pair of points whose coordinates are given. (Lesson 10-6)

62. (12, 3), (−8, 3)
63. (0, 0), (5, 12)
64. (6, 8), (3, 4)
65. (−8, −4), (−3, −8)

Simplify. (Lesson 10-2)

66. $\sqrt{20}$
67. $\sqrt{18}$
68. $\sqrt{2} \cdot \sqrt{8}$
69. $2\sqrt{32}$
70. $\sqrt{5} \cdot \sqrt{6}$
71. $\sqrt{40a^2}$
72. $\sqrt{\frac{t}{8}}$
73. $\sqrt{\frac{2}{7}} \cdot \sqrt{\frac{7}{3}}$

74. **FINANCIAL LITERACY** Determine the amount of an investment if \$250 is invested at an interest rate of 7.3% compounded quarterly for 40 years. (Lesson 9-7)

Skills Review

Find the greatest common factor for each set of monomials. (Lesson 8-1)

75. $2x, 8x^2$
76. $3y^2, 7y^3$
77. $7g, 10h$
78. $21c^2d^3, 14cd^2$
79. $9qt^2, 18q^2t^2, 27qt$
80. $10ab, 25a^2b^2, 30a^2b$

EXTEND 11-3

Graphing Technology Lab
Simplifying Rational Expressions

Math Online glencoe.com
- Other Calculator Keystrokes
- Graphing Technology Personal Tutor

When simplifying rational expressions, you can use a graphing calculator to support your answer. If the graphs of the original expression and the simplified expression overlap, they are equivalent. You can also use the graphs to see excluded values.

ACTIVITY Simplify a Rational Expression

Simplify $\frac{x^2 - 16}{x^2 + 8x + 16}$.

Step 1 Factor the numerator and denominator.

- $\frac{x^2 - 16}{x^2 + 8x + 16} = \frac{(x-4)(x+4)}{(x+4)(x+4)}$
  $= \frac{(x-4)}{(x+4)}$

When $x = -4$, $x + 4 = 0$. Therefore, x cannot equal -4 because you cannot divide by zero.

Step 2 Graph the original expression.

- Set the calculator to Dot mode.
- Enter $\frac{x^2 - 16}{x^2 + 8x + 16}$ as **Y1** and graph.

KEYSTROKES: 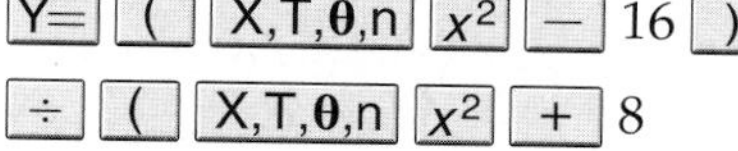

Y= (X,T,θ,n x² − 16)
÷ (X,T,θ,n x² + 8
X,T,θ,n + 16) Zoom 6

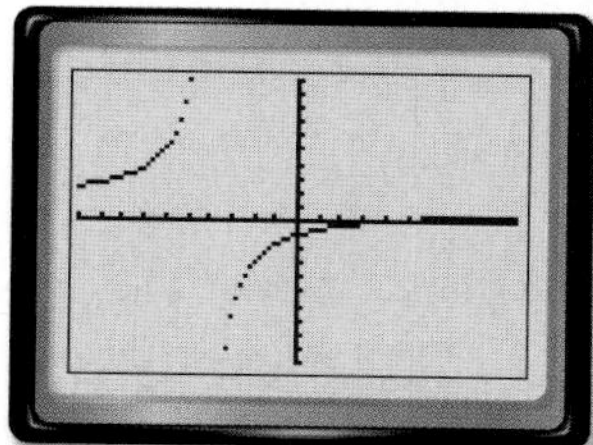

[−10, 10] scl: 1 by [−10, 10] scl: 1

Step 3 Graph the simplified expression.

- Enter $\frac{(x-4)}{(x+4)}$ as **Y2** and graph.

KEYSTROKES: Y= ▼ (X,T,θ,n − 4)
÷ (X,T,θ,n + 4)
Graph

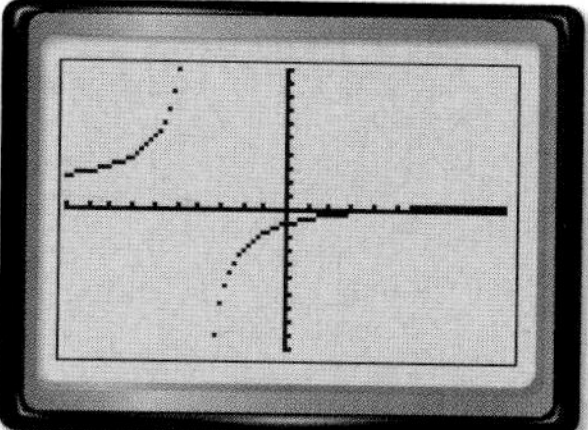

[−10, 10] scl: 1 by [−10, 10] scl: 1

Since the graphs overlap, the two expressions are equivalent.

Exercises

Simplify each expression. Then verify your answer graphically. Name the excluded values.

1. $\frac{5x + 15}{x^2 + 10x + 21}$

2. $\frac{x^2 - 8x + 12}{x^2 + 7x - 18}$

3. $\frac{2x^2 + 6x + 4}{3x^2 + 9x + 6}$

4. a. Simplify $\frac{3x - 8}{6x^2 - 16x}$.

b. How can you use the **TABLE** function to verify that the original expression and the simplified expression are equivalent?

c. How does the **TABLE** function show you that an x-value is an excluded value?

11-4 Multiplying and Dividing Rational Expressions

Then

You multiplied and divided polynomials. (Lesson 7-7 and 7-2)

Now

- Multiply rational expressions.
- Divide rational expressions.

Math Online

glencoe.com

- Extra Examples
- Personal Tutor
- Self-Check Quiz
- Homework Help

Why?

A recent survey showed 10- to 17-year olds talk on their cell phones an average of 3.75 hours per day during the summer. The expression below can be used to find the average number of minutes youth talk on their phones during summer, approximately 90 days.

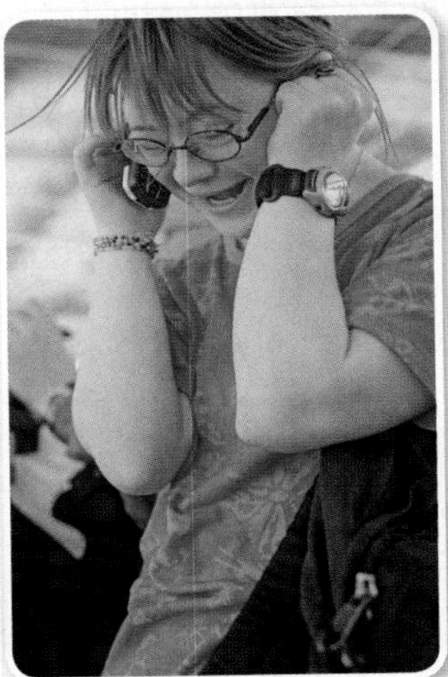

$$90\,\cancel{\text{days}} \cdot \frac{3.75\,\cancel{\text{hours}}}{\cancel{\text{day}}} \cdot \frac{60 \text{ minutes}}{1\,\cancel{\text{hour}}} = 20{,}250 \text{ minutes}$$

Multiply Rational Expressions To multiply fractions, you multiply numerators and multiply denominators. Use this same method to multiply rational expressions.

Key Concept — Multiplying Rational Expressions

For Your FOLDABLE

Words Let a, b, c, and d be polynomials with $b \neq 0$ and $d \neq 0$. Then, $\frac{a}{b} \cdot \frac{c}{d} = \frac{ac}{bd}$.

Example $\frac{x}{2x-3} \cdot \frac{4x^2}{5} = \frac{4x^3}{5(2x-3)}$

EXAMPLE 1 Multiply Expressions Involving Monomials

Find each product.

a. $\frac{r^2x}{9t^3} \cdot \frac{3t^4}{r}$

Divide by the common factors before multiplying.

$$\frac{r^2x}{9t^3} \cdot \frac{3t^4}{r} = \frac{\overset{r}{\cancel{r^2}}x}{\underset{3\ 1}{\cancel{9}\cancel{t^3}}} \cdot \frac{\overset{1\ t}{\cancel{3}\cancel{t^4}}}{\underset{1}{\cancel{r}}}$$ **Divide by the common factors 3, r, and t^3.**

$$= \frac{rxt}{3}$$ **Simplify.**

b. $\frac{a+4}{a^2} \cdot \frac{a}{a^2+2a-8}$

$$\frac{a+4}{a^2} \cdot \frac{a}{a^2+2a-8} = \frac{a+4}{a^2} \cdot \frac{a}{(a+4)(a-2)}$$ **Factor the denominator.**

$$= \frac{\overset{1}{\cancel{a+4}}}{\underset{a}{\cancel{a^2}}} \cdot \frac{\overset{1}{\cancel{a}}}{\underset{1}{\cancel{(a+4)}}(a-2)}$$ **The GCF is $a(a+4)$.**

$$= \frac{1}{a(a-2)} \text{ or } \frac{1}{a^2-2a}$$ **Simplify.**

Check Your Progress

1A. $\frac{3x}{16x^2} \cdot \frac{8x^2}{3}$ **1B.** $\frac{x+3}{x} \cdot \frac{5}{x^2+7x+12}$ **1C.** $\frac{y^2-3y-4}{y+5} \cdot \frac{y+5}{y^2-4y}$

Personal Tutor glencoe.com

StudyTip

Rational Expressions From this point on, assume that no denominator of a rational expression has a value of zero.

When you multiply fractions that involve units of measure, you can divide by the units in the same way that you divide by variables. Recall that this process is called *dimensional analysis*. You can use dimensional analysis to convert units of measure within a system and between systems.

Real-World EXAMPLE 2 Dimensional Analysis

SKI RACING Ann Proctor won the 2007 World Waterski Racing Championship race in her category when she finished the 88-kilometer course in 51.23 minutes. What was her average speed in miles per hour? (*Hint*: 1 km ≈ 0.62 mi)

$$\frac{88 \text{ km}}{51.23 \text{ min}} \cdot \frac{0.62 \text{ mi}}{1 \text{ km}} \cdot \frac{60 \text{ min}}{1 \text{ h}} = \frac{88 \cancel{\text{km}}}{51.23 \cancel{\text{min}}} \cdot \frac{0.62 \text{ mi}}{1 \cancel{\text{km}}} \cdot \frac{60 \cancel{\text{min}}}{1 \text{ h}}$$

$$= \frac{88 \cdot 0.62 \text{ mi} \cdot 60}{51.23 \cdot 1 \cdot 1 \text{ h}}$$ **Simplify.**

$$= \frac{3273.6 \text{ mi}}{51.23 \text{ h}}$$ **Multiply.**

$$\approx \frac{63.9 \text{ mi}}{\text{h}}$$ **Divide the numerator and the denominator by 51.23.**

Her average speed was 63.9 miles per hour.

Check Your Progress

2. SKI RACING What was Ann Proctor's speed in feet per second?

Personal Tutor glencoe.com

Divide Rational Expressions To divide by a fraction, you multiply by the reciprocal. You can use this same method to divide by a rational expression.

Key Concept Dividing Rational Expressions

For Your FOLDABLE

Symbols Let a, b, c, and d be polynomials with $b \neq 0$, $c \neq 0$, and $d \neq 0$. Then, $\frac{a}{b} \div \frac{c}{d} = \frac{a}{b} \cdot \frac{d}{c} = \frac{ad}{bc}$.

Example $\frac{x-3}{x} \div \frac{2x^2}{5} = \frac{x-3}{x} \cdot \frac{5}{2x^2} = \frac{5(x-3)}{2x^3}$

EXAMPLE 3 Divide by a Rational Expression

Find $\frac{4}{15n^3} \div \frac{12}{25n}$.

$$\frac{4}{15n^3} \div \frac{12}{25n} = \frac{4}{15n^3} \cdot \frac{25n}{12}$$ **Multiply by $\frac{25n}{12}$, the reciprocal of $\frac{12}{25n}$.**

$$= \frac{\overset{1}{\cancel{4}}}{\underset{3n^2}{\cancel{15n^3}}} \cdot \frac{\overset{5}{\cancel{25n}}}{\underset{3}{\cancel{12}}}$$ **Divide by common factors 4, 5, and n.**

$$= \frac{5}{9n^2}$$ **Simplify.**

Check Your Progress

Find each quotient.

3A. $\frac{15y^2}{4x} \div \frac{5y}{8x^3}$

3B. $\frac{12a^2}{5b} \div \frac{25a}{6b^2}$

Personal Tutor glencoe.com

ReadingMath

Rational Expressions In a rational expression, the fraction bar acts as a grouping symbol. In Example 5a, $\frac{4x+6}{x^2}$ is read *the quantity four x plus six divided by x squared.*

EXAMPLE 4 Divide by Rational Expressions and Polynomials

Find each quotient.

a. $\frac{2x+6}{x^2} \div (x+3)$

$$\frac{2x+6}{x^2} \div (x+3) = \frac{2x+6}{x^2} \div \frac{x+3}{1}$$ **Write the binomial as a fraction.**

$$= \frac{2x+6}{x^2} \cdot \frac{1}{x+3}$$ **Multiply by the reciprocal of $x + 3$.**

$$= \frac{2(x+3)}{x^2} \cdot \frac{1}{x+3}$$ **Factor $4x + 6$.**

$$= \frac{2\cancel{(x+3)}^{1}}{x^2} \cdot \frac{1}{\cancel{x+3}_{1}} \text{ or } \frac{2}{x^2}$$ **Divide out the common factor and simplify.**

b. $\frac{a-2}{4a+4} \div \frac{a+5}{a+1}$

$$\frac{a-2}{4a+4} \div \frac{a+5}{a+1} = \frac{a-2}{4a+4} \cdot \frac{a+1}{a+5}$$ **Multiply by the reciprocal.**

$$= \frac{a-2}{4\cancel{(a+1)}_{1}} \cdot \frac{\cancel{a+1}^{1}}{a+5}$$ **Factor $4a + 4$.**

$$= \frac{a-2}{4(a+5)}$$ **The GCF is $a + 1$ and simplify.**

Check Your Progress

4A. $\frac{4d-8}{2d-6} \div \frac{2d-4}{d-4}$

4B. $\frac{b+4}{3b+2} \div \frac{3b+12}{b+1}$

Personal Tutor glencoe.com

Sometimes you must factor a quadratic expression before you can simplify the quotient of rational expressions.

StudyTip

Canceling Remember that only factors can be canceled, not individual terms.

EXAMPLE 5 Expression Involving Polynomials

Find $\frac{y-3}{y^2-10y+16} \div \frac{y^2-9}{y-8}$.

$$\frac{y-3}{y^2-10y+16} \div \frac{y^2-9}{y-8}$$

$$= \frac{y-3}{y^2-10y+16} \cdot \frac{y-8}{y^2-9}$$ **Multiply by the reciprocal, $\frac{y-8}{y^2-9}$.**

$$= \frac{y-3}{(y-2)(y-8)} \cdot \frac{y-8}{(y-3)(y+3)}$$ **Factor $y^2 - 10y + 16$ and $y^2 - 9$.**

$$= \frac{\cancel{y-3}^{1}}{(y-2)\cancel{(y-8)}_{1}} \cdot \frac{\cancel{y-8}^{1}}{\cancel{(y-3)}_{1}(y+3)}$$ **The GCF is $(y - 3)(y - 8)$.**

$$= \frac{1}{(y-2)(y+3)}$$ **Simplify.**

Check Your Progress

Find each quotient.

5A. $\frac{p^2-4}{5p} \div \frac{p-2}{p+q}$

5B. $\frac{q^2+3q+2}{12} \div \frac{q+1}{q^2+4}$

Personal Tutor glencoe.com

Check Your Understanding

Example 1 p. 692

Find each product.

1. $\frac{2x^3}{7x} \cdot \frac{14}{x}$

2. $\frac{3ab}{4c^4} \cdot \frac{16c^2}{9b}$

3. $\frac{t^2}{(t-5)(t+5)} \cdot \frac{t+5}{6t}$

4. $\frac{8}{r+1} \cdot \frac{r^2-1}{2}$

Example 2 p. 693

5. **SLOTHS** The slowest land mammal is the three-toed sloth. It travels 0.07 mile per hour on the ground. What is this speed in feet per minute?

6. **EXERCISE** One hour of moderate inline skating burns approximately 330 Calories. If Nelia plans to do inline skating for 3 hours a week, how many Calories will she burn in a year from the skating?

Examples 3–5 pp. 693–694

Find each quotient.

7. $\frac{8}{3x^2} \div \frac{4}{x}$

8. $\frac{c^5}{2} \div \frac{c^3}{6d^2}$

9. $\frac{b^2+6b+5}{6b+6} \div (b+5)$

10. $\frac{2x+8}{x+3} \div \frac{x+4}{x^2+6x+9}$

Practice and Problem Solving

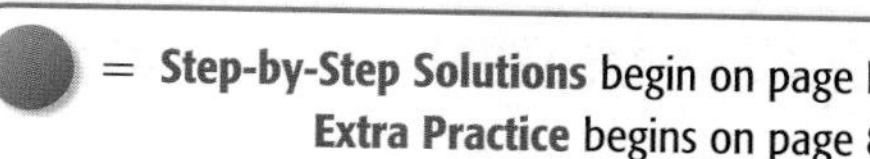
= **Step-by-Step Solutions** begin on page R12.
Extra Practice begins on page 815.

Example 1 p. 692

Find each product.

11. $\frac{10n^2}{4} \cdot \frac{2}{n}$

12. $\frac{12c^3}{21b} \cdot \frac{14b^2}{6c}$

13. $\frac{x^5y}{2z^3} \cdot \frac{18z^4}{xy}$

14. $\frac{5c^3d}{c^4d} \cdot \frac{f^2d^3c}{10cf^4}$

15. $\frac{9}{t-2} \cdot \frac{(t+2)(t-2)}{3}$

16. $\frac{(a+4)(a-5)}{a^2} \cdot \frac{6a}{a+4}$

17. $\frac{(k+6)(k-1)}{k+2} \cdot \frac{(k+1)(k+2)}{(k+1)(k-1)}$

18. $\frac{(r-8)(r+3)}{r} \cdot \frac{2r}{(r+8)(r+3)}$

19. $\frac{n^2+n-2}{n+2} \cdot \frac{4n}{n-1}$

20. $\frac{y^2-1}{y^2-49} \cdot \frac{y-7}{y+1}$

Example 2 p. 693

21. **FINANCIAL LITERACY** A scarf bought in Italy cost 18 Euros. The exchange rate at the time was 1 U.S. dollar = 0.73 Euro.
    a. How much did the scarf cost in U.S. dollars?
    b. If the exchange rate at the time was 1 Canadian dollar = 0.69 Euro, how much did the scarf cost in Canadian dollars?

22. **ROLLER COASTERS** A roller coaster has 6 trains. Each train has 3 cars, and each car seats 4 people. Write and simplify an expression including units to find the total number of people that can ride the roller coaster at one time.

Examples 3–5 pp. 693–694

Find each quotient.

23. $\frac{x^5}{y} \div \frac{x}{y^2}$

24. $\frac{3r^4}{k^2} \div \frac{18r^3}{k}$

25. $\frac{21b^3}{4c^2} \div \frac{7}{6c^2}$

26. $\frac{f^4g^2h}{x^2y} \div f^3g$

27. $\frac{6b-12}{b+5} \div (12b+18)$

28. $\frac{k+3}{k+2} \div \frac{k}{5k+10}$

29. $\frac{5x^2}{x^2-5x+4} \div \frac{10x}{x-1}$

30. $\frac{n^2+7n+12}{16n^2} \div \frac{n+3}{2n}$

31. $\frac{r+2}{r+1} \div \frac{4}{r^2+3r+2}$

32. $\frac{3a}{a^2+2a+1} \div \frac{a-1}{a+1}$

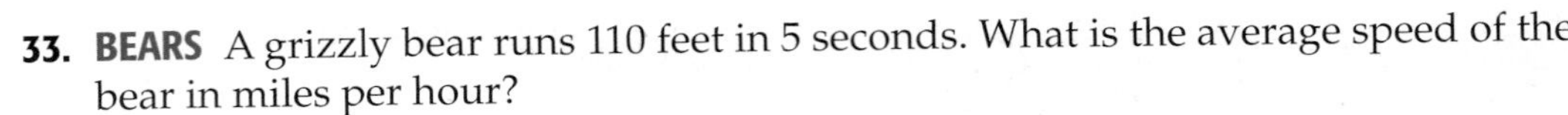

33. **BEARS** A grizzly bear runs 110 feet in 5 seconds. What is the average speed of the bear in miles per hour?

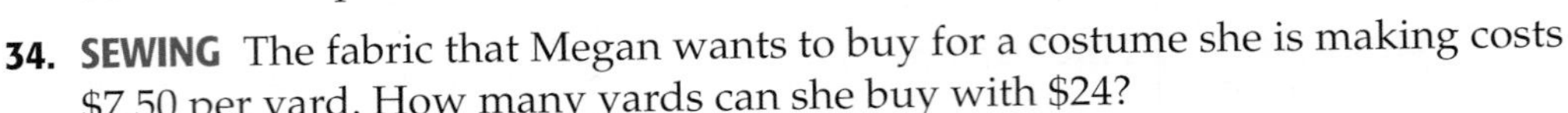

34. **SEWING** The fabric that Megan wants to buy for a costume she is making costs \$7.50 per yard. How many yards can she buy with \$24?

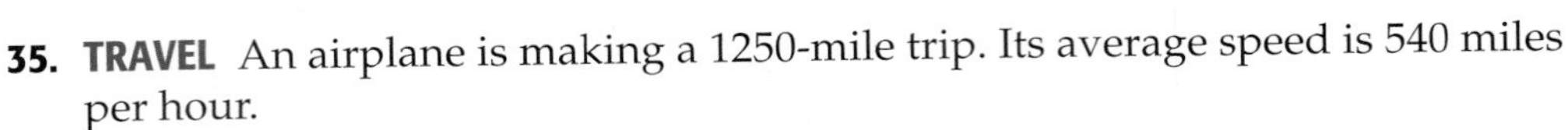

35. **TRAVEL** An airplane is making a 1250-mile trip. Its average speed is 540 miles per hour.

    a. Write a division expression you can use to find the number of hours that the trip will take. Include the units.

    b. Find the quotient. Round to the nearest tenth.

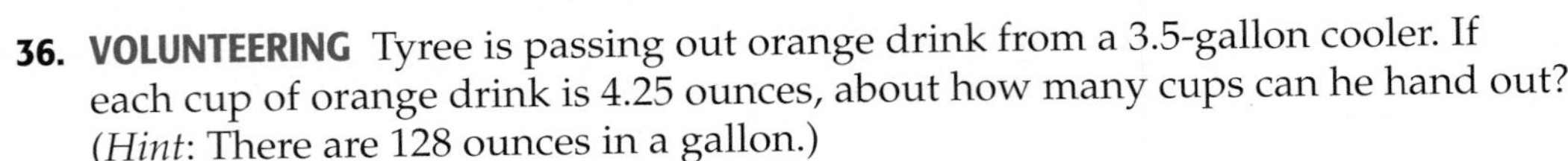

36. **VOLUNTEERING** Tyree is passing out orange drink from a 3.5-gallon cooler. If each cup of orange drink is 4.25 ounces, about how many cups can he hand out? (*Hint*: There are 128 ounces in a gallon.)

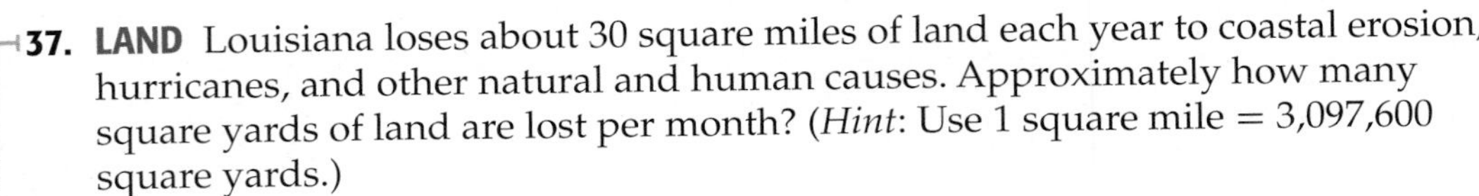

37. **LAND** Louisiana loses about 30 square miles of land each year to coastal erosion, hurricanes, and other natural and human causes. Approximately how many square yards of land are lost per month? (*Hint*: Use 1 square mile = 3,097,600 square yards.)

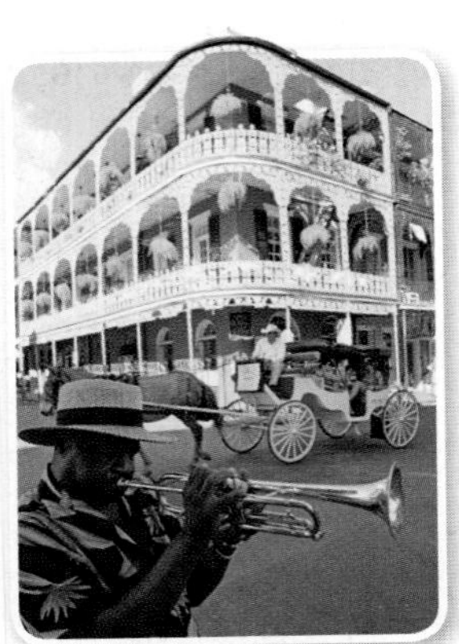

Real-World Link

Much of New Orleans sits 11 feet below sea level. Parts of the French Quarter have sunk as much as 2 feet in the past six decades.

Source: Imaginova

38. **GEOMETRY** Write an expression to represent the length of the rectangle.

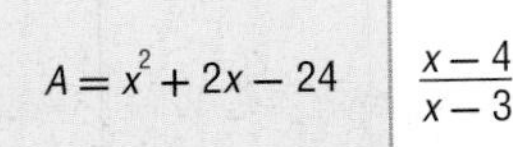

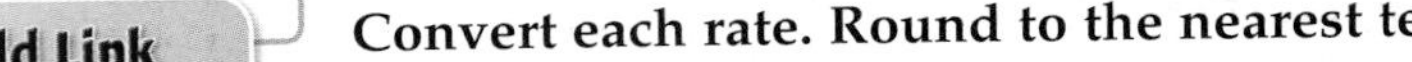

Convert each rate. Round to the nearest tenth.

39. 46 feet per second to miles per hour

40. 29.5 meters per second to kilometers per hour

41. 28 milliliters per second to cups per minute. (*Hint*: 1 liter ≈ 0.908 quarts)

42. 32.4 meters per second to miles per hour. (*Hint*: 1 mile ≈ 1.609 kilometers)

43. **LIFE SCIENCE** A human heart pumps about a cup of blood each time it beats. On average, a person's heart beats about 70 times a minute. Write and simplify an expression to find how many gallons of blood are pumped per hour.

44. **GEOMETRY** Refer to the prism at the right.

    a. Find the volume in cubic inches.

    b. Use the ratio $\frac{1 \text{ foot}^3}{1728 \text{ inches}^3}$ to write a multiplication expression to convert the volume to cubic feet. Then convert the volume.

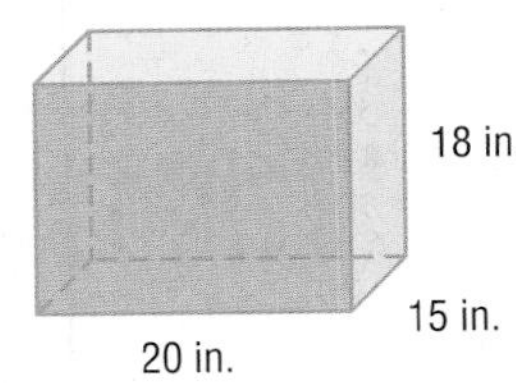

Find each product. Describe what the final answer represents.

45. $\frac{\$9.80}{1 \text{ hour}} \cdot \frac{15 \text{ hours}}{1 \text{ week}} \cdot \frac{52 \text{ weeks}}{1 \text{ year}}$

46. $\frac{\$2.85}{1 \text{ gallon of gasoline}} \cdot \frac{15 \text{ gallons of gasoline}}{1 \text{ fill-up}} \cdot \frac{3 \text{ fill-ups}}{1 \text{ month}} \cdot \frac{1 \text{ month}}{30 \text{ days}}$

47. $\frac{32 \text{ meters}}{1 \text{ second}} \cdot \frac{60 \text{ seconds}}{1 \text{ minute}} \cdot \frac{60 \text{ minutes}}{1 \text{ hour}} \cdot \frac{1 \text{ kilometer}}{1000 \text{ meters}} \cdot \frac{1 \text{ mile}}{1.609 \text{ kilometers}}$

48. $\frac{\$32{,}000}{1 \text{ year}} \cdot \frac{1 \text{ year}}{52 \text{ weeks}} \cdot \frac{1 \text{ week}}{40 \text{ hours}}$

49. **SPACE** The highest speed at which any spacecraft has ever escaped from Earth is 35,800 miles per hour by the *New Horizons* probe, which was launched in 2006. Convert this speed to feet per second. Round to the nearest tenth.

Real-World Link

The largest cylindrical aquarium ever built is located in a hotel lobby in Berlin. It is 25 meters high and contains 900,000 liters of seawater and 2500 tropical fish.

Source: South Travels

50. **ELECTRICITY** Simplify the expression below to find the cost of running a 3500-watt air conditioner for one week.

$$3500 \text{ watts} \cdot \frac{1 \text{ kilowatt}}{1000 \text{ watts}} \cdot \frac{168 \text{ hours}}{1 \text{ week}} \cdot \frac{10 \text{ cents}}{1 \text{ kilowatt} \cdot \text{hours}} \cdot \frac{1 \text{ dollar}}{100 \text{ cents}}$$

51. **AMUSEMENT PARKS** In an amusement park ride, riders stand along the wall of a circular room with a radius of 3.1 meters. The room completes 27 rotations per minute.

    a. Write an expression for the number of meters the room moves per second. (*Hint*: The circumference of a circle is $2\pi r$.)

    b. Simplify the expression you wrote in Part **a** and describe what it means.

52. **AQUARIUMS** An aquarium is a rectangular prism 30 inches long, 15 inches wide, and 18 inches high.

    a. Sketch and label a diagram of the aquarium. Then find the volume of the tank in cubic inches.

    b. Describe how to use the ratio $\frac{1 \text{ ft}^3}{1728 \text{ in}^3}$ to find the volume of the tank in cubic feet. Then find the volume. Round to the nearest tenth.

    c. Water weighs 62 pounds per cubic foot. How much would the water in the tank weigh if the tank were filled?

H.O.T. Problems

Use Higher-Order Thinking Skills

53. **FIND THE ERROR** Mei and Tamika are finding $\frac{2x+6}{x+5} \div \frac{2}{x+5}$. Is either of them correct? Explain.

Mei

$$\frac{2x+6}{x+5} \div \frac{2}{x+5}$$
$$= \frac{\cancel{2}x+6}{\cancel{x+5}} \cdot \frac{\cancel{x+5}}{\cancel{2}}$$
$$= x + 6$$

Tamika

$$\frac{2x+6}{x+5} \div \frac{2}{x+5}$$
$$= \frac{\cancel{2}(x+3)}{\cancel{x+5}} \cdot \frac{\cancel{2}}{\cancel{x+5}}$$
$$= \frac{4(x+3)}{(x+5)^2}$$

54. **REASONING** Find the missing term. Justify your answer.

$$\frac{?}{} \div \frac{10x^3}{21} = \frac{3}{2x}$$

55. **CHALLENGE** Find $\frac{x^2 - 3x - 10}{x^2 + 2x - 35} \cdot \frac{x^2 + 4x - 21}{x^2 + 9x + 14}$. Write in simplest form.

56. **WRITING IN MATH** Give an example and describe how you could use dimensional analysis to solve a real-world problem involving rational expressions.

57. **OPEN ENDED** Give an example of a real-world situation that could be modeled by the quotient of two rational expressions. Provide an example of this quotient.

58. **WRITE A QUESTION** A classmate found that the product of two rational expressions is $\frac{9x-3}{(x+3)(3x+1)}$. She wants to find the excluded values. Write a question to help her solve the problem.

59. **WRITING IN MATH** Describe how to use dimensional analysis to find the number of hours in one year.

Standardized Test Practice

60. GEOMETRY The perimeter of a rectangle is 30 inches. Its area is 54 square inches. Find the length of the longest side.

A 6 inches
B 9 inches
C 12 inches
D 30 inches

61. Find $\frac{c^2 - c - 6}{2c - 10} \div \frac{2c + 4}{3c - 15}$.

F $\frac{3(c-3)}{4}$
G $\frac{c+5}{c-3}$
H $\frac{4(c-3)}{3}$
J $\frac{c-3}{c-5}$

62. EXTENDED RESPONSE The weekly salaries of six employees at a fast food restaurant are \$140, \$220, \$90, \$180, \$140, \$200.

a. What is the mean of the six salaries?

b. What is the median of the six salaries?

c. What is the mode of the six salaries?

63. Tito has three times as many CDs as Dasan. Dasan has two thirds as many CDs as Brant. Brant has 27 CDs. How many CDs does Tito have?

A 54
B 27
C 18
D 32

Spiral Review

Simplify each expression. State the excluded values of the variables. (Lesson 11-3)

64. $\frac{20x^2y}{25xy}$

65. $\frac{14g^3h^2}{42gh^3}$

66. $\frac{64qt}{16q^2t^3}$

67. $\frac{y^2 + 10y + 16}{y + 2}$

68. $\frac{p^2 - 9}{p^2 - 5p + 6}$

69. $\frac{z^2 + z - 2}{z^2 - 3z + 2}$

Identify the asymptotes of each function. (Lesson 11-2)

70. $y = \frac{2}{x}$

71. $y = \frac{3}{x} + 5$

72. $y = \frac{1}{x - 5} - 4$

73. $y = \frac{1}{x + 3}$

74. $y = \frac{-1}{x + 6} + 7$

75. $y = \frac{2}{x - 8} - 3$

76. FORESTRY The number of board feet B that a log will yield can be estimated by using the formula $B = \frac{L}{16}(D^2 - 8D + 16)$, where D is the diameter in inches and L is the log length in feet. For logs that are 16 feet long, what diameter will yield approximately 256 board feet? (Lesson 8-6)

Find the degree of each polynomial. (Lesson 7-4)

77. 2

78. $-3a$

79. $5x^2 + 3x$

80. $d^4 - 6c^2$

81. $2x^3 - 4z + 8xz$

82. $3d^4 + 5d^3 - 4c^2 + 1$

83. DRIVING Tires should be kept within 2 pounds per square inch (psi) of the manufacturer's recommended tire pressure. If the recommendation for a tire is 30 psi, what is the range of acceptable pressures? (Lesson 5-5)

Skills Review

Factor each polynomial. (Lessons 8-3 and 8-4)

84. $x^2 - 18x - 40$

85. $x^2 - 5x + 6$

86. $x^2 - 2x - 24$

87. $3x^2 + 7x - 20$

88. $2x^2 + x - 15$

89. $8x^2 - 4x - 40$

Mid-Chapter Quiz

Lessons 11-1 through 11-4

Math Online glencoe.com
Math *in Motion,* Animation

1. Determine whether the table represents an inverse variation. Explain. (Lesson 11-1)

x	y
2	8
4	4
8	2
16	1

Assume that y varies inversely as x. Write an inverse variation equation that relates x and y. (Lesson 11-1)

2. $y = 5$ when $x = 10$

3. $y = -2$ when $x = 12$

Solve. Assume that y varies inversely as x. (Lesson 11-1)

4. If $y = 6$ when $x = 3$, find x when $y = 5$.

5. If $y = 3$ when $x = 2$, find y when $x = 4$.

State the excluded value for each function. (Lesson 11-2)

6. $y = \frac{2}{x}$

7. $y = \frac{1}{x - 6}$

Identify the asymptotes of each function. (Lesson 11-2)

8. $y = \frac{3}{2x + 4}$

9. $y = \frac{2}{x - 4}$

10. **MULTIPLE CHOICE** Jorge has $x^2 + 5x + 6$ square yards of carpet. He wants to carpet rooms that have areas of $x^2 + 8x + 15$ square yards. Write and simplify an expression to show how many rooms he can carpet. (Lesson 11-3)

A $\frac{x + 3}{x + 5}$

B $\frac{x + 2}{x + 5}$

C $\frac{x + 2}{x + 3}$

D $\frac{x + 6}{x + 5}$

Simplify each expression. State the excluded values of the variables. (Lesson 11-3)

11. $\frac{16x^2y^3}{8xy}$

12. $\frac{z - 5}{z^2 - 7z + 10}$

13. $\frac{3x - 15}{x^2 - 25}$

Find each product. (Lesson 11-4)

14. $\frac{(x + 5)(x - 3)}{x^3} \cdot \frac{5x}{x - 3}$

15. $\frac{a^2 + 2a + 1}{a + 1} \cdot \frac{a - 1}{a^2 - 1}$

16. $\frac{m}{m^2 + 3m + 2} \cdot \frac{m + 2}{m^2}$

17. **MULTIPLE CHOICE** Find the area of the rectangle. (Lesson 11-4)

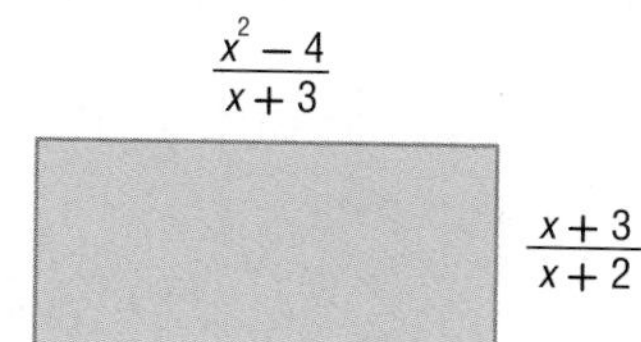

F $\frac{x + 2}{x - 2}$

G $\frac{x + 3}{x - 2}$

H 1

J $x - 2$

Find each quotient. (Lesson 11-4)

18. $\frac{x^4}{y^2} \div \frac{x}{y}$

19. $\frac{x + 3}{2x + 6} \div \frac{3x - 6}{4x - 8}$

20. $\frac{x^2 + 7x + 12}{x^2 - 25} \div \frac{x^2 - 9}{2x + 10}$

21. **MOTOR VEHICLES** In 2005, the U.S. produced 4,411,300 motor vehicles. This was 10% of the total motor vehicle production for the whole world. How many motor vehicles were produced worldwide in 2005?

11-5

Dividing Polynomials

Then
You divided rational expressions. (Lesson 11-5)

Now
- Divide a polynomial by a monomial.
- Divide a polynomial by a binomial.

Math Online
glencoe.com
- Extra Examples
- Personal Tutor
- Self-Check Quiz
- Homework Help
- Math in Motion

Why?

The equation below describes the distance d a horse travels when its initial velocity is 4 m/s, its final velocity is v m/s, and its acceleration is a m/s^2.

$$d = \frac{v^2 - 4^2}{2a}$$

There are different ways to simplify the expression.

Keep as one fraction.

$$\frac{v^2 - 4^2}{2a} = \frac{v^2 - 16}{2a}$$

Divide each term by $2a$.

$$\frac{v^2 - 4^2}{2a} = \frac{v^2}{2a} - \frac{4^2}{2a}$$
$$= \frac{v^2}{2a} - \frac{8}{a}$$

Divide Polynomials by Monomials To divide a polynomial by a monomial, divide each term of the polynomial by the monomial.

EXAMPLE 1 Divide Polynomials by Monomials

Find each quotient.

a. $(2x^2 + 16x) \div 2x$

$(2x^2 + 16x) \div 2x = \frac{2x^2 + 16x}{2x}$ Write as a fraction.

$= \frac{2x^2}{2x} + \frac{16x}{2x}$ Divide each term by $2x$.

$= \frac{\overset{x}{\cancel{2x^2}}}{\underset{1}{\cancel{2x}}} + \frac{\overset{8}{\cancel{16x}}}{\underset{1}{\cancel{2x}}}$ Divide out common factors.

$= x + 8$ Simplify.

b. $(b^2 + 12b - 14) \div 3b$

$(b^2 + 12b - 14) \div 3b = \frac{b^2 + 12b - 14}{3b}$ Write as a fraction.

$= \frac{b^2}{3b} + \frac{12b}{3b} - \frac{14}{3b}$ Divide each term by 3b.

$= \frac{\overset{b}{\cancel{b^2}}}{\underset{3}{\cancel{3b}}} + \frac{\overset{4}{\cancel{12b}}}{\underset{1}{\cancel{3b}}} - \frac{14}{3b}$ Divide out common factors.

$= \frac{b}{3} + 4 - \frac{14}{3b}$ Simplify.

Check Your Progress

1A. $(3q^3 - 6q) \div 3q$

1B. $(4t^5 - 5t^2 - 12) \div 2t^2$

1C. $(4r^6 + 3r^4 - 2r^2) \div 2r$

1D. $(6w^3 - 3w) \div 4w^2$

Personal Tutor glencoe.com

Math *in Motion*, Animation glencoe.com

Divide Polynomials by Binomials You can also divide polynomials by binomials. When a polynomial can be factored and common factors can be divided out, write the division as a rational expression and simplify.

EXAMPLE 2 Divide a Polynomial by a Binomial

Find $(h^2 + 9h + 18) \div (h + 6)$.

$$(h^2 + 9h + 18) \div (h + 6) = \frac{h^2 + 9h + 18}{h + 6}$$ **Write as a rational expression.**

$$= \frac{(h + 3)(h + 6)}{h + 6}$$ **Factor the numerator.**

$$= \frac{(h + 3)\cancel{(h + 6)}^{1}}{\cancel{h + 6}_{1}}$$ **Divide out common factors.**

$$= h + 3$$ **Simplify.**

Check Your Progress

Find each quotient.

2A. $(b^2 - 2b - 15) \div (b + 3)$

2B. $(x^2 + 11x + 24) \div (x + 8)$

Personal Tutor glencoe.com

If the polynomial cannot be factored or if there are no common factors by which to divide, you must use long division.

EXAMPLE 3 Use Long Division

Find $(y^2 + 4y + 12) \div (y + 3)$ by using long division.

Step 1 Divide the first term of the dividend, y^2, by the first term of the divisor, y.

$$\begin{array}{r} y \\ y + 3 \overline{) y^2 + 4y + 12} \\ \underline{(-)\ y^2 + 3y} \\ 1y + 12 \end{array}$$

$y^2 \div y = y$

Multiply y and $y + 3$

Subtract. Bring down the 12.

Step 2 Divide the first term of the partial dividend, $1y$, by the first term of the divisor, y.

$$\begin{array}{r} y + 1 \\ y + 3 \overline{) y^2 + 4y + 12} \\ \underline{(-)\ y^2 + 3y} \\ 1y + 12 \\ \underline{(-)\ y + 3} \\ 9 \end{array}$$

Subtract. Bring down the 12.

Multiply 1 and $y + 3$.

Subtract.

So, $(y^2 + 4y + 12) \div (y + 3)$ is $y + 1$ with a remainder of 9. This answer can be written as $y + 1 + \frac{9}{y + 3}$.

Check Your Progress

3A. $(3x^2 + 9x - 15) \div (x + 5)$

3B. $(n^2 + 6n + 2) \div (n - 2)$

Personal Tutor glencoe.com

Watch Out!

Polynomials When using long division, be sure the dividend is written in standard form. That is, the terms are written so that the exponents decrease from left to right.

$y^2 + 4y + 12$ yes

$4y + y^2 + 12$ no

Real-World EXAMPLE 4 Divide Polynomials to Solve a Problem

PARTIES The expression $5x + 250$ represents the cost of renting a picnic shelter and food for x people. The total cost is divided evenly among all the people except for the two who bought decorations. Find $(5x + 250) \div (x - 2)$ to determine how much each person pays.

$$\begin{array}{r} 5 \\ x - 2\overline{)5x + 250} \\ (-)\ 5x - 10 \\ \hline 260 \end{array}$$

So, $5 + \frac{260}{x - 2}$ represents the amount each person pays.

Check Your Progress

4. GEOMETRY The area of a rectangle is $(2x^2 + 10x - 1)$ square units, and the width is $(x + 1)$ units. What is the length?

Personal Tutor glencoe.com

When a dividend is written in standard form and a power is missing, add a term of that power with a coefficient of zero.

EXAMPLE 5 Insert Missing Terms

Find $(c^3 + 5c - 6) \div (c - 1)$.

$$\begin{array}{rl} c^2 + c + 6 & \\ c - 1\overline{)c^3 + 0c^2 + 5c - 6} & \text{Insert a } c^2\text{-term that has a coefficient of 0.} \\ (-)\ c^3 - c^2 & \text{Multiply } c^2 \text{ and } c - 1. \\ c^2 + 5c & \text{Subtract. Bring down the } 5c. \\ (-)\ c^2 - c & \text{Multiply } c \text{ and } c - 1. \\ 6c - 6 & \text{Subtract. Bring down the } -6. \\ (-)\ 6c - 6 & \text{Multiply 6 and } c - 1. \\ 0 & \text{Subtract.} \end{array}$$

So, $(c^3 + 5c - 6) \div (c - 1) = c^2 + c + 6$.

Check Your Progress Find each quotient.

5A. $(2r^3 + 2r^2 - 4) \div (r - 1)$

5B. $(x^4 + 2x^3 + 6x - 10) \div (x + 2)$

Personal Tutor glencoe.com

Check Your Understanding

Examples 1 and 2 pp. 700–701

Find each quotient.

1. $(8a^2 + 20a) \div 4a$
2. $(4z^3 + 1) \div 2z$
3. $(12n^3 - 6n^2 + 15) \div 6n$
4. $(t^2 + 5t + 4) \div (t + 4)$
5. $(x^2 + 3x - 28) \div (x + 7)$
6. $(x^2 + x - 20) \div (x - 4)$

Example 4 p. 702

7. **CHEMISTRY** The formula $y = \frac{400 + 3x}{50 + x}$ describes a mixture when x liters of a 25% solution are added to a 90% solution. Find $(400 + 3x) \div (50 + x)$.

Examples 3 and 5 pp. 701–702

Find each quotient. Use long division.

8. $(n^2 + 3n + 10) \div (n - 1)$
9. $(4y^2 + 8y + 3) \div (y + 2)$
10. $(4h^3 + 6h^2 - 3) \div (2h + 3)$
11. $(9n^3 - 13n + 8) \div (3n - 1)$

Practice and Problem Solving

● = **Step-by-Step Solutions** begin on page R12.
Extra Practice begins on page 815.

Examples 1 and 2 pp. 700–701

Find each quotient.

12. $(14x^2 + 7x) \div 7x$

13. $(a^3 + 4a^2 - 18a) \div a$

14. $(5q^3 + q) \div q$

15. $(6n^2 - 12n + 3) \div 3n$

16. $(8k^2 - 6) \div 2k$

17. $(9m^2 + 5m) \div 6m$

18. $(a^2 + a - 12) \div (a - 3)$

19. $(x^2 - 6x - 16) \div (x + 2)$

20. $(r^2 - 12r + 11) \div (r - 1)$

21. $(k^2 - 5k - 24) \div (k - 8)$

22. $(y^2 - 36) \div (y^2 + 6y)$

23. $(a^3 - 4a^2) \div (a - 4)$

24. $(c^3 - 27) \div (c - 3)$

25. $(4t^2 - 1) \div (2t + 1)$

26. $(6x^3 + 16x^2 - 60x + 39) \div (2x + 10)$

27. $(2h^3 + 8h^2 - 3h - 12) \div (h + 4)$

Example 4 p. 702

28. **GEOMETRY** The area of a rectangle is $(x^3 - 4x^2)$ square units, and the width is $(x - 4)$ units. What is the length?

29. **MANUFACTURING** The expression $-n^2 + 18n + 850$ represents the number of baseball caps produced by n workers. Find $(-n^2 + 18n + 850) \div n$ to write an expression for average number of caps produced per person.

Examples 3 and 5 pp. 701–702

Find each quotient. Use long division.

30. $(b^2 + 3b - 9) \div (b + 5)$

31. $(a^2 + 4a + 3) \div (a - 1)$

32. $(2y^2 - 3y + 1) \div (y - 2)$

33. $(4n^2 - 3n + 6) \div (n - 2)$

34. $(p^3 - 4p^2 + 9) \div (p - 1)$

35. $(t^3 - 2t - 4) \div (t + 4)$

36. $(6x^3 + 5x^2 + 9) \div (2x + 3)$

37. $(8c^3 + 6c - 5) \div (4c - 2)$

38. **GEOMETRY** The volume of a prism with a triangular base is $10w^3 + 23w^2 + 5w - 2$. The height of the prism is $2w + 1$, and the height of the triangle is $5w - 1$. What is the measure of the base of the triangle? (*Hint*: $V = Bh$)

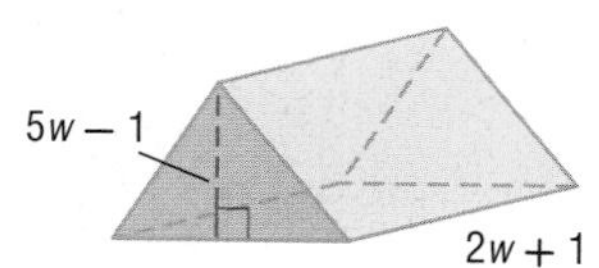

Use long division to find the expression that represents the missing length.

39.

40.

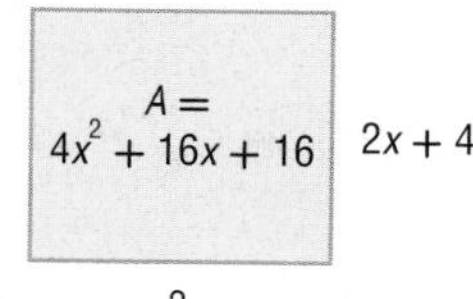

41. Determine the quotient when $x^3 + 11x + 14$ is divided by $x + 2$.

42. What is $14y^5 + 21y^4 - 6y^3 - 9y^2 + 32y + 48$ divided by $2y + 3$?

43. **FUNCTIONS** Consider $f(x) = \frac{3x + 4}{x - 1}$.

a. Rewrite the function as a quotient plus a remainder. Then graph the quotient, ignoring the remainder.

b. Graph the original function using a graphing calculator.

c. How are the graphs of the function and quotient related?

d. What happens to the graph near the excluded value of x?

Real-World Link

The higher the altitude, the lower the temperature at which water boils. Tibet is at a high altitude. The temperature of boiling water in Tibet is lower than that of boiling water in the United States. Tibetans drink their tea while it is still boiling.

Source: *Did You Know?*

44. **ROAD TRIP** The first Ski Club van has been on the road for 20 minutes, and the second van has been on the road for 35 minutes.

    a. Write an expression for the amount of time that each van has spent on the road after an additional t minutes.

    b. Write a ratio for the first van's time on the road to the second van's time on the road and use long division to rewrite this ratio as an expression. Then find the ratio of the first van's time on the road to the second van's time on the road after 60 minutes, 200 minutes.

45. **BOILING POINT** The temperature at which water boils decreases by about 0.9°F for every 500 feet above sea level. The boiling point at sea level is 212°F.

    a. Write an equation for the temperature T at which water boils x feet above sea level.

    b. Mount Whitney, the tallest point in California, is 14,494 feet above sea level. At approximately what temperature does water boil on Mount Whitney?

46. **MULTIPLE REPRESENTATIONS** In this problem, you will use picture models to help divide expressions.

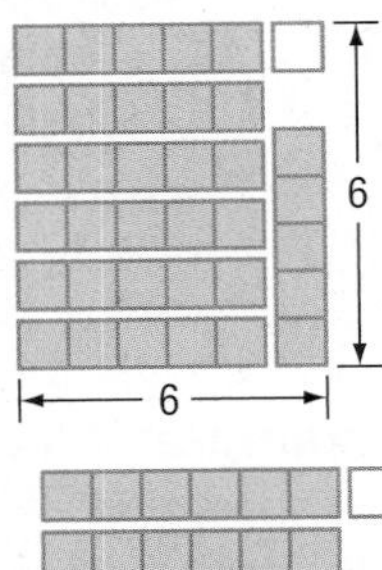

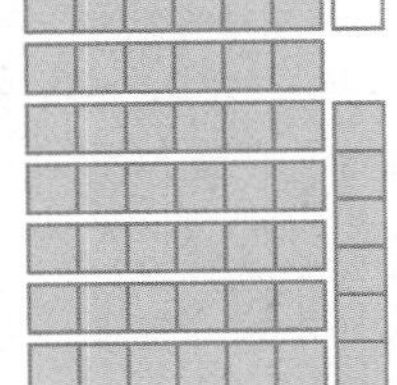

    a. **ANALYTICAL** The first figure models $6^2 \div 7$. Notice that the square is divided into seven equal parts. What are the quotient and the remainder? What division problem does the second figure model?

    b. **CONCRETE** Draw figures for $3^2 \div 4$ and $2^2 \div 3$.

    c. **VERBAL** Do you observe a pattern in the previous exercises? Express this pattern algebraically.

    d. **ANALYTICAL** Use long division to find $x^2 \div (x + 1)$. Does this result match your expression from part **c**?

H.O.T. Problems Use Higher-Order Thinking Skills

47. **FIND THE ERROR** Alvin and Andrea are dividing $c^3 + 6c - 4$ by $c + 2$. Is either of them correct? Explain your reasoning.

Alvin

$$
\begin{array}{r}
c^2 + 4c - 12 \\
c + 2 \overline{) c^3 + 6c - 4} \\
\underline{c^3 + 2c^2} \\
4c^2 - 4 \\
\underline{4c^2 + 8c} \\
-12c \\
\underline{-12c - 24} \\
24
\end{array}
$$

Andrea

$$
\begin{array}{r}
c^2 - 2c + 10 \\
c + 2 \overline{) c^3 + 0c^2 + 6c - 4} \\
\underline{c^3 + 2c^2} \\
-2c^2 + 6c \\
\underline{-2c^2 - 4c} \\
10c - 4 \\
\underline{10c + 20} \\
-24
\end{array}
$$

48. **CHALLENGE** The quotient of two polynomials is $4x^2 - x - 7 + \frac{11x + 15}{x^2 + x + 2}$. What are the polynomials?

49. **OPEN ENDED** Write a division problem involving polynomials that you would solve by using long division. Explain your answer.

50. **WRITING IN MATH** Describe the steps to find $(w^2 - 2w - 30) \div (w + 7)$.

Standardized Test Practice

51. Simplify $\frac{21x^3 - 35x^2}{7x}$.

A $3x^2 - 5x$
B $4x^2 - 6x$
C $3x - 5$
D $5x - 3$

52. EXTENDED RESPONSE The box shown is designed to hold rice.

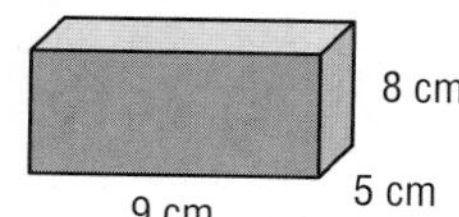

a. How much rice would fit in the box?

b. What is the area of the label on the box, if the label covers all surfaces?

53. Simplify $\frac{x^2 + 7x + 12}{x^2 + 5x + 6}$.

F $x + 4$
G $\frac{x+4}{x+2}$
H $x + 2$
J $\frac{x+2}{x+4}$

54. Susana bought cards at 6 for \$10. She decorated them and sold them at 4 for \$10. She made \$60 in profit. How many cards did she sell?

A 53
B 25
C 60
D 72

Spiral Review

Find each product. (Lesson 11-4)

55. $\frac{3x^3}{8x} \cdot \frac{16}{x}$
56. $\frac{3ad}{4c^4} \cdot \frac{8c^2}{6d}$
57. $\frac{t^2}{(t-4)(t+4)} \cdot \frac{t-4}{6t}$
58. $\frac{10}{r-2} \cdot \frac{r^2-4}{2}$

Find the zeros of each function. (Lesson 11-3)

59. $f(x) = \frac{x+2}{x^2 - 6x + 8}$
60. $f(x) = \frac{x^2 - 3x - 4}{x^2 - x - 12}$
61. $f(x) = \frac{x^2 + 6x + 9}{x^2 - 9}$

62. SHADOWS A 25-foot flagpole casts a shadow that is 10 feet long and a nearby building casts a shadow that is 26 feet long. How tall is the building? (Lesson 10-7)

Solve each equation. Check your solution. (Lesson 10-4)

63. $\sqrt{h} = 9$
64. $\sqrt{x+3} = -5$
65. $3 + 5\sqrt{n} = 18$
66. $\sqrt{x-5} = 2\sqrt{6}$

Solve each equation by using the Quadratic Formula. Round to the nearest tenth if necessary. (Lesson 9-5)

67. $v^2 + 12v + 20 = 0$
68. $3t^2 - 7t - 20 = 0$
69. $5y^2 - y - 4 = 0$
70. $2x^2 + 98 = 28x$
71. $2n^2 - 7n - 3 = 0$
72. $2w^2 = -(7w + 3)$

73. THEATER The drama club is building a backdrop using arches with a shape that can be represented by the function $f(x) = -x^2 + 2x + 8$, where x is the length of the arch in feet. The region under each arch is to be covered with fabric. (Lesson 9-2)

a. Graph the quadratic function and determine its x-intercepts.

b. What is the height of the arch?

Skills Review

Find each sum. (Lesson 7-5)

74. $(3a^2 + 2a - 12) + (8a + 7 - 2a^2)$
75. $(2c^3 + 3cd - d^2) + (-5cd - 2c^3 + 2d^2)$

Find the least common multiple for each set of numbers. (Concepts and Skills Bank Lesson 2)

76. 2, 4, 6
77. 3, 6, 8
78. 5, 12, 15
79. 14, 18, 24

11-6 Adding and Subtracting Rational Expressions

Then
You added and subtracted polynomials. (Lesson 7-5)

Now
- Add and subtract rational expressions with like denominators.
- Add and subtract rational expressions with unlike denominators.

New Vocabulary
least common multiple (LCM)
least common denominator (LCD)

Math Online
glencoe.com
- Extra Examples
- Personal Tutor
- Self-Check Quiz
- Homework Help

Why?

A survey asked families how often they eat takeout. To determine the fraction of those surveyed who eat takeout more than once a week, you can add. Remember that percents can be written as fractions with denominators of 100.

2–3 times a week	plus	daily	equals	more than once a week.
$\frac{30}{100}$	$+$	$\frac{8}{100}$	$=$	$\frac{38}{100}$

Thus, $\frac{38}{100}$ or 38% eat takeout more than once a week.

How Many Times a Week Families Eat Takeout

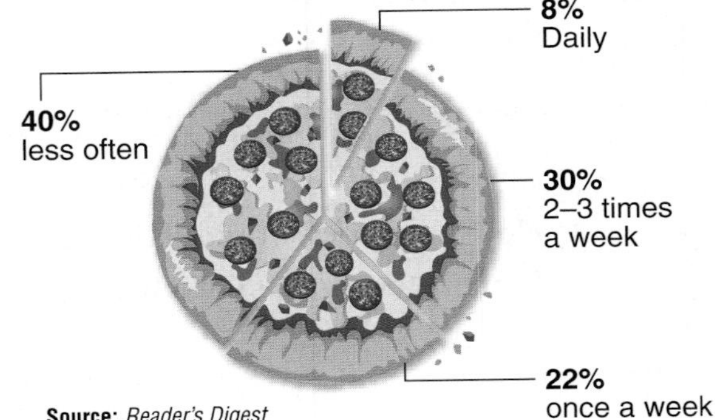

Source: *Reader's Digest*

Add and Subtract Rational Expressions with Like Denominators To add or subtract rational expressions that have the same denominator, add or subtract the numerators and write the sum or difference over the common denominator.

Key Concept — For Your FOLDABLE

Add or Subtract Rational Expressions with Like Denominators

Let a, b, and c be polynomials with $c \neq 0$.

$$\frac{a}{c} + \frac{b}{c} = \frac{a + b}{c} \qquad \frac{a}{c} - \frac{b}{c} = \frac{a - b}{c}$$

EXAMPLE 1 Add Rational Expressions with Like Denominators

Find $\frac{5n}{n+3} + \frac{15}{n+3}$.

$\frac{5n}{n+3} + \frac{15}{n+3} = \frac{5n + 15}{n + 3}$ **The common denominator is $n + 3$.**

$= \frac{5(n+3)}{n+3}$ **Factor the numerator.**

$= \frac{5\overset{1}{\cancel{(n+3)}}}{\underset{1}{\cancel{n+3}}}$ **Divide by the common factor, $n + 3$.**

$= \frac{5}{1}$ or 5 **Simplify.**

Check Your Progress

Find each sum.

1A. $\frac{8c}{6} + \frac{5c}{6}$

1B. $\frac{4t}{5xy} + \frac{7}{5xy}$

1C. $\frac{3y}{3+y} + \frac{y^2}{3+y}$

Personal Tutor glencoe.com

Watch Out!

Common Terms Remember that every term of the numerator and the denominator must be multiplied or divided by the same number for the fraction to be equivalent to the original.

EXAMPLE 2 Subtract Rational Expressions with Like Denominators

Find $\frac{3m-5}{m+4} - \frac{4m+2}{m+4}$.

$\frac{3m-5}{m+4} - \frac{4m+2}{m+4} = \frac{(3m-5)-(4m+2)}{m+4}$ The common denominator is $m + 4$.

$= \frac{(3m-5)+[-(4m+2)]}{m+4}$ The additive inverse of $(4m + 2)$ is $-(4m + 2)$.

$= \frac{3m-5-4m-2}{m+4}$ Distributive Property

$= \frac{-m-7}{m+4}$ Simplify.

Check Your Progress

Find each difference.

2A. $\frac{2h+4}{h+1} - \frac{5+h}{h+1}$

2B. $\frac{17h+4}{15h-5} - \frac{2h-6}{15h-5}$

Personal Tutor glencoe.com

You can sometimes use additive inverses to form like denominators.

EXAMPLE 3 Inverse Denominators

Find $\frac{3n}{n-4} + \frac{6n}{4-n}$.

$\frac{3n}{n-4} + \frac{6n}{4-n} = \frac{3n}{n-4} + \frac{6n}{-(n-4)}$ Rewrite $4 - n$ as $-(n - 4)$.

$= \frac{3n}{n-4} - \frac{6n}{n-4}$ Rewrite so the denominators are the same.

$= \frac{3n-6n}{n-4}$ or $-\frac{3n}{n-4}$ Subtract the numerators and simplify.

Check Your Progress

Find each sum or difference.

3A. $\frac{t^2}{t-3} + \frac{3}{3-t}$

3B. $\frac{2p}{p-1} - \frac{2p}{1-p}$

Personal Tutor glencoe.com

Add and Subtract with Unlike Denominators The **least common multiple (LCM)** is the least number that is a multiple of two or more numbers or polynomials.

EXAMPLE 4 LCMs of Polynomials

Find the LCM of each pair of polynomials.

a. $6x$ and $4x^3$

Step 1 Find the prime factors of each expression.

$6x = 2 \cdot 3 \cdot x$ $4x^3 = 2 \cdot 2 \cdot x \cdot x \cdot x$

Step 2 Use each prime factor, 2, 3, and x, the greatest number of times it appears in either of the factorizations.

$6x = 2 \cdot 3 \cdot x$ $4x^3 = 2 \cdot 2 \cdot x \cdot x \cdot x$

LCM $= 2 \cdot 2 \cdot 3 \cdot x \cdot x \cdot x$ or $12x^3$

Review Vocabulary

Factored Form A monomial is in factored form when it is expressed as the product of prime numbers and variables, and no variable has an exponent greater than 1. (Lesson 8-1)

b. $n^2 + 5n + 4$ **and** $(n + 1)^2$

$n^2 + 5n + 4 = (n + 1)(n + 4)$ **Factor each expression.**

$(n + 1)^2 = (n + 1)(n + 1)$

$(n + 1)$ is a factor twice in the second expression. $(n + 4)$ is a factor once.

LCM $= (n + 1)(n + 1)(n + 4)$ or $(n + 1)^2(n + 4)$

Check Your Progress

4A. $8m^2t$ and $12m^2t^3$

4B. $x^2 - 2x - 8$ and $x^2 - 5x - 14$

Personal Tutor glencoe.com

To add or subtract fractions with unlike denominators, you need to rename the fractions using the least common multiple of the denominators, called the **least common denominator (LCD)**.

Key Concept

For Your FOLDABLE

Add or Subtract Rational Expressions with Unlike Denominators

Step 1 Find the LCD.

Step 2 Write each rational expression as an equivalent expression with the LCD as the denominator.

Step 3 Add or subtract the numerators and write the result over the common denominator.

Step 4 Simplify if possible.

StudyTip

Checking Answers You can check whether you have simplified a rational expression correctly by substituting values, but this does not guarantee that the expressions are always equal. If the results are different, check for an error.

EXAMPLE 5 Add Rational Expressions with Unlike Denominators

Find $\frac{3t + 2}{t^2 - 2t - 3} + \frac{t + 1}{t - 3}$.

Find the LCD. Since $t^2 - 2t - 3 = (t - 3)(t + 1)$, the LCD is $(t - 3)(t + 1)$.

$\frac{3t + 2}{t^2 - 2t - 3} + \frac{t + 1}{t - 3} = \frac{3t + 2}{(t - 3)(t + 1)} + \frac{t + 1}{t - 3}$ **Factor $t^2 - 2t - 3$.**

$= \frac{3t + 2}{(t - 3)(t + 1)} + \frac{t + 1}{t - 3}\left(\frac{t + 1}{t + 1}\right)$ **Write $\frac{t+1}{t-3}$ using the LCD.**

$= \frac{3t + 2}{(t - 3)(t + 1)} + \frac{t^2 + 2t + 1}{(t - 3)(t + 1)}$ **Simplify.**

$= \frac{3t + 2 + t^2 + 2t + 1}{(t - 3)(t + 1)}$ **Add the numerators.**

$= \frac{t^2 + 5t + 3}{(t - 3)(t + 1)}$ **Simplify.**

Check Your Progress Find each sum.

5A. $\frac{4d^2}{d} + \frac{d + 2}{d^2}$

5B. $\frac{b + 3}{b} + \frac{b - 5}{b + 1}$

Personal Tutor glencoe.com

The formula time $= \frac{\text{distance}}{\text{rate}}$ is helpful in solving real-world applications.

Real-World Link

The distance a hang glider can travel is determined by its *glide ratio*, or the ratio of the forward distance traveled to the vertical distance dropped.

Source: HowStuffWorks

Real-World EXAMPLE 6 Add Rational Expressions

HANG GLIDING For the first 5000 meters, a hang glider travels at a rate of x meters per minute. Then, due to a stronger wind, it travels 6000 meters at a speed that is 3 times as fast.

a. Write an expression to represent how much time the hang glider is flying.

Understand For the first 5000 meters, the hang glider's speed is x. For the last 6000 meters, the hang glider's speed is $3x$.

Plan Use the formula $d = r \times t$ or $t = \frac{d}{r}$ to represent the time t of each section of the hang glider's trip, with rate r and distance d.

Solve Time to fly 5000 meters: $\frac{d}{r} = \frac{5000}{x}$ **$d = 5000$, $r = x$**

Time to fly 6000 meters: $\frac{d}{r} = \frac{6000}{3x}$ **$d = 6000$, $r = 3x$**

Total flying time: $\frac{5000}{x} + \frac{6000}{3x}$

$\frac{5000}{x} + \frac{6000}{3x} = \frac{5000}{x}\left(\frac{3}{3}\right) + \frac{6000}{3x}$ **The LCD is $3x$.**

$= \frac{15{,}000}{3x} + \frac{6000}{3x}$ **Multiply.**

$= \frac{\cancel{21{,}000}^{7000}}{\cancel{3}_{1}x}$ or $\frac{7000}{x}$ **Simplify.**

Check $\frac{5000}{x} + \frac{6000}{3x} = \frac{5000}{1} + \frac{6000}{3(1)}$ **Let $x = 1$ in the original expression.**

$= 5000 + 2000$ or 7000 **Simplify.**

$\frac{7000}{x} = \frac{7000}{1}$ or 7000 **Let $x = 1$ in the answer expression. Simplify.**

Since the expressions have the same value for $x = 1$, the answer is reasonable. ✓

b. If the hang glider is flying at a rate of 600 meters per minute for the first 5000 meters, find the total amount of time that the hang glider is flying.

$\frac{7000}{x} = \frac{7000}{600}$ **Substitute 600 for x in the expression.**

≈ 11.7 **Simplify.**

So, the hang glider is flying for approximately 11.7 minutes.

c. If the hang glider flew for approximately 15 minutes, find the rate the hang glider flew for the first 5000 meters.

$\frac{7000}{x} = 15$ **Set the expression equal to 15.**

$7000 = 15x$ **Multiply each side by x.**

$446.7 \approx x$ **Divide each side by 15 and simplify.**

The hang glider was flying at a rate of 466.7 meters per minute.

Check Your Progress

6. TRAINS A train travels 5 miles from Lynbrook to Long Beach and then back. The train travels about 1.2 times as fast returning from Long Beach. If r is the train's speed from Lynbrook to Long Beach, write and simplify an expression for the total time of the round trip.

Personal Tutor glencoe.com

To subtract rational expressions with unlike denominators, rename the expressions using the LCD. Then subtract the numerators.

EXAMPLE 7 Subtract Rational Expressions with Unlike Denominators

Find $\frac{5}{x} - \frac{2x+1}{4x}$.

$$\frac{5}{x} - \frac{2x+1}{3x} = \frac{5}{x}\left(\frac{4}{4}\right) - \frac{2x+1}{4x}$$ **Write $\frac{5}{x}$ using the LCD, 4x.**

$$= \frac{20}{4x} - \frac{2x+1}{4x}$$ **Simplify.**

$$= \frac{20 - (2x+1)}{4x}$$ **Subtract the numerators.**

$$= \frac{20 - 2x - 1}{4x} \text{ or } \frac{19 - 2x}{4x}$$ **Simplify.**

StudyTip

Simplifying Answers When simplifying a rational expression, you can leave the denominator in factored form, or multiply the terms.

Check Your Progress

Find each difference.

7A. $\frac{6}{t+3} - \frac{7}{t}$

7B. $\frac{y}{y-3} - \frac{2}{y^2+y-12}$

Personal Tutor glencoe.com

Check Your Understanding

Examples 1–3 pp. 706–707

Find each sum or difference.

1. $\frac{3}{7n} + \frac{2}{7n}$

2. $\frac{x+8}{2} + \frac{x}{2}$

3. $\frac{14r}{9-r} - \frac{2r}{r-9}$

4. $\frac{7}{5t} - \frac{3+t}{5t}$

Example 4 pp. 707–708

Find the LCM of each pair of polynomials.

5. $3t, 8t^2$

6. $5m + 15, 2m + 6$

7. $(x^2 - 8x + 7), (x^2 + x - 2)$

Examples 5 and 7 pp. 708 and 710

Find each sum or difference.

8. $\frac{6}{n^4} + \frac{2}{n^2}$

9. $\frac{3}{4x} + \frac{2}{5y}$

10. $\frac{4}{5n} - \frac{1}{10n^3}$

11. $\frac{8}{3c} - \frac{-5}{6d}$

12. $\frac{a}{a+4} + \frac{6}{a+2}$

13. $\frac{x}{x-3} - \frac{3}{x+2}$

Example 6 p. 709

14. **EXERCISE** Joseph walks 10 times around the track at a rate of x laps per hour. He runs 8 times around the track at a rate of $3x$ laps per hour. Write and simplify an expression for the total time it takes him to go around the track 18 times.

Practice and Problem Solving

● = **Step-by-Step Solutions** begin on page R12. **Extra Practice** begins on page 815.

Examples 1–3 pp. 706–707

Find each sum or difference.

15. $\frac{a}{4} + \frac{3a}{4}$

16. $\frac{1}{6m} + \frac{5m}{6m}$

17. $\frac{5y}{6} - \frac{y}{6}$

18. $\frac{11}{4r} - \frac{-1}{4r}$

19. $\frac{8b}{ab} + \frac{3a}{ab}$

20. $\frac{t+2}{3} + \frac{t+5}{3}$

21. $\frac{3c-7}{2c-1} + \frac{2c+1}{1-2c}$

22. $\frac{15x}{33x-9} + \frac{3}{9-33x}$

23. $\frac{n+6}{10} - \frac{n+1}{10}$

24. $\frac{5x+2}{2x+5} - \frac{x-8}{2x+5}$

25. $\frac{w+2}{8w} - \frac{2w-3}{8w}$

26. $\frac{3a+1}{a-1} - \frac{a+4}{a-1}$

Example 4
pp. 707–708

Find the LCM of each pair of polynomials.

27. x^3y, x^2y^2 **28.** $5ab, 10b$ **29.** $(3r - 1), (r + 2)$

30. $2n - 10, 4n - 20$ **31.** $(x^2 + 9x + 18), x + 3$ **32.** $(k^2 - 2k - 8), (k + 2)^2$

Examples 5 and 7
pp. 708 and 710

Find each sum or difference.

33. $\frac{5}{4x} + \frac{1}{10x}$ **34.** $\frac{6}{r} + \frac{2}{r^2}$ **35.** $\frac{3}{2a} + \frac{1}{5b}$

36. $\frac{6g}{g+5} - \frac{g-2}{2g}$ **37.** $\frac{7}{4k+8} - \frac{k}{k+2}$ **38.** $\frac{5}{2d+2} - \frac{d}{d+5}$

39. $\frac{-2}{7r} + \frac{4}{t}$ **40.** $\frac{n}{n-2} + \frac{n}{n+1}$ **41.** $\frac{d}{d+5} + \frac{7}{d-1}$

42. $\frac{4}{a} - \frac{1}{3a}$ **43.** $\frac{6}{5t^2} - \frac{2}{3t}$ **44.** $\frac{7}{4r} - \frac{3}{t}$

45. $\frac{w-3}{w^2-w-20} + \frac{w}{w+4}$ **46.** $\frac{n}{2n+10} + \frac{1}{n^2-25}$

47. $\frac{2x}{x^2+8x+15} - \frac{x+3}{x+5}$ **48.** $\frac{r-3}{r^2+6r+9} - \frac{r-9}{r^2-9}$

Example 6
p. 709

49. TRAVEL Grace walks to her friend's house 2 miles away and then jogs back home. Her jogging speed is 2.5 times her walking speed w.

a. Write and simplify an expression to represent the amount of time Grace spends going to and coming from her friend's house.

b. If Grace walks about 3.5 miles per hour, how many minutes did she spend going to and from her friend's house?

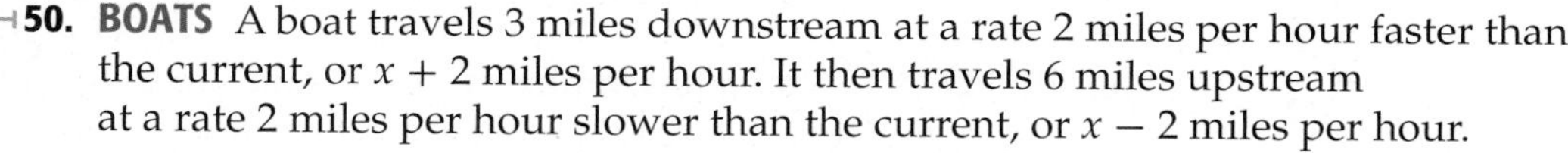

50. BOATS A boat travels 3 miles downstream at a rate 2 miles per hour faster than the current, or $x + 2$ miles per hour. It then travels 6 miles upstream at a rate 2 miles per hour slower than the current, or $x - 2$ miles per hour.

a. Write and simplify an expression to represent the total time it takes the boat to travel 3 miles downstream and 6 miles upstream.

b. If the rate of the current x is 4 miles per hour, how long did it take the boat to travel the 9 miles?

Real-World Link

In a recent year, there were nearly 12.8 million boats registered in the U.S.

Source: USCG Boating

51. SCHOOL Mr. Kim had 18 more geometry tests to grade than algebra tests. He graded 12 tests on Saturday and 20 tests on Sunday. Write an expression for the fraction of tests he graded if a represents the number of algebra tests.

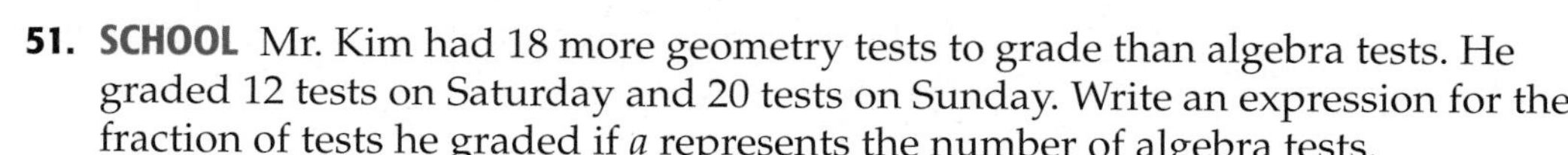

52. PLAYS A total of 1248 people attended the school play. The same number x attended each of the two Sunday performances. There were twice as many people at the Saturday performance than at both Sunday performances. Write an expression to represent the fraction of people who attended on Saturday.

Find each sum or difference.

53. $\frac{x+5}{x^2-4} - \frac{3}{x^2-4}$ **54.** $\frac{18y}{9y+2} - \frac{-4}{-2-9y}$

55. $\frac{k^2-26}{k-5} - \frac{1}{5-k}$ **56.** $\frac{8}{c-1} + \frac{c}{1-c}$

57. $\frac{2}{x-1} + \frac{3}{x+1} - \frac{4x-2}{x^2-1}$ **58.** $\frac{x^2-x-12}{x^2-11x+30} - \frac{x-4}{18-x}$

59. $\frac{a^2-5a}{3a-18} - \frac{7a-36}{3a-18}$ **60.** $\frac{8n-3}{n^2+8n+12} - \frac{5n-9}{n^2+8n+12}$

61. $\frac{x^2-16}{x^3} + \frac{x^3+1}{x^4}$ **62.** $\frac{x}{7x-3} + \frac{x+2}{15x+30}$

63. $\frac{5x}{3x^2+19x-14} - \frac{1}{9x^2-12x+4}$ **64.** $\frac{2x+7}{x^2-y^2} + \frac{-5}{x^2-2xy+y^2}$

Real-World Link

A regular triathlon includes a 1.5-kilometer swim, a 40-kilometer bike ride, and a 10-kilometer run. The first Olympic triathlons were held in 2000 in Sydney, Australia.

Source: USA Triathlon

65. TRIATHLONS In a sprint triathlon, athletes swim 400 meters, bike 20 kilometers, and run 5 kilometers. An athlete bikes 12 times as fast as she swims and runs 5 times as fast as she swims.

a. Simplify $\frac{400}{x} + \frac{20{,}000}{12x} + \frac{5000}{5x}$, an expression that represents the time it takes the athlete to complete the sprint triathlon.

b. If the athlete swims 40 meters per minute, find the total time it takes her to complete the triathlon.

GEOMETRY **Write an expression for the perimeter of each figure.**

66.

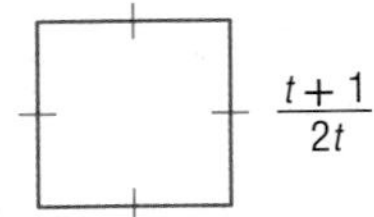

67.

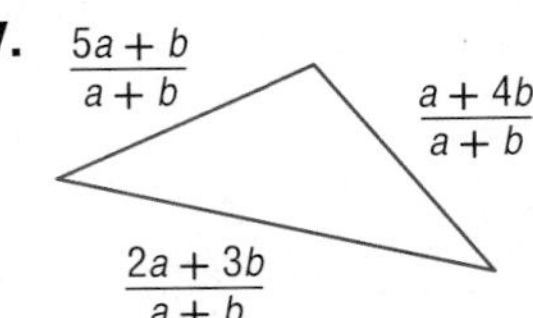

68.

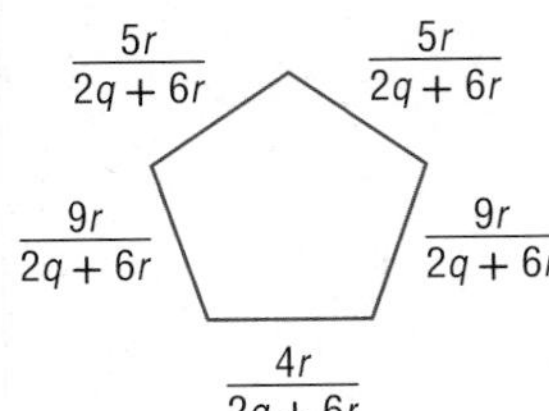

69. BIKES Marina rides her bike at an average rate of 10 miles per hour. On one day, she rides 9 miles and then rides around a large loop x miles long. On the second day, she rides 5 miles and then rides around the loop three times.

a. Write an expression to represent the total time she spent riding her bike on those two days. (*Hint:* Use $t = \frac{d}{r}$, where t is time, d is distance, and r is rate.) Then simplify the expression.

b. If the loop is 2 miles long, how long did Marina ride on those two days?

70. TRAVEL The Showalter family drives 80 miles to a college football game. On the trip home, their average speed is about 3 miles per hour slower.

a. Let x represent the average speed of the car on the way to the game. Write and simplify an expression to represent the total time it took driving to the game and then back home.

b. If their average speed on the way to the game was 68 miles per hour, how long did it take the Showalter family to drive to the game and back? Round to the nearest tenth.

H.O.T. Problems

Use **H**igher-**O**rder **T**hinking Skills

71. CHALLENGE Find $\left(\frac{4}{7y-2} + \frac{7y}{2-7y}\right)\left(\frac{y+5}{6} - \frac{y+3}{6}\right)$.

72. WRITING IN MATH Describe in words the steps you use to find the LCM in an addition or subtraction of rational expressions with unlike denominators.

73. CHALLENGE Is the following statement *sometimes*, *always*, or *never* true? Explain.

$$\frac{a}{x} + \frac{b}{y} = \frac{ay + bx}{xy}$$

74. OPEN ENDED Describe a real-life situation that could be expressed by adding two rational expressions that are fractions. Explain what the denominator and numerator represent in both expressions.

75. WRITING IN MATH Describe how to add rational expressions with denominators that are additive inverses.

Standardized Test Practice

76. SHORT RESPONSE An object is launched upwards at 19.6 meters per second from a 58.8-meter-tall platform. The equation for the object's height h, in meters, at time t seconds after launch is $h(t) = -4.9t^2 + 19.6t + 58.8$. How long after the launch does the object strike the ground?

77. Simplify $\frac{2}{5} + \frac{3}{25} + \frac{1}{10}$.

A $\frac{2}{5}$

B $\frac{3}{5}$

C $\frac{31}{50}$

D $\frac{5}{31}$

78. STATISTICS Courtney has grades of 84, 65, and 76 on three math tests. What grade must she earn on the next test to have an average of exactly 80 for the four tests?

F 84

G 80

H 98

J 95

79. Simplify $\frac{2}{x} + \frac{3}{x^2} + \frac{1}{2x}$.

A $\frac{3x + 2}{x^2}$

B $\frac{6}{2x^2}$

C $\frac{5x + 6}{2x^2}$

D $\frac{6 + x}{x^2}$

Spiral Review

Find each quotient. (Lesson 11-5)

80. $(6x^2 + 10x) \div 2x$

81. $(15y^3 + 14y) \div 3y$

82. $(10a^3 - 20a^2 + 5a) \div 5a$

Convert each rate. Round to the nearest tenth if necessary. (Lesson 11-4)

83. 23 feet per second to miles per hour

84. 118 milliliters per second to quarts per hour (*Hint*: 1 liter $\approx$ 1.06 quarts)

Find the length of the missing side. If necessary, round to the nearest hundredth. (Lesson 10-5)

85.
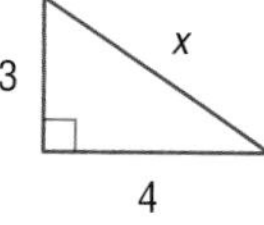

86.
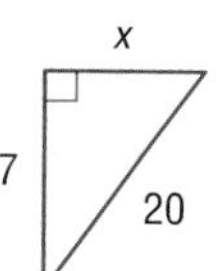

87.
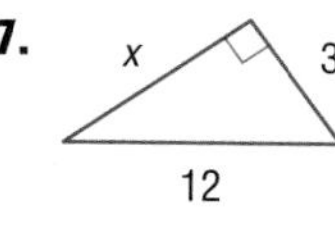

88. AMUSEMENT RIDE The height h in feet of a car above the exit ramp of a free-fall ride can be modeled by $h(t) = -16t^2 + s$. t is the time in seconds after the car drops, and s is the starting height of the car in feet. If the designer wants the ride to last 3 seconds, what should be the starting height in feet? (Lesson 8-6)

Express each number in scientific notation. (Lesson 7-3)

89. 12,300

90. 0.0000375

91. 1,255,000

92. FINANCIAL LITERACY Ruben has \$13 to order pizza. The pizza costs \$7.50 plus \$1.25 per topping. He plans to tip 15% of the total cost. Write and solve an inequality to find out how many toppings he can order. (Lesson 5-3)

Skills Review

Find each quotient. (Lesson 11-4)

93. $\frac{12}{3x^2} \div \frac{6}{x}$

94. $\frac{g^4}{2} \div \frac{g^3}{8d^2}$

95. $\frac{4y - 8}{y + 1} \div (y - 2)$

11-7 Mixed Expressions and Complex Fractions

Then

You simplified rational expressions. (Lesson 11-3)

Now

- Simplify mixed expressions.
- Simplify complex fractions.

New Vocabulary

mixed expression
complex fraction

Math Online

glencoe.com

- Extra Examples
- Personal Tutor
- Self-Check Quiz
- Homework Help

Why?

A Top Fuel dragster can cover $\frac{1}{4}$ mile in $4\frac{2}{5}$ seconds. The average speed in miles per second can be described by the expression below. It is called a *complex fraction*.

$$\frac{\frac{1}{4}\text{ mile}}{4\frac{2}{5}\text{ seconds}}$$

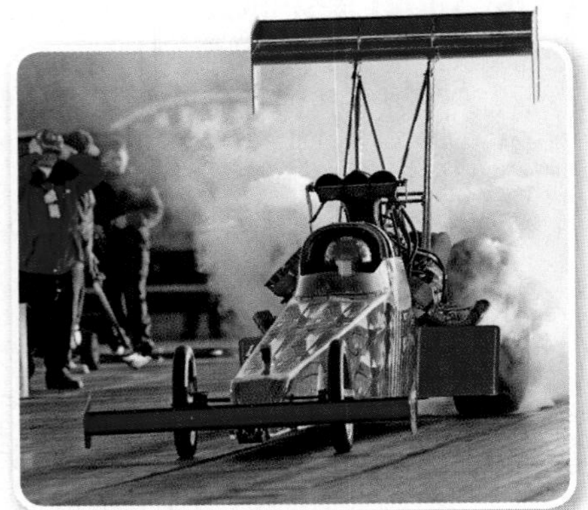

Simplify Mixed Expressions An expression like $2 + \frac{4}{x+1}$ is called a **mixed expression** because it contains the sum of a monomial, 2, and a rational expression, $\frac{4}{x+1}$. You can use the LCD to change a mixed expression to a rational expression.

EXAMPLE 1 Change Mixed Expression to Rational Expressions

Write $2 + \frac{4}{x-1}$ as a rational expression.

$2 + \frac{4}{x-1} = \frac{2(x-1)}{x-1} + \frac{4}{x-1}$ — The LCD is $x - 1$.

$= \frac{2(x-1)+4}{x-1}$ — Add the numerators.

$= \frac{2x-2+4}{x-1}$ — Distributive Property

$= \frac{2x+2}{x-1}$ — Simplify.

Check Your Progress

Write each mixed expression as a rational expression.

1A. $2 + \frac{5}{x}$

1B. $\frac{6y}{4y+8} + 5y$

Personal Tutor glencoe.com

Simplify Complex Fractions A **complex fraction** has one or more fractions in the numerator or denominator. You can simplify by using division.

numerical complex fraction

$$\frac{\frac{2}{3}}{\frac{5}{8}} = \frac{2}{3} \div \frac{5}{8}$$
$$= \frac{2}{3} \times \frac{8}{5}$$
$$= \frac{16}{15}$$

algebraic complex fraction

$$\frac{\frac{a}{b}}{\frac{c}{d}} = \frac{a}{b} \div \frac{c}{d}$$
$$= \frac{a}{b} \times \frac{d}{c}$$
$$= \frac{ad}{bc}$$

To simplify a complex fraction, write it as a division expression. Then find the reciprocal of the second expression and multiply.

Real-World EXAMPLE 2 Use Complex Fractions to Solve Problems

RACING Refer to the application at the beginning of the lesson. Find the average speed of the Top Fuel dragster in miles per minute.

$$\frac{\frac{1}{4}\text{ mile}}{4\frac{2}{5}\text{ seconds}} = \frac{\frac{1}{4}\text{ mile}}{4\frac{2}{5}\cancel{\text{ seconds}}} \times \frac{60\cancel{\text{ seconds}}}{1\text{ minute}}$$ **Convert seconds to minutes. Divide by common units.**

$$= \frac{\frac{1}{4} \times 60}{4\frac{2}{5}}$$ **Simplify.**

$$= \frac{\frac{60}{4}}{\frac{22}{5}}$$ **Express each term as an improper fraction.**

$$= \frac{\overset{15}{\cancel{60}} \times 5}{\underset{1}{\cancel{4}} \times 22}$$ **Use the rule $\frac{\frac{a}{b}}{\frac{c}{d}} = \frac{ad}{bc}$.**

$$= \frac{75}{22} \text{ or } 3\frac{9}{22}$$ **Simplify.**

So, the average speed of the Top Fuel dragster is $3\frac{9}{22}$ miles per minute.

Check Your Progress

2. RACING Refer to the information about the Jr. Dragster at the left. What is the average speed of the car in feet per second?

Personal Tutor glencoe.com

Real-World Link

A Jr. Dragster is a half-scale verson of a Top Fuel dragster. This car, which can go $\frac{1}{8}$ mile in $7\frac{9}{10}$ seconds, is designed to be driven by kids ages 8–17 in the NHRA Jr. Drag Racing League.

Source: NHRA

To simplify complex fractions, you can either use the rule as in Example 2, or you can rewrite the fraction as a division expression, as shown below.

EXAMPLE 3 Complex Fractions Involving Monomials

Simplify $\frac{\frac{8t^2}{v}}{\frac{4t}{v^3}}$.

$$\frac{\frac{8t^2}{v}}{\frac{4t}{v^3}} = \frac{8t^2}{v} \div \frac{4t}{v^3}$$ **Write as a division expression.**

$$= \frac{8t^2}{v} \times \frac{v^3}{4t}$$ **To divide, multiply by the reciprocal.**

$$= \frac{\overset{2t}{\cancel{8t^2}}}{\underset{1}{\cancel{v}}} \times \frac{\overset{2}{\cancel{v^3}}}{\underset{1}{\cancel{4t}}} \text{ or } 2tv^2$$ **Divide by the common factors 4*t* and *v* and simplify.**

Check Your Progress

Simplify each expression.

3A. $\frac{\frac{g^3h}{b}}{\frac{gh^3}{b^2}}$

3B. $\frac{\frac{-24m^3t^5}{p^2h}}{\frac{16pm^2}{t^4h}}$

Personal Tutor glencoe.com

Complex fractions may also involve polynomials.

EXAMPLE 4 Complex Fractions Involving Polynomials

Simplify each expression.

a. $\dfrac{\frac{2}{y+3}}{\frac{5}{y^2-9}}$

$$\dfrac{\frac{2}{y+3}}{\frac{5}{y^2-9}} = \frac{2}{y+3} \div \frac{5}{y^2-9}$$ **Write as a division expression.**

$$= \frac{2}{y+3} \times \frac{y^2-9}{5}$$ **To divide, multiply by the reciprocal.**

$$= \frac{2}{y+3} \times \frac{(y-3)(y+3)}{5}$$ **Factor $y^2 - 9$.**

$$= \frac{2}{\cancel{y+3}_{1}} \times \frac{(y-3)\cancel{(y+3)}^{1}}{5}$$ **Divide by the GCF, $y + 3$.**

$$= \frac{2(y-3)}{5}$$ **Simplify.**

StudyTip

Factoring When simplifying fractions involving polynomials, factor the numerator and the denominator of each expression if possible.

b. $\dfrac{\frac{n^2+7n-18}{n^2-2n+1}}{\frac{n^2-81}{n-1}}$

$$\dfrac{\frac{n^2+7n-18}{n^2-2n+1}}{\frac{n^2-81}{n-1}} = \frac{n^2+7n-18}{n^2-2n+1} \div \frac{n^2-81}{n-1}$$ **Write as a division expression.**

$$= \frac{n^2+7n-18}{n^2-2n+1} \times \frac{n-1}{n^2-81}$$ **Multiply by the reciprocal.**

$$= \frac{(n-2)(n+9)}{(n-1)(n-1)} \times \frac{n-1}{(n-9)(n+9)}$$ **Factor the polynomials.**

$$= \frac{(n-2)\cancel{(n+9)}^{1}}{(n-1)\cancel{(n-1)}_{1}} \times \frac{\cancel{n-1}^{1}}{(n-9)\cancel{(n+9)}_{1}}$$ **Divide out the common factors.**

$$= \frac{n-2}{(n-1)(n-9)}$$ **Simplify.**

Check Your Progress

4A. $\dfrac{\frac{a+7}{4}}{\frac{a^2-49}{10}}$

4B. $\dfrac{\frac{x+4}{x-1}}{\frac{x^2+6x+8}{2x-2}}$

4C. $\dfrac{\frac{c-d}{j+p}}{\frac{c^2-d^2}{j^2-p^2}}$

4D. $\dfrac{\frac{n^2+4n-21}{n^2-9n+18}}{\frac{n^2+3n-28}{n^2-10n+24}}$

Personal Tutor glencoe.com

Check Your Understanding

Example 1 p. 714

Write each mixed expression as a rational expression.

1. $\frac{2}{n} + 4$
2. $r + \frac{1}{3r}$
3. $6 + \frac{5}{t+1}$
4. $\frac{x+7}{2x} - 5x$

Example 2 p. 715

5. **ROWING** Rico rowed a canoe $2\frac{1}{2}$ miles in $\frac{1}{3}$ hour.
   a. Write an expression to represent his speed in miles per hour.
   b. Simplify the expression to find his average speed.

Examples 3 and 4 pp. 715–716

Simplify each expression.

6. $\dfrac{2\frac{1}{3}}{1\frac{2}{5}}$
7. $\dfrac{\frac{4}{5}}{6\frac{2}{3}}$
8. $\dfrac{\frac{a^2}{b^3}}{\frac{b^5}{a}}$
9. $\dfrac{\frac{y^4}{x^2}}{\frac{xy^2}{2x^2}}$
10. $\dfrac{\frac{6}{x-2}}{\frac{3}{x^2-x-2}}$
11. $\dfrac{\frac{r+s}{x^2-y^2}}{\frac{(r+s)^2}{x-y}}$
12. $\dfrac{\frac{2+q}{q^2-4}}{\frac{q+4}{q^2-6q+8}}$
13. $\dfrac{\frac{p+3}{p^2+p-6}}{\frac{p^2+4p+3}{p^2+6p+9}}$

Practice and Problem Solving

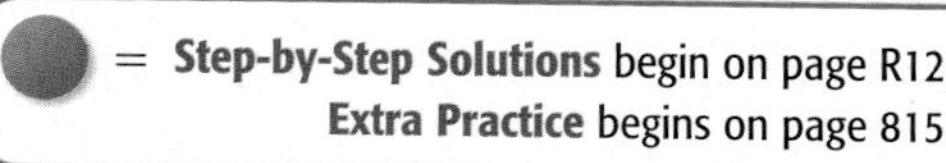
= **Step-by-Step Solutions** begin on page R12.
Extra Practice begins on page 815.

Example 1 p. 714

Write each mixed expression as a rational expression.

14. $10 + \frac{6}{f}$
15. $p - \frac{7}{2p}$
16. $5a - \frac{2a}{b}$
17. $3h + \frac{1+h}{h}$
18. $t + \frac{v+w}{v-w}$
19. $n^2 + \frac{n-1}{n+4}$
20. $(k+2) + \frac{k-1}{k-2}$
21. $(d-6) + \frac{d+1}{d-7}$
22. $\frac{h-3}{h+5} - (h+2)$

Example 2 p. 715

23. **READING** Ebony reads $6\frac{3}{4}$ pages of a book in 9 minutes. What is her average reading rate in pages per minute?

24. **HORSES** A thoroughbred can run $\frac{1}{2}$ mile in about $\frac{3}{4}$ minute. What is the horse's speed in miles per hour?

Examples 3 and 4 pp. 715–716

Simplify each expression.

25. $\dfrac{2\frac{2}{9}}{3\frac{1}{3}}$
26. $\dfrac{5\frac{3}{5}}{2\frac{1}{7}}$
27. $\dfrac{\frac{g^2}{h}}{\frac{g^5}{h^2}}$
28. $\dfrac{\frac{5n^4}{p^3}}{\frac{6n}{5p}}$
29. $\dfrac{\frac{2}{a}}{\frac{1}{a+6}}$
30. $\dfrac{\frac{t+5}{9}}{\frac{t^2-t-30}{12}}$
31. $\dfrac{\frac{j^2-16}{j^2+10j+16}}{\frac{15}{j+8}}$
32. $\dfrac{\frac{x-3}{x^2+3x+2}}{\frac{x^2-9}{x+1}}$

33. **COOKING** The Centralville High School Cooking Club has $12\frac{1}{2}$ pounds of flour with which to make tortillas. There are $3\frac{3}{4}$ cups of flour in a pound, and it takes about $\frac{1}{3}$ cup of flour per tortilla. How many tortillas can they make?

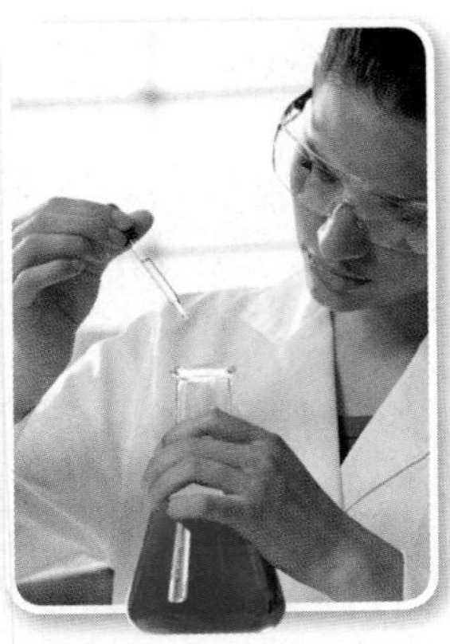

Real-World Career

Lab Technician
Lab technicians work with scientists, running experiments, conducting research projects, and running routine diagnostic samples. Lab technicians in any field need at least a two-year associate degree.

34. SCOOTER The speed v of an object spinning in a circle equals the circumference of the circle divided by the time T it takes the object to complete one revolution.

a. Use the variables v, r (the radius of the circle), and T to write a formula describing the speed of a spinning object.

b. A scooter has tires with a radius of $3\frac{1}{2}$ inches. The tires make one revolution every $\frac{1}{10}$ second. Find the speed in miles per hour. Round to the nearest tenth.

35. SCIENCE The *density* of an object equals $\frac{m}{V}$, where m is the mass of the object and V is the volume. The densities of four metals are shown in the table. Identify the metal of each ball described below. (*Hint*: The volume of a sphere is $V = \frac{4}{3}\pi r^3$.)

Metal	Density (kg/m³)
copper	8900
gold	19,300
iron	7800
lead	11,300

a. A metal ball has a mass of 15.6 kilograms and a radius of 0.0748 meter.

b. A metal ball has a mass of 285.3 kilograms and a radius of 0.1819 meter.

36. SIRENS As an ambulance approaches, the siren sounds different than if it were sitting still. If the ambulance is moving toward you at v miles per hour and blowing the siren at a frequency of f, then you hear the siren as if it were blowing at a frequency h. This can be described by the equation $h = \frac{f}{1 - \frac{v}{s}}$, where s is the speed of sound, approximately 760 miles per hour.

a. Simplify the complex fraction in the formula.

b. Suppose a siren blows at 45 cycles per minute and is moving toward you at 65 miles per hour. Find the frequency of the siren as you hear it.

Simplify each expression.

37. $15 - \frac{17x + 5}{5x + 10}$

38. $\frac{\frac{b}{b+3} + 2}{b^2 - 2b - 8}$

39. $\frac{1 + \frac{2c^2 - 6c - 10}{c + 7}}{2c + 1}$

40. $\frac{y - \frac{12}{y - 4}}{y - \frac{18}{y - 3}}$

41. $\frac{\frac{x^2 - 4x - 32}{x + 1}}{\frac{x^2 + 6x + 8}{x^2 - 1}}$

42. $\frac{\frac{r^2 - 9r}{r^2 + 7r + 10}}{\frac{r^2 + 5r}{r^2 + r - 2}}$

H.O.T. Problems

Use Higher-Order Thinking Skills

43. REASONING Describe the first step to simplify the expression below.

$$\frac{\left(\frac{y}{x} - \frac{x}{y}\right)}{\frac{x + y}{xy}}$$

44. REASONING Is $\frac{n}{1 - \frac{5}{p}} + \frac{n}{\frac{5}{p} - 1}$ *sometimes*, *always*, or *never* equal to 0? Explain.

45. CHALLENGE Simplify the rational expression below.

$$\frac{\frac{1}{t - 1} + \frac{1}{t + 1}}{\frac{1}{t} - \frac{1}{t^2}}$$

46. OPEN ENDED Write a complex fraction that, when simplified, results in $\frac{1}{x}$.

47. WRITING IN MATH Explain how complex fractions can be used to solve a problem involving distance, rate, and time. Give an example.

Standardized Test Practice

48. A number is between 44 squared and 45 squared. 5 squared is one of its factors, and it is a multiple of 13. Find the number.

A 1950

B 2000

C 2025

D 1975

49. SHORT RESPONSE Bernard is reading a 445-page book. He has already read 157 pages. If he reads 24 pages a day, how long will it take him to finish the book?

50. GEOMETRY Angela wanted a round rug to fit her room that is 16 feet wide. The rug should just meet the edges. What is the area of the rug rounded to the nearest tenth?

F 100.5 ft

G 804.2 ft²

H 50.3 ft

J 201.1 ft²

51. Simplify $7x + \frac{10}{2xy}$.

A $\frac{7x + 10}{2xy}$

B $\frac{7x^2y + 5}{xy}$

C $\frac{17x}{2xy}$

D $\frac{7xy + 5}{x^2y}$

Spiral Review

Find each sum or difference. (Lesson 11-6)

52. $\frac{6}{7x} - \frac{5 + x}{7x}$ **53.** $\frac{4}{d - 1} + \frac{d}{1 - d}$ **54.** $\frac{3q + 2}{2q + 1} + \frac{q - 5}{2q + 1}$

55. $\frac{2}{5m} - \frac{1}{15m^3}$ **56.** $\frac{10}{3g} - \frac{-3}{4h}$ **57.** $\frac{b}{b + 3} + \frac{6}{b - 2}$

Find each quotient. Use long division. (Lesson 11-5)

58. $(x^2 - 2x - 30) \div (x + 7)$ **59.** $(a^2 + 4a - 22) \div (a - 3)$ **60.** $(3q^2 + 20q + 11) \div (q + 6)$

61. $(3y^3 + 8y^2 + y - 7) \div (y + 2)$ **62.** $(6t^3 - 9t^2 + 6) \div (2t - 3)$ **63.** $(9h^3 + 5h - 8) \div (3h - 2)$

64. GEOMETRY Triangle ABC has vertices $A(7, -4)$, $B(-1, 2)$, and $C(5, -6)$. Determine whether the triangle has three, two, or no sides that are equal in length. (Lesson 10-6)

Graph each function. Determine the domain and range. (Lesson 10-1)

65. $y = 2\sqrt{x}$ **66.** $y = -3\sqrt{x}$ **67.** $y = \frac{1}{4}\sqrt{x}$

Factor each polynomial. If the polynomial cannot be factored, write *prime*. (Lesson 8-5)

68. $x^2 - 81$ **69.** $a^2 - 121$ **70.** $n^2 + 100$

71. $-25 + 4y^2$ **72.** $p^4 - 16$ **73.** $4t^4 - 4$

74. PARKS A youth group traveling in two vans visited Mammoth Cave in Kentucky. The number of people in each van and the total cost of the cave are shown. Find the adult price and the student price of the tour. (Lesson 6-3)

Van	Number of Adults	Number of Students	Total Cost
A	2	5	$77
B	2	7	$95

Skills Review

Solve each equation. (Lesson 2-2 and 2-3)

75. $6x = 24$ **76.** $5y - 1 = 19$ **77.** $2t + 7 = 21$

78. $\frac{p}{3} = -4.2$ **79.** $\frac{2m + 1}{4} = -5.5$ **80.** $\frac{3}{4}g = \frac{1}{2}$

11-8 Rational Equations

Then

You solved proportions. (Lesson 2-6)

Now

- Solve rational equations.
- Use rational equations to solve problems.

New Vocabulary

rational equation
extraneous solution
work problem
rate problem

Math Online

glencoe.com

- Extra Examples
- Personal Tutor
- Self-Check Quiz
- Homework Help
- Math in Motion

Why?

Oceanic species of dolphins can swim 5 miles per hour faster than coastal species of dolphins. An oceanic dolphin can swim 3 miles in the same time that it takes a coastal dolphin to swim 2 miles.

Dolphins			
Species	**Distance**	**Rate**	**Time**
coastal	2 miles	x mph	t hours
oceanic	3 miles	$x + 5$ mph	t hours

Since time $= \frac{\text{distance}}{\text{rate}}$, the equation below represents this situation.

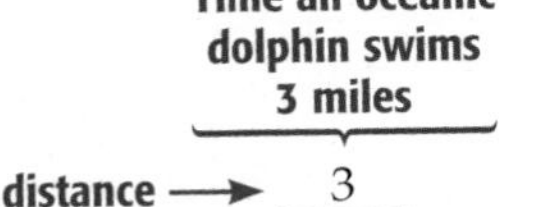

Time an oceanic dolphin swims 3 miles **equals** **time a coastal dolphin swims 2 miles.**

$$\frac{3}{x+5} = \frac{2}{x}$$

(**distance** → 3, **rate** → $x + 5$; 2 ← **distance**, x ← **rate**)

Solve Rational Equations A **rational equation** contains one or more rational expressions. When a rational equation is a proportion, you can use cross products to solve it.

Real-World EXAMPLE 1 Use Cross Products to Solve Equations

DOLPHINS Refer to the information above. Solve $\frac{3}{x+5} = \frac{2}{x}$ to find the speed of a coastal dolphin. Check the solution.

$\frac{3}{x+5} = \frac{2}{x}$	**Original equation**
$3x = 2(x + 5)$	**Find the cross products.**
$3x = 2x + 10$	**Distributive Property**
$x = 10$	**Subtract $2x$ from each side.**

So, a coastal dolphin can swim 10 miles per hour.

CHECK

$\frac{3}{x+5} = \frac{2}{x}$	**Original equation**
$\frac{3}{10+5} \stackrel{?}{=} \frac{2}{10}$	**Replace x with 10.**
$\frac{3}{15} \stackrel{?}{=} \frac{1}{5}$	**Simplify.**
$\frac{1}{5} = \frac{1}{5}$ ✓	**Simplify.**

Check Your Progress

Solve each equation. Check the solution.

1A. $\frac{7}{y-3} = \frac{3}{y+1}$

1B. $\frac{13}{10} = \frac{2f+0.2}{7}$

Personal Tutor glencoe.com

Another method that can be used to solve any rational equation is to find the LCD of all the fractions in the equation. Then multiply each side of the equation by the LCD to eliminate the fractions.

EXAMPLE 2 Use the LCD to Solve Rational Equations

Solve $\frac{4}{y} + \frac{5y}{y+1} = 5$. Check the solution.

Step 1 Find the LCD.

The LCD of $\frac{4}{y}$ and $\frac{5y}{y+1}$ is $y(y + 1)$.

Step 2 Multiply each side of the equation by the LCD.

$$\frac{4}{y} + \frac{5y}{y+1} = 5$$ **Original equation**

$$y(y+1)\left(\frac{4}{y} + \frac{5y}{y+1}\right) = y(y+1)(5)$$ **Multiply each side by the LCD, $y(y + 1)$.**

$$\left(\frac{\cancel{y}^{1}(y+1)}{1} \cdot \frac{4}{\cancel{y}_{1}}\right) + \left(\frac{y\cancel{(y+1)}^{1}}{1} \cdot \frac{5y}{\cancel{y+1}_{1}}\right) = y(y+1)(5)$$ **Distributive Property**

$$(y+1)4 + y(5y) = y(y+1)(5)$$ **Simplify.**

$$4y + 4 + 5y^2 = 5y^2 + 5y$$ **Multiply.**

$$4y + 4 + 5y^2 - 5y^2 = 5y^2 - 5y^2 + 5y$$ **Subtract $5y^2$ from each side.**

$$4y + 4 = 5y$$ **Simplify.**

$$4y - 4y + 4 = 5y - 4y$$ **Subtract $4y$ from each side.**

$$4 = y$$ **Simplify.**

CHECK

$$\frac{4}{y} + \frac{5y}{y+1} = 5$$ **Original equation**

$$\frac{4}{4} + \frac{5(4)}{4+1} \stackrel{?}{=} 5$$ **Replace y with 4.**

$$1 + 4 \stackrel{?}{=} 5$$ **Simplify.**

$$5 = 5 \checkmark$$ **Simplify.**

Check Your Progress

Solve each equation. Check your solutions.

2A. $\frac{2b-5}{b-2} - 2 = \frac{3}{b+2}$

2B. $1 + \frac{1}{c+2} = \frac{28}{c^2+2c}$

2C. $\frac{y+2}{y-2} - \frac{2}{y+2} = -\frac{7}{3}$

2D. $\frac{n}{3n+6} - \frac{n}{5n+10} = \frac{2}{5}$

Personal Tutor glencoe.com

Recall that any value of a variable that makes the denominator of a rational expression zero must be excluded from the domain.

In the same way, when a solution of a rational equation results in a zero in the denominator, that solution must be excluded. Such solutions are called **extraneous solutions**.

$$\frac{4+x}{x-5} + \frac{1}{x} = \frac{2}{x+1}$$ **5, 0, and −1 cannot be solutions.**

StudyTip

Solutions It is important to check the solutions of rational equations to be sure that they satisfy the original equation.

Vocabulary Link

extraneous
Everyday Use irrelevant or unimportant

extraneous solution
Math Use a result that is not a solution of the original equation

EXAMPLE 3 Extraneous Solutions

Solve $\frac{2n}{n-5} + \frac{4n-30}{n-5} = 5$. State any extraneous solutions.

$\frac{2n}{n-5} + \frac{4n-30}{n-5} = 5$	**Original equation**
$(n-5)\left(\frac{2n}{n-5} + \frac{4n-30}{n-5}\right) = (n-5)5$	**Multiply each side by the LCD, $n-5$.**
$\left(\frac{\overset{1}{\cancel{n-5}}}{1} \cdot \frac{2n}{\underset{1}{\cancel{n-5}}}\right) + \left(\frac{\overset{1}{\cancel{n-5}}}{1} \cdot \frac{4n-30}{\underset{1}{\cancel{n-5}}}\right) = (n-5)5$	**Distributive Property**
$2n + 4n - 30 = 5n - 25$	**Simplify.**
$6n - 30 = 5n - 25$	**Add like terms.**
$6n - 5n - 30 = 5n - 5n - 25$	**Subtract $5n$ from each side.**
$n - 30 = -25$	**Add 30 to each side.**
$n - 30 + 30 = -25 + 30$	**Add 30 to each side.**
$n = 5$	**Simplify.**

Since $n = 5$ results in a zero in the denominator of the original equation, it is an extraneous solution. So, the equation has no solution.

StudyTip

Solutions It is possible to get both a valid solution and an extraneous solution when solving a rational equation.

3. Solve $\frac{n^2 - 3n}{n^2 - 4} - \frac{10}{n^2 - 4} = 2$. State any extraneous solutions.

Personal Tutor glencoe.com

Use Rational Equations to Solve Problems You can use rational equations to solve **work problems**, or problems involving work rates.

Real-World EXAMPLE 4 Work Problem

JOBS At his part-time job at the zoo, Ping can clean the bird area in 2 hours. Natalie can clean the same area in 1 hour and 15 minutes. How long would it take them if they worked together?

Understand It takes Ping 2 hours to complete the job and Natalie $1\frac{1}{4}$ hours.

You need to find the rate that each person works and the total time t that it will take them if they work together.

Plan Find the fraction of the job that each person can do in an hour.

Ping's rate $\longrightarrow \frac{1 \text{ job}}{2 \text{ hours}} = \frac{1}{2}$ job per hour

Natalie's rate $\longrightarrow \frac{1 \text{ job}}{1\frac{1}{4} \text{ hours}}$ or $\frac{1 \text{ job}}{\frac{5}{4} \text{ hours}} = \frac{4}{5}$ job per hour

Since rate • time = fraction of job done, multiply each rate by the time t to represent the amount of the job done by each person.

StudyTip

Work Problems When solving work problems, remember that each term should represent the portion of a job completed in one unit of time.

Solve	Fraction of job Ping completes	plus	fraction of job Natalie completes	equals	1 job.
	$\frac{1}{2}t$	$+$	$\frac{4}{5}t$	$=$	1

$10\left(\frac{1}{2}t + \frac{4}{5}t\right) = 10(1)$ **Multiply each side by the LCD, 10.**

$10\left(\frac{1}{2}t\right) + 10\left(\frac{4}{5}t\right) = 10$ **Distributive Property**

$5t + 8t = 10$ **Simplify.**

$t = \frac{10}{13}$ **Add like terms and divide each side by 13.**

So, it would take them $\frac{10}{13}$ hour or about 46 minutes to complete the job if they work together.

Check In $\frac{10}{13}$ hour, Ping would complete $\frac{1}{2} \cdot \frac{10}{13}$ or $\frac{5}{13}$ of the job and Natalie would complete $\frac{4}{5} \cdot \frac{10}{13}$ or $\frac{8}{13}$ of the job. Together, they complete $\frac{5}{13} + \frac{8}{13}$ or 1 whole job. So, the answer is reasonable. ✓

Check Your Progress

4. RAKING Jenna can rake the leaves in 2 hours. It takes her brother Benjamin 3 hours. How long would it take them if they worked together?

Personal Tutor glencoe.com

Rational equations can also be used to solve **rate problems**.

Real-World EXAMPLE 5 Rate Problem

AIRPLANES An airplane takes off and flies an average of 480 miles per hour. Another plane leaves 15 minutes later and flies to the same city traveling 560 miles per hour. How long will it take the second plane to pass the first plane?

Record the information that you know in a table.

Plane	Distance	Rate	Time
1	d miles	480 mi/h	t hours
2	d miles	560 mi/h	$t - \frac{1}{4}$ hours

Plane 2 took off 15 minutes, or $\frac{1}{4}$ hour, after Plane 1

Since both planes will have traveled the same distance when Plane 2 passes Plane 1, you can write the following equation.

Distance for Plane 1 = Distance for Plane 2

$480 \cdot t = 560 \cdot \left(t - \frac{1}{4}\right)$ **distance = rate • time**

$480t = (560 \cdot t) - \left(560 \cdot \frac{1}{4}\right)$ **Distributive Property**

$480t = 560t - 140$ **Simplify.**

$-80t = -140$ **Subtract 560*t* from each side.**

$t = 1.75$ **Divide each side by −80.**

So, the second plane passes the first plane after 1.75 hours.

Check Your Progress

5. Lenora leaves the house walking at 3 miles per hour. After 10 minutes, her mother leaves the house riding a bicycle at 10 miles per hour. In how many minutes will Lenora's mother catch her?

Personal Tutor glencoe.com

Real-World Link

The longest nonstop commercial flight was 13,422 miles from Hong Kong Airport in China to London Heathrow in the United Kingdom. It took 22 hours and 42 minutes.

Source: *Guinness Book of World Records*

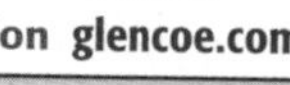

Math *in Motion*, Animation glencoe.com

Check Your Understanding

Examples 1–3 pp. 720–722

Solve each equation. State any extraneous solutions.

1. $\frac{2}{x+1} = \frac{4}{x}$
2. $\frac{t+3}{5} = \frac{2t+3}{9}$
3. $\frac{a+3}{a} - \frac{6}{5a} = \frac{1}{a}$
4. $4 - \frac{p}{p-1} = \frac{2}{p-1}$
5. $\frac{2t}{t+1} + \frac{4}{t-1} = 2$
6. $\frac{x+3}{x^2-1} - \frac{2x}{x-1} = 1$

Example 4 pp. 722–723

7. **WEEDING** Maurice can weed the garden in 45 minutes. Olinda can weed the garden in 50 minutes. How long would it take them to weed the garden if they work together?

Example 5 p. 723

8. **LANDSCAPING** Hunter is filling a 3.5-gallon bucket to water plants at a faucet that flows at a rate of 1.75 gallons a minute. If he were to add a hose that flows at a rate of 1.45 gallons per minute, how many minutes would it take him to fill the bucket? Round to the nearest tenth.

Practice and Problem Solving

= **Step-by-Step Solutions** begin on page R12.
Extra Practice begins on page 815

Examples 1–3 pp. 720–722

Solve each equation. State any extraneous solutions.

9. $\frac{8}{n} = \frac{3}{n-5}$
10. $\frac{6}{t+2} = \frac{4}{t}$
11. $\frac{3g+2}{12} = \frac{g}{2}$
12. $\frac{5h}{4} + \frac{1}{2} = \frac{3h}{8}$
13. $\frac{2}{3w} = \frac{2}{15} + \frac{12}{5w}$
14. $\frac{c-4}{c+1} = \frac{c}{c-1}$
15. $\frac{x-1}{x+1} - \frac{2x}{x-1} = -1$
16. $\frac{y+4}{y-2} + \frac{6}{y-2} = \frac{1}{y+3}$
17. $\frac{a}{a+3} + \frac{a^2}{a+3} = 2$
18. $\frac{12}{a+3} + \frac{6}{a^2-9} = \frac{8}{a+3}$
19. $\frac{3n}{n-1} + \frac{6n-9}{n-1} = 6$
20. $\frac{n^2-n-6}{n^2-n} - \frac{n-5}{n-1} = \frac{n-3}{n^2-n}$

Example 4 pp. 722–723

21. **PAINTING** It takes Noah 3 hours to paint one side of a fence. It takes Gilberto 5 hours. How long would it take them if they worked together?

22. **DISHWASHING** Ron works as a dishwasher and can wash 500 plates in two hours and 15 minutes. Chris can finish the 500 plates in 3 hours. About how long would it take them to finish all of the plates if they work together?

Example 5 p. 723

23. **ICE** A hotel has two ice machines in its kitchen. How many hours would it take both machines to make 60 pounds of ice? Round to the nearest tenth.

24. **CYCLING** Two cyclists travel in opposite directions around a 5.6-mile circular trail. They start at the same time. The first cyclist completes the trail in 22 minutes and the second in 28 minutes. At what time do they pass each other?

GRAPHING CALCULATOR For each function, a) describe the shape of the graph, b) use factoring to simplify the function, and c) find the zeros of the function.

25. $f(x) = \frac{x^2-x-30}{x-6}$
26. $f(x) = \frac{x^3+x^2-2x}{x+2}$
27. $f(x) = \frac{x^3+6x^2+12x}{x}$

28. **PAINTING** Morgan can paint a standard-sized house in about 5 days. For his latest job, Morgan hires two assistants. At what rate must these assistants work for Morgan to meet a deadline of two days?

29. **AIRPLANES** Headwinds push against a plane and reduce its total speed, while tailwinds push on a plane and increase its total speed. Let w equal the speed of the wind, r equal the speed set by the pilot, and s equal the total speed.

a. Write an equation for the total speed with a headwind and an equation for the total speed with a tailwind.

b. Use the rate formula to write an equation for the distance traveled by a plane with a headwind and another equation for the distance traveled by a plane with a tailwind. Then solve each equation for time instead of distance.

Real-World Link

Orange juice contains more than 10 times as much vitamin C as apple juice. Drinking an 8-ounce glass counts as one of the five recommended fruit and vegetable servings for the day.

Source: Ask Dr. Sears

30. **MIXTURES** A pitcher of fruit juice has 3 pints of pineapple juice and 2 pints of orange juice. Erin wants to add more orange juice so that the fruit juice mixture is 60% orange juice. Let x equal the pints of orange juice that she needs to add.

a. Copy and complete the table below.

Juice	Pints of Orange Juice	Total Pints of Juice	Percent of Orange Juice
original mixture		5	
final mixture	$2 + x$		0.6

b. Write and solve an equation to find the pints of orange juice to add.

31. **DORMITORIES** The number of hours h it takes to clean a dormitory varies inversely with the number of people cleaning it c and directly with the number of people living there p.

a. Write an equation showing how h, c, and p are related. (*Hint*: Include the constant k.)

b. It takes 8 hours for 5 people to clean the dormitory when there are 100 people there. How long will it take to clean the dormitory if there are 10 people cleaning and the number of people living in the dorm stays the same?

Solve each equation. State any extraneous solutions.

32. $\frac{4b + 2}{b^2 - 3b} + \frac{b + 2}{b} = \frac{b - 1}{b}$

33. $\frac{x^2 - x - 6}{x + 2} + \frac{x^3 + x^2}{x} = 3$

34. $\frac{y^2 + 5y - 6}{y^3 - 2y^2} = \frac{5}{y} - \frac{6}{y^3 - 2y^2}$

35. $\frac{x - \frac{6}{5}}{x} - \frac{x - 10\frac{1}{2}}{x - 5} = \frac{x + 21}{x^2 - 5x}$

H.O.T. Problems

Use **H**igher-**O**rder **T**hinking Skills

36. **CHALLENGE** Solve $\frac{2x}{x - 2} + \frac{x^2 + 3x}{(x + 1)(x - 2)} = \frac{2}{(x + 1)(x - 2)}$.

37. **REASONING** How is an excluded value of a rational expression related to an extraneous solution of a corresponding rational equation? Explain.

38. **OPEN ENDED** Write a problem about a real-world situation where work is being done. Write an equation that models the situation.

39. **REASONING** Find a counterexample for the following statement.

The solution of a rational equation can never be zero.

40. **WRITING IN MATH** Describe the steps for solving a rational equation that is not a proportion.

Standardized Test Practice

41. It takes Cheng 4 hours to build a fence. If he hires Odell to help him, they can do the job in 3 hours. If Odell built the same fence alone, how long would it take him?

A $1\frac{5}{7}$ hours
C 8 hours
B $3\frac{2}{3}$ hours
D 12 hours

42. In the 1000-meter race, Zoe finished 200 meters ahead of Taryn and 400 meters ahead of Evan. When Taryn finished, how far was she ahead of Evan?

F 400 m G 200 m H 150 m J 100 m

43. Twenty gallons of lemonade were poured into two containers of different sizes. Express the amount of lemonade poured into the smaller container in terms of g, the amount poured into the larger container.

A $g + 20$
C $g - 20$
B $20 + g$
D $20 - g$

44. GRIDDED RESPONSE The gym has 2-kilogram and 5-kilogram disks for weight lifting. They have fourteen disks in all. The total weight of the 2-kilogram disks is the same as the total weight of the 5-kilogram disks. How many 2-kilogram disks are there?

Spiral Review

Simplify each expression. (Lesson 11-7)

45. $\dfrac{\frac{c^2}{d}}{\frac{c^3}{d^2}}$

46. $\dfrac{\frac{5g^3}{h^2}}{\frac{6g}{5h}}$

47. $\dfrac{\frac{2}{b}}{\frac{4}{b-3}}$

48. $\dfrac{\frac{q-2}{9}}{\frac{q^2-6q+8}{12}}$

Find the LCM of each pair of polynomials. (Lesson 11-6)

49. $2h, 4h^2$

50. $5c^2, 12c^3$

51. $x - 4, x + 2$

52. $p - 7, 2(p - 14)$

Look for a pattern in each table of values to determine which kind of model best describes the data. (Lesson 9-9)

53.

x	0	1	2	3	4
y	4	5	6	7	8

54.

x	1	2	3	4	5
y	2	4	8	16	32

55.

x	−3	−2	−1	0	1
y	14	9	6	5	6

56.

x	3	4	5	6	7
y	3	5	7	9	11

57. GENETICS Brown genes B are dominant over blue genes b. A person with genes BB or Bb has brown eyes. Someone with genes bb has blue eyes. Mrs. Dunn has brown eyes with genes Bb, and Mr. Dunn has blue eyes. Write an expression for the possible eye coloring of their children. Then find the probability that a child would have blue eyes. (Lesson 7-8)

Solve each inequality. Check your solution. (Lesson 5-2)

58. $\frac{b}{10} \leq 5$

59. $-7 > -\frac{r}{7}$

60. $\frac{5}{8}y \geq -15$

Skills Review

Determine the probability of each event if you randomly select a marble from a bag containing 9 red marbles, 6 blue marbles, and 5 yellow marbles. (Lesson 0-11)

61. P(blue)

62. P(red)

63. P(not yellow)

Study Guide and Review

Math Online glencoe.com
- STUDY TO GO
- Vocabulary Review

Chapter Summary

Key Concepts

Inverse Variation (Lesson 11-1)

- You can use $\frac{x_1}{x_2} = \frac{y_2}{y_1}$ to solve problems involving inverse variation.

Rational Functions (Lesson 11-2)

- Excluded values are values of a variable that result in a denominator of zero.
- If vertical asymptotes occur, it will be at excluded values.

Rational Expressions (Lessons 11-3 and 11-4)

- Multiplying rational expressions is similar to multiplying rational numbers.
- Divide rational expressions by multiplying by the reciprocal of the divisor.

Dividing Polynomials (Lesson 11-5)

- To divide a polynomial by a monomial, divide each term of the polynomial by the monomial.

Adding and Subtracting Rational Expressions (Lesson 11-6)

- Rewrite rational expressions with unlike denominators using the least common denominator (LCD). Then add or subtract.

Complex Fractions (Lesson 11-7)

- Simplify complex fractions by writing them as division problems.

Solving Rational Equations (Lesson 11-8)

- Use cross products to solve rational equations with a single fraction on each side of the equals sign.

FOLDABLES® Study Organizer

Be sure the Key Concepts are noted in your Foldable.

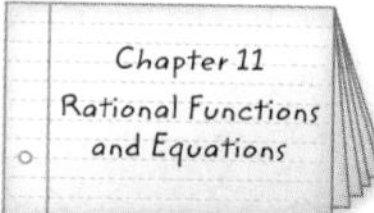

Key Vocabulary

asymptote (p. 679)
complex fraction (p. 714)
excluded value (p. 678)
extraneous solution (p. 721)
inverse variation (p. 670)
least common denominator (LCD) (p. 708)
least common multiple (LCM) (p. 707)
mixed expression (p. 714)
product rule (p. 671)
rate problems (p. 723)
rational equation (p. 720)
rational expression (p. 684)
rational function (p. 678)
work problems (p. 722)

Vocabulary Check

State whether each sentence is *true* or *false*. If *false*, replace the underlined word, phrase, expression, or number to make a true sentence.

1. The least common multiple for $x^2 - 25$ and $x - 5$ is $\underline{x - 5}$.
2. If the product of two variables is a nonzero constant, the relationship is an <u>inverse variation</u>.
3. If the line $x = a$ is a vertical <u>asymptote</u> of a rational function, then a is an excluded value.
4. A rational expression is a fraction in which the numerator and denominator are <u>fractions</u>.
5. The excluded values for $\frac{x}{x^2 + 5x + 6}$ are $\underline{-2 \text{ and } -3}$.
6. The equation $\frac{3x}{x - 2} = \frac{6}{x - 2}$ has an extraneous solution, $\underline{2}$.
7. A <u>rational expression</u> has one or more fractions in the numerator and denominator.
8. The expression $\frac{\frac{1}{2}}{\frac{3}{4}}$ can be simplified to $\underline{\frac{2}{3}}$.
9. A <u>direct variation</u> can be represented by an equation of the form $k = xy$, where k is a nonzero constant.
10. The rational function $y = \frac{2}{x - 1} + 3$ has a horizontal asymptote at $\underline{y = 3}$.

Study Guide and Review

Lesson-by-Lesson Review

11-1 Inverse Variation (pp. 669–676)

Solve. Assume that y varies inversely as x.

11. If $y = 4$ when $x = 1$, find x when $y = 12$

12. If $y = -1$ when $x = -3$, find y when $x = -9$

13. If $y = 1.5$ when $x = 6$, find y when $x = -16$

14. **PHYSICS** A 135-pound person sits 5 feet from the center of a seesaw. How far from the center should a 108-pound person sit to balance the seesaw?

EXAMPLE 1

If y varies inversely as x and $y = 28$ when $x = 42$, find y when $x = 56$.

Let $x_1 = 42$, $x_2 = 56$, and $y_1 = 28$. Solve for y_2.

$\frac{x_1}{x_2} = \frac{y_2}{y_1}$ **Proportion for inverse variation**

$\frac{42}{56} = \frac{y_2}{28}$ **Substitution**

$1176 = 56y_2$ **Cross multiply.**

$21 = y_2$

Thus, $y = 21$ when $x = 56$.

11-2 Rational Functions (pp. 677–683)

State the excluded value for each function.

15. $y = \frac{1}{x-3}$

16. $y = \frac{2}{2x-5}$

17. $y = \frac{3}{3x-6}$

18. $y = \frac{-1}{2x+8}$

19. **PIZZA PARTY** Katelyn ordered pizza and soda for her study group for \$38. The cost per person y is given by $y = \frac{38}{x}$, where x is the number of people in the study group. Graph the function and describe the asymptotes.

EXAMPLE 2

State the excluded value for the function $y = \frac{1}{4x + 16}$.

Set the denominator equal to zero.

$4x + 16 = 0$

$4x + 16 - 16 = 0 - 16$ **Subtract 16 from each side.**

$4x = -16$ **Simplify.**

$x = -4$ **Divide each side by 4.**

11-3 Simplifying Rational Expressions (pp. 684–691)

Simplify each expression.

20. $\frac{2xy^2}{16xyz}$

21. $\frac{x+4}{x^2+12x+32}$

22. $\frac{x^2+10x+21}{x^3+x^2-42x}$

23. $\frac{y^2-25}{y^2+3y-10}$

24. $\frac{3x^3}{3x^3+6x^2}$

25. $\frac{4y^2}{8y^4+16y^3}$

State the excluded values for each function.

26. $y = \frac{x}{x^2+9x+18}$

27. $y = \frac{10}{6x^2+7x-3}$

EXAMPLE 3

Simplify $\frac{a^2 - 7a + 12}{a^2 - 13a + 36}$.

Factor and simplify.

$\frac{a^2-7a+12}{a^2-13a+36} = \frac{(a-3)(a-4)}{(a-9)(a-4)}$ **Factor.**

$= \frac{a-3}{a-9}$ **Simplify.**

MIXED PROBLEM SOLVING
For mixed problem-solving practice, see page 855.

11-4 Multiplying and Dividing Rational Expressions (pp. 692–698)

Find each product or quotient.

28. $\frac{6x^2y^4}{12} \cdot \frac{3x^3y^2}{xy}$

29. $\frac{3x-6}{x^2-9} \cdot \frac{x+3}{x^2-2x}$

30. $\frac{x^2}{x+4} \div \frac{3x}{x^2-16}$

31. $\frac{3b-12}{b+4} \div (b^2-6b+8)$

32. $\frac{2a^2+7a-15}{a+5} \div \frac{9a^2-4}{3a+2}$

33. GEOMETRY Find the area of the rectangle shown. Write the answer in simplest form.

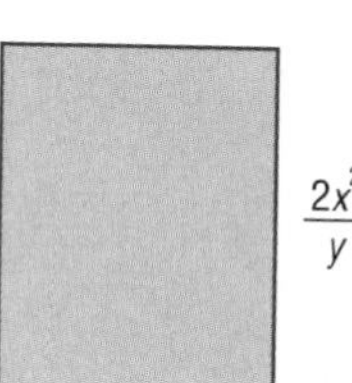

EXAMPLE 4

Find $\frac{7b^2}{9} \cdot \frac{6a^2}{b}$.

$\frac{7b^2}{9} \cdot \frac{6a^2}{b} = \frac{42a^2b^2}{9b}$ **Multiply.**

$= \frac{14a^2b}{3}$ **Simplify.**

EXAMPLE 5

Find $\frac{x^2-25}{x^2-9} \div \frac{x+5}{x-3}$.

$\frac{x^2-25}{x^2-9} \div \frac{x+5}{x-3} = \frac{(x+5)(x-5)}{(x+3)(x-3)} \div \frac{x+5}{x-3}$ **Factor.**

$= \frac{\overset{1}{\cancel{(x+5)}}(x-5)}{(x+3)\underset{1}{\cancel{(x-3)}}} \cdot \frac{\overset{1}{\cancel{x-3}}}{\underset{1}{\cancel{x+5}}}$ **Multiply by the reciprocal.**

$= \frac{x-5}{x+3}$ **Simplify.**

11-5 Dividing Polynomials (pp. 700–705)

Find each quotient.

34. $(x^3-2x^2-22x+21) \div (x-3)$

35. $(x^3+7x^2+10x-6) \div (x+3)$

36. $(5x^2y^2-10x^2y+5xy) \div 5xy$

37. $(48y^2+8y+7) \div (12y-1)$

38. GEOMETRY The area of a rectangle is $x^2+7x+13$. If the length is $(x+4)$, what is the width of the rectangle?

EXAMPLE 6

Find $(4x^2+17x-1) \div (4x+1)$.

$$\begin{array}{r} x+4 \\ 4x+1\overline{)4x^2+17x-1} \\ \underline{4x^2+x} \\ 16x-1 \\ \underline{16x+4} \\ -5 \end{array}$$

$4x^2 + x$: **Multiply x and $4x+1$.**
$16x - 1$: **Subtract, bring down −1.**
$16x + 4$: **Multiply 4 and $4x+1$.**
-5: **Subtract.**

The quotient is $x+4-\frac{5}{4x+1}$.

11-6 Adding and Subtracting Rational Expressions (pp. 706–713)

Find each sum or difference.

39. $\frac{5a}{b} - \frac{2a}{b}$

40. $\frac{-3}{2n-3} + \frac{2n}{2n-3}$

41. $\frac{3}{y+1} - \frac{y}{y-3}$

42. $\frac{1}{x+1} + \frac{3}{x-2}$

43. DESIGN Miguel is decorating a room that is $\frac{2x}{x+4}$ feet long and $\frac{8}{x+4}$ feet wide. What is the perimeter of the room?

EXAMPLE 7

Find $\frac{x^2}{x+1} + \frac{2x+1}{x+1}$.

$\frac{x^2}{x+1} + \frac{2x+1}{x+1} = \frac{x^2+2x+1}{x+1}$ **Add the numerators.**

$= \frac{(x+1)(x+1)}{x+1}$ **Factor.**

$= x+1$ **Simplify.**

CHAPTER 11 Study Guide and Review

11-7 Mixed Expressions and Complex Fractions (pp. 714–719)

Simplify each expression.

44. $\dfrac{\frac{a^2b^4}{c}}{\frac{a^3b}{c^2}}$

45. $\dfrac{x - \frac{35}{x+2}}{x + \frac{42}{x+13}}$

46. $\dfrac{\frac{x^2-25}{x+2}}{\frac{x-5}{x^2-4}}$

47. $\dfrac{y + 9 - \frac{6}{y+4}}{y + 4 + \frac{2}{y+1}}$

48. FABRICS Donna makes tablecloths to sell at craft fairs. A small one takes one-half yard of fabric, a medium one takes five-eighths yard, and a large one takes one and one-quarter yard.

a. How many yards of fabric does she need to make a tablecloth of each size?

b. One bolt of fabric contains 30 yards of fabric. Can she use the entire bolt of fabric by making an equal number of each type of tablecloth? Explain.

EXAMPLE 8

Simplify $\dfrac{\frac{x+3}{6}}{\frac{x^2-2x-15}{x}}$.

Write as a division expression.

$$\dfrac{\frac{x+3}{6}}{\frac{x^2-2x-15}{x}} = \frac{x+3}{6} \div \frac{x^2-2x-15}{x}$$

$$= \frac{x+3}{6} \cdot \frac{x}{x^2-2x-15}$$

$$= \frac{x+3}{6} \cdot \frac{x}{(x+3)(x-5)}$$

$$= \frac{x}{6(x-5)}$$

11-8 Rational Equations (pp. 720–727)

Solve each equation. State any extraneous solutions.

49. $\frac{5n}{6} + \frac{1}{n-2} = \frac{n+1}{3(n-2)}$

50. $\frac{4x}{3} + \frac{7}{2} = \frac{7x}{12} - 14$

51. $\frac{11}{2x} + \frac{2}{4x} = \frac{1}{4}$

52. $\frac{1}{x+4} - \frac{1}{x-1} = \frac{2}{x^2+3x-4}$

53. $\frac{1}{n-2} = \frac{n}{8}$

54. PAINTING Anne can paint a room in 6 hours. Oljay can paint a room in 4 hours. How long will it take them to paint the room working together?

EXAMPLE 9

Solve $\frac{3}{x^2+3x} + \frac{x+2}{x+3} = \frac{1}{x}$.

$$\frac{3}{x^2+3x} + \frac{x+2}{x+3} = \frac{1}{x}$$

$$x(x+3)\left(\frac{3}{x(x+3)}\right) + x(x+3)\left(\frac{x+2}{x+3}\right) = x(x+3)\left(\frac{1}{x}\right)$$

$$3 + x(x+2) = 1(x+3)$$

$$3 + x^2 + 2x = x + 3$$

$$x^2 + x = 0$$

$$x(x+1) = 0$$

$$x = 0 \text{ or } x = -1$$

The solution is -1, and there is an extraneous solution of 0.

Practice Test

Math Online glencoe.com
Chapter Test

Determine whether each table represents an inverse variation. Explain.

1.

x	y
2	10
4	12
8	14

2.

x	y
2	2
4	1
8	$\frac{1}{2}$

Find each product or quotient.

3. $\frac{(x+6)(x-2)}{x^3} \cdot \frac{7x^2}{x-3}$

4. $\frac{(x+3)}{y^2} \div \frac{x^2-9}{y}$

Solve. Assume that y varies inversely as x.

5. If $y = 3$ when $x = 9$, find x when $y = 1$.

6. If $y = 2$ when $x = 0.5$, find y when $x = 3$.

Simplify each expression. State the excluded values of the variables.

7. $\frac{z-6}{z^2-3z-18}$

8. $\frac{4x-28}{x^2-49}$

9. **MULTIPLE CHOICE** The area of a rectangle is $x^2 + 5x + 6$ square feet. If the width is $x + 2$, what is the length of the rectangle?

A $x + 2$

B $x + 3$

C 1

D 3

Simplify each expression.

10. $\frac{2\frac{1}{3}}{3\frac{1}{2}}$

11. $\frac{\frac{x^2-25}{x-2}}{\frac{x-5}{x-2}}$

12. $\frac{\frac{a-4}{a^2+6a+8}}{\frac{a^2-3a-4}{a^2-a-6}}$

13. $\frac{\frac{y^2+10y+24}{y^2-9}}{\frac{3y^2+17y-6}{2y^2-11y+15}}$

Find each quotient.

14. $(2x^2 + 10x) \div 2x$

15. $(4x^2 - 8x + 5) \div (2x + 1)$

16. $(3x^2 - 14x - 3) \div (x - 5)$

Assume that y varies inversely as x. Write an inverse variation equation that relates x and y.

17. $y = 2$ when $x = 8$

18. $y = -3$ when $x = 1$

Find each sum or difference.

19. $\frac{3}{x} + \frac{6}{x}$

20. $\frac{t-5}{t-6} + \frac{t+8}{t-6}$

21. $\frac{1}{x-6} + \frac{3}{x-2}$

22. $\frac{5}{x^2-2x-24} + \frac{x}{x-6}$

State the excluded value or values for each function.

23. $y = \frac{6}{x-1}$

24. $y = \frac{5}{x^2-5x-24}$

Identify the asymptotes of each function.

25. $y = \frac{2}{(x-4)(x+2)}$

26. $y = \frac{4}{x^2+3x-28} + 2$

27. **MULTIPLE CHOICE** Lee can shovel the driveway in 3 hours, and Susan can shovel the driveway in 2 hours. How long will it take them working together?

F 6 hours

G 5 hours

H $\frac{6}{5}$ hour

J 4 hours

28. **PAINTING** Sydney can paint a 60-square foot wall in 40 minutes. Working with her friend Cleveland, the two of them can paint the wall in 25 minutes. How long would it take Cleveland to do the job himself?

Model with an Equation

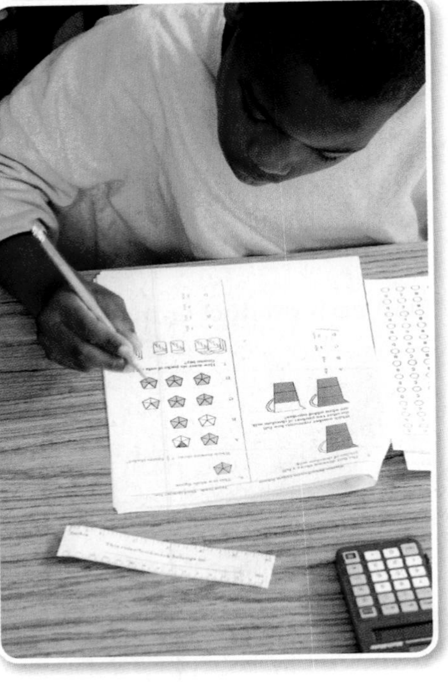

In order to successfully solve some standardized test questions, you will need to be able to write equations to model different situations. Use this lesson to practice solving these types of problems.

Strategies for Modeling with Equations

Step 1

Read the problem statement carefully.

Ask yourself:

- What am I being asked to solve?
- What information is given in the problem?
- What is the unknown quantity that I need to find?

Step 2

Translate the problem statement into an equation.

- Assign a variable to the unknown quantity.
- Write the word sentence as a mathematical number sentence.
- Look for keywords such as *is, is the same as, is equal to,* or *is identical to* that indicate where to place the equals sign.

Step 3

Solve the equation.

- Solve for the unknown in the equation.
- Check your answer to be sure it is reasonable and that it answers the question in the problem statement.

EXAMPLE

Read the problem. Identify what you need to know. Then use the information in the problem to solve.

It takes Craig 75 minutes to paint a small room. If Delsin can paint the same room in 60 minutes, how long would it take them to paint the room if they work together? Round to the nearest tenth.

A about 33.3 minutes
B about 38.4 minutes
C about 45.1 minutes
D about 50.3 minutes

Read the problem carefully. You know how long it takes Craig and Delsin to paint a room individually. Model the situation with an equation to find how long it would take them to paint the room if they work together.

Find the rate that each person works when painting individually.

Craig's rate: $\frac{1 \text{ job}}{75 \text{ minutes}} = \frac{1}{75}$ job per minute

Delsin's rate: $\frac{1 \text{ job}}{60 \text{ minutes}} = \frac{1}{60}$ job per minute

Let t represent the number of minutes it would take them to complete the job working together. Multiply each rate by the time t to represent the portion of the job done by each painter. Add these expressions and set them equal to 1 job. Then solve for t.

Portion that Craig completes	plus	portion that Delsin completes	equals	1 job.
$\frac{1}{75}t$	$+$	$\frac{1}{60}t$	$=$	1

Solve for t:

$\frac{1}{75}t + \frac{1}{60}t = 1$ **Original equation**

$300\left(\frac{1}{75}t + \frac{1}{60}t\right) = 1(300)$ **Multiply each side by the LCD, 300.**

$4t + 5t = 300$ **Simplify.**

$9t = 300$ **Combine like terms.**

$t \approx 33.3$ **Divide each side by 9.**

So, it would take Craig and Delsin about 33.3 minutes to paint the room working together. The correct answer is A.

Exercises

Read each problem. Identify what you need to know. Then use the information in the problem to solve.

1. Hana can finish a puzzle in 6 hours, while Eric can finish one in 5 hours. How long would it take them to finish a puzzle together? Round to the nearest tenth.

A about 1.8 hours

B about 2.4 hours

C about 2.5 hours

D about 2.7 hours

2. Roberto wants to print 500 flyers for his landscaping business. His printer can complete the job in 35 minutes, and his brother's printer can print them in 45 minutes. How long would it take to print the flyers using both printers? Round to the nearest whole minute.

F about 15 minutes

G about 18 minutes

H about 20 minutes

J about 23 minutes

CHAPTER 11

Standardized Test Practice

Cumulative, Chapters 1 through 11

Multiple Choice

Read each problem. Then fill in the correct answer on the answer document provided by your teacher or on a sheet of paper.

1. What is the inverse variation equation for the numbers shown in the table?

x	y
−8	16
−4	32
2	−64
8	−16
16	−8

A $y = -2x$

B $y = 8x$

C $xy = 24$

D $xy = -128$

2. Suppose a square has a side length given by the expression $\frac{x+5}{8x}$. What is the perimeter of the square?

F $\frac{4x + 20}{5x}$

G $\frac{2x + 10}{x}$

H $\frac{x + 5}{4x}$

J $\frac{x + 5}{2x}$

3. Find the distance between $(3, -6)$ and $(1, 4)$ on a coordinate grid. Round to the nearest tenth.

A 8.1

B 8.5

C 9.6

D 10.2

Test-TakingTip

Question 2 Sometimes you can eliminate answer choices as unreasonable because they are not in the proper form. Choices A and B show direct variation equations, so they can be eliminated.

4. In 1985, the population of a country was about 3.66 million people. By 2005, this number had grown to about 4.04 million people. What was the annual rate of change in population from 1985 to 2005?

F about 15,000 people per year

G about 19,000 people per year

H about 24,000 people per year

J about 38,000 people per year

5. Ricky's Rentals rented 12 more bicycles than scooters last weekend for a total revenue of $2,125. How many scooters were rented?

Item	Rental Fee
Bicycle	$20
Scooter	$45

A 26

B 29

C 37

D 41

6. The table shows the relationship between calories and fat grams contained in orders of french fries from various restaurants.

Calories	Fat Grams
240	14
280	15
310	16
260	12
340	16
350	18
300	13

Assuming the data can best be described by a linear model, how many fat grams would be expected to be contained in a 315-calorie order of french fries?

F 15 fat grams

G 16 fat grams

H 17 fat grams

J 18 fat grams

Short Response/ [illegible]dded Response

Record your answer[illegible] the answer sheet provided by your teacher or [illegible] a sheet of paper.

7. Suppose t[illegible]st term of a geometric sequence is 3 and t[illegible]h term is 192.
[illegible] the common ratio of the sequence?
   a. [illegible] an equation that can be used to find the [illegible] term of the sequence.
   [illegible]hat is the sixth term of the sequence?

[illegible] **GRIDDED RESPONSE** Peggy is having a cement walkway installed around the perimeter of her swimming pool with the dimensions shown below. Write an expression for the total area of the pool and the walkway. Then evaluate the expression for $x = 3$ to find the area, in square feet, of the pool and walkway.

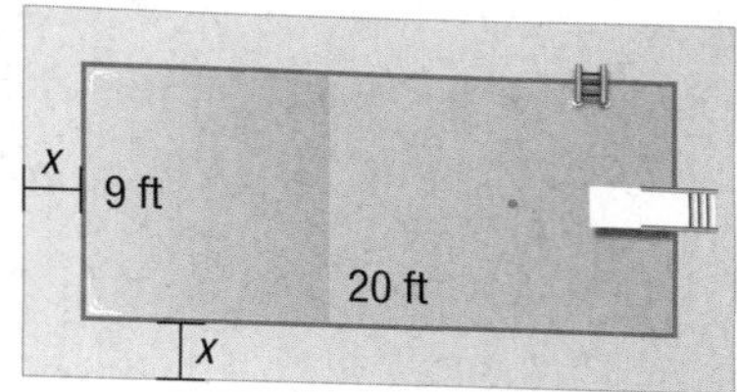

9. Use the equation $y = 2(4 + x)$ to answer each question.
   a. Complete the following table for the different values of x.
   b. Plot the points from the table on a coordinate grid. What do you notice about the points?

x	y
1	
2	
3	
4	
5	
6	

10. Jason received a \$50 gift certificate for his birthday. He wants to buy a DVD and a poster from a media store. (Assume that sales tax is included in the prices.) Write and solve a linear inequality to show how much he would have left to spend after making these purchases.

Weekend Blowout Sale
- All DVDs only **\$14.95**
- All CDs only **\$11.25**
- All posters only **\$10.99**

11. Simplify the complex fraction. Show your work.

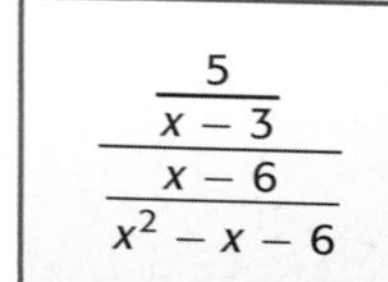

Extended Response

Record your answers on a sheet of paper. Show your work.

12. Carl's father is building a tool chest that is shaped like a rectangular prism. He wants the tool chest to have a volume of 30 cubic feet. The height of the chest will be 1 foot shorter than the width. The length will be 3 feet longer than the height.
   a. Sketch a model to represent the problem.
   b. Write a polynomial that represents the volume of the tool chest.
   c. What are the dimensions of the tool chest?

Need Extra Help?

If you missed Question...	1	2	3	4	5	6	7	8	9	10	11	12
Go to Lesson or Page...	11-1	11-6	10-6	3-3	6-2	4-5	9-8	7-7	1-4	5-1	11-7	8-6

CHAPTER 12 Statistics and Probability

Then

In chapter 0, you calculated simple probability.

Now

In Chapter 12, you will:

- Design surveys and evaluate results.
- Use permutations and combinations.
- Find probabilities of compound events.
- Design and use simulations.

Why?

RESTAURANTS A restaurant may ask their customers to complete a survey about their visit. The survey data can be analyzed using statistical methods. The restaurant staff can learn more about their customers and how to improve their experiences in the restaurant.

Statistics and Probability

Activity

Cathy's Café asks their customers to fill out the following survey.

Cathy's Cafe

We welcome your comments. Please fill out this questionnaire. Thank You!

1. Please rate the quality of the service you received from the server.
2. Please rate the quality of your main dish.
3. Please rate the quality of your Desserts.
4. Please rate the cleanliness of your booth.
5. Please rate your overall Dinner experience.

2/6

Math *in Motion*, Animation glencoe.com

EXPLORE 12-1

Algebra Lab
Survey Questions

For a survey to be valid, it should contain no bias or favoritism. Even though you may question people who are chosen randomly, questions may be worded to influence responses. These two different surveys on Internet sales tax had different results.

ACTIVITY

Analyze the difference between the two survey questions.

Question 1

Should there be sales tax on purchases made on the Internet?

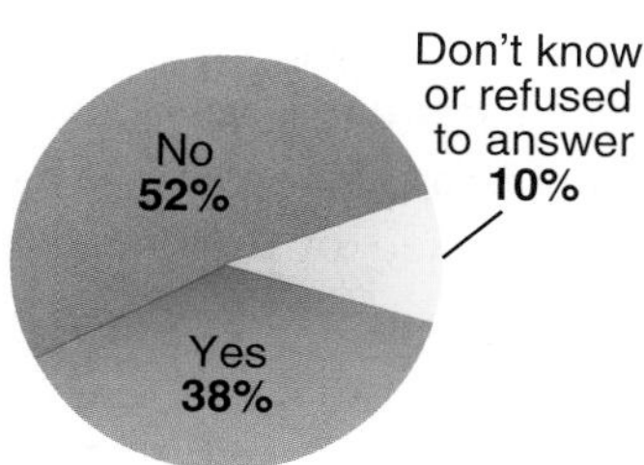

Question 2

Do you think people should or should not be required to pay the same sales tax for purchases made over the Internet as those made at a local store?

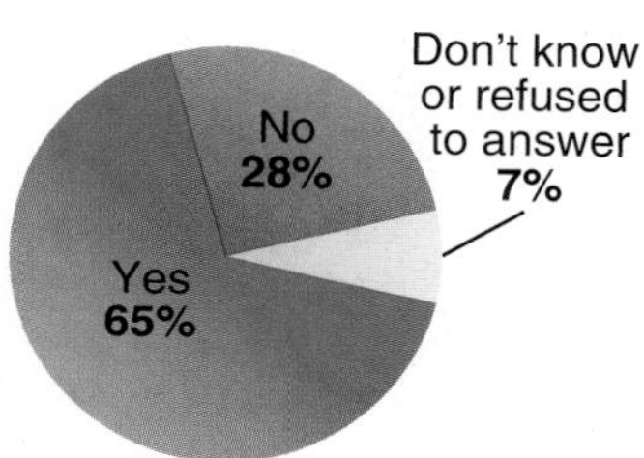

Notice that Question 2 includes more information. Pointing out that customers pay sales tax for items bought at a local store may give the people answering the survey a reason to say "yes."

Because they are random samples, the results of both of these surveys are accurate. However, the results could be used in a misleading way by someone with an interest in the issue. For example, an Internet retailer would prefer to state the results of Question 1.

Exercises

For Exercises 1 and 2, tell whether each question is likely to bias the results. Write *yes* or *no*. Explain your reasoning.

1. On a survey on environmental issues:
   - **a.** "Due to diminishing resources, should a law be made to require recycling?"
   - **b.** "Should the government require citizens to participate in recycling efforts?"
2. On a survey on education:
   - **a.** "Should schools fund extracurricular sports programs?"
   - **b.** "The budget of the River Valley School District is short of funds. Should taxes be raised in order for the district to fund extracurricular sports programs?"
3. You want to determine whether to serve hamburgers or pizza at a class party.
   - **a.** Write a survey question that would likely produce biased results.
   - **b.** Write a survey question that would likely produce unbiased results.

12-1

Designing a Survey

Then

You organized data by using matrices. (Lesson 6-7)

Now

- Design surveys.
- Identify various sampling techniques.

New Vocabulary

sample
population
survey
observational study
experiment
biased sample
simple random sample
stratified random sample
systematic random sample

Math Online

glencoe.com

- Extra Examples
- Personal Tutor
- Self-Check Quiz
- Homework Help

Why?

When manufacturing T-shirts, many steps and items must be checked for quality. These include fabrication, care labels, tags, trims, print artwork, and embroidery. It would be costly for a company to have each T-shirt inspected. Instead, they inspect a certain number of T-shirts.

All of the T-shirts that are made are a population, and the T-shirts that are inspected are a sample of the population. The inspectors draw conclusions about the sample and apply those conclusions to the entire population.

Design a Survey A **sample** is some portion of a larger group, called the **population**. Since it is impractical to examine every item in a population, a sample is selected to represent the population. After the sample is analyzed, conclusions can be drawn about the entire population. The larger the sample size, or the more samples taken, the more closely it approximates the population.

To accurately draw a conclusion from data received from a sample, you will need to first decide on the best method of collecting the data.

Key Concept Data Collection Techniques — For Your FOLDABLE

Type	Definition/Use	Example
survey	• Data are from responses given by a sample of the population. • To make a general conclusion about the population.	To determine whether the student body is happy with the spring dance theme, the dance committee asks a random sample of 50 students for their opinion.
observational study	• Data are recorded after just observing the sample. • To compare reactions and draw a conclusion about responses of the population.	A toy company watches some children play and notes the toys they play with the most. They conclude that the population of two-year-olds prefers toys that sing to toys that do not make noise.
experiment	• Data are recorded after changing the sample. • To make general conclusions about what will happen during an event.	A quality control manager runs the assembly machines 10 times at a certain rate. Each time the product is defective. She concludes that it would happen every time the machine runs at that pace.

StudyTip

Census A *census* is a survey in which every member of the population is questioned. So there is no sample.

EXAMPLE 1 Classify Data Collection Techniques

CHARITY A local charity is interested in finding out whether people are likely to give money to charity. They distributed 1000 questionnaires to people living in the neighborhood.

a. Identify the sample, and determine the population from which it was selected.

The sample is the 1000 people who received the questionnaires. The population is all the people in the neighborhood.

b. Classify the type of data collection used by this charity.

This is a survey. The data are from responses given by people in the sample.

Check Your Progress

Identify each sample, and suggest a population from which it was selected. Then classify the type of data collection used.

1A. RESEARCH A research facility analyzed two groups of rats to determine their reaction to sugar.

Group 1
Food with sugar

Group 2
Food with no sugar

1B. RECYCLING The city council wants to start a recycling program. They send out a questionnaire to 1000 random citizens asking what they would recycle.

Personal Tutor glencoe.com

There are factors that affect the collection of data and the conclusions drawn. If a sample favors one group over another, then the data are invalid because it is a **biased sample**. A sample is *unbiased* if it is random. Members of a **random sample** have an equal probability of being chosen.

EXAMPLE 2 Identify if the Sample is Valid

Identify each sample as *biased* or *unbiased*. Explain your reasoning.

a. MUSIC Every fifth person coming into a grocery store is asked to name a favorite radio station.

Unbiased; the sample is a random selection of people.

b. MUSIC Every fifth person at the Country Music Showcase is asked to name their favorite radio station.

Biased; because they are at a country music show, people may be more likely to select a country music station.

Check Your Progress

2A. POLITICS A journalist visits a senior center and chooses 10 individuals randomly to poll about various political topics.

2B. SHOES A shoe company conducts an observation study that involves 10 girls and 2 boys to see which shoes are the most popular.

Personal Tutor glencoe.com

Sampling Techniques Sample data are often used to estimate a characteristic within an entire population, such as voting preferences. A random sample of a population is selected so that it is representative of the entire population without any preference. Three common types of random samples are listed below.

Key Concept Random Samples *For Your FOLDABLE*

Type	Definition	Example
simple random sample	A sample that is equally likely to be chosen as any other sample from the population.	One hundred student ID numbers are randomly drawn from a hat, and those students are given a survey.
stratified random sample	The population is first divided in similar, nonoverlapping groups. A random sample is then selected from each group.	To reflect the diversity of the country, a candidate surveys citizens of various groups, based on their percent of the population.
systematic random sample	A sample in which the items in the sample are selected according to a specified time or item interval.	Every 10 minutes a toy is inspected. Or every 50th toy is inspected.

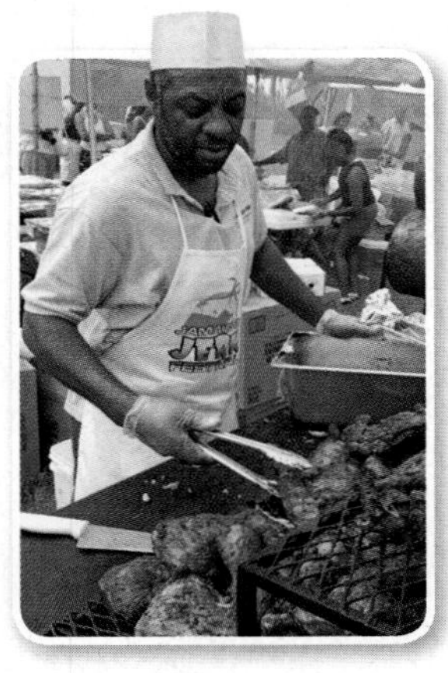

Real-World Link

Annually, in Pembroke Pines, Florida, the Jamaican Jerk Festival is held. A cooking competition is where each entrant must prepare any three dishes from the following categories: jerked pork, jerked chicken, jerked seafood, or other.

Source: Jamaican Jerk Festival

EXAMPLE 3 Classify a Random Sample

ZOOS Animals in a zoo are divided by continents. Then two animals were selected at random from each group to have their blood tested.

a. Identify the sample, and suggest a population from which it was selected.

The sample is the two animals selected from each continent. The population is the animals in the zoo.

b. Classify the sample as *simple*, *stratified*, or *systematic*. Explain your reasoning.

This is a stratified random sample. The animals are divided up into categories before there is a random selection.

Check Your Progress

Identify each sample, and suggest a population from which it was selected. Then classify the sample as *simple*, *stratified*, or *systematic*. Explain your reasoning.

3A. CONTESTS Refer to the information at the left. The cooks lined up randomly within their category, and every tenth cook in each category was selected to explain the rules.

3B. FOOD At a popular hamburger restaurant, the manager checks the quality of the burgers every 20 minutes, starting at a randomly selected time.

3C. SHOWER At a bridal shower, a sticker was placed on the bottom of three random plates. The guests who receive the starred plates will win a prize.

Personal Tutor glencoe.com

Check Your Understanding

Example 1 p. 741

Identify each sample, and determine a population from which it was selected. Then classify the type of data collection used.

1. **MUSIC** A record company wants to test five designs for an album cover. They randomly invite ten teens from a local high school to view the album covers.

2. **PARTIES** Federico is trying to decide on a theme and a color scheme for his party. He sends a survey in each invitation, asking guests for their opinions.

Example 2 p. 741

Identify each sample as *biased* or *unbiased*. Explain your reasoning.

3. **POLITICS** A group of students stands at the door of the school and asks every tenth student who they would vote for in the upcoming election and why.

4. **SHOPPING** Every fifteenth shopper at a clothing store is asked what they would want most for their birthday.

Example 3 p. 742

Identify the sample, and suggest a population from which it was selected. Then classify the sample as *simple*, *stratified*, or *systematic*. Explain your reasoning.

5. **SPORTS CARDS** Greg divides his rookie baseball cards by teams. Then he randomly selects cards and records the players' RBIs.

6. **TELEVISION** A nostalgia television network wants to conduct a cartoon marathon. To choose the episodes, they mail a questionnaire to people selected at random throughout the country.

Practice and Problem Solving

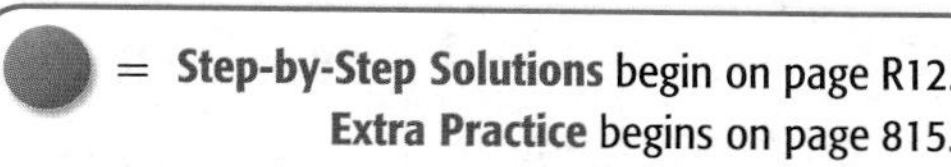
= **Step-by-Step Solutions** begin on page R12.
Extra Practice begins on page 815.

Example 1 p. 741

Identify each sample, and determine a population from which it was selected. Then classify the type of data collection used.

7. **FOOD** A frozen food company is considering creating frozen meals with tofu instead of meat. At a testing, they randomly give half of a group of 100 people the meals with meat and the other half the same meals with tofu and ask the people how they like the meals.

8. **PETS** The owners of dog care center want to know how many of each size crate they should order. They send flyers into the neighborhood to ask what size or breed of dog each person has.

9. **TRAVEL** A travel agency asks each of its customers for the past two years about their favorite and least favorite destinations.

Example 2 p. 741

Identify each sample as *biased* or *unbiased*. Explain your reasoning.

10. **MAGAZINES** A magazine publisher asks every tenth person at a fitness expo what magazines they have in their household.

11. **LIBRARY** The local library asks everyone who checks out a book if they also used the computers at the library.

12. **JEANS** A clothing chain gives its customers a card they can mail back that asks them questions about the customer's favorite brand of jeans.

13. **AMUSEMENT PARKS** An amusement park is deciding which rides to replace next year. As they leave the park, they ask teens what their least favorite ride is.

Example 3 p. 742

Identify the sample, and suggest a population from which it was selected. Then classify the sample as *simple, stratified,* or *systematic.* Explain your reasoning.

14. **TANNING** A tanning salon sorts its responses to a survey by the home states of the respondents. Then they are sorted to select teenagers.

15. **ART** Mitsu uses her blog about art to ask readers about their favorite medium and supply center. She then tabulates and publishes the results.

16. **CARS** The service manger at a car dealership inspects every fifth car to make sure cars are detailed after being serviced.

17. **MUSIC** A music store gives every fiftieth customer a free CD by a local artist.

18. **ELECTIONS** To estimate who the leading candidate is, the candidate's committee surveys a large group of people selected at random. The returns indicate that their candidate is leading 58% to 42%.

    a. Identify the sample. Suggest a population from which the sample was selected.
    b. State the method of data collection.
    c. Is the sample *biased* or *unbiased*? Explain.
    d. If unbiased, classify the random sample as *simple, stratified,* or *systematic.*

Real-World Link

The world's largest collection of Converse shoes is held by Joshua Mueller of Lakewood, Washington, consisting of 403 unique pairs.

Source: *The Guinness Book of World Records*

19. **SHOES** A shoe company surveys their customers about shoe design. This program keeps a count of styles and colors chosen by customers.

    a. Identify the sample. From what population was the sample selected?
    b. State the method of data collection.
    c. Is the sample *biased* or *unbiased*? Explain.
    d. If unbiased, classify the sample as *simple, stratified,* or *systematic.*

20. **MULTIPLE REPRESENTATIONS** Design and conduct your own survey.

    a. **WRITING** Write a question you would like to answer through a survey. The question should be meaningful to you. Describe the method you will use to gather the data, and explain why you chose that method.
    b. **ANALYTICAL** Devise a method to conduct your survey using an unbiased sample. Explain why you chose your sample.
    c. **CONCRETE** Conduct your survey.
    d. **TABULAR** Record your results in a table.
    e. **GRAPHICAL** Use a graph (line, circle, histogram, etc...) or other visual or graphic method to present your results to the class.

H.O.T. Problems Use Higher-Order Thinking Skills

21. **CHALLENGE** Consider the following survey proposal.

    Question: How do students feel about the new dress code?
    Method: Divide the student body by their four grade levels. Then, take a simple random sample from each of the four grades. Conduct the survey using this sample.

    Discuss the strengths and weaknesses of this survey.

22. **REASONING** Compare and contrast the three data collection techniques described in this lesson.

23. **OPEN ENDED** Describe a real-world example of an observational study.

24. **WRITING IN MATH** Explain why accurate surveys are important to companies, and how the companies use them.

Standardized Test Practice

25. GRIDDED RESPONSE The first stage of a rocket burns 28 seconds longer than the second stage. If the total burning time is 152 seconds, how many seconds is the first stage?

26. Ms. Brinkman invested \$30,000; part at 5%, and part at 8%. The total interest on the investment was \$2100 after one year. How much did she invest at 8%?

A \$10,000
B \$15,000
C \$20,000
D \$25,000

27. A pair of \$25 jeans is on sale for 15% off. What is the sale price?

F \$21.25
G \$24.25
H \$23.25
J \$22.25

28. GEOMETRY A piece of wire 42 centimeters long is bent into the shape of a rectangle with a width that is twice its length. Find the dimensions of the rectangle.

A 5 cm, 12 cm
B 7 cm, 14 cm
C 9 cm, 16 cm
D 11 cm, 18 cm

Spiral Review

Solve each equation. State any extraneous solutions. (Lesson 11-8)

29. $\frac{3}{c} = \frac{2}{c+2}$

30. $\frac{4}{f} = \frac{2}{f-3}$

31. $\frac{j}{j+2} = \frac{j-6}{j-2}$

32. $\frac{h-2}{h} = \frac{h-2}{h-5}$

33. $\frac{3m}{4} + \frac{1}{3} = \frac{3m+4}{6}$

34. $\frac{6}{5} + \frac{4p}{3} = \frac{8p}{5}$

35. $\frac{r-2}{r+2} - \frac{3r}{r-2} = -2$

36. $\frac{t-3}{t+3} - \frac{2t}{t-3} = -1$

37. $\frac{4v}{2v+3} - \frac{2v}{2v-3} = 1$

38. SPORTS When air is pumped into a ball, the pressure required can be computed by using the formula $P = \frac{3412.94}{\frac{4\pi r^3}{3}}$, where P represents the pressure in pound per square inch (psi), and r is the radius of the ball in inches.

a. Simplify the complex fraction.

b. Suppose the air pressure inside the ball is 8 psi. Approximate the radius of the ball to the nearest hundredth.

39. ROLLER COASTERS Suppose a roller coaster climbs 208 feet higher than its starting point, moving horizontally 360 feet. When it comes down, it moves horizontally 44 feet. (Lesson 10-5)

a. How far will it travel to get to the top of the ride?

b. How far will it travel on the downhill track?

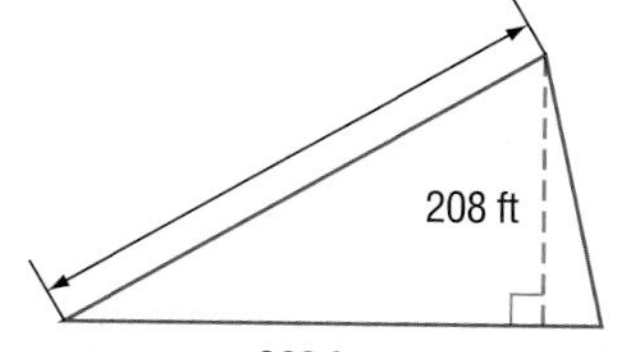

Skills Review

40. PHYSICAL SCIENCE Mr. Blackwell's students recorded the height of an object above the ground after it was dropped from a height of 5 meters. (Lesson 1-7)

Time (s)	0	0.2	0.4	0.6	0.8	1
Height (cm)	500	480	422	324	186	10

Draw a graph showing the relationship between the height of the object and time.

12-2 Analyzing Survey Results

Then
You designed surveys. (Lesson 12-1)

Now
- Summarize survey results.
- Evaluate survey results.

New Vocabulary
measures of central tendency
quantitative data
qualitative data

Math Online
glencoe.com
- Extra Examples
- Personal Tutor
- Self-Check Quiz
- Homework Help

Why?

Companies like to use surveys to get feedback on how they are doing in areas ranging from sales to their Web site.

A company recently received these results from a survey about their Web site.

What do these values mean? How were the data collected? Is the sample an accurate representation of their customers?

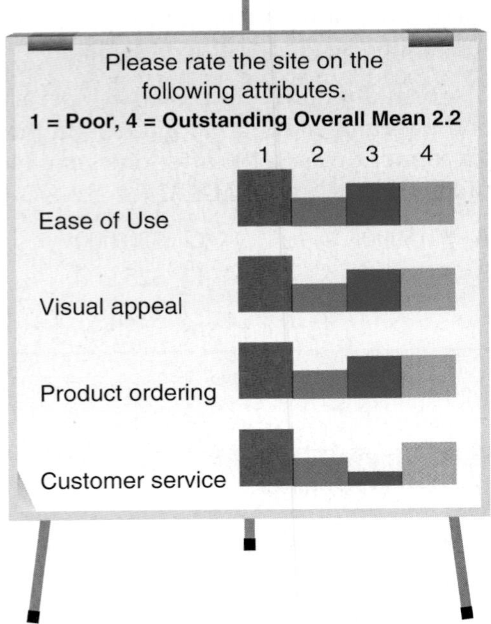

Summarize Survey Results Once data from a survey have been collected, they need to be summarized to be meaningful. We can summarize the data of a survey according to **measures of central tendency**.

Concept Summary Measures of Central Tendency — For Your FOLDABLE

Type	Description	When Best Used
mean	the sum of the data divided by the number of items in the data set	The data sets have no outliers.
median	the middle number of the ordered data, or the mean of the middle two numbers	The data set has outliers, but there are no big gaps in the middle of the data.
mode	the number or numbers that occur most often	The data set has many repeated numbers.

Some data cannot be analyzed using statistical methods. **Quantitative data** can be given and analyzed as numerical values. Some examples of these are test scores, hours that you have studied, or the weight of objects. **Qualitative data** cannot be given a numerical value. Some examples of these are gender, nationality, or television show preference.

It is also possible to have quantitative data and still not be able to find a measure of central tendency. This occurs when the data do not represent the same thing.

EXAMPLE 1 Select a Method to Summarize

Which measure of central tendency best represents the data, if any? Justify your answer. Then find the measure.

a. NUTRITION The table shows the number of Calories per serving of each vegetable.

Vegetable	Calories	Vegetable	Calories
asparagus	14	cauliflower	10
beans	30	celery	17
bell pepper	20	corn	66
broccoli	25	lettuce	9
cabbage	17	spinach	9
carrots	28	zucchini	17

List the values from least to greatest: 9, 9, 10, 14, 17, 17, 17, 20, 25, 28, 30, 66.

There is one value that is much greater than the rest of the data, 66. Also, there does not appear to be big gap in the middle of the data. There are only two sets of identical numbers. So, the median would best represent the data.

$$\{9, 9, 10, 14, 17, \underbrace{17, 17,}_{\text{The median is 17 Calories.}} 20, 25, 28, 30, 66\}$$

b. CONCERTS An amphitheater conducted a survey in which they asked 1000 adults the last time they attended a concert.

Results	
Response	**Percentage**
at least 3 years ago	8
1–3 years ago	15
6 months to 1 year ago	45
less than 6 months ago	32

A measure of central tendency cannot be calculated for this set of data. Each percentage in the table represents something different.

For example, 15% of the respondents attended a concert 1 to 3 years ago, while 32% saw a concert less than 6 months ago. So, the median value of 23.5% has no meaning in this situation.

Check Your Progress

1A. FINANCIAL LITERACY An electronics store records the number of customers it has during each hour of the day.

Number of Customers			
86	71	79	86
79	32	88	86
82	69	71	70
86	81	85	86

1B. BOOKS In a survey, students between the ages of thirteen and eighteen reported where they get their books. The responses were: teachers, 420; school library, 1320; public library, 1020; parents, 720; bookstore, 1020; Internet, 540; friends, 540; as a gift, 1020.

Watch Out!

Percents Always make sure that a survey that gives data in percents tells the size of the sample.

Personal Tutor glencoe.com

Evaluate Survey Results Once a survey has been conducted, data are summarized, a report of the findings and conclusions is made. However, bias can sometimes cause errors in the data, as well as how they are interpreted and reported.

You need to be able to judge the reliability of these survey reports. You can do this by making sure that the sample is random, large enough to be an accurate representation of the population, and that the source of the data is a reliable one.

Often newspaper, magazine, and television reports include the results of a survey. These surveys need to be judged for their validity before you make a decision based on them. Some questions that you may want to ask yourself are:

- What are the population and samples? Can I identify them easily? Are they biased?
- What is the source? Is the data source a reputable group? Could they be biased?
- Do the data actually support the conclusion?

Real-World Link

Generally, the larger the venue, the more you pay to get the best seats in the house. Arenas with 2000 to 4999 seats have a median price of $41.50 per seat.

Source: *USA TODAY*

EXAMPLE 2 Evaluate a Survey

YEARBOOKS Given the following portion of a survey report, evaluate the validity of the information and conclusion.

Question: **Should the school have an electronic yearbook this year?**

Sample: **Ballots were placed in random students' lockers.**

Conclusion: **The school should only offer an electronic yearbook this year.**

Results	
Choice	**Response**
electronic only	67%
traditional paper	22%
offer both	9%
no preference	2%

While the report states that students were chosen randomly, it does not say how many students were chosen. The results were given in percents. The 67% could mean about 34 out of 50. This may not be a large enough sample to represent a large school.

Check Your Progress

2. **CONCERTS** At a sold-out concert in a 5000-seat concert hall, every 10th attendee completed a survey.

Question: Did you feel that the price of the tickets was reasonable?

Conclusion: The prices for the tickets are reasonable and should remain the same.

Results	
Choice	**Response**
very reasonable	56
reasonable	185
somewhat reasonable	132
unreasonable	69
very unreasonable	58

Personal Tutor glencoe.com

The way in which results are displayed can influence how you interpret those results. Here are some factors.

- If the scale of a line graph, bar graph, or histogram is large, any changes may appear to be small, when they actually could be quite significant. If the scale of a graph is small, small changes or differences can be made to appear quite large.

Notice the scale in the graph at the right. This graph is misleading because it appears that gas prices are not increasing too much, when they are actually quite significant.

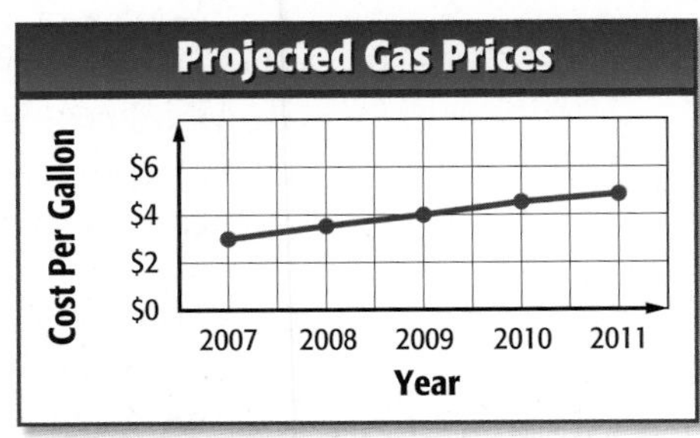

The following are features of a display that can influence the conclusion.

- The scales of graphs should be constant.
- Using percents rather than the actual numbers from a set of data can give a misleading result. However, if the numbers in the sample are large, percents best represent values.
- In a bar graph or histogram, all of the bars should have the same width. The heights of the bars represent the data values. Changing the width of a bar can exaggerate differences.
- If the colors on a circle graph, bar graph, or histogram are different shades of the same color, groups may visually blend together and influence how you interpret the results.

Real-World Link

Most public schools in the U.S. do not require uniforms, but have some kind of dress code. In 1994, a public school district in southern California made school uniforms mandatory that began a trend across the country.

Source: Schoolgirl Princess

EXAMPLE 3 Misleading Results

UNIFORMS A high school principal is considering whether to institute a school uniform policy. She sends out a survey to the students at her high school to get their opinions.

Question: How would you feel about having a school uniform policy?

Conclusion: It would not bother students if a uniform policy were instituted.

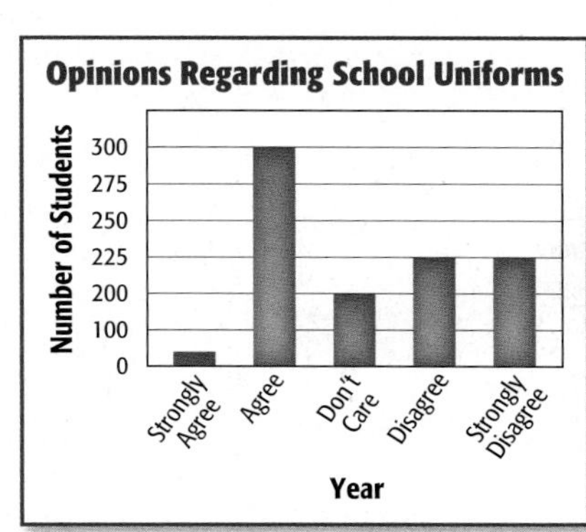

Determine whether the display gives an accurate picture of the survey results.

Upon first inspection, most of the students appear to agree with the uniforms. However, the scale in the graph is not constant. With a closer look, it appears that about 400 students either disagree or strongly disagree while a little more than 300 either agree or strongly agree.

In this case, the table is misleading, and the conclusion is invalid.

Check Your Progress

3. The city council wanted to see how local companies were donating resources to charities. The bar graph shows the results.

Question: How does your company contribute to local charities?

Conclusion: Donating money is the least popular contribution by local companies.

Determine whether the display gives an accurate picture of the survey results. Explain.

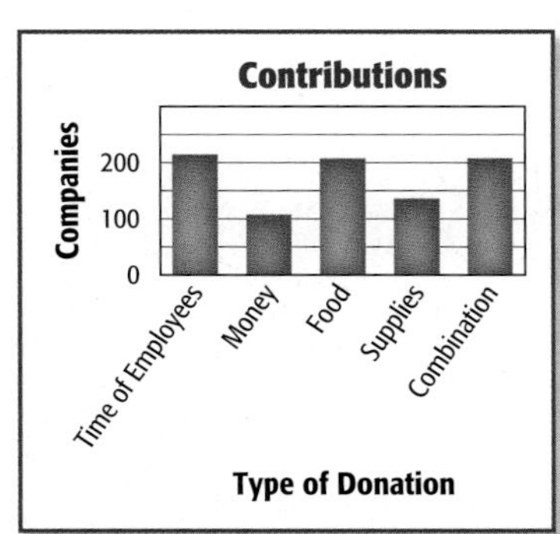

Personal Tutor glencoe.com

Check Your Understanding

Example 1 p. 747 **Which measure of central tendency best represents the data? Justify your answer. Then find the measure.**

1. **RECYCLING** Archmont High School is recycling newspapers as a fundraiser to buy some benches for the courtyard. The newspapers are gathered into 5-inch bundles. Ms. Sato counted the bundles received each Friday for the first quarter of the school year: 15, 12, 14, 15, 18, 15, 13, 14, 13, 12.

2. **TRAVEL** An online travel agency wants to design tours for families with teens. So, they surveyed students about their favorite vacation destinations. The top five responses were: beach, 25%; theme parks, 22%; lakes, 21%; historical sites, 17%; mountains, 15%.

Example 2 p. 748 **Given each survey report, evaluate the validity of the information and conclusion.**

3. **RADIO** A radio station is considering changing its format. It mails 1000 surveys to randomly selected houses within their listening area. They receive 750 responses.

Question: What type of radio station do you like?

Conclusion: The radio station should have a format of rap/hip-hop.

Results	
Choice	**Response**
talk/sports	26%
classic rock	32%
rap/hip-hop	39%
country	3%

4. **SHOPPING** A shopping mall owner wants to know during what hours the most people prefer to shop. One day every 10th person who entered the mall was asked what times he or she preferred to shop.

Conclusion: The mall should remain open from 9:00 A.M. until 9:00 P.M.

Results	
Choice	**Response**
before 9:00 A.M.	26%
9:00 A.M.–12:00 noon	12%
12:00 noon–3:00 P.M.	26%
3:00 P.M.–6:00 P.M.	27%
6:00 P.M.–9:00 P.M.	9%

5. **VOLUNTEERING** Sample: 21,700 6th- to 12th-grade students surveyed by *USA Weekend*, Youthnoise.com, and a volunteer organization called Key Club.

Question: Are youth interested in volunteering?

Conclusion: Youth are interested in volunteering.

How many hours a year do you volunteer?	
Number of Hours	**Percent (%)**
fewer than 20 hours	30
20 to 39 hours	35
40 to 59 hours	13
60 to 80 hours	7
more than 80 hours	15

6. **SPORTS** Sample: *Scholastic Magazine* asked about 3585 kids with online subscriptions about their favorite sports.

Question: What are kids' favorite sports to compete in?

Results: baseball/softball, 271; football, 436; basketball, 570; soccer, 279; hockey, 197; track, 209; swimming, 319; gymnastics, 197; skating, 289; bowling, 202; other, 616

Conclusion: Basketball is kids' favorite sport in which to compete.

Example 3 p. 749 **Determine whether each display gives an accurate picture of the survey results. Explain.**

7. **CONVERSATION** A nationwide survey was conducted on the time students in grades six through twelve spend talking one-on-one with a family member. The responses were divided by regions and are displayed in the graph.

   Conclusion: Students in the southeast talk with family members the most.

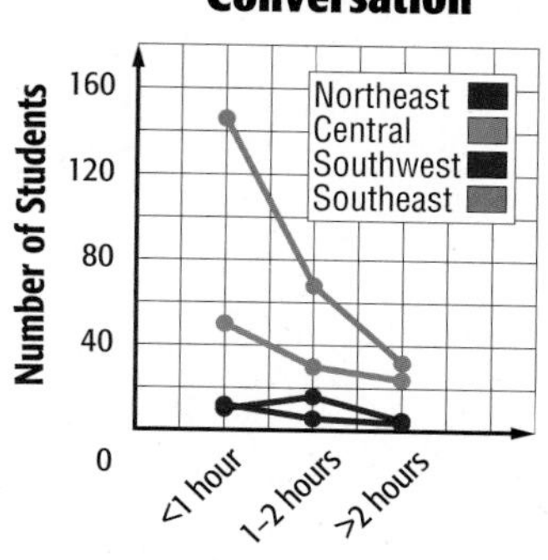

8. **TELEVISION** A survey conducted by a media network asked adults how many hours of television they watched each week. The results are displayed in the graph.

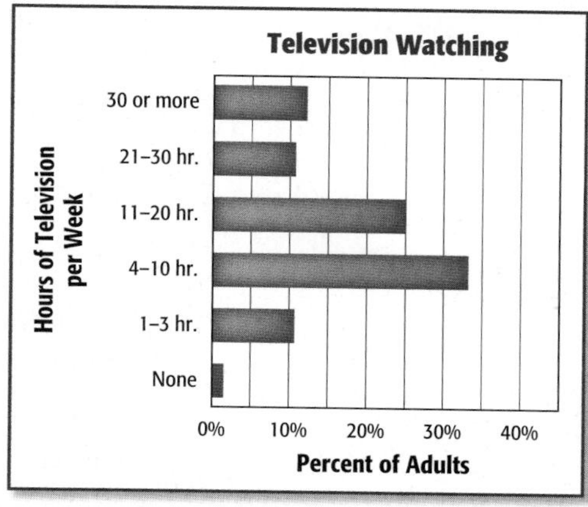

Practice and Problem Solving

= **Step-by-Step Solutions** begin on page R12.
Extra Practice begins on page 815.

Example 1 p. 747 **Which measure of central tendency best represents the data? Justify your answer. Then find the measure.**

9. **FOOD** A sub shop adds a new bread to its menu. To see if they should keep it, the manager counts how many orders of that bread type are taken each day: 10, 16, 14, 13, 17, 15, 18, 16, 19.

10. **MOVIES** A video store wants to order additional movies. They conducted a survey to find their members' favorite type of movie. The responses were: comedy, 21%; drama, 19%; horror, 12%; science fiction, 17%; action and adventure, 18%; mystery and suspense, 13%.

11. **MOTORCYCLES** A motorcycle dealership conducted a telephone survey of its customers from the last five years about customer satisfaction. The results are listed in the table.

Response	Weight	Total
very satisfied	5	182
satisfied	4	252
okay	3	365
dissatisfied	2	169
very dissatisfied	1	54

12. **CONTESTS** A beverage company introduced a contest in which winning codes were printed on bottle caps. One code awarded the winner a $1 million prize, two codes awarded each winner a new car worth $20,000, and 50,000 codes awarded each winner a free beverage worth $1.

13. **SUMMER ACTIVITIES** In a survey, students were asked about their favorite summer activity. The responses are listed in the table.

Summer Activities			
swimming	650	camping	432
travel	885	reading	281
sports	1123	other	514

Example 2
p. 748

Given each survey report, evaluate the validity of the information and conclusion.

14. **SERVICES** A salon wants to know which of its services gets used the most. It surveyed 1090 customers between January and March.

Question: For which service did you come in today?

Conclusion: The salon mostly does coloring and highlighting.

Results	
Choice	**Response**
haircut	294
styling	185
coloring/highlights	349
perm	153
combination of services	109

15. **FOOD** The freshman class decided to have a picnic on National Sandwich Day. They took a survey of all freshmen to decide which sandwich to serve.

Question: Which sandwich would you eat on National Sandwich Day?

Conclusion: They should serve peanut butter and jelly sandwiches.

Results	
Choice	**Response**
grilled cheese	10.6%
reuben	17.4%
hamburger	18.2%
hot dog	16.3%
peanut butter and jelly	37.5%

16. **NEWSPAPERS** To determine the popularity of the horoscope section, a newspaper sent at random a survey to 1000 of its subscribers.

Question: How often do you read your horoscope?

Results: every day, 9.9%; most days, 9.7%; not very often, 39.1%; never, 41.3%

Conclusion: The paper should eliminate the horoscope.

17. **MUSIC** A music store wanted to know where people hear about new music.

Question: How do you hear about new music being released?

Results: radio, 27%; TV, 24%; magazines, 19%; friends, 25%; the Internet, 5%

Conclusion: People hear about new music from a variety of sources.

18. **DRIVING** *The Canton Repository* polled 750 people.

Question: Have you ever talked on a cell phone while driving a car?

Results: never, 20.7%; a few times, 48.7%; not anymore, 5.1%; always, 25.5%

Conclusion: The people of Canton are careless drivers.

19. **ENTERTAINMENT** A national survey of students in first through twelfth grades was published in *Scholastic Magazine.* They received 5564 total votes.

Question: What type of TV shows and movies do you watch most?

Results: action, 1329; cartoons, 1115; comedy, 1423; drama, 358; horror, 332; music video, 1007

Conclusion: Students prefer comedies to other types of shows.

20. **BUSINESS** A poll was conducted by Junior Achievement.

Question: Is starting a business challenging?

Results: very easy, 2.5%; easy, 8.9%; somewhat challenging, 43.4%; difficult but possible, 44.1%; almost impossible, 1.1%

Conclusion: Students are aware of the challenges in starting a business.

21. **READING** A survey conducted by Smart Girl asked students why they read.

Results: just for fun, 25%; to learn new things, 24%; because they have to for school, 18%; because they get bored and have nothing else to do, 17%; their friends like to read and talk about books, 16%

Conclusion: Students read for a variety of reasons.

Example 3 p. 749

Determine whether each display gives an accurate picture of the survey results.

22. **MAGAZINES** The school library is ordering magazines. To make sure that they get ones that students will read, the librarian sends out a survey to 100 randomly chosen students from each grade.

Question: What type of magazine do you enjoy reading?

Conclusion: The library should order fashion, sports, music, and game magazines.

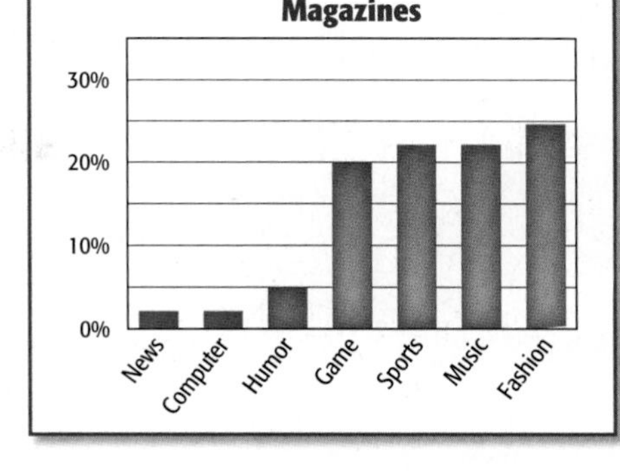

23. **ENVIRONMENT** A survey of 10,000 randomly chosen subscribers to an environmental magazine was conducted.

Question: What will be the biggest environmental challenge in the 21st century?

Conclusion: Finding places to put garbage is unimportant.

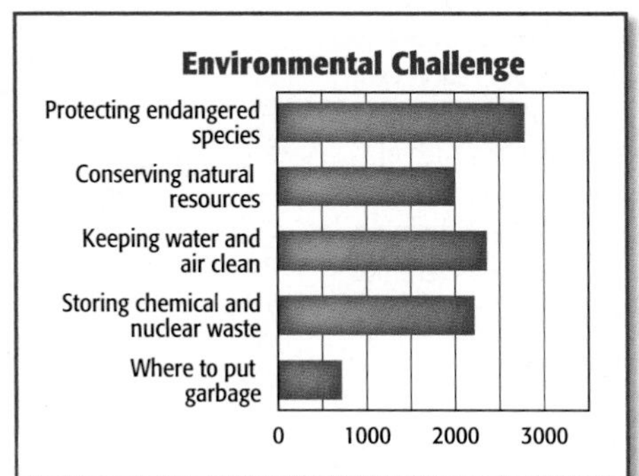

24. **ELECTIONS** A local elections board polled 498 high school seniors.

Question: Will elections ever be held online?

Conclusion: More seniors feel that elections will never be held online.

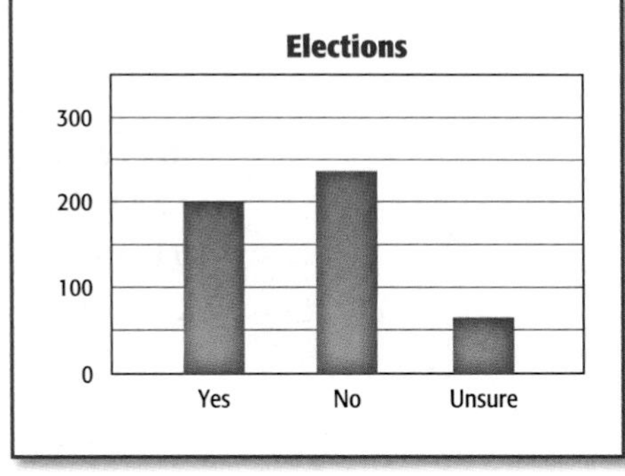

Real-World Link

First-time voters have become a focal point of political parties and interest groups. The number of new voters aged 18 to 24 has increased in recent years.

25. **SAFETY TRAINING** A chapter of the Red Cross offers classes designed for kids and teens in safety techniques. Of the participants, 74% take water safety class, 10% take babysitting classes, and 16% take first aid.

Question: Should the Red Cross continue babysitting classes? Write a valid conclusion using data to support your answer.

26. **BRACES** A journal for dentists conducted a nationwide survey of 5000 dentists. Does this graph accurately represent the data and conclusion? Justify your answer.

Question: What percent of patients do you refer to an orthodontist?

Percentage of Patients Referred

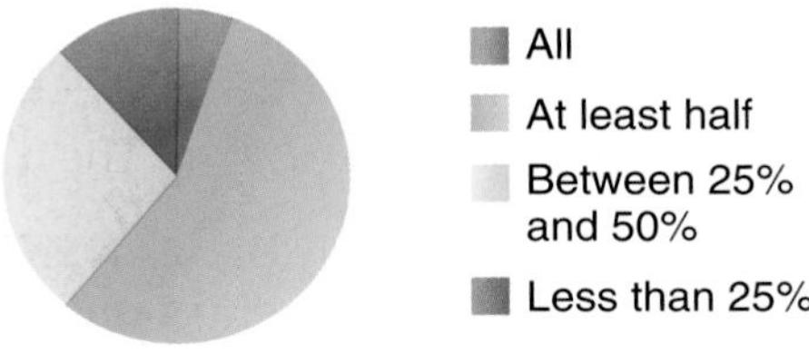

Conclusion: Many dentists refer more than half of their patients to orthodontists.

Real-World Link

Teenage girls are not that influenced by the behavior of celebrities. Of the 1700 teenage girls surveyed, only 5% thought that the excessive lifestyles of celebrities were cool. About 63% of the girls said that they were not influenced by celebrities.

Source: Yahoo News

27. **CELEBRITIES** A celebrity gossip magazine conducted a survey of their subscribers.

Question: Are you influenced by celebrities?

Conclusion: Their subscribers are not influenced by celebrities.

Does this graph accurately represent the data and conclusion? Justify your answer.

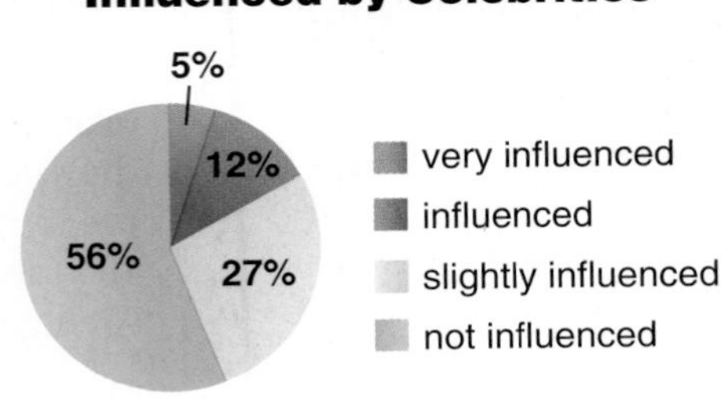

28. **MULTIPLE REPRESENTATIONS** In this problem, you will explore another way to analyze data.

a. **CONCRETE** Below is a distribution of coins in unequal stacks. Duplicate these stacks with coins.

b. **GRAPHICAL** Make a line plot of the stacks. Above each of these columns, record how much each stack differs from the mean (the number of coins per stack). Find the absolute value of each of these values.

c. **ANALYTICAL** Move the coins one at a time to make the stacks equal. Avoid unnecessary moves. Count the moves. The number of moves tells us how much the original set of stacks differs from the set of equal stacks.

d. **ANALYTICAL** Find the mean of the absolute values. Describe what this value is and what it means in these circumstances.

H.O.T. Problems

Use Higher-Order Thinking Skills

29. **FIND THE ERROR** Pepita and Ben are asked to decide which measure of central tendency to use given the data of test scores 84, 82, 80, 32, 87, 83, 85. Is either of them correct? Explain your reasoning.

Pepita	Ben
The mean is the best measure of central tendency because the data are clumped together, and there are no repeated values.	*The median is the best measure of central tendency because the data are clumped together, but there is an outlier, and there are no repeated values.*

30. **CHALLENGE** Find a set of numbers that satisfies each list of conditions.

a. The mean, median, and mode are all the same number.

b. The mean is greater than the median.

c. The mode is 10 and the median is greater then the mean.

d. The mean is 6, the median is 5.5, and the mode is 9.

31. **OPEN ENDED** Describe a survey you would like to conduct. Include the sample, population, method of questioning, and how you would display the results.

32. **WRITING IN MATH** Explain why a company may display survey results inaccurately. Give one example of how they might accomplish this.

Standardized Test Practice

33. At the county fair, 1000 tickets were sold. Adult tickets cost \$8.50, children's tickets cost \$4.50, and a total of \$7300 was collected. How many children's tickets were sold?

A 700 C 400
B 600 D 300

34. Edward has 20 dimes and nickels, which together total \$1.40. How many nickels does he have?

F 12 H 8
G 10 J 6

35. If 4.5 kilometers is about 2.8 miles, about how many miles is 6.1 kilometers?

A 3.2 miles C 3.8 miles
B 3.6 miles D 4.0 miles

36. EXTENDED RESPONSE Three times the width of a certain rectangle exceeds twice its length by three inches, and four times its length is twelve more than its perimeter.

a. Translate the sentences into equations.

b. Find the dimensions of the rectangle.

c. What is the area of the rectangle?

Spiral Review

Identify each sample and suggest a population from which it is selected. State whether the sample is *biased* or *unbiased*. If unbiased, classify the sample as *simple, stratified,* or *systematic*. (Lesson 12-1)

37. SCHOOL Twenty names were drawn from a container containing identical pieces of paper with the names of every member of the senior class. These seniors were then asked who they would choose for senior class president.

38. BOOKS To check the quality of the books being manufactured, an inspector checks every 50th book that comes off the line.

39. COMPUTERS Mayfield High School participated in a survey to find out how teens feel about certain issues involving social networks. The responses are divided into group by each age and then tallied for each question.

Find the zeros of each function. (Lesson 11-8)

40. $f(x) = \frac{x^2 - 8x + 15}{x^2 + 5x - 6}$

41. $f(x) = \frac{x^2 - x - 12}{x^2 - 6x + 8}$

42. $f(x) = \frac{x^2 - x - 30}{x^2 - 3x - 18}$

Find the values of the three trigonometric ratios for angle A. (Lesson 10-8)

43.

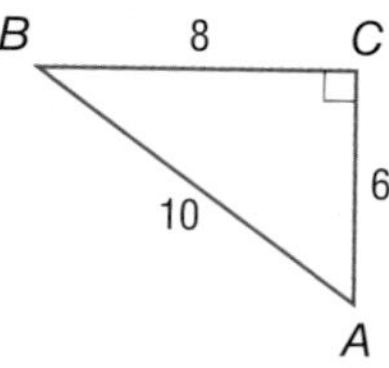

44.

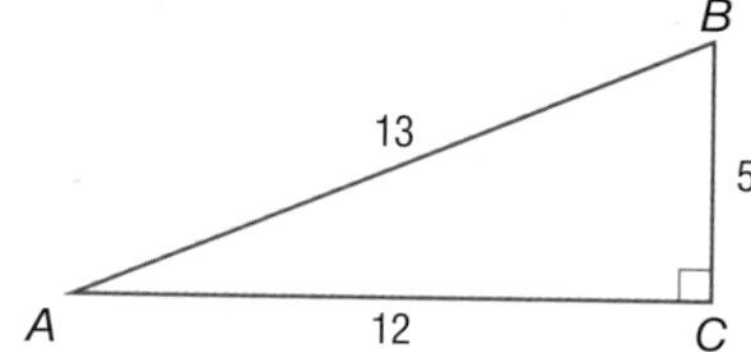

45.

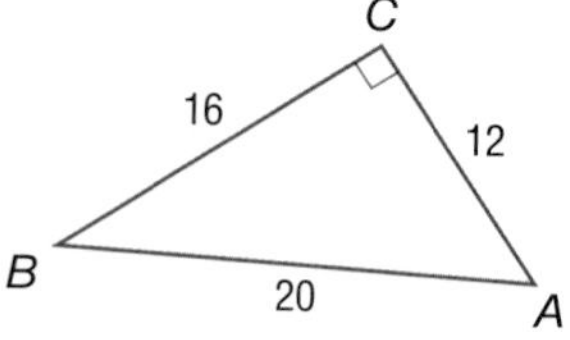

Skills Review

Find the mean, median, and mode to the nearest tenth for each set of data. (Lesson 0-12)

46. 100, 105, 100, 105, 100, 110

47. 12, 25, 14, 35, 42, 27, 31, 48

48. 90, 85, 92, 99, 78, 82, 92, 90

49. 1, 5, 3, 7, 6, 2, 9, 2, 5, 1, 9, 1

50. 55, 65, 45, 35, 65, 25, 85

51. 25, 28, 21, 26, 25, 27, 29, 30

12-3 Statistics and Parameters

Then

You organized and summarized survey results. (Lesson 12-2)

Now

- Use statistics to analyze survey results.
- Analyze data sets using statistics.

New Vocabulary

statistical inference
statistic
parameter
univariate data
measures of variation
mean absolute deviation
standard deviation
variance

Math Online

glencoe.com

- Extra Examples
- Personal Tutor
- Self-Check Quiz
- Homework Help

Why?

At the start of every class period for one week, each of Mr. Day's algebra students randomly draws 9 pennies from a jar of 1000 pennies. Each student calculates the mean age of the random sample of pennies drawn and then returns the pennies to the jar.

How does the mean age for 9 pennies compare to the mean age of all 1000 pennies?

Statistics and Parameters In this situation, the statistics of a sample are used to draw conclusions about the entire population. This is called **statistical inference**.

In the scenario above, each student takes a random sample of pennies from the jar. The jar of 1000 pennies represents the population. A **statistic** is a measure that describes a characteristic of a sample. A **parameter** is a measure that describes a characteristic of a population. Parameters are usually estimated values based on the statistics of a carefully chosen random sample. A statistic can and usually will vary from sample to sample. A parameter will not change, for it represents the entire population.

EXAMPLE 1 Identify Statistics and Parameters

Identify the sample and the population for each situation. Then describe the sample statistic and the population parameter.

a. At a local university, a random sample of 40 scholarship applicants is selected. The mean grade-point average of the 40 applicants is calculated.

Sample: the group of 40 scholarship applicants
Population: all applicants
Sample statistic: mean grade-point average of the sample
Population parameter: mean grade-point average of all applicants

b. A stratified random sample of registered nurses is selected from all hospitals in a three-county area, and the median salary is calculated.

Sample: randomly selected registered nurses from hospitals in three-county area
Population: all nurses at the hospitals in the same region
Sample statistic: median salary of nurses in the sample
Population parameter: median salary of all nurses in sampled hospitals

Check Your Progress

1. CEREAL Starting with a randomly selected box of Co-Co-Chunks from the manufacturing line, every 50th box of cereal is removed and weighed. The mode weight of a day's sample is calculated.

Personal Tutor glencoe.com

Review Vocabulary

bivariate data data that involve two variables (Lesson 4-5)

Statistical Analysis Data that involve only one variable are called **univariate data**. This kind of data can be represented by measures of central tendency, such as the mean, median, and mode. Univariate data can also be represented by **measures of variation**, such as range, quartiles, and interquartile range.

Concept Summary — Measures of Variation — For Your FOLDABLE

Type	Description	When Best Used
range	the difference between the greatest and least values	to describe which numbers are included in the data set
quartile	the values that divide the data set into four equal parts	to determine values in the upper or lower portions of a data set
interquartile range	the range of the middle half of a data set; the difference between the upper and lower quartiles	to determine what values lie in the middle half of the data set

The **mean absolute deviation** is the average of the absolute values of the differences between the mean and each value in the data set. Recall that absolute value is the distance from a number to zero on a number line.

Key Concept — Mean Absolute Deviation — For Your FOLDABLE

Step 1 Find the mean.

Step 2 Find the sum of the absolute values of the differences between each value in the set of data and the mean.

Step 3 Divide the sum by the number of values in the set of data.

Real-World Link

Recently, Japanese comics for girls, called "shojo," have become popular. These comics are available in three forms: comic books, graphic novels, and online comics.

Source: Disney Family

EXAMPLE 2 Use the Mean Absolute Deviation

MARKETING Each person that visited the Comic Book Shoppe's web site was asked to enter the number of times each month they buy a comic book. They received the following responses in one day: 2, 2, 3, 4, 14. Find the mean absolute deviation to the nearest tenth.

Step 1 The mean of this set of data is 5.

Step 2 Find the sum of the absolute values of the differences between each value and the mean.

$|2 - 5| + |2 - 5| + |3 - 5| + |4 - 5| + |14 - 5| = 3 + 3 + 2 + 1 + 9$ or 18

Step 3 Divide the sum by the number of values: $18 \div 5 = 3.6$.

Check Your Progress

2. **DANCES** The prom committee kept count of how many tickets it sold each day during lunch: 12, 32, 36, 41, 22, 47, 51, 33, 37, 49. Find the mean absolute deviation of these data.

Personal Tutor glencoe.com

StudyTip

Symbols The mean of a sample and the mean of a population are calculated in the same way. $\bar{x}$ usually refers to the mean of a sample, but in this text, it will refer to the mean of a population.

The **standard deviation** is a calculated value that shows how the data deviates from the mean of the data. The standard deviation is represented by the lower-case Greek symbol sigma, σ. The **variance** of the data is the square of the standard deviation. Use the method below to calculate the variance and standard deviation.

Key Concept Variance and Standard Deviation

For Your FOLDABLE

Step 1 Find the mean, $\bar{x}$.

Step 2 Find the square of the difference between each value in the set of data and the mean. Then sum the squares and divide by the number of values in the set of data. The result is the variance.

Step 3 Take the square root of the variance to find the standard deviation.

EXAMPLE 3 Find the Variance and Standard Deviation

Find the mean, variance, and standard deviation of 3, 6, 11, 12, and 13 to the nearest tenth.

Step 1 To find the mean, add the numbers and then divide by the number of values in the data set.

$$\bar{x} = \frac{3 + 6 + 11 + 12 + 13}{5} = \frac{45}{5} \text{ or } 9$$

Step 2 To find the variance, square the difference between each number and the mean. Then add the squares, and divide by the number of values.

$$\sigma^2 = \frac{(3-9)^2 + (6-9)^2 + (11-9)^2 + (12-9)^2 + (13-9)^2}{5}$$

$$= \frac{(-6)^2 + (-3)^2 + 2^2 + 3^2 + 4^2}{5}$$

$$= \frac{36 + 9 + 4 + 9 + 16}{5} \text{ or } \frac{74}{5}$$

Step 3 The standard deviation is the square root of the variance.

$\sigma^2 = \frac{74}{5}$ **Variance**

$\sqrt{\sigma^2} = \sqrt{\frac{74}{5}}$ **Take the square root of the variance.**

$\sigma \approx 3.8$ **Use a calculator.**

The mean is 9, the variance is $\frac{74}{5}$ or 14.8, and the standard deviation is about 3.8.

Check Your Progress

Find the mean, variance, and standard deviation of each set of data to the nearest tenth.

3A. 6, 10, 15, 11, 8

3B. 92, 84, 71, 83, 100

Personal Tutor glencoe.com

StudyTip

Categories of Data Quantitative data can also be called *measurement data*. Qualitative data is also known as *categorical data*.

The standard deviation illustrates the spread of a set of data. For example, when the mean is 75 and the standard deviation is 3, we know that almost all of the data values are very close to the mean. When the mean is 75 and the standard deviation is 15, then the data are more spread out and there may be an outlier.

StudyTip

Symbols The standard deviation of a sample *S* and the standard deviation of a population σ are calculated in different ways. In this text, you will calculate the standard deviation of a population.

Real-World EXAMPLE 4 Statistical Analysis

NUTRITION Caleb kept track of the Calories he ate each day. Find the standard deviation of the data set.

Day	Sun	Mon	Tues	Wed	Thurs	Fri	Sat
Calories	1800	2000	2100	2250	1900	2500	2000

Use a graphing calculator to find the standard deviation. Clear all lists. Press STAT ENTER, and enter each data value into L1, pressing ENTER after each value. To view the statistics, press STAT ▶ 1 ENTER. So, the standard deviation is about 216.9.

Check Your Progress

4. Caleb tracked his Calorie intake for another week: 1950, 2000, 2100, 2000, 1900, 2100, 2000. Find the standard deviation of his Calorie intake for this week.

Personal Tutor glencoe.com

Check Your Understanding

Example 1 p. 756

Identify the sample and the population for each situation. Then describe the sample statistic and the population parameter.

1. **POLITICS** A random sample of 1003 Mercy County voters is asked if they would vote for the incumbent for governor. The percent responding *yes* is calculated.
2. **BOOKS** A random sample of 1000 U.S. college students is surveyed about how much they spend on books per year.

Example 2 p. 757

Find the mean absolute deviation to the nearest tenth.

3. **FINANCIAL LITERACY** Iye is waiting tables at the Pizza Pan Restaurant. He is keeping track of the tips that he receives each hour: \$20, \$31, \$24, \$22, \$35, \$12.
4. **PARTIES** Dalila kept an account of what each cousin gave toward their grandmother's birthday party: \$25, \$24, \$36, \$28, \$34, \$25, \$17.

Example 3 p. 758

Find the mean, variance, and standard deviation of each set of data to the nearest tenth.

5. 3, 4, 18, 21, 17
6. 12, 15, 18, 21

Example 4 p. 759

7. **ELECTRONICS** Ed surveyed his classmates to find out how many electronic gadgets each person has in their home. Find the standard deviation of the data set to the nearest tenth: 3, 10, 11, 10, 9, 11, 12, 8, 11, 8, 7, 12, 11, 11, 5.

Practice and Problem Solving

● = **Step-by-Step Solutions** begin on page R12.
Extra Practice begins on page 815.

Example 1 p. 756

Identify the sample and the population for each situation. Then describe the sample statistic and the population parameter.

8. A stratified random sample of high school students from each school in the county was polled about the time spent each week on extracurricular activities.
9. A stratified random sample of 2500 high school students across the country was asked how much money they spent each month.

Example 2 p. 757

Find the mean absolute deviation to the nearest tenth.

10. DVDS Mr. Robinson asked his students to count the number of DVDs they owned.

Number of DVDs					
26	39	5	82	12	14
0	3	15	19	41	6
2	0	11	1	19	29

11. SALES An amusement park manager wanted to keep track of how many bags of cotton candy were sold each hour: 16, 24, 15, 17, 22, 16, 18, 24, 17, 13, 25, 21.

Example 3 p. 758

Find the mean, variance, and standard deviation of each set of data to the nearest tenth.

12. 3, 8, 7, 12

13. 76, 78, 83, 74, 75

14. 0.01, 0.03, 0.1, 0.5

15. 0.8, 0.01, 0.06, 0.02, 0.4, 0.8, 0.5

Example 4 p. 759

16. ONLINE AUCTIONS Scott makes keychains and sells them on an online auction site. He tracks the selling price of each keychain: \$3.25, \$4.50, \$5.00, \$5.75, \$2.25, \$8.50, \$6.00, \$3.50, \$4.50, \$5.00. Find the standard deviation to the nearest tenth.

17. PART-TIME JOBS Ms. Johnson asks all of the girls on the tennis team how many hours each week they work at part-time jobs: 10, 12, 0, 6, 9, 15, 12, 10, 11, 20. Find the standard deviation to the nearest tenth.

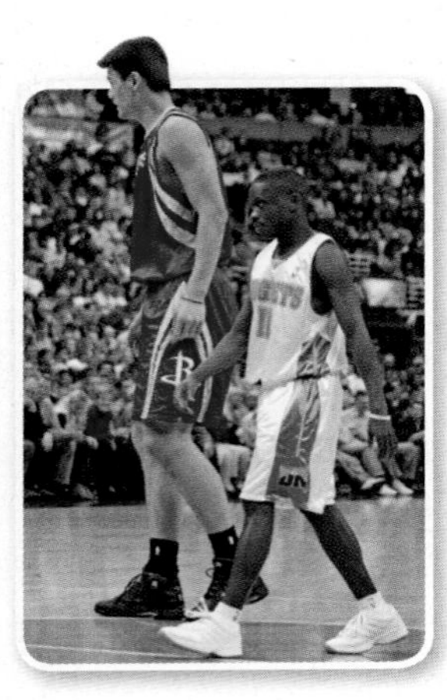

Real-World Link

One of the tallest players in the NBA was Yao Ming at 7 ft 5 in. One of the shortest NBA players was Earl Boykins at 5 ft 5 in.

Source: Inside Hoops

18. BASKETBALL The heights of players on an NBA team are shown.

Heights of Professional Basketball Players (in.)				
80	77	83	74	78
80	83	74	83	69
78	85	81	81	79

a. Find the mean and standard deviation.

b. Suppose the 5′9″ player was traded for Earl Boykins. Find the mean and the standard deviation. Describe the effect this trade has on the calculations.

c. Performing one or more operations on a data set transforms the data. If the operation can be written as a linear function, the transformation is a **linear transformation**. Convert each measure to centimeters then find the mean. Next, convert the mean to inches. How does this calculation compare to the mean found in part **a**?

19. PENNIES Mr. Day has another jar of pennies on his desk. There are 30 pennies in this jar. Theo looks at 5 pennies from the jar and replaces them. Lydia looks at 10 pennies and replaces them, and Peter looks at 20 pennies and replaces them.

Years of Pennies in Jar					
2001	1990	2000	1982	1991	1975
2007	1981	2005	2007	2003	2005
1997	1974	1992	1994	1991	1992
2000	1995	1999	2005	2006	2005
2004	2004	1998	2001	2002	2006

a. Identify the sample and the population for each situation. Then describe a statistic and a parameter.

b. The years of Theo's pennies are 1974, 1975, 1981, 1999, 1992. Find the mean and mean absolute deviation.

c. The years of Lydia's pennies are 2004, 1999, 2004, 2005, 1991, 2003, 2005, 2000, 2001, 1998. Find the mean and mean absolute deviation.

d. The years of Peter's pennies are 2007, 2005, 1975, 2003, 2005, 1997, 1992, 1994, 1991, 1992, 2000, 1999, 2005, 1982, 2005, 2004, 1998, 2001, 2002, 2006. Find the mean and mean absolute deviation.

e. Find the mean and mean absolute deviation for all of the pennies in the jar. Which sample was more similar to the full population? Explain.

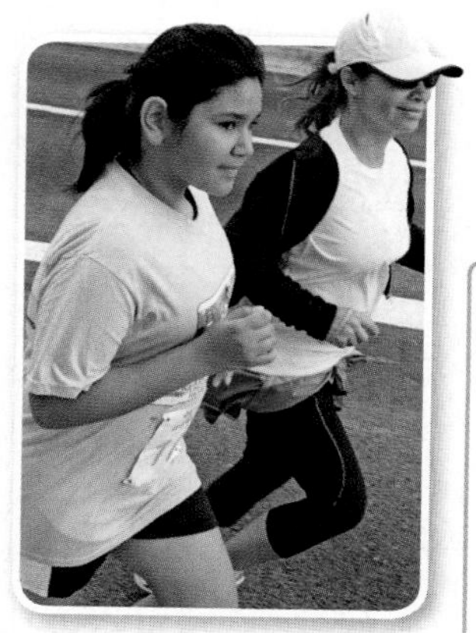

Real-World Link

A growing number of teens are completing marathons and 21-kilometer half-marathons at major races.

20. **BABYSITTING** Samantha wants to see if she is getting a fair wage for babysitting at \$8.50 per hour. She takes a survey of her friends to see what they charge per hour. The results are \$8, \$8.50, \$9, \$7.50, \$10, \$8.25, \$8.75. Find the mean absolute deviation of the data to the nearest tenth.

21. **RUNNING** The results of a 5K race are published in a local paper. Over a thousand people participated, but only the times of the top 15 finishers are listed.

15th Annual 5K Road Race					
Place	Time (min:s)	Place	Time (min:s)	Place	Time (min:s)
1	15:56	6	16:34	11	17:14
2	16:06	7	16:41	12	17:46
3	16:11	8	16:54	13	17:56
4	16:21	9	17:00	14	17:57
5	16:26	10	17:03	15	18:03

a. Find the mean and mean absolute deviation of the top 15 running times to the nearest tenth. (*Hint:* Convert each time to seconds.)

b. Identify the sample and population.

c. Analyze the sample. Classify the data as *qualitative* or *quantitative*. Can a statistical analysis of the sample be applied to the population? Explain.

H.O.T. Problems

Use Higher-Order Thinking Skills

22. **FIND THE ERROR** Amy and Esteban are describing one way to increase the accuracy of a survey. Is either of them correct? Explain your reasoning.

Amy	Esteban
The survey should include as many people in the population as possible.	The sample for the survey should be chosen randomly. Several random samples should be taken.

23. **CHALLENGE** Find the mean and standard deviation of the population of data represented by the histogram.

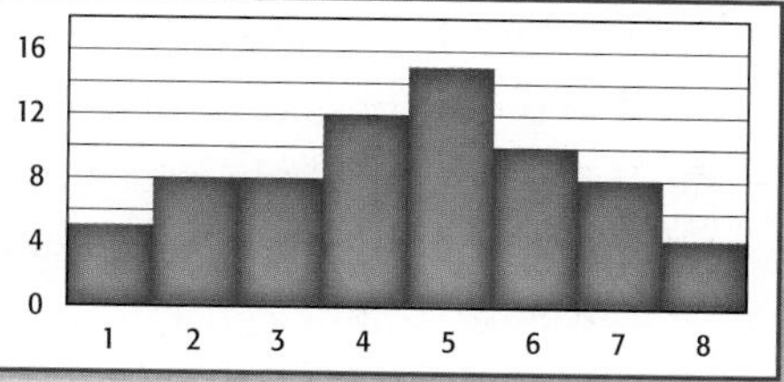

24. **REASONING** Determine whether the following statement is *sometimes, always,* or *never* true. Explain. *Two random samples taken from the same population will have the same mean and standard deviation.*

25. **WRITING IN MATH** Compare and contrast statistics and parameters. Include an example of each.

26. **OPEN ENDED** Describe a real-world situation in which it would be useful to use a sample mean to help estimate a population mean. Describe how you could collect a random sample from this population.

27. **WRITING IN MATH** Compare and contrast standard deviation and mean absolute deviation.

Standardized Test Practice

28. Melina bought a shirt that was marked 20% off for $15.75. What was the original price?

A $16.69 **C** $18.69
B $17.69 **D** $19.69

29. SHORT RESPONSE A group of students visited the Capitol building. Twenty students met with the local representative. This was 16% of the students. How many student ambassadors were there altogether?

30. The tallest 7 trees in a park have heights of 19, 24, 17, 26, 24, 20, and 18 meters. Find the median of their heights.

F 17 **H** 21
G 20 **J** 24

31. It takes 3 hours for a boat to travel 27 miles upstream. The same boat travels 30 miles downstream in 2 hours. Find the speed of the boat.

A 12 mph **C** 3 mph
B 14 mph **D** 5 mph

Spiral Review

Tell which measure of central tendency best represents the data. Justify your answer. Then find the measure. (Lesson 12-2)

32. FOOD DRIVE A high school is offering an incentive to classes that bring in the canned goods. The pounds of food brought by the classes are represented by the data set 8, 12, 18, 25, 21, 5, 10, and 14.

33. TEST SCORES The results of the math test are 78, 81, 85, 86, 88, 85, 90, 91, 85, 95, and 98.

Identify each sample as *biased* or *unbiased*. Explain your reasoning. (Lesson 12-1)

34. SHOPPING Every tenth person walking into the mall is asked to name their favorite store.

35. MUSIC Every fifth person at a rock concert is asked to name their favorite radio station.

36. GEOMETRY If the side length of a cube is s, the volume is represented by s^3, and the surface area is represented by $6s^2$. (Lesson 7-1)

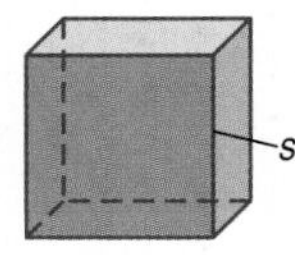

a. Are the expressions for volume and surface area monomials? Explain.

b. If the side of a cube measures 3 feet, find the volume and surface area.

c. Find a side length s such that the volume and surface area have the same measure.

d. The volume of a cylinder can be found by $V = \pi r^2 h$. Suppose you have two cylinders. Each dimension of the second is twice the measure of the first, so $V = \pi(2r)^2(2h)$. What is the ratio of the volume of the first cylinder to the second?

Skills Review

A bowl contains 3 red chips, 6 green chips, 5 yellow chips, and 8 orange chips. A chip is drawn randomly. Find each probability. (Lesson 0-11)

37. red **38.** orange **39.** yellow or green

40. not orange **41.** not green **42.** red or orange

CHAPTER 12

Mid-Chapter Quiz

Lessons 12-1 through 12-3

Math Online glencoe.com

Math *in Motion,* Animation

Identify each sample, and suggest a population from which it was selected. Then classify the type of data collection used. (Lesson 12-1)

1. **CEREAL** A cereal company invites 100 children and parents to test a new cereal.

2. **SCHOOL LUNCH** A school is creating a new lunch menu. They send out a questionnaire to all students with odd homeroom numbers to see what items should be on the new menu.

3. **MEDICINE** A research facility gave a new medicine to hamsters and determined that 1 out of every 50 hamsters that took the medicine lost its hair. They conclude that the same thing will happen to every 50 people who take the medicine.

4. **MASCOTS** The cheerleaders send out a flyer with pictures of options for the new mascot to all the girls in the school. The new mascot is chosen from the favorite from the survey.

Identify each sample as *biased* or *unbiased*. Explain your reasoning. (Lesson 12-1)

5. **ART** Every fifth person leaving the art museum is asked to name their favorite piece.

6. **SHOPPING** Each person leaving the Earring Pagoda is asked to name their favorite store in the mall.

7. **FOOTBALL** Every 10th student leaving the student union at Ohio State is asked to name their favorite college football team.

8. **CLASSES** Every 5th person leaving the school is asked to name their favorite class.

9. **MULTIPLE CHOICE** Every 10 minutes, Kaleigh writes down whether the TV is showing a commercial or a program. Which of the following best describes the sample? (Lesson 12-1)

   A simple **C** systematic

   B stratified **D** none of the above

Which measure of central tendency best represents the data? Justify your answer. Then find the measure. (Lesson 12-2)

10. **PLAY AREA** Ian listed the ages of the children playing at the play area at the mall.

    2, 3, 2, 2, 4, 2, 3, 2, 8, 3, 4, 2

11. **RECYCLING** Marielle is in charge of recycling cans at her school. She counts the number of cans recycled each week.

    22, 10, 23, 25, 24, 23, 25, 19

12. Does the following give an accurate picture of the survey results? (Lesson 12-2)

    A survey of 500 students was conducted.
    Question: What is the most important aspect of school?
    Conclusion: Preparing for the future is not important at all.

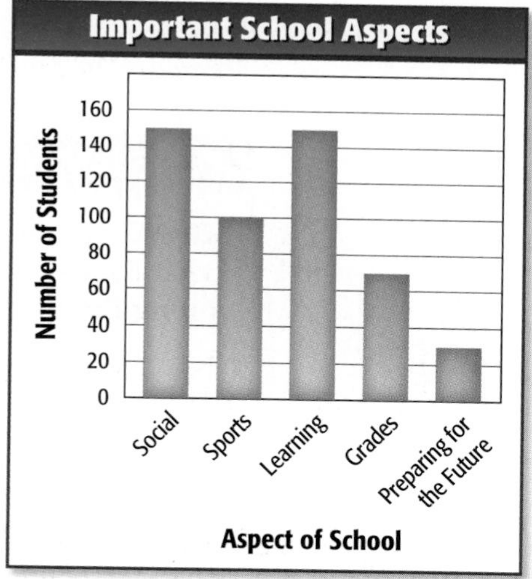

Find the mean, variance, and standard deviation to the nearest tenth for each set of data. (Lesson 12-3)

13. 2, 4, 5, 7, 7, 9

14. 13, 14, 18, 21

15. **MULTIPLE CHOICE** Several friends are chipping in to buy a gift for their teacher. Ignati is keeping track of how much each friend gives for the gift. Find the mean absolute deviation. (Lesson 12-3)

    \$10, \$5, \$3, \$6, \$7, \$8

    F 2.22 **H** 1.833

    G 6.5 **J** 2.4

12-4 Permutations and Combinations

Then

You used the Fundamental Counting Principle. (Lesson 0-11)

Now

- Use permutations.
- Use combinations.

New Vocabulary

sample space
permutation
factorial
combination

Math Online

glencoe.com

- Extra Examples
- Personal Tutor
- Self-Check Quiz
- Homework Help

Why?

Angie's coach told her that she would bat sixth in the softball game. When a coach decides on the team's lineup, the order in which she fills in the names determines the order in which the players will bat.

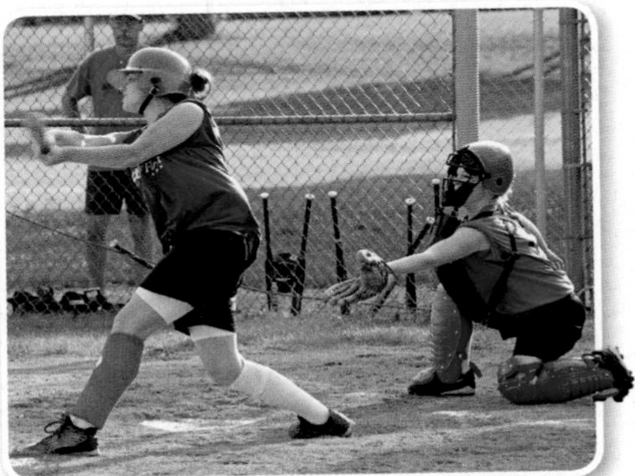

Permutations The list of all of the people or objects in a group is called the **sample space**. When the objects are arranged so that order is important and every possible order of the objects is provided, the arrangement is called a **permutation**.

Suppose Angie's coach has 4 players in mind for the first 4 spots in the lineup. The Fundamental Counting Principle can be used to determine the number of permutations. A batter cannot bat first and second, so once that player is chosen, she is not available for the next choice.

number of permutations		choices for 1st batter		choices for 2nd batter		choices for 3rd batter		choices for 4th batter
P	$=$	4	$\cdot$	3	$\cdot$	2	$\cdot$	1
	$= 24$							

There are 24 different ways to arrange the first four batters.

Real-World EXAMPLE 1 Permutation

TRAVEL A travel agency is planning a vacation package in which travelers will visit 5 cities around Europe. How many ways can the agency arrange the 5 cities along the tour?

$$\text{Number of ways to arrange the cities} = 5 \cdot 4 \cdot 3 \cdot 2 \cdot 1 = 120$$

There are 120 ways to arrange the cities.

Check Your Progress

1. **MOVIES** Lloyd and five friends go to a movie. In how many different ways can they sit together in a row of 6 empty seats?

Personal Tutor glencoe.com

The expression used in Example 1 to calculate the number of permutations of the five cities, $5 \cdot 4 \cdot 3 \cdot 2 \cdot 1$, can be written as $5!$, which is read *5 factorial*.

Key Concept Factorial

For Your FOLDABLE

Words The **factorial** of a positive integer n is the product of the positive integers less than or equal to n.

Symbols $n! = n \cdot (n - 1) \cdot (n - 2) \cdot \ldots \cdot 1$, where $n \geq 1$. Also, $0! = 1$.

Suppose Angie's coach has 5 players in mind for the top 3 spots in the lineup. The Fundamental Counting Principle can be used to determine the number of permutations.

choices for 1st batter		choices for 2nd batter		choices for 3rd batter	
5	$\cdot$	4	$\cdot$	3	= 60 permutations

Notice that $5 \cdot 4 \cdot 3$ is the same as $\frac{5 \cdot 4 \cdot 3 \cdot 2 \cdot 1}{2 \cdot 1}$. This relationship is expressed in the following formula.

ReadingMath

Notation The number of permutations $P(n, r)$ of n objects taken r at a time can also be written as nPr.

Key Concept Permutation Formula — For Your FOLDABLE

Words The number of permutations of n objects taken r at a time is the quotient of $n!$ and $(n - r)!$.

Symbols $P(n, r) = \frac{n!}{(n - r)!}$

Real-World EXAMPLE 2 Use the Permutation Formula

LIBRARY The librarian is placing 6 of 10 magazines on a shelf in a showcase. How many ways can she arrange the magazines in the case?

$$P(n, r) = \frac{n!}{(n - r)!}$$ **Permutation Formula**

$$P(10, 6) = \frac{10!}{(10 - 6)!}$$ **$n = 10$ and $r = 6$**

$$= \frac{10!}{4!}$$ **Simplify.**

$$= \frac{10 \cdot 9 \cdot 8 \cdot 7 \cdot 6 \cdot 5 \cdot \cancel{4} \cdot \cancel{3} \cdot \cancel{2} \cdot \cancel{1}}{\cancel{4} \cdot \cancel{3} \cdot \cancel{2} \cdot \cancel{1}}$$ **Divide by common factors.**

$$= 151{,}200$$ **Simplify.**

There are 151,200 ways for the librarian to arrange the magazines.

Check Your Progress

2. **FASHION** A designer has created 15 outfits and needs to select 10 for a fashion show. How many ways can the designer arrange the outfits for the show?

Personal Tutor glencoe.com

StudyTip

Permutations and Combinations If order matters in a group, the group is a *permutation*. If order does not matter in a group, the group is a *combination*.

Combinations A selection of objects in which order is not important is called a **combination**. To find all two-letter combinations of *A*, *B*, and *C*, you would list all of the arrangements of two letters, which are listed below.

AB *BA* *AC* *CA* *BC* *CB*

Because order does not matter, *AB* and *BA* are the same, so there are 2! ways to write the same letters. Divide $P(n, r)$ by 2! to remove the groups with identical objects.

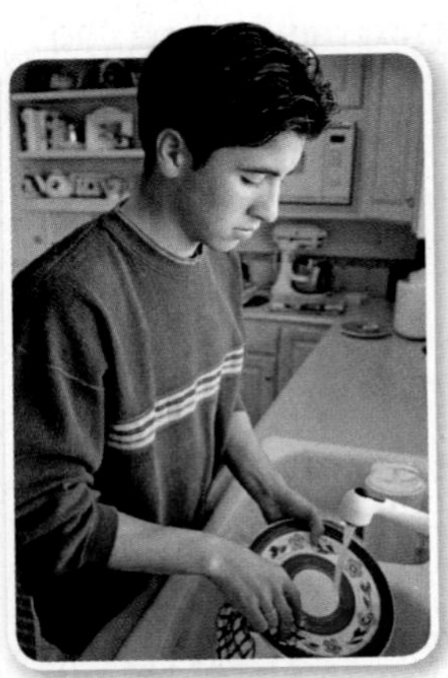

Real-World Link

About 27% of teens make dinner for their families at least some of the time.

Source: *American Demographics*

Real-World EXAMPLE 3 Combination

FAMILY Horacio has 2 brothers and 2 sisters. Their parents assign chores to them each week. How many ways can two children be chosen to wash the dishes?

Since the order in which the children are chosen does not matter, we need to find the number of combinations of 5 children taken 2 at a time.

$$C(n, r) = \frac{\text{number of permutations}}{\text{number of permutations with identical objects}}$$

First find the number of permutations.

$P(5, 2) = \frac{5!}{3!}$ or 20 $\quad$ **$n = 5$ and $r = 2$**

Because we are choosing 2, there are $2! = 2$ permutations with identical objects.

$C(n, r) = \frac{20}{2}$ or 10

There are 10 possible ways to choose 2 children.

Check Your Progress

3. **TEST** Louis is given the option of answering any 10 out of the 12 questions on his history test. How many ways can he complete the test?

Personal Tutor glencoe.com

We can now state a formula for combinations.

Key Concept Combination Formula

For Your FOLDABLE

Words The number of combinations of n objects taken r at a time is the quotient of $n!$ and $(n - r)!r!$.

Symbols $C(n, r) = \frac{n!}{(n - r)!r!}$

Real-World EXAMPLE 4 Use the Combination Formula

RETAIL Marques works part-time at a local department store. His manager asked him to choose for display 5 different styles of shirts from the wall of the store that has 8 shirts on it to put in a display. How many ways can Marques choose the shirts?

$C(n, r) = \frac{n!}{(n - r)!r!}$	**Combination Formula**
$= \frac{8!}{(8 - 5)!5!}$	**$n = 8$ and $r = 5$**
$= \frac{8!}{3!5!}$	**Simplify.**
$= \frac{8 \cdot 7 \cdot 6 \cdot \cancel{5} \cdot \cancel{4} \cdot \cancel{3} \cdot \cancel{2} \cdot \cancel{1}}{3 \cdot 2 \cdot 1 \cdot \cancel{5} \cdot \cancel{4} \cdot \cancel{3} \cdot \cancel{2} \cdot \cancel{1}}$	**Divide by common factors.**
$= \frac{336}{6}$ or 56	**There are 56 ways for Marques to select 5 shirts.**

Check Your Progress

4. **SPRING DANCE** A group of four students is selecting corsages and boutonnières to wear to the spring dance. They can choose from 18 different flowers which consist of 4 roses, 6 carnations, and 8 tulips. In how many ways can 4 flowers be chosen to wear?

Personal Tutor glencoe.com

StudyTip

Combination Formula The number of combinations is the quotient of the number of permutations of n objects taken r at a time and the number of permutations of r objects.

> **Watch Out!**
>
> **Combinations** Not all everyday uses of the word *combination* are descriptions of mathematical combinations. Notice that in Example 5, the combination of a lock is described by a permutation.

We can use permutations or combinations to find the probability of an event.

Real-World EXAMPLE 5 Probability Using a Permutation

BICYCLES A combination lock requires a three-digit code made up of the digits 0 through 9. No number can be used more than once.

a. How many different arrangements are possible?

Since the order of the numbers in the code is important, this is a permutation of 10 digits taken 3 at a time.

$P(n, r) = \frac{n!}{(n-r)!}$	**Permutation Formula**
$= \frac{10!}{(10-3)!}$	**$n = 10$ and $r = 3$**
$= \frac{10!}{7!}$	**Simplify.**
$= \frac{10 \cdot 9 \cdot 8 \cdot \cancel{7} \cdot \cancel{6} \cdot \cancel{5} \cdot \cancel{4} \cdot \cancel{3} \cdot \cancel{2} \cdot \cancel{1}}{\cancel{7} \cdot \cancel{6} \cdot \cancel{5} \cdot \cancel{4} \cdot \cancel{3} \cdot \cancel{2} \cdot \cancel{1}}$	**Divide by common factors.**
$= 720$	**Simplify.**

There are 720 possible codes.

b. What is the probability that all of the digits are odd?

Use the Fundamental Counting Principle to determine the number of ways for the three digits to be odd.

- There are five odd digits: 1, 3, 5, 7, and 9.
- The number of choices for the three digits, if they are odd, is $5 \cdot 4 \cdot 3$. So, the number of favorable outcomes is 60.

$P(\text{all digits odd}) = \frac{60}{720}$ ← **number of favorable outcomes** / ← **number of possible outcomes**

$= \frac{1}{12}$ **Simplify.**

The probability that all of the digits are odd numbers is $\frac{1}{12}$ or about 8%.

Check Your Progress

SPANISH CLUB The Spanish club is electing a president, vice president, secretary, and treasurer. Rebekah and Lydia are among the nine students who are running.

5A. How many ways can the Spanish club choose their officers?

5B. Assuming that the positions are chosen at random, what is the probability that either Rebekah or Lydia will be chosen as president or vice president?

Personal Tutor glencoe.com

Check Your Understanding

Example 1 p. 764

1. CHARITY A youth charity group is holding a raffle and wants to display a picture of the 6 prizes on a flyer. How many ways can they arrange the prizes in a row?

Examples 2–4 pp. 765–766

Evaluate each expression.

2. $P(7, 2)$ **3.** $P(9, 3)$ **4.** $C(6, 4)$ **5.** $C(5, 2)$

6. RECYCLING Juana is setting recycling bins out for pick-up. She has one bin each for glass, plastic, paper and aluminum. How many ways can she arrange the bins in a row?

7. **FOOD** Linda is preparing to bake a cake. She gets the ingredients out and sets them on the counter top. Six of the 14 ingredients are spices. How many ways can she arrange the spices in a row on the countertop?

Example 5 p. 767

8. **ICE CREAM** The Dairy Barn offers 5 varieties of chocolate ice cream, 4 varieties of candy-flavored ice cream, and 6 varieties of berry-flavored ice cream.
   a. In how many ways can a customer choose 3 different flavors of ice cream?
   b. Does the selection involve a *permutation* or a *combination*?
   c. If the ice cream flavors are chosen randomly, what is the probability that a customer will select all chocolate?

Practice and Problem Solving

 = **Step-by-Step Solutions** begin on page R12.
Extra Practice begins on page 815.

Example 1 p. 764

9. **PHOTOGRAPHY** The four captains of the football team are being arranged in a row for a newspaper photograph. How many ways can the photographer arrange the players for the photograph?

10. **SCIENCE FAIR** There are 8 finalists in a science fair competition. How many ways can they stand in a row on the stage?

11. **AMUSEMENT PARKS** Tino is entering an amusement park with 5 of his friends. At the gate they must go through a turnstile one by one. How many ways can Tino and his friends go in?

12. **JOBS** At a fast food restaurant there are 4 employees that are capable of running the cash registers. How many ways can the manager arrange the employees at the four front counter registers?

Examples 2–4 pp. 765–766

Evaluate each expression.

13. $P(6, 6)$
14. $P(5, 1)$
15. $P(4, 1)$
16. $P(7, 3)$
17. $C(7, 6)$
18. $C(5, 3)$
19. $C(5, 5)$
20. $C(3, 0)$

21. **DANCE** At the spring dance, Christy and 7 of her friends sit on one side of a table. How many ways can they fill the 10 empty seats?

22. **JEWELRY** Jewel works at the jewelry store in the mall. Her manager asks her to place 3 of the 12 birthstone necklaces in the front display case. How many ways can she arrange the necklaces in the display case?

Example 5 p. 767

23. **MARBLES** Fifteen marbles out of 20 must be randomly selected. There are 7 red marbles, 8 purple marbles, and 5 green marbles from which to choose. What is the probability that 5 of each color are selected?

24. **SCHOOL PLAY** Westerville High school is seeking volunteers to help decorate for the winter dance. In all, 4 freshmen, 5 sophomores, 6 juniors, and 8 seniors tried out for the 12 open spots.
   a. How many ways can the 12 spots be chosen?
   b. If the students are chosen randomly, what is the probability that at least one senior will be chosen?

Determine whether each situation involves a *permutation* or a *combination*.

25. choosing 3 different pizza toppings from a list of 12
26. selecting 4 different ingredients out of 8 for a salad
27. choosing team captains for a football team
28. choosing the first-, second-, and third-place winner of an art competition
29. selecting 5 books to read from a list of 8
30. an arrangement of the letters in the word *probability*

31. **GAMES** Tonisha is playing a game in which you make words to score points. There are 12 letters in the box, and she must choose 4. She cannot see the letters.
    a. Suppose the 12 letters are all different. In how many ways can she choose 4?
    b. She chooses A, T, R, and E. How many different arrangements of three letters can she make?
    c. How many of the three-letter arrangements are words? List them.

32. **PAGEANTS** The Teen Miss USA pageant has 51 contestants. The judges choose Teen Miss USA and four runners-up.
    a. Does the selection involve a *permutation* or a *combination*? Explain.
    b. In how many ways can the judges choose Teen Miss USA and four runners-up?

33. **BASKETBALL** The coach had to select 5 out of 12 players on the team to start the game. How many different groups of players could be selected as starters?

34. **LOCKER** Christopher cannot remember the order of his locker combination. He only remembers that it contains the numbers 5, 16, and 31. What is the maximum number of attempts Christopher could make?

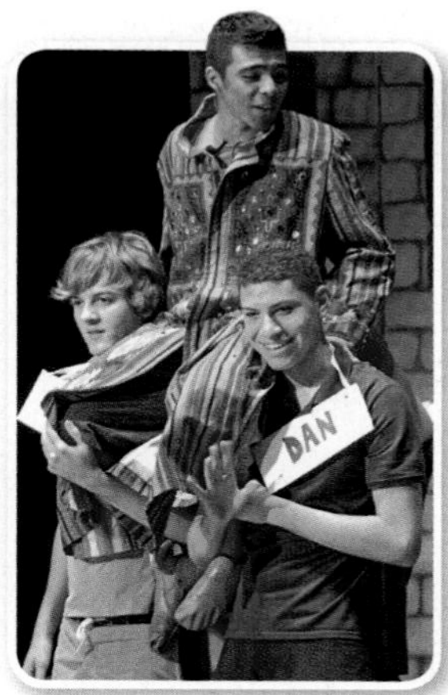

Real-World Link

A recent study found that schools with higher levels of student participation in the fine arts receive higher academic ratings and have lower dropout rates.

Source: The National Association for Music Education

H.O.T. Problems
Use **H**igher-**O**rder **T**hinking Skills

35. **FIND THE ERROR** Sydney and Ming want to form a 4-person committee to be in charge of decorations for the dance. They are determining how many committees are possible if 10 people are available. Is either of them correct?

Sydney	Ming
$P(10, 4) = \frac{10!}{(10-4)!} = 5040$	$C(10, 4) = \frac{10!}{(10-4)!4!} = 210$

36. **CHALLENGE** Seven identical mathematics books and 4 identical science books are to be stored on one shelf. In how many different ways can the books be arranged?

37. **WHICH ONE DOESN'T BELONG?** Determine which situation does not belong. Explain.

choosing 5 players on a quiz team	choosing 10 colored marbles from a bag
choosing 4 horses from 6 to run race	ranking students in a senior class

38. **REASONING** Determine whether the statement $P(n, r) = C(n, r)$ is *sometimes, always*, or *never* true. Explain your reasoning.

39. **WRITING IN MATH** Write a situation in which order is not important when 3 of 8 objects are being selected.

Standardized Test Practice

40. A gardener wants to plant 3 different types of flowers along a path. If she is choosing from 8 different types of flowers, how many ways can the 3 flowers be planted?

A 342 B 338 C 336 D 328

41. If Jack can tie 21 knots in 15 minutes, how many can he tie in 25 minutes if he continues at the same pace?

F 18 G 35 H 36 J 37

42. Shante has 30 coins, quarters and dimes, that total $5.70. How many quarters does she have?

A 12 C 18
B 15 D 20

43. SHORT RESPONSE There are 3 red candies in a bag of 20 candies. If you draw one without looking, what is the probability of drawing a red candy?

Spiral Review

Find the mean, variance, and standard deviation of each set of data to the nearest tenth. (Lesson 12-3)

44. 76, 47, 59, 47, 72, 89

45. 20, 30, 10, 40, 20, 12, 50

46. 44, 34, 64, 74, 94, 104, 55

47. 1, 5, 9, 4, 2, 4, 8, 4, 2, 1

48. SURVEY A soda manufacturing company surveyed customers to find the number of cans of soda they drank in a week. They received the following responses: 14, 7, 3, 0, 10, 12, and 10. Which measure of central tendency best represents the data? Justify your answer. Then find the measure. (Lesson 12-2)

49. PET CARE Kendra takes care of pets while their owners are away. One week she has three dogs that all eat the same dog food at the rates shown. How many bags of food should Kendra buy for that week? (Lesson 11-6)

Max
12 days/bag

Miles
15 days/bag

Stormy
16 days/bag

Find each product. (Lesson 11-4)

50. $\frac{8}{x^2} \cdot \frac{x^4}{4x}$

51. $\frac{10r^3}{6n^3} \cdot \frac{42n^2}{35r^3}$

52. $\frac{10y^3z^2}{6wx^3} \cdot \frac{12w^2x^2}{25y^2z^4}$

53. $\frac{(n-1)(n+1)}{(n+1)} \cdot \frac{(n-4)}{(n-1)(n+4)}$

54. $\frac{(x-8)}{(x+8)(x-3)} \cdot \frac{(x+4)(x-3)}{(x-8)}$

55. $\frac{3a^2b}{2gh} \cdot \frac{24g^2h}{15ab^2}$

56. COOKING The formula $t = \frac{40(25 + 1.85a)}{50 - 1.85a}$ relates the time t in minutes that it takes to cook an average-size potato in an oven to the altitude a in thousands of feet. (Lesson 11-3)

a. What is the value of t for an altitude of 4500 feet?

b. Calculate the time it takes to cook a potato at 3500 feet and at 7000 feet. How do your cooking times compare?

Skills Review

Ten red tiles, 12 blue tiles, 8 green tiles, 4 yellow tiles, and 12 black tiles are placed in a bag and selected at random. Find each probability. (Lesson 0-11)

57. P(blue)

58. P(red)

59. P(black or yellow)

60. P(green or red)

61. P(not blue)

62. P(not green)

12-5

Probability of Compound Events

Why?

Evita is flying from Cleveland to Honolulu. The airline reports that the flight from Cleveland to Honolulu has a 40% on-time record. The airline also reported that they lose luggage 5% of the time. What is the probability that both the flight will be on time and Evita's luggage will arrive?

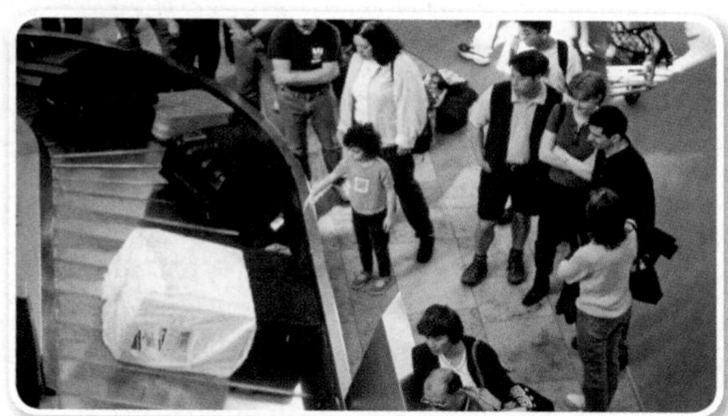

Then

You calculated simple probability. (Lesson 0-11)

Now

- Find probabilities of independent and dependent events.
- Find probabilities of mutually exclusive events.

Math Online

glencoe.com

- Extra Examples
- Personal Tutor
- Self-Check Quiz
- Homework Help
- Math in Motion

Independent and Dependent Events Recall that one event, like flying from Cleveland to Honolulu, is called a *simple event*. A **compound event** is made up of two or more simple events. So, the probability that the flight will be on time and the luggage arrives is an example of a compound event. The plane being on time may not affect whether luggage is lost. These two events are called **independent events** because the outcome of one event does not affect the outcome of the other.

Key Concept — **Probability of Independent Events** — For Your FOLDABLE

Words If two events, A and B, are independent, then the probability of both events occurring is the product of the probability of A and the probability of B.

Model

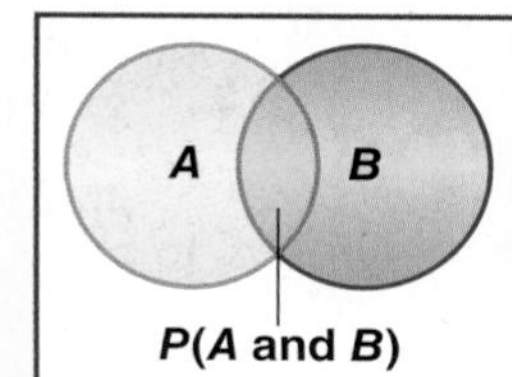

Symbols $P(A \text{ and } B) = P(A) \cdot P(B)$

Math *in Motion*, BrainPOP® glencoe.com

Real-World EXAMPLE 1 — Independent Events

MARBLES A bag contains 6 black marbles, 9 blue marbles, 4 yellow marbles, and 2 green marbles. A marble is selected, replaced, and a second marble is selected. Find the probability of selecting a black marble, then a yellow marble.

First marble: $P(\text{black}) = \frac{6}{21}$ ← number of black marbles / total number of marbles

Second marble: $P(\text{yellow}) = \frac{4}{21}$ ← number of yellow marbles / total number of marbles

$P(\text{black, yellow}) = P(\text{black}) \cdot P(\text{yellow})$ — **Probability of independent events**

$= \frac{6}{21} \cdot \frac{4}{21}$ or $\frac{24}{441}$ — **Substitution**

The probability is $\frac{24}{441}$ or about 5.4%.

Check Your Progress

Find each probability.

1A. $P(\text{blue, green})$

1B. $P(\text{not black, blue})$

Personal Tutor glencoe.com

When the outcome of one event affects the outcome of another event, they are **dependent events**. In Example 1, if the marble was not placed back in the bag, then drawing the two marbles would have been dependent events. The probability of drawing the second marble depends on what marble was drawn first.

Key Concept — Probability of Dependent Events

For Your FOLDABLE

Words If two events, A and B, are dependent, then the probability of both events occurring is the product of the probability of A and the probability of B after A occurs.

Symbols $P(A \text{ and } B) = P(A) \cdot P(B \text{ following } A)$

Recall that the complement of a set is the set of all objects that do *not* belong to the given set. In a standard deck of cards, the complement of drawing a heart is drawing a diamond, club, or spade. So, the probability of drawing a heart is $\frac{13}{52}$, and the probability of not drawing a heart is $\frac{52-13}{52}$ or $\frac{39}{52}$.

The sum of the probabilities for any two complementary events is 1.

Real-World Link

A standard deck of cards consists of 52 cards. There are 4 suits: hearts, diamonds, clubs, and spades. There are 13 cards of each suit. An ace, 2, 3, 4, 5, 6, 7, 8, 9, 10, jack, queen, and king. The hearts and diamonds are red, and the clubs and spades are black.

Real-World EXAMPLE 2 Dependent Events

CARDS Cynthia randomly draws three cards from a standard deck one at a time without replacement. Find the probability that the cards are drawn in the given order.

a. P(diamond, spade, diamond)

First card: $P(\text{diamond}) = \frac{13}{52}$ or $\frac{1}{4}$ ← number of diamonds / total number of cards

Second card: $P(\text{spade}) = \frac{13}{51}$ ← number of spades / number of cards remaining

Third card: $P(\text{diamond}) = \frac{12}{50}$ or $\frac{6}{25}$ ← number of diamonds remaining / number of cards remaining

$$P(\text{diamond, spade, diamond}) = P(\text{diamond}) \cdot P(\text{spade}) \cdot P(\text{diamond})$$
$$= \frac{1}{4} \cdot \frac{13}{51} \cdot \frac{6}{25} \text{ or } \frac{13}{850} \quad \textbf{Substitution}$$

The probability is $\frac{13}{850}$ or about 1.5%.

b. P(four, four, not a jack)

After Cynthia draws the first two fours from the deck of 52 cards, there are 50 cards left. Since neither of these cards are jacks, there are still four jacks left in the deck. So, there are 52 − 2 − 4 or 46 cards that are not jacks.

$$P(\text{four, four, not a jack}) = P(\text{four}) \cdot P(\text{four}) \cdot P(\text{not a jack})$$
$$= \frac{4}{52} \cdot \frac{3}{51} \cdot \frac{46}{50}$$
$$= \frac{552}{132{,}600} \text{ or } \frac{23}{5525}$$

The probability is $\frac{23}{5525}$ or about 0.4%.

Problem-Solving Tip

Act It Out Acting out the situation can help you understand what the question is asking. Use a deck of cards to represent the situation described in the problem.

Check Your Progress Find each probability.

2A. P(two, five, not a five) **2B.** P(heart, not a heart, heart)

Personal Tutor glencoe.com

Mutually Exclusive Events Events that cannot occur at the same time are called **mutually exclusive events**. Suppose you wanted to find the probability of drawing a heart or a diamond. Since a card cannot be both a heart and a diamond, the events are mutually exclusive.

StudyTip

and* and *or While probabilities involving *and* deal with independent and dependent events, probabilities involving *or* deal with mutually exclusive and non-mutually exclusive events.

Key Concept **Mutually Exclusive Events** For Your FOLDABLE

Words If two events, A and B, are mutually exclusive, then the probability that either A or B occurs is the sum of their probabilities.

Symbols $P(A \text{ or } B) = P(A) + P(B)$

Model

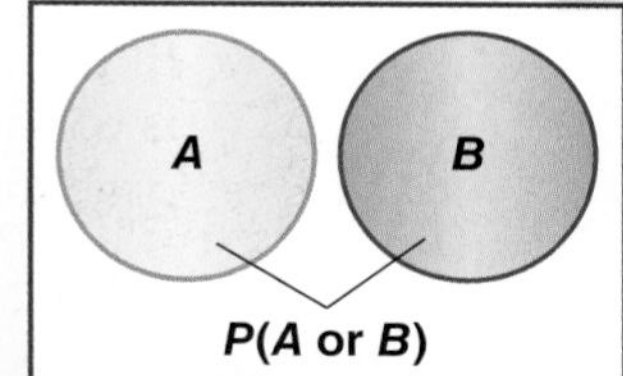

Real-World EXAMPLE 3 **Mutually Exclusive Events**

A die is being rolled. Find each probability.

a. P**(3 or 5)**

Since a die cannot show both a 3 and a 5 at the same time, these events are mutually exclusive.

$P(\text{rolling a 3}) = \frac{1}{6}$ ← number of sides with a 3 / total number of sides

$P(\text{rolling a 5}) = \frac{1}{6}$ ← number of sides with a 5 / total number of sides

$P(3 \text{ or } 5) = P(\text{rolling a 3}) + P(\text{rolling a 5})$ **Probability of mutually exclusive events**

$= \frac{1}{6} + \frac{1}{6}$ **Substitution**

$= \frac{2}{6} \text{ or } \frac{1}{3}$ **Add.**

The probability of rolling a 3 or a 5 is $\frac{1}{3}$ or about 33%.

StudyTip

Alternative Method In Example 3a, you could have placed the number of possible outcomes over the total number of outcomes.

$\frac{1+1}{6} = \frac{2}{6} \text{ or } \frac{1}{3}$

b. P**(at least 4)**

Rolling at least a 4 means you can roll either a 4, 5, or a 6. So, you need to find the probability of rolling a 4, 5, or a 6.

$P(\text{rolling a 4}) = \frac{1}{6}$ ← number of sides with a 4 / total number of sides

$P(\text{rolling a 5}) = \frac{1}{6}$ ← number of sides with a 5 / total number of sides

$P(\text{rolling a 6}) = \frac{1}{6}$ ← number of sides with a 6 / total number of sides

$P(\text{at least 4}) = P(\text{rolling a 4}) + P(\text{rolling a 5}) + P(\text{rolling a 6})$ **Mutually exclusive events**

$= \frac{1}{6} + \frac{1}{6} + \frac{1}{6}$ **Substitution**

$= \frac{3}{6} \text{ or } \frac{1}{2}$ **Add.**

The probability of rolling at least a 4 is $\frac{1}{2}$ or about 50%.

Check Your Progress

3A. $P(\text{less than 3})$ **3B.** $P(\text{even})$

Personal Tutor glencoe.com

ReadingMath

A* or *B Unlike everyday language, the expression *A* or *B* allows the possibility of both *A* and *B* occurring.

Suppose you want to find the probability of randomly drawing a 2 or a diamond from a standard deck of cards. Since it is possible to draw a card that is both a 2 and a diamond, these events are not mutually exclusive.

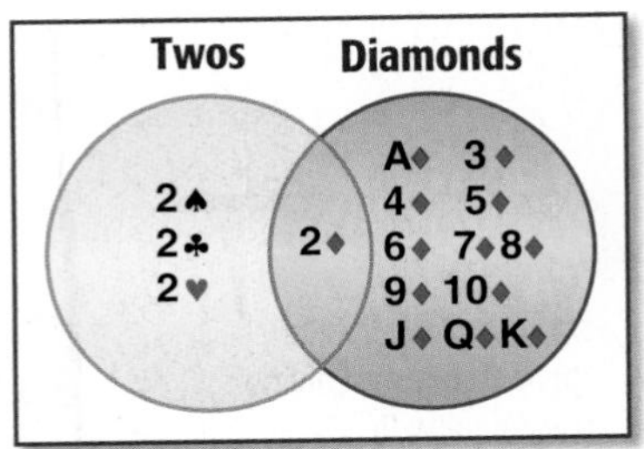

$P(2)$	$P(\text{diamond})$	$P(2, \text{diamond})$
$\frac{4}{52}$	$\frac{13}{52}$	$\frac{1}{52}$

In the first two fractions above, the probability of drawing the two of diamonds is counted twice, once for a two and once for a diamond. To find the correct probability, subtract $P(2 \text{ of diamonds})$ from the sum of the first two probabilities.

$$P(2 \text{ or a diamond}) = P(2) + \text{P(diamond)} - P(2 \text{ of diamonds})$$
$$= \frac{4}{52} + \frac{13}{52} - \frac{1}{52}$$
$$= \frac{16}{52} \text{ or } \frac{4}{13}$$

The probability is $\frac{4}{13}$ or about 31%.

Watch Out!

Intersection of Events When determining the probability of events that are not mutually exclusive, you may count the intersection of the events twice since it occurs in both events. It only actually occurs once.

Key Concept: Events that are Not Mutually Exclusive

For Your FOLDABLE

Words If two events, *A* and *B*, are not mutually exclusive, then the probability that either *A* or *B* occurs is the sum of their probabilities decreased by the probability of both occurring.

Model

$P(A \text{ or } B)$

Symbols $P(A \text{ or } B) = P(A) + P(B) - P(A \text{ and } B)$

Real-World EXAMPLE 4 Events that are Not Mutually Exclusive

STUDENT ATHLETES Of 240 girls, 176 are on the Honor Roll, 48 play sports, and 36 are on the Honor Roll and play sports. What is the probability that a randomly selected student plays sports or is on the Honor Roll?

Since some students play sports and are on the Honor Roll, the events are not mutually exclusive.

$P(\text{sports}) = \frac{48}{240}$ $\quad P(\text{Honor Roll}) = \frac{176}{240}$ $\quad P(\text{sports and Honor Roll}) = \frac{36}{240}$

$$P(\text{sports or Honor Roll}) = P(\text{sports}) + P(\text{HR}) - P(\text{sports and HR})$$
$$= \frac{48}{240} + \frac{176}{240} - \frac{36}{240} \quad \textbf{Substitution}$$
$$= \frac{188}{240} \text{ or } \frac{47}{60} \quad \textbf{Simplify.}$$

The probability is $\frac{47}{60}$ or about 78%.

Check Your Progress

4. PETS Out of 5200 households surveyed, 2107 had a dog, 807 had a cat, and 303 had both a dog and a cat. What is the probability that a randomly selected household has a dog or a cat?

Personal Tutor glencoe.com

Check Your Understanding

Examples 1 and 2
pp. 771–772

Determine whether the events are *independent* or *dependent*. Then find the probability.

1. **BABYSITTING** A toy bin contains 12 toys, 8 stuffed animals, and 3 board games. Marsha randomly chooses 2 items for the child she is babysitting. What is the probability that she chose 2 stuffed animals as the first two choices?

2. **FRUIT** A basket contains 6 apples, 5 bananas, 4 oranges, and 5 peaches. Drew randomly chooses one piece of fruit, eats it, and chooses another. What is the probability that he chose a banana and then an apple?

3. **MONEY** Nakos has 4 quarters, 3 dimes, and 2 nickels in his pocket. Nakos randomly picks two coins out of his pocket. What is the probability that he did not choose a dime either time, if he replaced the first coin before choosing a second coin?

4. **BOOKS** Joanna needs a book to prop up a table leg. She randomly selects a book, puts it back on the shelf, and selects another book. What is the probability that Joanna selected two math books?

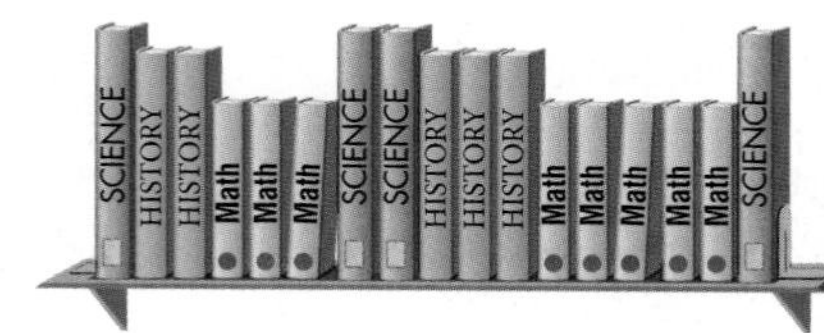

Examples 3 and 4
pp. 773–774

A card is drawn from a standard deck of playing cards. Determine whether the events are *mutually exclusive* or *not* mutually exclusive. Then find the probability.

5. *P*(two or queen)
6. *P*(diamond or heart)
7. *P*(seven or club)
8. *P*(spade or ace)

Practice and Problem Solving

 = **Step-by-Step Solutions** begin on page R12.
Extra Practice begins on page 815.

Examples 1 and 2
pp. 771–772

Determine whether the events are *independent* or *dependent*. Then find the probability.

9. **COINS** If a coin is tossed 4 times, what is the probability of getting tails all 4 times?

10. **DICE** A die is rolled twice. What is the probability of rolling two different numbers?

11. **CANDY** A box of chocolates contains 10 milk chocolates, 8 dark chocolates, and 6 white chocolates. Sung randomly chooses a chocolate, eats it, and then randomly chooses another. What is the probability that Sung chose a milk chocolate and then a white chocolate?

12. **DICE** A die is rolled twice. What is the probability of rolling the same numbers?

13. **PETS** Chuck and Rashid went to a pet store to buy dog food. They chose from 10 brands of dry food, 6 brands of canned food, and 3 brands of pet snacks. What is the probability that both chose dry food, if Chuck randomly chose first and liked the first brand he picked up?

14. **CARS** A rental agency has 12 white sedans, 8 gray sedans, 6 red sedans, and 3 green sedans for rent. Mr. Escobar rents a sedan, returns it because the radio is broken, and gets another sedan. Assuming the returned sedan remains in circulation, what is the probability that Mr. Escobar was given a green sedan and then a gray sedan?

Real-World Link

About 65% of pet owners acquire their pets free or at low cost.

Source: National Council or Pet Population Study and Policy

Examples 3 and 4
pp. 773–774

Determine whether the events are *mutually exclusive* or *not* mutually exclusive. Then find the probability.

Real-World Link

A bowling museum and hall of fame is located in St. Louis, Missouri. The museum spans 50,000 square feet and is 3 stories tall.

Source: International Bowling Museum and Hall of Fame

15. BOWLING Cindy's bowling records indicate that for any frame, the probability that she will bowl a strike is 30%, a spare 45%, and neither 25%. What is the probability that she will bowl either a spare or a strike for any given frame?

16. SPORTS CARDS Dario owns 145 baseball cards, 102 football cards, and 48 basketball cards. What is the probability that he randomly selects a baseball or a football card?

17. SCHOLARSHIPS 3000 essays were received for a $5000 college scholarship. 2865 essays were the required length, 2577 of the applicants had the minimum required grade-point average, and 2486 had the required length and minimum grade-point average. What is the probability that an essay selected at random will have the required length or the required grade-point average?

18. KITTENS Ruby's cat had 8 kittens. The litter included 2 orange females, 3 mixed-color females, 1 orange male, and 2 mixed-color males. Ruby wants to keep one kitten. What is the probability that she randomly chooses a kitten that is female or orange?

CHIPS A restaurant serves red, blue, and yellow tortilla chips. The bowl of chips Gabriel receives has 10 red chips, 8 blue chips, and 12 yellow chips. After Gabriel chooses a chip, he eats it. Find each probability.

19. P(red, blue)

20. P(blue, yellow)

21. P(yellow, not blue)

22. P(red, not yellow)

23. SOCKS Damon has 14 white socks, 6 black socks, and 4 blue socks in his drawer. If he chooses two socks at random, what is the probability that the first two socks are white?

Cards are being randomly drawn from a standard deck of cards. Once a card is drawn, it is not replaced. Find each probability.

24. P(heart or spade)

25. P(spade or club)

26. P(queen, then heart)

27. P(jack, then spade)

28. P(five, then red)

29. P(ace or black)

30. CANDY A bag contains 10 red, 6 green, 7 yellow, and 5 orange jelly beans. What is the probability of randomly choosing a red jelly bean, replacing, randomly choosing another red jelly bean, replacing, and then randomly choosing an orange jelly bean?

31. SPORTS The extracurricular activities in which the senior class at Valley View High School participate are shown in the Venn diagram.

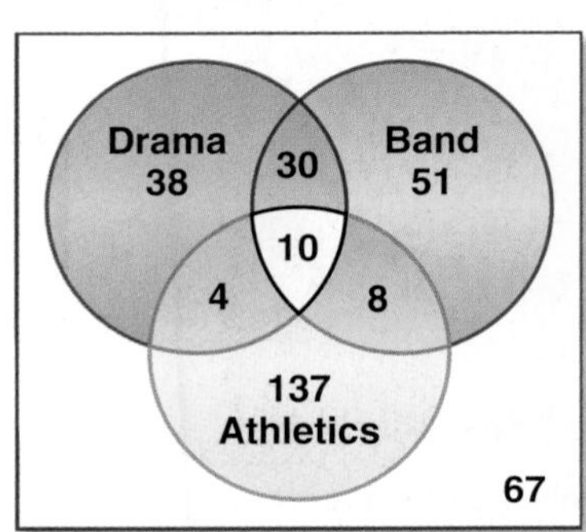

a. How many students are in the senior class?

b. How many students participate in athletics?

c. If a student is randomly chosen, what is the probability that the student participates in athletics or drama?

d. If a student is randomly chosen, what is the probability that the student participates in only drama and band?

32. **TILES** Kirsten and José are playing a game. Kirsten places tiles numbered 1 to 50 in a bag. José selects a tile at random. If he selects a prime number or a number greater than 40, then he wins the game. What is the probability that José will win on his first turn?

ReadingMath

Conditional Probability $P(B|A)$ is read the *probability* of *B, given A.*

33. **MULTIPLE REPRESENTATIONS** In this problem, you will explore conditional probability. **Conditional probability** is the probability that event B occurs given that event A has already occurred. It is calculated by dividing the probability of the occurrence of both events by the probability of the occurrence of the first event. The notation for conditional probability is $P(B \mid A)$.

   a. **GRAPHICAL** Draw a Venn diagram to illustrate $P(A \text{ and } B)$.

   b. **VERBAL** Write the formula for $P(B \mid A)$ given the Venn diagram.

   c. **ANALYTICAL** A jar contains 12 marbles, of which 8 marbles are red and 4 marbles are green. If marbles are chosen without replacement, find $P(\text{red})$ and $P(\text{red, green})$.

   d. **ANALYTICAL** Using the probabilities from part **c** and the Venn diagram in part **a**, determine the probability of choosing a green marble on the second selection, given that the first marble selected was red.

   e. **ANALYTICAL** Write a formula for finding a conditional probability.

   f. **ANALYTICAL** Use the definition from part **e** to answer the following: At a basketball game, 80% of the fans cheered for the home team. In the same crowd, 20% of the fans were waving banners and cheering for the home team. What is the probability that a fan waved a banner given that the fan cheered for the home team?

H.O.T. Problems

Use **H**igher-**O**rder **T**hinking Skills

34. **FIND THE ERROR** George and Aliyah are determining the probability of randomly choosing a blue or red marble from a bag of 8 blue marbles, 6 red marbles, 8 yellow marbles, and 4 white marbles. Is either of them correct? Explain.

George

$$P(\text{blue or red}) = P(\text{blue}) \cdot P(\text{red}) = \frac{8}{26} \cdot \frac{6}{26} = \frac{48}{676}$$

about 7%

Aliyah

$$P(\text{blue or red}) = P(\text{blue}) + P(\text{red}) = \frac{8}{26} + \frac{6}{26} = \frac{14}{26}$$

about 54%

35. **CHALLENGE** In some cases, if one bulb in a string of holiday lights fails to work, the whole string will not light. If each bulb in a set has a 99.5% chance of working, what is the maximum number of lights that can be strung together with at least a 90% chance that the whole string will light?

36. **REASONING** Suppose there are three events A, B, and C that are not mutually exclusive. List all of the probabilities you would need to consider in order to calculate $P(A \text{ or } B \text{ or } C)$. Then write the formula you would use to calculate it.

37. **OPEN ENDED** Describe a situation in your life that involves dependent and independent events. Explain why the events are dependent or independent.

38. **WRITING IN MATH** Explain why the subtraction occurs when finding the probability of two events that are not mutually exclusive.

Standardized Test Practice

39. In how many ways can a committee of 4 be selected from a group of 12 people?

A 48
B 483
C 495
D 11,880

40. A total of 925 tickets were sold for \$5925. If adult tickets cost \$7.50 and children's tickets cost \$3.00, how many adult tickets were sold?

F 700
G 600
H 325
J 225

41. SHORT RESPONSE A circular swimming pool with a diameter of 28 feet has a deck of uniform width built around it. If the area of the deck is 60π square feet, find its width.

42. The probability of heads landing up when you flip a coin is $\frac{1}{2}$. What is the probability of getting tails if you flip it again?

A $\frac{1}{4}$
B $\frac{1}{3}$
C $\frac{1}{2}$
D $\frac{3}{4}$

Spiral Review

43. SHOPPING The Millers have twelve grandchildren, 5 boys and 7 girls. For their anniversary, the grandchildren decided to pool their money and have three of them shop for the entire group. (Lesson 12-4)

a. Does this situation represent a *combination* or *permutation*?

b. How many ways are there to choose the three?

c. What is the probability that all three will be girls?

44. ECOLOGY A group of 1000 randomly selected teens were asked if they believed there was global warming. The results are shown in the table. Find the mean absolute deviation to the nearest tenth. (Lesson 12-3)

Teen Ecology Survey Results

Response	Number
Yes, strongly agree	312
Yes, mildly agree	340
No, I don't think so	109
No, absolutely not	116
Not sure	123

Solve each equation. State any extraneous solutions. (Lesson 11-8)

45. $\frac{4}{a} = \frac{3}{a-2}$

46. $\frac{3}{x} = \frac{1}{x-2}$

47. $\frac{x}{x+1} = \frac{x-6}{x-1}$

48. $\frac{2n}{3} + \frac{1}{2} = \frac{2n-3}{6}$

49. COOKING Hannah was making candy using a two-quart pan. As she stirred the mixture, she noticed that the pan was about $\frac{2}{3}$ full. If each piece of candy has a volume of about $\frac{3}{4}$ ounce, approximately how many pieces of candy will Hannah make? (*Hint:* There are 32 ounces in a quart.) (Lesson 11-3)

50. GEOMETRY A rectangle has a width of $3\sqrt{5}$ centimeters and a length of $4\sqrt{10}$ centimeters. Find the area of the rectangle. Write as a simplified radical expression. (Lesson 10-2)

Skills Review

Solve each equation. Check your solution. (Lesson 10-4)

51. $\sqrt{-3a} = 6$

52. $\sqrt{a} = 100$

53. $\sqrt{-k} = 4$

54. $5\sqrt{2} = \sqrt{x}$

55. $3\sqrt{7} = \sqrt{-y}$

56. $3\sqrt{4a} - 2 = 10$

12-6

Probability Distributions

Then

You found probabilities with permutations and combinations. (Lesson 12-4)

Now

- Find probabilities by using random variables.
- Solve real-world problems using distributions.

New Vocabulary

random variable
discrete random variable
probability distribution
probability histogram

Math Online

glencoe.com

- Extra Examples
- Personal Tutor
- Self-Check Quiz
- Homework Help

Why?

A gaming software company with five online games on the market is interested in how many games their customers play. They surveyed 1000 randomly chosen customers. The results of the survey are shown.

Number of Computer Games	Number of Customers
1	130
2	110
3	150
4	500
5	110

Random Variables and Probability A variable with a value that is the numerical outcome of a random event is called a **random variable**. A random variable with a finite number of possibilities is a **discrete random variable**. We can let the random variable G represent the number of different games. So, G can equal 1, 2, 3, 4, or 5.

EXAMPLE 1 Random Variables

A graduation supply company offers 5 items that can be purchased for graduation: a diploma frame, graduation picture, cap and gown, senior key ring, and class pin. The school takes a poll of the seniors to see how many of these items each senior is buying. The results are shown.

Number of Items Being Purchased	Number of Seniors
0	12
1	122
2	134
3	115
4	145
5	97

a. Find the probability that a randomly chosen senior is buying exactly 3 items.

Let X represent the number of items being purchased. There is only one outcome in which 3 items are being purchased, and there are 625 seniors.

$$P(X = 3) = \frac{\text{3 items being purchased}}{\text{seniors surveyed}}$$

$P(X = n)$ is the probability of X occurring n times.

$$= \frac{115}{625} \text{ or } \frac{23}{125}$$

The probability is $\frac{23}{125}$ or 18.4%

b. Find the probability that a randomly chosen senior buys at least 4 items.

There 145 + 97 or 242 seniors who are purchasing at least 4 items.

$$P(X \geq 4) = \frac{242}{625}$$

The probability is $\frac{242}{625}$ or about 38.7%.

Check Your Progress

GRADES After an algebra test, there are 7 students with As, 9 with Bs, 11 with Cs, 3 with Ds, and 2 with Fs.

1A. Find the probability that a randomly chosen student has a C.

1B. Find the probability that a randomly chosen student has at least a B.

Personal Tutor glencoe.com

StudyTip

Discrete and Continuous Data Data is discrete if the observations can be counted; for example, the number of kittens in a litter. Data is continuous if the data can take on any value within an interval. For example, the height of each person in a sample is continuous data.

Probability Distributions A **probability distribution** is the probability of every possible value of the random variable. A **probability histogram** is a histogram that displays a probability distribution.

Key Concept — Properties of Probability Distributions

For Your FOLDABLE

- The probability of each value of X is greater than or equal to 0 and is less than or equal to 1.
- The sum of the probabilities of all values of X is 1.

EXAMPLE 2 Probability Distribution

PIZZA The table shows the probability distribution of the number of times a customer orders pizza each month.

Pizzas Ordered Per Month	
X = Number of Pizzas	Probability
0	0.10
1	0.12
2	0.44
3	0.24
4+	0.10

a. Show that the distribution is valid.

- For each value of X, the probability is greater than or equal to 0 and less than or equal to 1.
- The sum of the probabilities, $0.10 + 0.12 + 0.44 + 0.24 + 0.10$, is 1.

b. What is the probability that a customer orders pizza fewer than three times per month?

The probability of a compound event is the sum of the probabilities of each individual event. The probability of a customer ordering fewer than 3 times per month is the sum of the probability of ordering 2 times per month plus the probability of ordering one time per month.

$P(X < 3) = P(X = 2) + P(X = 1) + P(X = 0)$ — **Sum of individual probabilities**

$= 0.44 + 0.12 + 0.10$ — **$P(X = 2) = 0.44$, $P(X = 1) = 0.12$, and $P(X = 0) = 0.10$**

$= 0.66$ — **Add.**

c. Make a probability histogram of the data.

Use the data from the probability distribution table to draw a histogram. Remember to label each axis and give the histogram a title.

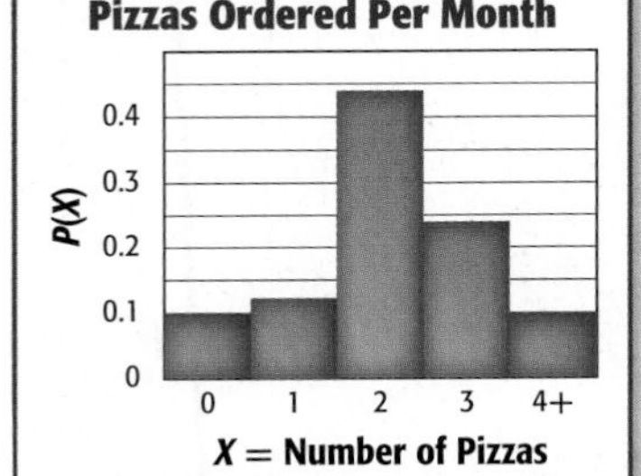

Real-World Link

In October of 2007, Joseph Jones, then a high school senior, ate 83 slices of pepperoni pizza within 10 minutes at an eating competition.

Source: About Pizza

Check Your Progress

The table shows the probability distribution of adults who play golf by age range.

Golfers By Age	
A = Ages	Probability
18–24	0.13
25–34	0.18
35–44	0.21
45–54	0.19
55–64	0.12
65+	0.17

2A. Show that the distribution is valid.

2B. What is the probability that an adult golfer is 35 years old or older?

2C. Make a probability histogram of the data.

Personal Tutor glencoe.com

Check Your Understanding

Example 1 p. 779

1. **GPS** A car dealership surveys 10,000 of its customers who have a GPS system to ask how often they have used the system within the past year. The results are shown.

Customers Using the GPS System	
Uses	**Customers**
0	1382
1–5	2350
6–10	2010
11–15	1863
16–20	1925
21+	470

   a. Find the probability that a randomly chosen customer will have used the GPS system more than 20 times.

   b. Find the probability that a randomly chosen customer will have used the GPS system no more than 10 times.

2. **JEANS** A fashion boutique ordered jeans with different numbers of stripes down the outside seams. The table shows the probability distribution of the number of each type of jean sold in a particular week.

Types of Jeans Sold	
X = Number of Stripes	**Probability**
0	0.15
1	0.19
2	0.26
3	0.22
4	0.18

   a. Show that the distribution is valid.

   b. What is the probability that a randomly chosen pair of jeans has fewer than 3 stripes?

   c. Make a probability histogram of the data.

Example 2 p. 780

3. **HOME THEATER** An electronics store sells the components and speakers for home theaters. The store surveyed its customers to see how many of the 10 components they bought. The results are shown.

Home Theater Components Purchased	
Components	**Customers**
0–2	26
3–4	42
5–6	33
7–8	24
9–10	40

   a. Find the probability that a randomly chosen customer bought 5 or 6 components.

   b. Find the probability that a randomly chosen customer bought fewer than 5 components.

Practice and Problem Solving

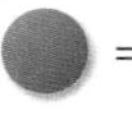 = **Step-by-Step Solutions** begin on page R12.
Extra Practice begins on page 815.

Example 1 p. 779

4. **FOOD DRIVE** Ms. Valdez's biology class held a food drive. The class kept track of the types of food donated.

Food Drive Donations Count	
Product	**Packages**
boxed dinner	36
pasta	22
juice	12
soup	45

   a. Find the probability that a randomly chosen product will be soup.

   b. Find the probability that a randomly chosen product will be a boxed dinner or pasta.

5. **SCHOOL SPIRIT** The student council wants to organize a spirit club to cheer at school sporting events. They surveyed the student body and asked students how many sporting events they typically attend each year.

Number of Sporting Events	0–5	6–10	11–15	16–20	21+
Number of Students	96	112	204	108	80

   a. Find the probability a randomly chosen student attended at most 10 events.

   b. Find the probability a randomly chosen student attended at least 16 events.

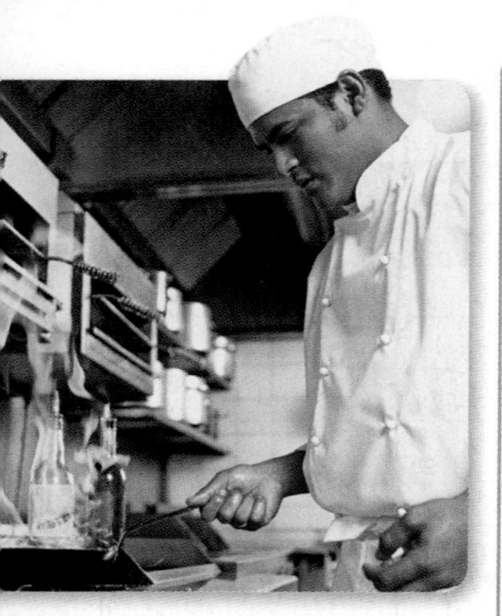

Real-World Career

Executive chefs plan the menu and oversee kitchen operations in a restaurant. They are required to have years of training and experience.

6. RESTAURANTS Kwan Chinese Restaurant has a delivery service. Mr. Kwan is keeping track of how many deliveries they have each week for a year. The results are shown.

a. Find the probability that there will be more than 20 deliveries in a randomly chosen week.

b. Find the probability that there will be fewer than 21 deliveries in a randomly chosen week.

Kwan Chinese Restaurant Deliveries	
Deliveries Per Week	**Weeks**
0–5	0
6–10	9
11–20	18
21–25	12
26+	13

7. PARTY Chrystal owns a company that plans parties for children. Throughout the year she has kept a count of each party theme she has used. The table shows the results of her tally.

Theme of Party	animal	circus	superhero	sports	music	other
Number of Parties	42	15	9	45	32	35

a. Find the probability that a randomly chosen theme will be animal or sports.

b. Find the probability that a randomly chosen theme will not be animal or sports.

Example 2
p. 780

8. MUSIC A Web site conducted a survey on the format of music teens listened to. The table shows the probability distribution of the results.

a. Show that the distribution is valid.

b. What is the probability that the type of format randomly chosen will be an MP3 or online?

c. Make a probability histogram of the data.

Formats for Music	
Format	**Probability**
CDs	0.35
radio	0.31
mini-disc	0.02
MP3	0.11
online	0.19
other	0.02

9. GRADES Mr. Rockwell's Algebra class took a chapter test last week. The table shows the probability distribution of the results.

a. Show that the distribution is valid.

b. What is the probability that a student chosen at random will have no higher than a B?

c. Make a probability histogram of the data.

Algebra Test Grades	
Grade	**Probability**
A	0.29
B	0.43
C	0.17
D	0.11
F	0

10. SKATE PARKS The park department asked the counties that had skate parks what equipment was allowed to be used in their park. The table shows the probability distribution of the results.

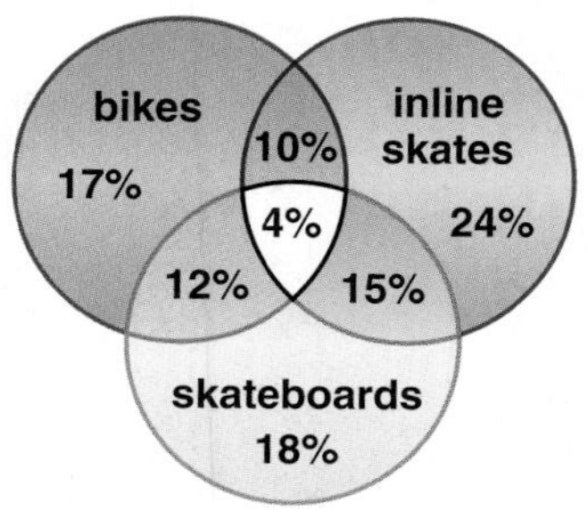

a. Show that the distribution is valid.

b. What is the probability that a park chosen at random allows bikes or skateboards?

c. Make a probability histogram of the data.

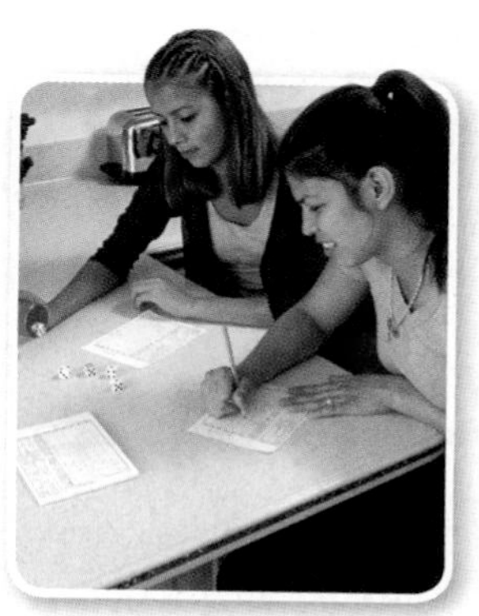

Real-World Link

Dice games have been in existence at least since 6000 BC. Found objects such as sea shells, pebbles, and nut shells were used before cubical dice became commonplace.

Source: NTL World

11. **MARKETING** A retail marketing group conducted a survey on teen shopping habits and asked where the teens did most of their holiday shopping. The table shows the probability distribution of the results

Types of Stores	malls	individual stores	online	catalogs	other
Probability	0.35	0.32	0.17	0.11	0.05

a. Show that the distribution is valid.

b. What is the probability that a shopper chosen at random will shop online or in a catalog?

c. Make a probability histogram of the data.

12. **SPORTS CARDS** Joshua mixed up all of his sports cards and placed them in a bag. Then he told his sister Drea that she could keep whatever card she randomly drew out of the bag.

Joshua's Sports Cards	
Sport	**Number Sold**
baseball	53
football	27
basketball	39
hockey	21

a. What is the probability that a randomly chosen card is hockey or football?

b. Make a probability distribution table for the data. Round to the nearest hundredth.

c. Is the distribution valid? Why or why not?

d. Make a probability histogram of the data.

13. **MULTIPLE REPRESENTATIONS** In this problem, you will explore the differences between a prediction and what actually happens.

a. **VERBAL** What is the probability of rolling a 2 on a die? What is the probability of rolling a 1 or a 6? What is the probability of rolling an odd number?

b. **ANALYTICAL** Roll the die 20 times. Record the value of the die after each roll.

c. **ANALYTICAL** Determine the probability distribution for X = value of the die.

d. **VERBAL** From your probability distribution, what is the probability of rolling a 2? What is the probability of rolling a 1 or a 6? What is the probability of rolling an odd number? Explain why the numbers may not be the same.

H.O.T. Problems

Use **H**igher-**O**rder **T**hinking Skills

14. **CHALLENGE** What is wrong with the probability distribution shown? Explain your reasoning.

15. **REASONING** Suppose two dice are rolled twelve times. Which sum is most likely to occur? Make a table to show the probability distribution. Then make a probability histogram to confirm your answer.

16. **REASONING** Explain why the sum of the probabilities in a probability distribution should always be 1. Include an example.

17. **OPEN ENDED** Write a real-world problem in which you could find a probability distribution. Create a probability histogram for your data.

18. **WRITING IN MATH** Write a real-world story in which you are the owner of a business. Explain how you could use a probability distribution to help you make a business decision.

Standardized Test Practice

19. A bucket contains 10 balls numbered 1, 1, 2, 3, 4, 4, 4, 5, 6, and 6. A ball is randomly chosen from the bucket. What is the probability of drawing a ball with a number greater than 6?

A $\frac{1}{5}$ **C** 1

B $\frac{3}{10}$ **D** 0

20. SHORT RESPONSE Mr. Bahn has $20,000 to invest. He invests part at 6% and the rest at 7%. He earns $1280 in interest within a year. How much did he invest at 7%?

21. Suppose there are 10 tickets in a box for a drawing numbered as follows: 1, 2, 2, 3, 4, 4, 6, 6, 9, and 9. A single ticket is randomly chosen from the box. What is the probability of drawing a ticket with a number less than 10?

F 0 **G** $\frac{1}{5}$ **H** $\frac{3}{10}$ **J** 1

22. GEOMETRY The height of a triangle is 5 inches less than the length of its base. If the area of the triangle is 52 square inches, find the base and the height.

A 15 in., 9 in. **C** 13 in., 8 in.

B 11 in., 7 in. **D** 17 in., 11 in.

Spiral Review

23. PET TOYS A pet store has a bin of clearance items that contains 6 balls, 5 tug toys, 8 rawhide chews, and 4 chew toys, all in equal-sized boxes. If Johnda reaches in the box and pulls out two items, what is the probability that she will pull out a tug toy each time? (Lesson 12-5)

A die is rolled and a spinner is spun like the one shown. Find the probability. (Lesson 12-4)

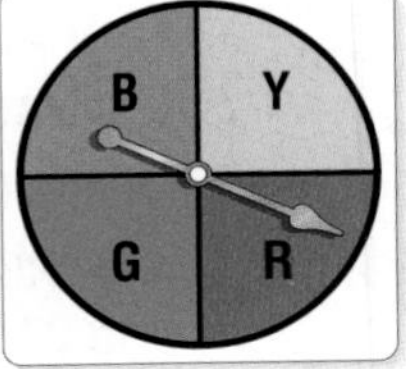

24. $P(3 \text{ and } Y)$

25. $P(\text{even and } G)$

26. $P(\text{prime number and } R \text{ or } B)$

27. $P(4 \text{ and not } Y)$

28. GAMES For a certain game, each player rolls four dice at the same time. (Lesson 12-3)

a. Do the outcomes of rolling the four dice represent permutations or combinations? Explain.

b. How many outcomes are possible?

c. What is the probability that four dice show the same number on a single roll?

Find each sum. (Lesson 11-6)

29. $\frac{4}{a^2} + \frac{6}{a}$

30. $\frac{3}{b^3} + \frac{7}{b^2}$

31. $\frac{4}{d+6} + \frac{5}{d-5}$

32. $\frac{f}{f+5} + \frac{4}{f-4}$

33. $\frac{8h}{h+6} + \frac{h}{h-3}$

34. $\frac{7k}{k-3} + \frac{k}{k+2}$

Skills Review

35. Write an expression to represent the probability of tossing a coin n times and getting n heads. Express as a power of 2. (Lesson 7-2)

36. Write an expression to represent the probability of rolling a die n times and getting 3 n times. Express as a power of 6. (Lesson 7-2)

37. Write an expression to represent the probability of rolling a die n times and getting a prime number n times. Express as a power. (Lesson 7-2)

Graphing Technology Lab
The Normal Curve

Math Online glencoe.com
- Other Calculator Keystrokes
- Graphing Technology Personal Tutor

When there are a large number of values in a data set, the frequency distribution tends to cluster around the mean of the set in a distribution (or shape) called a **normal distribution**. The graph of a normal distribution is called a **normal curve**. Since the shape of the graph resembles a bell, the graph is also called a *bell curve*.

Data sets that have a normal distribution include reaction times of drivers that are the same age, achievement test scores, and the heights of people that are the same age.

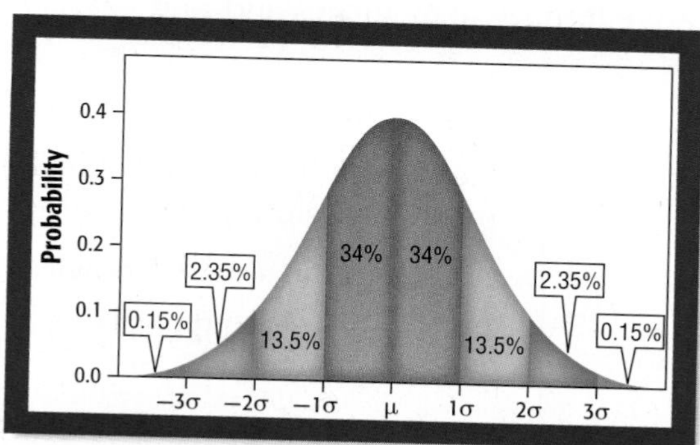

You can use a graphing calculator to graph and analyze a normal distribution if the mean and standard deviation of the data are known.

ACTIVITY 1 Graph a Normal Distribution

HEIGHT The mean height of 15-year-old boys in the city where Isaac lives is 67 inches, with a standard deviation of 2.8 inches. Use a normal distribution to represent these data.

Step 1 Set the viewing window.

- Xmin = 67 [−] 3 [×] 2.8 or 58.6
- Xmax = 67 [+] 3 [×] 2.8 or 75.4
- Xscl = 2.8
- Ymin = 0
- Ymax = 1 [÷] (2 [×] 2.8)
- Yscale = 1

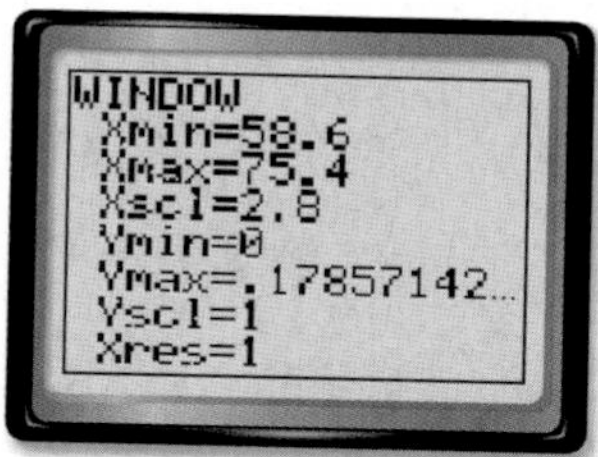

Step 2 By entering the mean and standard deviation into the calculator, we can graph the corresponding normal curve. Enter the values using the following keystrokes.

KEYSTROKES: [Y=] [2nd] [DISTR] [ENTER] [X,T,θ,n] [,] 67 [,] 2.8 [)] [Graph]

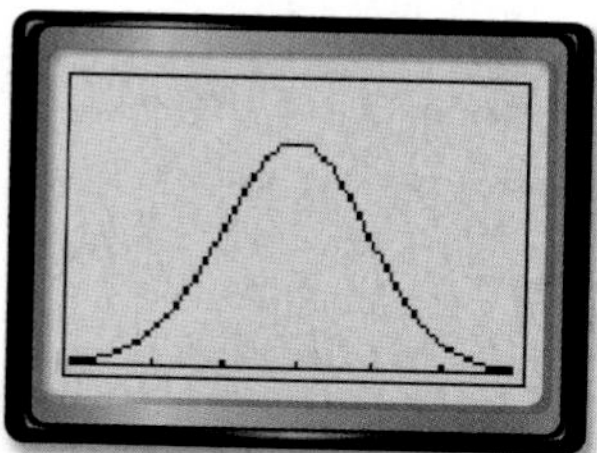

[58.6, 75.4] scl: 2.8 by [0, 0.17857142] scl: 1

The probability of a range of values is the area under the curve.

ACTIVITY 2 **Analyze a Normal Distribution**

Use the graph to answer questions about the data. What is the probability that Isaac will be at least 67 inches tall when he is 15?

The sum of all the y-values up to $x = 67$ would give us the probability that Isaac will be less than or equal to 67 inches. This is also the area under the curve. We will shade the area under the curve from negative infinity to 67 inches and find the area of the shaded portion of the graph.

Step 1 ShadeNorm Function

KEYSTROKES: 2nd [DISTR] ▶ ENTER

Step 2 Shade the graph.

Next enter the lowest value, highest value, mean, and standard deviation.

On the TI-84 Plus, 1×10^{-99} represents negative infinity.

KEYSTROKES: 1 2nd [EE] (–) 99 , 67 , 67 , 2.8) ENTER

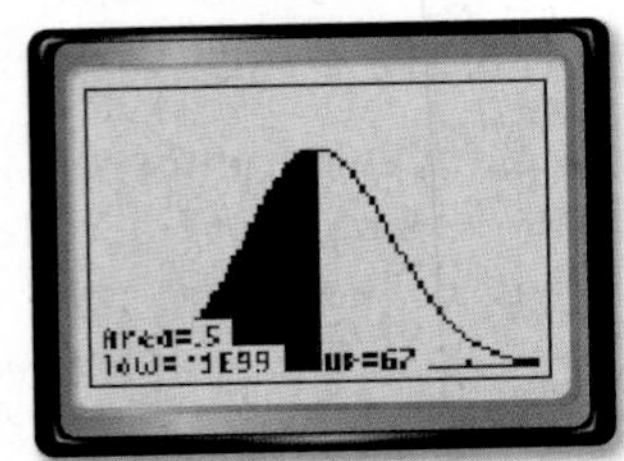

[58.6, 75.4] scl: 2.8 by [0, 0.17857142] scl: 1

The area is given as 0.5. The probability that Isaac will be 67 inches tall is 0.5 or 50%. Since the mean value is 67, we expect the probability to be 50%.

Exercises

1. What is the probability that Isaac will be at least 6 feet tall when he is 15?
2. What is the probability that Isaac will be between 65 and 68 inches?
3. If the mean height of 15-year-old girls in the same city is 64 inches with a standard deviation of 2.1, what is the probability that Isaac's sister, Maria, will be at least 64 inches tall when she is 15?
4. What is the probability that Maria will be no taller than 5 feet when she is 15 years old?

Extension

Refer to the curve at the right.

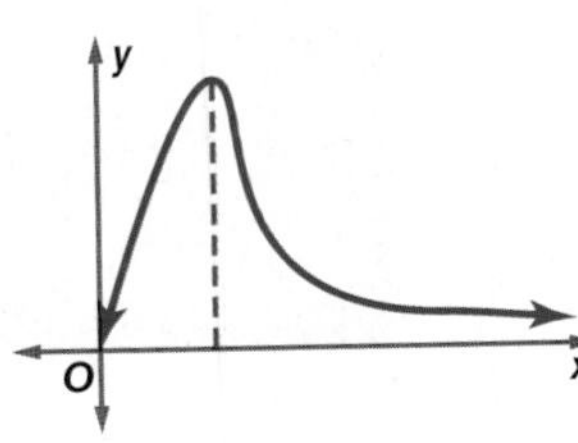

5. Compare this curve to the normal curve in Activity 1.
6. Describe where an outlier of the data set would be graphed on this curve.

12-7 Probability Simulations

Then

You used probability distributions. (Lesson 12-6)

Now

- Design simulations to estimate probabilities.
- Summarize data from simulations.

New Vocabulary

experimental probability
theoretical probability
relative frequency
empirical study
expected value
simulation

Math Online

glencoe.com

- Extra Examples
- Personal Tutor
- Self-Check Quiz
- Homework Help

Why?

Alex has been practicing his penalty kicks. He expects to be able to make at least 63% of his penalty kicks. To test this, he takes 50 penalty kicks, of which he makes 63% of them.

Experimental Probability and Expected Value

Experimental probability is determined using data from tests or experiments. Alex's experimental probability is 63%. Therefore, he *expects* to make 63% of his future kicks.

Experimental probability should not be confused with theoretical probability. **Theoretical probability** is the likeliness of an event happening. For example, when tossing a coin, the theoretical probability of it landing on heads is always 0.5, while an experiment tossing many coins may produce a different *experimental* probability.

Experimental probability is the ratio of the number of times an outcome occurs to the total number of events or trials. The ratio is also known as the **relative frequency.**

$$\text{experiment probability} = \frac{\text{frequency of an outcome}}{\text{total number of trials}}$$

It is often useful to perform an **empirical study**. In this study, an experiment is performed repeatedly, data are collected and combined, the results are analyzed, and an expected value can be calculated. The **expected value** is the average value that is expected for the outcome of one trial.

Real-World EXAMPLE 1 Experimental Probability

a. SOCCER What is the experimental probability that Alex successfully makes his goal kicks?

$$\text{experimental probability} = \frac{33}{50} \begin{matrix} \leftarrow \text{frequency of successes} \\ \leftarrow \text{total number of goal kicks} \end{matrix}$$

The experimental probability of the test is $\frac{33}{50}$ or 66%.

b. SOCCER Alex takes 50 kicks two more times. He makes 29 of the first 50 kicks and 34 of the second 50. What is the experimental probability of all three tests?

$$\text{experimental probability} = \frac{96}{150} \text{ or } \frac{16}{25}$$

The experimental probability of the three tests was $\frac{16}{25}$ or 64%.

Check Your Progress

1. **GAMES** Hakeem rolls a die 20 times. A 4 appears 5 times. What is the experimental probability of rolling a 4?

Personal Tutor glencoe.com

Performing Simulations A **simulation** allows you to find an experimental probability by using objects to act out an event that would be difficult or impractical to perform. You can conduct simulations using one or more objects such as dice, coins, marbles, or spinners. The theoretical probability of objects you choose should be identical to the experimental probability.

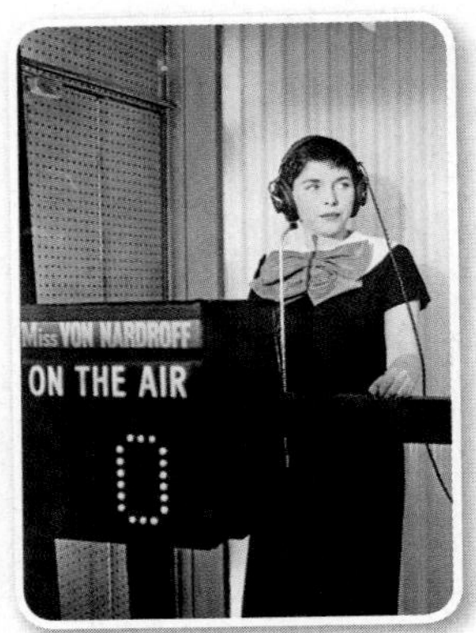

Real-World Link

Trivia quizzes and quiz shows have a great history in the U.S. Here are some dates in trivia quiz history.

1935: First radio quiz show

1950: First real TV quiz show "Truth or Consequences"

1955: First big money TV quiz show "The $64,000 Question"

Source: Trivia Wisdom

Real-World EXAMPLE 2 Simulation

In a trivia game, a player answers one of three questions on a card, two of which are multiple choice.

a. What could be used to simulate getting a multiple choice question? Explain.

You could use a die, where rolling 1 through 4 represents a multiple-choice question and rolling 5 or 6 does not.

b. Describe a way to conduct the simulation if 3 out of 5 are multiple-choice.

A spinner with five equally sized spaces, three of which are colored red, could be used. If the spinner lands on red, then the question is multiple choice.

Check Your Progress

2. Of the 48 games remaining on Bobbie's favorite basketball team's schedule, 24 will not be televised.

A. What could be used to simulate whether the next game will be televised?

B. Describe a simulation if 36 of the 48 games will not be televised.

Personal Tutor glencoe.com

Real-World EXAMPLE 3 Experimental Probability and Expected Value

QUALITY CONTROL Eloy inspects automobile frames as they come through the assembly line. From previous observations, he expects to find a weld defect in one out of ten of the frames each day.

a. What objects can be used to model the possible outcomes of the automobile inspection? Explain.

Use a simulation that has 10 objects, where 1 out of the 10 objects represents a defect. One possible simulation would be to place 10 marbles in a bag. Let 1 red marble represent the defects and 9 yellow marbles represent the good automobiles. One marble at a time can be drawn out of the bag, the results recorded, and the marble replaced in the bag. Repeat this 9 more times representing 10 automobiles.

b. What is the expected value that there is one automobile frame found with defects in a certain day?

On average, 1 out of 10 automobiles has a defect. So, the expected value is $\frac{1}{10}$ or 10%.

c. Run the simulation using Eloy's expectation of defects. What is the experimental probability of having a defect for every ten frames?

After 50 simulations, or 500 drawings, 57 had a defect, so the experimental probability of getting a defect is $\frac{57}{500}$ or 11.4%.

Defects	Frequency	Total
0	12	0
1	24	24
2	10	20
3	3	9
4	1	4
		57

Reading Math

Law of Large Numbers The *Law of Large Numbers* states that as the number of trial increases, the experimental probability gets closer to the theoretical probability.

d. How does the experimental probability compare to the expected value?

The expected value is 10%, and the experimental probability is 15%. They are relatively close. As the number of trials increased it would get closer to the expected value.

Check Your Progress

3. TELEVISION Anthony's favorite show is going to be replayed by a cable channel. They are going to play one episode each day in a random order. Anthony missed 3 of the 22 episodes.

A. What objects can be used to model the possible outcomes of one of the shows that he missed being aired on the first day?

B. What is the expected value that it will be a show he missed?

C. The results of a simulation Anthony performed are shown. What is the experimental probability that it will be a show he missed?

Show	Frequency
missed	5
watched	45

D. How does the experimental probability compare to the expected value?

Personal Tutor glencoe.com

Check Your Understanding

Example 1 p. 787

1. GAMES Games at the fair require the majority of players to lose in order for game owners to make a profit. New games are tested to make sure they have sufficient difficulty. The results of three test groups are listed in the table. The owners want a maximum of 33% of players to win. There were 50 participants in each test group.

a. What is the experimental probability that the participant was a winner in the second group?

Result	Group 1	Group 2	Group 3
Winners	13	15	19
Losers	37	35	31

b. What is the experimental probability of winning for all three groups?

2. BATTING AVERAGE In a computer baseball game, a baseball player has a batting average of 300. That is, he gets a hit 300 out of 1000, or 30%, of the times he is at bat. What could be used to simulate the player taking a turn at bat?

Example 2 p. 788

3. TEST On a true-false test, Marlene answered 16 out of the 20 questions correctly by guessing randomly.

a. What could be used to simulate her correctly answering a question? Explain.

b. Describe a way to simulate the next 20 questions.

Example 3 p. 788

4. LOTTERIES In a certain state, lottery numbers are five-digit numbers. Each digit can be 1, 2, 3, 4, 5, or 6. Once a week, a winning number is chosen randomly.

a. How many five-digit numbers are possible? Explain how you calculated the number of possible outcomes.

b. Perform a simulation for winning the lottery. Describe the objects you used.

c. According to your experiment, if you buy one ticket, what is the experimental probability of winning?

d. How does your experimental probability compare to the theoretical probability of winning?

Practice and Problem Solving

● = **Step-by-Step Solutions** begin on page R12.
Extra Practice begins on page 815.

Example 1 p. 787

5. **CARDS** Javier is drawing a card from a standard deck of cards, recording the suit, and then replacing the card in the deck. The table below shows his results.

Suit	clubs	diamonds	hearts	spades
Frequency	7	4	5	9

a. Find the experimental probability of drawing a heart.

b. Find the experimental probability of drawing a black card.

c. Javier repeated his test. The results are shown below. Find the experimental probability of drawing a spade for both tests.

Suit	clubs	diamonds	hearts	spades
Frequency	5	8	6	6

Example 2 p. 788

6. **CDs** There are 6 country CDs, 8 pop CDs, 3 rap CDs, and 7 rock CDs in a storage case. What could be used for a simulation to determine the probability of randomly selecting any one type of CD?

7. **TESTS** What could be used to simulate guessing on a multiple-choice test with 4 possible answers for each question?

Example 3 p. 788

8. **JEANS** Julie examines the stitching on pairs of jeans that are produced at a manufacturing plant. She expects to find defects in 1 out of every 16 pairs.

a. What can be used to model the possible outcome of a pair of jeans having defects? Explain.

b. What is the expected value that a random pair of jeans has a defect?

c. The results of simulations using Julie's expectations are shown. What is the experimental probability that a random pair of jeans will have a defect?

d. How does the experimental probability compare to the expected value?

Defects	Frequency
0	71
1	9
2	11
3	6
4	3

Real-World Link

Women between the ages of 16 and 24 have an average of 8 pairs of jeans.

Source: Cotton Incorporated

9. **DIE ROLL** Roll a die 25 times and record your results. Find each probability based on your results.

a. What is the probability of rolling a 2?

b. What is the probability of rolling a prime number?

c. What is the probability of rolling an even number or 3?

d. Compare your results to the theoretical probabilities.

10. **PRIZES** For its tenth anniversary, a video store randomly gives each customer a prize from the following choices: a free movie rental, a free video game rental, a free bag of popcorn, or a free pre-viewed movie. The chance of winning each prize is equal.

a. What could be used to perform a simulation of this situation? Explain.

b. How could you use this simulation to model the next 50 free items?

11. **COIN TOSS** Toss 4 coins, one at a time, 20 times, and record the number of heads and tails. Find each probability based on your results.
    a. P(any three coins will show tails)
    b. P(any two coins will show heads)
    c. P(the first coin will show heads and the fourth coin will show tails)

Math History Link

Ada Byron Lovelace (1815–1852)
Lovelace is known for generating a sequence of numbers using an early model of a computer. She is credited with being the first computer programmer. In 1980, the U.S. Department of Defense named a programming language after her.

12. **PRIZES** For a promotion, the concession stands at a football stadium are giving away free items. Every time a customer buys something, a wheel is spun to choose the customer's prize. Each prize is equally likely.

    a. Other than a spinner, what object could be used to simulate this situation? Explain.
    b. Perform the simulation until you have received at least one of each item.
    c. In your simulation, how many items must be bought to win every prize? Choose another representation to solve the problem. How do the solutions compare?

13. **GRAPHING CALCULATOR** With every purchase at a fast-food restaurant, you receive a scratchoff game card with two circles. You choose one circle to scratch. One reveals a prize, and the other reveals "Sorry. Try Again." The chance that the prize on a card is cash is 5%, a sandwich is 20%, a drink is 50%, and fries is 25%. Determine how many game cards you must scratch to win the cash.
    a. Generate a list of 100 random 0s and 1s in L1 of the calculator. 0 means you did not win a prize, and 1 means you won a prize.
    b. Generate a list of 100 random integers from 1 to 20 in L2. Let each 1 represent the cash prize.
    c. In L3, multiply the values in L1 and L2 together. What value represents that you won the cash prize? Explain.
    d. From your simulation, how many game cards had to be scratched to win the cash?

14. **ROLLING A DIE** Multiply each number on a die by the probability of rolling that number. Add these values.
    a. What is the expected value of one roll of a die?
    b. Find the expected value of the sum of the numbers on two dice.

H.O.T. Problems
Use Higher-Order Thinking Skills

15. **REASONING** The experimental probability of heads when a coin is tossed 15 times is *sometimes*, *never*, or *always* equal to the theoretical probability. Explain.

16. **CHALLENGE** Lenora tested her tennis ball machine by running 5 simulations. The experimental probability of the machine being accurate is 7% higher than the results of the 5th simulation. Determine the results of the 5th simulation.

Simulation	Accuracy
1	95%
2	85%
3	90%
4	85%

17. **REASONING** Find a counterexample to the following statement. Explain.

    It is possible for an experimental probability to be 0 if the theoretical probability is 1.

18. **OPEN ENDED** Describe a situation at your school that could be represented by a simulation. What could you use to simulate the situation?

19. **WRITING IN MATH** Compare and contrast experimental and theoretical probability.

Standardized Test Practice

20. GEOMETRY Suppose a covered water tank in the shape of a right circular cylinder is thirty feet long and eight feet in diameter. What is the surface area of the cylinder?

A 272π ft^2
C 286π ft^2
B 153π ft^2
D 248π ft^2

21. SHORT RESPONSE How many different ways can the letters P, Q, R, S be arranged?

22. In how many ways can the letters in the word STATISTICS be arranged?

F 50,400
H 15,400
G 20,800
J 3480

23. Two consecutive numbers have a sum of 91. What are the numbers?

A 41, 50
C 45, 46
B 44, 47
D 49, 42

Spiral Review

24. The table shows a class's grade distribution, where $A = 4.0$, $B = 3.0$, $C = 2.0$, $D = 1.0$, and $F = 0$. (Lesson 12-6)

G = Grade	0	1.0	2.0	3.0	4.0
Probability	0.05	0.05	0.30	0.35	0.25

a. Is the probability distribution valid? Explain.

b. What is the probability that a student passes the course?

c. What is the probability that a student chosen at random from the class receives a grade of C or better?

Review problems are color coded by lesson. Each student draws and then returns a colored ball from a bucket to see which lesson to review. There are 5 red, 10 yellow, 5 blue, and 2 green balls. Find each probability described. (Lesson 12-4)

25. P(yellow)

26. P(red)

27. P(red, blue)

28. P(yellow, green)

29. PARTIES Student Council is planning a party for the school volunteers. There are five 66-ounce unopened bottles of soda left from a recent dance. When poured over ice, $5\frac{1}{2}$ ounces of soda fills a cup. How many servings of soda do they have? (Lesson 11-7)

Write an inverse variation equation that relates x and y. Assume that y varies inversely as x. Then solve. (Lesson 11-1)

30. If $y = 8.5$ when $x = -1$, find x when $y = -1$.

31. If $y = 8$ when $x = 1.55$, find x when $y = -0.62$.

32. If $y = 6.4$ when $x = 4.4$, find x when $y = 3.2$.

33. TOPOGRAPHY To determine the mileage between landmarks, the U.S. military superimposes a coordinate grid over a map of the region. The units on this grid are approximately equal to 50,000 feet. So, a distance of 3 units on the grid equals an actual distance of 3(50,000) or 150,000 feet. Suppose the locations of two landmarks are at (132, 428) and (254, 105). Find the actual distance between these landmarks to the nearest mile. (Lesson 10-6)

Write each fraction as a percent rounded to the nearest whole number. (Lesson 0-6)

34. $\frac{26}{58}$

35. $\frac{55}{125}$

36. $\frac{14}{128}$

37. $\frac{82}{110}$

38. $\frac{76}{124}$

39. $\frac{23}{86}$

Study Guide and Review

Math Online glencoe.com
- STUDY TO GO
- Vocabulary Review

Chapter Summary

Key Concepts

Designing a Survey and Analyzing Results (Lessons 12-1 and 12-2)

- The three methods for collecting data are surveys, observational studies, and experiments.
- A sample is biased if one group is favored over another.
- Data can be organized by mean, median, mode, range, quartile and interquartile range.

Statistics (Lesson 12-3)

- A parameter is a characteristic of a whole population.
- The mean absolute value is the average of the absolute values of differences between the mean and each value and the data set.

Permutations and Combinations (Lesson 12-4)

- In a permutation, the order of objects is important, $P(n,r) = \frac{n!}{(n-r)!}$
- In a combination, the order of objects is not important, $C(n,r) = \frac{n!}{(n-r)!r!}$

Probability Distributions and Simulations (Lessons 12-6 and 12-7)

- For each value of X, $0 \leq P(X) \leq 1$. The sum of the probabilities for all values of X is 1.
- Theoretical probability describes expected outcomes, while experimental probability describes tested outcomes.
- Simulations are used to perform experiments that would be difficult or impossible to perform in real life.

FOLDABLES® Study Organizer

Be sure the Key Concepts are noted in your Foldable.

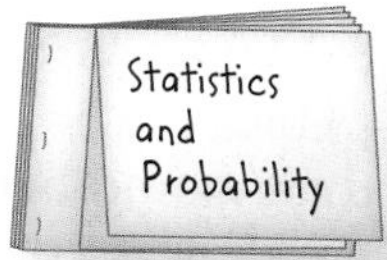

Key Vocabulary

biased sample (p. 741)
combination (p. 765)
complement (p. 772)
compound event (p. 771)
conditional probability (p. 777)
dependent events (p. 772)
discrete random variable (p. 779)
empirical study (p. 787)
experiment (p. 740)
experimental probability (p.787)
factorial (p. 764)
independent events (p. 771)
mean absolute deviation (p. 757)
mutually exclusive (p. 773)
parameter (p. 756)
permutation (p. 764)
population (p. 740)
probability distribution (p. 780)
qualitative data (p. 758)
quantitative data (p. 758)
random variable (p. 779)
relative frequency (p. 787)
sample (p. 740)
simple random sample (p. 742)
simulation (p. 788)
standard deviation (p. 758)
statistic (p. 756)
statistical inference (p. 756)
stratified random sample (p. 742)
survey (p. 740)
systemic random sample (p. 742)
theoretical probability (p. 787)
univariate data (p. 757)
variance (p. 758)

Vocabulary Check

Choose the word or term that best completes each sentence.

1. The arrangement in which order is important is called a (combination, permutation).
2. Rolling one die and then another are (dependent, independent) events.
3. The sum of probabilities of complements equals (0, 1).
4. Randomly drawing a marble from a jar and then drawing another marble are dependent events if the marbles (are, are not) replaced.
5. Events that cannot occur at the same time are (mutually exclusive, inclusive).

CHAPTER 12

Study Guide and Review

Lesson-by-Lesson Review

12-1 Designing a Survey (pp. 739–745)

6. **SCHOOL DANCE** The homecoming dance committee is trying to decide on a theme. They send out a questionnaire to all of the girls in the school. Identify the sample and suggest a population from which it was selected. Then, classify the type of data collection used.

7. **GOVERNMENT** To determine whether voters support a new trade agreement, 5 people from the list of registered voters in each state and in the District of Columbia are selected at random. Is the sample *biased* or *unbiased*?

8. **CANDY BARS** To ensure that all of the chocolate bars are the appropriate weight, every 50th bar on the conveyor belt in the candy factory is removed and weighed. Is the sample *simple*, *stratified*, or *systematic*?

EXAMPLE 1

For the situation, identify the sample and suggest a population from which it was selected. Then, classify the type of data collection used.

An artist is trying to choose a cover for a children's book. She sends out a flyer with the two covers to all of the students at one school. She asks them to check their favorite cover.

The sample is all of the students at the one school.
The population is all children who read books.
The type of data collection is a survey.

EXAMPLE 2

People listening to a country music radio station are asked to name their favorite type of music. Identify the sample as *biased* or *unbiased*.

The sample is biased because people listening to a country music station are more likely to vote for country music as their favorite.

12-2 Analyzing Survey Results (pp. 746–755)

Tell which measure of center best represents the data. Then find the measure of center.

9. **CLASSROOM** Sophia keeps track of the ages of the students in her class. She wants to best represent the ages of her classmates: 13, 14, 13, 13, 14, 13, 13, 15, 14, 13, 14, 14, 10.

10. **PETS** Jason conducts a survey about the number of pets his friends have. He wants to best represent the number of pets: 0, 2, 1, 2, 4, 0, 2, 3, 1, 2, 1, 2, 3, 2, 4, 3, 10.

11. **LUNCH PRICES** The cafeteria wants to show the best representation of how much students spend on lunch: 2, 3, 4, 3, 2, 4, 1, 3, 4, 2, 3, 3, 4, 3.

EXAMPLE 3

FINANCIAL LITERACY A company wants to show the best representation of an employee's salary. The salaries of the employees in the company are \$25,000, \$30,000, \$28,000, \$29,000, \$30,000, and \$65,000. Tell which measure of center best represents the data. Then find the measure of center.

List the values from least to greatest: \$25,000, \$28,000, \$29,000, \$30,000, \$30,000, \$65,000.

There is one value that is much greater than the rest of the data, \$65,000. Also, there does not appear to be a big gap in the middle of the data. The median would best represent the data.

The median is \$29,500.

MIXED PROBLEM SOLVING
For mixed problem-solving practice, see page 845.

12-3 Statistics and Parameters (pp. 756–762)

Find the mean absolute deviation to the nearest tenth.

12. **SHOVELING SIDEWALKS** Ben is shoveling sidewalks to raise money over break. He is keeping track of how many he shovels each day: 2, 4, 3, 5, 3.

13. **CANDY BARS** Luci is keeping track of the number of candy bars each member of the drill team sold.

20, 25, 30, 50, 40, 60, 20, 10, 42

Find the mean, variance, and standard deviation to the nearest tenth for each set of data.

14. 1, 1, 3, 4, 6

15. 10, 11, 11, 10, 12, 13, 14, 10

16. 3, 5, 6, 2, 1, 5

17. 10, 11, 10, 11, 12

18. 15, 16, 16, 15, 16, 17, 18

19. **FOOD** A fast food company polled a random sample of its customers to find how many times a month they eat out: 10, 3, 12, 15, 7, 8, 4, 12, 9, 14, 12. Find the mean absolute deviation of the data set to the nearest tenth.

EXAMPLE 4

GIFTS Joshua is collecting money from his family for his grandmother. He keeps track of how much was donated: 10, 5, 20, 15, 10. Find the mean absolute deviation.

First, find the mean of the data.

$$\frac{10 + 5 + 20 + 15 + 10}{5} = \frac{60}{5} = 12$$

Next, find the absolute value of the difference between the mean and each value.

$|10 - 12| = 2; |5 - 12| = 7; |20 - 12| = 8;$
$|15 - 12| = 3; |10 - 12| = 2$

Now, find the mean of the differences.

$$\frac{2 + 7 + 8 + 3 + 2}{5} = \frac{22}{5} = 4.4$$

The mean absolute deviation is 4.4.

EXAMPLE 5

Find the mean, variance, and standard deviation for 2, 4, 3, 5, and 6.

mean: $\frac{2 + 4 + 3 + 5 + 6}{5} = \frac{20}{5} = 4$

$$\sigma^2 = \frac{(2-4)^2 + (4-4)^2 + (3-4)^2 + (5-4)^2 + (6-4)^2}{5}$$

$$\sigma^2 = \frac{4 + 0 + 1 + 1 + 4}{5} \text{ or } 2$$

$$\sigma = \sqrt{2} \text{ or about } 1.4$$

12-4 Probability with Permutations and Combinations (pp. 764–770)

Evaluate each expression.

20. $C(10, 3)$

21. $C(9, 5)$

22. $P(6, 3)$

23. $P(5, 4)$

24. **PHOTOS** The Spanish teacher at South High School wants to arrange 7 students who traveled to Mexico for a yearbook photo.

a. Is this a permutation or combination?

b. How many ways can the students be arranged?

EXAMPLE 6

Find $C(8, 3)$.

$$C(8, 3) = \frac{8!}{(8-3)!3!} = \frac{8!}{5!3!} = \frac{8 \cdot 7 \cdot 6 \cdot 5 \cdot 4 \cdot 3 \cdot 2 \cdot 1}{5 \cdot 4 \cdot 3 \cdot 2 \cdot 1 \cdot 3 \cdot 2 \cdot 1}$$

$$= \frac{8 \cdot 7 \cdot 6}{6} = 56$$

EXAMPLE 7

Find $P(4, 2)$.

$$P(4, 2) = \frac{4!}{(4-2)!} = \frac{4 \cdot 3 \cdot 2 \cdot 1}{2 \cdot 1} = 4 \cdot 3 = 12$$

CHAPTER 12

Study Guide and Review

12-5 Probability of Compound Events (pp. 771–778)

A box contains 8 red chips, 6 blue chips, and 12 white chips. Three chips are randomly drawn from the box and are not replaced.

25. P(red, white, blue) **26.** P(red, red, red)

27. P(red, white, white) **28.** P(blue, blue)

One card is randomly drawn from a standard deck of 52 cards. Find each probability.

29. P(heart or red)

30. P(10 or spade)

EXAMPLE 8

A bag of colored paper clips contains 30 red clips, 22 blue clips, and 22 green clips. Find each probability if three clips are drawn randomly from the bag and are not replaced. Find P(blue, red, green).

First clip: $P(\text{blue}) = \frac{22}{74}$ Second clip: $P(\text{red}) = \frac{30}{73}$

Third clip: $P(\text{green}) = \frac{22}{72}$

$P(\text{blue, red, green}) = \frac{22}{74} \cdot \frac{30}{73} \cdot \frac{22}{72} = \frac{605}{16{,}206}$

12-6 Probability Distributions (pp. 779–786)

A local cable provider asked its subscribers how many television sets they had in their homes. The results of their survey are shown in the probability distribution.

X = Number of Televisions	Probability
1	0.18
2	0.36
3	0.34
4	0.08
5+	0.04

31. Show that the distribution is valid.

32. If a household is selected at random, what is the probability that it has fewer than 4 televisions?

EXAMPLE 9

The table shows the probability distribution for the number of activities in which students at Midpark High School participate.

X = Number of Activities	Probability
0	0.04
1	0.12
2	0.37
3	0.30
4+	0.17

What is the probability that a randomly chosen student participates in 1 to 3 activities?

$$P(1 \leq X \leq 3) = P(X = 1) + P(X = 2) + P(X = 3)$$
$$= 0.12 + 0.37 + 0.30$$
$$= 0.79 \text{ or } 79\%$$

12-7 Probability Simulations (pp. 787–792)

The results of a simulation of coin flipping are shown.

Outcome	Frequency
heads	25
tails	75

33. What is the experimental probability of heads?

34. What is the experimental probability of tails?

35. What is the theoretical probability of heads?

36. How can guessing randomly on a true-false question be simulated?

EXAMPLE 10

On a multiple-choice test with four choices, Maya answered 18 out of 20 correctly by guessing randomly. Describe a way to simulate this method.

A spinner with 4 equal sections could be spun 20 times with the results of each spin being recorded.

Practice Test

Math Online glencoe.com
Chapter Test

Identify each sample, and suggest a population from which it was selected. Then classify the type of data collection used.

1. **TOYS** A toy company invites 50 children in to test a new toy and records the reactions.

2. **FLOWERS** A nursery is sending out questionnaires to determine which flowers people like best. They send the questionnaires out to all people over 50 on their mailing list.

3. **MULTIPLE CHOICE** On a multiple-choice test with four choices, Zack answered 12 out of 20 questions correctly. What could be used to simulate his correctly answering a question?

   A tossing a coin

   B rolling a six-sided number cube

   C spinning a spinner with four equal sections

   D rolling a three-sided number cube

Evaluate each expression.

4. $P(7, 5)$
5. $C(10, 4)$
6. $C(7, 2)$
7. $P(6, 3)$

Which measure of central tendency best represents the data? Justify your answer. Then find the measure.

8. **VOTING** The polling place kept a list of all the ages of the people who voted: 21, 25, 32, 41, 32, 20, 65, 33, 30, 72.

9. **SHOPPING** A department store kept track of the number of items shoppers purchased on a given day: 3, 5, 4, 3, 4, 5, 5, 3, 2, 3, 2, 10.

Find the mean, variance, and standard deviation to the nearest tenth for each set of data.

10. 4, 5, 5, 6, 6, 8, 9, 10

11. 22, 25, 27, 30

12. 10, 10, 12, 14

13. **SALES** Nate is keeping track of how much people spent at the school bookstore in one day. Find the mean absolute deviation for the data to the nearest tenth: 1, 1, 2, 3, 4, 5, 12.

Identify each sample as *biased* or *unbiased*. Explain your reasoning.

14. **NEWSPAPERS** A survey is sent to all people who subscribe to *The Dispatch* to determine what newspaper people prefer to read.

15. **SHOPPING** Each person leaving the Maxtowne Mall is asked to name their favorite clothing store in the mall.

16. **PIZZA** How many ways can 3 different toppings be chosen from a list of 10 toppings?

17. What is the theoretical probability of tossing heads when a coin is tossed?

18. A die is rolled twice. What is the probability of getting a 2 then a 3?

19. **EDUCATION** Kristin surveys 200 people in her school to determine how many nights a week students do homework. The results are shown.

Number of Nights	Number of Students
0	10
1	30
2	50
3	90
4	10
5 or more	10

   a. Find the probability that a randomly chosen student will have studied more than 4 nights.

   b. Find the probability that a randomly chosen student will have studied no more that 3 nights.

20. **MULTIPLE CHOICE** The second graders are divided into boys and girls. Then 2 girls and 2 boys are chosen at random to represent the class at the Pride Assembly. Which of the following best describes the sample?

   F simple

   G stratified

   H systematic

   J none of the above

Organize Data

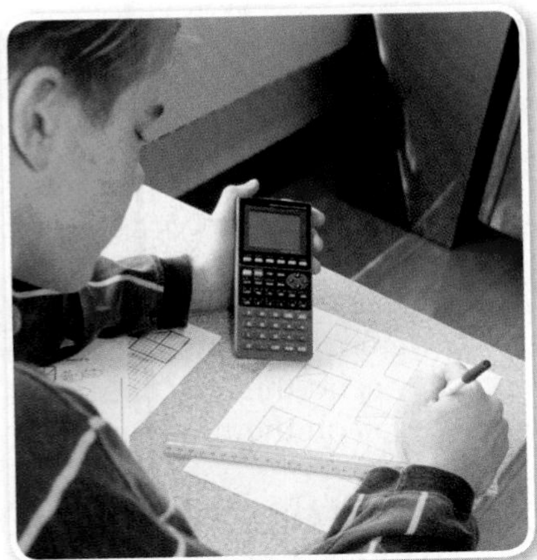

Sometimes you may be given a set of data that you need to analyze in order to solve problems on a standardized test. Use this lesson to practice organizing data to help you solve problems.

Strategies for Organizing Data

Step 1

When you are given a problem statement containing data, consider:

- **making a list** of the data.
- **using a table** to organize the data.
- **using a data display** (such as a bar graph, Venn diagram, circle graph, line graph, or box-and-whisker plot) to organize the data.

Step 2

Organize the data.

- Create your table, list, or data display.
- If possible, fill in any missing values that can be found by intermediate computations.

Step 3

Analyze the data to solve the problem.

- Reread the problem statement to determine what you are being asked to solve.
- Use the properties of algebra to work with the organized data and solve the problem.
- If time permits, go back and check your answer.

EXAMPLE

Read the problem. Identify what you need to know. Then use the information in the problem to solve. Show your work.

Of the 24 students in a music class, 10 play the flute, 14 play the piano, and 13 play the guitar. Two students play the flute only, 5 the piano only, and 7 the guitar only. One student plays the flute and the guitar but not the piano. Two students play the piano and guitar but not the flute. Three students play all the instruments. If a student is selected at random, what is the probability that he or she plays the piano and flute, but not the guitar?

Scoring Rubric	
Criteria	**Score**
Full Credit: The answer is correct and a full explanation is provided that shows each step.	2
Partial Credit: • The answer is correct, but the explanation is incomplete. • The answer is incorrect, but the explanation is correct.	1
No Credit: Either an answer is not provided or the answer does not make sense.	0

Read the problem carefully. The data is difficult to analyze as it is presented. Use a Venn diagram to organize the data and solve the problem.

Example of a 2-point response:

Use a Venn diagram to organize the data. Fill in all of the information given in the problem statement. There are 14 students who play the piano, so 14 – 5 – 2 – 3 or 4 students play the piano and the flute, but not the guitar. Find the probability.

$P(\text{piano and flute}) = \frac{4}{24}$ or $\frac{1}{6}$

So, the probability that a randomly selected student plays the piano and flute but not the guitar is $\frac{1}{6}$.

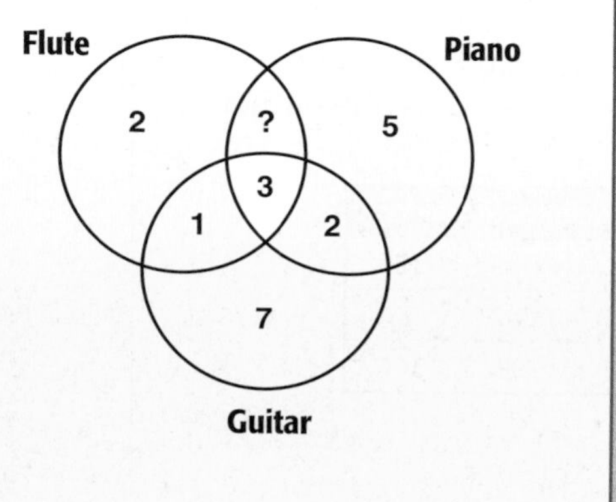

Exercises

Read the problem. Identify what you need to know. Then use the information in the problem to solve. Show your work.

1. There are 40 students, 9 camp counselors, and 5 teachers at Camp Kern. Each person is assigned to one activity this afternoon. There are 9 students going hiking and 17 students going horseback riding. Of the camp counselors, 2 will supervise the hike and 3 will help with the canoe trip. There are 2 teachers helping with the canoe trip and 2 going horseback riding. Suppose a person is selected at random during the afternoon activities. What is the probability that the one selected is a student on the canoe trip or a camp counselor on a horse? Express your answer as a fraction.

2. The table shows the number of coins in a piggy bank.

Coin	Number
Penny	16
Nickel	18
Dime	20
Quarter	10

   a. Find the probability that a randomly selected coin will be a dime.

   b. Find the probability that a randomly selected coin will be either a nickel or a quarter.

3. It takes Craig 40 minutes to mow his family's lawn. His brother Jacob can do the same job in 50 minutes. How long would it take them to mow the lawn together? Round your answer to the nearest tenth of a minute.

CHAPTER 12

Standardized Test Practice

Cumulative, Chapters 1 through 12

Multiple Choice

Read each question. Then fill in the correct answer on the answer document provided by your teacher or on a sheet of paper.

1. What are the excluded values of the variable in the expression $\frac{x^2 - x - 12}{x^2 - x - 2}$?

A $-1, 2$

B $-2, 2$

C $-2, 1$

D $-3, 4$

2. The table shows the number of Calories in twelve different snacks. Which measure of central tendency would be *most* affected by the outlier 342 Calories?

Number of Calories in Snacks			
122	87	149	121
64	138	342	72
179	105	99	114

F mean

G median

H mode

J range

3. Which of the following is *not* a factor of $x^4 - 6x^2 - 27$?

A $x^2 + 3$

B $x - 3$

C $x + 3$

D $x^2 - 3$

4. Eduardo has 20 CDs. He wants to choose 3 of them at random to take on a road trip. How many different ways can he do this if the order is *not* important?

F 60

G 84

H 1,140

J 1,482

5. Which of the following does *not* accurately describe the graph $y = -2x^2 + 4$?

A The parabola is symmetric about the y-axis.

B The parabola opens downward.

C The parabola has the origin as its vertex.

D The parabola crosses the x-axis in two different places.

6. The highest point in North Carolina is Mt. Mitchell at an elevation of 2,037 meters above sea level. Suppose the position of a hiker is given by the function $p(t) = -2.5t + 2{,}037$, where t is the number of minutes. Which of the following is the best interpretation of the slope of the function?

F The hiker's initial position was 2,037 feet below sea level.

G The hiker's initial position was 2,037 feet above sea level.

H The hiker is descending at a rate of 2.5 meters per minute.

J The hiker is ascending at a rate of 2.5 meters per minute

7. Jorge has made 39 out of 52 free throw attempts this season. What is the experimental probability that he makes a free throw?

A 54%

B 68%

C 75%

D 79%

8. Which equation passes through the points $(-1, -3)$ and $(-2, 3)$.

F $y = -6x - 9$

G $y = -\frac{1}{4}x + 3$

H $y = 4x - 5$

J $y = \frac{2}{3}x + 1$

9. At a museum, each child admission costs \$5.75 and each adult costs \$8.25. How much does it cost a family that consists of 2 adults and 4 children?

A \$34.50

B \$39.50

C \$44.50

D \$49.50

Test-TakingTip

Question 4 Since order is not important, you are looking for the number of combinations of CDs that can be chosen.

Short Response/Gridded Response

Record your answers on the answer sheet provided by your teacher or on a sheet of paper.

10. GRIDDED RESPONSE Suppose Colleen spins the spinner below 80 times and records the results in a frequency table. How many times should she expect to spin a vowel?

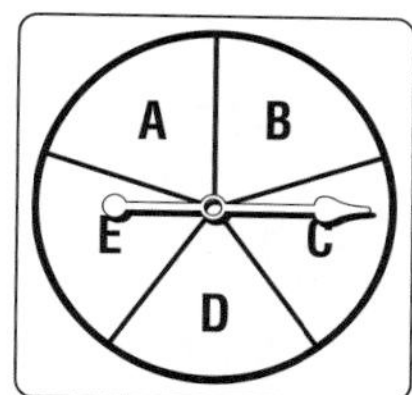

11. What is the value of sin B? Express your answer as a fraction.

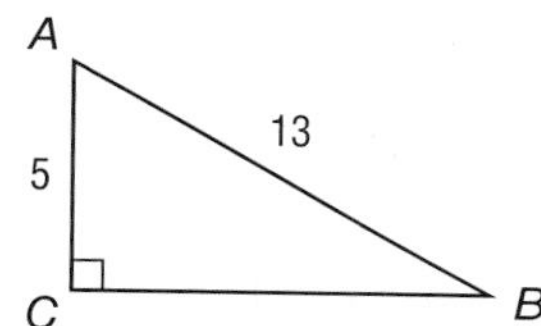

12. Graph $f(x) \geq |x - 2|$ on a coordinate grid.

13. GRIDDED RESPONSE Find the standard deviation of the set of data below. Show your work. Round to the nearest tenth if necessary.

14	11	9	6
10	16	15	13
9	12	19	10

14. Larissa has 5 peanut butter cookies, 7 chocolate chip cookies, 4 sugar cookies, and 9 oatmeal raisin cookies in a jar. If she picks two cookies at random without replacing them, what is the probability that she will choose a peanut butter cookie then a sugar cookie? Express your answer as a fraction.

15. Write an expression that describes the area in square units of a triangle with a height of $4c^3d^2$ and a base of $3cd^4$.

16. Casey made 84 field goals during the basketball season for a total of 183 points. Each field goal was worth either 2 or 3 points. How many 2-point and 3-point field goals did Casey make during the season?

17. GRIDDED RESPONSE The booster club pays \$180 to rent a concession stand at a football game. They purchase cans of soda for \$0.25 and sell them at the game for \$1.15. How many cans of soda must they sell to break even?

Extended Response

Record your answers on a sheet of paper. Show your work.

18. To predict whether or not an issue on a ballot will pass or fail, a committee randomly calls 250 houses with area codes that are inside the voting district and asks the opinions of registered voters. Based on these efforts, the committee determines that 71% (±2.5%) of the voting population supports the issue. The committee concludes that the issue will pass.

a. Identify the sample.

b. Describe the population.

c. What method of data collection did the committee use: survey, experiment, or observational survey? Explain.

d. Is the sample *biased* or *unbiased*. Explain.

e. If unbiased, classify the sample as *simple*, *stratified*, or *systematic*. Explain.

Need Extra Help?

If you missed Question...	1	2	3	4	5	6	7	8	9	10	11	12	13	14	15	16	17	18
Go to Lesson or Page...	11-2	12-5	8-5	12-4	9-3	3-3	7-3	12-7	1-3	12-7	10-8	5-6	12-3	12-5	7-1	6-5	2-4	12-1

Student Handbook

Built-In Workbooks

Reference

How to Use the Student Handbook

The Student Handbook is the additional skill and reference material found at the end of the text. This Handbook can help you answer these questions.

What If I Need Problem-Solving Practice?
You have probably used several different problem-solving strategies in previous math courses. The **Problem-Solving Handbook** section provides example and problems for refreshing your skills at using various strategies.

What If I Need More Practice?
You, or your teacher, may decide that working through some additional problems would be helpful. The **Extra Practice** section provides these problems for each lesson so you have ample opportunity to practice new skills.

What If I Have Trouble with Word Problems?
The **Mixed Problem Solving** portion of the book provides additional word problems that use the skills presented in each lesson. These problems give you real-world situations where math can be applied.

What if I Forget What I Learned Last Year?
Use the **Concepts and Skills Bank** section to refresh your memory about things you have learned in other math classes. Here's a list of the topics covered in your book.

1. Converting Units of Measure
2. Factors and Multiples
3. Prime Factorization
4. Measuring Angles
5. Venn Diagrams
6. Misleading Graphs

What If I Need to Check a Homework Answer?
The answers to odd-numbered problems are included in **Selected Answers and Solutions**. Check your answers to make sure you understand how to solve all of the assigned problems.

What If I Forget a Vocabulary Word?
The **English-Spanish Glossary** provides a list of important or difficult words used throughout the textbook. It provides a definition in English and Spanish as well as the page number(s) where the word can be found.

What If I Need to Find Something Quickly?
The **Index** alphabetically lists the subjects covered throughout the entire textbook and the pages on which each subject can be found.

What if I Forget a Formula?
Inside the back cover of your math book is a list of **Formulas and Symbols** that are used in the book.

Problem-Solving Handbook

Problem-Solving Strategy: Look for a Pattern

There are many problem-solving strategies in mathematics. One of the most common is to **look for a pattern**. To use this strategy, analyze the first few numbers or figures in a pattern and identify a rule that relates the first number or figure in the pattern to the second, and then to the third, and so on. Then use the rule to extend the pattern and find a solution.

EXAMPLE

Refer to the graph. Describe the pattern in the coordinates and predict the next point in the pattern in the positive direction.

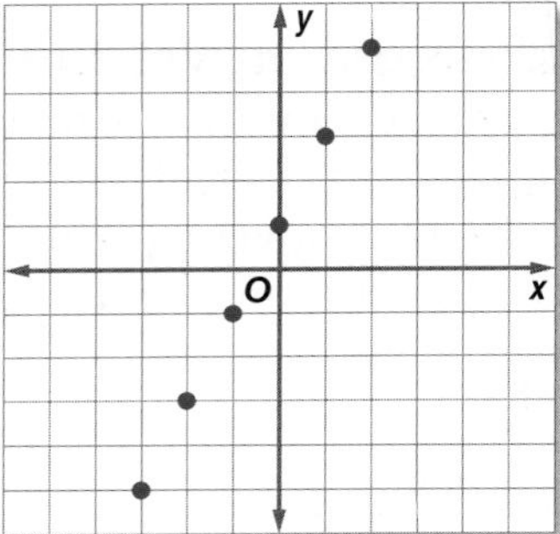

Step 1 List the coordinates of the points shown on the graph.

(−3, −5), (−2, −3), (−1, −1), (0, 1), (1, 3), and (2, 5)

Step 2 Identify the pattern in the x-coordinates and the y-coordinates.

As the x-coordinates increase by 1, the y-coordinates increase by 2 each time. So in the positive direction, the next point is $(2 + 1, 5 + 2)$ or $(3, 7)$.

Practice

Solve each problem by looking for a pattern.

1. The graph of a function passes through the points shown.

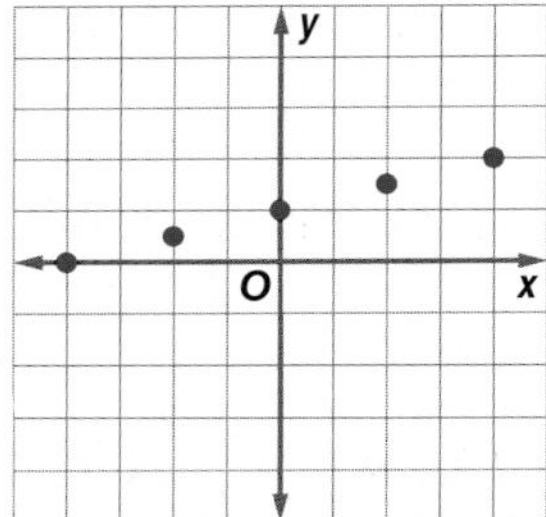

a. Describe the pattern in the coordinates, and predict the next point in the pattern in the positive direction.

b. Predict the next point in the pattern in the negative direction.

2. List the first five common multiples of 3, 4, and 6. Write an expression to describe all common multiples of 3, 4, and 6.

3. What is the perimeter of the twelfth figure?

Figure 1 Perimeter – 6 | Figure 2 Perimeter – 8 | Figure 3 Perimeter – 10

4. The football parent booster group sold hot chocolate at the game on Friday night. The table shows the total amount of money raised based on the number of cups of hot chocolate sold. Use this data table to determine how much money would be raised selling 75 cups of hot chocolate.

Number of Cups Sold	Amount of Money Raised ($)
15	11.25
30	22.50
45	33.75
60	45.00
75	?

Problem-Solving Strategy: Create a Table

One strategy for solving problems is to **create a table**. A table allows you to organize information in an understandable way.

Real-World EXAMPLE

A fruit machine accepts dollars, and each piece of fruit costs 65 cents. If the machine gives only nickels, dimes, and quarters, what combinations of those coins are possible as change for a dollar?

The machine will give back $1.00 − $0.65 or 35 cents in change in a combination of nickels, dimes, and quarters.

Make a table showing different combinations of nickels, dimes, and quarters that total 35 cents. Organize the table by starting with the combinations that include the most quarters.

Quarters	Dimes	Nickels
1	1	0
1	0	2
0	3	1
0	2	3
0	1	5
0	0	7

The total for each combination of the coins is 35 cents. There are 6 combinations possible.

Practice

Solve each problem by creating a table.

1. How many ways can you make change for a half-dollar using only nickels, dimes, and quarters?
2. A penny, a nickel, a dime, and a quarter are in a purse. How many amounts of money are possible if you grab two coins at random?
3. How many ways can you receive change for a quarter if at least one coin is a dime?
4. Johanna had a bag of four marbles. One marble is blue. Two marbles are green. One marble is orange. How many different ways are there to draw the marbles out of the bag one at a time?
5. At Midas High School, students are selling popcorn at a football game. Each small bag of popcorn is $1.25. Each large bag of popcorn is $2.25. Create a table to show the purchase price of up to five bags of each size of popcorn.
6. Make a table to show the ordered pairs that satisfy the equation $f(x) = 3x^3 - 4$. Use the domain of integers between 3 and 10.
7. The equation of a line is $y = \frac{1}{2}x - 2$. Create a table to show five ordered pairs with x-coordinates belonging to the set $\{-2, -1, 0, 1, 2\}$.
8. Aria asked her friends whether they used wrapping paper, gift bags, recycled paper, or no wrapping to wrap birthday presents. Create a table to show how many students preferred each method of the 24 students she asked. Then predict how many would choose each method if 120 students were asked.

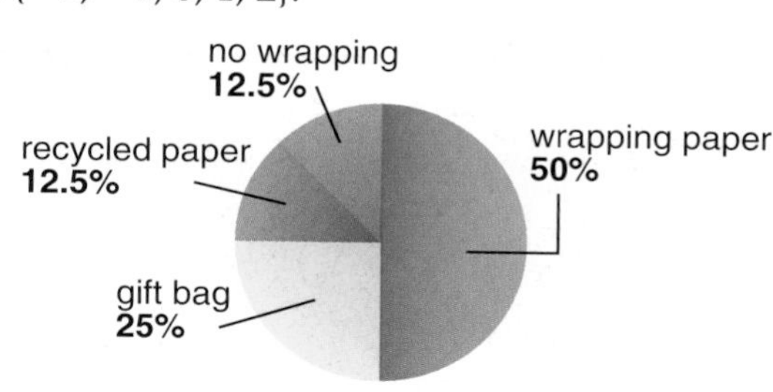

Problem-Solving Strategy: Make a Chart

Data presented in a problem can be organized by **making a chart**. This problem-solving strategy allows you to see patterns and relationships among data.

Real-World EXAMPLE

It takes an average driver 1.5 seconds to begin braking after they see an obstruction. The driver can safely decelerate a car or light truck with good tires on a dry street surface at the rate of about 15 feet per second (fps). The distance vehicle will travel while braking can be found by multiplying the initial velocity by the deceleration time and dividing by 2. Find the stopping distance for a car traveling at 45, 55, 65 and 75 miles per hour.

Step 1 Make a chart that includes the given information and the information to be found. Before setting up your chart, think about how speed, distance, and time are related.

Since the distance is listed in feet, find the rate of speed it takes to stop in feet per second. To convert from miles per hour to feet per second, multiply by a conversion factor of $\frac{3600 \text{ s}}{5280 \text{ ft}}$. Then find the deceleration time by dividing the initial velocity by 15 fps.

Initial Velocity (mph)	Initial Velocity (fps)	Deceleration Time (s)	Distance Traveled before Braking (ft)	Distance Traveled while Braking (ft)	Total Stopping Distance (ft)
45	66	4.4			
55	80.7	5.38			
65	95.3	6.35			
75	110	7.33			

Step 2 To find the distance traveled before braking, multiply the initial velocity by the reaction time, 1.5 seconds. Then find the distance traveled while braking. Add to find the total stopping distance.

Initial Velocity (mph)	Initial Velocity (fps)	Deceleration Time (s)	Distance Traveled before Braking (ft)	Distance Traveled while Braking (ft)	Total Stopping Distance (ft)
45	66	4.4	99	145.2	244.2
55	80.7	5.38	121.05	217.08	338.13
65	95.3	6.35	142.95	302.58	445.53
75	110	7.33	165	403.15	568.15

Practice

Solve each problem by making a chart.

1. As the length of a square doubles, the area increases by a scale factor. Using the squares in the diagram, make a chart of each length and each area. Then find the scale factor.

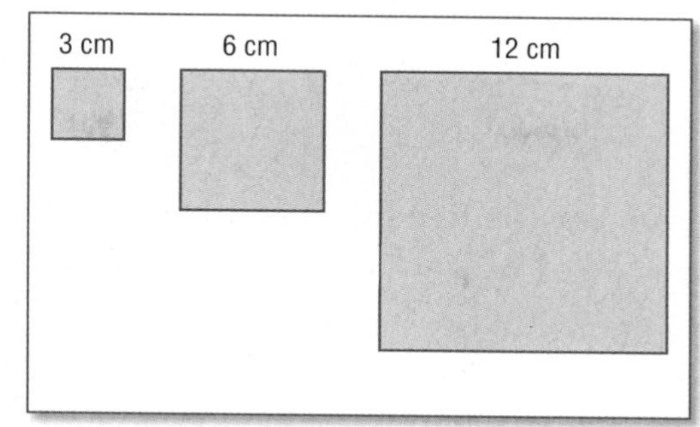

2. Given the functions $f(x) = 2x + 3$ and $g(x) = -x - 3$, use a table to find $f(x) - g(x)$ for all positive integers less than or equal to 6.

3. The chart shows at which point drivers determine when they are going to fill the gas tank. Make a chart to show how many people of 200 surveyed would be expected to have each response.

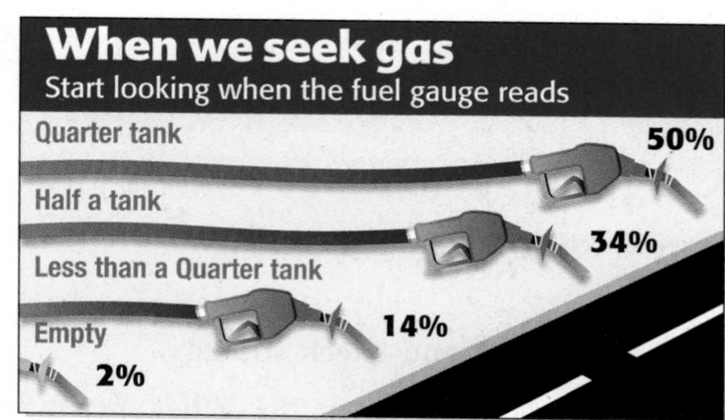

Source: *Bruskin/Goldring for Exxon*

4. The following table shows the official state reptile in each state. Make a tally chart that shows how many states have turtles (tortoise, terrapin), snakes, alligators, lizards, toads, or none. Find the ratio of states with alligators to states with no official state reptile.

AL	red-bellied turtle	**LA**	American alligator	**OH**	black racer (snake)
AK	none	**ME**	none	**OK**	collared lizard
AZ	ridge-nosed rattlesnake	**MD**	diamondback terrapin	**OR**	none
AR	none	**MA**	garter snake	**PA**	none
CA	desert tortoise	**MI**	painted turtle	**RI**	none
CO	none	**MN**	Blanding's turtle	**SC**	loggerhead turtle
CT	none	**MS**	American alligator	**SD**	none
DE	none	**MO**	three-toed turtle	**TN**	eastern box turtle
FL	American alligator	**MT**	none	**TX**	Texas horned lizard
GA	gopher tortoise	**NE**	none	**UT**	none
HI	none	**NV**	desert tortoise	**VT**	none
ID	none	**NH**	none	**VA**	none
IL	painted turtle	**NJ**	none	**WA**	none
IN	none	**NM**	whiptail lizard	**WV**	none
IA	none	**NY**	snapping turtle	**WI**	none
KS	ornate box turtle	**NC**	eastern box turtle	**WY**	horned toad
KY	none	**ND**	none		

Problem-Solving Strategy: Guess-and-Check

To solve some problems, you can make a reasonable guess and then check it in the problem. You can then use the results to improve your guess until you find the solution. This strategy is called **guess-and-check**.

EXAMPLE

The product of two even consecutive integers is close to 1000.

Make a guess. Let's try 24 and 26. → $24 \times 26 = 624$ ← This product is too low.

Adjust the guess upward.
Try 30 and 32. → $30 \times 32 = 960$ ← This product is still too low.

Adjust the guess upward again.
Try 34 and 36. → $34 \times 36 = 1224$ ← This product is too high.

Try between 30 and 34.
Try 32 and 34. → $32 \times 34 = 1088$ ← This is the correct product.

The integers are 32 and 34.

Practice

Solve each problem by using the guess-and-check strategy.

1. The product of two consecutive odd integers is 783. What are the integers?
2. Brianne is three times as old as Camila. Four years from now she will be just two times as old as Camila. How old are Brianne and Camila now?
3. Rafael is burning a CD for Selma. The CD will hold 35 minutes of music. Which songs should he select from the list to record the maximum time on the CD without going over?

Song	A	B	C	D	E	F	G	H	I	J
Time	5 min 4 s	9 min 10 s	4 min 12 s	3 min 9 s	3 min 44 s	4 min 30 s	5 min 0 s	7 min 21 s	4 min 33 s	5 min 58 s

4. Each hand in the human body has 27 bones. There are 6 more bones in the fingers than in the wrist. There are 3 fewer bones in the palm than in the wrist. How many bones are in each part of the hand?
5. The Science Club sold candy bars and soft pretzels to raise money for an animal shelter. They raised a total of \$62.75. They made 25¢ profit on each candy bar and 30¢ profit on each pretzel sold. How many of each did they sell?
6. The product of two consecutive even integers is 4224. Find the integers.
7. Odell has the same number of quarters, dimes, and nickels. In all he has \$4 in change. How many of each coin does he have?
8. Anita sold tickets to the school musical. She had 2 types of bills worth \$75 for the tickets she sold. If all the money that the club collected was in \$5 bills, \$10 bills, and \$20 bills, how many of each bill did Anita have?
9. Two angles of a triangle are shown. Find the third angle if the three angles have a sum of 180°.

38°
32°

Problem-Solving Strategy: Work Backward

On most problems, a set of conditions or facts is given and an end result must be found. However some problems start with the result and ask for something that happened earlier. The strategy of **working backward** can be used to solve problems like this. To use this strategy, start with the end result and *undo* each step.

Real-World EXAMPLE

Kendrick spent half of the money he had this morning on lunch. After lunch, he loaned his friend a dollar. Now he has \$1.50. How much money did Kendrick start with?

Start with end result, \$1.50, and work backward to find the amount Kendrick started with.

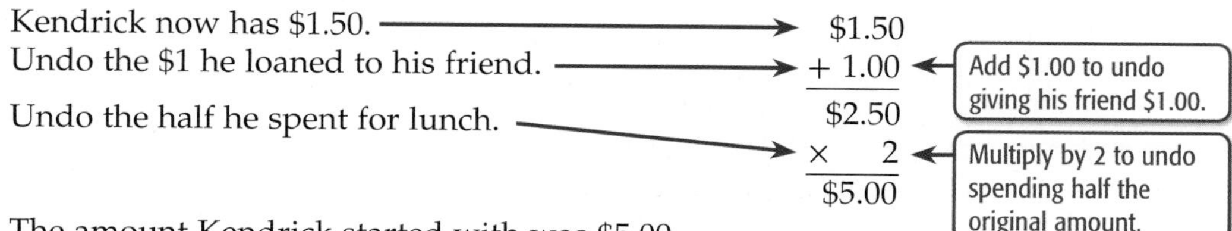

The amount Kendrick started with was \$5.00.

CHECK Kendrick started with \$5.00. If he spent half of that, or \$2.50, on lunch and loaned his friend \$1.00, he would have \$1.50 left. This matches the amount stated in the problem, so the solution is correct.

Practice

Solve each problem by working backward.

1. A certain number is multiplied by 3, and then 5 is added to the result. The final answer is 4. What is the number?
2. A certain bacteria doubles its population twice each day. After 3 full days, there are 1600 bacteria in a culture. How many bacteria were there at the beginning of the first day?
3. To catch a 7:30 A.M. bus, Don needs 30 minutes to get dressed, 30 minutes for breakfast, and 5 minutes to walk to the bus stop. What time should he wake up?
4. Find the length of the side of the quadrilateral if the perimeter is 83 meters.
5. Troy lives $1\frac{1}{8}$ miles from his school. If he has walked $\frac{1}{4}$ mile to meet a friend and then they walked another $\frac{1}{2}$ mile to meet another friend, how far do they still need to walk to get to school?

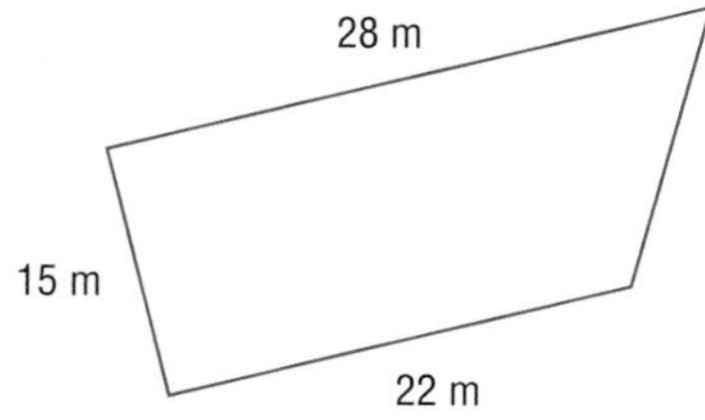

6. If a lizard in its cage weighs 23 pounds, find the weight of the lizard if the cage weighs 19 pounds and the sand weighs 2 pounds.
7. Mattie spent \$125.50 on three items at the mall. She bought one pair of socks for \$1.20, one pair of shoes for \$48.95. How much money did she spend on the jacket she purchased?

Problem-Solving Strategy: Solve a Simpler Problem

One of the strategies you can use to solve a problem is to **solve a simpler problem**. To use this strategy, first solve a simpler or more familiar case of the problem. Then use the same concept and relationships to solve the original problem.

EXAMPLE

Find the sum of the number 1 through 500.

Consider a simpler problem. Find the sum of the numbers 1 through 10. Notice that you can group the addends into partial sums as shown below.

$1 + 2 + 3 + 4 + 5 + 6 + 7 + 8 + 9 + 10 = 55$

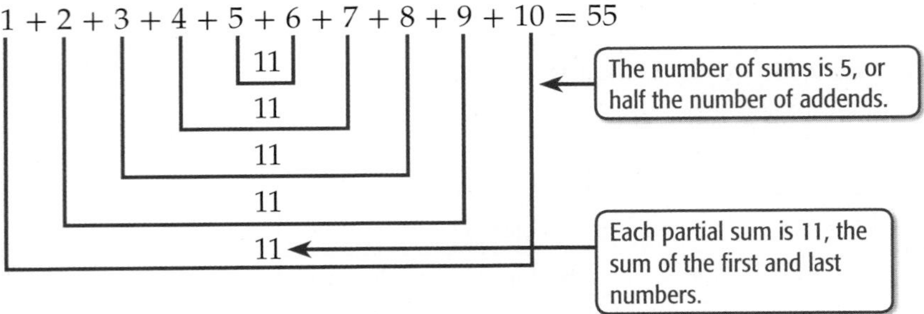

The sum is 5×11 or 55.

Use the same concepts to find the sum of the numbers 1 through 500.

$$1 + 2 + 3 + \ldots + 499 + 500 = 250 \times 501$$
$$= 125{,}250$$

Multiply half the number of addends, 250, by the sum of the first and last numbers, 501.

Practice

Solve each problem by solving a simpler problem.

1. Find the number of squares of any size in the game board shown at the right.

2. Find the sum of the numbers through 1000.
3. How many links are needed to join 30 pieces of chain into one long chain?
4. Three people can pick six baskets of apples in one hour. How many baskets of apples can 2 people pick in one-half hour?
5. A shirt shop has 112 orders for T-shirt designs. Three designers can make 2 shirts in 2 hours. How many designers are needed to complete the orders in 8 hours?
6. Find the area of the figure at the right.

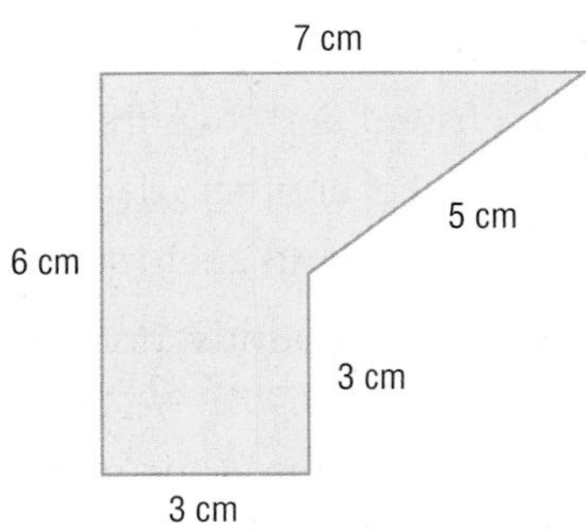

7. To add $\frac{3}{8}$ and $\frac{7}{12}$, how could you solve a simpler problem that leads to the answer?
8. If one fourth of the 136 freshmen at Bayridge High School packed their lunches, explain how to use a simpler problem to find the number of students.
9. Explain how to find the greatest common factor of 35 and 49 using a simpler problem.

Problem-Solving Strategy: Draw a Diagram

Another strategy for solving problems is to **draw a diagram**. There will be times when a sketch or diagram will give you a better picture of how to tackle a mathematics problem. Adding details like units, labels, and numbers to the drawing or sketch can help you make decisions on how to solve the problem.

Real-World EXAMPLE

Imani is trying to determine the number of 9-inch tiles needed to cover her patio. The rectangular patio measures 8-feet by 10-feet. What is the minimum number of 9-inch tiles Imani should purchase?

First, draw a diagram of the situation. Express the measurement of the patio in inches.

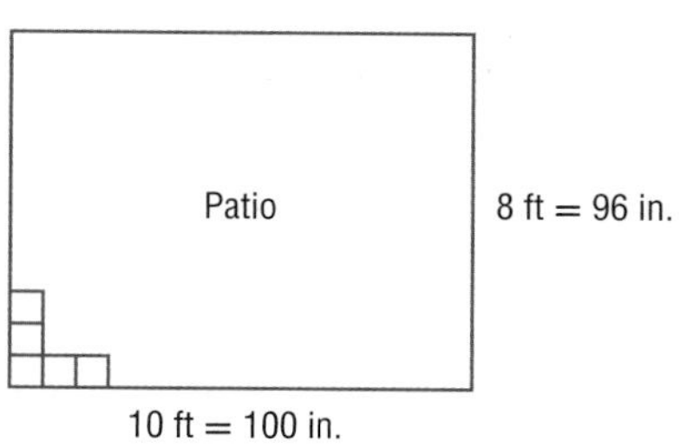

If each tile is 9 inches square, the minimum number of tiles for the width of the patio is $96 \div 9 \approx 10.7$ or 11 tiles.

The minimum number of tiles for the length of the patio is $120 \div 9 \approx 13.3$ or 14 tiles.

So the minimum number of tiles Imani needs to cover the patio is 11×14 or 154 tiles.

Practice

Solve each problem by drawing a diagram.

1. The area of a rectangular flower bed is 24 square feet. If the sides are whole number dimensions, how many combinations of lengths and widths are possible for the flower bed? List them.
2. Kevin was hired to paint a mural on a wall that measures 5 feet by 20 feet. Starting from the center of the wall, he will paint a square that measures 3 feet. The dimensions of the next square will be 0.5 times greater than and centered on the previous square. How many squares can Kevin paint on the wall?
3. It takes 42 minutes to cut a 2-inch by 4-inch piece of wood into 7 equally sized pieces. How long will it take to cut a similar 2-inch by 4-inch piece into 4 equally sized pieces?
4. Find the number of line segments that can be drawn between any two vertices of an octagon.

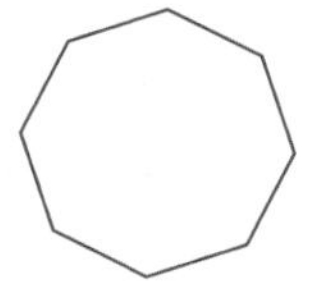

5. How many different teams of 3 players can be chosen from 8 players?
6. A right triangle has measures of 11, 60, and 61 millimeters. Find the area of the triangle.

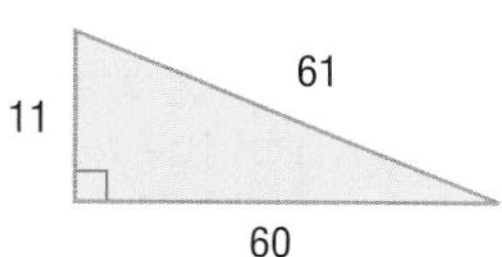

7. Nitarren is trying to decide how many 9 inch diameter pies will fit on her dessert table that measures 4 feet by 2 feet. How many pies could Nitarren fit on the table?

Problem-Solving Strategy: Use Estimation

When you need to make a decision on the basis of inexact information, a common strategy is to **use estimation**. Often estimation is used when an exact answer is not required or when mental math is used rather than a calculator or paper and pencil. You should use estimation to determine if your answer is reasonable.

Real-World EXAMPLE

In a recent year, 51 million international visitors came to the United States. Given the information in the table, estimate what percentage of the visitors were from Japan.

Source: Travel Industry Association

Step 1 Determine about how many international visitors came to the United States.

51 million or 51,000,000 would be an easier number to work with if it was rounded to only one digit in the ten millions place, so round 51,000,000 to 50,000,000.

Step 2 Determine from the chart the number of visitors who were from Japan.

According to the chart, 5,000,000 visitors to the United States were from Japan.

Step 3 To determine the percentage, divide.

$$\frac{\text{number of visitors from Japan}}{\text{total number of visitors}} = \frac{5{,}000{,}000}{50{,}000{,}000}$$

$$= 0.1 \text{ or } 10\%$$

About 10% of the visitors to the United States in 2000 were from Japan.

Practice

Solve each problem by using estimation.

1. Use the table at the right to find the total area of the Great Lakes.

Great Lakes	
Great Lake	**Area (mi²)**
Lake Superior	31,698
Lake Huron	23,011
Lake Michigan	22,316
Lake Erie	9922
Lake Ontario	7320

2. The length of Fun Center's go-kart track is 843 feet. If Nadia circled the track 9 times, about how many feet did she travel?

3. The Student Council is making pizzas to sell at the football game on Friday. Each pizza requires $2\frac{1}{4}$ cups of cheese. If Student Council members make 25 pizzas, how many cups of cheese will they need?

4. In Florida, National Parks, State Parks, and State Forests have acreage shown in the table. If the total land and water area is 65,754.59 square miles, about what percentage of the total area is parks and forests? Use 1 square mile = 640 acres.

Type of Park	Area (in acres)
National Parks	2,571,164.45
State Parks	890,000
State Forests	723,000

Source: RAND Florida

Problem-Solving Strategy: Eliminate Unnecessary Information

A useful problem-solving strategy is to learn how to **eliminate unnecessary information**. If there is a diagram, it is important to determine if all or some of the information is necessary to find a solution.

Real-World EXAMPLE

Twila is making a quilt that shows a house repeating on each block of the quilt. Which information is unnecessary to find the area of the white door of the house?

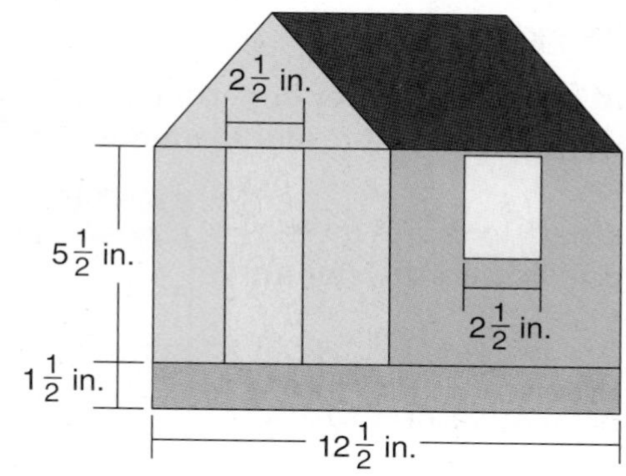

The dimensions needed to find the area of the door are the length and width of the door. So, the dimensions of $2\frac{1}{2}$ inches by $5\frac{1}{2}$ inches are needed. The other dimensions, such as the width of the window or the size of the square, are unnecessary.

Practice

Solve each problem by eliminating unnecessary information.

1. Which information is not necessary to find the temperature difference between the record lows in Alaska and Maine?

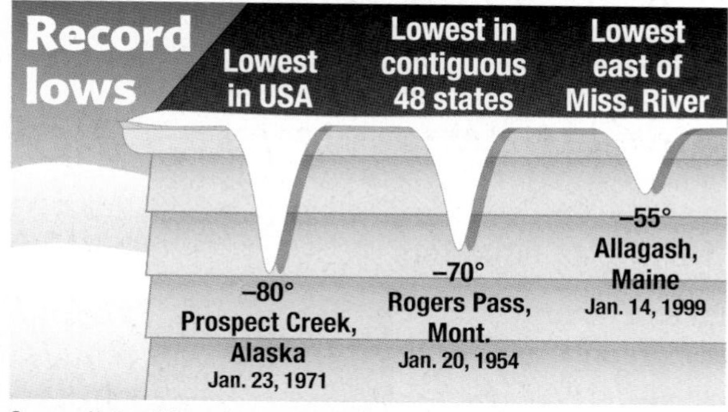

Source: National Oceanic and Atmospheric Administration

2. The Gemini North telescope was placed in Mauna Kea, Hawaii, in the year 2000. The Gemini South telescope was placed in Cerro Pachon, Chili, in the year 2001. Each of the twin telescopes are 8.1 meters in diameter. What information is not necessary to find the circumference of the base of the telescopes?

3. Which information is not necessary to find the difference in height between the tallest building and the 4th tallest building?

Rank	Building, city	Year	Stories	Height	
				m	ft
1.	Taipei 101, Taipei, Taiwan	2004	101	508	1,667
2.	Petronas Tower 1, Kuala Lumpur, Malaysia	1998	88	452	1,483
3.	Petronas Tower 2, Kuala Lumpur, Malaysia	1998	88	452	1,483
4.	Sears Tower, Chicago	1974	110	442	1,451
5.	Jin Mao Building, Shanghai	1999	88	421	1,381

Source: Council on Tall Building and Urban Habitat

Problem-Solving Strategy: Write an Equation

A natural outcome of recognizing mathematical patterns and organizing data is to **write an equation**. Look at a set of data or read a word problem to determine which values are constants and which values vary. Figure out the dependent and independent variables in order to write an equation to reflect the given situation.

Real-World EXAMPLE

For every $10 gift card sold, the theater department at Wallace High School earns $1.25. Write an equation to represent the amount raised based on the number of cards sold.

Number of Cards	Amount of Money ($)
1	1.25
2	2.50
3	3.75
4	5.00

Step 1 Make a table of data to represent the number of cards sold and the amount of money raised.

Step 2 Find the value that varies.The value that varies is the number of cards sold and the amount raised based on the number of cards sold.

Step 3 Find the value that is constant for each card sold. The value that is constant for each card sold is the amount of money raised per card, $1.25. This is the slope of the line.

Step 4 Write the equation that shows how the total amount changes based on the number of cards sold.

The equation is $y = 1.25x$.

Practice

Solve each problem by writing an equation.

Type of Veterans	Total Number of Veterans 23,425,051
Gulf War veterans	18.7%
Vietnam-era veterans	33.5%
Korean War veterans	13.3%
World War II veterans	13.9%

Source: U.S. Census Bureau

1. Write an equation that can be used to find the number of Korean War Veterans.

2. The land area of Alaska is 571,951 square miles. Montana's land area is 145,552 square miles. Write an equation to find how much larger Alaska's land area is compared to Montana.

3. Increasingly, Internet users are using broadband Internet in their homes. Write an equation for the slope of this data in the five years shown.

Broadband at Home	
Survey date	Users (in millions)
June 2000	6
May 2005	66

Source: Pew Internet Project

4. Iceland spent approximately 8.8% of its gross domestic product (GDP) on public health expenditures in 2006, the highest percentage of all countries. If Iceland had a GDP of $11,380,000,000 in 2006, write an equation to show how much money was spent on public health expenditures.

Extra Practice

Lesson 1-1 Variables and Expressions (pp. 5–9)

Write an algebraic expression for each verbal expression.

1. the sum of b and 21

2. the product of x and 7

3. the sum of 4 and 6 times a number z

4. the sum of 8 and -2 times n

5. one-half the cube of a number x

6. four-fifths the square of m

Write a verbal expression for each algebraic expression.

7. $2n$

8. 10^7

9. m^5

10. xy

11. $5n^2 - 6$

12. $9a^3 + 1$

13. $17 - 4m^5$

14. $\frac{12z^2}{5}$

15. $3x^2 - 2x$

Lesson 1-2 Order of Operations (pp. 10–15)

Evaluate each expression.

1. $3 + 8 \div 2 - 5$

2. $4 + 7 \cdot 2 + 8$

3. $5(9 + 3) - 3 \cdot 4$

4. $9 - 3^2$

5. $(8 - 1) \cdot 3$

6. $4(5 - 3)^2$

7. $3(12 + 3) - 5 \cdot 9$

8. $5^3 + 6^3 - 5^2$

9. $16 \div 2 \cdot 5 \cdot 3 \div 6$

10. $7(5^3 + 3^2)$

11. $\frac{9 \cdot 4 + 2 \cdot 6}{6 \cdot 4}$

12. $25 - \frac{1}{3}(18 + 9)$

13. 2^4

14. 10^2

15. 7^3

16. 20^3

17. 3^6

18. 4^5

19. 10^6

20. 3^5

21. 15^3

Evaluate each expression if $a = 2$, $b = 5$, $x = 4$, and $n = 10$.

22. $8a + b$

23. $48 + ab$

24. $a(6 - 3n)$

25. $bx + an$

26. $x^2 - 4n$

27. $3b + 16a - 9n$

28. $n^2 + 3(a + 4)$

29. $(2x)^2 + an - 5b$

30. $[a + 8(b - 2)]^2 \div 4$

31. $(3b^2)^3 + 2x$

32. $b^2 + 3n^3 - 4(x - 8)^2$

33. $[x + 2n(3a + 4)] - a^4$

Lesson 1-3 Properties of Numbers (pp. 16–22)

Evaluate each expression using the properties of numbers. Name the property used in each step.

1. $\frac{2}{3}[15 \div (12 - 2)]$

2. $\frac{7}{4}\left[4 \cdot \left(\frac{1}{8} \cdot 8\right)\right]$

3. $[(18 \div 3) \cdot 0] \cdot 10$

Use the properties of numbers to evaluate each expression.

4. $23 + 8 + 37 + 12$

5. $19 + 46 + 81 + 54$

6. $10.25 + 2.5 + 3.75$

7. $22.5 + 17.6 + 44.5$

8. $2\frac{1}{3} + 6 + 3\frac{2}{3} + 4$

9. $5\frac{6}{7} + 15 + 4\frac{1}{7} + 25$

10. $6 \cdot 8 \cdot 5 \cdot 3$

11. $18 \cdot 5 \cdot 2 \cdot 5$

12. $0.25 \cdot 7 \cdot 8$

13. $90 \cdot 12 \cdot 0.5$

14. $5\frac{1}{3} \cdot 4 \cdot 6$

15. $4\frac{5}{6} \cdot 10 \cdot 12$

Lesson 1-4 The Distributive Property (pp. 23–29)

Use the Distributive Property to rewrite each expression. Then evaluate.

1. $5(2 + 9)$
2. $8(10 + 20)$
3. $6(y + 4)$
4. $9(3n + 5)$
5. $32\left(x - \frac{1}{8}\right)$
6. $c(7 - d)$

Simplify each expression. If not possible, write *simplified*.

7. $13a + 5a$
8. $21x - 10x$
9. $8(3x + 7)$
10. $4m - 4n$
11. $3(5am - 4)$
12. $15x^2 + 7x^2$
13. $9y^2 + 13y^2 + 3$
14. $11a^2 - 11a^2 + 12a^2$
15. $6a + 7a + 12b + 8b$
16. $5a + 6b + 7a$
17. $8x + 4y + 9x$
18. $3a + 5b + 2c + 8b$

Lesson 1-5 Equations (pp. 31–37)

Find the solution of each equation if the replacement sets are $x = \{0, 2, 4, 6, 8\}$ and $y = \{1, 3, 5, 7, 9\}$.

1. $x - 4 = 4$
2. $25 - y = 18$
3. $3x + 1 = 25$
4. $5y - 4 = 11$
5. $14 = \frac{96}{x} + 2$
6. $0 = \frac{y}{3} - 3$

Solve each equation.

7. $x = \frac{27 + 9}{2}$
8. $\frac{18 - 7}{13 - 2} = y$
9. $n = \frac{6(5) + 3}{2(4) + 3}$
10. $\frac{5(4) - 6}{2^2 + 3} = z$
11. $\frac{7^2 + 9(2 + 1)}{2(10) - 1} = t$
12. $a = \frac{3^3 + 5^2}{2(3 - 1)}$

Lesson 1-6 Relations (pp. 38–44)

Describe what is happening in each graph.

1. The graph shows the average monthly high temperatures for a city over a one-year period.

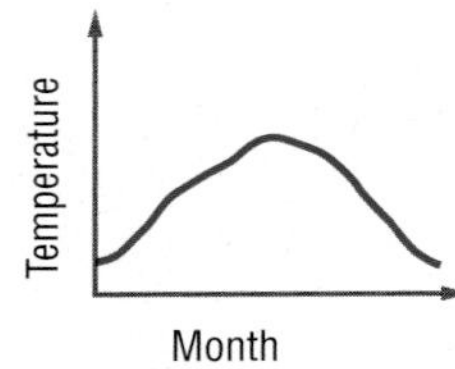

2. The graph shows the speed of a roller coaster car during a two-minute ride.

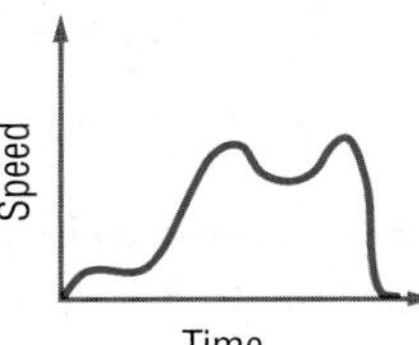

Express the relation shown in each table, mapping, or graph as a set of ordered pairs. Then describe the domain and range.

3.

x	y
1	3
2	4
3	5
4	6
5	7

4.

x	y
−4	1
−2	3
0	1
2	3
4	1

5.

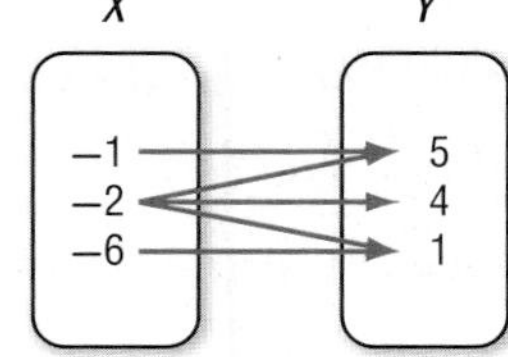

Lesson 1-7 Functions (pp. 45–52)

Determine whether each relation is a function. Explain.

1.

x	y
1	3
2	5
1	−7
2	9

2. 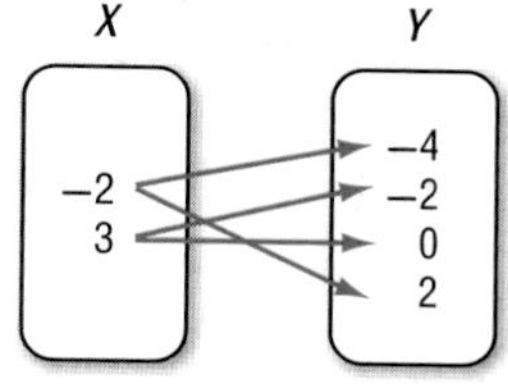

3. 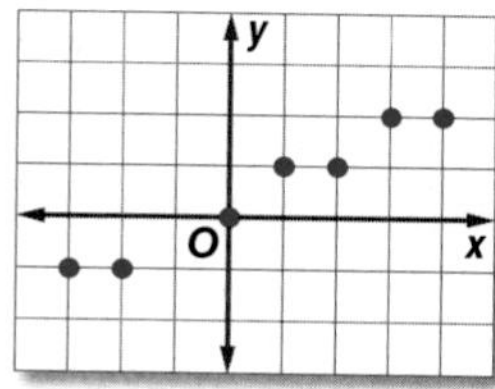

4. $x^2 + y = 11$

5. $y = 2$

6. $\{(-2, 4), (1, 3), (5, 2), (1, 4)\}$

If $f(x) = 2x + 5$ and $g(x) = 3x^2 - 1$, find each value.

7. $f(-4)$

8. $g(2)$

9. $f(3) - 5$

10. $g(a + 1)$

Lesson 1-8 Logical Reasoning and Counterexamples (pp. 54–59)

Identify the hypothesis and conclusion of each statement.

1. If an animal is a dog, then it barks.
2. If a figure is a pentagon, then it has five sides.
3. If $3x - 1 = 8$, then $x = 3$.
4. If 0.5 is the reciprocal of 2, then $0.5 \cdot 2 = 1$.

Identify the hypotheses and conclusion of each statement. Then write the statement in if-then form.

5. A square has four congruent sides.
6. $6a + 10 = 34$ when $a = 4$.
7. The video store is open every night.
8. The band will not practice on Thursday.

Find a counterexample for each conditional statement.

9. If the season is spring, then it does not snow.
10. If you live in Portland, then you live in Oregon.
11. If $2y + 4 = 10$, then $y < 3$.
12. If $a^2 > 0$, then $a > 0$.

Lesson 2-1 Writing Equations (pp. 75–80)

Translate each sentence into a formula.

1. A number z times 2 minus 6 is the same as m divided by 3.
2. The cube of a decreased by the square of b is equal to c.
3. Twenty-nine decreased by the product of x and y is the same as z.
4. The perimeter P of an isosceles triangle is the sum of twice the length of leg a and the length of the base b.
5. Thirty increased by the quotient of r and t is equal to v.
6. The area A of a rhombus is half the product of lengths of the diagonals a and b.

Translate each equation into a sentence.

7. $0.5x + 3 = -10$
8. $\frac{n}{-6} = 2n + 1$
9. $18 - 5h = 13h$
10. $n^2 = 16$
11. $2x^2 + 3 = 21$
12. $\frac{m}{n} + 4 = 12$

Extra Practice

Extra Practice

Lesson 2-2 Solving One-Step Equations (pp. 81–89)

Solve each equation. Check your solution.

1. $-2 + g = 7$
2. $9 + s = -5$
3. $-4 + y = -9$
4. $m + 6 = 2$
5. $t + (-4) = 10$
6. $v - 7 = -4$
7. $a - (-6) = -5$
8. $-2 - x = -8$
9. $d + (-44) = -61$
10. $b - (-26) = 41$
11. $p - 47 = 22$
12. $-63 - f = -82$
13. $7p = 35$
14. $-3x = -24$
15. $2y = -3$
16. $62y = -2356$
17. $\frac{a}{-6} = -2$
18. $\frac{c}{-59} = -7$
19. $\frac{7}{10} - a = \frac{1}{2}$
20. $f - \left(-\frac{1}{8}\right) = \frac{3}{10}$
21. $-4\frac{5}{12} = t - \left(-10\frac{1}{36}\right)$
22. $x + \frac{3}{8} = \frac{1}{4}$
23. $1\frac{7}{16} + s = \frac{9}{8}$
24. $17\frac{8}{9} = d + \left(-2\frac{5}{6}\right)$
25. $-\frac{5}{9}r = 7\frac{1}{2}$
26. $2\frac{1}{6}j = 5\frac{1}{5}$
27. $3 = 1\frac{7}{11}q$

Lesson 2-3 Solving Multi-Step Equations (pp. 90–96)

Solve each equation. Check your solution.

1. $2x - 5 = 3$
2. $4t + 5 = 37$
3. $7a + 6 = -36$
4. $47 = -8g + 7$
5. $-3c - 9 = -24$
6. $5k - 7 = -52$
7. $5s + 4s = -72$
8. $3x - 7 = 2$
9. $8 + 3x = 5$
10. $-3y + 7.569 = 24.069$
11. $7 - 9.1f = 137.585$
12. $6.5 = 2.4m - 4.9$
13. $\frac{n}{5} + 6 = -2$
14. $\frac{d}{4} - 8 = -5$
15. $-\frac{4}{13}y - 7 = 6$
16. $\frac{p + 3}{10} = 4$
17. $\frac{h - 7}{6} = 1$
18. $\frac{5f + 1}{8} = -3$
19. $\frac{4n - 8}{-2} = 12$
20. $\frac{-3t - 4}{2} = 8$
21. $4.8a - 3 + 1.2a = 9$

Lesson 2-4 Solving Equations with the Variable on Each Side (pp. 97–102)

Solve each equation. Check your solution.

1. $5x + 1 = 3x - 3$
2. $6 - 8n = 5n + 19$
3. $-3z + 5 = 2z + 5$
4. $\frac{2}{3}h + 5 = -4 - \frac{1}{3}h$
5. $\frac{1}{2}a - 4 = 3 - \frac{1}{4}a$
6. $6(y - 5) = 18 - 2y$
7. $-28 + p = 7(p - 10)$
8. $\frac{1}{3}(b - 9) = b + 9$
9. $-4x + 6 = 0.5(x + 30)$
10. $4(2y - 1) = -8(0.5 - y)$
11. $1.9s + 6 = 3.1 - s$
12. $2.85y - 7 = 12.85y - 2$
13. $2.9m + 1.7 = 3.5 + 2.3m$
14. $3(x + 1) - 5 = 3x - 2$
15. $\frac{x}{2} - \frac{1}{3} = \frac{x}{3} - \frac{1}{2}$
16. $\frac{6v - 9}{3} = v$
17. $\frac{3t + 1}{4} = \frac{3}{4}t - 5$
18. $0.4(x - 12) = 1.2(x - 4)$
19. $3y - \frac{4}{5} = \frac{1}{3}y$
20. $\frac{3}{4}x - 4 = 7 + \frac{1}{2}x$
21. $-0.2(1 - x) = 2(4 + 0.1x)$
22. $3.2(y + 1) = 2(1.4y - 3)$

Lesson 2-5 Solving Equations Involving Absolute Value (pp. 103–109)

Solve each open sentence.

1. $|c - 5| = 4$
2. $|e + 3| = 7$
3. $|4 - g| = 6$
4. $|10 - k| = 8$
5. $|2j + 4| = 12$
6. $|2r - 6| = 10$
7. $|6 - 3w| = 8$
8. $|7 + 2x| = 14$
9. $|4z + 6| = 12$

Evaluate each expression when $a = 7$, $b = 5$, and $c = -2$.

10. $|a - b|$
11. $|b - a|$
12. $|2a + c|$
13. $-|c| + |b|$
14. $-|b| + |a + c|$
15. $|4b - c|$

Lesson 2-6 Ratios and Proportions (pp. 111–118)

Solve each proportion. If necessary, round to the nearest hundredth.

1. $\frac{4}{5} = \frac{x}{20}$
2. $\frac{b}{63} = \frac{3}{7}$
3. $\frac{y}{5} = \frac{3}{4}$
4. $\frac{7}{4} = \frac{3}{a}$
5. $\frac{t-5}{4} = \frac{3}{2}$
6. $\frac{x}{9} = \frac{0.24}{3}$
7. $\frac{n}{3} = \frac{n+4}{7}$
8. $\frac{12q}{-7} = \frac{30}{14}$
9. $\frac{1}{y-3} = \frac{3}{y-5}$
10. $\frac{x}{8.71} = \frac{4}{17.42}$
11. $\frac{a-3}{8} = \frac{3}{4}$
12. $\frac{6p-2}{7} = \frac{5p+7}{8}$
13. $\frac{2}{9} = \frac{k+3}{2}$
14. $\frac{5m-3}{4} = \frac{5m+3}{6}$
15. $\frac{w-5}{4} = \frac{w+3}{3}$
16. $\frac{96.8}{t} = \frac{12.1}{7}$
17. $\frac{r-1}{r+1} = \frac{3}{5}$
18. $\frac{4n+5}{5} = \frac{2n+7}{7}$

Determine whether each pair of ratios are equivalent ratios. Write *yes* or *no*.

19. $\frac{3}{4}, \frac{7}{8}$
20. $\frac{3.8}{2}, \frac{4.1}{4}$
21. $\frac{8}{9}, \frac{17.6}{19.8}$
22. $\frac{5}{6}, \frac{20}{24}$
23. $\frac{5}{4}, \frac{30.5}{24.4}$
24. $\frac{1}{3}, \frac{3}{1}$

Lesson 2-7 Percent of Change (pp. 119–125)

State whether each percent of change is a percent of *increase* or a percent of *decrease*. Then find each percent of change. Round to the nearest whole percent.

1. original: \$100
   new: \$67
2. original: 62 acres
   new: 98 acres
3. original: 322 people
   new: 289 people
4. original: 78 pennies
   new: 36 pennies
5. original: \$212
   new: \$230
6. original: 35 mph
   new: 65 mph

Find the final price of each item.

7. television: \$299
   discount: 20%
8. book: \$15.95
   sales tax: 7%
9. software: \$36.90
   sales tax: 6.25%
10. boots: \$49.99
    discount: 15%
    sales tax: 3.5%
11. jacket: \$65
    discount: 30%
    sales tax: 4%
12. backpack: \$28.95
    discount: 10%
    sales tax: 5%

Extra Practice

Lesson 2-8 Literal Equations and Dimensional Analysis (pp. 126–131)

Solve each equation or formula for x.

1. $x + r = q$
2. $ax + 4 = 7$
3. $2bx - b = -5$
4. $\frac{x - c}{c + a} = a$
5. $\frac{x + y}{c} = d$
6. $\frac{ax + 1}{2} = b$
7. $d(x - 3) = 5$
8. $nx - a = bx + d$
9. $3x - r = r(-3 + x)$
10. $y = \frac{5}{9}(x - 32)$
11. $A = \frac{1}{2}h(x + y)$
12. $A = 2\pi r^2 + 2\pi rx$

Solve each equation or formula for the variable indicated.

13. $S = 2b(a + c) + 2ac$, for b
14. $g(h + 8) = -j$, for h
15. $\frac{11n + p}{t} = 8$, for p
16. $\frac{u - 10v}{7} = w$, for u
17. $\frac{9x - y}{z - 10} = 12$, for x
18. $\frac{2k - 3}{m + p} = 5$, for k
19. $14h + j = 2h$, for h
20. $-18k + m = 22k$, for k
21. $39k + 3j = -6m$, for j

Lesson 2-9 Weighted Averages (pp. 132–138)

1. **ADVERTISING** An advertisement for grape drink claims that the drink contains 10% grape juice. How much pure grape juice would have to be added to 5 quarts of the drink to obtain a mixture containing 40% grape juice?
2. **GRADES** In Ms. Pham's social studies class, a test is worth four times as much as homework. If a student has an average of 85% on tests and 95% on homework, what is the student's average?
3. **ENTERTAINMENT** At the Golden Oldies Theater, tickets for adults cost \$5.50 and tickets for children cost \$3.50. How many of each kind of ticket were purchased if 21 tickets were bought for \$83.50?
4. **FOOD** Wes is mixing peanuts and chocolate pieces. Peanuts sell for \$4.50 a pound and the chocolate sells for \$6.50 a pound. How many pounds of chocolate mixes with 5 pounds of peanuts to obtain a mixture that sells for \$5.25 a pound?
5. **TRAVEL** Missoula and Bozeman are 210 miles apart. Sheila leaves Missoula for Bozeman and averages 55 miles per hour. At the same time, Casey leaves Bozeman and averages 65 miles per hour as he drives to Missoula. When will they meet? How far will they be from Bozeman?

Lesson 3-1 Graphing Linear Equations (pp. 153–160)

Determine whether each equation is a linear equation. Write *yes* or *no*. If *yes*, write the equation in standard form.

1. $3x = 2y$
2. $2x - 3 = y^2$
3. $4x = 2y + 8$
4. $5x - 7y = 2x - 7$
5. $2x + 5x = 7y + 2$
6. $\frac{1}{x} + \frac{5}{y} = -4$

Graph each equation by using the x- and y-intercepts or by making a table.

7. $3x + y = 4$
8. $y = 3x + 1$
9. $3x - 2y = 12$
10. $2x - y = 6$
11. $2x - 3y = 8$
12. $y = -2$
13. $y = 5x - 7$
14. $x = 4$
15. $x + \frac{1}{3}y = 2$
16. $5x - 2y = 8$
17. $4.5x + 2.5y = 9$
18. $\frac{1}{2}x + 3y = 12$

Lesson 3-2 Solving Linear Equations by Graphing (pp. 161–168)

Solve each equation.

1. $-x + 6 = 0$

2. $8x + 2 = 0$

3. $4x + 3 = -2 + 4x$

4. $-2x + 5 = 5$

5. $3 = 4x - 1$

6. $\frac{1}{2}x + 5 = -1$

7. $6x - 4 = 8 + 6x$

8. $\frac{1}{2} = 3x - 4$

9. $7 = -x - 4$

Lesson 3-3 Rate of Change and Slope (pp. 170–178)

Find the slope of the line that passes through each pair of points.

1.

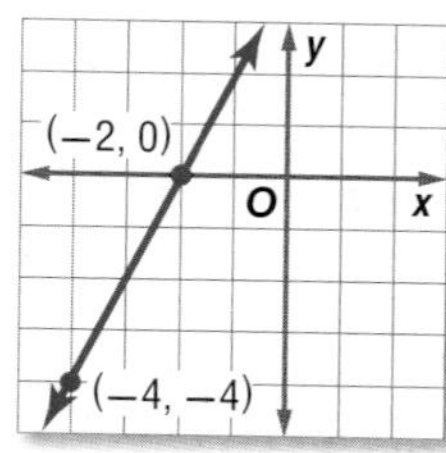

2.

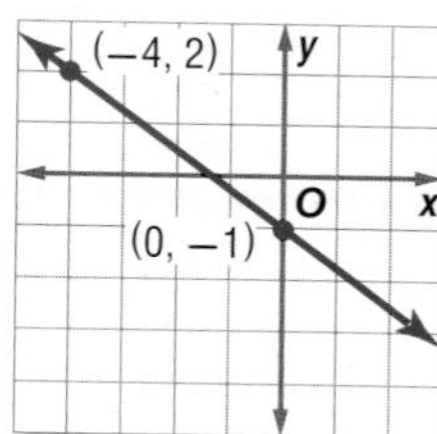

3. $(-2, 2), (3, -3)$

4. $(-2, -8), (1, 4)$

5. $(3, 4), (4, 6)$

6. $(-5, 4), (-1, 11)$

7. $(18, -4), (6, -10)$

8. $(-4, -6), (-4, -8)$

9. $(0, 0), (-1, 3)$

10. $(-8, 1), (2, 1)$

Find the value of r so the line that passes through each pair of points has the given slope.

11. $(-1, r), (1, -4), m = -5$

12. $(r, -2), (-7, -1), m = -\frac{1}{4}$

Lesson 3-4 Direct Variation (pp. 180–186)

Name the constant of variation for each equation. Then find the slope of the line that passes through each pair of points.

1.

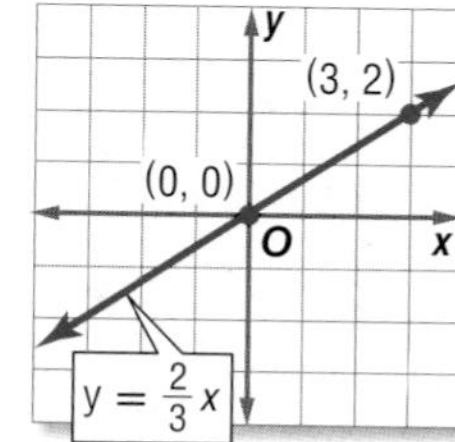

2.

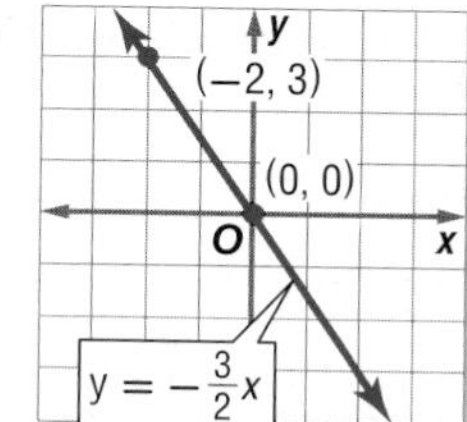

3.

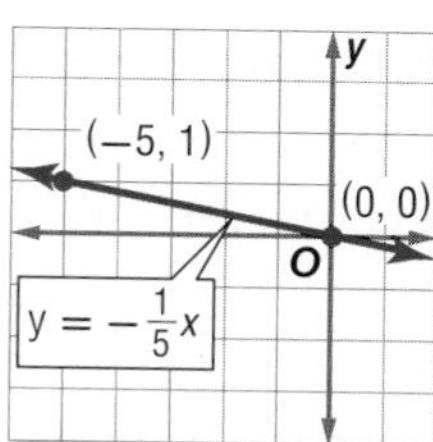

Graph each equation.

4. $y = 5x$

5. $y = -6x$

6. $y = -\frac{4}{3}x$

Suppose y varies directly as x. Write a direct variation equation that relates x and y. Then solve.

7. If $y = 45$ when $x = 9$, find y when $x = 7$.

8. If $y = -7$ when $x = -1$, find x when $y = -84$.

Extra Practice

Lesson 3-5 Arithmetic Sequences as Linear Functions (pp. 187–194)

Determine whether each sequence is an arithmetic sequence. Write *yes* or *no*. Explain.

1. $-2, -1, 0, 1, \ldots$

2. $3, 5, 8, 12, \ldots$

3. $2, 4, 8, 16, \ldots$

4. $-21, -16, -11, -6, \ldots$

5. $0, 0.25, 0.5, 0.75, \ldots$

6. $\frac{1}{3}, \frac{1}{9}, \frac{1}{27}, \frac{1}{81}, \ldots$

Find the next three terms of each arithmetic sequence.

7. $3, 13, 23, 33, \ldots$

8. $-4, -6, -8, -10, \ldots$

9. $-2, -1.4, -0.8, -0.2, \ldots$

10. $5, 13, 21, 29, \ldots$

11. $\frac{3}{4}, \frac{7}{8}, 1, \frac{9}{8}, \ldots$

12. $-\frac{1}{3}, -\frac{5}{6}, -\frac{4}{3}, -\frac{11}{6}, \ldots$

Write an equation for the *n*th term of the arithmetic sequence. Then graph the first five terms in the sequence.

13. $-3, 1, 5, 9, \ldots$

14. $25, 40, 55, 70, \ldots$

15. $-9, -3, 3, 9, \ldots$

16. $-3.5, -2, -0.5, \ldots$

Lesson 3-6 Proportional and Nonproportional Relationships (pp. 195–200)

Write an equation in function notation for each relation.

1.

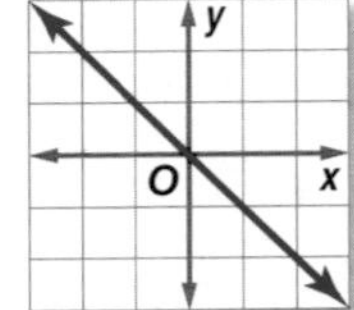

2.

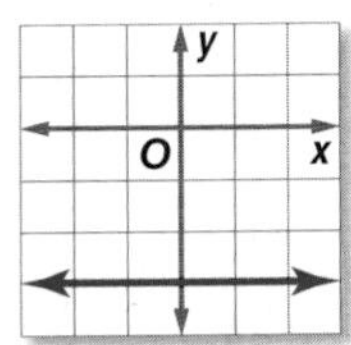

3.

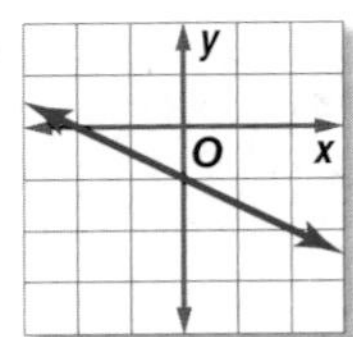

4.

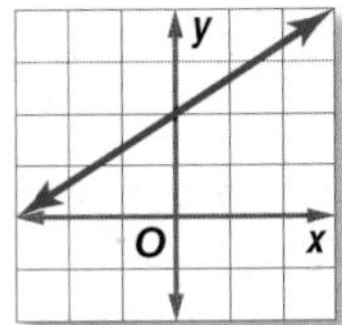

The table shows the number of cups of flour used to make batches of brownies.

Batches of brownies	1	2	3	4
Cups of flour	4	8	12	16

5a. Write an equation for the data given.

5b. Graph the equation.

5c. Find the number of cups of flour needed for 6 batches of brownies.

Lesson 4-1 Graphing Equations in Slope-Intercept Form (pp. 214–221)

Write an equation, in slope-intercept form, of the line with the given slope and *y*-intercept.

1. m: 5, y-intercept: -15

2. m: -6, y-intercept: 3

3. m: 0.3, y-intercept: -2.6

4. m: $-\frac{4}{3}$, y-intercept: $\frac{5}{3}$

5. m: $-\frac{2}{5}$, y-intercept: 2

6. m: $\frac{7}{4}$, y-intercept: -2

Write an equation in slope-intercept form for each graph shown.

7.

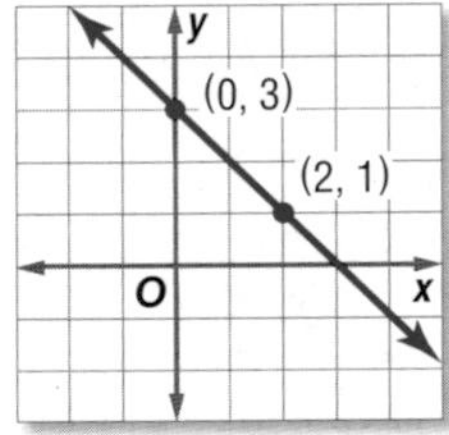

8.

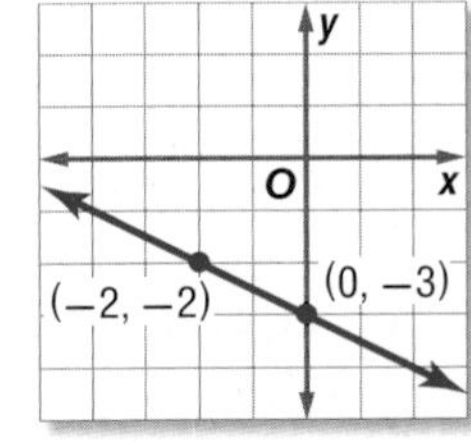

9.

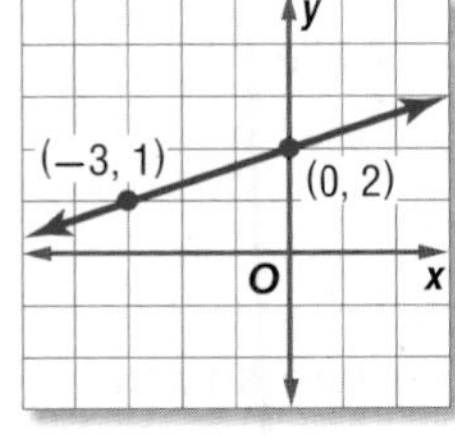

Graph each equation.

10. $y = 5x - 1$

11. $y = -2x + 3$

12. $3x - y = 6$

Lesson 4-2 Writing Equations in Slope-Intercept Form (pp. 224–230)

Write an equation of the line that passes through the given point and has the given slope.

1. $(0, 0)$; $m = -2$

2. $(-3, 2)$; $m = 4$

3. $(0, 5)$; $m = -1$

4. $(-2, 3)$; $m = -\frac{1}{4}$

5. $(1, -5)$; $m = \frac{2}{3}$

6. $\left(\frac{1}{2}, \frac{1}{4}\right)$; $m = 8$

Write an equation of the line that passes through each pair of points.

7. $(-1, 7), (8, -2)$

8. $(4, 0), (0, 5)$

9. $(8, -1), (7, -1)$

10. $(-2, 3), (1, 3)$

11. $(0, 0), (-4, 3)$

12. $\left(-\frac{1}{2}, \frac{1}{2}\right), \left(\frac{1}{4}, \frac{3}{4}\right)$

Lesson 4-3 Writing Equations in Point-Slope Form (pp. 231–236)

Write an equation in point-slope form for the line that passes through each point with the given slope. Then graph the equation.

1. $(5, -2)$, $m = 3$

2. $(0, 6)$, $m = -2$

3. $(-3, 1)$, $m = 0$

4. $(-2, -4)$, $m = \frac{3}{4}$

Write each equation in standard form.

5. $y + 3 = 2(x - 4)$

6. $y + 3 = -\frac{1}{2}(x + 6)$

7. $y - 4 = -\frac{2}{3}(x - 5)$

8. $y + 2 = \frac{4}{3}(x - 6)$

9. $y - 1 = 1.5(x + 3)$

10. $y + 6 = -3.8(x - 2)$

Write each equation in slope-intercept form.

11. $y - 1 = -2(x + 5)$

12. $y + 3 = 4(x - 1)$

13. $y - 6 = -4(x - 2)$

14. $y + 1 = \frac{4}{5}(x + 5)$

15. $y - 2 = -\frac{3}{4}(x - 2)$

16. $y + \frac{1}{4} = \frac{2}{3}\left(x + \frac{1}{2}\right)$

Lesson 4-4 Parallel and Perpendicular Lines (pp. 237–243)

Write the slope-intercept form of an equation for the line that passes through the given point and is parallel to the graph of each equation. Then write an equation in slope-intercept form for the line that passes through the given point and is perpendicular to the graph of each equation.

1. $(1, 6)$, $y = 4x - 2$

2. $(4, 6)$, $y = 2x - 7$

3. $(-3, 0)$, $y = \frac{2}{3}x + 1$

4. $(5, -2)$, $y = -3x - 7$

5. $(0, 4)$, $3x + 8y = 4$

6. $(2, 3)$, $x - 5y = 7$

Determine whether the graphs of each pair of equations are *parallel, perpendicular,* or *neither.*

7. $y = -2x + 11$
$y + 2x = 23$

8. $3y = 2x + 14$
$2x - 3y = 2$

9. $y = -5x$
$y = 5x - 18$

Lesson 4-5 Scatter Plots and Lines of Fit (pp. 245–252)

Determine whether each graph shows a **positive correlation**, *a* **negative correlation**, *or* **no correlation**. *If there is a correlation, describe its meaning in the situation.*

1. 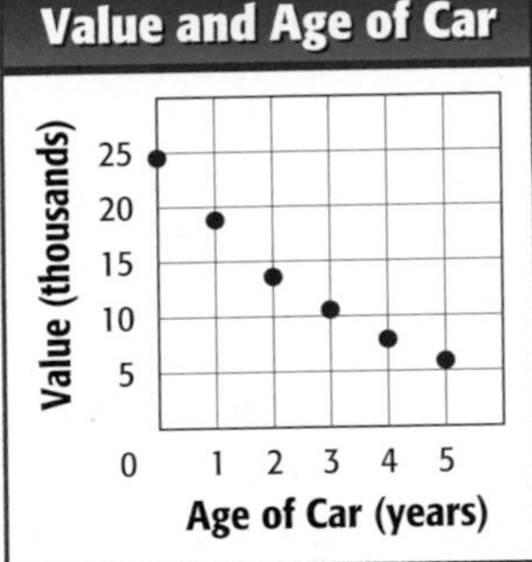

2. 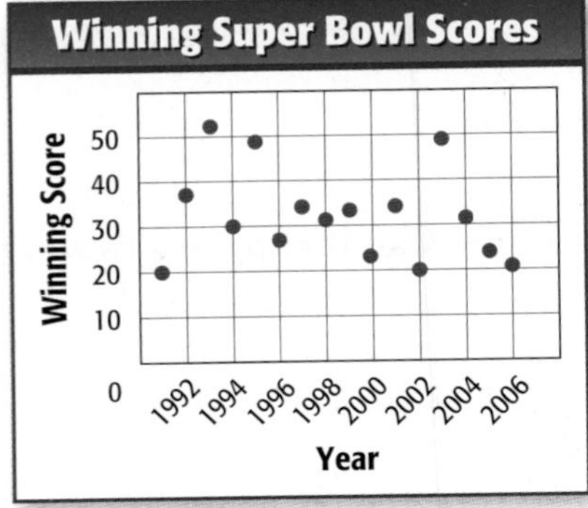

Source: ESPN Almanac

Lesson 4-6 Regression and Median-Fit Lines (pp. 253–260)

1. **COMPUTERS** The media center is keeping track of student computer use to determine if they need to purchase more computers. There are 6 computers in the center. The results of the 10-day tracking period are recorded in the table. Write an equation for the regression line for the data. Then find the correlation coefficient.

Student Computer Use										
Day	1	2	3	4	5	6	7	8	9	10
Number of Students	48	62	75	70	32	52	70	63	81	40

2. **STOCKS** For a class project, the students were asked to keep track of a stock price for a company of local interest. They were to record the closing price of the stock at the end of each Friday for 8 weeks. Marcus was out of town one week and missed the price for week 4. Find the equation for the data, and approximate the price for week 4.

Weekly Closing Stock Price for Barney's International							
Week	1	2	3	5	6	7	8
Closing Price	29	27	29	33	34	33	34

Lesson 4-7 Special Functions (pp. 261–269)

Graph each function.

1. $f(x) = -3[\![x]\!]$
2. $g(x) = [\![4x]\!]$
3. $h(x) = 2[\![x]\!] - 3$
4. $f(x) = |x + 2|$
5. $g(x) = |3x - 1|$
6. $f(x) = \begin{cases} -2x \text{ if } x < 3 \\ x + 1 \text{ if } x \geq 3 \end{cases}$

7. **PARKING** Short-term parking at the airport is $2 for the first half-hour and $1 for each half-hour after that. Draw a graph that represents this information.

Short-Term Parking Fees	
Hours	Fees
$0 < h \leq \frac{1}{2}$	$2
$\frac{1}{2} < h \leq 1$	$3
$1 < h \leq 1\frac{1}{2}$	$4

Extra Practice

Lesson 5-1 Solving Inequalities by Addition and Subtraction (pp. 283–288)

Solve each inequality. Check your solution, and then graph it on a number line.

1. $c + 9 \leq 3$
2. $d - (-3) < 13$
3. $z - 4 > 20$
4. $h - (-7) > -2$
5. $-11 > d - 4$
6. $2x > x - 3$
7. $2x - 3 \geq x$
8. $16 + w < -20$
9. $14p > 5 + 13p$
10. $-7 < 16 - z$
11. $1.1v - 1 > 2.1v - 3$
12. $\frac{1}{2}t + \frac{1}{4} \geq \frac{3}{2}t - \frac{2}{3}$
13. $9x < 8x - 2$
14. $-2 + 9n \leq 10n$
15. $a - 2.3 \geq -7.8$
16. $5z - 6 > 4z$

Define a variable, write an inequality, and solve each problem. Check your solution.

17. The sum of a number and negative six is greater than 9.
18. Negative five times a number is less than the sum of negative six times the number and 12.

Lesson 5-2 Solving Inequalities by Multiplication and Division (pp. 290–295)

Solve each inequality. Check your solution.

1. $7b \geq -49$
2. $-5j < -60$
3. $\frac{w}{3} > -12$
4. $\frac{p}{5} < 8$
5. $-8f < 48$
6. $-0.25t \geq -10$
7. $\frac{g}{-8} < 4$
8. $-4.3x < -2.58$
9. $4c \geq -6$
10. $6 \leq 0.8n$
11. $\frac{2}{3}m \geq -22$
12. $-25 > -0.05a$
13. $-15a < -28$
14. $-\frac{7}{9}x < 42$
15. $0.375y \leq 32$
16. $-7y \geq 91$

Define a variable, write an inequality, and solve each problem. Then check your solution.

17. Negative one times a number is greater than −7.
18. Three fifths of a number is at least negative 10.
19. Seventy-five percent of a number is at most 100.

Lesson 5-3 Solving Multi-Step Inequalities (pp. 296–301)

Solve each inequality. Check your solution.

1. $3y - 4 > -37$
2. $7s - 12 < 13$
3. $-5q + 9 > 24$
4. $-6v - 3 \geq -33$
5. $-2k + 12 < 30$
6. $-2x + 1 < 16 - x$
7. $15t - 4 > 11t - 16$
8. $13 - y \leq 29 + 2y$
9. $5q + 7 \leq 3(q + 1)$
10. $2(w + 4) \geq 7(w - 1)$
11. $-4t - 5 > 2t + 13$
12. $\left\{\frac{2t + 5}{3}\right\} < -9$
13. $\frac{z}{4} + 7 \geq -5$
14. $13r - 11 > 7r + 37$
15. $8c - (c - 5) > c + 17$
16. $-5(k + 4) \geq 3(k - 4)$
17. $9m + 7 < 2(4m - 1)$
18. $3(3y + 1) < 13y - 8$
19. $5x \leq 10(3x + 4)$
20. $3\left(a + \frac{2}{3}\right) \geq a - 1$

Extra Practice

Lesson 5-4 Solving Compound Inequalities (pp. 304–309)

Solve each compound inequality. Then graph the solution set.

1. $2 + x < -5$ *or* $2 + x > 5$
2. $-4 + t > -5$ *or* $-4 + t < 7$
3. $3 \leq 2g + 7$ *and* $2g + 7 \leq 15$
4. $2v - 2 \leq 3v$ *and* $4v - 1 \geq 3v$
5. $3b - 4 \leq 7b + 12$ *and* $8b - 7 \leq 25$
6. $-9 < 2z + 7 < 10$
7. $5m - 8 \geq 10 - m$ *or* $5m + 11 < -9$
8. $12c - 4 \leq 5c + 10$ *or* $-4c - 1 \leq c + 24$
9. $2h - 2 \leq 3h \leq 4h - 1$
10. $3p + 6 < 8 - p$ *and* $5p + 8 \geq p + 6$
11. $2r + 8 > 16 - 2r$ *and* $7r + 21 < r - 9$
12. $-4j + 3 < j + 22$ *and* $j - 3 < 2j - 15$
13. $2(q - 4) \leq 3(q + 2)$ *or* $q - 8 \leq 4 - q$
14. $\frac{1}{2}w + 5 \geq w + 2 \geq \frac{1}{2}w + 9$
15. $n - (6 - n) > 10$ *or* $-3n - 1 > 20$
16. $-(2x + 5) \leq x + 5 \leq 2x - 9$
17. **WIND SPEED** The Fujita Scale (F-scale) is the official classification system for tornado damage. One factor used to classify a tornado is wind speed. Use the information in the table to write an inequality for the range of wind speeds of an F3 tornado.

F-Scale	Wind Speed
F0	40–72 mph
F1	73–112 mph
F2	113–157 mph
F3	158–206 mph
F4	207–260 mph
F5	261–318 mph

Lesson 5-5 Inequalities Involving Absolute Value (pp. 310–314)

Solve each inequality. Then graph the solution set.

1. $|x + 4| < 10$
2. $|y - 3| \geq 3$
3. $|2x + 5| > 2$
4. $\left|\frac{2x - 3}{5}\right| \geq 4$
5. $\left|\frac{3m - 2}{2}\right| \geq 7$
6. $\left|\frac{2n + 8}{3}\right| < 2$
7. $|5v + 2| < 4$
8. $|w - 8| \geq 14$
9. $|3x + 2| > 8$
10. $\left|\frac{4m + 5}{3}\right| \geq 5$
11. $\left|\frac{y + 4}{5}\right| < 4$
12. $\left|\frac{3h - 5}{2}\right| > 5$

Lesson 5-6 Graphing Inequalities in Two Variables (pp. 315–320)

Determine which ordered pairs are part of the solution set for each inequality.

1. $x + y \geq 0$, $\{(0, 0), (1, -3), (2, 2), (3, -3)\}$
2. $2x + y \leq 8$, $\{(0, 0), (-1, -1), (3, -2), (8, 0)\}$

Graph each inequality.

3. $y \leq -2$
4. $x < 4$
5. $x + y < -2$
6. $3y - 2x \leq 2$
7. $y > 4x - 1$
8. $3x + y > 1$
9. **DELIVERIES** A delivery truck with a 4000-pound weight limit is transporting televisions that weigh 77 pounds each and microwaves that weigh 55 pounds each.
   a. Define variables and write an inequality for this situation.
   b. Will the truck be able to deliver 35 televisions and 25 microwaves at once? Explain.

Lesson 6-1 Graphing Systems of Equations (pp. 333–339)

Graph each system of equations. Then determine whether the system has no solution, one solution, or infinitely many solutions. If the system has one solution, name it.

1. $y = 3x$
   $4x + 2y = 30$
2. $x = -2y$
   $x + y = 1$
3. $y = x + 4$
   $3x + 2y = 18$
4. $x + y = 6$
   $x - y = 2$
5. $x + y = 6$
   $3x + 3y = 3$
6. $y = -3x$
   $4x + y = 2$
7. $2x + y = 8$
   $x - y = 4$
8. $\frac{1}{5}x - y = \frac{12}{5}$
   $3x - 5y = 6$
9. $x + 2y = 0$
   $y + 3 = -x$
10. $x + 2y = -9$
    $x - y = 6$
11. $x + \frac{1}{2}y = 3$
    $y = 3x - 4$
12. $\frac{2}{3}x + \frac{1}{2}y = 2$
    $4x + 3y = 12$

Lesson 6-2 Substitution (pp. 342–347)

Use substitution to solve each system of equations. If the system does not have exactly one solution, state whether it has no solutions or infinitely many solutions.

1. $y = x$
   $5x = 12y$
2. $y = 7 - x$
   $2x - y = 8$
3. $x = 5 - y$
   $3y = 3x + 1$
4. $3x + y = 6$
   $y + 2 = x$
5. $x - 3y = 3$
   $2x + 9y = 11$
6. $3x = -18 + 2y$
   $x + 3y = 4$
7. $x + 2y = 10$
   $-x + y = 2$
8. $2x = 3 - y$
   $2y = 12 - x$
9. $6y - x = -36$
   $y = -3x$
10. $\frac{3}{4}x + \frac{1}{3}y = 1$
    $x - y = 10$
11. $x + 6y = 1$
    $3x - 10y = 31$
12. $3x - 2y = 12$
    $\frac{3}{2}x - y = 3$
13. $2x + 3y = 5$
    $4x - 9y = 9$
14. $x = 4 - 8y$
    $3x + 24y = 12$
15. $3x - 2y = -3$
    $25x + 10y = 215$

Lesson 6-3 Elimination Using Addition and Subtraction (pp. 348–354)

Use elimination to solve each system of equations.

1. $x + y = 7$
   $x - y = 9$
2. $2x - y = 32$
   $2x + y = 60$
3. $-y + x = 6$
   $y + x = 5$
4. $s + 2t = 6$
   $3s - 2t = 2$
5. $x = y - 7$
   $2x - 5y = -2$
6. $3x + 5y = -16$
   $3x - 2y = -2$
7. $x - y = 3$
   $x + y = 3$
8. $x + y = 8$
   $2x - y = 6$
9. $2s - 3t = -4$
   $s = 7 - 3t$
10. $-6x + 16y = -8$
    $6x - 42 = 16y$
11. $3x + 0.2y = 7$
    $3x = 0.4y + 4$
12. $9x + 2y = 26$
    $1.5x - 2y = 13$
13. $x = y$
    $x + y = 7$
14. $4x - \frac{1}{3}y = 8$
    $5x + \frac{1}{3}y = 6$
15. $2x - y = 3$
    $\frac{2}{3}x - y = -1$

Lesson 6-4 Elimination Using Multiplication (pp. 355–360)

Use elimination to solve each system of equations.

1. $-3x + 2y = 10$
   $-2x - y = -5$
2. $2x + 5y = 13$
   $4x - 3y = -13$
3. $5x + 3y = 4$
   $-4x + 5y = -18$
4. $\frac{1}{3}x - y = -1$
   $\frac{1}{5}x - \frac{2}{5}y = -1$
5. $3x - 5y = 8$
   $4x - 7y = 10$
6. $x - 0.5y = 1$
   $0.4x + y = -2$
7. $x + 8y = 3$
   $4x - 2y = 7$
8. $4x - y = 4$
   $x + 2y = 3$
9. $3y - 8x = 9$
   $y - x = 2$
10. $x + 4y = 30$
    $2x - y = -6$
11. $3x - 2y = 0$
    $4x + 4y = 5$
12. $9x - 3y = 5$
    $x + y = 1$
13. $2x - 7y = 9$
    $-3x + 4y = 6$
14. $2x - 6y = -16$
    $5x + 7y = -18$
15. $6x - 3y = -9$
    $-8x + 2y = 4$

Extra Practice

Lesson 6-5 Applying Systems of Linear Equations (pp. 362–368)

Determine the best method to solve each system of equations. Then solve the system.

1. $y = 2x + 1$
   $y = -3x + 1$
2. $y = 5x - 8$
   $y = 3x$
3. $x + 2y = -6$
   $x = y + 3$
4. $2x - 3y = 5$
   $y = -6x$
5. $x = -1$
   $y = 8$
6. $4x + y = 5$
   $-4x - 2y = 9$
7. $-7x + 3y = -4$
   $2x + 3y = 5$
8. $4x - y = 11$
   $x + 2y = 5$
9. $2y - x = -7$
   $x + 3y = 5$
10. $-13x + 8y = -6$
    $3x - 4y = 2$
11. $-x + 7y = 9$
    $-4x + 6y = -8$
12. $2x + 5y = 7$
    $5x - 2y = 13$
13. $12x - 3y = 7$
    $x = 2 + 13y$
14. $6x = 5$
    $9y - 2x = 7$
15. $17x + 8y = -4$
    $-8y - 2x = 9$

Lesson 6-6 Organizing Data Using Matrices (pp. 369–375)

Perform the indicated matrix operations. If the matrix does not exist, write *impossible*.

1. $\begin{bmatrix} 3 & 5 \\ -7 & 2 \end{bmatrix} + \begin{bmatrix} -2 & 6 \\ 8 & -1 \end{bmatrix}$
2. $[0 \quad -1 \quad 3] + \begin{bmatrix} 5 \\ -2 \\ -3 \end{bmatrix}$
3. $\begin{bmatrix} 45 & 36 & 18 \\ 63 & 29 & 5 \end{bmatrix} - \begin{bmatrix} 45 & -2 & 36 \\ 18 & 9 & 10 \end{bmatrix}$
4. $4[-8 \quad 2 \quad 9] - 3[2 \quad -7 \quad 6]$
5. $5\begin{bmatrix} 6 & -2 \\ 5 & 4 \end{bmatrix} - 2\begin{bmatrix} 6 & -2 \\ 5 & 4 \end{bmatrix} + 4\begin{bmatrix} 7 & -6 \\ -4 & 2 \end{bmatrix}$
6. $1.3\begin{bmatrix} 3.7 \\ -5.4 \end{bmatrix} + 4.1\begin{bmatrix} 6.4 \\ -3.7 \end{bmatrix} - 6.2\begin{bmatrix} -0.8 \\ 7.4 \end{bmatrix}$

Use matrices *A*, *B*, *C*, *D*, and *E* to find the following.

$A = \begin{bmatrix} 1 & 0 \\ 0 & 1 \end{bmatrix}, B = \begin{bmatrix} -1 & 0 \\ 0 & -1 \end{bmatrix}, C = \begin{bmatrix} 2 & -2 \\ -3 & 3 \end{bmatrix}, D = \begin{bmatrix} -2 & 2 \\ 3 & -3 \end{bmatrix}, E = \begin{bmatrix} 5 & -3 \\ -2 & 4 \end{bmatrix}$

7. $A + B$
8. $C + D$
9. $A - B$
10. $4B$
11. $D - C$
12. $E + 2A$
13. $D - 2B$
14. $2A + 3E - D$

Lesson 6-7 Using Matrices to Solve Systems of Equations (pp. 376–381)

Write an augmented matrix equation for each system of equations.

1. $5a + 3b = 6$
   $2a - b = 9$
2. $3x + 4y = -8$
   $2x - 3y = 6$
3. $m + 3p = 1$
   $4m - p = -22$
4. $4c - 3d = -1$
   $5c - 2d = 39$
5. $x + 2y - z = 6$
   $-2x + 3y + z = 1$
   $x + y + 3z = 8$
6. $2a - 3b - c = 4$
   $4a + b + c = 15$
   $a - b - c = -2$

Solve each matrix equation or system of equations.

7. $\left[\begin{array}{cc|c} 3 & 4 & 33 \\ 2 & -5 & -1 \end{array}\right]$
8. $\left[\begin{array}{cc|c} -1 & 1 & 0 \\ 7 & -6 & 3 \end{array}\right]$
9. $\left[\begin{array}{cc|c} 1 & 0 & -29 \\ 0 & 1 & 52 \end{array}\right]$
10. $5x - y = 7$
    $8x + 2y = 4$
11. $3m + t = 4$
    $2m + 2t = 3$
12. $6c + 5d = 7$
    $3c - 10d = -4$
13. $3a - 5b = 1$
    $a + 3b = 5$
14. $2r - 7t = 24$
    $-r + 8t = -21$
15. $x + y = -3$
    $3x - 10y = 43$
16. $2m - 3p = 3$
    $-4m + 9p = -8$
17. $x + y = 1$
    $2x - 2y = -12$

Lesson 6-8 Systems of Inequalities (pp. 382–387)

Solve each system of inequalities.

1. $x \le 5$
   $y \ge -3$
2. $y < 3$
   $y - x \ge -1$
3. $x + y < 5$
   $x < 2$
4. $y + x < 2$
   $y \ge x$
5. $x + y \le 2$
   $y - x \le 4$
6. $y \le x + 4$
   $y - x \ge 1$
7. $y < \frac{1}{3}x + 5$
   $y > 2x + 1$
8. $y + x \ge 1$
   $y - x \ge -1$
9. $|x| > 2$
   $|y| \le 5$
10. $x \le 2$
    $y \le 3$
    $y \ge -\frac{3}{2}x + 3$
11. $y \le x$
    $y \le -x + 4$
    $y \ge -1$
12. $y \le -1$
    $3x - 2y \ge 6$

Lesson 7-1 Multiplying Monomials (pp. 401–407)

Determine whether each expression is a monomial. Write *yes* or *no*. Explain your reasoning.

1. $n^2 - 3$
2. 5^3
3. $9a^2b^3$
4. $15 - x^2y$

Simplify each expression.

5. $a^5(a)(a^7)$
6. $(r^3t^4)(r^4t^4)$
7. $(x^3y^4)(xy^3)$
8. $(bc^3)(b^4c^3)$
9. $(-3mp^2)(5m^3p^2)$
10. $[(3^3)^2]^2$
11. $(3n3t^2)(-4n^3t^2)$
12. $x^3(x^4y^3)$
13. $(1.1g^2h^4)^3$
14. $-\frac{3}{4}a(a^2b^3c^4)$
15. $\left(\frac{1}{2}w^3\right)^2(6w^4)^2$
16. $[(-2^3)^3]^2$

GEOMETRY **Express the volume of each solid as a monomial.**

17. 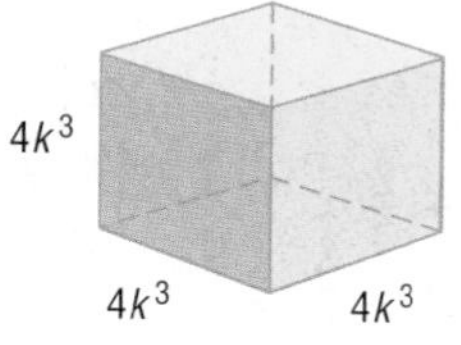

18. 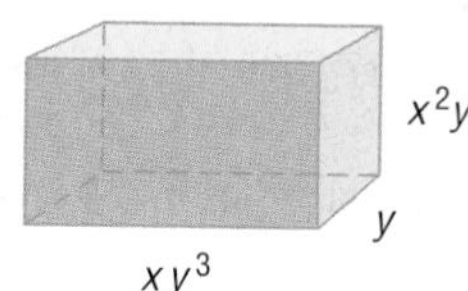

19. 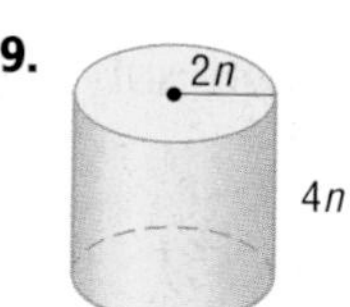

Extra Practice

Lesson 7-2 Dividing Monomials (pp. 408–415)

Simplify. Assume that no denominator is equal to zero.

1. $\frac{6^{10}}{6^7}$
2. $\frac{b^6c^5}{b^3c^2}$
3. $\frac{(-a)^4b^8}{a^4b^7}$
4. $\frac{(-x)^3y^3}{x^3y^6}$
5. $\frac{12ab^5}{4a^4b^3}$
6. $\frac{24x^5}{-8x^2}$
7. $\frac{-9h^2k^4}{18h^5j^3k^4}$
8. $\left(\frac{2a^2b^4}{3a^3b}\right)^2$
9. $a^5b^0a^{-7}$
10. $\frac{(-u^{-3}v^3)^2}{(u^3v)^{-3}}$
11. $\left(\frac{a^3}{b^2}\right)^{-3}$
12. $\left(\frac{2x}{y^{-3}}\right)^{-2}$
13. $\frac{(-r)s^5}{r^{-3}s^{-4}}$
14. $\frac{28a^{-4}b^0}{14a^3b^{-1}}$
15. $\frac{(j^2k^3m)^4}{(jk^4)^{-1}}$
16. $\left(\frac{-2x^4y}{4y^2}\right)^0$
17. $\left(\frac{-18x^0a^{-3}}{-6x^{-2}a^{-3}}\right)$
18. $\left(\frac{2a^3b^{-2}}{2^{-1}a^{-5}b^3}\right)^{-1}$
19. $\left(\frac{5n^{-1}m^2}{2nm^{-2}}\right)^0$
20. $\frac{(3ab^2c)^{-3}}{(2a^2bc^2)^2}$

Lesson 7-3 Scientific Notation (pp. 416–422)

Express each number in scientific notation.

1. 1,400,322
2. 134,490,000
3. 0.00009
4. 0.004500
5. 12,000,000
6. 0.0000233

Express each number in standard form.

7. 2.23×10^{-2}
8. 5.4×10^3
9. 3.334×10^{-9}
10. 4.7×10^{-6}
11. 5.22×10^4
12. 7.256×10^{-1}

Evaluate. Express the results in both scientific notation and standard form.

13. $(2.3 \times 10^{-2})(2.55 \times 10^3)$
14. $(5.23 \times 10^{-7})(8.2 \times 10^5)$
15. $\frac{3.344 \times 10^6}{4.2 \times 10^{-3}}$
16. $\frac{2.644 \times 10^{-5}}{3.2 \times 10^{-2}}$

Lesson 7-4 Polynomials (pp. 424–429)

State whether each expression is a polynomial. If so, identify it as a *monomial*, a *binomial*, or a *trinomial*.

1. $5x^2y + 3xy - 7$
2. 0
3. $\frac{5}{k} - k^2y$
4. $3a^2x - 5a$

Find the degree of each polynomial.

5. $a + 5c$
6. $14abcd - 6d^3$
7. $\frac{a^3}{4}$
8. 10
9. $-4h^5$
10. $\frac{x^2}{3} - \frac{x}{2} + \frac{1}{5}$
11. -6
12. $a^2b^3 - a^3b^2$

Arrange the terms of each polynomial so that the powers of x are in ascending order.

13. $2x^2 - 3x + 4x^3 - x^5$
14. $x^3 - x^2 + x - 1$
15. $2a + 3ax^2 - 4ax$
16. $-5bx^3 - 2bx + 4x^2 - b^3$
17. $x^8 + 2x^2 - x^6 + 1$
18. $cdx^2 - c^2d^2x + d^3$

Arrange the terms of each polynomial so that the powers of x are in descending order.

19. $5x^2 - 3x^3 + 7 + 2x$
20. $-6x + x^5 + 4x^3 - 20$
21. $5b + b^3x^2 + \frac{2}{3}bx$
22. $21p^2x + 3px^3 + p^4$
23. $3ax^2 - 6a^2x^3 + 7a^3 - 8x$
24. $\frac{1}{3}s^2x^3 + 4x^4 - \frac{2}{5}s^4x^2$

Lesson 7-5 Adding and Subtracting Polynomials (pp. 433–438)

Find each sum or difference.

1. $(3a^2 + 5) + (4a^2 - 1)$
2. $(5x - 3) + (-2x + 1)$
3. $(6z + 2) - (9z + 3)$
4. $(-4n + 7) - (-7n - 8)$
5. $(-7t^2 + 4ts - 6s^2) + (-5t^2 - 12ts + 3s^2)$
6. $(6a^2 - 7ab - 4b^2) - (2a^2 + 5ab + 6b^2)$
7. $(4a^2 - 10b^2 + 7c^2) + (-5a^2 + 2c^2 + 2b)$
8. $(z^2 + 6z - 8) - (4z^2 - 7z - 5)$
9. $(4d + 3e - 8f) - (-3d + 10e - 5f + 6)$
10. $(7g + 8h - 9) + (-g - 3h - 6k)$
11. $(9x^2 - 11xy - 3y^2) - (x^2 - 16xy + 12y^2)$
12. $(-3m + 9mn - 5n) + (14m - 5mn - 2n)$
13. $(6 - 7y + 3y^2) + (3 - 5y - 2y^2) + (-12 - 8y + y^2)$
14. $(-7c^2 - 2c - 5) + (9c - 6) + (16c^2 + 3) + (-9c^2 - 7c + 7)$

Lesson 7-6 Multiplying a Polynomial by a Monomial (pp. 439–444)

Find each product.

1. $-3(8x + 5)$
2. $3b(5b + 8)$
3. $1.1a(2a + 7)$
4. $\frac{1}{2}x(8x - 6)$
5. $7xy(5x^2 - y^2)$
6. $5y(y^2 - 3y + 6)$
7. $-ab(3b^2 + 4ab - 6a^2)$
8. $4m^2(9m^2n + mn - 5n^2)$
9. $4st^2(-4s^2t^3 + 7s^5 - 3st^3)$

Simplify each expression.

10. $-3a(2a - 12) + 5a$
11. $6(12b^2 - 2b) + 7(-2 - 3b)$
12. $x(x - 6) + x(x - 2) + 2x$
13. $11(n - 3) + 2(n^2 + 22n)$
14. $-2x(x + 3) + 3(x + 3)$
15. $4m(n - 1) - 5n(n + 1)$

Solve each equation.

16. $-6(11 - 2x) = 7(-2 - 2x)$
17. $11(n - 3) + 5 = 2n + 44$
18. $a(a - 6) + 2a = 3 + a(a - 2)$
19. $q(2q + 3) + 20 = 2q(q - 3)$
20. $w(w + 12) = w(w + 14) + 12$
21. $x(x - 3) + 4x - 3 = 8x + x(3 + x)$
22. $-3(x + 5) + x(x - 1) = x(x + 2) - 3$
23. $n(n - 5) + n(n + 2) = 2n(n - 1) + 1.5$

Lesson 7-7 Multiplying Polynomials (pp. 447–452)

Find each product.

1. $(d + 2)(d + 5)$
2. $(z + 7)(z - 4)$
3. $(m - 8)(m - 5)$
4. $(a + 2)(a - 19)$
5. $(c + 15)(c - 3)$
6. $(x + y)(x - 2y)$
7. $(2x - 5)(x + 6)$
8. $(7a - 4)(2a - 5)$
9. $(4x + y)(2x - 3y)$
10. $(7v + 3)(v + 4)$
11. $(7s - 8)(3s - 2)$
12. $(4g + 3h)(2g - 5h)$
13. $(4a + 3)(2a - 1)$
14. $(7y - 1)(2y - 3)$
15. $(2x + 3y)(4x + 2y)$
16. $(12r - 4s)(5r + 8s)$
17. $(-a + 1)(-3a - 2)$
18. $(2n - 4)(-3n - 2)$
19. $(x - 2)(x^2 + 2x + 4)$
20. $(3x + 5)(2x^2 - 5x + 11)$
21. $(4s + 5)(3s^2 + 8s - 9)$
22. $(5x - 2)(-5x^2 + 2x + 7)$
23. $(-n + 2)(-2n^2 + n - 1)$
24. $(x^2 - 7x + 4)(2x^2 - 3x - 6)$
25. $(x^2 + x + 1)(x^2 - x - 1)$
26. $(a^2 + 2a + 5)(a^2 - 3a - 7)$
27. $(5x^4 - 2x^2 + 1)(x^2 - 5x + 3)$

Lesson 7-8 Special Products (pp. 453–458)

Find each product.

1. $(t + 7)^2$
2. $(w - 12)(w + 12)$
3. $(q - 4h)^2$
4. $(10x + 11y)(10x - 11y)$
5. $(4p + 3)^2$
6. $(2b - 4d)(2b + 4d)$
7. $(a + 2b)^2$
8. $(3x + y)^2$
9. $(6m + 2n)^2$
10. $(3m - 7d)^2$
11. $(5b - 6)(5b + 6)$
12. $(1 + x)^2$
13. $(5x - 9y)^2$
14. $(8a - 2b)(8a + 2b)$
15. $\left(\frac{1}{4}x + 4\right)^2$
16. $\left(\frac{1}{2}x - 10\right)\left(\frac{1}{2}x + 10\right)$
17. $\left(\frac{1}{3}n - m\right)\left(\frac{1}{3}n + m\right)$
18. $(a - 1)(a - 1)(a - 1)$
19. $(x + 2)(x - 2)(2x + 5)$
20. $(4x - 1)(4x + 1)(x - 4)$
21. $(x - 5)(x + 5)(x + 4)(x - 4)$
22. $(a + 1)(a + 1)(a - 1)(a - 1)$
23. $(n - 1)(n + 1)(n - 1)$
24. $(2c + 3)(2c + 3)(2c - 3)(2c - 3)$
25. $(4d + 5g)(4d + 5g)(4d - 5g)(4d - 5g)$

Lesson 8-1 Monomials and Factoring (pp. 471–474)

Factor each monomial completely.

1. $240mn$
2. $-64a^3b$
3. $-26xy^2$
4. $-231xy^2z$
5. $44rs^2t^3$
6. $-756m^2n^2$

Find the GCF of each set of monomials.

7. 16, 60
8. 15, 50
9. 45, 80
10. 29, 58
11. 55, 305
12. 126, 252
13. 128, 245
14. $7y^2, 14y^2$
15. $4xy, -6x$
16. $35t^2, 7t$
17. $16pq^2, 12p^2q, 4pq$
18. 5, 15, 10
19. $12mn, 10mn, 15mn$
20. $14xy, 12y, 20x$
21. $26jk^4, 16jk^3, 8j^2$

Lesson 8-2 Using the Distributive Property (pp. 476–482)

Use the Distributive Property to factor each polynomial.

1. $10a^2 + 40a$
2. $15wx - 35wx^2$
3. $27a^2b + 9b^3$
4. $11x + 44x^2y$
5. $16y^2 + 8y$
6. $14mn^2 + 2mn$
7. $25a^2b^2 + 30ab^3$
8. $2m^3n^2 - 16mn^2 + 8mn$
9. $2ax + 6xc + ba + 3bc$
10. $6mx - 4m + 3rx - 2r$
11. $3ax - 6bx + 8b - 4a$
12. $a^2 - 2ab + a - 2b$
13. $8ac - 2ad + 4bc - bd$
14. $2e^2g + 2fg + 4e^2h + 4fh$
15. $x^2 - xy - xy + y^2$

Solve each equation. Check your solutions.

16. $a(a - 9) = 0$
17. $d(d + 11) = 0$
18. $z(z - 2.5) = 0$
19. $(2y + 6)(y - 1) = 0$
20. $(4n - 7)(3n + 2) = 0$
21. $(a - 1)(a + 1) = 0$
22. $10x^2 - 20x = 0$
23. $8b^2 - 12b = 0$
24. $14d^2 + 49d = 0$
25. $15a^2 = 60a$
26. $33x^2 = -22x$
27. $32x^2 = 16x$

Lesson 8-3 Quadratic Equations: $x^2 + bx + c = 0$ (pp. 485–491)

Factor each trinomial.

1. $x^2 - 9x + 14$
2. $a^2 - 9a - 36$
3. $x^2 + 2x - 15$
4. $n^2 - 8n + 15$
5. $b^2 + 22b + 21$
6. $c^2 + 2c - 3$
7. $x^2 - 5x - 24$
8. $n^2 - 8n + 7$
9. $m^2 - 10m - 39$
10. $z^2 + 15z + 36$
11. $s^2 - 13st - 30t^2$
12. $y^2 + 2y - 35$
13. $r^2 + 3r - 40$
14. $x^2 + 5x - 6$
15. $x^2 - 4xy - 5y^2$
16. $r^2 + 16r + 63$
17. $v^2 + 24v - 52$
18. $k^2 - 27kj - 90j^2$

Solve each equation. Check your solutions.

19. $a^2 + 3a - 4 = 0$
20. $x^2 - 8x - 20 = 0$
21. $b^2 + 11b + 24 = 0$
22. $y^2 + y - 42 = 0$
23. $k^2 + 2k - 24 = 0$
24. $r^2 - 13r - 48 = 0$
25. $n^2 - 9n = -18$
26. $2z + z^2 = 35$
27. $-20x + 19 = -x^2$
28. $10 + a^2 = -7a$
29. $z^2 - 57 = 16z$
30. $x^2 = -14x - 33$

Lesson 8-4 Quadratic Equations: $ax^2 + bx + c = 0$ (pp. 493–498)

Factor each trinomial, if possible. If the trinomial cannot be factored using integers, write *prime*.

1. $4a^2 + 4a - 63$
2. $3x^2 - 7x - 6$
3. $4r^2 - 25r + 6$
4. $2z^2 - 11z + 15$
5. $3a^2 - 2a - 21$
6. $4y^2 + 11y + 6$
7. $6n^2 + 7n - 3$
8. $5x^2 - 17x + 14$
9. $2n^2 - 11n + 13$
10. $5a^2 - 3a + 15$
11. $18v^2 + 24v + 12$
12. $4k^2 + 2k - 12$
13. $10x^2 - 20xy + 10y^2$
14. $12c^2 - 11cd - 5d^2$
15. $30n^2 - mn - m^2$

Solve each equation. Check your solutions.

16. $8t^2 + 32t + 24 = 0$
17. $6y^2 + 72y + 192 = 0$
18. $5x^2 + 3x - 2 = 0$
19. $9x^2 + 18x - 27 = 0$
20. $4x^2 - 4x - 4 = 4$
21. $12n^2 - 16n - 3 = 0$
22. $12x^2 - x - 35 = 0$
23. $18x^2 + 36x - 14 = 0$
24. $15a^2 + a - 2 = 0$
25. $14b^2 + 7b - 42 = 0$
26. $13r^2 + 21r - 10 = 0$
27. $35y^2 - 60y - 20 = 0$

Lesson 8-5 Quadratic Equations: Differences of Squares (pp. 499–504)

Factor each polynomial, if possible. If the polynomial cannot be factored, write prime.

1. $x^2 - 9$
2. $a^2 - 64$
3. $4x^2 - 9y^2$
4. $1 - 9z^2$
5. $16a^2 - 9b^2$
6. $8x^2 - 12y^2$
7. $a^2 - 4b^2$
8. $75r^2 - 48$
9. $x^2 - 36y^2$
10. $3a^2 - 16$
11. $9x^2 - 100y^2$
12. $49 - a^2b^2$
13. $5a^2 - 48$
14. $169 - 16t^2$
15. $8r^2 - 4$
16. $-45m^2 + 5$

Solve each equation by factoring. Check your solutions.

17. $4x^2 = 16$
18. $2x^2 = 50$
19. $9n^2 - 4 = 0$
20. $a^2 - \frac{25}{36} = 0$
21. $\frac{16}{9} - b^2 = 0$
22. $18 - \frac{1}{2}x^2 = 0$
23. $20 - 5g^2 = 0$
24. $16 - \frac{1}{4}p^2 = 0$
25. $\frac{1}{4}c^2 - \frac{4}{9} = 0$
26. $2q^3 - 2q = 0$
27. $3r^3 = 48r$
28. $100d - 4d^3 = 0$

Extra Practice

Lesson 8-6 Quadratic Equations: Perfect Squares (pp. 505–512)

Determine whether each trinomial is a perfect square trinomial. Write *yes* or *no*. If so, factor it.

1. $x^2 + 12x + 36$
2. $n^2 - 13n + 36$
3. $a^2 + 4a + 4$
4. $x^2 - 10x - 100$
5. $2n^2 + 17n + 21$
6. $4a^2 - 20a + 25$

Factor each polynomial, if possible. If the polynomial cannot be factored, write *prime*.

7. $3x^2 - 75$
8. $4p^2 + 12pr + 9r^2$
9. $6a^2 + 72$
10. $s^2 + 30s + 225$
11. $24x^2 + 24x + 9$
12. $1 - 10z + 25z^2$
13. $28 - 63b^2$
14. $4c^2 + 2c - 7$

Solve each equation. Check your solutions.

15. $x^2 + 22x + 121 = 0$
16. $343d^2 = 7$
17. $(a - 7)^2 = 5$
18. $c^2 + 10c + 36 = 11$
19. $16s^2 + 81 = 72s$
20. $9p^2 - 42p + 20 = -29$

Extra Practice

Lesson 9-1 Graphing Quadratic Functions (pp. 525–536)

Use a table of values to graph each function. State the domain and range.

1. $y = x^2 + 6x + 8$
2. $y = -x^2 + 3x$
3. $y = -x^2$

Find the vertex, the equation of the axis of symmetry, and the y-intercept.

4. $y = -x^2 + 2x - 3$
5. $y = 3x^2 + 24x + 80$
6. $y = x^2 - 4x - 4$
7. $y = 5x^2 - 20x + 37$
8. $y = 3x^2 + 6x + 3$
9. $y = 2x^2 + 12x$
10. $y = x^2 - 6x + 5$
11. $y = x^2 + 6x + 9$
12. $y = -x^2 + 16x - 15$

Consider each equation.

a. Determine whether the function has *maximum* or *minimum* value.
b. State the maximum or minimum value.
c. What are the domain and range of the function.

13. $y = 4x^2 - 1$
14. $y = -2x^2 - 2x + 4$
15. $y = 6x^2 - 12x - 4$
16. $y = -x^2 - 1$
17. $y = -x^2 + x + 1$
18. $y = -5x^2 - 3x + 2$

Lesson 9-2 Solving Quadratic Equations by Graphing (pp. 537–543)

Solve each equation by graphing.

1. $a^2 - 25 = 0$
2. $n^2 - 8n = 0$
3. $d^2 + 36 = 0$
4. $b^2 - 18b + 81 = 0$
5. $x^2 + 3x + 27 = 0$
6. $-y^2 - 3y + 10 = 0$

Solve each equation by graphing. If integral roots cannot be found, estimate the roots to the nearest tenth.

7. $x^2 + 2x - 3 = 0$
8. $-x^2 + 6x - 5 = 0$
9. $-a^2 - 2a + 3 = 0$
10. $2r^2 - 8r + 5 = 0$
11. $-3x^2 + 6x - 9 = 0$
12. $c^2 + c = 0$
13. $3t^2 + 2 = 0$
14. $-b^2 + 5b + 2 = 0$
15. $3x^2 + 7x = 1$

Lesson 9-3 Transformations of Quadratic Functions (pp. 544–551)

Describe how the graph of each function is related to the graph of $f(x) = x^2$.

1. $g(x) = x^2 - 5$
2. $h(x) = \frac{1}{2}x^2$
3. $h(x) = -x^2 + 5$
4. $g(x) = x^2 + 9$
5. $g(x) = -3x^2$
6. $h(x) = -x^2 - 6$
7. $g(x) = \frac{3}{4}x^2 + 4$
8. $h(x) = 1.2x^2 - 7.5$
9. $g(x) = -7 - \frac{4}{3}x^2$

List the functions in order from the most compressed to the least compressed graph.

10. $g(x) = -2.3x^2, h(x) = \frac{2}{3}x^2$
11. $g(x) = 5x^2, h(x) = \frac{1}{2}x^2$
12. $g(x) = -x^2, h(x) = \frac{5}{3}x^2, f(x) = -2.5x^2$
13. $g(x) = -6x^2, h(x) = 2x^2, f(x) = 0.4x^2$

Lesson 9-4 Solving Quadratic Equations by Completing the Square (pp. 552–557)

Solve each equation by taking the square root of each side. Round to the nearest tenth if necessary.

1. $x^2 - 4x + 4 = 9$
2. $t^2 - 6t + 9 = 16$
3. $b^2 + 10b + 25 = 11$
4. $a^2 - 22a + 121 = 3$
5. $x^2 + 2x + 1 = 81$
6. $t^2 - 36t + 324 = 85$

Find the value of c that makes each trinomial a perfect square.

7. $a^2 + 20a + c$
8. $x^2 + 10x + c$
9. $t^2 + 12t + c$
10. $y^2 - 9y + c$
11. $p^2 - 14p + c$
12. $b^2 + 13b + c$

Solve each equation by completing the square. Round to the nearest tenth if necessary.

13. $a^2 - 8a - 84 = 0$
14. $c^2 + 6 = -5c$
15. $p^2 - 8p + 5 = 0$
16. $2y^2 + 7y - 4 = 0$
17. $t^2 + 3t = 40$
18. $x^2 + 8x - 9 = 0$
19. $y^2 + 5y - 84 = 0$
20. $t^2 + 12t + 32 = 0$
21. $2x - 3x^2 = -8$
22. $2y^2 - y - 9 = 0$
23. $2z^2 - 5z - 4 = 0$
24. $8t^2 - 12t - 1 = 0$

Lesson 9-5 Solving Quadratic Equations by Using the Quadratic Formula (pp. 558–565)

Solve each equation by using the Quadratic Formula. Round to the nearest tenth if necessary.

1. $x^2 - 8x - 4 = 0$
2. $x^2 + 7x - 8 = 0$
3. $x^2 - 5x + 6 = 0$
4. $y^2 - 7y - 8 = 0$
5. $m^2 - 2m = 35$
6. $4n^2 - 20n = 0$
7. $m^2 + 4m + 2 = 0$
8. $2t^2 - t - 15 = 0$
9. $5t^2 = 125$
10. $t^2 + 16 = 0$
11. $-4x^2 + 8x = -3$
12. $3k^2 + 2 = -8k$
13. $8t^2 + 10t + 3 = 0$
14. $3x^2 - \frac{5}{4}x - \frac{1}{2} = 0$
15. $-5b^2 + 3b - 1 = 0$
16. $n^2 - 3n + 1 = 0$
17. $2z^2 + 5z - 1 = 0$
18. $3h^2 = 27$

State the value of the discriminant for each equation. Then determine the number of real solutions of the equation.

19. $3f^2 + 2f = 6$
20. $2x^2 = 0.7x + 0.3$
21. $3w^2 - 2w + 8 = 0$
22. $4r^2 - 12r + 9 = 0$
23. $x^2 - 5x = -9$
24. $25t^2 + 30t = -9$

Lesson 9-6 Exponential Functions (pp. 567–572)

Graph each function. State the y-intercept, and state the domain and range. Then use the graph to determine the approximate value of the given expression. Use a calculator to confirm the value.

1. $y = 7^x; 7^{1.5}$
2. $\left(\frac{1}{3}\right)^x; \left(\frac{1}{3}\right)^{5.6}$
3. $y = \left(\frac{3}{5}\right)^x; \left(\frac{3}{5}\right)^{-4.2}$

Graph each function. State the y-intercept.

4. $y = 3^x + 1$
5. $y = 2^x - 5$
6. $y = 2^{x+3}$
7. $y = 3^{x+1}$
8. $y = \left(\frac{2}{3}\right)^x$
9. $y = 5\left(\frac{2}{5}\right)^x$
10. $y = 5(3^x)$
11. $y = 4(5)^x$
12. $y = 2(5)^x + 1$
13. $y = \left(\frac{1}{2}\right)^{x+1}$
14. $y = \left(\frac{1}{8}\right)^x$
15. $y = \left(\frac{3}{4}\right)^x - 2$

Determine whether the data in each table display exponential behavior. Explain why or why not.

16.

x	−1	0	1	2
y	−5	−1	3	7

17.

x	1	2	3	4
y	25	125	625	3125

Lesson 9-7 Growth and Decay (pp. 573–579)

1. **FARMING** Mr. Rogers purchased a combine for \$175,000 for his farming operation. It is expected to depreciate at a rate of 18% per year. What will be the value of the combine in 3 years?
2. **REAL ESTATE** The Jacksons bought a house for \$65,000 in 1992. Houses in the neighborhood have appreciated at the rate of 4.5% a year. How much is the house worth in 2003?
3. **POPULATION** In 1950, the population of a city was 50,000. Since then, the population has increased by 2.25% per year. If it continues to grow at this rate, what will the population be in 2005?
4. **BEARS** In a particular state, the population of black bears has been decreasing at the rate of 0.75% per year. In 1990, it was estimated that there were 400 black bears in the state. If the population continues to decline at the same rate, what will the population be in 2010?
5. **INVESTMENTS** Determine the amount of an investment if \$500 is invested at an interest rate of 5.75% compounded monthly for 25 years.

Lesson 9-8 Geometric Sequences as Exponential Functions (pp. 580–585)

1. **MONEY** Marco deposited \$8500 in a 4-year certificate of deposit earning 7.25% compounded monthly. Write an equation for the amount of money Marco will have at the end of the four years. Then find the amount.
2. **TRANSPORTATION** Elise is buying a new car for \$21,500. The rate of depreciation on this type of car is 8% per year. Write an equation for the value of the car in 5 years. Then find the value of the car in 5 years.
3. **POPULATION** In 2000, the town of Belgrade had a population of 3422. For each of the next 8 years, the population increased by 4.9% per year. Find the projected population of Belgrade in 2008.

Lesson 9-9 Analyzing Functions with Successive Differences (pp. 586–591)

Graph each set or ordered pairs. Determine whether the ordered pairs represent a *linear* function, a *quadratic* function, or an *exponential* function.

1. $(-2, 1), (-1, -2), (0, -3), (1, -2), (2, 1)$
2. $(-4, -5), (-2, -4), (0, -3), (2, -2), (4, 0)$
3. $(-3, 1), (-2, 2), (-1, 4), (0, 8), (1, 16)$

Determine which model best describes the data. Then write an equation for the function that models the data.

4.

x	−2	−1	0	1	2
y	1	0.25	0	0.25	1

5.

x	−1	0	1	2	3
y	$\frac{2}{3}$	2	6	18	54

6.

x	−6	−5	−4	−3	−2
y	2	2.5	3	3.5	4

7. **COMPUTER GAME** Kylie started a game of computer "Tag, You're It". She tagged two friends and so there are three players after round 1. Each of them tagged three friends, and so on. The table shows the number of players after the first few rounds. Determine which function best models the number of players and then write a function that models the data.

Round	1	2	3	4	5
Players	3	9	27	81	243

8. **ROCKET** The table shows the height of a rocket that is launched from ground level after a period of 4 consecutive seconds. Determine which model best represents the height of the rocket with respect to time. Then write a function that models the data.

Times(s)	0	1	2	3	4
Height (ft)	0	1.2	4.8	10.8	19.2

Lesson 10-1 Square Root Functions (pp. 605–610)

Graph each function. Determine the domain and range of the function.

1. $y = \sqrt{x - 4}$
2. $y = \sqrt{x + 3} - 1$
3. $y = \frac{1}{3}\sqrt{x + 2}$
4. $y = \sqrt{2x + 5}$
5. $y = -\sqrt{4x}$
6. $y = 2\sqrt{x}$
7. $y = -3\sqrt{x}$
8. $y = \sqrt{x} + 5$
9. $y = \sqrt{2x} - 1$
10. $y = 5\sqrt{x} + 1$
11. $y = \sqrt{x + 1} - 2$
12. $y = 6 - \sqrt{x + 3}$

Lesson 10-2 Simplifying Radical Expressions (pp. 612–617)

Simplify.

1. $\sqrt{50}$
2. $\sqrt{200}$
3. $\sqrt{162}$
4. $\sqrt{700}$
5. $\frac{\sqrt{3}}{\sqrt{5}}$
6. $\frac{\sqrt{72}}{\sqrt{6}}$
7. $\sqrt{\frac{8}{7}}$
8. $\sqrt{\frac{7}{32}}$
9. $\sqrt{\frac{5}{8}} \cdot \sqrt{\frac{2}{6}}$
10. $\sqrt{\frac{2}{3}} \cdot \sqrt{\frac{3}{2}}$
11. $\sqrt{\frac{2x}{30}}$
12. $\sqrt{\frac{50}{z^2}}$
13. $\sqrt{10} \cdot \sqrt{20}$
14. $\sqrt{7} \cdot \sqrt{3}$
15. $6\sqrt{2} \cdot \sqrt{3}$
16. $5\sqrt{6} \cdot 2\sqrt{3}$
17. $\sqrt{4x^4y^3}$
18. $\sqrt{200m^2y^3}$
19. $\sqrt{12tx^3}$
20. $\sqrt{175a^4b^6}$
21. $\sqrt{\frac{54}{g^2}}$
22. $\sqrt{99x^3y^7}$
23. $\sqrt{\frac{32c^5}{9d^2}}$
24. $\sqrt{\frac{27p^4}{3p^2}}$

Lesson 10-3 Operations with Radical Expressions (pp. 619–623)

Simplify.

1. $3\sqrt{11} + 6\sqrt{11} - 2\sqrt{11}$
2. $6\sqrt{13} + 7\sqrt{13}$
3. $2\sqrt{12} + 5\sqrt{3}$
4. $9\sqrt{7} - 4\sqrt{2} + 3\sqrt{2}$
5. $3\sqrt{5} - 5\sqrt{3}$
6. $4\sqrt{8} - 3\sqrt{5}$
7. $2\sqrt{27} - 4\sqrt{12}$
8. $8\sqrt{32} + 4\sqrt{50}$
9. $\sqrt{45} + 6\sqrt{20}$
10. $2\sqrt{63} - 6\sqrt{28} + 8\sqrt{45}$
11. $14\sqrt{3t} + 8$
12. $7\sqrt{6x} - 12\sqrt{6x}$
13. $5\sqrt{7} - 3\sqrt{28}$
14. $7\sqrt{8} - \sqrt{18}$
15. $7\sqrt{98} + 5\sqrt{32} - 2\sqrt{75}$
16. $4\sqrt{6} + 3\sqrt{2} - 2\sqrt{5}$
17. $-3\sqrt{20} + 2\sqrt{45} - \sqrt{7}$
18. $4\sqrt{75} + 6\sqrt{27}$
19. $10\sqrt{\frac{1}{5}} - \sqrt{45} - 12\sqrt{\frac{5}{9}}$
20. $\sqrt{15} - \sqrt{\frac{3}{5}}$
21. $3\sqrt{\frac{1}{3}} - 9\sqrt{\frac{1}{12}} + \sqrt{243}$

Find each product.

22. $\sqrt{3}(\sqrt{5} + 2)$
23. $\sqrt{2}(\sqrt{2} + 3\sqrt{5})$
24. $(\sqrt{2} + 5)^2$
25. $(3 - \sqrt{7})(3 + \sqrt{7})$
26. $(\sqrt{2} + \sqrt{3})(\sqrt{3} + \sqrt{2})$
27. $(4\sqrt{7} + \sqrt{2})(\sqrt{3} - 3\sqrt{5})$

Lesson 10-4 Radical Equations (pp. 624–628)

Solve each equation. Check your solution.

1. $\sqrt{5x} = 5$
2. $4\sqrt{7} = \sqrt{-m}$
3. $\sqrt{t} - 5 = 0$
4. $\sqrt{3b} + 2 = 0$
5. $\sqrt{x} - 3 = 6$
6. $5 - \sqrt{3x} = 1$
7. $2 + 3\sqrt{y} = 13$
8. $\sqrt{3g} = 6$
9. $\sqrt{a} - 2 = 0$
10. $\sqrt{2j} - 4 = 8$
11. $5 + \sqrt{x} = 9$
12. $\sqrt{5y + 4} = 7$
13. $7 + \sqrt{5c} = 9$
14. $2\sqrt{5t} = 10$
15. $\sqrt{44} = 2\sqrt{p}$
16. $4\sqrt{x - 5} = 15$
17. $4 - \sqrt{x - 3} = 9$
18. $\sqrt{10x^2 - 5} = 3x$
19. $\sqrt{2a^2 - 144} = a$
20. $\sqrt{3y + 1} = y - 3$
21. $\sqrt{2x^2 - 12} = x$
22. $\sqrt{b^2 + 16} + 2b = 5b$
23. $\sqrt{m + 2} + m = 4$
24. $\sqrt{3 - 2c} + 3 = 2c$

Lesson 10-5 The Pythagorean Theorem (pp. 630–635)

If c is the measure of the hypotenuse of a right triangle, find each missing measure. If necessary, round to the nearest hundredth.

1. $b = 20, c = 29, a = ?$
2. $a = 7, b = 24, c = ?$
3. $a = 2, b = 6, c = ?$
4. $b = 10, c = \sqrt{200}, a = ?$
5. $a = 3, c = 3\sqrt{2}, b = ?$
6. $a = 6, c = 14, b = ?$
7. $a = \sqrt{11}, c = \sqrt{47}, b = ?$
8. $a = \sqrt{13}, b = 6, c = ?$
9. $a = \sqrt{6}, b = 3, c = ?$
10. $b = \sqrt{75}, c = 10, a = ?$
11. $b = 9, c = \sqrt{130}, a = ?$
12. $a = 9, c = 15, b = ?$

Determine whether each set of measures can be sides of a right triangle. Then determine whether they form a Pythagorean Triple.

13. $(14, 48, 50)$
14. $(20, 30, 40)$
15. $(21, 72, 75)$
16. $(5, 12, \sqrt{119})$
17. $(15, 39, 36)$
18. $(10, 12, \sqrt{22})$
19. $(2, 3, 4)$
20. $(\sqrt{7}, 8, \sqrt{71})$

Lesson 10-6 The Distance and Midpoint Formulas (pp. 636–641)

Find the distance between each pair of points with the given coordinates.

1. $(4, 2), (-2, 10)$
2. $(-5, 1), (7, 6)$
3. $(4, -2), (1, 2)$
4. $(-2, 4), (4, -2)$
5. $(3, 1), (-2, -1)$
6. $(-2, 4), (7, -8)$
7. $(-5, 0), (-9, 6)$
8. $(5, -1), (5, 13)$
9. $(3\sqrt{2}, 7), (5\sqrt{2}, 9)$
10. $(6, 3), (10, 0)$
11. $(3, 6), (5, -5)$
12. $(-4, 2), (5, 4)$

Find the possible values of a if the points with the given coordinates are the indicated distance apart.

13. $(0, 0), (a, 3); d = 5$
14. $(2, -1), (-6, a); d = 10$
15. $(1, 0), (a, 6); d = \sqrt{61}$

Lesson 10-7 Similar Triangles (pp. 642–647)

Determine whether each pair of triangles is similar. Justify your answer.

1.
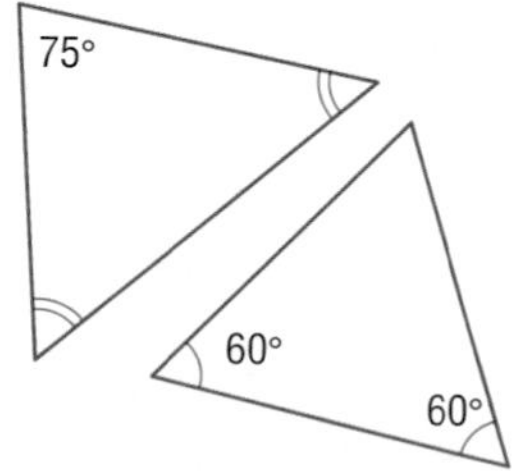

2.
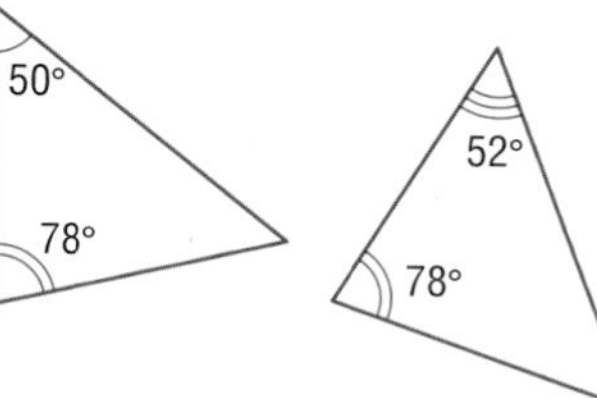

3.
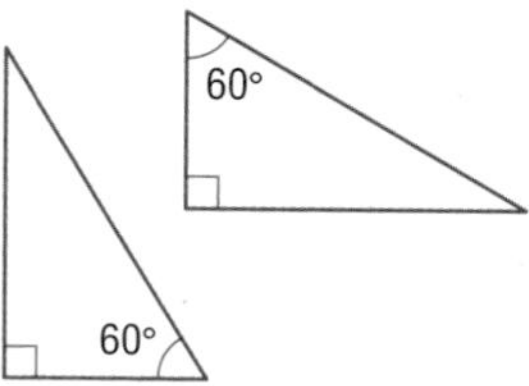

For each set of measures given, find the measures of the missing sides if $\triangle ABC \sim \triangle DEF$.

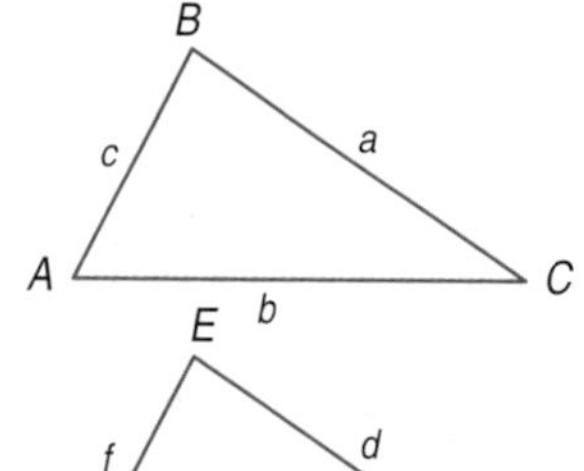

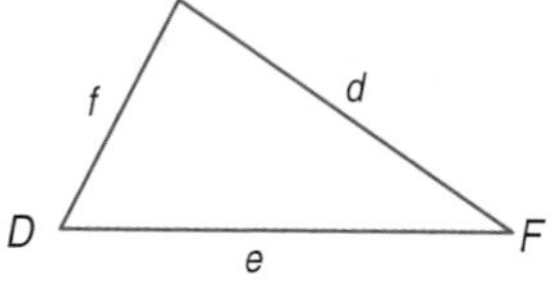

4. $a = 5, d = 10, b = 8, c = 7$
5. $a = 2, b = 3, c = 4, d = 3$
6. $a = 6, d = 4.5, e = 7, f = 7.5$
7. $a = 15, c = 20, b = 18, f = 10$
8. $f = 17.5, d = 8.5, e = 11, a = 1.7$

Lesson 10-8 Trigonometric Ratios (pp. 649–655)

Find the values of the three trigonometric ratios for angle X.

1.
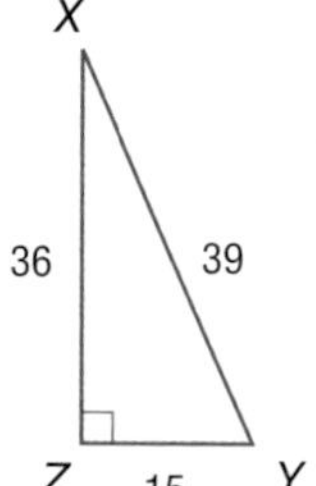

2.
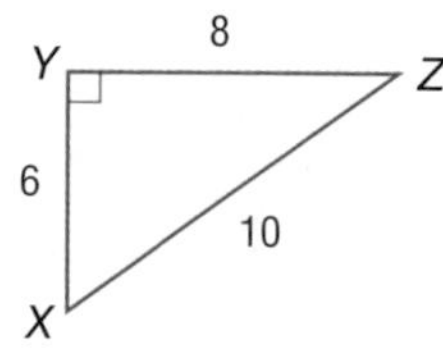

Use a calculator to find the values of each trigonometric ratio to the nearest ten-thousandth.

3. $\tan 42°$
4. $\sin 19°$
5. $\cos 78°$

Extra Practice

Extra Practice

Lesson 11-1 Inverse Variation (pp. 669–676)

Assume that y varies inversely as x. Write an inverse variation equation that relates x and y. Then graph the equation.

1. $y = 10$ when $x = 7.5$
2. $y = -5$ when $x = 3$
3. $y = -6$ when $x = -2$
4. $y = 1$ when $x = -0.5$
5. $y = -2.5$ when $x = 3$
6. $y = -2$ when $x = -1$

Solve. Assume that y varies inversely as x.

7. If $y = 54$ when $x = 4$, find x when $y = 27$.
8. If $y = 18$ when $x = 6$, find x when $y = 12$.
9. If $y = 12$ when $x = 24$, find x when $y = 9$.
10. If $y = 8$ when $x = -8$, find y when $x = -16$.
11. If $y = 3$ when $x = -8$, find y when $x = 4$.
12. If $y = 27$ when $x = \frac{1}{3}$, find y when $x = \frac{3}{4}$.
13. If $y = -3$ when $x = -8$, find y when $x = 2$.
14. If $y = -3$ when $x = -3$, find x when $y = 4$.
15. If $y = -7.5$ when $x = 2.5$, find y when $x = -2.5$.
16. If $y = -0.4$ when $x = -3.2$, find x when $y = -0.2$.

Lesson 11-2 Rational Functions (pp. 677–683)

Identify the asymptotes of each rational function.

1. $f(x) = \frac{1}{x + 4}$
2. $f(x) = \frac{x - 2}{x + 3}$
3. $f(x) = \frac{x}{x + 2}$
4. $f(x) = \frac{1}{x - 5}$
5. $f(x) = \frac{3x}{x + 1}$
6. $f(x) = \frac{x}{x - 6}$

State the excluded value for each function.

7. $f(x) = \frac{x + 2}{x + 3}$
8. $y = \frac{5}{x}$
9. $y = \frac{x}{x - 5}$
10. $y = \frac{2}{3x - 3}$
11. $y = \frac{x}{2x + 1}$
12. $y = \frac{3}{x + 2}$

Lesson 11-3 Simplifying Rational Expressions (pp. 684–691)

State the excluded values for each rational expression.

1. $\frac{x}{x + 1}$
2. $\frac{m}{n}$
3. $\frac{c - 2}{c^2 - 4}$
4. $\frac{b^2 - 5b + 6}{b^2 - 8b + 15}$

Simplify each expression. State the excluded values of the variables.

5. $\frac{13a}{39a^2}$
6. $\frac{38x^2}{42xy}$
7. $\frac{p + 5}{2(p + 5)}$
8. $\frac{a + b}{a^2 - b^2}$
9. $\frac{y + 4}{y^2 - 16}$
10. $\frac{c^2 - 4}{c^2 + 4c + 4}$
11. $\frac{a^2 - a}{a - 1}$
12. $\frac{x^2 + 4}{x^4 - 16}$
13. $\frac{r^3 - r^2}{r - 1}$
14. $\frac{4t^2 - 8}{4t - 4}$
15. $\frac{6y^3 - 12y^2}{12y^2 - 18}$
16. $\frac{5x^2 + 10x + 5}{3x^2 + 6x + 3}$

Lesson 11-4 Multiplying and Dividing Rational Expressions (pp. 692–698)

Find each product.

1. $\frac{a^2b}{b^2c} \cdot \frac{c}{d}$
2. $\frac{6a^2n}{8n^2} \cdot \frac{12n}{9a}$
3. $\frac{2a^2d}{3bc} \cdot \frac{9b^2c}{16ad^2}$
4. $\frac{10n^3}{6x^3} \cdot \frac{12n^2x^4}{25n^2x^2}$
5. $\frac{6m^3c}{10a^2} \cdot \frac{4a^2m}{9c^3}$
6. $\frac{5n-5}{3} \cdot \frac{9}{n-1}$
7. $\frac{(a-5)(a+1)}{(a+1)(a+7)} \cdot \frac{(a+7)(a-6)}{(a+8)(a-5)}$
8. $\frac{x-1}{(x+2)(x-3)} \cdot \frac{x+2}{(x-3)(x-1)}$
9. $\frac{a^2}{a-b} \cdot \frac{3a-3b}{a}$
10. $\frac{2a+4b}{5} \cdot \frac{25}{6a+8b}$

Find each quotient.

11. $\frac{5m^2p}{12a^2} \div \frac{30m^4}{18ap}$
12. $\frac{25g^7h}{28t^3} \div \frac{5g^5h^2}{42s^2t^3}$
13. $\frac{6a+4b}{36} \div \frac{3a+2b}{45}$
14. $\frac{x^2y}{18z} \div \frac{2yz}{3x^2}$
15. $\frac{p^2}{14tr^3} \div \frac{2r^2p}{7t}$
16. $\frac{5d-f}{5d+f} \div (25d^2 - f^2)$
17. $\frac{t^2-2t-15}{t-5} \div \frac{t+3}{t+5}$
18. $\frac{5x+10}{x+2} \div (x+2)$
19. $\frac{3d}{2d^2-3d} \div \frac{9}{2d-3}$

Lesson 11-5 Dividing Polynomials (pp. 700–705)

Find each quotient.

1. $(2x^2 - 11x - 20) \div (2x + 3)$
2. $(a^2 + 10a + 21) \div (a + 3)$
3. $(m^2 + 4m - 5) \div (m + 5)$
4. $(x^2 - 2x - 35) \div (x - 7)$
5. $(c^2 + 6c - 27) \div (c + 9)$
6. $(y^2 - 6y - 25) \div (y + 7)$
7. $(3t^2 - 14t - 24) \div (3t + 4)$
8. $(2r^2 - 3r - 35) \div (2r + 7)$
9. $\frac{12n^2+36n+15}{6n+3}$
10. $\frac{10x^2+29x+21}{5x+7}$
11. $\frac{4t^3+17t^2-1}{4t+1}$
12. $\frac{2a^3+9a^2+5a-12}{a+3}$
13. $\frac{27c^2-24c+8}{9c-2}$
14. $\frac{4b^3+7b^2-2b+4}{b+2}$

Lesson 11-6 Adding and Subtracting Rational Expressions (pp. 706–713)

Find each sum.

1. $\frac{4}{z} + \frac{3}{z}$
2. $\frac{a}{12} + \frac{2a}{12}$
3. $\frac{5}{2t} + \frac{-7}{2t}$
4. $\frac{y}{2} + \frac{y}{2}$
5. $\frac{k}{3} + \frac{2k}{7}$
6. $\frac{5}{2a} + \frac{-3}{6a}$
7. $\frac{6}{5x} + \frac{7}{10x^2}$
8. $\frac{4a}{2a+6} + \frac{3}{a+3}$
9. $\frac{3t+2}{3t-2} + \frac{t+2}{t^2-4}$
10. $\frac{-3}{a-5} + \frac{-6}{a^2-5a}$

Find each difference.

11. $\frac{5x}{24} - \frac{3x}{24}$
12. $\frac{7p}{3} - \frac{8p}{3}$
13. $\frac{8k}{5m} - \frac{3k}{5m}$
14. $\frac{8}{m-2} - \frac{6}{m-2}$
15. $\frac{y}{b+6} - \frac{2y}{b+6}$
16. $\frac{a+2}{6} - \frac{a+3}{6}$
17. $\frac{3z}{7w^2} - \frac{2z}{w}$
18. $\frac{p}{t^2} - \frac{r}{3t}$
19. $\frac{a}{a^2-4} - \frac{4}{a+2}$
20. $\frac{m}{m-p} - \frac{5}{m}$
21. $\frac{y+5}{y-5} - \frac{2y}{y^2-25}$

Find the LCM for each pair of polynomials.

22. $27a^2bc, 36ab^2c^2$
23. $3m - 1, 6m - 2$
24. $x^2 + 2x + 1, x^2 - 2x - 3$

Lesson 11-7 Mixed Expressions and Complex Fractions (pp. 714–719)

Write each mixed expression as a rational expression.

1. $4 + \frac{2}{x}$
2. $8 + \frac{5}{3t}$
3. $3b + \frac{b+1}{2b}$
4. $3z + \frac{z+2}{z}$
5. $\frac{2}{a-2} + a^2$
6. $3r^2 + \frac{4}{2r+1}$

Simplify each expression.

7. $\frac{3\frac{1}{2}}{4\frac{3}{4}}$
8. $\frac{\frac{x^2}{y}}{\frac{y}{x^3}}$
9. $\frac{\frac{t^4}{u}}{\frac{t^3}{u^2}}$
10. $\frac{\frac{x-3}{x+1}}{\frac{x^2}{y^2}}$
11. $\frac{\frac{y}{3} + \frac{5}{6}}{2 + \frac{5}{y}}$
12. $\frac{\frac{1}{x} + \frac{1}{y}}{\frac{1}{y} - \frac{1}{x}}$
13. $\frac{\frac{t-2}{t^2-4}}{t^2 + 5t + 6}$
14. $\frac{a + \frac{2}{a+1}}{a - \frac{3}{a-2}}$

Lesson 11-8 Rational Equations (pp. 720–727)

Solve each equation. State any extraneous solutions.

1. $\frac{k}{6} + \frac{2k}{3} = -\frac{5}{2}$
2. $\frac{2x}{7} + \frac{27}{10} = \frac{4x}{5}$
3. $\frac{18}{b} = \frac{3}{b+3}$
4. $\frac{3}{5x} + \frac{7}{2x} = 1$
5. $\frac{2a-3}{6} = \frac{2a}{3} + \frac{1}{2}$
6. $\frac{3x+2}{x} + \frac{x+3}{x} = 5$
7. $\frac{2b-3}{7} - \frac{b}{2} = \frac{b+3}{14}$
8. $\frac{2y}{y-4} - \frac{3}{5} = 3$
9. $\frac{2t}{t+3} + \frac{3}{t} = 2$
10. $\frac{5x}{x+1} + \frac{1}{x} = 5$
11. $\frac{r-2}{r+2} - \frac{2r}{r+9} = 6$
12. $\frac{m}{m+1} + \frac{5}{m-1} = 1$
13. $\frac{2x}{x-3} - \frac{4x}{3-x} = 12$
14. $\frac{14}{b-6} = \frac{1}{2} + \frac{6}{b-8}$
15. $\frac{a}{4a+15} - 3 = -2$
16. $\frac{5x}{3x+10} + \frac{2x}{x+5} = 2$
17. $\frac{2a-3}{a-3} - 2 = \frac{12}{a+2}$
18. $\frac{z+3}{z-1} + \frac{z+1}{z-3} = 2$

Lesson 12-1 Designing a Survey (pp. 739–745)

Identify each sample, suggest a population from which it was selected, and state whether it is *unbiased* (random) or *biased*. If unbiased, classify the sample as *simple*, *stratified*, or *systematic*. If biased, classify as *convenience* or *voluntary response*.

1. The sheriff has heard that many dogs in the county do not have licenses. He checks the licenses of the first ten dogs he encounters.
2. Every fifth car is selected from the assembly line. The cars are also identified by the day of the week during which they were produced.
3. A table is set up outside of a large department store. All people entering the store are given a survey about their preference of brand for blue jeans. As people leave the store, they can return the survey.
4. A community is considering building a new swimming pool. Every twentieth person on a list of residents is contacted for their opinion.
5. A group of custom chopper builders are interested in finding the type of art work people prefer on their bikes. They line up a group of each of their choppers at a show to see which designs people appear to prefer.

Lesson 12-2 Analyzing Survey Results (pp. 746–755)

Tell which measure of center best represents the data. Justify your answer. Then find the measure of center.

1. **COMPUTER USAGE** To determine if they need to add more computers for internet access at the Grandview Library, the staff kept track of the number of patrons who used the computers each evening between 5PM and 8PM during a two-week period. {32, 12, 45, 38, 26, 29, 31, 22, 40, 20, 24, 27}

2. **CHOCOLATE MILK** The preschool cafeteria staff wanted to decide if they should continue to order chocolate milk for the children. For two weeks they kept a count of the number of half-pints of chocolate milk requested by the children. {21, 17, 17, 21, 18, 18, 21, 18, 19, 20}

Given the following portion of a survey report, evaluate the validity of the information and conclusion.

3. **ONLINE SURVEYS** An online survey was conducted by an advertising firm to determine how many potential customers they might reach. The firm sent out 10,000 surveys and received 2,300 online responses. **Question:** How many hours per week, outside of work time, do you spend on your computer? **Results:** 0–2, 4%; 3–6, 20%; 7–12, 32%; 13–20, 20%; 21–30, 12%; more than 30, 12%. **Conclusion:** One-third of all adults spend between 7 and 12 hours each week on their home computers.

Lesson 12-3 Statistics and Parameters (pp. 756–762)

Identify the sample and the population for each situation. Then describe the sample statistic and the population parameter.

1. As part of their quality control program, the Venice Pizza Parlor called 125 of its delivery customers to see if they were satisfied with the delivery service. The percentage of those responding YES is calculated.

2. A random stratified sample of 1,200 high school band members from across the country is surveyed about how many hours they spend practicing each week during football season.

Find the mean, variance, and standard deviation of each set of data.

3. {7, 10, 22, 15, 11}
4. {11, 16, 20, 17}
5. {45, 40, 54, 67, 44}

Lesson 12-4 Permutations and Combinations (pp. 764–770)

Determine whether each situation involves a permutation or combination. Explain your reasoning.

1. three topping flavors for a sundae from ten topping choices
2. selection and placement of four runners on a relay team from 8 runners
3. five rides to ride at an amusement park with twelve rides
4. first, second, and third place winners for a 10K race

Evaluate each expression.

5. $P(5, 2)$
6. $P(7, 7)$
7. $C(10, 2)$
8. $C(6, 5)$

Extra Practice

Lesson 12-5 Probability of Compound Events (pp. 771–778)

1. **SIBLINGS** Perry took a survey of his classmates to see how many siblings each had. The results are in the table.
   a. Find the probability that a randomly chosen classmate will have 4 siblings.
   b. Find the probability that a randomly chosen classmate will have less than 3 siblings.

Classmates' Siblings	
Number of Siblings	Number of Classmates
0	3
1	8
2	4
3	5
4	3
5	1

2. **CARS** A customer visiting a new car lot was looking for a convertible. The Buyers Co. lot had 8 black, 6 white, 4 silver, 5 red and 7 green convertibles to consider.
   a. Find the probability that a randomly chosen car will be silver.
   b. Find the probability that a randomly chosen car will be red or black.

Lesson 12-6 Probability Distributions (pp. 779–786)

Toss 4 coins, one at a time, 50 times and record your results.

1. Based on your results, what is the probability that any two coins show tails?
2. Based on your results, what is the probability that the first and fourth coins show heads?
3. What is the theoretical probability that all four coins show heads?

Use the table that shows the results of a survey about household occupancy.

Number in Household	Number of Households
1	172
2	293
3	482
4	256
5 or more	148

4. Find the experimental probability distribution for the number of households of each size.
5. Based on the survey, what is the probability that a person chosen at random lives in a household with five or more people?
6. Based on the survey, what is the probability that a person chosen at random lives in a household with 1 or 2 people?

Lesson 12-7 Probability Simulations (pp. 787–792)

Determine whether the events are independent or dependent. Then find the probability.

1. A bag of colored candies contains 12 red, 5 brown, 7 orange, 6 blue, 4 green and 6 purple candies. Alex reaches in the bag and pulls out one piece of candy and then a second piece without replacement. What is the probability that the first two choices are blue?
2. Gretchen and her friends are having a sleepover and movie marathon. They have rented four movies: a comedy, a horror movie, a mystery and a romance. What is the probability that they will watch a mystery and then a romance?
3. For a special game, the sides on a number cube are colored red for the odd numbers, 1, 3, and 5, and green for the even numbers, 2, 4 and 6. What is the probability on two rolls of the cube of getting a green side, followed by a 4?

Mixed Problem Solving

Chapter 1 Expressions, Equations, and Functions (pp. 2–71)

1. **GEOMETRY** The area of a rectangle is the product of the length ℓ and the width w. (Lesson 1-1)

   a. Write an expression for the area of a rectangle.

   b. Write an expression in terms of ℓ for the area of a rectangle that is twice as wide as it is long.

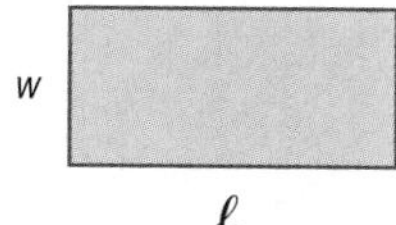

2. **SIGNS** A restaurant has a cylindrical shaped sign made to look like a bucket of their fried chicken. The volume of a cylinder can be found by the product of the radius squared, the height of the cylinder and π.

   a. Write an expression for the volume of a cylinder.

   b. Evaluate the volune of the sign if the radius is 3 feet and the height is 6 feet.

   c. For a bill board on the freeway, the dimensions of the sign were changed to a 9 foot diameter and a height of 9 feet. Find the volume of the bucket on the billboard. (Lesson 1-2)

3. **TRAVEL** Ticket prices for an amusement park are shown in the table below. There are two adults and three children in the Careen family. The children's ages are 11, 8, and 4. (Lesson 1-3)

   a. Write an expression for the cost c of the day at the amusement park for the family.

   b. Find the cost of the day at the amusement park.

   c. What is the cost per day of the vacation if the family stayed three days?

   d. What is the cost per day of the vacation if the family stayed seven days?

Length of Vacation	1-Day	3-Day	7-Day
Guests (Ages 10+)	$71	$203	$219
Guests (Ages 3–9)	$60	$171	$182

4. **FOOD** The cafeteria has the following menu.

Menu	
Items	Cost ($)
pizza	2.75
sandwich	1.50
fries	0.75
drink	1.25

Charlie gets lunch for himself and two friends. Charlie wants pizza and a drink. His friends each have a sandwich, fries, and a drink. (Lesson 1-4)

   a. Write and evaluate an expression to find the total cost.

   b. How much would it cost if Charlie bought the same thing as his friends?

5. **DISTANCE** In a given amount of time, Jena drove twice as far as Rita. Altogether they drove 120 miles. Find the number of miles driven by Jena. (Lesson 1-5)

6. **GRAPH** Describe what is happening in the graph. The graph represents Nikki's trip to the store. (Lesson 1-6)

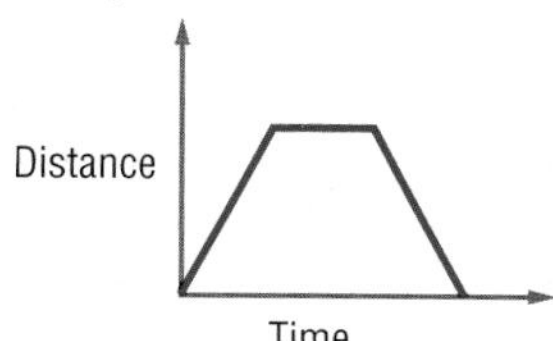

7. Is the following relation a function? Explain. (Lesson 1-7)

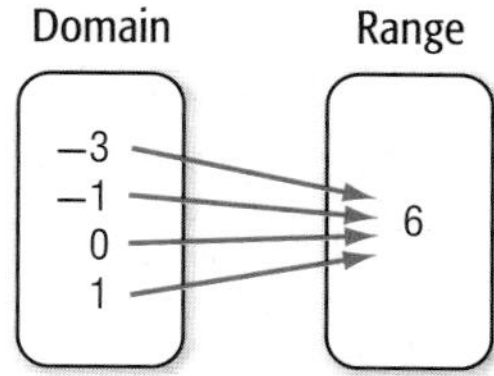

8. Find a counterexample for the following statement. *If you read music, you sing in a choir.* (Lesson 1-7)

9. Identify the hypothesis and conclusion of the following statement. Then write the statement in if-then form. *Mei and Janel will go to the mall after school.* (Lesson 1-7)

1. **BABY-SITTING** As a part-time job, Andrea babysits. She charges $4 per hour and $5 for gas money. Write an equation that could be used to find how much she will make if she babysits for 4 hours. (Lesson 2-1)

2. **WORLD RECORD** At 107 inches, Robert Pershing Wadlow was the tallest man to ever live. If the average 15-year-old male is 68 inches tall, write and solve an equation to find the number of inches taller Mr. Wadlow was than the average 15-year-old male. (Lesson 2-1)

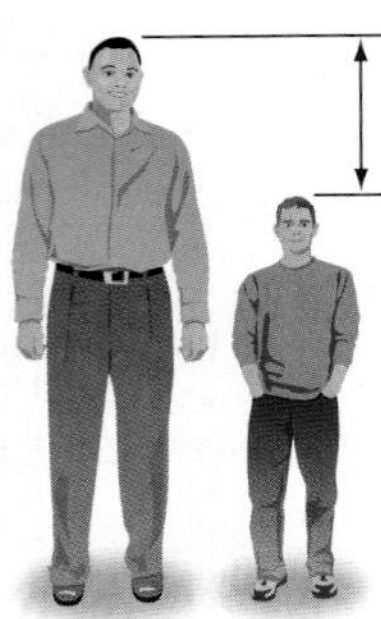

3. Three fifths of Marcus's classmates have brown eyes. There are 18 students who have brown eyes. Determine how many students are in the class. (Lesson 2-2)

4. **LOCKER NUMBERS** Andy, Kayla, and Samantha have 3 lockers in a row. The lockers are numbered consecutively and the sum of the locker numbers is 756. Write and solve an equation to find all three locker numbers. (Lesson 2-3)

5. **BEAD WORK** Sarah said, "I counted my beads, and I have three more than 4 times a certain mystery number." Jasmine had the same amount of beads as Sarah, but said she had 9 less than 5 times the mystery number. They challenged their art teacher to discover the mystery number. (Lesson 2-4)
   a. What equation could the art teacher use to find the number?
   b. What is the mystery number?
   c. How many beads does each girl have?

6. **DRIVING** A car has an average fuel economy of 23 miles per gallon. The car gets 4 miles per gallon fewer in city driving and 4 miles per gallon more in highway driving. Write and solve an absolute value equation to find the fuel economy of the car in city and highway driving. (Lesson 2-5)

7. **LAWN CARE** José's mom tells him she will pay him $20 if he mows the lawn in an hour, plus or minus 15 minutes. If it takes José 73 minutes to mow the lawn, will his mom pay him? Explain your reasoning. (Lesson 2-5)

8. **ARCHITECTURE** Isaac drew the following floor plan of his room. Write and solve a proportion to find the actual length of his room if the actual width is 12 meters. (Lesson 2-6)

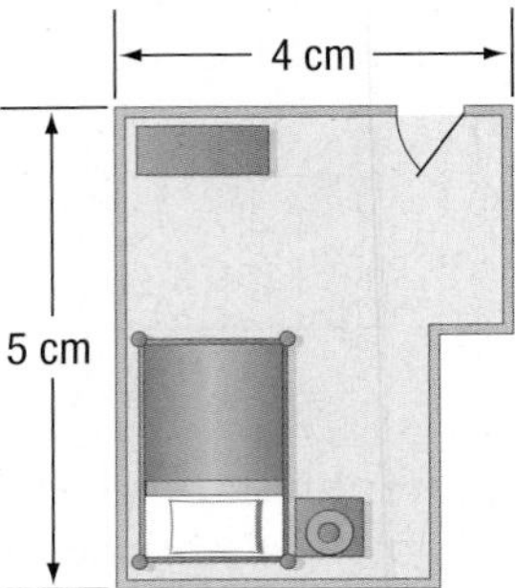

9. **CARS** Daniel's father purchased a vintage car when it was new. He paid $3334. Last year, he sold it for $28,250. (Lesson 2-7)
   a. Determine the percent of change to the nearest percent.
   b. Is this a percent of *increase* or *decrease*? Explain.

10. **WEATHER** Emily went to Canada for vacation. Listening to the weather report, she heard that the high for the day was going to be 23°C. She remembered that the formula for changing Fahrenheit to Celsius was $C = \frac{5}{9}(F - 32)$. (Lesson 2-8)
    a. Solve the formula for F.
    b. Use the new formula to find that day's high temperature in degrees Fahrenheit.
    c. If the low temperature for the day was 16°C, find the day's low in °F.

11. **TRANSPORTATION** Hector rode his bike to Matthew's house. It took him 10 minutes to ride the 3-mile trip. When it was time to leave, he saw that his tire had gone flat and he would need to walk home. It took him 45 minutes to walk home. What was his average speed for the round trip? Express your answer in miles per hour rounded to the nearest tenth. (Lesson 2-9)

1. **MOVIES** Tickets for the movies cost $5 for students and $8 for adults. The equation $5x + 8y = 80$ represents the number of students x and adults y who can attend a movie for $80. (Lesson 3-1)

   a. Use the x- and y-intercepts to graph the equation.

   b. Describe what these values mean in the context of this situation.

2. **FUNDRAISING** Shawn's class is selling candy to raise money for a class trip. They paid $55 for the candy, and they are selling each candy bar for $1.75. The function $y = 1.75x - 55$ represents their profit y for selling x candy bars. (Lesson 3-2)

   a. Find the zero of the function.

   b. Describe what the zero means in the context of this situation.

   c. Graph the function.

3. **PHOTOS** The average cost of using an online photo finisher decreased from $0.50 per print to $0.27 per print between 2002 and 2005. Find the average rate of change in the cost. Explain what the rate of change means. (Lesson 3-3)

4. **WAGES** The weekly salary Enrique earns w varies directly as the hours h that he works. (Lesson 3-4)

   a. Write this situation as an equation.

   b. Graph the equation if Enrique makes $11.50 an hour.

   c. How many hours does Enrique have to work to earn $690?

5. **MONEY** The table represents Tiffany's income. (Lesson 3-5)

Hours Worked	Income ($)
1	20.50
2	29.00
3	37.50
4	46.00

   a. Write an equation for this sequence.

   b. Use the equation to find Tiffany's income if she works 20 hours.

6. **SAVINGS** Payat has $680 in a savings account. He makes a deposit after he receives each paycheck. After one month he has $758 in the account. The next month the balance is $836. The balance after the third month is $914. (Lesson 3-5)

   a. Write a function to represent the arithmetic sequence.

   b. Graph the function.

   c. How much will Payat have in his savings account at the end of six months?

7. **SALES** Percy sells cell phones in the mall. In addition to his salary, he receives a bonus for each cell phone that he sells. The table shows the number of phones he sells and the amount of his bonus. (Lesson 3-6)

Number of Phones	Bonus Pay ($)
1	65
2	130
3	195
4	260
5	325

   a. Graph the data.

   b. Write an equation to describe this relationship.

   c. Determine the amount of his bonus if Percy sells 10 phones.

8. **STRAWBERRIES** The table below shows the cost of picking your own strawberries at a local farm. (Lesson 3-6)

Number of Pounds	Total Cost ($)
1	1.25
2	2.50
3	3.75
4	5.00

   a. Graph the data.

   b. Write an equation in function notation to describe this relationship.

   c. How much would 6 pounds of strawberries cost if you picked them yourself?

   d. How much would 9.5 pounds of strawberries cost if you picked them yourself?

1. **FINANCES** Brandon is saving money to purchase a new video game system. He has \$94 and plans to save \$7 each week for the next several weeks. (Lesson 4-1)
   a. Write an equation for the total amount S that he has saved after w weeks.
   b. Graph the equation.
   c. Find out how much Brandon will have saved after 8 weeks.

2. **PHOTOGRAPHY** An online photo processing company produces printed and bound photo albums. The standard price for an album includes up to 20 pages. There is a fee for every two additional pages. To make a 30-page book, you pay \$29.94. To make a 42-page book, you pay \$41.88. (Lesson 4-2)
   a. What is the cost for each additional 2 pages?
   b. What is the cost of a 20-page book?

3. **FITNESS** The graph shows the cost of a membership to the local gym. Write an equation in point-slope form to find the total price y for any number of months x. (Lesson 4-3)

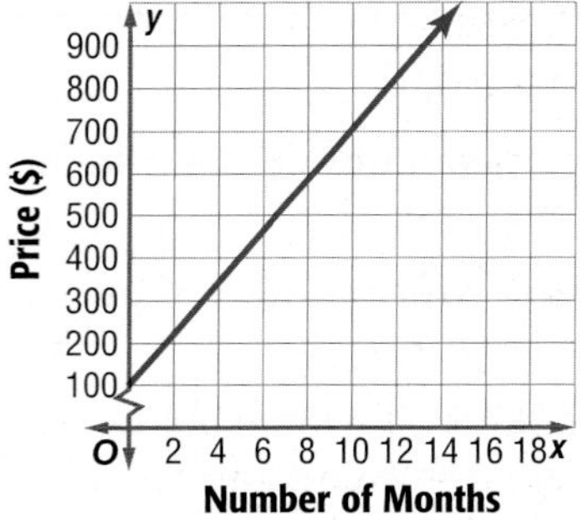

4. **MAPS** The map below shows Interstate Highways 80 and 70 through a portion of Wyoming and Colorado. (Lesson 4-4)

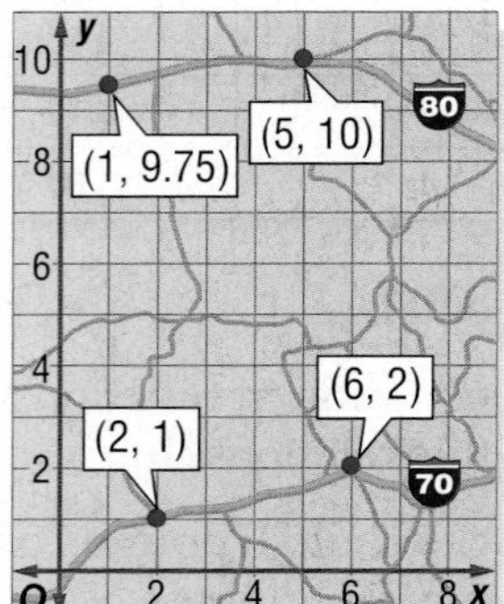

   a. Find the slope of the lines that approximate the path of each highway.
   b. Are the highways parallel?

5. **PLANETS** The table shows the distance between selected planets and the Sun and the diameters of the planets. Draw a scatter plot and determine what relationship exists, if any, in the data. (Lesson 4-5)

Planet	Distance from Sun (10^6 mi)	Diameter (mi)
Mercury	36.0	3032
Venus	67.2	7521
Earth	93.0	7926
Mars	141.6	2159

6. **ATTENDANCE** Rodeo Houston is an event that features rodeo events, a Bar-B-Que competition, concerts by top performers, art auction and contributes scholarships to youth organizations. The attendance for the rodeo championships are listed in the table below.

Year	2002	2004	2006	2008
Attendance	68,266	70,668	72,867	71,165

   a. Find an equation for the median-fit line.
   b. According to the equation, how many attended the rodeo's championship in 2009?

7. **PARKING** Mika was visiting the state library. The fees for parking are shown in the table. (Lesson 4-7)

Time	Cost
15 minutes or less	free
first hour	\$0.50
each addtional hour or partial hour	additional \$1.00
maximum daily rate	\$5.00

   a. Draw a graph that shows the parking fee y for the time spent parked x.
   b. If Mika visited the state library for three and a half hours, how much did she pay for parking?

1. **SHOPPING** Jeff is buying a new car but owes \$3000 on his old one. Jeff can spend no more than \$18,000 to pay off his old car and buy a new one. (Lesson 5-1)
   a. Write an inequality to show how much Jeff can spend on his new car.
   b. Solve the inequality.

2. **TOMATOES** There are more than 10,000 varieties of tomatoes. One seed company produces seed packages for 200 varieties of tomatoes. For how many varieties do they not provide seeds? (Lesson 5-1)

3. **SURVEYS** Of the students at Davidson High School, fewer than 92 students said they like strawberry ice cream. This is about one sixth of those surveyed. (Lesson 5-2)
   a. Write an inequality to represent this situation.
   b. How many students were surveyed?

4. **ARCHITECTURE** The dimensions of Lisle's room are shown below. If the area of the room is at least 96 square feet, what is the least width the room could have? (Lesson 5-2)

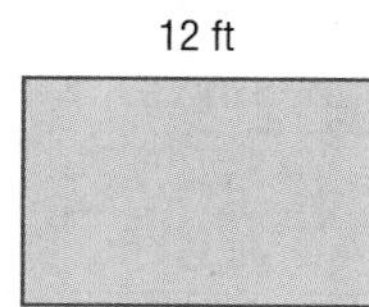

   a. Define a variable.
   b. Write an inequality.
   c. Solve the problem.

5. **SAVINGS** Ramone is raking yards for \$22 per yard to earn money for a car. So far he has \$2150 saved. The car that Ramone wants to buy costs at least \$8290. (Lesson 5-3)
   a. Write an inequality to show how many more yards Ramone still needs to rake to earn enough money to buy the car.
   b. Solve the inequality.

6. **CHARITY** Avery High School holds a walk-a-thon each fall to raise money for charity. This year they want to raise at least \$375. Each student earns \$0.75 for every half mile walked. How many miles will the students need to walk? (Lesson 5-3)

7. **SALES** A school buys 500 T-shirts. In addition to the price per shirt, there is a \$45 set-up fee. The school can afford to spend no more than \$2295. (Lesson 5-3)
   a. Write an inequality to show the relationship.
   b. What must the price be for the school to afford the shirts?

8. **FISH TANK** The temperature in a fish tank must be at least 77°F and at most 83°F. (Lesson 5-4)
   a. Write a compound inequality that describes acceptable temperatures for a fish tank.
   b. Graph the inequality.

9. **MOVIES** A recent survey showed that 73% of young adults had seen a movie recently. The margin of sampling error was within 5 percentage points. Find the range of young adults who saw a movie recently. (Lesson 5-5)

10. **SNOW** When the temperature in the clouds is 7°F plus or minus 3°F, star-shaped crystals of snow form. At 14°F plus or minus 4°F, plate-shaped crystals are formed. (Lesson 5-5)
    a. Find the range of temperature that produces star-shaped crystals.
    b. Find the range of temperature that produces plate-shaped crystals.

11. **JOBS** It takes a librarian 1 minute to renew a library card and 3 minutes to make a new card. Together, she can spend no more than 30 minutes renewing and making cards. Write an inequality to represent this situation, if x is the number of cards she renews and y is the number of new cards she makes. (Lesson 5-6)

12. **MOVING** A moving company charges \$95 an hour and \$0.08 per mile. If Brianna has only \$500 for moving expenses, can she afford to hire this moving company if it will take 5 hours and the houses are 75 miles apart? (Lesson 5-6)

13. **DELIVERIES** A delivery truck is transporting the items shown. (Lesson 5-6)

Item	Weight (lb)
television	48
computer	21

If the delivery truck has a 2500-pound weight limit, will the truck be able to deliver 30 televisions and 45 computers? Explain.

Mixed Problem Solving

1. **FUNDRAISERS** The French Club is selling heart-shaped lollipops to raise money for a trip to Quebec. They paid \$20 for the candy mold, and the ingredients for each lollipop cost \$0.50. They plan to sell the lollipops for \$1.00. (Lesson 6-1)

   a. Write an equation for the cost of supplies y for the number of lollipops sold x, and an equation for the income y for the number of lollipops sold x.

   b. Graph each equation.

   c. How many lollipops do they need to sell before they begin to make a profit?

2. **COMPUTER REPAIR** An in-home computer repair provider charges \$40 per hour for installations and \$60 per hour for troubleshooting and repair. Last week the busiest employee worked 40 hours and brought in \$2100. How many hours did the employee spend doing each type of service? (Lesson 6-2)

3. **TOURS** The Snider family and the Rollins family are traveling together on a trip to visit a candy factory. The number of people in each family and the total cost are shown in the table below. Find the adult admission price and the children's admission price. (Lesson 6-3)

Family	Number of Adults	Number of Children	Total Cost
Snider	2	3	\$58
Rollins	2	1	\$38

4. **GEOGRAPHY** In 2004, the state capital with the smallest population was Montpelier, Vermont. The state capital with the largest population was Phoenix, Arizona. The difference between the two populations was 1,410,006. The population of Phoenix was 3,881 more than 176 times as large as the population of Montpelier. Find the population of each city. (Lesson 6-4)

5. **CANOEING** A canoe travels 10 miles upstream in 2.5 hours. The return trip takes the canoe 2 hours. Find the rate of the boat in still water and the rate of the current. (Lesson 6-4)

6. **FAIR** At a county fair, the cost for 4 slices of pizza and 2 orders of French fries is \$21.00. The cost of 2 slices of pizza and 3 orders of French fries is \$16.50. To find out how much a single slice of pizza and an order of French fries costs, determine the best method to solve the system of equations. Then solve the system. (Lesson 6-5)

7. **OFFICE SUPPLIES** At a sale, Ricardo bought 24 reams of paper and 4 inkjet cartridges for \$320. Britney bought 2 reams of paper and 1 inkjet cartridge for \$50. The reams of paper were all the same price and the inkjet cartridges were all the same price. A system of equations can be used to represent this situation. Determine the best method to solve the system of equations. Then solve the system. (Lesson 6-5)

8. **ELEVATIONS** The highest and lowest elevations for several states are shown in the table below. (Lesson 6-6)

State	Highest Elevation (ft)	Lowest Elevation (ft)
Alaska	20,320	0
California	14,494	−282
Colorado	14,433	3,315
Hawaii	13,796	0
Louisiana	535	−8
Wyoming	13,804	3,099

   a. Write a matrix to organize the given data.

   b. What are the dimensions of the matrix?

   c. Which state has the highest elevation? the lowest elevation?

9. **KENNEDY SPACE CENTER** Admission to the Kennedy Space Center for 2 adults and 3 children costs \$160. Admission for 5 adults and 2 children is \$246. (Lesson 6-7)

   a. Write a system of linear equations to model the situation. Let a represent adult admission, and let c represent child admission.

   b. Write the augmented matrix.

   c. What is the price for adult and child admissions?

1. **POOLS** Find the area of the swimming pool below. (Lesson 7-1)

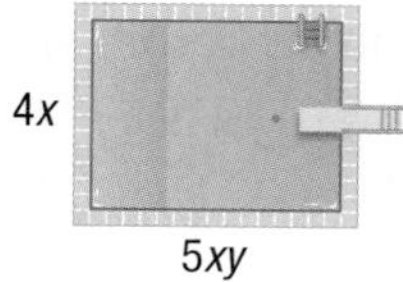

2. **GARDENING** Felipe is planting a flower garden that is shaped like a trapezoid. Use the formula $A = \frac{1}{2}h(b_1 + b_2)$ to find the area of Felipe's garden. (Lesson 7-1)

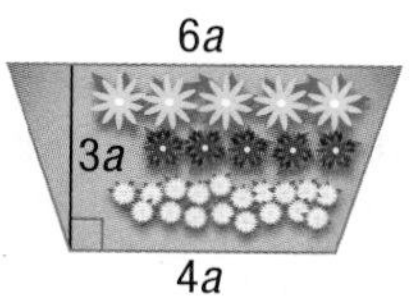

3. **GEOMETRY** The area of a rectangle is $36m^4n^6$ square meters. The length of the rectangle is $6m^3n^3$ meters. What is the width of the rectangle? (Lesson 7-2)

4. **ASTRONOMY** The order of magnitude of the mass of Earth is about 10^{24}. The order of magnitude of the mass of the Moon is about 10^{22}. How many orders of magnitude as big is the Earth as the Moon? (Lesson 7-2)

5. **MAMMALS** A blue whale has been caught that was 4.2×10^5 pounds. The smallest mammal is a bumblebee bat, which is about 0.0044 pound. (Lesson 7-3)

   a. Write the whale's weight in standard form.

   b. Write the bat's weight in scientific notation.

   c. How many orders of magnitude as big is a blue whale as a bumblebee bat?

6. **GARDENS** Megan had a square garden. She expanded this garden by 3 feet in one direction and 5 feet in the other. (Lesson 7-4)

   a. Write a polynomial to describe the area of the new garden if the old garden was x feet wide.

   b. If the original width was 12 feet, find the area of the new garden.

   c. Find the area of the new garden if the original width was 16.

7. **GEOMETRY** Write a polynomial that represents the area of the figure. (Lesson 7-5)

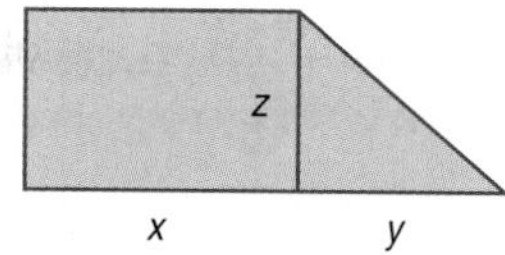

8. **GEOMETRY** Find the perimeter and area of the rectangle shown below. (Lesson 7-5)

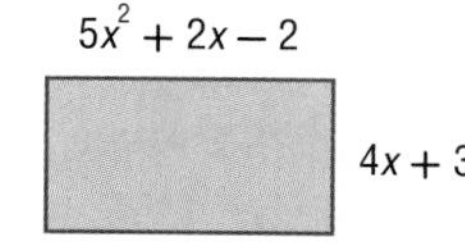

9. **SHOPPING** Nicole bought x shirts for \$15.00 each, y pairs of pants for \$25.72 each, and z belts for \$12.53 each. Sales tax on these items was 7%. Write an expression to find the total cost of Nicole's purchases. (Lesson 7-6)

10. **MANUFACTURING** A company is designing a box for dry pasta in the shape of a rectangular prism. The length is 2 inches more than twice the width, and the height is 3 inches more than the length. Write an expression for the volume of the box. (Lesson 7-7)

11. **GEOMETRY** Write an expression for the area of the trapezoid shown. (Lesson 7-11)

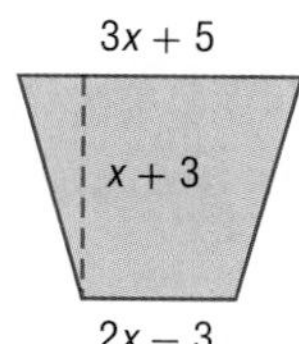

12. **CARPENTRY** Miguel's room is x feet on each side. He adds book shelves that are 2 feet deep to two adjacent walls. (Lesson 7-8)

    a. Show how the new area of the floor space can be modeled by the square of a binomial.

    b. Find the square of this binomial.

13. **TOYS** A flying disk shaped like a circle has a radius of $x - 4$ centimeters. (Lesson 7-8)

    a. Write an expression representing the area of the flying disk.

    b. If x is 15, what is the area of the flying disk?

1. **CANDY** Sada was packing candy into gift bags. She has 42 chocolate truffles, 96 pieces of taffy, and 108 jelly beans. Sada wants to package the same number of candies in each bag, and each bag should have every type of candy. (Lesson 8-1)

   a. If she puts the greatest possible number of candies in each bag, how many bags can she make?

   b. How many pieces of each type of candy will be in each bag?

2. **VOLUNTEERING** Catalina's class collected soap, washcloths, and toothbrushes to give to several homeless shelters in the area. The number of each item collected is shown in the table below. (Lesson 8-1)

Item	Number
soap	84
washcloth	24
toothbrush	72

   a. If she puts the greatest possible number of items in each box, how many boxes can she make?

   b. How many of each item will be in each box?

3. **SOCCER** Jorge is kicking a soccer ball straight up into the air. The height of the ball is described by the equation $h = -16r^2 + 20t$, where h is the height of the ball and t is the time in seconds. How long will it take the ball to hit the ground? (Leson 8-2)

4. **PHOTOGRAPHY** Kelsey has a 4-inch by 6-inch photograph she wants to frame with a mat. The area of the picture and mat is twice as large as the area of the picture itself. If she wants the mat to be the same width on all sides, what are the outside dimensions of the mat? (Lesson 8-3)

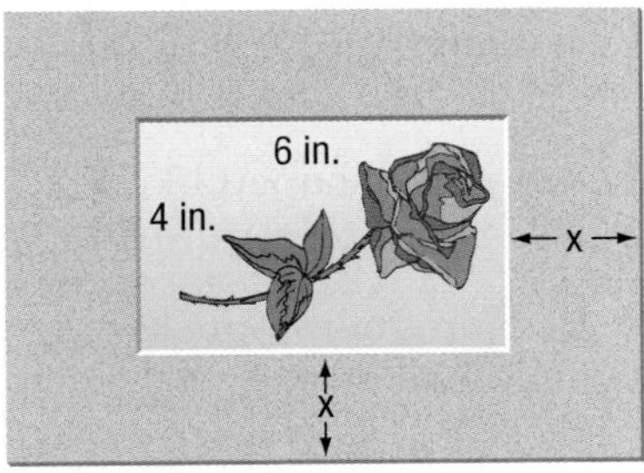

5. **BALLET** A ballet dancer is practicing a leap from the stage to an upper deck. If her initial velocity as she leaps is 18 feet per second, how long is she in the air before she lands on the upper deck, 5 feet above the lower deck? (Lesson 8-4)

6. **GRAPHICS** A computer cartoonist uses the equations $y = x^2 - 4$ and $y = 4 - x^2$ as the basis for his drawing of a fish. The graphs of the two equations intersect at the roots of the equations. Find the roots. (Lesson 8-5)

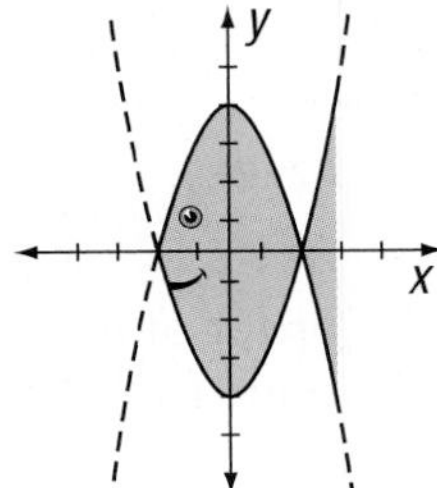

7. **STORAGE** Chase is building a box for the shelves in his father's garage. In order to fit the shelves as neatly as possible, the box needs to be 1 inch taller and 5 inches longer than it is wide. (Lesson 8-6)

   a. Let w represent the width of the box. Write an expression for each remaining dimensions in terms of w.

   b. If the volume of the box is 2112 cubic feet, what are the dimensions of the box?

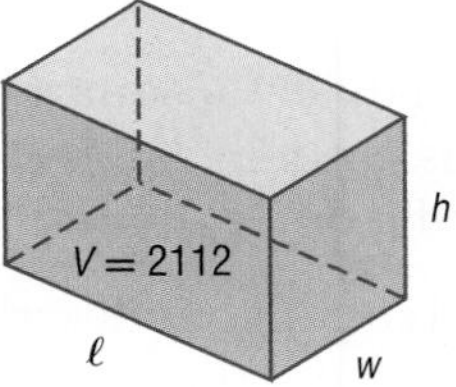

8. **RECREATION** Danille and Pete make a slingshot to shoot water balloons. If a water balloon shoots up with an initial velocity of 64 feet per second, how long is it until it hits the ground? (Lesson 8-6)

1. **PHYSICAL SCIENCE** A model rocket is launched with a velocity of 64 feet per second. The equation $h = -16t^2 + 64t$ gives the height of the rocket t seconds after it is launched. (Lesson 9-1)

   a. Graph the function.

   b. What is the maximum height that the rocket reaches?

   c. How many seconds is the rocket in the air?

2. **FIREWORKS** Some fireworks are fired vertically into the air from the ground at an initial velocity of 80 feet per second. The equation $h(t) = -16t^2 + 80t$ models the height h, in feet, of the fireworks after t seconds. Find the highest point reached by the projectile just as it explodes. (Lesson 9-1)

3. **BASEBALL** The International Space Agency has landed a robotic explorer on a planet. The robot launches a baseball directly upward at 147 ft/s. The equation for the path of the baseball is $h = -49t^2 + 147t$, where h is the height of the baseball in feet and t is time in seconds. Assuming no wind, how long will it take for the ball to land on the surface? (Lesson 9-2)

4. **DESIGN** For a design competition, Maggie dropped a device from the bleachers. The equation of the device's height in feet h after t second is $h = -16t^2 + 20$. Compare the graph of this function with its parent graph. (Lesson 9-3)

5. **PHYSICS** An object is launched at 17.2 m/s from a 25-meter tall platform. The equation for the object's height h at time t seconds after launch is $h(t) = -4.9t^2 + 17.2t + 25$, where h is in meters. (Lesson 9-4)

   a. Graph this equation.

   b. When does the object strike the ground?

6. **GEOMETRY** The area of a square can be tripled by increasing its length by 6 centimeters and increasing its width by 3 centimeters.

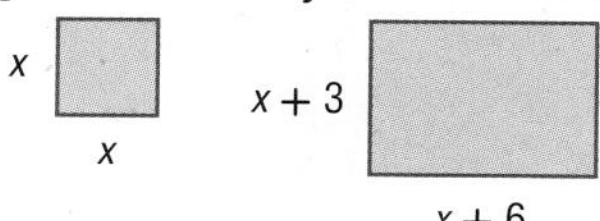

   What is the length of the side of the square? (Lesson 9-4)

7. **FENCE** Jackie and Ken are building a fence around a rectangular field. They want to enclose an area of 75 square feet. The width should be 3 feet longer than the length of the field. What are the dimensions of the field? Round to the nearest tenth, if necessary. (Lesson 9-5)

8. **VOLLEYBALL** A volleyball is thrown straight up. The equation that describes its motion is $h(t) = -16t^2 + 48t + 4$, where h represents the height in feet and t represents time in seconds. How long will it take for the ball to hit the ground? Round to the nearest hundredth, if necessary. (Lesson 9-5)

9. **BIOLOGY** The function $f(t) = 50(1.07)^t$ models the growth of a fly population, where $f(t)$ is the number of flies and t is time in days. After three weeks, approximately how many flies are in this population? (Lesson 9-6)

10. **INVESTMENT** Nicholas invested \$2000 with a 5.75% interest rate compounded monthly. How much money will Nicholas have after 5 years? (Lesson 9-7)

11. **RESTAURANTS** The total restaurant sales in the United States increased at an annual rate of about 5.2% between 1996 and 2004. In 1996, total sales were \$310 billion. (Lesson 9-7)

    a. Write an equation for the average sales per year t years after 1996.

    b. Predict the total restaurant sales in 2012.

12. **TENNIS** Each year a local country club sponsors a tennis tournament. Play starts with 256 participants. During each round, half of the players are eliminated. How many players remain after 6 rounds? (Lesson 9-8)

13. **CAR CLUB** The table shows the number of car club members for four consecutive years after it began. (Lesson 9-9)

Time (years)	0	1	2	3	4
Members	10	20	40	80	160

    a. Determine which model best represents the data.

    b. Write a function that models the data.

    c. Predict the number of car club members after 6 years.

1. **PENDULUMS** The period of a pendulum is the time in seconds it takes to swing from one side to the other and back. If the length of the pendulum ℓ is given in meters, the period T is given by $T = 2\pi\sqrt{\frac{\ell}{g}}$, where g is the gravitational constant, 9.8 meters per second squared. The Foucault Pendulum at the Pantheon in Paris, France, has a length of 67 meters. Find the period of the Foucault Pendulum. (Lesson 10-1)

2. **KINETIC ENERGY** The speed v of a ball in meters per second can be determined by the equation $v = \sqrt{\frac{2k}{m}}$, where k is the kinetic energy in Joules and m is the mass of the ball in kilograms. (Lesson 10-2)

   a. Simplify the formula if the mass of the ball is 2 kilograms.

   b. If the ball is traveling 10 meters per second, what is the kinetic energy of the ball in Joules?

3. **WALLPAPER** Ruby is designing a wallpaper border using a pattern of two triangles. If the measurements given are in inches, find the exact length of the segment shown. Then find the length to the nearest quarter inch. (Lesson 10-3)

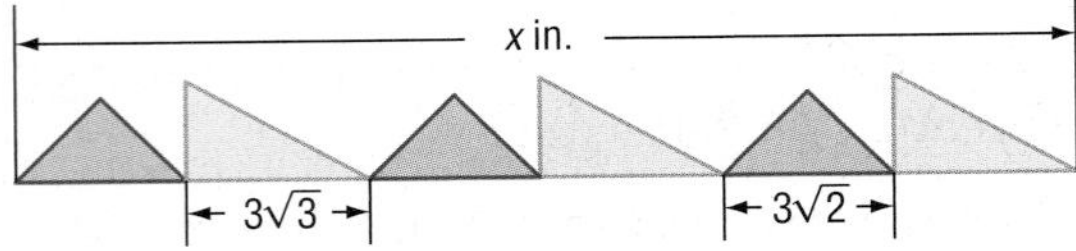

4. **SOUND** The speed of sound V in meters per second near Earth's surface is given by $V = 20\sqrt{t + 273}$, where t is the surface temperature in degrees Celsius. (Lesson 10-3)

   a. What is the speed of sound near Earth's surface at $-1°C$ and at $6°C$ in simplest form?

   b. Approximately how much faster is the speed of sound at $6°C$ than at $-1°C$?

5. **SKYDIVING** The approximate time t in seconds that it takes an object to fall a distance of d feet is given by $t = \sqrt{\frac{d}{16}}$. Suppose a parachutist falls 13 seconds before the parachute opens. How far does the parachutist fall during this time period? (Lesson 10-4)

6. **DELIVERY** Ben and Amado are delivering a freezer. The bank in front of the house is the same height as the back of the truck. They set up their ramp as shown. What is the length of the slanted part of the ramp to the nearest foot? (Lesson 10-5)

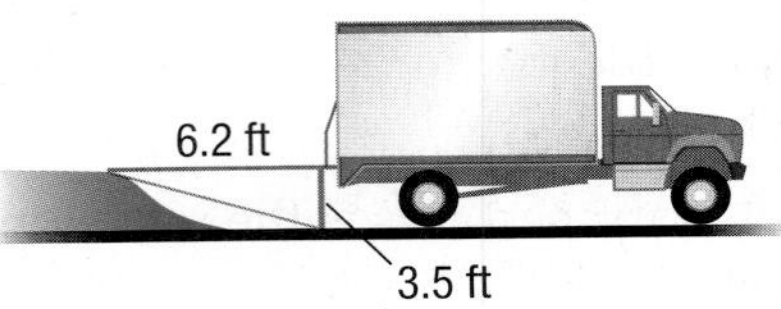

7. **LADDER** An 18-foot ladder is leaning against a building. For stability, the base of the ladder must be 3 feet away from the wall. How far up the wall does the ladder reach? (Lesson 10-5)

8. **SNOWMOBILES** Tai went snowmobiling with friends. His starting and ending points are shown on the graph. Use the Distance Formula to find how far he has to ride if he can ride straight home. (Lesson 10-6)

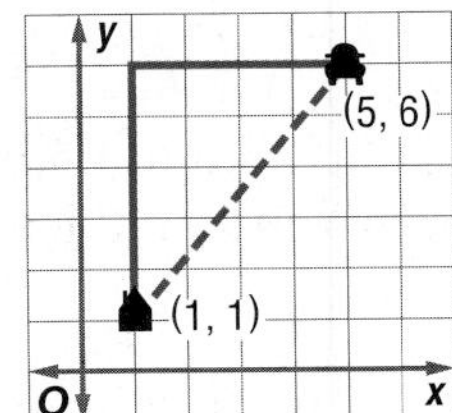

9. **FLAGS** Sarah needs to determine the height of the flag pole in her yard. On a sunny afternoon, Sarah's shadow is 2 feet 6 inches long. She is 5 feet 3 inches tall. The shadow of the flag pole is 6 feet long. How tall is the flag pole to the nearest tenth of a foot? (Lesson 10-7)

10. **SCHOOL** For a class project, Hailey needs to measure the height of her school. Hailey places a mirror 75 feet from the base of the school and backs 10 feet away from the mirror. If Hailey is 5 feet tall, how tall is the school? (Lesson 10-7)

11. **LIGHTHOUSE** How tall is the lighthouse? (Lesson 10-8)

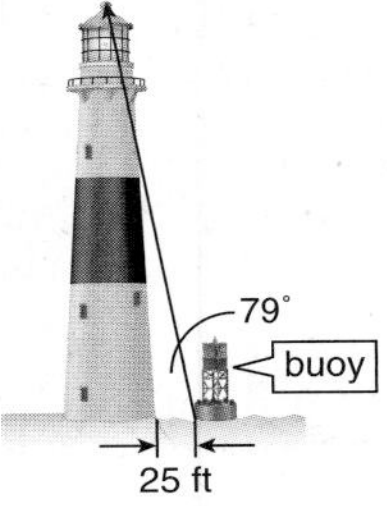

Mixed Problem Solving

Chapter 11 Rational Functions and Equations (pp. 666–735)

1. **GEOMETRY** The length of a rectangle with a constant area is inversely proportional to its width. A rectangle has a length of 10 inches and a width of 6 inches. Another rectangle, has the same area as the first rectangle, but its width is 3 inches. Find the length of the second rectangle. (Lesson 11-1)

2. **PHYSICS** Given a constant force, the acceleration of an object varies inversely with its mass. A shot putter uses a constant force to put a mass of 8 pounds with an acceleration of 10 feet per second squared. If the shot putter uses the same force on a 16-pound mass, what would be the acceleration? (Lesson 11-1)

3. **PHYSICS** A rectangle has an area of 36 square feet. The function $\ell = \frac{36}{w}$ shows the relationship between the length and width of the rectangle. Graph the function. (Lesson 11-2)

4. **RUNNING** A runner runs 12 miles each morning. Her average speed y is given by $y = \frac{12}{x}$, where x is the time it takes her to run 12 miles. (Lesson 11-2)

   a. Graph the function.

   b. Describe the asymptotes.

5. **PAINTING** The area of each wall in Isabel's room can be expressed as $2x^2 + 5x + 3$ square feet. A gallon of paint will cover an area that can be expressed as $x^2 - 3x - 4$ square feet. Write an expression that gives the number of gallons of paint that Isabel will need to buy to paint her room. (Lesson 11-3)

6. **TORTOISE** A giant tortoise can travel 0.17 mile per hour. What is this speed in feet per minute? (Lesson 11-4)

7. **EXCHANGE RATE** A pair of shoes bought in France cost 125 euros. The exchange rate at the time was 1 U. S. dollar = 0.68 euro. How much did the shoes cost in U. S. dollars? (Lesson 11-4)

8. **GAS** A motorcycle travels 225 kilometers with 5 liters of gas. How many liters of gas are needed to travel 135 kilometers? (Lesson 11-4)

9. **HOUSING** The total cost per month for a dorm is $2250 split equally among 15 students. If 10 more students join the dorm and each pays the same rate as each of the original 15 students, what will be the monthly expenditure? (Lesson 11-4)

10. **GEOMETRY** The area of a rectangle is $3x^2 - 6x - 24$ square units. What is the length? (Lesson 11-5)

    $A = 3x^2 - 6x - 24$ $x - 4$

11. **GARDENING** Trey planted a triangular garden. Write an expression for the perimeter of the triangle. (Lesson 11-6)

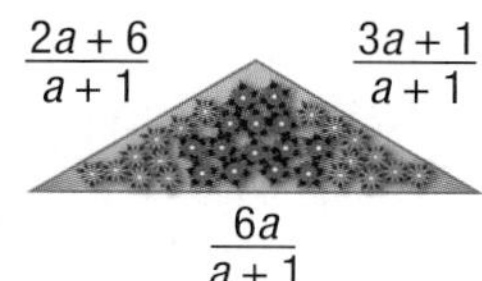

12. **TRACK** Trina walked 7 times around the track at a rate of t laps per hour. She ran around the track 10 times at a rate of $4t$ laps per hour. Write and simplify an expression for the total time it took her to go around the track 17 times. (Lesson 11-6)

13. **PLAY** A total of 1374 people attended three plays. Let p be the number of people who attended the third play. There were twice as many people at the second play than the third and three times as many people at the first play than the third. Write an expression to represent the fraction of people who attended the first and second plays. (Lesson 11-6)

14. **READING** Garcia reads $11\frac{1}{2}$ pages of a book in 8 minutes. What is his average reading rate in pages per minute? (Lesson 11-7)

15. **RACING** A drag racer drove $\frac{1}{4}$ mile in about $11\frac{3}{4}$ seconds. What is the speed in miles per hour? (Lesson 11-7)

16. **CAR REPAIR** Jack and Ivan are working on Ivan's car. Jack can complete the job in 6 hours, but Ivan will take 8 hours. They work together for two hours and Jack has to go. How long will it take Ivan to finish his car? (Lesson 11-8)

1. **CANDY** Alicia asked every fourth student that came in the classroom to choose their favorite from three types of candy. (Lesson 12-1)
   a. Identify the sample, and suggest a population from which it was selected.
   b. Classify the type of data collection used.
   c. Identify the sample as *biased* or *unbiased*. Explain your reasoning.
   d. Classify the sample as *simple*, *stratified*, or *systematic*. Explain your reasoning.

2. **HOUSING** A listing was made of the rents paid for apartments in a particular neighborhood: {$450, $590, $650, $520, $480, $800, $720, $1600, $600}. Which measure of central tendency best represents the data? Justify your selection, and then find the measure. (Lesson 12-2)

3. **CLUBS** Given the following portion of a survey report, evaluate the validity of the information and conclusion. (Lesson 12-2)

   Question: Should the media club meet before or after school?

   Sample: Students were randomly given an invitation to vote while waiting in the school commons for school to start.

   Results: 68% before school, 24% after school, 8% no preference

   Conclusion: The media club should meet before school.

4. **GRADES** Lynette took 5 tests in life science this grading period. Find the mean, variance, and standard deviation of the grades she earned: {76, 88, 82, 91, 78}. (Lesson 12-3)

5. **FOOTBALL** The following table shows information about the number of carries a running back had over a number of years. Find the mean absolute deviation of the number of carries. (Lesson 12-3)

Year	Number of Carries	Yards per Carry	Yards	Touchdowns
2002	40	3.7	148	0
2003	90	4.2	378	2
2004	105	4	420	3
2005	115	3.3	379.5	7
2006	140	4.3	602	9

6. **BAND** In a high school band, six girls and four boys play trumpets. Before auditions at the beginning of the year, the new band director randomly assigns chairs to the students. (Lesson 12-4)
   a. How many ways can the band director assign first chair, second chair, and third chair?
   b. What is the probability that the first three chairs will be assigned to boys?

7. **CARDS** If you have drawn and kept the 8 of spades and the 7 of hearts from a standard deck of cards, what is the chance you will draw another 7 or a spade next? (Lesson 12-5)

8. **COOKING** Nate has a shelf with 3 cans of green beans, 2 cans of corn, and 5 cans of peas and a shelf of 1 package each of egg noodles, rice and ziti. In the freezer, he has 2 pounds of chicken and 3 pounds of beef. If he randomly grabs one item from each area to put in a casserole, what is the probability that he makes a chicken-corn-ziti casserole? (Lesson 12-5)

9. **COATS** The table shows the probability distribution of the number of each type of coat that was sold in a particular week at a sports store. (Lesson 12-6)

Types of Coats Sold

X = Number of Zippered Pockets	Probability
0	0.03
1	0.08
2	0.08
3	0.10
4	0.30
5	0.41

   a. Show that the distribution is valid.
   b. What is the probability that a randomly chosen coat has fewer than 4 pockets?
   c. Make probability histogram of the data.

10. **BASKETBALL** Liam plays basketball. Last year, he made 80% of his attempted free throws. What objects can be used to model the possible outcomes of his next free throw? Explain. (Lesson 12-7)

1 Converting Units of Measure

There are three types of measurement: length, capacity, and mass. There are two measuring systems that we use: customary and metric.

The general rule when converting between units of measurement is:

- to convert from larger units to smaller units, multiply;
- to convert from smaller units to larger units, divide.

The diagram below shows the relationship between metric units of length

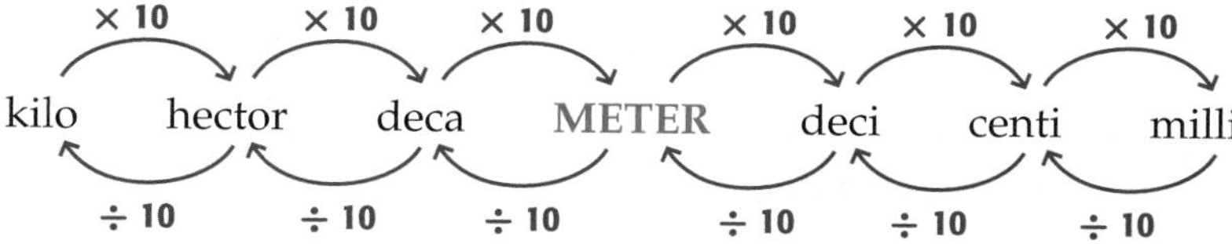

Units of Length	
Customary	**Metric**
1 foot (ft) = 12 inches (in.) 1 yard (yd) = 3 feet 1 mile(mi) = 1760 yards or 5280 feet	1 kilometer (km) = 1000 meters (m) 1 meter (m) = 100 centimeters (cm) 1 centimeter (cm) = 10 millimeters (mm)

Comparing Metric and Customary Units of Length	
Conversion Factors	**Examples**
1 millimeter (mm) ≈ 0.04 inch (in.)	height of a comma
1 centimeter (cm) ≈ 0.4 inch (in.)	half the width of a penny
1 meter (m) ≈ 1.1 yards or 3.3 feet (ft)	width of a doorway
1 kilometer (km) ≈ 0.6 mile (mi)	length of a city block

EXAMPLE 1 Units of Length

Complete each sentence.

1A. $5 \text{ m} = \underline{\ ?\ } \text{ cm}$

$5(100) = 500$ — **To convert from meters to centimeters, multiply by 100.**

$5 \text{ m} = 500 \text{ cm}$

1B. $63 \text{ m} = \underline{\ ?\ } \text{ km}$

$63 \div 1000 = 0.063$ — **To convert from meters to kilometers, divide by 1000.**

$63 \text{ m} = 0.063 \text{ km}$

Units of Capacity	
Customary	**Metric**
1 cup (c) = 8 fluid ounces (fl oz) 1 pint (pt) = 2 cups (c) 1 quart (qt) = 2 pints (pt) 1 gallon (gal) = 4 quarts	1 liter (L) = 1000 milliliters (mL)

Comparing Metric and Customary Units of Capacity	
Conversion Factors	**Examples**
1 milliliter (mL) ≈ 0.03 fluid ounce (fl oz)	drop of water
1 liter (L) ≈ 1 quart (qt)	bottle of ketchup

EXAMPLE 2 Units of Capacity

Complete each sentence.

2A. $8 \text{ gal} = \underline{\ ?\ }$ quarts

$8(4) = 32$ — **Larger → smaller, multiply.**

$8 \text{ gal} = 32 \text{ qt}$

2B. $24 \text{ fl oz} = \underline{\ ?\ }$ c

$24 \div 8 = 3$ — **Smaller → larger, divide.**

$24 \text{ fl oz} = 3 \text{ c}$

2C. $16 \text{ pt} = \underline{\ ?\ }$ gal

$16 \div 2 = 8 \text{ qt}$ — **Convert from pt to qt.**

$8 \div 4 = 2 \text{ gal}$ — **Convert from qt to gal**

$16 \text{ pt} = 2 \text{ gal}$

Customary Units of Weight	Metric Units of Weight
1 pound (lb) = 16 ounces (oz) 1 ton (T) = 2000 pounds (lb)	1 kilogram (kg) = 1000 grams (g) 1 gram (g) = 1000 milligrams (mg)

Comparing Metric and Customary Units of Mass	
Conversion Factors	**Examples**
1 gram (g) = 0.04 ounce (oz)	one raisin
1 kilogram (kg) = 2.2 pounds (lb)	a pineapple

EXAMPLE 3 Units of Mass

Complete each sentence.

3A. $6.8 \text{ kg} = \underline{\ ?\ }$ lb

$6.8(2.2) = 14.96$ — **Larger → smaller, multiply.**

$6.8 \text{ kg} = 14.96 \text{ lb}$

3B. $1.6 \text{ oz} = \underline{\ ?\ }$ g

$1.6 \div 0.04 = 40$ — **Smaller → larger, divide.**

$1.6 \text{ oz} = 40 \text{ g}$

3C. $1\text{T} = \underline{\ ?\ }$ kg

$1(2000) = 2000 \text{ lb}$ — **Convert from T to lb.**

$2000 \div 2.2 \approx 909.1 \text{ kg}$ — **Convert from lb to kg.**

$1\text{T} \approx 909 \text{ kg}$

Exercises

Complete each sentence.

1. $5 \text{ km} = \underline{\ ?\ }$ m
2. $3.5 \text{ cm} = \underline{\ ?\ }$ mm
3. $6\text{L} = \underline{\ ?\ }$ mL
4. $37 \text{ c} = \underline{\ ?\ }$ pt
5. $0.2 \text{ mi} = \underline{\ ?\ }$ ft
6. $400 \text{ yd} = \underline{\ ?\ }$ ft
7. $18 \text{ cm} = \underline{\ ?\ }$ in.
8. $0.75 \text{ L} = \underline{\ ?\ }$ qt
9. $93.5 \text{ lb} = \underline{\ ?\ }$ kg
10. $210 \text{ mm} = \underline{\ ?\ }$ cm
11. $65 \text{ g} = \underline{\ ?\ }$ kg
12. $20 \text{ mL} = \underline{\ ?\ }$ L
13. $52.9 \text{ kg} = \underline{\ ?\ }$ lb
14. $800 \text{ fl oz} = \underline{\ ?\ }$ mL
15. $9.05 \text{ yd} = \underline{\ ?\ }$ m

2 Factors and Multiples

Two or more numbers that are multiplied to form a **product** are called **factors**.

For example, because $6(7) = 42$, 6 and 7 are factors of 42.

Factor Rules		
This number is a factor of...	**Example**	**Reason**
2 if the ones digit is divisible by 2.	164	4 is divisible by 2.
3 if the sum of the digits is divisible by 3.	123	$1 + 2 + 3 = 6$, and 6 is divisible by 3.
5 if the ones digit is 0 or 5.	120	The ones digit is a 0.
6 if the number is divisible by 2 and 3.	48	48 is divisible by 2 and 3.
9 if the sum of the digits is divisible by 9.	189	$1 + 8 + 9 = 18$, and 18 is divisible by 9.
10 if the ones digit is 0.	1250	The ones digit is 0.

EXAMPLE 1

Determine whether 138 has a factor of 2, 3, 5, 6, 9, or 10.

Number	Factor?	Reason
2	yes	8 is divisible by 2.
3	yes	$1 + 3 + 8 = 12$, and 12 is divisible by 3.
5	no	The ones digit is not 0 or 5.
6	yes	138 has a factor of 2 and 3.
9	no	$1 + 3 + 8 = 12$, and 12 is not divisible by 9.
10	no	The ones digit is not 0.

So, 138 has factors of 2, 3, and 6.

You can also use the factor rules to find all of the factors of a number. Use division to find the other factor in each factor pair.

EXAMPLE 2

List all of the factors of 72.

Number	Factor?	Factor Pairs
1	yes	$1 \cdot 72$
2	yes	$2 \cdot 36$
3	yes	$3 \cdot 24$
4	yes	$4 \cdot 18$
5	no	...
6	yes	$6 \cdot 12$
7	no	...
8	yes	$8 \cdot 9$
9	yes	$9 \cdot 8$

You can stop finding factors when the numbers start repeating.

So, the factors of 72 are 1, 2, 3, 4, 6, 8, 9, 12, 18, 24, 36, and 72.

A multiple is the product of a specific number and any whole number. So, 64 is a multiple of 4 because $4(16) = 64$.

EXAMPLE 3

Determine whether 375 is a multiple of 15.

We can use the factor rules to help determine multiples. Since $15 = 3(5)$, check 3 and 5.

Number	Factor of 375?	Factor Pairs	Is 375 a Multiple?
3	yes	3 • 125	Yes, 375 is a multiple of 3.
5	yes	5 • 75	Yes, 375 is a multiple of 5.
15	yes	15 • 25	Yes, 375 is a multiple of 15.

Since $375 \div 15 = 25$ with no remainder, 375 is a multiple of 15.

Exercises

Determine whether each number has a factor of 2, 3, 5, 6, 9, or 10.

1. 39 **2.** 46 **3.** 35
4. 18 **5.** 44 **6.** 23
7. 22 **8.** 66 **9.** 212
10. 250 **11.** 118 **12.** 378
13. 995 **14.** 510 **15.** 5010
16. 1052 **17.** 32,460 **18.** 3039

List all of the factors of each number.

19. 28 **20.** 75 **21.** 14 **22.** 57
23. 81 **24.** 52 **25.** 42 **26.** 63
27. 60 **28.** 90 **29.** 114 **30.** 124
31. 102 **32.** 135 **33.** 365 **34.** 225

Determine whether the first number is a multiple of the second.

35. 49, 3 **36.** 64, 9 **37.** 63, 3
38. 60, 6 **39.** 135, 5 **40.** 102, 4
41. 121, 11 **42.** 905, 12 **43.** 364, 9
44. 536, 3 **45.** 657, 7 **46.** 234, 6
47. 282, 31 **48.** 3612, 12 **49.** 3585, 65

50. **MUSIC** Seventy-two members of the marching band will march in the Homecoming Parade. They will need to march in rows with the same number of students in each row.

a. Can the band be arranged in rows of 7? Explain.

b. How many different ways could the marching band members be arranged? Describe the arrangements.

51. **CALENDARS** Years that are multiples of 4, called *leap years*, are 366 days long. Use the rule given below to determine whether 2010, 2015, 2016, 2022, and 2032 are leap years.

If the last two digits form a number divisible by 4, then the number is divisible by 4.

Concepts and Skills Bank

3 Prime Factorization

A **prime number** is a whole number greater than 1, for which the only factors are 1 and itself. A whole number greater than 1 that has more than two factors is a **composite number**. The numbers 0 and 1 are *neither* prime *nor* composite. Notice that 0 has an endless number of factors and 1 has only one factor, itself.

EXAMPLE 1 Identify Numbers as Prime or Composite

Determine whether each number is *prime* or *composite*.

a. 27

The numbers 1, 3, and 9 divide into 27 evenly. So, 27 is a composite number.

b. 41

The only numbers that divide evenly into 41 are 1 and 41. So, 41 is a prime number.

A whole number expressed as the product of prime factors is called the **prime factorization**. The prime factors can be written in any order usually from the least prime factor to the greatest prime factor. Disregarding order, there is only one way to write the prime factorization of a whole number.

EXAMPLE 2 Prime Factorization of a Whole Number

Write the prime factorization of 120.

Method 1 Find the least prime factors.

$120 = 2 \cdot 60$ — **The least prime factor of 120 is 2.**

$= 2 \cdot 2 \cdot 30$ — **The least prime factor of 60 is 2.**

$= 2 \cdot 2 \cdot 2 \cdot 15$ — **The least prime factor of 30 is 2.**

$= 2 \cdot 2 \cdot 2 \cdot 3 \cdot 5$ — **The least prime factor of 15 is 3.**

Method 2 Use a factor tree.

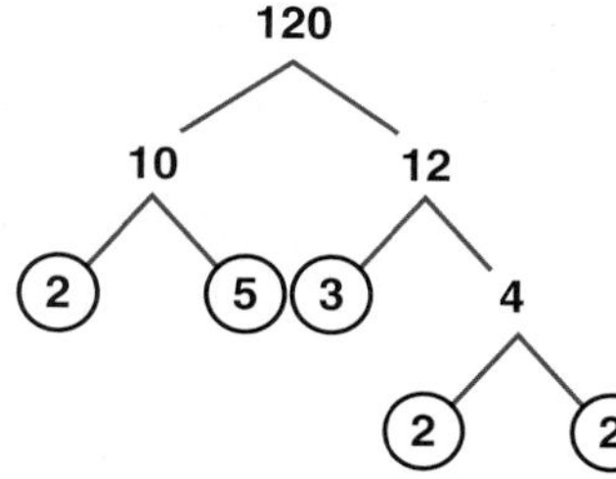

Choose any two factors of 120 to begin.
Keep finding factors until each branch ends in a prime factor.

The prime factorization is complete because 2, 3, and 5 are prime numbers. Thus, the prime factorization of 120 is $2 \cdot 2 \cdot 2 \cdot 3 \cdot 5$ or $2^3 \cdot 3 \cdot 5$.

Exercises

Determine whether each number is *prime* or *composite*.

1. 48 **2.** 29 **3.** 63 **4.** 75 **5.** 71 **6.** 43

Write the prime factorization of each number. Use exponents for repeated factors.

7. 20 **8.** 28 **9.** 64

10. 120 **11.** 140 **12.** 85

13. 144 **14.** 221 **15.** 84

16. 275 **17.** 210 **18.** 441

19. 351 **20.** 900 **21.** 1350

4 Measuring Angles

The most common measure of angles is the **degree** (°). You can use a **protractor** to measure angles in degrees.

EXAMPLE 1

Use a protractor to measure $\angle ABC$.

Step 1 Place the center point of the protractor's base on vertex B. Align the straight side with the side $\overline{AB}$ so that the marker for 0° is on the side.

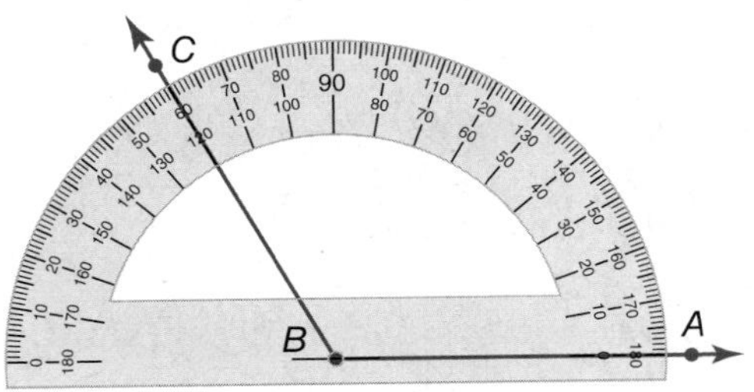

Step 2 Use the scale that begins with 0° at $\overline{AB}$. Read where the other side of the angle, $\overline{BC}$, crosses this scale.

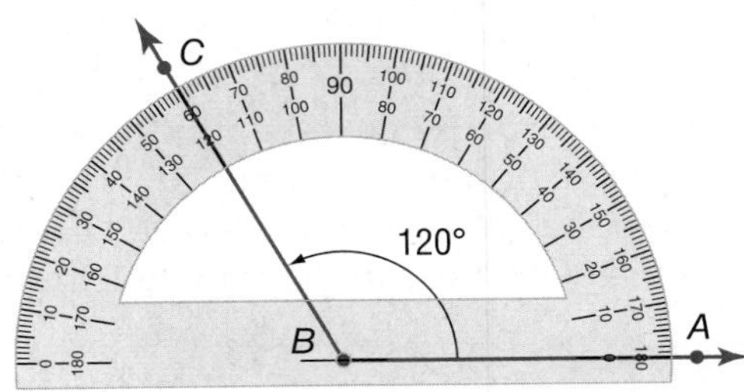

The measure of $\angle ABC$ is 120°. Use the symbols $m\angle ABC = 120°$.

If the sides of the angle are too small to reach the scale on the protractor, you can extend the sides until they are long enough.

Notice that there are measurements going in two directions on the protractor. One is on the outside of the other. You may use either scale for measuring, but you must use the one where one of your sides lines up with 0° to get the correct measurement.

Acute angles have measures less than 90°.
Right angles have measures equal to 90°.
Obtuse angles have measures between 90° and 180°.
Straight angles have measures equal to 180°.

So we can classify $\angle ABC$ as an obtuse angle.

A protractor can also be used to draw an angle of a certain measure.

EXAMPLE 2

Draw $\angle D$ with a measure of 85°.

Step 1 Draw a ray with an endpoint D. Make sure that the ray is long enough so that it crosses the edge of the protractor.

Step 2 Place the center point of the protractor on D. Align the mark labeled 0 with the ray you drew.

Step 3 Use the scale that begins with 0. Locate the mark labeled 85. Then draw the other side of the angle.

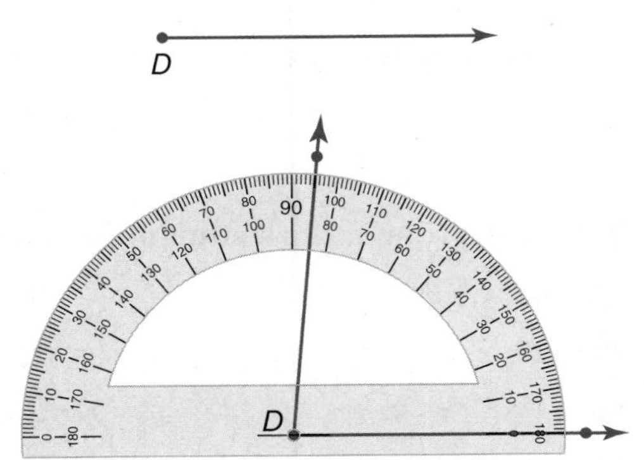

We can classify this angle as an acute angle.

Concepts and Skills Bank

Exercises

Use the protractor to find the measure of each angle. Then classify each angle as *acute, right, obtuse,* or *straight*.

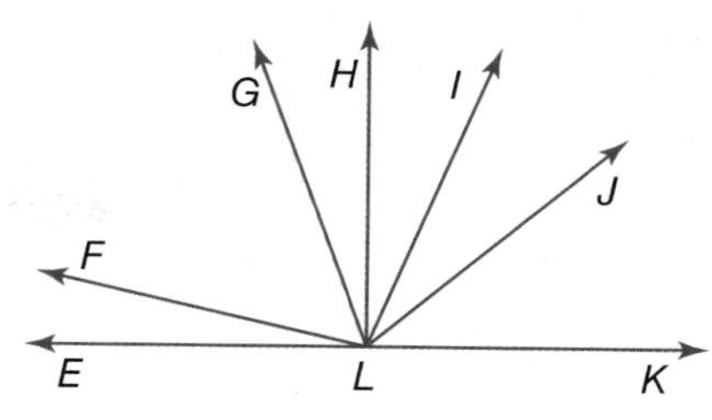

1. $\angle JLK$

2. $\angle ELF$

3. $\angle ELK$

4. $\angle GLJ$

5. $\angle FLI$

6. $\angle JLF$

7. $\angle GLF$

8. $\angle ILG$

9. $\angle ILK$

10. $\angle GLK$

11. Does $\angle FLI$ have the same measure as $\angle ELH$? Explain.

12. Which angle, if any, has the same measure as $\angle ELH$? Explain.

Use a protractor to measure each angle. Then classify each angle as *acute, right, obtuse,* or *straight*.

13.

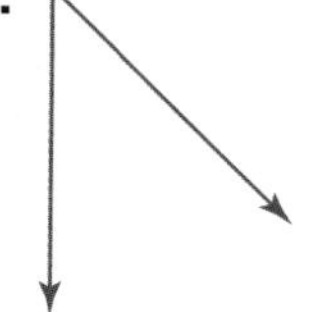

14.

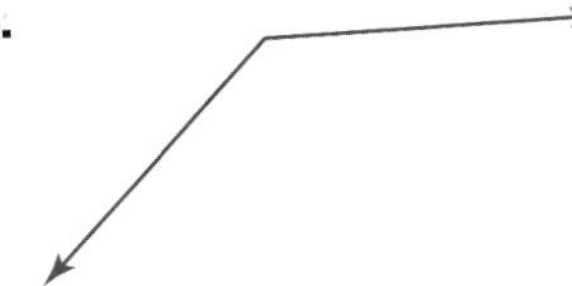

15.

16.

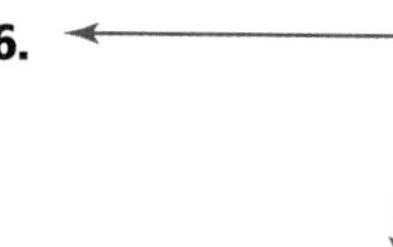

17.

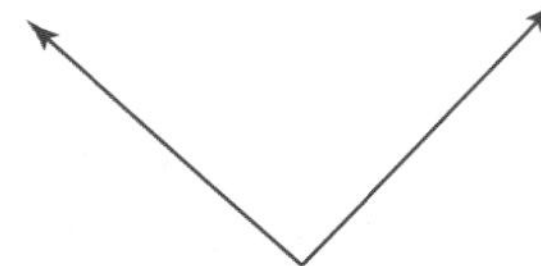

18.

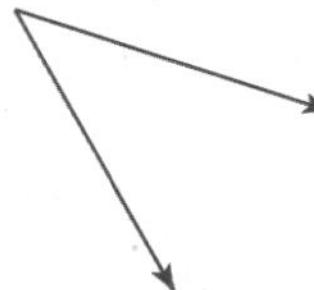

19.

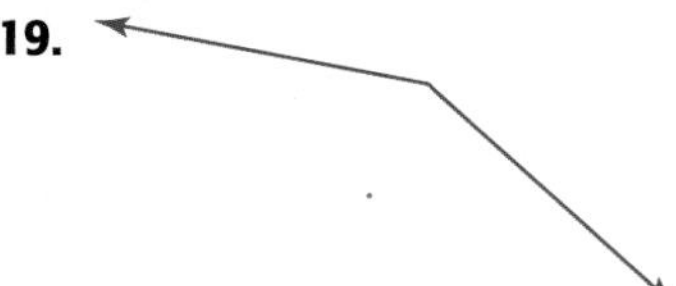

20.

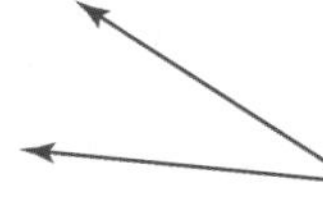

Use a protractor to draw an angle with each measurement. Then classify each angle.

21. 40°

22. 70°

23. 65°

24. 85°

25. 95°

26. 110°

27. 155°

28. 140°

29. 35°

30. 180°

31. 20°

32. 165°

5 Venn Diagrams

A Venn diagram shows the relationships among sets of numbers or objects by using overlapping circles in a rectangle.

The Venn diagram at the right shows the factors of 12 and 20. The common factors are in both circles.

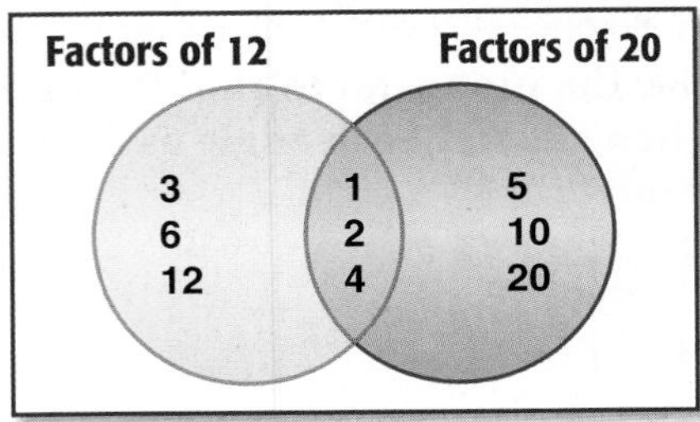

EXAMPLE 1

COLLEGE Members of a senior class were polled and asked what type of federal financial aid they are receiving for college.

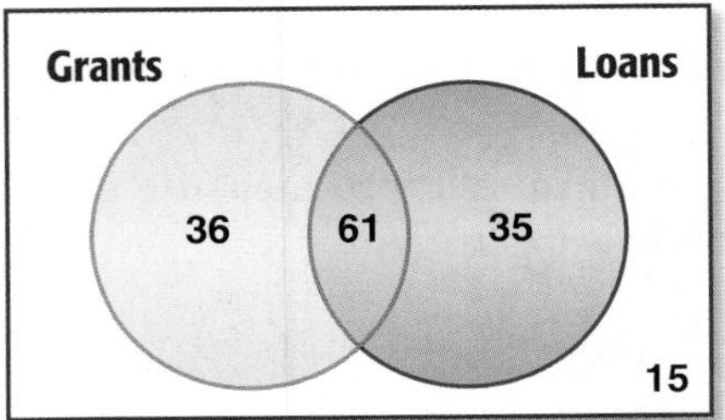

a. How many students are only receiving grants?

The number only in the grant circle is 36. So, 36 seniors are receiving only grants.

b. How many students are receiving both loans and grants?

The value in both circles is 61. So, 61 seniors are receiving both loans and grants.

c. How many students are not receiving any grants or loans?

The value outside of the circles is 15. So, 15 seniors are not receiving any grants or loans.

EXAMPLE 2

Refer to the Venn diagram at the right.

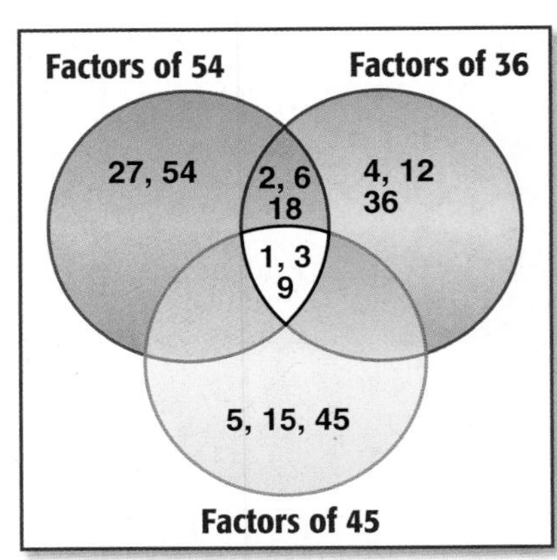

a. What are the factors in all three circles?

The factors in all three circles are 1, 3, and 9.

b. What is the greatest common factor of all three numbers?

The greatest common factor of all three numbers is 9.

c. What factors are shared only by 54 and 36?

The factors 2, 6, 18 are shared only by 54 and 36.

d. What is the greatest common factor of 54 and 36?

The greatest common factor between 54 and 36 is 18.

e. What factors are shared only by 54 and 45?

There are no factors shared by only 54 and 45. The factors that they have in common are shared by all three numbers.

f. What is the greatest common factor of 54 and 45?

The greatest common factor between 54 and 45 is 9.

Exercises

Refer to the Venn diagram at the right.

1. Which factors are in both circles?
2. What is the greatest common factor of 56 and 32?

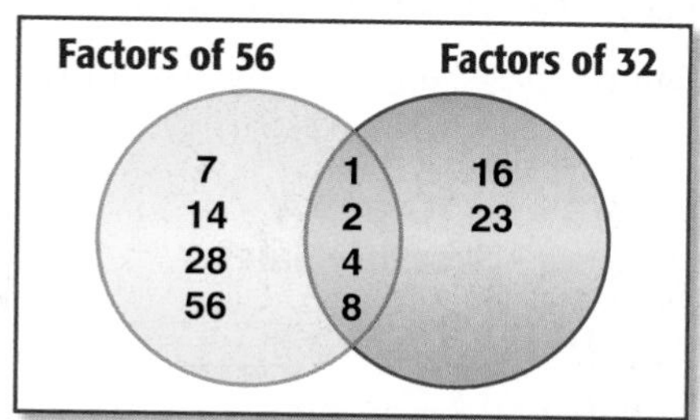

Refer to the Venn diagram at the right.

3. What are the factors in all three circles?
4. What is the greatest common factor of all three numbers?
5. What factors are shared only by 28 and 42?
6. What is the greatest common factor of 28 and 42?
7. What factors are shared only by 16 and 42?
8. What is the greatest common factor of 16 and 42?

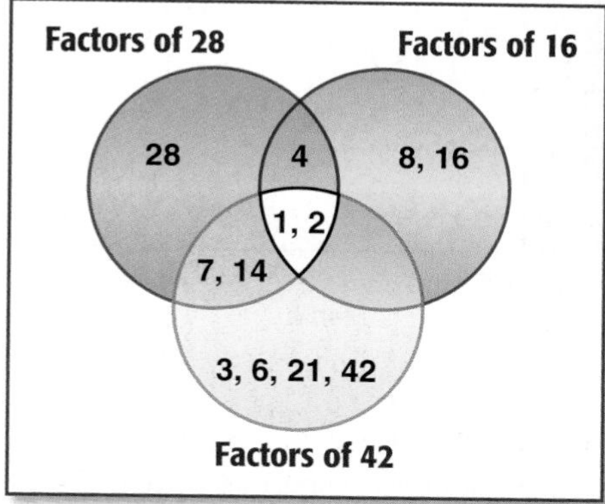

TRANSPORTATION **Twenty people who have traveled across the country in the last year were asked if they used land or air transportation. The results are shown in the Venn diagram.**

9. How many people traveled across the country by land?
10. Explain what the 5 in the diagram represents.

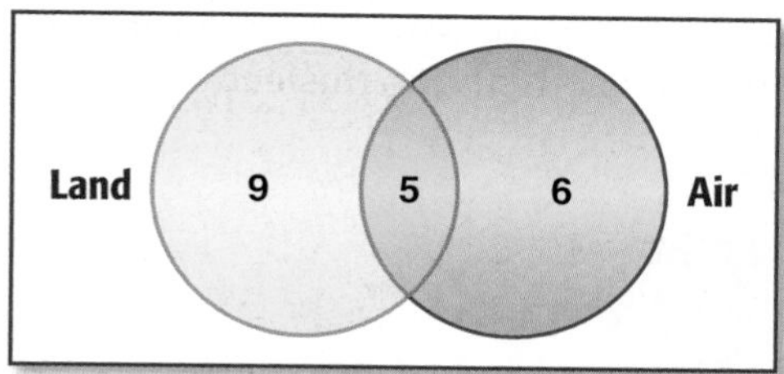

SPORTS **Ten students were asked which sports they played in the past month. Use the Venn diagram shown at the right to answer the following questions.**

11. Which student(s) have played basketball in the last month?
12. Who played only basketball and baseball?
13. Which student(s) have played only soccer in the last month?
14. Who played only soccer and baseball?
15. Which student(s) have played all three sports?
16. Who played at least two of the three sports?
17. What sports did Ted participate in the past month?

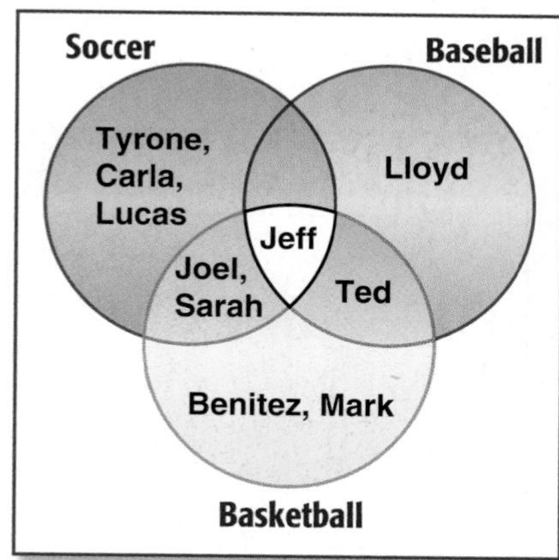

MEDIA **A hundred teens were polled and asked which of three news sources they use daily. The results are shown in the Venn Diagram at the right.**

18. How many teens use only television as a daily news source?
19. How many teens use television as a daily news source?
20. Which news source(s) was used by exactly 6 teens?
21. How many teens use only the Internet as a daily news source?
22. Which new source(s) was used by exactly 7 teens?
23. How many teens use all three as daily news sources?

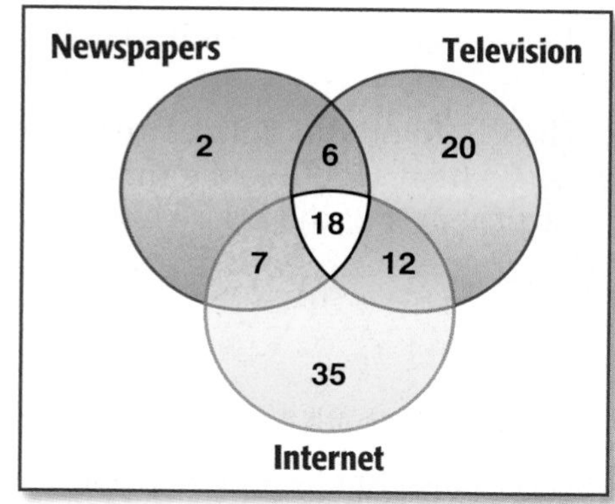

Concepts and Skills Bank

6 Misleading Graphs

The graphs below show the monthly sales of a local magazine.

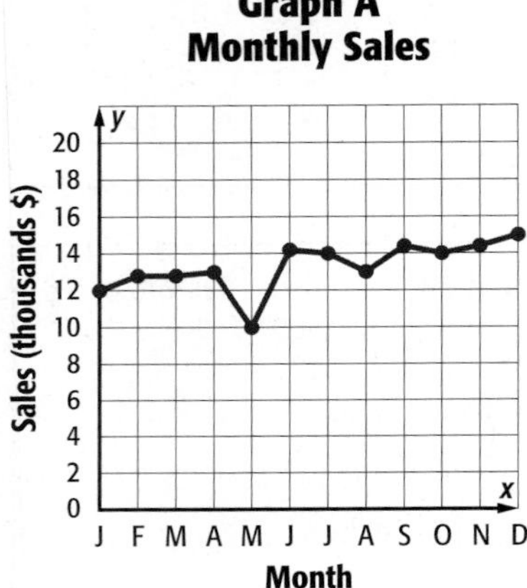

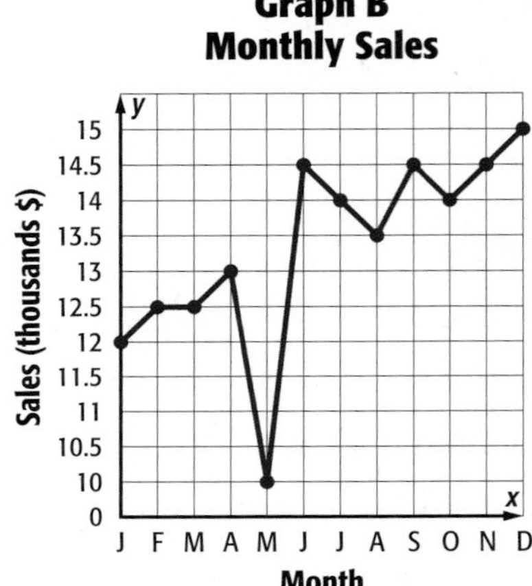

The graphs show the same information. However, while the first graph shows fairly steady sales, the second graph shows dramatic increases and decreases in sales. The graphs have the same horizontal axis, but notice that the scales are very different on the vertical axis. By changing the scale on a graph, the visual impression can be changed so that it is misleading.

EXAMPLE 1

The graphs show the number of visitors at a lodge in the Smoky Mountains each year. Which graph appears to show the greatest increase in the number of visitors?

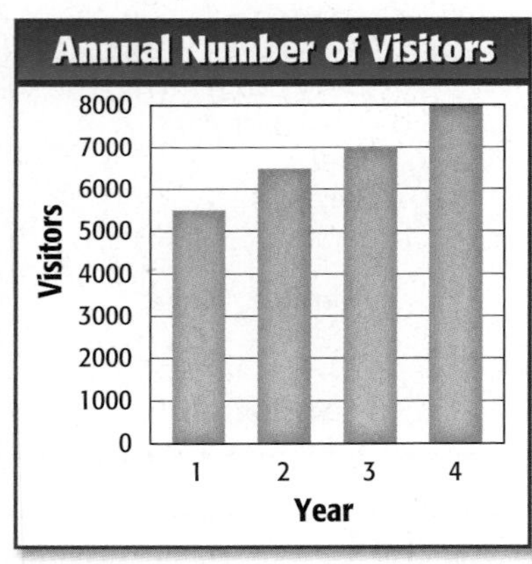

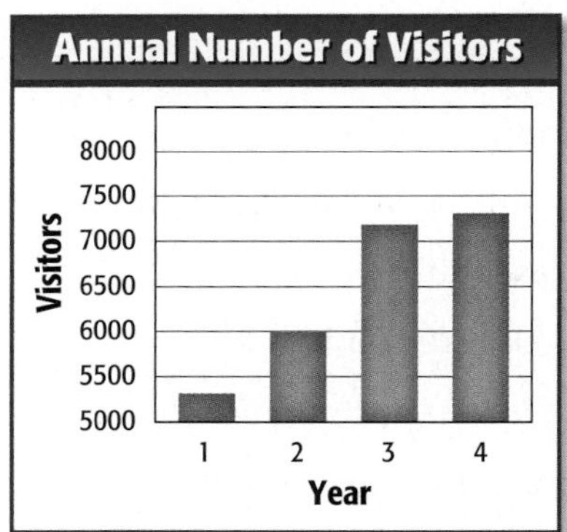

The graph on the right appears to show the greatest growth. The increments are 500 on the vertical axis, rather than 1000 like the first graph. The vertical axis of the graph on the left begins at 0, while the vertical axis of the graph on the right begins at 5000.

More Hints:

- Be sure that both axes have a scale that remains the same throughout the graph.
- Make sure that graphs with pictures, use the same picture throughout the graph.

Exercises

1. **TRAFFIC** The graphs display the number of cars that went through a tollbooth during the previous hour. Which graph appears to show the greatest increase in the number of vehicles? Explain.

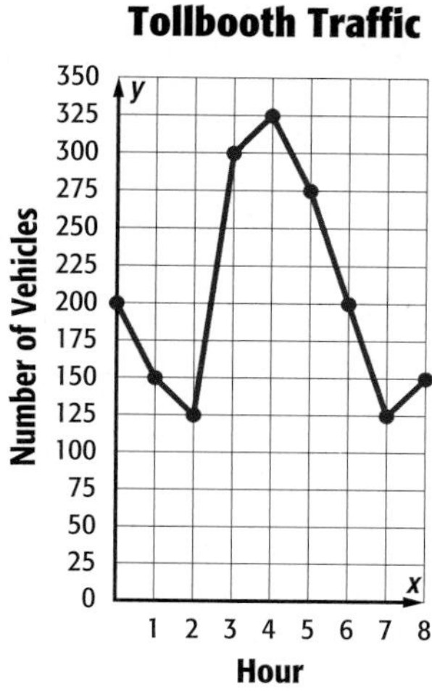

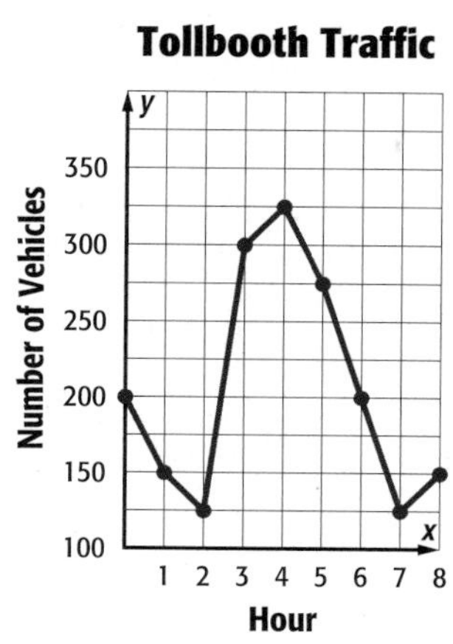

2. Which graph appears to have higher gas prices? Explain.

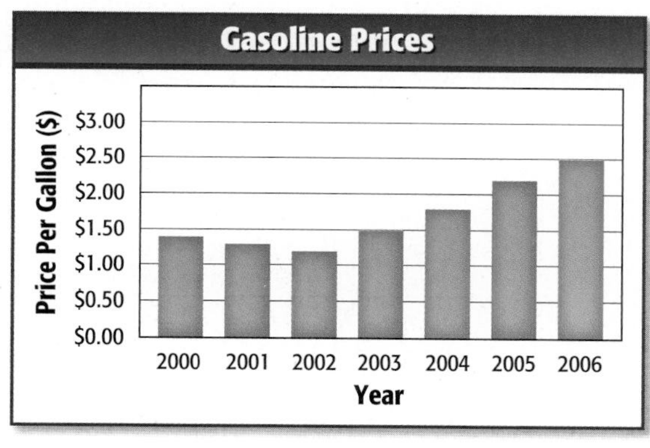

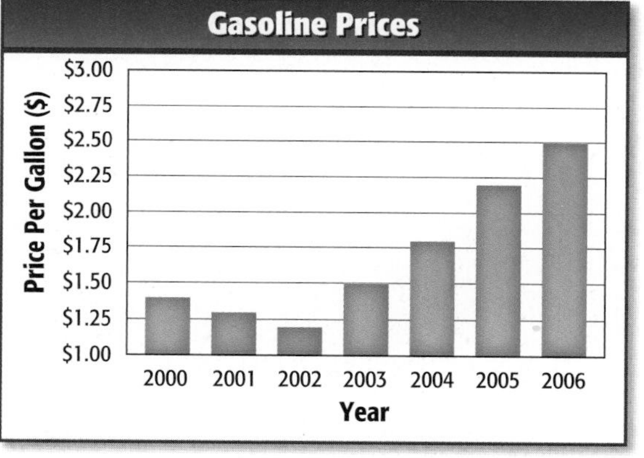

3. **BUSINESS** According to the information in the graph below, there were no visitors to the ski resort in 1999. Determine if this statement is accurate. Justify your response.

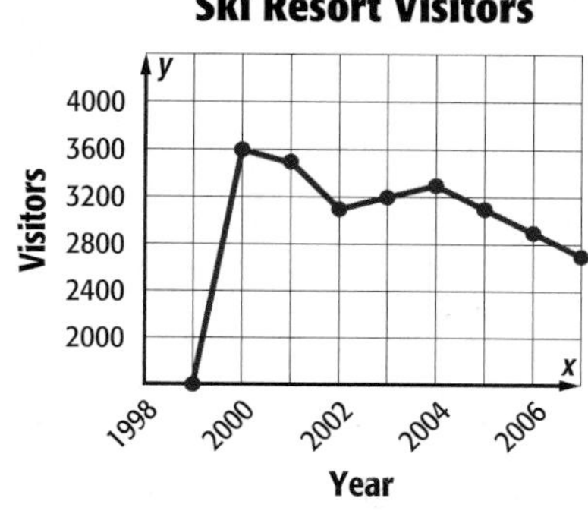

Concepts and Skills Bank

Key Concepts

Expressions, Equations, and Functions — Chapter 1

Translating Verbal to Algebraic Expressions (p. 6)
Addition more than, sum, plus, increased by, added to
Subtraction less than, subtracted from, difference, decreased by, minus
Multiplication product of, multiplied by, times, of
Division quotient of, divided by

Order of Operations (p. 10)
Step 1 Evaluate expressions inside grouping symbols.
Step 2 Evaluate all powers.
Step 3 Multiply and/or divide from left to right.
Step 4 Add and/or subtract from left to right.

Reflexive Property (p. 16)
Any quantity is equal to itself.

Symmetric Property (p. 16)
If one quantity equals a second quantity, then the second quantity equals the first.

Transitive Property (p. 16)
If one quantity equals a second quantity and the second quantity equals a third quantity, then the first quantity equals the third quantity.

Substitution Property (p. 16)
A quantity may be substituted for its equal in any expression.

Additive Identity (p. 16)
For any number a, the sum of a and 0 is a.

Additive Inverse (p. 16)
A number and its opposite are additive inverses of each other. The sum of a number and its additive inverse is 0.

Multiplicative Identity (p. 17)
For any number a, the product of a and 1 is a.

Multiplicative Property of Zero (p. 17)
For any number a, the product of a and 0 is 0.

Multiplicative Inverse (p. 17)
For every number $\frac{a}{b}$, where $a, b \neq 0$, there is exactly one number $\frac{b}{a}$ such that the product of $\frac{a}{b}$ and $\frac{b}{a}$ is 1.

Commutative Property (p. 18)
The order in which you add or multiply numbers does not change their sum or product.

Key Concepts

Associative Property (p. 18)
The way you group three or more numbers when adding or multiplying does not change their sum or product.

Distributive Property (p. 23)
For any numbers a, b, and c, $a(b + c) = ab + ac$ and $(b + c)a = ba + ca$ and $a(b - c) = ab - ac$ and $(b - c)a = ba - ca$.

Function (p. 45)
A function is a relation in which each element of the domain is paired with *exactly* one element of the range.

Chapter 2 Linear Equations

Addition Property of Equality (p. 83)
If an equation is true and the same number is added to each side of the equation, the resulting equivalent equation is also true.

Subtraction Property of Equality (p. 84)
If an equation is true and the same number is subtracted from each side of the equation, the resulting equivalent equation is also true.

Multiplication Property of Equality (p. 84)
If an equation is true and each side is multiplied by the same nonzero number, the resulting equation is equivalent.

Division Property of Equality
If an equation is true and each side is divided by the same number, the resulting equation is true.

Steps for Solving Equations (p. 99)

Step 1 Simplify the expressions on each side. Use the Distributive Property as needed.

Step 2 Use the Addition and/or Subtraction Properties of Equality to get the variables on one side and the numbers without variables on the other side. Simplify.

Step 3 Use the Multiplication or Division Property of Equality to solve.

Absolute Value Equations (p. 104)
When solving equations that involve absolute values, there are two cases to consider.

Case 1 The expression inside the absolute value symbol is positive.

Case 2 The expression inside the absolute value symbol is negative.

Means-Extremes Property of Proportion (p. 112)
In a proportion, the product of the extremes is equal to the product of the means.

Chapter 3 Linear Functions

Standard Form of a Linear Equation (p. 153)
The standard form of a linear equation is $Ax + By = C$, where $A \geq 0$, A and B are not both zero, and A, B, and C are integers with a greatest common factor of 1.

Linear Function (p. 161)
A linear function can be described by an equation of the form $f(x) = x$.

Rate of Change (p. 170)
If x is the independent variable and y is the dependent variable, then

$$\text{rate of change} = \frac{\text{change in } y}{\text{change in } x}.$$

Slope (p. 173)
The slope of a line is the ratio of the rise to the run.

Arithmetic Sequence (p. 187)
An arithmetic sequence is a numerical pattern that increases or decreases at a constant rate called the *common difference.*

***n*th Term of an Arithmetic Sequence** (p. 188)
The nth term of an arithmetic sequence with first term a_1 and common difference d is given by $a_n = a_1 + (n - 1)d$, where n is a positive integer.

Proportional Relationship (p. 195)
A relationship is proportional if its equation is of the form $y = kx$, $k \neq 0$. The graph passes through (0, 0).

Chapter 4 Linear Functions and Equations

Slope-Intercept Form (p. 214)
The slope-intercept form of a linear equation is $y = mx + b$, where m is the slope and b is the y-intercept.

Point-Slope Form (p. 231)
The linear equation $y - y_1 = m(x - x_1)$ is written in point-slope form, where (x_1, y_1) is a given point on a non-vertical line and m is the slope of the line.

Writing Equations (p. 232)

Given the Slope and One Point	Given Two Points
Step 1 Substitute the values of m, x, and y into the slope-intercept form and solve for b. Or, use the point-slope form. Substitute the value of m and let x and y be (x_1, y_1). **Step 2** Write the slope-intercept form using the values of m and b.	**Step 1** Find the slope. **Step 2** Choose one of the two points to use. **Step 3** Follow the steps for writing an equation given the slope and one point.

Parallel and Perpendicular Lines (p. 240)

Parallel Lines	Perpendicular Lines
Two nonvertical lines are parallel if they have the same slope.	Two nonvertical lines are perpendicular if the product of their slopes is −1.

Using a Linear Function to Model Data (p. 246)

Step 1 Make a scatter plot. Determine whether any relationship exists in the data.

Step 2 Draw a line that seems to pass close to most of the data points.

Step 3 Use two points on the line of fit to write an equation for the line.

Step 4 Use the line of fit to make predictions.

Greatest Integer Function (p. 261)

A greatest interger function can be described by an equation of the form $f(x) = [\![x]\!]$.

Absolute Value Function (p. 262)

$f(x) = |x|$, defined as $f(x) = \begin{cases} x \text{ if } x > 0 \\ 0 \text{ if } x = 0 \\ -x \text{ if } x < 0 \end{cases}$

Linear Inequalities — Chapter 5

Addition Property of Inequalities (p. 283)

If any number is added to each side of a true inequality, the resulting inequality is also true.

Subtraction Property of Inequalities (p. 284)

If any number is subtracted from each side of a true inequality, the resulting inequality is also true.

Multiplication Property of Inequalities (p. 290)

If a true inequality is multiplied by a positive number, the resulting inequality is also true. If a true inequality multiplied by a negative number, the direction of the inequality sign is changed to make the resulting inequality also true.

Division Property of Inequalities (p. 292)

If a true inequality is divided by a positive number, the resulting inequality is also true. If a true inequality is divided by a negative number, the direction of the inequality sign is changed to make the resulting inequality also true.

Graphing Linear Inequalities (p. 315)

Step 1 Graph the boundary. Use a solid line when the inequality contains $\leq$ or $\geq$. Use a dashed line when the inequality contains $<$ or $>$.

Step 2 Use a test points to determine which half-plane should be shaded.

Step 3 Shade the half-plane that contains the solution.

Systems of Linear Equations and Inequalities — Chapter 6

Solving by Substitution (p. 342)

Step 1 When necessary, solve at least one equation for one variable.

Step 2 Substitute the resulting expression from Step 1 into the other equation to replace the variable. Then solve the equation.

Step 3 Substitute the value from Step 2 into either equation, and solve for the other variable. Write the solution as an ordered pair.

Solving by Elimination Using Addition (p. 348)

Step 1 Write the system so like terms with the same or opposite coefficients are aligned.

Step 2 Add or subtract the equations eliminating one variable. Then solve the equation.

Step 3 Substitute the value from Step 2 into one of the equations and solve for the other variable. Write the solution as an ordered pair.

Solving by Elimination Using Multiplication (p. 355)

Step 1 Multiply at least one equation by a constant to result in two equations that contain opposite terms.

Step 2 Add or subtract the equations eliminating one variable. Then solve the equation.

Step 3 Substitute the value from Step 2 into one of the equations and solve for the other variable. Write the solution as an ordered pair.

Elementary Row Operations (p. 377)

The following operations can be performed on an augmented matrix.

- Interchange any two rows.
- Multiply all entries in a row by a nonzero constant.
- Replace one row with the sum of that row and a multiple of another row.

Polynomials — Chapter 7

Product of Powers (p. 402)

To multiply two powers that have the same base, add their exponents.

Power of a Power (p. 402)

To find the power of a power, multiply the exponents.

Power of a Product (p. 403)

To find the power of a product, find the power of each factor and multiply.

Simplify Expressions (p. 404)

To simplify a monomial expression, write an equivalent expression in which:

- each base appears exactly once,
- there are no powers of powers, and
- all fractions are in simplest form.

Quotient of Powers (p. 408)

To divide two powers with the same base, subtract the exponents.

Power of a Quotient (p. 409)

To find the power of a quotient, find the power of the numerator and the power of the denominator.

Zero Exponent Property (p. 410)

Any nonzero number raised to the zero power is equal to 1.

Negative Exponent Property (p. 410)

For any nonzero number a and any integer n, a^{-n} is the reciprocal of a^n. Also, the reciprocal of $a^{-n} = a^n$.

Standard Form to Scientific Notation (p. 416)

Step 1 Move the decimal point until it is to the right of the first nonzero digit. The result is a real number a.

Step 2 Note the number of places n and the direction that you moved the decimal point.

Step 3 If the decimal point is moved left, write the number as $a \times 10^n$. If the decimal point is moved right, write the number as $a \times 10^{-n}$.

Step 4 Remove the extra zeros.

Scientific Notation to Standard Form (p. 417)

Step 1 In $a \times 10^n$, note whether $n > 0$ or $n < 0$.

Step 2 If $n > 0$, move the decimal point n places right. If $n < 0$, move the decimal point $-n$ places left.

Step 3 Insert zeros, decimal point, and commas as needed to indicate place value

FOIL Method (p. 448)

To multiply two binomials, find the sum of the products of F the *First* terms, O the *Outer* terms, I the *Inner* terms, L and the *Last* terms.

Square of a Sum (p. 453)

The square of $a + b$ is the square of a plus twice the product of a and b plus the square of b.

Square of a Difference (p. 454)

The square of $a - b$ is the square of a minus twice the product of a and b plus the square of b.

Product of a Sum and a Difference (p. 455)

The product of $a + b$ and $a - b$ is the square of a minus the square of b.

Chapter 8 Factoring and Quadratic Equations

Factoring by Grouping (p. 477)

A polynomial can be factored by grouping only if all of the following conditions exist.

- There are four or more terms.
- Terms have common factors that can be grouped together.
- There are two common factors that are identical or additive inverses of each other.

Zero Product Property (p. 478)

If the product of two factors is 0, then at least one of the factors must be 0.

Factoring $x^2 + bx + c$ (p. 485)

To factor trinomials in the form $x^2 + bx + c$, find two integers, m and p, with a sum of b and a product of c. Then write $x^2 + bx + c$ as $(x + m)(x + p)$.

Factoring $ax^2 + bx + c$ (p. 493)

To factor trinomials in the form $ax^2 + bx + c$, find two integers, m and p, with a sum of b and a product of ac. Then write $ax^2 + bx + c$ as $ax^2 + mx + px + c$, then factor by grouping.

Difference of Squares (p. 499)
$a^2 - b^2 = (a + b)(a - b)$ or $(a - b)(a + b)$

Factoring Perfect Square Trinomials (p. 505)
$a^2 + 2ab + b^2 = (a + b)(a + b) = (a + b)^2$
$a^2 - 2ab + b^2 = (a - b)(a - b) = (a - b)^2$

Square Root Property (p. 508)
To solve a quadratic equation in the form $x^2 = n$, take the square root of each side.

Chapter 9 Quadratic and Exponential Functions

Quadratic Functions (p. 525)
A quadratic function can be described by an equation of the form $f(x) = x^2$.

Maximum and Minimum Values (p. 528)
The graph of $f(x) = ax^2 + bx + c$, where $a \neq 0$:

- opens up and has a minimum value when $a > 0$, and
- opens down and has a maximum value when $a < 0$.
- The range of a quadratic function is all real numbers greater than or equal to the minimum, or all real numbers less than or equal to the maximum.

Graph Quadratic Functions (p. 529)

Step 1 Find the equation of the axis of symmetry.

Step 2 Find the vertex, and determine whether it is a maximum or minimum.

Step 3 Find the y-intercept.

Step 4 Use symmetry to find additional points on the graph, if necessary.

Step 5 Connect the point with a smooth curve.

Vertical Translations (p. 544)
The graph of $f(x) = x^2 + c$ translates the graph of $f(x) = x^2$ vertically.
If $c > 0$, the graph of $f(x) = x^2$ is translated $|c|$ units up.
If $c < 0$, the graph of $f(x) = x^2$ is translated $|c|$ units down.

Dilations (p. 545)
The graph of $f(x) = ax^2$ stretches or compresses the graph of $f(x) = x^2$ vertically.
If $|a| > 1$, the graph of $f(x) = x^2$ is stretched vertically.
If $0 < |a| < 1$, the graph of $f(x) = x^2$ is compressed vertically.

Reflections (p. 546)
The graph of the function $-f(x)$ reflects the graph of $f(x) = x^2$ across the x-axis.
The graph of the function $f(-x)$ reflects the graph of $f(x) = x^2$ across the y-axis.

Completing the Square (p. 552)
To complete the square for any quadratic expression of the form $x^2 + bx$, follow the steps below

Step 1 Find one half of b, the coefficient of x.

Step 2 Square the result in Step 1.

Step 3 Add the result of Step 2 to $x^2 + bx$.

The Quadratic Formula (p. 558)

The solutions of a quadratic equation $ax^2 + bx + c = 0$, where $a \neq 0$ are given by the Quadratic Formula.

$$x = \frac{-b \pm \sqrt{b^2 - 4ac}}{2a}$$

Exponential Function (p. 567)

An exponential function is a function that can be described by an equation of the form $y = ab^x$, where $a \neq 0$, $b > 0$, and $b \neq 1$.

General Equation for Exponential Growth (p. 573)

$y = a(1 + r)^t$

General Equation for Compound Interest

$A = P\left(1 + \frac{r}{n}\right)^{nt}$

General Equation for Exponential Decay (p. 574)

$y = a(1 - r)^t$

***n*th term of a Geometric Sequence** (p. 582)

The nth term a_n of a geometric sequence with first term a_1 and common ratio r is given by the following formula, where n is any positive integer.

$$a_n = a_1 r^{n-1}$$

Chapter 10 Radical Functions and Geometry

Square Root Function (p. 605)

A square root function can be described by an equation of the form $f(x) = \sqrt{x}$.

Graphs of Square Root Functions (p. 606)

Step 1 Draw the graph of $y = a\sqrt{x}$. The graph starts at the origin and passes through the point at $(1, a)$. If $a > 0$, the graph is in quadrant I. If $a < 0$, the graph is reflected across the x-axis and is in quadrant IV.

Step 2 Translate the graph $|c|$ units up if $c > 0$ and down if $c < 0$.

Step 3 Translate the graph $|h|$ units left if $h > 0$ and right if $h < 0$.

Product Property of Square Roots (p. 612)

For any nonnegative real numbers a and b, the square root of ab is equal to the square root of a times the square root of b.

Quotient Property of Square Roots (p. 613)

For any real numbers a and b, where $a \geq 0$ and $b > 0$, the square root of $\frac{a}{b}$ is equal to the square root of a divided by the square root of b.

Power Property of Equality (p. 624)

If you square both sides of an equation, the resulting equation is still true.

The Pythagorean Theorem (p. 630)

If a triangle is a right triangle, then the square of the length of the hypotenuse is equal to the sum of the squares of the lengths of the legs.

Converse of the Pythagorean Theorem (p. 631)
If a triangle has side lengths a, b, and c such that $c^2 = a^2 + b^2$, then the triangle is a right triangle. If $c^2 \neq a^2 + b^2$, then the triangle is not a right triangle.

The Distance Formula (p. 636)
The distance d between any two points with coordinates (x_1, y_1) and (x_2, y_2) is given by $d = \sqrt{(x_2 - x_1)^2 + (y_2 - y_1)^2}$.

The Midpoint Formula (p. 638)
The midpoint M of a line segment with endpoints at (x_1, y_1) and (x_2, y_2) is given by $M = \left(\frac{x_1 + x_2}{2}, \frac{y_1 + y_2}{2}\right)$.

Similar Triangles (p. 642)
If two triangles are similar, then the measures of their corresponding angles are equal, and the measures of their corresponding sides are proportional.

Trigonometric Ratios (p. 649)

$$\text{sine of } \angle A = \frac{\text{leg opposite } \angle A}{\text{hypotenuse}}$$

$$\text{cosine of } \angle A = \frac{\text{leg adjacent to } \angle A}{\text{hypotenuse}}$$

$$\text{tangent of } \angle A = \frac{\text{leg opposite } \angle A}{\text{leg adjacent to } \angle A}$$

Inverse Trigonometric Functions (p. 651)
If $\angle A$ is an acute angle and the sine of A is x, then the inverse sine of x is the measure of $\angle A$.
If $\angle A$ is an acute angle and the cosine of A is x, then the inverse cosine of x is the measure of $\angle A$.
If $\angle A$ is an acute angle and the tangent of A is x, then the inverse tangent of x is the measure of $\angle A$.

Chapter 11 Rational Functions and Equations

Inverse Variation (p. 670)
y varies inversely as x if there is some nonzero constant k such that $y = \frac{k}{x}$ or $xy = k$, where $x \neq 0$ and $y \neq 0$.

Product Rule for Inverse Variations (p. 671)
If (x_1, y_1) and (x_2, y_2) are solutions of an inverse variation, then the products x_1y_1 and x_2y_2 are equal.

Rational Function (p. 678)
A rational function can be described by an equation of the form $y = \frac{p}{q}$, where p and q are polynomials and $q \neq 0$.
$f(x) = \frac{1}{x}$

Asymptotes (p. 679)

A rational function in the form $y = \frac{a}{x - b} + c$ has a vertical asymptote at the x-value that makes the denominator equal zero, $x = b$. It has a horizontal asymptote at $y = c$.

Simplifying Rational Expressions (p. 685)

Let a, b, and c be polynomials with $a \neq 0$ and $c \neq 0$.

$$\frac{ba}{ca} = \frac{b \cdot a}{c \cdot a} = \frac{b}{c}$$

Multiplying Rational Expressions (p. 692)

Let a, b, c, and d be polynomials with $b \neq 0$ and $d \neq 0$.

$$\frac{a}{b} \cdot \frac{c}{d} = \frac{ac}{bd}$$

Dividing Rational Expressions (p. 693)

Let a, b, c, and d be polynomials with $b \neq 0$, $c \neq 0$, and $d \neq 0$.

$$\frac{a}{b} \div \frac{c}{d} = \frac{a}{b} \cdot \frac{d}{c} = \frac{ad}{bc}$$

Add or Subtract Rational Expressions (p. 708)

Use the following steps to add or subtract rational expressions with unlike denominators.

Step 1 Find the LCD.

Step 2 Write each rational expression as an equivalent expression with the LCD as the denominator.

Step 3 Add or subtract the numerators and write the result over the common denominator.

Step 4 Simplify if necessary.

Statistics and Probability — Chapter 12

Data Collection Techniques (p. 740)

survery Data are from responses given by sample of the population.
observational study Data are recorded after just observing the sample.
experiment Data are recorded after changing the sample.

Random Samples (p. 742)

simple random sample A sample that is equally likely to be chosen as any other sample from the population.
stratified random sample The population is first divided in similar, nonoverlapping groups. A sample is then selected from each group.
systematic random sample Items in the sample are selected according to a specified time or item interval.

Measures of Central Tendency (p. 746)

mean the sum of the data divided by the number of items in the data set
median the middle number of the ordered data, or the mean of the middle two numbers
mode the number or numbers that occur most often

Measures of Variation (p. 757)

range the difference between the greatest and least values

quartile the values that divide the data set into four equal parts

interquartile range the range of the middle half of a data set; the difference between the upper and lower quartiles

Mean Absolute Deviation (p.757)

Step 1 Find the mean.

Step 2 Find the sum of the absolute values of the differences between each value in the set of data and the mean.

Step 3 Divide the sum by the number of values in the set of data.

Variance and Standard Deviation (p. 758)

Step 1 Find the mean, $\overline{x}$.

Step 2 Find the square of the difference between each value in the set of data and the mean. Then divide by the number of values in the set of data. The result is the variance.

Step 3 Take the square root of the variance.

Factorial (p. 764)

The factorial of a positive integer n is the product of the integers less than or equal to n.

Permutation Formula (p. 765)

The number of permutations of n objects taken r at a time is the quotient of $n!$ and $(n - r)!$.

Combination Formula (p. 766)

The number of combinations of n objects taken r at a time is the quotient of $n!$ and $(n - r)!r!$.

Probability of Independent Events (p. 771)

If two events, A and B, are independent, then the probability of both events occurring is the product of the probability of A and the probability of B.

Probability of Dependent Events (p. 772)

If two events, A and B, are dependent, then the probability of both events occurring is the product of the probability of A and the probability of B after A occurs.

Mutually Exclusive Events (p. 773)

If two events, A and B, are mutually exclusive, then the probability that either A or B occurs is the sum of their probabilities.

Events that are Not Mutually Exclusive (p. 774)

If two events, A and B, are not mutually exclusive, then the probability that either A or B occurs is the sum of their probabilities decreased by the probability of both occurring.

Properties of Probability Distributions (p. 780)

1. The probability of each value of X is greater than or equal to 0 and is less than or equal to 1.
2. The sum of the probabilities of all values of X is 1.

For Homework Help, go to Hotmath.com
Complete, step-by-step solutions of most odd-numbered exercises are provided free of charge.

Chapter 0 Preparing for Algebra

Pages P5–P6 Lesson 0-1

1. estimate; about 700 mi **3.** estimate; about 7 times **5.** exact; $98.75

Pages P7–P10 Lesson 0-2

1. integers, rationals **3.** irrationals **5.** irrationals **7.** rationals **9.** rationals **11.** irrationals

13. **15.** **17.**

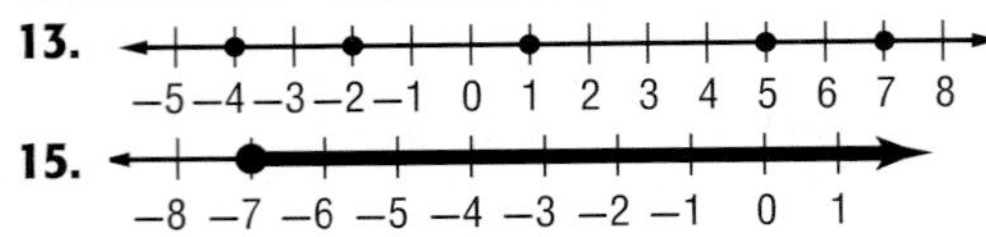

19. $\frac{5}{9}$ **21.** $\frac{13}{99}$ **23.** −5 **25.** $\pm\frac{4}{7}$ **27.** 7 **29.** −2.5 **31.** $\frac{5}{18}$ **33.** 5 **35.** 12

Pages P11–P12 Lesson 0-3

1. 5 **3.** −27 **5.** −22 **7.** −32 **9.** 22 **11.** 5 **13.** 8 **15.** −9 **17.** −115 **19.** 17° **21.** $150 **23.** $125

Pages P13–P16 Lesson 0-4

1. $<$ **3.** $<$ **5.** $=$ **7.** 3.06, $3\frac{1}{6}$, $3\frac{3}{4}$, 3.8 **9.** −0.5, $-\frac{1}{9}$, $\frac{1}{10}$, 0.11 **11.** $\frac{3}{5}$ **13.** $\frac{1}{16}$ **15.** 1 **17.** $2\frac{2}{3}$ **19.** $\frac{1}{9}$ **21.** $\frac{1}{6}$ **23.** $\frac{17}{30}$ **25.** $\frac{1}{4}$ **27.** −36.9 **29.** −19.33 **31.** 153.8 **33.** 93.3 **35.** $-\frac{5}{6}$ **37.** $\frac{9}{20}$ **39.** $\frac{2}{3}$ **41.** $\frac{3}{10}$

Pages P17–P19 Lesson 0-5

1. 0.85 **3.** −7.05 **5.** 60 **7.** −4.8 **9.** −1.52 **11.** $\frac{6}{35}$ **13.** $\frac{2}{33}$ **15.** $\frac{21}{4}$ or $5\frac{1}{4}$ **17.** $-\frac{1}{2}$ **19.** $-\frac{1}{8}$ **21.** $\frac{10}{11}$ **23.** $\frac{5}{2}$ or $2\frac{1}{2}$ **25.** $\frac{7}{6}$ or $1\frac{1}{6}$ **27.** $-\frac{23}{14}$ or $-1\frac{9}{14}$ **29.** $-\frac{3}{16}$ **31.** 2 **33.** 3 **35.** $-\frac{3}{10}$ **37.** $\frac{9}{2}$ or $4\frac{1}{2}$ **39.** $\frac{11}{20}$ **41.** $\frac{5}{18}$ **43.** 3 slices **45.** 34 uniforms **47.** 6 ribbons

Pages P20–P22 Lesson 0-6

1. $\frac{1}{20}$ **3.** $\frac{11}{100}$ **5.** $\frac{39}{50}$ **7.** $\frac{3}{500}$ **9.** $\frac{14}{1}$ **11.** 40% **13.** 160 **15.** 9.5 **17.** 48 **19.** 0.25% **21.** 24.5 **23.** 150% **25.** 90% **27.** 5% **29a.** 20 g **29b.** 2350 mg **29c.** 44% **31.** 6 animals

Pages P23–P25 Lesson 0-7

1. 20 m **3.** 90 in. **5.** 32 in. **7.** 29 ft **9.** 25.0 in. **11.** 31.4 in. **13.** 23.2 m **15.** 848.2 in. **17.** 13.4 cm **19.** 10.3 ft

Pages P26–P28 Lesson 0-8

1. 6 cm^2 **3.** 120 m^2 **5.** 81 ft^2 **7.** 9 ft^2 **9.** 14.1 in^2 **11.** 12.6 ft^2 **13.** 50.3 cm^2 **15.** 201.1 in^2 **17.** 1620 ft^2 **19.** 19.6 mi^2 **21.** 22.1 cm^2 **23.** 4.0 cm^2

Pages P29–P30 Lesson 0-9

1. 30 cm^3 **3.** 48 yd^3 **5.** 1404 ft^3 **7.** 20 m^3 **9.** 27 m^3 **11.** 2070 in^3 **13.** 1 ft **15.** 4 cm **17.** 2770.9 in^3 **19a.** 128 ft^3 **19b.** 80 ft^3 **19c.** 5 ft 4 in.

Pages P31–P32 Lesson 0-10

1. 68 in^2 **3.** 220 mm^2 **5.** 37 ft^2 **7.** 48 m^2 **9.** 216 in^2 **11.** 480.7 in^2 **13.** 24 m^2 **15.** 77 ft^2 **17.** 40.8 in^2

Pages P33–P36 Lesson 0-11

1. $\frac{4}{15}$ **3.** $\frac{1}{2}$ **5.** $\frac{5}{6}$ **7.** $\frac{1}{2}$ **9.** $\frac{2}{3}$ **11.** 20 **13.** 12 codes **15.** $\frac{11}{24}$ **17.** 1:5 **19.** 13:11 **21.** 16 orders

Pages P37–P39 Lesson 0-12

1. 5; 5; 5 **3.** 54; 53; 53 **5.** 9; 4; no mode **7.** 6; 6; no mode **9.** 28; 19; 7; 26 **11.** 16; 10; 5; 17 **13.** 6; 7; 5; 9 **15.** mean: $2.50; median: $2.25; mode: $2.00 **17.** 89 **19.** about 92.7 **21.** 118.5 min

Pages P40–P43 Lesson 0-13

1.

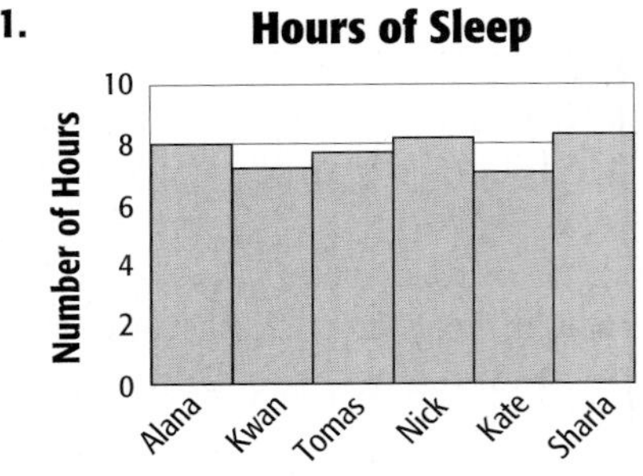

3.

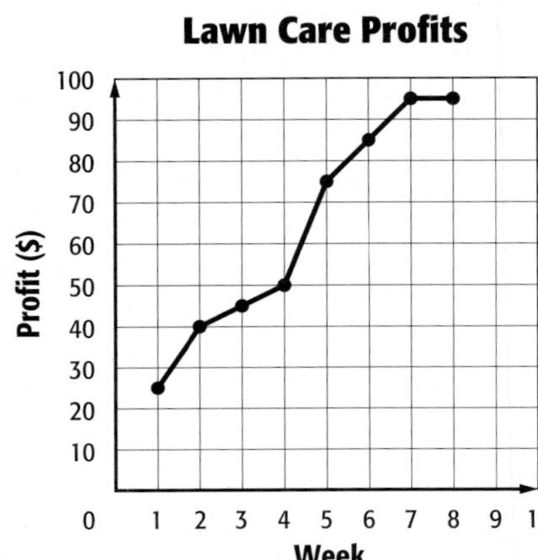

5.

Stem	Leaf
1	8 8
2	1 3 6 6 6 8
3	0 1 1 2 3 4
4	7

Key: 1|8 = 18

Removing 47 leaves Q_1 the same, changes Q_2 to 27 and Q_3 to 31.

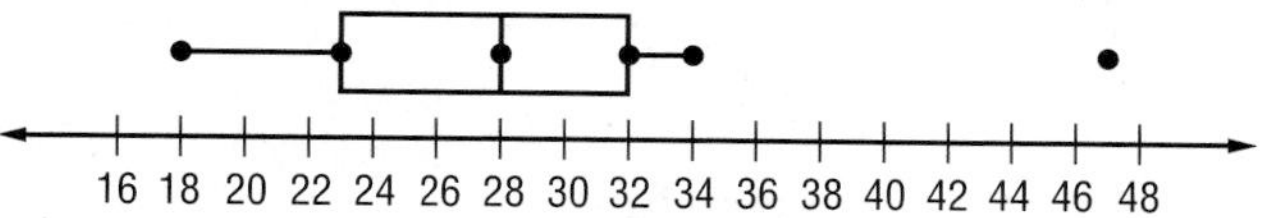

7.

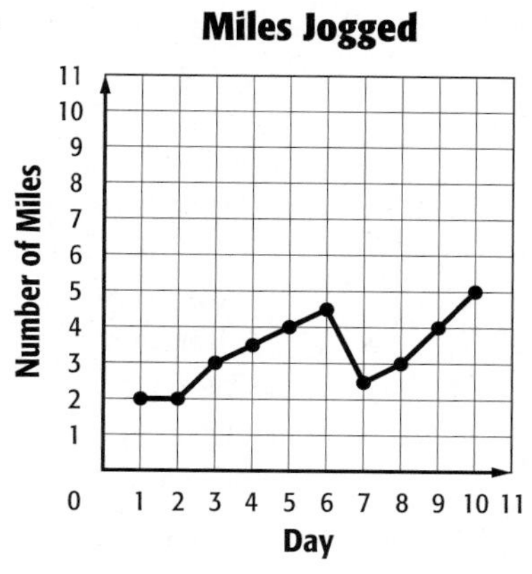

Chapter 1 Expressions, Equations, and Functions

Page 3 Get Ready

1. $\frac{2}{3}$ **3.** 3 **5.** *simplest form* **7.** 19 **9.** $\frac{8}{11}$
11. 8.2 cm **13.** 20 m **15.** 34.02 **17.** 1.9 **19.** 0.56

Pages 5–9 Lesson 1-1

1. Sample answer: the product of 2 and m
3. Sample answer: a squared minus 18 times b
5. $6 - t$ **7.** $1 - \frac{r}{7}$ **9.** $n^3 + 5$ **11.** Sample answer: four times a number q **13.** Sample answer: 15 plus r **15.** Sample answer: 3 times x squared
17 Sample answer: 6 more than the product 2 times a
19. $7 + x$ **21.** $5n$ **23.** $\frac{f}{10}$ **25.** $3n + 16$ **27.** $k^2 - 11$
29. $\pi r^2 h$ **31.** Sample answer: twenty-five plus six times a number squared **33.** Sample answer: three times a number raised to the fifth power divided by two
35 a. Words: $\frac{3}{4}$ of the number of dreams
Expression: $\frac{3}{4} \cdot d$
The expression is $\frac{3}{4}d$.
b. $\frac{3}{4}(28) = 21$ dreams

37a.

10^2	•	10^1	=	10 • 10 • 10	=	10^3
10^2	•	10^2	=	10 • 10 • 10 • 10	=	10^4
10^2	•	10^3	=	10 • 10 • 10 • 10 • 10	=	10^5
10^2	•	10^4	=	10 • 10 • 10 • 10 • 10• 10	=	10^6

37b. $10^2 \cdot 10^x = 10^{(2 + x)}$
37c. The exponent of the product of two powers is the sum of the exponents of the powers with the same bases. **39.** Sample answer: x is the number of minutes it takes to walk between my house and school. $2x + 15$ represents the amount of time in minutes I spend walking each day since I walk to and from school and I take my dog on a 15 minute walk. **41.** 6 **43.** D **45.** $\frac{31}{36}$

47.

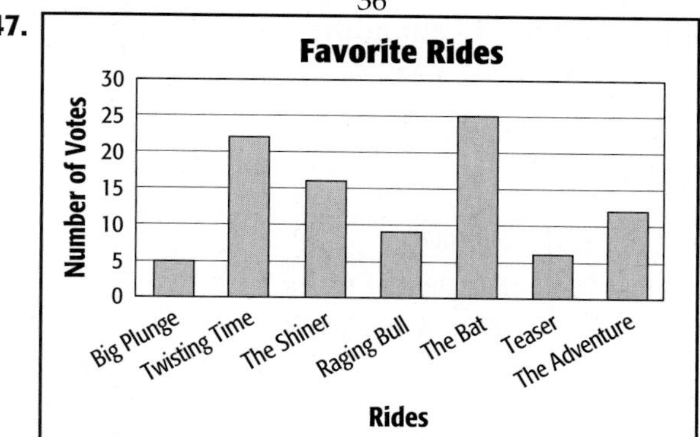

49. mean = 5.6; median = 6.5; mode = 7
51. mean = 15.25; median = 15.5; mode = 24
53. $\frac{21}{55}$ **55.** $\frac{20}{9}$ **57.** 1.46 **59.** 24.61 **61.** 21.16

Pages 10–15 Lesson 1-2

1. 81 **3.** 243
5 $5 \cdot 5 - 1 \cdot 3 = 25 - 3$
$= 22$
7. 28 **9.** 12 **11.** 20 **13.** $20 + 3 \times 4.95$; \$34.85
15. 49 **17.** 64 **19.** 14 **21.** 36 **23.** 14
25. 142 **27.** 36 **29.** 3 **31.** 1
33 $(2t + 3g) \div 4 = (2(11) + 3(2)) \div 4$
$= (22 + 6) \div 4$
$= (28) \div 4$
$= 7$
35. 149 **37.** $3344 - 148 = 3196$ **39.** 16 **41.** 729
43. 177 **45.** 324 **47.** 29 **49.** 4080 **51.** $\frac{97}{31}$ **53.** 0
55. $28(7) + 12(9.75) + 30(7) + 15(9.75)$; \$669.25
57 a.

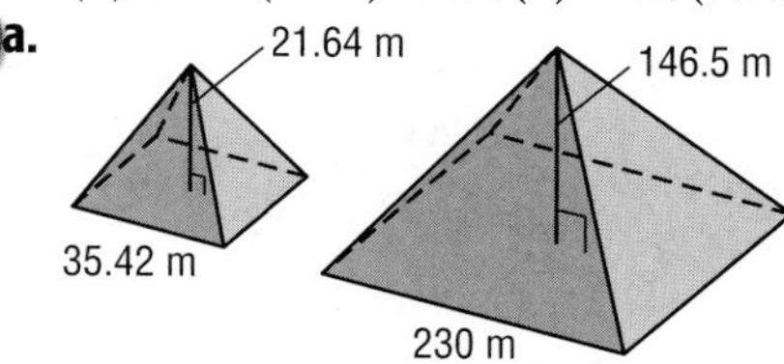

b. Words: one third times 230 squared times 146.5 m minus one third times 35.42 squared times 21.64 m

c. Expression: $\frac{1}{3}(230)^2(146.5) - \frac{1}{3}(35.42)^2(21.64)$
$\approx 2583283.33 - 9049.68$
$\approx 2574233.656 \text{ m}^3$
59. Curtis; Tara subtracted $10 - 9$ before multiplying 4 by 10. **61.** $5 + 4 - 3 - 2 - 1$
63. Sample answer: Area of a trapezoid: $\frac{1}{2}h(b_1 + b_2)$;

according to the order of operations, you have to add the lengths of the bases together first and then multiply by the height and by $\frac{1}{2}$. **65.** A

67. $\frac{5}{16}$; $\frac{1}{4}$; Experimental probability is what happens in the trials. In this case the experimental probability is the probability that the customers have actually received popcorn. Theoretical probability is what should happen theoretically.

69. 14 minus 9 times c **71.** the difference of 4 and v divided by w **73.** 9π units^2 **75.** $12b$ units^2

77. 2.57 **79.** 13.192 **81.** $\frac{2}{3}$

Pages 16–22 Lesson 1-3

1. $(1 \div 5)5 \cdot 14$
$= \frac{1}{5} \cdot 5 \cdot 14$ Substitution
$= (1) \cdot 14$ Multiplicative Inverse
$= 14$ Multiplicative Identity

3. $5(14 - 5) + 6(3 + 7) = 5(9) + 6(10)$ Substitution
$= 45 + 60$ Substitution
$= 105$ Substitution

5. $23 + 42 + 37$
$= 23 + 37 + 42$ Commutative (+)
$= (23 + 37) + 42$ Associative (+)
$= 60 + 42$ Substitution
$= 102$ Substitution

7. $3 \cdot 7 \cdot 10 \cdot 2$
$= 3 \cdot 2 \cdot 7 \cdot 10$ Commutative (×)
$= (3 \cdot 2) \cdot (7 \cdot 10)$ Associative (×)
$= 6 \cdot 70$ Substitution
$= 420$ Substitution

9 $3(22 - 3 \cdot 7) = 3(22 - 21)$ Multiply 3 and 7.
$= 3(1)$ $22 - 21 = 1$
$= 3$ Multiplicative Identity

11. $\frac{3}{4}[4 \div (7 - 4)]$
$= \frac{3}{4}[4 \div 3]$ Substitution
$= \frac{3}{4} \times \frac{4}{3}$ Substitution
$= 1$ Multiplicative Inverse

13. $2(3 \cdot 2 - 5) + 3 \cdot \frac{1}{3}$
$= 2(6 - 5) + 3 \cdot \frac{1}{3}$ Substitution
$= 2(1) + 3 \cdot \frac{1}{3}$ Substitution
$= 2 + 3 \cdot \frac{1}{3}$ Multiplicative Identity
$= 2 + 1$ Multiplicative Inverse
$= 3$ Substitution

15. $2 \cdot \frac{22}{7} \cdot 14^2 + 2 \cdot \frac{22}{7} \cdot 14 \cdot 7$
$= 2 \cdot \frac{22}{7} \cdot 196 + 2 \cdot \frac{22}{7} \cdot 14 \cdot 7$ Substitution
$= \frac{44}{7} \cdot 196 + \frac{44}{7} \cdot 14 \cdot 7$ Substitution
$= 1232 + 616$ Substitution
$= 1848$ Substitution
The surface area is about 1848 in^2.

17. $25 + 14 + 15 + 36$
$= 25 + 15 + 14 + 36$ Commutative(+)
$= (25 + 15) + (14 + 36)$ Associative(+)
$= 40 + 50$ Substitution
$= 90$ Substitution

19. $3\frac{2}{3} + 4 + 5\frac{1}{3} = 3\frac{2}{3} + 5\frac{1}{3} + 4$ Commutative (+)
$= \left(3\frac{2}{3} + 5\frac{1}{3}\right) + 4$ Associative (+)
$= 9 + 4$ Substitution
$= 13$ Substitution

21. $4.3 + 2.4 + 3.6 + 9.7$
$= 4.3 + 9.7 + 2.4 + 3.6$ Commutative (+)
$= (4.3 + 9.7) + (2.4 + 3.6)$ Associative (+)
$= 14 + 6$ Substitution
$= 20$ Substitution

23. $12 \cdot 2 \cdot 6 \cdot 5 = 12 \cdot 6 \cdot 2 \cdot 5$ Commutative (×)
$= (12 \cdot 6) \cdot (2 \cdot 5)$ Associative (×)
$= 72 \cdot 10$ Substitution
$= 720$ Substitution

25. $0.2 \cdot 4.6 \cdot 5 = (0.2 \cdot 4.6) \cdot 5$ Associative (×)
$= 0.92 \cdot 5$ Substitution
$= 4.6$ Substitution

27. $1\frac{5}{6} \cdot 24 \cdot 3\frac{1}{11} = 1\frac{5}{6}\left(24 \cdot 3\frac{1}{11}\right)$ Associative (×)
$= 1\frac{5}{6} \cdot \frac{216}{11}$ Substitution
$= 136$ Substitution

29a. Sample answer: $2(10.95) + 3(7.5) + 2(5) + 5(18.99)$; $2(10.95 + 5) + 3(7.5) + 5(18.99)$

29b. \$149.35

31 $4(-1) + 9(4) - 2(6) = -4 + 36 - 12$
$= 32 - 12$
$= 20$

33. -18 **35.** 192 **37.** Additive Identity; $35 + 0 = 35$ **39.** 0; Additive Identity **41.** 7; Reflexive Property **43.** 3; Multiplicative Identity **45.** 2; Commutative Property **47.** 3; Multiplicative Inverse **49a.** $4 + 5x + 4 + 5x + 3y$ **b.** 49 units

51. 88 units

53 **a.**

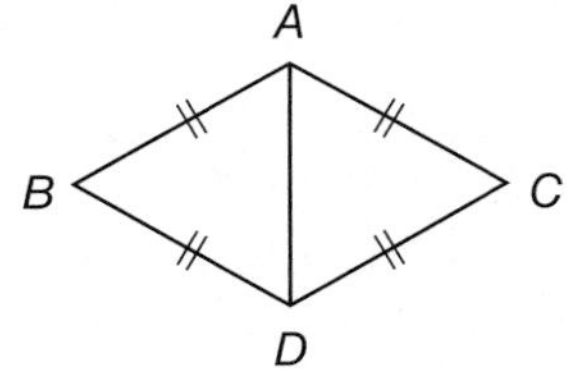

b. $\overline{AD} \cong \overline{AD}$ by the Reflexive Property. The Transitive Property shows that if $\overline{AB} \cong \overline{AC}$ and $\overline{AC} \cong \overline{DC}$, then $\overline{AB} \cong \overline{DC}$ and if $\overline{AB} \cong \overline{BD}$ and $\overline{AB} \cong \overline{AC}$, then $\overline{BD} \cong \overline{AC}$.

c. Since the sides are all congruent, each side has a length x. So, $P = x + x + x + x$.

55. Sample answer: You cannot divide by 0.

57. Sometimes; when a number is subtracted from itself then it holds but otherwise it does not.

59. $(2j)k = 2(jk)$; The other three sentences illustrate the Commutative Property of Addition or Multiplication. This equation represents the Associative Property of Multiplication. **61.** D **63.** C **65.** 14 **67.** 6 **69.** 26 ft; 40 ft^2

71. about 64.7 % **73.** $\frac{23}{2}$ **75.** $\frac{6}{35}$ **77.** $\frac{6}{11}$ **79.** 6

Pages 23–29 Lesson 1-4

1. 25(12 + 15); \$675 **3.** $\left(6 + \frac{1}{9}\right)9$; 55 **5.** $g(5) + (-9)(5)$; $5g - 45$ **7.** simplified

9. $4(2x + 6)$
$= 4(2x) + 4(6)$ Distributive Property
$= 8x + 24$ Multiply.

11 $4(5 + 3 + 4) = 4(8 + 4)$
$= 4(12)$
$= 48$ activities

13. 6(4) + 6(5); 54 **15.** 6(6) − 6(1); 30 **17.** 14(8) − 14(5); 42 **19.** 4(7) − 4(2); 20 **21.** 7(500 − 3); 3479

23. $36\left(3 + \frac{1}{4}\right)$; 117 **25.** $2(x) + 2(4)$; $2x + 8$

27. $4(8) + (-3m)(8)$; $32 - 24m$ **29.** $18r$ **31.** $2m + 7$

33. $34 - 68n$ **35.** $13m + 5p$ **37.** $4fg + 17g$

39. $7(a^2 + b) - 4(a^2 + b)$
$= 7a^2 + 7b - 4a^2 - 4b$ Substitution
$= 7a^2 - 4a^2 + 7b - 4b$ Commutative (+)
$= (7 - 4)a^2 + (7 - 4)b$ Distributive Prop.
$= 3a^2 + 3b$ Substitution

41 A hexagon has six sides so an expression for the perimeter is $6(3x + 5)$.
$6(3x + 5) = 6(3x) + 6(5)$
$= 18x + 30$ units

43. $14m + 11g$ **45.** $12k^3 + 12k$ **47.** $19x + 8$

49. $9 - 54b$ **51.** $12c - 6cd^2 + 6d$ **53.** $7y^3 + y^4$

55a. $2(x + 3)$

55b.

Area	Factored form
2x + 6	2(x + 3)
3x + 3	3(x + 1)
3x − 12	3(x − 4)
5x + 10	5(x + 2)

55c. Divide each term of the expression by the same number. Then write the expression as a product. **57.** It should be considered a property of both. Both operations are used in $a(b + c) = ab + ac$. **59.** You can use the Distributive Property to calculate quickly by expressing any number as a sum or difference of a more convenient number. Answers should include the following: Both methods result in the correct answer. In one method you multiply then add, and in the other you add then multiply. **61.** G **63.** about $\frac{1}{3}$ or 33%

65. $0.24 \cdot 8 \cdot 7.05 = (0.24 \cdot 8) \cdot 7.05$ Associative (×)
$= 1.92 \cdot 7.05$ Substitution
$= 13.536$ Substitution

67. $\frac{4[6(30) + 3(20)]}{60}$; 16 hours **69.** 21:48

71. 384 in^2 **73.** 15 **75.** 60 **77.** 192

Pages 31–37 Lesson 1-5

1. 13 **3.** 12 **5.** B **7.** 3 **9.** all real numbers **11.** {12} **13.** {5} **15.** {16} **17.** {3} **19.** 14 **21.** 2 **23.** 2 **25.** 5 **27.** no solution **29.** all real numbers

31 $(2^4 - 3 \cdot 5)q + 13 = (2 \cdot 9 - 4^2)q + \left(\frac{3 \cdot 4}{12} - 1\right)$
$(16 - 15)q + 13 = (18 - 16)q + (1 - 1)$
$1q + 13 = 2q + 0$
$1q + 13 = 2q$
$13 = 2q - 1q$
$13 = 1q$
$q = 13$

33. 41 students

35 Words: the number of calories equals 2836 plus 3091
Expression: $C = 2836 + 3091$ Solve: $C = 5927$

37.

x	3x − 2	y
−2	3(−2) − 2	−8
−1	3(−1) − 2	−5
0	3(0) − 2	−2
1	3(1) − 2	1
2	3(2) − 2	4

39. 20 **41.** 66 **43.** 5 **45.** $c = 15$ **47a.** $5 = \frac{1000}{r}$; 20

47b.

Intial Pressure p_1 (mm Hg)	Final Pressure p_2 (mm Hg)	Resistance r (mm Hg/L/min)	Blood Flow Rate F (L/min)
100	0	20	5
100	0	30	≈ 3.33
165	5	40	4
90	30	10	12

49. solution **51.** not a solution **53.** solution **55.** solution

57.

x	3x + 5	y
−2	3(−2) + 5	−1
−1	3(−1) + 5	2
0	3(0) + 5	5
1	3(1) + 5	8
2	3(2) + 5	11

59.

x	$\frac{1}{2}x + 2$	y
−2	$\frac{1}{2}(-2) + 2$	1
−1	$\frac{1}{2}(-1) + 2$	1.5
0	$\frac{1}{2}(0) + 2$	2
1	$\frac{1}{2}(1) + 2$	2.5
2	$\frac{1}{2}(2) + 2$	3

61a. w ; $2 + w$

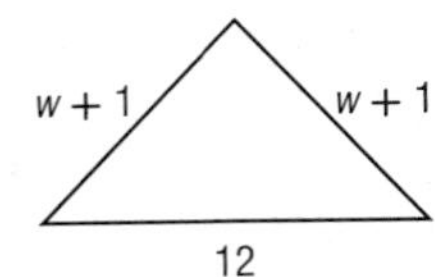

61b. perimeter of rectangle $= 2(2 + w) + 2w$ or $4 + 4w$; perimeter of triangle $= 2(w + 1) + 12 = 2w + 14$.
61c. $4 + 4w = 2w + 14$; $w = 5$ in.

63 b.

Layers	1	2	3	4	5	6	7
Cubes	4	8	12	16	20	24	28

c. From the table, we can tell that each layer adds 4 more cubes. Notice $8 - 4 = 4$; $12 - 8 = 4$; $16 - 12 = 4$; $20 - 16 = 4$; $24 - 20 = 4$; $28 - 24 = 4$

d. The number of cubes is 4 times the number of layers, or $c = 4L$.

65. Sample answer: $3x + 12 = 3(x + 4)$ **67.** Tom; Li-Cheng added 6 + 4 instead of dividing 6 by 8. She did not follow the order of operations. **69.** Sample answer: $3x - 2 = -23$ **71.** C **73.** G **75.** 30 (500 + 750) **77.** $p = \frac{1}{12}$; Multiplicative Inverse **79.** 1040 in^3 **81.** $\frac{3}{20}$ **83.** estimate; 10 gal **85.** 6.74 **87.** 1.65 **89.** $\frac{29}{28}$

Pages 38–44 *Lesson 1-6*

1.

x	y
4	3
−2	2
5	−6

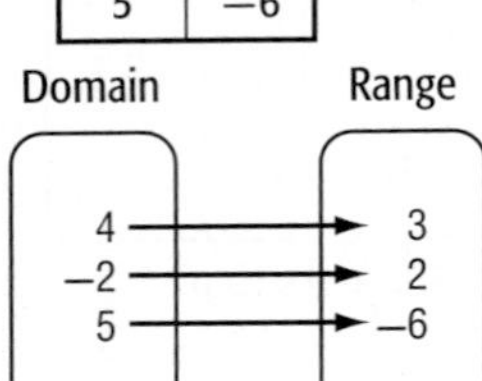

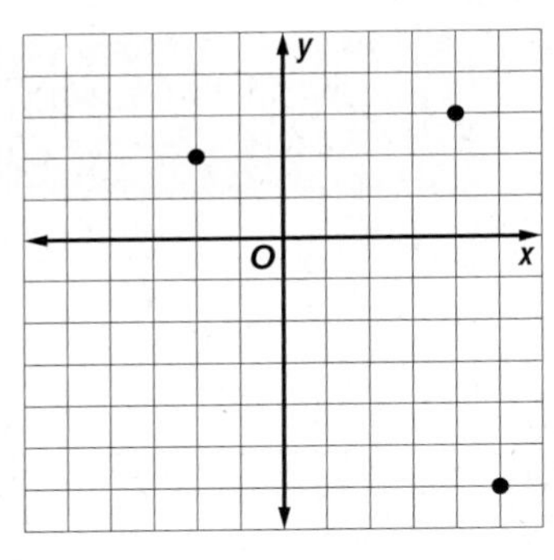

D = {−2, 4, 5}; R = {−6, 2, 3}

3. I: the temperature of the compound; D: the pressure of the compound **5.** I: number of concert tickers, D: cost of tickets **7.** The track team starts by running or walking, and then stops for a short period of time, then continues at the same pace. Finally, they run or walk at a slower pace.

9.

x	y
0	0
−3	2
6	4
−1	1

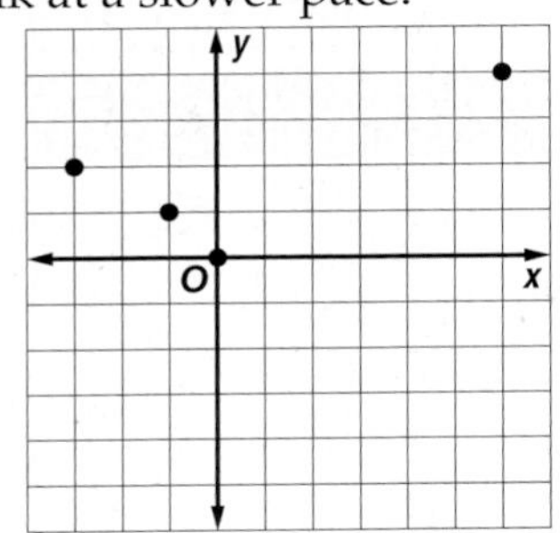

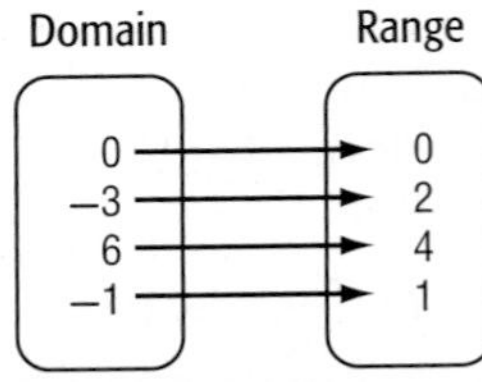

D = {0, −3, 6, −1}; R = {0, 2, 4, 1}

11.

x	y
6	1
4	−3
3	2
−1	−3

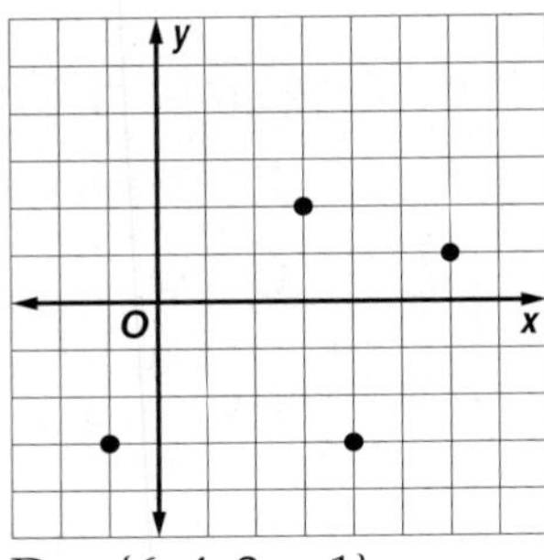

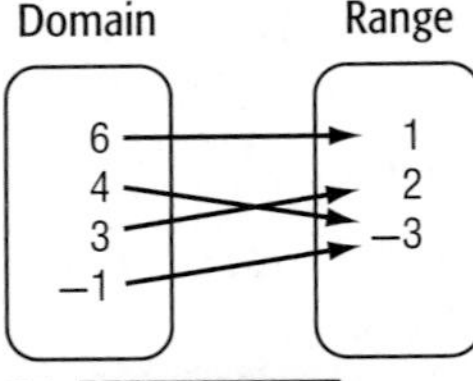

D = {6, 4, 3, −1}; R = {1, −3, 2}

13.

x	y
6	7
3	−2
8	8
−6	2
2	−6

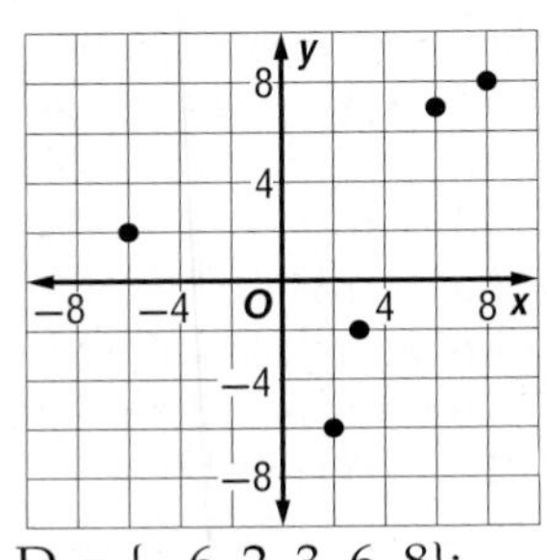

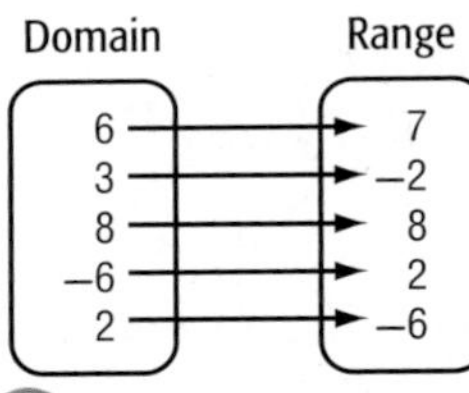

D = {−6, 2, 3, 6, 8}; R = {−6, −2, 2, 7, 8}

15 The number of students who attend is the independent variable because it does not depend on the amount of food there will be. The amount of food is the dependent variable because it depends on the number of students who attend.

17. The bungee jumper starts at the maximum height, and then jumps. After the initial jump, the jumper bounces up and down until coming to a rest.

19 Use the graph to determine what is happening to the value of the baseball card. The values are continually increasing.

21. (1, 5); The dog walker earns \$5 for walking 1 dog. **23.** I: number of dogs walked; D: amount earned **25.** (5, 6); In the year 2005, sales were about \$6 million. **27.** {(1, 2.50), (2, 5.50), (5, 10.00), (8, 18.75)}; D = {1, 2, 5, 8}; R = {2.50, 5.50, 10.00, 18.75} **29.** {(4, −1), (8, 9), (−2, −6), (7, −3)} **31.** {(4, −2), (−1, 3), (−2,−1), (1, 4)}

33. Sample answer:

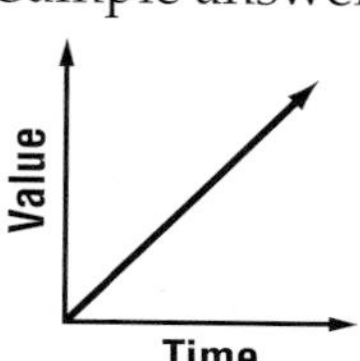

35. Sample answer:

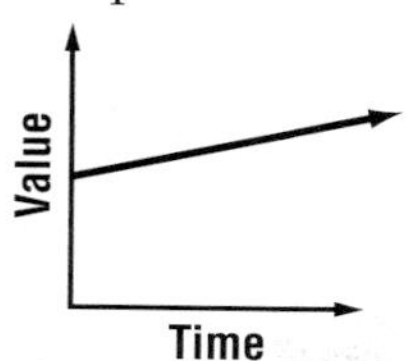

37 a.

b	$w = 2\left(\frac{b}{3}\right)$	w
100	$w = 2\left(\frac{100}{3}\right)$	66.7
105	$w = 2\left(\frac{105}{3}\right)$	70
110	$w = 2\left(\frac{110}{3}\right)$	73.3
115	$w = 2\left(\frac{115}{3}\right)$	76.7
120	$w = 2\left(\frac{120}{3}\right)$	80
125	$w = 2\left(\frac{125}{3}\right)$	83.3
130	$w = 2\left(\frac{130}{3}\right)$	86.7

b. The independent variable is the body weight b. The dependent variable is the water weight w.

c. The domain is the set of b values. D = {100, 105, 110, 115, 120, 125, 130}. The range is the set of all w values. R = {66.7, 70, 73.3, 76.7, 80, 83.3, 86.7}

Water Weight Per Body Weight

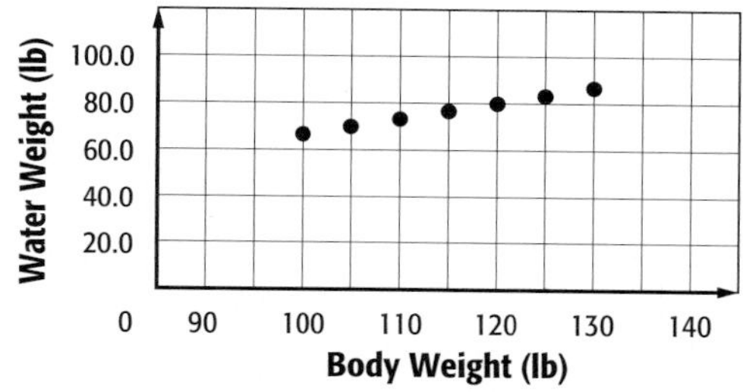

d. Graph the following ordered pairs: (66.7, 100), (70, 105), (73.3, 110), (76.7, 115), (80, 120), (83.3, 125), (86.7, 130).

Body Weight Per Water Weight

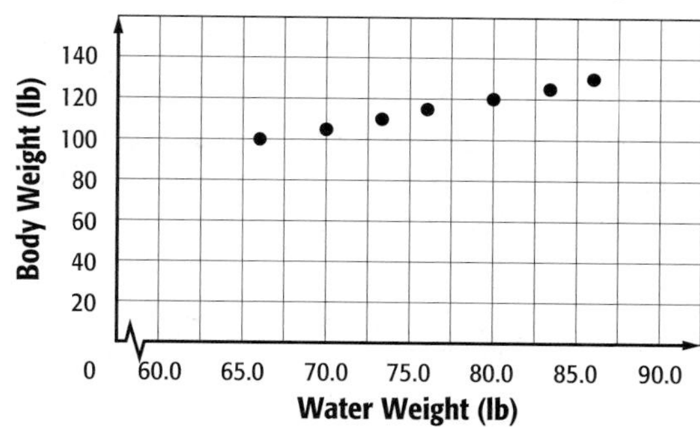

This graph shows what a person's body weight would be based on their water weight.

41. Reversing the coordinates gives (1, 0), (3, 1), (5, 2), and (7, 3).

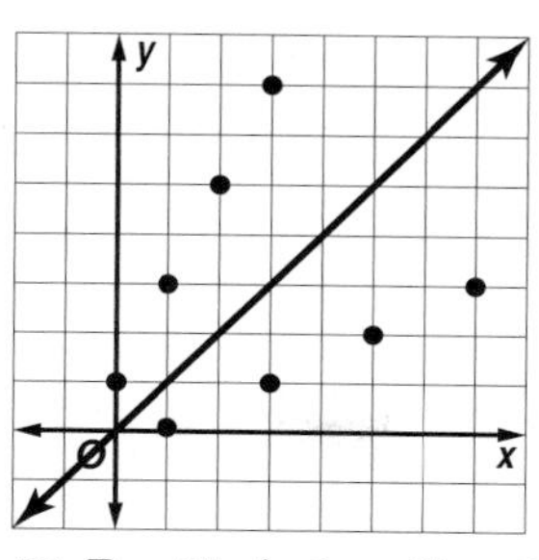

Each point in the original relation is the same distance from the line as the corresponding points of the reverse relation. The graphs are symmetric about the line $y = x$.

43. B **45.** (−1, −3) **47.** {2} **49.** {3} **51.** $\frac{1}{8}$ **53.** 50.27 cm **55.** 64 **57.** 6.25 **59.** 49

Pages 45–52 Lesson 1-7

1. Yes; for each input there is exactly one output. **3.** No; the domain value 2 is paired with both 2 and −4. **5.** no; when $x = 0$, $y = 1$ and $y = 6$ **7.** yes; its graph passes the vertical line test
9a. {(0, 48,560), (1, 48,710), (2, 48,948), (3, 49,091)}
9b.

School Enrollment

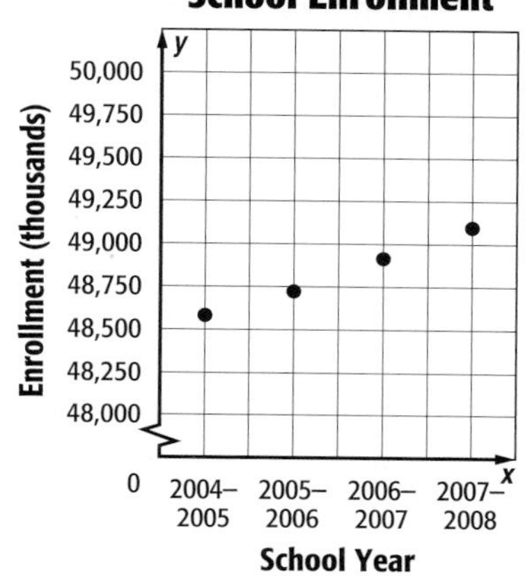

11 $f(-3) = 6(-3) + 7$
$= -18 + 7$
$= -11$

13. $6r - 5$ **15.** $a^2 + 5$ **17.** $6q + 13$ **19.** $b^2 - 7$ **21.** no **23.** yes **25.** yes **27.** yes **29.** yes **31.** yes **33.** −1 **35.** 14 **37.** −4 **39.** $-8y - 3$ **41.** $-2c + 7$ **43.** $-10d - 15$

45 a. Create a table using the rule given.

t	$0.8t + 72$	$f(t)$
0	0.8(0) + 72	72
10	0.8(10) + 72	80
20	0.8(20) + 72	88
30	0.8(30) + 72	96
40	0.8(40) + 72	104
50	0.8(50) + 72	112

Plot the ordered pairs on a coordinate plane.

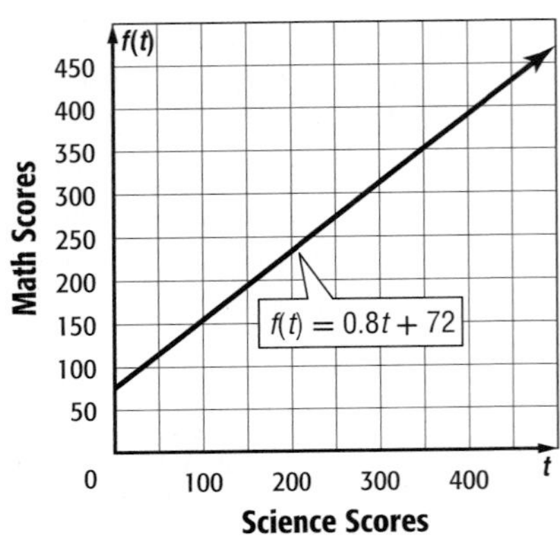

b. $308 = 0.8t + 72$
$236 = 0.8t$
$295 = t$

c. The domain is the set of science scores, the range is the set of math scores.

47 The graph represents a function because each x-value is paired with only one y-value.

49. Sample answer: {(−2, 3), (0, 3), (2, 5)}

Domain | Range
−2 → 3
0 → 3
2 → 5

51. $f(g + 3.5) = -4.3g - 17.05$ **53.** Sample answer: $f(x) = 3x + 2$ **55.** Sample answer: Functions can be used in traffic safety studies to determine the relationship between the speed of a car and the distance it takes to stop. This can help plan intersections and speed limits. This function can also help law enforcement officials to understand the cause of an accident. **57.** J **59.** her first game **61.** $\frac{13}{2}$ **63.** $4(1.99) + 10(0.25) + 4(1.85) = C$, where C is the cost of the items Tom needs. $C = 17.86$, so the cost is \$17.86, which is not less than \$10. **65.** sample answer: two thirds times x **67.** 38.016 cm^3 **69.** 288,000 mm^3 **71.** −1 **73.** 40 **75.** 65

Pages 54–59 Lesson 1-8

1. H: the game is on Saturday; C: Eduardo will play **3.** H: $52 - 4x = 28$; C: $x = 6$ **5.** H: two lines are perpendicular; C: they form right angles; If two lines are perpendicular, then they form right angles.

7 No valid conclusion. The last number could be a 5 instead of a 0.

9. No valid conclusion; the last digit is a 5.

11. A one-page paper is assigned. **13.** $y = -1$

15 Hypothesis: You are in a grocery store. Conclusion: you will buy food.

17. H: x equals y, and y equals z; C: x equals z

19. H: you play basketball; C: you are tall

21. H: it is after class; C: Joe will go to the mall; If it is after class, then Joe will go to the mall. **23.** H: $m < 12$; C: $5m - 8 < 52$; If $m < 12$, then $5m - 8 < 52$. **25.** H: two numbers are even; C: their sum is an even number; If two numbers are even, then their sum is an even number. **27.** H: you are a science teacher; C: you like to conduct experiments; If you are a science teacher, then you like to conduct experiments. **29.** No valid conclusion; the statement does not say that Belinda will not receive an A in the course if she scores lower than a 90% on the exam. **31.** Belinda did not score higher than 90% on the exam. **33.** You attend the banquet, but do not eat because you are feeling ill. **35.** $6 \div 3 = 2$ **37.** $(-1)^2 = 1$ **39.** The number is 2, an even number.

43 **a.** A, D **b.**

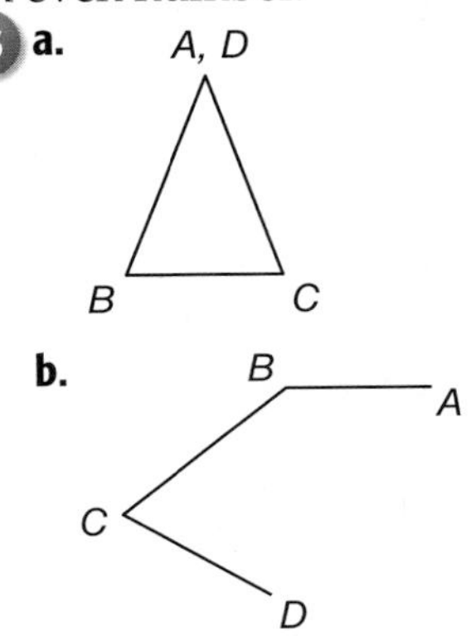

45. No; sample answer: Let $b = 4$ and $c = 5$; then $2 + (4 \times 5) \neq (2 + 4)(2 + 5)$. **47.** Sample answer: If you go swimming, then you get wet; H: you go swimming; C: you get wet. **49.** Sample answer: If you live in Ohio, then you live in Columbus. You do not have to live in Columbus, you could live in Canton. **51.** A **53.** G **55.** yes **57.** yes **59.** $\frac{1}{5}$ **61.** 38 cm **63.** 81 **65.** 150.0625 **67.** 625

Pages 62–66 Chapter 1 Study Guide and Review

1. true **3.** false; not in simplest form **5.** true **7.** false; multiplicative identity **9.** The product of 3 and x squared. **11.** $x + 9$ **13.** $4x - 5$ **15.** 216 **17.** $2.50 + 3.25g$ **19.** 18 **21.** 2 **23.** 3 **25.** 5 **27.** $2.75(3) + 4.25(2)$; \$16.75

29. $[5 \div (8 - 6)]$
$= [5 \div 2]\frac{2}{5}$ Substitution
$= \frac{5}{2} \cdot \frac{2}{5}$ Substitution
$= 1$ Multiplicative Inverse

31. $2 \cdot \frac{1}{2} + 4(4 \cdot 2 - 7)$
$= 2 \cdot \frac{1}{2} + 4(8 - 7)$ Substitution
$= 2 \cdot \frac{1}{2} + 4(1)$ Substitution
$= 1 + 4(1)$ Multiplicative Inverse
$= 1 + 4$ Multiplicative Identity
$= 5$ Substitution

Selected Answers and Solutions

33. $7\frac{2}{5} + 5 + 2\frac{3}{5}$
$= 7\frac{2}{5} + 2\frac{3}{5} + 5$ Commutative (+)
$= (7\frac{2}{5} + 2\frac{3}{5}) + 5$ Associative (+)
$= 10 + 5$ Substitution
$= 15$ Substitution

35. $5.3 + 2.8 + 3.7 + 6.2$
$= 5.3 + 3.7 + 2.8 + 6.2$ Commutative (+)
$= (5.3 + 3.7) + (2.8 + 6.2)$ Associative (+)
$= 9 + 9$ Substitution
$= 18$ Substitution

37. $(2 + 3)6$
$= (2)6 + (3)6$
$= 12 + 18$
$= 30$

39. $8(6 - 2)$
$= 8(6) - 8(2)$
$= 48 - 16$
$= 32$

41. $-2(5 - 3)$
$= -2(5) - -2(3)$
$= -10 + 6$
$= -4$

43. $3(x + 2)$
$= 3(x) + 3(2)$
$= 3x + 6$

45. $6(d - 3)$
$= 6(d) - 6(3)$
$= 6d - 18$

47. $(9y - 6)(-3)$
$= (9y)(-3) - (6)(-3)$
$= -27y + 18$

49. $4(3 + 5 + 4)$; 48 **51.** 7 **53.** 9 **55.** 5 **57.** 9

59.

x	y
1	3
2	4
3	5
4	6

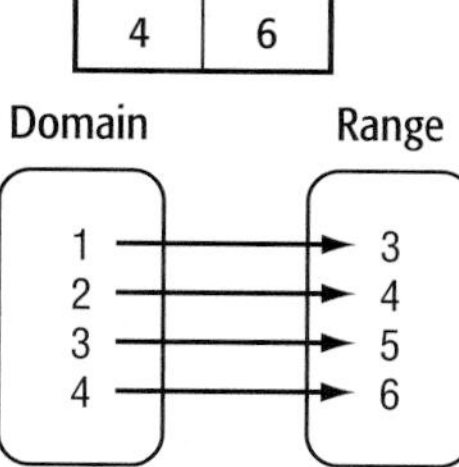

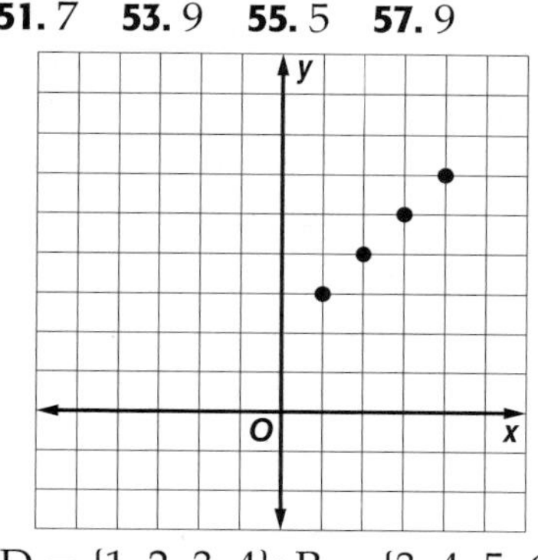

D = {1, 2, 3, 4}; R = {3, 4, 5, 6}

61.

x	y
−2	4
−1	3
0	2
−1	2

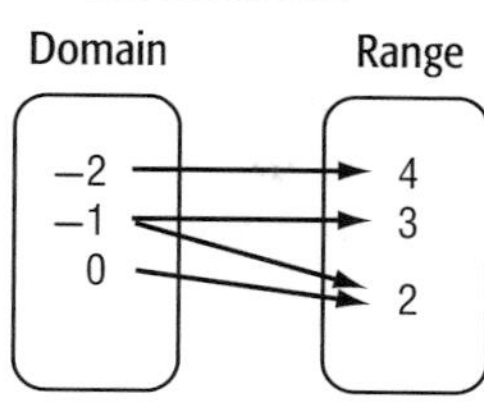

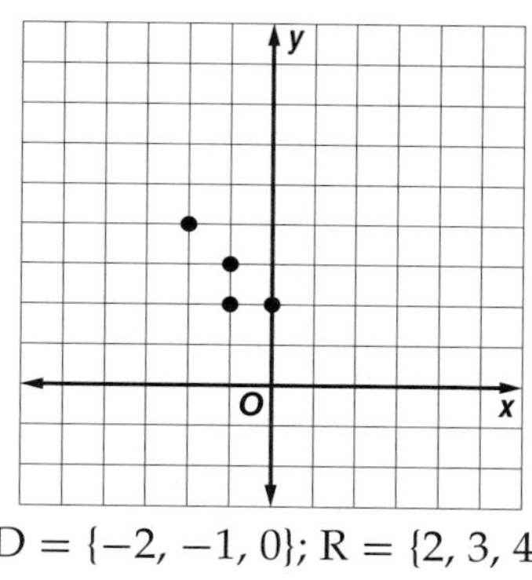

D = {−2, −1, 0}; R = {2, 3, 4}

63. {(−2, −2), (0, −3), (2, −2), (2, 0), (4, −1)}
65. function **67.** function **69.** −2 **71.** 4
73. $2m + 8$
75.

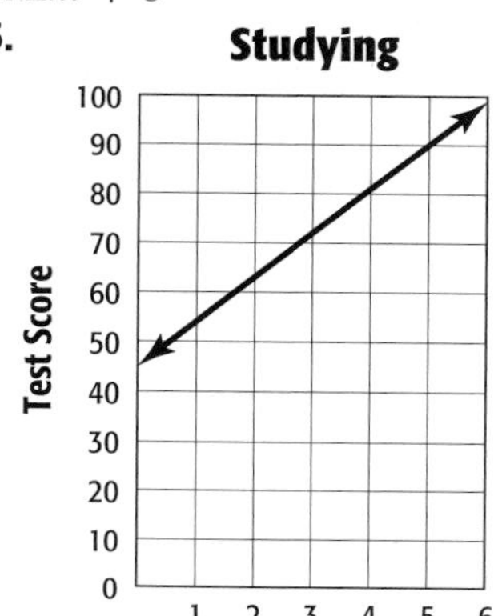

77. Hypothesis: $2x + 7 > 31$; Conclusion: $x > 12$
79. $x = 16$

Chapter 2 Linear Equations

Page 73 Get Ready

1. $3n - 4$ **3.** $2b - 11$ **5.** 2 **7.** 11 **9.** 11
11. \$28.40 **13.** 20% **15.** 21%

Pages 75–80 Lesson 2-1

1. $15 - 3r = 6$
3 Words: A number squared plus 12 is the same as the quotient of p and 4.
Equation: $n^2 = p \div 4$
The equation is $n^2 = p \div 4$.
5. $8 + 3k = 5k - 3$ **7.** $\frac{25}{t} + 6 = 2t + 1$ **9.** $1900 + 30w = 2500$; 20 **11.** $P = 5s$ **13.** $4\pi r^2 = S$
15. Sample answer: The product of seven and m minus q is equal to 23. **17.** Sample answer: Three times the sum of g and eight is the same as 4 times h minus 10. **19.** Sample answer: A team of gymnasts completed in a regional meet. Each member of the team won 3 medals. There were a total of 45 medals won by the team. How many team members were there? **21.** $f - 5g = 25 - f$
23. $4(14 + c) = a^2$ **25.** $3 \cdot 10 = 12f$; $2\frac{1}{2}$ flats

27 Words: C is five ninths times the difference of F and 32.
Equation: $C = \frac{5}{9} \cdot (F - 32)$
The equation is $C = \frac{5}{9}(F - 32)$.

29. $I = prt$ **31.** Sample answer: Four times m is equal to fifty-two. **33.** Sample answer: Fifteen less than the square of r equals the sum of t and nineteen. **35.** Sample answer: One third minus four fifths of z is four thirds of y cubed.
37. Sample answer: Ashley has a credit card that charges 12% interest on the principal balance. If Ashley's payment was \$224, what was the principal balance on the credit card? **39.** Sample answer: Fred was teaching his friends a new card game. Each player gets 5 cards, and 7 cards are placed in the center of the table. Since there are 52 cards in a deck, find how many players can play the game. **41.** C **43.** D

45 Words: the number of tent stakes + packets of drink mix + bottles of water = 17
$d = 3t$
$w = t + 2$
$$t + d + w = 17$$
$$t + 3t + (t + 2) = 17$$
$$5t + 2 = 17$$
$$5t + 2 - 2 = 17 - 2$$
$$5t = 15$$
$$t = 3$$
She brought 3 tent stakes.

47. Sample answer: My favorite television show has 30 new episodes each year. So far eight have aired. How many new episodes are left?
49. $\ell = \frac{P - 2w}{2}$ **51.** C **53.** 180 m **55.** I could have been born in Florida, moved to Kentucky, and still live in Kentucky. **57.** 10 is divisible by 2, but not by 4. **59a.** independent: number of sides; dependent: interior angle sum **59b.** Domain: all integers greater than or equal to 3; Range: all positive integer multiples of 180 **59c.** Discrete; sample answer: There cannot be a polygon with 3.5 sides, so the function cannot be continuous.
59d. 1440 **61.** 1,000,000 **63.** 125

Pages 83–89 *Lesson 2-2*

1. 28 **3.** $\frac{5}{6}$ **5.** 9 **7.** -4.1 **9.** $-3\frac{1}{4}$ **11.** 16
13. $\frac{10}{9}$ or $1\frac{1}{9}$ **15.** $-\frac{4}{7}$ **17.** \$22.75 **19.** 116 **21.** 22
23. -11

25
$$-16 - (-t) = -45$$
$$-16 + 16 - (-t) = -45 + 16$$
$$t = -29$$
Check: $-16 + (-29) = -45$
$$-45 = -45 \text{ Yes}$$

27. -32 **29.** -7 **31.** $1\frac{1}{8}$ **33.** $1\frac{2}{7}$ **35.** -708 **37.** 33
39. -2 **41.** $-1\frac{1}{9}$ **43.** $24.9 = 8.1 + t$; 16.8 hours

45. -77 **47.** $\frac{16}{3}$ **49.** -10 **51.** $-\frac{10}{7}$ or $-1\frac{3}{7}$
53. 18 **55.** 225 **57.** $\frac{2}{3} = -8n$; $n = -\frac{1}{12}$
59. $\frac{4}{5} = \frac{10}{16}n$; $n = \frac{32}{25}$

61 Words: Four and four fifths times a number is one and one fifth.
Equation: $4\frac{4}{5} \cdot n = 1\frac{1}{5}$
The equation is $4\frac{4}{5}n = 1\frac{1}{5}$.
Solve:
$$4\frac{4}{5}n = 1\frac{1}{5}$$
$$\frac{24}{5}n = \frac{6}{5}$$
$$5\left(\frac{24}{5}n\right) = 5\left(\frac{6}{5}\right)$$
$$24n = 6$$
$$n = \frac{6}{24}$$
$$n = \frac{1}{4}$$

63. $555 = 139 + p$; 416 **65.** $180 = t + 154$; 26 s
67. $1.6 - m = 0.8$; \$0.8 million

69 Words: 45 million fewer than 57 million is the number who have blogs.
Equation: $57 - 45$
Solve: $57 - 45 = 12$ million

71a. $350 + m = 1000$; \$650 **71b.** $350 + 225 + m = 1000$; \$425 **71c.** $6t = 1000$; 167 **73.** Sample answer: $12 + n = 25$; subtract 12 from each side or add -12 to each side.
75a. Sometimes; $0 + 0 = 0$ but $2 + 2 \neq 2$.
75b. Always; this is the Addition Identity Property. **77.** Sample answer: If we multiply each side of the first equation by 3 the result is the second equation. So, they have the same solution although they have different variables. **79.** C
81. F **83.** $2r + 3k = 13$ **85.** $m^2 - p^3 = 16$
87. Sample answer: If it is Monday, then the trash is picked up. **89.** Sample answer: If $x^2 - 3x = 40$, then $x = 8$. **91.** $12(5 + 18 + 12)$; 420 hours

Pages 91–96 *Lesson 2-3*

1
$$3m + 4 = -11$$
$$3m + 4 - 4 = -11 - 4$$
$$3m = -15$$
$$m = -5$$
Check: $3m + 4 = -11$
$$3(-5) + 4 = -11$$
$$-15 + 4 = -11$$
$$-11 = -11 \text{ Yes}$$

3. -55 **5.** 61 **7.** $12 - 2n = -34$; 23
9. $n + (n + 2) + (n + 4) = 75$; 23, 25, 27 **11.** -5
13. -5 **15.** 70 **17.** 27 **19.** 16 **21.** -61

23 Equation: $49.99 + 0.15m = 100$
$$49.99 - 49.99 + 0.15m = 100 - 49.99$$
$$0.15m = 50.01$$
$$m \approx 333$$

So, he can use the phone for 650 + 333 or 983 minutes.

25. $17 = 6x - 13; n = 5$ **27.** $n + (n + 2) + (n + 4) = 141$; 45, 47, 49 **29.** $n + (n + 1) + (n + 2) + (n + 3) = -142$; −37, −36, −35, −34 **31.** $-7\frac{3}{5}$ **33.** −72 **35.** 108 **37.** $\frac{4}{5}$ **39.** $\frac{33}{14}$ **41.** $7\frac{1}{4}$ yr or 7 yr 3 mo

43
$$3.7q + 26.2 = 111.67$$
$$3.7q + 26.2 - 26.2 = 111.67 - 26.2$$
$$3.7q = 85.47$$
$$q = 23.1$$

45. 31.6 **47.** −3.5 **49.** 5

51a. $5x + 275 = x(6 + 15 + 9)$; 11 visits

51b.

Visits	Cost for Members	Cost for Nonmembers
3	290	90
6	305	180
9	320	270
12	335	360
15	350	450

51c. **Park Costs**

Cost ($) — 0, 100, 150, 200, 250, 300, 350, 400, 450, 500

Number of Visits — 0, 2, 4, 6, 8, 10, 12, 14, 16, 18

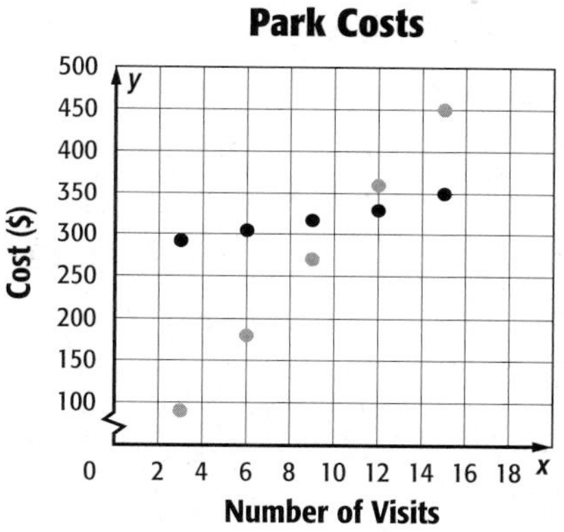

Both functions are linear. If a person is going to visit the park fewer than 11 times, it will be cheaper to be a nonmember. **53.** Sample answer: A pair of designer jeans costs $60. This is $40 more than twice the cost of a T-shirt. How much is the T-shirt? The T-shirt costs $10. **55.** 15 sides **57.** Sample answer: In order to solve the equation $4k + 20 = 236$, you would first subtract 20 from each side and then divide each side by 4. **59.** 84 **61.** B **63.** 1,379 **65.** Three times a number h is increased by 7 to equal 20. **67.** Three multiplied by a number p is the same as the difference of 8 times p and r. **69.** The product of $\frac{1}{2}$ and v is equal to the product of $\frac{2}{3}$ and v plus 4. **71.** 0; Additive Identity **73.** 4; Additive Inverse **75.** 53 **77.** 1000

Pages 97–102 Lesson 2-4

1. 4 **3.** −7 **5.** no solution **7.** all numbers **9.** A **11.** 4

13
$$6 + 3t = 8t - 14$$
$$6 + 3t - 3t = 8t - 3t - 14$$
$$6 = 5t - 14$$
$$6 + 14 = 5t - 14 + 14$$
$$20 = 5t$$
$$4 = t$$

Check:
$$6 + 3t = 8t - 14$$
$$6 + 3(4) = 8(4) - 14$$
$$6 + 12 = 32 - 14$$
$$18 = 18 \text{ Yes}$$

15. $2\frac{2}{5}$ **17.** 6 **19.** −5 **21.** 1 **23.** −4, −2 **25.** no solution **27.** all numbers **29.** −25 **31.** 15 **33.** 3 **35.** −2 **37.** −2, 0

39 Equation: $1500 + 0.80x = 1.59x$

Solve:
$$1500 + 0.80x = 1.59x$$
$$1500 + 0.80x - 0.80x = 1.59x - 0.80x$$
$$1500 = 0.79x$$
$$1899 \approx x$$

41a. Sample answer: $y = 2x + 4$

x	−2	−1	0	1	2
y	0	2	4	6	8

$y = -x - 2$

x	−2	−1	0	1	2
y	0	−1	−2	−3	−4

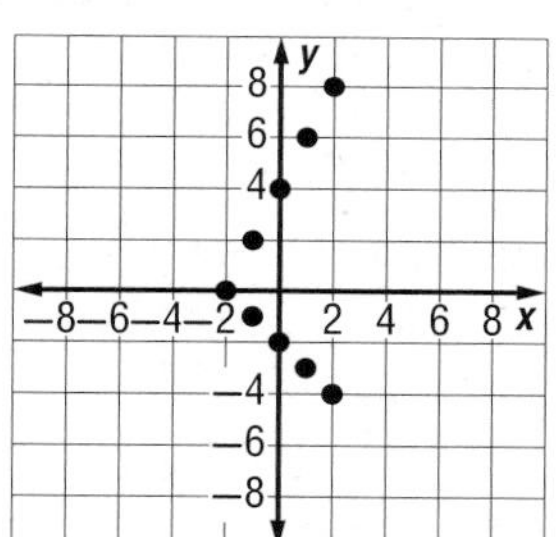

41b. −2 **41c.** Sample answer: The solution in part b is the x-coordinate for the point of intersection on the graph. **43.** Sample answer: $2x + 1 = \frac{3}{2}x - 2$; First I chose $\frac{3}{2}$ as the fractional coefficient. Then I chose 2 for the coefficient for the variable on the other side of the equation. After substituting −6 in for x on both sides, 1 must be added to the left and 2 must be subtracted from the right to balance the equation. **45a.** Incorrect; the 2 must be distributed over both g and 5; 6. **45b.** correct **45c.** Incorrect; to eliminate $-6z$ on the right side of the equal sign, $6z$ must be added to each side of the equation; 1. **47.** Sample answer: If the equation has variables on both sides of the equation, you must add or subtract so that the variable only appears on one side of the equation. After that step, solving the equation uses the same steps. **49.** J **51.** A **53.** $-2\frac{2}{3}$ **55.** −15 **57.** −15 **59.** $34 **61.** 2; Multiplicative Identity **63.** $\frac{2}{3}$; Additive Identity **65.** 7; Substitution **67.** $5(m + k) = 7k$ **69.** 5

71. −24 **73.** 11

Pages 103–109 Lesson 2-5

1. 15 **3.** −4

5. {4, −2}; 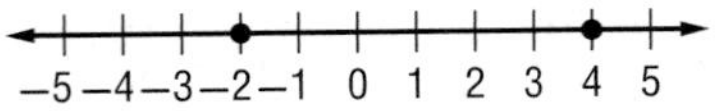

7. {−6, −2} 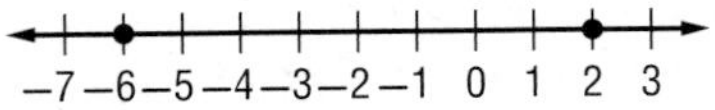

9. ∅ 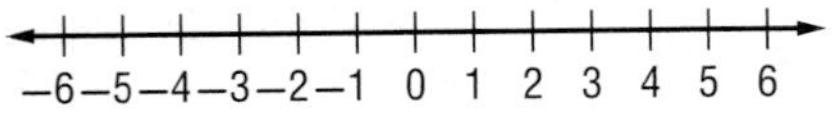

11 Find the point that is the same distance from −2 and 4. This is the midpoint between −2 and 4, which is 1.The distance from −2 to 1 is 3 units. The distance from 4 to 1 is 3 units. So, an equation is $|x - 1| = 3$

13
$$\begin{aligned}|2x + z| + 2y &= |2(2.1) + (-4.2)| + 2(3)\\ &= |4.2 + (-4.2)| + 6\\ &= |0| + 6\\ &= 0 + 6\\ &= 6\end{aligned}$$

15. −7.4 **17.** 8.4 **19.** −9.6 **21.** 0.4

23. {−11, −9}; 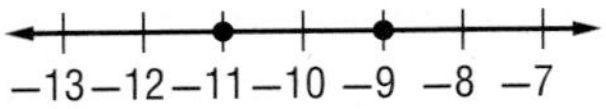

25. {7, −3}; 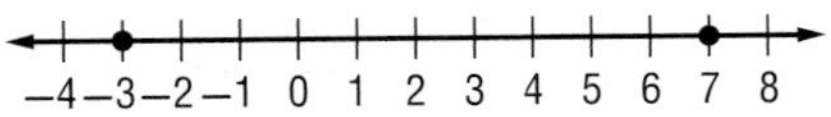

27. ∅ 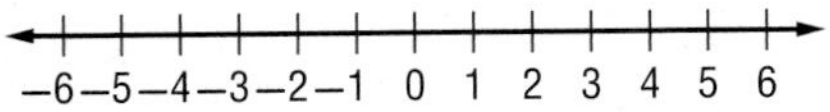

29. {0, 6} 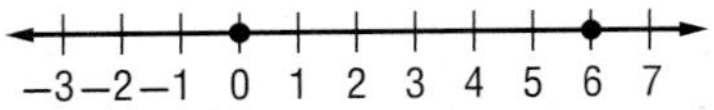

31. 11% to 19% **33.** $|x| = 4$ **35.** $|x - 1| = 4$

37. {−24, 16}

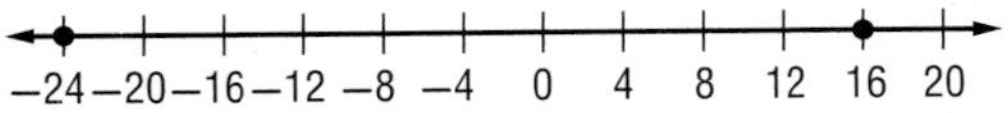

39. $\left\{3, -\frac{9}{5}\right\}$ 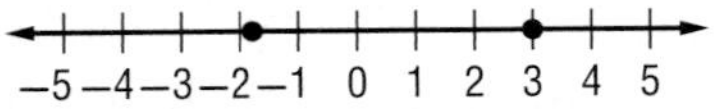

41. no solution 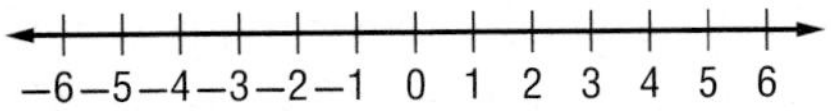

43a. $|x - 52| = 2$; {50, 54} **43b.** $|x - 53| = 1$; {52, 53} **43c.** 203 and 214 seconds **45a.** 47 to 53 mph **45b.** The slower the speed that the speedometer is calibrated at the more accurate the setting.

47. $|x| = 1\frac{1}{2}$ **49.** $\left|x - \frac{1}{4}\right| = \frac{1}{4}$ **51.** $\left|x + \frac{1}{3}\right| = 1$

53 **a.** Words: The number of people is 20,000, plus or minus 1,000.
Variable: Let h be the number of people who can clearly hear voices.
Equation: $|h - 20{,}000| = 1{,}000$

b. Solve the equation found in part **a**.

$|h - 20{,}000| = 1{,}000$

$h - 20{,}000 = 1{,}000$ or $h - 20{,}000 = -1{,}000$

$h - 20{,}000 + 20{,}000 = 1{,}000 + 20{,}000$ or

$h - 20{,}000 + 20{,}000 = -1{,}000 + 20{,}000$

$h = 21{,}000$ or $h = 19{,}000$

c. To find the range, find 21,000 − 19,000 = 2,000

55 **a.** Let p = the number of points awarded for each question; $|p| = 10$

b. Sample answer:

Number of questions correct	points
0	0
1	10
2	20
3	30
4	40
5	50

c. Sample answer: In science class, absolute values can be used for tolerance ranges of pollution on plants.

57. Sometimes; when $x = -1$, the value is 0. **59.** Sometimes; when c is a negative value whose absolute value is greater than x, the inequality is true. **61.** An absolute value represents a distance from zero on a number line. A distance can never be a negative number. **63.** Wesley; the absolute value of a number cannot be a negative number.

65. D **67.** A **69.** $\frac{1}{2}n + 16 = \frac{2}{3}n - 4$; 120 **71.** 10 in **73.** $\frac{2}{5}n = -24$; −60 **75.** $12 = \frac{1}{5}n$; 60

Pages 111–117 Lesson 2-6

1. no

3 $\frac{1.4}{2.1}$ is written in simplified form. $\frac{2.8}{4.4} = \frac{1.4}{2.2}$. Since the fractions are not equal, the ratios are not equivalent.

5. 5 **7.** about 253.3 min or 4 hours 13.3 min **9.** yes **11.** no **13.** yes **15.** 40 **17.** 29.25 **19.** 9.8 **21.** 1.32 **23.** 0.84

25
$$\frac{t}{0.3} = \frac{1.7}{0.9}$$
$$0.9t = 1.7(0.3)$$
$$0.9t = 0.51$$
$$t \approx 0.57$$

27. 6 **29.** 11 **31.** about $262.59 **33.** 150 mi **35.** 18 **37.** 0.8 **39.** 11 **41.** 130 students

43 **a.** Write each ratio.

for 2000, $\frac{\text{indoor theaters}}{\text{total theaters}} = \frac{35{,}567}{36{,}250}$

for 2001, $\frac{\text{indoor theaters}}{\text{total theaters}} = \frac{34{,}490}{35{,}173}$

for 2002, $\frac{\text{indoor theaters}}{\text{total theaters}} = \frac{35{,}170}{35{,}836}$

for 2003, $\frac{\text{indoor theaters}}{\text{total theaters}} = \frac{35{,}361}{35{,}995}$

for 2004, $\frac{\text{indoor theaters}}{\text{total theaters}} = \frac{36{,}012}{36{,}653}$

for 2005, $\frac{\text{indoor theaters}}{\text{total theaters}} = \frac{37{,}092}{37{,}740}$

for 2006, $\frac{\text{indoor theaters}}{\text{total theaters}} = \frac{37{,}776}{38{,}425}$

b. None of the ratios form a proportion.

45a.

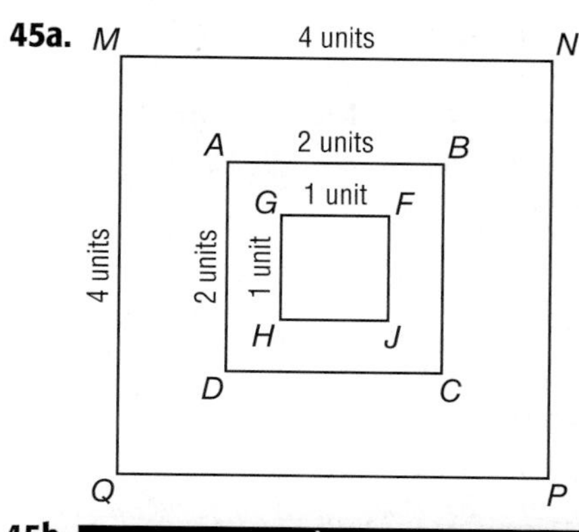

45b.

ABCD		MNPQ		GFHJ	
Side length	2	Side length	4	Side length	1
Perimeter	8	Perimeter	16	Perimeter	4

45c. If the length of a side is increased by a factor, the perimeter is also increased by that factor. If the length of the sides are decreased by a factor, the perimeter is also decreased by the same factor. **47.** Ratios and rates each compare two numbers by using division. However, rates compare two measurements that involve different units of measure. **49.** Neither; Tim inverted the woman-to-man comparison when he wrote the second ratio. Aisha wrote a proportion that is equivalent to Tim's incorrect proportion. **51.** C **53.** G **55.** ∅ **57.** {10, −7} **59.** 30 years **61.** −7 **63.** −48 **65.** 13 **67.** 5.5 **69.** 3.5

Pages 119–124 Lesson 2-7

1 It is an increase.

$125 - 78 = 47$

$47 \div 78 \approx 0.60.$

The percent of increase is about 60%.

3. inc.; 33% **5.** 146 mi **7.** \$38.42 **9.** \$53.07 **11.** \$17.21 **13.** \$22.10

15 It is a decrease.

$16 - 10 = 6$

$6 \div 16 \approx 0.38.$

The percent of decrease is about 38%.

17. dec.; 77% **19.** inc.; 127% **21.** inc.; 90% **23.** \$12,400 **25.** \$47.48 **27.** \$27.31 **29.** \$10.66 **31.** \$76.49 **33.** \$16.42 **35.** \$11.99 **37.** \$48.04 **39.** about 13.5% increase **41a.** First girl's dress - \$15; Second girl's dress = \$25.50 **41b.** The second girl. by \$0.50

43 Find the percent of increase or decrease for each grocery item. Ground beef had the biggest increase with a 41.1% increase.

45. Sample answer: A CD is on sale for \$9.99. If tax is 6.5%, what will the CD cost? **47.** Xavier; Maddie divided by the new amount instead of the original amount. **49.** Sample answer: To determine whether a percent of change is a percent of increase or decrease, compare the new amount with the old amount. If the new amount is greater, the change is an increase. If the new amount is less, the change is a decrease. To find the percent of change, subtract the original from the new amount. Then write a proportion, comparing the change to the original amount. The answer should be written as a percent. **51.** \$72 **53.** C **55.** 12 **57.** 4 **59.** 5.6 **61.** 3 **63.** −6 **65.** −7 **67.** If two lines are perpendicular, then they meet to form four right angles. If two lines meet to form four right angles, then they are perpendicular. **69.** Sample answer: Six more than twice a number *f* equals nineteen. **71.** Sample answer: The product of three and a number *a* when added to 5 is equal to the difference of 27 and two times *a*. **73.** Sample answer: The fourth power of a number *d* increased by sixty-four is three times that number *d* to the third power plus seventy-seven.

Pages 126–131 Lesson 2-8

1

$$5a + c = -8a$$
$$5a - 5a + c = -8a - 5a$$
$$c = -13a$$
$$-\frac{c}{13} = a$$

3. $k = -7n - m$ **5a.** $h = \frac{V}{\pi r^2}$ **5b.** 8 in. **7.** about 0.43875 ft

9

$$x = b - cd$$
$$x - b = -cd$$
$$\frac{x-b}{-d} = c$$

11. $m = \frac{-n+p}{10}$ **13.** $v = \frac{9}{5}(z - w)$ **15.** $f = \frac{6g - 10}{d}$ **17a.** $v_f = at + v_i$ **17b.** 10 ft/s^2 **19.** 49.8 L **21.** $t = \frac{w - 11v}{31}$ **23.** $c = \frac{-13 + f}{10 - d}$ **25.** 1.0 mm/s **27.** 3.9 km/s **29.** $t - 7 = r + 6$; $t = r + 13$ **31.** $\frac{9}{10}g = 7 + \frac{2}{3}k$; $k = \frac{3}{2}\left[\frac{9}{10}g - 7\right]$

33

$$S = 2w(\ell + h) + 2lh$$
$$214 = 2(6)(7 + h) + 2(7)h$$
$$214 = 12(7 + h) + 14h$$
$$214 = 84 + 12h + 14h$$
$$130 = 26h$$
$$5 = h$$

So, 5 inches.

35. about 396 in^3 **37.** Sandrea; she performed each step correctly; Fernando omitted the negative sign from $-5b$. **39a.** $x = \frac{y-1}{yn-1}$ **39b.** $y = -\frac{1}{3}x$ **41.** D **43.** 15 **45.** \$101.76 **47.** \$46.33 **49.** \$56.95 **51.** 1.67 **53.** 5.14 **55.** 50(7.50) + 90(5.00); \$825 **57.** −0.5 **59.** −1.5 **61.** 2

Pages 132–138 Lesson 2-9

1

	Weight	Price	Total Price
Soup	10	0.15	0.15(10)
Salad	x	0.20	$0.20x$
Total			3.30

$$0.15(10) + 0.20x = 3.30$$
$$1.50 + 0.20x = 3.30$$
$$0.20x = 1.80$$
$$x = 9$$

She bought 9 oz of salad.

3. 10 mph **5.** 2 hours

7 a.

	Amount	Percent	Total
Metallic Balloons	b	\$2.00	$2.00b$
Bunches of helium Balloons	$b - 36$	\$3.50	$3.50(b - 36)$

b. $2.00b + 3.50(b - 36) = 281$

c. $2.00b + 3.50(b - 36) = 281$
$$2b + 3.5b - 126 = 281$$
$$5.5b - 126 = 281 + 126$$
$$5.5b = 407$$
$$b = 74$$

There were 74 dozen metallic balloons sold.

d. $b - 36 = 74 - 36$
$$= 38$$

There were 38 dozen bunches of balloons sold.

9. about 16.67 gal **11.** about 22.2 mph **13.** $1\frac{1}{7}$hours or 1 h 8 min 34 s **15.** 10 gal **17.** 10.89 mph

19 a. $D = rt$
$$= 65(6)$$
$$= 390$$ He could drive 390 miles.

b. $D = rt$
$$\frac{D}{r} = t$$
$$\frac{625}{65} = t$$
$$9.62 \approx t$$ It will take about 9.62 hours.

21. 33 mi **23.** Sample answer: For a 50% solution being added to a 100% solution to produce a 75% resulting solution, the quantity of each must be the same. **25.** Sample answer: How many grams of salt must be added to 36 grams of a 15% salt solution to obtain a 50% salt solution? **27.** B **29.** C **31.** $\frac{-5 + b}{2b}$ **33.** $\frac{A}{2\pi r} - r$ **35.** Sample answer: The quotient of n and -6 is the same as the sum of two times n and one. **37.** Sample answer: The sum of three and twice x squared is equal to twenty-one. **39.** (4, 25); Sample answer: If four cars are washed, \$25 is earned. **41.** \$583.50 **43.** -2 **45.** -7 **47.** 24

Pages 139-144 Study Guide and Review

1. false, variable **3.** true **5.** false, ratio **7.** false, decrease **9.** $5x + 3 = 15$ **11.** $\frac{1}{2}m^3 = 4m - 9$ **13.** h squared minus five times h plus six is equal to zero. **15.** width: 8 ft, length: 19 ft **17.** -5 **19.** 2.1 **21.** 6 **23.** 14 **25.** 6 **27.** -11 **29.** 17 **31.** 2 **33.** 38.1 **35.** 19, 21, 23 **37.** 3 **39.** -2 **41.** 2 **43.** -8 **45.** 21 **47.** 28 **49.** -144 **51.** $\{-5, 17\}$

−8 −4 0 4 8 12 16 20

53. $\{-27, 63\}$

−60 −45 −30 −15 0 15 30 45 60 75

55. yes **57.** 20 **59.** 12 **61.** increase, 25% **63.** decrease, 17% **65.** \$52.19 **67.** \$55.20 **69.** \$33.75 **71.** $y = \frac{9 - 3x}{2}$ **73.** $m = \frac{15 - 9n}{-5}$ **75.** $y = \frac{5}{2}(m - n)$ **77.** $h = \frac{2A}{a + b}$ **79.** 52 mph

Chapter 3 Linear Functions

Page 151 Get Ready

1.

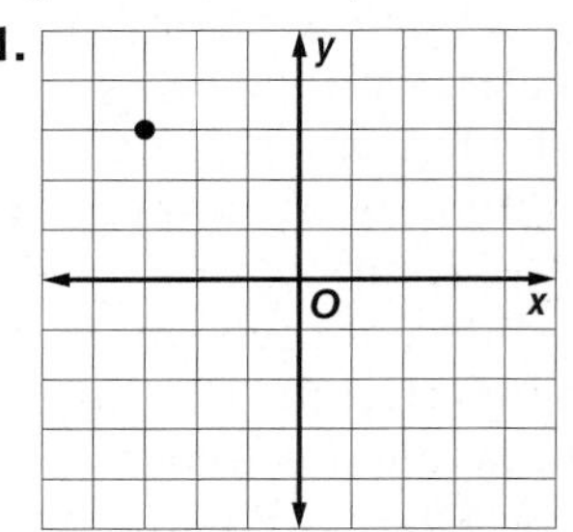

3.

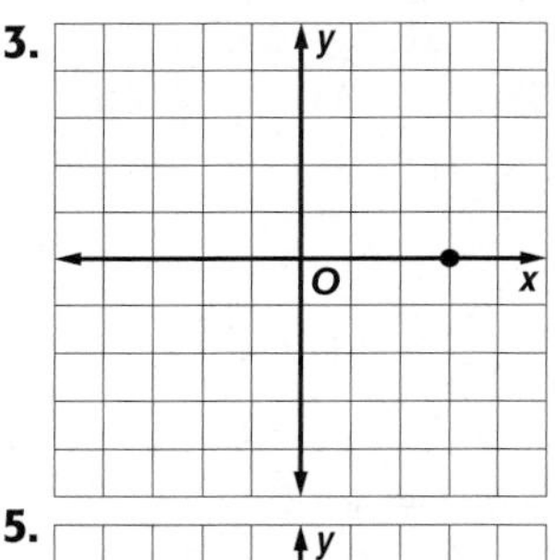

5.

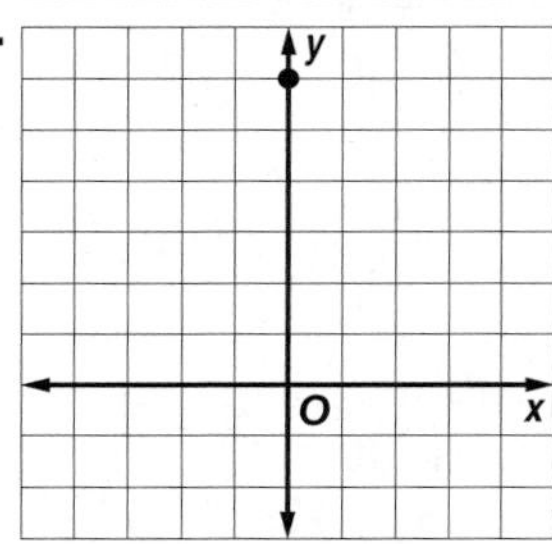

7. $(3, -1)$ **9.** $(3, 2)$ **11.** $(5, 0)$ **13.** $y = -3x + 1$ **15.** $y = \frac{5}{2}x - 6$ **17.** $y = -10x + 6$ **19.** $\frac{1}{4}$ **21.** 0 **23.** about \$13.5 million

Pages 153-160 Lesson 3-1

1. yes; $x - y = -5$ **3.** yes; $y = 1$ **5.** 25, -4; The x-intercept 25 means that after 25 minutes, the temperature is 0°F. The y-intercept -4 means that at time 0, the temperature is -4°F.

7.

9.

x	x + 2y = 4	y	(x, y)
−4	(−4) + 2y = 4	4	(−4, 4)
−2	(−2) + 2y = 4	3	(−2, 3)
0	(0) + 2y = 4	2	(0, 2)
2	(2) + 2y = 4	1	(2, 1)
4	(4) + 2y = 4	0	(4, 0)

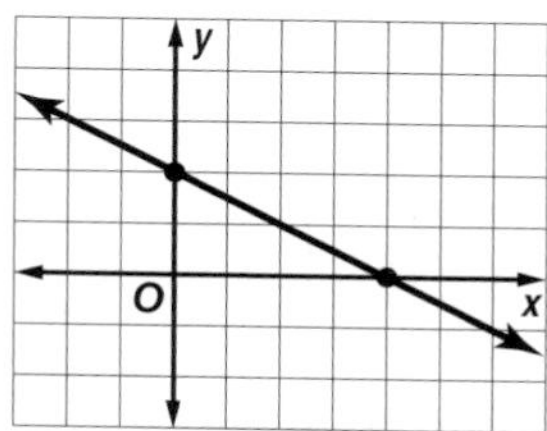

11.

x	y = 3	y	(x, y)
−2	y = 3	3	(−2, 3)
−1	y = 3	3	(−1, 3)
0	y = 3	3	(0, 3)
1	y = 3	3	(1, 3)
2	y = 3	3	(2, 3)

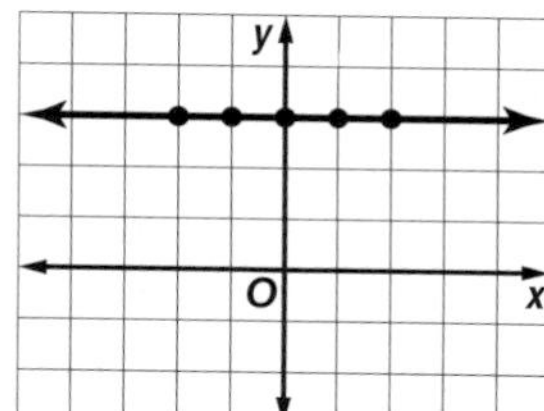

13 $5x + y^2 = 25$
Since the y term is squared, this equation cannot be written in the form $AX + BY = c$, so it is not a linear equation.

15. no **17.** yes; $4x + y = 0$ **19.** 3, 4 **21.** 6, 20; The x-intercept represents the number of seconds that it takes the Eagle to land. The y-intercept represents the initial height of the Eagle.

23.

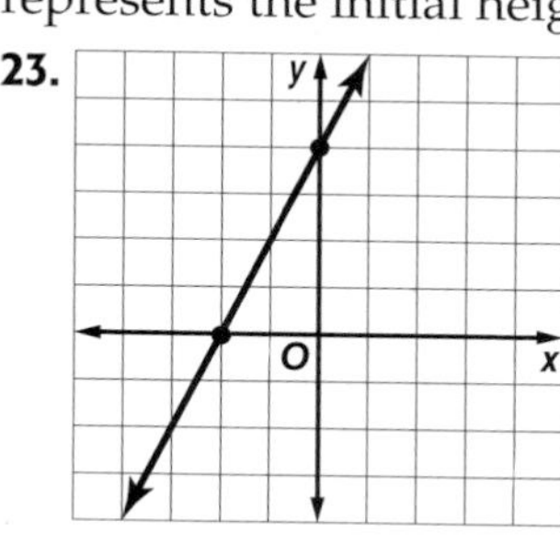

25.

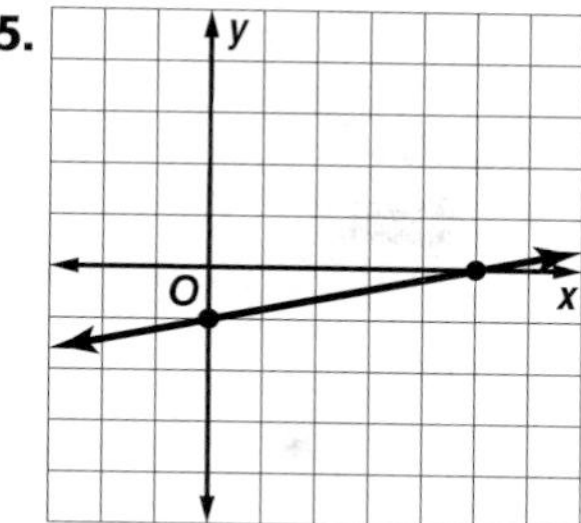

27.

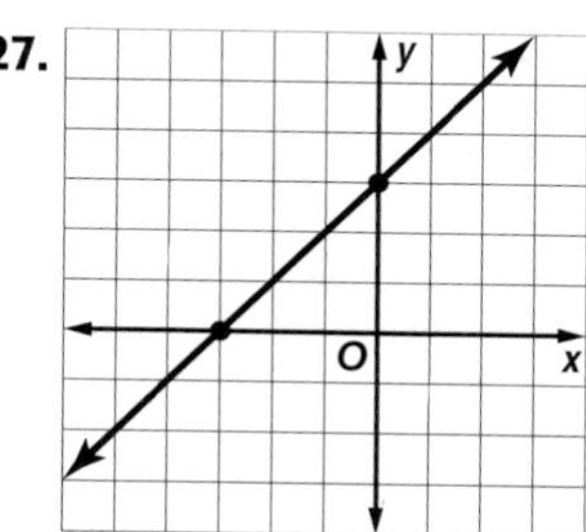

29.

x	y
−2	0
−2	1
−2	2

31.

x	y
−1	8
0	0
1	−8

33.

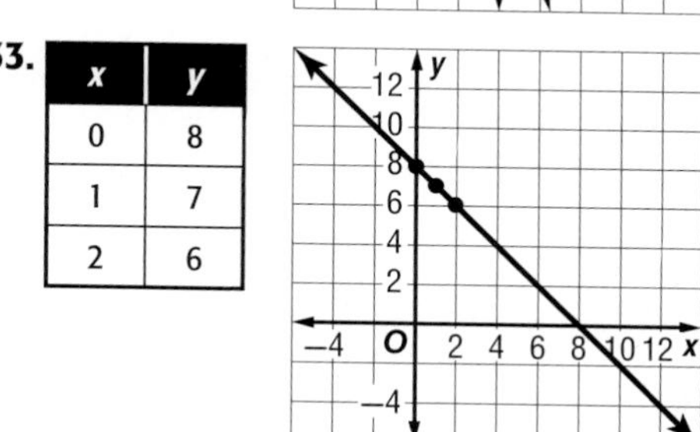

x	y
0	8
1	7
2	6

35 a. The domain is all real numbers so there are infinitely many solutions. Select values from the domain and make a table.

v	p = 0.15v	p	(v, p)
0	p = 0.15(0)	0	(0, 0)
2	p = 0.15(2)	0.3	(2, 0.3)
4	p = 0.15(4)	0.6	(4, 0.6)
6	p = 0.15(6)	0.9	(6, 0.9)
8	p = 0.15(8)	1.2	(8, 1.2)
10	p = 0.15(10)	1.5	(10, 1.5)

b. Create ordered pairs and graph them.

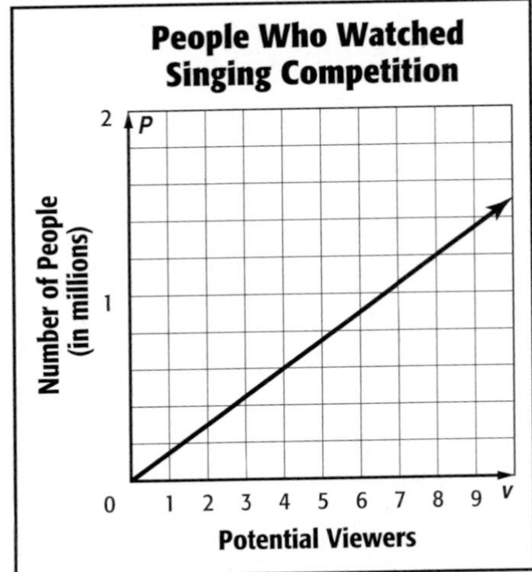

c. Using the graph, when there are 14 million potential viewers, there will about 2.1 million people who watch.

d. A negative does not make sense because you cannot have a negative number of viewers.

37. yes; $3x - 4y = 60$ **39.** yes; $3a = 2$

41. yes; $9m - 8n = -60$

43.

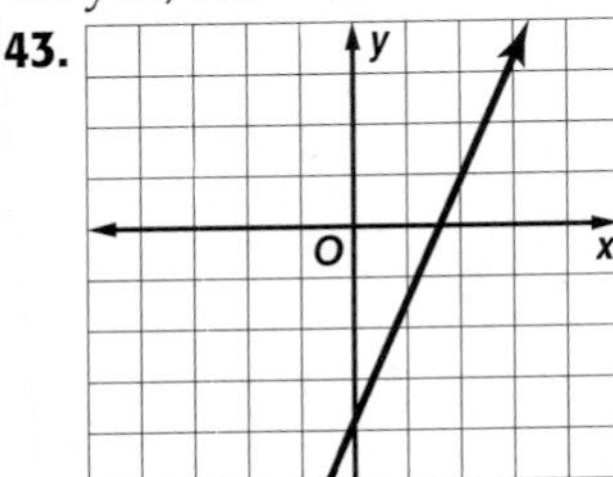

45.

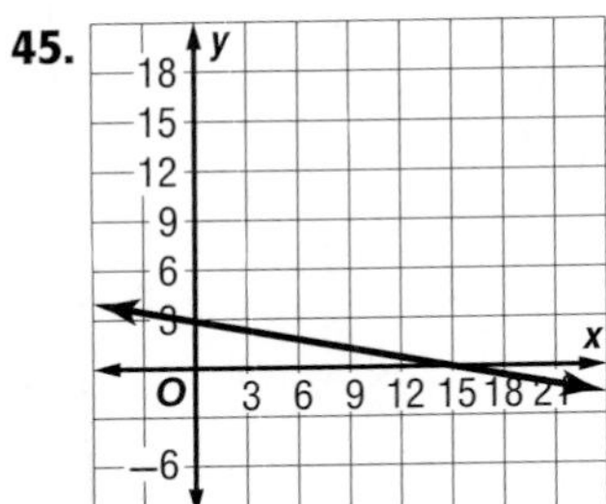

47.

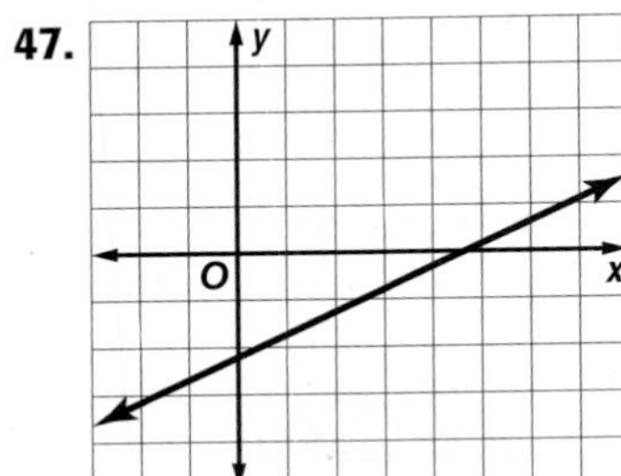

49. No; Sample answer: The rental car would cost \$176. Mrs. Johnson only has \$160 to spend.

51 $5x + 3y = 15$

To find the x-intercept, let $y = 0$.

$$5x + 3y = 15$$
$$5x + 3(0) = 15$$
$$5x = 15$$
$$x = 3$$

The x-intercept is 3. This means the graph intersects the x-axis at (3, 0).

To find the y-intercept, let $x = 0$.

$$5x + 3y = 15$$
$$5(0) + 3y = 15$$
$$3y = 15$$
$$y = 5$$

The y-intercept is 5. This means the graph crosses the y-axis at (0, 5).

53. $2\frac{1}{2}$; $-1\frac{2}{3}$ **55.** 12; −3

57a.

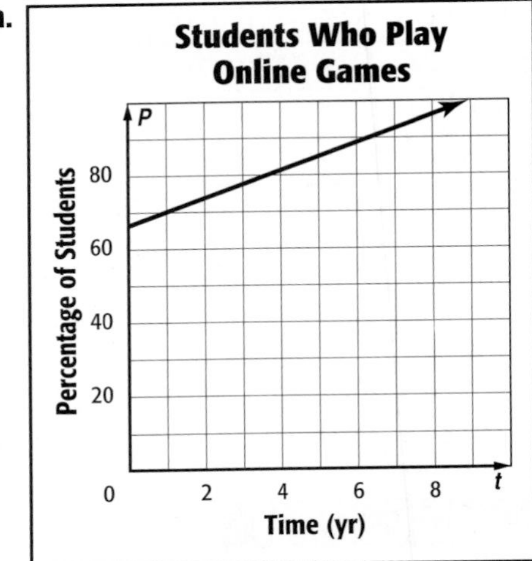

57b. 96%

59.

Perimeter of a Square	
Side Length	Perimeter
1	4
2	8
3	12
4	16

Sample answer: Yes; we used the formula $P = 4s$, which is linear.

Area of a Square	
Side Length	Area
1	1
2	4
3	9
4	16

Sample answer: No; we used the formula $A = s^2$, which is not linear.

Volume of a Cube	
Side Length	Volume
1	1
2	8
3	27
4	64

Sample answer: No; we used the formula $V = s^3$, which is not linear.

61. Sample answer: $y = 8$; horizontal line

63. Sample answer: $x - y = 0$; line through (0, 0)

65. D **67.** \$30 **69.** 270 rolls of solid wrap, 210 rolls of print wrap **71.** $g = \frac{5+m}{2+h}$ **73.** $z = \frac{c-b}{2}$ **75.** $-\frac{23}{14}$ **77.** -56

Pages 161–166 Lesson 3-2

1. 3 **3.** $\frac{1}{2}$ **5.** no solution **7.** no solution **9.** Tyrone must deliver 40 newspapers for the papers in his bag to weigh 0 pounds. **11.** -3

13. no solution **15.** $-\frac{10}{7}$ or $-1\frac{3}{7}$

17

$$\begin{aligned} 5x - 5 &= 5x + 2 \\ 5x - 5 + 5 &= 5x + 2 + 5 \\ 5x &= 5x + 7 \\ 5x - 5x &= 5x - 5x + 7 \\ 0 &= 7 \end{aligned}$$

There are no solutions.

19. no solution **21.** no solution **23.** 100; She can download a total of 100 songs before the gift card is completely used. **25.** -8 **27.** $\frac{10}{3}$ or $3\frac{1}{3}$ **29.** $-\frac{34}{13}$ or $-2\frac{8}{13}$ **31.** $\frac{17}{25}$ **33.** $\frac{15}{8}$ or $1\frac{7}{8}$ **35.** 3 **37.** 4:00 P.M. **39.** -3 **41.** -2 **43.** $\frac{9}{8}$ or $1\frac{1}{8}$

45 **a.** Sample answers given:

Number of Songs Downloaded	Total Cost (\$)	Total Cost / Number Songs Downloaded
2	4	$\frac{4}{2} = 2$
4	8	$\frac{8}{4} = 2$
6	12	$\frac{12}{6} = 2$
8	16	$\frac{16}{8} = 2$
10	20	$\frac{20}{10} = 2$

b. As the number of songs downloaded increases by 2, the cost increases by 4.

c. The value of the total cost divided by the number of songs downloaded represents the cost per song. It costs \$2 per song.

47. 3 **49.** Sample answer: $3 + 4x = 0$; $y = 3 + 4x$ or $f(x) = 3 + 4x$ **51.** A **53.** B **55.** $-5, 10$ **57.** $7, -2$ **59.** H: a number is divisible by 10; C: it is divisible by 5; If a number is divisible by 10, then it is divisible by 5. **61.** $\frac{5}{2}$ **63.** $-\frac{1}{2}$ **65.** $\frac{2}{3}$ **67.** 11

Pages 170–178 Lesson 3-3

1. $\frac{4}{3}$ **3a.** 2.005; There was an average increase in ticket price of \$2.005 per year. **3b.** Sample answer: 1998–2000; A steeper segment means a greater rate of change. **3c.** Sample answer: 1998–2000; Ticket prices show a sharp increase. **5.** No; the y-values do not decrease by a constant amount. **7.** -1

9. $\frac{7}{9}$ **11.** 0 **13.** -8

15 rate of change $= \frac{\text{change in } y}{\text{change in } x}$

$$\begin{aligned} &= \frac{9-15}{2-1} \\ &= \frac{-6}{1} \\ &= -6 \end{aligned}$$

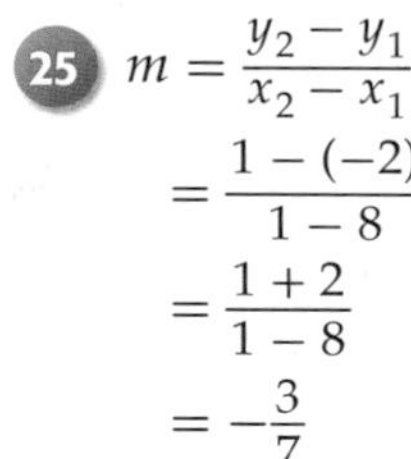

17. $\frac{1}{2}$ **19a.** Sample answer: $P = -1811.67t + 19548.30$ **19b.** The car value depreciates by \$1811.67 each year. **19c.** \$6866.61 **21.** No; the y-values do not increase at a constant rate. **23.** Yes; both the x-values and the y-values increase at a constant rate.

25

$$\begin{aligned} m &= \frac{y_2 - y_1}{x_2 - x_1} \\ &= \frac{1-(-2)}{1-8} \\ &= \frac{1+2}{1-8} \\ &= -\frac{3}{7} \end{aligned}$$

27. undefined **29.** $\frac{5}{17}$ **31.** 0 **33.** undefined **35.** $\frac{10}{3}$ **37.** $\frac{3}{4}$ **39.** 6 **41.** Sample answer: about 0.5 **43.** $\frac{15}{4}$ **45.** $-\frac{2}{3}$

47 **a.** Plot the ordered pairs on a coordinate plane. Connect the points with a line.

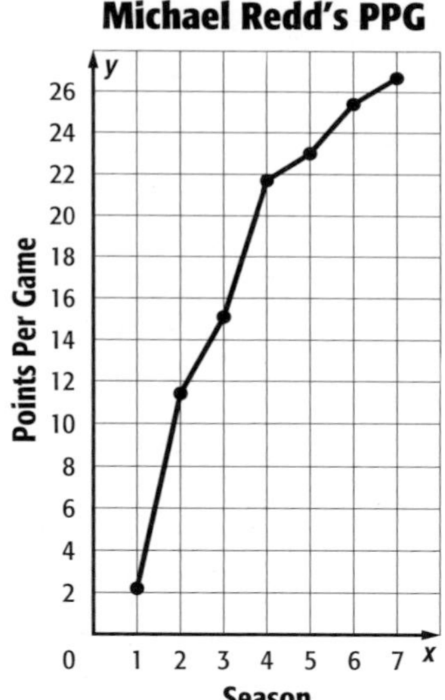

b. The steepest line is between 2000–2001 and 2001–2002, so that is when the PPG increased the most.

c. The rate of change was much more dramatic or steeper in the first four years, whereas it leveled off in the last few years.

49. The rate of change is $2\frac{1}{4}$ inches of growth per week. **51.** Sample answer: Slope can be used to describe a rate of change. Rate of change is a ratio that describes how much one quantity changes with respect to a change in another quantity. The slope of a line is also a ratio and it is the ratio of the change in the y-coordinates to the change in the x-coordinates. **53.** A **55.** \$4 **57.** -6 **59.** 4

61. $-1, 2$ **63.** 12 **65.** $\frac{5}{16}$ **67.** 5

Pages 180–186 Lesson 3-4

1. $-\frac{4}{5}; -\frac{4}{5}$

3.

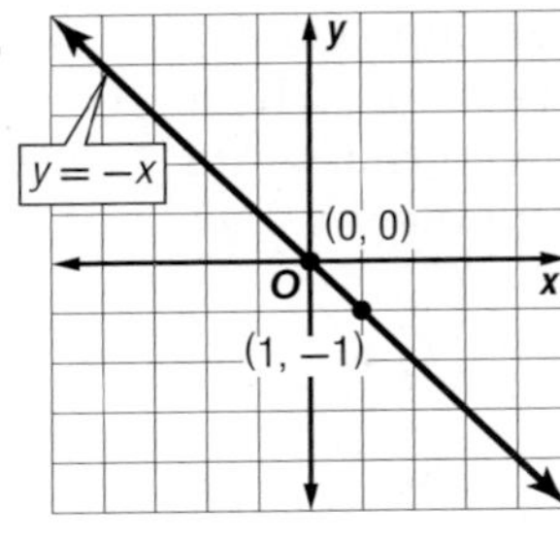

5.

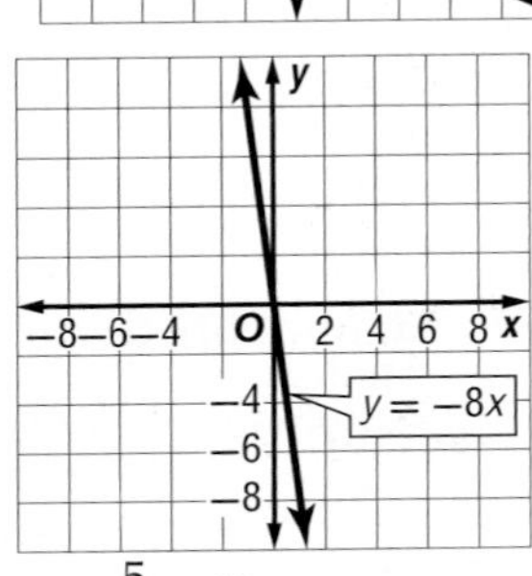

7. $y = \frac{5}{4}x$; 40

9a. $y = \frac{12}{5}x$;

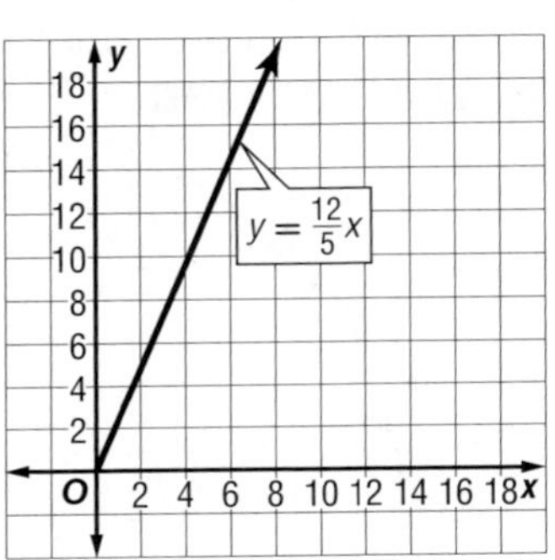

9b. 40

11 The constant of variation is -5.

$$m = \frac{5 - 0}{-1 - 0}$$
$$= \frac{5}{-1}$$
$$= -5$$

13. $-\frac{1}{5}; -\frac{1}{5}$ **15.** $-12; -12$

17.

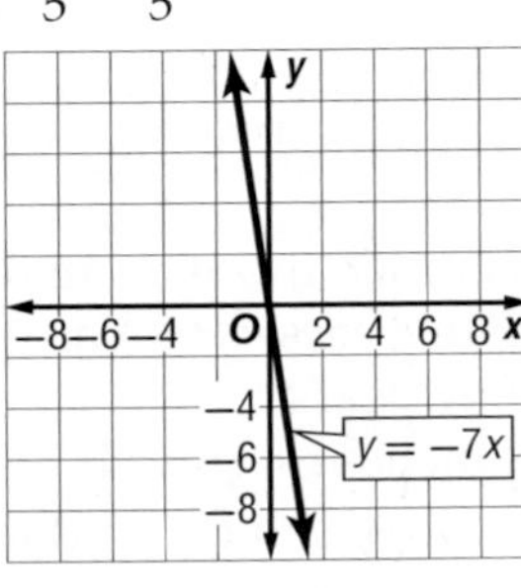

19.

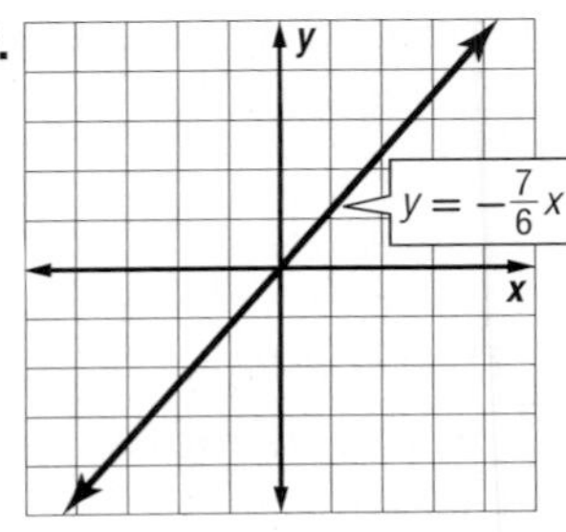

21.

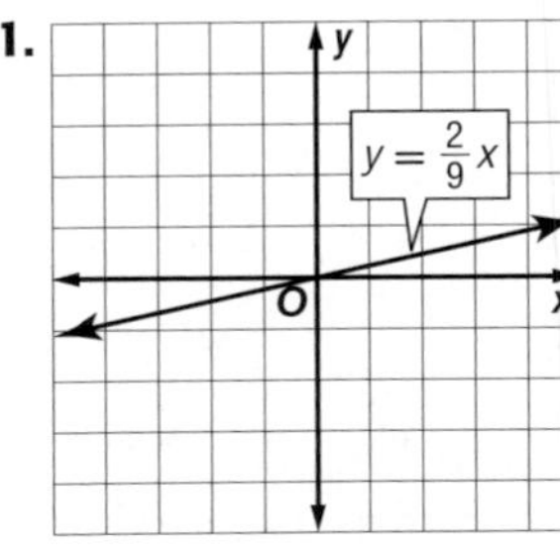

23.

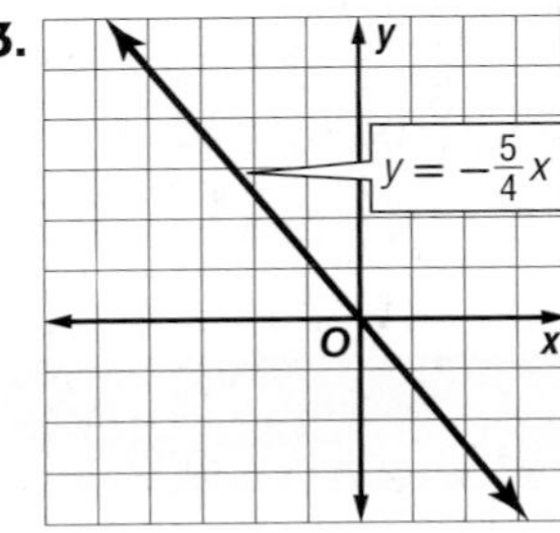

25

$y = kx$

$22 = k(8)$

$\frac{11}{4} = k$

Therefore, the direct variation equation is

$y = \frac{11}{4}x$.

$y = \frac{11}{4}x$

$y = \frac{11}{4}(-16)$

$y = 11(-4)$

$y = -44$

Therefore, $y = -44$ when $x = -16$.

27. $y = 14x; 1\frac{1}{7}$ **29a.** $y = 1800x$ **29b.** 7 yr 6 mo

31. $y = 20x; \frac{5}{4}$ **33.** $y = -3.75x; -30$

35. dark green **37.** lime green

39 Since the songs are \$0.99 each, the equation is $T = 0.99s$. The graph of the equation runs through the origin with a slope of 0.99.

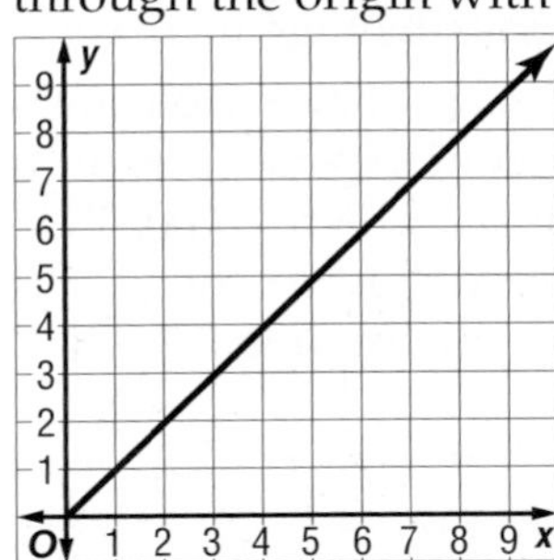

41a.

41b. Sample answer: The constant of variation, slope, and rate of change of a graph all have the same value. **41c.** Sample answer: Find the absolute value of k in each equation. The one with the great value of $|k|$ has the steeper graph.
43. $C = 9.95n$ **45.** $z = \frac{1}{9}x$; It is the only equation that is a direct variation. **47.** Sample answer: $y = 0.50x$ represents the cost of x apples.

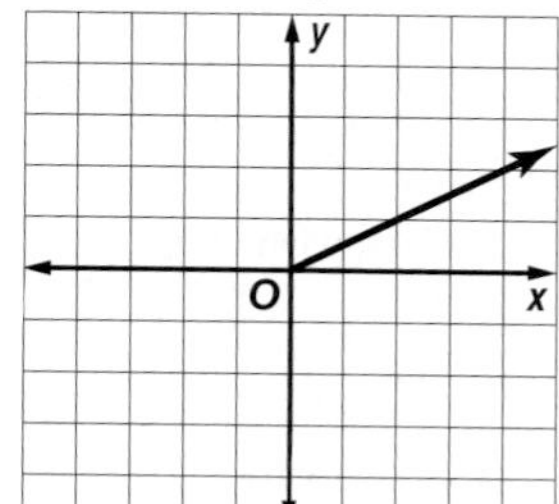

The rate of change, 0.50, is the cost per apple.

49. Neither; the slope is constant, but it is k.
51. A **53.** D **55.** 5.8; There was an average increase of 5.8 channels per year.

57.

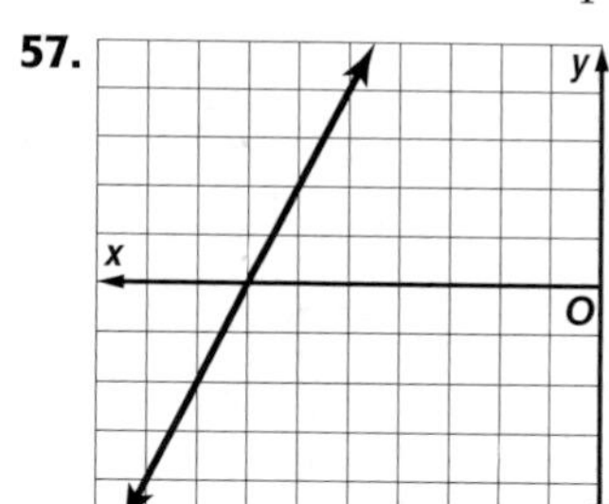

−7

59. −4

61.

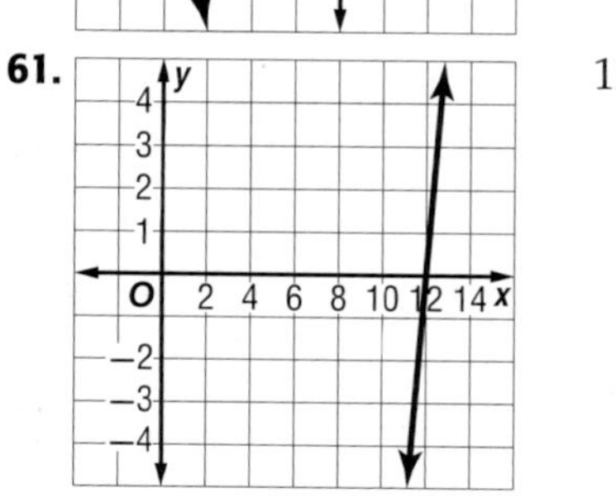

12

63. 12 **65.** −2 **67.** −28 **69.** −12 **71.** −6 **73.** −12

Pages 187–193 Lesson 3-5

1. No; there is no common difference. **3.** 0, −3, −6
5. $a_n = 17 - 2n$

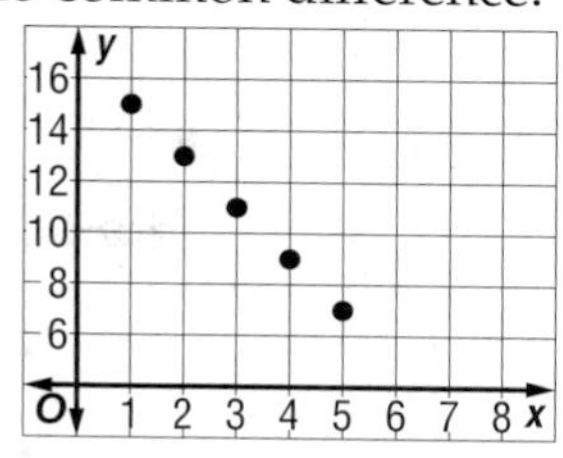

7. $f(n) = 55n + 525$

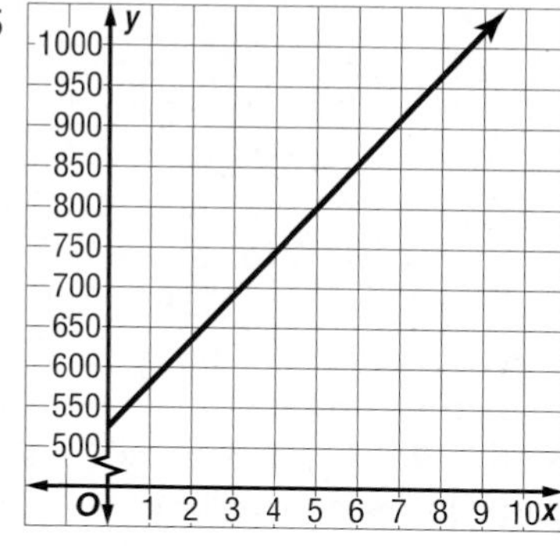

9. No; there is no common difference. **11.** Yes; the common difference is 2.6. **13.** 30, 36, 42
15 Step 1: Find the common difference by subtracting successive terms.

$-\frac{1}{2}, \quad 0, \quad \frac{1}{2}, \quad 1, \ldots$

$+\frac{1}{2} \quad +\frac{1}{2} \quad +\frac{1}{2}$

The common difference is $\frac{1}{2}$.

Step 2: Add $\frac{1}{2}$ to the last term of the sequence to get the next term.

$1 \quad 1\frac{1}{2} \quad 2 \quad 2\frac{1}{2}$

$+\frac{1}{2} \quad +\frac{1}{2} \quad +\frac{1}{2}$

The next three terms are $1\frac{1}{2}, 2, 2\frac{1}{2}$.
17. $3\frac{7}{12}, 4\frac{1}{3}, 5\frac{1}{12}$
19. $a_n = 5n - 7$

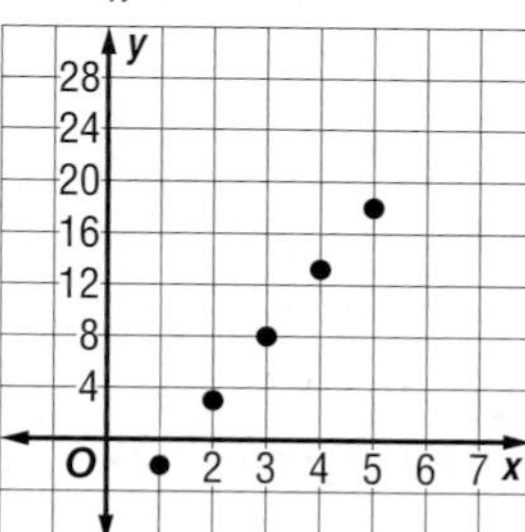

21. $a_n = 0.25n - 1$

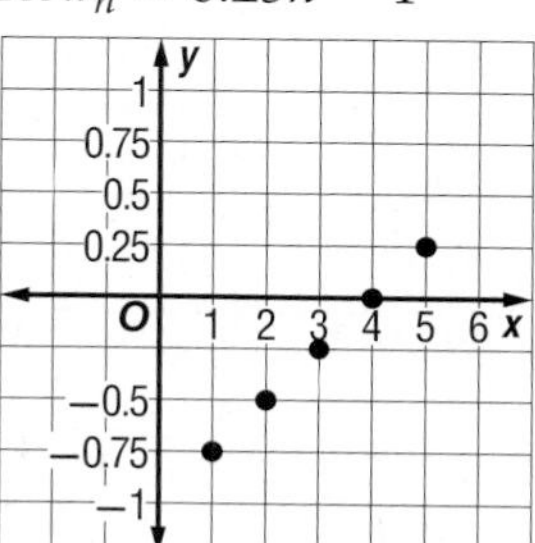

23a. $f(n) = 0.80n$
23b. D = {10, 20, 30, 40, ...}

Selected Answers and Solutions

25 The ordered pairs are (10, 7.50), (15, 8.75), (20, 10), (25, 11.25). So, the rate of change is $\frac{8.75 - 7.50}{15 - 10} = \frac{1.25}{5} = 0.25$.
The cost is \$0.25 per word.
$f(n) = 0.25n + b$
$7.50 = 0.25(10) + b$
$7.50 = 2.50 + b$
$5 = b$
So, the equation is $f(n) = 0.25n + 5$.
27. 77 **29.** 25,646 **31a.** Sample answer: the first two terms are both 1. Starting with the third term, the two previous terms are added together to get the next term; 5, 8, 13, 21, 34. **31b.** $a_n = a_{n-2} + a_{n-1}$ **31c.** 610 **31d.** There is no common difference. **33.** −1 **35a.** Yes; there is a common difference.; x; $5x + 1$, $6x + 1$, $7x + 1$.
35b. No; there is no common difference.
37. 8 yr **39.** H **41.** 3, 3 **43.** $-\frac{3}{7}$ **45.** 3 **47.** 2
49. Sample answer: $453{,}000 - d = 369{,}000$; 84,000
51–55.

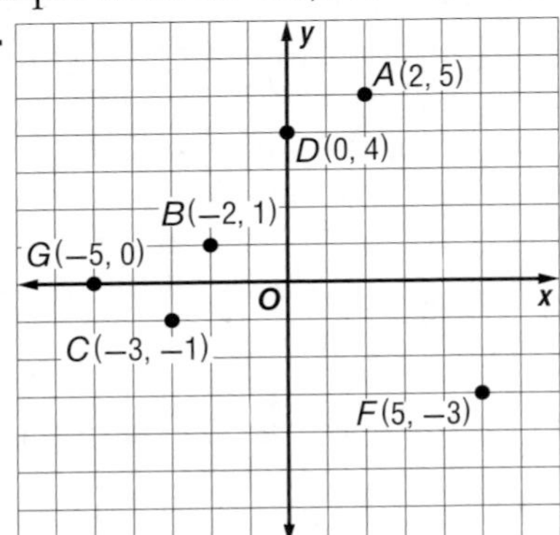

Pages 195–200 Lesson 3-6

1a.

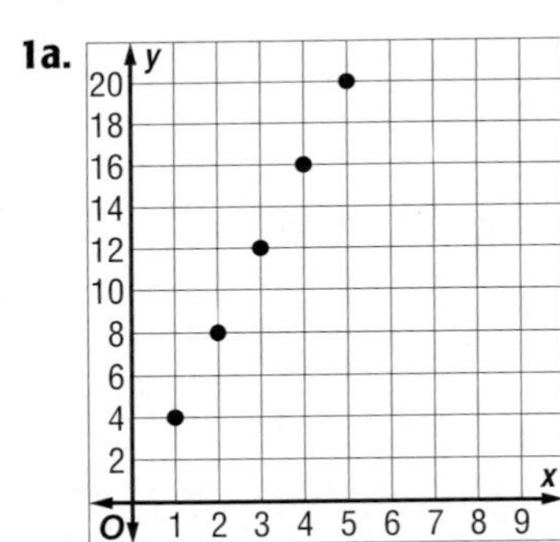

1b. $y = 4x$ **1c.** The perimeter is 4 times the length of the side.
3. $f(x) = -x + 3$

5 Select points from the graph and place them in a table.

x	0	1	2	3
y	0	2	4	6

The difference between the x-values is 1, while the difference between the y-values is 2. This suggests that $y = 2x$.
This works for each ordered pair, so the equation is $f(x) = 2x$.
7. $f(x) = 3x - 2$ **9.** $f(n) = 3n - 3$; Nonproportional; the function does not describe a direct variation.

11. $y = 2.25x + 2.50$
13 **a.** Sample answer:

Number of T-shirts ordered	5	10	15	20	25
Cost ($)	13	23	33	43	53

b. In functional notation, the equation is $C(t) = 2t + 3$.
c. The equation goes through (0, 3) and has a slope of 2.

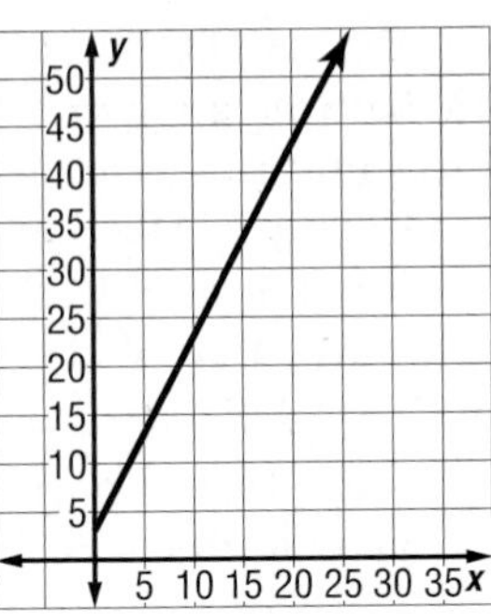

d. The relationship is nonproportional because $\frac{13}{5} \neq \frac{23}{10} \neq \frac{33}{15} \neq \frac{43}{20} \neq \frac{53}{25}$.
15. Sample answer: 4, 7, 10, 13; add a common difference of 3; $a_n = 3n + 1$. **17.** $f(n) = 3n + 2$ is the related function for the arithmetic sequence 5, 8, 11, 14, …, but it is not proportional. The line through (1, 5) and (2, 8) does not pass through (0, 0). **19.** D
21. H **23.** 43, 53, 63 **25.** $\frac{5}{4}, \frac{11}{8}, \frac{3}{2}$ **27.** $y = 7x$; −12 **29a.** $V = \frac{1}{3}\pi r^2 h$ **29b.** about 3142 cm^3
31. $y = 3x - 5$
33.

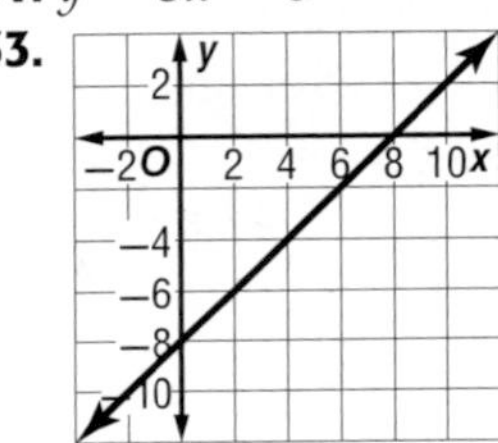

35.

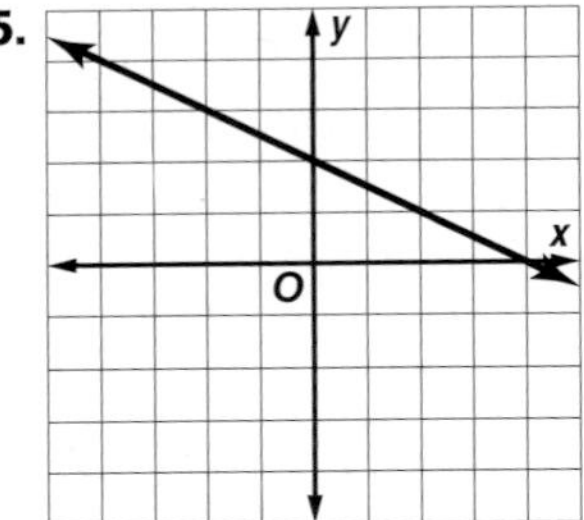

Pages 201–204 Study Guide and Review

1. true **3.** false; common difference **5.** true
7. false; The slope of $y = 5$ is 0. **9.** true
11. −8, 6
13.

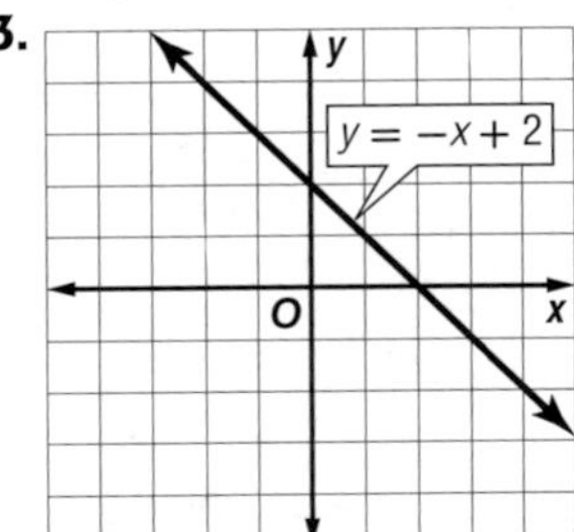

15.

17a. 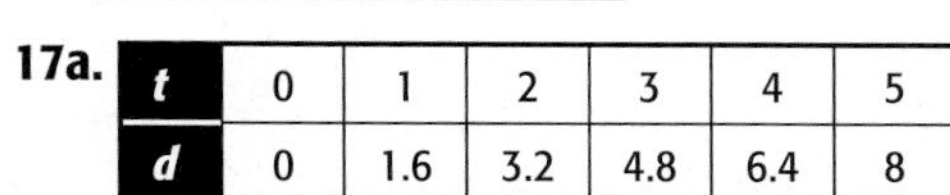

t	0	1	2	3	4	5
d	0	1.6	3.2	4.8	6.4	8

Speed of Sound

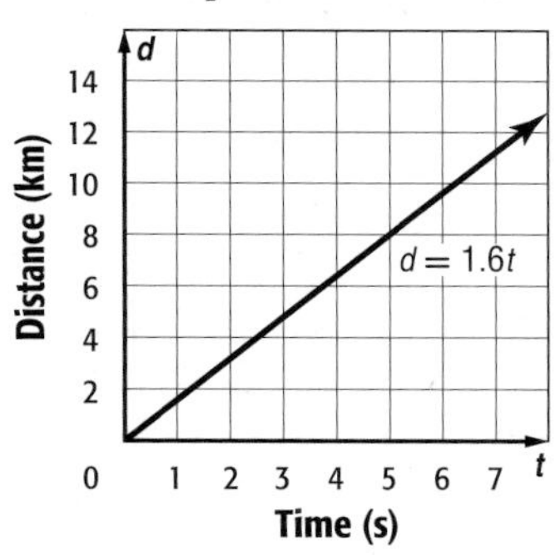

17b. about 11 km **19.** 6 **21.** $-\frac{1}{2}$

23.

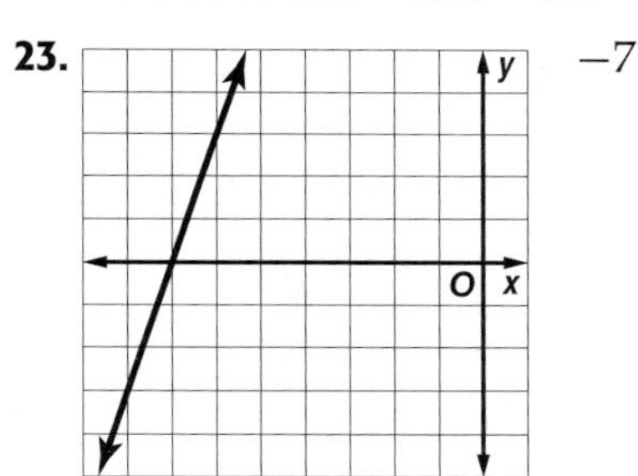

−7

25.

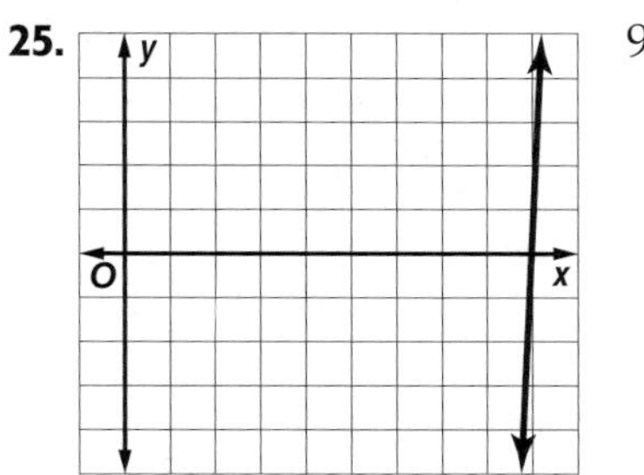

9

27. 3 **29.** $-\frac{1}{2}$ **31.** −0.046; an average decrease in cost of $0.046 per year

33.

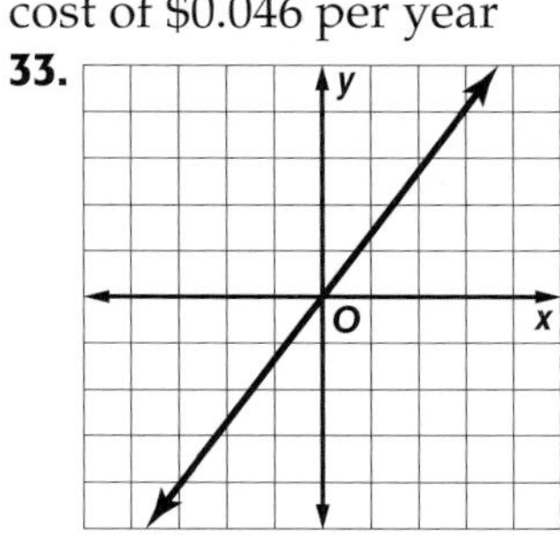

35. $y = 7.5x$; $y = 60$ **37.** $y = -x$; $y = -7$ **39.** 26, 31, 36 **41.** $a_n = 5n + 1$ **43.** $a_n = 4820n$; 15 s

45a.

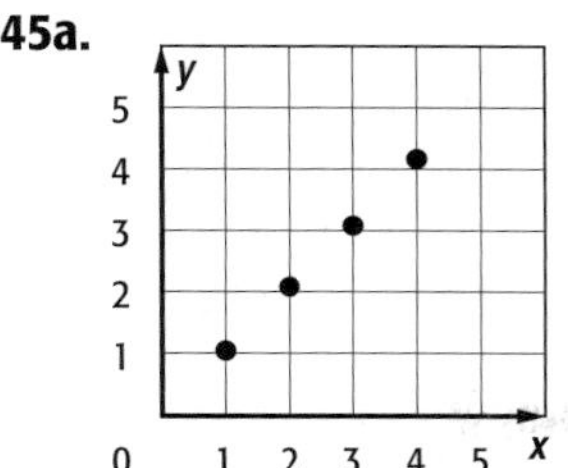

45b. $f(x) = 1.25x$ **45c.** $7.50

Chapter 4 Linear Functions and Relations

Page 211 Get Ready

1. 13 **3.** 14 **5.** $282.50 **7.** $x = 3 + 2y$ **9.** $x = \frac{3}{4}y + 3$ **11.** (4, 2) **13.** (2, −4) **15.** (−3, −3)

Pages 214–221 Lesson 4-1

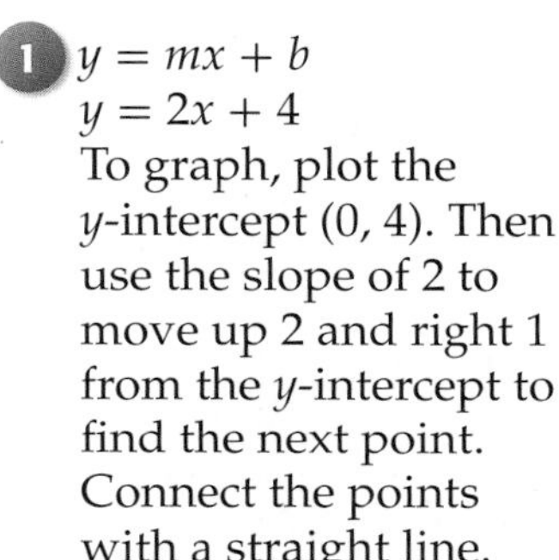

1 $y = mx + b$
$y = 2x + 4$
To graph, plot the y-intercept (0, 4). Then use the slope of 2 to move up 2 and right 1 from the y-intercept to find the next point. Connect the points with a straight line.

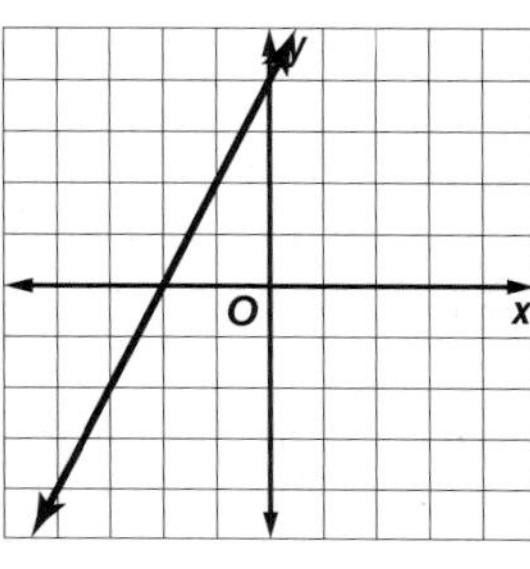

3. $y = \frac{3}{4}x - 1$

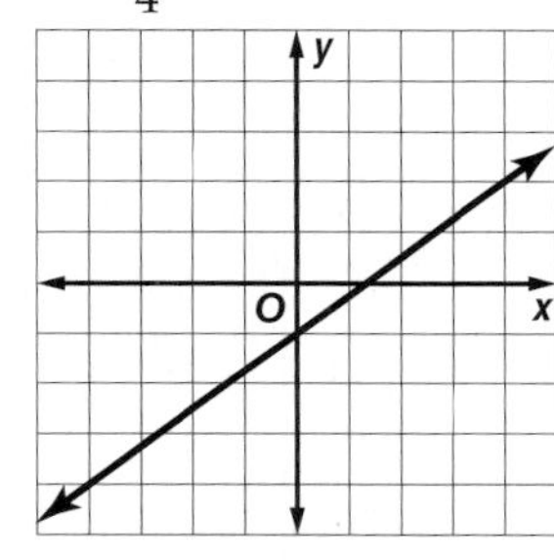

5.

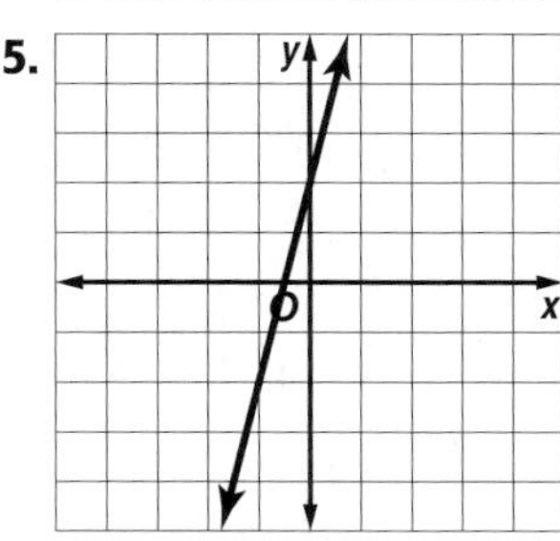

7.

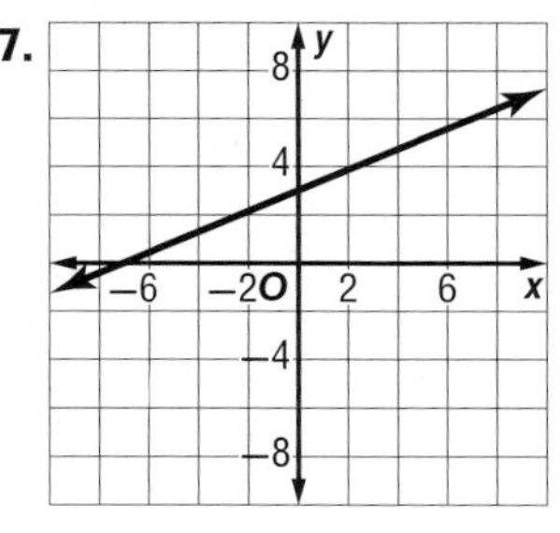

Selected Answers and Solutions

9.

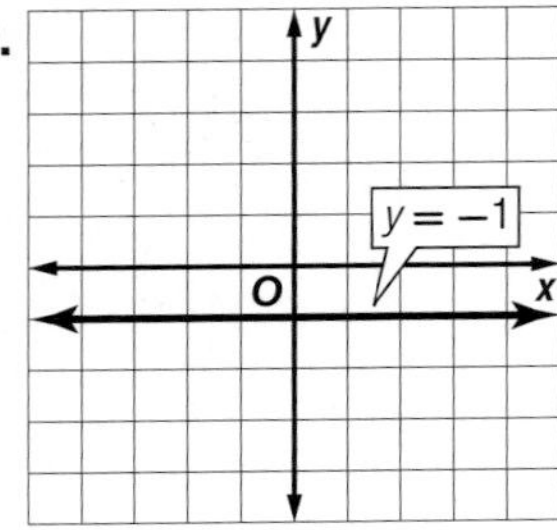

11. $y = \frac{2}{3}x + 2$ **13.** not possible **15a.** $S = 10w + 75$

15b.

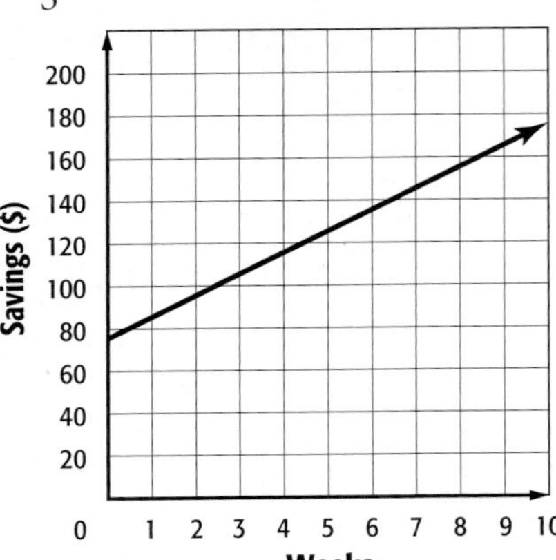

15c. $155

17 $y = mx + b$
$y = 5x + 8$
To graph, plot the y-intercept (0, 8). Then use the slope of 5 to move up 5 and right 1 from the y-intercept to find the next point. Connect the points with a straight line.

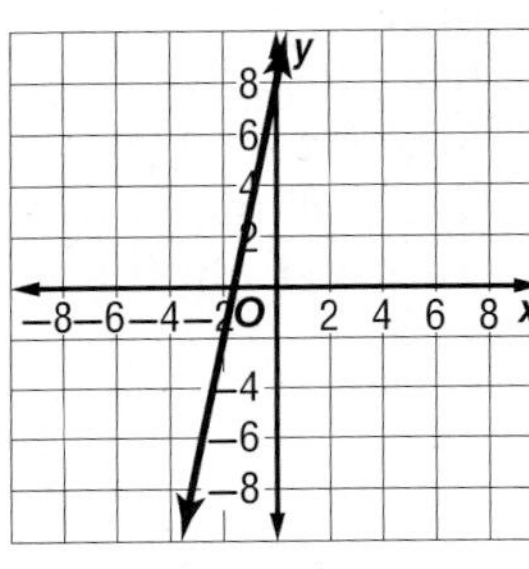

19. $y = -4x + 6$ **21.** $y = 3x - 4$

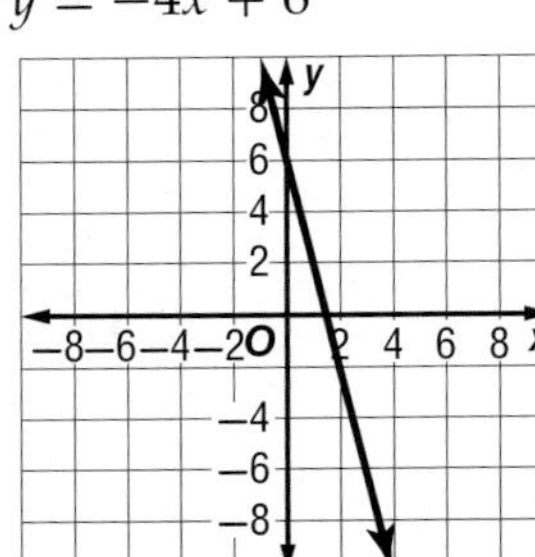

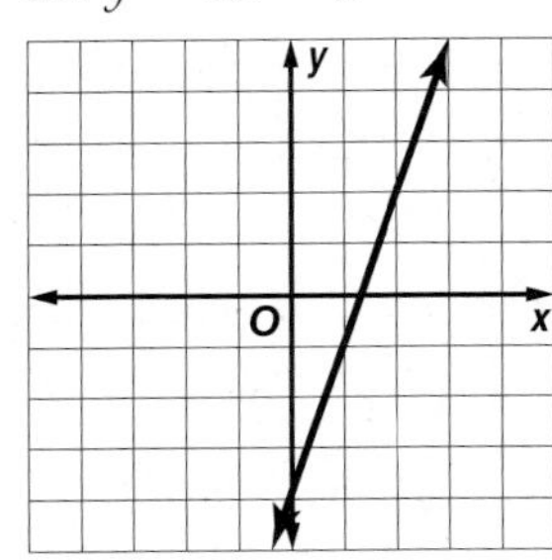

23.

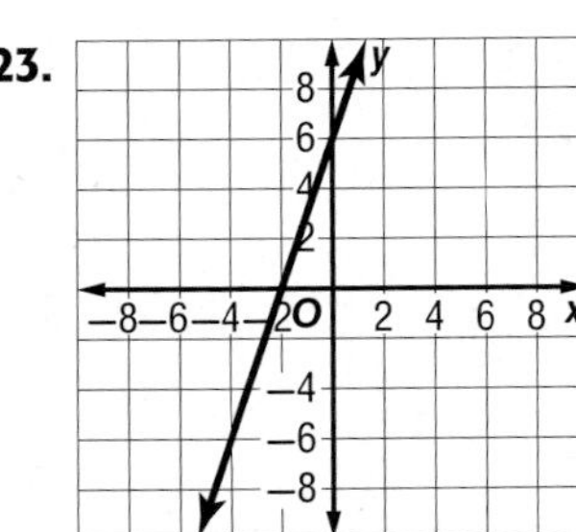

25.

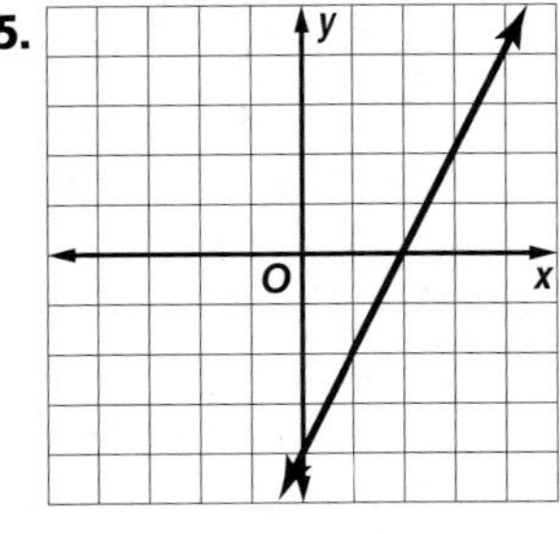

27.

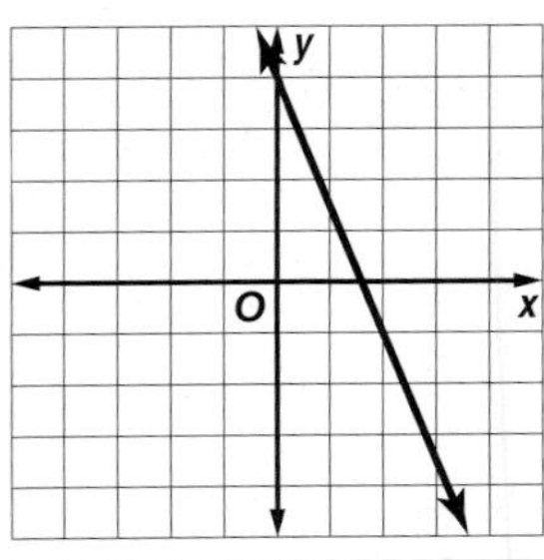

29.

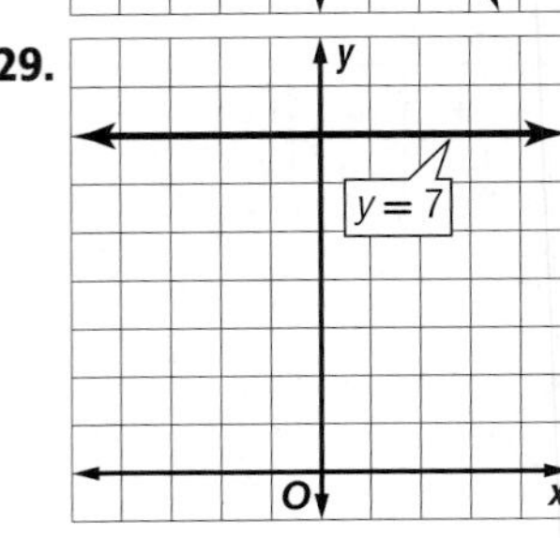

31.

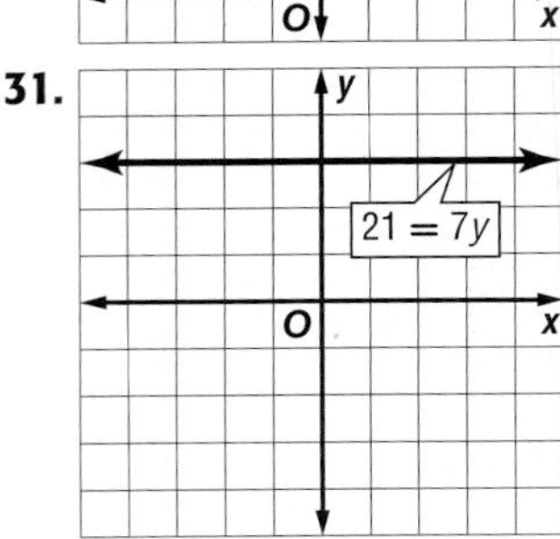

33. $y = -\frac{3}{5}x + 4$ **35.** $y = \frac{1}{2}x - 3$

37 a. Words: the population is 1267 plus 123 per year.
Equation: $P = 1267 + 123t$

b. Graph the equation by plotting the y-intercept of (0, 1267). Then use the slope of 123 to move up 123 up and 1 right.

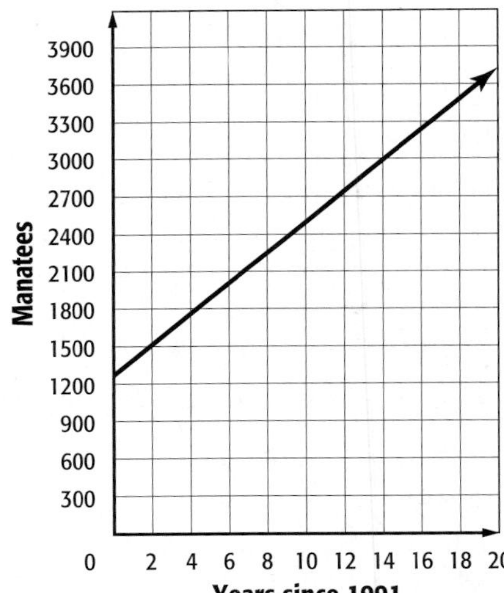

c. $P = 1267 + 123t$
$P = 1267 + 123(15)$
$P = 1267 + 1845$
$P = 3112$ manatees

39. $y = \frac{2}{3}x - 5$ **41.** $y = -\frac{3}{7}x + 2$ **43.** $y = 5$

45.

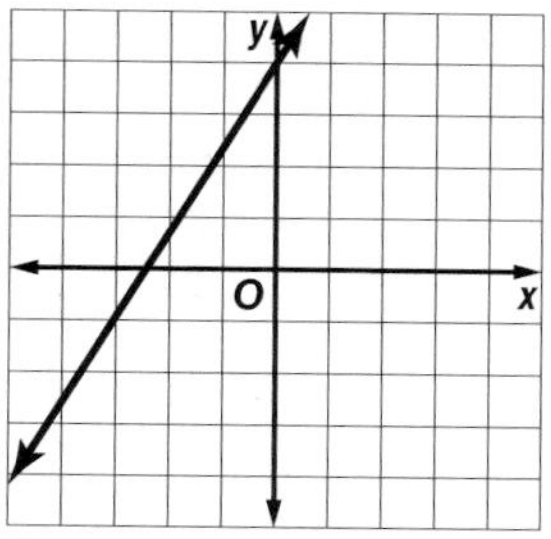

47.

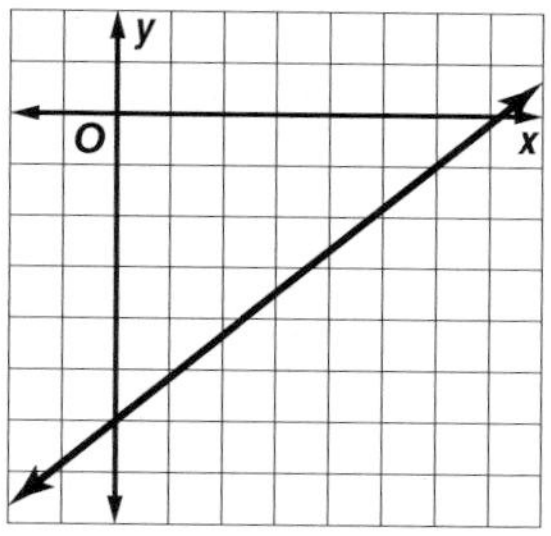

49.

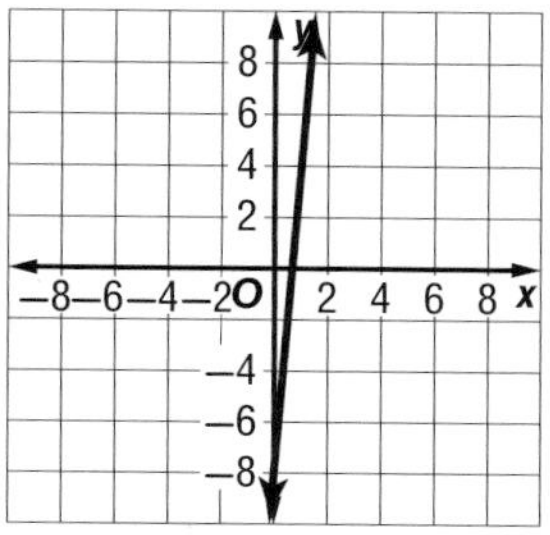

51a. $T = 157c + 218$ **b.** \$5242 **53.** $y = 0.5x + 7.5$ **55.** $y = -1.5x - 0.25$ **57.** $y = 3x$
59a. $C = 45m + 145$ **59b.** The cost per month to maintain the membership. **59c.** The start up fee **59d.** \$1225 **61a.** $C = 3.25 + 0.5256t$

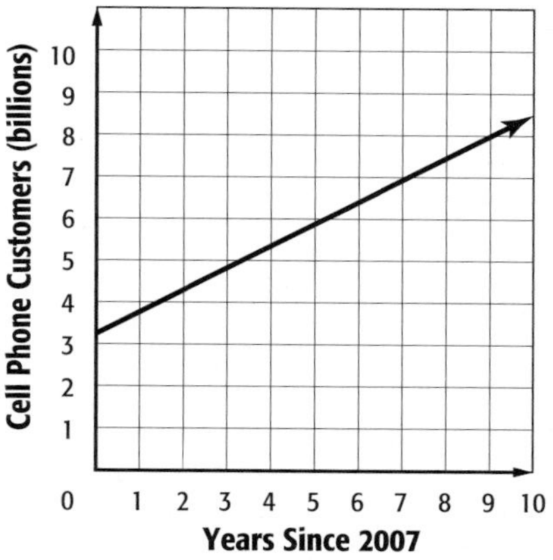

61b. 5.3524 billion **61c.** 2019 **63.** No; because a vertical line has no slope, it cannot be written in slope-intercept form. **65.** Sample answer: We would first have to rewrite the equation in slope-intercept form. The rate of change is also the slope, so, the coefficient for the x-variable is the rate of change. Assume that the coefficient of y is not 0. **67.** B **69.** C **71.** $a_n = 4_n - 1$; nonproportional, does not contain (0, 0) **73.** $a_n = 3_n + 10$; nonproportional, does not contain (0, 0) **75a.** \$25,500 **75b.** \$142,500 **77.** $y = -4x; -5$ **79.** $y = 0.8x; -7.5$ **81.** $-\frac{2}{5}$ **83.** 0

Pages 224–230 *Lesson 4-2*

1. $y = 3x - 12$ **3.** $y = -x + 6$ **5.** $y = -3x + 9$ **7.** $y = 5x + 8$ **9a.** $C = 35p + 75$ **9b.** \$600

11 $y = mx + b$
$4 = -1(-1) + b$
$4 = 1 + b$
$3 = b$
So, the equation is $y = -x + 3$.

13. $y = 8x - 55$ **15.** $y = 2x + 2$ **17.** $y = -x + 3$ **19.** $y = 7x - 16$ **21.** $y = 2x$ **23a.** $y = 9.575x + 337.1$ **23b.** 452 million **25.** $y = \frac{1}{2}x$ **27.** $y = -\frac{3}{4}x + 8\frac{1}{2}$ **29.** $y = \frac{2}{7}x - 2\frac{4}{7}$ **31a.** $G = 6.4t + 49.7$

31b.

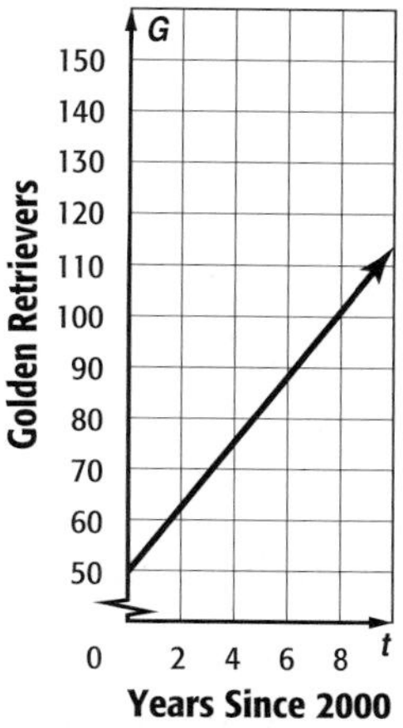

31c. 126,500 **33a.** \$2.75 **33b.** \$35.40

35 First, find the slope.

$$m = \frac{y_2 - y_1}{x_2 - x_1} = \frac{5 - (-3)}{2 - 5} = \frac{5 + 3}{2 - 5} = \frac{8}{-3} = -\frac{8}{3}$$

Next, use the slope-intercept formula.

$y = mx + b$
$5 = -\frac{8}{3}(2) + b$
$5 = -\frac{16}{3} + b$
$\frac{31}{3} = b$
$10\frac{1}{3} = b$
So, the equation is $y = -2\frac{2}{3}x + 10\frac{1}{3}$.

37. $y = -x - \frac{7}{12}$ **39.** Yes; substituting 6 and −2 for x and y, respectively, results in an equation that is true. **41.** B; x represents the number of raffle tickets sold, y represents the total amount of money in the treasury. **43a.** 605.2 **43b.** 2032; In that year the waste would be 0 tons. After that, the waste would be a negative amount, which is impossible.

45 a. $C = 52t + b$
$275 = 52(5) + b$
$275 = 260 + b$

$15 = b$

So, the equation is $C = 52t + 15$.

b.

Number of tickets	3	4	6	7
Cost ($)	171	223	327	379

c. Graph the equation by graphing the y-intercept (0, 15) and use the slope of 52 to find the next point.

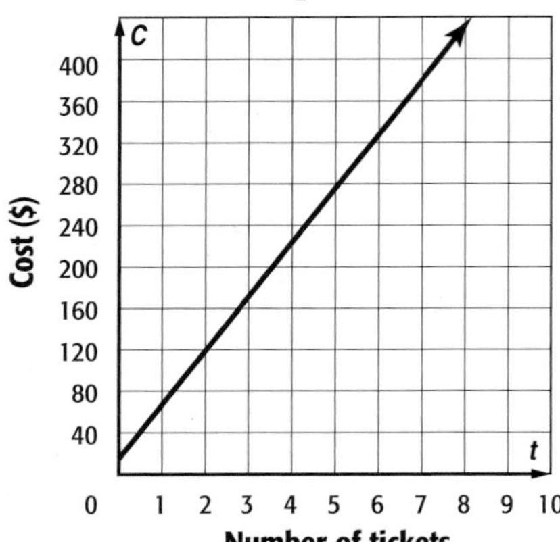

Eight tickets would be $431.

47. Jacinta; Teresa switched the x- and y-coordinates on the point that she entered in step 3.

49a. $y = -\frac{A}{B}x + \frac{C}{B}$ **49b.** slope $= -\frac{A}{B}$

49c. y-intercept $= \frac{C}{B}$ **49d.** no, $B \neq 0$ **51.** Sample answer: If the problem is about something that could suddenly change, such as weather or prices, the graph could suddenly spike up. You need a constant rate of change to produce a linear graph.

53. D **55.** B

57.

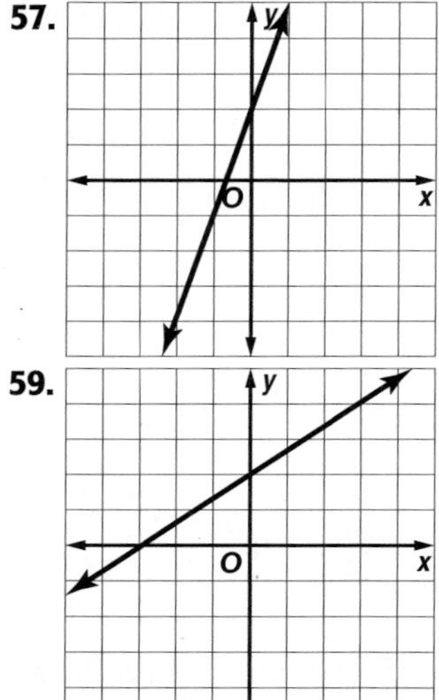

59.

61.

63. $f(x) = -2x$

65a.

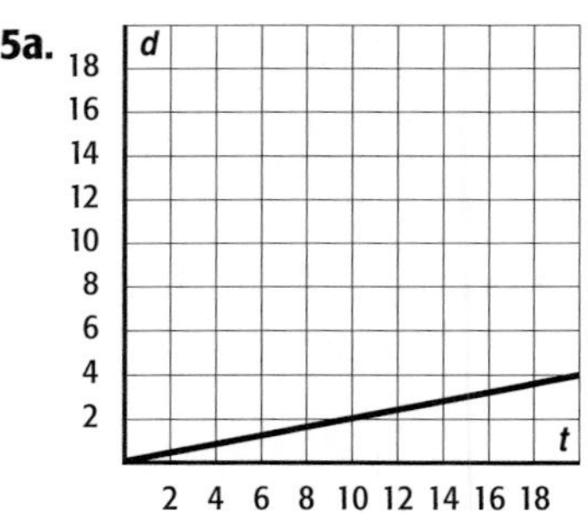

65b. about 14 seconds

67. −22 **69.** −207

71. 1.5 **73.** 7

75. −1 **77.** 1

Pages 231–236 Lesson 4-3

1 $y - y^2 = m(x - x^2)$

$(y - 5) = -6(x - (-2))$

$(y - 5) = -6(x + 2)$

Plot the point (−2, 5) and use the slope of −6 to find the next point.

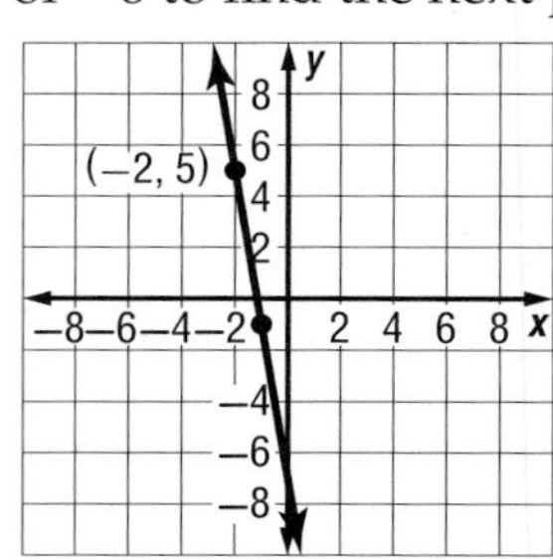

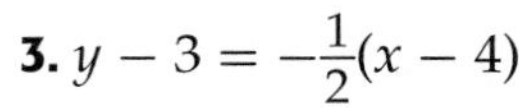

3. $y - 3 = -\frac{1}{2}(x - 4)$

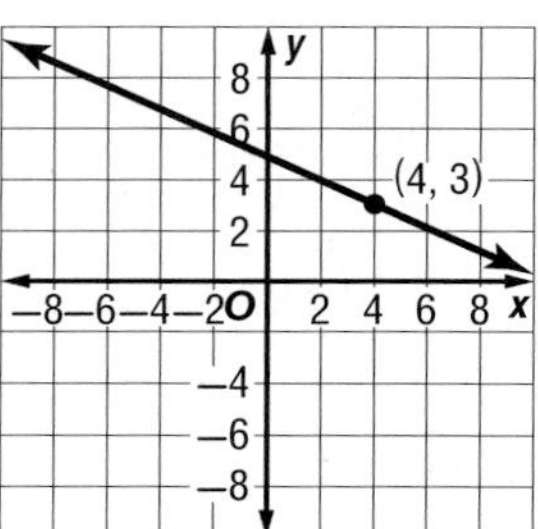

5. $5x + y = -22$ **7.** $y = 4x + 34$ **9.** $y = x + 13$

11. $y - 3 = 7(x - 5)$

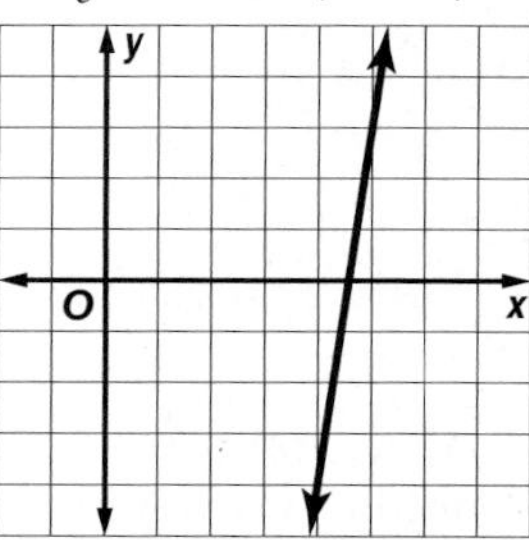

13. $y + 3 = -1(x + 6)$

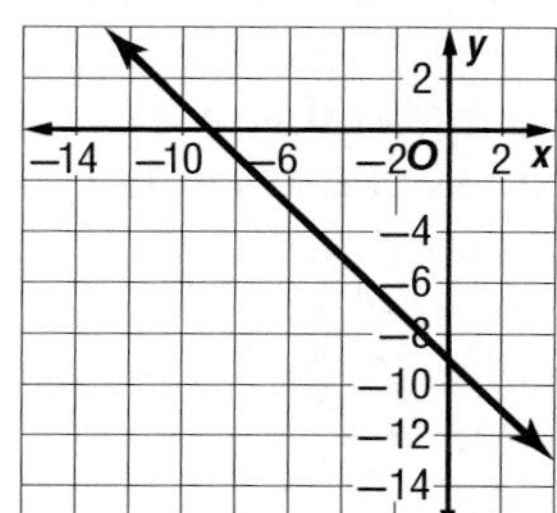

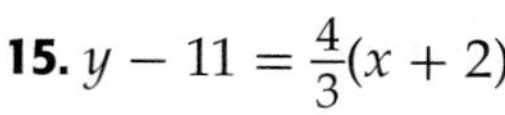

15. $y - 11 = \frac{4}{3}(x + 2)$

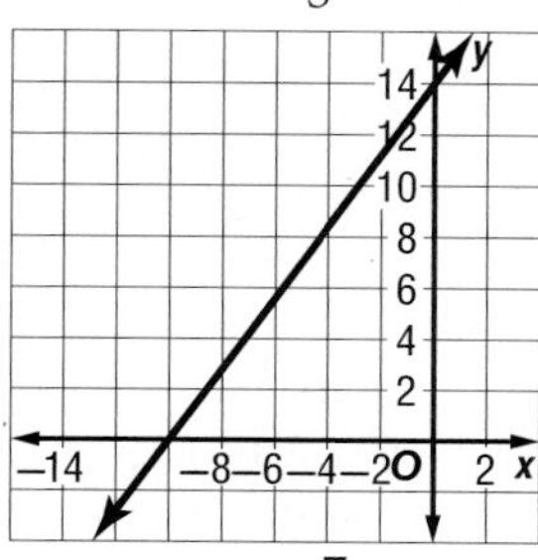

17. $y + 9 = -\frac{7}{5}(x + 2)$ **19.** $2x - y = 6$
21. $6x + y = -45$ **23.** $9x - 10y = 43$
25. $x + 6y = -7$ **27.** $y = -2x + 20$
29. $y = -6x - 47$ **31.** $y = \frac{1}{6}x - \frac{8}{3}$ **33.** $y = -\frac{2}{3}x - 5$

35 The slope is -5 and the point is $(6, 4)$. So, the equation is $y - 4 = -5(x - 6)$. Replace x with 2 to determine how many were rented the second week.

$$y - 4 = -5(2 - 6)$$
$$y - 4 = -5(-4)$$
$$y - 4 = 20$$
$$y = 24$$

So, 24 copies were rented.

37. $11x + 12y = 58$ **39.** $14x - 10y = 91$

41. $y + 1 = \frac{3}{2}(x + 4)$

43

$$y + \frac{3}{5} = x - \frac{2}{5}$$
$$y + \frac{3}{5} - \frac{3}{5} = x - \frac{2}{5} - \frac{3}{5}$$
$$y = x - \frac{5}{5}$$
$$y = x - 1$$

45. $y = \frac{5}{6}x$ **47.** $y - 4 = \frac{4}{7}(x + 9)$; $y = \frac{4}{7}x + \frac{64}{7}$; $4x - 7y = -64$ **49.** $y + 4 = 3(x + 1)$; The slope-intercept form is not $y = 3x + 2$. **51.** Sample answer: Jocari spent \$14 to go to an amusement park and ride ponies. The price she paid included admission. The 5 pony rides cost \$2 each; $y - 14 = 2(x - 5)$, $y = 2x + 4$. **53.** Sample answer: $y - g = \frac{j - g}{h - f}(x - f)$

55. B **57.** J **59.** $y = x - 2$ **61.** $y = -2x + 1$

63. $y = -2$ **65.** $y = -2x + 6$ **67.** $.y = \frac{1}{2}x + 3$

69. $y = 3$ **71.** Yes; there are only 364 seats.

73. $a = \frac{v - r}{t}$ **75.** $b = \frac{-t + 5}{4}$

Pages 237–243 Lesson 4-4

1. $y = \frac{1}{2}x + 2\frac{1}{2}$ **3.** Slope of $\overline{AC} = \frac{1 - 7}{-2 - 5}$ or $\frac{6}{7}$; slope of $\overline{BD} = \frac{-3 - 4}{3 - (-3)}$ or $-\frac{7}{6}$; the paths are perpendicular.

5 Graph each line on a coordinate plane. $y = -2x$ and $4y = 2x + 4$ are perpendicular to $2y = x$; $2y = x$ and $4y = 2x + 4$ are parallel.

7. $y = 2x + 7$ **9.** $y = \frac{3}{2}x$ **11.** $y = x - 5$

13. $y = -5x + 2$ **15.** $y = -\frac{3}{4}x + 1\frac{1}{2}$ **17.** Yes; the line containing $\overline{AD}$ and the line containing $\overline{BC}$ have the same slope, $\frac{1}{3}$. Therefore one pair of sides is parallel. The slope of $\overline{AB}$ is undefined and the slope of $\overline{CD}$ is $-\frac{5}{3}$. **19.** Yes; the slopes are -6 and $\frac{1}{6}$. **21.** $2x - 8y = -24$ and $4x + y = -2$ are perpendicular; $2x - 8y = -24$ and $x - 4y = 4$ are parallel.

23 The slope of the given line is -2. So, the slope of a line perpendicular is $\frac{1}{2}$.

$$y = mx + b$$
$$-2 = \frac{1}{2}(-3) + b$$
$$-2 = -\frac{3}{2} + b$$
$$-\frac{1}{2} = b$$

The equation is $y = \frac{1}{2}x - \frac{1}{2}$.

25. $y = -3x - 7$ **27.** $y = -\frac{1}{5}x + 8\frac{3}{5}$ **29.** $y = 2x + 16$ **31.** $y = -\frac{1}{5}x - \frac{3}{25}$ **33.** neither

35. perpendicular **37.** neither **39.** $y = 7x$

41 Find the slopes.

$$m = \frac{-1 - 6}{4 - 2} = \frac{-7}{2} = -\frac{7}{2}$$

$$m = \frac{12 - 10}{14 - 7} = \frac{2}{7}$$

Since the slopes are opposite reciprocals, the objects are perpendicular.

43a.

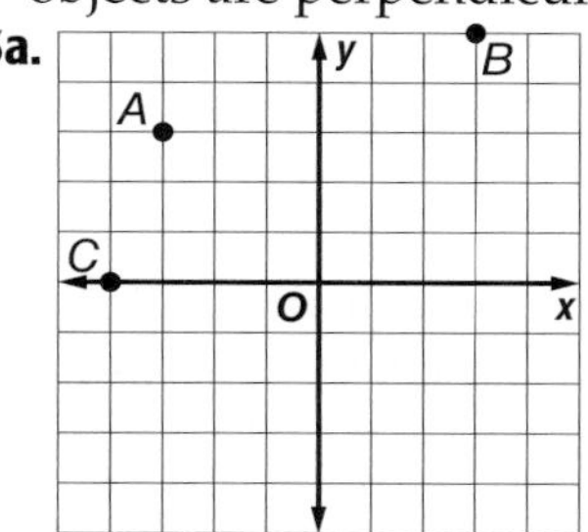

43b. Sample answer: $(2, 2)$; $\overline{AB}$ and $\overline{CD}$ both have slope $\frac{1}{3}$, and $\overline{AC}$ and $\overline{BD}$ both have slope 3.

43c. Two; sample answer: Move C to $(-2, 0)$ and move D to $(4, 2)$. Moving C changes the slpoe of $\overline{AC}$ to -3. This is the opposite reciprocal of the slope $\overline{AC}$, $\frac{1}{3}$. Moving D also changes the slope of $\overline{BD}$ so $\overline{BD}$ is perpendicular to $\overline{AB}$ and $\overline{CD}$ and it is parallel to $\overline{AC}$. **45.** Always; horizontal lines and vertical lines intersect at right angles. **47.** Carmen is correct; she correctly determined the slope of the perpendicular line. **49.** A **51.** B **53.** $-4x + y = 5$

55. $5x + y = -8$ **57.** $-5x + 6y = -14$
59a. $C = 10h + 15$ **59b.** \$95 **61.** $y = -5x - 21$
63. $y = 2x - 1$ **65.** $y = -5x - 6$ **67.** simplified
69a. $25(5) + 10(8.5) + 35(5) + 12(8.5)$ **69b.** \$487

71.

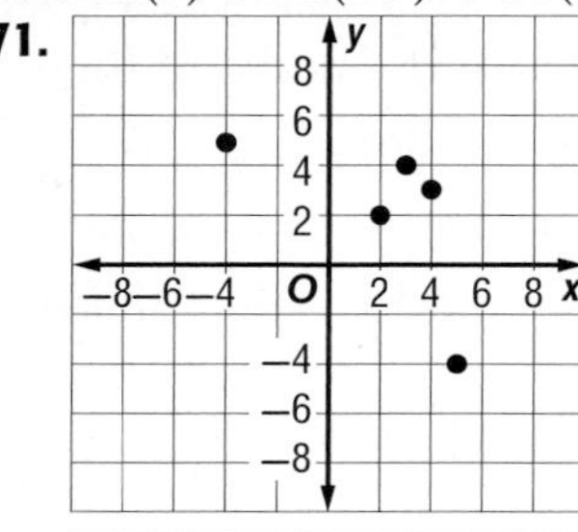

$D = \{3, 4, 2, 5, -4\}$;
$R = \{4, 3, 2, -4, 5\}$

73.

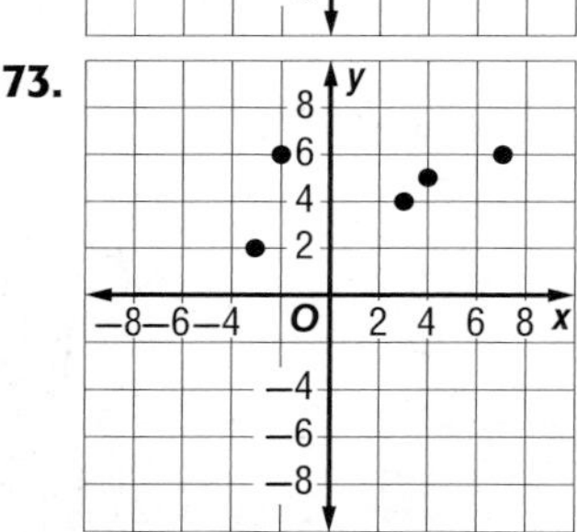

$D = \{7, 3, 4, -2, -3\}$;
$R = \{6, 4, 5, 6, 2\}$

Pages 245–251 Lesson 4-5

1. Positive; the longer you practice free throws, the more free throws you will make.

3a.

Median Age of Females When First Married

Age: 50, 45, 40, 35, 30, 25, 20, 15, 10, 5

Year: 1996, 1998, 2000, 2002, 2004, 2006, 2008, 2010

Positive; independent variable is year and dependent variable is median age of females when they were first married. **3b.** See above graph. **3c.** Sample answer: Using (1996, 24.8) and (2006, 25.9) and rounding, $y = 0.11x - 194.8$ **3d.** Sample answer: 27.0 **3e.** Yes, according to the equation, the median age would be 31.4, which is likely.

5 As the height increases, the percentage decreases. The graph shows a slight negative correlation. This correlation means that the taller a player is, the lower their percentage of 3-point shots made is.

7 There is not pattern to the graph, so there is no correlation between the speed of a vehicle and the miles per gallon.

9a. $y = -1783x + 78{,}349$ **9b.** 56,953 **9c.** No; the average attendance will fluctuate with other variables such as how good the team is that year.

11 a. The independent variable is the duration of the eruptions and the dependent variable is the interval of the eruptions. This is because the duration of the eruptions is not affected by the interval.

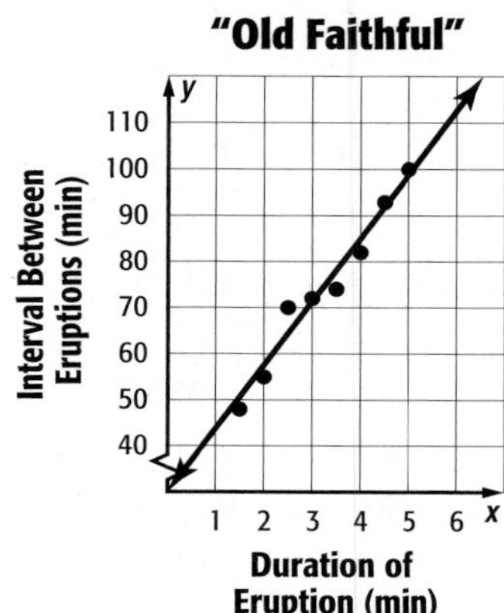

The interval increases as the duration increases, so there is a positive correlation between the independent and dependent variables.

b. Use (2, 55) and (4, 82)

$$m = \frac{y_2 - y_1}{x_2 - x_1}$$
$$= \frac{82 - 55}{4 - 2}$$
$$= \frac{27}{2}$$
$$= 13.5$$

So, the slope is 13.5.

$$y = 13.5x + b$$
$$55 = 13.5(2) + b$$
$$55 = 27 + b$$
$$28 = b$$
$$y = 13.5x + 28$$
$$y = 13.5(7.5) + 28$$
$$y = 101.25 + 28$$
$$y = 129.25 \text{ min}$$

c. Sample answer: The duration of an eruption is not dependent on the previous interval. Only the interval can be predicted by the length of the eruption.

13. Sample answer: The salary of an individual and the years of experience that they have; this would be a positive correlation because the more experience an individual has, the higher the salary would probably be. **15.** Neither; line g has the same number of points above the line and below the line. Line f is close to 2 of the points, but for the rest of the data there are 3 points above and 3 points below the line. **17.** Sample answer: You can visualize a line to determine whether the data has a positive or negative correlation. The following graph shows the ages and heights of people. To predict a person's age given his or her height, write a linear equation for the line of fit.

Then substitute the person's height and solve for the corresponding age. You can use the pattern in the scatter plot to make decisions.

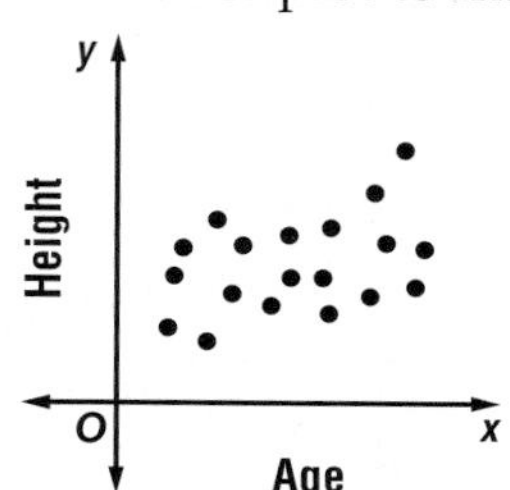

19. F **21.** 22 days **23.** neither **25.** perpendicular **27.** $2x + y = 1$ **29.** $-x + 2y = -12$ **31.** $2x + 5y = 26$

33.

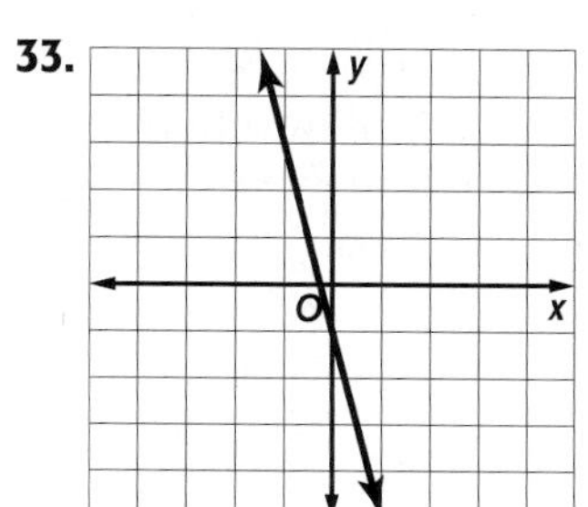

35. $\frac{4}{7}$ **37.** $\frac{3}{5}$

39. 16 **41.** 1.5 h

43.

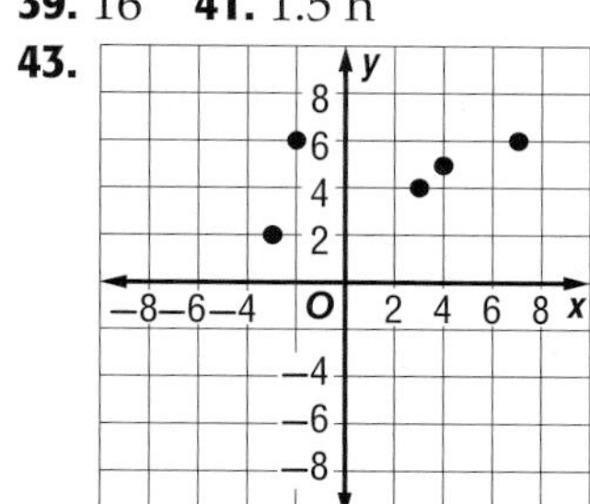

D = {7, 3, 4, −2, −3}; R = {6, 4, 5, 2}

Pages 253–260 Lesson 4-6

1. $y = 1.18x + 11$; 0.7181
3a. $y = -271.88x + 554.48$ **3b.** $78.69

Rental Properties

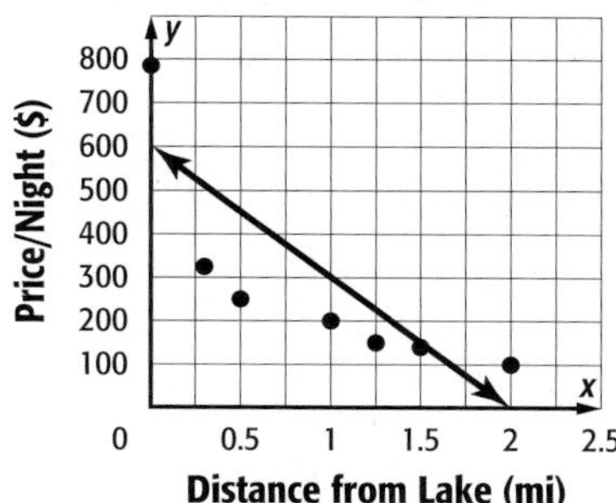

5 Step 1: Enter the data by pressing **STAT** and selecting the Edit option. Let the year 2000 be represented by 0. Enter the years since 2000 into List 1 (L1). These will represent the x-values. Enter the number of auditions into List 2 (L2). These will represent the y-values.
Step 2: Perform the regression by pressing STAT and selecting the **CALC** option. Scroll down to LinReg($ax + b$) and press ENTER.
The equation is $y = 3.54x + 19.68$.
The correlation coefficient is 0.9007.

7 a. Enter the data using 0 for 1975. Use med-med to find $y = 609.08x + 1680.8$.
b. $2003 - 1975 = 28$. Substitute 28 into the equation in **a** to get 18,734.32. There were about 18,735 entrants in 2003.

9a. $y = 0.095x - 94.58$

9b.

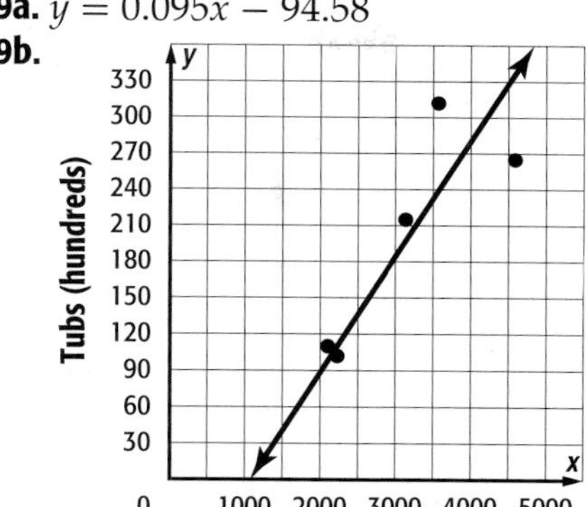

9c. about 48 tubs; about 380 tubs
11a. $y = 9.8x + 28.79$ **11b.** $146.39 **11c.** Sample answer: No; the number is not within a reasonable range of the other pairs of jeans.

13 Step 1: Enter the data by pressing **STAT** and selecting the Edit option. Let the year 1998 be represented by 0. Enter the years since 1998 into List 1 (L1). These will represent the x-values. Enter the amount raised into List 2 (L2). These will represent the y-values.
Step 2: Perform the regression by pressing **STAT** and selecting the **CALC** option. Scroll down to LinReg($ax + b$) and press ENTER.
The equation is $y = 420.17x + 1682.22$.
The correlation coefficient is 0.9464.

15a. $y = 87{,}390.5x + 4{,}018{,}431$ **15b.** about 5,591,460 **17a.** $y = 361.38x + 4840.6$

17b.

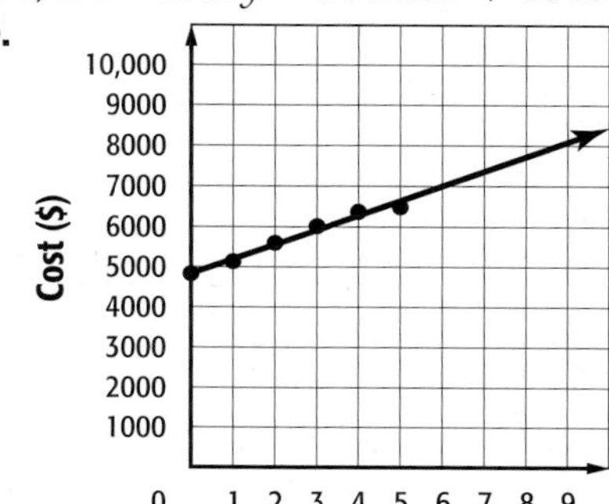

17c. $12,068.28

19.

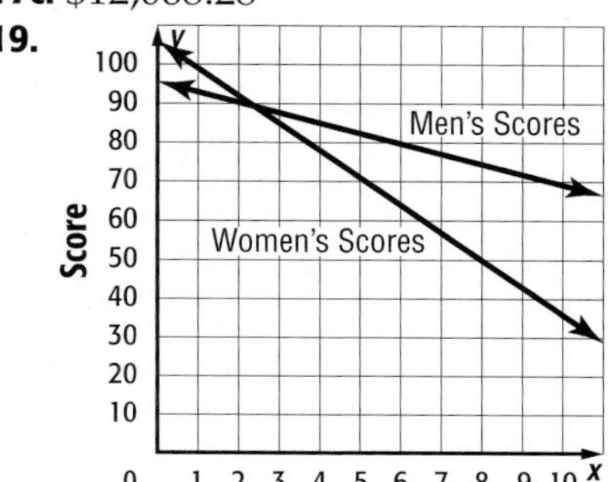

Sample answer: Men: $y = -2.92x + 95.92$; women: $y = -7x + 106$; women's scores have a steeper slope than men's.

Selected Answers and Solutions

23. C **25.** H **27a.** negative correlation **27b.** \$3600 **27c.** No; according to the line of fit, the cost would be \$0. **29.** $3x - y = 1$ **31.** $2x + y = 8$ **33.** $2x - 3y = -21$ **35.** $\frac{4}{7}$ **37.** $\frac{3}{5}$ **39.** 3 **41.** $a^2 - a + 1$

43.

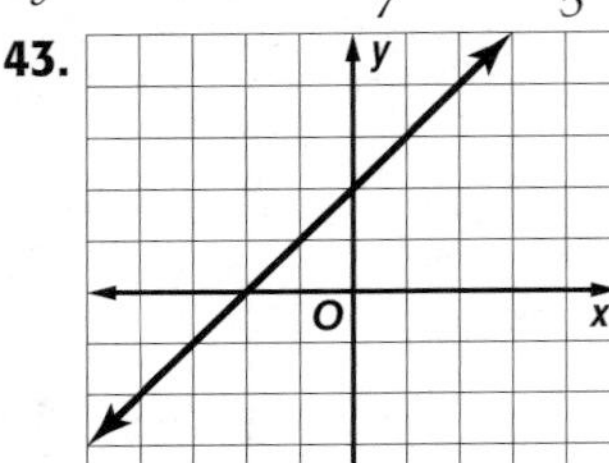

45.

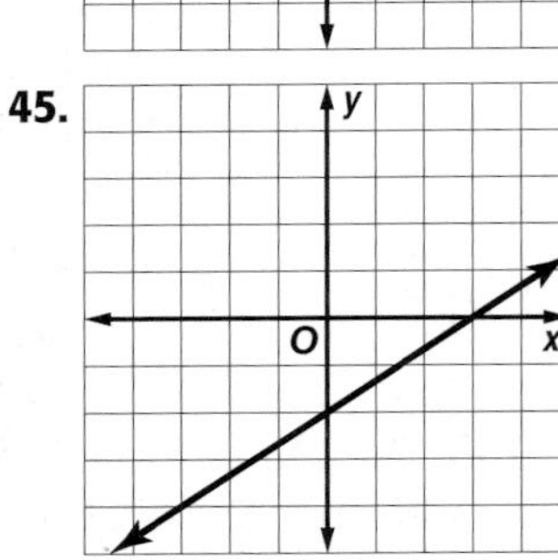

Pages 261–268 Lesson 4-7

1.

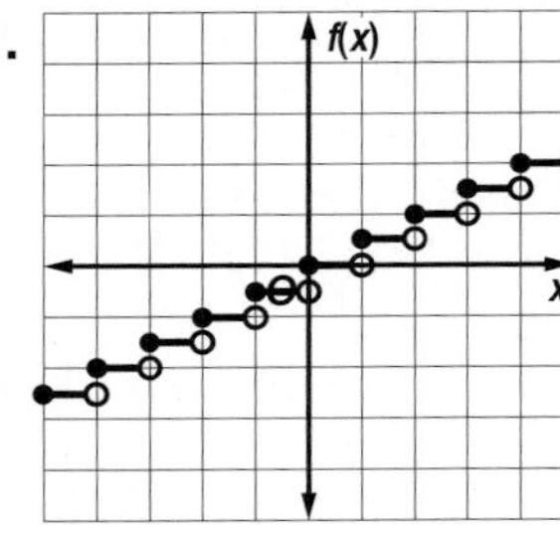

D = all real numbers, R = all integer multiples of 0.5

3.

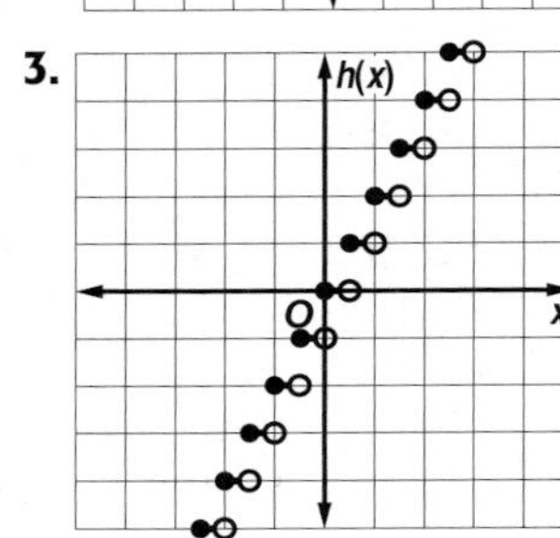

D = all real numbers, R = all integers

5.

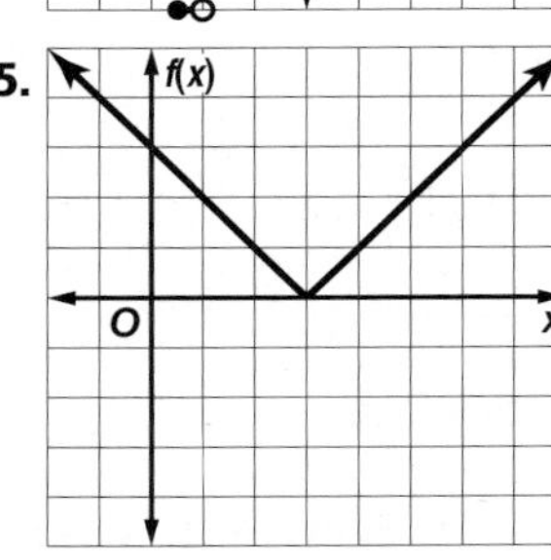

D = all real numbers, R = $f(x) \geq 0$

7.

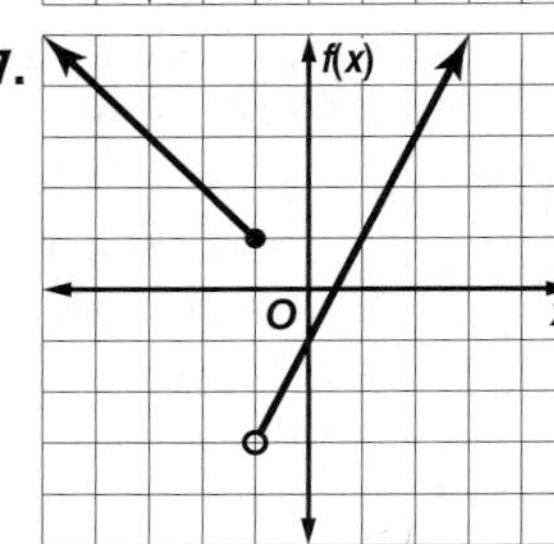

D = all real numbers, R = $f(x) > -3$

9 $f(x) = 3[\![x]\!]$

Create a table.

x	$3[\![x]\!]$	$f(x)$
0	$3[\![0]\!]$	0
0.25	$3[\![0.25]\!]$	0
0.5	$3[\![0.5]\!]$	0
1	$3[\![1]\!]$	3
1.25	$3[\![1.25]\!]$	3
1.5	$3[\![1.5]\!]$	3
2	$3[\![2]\!]$	6
2.25	$3[\![2.25]\!]$	6
2.5	$3[\![2.5]\!]$	6
3	$3[\![3]\!]$	9

Graph on a coordinate grid.

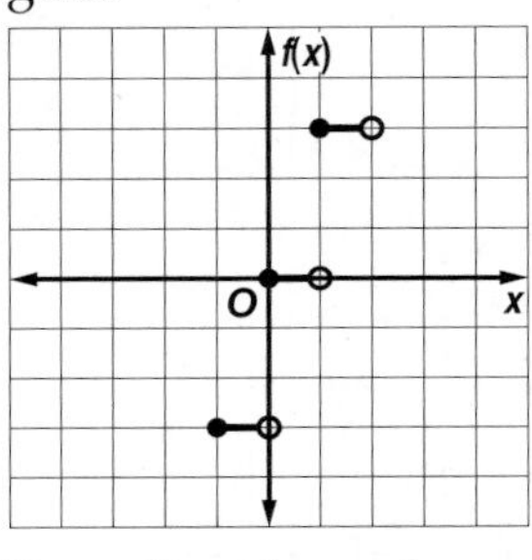

D = all real numbers, R = all integer multiples of 3

11.

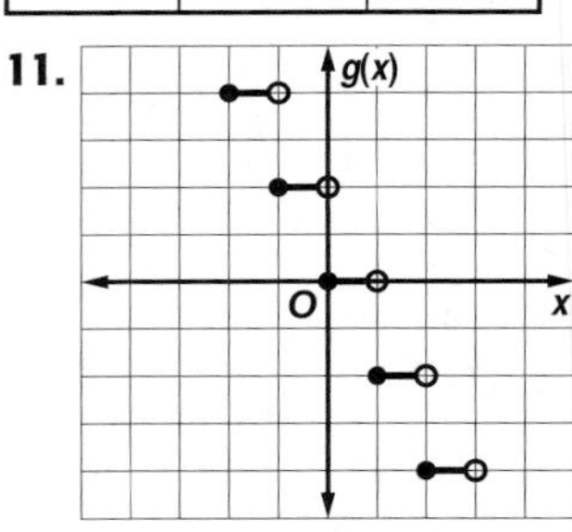

D = all real numbers, R = all even integers

13.

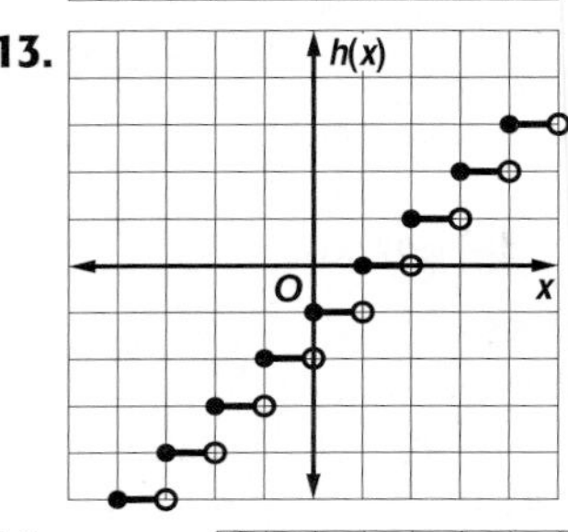

D = all real numbers, R = all integers

15a.

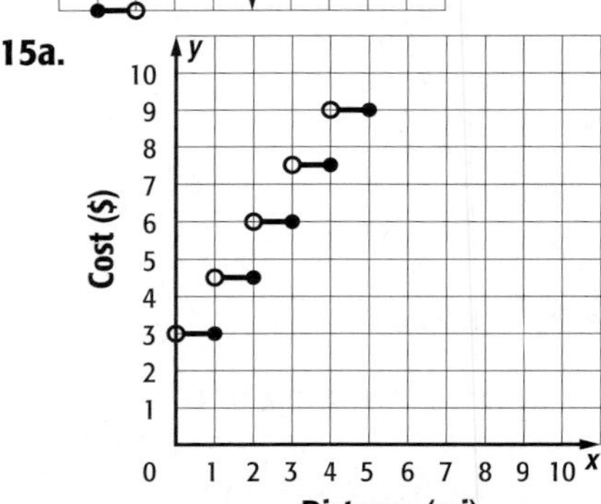

15b. \$16.50

17.

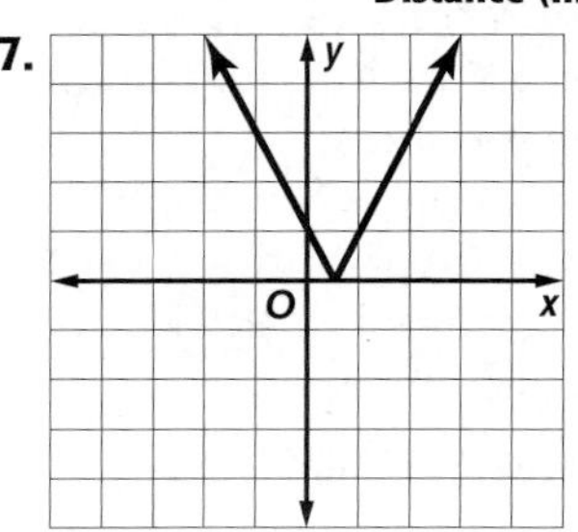

D = all real numbers, R = $y \geq 0$

19.

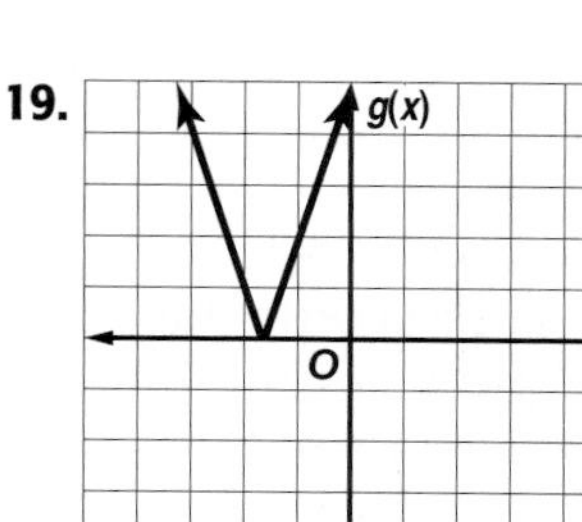

D = all real numbers,
R = $g(x) \geq 0$

21.

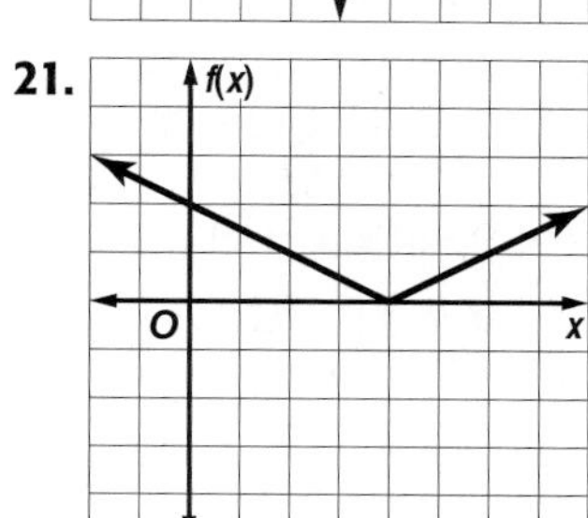

D = all real numbers,
R = $f(x) \geq 0$

23.

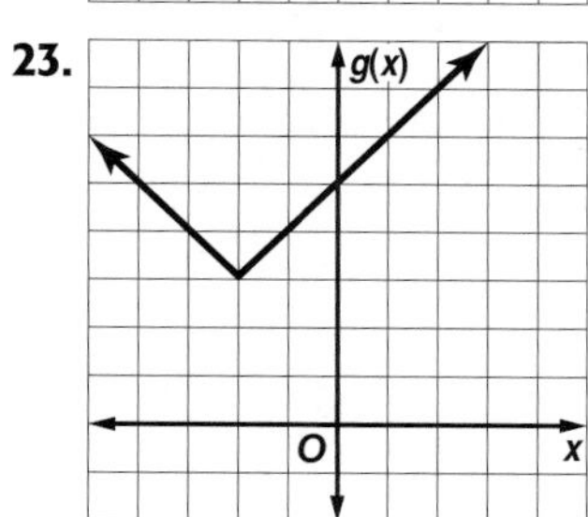

D = all real numbers,
R = $g(x) \geq 3$

25.

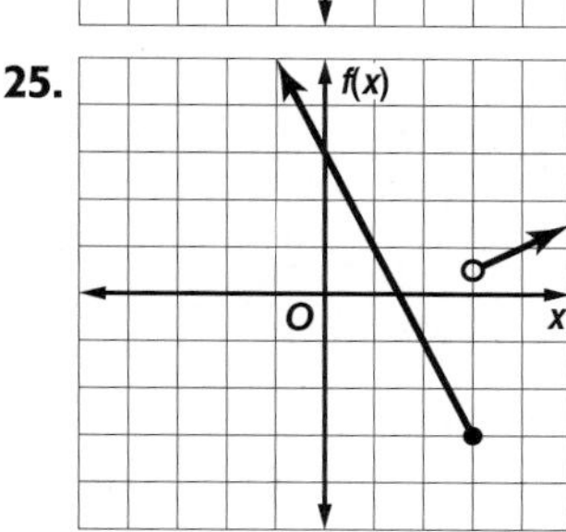

D = all real numbers,
R = all real numbers

27.

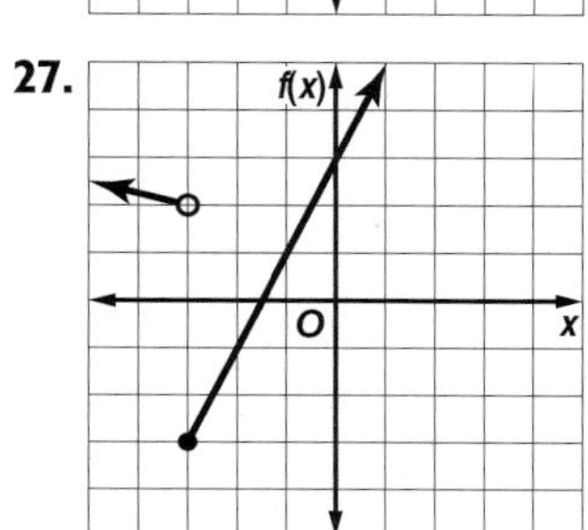

D = all real numbers,
R = $f(x) \geq -3$

29.

D = all real numbers,
R = $f(x) \geq -2.5$

31 The graph extends infinitely to the left and to the right, so the domain is all real numbers. The range is all real numbers greater than or equal to 4 because the line increases from both sides of (6, 4).

33. D = all real numbers; R = all integers

35. D = all real numbers; R = $y > -4$

37.

Number of People

Length of Boat (ft)

39. D **41.** B

43 a.

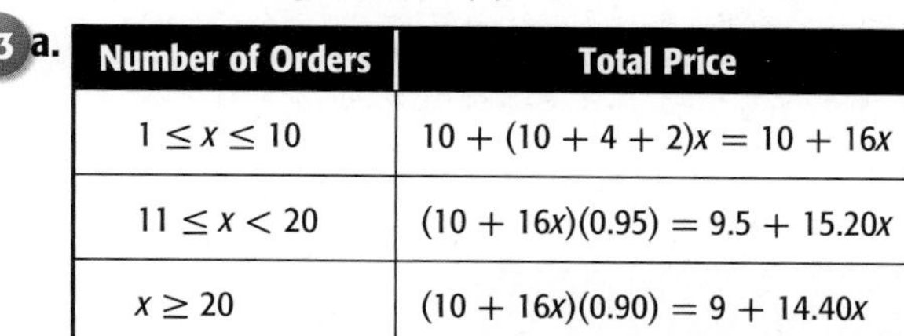

Number of Orders	Total Price
$1 \leq x \leq 10$	$10 + (10 + 4 + 2)x = 10 + 16x$
$11 \leq x < 20$	$(10 + 16x)(0.95) = 9.5 + 15.20x$
$x \geq 20$	$(10 + 16x)(0.90) = 9 + 14.40x$

b. $y = \begin{cases} 10 + 16x \text{ if } 1 \leq x \leq 10 \\ 9.50 + 15.20x \text{ if } 11 \leq x \leq 20 \\ 9 + 14.40x \text{ if } x \geq 20 \end{cases}$

c.

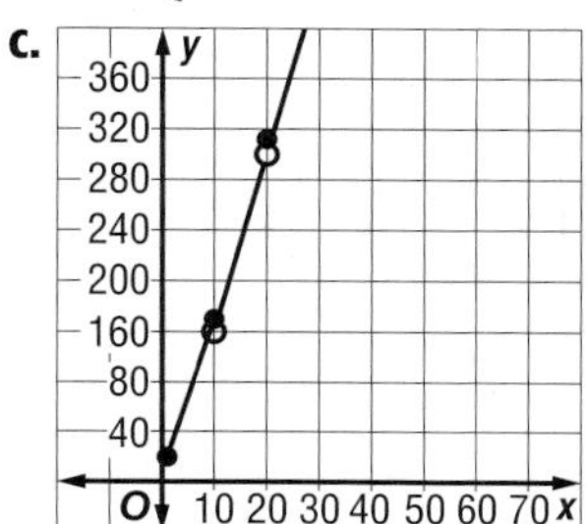

45a.

x	−5	−4	−3	−2	−1	0	1	2	3	4	5
$f(x)$	13	11	9	7	5	3	5	7	9	11	13

45b.

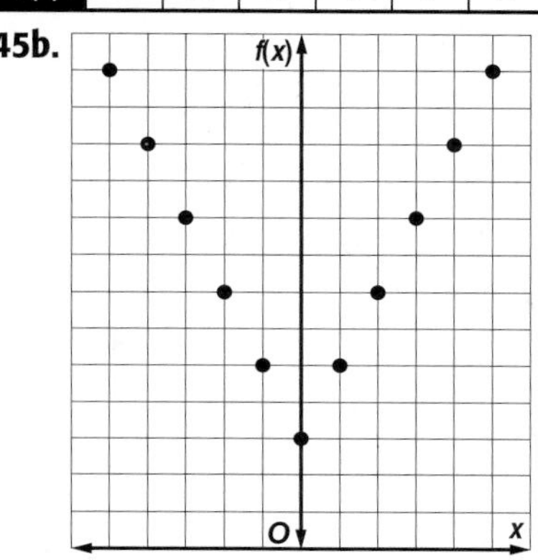

45c.

45d. The graph is shifted 1.5 units to the right and 3 units up.

47.

Number of Shows

9 8 7 6 5 4 3 2 1 0

200 400 600 800

Tickets Sold

49 $g(x) = \frac{1}{3}|x| + 4$

Set up a table.

x	$\frac{1}{3}\|x\| + 4$	$g(x)$
−6	$\frac{1}{3}\|-6\| + 4$	6
−3	$\frac{1}{3}\|-3\| + 4$	5
0	$\frac{1}{3}\|0\| + 4$	4
3	$\frac{1}{3}\|3\| + 4$	5
6	$\frac{1}{3}\|6\| + 4$	6

Plot on a coordinate grid.

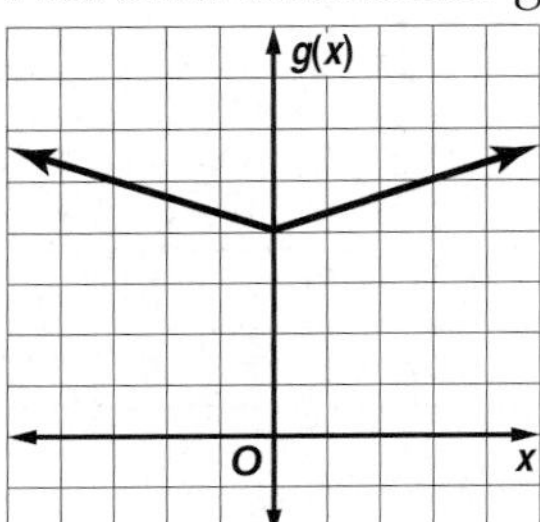

51.

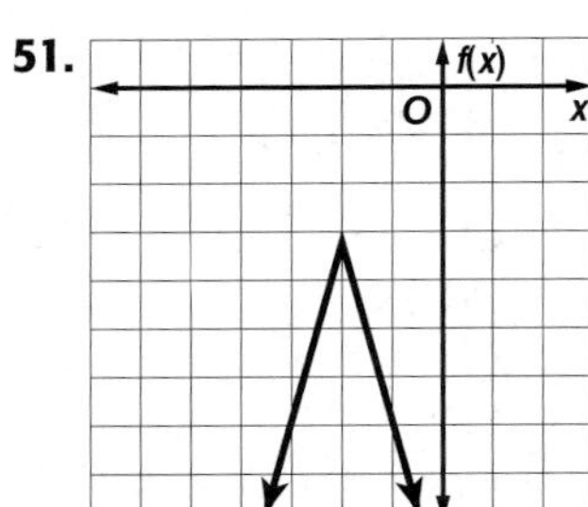

53.

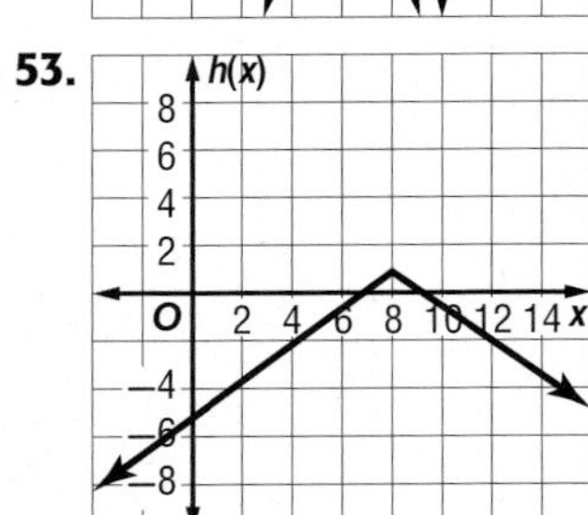

55. No; the pieces of the graph overlap vertically, so the graph fails the vertical line test.

57. $f(x) = \begin{cases} \frac{1}{2}x - 3 & \text{if } x > 6 \\ -\frac{1}{2}x - 1 & \text{if } x \leq 6 \end{cases}$

59. Sample answer: The cost of exceeding the allotted minutes on your cell phone plan is billed per minute. So, if you are 2.5 minutes over, you are charged for 3 minutes of overage. The graph of this is called a step function because you can't trace the graph without picking up your pencil. There is a range of x-values that have the same y-value on the graph. **61.** C **63.** B **65.** $y = 2.3x + 1.5$ **67.** $y = 10.7x + 10.1$ **69.** Positive; it means the more you study, the better your test score. **71.** −4.4 **73.** 1.2 **75.** −3 **77.** 21 **79.** $-\frac{1}{3}$

Pages 270–274 Study Guide and Review

1. T **3.** F; scatter plot **5.** T **7.** T **9.** F, greatest integer function

11.

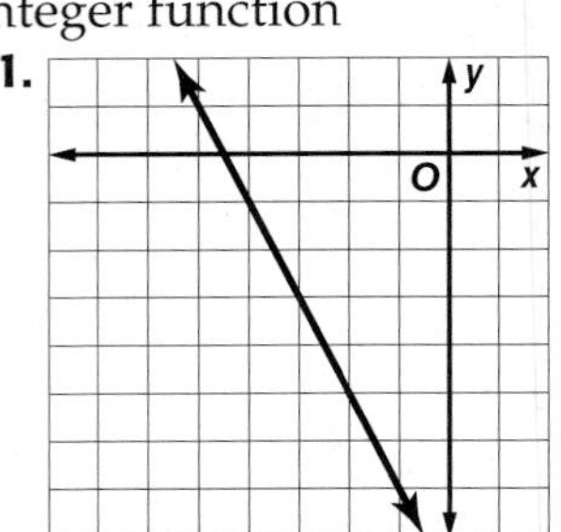

$y = -2x - 9$

13.

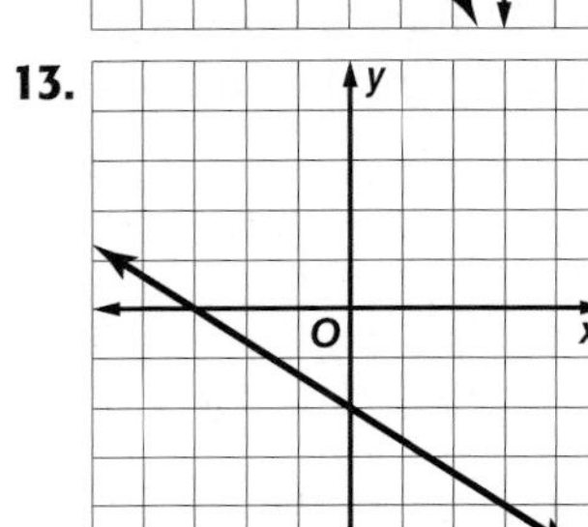

$y = -\frac{5}{8}x - 2$

15.

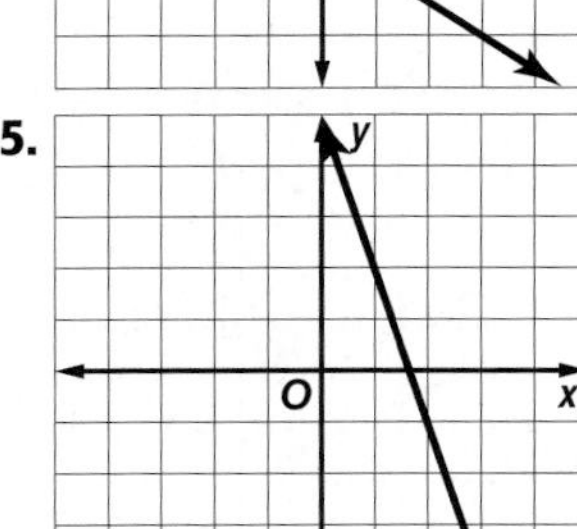

17.

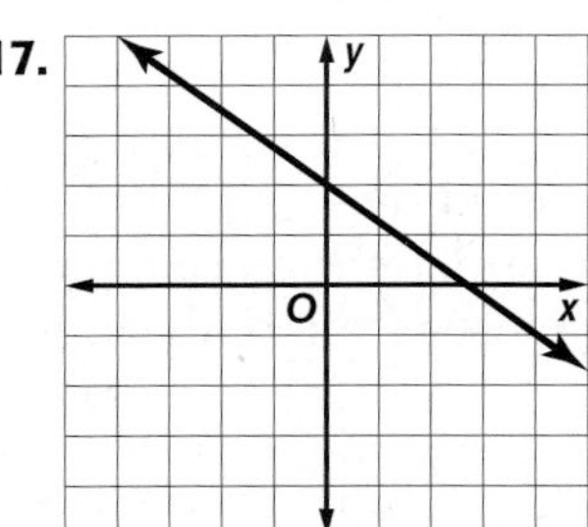

19. $y = 3x - 1$ **21.** $y = \frac{2}{5}x + \frac{1}{5}$ **23.** $y = x - 3$ **25.** $y = \frac{1}{2}x + \frac{7}{2}$ **27.** $y = 60x + 450$ **29.** $y - 1 = -3(x + 2)$ **31.** $5x - y = 7$ **33.** $x - 2y = 11$

35. $y = 3x - 13$ **37.** $y = 5x + 2$ **39.** $y = x + 3$ **41.** $y = -2x - 7$ **43.** $y = -\frac{1}{3}x + \frac{14}{3}$ **45.** $y = -3x - 13$ **47.** positive **49.** $y = 5.36x + 11$; 65

51. D = all real numbers; R = all integers

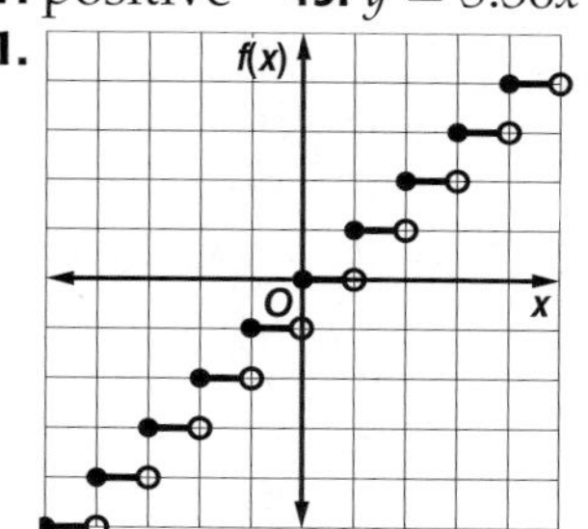

53. D = all real numbers; R = $f(x) \geq 0$

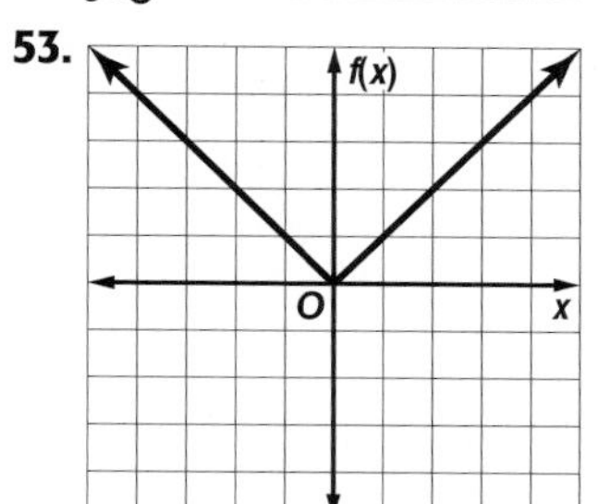

55. D = all real numbers; R = $f(x) < -1$ or $y \geq 3$

f(x)
O
x

57. D = all real numbers; R = $y \geq -2$

Chapter 5 Linear Inequalities

Page 281 Get Ready

1. −10 **3.** 24.6 **5.** −11 **7.** 21 **9.** 5 **11.** −4 **13.** {−29, 7} **15.** 34%, 30%

Pages 283–288 Lesson 5-1

1. $\{x \mid x > 10\}$

8 9 10 11 12 13 14 15 16 17 18 19

3. $\{g \mid g < -4\}$

−9 −8 −7 −6 −5 −4 −3 −2 −1 0 1

5. $\{n \mid n < 11\}$

7 8 9 10 11 12 13 14 15 16 17

7. $\{r \mid r > 6\}$

0 1 2 3 4 5 6 7 8

9. Sample answer: Let n = the number, $n + 4 \geq 10$; $\{n \mid n \geq 6\}$. **11.** no more than 92 ft

13
$$p - 6 \geq 3$$
$$p - 6 + 6 \geq 3 + 6$$
$$p \geq 9$$

Place a closed circle on 9 and an arrow to the right.

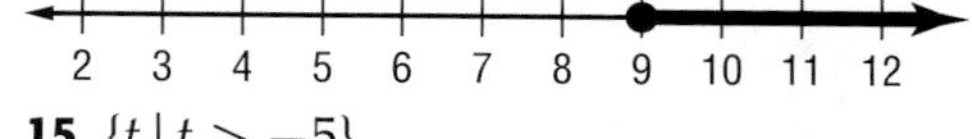

15. $\{t \mid t > -5\}$

−10 −9 −8 −7 −6 −5 −4 −3 −2 −1 0

17. $\{r \mid r < -5\}$

−7 −6 −5 −4 −3 −2 −1 0 1 2 3

19. $\{q \mid q \leq 7\}$

0 1 2 3 4 5 6 7 8

21. $\{h \mid h < 30\}$

26 27 28 29 30 31 32 33 34

23. $\{c \mid c < -27\}$

−31 −29 −27 −25 −23

25. $\{z \mid z \leq 4\}$

−5 −4 −3 −2 −1 0 1 2 3 4 5

27. $\{y \mid y \leq -6\}$

−11 −10 −9 −8 −7 −6 −5 −4 −3 −2 −1

29. $\{a \mid a > -9\}$

−13 −11 −9 −7 −5

31. Sample answer: Let n = the number, $n - 8 < 21$; $\{n \mid n < 29\}$. **33.** Sample answer: Let n = the number, $2n + 5 \leq n - 3$; $\{n \mid n \leq -8\}$. **35.** Sample answer: Let n = the number of online teens that do not use the Internet at school in millions; $n > 21 - 16$; $\{n \mid n > 5\}$, at least 5 million teens use the Internet but not at school. **37.** Sample answer: Let t = the original water temperature; $t + 4 < 81$; $\{t \mid t < 77\}$; the water temperature was originally less than 77°.

39 Let x represent the amount left on the card.
$$x + 32 + 26 < 75$$
$$x + 58 < 75$$
$$x + 58 - 58 < 75 - 58$$
$$x < 17$$
She will have no more than $17 left on the card.

41. $\{c \mid c \geq 3.7\}$

3.5 3.6 3.7 3.8 3.9 4.0 4.1 4.2 4.3 4.4 4.5

43. $\left\{k \mid k > -\frac{5}{12}\right\}$

−1 −0.8 −0.6 −0.4 −0.2 0

45a.

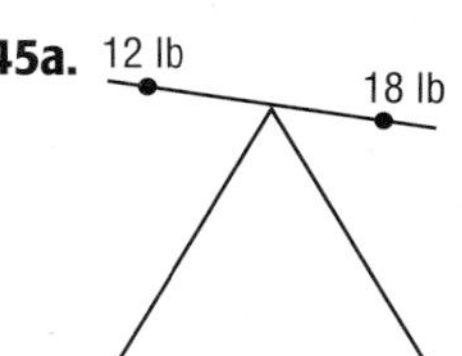

45b. 12 lb < 18 lb

Selected Answers and Solutions

45c.

	12	<	18
2	24	<	36
3	36	<	54
4	48	<	72
$\frac{1}{2}$	6	<	9
$\frac{1}{3}$	4	<	6
$\frac{1}{4}$	3	<	$4\frac{1}{2}$

45d. If a true inequality is multiplied by a positive number, the resulting inequality is also true. If a true inequality is divided by a positive number, the resulting inequality is also true. **47.** 10 **49.** 3 **51.** 26 **53.** $c < a < d < b$ **55.** Solving linear inequalities is similar to solving linear equations. You must isolate the variable on one side of the inequality. To graph, if the problem is a less than or a greater than inequality, an open circle is used. Otherwise a dot is used. If the variable is on the left hand side of the inequality, and the inequality sign is less than (or less than or equal to), the graph extends to the left; otherwise it extends to the right. **57.** C **59.** B

61.

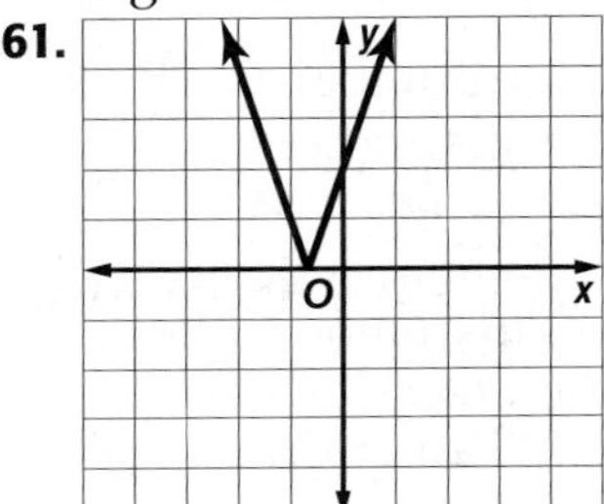

63.

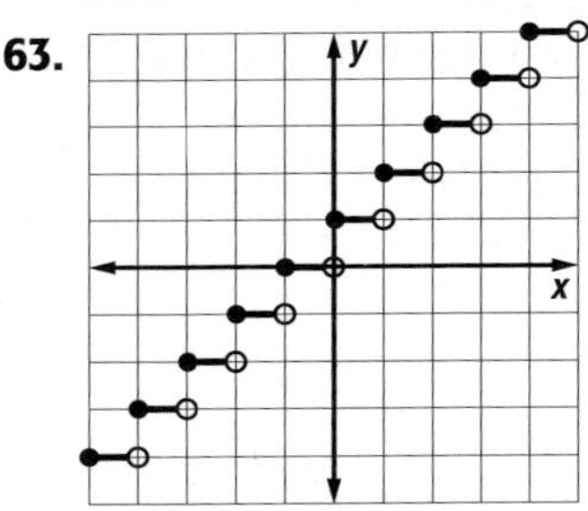

65. $y = -x - 2$ **67.** $y = -2x - 1$ **69.** blue **71.** 25 **73.** $y = 7x$; \$210 **75.** -30 **77.** $\frac{1}{10}$ **79.** 16 **81.** $-\frac{1}{9}$

Pages 290–295 Lesson 5-2

1. Let d = the number of DVDs sold; $15d > 5500$; $d > 366.67$; the band sold at least 367 DVDs. **3.** $r \geq 8$ **5.** $h < -10$ **7.** $v > -12$ **9.** $z \geq -8$ **11.** Let p = the number of pay periods for which Rodrigo will need to save; $25p \geq 560$; $p \geq 22.4$; Rodrigo will need to save for 23 weeks.

13
$$\frac{1}{2}a < 20$$
$$2\left(\frac{1}{2}a\right) < 2(20)$$
$$a < 40$$
Check by substituting values less than 40.

15. $d \geq 68$ **17.** $f < 432$ **19.** $j \leq -16$ **21.** $p \leq 16$ **23.** $y > -16$ **25.** $v < 12$ **27.** $b \leq -\frac{3}{4}$ **29.** $f < -\frac{5}{7}$ **31.** no more than 4 **33.** no more than 32 people **35.** b **37.** d

39
$$\frac{2}{3}x < 42$$
$$3\left(\frac{2}{3}x\right) < 3(42)$$
$$2x < 126$$
$$x < 63$$
fewer than 63 employees

41a.

h cm
b cm
b cm

41b. $h = \frac{216}{b^2}$

41c.

b	1	3	6	9	12
h	216	24	6	$\frac{8}{3}$	$\frac{3}{2}$

41d. $b < h$ when $0 < b < 6$; $b > h$ when $h < 6$.
43. $\frac{-96c}{-96} > \frac{12c}{-96}$; $c > -\frac{d}{8}$; $-96c \cdot -\frac{1}{96} > 12d \cdot -\frac{1}{96}$ $c > -\frac{d}{8}$
45. Sometimes; the statement is true when $a > 0$ and $b < 0$. **47.** Sample answer: The inequality symbol changes directions when multiplying or dividing by a negative number so that the inequality remains true. For example, dividing $-2x > 4$ by -2 results in $x < -2$. **49.** 10 in. **51.** C

53. $\left\{y \mid y \geq -\frac{11}{26}\right\}$

−5 −4 −3 −2 −1 0 1 2 3 4 5

55. D = {all real numbers}; R = $\{y \mid y \geq 0\}$
57. D = {all real numbers}; R = $\{y \mid y \leq 1\}$
59. 2 hours **61.** {1, 7} **63.** 2 **65.** $\frac{33}{8}$ **67.** 3

Pages 296–301 Lesson 5-3

1. $4n + 60 \leq 800$; $n \leq 185$; at most 185 lb per person

3
$$6h - 10 \geq 32$$
$$6h - 10 + 10 \geq 32 + 10$$
$$6h \geq 42$$
$$h \geq 7$$
$\{h \mid h \geq 7\}$

5. $\{x \mid x < -12\}$ **7.** Sample answer: Let n = the

number; $4n - 6 > 8 + 2n$; $\{n \mid n > 7\}$. **9.** $\{v \mid v \geq 0\}$
11. $\varnothing$

13
$$\begin{aligned} 21 &> 15 + 2a \\ 21 - 15 &> 2a \\ 6 &> 2a \\ 3 &> a \qquad \{a \mid a < 3\} \end{aligned}$$

15. $\{w \mid w > 56\}$ **17.** $\{w \mid w < -3\}$ **19.** $\left\{p \mid p > -\frac{24}{5}\right\}$
21. $\{h \mid h < -15\}$ **23.** Sample answer: Let n = the number; $\frac{2}{3}n + 6 \geq 22$; $\{n \mid n \geq 24\}$. **25.** Sample answer: Let n = the number; $8n - 27 \leq -n + 18$; $\{n \mid n \leq 5\}$. **27.** Sample answer: Let n = the number; $3(n + 7) > 5n - 13$; $\{n \mid n < 17\}$
29. $\left\{n \mid n > -\frac{1}{3}\right\}$ **31.** $\varnothing$ **33.** $\{t \mid t \geq -1\}$ **35.** Sample answer: Let s = the amount of sales made, $35{,}000 + 0.08s > 65{,}000$; $\{s \mid s > 375{,}000\}$; the sales must be more than \$375,000.

37.

$6(m - 3) > 5(2m + 4)$	Original inequality
$6m - 18 > 10m + 20$	Distributive Property
$6m - 18 - 6m > 10m + 20 - 6m$	Subtract 6*m* from each side.
$-18 > 4m + 20$	Simplify.
$-18 - 20 > 4m + 20 - 20$	Subtract 20 from each side.
$-38 > 4m$	Simplify.
$\frac{-38}{4} > \frac{4m}{4}$	Divide each side by 4.
$-9.5 > m$	Simplify.
$\{m \mid m < -9.5\}$	

39a. $5t + 565 \geq 1500$; $t \geq 187$
39b. [number line: 182 183 184 185 186 187 188 189 190 191 192]

41 a. Words: temperature can be greater than 104
$t > 104$
b. $F > 104$
$$\begin{aligned} \tfrac{9}{5}C + 32 &> 104 \\ \tfrac{9}{5}C &> 72 \\ 5\left(\tfrac{9}{5}C > 72\right) \\ 9C &> 360 \\ C &> 40 \end{aligned}$$

43. 1, 3, 5, 7; 3, 5, 7, 9; 5, 7, 9, 11; 7, 9, 11, 13
45. $\left\{x \mid x \geq \frac{1}{2}\right\}$ **47.** $\{m \mid m \geq 18\}$ **49.** $\{x \mid x \leq 8\}$
51. $\{x \mid x > -6\}$ **53.** $\{x \mid x \geq 1.5\}$ **55.** Add $3p$ and 2 to each side. The inequality becomes $9 \geq 3p$. Then divide each side by 3 to get $3 \geq p$. **57.** Sample answer: $2x + 4 > 2$ and $3x + 1 > -2$ both have the graph of $x > -1$. **59.** Sample answer: The solution set for an inequality that results in a false statement is the empty set, as in $12 < -15$. The solution set for an inequality in which any value of x results in a true statement is all real numbers, as in $12 \leq 12$. **61.** G **63.** D **65.** $\{b \mid b > -4\}$

67. $\{h \mid h < 14\}$

69. $\{m \mid m \geq 1\}$
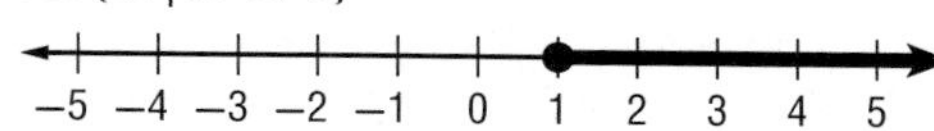
71. 4
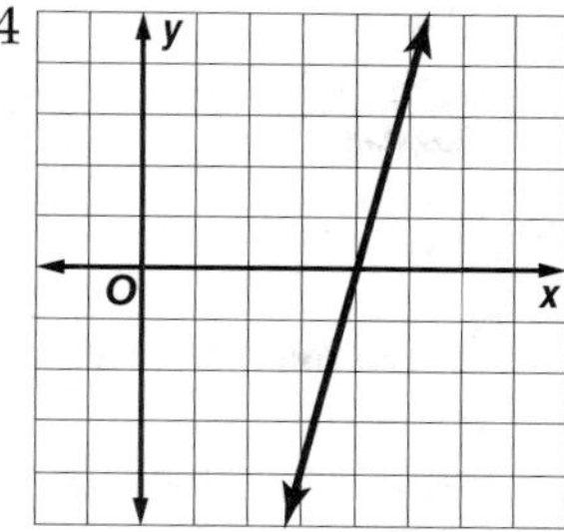

73. 118.0 million
75. 8 **77.** $12(29.95 + 4)$ or $12(29.95) + 12(4)$; \$407.40
79.
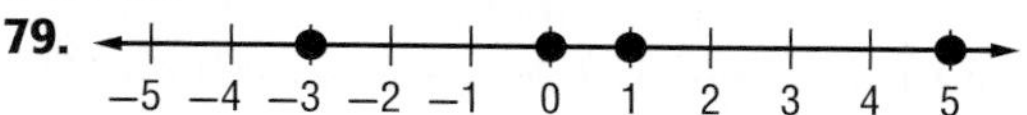
81.
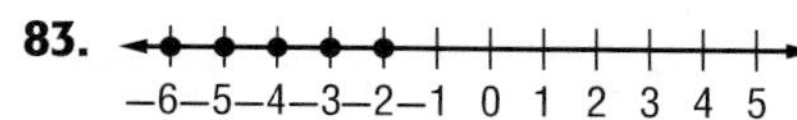
83. [number line: −6 −5 −4 −3 −2 −1 0 1 2 3 4 5]

Pages 304–309 *Lesson 5-4*

1. $\{p \mid 12 \leq p \leq 16\}$
3. $\{a \mid a > 5\}$
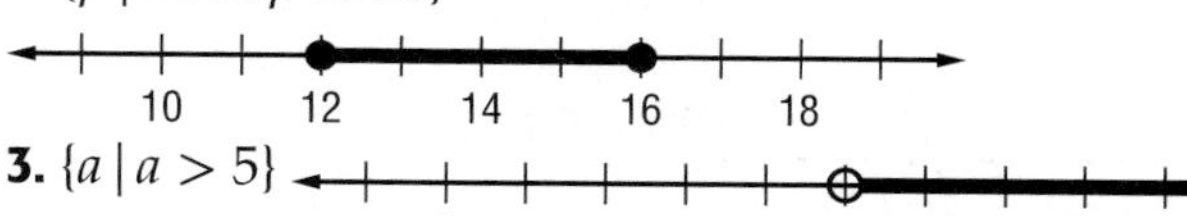
5. 11 psi $\leq x \leq$ 56 psi

7
$$\begin{aligned} n + 2 &\leq -5 \\ n + 2 - 2 &\leq -5 - 2 \\ n &\leq -7 \end{aligned} \quad \text{and} \quad \begin{aligned} n + 6 &\geq -6 \\ n + 6 - 6 &\geq -6 - 6 \\ n &\geq -12 \end{aligned}$$
The solution set is $\{n \mid -12 \leq n \leq -7\}$.
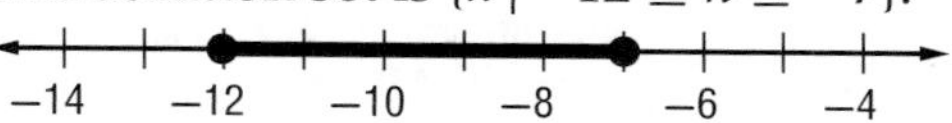

9. $\{t \mid t \geq 1 \text{ or } t < -1\}$
11. $\{c \mid -1 \leq c < 2\}$
13. $\{m \mid m$ is a real number.$\}$ 
15. $\{y \mid y < -3\}$
17. Sample answer: Let x = the smaller of two consecutive odd numbers, then $8 \leq 2x + 2 \leq 24$; $3 \leq x \leq 11$; 3, 5; 5, 7; 7, 9; 9, 11; 11, 13
19 The graph shows $x > -3$ and $x \leq 2$, so the inequality is $-3 < x \leq 2$.
21. $x < -4$ or $x > -3$ **23.** $x \leq -3$ or $x > 0$
25. $\left\{a \mid -3 < a \leq \frac{1}{2}\right\}$

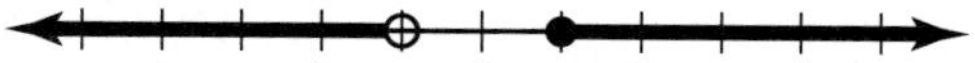

27. $\{n \mid n < -3 \text{ or } n > -3\}$

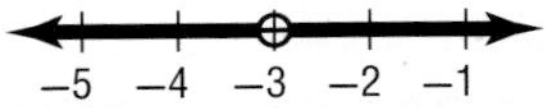

29. Sample answer: Let n = the number; $5 \le n - 8 \le 14$; $\{n \mid 13 \le n \le 22\}$. **31.** $-5n > 35$ or $-5n < 10$; $\{n \mid n < -7 \text{ or } n > -2\}$ **33.** $t < 75$ or $t > 90$
35. Sample answer: Let t = the temperature; $23 \le t \le 33$.

37 a. Category 3 has wind speeds between 111 and 130: $111 \le x \le 130$
Category 4 has wind speeds between 131 and 155: $131 \le x \le 155$

b. The union is all values between 111 and 155: $111 \le x \le 155$.
The intersection is all values they have in common, which is none. The intersection is the empty set, $\varnothing$.

39. Neither; Chloe did not add 5 to 3, and Jonas did not add 5 to 7. **41.** Sample answer: $x \le 2$ or $x \ge 4$ **43.** Sample answer: The speed at which a roller coaster runs while staying on the track could represent a compound inequality that is an intersection. **45.** H **47.** B **49.** at least 22 subscriptions **51.** 5.6 **53.** 3.94 **55.** 3.28 **57.** Ian will buy a DVD box set. **59.** The DVD box set cost \$70 or more.

61. $5 + (4 - 2^2)$
$= 5 + (4 - 4)$ Substitution
$= 5 + 0$ Substitution
$= 5$ Additive identity

63. $2(4 \cdot 9 - 3) + 5 \cdot \frac{1}{5}$
$= 2(36 - 3) + 5 \cdot \frac{1}{5}$ Substitution
$= 2(33) + 5 \cdot \frac{1}{5}$ Substitution
$= 66 + 5 \cdot \frac{1}{5}$ Substitution
$= 66 + 1$ Multiplicative Inverse
$= 67$ Substitution

65. 3 **67.** $12\frac{2}{3}$ **69.** 30 **71.** 10

Pages 310–314 Lesson 5-5

1. $\{a \mid 2 < a < 8\}$
3. $\varnothing$ 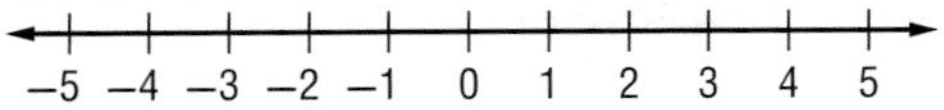
5. $\{n \mid n \le -8 \text{ or } n \ge -2\}$ 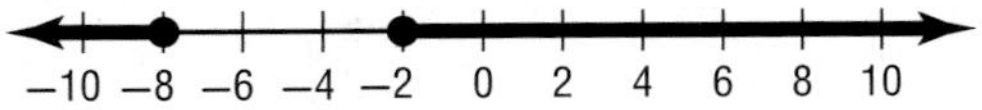
7. $\{m \mid 70.10 \le m \le 71.60\}$
9 $|r + 1| \le 2$

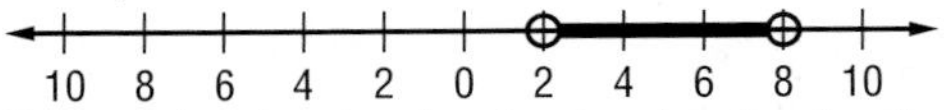

Case 1:		Case 2:
$r + 1$ is positive.	and	$r + 1$ is negative.
$r + 1 \le 2$	and	$r + 1 \ge -2$
$r \le 1$	and	$r \ge -3$

The solution set is $\{r \mid -3 \le r \le 1\}$.

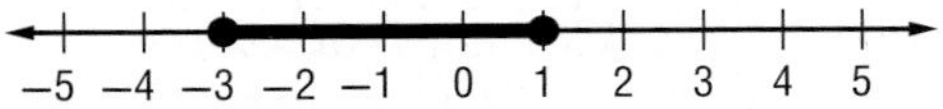

11. $\{h \mid -3 < h < 5\}$
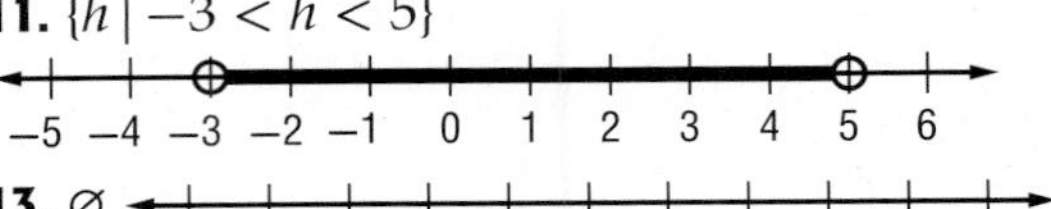
13. $\varnothing$
15. $\{k \mid k < 1 \text{ or } k > 7\}$ 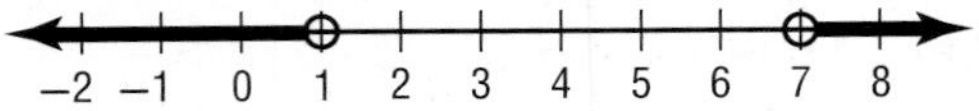
17. $\{p \mid p \le -3 \text{ or } p \ge 2\}$ 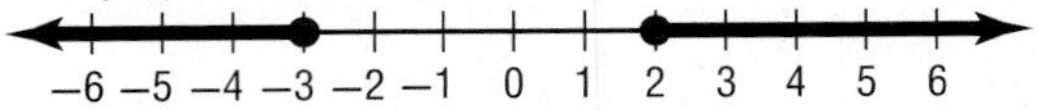
19. $\{c \mid c \text{ is a real number.}\}$ 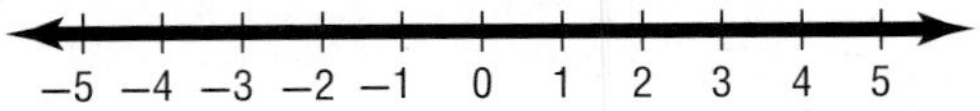
21. $\{n \mid n \le -5\frac{1}{4} \text{ or } n \ge 3\frac{3}{4}\}$ 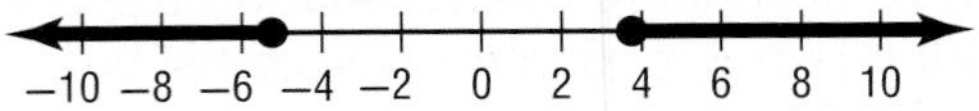
23. $\{h \mid -5\frac{2}{3} < h < 5\}$
25. $\varnothing$ 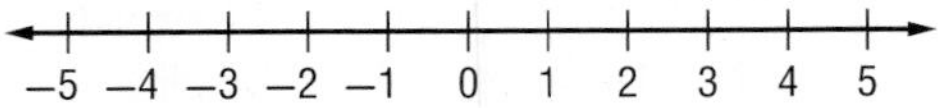
27. $\{r \mid -2 < r < \frac{2}{3}\}$
29. $\{h \mid -1.5 < h < 4.5\}$
31a. $\{t \mid t < 32 \text{ or } t > 212\}$
31b. 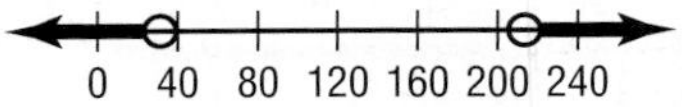
31c. $|t - 122| > 90$ **33.** $|x + 1| \le 4$
35. $|x - 5.5| > 4.5$
37 $|g - 52| \le 5$

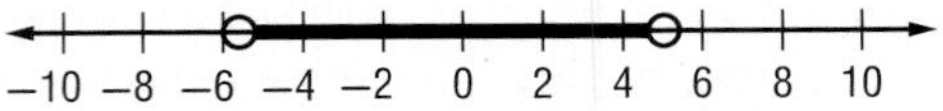

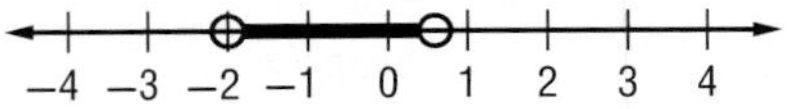

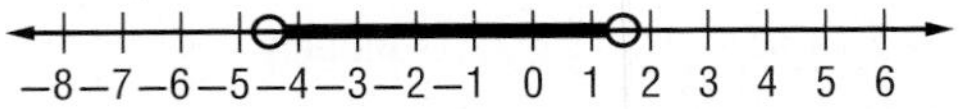

Case 1:		Case 2:
$g - 52$ is positive.	and	$g - 52$ is negative.
$g - 52 \le 5$	and	$g - 52 \ge -5$
$g \le 57$	and	$g \ge 47$

The solution is $\{g \mid 47 \le g \le 57\}$.

39. $|t - 38| \le 1.5$ **41.** $|c - 55| \le 3$ **43.** Sample answer: Lucita forgot to change the direction of the inequality sign for the negative case of the absolute value. **45.** Sample answer: If $t = 0$, then the absolute value is equal to 0, not greater than 0. **47.** Sample answer: When an absolute value is on the left and the inequality symbol is $<$ or $\le$, the compound sentence uses *and*, and if the inequality symbol is $>$ or $\ge$, the compound sentence used *or*. To solve, if $|x| < n$, then set up and solve the inequalities $x < n$ and $x > -n$, and if $|x| > n$, then set up and solve the inequalities $x > n$ or $x < -n$.
49. J **51.** B **53.** $\{t \mid 5 \le t \le 6\}$ 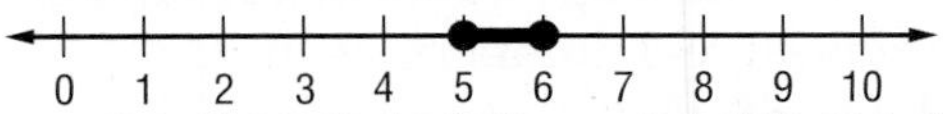
55. 681,654,568 \$2 bills and 1,175,919,377 \$50 bills
57. 18 **59.** -20

61. 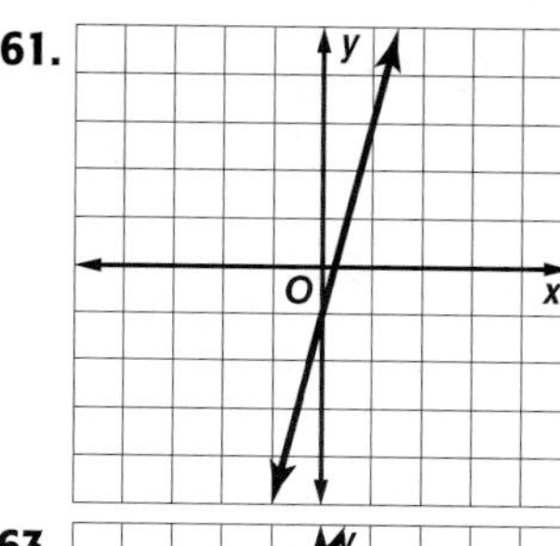

63. 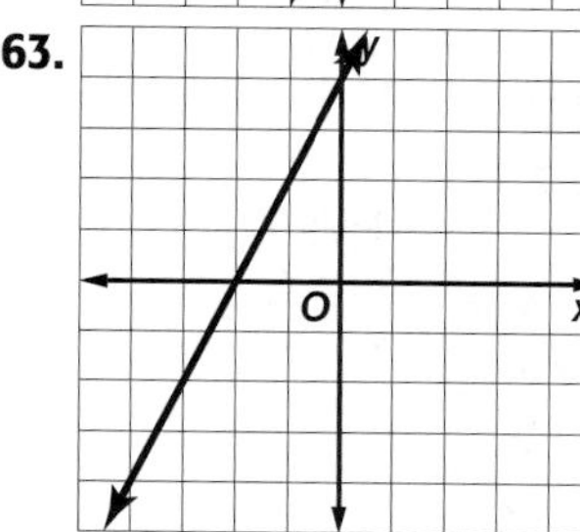

65. 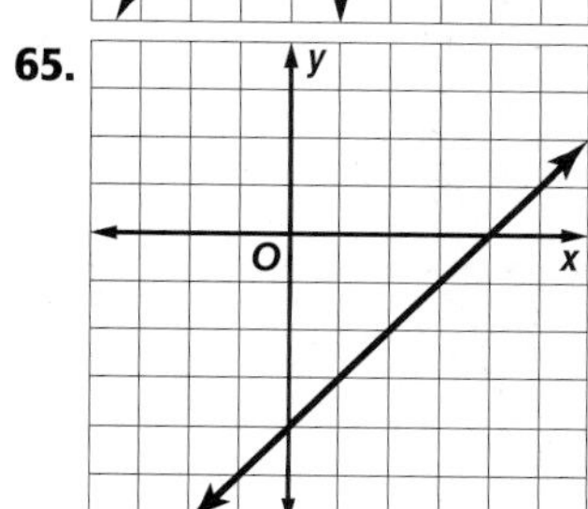

67. 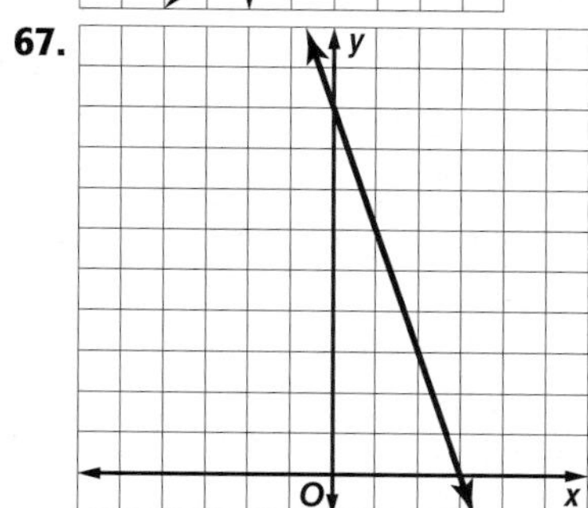

Pages 315–320 Lesson 5-6

1. 3. 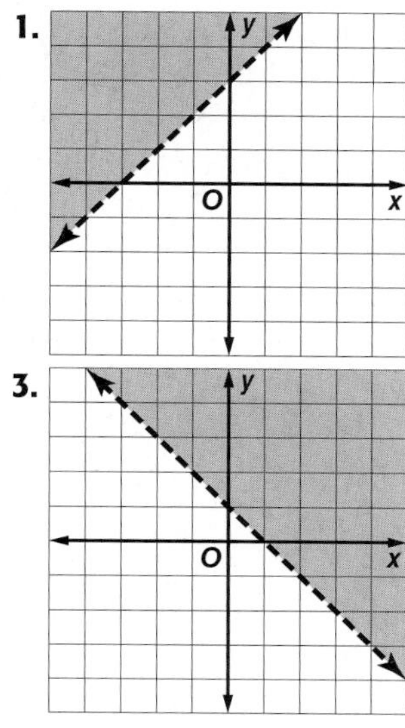

5. 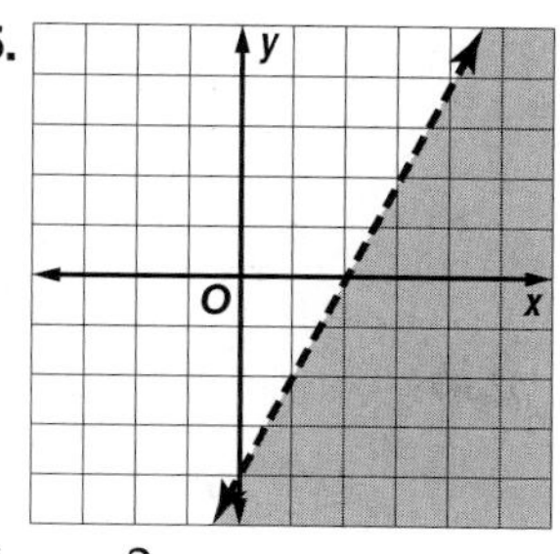

7. $x < 2$

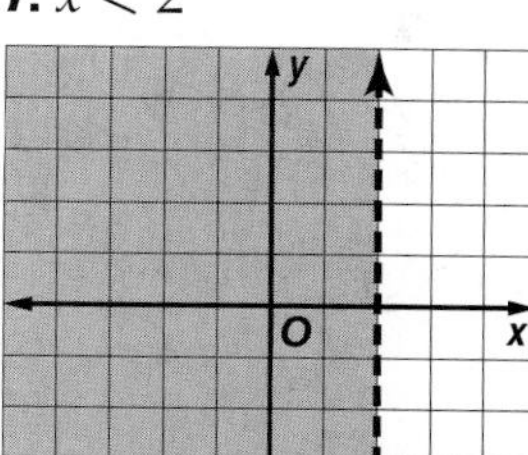

9. $y \leq 13$

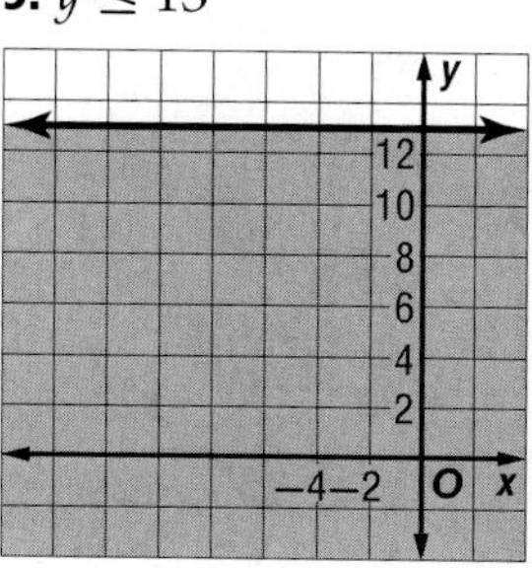

11a. $115x + 685y \geq 2300$ **11b.** Sample answer: 1 skim board and 4 surfboards

13. 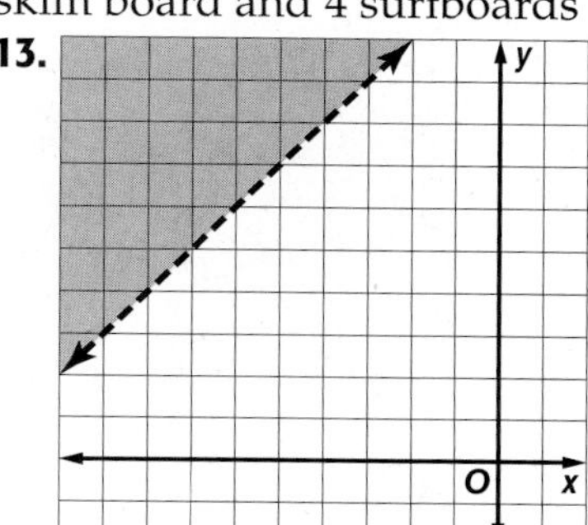

15. 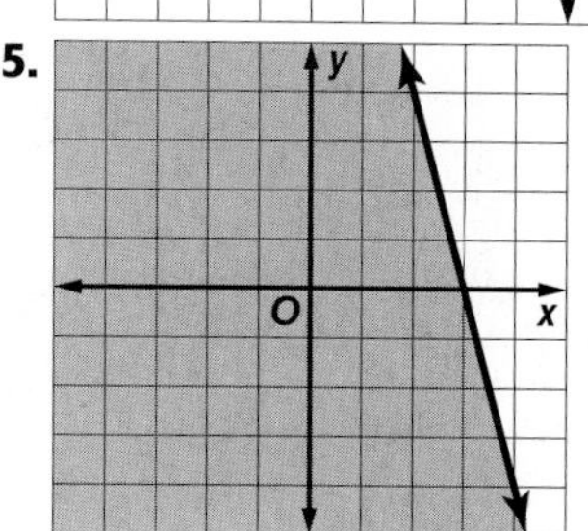

17. 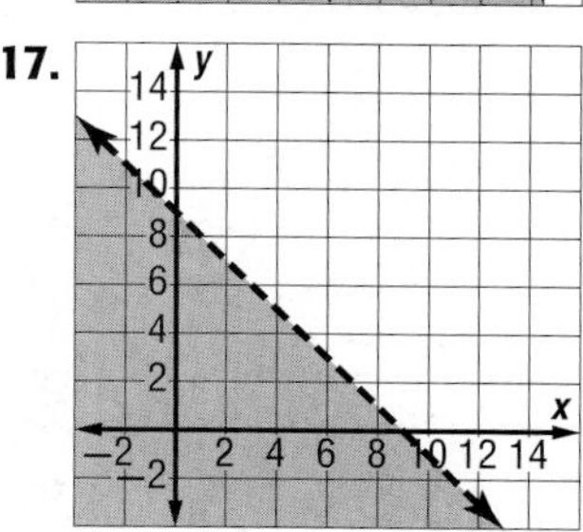

19. 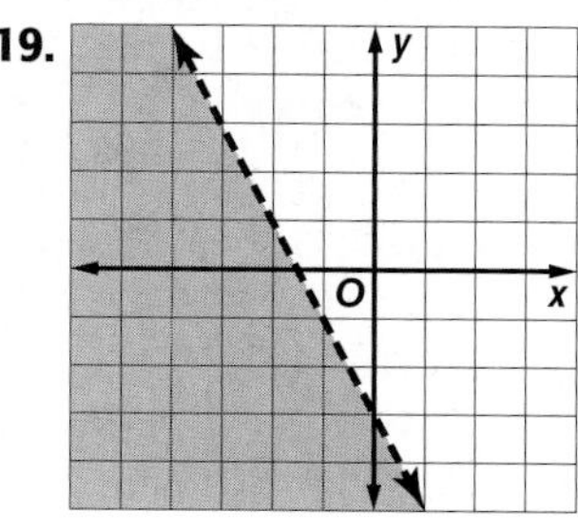

Selected Answers and Solutions

21.

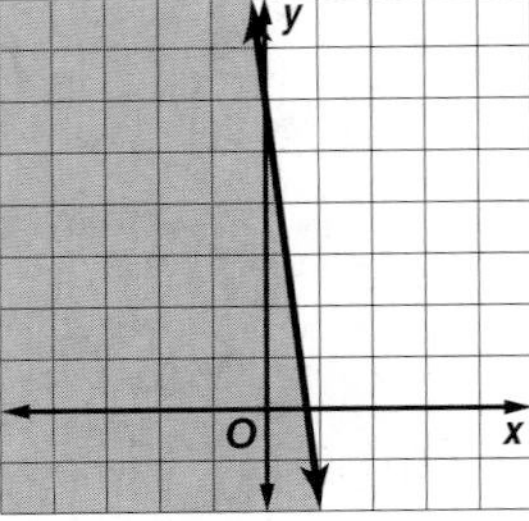

23.

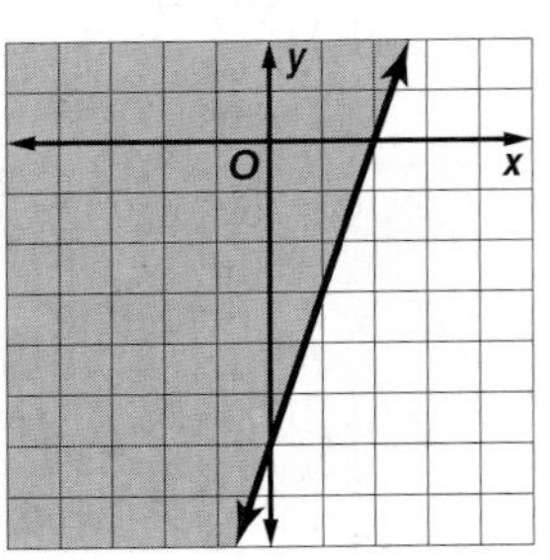

25. $x > 2$

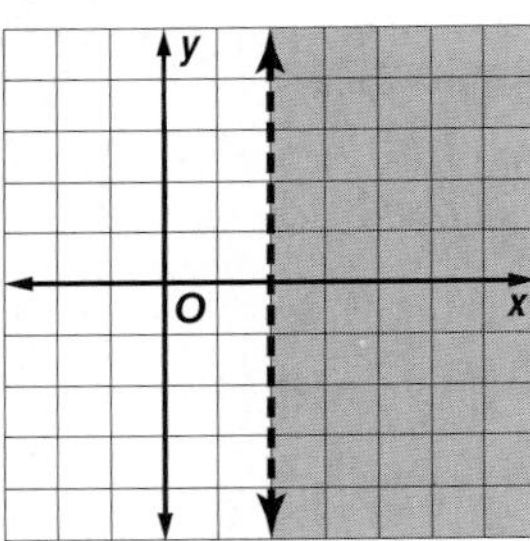

27. $y \leq 5$

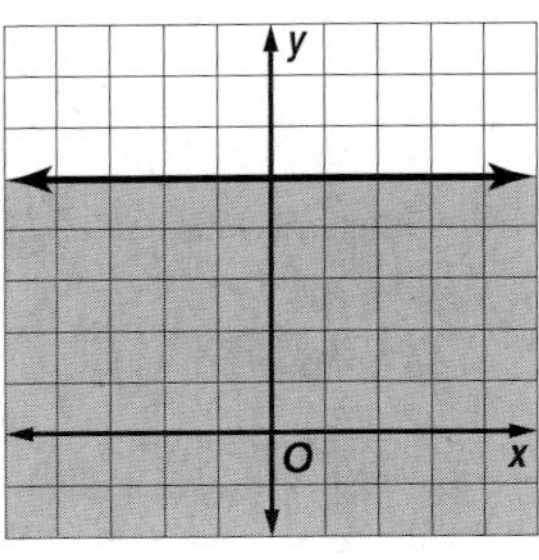

29. $x > -\frac{19}{14}$

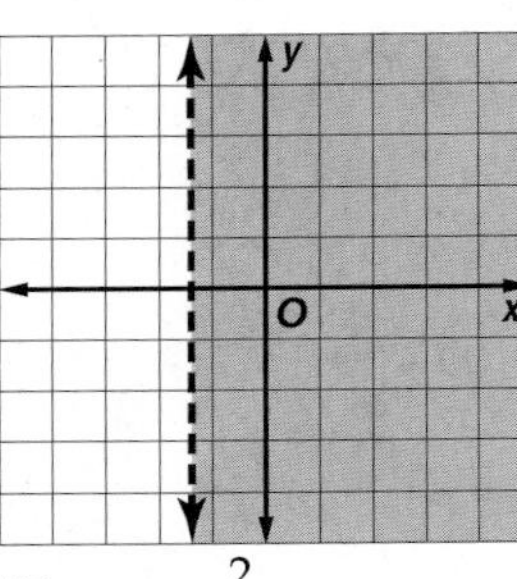

31. $x < -\frac{2}{3}$

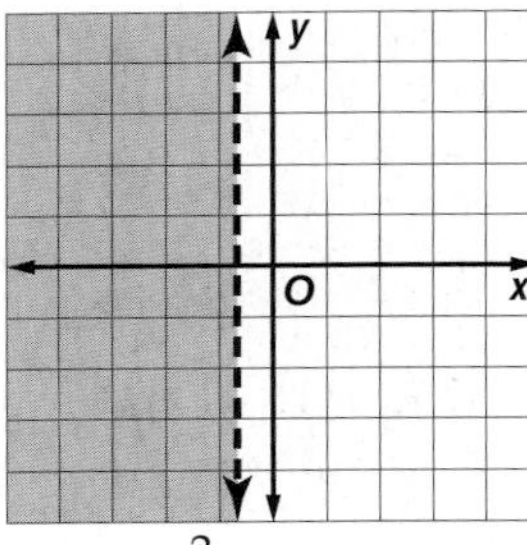

33. $x \leq -\frac{2}{3}$

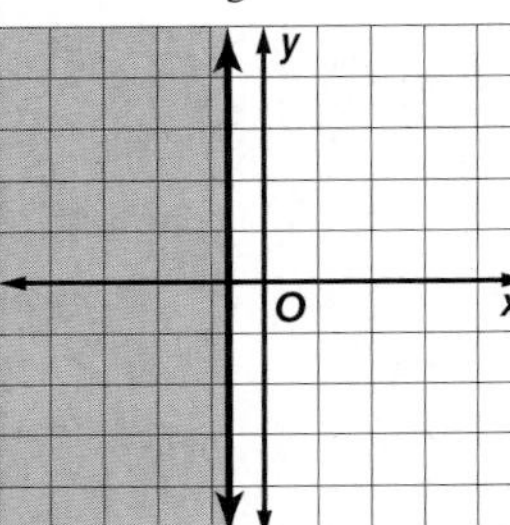

35. $x > \frac{3}{7}$

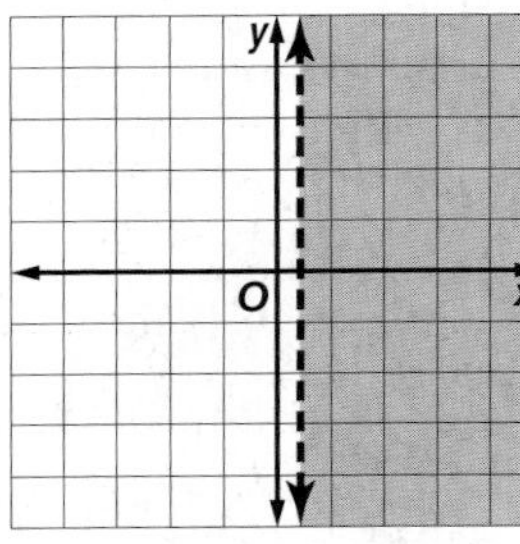

37 **a.** They need to make more than \$2000. Let x represent the number of hot dogs sold and let y represent the number of sodas sold.

$1x + 1.25y \geq 2000$

b. Step 1: First, solve for y in terms is x.

$x + 1.25y \geq 2000$

$1.25y \geq -x + 2000$

$y \geq 0.80x + 2000$

Then, graph $y = 0.80x + 2000$. Because the inequality involves $\geq$, graph the boundary with a solid line.

Step 2: Select a test point in either half-plane. A simple choice is (0, 0).

$x + 1.25y \geq 2000$

$0 + 1.25(0) \geq 2000$

$0 \geq 2000$

Step 3: Since this statement is not true, the half-place containing the origin is not the solution. Shade the other half-plane.

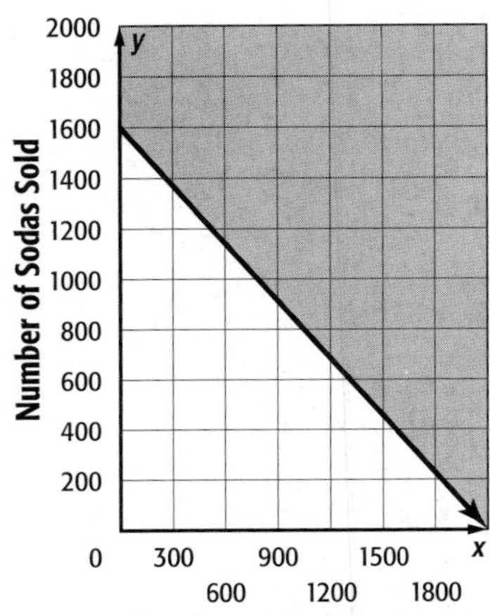

c. Sample answer: (400, 1600), (200, 1500), (300, 1400), (400, 1300), (1000, 1000)

d. Sample points should be in the shaded region of graph in part *b*.

39 $x < -4$

Step 1: Graph $x = 4$ with a dotted line since it involves $<$.

Step 2: Choose a test point in either half plane. A simple choice is (0, 0).

$0 < -4$

Step 3: Since this statement is not true, the half-place containing the origin is not the solution. Shade the other half-plane.

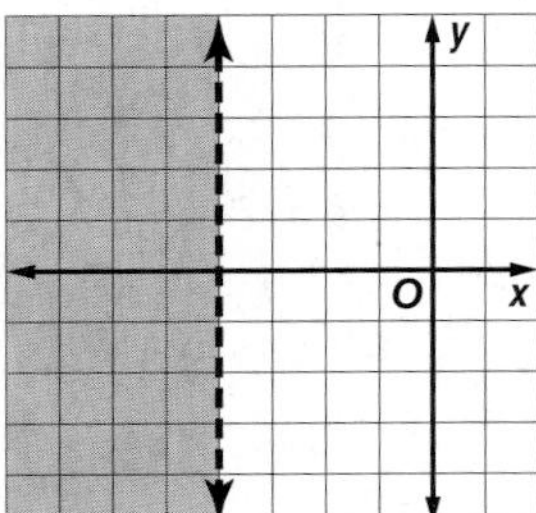

(2, 1): $2 < -4$ NO

(−3, 0): $-3 < -4$ NO

(0, −3): $0 < -4$ NO

(−5, −5): $-5 < -4$ YES

(−4, 2): $4 < -4$ NO

41. 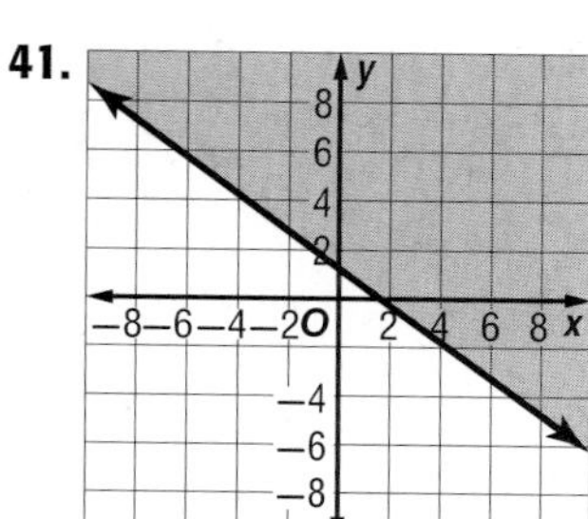

(1, 1), (2, 5), (6, 0)

43. 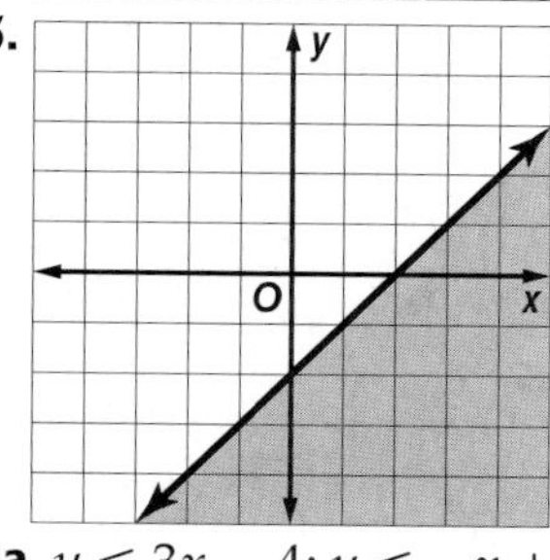

(7, 5), (5, 3), (2, −5)

45a. $y \leq 3x - 4$; $y \leq -x + 4$

45b. 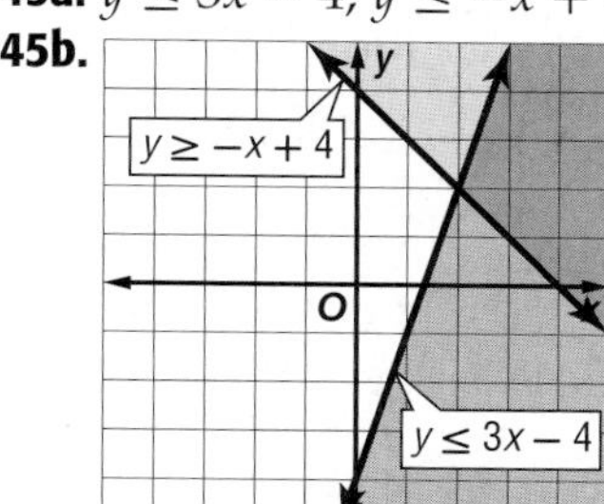

45c. The overlapping region represents the solutions that make both A and B true.

45d. 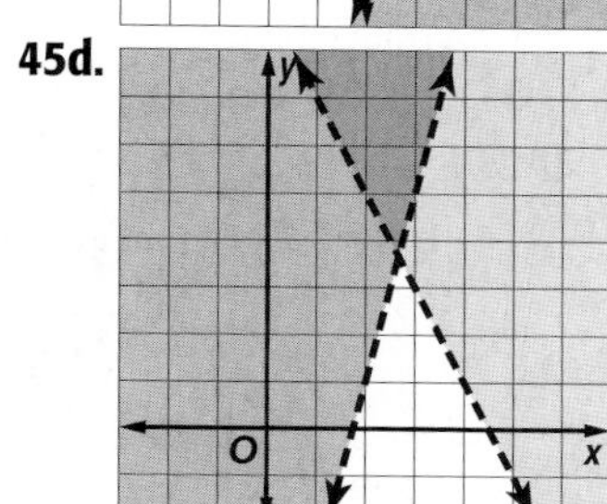

47.

49. Sample answer: The equation $y > 7x$ represents an employee earning $7 per hour plus the possibility of earning tips. Both the domain and range are nonnegative real numbers because hours worked and pay cannot be negative. **51.** B **53.** F **55.** $\{y \mid y > 6 \text{ or } y < -2\}$ **57.** ∅ **59.** $\{p \mid 4 < p < 10\}$ **61.** $y = 8x - 11$ **63.** $y = -\frac{3}{2}x - 17$ **65.** $r = \frac{w - sm}{10}$

Pages 322–324 Study Guide and Review

1. false; more **3.** false; intersection **5.** true **7.** true **9.** true

11. $w > 13$

13. $h < -5$

15. $p \leq -2$

17. no more than 9 **19.** $g \geq -20$ **21.** $w \leq 11$ **23.** $t < -72$ **25.** $h < 7$ **27.** $x \leq -5$ **29.** Sample answer: Let x = the number; $4x - 6 < -2$; $x < 1$.

31. $2 < m < 9$

33. $x \leq 3$ or $x > 6$

35. $-5 < x < 13$

37. $-7 \leq c \leq 4$

39. $-\frac{7}{3} \leq d \leq 3$

41. $t < -13$ or $t > 7$

43. $-20 \leq m \leq -18$

45. 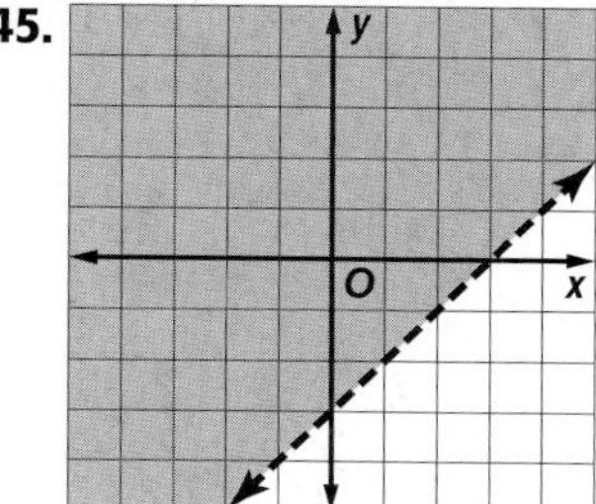

47. 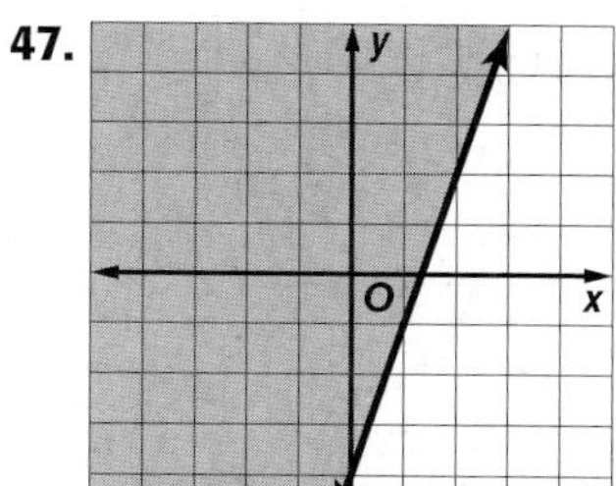

49. 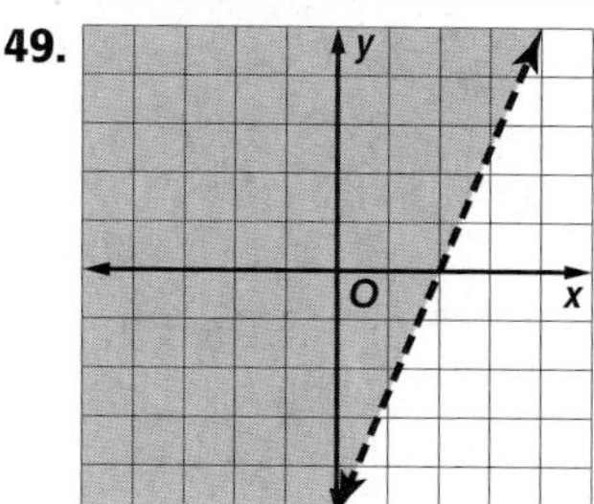

51. (1, 2), (3, −2)

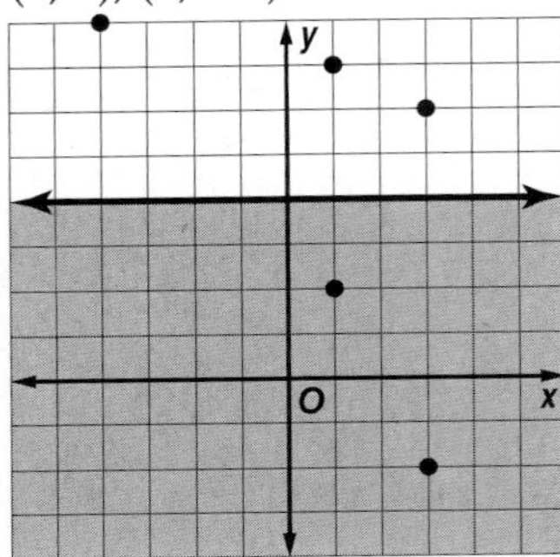

53. $2x + 3y \leq 24$

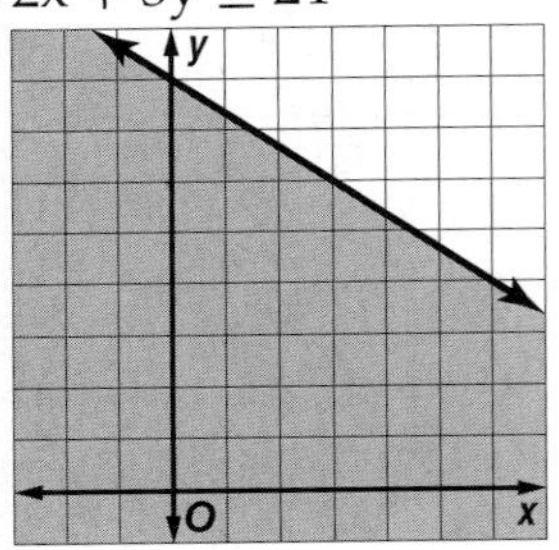

Selected Answers and Solutions

Chapter 6 Systems of Linear Equations and Inequalities

Page 331 Get Ready

1. (4, 0) **3.** (−2, −3) **5.** (−1, −1) **7.** $x = 6 - 2y$ **9.** $m = 2n + 6$ **11.** $l = \dfrac{P - 2w}{2}$ **13.** $b = \dfrac{2A}{h}$

Pages 333–339 Lesson 6-1

1. consistent and independent **3.** inconsistent **5.** consistent and independent

7. 1 solution, (−4, 0)

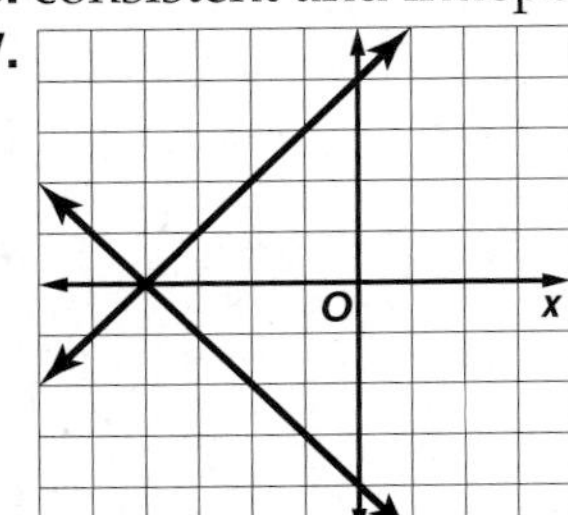

9a. Alberto: $y = 20x + 35$; Ashanti: $y = 10x + 85$

9b.

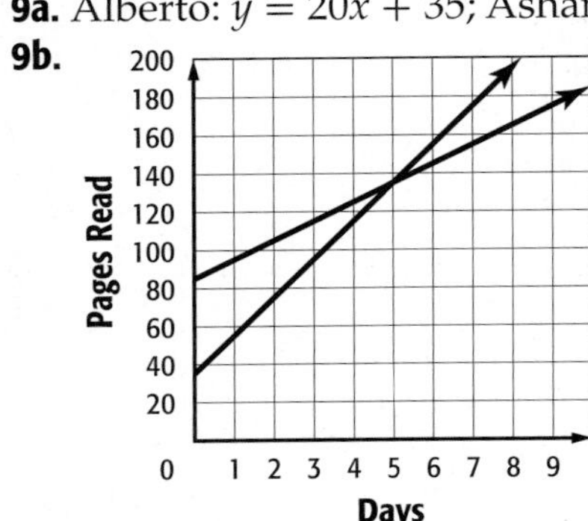

9c. (5, 135); Alberto will have read more after 5 days.

11. consistent and independent

13 Since these two graphs intersect at one point, there is exactly one solution. Therefore, the system is consistent and independent.

15. consistent and independent

17. 1 solution; $\left(-\frac{5}{6}, -1\frac{1}{3}\right)$

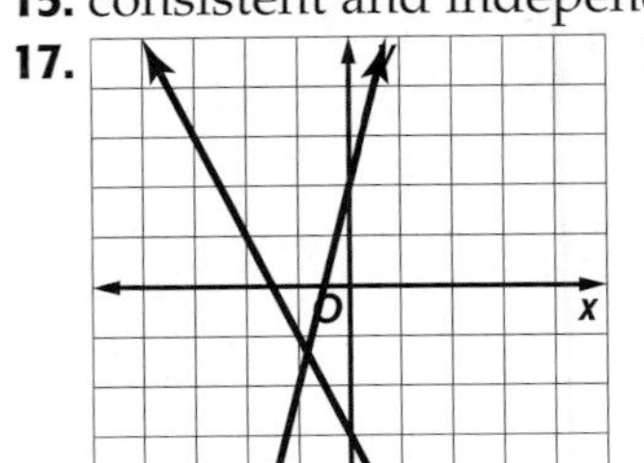

19. infinitely many

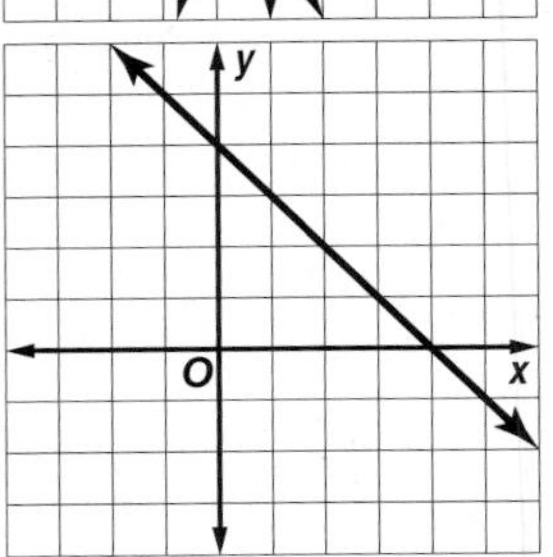

21. 1 solution; (5, −1)

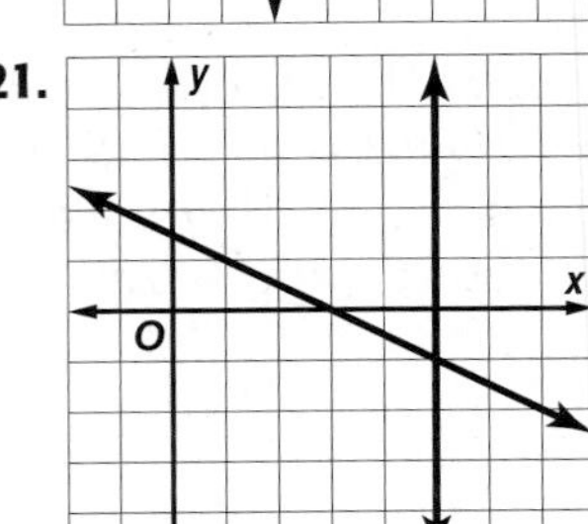

23. no solution

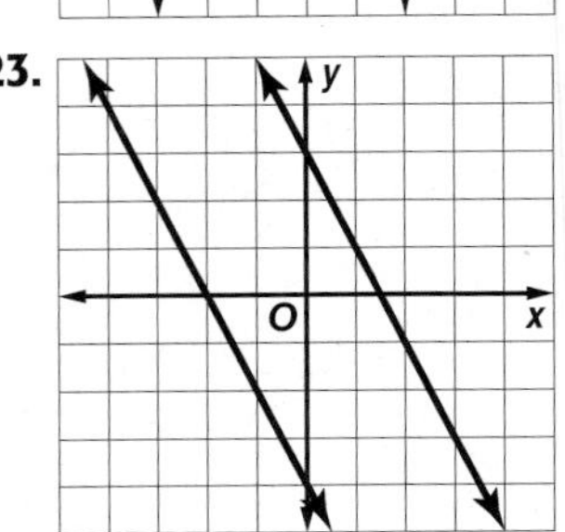

25a. Akira: $y = 30x + 22$; Jen: $y = 20x + 53$

25b.

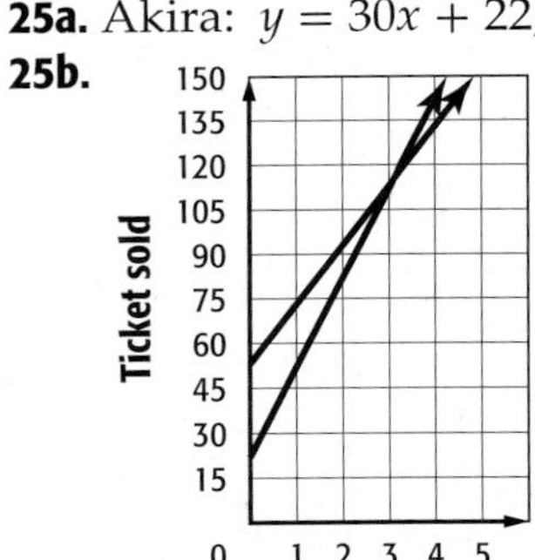

25c. (3.1, 115); After about 3 days Akira will have sold more tickets.

27 Graph the two equations on the same coordinate plane. The graphs appear to intersect at (−4, −2). Check by substituting into the equations.

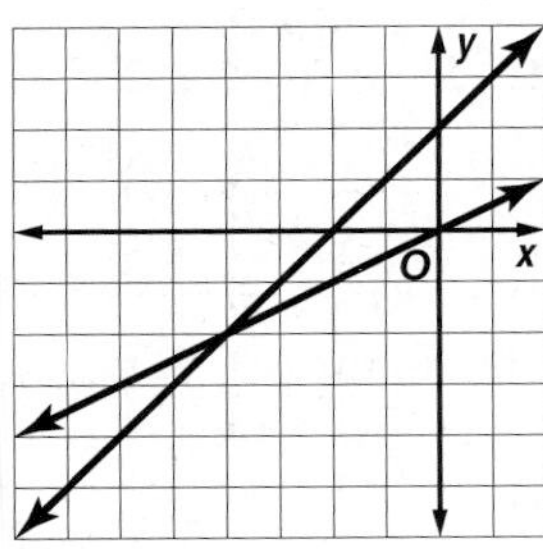

$y = \frac{1}{2}x$

$-2 = \frac{1}{2}(-4)$

$-2 = -2$ Yes

$y = x + 2$

$-2 = -4 + 2$

$-2 = -2$ Yes

So, the solution is $(-4, -2)$.

29. 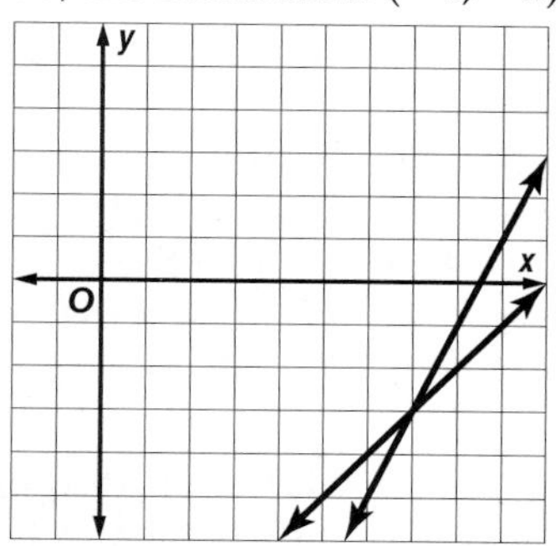

1 solution, $(7, -3)$

31. 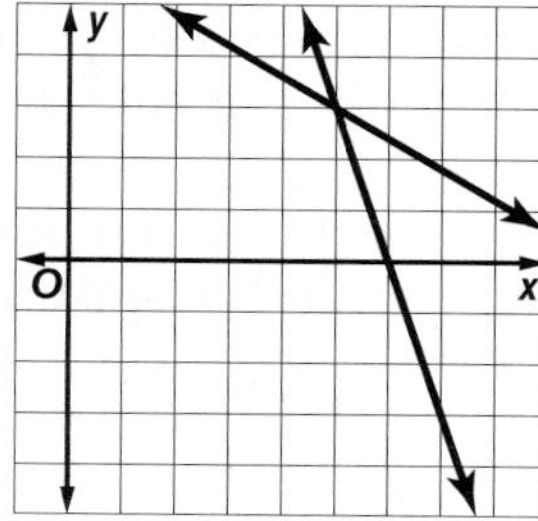

1 solution, $(5, 3)$

33. 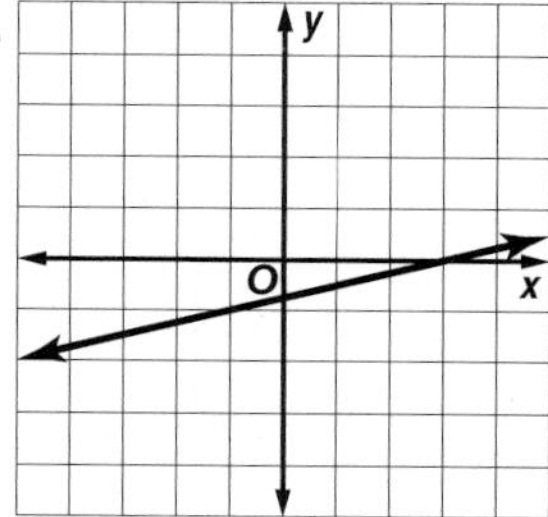

infinitely many

35. 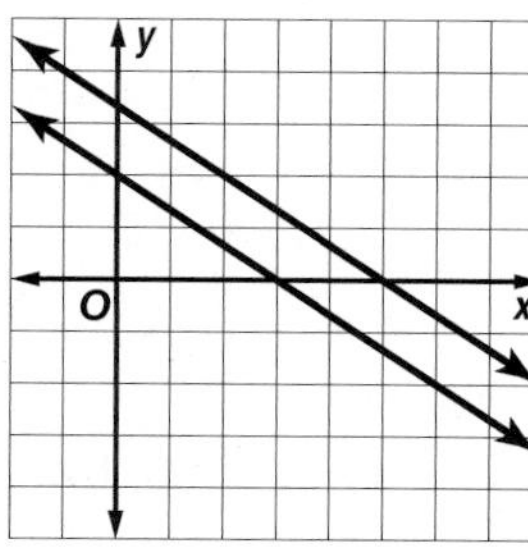

no solution

37. 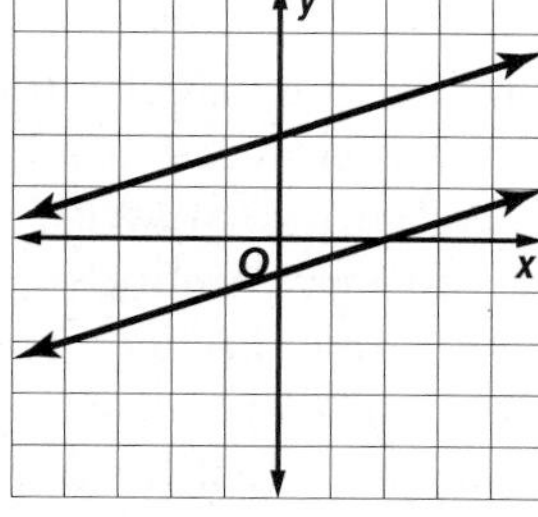

no solution

39. 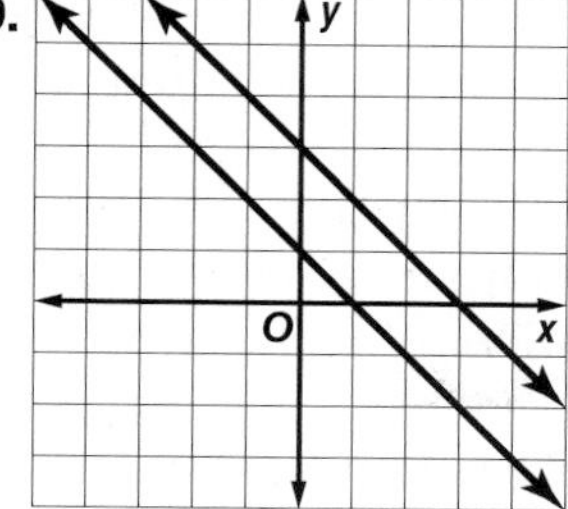

no solution

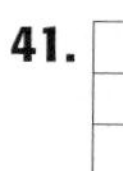

41. 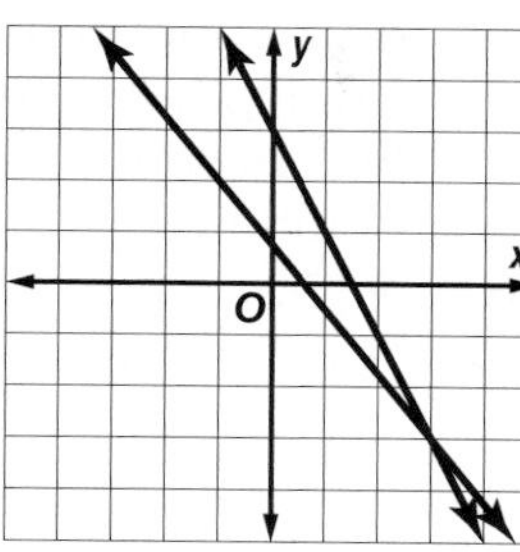

1 solution, $(3, -3)$

43. no solution

45 a. Lookatme:

Words: Started at 2.5 million and rose 13.1 million each year.

Expression: $2.5 + 13.1x$; $y = 13.1x + 2.5$

Buyourstuff:

Words: Started at 59 million and dropped by 2 million each year.

Expression: $59 - 2x$; $y = -2x + 59$

b.

Years Since 2005	Lookatme Vistors (mln)	Buyourstuff Vistors (mln)
0	2.5	59
1	15.6	57
2	28.7	55
3	41.8	53
4	54.9	51

c. Plot each equation on the same coordinate plane.

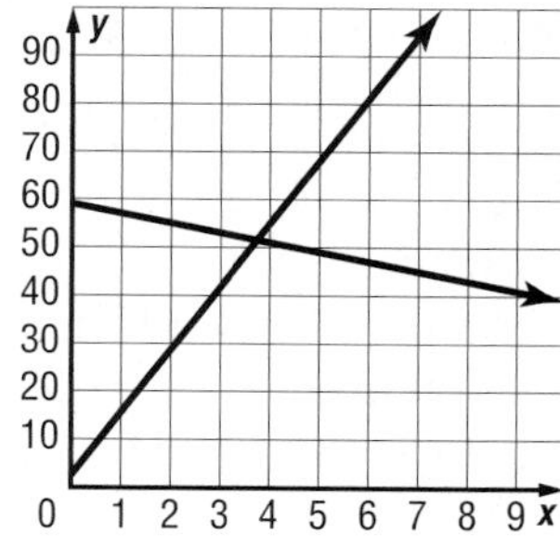

d. The graphs appear to intersect when $x = 3$, so they are about the same 3 years after 2005, or 2008.

e. The domain, or input values will be all values greater than 0 since negative values do not make sense for years. So, $D = \{x \mid x \geq 0\}$ The range, or output values will be all values greater than 0 since negative values do not make sense for the number of visitors. So, $R = \{y \mid y \geq 0\}$.

47. Francisca; if the item is less than \$100, then \$10 off is better. If the item is more than \$100, then the 10% is better. **49.** If the equations are linear and have more than one common solution, they must be consistent and dependent, which means that they have an infinite number of solutions in common. **51.** Sample answers: $y = 5x + 3$; $y = -5x - 3$; $2y = 10x - 6$ **53.** 14,745,600,000 bacteria **55.** H

57.

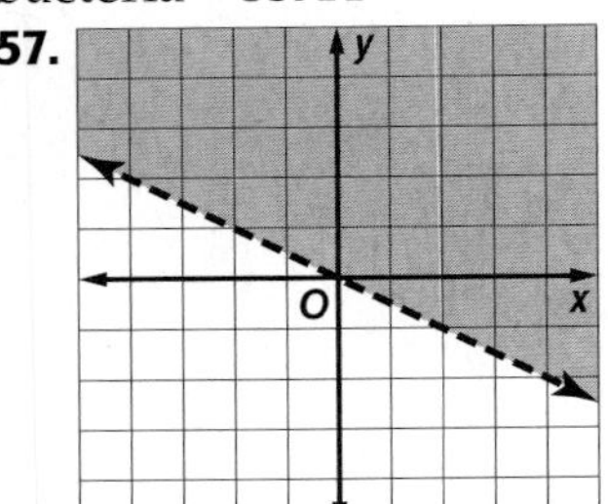

59.

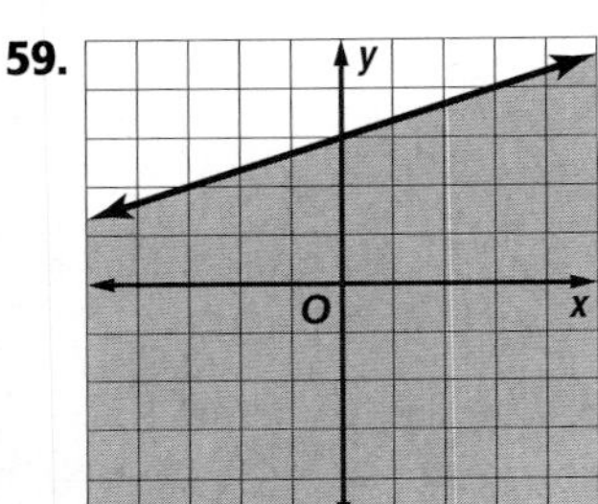

61.

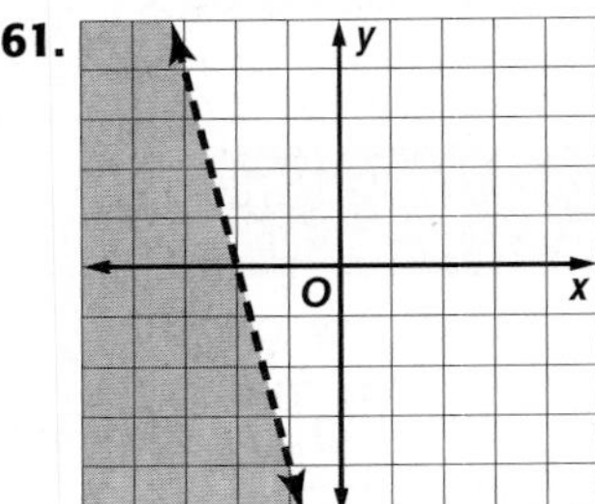

63. 1475 to 1525 books **65.** $y = -3x - 8$ **67.** $y = \frac{1}{2}x - 3$ **69.** 22 **71.** 92 **73.** −16 **75.** 7

Pages 342–347 Lesson 6-2

1. (5, 10) **3.** (2, 0) **5.** infinitely many **7a.** $x = m\angle X$, $y = m\angle Y$; $x + y = 180$, $x = 24 + y$ **7b.** $x = 102°$, $y = 78°$

9 Step 1: One equation is already solved for y.

$y = 4x + 5$

$2x + y = 17$

Step 2: Substitute $4x + 5$ for y in the second equation.

$$2x + y = 17$$
$$2x + 4x + 5 = 17$$
$$6x + 5 = 17$$
$$6x = 12$$
$$x = 2$$

Step 3: Substitute 2 for x in either equation to find y.

$y = 4x + 5$

$y = 4(2) + 5$

$y = 8 + 5$

$y = 13$

The solution is (2, 13).

11. (−3, −11) **13.** (−1, 0) **15.** infinitely many **17.** (2, 3) **19.** no solution **21.** (2, 0) **23a.** Let x = number of years since 2000, and let y = the number of nurses; supply, $y = -8000x + 1{,}890{,}000$; demand, $y = 82{,}000x + 2{,}000{,}000$ **23b.** during 1998

25 a. Men: 1:51:39 = 60 + 51 = 111, then round up because the number of seconds is greater than 30. So 1:51:39 rounds to 112.
1:49:31 = 60 + 49 = 109, then round up because the number of seconds is greater than 30. So, 1:49:31 rounds to 110.
Women: 1:54:33 = 60 + 54 = 114, then round up because the number of seconds is greater than 30. So, 1:54:31 rounds to 115.
1:58:03 = 60 + 58 = 118, then round down because the number of seconds is less than 30. So, 1:58:03 rounds to 118.

b. The y-intercept is (0, 112). Find the rate of change.

$$m = \frac{112 - 110}{0 - 5} = \frac{2}{-5} = -0.4$$

So, the equation is $y = -0.4x + 112$.
The y-intercept is (0, 115). Find the rate of change.

$$m = \frac{118 - 115}{5 - 0} = \frac{3}{5} = 0.6$$

So, the equation is $y = 0.6x + 115$.

c. never; If you graph the two equations, the graphs do not cross in the positive values of x. Negative values will not make sense in terms of the word problem.

27. Neither; Guillermo substituted incorrectly for b. Cara solved correctly for b, but misinterpreted the pounds of apples bought. **29.** Sample answer: The solutions found by each of these methods should be the same. However, it may be necessary to estimate using a graph. So, when a precise solution is needed, you should use substitution. **31.** An equation containing a variable with a coefficient of 1 can easily be solved for the variable. That expression can then be substituted into the second equation for the variable.

33. $\frac{5}{6}$ **35.** C

37. 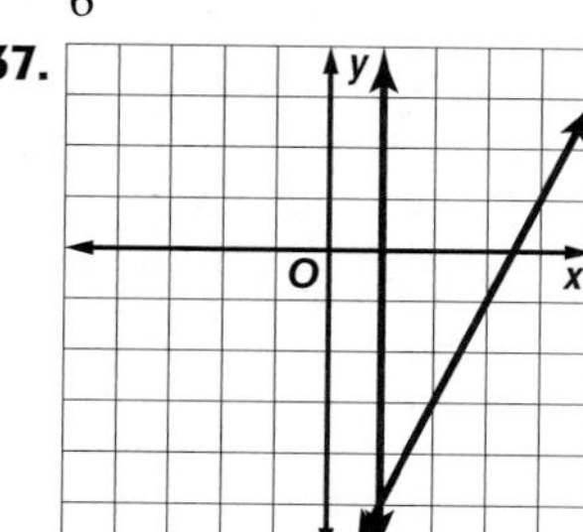 one solution; (1,–5)

39. 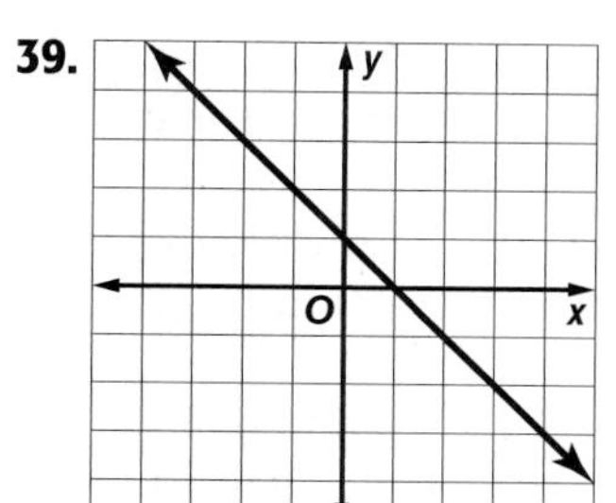 infinitely many solutions

41. $v \geq -2$ **43.** $q \leq -40$ **45.** $t \geq 3$
47. $55b + 15$ **49.** $11h^2 + 12h$

Pages 348–354 Lesson 6-3

1. (2, 3)

3 Step 1: The like terms are already aligned.

$7f + 3g = -6$
$7f - 2g = -31$

Step 2: Subtract the equations.

$$\begin{array}{r} 7f + 3g = -6 \\ (-)\ 7f - 2g = -31 \\ \hline 5g = 25 \\ g = 5 \end{array}$$

Step 3: Substitute 5 for g in either equation to find f.

$7f + 3(5) = -6$
$7f + 15 = -6$
$7f = -21$
$f = -3$

The solution is (–3, 5).

5. 6, 18 **7.** (–3, 4) **9.** (–3, 1) **11.** (4, –2)
13. (8, –7) **15.** (4, 7) **17.** (4, 1.5) **19.** 5, 17

21

Three times a number	minus	another number	is	–3.
$3x$	$-$	y	$=$	-3
The first number	plus	the second number	is	11.
x	$+$	y	$=$	11

Steps 1 and 2: Write the equations vertically and add.

$$\begin{array}{r} 3x - y = -3 \\ x + y = 11 \\ \hline 4x = 8 \\ x = 2 \end{array}$$

Step 3: Substitute 2 for x in either equation to find y.

$x + y = 11$
$2 + y = 11$
$y = 9$

The numbers are 2 and 9.

23. adult, \$5.95; children, \$3.95 **25.** (2, –1)
27. $\left(-\frac{5}{6}, 3\right)$ **29.** $\left(2\frac{7}{9}, 13\frac{1}{3}\right)$ **31a.** $x + y = 66$; $x = 30 + y$ **31b.** (48, 18) **31c.** There are 48 teams that are not from the U.S. and 18 teams that are from the U.S.

31d. 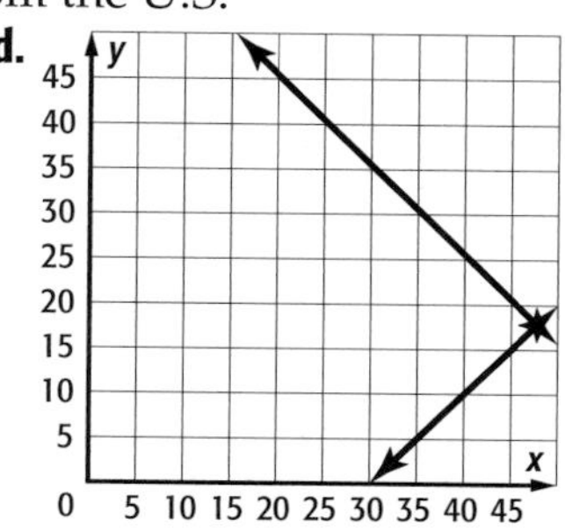

33 a. One way to get 15 points is to use 4 pennies and 3 paper clips.

$4(3) + 3 = 12 + 3$
$= 15$

b. The total number of objects is 9.

$p + c = 9$

Pennies are worth 3 points each and paper clips are worth 1 point each for a total of 15 points.

$3p + c = 15$

Solve:

$$\begin{array}{r} p + c = 9 \\ (-)\ 3p + c = 15 \\ \hline -2p = -6 \\ p = 3 \end{array}$$

Substitute 3 for p in either equation to find c,

$p + c = 9$
$3 + c = 9$
$c = 6$

So, $p = 3$ and $c = 6$.

c.

p	$c = 9 - p$	$3p + c$
0	9	$3(0) + 9 = 9$
1	8	$3(1) + 8 = 11$
2	7	$3(2) + 7 = 13$
3	6	$3(3) + 6 = 15$
4	5	$3(4) + 5 = 17$
5	4	$3(5) + 4 = 19$

Selected Answers and Solutions

d. Yes, since the pennies are 3 points each, 3 of them makes 9 points. Add the 6 points from 6 paper clips and you get 15 points.
35. The result of the statement is false, so there is no solution. **37.** Sample answer: $-x + y = 5$; I used the solution to create another equation with the coefficient of the x-term being the opposite of its corresponding coefficient. **39.** Sample answer: It would be most beneficial when one variable has either the same or opposite coefficients in each of the equations. **41.** A **43.** B **45.** (15, 5) **47.** (3, 11) **49.** (−2, 2) **51.** Yes; each pair of opposite sides have the same or an undefined slope, so they are parallel. **53.** −5 **55.** −20 **57.** $11w^2 - 9w$ **59.** $-2y - 35$

Pages 355–360 ***Lesson 6-4***

1. (3, 2)
3 Eliminate y:

$$\begin{aligned}(4x + 2y &= -14)(-3) \\ (5x + 3y &= -17)(2)\end{aligned} \qquad \begin{aligned}-12x - 6y &= 42 \\ 10x + 6y &= -34 \\ \hline -2x &= 8 \\ x &= -4\end{aligned}$$

Now, substitute −4 for x in either equation to find the value of y.

$$\begin{aligned}4x + 2y &= -14 \\ 4(-4) + 2y &= -14 \\ -16 + 2y &= -14 \\ 2y &= 2 \\ y &= 1\end{aligned}$$

The solution is (−4, 1).
5. 6 mph **7.** (−1, 3) **9.** (−3, 4) **11.** (−2, 3) **13.** (3, 5) **15.** (1, −5) **17.** (0, 1)
19

Seven times a number	plus	three times another number	equals	−1.
$7x$	$+$	$3y$	$=$	-1

The sum of the two numbers is −3.

$$x + y = -3$$

$$\begin{aligned}7x + 3y &= -1 \\ (x + y &= -3)(-3)\end{aligned} \qquad \begin{aligned}7x + 3y &= -1 \\ -3x - 3y &= 9 \\ \hline 4x &= 8 \\ x &= 2\end{aligned}$$

Now, substitute 2 for x in either equation to find y.

$$\begin{aligned}x + y &= -3 \\ 2 + y &= -3 \\ y &= -5\end{aligned}$$

The two numbers are 2 and −5.
21. (2.5, 3.25) **23.** $\left(3, \frac{1}{2}\right)$ **25a.** $120n + 180s = 1500$ **25b.** $90n + 120s = 1050$ **25c.** (5, 5); To be cost-effective, the robots must save the time of 5 nurses and 5 support staff.
27 a. Let x be the cost of a batting token and y be the cost of the miniature golf games.
For the first group, the equation is $16x + 3y = 30$.
For the second group, the equation is $22x + 5y = 43$.
b. Solve.

$$\begin{aligned}(16x + 3y &= 30)(5) \\ (22x + 5y &= 43)(-3)\end{aligned} \qquad \begin{aligned}80x + 15y &= 150 \\ -66x - 15y &= -129 \\ \hline 14x &= 21 \\ x &= 1.5\end{aligned}$$

Now, substitute 1.5 for x in either equation to find y.

$$\begin{aligned}16x + 3y &= 30 \\ 16(1.5) + 3y &= 30 \\ 24 + 3y &= 30 \\ 3y &= 6 \\ y &= 2\end{aligned}$$

A batting token costs $1.50 and a game of miniature golf costs $2.
29. One of the equations will be a multiple of the other. **31.** Sample answer: $2x + 3y = 6$, $4x + 9y = 5$ **33.** Sample answer: A variable that has a nonzero coefficient in each equation may be eliminated using multiplication. Calculations may be easier if a variable requiring only one equation to be multiplied or a variable with a smaller coefficient is eliminated. **35.** G **37.** D
39. (−1, −1) **41.** (9, 3) **43.** (0, 6)
45. $m \le 13$ and $m \ge -3$

47. $w > 1$ or $w < -10$

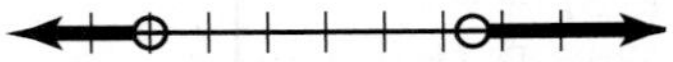

49. $A = \frac{1}{2}bh$ **51.** $V = \ell wh$ **53.** $A = \pi r^2$

Pages 362–367 ***Lesson 6-5***

1. elim (×); (2, −5) **3.** elim (+); $\left(-\frac{1}{3}, 1\right)$
5a. $4t + 3j = 181$; $t + 2j = 94$ **5b.** Substitution
5c. Each T-shirt cost $16 and each pair of jeans cost $39. **7.** subst.; (2, −2) **9.** elim. (−); $\left(1, -\frac{1}{2}\right)$
11

$$\begin{aligned}-5x + 4y &= 7 \\ -5x - 3y &= -14\end{aligned}$$

Since there are no coefficients of 1, elimination is the best method.

$$\begin{aligned}(-5x + 4y &= 7)(-1) \\ -5x - 3y &= -14\end{aligned} \qquad \begin{aligned}5x - 4y &= -7 \\ -5x - 3y &= -14 \\ \hline -7y &= -21 \\ y &= 3\end{aligned}$$

Now substitute 3 for y in either equation to find x,

$$\begin{aligned}-5x + 4y &= 7 \\ -5x + 4(3) &= 7 \\ -5x + 12 &= 7 \\ -5x &= -5 \\ x &= 1\end{aligned}$$

The solution is (1, 3).
13. $m + t = 40$ and $m = 3t - 4$; 29 movies, 11 television shows **15.** 880 books; If they sell this number, then their income and expenses both equal $35,200.

17 a. Let x be the cost per pound of the aluminum cans and y be the cost per pound of the newspapers.
For Mara, the equation is $9x + 26y = 3.77$.
For Ling, the equation is $9x + 114y = 4.65$.

b. Elimination is the best method for solving these equations.

$$\begin{array}{ll} (9x + 26y = 3.77)(-1) & -9x - 26y = -3.77 \\ 9x + 114y = 4.65 & \underline{\;\;9x + 114y = \;\;4.65} \\ & \qquad\quad 88y = \;\;0.88 \\ & \qquad\quad\;\; y = \;\;0.01 \end{array}$$

Now substitute 0.01 for y in either equation to find x.

$$\begin{aligned} 9x + 26y &= 3.77 \\ 9x + 26(0.01) &= 3.77 \\ 9x + 0.26 &= 3.77 \\ 9x &= 3.51 \\ x &= 0.39 \end{aligned}$$

The aluminum cans are \$0.39 per pound. This solution is reasonable.

19a. \$1.15 **19b.** \$9.15 **21.** Sample answer: $x + y = 12$ and $3x + 2y = 29$, where x represents the cost of a student ticket for the basketball game and y represents the cost of an adult ticket; substitution could be used to solve the system; (5, 7) means the cost of a student ticket is \$5 and the cost of an adult ticket is \$7.

23. Graphing: (2, 5)

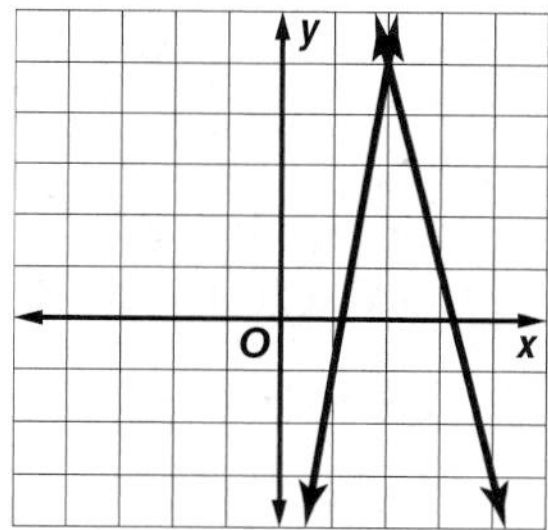

elimination by addition:

$$\begin{aligned} 4x + y &= 13 \\ 6x - y &= 7 \\ 10x &= 20 \\ x &= 2 \\ 4(2) + y &= 13 \\ y &= 5 \end{aligned}$$

substitution:

$$\begin{aligned} y &= -4x + 13 \\ 6x - (-4x + 13) &= 7 \\ 6x + 4x - 13 &= 7 \\ 10x &= 20 \\ x &= 2 \\ 4(2) + y &= 13 \\ y &= 5 \end{aligned}$$

25. The third system; this system is the only one that is not a system of linear equations. **27.** A **29.** 10 ft **31.** (0, 3) **33.** (2, 1)

35.

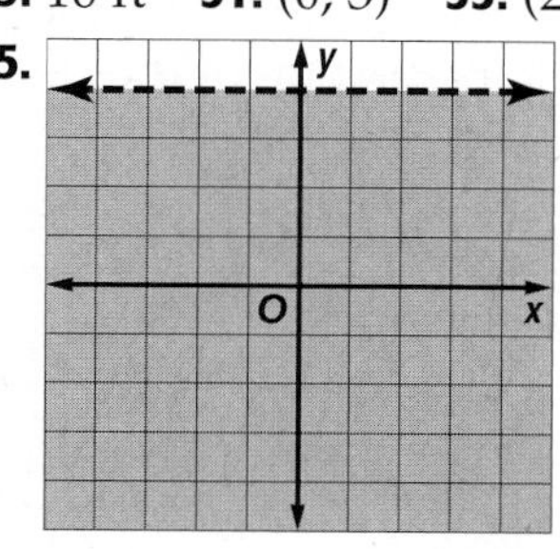

37.

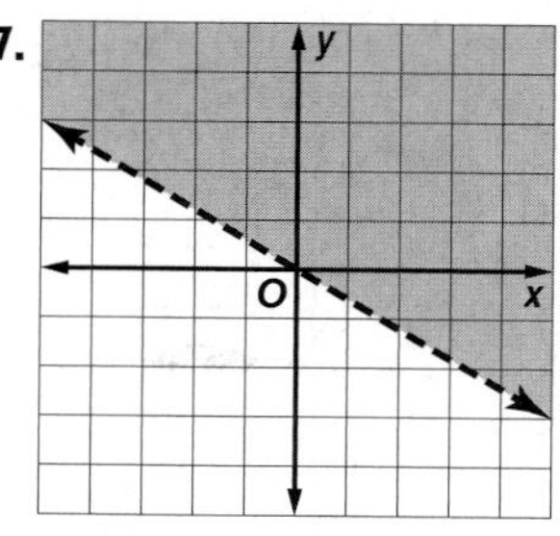

39. −12.31 **41.** 6.6 **43.** −93.19

Pages 369–375 Lesson 6-6

1. 2×4; second row and first column **3.** 1×4; first row and third column

5a.

$$\begin{array}{l} \\ \textit{Single} \\ \textit{Double} \\ \textit{Suite} \end{array} \begin{array}{c} \begin{array}{cc} \textit{Weekday} & \textit{Weekend} \end{array} \\ \begin{bmatrix} 69 & 89 \\ 79 & 109 \\ 99 & 139 \end{bmatrix} \end{array}$$

5b. 3×2 **5c.** suite on a weekend; single on a weekday

7. $\begin{bmatrix} 12 & 1 & 14 \\ -9 & 19 & 9 \\ -8 & -7 & -3 \end{bmatrix}$ **9.** impossible **11.** 5×2; second row and first column

13 The matrix has 4 rows and 6 columns, so the dimensions are 4×6. 7 is located in the third column and fourth row.

15. 6×4; second row and fourth column

17 a. Enter the data into a matrix.

$$\begin{array}{l} \\ \textit{Ohio} \\ \textit{Florida} \\ \textit{New York} \\ \textit{North Carolina} \end{array} \begin{array}{c} \begin{array}{cc} \textit{Land area} & \textit{People} \end{array} \\ \begin{bmatrix} 40{,}948 & 277.3 \\ 53{,}926 & 296.4 \\ 47{,}213 & 401.9 \\ 48{,}710 & 165.2 \end{bmatrix} \end{array}$$

b. The matrix has 4 rows and 2 columns, so the dimensions are 4×2.

c. New York has the most people per square mile, 401.9.
North Carolina has the least people per square mile, 165.2.

19. $\begin{bmatrix} -6 & 1 & 0 \\ 7 & -8 & 15 \end{bmatrix}$ **21.** $\begin{bmatrix} 10 & -5 & 0 \\ 5 & -15 & 25 \\ 35 & 50 & -55 \\ 40 & -45 & -20 \end{bmatrix}$

23. $[-3 \quad 12 \quad -11 \quad -10]$

25a. On Saturday at the store on Elm St., \$245 in glazed donuts were sold.

25b. $\begin{bmatrix} 95 & 205 & 70 & 51 \\ 105 & 245 & 79 & 49 \end{bmatrix}$; $\begin{bmatrix} 167 & 295 & 99 & 79 \\ 159 & 289 & 107 & 88 \end{bmatrix}$

25c. $\begin{bmatrix} 262 & 500 & 169 & 130 \\ 264 & 534 & 186 & 137 \end{bmatrix}$

Main Street: Chocolate 262, Glazed 500, Powered 169, Lemon filled 130
Elm Street: Chocolate 264, Glazed 534, Powdered 186, Lemon filled 137

25d. glazed **27.** [−16 25 9] **29.** impossible

31

$$\begin{bmatrix}-5&2\\12&-11\\9&0\\-1&7\\6&5\\-4&2\end{bmatrix}+4\begin{bmatrix}10&4\\-1&-3\\5&-8\\-9&0\\1&4\\-3&2\end{bmatrix}=\begin{bmatrix}-5&2\\12&-11\\9&0\\-1&7\\6&5\\-4&2\end{bmatrix}+\begin{bmatrix}40&16\\-4&-12\\20&-32\\-36&0\\4&16\\-12&8\end{bmatrix}$$

$$=\begin{bmatrix}-5+40&2+16\\12+(-4)&-11+(-12)\\9+20&0+(-32)\\-1+(-36)&7+0\\6+4&5+16\\-4+(-12)&2+8\end{bmatrix}$$

$$=\begin{bmatrix}35&18\\8&-23\\29&-32\\-37&7\\10&21\\-16&10\end{bmatrix}$$

33. Sample answer: $\begin{bmatrix}6&1&9\\1&3&2\end{bmatrix}$ and $\begin{bmatrix}1&3&2\\4&2&2\end{bmatrix}$

35. Sample answer: $\begin{bmatrix}4&5\\5&1\end{bmatrix}$ **37.** Sample answer:

The number of miles hiked on a 3-day hiking trip. On the first day, 5 miles were hiked. On the second day, 8 miles were hiked. On the third day, 10 miles were hiked.

$$\begin{array}{r}\\ \textit{Day 1}\\ \textit{Day 2}\\ \textit{Day 3}\end{array}\overset{\textit{Miles}}{\begin{bmatrix}5\\8\\10\end{bmatrix}}$$

39. C **41.** G **43.** $x + y = 500$, $0.25x + 0.5y = 170$; 320 gal of 25%; 180 gal of 50% **45.** (2, −7)
47. $1.5a - 0.3a \geq 75$; $a \geq 62.5$; at least 63 apples
49. 42, 48, 54 **51.** 25(28 + 18) = \$1150 **53.** −3
55. −153 **57.** 23

Pages 376–381 Lesson 6-7

1. $\left[\begin{array}{cc|c}-1&3&-10\\5&-2&7\end{array}\right]$ **3.** $\left[\begin{array}{cc|c}1&2&-1\\2&-2&-9\end{array}\right]$

5 Place the coefficients of the equations and the constants into a matrix.

$$\left[\begin{array}{cc|c}3&4&-5\\2&-1&6\end{array}\right]$$

7. (−1, −2) **9.** (−1, 6) **11a.** $3n + 4b = 22.25$ $3n + 10b = 29.75$ **11b.** $\left[\begin{array}{cc|c}3&4&22.25\\3&10&29.75\end{array}\right]$

11c. new: \$5.75; used: \$1.25

13. $\left[\begin{array}{cc|c}-4&-3&-8\\1&1&-12\end{array}\right]$ **15.** $\left[\begin{array}{cc|c}-6&1&-15\\1&-2&13\end{array}\right]$

17. $\left[\begin{array}{cc|c}1&-1&7\\9&-5&23\end{array}\right]$ **19.** (−5, 4) **21.** (−2, −4)

23. (3, −7) **25.** (−4, −8) **27.** (3, −4) **29.** $\left(\frac{3}{2}, \frac{1}{3}\right)$

31. $x = 16$, $y = -2$ **33.** $3x + 2y = 7$, $-x - 4y = 5$

35 Place the coefficients of the matrix into equations.
$-x + 9y = 12$
$2x + 3y = -7$

37 First, write the equations.
$x + y = 16$
$22x + 40y = 460$
Now, write the augmented matrix for the equations.

$$\left[\begin{array}{cc|c}1&1&16\\22&40&460\end{array}\right]$$

Step 1: To make the first element in row 2 zero, multiply row 1 by −22 and add to row 2.

$$\left[\begin{array}{cc|c}1&1&16\\0&18&108\end{array}\right]$$

Step 2: To make the second element in row 2 one, divide row 2 by 18.

$$\left[\begin{array}{cc|c}1&1&16\\0&1&6\end{array}\right]$$

Step 3: To make the second element in row 1 zero, multiply row 2 by −1 and add to row 2.

$$\left[\begin{array}{cc|c}1&0&10\\0&1&6\end{array}\right]$$

The solution is $x = 10$ and $y = 6$. So there are 10 boxes of notebooks and 6 boxes of mugs.

39. 5 movies and 3 games **41.** infinitely many **43.** no solution **45.** infinitely many **47.** Sample answer: The graphs of these two lines are parallel, so the system has no solution. **49.** (0, 2)
51. Sample answer: An augmented matrix consists of the coefficients and constant terms of a system. Row operations are used until the coefficient portion of the matrix is the identity matrix. The x-coordinate is the top number in the constant portion of the matrix, and the y-coordinate is the bottom number in the constant portion of the matrix. **53.** A **55.** B

57. $\begin{bmatrix}11&-18\\-3&18\\-12&-4\end{bmatrix}$ **59.** impossible

61. $\{p \mid 28 \leq p \leq 32\}$ **63.** $y = 2x - 5$

65. $y = \frac{2}{3}x + 1$ **67.** $y = -\frac{1}{2}x + \frac{3}{2}$ **69.** $-9 + 24x$

71. $-24m + 12$ **73.** $48c + 36b$

Pages 382–386 Lesson 6-8

1.

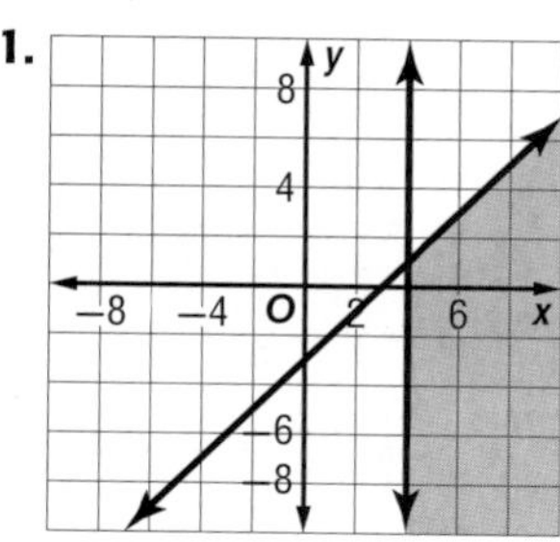

3.

5. no solution

7.

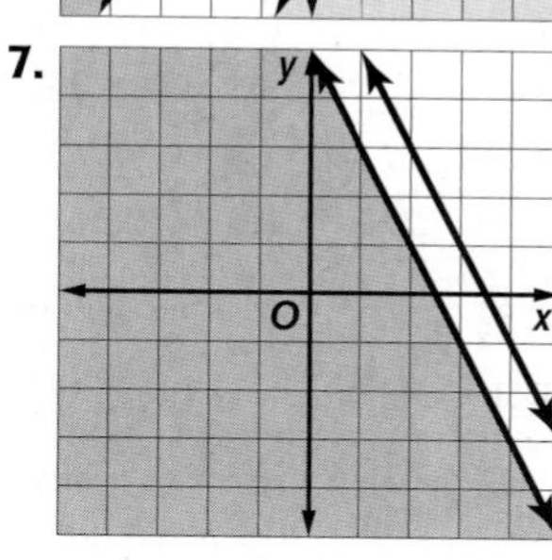

9a. Let h = the height of the driver in inches and w = the weight of the driver in pounds; $h < 79$ and $w < 295$.

Driving Requirements

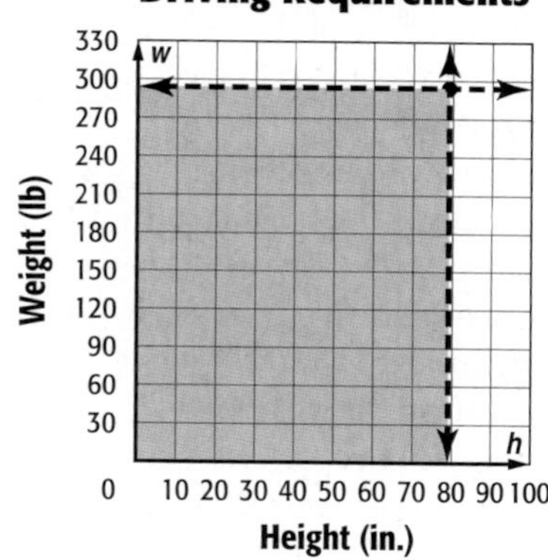

9b. Sample answer: 72 in. and 220 lb **9c.** Yes, the point falls in the overlapping region.

11 Graph both inequalities on the same coordinate plane.
$y \geq 0$ has a solid line.
$y \leq x - 5$ has a solid line.
The solution is the intersection of the shading.

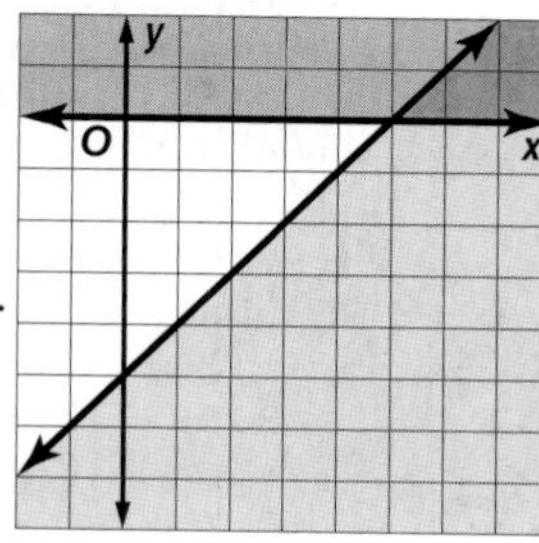

13.

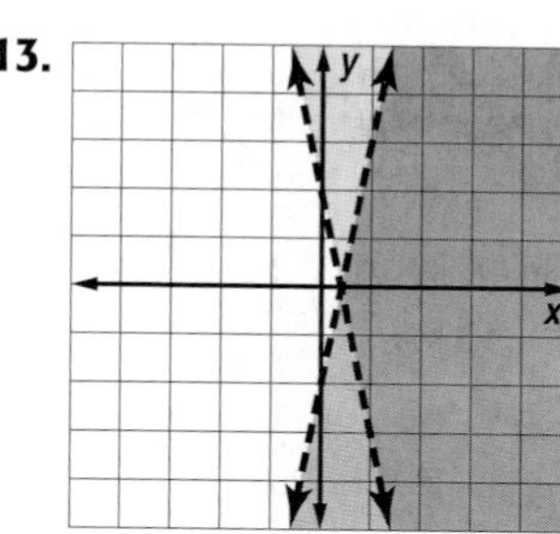

15.

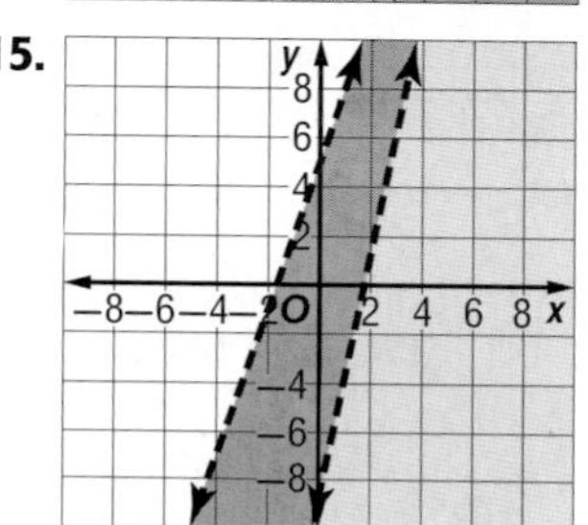

17. no solution

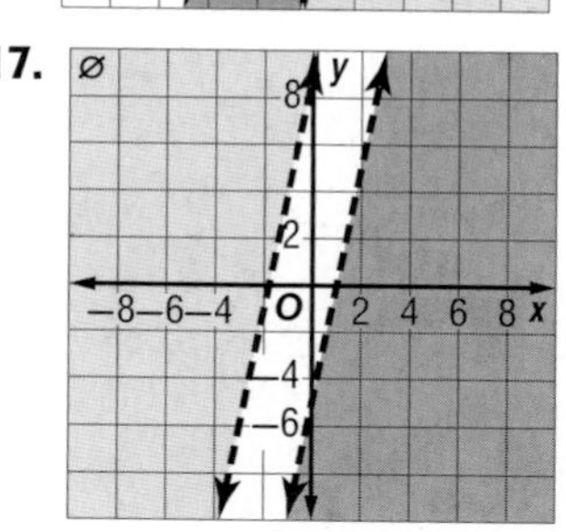
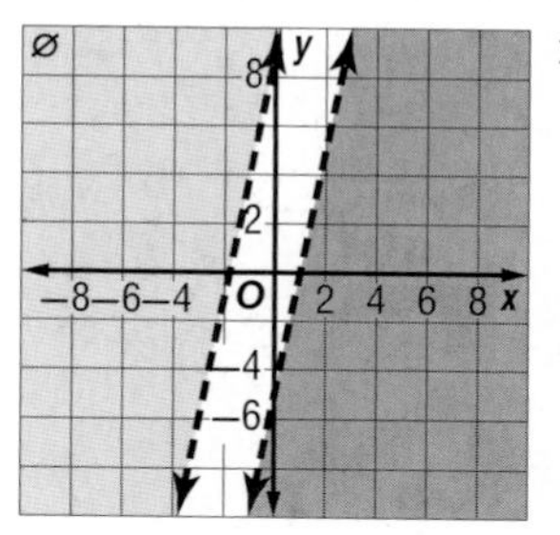

19.

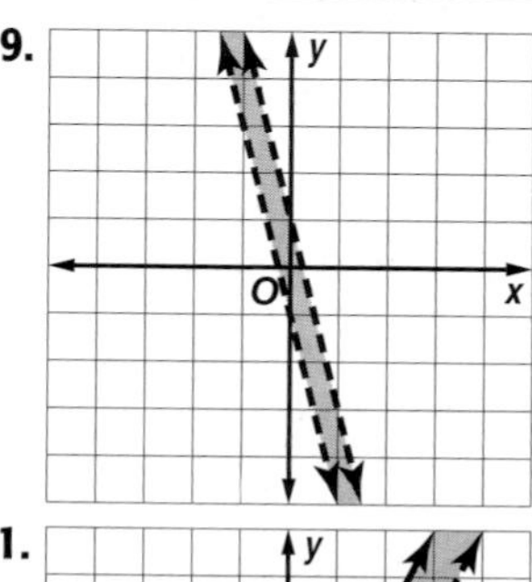

21.

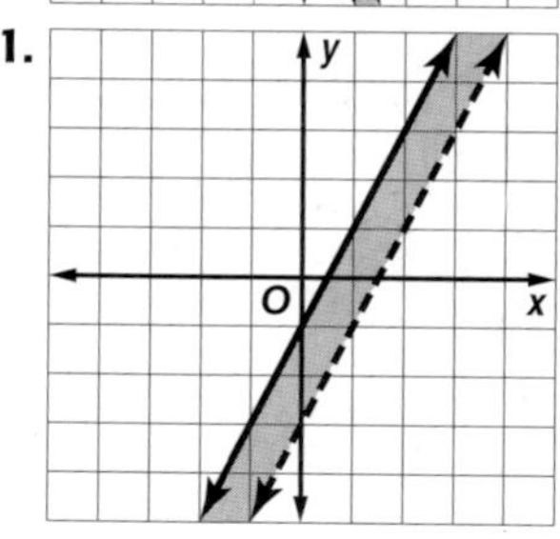

Selected Answers and Solutions

23.

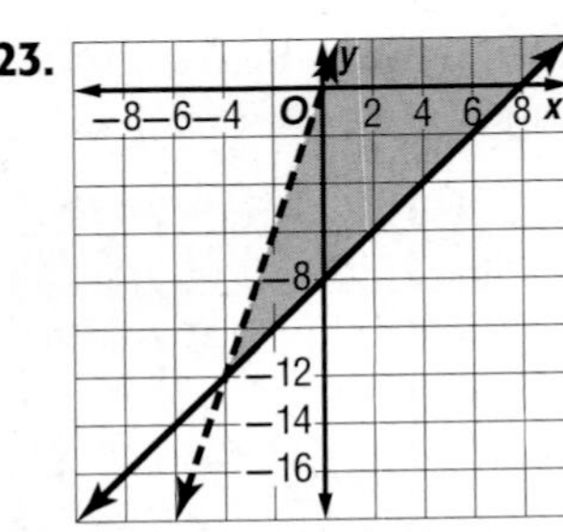

25a. Let f = square footage and let p = price; $1000 \leq f \leq 17{,}000$ and $10{,}000 \leq p \leq 150{,}000$

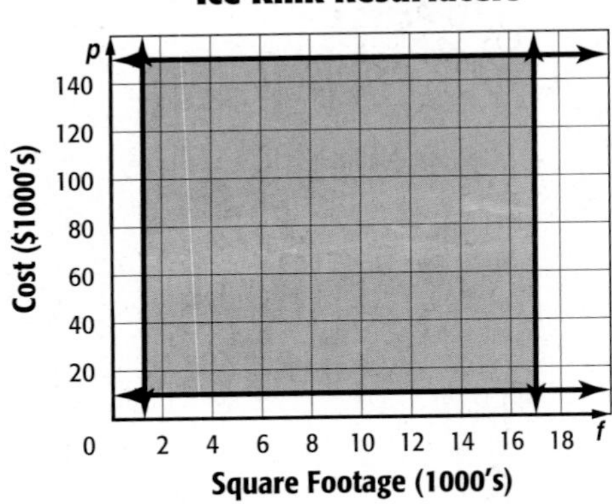

25b. Sample answer: an ice resurfacer for a rink of 5000 ft^2 and a price of \$20,000 **25c.** Yes; the point satisfies each inequality.

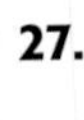

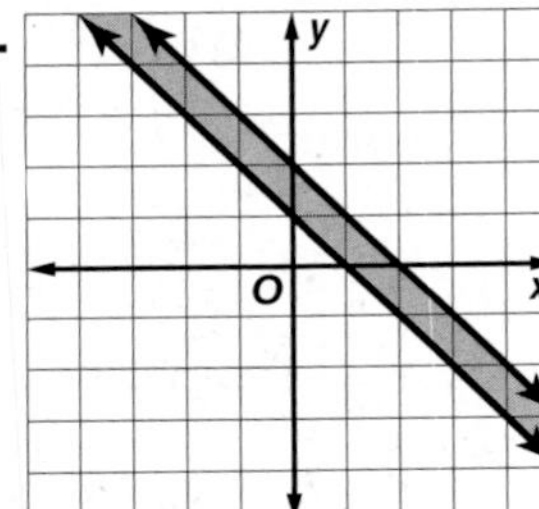

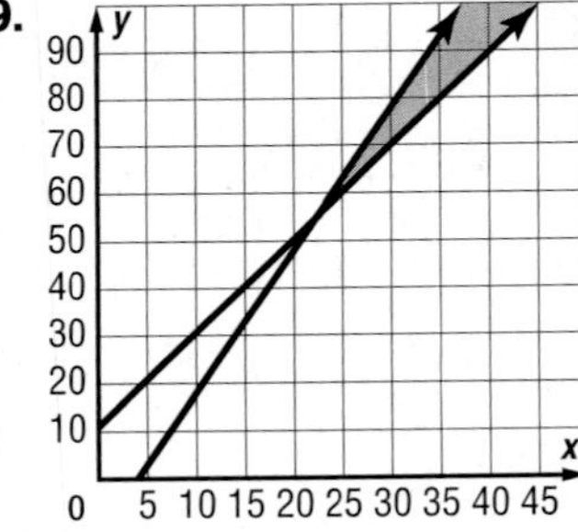

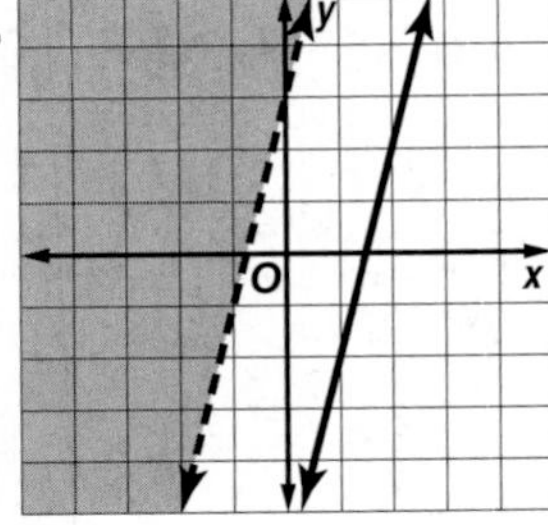

33.

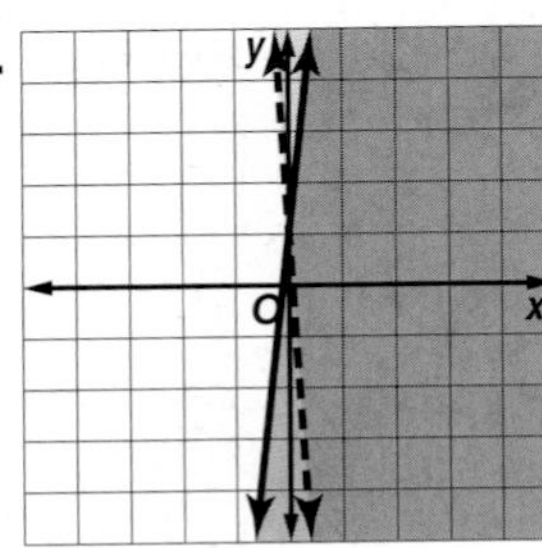

35.

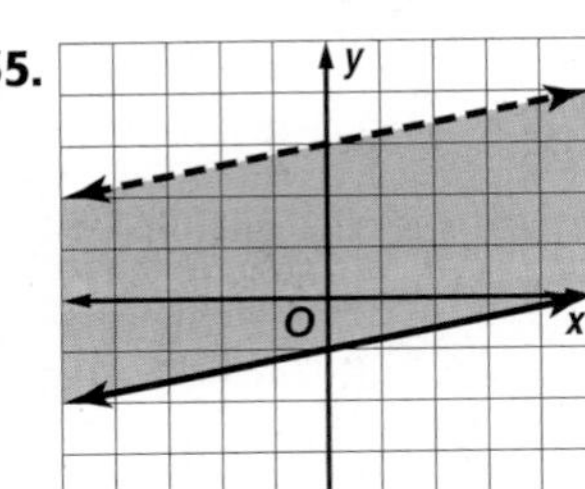

37 a. Let x be the number of hours she works for a photographer and y be the number of hours she works coaching.

$x + y \leq 20$

$15x + 10y \geq 90$

b. Graph both inequalities on the same grid.

$x + y \leq 20$ and $15x + 10y \geq 90$ have solid lines.

The solution is the intersection of the shading.

Earnings

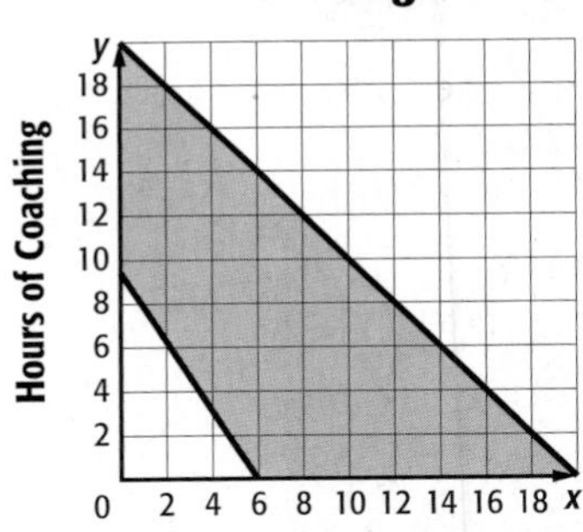

c. Two ordered pairs that are in the shaded area are (6, 10) and (8, 10). This means she could work for the photographer for 6 hours and coach for 10 or work for the photographer for 8 hours and coach for 10.

d. (2, 2) is not a solution because it does not fall in the shaded region. She would not earn enough money.

39. Sometimes; sample answer: $y > 3$, $y < -3$ will have no solution, but $y > -3$, $y < 3$ will have solutions. **41.** Sample answer: $3x - y < -4$

43. Sample answer: The yellow region represents the beats per minute below the target heart rate. The blue region represents the beats per minute above the target heart rate. The green region represents the beats per minute within the target heart rate. Shading in different colors clearly shows the overlapping solution set of the system of inequalities. **45.** D **47.** A **49.** $(-3, -3)$

51. $(-2, 8)$ **53.** $(6, 7)$ **55.** $\begin{bmatrix} 1 & -4 & 5 \\ 3 & -6 & 0 \\ 1 & -5 & 1 \end{bmatrix}$

57. $\begin{bmatrix} 7 & 14 & 7 \\ -3 & 4 & -6 \\ 3 & -1 & 13 \end{bmatrix}$ **59.** $\begin{bmatrix} -6 & -18 & -2 \\ 6 & -10 & 6 \\ -2 & -4 & -12 \end{bmatrix}$

61.

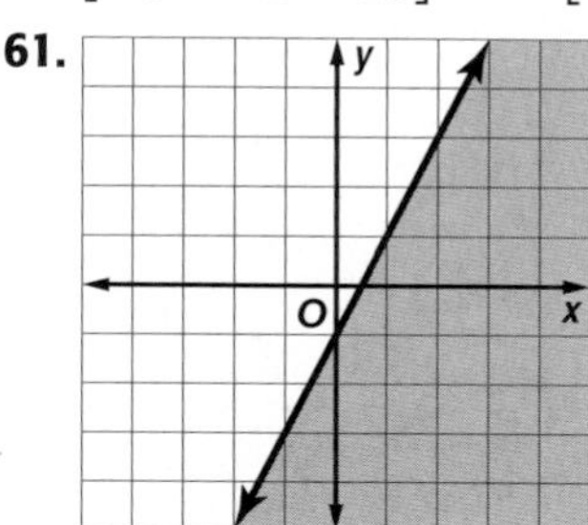

63.

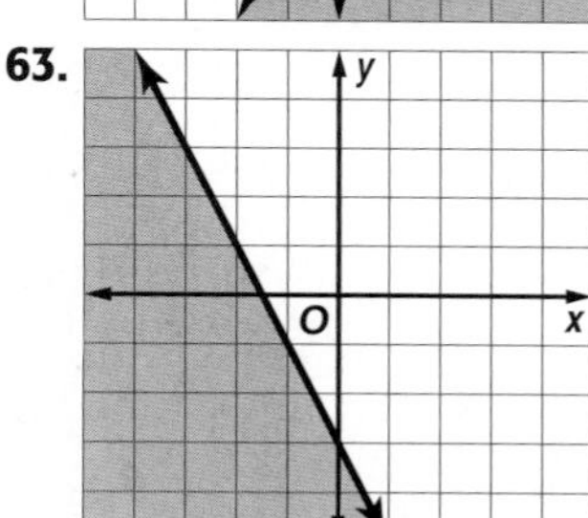

65. 16

Pages 388–392 Study Guide and Review

1. true **3.** false; dependent **5.** true **7.** false; element **9.** true

11.

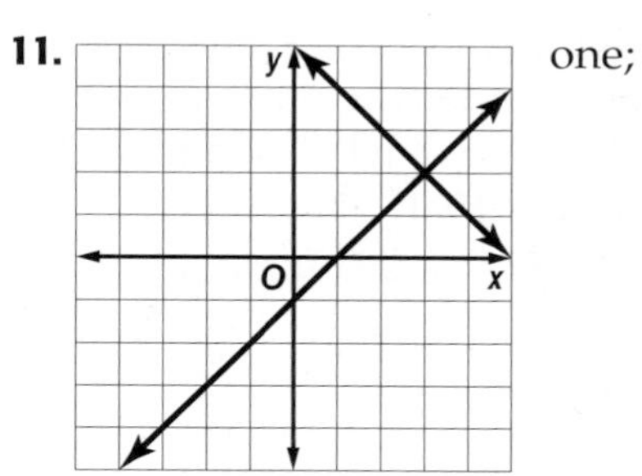

one; (3, 2)

13.

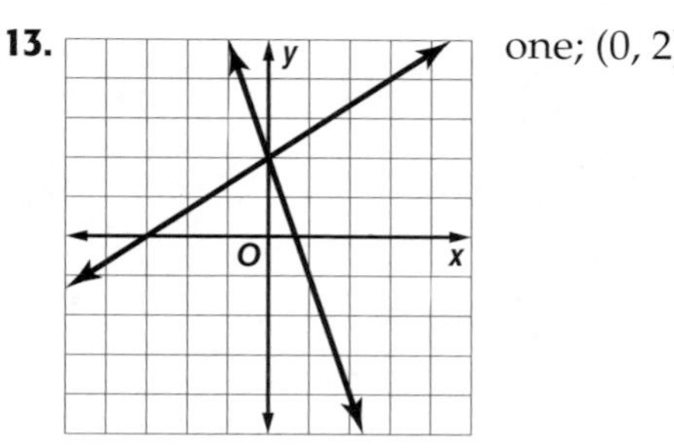

one; (0, 2)

15.

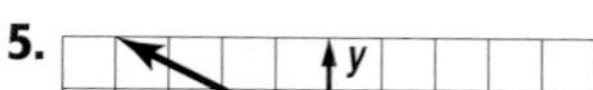

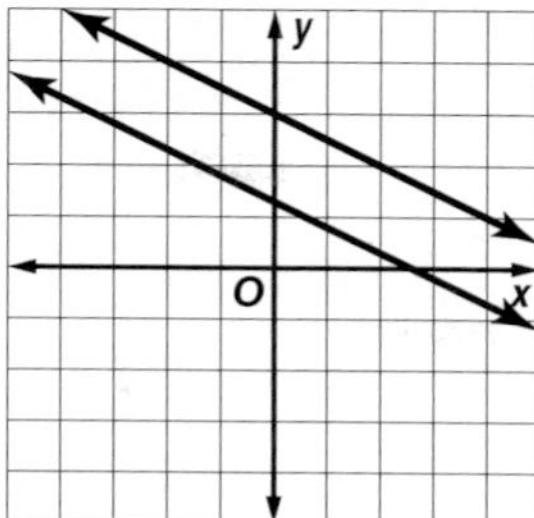

no solution

17. Sample answer: Let x be one number and y the other number; $x + y = 14$; $x - y = 4$; (9, 5)

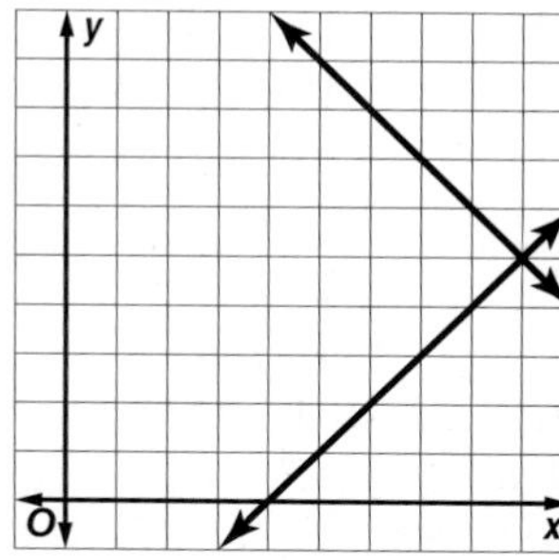

19. $(2, -10)$ **21.** $(2, -6)$ **23.** $(-3, 4)$ **25.** $(9, 4)$ **27.** $(4, -2)$ **29.** $\left(\frac{1}{2}, 6\right)$ **31.** $(-3, 5)$ **33.** Sample answer: Let f be the first type of card and let c be the second type of card; $f + c = 24$, $f + 3c = 50$; 11, \$1 cards and 13, \$3 cards. **35.** $(5, 7)$ **37.** $(2, 5)$ **39.** $(6, -1)$ **41.** $(1, -2)$ **43.** $(2, -6)$ **45.** $(24, -4)$ **47.** $(-2, 1)$ **49.** $(2, 5)$ **51.** Sample answer: Let d represent the dimes and let q represent the quarters; $d + q = 25$, $0.10d + 0.25q = 4$; 15 dimes, 10 quarters **53.** $\begin{bmatrix} -1 & -4 & 7 \\ -14 & 0 & -10 \end{bmatrix}$ **55.** 176 votes

57. $(3, -2)$ **59.** $(-4, -2)$ **61.** $(3, -2)$ **63.** $(3, -7)$ **65.** $(9, 1)$

67.

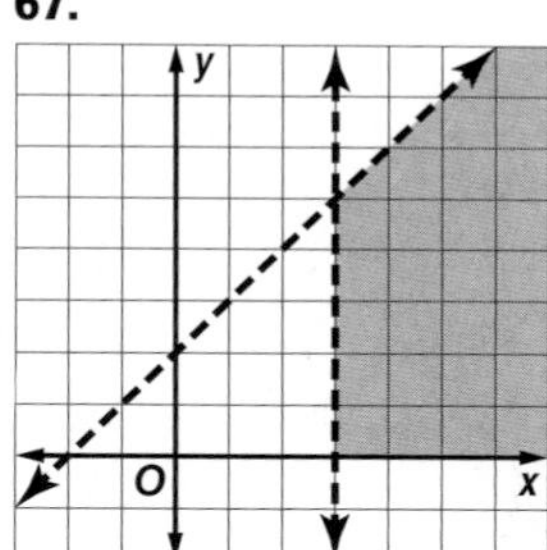

69.

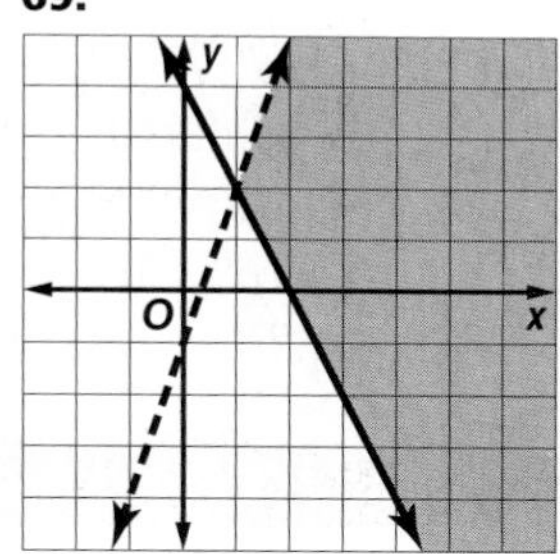

71.

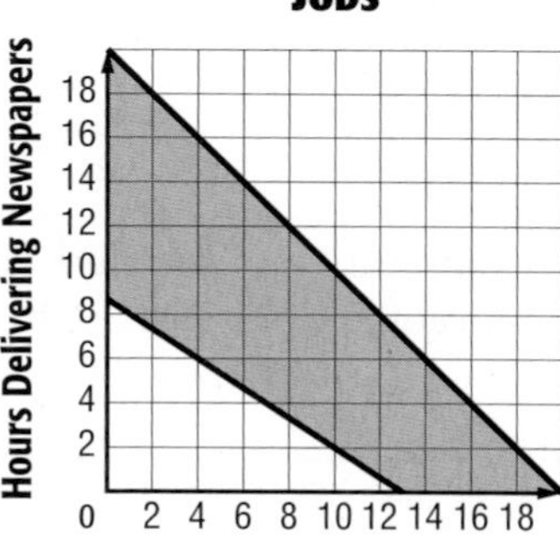

Chapter 7 Nonlinear Expressions, Equations, and Functions

Page 399 Get Ready

1. 4^5 **3.** 6^2 **5.** b^6 **7.** $\left(\frac{1}{3}\right)^8$ or $\frac{1}{3^8}$ **9.** 8 **11.** 27 **13.** $\frac{4}{9}$ **15.** $\frac{1}{32}$ **17.** 105 cm^3

Pages 404–407 Lesson 7-1

1. Yes; constants are monomials. **3.** No; there is a variable in the denominator. **5.** Yes; this is a product of a number and variables. **7.** k^4

9
$$\begin{aligned} 2q^2(9q^4) &= (2 \cdot 9)(q^2 \cdot q^4) \\ &= 18q^{2+4} \\ &= 18q^6 \end{aligned}$$

11. 3^8 or 6561 **13.** $16a^8b^{18}c^2$ **15.** $81p^{20}t^{24}$ **17.** $800x^8y^{12}z^4$ **19.** $-18g^7h^3j^{10}$ **21.** Yes; constants are monomials. **23.** No; there is addition and more than one term. **25.** Yes; this can be written as the product of a number and a variable.

27
$$\begin{aligned} (q^2)(2q^4) &= 2(q^2 \cdot q^4) \\ &= 2q^{2+4} \\ &= 2q^6 \end{aligned}$$

29. $9w^8x^{12}$ **31.** $7b^{14}c^8d^6$ **33.** $j^{20}k^{28}$ **35.** 28 *or* 256 **37.** $4096r^{12}t^6$ **39.** $20c^5d^5$ **41.** $16a^{21}$ **43.** $512g^{27}h^{18}$ **45.** $294p^{27}r^{19}$ **47.** $30a^5b^7c^6$ **49.** $0.25x^6$ **51.** $-\frac{27}{64}c^3$ **53.** $-9x^3y^9$ **55.** $2{,}985{,}984r^{28}w^{32}$ **57a.** $0.12c$ **57b.** \$280 **59.** $15x^7$

61 a.
$$\begin{aligned} V &= \pi r2h \\ &= \pi(2p^3)^2(4p^3) \\ &= \pi(22)(p^3)^2(4p^3) \\ &= \pi(4)(p^6)(4p^3) \\ &= \pi(4 \cdot 4)(p^6 \cdot p^3) \\ &= \pi(16)(p^{6+3}) \\ &= 16\pi p^9 \end{aligned}$$

b.

radius	height	Volume
$4p$	p^7	$16\pi p^9$
$4p^2$	p^5	$16\pi p^9$
$2p^3$	$4p^3$	$16\pi p^9$
$2p^4$	$4p$	$16\pi p^9$
$2p$	$4p^7$	$16\pi p^9$

c. If the height of the container is doubled, the volume of the container is doubled. So, the volume is $32\pi p^9$

63a.

Power	3^4	3^3	3^3	3^1	3^0	3^{-1}	3^{-3}	3^{-3}	3^{-4}
Value	81	27	9	3	1	$\frac{1}{3}$	$\frac{1}{9}$	$\frac{1}{27}$	$\frac{1}{81}$

63b. 1 and $\frac{1}{5}$ **63c.** $\frac{1}{a^n}$ **63d.** Any nonzero number raised to the zero power is 1.

65a.

Equation	Related Expression	Power of x	Linear or Nonlinear
$y = x$	x	1	linear
$y = x^2$	x^2	2	nonlinear
$y = x^3$	x^3	3	nonlinear

65b.

[−10, 10] scl: 1 by [−10, 10] scl: 1

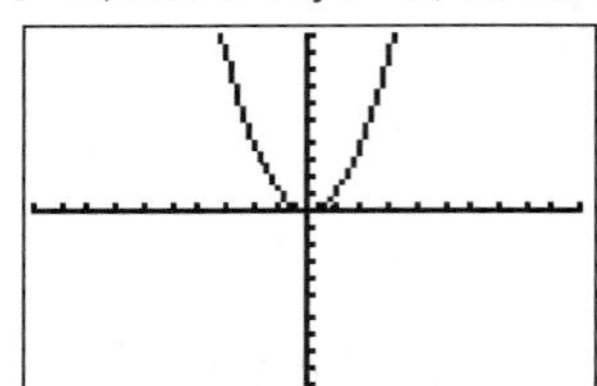

[−10, 10] scl: 1 by [−10, 10] scl: 1

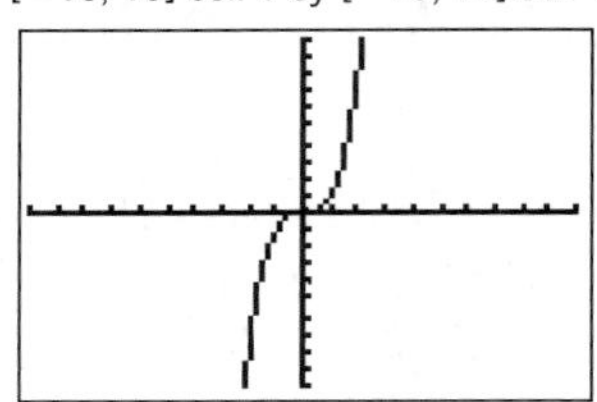

[−10, 10] scl: 1 by [−10, 10] scl: 1

65c. See chart for 65a. **65d.** If the power of x is 1, the equation or its related expression is linear. Otherwise, it is nonlinear. **67.** Sample answer: The area of a circle or $A = \pi r^2$, where r is the radius, can be used to find the area of any circle. The area of a rectangle or $A = w \times \ell$, where w is the width and ℓ is the length, can be used to find the area of any rectangle. **69.** F **71.** The x-intercept does not change.

73.

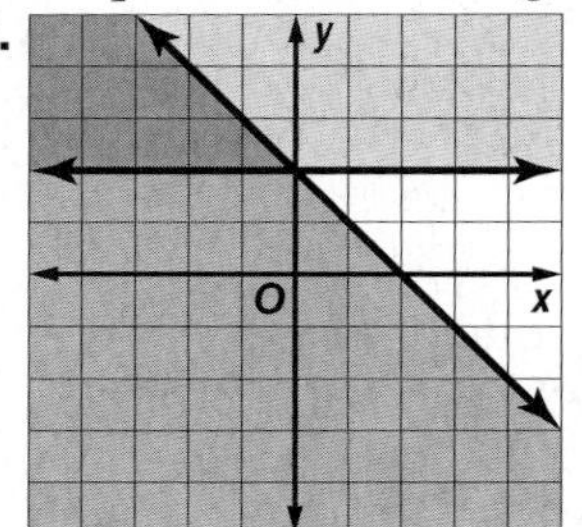

75.

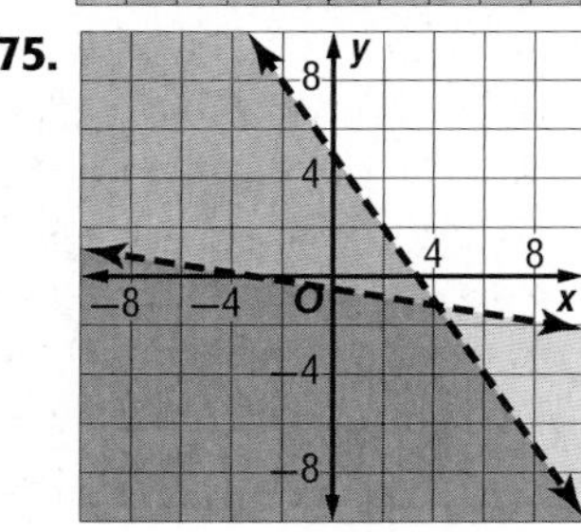

77. impossible **79.** $\begin{bmatrix} -2 & -5 & -16 \\ 11 & -1 & 9 \\ 18 & -3 & 12 \end{bmatrix}$ **81.** 8 **83.** −7.05 **85.** 13

1. t^3u^3

3 $\frac{m^6r^5p^3}{m^5r^2p^3} = \left(\frac{m^6}{m^5}\right)\left(\frac{r^5}{r^2}\right)\left(\frac{p^3}{p^3}\right)$
$= m^{6-5}r^{5-2}p^{3-3}$
$= m^1r^3p^0$
$= mr^3$

5. ghm **7.** xyz **9.** $\frac{4a^6b^{10}}{9}$ **11.** $\frac{32c^{15}d^{25}}{3125g^{10}}$ **13.** 1

15. $\frac{g^2h^4}{f^3}$ **17.** $\frac{a^5c^{13}}{3b^9}$ **19.** m^2p **21.** $\frac{r^4p^2}{4m^3t^4}$ **23.** $\frac{9x^2y^8}{25z^4}$

25. $\frac{p^6t^{21}}{1000}$ **27.** a^2b^7c

29 $\left(\frac{2r^3t^6}{5u^9}\right)^4 = \frac{2^4(r^3)^4(t^6)^4}{5^4(u^9)^4}$
$= \frac{16r^{12}t^{24}}{625u^{36}}$

31. 1 **33.** $\frac{p^4r^2}{t^3}$ **35.** $\frac{-f}{4}$ **37.** k^2mp^2 **39.** $\frac{3t^7}{u^6v^2}$

41. $\frac{r^3}{t^2x^{10}}$ **43.** 10^6; 10^8; about 10^2 or 100 times as many users as hosts **45.** $-\frac{w^9}{3}$

47. $1600k^{13}$ **49.** $\frac{5q}{r^6t^3}$ **51.** $\frac{4g^{12}}{h^4}$ **53.** $\frac{4x^8y^4}{z^6}$

55. $\frac{16z^2}{y^8}$ **57.** 100

59 **a.** the probability is $\frac{1}{6}$ multiplied d times, or $\left(\frac{1}{6}\right)^d$.

b. $\left(\frac{1}{6}\right)^d = (6^{-1})^d$
$= 6^{-d}$

61. Sometimes; sample answer: The equation is true when $x = 0$, $y = 2$, and $z = 3$, but it is false when $x = 1$, $y = 2$, and $z = 3$.

63. $\frac{1}{x^n} = \frac{x^0}{x^n} = x^{0-n} = x^{-n}$ **65.** The Quotient of Powers Property is used when dividing two powers with the same base. The exponents are subtracted. The Power of a Quotient Property is used to find the power of a quotient. You find the power of the numerator and the power of the denominator. **67.** J **69.** B

71.

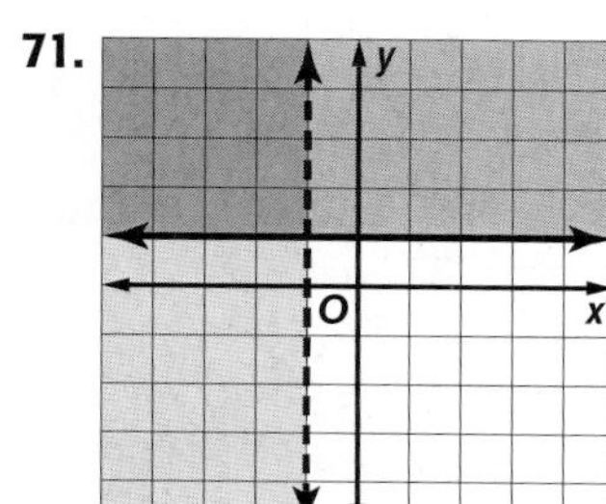

73.

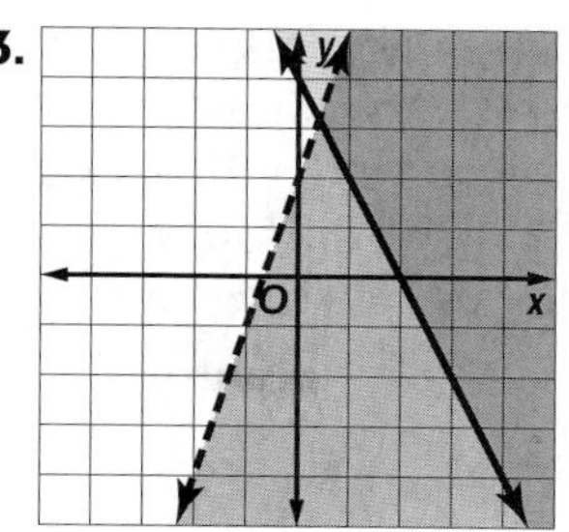

75. $h > 5$ **77.** $u \leq 35$ **79.** $n \geq -2$ **81.** 87 **83.** 121 **85.** 10,000 **87.** 125 **89.** 4096

1. 1.85×10^8 **3.** 5.64×10^{-4} **5.** 1.3×10^{10} **7.** 19,800,000 **9.** 0.00000003405 **11.** 1.74×10^{15}; 1,740,000,000,000,000 **13.** 4.7138×10^{-2}; 0.047138 **15.** 4.5×10^3; 4,500 **17.** 8.5×10^{-13}; 0.00000000000085 **19a.** 0.01, 0.000001 **19b.** 1×10^{-2}, 1×10^{-6} **19c.** 0.00000000001; 1×10^{-11}

21 58,600,000
Step 1: 58,600,000 ⟶ 5.8600000
Step 2: The decimal point moved 7 places to the left, so $n = 7$.
Step 3: 58,600,000 = 5.8600000×10^7
Step 4: 5.86×10^7

23. 1.3×10^{-6} **25.** 7.09×10^{-10} **27.** 5.5×10^9 **29.** 94,000,000 **31.** 0.0005 **33.** 0.00000622 **35.** 11,000,000 **37.** 8×10^7; 80,000,000

39 $(6.5 \times 10^7)(7.2 \times 10^{-2}) = (6.5 \times 7.2)(10^7 \times 10^{-2})$
$= 46.8 \times 10^5$
$= (4.68 \times 10^1) \times 10^5$
$= 4.68 \times 10^6$
$= 4{,}680{,}000$

41. 2.2×10^7; 22,000,000 **43.** 1.7889×10^{-6}; 0.0000017889 **45.** 6.89×10^5; 689,000 **47.** 4.7008×10^3; 4700.8 **49.** 5×10^{-6}; 0.000005 **51.** 2.448×10^{-10}; 0.0000000002448 **53.** 2.205×10^{-5}; 0.00002205 **55.** 2.325×10^5; 232,500 **57.** 6.1×10^{-8}; 0.000000061 **59.** 1.935×10^2; 193.5 **61.** 9×10^{-4}; 0.0009 **63.** 3×10^5

65.

Time	Kilometers Traveled
1 day	2.592×10^{10}
1 week	1.8144×10^{11}
1 month	7.776×10^{11}
1 year	9.4608×10^{12}

67 $(6.623 \times 10^9) \div (1.483 \times 10^8) = (6.623 \div 1.483) \times (10^9 \div 10^8)$
$\approx 4.47 \times 10^1$
There are about 44.7 persons per square kilometer.

69a. corn: 9.29×10^7, 92,900,000; soybeans: 6.41×10^7, 64,100,000; cotton: 1.11×10^7, 11,100,000

69b. about 1.4493×100; 1.4493 **69c.** about 8.3694×10^0; 8.3694 **71.** Pete is correct; Syreeta moved the decimal point in the wrong direction. **73.** Always; if the numbers are $a \times 10^m$ and $b \times 10^n$ in scientific notation, then $1 \le a < 10$ and $1 \le b < 10$. So $1 \le ab < 100$. **75.** Sample answer: Divide the numbers to the left of the × symbols. Then divide the powers of 10. If necessary, rewrite the results in scientific notation. To convert that to standard form, check to see if the exponent is positive or negative. If positive, move the decimal point to the right, and if negative, to the left. The number of places to move the decimal point is the absolute value of the exponent. Fill in with zeros as needed.

77. H **79.** B **81.** 6^2 or 36 **83.** $\frac{81a^{16}b^{16}}{4096c^8}$ **85.** $\frac{n^6p^3}{8}$

87. $y = 3x - 1$ **89.** $y = -2x - 12$

91. $y = \frac{2}{3}x + 7$ **93.** $13u$

95. simplified **97.** $65 + 52a$

Pages 426–429 Lesson 7-4

1. yes; trinomial **3.** yes; monomial **5.** yes; binomial **7.** 0 **9.** 1

11 Step 1: Find the degree of each term.

$12 \longrightarrow 0$

$-7q^2t \longrightarrow 2 + 1 = 3$

$8r \longrightarrow 1$

Step 2: The degree of the polynomial is the highest degree in the polynomial, 3

13. 4 **15.** $2x^5 + 3x - 12$; 2 **17.** $-5z^4 - 2z^2 + 4z$; -5 **19a.** 3300 students **19b.** 6000 students **21.** yes; monomial **23.** No; the exponent is a variable. **25.** yes; binomial **27.** 1 **29.** 3 **31.** 7 **33.** $7y^3 + 8y$; 7

35 Step 1: Find the degree of each term.

$-4d^4 \longrightarrow 4$

$1 \longrightarrow 0$

$-d^2 \longrightarrow 2$

Step 2: Write the terms in descending order:

$-4d^4 - d^2 + 1$

The leading coefficient is -4

37. $-r^3 + r + 2$; -1 **39.** $-b^6 - 9b^2 + 10b$; -1 **41.** quadratic trinomial **43.** quartic binomial **45.** quintic polynomial **47a.** $\frac{1}{3}\pi r^2h$ **47b.** about 9.42 in^3 **47c.** about 6.7 in. **49.** $6x^2$

51 **a.** Sample answer:

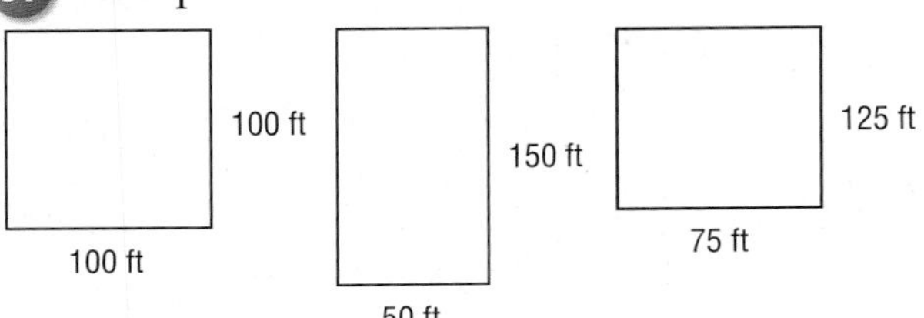

b.

Rectangle	Length	Width	Area
1	100 ft	100 ft	10,000 ft^2
2	50 ft	150 ft	7500 ft^2
3	75 ft	125 ft	9375 ft^2
4	x ft	$(200 - x)$ ft	$x(200 - x)$ ft^2

c.

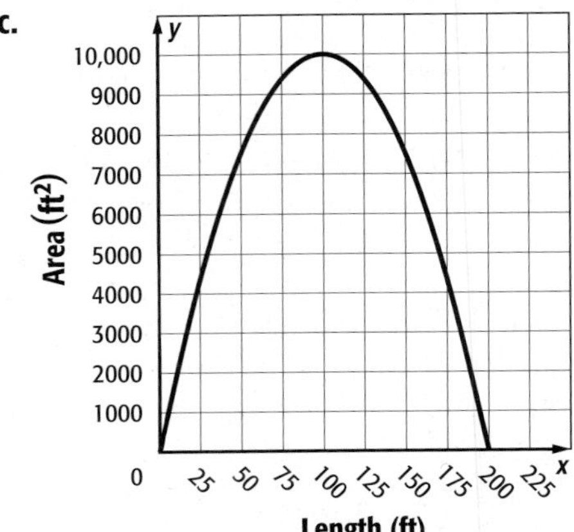

The highest point of the graph is (100, 10,000), so largest area possible is 10,000 ft^2.

d. The length and the width of the rectangle must be 100 feet each to have the largest area.

53. $2x + 1$, where x is an integer **55.** Sample answer: $x^3 - x^2 + 1$ **57.** D **59.** 500 **61.** 0.0000006 **63.** 810,000 **65.** 0.000000132 **67.** $\frac{1}{a^4}$ **69.** 1 **71.** $\frac{p^8}{k^4m^2}$ **73.** 6 two-seat, 11 four-seat **75.** $12t^2 - 3t$ **77.** simplified **79.** $\frac{5u}{2} + u^2$

Pages 435–438 Lesson 7-5

1. $4x^3 + 5$ **3.** $-a^2 + 6a - 3$ **5.** $-8z^3 - 3z^2 - 2z + 13$

7 $(2c^2 + 6c + 4) + (5c^2 - 7)$

$= [2c^2 + 5c^2] + 6c + [4 + (-7)]$

$= 7c^2 + 6c + (-3)$

$= 7c^2 + 6c - 3$

9a. $D = 6n + 14$ **9b.** 116,000 students **9c.** 301,000 student

11 $(2x + 3x^2) - (7 - 8x^2)$

$= (2x + 3x^2) + (-7 + 8x^2)$

$= [3x^2 + 8x^2] + 2x + (-7)$

$= 11x^2 + 2x - 7$

13. $2z^2 + z - 11$ **15.** $-2b^2 + 2a + 9$ **17.** $7x^2 - 2xy - 7y$ **19.** $3x^2 - rxt - 8r^2x - 6rx^2$ **21.** $-cd^2 + 6cd - 10$ **23.** $9x + 4y - 17z$ **25.** $2c^2 - c + 8$ **27.** $12x + 1\frac{1}{4}$ **29.** $4x$ **31.** $6x + 16y$ **33a.** $182 - 6x$ **33b.** 39, 97, 44

35 **a.** Words: \$15 plus \$0.15 per mile

Expression: 15 + $0.15m$

The expression is $15 + 0.15m$.

b. $15 + 0.15m = 15 + 0.15(145)$

$= 15 + 21.75$

$= 36.75$

The cost is \$36.75.

c. $4(15 + 0.15m) = 4[15 + 0.15(105)]$
$= 4(15 + 15.75)$
$= 4(30.75)$
$= 123$
The cost is \$123.

d. $7(15 + 0.15m) = 7[15 + 0.15(220)]$
$= 7(15 + 33)$
$= 7(48)$
$= 336$
The cost is \$336.

37. Sample answer: $3x^3 - 8x + 9$, $x^3 - x + 1$
39. Sample answer: $(2x - 3) - (4x - 3) = -2x$ but $(4x - 3) - (2x - 3) = 2x$ **41.** Sample answer: To add polynomials in a horizontal format, you combine like terms. For the vertical format, you write the polynomials in standard form, align like terms in columns, and combine like terms. To subtract polynomials in a horizontal format you find the additive inverse of the polynomial you are subtracting, and then combine like terms. For the vertical format you write the polynomials in standard form, align like terms in columns, and subtract by adding the additive inverse. **43.** $8x + 12$ units **45.** C **47.** 1 **49.** 5 **51.** 3 **53.** 5 **55.** \$80,000 **57.** no **59.** no **61.** yes; 0.5 **63.** $-2n^8$ **65.** $-40u^5z^9$ **67.** 64 **69.** $288x^8y^{10}z^6$

Pages 441–444 Lesson 7-6

1. $-15w^3 + 10w^2 - 20w$ **3.** $32k^2m^4 + 8k^3m^3 + 20k^2m^2$

5. $2ab(7a^4b^2 + a^5b - 2a) = 2ab(7a^4b^2) + 2ab(a^5b) + 2ab(-2a)$
$= 14a^5b^3 + 2a^6b^2 + (-4a^2b)$
$= 14a^5b^3 + 2a^6b^2 - 4a^2b$

7. $4t^3 + 15t^2 - 8t + 4$ **9.** $-5d^4c^2 + 8d^2c^2 - 4d^3c + dc^4$ **11.** 20 **13.** $-\frac{20}{49}$ **15.** 20 **17.** 1 **19.** $f^3 + 2f^2 + 25f$ **21.** $10j^5 - 30j^4 + 4j^3 + 4j^2$ **23.** $8t^5u^3 - 40t^4u^5 + 8t^3u$ **25.** $-8a^3 + 20a^2 + 4a - 12$ **27.** $-9g^3 + 21g^2 + 12$ **29.** $8n^4p^2 + 12n^2p^2 + 20n^2 - 8np^3 + 12p^2$

31. $7(t^2 + 5t - 9) + t = t(7t - 2) + 13$
$7t^2 + 35t - 63 + t = 7t^2 - 2t + 13$
$7t^2 + 36t - 63 = 7t^2 - 2t + 13$
$36t - 63 = -2t + 13$
$38t = 76$
$t = 2$

33. $\frac{43}{6}$ **35.** $\frac{30}{43}$ **37.** $20np^4 + 6n^3p^3 - 8np^2$
39. $-q^3w^3 - 35q^2w^4 + 8q^2w^2 - 27qw$ **41a.** $53.50 - 0.25h$ **41b.** \$50.50

43 **a.** $A = \ell w$
$= (1.5x + 24)x$
$= 1.5x^2 + 24x$
b. $x(x - 9) = x^2 - 9x$
c. $2(2.5x) = 2(2.5)(36)$
$= 180$ ft
$2(x + 6) = 2(36 + 6)$
$= 2(42)$
$= 84$ ft
Perimeter = 180 + 84 or 264 ft
Number of stepping stones = $264 \div 3$ or 88 stones

45. Ted; Pearl used the Distributive Property incorrectly. **47.** $8x^2y^{-2} + 24x^{-10}y^8 - 16x^{-3}$
49. Sample answer: $3n$, $4n + 1$; $12n^2 + 3n$ **51.** B
53. A **55.** $-3x^2 + 1$ **57.** $-9a^2 + 4a + 7$ **59.** $6ab + 2a + 4b$ **61.** 1 **63.** 2 **65.** 5 **67.** $y - 3{,}600{,}000 = 300{,}000(x - 1990)$; 9,600,000 people **69.** $f(x) = -0.5x$ **71.** $6y^3$ **73.** $15z^7 - 6z^4$ **75.** $-8p^5 + 10p^{10}$

Pages 450–452 Lesson 7-7

1. $x^2 + 7x + 10$ **3.** $b^2 - 4b - 21$ **5.** $16h^2 - 26h + 3$ **7.** $4x^2 + 72x + 320$ **9.** $16y^4 + 28y^3 - 4y^2 - 21y - 6$ **11.** $10n^4 + 11n^3 - 52n^2 - 12n + 48$
13. $2g^2 + 15g - 50$

15 $(4x + 1)(6x + 3) = 4x(6x) + 4x(3) + 1(6x) + 1(3)$
$= 24x^2 + 12x + 6x + 3$
$= 24x^2 + 18x + 3$

17. $24d^2 - 62d + 35$ **19.** $49n^2 - 84n + 36$
21. $25r^2 - 49$ **23.** $33z^2 + 7yz - 10y^2$ **25.** $2y^3 - 17y^2 + 37y - 22$ **27.** $m^4 + 2m^3 - 34m^2 + 43m - 12$
29. $6b^5 - 3b^4 - 35b^3 - 10b^2 + 43b + 63$ **31.** $2m^3 + 5m^2 - 4$ **33.** $4\pi x^2 + 12\pi x + 9\pi - 3x^2 - 5x - 2$

35 **a.** $A = \ell w$
$= (3y + 4)(6y - 5)$
$= 3y(6y) + 3y(-5) + 4(6y) + 4(-5)$
$= 18y^2 - 15y + 24y - 20$
$= 18y^2 + 9y - 20$
b. $3y + 4 = 31$
$3y = 27$
$y = 9$
So, the width is $6y - 5 = 6(9) - 5$
$= 54 - 5$
$= 49$
$A = \ell w$
$= (31)(49)$
$= 1519$ ft^2

37. $a^2 - 4ab + 4b^2$ **39.** $x^2 - 10xy + 25y^2$
41. $125g^3 + 150g^2h + 60gh^2 + 8h^3$ **43a.** $x > 4$; If $x = 4$, the width of the rectangular sandbox would be zero and if $x < 4$ the width of the rectangular sandbox would be negative. **43b.** square **43c.** 4 ft^2
45. Always; you can group two adjacent terms of a trinomial, treat the trinomial as the sum of two quantities, and apply the FOIL method. For example, $(2x + 3)(x^2 + 5x + 7) = (2x + 3)[x^2 + (5x + 7)] = 2x(x^2) + 2x(5x + 7) + 3(x^2) + 3(5x + 7)$. Then use the Distributive Property and simplify. **47.** Sample answer: $x - 1$, $x^2 - x - 1$. $(x - 1)(x^2 - x - 1) = x^3 - 2x^2 + 1$ **49.** The Distributive Property can be used with a vertical or horizontal format by

distributing, multiplying, and combining like terms. The FOIL method is used with a horizontal format. You multiply the first, outer, inner, and last terms of the binomials and then combine like terms. A rectangular method can also be used by writing the terms of the polynomials along the top and left side of a rectangle and then multiplying the terms and combining like terms.
51. F **53.** $\frac{3}{2}$ **55.** $4a^2 + 5$ **57.** $3n^3 - 6n^2 + 10$ **59.** $4b + c + 2$ **61.** $-7m^3 - 3m^2 - m + 17$ **63.** $-56t^{12}$ **65.** $50y^6 - 27y^9$

Pages 455–457 Lesson 7-8

1. $x^2 + 10x + 25$
3 $(2x + 7y)^2 = (2x)^2 + 2(2x)(7y) + (7y)^2$
$= 4x^2 + 28xy + 49y^2$
5. $g^2 - 8gh + 16h^2$ **7a.** $0.5Dy + 0.5y^2$ **7b.** 50% **9.** $x^2 - 25$ **11.** $81t^2 - 36$ **13.** $b^2 - 12b + 36$ **15.** $x^2 + 12x + 36$ **17.** $81 - 36y + 4y^2$ **19.** $25t^2 - 20t + 4$ **21a.** $(T + t)^2 = T^2 + 2Tt + t^2$
21b. *TT*: 25%; *Tt*: 50%; *tt*: 25%
23 $(b + 7)(b - 7) = b^2 - (7)^2$
$= b^2 - 49$
25. $16 - x^2$ **27.** $9a^4 - 49b^2$ **29.** $64 - 160a + 100a^2$ **31.** $9t^2 - 144$ **33.** $9q^2 - 30qr + 25r^2$ **35.** $g^2 + 10gh + 25h^2$ **37.** $9a^8 - b^2$ **39.** $64a^4 - 81b^6$ **41.** $\frac{4}{25}y^2 - \frac{16}{5}y + 16$ **43.** $4m^3 + 16m^2 - 9m - 36$ **45.** $2x^2 + 2x + 5$ **47.** $6x + 3$ **49.** $c^3 + 3c^2d + 3cd^2 + d^3$ **51.** $f^3 + f^2g - fg^2 - g^3$ **53.** $n^3 - n^2p - np^2 + p^3$
55 **a.** $A = 3.14(r + 9)^2$
$= 3.14(r^2 + 18r + 81)$
$\approx (3.14r^2 + 56.52r + 254.34)\text{ft}^2$
b. $38^2 - (3.14r^2 + 56.52r + 254.34)$
$= 1444 - 3.14r^2 - 56.52r - 254.34$
$\approx (1189.66 - 3.14r^2 - 56.52r)\text{ft}^2$
57. Sample answer: $(2c + d)(2c - d)$; The product of these binomials is a difference of two squares and does not have a middle term. The other three do. **59.** 81 **61.** Sample answer: To find the square of a sum, apply the FOIL method or apply the pattern. The square of the sum of two quantities is the first quantity squared plus two times the product of the two quantities plus the second quantity squared. The square of the difference of two quantities is the first quantity squared minus two times the product of the two quantities plus the second quantity squared. The product of the sum and difference of two quantities is the square of the first quantity minus the square of the second quantity. **63.** D **65.** C **67.** $2c^2 + 5c - 3$ **69.** $8h^2 - 34h + 21$ **71.** $40m^2 + 47m + 12$ **73.** $3c^2 - 2c$ **75.** $-13d^2 - 18d$ **77.** $19p^2 - 18p$ **79.** (2, 0) **81.** $t < 18$ or $t > 22$ **83.** $y = \frac{1}{5}x + 6$ **85.** 15 lb **87.** $2^3 \cdot 3 \cdot 5$ **89.** $3 \cdot 5 \cdot 11$

Pages 459–462 Chapter 7 Study Guide and Review

1. binomial **3.** trinomial **5.** power of a power **7.** scientific notation **9.** polynomial **11.** x^9 **13.** $20a^6b^6$ **15.** $64r^{18}t^6$ **17.** $8x^{15}$ **19.** $45\pi x^4$ **21.** $\left(\frac{27x^3y^9}{8z^3}\right)$ **23.** $\frac{c^6}{a^3}$ **25.** x^6 **27.** $\frac{6}{yx^3}$ **29.** 2.3×10^6 **31.** about 9.1×10^{-2} **33.** $-x^4 + 1$ **35.** $3x^5 + x^3 - 2x^2 + 6x - 2$ **37.** $-2x^3 - 3$ **39.** $-x^2 - x + 6$ **41.** $4x^2 + 4x + 8$ **43.** 1 **45.** 2 **47.** $x^2 + 4x - 21$ **49.** $6r^2 + rt - 35t^2$ **51.** $10x^2 + 7x - 12$ **53.** $9x^2 - 12x + 4$ **55.** $4x^2 - 9$ **57.** $9m^2 - 4$

Chapter 8 Factoring and Quadratic Equations

Page 469 Chapter 8 Get Ready

1. $a(a) + a(5)$; $a^2 + 5a$ **3.** $n(n) + n(-3n^2) + n(2)$; $n^2 - 3n^3 + 2n$ **5.** $5(9 + 3 + 6)$; \$90 **7.** $x^2 + 3x - 4$ **9.** $3x^2 + 11x - 20$ **11.** $54a^2 - 12ab - 2b^2$ **13.** $9 - 6a + a^2$ **15.** $9x^2 - 12xy + 4y^2$ **17.** $x^2 - 36$ in^2

Pages 472–474 Lesson 8-1

1. $2 \cdot 2 \cdot 3 \cdot g \cdot g \cdot h \cdot h \cdot h \cdot h$ **3.** $-1 \cdot 17 \cdot x \cdot x \cdot x \cdot y \cdot y \cdot z$ **5.** $24cd$ **7.** xy^3 **9.** 1 in.
11 $-35a^3c^2 = -1 \cdot 35a^3c^2$
$= -1 \cdot 5 \cdot 7 \cdot a \cdot a \cdot a \cdot c \cdot c$
13. $3 \cdot 3 \cdot 3 \cdot 3 \cdot n \cdot n \cdot n \cdot n \cdot n \cdot p$ **15.** $11 \cdot 11 \cdot a \cdot b \cdot c \cdot c \cdot c$ **17.** $2z$ **19.** $4r$ **21.** $5t$ **23.** height 1 in., base 56 in.; height 2 in., base 28 in.; height 4 in., base 14 in.; height 28 in., base 2 in.; height 14 in., base 4 in.; height 7 in., base 8 in.; height 56 in., base 1 in.; height 8 in., base 7 in.
25 $80 = 5 \cdot 16, 8 \cdot 10, 10 \cdot 8, 16 \cdot 5, 20 \cdot 4$
So, the DVDs can be arranged as 5 DVDs on 16 shelves, 8 DVDs on 10 shelves, 10 DVDs on 8 shelves, 16 DVDs on 5 shelves, 20 DVDs on 4 shelves.
27. 11, 13; 17, 19; 29, 31; 41, 43; 59, 61 **29.** 22 and 33 **31.** False; sample answer: The monomials $99x^5y^{11}z^{30}$ and $101abc$ have a GCF of 1. **33.** Sample answer: $6y^3$, $12y^4$, $18y^5$; 6 is the greatest numerical factor that all three monomials have in common, and y^3 is the highest power of y that they all have in common. **35.** A **37.** B **39.** $a^2 - 8a + 16$ **41.** $z^2 - 10z + 25$ **43.** $y^2 + 4y + 4$ **45.** $2m^2 + 5m - 12$ **47.** $t^2 + 11t + 18$ **49.** $p^2 + 6pq + 9q^2$ **51.** (2, 7) **53.** (−4, 3) **55.** $8x - 14$ **57.** $-6h^2 + h$ **59.** $5(y - 2)$

Pages 479–482 Lesson 8-2

1. $3(7b - 5a)$ **3.** $gh(10gh + 9h - g)$
5 $np + 2n + 8p + 16 = (np + 2n) + (8p + 16)$
$= n(p + 2) + 8(p + 2)$
$= (n + 8)(p + 2)$
7. $(b + 5)(3c - 2)$ **9.** 0, −10 **11.** 0, $\frac{3}{4}$

13a. 0, 2.08125 **13b.** 17.3 ft, 2.6 ft **15.** $8(2t - 5y)$
17. $2k(k + 2)$ **19.** $2ab(2ab + a - 5b)$

21 $fg - 5g + 4f - 20 = (fg - 5g) + (4f - 20)$
$= g(f - 5) + 4(f - 5)$
$= (g + 4)(f - 5)$

23. $(h + 5)(j - 2)$ **25.** $(9q - 10)(5p - 3)$
27. $(3d - 5)(t - 7)$ **29.** $(3t - 5)(7h - 1)$
31. $(r - 5)(5b + 2)$ **33.** $gf(5f + g + 15)$
35. $3cd(9d - 6cd + 1)$ **37.** $2(8u - 15)(3t + 2)$
39. 0, 3 **41.** $-\frac{1}{2}, -2$ **43.** 0, −3 **45a.** ab
45b. $(a + 6)(b + 6)$ **45c.** $6(a + b + 6)$

47a.

x	y
0	0
1	9
2	12
3	9
4	0

47b.

47c. 12 ft

49 $h = 64t - 16t^2$
$0 = 64t - 16t^2$
$0 = 16t(4 - t)$
$16t = 0$ or $4 - t = 0$
$t = 0$ or $4 = t$
So, the arrow hits the ground after 4 seconds.

51a. 3 and −2

51b.

x^2	$+3x$
$-2x$	-6

51c.

	x	$+3$
x	x^2	$+3x$
-2	$-2x$	-6

$(x + 3)(x - 2)$

51d. Sample answer: Place x^2 in the top left-hand corner and place −40 in the lower right-hand corner. Then determine which two factors have a product of −40 and a sum of −3. Then place these factors in the box. Then find the factor of each row and column. The factors will be listed on the very top and far left of the box. **53.** If $a = 0$ and $b = 0$, then all real numbers are solutions. If $a \neq 0$, then the solutions are $-\frac{b}{a}$ and $\frac{b}{a}$. **55.** Sample answer: $a = 0$ or $a = b$ for any real values of a and b. **57.** D **59.** 350 **61.** 5 **63.** $8x$ **65.** $8c^2d$ **67.** $0.5Bb + 0.5b^2; \frac{1}{2}$ **69.** p^7r^5 **71.** $81x^2y^{14}$ **73.** 16,777,216 **75.** $\{y \mid y > -11\}$ **77.** $\{k \mid k > -9\}$ **79.** $\{z \mid z \geq -48\}$

81. $a^2 + 7a + 10$ **83.** $z^2 - 9z + 8$ **85.** $x^2 - 13x + 42$

Pages 489–491 Lesson 8-3

1. $(x + 2)(x + 12)$ **3.** $(n + 7)(n - 3)$ **5.** −3, 7
7. 6, 9 **9.** −8, 9 **11.** 8 in. by 12 in. **13.** $(y - 9)(y - 8)$ **15.** $(n - 7)(n + 5)$ **17.** $(x - 2)(x - 20)$
19. $(m + 6)(m - 7)$

21 $y^2 + y = 20$
$y^2 + y - 20 = 0$
$(y + 5)(y - 4) = 0$
$y + 5 = 0$ or $y - 4 = 0$
$y = -5$ $y = 4$

23. −2, −9 **25.** 2, 16 **27.** −4, −14 **29.** 4, 12
31. $(x - 6)$ ft **33.** $(q + 2r)(q + 9r)$ **35.** $(x - y)(x - 5y)$

37 **a.** Sample answer: Let w represent the width of the swimming pool. So, the length of the pool is $w + 20$.
$A = \ell w$
$= (w + 20)w$
$525 = (w + 20)w$

b. $525 = (w + 20)w$
$525 = w^2 + 20w$
$0 = w^2 + 20w - 525$
$0 = (w + 35)(w - 15)$
$w + 35 = 0$ or $w - 15 = 0$
$w = -35$ $w = 15$

c. The solution of 15 means that the width is 15 feet and the length is 35 feet. The solution −35 does not make sense because length cannot be negative.

39. $4x - 26$ **41.** Charles; Jerome's answer once multiplied is $x^2 - 6x - 16$. The middle term should be positive. **43.** −15, −9, 9, 15 **45.** 4, 6 **47.** $x^2 + 19x - 20$; $(x - 1)(x + 20)$ **49.** Sample answer: Find factors m and n such that $m + n = b$ and $mn = c$. If b and c are positive, then m and n are positive. If b is negative and c is positive, then m and n are negative. When c is negative, m and n have different signs and the factor with the greater absolute value has the same sign as b.
51. 204 **53.** A **55.** $11x(1 + 4xy)$
57. $(2x + b)(a + 3c)$ **59.** $(x - y)(x - y)$
61. $\begin{bmatrix} 12 & -9 & 12 \\ 1 & -3 & -10 \end{bmatrix}$ **63.** impossible **65.** about 6 ft
67. $(3x - 4)(a - 2b)$

Pages 496–498 Lesson 8-4

1. $(3x + 2)(x + 5)$ **3.** prime **5.** $-\frac{3}{2}, -3$ **7.** $\frac{4}{3}, 2$
9a. 6 ft **9b.** 6 seconds

11 $2x^2 + 19x + 24$

factors of 48	sum of 19
3, 16	19

$2x^2 + 3x + 16x + 24 = (2x^2 + 3x) + (16x + 24)$

$= x(2x + 3) + 8(2x + 3)$
$= (x + 8)(2x + 3)$

13. $2(2x + 5)(x + 7)$ **15.** $(4x - 5)(x - 2)$ **17.** prime **19.** prime **21.** prime **23.** $\frac{3}{2}, -6$ **25.** $\frac{2}{3}, 8$ **27.** $-\frac{1}{3}, 2$ **29a.** $10 = -16t^2 + 20t + 6$ **29b.** 1 sec **29c.** Less; sample answer: It starts closer to the ground so the shot will not have as far to fall.

31 Words: 6 times the square of a number plus 11 times the number equals 2

Equation: $6 \cdot x^2 + 11 \cdot x = 2$

$6x^2 + 11x = 2$

$6x^2 + 11x - 2 = 0$

factors of −12	sum of 11
−1, 12	11

$$6x^2 + 11x - 2 = 0$$
$$6x^2 - 1x + 12x - 2 = 0$$
$$(6x^2 - 1x) + (12x - 2) = 0$$
$$x(6x - 1) + 2(6x - 1) = 0$$
$$(x + 2)(6x - 1) = 0$$
$$x + 2 = 0 \quad \text{or} \quad 6x - 1 = 0$$
$$x = -2 \qquad 6x = 1$$
$$x = \frac{1}{6}$$

The numbers are −2 and $\frac{1}{6}$.

33. $-(x + 2)(4x + 7)$ **35.** $-(2x - 7)(3x - 5)$ **37.** $-(3x - 4)(4x + 5)$ **39a.** a^2 and b^2 **39b.** $a^2 - b^2$ **39c.** width: $a - b$, length: $a + b$ **39d.** $(a - b)(a + b)$ **39e.** $(a - b)(a + b)$; the figure with area $a^2 - b^2$ and the rectangle with area $(a - b)(a + b)$ have the same area, so $a^2 - b^2 = (a - b)(a + b)$. **41.** $(12x + 20y)$ in.; The area of the square equals $(3x + 5y)(3x + 5y)$ in^2, so the length of one side is $(3x + 5y)$ in. The perimeter is $4(3x + 5y)$ or $(12x + 20y)$ in. **43.** Sample answer: $10x^2 + x - 3 = 0$; The polynomial factors into $(2x - 1)(5x + 3) = 0$, so the solutions are $\frac{1}{2}$ and $-\frac{3}{5}$. **45.** 6 **47.** J **49.** $(x - 2)(x - 7)$ **51.** $(x + 3)(x - 8)$ **53.** $(r + 8)(r - 5)$ **55.** {0, 9} **57.** {0, 2} **59.** {0, 4}

61.

Sample answers: 2 light, 8 dark; 6 light, 8 dark; 7 light, 4 dark

63. $\{d \mid d \leq 5 \text{ or } d > 7\}$

65. ∅

67. $\{y \mid 3 < y < 6\}$

69. 4 **71.** 8 **73.** 11

Pages 501–504 Lesson 8-5

1. $(x + 3)(x - 3)$ **3.** $9(m + 4)(m - 4)$ **5.** $(u^2 + 9)(u + 3)(u - 3)$

7 $20r^4 - 45n^4 = 5(4r^4 - 9n^4)$
$= 5\left((2r^2)^2 - (3n^2)^2\right)$
$= 5(2r^2 + 3n^2)(2r^2 - 3n^2)$

9. $(c + 1)(c - 1)(2c + 3)$ **11.** $(t + 4)(t - 4)(3t + 2)$ **13.** 36 mph **15.** $(q + 11)(q - 11)$ **17.** $6(n^2 + 1)(n + 1)(n - 1)$ **19.** $(r + 3t)(r - 3t)$ **21.** $h(h + 10)(h - 10)$ **23.** $(x + 9)(x - 9)(2x - 1)$ **25.** $7(h^2 + p^2)(h + p)(h - p)$ **27.** $6k^2(h^2 + 3k)(h^2 - 3k)$ **29.** $(f + 8)(f - 8)(f + 2)$ **31.** $10q(q + 11)(q - 11)$ **33.** $p^3r(r + 1)(r - 1)(r^2 + 1)$ **35.** $(r + 10)(r - 10)(r - 5)$ **37.** $(a + 7)(a - 7)$ **39.** $3(m^4 + 81)$ **41.** $2(a + 4)(a - 4)(6a + 1)$ **43.** $3(m + 5)(m - 5)(5m + 4)$ **45a.** $-0.5x(x - 9)$ **45b.** 9 ft **45c.** 10.125 ft

47 **a.** $S = -25m^2 + 125m$
$0 = -25m^2 + 125m$
$0 = -25m(m - 5)$
$0 = -25m \quad \text{or} \quad 0 = m - 5$
$0 = m \qquad 5 = m$

So, they will stop selling in month 5.

b. The peak will occur halfway between 0 and 5, or 2.5.

c. The peak amount is $S = -25(2.5)^2 + 125(2.5)$
$= -156.25 + 312.5$
$= 156.25$

The peak is 156,250 copies.

49 $100 = 25x^2$
$0 = 25x^2 - 100$
$0 = 25(x^2 - 4)$
$0 = 25(x + 2)(x - 2)$
$x + 2 = 0 \quad \text{or} \quad x - 2 = 0$
$x = -2 \qquad x = 2$

51. $\frac{3}{8}, -\frac{3}{8}$ **53.** −45, 45 **55.** $\frac{3}{16}, -\frac{3}{16}$ **57.** Lorenzo; sample answer: Checking Elizabeth's answer gives us $16x^2 - 25y^2$. The exponent on x in the final product should be 4. **59.** $(x^4 - 3)(x^4 + 3) \cdot (x^8 + 9)$ **61.** false; $a^2 + b^2$ **63.** When the difference of squares pattern is multiplied together using the FOIL method, the outer and inner terms are opposites of each other. When these terms are added together, the sum is zero. **65.** G **67a.** Car A, because it is traveling at 65 mph, and Car B is traveling at 60 mph. **67b.** $5t - 10$ **67c.** 2.5 mi **69.** prime **71.** {3, 6} **73.** {6, 16}

75. $\{t \mid t \geq 4\}$

77. $\{k \mid k > 4\}$

−1 0 1 2 3 4 5 6 7 8 9

79. $\{m \mid m \geq 3\}$

−1 0 1 2 3 4 5 6 7 8 9

81. the seventh week **83.** $x^2 - 4x + 4$
85. $4x^2 - 20x + 25$ **87.** $16x^2 + 40x + 25$

Pages 509–512 Lesson 8-6

1. yes; $(5x + 6)^2$ **3.** $(x - 4)(2x + 7)$ **5.** $4(x^2 + 16)$
7. ± 3 **9.** $\frac{3}{8}$ **11.** 0.6 second **13.** yes; $(4x - 7)^2$
15 Since the last term is not a perfect square, the trinomial is not a perfect square trinomial
17. prime **19.** $8(y - 5z)(y + 5z)$
21. $2m(2m - 7)(3m + 5)$ **23.** $3(2x - 7)^2$
25. $3p(2p + 1)(2p - 1)$ **27.** $2t(t + 6)(2t - 7)$
29. $2a(a - b)(b + 1)(b - 1)$ **31.** $3k(k - 4)(k - 4)$
33. prime
35 $(y - 4)^2 = 7$
$y - 4 = \pm\sqrt{7}$
$y = 4 \pm \sqrt{7}$
37. $\frac{3}{4}$ **39.** 6 **41.** $\frac{1}{3}$ **43.** $8 \pm \sqrt{6}$ **45.** 20 ft
47. $|4x + 5|$ **49a.** $w^3 + 14w^2 + 48w$ **49b.** 4 in. wide by 10 in. long by 12 in. high
51 **a.** 42 in. = 3.5 ft., width = w, length = $w + 5$, height = 3.5
Area of the surface = $\ell \cdot w$ or $(w + 5)(w)$
$V = \ell \cdot w \cdot h$
$1750 = (w + 5)(w)(3.5)$
$500 = (w + 5)(w)$
Area of the surface is 500 ft^2
b. $500 = (w + 5)(w)$
$= w^2 + 5w$
$0 = w^2 + 5w - 500$
$= (w + 25)(w - 20)$
$0 = w + 25$ or $0 = w - 20$
$-25 = w$ $20 = w$
The dimensions are 20 ft by 25 ft by 42 in.
c. Because the volume is doubled, we can double any one of the dimensions. 40 ft by 25 ft by 42 in., 20 ft by 50 ft by 42 in., or 20 ft by 25 ft by 84 in.
d. Because 2 of the dimensions are doubled the volume is increased by a factor of 4. The ratio is 1:4
53. Adriano; sample answer: Debbie did not factor the expression completely. **55.** Sample answer: $x^2 - 3x + \frac{9}{4} = 0; \left\{\frac{3}{2}\right\}$ **57.** First look for a GCF in all the terms and factor the GCF out of all the terms. Then, if the polynomial has two terms, check if the terms are the differences of squares and factor if so. If the polynomial has three terms, check if the polynomial is a perfect square polynomial and factor if so. If the polynomial has four or more terms, factor by grouping. If the polynomial does not have a GCF and cannot be factored, the polynomial is a prime polynomial. **59.** Sample answer: $x^4 - 1$; 1, −1
61. B **63.** H **65.** $(x - 4)(x + 4)$ **67.** $(1 - 10p)(1 + 10p)$ **69.** $(5n - 1)(5n + 1)$ **71.** $\{-2, 4\}$ **73.** $\{-2, 1\}$
75. $\{2, 3\}$ **77.** 10^1 or 10 **79.** $\frac{1}{10{,}000}$ **81.** $-\frac{2}{3}$ **83.** $\frac{3}{8}$
85. undefined

Pages 513–516 Chapter 8 Study Guide and Review

1. false; sample answer: $x^2 + 5x + 7$ **3.** true
5. true **7.** true **9.** false; difference of squares
11. $2 \cdot 2 \cdot 7 \cdot x \cdot x \cdot x$ **13.** $2 \cdot 2 \cdot 17 \cdot c \cdot d \cdot d \cdot d$
15. 11 **17.** $6ab$ **19.** 12 by 12 **21.** $7xy(2x - 3 + 5y)$
23. $(a + b)(a - 4c)$ **25.** $(3a + 5b)(8m - 3n)$ **27.** 0, 2
29. $0, \frac{5}{2}$ **31.** $(x - 5)(x - 3)$ **33.** $(x - 6)(x + 1)$
35. −10, 5 **37.** −8, −4 **39.** −10, −1
41. $2(2x - 1)(3x + 7)$ **43.** $3(x - 5)(x + 3)$
45. $\frac{3}{4}, -\frac{4}{5}$ **47.** $2, \frac{1}{4}$ **49.** $3x + 7$
51. $(8 + 5x)(8 - 5x)$ **53.** $3(x + 1)(x - 1)$
55. $\frac{5}{3}, -\frac{5}{3}$ **57.** −5, 5 **59.** $(x + 6)^2$ **61.** $(3y - 2)^2$
63. $(x^2 + 1)(x + 1)(x - 1)$ **65.** 16, −6 **67.** −4, 4
69. 2.5 ft

Chapter 9 Quadratic and Exponential Functions

Page 523 Chapter 9 Get Ready

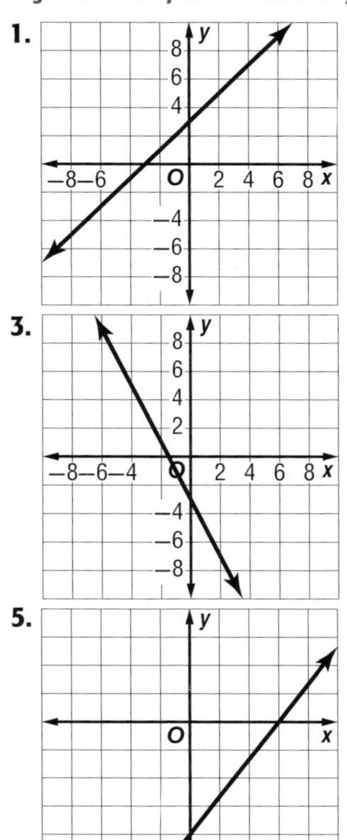

7.

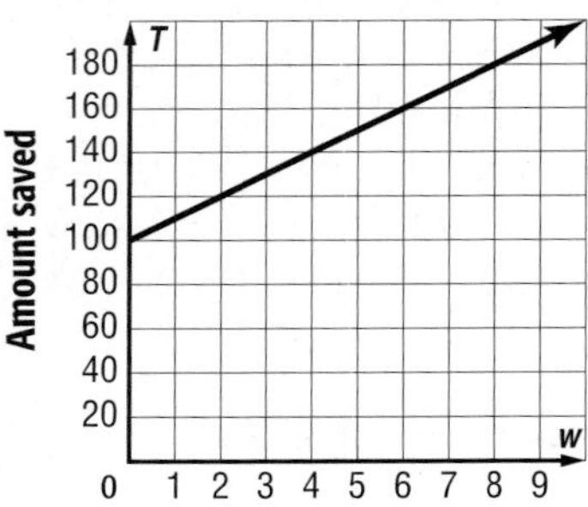

9. no **11.** yes; $(x + 10)^2$ **13.** yes; $(k - 8)^2$ **15.** no **17.** 34, 42, 50 **19.** 11, 13, 15

Pages 531–535 Lesson 9-1

1.

x	y
−3	0
−2	−6
−1	−8
0	−6
1	0
2	10

D = {all real numbers}; R = $\{y \mid y \geq -8\}$

3.

x	y
−1	4
0	−3
1	−8
2	−11
3	−12
4	−11
5	−8
6	−3
7	4

D = {all real numbers}; R = $\{y \mid y \geq -12\}$

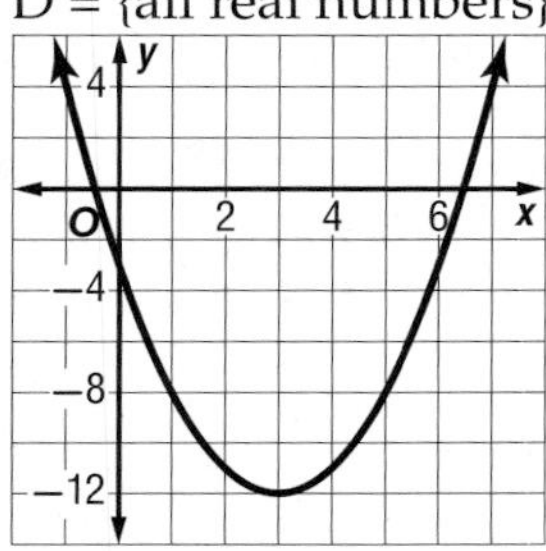

5. vertex (−1, 5), axis of symmetry $x = -1$, y-intercept 3 **7.** vertex (−2, −12), axis of symmetry $x = -2$, y-intercept −4 **9.** vertex (1, 2), axis of symmetry $x = 1$, y-intercept −1 **11.** vertex (2, 1), axis of symmetry $x = 2$, y-intercept 5

13 a. Since the a value is −1, the graph opens downward and has a maximum.

b. In this equation $a = -1$, $b = 4$, and $c = -3$.

$$x = \frac{-b}{2a}$$
$$= \frac{-4}{2(-1)}$$
$$= 2$$

To find the vertex, use the value you found for the x-coordinate of the vertex. To find the y-coordinate, substitute the value for x in the original equation.

$$y = -x^2 + 4x - 3$$
$$= -(2)^2 + 4(2) - 3$$
$$= -4 + 8 - 3$$
$$= 1$$

The maximum is at (2, 1).

c. The domain of the function is D = $\{x \mid x \text{ is all real numbers}\}$. The range is R = $\{y \mid y \leq 1\}$.

15a. maximum **15b.** 6 **15c.** D = {all real numbers}; R = $\{y \mid y \leq 6\}$

17.

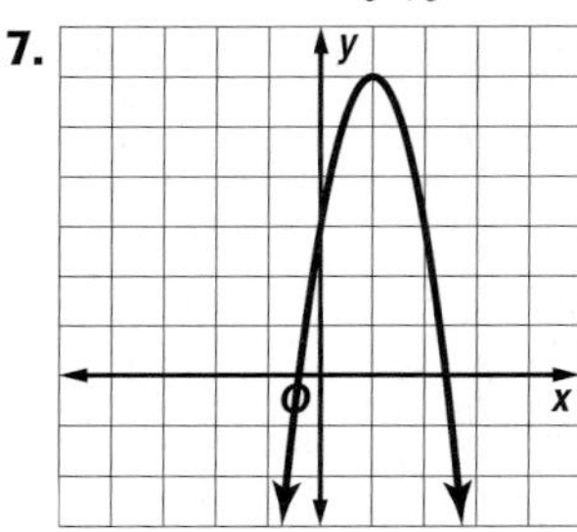

19.

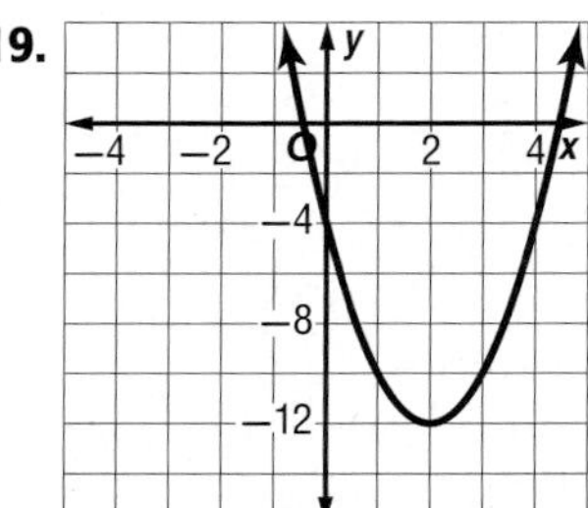

21a.

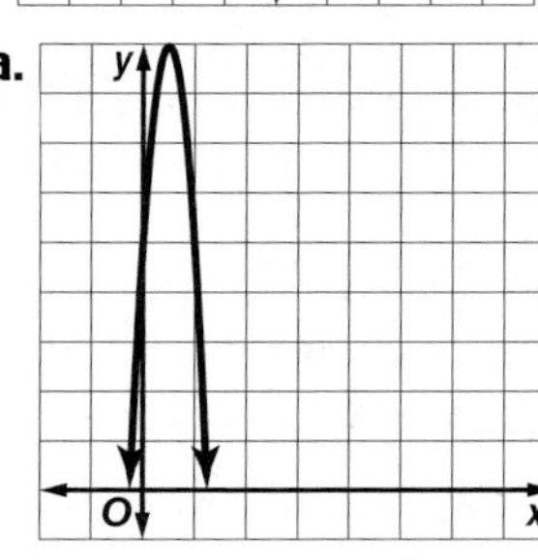

21b. 5 ft **21c.** 9 ft

23.

x	y
−3	13
−2	7
−1	5
0	7
1	13

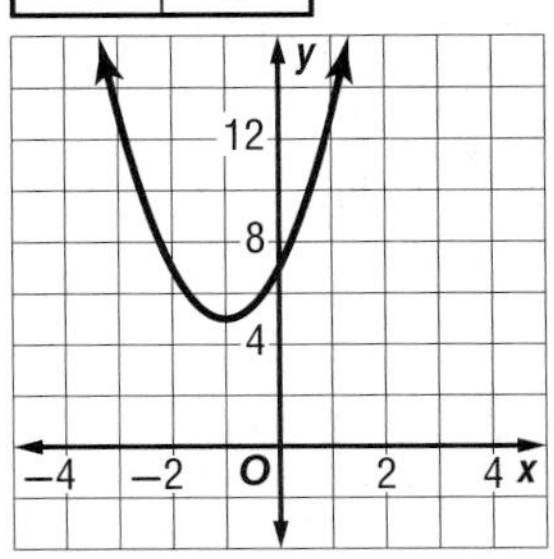

D = {all real numbers}; R = $\{y \mid y \geq 5\}$

25.

x	y
0	5
−1	−4
−2	−7
−3	−4
−4	5

D = {all real numbers}; R = $\{y \mid y \geq -7\}$

27.

x	y
3	2
2	−1
1	−2
0	−1
−1	2

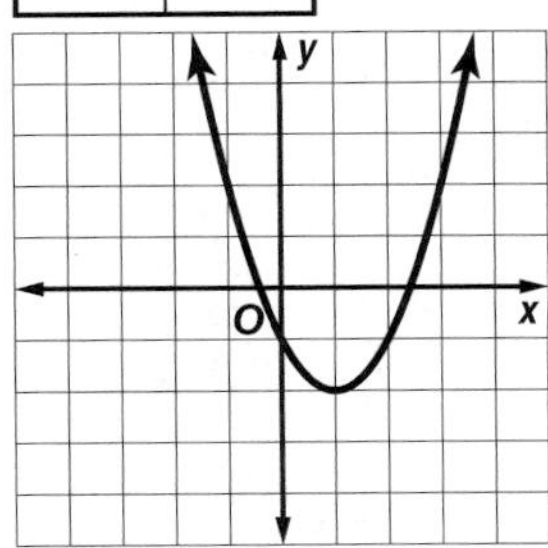

D = {all real numbers}; R = $\{y \mid y \geq -2\}$

29. vertex (0, 1), axis of symmetry $x = 0$, y-intercept 1 **31.** vertex (1, 1), axis of symmetry $x = 1$, y-intercept 4 **33.** vertex (0, 0), axis of symmetry $x = 0$, y-intercept 0

35 In this equation $a = 2$, $b = 12$, and $c = 10$.

$x = \frac{-b}{2a} \qquad = \frac{-12}{2(2)}$

The equation for the axis of symmetry is $x = -3$.

To find the vertex, use the value you found for the axis of symmetry as the x-coordinate of the vertex. To find the y-coordinate, substitute the value for x in the original equation.

$y = 2x^2 + 12x + 10$
$= 2(-3)^2 + 12(-3) + 10$
$= -8$

The vertex is at (−3, −8).

The y-intercept occurs at $(0, c)$. So, in this case, the y-intercept occurs is 10.

37. vertex (−3, 4), axis of symmetry $x = -3$, y-intercept −5 **39.** vertex (2, −14), axis of symmetry $x = 2$, y-intercept 14 **41.** vertex (1, −15), axis of symmetry $x = 1$, y-intercept −18 **43a.** maximum **43b.** 9 **43c.** D = {all real numbers}, R = $\{y \mid y \leq 9\}$ **45a.** minimum **45b.** −48 **45c.** D = {all real numbers}, R = $\{y \mid y \geq -48\}$ **47a.** maximum **47b.** 33 **47c.** D = {all real numbers}, R = $\{y \mid y \leq 33\}$ **49a.** maximum **49b.** 4 **49c.** D = {all real numbers}, R = $\{y \mid y \leq 4\}$ **51a.** maximum **51b.** 3 **51c.** D = {all real numbers}, R = $\{y \mid y \leq 3\}$

53.

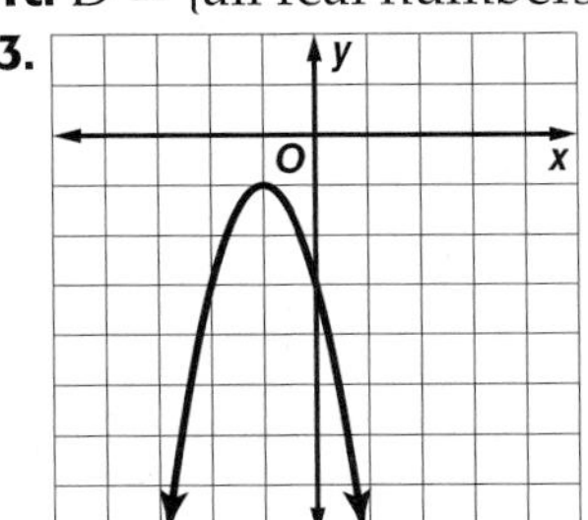

55.

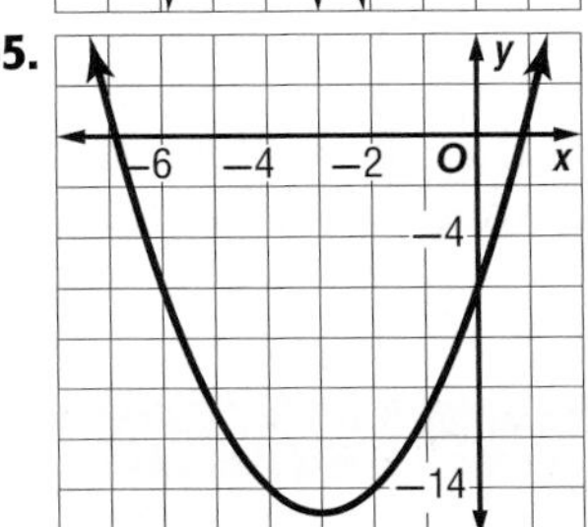

57.

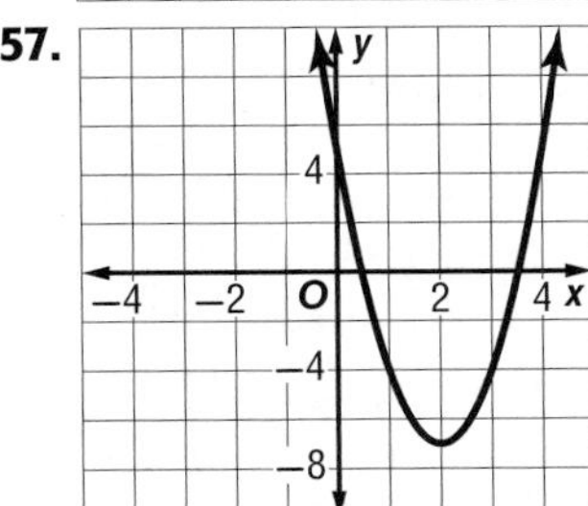

59.

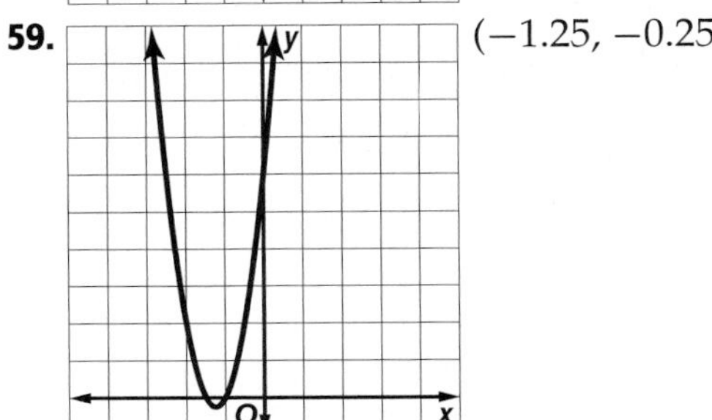

(−1.25, −0.25)

61. 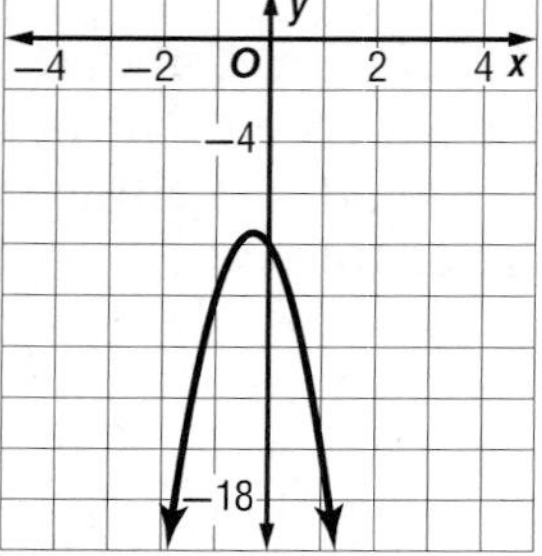

(−0.3, −7.55)

63a. 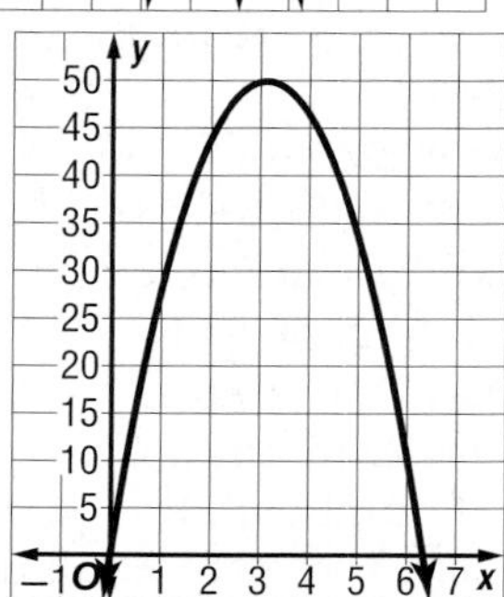

63b. 0 m **63c.** ≈ 50.0 m **63d.** ≈6.4 seconds
63e. D = $\{x \mid 0 \le x \le 6.4\}$; R = $\{y \mid 0 \le y \le 50.0\}$

65 a. $h = -16t^2 + 90t$
$= -16(1)^2 + 90(1)$
$= 74$ ft

b. $126 = -16t^2 - 90t$
$0 = -16t^2 - 90t - 126$
$0 = (t - 3)(-16t + 42)$
$t = 3$ and $t = 2.625$

c. $h = -16t^2 + 90t$
$0 = -16t^2 + 90t$
$0 = -16t(t - 5)$
$t = 0$ and $t = 5.625$

These represent the time that the ball leaves the ground initially and the time it returns to the ground.

67 a.

Equation	Related Function	Zeros	y-Values
$x^2 - x = 12$	$y = x^2 - x - 12$	−3, 4	−3: 8, −6; 4: −6, 8
$x^2 + 8x = 9$	$y = x^2 + 8x - 9$	−9, 1	−9: 11, −9; 1: −9, 11
$x^2 = 14x - 24$	$y = x^2 - 14x + 24$	2, 12	2: 11, −9; 12: −9, 11
$x^2 + 16x = -28$	$y = x^2 + 16x + 28$	−14, −2	−14: 13, −11; −2: −11, 3

b. Use a graphing calculator to graph.

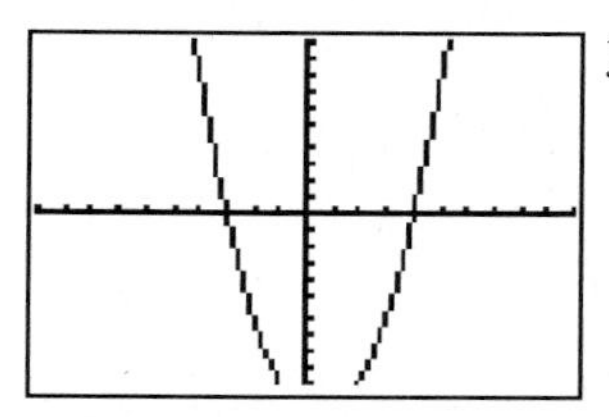
$y = x^2 - x - 12$

[−10, 10] scl: 1 by [−10, 10] scl: 1

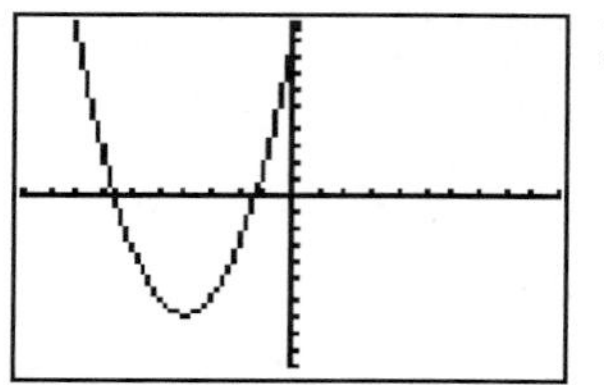
$y = x^2 - 8x - 9$

[−15, 15] scl: 2 by [−30, 30] scl: 5

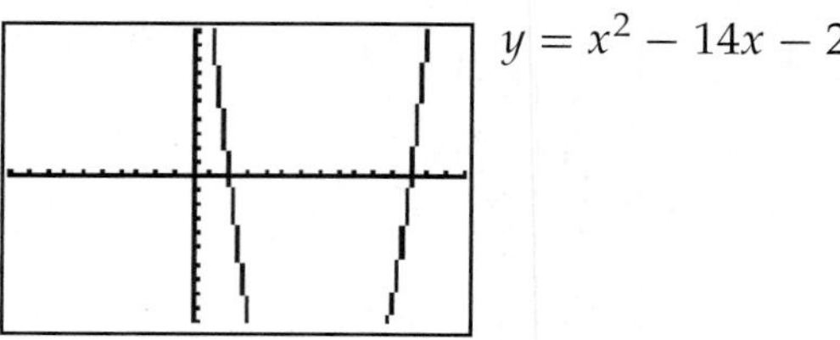
$y = x^2 - 14x - 24$

[−10, 15] scl: 1 by [−10, 10] scl: 1

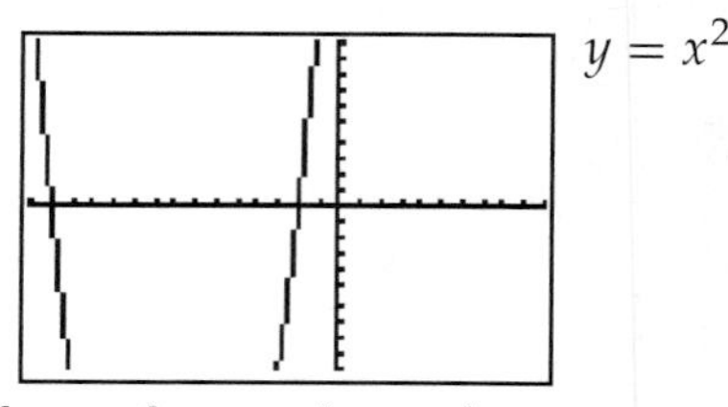
$y = x^2 - 16x - 28$

[−15, 10] scl: 1 by [−10, 10] scl: 1

c. Use the table function on the calculator to identify the zeros. The zeros are the values where y is 0. Also list the y-values that are to the left and right of the zeros.

d. The function values have opposite signs just before and just after the zeros.

69. Jade; Chase forgot the negative sign with the −4. **71.** (−1, 9); Sample answer: I graphed the points given, and sketched the parabola that goes through them. I counted the spaces over and up from the vertex to the other point and did the same on the opposite side. **73.** Sample answer: The function $y = -x^2 - 4$ has a vertex at (0, −4), but it is a maximum.

75. C **77.** D **79.** yes; $(2x + 1)^2$ **81.** no **83.** prime **85.** $b^2 - 4b - 21$ **87.** $2x^2 + 17x - 9$ **89.** (2, 1) **91.** (4, 2) **93.** 10 **95.** −6

Pages 540-542 Lesson 9-2

1. 2, −5

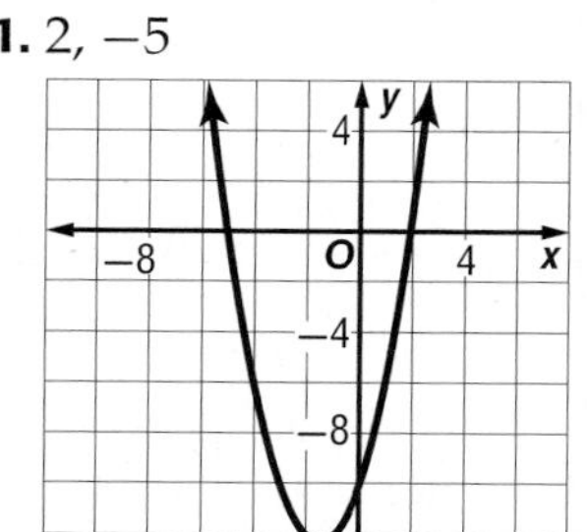

3. −2

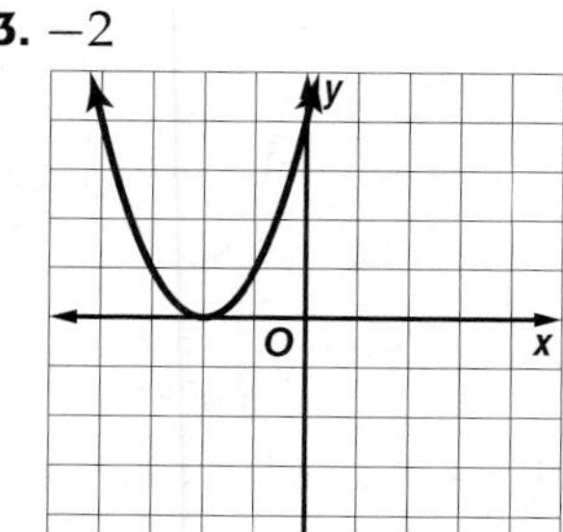

23.

x	y
−3	13
−2	7
−1	5
0	7
1	13

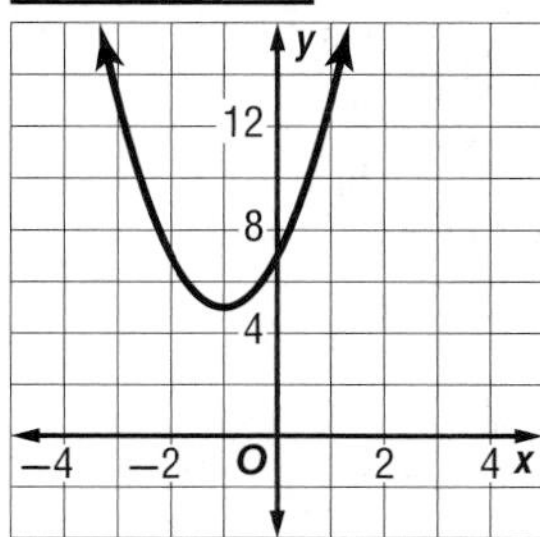

D = {all real numbers}; R = $\{y \mid y \geq 5\}$

25.

x	y
0	5
−1	−4
−2	−7
−3	−4
−4	5

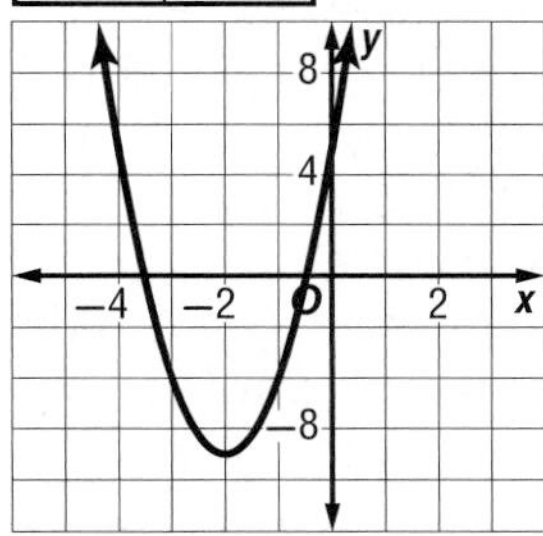

D = {all real numbers}; R = $\{y \mid y \geq -7\}$

27.

x	y
3	2
2	−1
1	−2
0	−1
−1	2

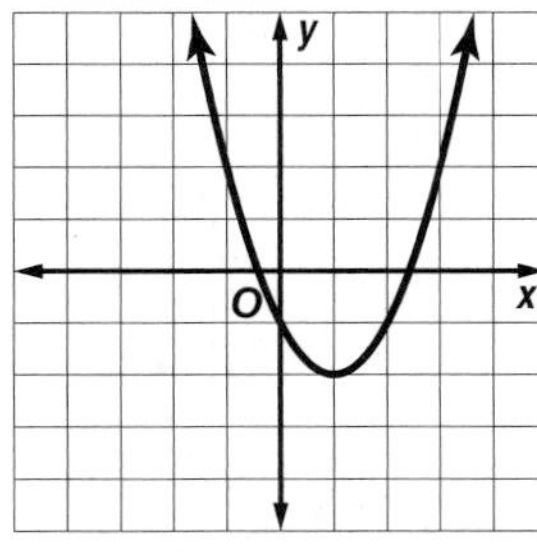

D = {all real numbers}; R = $\{y \mid y \geq -2\}$

29. vertex (0, 1), axis of symmetry $x = 0$, y-intercept 1 **31.** vertex (1, 1), axis of symmetry $x = 1$, y-intercept 4 **33.** vertex (0, 0), axis of symmetry $x = 0$, y-intercept 0

35 In this equation $a = 2$, $b = 12$, and $c = 10$.

$x = \frac{-b}{2a}$ $\qquad = \frac{-12}{2(2)}$

The equation for the axis of symmetry is $x = -3$.

To find the vertex, use the value you found for the axis of symmetry as the x-coordinate of the vertex. To find the y-coordinate, substitute the value for x in the original equation.

$y = 2x^2 + 12x + 10$
$= 2(-3)^2 + 12(-3) + 10$
$= -8$

The vertex is at (−3, −8).

The y-intercept occurs at (0, c). So, in this case, the y-intercept occurs is 10.

37. vertex (−3, 4), axis of symmetry $x = -3$, y-intercept −5 **39.** vertex (2, −14), axis of symmetry $x = 2$, y-intercept 14 **41.** vertex (1, −15), axis of symmetry $x = 1$, y-intercept −18 **43a.** maximum **43b.** 9 **43c.** D = {all real numbers}, R = $\{y \mid y \leq 9\}$ **45a.** minimum **45b.** −48 **45c.** D = {all real numbers}, R = $\{y \mid y \geq -48\}$ **47a.** maximum **47b.** 33 **47c.** D = {all real numbers}, R = $\{y \mid y \leq 33\}$ **49a.** maximum **49b.** 4 **49c.** D = {all real numbers}, R = $\{y \mid y \leq 4\}$ **51a.** maximum **51b.** 3 **51c.** D = {all real numbers}, R = $\{y \mid y \leq 3\}$

53.

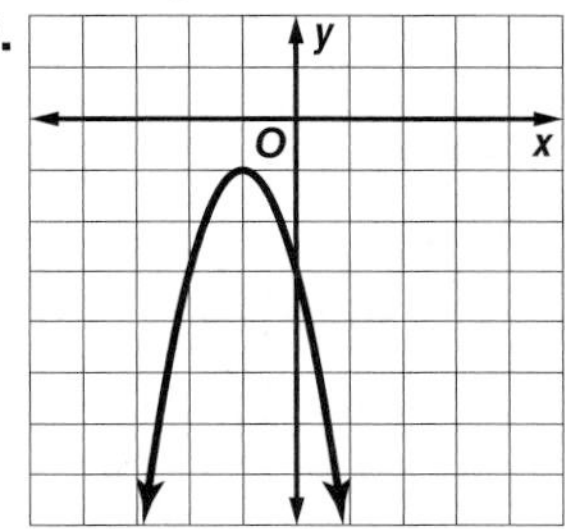

55.

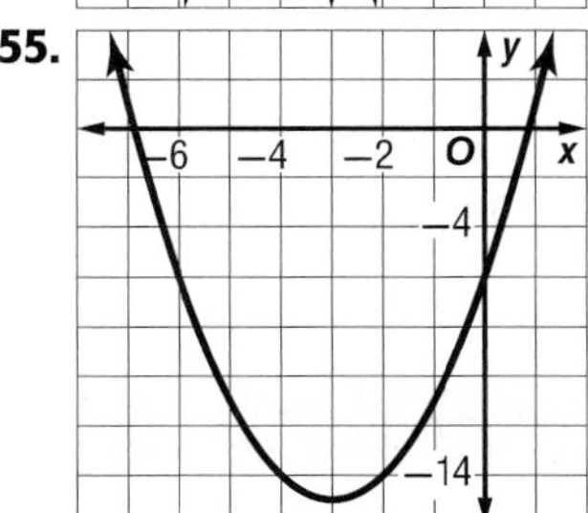

57.

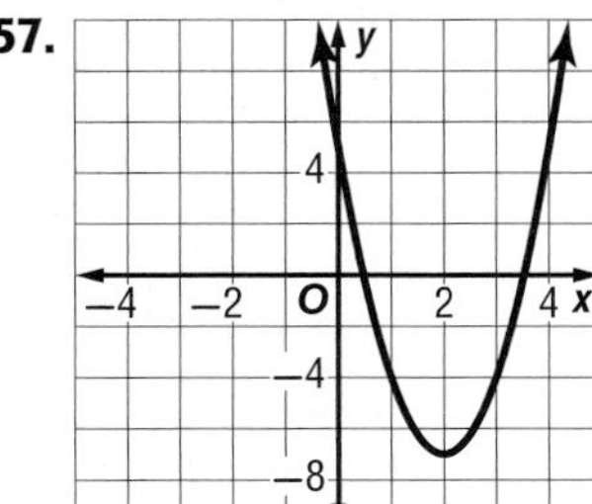

59.

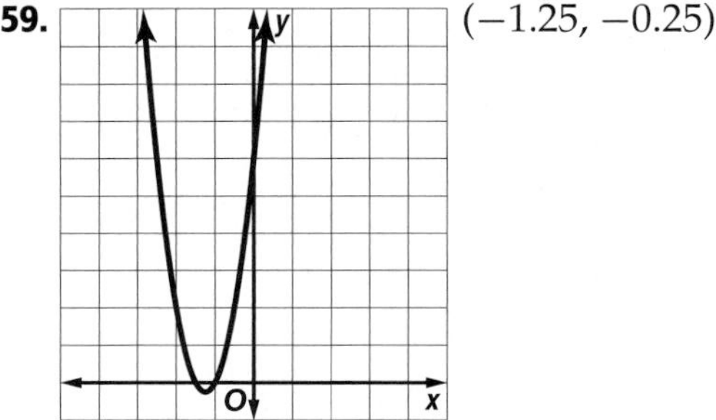

(−1.25, −0.25)

61. 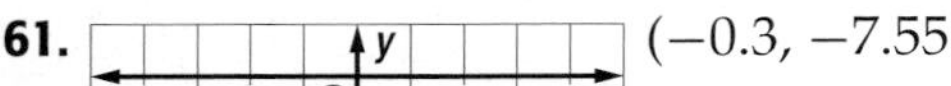
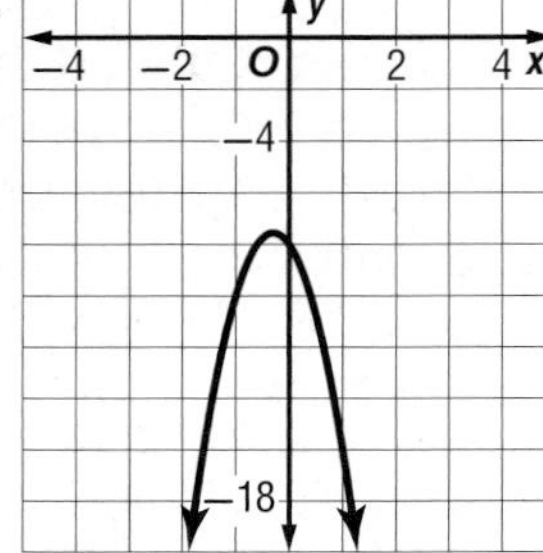
$(-0.3, -7.55)$

63a. 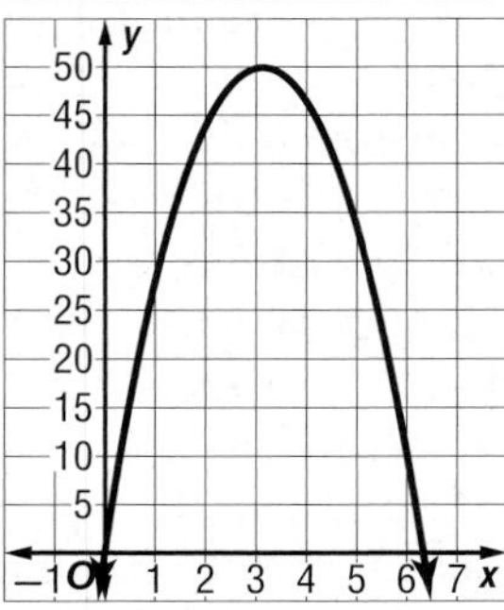

63b. 0 m **63c.** ≈ 50.0 m **63d.** ≈ 6.4 seconds
63e. $D = \{x \mid 0 \le x \le 6.4\}$; $R = \{y \mid 0 \le y \le 50.0\}$

65 a. $h = -16t^2 + 90t$
$= -16(1)^2 + 90(1)$
$= 74$ ft

b. $126 = -16t^2 - 90t$
$0 = -16t^2 - 90t - 126$
$0 = (t - 3)(-16t + 42)$
$t = 3$ and $t = 2.625$

c. $h = -16t^2 + 90t$
$0 = -16t^2 + 90t$
$0 = -16t(t - 5)$
$t = 0$ and $t = 5.625$

These represent the time that the ball leaves the ground initially and the time it returns to the ground.

67 a.

Equation	Related Function	Zeros	y-Values
$x^2 - x = 12$	$y = x^2 - x - 12$	−3, 4	−3: 8, −6; 4: −6, 8
$x^2 + 8x = 9$	$y = x^2 + 8x - 9$	−9, 1	−9: 11, −9; 1: −9, 11
$x^2 = 14x - 24$	$y = x^2 - 14x + 24$	2, 12	2: 11, −9; 12: −9, 11
$x^2 + 16x = -28$	$y = x^2 + 16x + 28$	−14, −2	−14: 13, −11; −2: −11, 3

b. Use a graphing calculator to graph.

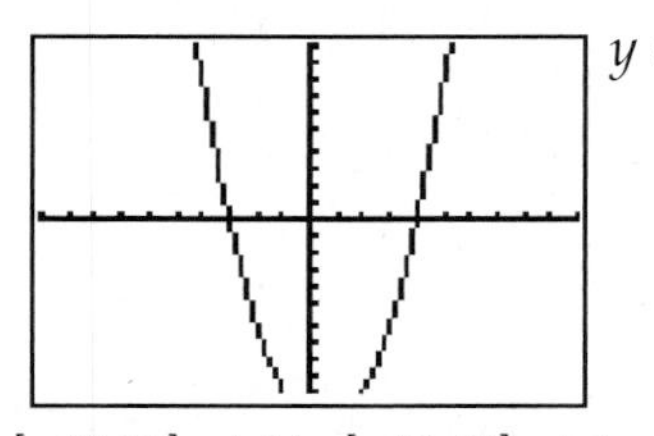
$y = x^2 - x - 12$
[−10, 10] scl: 1 by [−10, 10] scl: 1

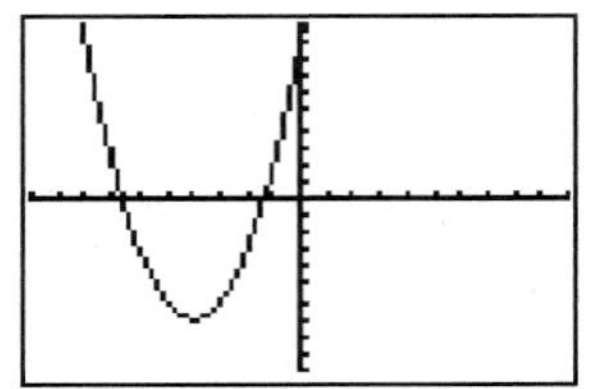
$y = x^2 - 8x - 9$
[−15, 15] scl: 2 by [−30, 30] scl: 5

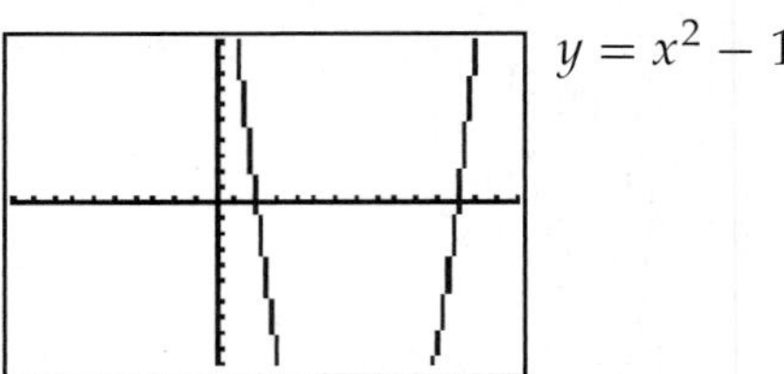
$y = x^2 - 14x - 24$
[−10, 15] scl: 1 by [−10, 10] scl: 1

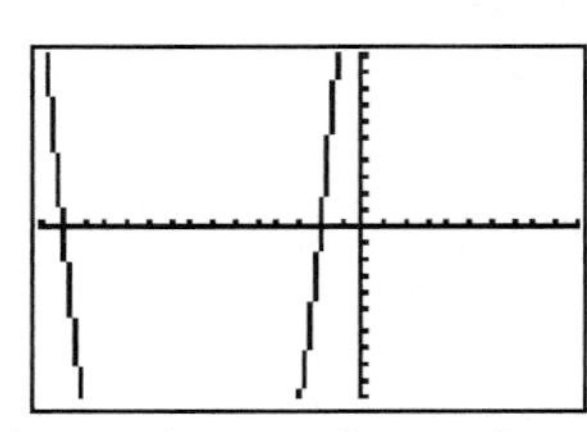
$y = x^2 - 16x - 28$
[−15, 10] scl: 1 by [−10, 10] scl: 1

c. Use the table function on the calculator to identify the zeros. The zeros are the values where y is 0. Also list the y-values that are to the left and right of the zeros.
d. The function values have opposite signs just before and just after the zeros.

69. Jade; Chase forgot the negative sign with the −4. **71.** $(-1, 9)$; Sample answer: I graphed the points given, and sketched the parabola that goes through them. I counted the spaces over and up from the vertex to the other point and did the same on the opposite side. **73.** Sample answer: The function $y = -x^2 - 4$ has a vertex at $(0, -4)$, but it is a maximum.
75. C **77.** D **79.** yes; $(2x + 1)^2$ **81.** no **83.** prime **85.** $b^2 - 4b - 21$ **87.** $2x^2 + 17x - 9$ **89.** (2, 1) **91.** (4, 2) **93.** 10 **95.** −6

Pages 540–542 Lesson 9-2

1. 2, −5

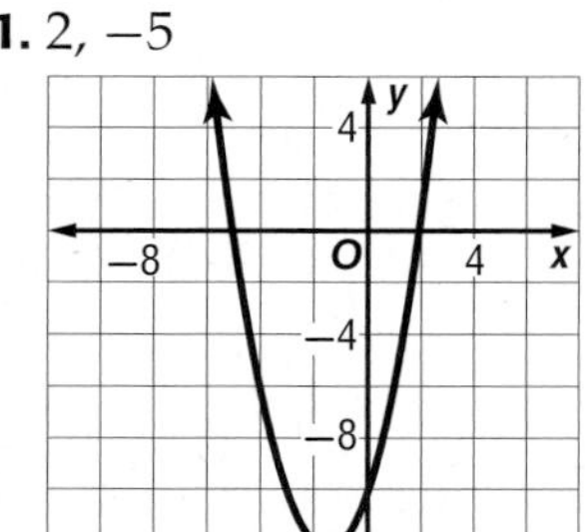

3. −2

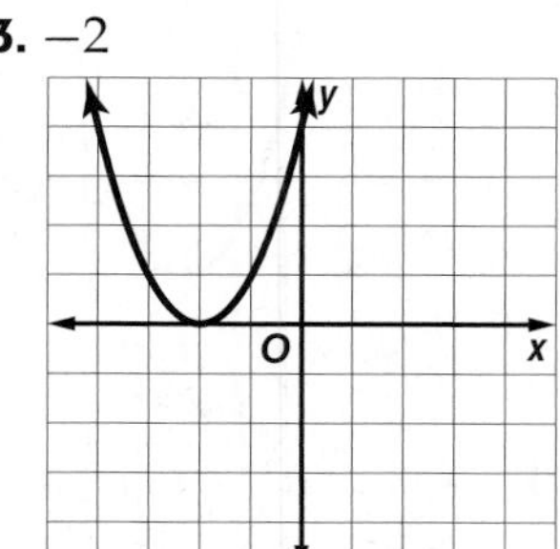

5. −5.2, 0.2

7. 5, −5 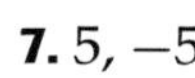

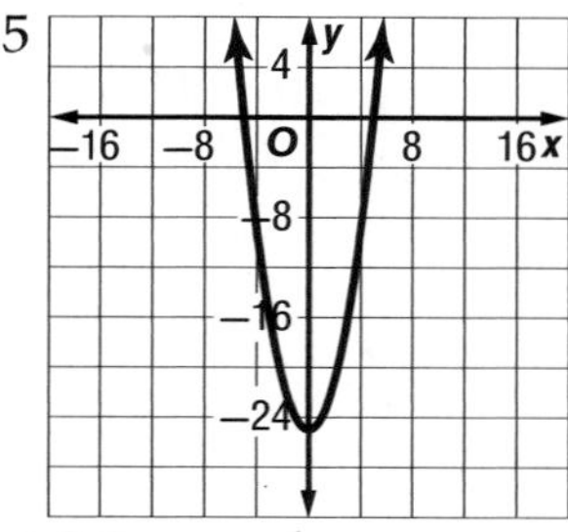

9. about 8.4 seconds

11 Step 1: Graph the related function of $f(x) = x^2 + 2x - 24 = 0$

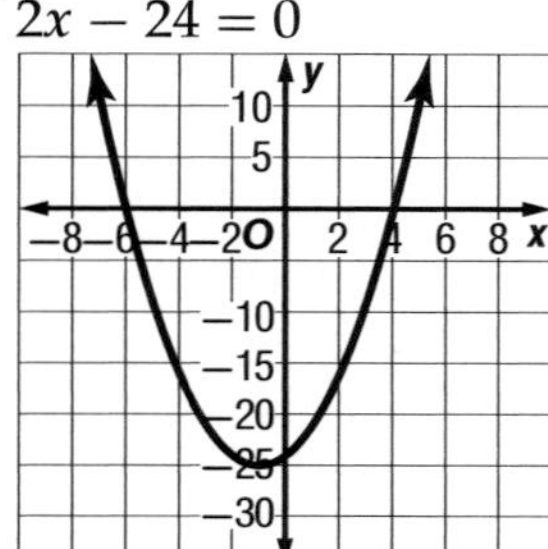

Step 2: The x-intercepts appear to be at −6 and 4, so the solutions are −6 and 4.

13. ∅

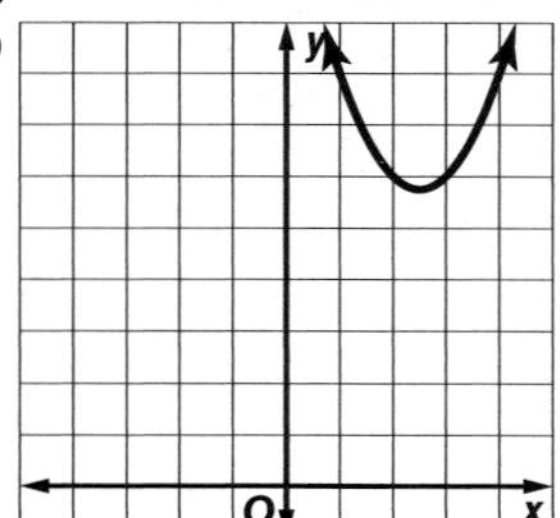

15. 1

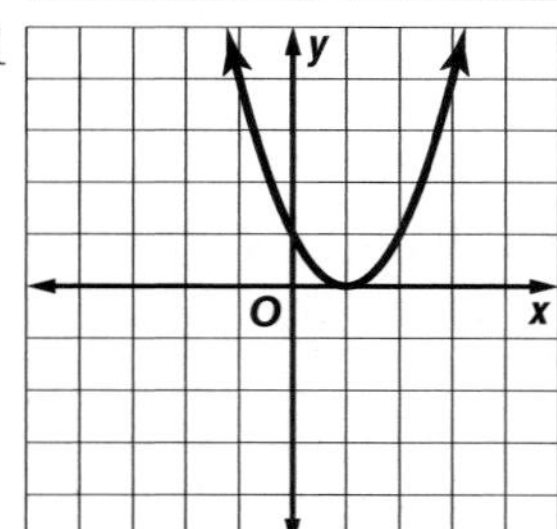

17. ∅

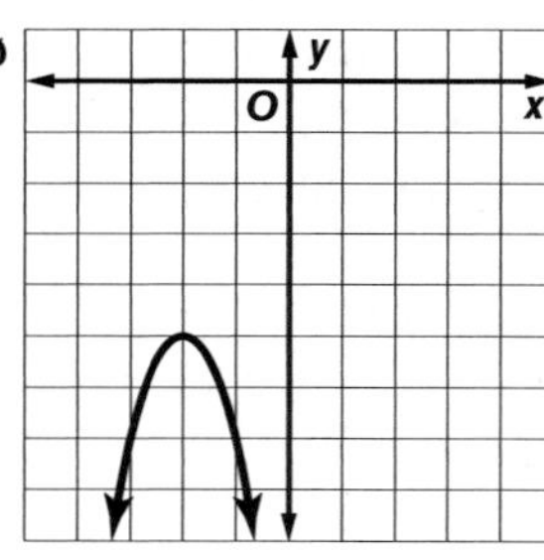

19. −6

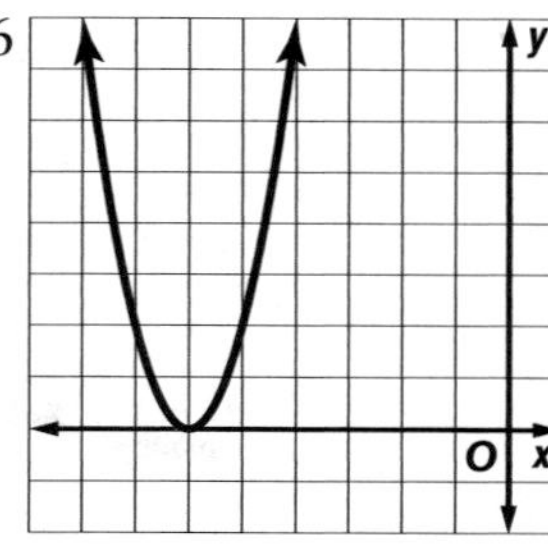

21. 8, −10

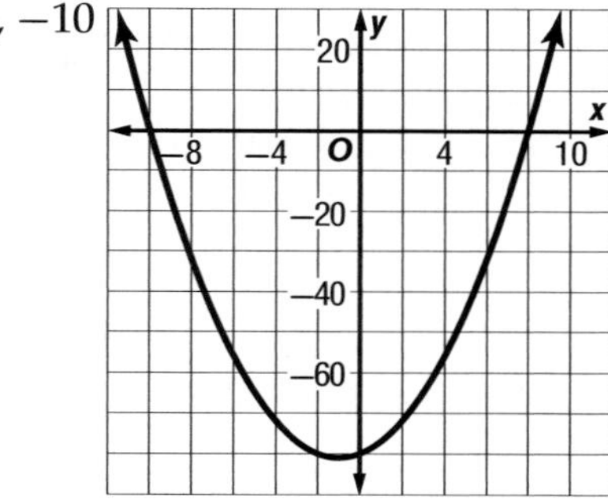

23. 6.9, −2.9

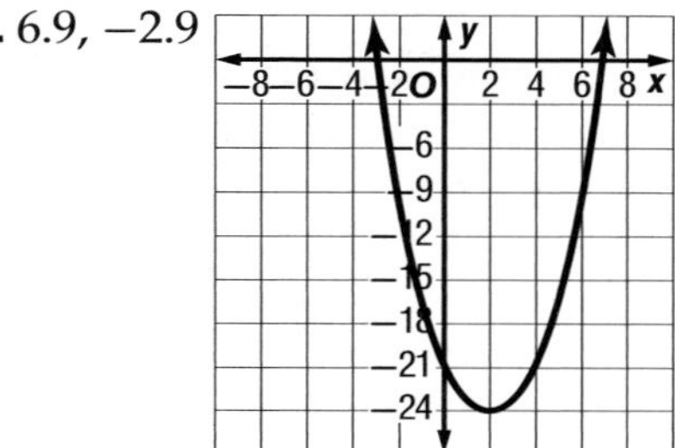

25. 3.3, 1.2

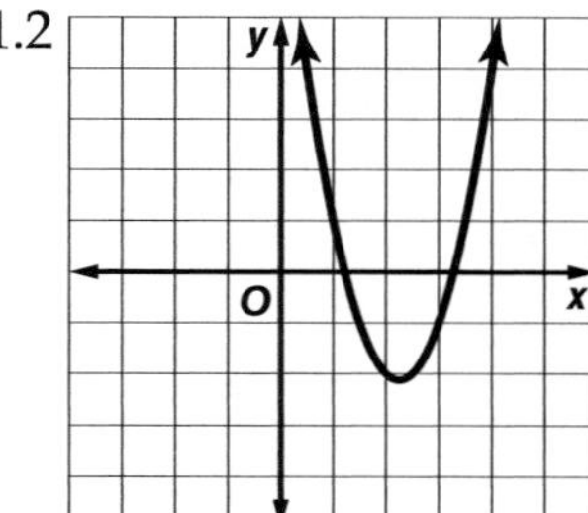

27. 3.1, −8.1

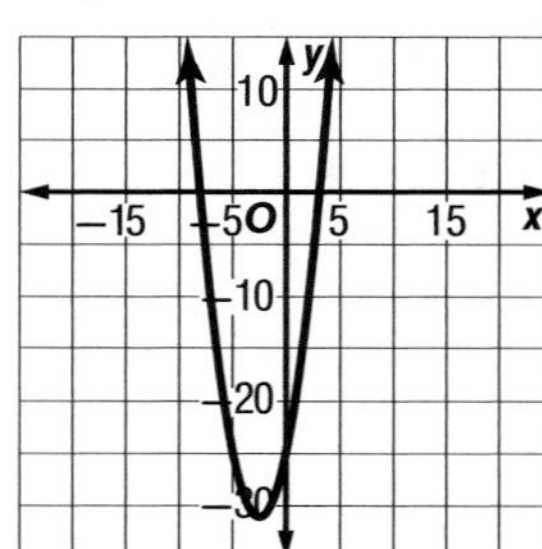

29. about 11.6 seconds **31.** 0; no real roots

33. 2; −4, −8 **35.** −3, 4

37 **a.** $h = -16t^2 + 30t + 10$

$0 = -16t^2 + 30t + 10$

Graph the equation and find the x-intercepts.

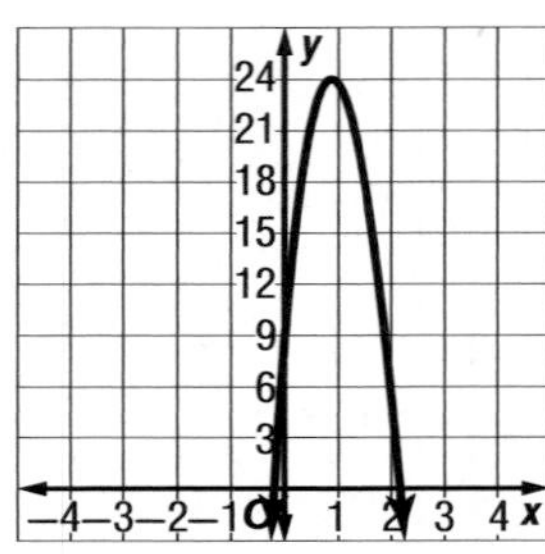

The positive x-intercept appears to be at 2.2, so she is in the air 2.2 seconds.

b. From the graph, she appears to hit a height of 15 feet at 0.2 seconds and 1.7 seconds.

c. $x = \frac{-b}{2a}$

$= \frac{-30}{2(-16)}$

$h = (-16)(0.9375)^2 + 30(0.9375) + 10$

≈ 24 feet

Her maximum height is about 24 feet, so she gets the bonus points.

39. $-2, 1, 4$ **41.** Iku; sample answer: The zeros of a quadratic function are the x-intercepts of the graph. Since the graph does not cross the x-axis, there are no x-intercepts and no real zeros. **43.** Sometimes; for (1, 3), the y-value is greater than 2, but for (1, −1), it is less than 2. **45.** 1.5 and −1.5; Sample answer: Make a table of values for x from −2.0 to 2.0. Use increments of 0.1. **47.** A

49. about 21.4 mi

Boat 1
5 mi
Dock
4 mi
12 mi
Boat 2
9 mi

51. $x = 0$; (0, 0); min

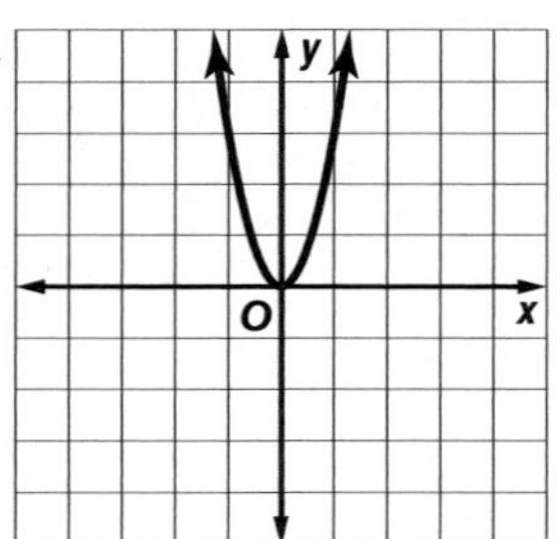

53. $x = 2$; (2, −3); max

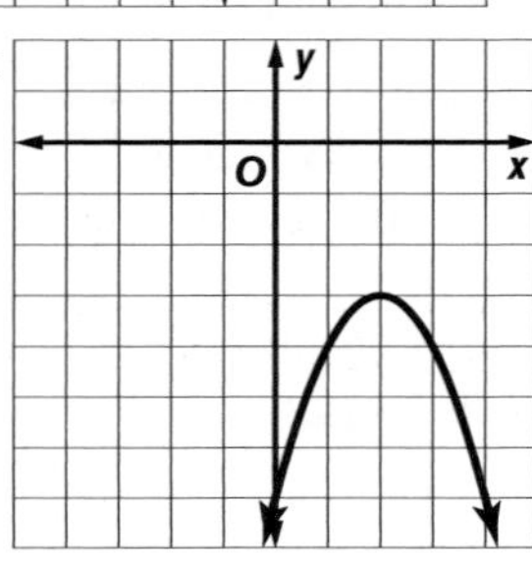

55. $x = -\frac{1}{3}$; $\left(-\frac{1}{3}, \frac{2}{3}\right)$; min

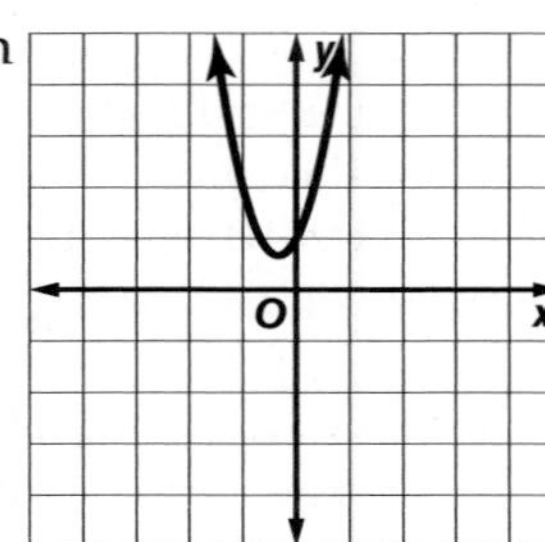

57. $-4, 4$ **59.** $-\frac{3}{2}, \frac{5}{2}$ **61.** $-3 \pm \sqrt{5}$ **63.** $7n^2 + 1$ **65.** $-3b^4 + 2b^3 - 9b^2 + 13$ **67.** $x + y = 180$; $x = y + 24$; 102°, 78° **69.** $y - 6 = -7(x + 3)$

71.

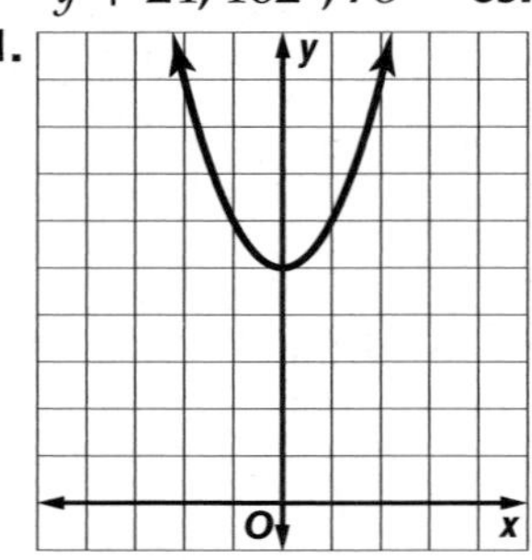

73.

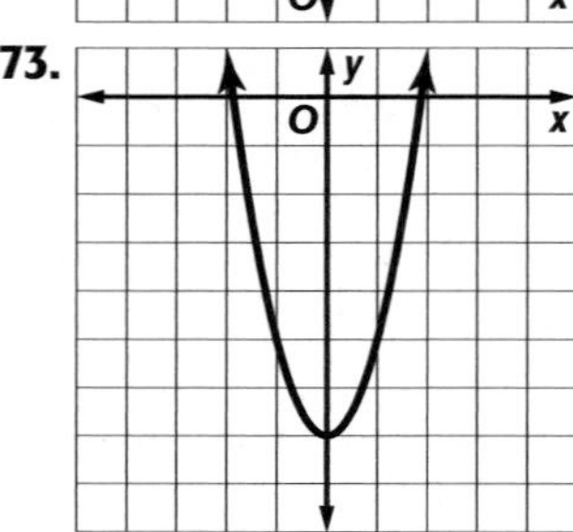

75.

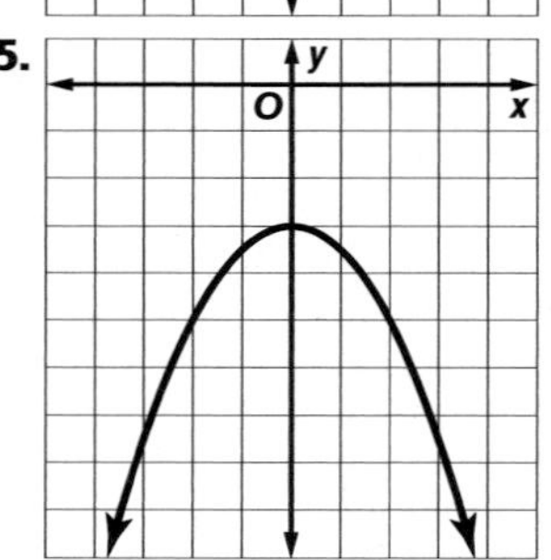

Pages 547–549 Lesson 9-3

1. translated down **3.** reflected across the x-axis, translated up **5.** reflected across the x-axis, stretched vertically **7.** C

9 The function can be written as $f(x) = ax^2 + c$ where $a = -1$ and $c = -7$. Since $-7 < 0$ and $-1 < 0$ the graph of $y = -x^2 - 7$ translates the graph of $y = x^2$ down 7 units and reflects it across the x-axis. **11.** compressed vertically, translated up **13.** stretched vertically, translated up **15.** stretched vertically, translated up **17.** stretched vertically, translated down **19.** A **21.** F **23.** E **25.** $g(x)$, $h(x)$ **27.** $h(x)$, $g(x)$, $f(x)$

29 **a.** The two equations are $h = -16t^2 + 20{,}000$ and $h = -16t^2 + 30{,}000$

b. The rock from the 20,000-inch cliff will hit the

ground first because it started lower and fell at the same rate as the other one.
31. Translate the graph of $f(x)$ down.
33. Compress vertically the graph of $f(x)$.
35a. Sometimes; this only occurs if $c = 0$. For any other value, the graph will be translated up or down. **35b.** Always; the negative sign only reflects the graph over the x-axis. Both graphs are dilated by a factor of a. **35c.** Never; the coefficient of x^2-term would have to be negative for the graph to open downward. **37.** Sample answer: Not all reflections over the y-axis produce the same graph. If the vertex of the original graph is not on the y-axis, the graph will not have the y-axis as its axis of symmetry and its reflection across the y-axis will be a different parabola. **39.** Sample answer: For $y = ax^2$, the parent graph is stretched vertically if $a > 1$ or compressed vertically if $0 < a < 1$. The y-values in the table will all be multiplied by a factor of a. For $y = x^2 + c$, the parent graph is translated up if c is positive and moved down if c is negative. The y-values in the table will all have the constant c added to them or subtracted from them. For $y = ax^2 + c$, the graph will either be stretched vertically or compressed vertically based upon the value of a and then will be translated up or down depending on the value of c. The y-values in the table will be multiplied by a factor of a and the constant c added to them.
41. D **43.** B **45.** 4, 6

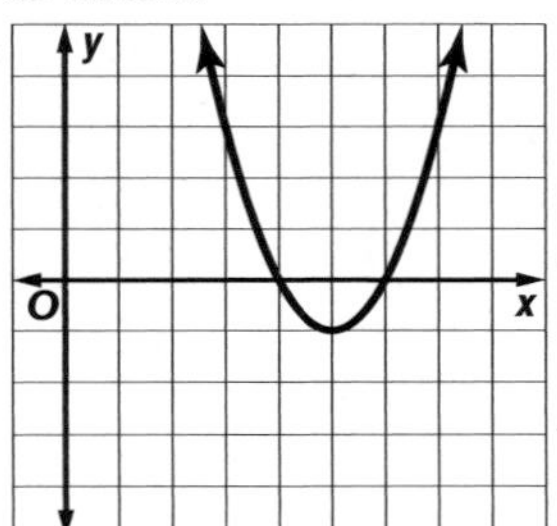

47. $-1, \frac{3}{2}$

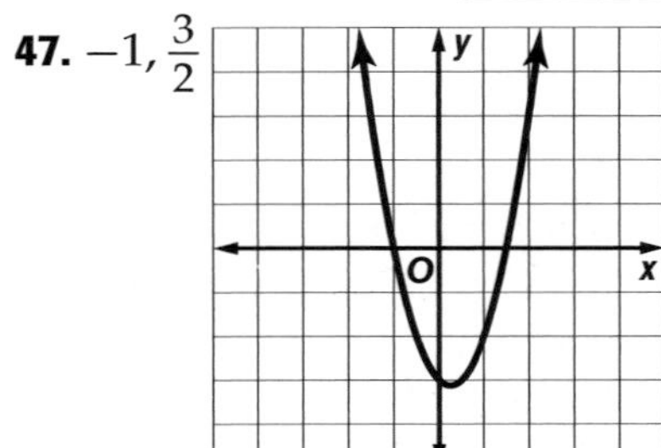

49. $-\frac{5}{3}, \frac{3}{4}$

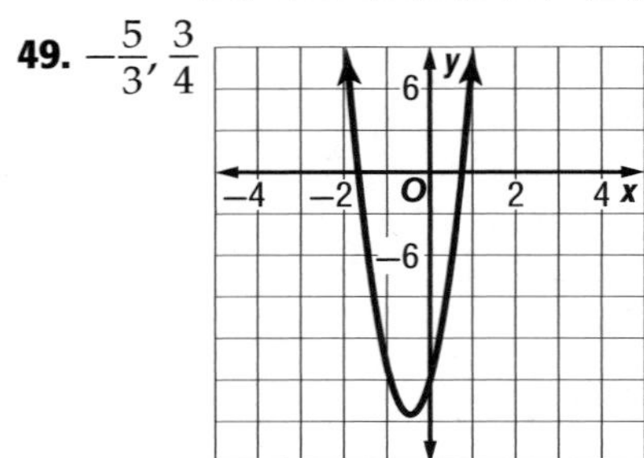

51. $(2, 2)$; $x = 2$; 6 **53.** $1.50t + 1.25mt$
55. $\{f \mid f > -8\}$ **57.** $(4x - 3)^2$ **59.** $(5x - 6)^2$ **61.** $(6x - 7)^2$

Pages 554-557 Lesson 9-4

1 Step 1: Find $\frac{1}{2}$ of $-18 = -9$
Step 2: Square the result in step 1: $(-9)^2 = 81$
Step 3: Add the result of step 2 to $x^2 - 18x$: $x^2 - 18x + 81$
Thus, $c = 81$.
3. $\frac{81}{4}$ **5.** −5.2, 1.2 **7.** −2.4, 0.1 **9.** 8 ft by 18 ft
11. 144 **13.** $\frac{289}{4}$ **15.** $\frac{169}{4}$ **17.** $\frac{225}{4}$
19
$$x^2 + 6x - 16 = 0$$
$$x^2 + 6x = 16$$
$$x^2 + 6x + 9 = 16 + 9$$
$$(x + 3)^2 = 25$$
$$x + 3 = \pm 5$$
$$x = -3 \pm 5$$
The solutions are 2 and −8.
21. −1, 9 **23.** −0.2, 11.2 **25.** ∅ **27.** −2.6, 1.1 **29.** −1.1, 6.1 **31.** on the 30th and 40th days after purchase **33.** 5.3 **35.** −21 and −23 **37.** −1, 2 **39.** 0.2, 0.9 **41.** −8.2, 0.2
43 **a.** The object on Earth will reach the ground first because it is falling at a faster rate.
b. Mars: $0 = -1.855t^2 + 120$
$-120 = -1.855t^2$
$64.69 \approx t^2$
$\pm 8.0 \approx t$
So, $t \approx 8.0$ seconds
Earth: $0 = -4.9t^2 + 120$
$-120 = -4.9t^2$
$24.49 = t^2$
$\pm 4.9 \approx t$
So, $t = 4.9$ seconds.
c. Sample answer: Yes, the acceleration due to gravity is much greater on Earth than on Mars, so the time to reach the ground should be much less.
45. −30 and 30
47a–b.

Trinomial	$b^2 - 4ac$	Number of Roots
$x^2 - 8x + 16$	0	1
$2x^2 - 11x + 3$	97	2
$3x^2 + 6x + 9$	−72	0
$x^2 - 2x + 7$	−24	0
$x^2 + 10x + 25$	0	1
$x^2 + 3x - 12$	57	2

47c. If $b^2 - 4ac$ is negative, the equation has no solutions. If $b^2 - 4ac$ is zero, the equation has one solution. If $b^2 - 4ac$ is positive, the equation has 2 solutions. **47d.** 0 because $b^2 - 4ac$ is negative. The equation has no real solutions because taking the square root of a negative number does not produce a real number. **49.** None; sample answer: If you add $\left(\frac{b}{2}\right)^2$ to each side of the equation and each side of the inequality, you get $x^2 + bx + \left(\frac{b}{2}\right)^2 = c + \left(\frac{b}{2}\right)^2$ and $c + \left(\frac{b}{2}\right)^2 < 0$. Since the left side of the last equation is a perfect square, it cannot equal the negative number $c + \left(\frac{b}{2}\right)^2$. So, there are no real solutions. **51.** Sample answer: $x^2 - 8x + 16 = 0$ **53.** B **55.** 32 **57.** translated down **59.** expanded vertically, translated up **61.** expanded vertically, translated up **63.** $40 = -16t^2 + 250$; about 3.6 s **65.** 16 **67.** 1 **69.** $\frac{1}{m^3b^3}$ **71.** $\{z \mid -8 < z < -2\}$ **73.** $\{y \mid y \geq 5.5 \text{ or } y \leq -2.5\}$ **75.** $\{c \mid -2.2 \leq c \leq 3\}$ **77.** ±10 **79.** ± 7.8 **81.** not a real number

Pages 562-564 Lesson 9-5

1. −3, 5 **3.** 6.4, 1.6 **5.** 0.6, 2.5 **7.** $-6, \frac{1}{2}$ **9.** $\pm\frac{5}{3}$ **11.** −3; no real solutions **13.** 0; one real solution **15.** The discriminant is −890.24, so the equation has no solutions. Thus, Eva will not reach a height of 20 feet.

17 $x^2 + 16 = 0$

For this equation, $a = 1$, $b = 0$, and $c = 16$

$$x = \frac{-b \pm \sqrt{b^2 - 4ac}}{2a}$$
$$= \frac{0 \pm \sqrt{(0)^2 - 4(1)(16)}}{2(1)}$$
$$= \frac{\sqrt{-64}}{2}$$

So, there is no real solution. The solution can be written ∅.

19. 2.2, −0.6 **21.** $-3, -\frac{6}{5}$ **23.** 0.5, −2 **25.** 0.5, −1.2 **27.** 3 **29.** −1.2, 5.2 **31.** −2, 5 **33.** −6.2, −0.8 **35.** −0.07; no real solution **37.** 12.64; two real solutions **39.** 0; one real solution **41a.** in 1993 and 2023 **41b.** Sample answer: No; the parabola has a maximum at about 66, meaning only 66% of the population would ever have high-speed Internet. **43.** 0 **45.** 1 **47.** −1.4, 2.1

49 **a.** $(20 - 2x)(25 - 7x) = 375$

b. $500 - 50x - 140x + 14x^2 = 375$

$14x^2 - 190x + 125 = 0$

$$x = \frac{-b \pm \sqrt{b^2 - 4ac}}{2a}$$
$$= \frac{190 \pm \sqrt{(190)^2 - 4(14)(125)}}{2(14)}$$
$$= \frac{190 \pm \sqrt{29100}}{28}$$
$$\approx 0.7 \text{ and } 12.9$$

c. The margins should be 0.7 in. on the sides and 4(0.7) or 2.8 in. on the top and 3(0.7) or 2.1 in. on the bottom.

51. $k < \frac{9}{40}$ **53.** none **55.** two **57.** Sample answer: positive discriminant: $f(x) = x^2 - 4$, negative discriminant: $f(x) = x^2 + 4$, zero discriminant: $f(x) = x^2 - 8x + 16$ **59.** D **61.** G **63.** $\frac{4}{3}, \frac{3}{2}$ **65.** $\frac{5}{2}$

67. Translate down 6. **69.** Positive; as time goes on, more people use electronic tax returns. **71.** $12x + 3y \leq 60$

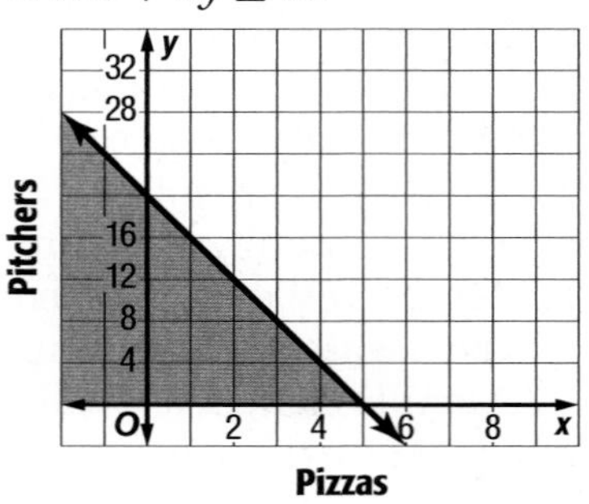

73. 4 **75.** 0 **77.** −75

Pages 570–572 Lesson 9-6

1.

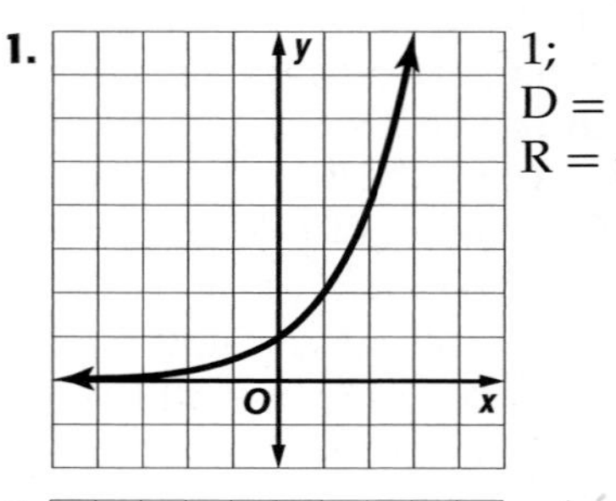

1; D = {all real numbers}; $R = \{y \mid y > 0\}$; $2^{1.5} \approx 2.8$

3.

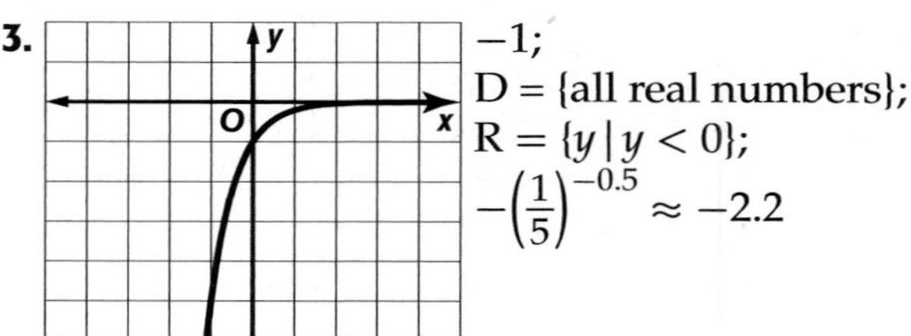

−1; D = {all real numbers}; $R = \{y \mid y < 0\}$; $-\left(\frac{1}{5}\right)^{-0.5} \approx -2.2$

5.

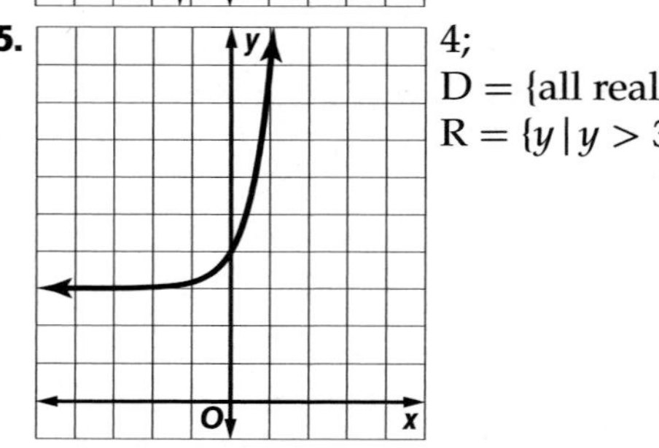

4; D = {all real numbers}; $R = \{y \mid y > 3\}$

7a. $D = \{d \mid d \geq 0\}$, the number of days is greater than or equal to 0; $R = \{y \mid y \geq 100\}$, the number of fruit flies is greater than or equal to 100. **7b.** about 198 fruit flies **9.** Yes; the domain values are at regular intervals, and the range values have a common factor of 4.

11.

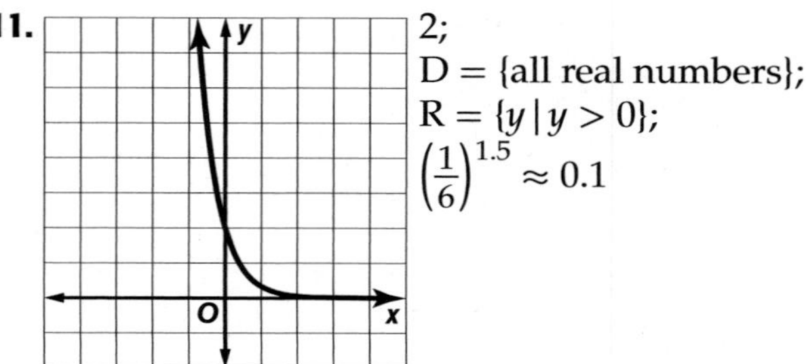

2; D = {all real numbers}; $R = \{y \mid y > 0\}$; $\left(\frac{1}{6}\right)^{1.5} \approx 0.1$

13. 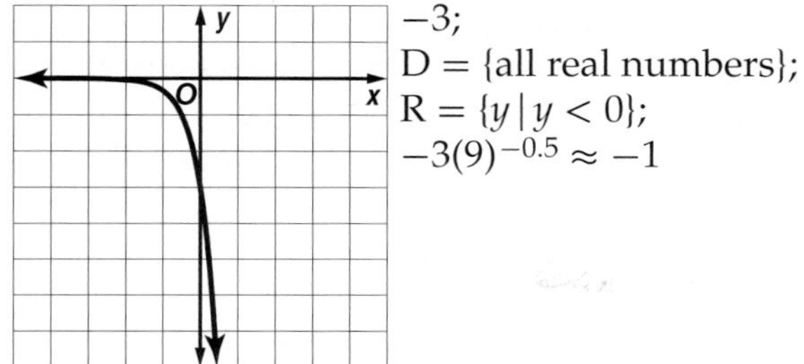

-3;
D = {all real numbers};
R = $\{y \mid y < 0\}$;
$-3(9)^{-0.5} \approx -1$

15. 3;
D = {all real numbers};
R = $\{y \mid y > 0\}$;
$3(11)^{-0.2} \approx 1.9$

17. 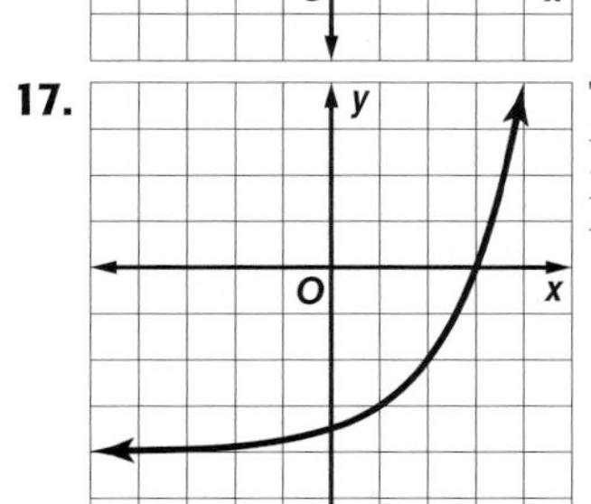

The y-intercept is -3.5;
D = {all real numbers};
R = $\{y \mid y > -4\}$

19. 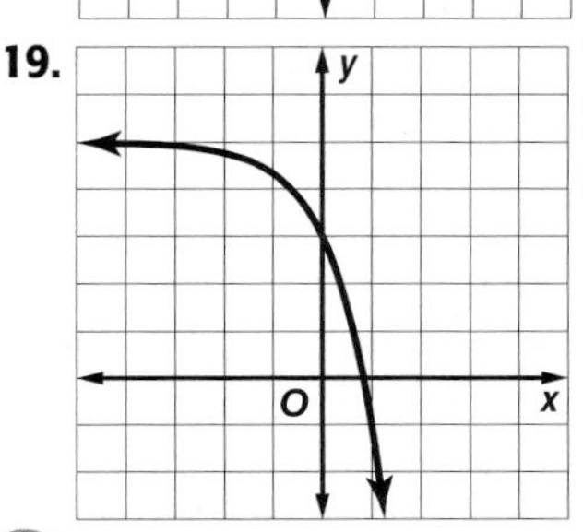

The y-intercept is 3;
D = {all real numbers};
R = $\{y \mid y < 5\}$

21 No; the domain values are at regular intervals of 4.
$2 \times (-2) = 4$
$-4 \times (-2) = 8$
$8 \times (-2) = -16$
$-16 \times (-2) = 32$
The range values differ by the common factor of -2. The range values do not have a positive common factor.

23. Yes; the domain values are at regular intervals, and the range values have a common factor of 2.

25 $P = 1.5^x$
$= 1.5^4$
≈ 5.06 or 506%
This enlargement is about 506% bigger than the original.

27. exponential **29.** linear **31.** quadratic **33.** about 198 students **35.** a vertical stretch by a factor of 3 **37.** a translation down 3 units **39.** a vertical stretch by a factor of 5 and a reflection over the x-axis. **41.** $f(x) = 3(2)^x$ **43.** Sample answer: The number of teams competing in a basketball tournament can be represented by $y = 2^x$, where the number of teams competing is y and the number of rounds is x.

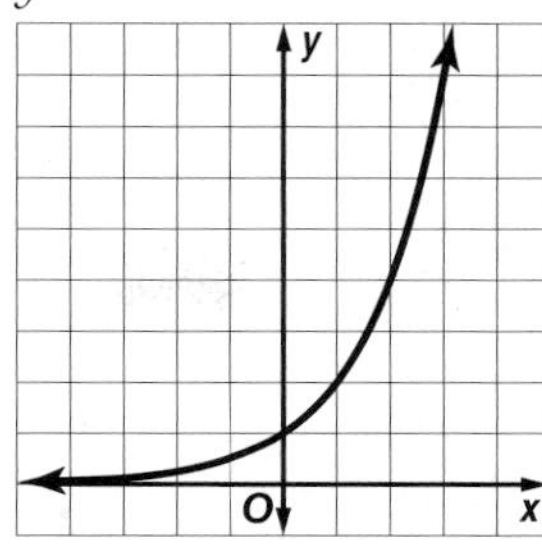

The y-intercept of the graph is 1. The graph increases quickly for $x > 0$.

45. Sample answer: First, look for a pattern by making sure that the domain values are at regular intervals and the range values differ by a common factor. **47.** B **49.** A **51.** $-5, 0.5$ **53.** ± 5 **55.** 3.1, 10.9 **57.** 2.52×10^2, 252 **59.** 32 km^2/min^2 **61.** $(-5, 20)$ **63.** 9, 11, 13 **65.** 16.5, 19, 21.5 **67.** $\frac{7}{2}, \frac{17}{4}, 5$

Pages 575-577 Lesson 9-7

1. about $37,734.73 **3a.** $y = 2200(0.98)^t$ **3b.** about 1624 **5.** about 92,095,349

7 $A = P\left(1 + \frac{r}{n}\right)^{nt}$
$= 6600\left(1 + \frac{0.045}{12}\right)^{12(4)}$
$= 6600(1.00375)^{48}$
≈ 7898.97 or about $7898.97

9. Sample answer: No; she will have about $199.94 in the account in 4 years.

11 $y = a(1 + r)^t$
$= 1{,}211{,}537(1 + 0.0106)^{20}$
$= 1{,}211{,}537(1.0106)^{20}$
$\approx 1{,}495{,}969$

13a. $I = 221{,}000(1.086)^t$ **13b.** about $645,922 **15a.** $316.82 **15b.** $19,009.20 **17.** about 9.2 yr **19.** A population is 200 and increasing at a rate of 5% annually. **21.** C **23.** D

25. 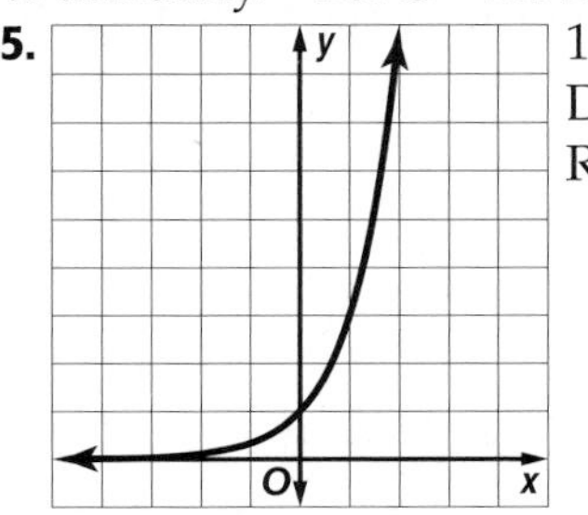

1;
D = {all real numbers},
R = $\{y \mid y > 0\}$

27. 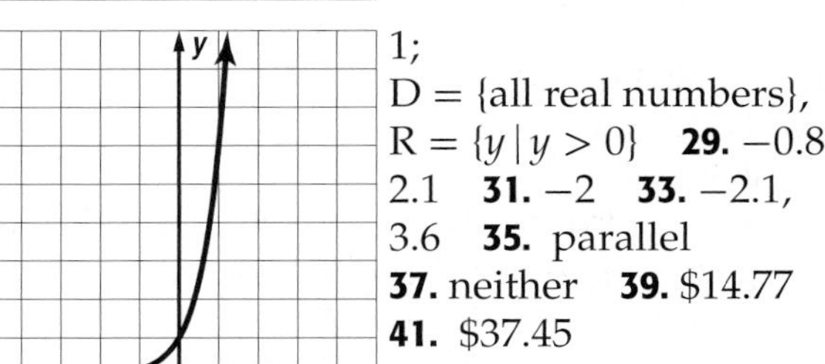

1;
D = {all real numbers},
R = $\{y \mid y > 0\}$ **29.** -0.8, 2.1 **31.** -2 **33.** -2.1, 3.6 **35.** parallel **37.** neither **39.** $14.77 **41.** $37.45

Selected Answers and Solutions

43.

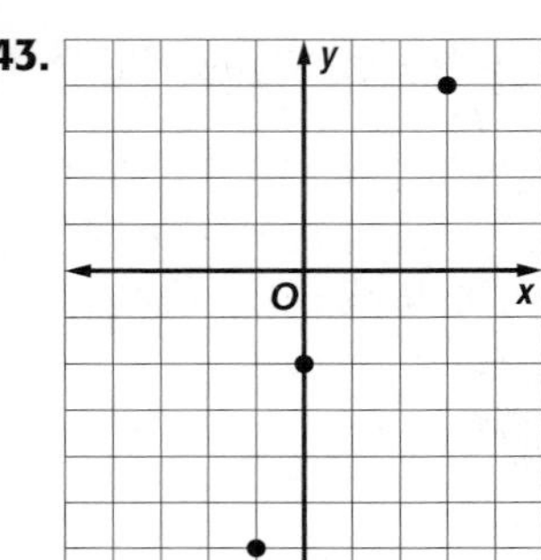

Pages 583-585 Lesson 9-8

1. Geometric; the common ratio is $\frac{1}{5}$.
3. Arithmetic; the common difference is 3.
5. 160, 320, 640 **7.** $-\frac{1}{16}, -\frac{1}{64}, -\frac{1}{256}$ **9.** $a_n = -6 \cdot (4)^{n-1}; -1536$ **11.** $a_n = 72 \cdot \left(\frac{2}{3}\right)^{n-1}; \frac{4096}{2187}$
13.

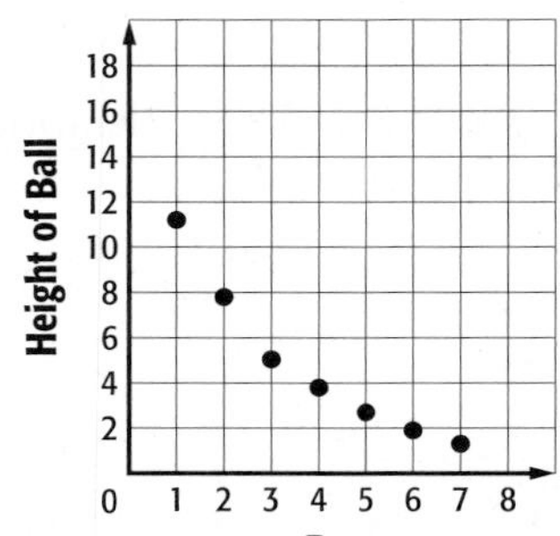

15. Arithmetic; the common difference is 10.
17. Geometric; the common ratio is $\frac{1}{2}$.
19. Neither; there is no common ratio or difference.
21 Step 1: Find the common ratio.
$36 \times \frac{1}{3} = 12$
$12 \times \frac{1}{3} = 4$
The common ratio is $\frac{1}{3}$.
Step 2: Multiply each term by the common ratio to find the next three terms.
$4\left(\frac{1}{3}\right) = \frac{4}{3}$
$\left(\frac{4}{3}\right)\left(\frac{1}{3}\right) = \frac{4}{9}$
$\left(\frac{4}{9}\right)\left(\frac{1}{3}\right) = \frac{4}{27}$
So, the next three terms are $\frac{4}{3}, \frac{4}{9}$, and $\frac{4}{27}$.
23. $\frac{25}{4}, \frac{25}{16}, \frac{25}{64}$ **25.** $2, \frac{1}{4}, \frac{1}{32}$ **27.** 134,217,728
29. −1,572,864 **31.** 19,683
33 **a.** $1 \times 2 = 2$
$2 \times 2 = 4$
The common ratio is 2, so the second option forms a geometric sequence.
b. first option: She will receive 30(9) = \$270
second option: In the 9th week, she will get $a_n = a_1 r^{n-1}$.
$a_n = 1(2)^{9-1}$
$= 1(2)^8$
$= 256$
In the 8th week, she will get $a_n = a_1 r^{n-1}$.
$a_n = 1(2)^{8-1}$
$= 1(2)^7$
$= 128$
Over the nine weeks she will earn: 1 + 2 + 4 + 8 + 16 + 32 + 64 + 128 + 256 or \$511.
She should choose the second option.
35. $9; \frac{1}{3}$
37a.

Richter Number (x)	Increase in Magnitude (y)	Rate of Change (slope)
1	1	–
2	10	9
3	100	90
4	1,000	900
5	10,000	9000

37b.

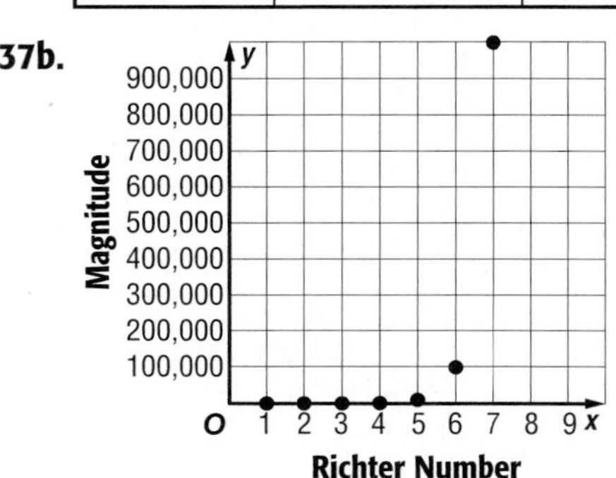

37c. The graph appears to be exponential. The rate of change between any two points does not match any others. **37d.** $1 \cdot (10)^{x-1} = y$ **39.** Neither Haro calculated the exponent incorrectly; Matthew did not enclose the common ratio in parentheses.
41. $1, \frac{3}{4}, \frac{9}{16}, \ldots$ **43.** B **45.** 15 dimes and 20 quarters **47.** 162, 486, 1458 **49.** $\frac{1}{16}, -\frac{1}{32}, \frac{1}{64}$ **51.** 0.1296, 0.07776, 0.046656
53.

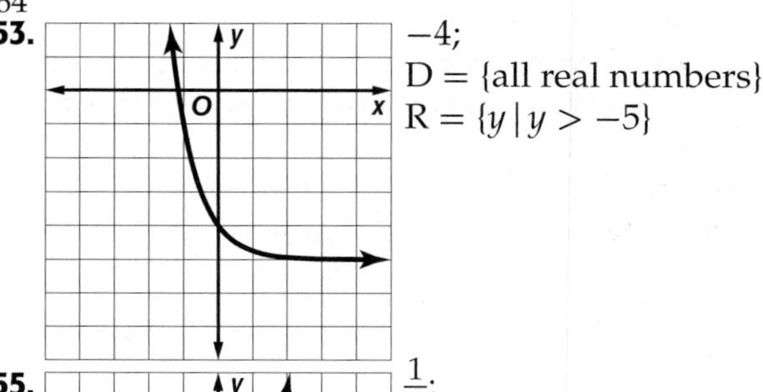

−4;
D = {all real numbers};
R = $\{y \mid y > -5\}$

55.

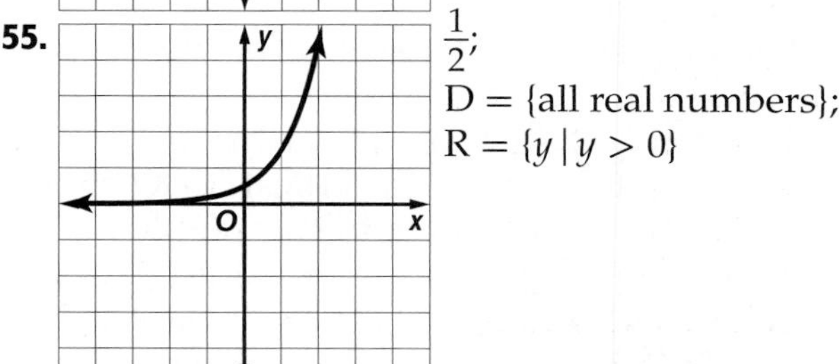

$\frac{1}{2}$;
D = {all real numbers};
R = $\{y \mid y > 0\}$

57. at least \$3747 **59.** $y = -3x - \frac{2}{3}$ **61.** $y = \frac{1}{2}x - 9$ **63.** $y = -6x - 7$ **65.** 7412.01 **67.** 371.50 **69.** 96,150.24

Pages 589-591 Lesson 9-9

1. 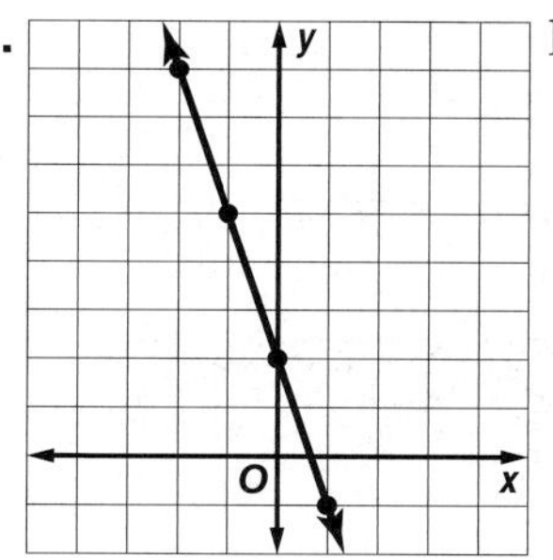 linear

3. exponential

5. quadratic **7.** exponential **9.** exponential; $y = 3 \times 3^x$ **11.** linear; $y = \frac{1}{2}x + \frac{5}{2}$ **13.** linear: $y = 0.5x + 3$

15. linear

17. quadratic

19. exponential

21 Look for a pattern in the y-values. Start with comparing first differences.

10 2.5 0 2.5 10

−7.5 −2.5 2.5 7.5

The first differences are not all equal. So, the table of values does not represent a linear function. Find the second differences and compare.

−7.5 −2.5 2.5 7.5

+5 +5 +5

The second differences are all equal, so the table of values represents a quadratic function.
Write an equation for the function that models the data.
The equation has the form $y = ax^2$. Find the value of a by choosing one of the ordered pairs from the table of values. Let's use (2, 10).

$y = ax^2$

$10 = a(2)^2$

$10 = 4a$

$\frac{5}{2} = a$

$2.5 = a$

An equation that models the data is $y = 2.5x^2$.

23. exponential; $y = 0.2 \cdot 5^x$

25. linear; $y = -5x - 0.25$

27 a. Graph the ordered pairs on a coordinate plane.

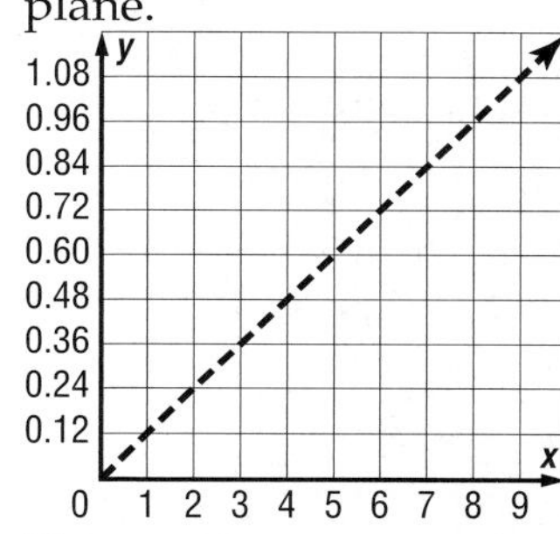

The graph appears to be linear.

b. Look at the first differences of the y-values.

0.12 0.24 0.36 0.48 0.60 0.72

+0.12 +0.12 +0.12 +0.12 +0.12

The common difference is 0.12.
The equation is $y = 0.12x$.

c. $y = 0.12x$
$= 0.12(10)$
$= \$1.20$

29a.

Time (hour)	0	1	2	3	4
Amount of Bacteria	12	36	108	324	972

29b. exponential **29c.** $b = 12 \times 3^t$ **29d.** 78,732

31. Sample answer: $y = 2x^2 - 5$ **33.** $y = 4x + 1$

35. The data can be graphed to determine which function best models the data. Also, differences and ratios of the y-values can be used. If the first differences are constant, the data are modeled by a linear function. If the second differences are constant, but the first difference are not, the data can be modeled by a quadratic function. If the ratios are constant, then the data can be modeled by an exponential function.

37. A **39.** B **41.** $a_n = 1(2)^{n-1}$; 64 **43.** $a_n = 4(-3)^{n-1}$; 2916 **45.** $a_n = 22(2)^{n-1}$; 1408 **47.** $x^2 - 8x + 16$ **49.** $16x^2 - 56x + 49$ **51.** $25x^2 - 36y^2$ **53.** $C = 10h + 15$; \$95 **55.** yes; $2x + y = 6$ **57.** yes; $y = -5$ **59.** no

61.

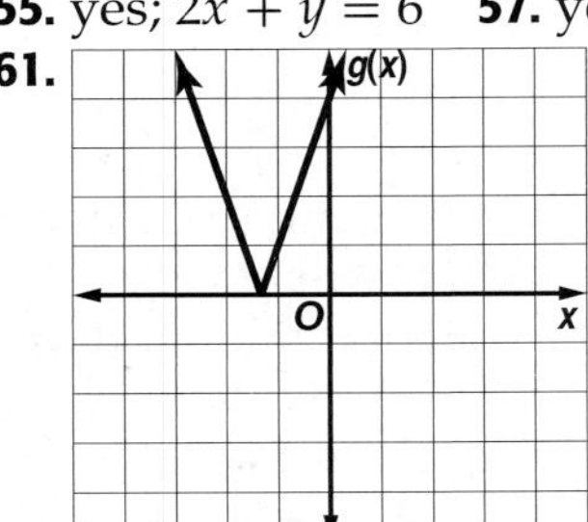

Pages 592–596 Chapter 9 Study Guide and Review

1. true **3.** false; parabola **5.** false; two **7.** true **9.** true **11a.** minimum **11b.** 0 **11c.** D = all real numbers; R = $y \mid y \geq 0\}$ **13a.** minimum **13b.** −4 **13c.** D = all real numbers; R = $\{y \mid y \geq -4\}$ **15a.** maximum **15b.** 16 **15c.** D = all real numbers; R = $\{h \mid h \leq 16\}$ **17.** 3 **19.** −4.6, 0.6 **21.** −0.8, 3 **23.** shifted up 8 units **25.** vertical stretch **27.** vertical compression **29.** $y = 2x^2 - 3$ **31.** 1, −7 **33.** 10, −2 **35.** −0.7, 7.7 **37.** −8, 6 **39.** −0.7, 0.5 **41.** −5, 1.5 **43.** −2.5, 1.5

45. y-intercept 1; D = all real numbers; R = $\{y \mid y > 0\}$

47. y-intercept 3; D = all real numbers; R = $\{y \mid y > 0\}$

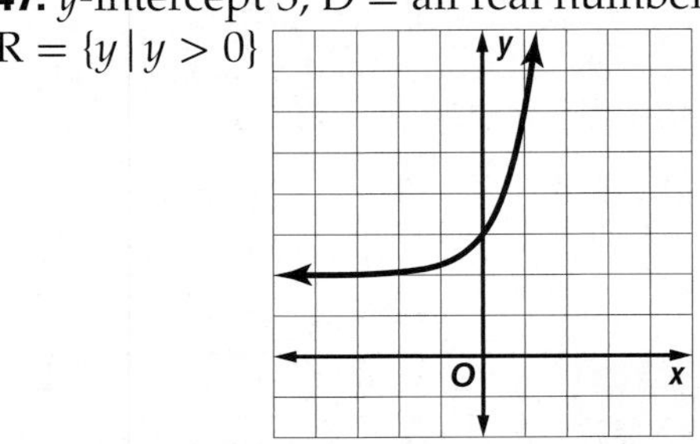

49. about 568 **51a.** $1200(1 - 0.03)^5$ **51b.** \$1030.48 **53.** 81, 243, 729 **55.** $a_n = -1(-1)^{n-1}$ **57.** $a_n = 256\left(\frac{1}{2}\right)^{n-1}$ **59.** quadratic; $y = 3x^2$ **61.** quadratic; $y = -x^2$ **63.** linear; $y = 1.50x + 1$

Chapter 10 Advanced Functions and Equations

Page 603 Chapter 10 Get Ready

1. 9.06 **3.** 3.87 **5.** 10 ft **7.** $13x - 3y$ **9.** $3m + 3n + 10$ **11.** 0, 2 **13.** 2, 5 **15.** 10 **17.** yes

Pages 608–610 Lesson 10-1

1.

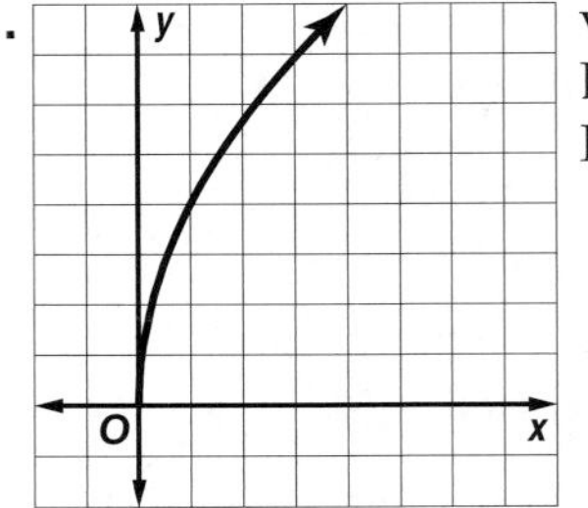

vertical stretch of $y = \sqrt{x}$; D = $\{x \mid x \geq 0\}$, R = $\{y \mid y \geq 0\}$

3.

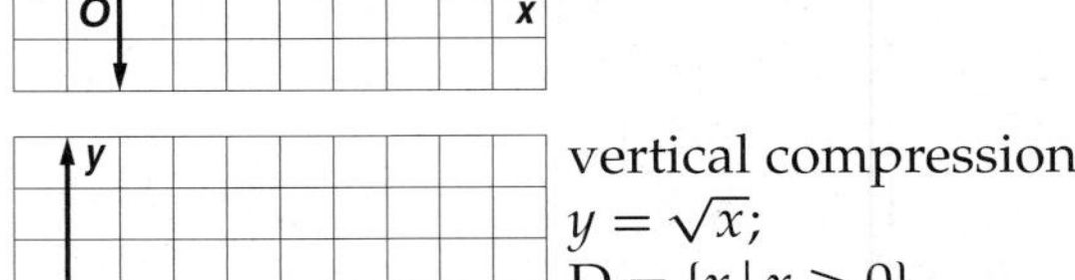

vertical compression of $y = \sqrt{x}$; D = $\{x \mid x \geq 0\}$, R = $\{y \mid y \geq 0\}$

5. translated up 3; D = $\{x \mid x \geq 0\}$, R = $\{y \mid y \geq 3\}$

7.

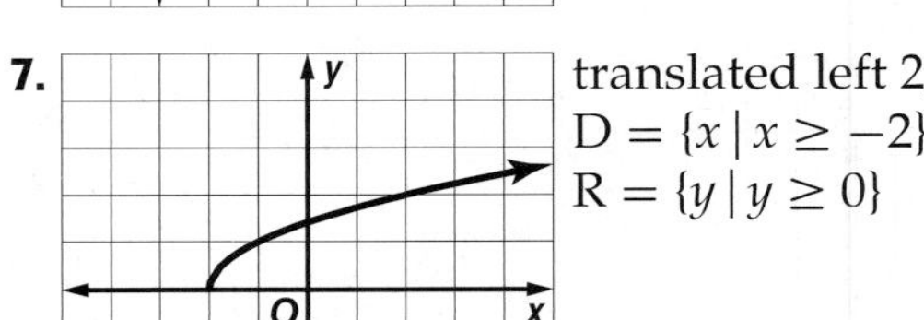

translated left 2; D = $\{x \mid x \geq -2\}$, R = $\{y \mid y \geq 0\}$

9.

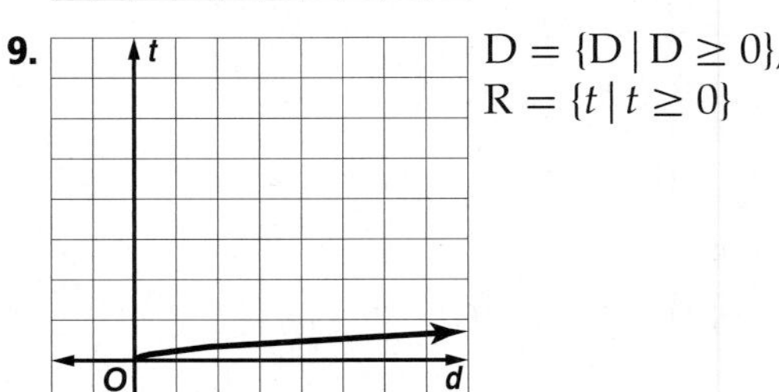

D = $\{D \mid D \geq 0\}$, R = $\{t \mid t \geq 0\}$

11. 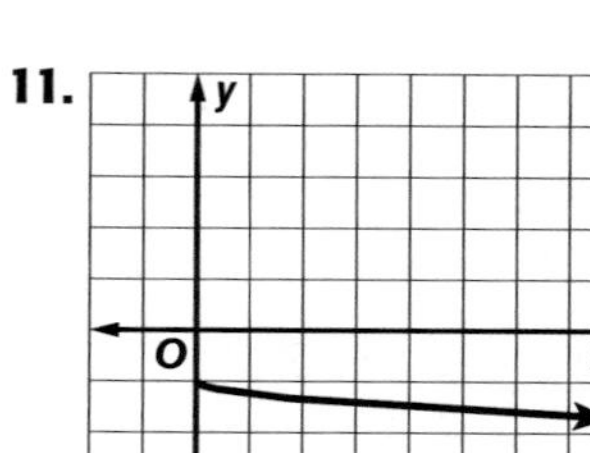

vertical compression of $\sqrt{x}$, reflected across the x-axis and translated down 1;
$D = \{x \mid x \geq 0\}$,
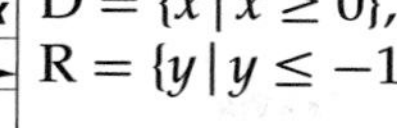
$R = \{y \mid y \leq -1\}$

13. 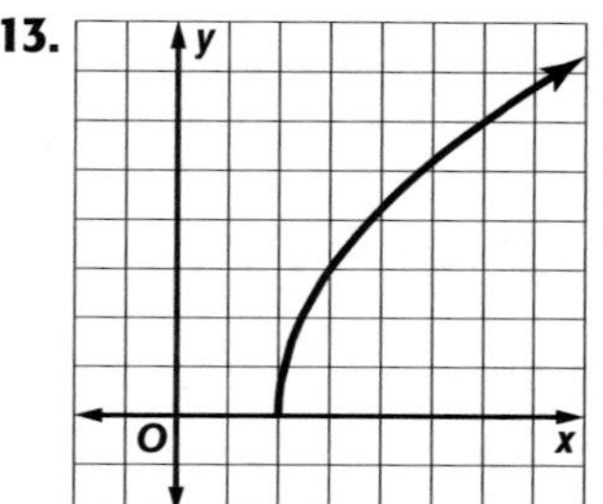

translated right 2 and vertical stretch of $\sqrt{x}$;
$D = \{x \mid x \geq 2\}$,
$R = \{y \mid y \geq 0\}$

15 Step 1: Make a table. Choose nonnegative values for x.

x	y
0	0
0.5	≈0.35
1	0.5
2	0.71
3	0.87
4	1

Step 2: Plot the points and draw a smooth curve.

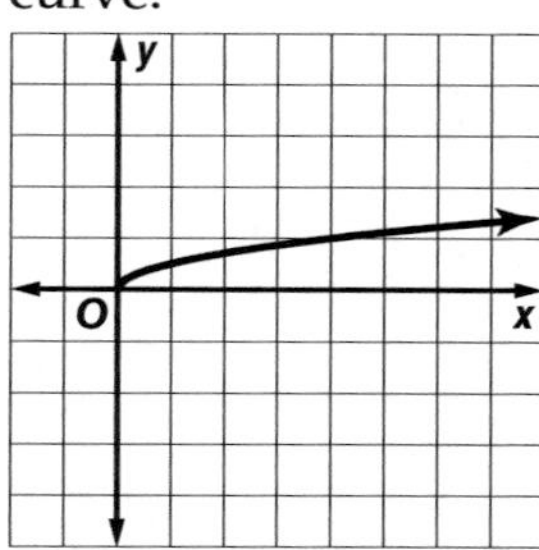

The graph is a vertical compression of $\sqrt{x}$. The domain is $\{x \mid x \geq 0\}$. The range is $\{y \mid y \geq 0\}$.

17. 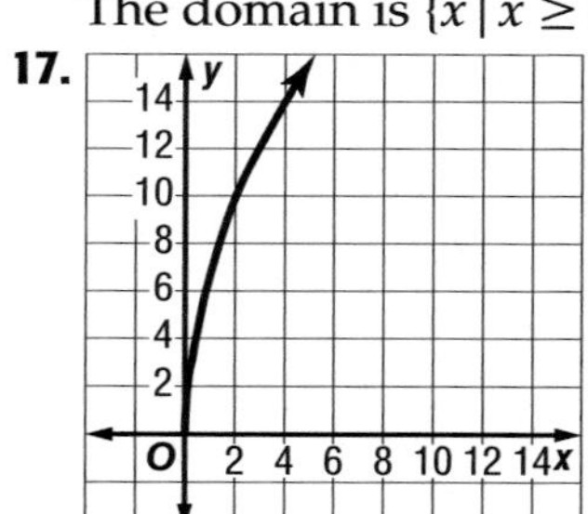

vertical stretch of $\sqrt{x}$;
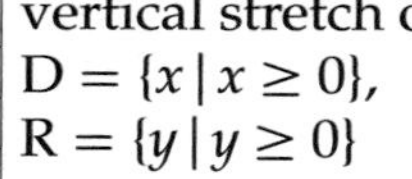
$D = \{x \mid x \geq 0\}$,
$R = \{y \mid y \geq 0\}$

19. 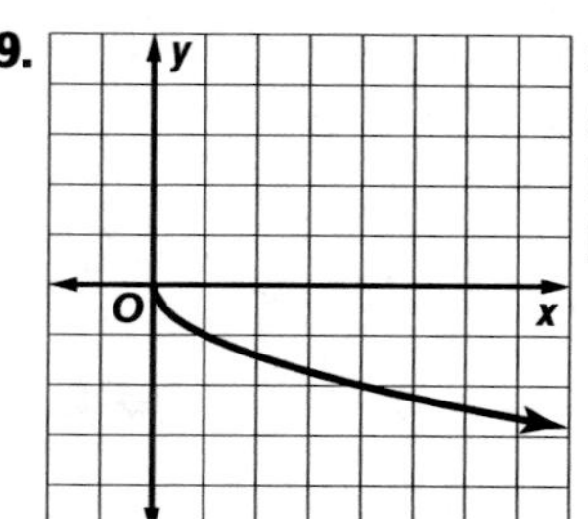

reflected across the x-axis;
$D = \{x \mid x \geq 0\}$,
$R = \{y \mid y \leq 0\}$

21. 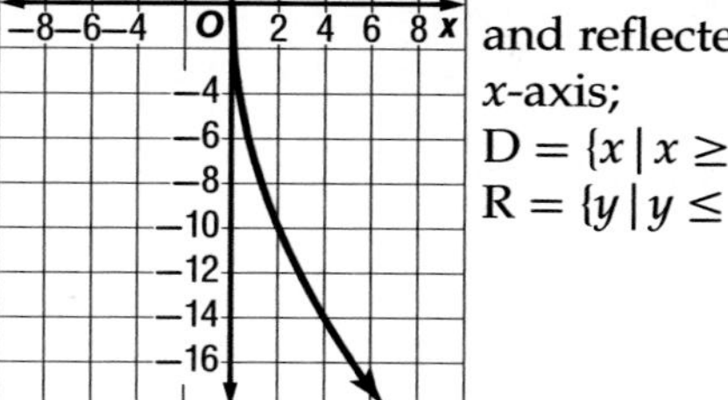

vertical stretch of $\sqrt{x}$ and reflected across the x-axis;
$D = \{x \mid x \geq 0\}$,
$R = \{y \mid y \leq 0\}$

23. 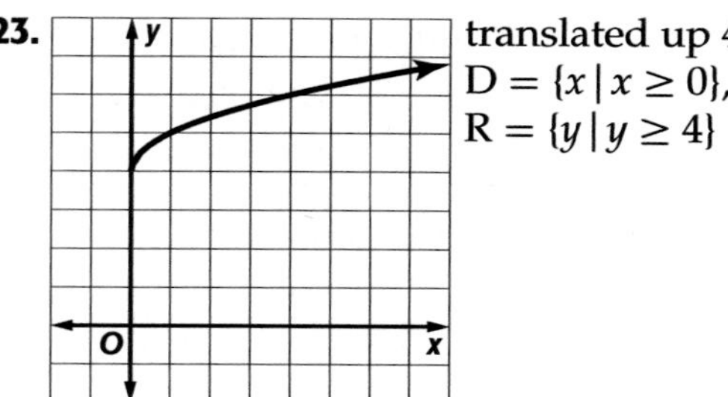

translated up 4;
$D = \{x \mid x \geq 0\}$,
$R = \{y \mid y \geq 4\}$

25. 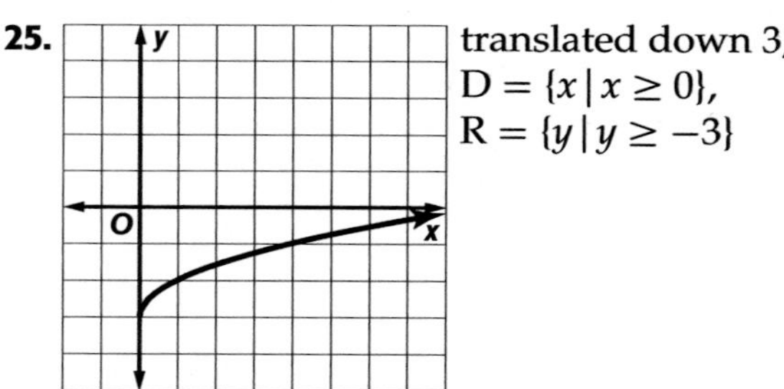

translated down 3;
$D = \{x \mid x \geq 0\}$,
$R = \{y \mid y \geq -3\}$

27. 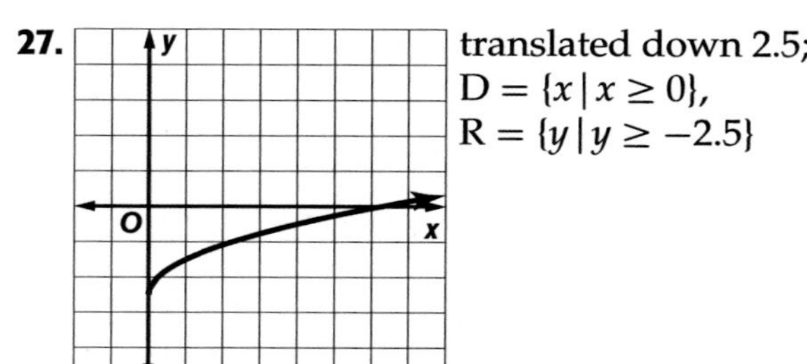

translated down 2.5;
$D = \{x \mid x \geq 0\}$,
$R = \{y \mid y \geq -2.5\}$

29. 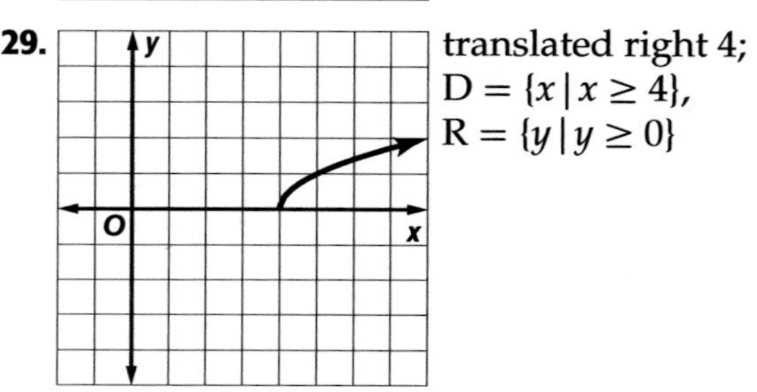

translated right 4;
$D = \{x \mid x \geq 4\}$,
$R = \{y \mid y \geq 0\}$

Selected Answers and Solutions

31. 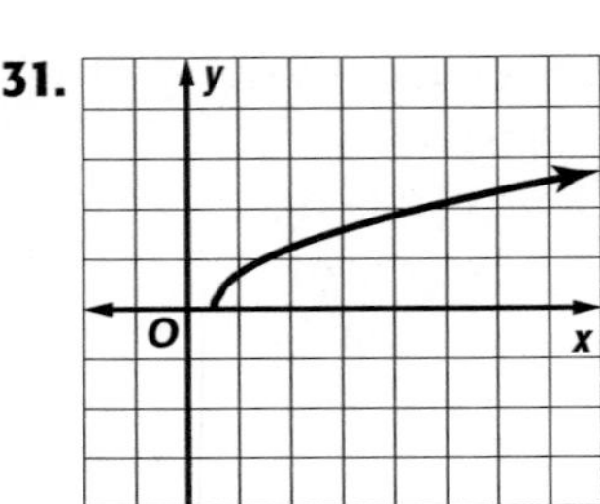

translated right 0.5; $D = \{x \mid x \geq 0.5\}$, $R = \{y \mid y \geq 0\}$

33. 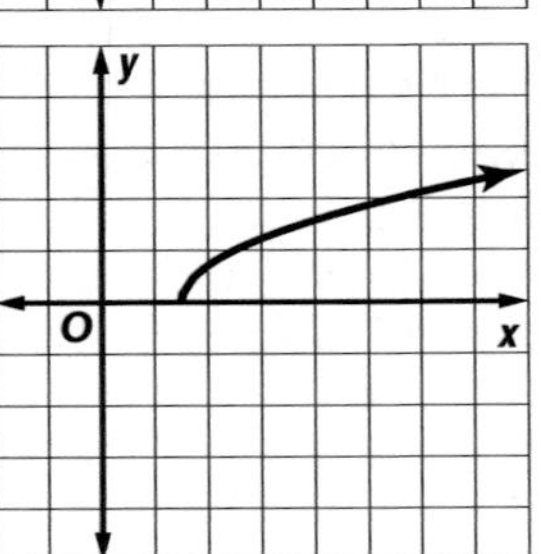

translated right 1.5; $D = \{x \mid x \geq 1.5\}$, $R = \{y \mid y \geq 0\}$

35. 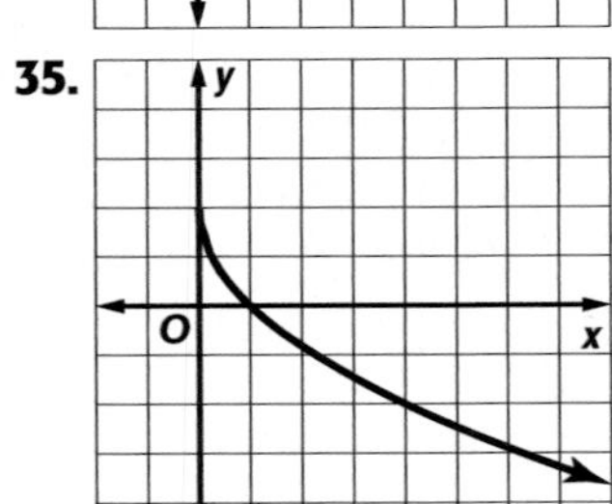

vertical sretch of $y = \sqrt{x}$, reflected across the x-axis, and translated up 2; $D = \{x \mid x \geq 0\}$, $R = \{y \mid y \leq 2\}$

37. 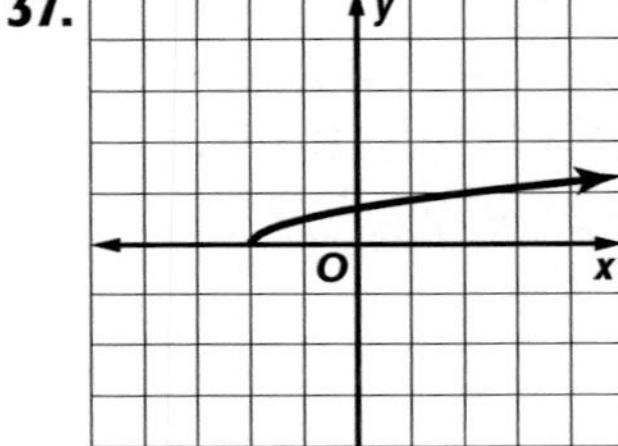

vertical compression of $y = \sqrt{x}$ and translated left 2; $D = \{x \mid x \geq -2\}$, $R = \{y \mid y \geq 0\}$

39. 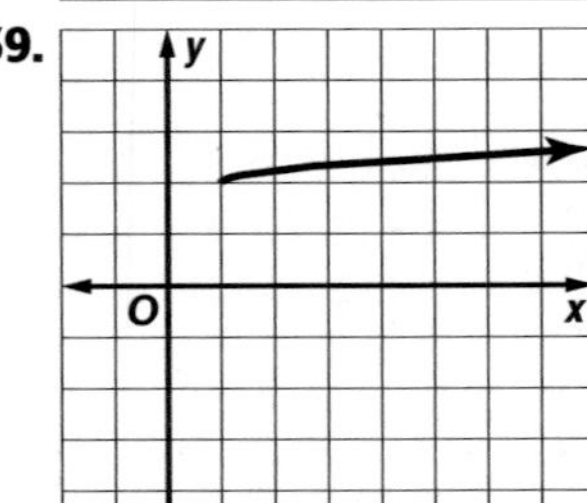

vertical compression of $y = \sqrt{x}$ and translated up 2 and right 1; $D = \{x \mid x \geq 1\}$, $R = \{y \mid y \geq 2\}$

41. 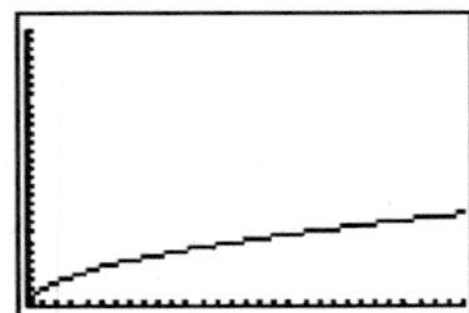

[0, 28] scl: 1 by [0, 28] scl: 1

43 **a.** Use a graphing calculator to graph the equation.

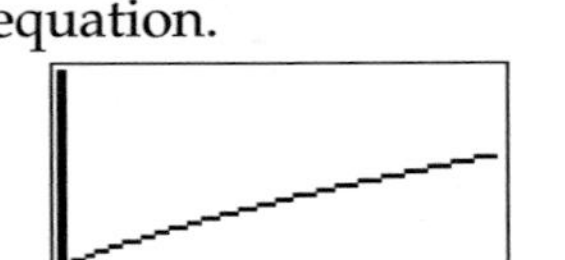

[0, 1000] scl: 20 by [0, 1000] scl: 0.1

b. $c = 331.5\sqrt{1 + \frac{t}{273.15}}$
$= 331.5\sqrt{1 + \frac{55}{273.15}}$
$= 331.5\sqrt{1.201}$
≈ 363.3 m/s

c. When $t = 65°$C:
$c = 331.5\sqrt{1 + \frac{t}{273.15}}$
$= 331.5\sqrt{1 + \frac{65}{273.15}}$
$= 331.5\sqrt{1.238}$
≈ 368.8 m/s

So, an increase of 10°C increases the speed by about 5.5 m/s.

45. False; sample answer: The domain of $y = \sqrt{x + 3}$ includes −1, −2, and −3.

47. $y = \sqrt{x - 4}$

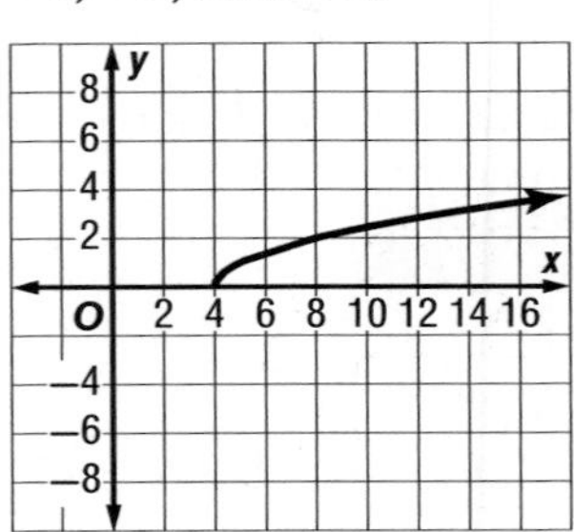

49. $y = \sqrt{x} + 3$; it is a translation of $y = \sqrt{x}$; the other equations represent vertical stretches or compressions. **51.** The value of a is negative. For the function to have negative y-values, the value of a must be negative.

53. A **55.** D

57. 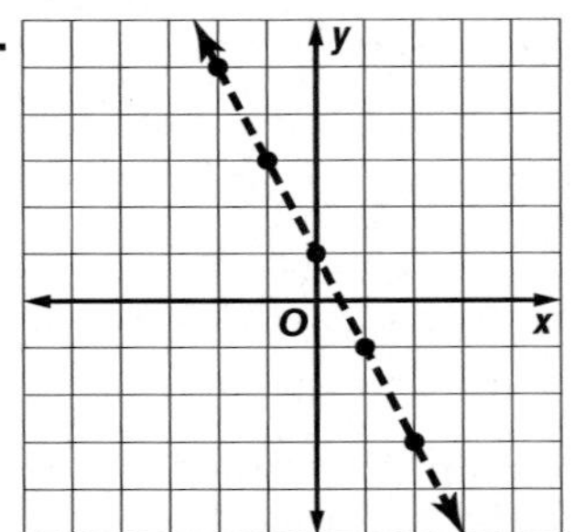

linear

59. 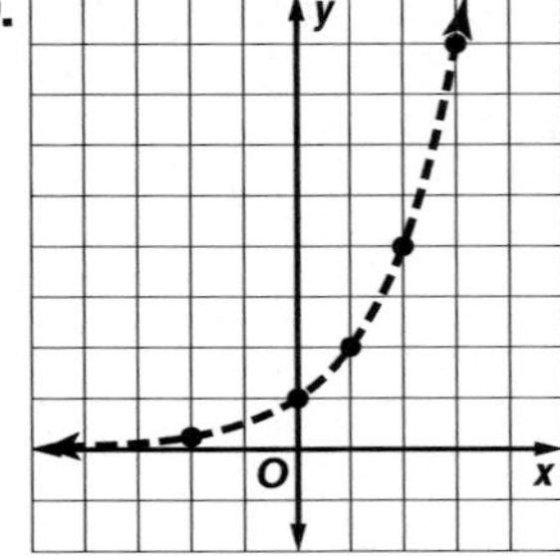

exponential

61. 1280; 5120; 20,480 **63.** 2, 4, 8 **65.** Positive; as the number of grams of fat increases, the amount of Calories increases. **67.** $-3 \cdot 11 \cdot a \cdot a \cdot b$
69. $-2 \cdot 3 \cdot 3 \cdot 3 \cdot 7 \cdot n \cdot q \cdot q \cdot r \cdot r$ **71.** $-2 \cdot 2 \cdot 2 \cdot 2 \cdot 2 \cdot 5 \cdot x \cdot x \cdot y \cdot y \cdot y \cdot y$

Pages 615–617 Lesson 10-2

1. $2\sqrt{6}$ **3.** 10 **5.** $3\sqrt{6}$ **7.** $2x^2|y^3|\sqrt{15y}$

9. $3b^2|c|\sqrt{11ab}$ **11.** $\frac{9-3\sqrt{5}}{4}$ **13.** $\frac{2+2\sqrt{10}}{-9}$

15. $\frac{24+4\sqrt{7}}{29}$ **17.** $2\sqrt{13}$ **19.** $6\sqrt{2}$ **21.** $9\sqrt{3}$

23. $5\sqrt{2}$ **25.** $12\sqrt{14}$ **27.** $15|t|$ **29.** $2|a||b|\sqrt{7b}$

31. $21|m|\sqrt{7mp}$ **33.** $2|a^3||b|\sqrt{5b}$

35 a. $v = \sqrt{64h}$
$= \sqrt{8 \cdot 8h}$
$= \sqrt{8^2} \cdot \sqrt{h}$
$= 8\sqrt{h}$

b. $v = 8\sqrt{h}$
$= 8\sqrt{134}$
≈ 92.6 ft/s

37 $\sqrt{\frac{32}{t^4}} = \frac{\sqrt{32}}{\sqrt{t^4}}$
$= \frac{\sqrt{16 \cdot 2}}{t^2}$
$= \frac{\sqrt{16} \cdot \sqrt{2}}{t^2}$
$= \frac{4\sqrt{2}}{t^2}$

39. $\frac{2|c|\sqrt{51ac}}{9|a|}$ **41.** $\frac{3\sqrt{15}}{20}$ **43.** $\frac{35-7\sqrt{3}}{22}$

45. $\frac{6\sqrt{3}+9\sqrt{2}}{2}$ **47.** $\frac{5\sqrt{6}-5\sqrt{3}}{3}$ **49a.** $I = \frac{\sqrt{PR}}{R}$

49b. 3.9 amps

51.

Distance	3	6	9	12	15
Height	6	24	4	96	150

53. $\pm\frac{\sqrt{3}}{3}$ **55.** Sample answer: $1 + \sqrt{2}$ and $1 - \sqrt{2}$; $(1 + \sqrt{2})(1 - \sqrt{2}) = 1 - 4 = -3$ **57.** No radicals can appear in the denominator of a fraction. So, rationalize the denominator to get rid of the radicand in the denominator. Then check if any of the radicands have perfect square factors other than 1. If so, simplify. **59.** H **61.** 507.50

63.

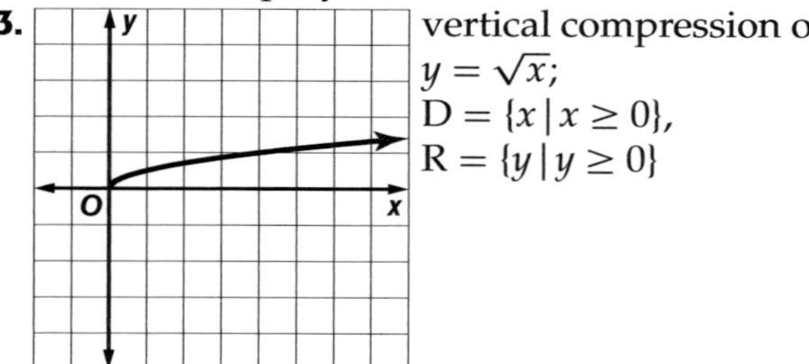

vertical compression of $y = \sqrt{x}$; $D = \{x \mid x \geq 0\}$, $R = \{y \mid y \geq 0\}$

65.

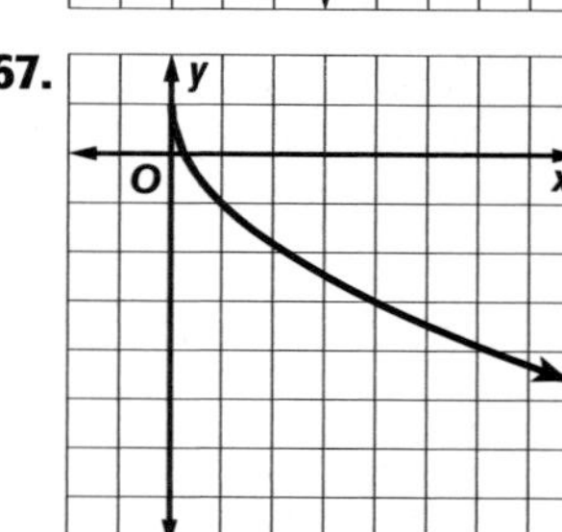

reflected across the x-axis and translated left 1; $D = \{x \mid x \geq -1\}$, $R = \{y \mid y \leq 0\}$

67. stretched vertically, reflected across the x-axis, and translated up 1 $D = \{x \mid x \geq 0\}$, $R = \{y \mid y \leq 1\}$

69. quadratic **71.** about 2,050,422 **73.** ∅ **75.** −2.9, 2.4 **77.** −0.5, 0.6 **79.** $(2 - 3a)(2 + 3a)$ **81.** $2(4x^2 + y^2)(2x - y)(2x + y)$ **83.** $(x + 3)(x - 3)(x - 3)$ **85.** $2^3 \cdot 3$ **87.** $2^2 \cdot 3^2 \cdot 5$ **89.** $2^2 \cdot 3 \cdot 5$

Pages 621–623 Lesson 10-3

1 $3\sqrt{5} + 6\sqrt{5} = (3 + 6)\sqrt{5}$
$= 9\sqrt{5}$

3. $-5\sqrt{7}$ **5.** $8\sqrt{5}$ **7.** $5\sqrt{2} + 2\sqrt{3}$ **9.** $72\sqrt{3}$ **11.** $\sqrt{21} + 3\sqrt{6}$ **13.** $14.5 + 3\sqrt{15}$ **15.** $11\sqrt{6}$ **17.** $3\sqrt{2}$ **19.** $5\sqrt{10}$ **21.** $60 + 32\sqrt{10}$ **23.** $3\sqrt{5} + 6 - \sqrt{30} - 2\sqrt{6}$ **25.** $5\sqrt{5} + 5\sqrt{2}$ **27.** $\frac{-4\sqrt{5}}{5}$ **29.** $\sqrt{2}$ **31.** $14 - 6\sqrt{5}$

33 a. $v_0 = \sqrt{v^2 - 64h}$
$= \sqrt{(120)^2 - 64(225)}$
$= \sqrt{0}$
$= 0$ ft/s

b. Sample answer: In the formula, we are taking the square root of the difference, not the square root of each term.

35. $\sqrt{170}$; about 13 amps **37.** $\left(a\sqrt{b} + c\sqrt{f}\right) \cdot \left(a\sqrt{b} - c\sqrt{f}\right) = a^2\sqrt{b^2} - ac\sqrt{bf} + ac\sqrt{bf} - c^2\sqrt{f^2} = a^2b - c^2f$; Sample answer: This pattern is the difference of squares. **39.** Sample answer: You can use the FOIL method. You multiply the first terms within the parentheses. Then you multiply the outer terms within the parentheses. Then you would multiply the inner terms within the parentheses. And, then you would multiply the last terms within each parentheses. Combine any like terms and simplify any radicals. Sample answer: $\left(\sqrt{2} + \sqrt{3}\right)\left(\sqrt{5} + \sqrt{7}\right) = \sqrt{10} + \sqrt{14} + \sqrt{15} + \sqrt{21}$. **41.** C **43.** C **45.** $2\sqrt{6}$ **47.** $5|a|b^2\sqrt{2ab}$ **49.** $3|c|d^2f^2\sqrt{7cf}$

51. 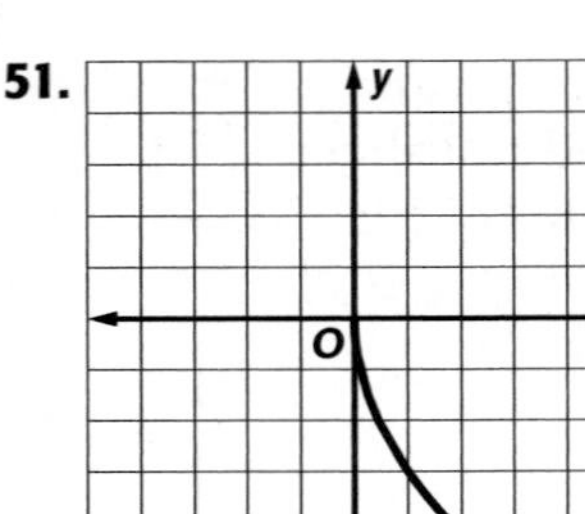
stretched vertically and reflected across the x-axis
$D = \{x \mid x \geq 0\}$,
$R = \{y \mid y \leq 0\}$

53. translated right 4;
$D = \{x \mid x \geq 4\}$,
$R = \{y \mid y \geq 0\}$

55. translated down 2;
$D = \{x \mid x \geq 0\}$,
$R = \{y \mid y \geq -2\}$

57. $(x + 3)(x + 9)$ **59.** $(p - 8)(p - 9)$
61. $(y - 7)(y + 6)$ **63.** -0.5 **65.** 18.7 **67.** 24

Pages 626–628 Lesson 10-4

1. $r = \dfrac{\sqrt{\pi x}}{2\pi}$ **3.** 2 **5.** 10 **7.** 6

9

$$\begin{aligned} \sqrt{a} + 11 &= 21 \\ \sqrt{a} &= 10 \\ (\sqrt{a})^2 &= (10)^2 \\ a &= 100 \end{aligned}$$

11. 39 **13.** 17 **15.** 3 **17.** 6 **19.** 7 **21a.** 52 ft
21b. Increases; sample answer: If the length is longer, the quotient and square root will be a greater number than before. **23.** no solution
25. 235.2 **27.** 3

29 a.

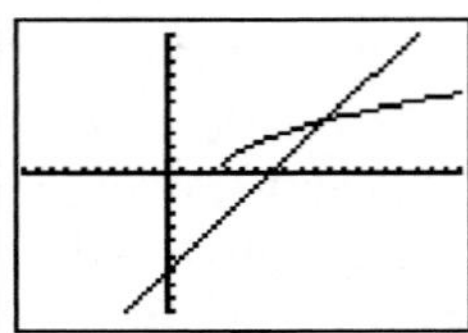

[−10, 20] scl: 1 by [−10, 10] scl: 1

c.

Intersection
X=10.828427 Y=3.8284271

[−10, 20] scl: 1 by [−10, 10] scl: 1

d.

$$\begin{aligned} \sqrt{2x - 7} &= x - 7 \\ \left(\sqrt{2x - 7}\right)^2 &= (x - 7)^2 \\ 2x - 7 &= x^2 - 14x + 49 \\ 0 &= x^2 - 16x + 56 \\ x &= \frac{-b \pm \sqrt{b^2 - 4ac}}{2a} \\ &= \frac{16 \pm \sqrt{(16)^2 - 4(1)(56)}}{2(1)} \\ &= \frac{16 \pm \sqrt{32}}{2} \\ &\approx 10.83 \text{ and } 5.17 \end{aligned}$$

When checking, 5.17 does not work, so the answer is about 10.83 which is the same as we got using the calculator.

31. Jada; Fina had the wrong sign for $2b$ in the fourth step. **33.** Sample answer: In the first equation, you have to isolate the radical first by subtracting 1 from each side. Then square each side to find the value of x. In the second equation, the radical is already isolated, so square each side to start. Then subtract 1 from each side to solve for x. **35.** Sometimes; the equation is true for $x \geq 2$, but false for $x < 2$. **37.** Sample answer: Add or subtract any expressions that are not in the radicand from each side. Multiply or divide any values that are not in the radicand to each side. Square each side of the equation. Solve for the variable as you did previously. **39.** C **41.** D

43. $4\sqrt{3}$ **45.** $42\sqrt{2}$ **47.** $\dfrac{c^2\sqrt{5cd}}{2|d^3|}$ **49.** about 1.3 s and 4.7 s **51.** $(2p + 3)(3p - 2)$ **53.** prime
55. $(2a + 3)(a - 6)$ **57.** Yes; $4x^3$ is the product of a number and three variables. **59.** No; $4n + 5p$ shows addition, not multiplication alone of numbers and variables. **61.** Yes; $\frac{1}{5}abc^{14}$ is the product of a number, $\frac{1}{5}$, and several variables.
63. 1,000,000 **65.** $64v^2$ **67.** $1000y^6$

Pages 632–635 Lesson 10-5

1. 5 **3.** 18.03 **5a.** about 127 ft **5b.** about 127 ft
5c. about 132 ft **7.** yes **9.** no

11

$$\begin{aligned} a^2 + b^2 &= c^2 \\ (2)^2 + b^2 &= 12^2 \\ 4 + b^2 &= 144 \\ b^2 &= 140 \\ b &\approx 11.83 \end{aligned}$$

13. 29.66 **15.** 5.29 **17.** 7.21

19

$$\begin{aligned} a^2 + b^2 &= c^2 \\ (20)^2 + (26)^2 &= c^2 \\ 400 + 676 &= c^2 \\ 1076 &= c^2 \\ 2.8 &\approx c \end{aligned}$$

The diagonal of the TV stand is about 32.8 inches which is larger than 27 inches, so the TV will fit.

21. no; no **23.** no; no **25.** no, no **27.** yes; yes
29a. about 20.20 **29b.** 111.1 units^2 **31.** a 30-ft ladder **33.** 8.06 **35.** about 4.24 m **37.** $5\sqrt{3}$ in. or

about 8.66 in. **39.** about 6.7 ft **41.** about 42.5 in.

43 $a^2 + b^2 = c^2$
$(8)^2 + x^2 = (x + 2)^2$
$64 + x^2 = x^2 + 4x + 4$
$64 = 4x + 4$
$60 = 4x$
$15 = x$
So, $b = 15$ and $c = 15 + 2$, or 17

45. $a = 65$; $b = 72$ **47.** $a = 9$; $b = 40$; $c = 41$
49. about 3.29 cm; about 7.29 cm **51.** $8\sqrt{2}$
53. Sample answer:

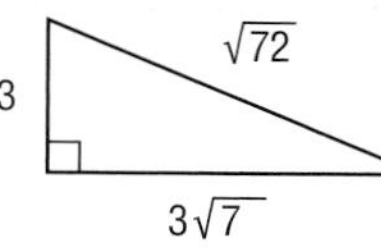

55. C **57.** \$88 **59.** 256 **61.** 10 **63.** 3 **65.** $7\sqrt{3}$
67. $10\sqrt{7}$ **69.** $12\sqrt{5} - 5\sqrt{3}$ **71.** about \$37.27 million; about \$41.74 million; about \$46.75 million
73. compressed vertically **75.** translated up 10
77. reflected across the x-axis, translated down $\frac{4}{3}$
79. $b^2 + 10b + 16$ **81.** $y^2 - 4y - 32$
83. $2w^2 + 9w - 35$ **85.** 20 **87.** 8

Pages 638–641 Lesson 10-6

1. $2\sqrt{34}$

3 $d = \sqrt{(x_2 - x_1)^2 + (y_2 - y_1)^2}$
$= \sqrt{(6 - 3)^2 + (-2 - 0)^2}$
$= \sqrt{(3)^2 + (-2)^2}$
$= \sqrt{9 + 4}$
$= \sqrt{13}$

5a. $2\sqrt{29}$ or about 10.77 ft **5b.** $\sqrt{5}$ or about 2.24 ft
7. −4 or 4 **9.** −18 or 6 **11.** (4, 0) **13.** $\left(-\frac{1}{2}, 0\right)$
15. (1, 6) **17.** $\left(-\frac{19}{2}, 14\right)$ **19.** 3 **21.** $\sqrt{34}$ **23.** $\sqrt{106}$
25. $\sqrt{365}$ **27.** $\sqrt{29}$ **29.** $8\sqrt{2}$ **31.** −9 **33.** 2 or 4
35. 3 or 7 **37.** $\left(\frac{7}{2}, \frac{5}{2}\right)$ **39.** (−2, 7)

41 $M = \left(\frac{x_1 + x_2}{2}, \frac{y_1 + y_2}{2}\right)$
$= \left(\frac{-5 + 3}{2}, \frac{5+(-3)}{2}\right)$
$= \left(\frac{-2}{2}, \frac{2}{2}\right)$
$= (-1,1)$

43. $\frac{10}{3}$ or $3\frac{1}{3}$ **45.** $2\sqrt{14}$ **47.** $\sqrt{68} + \sqrt{26} + \sqrt{104} + \sqrt{82}$; 32.6 units

49 **a.** $d = \sqrt{(x_2 - x_1)^2 + (y_2 - y_1)^2}$
$= \sqrt{(10)^2 + (8)^2}$
$= \sqrt{100 + 64}$
$= \sqrt{164}$
≈ 12.80 miles

b. $d = \sqrt{(x_2 - x_1)^2 + (y_2 - y_1)^2}$
$= \sqrt{(6)^2 + (12)^2}$
$= \sqrt{36 + 144}$
$= \sqrt{180}$
≈ 13.42 miles

c. $d = \sqrt{(x_2 - x_1)^2 + (y_2 - y_1)^2}$
$= \sqrt{(16)^2 + (4)^2}$
$= \sqrt{256 + 16}$
$= \sqrt{272}$
$= 4\sqrt{17}$
≈ 16.49 miles

51. (3.375, −0.25) **53.** $\left(\frac{11}{30}, \frac{23}{20}\right)$ **55.** Sample answer: The Distance Formula requires values to be squared. Once the coordinates and a are substituted into the formula and simplified, the result is a quadratic equation that can result in two possible values for a once solved.
57. Sample answer: (0, 2)

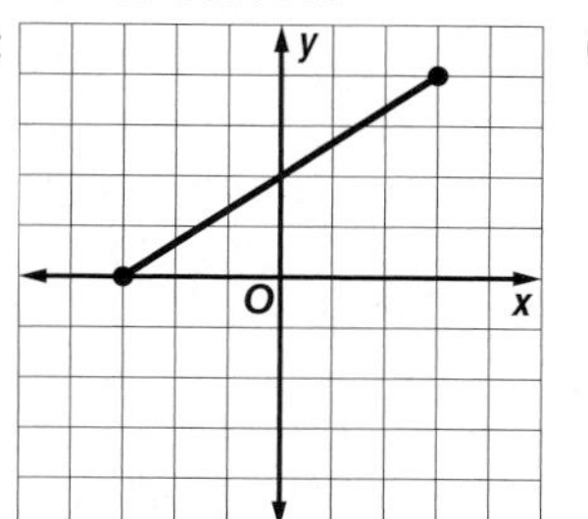

59. 15 mi **61.** H **63.** 65 **65.** 10.72
67. 8.19 **69.** 0.06 **71.** 18 **73.** 8.17 **75.** 1.32

Pages 644–647 Lesson 10-7

1. Yes; the angle measures are equal.

3 $\frac{a}{x} = \frac{b}{y} = \frac{c}{z}$
$\frac{4}{6} = \frac{6}{y} = \frac{8}{z}$
$\frac{4}{6} = \frac{6}{y}$
$4y = 6(6)$
$4y = 36$
$y = 9$
$\frac{4}{6} = \frac{8}{z}$
$4z = 6(8)$
$4z = 48$
$z = 12$

5. $c = 6$; $y = 25$ **7.** 17 ft **9.** No; the sides are not proportional. **11.** No; the angle measures are not equal. **13.** Yes; the angle measures are equal.
15. $m = 3$; $k = 8$ **17.** $k = 4.2$; $r = 0.28$ **19.** $h = \sqrt{5}$; $r = 2$ **21.** 20 in. from S

23 $\frac{1}{10} = \frac{x}{151}$
$1(151) = 10x$
$151 = 10x$
$15.1 = x$
$15\frac{1}{10}$ in. $= x$

25 **a.** first pair: 3 to 6 or 1 to 2
second pair: 4 to 12 or 1 to 3
third pair: 6 to 24 or 1 to 4

b.

Similar Triangles		Ratios	Perimeter	Ratio of perimeters
Pair 1	smaller triangle	1:2	$2 + 3 + 4 = 9$	1:2
	larger triangle		$4 + 6 + 8 = 18$	
Pair 2	smaller triangle	1:3	$3 + 4 + 5 = 12$	1:3
	larger triangle		$9 + 12 + 15 = 36$	
Pair 3	smaller triangle	1:4	$4 + 6 + 8 = 18$	1:4
	larger triangle		$16 + 24 + 32 = 72$	

c. The perimeters are in the same proportion as the side measures of the two similar triangles.

d. It would be the same or 1:6.

27. $\triangle XYZ \sim \triangle XZW$, $\triangle XYZ \sim \triangle ZYW$, $\triangle XZW \sim \triangle ZYW$; the triangles are similar to each other because the angle measures are equal. **29.** $\triangle PQR$ has a base that is twice the base and twice the height of $\triangle ABC$. The triangles are similar because their corresponding angles are congruent. The area of $\triangle PQR$ is four times the area of $\triangle ABC$.

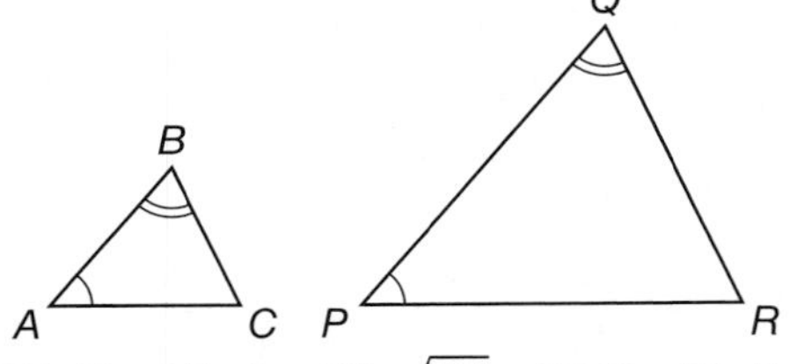

31. D **33.** A **35.** $\sqrt{37}$ **37.** 2 **39.** $5\sqrt{2}$ **41.** yes **43.** yes **45.** no **47.** about 2017 **49.** $4(a - 3b)(a + 3b)$ **51.** $2(5g + 2)^2$ **53.** $2(5n + 1)(2n + 3)$ **55.** $-\frac{1}{3}$ **57.** 2.25 **59.** 8

Pages 652–655 Lesson 10-8

1. $\sin A = \frac{24}{25}$; $\cos A = \frac{7}{25}$; $\tan A = \frac{24}{7}$

3 $\sin A = \frac{\text{opposite}}{\text{hypotenuse}} = \frac{5}{13}$

$\cos A = \frac{\text{adjacent}}{\text{hypotenuse}} = \frac{12}{13}$

$\tan A = \frac{\text{opposite}}{\text{adjacent}} = \frac{5}{12}$

5. 0.6018 **7.** 0.2493 **9.** $m\angle X \approx 51°$; $XY \approx 4.4$; $YZ \approx 5.4$ **11.** $m\angle Q \approx 60°$; $RQ \approx 2.9$; $PQ \approx 5.8$ **13.** about 11,326.2 ft **15.** 66° **17.** 33° **19.** $\sin B = \frac{5}{13}$; $\cos B = \frac{12}{13}$; $\tan B = \frac{5}{12}$ **21.** 0.0349 **23.** 0.7193 **25.** 0.9563 **27.** 0.5

29 Step 1: Find the measure of $\angle Y$.
$180° - (90° + 47°) = 43°$
Step 2: Find $\overline{XY}$ or z. Since you are given the measure of the side opposite $\angle X$, and are finding the measure of the hypotenuse, use the sine ratio.

$\sin 47° = \frac{16}{z}$
$z\sin 47° = 16$
$z = \frac{16}{\sin 47°}$
$z \approx 21.9$

Step 3: Find $\overline{XZ}$ or y. Since you are given the measure of the side opposite $\angle X$, and are finding the measure of the side adjacent to $\angle X$, use the tangent ratio.

$\tan 47° = \frac{16}{y}$
$y\tan 47° = 16$
$y = \frac{16}{\tan 47°}$
$y \approx 14.9$

31. $\angle R = 76°$; $QR = 7.2$; $\overline{PR} = 1.7$ **33.** $\angle Y = 39°$; $\overline{WU} = 11.3$; $\overline{UY} = 18.0$ **35.** about 53 ft **37.** 62° **39.** 31° **41.** 50°

43 $\tan 8° = \frac{5000}{x}$
$x\tan 8° = 5000$
$x = \frac{5000}{\tan 8°}$
$x \approx 35{,}577$ ft

45. $\sin A = \frac{\sqrt{7}}{4}$; $\tan A = \frac{\sqrt{7}}{3}$ **47.** $\cos A = \frac{\sqrt{15}}{4}$; $\tan A = \frac{\sqrt{15}}{15}$ **49.** about 0.5 mi **51.** $a = 5$; $c = 5$ **53.** Sample answer: Find the measure of $\angle A$ in the following triangle; $m\angle A \approx 56°$.

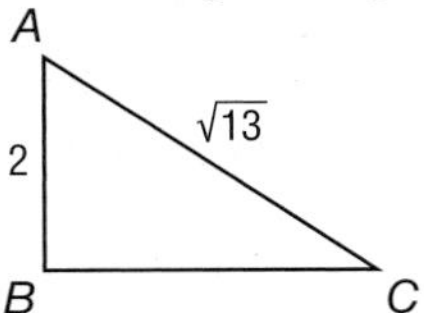

55. Use the angle given and the measure of the known side to set up one of the trigonometric ratios. The sine ratio uses the opposite side and hypotenuse of the triangle. The cosine ratio uses the adjacent side and hypotenuse of the triangle. The tangent ratio uses the opposite and adjacent sides of the triangle. Set up the ratio and solve for the unknown measure. **57.** F **59a.** increase **59b.** Sample answer: The sum of their squares is 16^2 or 256. **59c.** about 15.7 ft **61.** $a = 27$; $c = 12$ **63.** $a = 5$; $f = 8$ **65.** (7, 3) **67.** The amount of sales must be more than \$260,000. **69.** 8 **71.** 4.62

Pages 656–660 Chapter 10 Study Guide and Review

1. false; Sample answer: 3, 4, and 5 **3.** true **5.** false; longest **7.** false; $\{x \mid x \geq 0\}$ **9.** true

11.

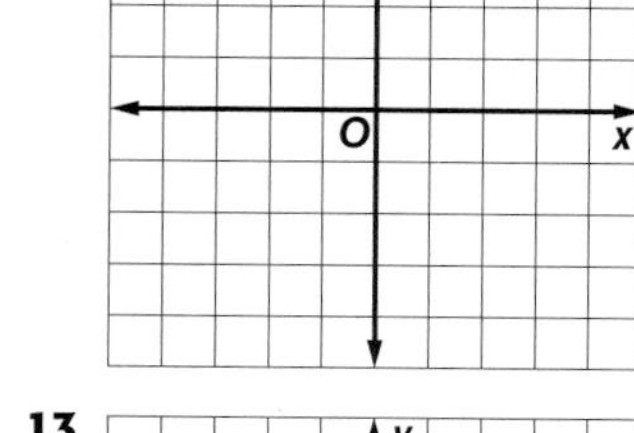

translated up 3;
$D = \{x \mid x \geq 0\}$,
$R = \{y \mid y \geq 3\}$

13. stretched vertically and reflected across the x-axis;
$D = \{x \mid x \geq 0\}$,
$R = \{y \mid y \leq 0\}$

15.

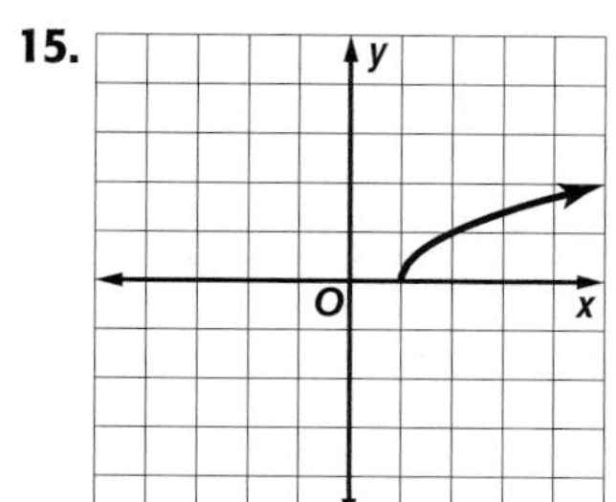

translated right 1;
$D = \{x \mid x \geq 1\}$,
$R = \{y \mid y \geq 0\}$

17. 9.5 in. **19.** $2|b|\sqrt{5ab}$ **21.** 36 **23.** $2\sqrt{2} + 3$ **25.** $\frac{\sqrt{30}}{10}$ **27.** $\frac{5\sqrt{7} - 30}{-29}$ **29.** $-2\sqrt{6} + 11\sqrt{3}$ **31.** $4\sqrt{3x}$ **33.** $5\sqrt{2} + 3\sqrt{6}$ **35.** $24\sqrt{10} + 8\sqrt{2} + 6\sqrt{15} + 2\sqrt{3}$ **37.** no solution **39.** 32 **41.** 12 **43.** 1600 ft **45.** yes **47.** no **49.** yes **51.** no **53.** 5; $\left(-\frac{1}{2}, 4\right)$ **55.** 9.2; $\left(-3, -\frac{7}{2}\right)$ **57.** 18 or −8 **59.** about 4.1 ft **61.** $e = 6, f = 7.5$ **63.** $d = 10$, $e = 14$ **65.** $\cos A = \frac{5}{13}$, $\sin A = \frac{12}{13}$, $\tan A = \frac{12}{5}$ **67.** 6 ft

Chapter 11 Rational Functions and Equations

Page 667 Chapter 11 Get Ready

1. $\frac{8}{3}$ **3.** $\frac{21}{2}$ **5.** $\frac{10}{3}$ or $3\frac{1}{3}$ inches **7.** $5cd$ **9.** $4xy$ **11.** $2x(x - 2)$ **13.** $3x(2y + 5)$ **15.** $(x + 3), (x + 2)$

Pages 673–676 Lesson 11-1

1. Direct; the data in the table can be represented by the equation $y = 2x$. **3.** Inverse; $xy = 4$.

5.

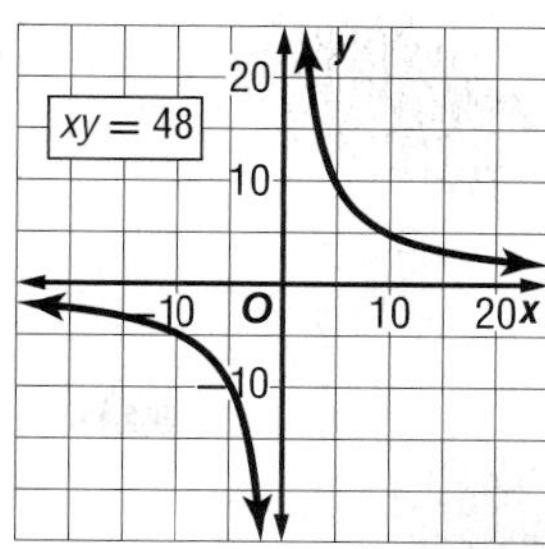

7.

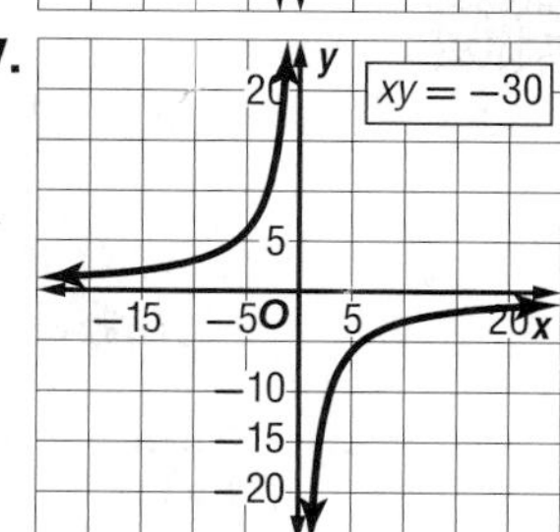

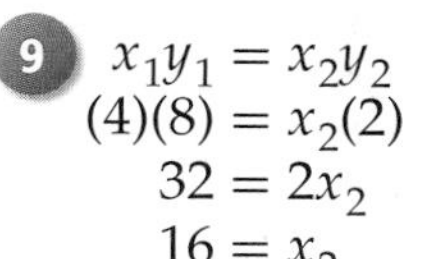

9 $x_1y_1 = x_2y_2$
$(4)(8) = x_2(2)$
$32 = 2x_2$
$16 = x_2$

11. −7.5

13a.

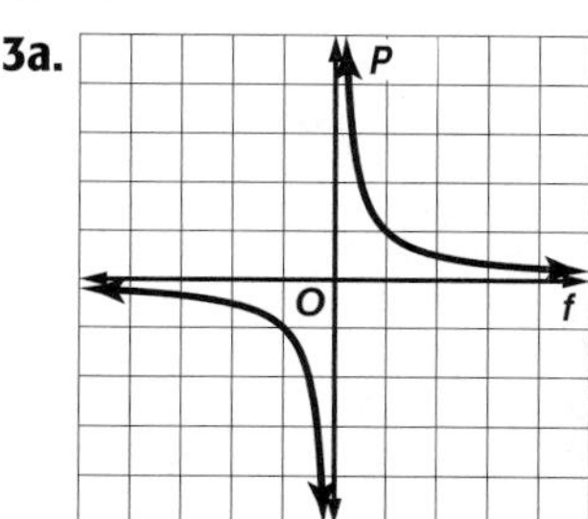

13b. 5 to −2.5 diopters **15.** Direct; $y = -3x$ **17.** Inverse; $xy = -40$ **19.** Inverse; $xy = \frac{1}{4}$ **21.** Direct; $y = 9x$

23.

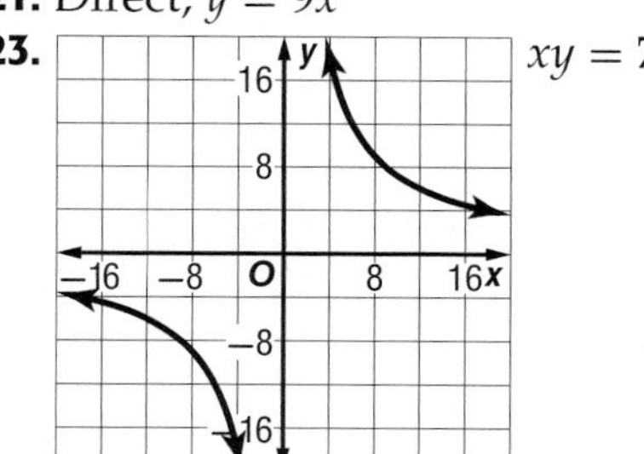

$xy = 72$

25.

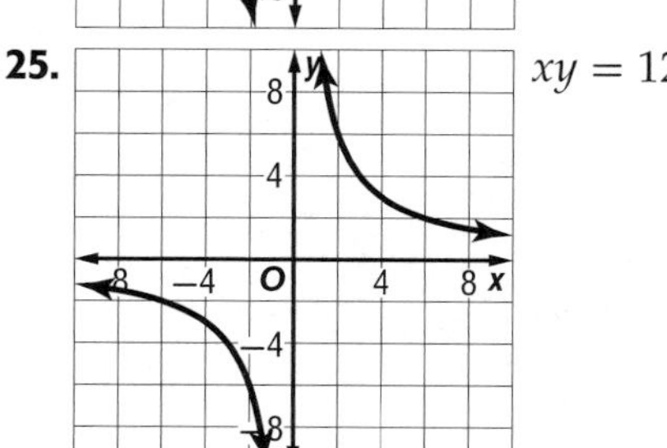

$xy = 12$

27. 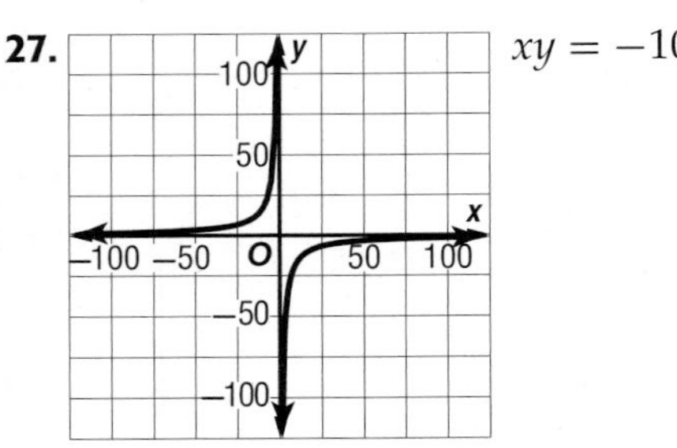

$xy = -108$

29. 15 **31.** −3 **33.** 9.6

35
$$x_1y_1 = x_2y_2$$
$$(420)(523) = (707)y_2$$
$$219{,}660 = 707y_2$$
$$311 \approx y_2$$
approximately 311 cycles per second

37. Direct; the number of lemonades times the cost per lemonade equals the total cost. So the ratio $\frac{\text{total cost}}{\text{number of lemonades}}$ is a constant \$1.50.

39. Inverse; the number of friends times the number of tokens per person equals the constant 30. **41.** Inverse; $xy = 21$ **43.** Direct; $y = \frac{1}{2}x$

45
$$x_1y_1 = x_2y_2$$
$$(9.2)(6) = x_2(3)$$
$$55.2 = 3x_2$$
$$18.4 = x_2$$

47. 2.5 **49.** \$4

51a.

Hours per Week *h*	Number of Weeks *w*
1	40
2	20
4	10
5	8
8	5
10	4

51b. The number of weeks decreases.
51c. $hw = 40$ or $w = \frac{40}{h}$

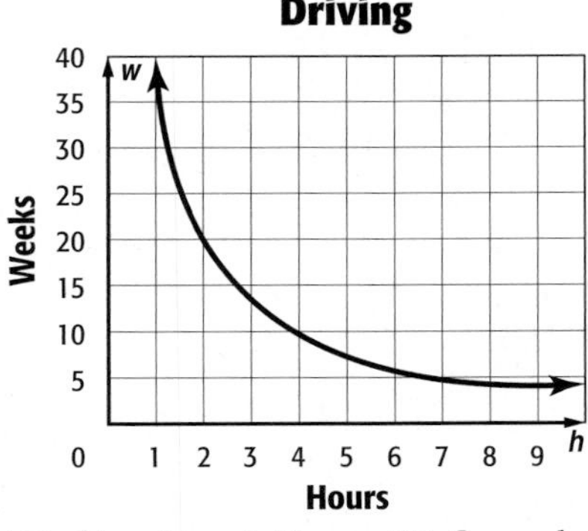

53. direct variation **55.** Sample answer: Newton's Law of Gravitational Force is an example of an inverse variation that models real-world situations. The gravitational force exerted on two objects is inversely proportional to the square of the distances between the two objects. The force exerted on the two objects, times the square of the distance between the two objects, is equal to the gravitational constant times the masses of the two objects. **57.** B **59.** C **61.** $\sin A = 0.7241$, $\cos A = 0.6897$, $\tan A = 1.05$ **63.** $\sin A = 0.8182$, $\cos A = 0.5750$, $\tan A = 1.4230$ **65.** $\frac{9}{10}$ **67.** 1 **69.** no solution **71.** 7^2 or 49 **73.** $\frac{q^4}{2p^5}$ **75.** $\frac{4a^4b^2}{c^6}$ **77.** $\frac{1}{mn}$

Pages 681–683 Lesson 11-2

1. $x = 0$ **3.** $x = 1$

5. 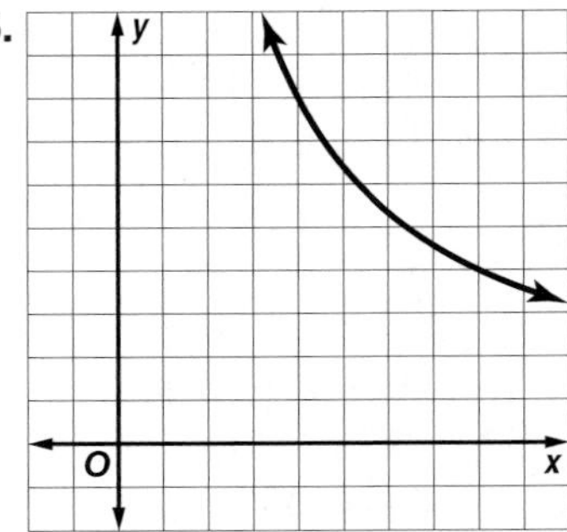

7. $x = 0$; $y = -1$ 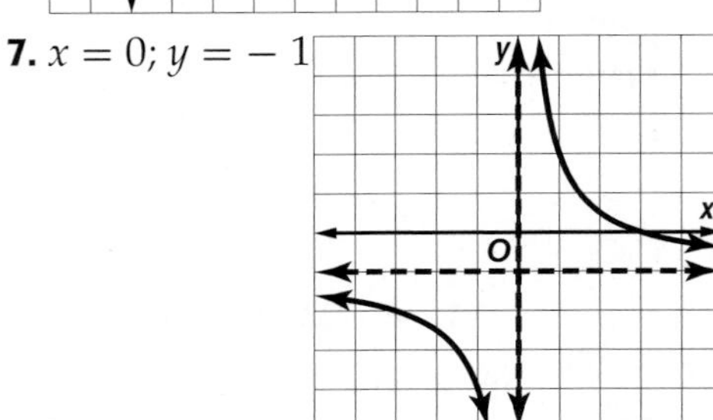

9. $x = -2$; $y = 0$ 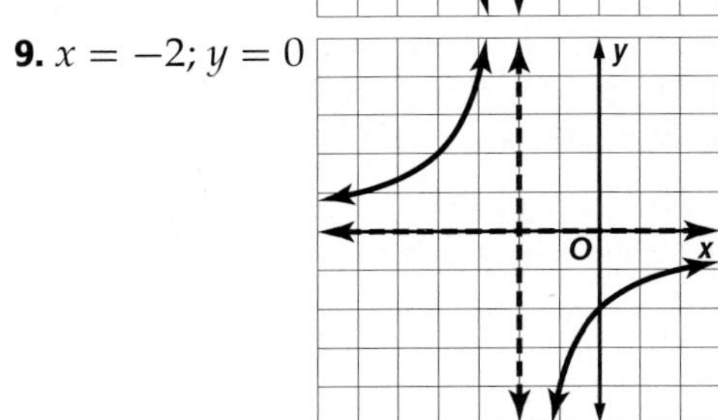

11. $x = -1$; $y = -5$ 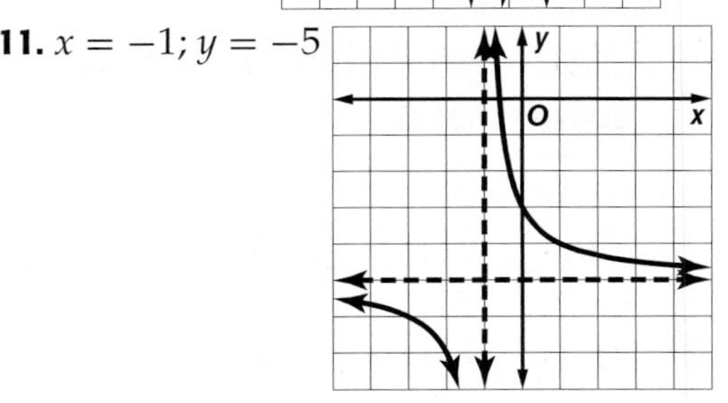

13. $x = 8$ **15.** $x = -6$ **17.** $x = -5$ **19.** $x = -7$

21. 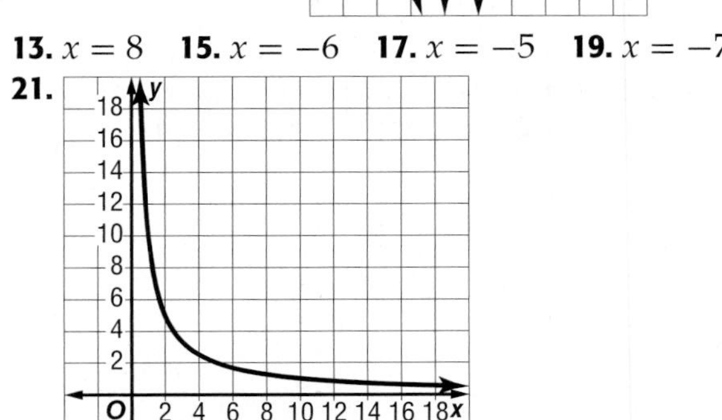

Selected Answers and Solutions

23 Step 1: Identify and graph the asymptotes using dashed lines.

vertical asymptote: $x = 0$
horizontal asymptote: $y = 0$

Step 2: Make a table of values and plot the points. Then connect them.

x	-2	-1	1	2
y	$\frac{3}{2}$	3	3	$\frac{3}{2}$

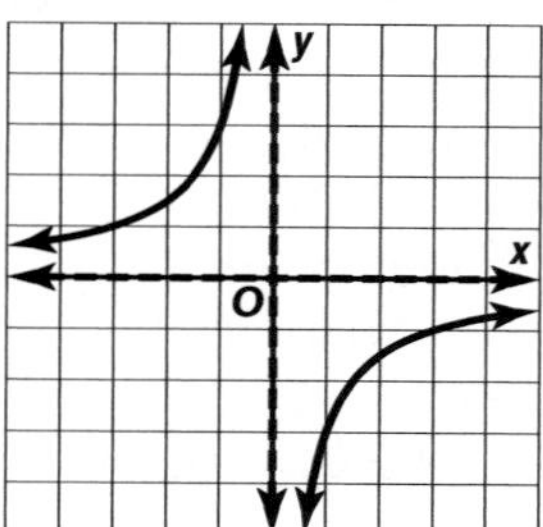

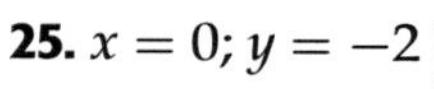

25. $x = 0;\ y = -2$

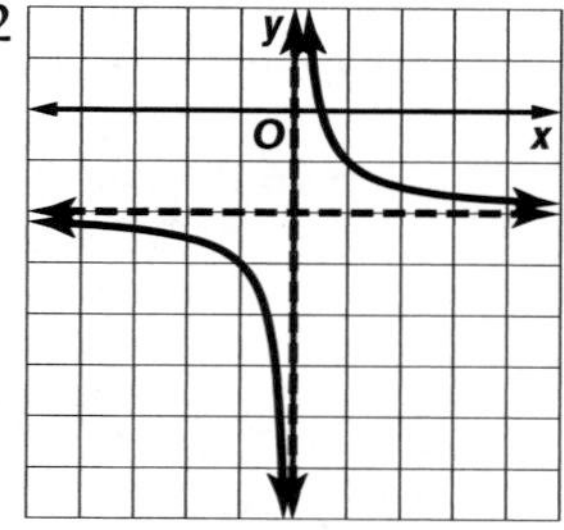

27. $x = 2;\ y = 0$

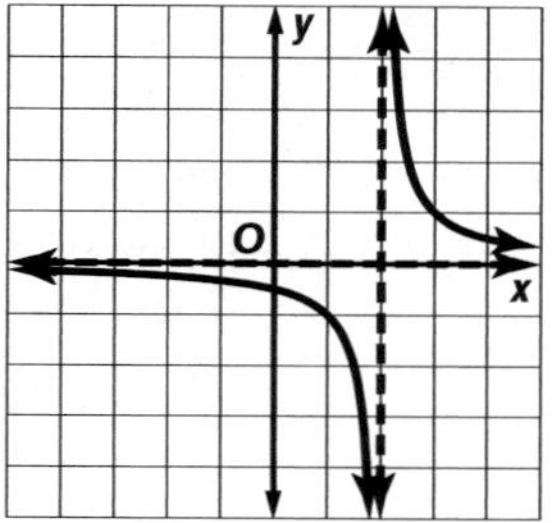

29. $x = 1;\ y = 0$

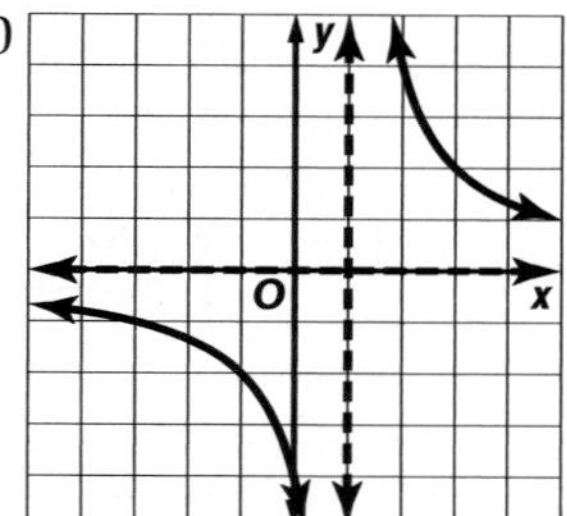

31. $x = 1;\ y = -2$

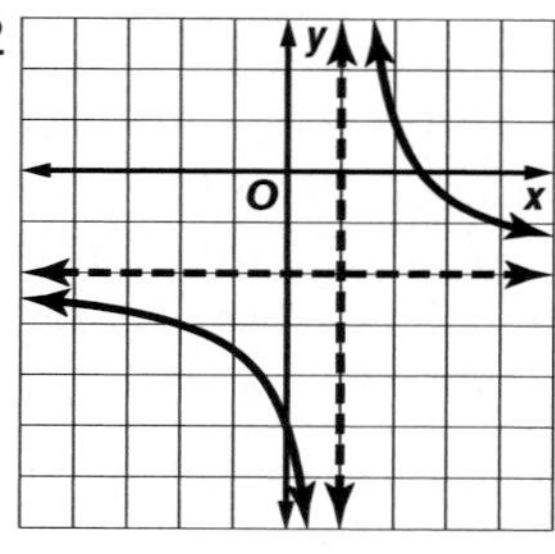

33. $x = -4;\ y = 3$

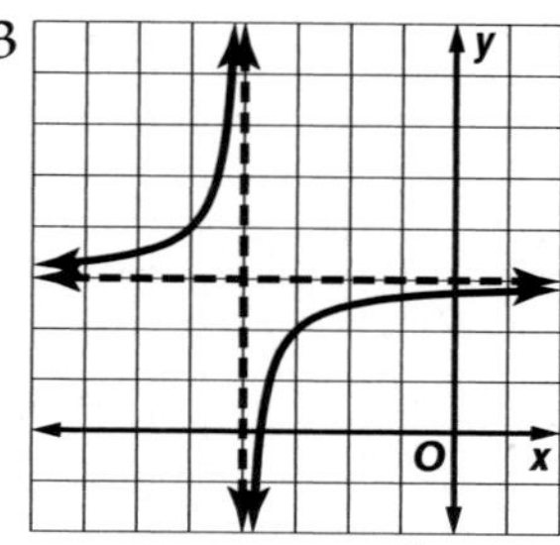

35a. $x = 3$ and $y = 2$ **35b.** $y = \frac{1}{x-3} + 2$

37a. Sample answer: The total cost of the trip equals the cost of a ticket plus the cost of the star-naming package divided by the number of people. **37b.** $y = \frac{95}{p} + 8.50$

37c.

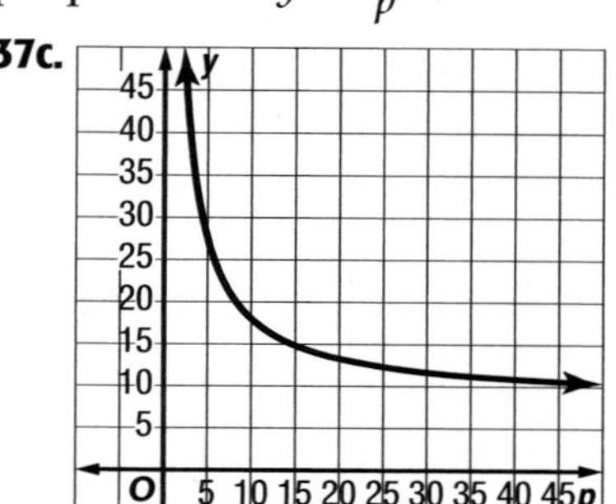

37d. Sample answer: 15 people

39. $x = -1, x = 1;\ y = 1$

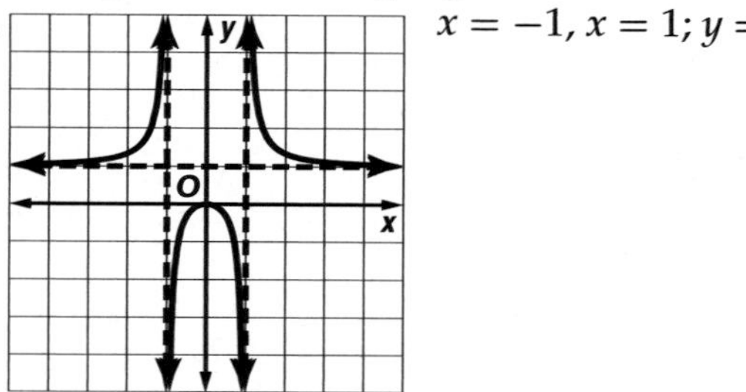

41 **a.** The domain will be positive values since negative values do not make sense for the base of a trapezoid.
The range must also be positive values since negative values do not make sense for the height.
D = {all positive real numbers}
R = {all positive real numbers}

b. Step 1: Identify and graph the asymptotes using dashed lines.
vertical asymptote: $b_1 = -8$
horizontal asymptote: $h = 0$
Step 2: Make a table of values and plot the points. Then connect them.

b_1	2	4	6	8
h	12.8	10.7	9.1	8

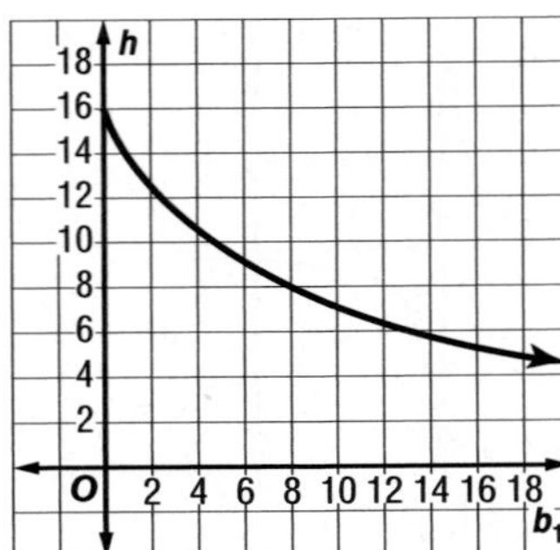

c. When $b_1 = 10$, h will be about 7 units.
43. The graph of $y = \frac{1}{x+5} - 2$ is the graph of $y = \frac{1}{x}$ translated 5 units to the left and 2 units down.
45. False; sample answer: The graph of $y = \frac{1}{x}$ has no x- or y-intercepts. **47.** The vertical asymptote of a rational function occurs for values of x that make the denominator zero. **49.** 45 seconds
51. D **53.** 78° **55.** 40° **57.** 60° **59.** 61.85
61. 10.54 **63.** 9.22 **65.** $\sqrt{465}$ or about 21.56 mi
67. $(w + 16)(w - 3)$ **69.** $(3 + a)(24 + a)$
71. $(d - 2)(d - 5)$ **73.** $(n + 9)(n - 6)$
75. $(4b - 3)(6b + 1)$ **77.** $2(x - 3)(3x + 2)$

Pages 687–690 *Lesson 11-3*

1. 4, −4 **3.** 16.9 units of force

5 $\frac{(-3r)(10r^4)}{6r^5} = \frac{-30r^5}{6r^5} = -5r^0 = -5$

The excluded value is when $6r^5 = 20$, or when $r = 0$.

7. $x + 7$; −4 **9.** $-\frac{3}{9+y}$; −9, 9 **11.** 3 **13.** 1, −1
15. 3, −8 **17.** $\frac{2\pi(5t)}{\pi(5t)^2} = \frac{2}{5t}$; 0 **19.** $\frac{16}{n^2}$; $n \neq 0, p \neq 0$
21. $-2c$; 0 **23.** $\frac{a}{a-6}$; 6, −3 **25.** $x - 4$; −8
27. $\frac{2p}{p+7}$; 7, −7 **29.** $\frac{-(8+c)}{c+1}$; 8, −1

31 $f(x) = \frac{x^2 + 3x - 4}{x^2 + 9x + 20} = \frac{(x+4)(x-1)}{(x+4)(x+5)} = \frac{x-1}{x+5}$

When $x = 1$, $f(x) = 0$, so 1 is the zero of $f(x)$.

33. 2, 4 **35.** 0, 6 **37a.** $\frac{250\pi}{t}$ **37b.** about 157 ft/min
39. $\frac{4x^4 - 5y^2}{y^3}$; $x, y \neq 0$

41 **a.** Surface area: $2(2x^2) + 2(x^2) + 2(2x^2) = 4x^2 + 2x^2 + 4x^2 = 10x^2$

Volume: $2x(x)(x) = 2x^3$

So, the ratio is $\frac{10x^2}{2x^3} = \frac{5}{x}$.

The excluded value is when $2x^3 = 0$, or when $x = 0$.

b. Surface area: $2\pi a^2 + 2\pi ab$

Volume: $\pi a^2 b$

So, the ratio is $\frac{2\pi a^2 + 2\pi ab}{\pi a^2 b} = \frac{2\pi a^2}{\pi a^2 b} + \frac{2\pi ab}{\pi a^2 b} = \frac{2a + 2b}{ab}$

The excluded value is when $\pi a^2 b = 0$, or when $a = 0$ or $b = 0$.

43. No; Colleen did not show the simplified expression, and Sanson used the simplified expression to find the excluded value. **45.** Every polynomial P can be written as $\frac{P}{1}$, where the numerator and denominator are polynomials; hence every polynomial is also a rational expression. **47.** No; the numerator and denominator have $x - 2$ as a common factor.
49. C **51.** G **53.** 0 **55.** 3 **57.** 20 **59.** 5
61. 8.4 in. **63.** 13 **65.** $\sqrt{41}$ **67.** $3\sqrt{2}$ **69.** $8\sqrt{2}$
71. $2|a|\sqrt{10}$ **73.** $\frac{\sqrt{6}}{3}$ **75.** $2x$ **77.** 1 **79.** $9qt$

Pages 695–698 *Lesson 11-4*

1. $4x$ **3.** $\frac{t}{6(t-5)}$ **5.** 6.16 ft/min **7.** $\frac{2}{3x}$ **9.** $\frac{1}{6}$

11 $\frac{10n^2}{4} \cdot \frac{2}{n} = \frac{5n}{1} \cdot \frac{1}{1} = 5n$

13. $9x^4z$ **15.** $3(t + 2)$ **17.** $k + 6$ **19.** $4n$
21a. about \$24.66 **21b.** about \$26.09 **23.** x^4y
25. $\frac{9b^3}{2}$ **27.** $\frac{b-2}{(b+5)(2b+3)}$ **29.** $\frac{x}{2(x-4)}$
31. $\frac{(r+2)^2}{4}$ **33.** 15 mi/h **35a.** 1250 mi ÷ 540 mi/h
35b. 2.3 h **37.** 7,744,000 yd^2/mo **39.** about 31.4 mi/h **41.** 6.1 c/min **43.** $\frac{1 \text{ cup}}{1 \text{ beat}} \cdot \frac{70 \text{ beats}}{1 \text{ minute}} \cdot \frac{1 \text{ gallon}}{16 \text{ cups}} \cdot \frac{60 \text{ minutes}}{1 \text{ hour}} = 262.5$ gal/h

45 $\frac{\$9.80}{1 \text{ hour}} \cdot \frac{15 \text{ hours}}{\text{week}} \cdot \frac{52 \text{ weeks}}{\text{year}} = \frac{\$7644}{\text{year}}$

This answer represents earning per year.

47. about 71.6 mi/h; converting 32 meters per second to miles per hour **49.** about 52,506.7 ft/s

51 **a.** $\frac{27 \text{ rotations}}{1 \text{ minute}} \cdot \frac{1 \text{ minute}}{60 \text{ seconds}} \cdot \frac{2\pi(3.1) \text{ meters}}{\text{rotation}}$

b. $\frac{27 \text{ rotations}}{1 \text{ minute}} \cdot \frac{1 \text{ minute}}{60 \text{ seconds}} \cdot \frac{2\pi(3.1) \text{ meters}}{\text{rotation}} \approx \frac{8.8 \text{ meters}}{\text{second}}$

This means that the room moves about 8.8 meters per second.

53. Neither; Tamika did not multiply by the reciprocal, and Mei incorrectly factored out the 2 in $2x + 6$. **55.** $\frac{x-3}{x+7}$ **57.** Sample answer: The height of a cylinder when you know an expression for the volume and radius; $\frac{V}{\pi r^2} = h$; $V = \pi(x^3 - 6x^2 + 9x)$, $r = (x - 3)$. **59.** Sample answer: Write ratios comparing the number of days in one year and the number of hours in one day. Then multiply the ratios: $\frac{365 \text{ days}}{1 \text{ year}} \cdot \frac{24 \text{ hours}}{1 \text{ day}}$; 8760 hours. **61.** F **63.** A **65.** $\frac{g^2}{3h}$; $g \neq 0, h \neq 0$ **67.** $y + 8$; -2 **69.** $\frac{z+2}{z-2}$; 2, 1 **71.** $x = 0, y = 5$ **73.** $x = -3, y = 0$ **75.** $x = 8, y = -3$ **77.** 0 **79.** 2 **81.** 3 **83.** $\{p \mid 28 \leq p \leq 32\}$ **85.** $(x-2)(x-3)$ **87.** $(x+4)(3x-5)$ **89.** $4(x+2)(2x-5)$

Pages 702–705 ***Lesson 11-5***

1 $(8a^2 + 20a) \div 4a = \frac{8a^2}{4a} + \frac{20a}{4a}$
$= 2a + 5$

3. $2n^2 - n + \frac{5}{2n}$ **5.** $x - 4$ **7.** $3 + \frac{250}{x+50}$ **9.** $4y + \frac{3}{y+2}$ **11.** $3n^2 + n - 4 + \frac{4}{3n-1}$ **13.** $a^2 + 4a - 18$ **15.** $2n - 4 + \frac{1}{n}$ **17.** $\frac{3}{2}m + \frac{5}{6}$ **19.** $x - 8$

21 $(k^2 - 5k - 24) \div (k - 8) = \frac{(k^2 - 5k - 24)}{(k-8)}$
$= \frac{(k-8)(k+3)}{k-8}$
$= k + 3$

23. a^2 **25.** $2t - 1$ **27.** $2h^2 - 3$ **29.** $-n + 18 + \frac{850}{n}$ **31.** $a + 5 + \frac{8}{a-1}$ **33.** $4n + 5 + \frac{16}{n-2}$ **35.** $t^2 - 4t + 14 - \frac{60}{t+4}$ **37.** $2c^2 + c + 2 - \frac{1}{4c-2}$ **39.** $x + 3$ **41.** $x^2 - 2x + 15 - \frac{16}{x+2}$

43a.

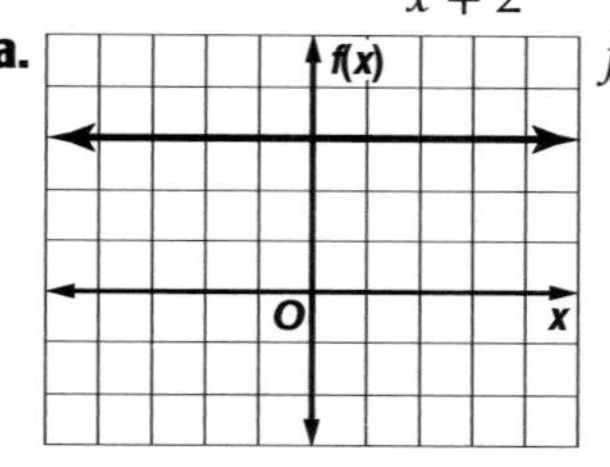

$f(x) = 3 + \frac{7}{x-1}$

43b.

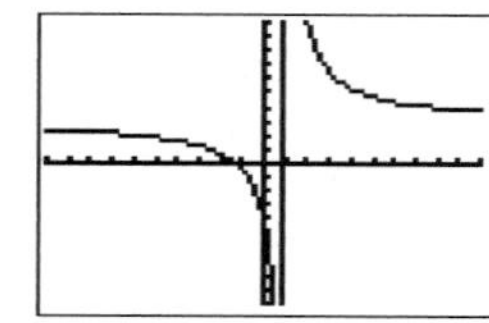

[−10, 10] scl: 1 by [−10, 10] scl:1

43c. The graph of the quotient ignoring the remainder is an asymptote of the graph of the function. **43d.** As x approaches 1 from the left, y approaches negative infinity. As x approaches 1 from the right, y approaches positive infinity.

45 **a.** Words: 212° decreases by 0.9 for every 500 feet above sea level.

Equation: $T = 212 - \left(\frac{0.9}{500}\right)x$

b. $212 - \left(\frac{0.9}{500}\right)x = 212 - \left(\frac{0.9}{500}\right)(14{,}494)$
$= 212 - 26.1$
$= 185.9°$ F

47. Andrea; Alvin did not take into account the missing term. **49.** Sample answer: $(a^2 + 4a - 22) \div (a - 3)$; The polynomial $a^2 + 4a - 22$ is prime, so the problem can be solved by using long division.

51. A **53.** G **55.** $6x$ **57.** $\frac{t}{6(t+4)}$ **59.** -2 **61.** -3 **63.** 81 **65.** 9 **67.** $-10, -2$ **69.** $-0.8, 1$ **71.** $-0.4, 3.9$ **73a.** -2 and 4

73b. 9 ft **75.** $-2cd + d^2$ **77.** 24 **79.** 504

Pages 710–713 ***Lesson 11-6***

1. $\frac{5}{7n}$ **3.** $\frac{16r}{9-r}$ **5.** $24t^2$ **7.** $(x-7)(x-1)(x+2)$ **9.** $\frac{15y+8x}{20xy}$ **11.** $\frac{16d+5c}{6cd}$ **13.** $\frac{x^2-x+9}{x^2-x-6}$

15 $\frac{a}{4} + \frac{3a}{4} = \frac{a+3a}{4}$
$= \frac{4a}{a}$
$= a$

17. $\frac{2y}{3}$ **19.** $\frac{8b+3a}{ab}$ **21.** $\frac{c-8}{2c-1}$ **23.** $\frac{1}{2}$ **25.** $\frac{-w+5}{8w}$ **27.** x^3y^2 **29.** $(3r-1)(r+2)$ **31.** $(x+6)(x+3)$ **33.** $\frac{27}{20x}$ **35.** $\frac{15b+2a}{10ab}$ **37.** $\frac{7-4k}{4(k+2)}$ **39.** $\frac{-2t+28r}{7rt}$ **41.** $\frac{d^2+6d+35}{(d+5)(d-1)}$ **43.** $\frac{18-10t}{15t^2}$ **45.** $\frac{w^2-4w-3}{(w+4)(w-5)}$ **47.** $\frac{-x^2-4x-9}{(x+3)(x+5)}$ **49a.** $\frac{2}{w} + \frac{2}{2.5w}$; $\frac{7}{2.5w}$ **49b.** 48 min **51.** $\frac{32}{2a+18}$

53 $\frac{x+5}{x^2-4} - \frac{3}{x^2-4} = \frac{x+5-3}{x^2-4}$
$= \frac{x+2}{(x+2)(x-2)}$
$= \frac{1}{x-2}$

55. $k + 5$ **57.** $\frac{1}{x-1}$ **59.** $\frac{a-6}{3}$ **61.** $\frac{2x^3-16x+1}{x^4}$ **63.** $\frac{15x^2-11x-7}{9x^3+51x^2-80x+28}$ **65a.** $\frac{9200}{3x}$ **65b.** about 76 min 40 seconds **67.** 8

69 **a.** $t = D/r$
Day 1: $t = \frac{9+x}{10}$
Day 2: $t = \frac{5+3x}{10}$
So, the total for the 2 days was
$\frac{9+x}{10} + \frac{5+3x}{10} = \frac{9+x+5+3x}{10}$
$= \frac{14+4x}{10}$
$= \frac{7+2x}{5}$

b. $\frac{7+2x}{5} = \frac{7+2(2)}{5}$
$= \frac{7+4}{5}$
$= \frac{11}{5}$ or 2.2 Marina rode her bike 2.2 hours on those 2 days.

71. $\frac{4-7y}{21y-6}$ **73.** always; $\frac{a}{x}+\frac{b}{y}=\frac{a}{x}\cdot\frac{y}{y}+\frac{b}{y}\cdot\frac{x}{x}=\frac{ay}{xy}+\frac{bx}{yx}=\frac{ay+bx}{xy}$, $x, y \neq 0$ **75.** First, factor -1 out of one of the denominators so that it is like the other. Then rewrite the denominator without parentheses. Finally, add or subtract the numerators and write the result over the like denominator. **77.** C **79.** C **81.** $5y^2+\frac{14}{3}$ **83.** about 15.7 mi/h **85.** 5 **87.** 11.62 **89.** 1.23×10^4 **91.** 1.255×10^6 **93.** $\frac{2}{3x}$ **95.** $\frac{4}{y+1}$

Pages 171–719 *Lesson 11-7*

1. $\frac{2+4n}{n}$ **3.** $\frac{6t+11}{t+1}$ **5a.** $\frac{2\frac{1}{2}\text{ mi}}{\frac{1}{3}\text{ h}}$ **5b.** $\frac{15}{2}$ or $7\frac{1}{2}$ mi/h **7.** $\frac{3}{25}$ **9.** $\frac{2y^2}{x}$ **11.** $\frac{1}{(x+y)(r+s)}$ **13.** $\frac{p+3}{p^2-p-2}$

15
$$\frac{p-7}{2p}=\frac{p(2p)}{2p}-\frac{7}{2p}$$
$$=\frac{2p^2}{2p}-\frac{7}{2p}$$
$$=\frac{2p^2-7}{2p}$$

17. $\frac{3h^2+h+1}{h}$ **19.** $\frac{n^3+4n^2+n-1}{n+4}$ **21.** $\frac{d^2-12d+43}{d-7}$ **23.** $\frac{3}{4}$ page/min **25.** $\frac{2}{3}$ **27.** $\frac{h}{g^3}$ **29.** $\frac{2a+12}{a}$ **31.** $\frac{(j-4)(j+4)}{15(j+2)}$ **33.** about 140

35 $D=\frac{m}{V}$; $V=\frac{4}{3}\pi r^3$
$$D=\frac{m}{\frac{4}{3}\pi r^3}$$
$$D=\frac{3m}{4\pi r^3}$$

a. $D=\frac{3(15.6)}{4\pi\,(0.0748)^3}\approx 8898.794862$ or 8900
The metal is copper.

b. $D=\frac{3(285.3)}{4\pi\,(0.1819)^3}\approx 11{,}316.57654$ or 11,300
The metal is lead.

37. $\frac{58x+145}{5x+10}$ **39.** $\frac{c-3}{c+7}$ **41.** $\frac{(x-8)(x-1)}{x+2}$

43. Find the lowest common denominator for the fractions in the numerator and simplify to $\frac{y^2-x^2}{xy}$.

45. $\frac{2t^3}{(t-1)(t^2-1)}$ **47.** Sample answer: Time equals distance divided by rate or $\frac{d}{r}$. When the distance or the rate is given as a fraction or mixed number, the expression $\frac{d}{r}$ becomes a complex fraction. Example: Someone walks $\frac{3}{4}$ mile in $10\frac{1}{2}$ minutes; the time in miles per minute is $\frac{\frac{3}{4}}{10\frac{1}{2}}$, which simplifies to $\frac{1}{14}$ mi/min. **49.** 12 days **51.** B **53.** $\frac{-d+4}{d-1}$ **55.** $\frac{6m^2-1}{15m^3}$ **57.** $\frac{b^2+4b+18}{(b+3)(b-2)}$ **59.** $a+7-\frac{1}{a-3}$ **61.** $3y^2+2y-3-\frac{1}{y+2}$

63. $3h^2+2h+3-\frac{2}{3h-2}$

65.

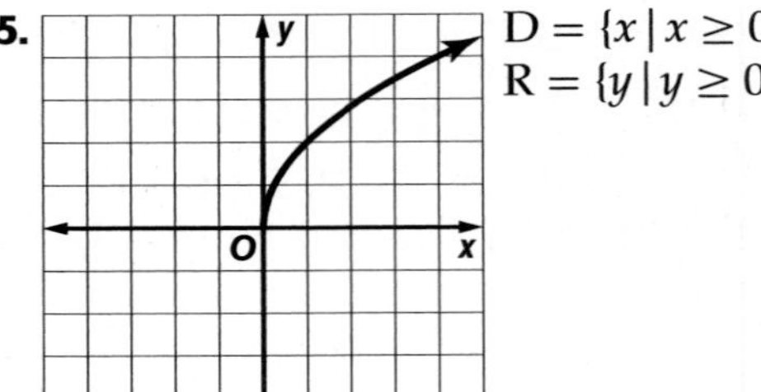

D = {x | x ≥ 0}; R = {y | y ≥ 0}

67. D = {x | x ≥ 0}; R = {y | y ≥ 0}

69. $(a-11)(a+11)$ **71.** $(2y-5)(2y+5)$ **73.** $4(t-1)(t+1)(t^2+1)$ **75.** 4 **77.** 7 **79.** -11.5

Pages 724–726 *Lesson 11-8*

1. -2 **3.** $-\frac{4}{5}$ **5.** -3 **7.** $\frac{15}{38}$ hour or about 0.4 hour

9
$$\frac{8}{n}=\frac{3}{n-5}$$
$$8(n-5)=3n$$
$$8n-40=3n$$
$$-40=-5n$$
$$8=n$$

11. $\frac{2}{3}$ **13.** -13 **15.** 0 **17.** $-2, 3$ **19.** no solution; extraneous: 1 **21.** $\frac{15}{8}$ hours or $1\frac{7}{8}$ hours **23.** 26.2 hours **25a.** line **25b.** $f(x)=\frac{(x+5)(x-6)}{x-6}=x+5$ **25c.** -5 **27a.** parabola **27b.** $f(x)=x^2+6x+12$ **27c.** no real zeros **29a.** $s=r-w$; $s=r+w$ **29b.** $d=t(r-w)$, $d=t(r+w)$; $t=\frac{d}{r-w}$, $t=\frac{d}{r+w}$

31 **a.** $h=\frac{kp}{c}$

b. $8=\frac{\frac{k(100)}{5}}{}$
$$8=\frac{k(100)}{5}$$
$$8=20k$$
$$\frac{2}{5}=k$$
$$h=\frac{\frac{2}{5}p}{c}$$
$$=\frac{\frac{2}{5}(100)}{10}$$
$$=4\text{ hours}$$

33. $-1\pm\sqrt{7}$; extraneous: -2 and 0 **35.** $\frac{50}{11}$

37. The extraneous solution of a rational equation is the excluded value of one of the expressions in the equation. **39.** Sample answer: $\frac{x}{8}=0$ **41.** D **43.** D **45.** $\frac{d}{c}$ **47.** $\frac{b-3}{2b}$ **49.** $4h^2$ **51.** $(x-4)(x+2)$ **53.** linear **55.** quadratic **57.** $0.5Bb+0.5b^2$; $\frac{1}{2}$

59. $\{r \mid r > 49\}$ **61.** 0.3 **63.** 0.75

Pages 727–731 Chapter 11 Study Guide and Review

1. false; $x^2 - 25$ **3.** true **5.** true **7.** false; complex fraction **9.** false; inverse variation **11.** $\frac{1}{3}$ **13.** $-\frac{9}{16}$ **15.** 3 **17.** 2

19. 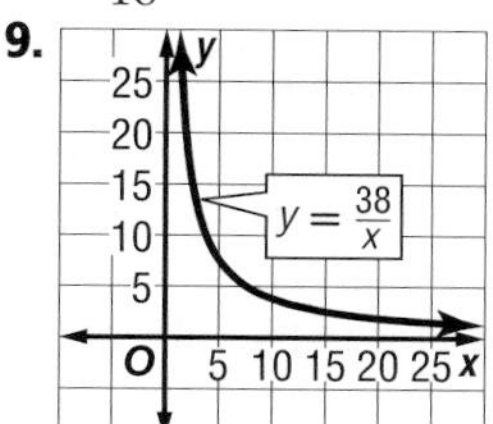

The vertical asymptote is at $x = 0$ and the horizontal asymptote is at $y = 0$.

21. $\frac{1}{x+8}$ **23.** $\frac{y-5}{y-2}$ **25.** $\frac{1}{2y(y+2)}$ **27.** $-\frac{3}{2}, \frac{1}{3}$ **29.** $\frac{3}{x^2 - 3x}$ **31.** $\frac{3}{(b+4)(b-2)}$ **33.** xy **35.** $x^2 + 4x - 2$ **37.** $4y + 1 + \frac{8}{12y - 1}$ **39.** $\frac{3a}{b}$ **41.** $\frac{-y^2 + 2y - 9}{(y+1)(y-3)}$ **43.** 4 ft **45.** $\frac{x^2 + 8x - 65}{x^2 + 8x + 12}$ **47.** $\frac{y^2 + 11y + 10}{y^2 + 6y + 8}$ **49.** $\frac{2}{5}$, extraneous: 2 **51.** 24 **53.** −2, 4

Chapter 12 Probability and Statistics

Page 737 Chapter 12 Get Ready

1. $\frac{3}{7}$ **3.** $\frac{4}{7}$ **5.** $\frac{1}{6}$ **7.** $\frac{5}{6}$ **9.** $\frac{7}{256}$ **11.** $\frac{84}{625}$ **13.** 82.4% **15.** 85.6% **17.** 35%

Pages 743–745 Lesson 12-1

1. Sample: the 10 teens; population: all teens; observational study **3.** Unbiased; each student is equally likely to be the tenth student. **5.** Sample: the rookie cards; population: all of the cards; stratified, because the cards are divided before the sample is selected.

7 Sample answer: The sample is the 100 people the test. The population is all people. The type of data collection is an experiment because they give the people the food and ask what they think.

9. Sample: the customers for the past two years; population: all past customers; survey **11.** Biased; they only poll customers who check out books.

13. Biased; because the park only asked teens, the respondents are more likely to select certain rides.

15 Sample answer: The sample is the blog readers. The population is all artists. This sample is simple because the sample is equally likely to be chosen.

17. Sample: every fiftieth customer; population: all of the customers; systematic: a customer was selected at a regular interval. **19a.** the people that they surveyed; all customers **19b.** observational study **19c.** Unbiased; each person is equally likely to be chosen. **19d.** systematic **21.** Sample answer: This method of selecting a sample is valid. Each student has an equally likely chance of being selected for the sample. A weakness may be that this would not reflect that one grade may feel more strongly about the dress code than another. **23.** Sample answer: A video game company wants to know how their game compares with its competitors. So, they set up a room with the game and three games of their competitors and observe which game the people in the sample prefer. **25.** 90 **27.** F **29.** −6 **31.** −6 **33.** $\frac{4}{3}$ **35.** $-\frac{2}{5}$ **37.** $\frac{1}{2}$ **39a.** ≈ 415.8 ft **39b.** ≈ 212.6 ft

Pages 750–755 Lesson 12-2

1. Mode; there are repeating values in the data; 15.

3 Sample answer: Not valid; there is no mention of what kind of format the station currently has. People are more likely to respond if they already listen to that station or that format.

5. Sample answer: The Key Club is volunteer organization, so the data may be biased.

7. The data in the graph support the conclusion. The display is accurate.

9 List the values from least to greatest: 10, 13, 14, 15, 16, 16, 17, 18, 19.
There is no one value that is much greater or less than the rest of the data. Also, there does not appear to be a big gap in the middle of the data. So, the mean would best represent the data. The mean is approximately 15.33.

11. Mean; the data are weighted and the weighted average needs to be calculated in ordered to determine customer response. **13.** Cannot be calculated; the data values represent different things. **15.** Sample answer: The data seem to be unbiased. The survey is valid.

17 This is not valid because the sample is not mentioned and there is no way to determine the sample size.

19. Sample answer: The data seem unbiased, the source seems reputable, the data support the conclusion. The survey is valid. **21.** Sample answer: Because the source of the survey is Smart Girl, the sample is most likely to be girls and therefore, you are unable to make an unbiased conclusion about all teens.

23 Sample answer: The conclusion is too broad for the data presented.

25. Sample answer: The Red Cross should continue to offer the babysitting class. While only 10% of its participants are in the class, that is still one of the higher percents, and the service is an important one. **27.** Yes; there are no similar colors and there is numerical data for each section. **29.** Ben; there were no large gaps, no repeated values, but there was an outlier of 32. **31.** Sample answer: I would like to conduct a survey about chewing gum in school. I would use a sample selected by randomly picking 50 names from each grade to represent the student body. I would give a series of questions that would evaluate the students' attitudes about the gum chewing policy. I would display the results in a series of bar graphs, one for each question. **33.** D **35.** C **37.** 20 seniors; senior class; unbiased; simple **39.** Mayfield High School; everyone who participated in survey; stratified **41.** −3 **43.** $\sin A = \frac{4}{5}$; $\cos A = \frac{3}{5}$; $\tan A = \frac{4}{3}$

45. $\sin A = \frac{4}{5}$; $\cos A = \frac{3}{5}$; $\tan A = \frac{4}{3}$

47. 29.3; 29; no mode **49.** 4.3; 4; 1 **51.** 26.4; 26.5; 25

Pages 759–762 Lesson 12-3

1. sample: 1003 voters in Mercy County; population: all voters in Mercy County; sample statistic: the number of people in the sample who would vote for the incumbent candidate; population parameter: the number of people in the county who would vote for the incumbent candidate **3.** 6 **5.** 12.6, 57.0, 7.6 **7.** 2.5

9 Sample: stratified random sample from schools in the county
Population: all high shool students in the county
Sample Statistic: time spent each week on extracurricular activites by the sample
Population Parameter: time spent each week on extracurricular activites by all students in the county

11. 3.5

13 Step 1: To find the mean, add the numbers and then divide by how many numbers are in the data set.

$$x = \frac{76 + 78 + 83 + 74 + 75}{5} = \frac{386}{5} = 77.2$$

Step 2: To find the variance, square the difference between each number and the mean. Then divide by the number of values.

$$\sigma^2 = [(76 - 77.2)^2 + (78 - 77.2)^2 + (83 - 77.2)^2 + (74 - 77.2)^2 + (75 - 77.2)^2] \div 5 = \frac{50.8}{5} \approx 10.2$$

Step 3: The standard deviation is the square root of the variance.

$$\sigma^2 = 10.2$$
$$\sqrt{\sigma^2} = \sqrt{10.2}$$
$$\sigma \approx 3.2$$

15. 0.4, 0.1, 0.3 **17.** 5.0 **19.a.** Sample answer: The pennies chosen by Tyrone, Lydia, and Peter each represent a sample. The 30 pennies in the jar is the population. The sample statistic is the mean year of the pennies in the sample. The population parameter is the mean year of the pennies in the population. **19b.** ≈ 1984, ≈ 9.0
19c. ≈ 2001, ≈ 3.2 **19d.** ≈ 1998, ≈ 6.4
19e. ≈ 1997, ≈ 7.4; Peter's sample was the most accurate. The mean year of his sample was 1 year off from the actual mean year. The samples that had more pennies were more accurate.

21 a. Step 1: To find the mean, add the numbers and then divide by how many numbers are in the data set.

$x = 956 + 966 + 971 + 981 + 986 + 991 + 1001 + 1014 + 1020 + 1023 + 1034 + 1066 + 1076 + 1077 + 1083/15$

$= \frac{15245}{15}$

$= 1016.3$ seconds or 16.9 minutes

Step 2: To find the mean absolute deviation, first find the sum of the absolute values of the differences between each value in the set of data and the mean.

$|956 - 1016.5| + |966 - 1016.5| + |971 - 1016.5| + |981 - 1016.5| + |986 - 1016.5| + |994 - 1016.5| + |1001 - 1016.5| + |1014 - 1016.5| + |1020 - 1016.5| + |1023 - 1016.5| + |1034 - 1016.5| + |1066 - 1016.5| + |1076 - 1016.5| + |1077 - 1016.5| + |1083 - 1016.5| = 499.5$

Step 3: Then divide the sum by the number of values in the data set.

$\frac{499.5}{15} \approx 33.3$ seconds

b. The sample is the top 15 runners. The population is all the people who ran.
c. The data is quantitative. No, since the sample is the top 15 runners in the race, it is not random. So, it would not be accurate to apply the mean and standard deviation of the running times to the population.

23. 8.75, 3.6 **25.** A statistic is a characteristic that is computed on a sample of the population. A parameter is a characteristic of the entire population. Sample answer: To determine the average height of a student at your high school, you can measure the heights of a random sample of students at your school. The mean height of the sample is a statistic; the actual mean height of the students at your school is a parameter.
27. Both are calculated statistical values that show how each data value deviates from the mean of the data set. The mean absolute deviation is calculated by taking the mean of the absolute values of the differences between each number and the mean of the data set. To find the standard deviation, you

square each difference and then take the square root of the mean of the squares. **29.** 125 **31.** A **33.** Mode; one value is repeated; 85. **35.** Biased; because they are at a rock concert, they are more likely to select a rock music station. **37.** $\frac{3}{22}$ **39.** $\frac{1}{2}$ **41.** $\frac{8}{11}$

Pages 767–770 Lesson 12-4

1. 720

3 $P(n, r) = \frac{n!}{(n-r)!}$

$P(9, 3) = \frac{9!}{(9-3)!}$

$= \frac{9!}{6!}$

$= \frac{9 \cdot 8 \cdot 7 \cdot 6 \cdot 5 \cdot 4 \cdot 3 \cdot 2 \cdot 1}{6 \cdot 5 \cdot 4 \cdot 3 \cdot 2 \cdot 1}$

$= 504$

5. 10 **7.** 720

9 The number of ways is $4 \cdot 3 \cdot 2 \cdot 1 = 24$ ways

11. 720 **13.** 720 **15.** 4 **17.** 7 **19.** 1 **21.** 1,814,400 **23.** $\frac{49}{646}$ or about 7.59%

25 This is a combination because the order of the toppings put on the pizza does not matter.

27. combination **29.** combination **31a.** 495 **31b.** 24 arrangements **31c.** 9; ART, ATE, ARE, TAR, TEA, RAT, EAT, EAR, ERA **33.** 792 groups **35.** Ming; since order is not important, combinations should have been used.
37. Determining class rank in a senior class; this is the only situation in which order matters.
39. Sample answer: choosing 3 clubs out of 8 to join **41.** G **43.** $\frac{3}{20}$ **45.** 26; 187.4, 13.7 **47.** 4, 6.8, 2.6 **49.** 2 bags **51** $\frac{2}{n}$ **53.** $\frac{n-4}{n+4}$ **55.** $\frac{12ag}{5b}$ **57.** $\frac{6}{23}$ **59.** $\frac{8}{23}$ **61.** $\frac{17}{23}$

Pages 775–778 Lesson 12-5

1. dependent; $\frac{28}{253}$ or about 11% **3.** independent; $\frac{4}{9}$ or about 44% **5.** mutually exclusive; $\frac{2}{13}$ or about 15% **7.** not mutually exclusive; $\frac{4}{13}$ or about 31% **9.** independent; $\frac{1}{16}$ or about 6%

11 These events are dependent because the chocolate is not being replaced.

First, milk chocolate: $P = \frac{10}{24}$

Second, white chocolate: $P = \frac{6}{23}$

$P(\text{milk, then white}) = \left(\frac{10}{24}\right)\left(\frac{6}{23}\right)$

$= \frac{60}{552}$

$= \frac{5}{46}$ or about 11%

13. dependent; $\frac{5}{19}$ or about 26% **15.** mutually exclusive; $\frac{3}{4}$ or 75% **17.** not mutually exclusive; $\frac{739}{750}$ or about 98.5% **19.** $\frac{8}{87}$ or about 9% **21.** $\frac{42}{145}$ or about 29%

23 choose white sock: $P = \frac{14}{24}$

choose white sock: $P = \frac{13}{23}$

$P(\text{milk, then white}) = \left(\frac{14}{24}\right)\left(\frac{13}{23}\right)$

$= \frac{182}{552}$

$= \frac{91}{276}$ or about 33%

25. $\frac{1}{2}$ or 50% **27.** $\frac{4}{221}$ or about 19% **29.** $\frac{7}{13}$ or about 54% **31a.** 345 **31b.** 159 **31c.** $\frac{227}{345}$ or about 66% **31d.** $\frac{2}{23}$ or about 9%

33 **a.**

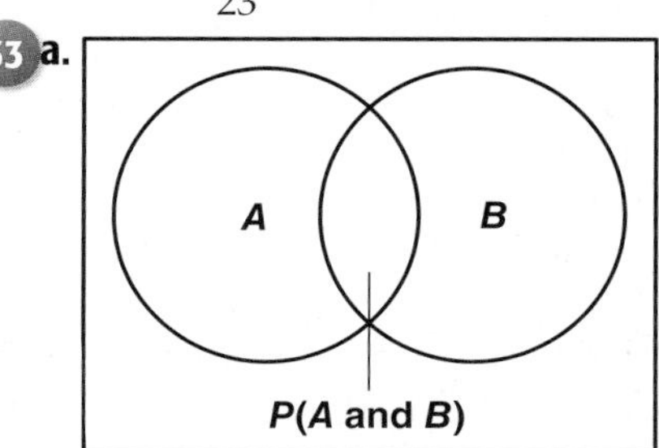

b. Divide the overlap of the two circles by the $P(A)$ circle; $P(B/A) = \frac{P(A \text{ and } B)}{P(A)}$.

c. $P(\text{red}) = \frac{8}{12} = \frac{2}{3}$

$P(\text{red, green}) = \left(\frac{2}{3}\right)\left(\frac{4}{11}\right)$

$= \frac{8}{33}$

d. $P(\text{green/red}) = \frac{\frac{8}{33}}{\frac{2}{3}}$

$= \frac{8}{33} \cdot \frac{3}{2}$

$= \frac{4}{11}$

e. $P(B \mid A) = \frac{P(A \text{ and } B)}{P(A)}$

f. $P(B \mid A) = \frac{P(A \text{ and } B)}{P(A)}$

$= \frac{0.20}{0.80}$

$= 0.25$ or 25%

35. 21 **37.** Sample answer: Choosing a CD to listen to, putting it back, and then choosing another CD to listen to would be an independent event since the CD was placed back before the second CD was chosen. Choosing a pair of jeans to wear would be a dependent event if I did not like the first pair chosen, and I did not put them back. **39.** C **41.** 2 ft **43a.** combination **43b.** 220 **43c.** $\frac{210}{1320}$ or 16% **45.** 8 **47.** $-\frac{3}{2}$ **49.** about 57 pieces **51.** −12 **53.** −16 **55.** −63

Selected Answers and Solutions

1a. $\frac{47}{1000} = 0.047 = 4.7\%$ **1b.** $\frac{2871}{5000} = 0.5742 = 57.4\%$ **3a.** $\frac{1}{5} = 0.2 = 20\%$ **3b.** $\frac{68}{165} \approx 0.4121 \approx 41.2\%$

5 a. $P = \frac{96}{100} + \frac{112}{600}$
$= \frac{208}{600}$
$= \frac{26}{75}$ or about 34.7%

b. $P = \frac{108}{600} + \frac{80}{600}$
$= \frac{188}{600}$
$= \frac{47}{150}$ or about 31.3%

7a. $\frac{87}{178} \approx 0.4888 \approx 48.9\%$ **7b.** $\frac{91}{178} \approx 0.5112 \approx 51.1\%$

9 a. All of the values in data are between 0 and 1. $0.29 + 0.43 + 0.17 + 0.11 + 0 = 1$ so the distribution is valid.

b. $P = 0.43 + 0.17 + 0.11 + 0$
$= 0.71$ or 71%

c. Use the data from the probability distribution table to draw a histogram. Remember to label each axis and give the histogram a title.

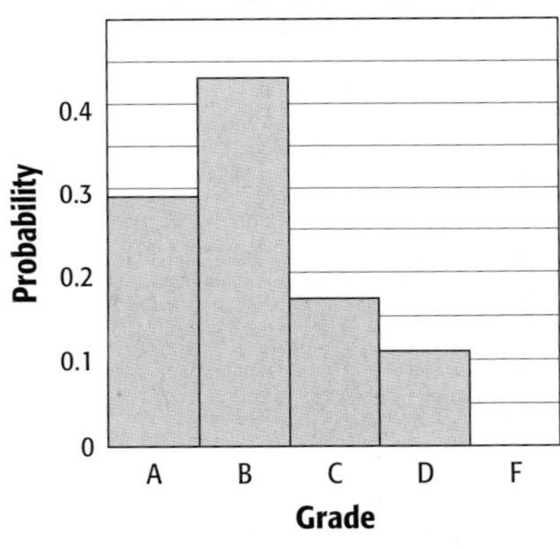

11 a. $0.35 + 0.32 + 0.17 + 0.11 + 0.05 = 1$ so the distribution is valid.

b. $P = 0.17 + 0.11$
$= 0.28$ or 28%

c. Use the data from the probability distribution table to draw a histogram. Remember to label each axis and give the histogram a title

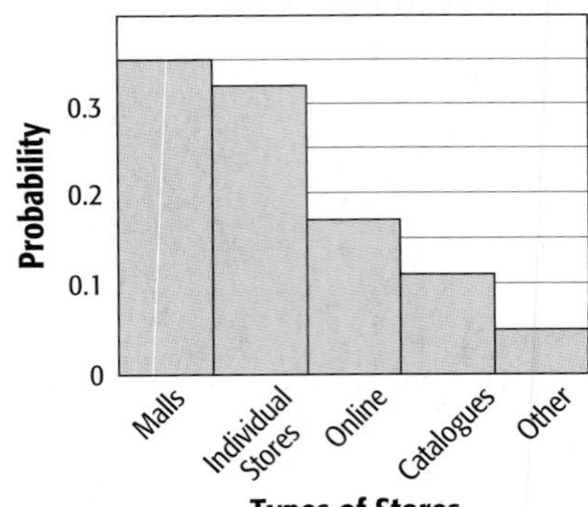

13a. $\frac{1}{6}, \frac{1}{3}, \frac{1}{2}$

13b. Sample answer: 1: 6; 2: 3; 3: 7; 4: 0; 5: 2; 6: 2

13c. Sample answer:

1	0.3
2	0.15
3	0.35
4	0
5	0.1
6	0.1

13d. $0.15 = \frac{3}{20} = 15\%$; $0.4 = \frac{2}{5} = 40\%$; $0.75 = \frac{3}{4} = 75\%$; The first set of probabilities were predictions based on everything being equal. The second set were based on what actually happened.

15. The sum 7 is most likely to happen.

X = Sum of Dice	2	3	4	5	6	7	8	9	10	11	12
Probability	$\frac{1}{36}$	$\frac{1}{18}$	$\frac{1}{12}$	$\frac{1}{9}$	$\frac{5}{36}$	$\frac{1}{6}$	$\frac{5}{36}$	$\frac{1}{9}$	$\frac{1}{12}$	$\frac{1}{18}$	$\frac{1}{36}$

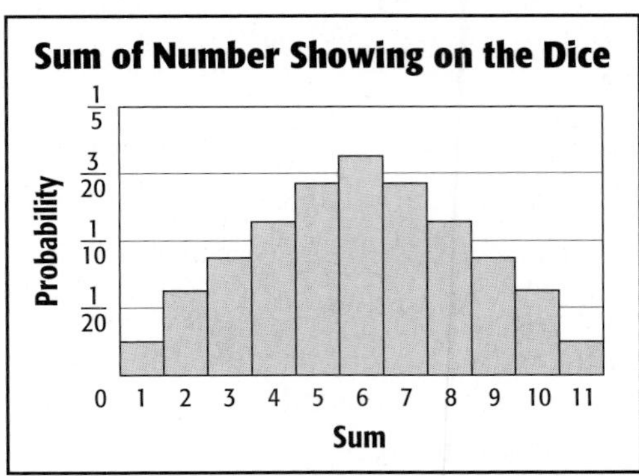

17. Sample answer: There are 870 students in a school: 179 freshmen, 215 sophomores, 211 juniors, and 265 seniors. Find the probability distribution for each class. What is the probability that a randomly chosen student is a sophomore?

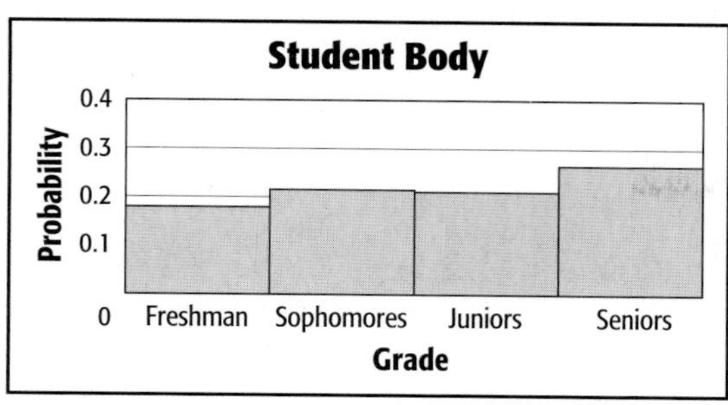

19. D **21.** J **23.** $\frac{10}{253}$ or about 4% **25.** $\frac{1}{8}$ **27.** $\frac{1}{8}$
29. $\frac{6a+4}{a^2}$ **31.** $\frac{9d+10}{(d+6)(d-5)}$ **33.** $\frac{9h(h-2)}{(h+6)(h-3)}$
35. $\frac{1}{2^n}$ **37.** $\frac{1}{2^n}$

Pages 789–792 Lesson 12-7

1a. $\frac{3}{10}$ or 30% **1b.** $\frac{47}{150}$ or about 31%

3a. You could flip a coin where heads represents true and tails represents false.
b. Flip a coin 20 times and record the results using heads for true and tails for false.

5a. $\frac{1}{5}$ or 20% **5b.** $\frac{16}{25}$ or 64% **5c.** $\frac{3}{10}$ or 30%

7 You could spin a spinner with 4 equal sections. The first section could represent the first choice, the second section could represent the second choice, the third section could represent the third choice and the fourth section could represent the fourth choice.

11 This will depend on students work. Sample answers are given:
a. $P(3 \text{ heads}) = \frac{1}{4}$
b. $P(2 \text{ heads}) = \frac{3}{8}$
c. $P(1\text{st and last heads}) = \frac{1}{16}$

13a. Sample answer:

L1	L2	L3 1
0	------	------
0		
0		
0		
1		
0		
0		

L1(1)=0

13b. Sample answer:

L1	L2	L3 2
0	12	------
0	8	
0	18	
0	6	
1	9	
0	19	
0	14	

L2(1)=12

13c. A value of 1 in the third list means you won the cash prize because a 1 in the first list represents you won a prize and a 1 in the second list represents the cash prize.

13d. Sample answer:

L1	L2	L3 3
1	20	20
0	1	0
1	9	9
1	18	18
1	11	11
1	13	13
1	1	1

L3(33)=1

33 cards would have to be scratched in order to win the cash prize.

15. Never; the theoretical probability of getting tails from a coin toss is one half, or 7.5 times of 15 tosses. However, in an experiment you can never achieve 7.5 tails, only whole numbers. **17.** The theoretical probability of tossing a coin and showing heads or tails is 1, since it will always show one or the either. So when tossing a coin in an experiment, showing heads or tails will always happen, making the experimental probability always greater than 0. **19.** Theoretical probabilities are based on actual results, but experimental probabilities are based on trial results. The probabilities become closer and closer as more trials are done in the experiments. **21.** 24
23. C **25.** $\frac{10}{22}$ or about 45.5% **27.** $\frac{25}{484}$ or about 5.2% **29.** 60 **31.** $xy = 12.4$; -20
33. 17,263.61781 ft **35.** 44% **37.** 75% **39.** 27%

Pages 793–796 Chapter 12 Study Guide and Review

1. permutation **3.** 1 **5.** mutually exclusive
7. unbiased **9.** median; 13 **11.** median; 3
13. 13.3 **15.** 11.4; 2; 1.4 **17.** 10.8, 0.6, 0.8
19. 3.1 **21.** 126 **23.** 120 **25.** $\frac{12}{325}$ **27.** $\frac{22}{325}$ **29.** $\frac{1}{2}$
31. For each X, the probability is greater than or equal to 0 and less than or equal to $0.18 + 0.36 + 0.34 + 0.08 + 0.04 = 1$, so the sum of the probabilities is 1. **33.** 0.25 **35.** 0.50

Selected Answers and Solutions

Photo Credits

Cover: Michael Wong/CORBIS; **P6** altrendo images/Altrendo/Getty Images; **P16** George Disario/CORBIS; **P28** Skyscan Photolibrary/Alamy; **P33** Eyewire/Getty Images; **P41** Martin Harvey/Gallo Images/Getty Images; **002–003** JUPITERIMAGES/Comstock Images/Alamy; **005** Gabe Palmer/CORBIS; **006** Masterfile; **008** Image Source/CORBIS; **010** Ken Biggs/Stone/Getty Images; **012** NOAA Photo by Will von Dauster; **014** guichaoua/Alamy; **016** JUPITERIMAGES/Brand X/Alamy; **018** Stewart Cohen/Blend Images/Getty Images; **020** Ariel Skelley/CORBIS; **023** SuperStock/AGE Fotostock; **024** Gail Mooney/Masterfile; **028** age fotostock/SuperStock; **031** Paul Rees/Imagestate; **035** Hola Images/Getty Images; **036** Matthew Mawson/Alamy; **038** Jeff Hunter/Photographer's Choice/Getty Images; **040** Masterfile; **042** blickwinkel/Alamy; **043** Robert Michael/CORBIS; **046** Rainer Raffalski/Alamy; **051** David P. Hall/CORBIS; **054** Chuck Pefley/Alamy; **057** AP Images; **058** AP Images; **068** fStop/Alamy; **072–073** Bob Daemmrich/PhotoEdit; **075** Doug Murray/Reuters/CORBIS; **076** SUNNYphotography.com/Alamy; **077** Art Resource, NY; **079** Ty Allison/Taxi/Getty Images; **083** Brian Bailey/Stone/Getty Images; **085** Ronnie Kaufman/Photolibrary; **087** Jim West/Alamy; **088** Stockbyte/Alamy; **092** William Thomas Cain/Getty Images; **095** YOSHIKAZU TSUNO/AFP/Getty Images; **097** Gen Nishino/Taxi/Getty Images; **101** David P. Hall/CORBIS; **104** Bill Aron/PhotoEdit; **107** Masterfile; **108** JUPITERIMAGES/Creatas/Alamy; **111** Digital Vision/Alamy; **113** Blend Images/SuperStock; **114** Rab Harling/Alamy; **115** Gary Crabbe/Alamy; **116** Gail Mooney/Masterfile; **118** Image Source/CORBIS; **120** Lloyd Sutton/Masterfie; **122** Jed Jacobsohn/Allsport/Getty Images; **123** ACE STOCK LIMITED/Alamy; **126** Stockbyte/SuperStock; **127** Erika Nelson/World's Largest Things, Lucas, KS; **129** Lisette Le Bon/SuperStock; **130** AP Images; **132** Photodisc/Getty Images; **134** David Young-Wolff/PhotoEdit; **137** Purcell Team/Alamy; **150–151** Wolfgang Kaehler; **153** Wesley Hitt/Alamy; **158** CORBIS; **159** Bill Aron/PhotoEdit; **163** Comstock/PunchStock; **164** Ausloser/zefa/CORBIS; **169 (cl)** Ed Imaging; **169 (cr)** Ed Imaging; **169 (bl)** PHOTOSPORT; **169 (tr)** The McGraw-Hill Companies; **170** J Silver/SuperStock; **176 (bc)** CORBIS; **176 (br)** Laura Hinshaw/Index Stock Imagery; **177** Artiga Photo/Masterfile; **180** NOVASTOCK/PhotoEdit; **182** Alamy Images; **184** Alamy Images; **185** Tom Grill/CORBIS; **187** Alamy Images; **188** Bettmann/CORBIS; **190** Stewart Cohen/Getty Images; **192** D. Hurst/Alamy Images; **194** Brooklyn Production/CORBIS; **196** CORBIS Super RF/Alamy; **199** Mark A. Johnson/Alamy; **206** Steven May/Alamy; **210–211** Alex Segre/Alamy; **213** Ed Imaging; **214** Image Source Pink/Getty Images; **217** Brand X Pictures/PunchStock; **219** Scott Tysick/Masterfile; **220** Sugar Gold Images/Photolibrary; **224** Blend Images/Alamy; **226** imagebroker/Alamy; **228** CORBIS; **229** Juliet Brauner/Alamy; **231** Chuck Savage/CORBIS; **235** Hermann Erber/LOOK/Getty Images; **237** Purestock/Getty Images; **238** Arcaid/Alamy; **242** Richard T. Nowitz/CORBIS; **246** AFP PHOTO/Stan HONDA/Getty Images; **249** George Rose/Getty Images; **250** Chuck Pefley/Alamy; **252** John Giustina/Photodisc/Getty Images; **254** AP Images; **255** Alan Thornton/Stone/Getty Images; **257** JUPITERIMAGES/Thinkstock/Alamy; **258** Courtesy of American Statistical Association; **259** Tracy Frankel/WireImage/Getty Images; **261** David J. Green/Alamy; **262** Yellow Dog Productions/Getty Images; **266** UpperCut Images/Alamy; **267** Visions of America, LLC/Alamy; **276** Chuck Franklin/Alamy; **280–281** Digital Vision Ltd./SuperStock; **283** JUPITERIMAGES/PIXLAND/Alamy; **285** IT Stock/PunchStock; **287** Jeff Greenberg/PhotoEdit; **290** Inti St Clair/Digital Vision/Getty Images; **291** blickwinkel/Alamy; **293** Brand X Pictures/PunchStock; **294** Chantelle Gorham - Northwest Fudge Factory; **296** Design Pics Inc./Alamy; **299** Masterfile; **300** Martin Harvey/CORBIS; **303 (cl)** George Grall/National Geographic/Getty Images; **303 (c)** Mike Randolph/Masterfile; **304** Joe Schwartz; **307** Larry Williams/CORBIS; **308** Laboratory for Atmospheres, NASA Goddard Space Flight Center; **310** David L. Moore/Alamy; **311** Stockbyte/Alamy; **313** Getty Images; 315 David Young-Wolff/PhotoEdit; **317** Spencer Grant/PhotoEdit; **319** David Young-Wolff/PhotoEdit; **326** Rob Gage/Taxi/Getty Images; **330–331** Masterfile; **335** Brand X Pictures/PunchStock; **337** Jason Horowitz/zefa/CORBIS; **338** JUPITERIMAGES/BananaStock/Alamy; **342** Colin Young-Wolff/PhotoEdit; **344** Michael Newman/PhotoEdit; **345** Rolf Bruderer/Masterfile; **346** WireImageStock/Masterfile; **348** Miles Ertman/Masterfile; **350** Photodisc/Alamy; **352** AP Images; **353** Science Source/Photo Researchers, Inc.; **355** Photodisc/SuperStock; **358** David M. Grossman/Phototake; **359** Wolfgang Kaehler/Alamy; **362** Mike Powell/Getty Images; **364** Tony Camacho/Photo Researchers, Inc.; **365** Spencer Grant/PhotoEdit; **366** JUPITERIMAGES/Creatas/Alamy; **368** David Young-Wolff/PhotoEdit; **370** Mark Richards/PhotoEdit; **373** foodfolio/Alamy; **374** AP Images; **376** Jonathan Nourok/PhotoEdit; **378** Jeff Henry/CORBIS; **380** JUPITERIMAGES/Brand X/Alamy; **383** Michael Newman/PhotoEdit; **385** Patrick Olear/PhotoEdit; **394** CORBIS Premium RF/Alamy; **398–399** Roger Ressmeyer/CORBIS; **401** Zoomstock/Masterfile;

405 Steven Widoff/Alamy; **406** Deborah Betz Collection/CORBIS; **408** Taxi/Getty Images; **411** Blend Images/SuperStock; **412** Holt Studios International Ltd/Alamy; **414** Roger Ressmeyer/CORBIS; **416** StockTrek/Getty Images; **418** Michael Ochs Archives/CORBIS; **420** Bill Frymire/Masterfile **421** Nevada Wier/CORBIS; **424** JUPITERIMAGES/Creatas/Alamy; **426** FOUNDATION SKATEBOARD COMPANY; **428** CORBIS; **433** Jeff Greenberg/PhotoEdit; **435** Digital Vision/Alamy; **436** David R. Frazier Photolibrary, Inc./Alamy; **437** Marcel Mettelsiefen/Getty Images; **439** David Schmidt/Masterfile; **440** Noriyuki Yamagashira/ANYONE/amana images/Getty Images; **443** David Ashdown/Keystone/Getty Images; **447** Jo Katanigra/Alamy; **449** Roger Charity/Taxi/Getty Images; **451** Kevin C. Cox/Getty Images; **453** Alex Segre/Alamy; **456** Paul Sutherland/Digital Vision/Getty Images; **457** Yves Herman/Reuters/CORBIS; **464** Bob Daemmrich/PhotoEdit; **468–469** Jon Arnold Images Ltd/Alamy; **471** Spencer Grant/PhotoEdit; **473** Flying Colours Ltd/Digital Vision/Getty Images; **476** Dan Galic/Alamy; **479** Jeff Greenberg/PhotoEdit; **481** King's Dominion; **485** Image Source Black/Getty Images; **488** GLG3/Photodisc/Getty Images; **490** Patrick B. Kraemer/epa/CORBIS; **493** Kim Karpeles/Alamy; **495** Gallo Images/CORBIS; **497** LWA/Dann Tardif/Blend Images/Getty Images; **499** David Young-Wolff/PhotoEdit; **502** David Young-Wolff/PhotoEdit; **505** Michael Newman/PhotoEdit; **509** Pictorial Press Ltd/Alamy; **510** JEFF HAYNES/Staff/AFP/Getty Images; **511** Joseph Barnell/SuperStock; **518** Michael Newman/PhotoEdit; **522–523** Stephen Chernin/Getty Images; **525** Bob Turner/Alamy; **530** Andy Lyons/Getty Image; **533** Tony Freeman/PhotoEdit; **536** Tsolo/Science Faction/Getty Images; **537** N.C. Department of Agriculture and Consumer Services; **539** moodboard/Alamy; **541** AP Images; **547** DOMINIQUE BRAUD/Animals Animals Earth Scenes; **548** Daniel Dempster Photography/Alamy; **552** Buzz Pictures/Alamy; **554** AP Images; **555** Raimund Koch/Photonica/Getty Images; **556** Stocktrek/age fotostock; **558** Creatas/PunchStock; **563** Digital Vision/Alamy; **568** Jonathan Nourok/PhotoEdit; **571** The Great Picture: Copyright 2006, The Legacy Project; **574** Digital Vision/Alamy; **576** Digital Vision/Alamy; **578** Karen Moskowitz/Taxi/Getty Images; **579** The Granger Collection, New York.; **580** Matt Sullivan/Reuters/CORBIS; **582** JUPITERIMAGES/Brand X/Alamy; **584** Tony Freeman/PhotoEdit; **586** PhotoAlto/SuperStock; **588** rubberball/Rubberball Productions/Royalty-Free/Getty Images; **598** Wright Group Stockdisc/PunchStock; **602–603** Stockbyte/PunchStock; **605** Louie Psihoyos/CORBIS; **607** Didier Dorval/Masterfile; **609** Glen Allison/Getty Images; **612** Cameron Davidson/Alamy; **615** Comstock/PunchStock; **616** V. Spencer Grant/PhotoEdit; **622** AP Images; **624** Digital Vision/Alamy; **626** Photo Courtesy of Silver Dollar City; **627** Johannes Rodach/StockFood Creative/Getty Images; **630** HANNSdribbling 10; **631** Gerry Walden/Alamy; **633** Blue Lantern Studio/CORBIS; **634** wendy connett/Alamy; **637** william casey/Alamy; **640** Jeff Greenberg/Alamy; **642** Jim Craigmyle/CORBIS; **645** Tommy Hindley/NewSport/CORBIS; **649** M Stock/Alamy; **651** Masterfile; **654** Comstock/JupiterImages; **662** JUPITERIMAGES/Polka Dot/Alamy; 666–667 Doug Pensinger/Getty Images; **670** Kim Karpeles/Alamy; **672** Trevor Lush/UpperCut Images/Getty Images; **674** Digital Vision; **675** David Young-Wolff/PhotoEdit; **678** Image Source Pink/Alamy; **679** Image Source/PunchStock; **680** Courtesy of Evelyn B. Granville; **682** Stockbyte; **684** Tatsuyuki Tayama Fujifotos/The Image Works; **688** Bettmann/CORBIS; **689** Will & Deni McIntyre/Photo Researchers, Inc.; **692** Mika/zefa/CORBIS; **696** Philip Gould/CORBIS; **697** GmbH & Co.KG/Alamy; **700** Juniors Bildarchiv/Alamy; **704** Tiziana and Gianni Baldizzone/CORBIS; **709** Gerolf Kalt/zefa/CORBIS; **711** Ingram Publishing/AGE Fotosearch; **712** Robert Michael/CORBIS; **714** David Allio/Icon SMI/CORBIS; **715** AP Images; 718 Rubberball Productions/Getty Images; **720** Creatas/PunchStock; **723** George Hall/CORBIS; **725** Glow Images/Alamy; **732** Will & Deni McIntyre/CORBIS; **736** Dex Image/Alamy; **740** Steve Prezant/CORBIS; **742** Jeff Greenberg/PhotoEdit; **744** Christian Schmidt/zefa/CORBIS; **748** Ryan McVay/Stone/Getty Images; **749** Robin Nelson/PhotoEdit; **753** Bob Daemmrich/PhotoEdit; **754** image100/CORBIS; **756** Halfdark/fStop/Getty Images; **757** Geoff A Howard/Alamy; **760** GARY CASKEY/Reuters/CORBIS; **761** Jeff Greenberg/Alamy; **764** Allan Munsie/Alamy; **766** Photodisc/Alamy; **769** Bob Daemmrich/PhotoEdit; **771** Spencer Grant/PhotoEdit; **772** Stock 4B RF; **775** JUPITERIMAGES/Thinkstock/Alamy; **776** A. Ramey/PhotoEdit; **780** Brand X Pictures; **782** Stockbyte/Getty Images; **783** David Young-Wolff/PhotoEdit; **787** Rana Faure/DigitalVision/Getty Images; **788** Bettmann/CORBIS; **790** Stewart Cohen/Taxi/Getty Image; **791** The Bridgeman Art Library International; **798** Jeff Greenberg/Alamy; **802** Eclipse Studios; **810** Judith Collins/Alamy

Glossary/Glosario

Math Online

A mathematics multilingual eGlossary is available at **glencoe.com**. The glossary includes the following languages:

Arabic	English	Portuguese	Urdu
Bengali	Haitian Creole	Russian	Vietnamese
Brazilian	Hmong	Spanish	
Cantonese	Korean	Tagalog	

Cómo usar el glosario en español:

1. Busca el término en inglés que desees encontrar.
2. El término en español, junto con la definición, se encuentran en la columna de la derecha.

English | Español

A

absolute value (p. P11) The distance a number is from zero on the number line.

valor aboluto Es la distancia que dista de cero en una recta numerica.

absolute value function (p. 242) A function written as $f(x) = |x|$, in which $f(x) \geq 0$ for all values of x.

función del valor absoluto Una función que se escribe $f(x) = |x|$, donde $f(x) \geq 0$, para todos los valores de x.

additive identity (p. 10) For any number a, $a + 0 = 0 + a = a$.

identidad de la adición Para cualquier número a, $a + 0 = 0 + a = a$.

additive inverse (p. P11) Two integers, x and $-x$, are called additive inverses. The sum of any number and its additive inverse is zero.

inverso aditivo Dos enteros x y $-x$ reciben el nobre de inversos aditivos. La suma de cualquier número y su inverso aditivo es cero.

algebraic expression (p. 5) An expression consisting of one or more numbers and variables along with one or more arithmetic operations.

expresión algebraica Una expresión que consiste en uno o más números y variables, junto con una o más operaciones aritméticas.

area (p. P26) The measure of the surface enclosed by a geometric figure.

área La medida de la superficie incluida por una figura geométrica.

arithmetic sequence (p. 187) A numerical pattern that increases or decreases at a constant rate or value. The difference between successive terms of the sequence is constant.

sucesión aritmética Un patrón numérico que aumenta o disminuye a una tasa o valor constante. La diferencia entre términos consecutivos de la sucesión es siempre la misma.

asymptote (p. 679) A line that a graph approaches but never crosses.

asíntota Una línea a que un gráfico acerca pero nunca cruza.

augmented matrix (p. 376) A coefficient matrix with an extra column containing the constant terms.

matriz aumentada una matriz del coeficiente con una columna adicional que contiene los términos de la constante

axis of symmetry (p. 525) The vertical line containing the vertex of a parabola.

eje de simetría La recta vertical que pasa por el vértice de una parábola.

B

bar graph (p. P40) A graphic form using bars to make comparisons of statistics.

gráfico de barra Forma gráfica usando barras para comparar estadísticas

base (p. 5) In an expression of the form x^n, the base is x.

base En una expresión de la forma x^n, la base es x.

best-fit line (p. 253) The line that most closely approximates the data in a scatter plot.

recta de ajuste óptimo La recta que mejor aproxima los datos de una gráfica de dispersión.

biased sample (p. 741) A sample in which one or more parts of the population are favored over others.

muestra sesgada Muestra en que se favorece una o más partes de una población en vez de otras partes.

binomial (p. 424) The sum of two monomials.

binomio La suma de dos monomios.

bivariate data (p. 245) Data with two variables.

datos bivariate Datos con dos variables.

boundary (p. 315) A line or curve that separates the coordinate plane into regions.

frontera Recta o curva que divide el plano de coordenadas en regiones.

box-and-whisker plot (p. P42) A diagram that divides a set of data into four parts using the median and quartiles. A box is drawn around the quartile values and whiskers extend from each quartile to the extreme data points.

diagrama de caja y patillas Diagram que divide un conjunto de datos en cuatro partes usando la mediana y los cuartiles. Se dibuja una caja alrededor de los cuartiles y se extienden patillas de cada uno de ellos a los valores extremos.

C

center (p .P24) The given point from which all points on the circle are the same distance.

centro Punto dado del cual equidistan todos los puntos de un circulo.

circle (p. P24) The set of all points in a plane that are the same distance from a given point called the center.

círculo Conjunto de todos los puntos del plano que están a la misma distancia de un punto dado del plano llamado centro.

circle graph (p. P41) A type of statistical graph used to compare parts of a whole.

gráfico del círculo Tipo de gráfica estadística que se usa para comparar las partes de un todo.

circumference (p. P24) The distance around a circle.

circunferencia Longitud del contorno de un círculo.

closed half-plane (p. 315) The solution of a linear inequality that includes the boundary line.

mitad-plano cerrado La solución de una desigualdad linear que incluye la línea de límite.

coefficient (p. 26) The numerical factor of a term.

coeficiente Factor numérico de un término.

combination (p. 765) An arrangement or listing in which order is not important.

combinación Arreglo o lista en que el orden no es importante.

common difference (p. 187) The difference between the terms in a sequence.

diferencia común Diferencia entre términos consecutivos de una sucesión.

common ratio (p. 580) The ratio of successive terms of a geometric sequence.

razón común El razón de términos sucesivos de una secuencia geométrica.

complements (p. P33 and 772) One of two parts of a probability making a whole.

complementos Una de dos partes de una probabilidad que forma un todo.

completing the square (p. 552) To add a constant term to a binomial of the form $x^2 + bx$ so that the resulting trinomial is a perfect square.

completar el cuadrado Adición de un término constante a un binomio de la forma $x^2 + bx$, para que el trinomio resultante sea un cuadrado perfecto.

complex fraction (p. 714) A fraction that has one or more fractions in the numerator or denominator.

fracción compleja Fracción con una o más fracciones en el numerador o denominador.

composite number (p. 420) A whole number, greater than 1, that has more than two factors.

número compuesto Número entero mayor que 1 que posee más de dos factores.

compound event (p. 771) Two or more simple events.

evento compuesto Dos o más eventos simples.

compound inequality (p. 304) Two or more inequalities that are connected by the words *and* or *or*.

desigualdad compuesta Dos o más desigualdades que están unidas por las palabras y u o.

compound interest (p. 574) A special application of exponential growth.

interés compuesto Aplicación especial de crecimiento exponencial.

conclusion (p. 54) The part of a conditional statement immediately following the word *then*.

conclusión Parte de un enunciado condicional que sigue inmediatamente a la palabra entonces.

conditional probability (p. 777) The probability of an event under the condition that some preceding event has occurred.

probabilidad condicional La probabilidad de un acontecimiento bajo condición que ha ocurrido un cierto acontecimiento precedente.

conditional statements (p. 54) Statements written in the form *If A, then B.*

enunciados condicionales Enunciados de la forma Si A, entonces B.

conjugates (p. 614) Binomials of the form $a\sqrt{b} + c\sqrt{d}$ and $a\sqrt{b} - c\sqrt{d}$.

conjugados Binomios de la forma $a\sqrt{b} + c\sqrt{d}$ and $a\sqrt{b} - c\sqrt{d}$.

consecutive integers (p. 92) Integers in counting order.

enteros consecutivos Enteros en el orden de contar.

consistent (p. 333) A system of equations that has at least one ordered pair that satisfies both equations.

consistente Sistema de ecuaciones para el cual existe al menos un par ordenado que satisface ambas ecuaciones.

constant (p. 153, 401) A monomial that is a real number.

constante Monomio que es un número real.

constant of variation (p. 180) The number k in equations of the form $y = kx$.

constante de variación El número k en ecuaciones de la forma $y = kx$.

continuous function (p. 46) A function that can be graphed with a line or a smooth curve.

función continua Función cuya gráfica puedes ser una recta o una curva suave.

convenience sample (p. 643) A sample that includes members of a population that are easily accessed.

muestra de conveniencia Muestra que incluye miembros de una población fácilmente accesibles.

converse (p. 631) The statement formed by exchanging the hypothesis and conclusion of a conditional statement.

recíproco Enunciado que se obtiene al inter cambiar la hipótesis y la conclusión de un enucnciado condicional dado.

coordinate (p. P8) The number that corresponds to a point on a number line.

coordenada Número que corresponde a un punto en una recta numérica.

coordinate plane (p. 53) The plane containing the x- and y-axes.

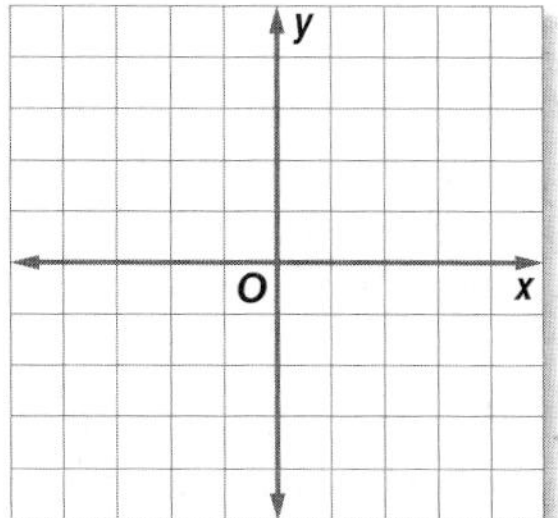

plano de coordenadas Plano que contiene los ejes x y y.

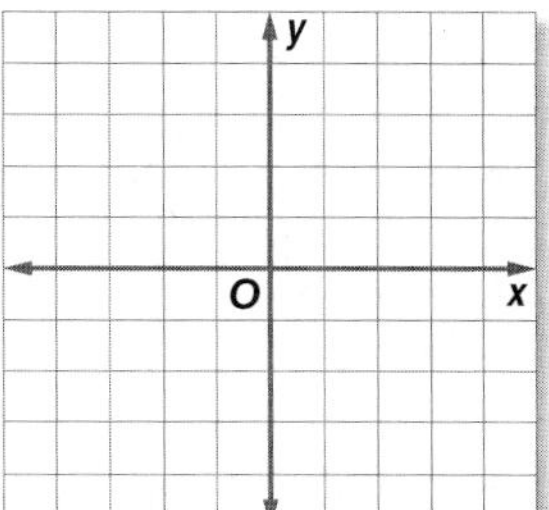

coordinate system (p. 38) The grid formed by the intersection of two number lines, the horizontal axis and the vertical axis.

sistema de coordenadas Cuadriculado formado por la intersección de dos rectas numéricas: los ejes x y y.

correlation coefficient (p. 253) A value that shows how close data points are to a line.

coeficiente de correlación Un valor que demostraciones cómo los puntos de referencias cercanos están a una línea.

cosine (p. 649) For an acute angle of a right triangle, the ratio of the measure of the leg adjacent to the acute angle to the measure of the hypotenuse.

coseno Para un ángulo agudo de un triángulo derecho, el razón de la medida de la pierna adyacente al ángulo agudo de la medida de la hipotenusa.

counterexample (p. 56) A specific case in which a statement is false.

contraejemplo Ejemplo específico de la falsedad de un enunciado.

D

deductive reasoning (p. 55) The process of using facts, rules, definitions, or properties to reach a valid conclusion.

razonamiento deductivo Proceso de usar hechos, reglas, definiciones o propiedades para sacar conclusiones válidas.

defining a variable (p. P5) Choosing a variable to represent one of the unspecified numbers in a problem and using it to write expressions for the other unspecified numbers in the problem.

definir una variable Consiste en escoger una variable para representar uno de los números desconocidos en un problema y luego usarla para escribir expresiones para otros números desconocidos en el problema.

degree of a monomial (p. 424) The sum of the exponents of all its variables.

grado de un monomio Suma de los exponentes de todas sus variables.

degree of a polynomial (p. 424) The greatest degree of any term in the polynomial.

grado de un polinomio El grado mayor de cualquier término del polinomio.

dependent (p. 333) A system of equations that has an infinite number of solutions.

dependiente Sistema de ecuaciones que posee un número infinito de soluciones.

dependent events (p. 771) Two or more events in which the outcome of one event affects the outcome of the other events.

eventos dependientes Dos o más eventos en que el resultado de un evento afecta el resultado de los otros eventos.

dependent variable (p. 40) The variable in a relation with a value that depends on the value of the independent variable.

variable dependiente La variable de una relación cuyo valor depende del valor de la variable independiente.

diameter (p. P24) The distance across a circle through its center.

diámetro La distancia a través de un círculo a través de su centro.

difference of two squares (p. 499) Two perfect squares separated by a subtraction sign.
$a^2 - b^2 = (a + b)(a - b)$ or
$a^2 - b^2 = (a - b)(a + b)$.

diferencia de cuadrados Dos cuadrados perfectos separados por el signo de sustracción.
$a^2 - b^2 = (a + b)(a - b)$ or
$a^2 - b^2 = (a - b)(a + b)$.

dilation (p. 24) A transformation that alters the size of a figure but not its shape.

dilatación Transformacióñ que altera el tamaño de una figure, pero no su forma.

dimension (p. 369) The number of rows, m, and the number of column, n, of a matrix written as $m \times n$.

dimension El número de filas, de m, y del número de la columna, n, de una matriz escrita como $m \times n$.

dimensional analysis (p. 128) The process of carrying units throughout a computation.

análisis dimensional Proceso de tomar en cuenta las unidades de medida al hacer cálculos.

direct variation (p. 180) An equation of the form $y = kx$, where $k \neq 0$.

variación directa Una ecuación de la forma $y = kx$, donde $k \neq 0$.

discrete function (p. 46) A function of points that are not connected.

función discreta Función de puntos desconectados.

discrete random variable (p. 779) A variable with a value that is a finite number of possible outcomes.

variable aleatoria discreta Variable cuyo valor es un número finito de posibles resultados.

discriminant (p. 562) In the Quadratic Formula, the expression under the radical sign, $b^2 - 4ac$.

discriminante En la fórmula cuadrática, la expresión debajo del signo radical, $b^2 - 4ac$.

Distance Formula (p. 636) The distance d between any two points with coordinates (x_1, y_1) and (x_2, y_2) is given by the formula

$$d = \sqrt{(x_2 - x_1)^2 + (y_2 - y_1)^2}.$$

Fórmula de la distancia La distancia d entre cualquier par de puntos con coordenadas (x_1, y_1) y (x_2, y_2) viene dada por la fórmula

$$d = \sqrt{(x_2 - x_1)^2 + (y_2 - y_1)^2}.$$

domain (p. 39) The set of the first numbers of the ordered pairs in a relation.

dominio Conjunto de los primeros números de los pares ordenados de una relación.

double root (p. 538) The roots of a quadratic function are the same number.

raíces dobles Las raíces de una función cuadrática son el mismo número.

E

element (p. 369) **1.** Each object or number in a set. (p. 715) **2.** Each entry in a matrix.

elemento **1.** Cada número u objeto de un conjunto. **2.** Cada entrada de una matriz.

elimination (p. 348) The use of addition or subtraction to eliminate one variable and solve a system of equations.

eliminación El uso de la adición o la sustracción para eliminar una variable y resolver así un sistema de ecuaciones.

empirical study (p. 787) Performing an experiment repeatedly, collecting and combining data, and analyzing the results.

estudio empírico Ejecución repetida de un experimento, recopilación y combinación de datos y análisis de resultados.

equally likely (p. P33) The outcomes of an experiment are equally likely if there are n outcomes and the probability of each is $\frac{1}{n}$.

igualmente probablemente Los resultados de un experimento son igualmente probables si hay resultados de n y la probabilidad de cada uno es $\frac{1}{n}$.

equation (p. 31) A mathematical sentence that contains an equals sign, =.

ecuación Enunciado matemático que contiene el signo de igualdad, =.

equivalent equations (p. 83) Equations that have the same solution.

ecuaciones equivalentes Ecuaciones que poseen la misma solución.

equivalent expressions (p. 16) Expressions that denote the same number.

expresiones equivalentes Expresiones que denotan el mismo número.

evaluate (p. 10) To find the value of an expression.

evaluar Calcular el valor de una expresión.

event (p. 650) Any collection of one or more outcomes in the sample space.

evento Cualquier colección de uno o más resultados de un espacio muestral.

excluded values (p. 678) Any values of a variable that result in a denominator of 0 must be excluded from the domain of that variable.

valores excluidos Cualquier valor de una variable cuyo resultado sea un denominador igual a cero, debe excluirse del dominio de dicha variable.

expected value (p. 791) The weighted average of all outcomes.

valor previsto El promédio cargado de todos los resultados.

experiment (p. 740) Data are recorded from outcomes involving characteristics of a sample.

experimento Los datos se registran de los resultados que implican características de una muestra.

experimental probability (pp. P35 and 787) What actually occurs when conducting a probability experiment, or the ratio of relative frequency to the total number of events or trials.

probabilidad experimental Lo que realmente sucede cuando se realiza un experimento probabilístico o la razón de la frecuencia relativa al número total de eventos o pruebas.

exponent (p. 5) In an expression of the form x^n, the exponent is n. It indicates the number of times x is used as a factor.

exponente En una expresión de la forma x^n, el exponente es n. Éste indica cuántas veces se usa x como factor.

exponential decay (p. 574) When an initial amount decreases by the same percent over a given period of time.

desintegración exponencial La cantidad inicial disminuye según el mismo porcentaje a lo largo de un período de tiempo dado.

exponential function (p. 567) A function that can be described by an equation of the form $y = a^x$, where $a > 0$ and $a \neq 1$.

función exponencial Función que puede describirse mediante una ecuación de la forma $y = a^x$, donde $a > 0$ y $a \neq 1$.

Glossary/Glosario

exponential growth (p. 573) When an initial amount increases by the same percent over a given period of time.

crecimiento exponencial La cantidad inicial aumenta según el mismo porcentaje a lo largo de un período de tiempo dado.

extraneous solutions (pp. 721, 624) Results that are not solutions to the original equation.

soluciones extrañas Resultados que no son soluciones de la ecuación original.

extremes (p. 112) In the ratio $\frac{a}{b} = \frac{c}{d}$, a and d are the extremes.

extremos En la razón $\frac{a}{b} = \frac{c}{d}$, a y d son los extremos.

F

factored form (p. 471) A monomial expressed as a product of prime numbers and variables in which no variable has an exponent greater than 1.

forma reducida Monomio escrito como el producto de números primos y variables y en el que ninguna variable tiene un exponente mayor que 1.

factorial (p. 764) The expression $n!$, read n factorial, where n is greater than zero, is the product of all positive integers beginning with n and counting backward to 1.

factorial La expresión $n!$, que se lee n factorial, donde n que es mayor que cero, es el producto de todos los números naturales, comenzando con n y contando hacia atrás hasta llegar al 1.

factoring (p. 476) To express a polynomial as the product of monomials and polynomials.

factorización La escritura de un polinomio como producto de monomios y polinomios.

factoring by grouping (p. 477) The use of the Distributive Property to factor some polynomials having four or more terms.

factorización por agrupamiento Uso de la Propiedad distributiva para factorizar polinomios que poseen cuatro o más términos.

factors (p. 6) In an algebraic expression, the quantities being multiplied are called factors.

factores En una expresión algebraica, los factores son las cantidades que se multiplican.

family of functions (p. 161) A group of functions that have one or more similar characteristics.

familia de funciones Un grupo de las funciones que tienen unas o más características similares.

family of graphs (pp. 197, 478) Graphs and equations of graphs that have at least one characteristic in common.

familia de gráficas Gráficas y ecuaciones de gráficas que tienen al menos una característica común.

FOIL method (p. 448) To multiply two binomials, find the sum of the products of the First terms, the Outer terms, the Inner terms, and the Last terms.

método FOIL Para multiplicar dos binomios, busca la suma de los productos de los primeros (First) términos, los términos exteriores (Outer), los términos interiores (Inner) y los últimos términos (Last).

formula (p. 76) An equation that states a rule for the relationship between certain quantities.

fórmula Ecuación que establece una relación entre ciertas cantidades.

four-step problem-solving plan (p. P5)
Step 1 Explore the problem.
Step 2 Plan the solution.
Step 3 Solve the problem.
Step 4 Check the solution.

plan de cuatro pasos para resolver problemas
Paso 1 Explora el problema.
Paso 2 Planifica la solución.
Paso 3 Resuelve el problema.
Paso 4 Examina la solución.

frequency table (p. P40) A chart that indicates the number of values in each interval.

Tabla de frecuencias Tabla que indica el número de valores en cada intervalo.

function (p. 45) A relation in which each element of the domain is paired with exactly one element of the range.

función Una relación en que a cada elemento del dominio le corresponde un único elemento del rango.

function notation (p. 48) A way to name a function that is defined by an equation. In function notation, the equation $y = 3x - 8$ is written as $f(x) = 3x - 8$.

notación funcional Una manera de nombrar una función definida por una ecuación. En notación funcional, la ecuación $y = 3x - 8$ se escribe $f(x) = 3x - 8$.

Fundamental Counting Principle (p. 764) If an event M can occur in m ways and is followed by an event N that can occur in n ways, then the event M followed by the event N can occur in $m \times n$ ways.

Principio fundamental de contar Si un evento M puede ocurrir de m maneras y lo sigue un evento N que puede ocurrir de n maneras, entonces el evento M seguido del evento N puede ocurrir de $m \times n$ maneras.

G

general equation for exponential decay (p. 511) $y = C(1 - r)^t$, where y is the final amount, C is the initial amount, r is the rate of decay expressed as a decimal, and t is time.

ecuación general de desintegración exponencial $y = C(1 - r)^t$, donde y es la cantidad final, C es la cantidad inicial, r es la tasa de desintegración escrita como decimal y t es el tiempo.

general equation for exponential growth (p. 510) $y = C(1 + r)^t$, where y is the final amount, C is the initial amount, r is the rate of change expressed as a decimal, and t is time.

ecuación general de crecimiento exponencial $y = C(1 + r)^t$, donde y es la cantidad final, C es la cantidad inicial, r es la tasa de cambio del crecimiento escrita como decimal y t es el tiempo.

geometric sequence (p. 580) A sequence in which each term after the first is found by multiplying the previous term by a constant r, called the common ratio.

secuencia geométrica Una secuencia en la cual cada término después de que la primera sea encontrada multiplicando el término anterior por un r constante, llamado el razón común.

graph (p. P8) To draw, or plot, the points named by certain numbers or ordered pairs on a number line or coordinate plane.

graficar Marcar los puntos que denotan ciertos números en una recta numérica o ciertos pares ordenados en un plano de coordenadas.

greatest common factor (GCF) (p. 471) The product of the prime factors common to two or more integers.

máximo común divisor (MCD) El producto de los factores primos comunes a dos o más enteros.

greatest integer function (p. 261) A step function, written as $f(x) = [\![x]\!]$, where $f(x)$ is the greatest integer less than or equal to x.

La función más grande del número entero Una función del paso, escrita como $f(x) = [\![x]\!]$, donde está el número entero $f(x)$ es el número más grande menos que o igual a x.

H

half-plane (p. 315) The region of the graph of an inequality on one side of a boundary.

semiplano Región de la gráfica de una desigualdad en un lado de la frontera.

histogram (p. P40) A graphical display that uses bars to display numerical data that have been organized into equal intervals.

histograma Una exhibición gráfica que utiliza barras para exhibir los datos numéricos que se han organizado en intervalos iguales.

hypotenuse (p. 630) The side opposite the right angle in a right triangle.

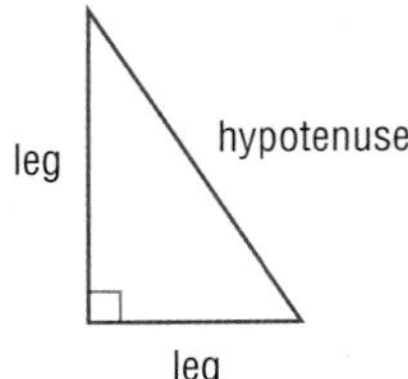

hipotenusa Lado opuesto al ángulo recto en un triángulo rectángulo.

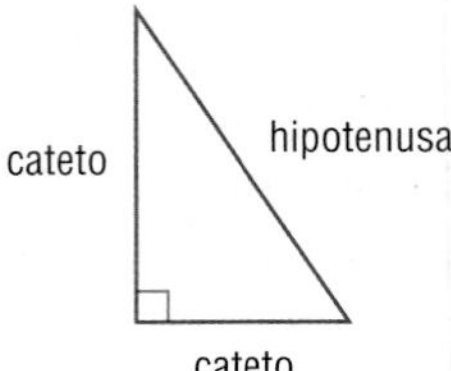

hypothesis (p. 54) The part of a conditional statement immediately following the word if.

hipótesis Parte de un enunciado condicional que sigue inmediatamente a la palabra si.3

I

identity (pp. 33, 98) An equation that is true for every value of the variable.

identidad Ecuación que es verdad para cada valor de la variable.

identity matrix (p. 377) A square matrix that, when multiplied by another matrix, equals that same matrix. If A is any $n \times n$ matrix and I is the $n \times n$ identity matrix, then $A \cdot I = A$ and $I \cdot A = A$.

matriz de la identidad una matriz cuadrada que, cuando es multiplicada por otra matriz, iguala que la misma matriz. Si A es alguna de la matriz $n \times n$ e I es la matriz de la identidad de $n \times n$, entonces $A \cdot I = A$ e $I \cdot A = A$.

if-then statements (p. 54) Conditional statements in the form *If A, then B*.

enunciados si-entonces Enunciados condicionales de la forma *Si A, entonces B*.

inclusive (p. 666) Two events that can occur at the same time.

inclusivos Dos eventos que pueden ocurrir simultáneamente.

inconsistent (p. 333) A system of equations with no ordered pair that satisfy both equations.

inconsistente Un sistema de ecuaciones para el cual no existe par ordenado alguno que satisfaga ambas ecuaciones.

independent (p. 333) A system of equations with exactly one solution.

independiente Un sistema de ecuaciones que posee una única solución.

independent events (p. 771) Two or more events in which the outcome of one event does not affect the outcome of the other events.

eventos independientes El resultado de un evento no afecta el resultado del otro evento.

independent variable (p. 40) The variable in a function with a value that is subject to choice.

variable independiente La variable de una función sujeta a elección.

inductive reasoning (p. 195) A conclusion based on a pattern of examples.

razonamiento inductivo Conclusión basada en un patrón de ejemplos.

inequality (p. 16) An open sentence that contains the symbol $<, \leq, >$, or $\geq$.

desigualdad Enunciado abierto que contiene uno o más de los símbolos $<, \leq, >$, o $\geq$.

integers (p. P7) The set $\{\ldots, -2, -1, 0, 1, 2, \ldots\}$.

enteros El conjunto $\{\ldots, -2, -1, 0, 1, 2, \ldots\}$.

interquartile range (p. P42) The range of the middle half of a set of data. It is the difference between the upper quartile and the lower quartile.

amplitud intercuartílica Amplitude de la mitad central de un conjunto de datos. Es la diferenccia entre el cuartil superior y el inferior.

intersection (p. 304) The graph of a compound inequality containing and; the solution is the set of elements common to both inequalities.

intersección Gráfica de una desigualdad compuesta que contiene la palabra y; la solución es el conjunto de soluciones de ambas desigualdades.

inverse (p. 145) The inverse of any relation is obtained by switching the coordinates in each ordered pair.

inversa La inversa de una relación se halla intercambiando las coordenadas de cada par ordenado.

inverse cosine (p. 651) If $\angle A$ is an acute angle and the cosine of A is x, then the inverse cosine of x is the measure of $\angle A$.

coseno inverso Si el $\angle A$ es un ángulo agudo y el coseno de A es x, entonces el coseno inverso de x es la medida de $\angle$A.

inverse sine (p. 651) If $\angle A$ is an acute angle and the sine of A is x, then the inverse sine of x is the measure of $\angle A$.

seno inverso Si $\angle A$ es un ángulo agudo y el seno de A es x, entonces el seno inverso de x es la medida de $\angle A$.

inverse tangent (p. 651) If $\angle A$ is an acute angle and the tangent of A is x, then the inverse tangent of x is the measure of $\angle A$.

tangente inverso Si el $\angle A$ es un ángulo agudo y la tangente de A es x, entonces la tangente inversa de x es la medida de $\angle A$.

inverse variation (p. 670) An equation of the form $xy = k$, where $k \neq 0$.

variación inversa Ecuación de la forma $xy = k$, donde $k \neq 0$.

irrational numbers (p. P7) Numbers that cannot be expressed as terminating or repeating decimals.

números irracionales Números que no pueden escribirse como decimales terminales o periódicos.

L

leading coefficient (p. 425) The coefficient of the term with the highest degree.

coeficiente inicial El coeficiente del término con el grado más alto (el primer coeficiente inicial).

least common denominator (LCD) (p. 708) The least common multiple of the denominators of two or more fractions.

mínimo denominador común (mcd) El mínimo común múltiplo de los denominadores de dos o más fracciones.

least common multiple (LCM) (p. 707) The least number that is a common multiple of two or more numbers.

mínimo común múltiplo (mcm) El número menor que es múltiplo común de dos o más números.

legs (p. 549) The sides of a right triangle that form the right angle.

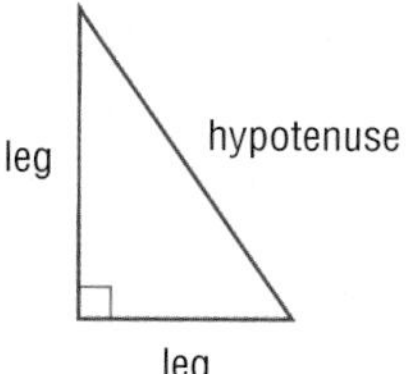

catetos Lados de un triángulo rectángulo que forman el ángulo recto del mismo.

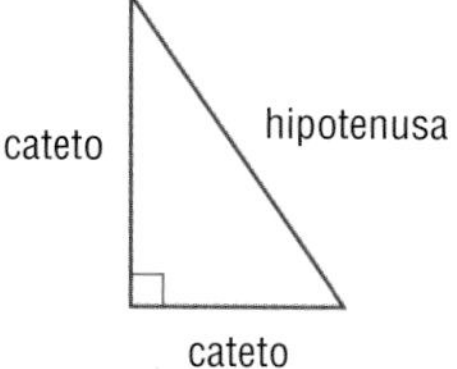

like terms (p. 25) Terms that contain the same variables, with corresponding variables having the same exponent.

términos semejantes Expresiones que tienen las mismas variables, con las variables correspondientes elevadas a los mismos exponentes.

linear equation (p. 153) An equation in the form $Ax + By = C$, with a graph that is a straight line.

ecuación lineal Ecuación de la forma $Ax + By = C$, cuya gráfica es una recta.

linear extrapolation (p. 226) The use of a linear equation to predict values that are outside the range of data.

extrapolación lineal Uso de una ecuación lineal para predecir valores fuera de la amplitud de los datos.

linear function (p. 155) A function with ordered pairs that satisfy a linear equation.

función lineal Función cuyos pares ordenados satisfacen una ecuación lineal.

linear interpolation (p. 247) The use of a linear equation to predict values that are inside of the data range.

interpolación lineal Uso de una ecuación lineal para predecir valores dentro de la amplitud de los datos.

linear regression (p. 253) An algorithm to find a precise line of fit for a set of data.

regresión linear Un algoritmo para encontrar una línea exacta del ajuste para un sistema de datos.

linear transformation (p. 760) One or more operations performed on a set of data that can be written as a linear funciton.

transformación lineal Una o más operaciones que se hacen en un conjunto de datos y que se pueden escribir como una función lineal.

line of fit (p. 246) A line that describes the trend of the data in a scatter plot.

recta de ajuste Recta que describe la tendencia de los datos en una gráfica de dispersión.

literal equation (p. 127) A formula or equation with several variables.

ecuación literal Un fórmula o ecuación con varias variables.

look for a pattern (p. 172) Find patterns in sequences to solve problems.

buscar un patrón Encontrar patrones en sucesiones para resolver problemas.

lower quartile (p. P38) Divides the lower half of the data into two equal parts.

cuartil inferior Éste divide en dos partes iguales la mitad inferior de un conjunto de datos.

M

mapping (p. 38) Illustrates how each element of the domain is paired with an element in the range.

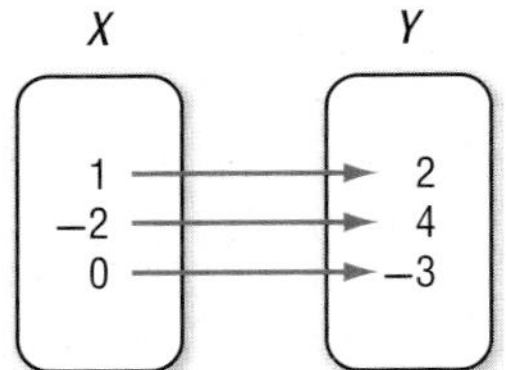

aplicaciones Ilustra la correspondencia entre cada elemento del dominio con un elemento del rango.

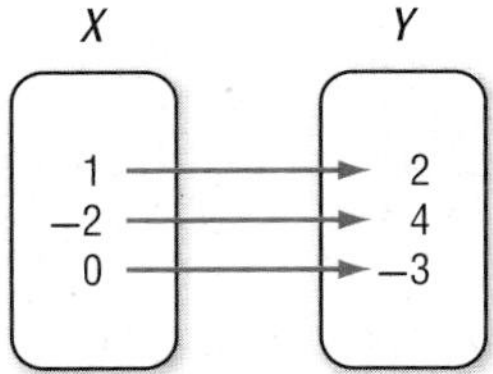

measures of central tendency (pp. P37, 746) Numbers or pieces of data that can represent the whole set of data.

medidas de tendencia central Números o fragmentos que pueden representar el conjunto de datos total de datos.

matrix (p. 369) Any rectangular arrangement of numbers in rows and columns.

matriz Disposción rectangular de numeros colocados en filas y columnas.

maximum (p. 525) The highest point on the graph of a curve.

máximo El punto más alto en la gráfica de una curva.

mean (p. P37) The sum of numbers in a set of data divided by the number of items in the data set.

media La suma de los números de un conjunto de datos dividida entre el numero total de artículos.

mean absolute deviation (p. 757) The average of the absolute values of differences between the mean and each value in a data set. It is used to predict errors and to judge equality.

desviación absoluta media El promedio de los valores absolutos de diferencias entre el medio y cada valor de un conjunto de datos. Ha usado para predecir errores y para juzgar igualdad.

means (p. 112) The middle terms of the proportion.

medios Los términos centrales de una proporción.

measures of central tendency (pp. P37, 746) Numbers or pieces of data that can represent the whole set of data.

medidas de tendencia central Números o fragmentos que pueden representar el conjunto de datos total de datos.

measures of variation (p. P38) Used to describe the distribution of statistical data.

medidas de variación Números que se usan para describir la distribución o separación de un conjunto de datos.

median (p. 37) The middle number in a set of data when the data are arranged in numerical order. If the data set has an even number, the median is the mean of the two middle numbers.

mediana El número central de conjunto de datos, una vezque los datos han sido ordenados numéricamente. Si hay un número par de datos, la mediana es el promedio de los datos centrales.

median fit line (p. 255) A type of best-fit line that is calculated using the medians of the coordinates of the data points.

línea apta del punto medio Tipo de mejor-cupo la línea se calcula que usando los puntos medios de los coordenadas de los puntos de referencias.

midpoint (p. 638) The point halfway between the endpoints of a segment.

punto medio Punto que divide a un segmento separándolo en dos segmentos congruentes.

minimum (p. 525) The lowest point on the graph of a curve.

mínimo El punto más bajo en la gráfica de una curva.

mixed expression (p. 714) An expression that contains the sum of a monomial and a rational expression.

expresión mixta Expresión que contiene la suma de un monomio y una expresión racional.

mixture problems (p. 132) Problems in which two or more parts are combined into a whole.

problemas de mezclas Problemas en que dos o más partes se combinan en un todo.

mode (p. P37) The number(s) that appear most often in a set of data.

moda El número(s) que aparece más frecuencia en un conjunto de datos.

monomial (p. 401) A number, a variable, or a product of a number and one or more variables.

monomio Número, variable o producto de un número por una o más variables.

multiplicative identity (p. 17) For any number a, $a \cdot 1 = 1 \cdot a = a$.

identidad de la multiplicación Para cualquier número $a \cdot 1 = 1 \cdot a = a$.

multiplicative inverses (pp. P18, 17) Two numbers with a product of 1.

inversos multiplicativos Dos números cuyo producto es igual a 1.

multi-step equation (p. 91) Equations with more than one operation.

ecuaciones de varios pasos Ecuaciones con más de una operación.

mutually exclusive (p. 773) Events that cannot occur at the same time.

mutuamente exclusivos Eventos que no pueden ocurrir simultáneamente.

Glossary/Glosario

N

natural numbers (p. P7) The set {1, 2, 3, …}.

números naturales El conjunto {1, 2, 3, …}.

negative correlation (p. 227) In a scatter plot, as x increases, y decreases.

correlación negativa En una gráfica de dispersión, a medida que x aumenta, y disminuye.

negative exponent (p. 410) For any real number $a \neq 0$ and any integer n, $a^{-n} = \frac{1}{a^n}$ and $\frac{1}{a^{-n}} = a^n$.

exponiente negativo Para números reales, si $a \neq 0$, y cualquier número entero n, entonces $a^{-n} = \frac{1}{a^n}$ and $\frac{1}{a^{-n}} = a^n$.

negative number (p. P7) Any value less than zero.

número negativo Cualquier valor menor que cero.

nonlinear function (pp. 48 and 525) A function with a variable term that has an exponent other than 1 or 0.

función no lineal Una función con un término variable que tiene un exponente con excepción de 1 o de 0.

number theory (p. 92) The study of numbers and the relationships between them.

teoría del número El estudio de números y de las relaciones entre ellas.

O

observational study (p. 740) Data are recorded and made regarding certain activities within a sample.

estudio de observación Datos que se registran y se hacen con respecto a ciertas actividades dentro de una muestra.

odds (p. P35) The ratio of the probability of the success of an event to the probability of its complement.

probabilidades El cociente de la probabilidad del éxito de un acontecimiento a la probabilidad de su complemento.

open half-plane (p. 315) The solution of a linear inequality that does not include the boundary line.

abra el mitad-plano La solución de una desigualdad linear que no incluya la línea de límite.

open sentence (p. 31) A mathematical statement with one or more variables.

enunciado abierto Un enunciado matemático que contiene una o más variables.

opposites (p. P11) Two numbers with the same absolute value by different signs.

opuestos Dos números que tienen el mismo valor absolute, pero que tienen distintos signos.

ordered pair (p. 38) A set of numbers or coordinates used to locate any point on a coordinate plane, written in the form (x, y).

par ordenado Un par de números que se usa para ubicar cualquier punto de un plano de coordenadas y que se escribe en la forma (x, y).

order of magnitude (p. 411) The order of magnitude of a quantity is the number rounded to the nearest power of 10.

orden de magnitud de una cantidad Un número redondeado a la potencia más cercana de 10.

order of operations (p. 10)

1. Evaluate expressions inside grouping symbols.
2. Evaluate all powers.
3. Do all multiplications and/or divisions from left to right.

orden de las operaciones

1. Evalúa las expresiones dentro de los símbolos de agrupamiento.
2. Evalúa todas las potencias.
3. Multiplica o divide de izquierda a derecha.

4. Do all additions and/or subtractions from left to right.

4. Suma o resta de izquierda a derecha.

origin (p. 38) The point where the two axes intersect at their zero points.

origen Punto donde se intersecan los dos ejes en sus puntos cero.

outliers (p. P42) Data that are more than 1.5 times the interquartile range beyond the quartiles.

valores atípicos Datos que distan de los cuartiles más de 1.5 veces la amplitude intercuartílica.

P

parabola (p. 525) The graph of a quadratic function.

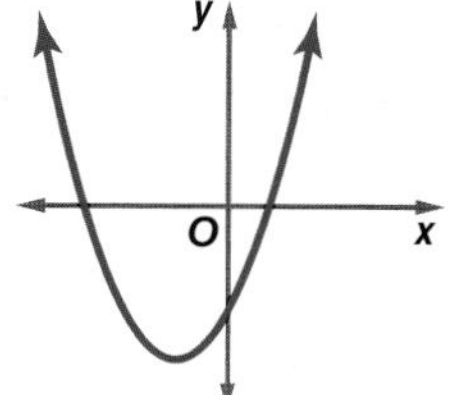

parábola La gráfica de una función cuadrática.

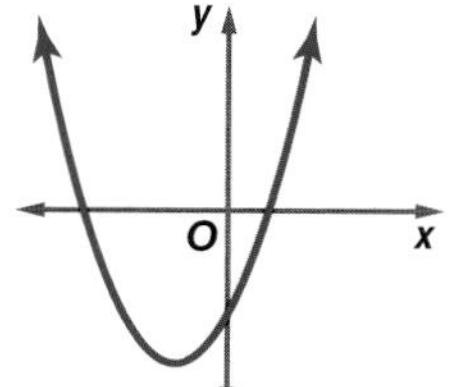

parallel lines (p. 237) Lines in the same plane that never intersect and have the same slope.

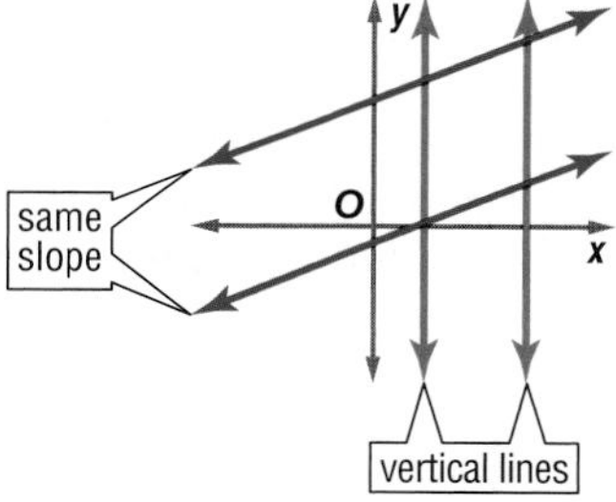

rectas paralelas Rectas en el mismo plano que no se intersecan jamás y que tienen pendientes iguales.

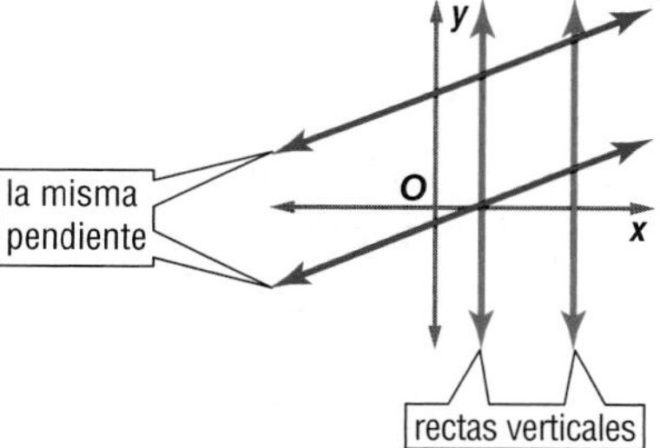

parameter (p. 756) A characteristic of the population as a whole.

parámetro Una característica de la población en su totalidad.

parent function (p. 161) The simplest of functions in a family.

función basíca La función más fundamental de un familia de funciones.

parent graph (p. 197) The simplest of the graphs in a family of graphs.

gráfica madre La gráfica más sencilla en una familia de gráficas.

percent (p. P20) A ratio that compares a number to 100.

porcentaje Razón que compara un numero con 100.

percent of change (p. 119) When an increase or decrease is expressed as a percent.

porcentaje de cambio Cuando un aumento o disminución se escribe como un tanto por ciento.

percent of decrease (p. 119) The ratio of an amount of decrease to the previous amount, expressed as a percent.

porcentaje de disminución Razón de la cantidad de disminución a la cantidad original, escrita como un tanto por ciento.

percent of increase (p. 119) The ratio of an amount of increase to the previous amount, expressed as a percent.

porcentaje de aumento Razón de la cantidad de aumento a la cantidad original, escrita como un tanto por ciento.

percent proportion (p. P20)
$\frac{\text{part}}{\text{whole}} = \frac{\text{percent}}{100}$ or $\frac{a}{b} = \frac{P}{100}$

perimeter (p. P23) The distance around a geometric figure.

perfect square (p. P7) A number with a square root that is a rational number.

perfect square trinomial (p. 505) A trinomial that is the square of a binomial.
$(a + b)^2 = (a + b)(a + b) = a^2 + 2ab + b^2$ *or*
$(a - b)^2 = (a - b)(a - b) = a^2 - 2ab - b^2$

permutation (p. 764) An arrangement or listing in which order is important.

perpendicular line (p. 238) Lines that intersect to form a right angle.

piecewise function (p. 261) A function written using two or more expressions.

piecewise-defined function (p. 262) A function that is written using two or more expressions.

point-slope form (p. 231) An equation of the form $y - y_1 = m(x - x_1)$, where m is the slope and (x_1, y_1) is a given point on a nonvertical line.

polynomial (p. 424) A monomial or sum of monomials.

population (p. 740) A large group of data usually represented by a sample.

positive correlation (p. 227) In a scatter plot, as x increases, y increases.

positive number (p. P7) Any value that is greater than zero.

power (p. 5) An expression of the form x^n, read *x to the nth power*.

prime factorization (p. 420) A whole number expressed as a product of factors that are all prime numbers.

prime number (p. 420) A whole number, greater than 1, with only factors that are 1 and itself.

prime polynomial (p. 495) A polynomial that cannot be written as a product of two polynomials with integral coefficients.

proporción porcentual
$\frac{\text{parte}}{\text{todo}} = \frac{\text{por ciento}}{100}$ or $\frac{a}{b} = \frac{P}{100}$

perímetro Longitud alrededor una figura geométrica.

cuadrado perfecto Número cuya raíz cuadrada es un número racional.

trinomio cuadrado perfecto Un trinomio que es el cuadrado de un binomio.
$(a + b)^2 = (a + b)(a + b) = a^2 + 2ab + b^2$ *or*
$(a - b)^2 = (a - b)(a - b) = a^2 - 2ab - b^2$

permutación Arreglo o lista en que el orden es importante.

recta perpendicular Recta que se intersecta formando un ángulo recto

función por partes Función que se escribe usando dos o más expresiones.

función por partes Función que se escribe usando dos o más expresiones.

forma punto-pendiente Ecuación de la forma $y - y_1 = m(x - x_1)$, donde m es la pendiente y (x_1, y_1) es un punto dado de una recta no vertical.

polinomio Un monomio o la suma de monomios.

población Grupo grande de datos, representado por lo general por una muestra.

correlación positiva En una gráfica de dispersión, a medida que x aumenta, y aumenta.

número positivos Cualquier valor mayor que cero.

potencia Una expresión de la forma x^n, se lee *x a la enésima potencia*.

factorización prima Número entero escrito como producto de factores primos.

número primo Número entero mayor que 1 cuyos únicos factores son 1 y sí mismo.

polinomio primo Polinomio que no puede escribirse como producto de dos polinomios con coeficientes enteros.

Glossary/Glosario

principal square root (p. 49) The nonnegative square root of a number.

raíz cuadrada principal La raíz cuadrada no negativa de un número.

probability (p. P33) The ratio of the number of favorable equally likely outcomes to the number of possible equally likely outcomes.

probabilidad La razón del número de maneras en que puede ocurrir el evento al numero de resultados posibles.

probability distribution (p. 780) The probability of every possible value of the random variable x.

distribución de probabilidad Probabilidad de cada valor posible de una variable aleatoria x.

probability histogram (p. 780) A way to give the probability distribution for a random variable and obtain other data.

histograma probabilístico Una manera de exhibir la distribución de probabilidad de una variable aleatoria y obtener otros datos.

product (p. 5) In an algebraic expression, the result of quantities being multiplied is called the product.

producto En una expresión algebraica, se llama producto al resultado de las cantidades que se multiplican.

product rule (p. 671) If (x_1, y_1) and (x_2, y_2) are solutions to an inverse variation, then $y_1x_1 = y_2x_2$.

regla del producto Si (x_1, y_1) y (x_2, y_2) son soluciones de una variación inversa, entonces $y_1x_1 = y_2x_2$.

proportion (p. 111) An equation of the form $\frac{a}{b} = \frac{c}{d}$ stating that two ratios are equivalent.

proporción Ecuación de la forma $\frac{a}{b} = \frac{c}{d}$ que afirma la equivalencia de dos razones.

Pythagorean Theorem (p. 549) If a and b are the measures of the legs of a right triangle and c is the measure of the hypotenuse, then $c^2 = a^2 + b^2$.

Teorema de Pitágoras Si a y b son las longitudes de los catetos de un triángulo rectángulo y si c es la longitud de la hipotenusa, entonces $c^2 = a^2 + b^2$.

Pythagorean triple (p. 631) Whole numbers that satisfy the Pythagorean Theorem.

Triple pitagórico Números enteros que satisfacen el Teorema de Pitágoras.

Q

quadratic equation (pp. 485, 537) An equation of the form $ax^2 + bx + c = 0$, where $a \neq 0$.

ecuación cuadrática Ecuación de la forma $ax^2 + bx + c = 0$, donde $a \neq 0$.

quadratic expression (p. 448) An expression in one variable with a degree of 2 written in the form $ax^2 + bx + c$.

expression cuadratica Una expresión en una variable con un grado de 2, escritos en la forma $ax^2 + bx + c$.

Quadratic Formula (p. 558) The solutions of a quadratic equation in the form $ax^2 + bx + c$, where $a \neq 0$, are given by the formula $x = \frac{-b \pm \sqrt{b^2 - 4ac}}{2a}$.

Fórmula cuadrática Las soluciones de una ecuación cuadrática de la forma $ax^2 + bx + c$, donde $a \neq 0$, vienen dadas por la fórmula $x = \frac{-b \pm \sqrt{b^2 - 4ac}}{2a}$.

quadratic function (p. 525) An equation of the form $y = ax^2 + bx + c$, where $a \neq 0$.

función cuadrática Función de la forma $y = ax^2 + bx + c$, donde $a \neq 0$.

qualitative data (p. 746) Data that can not be given a numerical value

datos cualitativos Datos que no se pueden dar un valor numérico.

quantitative data (p. 746) Data that can be given as a numerical value

datos cuantitativos Datos que se pueden dar como numérico

Glossary/Glosario

quartile (p. P38) The values that divide a set of data into four equal parts.

cuartile Valores que dividen en conjunto de datos en cuarto partes iguales.

R

radical equations (p. 624) Equations that contain radicals with variables in the radicand.

ecuaciones radicales Ecuaciones que contienen radicales con variables en el radicando.

radical expression (p. 612) An expression that contains a square root.

expresión radical Expresión que contiene una raíz cuadrada.

radical function (p. 605) A function that contains radicals with variables in the radicand.

ecuaciones radicales Ecuaciones que contienen radicales con variables en el radicando.

radical sign (p. 49) The symbol $\sqrt{\ }$, used to indicate a nonnegative square root.

signo radical El símbolo $\sqrt{\ }$, que se usa para indicar la raíz cuadrada no negativa.

radicand (p. 605) The expression that is under the radical sign.

radicando La expresión debajo del signo radical.

radius (p. P24) Distance from the center to any point on the circle.

radio Distancia del centro cualquier punto de un círculo.

random sample (p. 741) A sample that is chosen without any preference, representative of the entire population.

muestra aleatoria Muestra tomada sin preferencia alguna y que es representativa de toda la población.

random variable (p. 779) A variable with a value that is the numerical outcome of a random event.

variable aleatoria Una variable cuyos valores son los resultados numéricos de un evento aleatorio.

range (p. 39) The set of second numbers of the ordered pairs in a relation.

rango Conjunto de los segundos números de los pares ordenados de una relación.

rate (p. 113) The ratio of two measurements having different units of measure.

tasa Razón de dos medidas que tienen distintas unidades de medida.

rate of change (p. 170) How a quantity is changing over time.

tasa de cambio Cómo cambia una cantidad con el tiempo.

rate problems (pp. 134, 725) Rational equations are used to solve problems involving transportal rates.

problemas de tasas Ecuaciones racionales que se usan para resolver problemas de tasas de transportación.

ratio (p. 111) A comparison of two numbers by division.

razón Comparación de dos números mediante división.

rational approximation (p. 50) A rational number that is close to, but not equal to, the value of an irrational number.

aproximación racional Número racional que está cercano, pero que no es igual, al valor de un número irracional.

rational equations (p. 720) Equations that contain rational expressions.

ecuaciones racionales Ecuaciones que contienen xpresiones racionales.

rational expression (p. 684) An algebraic fraction with a numerator and denominator that are polynomials.

expresión racional Fracción algebraica cuyo numerador y denominador son polinomios.

rational function (p. 678) An equation of the form $f(x) = \frac{p(x)}{q(x)}$, where $p(x)$ and $q(x)$ are polynomial functions, and $q(x) \neq 0$.

función racional Ecuación de la forma $f(x) = \frac{p(x)}{q(x)}$, donde $p(x)$ y $q(x)$ son funciones polinomiales y $q(x) \neq 0$.

rationalizing the denominator (p. 612) A method used to eliminate radicals from the denominator of a fraction.

racionalizar el denominador Método que se usa para eliminar radicales del denominador de una fracción.

rational numbers (p. P7) The set of numbers expressed in the form of a fraction $\frac{a}{b}$, where a and b are integers and $b \neq 0$.

números racionales Conjunto de los números que pueden escribirse en forma de fracción $\frac{a}{b}$, donde a y b son enteros y $b \neq 0$.

real numbers (p. P7) The set of rational numbers and the set of irrational numbers together.

números reales El conjunto de los números racionales junto con el conjunto de los números irracionales.

reciprocal (pp. P18, 17) The multiplicative inverse of a number.

recíproco Inverso multiplicativo de un número.

recursive formula (p. 188) Each term is formulated from one or more previous terms.

fórmula recursiva Cada tórmino proviene de uno o más terminos anteriores.

reflection (p. 545) A transformation where a figure, line, or curve, is flipped across a line.

reflexión Transformación en que cadapunto de una figura se aplica a través de una recta de simetría a su imagen correspondiente.

relation (p. 38) A set of ordered pairs.

relación Conjunto de pares ordenados.

relative frequency (p. 787) The number of times an outcome occurred in a probability experiment.

frecuencia relativa Número de veces que aparece un resultado en un experimento probabilístico.

replacement set (p. 31) A set of numbers from which replacements for a variable may be chosen.

conjunto de sustitución Conjunto de números del cual se pueden escoger sustituciones para una variable.

root (p. 161) The solutions of a quadratic equation.

raíces Las soluciones de una ecuación cuadrática

row reduction (p. 377) The process of performing elementary row operations on an augmented matrix to solve a system.

reducción de la fila El proceso de realizar operaciones elementales de la fila en una matriz aumentada para solucionar un sistema.

S

sample (p. 740) Some portion of a larger group selected to represent that group.

muestra Porción de un grupo más grande que se escoge para representarlo.

sample space (pp. P33, 764) The list of all possible outcomes.

espacio muestral Lista de todos los resultados posibles.

scalar (p. 371) A constant that is multiplied by a matrix.

escalar Una constante ques es multilicado por una matriz.

scalar multiplication (p. 371) Multiplication of a vector by a scalar.

multiplicación por escalare Multiplicación de un vector por una constante que es un vector.

scale (p. 108) A ratio or rate used when making a model of something that is too large or too small to be conveniently shown at actual size.

escala Razón o tasa que se usa al construir un modelo de algo que es demasiado grande o pequeño como para mostrarlo de tamaño natural.

scale (p. 114) The relationship between the measurements on a drawing or model and the measurements of the real object.

escala Relación entre las medidas de un dibujo o modelo y las medidas de la figura verdadera.

scale model (p. 114) A model used to represent an object that is too large or too small to be built at actual size.

modelo a escala Modelo que se usa para representar un figura que es demasiado grande o pequeña como para ser construida de tamaño natural.

scatter plot (p. 245) Two sets of data plotted as ordered pairs in a coordinate plane.

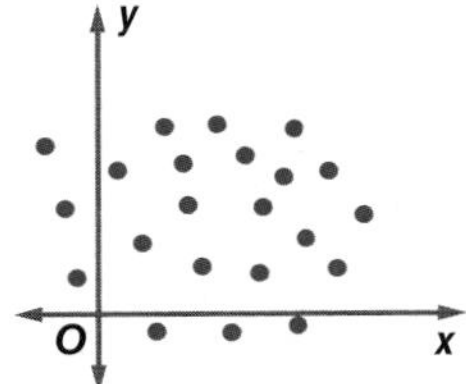

gráfica de dispersión Dos conjuntos de datos graficados como pares ordenados en un plano de coordenadas.

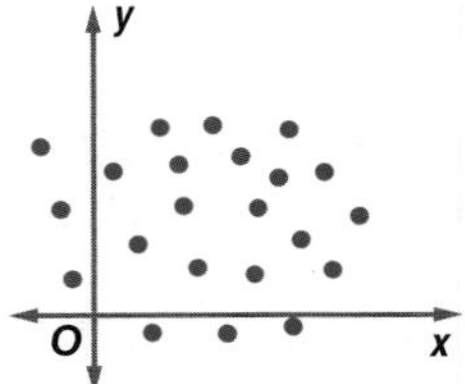

scientific notation (p. 416) A number in scientific notation is expressed as $a \times 10^n$, where $1 \leq a < 10$ and n is an integer.

notación científica Un numero en notación científica se escribe con $a \times 10^n$, donde $1 \leq a < 10$ y n es un número entero.

sequence (p. 187) A set of numbers in a specific order.

sucesión Conjunto de números en un orden específico.

set (p. 15) A collection of objects or numbers, often shown using braces { } and usually named by a capital letter.

conjunto Colección de objetos o números, que a menudo se exhiben usando paréntesis de corchete { } y que se identifican por lo general mediante una letra mayúscula .

set-builder notation (p. 284) A concise way of writing a solution set. For example, $\{t \mid t < 17\}$ represents the set of all numbers t such that t is less than 17.

notación de construcción de conjuntos Manera concisa de escribir un conjunto solución. Por ejemplo, $\{t \mid t < 17\}$ representa el conjunto de todos los números t que son menores o iguales que 17.

similar triangles (p. 642) Triangles having the same shape but not necessarily the same size.

semejantes Que tienen la misma forma, pero no necesariamente el mismo tamaño.

simple event (pp. 98, 663) A single event.

evento simple Un sólo evento.

simple random sample (p. 742) A sample that is as likely to be chosen as any other from the population.

muestra aleatoria simple Muestra de una población que tiene la misma probabilidad de escogerse que cualquier otra.

simplest form (p. 25) An expression is in simplest form when it is replaced by an equivalent expression having no like terms or parentheses.

forma reducida Una expresión está reducida cuando se puede sustituir por una expresión equivalente que no tiene ni términos semejantes ni paréntesis.

simulation (p. 788) Using an object to act out an event that would be difficult or impractical to perform.

sine (p. 649) For an acute angle of a right triangle, the ratio of the measure of the leg opposite the acute angle to the measure of the hypotenuse.

slope (p. 172) The ratio of the change in the y-coordinates (rise) to the corresponding change in the x-coordinates (run) as you move from one point to another along a line.

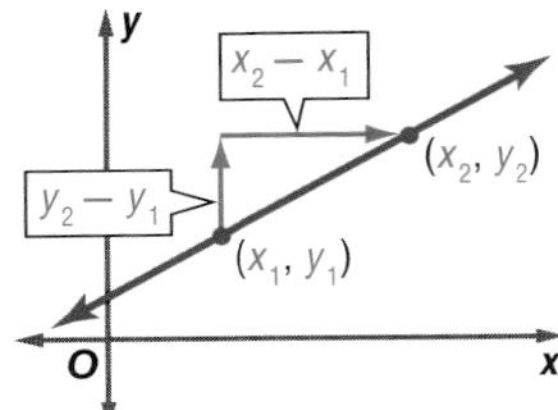

slope-intercept form (p. 214) An equation of the form $y = mx + b$, where m is the slope and b is the y-intercept.

solution (p. 31) A replacement value for the variable in an open sentence.

solution set (p. 31) The set of elements from the replacement set that make an open sentence true.

solve an equation (p. 83) The process of finding all values of the variable that make the equation a true statement.

solving an open sentence (p. 31) Finding a replacement value for the variable that results in a true sentence or an ordered pair that results in a true statement when substituted into the equation.

solving the triangle (p. 650) Finding the measures of all of the angles and sides of a triangle.

square root (p. P7) One of two equal factors of a number.

square root function (p. 605) Function that contains the square root of a variable.

standard deviation (p. 757) Is the square root of the variance.

simulación Uso de un objeto para representar un evento que pudiera ser difícil o poco práctico de ejecutar.

seno La razón entre la medida del cateto opuesto al ángulo agudo y la medida de la hipotenusa de un triángulo rectángulo.

pendiente Razón del cambio en la coordenada y (elevación) al cambio correspondiente en la coordenada x (desplazamiento) a medida que uno se mueve de un punto a otro en una recta.

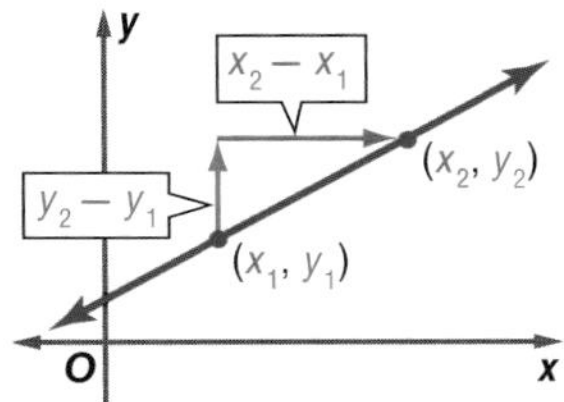

forma pendiente-intersección Ecuación de la forma $y = mx + b$, donde m es la pendiente y b es la intersección y.

solución Valor de sustitución de la variable en un enunciado abierto.

conjunto solución Conjunto de elementos del conjunto de sustitución que hacen verdadero un enunciado abierto.

resolver una ecuación Proceso en que se hallan todos los valores de la variable que hacen verdadera la ecuación.

resolver un enunciado abierto Hallar un valor de sustitución de la variable que resulte en un enunciado verdadero o un par ordenado que resulte en una proposición verdadera cuando se lo sustituye en la ecuación.

resolver un triángulo Hallar las medidas de todos los lados y todos los ángulos de un triángulo.

raíz cuadrada Uno de dos factores iguales de un número.

función radical Función que contiene la raíz cuadrada de una variable.

desviación tipica Calculada como la raíz cuadrada de la varianza.

upper quartile (p. P38) The median of the upper half of a set of data.

cuartil superior Mediana de la mitad superior de un conjunto de datos.

V

variable (p. 5) Symbols used to represent unspecified numbers or values.

variable Símbolos que se usan para representar números o valores no especificados.

variance (p. 757) The mean of the squares of the deviations from the arithmetic mean.

varianza Media de los cuadrados de las desviaciones de la media aritmética.

vertex (p. 525) The maximum or minimum point of a parabola.

vértice Punto máximo o mínimo de una parábola.

vertical line test (p. 47) If any vertical line passes through no more than one point of the graph of a relation, then the relation is a function.

prueba de la recta vertical Si cualquier recta vertical pasa por un sólo punto de la gráfica de una relación, entonces la relación es una función.

volume (p. P29) The measure of space occupied by a solid region.

volumen Medida del espacio que ocupa un solido.

voluntary response sample (p. 643) A sample that involves only those who want to participate.

muestra de respuesta voluntaria Muestra que involucra sólo aquellos que quieren participar.

W

weighted average (p. 132) The sum of the product of the number of units and the value per unit divided by the sum of the number of units, represented by M.

promedio ponderado Suma del producto del número de unidades por el valor unitario dividida entre la suma del número de unidades y la cual se denota por M.

whole numbers (p. P7) The set {0, 1, 2, 3, ...}.

números enteros El conjunto {0, 1, 2, 3, ...}.

work problems (p. 722) Rational equations are used to solve problems involving work rates.

problemas de trabajo Las ecuaciones racionales se usan para resolver problemas de tasas de trabajo.

X

x-axis (p. 38) The horizontal number line on a coordinate plane.

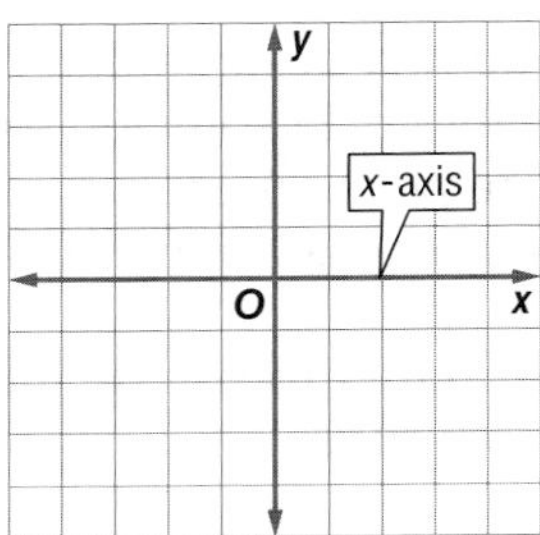

eje x Recta numérica horizontal que forma parte de un plano de coordenadas.

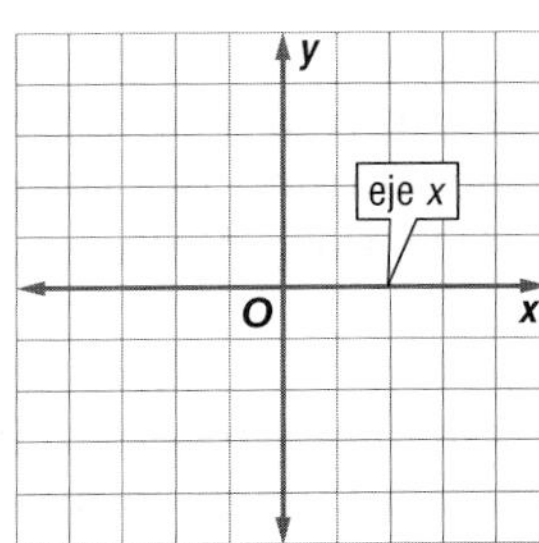

x-coordinate (p. 38) The first number in an ordered pair.

coordenada x El primer número de un par ordenado.

x-intercept (p. 154) The x-coordinate of a point where a graph crosses the x-axis.

intersección x La coordenada x de un punto donde la gráfica corte al eje de x.

Y

y**-axis** (p. 38) The vertical number line on a coordinate plane.

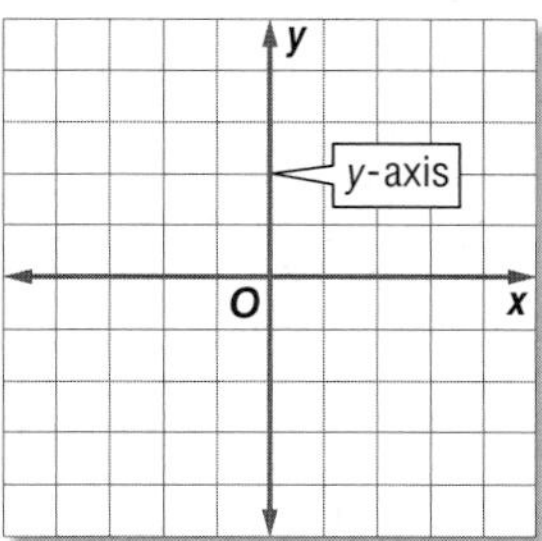

y**-coordinate** (p. 38) The second number in an ordered pair.

y**-intercept** (p. 154) The y-coordinate of a point where a graph crosses the y-axis.

eje y Recta numérica vertical que forma parte de un plano de coordenadas.

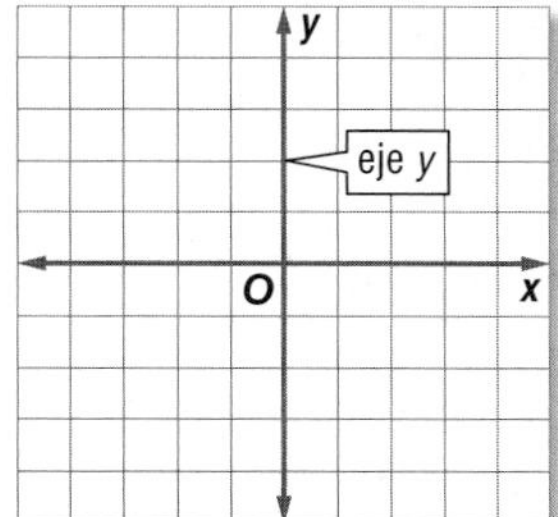

coordenada y El segundo número de un par ordenado.

intersección y La coordenada y de un punto donde la grafica corta al eje de y.

Z

zero (p. 161) The roots, or x-intercepts, of a quadratic function.

zero exponent (p. 410) For any nonzero number a, $a^0 = 1$. Any nonzero number raised to the zero power is equal to 1.

cero Las raíces o intersecciones x de una función cuadrática.

exponente cero Para cualquier número distinto a cero a, $a^0 = 1$. Cualquier número distinto a cero levantado al potente cero es igual a 1.

Index

A

B

C

Empirical study, 787

Which One Doesn't Belong?

Extended Response. *See* Assessment

Index

K

Index

L

M

Index

Q

R

S

Index

T

Index

Symbols

$\neq$	is not equal to	AB	measure of $\overline{AB}$
$\approx$	is approximately equal to	$\angle$	angle
$\sim$	is similar to	$\triangle$	triangle
$>, \geq$	is greater than, is greater than or equal to	$^\circ$	degree
$<, \leq$	is less than, is less than or equal to	π	pi
$-a$	opposite or additive inverse of x	$\sin x$	sine of x
$\lvert a \rvert$	absolute value of a	$\cos x$	cosine of x
$\sqrt{a}$	principal square root of a	$\tan x$	tangent of x
$a : b$	ratio of a to b	!	factorial
(x, y)	ordered pair	$P(a)$	probability of a
$f(x)$	f of x, the value of f at x	$P(n, r)$	permutation of n objects taken r at a time
$\overline{AB}$	line segment AB	$C(n, r)$	combination of n objects taken r at a time

Algebraic Properties and Key Concepts

Identity	For any number a, $a + 0 = 0 + a = a$ and $a \cdot 1 = 1 \cdot a = a$.
Substitution (=)	If $a = b$, then a may be replaced by b.
Reflexive (=)	$a = a$
Symmetric (=)	If $a = b$, then $b = a$.
Transitive (=)	If $a = b$ and $b = c$, then $a = c$.
Commutative	For any numbers a and b, $a + b = b + a$ and $a \cdot b = b \cdot a$.
Associative	For any numbers a, b, and c, $(a + b) + c = a + (b + c)$ and $(a \cdot b) \cdot c = a \cdot (b \cdot c)$.
Distributive	For any numbers a, b, and c, $a(b + c) = ab + ac$ and $a(b - c) = ab - ac$.
Additive Inverse	For any number a, there is exactly one number $-a$ such that $a + (-a) = 0$.
Multiplicative Inverse	For any number $\frac{a}{b}$, where $a, b \neq 0$, there is exactly one number $\frac{b}{a}$ such that $\frac{a}{b} \cdot \frac{b}{a} = 1$.
Multiplicative (0)	For any number a, $a \cdot 0 = 0 \cdot a = 0$.
Addition (=)	For any numbers a, b, and c, if $a = b$, then $a + c = b + c$.
Subtraction (=)	For any numbers a, b, and c, if $a = b$, then $a - c = b - c$.
Multiplication and Division (=)	For any numbers a, b, and c, with $c \neq 0$, if $a = b$, then $ac = bc$ and $\frac{a}{c} = \frac{b}{c}$.
Addition (>)*	For any numbers a, b, and c, if $a > b$, then $a + c > b + c$.
Subtraction (>)*	For any numbers a, b, and c, if $a > b$, then $a - c > b - c$.
Multiplication and Division (>)*	For any numbers a, b, and c, 1. if $a > b$ and $c > 0$, then $ac > bc$ and $\frac{a}{c} > \frac{b}{c}$. 2. if $a > b$ and $c < 0$, then $ac < bc$ and $\frac{a}{c} < \frac{b}{c}$.
Zero Product	For any real numbers a and b, if $ab = 0$, then $a = 0$, $b = 0$, or both a and b equal 0.
Square of a Sum	$(a + b)^2 = (a + b)(a + b) = a^2 + 2ab + b^2$
Square of a Difference	$(a - b)^2 = (a - b)(a - b) = a^2 - 2ab + b^2$
Product of a Sum and a Difference	$(a + b)(a - b) = (a - b)(a + b) = a^2 - b^2$

* ***These properties are also true for $<$, $\geq$, and $\leq$.***

Formulas

Slope	$m = \frac{y_2 - y_1}{x_2 - x_1}$
Distance on a coordinate plane	$d = \sqrt{(x_2 - x_1)^2 + (y_2 - y_1)^2}$
Midpoint on a coordinate plane	$M = \left(\frac{x_1 + x_2}{2}, \frac{y_1 + y_2}{2}\right)$
Pythagorean Theorem	$a^2 + b^2 = c^2$
Quadratic Formula	$x = \frac{-b \pm \sqrt{b^2 - 4ac}}{2a}$
Perimeter of a rectangle	$P = 2\ell + 2w$ or $P = 2(\ell + w)$
Circumference of a circle	$C = 2\pi r$ or $C = \pi d$

Area

rectangle	$A = \ell w$	**trapezoid**	$A = \frac{1}{2}h(b_1 + b_2)$
parallelogram	$A = bh$	**circle**	$A = \pi r^2$
triangle	$A = \frac{1}{2}bh$		

Surface Area

cube	$S = 6s^2$	**regular pyramid**	$S = \frac{1}{2}P\ell + B$
prism	$S = Ph + 2B$	**cone**	$S = \pi r\ell + \pi r^2$
cylinder	$S = 2\pi rh + 2\pi r^2$		

Volume

cube	$V = s^3$	**regular pyramid**	$V = \frac{1}{3}Bh$
prism	$V = Bh$	**cone**	$V = \frac{1}{3}\pi r^2 h$
cylinder	$V = \pi r^2 h$		

Measures

Metric	Customary
Length	
1 kilometer (km) = 1000 meters (m)	1 mile (mi) = 1760 yards (yd)
1 meter = 100 centimeters (cm)	1 mile = 5280 feet (ft)
1 centimeter = 10 millimeters (mm)	1 yard = 3 feet
	1 foot = 12 inches (in.)
	1 yard = 36 inches
Volume and Capacity	
1 liter (L) = 1000 milliliters (mL)	1 gallon (gal) = 4 quarts (qt)
1 kiloliter (kL) = 1000 liters	1 gallon = 128 fluid ounces (fl oz)
	1 quart = 2 pints (pt)
	1 pint = 2 cups (c)
	1 cup = 8 fluid ounces
Weight and Mass	
1 kilogram (kg) = 1000 grams (g)	1 ton (T) = 2000 pounds (lb)
1 gram = 1000 milligrams (mg)	1 pound = 16 ounces (oz)
1 metric ton (t) = 1000 kilograms	